MINNESOTA COUNTIES

The Breeding Birds of Minnesota

Male Common Yellowthroat. Photo by Michael Furtman.

The Breeding Birds of Minnesota

HISTORY, ECOLOGY, AND CONSERVATION

Lee A. Pfannmuller, Gerald J. Niemi, and Janet C. Green

FOREWORD BY Francesca J. Cuthbert

Kim Rewinkel, TECHNICAL EDITOR

University of Minnesota Press
Minneapolis
London

Portions of chapter 1 were published in "The Making of Minnesota's New Atlas of Breeding Birds" by Lee Pfannmuller in *Agate Magazine*.

Published by the University of Minnesota Press
111 Third Avenue South, Suite 290
Minneapolis, MN 55401-2520
http://www.upress.umn.edu

ISBN 978-1-5179-0679-5 (hc)

A Cataloging-in-Publication record for this book is available from the Library of Congress.

Printed in China on acid-free paper.

31 30 29 28 27 26 25 24 10 9 8 7 6 5 4 3 2 1

PRODUCED BY WILSTED & TAYLOR PUBLISHING SERVICES
Art direction Christine Taylor
Copy editing Nancy Evans
Design and composition Nancy Koerner, Laurel Muller, and Michael Starkman
Proofreading Jennifer Brown, Sophia Fox, and Melody Lacina
Color management Evan Winslow Smith

ENDSHEETS: County data derived from 2010 TIGER files prepared by the U.S. Census Bureau.

PAGE X: Ruffed Grouse (David Brislance), Horned Grebe (Sparky Stensaas), Ruby-throated Hummingbird (David Brislance), Sora (Rebecca Field), Wilson's Phalarope (Rebecca Field), Ring-billed Gull (Sparky Stensaas).

PAGE XI: American White Pelican (Rebecca Field), Green Heron (Rebecca Field), Bald Eagle (David Brislance), Northern Hawk Owl (David Brislance), Belted Kingfisher (Mike Lentz), Northern Flicker (David Brislance).

PAGE XII: Yellow-bellied Flycatcher (David Brislance), Yellow-throated Vireo (Gerald Hoekstra), Blue Jay (David Brislance), Black-capped Chickadee (David Brislance), Bank Swallow (Karl Bardon), Cedar Waxwing (Mike Lentz).

PAGE XIII: House Wren (Mike Lentz), Northern Mockingbird (Sparky Stensaas), Eastern Bluebird (Rebecca Field), Purple Finch (David Brislance), Clay-colored Sparrow (Jean Brislance), Eastern Towhee (David Brislance).

PAGE XIV: Yellow-headed Blackbird (Mike Lentz), Baltimore Oriole (Mike Lentz), Black-and-white Warbler (David Brislance), American Redstart (Gerald Hoekstra), Yellow Warbler (Mike Lentz), Northern Cardinal (Jean Brislance).

PAGE XV: Chukar, Passenger Pigeon (Mark Catesby), Black-necked Stilt (Don Edwards), Barn Owl (Vaclav Matous), Rock Wren (Travis Maher), Mountain Bluebird (Danita Delimont).

T*his book* is for the long-term survival and conservation of Minnesota birds and for anyone with a desire to know something about them. It is dedicated to the hundreds of atlas participants, partners, and contributors whose outstanding efforts made Minnesota's first breeding bird atlas a success and made this book possible. Their work, coupled with the dedication of thousands of bird watchers, ornithologists, nature enthusiasts, resource managers, and conservationists who have made past and present contributions to our knowledge of Minnesota birds, serves as the very foundation of all efforts to conserve our remarkable avifauna.

The University of Minnesota Press thanks the following individuals for their generous contributions to the publication of this book:

David Hartwell • Gene Merriam and Sara Meyer
Manley and Ann Olson • Deborah Reynolds
Steve Wilson and Mary Shedd • Fred and Eleanor Winston

This publication has also been supported by grants from the following organizations:

Audubon Chapter of Minneapolis
Minnesota Ornithologists' Union—Savaloja Grant
The Saint Paul Audubon Society
Duluth Superior Area Community Foundation—Biodiversity Fund

DULUTH SUPERIOR AREA COMMUNITY FOUNDATION

The University of Minnesota Press also gratefully acknowledges financial assistance provided for the publication of this book by Audubon Minnesota and the University of Minnesota's Natural Resources Research Institute.

Audubon
MINNESOTA, IOWA, MISSOURI

Natural Resources Research Institute
UNIVERSITY OF MINNESOTA DULUTH
Driven to Discover

This publication was made possible in part by the people of Minnesota through a grant funded by an appropriation to the Minnesota Historical Society from the Minnesota Arts and Cultural Heritage Fund. Any views, findings, opinions, conclusions, or recommendations expressed in this publication are those of the authors and do not necessarily represent those of the State of Minnesota, the Minnesota Historical Society, or the Minnesota Historic Resources Advisory Committee.

CLEAN WATER LAND & LEGACY AMENDMENT

< Adult Bald Eagle. Photo by David Brislance.

CONTENTS

< Female Common Merganser with young. Photo by Michael Furtman.

FLYCATCHERS

VIREOS

SHRIKES

JAYS, MAGPIES, CROWS, AND RAVENS

CHICKADEES AND TITMICE

LARKS

SWALLOWS

KINGLETS

WAXWINGS

NUTHATCHES

CREEPERS

GNATCATCHERS

CHATS

BLACKBIRDS

WARBLERS

TANAGERS, CARDINALS, AND GROSBEAKS

II. FORMER AND INCIDENTAL NESTING SPECIES

FOREWORD

Francesca J. Cuthbert

< Sora. Photo by Mike Lentz.

I have lived in Minnesota for more than forty years, and my passion for birds has taken me to all of the biomes in the state for my own research or that of my graduate students. Until the publication of the online version of the Minnesota Breeding Bird Atlas (www.mnbirdatlas.org) in 2017, my "go-to" critical resources on the state's birds have always been Thomas S. Roberts's *The Birds of Minnesota* (1932) as well as the publications by Green and Janssen (1975) and Janssen (1987, 2019). I have often reflected on which book I would grab first from my extensive office library if the building were on fire; without a doubt, it would be my tattered and yellowed copies of Roberts's two volumes. They are such remarkable treasures of the deep history of Minnesota's birds that they stand as the best record of avian relative abundance and distribution in the early history of our state up until the early 1930s. For anyone attempting to understand current species status or seeking insights on the distribution of Minnesota's breeding birds, the Roberts–Green–Janssen route through the literature is essential. But today we also have this truly exceptional book, *The Breeding Birds of Minnesota: History, Ecology, and Conservation*. Using the groundbreaking data gathered by the Minnesota Breeding Bird Atlas as a foundation, the book complements the work of earlier authors and transitions bird knowledge from their early contributions into a discrete window of time (2009–13) in the present. It establishes a permanent, detailed record of Minnesota's breeding birds from years past to the present while also identifying issues that many species will confront in the years ahead.

What is unique about the Minnesota Breeding Bird Atlas? It is much more than descriptions of change since the previously published efforts. Of note, it was the first systematic data collection endeavor for the entire state using a repeatable process for all observed species. It therefore serves as a foundation for future atlas editions by

A male Northern Flicker extends his impressively long tongue while foraging on the ground. All woodpeckers have long tongues to help them access food that other birds cannot reach, but the Northern Flicker takes top prize in having the longest tongue of any North American woodpecker. Woodpecker tongues are so long that they wrap around the back of the skull to fit in their head when not in use. Photo by David Brislance.

using a method that is comparable and quantifies change. Previous attempts to document bird distribution in Minnesota were limited in geography (the North American Breeding Bird Survey), focused primarily on rare species and native habitats (Minnesota Biological Survey), or used protocols that are not repeatable. Also, the Minnesota atlas relied on survey design using observations to indicate confidence that a species was breeding within a limited area (9 square miles [2,331 ha]).

Incorporating the outcomes of the atlas, one of the truly valuable contributions of *The Breeding Birds of Minnesota* that will definitely appeal to readers is the detailed information within each species account. The general topics covered include historical and current breeding distribution, habitat, population abundance, and conservation; the thoroughness of coverage is overwhelmingly impressive. Although the reader should always check Roberts, the species accounts include his key information along with details from other relevant authors. For example, the account for the Canada Goose cites information on history, biology, management, and trends from 19 sources. Finally, it is important to note that the content of this book and the online atlas is a tribute to the hundreds of Minnesota atlas participants and organizations who contributed their time and data to the initiative. Their data were meticulously reviewed by three of Minnesota's premier ornithologists: Janet Green, Gerald Niemi, and Lee Pfannmuller. With more than a collective century of experience and knowledge about avian natural history and conservation, these individuals have incorporated, analyzed, and summarized the atlas findings into a new, comprehensive assessment of Minnesota's breeding avifauna that is a valuable reference for anyone interested in birds.

What do we learn from the book? Despite the recent concern about significant bird declines across North America (Rosenberg et al. 2019), not all of the news is gloomy. When we examine species that Roberts identified as in serious trouble or on the verge of extinction in the state, we discover some surprises to celebrate. For example, Roberts (1932) stated the American White Pelican was "once common [but] disappeared many years ago as a breeding bird"; the Sandhill Crane has "about vanished" as a breeding bird, and the Common Raven "is apparently nearing extinction." If Roberts were alive today, he would be amazed at how these species have increased in abundance and expanded their distribution in Minnesota owing to creative management, protection, and, in some cases, shifting land-use policies. These examples represent just a few among other positive changes. In contrast, he would be dismayed to discover species he once described as abundant and widespread that are now in serious decline, both in Minnesota and elsewhere in North America. Two examples include the Black Tern (a species Roberts described as an "abundant summer resident, breeding throughout the state") and Eastern Meadowlark (described as an "abundant summer resident in the eastern part of the state"), as well as multiple other species, especially those associated with Minnesota's grasslands. A major message from all works, from Roberts to the present, is that bird populations are ever changing: they increase, decrease, expand, and contract—most often in response to human activities on the landscape.

Why are the Minnesota Breeding Bird Atlas and the publication of *The Breeding Birds of Minnesota* such important accomplishments? "If you take care of birds, you take care of most of the environmental problems in the world." This quote, attributed to Dr. Thomas Lovejoy, was used to spearhead the National Geographic Society's 2018 Year of the Bird, a wide United States effort to commemorate the centennial of the Migratory Bird Treaty Act. These words are just as timely and relevant when celebrating the work and documentation of Minnesota's inaugural Breeding Bird Atlas. The health of bird populations is tied closely to the health of ecosystems from local to global levels. Many studies have demonstrated that birds are important indicators of the condition of critical ecosystem components, including air, land, and water. Where birds decline or struggle to survive, we know that human well-being is affected too. The message from *The Breeding Birds of Minnesota* is that many species of birds still have a wide distribution in the state, but key efforts that track national trends also show that a number of these species, notably grassland birds and those already affected by climate change, are in serious decline. This book is ultimately a call for action by state and federal agencies and environmental organizations that develop policies, enforce legislation, and support on-the-ground action to address conservation issues. It also is a source of endless ideas for high-priority research by Minnesota's avian biologists and their graduate students who include birds in their research agendas. And finally, citizen scientists will recognize their critical role in bird conservation and be motivated to continue to participate in important monitoring efforts such as the Breeding Bird Survey or Christmas Bird Count, as well as future versions of the atlas.

Atlas data presented in *The Breeding Birds of Minnesota* ultimately set the stage for questions that cannot be answered until the next effort to summarize the distribution and status of birds in the state. How will today's concerns impact bird populations in the future? To the readers and everyone picking up this book to review one species or many, it is a stimulus for conservation. By presenting these data in a historical context, the book becomes a timely and essential reference in this age of environmental crises; its value in promoting conservation is unquestionably critical. I hope that, a century or more from now, ornithologists, birders, and emerging young scholars will echo my excitement as they blow the dust off Roberts's two volumes and do the same for the 2023 edition of *The Breeding Birds of Minnesota*. Despite an unknown future for birds, our environment, and the storage of historical records, I predict all three will be protected as valued resources for future generations of Minnesota's citizens.

INTRODUCTION

The Historical Legacy of Minnesota's Avifauna

< Cedar Waxwing eating crabapples.
Photo by David Brislance.

For the first time in Minnesota's history, we have an assessment of the current distribution of 250 bird species that breed in the state nearly every summer, completed at a scale of resolution much finer than has ever been attempted in the past. And for many of these species we have a more accurate estimate of their statewide abundance. This extraordinary outcome reflects the efforts of hundreds of enthused field surveyors who combed all corners of the state to document Minnesota's avifauna during the course of five breeding seasons, from 2009 through 2013. The treasure chest of data collected by Minnesota's first Breeding Bird Atlas will help guide bird conservation efforts in the twenty-first century and spark the curiosity of bird enthusiasts across the state.

When placed in a historical context, the data gathered during the atlas project provide a fascinating and remarkable portrait of the present health of our state's avifauna. Some indicators suggest a strong and healthy system, while others signal trouble ahead. Minnesota's landscape, for example, continues to support one of the most diverse avifaunas in North America. Notable are the breeding populations of 21 native waterfowl species, 32 waterbirds, 10 owls, 9 woodpeckers, 12 flycatchers, 17 sparrows, and 31 wood warblers! Not only do native forests, grasslands, wetlands, lakes, and streams provide suitable habitat for these and many more species but also restored grasslands provide habitat for native prairie species such as the Short-eared Owl and Grasshopper Sparrow; impoundments designed to control spring flooding provide habitat for such rare wetland species as the Eared Grebe and King Rail; and restored stream channels provide critical habitat for the Belted Kingfisher and Northern Rough-winged Swallow.

But not all systems are in good working order. Substantial numbers of species are in decline and many native species, such as the Chestnut-collared Longspur, Cerulean Warbler, Common Tern, Connecticut

Great Gray Owl. Photo by Mike Lentz.

Warbler, Horned Grebe, and Piping Plover, are barely surviving in the state. Minnesota is not alone in witnessing such changes. Rosenberg and his colleagues (2019) documented a nearly 30 percent decline in wild bird populations across the continental United States and Canada since 1970. That is a staggering loss of nearly 3 billion birds! In many respects, it is surprising that Minnesota has not had more extinctions like the Passenger Pigeon or statewide extirpations like the Swallow-tailed Kite or Thick-billed Longspur. This is a strong testament to the conservation ethic that exists among the people and organizations in the state. The population recoveries of many species, including Canada Geese, Bald Eagles, Osprey, Peregrine Falcons, Sandhill Cranes, and Trumpeter Swans, are astounding, a tribute to well-organized efforts, and an excellent illustration of the value of conservation work.

In the past and present, Minnesota has thousands of outstanding bird-watchers, nature enthusiasts, and professionals who have gathered exceptional historical information on Minnesota's avifauna. The legacy begins with Native American knowledge and sacred stories. Ogimaagiizhig Odoodeman Adikwan (a.k.a. Charles Grolla), an Ojibwe from the Red Lake Indian Reservation, has published a wonderful compilation of stories about northern Minnesota birds in an effort to preserve his community's oral history (Adikwan 2019). Each story reflects an intimate understanding of bird behavior and a deep respect for each bird and its cultural significance to the tribe. *Naanokaashiinh,* for example, the Ojibwe name for the Ruby-throated Hummingbird, is a reference to the bird's "fighting nature" and aggressive defense of its young. Given its personality, the tribe associates the bird with strong warriors and strong healing properties. *Diindiisi* "is known for being the gossip and tattletale" who is responsible for alerting "everything in the forest of your presence." *Diindiisi* is, of course, the Blue Jay.

Later, the journals of explorers such as Henry Schoolcraft (1821, 1834, 1854), Stephen Long (Keating 1824), and George Featherstonhaugh (1847) provided important observational data as their journeys took them across the lands and rivers that would soon become Minnesota. Large game animals, including bison, elk, cranes, swans, and grouse, were most frequently mentioned as they also served as essential provisions for the explorers. Keating wrote the most detailed account of the birds encountered. Indeed, Finseth (1958) declared that 1823 marked the year that written documentation of Minnesota ornithology began. This was the year that Major Stephen Long traveled from Fort Snelling to the source of the Red River. Keating's narrative of the expedition included a list of 50 different bird species observed during the journey, including Passenger Pigeons, Upland Plovers (now known as Upland Sandpipers), and American Woodcock.

Premier among the early contributors to our knowledge of Minnesota's avifauna is Thomas Sadler Roberts, whose two-volume set of books, *The Birds of Minnesota,* published in 1932 and revised in 1936, is an absolute classic that set the stage on Minnesota birds from the late 1800s to the early 1900s. As Lotus D. Coffman, then president of the University of Minnesota, stated in the Foreword, "During a half century of investigation and first-hand observation he [Roberts] has assembled a vast body of information on the birds of this region." He emphasized the "years of painstaking labor" by Roberts as he "accumulated stores of knowledge for the enrichment of this and future generations." Indeed, enough superlatives cannot be attributed to the importance of Roberts's contributions and attention to detail. We know so much more about the ornithological history of birds in Minnesota because of this tome.

Another important work that tracked Minnesota wildlife from 1850 to 1900 was a thesis by Evadene Burris Swanson (1940, reprinted in 2007), "The Use and Conservation of Minnesota Wildlife 1850–1900." This thesis is loaded with details on Minnesota wildlife during this period, especially for larger game animals that were better documented than the more diminutive wildlife species, but it also includes many details on the Trumpeter Swan, the Sandhill and Whooping Cranes, and the Passenger Pigeon. Both Roberts (1932) and Finseth (1958) note other individuals who provided important notes on the state's avifauna. Roberts (1932), however, stated that the "first real report on Minnesota birds, as such was an annotated list of 138 species by Thomas Martin Trippe, published in the Proceedings of the Essex Institute for 1871." This was later followed by Dr. Philo Louis Hatch, Minnesota's first state ornithologist, who published a bound 487-page report in 1892 titled *First Report: Accompanied with Notes on the Birds of Minnesota*.

Since the 1930s and 1940s, considerably more information has been accumulated on Minnesota's birds. The Minnesota Ornithologists' Union (MOU) was established in 1938. *The Flicker*, initially a publication of the Minnesota Bird Club, one of six original local organizations focused on Minnesota birds, was published in 1929; in 1964 it became *The Loon*. The MOU has compiled a vast amount of information on Minnesota's birds, and the journal issues have continued to build our knowledge on their distribution, abundance, and natural history. A prominent figure during this period was Dr. Walter Breckenridge, professor at the University of Minnesota and director of the Minnesota Museum of Natural History. Breckenridge published hundreds of peer-reviewed papers and popular articles on Minnesota birds and wildlife for seventy years, from 1929 to 1999 (Breckenridge 2009). For example, his summary *A Century of Minnesota Wild Life* (Breckenridge 1949) highlighted the extreme variability in game populations and their recovery once game laws were implemented. He also lamented the reduced populations or loss of many species, including prairie birds and birds of prey like the Chestnut-collared Longspur and Swallow-tailed Kite. He closed the article with these words of wisdom: "Minnesotans of the present day should be educated to a feeling of stewardship in relation to this resource [wildlife], for it played a glamorous role in the state's romantic past, and it can and should figure largely in affording a fuller and richer life for those who build Minnesota's future."

The books by Green and Janssen (1975) and Janssen (1987, 2019) were especially important in filling in the gaps of knowledge of breeding distribution from the 1930s to the present. These documents are repeatedly cited throughout this book. Similarly, the wealth of information that the Minnesota Department of Natural Resources has accumulated on Minnesota's breeding birds with such programs as the Minnesota Biological Survey, the Nongame Wildlife Program, and several targeted game bird programs, as well as the vast resources available on their websites, were essential in expanding details on distribution and relative abundance, often in remote places not necessarily visited by most Minnesotans.

Based on the trajectory of the past and the predictions for the future, the probability of species extinctions and an overall loss of biodiversity are projected to be among our most important challenges in the twenty-first century. Because of the massive number of global extinctions already occurring, a new geological epoch has been named, the Anthropocene. This period was defined because human activity has become the dominant influence on climate and the environment. The consequences of these changes to the basic structure and functioning of our natural ecosystems are now unclear.

Given this current status, it is interesting to note Roberts's words from his book in 1932, when he was reflecting on how conditions had changed in the past "forty or fifty years." "Birds as a whole, non-game- as well as game-birds, have, during that interval, undergone a very great diminution in numbers. Some species, formerly abundant, are now represented by a mere remnant, while a few have disappeared entirely." Roberts goes on to discuss many of the causes of his day that led to these changes. They include "undesirable species" like the "Grackle, and the Crow," the "prairies are plowed," the "forests are cut down," "marshes and shallow lakes are drained," "many kinds have been and still are being killed by the carrier of gun or rifle," and the domestic cat when returned to a feral state "has become one of the most destructive enemies of our wild birds." Most of these problems have continued to the present, although protection of raptorial and migratory birds has improved since laws have been implemented, and game laws have been effective in maintaining healthy populations of most game species.

Besides these continuing problems facing Minnesota wildlife, today we have two additional threats, climate change and invasive species, and a reemerging one, xenobiotic chemicals. All of these threats affect birds in insidious ways; they proceed gradually and subtly over long time frames and ultimately will have large effects on the distribution and abundance of many species.

Amid all these challenges, Minnesota's first Breeding Bird Atlas project (MNBBA) continues the long and remarkable legacy of documenting the state's breeding avifauna. From 2009 through 2013, volunteers and experienced surveyors contributed nearly 400,000 records to establish the most up-to-date and accurate assessment of Minnesota's breeding birds since Roberts's tome was first published in 1932. Once the fieldwork was completed, a website (www.mnbirdatlas.org) was launched in November 2017 to present the atlas results. This online, interactive website is an excellent resource for naturalists, bird-watchers, researchers, and historians, often providing even more in-depth information beyond the accounts in the following pages.

This male Chestnut-sided Warbler, like most insectivorous breeding birds, is astonishingly effective at catching insects to feed its young during the breeding season. Photo by David Brislance.

The goal of this book has been to condense, update, and expand on the website content. The organization of the book and the website are parallel in many respects. In the species accounts, we retained the information on life history, distribution, breeding habitat, population abundance, and conservation. We have included an introduction where we provide more particular information on each species, such as where it may be encountered in Minnesota, identification tips, why it has such a bizarre name, and a broad description of its breeding range. We also updated much of the trend information to what was available as of February 2021 (e.g., Sauer et al. 2020) as well as the climate change prediction models that became available in October 2019

(NAS 2019). Our foremost goal was to focus on Minnesota-centric information when available. The taxonomic order in which we present Minnesota breeding birds reflects the changes outlined in the 62nd supplement to the American Ornithological Society's *Check-list of North American Birds,* published July 1, 2021 (Chesser et al. 2021). The order in which birds are presented, as well as the species' scientific names, can change as new research improves our understanding of the relationship between different species. The application of DNA testing has significantly increased the rate at which these changes are occurring. Indeed, the 62nd supplement includes several adjustments to the order in which many of the songbird families are presented. A handful of scientific names have also been altered. The species accounts here reflect these changes.

Because breeding bird atlases may be unfamiliar to some readers, the book begins with an overview of the atlasing phenomenon: how it began and the history of its launch in Minnesota (chapter 1, "Minnesota's Breeding Bird Atlas"). A brief description of how data were collected in Minnesota follows the overview and concludes with brief summaries and highlights of the MNBBA. Chapter 2, "Biogeography of Minnesota Birds," describes the distribution of Minnesota's breeding birds in response to the state's ecological landscape. Chapter 3, "Reading the Species Accounts," provides a detailed summary and explanation of the information and source material included in each species account. This is followed by the heart of the book, the individual species accounts. Part I comprises accounts for all 250 species that were documented during the 2009–13 breeding seasons, followed by Part II, abbreviated accounts for 19 species that formerly bred in the state.

We end with our best effort to recognize all the organizations and countless individuals whose love of birds made this book possible. Hundreds of birders from all walks of life participated in Minnesota's first breeding bird atlas, the cornerstone of this publication. All share a love of Minnesota's avifauna and a commitment to its conservation. Some reported on birds in their backyards or local parks, while others, fortified with mosquito netting, hiked into remote areas before sunrise to document the morning chorus or traveled hundreds of miles to amass thousands of records. Each and every contribution was important.

We would like to make particular note of three special participants who have since passed away. First was birder extraordinaire Warren E. Nelson. Bird guide, photographer, and author, Warren was known to all in the Minnesota birding community as the go-to expert on boreal birds, especially in Aitkin County. Thousands of atlas records from sites not far from his hometown of Aitkin are attributed to his efforts. Second was a well-known science teacher in the Brainerd Lakes area, Kenneth R. Perry, who enthusiastically shared his love of birds and the outdoors with his students. Ken and his wife, Pam, were co-regional coordinators of the MNBBA in the Brainerd Region. Not only did they contribute several thousand records but one only needs to look at a handful of MNBBA species distribution maps to appreciate the outstanding job they did in recruiting volunteers in their region. Finally, a gifted birder and wetland biologist with the U.S. Fish and Wildlife Service, Robert P. Russell Jr. maintained a contagious enthusiasm for birds and most everything else in life. As one of the very top contributors to the MNBBA, Robert traveled to every corner of the state during the MNBBA's five field seasons. He was the only participant to collect records in every one of the state's 32 atlas regions, including his beloved Stearns County. We will be forever grateful for their efforts.

Finally, our hope is that this tome follows in the tradition of all the wonderful manuscripts and sacred stories about Minnesota's avifauna that have preceded it. It would be an honor to all who contributed to Minnesota's Breeding Bird Atlas and to *The Breeding Birds of Minnesota* if these pages become a critical resource that helps continue the state's conservation legacy. Indeed, birds have the power to galvanize us to address the conservation needs of today and to tackle the unknown challenges of tomorrow. May their beauty and music inspire such action.

1

MINNESOTA'S BREEDING BIRD ATLAS

< Belted Kingfisher pair (female on the left and male on the right). Photo by Mike Lentz.

Atlasing and Initiation of Minnesota's Breeding Bird Atlas

The word "atlas" typically refers to a collection of maps, often depicting roads, cities, and land ownership, to help guide our travels. More recently, the verb "atlasing" has come to embrace the effort to geographically catalog the distribution and abundance of elements of biodiversity during an established time period. Made possible by engaging hundreds of knowledgeable volunteers, "atlasing" had its origins in Great Britain in the 1950s when volunteers helped botanists document and map the diversity and distribution of the country's flora using a 10 km × 10 km grid. Prior to this effort, the presence or absence of species was recorded at the scale of vice-counties, a subdivision of the country's 86 counties. Mapping at the much finer scale of 100 square kilometers not only provided considerably more detail but also established a meticulous baseline for monitoring future changes. It also provided comprehensive data to better inform local conservation and land-use planning efforts. Published in 1962, the *Atlas of the British Flora* became a sentinel publication that prompted others to initiate similar efforts (Perring and Walters 1962).

Given the popularity of bird-watching, it was only natural that an atlas of birds soon followed, with the publication of the *Atlas of Breeding Birds of the West Midlands* (Lord and Munns 1970), an area encompassing three counties in Great Britain. This was followed by the 1976 publication of the *Atlas of Breeding Birds in Britain and Ireland* (Sharrock 1976). The movement quickly traveled overseas, and by 1974 Massachusetts was the first state to launch a breeding bird atlas, followed by Vermont in 1976, Maine in 1978, and New York in 1980. In the ensuing years, breeding bird atlas efforts continued to sprout up across the United States and Canada. Indeed, by the year 2000 Minnesota was the only state in the Mississippi River valley and one of only seven states in the nation that had not launched a similar initiative.

The atlasing movement, however, did not go unnoticed in Minnesota. Staff at the Department of Natural Resources (MNDNR) attended conferences that focused on national atlas efforts and protocols, engaged in conversations with natural resources staff in neighboring states that had conducted atlases, and met with individuals active in the statewide birding community. These efforts were concurrent with a growing national awareness of the plight of Neotropical migrant songbirds. Spurred by John Terborgh's provocative book, *Where Have All the Birds Gone?,* published in 1989, researchers and resource managers saw the need to broaden traditional resource management beyond its historical focus on game birds and direct more resources to what was commonly referred to as "all-bird" conservation. The book led to two critical symposia held at Manomet Bird Observatory in 1989 and in Atlanta in 1990. Attended by MNDNR staff, these gatherings blossomed into a network of more than 150 organizations now formally recognized as Partners in Flight (PIF), an international collaborative dedicated to keeping common birds common and helping species at risk throughout the Americas. Although the original focus was on landbirds, similar initiatives that focused on waterbirds and shorebirds soon followed. Together with a long-standing group of resource managers dedicated to waterfowl management, all four initiatives were embraced under the umbrella of the North American Bird Conservation Initiative (NABCI).

Several years earlier, in 1977, the creation of the MNDNR's Nongame Wildlife Program had already initiated a new effort to devote resources to the conservation of all Minnesota wildlife species. Many birds, including the Peregrine Falcon, Trumpeter Swan, and Eastern Bluebird, were among the initial beneficiaries of the broadened approach. Launched in 1987, the Minnesota County Biological Survey, now known as the Minnesota Biological Survey (MBS), focused even more attention on a county-by-county inventory of rare features, including rare birds, further expanding our knowledge of the distribution and abundance of Minnesota's avifauna. As important as these efforts were, the Partners in Flight collaborative reinforced the notion that effective bird conservation depended on, first, an integrated partnership of agencies, conservation organizations, and landowners combining resources and working together to achieve mutual goals; and, second, a solid foundation of baseline data on the distribution and abundance of birds that would allow a rigorous approach to monitoring changes over time.

Enter Bird Conservation Minnesota (BCM), a collaborative of individuals, organizations, and institutions organized in 2003 to expand the scope of bird conservation efforts in the state. Undeterred by the numerous challenges of initiating an atlas, including the thousands of hectares of inaccessible lands in northern Minnesota, BCM sought funding to conduct a statewide breeding bird atlas. After numerous meetings and workshops, Audubon Minnesota, working with other BCM partnership organizations, successfully secured a major state appropriation from the Minnesota Environment and Natural Resources Trust Fund, as recommended by the Legislative-Citizen Commission on Minnesota Resources (LCCMR), to launch Minnesota's first statewide atlas. With additional support provided by Audubon Minnesota, the Minnesota Department of Natural Resources, the Minnesota Ornithologists' Union, the Natural Resources Research Institute of the University of Minnesota–Duluth (UMD), and the United States Fish and Wildlife Service, preparatory work began in 2008, and the atlas was officially launched during the 2009 breeding season under the direction of Audubon's newly hired atlas coordinator, Bonnie Sample. Data collection continued for five summers through the end of the 2013 breeding season. Then, from the fall of 2013 through the fall of 2017, the data collected by nearly 700 volunteers, paid survey staff, and numerous resource organizations were compiled, reviewed, analyzed, and summarized. The results of that work were published as a website in November 2017 (https://mnbirdatlas.org) (Pfannmuller et al. 2017). Once the website was completed, work immediately began on this book.

Data Collection by the Minnesota Breeding Bird Atlas (MNBBA)

Atlas surveys were conducted in predetermined areas to ensure that the atlas covered the entire state. Most atlases use a grid system to determine the survey area. The MNBBA technical committee chose the US Public Land Survey System to select these areas. The Public Land Survey System is based on a 36-square-mile (9,324 ha) area called a township.

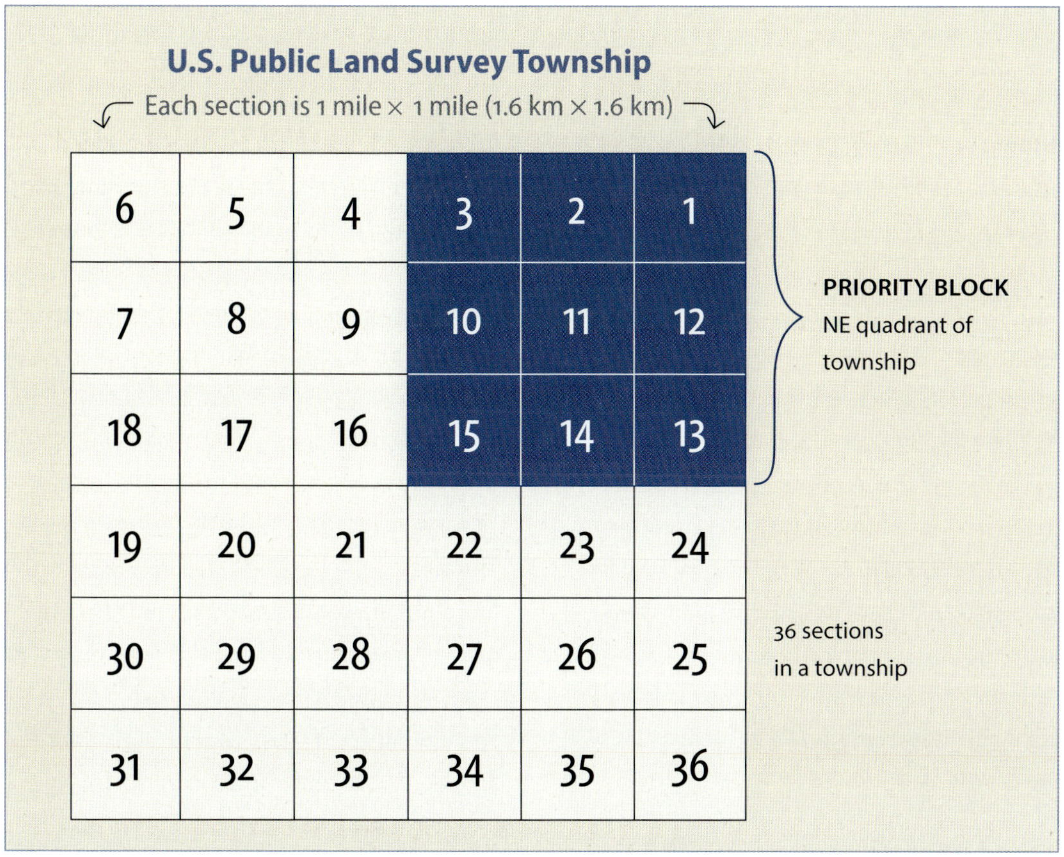

Figure 1.1. The northeast (NE) quadrant (4.8 km × 4.8 km [3 miles × 3 miles]) of each township was selected as a "priority block" to focus recording of breeding evidence for the Minnesota Breeding Bird Atlas. Volunteers could adopt a priority block or were assigned a priority block by a regional coordinator. Priority blocks also served as the area for focused sampling for point counts.

Townships are subdivided into 36 sections (each 1.6 km × 1.6 km [1 mile × 1 mile]). To reduce survey bias and ensure systematic coverage across the state, the MNBBA randomly selected the northeast quadrant (4.8 km × 4.8 km [3 miles × 3 miles]) of every township in the state to survey. These predetermined areas were called "priority blocks" (Figure 1.1).

Rivers, state and county borders, or occasional variations in the grid system can create townships less than 36 square miles (9,324 ha). In cases where the northeast priority block was smaller than 8 square miles (2,072 ha), the southeast quadrant was selected, or the process continued clockwise until a complete quadrant was identified. If the township had no quadrants that met the size criteria, no priority blocks were designated in the township. Blocks were identified by the township and range number plus a one-letter code that described whether the block was the northeast (a), southeast (b), southwest (c), or northwest (d) quadrant in the township. For example, a block labeled T123R15a is in Township 123, Range 15, in the northeast quadrant of the township.

VOLUNTEER DATA COLLECTION

Volunteers adopted blocks for coverage or were recruited by regional coordinators or others across the state to survey atlas blocks using standard atlas techniques. The primary purpose of the volunteer network was to identify the breeding bird species present and their breeding status within the selected northeast quadrant (priority block) for each of 2,443 possible townships in the state of Minnesota. Volunteers could also gather data from any block within the state. Although details and guidance provided to the volunteers emphasized that as many priority blocks as possible be covered, many volunteers also covered secondary blocks (e.g., b, c, or d quadrants of a township), especially when it was convenient to do so. This was particularly true in population centers across the state. To ensure coverage in more remote regions of the state, during several breeding seasons field personnel were hired to specifically sample these regions and augment the volunteer data. A complete list of regional coordinators and atlas participants can be found in the Acknowledgments.

POINT COUNTS DATA COLLECTION

In addition to the traditional, volunteer-based sampling used by atlases, the MNBBA used another complementary method for gathering data on Minnesota's breeding birds: the one-time sampling of three breeding bird point counts established within each priority block. With volunteer sampling, it is difficult to standardize the gathering of data on time of sampling, level of effort, and observer ability. Each of these requires a more rigorous, systematic approach to better assess habitat relationships and relative abundances, to ensure statewide coverage, and to detect bird population changes over time. Each individual who was gathering point count data was also expected to record their observations outside of the point count periods both within the priority blocks and in transit between priority blocks. These observations were entered in the same fashion as standard data from the volunteers.

Two examples highlight the importance of these complementary approaches. Volunteer efforts are of immense importance for an atlas because volunteer work results in extensive coverage of the state by participants with high skill levels. They also allow individuals to gather data close to their homes and in their backyards. Nevertheless, volunteer efforts can result in uneven coverage of the state because they are focused on urban areas where there are more people; and in an abundance of data gathered in the "best," well-known birding places.

The goal for including a network of point counts to gather breeding bird atlas data was to ensure equal and consistent coverage across the state of Minnesota. This included a random sampling approach and standardization of effort in gathering data within each township. Specific objectives to be addressed by the point count protocol for the MNBBA included the following:

> Estimate the relative abundance of breeding bird species in Minnesota.
>
> Identify where in the state breeding bird species were most likely to be detected in terms of habitat, landscape context, and climate regime.

Further details on the point count methodology, training of point count surveyors, point count selection and sampling design, and the review and analysis of the MNBBA data can be found in Appendix A.

CONTRIBUTED DATA

More than 20 agencies and organizations contributed nearly 80 sets of data during the relevant window (2009–13) that considerably supplemented the ability to accurately represent the status of all breeding birds in the state. An emphasis was placed on datasets that provided information on rare species or species that are difficult for volunteers to survey or detect. A complete list of all contributed data is provided in Appendix B.

ATLAS MATERIALS

The MNBBA website contains substantial background information that was available to all atlas volunteers to aid and guide their field efforts. These included the MNBBA Handbook; a letter requesting permission to survey private property; a vehicle placard; various field forms to document observations; bird identification resources; a master species list with safe dates for nesting and general habitat information; suggestions for surveying an atlas block; and guidelines for nocturnal and marsh bird surveys.

MNBBA Data Summaries

Even though 9,774 quarter township blocks (4.8 km × 4.8 km [3 miles × 3 miles]) were possible, at least 10 detections of birds were recorded in 4,733 blocks (48 percent) (Table 1.1). Although not all priority blocks were adopted by observers, out of 2,352 priority blocks, 2,337 (99 percent) had at least 10 records or more, indicating that observers incidentally recorded birds in these blocks even though they did not necessarily adopt them. The distribution and coverage of the point counts across the state were also impressive, with 6,919 point counts sampled by atlas surveyors, which represented coverage for over 98 percent of the priority blocks (Figure 1.2). The spatial distribution of the point counts illustrates that all corners of the state were adequately covered. The major exception was the Red Lake Peatland, where even the Marsh Master® vehicle was unable to traverse the hydric regions in the center of this vast ecosystem.

Breeding bird records are reported in four broad categories (Observed, Possible, Probable, or Confirmed) that represent 20 different types of breeding evidence (Table 1.2). Types of evidence vary from simple observations of a bird flying overhead, which does not provide breeding evidence, to the best confirmed evidence,

Table 1.1 Number of atlas blocks and priority blocks sampled during the MNBBA

	ALL BLOCKS	PRIORITY BLOCKS
Total number of blocks in Minnesota	9,774	2,352
Total number of blocks with ≥ 10 records	4,733	2,337
Total number of species detected	250	243
Total number of species confirmed breeding	231	218

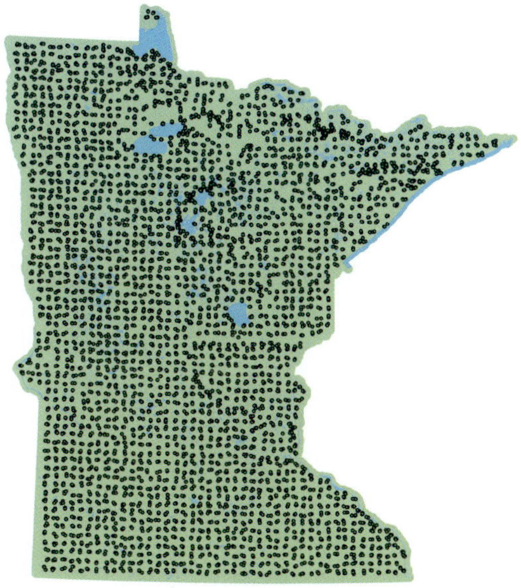

Figure 1.2. Distribution of more than 7,000 point counts sampled and included as part of the Minnesota Breeding Bird Atlas from 2009 to 2013. The clumping of points in northeast and north-central Minnesota is due to the inclusion of the point count samples from the bird monitoring program in the Superior and Chippewa National Forests, respectively. Clumping of the point count locations also occurs near large bodies of water, such as the Mississippi River, Upper and Lower Red Lake, Leech Lake, and Mille Lacs Lake.

a nest with young seen or heard. It is important to note that the definition of confirmed nesting of a species that is used by Breeding Bird Atlas projects in the United States, Canada, and many other countries, including the Minnesota Breeding Bird Atlas, differs from a more restrictive definition used by the Minnesota Ornithologists' Union (MOU) (e.g., Hertzel and Janssen 1998).

Species accounts summarize data for all 250 breeding bird species detected during the MNBBA. A complete summary of the number of species detections and the percent of the total detections among all the blocks and in the priority blocks is in Appendix C. This appendix also includes the total number of species detections in all the point counts and in the first, randomly selected point count in each block. Collectively, these random points provide an estimate of bird species diversity across the state of Minnesota. We also summarize the 25 most abundant species and their frequency of encounter on the first, randomly selected point counts (Table 1.3). The methodology for the Minnesota population estimates was based on the Breeding Bird Probability Distribution Modeling Strategy 1 (see Appendix A under Data Analysis).

It may come as a surprise that the most abundant species in Minnesota based on the MNBBA was the Common Yellowthroat (Table 1.3). However, its high abundance estimate in the state is due to several factors: first, it is indeed a frequently encountered species that was found on over 50 percent of the random point counts gathered throughout the state; second, it occurs in nearly all of the cover types of the state and often in edges such as ditches, even in highly agricultural areas; third, it has a short detection distance; and fourth, in suitable habitat it can achieve relatively high densities. Similarly, the high population estimate for the second most abundant species, the Horned Lark, was also surprising. However, Horned Larks were highly associated with cropland, the most abundant habitat type in Minnesota, and their effective detection distance is relatively short. Find a field of young corn or soybeans in Minnesota and you are bound to find multiple pairs of breeding Horned Larks. Comprising more than 75 percent of the landscape in one-third of Minnesota's 87 counties (Gunderson et al. 2019), croplands provide a bounty of nesting opportunities for species that find this rather inhospitable habitat suitable. Indeed, the Horned Lark excels in this regard.

In general, most of the species that have the highest estimated populations in the state are small species with associated small breeding territories; are frequently encountered; and occur in many of the most prevalent habitat types. Among the larger species in the top 25 are the Red-winged Blackbird, American Robin, Gray Catbird, and Baltimore Oriole. It is likely no surprise to any observer of birds that the Red-winged Blackbird and American Robin would have high populations. The American Robin was the most frequently recorded of all the bird species identified; individuals were recorded in over 58 percent (1,353) of the 2,314 randomly located point counts (Table 1.3; Appendix C). The next five most frequent species recorded were the Common Yellowthroat (1,168, 50.5 percent), Red-winged Blackbird (1,127, 48.7 percent), Song Sparrow (1,066, 46.1 percent), American Crow (988, 42.7 percent), and American Goldfinch (872, 37.7 percent).

Unfortunately, we cannot dismiss the potential roadside bias in the population estimates and frequency calculations. Even though larger, busy roads such as interstates, US, and state highways were avoided for sampling, complete coverage of a state the size of Minnesota was impossible without using the existing road networks. The estimates for Common Yellowthroat, Horned Lark, Song

Table 1.2 Breeding categories, codes, and evidence descriptions used by the MNBBA, 2009–13

CATEGORY	CODE	EVIDENCE	CATEGORY	CODE	EVIDENCE
Observed	O	Species observed in a block within safe dates, but no evidence of breeding. Use this code for species not in suitable nesting habitat. For example, this code applies to raptors or gulls flying over, herons and other colonial nesters foraging with no observed rookery, ducks summering on a pond, or pelicans loafing. Check Evidence Reporting Tips for other examples if you have questions.	Confirmed	CN	Bird seen carrying nesting material, such as sticks, grass, mud, cobwebs, etc.
Possible	X	Species encountered in suitable nesting habitat within safe dates, such as a Virginia Rail in a marsh. Use this code when you hear a male singing in a likely nesting area on only one occasion. For raptors, "X" is not a reliable code owing to extensive home ranges; use "O." Also, use caution with northern irruptive owls (Northern Hawk Owl, Great Gray Owl, Boreal Owl). Check Evidence Reporting Tips if you have questions.	Confirmed	NB	Nest building seen at the actual nest site, including wrens and woodpeckers.
Probable	M	Multiple (five or more) singing or territorial birds of a species detected in suitable habitat within the block on the same day. Most species listed as "M" can be upgraded during a later visit. Check Evidence Reporting Tips for species-specific examples.	Confirmed	DD	Distraction displays, defense of unknown nest or young, or injury feigning. Killdeers may give a "broken wing" act; a Red-winged Blackbird may dive at you near the nest site; or an Ovenbird may run about with wings fluttering.
Probable	S	Singing male in suitable habitat at same location at least five or more days apart. Be aware that some species do not sing to demonstrate territoriality or to attract a mate. Check Evidence Reporting Tips for species-specific examples.	Confirmed	UN	Used nest or eggshells found. Unless carefully identified, use this code only for unmistakable eggshells and nests that were used during the atlas period.
Probable	P	Pair observed in suitable nesting habitat. Geese, swans, ducks: "P" is not reliable as breeding evidence; code as "X." For more information, check the Evidence Reporting Tips.	Confirmed	ON	Occupied nest indicated by adult entering or leaving nest site in circumstances indicating an occupied nest, including those in high trees, cliffs, cavities, and chimneys where the contents of the nest and incubating or brooding adult cannot be seen.
Probable	T	Permanent territory presumed through defense of breeding territory by fighting or chasing individuals of same species. Geese, swans, ducks: "T" is not reliable as breeding evidence; code as "X." For more information, check the Evidence Reporting Tips.	Confirmed	FY	Adult bird carrying food for young or feeding long-tail young. This code includes this year's birds with adults (e.g., duck broods not full grown). Be aware that some adults carry food a long distance or may be engaged in courtship feeding.
Probable	C	Courtship or copulation behavior, including displays or transfer of food. Geese, swan, ducks: "C" is not reliable as breeding evidence; code as "X." For more information, check the Evidence Reporting Tips.	Confirmed	FL	Recently fledged short-tail young or downy young. This code includes dependent young only. Be cautious of species that range widely soon after fledging.
Probable	N	Bird observed visiting the same likely nest site repeatedly, but insufficient behavior for upgrading to Confirmed. Especially useful for cavity nesters or for a shrub-nesting species that flies into the same thicket and disappears several times.	Confirmed	FS	Adult bird seen carrying fecal sac. Many passerine adults keep their nests clean by carrying membranous, white fecal sacs away from the nest.
Probable	A	Agitated behavior or anxiety calls from adults usually indicate a nest site or young in the vicinity.	Confirmed	NE	Nest with eggs.
Confirmed	PE	Physiological evidence of breeding based on bird in the hand, such as a highly vascularized swollen incubation (brood) patch or an egg in the oviduct. For bird-banders only, using expert judgment.	Confirmed	NY	Nest with young seen or heard.

Note: Tips on the use of these evidence codes can be found at https://mnbirdatlas.org/wp-content/uploads/2016/11/BreedingEvidenceCodes_Tips.pdf.

Table 1.3 Median population estimates of breeding adults for the 25 most abundant bird species in Minnesota

BIRD SPECIES	NUMBER OF BREEDING ADULTS[a]	NUMBER OF DETECTIONS[b]	% DETECTED[c]	BIRD SPECIES	NUMBER OF BREEDING ADULTS[a]	NUMBER OF DETECTIONS[b]	% DETECTED[c]
Common Yellowthroat	14,700,000	1,168	50.5	House Wren	4,900,000	524	22.6
Horned Lark	12,200,000	518	22.4	Sedge Wren	4,750,000	238	10.3
Song Sparrow	10,400,000	1,066	46.1	Chipping Sparrow	4,500,000	661	28.6
Nashville Warbler	9,210,000	416	18.0	American Goldfinch	4,430,000	872	37.7
Red-winged Blackbird	8,850,000	1,127	48.7	Swamp Sparrow	3,740,000	265	11.5
Red-eyed Vireo	8,540,000	861	37.2	Least Flycatcher	3,580,000	279	12.1
Yellow Warbler	8,100,000	438	18.9	American Robin	3,520,000	1,353	58.5
American Redstart	7,730,000	389	16.8	Vesper Sparrow	3,510,000	491	21.2
Savannah Sparrow	6,850,000	470	20.3	Clay-colored Sparrow	3,370,000	374	16.2
Bobolink	6,510,000	326	14.1	Gray Catbird	3,230,000	312	13.5
Chestnut-sided Warbler	6,140,000	420	18.2	Baltimore Oriole	3,230,000	211	9.1
Black-and-white Warbler	5,620,000	292	12.6	White-throated Sparrow	2,820,000	417	18.0
Ovenbird	5,590,000	615	26.6				

[a] Median population estimates (ranked highest to lowest), calculated from point count data gathered during the Minnesota Breeding Bird Atlas.
[b] The number of times a species was detected on the first randomly selected point counts (n = 2,314).
[c] The percentage (frequency) of detections on the 2,314 first randomly selected point counts.

Sparrow, Red-winged Blackbird, American Robin, and Vesper Sparrow all suggest a potential roadside bias, because these species are frequently detected in roadside ditches, along habitat edges, on telephone or power lines, and frequently on fences or fence posts. In contrast, many of the species on the top 25 list—such as the Nashville Warbler, Red-eyed Vireo, American Redstart, Savannah Sparrow, Chestnut-sided Warbler, Black-and-white Warbler, Ovenbird, Swamp Sparrow, Least Flycatcher, Clay-colored Sparrow, Gray Catbird, or White-throated Sparrow—have little association with roads. Perhaps in the future, with drones, digital audio recorders, or other new technology, we can achieve a better random sample of our Minnesota breeding birds.

MNBBA Highlights

In addition to the data summaries mentioned above, atlas participants are often asked, "What did we learn about Minnesota's breeding birds? In particular, what surprised us, what didn't we know before, and how have things changed in the past 150 years?" The following is a brief sample of the many MNBBA highlights and outcomes.

- ***A total of 250 bird species was detected during the MNBBA, and 231 were confirmed nesting.*** The total number of species detected in Minnesota is similar to the number that was documented in many neighboring states during their first atlas (Iowa, 199; Michigan, 233; South Dakota, 219; Wisconsin, 237). Neighboring provinces, which are much larger in size, detected significantly more species (Manitoba, 312; Ontario, 292).

  Among the 250 species detected were 243 species classified as Regular in Minnesota by the Minnesota Ornithologists' Union (present at least nine years in the past ten years either as migrants, winter visitants, or breeding species); 6 species classified as Casual (present in three to eight of the past ten years); and one Extirpated species, the Northern Bobwhite (MOU 2020). Although the latter was formerly a permanent resident in southern Minnesota, ornithologists believe that a natural Bobwhite population no longer exists in the state. Birds that are seen or heard are considered to be intentionally released or escaped birds from game farms.

Excluding species that were formerly present in Minnesota but are now extinct (Passenger Pigeon) and extirpated species that are gone from Minnesota but still present elsewhere (Northern Bobwhite and Eskimo Curlew), a total of 443 species has been reported in the state as of 2020 (MOU 2020). A large number of these are vagrants, migrants, and winter visitors.

- ***At least two 9-square-mile (2,331 ha) atlas blocks each reported 101 potentially nesting bird species: one in central Aitkin County and one along the Minnesota River bordering both Lac qui Parle and Big Stone Counties.***

As discussed in chapter 2, "Biogeography of Minnesota Birds," Minnesota is uniquely situated within four different ecological provinces (the Eastern Broadleaf Forest, the Prairie Parkland, the Laurentian Mixed Forest, and the Tallgrass Aspen Parklands), providing a rich matrix of vegetation types that attract a wide diversity of birds. Located in the Laurentian Mixed Forest, the Aitkin County block encompassed lakes, streams, extensive lowland and upland forests, and a few small farmsteads. Among the 101 bird species were 7 waterfowl, 3 upland game birds, 7 woodpeckers, 5 flycatchers, 5 swallows, 8 sparrows, and 9 wood warblers. The block along the Minnesota River contained the river's floodplain, numerous back channels, outcrops, and grasslands. Located partially within the Lac qui Parle Wildlife Management Area, the block was devoid of development. Among the 101 species reported there were 10 waterfowl, 4 upland game birds, 5 shorebirds, 6 flycatchers, 6 swallows, 9 sparrows, 9 blackbirds, and 4 wood warblers. No doubt a concerted effort by atlas participants in these blocks contributed to these impressive results.

- ***Data gathered during the MNBBA proved the effectiveness of conservation efforts and the ultimate recovery of the Bald Eagle, Osprey, Peregrine Falcon, and Trumpeter Swan.***

When Roberts wrote his definitive work on Minnesota birds in 1932, three of these species had declined precipitously; the settler's gun had taken its toll on each one. Roberts would be astonished to learn that less than one hundred years later MNBBA participants would report 2,113 Bald Eagle records in all but two of Minnesota's 87 counties (Lincoln and Pipestone), 938 Osprey records in 49 counties, and 1,173 Trumpeter Swan records in 72 counties!

The Peregrine Falcon's demise was to come along decades after Roberts's publication, coinciding instead with the widespread use of organochlorine pesticides, especially DDT. By the late 1960s, the Falcon was rarely seen in the state. Aggressive reintroduction efforts that began in the 1980s would eventually lead to a total of 132 atlas records from 25 counties, primarily in northeastern and southeastern Minnesota.

- ***New breeding sites were documented for rare species like the Bufflehead and Common Tern.***

Many factors can influence a bird's selection of breeding habitat each summer, including weather, disturbance, and prey abundance. The result may lead to some unexpected observations. During the atlas, a Bufflehead pair was confirmed breeding in Cottonwood County, in a small wetland more than 560 km south of the species' long-established breeding range in northwestern Minnesota. And, in the heart of the Brainerd Lakes Area, observers were surprised to find a small Common Tern colony on Pelican Lake in Crow Wing County. Prior to its detection, the majority of the state's tern population was confined to four relatively remote sites. New discoveries are always possible for those who explore off the beaten track!

- ***Further details were provided on the range expansion of species that are relatively new to Minnesota, such as the Eurasian Collared-Dove and Great-tailed Grackle.***

After it first appeared in Florida in 1986, the non-native Eurasian Collared-Dove quickly spread across the continent, moving in a northwesterly direction. The first documented report in Minnesota was only twelve years later, in 1998; the first confirmed breeding occurred in Houston County in 2001. When the atlas was completed in 2013, 133 Collared-Dove records were reported from 44 Minnesota counties, stretching as far north as Polk County in northwestern Minnesota and Itasca County in the north-central region. Nesting was confirmed in 10 counties.

The Great-tailed Grackle also entered Minnesota from the south. Formerly native to Mexico and Central America, the species was found only as far north as southern Texas by 1900. Decades passed before it was first reported in Minnesota in 1982. The Grackle's rate of spread has been considerably slower than that of the Dove's. It was the summer of 2000 before the state's first nesting record was documented in Jackson County, and by the end of the atlas only 11 records were reported from five counties. Jackson County remained the only known nesting site.

- ***Evidence documented the continued range expansion and growing abundance of many breeding species, such as the Wild Turkey, Sandhill Crane, American White Pelican, Red-bellied Woodpecker, Blue-gray Gnatcatcher, and Northern Cardinal.***

Each of these species has a unique story, but all were very uncommon to absent in Minnesota approximately one hundred years ago (Roberts 1932). Since then, they have expanded far beyond even Roberts's expectations. The Wild Turkey has proven to be an especially adaptable species. Although their status as a native Minnesota species has been disputed, Turkeys found Minnesota's diverse landscapes to their liking following successful introduction efforts in the 1970s. During the atlas, they were reported in 79 Minnesota counties stretching as far north as Itasca, Pennington, and St. Louis.

On the other hand, both the Sandhill Crane and the American White Pelican were abundant in the 1800s but quite rare by the end of the nineteenth century. Hunting had taken its toll on these large, conspicuous birds. Once hunting regulations were enacted and enforced and habitat conservation efforts were underway, populations of both species began to rebound. During the atlas, Sandhill Cranes were reported in 66 counties with breeding confirmed in 52 counties; American White Pelicans were confirmed breeding in 11 counties.

The stories of the Red-bellied Woodpecker, Blue-gray Gnatcatcher, and Northern Cardinal are markedly different than those described above. All three species expanded into Minnesota by following the Mississippi River valley northward from the core of their more southern breeding distributions. First documented in the state in the late 1800s, by 1932 all three had reached Minneapolis but remained most common from Red Wing south to the Iowa border. By the end of the atlas, the Red-bellied Woodpecker had been documented in 80 counties, the Blue-gray Gnatcatcher in 54 counties, and the Northern Cardinal in 79 counties!

- ***Evidence revealed the decline of several secretive, nocturnal species, such as the Eastern Whip-poor-will and Eastern Screech-Owl.***

Poorly monitored because of their crepuscular and nocturnal habits, these once abundant species are quite uncommon today. As recently as the 1970s and 1980s, the Eastern Whip-poor-will was considered abundant along the Mississippi River and in the Anoka Sand Plain. Atlas participants, however, reported the species in only 16 blocks in these two regions, compared to a total of 139 blocks in northern counties. And the little Eastern Screech-Owl, once considered by Roberts to be the most abundant owl in Minnesota, was outnumbered by the Barred Owl, Great Horned Owl, and Northern Saw-whet Owl during the atlas. Although the large Barred Owl and Great Horned Owl are considerably more conspicuous and easier to detect, the Northern Saw-whet Owl is actually a little smaller than the Screech-Owl.

- ***Significant population declines of the Common Nighthawk, Black Tern, Purple Martin, and Bobolink were documented, even though their statewide distribution has changed very little in the past one hundred years.***

Sometimes, species are so widespread that their decline is not easily noted unless biologists specifically document their abundance. To assess how populations have changed since the 1800s we have only the words of early explorers, naturalists, and ornithologists for comparison. Roberts (1932), for example, described the Black Tern as more abundant and widely distributed in Minnesota than any other waterbird. Although atlas participants confirmed its wide distribution (present in 72 of Minnesota's 87 counties), their records also illustrate that it is no longer the most common waterbird. Indeed, the Ring-billed Gull, Common Loon, Double-crested Cormorant, American White Pelican, Great Blue Heron, Great Egret, and Green Heron were each detected in more atlas blocks than the Black Tern.

Fortunately, biologists now have a more accurate tool for detecting changes in population abundance. Since 1966, volunteers have been collecting bird population data along 4,100 roadside survey routes, each 40 km (25 miles) in length. Known as the North American Breeding Bird Survey (commonly known as the BBS), the data collected by these volunteers are included in the accounts that follow and also document significant population declines for all four of these species, as well as for many more that regularly breed in Minnesota.

- ***The perilous states of the Black-crowned Night-Heron, Piping Plover, Sharp-tailed Grouse, Loggerhead Shrike, Cerulean Warbler, and Connecticut Warbler were reported.*** In the 1970s, there was widespread concern for the future of two well-known Minnesota species, the Bald Eagle and Peregrine Falcon. Both were subsequently listed under the State and Federal Endangered Species Acts; both, as noted above, have since experienced spectacular recoveries.

Unlike these well-known and charismatic raptors, smaller and less conspicuous species now dominate the concerns of state natural resource biologists. Although the diminutive Piping Plover was state listed in 1984 and federally listed in 1985, MNBBA results reinforced the need to draw attention to this and many other species in decline. The Plover, for example, was reported in only two of the 6,987 atlas blocks visited; both were located on the remote shores of Lake of the Woods. Not only does this outcome reinforce the federal status of the species, it clearly suggests that its future as a Minnesota breeding bird is in doubt.

The outlook for the Black-crowned Night-Heron is also troubling. Roberts (1932) described the "numerous large colonies of breeding birds" located throughout the "prairie and semiprairie portions of the state." Restricted primarily to southern Minnesota, what little data was available in the late 1900s suggested numbers were declining. Indeed, atlas results suggest that the toll on this crepuscular species may be much greater than natural resource managers understood and warrants close attention. The same is true for the Loggerhead Shrike. Nearly one-quarter (8) of all atlas records (33) were restricted to one small area in Dakota County, leaving biologists uncertain about the factors driving its decline. Similar stories appear in the following pages for many species, including the Cerulean Warbler and Connecticut Warbler.

- ***The devastating impacts of the loss of over 98 percent of Minnesota's native prairies were registered for many grassland birds, including the Chestnut-collared Longspur, Grasshopper Sparrow, and Sprague's Pipit.*** One can only imagine the vista of tall prairie grasses that once stretched for miles in all directions, originally covering nearly 20 million acres of western Minnesota. As Roberts (1932) noted during his 1879 trip to Grant and Traverse Counties, the "whole region teemed with nesting birds." Waterfowl and waterbirds filled the marshes and lakes, while the calls and songs of Greater Prairie-Chickens, Short-eared Owls, Upland Sandpipers, Marbled Godwits, and grassland sparrows rang across the landscape. As the virgin prairies were lost to the plow, many species retreated to the few remaining tracts, slowly disappearing as the tracts became too small and fragmented to support sustainable populations. Among the victims were the Chestnut-collared Longspur, Sprague's Pipit, Thick-billed Longspur, and Baird's Sparrow. Some are now gone entirely; others are barely hanging on. Others still, like the Grasshopper Sparrow and Upland Sandpiper, were able to adapt to restored grasslands and agriculture's grazed pastures and hayfields, albeit at reduced numbers, as the atlas results illustrate.

- ***Participants provided strong evidence of the southern movements of two species, the Merlin and Common Raven.***
When most Minnesota birds extend their breeding range within the state, they expand from the southern regions northward. This was especially true for many species found in the state's deciduous woodlands. As the pineries and spruce-fir forests of the northern boreal region were logged, they were replaced by deciduous stands of aspen and birch that provided suitable habitat for many of the deciduous-loving southern species, including the Wood Thrush, Yellow-throated Vireo, and White-breasted Nuthatch. In the future, even more species are predicted to move northward in the state as temperatures continue to rise and change the vegetative composition of our forests (NAS 2019).

Two notable species, however, have moved southward. As recently as the 1970s, Merlins were largely confined to the northern counties bordering Canada. Forty years later, atlas participants observed the birds as far south as Wabasha and Winona Counties and confirmed breeding in four metropolitan counties: Dakota, Hennepin, Ramsey, and Washington. The Merlin's recent adaptability to highly urbanized areas has been noted throughout the Great Lakes region.

Another northern species, the Common Raven, was so rare during Roberts's days that he feared it was nearing extirpation in the state. The Raven was known primarily as a winter visitant in northern Minnesota; neither Hatch (1892) nor Roberts cites any early evidence that it was more broadly distributed during the breeding season beyond the Canadian border. Indeed, it was 1929 before nesting was first documented in the state along the Rainy River (Roberts 1936). Cutright and colleagues (2006), however, note that Ravens were widespread throughout Wisconsin at the time of European settlement, and Brewer and colleagues (1991) cite nesting populations in the southern Lower Peninsula of Michigan in the late 1800s, suggesting that the species may have occurred farther south in Minnesota as well. Unfortunately, during those days many Ravens were shot, poisoned, or unintentionally caught in wolf traps. Slowly, however, the species has increased in abundance and spread both to the west and south throughout the Great Lakes region. In Minnesota, by the end of the twentieth century, nesting was reported as far south as Isanti County and as far west as Marshall County. During the MNBBA, breeding was confirmed even farther south, including Becker and Norman Counties in the west, Morrison and Todd Counties in central Minnesota, and Washington County in the east. A variety of factors are likely contributing either to its recovery or to its range expansion, including greater protections and more feeding and nesting opportunities.

- ***MNBBA registered the gradual reoccupation of southern Minnesota by species formerly documented in the region but gone for many years, including the Common Loon, Northern Harrier, and Clay-colored Sparrow.***
The rich prairie soils of southern Minnesota were among the first to be converted by European settlers into croplands, hayfields, and pastures. Ditches also crisscrossed the landscape, draining streams, wetlands, and shallow lakes. Birds that originally flourished in the region were bound to be impacted. Early accounts indicate that many species, such as the Common Loon, Northern Harrier, and Clay-colored Sparrow, were among those whose breeding ranges in the state retreated northward. Only recently, as conservation efforts have helped to improve water quality, restore wetlands, and establish grassland cover, have some of these species begun to reoccupy the southern counties. The Northern Harrier, for example, had essentially disappeared from southern Minnesota by the 1970s. Its gradual reoccupation of the region began in the 1980s, and by 1998 it was found breeding in five counties south of the Minnesota River (Hertzel and Janssen 1998). By the time the atlas was completed in 2013, Harriers had been reported breeding in nine counties south of the river and were reported in all 87 Minnesota counties except Watonwan in south-central Minnesota.

- ***Population analyses revealed that populations of many Minnesota birds are much larger than previous estimates suggested, including the Winter Wren, Golden-winged Warbler, Wood Thrush, Clay-colored Sparrow, and Scarlet Tanager.*** One key to assessing conservation priorities among North America's rich diversity of bird species is to accurately calculate their population size. Since the early 2000s, Partners in Flight (PIF) biologists have relied on abundance data collected by the North American Breeding Bird Survey (BBS) to assess population abundance. Although these data are helpful in many regards, the number of BBS routes in any given state is relatively low, and those that are established only reliably survey road-accessible areas. Because the MNBBA collected abundance data across the state at a much finer scale, including in areas that were inaccessible to BBS's roadside surveys, MNBBA research biologists were able to establish more accurate estimates. Not surprisingly, these were often orders of magnitude larger than those established by PIF. For example, PIF (2021) estimated Minnesota's Winter Wren population to number 180,000 breeding adults, while the MNBBA estimated a state population of 588,000 adults. PIF estimated the Minnesota Clay-colored Sparrow population to number 1.9 million, while the MNBBA estimate was 3.4 million!

- ***The ecological importance of vast wetlands in northwestern Minnesota to numerous wetland-dependent species was documented, including the Bufflehead, American Wigeon, American Bittern, Yellow Rail, Alder Flycatcher, and Swamp Sparrow.*** Although many Minnesotans consider the Boundary Waters Canoe Area Wilderness to be the state's premier wilderness area, natural resource biologists recognize that the vast peatlands of north-central Minnesota and the expansive wetlands in the far northwestern counties are among the state's most unique and inaccessible terrains. Indeed, they certainly are not areas that birders visit on a regular basis. This illustrates one of the most valuable aspects of the Breeding Bird Atlas. It is based on the premise that every corner of the state is surveyed, no matter how difficult the challenge. Atlas surveys conducted in these remote regions revealed how significant they are to a host of wetland species, not only from a Minnesota perspective but nationally as well. The critical data collected in these wild areas is a tribute to many dedicated participants who trekked through these remote and difficult landscapes.

These highlights are just a sample of the many tales of Minnesota birds that can be found in the species accounts. In these stories, the reader will find the first comprehensive account of the status and distribution of species that breed in Minnesota since the 1932 publication and 1936 revised edition of Roberts's tome, *The Birds of Minnesota*. Not only do these new accounts look to the past to establish the historical record of each species in Minnesota but they also look forward and include predictions about the potential impacts of current and future threats, including climate change, continued habitat loss, interspecific competition, and declines in prey abundance. As readers explore the following pages, we hope that they learn more about Minnesota's surprisingly rich and diverse community of breeding birds, their unique stories, and the importance of their conservation.

< Black-capped Chickadee.
Photo by David Brislance.

2 BIOGEOGRAPHY OF MINNESOTA BIRDS

< The matrix of grasslands, wetlands, and woody habitat on the Kandi WMA in Kandiyohi County is typical of Minnesota's prairie forest border. Photo by Lee A. Pfannmuller.

Minnesotans are fortunate to have such a wealth of bird species. This is due to many factors, but most notable is the smorgasbord of habitats that occur in the state. We have many different kinds of forests, wetlands, grasslands, and prairies, all of which contribute to this magnificent species diversity. These habitats were formed long ago as a result of the receding glaciers and the subsequent recolonization by plants, and they have been continually altered by natural disturbances such as fire, wind, and insects. Many habitats have been permanently altered by human activity, such as the conversion of land and wetlands to agriculture, human settlements, industrial sites, and recreational activities (Table 2.1). These changes are well illustrated by comparing Marschner's map, which documented the vegetation around the time of European settlement in 1890, with the vegetation map of 1990 (Figure 2.1). For instance, from 1890 to 1990, over 98 percent of the native prairie has been converted, primarily to agricultural land uses. In comparison, about 50 percent of open wetlands has been converted to other land uses, but only 19 percent of lowland conifer/shrubland has been lost.

Invasive species are also transforming the character of many habitats, for example, European buckthorn (*Rhamnus cathartica*) or non-native earthworms in forested areas, purple loosestrife (*Lythrum salicaria*) and *Phragmites* in wetlands, or zebra mussels (*Dreissena polymorpha*) in lakes. Finally, the warming climate is changing many of our habitats; deciduous vegetation is slowly colonizing Minnesota's northern coniferous forests; an increase in the number of forest fires is predicted to expand open habitats and further encroach on our forested areas; and aquatic areas like lakes, ponds, and wetlands are changing due to greater variability in precipitation patterns.

Our goal here is to describe and illustrate the current and general biogeographic patterns of Minnesota's habitats and associated avifauna. Minnesotans have a vast array of resources that provide

Table 2.1 Change in cover types between circa 1848–1907 and 1990

COVER TYPE	ACRES (CIRCA 1848–1907)	ACRES (1990)	% CHANGE
Cropland	0	23,981,079	100
Grassland	0	5,109,924	100
Developed	0	599,675	100
Open wetland	4,163,031	2,074,773	−50
Lowland conifer/shrubland	6,639,649	5,350,747	−19
Prairie	15,677,426	27,632	−99
Upland shrub/woodland	6,383,580	1,031,659	−84
Upland deciduous (aspen–birch)	8,362,227	7,053,315	−16
Upland deciduous (hardwoods)	4,388,564	2,179,753	−50
Total acres	**48,774,203**	**49,073,973**	

Note: Dates are based on the General Land Office Survey (circa 1848–1907, depending on the region of the state) and the 1990 Gap Analysis Project (University of Minnesota Institute on the Environment 2008).

Total acreage amounts differ primarily because of increased mapping accuracy and/or change in the amount of open water area; open water acreages are not included in the cover type data.

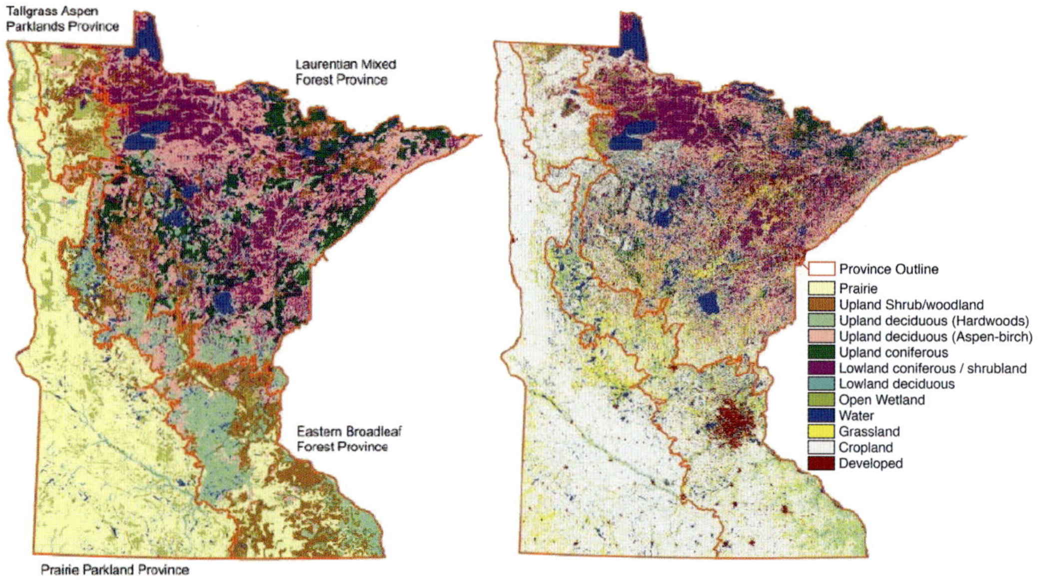

Figure 2.1. Marschner's map of vegetation around the time of European settlement and accompanying landcover based on 1990 GAP data. Reproduced from University of Minnesota Institute on the Environment 2008.

basic information on Minnesota. Examples include summaries of Minnesota's climate (Seeley 2015), geology (Ojakangas and Matsch 1982), natural heritage (Tester et al. 2020), soils (Cummins and Grigal 1981), landscapes (Hart and Ziegler 2008), and many others, especially with the wealth of additional information on the Internet. For a background summary on relevant information on Minnesota's birds, we suggest, first, the introductory chapters of Roberts (1932 or 1936); second, the chapter on "Geography and Ecology of Minnesota's Birdlife" (Green and Janssen 1975); third, the "Introduction" of Coffin and Pfannmuller (1988); and finally, the Introduction chapter on "The Unique Geography, Diverse Habitats, and Rich Avifauna of Minnesota" (Janssen 2019). For more detailed reading, the two wildlife action plans, *Tomorrow's Habitat for the Wild and Rare* (MNDNR 2006c) and the updated *Minnesota's Wildlife Action Plan, 2015–2025* (MNDNR 2015b), contain a wealth of information; in addition, Minnesota's Biological Survey provides substantial details on birds and conservation (https://www.dnr.state.mn.us/mbs).

Minnesota is situated in a unique part of North America. It is positioned in the center of this continent where four major ecological provinces meet: the Eastern Broadleaf Forest, the Prairie Parkland, the Laurentian Mixed Forest, and the Tallgrass Aspen Parklands (Figure 2.2; MNDNR 2019f). On the array of these provinces, we can overlay the remnants of the topography left by the receding glaciers and the complex hydrology of three major watersheds that flow to the south (Mississippi, Minnesota, and St. Croix Rivers), to the north (Red and Rainy Rivers), and to the east (Pigeon and St. Louis Rivers), which results in a tapestry of habitats that drives much of the resultant high breeding bird diversity of Minnesota.

Nothing exemplifies the high breeding bird diversity of Minnesota within the United States and Canada better than the species richness map that results from the thousands of volunteer North American Breeding Bird Survey (BBS) routes (Figure 2.3). Minnesota

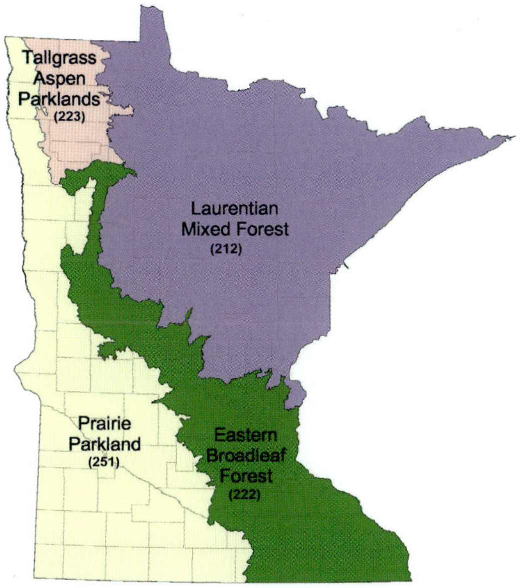

Figure 2.2. Minnesota's four ecological provinces, based on the Minnesota Department of Natural Resources Ecological Classification System (https://www.dnr.state.mn.us/ecs/index.html).

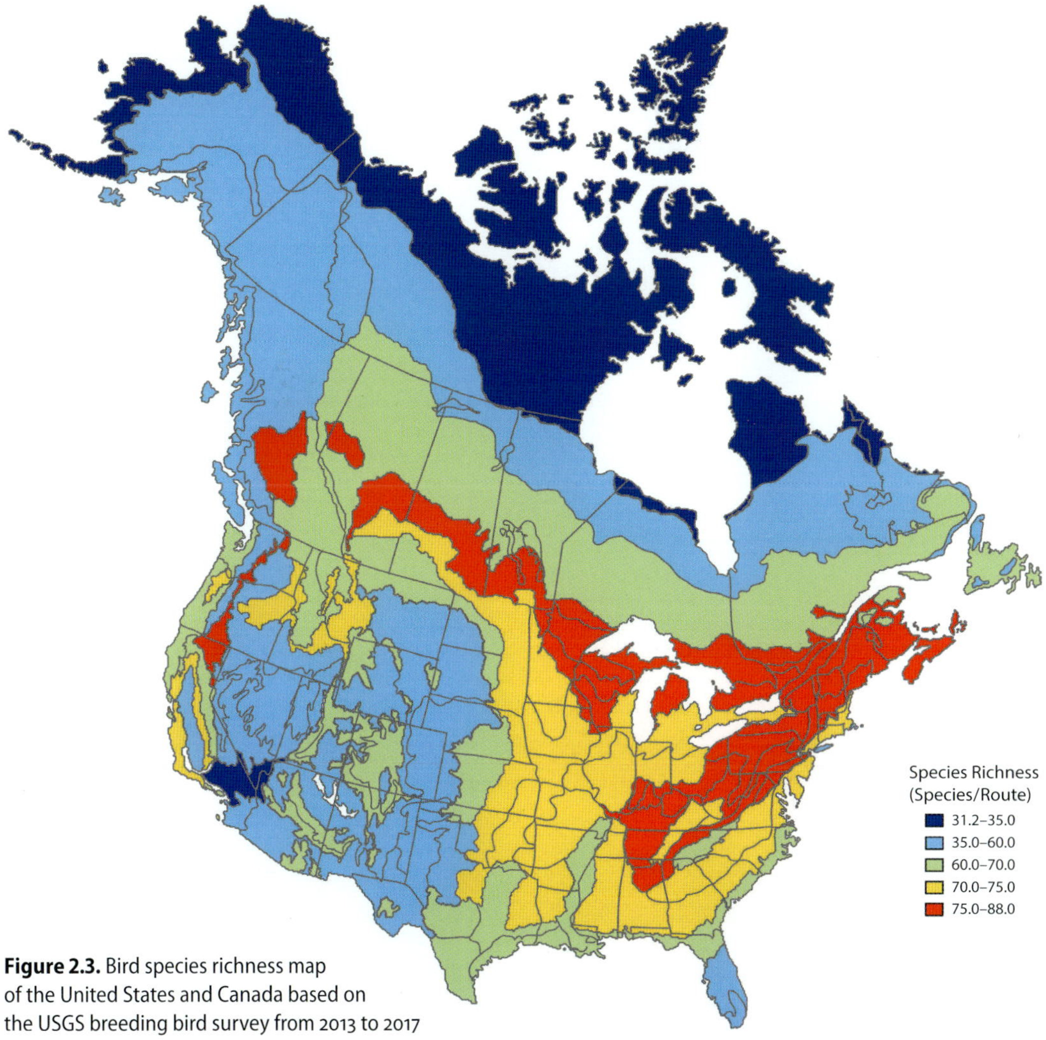

Figure 2.3. Bird species richness map of the United States and Canada based on the USGS breeding bird survey from 2013 to 2017 (Sauer et al. 2020). Species richness was calculated as the total number of species seen on each route between 2013 and 2017 within the strata defined in Robbins et al. (1986). Note that species richness values are higher than values presented in Robbins et al. (1986, 126) because their map was based on species richness for a single year (1979).

and many of the states and provinces surrounding the Laurentian Great Lakes are part of the region with the highest breeding bird diversity anywhere in the United States and Canada! Within Minnesota, the northeastern and north-central portions of the state fall within this zone of high species richness, where approximately 75–88 bird species on average are observed on BBS routes within this region (red area). Contrast this with the lower average numbers detected on the same type of routes in the central United States (70–75 bird species in orange) or a range of 35–70 species for most of California or Florida (blue, orange, and green colors).

Within each of the major biomes of Minnesota, the MNDNR (2019f) has developed an Ecological Classification System (ECS) that allows progressively smaller areas of land and water with similar physiographic features to be mapped at finer scales within the ecological provinces, called ecological subsections (Figure 2.4). Bird species that require similar habitats can be grouped together because of their affinities to common features such as climate, geology, hydrology, and topography. Both the ecological provinces and subsections were extensively used in describing the breeding distributions of species in "Current Breeding Species."

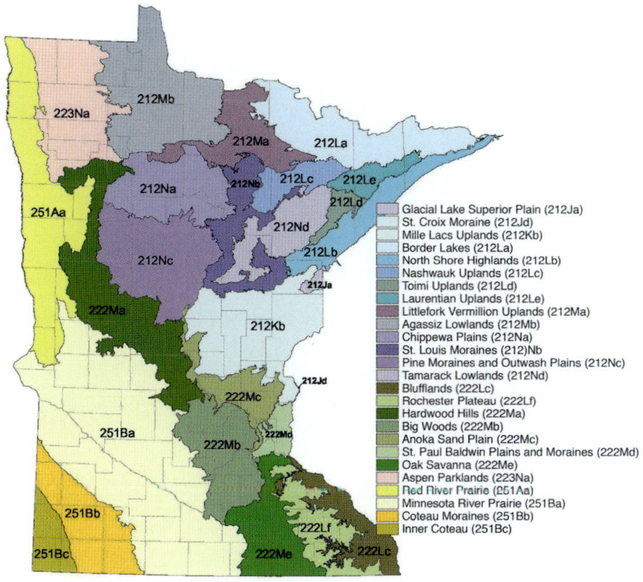

Figure 2.4. Minnesota's ecological subsections based on the Minnesota Department of Natural Resources Ecological Classification System (https://www.dnr.state.mn.us/ecs/index.html).

(A)   (B)

Figure 2.5. Examples of cropland habitats in Minnesota. **A.** The mix of crops, grass cover, and the fencerow is attractive to a variety of grassland species. The large trees around the farmstead in the upper right also provide habitat for many species like the Baltimore Oriole, American Robin, and Downy Woodpecker. **B.** Grassy ditches are important habitats used by birds such as the Gray Partridge, Ring-necked Pheasant, and Vesper Sparrow. Photos by Gerald J. Niemi.

Major Habitat Types in Minnesota

The state is composed of many different habitat types in addition to those delineated by the ECS and could include literally hundreds of different cover types depending on the detail needed. Here we summarize and illustrate 13 major habitat classes, each of which represents over 1 percent of the landcover in Minnesota. Collectively, these 13 cover types represent about 99 percent of the landcover in the state. We used the published literature and our collective knowledge to identify the common or characteristic birds that nest in these 13 cover types. These designations were also supported by the quantitative values (percent perfect indicator [PPI]) described in Appendix A and presented in Appendix D. The remaining cover types include many rare habitats that are important to selected species of birds (e.g., sandy-cobble beaches for the Piping Plover and Common Tern); those rare types are identified in the Breeding Habitat section of each species account. In addition, the entire state includes a substantial amount of open water, which is not included here, primarily because species do not nest in open water without any suitable substrate to support a nest. For more detailed information on these major habitat types, see Coffin and Pfannmuller (1988), Tester et al. (2020), or the *Field Guides to the Native Plant Communities of Minnesota* (MNDNR 2003a, 2005a, 2005b).

CROPLAND

Cropland habitat represents about 36 percent of Minnesota's landcover and is associated with Minnesota's substantial agricultural industry in the southern and western portions of the state (Figure 2.5, Figure 2.6). This habitat has replaced much of the native prairie found in the Prairie Parkland Province. The major crops include corn (*Zea mays*), soybeans (*Glycine max*), and wheat (*Triticum* spp.). The primary habitats for birds within these areas include orchards, shelterbelts, ditches, remnant patches of native prairie, small wetlands, and other non-crop areas.

Figure 2.6. Spatial distribution of cropland habitat in Minnesota.

Croplands are relatively depauperate in species diversity and abundance. Common or characteristic species of the cropland landscape include the ubiquitous Red-winged Blackbird and Common Grackle in wetlands, shrubby areas, and shelterbelts; the American Kestrel, Brewer's Blackbird, and Vesper Sparrow along

(A)

(B)

Figure 2.7. Grassland habitat examples in Minnesota. **A.** The Felton Prairie in western Minnesota, a native prairie and one of the primary breeding areas for the Chestnut-collared Longspur and Upland Sandpiper. **B.** A mesic grassland in Yellow Medicine County that provides habitat for a variety of birds, such as Brewer's Blackbird, Bobolink, Western Meadowlark, Sedge Wren, Ring-necked Pheasant, and Grasshopper Sparrow. Photos by Gerald J. Niemi (A) and Lee A. Pfannmuller (B).

roadsides; Brown Thrasher in shrubby habitats; Killdeer, Mourning Dove, Red-tailed Hawk, Ring-necked Pheasant, Rock Pigeon, Savannah Sparrow, Upland Sandpiper, and Western Meadowlark in a variety of open areas with moderate vegetation; and Horned Larks within the croplands. Many of them are easiest to observe on fences and fence posts or on power or telephone lines.

UPLAND GRASSLAND

Upland grassland habitat constitutes 11 percent of Minnesota's landcover; it comprises a mix of open areas dominated by grasses, herbaceous vegetation, or low shrubs, including restored grasslands, such as those enrolled in the Conservation Reserve Program, and the very small percentage of native prairie that remains (Figure 2.7). It includes sites that vary from dry upland grasslands to low-lying wet to wet-mesic sites. These habitats are widely scattered throughout Minnesota, except in the northeast, and are very prominent in the northwestern region within the Tallgrass Aspen Parklands Province (Figure 2.8). This habitat also includes many pastures, hayfields, and areas with scattered or recently burned or logged areas that are slow to regenerate.

The avifauna of these upland grassland habitats can be amazingly diverse, depending on the overall structure and diversity of the vegetation. Prominent bird species found in this habitat include the American Kestrel, Bobolink, Brewer's Blackbird, Brown Thrasher, Clay-colored Sparrow, Eastern Bluebird, Eastern Kingbird, Eastern Meadowlark (east primarily), Mourning Dove, Red-winged Blackbird, Ring-necked Pheasant, Sandhill Crane, Savannah Sparrow, Sedge Wren, Song Sparrow, Tree Swallow, Warbling Vireo (where trees are present), and Western Meadowlark. Especially interesting species that also occur include the Black-billed Magpie (near farmsteads in the north-central and northwest regions), Dickcissel, Grasshopper Sparrow, LeConte's Sparrow, Orchard Oriole (with shrubs and trees), Sharp-tailed Grouse, and Upland Sandpiper.

Figure 2.8. Spatial distribution of upland grassland habitat in Minnesota. The dark area in northeastern Minnesota near the Canadian border represents a large recent burn area.

NORTHERN HARDWOODS

This forested habitat comprises 8 percent of Minnesota's landcover and is primarily found in the central regions of the state within the Eastern Broadleaf Forest and

(A) (B)

Figure 2.9. Examples of the northern hardwood forests of Minnesota. **A.** Large trees from a hardwood forest in Lake Maria State Park provide breeding habitat for many forest-dwelling species, such as the Yellow-throated Vireo, Least Flycatcher, White-breasted Nuthatch, and many species of woodpeckers, while the open understory is preferred by the Wood Thrush and Ovenbird. **B.** A typical aspen forest in northern Minnesota with a complex understory that is an excellent breeding habitat for many deciduous tree-dwelling species, such as the Yellow-bellied Sapsucker, Veery, and Red-eyed Vireo. Photos by Lee A. Pfannmuller (A) and Gerald J. Niemi (B).

Figure 2.10. Spatial distribution of the northern hardwood forest habitat in Minnesota.

the Laurentian Mixed Forest Provinces (Figure 2.9, Figure 2.10). These are closed-canopy forests dominated by deciduous trees such as sugar maple (*Acer saccharum*), basswood (*Tilia americana*), red (*Quercus rubra*), bur (*Q. macrocarpa*), or white oak (*Q. alba*), and green ash (*Fraxinus pennsylvanica*) in the more southern regions where it occurs and aspen (*Populus* spp.), birch (*Betula* spp.), or black ash (*Fraxinus nigra*) in the more northern regions; the latter are often mixed with coniferous trees such as balsam fir (*Abies balsamea*), pine (*Pinus* spp.), and white spruce (*Picea glauca*). The habitat type can consist of many different age classes, depending on the disturbance history of a stand and whether it has been logged or disturbed in other ways such as by wind. In addition, lowland hardwood forests are included within this forest cover type because they are often difficult to distinguish from upland forests with remote sensing data. These forests have also been subjected to substantial fragmentation owing to conversion to agriculture, urbanization, exurbanization, or other forms of parcelization.

Depending on the diversity and structure of the vegetation within these forests, bird species diversity can be substantial. The most abundant species within this forest type include the American Redstart, Downy Woodpecker, Hairy Woodpecker, Least Flycatcher, Ovenbird, Pileated Woodpecker, Red-eyed Vireo, Veery, White-breasted Nuthatch, Wood Thrush, Yellow-bellied Sapsucker, and Yellow-throated Vireo. However, almost any of the bird species found in the oak forest, the northern mixed forest, or pine forest habitats not specifically mentioned here could also be found in the northern hardwoods forest.

NORTHERN MIXED FORESTS

The northern mixed forest habitat is similar to the northern hardwoods forest, but has a greater coniferous tree component (Figure 2.11). These habitats are largely associated with the Laurentian Mixed Forest Province in the northeastern and north-central regions of the state and constitute 7.7 percent of Minnesota's landcover (Figure 2.12). These forests occur in a variety of age classes, from regenerating young forests to mature old growth, and are a major focus

(A) (B)

Figure 2.11. Examples of northern mixed forest habitat in Minnesota have a mixture of both deciduous and coniferous tree species and varied understories. **A.** A hint of damage by spruce budworm is shown on the right along with a mix of canopy tree species. **B.** A deciduous canopy dominates, with extensive conifers in the understory. Mixed forests typically have a high diversity and density of birds with affinities to both deciduous and coniferous tree species, such as the common Red-eyed Vireo and Ovenbird, and also the Broad-winged Hawk, Rose-breasted Grosbeak, Ruffed Grouse, Purple Finch, Black-throated Green Warbler, and White-throated Sparrow. Photos by Gerald J. Niemi.

Figure 2.12. Spatial distribution of northern mixed forest habitat in Minnesota.

of the timber industry. Hence, many stands are second-growth forests that have been logged or subjected to other disturbances such as fire, wind, or insects. They are largely dominated by aspen and birch, but red maple (*Acer rubrum*), white spruce, balsam fir, and white (*Pinus strobus*) or red pine (*P. resinosa*) frequently occur as well. Lowland deciduous forests dominated by black ash are also frequently combined with this forest type. The understory often has a substantial shrub layer, including hazel, dogwood, and many other shrub species or saplings of the canopy trees.

Many of the prominent bird species found in the northern hardwoods forests are also found in the northern mixed forest habitat. This is especially true of abundant species like the Ovenbird, Red-eyed Vireo, Veery, and Yellow-bellied Sapsucker. Other common species in this habitat include the Black-throated Green Warbler, Broad-winged Hawk, Purple Finch, Rose-breasted Grosbeak, Ruffed Grouse, and White-throated Sparrow.

LOWLAND CONIFEROUS FOREST

The lowland coniferous forest is a habitat that occurs in wet or occasionally in transitions between wet to dry places such as in riparian areas (Figure 2.13). This cover type is found exclusively in the northeastern and north-central parts of Minnesota and represents 7.3 percent of Minnesota's landcover, especially in the Agassiz Lowlands and Tamarack Lowlands Subsections (Figure 2.14). Dominant trees are black spruce (*Picea mariana*), tamarack (*Larix laricina*), and white cedar (*Thuja occidentalis*), but they are often mixed with balsam fir and black ash. This habitat type includes many age classes, including old-growth stands that can exceed 100–200 years in age.

Some of the most distinctive species of the northern forested regions of Minnesota are almost exclusively found in lowland coniferous forest. These include the Boreal Chickadee, Canada Jay, Connecticut Warbler, and Yellow-bellied Flycatcher. Other species include the Bay-breasted Warbler, Dark-eyed Junco, Golden-crowned Kinglet, Hermit Thrush, Nashville Warbler, Purple Finch, Ruby-crowned Kinglet, Swainson's Thrush, White-throated Sparrow, Winter Wren, and Yellow-rumped Warbler.

Figure 2.13. An example of a typical lowland coniferous forest in Minnesota with a mixture of black spruce (*Picea mariana*) and tamarack (*Larix laricina*) trees and a thick ground layer of *Sphagnum* moss and ericaceous heath vegetation. Lowland coniferous forests are critical habitats for many bird species that are primarily found in these forests such as the Boreal Chickadee, Connecticut Warbler, and Yellow-bellied Flycatcher, but are also important for the Winter Wren, Dark-eyed Junco, Golden and Ruby-crowned Kinglets, Hermit and Swainson's Thrushes, and Nashville Warbler. Photo by Gerald J. Niemi.

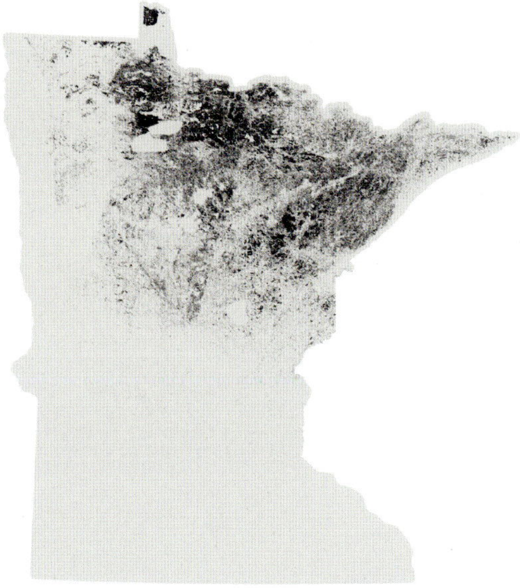

Figure 2.14. Spatial distribution of lowland coniferous forest in Minnesota.

Figure 2.15. Bog habitats in Minnesota are dominated by a ground layer of *Sphagnum* moss and ericaceous heath vegetation, along with a highly variable tree layer, from no trees to low-statured, stunted black spruce (*Picea mariana*) and tamarack (*Larix laricina*). The avifauna of bogs have few species and low densities, but nevertheless can be interesting, especially because of the presence of species such as Sharp-tailed Grouse, Short-eared Owl, Lincoln's Sparrow, and Palm Warbler. Photo by Gerald J. Niemi.

Figure 2.16. Spatial distribution of bog habitat in Minnesota.

BOG

Substantial areas of open bog or muskeg occur in the northeastern and north-central regions of Minnesota and represent about 6 percent of the Minnesota landscape (Figure 2.15, Figure 2.16). As with the lowland coniferous forest, these habitats are well-represented in the Agassiz Lowlands and Tamarack Lowlands Subsections. Habitats are wet, acidic areas dominated by *Sphagnum* mosses and a host of ericaceous heath-type plant species. The areas often have scattered but small black spruce or tamarack trees present. Despite their low stature, these trees may be more than one hundred years old. Bogs are often adjacent to or intermixed with lowland coniferous forests. Bogs can also be remnants of lowland coniferous forests by virtue of a successional process known as paludification.

Several bird species that are common in these open bog habitats include the Clay-colored Sparrow, Common Yellowthroat, Lincoln's Sparrow, Savannah Sparrow, Sedge Wren, and Swamp Sparrow. In bogs where some trees are present, the Boreal Chickadee, Canada Jay, Connecticut Warbler, Dark-eyed Junco, Hermit Thrush, Nashville Warbler, Palm Warbler, Tennessee Warbler, White-throated Sparrow, Yellow-bellied Flycatcher, Yellow-rumped Warbler, and other species of the lowland conifer habitat type can be found. In the massive area of the Red Lake Peatland, the Bobolink and LeConte's Sparrow are especially common, while Sharp-tailed Grouse and Short-eared Owls can also be observed in this habitat type.

MARSH AND WET MEADOW

The marsh and wet meadow habitat is composed of low-statured wetland vegetation dominated by grasses, sedges, cattails (*Typha* spp.), and other types of aquatic plants (Figure 2.17). This habitat includes shallow and deep marsh areas, such as hemi-marshes and fens; many have patchy shrub cover. This habitat type represents 5.6 percent of Minnesota's landcover and is widely scattered throughout the state but is especially prominent in the Tallgrass Aspen Parklands Province (Figure 2.18). A substantial proportion of this habitat type has been converted to agricultural or urbanized areas, especially in western and southern Minnesota.

Bird species within this habitat are distinctive from many of those previously described but can overlap considerably with species in shrub wetlands. Depending on the location of this habitat and the specific vegetation present, characteristic breeding species can include the American Bittern, Black Tern, Common Yellowthroat, Marsh Wren, Red-winged Blackbird, Sandhill Crane, Sedge Wren, Song Sparrow, Sora, Swamp Sparrow, Tree Swallow, Virginia Rail, Wilson's Snipe, Yellow Warbler, and Yellow-headed Blackbird. In the large open

(A)

(B)

Figure 2.17. Marsh and wet meadow habitats in Minnesota can vary considerably. **A.** Extensive sedge, grass, and cattail vegetation found at Carlos Avery State Wildlife Management Area. **B.** The extensive wetland complex in Lac qui Parle County in western Minnesota. These productive wetlands provide breeding habitat, cover, and foraging area for numerous waterfowl, grebes, terns, rails, herons, bitterns, raptors, and a host of songbirds. Notable species include the Trumpeter Swan, Horned Grebe, King Rail, Sandhill Crane, and Northern Harrier. Photos by Lee A. Pfannmuller (A) and Gerald J. Niemi (B).

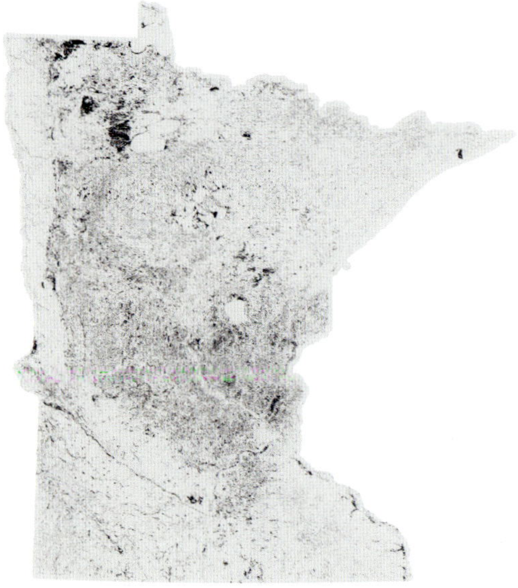

Figure 2.18. Spatial distribution of marsh and wet meadow habitat in Minnesota.

Figure 2.19. A typical upland coniferous forest in Minnesota consists of a variety of pine (*Pinus* spp.), spruce (*Picea* spp.), or balsam fir (*Abies balsamea*) with varied understories of shrubs and ground vegetation. The variability in the age, density, and structure of these forests provides different opportunities for nesting by conifer-dependent bird species such as the Blue-headed Vireo, Black-throated Blue Warbler, Canada Warbler, Cape May Warbler, Evening Grosbeak, Magnolia Warbler, Northern Parula, Red-breasted Nuthatch, and Yellow-rumped Warbler. Photo by Gerald J. Niemi.

fens of the Red Lake Peatland in northwestern Minnesota, the Bobolink, LeConte's Sparrow, and Sedge Wren are especially abundant, and the Sharp-tailed Grouse also can be found.

UPLAND CONIFEROUS FOREST

Representing 4.4 percent of Minnesota's landcover, coniferous forests are dominant in the northernmost areas of northeastern Minnesota, especially in the region known as the Arrowhead (Figure 2.19, Figure 2.20). These are closed-canopy forests dominated by white spruce, balsam fir, and three species of pine (red, white, and jack); in the far northern regions, black spruce can be found in the uplands. This habitat occurs in many age classes because the trees are subject to logging, forest fire, wind, and insect damage (e.g., spruce budworm). Deciduous trees, such as aspen and birch, often occur in low abundance.

These conifer forests house a wide variety of breeding bird species, many of which are shared with the northern mixed forest, pine forest, and lowland coniferous forest cover types. Common breeding species include the Blackburnian Warbler, Black-throated Blue Warbler, Black-throated Green Warbler, Blue-headed Vireo, Canada Jay, Canada Warbler, Cape May Warbler, Evening Grosbeak, Golden-crowned Kinglet, Hermit Thrush, Magnolia Warbler, Nashville Warbler, Northern Parula, Ovenbird, Pileated Woodpecker, Purple Finch, Red-breasted Nuthatch, Red-eyed Vireo, Rose-breasted Grosbeak, Ruffed Grouse, Swainson's Thrush, Tennessee Warbler, White-throated Sparrow, Winter Wren, Yellow-bellied Sapsucker, and Yellow-rumped Warbler.

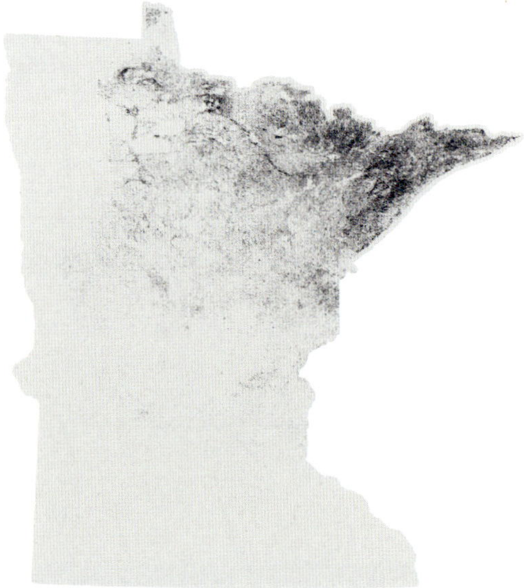

Figure 2.20. Spatial distribution of upland coniferous forest habitat in Minnesota.

(A)   (B)

Figure 2.21. Shrub wetlands are a common feature across much of the Minnesota landscape. As the name implies, the predominant feature is the presence of alder (*Alnus* spp.) or willow (*Salix* spp.) shrubs with varying densities, a ground layer of sedges, grasses, or herbaceous vegetation, and often a few scattered dead or live trees. **A.** A typical high-density shrub wetland; **B.** A riparian floodplain dominated by shrubs. Both areas provide suitable habitat for the American Woodcock, Common Yellowthroat, Swamp Sparrow, Yellow Warbler, and, depending on the location in Minnesota, the Alder or Willow Flycatcher. Photos by Lee A. Pfannmuller (A) and Gerald J. Niemi (B).

SHRUB WETLAND

This wetland type has some overlap with the marsh and wet meadow habitat type, with the primary distinction being the extensive presence of woody shrubs, higher-stature vegetation (e.g., 1–4 m high), and less wet (Figure 2.21). Shrub wetlands comprise 4 percent of Minnesota's landcover and primarily occur in central and north-central Minnesota, especially in the Agassiz Lowlands Subsection (Figure 2.22). In many areas they also comprise the riparian zone, bordering lakes, rivers, streams, and wetlands. Dominant shrub vegetation includes several species of willow (*Salix* spp.) and alder (*Alnus* spp.), plus bog birch (*Betula pumila*) in the northern regions.

As previously mentioned, bird species in this habitat often can overlap with the marsh and wet meadow habitat type, depending on the extent of the shrubs and the depth of water in the wetland. Typical species of this habitat include the Alder Flycatcher (north), American Woodcock, Common Yellowthroat, Sandhill Crane, Sedge Wren, Song Sparrow, Swamp Sparrow, Veery, Virginia Rail, Willow Flycatcher (south and northwest), Wilson's Snipe, and Yellow Warbler. Shrub wetlands are an important habitat type for the Golden-winged Warbler; its breeding distribution in north-central Minnesota corresponds well with the distribution of this habitat.

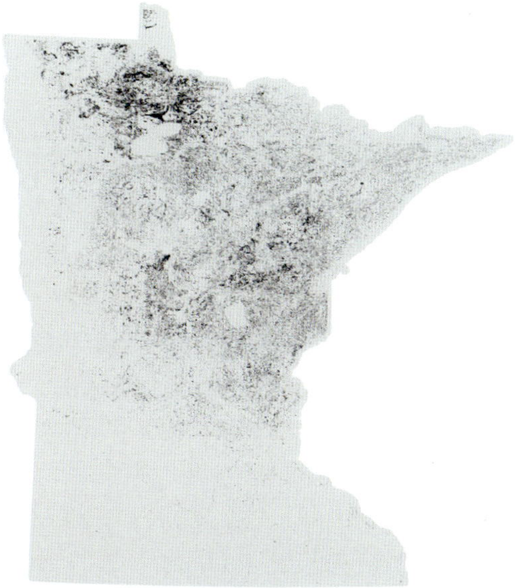

Figure 2.22. Spatial distribution of shrub wetland habitat in Minnesota.

DEVELOPED

A habitat type most are familiar with is developed land, which comprises 5 percent of Minnesota's landcover (Figure 2.23). This habitat is quite diverse and includes low to high human residential development, urban herbaceous and shrub cover, open or barren areas (such as golf courses, brownfields, gravel pits, and mine lands), as well as industrial sites and any habitat associated with the human built environment. The Twin Cities metropolitan area has the greatest share of developed land, but many other moderate to large cities and mine lands contribute substantially to this habitat type (Figure 2.24). Even though the percentage of this habitat type in the state seems small, many other habitat types are intermixed with developed land, such as cropland, wetlands, and various forested habitats.

(A)

(B)

Figure 2.23. Developed habitat can vary from industrial and high-residential areas to exurban landscapes with small farms and orchards. **A.** A panorama of the city of Redwing in southeastern Minnesota illustrating many features of development, such as industrial areas used by the House Sparrow, European Starling, Rock Pigeon, and Chimney Swift to cliffs used by the Belted Kingfisher and Bank Swallow to suburban developments with grassy areas, shrubbery, gardens, and trees. **B.** An exurban landscape in the western Twin Cities region with many features that are attractive as breeding habitats. Prominent bird species breeding in suburban areas and exurban areas include the House Wren, American Robin, Black-capped Chickadee, Northern Cardinal, House Finch, Chipping Sparrow, Mourning Dove, Common Grackle, and Ruby-throated Hummingbird. Photos by Lee A. Pfannmuller (A) and Gerald J. Niemi (B).

Breeding bird species of developed land are among the most familiar birds to the public. They include the American Robin, Barn Swallow, Black-capped Chickadee, Canada Goose, Cedar Waxwing, Chimney Swift, Chipping Sparrow, Common Grackle, European Starling, House Finch, House Sparrow, Mallard, Mourning Dove, Northern Cardinal, Purple Martin, Rock Pigeon, Ruby-throated Hummingbird, Tree Swallow, and White-breasted Nuthatch.

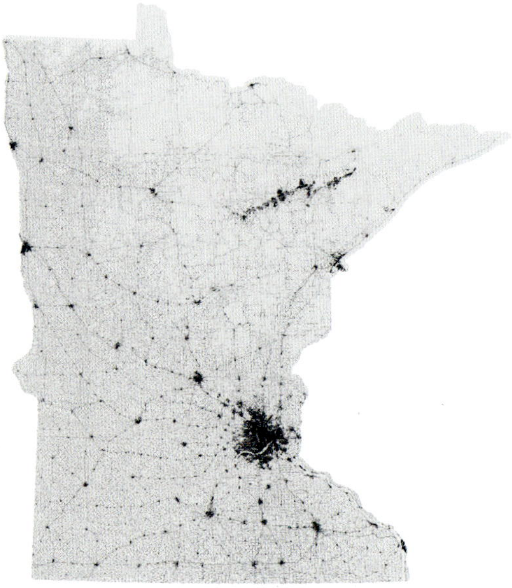

Figure 2.24. Spatial distribution of developed habitat in Minnesota.

OAK FOREST

Oak forest habitat represents about 1.9 percent of Minnesota's landcover and is prominent in the Eastern Broadleaf Forest Province, especially in southeastern Minnesota (Figure 2.25, Figure 2.26). Oak forests are in uplands, where they are often found in transitional regions with adjacent prairie ecosystems. Drier soil conditions and frequent fires can play a major role in the presence and perpetuation of this forest type. This cover type has declined substantially from its historical presence, partly owing to fire suppression and conversion to agricultural or developed land. Bur and pin oak (*Quercus palustris*) are usually dominant tree species, while white and black oak (*Q. velutina*) also are common in the southeastern part of the state.

As with many of the forest types, many bird species of the northern mixed forest and the pine–oak barrens cover types can also be found in this habitat. However, the following species tend to be prominent: American Redstart, Baltimore Oriole, Blue Jay, Blue-gray Gnatcatcher, Downy Woodpecker, Eastern Bluebird, Eastern Towhee, Eastern Wood-Pewee, Great Crested Flycatcher, Indigo Bunting, Northern Cardinal, Pileated Woodpecker, Red-bellied Woodpecker, Scarlet Tanager, White-breasted Nuthatch, and Wild Turkey.

PINE–OAK BARRENS AND OAK SAVANNA

Pine-oak barrens and oak savannas are upland forests where the canopy is open due to the low density of trees. This cover type is dominated by oak (bur and pin) or a combination of oak and jack pine (*Pinus banksiana*) (Figure 2.27). Historically, this

Figure 2.25. As the name implies, oak forests are dominated by oaks (*Quercus* spp.), but the density of trees can vary considerably. The photo shows a typical closed-canopy oak forest with large trees and a variable understory. The avifauna attracted to this habitat includes Red-headed and Red-bellied Woodpeckers, Blue-gray Gnatcatcher, American Redstart, Eastern Towhee, and Indigo Bunting, as well as many species found in other deciduous forest cover types. Photo by Gerald J. Niemi.

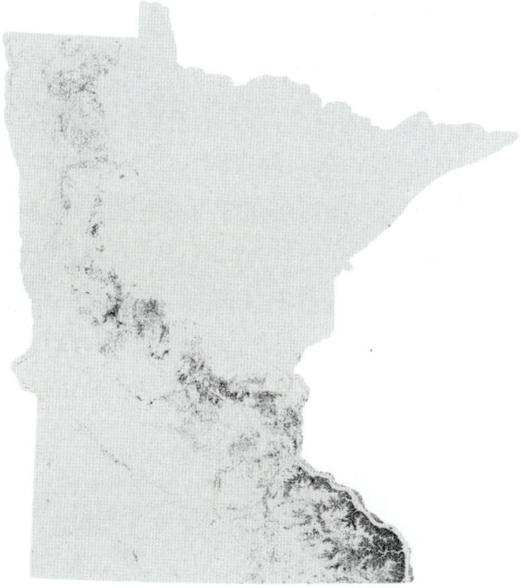

Figure 2.26. Spatial distribution of oak forest habitat in Minnesota.

was a fire-dependent forest type where it occurred in xeric soil conditions and it was often a transition forest type with prairies to the west. The main features distinguishing it from oak forests are the lower tree densities and lower stature of trees. The cover type is most common in sandy soils of west-central Minnesota and in the St. Croix Moraine Subsection in east-central Minnesota (Figure 2.28). It constitutes 1.2 percent of Minnesota's landcover.

Most of the bird species found in the oak forest cover type are also found in this habitat, although they are often in lower densities because tree densities are much lower.

PINE FOREST

This closed-canopy forest habitat formerly occupied a substantially larger area in northeastern and north-central Minnesota and has been greatly reduced owing to extensive logging that began in the mid-1800s. It currently occupies about 1 percent of Minnesota's landcover and remains within its historical distribution (Figure 2.29, Figure 2.30). Two species of pine are generally dominant in this forest type, red and white, but jack pine is also prominent on sandy soils. Included within this cover type are plantations of red pine. Large tracts of pine forests are still found in the Chippewa National Forest, and many larger tracts of old growth occur in the Boundary Waters Canoe Area Wilderness, Voyageurs National Park, and within the Superior National Forest.

Pine forests share many of the same bird species found in the northern mixed forest and the upland coniferous forest cover types. However, among the common breeding bird species found in this habitat are the Blackburnian Warbler, Black-throated Green Warbler, Brown Creeper, Eastern Wood Pewee, Least Flycatcher, Ovenbird, Pileated Woodpecker, Pine Warbler (especially prominent), Red-breasted Nuthatch, Red-eyed Vireo, Rose-breasted Grosbeak, Scarlet Tanager, Veery, Yellow-bellied Sapsucker, Yellow-rumped Warbler, and Yellow-throated Vireo (western).

(A)

(B)

Figure 2.27. The pine–oak barrens and oak savanna differ from oak forests in the lower density and stature of trees as well as the openness of the canopy. **A.** The presence of jack pine (*Pinus banksiana*) is a distinguishing feature of the pine–oak barrens. **B.** The Helen Allison Savanna Scientific and Natural Area in east-central Minnesota is an excellent example of an oak savanna. Both the barrens and savanna have an open canopy interspersed with grasses and herbaceous vegetation. Historically, both habitats were more common in Minnesota because they were maintained by frequent fires. Their breeding avifauna, however, includes the same suite of bird species as those of the oak forests. Photos by Gerald J. Niemi (A) and Lee A. Pfannmuller (B).

Figure 2.28. Spatial distribution of pine–oak barrens and oak savanna habitat in Minnesota.

(A)

(B)

Figure 2.29. Pine forests formerly were a more common feature of the Minnesota landscape but have been greatly reduced by logging since the mid-1800s. **A.** A typical mature pine forest with an open understory at Minnesota Point Pine Forest Scientific and Natural Area in Duluth. **B.** A canopy perspective of typical majestic pines in northern Minnesota. Large pine trees are especially attractive to the Blackburnian Warbler, Pine Warbler, Brown Creeper, Eastern Wood Pewee, Pileated Woodpecker, and Yellow-rumped Warbler. Photos by John C. Green (A) and Gerald J. Niemi (B).

Figure 2.30. Spatial distribution of pine forest habitat in Minnesota.

Biogeographic Patterns of Minnesota Birds

The Minnesota Breeding Bird Atlas provided an opportunity to present the contemporary biogeographic patterns of birds during the breeding season in Minnesota. The broad coverage of point counts gathered in Minnesota's townships during the atlas period delivered the standardized information necessary to produce maps of these patterns. Additional information on these patterns can be found in Green and Jansen (1975), Green (1991, 1995), Niemi and Hanowski (1992b), Mladenoff et al. (1997), and Niemi et al. (1998).

BIRD SAMPLING

Producing a statewide biogeographic map requires that both the effort (time) by observers and the area being sampled be equivalent from one place to the next. We explained the overall bird sampling methods in the Data Collection section in chapter 1 with additional details provided in Appendix A. Briefly, standardization of data gathering was completed by the point count samples of each township where each township was sampled by three point counts in which the total time of sampling was 30 minutes and the area sampled by the three point counts was 30 hectares when detections of birds were limited to 100 m. On the surface, this may seem like a small area, but gathering these data over the entire state and for 2,352 priority blocks, many of which were inaccessible, was a substantial effort. The BBS also controls the amount of time sampling on a route (50 stops each for 3 minutes = 150 minutes), but the area of sampling is not controlled because counts use unlimited distance. For example, an observer can detect a much larger area in open habitats compared with the more limited detection in a forest environment. In addition, the BBS routes are long and linear, covering 40.2 km (25 miles), while the MNBBA was restricted to covering a single priority block (4.8 km × 4.8 km [3 miles × 3 miles]). Therefore, the species richness maps of the BBS and the MNBBA look different because of the different spatial scales upon which birds were counted and how species richness is plotted. The spatial scale for point counts in the MNBBA were highly controlled in the area sampled and the effort expended.

BIOGEOGRAPHIC MAPPING ANALYSIS

We calculated richness at the township scale using MNBBA point count data limited to 100 m and excluded flyover detections. All three point counts were included in the calculation. The number of species detected in the priority block was used as the sample for the entire township; in rare cases, data from the alternative block was used. In addition to total species richness, we also calculated species richness for the following groups: flycatchers (family Tyrannidae), thrushes (family Turdidae), warblers (family Parulidae), and sparrows (family Passerellidae). All calculations were made using the R statistical software.

Overall Bird Species Richness

In contrast with the overall BBS species richness map (Figure 2.3), species richness for Minnesota based on the MNBBA illustrated a wider range of diversity across the state. Generally, there was a band of high species richness (average greater than 25 species/sampled township) in the Eastern Broadleaf Forest Province (Figure 2.31). In Minnesota, this province, which is the dominant biome of the Eastern United States, represents an ecotone between the Laurentian Mixed Forest Province and the Prairie Parkland Province (Niemi et al. 1998). Ecotones are notorious for being areas of high species diversity because they contain a mixture of spatial scales of land and a variety of habitat types. In this case, the Eastern Broadleaf Forest Province is generally dominated by deciduous trees and, in Minnesota, is greatly influenced by the boreal nature of the Laurentian Mixed Forest Province (e.g., more coniferous trees) and the open landscape of the Prairie Parkland Province (e.g., open non-forested habitats). Furthermore, with human development and aquatic habitats such as wetlands, lakes, and flowing water, it

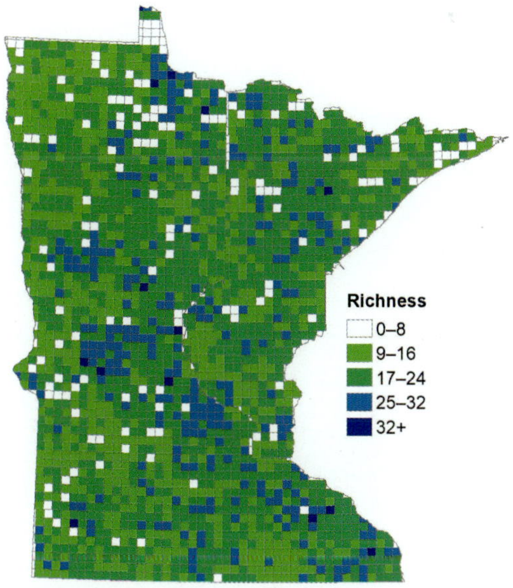

Figure 2.31. Bird species richness map of Minnesota based on the point count data collected within each township.

results in a diversity of habitat types within this broad region.

Note that many townships are white with no or few species detected. These areas are large bodies of water (such as Upper and Lower Red Lakes or Mille Lacs Lake), highly industrialized areas, inaccessible blocks, or they simply were not visited by any observers. However, most townships would have bird species present, such as gulls or waterfowl in open water, European Starlings, Rock Pigeons, or House Sparrows in industrialized areas, and many species in those areas that were too inaccessible.

The northeastern and north-central regions of the state also had many townships with a large number of species. These are likely areas with diverse habitats, including forests, wetlands, and open areas. Areas with the lowest species richness were primarily found in western regions of the state, which are dominated by cropland—a habitat that provides limited plant and landscape diversity for bird species. Exceptions, however, are noted such as near Big Stone National Wildlife Refuge in Big Stone County, in Pipestone and Rock Counties in the extreme southwest, and in eastern Clay County, an area with a substantial number of wildlife and waterfowl management areas.

FLYCATCHER SPECIES RICHNESS

The distribution of the 12 flycatcher (Tyrannidae) species found in Minnesota had their highest species richness in the Laurentian Mixed Forest Province and to some extent in the transition region with the Eastern Broadleaf Forest Province (Figure 2.32). The 12 flycatcher species in the state vary from those highly associated with coniferous forests (Yellow-bellied and Olive-sided Flycatcher) to those associated with mixed or deciduous forests (Least Flycatcher, Eastern Wood-Pewee, Great Crested Flycatcher, and Acadian Flycatcher) and shrubby habitats (Alder and Willow Flycatcher) to those of open habitats (Eastern and Western Kingbird, and Say's Phoebe). The Eastern Phoebe is widespread in the state, primarily associated with riparian habitats and available nest sites. Hence, those regions that supply a combination of these habitats tend to have the most flycatcher species.

THRUSH SPECIES RICHNESS

Five of the six species of thrushes (Turdidae) found in Minnesota tend to be forest-associated species and, therefore, their species richness is highest in the northeastern and central regions of the state (Figure 2.33). Undoubtedly many of the areas in the western and southern portions of the state with one to two species include the American Robin or Eastern Bluebird, species that occur in open or semi-open areas. The American Robin is ubiquitous statewide and joins the woodland thrushes (Veery, Swainson's Thrush, Hermit Thrush, and Wood Thrush) in the central, north-central, and northeastern regions. In regions dominated by cropland and open wetlands, thrushes, including the American Robin and Eastern Bluebird, are scarce because they both require trees or (the latter) nest boxes for breeding habitat. In addition, both the American Robin and Wood Thrush decline in abundance in the heavily forested regions in extreme northeastern Minnesota.

WARBLER SPECIES RICHNESS

Minnesota has 31 species of warblers (Parulidae) that have been found nesting in the state. Since most are highly associated with both deciduous and coniferous

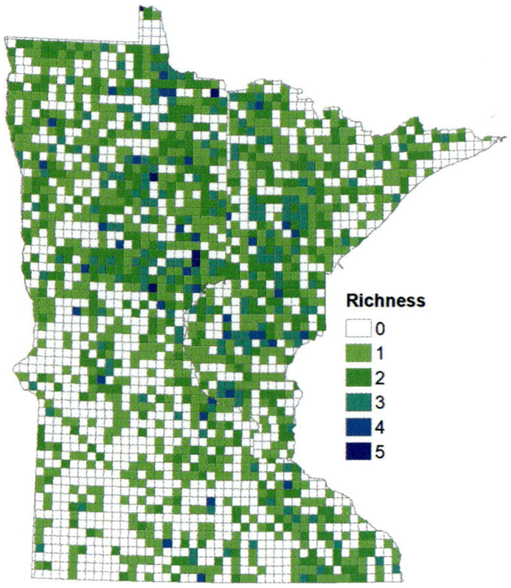

Figure 2.32. Species richness of flycatchers (Tyrannidae) in Minnesota based on the point count data collected within each township.

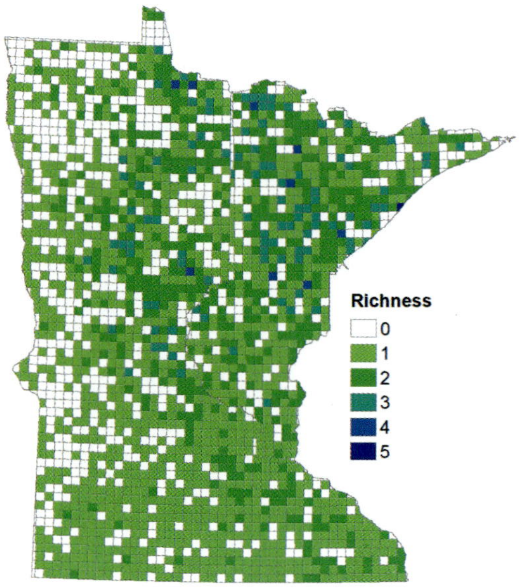

Figure 2.33. Species richness of thrushes (Turdidae) in Minnesota based on the point count data collected within each township.

forests, their greatest species richness is found in the north-central and northeastern regions (Figure 2.34). Most warbler species do not occur in the southern and western counties of Minnesota, or they are found in very low densities. Those that do occur in these regions are the Common Yellowthroat and Yellow Warbler, the former a very ubiquitous species in wetlands and shrubby habitats and the latter also in shrubby areas. The diversity and resultant colorful tapestry of these warblers in the forested areas of Minnesota are among the highest anywhere in North America (Price et al. 1995). In numerous townships of these regions, from 9 to 14 warbler species were detected during three 10-minute point counts in the breeding season. Volunteer counts in priority blocks in these forested regions could easily exceed 15 or more warbler species.

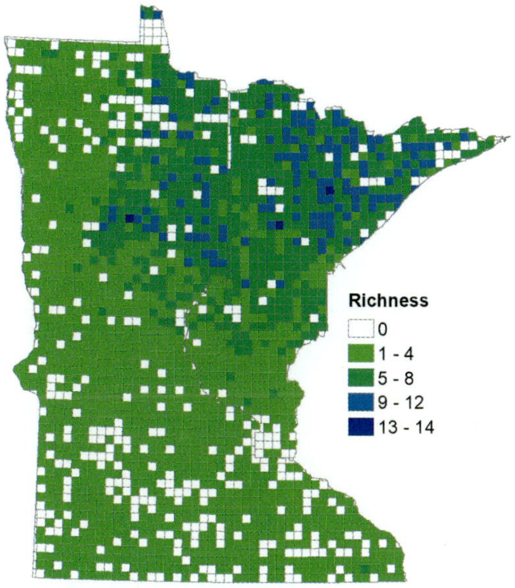

Figure 2.34. Species richness of warblers (Parulidae) in Minnesota based on the point count data collected within each township.

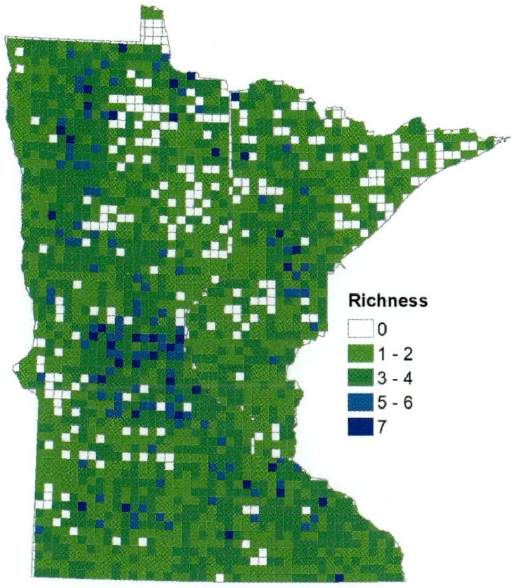

Figure 2.35. Species richness of sparrows (Passerellidae) in Minnesota based on the point count data collected within each township.

SPARROW SPECIES RICHNESS

The sparrows in the family Passerellidae represent a relatively large group of 17 species with a mix of ecological characteristics. In contrast with the warblers, many of the sparrows tend to be found in open country and in lower-stature vegetation consisting of shrubs, grasses, and sedges. However, there are many exceptions, such as the Chipping Sparrow and Dark-eyed Junco, which occur in forested environments. Patches of high species richness can be observed in numerous places throughout the state (Figure 2.35), including the northern portions of the Minnesota River Prairie Subsection and portions of the Hardwood Hills Subsection. Low species richness of sparrows occurs in the heavily forested regions of north-central and northeastern Minnesota; however, the White-throated Sparrow is a very ubiquitous species in these regions.

3 READING THE SPECIES ACCOUNTS

< Male Ruby-crowned Kinglet.
Photo by David Brislance.

The Internet has created an explosion of information on birds, and Minnesota's avifauna proves no exception. With a simple click of the mouse or the touch of a screen, a vast plethora of images, text, and data on any bird species is available. Managing and selecting the appropriate material from this glut of information has been a challenge. For each species, we provide a brief summary of basic life history information, including its Minnesota seasonal status, migration, foods consumed, nest structure, and level of conservation concern. This is followed by more in-depth text on the species: first, an introduction that summarizes identification hints, primary vocalizations, the origin of its name, unique features of its distribution or behavior, and its overall breeding distribution; second, a historical and current account of its Minnesota breeding distribution; third, details on its breeding habitat; fourth, its population abundance and trend; and, finally, conservation, including major threats to its ecology and possible management solutions.

Owing to the massive amount of data available for each species, we have restricted most species accounts to approximately 1,500 words, excluding figures or tables. A few species receive more extensive coverage, with up to 3,000 words, especially those designated as state *Endangered, Threatened,* or *Special Concern*; those that are managed; or those that figure prominently in Minnesota. This book includes accounts for all 250 species reported during the Minnesota Breeding Bird Atlas (MNBBA) project. The extensive records included in the atlas reflect the passion that Minnesotans have for birds. This first atlas provides a snapshot in time on the nesting status of Minnesota's breeding birds, and we strongly recommend that an atlas be repeated at twenty-year intervals like the second-generation atlases that have occurred in our neighboring states and provinces of Iowa, Manitoba, Michigan, Ontario, South Dakota, and Wisconsin.

If more information is desired for a species, more is provided on the MNBBA website (www.mnbirdatlas.org) or in the extensive references included in each account. We especially direct readers to the free information on Minnesota birds available at the Minnesota Ornithologists' Union (MOU) (http://moumn.org/) or at the Minnesota Department of Natural Resources (MNDNR) (http://www.dnr.state.mn.us/nr/index.html). Additional excellent resources of general information on birds include the Cornell Laboratory of Ornithology (https://www.allaboutbirds.org/) or the Birds of the World (https://birdsoftheworld.org), a cooperative endeavor of the American Ornithological Society, the Cornell Laboratory of Ornithology, and nine other conservation organizations. The latter is the most comprehensive summary of what is known about North American bird species, but full access requires a paid account. We apologize to those individuals who have published research results for any species that have not been included.

We truly hope these species accounts will motivate the collection of more quantitative data on the nesting activities, habitats, and status of Minnesota birds. We have learned a lot about the distribution of Minnesota's birds based on observations, but we know too little about the actual nesting of these birds. The determination of nesting success is absolutely critical to our long-term understanding of whether a population is sustainable. For example, what are the important micro and macro characteristics that determine where species nest? How many birds are actually nesting in an area? Was the nesting attempt successful, and if not, why not? We know a fair amount about many species, especially the common ones, but there is still much to be discovered about most bird species nesting in our state.

Below, we provide a more detailed summary and explanation of the primary categories of information included in the individual species accounts.

Minnesota Seasonal Status: The seasonal status of each species in Minnesota is briefly described in terms of its frequency and season of occurrence. We also note if a species is non-native. Two species of birds are officially listed by the MNDNR (2020b) as "invasive terrestrial animals": the European Starling and the Mute Swan. Invasive species are those that are not native to Minnesota and cause economic or environmental harm or harm to human health.

We use the following definitions for describing the species occurrence during the breeding season:

Regular: The species is found every year in the state of Minnesota. Regular species are further classified as either:
- Permanent, year-round residents; or
- Breeding residents present primarily during the summer season.

Visitant: A non-resident species whose occurrence in Minnesota is outside the species' expected distribution range; it occurs in Minnesota five to ten years each decade.

Vagrant: A non-resident species whose occurrence in Minnesota is well outside the species' normal distribution range; it occurs in Minnesota one to four years each decade.

Former Breeding Species: Species that formerly bred in Minnesota prior to the year 2000 (although extremely rare, documented incidents of more recent breeding records for a few of these species are noted—e.g., King Rail, Barn Owl, and Black-necked Stilt).

Each species has also been assigned to a relative abundance category based on the total number of records included in the atlas database. Several metrics of relative abundance could be used, such as the total number of blocks where breeding evidence was detected, the total number of priority blocks where breeding evidence was detected, or the total number of detections from the randomly selected points. However, the correlation among all of these metrics yields an $r > 0.88$, which indicates that each would provide similar information about its relative abundance. Total records are also most useful for such birds as aquatic species, raptors, and nocturnal species, which are not easily detected except with specific observational techniques, such as night surveys or searches within habitats that are difficult to access. We recognize that this is, at best, a relative index. For highly colonial species, such as many herons, egrets, gulls, terns, cormorants, and pelicans, the index is not a true reflection of abundance. Each atlas record can often represent a breeding colony that may consist of hundreds or even thousands of breeding individuals.

Categories of abundance are designated as follows (the number of species in each category is in parentheses): very abundant, more than 6,000 records (9 species); abundant, 4,000–5,999 records (22); common, 1,000–3,999 records (71); uncommon, 100–999 records (82); rare, 10–99 records (41); and very rare, 1–9 records (25).

Migration: For each species, we provide a brief summary of its migration strategy. We identify whether it is a permanent resident, partial migrant (a portion of the population migrates out of Minnesota while others stay all year, such as the Belted Kingfisher and Red-breasted Nuthatch), short-distance migrant (winters in the United States), medium-distance migrant (winters in Mexico, Central America, and the Caribbean), or long-distance migrant (winters in South America), and where it can be found in winter.

Food: A list of primary food items and foraging strategies, with an emphasis on the breeding season.

Nest: Physical aspects of the nest and its placement.

Conservation Concern: For each species, we summarize its conservation status by using a few key rankings. A universal, global ranking for all waterfowl, waterbirds, shorebirds, and landbirds was recently developed by Partners in Flight (PIF) (2021). Data for many of the landbirds are also presented in the PIF 2016 Landbird Conservation Plan (Rosenberg, Kennedy, et al. 2016). Although the majority of scores for landbirds remain unchanged, readers familiar with the earlier document may notice a few updates. These scores are based on a combination of the species' global population size, global breeding and non-breeding distribution, threats to its breeding and non-breeding range in North America, and its overall population trend in North America. They give an overall perspective on the North American conservation status of the species. The lowest score possible for a species is 4 (the lowest Continental Concern), and the highest score possible is 20 (the highest Continental Concern). (*Note:* While we identify the PIF scores as Continental Concern Scores, PIF refers to them as Continental Combined Scores, or CCS.) These scores will change over time as more knowledge is gained on the population, distribution, threats, and trends for each species.

The delineation of "Low," "Moderate," and "High" scores is taken from the 2016 Landbird Conservation Plan (Rosenberg, Kennedy, et al. 2016); scores from 4 to 8 are "Low," scores from 9 to 13 are "Moderate," and scores from 14 to 20 are "High." The scores for Minnesota species range from a low of 5 (e.g., House Wren, American Robin, and Northern Cardinal) to a high of 18 (Piping Plover).

We have chosen to highlight the PIF Continental Concern Scores because they have recently been developed for all birds found in North America south to Panama and they reflect important conservation data (e.g., total population size) for those species whose breeding ranges extend beyond North America (e.g., holarctic species). We also note if a species is federally listed (USFWS 2021) and/or state listed (MNDNR 2013), and if it is a game species in Minnesota (2021 Minnesota Statutes 97A.015, Subd. 24, Game Birds). Finally, we identify species on the Red List of Threatened Species established by the International Union for Conservation of Nature (IUCN 2021). The Red List assigns species to one of nine categories. We have chosen to identify species classified in the two highest categories of concern that pertain to current breeding species in the state: Vulnerable and Endangered. The list is updated regularly, and we selected the most recent list available when this manuscript was completed (Version 1 published in the spring of 2021). Further details are often discussed in the Conservation section of the account.

MINNESOTA BREEDING DISTRIBUTION

To describe the breeding status of species in Minnesota, a historical path has been followed, which fortunately is available for Minnesota birds. For instance, a general chronological sequence includes P. L. Hatch (1892) in his *First Report: Accompanied with Notes on the Birds of Minnesota*; the monumental two-volume tome of Thomas S. Roberts, *Birds of Minnesota* (1932, 1936); Green and Janssen's 1975 book, *Minnesota Birds: Where, When, and How Many*; Janssen's 1987 book, *Birds in Minnesota*; Hertzel and Janssen's 1998 publication, *County Nesting Records of Minnesota Birds*; and Janssen's 2019 revision and expansion of his 1987 book. Note that Hatch is not always used, because his observations are often very limited in detail and coverage in Minnesota. Important distributional records and compilations published in *The Loon,* the quarterly publication of the Minnesota Ornithologists' Union, are also frequently cited.

These materials are generally followed with a summary of the extensive coverage of the Minnesota Biological Survey (MNDNR 2020c) and their bird inventories of Minnesota counties. Although the focus of the Survey's bird work was on rare birds, data were collected incidentally on all birds when point counts were conducted. Survey work was completed from 1987 to 2014 in 82 of Minnesota's 87 counties. The results of surveys conducted in all or portions of five northern counties (Koochiching, Lake of the Woods, and northern Beltrami, Clearwater, and St. Louis Counties) were not available at the time the MNBBA species accounts were prepared. Finally, we present the results of the MNBBA data gathered from 2009 to 2013.

Each of the historical documents summarized in the MNBBA species accounts represents a summary of the status of breeding evidence for the primary species that were known to nest in Minnesota at the time of the publication. Each varies considerably in the level of detail presented, especially on what constitutes evidence of nesting. Many of the earlier accounts by Hatch and by Roberts could be vague on the level of detail they provided and what constituted confirmed nesting. For instance, Roberts (1932, 1936) often used the terms Cass Lake, Leech Lake, Itasca Park, and Mille Lacs. These areas are generally well-known to most Minnesotans, but they may not be specific to a county, because many of these locations encompass more than one county. When the county was not clear from the document, we simply refer to the place name, such as Mille Lacs. Minneapolis and St. Paul were also used as place names by Roberts, but we interpret these as Hennepin County and Ramsey County, respectively.

Brief details on what Hatch or Roberts stated in their summary accounts regarding nesting activity are often reported for each species (see Appendix E). Green and Janssen (1975) summarized a species' breeding evidence as either positive (e.g., nest with eggs, or adult sitting on nest constantly, or eggshells near nest; young in nest seen or heard; downy young, or young still unable to fly, seen away from nest) or inferred nesting (nest-building or excavation of nest-hole in progress; adults seen in distraction display or feigning injury; used nest found; recently fledged young seen; adult seen carrying fecal sac; adult seen carrying food for young; adult seen entering or leaving nest site in circumstances indicating occupied nest). Two additional symbols were used to indicate former positive nesting or former inferred nesting on their distribution maps. Breeding evidence from Roberts (1932, 1936) was also included in their breeding evidence maps.

Janssen (1987) summarized breeding evidence since 1970. He included only confirmed nesting records for a county using the same positive definition used by Green and Janssen (1975). Hertzel and Janssen (1998) also summarized breeding evidence since 1970 and they defined a confirmed nesting record as a nest with eggs, an adult on a nest, identifiable eggshells near a nest, downy young in a nest or nearby, or flightless young. Updates delineated by the Hertzel and Janssen (1998) maps that differ from the Janssen (1987) maps are identified in the MNBBA species accounts. Updates by Janssen (2019) at the county level through 2013 include results of the MNBBA, but here we present results at the township level. The Minnesota Biological Survey identified locations where species were reported during the breeding season but did not have a focused effort on finding nesting evidence except through incidental observations. Each record represents "one or more individuals, observed (seen or heard) at a specific location" (MNDNR 2020a).

The MNBBA used definitions for breeding evidence that are standard among breeding bird atlases completed in the United States and throughout many parts of the world (see Table 1.2). For instance, the definitions used by the MNBBA are consistent with the atlases completed in the states and provinces surrounding Minnesota, including Iowa, Manitoba, Michigan, Ontario, South Dakota, and Wisconsin. These definitions have some overlap with those used by Green and Janssen (1975), Janssen (1987), and Hertzel and Janssen (1998), but also many differences and additional categories of breeding evidence. For example, confirmed nesting with the MNBBA included all the definitions used by Green and Janssen, Janssen, and Hertzel and Janssen, but included confirmed nesting for several behaviors that Green and Janssen (1975) classified as inferred nesting, such as nest building seen at the actual site (code NB), bird seen carrying nesting material such as sticks, grass, mud, cobwebs, etc. (code CN), and an adult seen carrying a fecal sac (code FS). For confirmed nesting, Janssen (2019) used a combination of Janssen (1987) and the MNBBA definitions, but evidence was restricted to the 2001–17 period. These differences in definitions require caution when making comparisons with previous confirmed breeding evidence in Minnesota. Because there are differences in definition of confirmed nesting, the MNBBA often documented nesting in more counties than earlier accounts because the MNBBA definitions were less restrictive.

The word "records" is generally used to describe entries by the MNBBA participants for each species. A record refers to an entry into the MNBBA database for a specific block. A record may represent a single detection of a species within a block or it may include a number of detections of a species within a block. Detections refer to both visual and aural observations of a species. Two of the breeding evidence codes (codes M and S) specifically reference multiple detections of a species and are assigned as "Probable" breeding evidence. The multiple detections, however, may constitute only one record in the MNBBA database. Hence, the number of detections of a species during the MNBBA may greatly exceed the official number of records reported in a species account. The word "observations" is used sparingly in the following pages because it is often interpreted as a visual detection only.

MNBBA Results

Major results from the MNBBA were summarized in each species account, such as the number of blocks or priority blocks where a species was reported and general statements about the species' distribution or breeding evidence. Breeding evidence is summarized to the highest evidence code reported for a given block during the MNBBA. For example, a species may have been reported several times within a particular block (i.e., multiple detections) but only that record with the highest breeding evidence code is shown on the statewide breeding distribution map. Note that although many species appear to have their center of abundance in the Twin Cities and the Brainerd Lakes region (e.g., Mallard, Mourning Dove, and House Wren), this usually reflects the number of volunteers and their remarkable atlas field efforts in each region.

Two different base maps of Minnesota were used to display the MNBBA results. The majority of maps display six of Minnesota's largest lakes: Lake of the Woods, Lake Winnibigoshish, Leech Lake, Mille Lacs Lake, Upper Red Lake, and Lower Red Lake. However, for a few species that nest or are frequently observed on islands on these lakes, the lakes are not shown. This omission allowed us to clearly delineate atlas blocks within the lake's perimeter where the species was documented. The text also may identify counties with confirmed breeding evidence that is not easy to see on the small MNBBA distribution maps. This is especially true along the North Shore of Lake Superior or along the perimeter of large lakes, where many atlas blocks contain more water than lakeshore. Readers can easily find more detailed information at a smaller scale on the interactive maps available at www.mnbirdatlas.org.

When suitable data were available for a species, a probability-based predictive model was derived using the point counts gathered throughout the state of Minnesota. These models use habitat at a point count, landscape information surrounding the point, and climate data associated with the point to make predictions on the probability of observing a species within its breeding range in Minnesota. A brief overview of the point counts can be found in chapter 1 under Point Counts Data Collection; details regarding the collection and analysis of data at point counts can be found in Appendix A.

For the predictive species distribution maps, we use different expressions of the relative abundance of a species based on their detectability and behavior. These expressions fall into three different categories. First, for species that are relatively common or abundant and have strong territorial behavior, we express densities as pairs per 40 ha (100 acres). These are largely based on singing behavior of territorial males, and it is commonly assumed there is a 1:1 sex ratio of males to females (see Appendix A under Breeding Bird Probability Distribution Modeling Strategy 1). Second, the number of individuals observed per 10-minute unlimited distance point count is used to express relative abundance for species that are wide-ranging (e.g., Chimney Swift, swallows); heard from a great distance making estimates of detection distance difficult (e.g., Ring-necked Pheasant, American Bittern, Sandhill Crane, or Pileated Woodpecker); or landbird species that are frequently detected by vocalization but generally not as territorial males (e.g., nuthatches, Ruby-throated Hummingbird, and many woodpeckers) (see Appendix A under Breeding Bird Probability Distribution Modeling Strategy 2). And third, we use landcover suitability (low to high) for species that are rare or not amenable to quantitative sampling of their populations (e.g., raptors, waterfowl, or many landbirds that are uncommon or rare) (see Appendix A under Breeding Bird Probability Distribution Modeling Strategy 3).

Both the breeding evidence and

predictive species distribution maps could be useful for conservation planning, to identify places where a particular species might be observed, to identify places where nest sites might be found in future atlases, or to gain a better understanding of a species' Minnesota distribution and abundance.

Surveyed Blocks

As discussed in chapter 1 (under MNBBA Data Summaries), Minnesota encompasses 9,774 atlas blocks, and 2,352 of these are priority blocks. Theoretically, approximately one-quarter of all the blocks should be priority blocks (2,443), but many townships on the state boundary are only partial in size. As noted in the discussion on Minnesota's BBA grid system (chapter 1, under Data Collection by the Minnesota Breeding Bird Atlas [MNBBA]), in cases where the northeast block was smaller than 8 square miles, we selected the southeast block or continued clockwise to select a complete block. If the township had no blocks that met the size criteria, the township was excluded from the MNBBA's list of priority blocks that were the focus of the MNBBA. Volunteers, however, could still collect data from those smaller blocks.

In the end, a total of 4,733 blocks (48 percent of possible atlas blocks) and 2,337 priority blocks (99 percent of priority blocks) had at least 10 records during the atlas. We use these "surveyed" block numbers to calculate the summary statistics mentioned in the species accounts and detailed in Appendix C. However, we include additional blocks when a species was reported from one or more blocks with fewer than 10 records. For example, the Alder Flycatcher was reported from 11 blocks with fewer than 10 records. So, to calculate the proportion of blocks where the species was probable, we divide 827 (the number of blocks where it was probable) by 4,744 (i.e., 4,733 + 11). This is also the reason for some blocks in the block maps being coded as "not surveyed" for most species but "observed" or greater for a handful of other species. Because of this, an astute reader may notice the denominator change from species to species when calculating the proportion of blocks where a species was detected.

In our summary of the MNBBA data for a species, we primarily use counties or common place names in Minnesota (e.g., North Shore of Lake Superior, Twin Cities metropolitan area) to describe distributions of the species. Many atlas blocks include two counties; a small number even include three counties. As a result, when the atlas data are summarized by county, both counties that were covered by the block are included in the summaries. We have used a rule that if 5 percent of the block was in a county, then the county was included for that respective breeding code. Species accounts often note which counties are included merely because blocks straddled county lines. This was particularly true for many riparian species that occurred along the Minnesota River valley, where a large number of blocks along the river's floodplain encompass two counties. In addition, we often searched maps or satellite imagery to identify suitable habitats within a block for a species, but often suitable habitats (e.g., large wetlands) crossed county lines.

We also use the Ecological Classification System (ECS) that is commonly used by the MNDNR (2019f). The ECS groups regions of the state into broader categories that can be more useful for management, because the groupings have similar climate, geology, topography, soils, hydrology, and vegetation. In our species accounts we have confined our use of the ECS to three levels: provinces, sections, and subsections. A description of the ECS system can be found at http://www.dnr.state.mn.us/ecs/index.html; illustrations of the provinces and subsections can be found in chapter 2 (Figures 2.2 and 2.4, respectively).

For most species, we have summarized major historical changes that have occurred in their breeding distribution in Minnesota, in the region, or in North America, plus potential reasons for these changes. The basis for this evidence was primarily derived from the historical accounts, the published literature, breeding bird atlas data available for Iowa, Manitoba, Michigan, Ontario, South Dakota, and Wisconsin, and especially from the Birds of the World series mentioned earlier.

BREEDING HABITAT

Breeding habitats are broadly described and, when possible, specific descriptions from Minnesota or the surrounding region are used, especially when sample sizes were large and completed in a probability-based design. We also provide a photograph of "typical breeding habitat in Minnesota." We usually use only the common name of the tree or shrub species used in the text, but scientific names are occasionally incorporated. We provide a perspective on whether the species has been reported to be sensitive to landscape issues such as habitat fragmentation of forest, grassland/prairie, or wetland habitats.

For bird species that were detected with a reasonable sample size on the MNBBA point counts, a habitat profile of the types of habitat where the species was detected is provided in Appendix D. The point counts were documented with a global positioning system (GPS) so the detection of a species could be related to overlays of their habitat from remote sensing imagery.

For these habitat profiles we use 13 different habitat types that represent the majority of major habitats found in Minnesota (chapter 2, under Major Habitat Types in Minnesota). Many habitat classifications for Minnesota, however, have 40 or more different categories. The species' affinity to a habitat type uses a percent perfect indicator (PPI) index that expresses how strongly a species is related with a habitat type. PPI values are calculated using both how abundant a species is in the habitat category and its frequency of occurrence in the habitat. PPI values range as high as 1.0 if a species is always found in the same habitat type (Dufrêne and Legendre 1997). See Appendix A under Data Analysis: Habitat Associations for more details on the calculation of PPI values.

When covering such a massive area as an entire state, documenting the habitat for rare species or rare habitats can become problematic and result in spurious or inappropriate habitat representations. We reviewed many iterations of habitat classifications with more and fewer habitat types. We concluded that the 13 habitat types used here best represent the most bird species across the entire state. Many bird species, such as those found in wetlands, lowland deciduous forests, or in upland open areas (e.g., grasslands, croplands, or shrublands), were broadly classified into a general habitat type. In these situations, we urge the reader to gain a better understanding of the species' breeding habitat from the narrative statements in the text.

POPULATION ABUNDANCE

In recent years, many sophisticated models have been developed nationally to estimate the breeding populations of the majority of landbirds that occur in North America. Several programs, such as PIF (e.g., Blancher et al. 2013; Rosenberg, Kennedy, et al. 2016; PIF 2021), the North American Bird Conservation Initiative (NABCI 2016), Environment Canada (2011), and Environment and Climate Change Canada (Environment Canada 2014b), have estimated breeding populations using the North American Breeding Bird Survey (BBS) as a major source of information for these models. We caution, however, that the estimates from Environment Canada (2011) and Environment and Climate Change Canada (Environment Canada 2014b) are very conservative, with broad population range estimates. When available, we always cite the most recent data provided by PIF (2021) and/or other relevant data sources, especially for species not well surveyed by the BBS (e.g., most waterfowl, waterbirds, and shorebirds). For landbirds, PIF (2021) included the most recent global population estimates as well as the most recent population estimates for the United States and Canada and for the state of Minnesota when sufficient BBS data are available.

The BBS is a North American network of randomly selected roadside routes that are annually censused by volunteers. Routes are on secondary roads and consist of 50 stops, each 0.5 mile (0.8 km) apart, where all birds seen or heard in a 0.25 mile (0.4 km) radius around each point are counted over a 3-minute period (Sauer et al. 2020). The national program was initiated in 1966 but did not begin in Minnesota and states west of the Mississippi River until 1967. Thousands of routes are sampled each year. The BBS was primarily developed to estimate breeding bird population trends in the United States and Canada. We report and illustrate BBS population trends only for those species that had sufficient sample sizes over the long term with moderate precision and moderate abundance on routes. Depending on their reliability, trends may be reported and illustrated for the core BBS survey region (southern Canada and the US), for the US only, for a specific Bird Conservation Region (see Figure 3.1), or for the state of Minnesota.

Reliable trends were indicated by color codes of dark blue at the BBS website for data spanning 1966–2019 (Sauer et al. 2020). The color codes of yellow or red were considered deficient primarily because the species is uncommon and/or the sample size is small. When the yellow or red color-coded trend data are used, we state that the trend should be used with caution and we use it for a specific purpose, such as when the trend was particularly convincing for Minnesota, a region, or nationally (Sauer et al. 2020).

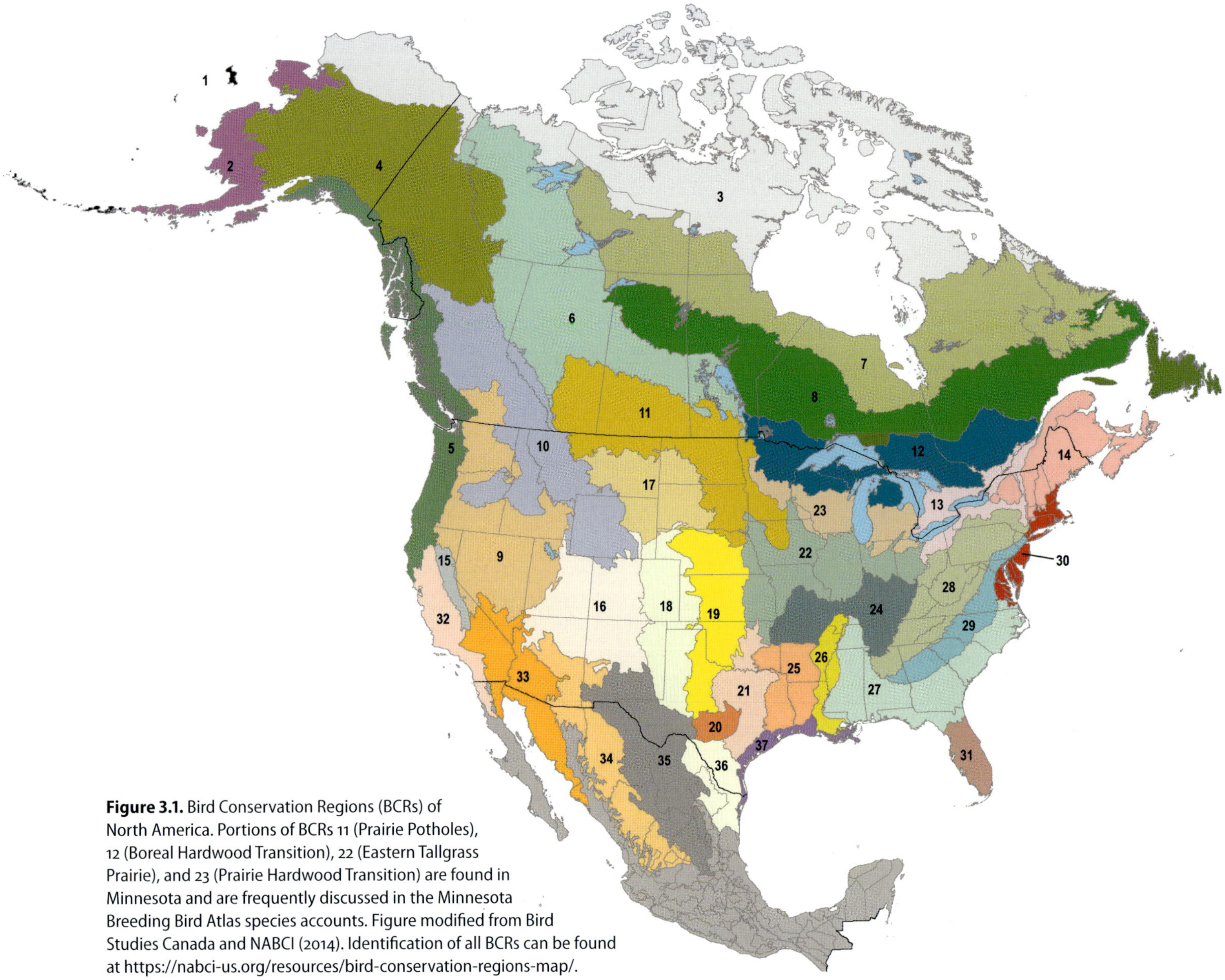

Figure 3.1. Bird Conservation Regions (BCRs) of North America. Portions of BCRs 11 (Prairie Potholes), 12 (Boreal Hardwood Transition), 22 (Eastern Tallgrass Prairie), and 23 (Prairie Hardwood Transition) are found in Minnesota and are frequently discussed in the Minnesota Breeding Bird Atlas species accounts. Figure modified from Bird Studies Canada and NABCI (2014). Identification of all BCRs can be found at https://nabci-us.org/resources/bird-conservation-regions-map/.

BBS trends include 95 percent credible intervals because the trends are based on a Bayesian statistical analysis. The more familiar 95 percent confidence intervals are included for some of the waterfowl species because they are based on more well-known statistical techniques such as regression. Regardless of the method, the interval width displayed by the upper and lower lines on a graph of the species population trend provides a perspective on the uncertainty of the trend estimate; narrower intervals increase confidence in the trend.

BBS analyses from 1966 to 2019 also include an option to calculate the trend using core and/or non-core BBS survey areas. Core BBS routes include those originally established in the United States and southern Canada by 1970 (Link et al. 2020). However, after 1970 additional routes were added in Alaska and northern Canada and are identified as non-core areas. Because these non-core areas have received sporadic coverage, data only were determined to be sufficient for analysis beginning in 1993. Therefore, accounts for a few bird species that occur in the far north (e.g., Palm Warbler) included trends from 1993 to 2019 in the core and non-core areas. Many Minnesota breeding species also have summer distributions that extend south of the US border. Although BBS routes are being established in

Mexico, currently data are too limited to allow their incorporation into the statistical analysis.

The BBS population trends displayed in the species accounts, be they for the core BBS survey region, an individual Bird Conservation Region (BCR), or the state of Minnesota, were standardized so each figure displays the trend data beginning with the year 1967 and ending in 2019. These dates are noted in each figure caption. The figures were prepared for this publication by US Geological Survey staff at the Patuxent Wildlife Research Center in Laurel, Maryland. The calculation of the trend (i.e., the average annual percent change per year) discussed in the text may span either the years 1966–2019 or 1967–2019. Population trends reported for the core BBS survey area (i.e., "surveywide"), the Boreal Hardwood Transition BCR, and the Prairie Hardwood Transition BCR begin with the date 1966; population trends for the state of Minnesota and the Prairie Potholes BCR begin with the year 1967. The 1966 or 1967 distinction was simply due to when the BBS routes were initiated; in 1966 for BBS routes east of the Mississippi River or 1967 for routes west of the Mississippi River.

Besides their use in estimating population trends, BBS data have been used to estimate breeding populations in specific US states or Canadian provinces if a species occurs within that jurisdiction (PIF 2021). Population estimates were also developed using the MNBBA data (see details in Appendix A under Species Distribution Models and Population Estimates). The MNBBA statewide estimates usually varied considerably from those presented by PIF (2021) and were frequently much larger. Underestimates of population sizes using the BBS roadside counts have been identified previously, such as by Sólymos and colleagues (2018), who stated that population estimates "are most likely to be underestimates for most species." PIF (Will et al. 2019) has clearly recognized the issues associated with population estimates using the BBS data but emphasizes that the BBS data represent the only North American data capable of providing population estimates for most breeding landbirds. The reasons for differences in population estimates between the MNBBA and PIF (2021) include several factors. First, the MNBBA estimates were based only on data gathered in Minnesota. Second, the methods for population estimates used in the MNBBA modeling approach were similar to those used by Sólymos and colleagues (2018), which incorporated species detectability and distance to detection in population estimates (Thogmartin 2010). Third, MNBBA sampled and used point counts in roadless areas such as in the Boundary Waters Canoe Area Wilderness and in the Red Lake Peatland, which are neither sampled by the BBS nor incorporated in population estimates by PIF (2021). Fourth, MNBBA used the extensive data gathered during the National Forest Bird Monitoring Program in the Chippewa and Superior National Forests (Niemi, Howe, et al. 2016). And, finally, MNBBA assumed a 1:1 ratio of males:females for territorial bird species that were primarily detected by their territorial song. Note that PIF (2021) also included adjustments in their species population estimates such as detection area and regional area modifications (e.g., large lakes), and detectability adjustments such as pairing ratios and time of day (Will et al. 2019). Estimating bird species populations will continue to be a challenging and evolving topic of research well into the future (Stanton et al. 2019; Edwards et al. 2023).

The primary purpose of the BBS was to estimate population trends for North American birds (Sauer et al. 2020). The BBS also provides a wealth of other information used for such items as relative abundance maps in the United States and Canada; trend maps depicting where a species is stable, increasing, or decreasing; and specific estimates of abundance for specific areas (e.g., average species detections per route per year in Minnesota) (see Sauer et al. 2020). The BBS data set has been used extensively in the species accounts to present data on population estimates, population trends, and relative abundance of the species.

CONSERVATION

Our final section provides an overall context on the status of the species in Minnesota and nationally. We cite the Continental Concern Score that was developed by PIF (2021) to provide this overall perspective on the species status. We also identify other conservation designations of the species, especially those in Minnesota. Because these change frequently, Minnesota birds designated as *Species in Greatest Conservation Need* are only occasionally pointed out. We summarize some of the major threats that are known to affect the species, such as habitat change and loss, climate change, risk of migration such as collisions with towers and buildings, or other factors that appear to be affecting the species distribution or abundance. Finally, we include conservation actions that have been implemented or potentially would be useful if implemented to reduce threats to the species.

The future of climate change–related threats for Minnesota's breeding birds are included based on an analysis of bird species range changes in North America completed by scientists at the National Audubon Society (NAS 2019; Wilsey, Bateman, et al. 2019). The study used more than 70 data sources and more than 140 million bird records for a total of 604 bird species gathered across the North American continent. The analysis examined how the current breeding and winter ranges for birds were predicted to change in North America in response to three global warming scenarios; vegetation, including human land use in agriculture and urbanization; and several localized impacts, such as sea-level rise, extreme weather, drought, and lake level changes. The three global warming scenarios were based on the 2014 Intergovernmental Panel on Climate Change Fifth Assessment Report. The climate scenarios included, first, a 1.5°C (2.7°F) of warming above preindustrial levels, which is considered imminent between 2030 and 2052 if we do nothing to mitigate climate change; second, a 2°C (3.6°F) of warming, which could happen as soon as 2050; and, third, a 3°C (5.4°F) of warming, which could occur by 2080 (Wilsey, Bateman, et al. 2019). Additional papers with analyses using the results of the study can be found in Bateman and colleagues (2019) and Wilsey, Taylor, and colleagues (2019).

We primarily use the National Audubon Society (NAS) interactive website (Survival by Degrees: 389 Birds Species on the Brink) to summarize the overall changes in the breeding range and, when relevant, the wintering range changes for the breeding bird species of Minnesota (NAS 2019). We generally concentrate on the 2°C (3.6°F) warming scenario in our interpretations for future potential changes in North America and in Minnesota because the predictions are relevant within a near-term time frame (2050). However, we frequently comment on the other two scenarios if we deem them particularly relevant. Several species that clearly nest in Minnesota but are either very rare and/or on the very periphery of their breeding range in the state (e.g., Horned Grebe, Solitary Sandpiper, and Little Blue Heron) are omitted within the "current breeding range" of the analysis presented in the NAS website. In these situations, we comment on the overall changes predicted in North America. Under all three scenarios, the models for Minnesota's regular breeding species depicting those with the highest vulnerability are those largely confined to the Boreal Forest during the breeding season. Highly vulnerable species are defined as "those projected to experience the highest percent of current range loss with limited opportunity for future range gains" (NAS 2019). Species classified as habitat generalists are projected to be the least vulnerable and include such species as the Mourning Dove, Red-tailed Hawk, Great Horned Owl, Downy Woodpecker, Warbling Vireo, and Red-winged Blackbird.

We freely admit that we were surprised or alarmed by many of the climate projections even for the imminent 1.5°C warming scenario. Many species that nest in the northeastern and north-central portion of the state were projected to be extirpated under the 1.5°C climate scenario by 2050! Included were such classic boreal species as the Spruce Grouse, Northern Hawk Owl, Great Gray Owl, Boreal Owl, Black-backed Woodpecker, Olive-sided Flycatcher, Boreal Chickadee, Golden-crowned and Ruby-crowned Kinglets, Hermit Thrush, Connecticut Warbler, and Palm Warbler. As alarming as the total list of 27 species is, it is less surprising when one considers that "Minnesota is the second-fastest warming state in the nation" (NAS 2019). Appendix F provides a complete list of all Minnesota regular breeding species that are predicted to be extirpated in the state under each of NAS's modeled climate scenarios.

Finally, a major concern in NAS's projection of breeding range gains to the north of Minnesota regards the minimal evidence that the vegetation can migrate to the far north of Canada and Alaska within the time frames of 2050 or even 2080. If the lag times are too slow for the vegetation to respond, then species will need to adapt or face even higher probabilities of extirpation or extinction. Moreover, many long-distance migrants will face even greater flight distances to Central and South America as well as more hazards en route between breeding and wintering areas.

Abbreviations Commonly Used in Species Accounts

To reduce the length of the book and make the accounts more efficient to read, we abbreviated the frequently used names of many organizations.

AOU American Ornithologists' Union, now known as the American Ornithological Society (AOS) because of the merger of the AOU and the Cooper Ornithological Society in 2016 (https://americanornithology.org/). The AOS is a diverse, global network of empowered professionals, working together to advance the scientific study and conservation of birds.

BBS BBS is an abbreviation for the North American Breeding Bird Survey, a cooperative effort between the United States Geological Survey and the Canadian Wildlife Service to monitor the status and trends of North American bird populations. The sources used in this book can be found at https://www.mbr-pwrc.usgs.gov/; additional descriptions can be found in Sauer and colleagues (2020). BBS data may be summarized at different geographical levels. For example, "Minnesota BBS" refers to data collected along the BBS routes in Minnesota, while "surveywide BBS" refers to data collected on BBS routes established in the core region of the United States and southern Canada.

CBC The Audubon Christmas Bird Count is a count of winter birds conducted by volunteers across North America and south through Mexico and northern South America. Data are collected on one day from mid-December through the first week of January. Initiated more than a century ago, a summary of trends is available at: https://www.audubon.org/conservation/where-have-all-birds-gone.

CLO Cornell Laboratory of Ornithology, information used from their website at https://www.allaboutbirds.org/.

MBS Minnesota Biological Survey, a program of the MNDNR to systematically collect, interpret, monitor, and deliver data on plants and animals in Minnesota. As of the time of writing (February 2021), 82 of 87 Minnesota counties have been completed and summarized, but results are not complete for Koochiching and Lake of the Woods Counties or for northern portions of Beltrami, Clearwater, and St. Louis Counties. Data for some species may be limited in these counties; see https://www.dnr.state.mn.us/eco/mcbs/bird-map-list.html.

MNBBA Minnesota Breeding Bird Atlas, the breeding bird atlas project and data gathered from 2009 to 2013 in Minnesota, as summarized in the website https://mnbirdatlas.org/.

MNDNR Minnesota Department of Natural Resources.

MOU Minnesota Ornithologists' Union, the state ornithological society of Minnesota.

NABCI North American Bird Conservation Initiative, a program that brings together partners that work across state and country boundaries to ensure that North American bird conservation is coordinated at an appropriate scale for birds; see the Bibliography for specific documents cited in the text.

NWR National Wildlife Refuge, a system of public lands protected and managed by the United States Fish and Wildlife Service.

PIF Partners in Flight is an international program of more than 150 partner organizations distributed throughout the Western Hemisphere. The program is focused on all aspects of landbird conservation, from science to land management and outreach. The primary sources used are PIF Science Committee (2013), Rosenberg, Kennedy, and colleagues (2016), PIF (2019), and PIF (2021).

SNA Scientific and Natural Area, public lands managed by the MNDNR that protect outstanding ecological resources.

USFS United States Forest Service.

USFWS United States Fish and Wildlife Service.

USGS United States Geological Survey.

WMA Wildlife Management Area.

OVERLEAF: Male Ruby-throated Hummingbird taking flight. >
Photo by David Brislance.

SPECIES ACCOUNTS

I. CURRENT BREEDING SPECIES

NOTE: In the species accounts, illustrations are referenced with bold letters in the text and captions.

CANADA GOOSE

Branta canadensis maxima

Minnesota Seasonal Status: Regular breeding resident and migrant; regularly observed during the winter. The Canada Goose was abundant during the MNBBA.
Migration: Resident to short-distance migrant. In early summer, young birds that do not breed and failed breeders migrate to northern and subarctic areas, where they molt their flight feathers and spend the summer.
Food: Feeds on submerged aquatic vegetation in wetlands and lakes and grazes on grasses, sedges, and grains in uplands.
Nest: A depression located on elevated sites within or adjacent to wetlands; muskrat houses and artificial platforms are frequently used.
Conservation Concern: A game species; assigned a low PIF Continental Concern Score of 5/20.

Photo by Stephen Maxson.

Few wildlife species are as well-known as the Canada Goose. "Once seen, its grandeur creates an impression on the mind which even the casual observer never forgets" (Bent 1925). Familiar even in flight, this impressive bird is a complex species, demonstrating extensive morphological variations across its breeding range. In addition to differences in plumage and body proportions, Canada Geese range in size from the smallest Cackling Goose, which is just a little larger than a Mallard, to the Giant Canada Goose, with a wingspan of nearly 2 m. Eleven subspecies that were initially described were split into two different species in 2004: the large Canada Goose (*Branta canadensis*, with seven subspecies) and the smaller Cackling Goose (*Branta hutchinsii*, with four subspecies) (Banks et al. 2004). It is the Giant Canada Goose, specifically the subspecies *B. c. maxima,* that breeds in Minnesota. Formerly common "from south-central Manitoba and eastern North Dakota south to northern Arkansas" (Baldassarre 2014), the subspecies now ranges across the southern regions of eastern Canada, south through the eastern Dakotas to the Gulf coast, and east to the Atlantic coast. Breeding densities are especially high in the northern Great Plains. Cackling Geese breed in the Arctic tundra but are common migrants in Minnesota (Janssen 2019; S. P. Stucker 2004).

The return of the Giant Canada Goose from the brink of extinction in the early twentieth century to a bird so abundant that population control measures are now necessary is quite remarkable. Indeed, Minnesota is at the very heart of this extraordinary story.

MINNESOTA BREEDING DISTRIBUTION

Although the Canada Goose was a common summer resident throughout Minnesota in the early 1800s, its numbers had already "materially diminished" in the rapidly settled southern counties by the late 1800s (Hatch 1892). Forty years later, Roberts (1932) noted that even in the remote counties of northwestern Minnesota populations had been decimated by extensive wetland drainage, overharvesting, and egg collection. The last documented nesting in the state was in 1929 on Swan Lake in Nicollet County.

Coincidentally, the Giant Canada Goose disappeared from many of its traditional breeding grounds in North America, leading biologists to assume that the bird was either extirpated throughout much of its range or totally extinct. Given its plight, the USFWS began transplanting birds from captive flocks onto many national wildlife refuges, including the Agassiz NWR in northwestern Minnesota in the early 1950s (Dokken 2012).

But it was a discovery in the winter of 1962 in southern Minnesota that truly changed the bird's future (H. C. Hanson 1965). For years a small flock of large Canada Geese had resided in the city of Rochester. The first flock, secured from a private breeder in North Dakota, had been introduced by one of the Mayo brothers onto his private estate in 1924; in 1936, the city also introduced a small flock on Silver Lake (Eckberg 2010).

When Dr. Mayo stopped feeding his birds in 1939, the flock dispersed to nearby sites, including Silver Lake. This small population persisted and grew further when another dozen birds were purchased. But the real game changer was the construction of a power plant on the lakeshore in 1948. Its warm-water discharge kept Silver Lake open throughout the coldest months, enticing migrant geese from unknown breeding grounds in Canada to stay throughout the winter. By 1960 the winter flock had grown to 4,000 geese, and residents noticed that some of the wild birds were extremely large. Banding data documented that these large migrant geese nested in the Interlake Region of Manitoba (H. C. Hanson 1965).

The "rediscovery" of a wild Giant Canada Goose population led to a surge of new reintroduction efforts in Minnesota and surrounding states. By 1975, it had become an established breeding bird in all but the northeastern region of the state (Green and Janssen 1975), and populations continued to expand through the late 1900s, especially in urban areas, where the birds were protected and where they found abundant nesting and foraging opportunities. By 1998 they were confirmed nesting in 65 of Minnesota's 87 counties (Hertzel and Janssen 1998).

Today the Canada Goose remains a widely distributed and abundant waterfowl species, second only to the Mallard in the total number of records (4,877) reported during the MNBBA. Documented in 53.3 percent (2,618/4,916) of the surveyed atlas blocks, breeding was confirmed in all 87 Minnesota counties **(A)**. The abundance of records in the Twin Cities and Brainerd Lakes region is a testament to local volunteer efforts as well as the adaptability of

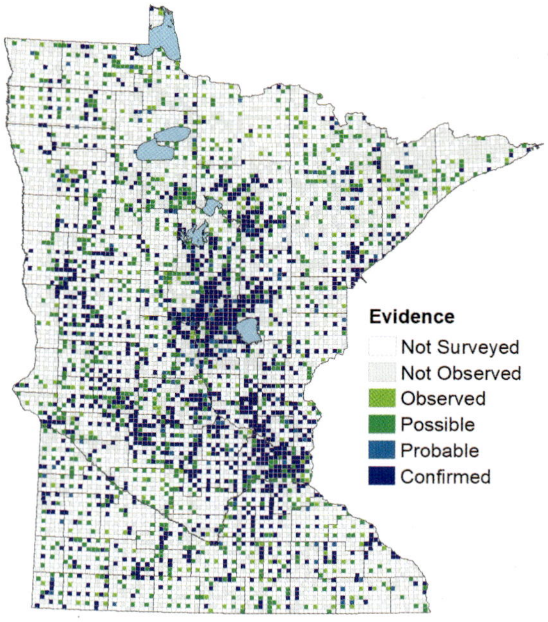

A. Breeding distribution of the Canada Goose in Minnesota based on the Breeding Bird Atlas (2009–13).

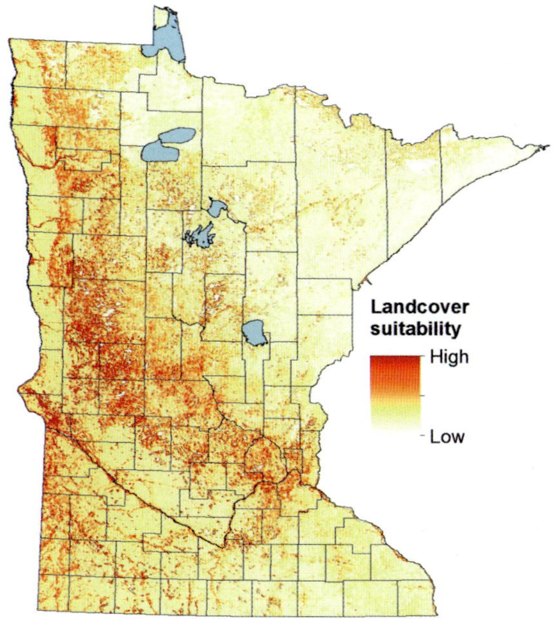

B. Landcover suitability map for the Canada Goose in Minnesota based on habitat, landscape context, and climate data gathered during the Minnesota Breeding Bird Atlas (2009–13) using the MaxEnt modeling approach.

the species to residential landscapes. The landcover suitability map, however, predicts that the core of the species' breeding range lies a little farther west of these two centers of MNBBA abundance **(B)**.

Now the most abundant and widely distributed subspecies of the Canada Goose in North America, *B. c. maxima* is well established and abundant throughout its former breeding range; in addition, birds have been translocated to sites beyond their traditional range into the southern United States (Mowbray et al. 2020). From these relocation sites, the population has continued to expand and increase in numbers.

BREEDING HABITAT

A wide range of habitats is selected during the breeding season, including forest and prairie wetlands, large reservoirs, and small urban ponds **(C)**. Two features are universal to these diverse sites: high-quality foraging habitat and dry, elevated sites in or near water for nesting. Geese are grazing herbivores that feed on aquatic plants and on grasses, sedges, and cultivated grains, so upland foraging habitats need to be located close to water (Mowbray et al. 2020; Baldassarre 2014).

POPULATION ABUNDANCE

The Giant Canada Goose (*B. c. maxima*) population has grown exponentially since its near extinction nearly one hundred years ago. Although a precise population estimate is not available, the birds "now represent a large proportion of all Canada Geese in the Mississippi Flyway," which totaled 1.5 million birds in 2019 (USFWS 2019). That same year, the entire North American Canada Goose/Cackling Goose population was estimated to number 8.1 million. Although *B. c. maxima* occurs in other flyways, the Mississippi Flyway represents the core of its breeding population. Its successful recovery is attributed to many factors, including reintroduction efforts, the species' adaptability to diverse habitats, its tolerance of urban/suburban landscapes, ample food supplies, minimal harvest pressure in urban environments, and the absence of natural predators (Baldassarre 2014).

In Minnesota, two long-term monitoring efforts document the species' spectacular recovery. The first, the BBS, shows an average annual increase of 14.71 percent across the state from 1967 to 2019 (Sauer et al. 2020)! Although these data lack some reliability, the trend line mirrors the statistically significant population increase observed throughout the BBS survey region **(D)**. The second, conducted in western and central Minnesota by state waterfowl biologists, demonstrates a steady increase in numbers since 1968 until the first decade of the twenty-first century, when numbers dropped and stabilized at approximately 150,000 birds (Cordts 2019).

CONSERVATION

As Giant Canada Goose populations recovered, management transitioned from reintroduction efforts to nuisance control. Urban residents, for example, are well aware of the challenges posed by the birds' excrement and overgrazing. They also pose a dangerous air safety hazard, well-documented when US Airways Flight 1549 struck a flock of Geese shortly after takeoff from New York City's LaGuardia Airport in 2009.

C. A large reservoir in west-central Minnesota, embedded in grasslands that provide an uninterrupted view in all directions, is among the many habitats where the Canada Goose can be found during the breeding season. Photo by Lee A. Pfannmuller.

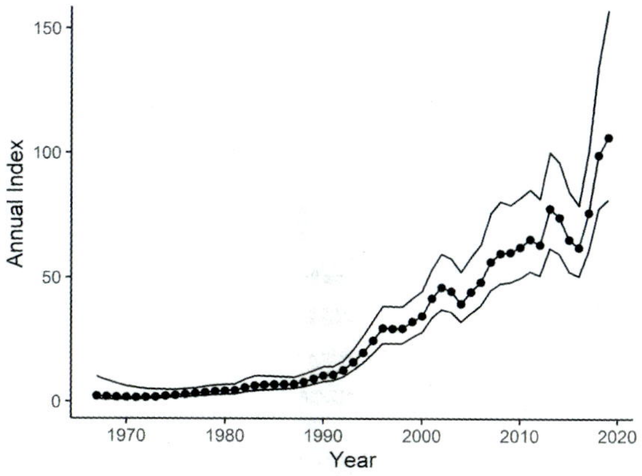

D. Surveywide breeding population trend and 95 percent credible interval (solid black lines) for the Canada Goose based on the 1967–2019 Breeding Bird Survey (Sauer et al. 2020).

Resident Geese became such a nuisance in the Twin Cities that relocation efforts were begun in the 1980s. Geese were rounded up from city lakes and transported to southern states to facilitate recovery efforts. It was not long, however, before southern populations grew to such a level that states would no longer accept additional birds. Today eggs are oiled to prevent hatching, shorelines are managed to make them less attractive to Geese, and birds are collected in the fall, processed, and donated to food pantries. In rural areas, the MNDNR also issues kill permits to farmers and uses special hunting seasons or conservation actions to reduce populations in areas where crop damage and other problems occur.

The Canada Goose is a popular game bird; the 2019 continental harvest totaled approximately 2.8 million birds, while nearly 200,000 were harvested in Minnesota (Raftovich et al. 2020). Many states, including Minnesota, have established early-season hunts to reduce the population of resident Geese before the northern populations migrate through (MNDNR 2015c).

Ultimately, aggressive harvest and control efforts may slow the species' population growth. Climate change could also play a role. The species is moderately vulnerable under a 2°C warming scenario; NAS (2019) projected a loss of 17 percent of the species' current breeding range by 2050 with only a 1 percent gain northward. Some of the loss is predicted to occur in southern Minnesota. Nevertheless, the adaptable Giant Canada Goose, once thought to be gone forever from Minnesota's wetlands, will likely remain in the state for many years to come.

MUTE SWAN

Cygnus olor

Minnesota Seasonal Status: Non-native visitant all seasons; documented nesting four times, but no subsequent viable breeding population. The Mute Swan was very rare during the MNBBA.
Migration: The birds are sedentary; year-round populations are established where there is open water during the winter or in urban areas where they are fed.
Food: Primarily submerged aquatic plants; small amounts of aquatic animals are consumed incidentally with the plants.
Nest: A depression within the center of a mound constructed largely of aquatic emergent vegetation; the nest is usually placed along the shoreline of a peninsula or island.
Conservation Concern: Designated an unprotected, regulated invasive species in Minnesota.

Photo by Larry Sirvio.

Large, white, and graceful—it is easy to understand the aesthetic appeal of the Mute Swan and why this Eurasian species was introduced to North America. Although some birds were intentionally introduced into the wild, the majority of Mute Swans acquired in the mid-1800s and early 1900s were for ornamental display in zoos, parks, and private estates. They were usually kept captive, with their wings clipped to prevent flight. Yet, despite intentions to keep the birds confined, feral breeding populations were eventually established in many states, especially along the northeastern Atlantic coast from Maine to South Carolina, and throughout the Great Lakes states. Smaller populations were established in the Pacific Northwest. Early introductions of birds that eventually spread were in Maryland, Massachusetts (1909 and 1916, Boston), Michigan (1919, Charlevoix County), and New York (1910 and 1912, Hudson River and Long Island).

Approximately the same size as Minnesota's native breeding resident the Trumpeter Swan, the adult Mute Swan is readily distinguished by its orange bill, with a black tooth at the tip and a fleshy black knob at the base. Its wings are often held slightly above its back, and its neck is usually held in a graceful, sinuous curve. Both the Trumpeter Swan and the similar Tundra Swan, which migrates through Minnesota, hold their necks erect. Although its common name implies otherwise, the Mute Swan's vocal repertoire includes hisses, grunts, snorts, and a rather quiet bugle call.

Fiercely territorial with voracious appetites that result in uprooting far more aquatic vegetation than they consume, Mute Swans can have a devastating impact on wetland communities. On the Atlantic Seaboard, large invasive populations are now a problem in New York, southern New England, and the Chesapeake Bay. Several states in the Great Lakes region have viable populations that are expanding and are likely the origin of many birds observed in Minnesota. In addition to Michigan's large population, which originated in the early 1900s, Mute Swans were reintroduced in or expanded to Ohio, Indiana, Illinois, and Wisconsin in the 1960s and 1970s, mostly at sites along the Great Lakes shorelines.

MINNESOTA BREEDING DISTRIBUTION

The first record of Mute Swans in Minnesota was on May 27, 1974, when three birds were seen on Lake Superior at Duluth, and later in June on the St. Louis River (Green 1975). Publicity surrounding those records turned up several more observations on Lake Superior from Cook County (January 1975) and the Thunder Bay area in Ontario (1974) to Ashland, Wisconsin (1973, 1974). It was later discovered that a feral breeding population was established in 1971–75 in northwestern Wisconsin (S. D. Robbins 1991).

In the 1980s there were yearly, mostly spring, reports from Lake Superior at Duluth. Although it was assumed that these birds originated from the Ashland population, the MOU Records Committee considered Lake Superior Mute Swan reports acceptable as records of wild birds (Eckert 1992). During this period, the first Minnesota breeding record was documented in Duluth in 1988, when two adults fledged one young (Wiens 1989). No viable population was established.

It has been more difficult to consider subsequent Minnesota records as reports of wild birds. Most are classified as origin "unknown." A compilation of Mute Swan records from the MOU (2021b) shows that observations were reported every year from 1980 to 2020. They occur during all seasons: spring, forty-eight years; summer, twenty-five years; fall, thirty-four years; and winter, thirty years. Reported from 57 of Minnesota's 87 counties, all but 14 of these county records are in the southern half of the state.

In addition to the Duluth (1988) nesting record, nesting has been documented in Steele (1989), Washington (2005), and Anoka (2006) Counties. Nonbreeding records are concentrated in the Twin Cities (many observers), Goodhue County (captive flock, winter open water), and Sherburne/Stearns Counties (winter open water from power plant). A wild breeding population, however, does not exist in Minnesota. Birds are assumed to be escapees or releases from captive flocks or wanderers from other states. In Minnesota, most open water freezes every winter, which constrains a sedentary species from establishing a viable population. Another constraint is their unprotected status under both state and federal authorities. Birds found nesting in natural habitats can be removed as a nuisance.

MNBBA records were from only two locations: McLeod County (July 2012) and the Minnesota River

Valley NWR on the border of Carver and Scott Counties (August 2009) **(A)**. Additional information about these records was reported to the MOU: the bird in McLeod County was with Trumpeter Swans and was seen multiple times; the bird on the refuge had been observed the previous two years, and there were multiple reports in 2009 on June 20 and from August 1 through October 18. Several Mute Swan observations were reported to the MOU during the course of the MNBBA but were outside of the breeding season. Like most observations, all were incidental reports from ponds, lakes, wastewater treatment plants, and near the casinos in Goodhue and Pine Counties.

BREEDING HABITAT

The Mute Swan's breeding habitat is described as ponds, lakes, inland and coastal marshes, and slow-flowing rivers (Kaufman 1996). For this semi-domesticated bird, any site with water and abundant food is likely to be sufficiently suitable, including sewage lagoons. In Minnesota, incidental records range from Lake Superior to small lakes and ponds **(B)**. Mute Swans are also reported on large rivers, such as the Mississippi and Minnesota, especially where open water is available during the winter months.

When Mute Swans remain in an area with good feeding habitat, their foraging can be very destructive to submerged aquatic vegetation. These large birds require 1.8–3.6 kg of food a day, consuming many aquatic plants like pondweeds, common waterweed, wild celery, and wild rice. Their method of feeding, by uprooting much of the vegetation, can damage the site's productivity (Baldassarre 2014).

B. The general location and habitat of the Mute Swan nest in the St. Louis River in 1988. Photo by Gerald J. Niemi.

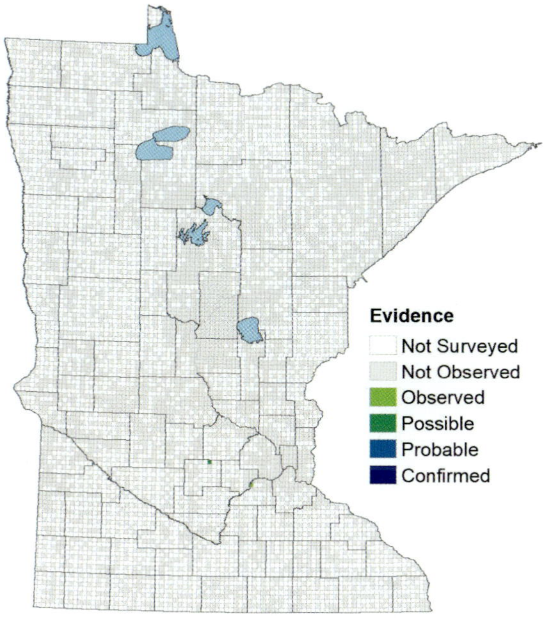

A. Breeding distribution of the Mute Swan in Minnesota based on the Breeding Bird Atlas (2009–13).

POPULATION ABUNDANCE

Although Mute Swans were first reported breeding in the wild in 1910 along the lower Hudson River (Baldassarre 2014), the wild population did not begin to expand dramatically until the early 1990s, when it increased on the East Coast and in the Great Lakes, particularly in Michigan. The continental population is now estimated at approximately 22,000 to 25,000 birds (Baldassarre 2014). Recent population increases have been extraordinary. In Maryland, for example, records from midwinter surveys increased from 2,145 in 2002 to 4,345 in 2006 (Ellison 2010); in Michigan, estimated numbers increased from 500 birds in 1972 to 15,500 in 2010 (Michigan DNR 2012).

CONSERVATION

Several factors have prompted state and federal agencies to increase management of Mute Swans. Contrary to most conservation plans, these policies focus on control rather than on protection. First, federal agencies changed their strategy of ignoring this feral exotic species following a 2001 lawsuit that resulted in a ruling that the Mute Swan is not a protected species under the Migratory Bird Treaty Act. As a result, the federal government does not have regulatory authority over the Mute Swan and management control is the responsibility of the states. Second, the desirability of control was made obvious by field observations that showed that increasing Mute Swan populations were having a detrimental effect on wetland ecology, and their aggressiveness was a threat not only to other waterfowl but also to humans. Lastly, state wildlife agencies recognized their responsibility and established a structure to implement controls through organized working groups, management plans, and cooperative agreements.

Because of the public appeal of this beautiful and partly tame bird, direct control methods like shooting birds cause controversy. Most acceptable control activities, such as destroying eggs or small young, although used, are not very effective in stopping population expansions because the adults are long-lived. Another approach has been public educational efforts to demonstrate that this invasive Swan causes detrimental effects to native waterfowl. It is necessary to make it clear that management agencies have the authority for control. Some physical control activities have been implemented where appropriate. Minnesota has taken the approach that preventing the establishment of a Mute Swan breeding population is critical. In 1996, the Minnesota Legislature defined this bird as "Unprotected" (2021 Minnesota Statutes: Chapter 97A.015, Subd. 24, Unprotected birds). It has been classified as "a regulated invasive species whose introduction into the wild is prohibited. Possession of captive birds requires a state game farm license and fencing to contain them" (Minnesota Administrative Rules: Chapter 6216.0260, Subp. 5, Birds). There is no explicit Minnesota management plan for the Mute Swan.

TRUMPETER SWAN

Cygnus buccinator

Photo by Laura Erickson.

Minnesota Seasonal Status: Regular breeding resident widely distributed throughout the state but most common in east-central and north-central Minnesota, and in the southern counties of northwestern Minnesota. Most birds remain through the winter months where open water is available. Trumpeter Swans were common during the MNBBA.
Migration: Short-distance migrant. The majority of Minnesota's breeding population migrates only as far south as central and southern Minnesota to sites where there is open water and an abundant food supply. Two power plant sites now offer some of the best opportunities for Minnesota's wintering swans: Xcel Energy's Monticello Nuclear Generating Plant along the Mississippi River, and Otter Tail Power Company's coal-fired power plant in Fergus Falls, located along the Otter Tail River. Indeed, the former has become a winter destination site for viewing nearly 2,000 Trumpeter Swans, which are attracted by the site's open water and local efforts to feed the birds.
Food: Largely herbivorous, feeding on submerged and emergent aquatic vegetation during the summer; during the winter, birds may forage in agricultural fields for waste grain.
Nest: Large platform of aquatic vegetation built on a variety of wetland structures, including muskrat houses, beaver dens, floating mounds of vegetation, and islands.
Conservation Concern: Assigned a moderate PIF Continental Concern Score of 11/20. Formerly extirpated, but now classified as *Special Concern* in Minnesota.

Recovered from the brink of extinction, the Trumpeter Swan is now thriving amid Minnesota's grasslands and forests, where its unmistakable bugle-like call had fallen silent for nearly a century. Throughout recorded history, swans have always had a special place in our hearts and have been inextricably woven into our lore. So it is not surprising to find naturalists in the new world waxing poetic about the magnificence and grace of this bird. As John James Audubon recounted, "To form a perfect conception of the beauty and elegance of these swans, you must observe them when they are not aware of your proximity." He went on to describe watching a small flock of swans gliding through the air with "surprising agility and grace." The experience left him feeling "more happy and void of care than I can describe" (Audubon 1827–38).

Nearly 1.4 to 1.6 m (4 to 6 feet) in length and weighing between 9 and 13.5 kg (20 to 30 pounds), the Trumpeter Swan is the largest waterfowl species in North America. Its size requires a runway of at least 100 m of open water for the bird to become airborne. Its elongated windpipe is nearly three times larger in volume than that found in other birds its size (Hinds and Calder 1971) and helps produce the unique trumpet-like call. Indeed, the species' Latin name, *Cygnus buccinator,* means "swan with a trumpet."

Despite its size and conspicuous white plumage, two other large white swans that occur in Minnesota might challenge identification efforts. The Mute Swan, an invasive species, is quite rare and can be easily distinguished by its bright orange bill with a large black knob at the base, near the forehead. Distinguishing the Trumpeter Swan from the Tundra Swan, a common spring and fall migrant in the state, is more difficult. Although they can be easily differentiated by their calls (the Tundra has a higher-pitched, more mellow call), positively identifying birds far out on the water or in a field when they are not calling is an altogether different matter. Most birders rely on facial features, including the prominence of their eyes, the color of the lores, and the profile of the bill between the eyes, but even these can be confusing without help from those most experienced with the birds **(A)**. Luckily, during the breeding season there is no confusion!

Unfortunately, the Trumpeter Swan was an easy target for the region's early European settlers and

A. Features of the Trumpeter Swan's head that aid identification include the red "lip-stick" line on the lower mandible; the absence of any yellow color at the base of the bill near the eye; and the broad, flat bill that makes the black eye look indistinguishable. Photo by Laura Erickson.

market hunters. Although the older adults were considered "about as tough and unfit for food as an old horse" (Forbush 1912), the young birds were quite tender and tasty. But the adults were not spared from the gun. Often shot just for sport, they were prized for their skin and feathers, which were used for women's powder puffs, fashionable hats, and writing quills (CLO 2017a).

The swan's demise was so rapid that, in the early 1900s, this once-abundant species, originally found throughout Canada and the central United States, was declared a vanishing species, on the brink of extinction. By 1932, only 69 Trumpeter Swans were known to remain in the remote mountain valleys of Montana, Idaho, and Wyoming (Matteson et al. 1988). In 1935, the core of this breeding area, in the Centennial Valley of southwestern Montana, was protected as the Red Rock Lakes NWR. Located just 50 miles west of Yellowstone National Park, the region was an ideal site for the birds. In addition to stable water levels and abundant forage, hot springs provided open water year-round, muskrat houses and islands provided nesting sites, and the region's remoteness provided a safe and secure environment. With the protection of the Migratory Bird Treaty Act of 1918, all the ingredients were present to establish a foundation for future recovery efforts (Ross 1998). Bolstered further by the discovery of a small population in Alberta and an even larger one in Alaska's Copper River Delta, efforts to restore the population soon began in earnest.

Population recovery is a long, slow process for this long-lived species. Young swans do not begin breeding until they are at least four years old. Although the birds may live more than twenty years, they raise only one brood of four to six young per year. Wet and cold weather, unstable water levels, predators, and other threats can result in significant losses. Regardless, the hard work of local, state, and federal resource agencies, tribes, and conservation organizations has resulted in one of the most successful wildlife conservation stories in North America. In Minnesota, this majestic bird, once considered long-gone from the state, is now trumpeting its return from wetlands along the Iowa border north all the way to Kittson County in the northwest and Cook County in the northeast.

Although the original extent of its breeding range is not entirely known, the Trumpeter Swan formerly occupied much of the north-central United States and Canada, from the Great Lakes west to British Columbia and southern Alaska. Today its distribution is quite fragmented, with populations in Alaska and western Canada, the Rockies, the Central Plains, and the Upper Midwest **(B)**.

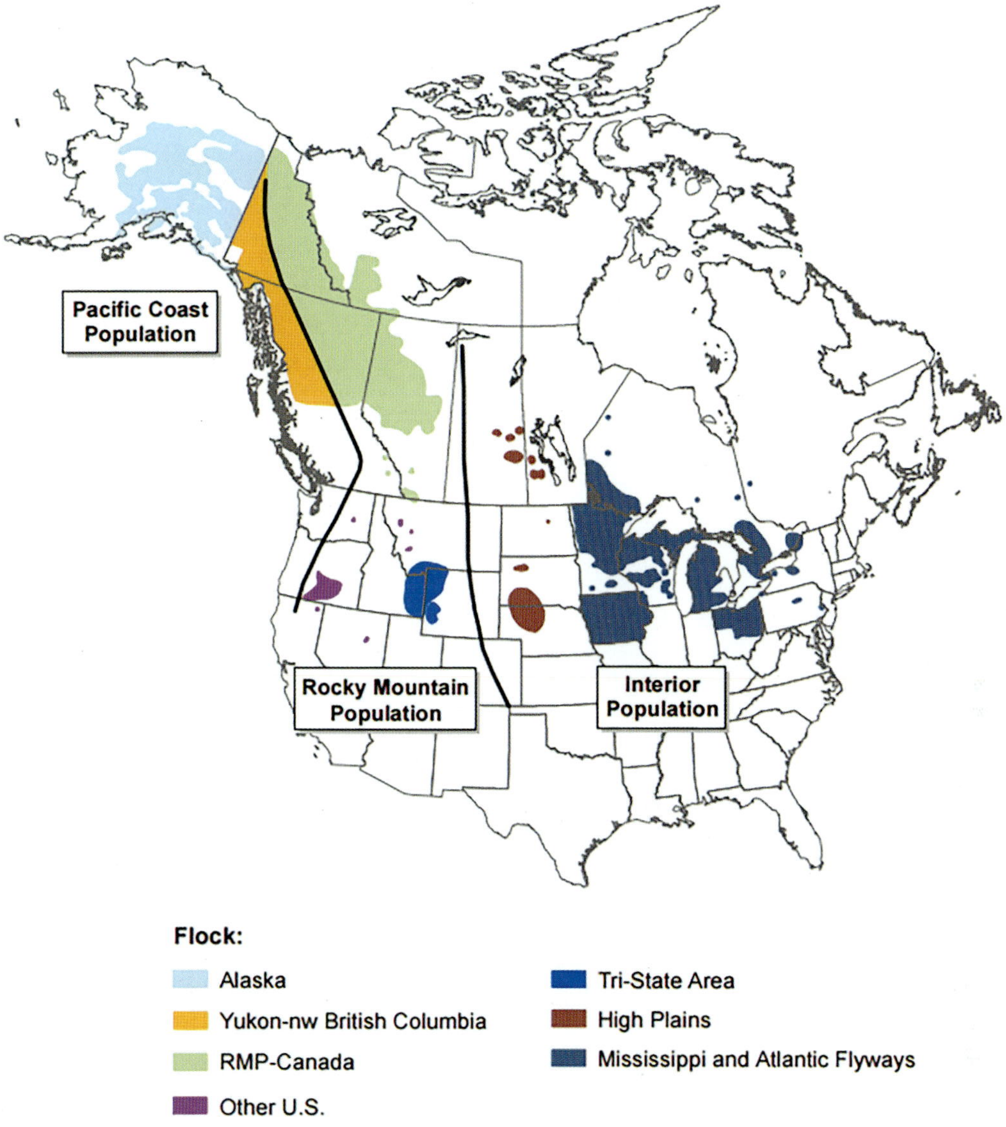

B. Current distribution of the Trumpeter Swan in North America. Reproduced from Groves (2017, fig.1).

MINNESOTA BREEDING DISTRIBUTION

When Roberts compiled his treatise on Minnesota birds in 1932, he wrote, "The days of the Trumpeter Swan as a bird of Minnesota have long since passed." He believed that, prior to the late 1800s, it was a breeding resident throughout the grasslands and sparsely wooded regions of the state. Breeding records were available from Heron Lake in Jackson County (1883), Everson Lake in Meeker County (1884 or 1885), Pike Lake in central Minnesota (1853), and Swan Lake in Nicollet County (1823). The birds on Swan Lake were reported by G. C. Beltrami, who was part of the Long Expedition to the territory in 1823. Keating, the expedition's historian, reported that the Sioux name for the lake, *Manha tanka otamenda,* meant Lake-of-the-Many-Large-Birds, no doubt in reference to the longtime tradition of Trumpeter Swans nesting there (Roberts 1932).

Any further accounts of the species' presence in the state were absent until 1965, when a provocative article by Robert Turner, titled "Will Trumpeter Swans Return to Minnesota?," appeared in the statewide ornithological journal *The Loon* (Turner 1965). Turner questioned whether some recent summer records of Whistling Swans (now known as Tundra Swans) in Minnesota were actually Trumpeter Swans. Trumpeter Swans had been introduced to Lacreek NWR in southwestern South Dakota, and he argued that vagrants from that population could already be showing up in Minnesota. Lacreek NWR was the first location in the Great Plains to receive 20 cygnets from the Red Rock Lakes NWR in Montana in 1960. Additional releases in 1961 and 1962 resulted in

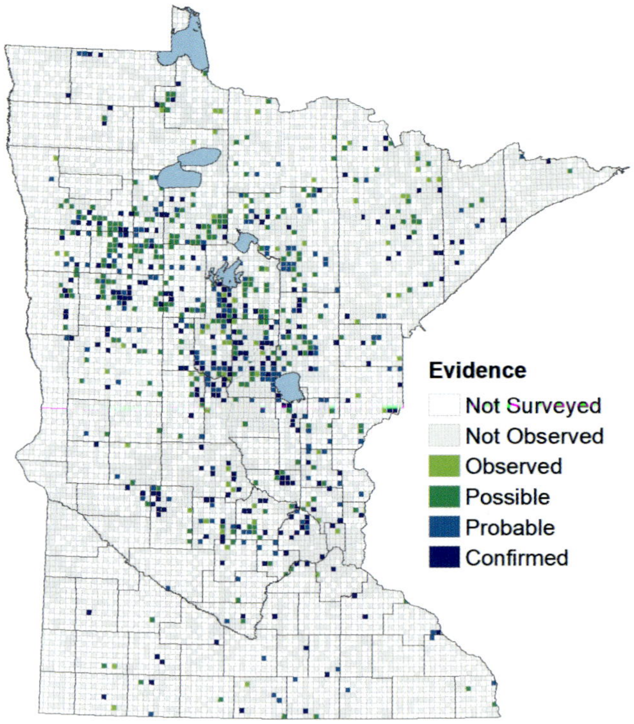

C. Breeding distribution of the Trumpeter Swan in Minnesota based on the Breeding Bird Atlas (2009–13).

a steadily growing population on the refuge and in the surrounding vicinity. Turner urged the USFWS to establish a breeding population in Minnesota.

In 1966, the Hennepin County Park Reserve District (now the Three Rivers Park District) began experimenting with the reintroduction of Trumpeter Swans to the county's park reserves. Their long-term goal was to establish a free-flying population of at least 100 birds with 15 breeding pairs within the district and throughout south-central Minnesota (Gillette 1986). Forty swans were secured from the Red Rock Lakes NWR, but initial efforts to simply release the birds proved unsuccessful (Gillette 1974). The tide turned when the district initiated its own captive breeding program, caring for the young cygnets and not releasing them until they were two years old. The first release of young raised by this program occurred in 1978, and the first successful nesting occurred in 1979 (MNDNR 2019l). By 1996, more than 245 young birds had been raised and were successfully nesting within the park district and beyond. One pair released from Lake Rebecca Park Reserve in 1983 even nested just north of Grand Rapids in Itasca County in 1986 (Kittelson 1996; Gillette 1986). The park district's successful program continued for another thirteen years, until it was retired in 2009.

In 1982, the MNDNR outlined an initiative to build on the work of the park district by reintroducing birds in northwestern Minnesota, working collaboratively with state and federal agencies, provinces, universities, tribes, and the Trumpeter Swan Society. Kittelson (1996) and Henderson (2017) provide comprehensive overviews of the initiative. Briefly, eggs were acquired from Red Rock Lakes and Lacreek NWRs, from the Minnesota and Brookfield Zoos, from private propagators, and from the state of Alaska. The birds were reared at the Carlos Avery WMA in Anoka County until they were two years old. Releases began in 1987 near Tamarac NWR in Becker County and continued for more than twenty years. Birds were also released at other sites in northwestern Minnesota, farther east in Itasca and St. Louis Counties, on Swan Lake in Nicollet County, and on Heron Lake in Jackson County. By 1998, Hertzel and Janssen had confirmed nesting in six counties since 1970: Aitkin, Becker, Carver, Hennepin, Hubbard, and Polk. The 1986 Itasca County nesting was not included because it was an observation of an adult pair with four cygnets, which did not meet the definition of a confirmed nesting record by the MOU (Hertzel and Janssen 1998).

By 1999, more than 75 breeding pairs had become established in the state (MNDNR 2000). A winter survey in 2011 documented a statewide population of more than 5,500 birds (MNDNR 2011), and a comprehensive breeding-season survey in 2015 found more than 17,000 individuals (MNDNR 2016; Henderson 2017). Owing to similar restoration efforts elsewhere in the United States, Trumpeter Swans are now found in many portions of their former breeding range, including in Wisconsin, Iowa, and Michigan.

From 2009 to 2013, when the MNBBA was conducted, participants reported 1,173 Trumpeter Swan records in 16.0 percent (772/4,832) of the surveyed atlas blocks. Breeding was confirmed in 244 (5.0 percent) of the blocks **(C)**. The birds were most abundant in the western portions of the Laurentian Mixed Forest Province, south of the expansive peatlands, and were least abundant in the Prairie Parkland and Tallgrass Aspen Parklands Provinces of western Minnesota. They were reported in 72 of Minnesota's 87 counties and were confirmed breeding in 55 counties. It has been a remarkable recovery for a species on the brink of extinction in the early 1900s.

BREEDING HABITAT

Throughout their breeding range, swans utilize a variety of freshwater wetland habitats, including marshes, ponds, shallow lakes, and, occasionally, slow-moving rivers. Quiet bays of large lakes are also used. The actual size of the site seems less important than the following criteria: sufficient open water to enable the birds to become airborne (approximately 100 m or more), abundant aquatic plants for foraging, stable water levels, good water quality, suitable structures or vegetation for nesting (e.g., muskrat or beaver houses), and limited human disturbance (Baldassarre 2014; C. D. Mitchell and Eichholz 2020) **(D)**. A field study conducted in northern Minnesota in 2007 and 2008 found that swans actually preferred smaller wetlands and lakes (less than 24 ha) with extensive forest cover along at least 75 percent of the shoreline and shallow beds of cattails, which provide tubers for feeding. Such sites were typically less disturbed by shoreline development and recreational boat traffic (Henderson 2017).

Productivity of nesting swans is often higher in sites that have irregular shorelines, are shallow (less than 1.2 m deep), and support a diverse and abundant community of aquatic plants, especially submerged macrophytes. Studies suggest sites that provide multiple nesting opportunities and open water early in the season are preferred (C. D. Mitchell and Eichholz 2020). In the central United States, including Minnesota, wetlands usually support only one breeding pair; farther north and west they may support multiple pairs. As the population size continues to increase in the interior, lakes may eventually support more pairs. Territories range in size from as small as 1.5 ha to more than 100 ha (Baldassarre 2014).

D. Shallow lakes with natural, undeveloped shorelines and a rim of emergent vegetation provide breeding habitat for the Trumpeter Swan in northwestern Minnesota. Photo by Lee A. Pfannmuller.

POPULATION ABUNDANCE

Since 1968, the USFWS has coordinated the North American Trumpeter Swan Survey. Following the initial survey in 1968, the second one was conducted in 1975, and it has been repeated every five years since. Cooperators include federal, state, and provincial resource agencies and individuals across both the United States and Canada. Regions can establish their own survey design but are encouraged to address several overarching goals, including providing an estimate of annual reproductive success. The most recent survey data are available for 2015, when the continental population of Trumpeter Swans was estimated at 63,016 birds, an increase of 63 percent since the previous survey in 2010 **(E)**. The estimated average annual growth rate since 1968 was 6.6 percent, an exponential rate of population growth (Groves 2017). Range-wide, the population has expanded so quickly that the expense of the survey is now a challenge for many of the cooperating agencies.

In Minnesota, the 2019 Minnesota Waterfowl Breeding Population Survey generated a statewide population estimate of 23,200 swans (Cordts 2019). This represents more than a 300 percent increase over the 2010 estimate of 5,550 birds.

CONSERVATION

The recovery of the Trumpeter Swan is truly a conservation success story. In Minnesota, the species was deemed extirpated in 1984, when the first state list of *Endangered, Threatened,* and *Special Concern* species was published. In 1996 it was downgraded to *Threatened* status, and in 2013 downgraded again to *Special Concern* status (MNDNR 2012). Because Minnesota supports nearly 30 percent of the North American population, Audubon Minnesota designated it a Minnesota Stewardship Species (Pfannmuller 2012). Although the continental population is growing at an unprecedented rate, the number of breeding adults is still exceptionally small. As a result, nationally it is considered a moderate conservation priority (PIF 2021).

During the course of the recovery efforts, there were numerous setbacks that clearly highlight many of the risks that still confront the species. The most significant is lead poisoning. When the birds tip up and forage for tubers and roots among the bottom sediments, they can easily ingest lead shot and lead sinkers. Although lead shot has been banned nationally for waterfowl hunting since 1991, wetlands that were popular waterfowl hunting spots may have decades of accumulated spent shot in their sediments. From 1986 to 1989, the University of Minnesota's Raptor Center documented that approximately 40 percent of the Trumpeter Swan fatalities in the state were the result of lead poisoning (Degernes and Frank 1991). Whether that level of mortality persists is unknown. The Raptor Center still receives dead swans, but only about 10 birds per year are submitted (D. Larson, pers. comm.).

Other risks to local populations include collisions with power lines, illegal shooting from vandalism or misidentification with snow geese, human disturbance during the nesting season, and the loss and degradation of wetland habitats.

Because the Trumpeter Swan is a wetland-dependent species, the potential impacts of climate change are a concern. Wetlands, particularly in the continent's interior, are considered at risk as temperatures continue to rise. A recent assessment by NAS (2019) classified the species as highly vulnerable under a 2°C warming scenario that projects that 81 percent of its current breeding range, including much of Minnesota, would be lost by the year 2050, while the amount of habitat gained in Canada would only increase by 52 percent. Interestingly, should the climate increase by 3°C, the amount of habitat gained in the arctic regions could compensate for the projected loss, reducing the species' vulnerability to a moderate ranking.

Although Minnesota's Trumpeter Swan population is growing at an unprecedented rate, there are clearly threats that could impact current and future populations. History has shown repeatedly that we should never take a species' abundance for granted, even one as large and conspicuous as the elegant Trumpeter Swan.

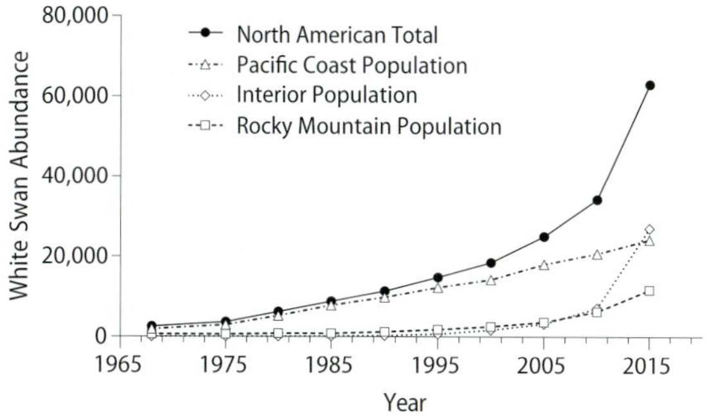

E. Trends in the abundance of adult and subadult Trumpeter Swans ("white swans") in the Pacific coast, Rocky Mountain, Interior, and total North American populations, 1968–2015. Reproduced from Groves (2017, fig. 2).

WOOD DUCK

Aix sponsa

Minnesota Seasonal Status: Regular breeding resident and migrant; occasionally observed during the winter. The Wood Duck was common during the MNBBA.
Migration: Short-distance migrant that winters primarily in the United States.
Food: Dabbling duck that feeds on aquatic and terrestrial invertebrates, seeds, and fruits.
Nest: Secondary cavity nester that selects large, mature trees in riparian habitats; also uses nest boxes.
Conservation Concern: A game species; assigned a low PIF Continental Concern Score of 7/20.

Superlatives abound when describing the male Wood Duck in his full nuptial plumage. Even his Latin name, which means a waterbird in bridal dress, is a pronouncement of the bird's stylish attire. Roberts (1932), one of the bird's many admirers, wrote that "the brilliancy, variety, and exquisite pattern of its colors is rivalled by few, if any, of its kind in the world." The male's iridescent green head crest, the white "bridle" on its black face, and its large red eye easily distinguish it from every other waterfowl species. Even the female has a subtle elegance, with a teardrop patch of white around the eye and a white throat that contrast with her grayish-brown plumage. The common vocalization we associate with the Wood Duck, *oo-eek, oo-eek,* is actually the female's raspy, high-pitched call, often given as she takes flight.

Supremely adapted to its tree-dwelling habits, the Wood Duck's slender body enables it to squeeze into narrow nesting cavities. Toes are equipped with a sharp claw that helps the hatchlings to climb out of the nesting cavity and the adults to perch on tree branches. Its eyes, larger than that of any other dabbling duck, help it navigate through the forest canopy, while its short wings and broad tail assist with aerial maneuverability.

The Wood Duck's primary breeding range extends across the eastern United States and southeastern Canada. Farther west, a population occurs from Montana west to the Pacific coast and south through California; a small population is also present in Cuba. Some of the highest breeding densities occur in west-central Minnesota.

Photo by David Brislance.

MINNESOTA BREEDING DISTRIBUTION

One hundred years ago the Wood Duck's extinction seemed imminent. Like so many waterfowl, prior to the enactment of hunting regulations the birds were harvested year-round, severely depleting populations (Bellrose 1976). At the same time, the loss of cavity trees due to extensive logging also likely contributed to population declines (J. Lawrence, pers. comm.). In Minnesota, Roberts (1932) noted that Wood Ducks were an abundant summer resident in the 1800s but "greatly reduced in numbers" by the early 1900s. Nesting was only confirmed in Grant County, while young broods were reported in just five counties: Aitkin, Clearwater, Hennepin, St. Louis, and Stearns.

When the federal Migratory Bird Treaty Act was signed in 1918, the Wood Duck hunting season was immediately closed and remained so until 1941. The response was nearly immediate. Only a few years later, Roberts (1932) noted it was nesting "in fair numbers" throughout the state.

When Green and Janssen (1975) published an updated account of the species' status forty years later, the Wood Duck had become the fourth most abundant nesting duck in the state (Lee et al. 1964). It was scarce, however, in the northeast and was absent in the Red River valley and southwestern Minnesota. Twelve years later, Janssen (1987) reported its breeding range was continuing to expand, with nesting confirmed in 68 counties since 1970. By 1998, Hertzel and Janssen (1998) documented Wood Duck nesting records in all but five counties: Beltrami, Benton, Otter Tail, Steele, and Wilkin. Likewise, the MBS reported the species was widely distributed across the state.

The Wood Duck was the second most abundant duck reported during the MNBBA; participants reported a total of 2,828 Wood Duck records from 36.1 percent (1,738/4,821) of the surveyed atlas blocks **(A)**. Reported from each of Minnesota's 87 counties, breeding was confirmed in all but two counties: Traverse and Red Lake. The density of reports in the Twin Cities and Brainerd Lakes regions reflects, in part, the number of atlas volunteers concentrated in these areas. Despite their statewide distribution, Wood Ducks remained least abundant in the heavily cultivated regions of the Red River valley and in far north-central Minnesota.

The land suitability map predicts that the most suitable Wood Duck habitat occurs across central Minnesota, north to the Hardwood Hills Subsection and south throughout southwest and south-central Minnesota **(B)**. Although the birds are sparse throughout the Red River valley and in the southeast, the rivers and streams in these regions still provide highly suitable habitat for this riparian species.

More than one hundred years after many predicted the species' demise, the Wood Duck not only has recovered to the point of reoccupying most of its original breeding range but also has expanded its breeding populations north in Canada and west in the Central Plains (Hepp and Bellrose 2020).

BREEDING HABITAT

Wood Ducks select mature woodlands that border lakes, streams, rivers, and small wetland pools **(C)**. Old or dead and dying trees provide nesting cavities, while shallow wetlands provide foraging sites for the adults and young broods. When natural cavities are

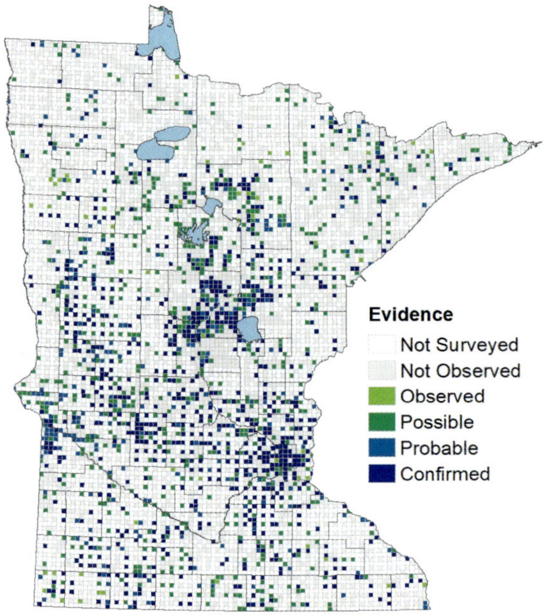

A. Breeding distribution of the Wood Duck in Minnesota based on the Breeding Bird Atlas (2009–13).

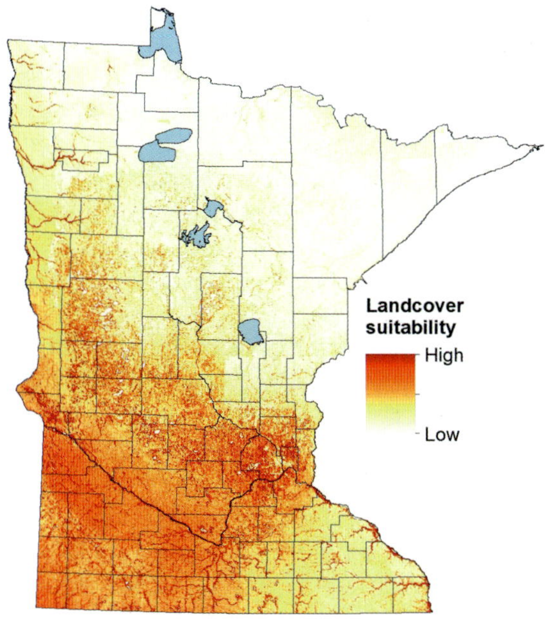

B. Landcover suitability map for the Wood Duck in Minnesota based on habitat, landscape context, and climate data gathered during the Minnesota Breeding Bird Atlas (2009–13) using the MaxEnt modeling approach.

scarce, the birds take readily to nest boxes. Wetland cover is important, particularly during brood rearing, and may be furnished by overhanging trees and shrubs, by wooden debris, or by emergent aquatic vegetation. Large expanses of open water are generally avoided. Although nesting cavities adjacent to water are usually preferred, Wood Ducks may nest up to 2 km from the nearest body of water (Hepp and Bellrose 2020). A Minnesota study demonstrated that 70 percent of females stayed within 1 km of the nesting cavity (Baldassarre 2014).

C. The Wood Duck is common along quiet backwaters of the Mississippi River that provide plenty of cover, foraging opportunities, and mature trees for nesting. Photo by Lee A. Pfannmuller.

POPULATION ABUNDANCE

The Wood Duck's forest habitat makes it difficult to effectively monitor populations and assess population abundance. The North American Waterfowl Breeding Population and Habitat Survey, the gold standard for monitoring waterfowl populations, more effectively monitors species that nest in prairie wetlands than birds that nest in woodlands. As a result, waterfowl biologists rely on other survey tools. The BBS, for example, was designed to monitor songbirds but provides a reasonable population index for other common species like the Wood Duck. Indeed, since the BBS began in 1966, Wood Ducks have shown a significant, surveywide population increase, averaging 1.16 percent per year through 2019 (Sauer et al. 2020). In Minnesota, the BBS trend is similar, with a significant annual increase of 1.42 percent per year from 1967 to 2019 **(D)**.

Factors responsible for the increase include strictly managed harvest regulations, protection and restoration of riparian wetlands, and the gradual maturation of forests in the northeastern United States following the abandonment of small farmsteads. In some areas, nest boxes are helping to restore local populations.

Minnesota waterfowl biologists have another tool to assess Wood Duck populations. The state Waterfowl Breeding Population Survey monitors waterfowl populations in western and central Minnesota. Data collected by this survey, however, illustrate a steep decline in Wood Duck numbers, particularly beginning in the year 2001 (Cordts 2019). Earlier leaf-out dates that make it more difficult to detect the forest-inhabiting Wood Duck might explain some of the decline. Nevertheless, the same survey also illustrates declines for two other abundant waterfowl species, the Mallard and Blue-winged Teal.

The different population trends documented by the Minnesota BBS and the Minnesota Waterfowl Survey may indicate that neither monitoring tool is entirely effective for this woodland-dependent species, or they may signal that Wood Ducks are doing well at the statewide level but are experiencing local declines in areas more intensively surveyed by the Waterfowl Survey. Poor wetland conditions are widely cited as factors negatively impacting waterfowl populations in western Minnesota.

Regardless, across North America, Wood Duck populations have certainly increased from approximately 2.8 million birds in the early 1990s (Bellrose and Holm 1994) to a current estimate of 4.6 million more than twenty years later (North American Waterfowl Management Plan Committee 2012). In Minnesota, the statewide population is believed to exceed 100,000 birds (MNDNR 2019n).

CONSERVATION

The Wood Duck's recovery is one of the most successful wildlife recovery efforts in North America. Other than the Eastern Bluebird, there is no other bird species for which the general public has been so actively engaged in its restoration. Since its acceptance of artificial nesting structures was first documented in the 1930s, conservation volunteers across the United States have built and maintained thousands of Wood Duck nest boxes. Although habitat remains the most important element in Wood Duck conservation, nest boxes have established or increased populations in many areas (Baldassarre 2014).

A popular species among waterfowl hunters, in the fall of 2019 Wood Ducks comprised approximately 9.7 percent of the estimated duck harvest in the United States. In Minnesota, Wood Ducks are second only to Mallards in the number of birds harvested, comprising 18.1 percent (80,716 birds) of the estimated state harvest in 2019 (Raftovich et al. 2020).

Climate change, however, may pose some risk to the species' future in Minnesota. Under a 2°C warming scenario, NAS (2019) projected a loss of 20 percent of its current breeding range by 2050, including in portions of southern and northwestern Minnesota, and a 37 percent gain farther north and west. Although it was assigned a low vulnerability ranking, management practices that protect and restore Minnesota's riparian forests are essential to ensure that this elegant species can adapt to such change.

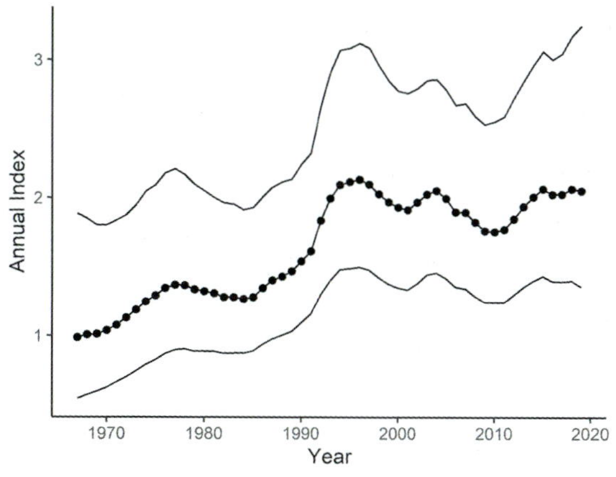

D. Breeding population trend and 95 percent credible interval (solid black lines) for the Wood Duck in Minnesota based on the 1967–2019 Breeding Bird Survey (Sauer et al. 2020).

BLUE-WINGED TEAL

Spatula discors

Minnesota Seasonal Status: Regular breeding resident and migrant. The Blue-winged Teal was common during the MNBBA.
Migration: Long-distance migrant, wintering from the southern United States south to northern South America.
Food: Dabbler that feeds on aquatic invertebrates and aquatic plants; rarely feeds away from water except during fall migration.
Nest: Ground nester.
Conservation Concern: A game species; assigned a low PIF Continental Concern Score of 7/20.

This handsome little dabbling duck is known for its swift and agile flight, its quiet nature, and its sensitivity to variable wetland conditions on its nesting grounds. Audubon himself reflected on the speed and beauty of the bird: "The flight of the blue-winged teal is extremely rapid and well sustained. Indeed, I have thought that, when traveling, it passes through the air with a speed equal to that of the passenger pigeon. When flying in flocks in clear sunny weather, the blue of their wings glistens like polished steel" (quoted in Bent 1923).

The blue speculum provides the species' namesake. On the water, however, it is the male's shiny gray head and the white crescent between each eye and the bill that are the most distinguishing features. In contrast, the mottled gray-brown female has just a small area of white near the bill and a faint dark line behind the eye. The species' scientific name, *Spatula discors,* refers to the bird's broad, flat bill and the discordant sounds of the male's high-pitched whistle and its lower-pitched, nasal bleat.

One of the most abundant waterfowl species in Minnesota, the Blue-winged Teal has a breeding range that stretches across eastern Alaska, Canada, and the northern United States, south through the Great Plains. Population numbers may fluctuate dramatically depending on year-to-year variations in wetland conditions, especially in the northern Great Plains and Canada's Prairie Parklands, the core of the species' distribution.

MINNESOTA BREEDING DISTRIBUTION

Roberts (1932) made the following observation regarding the Blue-winged Teal's abundance in Minne-

Photo by Mike Lentz.

sota in the early 1900s: "While the Blue-winged Teal is far less abundant than it was fifty years ago, it is today the commonest of the summer resident Ducks, having endured the destructive changes attendant upon the settling of the state better than any other species."

Indeed, when Walter Breckenridge surveyed western Minnesota in 1929, he found the species to be the most common duck in the region (Roberts 1932). At the time, confirmed nesting records were available from Grant, Jackson, Kittson, and Polk Counties in western Minnesota and from Hennepin County in east-central Minnesota. Nesting was inferred in Big Stone, Cook, and Lincoln Counties.

Green and Janssen (1975) described the Blue-winged Teal as a summer resident throughout the state and, once again, as the most abundant breeding duck in the state. When Janssen (1987) prepared an updated account just twelve years later, the Teal had moved to second place, outnumbered by the Mallard, which continues to be the most common waterfowl species in Minnesota to this day. Although breeding records were available statewide, Janssen noted that the species was least abundant in northeastern Minnesota. Hertzel and Janssen (1998) reported nesting had been confirmed in 69 of Minnesota's 87 counties since 1970; 11 of the 18 counties where it was not documented nesting were in the southern third of the state.

By 2014, MBS field staff documented the species' abundance across the Prairie Parkland Province, with 470 breeding-season observations in all but the most intensively cultivated counties. Farther to the north and east, observations were more sparsely distributed.

The MNBBA's 1,524 Blue-winged Teal detections documented a distribution similar to that delineated by the MBS but with considerably more occurrences in east-central Minnesota, including from Benton, Isanti, Sherburne, and Wright Counties. Reported from 20.3 percent (967/4,768) of the surveyed atlas blocks, breeding was confirmed in 159 blocks and deemed probable in an additional 275 blocks **(A)**. The species was observed in all but one of Minnesota's 87 counties (absent in Pine County) and was confirmed breeding in 51 counties. It was most abundant in the Prairie Parkland Province, especially in west-central Minnesota, and in the northern half of the Eastern Broadleaf Forest Province; it was least abundant in the Laurentian Mixed Forest Province. A strong reflection of its MNBBA distribution, the Teal's land suitability model emphasizes the importance of wetlands stretching from Lac qui Parle County to the Canadian border **(B)**.

The Blue-winged Teal was the third most common duck species reported during the MNBBA, outnumbered by the Mallard and Wood Duck. Although its relative abundance has changed since Roberts's 1932 account, the core of its Minnesota breeding distribution has likely changed very little.

Rohwer and his colleagues (2020) noted several historical changes in the species' breeding distribution. Most significant may be the species' increase in abundance along the East Coast, where the creation of waterfowl impoundments has improved habitat opportunities. Increases have also been noted in the western states. In contrast, in the Upper Midwest, where wetland losses have been significant, the species is likely less common than it was prior to European settlement.

BREEDING HABITAT

The Blue-winged Teal prefers temporary and seasonal wetlands that are teeming with aquatic invertebrates **(C)**. As the breeding season progresses and shallow wetlands dry up, it often shifts to more permanent wetland basins (G. A. Swanson et al. 1974). Large permanent wetlands are especially important

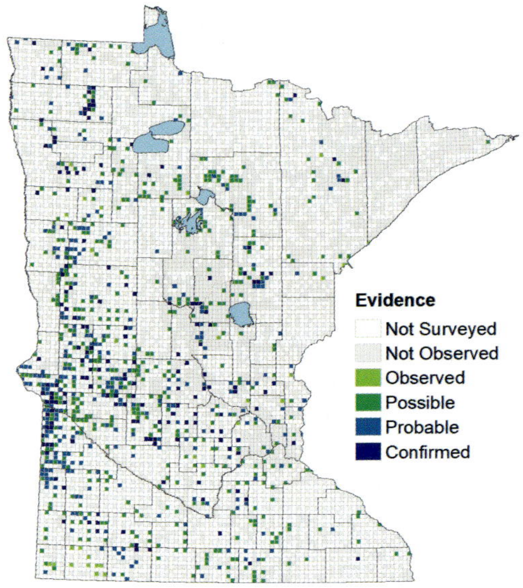

A. Breeding distribution of the Blue-winged Teal in Minnesota based on the Breeding Bird Atlas (2009–13).

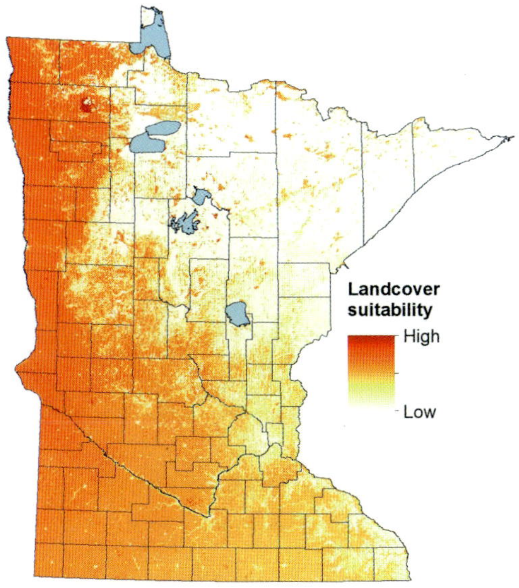

B. Landcover suitability map for the Blue-winged Teal in Minnesota based on habitat, landscape context, and climate data gathered during the Minnesota Breeding Bird Atlas (2009–13) using the MaxEnt modeling approach.

in drought years. When conditions are extremely poor, Teal may simply move to an entirely different region of their breeding range.

Nesting occurs in upland grasslands or wet meadows. Nests usually are placed within 100–150 m of water, though distances up to 1.6 km away have been reported (Baldassarre 2014). Residual cover from the previous growing season is important. Burning, mowing, or grazing activities late in the summer can reduce a site's suitability for nesting the following spring. At the landscape scale, the availability of grassland cover can be as important as the density of wetlands in determining habitat suitability (Rohwer et al. 2020).

C. Small wetlands that provide abundant populations of aquatic invertebrates and nearby grassland nesting cover are frequented by Blue-winged Teal during the breeding season. Photo by Lee A. Pfannmuller.

POPULATION ABUNDANCE

In 2019, the North American Waterfowl Breeding Population and Habitat Survey estimated a population of 5.4 million Blue-winged Teal in the traditional survey area of Alaska, Canada, and the northern Great Plains **(D)**. The number was just above the goal of 4.9 million birds established by the North American Waterfowl Management Plan Committee (2014). Since monitoring began in 1955, Blue-winged Teal numbers hit an all-time low of 2.8 million birds in 1990, at the peak of a widespread drought in the Great Plains. Fortunately, higher precipitation levels, coupled with focused conservation efforts, soon paid dividends in the Prairie Pothole Region as their numbers steadily increased. Nevertheless, wetland conditions remain a strong predictor of breeding population size, as evidenced by the fluctuations shown. Currently, the Blue-winged Teal is the second most abundant duck species in North America, following the Mallard.

In Minnesota, the Waterfowl Breeding Population Survey has tracked Teal numbers in central and western Minnesota since 1968. Although numbers have fluctuated widely over the years, in 2019 they were 7 percent above the long-term average of 208,000 birds (Cordts 2019). Unfortunately, the BBS, which is less intensive but statewide in scope, documents a significant long-term decline in Minnesota, averaging −2.03 percent per year from 1967 to 2019 **(E)**. Habitat loss and the weak protection afforded to many temporary and seasonal wetlands are considered the primary culprits (MNDNR 2006a).

CONSERVATION

Although the Blue-winged Teal is considered a low conservation priority (PIF 2021), it is an important game species and one of the most common ducks harvested in the United States. Harvest statistics for the Blue-winged Teal and the rare Cinnamon Teal are combined and average approximately 1 million birds per year; 65,000–120,000 of the birds are taken in Minnesota (e.g., Raftovich et al. 2020). Teal migration actually is well underway when the waterfowl season opens each fall, which effectively reduces the overall impact of hunting mortality. With such a long distance to travel, this neotropical migrant makes an early departure. Even though some states offer an early teal season, hunting mortality remains relatively low (Baldassarre 2014).

While climate change is a concern, the species is less vulnerable than many waterfowl that breed in Minnesota. Under a 2°C warming scenario, NAS (2019) projected the loss of 17 percent of the Teal's

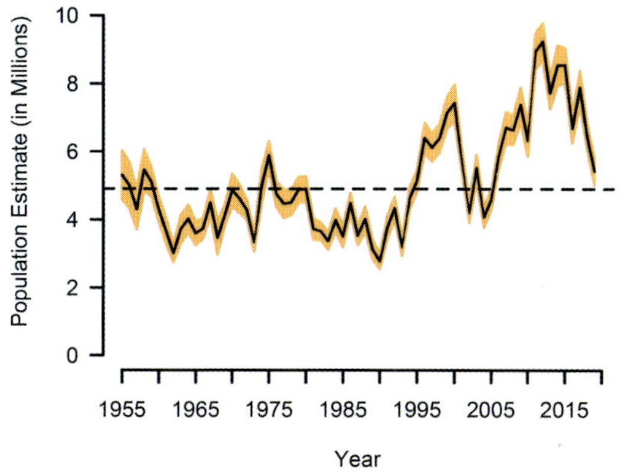

D. Blue-winged Teal breeding population estimates, 90 percent confidence intervals (orange band), and North American Waterfowl Management Plan population goals (dashed line; North American Waterfowl Management Plan Committee 2014) in the traditional survey area (strata 1–18, 20–50, 75–77). Reproduced from USFWS (2019, fig. 4).

current breeding range by 2050 and a 28 percent gain northward, resulting in a rank of low vulnerability. In Minnesota, the core of its breeding population in the west-central and northwestern counties would be spared, but significant losses are predicted in southern and east-central regions of the state. The same models projected that the species' winter habitat in the United States would remain stable, but Baldassarre (2014) noted concern regarding habitat loss along the Gulf coast. Unfortunately, little is known about critical wintering sites in South America.

The Blue-winged Teal has benefited from many habitat conservation initiatives since the 1990s, but additional work is still needed. Its future depends on the success of such efforts and on the support of duck hunters and conservationists who support wetland conservation through such well-respected organizations as Ducks Unlimited.

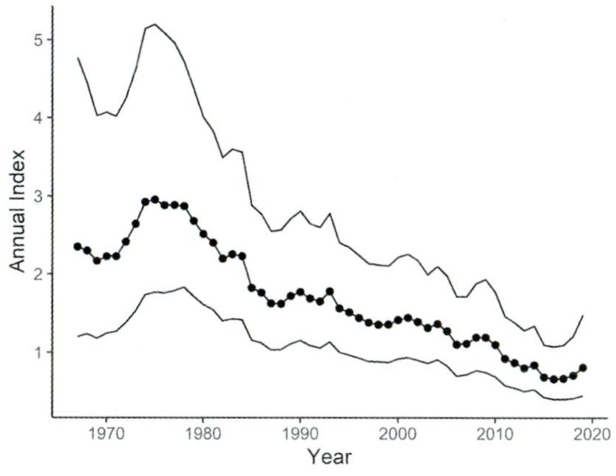

E. Breeding population trend and 95 percent credible intervals (solid black lines) for the Blue-winged Teal in Minnesota based on the 1967–2019 Breeding Bird Survey (Sauer et al. 2020).

NORTHERN SHOVELER

Spatula clypeata

Photo by Mike Lentz.

Minnesota Seasonal Status: Regular breeding resident and migrant; occasionally overwinters in southern Minnesota. The Northern Shoveler was uncommon during the MNBBA.
Migration: Most populations are medium-distance migrants wintering from the southern United States to Mexico, the Pacific coast of Central America, and the Caribbean.
Food: Aquatic invertebrates and seeds; feeds by straining water through its shovel-like bill as it skims the water's surface.
Nest: Ground nester.
Conservation Concern: A game species; assigned a low PIF Continental Concern Score of 7/20.

The male Northern Shoveler is unmistakable. From afar, the broad chestnut patch on its belly and flanks contrasts sharply with the white breast. This, combined with its iridescent dark-green head, immediately draws the observer's attention. It is only upon closer inspection that its long, broad, spatula-shaped bill becomes apparent. Often referring to it as a "Spoonbill," Roberts (1932) lamented the Shoveler's low status among waterfowl: "This, perhaps, is due to the big, ungainly bill, which imparts a clumsy appearance to the head and diverts attention from the handsome dress that he wears."

As awkward as it may appear, the spoon-shaped bill is a unique adaptation. Along the edges are small, comb-like projections, or lamellae. These act as a sieve that allows the Shoveler to strain water through its bill to retrieve its favorite food item, water fleas (Baldassarre 2014). Lamellae are also present in other dabbling ducks. Most have 50–70 along each mandible; teal may have up to 120–130. But the Shoveler takes the prize. The lower mandible is densely packed with about 220 lamellae, while the upper mandible sports approximately 180 (Davis 2019). This efficient feeding machine also enables Shovelers to keep their eyes above water and alert for predators that approach from above, unlike other dabbling ducks who more frequently tip their bottoms up to feed underwater (Baldassarre 2014). Not only does the genus name, *Spatula,* reference the bill, the Latin name *clypeata* does as well, meaning "armed with a shield."

The mottled grayish-brown female is also easily identified by her spatulate bill. In flight, a green speculum, bordered above by a white stripe and blue upper wing coverts, is visible in both sexes, though the colors are quite dull in the female. The male has an unusual *took-took* call during courtship; a rattling sound made by the wings when the birds take flight also aids identification.

A Holarctic species, in North America the Northern Shoveler breeds from Alaska south through central Canada and the north-central states. Smaller, year-round populations are found from the Pacific Northwest southeast to New Mexico. The species is most abundant in the prairie potholes of the northern Great Plains and Prairie Provinces.

MINNESOTA BREEDING DISTRIBUTION

Roberts (1932) described the Northern Shoveler as "one of the commonest nesting ducks" found "throughout the state." Confirmed nesting records, however, were available from only two counties, Hennepin and Polk, while other evidence of breeding was reported from Cook, Jackson, and Marshall Counties. Despite Roberts's statement about the species' broad distribution, one has to wonder how prevalent it was in the northern forests, despite the inferred nesting report from Devil's Track Lake in Cook County (1922).

Like so many other waterfowl, by 1932 the Shoveler population had already begun to rebound from a dramatic decline in the late 1800s caused largely by unregulated harvesting. In 1898, a longtime observer at Heron Lake wrote, "Once very common and breeding everywhere here but now rarely found nesting in this locality." Nearly twenty years would pass before the birds were again nesting at this renowned waterfowl lake. The species' recovery, however, may have been rather slow. Compilers for the MOU's summer seasonal report noted that despite being one of the state's most common nesting ducks, the Northern Shoveler had only been reported twice from 1940 to 1945 (Longley and Harrell 1946).

Years later, Green and Janssen (1975) described the Shoveler as a summer resident restricted primarily to western and south-central Minnesota. The species was scarce in the southeast and in the northern forests. Although two more nesting records had been documented in northeastern Minnesota (Itasca County in 1949 and Aitkin County in 1972), the records represented single occurrences of breeding pairs. Janssen described the same breeding range in his 1987 update. He delineated 13 counties where nesting had been confirmed since 1970; four were east of the species' primary breeding range (Aitkin, Cook, Itasca, and St. Louis). Hertzel and Janssen (1998) later deleted Cook and Itasca Counties from Janssen's original map and added three western counties (Kittson, Polk, and Stevens). By 2014, all but a handful of the 85 breeding-season locations identified by the MBS were confined to western and south-central Minnesota.

During the MNBBA, the Northern Shoveler's distribution remained similar to that described by earlier accounts. Participants reported 314 records from 5.1 percent of the surveyed atlas blocks (244/4,740); breeding was confirmed in 11 blocks **(A)**. The species was reported from 57 of Minnesota's 87 counties and was confirmed breed-

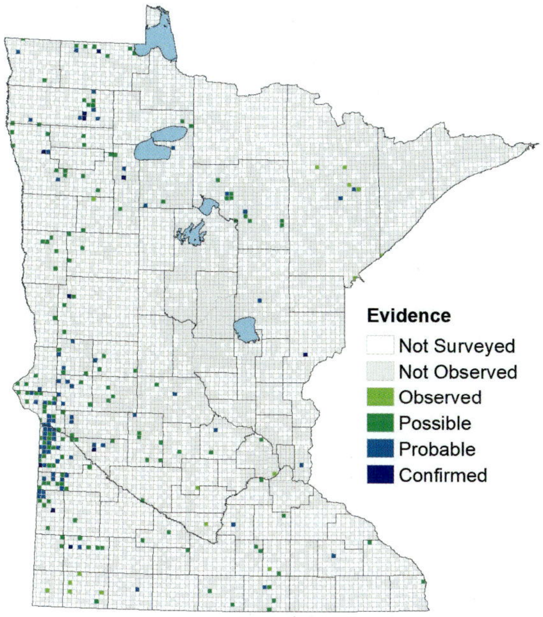

A. Breeding distribution of the Northern Shoveler in Minnesota based on the Breeding Bird Atlas (2009–13).

ing in 10 counties; all but one (Pine) were located in western Minnesota. Breeding was deemed probable, however, in several northern counties, including Aitkin, Beltrami, Itasca, and St. Louis. Farther south, probable breeding was reported as far east as Washington and Olmsted Counties. Seven of the counties with confirmed breeding were new to the list published by Hertzel and Janssen in 1998: Chippewa, Clearwater, Lac qui Parle, Murray, Otter Tail, Pine, and Roseau. The abundance of records in west-central Minnesota represents impressive survey efforts by local atlas participants.

Although records in the Laurentian Mixed Forest Province were sparse, the region did report 29 records, including seven in St. Louis County and 10 in Itasca County. These records may represent a gradual reoccupation of areas where the species was formerly more common, as Roberts (1932) implied. Indeed, Dubowy and colleagues (2020) noted that the Shoveler may have bred east of its current distribution in years past. But the Shoveler has also demonstrated a tolerance for wastewater treatment ponds. Of the seven atlas blocks where it was reported in St. Louis County, three had wastewater treatment ponds and one included a former mine pit filled with water. Atlas blocks where the Northern Shoveler was reported in Itasca County, however, included only natural wetlands. The species' recent presence as a breeding species in eastern Canada is also attributed, in part, to the "proliferation of sewage-treatment plants" (Dubowy et al. 2020).

BREEDING HABITAT

In the prairie potholes, the Northern Shoveler prefers small seasonal or semipermanent wetlands **(B)**. North and east of the prairies, forested wetlands are selected (Baldassarre 2014). As noted above, wastewater treatment ponds also provide suitable habitat. Nests are usually placed within 50 m of water in short-cover vegetation, including pastures, hayfields, and grasslands (Dubowy et al. 2020).

POPULATION ABUNDANCE

In 2019, the North American Waterfowl Breeding Population and Habitat Survey estimated that 3.6 million birds resided in the traditional survey area of Alaska, Canada, and the northern Great Plains **(C)**, a number significantly higher than the goal of 2.5 million birds established by the North American Waterfowl Management Plan Committee (2014). After remaining stable for nearly forty years, from 1955 to the mid-1990s, Shoveler numbers began to surge upward. This upward trend is similar to that observed for many waterfowl species that are abundant in the northern Great Plains. Improved precipitation patterns in the region in the 1990s, along with an increased focus on conservation, are likely responsible.

Despite this hopeful outlook for the continental population, the Minnesota Waterfowl Breeding Population Survey demonstrates wide fluctuations with an overall downward trend in numbers. Since 1987, peak numbers were reported in 1990 (26,545) and in 2000 (26,175) but have remained relatively low since then with an average of 5,490 birds per year from 2010 to 2019 (Cordts 2019). The BBS also shows a nonsignificant declining population in Minnesota, although the statistical precision of the data is poor (Sauer et al. 2020). Loss of wetland habitat is likely the primary cause.

B. In western Minnesota, the Northern Shoveler prefers shallow wetland basins with submerged vegetation for foraging and adjacent grassland cover for nesting. Photo by Gary T. Seim.

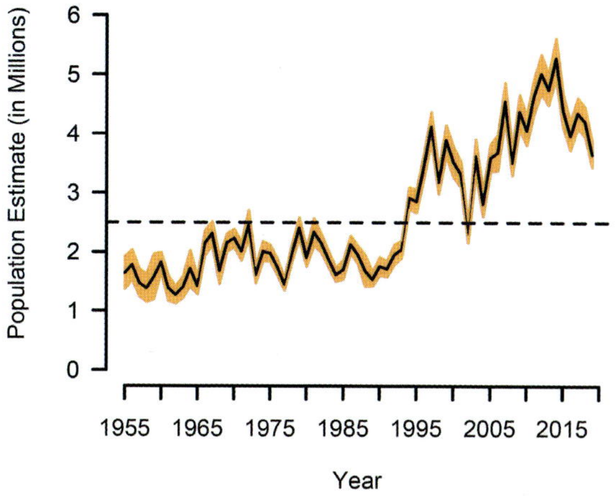

C. Northern Shoveler breeding population estimates, 90 percent confidence intervals (orange band), and North American Waterfowl Management Plan population goals (dashed line; North American Waterfowl Management Plan Committee 2014) in the traditional survey area (strata 1–18, 20–50, 75–77). Reproduced from USFWS (2019, fig. 4).

CONSERVATION

Given its robust continental population, the Northern Shoveler is considered a low conservation priority (PIF 2021). A harvested species, it is not highly pursued by waterfowl hunters. In 2019, it comprised 9.4 percent of the monitored duck population in North America's traditional survey area (USFWS 2019) and accounted for only 4.4 percent of the total waterfowl harvest (Raftovich et al. 2020). That same year Minnesota's Northern Shoveler harvest (6,356 birds) comprised less than 1.4 percent of the total Shoveler harvest in North America (469,340 birds).

Like many waterfowl dependent on the Great Plains, the impact of climate change is a concern. Warming temperatures are predicted to decrease the region's wetlands by more than 60 percent (NABCI 2010). Nevertheless, an analysis of the potential impacts of climate change by NAS (2019) classified the Shoveler's vulnerability as low under all three warming scenarios modeled, ranging from an increase of 1.5°C to 3°C. Losses were predicted to be scattered throughout Minnesota, with gains predicted in the northwest and north-central counties by 2050.

Although several toxic chemicals, including selenium, DDT, and lead, have been detected in Shovelers, their impacts on population numbers are unknown (Dubowy et al. 2020). Overall, it appears that the most important conservation measure is to preserve and restore critical breeding habitat. Ambitious interagency efforts designed to reverse the decline of all waterfowl populations are essential to ensure a sustainable future for Minnesota's diverse waterfowl community, including the unique Spoonbill.

GADWALL

Mareca strepera

Minnesota Seasonal Status: Regular breeding resident and migrant; occasionally seen during the winter in southern Minnesota. The Gadwall was uncommon during the MNBBA.
Migration: Medium-distance migrant that winters in the southern United States and Mexico; populations in the western United States are year-round residents.
Food: A dabbling duck that feeds primarily on submerged aquatic plants.
Nest: Ground nester.
Conservation Concern: A game species; assigned a low PIF Continental Concern Score of 8/20.

Photo by Mike Lentz.

When the winter ice begins to melt on Minnesota's wetlands in early spring, the first wave of migrant dabblers arrives. The males look quite dapper in their breeding plumage, displaying a stunning array of colors that sharply contrast with the grays and browns of the surrounding landscape. Concealed among the rafts of waterfowl, the Gadwall, an uncommon nesting bird in Minnesota, might be easily overlooked.

From a distance, their gray and brown plumage makes it easy to mistake the birds for Mallard hens. Indeed, the female Gadwall looks like a slightly smaller and grayer female Mallard. The male's most diagnostic feature is his black rump. His buffy head contrasts with a somewhat darker, patterned breast and gray back. If the lighting is just right, the subtlety of his elegant plumage is striking. Silvery secondary wing feathers are overlaid by tawny-colored wing scapulars, while an intricate, wavy pattern of white and black on the upper chest grades into a white belly. In flight, they are the only dabbler with a patch of white on the speculum. *Mareca,* the scientific name of the Gadwall's genus, originates from the Brazilian term *marreco,* meaning "smaller duck." The Latin word *strepere* means "to make a loud noise," a reference to the female's rather loud quack call, which is similar to the female Mallard's but higher in pitch.

The Gadwall's "meteoric rise" since the late twentieth century to become one of the most widely distributed and common ducks in North America is a story unique among the continent's waterfowl (Young 2017). A Holarctic species, in North America it is broadly distributed throughout the western United States, including the Great Plains, and the Canadian Prairie Provinces; scattered populations are found in southern Alaska, northwestern Canada, and throughout the Great Lakes region. Some of the densest populations occur in North Dakota.

MINNESOTA BREEDING DISTRIBUTION

There was a time, wrote Roberts (1932), when the Gadwall "was one of the commonest species in the prairie." During a visit to Grant County in 1879, he described it as common, if not more so, than the ubiquitous Mallard. Colleagues also wrote of the species' abundance in Otter Tail County in 1881 and at Heron Lake in Jackson County and Mud Lake in Marshall County in 1901. Yet by the early 1900s, Roberts noted that the species was experiencing a significant decline: "It was one of the Ducks that suffered most severely from the settling of the country, probably as much from the breaking-up of the prairie, where it commonly nested, as from the hunters."

Beginning in the 1920s, numbers began to rebound in response to enactment of harvest regulations and passage of the Migratory Bird Treaty Act in 1918. By 1932, Roberts described the Gadwall as a breeding resident throughout "much of the prairie and open woodland regions of the state." Confirmed nesting records, however, were available only from Jackson and Grant Counties. Several broods of small young were observed in Marshall County.

In 1975, Green and Janssen reaffirmed the Gadwall's breeding distribution across western and southern Minnesota, noting it was most abundant at the Agassiz NWR in Marshall County. A few years later, Janssen (1987) cited scattered breeding records farther east, specifically in Aitkin, Crow Wing, and Mille Lacs Counties (his map identifies Mille Lacs County, but his written account mentions Cass County) as well as records from eight western counties where nesting had been confirmed since 1970: Big Stone, Lac qui Parle, Le Sueur, Lyon, Marshall, Pennington, Pope, and Yellow Medicine. Hertzel and Janssen (1998) later deleted the Aitkin County record and added Lake of the Woods. The MBS subsequently reported 91 breeding-season locations. All but two of the records were in the western and south-central regions of the state. One report was from central St. Louis County, and the second, from Carlton County.

During the MNBBA, the majority of the 280 Gadwall records collected were from far west-central Minnesota, followed by another population center in and around the Agassiz NWR. Although still rare, there were more reports of birds farther east than reported by the MBS, including observed and possible records in Aitkin, Cass, Crow Wing, Hennepin, Lake, and Wabasha Counties. Reported from 4.2 percent (201/4,737) of the surveyed atlas blocks, the birds were detected in 41 of Minnesota's 87 counties and were confirmed breeding in eight counties **(A)**. Five of the counties where breeding was confirmed were new to the list published by Hertzel and Janssen (1998): Grant, Lincoln, Polk, Stearns, and Stevens.

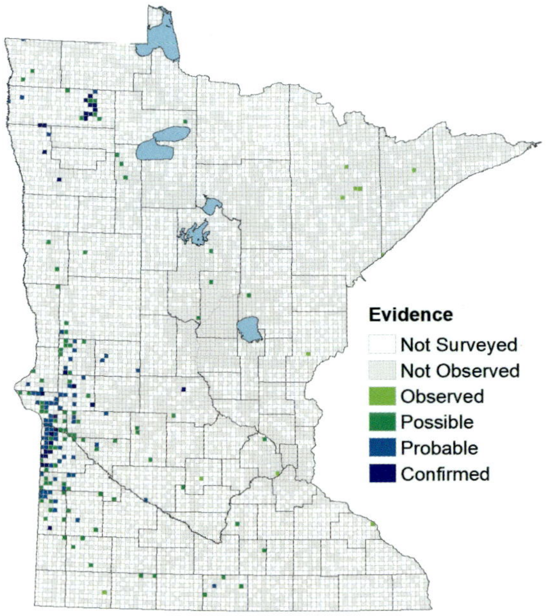

A. Breeding distribution of the Gadwall in Minnesota based on the Breeding Bird Atlas (2009–13).

The increase in the number of reports in eastern Minnesota may reflect changes that are being documented across North America, where the species has gradually expanded east through the Great Lakes and New England. Some of this expansion has been aided by local introductions, but the majority is attributed to the establishment of state and national wildlife refuges and the creation of impoundments that provide suitable habitat (Henny and Holgerson 1974). In their comprehensive review of the species, LeSchack and his colleagues (2020) note that the species' breeding range changed more during the twentieth century than that of any other duck in North America. In addition to its expansion in the East, it has also expanded in Oregon, Washington, and British Columbia.

B. Wetlands with abundant submerged and floating aquatic plants, a dense border of emergent vegetation, and located adjacent to grassland cover meet all the requirements for the breeding Gadwall. Photo by Lee A. Pfannmuller.

BREEDING HABITAT

In the Great Plains, Gadwalls prefer semipermanent and permanent wetlands that have an abundance of submerged vegetation for feeding and emergent vegetation for cover **(B)**. Both freshwater and alkaline wetlands are used. When natural wetland basins are limited, stock ponds and dugouts may be selected (LeSchack et al. 2020). Suitable upland nesting habitat may be dry or mesic prairie, restored grasslands, cultivated fields, hay meadows, or brushlands. The cover is usually dense, with 25 percent cover or greater. Although Gadwalls nest later than other dabblers, the presence of residual vegetation from the previous growing season may influence nesting habitat selection (Sousa 1985).

POPULATION ABUNDANCE

Globally, the total Gadwall population numbers approximately 4.4 million birds (PIF 2021). In 2019, the North American Waterfowl Breeding Population and Habitat Survey estimated a breeding population of 3.3 million Gadwalls in the traditional survey area located in portions of Alaska, Canada, and the north-central United States (USFWS 2019). The USFWS estimate is 73 percent higher than the goal of 1.9 million birds established for the traditional survey area by the North American Waterfowl Management Plan Committee (2014). With the exception of one major decline observed since 2000, the continental population has been steadily increasing for many years **(C)**. This long-term trend is corroborated by surveywide BBS data (Sauer et al. 2020).

A statewide population estimate is not available. Gadwalls have demonstrated wide fluctuations in numbers in the spring when Minnesota's Breeding Waterfowl Survey is conducted. In the past ten years (2010–19), numbers have ranged from a high of 11,855 birds in 2016 to a low of 2,703 birds in 2014 (Cordts 2019). Minnesota's Waterfowl Survey is not conducted statewide; the most intensively surveyed areas exclude some of the more densely populated Gadwall regions in the state. Although it lacks statistical precision, BBS data in Minnesota suggest a stable population.

The continental increase in Gadwall numbers largely reflects conditions in the Prairie Pothole Region, the core of the species' continental breeding distribution. Improved precipitation patterns that began in the region in the 1990s are a major factor driving higher population numbers in recent years, in concert with more focused conservation efforts.

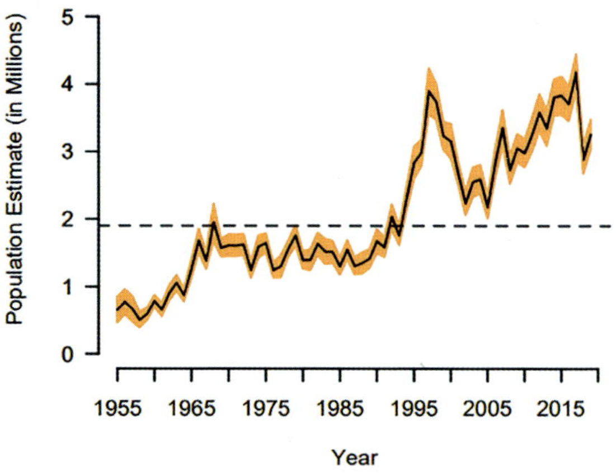

C. Gadwall breeding population estimates, 90 percent confidence intervals (orange band), and North American Waterfowl Management Plan population goals (dashed line; North American Waterfowl Management Plan Committee 2014) in the traditional survey area (strata 1–18, 20–50, 75–77). Reproduced from USFWS (2019, fig. 4).

The Gadwall is unique in having dodged the large-scale population declines that plagued many grassland-nesting ducks from the 1960s through the early 1980s. Gadwall numbers remained steady during these turbulent years, followed by an increase in numbers and the gradual expansion of its breeding range as conditions improved (Baldassarre 2014). Its late nesting habits and propensity to nest on islands are adaptive strategies that help maintain nesting success even when conditions are poor.

CONSERVATION

A harvested species, the Gadwall comprised nearly 11 percent (1.05 million birds) of the United States duck harvest in the fall of 2019 (USFWS 2019). Nationally it is one of the most common ducks bagged, usually ranking third behind the Mallard and Green-winged Teal. Nearly 3 percent (29,447 birds) of the species' harvest occurred in Minnesota (Raftovich et al. 2020).

Deemed a low conservation priority (PIF 2021), even a warming climate may pose relatively few concerns for the growing Gadwall population. Under a 2°C warming scenario, NAS (2019) assigned it a low vulnerability when models projected a 21 percent loss in the species' southern breeding range by 2050 and a 26 percent gain to the north. In Minnesota, range declines are projected in the southwest while the species expands into counties east of the Red River valley. Although these potential gains are promising, it is critical that aggressive wetland and grassland conservation efforts continue to help ensure the future of this elegant but often overlooked waterfowl species.

AMERICAN WIGEON

Mareca americana

Minnesota Seasonal Status: Regular breeding resident and migrant; occasionally observed during the winter. The American Wigeon was rare during the MNBBA.

Migration: Short- to medium-distance migrant that winters along the Pacific coast, in the Caribbean, and across the southern United States south to Panama.

Food: Dabbler and upland grazer feeding primarily on aquatic and terrestrial plants as well as seeds, waste grains, and invertebrates.

Nest: Ground nester.

Conservation Concern: A game species; assigned a moderate PIF Continental Concern Score of 10/20.

Photo by Mike Lentz.

The American Wigeon has been described as "a distinctive species with a perky attitude" (Mini et al. 2020); the male is easily identified by his creamy white forehead and crown, features that gave rise to its second common name, Baldpate. Its dark rump, bordered by white, is another good field mark coupled with the dark, iridescent green face patch and, in flight, the white upper wing coverts. The female can be easily confused with the females of other dabbling ducks, but her rounded head, small bill, and rusty flanks are good diagnostic features. Considered one of the more vocal waterfowl species, the male's high-pitched whistles (*whew-whew-whew*) are heard throughout the year.

In many ways the American Wigeon is more similar to a goose than to other dabbling ducks. The small and compact shape of its bill limits its effectiveness for filter feeding, a feeding strategy used by most dabblers, but enables the bird to apply more force to snap off the tops of succulent upland and aquatic vegetation. Its legs are also positioned relatively far forward on its body, allowing it to walk efficiently on land. Indeed, during the winter, Wigeons are commonly seen feeding on lawns and golf courses, similar to Canada Geese (Baldassarre 2014).

Giving it the nickname of the "poacher" or "robber," Bent (1923) described another interesting foraging strategy. "The baldpate, being a poor diver and yet extravagantly fond of the succulent roots of the, so-called, wild celery, has to be content with what small bits of this delicacy the canvasback lets drop or what it can steal from this expert diver on its return to the surface." The Canvasback is not the Wigeon's only target; it frequently robs aquatic plants from other waterfowl species as well.

Found during the breeding season from Alaska south through the northern Great Plains of the United States and Canada, small numbers also occur across the northern Great Lakes and the St. Lawrence River valley. Largely a boreal species, less than 16 percent of the population occurs in the United States (Mini et al. 2020).

MINNESOTA BREEDING DISTRIBUTION

In the early 1900s, Roberts (1932) knew this handsome bird only as a spring and fall migrant in Minnesota. There were no confirmed nesting records. Nevertheless, the occasional occurrence of pairs in late spring and early summer led him to speculate that it might be a rare summer resident. Most summer observations were in southwestern and west-central Minnesota, including a pair at Heron Lake in Jackson County (1898), two pairs at Lake Benton in Lincoln County (1928), and two pairs in Grant County (1929). Breeding was finally confirmed in 1932 when a female with 12 downy young was observed in Koochiching County by Gus Swanson (Roberts 1936).

Years later, Green and Janssen (1975) noted that the species was primarily a summer resident in north-central Minnesota and adjacent counties, despite scattered breeding records across the state. It was most common in Itasca and Beltrami Counties. Janssen (1987) described a similar distribution and identified 11 counties where nesting had been confirmed since 1970. Eight of the counties (Aitkin, Beltrami, Cass, Itasca, Lake of the Woods, Marshall, Polk, and St. Louis) were located within the core of the species' breeding distribution; three counties were located farther south (Lac qui Parle, Stearns, and Todd). Hertzel and Janssen (1998) later added Grant County to the list. Meanwhile, by 2014, the MBS had identified 27 American Wigeon breeding-season locations widely dispersed throughout much of the state.

The American Wigeon's rather sporadic breeding distribution was confirmed again by the MNBBA. Participants reported 55 records in 22 of Minnesota's 87 counties **(A)**. Breeding was confirmed in only three blocks in two counties: Marshall and Lac qui Parle. More than half of the records were located in the Laurentian Mixed Forest Province, followed by the Tallgrass Aspen Parklands Province. Scattered records, including the Lac qui Parle County breeding record, were reported in southern Minnesota.

It is difficult to know if the American Wigeon's distribution has changed over the past one hundred years or if it was simply overlooked by early observers amid the smaller wetlands of Minnesota's northern forests. The absence of breeding records in the late 1800s and early 1900s, however, compared with the relative prominence of records in north-central Minnesota beginning in the mid-1900s may reflect broader changes in the species' distribution. Gradually Wigeons have decreased in abundance in North America's Prairie Potholes, while their numbers have increased significantly in Alaska, and their range has expanded east across the northern Great Lakes region and the Maritime Provinces. Factors responsible for these changes may be related to the

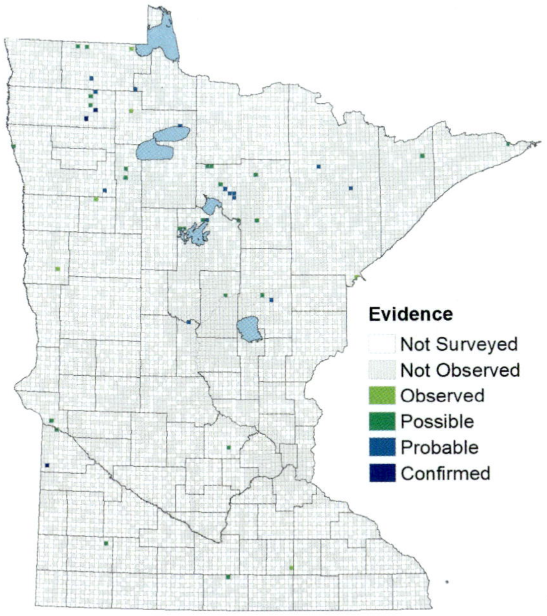

A. Breeding distribution of the American Wigeon in Minnesota based on the Breeding Bird Atlas (2009–13).

loss of suitable habitat in the Prairie Potholes coupled with the establishment of more suitable habitat in other regions, especially the creation of impoundments managed for waterfowl populations (Mini et al. 2020).

BREEDING HABITAT

The American Wigeon is found in shallow wetlands adjacent to upland grass and brushy vegetation **(B)**. Such features are common throughout North America's Prairie Potholes, Canada's boreal forests and parklands, and Alaska's coastal tundra. Regardless of location, the availability of dense upland cover and suitable wetland conditions are paramount. If drought conditions prevail in the Prairie Potholes Region, Wigeons bypass these areas and fly farther north to breed in Alaska and northwestern Yukon (Baldassarre 2014).

POPULATION ABUNDANCE

In 2019, the North American Waterfowl Breeding Population and Habitat Survey estimated a population of 2.8 million American Wigeons within the traditional survey area located in portions of Alaska, Canada, and the north-central United States (USFWS 2019). The number was slightly higher than the region's population objective of 2.6 million birds established by the North American Waterfowl Management Plan Committee (2014). A small number of Wigeons also occur in the eastern Canadian provinces and the state of Maine (an average of 18,000 per year from 1990 to 2010; Baldassare 2014).

Although population numbers have been relatively stable since monitoring began in 1955 **(C)**, significant regional shifts have been masked by the overall trend. From 1955 to the late 1990s, the number of Wigeons in Alaska grew from a low of 100,000 to more than 1 million. At the same time, the numbers of birds in the Prairie Potholes Region decreased; from 1997 to 2006, numbers were approximately 60 percent lower than numbers reported during the 1970s (Mini et al. 2020). More recently, the trend has slightly reversed: the number of birds in Alaska has decreased, while increases have been observed in Montana and the western Dakotas (USFWS 2019). Some of these changes likely reflect fluctuations in wetland conditions. The long-term increase in numbers of Wigeons in Alaska may also be due to higher rates of adult survival on the northern breeding grounds and/or on the Pacific coast wintering grounds, coupled with less suitable, long-term conditions in the Great Plains (Mini et al. 2020).

BBS data, however, suggest a population decline throughout the United States and Canada averaging −1.52 percent per year from 1966 to 2019 (Sauer et al. 2020). Although the trend is statistically significant, the Wigeon's relatively low abundance reduces its reliability. Given that BBS routes are also sparsely distributed in Alaska and northern Canada, where significant breeding concentrations occur, the decline is likely a more accurate depiction of population numbers in the more southern Prairie Parklands and Potholes.

American Wigeon numbers in Minnesota are quite variable. Unfortunately, the state's aerial waterfowl survey is not statewide in coverage and ex-

B. In northwestern Minnesota, the Thief Lake WMA in Marshall County provides an abundance of shallow wetlands and adjacent grassy nesting cover for the American Wigeon. Photo by Lee A. Pfannmuller.

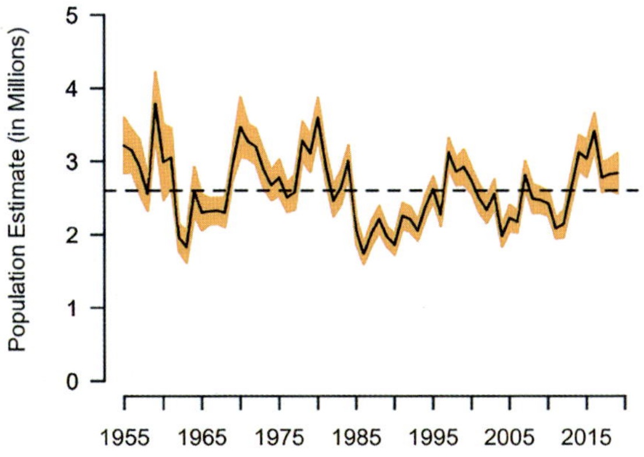

C. American Wigeon breeding population estimates, 90 percent confidence intervals (orange band), and North American Waterfowl Management Plan population goals (dashed line; North American Waterfowl Management Plan Committee 2014) in the traditional survey area (strata 1–18, 20–50, 75–77). Reproduced from USFWS (2019, fig. 4).

cludes the core of the species' breeding distribution (Cordts 2019). Minnesota also lies along the eastern periphery of the Wigeon's North American breeding range, so population fluctuations are expected.

CONSERVATION

The Wigeon's sensitivity to changing wetland conditions in the Prairie Potholes is a major concern. Indeed, the species responds more negatively to drought conditions in the region than any other waterfowl species. This, coupled with its relatively small population size, led to its assignment as a moderate conservation priority (PIF 2021).

Compounding these challenges are the long-term impacts of climate change. Should the climate increase by 2°C, NAS (2019) predicts the loss of 34 percent of the Wigeon's current breeding range by 2050 but a gain of 35 percent farther north. Much of the loss is concentrated in Canada's boreal forest. Unfortunately, even under a 1.5°C warming scenario, the Wigeon is predicted to retreat almost entirely from Minnesota and a large portion of the United States. Even a continued focus on wetland protection and restoration may not be sufficient to prevent its statewide extirpation.

A game species, Wigeons represented 5.5 percent of the United States duck harvest in 2019, with a total of 536,291 birds harvested; 11,652 birds were taken in Minnesota (Raftovich et al. 2020). Strictly managed, harvest levels are not considered a major concern. Nevertheless, species-specific management recommendations for this "perky" boreal inhabitant are lacking, which underscores the importance of carefully monitoring the population and its oscillating occupancy of critical breeding grounds across North America, especially in the Prairie Potholes and Alaska.

MALLARD

Anas platyrhynchos

Minnesota Seasonal Status: Regular breeding resident and migrant; observed during the winter throughout most of the state. The Mallard was a common species during the MNBBA.
Migration: Resident to medium-distance migrant; northern breeding populations winter along the coast of Canada and throughout the United States and northern Mexico.
Food: Dabbling duck that feeds on aquatic plants and invertebrates; also feeds on waste grain in uplands.
Nest: Ground nester.
Conservation Concern: A game species; assigned a low PIF Continental Concern Score of 7/20.

The most well-known duck in the world, the Mallard elicits an array of affections. Perhaps it is our memories of the classic childhood book *Make Way for Ducklings,* or our memories of frosty fall mornings sitting in a duck blind waiting for the first flight. Perhaps it is simply the birds' audacious attitude as they waddle across our city lawns. Endeared to all, the Mallard is one of the most successful birds in North America.

With his bold green head, yellow bill, and chestnut breast, the male is unmistakable. Even when he is feeding with his tail tipped up, the curly black tail feathers are diagnostic. The brownish-gray female may be confused with the similar but darker-plumaged American Black Duck. Her orange bill, with a dark center, is a helpful clue. The male and female Black Duck display a greenish-yellow and olive-colored bill, respectively. In flight, the Mallard has a violet-blue wing speculum with a white border. The familiar *quack* that we associate with the species is attributed to the female. The male gives a quieter, raspy call. *Platyrhynchos,* the Latin name for the species, refers to the bird's broad bill.

The Mallard exceeds all other ducks in abundance. Found throughout North America, it reaches its greatest abundance in the Prairie Potholes. Here it numbered more than 9 million birds in 2019, nearly 4 million more than the second most common species, the Blue-winged Teal (USFWS 2019). During the fall harvest, nearly one in three ducks taken are Mallards (Raftovich et al. 2020). It is a darling of the duck hunter; Bent (1923) expressed the sentiment of many when he wrote: "In my estimation there is no duck quite equal to a fat, grain-fed mallard, not even the far-famed canvasback."

Photo by Mike Lentz.

The Mallard's success can be attributed to many factors. A generalist in all matters, the bird is extremely adaptable in finding suitable habitats that range from pristine wetlands and grasslands to roadside ditches and city parks. It consumes nearly any available plant or small invertebrate and settles almost anywhere near open water. Despite its cosmopolitan nature, the Mallard is the bellwether for the status of all waterfowl that reside in the Prairie Potholes. When conditions are favorable for Mallards, nearly all waterfowl in the region benefit.

MINNESOTA BREEDING DISTRIBUTION

Although its numbers have waxed and waned over time, the Mallard has always been one of the most common waterfowl species in the state. Roberts (1932) reminisced of earlier times when the Mallard was so abundant "that almost every pot-hole had its breeding pair and the prairie sloughs were full of them." But by the early 1900s, the absence of harvest regulations had taken its toll. Luckily, the intrepid Mallard survived these devastating years, recovering to the point that it was once again a common breeding resident by the early 1930s. Roberts (1932) reported that among waterfowl only the Blue-winged Teal was more abundant.

By the 1970s, the Mallard was still second in abundance to the Blue-winged Teal (Green and Janssen 1975), but it surpassed the Teal when Janssen (1987) again updated the species' status just a few years later. Green and Janssen (1975) described the Mallard as statewide in distribution but "most numerous in the western regions." Confirmed nesting records compiled since 1970 documented records from all but 10 of Minnesota's 87 counties (Hertzel and Janssen 1998). The MBS further confirmed its statewide distribution.

MNBBA participants reported 5,229 Mallard records from 56.6 percent (2,766/4,888) of the surveyed atlas blocks and confirmed breeding in 21 percent (1,028) of the blocks **(A)**. Reported from every county in the state, the species was documented breeding in all but two counties, Fillmore and Yellow Medicine. Probable nesting, however, was reported in both counties. The high density of reports in the Brainerd Lakes region and the Twin Cities largely reflects the high number of MNBBA participants in those areas. The Mallard's landcover suitability model emphasizes the importance of the Prairie Parkland Province and the prairie–forest ecotone **(B)**. Wetlands in the Twin Cities also provide highly suitable habitats.

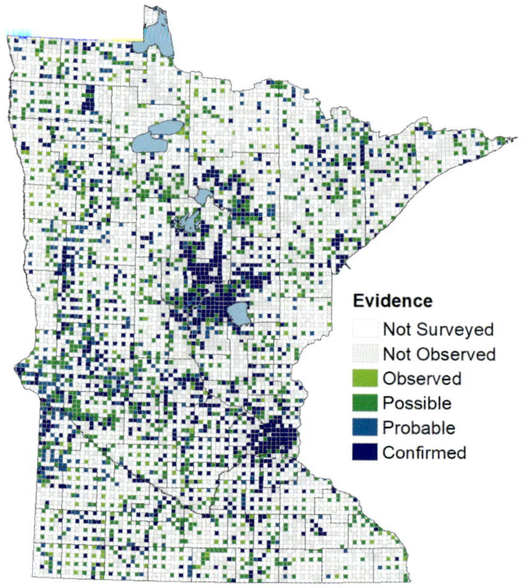

A. Breeding distribution of the Mallard in Minnesota based on the Breeding Bird Atlas (2009–13).

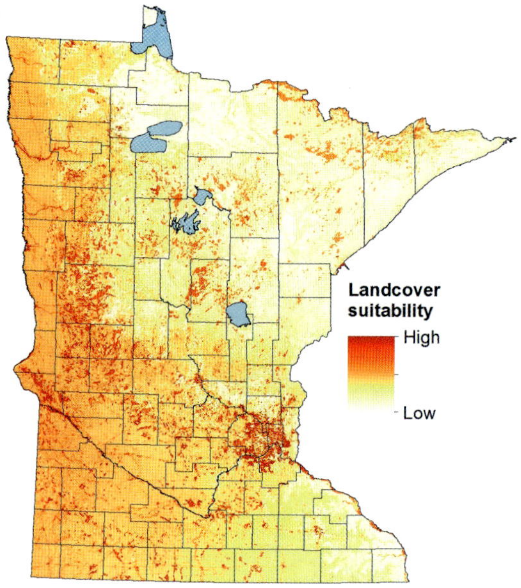

B. Landcover suitability map for the Mallard in Minnesota based on habitat, landscape context, and climate data gathered during the Minnesota Breeding Bird Atlas (2009–13) using the MaxEnt modeling approach.

Prior to the 1900s, the Mallard was largely restricted to western and central North America. Its expansion east was owing largely to the release of captive-reared birds for hunting. The species, for example, was "virtually absent" from Ontario prior to the 1950s but is now "the most widely distributed duck in the province" (Cadman et al. 2007). It is now common in Quebec and the Maritime Provinces as well.

BREEDING HABITAT

Open water and adjacent upland cover are the primary habitat requirements of the ubiquitous Mallard **(C)**. Although such basic needs can be met in a wide range of habitats, from boreal forests to manicured city parks, the prairie wetlands and grasslands of the northern Great Plains provide ideal conditions. When suitable nesting cover is limited, the bird may nest nearly 1.0 km away from open water but more commonly nests within 2 to 300 m (Drilling, Titman, et al. 2020). Occasionally it will nest on wetlands within dense stands of emergent vegetation.

C. A habitat generalist, the Mallard is typically found in productive, shallow wetland basins with abundant emergent vegetation along the shoreline and adjacent nesting cover in the uplands. Photo by Gary T. Seim.

POPULATION ABUNDANCE

In 2019 the Mallard's breeding population was estimated at 9.4 million birds in the mid-continent region of the North American Waterfowl Breeding Population and Habitat Survey. The estimate is 22 percent higher than the regional population goal of 7.7 million birds established by the North American Waterfowl Management Plan Committee (2014). The number of birds counted in the survey's eastern region was 1.0 million, yielding a total population estimate of 10.4 million (USFWS 2019).

Strongly influenced by wetland conditions in the Prairie Potholes, the population has declined significantly twice since 1955: first during the 1960s and second during the 1980s and early 1990s **(D)**. Both periods were times of notable drought that impacted the availability of wetlands and the quality of upland cover for all waterfowl.

Data collected on BBS routes in Minnesota document a significant long-term population increase averaging 1.57 percent per year from 1967 to 2019 **(E)**. Numbers gradually increase until the early 1990s, stabilize until 2000, and then decline, followed by years of wider fluctuations. Minnesota waterfowl biologists use a more targeted tool to monitor waterfowl populations in a region of western and central Minnesota that supports the highest density of lake basins outside of the northern forests. Begun in 1968, the survey documents a trend very similar to that demonstrated by BBS data (Cordts 2019).

In terms of overall abundance, the state waterfowl survey confirms that the Mallard is the most abundant duck species in the state. Since about 2005, numbers have fluctuated at a level just above the long-term average (1968–2019) of 228,000 birds. The survey's 2019 population estimate was 286,000 birds (Cordts 2019).

CONSERVATION

Although the Mallard is ranked a low conservation priority by PIF (2021), it is considered a high priority by the North American Waterfowl Management Plan Committee (2004). This reflects the species' ranking as the most abundant and important waterfowl species in North America. Indeed, as a game species, the Mallard is king. During the 2019 hunting season, 2.9 million birds were harvested in the United States, comprising 29.8 percent of the entire duck harvest. In Minnesota, 98,723 birds were harvested, or 22.2 percent of the state's waterfowl harvest (Raftovich et al. 2020).

Given its abundance, it is easy to understand why the Mallard drives most waterfowl management decisions in the United States, from the design and implementation of conservation measures to the framework of annual harvest regulations. Many of these measures were implemented in earnest following the disastrous drought of the 1980s and early 1990s. Some debate remains as to whether these conservation actions or the recent increase in precipitation have been responsible for the species' recovery (Drilling, Titman, et al. 2020). No doubt both have played an important role.

Unfortunately, the potential impacts of climate change could be formidable. The Prairie Potholes encompass only 10 percent of the continent's wetlands but produce 50–80 percent of the continent's ducks. Models predict that the number of wetlands in the region could decline by two-thirds (NABCI 2010). The Minnesota Interagency Climate Adaptation Team (2013) concluded that without major wetland restoration efforts, the state's "prairie potholes 'duck factory' could largely disappear by the end of the century." The widespread and highly adaptable Mallard, however, may be well suited to overcome such challenges. NAS (2019) models classified the species as having a low vulnerability to climate change under all three warming scenarios, ranging from 1.5°C to 3°C. Projections suggest that it will remain our most common and widespread waterfowl species in Minnesota.

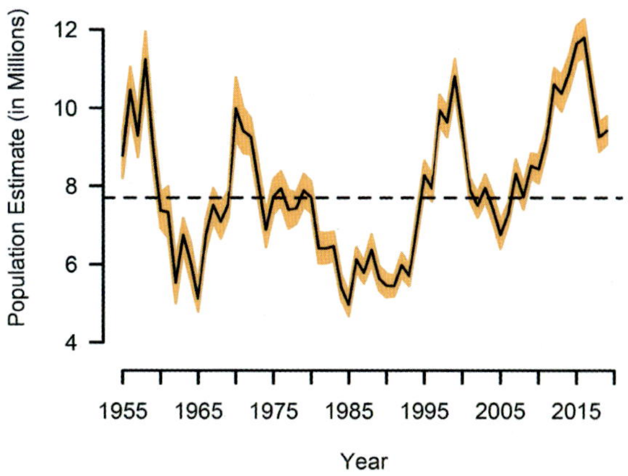

D. Mallard breeding population estimates, 90 percent confidence intervals (orange band), and North American Waterfowl Management Plan population goals (dashed line; North American Waterfowl Management Plan Committee 2014) in the traditional survey area (strata 1–18, 20–50, 75–77). Reproduced from USFWS (2019, fig. 4).

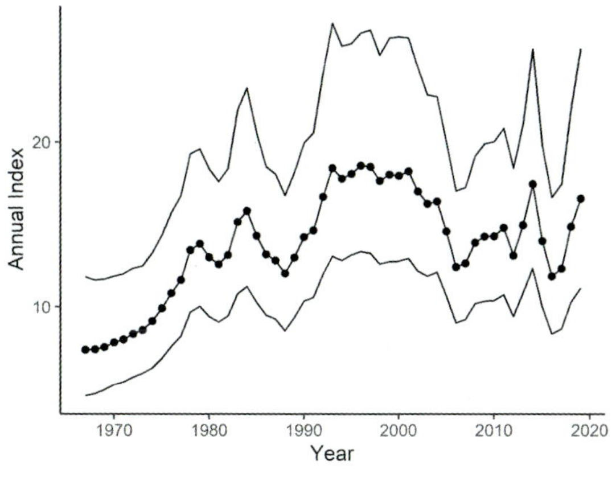

E. Breeding population trend and 95 percent credible interval (solid black lines) for the Mallard in Minnesota based on the 1967–2019 Breeding Bird Survey (Sauer et al. 2020).

AMERICAN BLACK DUCK

Anas rubripes

Minnesota Seasonal Status: Regular breeding resident and migrant; occasionally observed during the winter. The American Black Duck was rare during the MNBBA.
Migration: Short-distance or partial migrant; some populations are permanent residents, but northern breeding populations migrate to the southern United States.
Food: Dabbler and upland grazer feeding primarily on aquatic plants and aquatic invertebrates; grazes on waste corn and other grains in the fall and winter.
Nest: Ground nester; nests are usually located within dense shrub or herbaceous vegetation and occasionally in tree stumps or open tree cavities.
Conservation Concern: A game species; assigned a moderate PIF Continental Concern Score of 12/20.

Photo by Mike Lentz.

Unlike most waterfowl species, the male American Black Duck does not impress potential mates with a vibrant array of colors but relies solely on his persuasive courtship behaviors to attract their attention. Adorned in dark brown hues, he and his mate actually resemble the female of the closely related Mallard. Easily overlooked and relatively uncommon in the Great Lakes region, in the early 1900s the American Black Duck reigned supreme among waterfowl in the east, where it far outnumbered "all other species of fresh water ducks" (Bent 1923). One hundred years later, it is the Mallard that rules the roost (USFWS 2019).

Black Ducks are not truly black, but their brown plumage is considerably darker than that of the female Mallard. Sitting on the water, the male's dark brown body and gray-brown neck and head are diagnostic. The female is paler but still considerably darker than the female Mallard. The Latin term *rubripes* refers to the Black Duck's reddish-orange legs and feet. In flight, its purple speculum lacks the white borders of the Mallard's blue speculum. Unfortunately, identification can be complicated, because the two species easily hybridize with each other. They are so closely related that hybrids are usually fertile.

The American Black Duck's primary breeding range stretches from eastern Manitoba and northeastern Minnesota east across Canada, the northern Great Lakes, and New England, and south to the coastal regions of Virginia and North Carolina. Breeding densities are highest in northern New England and northeastern Canada.

MINNESOTA BREEDING DISTRIBUTION

Today Minnesota lies on the western periphery of the American Black Duck's breeding range. Before the early 1900s, however, the species did not breed west of Lake Erie and central Ontario (Palmer 1976). Roberts (1932) described it as a "straggler" in Minnesota in the late 1800s and early 1900s. It was a relatively rare bird in the hunter's bag; in 1893 a market hunter near Heron Lake reported that he had never seen it but he knew it was occasionally taken. From 1896 to 1910, only 14 were reported shot at the Long Meadow Gun Club south of Minneapolis.

Summer reports from northern Minnesota became more common during the 1920s. Roberts received a letter in 1922 from Carlos Avery, the Minnesota Game and Fish commissioner, commenting that his game wardens found the bird in "considerable numbers in the marshes and bogs between Red Lake and Rainy River" and that it appeared to be nesting in several areas. That same year Thaddeus Surber found several American Black Ducks on the Cascade River in Cook County and presumed from their behavior that nests or young were nearby. He also found several nesting pairs in northeastern Lake County and northwestern Cook County in 1926.

Green and Janssen (1975) later described the species' primary breeding range as northeastern Minnesota, including Koochiching County in the north and northern Pine County in the south. Beyond its primary range, there were confirmed nesting records from eight widely dispersed counties: Clay, Freeborn, Kittson, Marshall, Pope, Ramsey, Washington, and Wilkin. Several years later, Janssen (1987) delineated a breeding range that stretched west as far as eastern Roseau and Marshall Counties, south to Becker and Aitkin Counties, and east through the Arrowhead region. Since 1970, he reported confirmed nesting records from five counties: Aitkin, Becker, Cook, Lake, and Marshall. Summer observations also continued to be reported from southern and western Minnesota. In 1998, Hertzel and Janssen added eight counties with confirmed nesting to Janssen's original list. Four of the counties were in northern Minnesota (Cass, Koochiching, Lake of the Woods, and St. Louis), and four were in east-central and southeastern Minnesota (Goodhue, Hennepin, Ramsey, and Washington). Roseau County was later added by Janssen (2019).

By 2014, the MBS reported 31 breeding-season locations for the American Black Duck. All but one record (from Martin County) were restricted to northern counties, and the majority were reported from northern St. Louis and Lake Counties and from Cook County. Survey work had yet to be reported from far north-central Minnesota.

During the MNBBA, confirmed breeding evidence was restricted to 12 blocks in east-central

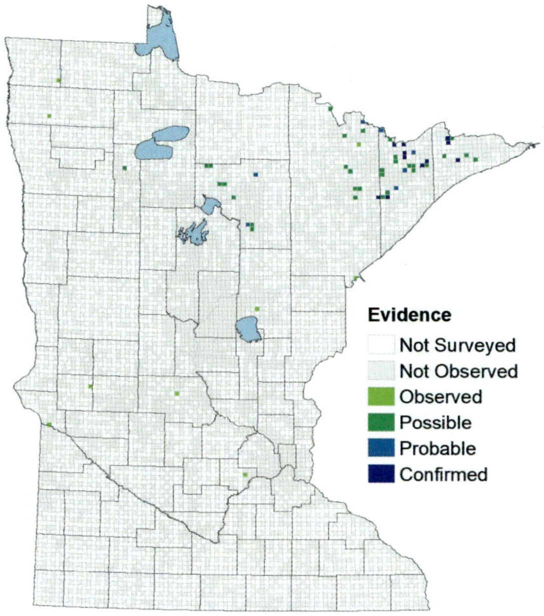

A. Breeding distribution of the American Black Duck in Minnesota based on the Breeding Bird Atlas (2009–13).

St. Louis County, northern Lake County, and Cook County, where the majority of the American Black Duck's 73 records were reported **(A)**. Probable breeding evidence was documented in Itasca County. With reports from a total of 13 counties, no birds were reported south of Carver County in the east and the Upper Minnesota River valley in the west. Breeding has not been documented in southern Minnesota since 1978 (Wabasha County). In the Twin Cities metropolitan area, the birds were last reported nesting in Hennepin County in 1976, in Ramsey County in 1977, and in Washington County in 1978 (MOU 2021b).

As the American Black Duck expanded its range into northern Minnesota in the early 1900s, it also expanded into western Ontario, North Dakota, Manitoba, Saskatchewan, and Alberta. This westward expansion occurred at a time when populations were declining in the East owing to a number of factors, including excessive hunting. Birds were also introduced in Washington and British Columbia (Longcore et al. 2020).

BREEDING HABITAT

Habitat generalists, American Black Ducks are found in such divergent sites as coastal marshes, estuaries, shallow lakes, and boreal bogs. Their lack of habitat specificity prompted Palmer (1976) to comment that the only singular requirement appeared to be the presence of water. In the Great Lakes region, the species is primarily dependent upon small lakes with stands of emergent vegetation, small impoundments established by beaver dams, and small wetland openings in bogs and lowland hardwood forests (Longcore et al. 2020) **(B)**.

POPULATION ABUNDANCE

Long-term population data on the American Black Duck have been fragmented for many years. One of the longest standing surveys is the Mid-Winter Inventory, which began in 1955 and monitors waterfowl in major winter concentration areas located in the United States along principal flyways. From 1955 to 2015, these data show a dramatic 62 percent population decline **(C)**. The rate of decline has slowed since the 1980s, but overall numbers remain relatively low (USFWS 2016a).

In 1990, the North American Waterfowl Breeding Population and Habitat Survey, which previously limited its focus to portions of Alaska and the Prairie Potholes Region, was expanded to include a large region of eastern Canada and New England. Further expanded in 1998, the survey now covers much of the Black Duck's breeding range. As more data are collected, this survey will eventually replace the Mid-Winter Inventory as the more reliable monitoring tool. Since 1998, however, American Black Duck populations in the eastern survey region have been relatively stable.

In 2019, the number of adults tallied by the survey was estimated at 729,400 (USFWS 2019). This is above the population objective of 628,000 birds established by the North American Waterfowl Management Plan Committee (2014) but below the 1998–2018 average of 860,000. Unfortunately, the size of Minnesota's population is unknown.

Overall, biologists agree that no single factor is responsible for the species' range-wide decline. Overharvesting, habitat loss and degradation, contaminants, and competition and hybridization with a growing Mallard population have all contributed to the loss of nearly 50 percent of the historical population (Baldassarre 2014).

B. A small forested wetland embedded in a northern Minnesota bog has an abundance of emergent vegetation and cover for the nesting American Black Duck. Photo by Courtney Celley/USFWS.

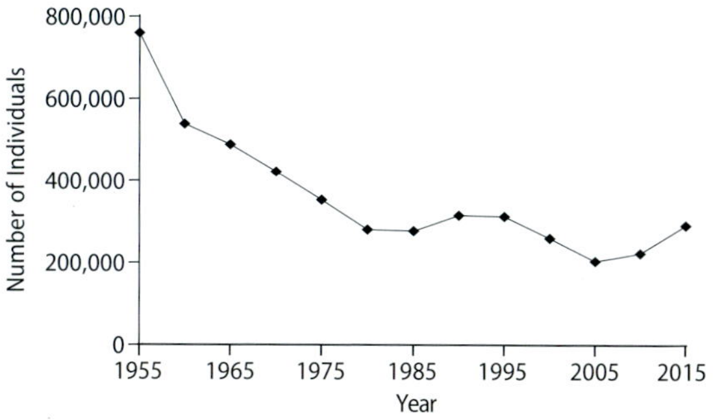

C. Population trend for the American Black Duck based on the Mid-Winter Inventory in the United States from 1955 to 2015 (USFWS 2016a).

CONSERVATION

Revered as a game bird in the eastern United States and Canada, the American Black Duck is recognized by many resource agencies and conservation initiatives as a priority species, including PIF (2021), which assigned it a moderate Continental Concern Score. In 1989, the Black Duck Joint Venture was launched specifically to gather scientific data that would guide the species' recovery.

The contribution of high harvest levels to the bird's demise has been intensely debated for many years. In 1982, a lawsuit alleging that harvest regulations were too permissive led to the establishment of very restrictive regulations, which helped slow the rate of decline (Longcore et al. 2020; Baldassarre 2014). In 2019, the annual harvest in North America was 132,644 birds, only 1.2 percent of the total duck harvest. That same year only 636 American Black Ducks were harvested in Minnesota (Raftovich et al. 2020).

In addition to all the other threats the species contends with, warming temperatures are also a concern. NAS (2019) classified the species as moderately vulnerable under all three warming scenarios, ranging from 1.5°C to 3°C. Significant losses were predicted along the western and southern borders of its breeding range, including most of northeastern Minnesota. As significant as this threat may be, biologists contend that the principal challenge may be the increasing population of Mallards, which outcompete and hybridize with the American Black Duck (Baldassarre 2014). Although the debate continues about whether the Black Duck and Mallard are indeed separate species, the future of the American Black Duck seems tenuous at best.

NORTHERN PINTAIL

Anas acuta

Minnesota Seasonal Status: Regular breeding resident and migrant; occasionally observed during the winter. The Northern Pintail was rare during the MNBBA.
Migration: Ranges from a year-round resident to a medium-distance migrant; birds breeding in the Prairie Potholes winter primarily along the Gulf coast of the United States.
Food: Dabbling duck that feeds on seeds, aquatic plants, and invertebrates; during the nonbreeding season it also forages in agricultural fields.
Nest: Ground nester; nests are usually located in short, sparse herbaceous cover, including croplands, or in grassland brush.
Conservation Concern: A game species; assigned a moderate PIF Continental Concern Score of 12/20.

Photo by Mike Lentz.

Considered the "vanguard of the spring migration" (Bent 1923), this slender and graceful bird is a rare sight in Minnesota during the breeding season. But long before the ice is out on the state's wetlands and small lakes, the first wave of puddle ducks appears, stopping briefly to rest in fields where the snow has melted just enough to create temporary pools. Invariably, these birds "are the advance couriers of the hardy Pintail Duck pushing northward, in answer to the spring urge, into a still wintery land that offers but indifferent accommodation" (Roberts 1932). Most are en route to their breeding grounds in Canada's Prairie Parklands.

Male Pintails are especially striking in their nuptial plumage. The elongated central tail feathers extend well beyond the body, giving rise to the species' common and scientific names. Their chocolate-brown heads contrast with their gray bodies and with a streak of white that extends from the breast up each side of the neck. In flight, a green speculum, bordered above by cinnamon and below by white, is visible. The grayish-brown female, however, is difficult to distinguish from the females of other dabbling ducks. Although her tail is considerably shorter than the male's, experienced observers look for her pointed tail, long neck, and buffy speculum when the birds are in flight.

Circumpolar in distribution, the Pintail's North American breeding range stretches from Alaska east to Newfoundland and Nova Scotia and south through the northwestern and north-central states. The core of its breeding range is in the Prairie Parklands of central Canada and in the northern Great Plains states.

MINNESOTA BREEDING DISTRIBUTION

The Northern Pintail likely reached its peak abundance before the majority of Minnesota's wetlands were drained and the unregulated hunting of waterfowl ceased with passage of the 1918 Migratory Bird Treaty Act. By the 1930s, Roberts (1932) wrote that the species still nested in "considerable numbers" in southern and western Minnesota and occasionally in forested areas, but it was "by no means, as abundant a summer resident as formerly." Prior to 1898, for example, it was quite common at Heron Lake in Jackson County and abundant at Mud Lake in Marshall County. Roberts (1932, 1936) identified confirmed nesting records in Anoka, Cass, Jackson, Lincoln, and Marshall Counties and inferred nesting records in Big Stone, Otter Tail, and Polk Counties.

Unfortunately, the Pintail never recovered to its former level of abundance. In the early 1960s, it comprised less than 5 percent of Minnesota's breeding duck population (Lee et al. 1964). Green and Janssen (1975) described it as a summer resident "primarily in the western and southern prairie regions" but also noted confirmed breeding records farther east in Aitkin, Beltrami, Cass, and Washington Counties. They depicted the species' breeding range as including all of northwestern Minnesota and most of southern Minnesota, with the exception of the southeast region. Janssen (1987) delineated 15 counties where breeding had been confirmed since 1970; only one, Aitkin County, was in the northern forest region. The species was not, Janssen wrote, a common breeder in the state. Hertzel and Janssen (1998) added four counties to the list with confirmed nesting since 1970; only Lake of the Woods was outside of the Pintail's primary breeding range. Meanwhile, the MBS identified just 32 breeding-season locations for the species. Restricted to the Prairie Parkland Province, nearly one-third of the records were located in just two counties: Marshall and Lac qui Parle.

The number of records reported during the MNBBA was only marginally higher. Indeed, among Minnesota's nine dabbling ducks, the Pintail and American Wigeon had the fewest detections, each with just 55. Reported from 1.0 percent of the surveyed atlas blocks (45/4,734), Pintails were distributed across 25 of Minnesota's 87 counties **(A)**. Breeding was confirmed in seven blocks that spanned seven counties, including the first documented nesting (fledged young) in Swift County. With the exception of confirmed breeding in Sherburne County, all confirmed and probable nesting records were in the Prairie Parkland Province. Possible breeding and observed records, however, were reported from several northern counties, specifically Aitkin, Beltrami, Cass, Lake, and St. Louis. The first confirmed nesting record for St. Louis County was logged two years after the atlas was completed, when a female with five young was found on an old tailings basin in 2015. One year later, nesting was also reported in Cook County, when a female with six young was observed at the Grand Portage wastewater treatment pond (MOU 2021b).

Overall, the Northern Pintail's distribution across Minnesota remains similar to its distribution nearly one hundred years ago, but its abundance is likely magnitudes smaller. The same appears true across its North American breeding range (Clark et al. 2020).

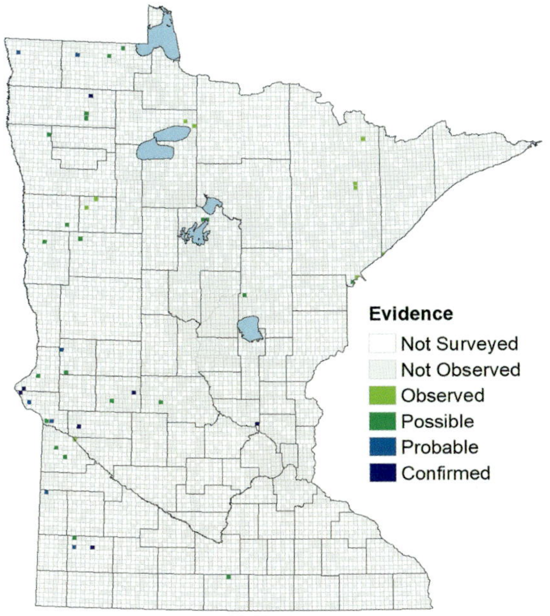

A. Breeding distribution of the Northern Pintail in Minnesota based on the Breeding Bird Atlas (2009–13).

BREEDING HABITAT

The Northern Pintail relies on ephemeral and semipermanent wetlands that support emergent aquatic vegetation and are embedded in a landscape with short and sparse upland cover **(B)**. Permanent wetland basins or impoundments are occasionally used (Clark et al. 2020). The species has a strong tendency to nest farther away from water (often 1–2 km) (Duncan 1987) and in more open cover than any other dabbling duck (Baldassarre 2014).

The Pintail also demonstrates a higher preference for nesting in croplands and grain stubble than other ducks, with estimates of 34–57 percent of the nests located in cultivated fields (Klett et al. 1988; Reynolds et al. 2001). Unfortunately, nesting success in such habitat is dismally low (5–11 percent). Because Pintails nest early in the breeding season, many nests are destroyed by agricultural activities as the season unfolds, while the low vegetative cover increases their vulnerability to predators.

Other nesting habitats include idle grasslands, planted cover, fall-seeded croplands, and hayfields, where nesting success is generally higher but variable (9–47 percent) (Clark et al. 2020). Sensitive to local drought conditions, Pintails will readily move farther north to seek wetlands that are more suitable (Baldassarre 2014).

POPULATION ABUNDANCE

In 2019 the North American Waterfowl Breeding Population and Habitat Survey (USFWS 2019) estimated a population of 2.3 million adults in the traditional survey area of Alaska, Canada, and the northern Great Plains **(C)**, a dramatic decline from more than 10 million birds reported in the mid-1950s and well below the goal of 4.0 million established by the North American Waterfowl Management Plan Committee (2014). BBS data also show a significant, surveywide population decline averaging −2.60 percent per year from 1966 to 2019 (Sauer et al. 2020).

It would be easy to ascribe the Pintail's steep decline entirely to habitat loss and drought conditions in the Great Plains. But, as precipitation patterns improved and wetland protection efforts began in earnest, the species recovery remained slow despite increases in other waterfowl populations. Poor nesting success is the principal culprit (M. R. Miller et al. 2001). As noted above, nests established in croplands early in the season are frequently lost to agricultural operations, while exposed nests that survive are an easy target for predators. Pintails are also less inclined than other species to renest when their initial efforts fail (Baldassarre 2014). These problems are especially acute in Canada, where nearly 75 percent of the prairie grasslands were converted to croplands during the 1900s (Trottier 1992). Compounding the challenge, Canadian farmers that previously allowed their fields to go fallow between harvest years are now planting crops every year (Carlyle 1997).

In Minnesota, Pintails comprise less than 1 percent of the total dabbling duck population. Only 172 birds were counted during the state's 2019 waterfowl survey, a number well below the 10-year average of 1,832 birds documented from 1987 to 1996 (Cordts 2005, 2019). The loss of more than 90 percent of Minnesota's wetlands and the weak protection afforded many temporary and seasonal wetlands are considered largely responsible for the decline (MNDNR 2006a).

B. Although the Northern Pintail often selects wetlands adjacent to cropland cover for nesting, this shallow wetland surrounded by grassland cover in west-central Minnesota also provided suitable habitat for a nesting pair.
Photo by Lee A. Pfannmuller.

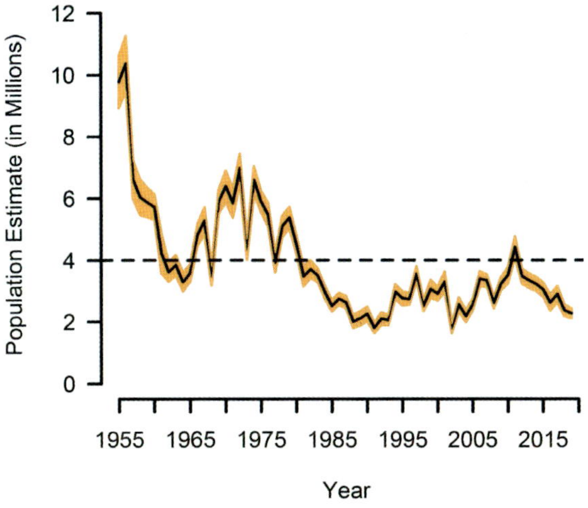

C. Northern Pintail breeding population estimates, 90 percent confidence intervals (orange band), and North American Waterfowl Management Plan population goals (dashed line; North American Waterfowl Management Plan Committee 2014) in the traditional survey area (strata 1–18, 20–50, 75–77). Reproduced from USFWS (2019, fig. 4).

CONSERVATION

Given its small, declining population, the impacts of hunting mortality on Northern Pintails were a concern for many years. Stricter harvest regulations enacted in 1985, however, reduced the United States harvest from approximately 1 million birds per year (Clark et al. 2020) to 317,720 in 2019 (Raftovich et al. 2020). That same year, 5,084 birds were harvested in Minnesota.

Another concern is the Pintail's reliance on temporary and semipermanent wetlands, habitats that are especially vulnerable as temperatures rise. Fortunately, the species vulnerability was projected to be low under a 2°C warming scenario (NAS 2019). Models actually predicted minor losses in western Minnesota coupled with substantial gains in northwestern and north-central counties. Wetland restoration and protection efforts in those regions are therefore critical.

Many of the challenges facing the species, however, can be resolved only by waterfowl biologists working side-by-side with farmers. On cultivated lands in Canada, Pintail conservation efforts now are focused on "the promotion of fall-seeded crops (winter wheat, fall rye), conversion of cropland to perennial forages and pasture," and establishing conservation easements on non-tilled lands (Clark et al. 2020). Indeed, integrating the needs of farmers with Pintail conservation measures may ultimately ensure the future of this elegant species.

GREEN-WINGED TEAL

Anas crecca

Minnesota Seasonal Status: Regular breeding resident and migrant; occasionally reported during the winter, principally in southern Minnesota. The Green-winged Teal was uncommon during the MNBBA.
Migration: Year-round resident in portions of the western United States; elsewhere the Green-winged Teal migrates to the coast of southern Alaska and western British Columbia and south to the western and southern United States, the Caribbean, and Mexico.
Food: Dabbler that forages on mudflats and in shallow water for seeds of aquatic plants, grasses, and sedges; also consumes aquatic invertebrates and occasionally waste grains.
Nest: Ground nester.
Conservation Concern: A game species; assigned a low PIF Continental Concern Score of 8/20.

Photo by Mike Lentz.

North America's smallest dabbling duck, the little Green-winged Teal makes up for its small size in its elegant plumage. Roberts (1932) gave it high praise in the following account: "The handsomely adorned, little male Green-winged Teal belongs to a group composed of the drakes of a few of our Ducks which, in full nuptial plumage, vie with one another in gorgeous colorings and striking patterns. Among these the Wood Duck is usually accorded first place, and the dapper, gay little Green-wing seems a fair candidate for the second place." Smitten, in general, with the vibrant attire of Minnesota's waterfowl, Roberts made very similar remarks of the Hooded Merganser.

On the water, the male's chestnut head, with a prominent green ear patch that stretches from the eye down the back of the neck, is a clear identification marker. In poor light conditions, particularly when the birds are at a distance, the white vertical shoulder patch is the best diagnostic feature. Like other dabblers, the female is a mottled gray-brown, but her creamy white undertail coverts are unique. The iridescent green feathers within the speculum, which provide the bird's namesake, are visible only in flight, which Roberts (1932) likened to that of a swallow because of its swift and erratic movements. *Anas crecca,* the species' Latin name, sheds little light on the species' unique features, literally translating to "duck with a quack"!

The Green-winged Teal has a global distribution spanning the northern reaches of North America, from Alaska and boreal Canada south into the north-central and northeastern states, as well as across northern Europe, Iceland, Greenland, and Asia. Since 2000, biologists in Europe have recognized two distinct species: the Common Teal (*A. c. crecca*) in Europe and the Green-winged Teal (*A. c. carolinensis*) in North America. The American Ornithological Society, the purveyor of all bird systematics in North America, has yet to make a similar decision. Prior to 1931, however, they considered the two populations to be separate species.

MINNESOTA BREEDING DISTRIBUTION

The Green-winged Teal has always been a common spring and fall migrant in Minnesota but has a spotty nesting history. In the late 1800s, Hatch (1892) reported it was a frequent breeder in Hennepin County, and Tom Miller, a hunting guide at Heron Lake in Jackson County, reported he found a nest at the lake in 1888 (Roberts 1932). By 1932, however, Roberts wrote that there had been no new nesting records since Hatch's and Miller's accounts and, although still a common migrant, the bird was "much reduced in numbers." Like most waterfowl, once the Migratory Bird Treaty Act was passed in 1918, numbers began to rebound. Pairs were observed during the summer months at a few widespread locations, ranging from Jackson County (1919) in the south, to Grant County (1930) in the north, and east to Isanti County (1927), leading Roberts to suspect that the bird might still be nesting in the state.

Green and Janssen's (1975) account of the species described the Green-winged Teal as a "resident throughout most of the state but very scarce." Nesting had been documented in 14 wide-ranging counties, including Marshall and Polk in the northwest, Nobles and Yellow Medicine in the southwest, Rice and Wabasha in the southeast, and Hennepin and Washington in east-central Minnesota. Several years later, Janssen (1987) noted that "scattered individuals or pairs can be found in almost any part of the state during the breeding season." He identified nine counties where nesting had been documented since 1970: Aitkin, Anoka, Beltrami, Big Stone, Clearwater, Lake, Roseau, Winona, and Wright. Eleven years later, Hertzel and Janssen (1998) added three more counties to the list: Crow Wing, Hennepin, and Marshall. By 2014, the MBS had reported 78 breeding-season locations for the Green-winged Teal since the late 1980s. Widely scattered throughout the state, the species was notably absent from east-central and southeastern Minnesota. The largest number of records was reported from Clearwater County in north-central Minnesota.

Like earlier accounts, the Green-winged Teal's 154 MNBBA records were widely dispersed across the state. Absent only in the southeast region, the species was found in 2.8 percent (133/4,738) of the surveyed atlas blocks. Breeding was confirmed in nine blocks and deemed probable in 26 blocks **(A)**. Observations were reported from 47 of Minnesota's 87 counties and breeding was confirmed in five counties (Anoka, Lac qui Parle, Mahnomen, Marshall, and St. Louis). The majority of records for this

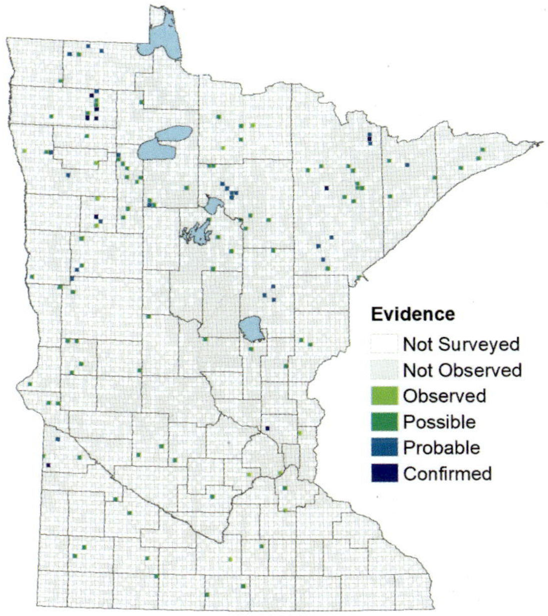

A. Breeding distribution of the Green-winged Teal in Minnesota based on the Breeding Bird Atlas (2009–13).

boreal-nesting species were in the Laurentian Mixed Forest and Tallgrass Aspen Parklands Provinces. As is the case for so many of Minnesota's waterfowl, Agassiz NWR, Thief Lake WMA, and the surrounding wetlands in Marshall County are extremely important for Green-winged Teal. Although it is certainly more abundant today than in the late 1800s, the species remains an uncommon breeding resident. Nevertheless, its 154 atlas records were greater than those reported for seven other waterfowl species: Bufflehead (27), Red-breasted Merganser (37), Lesser Scaup (49), Northern Pintail (55), American Wigeon (55), American Black Duck (73), and Canvasback (94).

In his comprehensive overview of the species, K. Johnson (2020) noted that there were no documented changes in the Teal's distribution. In Ontario, however, the species may have nested only in the far northern regions historically but today can be found province-wide (Cadman et al. 2007). Farther south, atlas reports from several states located near the southern periphery of the Teal's breeding range commented on its widely dispersed breeding distribution, including Wisconsin (Cutright et al. 2006), Michigan (Chartier et al. 2013), and South Dakota (Drilling, Sparks, et al. 2016).

BREEDING HABITAT

The Green-winged Teal is a northern species, adapted primarily to the boreal forests and aspen parklands of the Canadian Prairie Provinces and the broad river deltas of Alaska. Although it nests in the prairie potholes of the United States as well, it reaches its highest nesting densities in Canada's forests and open woodlands (K. Johnson 2020).

Few studies have delineated specific habitat requirements during the breeding season. The birds seem to prefer seasonal and semipermanent wetlands, although they also use dugouts and stock ponds. Wetlands usually have a high percentage of emergent cover and have shallow areas measuring less than 1 m in depth (Baldassarre 2014) **(B)**. Nests are placed in adjacent grasslands, brushy thickets, or aspen parklands.

POPULATION ABUNDANCE

The North American Waterfowl Breeding Population and Habitat Survey generated a 2019 continental population estimate of 3.5 million birds (USFWS 2019), down significantly from a high of 4.5 million in 2016. Over 90 percent of the birds (3.2 million) occurred in the survey's traditional survey area, which stretches from the Dakotas north to Alaska. Despite the recent decline, which was experienced by many waterfowl species, this estimate was still significantly higher than the 2.06 million population goal established for the traditional survey region by the North American Waterfowl Management Plan Committee (2014). Approximately 20 percent of the population occurs in the United States and 80 percent in Canada.

From a low of approximately 723,000 Green-winged Teal in 1962, populations have grown steadily **(C)**. Baldassarre (2014) speculates that Teal have prospered because they depend on habitats in the boreal region of Canada rather than the more southern Prairie Potholes Region, which has experienced extensive habitat loss and degradation. The bird's well-concealed nests may also contribute to higher reproductive success.

B. Extensive and productive wetland complexes in Marshall County, in the heart of Minnesota's Aspen Parklands, are one of the few sites in the state where the Green-winged Teal is regularly confirmed nesting. Photo by Lee A. Pfannmuller.

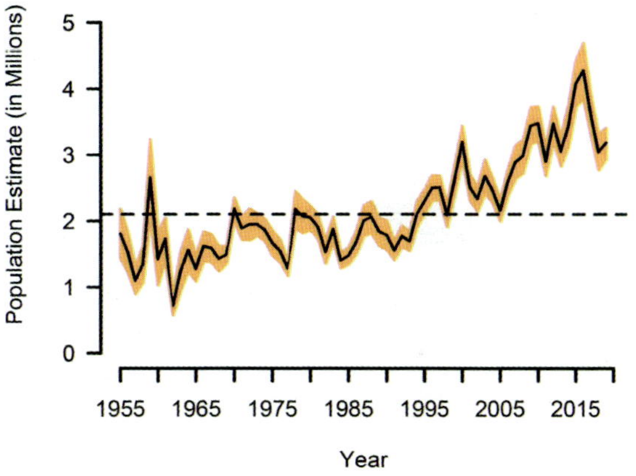

C. Green-winged Teal breeding population estimates, 90 percent confidence intervals (orange band), and North American Waterfowl Management Plan population goals (dashed line; North American Waterfowl Management Plan Committee 2014) in the traditional survey area (strata 1–18, 20–50, 75–77). Reproduced from USFWS (2019, fig. 4).

In Minnesota, where the annual Waterfowl Breeding Population Survey is restricted in geographical scope, Green-winged Teal comprise less than one percent of all dabbling ducks reported during the Minnesota survey (Cordts 2019). Since the mid-1980s, numbers have fluctuated from a low of zero in 1992 and 1995 to a high of only 2,170 in 1996. Wide fluctuations on the edge of the species' range, south of its prime breeding habitat, are to be expected.

CONSERVATION

A popular game bird, Green-winged Teal comprised 12.0 percent of the United States duck harvest in 2019 (1.2 million birds), second only to the Mallard (2.9 million birds). A total of 22,668 birds were taken in Minnesota, or 5.1 percent of the total state waterfowl harvest (Raftovich et al. 2020).

Given the healthy status of its population, little conservation attention has been focused on the Green-winged Teal, and it is ranked a low conservation priority (PIF 2021). Unfortunately, its future depends on Canada's boreal landscape, one of North America's ecosystems most vulnerable to climate change. Although habitat gains may occur across the Arctic, under a 2°C warming scenario, this moderately vulnerable species is projected to lose 30 percent of its current breeding range by 2050. Some of the heaviest losses are predicted in eastern Canada (NAS 2019). Under the same scenario, this stunning species is projected to become even less common across Minnesota.

CANVASBACK

Aythya valisineria

Minnesota Seasonal Status: Regular breeding species and migrant; occasionally observed in central and southern Minnesota during the winter. The Canvasback was rare during the MNBBA.
Migration: Short- to medium-distance migrant that winters along the United States coastline, in many central and southern states, and in northern Mexico.
Food: Diver that feeds primarily on aquatic plants and aquatic invertebrates.
Nest: Platform constructed in emergent vegetation over shallow water; occasionally placed on the ground.
Conservation Concern: A game species; assigned a moderate PIF Continental Concern Score of 10/20.

Photo by Mike Lentz.

Known as the "undisputed king of ducks" among waterfowl hunters (Washburn 2007), the stately Canvasback has been described with countless expressions and eulogies that pay homage to its royal status. McIntosh (1997), for example, praised the wading aristocrat as "America incarnate." Duck hunting, he believed, simply could not be any better than watching large flocks of these gregarious birds alight on a nearby wetland, "swarming a spread of decoys in the open water." Aldo Leopold and his colleagues Miles Pirnie and William Rowan (1944) wrote: "We offer no apology for thus elevating the lordly canvasback to a classical role, for among duck-minded people he has long been the 'gold standard' against which all lesser fry are weighed and measured."

Sitting low on the water, the male's chestnut-red head, black chest, and white body are striking. The sloping head profile is diagnostic on both the colorful male and the gray-brown female. Its common name originates from the habit of early European settlers comparing its white body to the white color of canvas tarps (Baldassarre 2014). The Latin name, *valisineria,* is a reference to one of its primary food items during the nonbreeding season, the winter buds of the aquatic plant wild celery, *Vallisneria americana.* A diet of wild celery imparts a "delicate" flavor to the Canvasback that has endeared it to waterfowl hunters for centuries.

The Canvasback's history is legendary. When Europeans first arrived, nearly 1 million birds were believed to winter in Chesapeake Bay, attracted by its vast beds of wild celery. But they soon became an easy target for market hunters who were supplying the insatiable appetites of eastern markets. Although the 1918 Migratory Bird Treaty Act brought an end to the carnage, the bay's water quality declined, causing a dramatic reduction in the entire community of submerged aquatic plants, including wild celery (Haramis et al. 2001). By the winter of 2017–18, Chesapeake Bay was a winter home to only 60,000 birds (Dybas 2018). Responding to these changes, Canvasbacks have actually "altered their traditional migration routes" and "modified [their] diets" (Mowbray 2020a).

During the breeding season, Canvasbacks can be found from central Alaska south across Canada's Prairie Provinces, the northern Great Plains of the United States, and in scattered locations farther to the east and west. The highest population densities occur in southern Saskatchewan, southern Manitoba, and western North Dakota.

MINNESOTA BREEDING DISTRIBUTION

In 1932, Roberts wrote that the Canvasback was formerly "a common summer resident" throughout western Minnesota. At the time, he lamented that it had "almost entirely disappeared as a breeding bird," although it continued to migrate through the state each spring and fall. He referenced a June 1929 field survey by Walter Breckenridge that began in Lincoln County in southwestern Minnesota and ended in Kittson County in the northwest. Not a single Canvasback was sighted. Yet, in the late 1800s, this handsome species was a regular summer resident at Heron Lake in southwestern Minnesota and at some of the large, shallow wetland basins in northwestern Minnesota, such as Thief Lake and Mud Lake in Marshall County. Nesting was confirmed at all three of these sites between 1896 and 1900. Roberts observed the species in Minneapolis in June 1925, which suggested it may be breeding in eastern Minnesota as well.

More than forty years later, the Canvasback's plight had improved considerably. Green and Janssen (1975) described it as a regular summer resident in northwestern Minnesota, stretching from northwestern Becker County north to the Canadian border. Nesting birds were occasionally documented as far south as Le Sueur, Lincoln, and Lyon Counties, and as far east as Hennepin County. Indeed, just a few years later, Jessen and Henderson (1979) reported that the summer of 1978 was the "best year for Canvasbacks breeding in Minnesota in recent history." The authors reported 699 young from 46 broods located throughout west-central and northwestern Minnesota, with one outlier in Faribault County, in south-central Minnesota.

In 1987, Janssen identified the species' breeding range as encompassing all of western Minnesota, stretching east to Hennepin, Sherburne, and Wright Counties in the north, and to Faribault County in the south. Northwestern Minnesota, however, still remained the core of the species' distribution. Since 1970, nesting had been confirmed in 23 counties. Hertzel and Janssen (1998) later added four counties to the list, all within the primary breeding range delineated by Janssen (1987). The MBS documented a similar distribution. By 2014, 46 breeding-season locations had been identified across the grasslands and aspen parklands of western Minnesota. The most eastern location was in northern Freeborn County.

MNBBA participants documented 94 Canvasback records from 1.6 percent of the surveyed atlas blocks (77/4,741). The records spanned 21 counties; all but one were located within the breeding range identified by earlier accounts **(A)**. The one outlier

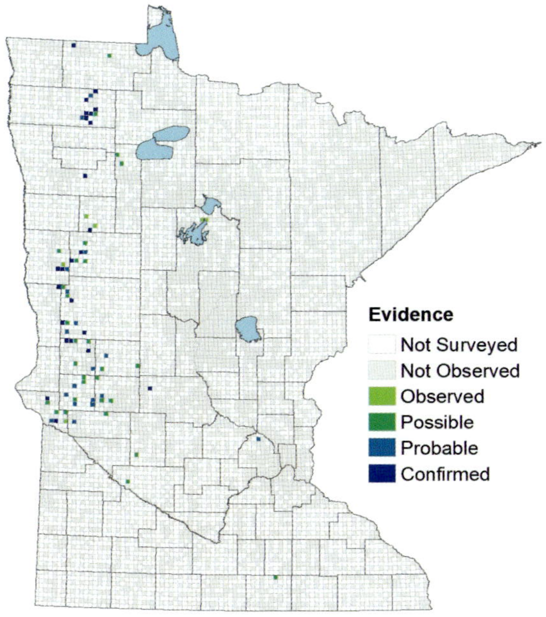

A. Breeding distribution of the Canvasback in Minnesota based on the Breeding Bird Atlas (2009–13).

was Cass County, where Canvasbacks were observed in a shallow bay in Leech Lake. Only one record was located south of the Minnesota River, in Freeborn County. Breeding was confirmed in 24 blocks and 11 counties. Seven of the 24 confirmed breeding records were from Marshall County. Grant County, where fledglings were observed in three blocks, was new to the list of counties with confirmed breeding compiled by Hertzel and Janssen (1998). A few years later, Janssen (2019) extended the species' breeding range farther east with new nesting records in Anoka and Aitkin Counties.

Although their distribution appears to have receded from southwestern Minnesota, Canvasbacks have made a significant comeback from the late 1800s, when Roberts (1932) considered them on the brink of extirpation. Beyond Minnesota, their breeding distribution is shifting farther to the northwest in Canada and Alaska, while their population is decreasing in abundance in eastern Canada (Mowbray 2020a).

BREEDING HABITAT

Canvasbacks select a wide variety of wetland basins for nesting, ranging from small potholes to sheltered bays of very large lakes **(B)**. Field studies in northwestern Minnesota found them using shallow and deepwater marshes, shrub wetlands, and shallow impoundments (Korschgen et al. 1996; Maxson and Riggs 1996). As diving birds, however, they prefer permanent, deepwater marshes or large lakes. A border of dense emergent vegetation is required for nesting, and a rich community of submerged aquatic vegetation, especially sago pondweed, for foraging (Mowbray 2020a).

POPULATION ABUNDANCE

In 2019, the North American Waterfowl Breeding Population and Habitat Survey estimated a breeding population of 651,900 birds in the survey's traditional survey area, which stretches from the Dakotas north to Alaska (USFWS 2019). This exceeded the goal of 581,000 birds established by the North American Waterfowl Management Plan Committee (2014). Although numbers have fluctuated since the survey began in 1955, they demonstrate a largely stable population trend **(C)**. BBS data corroborate this stable trend and even suggest a small increase since the year 2000 across the survey region (Sauer et al. 2020).

Canvasback population data collected by the Minnesota Waterfowl Breeding Population Survey largely mirror that at the national level with an overall stable population punctuated by wide fluctuations (Cordts 2019). Although Minnesota BBS data are not entirely reliable for this uncommon breeding species, they suggest a slowly increasing population since 1967 (Sauer et al. 2020).

CONSERVATION

The Canvasback's small population, coupled with concerns regarding threats to its wetland habitats, have prompted its assignment as a moderate conservation priority (PIF 2021). Locally, navigational pools 7 and 8 along the Mississippi River near La Crosse, Wisconsin, are of particular interest. Because they are a significant fall staging area for birds migrating from the northern Great Plains (Baldassarre 2014), efforts to minimize recreational disturbances in the pools during "this critical period of weight increase" are vital to the species' conservation (Mowbray 2020a).

B. A diving duck, the Canvasback usually prefers permanent, deepwater marshes with abundant emergent and submerged aquatic vegetation for breeding.
Photo by Lee A. Pfannmuller.

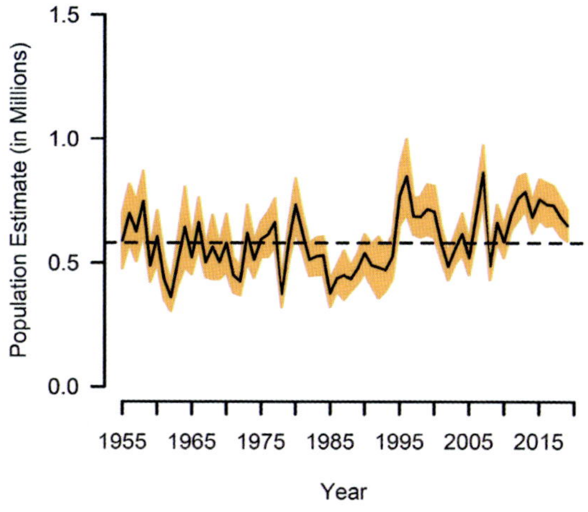

C. Canvasback breeding population estimates, 90 percent confidence intervals (orange band), and North American Waterfowl Management Plan population goals (dashed line; North American Waterfowl Management Plan Committee 2014) in the traditional survey area (strata 1–18, 20–50, 75–77). Reproduced from USFWS (2019, fig. 4).

Highly valued by waterfowl hunters, Canvasbacks comprised just 0.7 percent (72,900 birds) of the 2019 continental harvest (Raftovich et al. 2020). That same year, 4,661 birds were harvested in Minnesota. Because hunting is a major source of Canvasback mortality, biologists track population levels closely, making frequent modifications to the harvest framework, including season closures when necessary (Mowbray 2020a).

The impact of warming temperatures is another concern. The potential loss of nearly two-thirds of the prairie pothole wetlands could have a devastating impact on wetland-dependent species (NABCI 2010). As a result, close monitoring is warranted, even though warming scenarios ranging from 1.5°C to 3°C projected the Canvasback's vulnerability to climate change was low (NAS 2019). These same models, however, also predict this western species' gradual disappearance from most of Minnesota.

Finally, during the nonbreeding season, the Canvasback's diet of aquatic plant rhizomes and tubers, secured by probing wetland sediments, increases its ingestion of many contaminants, including spent lead pellets. Although lead shot is now prohibited for waterfowl hunting, its persistence in wetlands that are popular hunting spots harmed Canvasback populations for many years. Fortunately, recent studies suggest the incidence of lead shot ingestion is decreasing, thereby contributing to the species' recovery (Baldassarre 2014). Aggressive efforts to protect and restore wetland habitats have been equally important and will continue to be critical to sustaining populations of the regal Canvasback.

REDHEAD

Aythya americana

Minnesota Seasonal Status: Regular breeding resident and migrant; occasionally overwinters, primarily in southern Minnesota. The Redhead was uncommon during the MNBBA.
Migration: Short- to medium-distance migrant that winters from the central United States south through Mexico.
Food: Diver that feeds primarily on aquatic plants and aquatic invertebrates.
Nest: Platform constructed in emergent wetland vegetation; occasionally nests on muskrat houses.
Conservation Concern: A game species; assigned a low PIF Continental Concern Score of 8/20.

Photo by Mike Lentz.

True to its name, the Redhead does indeed have a red head. One might think this feature alone would distinguish this diving duck from other waterfowl species. Early ornithologists, however, confused it with a remarkably similar European species, the Common Pochard, and novice birders today may mistake it for another Minnesota species, the Canvasback. All three species have a red head, a black breast, and a nearly uniform light plumage on the back and sides. For many years the Redhead actually was known as the American Pochard.

Today Minnesotans need to distinguish the Redhead, now recognized as a distinct species from the Pochard (*A. ferina*), only from the Canvasback. Two important features are, first, the Redhead's high, boxy forehead, compared to the Canvasback's sloping head profile; and, second, the Redhead's gray plumage on its back and flanks, compared to the Canvasback's white plumage. Males also have a unique, catlike *meow* vocalization during the breeding season. Like most divers, the female is a light grayish brown; the absence of any white near the base of the bill can help distinguish her from many other female divers.

Redheads are a favorite among waterfowl hunters. Once the breeding season is over, they are a very social bird, often flocking in large rafts that can number in the thousands. This very trait makes them easily attracted to decoys, leading to the nickname "fool duck" among some hunters. The behavior contributed to the Redhead's demise in the late 1800s, when the absence of hunting regulations made it easy for hunters to bag 100 or more birds in a day.

A species of the western prairie potholes and prairie parklands, smaller Redhead populations also occur elsewhere, including Alaska, the Great Lakes region, and the St. Lawrence River valley. The highest breeding densities are found in North Dakota.

MINNESOTA BREEDING DISTRIBUTION

In 1932, when Roberts wrote his account of the Redhead's status in the early 1900s, the species was just beginning to recover from the widespread carnage of earlier years. Formerly abundant throughout the western prairies, the birds were especially numerous at Heron Lake, in Jackson County, where they bred in "great numbers." But "with the settling of the country and the excessive spring market-shooting that was soon inaugurated, the nesting birds decreased rapidly." Thomas Miller, a longtime hunting guide on Heron Lake, reported that "from 1883 to 1893 several thousand Redheads were killed [on the lake] every year" (Roberts 1932). Large numbers had been reported nesting in Marshall County, but "disastrous draining of the whole region rendered it unfit as a breeding ground." By the early 1900s, scattered reports of nesting pairs represented mere fragments of Minnesota's once-abundant population.

Despite the decimation, by the late 1920s there were hopeful signs that numbers were beginning to rebound with passage of the 1918 Migratory Bird Treaty Act. When Walter Breckenridge surveyed the western counties in May and June 1929, he observed an "unexpectedly large" number of Redheads stretching from Lincoln County in the southwest north to the Canadian border (Roberts 1932).

By the time Green and Janssen (1975) published their account of the species' status, it was well established as a summer resident throughout western and central Minnesota, with records extending as far east as the Twin Cities. Janssen's (1987) updated account further expanded the species' breeding range to encompass the entire western two-thirds of the state. He identified 20 counties where nesting had been confirmed since 1970. The counties were widely distributed across the state, with the exception of the north-central, northeastern, and southeastern regions. Hertzel and Janssen (1998) later added six counties to the list, all within the species' primary breeding range. By 2014, the MBS documented 112 breeding-season locations. The records were largely confined to the Prairie Parkland and Tallgrass Aspen Parklands Provinces but included scattered reports from a few eastern localities, including Lake County.

The species' breeding distribution was similarly defined by the 274 records collected during the MNBBA **(A)**. Detected in 4.3 percent (206/4,740) of the surveyed atlas blocks, breeding was confirmed in 26 blocks (0.5 percent) and deemed probable in 88 (1.9 percent). Observed in 45 of Minnesota's 87 counties, breeding was confirmed in 14 counties, including four that were new to the 1998 list published by Hertzel and Janssen (Cottonwood, Stearns, Swift, and Yellow Medicine). The birds were most abundant in the west-central region of the state, in part due to intensive search efforts, but important breeding habitat was also provided by the Agassiz NWR and the Thief Lake WMA in Marshall County. Although the core of the species' breeding distribution remains in western Minnesota, Janssen (2019) recently reported an expansion farther east with

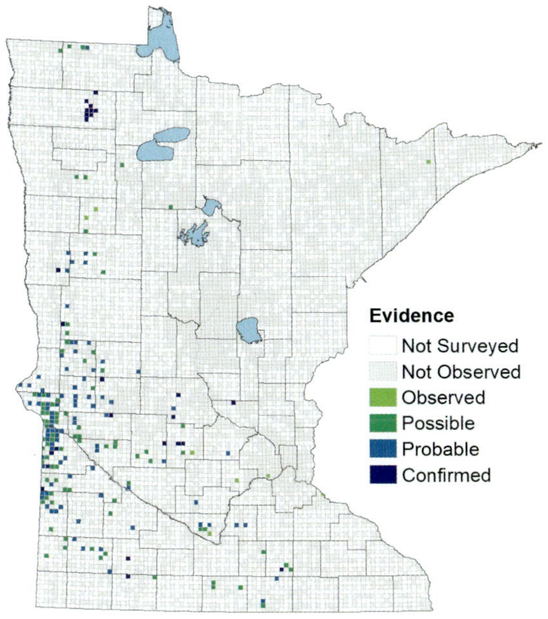

A. Breeding distribution of the Redhead in Minnesota based on the Breeding Bird Atlas (2009–13).

single nesting records each in Isanti, Itasca, and Koochiching Counties.

Elsewhere within its breeding range, the Redhead expanded its range into Alaska in the mid-1900s and established small, localized populations farther east along the Great Lakes and in the northeastern United States and southeastern Canada. Some of these eastern populations may have resulted from the release of captive-reared birds, although some natural pioneering eastward also occurred (Woodin and Michot 2020).

BREEDING HABITAT

Considered a habitat generalist, the Redhead uses a variety of seasonal and semipermanent wetlands with dense stands of emergent vegetation **(B)**. The most suitable wetlands either have a mix of open water and vegetation or are completely covered with emergents. Basins are often larger than 4 ha but may be much smaller in the Prairie Potholes Region (less than 1 ha) when wetland conditions are favorable (Woodin and Michot 2020).

B. The Redhead may select small, temporary wetland basins or larger, semipermanent basins with an abundance of emergent vegetation, such as this site in western Minnesota. Photo by Lee A. Pfannmuller.

POPULATION ABUNDANCE

The 2019 North American Waterfowl Breeding Population and Habitat Survey reported a population estimate of 732,200 birds across portions of Alaska, Canada, and the north-central United States (USFWS 2019), just above the population goal of 701,000 established by the North American Waterfowl Management Plan Committee (2014). Although it is a significant drop from the previous ten-year average of nearly 1.2 million birds from 2009 to 2018, many waterfowl species in the Prairie Potholes Region experienced declines in 2019 because of poor wetland conditions in Canada **(C)**.

In Minnesota, the spring Redhead count from 1987 to 1999 averaged 6,375 birds per year; the average from 2000 to 2019 was only 3,263 birds. The birds comprised 6 percent of the surveyed ducks reported in the late 1950s and early 1960s (Lee et al. 1964); since 2000 they comprised an average of only 1.4 percent of the surveyed ducks (Cordts 2019). Although not statistically robust at the state level, BBS data (1966–2019) reflect similar trends: stable to recently increasing numbers surveywide and decreasing populations in Minnesota (Sauer et al. 2020).

Factors that impact Redhead population numbers are speculative. In Minnesota, their declining population is similar to many members of the state's waterfowl community that have been impacted by the loss and degradation of wetlands. Farther west, the species is benefiting from higher precipitation in recent years and aggressive conservation efforts. Sensitive to wetland conditions, when conditions are favorable, Redheads maximize reproductive success by parasitizing neighboring nests and raising their own clutch. When conditions are poor, they reduce their energy expenditure by only parasitizing other nests or totally abandoning any reproductive effort. The result can be widely fluctuating numbers from year to year (Baldassarre 2014).

Conditions on the wintering grounds also influence population size. The Redhead's significant reliance on shoalgrass (*Halodule wrightii*) rhizomes during the winter months increases its vulnerability. In the Texas portion of the Laguna Madre, the Redhead's principal wintering ground, heavy grazing of these beds one year reduces the biomass available

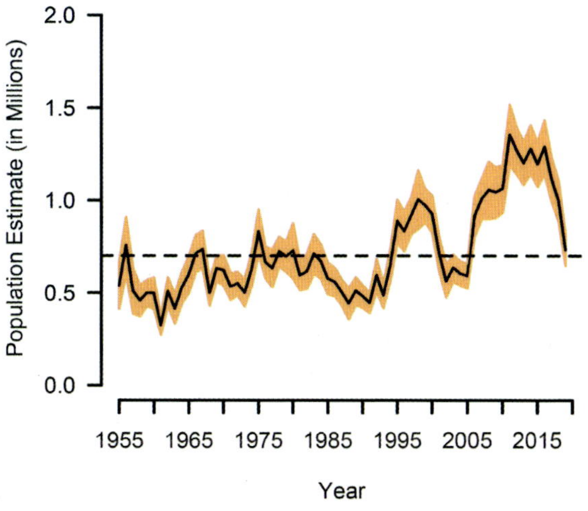

C. Redhead breeding population estimates, 90 percent confidence intervals (orange band), and North American Waterfowl Management Plan population goals (dashed line; North American Waterfowl Management Plan Committee 2014) in the traditional survey area (strata 1–18, 20–50, 75–77). Reproduced from USFWS (2019, fig. 4).

the following year (C. A. Mitchell et al. 1994). This process is further complicated by a slow decline in the size of the shoalgrass beds throughout the coastal lagoons (Baldassarre 2014).

CONSERVATION

The Redhead has been slow to recover from the slaughter in the late 1800s; regulations that tightly manage the harvest of Redheads have been an important factor in protecting the species. During the fall of 2019, Redheads comprised only 1.9 percent (188,793 birds) of the total duck harvest in the United States. Approximately 6 percent of the harvest (11,016 birds) occurred in Minnesota (Raftovich et al. 2020).

On the breeding grounds, the most important conservation measures are the continued emphasis on habitat protection and restoration. These measures are even more critical given the potential loss of wetlands in the Prairie Potholes Region as temperatures rise (NABCI 2010). Fortunately, under a 2°C warming scenario, NAS (2019) predicted a low vulnerability for the Redhead, with a loss of 18 percent of its current breeding range offset by a 36 percent gain. In Minnesota, losses would be more than compensated by predicted gains. A rise in sea levels, however, could severely impact a high percentage of the continental population that winters along the Gulf coast of southern Texas and northeastern Mexico (NAS 2019). Clearly, the Redhead warrants close monitoring in the years ahead, despite its increasing population and low conservation ranking (PIF 2021).

RING-NECKED DUCK

Aythya collaris

Minnesota Seasonal Status: Regular breeding resident and migrant; occasionally observed during the winter in southern Minnesota. The Ring-necked Duck was uncommon during the MNBBA.
Migration: Short- to medium-distance migrant that winters in the southern and western United States, the Caribbean, and Mexico.
Food: Diving duck that feeds on aquatic plants and aquatic invertebrates.
Nest: Shallow bowl of aquatic vegetation placed on a wetland hummock or mat of floating vegetation.
Conservation Concern: A game species; assigned a low PIF Continental Concern Score of 8/20.

Photo by Mike Lentz.

Minnesota's Ring-necked Duck is a bit of an anomaly. Even though its common name and Latin name, *Aythya collaris,* suggest that a prominent ring or collar is visible around the bird's neck, such a feature is "rarely seen in life" (Roberts 1932). When held, however, a well-hidden ring of chestnut feathers can be found at the base of the male's neck. The species' discovery is equally curious. Endemic to North America, the duck was first described by an English ornithologist who found a male specimen in a London market in 1801 (E. Donovan 1820).

The male is an elegant bird on the water. His diagnostic features include a tuft of crown feathers that give his head an angular appearance and a white crescent that separates the black breast from the gray flanks. A blue-gray bill with bold bands of white and black at the tip and a ring of white at the base is also unique. Many have even suggested that "Ring-billed Duck" would be a better descriptive name for the species. In the sunlight, a purple sheen is evident on his black head and neck. The brownish-gray female has a dark head, gray cheeks, and white eye-ring. The markings on her bill are less distinct, while the crown of her head is slightly raised. In flight, both sexes display a gray speculum in their wings.

The Ring-necked Duck's breeding distribution is primarily restricted to Canada but dips south into the northern Rockies, eastern North Dakota, the Great Lakes states, and northern New England. Breeding densities are highest in the taiga plains of north-central Canada and in the boreal forests of southeastern Canada.

MINNESOTA BREEDING DISTRIBUTION

In the early 1900s, unregulated hunting had taken its toll on Minnesota's Ring-necked Duck population, and it had largely disappeared as a breeding species. Reminiscing about earlier times, Roberts (1932) wrote that it was once so abundant in the fall that "it fairly swarmed at times about favored lakes." At the Long Meadow Gun Club on the Minnesota River, large numbers were harvested even by novice hunters, who "could not fail to get a good bag, shooting into big, compact, steady-flying flocks."

Prior to its near demise, Roberts described the bird as a common, statewide summer resident. Confirmed nesting records were few and limited to counties in east-central Minnesota: Carver (1898), Hennepin (1876), Isanti (1914), and Sherburne (1917). Records were scarce in the 1920s, but Roberts made special note of summer observations in the following counties as an indication that "it still breeds sparingly in the state": Cook (1922), Hennepin (1927), Kittson (1928, 1929), McLeod (1929), Otter Tail (1928), and Pennington (1928).

Once hunting regulations were enacted following passage of the 1918 Migratory Bird Treaty Act, the Ring-necked Duck made a steady comeback. By 1964, it was the third most common nesting duck in the state, trailing the Mallard and Blue-winged Teal (Lee et al. 1964). Green and Janssen (1975) described its distribution as spanning Minnesota's forested and forest–prairie transition regions, stretching south to Douglas County in the west and Washington County in the east. Occasionally it nested farther south, especially along the lower Minnesota River valley. Janssen (1987) expanded its breeding range to include all counties north and immediately south of the Minnesota River. He identified 27 counties where nesting had been confirmed since 1970; only one (Cottonwood County) was located south of the Minnesota River. Hertzel and Janssen (1998) later reported confirmed nesting in 31 counties since 1970, all within the breeding range identified by Janssen (1987).

By 2014, the MBS reported 290 Ring-necked Duck breeding-season locations. Absent from east-central and southeastern Minnesota, most records were in the northern forest, stretching from Cook County west to Becker, Beltrami, and Clearwater Counties. Scattered records were found farther south, including several in southwestern and south-central Minnesota.

The MNBBA's 820 Ring-necked Duck records documented a distribution similar to that delineated by the MBS but found the species was far more abundant in west-central Minnesota and in the western half of the Laurentian Mixed Forest Province, south of the peatlands **(A)**. Detected in 12.5 percent (601/4,797) of the surveyed atlas blocks, breeding was confirmed in 126 blocks (2.6 percent). The birds were observed in 56 of Minnesota's 87 counties and were confirmed breeding in 29 counties. Although the MNBBA confirmed breeding in only one county south of the Minnesota River (Lac qui Parle), Janssen's (2019) recent account noted a record in Redwood County as well. Regardless, the MNBBA landcover suitability map emphasizes the species' strong affiliation with the Laurentian Mixed Forest Province **(B)**. Large areas of highly suitable habitat are also found farther west in the Hardwood Hills Subsection and in the Tallgrass Aspen Parklands Province.

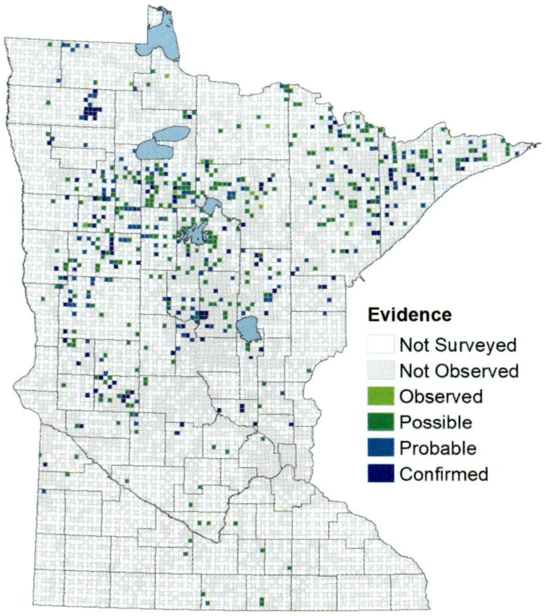

A. Breeding distribution of the Ring-necked Duck in Minnesota based on the Breeding Bird Atlas (2009–13).

The Ring-necked Duck made an impressive comeback during the past century. It was the fifth most common duck reported during the atlas, preceded by the Mallard, Wood Duck, Blue-winged Teal, and Hooded Merganser. As it was recovering in Minnesota, the species also expanded east across the Great Lakes and, more recently, across portions of the western United States and Canada (Roy et al. 2020). Factors attributed to its expansion across southern Ontario were likely at play elsewhere in eastern Canada and included natural pioneering eastward of western populations and an increase in beaver populations, which created suitable wetlands for the species in extensively forested landscapes (Cadman et al. 2007).

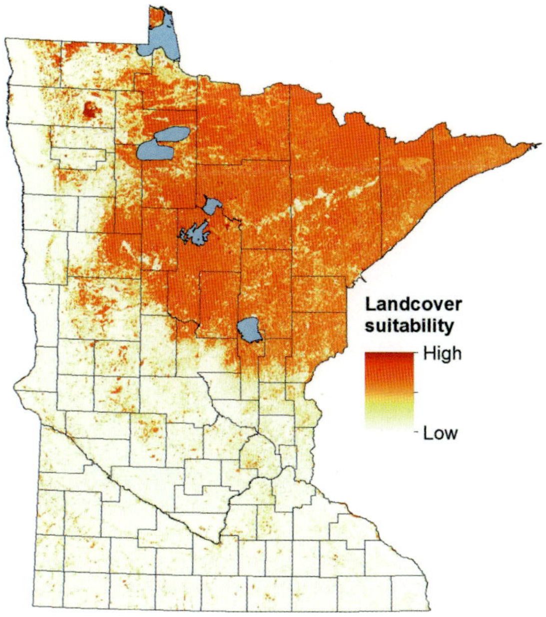

B. Landcover suitability map for the Ring-necked Duck in Minnesota based on habitat, landscape context, and climate data gathered during the Minnesota Breeding Bird Atlas (2009–13) using the MaxEnt modeling approach.

BREEDING HABITAT

Largely dependent on permanent wetlands in forested habitats, the Ring-necked Duck favors wetlands that are less than 1.5 m in depth, maintain stable water levels during the breeding season, and support abundant stands of emergent, submerged, and floating vegetation. Such sites include beaver ponds, bogs, and sedge fens, as well as prairie wetlands and impoundments (Goodwin 1957; Roy et al. 2020) **(C)**. The wetlands selected are generally fishless and are more acidic than most wetland basins (Roy et al. 2020). Baldassarre (2014) believes this selectivity enabled the species to expand its range because such sites are seldom used by other waterfowl species.

POPULATION ABUNDANCE

Ring-necked Ducks have increased significantly in the North American Waterfowl Breeding Population and Habitat Survey's Prairie Pothole Region, increasing from approximately 463,000 birds per year in the 1960s and 1970s to approximately 1.1 million birds per year between 1997 and 2010 (Baldassarre 2014). In 2015, numbers continued to increase to approximately 1.5 million (USFWS 2016b). Populations in the survey's eastern region also grew, but estimates are not strictly comparable between years owing to different sampling strategies. Nevertheless, by 2015, the eastern population numbered nearly 505,000 birds (USFWS 2016c), for a continental population estimate of 2.0 million. Although the trend line lacks some reliability, BBS data also demonstrated a significant population increase across the survey region from 1966 to 2019 (Sauer et al. 2020).

In Minnesota, a targeted population survey of the species was conducted from 2004 to 2013 (Zicus et al. 2008; J. S. Lawrence and Giudice 2013). The results documented 25,000 to 30,000 Ring-necked Ducks in key breeding areas in northwest, north-central, and northeastern Minnesota, with the highest densities in the north-central region. The state's annual Waterfowl Breeding Population Survey, however, is restricted to western and central Minnesota, where numbers have fluctuated widely since 1987 but do demonstrate a slight overall increase (Cordts 2019). Minnesota BBS data also corroborate an upward trend (Sauer et al. 2020).

C. The Ring-necked Duck prefers shallow wetlands surrounded by forest habitats that provide an abundance of emergent sedges along the shoreline and submerged and floating aquatic vegetation for foraging.
Photo by Debbie Petersen.

CONSERVATION

The species was designated a low conservation priority by PIF (2021); Baldassarre (2014) highlighted its healthy status, noting that it "has prospered more than any other species of North American diving duck." A popular species among many waterfowl hunters, the Ring-necked Duck comprised 3.8 percent (374,088 birds) of the 2019 United States duck harvest and 15.0 percent (66,945 birds) of the Minnesota harvest (Raftovich et al. 2020).

Although there are no immediate threats to the species, the ingestion of lead shot remains a concern. The use of lead shot was banned for waterfowl hunting in the United States in 1991 and in Canada in 1997, but this shallow diver is particularly prone to consuming spent shot that remains in the substrates of many wetlands (Roy et al. 2020). Because previous studies have demonstrated especially high levels of lead in both adults and young, continued monitoring is warranted (Baldassarre 2014).

The future of the Ring-necked Duck, however, is invariably tied to the future of North America's boreal forest, where nearly 70 percent of the population resides. Among the many concerns in the region are the expansion of natural resource extraction industries, such as oil and gas development, and the impacts of climate change. Boreal forests and their inhabitants are especially vulnerable to rising temperatures. Although Minnesota's population would remain relatively secure, under a 2°C warming scenario NAS (2019) projected a 26 percent loss of the species' current breeding range by 2050 and a 21 percent gain farther north. Classified as moderately vulnerable to warming temperatures, this northern forest inhabitant may eventually require focused conservation efforts.

LESSER SCAUP

Aythya affinis

Minnesota Seasonal Status: Regular breeding resident and migrant; occasionally observed during the winter. The Lesser Scaup was rare during the MNBBA.
Migration: Short- to medium-distance migrant that winters along the United States coastline and in the southern states, Mexico, the Caribbean, and Central America.
Food: Omnivorous diver feeding on aquatic invertebrates and aquatic plants.
Nest: Shallow bowl on the ground or in a mound of aquatic vegetation over water.
Conservation Concern: A game species; assigned a moderate PIF Continental Concern Score of 11/20.

Photo by Travis Bonovsky.

One of the most abundant waterfowl species in North America, the Lesser Scaup is one of two scaup species that occur in Minnesota. The other, not surprisingly, is the Greater Scaup. Indeed, the scientific name, *Aythya affinis,* means a diver related to the Greater Scaup. Both species are known as "blue-bills" for their pale blue-gray bills. Both species also migrate through Minnesota in the spring and fall. The tundra-loving Greater Scaup, however, breeds across the far northern regions of North America, Siberia, and northern Europe. The Lesser Scaup is restricted to Canada, dipping south into the northwestern and north-central United States, including Minnesota. The core of its breeding population occurs in a band stretching from the Northwest Territories south to the northern Great Plains.

The male's distinctive features include an iridescent black head, a black breast and rump, white flanks, and a variegated black-and-white back. The brownish-gray female has a dark brown head with a patch of white at the base of the bill. Unfortunately, this description is equally apt for the Greater Scaup.

Distinguishing the two species can challenge the most experienced field biologists. Even prominent ornithologists John James Audubon and Alexander Wilson were unaware that there were two species. Familiar with the Lesser Scaup, Audubon criticized Wilson's account of the species, who most likely was writing about the Greater Scaup (Bent 1925). In general, the Greater Scaup has a more rounded contour to the back of the head (compared to a small peak on the crown of the Lesser) and has more white in the trailing edge of the wing. Rather than agonize over these subtle differences, many observers simply refer to all blue-bills as "scaup."

MINNESOTA BREEDING DISTRIBUTION

Roberts's (1932) account of the Lesser Scaup emphasized its status as a common migrant, particularly in western Minnesota. His comments regarding its breeding status were limited. "It seems improbable," he wrote, "that the Scaup ever nested in considerable numbers anywhere in Minnesota, except, perhaps, in the northwest corner." Confirmed nesting records were limited to Marshall, Meeker, and Otter Tail Counties, and Itasca State Park. A June 1901 report from Mud Lake in Marshall County noted "many" Lesser Scaup were present and "evidently breeding," although no young or nests were seen. When Breckenridge surveyed western Minnesota in 1929, however, the largest concentration of birds he encountered was just three to four pairs at Twin Lakes in Kittson County (Roberts 1932).

Green and Janssen (1975) described the Lesser Scaup as Minnesota's rarest breeding duck. It was a summer resident in the northwest; one new nesting report was identified in central Minnesota (Kandiyohi County, 1942) while nonbreeding birds were regularly observed "in the northern regions and also as far south as the Iowa border." Janssen (1987) identified only one county where nesting had been confirmed since 1970: Marshall. No additional counties were added by Hertzel and Janssen (1998). The species' status changed little in the following years. By 2014, the MBS reported just 13 breeding-season locations in six counties: Clearwater, Kittson, Marshall, Pennington, Pope, and St. Louis.

The species' scattered summer distribution was evident when MNBBA participants reported only 49 records from 30 widely dispersed counties **(A)**. The majority of records were from western Minnesota, but records were also located in the north-central, northeastern, and south-central regions. Detected in 1.0 percent (46/4,734) of the surveyed atlas blocks, no evidence of confirmed breeding was found. Although two confirmed nesting reports were submitted by observers from Lac qui Parle and Anoka Counties, both were invalidated because they lacked documentation, a requirement for such a rare species. Probable breeding, however, was reported

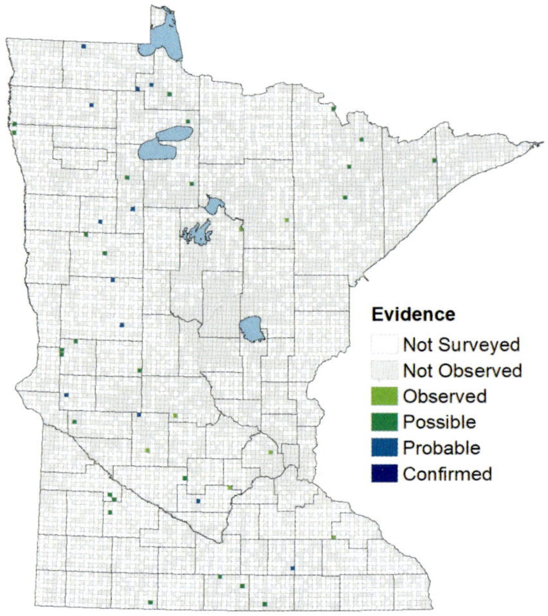

A. Breeding distribution of the Lesser Scaup in Minnesota based on the Breeding Bird Atlas (2009–13).

in 12 blocks ranging from Steele County in south-central Minnesota north to Roseau and Lake of the Woods Counties. Many of the observed and possible records may represent young, nonbreeding individuals. Because the population has a skewed sex ratio in favor of males, not all males can find mates. Indeed, males may comprise 55–87 percent of the population. In addition, females may delay breeding until their second or third year, especially if wetland conditions are poor (Baldassarre 2014).

The large number of counties where Lesser Scaup were reported during the MNBBA appears to continue a long-term trend. Since the 1960s, there has been a steady increase in the number of counties where the birds are reported each summer (MOU 2021b). This may reflect an influx of birds from the Dakotas, especially North Dakota, where breeding populations are gradually expanding (Sauer et al. 2020), or an increase in observers.

Overall, however, the species status in Minnesota has changed very little in the past century. Likewise, Anteau and his colleagues (2020) commented that "there is no compelling evidence of historical changes in the range of Lesser Scaup. However, there is new evidence that the breeding range is larger than was previously thought (Schummer et al. 2013)." Areas along the southern periphery of its breeding range, where the species is thought to have decreased in abundance, are now being reoccupied, including in North and South Dakota (Anteau et al. 2020).

BREEDING HABITAT

Typical breeding habitat consists of seasonal, semipermanent, and permanent wetlands with stands of emergent vegetation **(B)**. Such wetlands are usually embedded in a landscape surrounded by native prairie, restored grasslands, and hayfields. In North Dakota, Lesser Scaup used seasonal wetlands to a greater extent than any other diving duck (Kantrud and Stewart 1977). In Manitoba, the birds selected small wetlands, averaging less than 0.8 ha in size (Hammell 1973). As the young broods age, the female moves them to larger, more permanent water bodies (Sugden 1973).

POPULATION ABUNDANCE

Given the difficulty of distinguishing Greater Scaup and Lesser Scaup from the air, the North American Waterfowl Breeding Population and Habitat Survey combines data for the two species. In 2019, the combined estimate was 3.6 million (USFWS 2019), below the population objective established by the North American Waterfowl Management Plan Committee (2014) of 5.03 million and well below the original goal of 6.3 million established by the committee in 2004. Population numbers have gradually declined since reaching a peak in the 1970s and 1980s **(C)**.

B. Large, shallow wetland complexes in northwestern Minnesota are the mainstay of Minnesota's small breeding population of the Lesser Scaup. Nests are often placed in uplands, but, in large complexes like this, they may be constructed over water or in emergent vegetation.
Photo by Lee A. Pfannmuller.

Because Lesser Scaup comprise nearly 90 percent of the continental scaup population monitored by the survey, the downward trend largely reflects changes in its breeding population (Bellrose 1980). The decline was also most prevalent in Alaska, the Yukon, and the Northwest Territories, where Lesser Scaup are more abundant than Greater Scaup. Although populations in the northern Great Plains have increased, they contribute a small percentage to the continental population (Anteau et al. 2020). BBS data for the northern Great Plains (i.e., Prairie Potholes Region) also indicate a stable to slightly increasing population, while most data outside the region lack statistical reliability (Sauer et al. 2020).

The Lesser Scaup's decline represents "a biological conundrum" to waterfowl biologists (Baldassarre 2014). Although it has received little research attention in the past, studies are now examining a multitude of possible factors, including habitat loss, climate change, contaminants, declining recruitment, and diminished food resources at important migratory stopovers (Baldassarre 2014). A Minnesota study, for example, found 94 percent fewer amphipods, an important food item for Lesser Scaup, in wetlands that were sampled in 2000–2001 than in the same wetlands almost twenty years earlier (Anteau and Afton 2006). Minnesota waterfowl biologists are also investigating the death of thousands of migrant scaup in 2007 and 2008 caused by their ingestion of nonnative faucet snails, a host for several intestinal trematodes (Roy 2013).

CONSERVATION

Once a popular game bird, Lesser Scaup were the third most commonly harvested duck in Minnesota in the 1950s (Green and Janssen 1975) but ranked only 11th in 2019, comprising 1.4 percent (6,356 birds) of the state harvest. Nationally, they comprised less than 2 percent (180,474 birds) of the 2019 duck harvest (Raftovich et al. 2020). Declining population numbers have contributed to the decrease, as have special Scaup harvest regulations imposed from 1999 to 2011. The regulations not only reduced the overall take but may also have triggered a drop in hunter interest (Anteau et al. 2020).

As biologists unravel the mystery of the Lesser Scaup's decline, wetland protection and restoration efforts in Minnesota should ensure that suitable breeding habitat is available for birds dispersing from the neighboring Dakotas. But the species' long-term survival is tied to the protection of critical breeding habitat in Canada, where more than 65 percent of the population resides. Development in the region is a concern as habitat is lost to logging, agriculture, and a rapid expansion in the oil and gas industry. Warming temperatures may exacerbate these challenges. NAS (2019) classified the species as moderately vulnerable under a 2°C warming scenario that projected the loss of 45 percent of the Scaup's current breeding range, including its virtual disappearance from Minnesota, compensated by only a 10 percent gain to the far north. These concerns, coupled with the Lesser Scaup's population status, prompted its designation as a moderate conservation priority (PIF 2021).

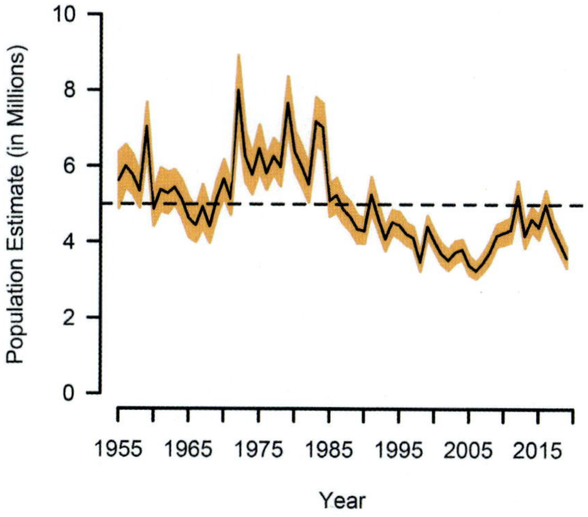

C. Scaup (Greater and Lesser) breeding population estimates, 90 percent confidence intervals (orange band), and North American Waterfowl Management Plan population goals (dashed line; North American Waterfowl Management Plan Committee 2014) in the traditional survey area (strata 1–18, 20–50, 75–77). Reproduced from USFWS (2019, fig. 4).

BUFFLEHEAD

Bucephala albeola

Minnesota Seasonal Status: Regular breeding species and migrant; occasionally observed during the winter, especially on Lake Superior. The Bufflehead was rare during the MNBBA.
Migration: Medium-distance migrant that winters along the coastlines of Canada and the United States and in the Great Lakes, the southern states, and northern Mexico.
Food: Diver that feeds primarily on aquatic insects in freshwater habitats and on mollusks and crustaceans in saltwater habitats.
Nest: Secondary cavity nester that occasionally uses nest boxes.
Conservation Concern: A game species; assigned a low PIF Continental Concern Score of 8/20.

Photo by Mike Lentz.

Stocky and compact, this handsome little diver is Minnesota's smallest duck. The name Bufflehead originates from the "odd, puffy" shape of the bird's head that has been likened to that of a buffalo (Kaufman 1996). Early naturalists often referred to it as the "Butterball," "Marionette," or "Dipper." Bent (1923) argued that "Spirit Duck" may be the most appropriate nickname. "The propriety of applying the name 'spirit duck' to this sprightly little duck will be appreciated by anyone who has watched it in its natural surroundings . . . as it plunges, quickly beneath the surface and bursts out again in full flight, disappearing in the distance with a blur of whirring wings."

A broad white patch that extends across the back of its iridescent black head distinguishes the male Bufflehead from all other male diving ducks with black-and-white plumages. Indeed, *Bucephala albeola* means "bullhead with a little white patch." The female, by contrast, has a brownish-gray body with a patch of white behind each eye. Often seen flying low over the water, the birds rarely travel in large flocks but prefer instead to move as pairs or in small groups. When diving, they first compress their plumage to reduce the volume of air and then propel themselves with a little leap out of the water before plunging downward (Gauthier 2020). Like the Common Goldeneye, they return to the surface some distance away, bobbing lightly upon the water.

The Bufflehead's boreal breeding distribution stretches from Alaska east to western Quebec, dipping south into the northwest and north-central states, including Minnesota. The highest breeding densities are restricted to regions of Alaska and western Canada.

MINNESOTA BREEDING DISTRIBUTION

For many years, Minnesota was considered just a migratory stopover for this unique little duck as it traveled to and from its breeding grounds in the Canadian boreal forest. In the early 1900s, Roberts (1932) described it as a common spring and fall migrant that also "occurs rarely in winter." He remarked that it was formerly much more abundant, "passing through the state in flocks of considerable size, while now it is usually seen in pairs or little parties of a half-dozen or so." Although there were no confirmed breeding reports, Roberts did speculate that the species might occasionally nest in the Northwest Angle. Based on a single bird observed in July 1928 in Cook County, he thought that it might occasionally nest in that region of the state as well.

Forty years later, the Bufflehead still had not been confirmed nesting in Minnesota. Green and Janssen (1975) classified it as a "casual visitant in the northern half of the state." Single birds had been reported during the summer months in Cass, Clearwater, Cook, Douglas, Kittson, Mahnomen, Marshall, and St. Louis Counties. There was also a report of three birds in August 1963 in Chisholm, St. Louis County.

Nesting was finally confirmed in 1978 when MNDNR wildlife manager George Davis observed a brood on the East Park WMA in Marshall County (G. Davis 1978). Davis's report also included a note that a colleague had observed Bufflehead broods on the Roseau River WMA in Roseau County. Seven years later, nesting was documented a little farther south, again in Marshall County, on the Agassiz NWR (Bell 1985).

Since then, an increasing number of summer observations have been reported from northern Minnesota and from several counties in central and south-central Minnesota (MOU 2021b), including two breeding-season locations identified by the MBS: one each in Clearwater County and Marshall County. Confirmed nesting reports, however, continued to be primarily limited to Marshall County. The one notable exception was a 1983 report of a brood observed on the western outskirts of the Twin Cities in Carver County (Heidel 1983). Then, in 1998, Beltrami was added to the short list of counties where nesting has been confirmed (Hertzel and Janssen 1998).

The species' dependence on wetlands in eastern Marshall County was again confirmed during the MNBBA. Participants reported just 27 Bufflehead records from 23 of the surveyed atlas blocks; breeding was confirmed in 13 blocks **(A)**. Fourteen of the 23 atlas blocks, including 11 blocks with confirmed nesting records, were from eastern Marshall County, in or adjacent to the Agassiz NWR and the Thief Lake WMA. The two other nesting records were north in Roseau County and, most surprisingly, more than 560 km to the south, in Cottonwood County! The

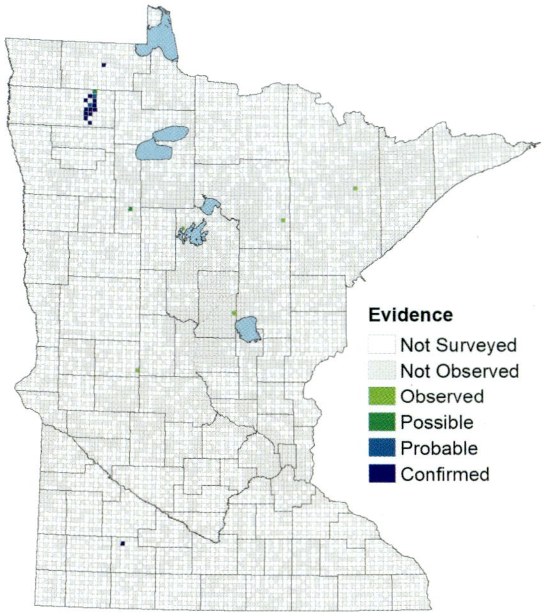

A. Breeding distribution of the Bufflehead in Minnesota based on the Breeding Bird Atlas (2009–13).

latter record was of a hen and brood on the Hurricane WMA and was confirmed with a photo. All remaining records were in the northern half of the state. Indeed, since the atlas was completed, nesting has also been confirmed in Kittson County (2014, 2020) and St. Louis County (2015, 2016; MOU 2021a).

The MNBBA's Cottonwood County breeding record and the 1983 Carver County record may not have been such a rarity in the 1800s. Gauthier (2020) references old breeding records from New Brunswick, Maine, and the northern Great Plains that suggest the species may have ranged farther east and south in the late 1800s and early 1900s. On the other hand, the species' presence in southern Minnesota would surely have been noticed even in Roberts's time if it were a regular breeding resident.

Located on the southern periphery of its North American breeding range, Minnesota's Bufflehead population in Marshall County is somewhat distant from other breeding localities. The nearest significant population occurs in southwestern Manitoba, more than 300 km to the northwest (Taylor 2018). Elsewhere, they are accidental to sparse breeders in southwestern Ontario (Cadman et al. 2007), eastern South Dakota (Drilling, Sparks, et al. 2016), and north-central North Dakota (Gauthier 2020).

BREEDING HABITAT

A species of the boreal forest and aspen parklands, Buffleheads are found in mixed coniferous and deciduous forests dominated by small lakes and wetlands **(B)**. Waters with extensive beds of emergent and submerged vegetation are usually avoided (Gauthier 2020). Nevertheless, like many boreal nesting species, detailed studies describing their habitat requirements are scarce.

Buffleheads frequently select nesting cavities that were excavated by Northern Flickers and are located close to water (Gauthier and Smith 1987). The openings are small enough to prevent competition with larger cavity-nesting waterfowl species. Yet, in Ontario the probability of observing Buffleheads "increased several-fold" between the first (1981–85) and second (2001–5) atlases, despite a documented decline in Northern Flickers (Cadman et al. 2007). The authors noted, however, that the Pileated Woodpecker population had increased.

POPULATION ABUNDANCE

Approximately 90 percent of the Bufflehead's population occurs from Manitoba west through Canada and Alaska, a region that falls within the traditional survey area of the North American Waterfowl Breeding Population and Habitat Survey (Baldassarre 2014). Bufflehead numbers within the traditional survey area increased from an average of approximately 471,000 in the 1960s to an average of 1.44 million from 2011 to 2015 **(C)**. In 2012, the North American Waterfowl Management Plan Committee estimated the continental population across its entire breeding range at 1.67 million birds. Few other waterfowl species have demonstrated such a significant, long-term population increase since the survey began. Reduced hunting pressure, particularly in Canada, may be partly responsible for the increase (Gauthier 2020).

B. The mainstay for the Bufflehead breeding population in Minnesota is the Agassiz NWR. The refuge's extensive array of wetlands, surrounded by scattered trees that provide nesting cavities, provides ideal nesting habitat. Photo by Gary T. Seim.

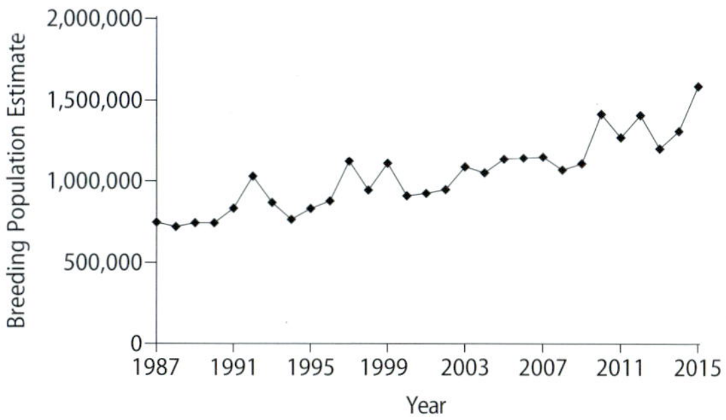

C. Bufflehead breeding population estimates for the North American Waterfowl Breeding Population and Habitat Survey's traditional survey area from 1987 to 2015 (USFWS 2016b).

Buffleheads are included in Minnesota's annual Waterfowl Breeding Population Survey. However, the late spring survey is more accurately an index of the migratory population traveling north to Canadian breeding grounds than an index of the state's small breeding population.

CONSERVATION

Harvest numbers for this diving duck are relatively small. During the 2019 hunting season, a total of 210,846 birds were harvested in the United States and 16,313 birds in Minnesota (Raftovich et al. 2020). Although hunting pressure appears to have declined, Gauthier (2020) expressed concern regarding the Bufflehead's vulnerability to overharvesting. Because it uses easily accessible, near-shore habitats during the fall and winter, high hunting mortality is possible. The bird's strong fidelity to these sites means that "recolonization of areas where local populations are overharvested is slow." Their delayed maturity, postponing breeding until their second full summer, also "hinders" recovery.

A recent analysis by NAS (2019) demonstrated the species' vulnerability to warming temperatures as well. Classified as highly vulnerable under a 2°C warming scenario that would nearly eliminate the species from Minnesota by 2050, as well as from 53 percent of its current breeding range, only a 20 percent gain was projected in the arctic north. The decrease would be most significant in western Canada, the core of the species' breeding population.

Habitat loss, however, remains the principal threat to the Bufflehead, especially in Canada's Prairie Parklands. In Minnesota, the aspen parklands are central to the species' continued presence, especially the region's public wetlands. Targeted wetland protection and restoration efforts in the area are critical to many wetland-dependent species, including the handsome little "spirit duck."

COMMON GOLDENEYE

Bucephala clangula

Minnesota Seasonal Status: Regular breeding resident and migrant; regularly observed in winter, especially on Lake Superior and the Mississippi River. The Common Goldeneye was uncommon during the MNBBA.
Migration: Short- to medium-distance migrant that winters along the coastlines of Canada and the United States and on large lakes and rivers from southern Canada to northern Mexico.
Food: Diving duck that feeds primarily on aquatic invertebrates; small fish are occasionally consumed.
Nest: Secondary cavity nester; nest boxes are also used.
Conservation Concern: A game species; assigned a moderate PIF Continental Concern Score of 9/20.

Photo by Mike Lentz.

The Common Goldeneye is a boreal species with a global distribution. In North America, its breeding range extends as far south as the northern Rockies, northern New England, and the northern Great Lakes. In Minnesota, it is one of the few ducks that is regularly encountered on a northern lake in June and on a Christmas Bird Count in December. As long as open water is available, the Common Goldeneye tolerates the most inhospitable conditions Minnesota can muster in the depth of winter. Active birds on the water and swift, rapid fliers in the air, Goldeneyes are entertaining to watch. When they are feeding, they plunge below the water's surface and reemerge some distance away, bouncing "up to the surface like a cork" (Millais 1913).

A handsome bird, the male is easily identified by its dark, iridescent green head, its white chest and belly, and its black back. An oval spot of white, just below the eye, is diagnostic. Its common name is a reference to its bright yellow eye, a trait shared with many other waterfowl species. The female has a rich chocolate brown head and a largely gray body; the male's white face patch is missing. Unfortunately, the species can be easily confused with the Barrow's Goldeneye, a rare Minnesota visitor during the nonbreeding season. Distinctions include a crescent-shaped, white patch on the face of the male Barrow's Goldeneye, more black feathers on the wings, and a short black spur on the sides of the breast.

Known as the "Pied Duck" among some Native American tribes (Bent 1925), the Goldeneye's most common colloquial name is "Whistler," a name earned from the whistling sound made by its rapid wingbeat. *Bucephala,* the Latin name for the genus, means "bull-headed" and refers to the relatively large size of the Goldeneye's head compared to its small body; *clangula* means "a little noise," likely a reference to the sound made by its wings.

MINNESOTA BREEDING DISTRIBUTION

Since the early 1900s, when Roberts (1932) published his account of the species, the Common Goldeneye has been considered a breeding resident across the state's northern forests. The species was especially numerous, he wrote, "in the lakes and streams of Cook and Lake Counties, north of Lake Superior." Surber, who visited the region in 1922, noted that "every lake and stretch of dead-water in the rivers [has] its quota of one to as many as half a dozen broods" (Roberts 1932). The birds were also common in northern Itasca County and in Itasca State Park. Nesting was only confirmed in Becker County (1883), but there were multiple reports of females with broods in Cook, Itasca, and Lake Counties and from Leech Lake and Itasca State Park.

Little had changed when Green and Janssen (1975) reported Goldeneyes were summer residents across north-central and northeastern Minnesota with the highest breeding densities in Beltrami and Itasca Counties. Janssen (1987) identified 14 northern counties where breeding had been confirmed since 1970. Aitkin, Cass, and Crow Wing Counties established the southern boundary of its breeding range; Becker and Roseau Counties established the western boundary. By 1998, when Hertzel and Janssen published an updated map of confirmed nesting records, nothing had changed.

The 118 breeding-season locations identified by the MBS revealed two major population centers: one stretching across the northern Arrowhead region, and one in north-central Minnesota, including Beltrami, Clearwater, Hubbard, and Itasca Counties. A few records were found farther south in Crow Wing, Morrison, and Otter Tail Counties.

The MNBBA tallied 472 Common Goldeneye records from 6.8 percent (322/4,746) of the surveyed atlas blocks and confirmed breeding in 142 blocks (3.0 percent) **(A)**. Most records were again in one of two population centers: the northern Arrowhead region, including all of Cook County and northern Lake and St. Louis Counties; and north-central Minnesota, where the majority of reports were in western Itasca, southern Beltrami, and northern Cass and Crow Wing Counties. The two centers are separated by the Iron Range and exten-

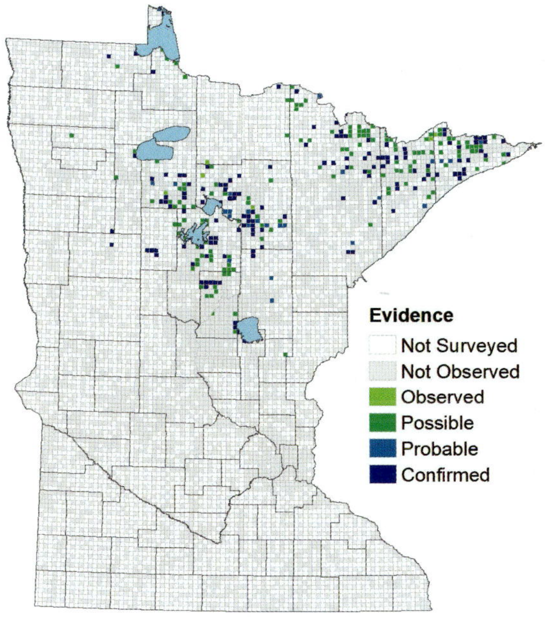

A. Breeding distribution of the Common Goldeneye in Minnesota based on the Breeding Bird Atlas (2009–13).

sive peatlands. Other records were scattered north to the Northwest Angle, west to Pennington County, and south to Kanabec County. Goldeneyes were observed in 18 of Minnesota's 87 counties and were confirmed breeding in 16 counties. Mille Lacs and Mahnomen Counties were additions to Hertzel and Janssen's (1998) map of counties where breeding had been confirmed since 1970. Since the atlas was completed, new nesting reports from Kittson County (2017, 2020) and Grant County (2018) have extended the western periphery of the Goldeneye's Minnesota breeding range (MOU 2021a).

The atlas findings are similar to those of the MBS and to the outcomes of earlier work by Zicus and Hennes (1988), who conducted an extensive survey of Minnesota's cavity-nesting waterfowl. With the exception of a few records just beyond the species' primary breeding range, little has changed in the past one hundred years. Elsewhere, no major historical changes to the species' breeding distribution have been reported (Eadie et al. 2020).

BREEDING HABITAT

Common Goldeneyes can be found in wetlands, large lakes, rivers, ponds, and streams. All are embedded in forested habitats where there are suitable nest cavities and an abundance of aquatic invertebrates. Cavities may be in live or dead trees that are located near the shoreline or up to 1.3 km away. Both coniferous and deciduous trees are used (Eadie et al. 2020). Although Goldeneyes may successfully nest on lakes with highly productive fish populations (Zicus et al. 1995), the fish compete with Goldeneyes for the same invertebrate forage base. Overall, numerous studies have shown that "both breeding pairs and broods exhibit a preference for lakes without fish" (Baldassarre 2014).

Many studies also cite the species' preference for clear, deep, oligotrophic lakes where stands of emergent and submerged vegetation are limited (Eadie et al. 2020) **(B)**. Zicus and his colleagues (1995) demonstrated use of a much wider range of habitats in Minnesota, including small ponds (16 ha) with extensive stands of bulrush and pondweeds as well as large lakes (100 to 1,250 ha), some with extensive stands of emergent vegetation and some with absolutely none.

POPULATION ABUNDANCE

Because it is difficult to distinguish Common and Barrow's Goldeneyes from the air, the North American Waterfowl Breeding Population and Habitat Survey combines data for the two species. Most birds encountered are Common Goldeneyes, but an estimate of their proportional abundance is not available. In 2015, the survey estimated 988,000 Goldeneyes across North America (USFWS 2016b), a number below the North American Waterfowl Management Plan Committee's (2004) estimate of 1.35 million birds approximately ten years earlier. However, data across the BBS survey region specifically for Common Goldeneyes suggest a relatively stable to slowly increasing continental population from 1966 to 2019 (Sauer et al. 2020).

The Minnesota Waterfowl Breeding Population Survey covers only the western portion of the Common Goldeneye's breeding range in the state. Although survey numbers have fluctuated widely, the 10-year average of 1,830 birds from 2010 to 2019 was just slightly higher than the previous 10-year average of 1,640 birds (Cordts 2019).

B. Clear, deep, oligotrophic lakes are typical of the breeding habitat of the Common Goldeneye in northeastern Minnesota. Mature trees along the shoreline provide critical nesting cavities. Photo by Debbie Petersen.

When Goldeneye populations decline, recovery can be slow. The birds usually postpone breeding until their second year and, when nesting attempts fail, they rarely renest. Overall, however, availability of nest cavities may be the most important factor regulating population numbers (Eadie et al. 2020).

CONSERVATION

Given its relatively small population and threats to its lake and riverine habitats, the Common Goldeneye is considered a moderate conservation priority (PIF 2021). Harvest pressure on the species is low. Together, Common and Barrow's Goldeneyes comprised less than one percent of the 2019 duck harvest in the United States (80,902 birds). That same year, 5,508 Common Goldeneyes were harvested in Minnesota (Raftovich et al. 2020).

Ultimately, the Goldeneye's survival is tied to the long-term health of the boreal forest. Habitat loss in the region is a major concern. Where forests remain, management practices that retain mature trees and snags, especially in riparian areas, should be a priority. Where natural cavities are limited, nest boxes are an effective conservation tool.

Climate change and industrialization are also significant threats. By 2050, under a 2°C warming scenario, NAS (2019) projected a moderate vulnerability for the Goldeneye, including its virtual disappearance as a breeding species in Minnesota, the loss of 35 percent of its continental breeding range, and a northern range gain of only 11 percent. On the other hand, not only does industrialization result in habitat loss but industrial emissions can also lead to long-term consequences, including acid rain. In the short term, acid precipitation reduces fish populations, which temporarily increases the availability of invertebrate food for Goldeneyes, but it may have long-term, negative impacts "through chronic, sublethal effects on physiology and survival" (Mallory et al. 1994). Fortunately, acid rain deposition has been gradually decreasing in many regions.

Contaminants pose another concern. Although Minnesota biologists detected DDT, PCBs, and mercury in Goldeneye eggs collected in the early 1980s, the levels were not considered harmful (Zicus et al. 1988). More recently, potentially harmful levels of mercury and selenium have been found in birds wintering in Utah (Vest et al. 2009). Although the "Whistler" population appears relatively secure in the short term, continued vigilance on many fronts is paramount.

HOODED MERGANSER

Lophodytes cucullatus

Minnesota Seasonal Status: Regular breeding resident and migrant; regularly observed during the winter. The Hooded Merganser was common during the MNBBA.
Migration: North-central populations are short-distance migrants that winter in the southern United States; farther south and in the Pacific Northwest, populations are nonmigratory.
Food: Diver that feeds primarily on aquatic insects, small fish, and crustaceans.
Nest: Secondary cavity nester; will use artificial nest boxes.
Conservation Concern: A game species; assigned a low PIF Continental Concern Score of 8/20.

Photo by Mike Lentz.

Although the Wood Duck may be the most colorful waterfowl species in Minnesota, the Hooded Merganser surely is the most elegant. Roberts (1932) used glowing praise to describe this smallest member of the merganser family: "The male Hooded Merganser is attired in but three simple colors—black, white, and pale rufous—but these are so disposed that he vies in beauty with the gay and many-hued Wood Duck drake. His great, fan-shaped crest, with jet-black border and brilliant white center, is an adornment equalled by no other member of the tribe."

The male's beautiful crest can be opened and closed quickly for an even more dramatic effect during courtship. By contrast, the female's brownish-gray plumage is quite subtle. Similar in appearance to the female Common and Red-breasted Mergansers, she is best distinguished by her smaller size, a bushier crest, and a yellow and brown bill compared to the red bill found on the larger mergansers. Like all mergansers, the "little sawbill" has a thin serrated bill for grabbing its underwater prey. The species' scientific name means a crested diver (*Lophodytes*) with a hood (*cucullatus*), the latter referring to the cowl-like black feathers of the neck and head. Generally a silent bird, the male makes a unique and very deep frog-like sound during courtship.

A forest inhabitant, the Hooded Merganser is broadly distributed across the eastern half of the United States and southeastern Canada, with a disjunct population in the Pacific Northwest. The core of the species' breeding range is in the Great Lakes states.

MINNESOTA BREEDING DISTRIBUTION

Roberts's (1932) brief history of the Hooded Merganser's status in Minnesota is nearly identical to that of every other waterfowl species in the state: "formerly an abundant migrant and common summer resident throughout the state but now much less numerous and a rather infrequent summer resident." Overhunting was common in the late 1800s and early 1900s and certainly contributed to the species' initial decline, but the wide-scale conversion of forests to agricultural lands in the south, coupled with logging of old-growth forests in the north, likely had a major impact on this forest-dwelling cavity nester as well. Roberts's account included numerous summer reports of birds that ranged from the cold, rapid streams of the northeast to the large wetlands and shallow lakes of the prairie. However, there were only two confirmed nesting records: one in Becker County (1886), and one in Kittson County (1899). An inferred breeding record was also reported in Lake County.

The species eventually rebounded from its early demise and was reported as a regular summer resident in most localities by the late 1900s. The birds were most abundant in the southeast and north-central counties, "very scarce" in the west-central region, and absent from the southwest (Green and Janssen 1975). Janssen's (1987) breeding distribution map excluded only the southwestern counties, from Lac qui Parle County southeast to central Martin County. He identified 29 counties where nesting had been confirmed since 1970. By 1998, Hertzel and Janssen confirmed nesting in 13 additional counties. Apart from a breeding record in Lac qui Parle County, the species was still not confirmed nesting in southwest or south-central Minnesota, despite summer observations in the region. By 2014, MBS field staff had tallied 180 Hooded Merganser breeding-season locations. The majority of records were from northeastern Minnesota and the north-central counties of Beltrami and Clearwater. Unlike earlier accounts, the species was absent from many southeastern counties but widely distributed across the southwest region.

During the MNBBA, Hooded Mergansers were reported in all but three of Minnesota's 87 counties (Chisago, Mower, and Wilkin) while breeding was confirmed in 73 counties. It is the most abundant of the three merganser species that occur in Minnesota; MNBBA participants reported 1,180 Hooded Merganser records in 18.8 percent (898/4,764) of the surveyed atlas blocks and confirmed breeding in 8.1 percent (386) of the blocks **(A)**. Abundant throughout the central core of the state, especially in far west-central Minnesota, the species was present in the southeast but only sparsely distributed.

The landcover suitability map for the Hooded Merganser **(B)** predicts that the only extensive region of unsuitable habitat is the peatlands of north-central Minnesota. Suitable habitat is present elsewhere throughout much of the state but is especially abundant in the outer rim of the Twin Cities and in riparian floodplains throughout western and southern Minnesota.

The Hooded Merganser was the fifth most frequently reported waterfowl species in the state during the atlas, following the Canada Goose, Mallard, Wood Duck, and Blue-winged Teal. Clearly this

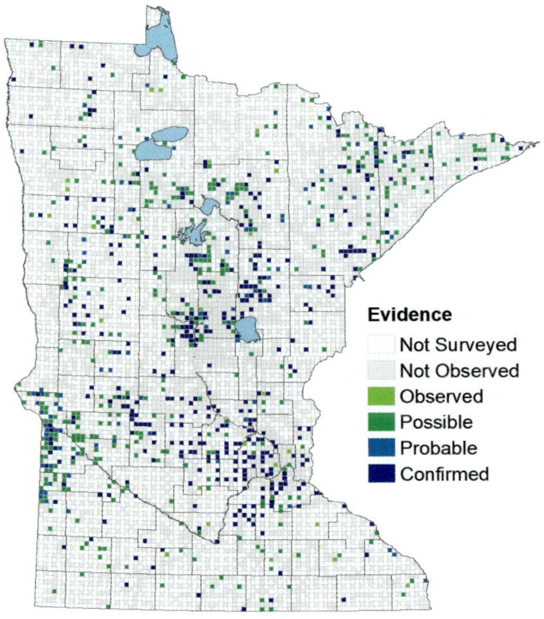

A. Breeding distribution of the Hooded Merganser in Minnesota based on the Breeding Bird Atlas (2009–13).

spectacular little duck is faring quite well and has found suitable wetlands and nesting opportunities even in small woodlots scattered throughout Minnesota's agricultural landscape. The widespread placement of Wood Duck nest boxes in this region has likely benefited the species as well.

Historically, Hooded Mergansers occurred throughout North America's eastern deciduous forest and in the mountainous regions of the western states. Small breeding populations still occur in many of these areas, but populations are now more localized in distribution and smaller in size (Dugger et al. 2020).

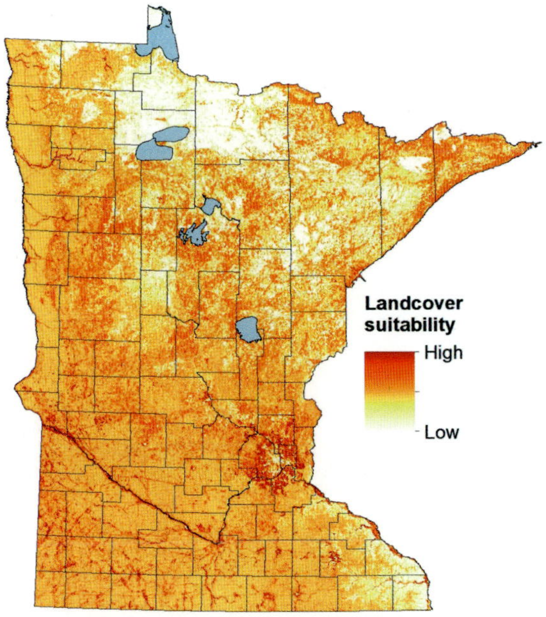

B. Landcover suitability map for the Hooded Merganser in Minnesota based on habitat, landscape context, and climate data gathered during the Minnesota Breeding Bird Atlas (2009–13) using the MaxEnt modeling approach.

BREEDING HABITAT

Hooded Mergansers use a variety of forested wetlands, including the riparian shorelines of lakes, rivers, and streams in otherwise open habitats **(C)**. The limiting factor is the availability of nesting cavities or nest boxes (Baldassarre 2014). In Minnesota's northern forests, the birds are most frequently found on relatively quiet lakes and streams (Danz et al. 2007). In the agricultural region, they use large, shallow wetlands and lakes that are surrounded by woodlots with appropriate cavities or where nest boxes have been erected. Natural cavities may be located in dead or live trees, in broken-off limbs, or atop broken-off tree trunks (Dugger et al. 2020). Old cavities excavated by Pileated Woodpeckers are commonly used (Baldassarre 2014).

POPULATION ABUNDANCE

The Hooded Merganser population appears to be stable and possibly increasing across North America (Baldassarre 2014). Unfortunately, many surveys do not distinguish among the three merganser species (Hooded, Common, and Red-breasted), reporting instead on the combined number of all mergansers. Using harvest statistics and survival rates to estimate population size, Dugger and his colleagues (2020) estimated the Hooded Merganser population to number from 329,000 to 414,000 adults from 1999 to 2006. In contrast, the North American Waterfowl Management Plan Committee (2012) estimated the continental population at approximately 1.1 million birds, noting the estimate was somewhat speculative and based on few data. BBS data show an increasing population trend surveywide from 1966 to 2019, but the results are compromised by the species' low abundance (Sauer et al. 2020).

In Minnesota, the annual Waterfowl Breeding Population Survey does distinguish Hooded Mergansers from Common and Red-breasted Mergansers. Since 1987, the numbers demonstrate a slow, increasing trend that rose significantly from 2016 to 2019 (Cordts 2019). The survey, however, is not conducted statewide. Minnesota BBS data, which are collected throughout the state, also suggest a population increase, but the results are still compromised by the species' low abundance and dependence on small, forested wetlands that are not adequately surveyed by roadside counts (Sauer et al. 2020).

C. A quiet, permanent wetland bordered by small stands of mature trees with nesting cavities is typical of the nesting habitat selected by the Hooded Merganser.
Photo by Lee A. Pfannmuller.

CONSERVATION

Most of the focus on waterfowl populations is on species that nest in the Prairie Potholes Region. Because the Hooded Merganser is a forest species, found largely east of the prairie grasslands, less attention has been directed at its life history and, consequently, its conservation and management requirements. Even though it is a harvested species, the Hooded Merganser is not an important game bird. Dependent on small fish, the birds are not particularly tasty (Bent 1925). In 2019, only 92,869 birds were harvested in North America and 11,228 birds, or 12 percent of the continental harvest, in Minnesota (Raftovich et al. 2020).

Factors negatively impacting populations include the availability of suitable nest cavities and, in some areas, competition with two other cavity-nesting waterfowl, the Common Goldeneye and Wood Duck (Dugger et al. 2020). Contaminants can pose another concern. Thin eggshells and detectable levels of DDE, PCBs, and mercury were previously found in Hooded Merganser eggs (e.g., Zicus et al. 1988). Fortunately, bans on the use of these chemicals have resulted in a reduction in contaminant levels and an increase in eggshell thickness (Rave et al. 2014). Finally, although the species' North American distribution has been classified as climate stable under a 2°C warming scenario, losses are projected across the southern periphery of its current breeding range by 2050 (NAS 2019).This includes Minnesota, where, with the exception of the northeastern region, significant losses are predicted statewide.

Until more is understood about the population status of this elegant little member of the merganser family, the best conservation approach is to focus on best-management practices that protect riparian woodlands and forests and improve wetland quality.

COMMON MERGANSER

Mergus merganser

Minnesota Seasonal Status: Regular breeding species and migrant; observed during the winter, especially on large rivers in central and southern Minnesota. The Common Merganser was uncommon during the MNBBA.
Migration: Short- to medium-distance migrant that winters along the northern coasts and inland across the central and southwestern United States, south to northern Mexico.
Food: Diver that feeds primarily on small fish but occasionally consumes other small vertebrates, such as frogs and birds, and aquatic invertebrates.
Nest: Secondary cavity nester but may nest in other crevices, including burrows, hollow logs, or nest boxes.
Conservation Concern: A game species; assigned a moderate PIF Continental Concern Score of 9/20.

The male Common Merganser is a handsome bird. On the water, its bright red bill, dark green head, black back, and white breast and flanks are prominent features. In flight, his bright orange-red legs and feet are a striking contrast to his white belly. The female, however, can challenge the best of birders. Her silhouette, her gray body, and her rufous-brown head, with a crest of feathers at the back, easily identify her as a female Merganser. Distinguishing her from a female Red-breasted Merganser is more difficult. A well-defined white chin, which is considerably less conspicuous on the female Red-breasted Merganser, is diagnostic, as is the sharp separation between the red head and grayish-white breast. Unfortunately, these features can be difficult to discern at a distance or in poor light conditions.

A common breeding species in northern Minnesota, the Common Merganser's status as a migrant is one for the record books. Each fall, thousands of birds concentrate on Lake Pepin, a widening of the Mississippi River south of Lock and Dam No. 3 at Red Wing, Minnesota. Counts of several thousand birds are not uncommon, including a record of 70,000 on November 26, 1995 (MOU 2021b). This spectacular migratory stopover, the largest in the world, contributed to Lake Pepin's designation as one of Minnesota's Important Bird Areas.

Also known by several colloquial names, including "Sheldrake," "Goosander," and "Sawbill," the

Photo by Michael Furtman.

Common Merganser's Latin name means "plunging goose," a reference to its diving behavior. Globally distributed, its North American breeding range extends from Alaska, across Canada's boreal forest, and as far south as the northwestern United States, the Great Lakes, and New England. Sparsely distributed, the highest breeding densities are observed in several western states.

MINNESOTA BREEDING DISTRIBUTION

In the first comprehensive account of Minnesota's avifauna, Hatch (1892) described the Common Merganser as a breeding resident in the state's "remoter districts." In earlier years, however, he reported finding them in Minneapolis, where he allegedly observed young "on the ponds and small grassy lakes." T. S. Roberts (1932) included this note in his own Common Merganser account. He also referenced Dr. Hvoslef's 1888 observation of a female and two small young in Fillmore County, an 1887 report from Heron Lake in Jackson County, and another report from northern Iowa. Collectively, these led him to suspect that the species once nested in limited numbers as far south as Iowa. At the time of his writing, however, there were no recent records from southern Minnesota. Instead, the species was restricted to the northern forests, "especially in the triangle formed by the Rainy and Pigeon rivers on the north and Lake Superior on the south" (Roberts 1932).

In their updated account, Green and Janssen (1975) described the Common Merganser as a breeding resident in north-central and northeastern Minnesota with recent summer observations as far west as Clearwater and Otter Tail Counties. They noted the nineteenth-century records from southern Minnesota but remarked that additional information to suggest that the species nested that far south was lacking. Janssen (1987) identified six northern counties where breeding had been confirmed since 1970: Beltrami, Cook, Koochiching, Lake, Lake of the Woods, and St. Louis. Hertzel and Janssen (1998) later added Cass and Itasca Counties to the list. By 2014, the MBS had identified 48 Common Merganser breeding-season locations. The most southern record was from southern Pine County.

MNBBA participants reported 397 Common Merganser records from 6.2 percent (294/4,740) of the surveyed atlas blocks; all but 14 were confined to the Laurentian Mixed Forest Province **(A)**. Breeding was confirmed in 88 blocks (1.9 percent). In addition to the eight counties identified by Hertzel and Janssen (1998), breeding was confirmed in Aitkin, Hubbard, Morrison, Sherburne, Stearns, and Wabasha. Two nesting reports submitted from Lac qui Parle and Ramsey Counties, however, were invalidated because they lacked sufficient details to document records so far beyond the species' primary breeding range.

The Merganser's northern breeding range was largely bisected by the Mesabi Iron Range in St. Louis and Itasca Counties and by extensive peatlands in southern St. Louis and northern Aitkin Counties. Although Merganser habitat is present throughout northern Minnesota, the MNBBA land-cover suitability map **(B)** suggests scattered pockets of suitable habitat are also available in the lakes region of west-central Minnesota, the northwest aspen parklands, and the larger lakes and rivers of southern Minnesota.

Indeed, some of the most interesting atlas records are scattered in east-central and southern Minnesota. Most notable are the confirmed breeding records in Sherburne, Stearns, and Wabasha

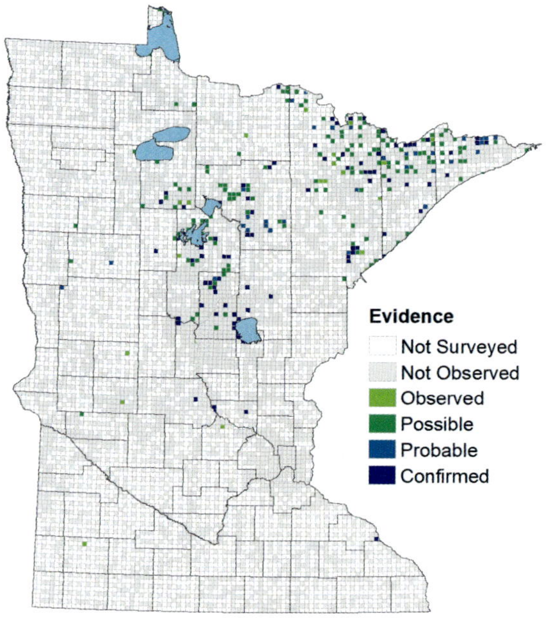

A. Breeding distribution of the Common Merganser in Minnesota based on the Breeding Bird Atlas (2009–13).

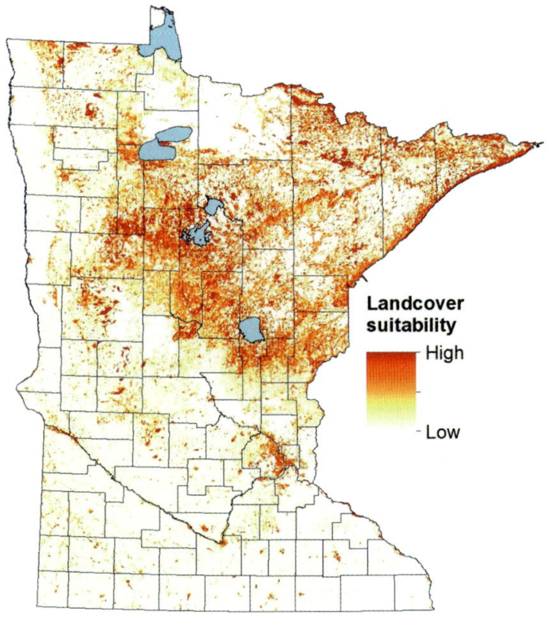

B. Landcover suitability map for the Common Merganser in Minnesota based on habitat, landscape context, and climate data gathered during the Minnesota Breeding Bird Atlas (2009–13) using the MaxEnt modeling approach.

Counties. From 1930 to 1979, there were summer MOU records beyond the Merganser's core breeding range in only six years; from 2000 to 2020 there were summer records beyond the core breeding range in eighteen of the twenty-one years, including every year since 2006 (MOU 2021b). Although some of these birds may be late spring migrants, there is a clear trend of increasing summer reports in southern Minnesota, where the Merganser may be expanding or, as Roberts (1932) originally speculated, reoccupying portions of its former breeding range. Careful documentation of future nesting reports in this region is important to help shed light on its status.

Minnesota is not the only state where the Merganser's distribution has changed. The species is recolonizing portions of its former range in several eastern states, including New York, Pennsylvania, and Vermont (Pearce et al. 2020), and expanding southward in the Great Lakes region. Michigan's second atlas (2002–9), for example, reported an increase in the number of townships in the southern Lower Peninsula with possible, probable, and confirmed Merganser breeding reports (Chartier et al. 2013).

BREEDING HABITAT

In the Great Lakes region, Mergansers are found on clear, freshwater rivers, lakes, and reservoirs that are bordered by mature coniferous or mixed coniferous–deciduous forests **(C)**. Abundant fish populations and suitable nest cavities are essential. Although they are often found on deep waters during the winter, during the summer most foraging occurs in waters less than 4 m deep (Baldassarre 2014). Recent breeding records south of the northern forest landscape, however, suggest the species has a broader habitat tolerance than previously described.

POPULATION ABUNDANCE

Population data for the Common Merganser are sparse. Because the North American Waterfowl Breeding Population and Habitat Survey doesn't distinguish between the three Merganser species (Common, Red-breasted, and Hooded), Baldassarre (2014) used a variety of data to generate a continental estimate of 943,000 birds between 1994 and 2007. The North American Waterfowl Management Plan Committee (2012) provided a slightly higher estimate of 1.2 million birds. Estimates are not available for Minnesota.

An accurate assessment of population trends is equally difficult to discern. The combined Merganser count for eastern North America shows relatively stable populations from 1998 to 2019 **(D)**. The BBS, however, tracks the three species separately. Although the Common Merganser data are not statistically robust, they demonstrate a small, surveywide decline averaging −0.40 percent per year from 1966 to 2019 (Sauer et al. 2020). In Minnesota, BBS data show a steady increase in Merganser numbers, but these data are quite limited.

Factors responsible for the possible decline in the mid- to late 1900s are not entirely known, but contaminants have been a threat to this fish consumer. In particular, two studies documented significant levels of organochlorines, mercury, and other heavy metals in Common Merganser eggs collected in the Great Lakes in the 1960s and 1970s (Faber and Hickey 1973; Haseltine et al. 1981).

C. In the Boundary Waters Canoe Area Wilderness, the Common Merganser breeds along clear-water lakes and rivers with large-boled trees adjoining mature boreal forests that provide suitable nesting cavities.
Photo by Jarad B. Niemi.

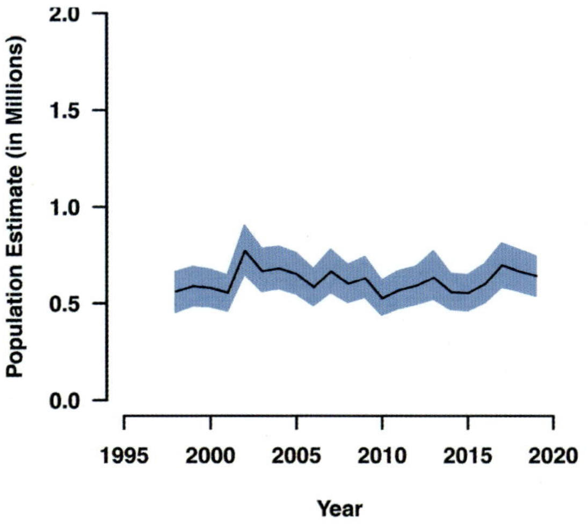

D. Breeding Merganser (Common, Hooded, and Red-breasted) population estimates and 90 percent credible intervals from Bayesian hierarchical models for the North American Waterfowl Breeding Population and Habitat Survey's eastern survey area from 1998 to 2019. Reproduced from USFWS (2019, fig. 5).

CONSERVATION

The Common Merganser is considered a moderate conservation priority, largely because of the uncertainty regarding its population status (PIF 2021). Harvest pressure is negligible. From 1999 to 2008 the United States harvest averaged only 19,797 birds per year, while the annual Canadian harvest averaged 7,127 birds (Baldassarre 2014).

Specific conservation measures for the species are lacking. Most management efforts have been directed at reducing the impact of Mergansers on local fish populations, primarily salmonids, and fish hatcheries. Although there have been some targeted control efforts, their effectiveness requires more careful study (Baldassarre 2014).

Today the primary threats to the species appear to be chemical contaminants, owing to its position high within the food chain, and habitat degradation, resulting from poor water quality and acid precipitation (Pearce et al. 2020; Baldassarre 2014). Warmer temperatures also pose a concern. NAS (2019) projected the species' North American breeding range was moderately vulnerable under all three climate scenarios they examined. Unfortunately, models for all three scenarios also predicted significant habitat losses across northeastern and north-central Minnesota. Two measures that may help buffer impacts to the Common Merganser are to maintain good water quality in Minnesota's northern lakes and rivers and to implement forest management practices that focus on retaining mature trees and snags for nesting.

RED-BREASTED MERGANSER

Mergus serrator

Photo by Mike Lentz.

Minnesota Seasonal Status: Regular breeding resident and migrant; occasionally observed during the winter. The Red-breasted Merganser was rare during the MNBBA.
Migration: Short- to medium-distance migrant that winters along the coastlines of Canada, the United States, northern Mexico, and the Great Lakes.
Food: Diver that feeds primarily on small fish and crustaceans.
Nest: Small, sheltered depression on the ground near water; occasionally in a hollow stump. May nest in small colonies and with other waterbirds.
Conservation Concern: A game species; assigned a moderate PIF Continental Concern Score of 10/20.

In Minnesota this boreal species is restricted to the cold waters of Lake Superior's northern shoreline and to large, deep lakes in the northern forest. It is the least common of our three Merganser species; the male is unmistakable in his spectacular breeding plumage. Bent (1923) was bursting with praise when he wrote the following: "The drake in his newly acquired nuptial plumage is resplendent with a metallic green headdress and waving crest, the whole set off by a long coral-red bill. The white ring about his neck, the reddish brown and speckled breast, the snowy flanks and wing patches, and the dark back all go to create a picture of great beauty as he swims or dives or restlessly flies to and from among the breakers."

The female, with her gray body and rufous head, is similar in appearance to the female Common Merganser. Her bill, however, is thinner; her rufous head is paler; her white throat blends into the adjacent feathers rather than appearing as a well-defined white patch seen on the female Common Merganser; and the distinction between her red head and gray breast is diffuse. Unfortunately, these differences can be difficult to see, especially in poor light or from afar.

Although these two Merganser species share much in common, there are two primary differences. First, unlike the cavity-nesting Common Merganser, the Red-breasted Merganser is a ground nester and, second, the Red-breasted Merganser occurs more frequently in salt water and estuaries than the more freshwater-loving Common Merganser. Like all Mergansers, the Red-breasted Merganser is also known as a "sawbill," for its thin, serrated bill.

Global in distribution, in North America the species' breeding range extends from Alaska south and east across Canada, and north to the southern coastal waters of Greenland. In the United States, it breeds locally within the Great Lakes and in northern New England.

MINNESOTA BREEDING DISTRIBUTION

In the late 1800s, Hatch (1892) wrote that the Red-breasted Merganser was rare throughout northern Minnesota and bred "sparingly" as far south as Minneapolis. T.S. Roberts (1932), however, seriously questioned Hatch's assertion that the species occurred beyond the northern forests. His files included no evidence that the birds ever occurred south of Mille Lacs Lake. Even to the north, reports were scarce, leading Roberts to describe the species as an uncommon summer resident.

Most reports came from Lake Superior, where birds were usually observed at the mouths of fast-moving rivers stretching from the St. Louis Bay in Duluth (1905) northeast to a site 12 miles north of Grand Marais (1928). All but one account presumed nesting but lacked positive confirmation. The exception was in June 1915, when Roberts and a colleague visited Spirit Island in Mille Lacs Lake. They observed a pair and saw one egg nestled within a rock crevice. His account also included his 1915 observation of a pair "evidently nesting" at Big Oak Island in Lake of the Woods.

Decades later, Green and Janssen (1975) reported that the Red-breasted Merganser remained a summer resident primarily along Lake Superior's shoreline. Although they were nesting on Mille Lacs Lake and on Big Sandy Lake in Aitkin County, they were quite scarce in both localities. Published nesting reports were also available from Beltrami County, inland St. Louis County, and as far south as Houston County. Unfortunately, these accounts lacked details and left open the possibility that females were observed and were misidentified.

Then, from 1975 to 1977, Oring (1978) reported observing multiple Red-breasted Merganser broods on Leech Lake in Cass County. No nests were found, but the species' identification was confirmed when a bird was collected. When Janssen (1987) provided another updated account, he identified five counties where nesting had been confirmed since 1970: Aitkin, Cass, Cook, Lake, and St. Louis. Hertzel and Janssen (1998) later deleted Oring's Cass County record, given that breeding was only inferred and not confirmed. The species' rarity was further corroborated by the MBS, which reported only one breeding-season location in central St. Louis County.

During the MNBBA, participants reported 36 Red-breasted Merganser records from 27 (0.6 percent) atlas blocks. Breeding was confirmed in five blocks; four of the blocks were along the North Shore of Lake Superior and one was located in Crow Wing County **(A)**. All breeding records were of females with broods.

Given the possible misidentification of female Mergansers, each MNBBA Red-breasted Merganser record was examined closely, particularly those distant from the species' traditional Lake Superior habitats. Four records were invalidated, including one from Lyon County in southwestern Minnesota. The confirmed breeding record from Pelican Lake in Crow Wing County was reported in July 2010 (a hen with four young). That same summer there were

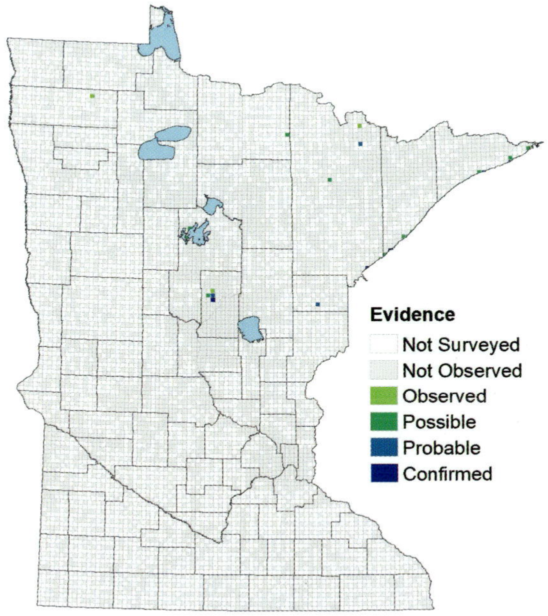

A. Breeding distribution of the Red-breasted Merganser in Minnesota based on the Breeding Bird Atlas (2009–13).

three other atlas reports of both male and female Red-breasted Mergansers on the lake, giving credence to the nesting report. Red-breasted Mergansers had not previously been documented nesting in the county.

Although Red-breasted Mergansers occasionally find suitable habitat in Minnesota's northern lakes west of Lake Superior, they remain a rare bird throughout the state. Their range and abundance have changed little in Minnesota and throughout North America over the past one hundred years (Craik et al. 2020).

BREEDING HABITAT

Breeding pairs are found in freshwater lakes and rivers, in addition to fresh, brackish, and saltwater wetlands near coastal waters (Baldassarre 2014). In Minnesota, they prefer large, deep lakes and rivers bordered by rocky shorelines, gravel outwash, or sandbars **(B)**. Shorelines with dense, low vegetative cover provide the shelter needed for ground nests, while an abundant supply of small fish provides the necessary forage. Smaller lakes are selected for nesting only when they are connected to large, deep bodies of water (Craik et al. 2020; Kaminski and Weller 1992).

POPULATION ABUNDANCE

Because it is difficult to distinguish the female Merganser species from one another, the North American Waterfowl Breeding Population and Habitat Survey simply reports the total number of all Mergansers. As a result, biologists have used other information to establish a rough estimate of the Red-breasted Merganser's continental population as ranging from 180,000 to 670,000 individuals (Baldassarre 2014). PIF's (2021) estimate of the global population, however, was only 600,000 birds, with approximately 400,000 in North America. No estimates are available for Minnesota.

Population trends are equally uncertain. The North American Waterfowl Breeding Population and Habitat Survey demonstrates a relatively stable population of all Merganser species in the eastern survey area since 1998 **(C)**. Species-specific data are collected by the BBS, but the continental data are quite limited, given that the survey's roadside counts do not adequately sample the Red-breasted Merganser's large lakes and coastal waters. Nevertheless, what data are available suggest a sharp, surveywide decline from the mid-1960s until the mid-1980s, followed by relatively stable numbers (Sauer et al. 2020). Minnesota BBS data are equally unreliable.

Although in-depth studies are limited, contaminants are likely responsible for declines observed in the late 1900s. High levels of DDE, PCBs, mercury, and other chemicals were found in the birds, especially in the 1970s before their widespread use was halted (D. H. White and Cromartie 1977; Heinz et al. 1983). Changes in prey availability and habitat conditions also may impact population numbers (Craik et al. 2020).

CONSERVATION

Although population data are sparse, the Red-breasted Merganser appears relatively secure in its remote, boreal breeding habitats. Hunting pressure is especially low. As a consumer of fish, the species is not a favorite among waterfowl hunters, and its preference for open, coastal waters in the fall makes it a difficult target. From 1999 to 2008 the annual continental harvest averaged only 20,000 birds (Baldassarre 2014). Nevertheless, PIF (2021) assigned it a moderate conservation score because of its small population and uncertain population trend.

B. Lake Superior's rocky shoreline, with a narrow fringe of low-growing trees and shrubs that provide nesting cover, is ideal breeding habitat for the Red-breasted Merganser. Photo by Gerald J. Niemi.

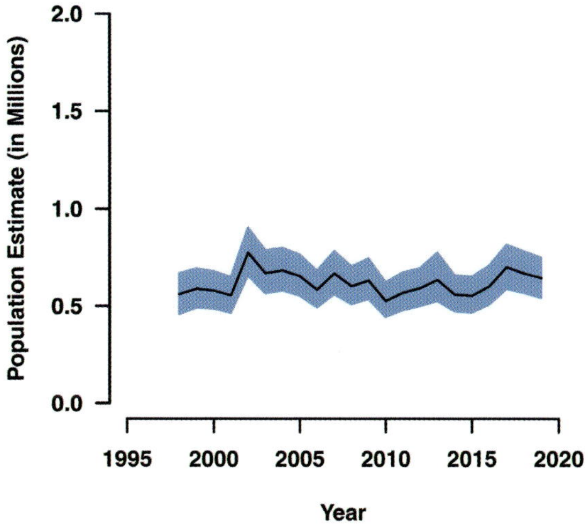

C. Breeding Merganser (Common, Hooded, and Red-breasted) population estimates and 90 percent credible intervals from Bayesian hierarchical models for the North American Waterfowl Breeding Population and Habitat Survey's eastern survey area from 1998 to 2019. Reproduced from USFWS (2019, fig. 5).

Like other boreal nesting waterfowl, the Red-breasted Merganser has not been the focus of specific conservation initiatives. The only targeted management measures have taken aim at the species' consumption of game fish. Efforts to control both Red-breasted and Common Mergansers in localities where young game fish populations are abundant have been largely ineffective, particularly in the Pacific Northwest. Some efforts on the Canadian Atlantic coast, however, have been locally effective in reducing the mortality of young salmon. Nevertheless, biologists caution that such measures warrant careful consideration regarding their cost and long-term effectiveness (Baldassarre 2014; Craik et al. 2020).

Finally, because the size of the boreal forest is predicted to decrease with warming temperatures, climate change is a significant concern for the Red-breasted Merganser. An analysis by NAS (2019) projected the species' breeding range in North America was moderately vulnerable under all three warming scenarios that were examined. At the southern periphery of its breeding range in Minnesota, the Red-breasted Merganser may have a secure future in the short term, but its long-term future is in doubt. Under a 3°C warming scenario, models predict its near extirpation from the state by 2080.

RUDDY DUCK

Oxyura jamaicensis

Minnesota Seasonal Status: Regular breeding species and migrant; occasionally observed during the winter. The Ruddy Duck was uncommon during the MNBBA.
Migration: Most northern populations winter in the central and southern United States and in Mexico; elsewhere populations are year-round residents.
Food: Diver that feeds primarily on aquatic invertebrates and occasionally on aquatic vegetation and seeds.
Nest: Platform nest placed over shallow water; sometimes constructed atop a muskrat house or American Coot nest.
Conservation Concern: A game species; assigned a low PIF Continental Concern Score of 8/20.

Photo by Mike Lentz.

From "cartoonishly bold" to "gaudy" and "bizarre," an amusing array of phrases has been used to describe the small Ruddy Duck. During the breeding season the male is striking: "his back glowing with the rich, red brown of his nuptial attire, offset by the pure white of his cheeks, his black crown, and above all his wonderful bill of the brightest, living, glowing sky blue" (Bent 1923). The female, by contrast, is a mottled gray-brown with only the hint of a cheek patch that is interrupted by a dark stripe below the eye. Her bill is dark gray in color.

The Ruddy Duck is unique among Minnesota waterfowl. *Oxyura,* the genus name, means "sharp-pointed tail." Composed of stiff, elongated feathers, it functions as an underwater rudder. Above water, it is prominent in many of the male's behavioral displays, often "held erect or even pointed forward till it nearly meets his upturned head" (Bent 1923). Like those of grebes and loons, the Ruddy Duck's legs are also located far back on the body, "increasing propulsion efficiency during diving but rendering the birds nearly helpless on land" (Johnsgard 2010). Other unusual features include an inflatable air sac in the male's neck that helps produce a guttural "belching" sound during courtship, large feet, and eggs that are nearly equal in size to those of a Great Blue Heron (Bent 1923). And, while other waterfowl compress the air in their feathers to partially submerge below the water's surface, like a submarine the Ruddy Duck is able to sink completely without diving. Many of these traits are so unusual that the taxonomic relationship of the Ruddy Duck and its "stiff-tailed" relatives to other waterfowl remains a matter of debate (Brua 2020).

Although the species' breeding range stretches from central Canada south through much of the western United States, northern Mexico, and the Caribbean, the core of its distribution is restricted to the northern Great Plains. The bird is considered a pest in Europe, where populations were accidentally introduced and now pose a threat to the rare, native White-headed Duck (*Oxyura leucocephala*).

MINNESOTA BREEDING DISTRIBUTION

Roberts (1932) described the Ruddy Duck as formerly "an abundant breeding bird in all the larger rush-grown and weed-grown sloughs of southern and western Minnesota." By 1900, however, its numbers were notably reduced. Unlike many waterfowl that were considered tasty treats by the early settlers, the Ruddy Duck was largely ignored. As a result, Roberts was unsure what caused its dramatic decline. But as the populations of other, more favored species diminished, the Ruddy Duck became more popular table fare. Indeed, the species' habit of diving rather than flying when disturbed made it easy to harvest large numbers during migration, likely contributing to the speed at which populations plummeted (Brua 2020).

Following passage of the Migratory Bird Treaty Act in 1918, the species began to slowly recover. In 1929, a game warden in west-central Minnesota reported that "he never saw the Ruddy when he first went to Grant County years ago, but that now it is everywhere" (Roberts 1932). Nevertheless, confirmed breeding records were limited to Hennepin, Jackson, and Kittson Counties.

More than forty years later, Green and Janssen (1975) summarized the species' distribution as occurring throughout western Minnesota and east as far as the Twin Cities. In northern Minnesota, nesting was even confirmed in Cass and Clearwater Counties. Nevertheless, referencing a 1964 report that summarized results of state waterfowl surveys conducted in the early 1960s (Lee et al. 1964), they noted the Ruddy Duck comprised only 2 percent of the state's breeding waterfowl population.

A few years later, Janssen (1987) outlined a western breeding distribution that stretched from eastern Roseau County, south to Stearns County, east to Anoka and Ramsey Counties, and south to Freeborn County. Since 1970, nesting had been confirmed in 26 counties, including two counties east of the primary breeding range (Cass and Carlton). Hertzel and Janssen (1998) later identified 31 counties where nesting had been confirmed since 1970 but excluded two counties identified by Janssen: Clearwater and Cass, leaving Carlton County as the easternmost nesting locality. By 2014, the MBS documented 134 breeding-season locations for the Ruddy Duck.

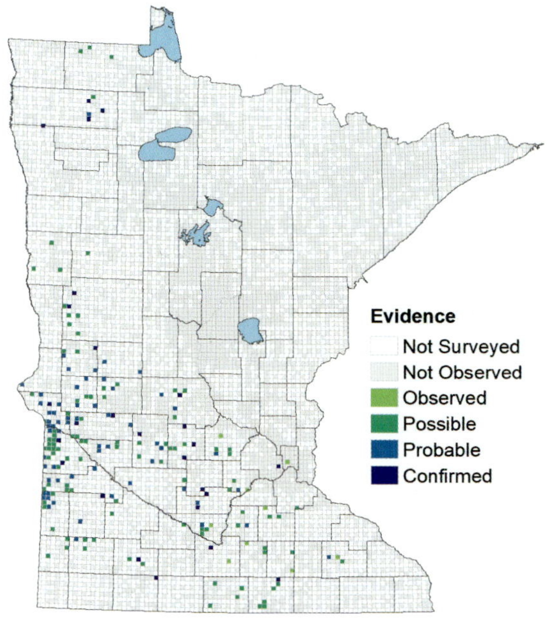

A. Breeding distribution of the Ruddy Duck in Minnesota based on the Breeding Bird Atlas (2009–13).

All but two of the records were within the species' primary range. The two outliers were records in Winona and Cass Counties.

MNBBA participants reported 257 Ruddy Duck records from 4.2 percent (199/4,741) of the surveyed atlas blocks found in 46 of Minnesota's 87 counties **(A)**. Breeding was confirmed in 26 blocks located in 18 counties. Nearly all of the records were within the species' primary range. The exceptions were three records in Olmsted County. No records were documented in northeastern or north-central Minnesota.

Since the Ruddy Duck recovered from its decline in the late 1800s and early 1900s, its Minnesota breeding distribution has remained similar to that reported in the 1930s. Still an uncommon species, it accounted for only 1.8 percent of the MNBBA records for all dabbling and diving ducks. Elsewhere, the species' breeding distribution has shown few changes. Although Brua (2020) suggested the extent of its breeding range may be declining in the Great Lakes, accounts in Wisconsin (Cutright et al. 2006), Michigan (Chartier et al. 2013), and Ontario (Cadman et al. 2007) indicate stable or increasing populations. Cadman and his colleagues (2007) attributed much of the increase in Ontario to the species' use of sewage lagoons that support abundant populations of midge larvae, one of the Ruddy Duck's favorite food items (Brua 2020).

BREEDING HABITAT

An occupant of many wetland habitats, including large wetland complexes, reservoirs, stock ponds, and sewage lagoons, Ruddy Ducks prefer semipermanent wetlands in the core of their breeding range in North and South Dakota (Brua 2020). Overall, their primary requirements are extensive beds of emergent vegetation where nests are well concealed and open-water areas that allow room for them to become airborne **(B)**. This chunky little bird needs a relatively long distance to run on the water and gain enough momentum to take flight (Brua 2020).

POPULATION ABUNDANCE

The North American Waterfowl Management Plan Committee (2012) estimated the total continental population of Ruddy Ducks at 1.2 million birds. In the prairie potholes and prairie parklands, monitoring data show a slow but steady increase from approximately 400,000 breeding adults in 1987 to more than 800,000 in 2015 (USFWS 2016b). Surveywide, BBS data also show a reliable but nonsignificant population increase averaging 0.43 percent per year from 1966 to 2019 **(C)**.

In Minnesota, waterfowl survey efforts show wide fluctuations in Ruddy Duck numbers and even suggest a downward trend. The species is one of the last waterfowl to arrive on its breeding grounds each spring (Baldassarre 2014). Because waterfowl surveys are timed to correspond with the nesting phenology of more favored species, such as the Mallard and the Blue-winged Teal, many Ruddy Ducks may still be en route to their breeding grounds when the survey is conducted. Wide fluctuations, therefore, may reflect the vagaries of migration more than the status of nesting populations.

B. Extensive beds of emergent vegetation and large open-water areas, which enable the Ruddy Duck to run along the water to become airborne, are the major habitat requirements of this unique waterfowl species. Photo by Lee A. Pfannmuller.

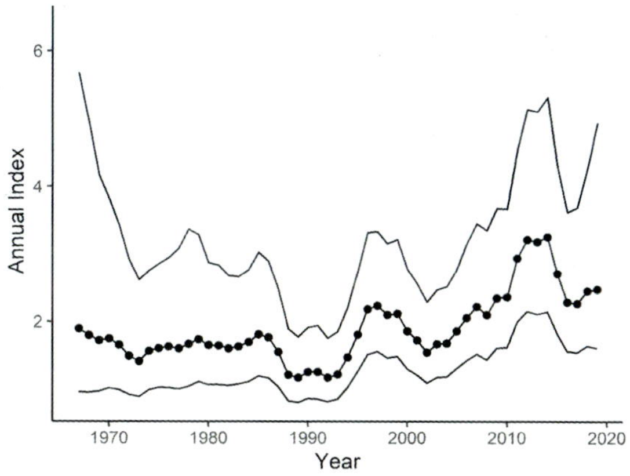

C. Surveywide breeding population trend and 95 percent credible interval (solid black lines) for the Ruddy Duck based on the 1967–2019 Breeding Bird Survey (Sauer et al. 2020).

CONSERVATION

Because populations are relatively stable and the Ruddy Duck is not a highly valued game species, the North American Waterfowl Management Plan Committee (2014) has not established a population goal for the species. During the 2019 season, 38,559 birds were harvested in the United States, comprising less than 1 percent of the total duck harvest (Raftovich et al. 2020). That same year only 847 birds were harvested in Minnesota.

Although it is assigned a low Continental Concern Score (PIF 2021), the Ruddy Duck faces threats to its breeding and wintering habitats. The majority of birds, for example, winter at a limited number of sites where mortality can negatively influence populations nationwide. Indeed, high levels of several contaminants—including mercury, cadmium, lead, and zinc—have been found in birds at two of these sites, the Chesapeake Bay and the San Francisco Bay (Baldassarre 2014; Brua 2020). Oil spills in these coastal areas also pose a concern. And, as with all waterfowl that nest in the prairie potholes, warmer temperatures may impact breeding populations (NABCI 2010). Nevertheless, NAS (2019) projected a low vulnerability for the Ruddy Duck's breeding distribution under all three warming scenarios they examined. The greatest losses were predicted in Mexico and in Canada's Boreal Taiga Plains northwest of the prairie potholes. Losses were predicted to be minimal in Minnesota.

The future of this unique member of Minnesota's waterfowl community should not be taken for granted. Although the species may benefit from initiatives to protect and restore wetlands, the little "stiff-tail" is not the focus of any targeted conservation efforts.

NORTHERN BOBWHITE

Colinus virginianus

Minnesota Seasonal Status: Former permanent resident; now introduced; no known wild populations. The Northern Bobwhite was very rare during the MNBBA.
Migration: Sedentary.
Food: Consumes plants and seeds with varying amounts of arthropods.
Nest: On the ground in a shallow depression lined with grass or leaves and with protective vegetation above.
Conservation Concern: Formerly a game bird; assigned a moderate PIF Continental Concern Score of 12/20.

Photo by Sparky Stensaas.

The Northern Bobwhite is beloved by hunters, rural residents, naturalists, and bird-watchers. Forbush and May (1955) reminisced about years past on a rural farm: "During my boyhood the cheery, heartening call of the Quail was one of the most common and welcome sounds of spring and summer." This experience was an uncommon occurrence in Minnesota because the species was extremely rare. As a game bird, it has been well studied in the United States and 22 subspecies have been described. The Bobwhite gets its name from its call, a clear whistled and unmistakable *bob-white.* Both sexes are chunky reddish-brown birds with gray tails. The male has a white throat and eye-stripe, while in the female the white is replaced with a buffy color.

The range of the Northern Bobwhite is widespread in the East; Minnesota is at or mostly beyond the northern edge of its traditional range. It occurs in the Great Plains, in the eastern tallgrass prairie, and throughout the southeastern United States. Its midwestern population is sparse north of southern Iowa, and it is rare to absent along the Appalachian Mountains. Its highest densities have been observed in western Oklahoma and southern Texas. Populations also are found in the Caribbean and south through Mexico.

MINNESOTA BREEDING DISTRIBUTION

Roberts (1932) summarized the absence of quail during the presettlement period as, "The various explorers into the country that is now Minnesota in the early part of the nineteenth century (Keating and Thomas Say of the Long Expedition, Schoolcraft, Featherstonhaugh, and others), who mentioned both game- and non-game birds, have nothing to say about seeing Quail." The lack of information about the Bobwhite was expressed by Gustav Swanson (1929): "Frequently we find in the periodicals or in the daily press references to the introduction of Bob-white . . . into Minnesota, and it seems to be quite generally known that the bird is not native to the state, but that it has been brought in by interested sportsmen. However, when a search is made for actual records of birds introduced, these references are for the most part found to be conspicuous for their lack of definiteness and real value."

Roberts (1932) reported on Bobwhite in the 1880s and 1890s in notes from Dr. Hvoslef of Lanesboro, Fillmore County: "Last winter, 1892–1893, I again saw a few flocks of Quail after it had been almost entirely absent for several years, and during the summer of 1897 I saw quite a number." Roberts also reported Bobwhite records during that time from Houston, Lincoln, McLeod, and Waseca Counties. He further stated that "from this time on the Bobwhite became a regular, established bird in southern Minnesota." It was assumed these birds were from an expanding native population in northern Iowa that resulted from favorable land-use changes from farming and spells of good weather.

Roberts also mentioned there were "various attempts of the Quail to gain a foothold in northern Minnesota" from the late nineteenth century up until 1915 in the Red River valley, near Mille Lacs Lake, and in Otter Tail and Todd Counties. He concluded that "it is not likely that a half-hardy, non-migrating bird like the Quail can ever become established in this land of deep snows and continued low temperatures." It is unknown whether these birds were a native expansion or from reintroductions. Interestingly, Aldo Leopold (1931) did game surveys in several midwestern states and mentioned that Pine County had many quail in 1896–98 after the Hinckley fire of 1894. These birds appear to have expanded from Wisconsin.

Regardless of how Northern Bobwhites came to occupy Minnesota in these early years, the population was sufficient to establish a hunting season from the 1920s through the early 1950s, when hundreds of birds were estimated to have been killed each year (Chesness 1964). According to the MNDNR (2015a), the Bobwhite had been hunted since 1858, and the first recorded harvest, estimated at 6,100 birds, was in 1919. The report concluded that during the 1920s, the estimated harvest peaked in 1927 (13,000 birds) and by 1932 it was confined to the southeastern area. Most of the nesting records given by Roberts were also from the 1920s. During this time period, the MNDNR was also releasing the Bobwhite, but where, when, and how many are not known. These releases were terminated in 1952, and, as harvested birds declined, hunting was stopped in 1958.

Reviewing the most recent period, Hertzel (2003) concluded that "most Northern Bobwhite reported in Minnesota during the past twenty-five years are probably released or escaped birds." He said there are 75 game farms that have Northern Bobwhites, as well as private breeders, citizens, and farmers who could be releasing hundreds of birds. There are no state requirements to report where these releases are made. In the yearly observations summarized in the MOU seasonal reports from 1970 to 2005, there is an indication that the species was present but becoming scarce, and some were tagged as "escapes." Breeding evidence was mentioned from Castle Rock, Dakota County, and reports from Blue Mounds State Park, Rock County, possibly indicated dispersing wild birds from South Dakota (Millard 1981).

Many of the MOU records in the latter years of the twentieth century appear to be single, random observations from 20 different counties. Most of the observations in those decades were from traditional places in Fillmore, Houston, Wabasha, and Winona Counties. No more MOU records were posted after 2004, and the species was removed from the MOU Regular List and reclassified as extirpated, "because all records are presumed to be escaped or released individuals, unless there is documentation with a record to suggest wild origin" (Eckert 2004).

The participants in the MNBBA recorded only four records and all were close to the Mississippi River in Houston, Wabasha, and Winona Counties **(A)**. Three of the records are confirmed nesting records. The other record was possible breeding, and the observer guessed it was a released bird. Since the MNBBA period, there have been scattered observations, including two records of fledged young, from Carver, Dakota, Hennepin, Houston, and Winona Counties (MOU 2021a). Without better monitoring, reporting, and research about the origin of these Bobwhite quail and others in the southeastern region of Minnesota, there is no way of knowing if they are native or released.

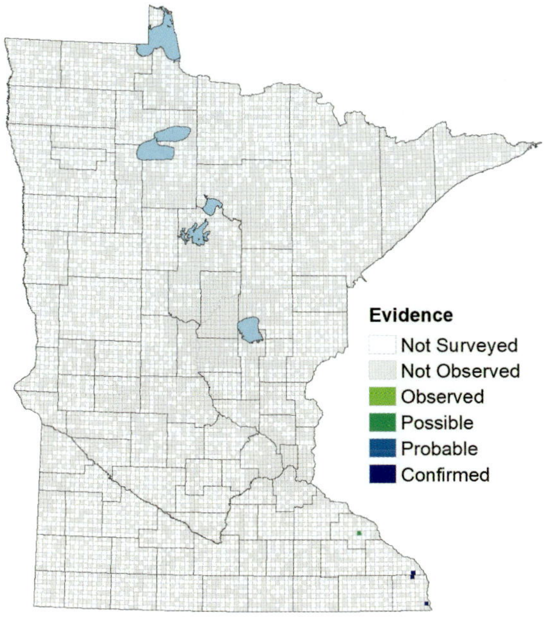

A. Breeding distribution of the Northern Bobwhite in Minnesota based on the Breeding Bird Atlas (2009–13).

B. Northern Bobwhites are attracted to areas that provide a diversity of vegetation, including open agricultural lands, shrubby fencerows, and mixed forests.
Photo by Lee A. Pfannmuller.

BREEDING HABITAT

Bobwhites use early-successional habitat of open, diverse patchy vegetation in agricultural lands, mixed forests, plantations, brushy savannas, and seminatural suburban landscapes (Brennan et al. 2020). Large open prairies or dense planted grasslands are not suitable, nor are closed-canopy forests. Where populations are persistent, a landscape of patchy fragmented vegetation is most important **(B)**.

POPULATION ABUNDANCE

The US and Canadian population of the Northern Bobwhite is estimated at 4.4 million birds, and a global population at 5.8 million (PIF 2021). This is a considerably larger population than any of the other five western quail species, which also have much smaller ranges. Trends from the BBS show a significant and drastic decline surveywide that averages −3.10 percent per year from 1966 to 2019 **(C)**. Rosenberg, Kennedy, and colleagues (2016) calculated a cumulative −83 percent decline in the population in the United States and Canada from 1970 to 2014. The Northern Bobwhite was designated a Common Bird in Steep Decline by PIF (2021). These severe population declines have given wildlife management agencies a strong cause for alarm. Multiple reasons are postulated for the decline, most prominently the change in agricultural land-use practices from small farms to large agribusinesses. The patchy landscape with hedgerows, brushy edges, and native grasslands created by small farms has been lost and replaced by efficient large-field monocultures (Brennan et al. 2020).

CONSERVATION

Because of severe population declines and threats to the Bobwhite breeding population, there are many concerns with its future. Fortunately, climate change does not appear to be a concern. NAS (2019) predicted a stable breeding and winter range in the United States for all three warming scenarios they examined, including increasing favorable conditions along its entire northern range. Restoring Bobwhite populations requires coordinated habitat management on both public and private lands. Two conservation organizations, Quail Forever and the National Bobwhite Conservation Initiative, have been working on quail restoration in the United States.

In Minnesota, the legal classification of the Northern Bobwhite is as a game bird. Since there has been no hunting season since 1958, there is essentially no management, nor is there a state program that regulates, surveys, or monitors the Bobwhite. In 2014, the state legislature, prompted by the two Quail Forever state chapters, required that the MNDNR do a *Feasibility Study for the Restoration of Wild Northern Bobwhite in Minnesota* (MNDNR 2015a). The report includes restoration challenges and highlights the lack of a genetic source population of wild birds. It concludes that "Minnesota's wild bobwhite population is likely extirpated, or nearly extirpated. Game farm raised domestic quail may persist in the wild for brief periods of time."

The future status of the Bobwhite in Minnesota is unclear. Funding for management or reintroductions is limited; however, the Quail Forever program has used state funding to acquire habitat in southeast Minnesota for the Bobwhite. Consequently, management of the Northern Bobwhite in Minnesota is an activity in progress.

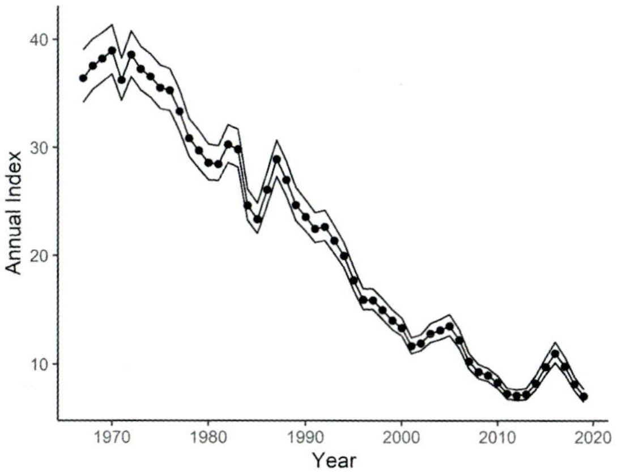

C. Surveywide breeding population trend and 95 percent credible interval (solid black lines) for the Northern Bobwhite based on the 1967–2019 Breeding Bird Survey (Sauer et al. 2020).

WILD TURKEY

Meleagris gallopavo

Minnesota Seasonal Status: Permanent resident; introduced native species; distribution statewide except farthest northern coniferous forest. The Wild Turkey was common during the MNBBA.
Migration: Sedentary.
Food: Omnivorous and highly varied diet by season, including arthropods, seeds, vegetation, tree nuts, berries, waste grain, and small amphibians and reptiles.
Nest: Bowl on the ground in dead leaves, often near the base of trees, brush piles, or in thick vegetation.
Conservation Concern: A game species; assigned a low PIF Continental Concern Score of 7/20.

Photo by Mike Lentz.

The Wild Turkey is an indigenous North American bird. Many Native American tribes in the eastern United States utilized Wild Turkeys for food and probably had semidomesticated flocks. The Pilgrims settling at Plymouth in 1620 encountered many Wild Turkeys in the open forests of eastern Massachusetts, and Turkeys became part of their fall Thanksgiving feast with the Wampanoag. There is a much-told tale that Benjamin Franklin wanted the Wild Turkey rather than the Bald Eagle to be the emblem on the Great Seal that was adopted by Congress in 1782.

There is no evidence that the Wild Turkey was ever a native species in Minnesota, but its rapid population increase and northward expansion in recent years renders this difficult to comprehend. During colonial times it was aggressively hunted and suffered massive habitat losses. Forbush and May (1955) summarized its precarious status from the mid-1800s and end with a strong quotation: "It is destined to vanish forever from the earth unless our people begin at once to protect it."

The largest North American game bird, the male Wild Turkey is unmistakable with its massive size, dark body, pink-and-blue naked head, red wattle, and blackish breast tuft. The female and immature birds are smaller, duller in coloration, and usually lack the breast tuft. The gobble call of the male is unique and can be heard a mile away.

The Wild Turkey ranges from the southern part of the Canadian provinces to the Mexican Yucatan and has six subspecies (McRoberts et al. 2020). Its abundance varies considerably, but its highest densities are in the southern Great Plains.

MINNESOTA BREEDING DISTRIBUTION

The question of whether the Wild Turkey was a resident in Minnesota in presettlement times has long been disputed. Roberts (1932) thoroughly examined reports from early explorers, including those by the Major Long expedition with Thomas Say as naturalist (1823), by Featherstonhaugh (1835) on the Minnesota River, and on the Mississippi River by the Zebulon Pike expedition (1805–1806). He concluded that "no eye-witness has left a written record so far as can be found, and no Minnesota specimen is in existence." He found writings of Father Hennepin in 1680 that mention Wild Turkey at Lake Pepin and those of Jonathan Carver from the same place in 1766 that were deemed not credible. These two records were also considered in error by A. W. Schorger, who looked at the names used in these old accounts (*Outardes, Coqs d'Inde,* Bustard) and found they were used for Sandhill Crane, Turkey, or Canada Goose. The latter species was most likely, given the context of the records (Schorger 1942).

A report by Aldo Leopold (1931) is also cited as evidence of the presettlement presence of the Turkey in Minnesota: "The ancestral range of eastern wild turkey . . . is believed to have included extreme southern Minnesota" (MNDNR 2006b). His report included a map with two records: Rock County (1871) and Blue Earth County (1773). The former is attributed to Hatch (1892) and is the same record that Roberts dismissed as "hearsay" and not supported by evidence. The latter is from Peter Pond, an important but semiliterate fur trader and not a naturalist, who made one mention of Turkey in his memoir about his travels as a trader (1765–90) in the upper Great Lakes region (see Gates 1933). Writing in New England (where there are Turkeys), decades after his trip on the St. Peter River (now the Minnesota River), he described the countryside along the St. Peter River: "the Wood & Meadowes abundans of Annamels, Sum turkeas, Buffeloes are Verey Plentey, the Common Dear are Plentey, and Larg, the Read and Moose Dear are Plentey hear, Espeshaley the former." These two records from Leopold (1931) do not support evidence of the presettlement presence of the Wild Turkey in Minnesota. To his credit, Leopold wrote that no "single statement in this report is offered as final or sufficient fact." The conclusion of these old records is that there is "no positive evidence that the species had ever existed in Minnesota" before introductions began (Green and Janssen 1975).

It is reasonable to speculate that Turkeys might have occurred in Minnesota, especially since Schorger found them on the Mississippi River southward from Prairie du Chien, and Henry Schoolcraft recorded them as common there in 1820 (Schorger 1942). Even without nineteenth-century evidence of occurrence in Minnesota, the habitat appears suitable, and introductions were attempted first in the early 1920s (Roberts 1932), in 1936, and from 1955 to 1959 (Kopischke and Johnson 1973). These releases failed, probably because pen-reared Turkeys were used. In 1964, 1965, and 1968 wild-trapped Merriam's Turkeys from several central states were released at the Whitewater WMA, but the MNDNR determined that this subspecies was not well adapted to Minnesota forests. Hence, in the early 1970s birds from the eastern subspecies (*M. g. silvestris*) were released in Houston County (MNDNR 2006b). This successful introduction led to permit hunting in three southeastern counties in 1978 (Houston, Wabasha, and Winona). Introductions continued through the 2000s in 62 counties in the south, the northwest, and in three northeastern counties: Aitkin, Crow Wing, and Pine. These introductions and subsequent dispersal created Turkey populations by 2010 throughout much of the state, except in five counties in the northeast. It was assumed that habitat in the mixed coniferous forest with heavy winter snows was not suitable.

However, the Wild Turkey's expansion throughout most of Minnesota has contradicted that assumption. The MNBBA found breeding birds in Aitkin, Carlton, western Itasca, and southwestern St. Louis Counties **(A)**. Participants reported 1,384 records for the Wild Turkey, including 19.9 percent (952/4,795) of the blocks surveyed, and confirmed nesting in 338 blocks in 68 counties. This represented a substantial expansion from the nine counties with confirmed nesting records reported by Hertzel and Janssen (1998) since 1970. Since the atlas, fledged young were also reported in Kittson County in 2019 (MOU 2021a). The landcover suitability model predicted that the most suitable habitat occurs in pockets throughout the Eastern Broadleaf Forest Province and along the Minnesota River valley **(B)**. Elsewhere, scattered pockets can be found in southern Minnesota, along the North Shore of Lake Superior, and in the western and southern regions of the Laurentian Mixed Forest Province.

BREEDING HABITAT

The Wild Turkey is found in open wooded areas, brush grasslands, and river bottoms, especially where wooded areas are adjacent to farm fields and pastures. During the MNBBA, it was primarily detected in oak forests and, to a lesser extent, in upland grassland, shrub wetland, northern hardwoods, and cropland **(C)**.

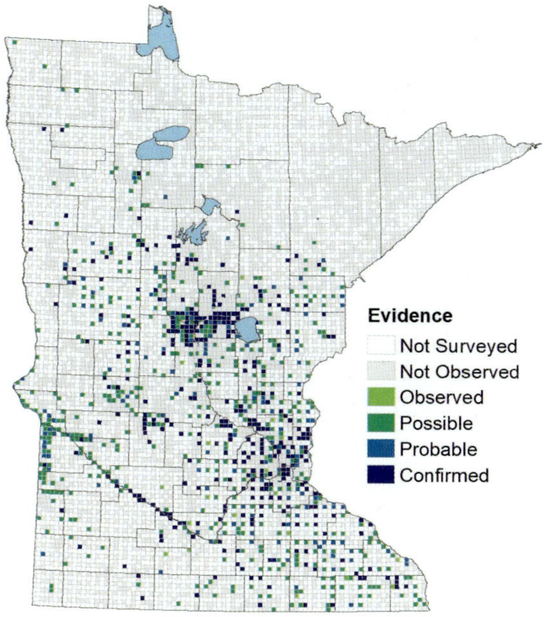

A. Breeding distribution of the Wild Turkey in Minnesota based on the Breeding Bird Atlas (2009–13).

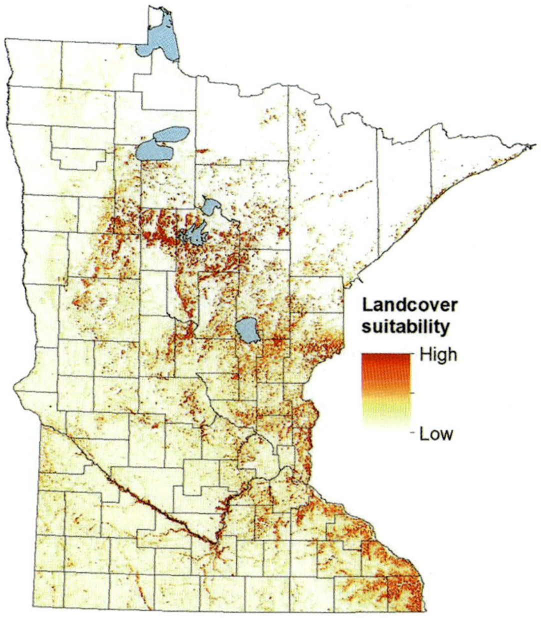

B. Landcover suitability of the Wild Turkey in Minnesota based on habitat, landscape context, and climate data gathered during the Minnesota Breeding Bird Atlas (2009–13) using the MaxEnt modeling approach.

POPULATION ABUNDANCE

The population in the United States and Canada was estimated by PIF (2021) as 6.7 million birds but an estimate for Minnesota was not attempted. The BBS roadside counts were unreliable for detecting population trends for the Wild Turkey in Minnesota, but trends in the United States from 1966 to 2019 have been significantly increasing by 4.98 percent per year (Sauer et al. 2020). Spring harvest levels in Minnesota from 1978 to 2020 reflect the substantial increase in the population up to the 2010 season and a slightly increasing harvest from around 10,000 birds in 2011 to nearly 14,000 in 2020 (Lyons 2020a) **(D)**. Fall harvest levels are substantially lower and birds per year from 2010 to 2019 ranged from 834 in 2018 to 1,752 in 2012 (Lyons 2020a).

CONSERVATION

The Wild Turkey is of minimal conservation concern because of its wide breeding distribution and recent population increase. It has gone from an original native population of millions to being almost extirpated by 1900, a decrease caused by habitat change and unregulated subsistence and market hunting. Now hunted in all states except Alaska, it is a great wildlife management success story. However, many conflicts have emerged with its increased population. These include conflicts when flocks roost on buildings or decks; they can pose threats to people, especially when young children are present; and they can cause damage to gardens and crops. Remedies include making bird feeders inaccessible, placing netting around gardens, and using dogs or loud noises to keep Turkeys away. The release or escape of pen-reared birds can contaminate the gene pool of wild birds. In Minnesota, releases without a permit are prohibited by law (2021 Minnesota Statutes: Chapter 97A.438), but the statute has been difficult to enforce. Finally, concerns about the potential impacts of Turkeys on native wildlife populations have also been raised.

C. In Minnesota, Wild Turkeys prefer oak forests that are often adjacent to grassland, shrub wetlands, and cropland. Photo by Lee A. Pfannmuller.

The future of the Wild Turkey in Minnesota is optimistic, especially if suitable habitat is maintained, harvest levels are managed appropriately, and the climate of the state continues to warm as current models predict. Indeed, NAS (2019) predicted substantial range expansions into northern Minnesota under all three of its climate change scenarios for both its breeding and winter ranges.

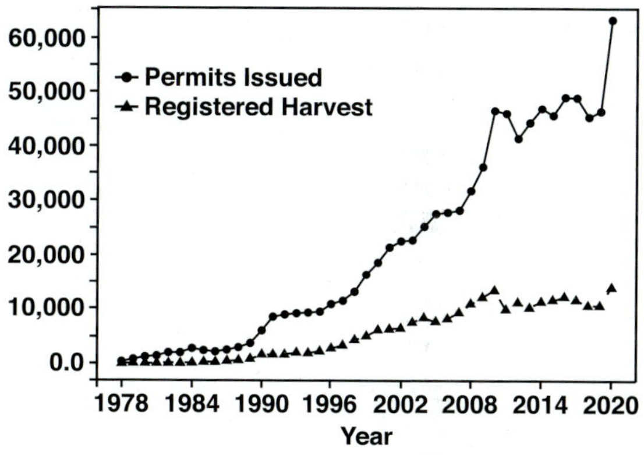

D. Permits issued and registered harvest for spring Wild Turkey seasons from 1978 to 2020 in Minnesota (Lyons 2020a, fig. 3).

RUFFED GROUSE

Bonasa umbellus

Minnesota Seasonal Status: Permanent resident. The Ruffed Grouse was a common species during the MNBBA.
Migration: Sedentary, known to have seasonal habitat movements within its home range.
Food: Primarily vegetarian, but diet varies by season. In the breeding season, consumes leaves of many herbaceous plants, fruits, and arthropods; the latter are important to chicks and adults. In other seasons, consumes buds, twigs, catkins, flowers, leafy greens, and acorns. Fond of aspen buds, catkins, and leaves but also uses hazel, willow, and birch.
Nest: On the ground in a depression with dead leaves, usually near the base of a tree, stump, log, or boulder.
Conservation Concern: A game species; the Ruffed Grouse was assigned a low PIF Continental Concern Score of 8/20.

Photo by David Brislance.

The Ruffed Grouse is the quintessential upland game bird of Minnesota. In 1892 Hatch, in his *First Report of the State Zoologist,* exclaimed, "Nowhere was the Ruffed Grouse more abundant than in all the deciduous forests of this State." He further said he had reliable reports that its distribution was "[f]rom the southern line of the state to the Lake of the Woods in the extreme north." T. S. Roberts (1932), in his treatise *Birds of Minnesota,* glorified the bird when he said, "Sportsmen will probably agree that the Ruffed Grouse, from every point of view, is the finest upland game-bird belonging to the state of Minnesota, if not to the whole of North America." Despite a potentially statewide distribution in earlier times, the species is now largely confined to the northern and north-central forest regions of Minnesota with scattered populations in the southeastern forests.

Colloquially, the Ruffed Grouse is often referred to simply as a *partridge.* It is a medium-sized game bird that is well camouflaged with its brown or reddish-brown and white barred, spotted plumage. Its small crest is not always visible, but its black *ruffed* feathers on the neck are diagnostic and form the basis of its common name. The explosive flight when startled in the forest was eloquently described by Forbush and May (1955): "thundering pinions seemed to fill the air with their sudden breath-taking roar, rose almost beneath my feet and sped away through flying leaves and bending twigs into the shadowy recesses of the forest." The Ruffed Grouse male is also well-known for its characteristic drumming, especially in the spring, which is usually delivered from a fallen log, stump, or rock. The deep thumping of the drumming starts slowly and increases in intensity, reminiscent of an engine starting.

Ruffed Grouse have a wide distribution across the temperate and boreal forest from the Maritime Provinces and the eastern United States to British Columbia and the Pacific Northwest. Southern limits in the Northeast extend into the Appalachian Mountains to northern Georgia; in the Midwest to southern Michigan, Minnesota, and Wisconsin; in the Rocky Mountains to central Utah; and along the Pacific coast to northern California. To the chagrin of Minnesotans, the highest densities of the species are found in western British Columbia.

MINNESOTA BREEDING DISTRIBUTION

As stated by Roberts, the Ruffed Grouse was breeding in all woodlands of the state; he noted that it was highly adaptable to agricultural and urbanized expansions in the early 1900s. Furthermore, he said, "Even with the cutting of the forests and the occupancy of the land, it has maintained a foothold in all available places in the settled portions of the state. Nearly every considerable bit of woodland in the southern part of Minnesota, with anything like wild conditions, still has its breeding Grouse, and scattered pairs may be found even in the environs of our larger cities." Roberts (1932) documented confirmed breeding from 1877 to 1924 in Anoka, Becker, Hennepin, Isanti, Marshall, Morrison, Polk, Stearns, and Wright Counties, as well as at Itasca State Park.

He commented on the "lean" and "fat" periods of its population cycles, which occurred as far back as 1879 without any "special connection with hunting." He devoted considerable text to the possibility that cycles were related to tularemia (*Francisella tularensis*), a common disease associated with the snowshoe hare. More recent evidence disputes the role of disease in the mortality of wild populations.

Forty years later and in contrast to Roberts's optimistic viewpoint of the Ruffed Grouse's occupancy throughout woodlands of the state, Green and Janssen (1975) limited its range to the "forested portions of the state." They outlined a distribution that eliminated the southern and southwestern counties west of a line from central Fillmore County to the Twin Cities metropolitan area, west to Swift County and north to central Kittson County. Most of the western tier of counties in Minnesota were not included in the Ruffed Grouse's distribution. Green and Janssen stated that there had been no reports of the species from Hennepin County for twenty-five years. They documented confirmed nesting in 31 counties, all within forested areas of the state.

Janssen (1987) described a Ruffed Grouse distribution similar to that reported by Green and Janssen (1975) but with potential extensions west into Rice County (Nerstrand Big Woods State Park) and eastern Steele County. He also reported a reintroduction of the species into the Three Rivers Park District (formerly Hennepin County Park District) in 1972, which produced some success, as documented by Sivets (1975) in Morris T. Baker County Park in 1974 and by Gillette (1976) in Elm Creek, Baker, and Crow-Hassan Park Reserves from 1974 to 1976. Janssen reported confirmed nesting records in 36 counties since 1970. The counties encompassed the same general range as Green and Janssen's (1975) distribution but extended the western boundary to include eastern Clay and eastern Norman Counties as well as south to northeastern Grant County. Hertzel and Janssen (1998) reported 42 counties with confirmed nesting records since 1970, when they added Fillmore, Kandiyohi, Pennington, Roseau, and Winona Counties to Janssen's 1987 list.

The MBS reported 304 breeding-season locations, which were consistent with the ranges described by both Green and Janssen (1975) and Janssen (1987), except for one location in Nicollet County in southern Minnesota. Locations were numerous in many counties, including in Aitkin, southern Beltrami, southern Cass, Clearwater, Cook, Houston, Itasca, northern Lake, Marshall, northwestern Morrison, Otter Tail, Pine, northern St. Louis, and Winona Counties.

The participants in the MNBBA reported 2,239 records, which corresponded well with what had been previously described by Green and Janssen (1975) and Janssen (1987) **(A)**. Ruffed Grouse breeding evidence was found in 26.0 percent (1,272/4,899) of all blocks surveyed. By far, most of the breeding evidence records were of probable nesting within 51.2 percent (651 of 1,272) of all blocks with records. Confirmed nesting was found in 227 blocks and 32 counties, including two counties (Chisago and Todd) that were not previously reported by Hertzel and Janssen (1998). Breeding records were particularly dense in the northeastern and northern regions of the state, basically north from northern Mille Lacs County to central Becker County and east from eastern Marshall and eastern Roseau Counties. Detections were lacking from the Twin Cities area, including Carver, Hennepin, Ramsey, and Wright Counties.

The landcover suitability map for the Ruffed Grouse illustrates that the most suitable habitat for the species is northeastern Minnesota, with suitability decreasing to the west and to the south **(B)**. Sparse populations are predicted in southeastern Minnesota and in extreme northwestern Minnesota.

Changes in the Minnesota distribution of the Ruffed Grouse have primarily been retractions from the southwest, south, and western margins of its former range reported by Roberts (1932). In their review of the Ruffed Grouse in North America, Rusch and colleagues (2020) said these retractions were widespread throughout many areas, including riparian woodlands of the Dakotas, eastern Iowa, eastern Kansas, and Nebraska. Further, Ruffed Grouse once occupied all of Ohio, Indiana, Illinois, Missouri, northwest Arkansas, western Tennessee, western Kentucky, eastern Maryland, and eastern Virginia. In many of these areas, Ruffed Grouse have been successfully reintroduced (Rusch et al. 2020). Roberts (1932) reported individuals or populations being maintained in fragmented woodlands in the southwest and in the Twin Cities metropolitan area, often following reintroductions.

Historical changes in the Ruffed Grouse's distribution were largely due to land clearing for agriculture and settlements, the consequent reduction in contiguous forest habitat, and market hunting (e.g., Rusch et al. 2020). These land use changes in the late 1800s and early 1900s resulted in reduced forest area in southern areas of both Minnesota and Wisconsin. Increased logging in the northern forests, formerly dominated by conifers and northern hardwoods, were regenerated with increased aspen and birch cover types, which provided a better food source for Grouse. Similar retractions of the species' range from the southern portions of Ontario (Cadman et al. 1987) and Michigan (R. Brewer et al. 1991) have been documented.

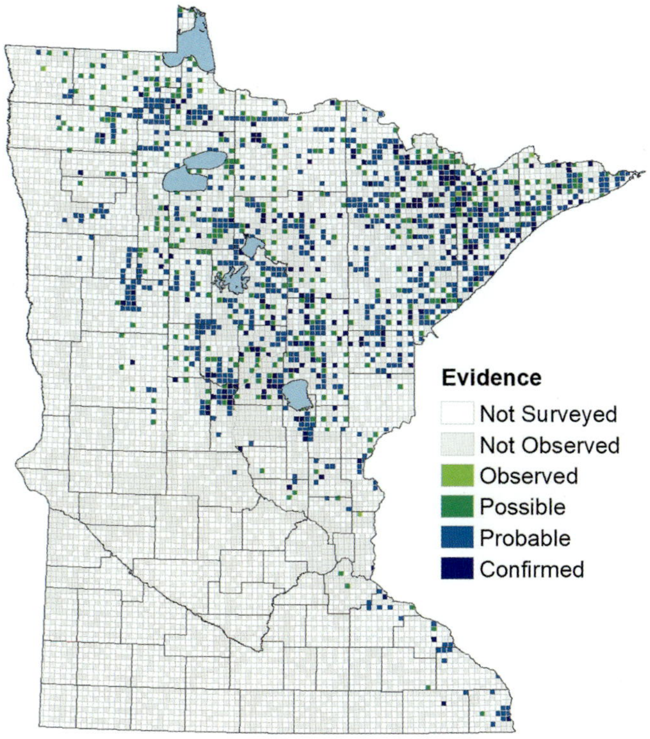

A. Breeding distribution of the Ruffed Grouse in Minnesota based on the Breeding Bird Atlas (2009–13).

BREEDING HABITAT

Substantial literature has accumulated on the breeding habitat for the Ruffed Grouse. Rusch and colleagues (2020) summarized that the highest densities in the midwestern states are in aspen-dominated forests with a mixture of young and older forests that provide cover and food. Vegetation structure around nest sites and drumming sites are generally described as hardwood or aspen stands with open understories that provide good visibility to detect ground predators, and dense overstory cover for protection from aerial predators (Rusch et al. 2020).

In Minnesota, Ruffed Grouse were also found in mixed deciduous–coniferous forests but in lower densities (Gullion and Marshall 1968; Gullion and Alm 1983). The MNDNR (2019j) identified that young to middle-aged aspen forests were the best habitat, and that alder lowlands and patches of gray dogwood were attractive in the summer and fall.

Gullion (1970) quantified drumming sites in Minnesota. Densities of understory stems were high and ranged from 30,000 to 33,000 stems per ha.

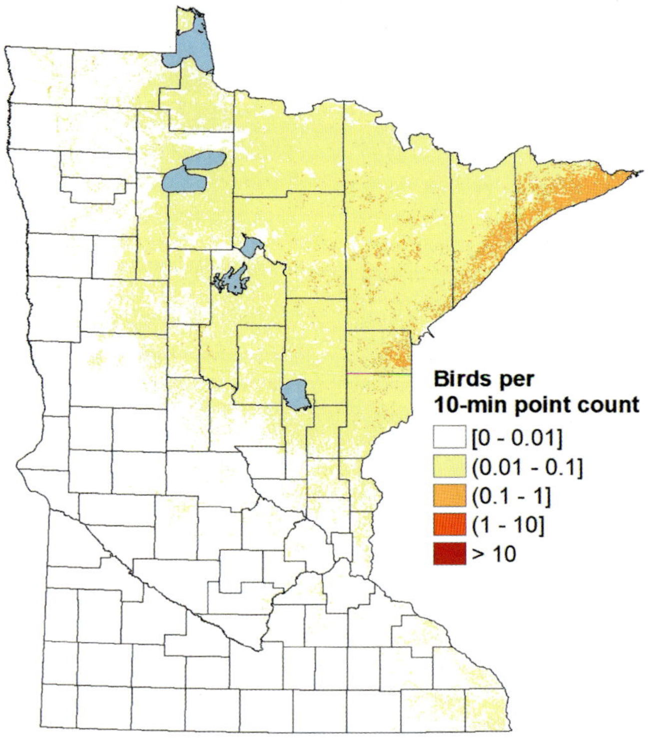

B. Predicted breeding distribution (birds detected per 10-minute point count) of the Ruffed Grouse in Minnesota based on habitat, landscape context, and climate data gathered during the Minnesota Breeding Bird Atlas (2009–13) using the Generalized Linear Modeling method without an adjustment for detectability.

Drumming logs were typically 20 cm in diameter and greater than or equal to 2 m in length (Gullion 1967). There were generally one to two logs per territory. Other objects, such as large boulders, mounds of dirt, exposed tree roots, and other structures that elevate Grouse above the ground were also used. Rusch and colleagues (2020) concluded that the lack of drumming logs or platforms was not likely to limit the distribution of Grouse.

Green (1995) discussed the Ruffed Grouse's need for edges and maintained that labeling it as an "edge species" was erroneous and the result of poor definition and misunderstanding. She quoted the following from Gullion (1984):

> Ruffed grouse have usually been included among the species for whom the development of a great deal of "edge" is supposedly beneficial. However, when we take a closer look at what happens at the edges where ruffed grouse concentrate, and why the birds are there, the story takes a different slant. Actually, extensive use of forest edges by ruffed grouse provides the best indication of how unsatisfactory a forest habitat has become for these birds. When grouse must depend upon edges to find the resources they need it means that the rest of the forest is deficient in those resources, and the quality of the habitat has deteriorated to such a state that only a small portion remains acceptable.

Rusch and colleagues (2020) also pointed out that large, contiguous blocks of upland aspen and oak forest are preferred to smaller, isolated, or fragmented woodlots surrounding agricultural fields. The effect of landscape fragmentation on Ruffed Grouse populations was also raised by North and Faber (2014). In their evaluation of the effect of the Twin Cities metropolitan area on regional distributions of breeding birds, they noted the lack of Ruffed Grouse populations in the metropolitan area. They were concerned that the region might serve as a bottleneck to dispersing Ruffed Grouse and prohibit gene flow between populations in the north and those in southeastern Minnesota. Small and colleagues (1991) had previously raised this issue about Ruffed Grouse and fragmented habitat in Wisconsin based on a study of 381 radio-marked grouse. They found that immigration from private lands was important to maintain populations of grouse on public lands, because mortality from both predation and hunting were high on public lands but only predation was high on private lands. Fragmented landscapes and the Ruffed Grouse's low ability to disperse could increasingly have substantial effects on populations by reducing colonization.

Habitat analysis in the Chippewa and Superior National Forests of Minnesota indicated that the species was found in many forest cover types during the breeding season, including aspen–birch, aspen–spruce–fir, lowland hardwoods, mixed swamp conifers, and upland hardwoods (Niemi, Howe, et al. 2016) **(C)**. The analysis was primarily based on drumming counts but pointed out the species' wide breeding habitat distribution and used a large, probability-based sample (n = 1,128; detections at 486 sample points).

POPULATION ABUNDANCE

PIF (2021) estimated a United States and Canadian population of 18 million adults but did not provide an estimate for Minnesota. Rosenberg, Kennedy, and colleagues (2016) estimated that the population had increased by 31 percent from 1970 to 2014. The North American BBS (Sauer et al. 2020) was unreliable for detecting population trends, because most counts are completed after the primary drumming season is over.

The MNDNR has completed spring roadside Ruffed Grouse drumming counts in Minnesota since 1949. Routes consist of 10 four-minute listening stops at 1.6-km intervals. The counts are completed in April or May. Statewide trends have been relatively stable from 1949 to 2020. Peaks generally occur about every ten years, with the last one in 2017 (Roy 2020c) **(D)**. The population index, as measured by drums per stop, for the Ruffed Grouse in southeastern Minnesota has been relatively low since 1991. The population cycles are less pronounced southward from the northeastern region, where they are most evident, and they are nonexistent in the southeast (Bergerud 1988; Roy 2020c). The annual harvest of Ruffed Grouse in Minnesota averages slightly more than 500,000 birds (range of 150,000 to 1.4 million) (Roy 2016b), but recent estimated harvest levels from 2009–10 to 2019–20 ranged from 195,515 to 465,580 (Davros and Dexter 2019).

Rusch and colleagues (2020) stated that average recorded densities are about three drumming males or nine adults per 40 ha in spring but suggested that these are overestimates because studies have been completed primarily in good habitat. They also surmised that densities during the peak of the cycles in Minnesota and Wisconsin were likely about 20

C. A habitat frequently used by Ruffed Grouse in a mixed deciduous–coniferous forest of northeastern Minnesota. This habitat had extensive balsam fir cover and was surrounded by extensive stands of aspen, birch, and alder. Photo by Gerald J. Niemi.

birds per 400 ha. Dessecker and colleagues (2006) estimated male drumming densities by habitat type in the Great Lakes region, which varied from 0.5 drumming males per 40 ha in pine forests to 3.5 drumming males per 40 ha in aspen–birch forest types. Home ranges of the Ruffed Grouse based on telemetry of 13 hens in Minnesota ranged from 2 to 14 ha (Maxson 1989).

CONSERVATION

Overall, there are limited conservation concerns for the Ruffed Grouse because it still has a relatively large population, wide distribution, and a stable or possibly increasing population in North America. Management efforts for Ruffed Grouse have chiefly focused on maintaining or improving habitat conditions to ensure a reasonable population for hunting. Strategic efforts to improve populations of the Ruffed Grouse have been developed by the Ruffed Grouse Society (Dessecker et al. 2006) and specifically for Bird Conservation Region 12, which includes Minnesota. The plan calls for restoring its population to levels observed in 1980 by 2025. The year 1980 was selected because it represents a point in time when population levels were considered *normal.* Prior to this time, in the early to mid-twentieth century, populations were considered to be higher than historically normal because of the creation of excessive early-successional vegetation and abandonment of farms during the early 1900s.

Regulated hunting has not been considered to be a problem in maintaining population levels of the species (Rusch et al. 2020). Historically, however, hunting did decrease populations in the late 1800s and early 1900s. In Minnesota, Hatch (1892) described the Ruffed Grouse as "mercilessly slaughtered by the pot-hunters," and he wrote that "their [Ruffed Grouse] 'glorious day is passing away' as fast as about 300 dogs and 700 double-barreled breech-loading shotguns can accomplish their annihilation." Schorger (1945) in Wisconsin described that shooting of Grouse for subsistence and sale was common through the early 1900s. For instance, 600 Grouse per day were shipped to markets in Milwaukee in 1898.

Considerable land is available for hunting Ruffed Grouse in Minnesota, and many conservation efforts have been implemented to maintain population levels. The MNDNR (2019g) states that there are 4.5 million ha (11 million acres) of public hunting land, 528 designated hunting areas covering nearly 405,000 ha (1 million acres), more than 40 designated Ruffed Grouse management areas, and 1,000 km (600 miles) of hunter walking trails. The Ruffed Grouse Society identified more than 100 management areas and special projects dedicated to Ruffed Grouse restoration and management between 1985 and 2010. Clearly, many of these areas provide extensive benefits to many other wildlife species as well.

Rusch and colleagues (2020) summarized the major issues for the future of the Ruffed Grouse as the following: maturation of the forests, especially in the eastern United States; forest fragmentation and lack of connectivity among forest tracts; management of hunting of small, isolated populations of Ruffed Grouse; future effects of pesticide use in agricultural regions and in forestry; and the need for an improved understanding of the genetic structure of Ruffed Grouse populations.

In 2020, when Rusch and colleagues published their review, no mention was made of climate change. NAS (2019), in their review on the effects of a 2°C change in climate in North America, identified the Ruffed Grouse as having low vulnerability, because 34 percent of its range would be gained farther north in Canada and Alaska but lost in its southern fringes by 2050. However, in Minnesota the species was predicted to be extirpated by 2080 under a 3°C warming scenario.

Like so many other bird species, the Ruffed Grouse has an unclear future in the long term in Minnesota. Furtman (1999) suggested its future is bright and that "Enlightened management of forests will, if we can keep our own population in check, ensure healthy habitats within which it can prosper." Yet his optimistic forecast was made prior to our understanding of climate change, especially in terms of how it may affect Minnesota. Indeed, its future will depend on many decisions that society makes, including the use of natural resources.

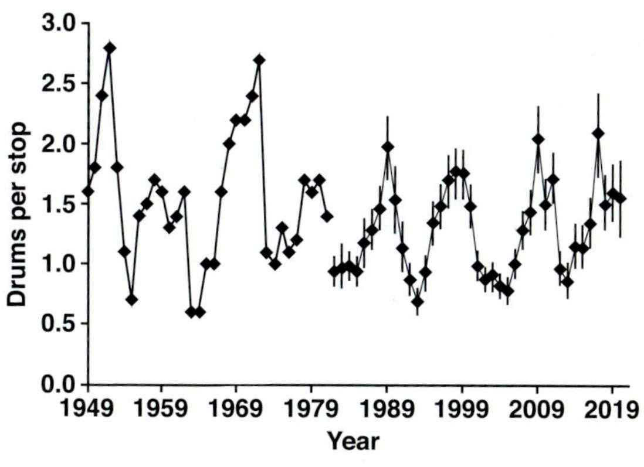

D. Statewide population index values for Ruffed Grouse in Minnesota from 1949 to 2020 (95 percent bootstrapped confidence intervals have been used since 1981; different methods were used prior to 1981; Roy 2020c, fig. 3).

SPRUCE GROUSE

Canachites canadensis

Minnesota Seasonal Status: Permanent resident. The Spruce Grouse was rare during the MNBBA.
Migration: Primarily sedentary but known to move short distances between breeding and wintering areas.
Food: Primarily conifer needles; some seeds, arthropods, and fruit depending on the season; forages in trees and on the ground.
Nest: On the ground, often in depressions of *Sphagnum* moss at the base of a tree.
Conservation Concern: A game species; assigned a low PIF Continental Concern Score of 7/20.

Photo by David Brislance.

The Spruce Grouse is a rare grouse restricted to the northern coniferous forests of Minnesota. Forbush and May (1955) vividly say it occurs "where the thick carpet of green sphagnum moss deadens every footfall, where tiny-leaved vinelets radiate over their mossy beds, there we may find this wild bird as tame as a barnyard fowl." The Spruce Grouse population has declined as human populations have encroached on its coniferous forest habitat. As Roberts (1932) concluded almost one hundred years ago, "its disappearance was undoubtedly [due to] the fact that it is naturally such a tame and indifferent bird in the face of danger that it was easily killed by whatever means happened to be at hand, even a club or a stone serving in the absence of a gun." This unwary behavior has provided it with an alternative common name, "Fools Hen."

The Spruce Grouse male has a dark plumage with a black throat, breast, and tail. The female superficially may be similar to a Ruffed Grouse, but it has heavy dark barring with white tips on the breast. Compared with the Ruffed Grouse, both sexes are more compact and lack a crest, black ruffs on the neck, and a black terminal band on the tail. Its species name, *canadensis,* simply denotes its primary distribution in Canada.

Spruce Grouse males have an elaborate courtship display and drum like a Ruffed Grouse, but their drumming is not nearly as loud. The drumming is delivered both in a standing position and often in flight. Robinson (1980) noted its drumming is heard for only about 100 m while the Ruffed Grouse can be heard for a much longer distance.

Spruce Grouse are widely distributed across the forested areas of the boreal forest in Canada from the Maritime Provinces to British Columbia and Alaska. In the continental United States, resident populations extend southward into western Montana, Idaho, Washington, and Oregon. Resident populations are also found in northern Minnesota, Wisconsin, and Michigan, and the northern portions of the Northeast, including Maine and New Hampshire. It is not found abundantly anywhere within its range.

MINNESOTA BREEDING DISTRIBUTION

Hatch (1892) stated that "Canada Grouse," as he knew the species, were "permanent residents of the northern half of the State, scattering individuals reaching a south line of its habitat about sixty miles northeast of Minneapolis." Roberts (1932) was more descriptive and reported the Spruce Grouse was found throughout the coniferous forest region of the state as far south as Mille Lacs and Carlton Counties, Wadena County in central Minnesota, and to eastern Marshall County in the northwest. He noted it was formerly "common, perhaps abundant, throughout the evergreen forests of the state."

Roberts referred to notes made from 1865 to 1868 by Dr. Frederick Washburn, who was a professor at the University of Minnesota and state entomologist from 1902 to 1918, who stated the bird was quite common about Mille Lacs Lake, where observers were constantly running across coveys of six or more on walks. They nested in mossy hummocks among the spruce. He also cited observations from Thomas Martin Trippe, whom Roberts credits with the "first real report on Minnesota birds" in 1871. Trippe said "it was abundant and breeding in the central part of the state from Carlton County west to the Mississippi River in 1870" (Roberts 1932). However, the human population in the state continued to expand, and Roberts labeled it a "disappearing bird" by the early 1900s. He stated that, as of 1932, it "survived in numbers only in the wild, largely uninhabited region south of the International Boundary from Lake of the Woods to Lake Superior, which includes the Superior National Forest."

Besides the observations of nesting in the Mille Lacs area in the 1800s, Roberts (1932) reported confirmed nesting only in Cook, Lake, and Roseau Counties; all were observations of hens with chicks or broods. In Roberts's 1936 revision of his treatise on Minnesota birds, he stated that he felt compelled "to place on record all definite information in regard to this steadily disappearing species." He included several additional observations from Lake of the Woods County, such as a survey made by Gustav Swanson in 1932 near International Falls on the northern border with Canada. Unfortunately, Swanson did not encounter a single Spruce Grouse but was told that the species had been seen near Ericsburg, Koochiching County, and that they were fairly common in the Northwest Angle of Lake of the Woods County. Roberts did add nesting activity from a brood of young that were just hatched near Ely, Minnesota, on July 12, 1933.

Years later, Green and Janssen (1975) reinforced the fact that the Spruce Grouse was a permanent resident in the northeastern and north-central regions of the state. They reported confirmed nesting in Hubbard, Itasca, and Lake of the Woods Counties, and in those previously reported by Roberts (1932, 1936). They also noted populations of the species that had been observed between 1960 and 1972 in northern Hubbard County. Janssen (1987) identified the species' primary distribution as northeastern and northern Minnesota, with observations from as far west as Roseau County and confirmed nesting to Hubbard and Beltrami Counties. He reported confirmed

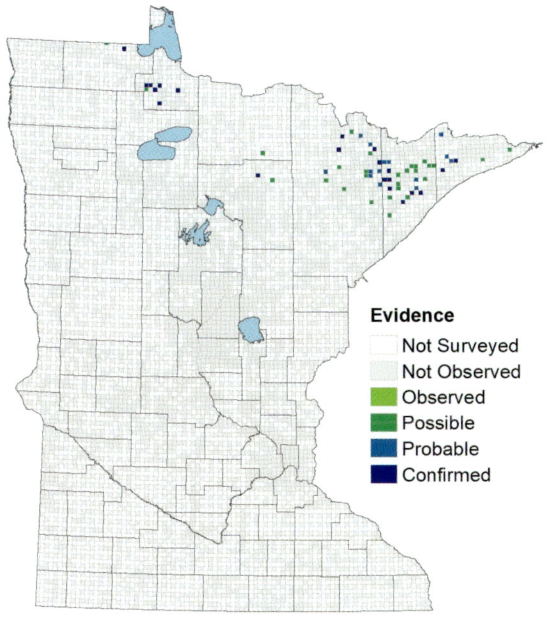

A. Breeding distribution of the Spruce Grouse in Minnesota based on the Breeding Bird Atlas (2009–13).

nesting in eight counties since 1970; besides the two aforementioned confirmed nesting records, others were reported from Lake of the Woods and northern Cook, Itasca, Koochiching, Lake, and St. Louis Counties. Later, Hertzel and Janssen (1998) added Roseau County but excluded Hubbard County from the list of confirmed nesting since 1970. MBS reported 17 breeding-season locations of the Spruce Grouse, including multiple locations in Cook, Lake, and St. Louis Counties and one location in Roseau County.

The MNBBA included 87 records of the Spruce Grouse, including 21 blocks with confirmed nesting **(A)**. These records included only 1.3 percent (63/4,735) of all surveyed blocks. Confirmed nesting was from six counties, all of which were previously identified by Janssen (1987) and Hertzel and Janssen (1998). No observations were made from the potential extant population noted by Janssen from northern Hubbard County. The records reported were densest in northern St. Louis County and central Lake County. One record was included from southern Koochiching County.

Roberts (1932) clearly documented that Spruce Grouse were formerly found as far south as the Mille Lacs Lake region. Therefore, based on the most recent distributional data, its range in Minnesota has retracted considerably over the past one hundred years, but has stabilized in the northern Minnesota counties from Cook to Roseau County and south to central St. Louis County and northern Itasca County. Zumeta and Cincotta (1982) identified a female Spruce Grouse in Pine County, but since that time, no other observations have been noted from that region.

Stenlund and Magnus (1951) attribute the substantial reduction in the Spruce Grouse distribution to intense logging and the many forest fires that occurred from 1890 to 1915. They argued that the recovery of the species began in the early 1940s, when regenerating forests became mature enough to support understories of evergreen trees, such as jack pine, spruce, and balsam fir. The historical range retractions of Spruce Grouse from the early 1900s have been noted in Michigan, Ontario, and Wisconsin (Cadman et al. 1987; Cutright et al. 2006; Chartier et al. 2013). Robinson (1980) stated that the population and range began to decline around 1912 in Michigan, but since 1940 the species has been maintaining its numbers and even increasing in some parts of the Upper Peninsula. The range retractions and decline in populations were largely attributed to hunting, logging, forest fires, and human settlement. Hence, the timing of distributional changes in Michigan closely parallel those observed in Minnesota.

Despite the dramatic changes that have historically occurred in the Spruce Grouse distribution in the upper Midwest, the recent revision of the Spruce Grouse's status in North America only vaguely mentions a 1943 publication by Scott, that "In the upper Midwestern states such as Wisconsin, the range of Spruce Grouse has retracted to the north substantially" (Schroeder et al. 2020).

Roy, Giudice, and Scharenbroich (2017), in their development of methodology to better survey Spruce Grouse, found the species relatively rare in the Northern Minnesota Drift and Lake Plains Section, and more abundant in the Northern Superior Uplands and the Northern Minnesota and Ontario Peatlands Sections. Generally, relative abundance of the Spruce Grouse increased from the southwest to the northeast in the north-central and northeastern regions of Minnesota.

BREEDING HABITAT

Spruce Grouse are primarily found in short-needled conifer-dominated forests and not always in spruce (Williamson et al. 2008). Besides spruce, they commonly can be found in forests dominated by pine, balsam fir, or cedar **(B)**. Schroeder and colleagues (2020) elaborate extensively on the wide variation of habitats used by Spruce Grouse across their breeding range and emphasize their association with regenerating areas following forest fires, especially among jack pine. They also suggest that Spruce Grouse appear to prefer relatively young conifer stands over more mature stands where forests are maintained by fire, insects, and some forest practices. Yet they also acknowledge many examples for variations of this habitat preference, such as the use of old-growth forests and bogs, use of upland and lowland conifer forests, and avoidance of clear-cuts and second growth.

The descriptions of breeding habitat used by Spruce Grouse in Minnesota are also confusing. Roberts (1932) included a long and detailed but eloquent description of the "Courting of the Spruce Grouse" abstracted from notes by his protégé, Walter Breckenridge. He described the breeding habitat as "spruce swamps among the Laurentian hills that lie along the Canadian boundary lakes." Furthermore, he noted "typical black spruce and cedar swamp flora with its ground-covering of sphagnum, Labrador tea, leatherleaf, and associated northern plants" and that "rocky outcroppings in and about the swamp were covered largely with spruce and jack-pine." Stenlund and Magnus (1951) emphasized jack pine, followed by black spruce, balsam, and tamarack, in that order, in their description of Spruce Grouse habitat. In contrast, L. Anderson (1973) and Haas (1974) both described breeding habitat as black spruce forest types. Eckert (2017) stated the species was "more associated with jack pines, at least in this part of the continent." Zlonis and colleagues (2017), in a recent modeling effort, found Spruce Grouse in the Agassiz Lowlands Subsection of northern Minnesota highly associated with large patches of black spruce habitat. Haas (1974) stated the species prefers relatively young, dense successional stands (2,500–3,500 stems per ha), 7–14 m in height, and with a relatively well-developed middle story.

The confusion on what forest types are used by Spruce Grouse was highlighted by Roy, Giudice, and Scharenbroich (2017) in their study of the species in Minnesota. They recognized the conflicts in the

B. Spruce Grouse in Minnesota prefer dense conifer forests, especially those with spruce or pine trees. This picture illustrates a black spruce forest with extensive cover of Labrador tea (*Ledum groenlandicum*) and *Sphagnum* moss. Photo by Gerald J. Niemi.

literature describing Spruce Grouse breeding habitat in Minnesota, especially forest ages used by the species. Therefore, when they began their study in 2014, they initially included jack pine, black spruce, tamarack, and white cedar and age classes within all stands. White cedar was added because of specific interest by managers in this forest cover type. However, during their investigations they also noted detections in forest cover types dominated by other coniferous tree species, including balsam fir, red pine, and white spruce. Hence, their research has broadened to identify the relative use of all conifer forest cover types, and their future research will be of considerable importance in deciphering the breeding habitat used by this elusive species.

In Michigan, Chartier and colleagues (2013) emphasized that landscape heterogeneity is important to Spruce Grouse because they historically were found in a disturbance matrix created by insect pests like spruce budworm and forest fire. They recommended that conservation efforts be directed toward maintaining coniferous habitats affected by natural disturbances.

POPULATION ABUNDANCE

Based on the Spruce Grouse Continental Conservation Plan (Williamson et al. 2008), PIF (2021) estimated a United States and Canadian population of 11 million breeding adults. Williamson and colleagues (2008) also provided a continental population range of 5 to 15 million individuals that was extrapolated from forest inventory data, total area of potential habitat types, and from the literature of densities by habitat type. From these data, they estimated the population range of the Spruce Grouse in Minnesota was between 22,022 and 88,090 individuals, which was the second highest state estimate in the eastern and midwestern United States. Maine had the largest population. Environment Canada (2014b) estimated an even wider population that ranged from 5 to 50 million adults for Canada, where the majority of the population exists. Dexter (2017) and Davros and Dexter (2019) recorded that between 7,081 and 26,568 Spruce Grouse were harvested each year from 2006–7 to 2019–20. The higher harvest potentially would represent a high proportion of the species population in the state. This compares with an average of about 500,000 Ruffed Grouse harvested each year.

Population trends of Spruce Grouse are difficult to measure because they are secretive and found in remote areas. Based on a combination of Christmas Bird Counts and the BBS, Environment Canada (2014b) suggested that Spruce Grouse populations in Canada have increased since 1970. Rosenberg, Kennedy, and colleagues (2016) estimated a greater than 200 percent increase in the Spruce Grouse population in the United States and Canada from 1970 to 2014. These estimates are generally based on the BBS roadside counts, but they are not reported by Sauer and colleagues (2020), so it is unclear what the basis is for this estimated substantial increase in the Spruce Grouse population. Reasons for population increases are unknown, but Spruce Grouse populations are primarily impacted by mortality due to depredation of adults, young, and eggs; human harvest where the species is legally or illegally hunted; and the amount of available habitat, which is interconnected with forest management, land ownership, and the extent and frequency of natural disturbances (Schroeder et al. 2020).

Schroeder and colleagues (2020) reported population densities ranging from 0 to 83 birds per 100 ha in Ontario and from 0 to 50 birds per 100 ha in Alaska, but they state that there are many reports of densities of fewer than 10 birds per 100 ha. Hence, populations can vary considerably, but they generally are quite low, and the species is hard to census because they seldom vocalize, nor do they flush easily. An example of the low density of the species in potentially good habitat of northern Minnesota was described by Abraham (2014) in a study of Spruce Grouse in the Red Lake WMA in extreme northern Minnesota. In 2012, only 32 Spruce Grouse were counted in 1,314 points in 83 randomly selected plots. Audio playbacks, line transects with dogs, and fecal droppings have been used to estimate population sizes (Schroeder et al. 2020). Roy, Giudice, and Scharenbroich (2020a) evaluated the use of playbacks of the Spruce Grouse *cantus* call, nasal clucks given by females on a perch, and fecal pellet counts to provide reliable information on population status, trends, and distribution in Minnesota. They found that fecal pellet counts provided more detections and were easier to implement than the *cantus* calls. The MNDNR has initiated the fecal pellet count protocol from 2018 to 2020 and expects the survey to detect meaningful changes over a ten-year period (Roy, Giudice, and Scharenbroich 2020b).

CONSERVATION

Overall, the Spruce Grouse is not considered to be a species of concern in North America because of its large population and its wide distribution across the Canadian boreal forest. It is considered the most widely distributed grouse species in North America (Robinson 1980). The MNDNR (2015b) classified it as a *Species in Greatest Conservation Need* primarily because of the loss of coniferous forests and their fragmentation in Minnesota. Environment Canada (2014b) considered the species to be secure in Canada. However, climate change may pose a substantial threat to this species.

The Spruce Grouse has been protected in Wisconsin since 1929. In Michigan, it has not been hunted since 1914. In the Superior National Forest, the species' primary habitats of spruce–fir and jack pine were considered abundant. Gregg and colleagues (2004) noted that the Superior National Forest management plan focuses on a range of natural variation in the landscape, better range of age classes, and larger harvest units in the forest, which should provide better habitat in the future.

Schroeder and colleagues (2020) emphasized that advanced succession due to the loss of disturbances such as fire, insects, or appropriate forest management can lead to long-term declines in Spruce Grouse populations. In addition, conversion of conifer-dominated forests to other stand types or open clear-cuts and single-species plantations can eliminate Spruce Grouse populations entirely. Szuba and Bendell (1983) suggested that forest practices can be beneficial to Spruce Grouse by mimicking fire and by keeping clear-cut areas small and interspersed with optimal habitat.

Environment Canada (2014b) has identified a number of Bird Conservation Regional Strategies, many of which provide specific benefits for Spruce Grouse. Similarly, the MNDNR (2015b), in their State Wildlife Action Plan, have identified many Conservation Focus Areas that have the potential to provide improved habitat conditions for the species.

Even though it has not recovered much of its former range in Minnesota or in many other areas of North America, by most accounts the Spruce Grouse is currently secure and its population is increasing. There is some optimism that this emblematic species of the dense north woods has a reasonable future in Minnesota. Yet the reduction in coniferous forests and a reduced optimal thermal regime in the state with future climate change renders its future more precarious. NAS (2019) predicted that 63 percent of the species' breeding range will be lost under a 2°C warming scenario of climate change by 2050. They consider the species as highly vulnerable in North America. In Minnesota, NAS (2019) forecasts that the species will be extirpated from the state by 2050 even under our imminent 1.5°C warming scenario. Dr. Roberts's (1932) prognosis that the species "will inevitably disappear before many years have passed" fortunately has not yet come to pass, but future predicted changes in land use and climate have certainly increased the odds of this diagnosis.

SHARP-TAILED GROUSE

Tympanuchus phasianellus

Minnesota Seasonal Status: Permanent resident. The Sharp-tailed Grouse was an uncommon species during the MNBBA.
Migration: Nonmigratory.
Food: Grains, seeds, buds, foliage, fruit, and insects, especially grasshoppers; variable depending on season and availability.
Nest: Sparsely lined depression on the ground, usually concealed with overhanging vegetation.
Conservation Concern: A game species; assigned a moderate PIF Continental Concern Score of 10/20.

The Sharp-tailed Grouse, formerly known as the Prairie Sharp-tailed Grouse, is the only indigenous prairie grouse in Minnesota (Roberts 1932; Berg 1997). Historically, there were seven known subspecies in North America, and the Minnesota subspecies had been designated the "prairie race" (*Tympanuchus phasianellus campestris*) (Hoffman and Thomas 2007; Connelly et al. 2020). Recent genetic evidence suggests that many of the formerly named subspecies have been found to be similar, with little differentiation between the Sharptails of the Great Plains (*T. p. jamesi*) and Minnesota's *T. p. campestris* (Spaulding et al. 2006). Berg (1997) emphasized that even though it is called the prairie race, the species occupies grass, brush, savanna, and boreal peatlands in Minnesota.

The Sharp-tailed Grouse is a lekking species where the males gather in a small arena during the spring to outshine other males and court females with elaborate plumages and display behaviors. The number of males gathered at a lek can vary considerably; reported numbers range from a pair of males up to 57. The territory size of their leks, though relatively small, has ranged from 46 to 258 square m (Connelly et al. 2020).

The Sharp-tailed Grouse does indeed have a pointed *sharp-tail,* which is diagnostic in distinguishing the species from the Greater Prairie-Chicken. Sharptails also have scaled and spotted underparts, while the Prairie-Chicken is barred. The genus name of the species, *Tympanuchus,* derives from the Greek for "drum bearing," which refers to two air sacs on either side of the neck that provide a booming sound heard during courtship. Its species name, *phasianellus,* is Greek for "little

Photo by David Brislance.

pheasant," because of its similarity to the larger Ring-necked Pheasant.

The Sharp-tailed Grouse is primarily distributed in the northern Great Plains. Its distribution spans from Nebraska and Colorado north to Saskatchewan and Alberta, west to Montana and Idaho, and east to Michigan and Ontario. Populations also exist in central Alaska, the southwestern Yukon Territory, parts of British Columbia, and northeastern Washington State. The highest densities are found in eastern Montana.

MINNESOTA BREEDING DISTRIBUTION

Roberts (1932) described its historical distribution as "the prairies in the summer and . . . the brushlands and open forests in the winter, and the early explorers and first settlers found it abundant everywhere." In his summary of the species in Minnesota, Berg (1997) described its presettlement range in Minnesota as "occupied" throughout the state with the exception of western Minnesota, primarily from Pipestone County in the south to Wilkin and Otter Tail Counties in the north.

In the 1930s, Roberts emphasized that it was "now confined largely to the northern part of the state from Pine County northward and westward in cut-over lands, in natural openings, and in brush areas, to the settled prairies of the west-central counties and the Red River Valley." This is largely the same area still occupied by the Sharp-tailed Grouse today, though the population has declined even more within this range.

Roberts (1932) documented confirmed nests with eggs in the Pembina valley on the Red River in 1873 and in Polk County in 1918 and 1920. Since the heading by Roberts's account was labeled "Minnesota Nesting," the Pembina valley breeding activity is assumed to be in Minnesota and not North Dakota.

In 1975, Green and Janssen emphasized the Sharp-tailed Grouse's distribution as primarily in the northern regions and most numerous in the northwestern part of the state in Beltrami, Lake of the Woods, Marshall, and Roseau Counties. Both Roberts (1932) and Green and Janssen (1975) described its distribution as changed after prairies began to be converted to agricultural land and then colonized by the Greater Prairie-Chicken. The Sharp-tailed Grouse retreated to the north. Green and Janssen described confirmed nesting in Aitkin, Lake of the Woods, Pine, Polk, and Roseau Counties, plus a 1942 nesting in Swift County in the west-central part of the state.

Janssen (1987) reinforced a distribution similar to that of Green and Janssen (1975) and noted that the Sharp-tailed Grouse had been recorded in 1968 in Cook County in extreme northeastern Minnesota. He suggested these were "vagrants" from a population in Thunder Bay, Ontario. Janssen documented confirmed nesting in five counties since 1970, including Aitkin, Beltrami, Lake of the Woods, Marshall, and Polk. A few years later, Hertzel and Janssen (1998) would add three more counties with confirmed nesting since 1970: Kittson, Pennington, and Roseau. Since 1998, young chicks were confirmed at the Kroschel WMA, Kanabec County, in May 2006 and flightless young in Big Stone NWR in July 2007 (MOU 2021a).

Berg (1997) summarized the breeding distribution of the Sharp-tailed Grouse in Minnesota as two disjunct populations. One population was highly associated with the Agassiz glacial lake bed in northwest Minnesota and the second with a combination of Aitkin, St. Croix, and Upham glacial lake beds in east-central and northeastern Minnesota. The latter also included the taconite ore tailing basins associated with iron mining operations.

The MBS recorded 55 breeding-season locations during its counts, which began in the late 1980s. These locations are largely consistent with the two disjunct populations identified by Berg (1997). One population in northwest Minnesota had the most locations in Clearwater, Kittson, Marshall, Pennington, Polk, Red Lake, and Roseau Counties. The other population was more sparsely distributed and included locations in Aitkin, Carlton, northeastern Kanabec, and St. Louis Counties.

The participants in the MNBBA reported 566 records for the Sharp-tailed Grouse and again reinforced the identification of two disjunct populations in the state **(A)**. The species was found in 6.9 percent (337/4,857) of the surveyed blocks. A total of 16 confirmed nests were reported, and

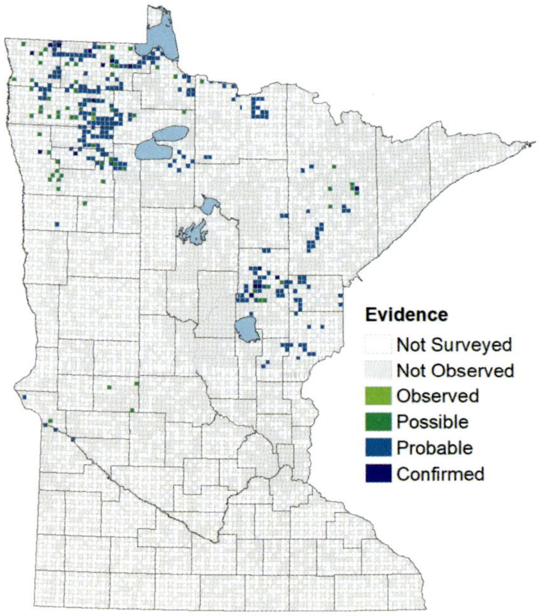

A. Breeding distribution of the Sharp-tailed Grouse in Minnesota based on the Breeding Bird Atlas (2009–13).

probable nesting was included from 253 blocks. A total of seven counties had confirmed nesting records, with new county records for Red Lake and St. Louis Counties, the latter in the former Iron Range mining lands. A new confirmed county nesting record was added in Lac qui Parle County in 2010 (MOU 2021a).

The distribution of the MNBBA records illustrated a triangular area of northwestern Minnesota from Norman County to Koochiching County to Kittson County. The northeastern and east-central population in Minnesota included Aitkin, Carlton, Itasca, Kanabec, Pine, and St. Louis Counties. Several additional probable nesting observations were recorded from Big Stone, Lac qui Parle, and Chippewa Counties as well as possible records from Pope County in west-central Minnesota. No observations were recorded from the outlying counties of Cook and Swift, where the species had been reported earlier. The latter, however, is adjacent to the probable MNBBA records in Big Stone, Lac qui Parle, and Chippewa Counties. Berg (1997) had previously identified the regions including Lac qui Parle, Pope, and Swift Counties as not occupied during presettlement times.

Berg (1997) emphasized that the species occupied less than one-third of its historic range in Minnesota. Similar retractions of the species' range were described by Berg (1990) in Wisconsin and in Michigan. The Wisconsin breeding bird atlas from 1995 to 2000 included 53 records of the Sharp-tailed Grouse (Cutright et al. 2006). The authors emphasized that the populations are still a fraction of what they were historically. Michigan recorded a slight increase in confirmed nesting of the Sharp-tailed Grouse from 13 townships during its 2002–8 breeding bird atlas, compared with 12 townships from its first atlas in 1983–88 (Chartier et al. 2013). They emphasized that the species' range is still considerably smaller than during presettlement times.

In their review of the Sharp-tailed Grouse in North America, Connelly and colleagues (2020) stated that the species originally occupied twenty-one states and eight Canadian provinces. They identified extirpations of the species from the early 1900s to 1970 from California, Illinois, Iowa, Kansas, Nevada, New Mexico, and Oregon. The species' distribution has also been greatly reduced in the southern regions of its range as well as in British Columbia, Colorado, and Utah. The primary reasons given by Connelly and colleagues were the conversion of native ranges to agricultural fields and habitat loss.

BREEDING HABITAT

The haunts of the Sharp-tailed Grouse are broadly described as grasslands, brushlands, savannas, barrens, parklands, open and shrubby peatlands, cutover and burned-over areas, and large disturbed open areas such as from mining operations (Roberts 1932; H. C. Hanson 1953; Berg 1990, 1997). The species is typically found in large, contiguous open areas in these habitat types **(B)**. Niemuth and Boyce (2000) suggested that aggregating clear-cuts over a large area can provide suitable habitat. As trees begin to grow in open habitats, the species will abandon these areas as breeding habitat. Sharpies use brush, riparian, and forested habitats for cover during winter (Connelly et al. 2020).

B. Sharp-tailed Grouse in the breeding season prefer wide open landscapes without trees, such as this area of grasses, sedges, and low brush in its east-central range in Minnesota. Photo by Gerald J. Niemi.

Because Sharptails are a lekking species, the habitat and landscape context of these sites are critically important. Lek sites have been described as occurring on a slightly elevated knoll and characterized by low, sparse vegetation with limited woody cover within 800 m of the lek site (Connelly et al. 2020). Prose (1987) stated that females require appropriate grassy, brush, or woody vegetation within 3.2 km of the lek site. Johnsgard (1983) suggested that shrub cover is important during the early brood stages. Hanowski, Christian, and Niemi (2000) quantified the landscapes surrounding occupied and unoccupied lek sites in the species' east-central range of Minnesota. They found that active lek sites had more bare ground, emergent aquatic vegetation, and shrubby bogs than random sites. They emphasized the importance of the landscape context (e.g., 3,000 m radius) surrounding lek sites and the negative effects of woody cover, especially trees, in the horizon around these sites. A predictive model estimated that about 8 percent of the species' east-central range in Minnesota had appropriate lek habitat.

Berg (1997) summarized that an optimum mix for nesting and brood habitat included 35 percent grass-legumes, 15 percent cropland, 7 percent sedge marsh, 25 percent lowland brush, and 13 percent young aspen and birch. Svedarsky (1977) also suggested that grassy roadsides in northwest Minnesota could provide suitable nesting habitat for the Sharp-tailed Grouse.

The MNDNR has recently initiated a series of experiments on the differential effects of prescribed fire and mowing on the Sharp-tailed Grouse within its two primary ranges in Minnesota (Roy and Shartell 2017). The response is anticipated to be measured for five to ten years to detect Sharp-tailed Grouse responses to these management treatments.

POPULATION ABUNDANCE

PIF (2021) estimated a Sharp-tailed Grouse population of 760,000 breeding adults in the United States and Canada and a Minnesota population of 44,000 individuals. The species is the third most abundant upland game bird in the state, behind the Ruffed Grouse and Ring-necked Pheasant (Berg 1997). North Dakota estimated a population of 170,000 adults (Dyke et al. 2015). Environment Canada (2011) suggested that the Sharp-tailed Grouse population in Canada ranged from 50,000 to 500,000 adults, with a moderate decrease since 1970.

Trend estimates from the BBS were unreliable for Minnesota and surveywide. However, trend estimates were reliable for the Badlands and Prairies

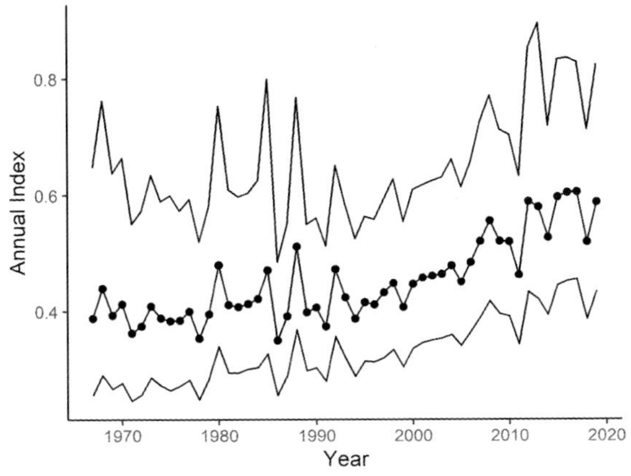

C. Surveywide breeding population trend and 95 percent credible interval (solid black lines) for the Sharp-tailed Grouse based on the 1967–2019 Breeding Bird Survey (Sauer et al. 2020).

BBS region that includes the species' primary breeding range in the Dakotas and eastern Montana. The trend from this region was stable from 1966 to 2019 (Sauer et al. 2020). The species' BBS trend in North Dakota was also credible and stable over the same time frame. Although not significant, the BBS surveywide trend is indicative of the pattern of population change in the Sharp-tailed Grouse in the United States and Canada from 1966 to 2019—a slow but highly variable pattern of increase **(C)**. The high variability renders its overall population trend as inconclusive.

Historically, the Sharp-tailed Grouse was pronounced by early explorers as "abundant everywhere" (Roberts 1932) and "numerous" in their northwestern range (Berg 1997), with flocks "so large that pioneers said they sometimes blocked the sun" (MNDNR 2019k). Berg (1997) described the decline continuing into the 1940s, when mail carrier "surveys in the spring of 1962 found birds in 21 northern counties," with most having 0.9–2.7 Sharptails/100 km of roads in the northwest and 0.1–0.9 Sharptails/100 km of roads in the east-central range. By the late 1960s, adult male densities were estimated to exceed 0.2 birds per square kilometer in only nine northwestern and two east-central counties. Green and Janssen (1975) stated that the Sharp-tailed Grouse reached its lowest population level in the mid-1960s. Spring lek count surveys have been completed by the MNDNR from 1980 to 2018 and, even though the number of leks surveyed each year is not consistent, the average number of Sharptails counted per lek has remained relatively stable at approximately 7 to 13 birds per lek (Roy 2018).

Berg (1990) illustrated the changes in hunter harvests from 1949 to 1988. In 1949, more than 150,000 birds were killed, but in 1965 this total reached a low of fewer than 10,000 birds. This number increased to approximately 50,000 in the late 1970s but has since declined. For example, during the more recent period from 2004–5 to 2019–20, the estimated hunter harvest has ranged from 5,681 to 16,820 (Dexter 2017; Davros and Dexter 2019). Hunter harvests, however, are difficult to compare among years because they are related to many factors, such as the population of Sharp-tailed Grouse, the number of hunters, the number of days hunted, weather conditions, and the length of the hunting season. Regardless, the prevailing evidence illustrates that both Sharp-tailed Grouse populations and hunter harvest rates have substantially declined compared with those in the 1950s and early 1960s, but populations now appear to be relatively stable.

Berg (1990) emphasized the status of the Sharp-tailed Grouse in Minnesota as perilous because of its massive decline since 1963. The primary causes have been the loss of appropriate habitat due to vegetational succession of the grassland-brushland ecosystem, improperly placed conifer plantations, intensive agricultural development, and fire suppression (Berg 1997; Connelly et al. 2020; Cutright et al. 2006). In addition, as an area-sensitive species, Sharp-tailed Grouse need large expanses of suitable habitat that have become fragmented into smaller parcels over time. Roy and Gregory (2019) used genetic analysis to assess the movements of Sharp-tailed Grouse between their two primary disjunct populations in the northwest and east-central portions of Minnesota. They found high connectivity of Sharpies within each of these disjunct populations but limited connectivity between the two. They emphasized a concern with the low population size in the east-central region, which renders the population vulnerable, and stressed the need for an increase in suitable habitat quantity within that region.

CONSERVATION

Even though the species has had a massive decline since the 1800s, the population has stabilized and, as reported above, was estimated to have increased over the past forty-five years. The PIF Science Committee (2013) had identified the Sharp-tailed Grouse as a Stewardship Species of regional concern. Dyke and colleagues (2015) labeled the Sharptail as a Level II Conservation Priority in North Dakota, primarily because of a historical decline in its population and because an estimated one-third of the entire North American Sharp-tailed Grouse population resides in the state. Environment Canada (2011) considered the species to be secure at present.

Under a 2°C warming scenario, NAS (2019) projected low vulnerability for the Sharp-tailed Grouse's breeding distribution in North America, with a 27 percent loss and 27 percent gain of its breeding range by 2050. Losses were predicted primarily in the Great Plains from Nebraska and Wyoming to southwestern Montana, while gains were in the far north in Canada and Alaska. In Minnesota, modest breeding range retractions were forecast along its eastern range from central to northeastern Minnesota, but western and north-central breeding ranges were spared. However, predictions of changes in its winter distribution indicated moderate vulnerability under the 2°C warming scenario, including even greater retractions of its range in central and northeastern Minnesota by 2050, which would further stress an already vulnerable population.

The goal of the Minnesota Sharp-tailed Grouse Management Plan (MNDNR 1990) was to restore the Sharptail to a level that would sustain an annual harvest of 40,000 birds. The plan has been hampered by funding and implementation issues and, therefore, the goal has not been met (Berg 1997). Management of the Sharp-tailed Grouse has also been included as a priority in the Minnesota Prairie Conservation Plan (Minnesota Prairie Plan Working Group 2011). The recent State Wildlife Action Plan (MNDNR 2015b) identified a number of conservation focus areas, such as the Aspen Parklands and Glacial Ridge, that could benefit Sharptails. Roy (2018) emphasized that Sharp-tailed Grouse rely on ongoing management of their habitat—such as prescribed fire, mowing, and shearing—but obstacles include funding, shortages in staffing, equipment, and weather. Moreover, most of these habitat management activities require coordination and permission from private landowners and government agencies, which in themselves can be challenging because of multiple objectives and opinions among these entities.

The habitat and landscape issues for this species are shared with many other species found within its habitat complex, including the Yellow Rail, Short-eared Owl, Bobolink, Eastern and Western Meadowlark, Nelson's Sparrow, and many others. Ultimately, the conservation status of the Sharp-tailed Grouse will depend on a commitment of many organizations to large-scale habitat and landscape management of the grassland, brushland, and peatland ecosystems of the state. Indeed, the future of the Sharp-tailed Grouse in Minnesota is in a very risky state.

GREATER PRAIRIE-CHICKEN

Tympanuchus cupido pinnatus

Minnesota Seasonal Status: Permanent resident. The Greater Prairie-Chicken was an uncommon species during the MNBBA.
Migration: Historically, Prairie-Chickens were short-distance migrants. Roberts (1932) included accounts of large flocks from Minnesota and northern Iowa moving to Missouri for the winter. Females were more likely than males to move large distances, and the movements were more pronounced in severe winters. Today the birds are largely nonmigratory (Svedarsky and Van Amburg 1996). Researchers have suggested that, as numbers declined, there was less competition for food during the winter months. Grasslands also have become more fragmented by croplands, which can provide abundant winter food resources and reduce the need to move south as temperatures fall (J. A. Johnson et al. 2020).
Food: Omnivorous; consumes a wide variety of insects, leaves, fruits, and seeds; waste grains are important during the winter months.
Nest: Ground nest built in a shallow depression.
Conservation Concern: A game species; assigned a high PIF Continental Concern Score of 16/20 and officially listed as *Special Concern* in Minnesota.

Photo by Dan Svedarsky.

No other species is more closely tied to the importance of grassland conservation than the Greater Prairie-Chicken. Its dependence on grassland habitats, coupled with its captivating breeding behavior, has endeared it to resource professionals, amateur birders, and nature enthusiasts throughout the northern Great Plains. Spend the pre-dawn hours of a frosty April morning in a Prairie-Chicken blind, quietly awaiting their appearance on the communal lek, and you will be treated to an unforgettable performance. Roberts (1932) wrote that the Prairie-Chicken's courting display "is one of the most interesting and striking episodes in its life history. In the olden days, when the birds were plentiful, the prairies and open spaces resounded on all sides with these weird, early morning performances."

Nearly identical in appearance, the male and female's brown, heavily barred plumage provides a degree of camouflage for these relatively large, plump birds amid the tall prairie grasses. But on the lek, the male is transformed as he lowers his head, inflates his orange "eyebrows" and air sacs, raises his neck feathers, drops his wings, and struts about with his upturned tail. With a "singularly devilish look" (Chapman 1908), he jerks his head and emits an unmistakable booming call in an effort to entice the onlooking females and deter the competing males. The species Latin name, *Tympanuchus cupido,* is a direct reference to this courtship performance. *Tympanuchus* means "drum-bearing" and alludes to the male's booming; *cupido* refers to his raised neck feathers, which were likened to Cupid's wings.

Formerly found throughout the grasslands of the Great Plains and Upper Midwest, the Greater Prairie-Chicken is now largely restricted to North Dakota, South Dakota, Nebraska, Kansas, and northwestern Minnesota. Small, scattered populations are found farther south and east. Three subspecies are recognized: the Heath Hen (*T. c. cupido*), originally found along the mid-Atlantic coast and now extinct; the Attwater's Prairie-Chicken (*T. c. attwateri*), which is now restricted to the central Gulf coast of Texas and is federally *Endangered*; and the Greater Prairie-Chicken (*T. c. pinnatus*). Although similar in appearance, the smaller Lesser Prairie-Chicken, *T. pallidicinctus,* is an entirely different species that occurs in the southern Great Plains.

MINNESOTA BREEDING DISTRIBUTION

The full extent of the Greater Prairie-Chicken's distribution prior to European settlement is unknown, but it likely included the tallgrass prairies of Illinois and Missouri stretching west to the central plains states of Nebraska, Kansas, and Oklahoma and stretching east to Indiana, Ohio, and Kentucky (J. A. Johnson et al. 2020). Its early status in Minnesota is unclear. Absent in the days of the earliest European explorers, it may have been a rare species along the state's very southern border in the early 1800s (Partch 1973).

What is undisputed is the species' rapid spread across Minnesota as settlers turned the prairie sod into agricultural fields and began clearing the northern forest for its valued timber. Although the Greater Prairie-Chicken would become the symbol of prairie conservation efforts in the state, it is a relative newcomer to Minnesota's western grasslands. The Sharp-tailed Grouse was the state's first truly native grassland grouse, broadly distributed throughout the state in presettlement times. The Sharp-tail's primary habitat is "open grass-brushland dominated by various grasses, sedges (*Carex* spp.), and willows (*Salix* spp.)" (Berg 1997). Even "open bogs associated with glacial lake beds" are used. The Prairie-Chicken, on the other hand, favors more open country and was well suited to the mosaic of small farms, croplands, pastures, and hayfields that followed the plow in the nineteenth century. As agricultural settlements spread across Minnesota's open landscape and Prairie-Chicken populations expanded, the Sharp-tailed Grouse retreated northward.

Documentation of the Prairie-Chicken's undisputed presence in Minnesota begins with Hatch's comprehensive report on Minnesota birds published in 1892. His good acquaintance, the Reverend E. G. Gear, was a chaplain stationed at Fort Snelling beginning in 1839. A natural history buff, the reverend remarked that Prairie-Chickens "were seldom seen at the first, but after the country began to become settled considerably, they increased in numbers perceptibly from year to year" (in Hatch 1892).

From that point forward, the species' progression across the state was well-documented. Advancing approximately 16 km per year (Partch 1973), the birds moved west from Fort Snelling, reaching St. Cloud by 1860 and the North Dakota border near Moorhead by 1879 (Partch 1970; Hoch 2011). Just two years later, Roberts (1932) found them in Kittson County. Although local residents reported the Chickens had only recently appeared in the region, by 1885 they were reported as "abundant" throughout much of the Red River valley.

Their progression into the state's forest region was slower. Nevertheless, logging, fires, wetland drainage, and the establishment of small farms enabled the species to expand into north-central and northeastern Minnesota and into forested areas of Wisconsin, Michigan, and Ontario (Svedarsky et al. 1997; Svedarsky et al. 2000). Prairie-Chickens were established in forest clearings in the Mille Lacs region as early as 1885 and reached Duluth and the North Shore of Lake Superior in the late 1880s or early 1900s. They were observed along the Iron Range in west-central St. Louis County in 1906, and in Itasca County in 1921 (Roberts 1932). By the early 1900s, their range covered nearly 92 percent of the state (Hoch 2011). Yet, as widespread as the species was, Roberts reported confirmed nesting records from only six counties: Grant, Kandiyohi, Kittson, Meeker, Otter Tail, and Pipestone. An adult with a young brood was also reported from Polk County.

The Greater Prairie-Chicken's heyday, however, was short-lived. As Roberts (1932) noted, it is an easy bird to shoot. When flushed, "it flies off in a direct course" and makes an easy target for the hunter's gun. Despite the enactment of hunting regulations as early as 1858, most restrictions were blatantly ignored. Even in the late 1800s, it was not unusual for hunters to take 50 to 100 birds per day (E. Swanson 1940). Roberts (1932) publicly admonished hunters for the excessive take: "Unless . . . sportsmen's organizations become less selfish and more willing to practice instead of vainly preach[ing] conservation, the days of the Prairie Chicken are numbered." In just ten years, the estimated Minnesota harvest declined from 328,914 birds in 1923 to only 29,216 in 1933 (A. B. Erickson and Petraborg 1952). During the 1935 harvest season, Prairie-Chickens "were found to be so scarce that the [Game and Fish] Department appealed to the hunters, by radio and press, to desist from killing the birds" (Roberts 1936). By 1942 the season was closed.

Hunting pressure certainly reduced population numbers significantly, but other changes were underway in the early twentieth century that influenced the Prairie-Chicken's distribution. Most important was the intensification of agricultural practices. New machines made it more efficient to farm large, uniform fields, which replaced the original mosaic of pastures, croplands, and native grasslands that provided ideal year-round habitat. Overall, when croplands comprised 20–30 percent of the landscape, the birds did well, but when croplands exceeded that percentage, they declined (Svedarsky et al. 1997; Svedarsky et al. 2003). Rather than retreating to southern Minnesota, their first gateway to the state, the dwindling numbers retreated to the unplowed prairies remaining along the beach ridges of former Glacial Lake Agassiz in northwestern Minnesota. A small, isolated population also remained farther east, near the town of Wadena.

By 1975, Green and Janssen described the species' distribution as restricted largely to the grasslands of northwestern Minnesota, from Wilkin County north to Polk and Red Lake Counties. The population decline that Roberts (1932) witnessed in the early 1900s had continued through the 1960s. Resident flocks that "were present in Fillmore and Mower Counties until the late 1930s" were long gone. Birds were last seen near Duluth in 1952 and "an outlying population in Morrison County" was last seen in 1965. The small eastern population in Wadena and Cass Counties, however, persisted.

Several years later, Janssen (1987) reported that a 1983 census estimated 1,500 males on booming grounds. The population, he noted, "has remained fairly stable over the past five years with some increases noted; this is probably due to more accurate censusing than to [an] actual increase in birds." Although small numbers of birds had been discovered in Hubbard and Marshall Counties, since 1970 confirmed nesting reports were limited to Clay and Chippewa Counties. Their presence in Chippewa County was the result of reintroduction efforts by the MNDNR in the early 1980s. Little had changed when the MBS was working in western Minnesota in the late 1980s and 1990s. A total of 60 breeding-season locations were identified; none was found in west-central Minnesota, but a few records were located in southwestern Otter Tail County. By 1998, Hertzel and Janssen were able to add Mahnomen and Wilkin Counties to the list of counties with confirmed nesting since 1970.

Since the latter half of the 1900s, a concerted effort has been made by state and federal resource agencies and conservation organizations to protect and restore the state's critical grassland resource. Now closely monitored, the status of the Prairie-Chicken population has improved to the point that hundreds of birds were used for at least four reintroduction projects, including in the states of Illinois and North Dakota and in Minnesota's upper Minnesota River valley (Svedarsky et al. 1997; Roy 2016a). A limited hunting season was even established in the fall of 2003. By 2020, the annual take averaged 98 birds (Roy 2020a).

By the time the MNBBA was conducted, participants reported 394 Greater Prairie-Chicken records from 3.0 percent (142/4,784) of the surveyed atlas blocks. Breeding was confirmed in 10 blocks **(A)**. The birds were documented in 20 of Minnesota's 87 counties and were confirmed breeding in four counties (Big Stone, Norman, Pipestone, and Polk). The Glacial Lake Agassiz beach ridges remained the core

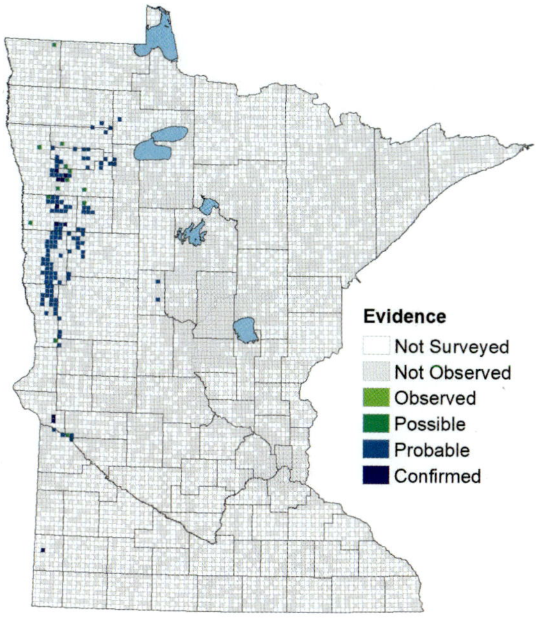

A. Breeding distribution of the Greater Prairie-Chicken in Minnesota based on the Breeding Bird Atlas (2009–13).

of the species' range, while small numbers were still present in Wadena County and in the upper Minnesota River valley. Outside of these known populations, isolated observations were documented in northern Kittson County and, to the south, in Pipestone County. The latter was a breeding record where a hen with four young was found in a location where a male had been booming in the spring. Additional reports of Prairie-Chickens in this same area were detailed by Schladweiler (2016). Atlas efforts were aided immeasurably by the contribution of survey data from the MNDNR.

Although many predicted the demise of the Prairie-Chicken nearly one hundred years ago, this adaptable bird has become, with considerable help from conservation organizations and resource agencies, a healthy and beloved representative of Minnesota's grassland resource. Unfortunately, beyond several states in the Great Plains and Midwest, the species is only a distant memory (J. A. Johnson et al. 2020).

B. Large, open grasslands, like those present at the Felton Prairie Complex in Clay County, are a critical feature of Greater Prairie-Chicken habitat in northwestern Minnesota during several stages of their life cycle. Photo by Lee A. Pfannmuller.

BREEDING HABITAT

Typically classified as an obligate grassland species, the Prairie-Chicken has complex habitat requirements. Dan Svedarsky and his students and colleagues at the University of Minnesota Crookston have spent decades documenting the conditions required during each stage of its life cycle. During the spring, the breeding season is officially launched when the birds gather at traditional display grounds, or leks. With the male's energy focused on his legendary performance, habitats that provide good visibility are essential. Not only must the females have full sight of the eligible males but also the males must be able to easily spot potential predators. As a result, patches of bare ground or grasslands with very low vegetation (less than 15 cm tall) are critical (Svedarsky et al. 2003). Sites that are on a small rise are often selected (J. A. Johnson et al. 2020).

While the male needs habitat that ensures visibility, the nesting female requires cover that provides concealment yet still enables her to detect predators and to respond quickly when disturbed. The structural features of the nesting cover generally are more important than the vegetation's composition. Planted cover on Conservation Reserve Program (CRP) lands is used extensively (Svedarsky et al. 2003).

Habitat used by the broods often differs from the female's nesting habitat. Brood habitat provides good food resources (insects for the young chicks) and concealment but enables movement. In Minnesota, nearly 70 percent of the brood habitat was located in sites recently disturbed by mowing, grazing, or burning (Svedarsky 1979).

The Greater Prairie-Chicken is also considered an area-sensitive species (e.g., Winter et al. 2001). A study in northwestern Minnesota found 95 percent of the birds were on grassland tracts larger than 200 ha; no individuals occurred on tracts smaller than 50 ha. Landscapes surrounding suitable habitat also supported little woody vegetation **(B)**.

POPULATION ABUNDANCE

From 1968 to 1997, Svedarsky and his colleagues (2000) compiled statewide estimates that suggested North America's Prairie-Chicken population peaked in the late 1970s at approximately 1.2 million birds, followed by a significant decline **(C)**. Three states supported more than 80 percent of the entire US population: Kansas, Nebraska, and South Dakota. Throughout this period, Minnesota's population was consistently less than 1 percent of the US population, ranging from an estimated high of 5,000 birds in 1968 to a low of 1,600 in 1985. More recently, PIF (2021) estimated the North American population at just 360,000 birds. Although the BBS is not a robust monitoring tool for Prairie-Chickens, the data demonstrate a small upward population trend across the survey region from 1990 to 2019 (Sauer et al. 2020).

In Minnesota, many booming grounds were surveyed as early as the 1950s and 1960s. But in 1974, members of the Minnesota Prairie-Chicken Society launched a more comprehensive survey effort, counting males on every known booming ground during the spring display season (Svedarsky et al. 1997). This effort served as a benchmark for the status of Minnesota's population for forty years. Once a limited hunting season was established in 2003, the MNDNR launched a collaborative effort with the Society to design and implement a standardized survey in 17 designated blocks. Volunteer observers also submit observations that occur outside of these blocks. Collectively, in 2020, a total of 1,004 males (which includes 84 birds of unknown sex) were located on 98 booming grounds (Roy 2020b). Although field methods are not strictly comparable, this is below the average of 1,302 males found on an average of 128 booming grounds from 1980 to 1984 (Wolfe 1984).

When data from the new standardized survey are examined separately, the results point to a significant decline since the survey began. In its first year, 2004, the survey tallied a total of 1,566 birds on 95 booming grounds; in 2020, surveyors tallied only 452 birds on 39 booming grounds.

The downturn coincides with a significant drop in CRP enrollment (Roy 2020b). More than 676,000 acres were lost from 2007 to 2016 in Minnesota alone (USDA 2017). Given the importance of CRP lands to Prairie-Chickens, this loss of habitat is staggering. Similar losses occurred throughout the species'

C. Population estimates (individuals) for the Greater Prairie-Chicken in the United States, 1968–97.

	1968	1979	1985	1989	1997
US population	778,860	1,217,672	756,054	756,268	372,742
Minnesota population	5,000	2,000	1,600	2,000	1,868
(% of total population)	(0.64%)	(0.16%)	(0.21%)	(0.26%)	(0.50%)

Source: Adapted from Svedarsky and colleagues (2000, table 1).

range. But there is historical and recent evidence suggesting the species experiences 10-year population cycles not unlike those observed in Ruffed Grouse (Roberts 1932; C. E. Carlson 1942; Toepfer 2007). The relative significance of the loss of suitable habitat, natural fluctuations in the population cycle, recent allowable harvests, and other factors, such as significant weather events and documented parasitism of Greater Prairie-Chicken nests by Ring-necked Pheasants (Toepfer 2007), is unknown.

CONSERVATION

Given its restricted distribution and small population size, PIF (2021) has rated the Greater Prairie-Chicken a high conservation priority, with the goal of restoring a portion of the population to safer levels within thirty years (Rosenberg, Kennedy, et al. 2016). In Minnesota, the Greater Prairie-Chicken has been officially classified as *Special Concern* since 1984 (Coffin and Pfannmuller 1988) and a game species since 2003. Now an enduring symbol of grassland conservation in the state, the species even has its own 4,100-kilogram, 4-meters-high statue in the west-central Minnesota town of Rothsay, the self-proclaimed Prairie-Chicken capital of Minnesota **(D)**!

Because the Greater Prairie-Chicken is an adaptable species that utilizes restored grasslands and native prairies, its numbers serve as an important ecological indicator of the health of grassland communities as well as a bellwether of farmland policies and prairie conservation efforts. Interagency initiatives designed to protect and restore grassland habitats are essential to the future conservation of the species; equally important are initiatives with annual payments that encourage farmers to set aside grasslands from cropland production. When farm commodity prices are high, the incentives often cannot compete with the high prices available on the commodity markets.

D. An enormous Greater Prairie-Chicken greets visitors to the small town of Rothsay, the Greater Prairie-Chicken capital of Minnesota. Photo by minnemom/flickr.

Although valiant efforts enabled the species to recover to the point where a limited harvest was reinstated in 2003, the future of the Greater Prairie-Chicken in Minnesota is not guaranteed. Prairie-Chickens depend on large grassland landscapes, and the pressure on farmers and developers to convert grasslands to other uses is a constant challenge. Sustaining local populations depends on many factors. Svedarsky and his colleagues (2000) recommend that 30 percent of the grassland cover within 1.6 km of an active booming ground be managed to provide good nesting cover (dense, short cover at 20 cm height) and the area be readily accessible to quality brood cover. Maintaining approximately 25 percent of the grassland landscape in croplands is equally important, as they provide an important food source, particularly during the winter.

Not only do development pressures threaten the ability to sustain these large landscapes, but also high precipitation can increase woody encroachment, particularly in the eastern portion of the species' range, including Minnesota (Svedarsky et al. 2000). Indeed, warmer temperatures are predicted to increase precipitation and storm events. Nevertheless, an analysis by NAS (2019) projected conditions in the northern Great Plains would gradually improve for Greater Prairie-Chickens as temperatures rise. The most notable range gains were projected north in Canada's central Prairie Provinces and west in eastern Montana and eastern Wyoming. The models also predicted gradually improving habitat conditions for the species across northwestern, north-central, and southwestern Minnesota until, under the 3°C warming scenario, conditions have improved throughout the entire Prairie Parkland, Tallgrass Aspen Parklands, and Eastern Broadleaf Forest Provinces. Although its habitats are significantly more restricted today, a warmer climate might certainly improve habitat and climate conditions for the Prairie-Chicken in some of Minnesota's transitional prairie–forest habitats, but a substantial amount of cropland would likely need to be restored to grassland for the species to attain the level of widespread distribution NAS projected. In the interim, the highest priority is to protect existing grasslands and manage them with appropriate burning, mowing, and/or grazing regimes to maintain quality habitat, reduce woody encroachment, and ensure that the Greater Prairie-Chicken remains a visible icon of Minnesota's grassland communities for generations to come.

GRAY PARTRIDGE

Perdix perdix

Minnesota Seasonal Status: Regular permanent resident in southwestern and western agricultural regions. The Gray Partridge was rare during the MNBBA.
Migration: Nonmigratory.
Food: Insects, seeds of domestic crops and weeds, and leafy greens.
Nest: Ground-scrape, usually lined with grasses in hayfields, pastures, fencerows, shelterbelts, and roadsides. Large clutches with an average of 17 eggs in the first clutch (range 10–22).
Conservation Concern: Introduced, game species; native to Europe and Asia.

Photo by Terry L. Johnson.

The Gray Partridge was introduced in the United States from Europe in the early 1900s. Roberts (1932) said it was first introduced in Minnesota from northern Iowa around 1913 in the southern tier of counties from Martin County westward. He surmised that the successful introduction came from about 6,000 birds released in several places in Iowa, where the greatest success was achieved in the northwestern portion of that state (Spiker 1929). Roberts further documented 89 introductions of 8,556 birds primarily in the southern portions of Minnesota from 1926 to 1929. Apparently, introductions were successful in most cases. However, this was not an inexpensive process; the 1,000 birds received from Czechoslovakia in 1929 cost $6.25 each (or $6,250 total), a considerable sum at that time.

The species is also known as the Hungarian Partridge, European Partridge, Bohemian Partridge, or by hunters simply as "Huns." Its modern common name, Gray Partridge, represents the dominant color of the bird, while its scientific name is a tautonym, meaning its genus and species name are the same. *Perdix* is both Greek and Latin for partridge. The plumages of both sexes are similar, with rusty stripes down the sides and orange faces, but the male has more orange on the face and a dark patch on its upper belly. Their primary vocalization is a harsh *kee-uck* call, which is reminiscent of a rusty gate. Both sexes are similar in size, larger than Northern Bobwhite and smaller than Ruffed Grouse. Hunters find them a challenging species to hunt because of their small size and their explosive flushing behavior.

Gray Partridge are primarily found in the upper Great Plains from Minnesota and Iowa west to Idaho, western Oregon and Washington, and north to southern Alberta, Saskatchewan, and Manitoba.

MINNESOTA BREEDING DISTRIBUTION

Roberts (1932) reported the Gray Partridge appeared to be a "hardy bird." He speculated that "if placed in suitable surroundings in a north-temperate climate," then it is likely to thrive in the southern counties of Minnesota. He wondered about the consequences of such an introduction. "Will this alien Partridge displace the native Bob-white and the Quails of the western United States? . . . [From the] average sportsman's standpoint such a result—a calamity in the opinion of the nature-lover—is viewed with equanimity if an equal or better game-bird be substituted, even though it be an interloper."

By 1936, Roberts (1936) noted the rapidity in the expansion of the Gray Partridge in Minnesota. He comments on an observation from Dr. G. H. Luedtke in 1934 from Martin County and vicinity where there was "one or more coveys on every 80-acre patch." Furthermore, he cites P. O. Fryklund and A. Torgerson, from Lake of the Woods County and Polk County respectively, who "report that implantations in that part of the state have given satisfactory results."

More than forty years later, Green and Janssen (1975) documented that it was most numerous in the southwestern part of the central region of Minnesota. It is "scarce north of Polk County, southeast of Olmsted County, and along the eastern margins of the central region." They stated that hunting began in 1939 and peak numbers were harvested in 1958. They described confirmed nesting from 17 counties, ranging in a triangle from Nobles and Pipestone Counties in the southwest, northward to Polk County, and southeast to Olmsted County. It was not found northeastward from Chisago County to Roseau County in the heavily forested regions of the state.

Janssen (1987) described a similar distribution in Minnesota and emphasized the high annual variability in its population, with peak numbers in the late 1970s and early 1980s; however, the population numbers have declined since 1984. He also pointed out eastward expansions into Houston County in the extreme southeast, into Anoka and southern Mille Lacs Counties in the central region, and into Beltrami and Hubbard Counties in the northwest regions. Janssen included confirmed nesting from 28 counties since 1970, including confirmed eastward nesting in Benton, Houston, and Sherburne Counties not previously reported by Green and Janssen (1975). Hertzel and Janssen (1998) completely documented confirmed nesting of the Gray Partridge throughout its breeding range of 64 counties. These included positive nesting in Morrison, Roseau, and Washington Counties, all on the margins of its breeding range.

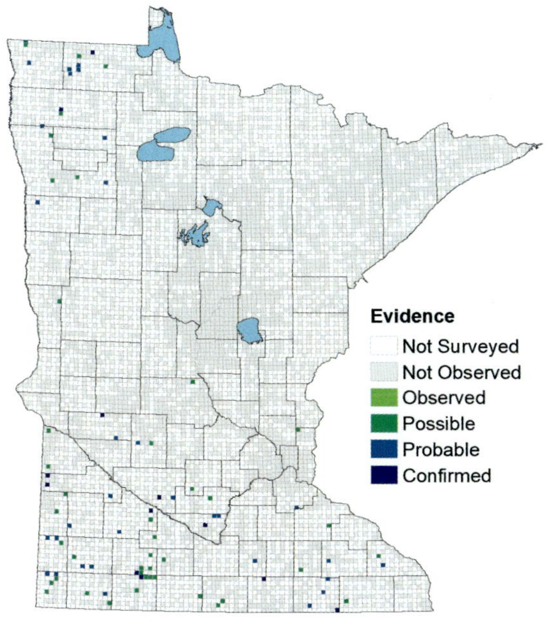

A. Breeding distribution of the Gray Partridge in Minnesota based on the Breeding Bird Atlas (2009–13).

The MNBBA recorded a total of 96 records in 1.8 percent of the blocks (86/4,755) **(A)**. The distribution included the same regions described above. Most observations were restricted to the southwest and southern regions or the northwestern regions of the state. Confirmed nesting was reported from 12 blocks in 11 counties: Cottonwood, Freeborn, Jackson, Marshall, Mower, Nicollet, Redwood, Renville, Roseau, Swift, and Yellow Medicine. All were previously reported by Hertzel and Janssen (1998).

In their review of the Gray Partridge in North America, Carroll and colleagues (2020) stated the species had been released into more than thirty states and provinces during the late 1800s and early 1900s. The most successful were in Alberta, where they spread into Saskatchewan and Montana. Since the 1950s, its range has retracted in the eastern United States and Canada, while it has gradually expanded in the prairie states, including southward into Nebraska and Missouri.

Mettler (1977) summarized that Gray Partridge populations and their range in Minnesota have expanded from relatively low populations in the 1950s and 1960s. She explained that, as the Ring-necked Pheasant population declined in the 1960s, the Gray Partridge population began to increase in the 1970s. These reciprocal changes are not completely known but may have been due to interspecific competition or weather-related issues. For instance, Gray Partridge prefer drier weather conditions, while Ring-necked Pheasants prefer wetter conditions. However, since the 1970s, Minnesota's Gray Partridge distribution has not changed extensively, but populations have declined since the late 1980s (Davros 2015).

BREEDING HABITAT

The Gray Partridge can be found in large tracts of grasslands, fields with cereal grains and row crops, shelterbelts, and roadsides (Carroll et al. 2020). In his habitat review, Arthur W. Allen (1984) repeatedly mentioned the importance of idle cover types—those not subject to grazing or crop or hay production. These included rights-of-way, idle and active farmsteads, shelterbelts, fence lines, irrigation ditches, and field edges **(B)**. Mettler (1977) emphasized that wildlife biologists know the Gray Partridge as a "dry land bird" and a "bird of open country," but it preferred higher vegetation for nesting than the Ring-necked Pheasant. Habitat suitability models indicate that it preferred to nest in vegetation at least 46 cm tall (A. W. Allen 1984).

B. Breeding habitat of the Gray Partridge is characterized by large, dry grasslands, especially those found in southern and northwestern Minnesota, as in this photo from Swift County. Photo by Lee A. Pfannmuller.

POPULATION ABUNDANCE

PIF (2021) estimated a United States and Canadian population of 820,000 birds, with 12,000 in the state of Minnesota. Carroll and colleagues (2020) surmised that the species' greatest populations existed prior to the 1960s. Mettler (1977) reported that the statewide average was 18 birds per 100 miles (161 km) based on roadside counts in 1941. Counts were low in the 1950s and 1960s, with generally fewer than 10 birds per 100 miles reported; however, they began to rise in the late 1960s (Lyons 2021). High counts of 20–45 birds per 100 miles were recorded from the late 1970s to the late 1980s, but counts of fewer than 5 birds per 100 miles have been reported since 2008 **(C)**. The BBS trends for Minnesota were unreliable but showed a tendency toward a decline of –0.44 percent per year from 1967 to 2019 (Sauer et al. 2020).

CONSERVATION

As an introduced species, the Gray Partridge is of minimal conservation concern, but of interest as a game species. However, NAS (2019) considered it as moderately vulnerable to climate change in both a 1.5°C and a 2°C warming scenario. Intensive agricultural management that reduces the availability of cover, food, and nesting sites is detrimental (Carroll et al. 2020). Messinger and Davros (2018) state that intensified agricultural land use, such as corn and soybeans, has reduced the amount of suitable habitat for the species in Minnesota. Besides available habitat, agricultural pesticides have caused direct mortality to the species (Carroll et al. 2020). Indirectly, Carroll and colleagues (2020) point out that pesticides affect partridge mortality because of the reduced availability of weeds and insect foods during the nesting season. Hay mowing reduces available habitat as well as causing mortality, especially if performed early in the season or too frequently. Carroll and colleagues (2020) emphasized that mortality primarily occurs during nesting and brood rearing owing to predation but can also occur in winter with food shortages or in severe weather, especially with deep and crusted snow.

Mettler (1977) suggested that population levels of the Gray Partridge are affected by multiple factors. She stated that the Partridge's hardiness in winter and its large clutch size were both positives compared with the Ring-necked Pheasant. Dry summers are positive because wet summers cause "mud-balling" on the legs of young and reduce their mobility. Neither Carroll and colleagues (2020) nor Mettler (1977) considers hunting a major factor because the Gray Partridge is a limited quarry, usually only incidentally hunted while hunting other species.

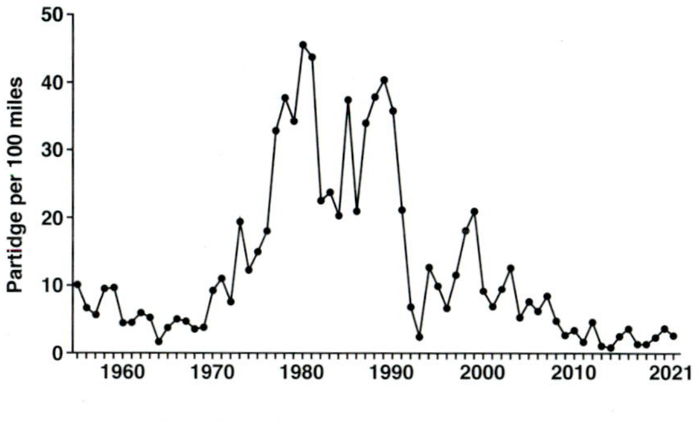

C. Range-wide index of the number of Gray Partridge observed per 100 miles (161 km) driven in Minnesota, 1955–2021, during August roadside surveys. The northwest region is not included. Adapted from Lyons 2021 (fig. 2b).

RING-NECKED PHEASANT

Phasianus colchicus

Minnesota Seasonal Status: An introduced species; now a permanent resident. The Ring-necked Pheasant was common during the MNBBA.
Migration: Relatively sedentary, seasonal; short distance movements of several kilometers are known in response to weather, food availability, and the need for cover.
Food: Grains, fruit, seeds, and arthropods are consumed on the ground.
Nest: On the ground, usually in tall vegetation in a natural depression; a rare nest parasite on grouse or ground-nesting waterfowl.
Conservation Concern: Introduced game species.

Photo by Michael Furtman.

The Ring-necked Pheasant is a very popular game bird and a species that is nearly impossible to misidentify, especially the male. The male is a spectacular specimen with red scarlet patches on the side of its face, a long barred tail, and an iridescent plumage that is a combination of bronze, tan, green, white, black, and gray. A distinctive white neck collar is also diagnostic and responsible for its common name. The female is smaller and buff-colored with various brown and black spots and bars. She is a perfectly camouflaged specimen to avoid detection from predators while nesting in thick cover (Wechsler 1985). The distinctive call of the male has been described as a two-syllable *kok cack,* often followed by a whirring of the wings.

A native of Asia, the Ring-necked Pheasant has been one of the most successfully introduced game birds throughout the globe. There are more than 30 varieties of Pheasants, and they have been the subject of numerous artists and gourmet cooks for centuries. Its genus name, *Phasianus,* literally means "Pheasant" in Greek, while its species name, *colchicus,* translates from Latin to "an ancient country through which the river Phasis flows . . . the place where a large number of pheasants were reported in ancient times" (Sandrock and Prior 2014).

Ring-necked Pheasants are widely distributed across the United States and portions of southern Canada, from Nova Scotia to California, primarily in open agricultural regions and grasslands. Its highest densities are in the Great Basin from Kansas to North Dakota. It is considered to be one of the most widely introduced bird species worldwide.

MINNESOTA BREEDING DISTRIBUTION

Roberts (1932) provided extensive details on the introduction of this species in Minnesota. The first documented attempts began in 1905 at the St. Paul Fish Hatchery. There may have been attempts before that time, but he could not locate any records. Additional introductions were made from 1909 to 1914. Roberts summarized the status of these introductions with a statement by Mr. Rider, the "executive agent" of the Minnesota Game and Fish Commission, who said, "We regret to state that the propagation of Chinese Pheasants has not proven a success in this state."

The pursuit of introducing Pheasants to Minnesota continued from 1915 to 1928, when 25,000 Pheasants and 60,000 eggs were distributed to farmers and others throughout the state by the State Game Farm. The farm was originally located near Lake Minnetonka and was moved to Mound in 1920. The farm was discontinued and reestablished near Madelia, Watonwan County, in 1928 (Roberts 1932).

Giudice and Ratti (2020) documented that the Ring-necked Pheasant was first introduced to North America in New York between 1730 and 1733, but the introductions failed. Despite many introductions throughout the United States, the first successful introduction was in 1882 in the Willamette Valley, Oregon. In contrast to earlier attempts, the MNDNR (2019i) reported that the Ring-necked Pheasant was first successfully released in Minnesota in 1916. Although Pheasants did not make their appearance in the wooded valleys of southeastern Minnesota until about 1925, Roberts (1932) stated that from about 1920, it "rapidly expanded in numbers in the southern, more open, and more thickly settled parts of the state." Minnesota benefited from an expanding population in South Dakota, where they "had multiplied enormously." Roberts noted that in the spring of 1930 "it was possible to see 150 to 200 birds by the roadside in a day's drive" near Windom in Cottonwood County.

As of 1932, Roberts said that "[p]heasants have been liberated in probably every county in the state" and "are present in small numbers here and there as far north as the Canadian boundary, but their continued existence in the northern counties will probably depend upon frequent restocking." The first Minnesota hunting season was in 1924. In his revision of his treatise on Minnesota birds, Roberts (1936) stressed that the Pheasant had "continued to increase greatly in the southern part of the state, and is apparently becoming well established and fairly numerous in the northern counties." He reported that it was increasing in Roseau County and wintering and nesting in the vicinity of Virginia, St. Louis County.

Roberts (1932, 1936) did not document many confirmed nests, but included two nests with eggs

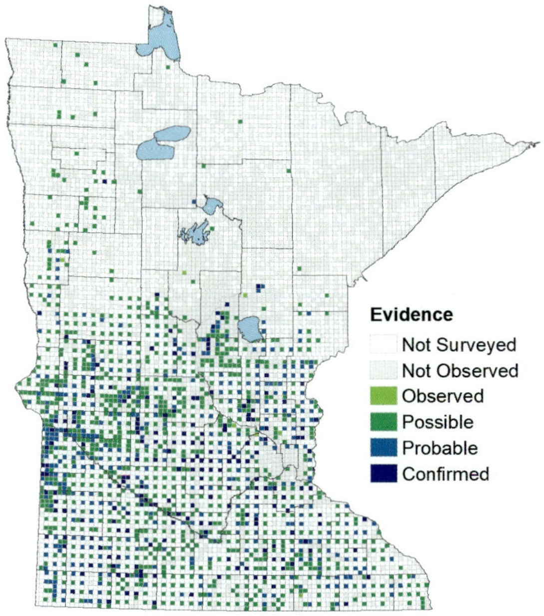

A. Breeding distribution of the Ring-necked Pheasant in Minnesota based on the Breeding Bird Atlas (2009–13).

at Lake Minnetonka and many in Minneapolis, all primarily in the 1920s. Given the rapid population increase and widespread distribution of the Ring-necked Pheasant in Minnesota, the Pheasant was likely successfully nesting in many areas of the state.

Green and Janssen (1975) reported the distribution of the species as southern and central in open-country regions as far north as Polk County in the northwest, and southern Aitkin, Carlton, Crow Wing, and Hubbard Counties. They included a map documenting confirmed nesting from 65 counties. The northern limits of confirmed nests included Kanabec, Mille Lacs, Morrison, Norman, Otter Tail, Pine, and Todd Counties, and Duluth, St. Louis County, where there was one confirmed nest.

By 1987, Janssen (1987) reported a distribution throughout most of the state, except in the northeast (though regular in Duluth), the north-central region, where it is absent, and the northwest, where it is rare. He stated it was most numerous in the Twin Cities area and north to the Mille Lacs Lake area. Janssen documented nesting in 31 counties since 1970, with a northern limit of Mille Lacs, Morrison, and Todd Counties. Hertzel and Janssen (1998) expanded confirmed nesting since 1970 to 66 counties, including all counties in southern Minnesota and most in central Minnesota. The northern limits included Becker and Norman Counties in the northwest, Crow Wing in the north-central region, and St. Louis County in the northeast.

The MBS, which began in the late 1980s, recorded 813 breeding-season locations during its county surveys. Their records included virtually all counties in the southern and central regions of the state as far north as Morrison County and northwest to Marshall County. They reported one location in northwestern Itasca County but no locations northeast from Aitkin and Pine Counties.

The participants in the MNBBA included 2,977 records for the Ring-necked Pheasant, which were distributed as far north as Kittson, Roseau, Lake of the Woods, Koochiching, and southeastern St. Louis Counties **(A)**. The species was recorded in 32.1 percent (1,534/4,772) of all surveyed blocks, and 45.4 percent (1,062/2,337) of priority blocks. Nesting was confirmed in 200 blocks and in 59 counties, including as far north as Aitkin, Polk, and Wadena Counties.

The breeding distribution of the species in the state was predominantly in the central and southern counties of Minnesota, where confirmed or probable nesting was reported from all counties south of a line from Aitkin and Pine Counties to Clay County, excluding Becker and Cass Counties. All breeding evidence in the northern counties was reported as possible nesting, which is defined as "species encountered in suitable nesting habitat within safe dates." The code for possible nesting is used when the species is heard calling or is visually observed. It records the possibility of nesting activity and a judgment that suitable habitat for the species to nest is present.

Overall, the breeding distribution of the Ring-necked Pheasant has not changed substantially since its initial expansion occurred in the 1920s and 1930s (Roberts 1932, 1936). The species' stronghold has been southern Minnesota, northeast to Pine County, to Crow Wing County in the central region, and northwest to Polk County. Scattered breeding records and detections have been reported sporadically in many northern regions of the state, but most or many could be birds released from game farms or by individuals.

Ring-necked Pheasant populations had been established throughout most of their North American range by the late 1930s and stable since 1941; however, their relative abundance has declined throughout much of their range (Giudice and Ratti 2020).

BREEDING HABITAT

Ring-necked Pheasants are primarily found in croplands that are interspersed with ditches, grasslands, pasture, wetlands, brushlands, and woodland borders (Giudice and Ratti 2020). Habitat and landscape features that provide for many needs of the species are preferred. The MNDNR (2019i) emphasizes that they "breed in grasslands but spend other parts of the year in both grasslands and cattail marshes near grain fields." MNBBA data gathered during 1,658 detections of the Ring-necked Pheasant during point counts identified cropland, upland grassland, and marsh/wet meadows as primary habitats **(B)**.

POPULATION ABUNDANCE

Based on the BBS, PIF (2021) estimated a United States and Canadian population of 17 million adults and a whopping global population of 56 million Ring-necked Pheasants. In Minnesota, their estimate was 780,000 individuals. The MNBBA did not estimate a population for this species. Ring-necked Pheasants can be heard for a great distance, especially in open country, which renders population estimates difficult and highly variable.

Ring-necked Pheasants have been monitored by roadside counts in Minnesota from 1955 to 2021 (Lyons 2021). Counts have varied from a peak in 1958 of more than 400 Pheasants per 100 miles (161 km), to lows since the mid-1960s of fewer than 150 Pheasants per 100 miles and often much lower **(C)**. In general, harvest levels closely follow the Pheasant counts. Lyons (2021) reported that recent Pheasant harvests (2011 to 2020) were under 300,000 per year **(C)**. For instance, 152,800 Pheasants were harvested in 2014–15, which was dramatically down from a high of 655,443 Pheasants taken in 2007–8 (Davros 2015). During this same period, the number of hunters varied from a low of 45,263 in 2017–18 to a

B. Breeding habitat of the Ring-necked Pheasant typically includes a variety of open areas that provide suitable food supplies and cover for nesting and for protection from inclement weather. Photo by Gerald J. Niemi.

high of 118,703 in 2006–7 (Davros 2015; Davros and Dexter 2019).

Giudice and Ratti (2020) suggested that Ring-necked Pheasant populations in North America probably peaked between the 1930s and 1950s. The surveywide BBS counts in the United States and Canada from 1966 to 2019 indicated a highly variable population that is trending downward **(D)**. Low counts have occurred since 2010. Declines in the species from 1966 to 2015 have been widespread across the United States, especially in the western states and throughout much of the Midwest **(E)**. In contrast, during this same period, increasing populations have been observed from Ne-

D. Surveywide breeding population trend and 95 percent credible interval (solid black lines) for the Ring-necked Pheasant based on the 1967–2019 Breeding Bird Survey (Sauer et al. 2020).

braska and the Dakotas to southern Manitoba and Saskatchewan.

The BBS roadside counts also provide a general measure of the relative abundance of the Ring-necked Pheasant in North America. In Minnesota, an average of 27 Pheasants was detected on BBS routes per year from 1967 to 2019 (Sauer et al. 2020). Highest detection rates were found in Kansas, with a mean of 155 per route per year. In Nebraska the mean is 49; in Iowa, 117; in South Dakota, 143; and in North Dakota, 79.

Declines in the population of Ring-necked Pheasants have primarily been attributed to changes in agricultural practices to large-scale "clean farming" and subsequent reductions in nesting and winter cover habitat (Clark and Bogenschutz 1999; Giudice and Ratti 2020). Messinger and Davros (2018) also summarize the fact that grassland habitat had declined over this same period, especially in Conservation Reserve Program (CRP) lands and that "losses correlate well with an overall decline in the pheasant population and harvest since the mid-2000s." The recent high of nearly 81,000 ha (2 million acres) enrolled in CRP in Minnesota in 2007 has declined to fewer than 60,750 ha (1.5 million acres) from 2015 to 2018.

Giudice and Ratti (2020) identified the critical issues affecting the population of Ring-necked Pheasants as available habitat and appropriate landscape configurations. Key factors in population regulation include reproductive success, predation, and overwinter survival. The MNDNR (2019i) notes that permanent protection of grasslands and wetlands is important to maintain populations of Pheasants, because there are not enough idle grasslands for hens to nest in and not enough cattail marshes where they can hide from cold

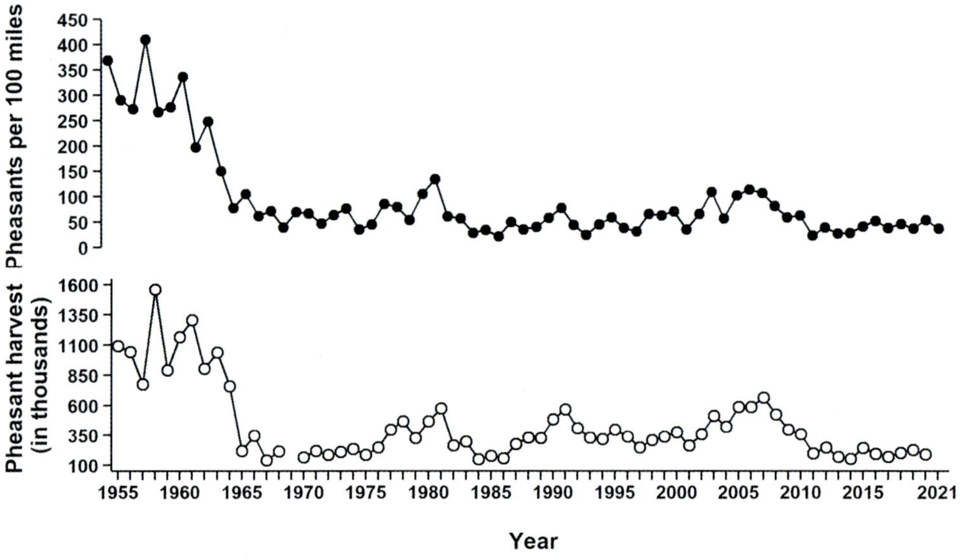

C. Roadside counts and associated harvest levels of the Ring-necked Pheasant in Minnesota from 1955 to 2021 or 2020, respectively. Adapted from Lyons (2021, fig. 2a).

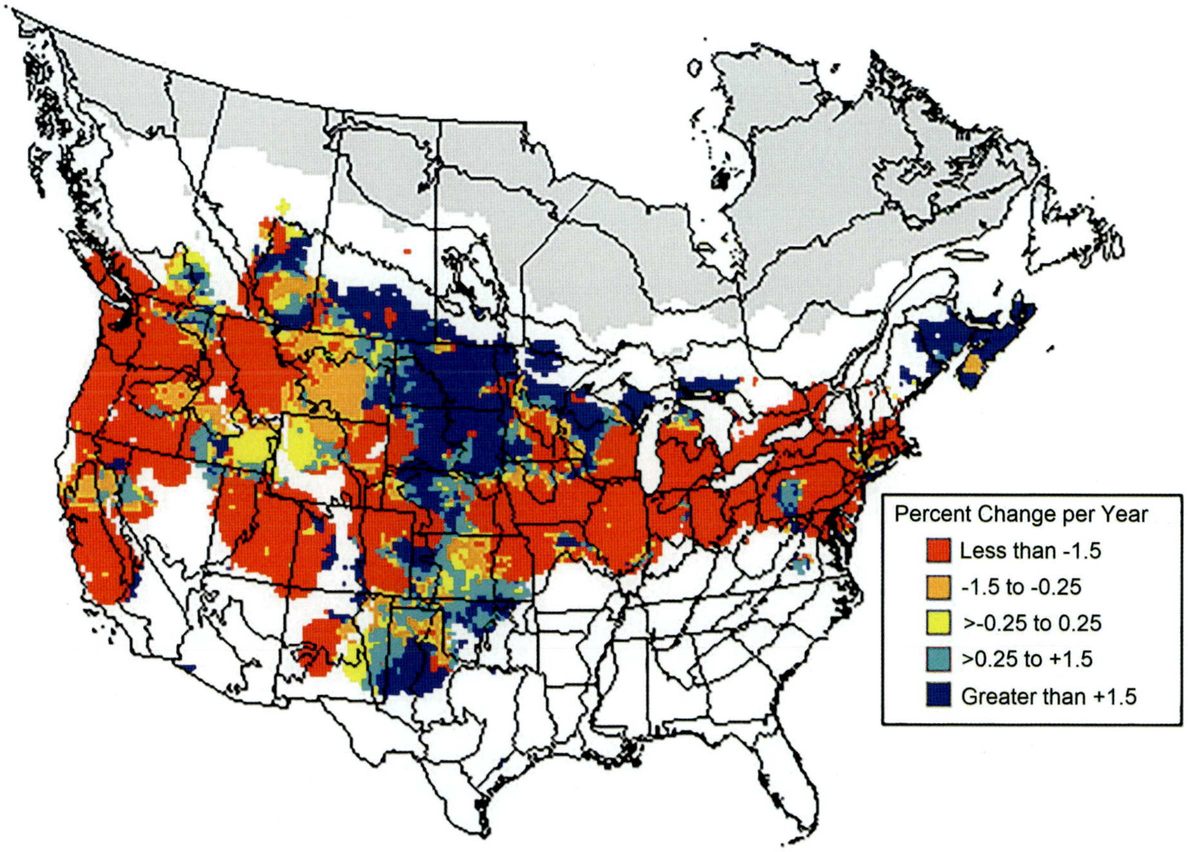

E. Breeding population trend map for the Ring-necked Pheasant in North America for 1966–2015 based on the Breeding Bird Survey (Sauer et al. 2020).

weather. Pheasant die-offs occur during extreme winter conditions such as blizzards.

Green and Janssen (1975) stated that peak populations of the species occurred in Minnesota in the early 1940s, and Janssen (1987) added peaks in the late 1950s. Both documents emphasized that "clean farming" in the south and west-central portions of the state resulted in substantial reductions in the species' population.

CONSERVATION

The Ring-necked Pheasant is a highly popular game bird throughout much of its range, including Minnesota. Therefore, populations are managed to maintain adequate levels. Roosters are the primary hunted sex, and hunting can significantly contribute to the annual mortality of male Pheasants. However, the Ring-necked Pheasant mating system is described as a "harem-defense polygyny" system. It is a compensatory mechanism in which groups of females will select one male or one male territory. Roosters can defend the females from other male intruders, and fewer males are necessary to successfully propagate the Pheasant population. Giudice and Ratti (2020) suggest that partly because of this system, the hunting mortality of male Pheasants is probably not an important limiting factor in most cases.

Most Pheasant conservation efforts are directed toward habitat and landscape management to provide appropriate nesting cover, thus reducing nesting losses, and to provide adequate winter cover. Historically, additional Pheasant stocking and predator control methods have been carried out, but these options are controversial and expensive. They may also have limited success in increasing Pheasant populations. Programs that increase available habitat, such as the CRP and conservation easement programs, have proved effective. Delaying mowing of roadsides and fields until late July has also been effective (Giudice and Ratti 2020). To this end, the MNDNR (2021) has a Pheasant Action Plan 2020–23 with the goals to increase, maintain, and enhance grassland habitat for pheasants; increase outdoor recreation related to pheasants and their habitat; and increase public awareness of grassland conservation for pheasants and people.

Regarding climate change, NAS (2019) projected the loss of 19 percent of the Pheasant's breeding range in North America by 2050 but with a 39 percent gain northward under a 2°C warming scenario, resulting in a stable breeding range. In Minnesota, the core of its breeding population in southern and western counties would be spared but range gains would continue to advance farther into northern regions.

Ring-necked Pheasants are, however, a nonnative introduced species that has been documented to have potentially negative effects on native grouse species, such as the Greater Prairie-Chicken (Svedarsky et al. 1982). Richard Kimmel (1988) cited nest parasitism, competition for habitat, transmission of disease, and aggressive behavior as harmful impacts associated with Ring-necked Pheasants. Climate change may exacerbate the negative interactions with native grouse in the northern portions of Minnesota.

Ring-necked Pheasants not only are a popular game bird but also elicit excitement from all wildlife enthusiasts, who marvel at the sight of the stoic and colorful male. Its popularity and its populations will likely be maintained for many years to come, but achieving the population levels of earlier years will be challenging for management agencies because of the many demands for the use of Minnesota's open lands.

PIED-BILLED GREBE

Podilymbus podiceps

Minnesota Seasonal Status: Regular breeding species and migrant; occasionally observed during the winter in central and southeastern Minnesota. The Pied-billed Grebe was common during the MNBBA.
Migration: Resident to medium-distance migrant; birds in the northern regions, including Minnesota, usually migrate to the southern United States and Mexico.
Food: Largely aquatic prey, including small frogs, fish, snails, crayfish, and aquatic insects secured by diving.
Nest: Floating platform constructed with aquatic vegetation and usually anchored to floating vegetation.
Conservation Concern: Assigned a low PIF Continental Concern Score of 7/20.

Photo by Mike Lentz.

If you spot a small, brown bird swimming alone on a nearby marsh or lake, chances are it is a Pied-billed Grebe. And it is likely that as soon as you see it, it will quickly dive and disappear, only to reappear some distance away. At times it may also sink below the surface of the water with only its head and neck exposed, making it nearly invisible. By expelling the air trapped between its body and its feathers, the bird reduces its buoyancy, allowing it to sink and conceal its presence. These seemingly mysterious ways have earned this little grebe such vernacular names as "Hell-Diver" and "Water-Witch."

During the breeding season, the bird's primary diagnostic trait is its thick, bluish-white bill with a prominent vertical black bar in the middle. The male and female are nearly identical, and both lose the coloration and markings on the bill during the nonbreeding season. *Podilymbus,* the genus name, translates to "rump-foot diver"; the species name, *podiceps,* also means "rump-foot," a reference to the placement of the Grebe's feet near the rear of its plump little body. Common in diving birds, the rear placement allows the feet to act as efficient propellers.

The Pied-billed Grebe's loud and distinctive call belies its small size. Roberts (1932) described the unique sound: "The loud and resonant spring notes or 'love-song' of the Little Diver is one of the most striking and characteristic bird utterances of all our marshlands. It has a far-reaching and vibrant quality like the call of the Cuckoo, . . . well expressed . . . by the syllables *cow-cow-cow-cow-cow-cow-cow-cow-cow-uh, cow-uh, cow-uh-cow-uh,* slow at first and more rapid toward the finish."

Found in a variety of wetland habitats, the Pied-billed Grebe has a breeding range that stretches across southern Canada, the United States, and farther south into Mexico, Central America, and the Caribbean Islands. The highest breeding densities are seen in the Northern Great Plains of the United States and Canada's Prairie Provinces.

MINNESOTA BREEDING DISTRIBUTION

Roberts's account of the species in 1932 described the Pied-billed Grebe as an "abundant summer resident, breeding throughout the state wherever there are suitable water and marsh conditions." He noted it was less common in the northeastern counties, where shallow wetlands were replaced by "rock-bound lakes and rivers." Unlike members of Minnesota's diverse waterfowl community, the Pied-billed Grebe seems to have gone largely unnoticed by the cadre of observers providing Roberts with frequent accounts of the distribution and abundance of waterfowl in the late 1800s and early 1900s. Nevertheless, the species was likely far more abundant prior to the loss of nearly 50 percent of Minnesota's wetlands (J. Anderson and Craig 1984). Despite its abundance, Roberts compiled confirmed nesting records from just seven counties that he and his colleagues frequently visited: Hennepin, Isanti, Jackson, Kittson, Marshall, Polk, and Stearns.

Forty years later, Green and Janssen (1975) described the species as a summer resident throughout the state. The exception, as in Roberts's time, was the Arrowhead region, where its preferred marshland habitat was far less common. Janssen (1987) described it as a casual breeder in these northeastern counties and identified 57 counties across the state where breeding had been confirmed since 1970. Later, Hertzel and Janssen (1998) added another seven counties to the list.

Survey work conducted by the MBS identified 357 breeding-season locations for the Pied-billed Grebe. The species was common throughout much of western Minnesota, but records were particularly scarce in the southeastern counties and in the northeast.

During the MNBBA, Pied-billed Grebes were detected in 15.0 percent (716/4,769) of the surveyed atlas blocks that were dispersed across all but two of Minnesota's 87 counties (Rock and Dodge). Breeding was confirmed in 190 blocks and 65 counties, including one record in far northeastern Cook County **(A)**. Twelve of the counties were additions to the list published by Hertzel and Janssen (1998).

Although there is no barometer of its former abundance in the late 1800s and early 1900s, the Pied-billed Grebe's current breeding range appears remarkably similar to that described by earlier accounts. The core of its distribution is in central Minnesota, from Dakota and Washington Counties in the east to Big Stone and Lac qui Parle Counties

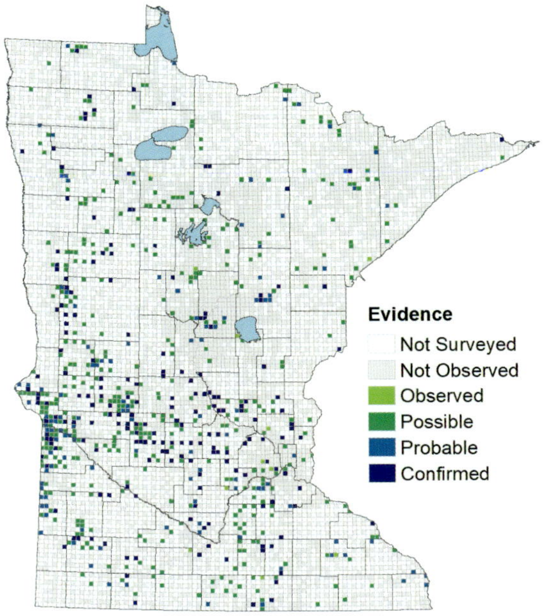

A. Breeding distribution of the Pied-billed Grebe in Minnesota based on the Breeding Bird Atlas (2009–13).

in the west, a region where productive shallow lake basins are prevalent. It remains scarce in the far northeast and north-central counties, particularly in Cook, Lake, and Koochiching Counties, as well as in the far southeastern corner of the state. Elsewhere in its breeding range, there have been few major changes in the species' overall distribution other than local responses to the loss of wetland habitat (Muller and Storer 2020).

One of six grebe species found in Minnesota during the MNBBA, the Pied-billed Grebe is the most widely distributed and most common. Nearly three of every four grebes reported during the MNBBA were Pied-billed Grebes.

BREEDING HABITAT

An inhabitant of small wetlands and shallow lakes, the Pied-billed Grebe tolerates a variety of landscapes, including agricultural, forested, and urban/suburban **(B)**. In North Dakota, the heart of its breeding range, 91 percent of the breeding pairs were found on seasonal and semipermanent wetlands. In Minnesota, 88.5 percent were found on semipermanent and permanent ponds and lakes (Muller and Storer 2020).

The Grebe's principal nesting requirements include stands of emergent vegetation and access to open water. The latter enables a quick retreat from predators or other disturbances and provides open areas for this diving forager to feed. Water depths greater than 25 cm are preferred (Muller and Storer 2020). Nests are usually nestled within stands of emergent vegetation but are occasionally placed in open water (Muller and Storer 2020).

POPULATION ABUNDANCE

Population data for the Pied-billed Grebe are scant; few demographic studies have been conducted. Based on expert opinion and reports dating from 2006 to 2008, Wetlands International (2012) estimated the entire population in the Western Hemisphere as approximately 126,000 to 255,000 birds. More recent assessments by PIF (2021) estimate the global population at 3.1 million birds and the population in the United States and Canada at 1.1 million. These numbers seem small given the species' broad distribution, but even winter totals from Christmas Bird Counts (CBC) in North America conducted from 2000 to 2015 ranged from only 20,500 to 38,000 individuals each year (NAS 2018). Despite its wide distribution, the Pied-billed Grebe is not an abundant species anywhere within its breeding range.

Although it is not a robust tool for monitoring wetland species, the BBS provides some indication of long-term population trends. Unfortunately, the Grebe's low relative abundance compromises the statistical analyses in most ecological regions, states, and provinces. The exception is the Prairie Potholes Region, where, despite wide fluctuations, the trend has remained relatively stable from 1967 to 2019 **(C)**. The data suggest a similar trend across the United States and Canada but indicate a small downward trend in Minnesota (Sauer et al. 2020).

Although this diminutive wetland inhabitant has not received as much attention as the waterfowl species it shares its habitat with, its population trend seems to mirror that of many of its co-inhabitants.

B. Pied-billed Grebe nests were abundant on this small, permanent lake basin in Anoka County, with their platform nests anchored to the short emergent vegetation and to the floating leaves of the water lilies. Photo by Lee A. Pfannmuller.

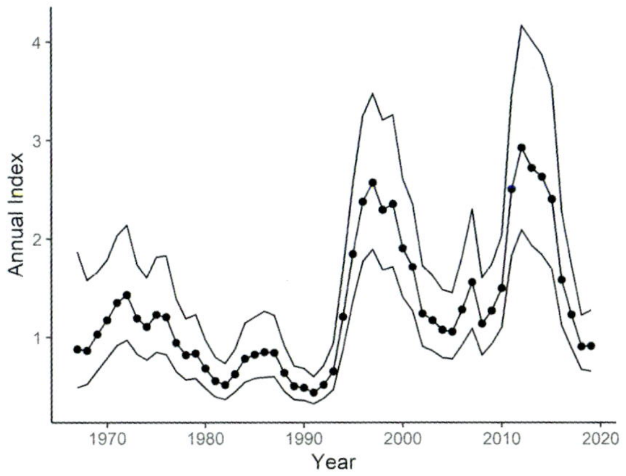

C. Breeding population trend and 95 percent credible interval (solid black lines) for the Pied-billed Grebe in the Prairie Potholes Bird Conservation Region based on the 1967–2019 Breeding Bird Survey (Sauer et al. 2020).

At the national scale, populations are stable, even increasing in some regions, but Minnesota populations appear to be declining.

CONSERVATION

Despite a possible downward trend in numbers, the Pied-billed Grebe remains a well-distributed species in regions of the state that provide suitable habitat. Because it is a low conservation priority (PIF 2021), there has been little focus on specific management needs, but like all wetland species, it will benefit from conservation initiatives designed to improve the quantity and quality of wetland habitats. Such efforts are particularly important given the predicted impacts of warming temperatures on wetlands across the northern Great Plains. Increased precipitation and more intense storm events will decrease the likelihood that water levels will be stable during the critical nesting season. Nevertheless, NAS (2019) projected the species' North American breeding distribution would remain stable under all three warming scenarios they examined. Even in Minnesota, breeding range losses were predicted to be quite small.

Surprisingly, however, communication tower collisions are a significant mortality factor for this nocturnal migrant. The annual mortality was estimated to be 5.6 percent of the North American population, third only behind the Yellow Rail and Swainson's Warbler (Longcore et al. 2013).

Next time you hear a loud and unearthly sounding call emanating from a nearby wetland, you can be assured that it is not coming from the vocally subdued waterfowl in the neighborhood but rather from the petite, often invisible, and relatively common little "Hell-Diver."

HORNED GREBE

Podiceps auritus

Minnesota Seasonal Status: Rare breeding species and a common migrant; occasionally seen during the winter on Lake Superior and in central and southeastern Minnesota. The Horned Grebe was very rare during the MNBBA.
Migration: Short-distance migrant that winters in coastal waters or inland lakes in the southeastern United States.
Food: Diver that feeds on aquatic insects, small fish, and crustaceans.
Nest: Floating platform constructed with aquatic vegetation; usually nests as solitary pairs but occasionally nests in loose colonies numbering fewer than 20 pairs.
Conservation Concern: Assigned a moderate PIF Continental Concern Score of 11/20 and officially listed as *Endangered* in Minnesota. Classified as *Vulnerable* on the IUCN Red List (IUCN 2021).

Photo by Sparky Stensaas.

Among the six grebe species that occur in Minnesota, the colorful Horned Grebe is the most restricted in its breeding distribution, confined largely to a few wetlands in the northwestern corner of the state. Slightly larger than the abundant Pied-billed Grebe, the Horned Grebe is more familiar as a spring and fall migrant, when it frequently congregates in large numbers, including on some of the larger lakes within the Twin Cities and along the North Shore of Lake Superior.

The species' Latin name, *auritus,* is a reference to its namesake: two ear tufts, or "horns," of yellow-gold feathers that are raised during courtship displays and aggressive encounters. Richly imbued, the black head feathers contrast with a chestnut-colored neck, upper breast, and flanks, and a gray-speckled back. In flight, white wing patches are visible. Males and females have identical plumages. Although its rufous neck easily distinguishes it from its close relative, the Eared Grebe, during the nesting season, observers need a more discerning eye to distinguish these two species in their gray-and-white winter plumage. Vocally, the Horned Grebe's nasal call descends in pitch and ends with a harsh chatter or trill. Bent (1919) described it as "a wonderful combination of weird, loud, striking notes" that was difficult to describe but hard to forget.

In North America, this Holarctic species' breeding distribution spans the interior of western Canada and Alaska, with small extensions south into Montana, the Dakotas, and northern Minnesota. The Horned Grebe is sparsely distributed throughout most of its breeding range; the core of its distribution is in the Canadian Prairie Provinces.

MINNESOTA BREEDING DISTRIBUTION

In the early 1900s, the Horned Grebe was an abundant spring and fall migrant in the state. Roberts (1932) commented that the numbers passing through southern Minnesota in the spring of 1928 were "truly astonishing." But their abundance during the summer months was less clear. Although Roberts described the Grebe as a statewide breeding resident, he wrote, "It is apparently rare during the nesting-season in the southern counties, but northward it is common in many suitable localities." Indeed, one observer reported Horned Grebes were more common than Pied-billed Grebes at Mud Lake in Marshall County. Nests were documented in just a few widely scattered localities: Heron Lake (1885) and Loon Lake (1900) in Jackson County, Cedar Lake in McLeod County (1926), and Mud Lake in Marshall County (1901). A female with young was also reported in Lincoln County (1897); two immature birds were reported on Silver Island Lake in Lake County (1930); and a number of adults were suspected nesting on Twin Lakes in Kittson County (1929). Whether these were widely dispersed breeding pairs or represented a more abundant nesting population, as Roberts contended, is not clear.

Regardless, when Green and Janssen (1975) provided an updated account forty years later, nesting was confined to northwestern Minnesota, principally Kittson, Mahnomen, Marshall, and Roseau Counties. The paucity of records, confined to such a small area of the state, led to its classification as a state *Special Concern* species in 1984 (Coffin and Pfannmuller 1988).

A few years later, Janssen (1987) reported that nesting had been confirmed in Pennington County and noted there were reports in the late 1970s and early 1980s of adults and broods on Amelia Lake in Pope County and adults in Becker and Cook Counties. Unlike earlier accounts, he also mentions a 1931 nesting record in Wright County and a 1944 record from Kandiyohi County, but no additional details were noted. Nesting had been confirmed in four counties since 1970: Marshall, Pennington, Pope, and Roseau. Hertzel and Janssen (1998) later added Kittson County to the list, but details regarding confirmed nesting in this county beyond 1966 could not be found.

During the summer of 1991, Boe (1992a) undertook a comprehensive review of all Horned Grebe nesting records and found only 17 documented nesting attempts from 1931 to 1989. She also conducted an extensive field survey of potential nesting habitat in northwestern Minnesota but located only one bird, on the Roseau River WMA in Roseau County. This led to the species' reclassification as a state *Threatened* species in 1996 (MNDNR 1995). Even extensive fieldwork conducted by the MBS identified

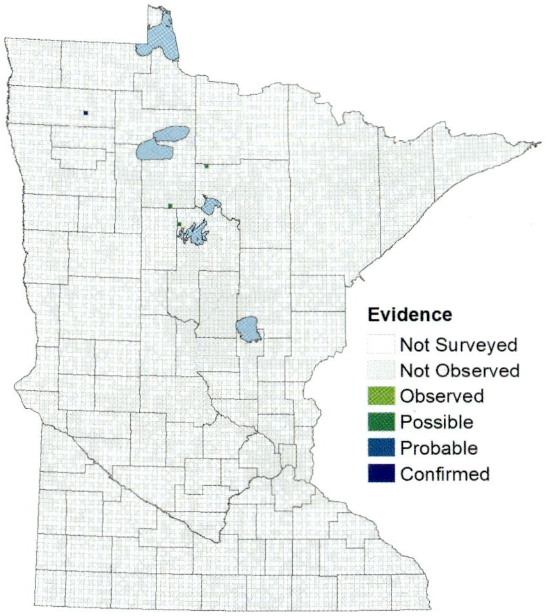

A. Breeding distribution of the Horned Grebe in Minnesota based on the Breeding Bird Atlas (2009–13).

only three breeding-season locations: two in Roseau County and one in Beltrami County.

Beginning in 2008, there were a handful of reports of adults in northeastern Minnesota, including a report of juveniles in St. Louis County, but there were few additional details (MOU 2021a). These were all late-summer reports. Because Horned Grebes begin moving from their nesting grounds in late summer, the observations may have been early migrants that nested elsewhere (J. Green, pers. comm.; MNDNR 2012). Reports are so rare that any account of possible nesting in the state should be accompanied with detailed documentation.

During the MNBBA, Horned Grebes were reported in only four of the surveyed atlas blocks, including one priority block. Breeding was confirmed in one block within the Agassiz NWR, where approximately four breeding pairs and one brood were observed in 2013 **(A)**. Three other possible breeding records were found in southern Beltrami County, northwestern Cass County, and northwestern Itasca County. Owing to the continuing rarity of records, the Horned Grebe was again reclassified in 2013 to a state *Endangered* species (MNDNR 2012).

Although it is clear that the species has declined in abundance and distribution in Minnesota in the past fifty to sixty years, the true extent of its distribution in the late 1800s and early 1900s is not well understood. Commenting on historical changes to its breeding range in North America, Stedman (2020) noted the species may have bred farther south and east in earlier times, although "some areas of [its] alleged eastern breeding range (e.g., New York and most of Ontario) actually hosted few or no birds, at least during the 1990s."

BREEDING HABITAT

Few descriptions of the Horned Grebe's breeding habitat are available. In general, it selects small freshwater wetlands (fewer than 10 ha) with extensive beds of emergent vegetation. Hemi-marshes with open-water areas amid dense beds of cattails, rushes, and sedges are preferred (Stedman 2020). Occasionally larger bodies of water are used (Butler et al. 2013), including in northwestern Minnesota **(B)**.

POPULATION ABUNDANCE

Population data for the Horned Grebe are sparse throughout its breeding range. A 2009 report summarizing the species' status in Canada provides the most recent population estimate (COSEWIC 2009b). Here, biologists estimate the population numbers from 200,000 to 500,000 breeding adults, with the majority of birds located in Saskatchewan and Alberta. Although this is an estimate only for Canada, approximately 90 percent of the species' breeding range occurs in the Canadian provinces.

Regarding long-term population trends, both the BBS and the annual Christmas Bird Count include Horned Grebes in their efforts to monitor bird populations. The former (Sauer et al. 2020) documents a declining continental population, while the latter documents a relatively stable population that includes both a downward trend in Canada and an upward trend in the United States (Meehan et al. 2020). Each protocol, however, has limitations that impede the ability to accurately assess and monitor this rare wetland species. Nevertheless, most experts agree that the Horned Grebe has declined in abundance since the 1960s and that its breeding range has contracted (Stedman 2020).

Horned Grebes were among the colonial waterbirds whose plumage was valued for the millinery trade in the nineteenth century, but populations declined long after shooting was a major source of mortality. Speculation regarding factors responsible for the current decline includes concern about the species' high vulnerability to predation during the summer and oil spills during the winter, wetland loss and degradation, and climate change (Butler et al. 2013; Stedman 2020).

CONSERVATION

The Horned Grebe has been assigned a moderate Continental Concern Score (PIF 2021); proposed conservation measures for this rare species are general at best. The only sites where the species is consistently reported in Minnesota are already protected wetlands managed by the USFWS and the MNDNR. Continued focus elsewhere on the protection and restoration of wetlands should be beneficial.

Unfortunately, warming temperatures may pose a significant threat to the Grebe's future. Under a 2°C warming scenario, NAS (2019) projected a loss of 61 percent of its current North American breeding range by 2050 and a gain of only 24 percent. The southern limit of its breeding range would move nearly 1,000 km north of where it currently breeds in northwestern Minnesota. Clearly actions to stem the tide of a warmer environment are critical for this highly vulnerable species. Although Minnesota birders may still enjoy this state *Endangered* species during migration, it is an extremely rare and elusive breeding species whose future appears even more tenuous.

B. This large wetland complex on the Agassiz NWR, with its extensive stands of emergent vegetation interspersed with open water, is the only site where the Horned Grebe has been reported nesting in Minnesota in recent years. Photo by Lee A. Pfannmuller.

RED-NECKED GREBE

Podiceps grisegena

Minnesota Seasonal Status: Regular breeding species and migrant; occasionally observed during the winter months, especially on Lake Superior. The Red-necked Grebe was uncommon during the MNBBA.
Migration: Short-distance migrant that winters primarily along northern coastlines, including from Alaska south along California; New England; and southeastern Canada.
Food: Diver that feeds on aquatic insects, fish, and crustaceans.
Nest: Floating platform constructed with aquatic vegetation and often anchored to emergent vegetation; may nest singly or in loose colonies.
Conservation Concern: Assigned a moderate PIF Continental Concern Score of 9/20.

Photo by Mike Lentz.

Known as Holboell's Grebe for many years, the Red-necked Grebe is midway in size between the larger Western Grebe and the smaller Pied-billed Grebe. The Latin name *grisegena* refers to one of the bird's most distinctive features, its gray (*griseus*) cheek (*gena*) patches. These, coupled with the rusty-red plumage on the front of its long neck and its black crown, clearly distinguish the male and female from any other wetland species during the breeding season. During the nonbreeding season, however, these features fade, and one must rely on size and other more subtle characteristics for correct identification.

The Red-necked Grebe shares many behavioral traits with other members of the grebe family. Its quick diving retreat from human intruders belies its aggressive behavior toward other marsh inhabitants that venture into its breeding territory. Slowly sinking below the water surface, the Red-necked Grebe swims toward the unwelcome intruder and attacks the unsuspecting bird on its underside with its bill (Stout and Nuechterlein 2020). Like its relatives, the Red-necked Grebe also has a beautiful and elaborate courtship display. Many birds, however, arrive on their breeding grounds already paired. Roberts (1932), for example, made no mention of its courtship behavior in Minnesota and neither he nor his colleague Walter Breckenridge appeared to have any personal experience with its vocalizations. But to the north, as he observed the birds on a moonlit summer night in Manitoba, Bent (1919) described the most common breeding-season vocalization as a series of "loon-like wailing cries" followed by a series of "short, plaintive, vibrating wails."

A Holarctic species whose distribution also spans Europe, western Asia, and Siberia, the Red-necked Grebe can be found breeding in North America from southern Alaska east to central Canada. In the United States, its range dips south into Minnesota, Montana, Idaho, and Washington; isolated populations are also scattered across a few other northern states. It is most abundant in north-central North Dakota.

MINNESOTA BREEDING DISTRIBUTION

In the early 1900s Roberts (1932) described the Red-necked Grebe's breeding range as primarily restricted to the northern and west-central regions of Minnesota: "It is rather common locally in the northern half of the state but is less frequent farther south." Confirmed nesting records were reported from six counties (Cass, Hennepin, Jackson, Meeker, Otter Tail, and Wright) and Itasca State Park; inferred nesting records were available from McLeod and Nicollet Counties. Noting its less frequent occurrence in southern Minnesota, Roberts speculated that the Red-necked Grebe may have been a more common breeder in the region in former times. In the 1936 revision to his book, confirmed nesting records from Ramsey and Wright Counties were also added (Roberts 1936).

When Green and Janssen (1975) published their updated account of Minnesota birds, new nesting records confirmed the grebe's breeding range as extending from northwestern Minnesota south through central Minnesota, east through the Twin Cities, and south again to Swan Lake in Nicollet County. Outside of this core area, nesting was confirmed as far east as Lake Vermilion in northern St. Louis County and as far south as Freeborn and Jackson Counties along the Iowa border. Along the upper Minnesota River valley, confirmed nesting was restricted to Big Stone County.

Little had changed when Janssen provided another update in 1987. There were two new nesting reports in southern Minnesota: one in Steele County (1982) and one in Lyon County (1986). Janssen identified 24 counties with confirmed nesting since 1970; Hertzel and Janssen (1998) later added another four counties to the list. All were largely within the species' established breeding range.

By 2014, the MBS documented 130 breeding-season locations for the Red-necked Grebe. No birds were reported in the greater Twin Cities metropolitan region, or in northeastern and southeastern Minnesota. Nearly 20 records south of the Minnesota River were confined to southwestern and south-central Minnesota.

During the MNBBA, observers reported 296 Red-necked Grebe records in 4.1 percent (197/4,749) of the surveyed atlas blocks. Nesting was confirmed in 86 blocks (1.8 percent) and deemed probable or possible in an additional 109 blocks (2.3 per-

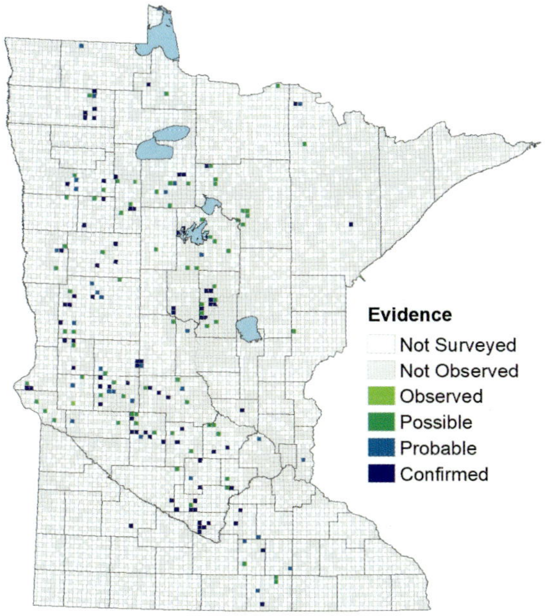

A. Breeding distribution of the Red-necked Grebe in Minnesota based on the Breeding Bird Atlas (2009–13).

cent) **(A)**. Detected in 45 of Minnesota's 87 counties, breeding was confirmed in 32 counties. Fourteen of these counties with confirmed breeding records were additions to the list published by Hertzel and Janssen in 1998; four of the new records were south of the Minnesota River valley.

Overall, there have been few changes to the Red-necked Grebe's breeding distribution in the past one hundred years. The species appears to be more common across central Minnesota than suggested by earlier accounts. It was largely absent, however, from the southwestern corner of the state. Records from south-central Minnesota, especially south of the Minnesota River, have become more common, suggesting the species is either expanding to the south or reoccupying areas that Roberts (1932) suspected were part of its former breeding range.

Few data are available to ascertain if the species' distribution has changed elsewhere in North America. Stout and Nuechterlein (2020) suggested that its breeding range may have extended farther east to Quebec and New Brunswick, but documentation is poor. In Wisconsin, the first large colony was reported in 1969 at Rush Lake in Winnebago County, where the birds have continued to nest as recently as 2017 (Cutright et al. 2006; WSO 2021).

BREEDING HABITAT

The summer haunts of the Red-necked Grebe are permanent wetlands with abundant emergent vegetation and the protected bays and shorelines of larger, deeper lakes where small stands of emergent bulrushes, sedges, cattails, or wild rice are still present **(B)**. Occasionally, the birds nest in bogs or in open-water areas with thick mats of submerged vegetation. Other important habitat features include protection from wind and waves, availability of nesting material, access to open water for diving, and suitable cover located a sufficient distance from shore to minimize access by predators (Stout and Nuechterlein 2020). In the southern portion of their breeding range, including Minnesota, Red-necked Grebes are rarely found on waters that are fewer than 2 ha in size (Riske 1976; De Smet 1983).

POPULATION ABUNDANCE

Because the Red-necked Grebe is not well monitored by roadside surveys such as the BBS, population numbers are only estimates by biologists familiar with the species. Numbers have ranged widely. Using data from 1995 to 2005, biologists with Wetlands International (2012) suggested that the species' global population ranged from 193,000 to 287,000 breeding adults; their estimate for the North American subspecies, *P. g. holboellii,* was 45,000. Stout and Nuechterlein (2020), however, believed this estimate was too low, as birds recorded in aerial surveys in the western Northwest Territories in the 1990s alone exceeded 20,000. Indeed, PIF (2021) recently projected a global population of 2.5 million birds and a population of 740,000 in the United States and Canada.

Unfortunately, owing to its sparse distribution and low abundance, accurately assessing the Red-necked Grebe's population trend is also challenging. Although surveywide BBS data lack statistical precision, population numbers were fairly stable from 1966 to approximately 2005, when a notable increase occurred through 2019 (Sauer et al. 2020). The increase appears to have been driven largely by increases in the Prairie Potholes Bird Conservation Region, especially in Manitoba, North Dakota, and Saskatchewan. Data collected by Audubon's Christmas Bird Count also suggest relatively stable to slowing increasing numbers since 1967 (Meehan et al. 2020). Reliable data on population numbers and trends are not available for Minnesota.

B. Although the Red-necked Grebe is commonly found in larger permanent wetlands and lakes in northern counties, this shallow wetland basin in west-central Minnesota, bordered by a dense stand of emergent vegetation, provided suitable habitat for a nesting pair. Photo by Lee A. Pfannmuller.

CONSERVATION

Site-level management recommendations for the Red-necked Grebe focus on maintaining a diverse community of emergent vegetation that provides a minimum cover of 25 percent (W. Brooks 2006) and protecting and restoring wetlands that are at least 4 ha in size and more than 20 cm deep. Maintaining stable water levels throughout the nesting season is critical (W. P. Mueller 2013b). Because Red-necked Grebes frequently nest in the shallow bays of large lakes that receive heavy recreational use, minimizing boat traffic in nesting areas during the breeding season can reduce disturbance and the impacts of boat-induced wakes that can swamp nests (Stout and Nuechterlein 2020).

Although populations are considered stable, more accurate population data are needed to assess trends and guide management. Much remains to be known about the species' habits, especially during the winter. There is concern, for example, that the coastal waters where they spend the winter are the major source of contaminants that have been found in adults, young, and eggs (Riske 1976; De Smet 1987; Forsyth et al. 1994). Oil spills in coastal waters can also significantly impact adult mortality. These concerns, coupled with threats to wetlands in the Prairie Potholes Region, contributed to the Grebe's assignment as a moderate conservation priority (PIF 2021).

On their breeding grounds, warmer temperatures pose another potential threat. NAS (2019) classified the Red-necked Grebe as a moderately vulnerable species under all three warming scenarios they examined. Unfortunately, the 2°C warming scenario projects the near extirpation of the species from Minnesota as the southern boundary of its breeding range moves north to southeastern Manitoba and southwestern Ontario. Monitoring the status of this uncommon and potentially vulnerable species in Minnesota is certainly warranted.

EARED GREBE

Podiceps nigricollis

Minnesota Seasonal Status: Regular breeding species and migrant. The Eared Grebe was rare during the MNBBA.
Migration: Short- to medium-distance migrant; most birds winter along the California coast.
Food: A diver that forages primarily on aquatic invertebrates; small fish and frogs are also taken.
Nest: Loosely constructed platform of aquatic vegetation that may be floating in shallow water or anchored to other vegetation. Strongly colonial but may nest as solitary pairs in small wetlands.
Conservation Concern: Assigned a moderate PIF Continental Concern Score of 10/20.

Photo by Mike Lentz.

The Eared Grebe is a striking little bird. The same size as the common Pied-billed Grebe, the upper body is black with rufous flanks visible beneath the folded black wings. Bright red eyes contrast with a tuft of black feathers on the top of the head and with long, golden ear tufts; the slender bill often appears to be curved slightly upward. Males and females are indistinguishable. A common summer vocalization is a high-pitched slur followed by an abrupt *ee-chk* (Cullen et al. 2020). In a large nesting colony, the sound was likened by one of T. S. Roberts's (1932) contemporaries, Ralph Hoffmann, to "the peeping of frogs."

The Eared Grebe may be confused with another Minnesota grebe with golden ear tufts, the Horned Grebe. During the summer, the Eared Grebe's black neck (i.e., *nigricollis*) is diagnostic; the Horned Grebe has a rufous neck. But during the winter, their gray-and-white plumages are so similar that observers must rely on the fine subtleties of head shape and color for correct identification.

The most abundant grebe in the world, the remarkable Eared Grebe is also one of the most gregarious. It has the longest flightless period of any bird capable of flight and it migrates later than any bird in North America. Eared Grebes are also capable of changing the size of their internal organs repeatedly during migration as they gorge their bellies with food and reduce the size of their breast muscles at each staging area, a process that is reversed just prior to departure. The extent of this transformation exceeds that known for any other North American bird (Cullen et al. 2020).

Present on every continent except Australia and Antarctica, in North America the Eared Grebe is restricted to the northern Great Plains of the United States, the Canadian Prairie Provinces, and to more arid landscapes farther west. Breeding densities are high wherever large nesting colonies occur.

MINNESOTA BREEDING DISTRIBUTION

In the early 1900s, Roberts (1932) described the Eared Grebe as a summer resident in Minnesota's western grasslands, breeding as far east as Renville County and as far north as eastern Kittson and Marshall Counties. Although he described Swan Lake in Nicollet County as a known breeding locality, no confirmed nesting report from the site was included in his account. He did, however, cite his own 1899 observation of a colony of 100 or more birds nesting on Heron Lake in Jackson County, as well as reports from many other observers of breeding concentrations, ranging from just a dozen birds to several hundred, in west-central and southwestern Minnesota. Large colonies that were once present in Marshall County, however, "disappeared when the whole region was drained." Overall, confirmed nesting records were limited to Jackson, Kittson, Marshall, and Renville Counties. An inferred breeding record was reported in Lyon County. Nearly thirty years would pass before nesting was documented at Swan Lake (R. Huber 1961a).

Green and Janssen's (1975) updated description of the species' breeding range was similar to that described by Roberts. Breeding had now been confirmed in 11 western and south-central counties. Unfortunately, in the twenty-five years preceding their book's publication, Eared Grebes had been reported nesting at only three sites: near the town of Alberta in Stevens County, at Swan Lake in Nicollet County, and at the Agassiz NWR in Marshall County.

In the late 1970s and early 1980s, the MNDNR's Nongame Wildlife Program assessed the status of the state's colonial waterbirds by conducting surveys and collecting reports from other natural resource managers and field biologists. Fourteen Eared Grebe colonies were reported active at least once between 1981 and 1983 (Guertin and Pfannmuller 1985), including one record as far east as Anoka County. Only two of the colonies supported more than 50 breeding pairs; the largest colony, numbering more than 100 pairs, was on the Agassiz NWR. An addendum also noted the identification of two new colonies in 1984 in Yellow Medicine County, one of which supported 103 nesting pairs. The growing number of colonies was believed to reflect reports of small colonies that were previously undetected rather than a true population increase. A few years later, a more intensive field study identified 15 active colonies in 1986, including another site in the metropolitan region in Carver County (Boe and Prahl 1987).

Janssen's 1987 account identified 16 counties where nesting had been confirmed since 1970, including two eastern counties: the Carver County record cited above and a new record in Freeborn County in south-central Minnesota. The Anoka County report was not included. Janssen (1987) speculated that the Grebe's use of sewage treatment

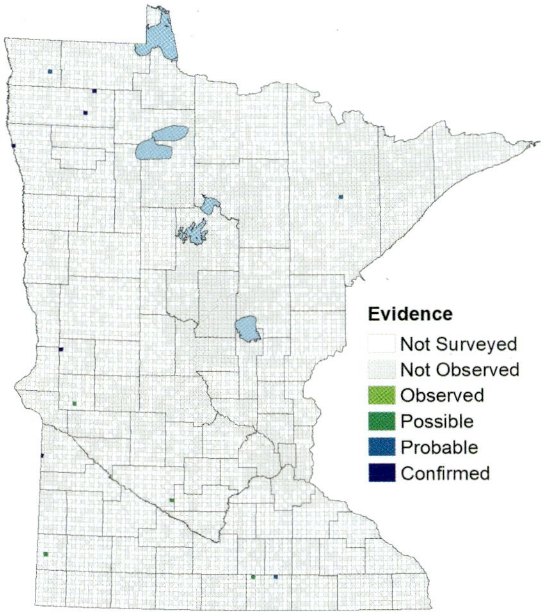

A. Breeding distribution of the Eared Grebe in Minnesota based on the Breeding Bird Atlas (2009–13).

ponds might explain some of the species' eastward range expansion. Meanwhile, the MBS identified 20 Eared Grebe breeding-season locations. Most notable was a report from central St. Louis County. With all the new fieldwork underway, by 1998 Hertzel and Janssen identified 30 counties where nesting had been confirmed since 1970, including Anoka, Hennepin, and Carver Counties in the greater Twin Cities metropolitan region.

During the MNBBA, however, Eared Grebes were reported from only 12 atlas blocks located in 10 widely scattered counties. The reports included the MBS's St. Louis County record noted above **(A)**. Confirmed breeding was documented in five blocks found in Grant, Lac qui Parle, Marshall, and Polk Counties. No birds were reported in the Twin Cities metropolitan region. Among the 12 blocks where Eared Grebes were reported, five had sewage treatment ponds within the block, including the St. Louis County record.

This small grebe remains a rare member of Minnesota's wetland bird community, while the paucity of MNBBA records suggests a possible decline in recent years. Elsewhere, however, the species' distribution is remarkably similar to that documented in years past (Cullen et al. 2020).

BREEDING HABITAT

Besides their occasional use of sewage ponds, Eared Grebes commonly select shallow wetland complexes with emergent vegetative cover **(B)**. In the west, the species is closely associated with saline wetlands. In Minnesota, Boe (1992b) found suitable wetlands were larger than 30 ha, had minimal human disturbance, and usually supported more submerged vegetation than unsuitable wetlands. Nests were usually placed within sparsely distributed emergent vegetation. Like many species dependent on shallow wetlands, Eared Grebes adapt to changing water levels by frequently shifting colony locations.

POPULATION ABUNDANCE

Systematic surveys of breeding colonies are rare and insufficient to provide solid census data. However, en route to their coastal wintering grounds, the majority of Eared Grebes stop at large saline lakes in the Great Basin, where they gorge on brine shrimp. An aerial census conducted at major staging areas in the fall of 1997 generated a North American population estimate of 4.1 million birds (Cullen et al. 2020). More recently, PIF (2021) estimated a global population of 3.1 million birds, with 2.0 million residing in the United States and Canada. The size of Minnesota's population is unknown, but approximately 650–1,050 nesting pairs were present each year from 1986 to 1989 (Boe and Nuechterlein 1990).

Although the utility of the BBS data for wetland species like the Eared Grebe is questionable, the data do reveal statistically robust trend lines in North America's Prairie Potholes Region and in the United States. In the Prairie Potholes, BBS data document a nonsignificant increase averaging 1.94 percent per year from 1967 to 2019 **(C)**; a slightly higher, nonsignificant increase of 2.06 percent per year is evident across the United States (Sauer et al. 2020). In the long term, aerial census data collected at staging areas may prove more useful. Since aerial censusing began in the late 1990s, "[n]o demonstrable trends in size or distribution" of North America's Eared Grebe population have been detected (Cullen et al. 2020).

B. The North Ottawa Impoundment, a flood-water reduction project in Grant County, provided the shallow waters and dense emergent vegetation required for an Eared Grebe nesting pair in 2017. Photo by Lee A. Pfannmuller.

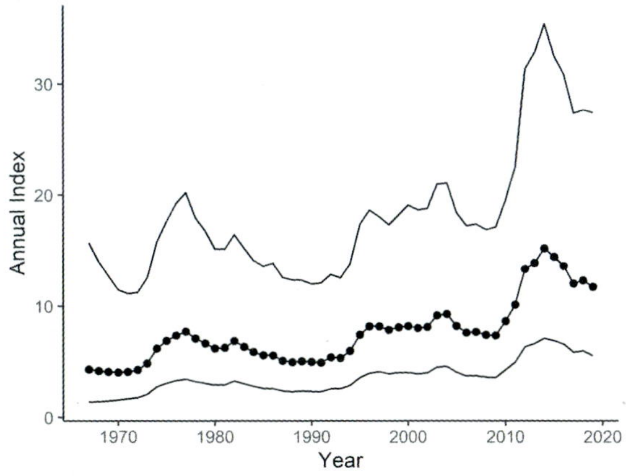

C. Breeding population trend and 95 percent credible interval (solid black lines) of the Eared Grebe in the Prairie Potholes Bird Conservation Region based on the 1967–2019 Breeding Bird Survey (Sauer et al. 2020).

CONSERVATION

Among the species targeted for the millinery trade, the Eared Grebe declined in numbers in the late 1800s. Regulatory protections eventually aided their recovery, but the loss of wetlands as European settlers moved west likely contributed to additional declines. The Grebe's preference for large, shallow wetland complexes with minimal disturbance made it particularly vulnerable to the large-scale landscape conversion that took place. Today, its uncertain population status, especially in Canada, coupled with threats to its breeding and nonbreeding habitat, has prompted the species' designation as a moderate conservation priority (PIF 2021). Its highly gregarious behavior exacerbates these concerns. Because hundreds of thousands of birds concentrate at a limited number of staging areas and wintering sites, a catastrophic event can negatively impact a significant portion of the population.

Like for all wetland-dependent species, aggressive wetland conservation policies and protection efforts are critical, particularly given the potential impacts of warming temperatures. Indeed, NAS (2019) classified the Eared Grebe as moderately vulnerable under a 2°C warming scenario with the potential loss of 39 percent of its current North American breeding range by 2050. Most losses would occur in the southwestern states and Mexico with only a 15 percent range gain farther north. Although the species' distribution in Minnesota seems secure under this scenario, losses are significant when temperatures rise 3°C. Given its rarity in the state, nothing should be taken for granted.

WESTERN GREBE

Aechmophorus occidentalis

Minnesota Seasonal Status: Regular breeding species and migrant. The Western Grebe was rare during the MNBBA.
Migration: Short- to medium-distance migrant; many northern birds winter along the Pacific coast. Southern and West Coast populations are permanent residents.
Food: Diver that feeds primarily on fish.
Nest: Platform of aquatic vegetation that is usually floating in open water but anchored to emergent or submerged vegetation. May nest as single pairs or in colonies.
Conservation Concern: Assigned a moderate PIF Continental Concern Score of 13/20.

Photo by Mike Lentz.

One of North America's most elegant waterbirds, the Western Grebe has sleek looks and a tall, curving neck that earned it the nickname "swan-grebe." Watching it gracefully glide along the waters of a Canadian wetland, Bent (1919) reflected on its "long white neck . . . the glossy black crown, the fiery red eyes, and the javelin-like beak. Who could help admiring such a picture of aquatic grace, such specialized mastery of its native element?" Nearly identical to the male, the female is a little smaller in size and her smaller bill appears slightly upturned.

The only bird the Western Grebe can be confused with is the Clark's Grebe (*A. clarkii*); the two were considered the same species until 1985. The primary distinction is that the Western Grebe's black crown extends below the eye; on the Clark's Grebe it stops above the eye. The latter's bill is also more orange and, when visible, the flanks show more white than the sooty gray seen on the Western. Both species occur in western Minnesota, so discriminating eyes are needed for positive identification. Good ears can also help: the Western Grebe's most common call is a two-syllable *kree-eeek*; the Clark's call is a single-syllable *kreek*. The species' scientific name refers to its "javelin-like beak" (*Aechmophorus*) and western (*occidentalis*) distribution.

All grebes have complex courtship rituals, but none rivals the elaborate ceremonies of the Western Grebe. Two, in particular, are quite spectacular. Carrying vegetation in their bills during the "Weed Ceremony," the birds stand erect in the water, facing each other as they touch bills and "dance" in a circle. Perhaps better-known is the "Rushing Ceremony," where, positioned side-by-side, the birds thrust their bodies upward out of the water and rapidly paddle forward with their feet (LaPorte et al. 2020b).

The Western Grebe is confined to the western United States, northern Mexico, and the southern half of the western Canadian Provinces. Minnesota lies on the eastern periphery of its breeding range. Scattered pockets of high abundance occur in several localities, but the highest densities are found in northern California and southern Oregon.

MINNESOTA BREEDING DISTRIBUTION

Roberts's 1932 account of the Western Grebe is brief. He considered it a "rare summer resident" that occurred "irregularly and in limited numbers" along the state's western border. At the time, there were only two nesting records from the late 1800s, both from Heron Lake in Jackson County. Although the birds continued to be sighted on the lake until 1923, there have been only intermittent reports since then (e.g., Koster 1995).

It was 1935 before the species was observed nesting at a new location, Lake Traverse in Traverse County. Additional reports later surfaced from Nicollet County (1961), Stevens County (1962), Marshall County (1969), and Kandiyohi County (1970). Reports of migrants also began increasing. In her status summary of the state's population, Burger (1971) concluded that Western Grebes were expanding eastward.

Little had changed when Green and Janssen published another update in 1975. Ten years later, Guertin and Pfannmuller (1985) summarized the Minnesota Nongame Wildlife Program's efforts to assess the status and distribution of all colonial waterbirds nesting in the state. The effort identified 29 Western Grebe colonies in 17 counties that were active at least once between 1981 and 1983. Colonies stretched as far east as Freeborn County in the south and Wright County in east-central Minnesota. Seven of the colonies were located in Big Stone County, and all but four of the colonies numbered 50 or fewer nesting pairs. Lake Osakis in Todd/Douglas Counties supported the largest colony, with more than 200 nesting pairs in 1981 (Guertin and Pfannmuller 1985).

When Janssen (1987) published his updated account, notable extensions to the species' distribution included Hennepin County in east-central Minnesota and Roseau County in the northwest. Since 1970, nesting had been documented in 24 counties. By 1998, Hertzel and Janssen had added another 11 counties to the list, again expanding the species' range eastward to Anoka County, north of the Twin Cities, and to Crow Wing County in north-central Minnesota.

By 2014, the MBS had identified 48 Western Grebe breeding-season locations. All but three of the locations were confined to the Prairie Parkland and Tallgrass Aspen Parklands Provinces. The remaining sites were located on the western border of the Eastern Broadleaf Forest Province, suggesting the species might be retreating westward.

MNBBA participants documented Western Grebes in 53 of the surveyed atlas blocks spanning 27 counties. Breeding was confirmed in only seven blocks and deemed probable in 17 **(A)**. Given the immediate adjacency of many blocks, the birds were

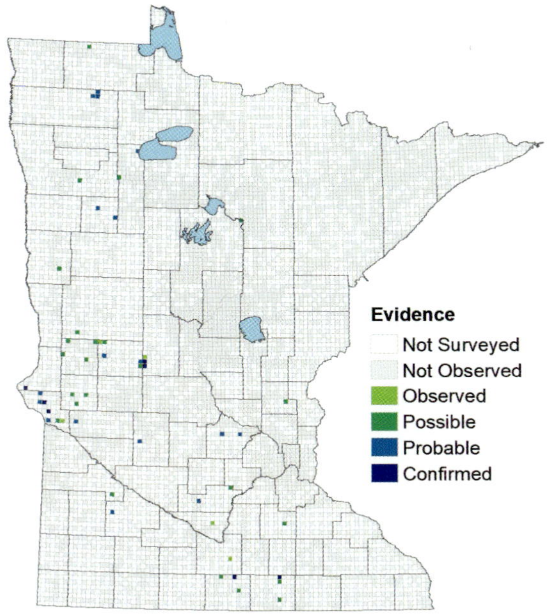

A. Breeding distribution of the Western Grebe in Minnesota based on the Breeding Bird Atlas (2009–13).

either confirmed (6) or probable (12) nesters in 18 locations. Three of the atlas reports came from the outlying metropolitan counties of Isanti and Wright; there was also one possible report on the border of Cass and Itasca Counties. The species was locally common in west-central Minnesota but absent from much of southwestern, southeastern, north-central, and northeastern Minnesota. Of the six grebe species observed during the MNBBA, Western Grebes comprised only 4.7 percent of the records. The Pied-billed Grebe and Red-necked Grebe together accounted for 93.4 percent of all detections.

Although the Western Grebe was more widely distributed than Roberts documented in the early 1900s, the small number of MNBBA confirmed and probable breeding localities (18), compared to the 29 colonies identified in the 1980s, is cause for concern. Elsewhere within its breeding range, the only notable change in its distribution was an expansion in the lower Colorado River valley, where the species is nesting on large reservoirs (LaPorte et al. 2020b).

BREEDING HABITAT

Western Grebes utilize large semipermanent or permanent wetlands with extensive beds of emergent vegetation interspersed with open water **(B)**. Bulrushes and phragmites are generally preferred to cattails (Nuechterlein 1975). Dense stands of submerged vegetation, such as milfoil and pondweed, also may be used. In Minnesota, Burger (1971) found the birds preferred nesting in inlets where the vegetation was more sparsely distributed.

POPULATION ABUNDANCE

As with many grebes, a reliable assessment of the species' population status is needed. One of the earliest assessments of the Western Grebe's population following recognition of the Clark's Grebe as a separate species was in 2002, when Kushlan and his colleagues projected a North American population of at least 110,000 Western Grebes and 10,000–20,000 Clark's Grebes. Nearly two decades later, PIF (2021) biologists generated significantly higher estimates of 1 million Western Grebes and 81,000 Clark's Grebes.

Despite these differences, in the United States BBS data illustrate a reliable and stable breeding population in the survey's early years and a small, albeit erratic, increase in recent years **(C)**. Unfortunately, BBS road surveys are not an effective tool for monitoring clustered populations of wetland species and they continue to combine data for both Western and Clark's Grebes. National Audubon's CBC, on the other hand, suggests a significant population decline over the same time period (Meehan et al. 2020). Statistically reliable data are not available for Minnesota.

Explanations for these conflicting trends are not readily apparent. The CBC data, however, may be more reliable. An in-depth analysis of these data from 1975 to 2010 along the Pacific coastline, the Western Grebe's primary wintering range, revealed a 52 percent decline in the continental population and a 900 km shift in the core of the species' winter distribution from British Columbia and northern Washington south to California (Wilson et al. 2013). Habitat losses and human disturbance on the breeding grounds, coupled with mortality due to oil spills and gill nets in coastal waters, were identified as major factors responsible for the population decline.

B. Although single pairs may nest on smaller wetlands, Western Grebe colonies select large wetlands with extensive beds of emergent vegetation or large lakes with protected bays. Photo by Lee A. Pfannmuller.

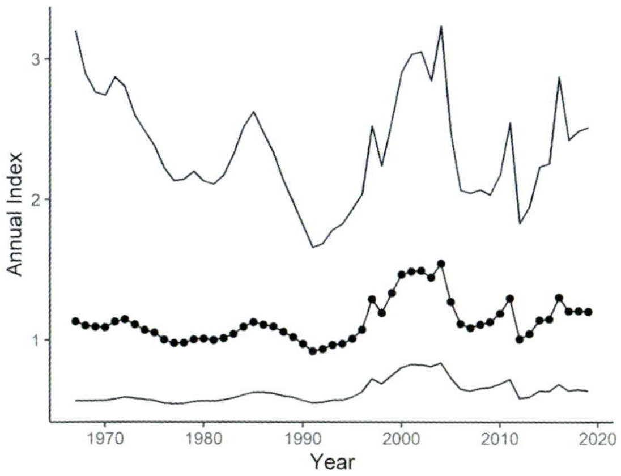

C. Combined breeding population trend and 95 percent credible interval (solid black lines) for the Western Grebe and Clark's Grebe in the United States based on the 1967–2019 Breeding Bird Survey (Sauer et al. 2020).

The southern shift in the core of the Grebe's winter range was attributed to a major shift in the abundance of two primary food items, Pacific herring (*Clupea pallasii*) and Pacific sardines (*Sardinops sagax*). It was not clear, however, if the latter also contributed to the cumulative population decline.

CONSERVATION

The Western Grebe was massacred by the thousands for its silky white feathers, and many colonies were decimated in the late 1800s and early 1900s. Although the population recovered, numbers remain relatively small. This, coupled with the concerns mentioned above, contributed to the species' designation as a moderate conservation priority (PIF 2021).

Management recommendations include maintaining stable water levels during the breeding season and reducing disturbances to nesting colonies. Colonies on lakes that are heavily used for water-based recreation are particularly vulnerable. Closing stands of emergent vegetation during the nesting season could help reduce disturbance and nesting failures. Periodic counts at some of the largest Minnesota colonies, such as Lake Osakis, would also provide essential data on this rare wetland species.

Warming temperatures are predicted to have their greatest impact on Western Grebe populations in the western states and in northwestern Mexico. Even then, under all three warming scenarios examined, NAS (2019) ranked the species with a low vulnerability to climate change, with the potential gains in habitat exceeding the potential losses. At least in the northern Great Plains, the future of the elegant "swan-grebe" may be secure in the near term.

CLARK'S GREBE

Aechmophorus clarkii

Minnesota Seasonal Status: Summer visitant; 10 records of presumed hybrid pairings; regular migrant. The Clark's Grebe was very rare during the MNBBA.
Migration: Available data suggests that most birds winter along the Pacific coast; populations in the southwestern United States and Mexico appear to be permanent residents.
Food: A diver that feeds primarily on fish as well as crustaceans and, in coastal waters, bristle worms.
Nest: Not well-known but considered similar to the Western Grebe's platform of aquatic vegetation.
Conservation Concern: Assigned a high PIF Continental Concern Score of 14/20.

Photo by Gregory Lis.

The type specimen for the Clark's Grebe was collected on its wintering grounds in the 1850s by John H. Clark, a zoologist working on the United States/Mexican Border Survey. The scientific description was published in 1858 by S. F. Baird, J. Cassin, and G. N. Lawrence (Mearns and Mearns 1992). This grebe languished for more than one hundred years as a color phase of the Western Grebe. In 1985, it obtained full species status after the publication of several genetic studies and formal acceptance by the AOU (1985).

The Clark's Grebe is remarkably similar to the Western Grebe in appearance and also shares its spectacular water-dancing breeding displays. The most notable physical distinctions are the Clark's Grebe's orange bill (yellowish-green in the Western Grebe), white lores (dark in the Western Grebe), and the black crown only extending above the red eyes (in the Western Grebe the black crown extends just below the eyes). Other more subtle distinctions pertain to the extent of the white plumage in the wings and on the flanks (more in the Clark's Grebe) and finer points regarding the shape of the crown's black cap and the plumage on the back. Overall, the greater amount of white showing on the head and flanks led to the Clark's Grebe long-standing identification as a "white morph" of the Western Grebe. Unfortunately, as Eckert (1989a) points out, there is considerable variability in all these field marks. Not only do the two grebes look alike to the casual observer, they also appear to share similar habits and habitats. Hybrids also occur, so a positive identification can be difficult.

During the breeding season, Clark's Grebe is found throughout the West from British Columbia to southern California, then eastward to the Dakotas, Nebraska, and northwestern Texas. During the winter the species is most common from southern California to southern Baja California (LaPorte et al. 2020a). In its breeding range, it commonly occurs as a rare resident in Western Grebe colonies. Recent surveys in adjacent states classified the Clark's Grebe as "observed" at three locations in North Dakota (Drilling 2015) and a "confirmed" breeder in several northern counties in South Dakota (Drilling, Sparks, et al. 2016). It has not been reported in Iowa. A Mexican subspecies (*A. c. clarkii*) also resides in portions of central Mexico, where it is a permanent resident.

MINNESOTA BREEDING DISTRIBUTION

When the taxonomic status of the Clark's Grebe was changed to a full species, field observers in Minnesota were alert for observations of the species in the state's scattered Western Grebe colonies. The first record approved by MOU was at Ash Lake, Grant County, on May 10, 1986 (Janssen 1986b). Given the uncertainty about how to separate it from the Western Grebe, MOU requirements were conservative, emphasizing detailed field notes over ambiguous photographs (Eckert 1990b). Three more migration records during the 1980s were accepted by MOU: Heron Lake, Jackson County; Moorhead, Clay County; and, most interestingly, one in the Duluth estuary, St. Louis County, on May 7, 1987. This latter record was well-documented (Hendrickson 1987). It was in a small flock of Western Grebes, which could be seen from both sides of the river, and consequently became the first record for Wisconsin (S. D. Robbins 1991). Seven other 1980s records were not accepted (Eckert 1989a). However, upon reviewing all these reports and the challenges of distinguishing the two species, Eckert concluded, "Early indications suggest that this species may actually be rare but Regular in Minnesota." Finally, in 1991, its presence during the summer months was firmly established at Thielke Lake in Big Stone County (Svingen 1991; Eckert 1991).

From 1986 through 2019, the Clark's Grebe has been recorded during one or more seasons in all years but three (MOU 2021a); its designated status by the MOU is "regular, rare migrant." Its summer occurrence is less regular, occurring eight years in the 1990s, seven years in the 2000s, and five years in the 2010s. Most of the summer records are concentrated in southwestern counties, outlined by Yellow Medicine to Faribault and Todd to Traverse. Migrant records are single occurrences scattered along the northwestern border counties (Clay, Kittson, and Marshall); southern border counties (Faribault, Freeborn, Jackson, and Martin); and the Twin Cities (Anoka, Dakota, and Hennepin). There is only one winter record: a bird found alive on December 7, 2015, along Highway 2, Lake County. Uninjured, it was released in Lake Superior (Fall et al. 2016).

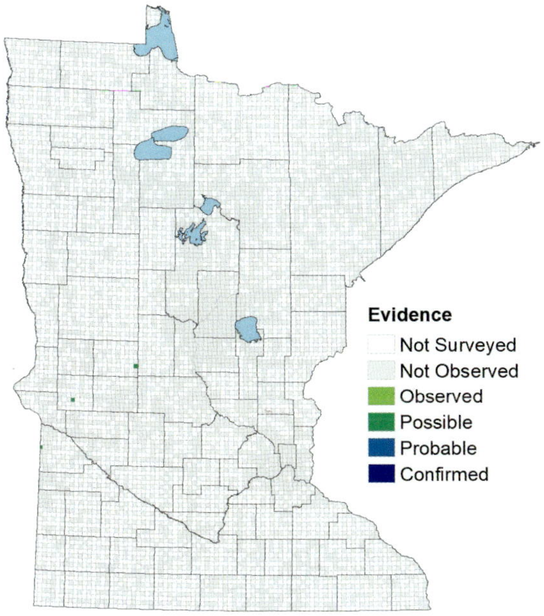

A. Breeding distribution of the Clark's Grebe in Minnesota based on the Breeding Bird Atlas (2009–13).

Because of identification subtleties and observational difficulties in Western Grebe colonies where the Clark's Grebe is found, there are no records of a mated pair of Clark's Grebe producing young. There are numerous records of young being fed or carried on the back of an adult Clark's Grebe, but a Clark's Grebe mate has never unequivocally been identified. Consequently, the observations from ten years with breeding evidence are considered hybrid pairings. The best published details of breeding activity are early records: Lake Traverse, Traverse County (Eckert and Buer 1991); Minnesota Lake, Faribault County (Janssen 1994); and Lake Osakis, Todd/Douglas Counties (Bardon 1997). Even though hybrid pairings are reported, LaPorte and his colleagues (2020a) stated, "Although there are field observations that suggest that hybrids between *A. occidentalis* [Western Grebe] and *A. clarkii* [Clark's Grebe] are fertile, this has yet to be fully examined." Indeed, the authors' general consensus was that, although there is a broad overlap in the breeding distribution of the Western and Clark's Grebe, mixed pairings and hybrids are generally rare. Clearly there is a need for in-depth field studies.

Unfortunately, recent (2000 to 2020) breeding records for Clark's Grebe in Minnesota are only short summaries in the Seasonal Reports of *The Loon,* with single observations that make it difficult to assess the species' overall status. Two places with multiple-year and possible hybrid breeding evidence that stand out are Thielke Lake, Big Stone County (2000, 2002, 2006) and Lake Osakis (2005, 2012, 2017) (MOU 2021a, 2021b). Indeed, the 2012 Lake Osakis record occurred during the MNBBA (adult Clark's Grebe with a flightless juvenile) as did breeding evidence gathered at Gorder Lake, Stevens County (2010; hybrid pair feeding young). Details on both records can be found in Janssen (2019) and MOU (2021a). One lone individual was also reported during the 2011 MNBBA field season at Pegg Lake in Lac qui Parle County (MOU 2021b). Although the MNBBA data entry system in use at the time did not recognize Clark's Grebe as a Minnesota breeding species, preventing data on the species to be entered, the records at Gorder Lake, Pegg Lake, and Lake Osakis (2012) place the Clark's Grebe on the list of species present in Minnesota during the atlas period. All three reports were deemed *Possible* breeding records **(A)**.

BREEDING HABITAT

In Minnesota, the Clark's Grebe has frequently been reported within Western Grebe colonies, which are found on freshwater lakes and wetlands with abundant emergent vegetation and open water **(B)**. Nevertheless, LaPorte and colleagues (2020a) noted that although "mixed colonies and flocks [are] common," the Clark's Grebe "[a]ssociates significantly more frequently with [its] own species than with Western Grebe[s] on the breeding grounds." Further research is needed to identify any unique differences in their habitat requirements.

POPULATION ABUNDANCE

PIF (2021) provided a breeding population estimate of 72,000 adults in the United States and Canada, with an additional 9,000 residing in Mexico. The total population of Western Grebes, which have a similar distribution to the Clark's Grebe, is more than 10 times larger, at 1 million birds. Unfortunately, until the species is studied more intensively, population trends are unknown. Indeed, since the BBS began its survey work in 1966, data on the Clark's Grebe has always been combined with the Western Grebe, which shows an overall stable population trend through 2019 throughout the survey region (Sauer et al. 2020).

CONSERVATION

Ranked as a high conservation priority (PIF 2021), largely because of its small population size and threats to its summer and winter wetland habitats, Clark's Grebe remains a species unfamiliar to most amateur birders, field biologists, and professional ornithologists. Indeed, not a single in-depth study on the species has yet been undertaken (LaPorte et al. 2020a). Much of the literature available focused on distinguishing the color morphs of the Western Grebe from one another until the "white morph" (i.e., the Clark's Grebe) was given full species status.

Given that the species appears to share the same wetland and coastal habitat utilized by Western Grebes, management concerns include human disturbance on the breeding grounds, changes in water levels during the critical breeding season, and predator access to nests. Oil spills, fishing nets, and contaminants are principal concerns in coastal winter habitats (LaPorte et al. 2020a). Climate change models developed by NAS (2019) predicted that warming temperatures will impact Clark's Grebe breeding populations in southern California the most, with a significant loss of habitat. In the Northern Great Plains, however, populations are predicted to expand eastward in both the Canadian Prairie Provinces of Saskatchewan and Manitoba and in eastern North Dakota, eastern South Dakota, and northwestern Minnesota.

Meanwhile, as we await potential opportunities for enjoying this new member of our avifauna even more, the most important need to guide future conservation and management is the initiation of field studies that further illuminate how it can be identified, its unique life history, and its preferred habitats.

B. Emergent vegetation along the shoreline of Lake Osakis, located in Todd and Douglas Counties, supports a large nesting colony of Western Grebes and frequently attracts the similar-looking Clark's Grebe as well. Photo by Gary T. Seim.

ROCK PIGEON

Columba livia

Minnesota Seasonal Status: A permanent resident found statewide. The Rock Pigeon was common during the MNBBA.
Migration: Year-round, nonmigratory resident.
Food: Ground forager that consumes seeds and fruits. In cities, the birds forage on food litter and are often fed popcorn, bread, peanuts, and various other food items.
Nest: Natural nest site is a rocky cliff ledge; in urban and rural landscapes it nests on a variety of protected ledges, including under bridges, on tall buildings, and on barn lofts.
Conservation Concern: As an introduced species, the Rock Pigeon is not a conservation concern and is an unprotected species in Minnesota.

Photo by Laura Erickson.

The common Rock Pigeon is one of the few wildlife species that urban inhabitants regularly encounter. Proudly strutting along the city's busy streets and parks, the birds readily consume our handouts and refuse while they roost and nest on the city's ubiquitous ledges, rooftops, bridges, and other erected structures. Equally at home in agricultural landscapes, they are a frequent fixture perched atop a farmstead's silos and barns or feeding on waste grain in the nearby fields. We so closely associate Pigeons with human habitation and take their presence for granted that their fascinating story is often overlooked.

The bird is originally a native of Europe, northern Africa, and India. Its common name is a reference to its native habitat of rocky ledges and cliffs. Domesticated Pigeons were introduced to North America by immigrants in the early 1600s and were the source of today's feral populations (Lowther and Johnston 2020). Although they display remarkable variation in color, most birds are bluish gray with iridescent throat feathers and two dark wing bars. They are commonly observed in flocks; their primary vocalization is a soft *coo.*

Today this adaptable species can be found in southern Alaska, across southern Canada, throughout the United States, and south to Mexico, Central America, the Caribbean, and northern South America. It is among the most intensively studied birds in the world, yielding new insights on a diversity of topics, including avian flight, navigation, and sensory perception (Lowther and Johnston 2020). One of the most intelligent and agile wildlife species, it possesses legendary homing abilities and speed of flight, which earned the birds high accolades during World War I and World War II. During an era when technical communication systems were still rudimentary and often undependable, carrier Pigeons were a reliable method for communicating messages between ships, ground troops engaged in battle behind enemy lines, and distant headquarter operations. Selectively bred for their homing skills, the birds had the ability to find their way back home from sites nearly 1,000 km away. Approximately 100,000 carrier Pigeons were used in World War I and successfully reached their destinations 95 percent of the time. Thousands more were employed during World War II. Indeed, 32 "war pigeons" were awarded medals from the United Kingdom for their contributions to the war effort and for saving hundreds of lives as military messengers (BBC 2019).

MINNESOTA BREEDING DISTRIBUTION

For a species that virtually surrounds anyone who lives in one of Minnesota's large cities or on a rural farmstead, we know remarkably little about the Rock Pigeon's first appearance in the state. It was 1960 before a specimen was even added to the collections at the University of Minnesota's Bell Museum of Natural History.

Throughout North America, early naturalists and ornithologists essentially ignored the species for more than two centuries. When people spoke of pigeons in the 1700s and 1800s, they invariably were referring to the country's most abundant member of the pigeon family, the Passenger Pigeon (*Ectopistes migratorius*). Years later, Arthur C. Bent's encyclopedic 21-volume series *The Life Histories of North American Birds* did not include an account of the Rock Pigeon, even though he addressed in detail many other nonnative species, including the European Starling and House Sparrow. In Minnesota, neither Hatch's 1892 account of Minnesota birds nor Roberts's 1932 two-volume treatise on the state's avifauna makes mention of the bird. In fact, it was 1974 before participants in National Audubon's annual Christmas Bird Counts regularly tallied the number of Rock Pigeons they observed (R. Brewer et al. 1991). Up until then, the birds were thought of more as urban pests than as truly "wild" members of the local avifauna.

Upon the introduction of Rock Pigeons in the early 1600s, early colonists domesticated them as a common farm animal, raised solely for the purpose of food. Escaped birds eventually gave rise to today's widely distributed feral populations. No doubt some of Minnesota's earliest settlers traveled with Pigeons in tow or had them shipped in at a later date. Unfortunately, we can only conjecture about the species' appearance based on accounts from other midwestern states. Even there the records are scanty. Temple remarked on the sparse records available during the nineteenth century in Wisconsin, commenting, "Perhaps these early pigeons were dismissed as captives and not recognized as truly feral birds, or perhaps birds that escaped were soon shot for pigeon pie!" (in Cutright et al. 2006). Farther east, in Ohio, there is a somewhat better account of the species' appearance. A feral population first became established on Kelleys Island in Lake Erie in 1906; then birds appeared in Cleveland in the mid-1920s and were widespread in the state by the 1930s (Rodewald et al. 2016). So it certainly seems possible that some feral birds were already present in Minnesota in the early 1900s.

The first written account of the species' presence in Minnesota was in the state ornithological journal, *The Loon,* in the summer of 1944. A summary of nesting records for the previous year included a small note that on May 26, 1943, young Pigeons were observed at the state capitol in St. Paul (Chambers 1944). Just thirty years later, when Green and Janssen (1975) published their updated status account, the species was already considered statewide in distribution, present wherever farms and cities dotted the landscape. Janssen (1987) later reported that nesting had been confirmed in five counties since 1970; Hertzel and Janssen (1998) added an additional six counties to the list. No doubt it was more widespread as a nesting species but simply not a priority for documentation among birding enthusiasts.

During the MNBBA, observers reported a total of 2,896 Rock Pigeon records in 37.7 percent (1,824/4,841) of the surveyed atlas blocks and in 47.2 percent (1,103/2,337) of the priority blocks. Breeding was confirmed in 418 blocks **(A)**. The species was reported in all 87 Minnesota counties and was confirmed breeding in 71 counties. Records were least abundant in the far northeastern counties of Cook, Lake, and northern St. Louis and in the far north-central region of the state.

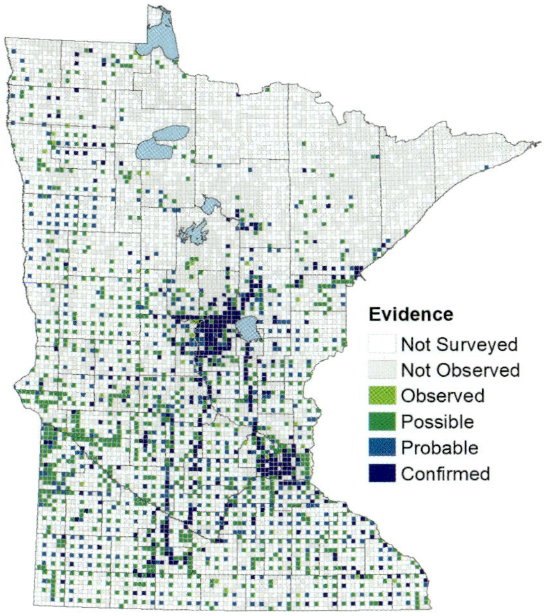

A. Breeding distribution of the Rock Pigeon in Minnesota based on the Breeding Bird Atlas (2009–13).

B. Farmsteads, with their abundance of food and nesting opportunities, provide ideal habitat for the Rock Pigeon across Minnesota's agricultural landscapes. Photo by Lee A. Pfannmuller.

BREEDING HABITAT

Wild Rock Pigeons in their native habitat nest on the ledges of rocky cliffs. Today, some feral birds will nest in such habitats when they are available and are close to a dependable food source, but they are more often found in close association with people, occupying small cities, dense urban areas, and rural farmsteads **(B)**. Any suitable ledge may be deemed appropriate nesting habitat, from small ledges on tall buildings, to large highway overpasses and small bridges, to grain silos and empty buildings. The most important requirements are simply a flat surface and some cover (Lowther and Johnston 2020). Today, feral populations of the Rock Pigeon are almost entirely dependent on human landscapes for roosting sites, nesting sites, and food.

POPULATION ABUNDANCE

PIF (2021) estimated the global population of Rock Pigeons at 140 million birds, of which approximately 16 million occur in the United States and Canada. The statewide estimate for Minnesota is 690,000.

Surprisingly, as adaptable as this species is, it has experienced a steady and significant population decline throughout its range. BBS data collected across the survey region since 1966 illustrate that populations have declined at a rate of −0.71 percent per year through 2019 (Sauer et al. 2020). Some of the biogeographical regions with the greatest average annual declines occur in the central and eastern United States, including the Eastern Tallgrass Prairie (−2.49 percent), the Central Hardwoods (−2.30 percent), the southeastern Piedmont (−4.63 percent), and the New England and Mid-Atlantic coast (−2.94 percent). BBS data collected in Minnesota demonstrate a slower but statistically significant decline for the species, averaging −0.64 percent per year from 1967 to 2019 **(C)**.

Factors responsible for the decline have not been closely examined but may include changes in agricultural operations and increased predator populations in urban areas. Focused efforts in some rural areas and cities to reduce local populations are also not uncommon (Lowther and Johnston 2020).

CONSERVATION

Despite declining populations, the feral Rock Pigeon remains a highly adaptable species that is widespread and relatively abundant across North America. A warming climate should pose no threat to this species. Since the bird is considered a pest in cities and a potential vector for livestock and poultry diseases in farming communities, management is focused primarily on reducing its use of urban and agricultural buildings for roosting and nesting with the use of poisoned bait, traps, and other creative tactics (Lowther and Johnston 2020).

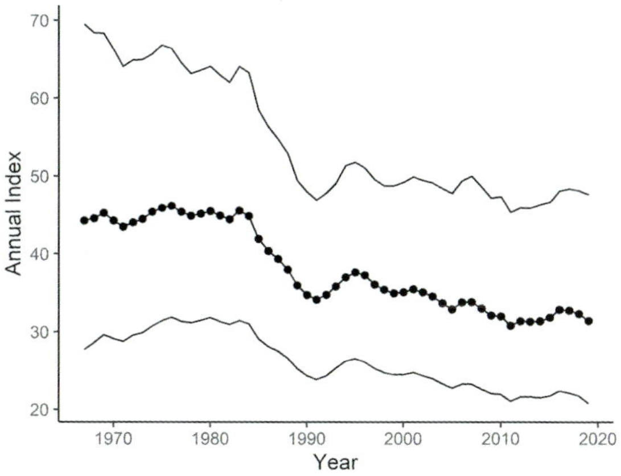

C. Breeding population trend and 95 percent credible interval (solid black lines) for the Rock Pigeon in Minnesota based on the 1967–2019 Breeding Bird Survey (Sauer et al. 2020).

EURASIAN COLLARED-DOVE

Streptopelia decaocto

Photo by Gerald Hoekstra.

Minnesota Seasonal Status: Regular breeding resident primarily in southern and western Minnesota, although its range is continuing to expand northward; winter observations are increasingly common. The Eurasian Collared-Dove was uncommon during the MNBBA.
Migration: Migratory movements unclear; appears to be a permanent resident but may move in response to food and roost site availability.
Food: Ground forager that feeds primarily on seeds as well as some insects and berries.
Nest: Stick platform usually placed within a tree or shrub but sometimes found on man-made structures.
Conservation Concern: As an introduced species, the Eurasian Collared-Dove is not a conservation concern and is an unprotected species in Minnesota. Its impact on native species, as the population continues to expand, remains uncertain.

The sights and sounds of the ubiquitous Mourning Dove and Rock Pigeon are so familiar that one might easily bypass another member of the dove family that is a new arrival to Minnesota, the Eurasian Collared-Dove. First seen in 1998 in Big Stone County, this introduced species has rapidly expanded across the state and has been documented nesting in several southern counties.

The story of the Eurasian Collared-Dove is an all-too-familiar account of the introduction of an exotic species. Oftentimes our knowledge of how a species was introduced is sketchy and speculative, but, in the case of the Collared-Dove, a very detailed account was documented by P. William Smith in 1987.

The story begins in the early 1970s. A small pet shop owner in the Bahamas asked a colleague traveling to Europe to acquire 10 pairs of Ringed Turtle-Doves (*Streptopelia capicola*) so he could build a local breeding stock for his business. Ringed Turtle-Doves (now known as African Collared-Doves) are a popular cage bird in the pet industry and have been domesticated for thousands of years. Thinking he had acquired what his friend requested, the buyer returned with "Indian Ring-necks" that, to the untrained eye, look nearly identical to Ring-necked Doves. But the buyer had acquired an entirely different species, the Eurasian Collared-Dove (*S. decaocto*). The latter were often referred to as Indian Ring-necks because it was believed that they were restricted largely to the Indian subcontinent before they expanded westward in the sixteenth century. Others, however, have pointed out that the species was known to the ancient Greeks and was likely present in a "relatively small area around the Black and Aegean seas" for at least four hundred years and "perhaps much longer" before expanding farther across Europe (P. Smith 1987).

A few years after receiving his new acquisition, the Bahama breeder's aviary was burglarized in 1974 and several Collared-Doves escaped. The owner decided to simply release the remaining birds and abandon his original plans to establish a local breeding stock. The number of Collared-Doves released is unknown but likely was fewer than 50.

P. Smith (1987) believes these released birds are the main source of the current North American population. There may have been a few other releases on the same island by those who acquired some of the original stock. And at least one other release is known to have occurred nearly 1,900 km to the southeast on the island of Guadaloupe, when a volcano was threatening to erupt after the Collared-Doves were originally acquired in 1976 (CLO 2017d; Lever 2005). P. Smith (1987) carefully documents the Dove's dispersal to adjacent islands, and he makes a strong case for the absence of any released birds in Florida or anywhere else in North America that could have been the original source of the introduction.

The Ringed Turtle-Dove and the Eurasian Collared-Dove are so similar that the exact date of the Collared-Dove's arrival in Florida is unknown, because the birds were initially dismissed as being escaped Ringed Turtle-Doves. P. Smith and his colleague Herbert W. Kale II irrefutably identified the increasing number of Collared-Doves in Florida as *S. decaocto* in the spring of 1986 (Smith and Kale 1986).

The similarity of the two species is the foundation of the confusion and the need to carefully discern the diagnostic features that separate the two species. Escaped, or intentionally released, Ringed Turtle-Doves always remain a possibility. Overall, the Eurasian Collared-Dove is a sandy gray bird with a distinct black collar on the back of the neck. *Streptopelia,* the Latin name for the genus, means collared (*streptos*) dove (*pelia*). The Ringed Turtle-Dove may be slightly lighter in color, but the best diagnostic tool is the bird's voice. The Eurasian Collared-Dove has a three-syllable *koo-KOO-kook* call, compared to the softer, usually two-syllable *kooeek-krrroooo(aw)* call of the Turtle-Dove (P. Smith 1987). In Minnesota, the primary challenge is to distinguish the bird from our ubiquitous Mourning Dove, which is darker in color, lacks the black neck collar, and has a pointed tail rather than the Collared-Dove's rounded tail. Unlike the rapid three-syllable call of the Collared-Dove, the Mourning Dove's slower, five-syllable call starts softly followed by louder coos.

Following their arrival in Florida, the Collared-Dove's range rapidly expanded across the United States with reports documented as far north as southern Alaska. Dispersal to the northwest is characteristic of the species' general movements. Along this gradient, its expansion tends to follow human-altered landscapes (Fujisaki et al. 2010). Although most of the expansion is presumed to have originated from wild birds dispersing from Florida, some caged birds have also been released in other localities. Today the species is broadly distributed across the western, central, and southeastern United States. It is largely absent from the Great Lakes region, New England, and the mid-Atlantic and north-central states.

MINNESOTA BREEDING DISTRIBUTION

It was only twelve years after the Eurasian Collared-Dove's appearance in Florida was confirmed that the first bird was observed in Minnesota, in the spring of 1998 in Big Stone County (Eckert 1998a). During the next two years, confirmed and well-documented reports were submitted from other southwestern counties, including Lyon, Martin, Mower, Pipestone, and Rock, as well as one record farther north, in Kandiyohi County (Svingen and Kinstler 2002). Several of the reports were of multiple birds (two to four). Then, in 2001, the first documented nesting report came from a municipal park in the town of Caledonia in Houston County (Svingen and Kinstler 2002). That same year, one bird was reported as far north as Roseau County. Enough reports continued to accumulate over the next few years that in 2004 the official Records Committee of the Minnesota Ornithologists' Union changed the statewide status of the Eurasian Collared-Dove from Accidental to Regular (Eckert 2004).

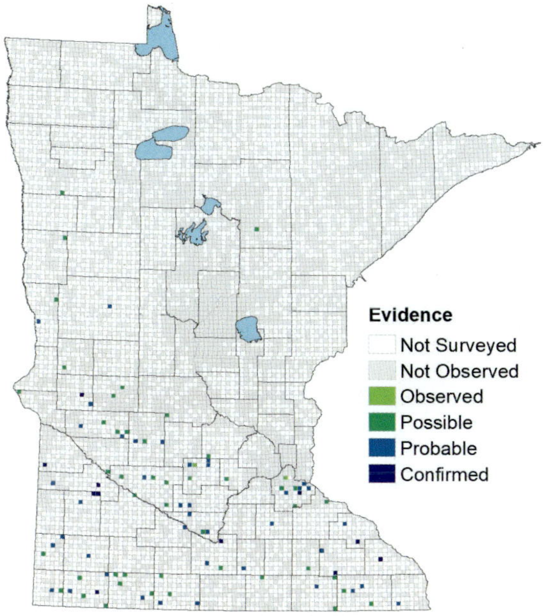

A. Breeding distribution of the Eurasian Collared-Dove in Minnesota based on the Breeding Bird Atlas (2009–13).

B. Spillage from grain silos and scattered trees for nesting meet the habitat requirements for an expanding Eurasian Collared-Dove population in small agricultural communities across Minnesota. Photo by Gary T. Seim.

The MNBBA documented 133 Eurasian Collared-Dove records in 2.1 percent of the atlas blocks that were sampled and in 2.5 percent of the priority blocks. Breeding was confirmed in 11 blocks covering 10 counties **(A)**. Only one county with confirmed breeding (Stevens) was north of the Minnesota River. The species was most abundant in the southern half of the state, with scattered records as far north as Polk County in the northwest and Itasca County in the northeast. A few years later, Janssen (2019) noted two additional nesting records north of the Minnesota River (McLeod and Stearns Counties) and two in southeastern Minnesota (Fillmore and Steele Counties). During the summer of 2018, nesting was also confirmed in Grant County (MOU 2021b).

BREEDING HABITAT

The primary requirements of the Eurasian Collared-Dove's habitat are trees, open ground, and available sources of grain. A variety of habitats meet these criteria, including city parks, residential areas, farmsteads, and large grain storage facilities **(B)**. Heavily forested areas are usually avoided.

POPULATION ABUNDANCE

PIF (2021) provided a global population estimate of 85 million birds. They estimated that approximately 10.2 percent of the global population, or 8.7 million birds, occur in the United States and Canada. The statewide estimate for Minnesota is 1,800 individuals. The BBS sample size is still too small at both the national and state level to conduct a robust statistical assessment of population trends. Nevertheless, the rapid growth of the population is clearly evident in the limited data that are available (Sauer et al. 2020).

Despite its rapid range expansion, the species remains a very uncommon bird on the northern periphery of its range in Minnesota. Across North America the highest breeding densities are encountered farther south and west, with particularly high numbers in western Nebraska and western Texas.

CONSERVATION

As an introduced, nonnative species, the Eurasian Collared-Dove is not a protected species. Having originated from Eurasia, it will find little that is likely to impede its rapid expansion northward, including an increasingly warmer climate. The bird has even become a popular bird to hunt in some regions of the southern United States, where population numbers are the highest.

There is some concern that the Collared-Dove may become so abundant that it will be the equivalent of the "gray starling," competing for available resources with other native North American birds (P. Smith 1987). Some have speculated, for example, that the bird's larger size could place the smaller-sized Mourning Dove at a competitive disadvantage, particularly when securing food resources. A study by Poling and Hayslette (2006) suggested, however, that Collared-Doves were not more aggressive or competitive than Mourning Doves in experimental trials. Regardless, because the real impact of the Eurasian Collared-Dove's rapid invasion is unknown at this early stage, populations of doves native to North America should be closely monitored. Meanwhile, Minnesotans should keep an eye out for this new member of our state's avifauna.

MOURNING DOVE

Zenaida macroura

Minnesota Seasonal Status: A regular breeding resident and migrant; regularly observed during the winter, especially in the Twin Cities region and southeastern Minnesota. The Mourning Dove was very abundant during the MNBBA.
Migration: Resident to medium-distance migrant; most north-central and northeastern breeding populations winter in the Gulf coast states south into Mexico and Central America.
Food: A ground forager, feeding almost entirely on the small seeds of cultivated grains as well as many grasses and forbs.
Nest: A loose platform of twigs and grass, usually placed in a tree or shrub; may also nest on the ground.
Conservation Concern: A game species; assigned a low PIF Continental Concern Score of 6/20.

Photo by Laura Erickson.

Take a summer drive through Minnesota's farm country and you are sure to spot the striking silhouette of numerous Mourning Doves. When they are perched atop telephone wires, their small head, long, tapered tail, and sleek appearance are unmistakable. Although it is the most abundant and widely distributed game bird in North America, "[i]t is quiet in voice, neutral in color, and so unobtrusive in deportment that it seems little more than a part of the background; a quiet, pastoral bird" (Bent 1932).

Their distinctive mournful call, *coo-oo, oo, oo, oo* is soft and easily overlooked amid the boisterous dawn chorus of birds that share their open-country habitat. Equally unique is the whistling sound made by their wings when they take flight, a trait that distinguished Doves from the now extinct but somewhat similar Passenger Pigeon. In the late 1800s, when the enormous flights of Passenger Pigeons in southern Minnesota were becoming but a faint memory, Roberts (1932) often received reports from observers who believed they had seen the extinct bird only to be told it was the common Mourning Dove.

Other distinguishing characteristics of this pale, grayish-brown bird are the black dot below the eye, a black bill, and white-tipped tail feathers that are evident in flight. During the breeding season, the males can be distinguished by their light, rosy breast feathers and a light blue cast on their crown and nape. The Latin name, *Zenaida macroura,* means "dove with a long tail."

Most Minnesota residents are familiar with the Mourning Dove. It has adapted well to our urban and agricultural landscapes and is a common visitor to our backyard feeders throughout the year. As prevalent as Mourning Doves are in the state, the core of their breeding range is farther south and west. Their entire range extends from southern Canada to the south across the United States and Mexico.

MINNESOTA BREEDING DISTRIBUTION

The Mourning Dove has always been recognized as a common and abundant summer resident throughout Minnesota, with the exception of the far northeastern and north-central counties, where it is uncommon. In 1932, Roberts noted that the Mourning Dove was one of the few birds in the state that had actually increased in abundance. "It was always common but it is now abundant and getting more so each year. With the clearing and settling of the northern forests it has appeared in many localities where it was formerly absent." He speculated that a decrease in the number of predatory hawks and owls due to persecution may have also contributed to the Dove's improved status. At the time, nesting was documented in 12 counties, ranging from Rock, Pipestone, and Jackson in the southwest, east to Waseca and Goodhue, north to the east-central counties of McLeod, Hennepin, Anoka, Isanti, and Sherburne, and then northwest to Polk and Kittson.

Little changed regarding the Dove's status in the following years. Both Green and Janssen (1975) and Janssen (1987) noted the species was absent from Cook and Lake Counties and from the extensively forested portions of Itasca, Koochiching, and St. Louis Counties. It reached its greatest abundance in the southern and western agricultural regions of the state. Janssen (1987) identified 48 counties where nesting had been confirmed since 1970. Hertzel and Janssen (1998) added another 11 counties to the list, including St. Louis County. The MBS further confirmed the species' common occurrence outside of the Laurentian Mixed Forest Province. Although the Mourning Dove was uncommon in the northern forest, numerous locations were documented in Itasca and St. Louis Counties, as well as one record in northern Lake County.

During the MNBBA, observers reported a total of 6,125 Mourning Dove records in 60.9 percent (2,948/4,842) of the surveyed atlas blocks and in 75.9 percent (1,773/2,337) of the priority blocks. Breeding evidence was documented in 493 blocks **(A)**. Although least abundant in the northern and eastern regions of the Laurentian Mixed Forest Province, the birds were observed in all 87 Minnesota counties. Breeding was

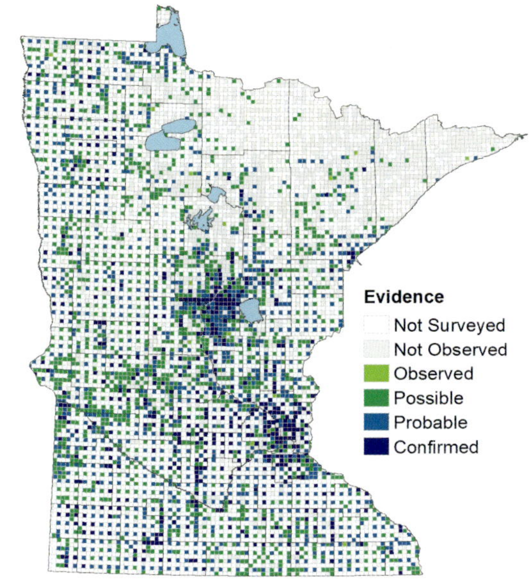

A. Breeding distribution of the Mourning Dove in Minnesota based on the Breeding Bird Atlas (2009–13).

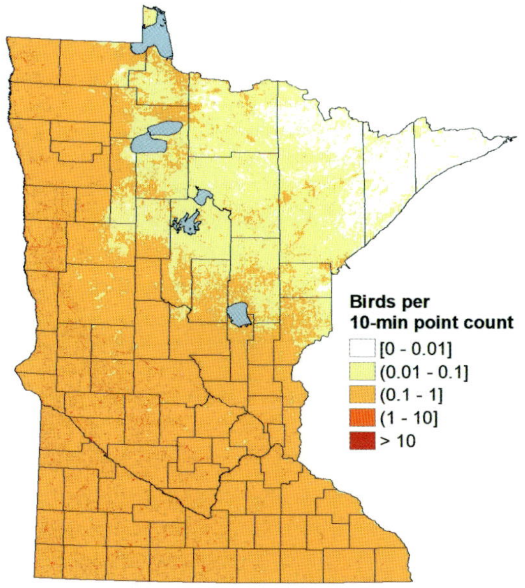

B. Predicted breeding distribution (birds detected per 10-minute point count) of the Mourning Dove in Minnesota based on habitat, landscape context, and climate data gathered during the Minnesota Breeding Bird Atlas (2009–13) using the Generalized Linear Modeling method without an adjustment for detectability.

confirmed in 78 counties; eight of the nine counties where breeding was not confirmed were in northeastern and north-central Minnesota. Twenty of the counties with confirmed breeding were additions to the list published in 1998 by Hertzel and Janssen.

The MNBBA predicted distribution map emphasizes the Mourning Dove's abundance and widespread distribution throughout the Eastern Broadleaf Forest, Prairie Parkland, and Tallgrass Aspen Parklands Provinces **(B)**. In the Laurentian Mixed Forest Province, it is predicted to be most abundant in areas of urban, industrial, and agricultural development, including in Duluth and along the Iron Range.

Prior to European settlement, Mourning Doves were likely present in scattered locations across much of the United States and southern Canada (Otis et al. 2020). As human settlement progressed to the north and west, their numbers increased, and their range expanded even farther, not only in Minnesota but throughout New England and southern Canada. Although clearing the eastern deciduous forest created extensive habitat for Doves, the two most important factors that favored the species were plowing the central grasslands for agriculture and introducing irrigation in the central and western states for cereal production (Tacha and Braun 1994).

BREEDING HABITAT

A bird of open country, the Mourning Dove is most frequently associated with farmsteads, open woodlands, edges of agricultural fields, suburbs, and forest clearings **(C)**. The only areas the Doves strictly avoid are extensively forested landscapes. They reach their highest abundance in areas that provide trees for nesting, perches for calling, and open ground for foraging (Kaufman 1996).

POPULATION ABUNDANCE

BBS data generated a North American population estimate of 150 million Mourning Doves (PIF 2021). Minnesota supports approximately 2.1 million birds. BBS data also demonstrate a significant and consistent downward trend in Mourning Dove populations across the BBS survey region, averaging a statistically significant decline of −0.42 percent per year from 1966 through 2019 (Sauer et al. 2020). The downward trend has been steeper in Minnesota, where the population has declined an average of −0.95 percent per year since 1967 **(D)**. August roadside surveys in Minnesota corroborate the statewide BBS data (Lyons 2020b).

Although annual mortality is known to be high due to a variety of factors, including hunting, predation, and disease, their relative contribution to the species' decline is speculative at best (Otis et al. 2020). Long-term habitat loss is also a concern. Not only is habitat lost to expanding residential development but also more intensive agricultural practices often eliminate shelterbelts and windbreaks that provide important nesting habitat (Tacha and Braun 1994).

CONSERVATION

The most popular game bird in the United States, the Mourning Dove is legally harvested in forty states. Intensively managed, the annual harvest is regulated at the federal level and guided by a national harvest management plan (USFWS 2005b). In 2004, following years of public debate, Minnesota reinstated a hunting season on Doves, nearly six decades after the species was removed from the list of game birds in 1947 (Henderson 1980). Since then, an average of 90,150 birds has been harvested in the state each fall from 2004 through 2019 (Davros and Dexter 2019), a mere fraction of the nearly 10 million birds harvested in the United States in the 2019–20 season (Seamans 2020).

C. In Minnesota's farm country, the Mourning Dove's breeding habitat typically includes a shelterbelt of trees and shrubs for nesting and roosting and adjacent fields for foraging. Photo by Lee A. Pfannmuller.

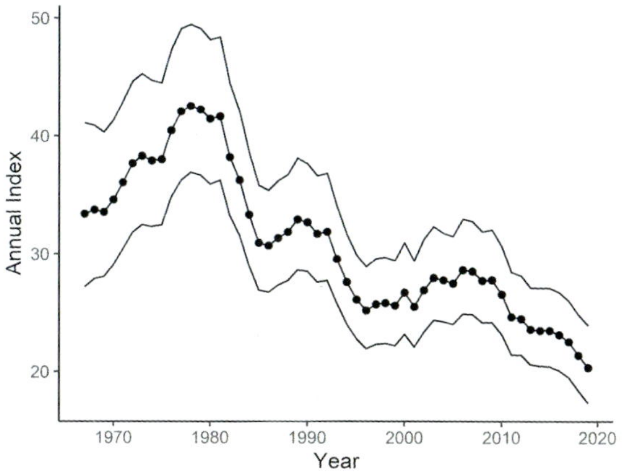

D. Breeding population trend and 95 percent credible interval (solid black lines) for the Mourning Dove in Minnesota based on the 1967–2019 Breeding Bird Survey (Sauer et al. 2020).

Because the numbers harvested are so high, especially in southern states, attention has focused on the Dove's ingestion of spent lead shot. Some agricultural fields are managed specifically to attract flocking birds in the fall for harvest. When these sites receive a lot of hunting pressure, lead shot can accumulate in the fields and be ingested by Doves and other wildlife. Very small numbers of shot can result in Dove mortality, so the overall impact on the population is a primary concern (Franson et al. 2009). One study suggested that the annual losses approximate the losses attributed to the annual harvest, or approximately 3 percent of the population (Schulz et al. 2006). In Minnesota the use of nontoxic shot is not required when hunting Doves unless the hunting takes place on federal wildlife refuges and waterfowl production areas (MNDNR 2015d).

In the long term, despite its declining population trend, the Mourning Dove remains an abundant species, well adapted to human-modified landscapes and with a low conservation priority (PIF 2021). Even warmer temperatures are unlikely to pose a significant threat. As temperatures rise, the most significant change is likely to be a broad range expansion along the northern edge of the Dove's current breeding distribution, including in northeastern Minnesota (NAS 2019). This ubiquitous inhabitant of Minnesota's open country will likely become even more widespread in coming years.

YELLOW-BILLED CUCKOO

Coccyzus americanus

Minnesota Seasonal Status: Regular breeding resident and migrant in Minnesota. The Yellow-billed Cuckoo was uncommon during the MNBBA.
Migration: Long-distance migrant that overwinters in South America, primarily east of the Andes Mountains.
Food: Large insects, such as hairy caterpillars, cicadas, grasshoppers, and crickets, especially eastern tent caterpillars; occasionally bird eggs and young, lizards, and small frogs.
Nest: Loose platform of twigs between two branches or close to the trunk of a tree or large shrub; generally, in thick vegetation and less than 6 m from the ground. Rarely a brood parasite.
Conservation Concern: Assigned a moderate PIF Continental Concern Score of 12/20.

A graceful, elegant bird that glides across open areas or threads its way quietly through the branches of trees, the Yellow-billed Cuckoo is quickly recognized as a cuckoo with its olive-brown upperparts, white underparts, sleek profile, and long tail with black-and-white markings (Bent 1940). Closer examination on whether it has rufous coloration in the wings or a yellow lower bill is required to distinguish it from that of the Black-billed Cuckoo. However, as Bent (1940) observed, "we hear the voice of the cuckoo much oftener than we see the bird." It has a series of call notes, *cu cu cu,* that is difficult to distinguish from that of the Black-billed, but its more complex vocalization is quite different and consists of 8–12 guttural syllables, described as *ka-ka-ka-ka-ka-kow-kow-kowlp-kowlp-kowlp-kowlp* (Hughes 2020b).

The Yellow-billed Cuckoo is a voracious consumer of caterpillars and, therefore, a highly beneficial species. Populations in localized areas can fluctuate substantially. In many years they can be very rare or absent, but in years of caterpillar outbreaks, especially by the eastern or western tent caterpillars (*Malacosoma americanum* or *M. californicum*), either or both cuckoo species can be common or abundant. Consumption rates of 40 gypsy moth (*Lymantria dispar*) caterpillars eaten in fifteen minutes and 47 tent caterpillars in six minutes have been reported (Hughes 2020b).

Photo by Paul Sparks.

The Yellow-billed Cuckoo is found primarily in the Eastern Temperate Deciduous Biome as far west as South Dakota, Nebraska, and Texas, but also has breeding populations in the Greater Antilles and Mexico. Highest densities occur in the West Indies and south-central United States, especially in parts of Arkansas, Mississippi, Missouri, and Oklahoma. The common name of the Yellow-billed Cuckoo is derived from its distinctive bill color, while its scientific name, *americanus,* describes its broad range in the Western Hemisphere.

MINNESOTA BREEDING DISTRIBUTION

The historical distribution of this species is confusing based on the descriptions of Roberts (1932). He emphasized it was more common in the southern portions of Minnesota than the Black-billed Cuckoo, the latter being more common in the north. Yet he cites observations by O.J. Murie from Moorhead, Clay County, in 1920, who stated, "Both species of Cuckoos are found here, but I am inclined to think that the Yellow-billed is the more common." Roberts was more definitive about its distribution in the northeastern forests, where it was found sparingly. For the southern forests, he quotes Dr. Hvoslef, who found it "rare until 1898" in Fillmore County along the Root River. Yet, in 1897, "Knox reported it as common in Jackson County" and Gleason said it was "usually common" at Lake Benton in Lincoln County. Roberts's only reported confirmed nests were from Minneapolis and Stearns County in 1892 and 1896, respectively.

More than forty years later, Green and Janssen (1975) described the Yellow-billed Cuckoo as a summer resident throughout the state but casual in the northeast and north-central regions. They identified most breeding records as being from the southern half of the state, with some breeding evidence from Clearwater, Pine, and Wadena Counties. Janssen (1987) emphasized its summer range as "southern and central portions of the state and up the western edge" to the northwestern region. He included confirmed nesting records since 1970 from Brown, Clay, Goodhue, Hennepin, Houston, Le Sueur, Olmsted, St. Louis, and Wright Counties. Hertzel and Janssen (1998) would later add confirmed nesting since 1970 in Rice and Scott Counties.

MBS included 167 breeding-season locations that were widely scattered in Minnesota. Most observations were from the southern tier of counties northward to Aitkin, Becker, and Hubbard. Single locations were included northwest to western Roseau County and northeast to southern St. Louis County. Clusters of breeding-season locations were from northwestern Otter Tail, northwestern Morrison, and throughout Douglas, northern Kandiyohi, and Pope Counties. Numerous locations were included from southwestern Minnesota.

The MNBBA reported 160 records that represented just 2.7 percent (130/4,735) of the surveyed atlas blocks **(A)**. The observations were amazingly widespread throughout most of the state, except in the northeast. No observations were recorded from Cook, Koochiching, Lake, and Lake of the Woods Counties, plus only single possible observations from Beltrami, Carlton, Kittson, Roseau, and St. Louis Counties. Many other counties throughout the state also had no breeding-season detections.

Most of the recorded breeding-season records were of possible nesting, which is a strong testament to the secretive and elusive behavior of the Yellow-billed Cuckoo. The only confirmed nesting was from west-central Douglas County. There were 18 probable nesting records from the southeastern region and in the suburban areas of the Twin Cities, as well as northwestward to Becker and Otter Tail Counties. Since the MNBBA period, additional con-

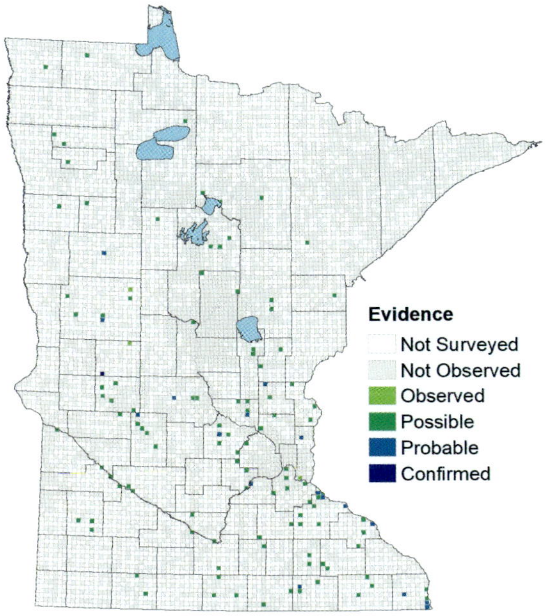

A. Breeding distribution of the Yellow-billed Cuckoo in Minnesota based on the Breeding Bird Atlas (2009–13).

firmed nests have been found in Steele County in 2014, in Chisago and Faribault Counties in 2019, and in Redwood County in 2020 (MOU 2021a). The Chisago County records were of adults feeding young in early September, which are late dates, but Cuckoos often nest late if food is available, and these may be second broods (Hughes 2020b).

Minnesota is on the northwestern periphery of the North American breeding distribution of the Yellow-billed Cuckoo. The historical distribution of the Yellow-billed Cuckoo was described by Roberts (1932) as confusing because of the high variability in population levels of both cuckoo species. One or both species may appear in a Minnesota locality as common one year and not present the next. Their attraction to outbreaks of caterpillars, especially tent caterpillars, gypsy moths, or other variable insect populations, is likely an underlying factor (Hughes 2020b). Identification of the two species may also contribute to confusion because the *cu cu cu* calls emitted by both species can be difficult to distinguish. However, the long and complex song of the Yellow-billed Cuckoo is very different and should not be confused with that of the Black-billed Cuckoo.

BREEDING HABITAT

Preble (1957), DeGraaf and colleagues (1991), and Hughes (2020b) emphasized the species' use of open woodlands with clearings, dense shrubby vegetation such as willow and alder, second-growth woodlands, and a strong association with watercourses **(B)**. In Minnesota's Chippewa National Forest, Niemi, Howe, and colleagues (2016) primarily found the species in lowland shrub cover types and in oak forests. In Michigan and Wisconsin, the following breeding habitats are frequently mentioned: dense shrubby vegetation, wet thickets, brushy fields, young northern hardwoods, backwaters of major rivers, and overgrown orchards (R. Brewer et al. 1991; Cutright et al. 2006).

POPULATION ABUNDANCE

On the basis of the BBS, the global population of the Yellow-billed Cuckoo was estimated by PIF (2021) as 9.6 million breeding adults, with a Minnesota population of 9,700. MNBBA did not estimate a breeding population in Minnesota, and BBS trend data in Minnesota were unreliable because of too few detections. Surveywide, the BBS revealed a significantly declining population that averaged −0.79 percent per year from 1966 to 2019 **(C)**.

Reductions in populations of the Yellow-billed Cuckoo have mostly been linked with the loss of riparian woodlands, which has led to extirpation and large reductions of its populations in the western United States. Its western subspecies is now listed as *Threatened* under the Endangered Species Act. The species is also sensitive to pesticides, which are often directed at its primary food sources (e.g., caterpillars). Hughes (2020b) noted its populations were affected in the past by DDT applications and provided evidence of current effects in the use of modern insecticides directed at control of gypsy moths and other insects. As with many other species, Yellow-billed Cuckoos are frequently found dead following collisions with towers, residences, and buildings during migratory periods. In addition, high predation rates at nests have been reported (Hughes 2020b).

B. Riparian thickets with a dense growth of wetland shrubs and small trees provide the cover that Yellow-billed Cuckoos require during the nesting season. Photo by Lee A. Pfannmuller.

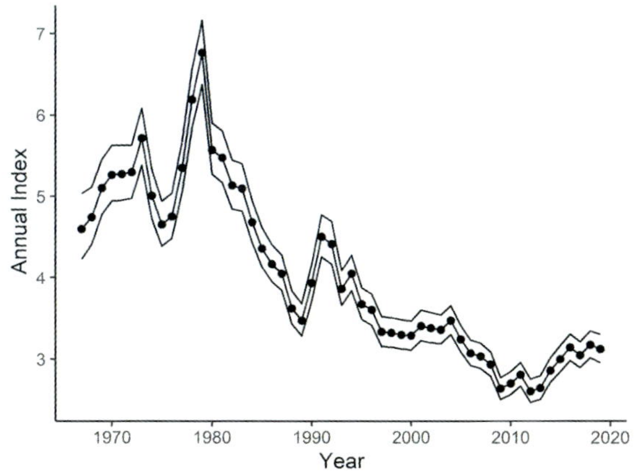

C. Surveywide breeding population trend and 95 percent credible interval (solid black lines) for the Yellow-billed Cuckoo based on the 1967–2019 Breeding Bird Survey (Sauer et al. 2020).

CONSERVATION

PIF (2021) listed the Yellow-billed Cuckoo as a Common Species in Steep Decline; however, Minnesota is not included as a critical area of conservation action in the Upper Mississippi/Great Lakes or in the Prairie Pothole Joint Venture Regions by the USFWS because it is on the periphery of its range in the state. The primary focus of conservation efforts is directed at maintaining suitable habitat, especially riparian woodlands that have been lost to agricultural and residential development, plus flood control management (Hughes 2020b).

The Yellow-billed Cuckoo was considered to have a stable breeding population under all three climate scenarios considered by NAS (2019). In Minnesota, it was also predicted to have an increase in its breeding range into central and northeastern Minnesota under all three climate scenarios. Anders and Post (2006) had previously documented broad-scale declines in population densities of the Yellow-billed Cuckoo following increased temperatures associated with North Atlantic Oscillation and El Niño Southern Oscillation events. The species' winter range is poorly known.

Like for many bird species of Minnesota, suitable breeding habitat, use of pesticides, and migratory hazards appear to be key issues for the long-term future of this distinctive member of Minnesota's avifauna. The decline in the Yellow-billed Cuckoo, along with the loss of its ravenous appetite for caterpillars, is cause for future concern about the subsequent reduction in ecological processes that maintain healthy, natural forests in our state.

BLACK-BILLED CUCKOO

Coccyzus erythropthalmus

Minnesota Seasonal Status: Regular breeding resident and migrant. The Black-billed Cuckoo was uncommon during the MNBBA.
Migration: Long-distance migrant; winters in South America, but distribution is not well known.
Food: Large arthropods, especially caterpillars and cicadas during outbreaks, grasshoppers, crickets, butterflies, and occasionally bird eggs.
Nest: Crude, platform-like nest between two branches or close to trunk; relatively low (1–2 m) in thick shrubbery, saplings, or vines; often described as flimsy, but more solidly constructed than the Yellow-billed Cuckoo's nest.
Conservation Concern: Assigned a moderate PIF Continental Concern Score of 13/20.

Photo by Larry Sirvio.

The Black-billed Cuckoo is a highly beneficial species because of its voracious appetite for caterpillars, especially those that can be highly injurious to trees, such as the eastern tent caterpillar (*Malacosoma americanum*), forest tent caterpillar (*M. californicum*), and gypsy moth (*Lymantria dispar*). During outbreaks, cuckoos are known to consume 10 to 15 caterpillars per minute. Hughes (2020a) stated that Black-billed Cuckoos frequently have more than 100 large caterpillars, or several hundred small caterpillars, in their stomach.

In Minnesota, the Black-billed Cuckoo tends to be more common than its sister species, the Yellow-billed Cuckoo, especially in the northern regions. As Bent (1940) suggested, "In appearance and habits our two cuckoos are very much alike, and their haunts are similar; both are often found together . . . though the black-billed is rather more of a woodland bird and rather more retiring than the yellow-billed." The Black-billed has the same slender appearance, with a long tail and graceful flight. It is distinguished from the Yellow-billed by its lack of reddish color in its wings, smaller black-and-white spots on the tail, and the lack of yellow on its bill. On close examination, it has a red orbital ring around its eye, but be careful, because juvenile Black-bills have a yellow orbital ring (Hughes 2020a).

The species' scientific name, *erythropthalmus,* is Greek for "red-eye." Like many bird species, the Black-billed is more often heard than seen. Its call can be difficult to distinguish from that of the Yellow-billed. Both species emit a *cu-cu-cu* type call, but the Black-billed call is less harsh and is usually given in phrases of two to five with a pause in between phrases. It also lacks the longer, more vibrant vocalization of the Yellow-billed, especially the *kwolp-kwolp-kwolp* ending.

Black-billed Cuckoos are found throughout the upper midwestern and northeastern United States, and south in the Appalachian Mountains to Tennessee. In Canada, it occurs from the Maritime Provinces to central Alberta. Highest densities are found from western Wisconsin, through central and northwest Minnesota, to southern Manitoba.

MINNESOTA BREEDING DISTRIBUTION

The Black-billed Cuckoo was described by Roberts (1932) as a "common summer resident" that was present throughout the state, but more evenly distributed and more northerly compared with the Yellow-billed Cuckoo. Before the 1930s, nesting was confirmed in Anoka, Grant, Hennepin, McLeod, Roseau, Sherburne, and St. Louis Counties and at Itasca State Park.

Forty years later, Green and Janssen (1975) described the Black-billed Cuckoo as a resident throughout the state, but numbers varied considerably in northern regions because of their dependence on outbreaks of tent caterpillars. Janssen (1987) stated it was widespread but never a numerous resident. He reported confirmed nesting records since 1970 from 28 counties, ranging from the northeast and northwest to southwestern Minnesota, but lacking from the southeastern portion of the state. Hertzel and Janssen (1998) added six additional counties to the confirmed nesting of the species since 1970, including three from the southeast.

MBS recorded 514 locations from virtually all of the counties that have been surveyed. No observations were recorded from Cook County, nor from several counties in the south-central, western, and southern tier of counties in the state, where less suitable habitat exists and tent caterpillars are rare.

MNBBA participants included 739 records and substantiated the Black-billed Cuckoo's wide distribution across the state. Participants submitted records for 13.2 percent (625/4,741) of the surveyed blocks **(A)**. Breeding was confirmed in 29 blocks and included 11 counties without previous confirmed nesting records. Records were detected in every county in the state except seven counties in southern Minnesota.

The probability distribution map illustrates that highest densities were predicted in northwestern Minnesota, primarily in the Tallgrass Aspen Parklands Province **(B)**. Additional areas with higher densities were in west-central Minnesota and scattered areas along the Mississippi River floodplain of southeastern Minnesota. Low densities were predicted throughout much of Minnesota.

Hughes (2020a), in a review of the Black-billed Cuckoo in North America, described historical northward movements into New York State, Quebec, Ontario, and the Maritime Provinces in the late 1800s and early 1900s. The suggested reason for these expansions was the increase in edge habitats resulting from conversion of primeval forests to agricultural land. During breeding bird atlases, both Wisconsin (Cutright et al. 2006) and Michigan

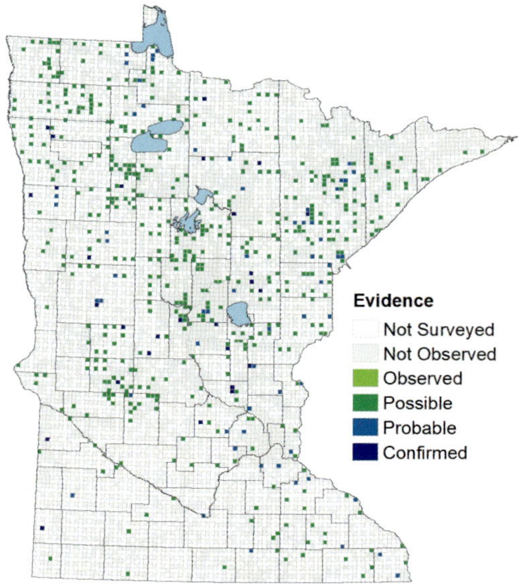

A. Breeding distribution of the Black-billed Cuckoo in Minnesota based on the Breeding Bird Atlas (2009–13).

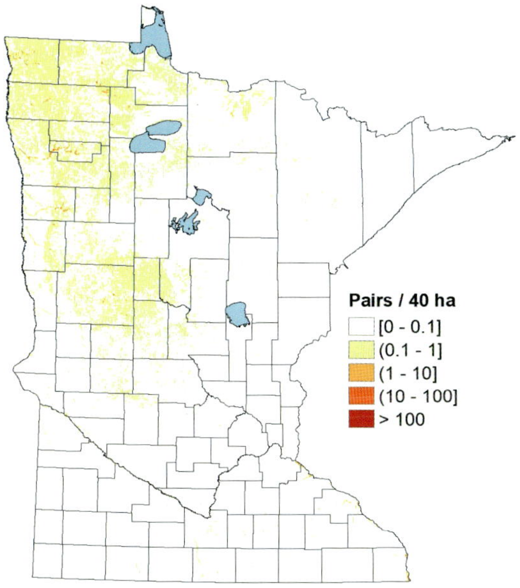

B. Predicted breeding distribution (pairs per 40 ha) of the Black-billed Cuckoo in Minnesota based on habitat, landscape context, and climate data gathered during the Minnesota Breeding Bird Atlas (2009–13) using the Generalized Linear Modeling method with an adjustment for detectability.

(Chartier et al. 2013) reported that the species was more common in the northern portions of these states than in the south.

Black-billed Cuckoos were known in the late 1800s to occur in large flocks during eruptions of food sources, but those flocks are a thing of the past (Hughes 2020a). Historical records from the midwestern United States are limited on whether the species has expanded northward over the past 150 years or if its current distribution is typical. The species' considerable variation in populations can obscure its distribution. Hence, suppositions about historical changes in the distribution of this species in Minnesota remain inconclusive.

BREEDING HABITAT

Black-billed Cuckoos are broadly described as occurring in groves, forest edges, pastures, abandoned farmlands, shrubby wetlands, and deciduous and mixed deciduous forests, especially associated with aspen, birch, and wet areas (S.D. Robbins 1991; Hughes 2020a). A common word in many habitat descriptions is "thicket," because the cuckoos are often found in dense vegetation, such as shrubbery, vines, and young trees **(C)**.

POPULATION ABUNDANCE

Based on the BBS, PIF (2021) estimated a global population of 880,000 individuals and a Minnesota population estimate of 120,000 or 14 percent of its global population. MNBBA estimated a population of 73,000 individuals; however, the 95 percent confidence interval was 52,000–446,000 individuals and within the range predicted by PIF. The wide confidence interval is due to the high annual variability of the species' population and is consistent with a species associated with insect outbreaks.

BBS estimates for Minnesota from 1967 to 2019 indicated a nonsignificant but negative population trend and a cyclical pattern of fluctuations **(D)**. Peaks in detections occurred in 1971, 1979, 1988, 1989, 2001, and 2017. The patterns in the peaks observed are likely due to outbreaks of tent caterpillars or, in some situations, of cicadas. The declining population trends in Black-billed Cuckoos were widespread across the United States and Canada, including a surveywide decline in North America that averaged −2.07 percent per year from 1966 to 2019. However, the BBS emphasized the estimate was highly variable and should be interpreted with caution. Rosenberg, Kennedy, and colleagues (2016) estimated a 68 percent decline in the North American population from 1970 to 2014.

Causes for the species' decline are thought to be related to available food supplies, including the effects of pesticides on caterpillars as well as habitat changes, such as reductions in forested areas with conversion to agriculture and residential areas. Despite its use of edges, the species has been identified as sensitive to habitat fragmentation (Galli et al. 1976; T. Martin 1981). Hughes (2020a) commented that it is frequently killed at television towers and tall buildings during nocturnal migration and its winter distribution in South America is poorly known.

CONSERVATION

PIF (2021) designated the Black-billed Cuckoo as a Yellow Watch List species based on its substantial population decline, its overall small population size,

C. Breeding habitats of the Black-billed Cuckoo in Minnesota are usually thickets of shrubs or vines along edges of forests. Note the tent of the eastern tent caterpillar (*Malacosoma americanum*) on the right. Photo by Gerald J. Niemi.

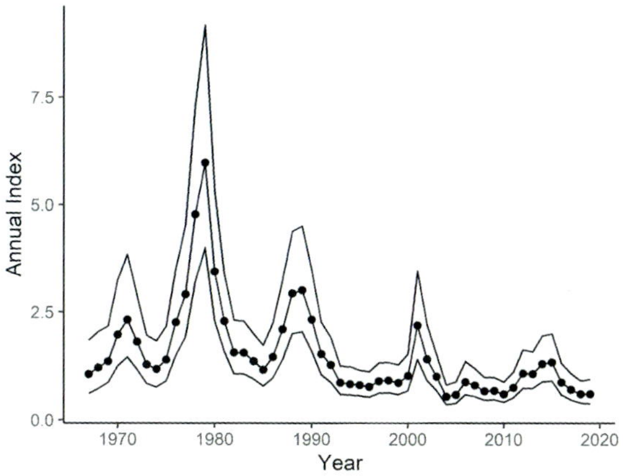

D. Breeding population trend and 95 percent credible interval (solid black lines) for the Black-billed Cuckoo in Minnesota based on the 1967–2019 Breeding Bird Survey (Sauer et al. 2020). Trend should be interpreted with caution because of the variability in the species' cyclic populations.

and threats to its breeding and wintering populations. Minnesota listed it as a *Species in Greatest Conservation Need* because of its population decline and suggested its population is no longer within the normal range of variation (MNDNR 2015b). Minnesota considers that it has a substantial responsibility toward conservation because over 10 percent of its breeding population occurs in the state (Pfannmuller 2012). The Minnesota State Wildlife Action Plan (MNDNR 2015b) has identified Conservation Focus Areas that could potentially benefit the species, such as the Aspen Parklands in northwestern Minnesota and the Pine-Sands South areas in Hubbard County.

Under a 2°C warming scenario, NAS (2019) projected low vulnerability for the Black-billed Cuckoo's breeding distribution in North America, including a 43 percent gain of its breeding range northward but a 31 percent loss in the United States, primarily along its southern breeding range, by 2050. Substantial losses of its breeding range were predicted to occur in the western regions in Minnesota.

Unfortunately, few specific management or conservation actions have been identified for this secretive and elusive species, possibly because the Black-billed Cuckoo remains an enigma. Little is known about the status of its food supplies other than that it eats caterpillars and cicadas when available. It may be sensitive to habitat fragmentation and is potentially vulnerable to collisions with towers and windows during migration. Wintering areas are poorly known, but eBird observations from December to February range from Guatemala to Paraguay, primarily along the western countries of South America (eBird 2020). Along with the effects of climate change, many questions about the future of this unique species remain in Minnesota and beyond.

COMMON NIGHTHAWK

Chordeiles minor

Minnesota Seasonal Status: Regular breeding resident and migrant. The Common Nighthawk was uncommon during the MNBBA.
Migration: Long-distance migrant that winters in South America.
Food: Aerial insectivore.
Nest: No nest is built; eggs are laid on gravel beaches, outcrops, forest floors, and flat gravel roofs.
Conservation Concern: Assigned a moderate PIF Continental Concern Score of 11/20.

Stroll along the city sidewalks during a summer evening and you may hear the unmistakable buzzy *peent* call of the Common Nighthawk. Indeed, *Chordeiles,* the scientific name for the genus, translates to "twilight music." Flying high above the cityscape in its aerial pursuit of insects, the Nighthawk is easily recognized from below by the white bands near the tips of its long, pointed wings **(A)**. A white triangular throat patch may also be visible. When it is not in flight, the bird's dark, mottled plumage provides perfect camouflage against the pebbly beaches, outcrops, and patches of bare ground where it nests. The Nighthawk's association with urban environments began in the late 1800s, when the construction of flat, gravel rooftops furnished a suitable replica of its natural habitats.

Unfortunately, the bird's common name belies its true nature. It is not nocturnal but crepuscular, being most active at dawn and dusk; and it is not a hawk. Like the Eastern Whip-poor-will, the Nighthawk belongs to a family of birds with the unlikely vernacular name of Goatsuckers. Birds in this family have such enormously wide mouths that ancient lore maintained they actually suckled the teats of goats. Although it was once a common species, the Common Nighthawk is now significantly less abundant for reasons that remain somewhat speculative.

A. In flight, the Common Nighthawk's white wing bands and white throat patch are visible from below.
Photo by Laura Erickson.

Photo by Laura Erickson.

Broadly distributed across North America, from the Yukon Territory and British Columbia east to the Atlantic coast and south throughout the United States and portions of Mexico and Central America, the Common Nighthawk reaches its highest breeding densities in the southern Great Plains.

MINNESOTA BREEDING DISTRIBUTION

In the early 1900s, Roberts (1932) compiled confirmed nesting reports of the Common Nighthawk from eight widely dispersed counties: Anoka, Grant, Hennepin, Isanti, Marshall, Meeker, Pipestone, and Polk. Reports of recently fledged young were also available from Cass, Lake, and Yellow Medicine Counties. He described the species as a common summer resident yet noted its recent decline: "It was never shot here for sport nor for eating, as it was in the south, and suitable nesting-places are just as plentiful as ever they were, but it has steadily grown scarcer and scarcer all over the state." Although Roberts considered hunting mortality insignificant in Minnesota, Bent (1940) included accounts from the late 1800s of widespread persecution due to the misbelief that the bird was indeed a hawk and capable of taking young poultry.

The decline in nesting birds may have gone unnoticed by many observers owing to the spectacular numbers seen during fall migration. Roberts described the magnitude of the flight from several locations, including Itasca State Park. Rarely seen in the park during the summer, from "about the middle of August and for two or three weeks thereafter there are days in which the air is literally filled with them."

Decades later, Green and Janssen (1975) and Janssen (1987) described the Nighthawk as a regular summer resident throughout the state. Janssen identified nesting in 13 counties since 1970. The counties were primarily restricted to southwestern, east-central, and northeastern Minnesota. Hertzel and Janssen (1998) later added five more counties to the compiled list. Meanwhile, the MBS reported just 84 breeding-season locations from widely scattered locations. Urban environments, however, were not targeted during the survey efforts. Coincidentally, concern regarding declining Common Nighthawk numbers in the Twin Cities region prompted the Three Rivers Park District to conduct several surveys from 1989 to 2001 that documented "dramatic declines" (Carter and Gillette 2002). The authors concluded that Nighthawks could become extirpated from the region within the coming decade.

MNBBA participants reported 274 Common Nighthawk records from 5.0 percent (235/4,747) of the surveyed atlas blocks **(B)**. Breeding was confirmed in 15 blocks; several straddled two counties. Although sparsely distributed, records came from all regions of the state except the Red River valley south of Marshall County. Reported from all seven metropolitan counties, the birds were noticeably absent from exurban areas, such as western Hennepin County and most of Anoka County. Overall, Nighthawks were observed in 70 of Minnesota's 87 counties and were confirmed nesting in 17 counties. Twelve counties were additions to the list published by Hertzel and Janssen (1998).

The MNBBA landcover suitability model predicts that highly suitable habitat occurs in the Twin Cities, in small towns and cities statewide, and across much of the northern forest. Even the grasslands and agricultural fields of western and southern Minnesota are predicted to provide moderately suitable habitat **(C)**.

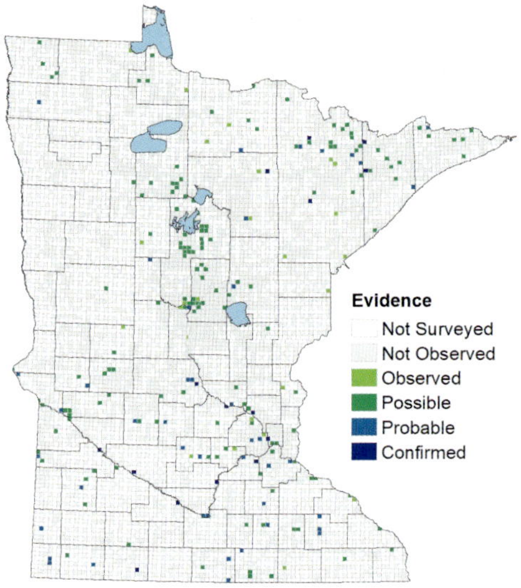

B. Breeding distribution of the Common Nighthawk in Minnesota based on the Breeding Bird Atlas (2009–13).

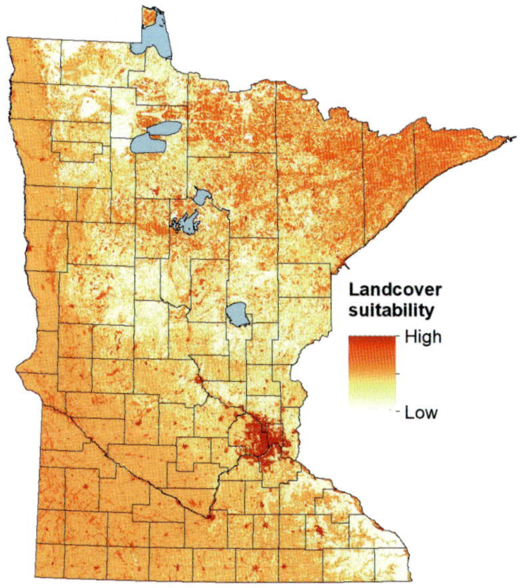

C. Landcover suitability map for the Common Nighthawk in Minnesota based on habitat, landscape context, and climate data gathered during the Minnesota Breeding Bird Atlas (2009–13) using the MaxEnt modeling approach.

Although the Nighthawk's distribution appears similar to that described nearly one hundred years ago, it was not a common species during the atlas. The same is true elsewhere within the species' breeding range; its distribution has changed little, but the overall population has declined significantly. Brigham and colleagues (2020) noted that five northeastern states had classified the species as "critically imperiled or imperiled," with notable declines also documented in New York and Florida.

BREEDING HABITAT

Open habitats used by Nighthawks include beaches, rock outcrops, grasslands, cultivated fields, and forest openings created by timber harvesting and natural disturbances **(D)**. In Michigan and Wisconsin, the birds are often associated with drier soils that support more open pine or pine–oak barren communities (Chartier et al. 2013; Cutright et al. 2006). Although open ground is required for nesting, foraging may occur in a number of habitats. Several studies, especially in grasslands, have emphasized the importance of nearby bodies of water for foraging (e.g., Ng 2009; Brigham et al. 2020); others suggest that in the boreal forest foraging occurs in all open habitats (e.g., Farrell et al. 2017; Knight et al. 2018; Knight and Bayne 2017).

When flat, graveled rooftops became a common form of construction, Nighthawks began inhabiting towns and cities. As early as 1869 they were found nesting on warehouse rooftops in Philadelphia (Bent 1940). This habit prevailed for decades until construction standards changed to flat, smooth, or rubberized roofs, which lack the same appeal (Marzilli 1989).

POPULATION ABUNDANCE

The global population of Common Nighthawks is estimated at 23 million birds, with 22 million residing in the United States and Canada (PIF 2021). In Minnesota, the statewide estimate is only 9,100 birds. Unfortunately, from 1966 to 2019, the species has demonstrated a statistically significant population decline across the BBS survey region averaging −1.23 percent per year **(E)**. With a cumulative decline of 58 percent from 1970 to 2014, the Nighthawk has been designated a Common Species in Steep Decline (Rosenberg, Kennedy, et al. 2016). Minnesota BBS data also suggest a downward trend (Sauer et al. 2020).

Given the difficulty of monitoring the crepuscular Nighthawk during the breeding season, confidence in these population numbers and trends is tentative (Kolbe 2018). A more accurate monitoring tool may be fall migration counts. Some of the largest concentrations of migrating Nighthawks in North America have been observed along Lake Superior's North Shore. The largest flight on record occurred on the evening of August 26, 1990, when 43,690 Nighthawks were counted passing over Duluth (Hendrickson and Eckert 1991). Since 2007, Hawk Ridge Bird Observatory has conducted systematic fall counts of all migrating birds, including Nighthawks, to complement their long-standing work on raptors (Bardon 2012). As the years progress, these annual counts may become a more robust index of Nighthawk population trends (Kolbe 2018).

Factors responsible for the species' decline are not well understood. Declines in urban areas may be attributed to changes in roof construction. Marzilli (1989) speculated that new, uniformly colored

D. A forest opening created by a rocky escarpment immediately adjacent to a lake is an example of the type of natural habitat utilized by the Common Nighthawk in northern Minnesota. Photo by Jarad B. Niemi.

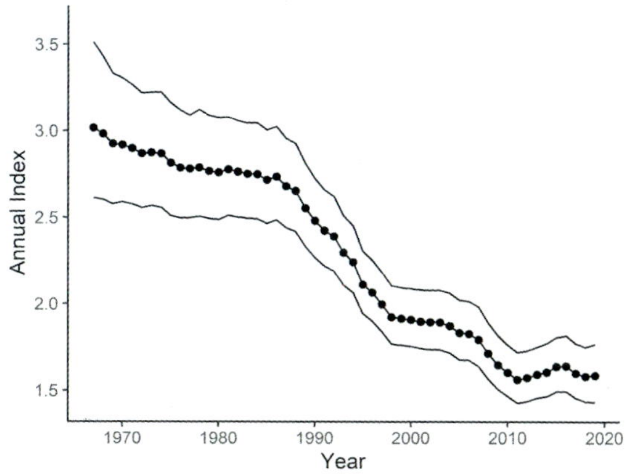

E. Surveywide breeding population trend and 95 percent credible interval (solid black lines) for the Common Nighthawk based on the 1967–2019 Breeding Bird Survey (Sauer et al. 2020).

rubber rooftops may not provide the camouflaged background preferred by the birds. The smooth surface also makes it difficult for the eggs to stay in place, while the rubber's hotter surface may be too warm to safely incubate eggs.

Attention has also focused on the Nighthawk's prey base. Population declines among aerial insectivores are significantly steeper than those reported in other birds, especially in the northeastern United States, where human population densities are high and industrial development is widespread (Nebel et al. 2010). The resulting increase in atmospheric pollutants and their negative impact on aerial insects could be contributing to the species' decline. Competition among aerial insectivores may also be at play. Studies suggest that bats may out-compete Nighthawks for food resources, placing additional pressures on the birds when prey populations are low (Boyce 1980; Shields and Bildstein 1979).

CONSERVATION

Despite its well-documented decline and status as a moderate conservation priority (PIF 2021), conservation measures for the Nighthawk are limited. The only management tool that has proved successful is the use of gravel pads on new, nongravel roof surfaces to attract nesting birds (Marzilli 1989).

Otherwise, biologists have noted the urgent need for research. In addition to the concerns related to the impacts of pollutants on insect abundance, studies are also beginning to draw attention to the potential impact of warmer temperatures on the Nighthawk's prey (e.g., Sánchez-Bayo and Wyckhuys 2019). Although NAS (2019) classified the Nighthawk as a climate-stable species under all three warming scenarios examined, this denizen of Minnesota's evening sky clearly merits more attention.

EASTERN WHIP-POOR-WILL

Antrostomus vociferus

Minnesota Seasonal Status: Regular breeding species and migrant. The Eastern Whip-poor-will was uncommon during the MNBBA.
Migration: Medium-distance migrant that winters in Florida and south along the Gulf coast through eastern Mexico and Central America.
Food: Aerial insectivore.
Nest: Eggs are laid on the forest floor.
Conservation Concern: Assigned a high PIF Continental Concern Score of 14/20.

Photo by Cynthia Bridge.

Like many nocturnal creatures, the Eastern Whip-poor-will is shrouded in mystery. Not only do its nighttime habits make it a rare sighting, but its cryptic plumage provides the perfect camouflage during daylight hours. As the bird roosts in the forest floor's leaf litter, or perches on a low-lying branch, its mottled, grayish-brown feathers blend perfectly with its surroundings. Males and females are nearly identical. How many people, one writer mused, who "know the whippoorwill's name ever saw the bird, or would recognize it if they did see it?" (Bent 1940).

Despite the rarity of sightings, the Eastern Whip-poor-will is well-known in areas where it commonly resides. Considered by some Native American legends to be an omen of approaching storms, future misfortune, or even imminent death, the bird's incessant, repetitive *whip-poor-will* call is unmistakable and may be given a "hundred times or more without a pause" (Bent 1940).

If you are lucky to spot a roosting bird during the daytime, a few features will help distinguish it from its more abundant relative, the Common Nighthawk, which is similar in size. The Nighthawk's long, pointed wings extend just beyond the tip of its tail when the bird is perched, while the Whip-poor-will's round wings are shorter than its tail. In flight, the Whip-poor-will's rounded wings not only are diagnostic but also lack the Nighthawk's white wing bars. *Antrostomus vociferus,* the Whip-poor-will's scientific name, translates to "a cavern-mouthed, noisy bird," a tribute to its persistent singing and to its enormous mouth, a characteristic of all members of the Goatsucker family.

Sparsely distributed throughout its breeding range, the Eastern Whip-poor-will is largely restricted to the Eastern Deciduous Forest Biome. It is found in southern Canada, from central Alberta east to the Maritime Provinces, and in the north-central United States south to northern Arkansas and east to South Carolina.

MINNESOTA BREEDING DISTRIBUTION

Because the Eastern Whip-poor-will's call is so distinctive, Roberts (1932) found many accounts from early explorers and settlers commenting on the species' presence and abundance in the late 1700s and early 1800s. Even in the late 1800s it was described as common or abundant in the Red River valley (1873), central Minnesota (1870), and the Twin Cities (1874). By the early 1900s, however, Roberts was lamenting its demise near Minneapolis as woodlands gave way to expanding development: "now it is rarely heard, only an occasional passing bird, spring and fall, stopping for a night or two to utter a mournful requiem for its numerous, long-since-departed fellows."

Although the species retreated from the Twin Cities, it persisted in areas where forests remained intact. Roberts reported five nesting records, which stretched from Wabasha and Goodhue Counties in southeastern Minnesota, to Hennepin and Stearns Counties in central Minnesota, and north to Lake of the Woods County.

Years later, Green and Janssen (1975) wrote that the Whip-poor-will was a summer resident throughout the state's forested regions, although very scarce in northeastern Minnesota and the adjacent counties of Itasca and Koochiching. Absent from the western grasslands, it was most abundant in southeastern Minnesota and in the northern counties of Beltrami, Clearwater, Lake of the Woods, Marshall, and Roseau.

Janssen (1987) later remarked that Whip-poor-wills were "numerous" along the Mississippi River and "well-represented" in the sand plains of Anoka and Isanti Counties and in northwestern and north-central Minnesota. Nevertheless, since 1970, nesting had been confirmed in only six counties: Anoka, Lake, Marshall, Polk, Sherburne, and Washington. Hertzel and Janssen (1998) later added Winona County to the list.

By 2014, the MBS had reported 67 breeding-season locations. Most records were confined to the Laurentian Mixed Forest Province, where it was especially common in northern Wadena County, and to the Tallgrass Aspen Parklands Province. Scattered records were found farther south, with a small concentration in the upper Minnesota River valley.

A somewhat similar distribution was reported by the MNBBA. Participants tallied 214 Eastern Whip-poor-will records in 3.3 percent (155/4,747) of the surveyed atlas blocks. Since the birds are more easily heard than seen, it is not surprising that breeding was confirmed in only six blocks but deemed probable or possible in 148 **(A)**. In total, the birds were reported in 28 of Minnesota's 87 counties and were reported nesting in seven counties (one block straddled two counties, Renville and Redwood).

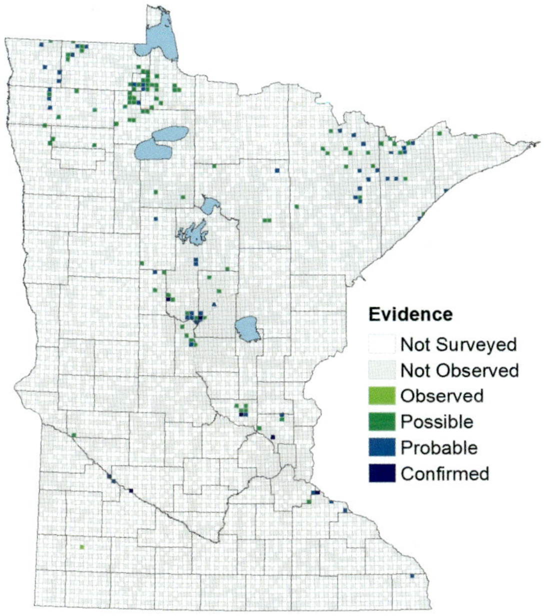

A. Breeding distribution of the Eastern Whip-poor-will in Minnesota based on the Breeding Bird Atlas (2009–13).

Despite the difficulty of surveying this cryptic species, atlas records reaffirmed its primary but patchy distribution within the Laurentian Mixed Forest and the Tallgrass Aspen Parklands Provinces. Records in southern Minnesota were largely limited to the Anoka Sand Plain, southeastern Dakota County, eastern Goodhue County, and the upper Minnesota River valley. The extent to which the species' range changed in the past century is not entirely clear, but southeastern Minnesota no longer represents a center of abundance.

Although many states have documented major population declines and local extirpations, no major change in the Whip-poor-will's overall breeding distribution has been documented. In Ontario, Ohio, and Michigan, for example, the number of townships and/or blocks reporting Whip-poor-wills declined 50–58 percent between completion of their first and second atlases, but their overall distribution changed very little (Cadman et al. 2007; Rodewald et al. 2016; Chartier et al. 2013).

BREEDING HABITAT

Open woodlands are the Whip-poor-will's preferred habitat. Although the birds select a variety of cover types, a degree of openness in the canopy and understory is a common feature of the site or its immediately adjacent habitat **(B)**. The birds are absent from extensive stands of uninterrupted forest unless natural or created openings exist, including open wetlands, sparsely vegetated rock outcrops, and recent clear-cuts (Farrell et al. 2017; Wilson 1985). Because they forage at sunrise and sunset and during moonlit nights, open forests and forest edges provide more visibility and enable this aerial insectivore to adeptly maneuver on the wing as it feeds.

Unfortunately, many areas that appear to provide suitable habitat are unoccupied. Wilson (1985) offered three explanations for this patchy distribution. First, many woodlands are heavily grazed and are not suitable for this ground-nesting species. Second, many woodlots are adjacent to agricultural fields where pesticide use can reduce aerial insect populations. Finally, competition with other aerial insectivores, such as bats, may be a limiting factor (Shields and Bildstein 1979; Boyce 1980).

POPULATION ABUNDANCE

Although the BBS is not well suited to monitoring nocturnal species, it provides the only long-term monitoring data for the Eastern Whip-poor-will. These data have been used to generate a North American population estimate of 1.8 million breeding adults. Minnesota supports approximately 15,000 birds, or less than 1 percent of the North American population (PIF 2021).

Since the BBS began in 1966, the Whip-poor-will has shown a steady, surveywide decline averaging −1.75 percent per year through 2019 (Sauer et al. 2020) and a cumulative decline of 69 percent from 1970 to 2014 (Rosenberg, Kennedy, et al. 2016). Even larger declines are seen throughout the central United States, ranging from Missouri east to Delaware and Maryland and south to Alabama and Georgia. The trend has also been downward in Minnesota. Population increases are evident only along the southern periphery of the species' breeding range in eastern Oklahoma and southern Arkansas. Unfortunately, because of the Whip-poor-will's sparse distribution and nocturnal habits, all BBS trends must be interpreted with caution (Sauer et al. 2020).

In addition to the explanations postulated by Wilson (1985) for its patchy distribution, other explanations for the species' decline include closure of forest openings due to forest succession, invasion of open woodlots by aggressive understory species such as buckthorn, and loss of habitat due to agriculture and development (Cink et al. 2020; Cutright et al. 2006). A recent analysis of the influence of prey availability and suitable habitat on the distribution and abundance of Whip-poor-wills concluded that food availability was a leading factor in the species' decline (English et al. 2017).

CONSERVATION

Specific management recommendations for the Whip-poor-will are lacking despite its high conservation ranking (PIF 2021). Nevertheless, protecting naturally open woodlands, such as the dry oak and aspen stands in Minnesota's aspen parklands, and the oak savannas on the Anoka Sand Plain, appears to be a critical requirement for the species.

Unfortunately, under a 2°C warming scenario, NAS (2019) projected the Whip-poor-will's continental breeding distribution is moderately vulnerable, with a 58 percent range loss by 2050 and only a 39 percent gain northward. Although their models failed to recognize the species' significant presence in northwest Minnesota, they do suggest further expansion into the state's northeastern counties. Extensive losses predicted in southern Minnesota, however, will have little impact given the Whip-poor-will's current scarcity in the region. Overall, this uncommon and unique species warrants our attention.

B. The Eastern Whip-poor-will prefers open woodlands, like this oak savanna habitat in the Sherburne NWR, that provide an open canopy and adjacent grasslands for foraging. Photo by Lee A. Pfannmuller.

CHIMNEY SWIFT

Chaetura pelagica

Minnesota Seasonal Status: Regular breeding resident and migrant. The Chimney Swift was uncommon during the MNBBA.
Migration: Long-distance migrant that winters in northwestern South America.
Food: Aerial insectivore.
Nest: Chimney or tree hollow.
Conservation Concern: Assigned a moderate PIF Continental Concern Score of 13/20. Classified as *Vulnerable* on the IUCN Red List (IUCN 2021).

Often referred to as a "flying cigar," the Chimney Swift is an unusual bird. Similar in size to a chickadee, the perpetually moving Swift rarely permits one to get as close as the curious chickadee allows. Indeed, these sooty-gray birds never stop their aerial pursuits unless they are sitting on a nest or roosting in a chimney or hollow tree. Incapable of perching on a tree branch, they bathe by gliding low over water, "smacking" the surface with their belly and shaking the water off as they bounce up and fly away. Nesting materials are gathered by grabbing small twigs with their feet as they dart through the treetops. Though infrequent, even mid-air copulations have been reported (Steeves et al. 2020).

With their stocky body, short tail, and long, pointed wings, the Swift's comparison to a cigar is readily apparent as it dashes overhead, delivering sharp *tsik* notes that sound like a high-pitched chatter. Similar in appearance, both sexes have stiff spines at the tip of their tail feathers and sharp, curved claws that enable them to cling for hours on vertical surfaces. The genus name, *Chaetura,* means "bristle or spine tail"; the species name, *pelagica,* means "of the sea" and likely refers to the bird's nomadic, airborne habits.

But it is the evening ritual of Chimney Swifts during migration that has captured the imagination of bird enthusiasts. Roosting communally, the "birds come in little groups and for a few minutes . . . dart and circle about the chimney until finally, all rising at once above the inky black opening, the tiny, ebony silhouettes drop with upturned wings in a rapid, continuous stream, until the last are settled for the night" (Roberts 1932).

Distributed across the eastern United States and southeastern Canada, Chimney Swifts attain high breeding densities in portions of North Carolina, Georgia, western Pennsylvania, and eastern Ohio.

Photo by Ryan Sanderson.

MINNESOTA BREEDING DISTRIBUTION

Roberts (1932) considered the Chimney Swift to be an abundant summer resident throughout much of Minnesota in the early 1900s. The one exception was the large, open stretches of prairie grasslands where trees and farmsteads were absent. The species could be found nesting in hollow trees in old-growth forests, in dilapidated cabins in northern Minnesota, and in chimneys and old farm buildings in southern Minnesota. Confirmed nesting records were available from Hennepin and Sherburne Counties in the east-central region and Cass and Cook Counties in the north.

The Swift's distribution changed little in the forty years that followed. Green and Janssen (1975) noted that the species occurred statewide but was most abundant near human habitations, where chimneys and abandoned buildings provided nest sites. It was still present, however, in "heavily forested areas, following its natural pattern of nesting in hollow trees."

A few years later, Janssen (1987) commented on the species' rarity in northwestern Minnesota. Despite its broad distribution, nesting had been confirmed in just five counties since 1970: Aitkin, Clearwater, Hennepin, Morrison, and Rice. Hertzel and Janssen (1998) later identified six counties with confirmed nesting since 1970; Clearwater County was not included on their map, but Kandiyohi and Watonwan Counties were added. Meanwhile, the MBS documented the species' sparse distribution, with only 168 breeding-season locations widely scattered across all regions except the aspen parklands.

MNBBA participants subsequently reported 836 Chimney Swift records from 12.8 percent (612/4,774) of the surveyed atlas blocks. Breeding was confirmed in only 82 blocks (1.7 percent) but was considered probable or possible in 494 blocks **(A)**. Confirmation of breeding occurred when participants observed the Swift's unique synchronized courtship flight or observed adults and/or recently fledged young enter or leave chimneys. Reported in all but one of Minnesota's 87 counties (Isanti), the Swifts were confirmed breeding in 42 counties, including both Polk and Roseau in the northwest. The Swift's land suitability model emphasizes its strong association with developed communities that provide nesting chimneys **(B)**.

The major change in the past one hundred years is the sparse but widespread occurrence of the species throughout the agricultural regions of the state, from southern Minnesota north through the Red River valley and aspen parklands. The small towns and cities that dot the landscape now provide more man-made opportunities for nesting. Indeed, as settlements spread across the Great Plains, the western periphery of the Swift's entire breeding range may have expanded westward (Steeves et al. 2020).

BREEDING HABITAT

Although hollow trees or tree cavities are still used for nesting where extensive mature forests remain, chimneys in cities and small towns have become the species' preferred breeding habitat **(C)**. Chimneys were adopted as alternate nesting and roosting sites as early as the late 1600s and have been used almost exclusively in most regions since the late 1700s (Steeves et al. 2020). When chimneys are in short supply, other building structures may be used,

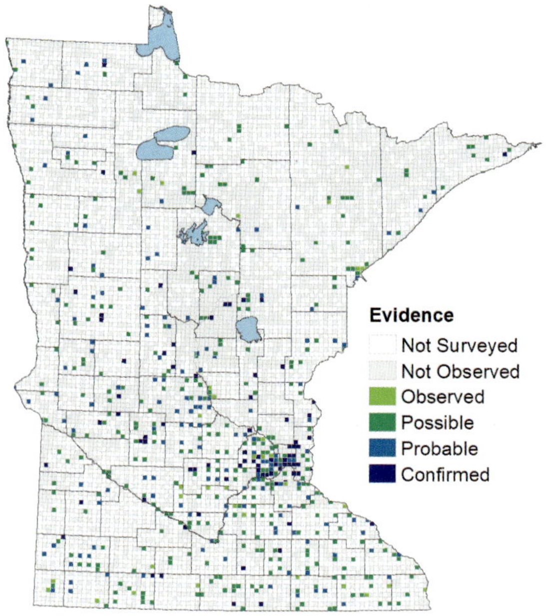

A. Breeding distribution of the Chimney Swift in Minnesota based on the Breeding Bird Atlas (2009–13).

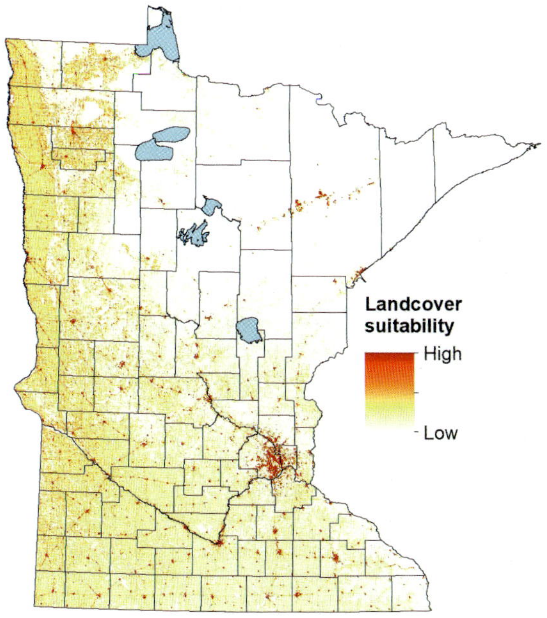

B. Landcover suitability map for the Chimney Swift in Minnesota based on habitat, landscape context, and climate data gathered during the Minnesota Breeding Bird Atlas (2009–13) using the MaxEnt modeling approach.

including open wells, abandoned buildings, and air vents. Overall, darkness and shelter are the primary features characterizing selected sites. Foraging occurs in a variety of habitats, sometimes several kilometers (3–6 km) from the nest (Steeves et al. 2020).

POPULATION ABUNDANCE

The North American Chimney Swift population is estimated at 8.8 million birds, with approximately 85,000 birds residing in Minnesota (PIF 2021). Although it would seem that chimneys are in plentiful supply, from 1966 to 2019 Swifts experienced a statistically significant decline across the BBS survey region, averaging −2.09 percent per year (Sauer et al.

C. The chimney on this old, rural building provides a dark, sheltered enclosure that is ideal nesting habitat for the Chimney Swift. Photo by Gerald J. Niemi.

2020). With a cumulative decline of 67 percent from 1970 to 2014, they are now recognized as a Common Species in Steep Decline (Rosenberg, Kennedy, et al. 2016). Minnesota BBS data demonstrated a similar, albeit slower, rate of loss averaging −1.55 percent per year from 1967 to 2019 **(D)**.

Given the unabated growth in industrial and urban development in the eastern United States, the species population decline seems surprising. Yet some have attributed the decline to a decrease in the availability of chimneys. Home and business owners are increasingly capping or caging the tops of old masonry or clay flue-tile chimneys, making them unavailable to Swifts and other wildlife that occasionally enter and are unwanted. As new homes are built, brick and stone chimneys are being replaced with insulated sheet-metal flues for furnaces and fireplaces that are too narrow for Swifts to use. Nevertheless, a few recent studies concluded that more suitable chimneys remain available than are actually being used by Swifts, suggesting that other factors may be at play (Steeves et al. 2020).

The species' strict dependence on flying insects, however, is a major concern. Nebel and his colleagues (2010) pointed out that aerial insectivores have declined at a much faster pace than any other guild of passerine birds. The declines were most prevalent in the densely populated northeastern United States, where high levels of atmospheric pollutants may be negatively impacting aerial insects. Indeed, the bird's sensitivity to pollutants was demonstrated by a unique study that examined guano deposits that had accumulated for five decades at a major roost site in Ontario. Years of peak DDT use correlated with a significant change in the Swift's diet as they decreased their consumption of beetles and increased their consumption of true bugs (Nocera et al. 2012). Unfortunately, although DDT use has abated, the Swift's decline has continued.

CONSERVATION

Chimney Swift enthusiasts have launched several conservation initiatives focused on public education, preservation of existing habitat, and creation of new nesting and roosting sites. Two of their most popular projects have been monitoring the number of birds inhabiting significant roost sites during migration and promoting the construction of nesting towers. Many organizations, including Audubon Minnesota, have rallied behind these initiatives. Working with volunteers and local civic groups, 68 nesting towers were built and erected in the Twin Cities metropolitan area and in several outstate locations (Audubon Minnesota 2016) **(E)**. Although their

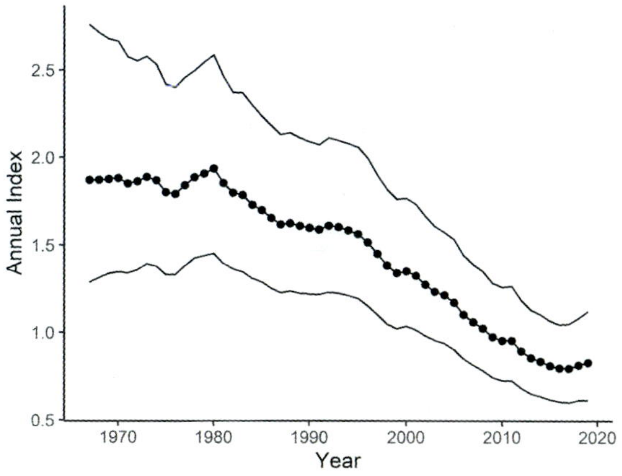

D. Breeding population trend and 95 percent credible interval (solid black lines) for the Chimney Swift in Minnesota based on the 1967–2019 Breeding Bird Survey (Sauer et al. 2020).

occupancy in Minnesota has not been monitored, nesting towers in some states have had mixed success. Low residency rates may reflect inappropriate design features or the fact that chimneys are not the limiting resource for local populations (Steeves et al. 2020).

Much remains to be learned about Chimney Swifts and the factors responsible for their decline. The recent attention focused on the imperiled status of insect populations especially requires close examination (e.g., Sánchez-Bayo and Wyckhuys 2019). Fortunately, warming temperatures are predicted to have little impact on the Swift (NAS 2019). In the interim, it is important to promote the value of retaining existing chimneys. Education needs to be directed not only to home and building owners but also to chimney sweep companies, who are frequently called upon to cap chimneys, making them unavailable to nesting and roosting birds.

E. Chimney Swift nesting tower in south Minneapolis. Photo by Lee A. Pfannmuller.

RUBY-THROATED HUMMINGBIRD

Archilochus colubris

Minnesota Seasonal Status: Regular breeding resident and migrant. The Ruby-throated Hummingbird was common during the MNBBA.
Migration: Medium-distance migrant; most birds winter in Central America.
Food: Nectar, tree sap, and insects acquired by hovering flight, hawking, and foliage gleaning.
Nest: Cup nest usually placed in deciduous trees and shrubs; often near or directly above water.
Conservation Concern: Assigned a low PIF Continental Concern Score of 8/20.

The smallest member of Minnesota's avifauna, the Ruby-throated Hummingbird has inspired nature writers for centuries. Its iridescent feathers, hovering flight, and remarkably long bill are marvels of natural adaptation. An early admirer, John James Audubon (1827–38), wrote: "Where is the person who, on seeing this lovely little creature moving on humming winglets through the air, suspended as if by magic in it, flitting from one flower to another . . . would not pause, admire, and instantly turn his mind with reverence toward the Almighty Creator."

This little jewel is even more beloved for its attraction to colorful garden flowers, including the red and orange blossoms of such species as columbine, honeysuckle, jewelweed, and bee balm. Its special attraction to these tubular flowers is an example of coevolution, each species adapting to the traits of the other. The flowers produce nectar in blossoms that are uniquely accessible to the Hummingbird, while the Hummingbird serves the role of pollinator.

With a weight of less than 4 grams (0.14 ounce), the bird's tiny size and unique habits inspired many legends regarding its behavior and life history. Since the Hummingbird was long thought incapable of extended migratory flights, many believed it hitched a ride on the backs of larger birds. But this little creature, with incredibly strong pectoral muscles that comprise nearly 25–30 percent of its total weight, nearly doubles its weight at migratory stopovers before choosing one of two migratory pathways: either a direct, non-stop flight of at least 1,000 km across the Gulf of Mexico; or a flight that avoids the open water and follows the Gulf's coastline. Weather, winds, and age may all influence the pathway chosen (Terres 1980). Zenzal and Moore

Photo by David Brislance.

(2016) estimated that the average Hummingbird even has a much longer flight range of approximately 2,200 km! The bird's unique flight was also a matter of speculation until high-speed photography enabled closer analysis. It is capable of hovering in place as well as moving up, down, forward, backward, and side to side; the bird's wings beat at a rate of 55 to 75 beats per second, depending on the exact maneuver underway.

With its glittering green back and head and the male's ruby throat (absent in the female), the Ruby-throated Hummingbird is unmistakable. It is the only hummingbird species that occupies the eastern United States and southern Canada, its range closely mirroring the distribution of the eastern deciduous forest. Several hummingbird species that nest in the western United States and Mexico occasionally visit Minnesota in late summer and fall, but only the Ruby-throat is a summer resident. Although the humming sound of its wings often provides our first clue to the bird's presence, the male emits a continuous series of single-pitched chips at daybreak, while both sexes give a more common, squeaky-sounding *chee-dit* call.

MINNESOTA BREEDING DISTRIBUTION

"This gem among Minnesota birds," Roberts (1932) wrote, "is a common, at times abundant, summer resident in all parts of the state. It is present alike in the groves of the prairie region and the open glades and the burned-over areas of the heavy forests." Confirmed nesting records were available from six counties: Fillmore, Hennepin, Kittson, Ramsey, Scott, and Sherburne. Inferred breeding reports were available from Cass and Goodhue Counties and Itasca State Park. Roberts extolled the rapid beat of the Hummingbird's wings, the "fantastic performance" of the male's courtship flight, and the "marvelous and exquisite little nest" built by the architecturally accomplished female.

Forty years later, Green and Janssen (1975) confirmed the Ruby-throated Hummingbird's status as a statewide resident, reporting that the species could be found breeding in all regions of the state. Janssen (1987) would later add that it was most abundant in northern Minnesota, gradually becoming less common as one moved farther south, and becoming "rare to absent over much of the south-central and southwest regions." Nesting was confirmed in 18 counties since 1970. Although most counties stretched from the southeast corner of the state northwest to Koochiching and Roseau Counties, two counties in the southwest were also included: Brown and Cottonwood. Hertzel and Janssen (1998) added seven more counties to the list of confirmed nesting, all within the species' primary breeding range.

By 2014, field biologists working with the MBS reported a total of 276 Ruby-throated Hummingbird breeding-season locations. The majority of reports were in the Laurentian Mixed Forest and Eastern Broadleaf Forest Provinces. Although it was absent from the Red River valley, there were scattered reports in the Tallgrass Aspen Parklands Province, along the Minnesota River valley, and in southwestern Minnesota.

During the MNBBA, observers reported 1,575 Ruby-throated Hummingbird records in 24.2 percent (1,146/4,744) of the surveyed atlas blocks and in 32.9 percent (769/2,337) of the priority blocks. Breeding was confirmed in 114 blocks **(A)**. The birds

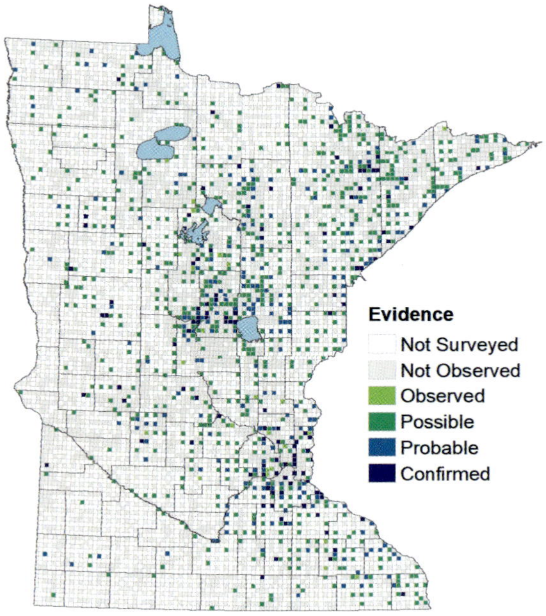

A. Breeding distribution of the Ruby-throated Hummingbird in Minnesota based on the Breeding Bird Atlas (2009–13).

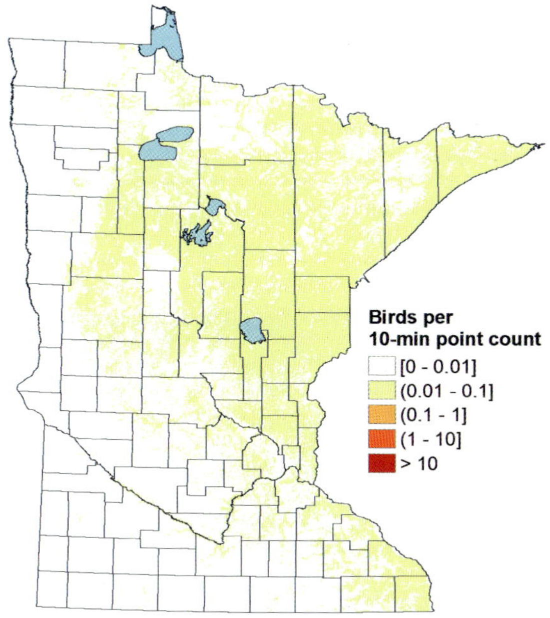

B. Predicted breeding distribution (birds detected per 10-minute point count) of the Ruby-throated Hummingbird in Minnesota based on habitat, landscape context, and climate data gathered during the Minnesota Breeding Bird Atlas (2009–13) using the Generalized Linear Modeling method without an adjustment for detectability.

were found in 81 of Minnesota's 87 counties and were documented breeding in 38 counties. With the exception of Lac qui Parle County, the counties with confirmed nesting were all located in southeastern Minnesota or north of the Minnesota River valley.

Once again, the Hummingbird was most abundant in the Laurentian Mixed Forest and Eastern Broadleaf Forest Provinces. There were more records in the Tallgrass Aspen Parklands and in the Prairie Parkland Provinces, particularly in west-central Minnesota, than reported by the MBS. Although the extensive peatlands of north-central Minnesota and the agricultural landscapes of western Minnesota provide less suitable habitat, small, scattered woodlands and residential areas provide a haven for the species in these landscapes, especially when backyard feeding stations are provided. The MNBBA probability distribution map depicts where the species is likely to be encountered throughout the state **(B)**. Although it is a reasonable portrayal of where the atlas reported many of its observations, it emphasizes the species' relatively low breeding density throughout its summer range.

Nearly a century ago, Roberts (1932) specifically noted the Hummingbird's presence in the prairie regions of the state; fifty years later, Janssen (1987) commented on its rarity and near absence in the southwest and south-central regions. Today, it again appears to be a regular but uncommon species in the region. Small-scale farming operations of the early twentieth century may have provided limited but suitable habitat for the species that disappeared as the intensity of agricultural practices increased. The reappearance of the species in these regions may reflect more interest on the part of landowners to provide Hummingbird feeding stations.

Throughout its breeding range, local populations likely have benefited from landscape plantings that attract Hummingbirds and the placement and maintenance of Hummingbird feeders. Apart from small-scale changes in the species' abundance and distribution, however, there is little evidence of any large-scale distributional change (Weidensaul et al. 2020).

BREEDING HABITAT

Generally characterized as an inhabitant of open woodlands, the Hummingbird occupies a variety of habitats, ranging from mature deciduous or mixed forests to more open woodlands, parks, orchards, and gardens **(C)**. It is also found in woodland groves in predominantly agricultural regions. Proximity to water may be an important habitat requirement (Natural Resources Conservation Service 1999). In the extensively forested landscapes of northern Minnesota and Wisconsin, the Hummingbird was most closely associated with small towns and communities, followed by open, wet habitats and dry, open oak forests (Niemi, Howe, et al. 2016).

POPULATION ABUNDANCE

Data collected by the BBS helped to generate a North American population estimate of 36 million breeding adults with a statewide estimate for Minnesota of 690,000 individuals (PIF 2021). Because Hummingbirds are rarely encountered on BBS routes, most population trendlines are unreliable. Nevertheless, BBS data collected across the survey region suggest an increasing population from 1966 to 2019, with an average increase of 0.95 percent per year (Sauer et al. 2020). A nearly identical trend is evident in Minnesota. Indeed, Hummingbird numbers are stable or increasing everywhere except in the Florida Peninsula, the Edwards Plateau in southern Texas, and the Mississippi Alluvial Plain, which stretches along the lower Mississippi River from southern Missouri to the Gulf coast. Greater homeowner interest in maintaining Hummingbird feeding stations and landscaping yards to attract a variety of birds is believed to be contributing to the increases.

CONSERVATION

Given its relatively stable population, the Ruby-throated Hummingbird is not the focus of any concerted conservation effort. Predictions regarding the species' response to warming temperatures are also favorable (NAS 2019). Losses predicted along the western edge of its breeding range, including in Minnesota, are offset by substantial gains to the north under all three warming scenarios examined.

Beloved by backyard bird enthusiasts, this smallest member of Minnesota's breeding bird community is likely to remain a common member of our avifauna for years to come. Indeed, many resources are available to encourage and instruct homeowners on methods to attract Hummingbirds to their yards and feeding stations (e.g., Natural Resources Conservation Service 1999; Henderson 1994; Stokes and Stokes 1989).

C. An abundance of flowering plants and widely spaced trees in the Noerenberg Memorial Gardens, located on the shore of Lake Minnetonka, provide ample habitat opportunities for the Ruby-throated Hummingbird. Photo by Lee A. Pfannmuller.

VIRGINIA RAIL

Rallus limicola

Minnesota Seasonal Status: Regular breeding resident and migrant. The Virginia Rail was uncommon during the MNBBA.
Migration: Short- to medium-distance migrant wintering along the southern coast of the United States and in northern Mexico.
Food: Probes mud and shallow waters to feed on aquatic invertebrates, small fish, frogs, and aquatic plants.
Nest: Woven platform and overhanging canopy of cattails and bulrushes; a small ramp may also be constructed. Several dummy nests are built that may serve as sites for feeding, resting, and brooding the young.
Conservation Concern: Legally harvested; assigned a moderate PIF Continental Concern Score of 9/20.

Photo by Travis Bonovsky.

Like many marsh birds, the Virginia Rail is rarely seen and poorly studied. Although it is about the size of a Robin, it is hidden by the dense emergent vegetation in its favorite wetland haunts. As one writer recounted: "The best way to observe these shy birds is to take up one's station near a pond or marsh frequented by them and watch patiently, silently, and immobile, and in so doing the watcher is often well rewarded" (Townsend 1926).

If you are lucky enough to catch a glimpse, the Virginia Rail's long, slightly down-curved red bill is diagnostic and distinguishes it from the more common Sora with its short yellow bill. The Virginia Rail is also adorned with reddish legs, gray cheeks, a rufous breast and belly, and striped flanks. Its plumage is similar to that of the very rare King Rail, which birders are always hopeful of finding, but the King Rail is considerably larger and has an orange bill and pale legs.

This bird is more commonly heard than seen; most field guides highlight the Virginia Rail's characteristic *kid-dik, kid-dik* call. Unfortunately, this vocalization is given only by the male for a short time in the spring (Conway 2020). Learning the Rail's wide array of other vocalizations will pay dividends in the long run, including the *grunt* given by both the male and female and the high-pitched *kiu* call.

Like other members of the Rallidae bird family, the Virginia Rail is well adapted to its marsh environment. Its strong feet and large toes enable it to move efficiently through dense emergent vegetation, walk atop the floating leaves of water lilies, or cling to the stalks of cattails and rushes. By laterally compressing its body, it increases its maneuverability in an environment that most of us struggle to walk through. Even the head feathers have modified tips that lessen wear on the plumage as the birds push their way through brittle stalks (Conway 2020).

Sparsely distributed throughout its breeding range, the Virginia Rail is found throughout most of southern Canada and the northern United States, extending as far south as New England in the east, the Great Lakes in the Midwest, and California, northern New Mexico, and Arizona in the west. Populations are present year-round along the Pacific coast and in scattered areas in the southwestern states, Mexico, and Central America.

MINNESOTA BREEDING DISTRIBUTION

As early as 1892, Hatch considered the Virginia Rail to be "generally distributed across the country" where suitable wetlands were present. Roberts (1932) also described the species as a statewide summer resident, noting that it wasn't nearly as abundant as the Sora. Although the Virginia Rail's distribution was presumed to be widespread, nesting records for this secretive marshland bird were limited to five counties: Hennepin, Isanti, Jackson, Kittson, and Polk.

When Green and Janssen (1975) wrote their updated account of the species' status, the Virginia Rail had been documented breeding as far north as the city of Virginia in St. Louis County but was not found farther east in Lake or Cook Counties. It was also considered scarce in the far northern counties of north-central Minnesota. Several years later, Janssen (1987) reported the first breeding record from Lake County. The species, however, remained uncommon throughout most of the forested regions of the state. Janssen's distribution map excluded northeastern Koochiching County, northern and eastern St. Louis County, and all of Lake and Cook Counties. Since 1970, nesting had been confirmed in 21 widely dispersed counties. By 1998, Hertzel and Janssen had added another three counties to the list.

MBS field biologists recorded 145 breeding-season locations for the Virginia Rail dispersed across 50 counties, including two locations in Cook County. The bird was largely absent, however, from four regions: southeastern Minnesota, except in the Weaver Bottoms area of the Mississippi River; most counties bordering the Minnesota River valley; large portions of east-central Minnesota, including

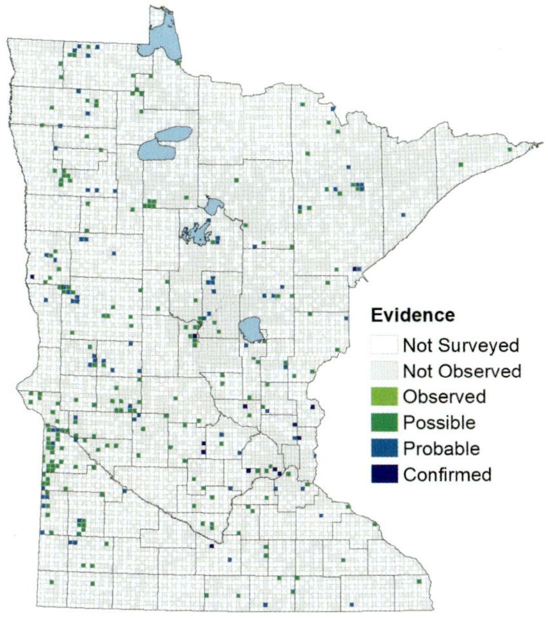

A. Breeding distribution of the Virginia Rail in Minnesota based on the Breeding Bird Atlas (2009–13).

Aitkin, Carlton, Cass, and Crow Wing Counties; and the Arrowhead region.

The 377 Virginia Rail detections reported by MNBBA participants were more widely dispersed than those reported by the MBS with records in 70 counties **(A)**. These included nearly all counties in east-central Minnesota, except Ramsey, and many counties bordering the Minnesota River, especially in western Minnesota. Documented from 6.5 percent (307/4,758) of the surveyed atlas blocks, breeding was more difficult to confirm for this secretive species. Confirmed evidence was reported from only 14 blocks spanning 12 counties, but breeding was deemed probable in an additional 36 counties. New confirmed breeding records were reported for Blue Earth, Carver, Chisago, Clay, Meeker, Sherburne, and Washington Counties. Despite its wide distribution, the species remains sparsely distributed, with pockets of high abundance in a few localities, most notably in western Lac qui Parle, Lincoln, and Yellow Medicine Counties, where survey efforts by atlas participants were quite impressive.

Although Roberts (1932) presumed the Virginia Rail occurred statewide, it was years before his supposition was proved true. New records in northeastern Minnesota most likely do not represent a range expansion but are, instead, an outcome of more intensive inventory efforts. Other atlas efforts in the upper Midwest, including Ontario, have noted that the species' distribution remained similar between atlases and unchanged from historical accounts (Cadman et al. 2007; Chartier et al. 2013; Rodewald et al. 2016).

BREEDING HABITAT

The Virginia Rail is predominantly an inhabitant of freshwater wetlands during the breeding season **(B)**. Suitable habitats provide shallow water (0–15 cm deep) and patches of dense emergent cover (30–70 percent) interspersed with open water and muddy substrates with a plentiful supply of aquatic invertebrates (Conway and Eddleman 1994). *Rallus limicola,* the species' scientific name, actually refers to a mud-dwelling (*limicola*) rail.

Floating residual vegetation or water lilies are often present and provide an additional surface for foraging, particularly in areas where the water is deep. Although dense cover is important, wetlands choked with emergents are usually avoided because movement through them is too difficult; wetlands with younger growth are preferred (Zimmerman, Jamison, et al. 2002b). Virginia Rails generally occupy drier areas of a wetland than the Sora (Conway 2020).

POPULATION ABUNDANCE

Given its reclusive habits, the Virginia Rail is extremely difficult to monitor. National population estimates have not been available until recently, when PIF (2021) provided a global estimate of 240,000 birds and an estimate of 230,000 in the United States and Canada. Even the BBS is not a robust tool for monitoring population trends. Virginia Rail detections are so low that these data have low credibility in every state, province, and bird conservation region in the United States and southern Canada, except North Dakota. Despite their lack of statistical rigor, BBS data suggest a stable to increasing population trend across the North America survey region but a decreasing trend in Minnesota (Sauer et al. 2020).

B. The Virginia Rail prefers shallow wetlands that supply an abundance of aquatic invertebrates for foraging. Photo by Lee A. Pfannmuller.

Another source of monitoring data comes from the Great Lakes Marsh Monitoring Program, coordinated by Bird Studies Canada (Tozer 2016). From 1996 to 2013, Virginia Rails demonstrated a statistically significant decline averaging −2.6 percent per year. Loss of wetlands coupled with environmental stress on those coastal wetlands that remain are believed to be the primary factors responsible for the decline.

CONSERVATION

The Virginia Rail is legally hunted in thirty-seven states, including Minnesota, and in Ontario. Although hunting pressure in the Midwest is considered minimal, annual statistics on the number of birds harvested are difficult to find (Conway and Eddleman 1994). Minnesota is one of the few states that collect harvest data, but data for all rails, except the American Coot, are combined. The average annual harvest from 2010 through 2019 was only 344 birds; this included three seasons where no birds were taken (Davros and Dexter 2019). Soras likely comprised the majority of the reported harvest.

Management recommendations for the Virginia Rail focus on the protection and restoration of wetland basins. Because it is less responsive than other wetland species to landscape-level features, Saunders et al. (2019) recommended that "setting aside individual wetlands or small wetland complexes" is likely sufficient to benefit the species. Wetlands overgrown with dense emergent cover, however, need to be periodically burned, mowed, plowed, or drawn down to set back succession and restimulate the aquatic invertebrate community (Zimmerman, Jamison, et al. 2002b).

Warmer temperatures may not pose a major threat to the Virginia Rail, which was designated a species with low vulnerability under a 2°C warming scenario. NAS (2019) projected a loss of 36 percent of its continental breeding range by 2050 but a gain of 65 percent to the north. Losses in Minnesota would occur primarily in the southwest region. Unfortunately, rising temperatures could increase drought conditions and extreme rainfall events, impacting breeding birds that are dependent on stable water levels. These and other challenges increase the importance of being able to accurately monitor population trends. As the bird is deemed a moderate conservation priority (PIF 2021), until more data are available, the status of this uncommon wetland inhabitant is uncertain.

SORA

Porzana carolina

Minnesota Seasonal Status: Regular breeding resident and migrant. The Sora was uncommon during the MNBBA.
Migration: Short- to long-distance migrant that winters from the southern United States south to northern South America.
Food: Probes the vegetation and wetland substrate for aquatic invertebrates and seeds.
Nest: Woven basket of aquatic vegetation, often with an overhanging canopy, either attached to surrounding bulrushes and cattails or lying atop mounds of decadent vegetation; a small ramp may lead up to the nest.
Conservation Concern: A game species; assigned a moderate PIF Continental Concern Score of 9/20.

Photo by Rebecca Field.

A secretive inhabitant of wetlands, the Sora is a common and widespread species that may be difficult to observe without hip waders or a canoe. Even then, upon sensing an intruder, this member of the rail family is more apt to quietly run unseen through the rushes and cattails than it is to flush and reward one with a quick glance. Wetlands bordering rural roads, however, can provide good opportunities to view Soras as they cautiously emerge from dense cover to forage in the open atop lily pads and scattered vegetative debris.

The thick, yellow bill is the most diagnostic feature on this stocky gray-brown bird, which is a little smaller than a robin. Its bill is bordered by black feathers that extend down the throat. Gray feathers cover the sides of the head and upper breast; the brown flanks are heavily barred with black and white, while the back displays similar colors. Long, greenish legs and large feet are adapted for walking on the marsh's floating vegetation. A little smaller than the male, the female has less black on the face, and her yellow bill is a little duller (Melvin and Gibbs 2020). *Porzana,* the genus name, comes from the Italian word *sporanza,* or "little rail."

Luckily, the Sora's vocalizations are unmistakable. The most common call is the descending whinny that begins with eight or ten notes that "are uttered very rapidly in an evenly descending scale," followed by several that are delivered more deliberately (Brewster 1902). Another common call is a two-syllable *ker-wee* that often precedes the whinny.

Once considered weak fliers, these plump little birds were thought to be incapable of long migratory flights. "Superstitious notions and absurd fancies" were related to the belief that "Soras buried themselves in the mud at the approach of cold weather," spending the winter in a torpid state (Audubon 1827–38). In time, biologists learned that Soras, like most birds, were indeed capable of long migrations, some traveling several thousand kilometers to reach their southern wintering grounds.

The Sora is widely distributed throughout Canada and the northern and western United States. The core of its breeding range is located in the northern Great Plains of the United States and the Canadian Prairie Provinces.

MINNESOTA BREEDING DISTRIBUTION

In the late 1800s, Hatch (1892) described the Sora as Minnesota's most abundant rail, breeding in "nearly all parts" of the state. It was so common, he noted, that in Minneapolis one had to walk only fifteen minutes in nearly any direction at sunset to encounter the bird. Forty years later, the species' abundance appeared to have changed very little. Roberts (1932) commented that "every marsh and grass-bordered lake harbors a considerable number of these birds." Nevertheless, the nesting of this secretive marsh bird was confirmed in only six counties: Cass, Crow Wing, Hennepin, Isanti, Kittson, and Polk.

Green and Janssen (1975) and Janssen (1987) both described the Sora as a statewide resident, noting that it was rare or absent in heavily forested regions of northern Minnesota. Janssen identified 19 counties where nesting had been confirmed since 1970, only three of which were south of the Minnesota River (Lac qui Parle, Le Sueur, and Wabasha). Later, Hertzel and Janssen (1998) would add another six counties to the list, including just one, Olmsted, south of the river. By 2014, the MBS had identified 408 Sora breeding-season locations. Records were least abundant in northeastern Minnesota and, with the exception of the southwestern Prairie Coteau, were rare in all counties south of a line between Washington County in the east and Lac qui Parle County in the west.

Although MNBBA participants documented a similar distribution, their impressive sampling efforts in west-central Minnesota resulted in a notable increase in records from Big Stone County south through northern Lincoln County. Detections were also more abundant from the Twin Cities metropolitan region north to Cass, Crow Wing, and St. Louis Counties **(A)**. The species was absent only from Blue Earth, Dodge, and Watonwan Counties in south-central Minnesota. A total of 839 Sora records were detected in 13.5 percent (644/4,762) of the surveyed blocks. Breeding was confirmed in just 13 blocks spread across 12 counties, all north of the Minnesota River, including one record in Cook County. Six counties were additions to the list published by Hertzel and Janssen (1998). Breeding was also deemed probable in 155 blocks, limited primarily to the northern two-thirds of the state. The species occurred in more than twice as many blocks (644) as its relative the Virginia Rail (307).

The statewide probability map generated with MNBBA data predicted low breeding densities throughout western and central Minnesota **(B)**. The

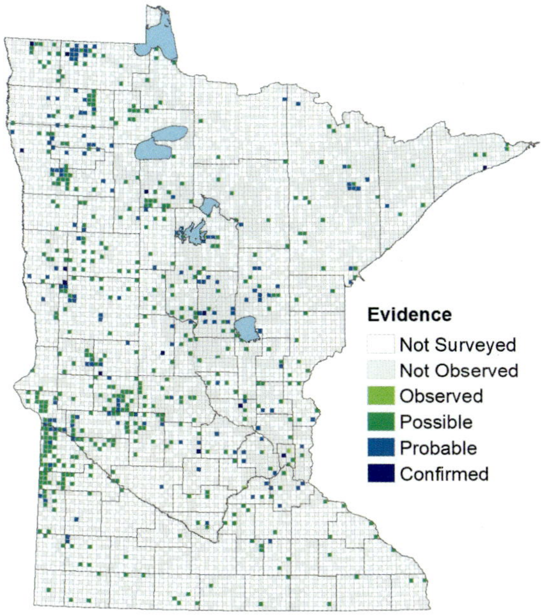

A. Breeding distribution of the Sora in Minnesota based on the Breeding Bird Atlas (2009–13).

region of highest abundance was predicted to occur in the extensive wetlands located immediately west of Red Lake and in the glacial outwash plains southwest of Leech Lake.

Although the Sora remains a widely distributed species in Minnesota, gone are the days when one could walk out the front door and be assured of encountering the bird. Beyond Minnesota, the extent of the Sora's distribution has changed little in the past one hundred years. Still widespread, its occurrence has become more localized in response to the extensive loss of wetland habitats (Melvin and Gibbs 2020).

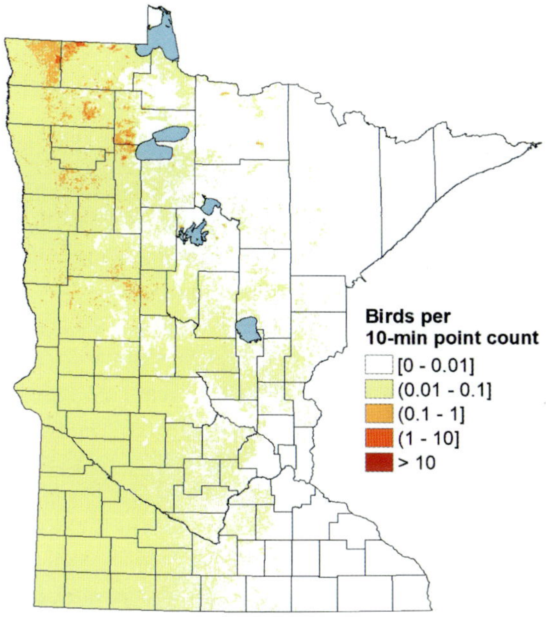

B. Predicted breeding distribution (birds detected per 10-minute point count) of the Sora in Minnesota based on habitat, landscape context, and climate data gathered during the Minnesota Breeding Bird Atlas (2009–13) using the Generalized Linear Modeling method without an adjustment for detectability.

BREEDING HABITAT

The Sora prefers semipermanent wetlands dominated by cattails, sedges, and bulrushes **(C)**. Wetlands where the emergent vegetation is interspersed with small areas of open water are preferred; wetlands with large expanses of open water are often avoided (Soulliere et al. 2007; Zimmerman, Jamison, et al. 2002a). Water levels vary widely; in deeper marshes, floating and submerged aquatic plants enable the birds to walk atop the vegetation and cross open water (Zimmerman, Jamison, et al. 2002a). Often found nesting in isolated marshes as small as 0.5 ha (Melvin and Gibbs 2020), Soras are more likely to be present when wetlands are abundant throughout the local landscape (Saunders et al. 2019).

POPULATION ABUNDANCE

Despite its being the most abundant rail in North America, information regarding the Sora's population size has been virtually nonexistent. Biologists have only recently provided a rough North American population estimate of 4.4 million birds (PIF 2021). Estimates for the Boreal Hardwood Transition Bird Conservation Region, which stretches across northern Minnesota, Wisconsin, and Michigan, range from 1,190 to 6,620 breeding pairs (Soulliere et al. 2007).

Given the Sora's relatively low detectability on BBS routes, trend data for most regions and states are not statistically robust. Nevertheless, the surveywide data suggest a relatively stable to slowly increasing population. In Minnesota, the population was stable from 1967 to about 1995 and then slowly declined through 2019 (Sauer et al. 2020). A downward trend was also documented by the Great Lakes Marsh Monitoring Program. Its survey of marshland inhabitants in more than 5,000 marshes in the Great Lakes basin demonstrated an average annual decline of −3.8 percent from 1996 to 2013 (Tozer 2016).

Not only is wetland loss a major concern, but the degradation of those wetlands that remain is a primary threat to the species. The infiltration of aquatic invasive species that choke out native wetland plants, sudden changes in water levels due to storm events, and water contamination from pesticides and sediments are all considered responsible for local population declines (Soulliere et al. 2007).

C. The Sora selects small, semipermanent wetlands dominated by emergent vegetation with scattered areas of open water. This particular site in Becker County supports both Soras and Virginia Rails. Photo by Lee A. Pfannmuller.

CONSERVATION

A harvested species, the Sora is legally hunted in thirty-one states, including Minnesota, and in Manitoba and Ontario (Melvin and Gibbs 2020). As the bird is targeted by relatively few hunters, the state's average annual rail harvest from 2010 to 2019 was only 344 birds (Davros and Dexter 2019). The reported total combines the harvest of Soras, Virginia Rails, and Common Gallinules, though Soras surely comprised the majority of birds taken. The practice of reporting combined totals occurs in most other states as well (Melvin and Gibbs 2020).

The species is designated a moderate conservation priority (PIF 2021); however, warmer temperatures could compound the challenges of habitat loss and degradation. The Sora is moderately vulnerable under all three warming scenarios examined; NAS (2019) predicted extensive range losses in Minnesota with near extirpation projected when temperatures rise by 3°C. Reproductive success may also be impacted by drought conditions and more extreme rainfall events brought on by warmer temperatures.

Conservation measures focus on protecting and restoring wetlands that both support diverse stands of emergent, floating, and submerged aquatic vegetation and are well interspersed with open water. Periodic drawdowns to rejuvenate the growth of native vegetation should improve habitat quality. Wetlands with dense stands of emergents, especially cattails, should be periodically disturbed by burning, plowing, mowing, or disking to increase cover diversity and create more open-water conditions (Zimmerman, Jamison, et al. 2002a). Such efforts should have multiple benefits for the entire community of secretive wetland inhabitants, including its most common member, the Sora.

COMMON GALLINULE

Gallinula galeata

Minnesota Seasonal Status: Regular breeding resident and migrant. The Common Gallinule was rare during the MNBBA.
Migration: Northern populations are short-distance migrants that winter in the southeastern United States; the migratory status of southern populations in the United States is less clear.
Food: Gleans food from the water surface and occasionally dives, feeding on the seeds and roots of aquatic vegetation and on small invertebrates.
Nest: Platform anchored to stems of emergent vegetation with a ramp leading down to the water's surface. Multiple nests are often constructed and used for resting by the adults and young broods.
Conservation Concern: A game species; assigned a moderate PIF Continental Concern Score of 9/20 and officially classified as *Special Concern* in Minnesota.

Photo by Michael Furtman.

Just a brief glimpse of this rail's bright red frontal plate immediately reveals its identity. Together with its rusty wings and rump, its long yellow legs, and the white stripe along the flanks and the underside of the tail, the Common Gallinule looks like an American Coot dressed for a party. Nearly the size of the Coot, but a much rarer inhabitant of Minnesota's wetlands, both the male and female Gallinule are similarly clad. If spotted, they are usually seen stalking among a marsh's dense reeds and cattails, rather than paddling about in the open water. The bird's vocal repertoire of loud and harsh hen-like grunts, clucks, and squawks has been described as "so varied and complex that it seems hopeless to attempt a full description of them" (Brewster 1891). Bannor and Kiviat (2020) observed that "concealed gallinules may emit sudden, loud calls that startle the unfamiliar observer into thinking there is a very large, dangerous, intoxicated chicken just around the bend."

Known by many names, including Common Moorhen and Red-billed Mud Hen, the Common Gallinule was considered a broad-ranging species present on every continent except Antarctica and Australia. Recent studies, however, have concluded that there are actually two species: the Common Gallinule (*G. galeata*) in North, Central, and South America, and the Eurasian Moorhen (*G. chloropus*) in Europe, Asia, and Africa (Chesser, Banks, et al. 2011).

Found in the eastern United States and southeastern Canada, the Gallinule has a breeding range that extends south through the Caribbean and from Mexico south to Chile and Argentina. Scattered populations also occur in several southwestern states and California. In the United States, the species is most abundant in southern Louisiana and Florida.

MINNESOTA BREEDING DISTRIBUTION

As early as 1892, Hatch recognized that the Common Gallinule, then known as the Florida Gallinule, was a regular but uncommon breeding species in Minnesota. He described it as common, however, in the upper Minnesota River floodplains near Big Stone Lake and along the Red River. Roberts (1932) later described it as a common summer resident in southern Minnesota, with records as far north as Otter Tail and Becker Counties. Within this range, it could be found breeding "in all the larger sloughs and shallow, grass-grown lakes." He also observed the birds in abundance at the Long Meadow Gun Club along the Minnesota River, writing in his journal: "It was no uncommon thing for ten or a dozen to be visible at the same time. . . . Their hen-like cackling, interspersed with loud, often shrill, cries usually kept up all through the night."

Confirmed nesting records, however, were limited to Heron Lake in Jackson County, Lake Shetek in Murray County, and Hennepin County. A nest reported in Otter Tail County lacked details. The absence of records in the Red River valley calls into question Hatch's assertion that the species was common in the region.

Nearly all reports of nesting pairs continued to come from southern Minnesota until, in July 1947, five young Gallinules were observed in Pope County (Chambers 1947). Then, between 1964 and 1966, there were four spring and summer reports from northern Minnesota: Marshall County (Ellis 1965), Stevens County (Strubbe 1965), Duluth (Green 1966), and Cass County (Egeland 1966). In light of these new records, Green and Janssen's (1975) updated species account suggested the Gallinule might breed farther north.

A few years later, however, Janssen (1987) reassessed the species' status and concluded that numbers had declined significantly in recent years. The Common Gallinule was now restricted largely to southeastern Minnesota, from the Twin Cities south through the backwaters of the Mississippi River and its tributaries. Sporadic reports continued from the central counties of Kandiyohi, Stearns, Todd, and Watonwan, and from as far north as Marshall County in the west and Aitkin County in the east, suggesting "the species may breed rarely in these areas." Since 1970, however, breeding had been confirmed in only nine central and southern counties: Anoka, Brown, Chisago, Hennepin, Houston, Ramsey, Sherburne, Stearns, and Washington. Hertzel and Janssen (1998) later deleted Chisago County from Janssen's 1987 map and added Morrison County. The decline discussed by Janssen (1987) also led to the species' official classification as a state *Special Concern* species (Coffin and Pfannmuller 1988).

When the MBS began its survey work in the late 1980s, the Common Gallinule appeared to have become even rarer. Only 15 breeding-season locations were reported. Several were concentrated along the Mississippi River valley from Wabasha County southward; others were located in Anoka, Becker, Nobles, and Watonwan Counties.

MNBBA participants reported 35 Gallinule records from 12 surveyed atlas blocks spanning 10 southern counties **(A)**. Breeding was confirmed in seven blocks and deemed probable in one block. Although a few records occurred along the Mississippi River south of the Twin Cities, several more were documented in south-central Minnesota, from southern Wright County south to northern Blue Earth County. Among the seven counties where nesting was confirmed, five were new to the list published by Hertzel and Janssen (1998): Blue Earth, Dakota, McLeod, Nicollet, and Wabasha.

Finding secretive marsh birds is challenging; finding rare, secretive marsh birds like the Gallinule is especially difficult. Although the species is likely more common than the MNBBA records suggest, the core of the Gallinule's breeding range now appears concentrated in central, south-central, and southeastern Minnesota. A few recent nesting records reported by Janssen (2019) and MOU (2021b) farther west in central and west-central Minnesota suggest that the rail may be slowly reoccupying portions of its former breeding range when wetland conditions are suitable.

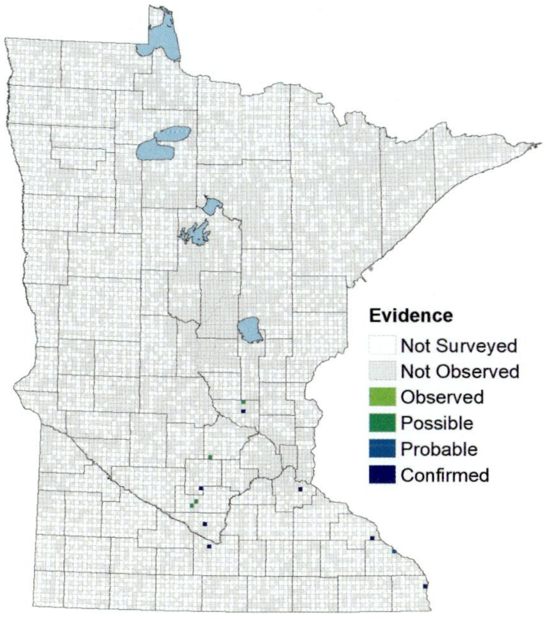

A. Breeding distribution of the Common Gallinule in Minnesota based on the Breeding Bird Atlas (2009–13).

During the twentieth century, the Common Gallinule gradually expanded its North American breeding range to the north and west. Reports from the Maritime Provinces in Canada, for example, were first documented in the 1960s, the same time that reports in northern Minnesota surfaced; reports from New Mexico were first documented in the late 1920s (Bannor and Kiviat 2020). Eventually, the range retraction observed in Minnesota beginning in the late 1970s was also observed in Ontario, where the Gallinule nearly disappeared from the most northern portion of its range (Cadman et al. 2007).

BREEDING HABITAT

Deep, permanent wetland basins are the Common Gallinule's preferred habitat. Hemi-marshes that contain dense stands of emergent vegetation interspersed with open water provide the best conditions **(B)**. Other important features include a dense cover of duckweed; floating mats of organic material that are used for preening, courtship, and resting; and muskrat lodges and runways. Water depths of 15–20 cm are preferred (Bannor and Kiviat 2020).

POPULATION ABUNDANCE

PIF (2021) provided a population estimate of 500,000 birds within the United States and Canada and a global population estimate of 5.1 million. Although BBS data lack statistical rigor given the Gallinule's low abundance, they suggest a relatively stable population trend across the survey region, with numbers increasing from 1966 to the early 1980s and then slowly declining through 2019 (Sauer et al. 2020). Regionally, the binational Great Lakes Marsh Monitoring Program documented a decline in Gallinule numbers from 1996 to 2013, with an average decrease of −4.5 percent per year (Tozer 2016). Population numbers or trends are not available for Minnesota.

Like that of all wetland-dependent species, the fate of the Common Gallinule is tied to the loss and degradation of wetlands, at least on the northern periphery of its range. Populations farther south appear to have adapted to more modified landscapes, including reservoirs and sewage treatment ponds (Bannor and Kiviat 2020).

B. Deepwater wetlands with an abundance of emergent vegetation scattered in small islands throughout the wetland provide breeding habitat for the Common Gallinule.
Photo by Lee A. Pfannmuller.

CONSERVATION

Considered a moderate conservation priority (PIF 2021), the Common Gallinule is a game species in thirty-one states, including Minnesota (Bannor and Kiviat 2020). The impact of the annual harvest is difficult to assess. Most states, including Minnesota, simply report the collective harvest of all rails and do not itemize the harvest per species. At least one research biologist who studied Gallinules for many years suggested that hunting pressure may be too high (Greij 1994). Given the species' rarity in Minnesota, few if any birds are likely harvested, but high harvest levels in southern states during migration and winter may impact northern breeding populations.

Despite apparent declines in the Common Gallinule's abundance, little focus has been directed at its conservation needs. Across its range, warming temperatures are actually projected to result in greater habitat gains than losses, including in Minnesota (NAS 2019). Unfortunately, few specific management recommendations are available, apart from the necessary protection, restoration, and management of wetland habitats. Indeed, the presence of nesting Gallinules in a newly restored wetland in Kandiyohi County (MOU 2021a) is a testament to the importance of such efforts to help ensure that this rare and seldom-seen inhabitant remains part of Minnesota's natural heritage.

AMERICAN COOT

Fulica americana

Photo by Michael Furtman.

Minnesota Seasonal Status: Regular breeding resident and migrant; occasionally observed during the winter. The American Coot was uncommon during the MNBBA.
Migration: Northern populations are medium- to long-distance migrants; most Coots winter along the southern Atlantic coast, the United States Gulf coast, the lower Mississippi River valley, and in the Caribbean, Mexico, and Central America.
Food: Pecks food from the water's surface or just below and occasionally dives, feeding on a variety of aquatic plants, invertebrates, and small fish. A kleptoparasite, the Coot also steals food from neighboring ducks.
Nest: Platform nest anchored to emergent vegetation, occasionally with a ramp leading to the water's surface. Numerous nests are constructed during courtship; one is usually selected for the eggs. A larger nest platform for the brood may be constructed after the eggs have hatched.
Conservation Concern: A game species; assigned a low PIF Continental Concern Score of 8/20.

Swimming among the teal, mallards, and shovelers, the American Coot might easily be mistaken as another member of the marsh's diverse waterfowl community. Yet, as duck-like as it appears in the water, on land its compact, hen-like body and its long legs and large feet reveal its true identity as a member of the rail family. Compared to those of the more reserved and relatively quiet waterfowl, the Coot's behavior and vocalizations are also notably different. Bent (1926) described it as a "conspicuous, noisy, and amusing clown" emitting a varied repertoire of grunts and squawks, raucously defending its nest when approached, and initiating spectacular territorial combats that are occasionally fatal to one of the participants (Bent 1926; Brisbin and Mowbray 2020). Often referred to as Mud Hens, Coots also bob their heads like a chicken when they are in motion, swimming among the cattails or walking along the shallow mudflats.

Since the Coot bears a black head, red eyes, and a slate-gray body, its most diagnostic feature is the white bill and frontal plate. Considered an adaptation that helps protect its face as it moves through the dense, hard stalks of cattails and bulrushes, it also serves a role in aggressive and sexual interactions. Often hidden from view, the wings have a trailing edge of white in the secondaries, and the undertail feathers have two patches of white; the latter are prominently displayed during courtship and aggressive encounters. Males and females are indistinguishable, although the male is slightly larger. Their most common vocalization is a guttural *krrp.*

The core of the American Coot's breeding range is the northern Great Plains of the United States and the Prairie Parklands of Canada. Although far less common elsewhere, the species is a year-round resident across much of the western United States and farther south through Mexico, Central America, and the Caribbean.

MINNESOTA BREEDING DISTRIBUTION

Roberts (1932) recognized the American Coot as an abundant breeding resident wherever suitable wetland habitat was available. He wrote, "During the month of June hundreds of Coots' nests may be found in practically all the sloughs and shallow lakes of the state." Despite the species' abundance, confirmed nesting reports were limited to a handful of sites that Roberts and his colleagues often visited: Heron Lake in Jackson County, numerous sites in Minneapolis and Hennepin County, and the northwest counties of Marshall and Kittson.

Green and Janssen (1975) and Janssen (1987) clarified that although the species was a statewide breeding resident, it was rare in the northeast and far north-central regions. Janssen described its primary range as occurring "west of the Mississippi River in the northern and central regions and west of Goodhue, Dodge, and Mower Counties in the south." Data collected by the MBS further confirmed this assessment. The 189 breeding-season locations the survey documented were confined primarily to the Prairie Parkland and Aspen Parklands Provinces. Janssen (1987) identified 46 counties where nesting had been confirmed since 1970; Hertzel and Janssen (1998) later added an additional 13 counties to the list.

A similar distribution to that described by Janssen (1987) was documented by the 490 American Coot records collected by MNBBA participants **(A)**. Found in 7.3 percent (349/4,749) of the surveyed atlas blocks, breeding was confirmed in 85 blocks (1.8 percent). Only 13 MNBBA records, including one breeding record in Aitkin County, were within the Laurentian Mixed Forest Province of northern Minnesota. Most of the records were within the southern half of the Prairie Parkland Province. The Coot was reported in 72 of Minnesota's 87 counties and was confirmed breeding in 35 counties. Seven counties were additions to the list published by Hertzel and Janssen (1998).

Although the American Coot's current distribution appears similar to that described by earlier accounts, the species' abundance has clearly declined since the days of Roberts (1932), when hundreds of birds could be found in any given wetland. Elsewhere, declines in both abundance and distribution have been noted since the early 1900s. While the American Coot's distribution has expanded westward, it has retracted and become quite scarce

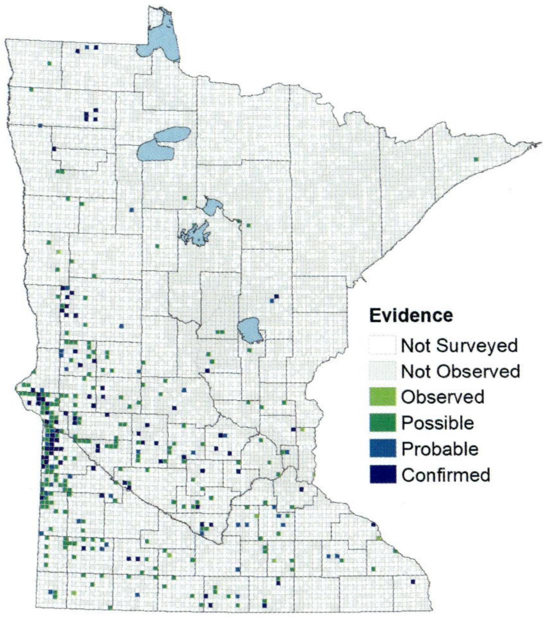

A. Breeding distribution of the American Coot in Minnesota based on the Breeding Bird Atlas (2009–13).

in the east, no doubt due largely to the extensive loss of wetland habitat (Brisbin and Mowbray 2020).

BREEDING HABITAT

The American Coot breeds in a variety of wetlands, ranging from prairie marshes and lakes to sewage ponds and canals **(B)**. Two primary characteristics define suitable sites: dense stands of emergent vegetation and standing water (less than 1.2 m) (Gullion 1954). As a result, the species is more closely associated with semipermanent and permanent wetlands than with seasonal or temporary wetlands. The latter are more frequently used during years with high water. Brackish wetlands are occasionally used, but freshwater sites are preferred (Tacha and Braun 1994; Brisbin and Mowbray 2020).

POPULATION ABUNDANCE

American Coot population data are provided by the North American Waterfowl Breeding Population and Habitat Survey. Within the survey's Traditional Survey area, which covers the core of the Coot's breeding population in the northern Great Plains and Canadian Prairie Provinces, 3.06 million American Coots were counted in 2015 (USFWS 2016b). Using BBS data, PIF (2021) provided a global estimate of 7.1 million birds and an estimate of 5.5 million in the United States and Canada. A statewide estimate for Minnesota is not available.

Historically, the American Coot was one of the most abundant inhabitants of the Prairie Potholes wetlands (Brisbin and Mowbray 2020). Yet, as the prairie sod became a productive agricultural landscape, the species' population plummeted owing to wetland drainage and intense harvesting. The full extent of these impacts is unknown. Fortunately, population estimates in the region have steadily increased since 1955, when the North American Waterfowl Survey began. Nonetheless, their reliability, especially in the survey's early years, has been questioned (Alisauskas and Arnold 1994). The focus of these surveys is waterfowl; accurately assessing Coot populations was a much lower priority. The BBS also provides an index of population trends. Unfortunately, from 1966 to 2019, the Coot has not demonstrated statistically significant increases anywhere within its breeding range; surveywide data show a nonsignificant, average annual decline of −0.96 percent. Numbers also illustrate a nonsignificant decline in the Prairie Potholes Bird Conservation Region, the core of the species' breeding range, averaging −0.86 percent per year **(C)**.

Although it fluctuates less, the downward trend line is similar in Minnesota, with a pronounced decline following the early 1980s. While the state's BBS data for the Coot lack statistical reliability, from 1967 to 2019 they illustrate an average annual decline of −1.03 percent. The downward trend is corroborated by Minnesota's annual Waterfowl Breeding Population Survey (Cordts 2019).

Brisbin and Mowbray (2020) note that habitat availability is the primary factor regulating American Coot populations but point out that "little information" is really available regarding the range of factors that influence the species. More research is needed to understand its true population status and why it is doing well in some areas and so poorly elsewhere. Unfortunately, the Coot's downward population trend in Minnesota is not unlike that observed for many waterfowl species in the state and likely reflects the impacts of large-scale wetland loss and degradation.

B. Although the American Coot is found in a variety of wetlands, lakes, and small ponds, most basins have deeper water than sites preferred by other members of the rail family, with extensive beds of emergent vegetation along the shoreline. Photo by Lee A. Pfannmuller.

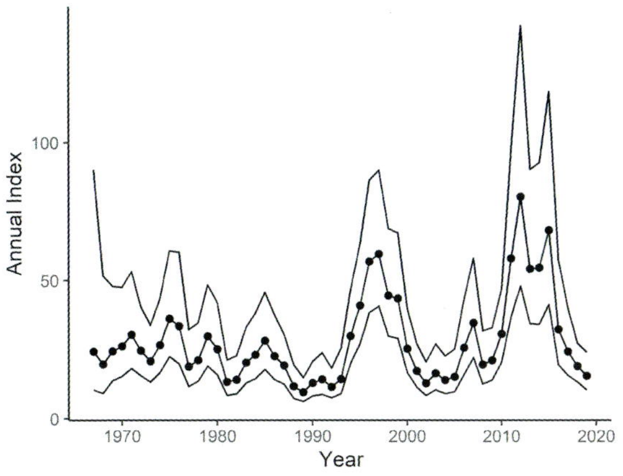

C. Breeding population trend and 95 percent credible interval (solid black lines) for the American Coot in the Prairie Potholes Bird Conservation Region based on the 1967–2019 Breeding Bird Survey (Sauer et al. 2020).

CONSERVATION

Like most rails, the American Coot is a harvested game species. Maligned and considered inedible by many northern hunters, the bird receives more admiration as one travels farther south. "Continue all the way to the Louisiana delta," one writer wrote, "and you've arrived in the Coot Appreciation Capital of the World" (Washburn 2014).

In Minnesota, hunting pressure is relatively low. From 2010 to 2019, the number of Minnesotans hunting Coots averaged 3,700 per year, and the annual harvest averaged approximately 17,000 birds (Davros and Dexter 2019). During the 2019 hunting season, the total Coot harvest in the United States was 242,600 (Raftovich et al. 2020). Overall, hunter interest in the bird has been declining in both Canada and the United States (Brisbin and Mowbray 2020).

Conservation measures are limited, but actions designed to benefit Minnesota's diverse waterfowl community should also provide lasting benefits to American Coots, especially aggressive wetland protection and restoration efforts. Although NAS (2019) assigned the Coot a low vulnerability under a 2°C warming scenario, the core of its population occurs in the Prairie Potholes, a region projected to experience more drought conditions and more severe storm events, posing even more concern for the species' long-term future.

YELLOW RAIL

Coturnicops noveboracensis

Minnesota Seasonal Status: Regular breeding resident and migrant. The Yellow Rail was rare during the MNBBA.
Migration: Short-distance migrant, wintering along the Gulf and southeast Atlantic coasts of the United States.
Food: Small freshwater snails supplemented with aquatic insects, terrestrial arthropods, and seeds.
Nest: Eggs are placed in a canopy-covered nest of sedges and grasses placed on the ground or on top of small tussocks of vegetation; a second nest without a canopy is constructed for brooding the young.
Conservation Concern: Assigned a high PIF Continental Concern Score of 15/20; officially classified as *Special Concern* in Minnesota.

The Yellow Rail is Minnesota's smallest and rarest member of the rail family. It is secretive, skulking, and "shrouded in mystery" (Bent 1926); its habits have been likened to those of a small mammal, slipping "silently through the grasses like a phantom" (Sprunt 1970). Peabody (1922) studied the species for twenty years and wrote, "Never but once, in the 20 odd years, did I ever actually see a yellow rail come out into the open."

This plump little Rail is distinguished by its buffy-yellow plumage, its short neck and bill, its short tail, and its heavily streaked back and flanks. The latter display a distinct pattern of thin, white bars. During the breeding season, the male's bill turns yellow. Although the bird is rarely seen in flight, white wing patches are diagnostic. The Latin name *Coturnicops* refers to its quail-like appearance; *noveboracensis* refers to New York, the locality of the type specimen.

Luckily, the male's simple courtship call is easily recognized. A series of *clicks* that sound like two coins repeatedly striking each other, the sequence is nearly always the same: two clicks followed by three: *click, click . . . click, click, click.* Adventurous birders carefully pursue the caller, which is heard most frequently after sunset, trying to catch a fleeting glimpse of one of the most sought-after birds in North America.

The Yellow Rail's breeding range stretches from Alberta and the southern Northwest Territories east to New Brunswick and northern Maine and south into northeastern Montana, North Dakota, and the northern Great Lakes states. A small population also persists in southern Oregon and northern California. Areas of abundance are poorly known, but Minnesota is considered the core of the United States' breeding population (Wires, Lewis, et al. 2010).

Photo by Chris Sloan.

MINNESOTA BREEDING DISTRIBUTION

As early as 1892, Hatch declared the Yellow Rail a rare summer resident based simply on five specimens in the possession of local taxidermists and private collectors. Having never observed the bird himself, he knew nothing of its habits and distribution. Even forty years later, it remained an elusive bird when Roberts (1932) summarized what little was known about its Minnesota status. There were summer observations from the late 1800s in Becker and Sherburne Counties, from 1928 in Marshall County, and one nesting record in Murray County in 1917 (Lake Wilson; adult, nest, and eight eggs were observed by Albert S. Peters [2018]). Then, in 1929, two colleagues visited Marshall and Kittson Counties in search of Sharp-tailed Sparrows (now known as Nelson's Sparrow) and found Yellow Rails at nearly every site where they found the Sparrow. Roberts concluded that the Rail was a summer resident in western Minnesota, and perhaps statewide.

As the years passed, scattered reports were slowly gathered from other localities. In 1952, a second nest was found and photographed by Nestor Hiemenz in a wetland just east of St. Cloud in Sherburne County (Janssen 1987), and in 1959 members of the Avifaunal Club discovered nine calling males at the Waubun Marsh, a wetland that straddles Mahnomen and Becker Counties (Huber 1959b). Nesting was confirmed at the site in 1962 and 1964 (Huber 1962b, 1964b). Because the Yellow Rail was consistently reported from the marsh in all but two years from 1959 to 1972, it became the go-to site for observing this rare species. Unfortunately, the site's dependability may have precluded wider investigation of other suitable nesting areas.

By the time Green and Janssen (1975) prepared their updated account on the status of Minnesota birds, recent nesting had been documented not only in Sherburne County and at the Waubun Marsh but also near McGregor in Aitkin County (1973). There were scattered summer reports from other counties as well (Clearwater, Crow Wing, Kittson, Marshall, and Traverse), leaving the authors to conclude that the Yellow Rail was most common in the northwestern and north-central regions of the state.

Gradually, during the mid-1970s, reports of the Rail's presence at the Waubun Marsh began to wane and were replaced by regular observations at the McGregor Marsh SNA. The marsh continues to be one of the most popular sites in North America to find Yellow Rails. Nevertheless, in nearly fifty years of active birding, the only documentation that the birds actually nest at the site was the 1973 report noted above, which was submitted by an exceptional birder, Terry Savaloja (1974). On June 2, Savaloja found an eggshell in the marsh that he identified as belonging to the species. Given its significance, he submitted it to the Bell Museum of Natural History. Then, on July 6, he "saw a downy rail which could have been a young Yellow Rail." Today this short note remains the only positive nesting evidence for Yellow Rails in Minnesota since nesting was confirmed at Waubun Marsh in the early 1960s. It is only rare observations of individual birds, and the birds' persistent calling at the site, which suggest that the McGregor Marsh remains a viable breeding location.

In 1980, new breeding-season localities were documented farther west when Yellow Rails were found on two grassland tracts: Anna Gronseth Prairie in Wilkin County (six calling males) and Pembina Trail Preserve in Pennington County (two calling males) (Eckert 1980). Janssen's 1987 update a few years later summarized additional summer reports gathered from counties east of the intensively cultivated Red River valley, including Beltrami, Cass, Clearwater, and Crow Wing, as well as from Kittson, Marshall, Norman, Polk, and Traverse. Regardless of the growing list of new locations, confirmed nesting reports of this small, secretive species had not been documented in any county since 1973. Janssen concluded that the species' breeding distribution was still "poorly known." Hertzel and Janssen (1998) were also unable to identify new confirmed nesting records.

More summer records were documented as the MBS began its systematic county-by-county surveys in northwestern Minnesota in the early 1990s. By 2014, the survey reported a total of 155 breeding-season locations, all within the northern third of the state, bounded by Kittson County in the northwest and Todd, Morrison, Mille Lacs, Kanabec, and Pine Counties in the south. The latter records extended the species' range farther south and east into central Minnesota than previous accounts had done since the 1970s.

Although the MBS did not uncover records of the species in northeastern Minnesota, in June 1993, when survey staff were working in western Minnesota, accomplished birder Kim Risen reported the Yellow Rail at two unusual locations much farther to the east: Cook County (two birds at one location) and southern St. Louis County (multiple birds found throughout the Sax-Zim Bog). As with the MBS, he also found birds in southern Pine County (several individuals at one location) (Risen 1993). These new records, coupled with the MBS's records in other central Minnesota counties, began to raise questions regarding the species' fidelity to breeding sites. Perhaps its dependence on suitable water levels meant that its presence at a given site is ephemeral and largely influenced by local wetland conditions. More research was underway in other parts of the species' range that would begin to address this issue.

Unfortunately, by the time the MNBBA was launched in 2009, the Yellow Rail's overall abundance seemed to have declined and its range receded to what appears to be its primary range in northwestern and north-central Minnesota. During the five-year atlas sampling period, observers tallied a total of 84 Yellow Rail records in 1.6 percent (76/4,748) of the surveyed atlas blocks. Surveyors were unable to confirm nesting anywhere, but it was deemed probable in 11 northern counties: Aitkin, Becker, Beltrami, Cass, Kittson, Koochiching, Lake of the Woods, Marshall, Polk, Roseau, and Wilkin **(A)**. The species was reported from 17 of Minnesota's 87 counties; one block straddled both Red Lake and Pennington Counties.

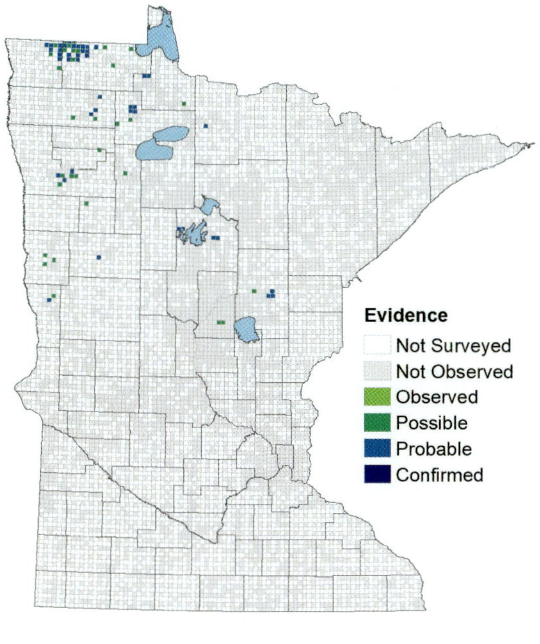

A. Breeding distribution of the Yellow Rail in Minnesota based on the Breeding Bird Atlas (2009–13).

Approximately 63 percent of all the blocks where the Yellow Rail was reported were located within the Tallgrass Aspen Parklands Province. Over 60 percent of all the records were contributed by the USFWS from targeted Yellow Rail and marsh bird surveys on state and federal lands. These surveys suggest that one of the most important sites for the Yellow Rail in Minnesota is the complex of state wildlife management areas (WMAs) in northern Kittson and Roseau Counties, including the Beaches Lake WMA, Caribou WMA, Roseau River WMA, and Skull Lake WMA. The McGregor Marsh SNA remains the site most frequently visited by birders interested in finding the Yellow Rail, yet these large northwestern wetland complexes may be more important to the species' long-term sustainability.

Although there is evidence that the Yellow Rail once nested as far south as Murray County (1917) and Sherburne County (1952), such southern locales appear to have been abandoned for many years. This, no doubt, is a consequence of the extensive loss of wetlands in central and southern Minnesota. More recently, the most southern extent of the Yellow Rail's breeding range in Minnesota is defined by breeding-season locations documented by the MBS in the central and east-central counties of Kanabec, Mille Lacs, Morrison, Pine, and Todd. Although the species was never found as far south during the MNBBA, studies have shown it regularly shifts its distribution depending on wetland conditions (Leston and Bookhout 2020; COSEWIC 2009a). Elsewhere within the species' breeding range, few wide-scale distributional changes have been documented recently other than local changes due to loss of wetland habitat (Cadman et al. 2007; Chartier et al. 2013).

Nevertheless, given the species' secrecy and rarity and the relative inaccessibility of its wetland habitat, documenting historical changes to its distribution is difficult. Leston and Bookhout (2020) note that "Historical and unconfirmed reports suggest that the species bred in the ne. U.S. south to about 40° N," which is as far south as southern New York. John James Audubon even reported the Yellow Rail was a year-round resident as far south as Florida and

B. Large, wide-open sedge wetlands are the preferred breeding habitat of the Yellow Rail. The absence of cattails is an important feature. Photo by Gerald J. Niemi.

Louisiana in the early nineteenth century, although "no subsequent confirmation of breeding at this latitude" is available (Leston and Bookhout 2020).

BREEDING HABITAT

Large sedge- and grass-dominated wetlands are the predominant breeding habitats utilized by the Yellow Rail **(B)**. Wet prairies and rich fens comprised of narrow-leaved sedges and wet meadows dominated by wide-leaved sedges and grasses are particularly important (MNDNR 2019o). An overlaying mat of dead vegetation from the previous growing season is critical. Not only does it help cover and conceal the nest but it also provides an overstory for runways that enable the species to quickly move through the densely vegetated site (COSEWIC 2009a). The birds are rarely present in wetlands dominated by cattails. Once woody shrubs begin to invade a site, its overall quality for the Yellow Rail diminishes.

Although water levels are an important factor guiding habitat selection, varying from saturated soils to standing water 15 cm in depth (COSEWIC 2009a; Hanowski and Niemi 1990), the abundance of the species' primary food item, snails, may be the ultimate predictor of suitable habitat, especially early in the breeding season. As the summer progresses and water levels subside, the Rails may move to the wetland's drier edges and feed more on insects and seeds (Pieper 1963a).

Locally, numbers of nesting birds fluctuate dramatically in response to changing wetland conditions at the beginning of each nesting season (COSEWIC 2009a). These changes may be the most dramatic in small, isolated wetlands compared with large wetland complexes. At the landscape scale, the species is more likely to be present in a given wetland if there is a landscape dominated by marshes and fens within 3 km of the occupied site (K. Martin 2012). This is borne out by the abundance of records in northern Kittson and Roseau Counties, where large, expansive wetland complexes are common.

POPULATION ABUNDANCE

Given its secretive nature, nighttime calling, and inaccessible habitat, the Yellow Rail is poorly surveyed by well-established monitoring programs, including the BBS, the Christmas Bird Count, and several new marsh bird monitoring initiatives. Canadian biologists, however, combined densities of Yellow Rails at known breeding sites with estimates of available habitat to generate a 1999 population estimate of 10,000 breeding adults in Canada and 1,200–1,500 in the United States (Alvo and Robert 1999). Ten years later, reports of local declines, coupled with known habitat loss in important nesting areas, led them to speculate that population numbers may have declined by less than 30 percent in the decade from 1999 to 2009 (COSEWIC 2009a). Recently, PIF (2021) also reported an estimated North American population size of 12,000 birds. Although no population estimate is available for Minnesota, the state appears to be the "mainstay" of the US population (Wires, Lewis, et al. 2010).

The primary threat to the species is the loss and degradation of wetland habitats on both its breeding and wintering grounds. Degradation can include agricultural runoff, woody invasion, and untimely manipulation of water levels in an effort to manage the habitat for other wetland nesting birds (Soulliere et al. 2007; Niemuth 2005). Indeed, for many years, waterfowl managers were eager to impound wet prairies and meadows, converting them to deepwater marshes that were attractive to nesting waterfowl. Today, wildlife managers are more sensitive to the needs of all wetland-dependent birds. Large wetland complexes are managed to provide a variety of conditions that benefit an array of species, including rails and shorebirds.

CONSERVATION

The Yellow Rail is a high conservation priority throughout its breeding range (PIF 2021) and has been ranked a High Concern by the North American Waterbird Conservation Plan (2006). It is federally listed as *Special Concern* in Canada and has been given legal status in many states and provinces (COSEWIC 2009a), including Minnesota, where it is listed as *Special Concern* (Coffin and Pfannmuller 1988). Minnesota has also designated the Rail a *Species in Greatest Conservation Need* in response to its rare and vulnerable breeding habitat (MNDNR 2015b). Fortunately, the harvest of Yellow Rails has been prohibited since 1968 (Leston and Bookhout 2020).

The continuing loss of shallow, graminoid wetlands poses the biggest threat to the Yellow Rail. Protection of public lands in northwestern Minnesota provides a solid foundation but needs to be augmented by continuing efforts to protect, restore, and manage the species' unique wetland habitat. Maintaining naturally fluctuating water levels is especially critical. This allows sites to be burned to reduce woody shrubs in dry years while still providing quality nesting habitat in wet years. Management plans also need to accommodate the bird's dependence on senescent vegetation because burned sites are not used for at least one year (Goldade et al. 2002).

Additional factors may impact population numbers. Annual mortality from collisions with communication towers, for example, averages 1 percent, a relatively high percentage given its small population (Longcore et al. 2013). In addition, rising water levels due to warmer temperatures may threaten coastal wetlands used by wintering birds, while breeding birds could lose a significant portion of their nesting habitat. NAS (2019) projected a high vulnerability of the species' continental breeding distribution under all three warming scenarios they examined. An increase of 2°C could result in the loss of 72 percent of its breeding range by 2050 and a gain of only 17 percent. Indeed, an increase of 1.5°C, which climate change biologists consider imminent, could cause the extirpation of the species in Minnesota.

Minnesota has a unique responsibility to protect this very rare and special member of our state's wetland avifauna. Although one's chances of actually seeing a Yellow Rail are slim, its unmistakable call is easy to identify and can be heard from quite some distance away. Warren Nelson (1991) provided an engaging account of his experience guiding birders from all over the world to the McGregor Marsh to see and hear the species. After straining to hear the Rail's first courtship clicks when darkness falls, and then wading through the cold wetland waters in search of the caller's location, the excitement of finally spotting a bird is indescribable. "The adrenaline," he wrote, "is flowing so fast you almost fly out of the marsh." Such is the thrill of seeing one of Minnesota's and North America's most elusive breeding birds.

SANDHILL CRANE

Antigone canadensis

Minnesota Seasonal Status: Regular breeding resident and migrant; some birds may linger into early winter before migrating south. The Sandhill Crane was common during the MNBBA.
Migration: Minnesota birds are short-distance migrants. Birds in eastern Minnesota winter in Georgia and Florida; birds in northwestern Minnesota winter along the Gulf coast of Texas.
Food: Omnivorous and an opportunistic feeder that consumes a wide variety of plants and animals from the ground and shallow waters.
Nest: A platform constructed of vegetation pulled from the nest site that is floating in shallow water, anchored to emergent vegetation, or placed in uplands immediately adjacent to the wetland.
Conservation Concern: Assigned a low PIF Continental Concern Score of 8/20; a game species in portions of its range, including northwestern Minnesota.

Photo by Stephen Maxson.

Watching a flock of Sandhill Cranes cross the open sky with their slow, deep wingbeats invokes images of earlier times, when grasslands were teeming with wildlife species long since gone. Aided by a long, coiled windpipe that can project their voice more than 3 km, their rich, bugle-like calls fill the sky long before the first birds are visible. One of "the world's oldest living birds," members of the crane family have "outlasted millions of species." Indeed, the "particularly successful Sandhill Crane of North America has not changed appreciably in ten million years" (Schoumatoff 2014). Remarking on their evolutionary history, Aldo Leopold (1949) wrote, "When we hear his call, we hear no mere bird. We hear the trumpet in the orchestra of evolution."

On the ground, Cranes stand nearly 1.5 m in height and are easily identified by their long legs, gray plumage, and red crown. Juveniles are a rich cinnamon brown their first year, but many of the adult's gray feathers also appear to be various shades of brown. The feathers become stained when the Cranes preen them with dirt or mud; soils rich in iron result in more of a reddish-brown coloration. Nesbitt (1975) speculated that the behavior may help to camouflage the Crane's large silhouette during the nesting season or when they are flightless during periods of molt.

Symbols of long life and eternal bliss, members of the crane family are revered by cultures throughout the world, including Native American tribes. Their exceptional vocalizations and their remarkable courtship dances are legendary. For many wildlife enthusiasts, witnessing the annual spring gathering of Sandhill Cranes in Nebraska is one of North America's most extraordinary wildlife spectacles. Nearly half a million birds stop along an 80-mile stretch of the Platte River for several weeks to feed on waste grain before continuing their trek northward.

An easy target for North America's early settlers, Sandhill Cranes have made a remarkable recovery from their earlier demise. Today they are widely distributed from Alaska, across Canada, and throughout the northern United States. Small, nonmigratory populations also occur in portions of Georgia, Florida, and Cuba. Some of the largest breeding populations are found in southern Manitoba and south through Minnesota and Wisconsin.

MINNESOTA BREEDING DISTRIBUTION

Given their size, it is no surprise that early explorers and fur traders took note of the bird's presence. When the fur trader Alexander Henry visited the Red River valley in 1804, he noted how common the "gray" cranes were near the town of Pembina, North Dakota, located directly across the Red River from Minnesota. William Keating's record of the Long Expedition up the Minnesota and Red River valleys in 1823 included an account of a large flock of Sandhill Cranes seen along the Minnesota River that was initially mistaken for a herd of elk. Members of the party could see so far across the open landscape that it was difficult "to determine with accuracy the nature or size of objects seen at a distance" (E. Swanson 1940; Nord and Breckenridge 1941).

In his treatise on Minnesota birds, Roberts (1932) included several accounts from early settlers during the late 1800s who commented on the species' abundance. Evadene Swanson (1940) compiled additional observations from published reports and newspaper stories of the same period. Most accounts, however, are descriptions of birds observed during the spring and fall migration, when thousands of Cranes nesting in Canada were passing through and hunters were proud to proclaim their success in securing the highly prized quarry. Only a handful of accounts pertain to the species' breeding status in the state. Indeed, Roberts (1932) could document only three confirmed breeding reports: one from Jackson County in 1883, and two from Polk County in 1919 and 1920. Despite the lack of nesting evidence, a few observers commented on the species' abundance during the summer, including in central Minnesota, in the vicinity of Heron Lake in Jackson County, and throughout the Red River valley. Their accounts led Roberts to declare that "while definite records of nests are largely lacking, there is every reason to believe that the Sandhill Crane nested commonly throughout Minnesota, south and west of the more heavily wooded area, until about 1875 or 1880."

But the intense subsistence and market hunting pressure on such a large and conspicuous bird, followed by widespread conversion of the prairie sod to agricultural cropland, led to a dramatic decline in numbers. Evadene Swanson (1940) cited a report from a game warden in Kittson County who noted that Cranes were still present in the county in 1892, but "their importance as a game bird ended in the 1880s." By the turn of the century, they were considered rare. Roberts (1932) described them as "much reduced in numbers and nesting only locally in a few places."

Drought conditions in the 1930s coupled with continued intensification of agricultural practices left few opportunities for this majestic bird to make a comeback, despite the closure of hunting seasons in

1916 under the newly signed Migratory Bird Treaty (Gerber et al. 2020) and subsequent passage of the 1918 Migratory Bird Treaty Act. These conditions, combined with the species' low annual productivity and delayed sexual maturity, slowed the pace of recovery. By 1944, Walkinshaw (1949) estimated that only 10 to 25 nesting pairs remained in Minnesota's remote northwestern counties. The Duxby Bog, now the Roseau River WMA, was considered one of the last strongholds (Henderson 1979a).

Then, in the 1950s, reports of nesting birds in the northwest began to increase (e.g., Magnus 1955; Jensen 1959; and Bremer 1966). More than 300 km to the southeast, accounts were also being reported from east-central Minnesota, including from Morrison, Aitkin, and Anoka Counties (e.g., Grewe 1968; Angell 1961; and Longley 1970). It was unclear if the eastern birds were remnants of a former population or if they originated from an expanding breeding population located in the Crex Meadows Wildlife Area in Grantsburg, Wisconsin (Henderson 1979b).

Several years later, Jay Johnson (1976) solicited data on the location and abundance of Cranes from natural resource agency personnel throughout the state from 1974 to 1975. Reported in 12 counties, 20 to 25 breeding pairs were estimated to occur in east-central Minnesota, and 50 to 60 pairs in the northwest. MNDNR's Nongame Wildlife Program conducted a similar assessment from 1977 to 1979, which yielded comparable population estimates and led to speculation that the statewide population might be as large as 150 to 300 nesting pairs (Henderson 1979a, 1979b).

These preliminary assessments unleashed a myriad of technical assessments and field studies. In 1985, a comprehensive assessment of the species' status and distribution in the state concluded that the population might be much larger than originally thought, consisting of at least 87 to 109 breeding pairs in the east-central region and 760 to 1,160 nesting pairs in the northwest (Tacha and Tacha 1985).

At the same time, the two population centers in the widely separated northwest and east-central regions of the state were beginning to merge. Resident Cranes were being reported throughout north-central Minnesota, with reports in Todd, Cass, Hubbard, and Crow Wing Counties. From a conservation and management standpoint, this was particularly interesting, since the birds in northwestern Minnesota were known to migrate to the Gulf coast of Texas, while the birds in east-central Minnesota were known to follow a different course, heading southeast to Georgia and Florida. The birds were also expanding their range farther south in the state, as evidenced by reports from Stevens, Cottonwood, and Houston Counties (Tacha and Tacha 1985), and north into St. Louis County. By 1998, Hertzel and Janssen had delineated 18 counties where nesting had been confirmed since 1970, ranging from Houston County in the southeast corner to Kittson County in the far northwest. And, by 2014, MBS field biologists reported 633 Sandhill Crane breeding-season locations, including one location in southwestern Minnesota.

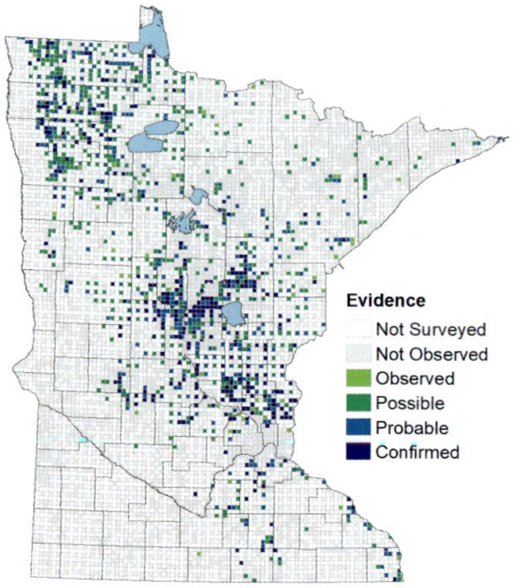

A. Breeding distribution of the Sandhill Crane in Minnesota based on the Breeding Bird Atlas (2009–13).

During the MNBBA, observers reported 1,943 Sandhill Crane records in 24.4 percent (1,186/4,856) of the surveyed atlas blocks and gathered confirmed breeding evidence in 6.0 percent (290) of the blocks **(A)**. The species was reported in 66 of Minnesota's 87 counties, while nesting was confirmed in 52 counties. Southwestern Minnesota and much of the Red River valley were the only regions lacking reports. Cranes were particularly abundant in the northwest Tallgrass Aspen Parklands Province and throughout central and east-central Minnesota. The species' high abundance in these regions is clearly delineated by the MNBBA landcover suitability model **(B)**. Small pockets of highly suitable habitat, however, are also predicted throughout southern Minnesota.

Based on Roberts's original account, it appears that intensively cultivated regions of the Red River valley and southwestern Minnesota are the only regions of the Sandhill Crane's former breeding range that it has not reoccupied. As Henderson (2016) pointed out, this impressive bird is a resilient species that has adapted well to a landscape that has changed in countless ways in the past 100 to 125 years. The Crane's recovery in Minnesota has also been repeated in other regions of its breeding range. Nevertheless, as in the case in Minnesota, its current distribution is still believed to be less extensive than it was in the early 1800s (Gerber et al. 2020).

Over the years, numerous Minnesota studies also examined the birds' breeding biology, migratory behavior, nest success, and habitat use (Maxson, Provost, and Davis 1990; Maxson, Fieberg, and Riggs 2008; Herr 1992; Provost 1991; DiMatteo 1992). More recent studies have focused on explicit management issues associated with their burgeoning population (e.g., Krapu et al. 2011; Fronczak 2014; Fronczak et al. 2015).

BREEDING HABITAT

The Sandhill Crane is an inhabitant of open landscapes. Tacha and Tacha (1985) summarized its suitable breeding habitat as "any large, shallow, relatively open, isolated wetland adjacent to open, upland foraging areas" **(C)**. Within Minnesota, wetlands can be as diverse as the extensive open peatlands and wet meadows of northwestern Minnesota to the backwaters of the Lower Mississippi River floodplain in southeastern Minnesota.

Despite their variability, suitable breeding areas always contain four primary elements: water, nesting cover, isolation, and a nearby foraging area (Provost 1991). Water depth is critical. Nest depredation in northwestern Minnesota occurred significantly more often when nests were placed on sites where the water was less than about 10 cm deep (Maxson, Fieberg, and Riggs 2008). Shallow water depths allow easier access to nests by mammalian predators and increase the nests' vulnerability to flooding during heavy rain events.

Moderately dense stands of cattails, sedges, phragmites, and bulrushes help conceal the nest. Low shrub densities are also preferred (Tacha and Tacha 1985). Sites where shrubs and emergent vegetation are so dense that they restrict access and the

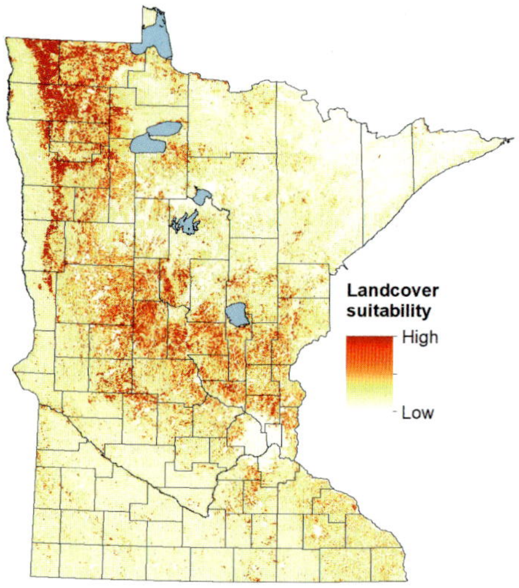

B. Landcover suitability map for the Sandhill Crane in Minnesota based on habitat, landscape context, and climate data gathered during the Minnesota Breeding Bird Atlas (2009–13) using the MaxEnt modeling approach.

C. The large managed wetland complex at the Carlos Avery WMA in east-central Minnesota provides the cover and relative isolation that the Sandhill Crane prefers for nesting. Photo by Lee A. Pfannmuller.

ability of the birds to view their surroundings are usually avoided (Maxson, Fieberg, and Riggs 2008). Although most nest sites are isolated, the birds do not appear to be restricted to large wetland basins. In northwestern Minnesota, for example, Provost (1991) found birds nesting in wetlands as small as 0.6 ha and as large as 1,862 ha.

Once the young have hatched, the family forages in shallow wetlands as well as in adjacent grasslands and agricultural fields. During the summer, nonbreeding adults are also found in these upland habitats (Provost 1991).

POPULATION ABUNDANCE

In 1975, the North American Sandhill Crane population was broadly estimated at 285,000 to 320,000 birds (Drewien et al. 1975). Recently, state and federal agencies conducted surveys that resulted in an estimate of 507,000 individuals (Gerber et al. 2020). Given the population's growth, even this number is likely too small.

Apart from general approximations of statewide numbers in the 1970s and 1980s, there has never been an in-depth assessment or survey of Minnesota's Sandhill Crane population. As the population expanded and interest in initiating a hunting season in northwestern Minnesota grew, the MNDNR added Cranes to their annual August roadside pheasant survey in 2009. In 2020, the number of Cranes seen statewide averaged 13.6 birds per 160 km. Regionally, the greatest numbers were seen in the east-central survey region (48.6 birds per 160 km), followed by the northwest region (39.7 birds per 160 km; Lyons 2020b).

Once a limited Sandhill Crane harvest season was launched in the fall of 2010 in northwestern Minnesota, the MNDNR also initiated a more intensive regional survey, conducted each May since 2012, to properly track potential impacts of the harvest on the local breeding population (Lawrence, Giudice, and Wright 2016). The population has ranged from a high of 4,469 birds in 2012 to a low of 1,952 birds in 2014. Limited to the far northwest region, the numbers represent only a fraction of the statewide population. Both this and the August roadside survey will need to continue for many years before long-term population trends are discernible.

The BBS, however, does provide a strong indication of long-term population trends across the North American survey region and in Minnesota. Surveywide, from 1966 to 2019, Cranes demonstrated a statistically significant increase, averaging 4.07 percent per year **(D)**. Although the data are less rigorous, the upward trend is even more dramatic in Minnesota, where the birds have increased an average of 6.72 percent per year (Sauer et al. 2020). Indeed, northwestern and east-central Minnesota now support some of the densest breeding concentrations across North America.

CONSERVATION

Given its spectacular recovery, PIF (2021) has designated the Sandhill Crane a low conservation priority. In Minnesota, the species was officially listed as *Special Concern* in 1984 (Coffin and Pfannmuller 1988), prompted largely by concerns regarding the availability of suitable habitat as agricultural practices continued to intensify and wetland acres continued to decline. Despite these concerns, the species prospered, and the Crane was delisted in 1996 (MNDNR 1995). Wildlife managers and research biologists all agree, however, that the Crane's future rests entirely on habitat availability. Indeed, at the heart of the Crane's remarkable recovery have been aggressive efforts since the mid-1900s to protect and restore shallow wetland habitats.

Success, however, begets both opportunities and challenges. Hunting Cranes, for example, is a long-established tradition and one that nearly brought the species to its demise. As a result, when the Migratory Bird Treaty was first signed in 1916, hunting seasons were closed in both Canada and the United States. Forty years passed before numbers began to rebound, prompting the reinstatement of hunting opportunities in Alaska, New Mexico, and Texas in 1961 and in Saskatchewan and Manitoba in 1964. As populations have continued to increase, hunting opportunities have expanded.

The harvest and management of North America's Sandhill Crane populations are determined by their geographical location. Minnesota's northwestern cranes belong to the Mid-Continent Population (MCP). Most states that harbor migrant birds from the MCP reinstated hunting seasons many years ago. Minnesota, however, delayed reopening the season until 2010, when it allowed hunting in all or portions of eight counties in the northwest region.

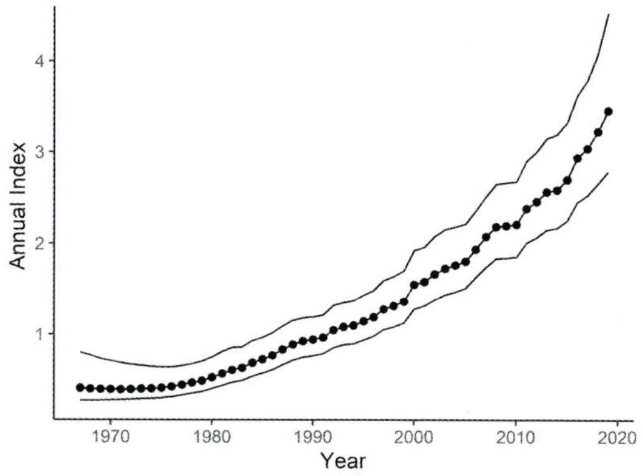

D. Surveywide breeding population trend and 95 percent credible interval (solid black lines) for the Sandhill Crane based on the 1967–2019 Breeding Bird Survey (Sauer et al. 2020).

Because these counties support resident breeding Cranes as well as migrants heading south from Canada, some biologists have expressed concern regarding the hunting season's impact on the local breeding population. Harvest levels, however, have steadily declined since the season first opened, with 830 birds taken in 2010 and only 129 in 2018 (Dubovsky 2019). The initial interest seems to have waned as hunters learn the challenges of hunting such wary targets. Nevertheless, adjustments to the hunting season have been made to ensure that the harvest is primarily focused on migrants and the local breeding population is managed sustainably.

Cranes that breed in eastern and central Minnesota, however, belong to the Eastern Population (EP) for management purposes. As of 2021, Alabama, Kentucky, and Tennessee are the only states that permit the fall harvest of EP migrants.

Meanwhile, as populations increase statewide, so do crop depredation complaints by farmers. The problem is particularly troublesome in the spring, when newly planted corn provides a tasty treat to migrating Cranes. A nonlethal repellent, Avipel, is available to help discourage depredation, but continuing complaints are also fueling a growing interest in opening a hunting season in eastern and central Minnesota. Studies are now underway to examine these populations more closely.

Finally, the influence of warming temperatures on current populations is a growing concern. A modeling exercise by NAS (2019) projected a moderate vulnerability for the Sandhill Crane under a 2°C warming scenario with the loss of 27 percent of its North American breeding range by 2050 and a gain of only 18 percent farther north. Under the same scenario, the species would be gone from all but the northeastern and east-central counties of Minnesota. Although we can marvel at the remarkable recovery that Sandhill Cranes have made in the state, work to ensure that their future remains secure is critical.

AMERICAN AVOCET

Recurvirostra americana

Minnesota Seasonal Status: Summer visitant, occasionally breeds; regular spring and fall migrant. The American Avocet was very rare during the MNBBA.
Migration: Medium-distance migrant that winters from the coasts of the Carolinas and southern California south along the Mexican coastlines. Winter populations in central California and Mexico and along the Texas Gulf coast may include permanent residents.
Food: Forages in shallow waters for aquatic invertebrates.
Nest: Ground scrape lined with vegetation or pebbles or completely unlined.
Conservation Concern: Assigned a moderate PIF Continental Concern Score of 11/20.

Photo by Rebecca Field.

This beautiful long-legged wading bird is emblematic of Pacific Ocean beaches and the inland alkaline lakes and wetlands of the Great Basin and Great Plains. Ryser (1985) commented that its presence "enlivens the bleakest yet most characteristic of Great Basin summer landscapes—the shallow foul bodies of alkaline or brackish water and their fringing flats of mud, alkali, or salt." Surprisingly, this elegant bird also occasionally breeds in western Minnesota.

There really is no other bird quite like the American Avocet. Standing nearly 46 cm tall, it is a striking contrast to the surrounding grassland landscape. During the breeding season, the rusty plumage on its head and neck is prominent, along with a white belly and a bold black-and-white pattern on its back and wings. As it wades through the shallow waters with its bluish-gray legs, the most conspicuous feature is its very thin, long, black bill that curves upward—hence the genus name, *Recurvirostra.* The sexes are identical, and both emit an oft-repeated, high-pitched *kleet* call.

The Avocet's breeding range is centered in the Great Basin and Great Plains and other areas of the arid west. It also breeds locally along the California coast, in the Central Valley, and in inland basins from Washington south to the Texas Panhandle. Another year-round population occurs along the Texas Gulf coast. To the north, the Avocet's breeding range includes the Canadian Prairie Provinces. The eastern margin of its range includes persistent breeding populations in North Dakota (Drilling 2015) and South Dakota (Drilling, Sparks, et al. 2016) and sporadic nesting in western Minnesota. It appears to be just a rare migrant in Iowa (Kent and Dinsmore 1996). High breeding densities are found in northern Utah. Historically, Avocets nested along the Atlantic coast as far north as New Jersey in the 1800s, but commercial plume hunting extinguished this population by 1900 (Ackerman et al. 2020).

MINNESOTA BREEDING DISTRIBUTION

In 1932, Roberts wrote, "Although still a common bird in western North Dakota, the Avocet has long been extinct in Minnesota and comparatively little is on record in regard to its former presence in the state. . . . There is reason, however, to believe that during the early years of the settlement of the state it was a summer resident and nested on the prairies of the west." However, it was not common, "as the early explorers . . . make no mention of it." Roberts cited records from the Dakotas and an anecdotal record from Jackson County, but during his own two-week trip to the prairie in Grant and Traverse Counties in June 1879, he "saw nothing of the Avocet." The only other record from the nineteenth century is an unknown number of birds collected near Browns Valley in Traverse County between June 21 and July 8, 1887, by Vernon Bailey (Roberts 1932).

Fortunately, Roberts's prediction of extinction in Minnesota did not come to pass. The first twentieth-century observation, a sighting in Murray County in 1934, was published in the second edition (1936) of *The Birds of Minnesota*. Other early records include single observations in August 1937 from Lincoln and Big Stone Counties (Rysgaard 1937b). Surprisingly, the next record was from May 21, 1944, of a bird observed on Minnesota Point, Duluth (Lakela 1944). When bird-watching on the western Minnesota prairie became more common in the 1950s, the Avocet was considered a regular migrant and irregular summer resident (Janssen 1987). The first nesting record was in Lyon County in 1956 (Southwick 1956). Subsequently, 23 nest records from 1956 to 1993 were published in *The Loon*. They were from ten western border counties inland to Clearwater, Lyon, and Todd, with an outlier in Faribault County. These were compiled and well mapped in the article "The American Avocet in Minnesota" by Peder Svingen (1998a).

After Svingen's article was published, seven more records with evidence of breeding were documented: Big Stone County (1999, 2000, 2003, 2005; MOU 2021b); Traverse County (Marengo 2000); Hamden Slough, Becker County (Wiens 2003); and Marshall County (Wiens 2004). Most of the breeding records and summer records in the twenty-first century were centered in Big Stone (Toqua Township), Lac qui Parle, and Traverse Counties. Considering how scattered in time and space these records have been and the few observers present early in the nineteenth century, it is probable that Avocets have always been rare nesters in Minnesota. The birds are likely outliers from the more substantial breeding populations in North and South Dakota.

Most American Avocet records in Minnesota, including hundreds documented throughout the twentieth century (Janssen 1987), are spring and fall migrants, including those in early June or late July through August. At least half of the migration observations were in counties near the borders of North and South Dakota. All but 18 of the 87 Minnesota counties have one or more records of the American Avocet (see map in Svingen 1998a), and more counties have been added in recent years. Eighteen records from 1934 to 2014 are from the city of Duluth, a migration hot spot, as well as from cities along the Iron Range in St. Louis County. If there are bird-watchers and open habitat in rural areas, it is likely the species will be spotted at some point in time.

During the summer breeding season, participants in the MNBBA reported four American Avocet records in four atlas blocks (one confirmed, one probable, and two observed). All four records were in Big Stone County **(A)**. This distribution is a small echo of the breeding distribution described above by Peder Svingen. Because bird records in recent years are mostly county listings without behavioral information separating migrants from breeders, it is difficult to judge if the range of the Avocet delineated by the MNBBA is indeed confined to just a few managed wildlife areas and their surroundings.

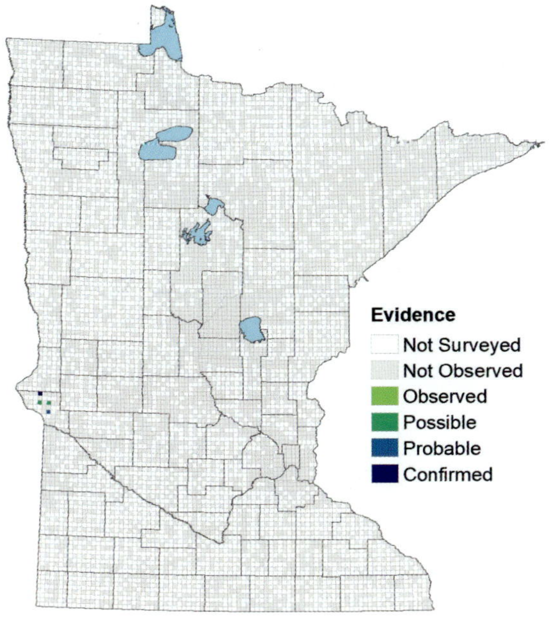

A. Breeding distribution of the American Avocet in Minnesota based on the Breeding Bird Atlas (2009–13).

B. American Avocet breeding habitat is characterized by mudflats with sparse vegetative cover for nesting, embedded in shallow wetlands for foraging, like this site at the North Ottawa Impoundment in Grant County. Photo by Lee A. Pfannmuller.

In recent years, from 2014 to 2020, a cadre of active birders has been visiting the new Red River valley impoundments, which are restored shallow wetlands created as part of large-scale flood control projects. Included among the many waterbirds observed were two new Avocet confirmed breeding records: from the Ulen Wetland Project, Clay County, a pair with two chicks, June 12, 2017; and from the North Ottawa Impoundment, Grant County, three pairs, one with new chick, June 13, 2018 (MOU 2021a).

BREEDING HABITAT

In North Dakota, breeding pairs are "largely restricted to ponds and lakes with exposed, sparsely vegetated shoreline or mud flats that are adjacent to expanses of shallow water **(B)**. These conditions occur naturally on alkali ponds and lakes and on sub-saline, semi-permanent ponds and lakes. During low-water conditions, exposed shorelines or mud flats are created in other types of wetlands by certain types of land-use practices, particularly overgrazing and tillage" (Stewart 1975). In Minnesota, the breeding habitat has seldom been described; sites mentioned included sewage treatment lagoons, impoundments, islands in a marsh, and bare edges of lakes, most notably at Salt Lake WMA, Lac qui Parle County.

POPULATION ABUNDANCE

The global population of the American Avocet (Canada, Mexico, and United States) is estimated at 460,000 birds (PIF 2021). A high proportion of the population occurs in the Prairie Pothole Region of North Dakota (Beyersbergen et al. 2004). Survey-wide, BBS data demonstrate a reliable, relatively stable population trend with a nonsignificant average annual increase of 0.33 percent per year from 1966 to 2019 (Sauer et al. 2020). A positive, but nonsignificant population trend (0.74 percent per year) has also been documented in the Prairie Pothole Region of the northern Great Plains **(C)**.

CONSERVATION

The primary threat to the American Avocet is the loss of habitat on its breeding and wintering grounds due to habitat destruction and wetland drainage. This, coupled with its relatively small population size, led to its assignment as a moderate conservation priority (PIF 2021). Pesticides and other toxins are another concern because both selenium and methylmercury can be high in the species' preferred habitats; in some circumstances, the viability of eggs and the survival of chicks can be affected by the high bioaccumulation levels of these substances (Ackerman et al. 2020).

Although it is a Great Plains inhabitant, warmer temperatures may not be a concern. NAS (2019) classified the Avocet's breeding distribution as stable under all three warming scenarios they examined. All of their models consistently compensated small range losses with significant gains in the western states and Mexico. Indeed, the range expansions they predicted in the eastern Dakotas increases the likelihood that this stunning, long-legged wading bird will be seen more often in western Minnesota as temperatures rise.

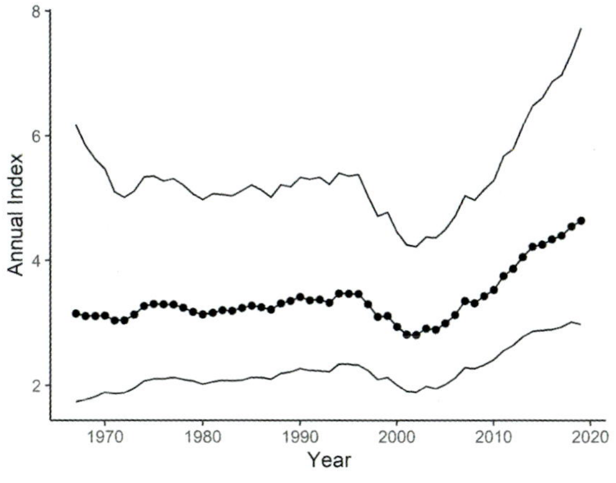

C. Breeding population trend and 95 percent credible interval (solid black lines) for the American Avocet in the Prairie Potholes Bird Conservation Region based on the 1967–2019 Breeding Bird Survey (Sauer et al. 2020).

KILLDEER

Charadrius vociferus

Minnesota Seasonal Status: Regular breeding resident and migrant; occasionally late fall or early spring migrants are observed during the winter months in southern Minnesota. The Killdeer was abundant during the MNBBA.
Migration: Short-distance migrant that winters along the Gulf coast and in the southeastern United States.
Food: Probes with its bill and pats the ground with its feet to find a smorgasbord of terrestrial and aquatic invertebrates that it consumes; occasionally feeds on frogs and small fish.
Nest: Small depression in the ground, usually lined with pebbles, twigs, and other debris.
Conservation Concern: Assigned a moderate PIF Continental Concern Score of 10/20.

Photo by David Brislance.

In the waning days of winter, when it seems that spring may never arrive, the familiar call of the Killdeer is the long-awaited sign that warmer days are ahead. Minnesota's most abundant shorebird arrives early, along with two other harbingers of spring, the Eastern Bluebird and American Robin. Each has an affinity for city parks, lawns, and other manicured landscapes that afford us ample opportunities to observe their habits at close range.

A little larger than a Robin, the Killdeer boldly announces its presence with its striking plumage and exaggerated behaviors. The sharp contrast between its brownish-gray wings and back, its white belly, and the two black bands across its chest is sufficient to identify the identically plumaged male and female. In flight, a rufous rump and upper tail coverts are clearly visible. Should you wander close to this *vociferous* shorebird, you are likely to hear its alarm calls long before you see it. The slightest disturbance will unveil an incessant *dee-ee* call that can start out slowly and progress into a rapid trill as it either chases the intruder or feigns injury. Its well-known *kill-dee* call is most frequently given in flight but may also be used as an alarm call (B. Jackson and Jackson 2020).

The Killdeer's spectacular efforts to protect its eggs and young are at the root of its notoriety. Roberts (1932) described its broken wing display as follows: "There is no bird that can display a more perfect imitation of severe injury than can the Killdeer when its nest and young are threatened. A few feet from the nest it falls flat on the ground all spread out as though hopelessly crushed, and crying piteously. If followed it revives sufficiently to limp along, dragging first one wing and then the other, tail spread widely, displaying the showy lateral feathers and the tawny rump in a most alluring way."

Broadly distributed from eastern Alaska across Canada and throughout the entire United States, south to Mexico and the Caribbean, the Killdeer is also a common summer inhabitant of open lands throughout Minnesota.

MINNESOTA BREEDING DISTRIBUTION

"The Killdeer," Roberts (1932) wrote, "is too common and too well known to need more than a general notice. There is no place in the state, except in deep forests, where its loud and penetrating notes, *kill-dee, kill-dee,* can not be heard." Although he described the species' nesting distribution as statewide, confirmed nesting records were limited to eight counties, stretching from Dakota and Goodhue in the southeast through the central counties of Hennepin, Isanti, McLeod, and Sherburne; west to Grant; and north to Polk. Other evidence of breeding was reported from Kittson, Lake of the Woods, and Marshall Counties.

Both Green and Janssen (1975) and Janssen (1987) considered the species common and widespread wherever suitable habitat was present. Janssen (1987) delineated 56 counties dispersed across all regions of the state where nesting had been documented since 1970. Hertzel and Janssen (1998) would later add an additional 11 counties to the list.

The MBS reported 799 breeding-season locations throughout the state. Despite their wide distribution, records were sparse in the Paleozoic Plateau of southeastern Minnesota, in portions of the northern forest, in east-central Minnesota, and in the heavily cultivated landscapes of western Minnesota where survey efforts were less intense.

The species' statewide distribution was reaffirmed by the MNBBA. Participants reported 4,190 Killdeer records in 49.5 percent (2,380/4,805) of the surveyed atlas blocks and in 66.8 percent (1,561/2,338) of the priority blocks. Breeding was confirmed in 618 blocks (12.9 percent) **(A)**. Although the Killdeer was reported from all 87 Minnesota counties, records were least abundant in northeastern and north-central Minnesota. Nesting was confirmed in 86 counties, 19 more than delineated by Hertzel and Janssen (1998). Brown County, along the Minnesota River in southwestern Minnesota, was the lone county without a confirmed nesting record during the atlas.

As broadly distributed as the Killdeer is outside of north-central and northeastern Minnesota, including the impressive survey results in the Twin

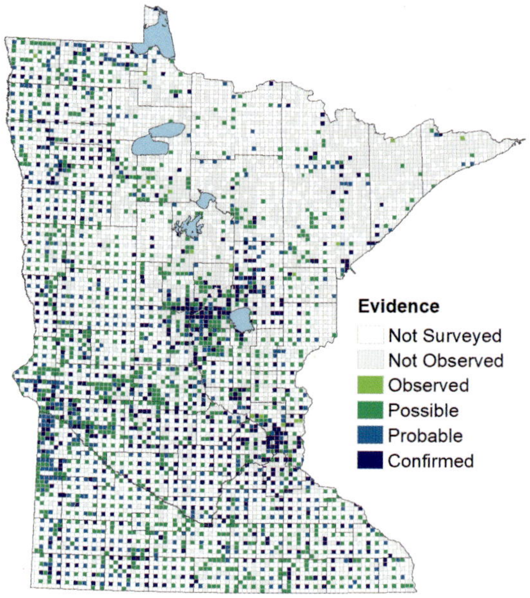

A. Breeding distribution of the Killdeer in Minnesota based on the Breeding Bird Atlas (2009–13).

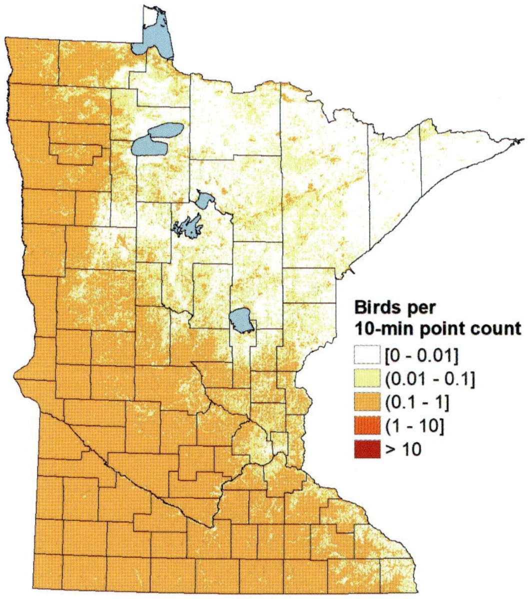

B. Predicted breeding distribution (birds detected per 10-minute point count) of the Killdeer in Minnesota based on habitat, landscape context, and climate data gathered during the Minnesota Breeding Bird Atlas (2009–13) using the Generalized Linear Modeling method without an adjustment for detectability.

Cities and the Brainerd region, the core of the species' range is clearly in the agricultural region of the state. The MNBBA probability map illustrates that predicted breeding densities gradually decrease east across the state's Eastern Broadleaf Forest and Laurentian Mixed Forest Provinces **(B)**. Only scattered pockets of birds are predicted to occur in Lake and Cook Counties. The one area of abundance in northern Minnesota is in the open habitat along the Iron Range in St. Louis and Itasca Counties.

Nearly one hundred years after Roberts wrote his first account of the Killdeer's status in Minnesota, little appears to have changed. No doubt, as the northern forest landscape became more fragmented by urban, industrial, and small-scale agricultural development, open habitats provided more suitable opportunities for the bird. Significant range expansions have also been documented across North America during the twentieth century, with birds expanding from interior breeding grounds into more coastal areas. Included are expansions northeast into Quebec, Nova Scotia, and Newfoundland, south into the Florida peninsula, and northwest along the coast of British Columbia (B. Jackson and Jackson 2020). Some states and provinces where the species originally expanded are now witnessing a retraction in the species' breeding range. Ontario, Michigan, and Vermont, for example, documented retractions from the northern regions of the state or province. These were believed to be partly due to abandonment and reforestation of marginal farmlands (Cadman et al. 2007; Chartier et al. 2013; Renfrew 2013).

BREEDING HABITAT

Historically, this member of the shorebird family was likely most abundant on open, sparsely vegetated areas that followed fires and floods or in gravel and sandbar habitats bordering rivers, lakes, and estuaries. Although it continues to use these habitats, the Killdeer has adapted to a wide range of modified landscapes that provide the most critical features: low, sparse vegetation (usually less than 0.1 m in height) and an open vista in all directions (B. Jackson and Jackson 2020). Rocks or pebbles may be present but are not essential. Nests located close to water are usually more successful (Kaufman 1996), although, as B. Jackson and Jackson (2020) note, a water sprinkler may be all that is needed. As a result, lawns, ball fields, farmsteads, golf courses, parking lots, city parks, and gravel pits are among the wide range of habitats now utilized by the Killdeer **(C)**.

POPULATION ABUNDANCE

Morrison and his colleagues (2001) extrapolated BBS data to generate a population estimate of 2 million Killdeer in the United States, Canada, and Mexico. Acknowledging that BBS data indicated widespread population declines, they suggested "adopting an intermediate value of 1,000,000 for the estimate, with a range of 200,000 to 2,000,000." Their estimate remained unchanged in 2006 (Morrison et al. 2006) but was doubled six years later to an estimate of 2 million birds (Andres et al. 2012). The authors commented that even this estimate may be conservative. Nevertheless, PIF (2021) recently placed the total population at 2.3 million and the number in the United States and Canada at 1.8 million. A statewide estimate is not available for Minnesota.

C. A low-lying portion of a cropland provides the open mudflats, low-lying vegetation, water, and open vista that the Killdeer prefers for breeding habitat. Photo by Lee A. Pfannmuller.

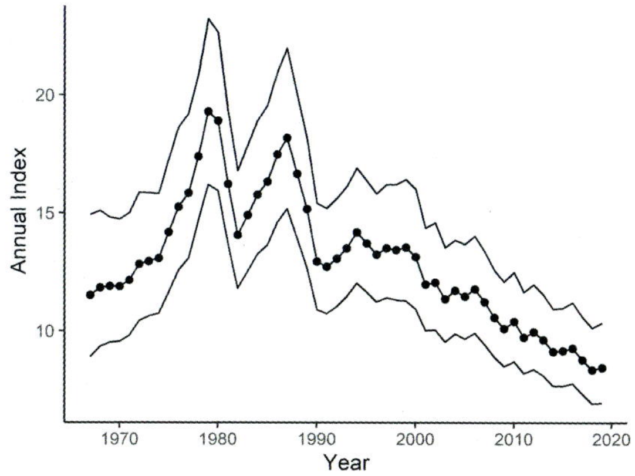

D. Breeding population trend and 95 percent credible interval (solid black lines) for the Killdeer in Minnesota based on the 1967–2019 Breeding Bird Survey (Sauer et al. 2020).

From 1966 to 2019, BBS data demonstrated a surveywide, statistically significant population decline averaging −0.57 percent per year (Sauer et al. 2020). The decline was considerably steeper in Canada compared to a relatively stable population in the United States. BBS data in Minnesota mirrored the surveywide trend, with a significant annual decrease of −0.59 percent since 1967 **(D)**.

Factors responsible for the overall decline are not well understood. The use of pesticides in many modified habitats utilized by the Killdeer, such as lawns, cultivated fields, and golf courses, may be responsible for some local declines. Several studies, for example, have demonstrated the species' susceptibility to a variety of chemicals used to control insects. Indeed, B. Jackson and Jackson (2020) suggested that many habitats that appear to be optimum sites for foraging and nesting (such as golf courses, airports, parking lots, and gravel roads) may actually "function as ecological sinks" because of the negative impacts of pesticides or the inadvertent destruction of eggs and chicks. Russell and his colleagues (2016) also cite the loss of isolated wetlands due to more intensive agricultural practices, such as tiling.

CONSERVATION

Given its broad distribution across North America and even portions of Central and South America, and its relatively large population size, the Killdeer is not a high conservation priority. Although the BBS documents a declining population trend, it is still considered a relatively common species. Even warmer temperatures are not predicted to pose a challenge to the species (NAS 2019). The Killdeer's persistent vocal proclamations are not likely to disappear anytime soon.

PIPING PLOVER

Charadrius melodus

Photo by Michael Furtman.

Minnesota Seasonal Status: Regular breeding species and migrant. The Piping Plover was a very rare species during the MNBBA.
Migration: Short-distance migrant that winters along the south Atlantic and Gulf coasts.
Food: Terrestrial and aquatic invertebrates secured by pecking and probing marine and freshwater shorelines as well as by gleaning prey from beach vegetation.
Nest: Shallow scrape usually lined with small pebbles and shell fragments.
Conservation Concern: The Piping Plover has been assigned a high PIF Continental Concern Score of 18/20. The Great Lakes population (which includes the Duluth-Superior Harbor) is federally *Endangered*; the northern Great Plains population (which includes Lake of the Woods) and the Atlantic coast population are federally *Threatened*. In Minnesota, the Piping Plover is officially classified as state *Endangered*.

The extremely rare diminutive Piping Plover is one of the most difficult Minnesota breeding species to observe. Recent breeding-season observations have been restricted to sandy beaches in Lake of the Woods, a scarce resource in one of the most remote regions of the state. Adding to the challenge, the light gray-brown plumage on the small Plover's back and wings provides a perfect camouflage against the sandy shoreline. Only during the breeding season are the narrow black neckband and strip of black on the forehead clear distinctive features on both the male and female.

The Piping Plover's Latin name, *Charadrius melodus,* is a direct reference to its soft, melodic voice. Described as having a "ventriloquial quality," the *peep peep* call seems to come from all directions, making the bird's location difficult to pinpoint (Bent 1927). The bird itself may not be moving; instead, it often crouches down into the sand in an effort to make itself even more invisible when disturbed.

Unbeknownst to most, the Piping Plover is Minnesota's only federally *Endangered* bird species. Nearly hunted to extinction in the nineteenth century, populations rebounded once the Migratory Bird Treaty Act was enacted in 1918. Today, however, the Plover's dependence on sandy beaches, a habitat coveted for human recreation and development, has once again placed it in peril. Now restricted to three disjunct populations, in 2011 approximately 59 percent of the continental breeding population was located along the north Atlantic coast, 39 percent in the northern Great Plains, and 2 percent along the Great Lakes (Elliott-Smith, Bidwell, et al. 2015).

MINNESOTA BREEDING DISTRIBUTION

One hundred years ago, Roberts (1932) described this endearing little shorebird as "an infrequent migrant" in Minnesota, with several May reports along the Mississippi River in southeastern Minnesota and a few scattered May reports from Waseca, Lake, and Lincoln Counties. One has to wonder whether his report of an adult collected in Duluth in early July 1923 for the American Museum of Natural History was an early fall migrant or actually a member of a very small breeding population that was discovered in the harbor sixteen years later. When Roberts himself saw four adults on July 15, 1915, "on the tip of the long sandy point at the entrance of the Rainy River into Lake of the Woods," he was less than 5 km from Pine and Curry Island. This site would later be documented as the primary Piping Plover breeding site in Minnesota, supporting as many as 20 to 30 nesting pairs during its peak in the 1970s and 1980s. Just a stone's throw from Pine and Curry Island is an even larger island in Ontario, Sable Island, which also has a history of supporting a small Piping Plover breeding population (Cadman et al. 2007).

Although breeding records had yet to be documented when Roberts published his treatise on Minnesota birds in 1932, he noted that Minnesota was "well within the assigned breeding-range of the Piping Plover." Indeed, that same year a young graduate student, Gustav Swanson, documented a nesting colony on what was then known as Curry Island in Lake of the Woods (Risser 1932). Four years later, in 1936, Olga Lakela (1940) found the first nesting pair on Minnesota Point in Duluth. In the 1930s, breeding records were also documented in Otter Tail County (1933), Douglas County (1936, 1937), and Mille Lacs County (1938), all notably during the drought years, when more beach areas were likely exposed (Green and Janssen 1975). Three young were also observed in 1941 in Roseau County, presumably along the shoreline of Lake of the Woods (Reif 1941). The only other confirmed nesting attempt is a report of two nests and two broods at Agassiz NWR in Marshall County in 1980 (Janssen 1987). Apart from these records, the bird has never been confirmed nesting anywhere other than in the Duluth-Superior Harbor and on Lake of the Woods (Green and Janssen 1975). A brief history of these two breeding populations follows.

Duluth-Superior Harbor

Following Lakela's initial report in 1936, Piping Plovers continued to nest on Minnesota Point through 1941. Only a handful of nesting pairs (two to eight) found suitable habitat in a location that had recently been augmented with fresh dredge spoils (MOU 2021b; Lakela 1940). Four seasons after Lakela's initial discovery, she lamented the fact that encroaching vegetation and the recreational popularity of Minnesota Point might eventually make the site unsuitable.

Her misgivings would prove true as the birds began a succession of moves to other dredge-spoil sites, which Common Terns, their frequent co-nester, also found attractive. These included Hearding Island, Barker's Island (Wisconsin), and the Port Terminal (Niemi and Davis 1979). Despite high levels of disturbance, several pairs occasionally continued to nest on Minnesota Point and on nearby Wisconsin Point.

Perhaps the most thorough documentation of the Plover's nesting history in the harbor was a series of observations by Lakela (Lakela 1946; Longley and Harrell 1946) and the Duluth Bird Club (e.g., Bronoel 1948, 1953, 1954a, 1955; Hofslund 1950a, 1952) on the nesting birds of Hearding Island, formerly known as Harbor Island. The club documented Plovers nesting on the site in eight of the eleven years from 1945 through 1955. Most years only one to three nests were found, but in 1948 they reported a record of ten nesting pairs.

Documentation of the Plover's nesting attempts in the harbor from the late 1950s through the 1960s is scanty, and most reports note only that the birds nested "as usual" in Duluth. By the early 1970s, the last site where they consistently were observed was the Port Terminal, another dredge-spoil site that was heavily disturbed by vehicle and people traffic. One to eight pairs attempted nesting at the terminal through 1985, the last year that nesting Piping Plovers were observed in the harbor (MOU 2021b).

Several efforts were made to provide a more secure breeding site elsewhere in the harbor for both Common Terns and Piping Plovers, but these attempts succeeded in relocating the terns to only one site, Interstate Island (Schimpf and Pfannmuller 1984). Today, Plovers are occasionally seen migrating through Duluth during the spring months, but sightings during the summer breeding season have been extremely rare.

Lake of the Woods

In 1932, when Swanson visited Curry Island (now Pine and Curry Island), he did not estimate the size of the Piping Plover colony but noted that he saw many young birds as well as one nest with two eggs (Risser 1932). The site was visited on five more occasions between 1934 and 1940 by an assortment of curious birders. The largest number of nests was reported when Swanson revisited the island in 1938, documenting a total of 15 nests (A. B. Erickson and Upson 1938).

Thirty-seven years passed until the next report of the Plover's activities on the island. In the summer of 1977, a total of three birds were observed, "two of which appeared in territorial conflict" (Eckert 1978a). Most interesting was a report two years later, when an estimated 20 to 30 pairs were observed (B. Anderson 1979). Then, in 1981, not only were birds continuing to nest at Pine and Curry Island (16 pairs plus five individuals; active nests located for eight of the 16 pairs), but two pairs (one nest with four eggs and one pair with three chicks) and nine individuals were found at another site on the lake's southern shoreline (Rocky Point), and one pair (a territorial pair with a broken egg) was found many miles farther north on a small islet just off the shore of the Northwest Angle (Hirsch 1981).

These observations, coupled with knowledge of the species' rarity and decline throughout North America, launched a long-term initiative to study and monitor the Piping Plover population on Lake of the Woods. Unfortunately, the late 1970s and early 1980s appeared to be the birds' heyday on the lake, when there were between 44 and 50 adult Plovers in 1982, 1984, and 1985 (Wiens and Cuthbert 1984; Wiens 1986). Today, the habitat available on Pine and Curry Island has diminished significantly. The island was nearly 6.4 km long in the 1980s but only 4 km remained in 2014 (Herwig 2014). The loss is largely a consequence of high water levels that erode the low-lying sandy beaches and the lack of sediment to subsequently replenish them. From 2010 to 2018, only one nest was reported in five of the nine years either on Pine and Curry Island or on Morris Point, immediately to the west of the island (Herwig, pers. comm.; MOU 2021b). On rare occasions, birds have been observed on islands farther north by the Northwest Angle (e.g., Garden Island, 2006; Kessen and Svingen 2007).

MNBBA Results

During the MNBBA, Piping Plovers were observed in only two atlas blocks: one nest with three eggs was found in 2013 on Pine and Curry Island; farther west, one pair was observed at Rocky Point **(A)**. Clearly, the contribution of Minnesota to the Plover's national population has diminished significantly. Indeed, the Piping Plover's status as a regular breeding species in the state is in jeopardy.

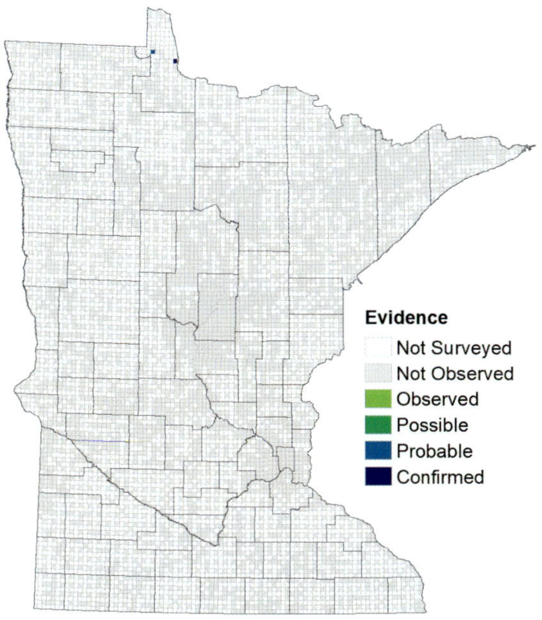

A. Breeding distribution of the Piping Plover in Minnesota based on the Breeding Bird Atlas (2009–13).

Elsewhere within its breeding range, the Piping Plover experienced similar declines in abundance and distribution. This was especially true in the Great Lakes region, where nesting Plovers had disappeared from everywhere but Michigan by the time they were listed as federally *Endangered* (Haig et al. 1988). Fortunately, the Great Lakes population is slowly recovering owing to aggressive conservation efforts. By 2017, Plovers had recolonized locations in Wisconsin, Illinois, Michigan, Pennsylvania, New York, and Ontario (Cuthbert and Saunders 2017).

BREEDING HABITAT

Throughout their breeding range, Piping Plovers utilize a wide variety of open, sandy habitats, ranging from alkali lakes, wetlands, and river floodplains in the northern Great Plains, to beaches and islands along the shorelines of large inland bodies of water **(B)**, to coastal sand dunes along the north Atlantic coast. In addition to these natural habitats, they are found in gravel mines, sand pits, industrial ponds, and on dredge-spoil islands (Elliott-Smith and Haig 2020).

The features shared by all these habitats are open spans of sand, gravel, or shell-covered beaches that are relatively flat and sparsely vegetated **(C)**. At the

B. The expansive beaches at Sleeping Bear Dunes in northern Michigan provide ideal habitat for a growing Piping Plover population along the Lake Michigan shoreline. The area behind the distant red fencing is closed to the public to provide safe areas for the nesting birds. Similar habitat is a rare commodity in Minnesota. Photo by Lee A. Pfannmuller.

C. Throughout its breeding range, Piping Plover nests are placed in sandy shoreline habitat largely devoid of vegetation. Photo by Laura Erickson.

Port Terminal in Duluth, where the birds nested for nearly fifteen years, investigators found the birds preferred sites with approximately 5 percent vegetation cover that averaged 13 cm in height (Niemi and Davis 1979). Nests are often located near large objects, such as stones or driftwood, and away from the water's edge (Elliott-Smith and Haig 2020).

Although Piping Plovers occasionally nest as single pairs in isolated locations, it is common to find clusters of birds nesting in a semicolonial manner (Elliott-Smith and Haig 2020). It is also common to find them nesting within tern colonies, particularly those of Least Terns and Common Terns. In Minnesota's two major breeding populations, Plover nests were frequently located within Common Tern nesting colonies. In such situations, their nests are often closer to the terns' nests than to those of other Plovers. Niemi and Davis (1979) speculated that the Plovers may benefit from the aggressive behavior of the Common Tern. Although the terns tolerate the diminutive Piping Plover as a neighbor, they are quite aggressive to another common co-nesting species, the Ring-billed Gull. Such behavior may impart a degree of nest protection to the neighboring Piping Plover as well.

POPULATION ABUNDANCE

When the Piping Plover was federally listed in 1985, biologists estimated that the entire North American population numbered approximately 2,200 breeding pairs (Haig et al. 1988). The situation was particularly dire for the endangered Great Lakes population, which was estimated at just a dozen pairs (Cuthbert 2016). Given the precarious status of the Piping Plover, the US Geological Survey began an International Piping Plover Census in 1991 to help assess the effectiveness of intensive management efforts, which were soon underway. Conducted every five years, the survey is conducted on the Plovers' breeding grounds and wintering grounds **(D)**.

Excluding the summer of 2006, from the summer of 1991 through 2011 the population averaged 5,770 breeding adults. In 2006, an unprecedented 8,092 birds were observed, with nearly twice as many adults observed in the northern Great Plains than were observed in 2011. The census compilers noted that the subsequent decrease in 2011 might be real, or it might be "an artifact of extreme flooding in the region in 2011 which rendered much traditional habitat unsuitable" (Elliott-Smith, Bidwell, et al. 2015). Only future years will shed light on the

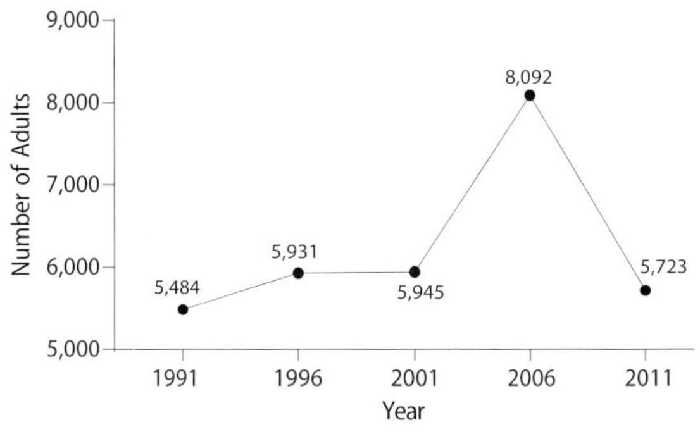

D. Number of Piping Plover breeding adults in North America from 1991 to 2011 (adapted from Elliott-Smith, Haig, and Powers 2009; Elliott-Smith, Bidwell, et al. 2015; Elliott-Smith and Haig 2020).

long-term trends. In the interim, although data from the 2016 breeding census are not yet available, PIF's (2021) recent estimate of 8,400 birds is a hopeful sign, documenting a nearly 90 percent increase from the total population estimate of approximately 4,400 breeding adults in the early 1980s (Haig et al. 1988). Intensive management efforts have been especially successful in the Great Lakes, where, from a low of just a dozen breeding pairs in the 1980s, the endangered population now numbers approximately 75–80 nesting pairs (Boissoneault 2021).

Unfortunately, the Piping Plover's status in Minnesota is quite tenuous; breeding pairs have been absent from the Duluth-Superior Harbor since 1985 and in recent years have nested only sporadically at Lake of the Woods. The rapid decline of the nesting population on the lake's Pine and Curry Island is illustrated **(E)**.

The primary threats to Piping Plovers are habitat loss and human disturbance. The open, sandy habitat it depends on during both the breeding and wintering seasons is highly sought after for recreational pursuits and coastline development. Even when habitat remains, recreational use of the beaches—including dogs running off-leash, driving by cars and recreational vehicles, beach grooming, and dense concentrations of people—causes serious problems (Cuthbert, pers. comm.). Predation is also a serious concern: the numerous predators common in such areas include dogs, cats, raccoons, gulls, and crows (Bracey, Chatterton, and Niemi 2016). In more remote regions, such as Lake of the Woods, high water levels may be managed to balance multiple competing needs, such as hydropower generation, recreation, and flooding. Unfortunately, high water levels jeopardize the low-lying sandy beaches where Plovers nest. Lack of sufficient sediments to replenish eroding beach lines can also result in habitat loss. In Lake of the Woods, the sediment load of two primary tributaries along the lake's south shoreline, the Rainy River and the Little Fork River, has notably declined in the past forty years (Herb et al. 2004).

CONSERVATION

Federally listed in 1985 as *Endangered* in the Great Lakes and *Threatened* in the northern Great Plains and along the Atlantic coast, the Piping Plover is also listed by states that support breeding and wintering populations. It was designated a Highly Imperiled species by the US Shorebird Conservation Plan Partnership (2016), and PIF (2021) assigned it a high Continental Concern Score. In Minnesota, it has been state-listed as *Endangered* since 1984 (Coffin and Pfannmuller 1988).

The species' recent recovery is due largely to intensive management efforts on its breeding grounds. Conservation organizations, resource professionals, and researchers cooperate to close nesting beaches during the breeding season, control avian and mammalian predators, and erect predator exclosures around active nests. When nests are at high risk, even translocation and captive rearing techniques have been implemented (Elliott-Smith and Haig 2020).

Habitat protection and acquisition are also important conservation tools. In Minnesota, Pine and Curry Island was acquired and designated as a state Scientific and Natural Area (SNA) specifically to aid the protection and management of this once-significant population. In the Duluth-Superior Harbor, Interstate Island was acquired as a wildlife management area in the hopes that it too would provide a secure nesting site not only for Common Terns but also for Piping Plovers. Unfortunately, Ring-billed Gulls have also colonized Interstate Island and Piping Plovers have never nested there.

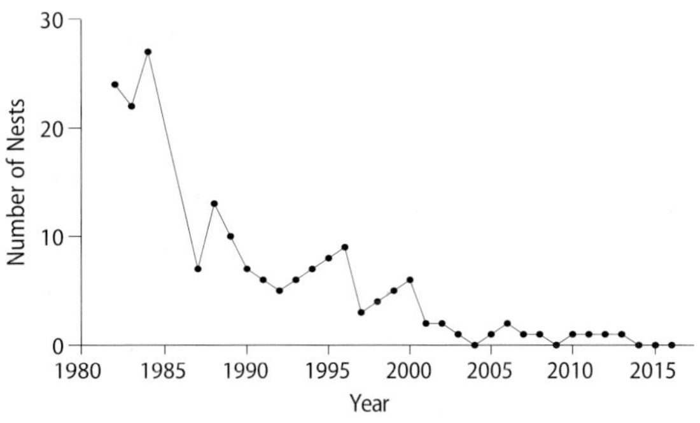

E. Number of Piping Plover nests on Pine and Curry Island, 1982–2016 (adapted from Haws 2004, 2008; Herwig, pers. comm.).

The dwindling population on Pine and Curry Island has been intensively managed and monitored since the early 1980s. Efforts have included predator control, the erection of predator exclosures, and closing beaches during the nesting season. Unfortunately, they have not offset the impacts of diminishing habitat.

Warming temperatures pose additional threats to species like the Piping Plover that are dependent on coastal or inland shoreline habitats. Rises in sea level will threaten both breeding and wintering areas, while more intense rain events in the northern Great Plains may result in more frequent flooding of shoreline habitats. An assessment of the potential impacts of climate change on the Piping Plover by NAS (2019) classified the species with moderate vulnerability under a 2°C warming scenario. Models predicted a loss of 74 percent of its North American breeding distribution and only a 39 percent range gain by 2050. Interestingly, some of the gains were predicted to occur in Minnesota. Under the same scenario, approximately 21 percent of the Plover's coastal winter habitat would be lost and compensated by only a 12 percent gain along the Atlantic seaboard.

In the meantime, Minnesota's best hope for reestablishing a sustainable breeding population of Piping Plovers may lie in the success underway to recover the Great Lakes and northern Great Plains populations. If managers are able to maintain suitable breeding habitat in key locations, as nearby populations recover more birds may attempt to colonize former breeding areas where suitable habitat still exists. In anticipation of that event, Price and Cuthbert (2002) prepared a Piping Plover habitat and recovery assessment plan for the St. Louis Estuary. The plan identified the habitat suitability and restoration potential of eight sites for harboring future Piping Plover populations. The St. Louis River Alliance, a conservation group in Duluth dedicated to restoring, protecting, and enhancing the St. Louis River, is also working to protect critical nesting beaches that might entice birds migrating through the harbor to remain and nest once again. The work of these dedicated ornithologists and volunteers just might result in Piping Plovers returning and thriving in the Duluth-Superior Harbor. That would truly be a conservation story for the ages!

UPLAND SANDPIPER

Bartramia longicauda

Minnesota Seasonal Status: Regular breeding resident and migrant. The Upland Sandpiper was uncommon during the MNBBA.
Migration: Long-distance migrant that winters in southern South America.
Food: Primarily arthropods and some seeds gleaned from the ground.
Nest: Shallow ground scrape lined with vegetation; surrounding vegetation may be shaped to conceal the nest from above.
Conservation Concern: Assigned a moderate PIF Continental Concern Score of 11/20.

Photo by Sparky Stensaas.

Perched atop boulders that were relicts of the landscape's glacial past, the Upland Sandpiper's 30 cm frame was both conspicuous and abundant across the Great Plains in the 1800s. Today the mournful call of a diminished population still heralds spring's arrival. A small head and thin neck lend a unique profile to the Sandpiper's speckled brown body, while its habit of teetering up and down is a behavior distinct among many members of the sandpiper family. After alighting on a perch, the birds raise both wings above their bodies "momentarily, ballerinalike, before inserting them delicately inside their flank feathers" (Johnsgard 1995). The performance inspired Leopold (1949) to write, "Whoever invented the word 'grace' must have seen the wing-folding of the plover [Sandpiper]." Graceful in its execution, the elevation of the wings functions as both a courtship and threat display (Houston et al. 2020).

Often given in flight, the Sandpiper's "wolf whistle" has a mystical quality that has enthralled listeners for centuries. Described as a "long-drawn, melodious *quai-le-e-e-e-e-e-e*" call, it is "a sweet, tremulous, plaintive whistle, rising slowly and descending again through the long *e-e-e-e* in a far reaching curve" (Roberts 1932). Given by both sexes, it is delivered only by the male once nesting commences (Houston et al. 2020).

Named in honor of the American naturalist William Bartram, the Upland Sandpiper was known for many years as the Upland Plover (Houston et al. 2020). Today its breeding range extends from the Canadian Prairie Provinces south to Oklahoma and east through the Ohio River valley and the Great Lakes states. Scattered populations are also present in Alaska, northwestern Canada, and western New England. The core of its distribution is in the northern Great Plains.

MINNESOTA BREEDING DISTRIBUTION

The Sandpiper was formerly abundant across Minnesota's grasslands. Roberts (1932) provides a compelling narrative on the loss of one of the prairie's most emblematic species: "To recite the history of the Upland Plover [Sandpiper] in Minnesota is to tell a sad tale of the wanton destruction of a valuable and once abundant bird that resulted in its almost complete extermination. Sixty years ago it was present all through the summer everywhere in the open country in countless thousands. Now it is a question whether the remnant left can be saved even with careful protection."

The decline of the Sandpiper, a tasty bird, was hastened by the Passenger Pigeon's demise in the 1880s. Anxious to continue satisfying the insatiable demands of eastern markets for game birds, hunters aimed their guns instead at Upland Sandpipers (Forbush 1912). Stories of market hunters shooting "great numbers" and shipping them east were common even in Minnesota (Roberts 1932). This killing, coupled with the loss of native prairie, led to the Sandpiper's virtual disappearance from southern Minnesota between 1895 and 1900. Although populations persisted in the northwest, their numbers were greatly reduced. Roberts reported confirmed nesting records from just six counties: Anoka, Grant, Hennepin, Lac qui Parle, Pennington, and Polk; nesting was inferred in Kandiyohi County.

The Upland Sandpiper, however, is an adaptable bird. Once the carnage stopped with passage of the 1918 Migratory Bird Treaty Act, it found the landscape's new hayfields and pastures a fair substitute for the disappearing prairies. It wasn't long before Roberts witnessed the species' slow recovery. It even occupied small areas of northern Minnesota where forests had been cleared for farming operations.

Accounts of the Upland Sandpiper's distribution by Green and Janssen (1975) and Janssen (1987) further documented the species' growing abundance. Although it reoccupied suitable habitat throughout much of the state, the species was most abundant in western Minnesota and "very scarce to absent" in far north-central and northeastern Minnesota. Janssen identified 25 counties where nesting had been confirmed since 1970. Hertzel and Janssen (1998) later identified 28 counties with confirmed nesting. The majority of the 545 breeding-season locations reported by the MBS were also located in western Minnesota but stretched as far east as Winona County in the south and Pine and St. Louis Counties in the north.

MNBBA participants reported 349 Sandpiper records in 5.3 percent (251/4,760) of the surveyed atlas blocks located in 54 of Minnesota's 87 counties. Breeding was confirmed in 20 blocks covering 16 counties **(A)**. Nine of the 16 counties were additions to Hertzel and Janssen's 1998 map, including seven counties in southwestern Minnesota. The species was most abundant in western Minnesota, especially from southern Traverse County to the Iowa border. The MNBBA probability map, however, emphasizes the low breeding densities that occur even in the most suitable grasslands **(B)**.

The most notable change in recent years is the Sandpiper's restricted distribution in northwestern

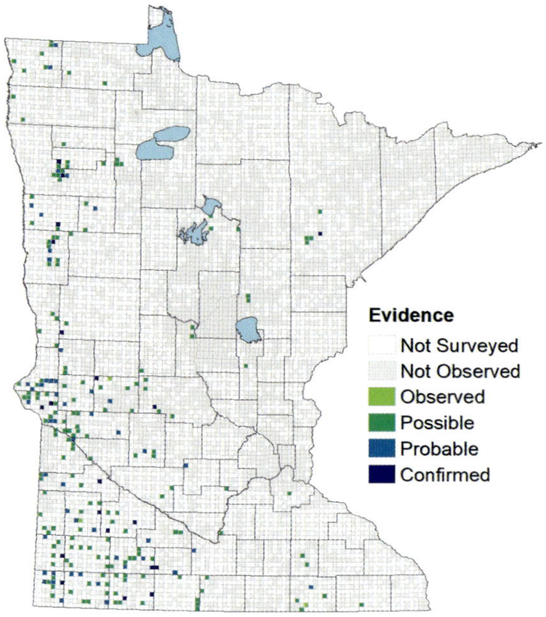

A. Breeding distribution of the Upland Sandpiper in Minnesota based on the Breeding Bird Atlas (2009–13).

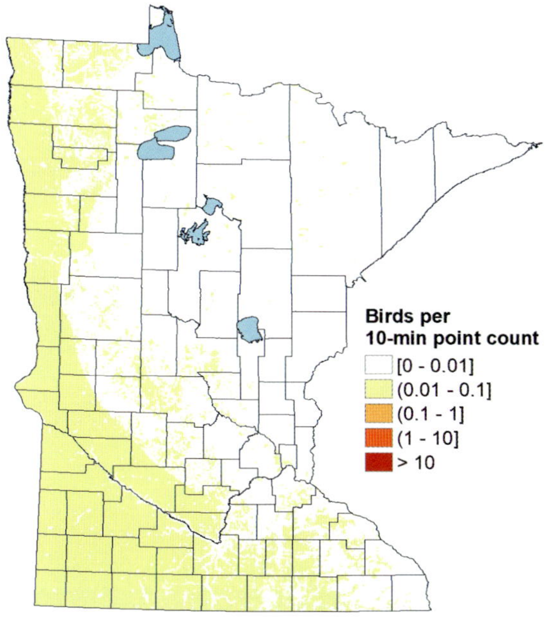

B. Predicted breeding distribution (birds detected per 10-minute point count) of the Upland Sandpiper in Minnesota based on habitat, landscape context, and climate data gathered during the Minnesota Breeding Bird Atlas (2009–13) using the Generalized Linear Modeling method without an adjustment for detectability.

Minnesota. Abundant in Kittson, Marshall, and Roseau Counties during the MBS, twenty years later it was restricted primarily to the glacial beach ridges of central Polk and Clay Counties. Changes in land use are likely responsible. For years, the region's extensive grasslands were marginally productive and commodity prices too low to warrant large investments to improve productivity, especially the installation of tiling systems. But as corn prices increased in the early 2000s, converting these grasslands to agricultural fields became quite economical.

Although the bird is less abundant than a century ago, the heart of the Sandpiper's breeding range remains in the Great Plains. Except for scattered remnants, it is largely absent from the eastern portion of its former range, including Ohio, Indiana, New York, and New England. In some of these areas, grass cover bordering airport runways provides the only remaining habitat (Houston et al. 2020).

BREEDING HABITAT

Originally confined to native prairies, the Upland Sandpiper has adapted to restored grasslands, grazed pastures, hayfields, croplands, and planted cover, including Conservation Reserve Program (CRP) fields **(C)**. Suitable sites are characterized by low to moderate herbaceous cover, minimal woody vegetation, and moderate grass cover, with little bare ground (Dechant, Dinkins, et al., "Upland Sandpiper," [1999] 2002; Hanowski and Niemi 1985). The birds usually avoid extensive areas of tall vegetation, preferring instead a matrix of different heights and cover densities. The ability to see over or through the vegetation is important, as is the availability of fence posts or other display perches. Grasslands larger than 100 ha are usually preferred; the species is rarely found in sites smaller than 50 ha (Dechant, Dinkins, et al., "Upland Sandpiper," [1999] 2002).

POPULATION ABUNDANCE

The North American population totals approximately 750,000 breeding adults (Andres et al. 2012; PIF 2021). A statewide estimate is not available. The BBS trends from 1966 to 2019 for the Sandpiper document a surveywide significant increase averaging 0.62 percent per year **(D)**. The increase that began in the mid-1980s may be attributed to the popularity of the CRP, which began in 1985 and provides payments to farmers for retiring highly erodible or sensitive agricultural lands for ten to fifteen years. Indeed, the small decline that began in the mid-1990s is coincident with a small drop in national CRP enrollment from 1995 through 2003 (USDA 2020).

Contrary to the surveywide trend, Minnesota BBS data document an average annual decline of −1.44 percent from 1967 to 2019 (Sauer et al. 2020). Because the Sandpiper is uncommon in Minnesota, the trend results are not statistically robust. Nevertheless, the significant reduction in records from the MBS (545) to the MNBBA (349) certainly suggests a drop in abundance. More intensive agricultural practices, coupled with a recent decline in CRP acres, are likely affecting Minnesota populations.

CONSERVATION

Dependent on grasslands throughout its life cycle, the Upland Sandpiper was listed as a Minnesota *Special Concern* species in 1984 because of low

C. Large grasslands with minimal woody cover provide nesting cover for the Upland Sandpiper, while boulders provide perches. Photo by Gerald J. Niemi.

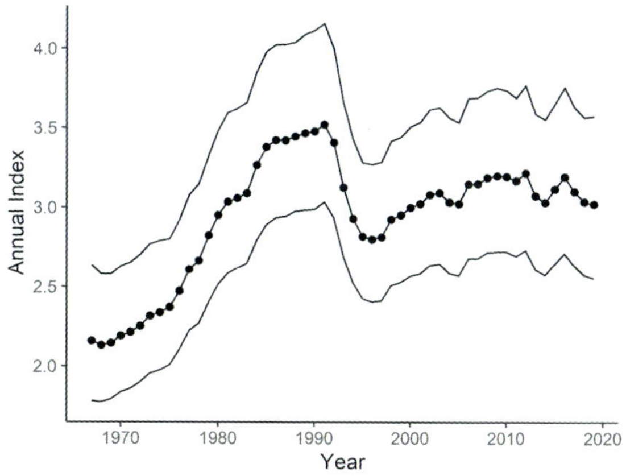

D. Surveywide breeding population trend and 95 percent credible interval (solid black lines) for the Upland Sandpiper based on the 1967–2019 Breeding Bird Survey (Sauer et al. 2020).

numbers and the continuing loss of critical habitat. But given the relative abundance of MBS records collected by the early 1990s, the Sandpiper was delisted in 1996 (MNDNR 1995). Continued concern, however, is warranted. Although the species utilizes CRP grasslands, when commodity prices are high there is little incentive to protect these set-aside acres. Coupled with cuts in federal appropriations for the program, Minnesota farmers enrolled nearly 800,000 fewer acres in CRP in 2019 than in 2007 (Stanley 2019). Nationally, fewer acres were enrolled in 2020 than in the previous thirty-three years (USDA 2020).

Habitat loss is not the only challenge threatening Upland Sandpipers. Like all shorebirds that winter in South America, the species is still hunted, even providing table fare in some Argentinian restaurants (Houston et al. 2020). Fortunately, although grasslands are predicted to become drier and be subjected to more variability in precipitation as temperatures rise, NAS (2019) ranked the Sandpiper's breeding distribution as climate stable, even noting substantial gains in breeding habitat throughout eastern Minnesota.

Management recommendations designed to ensure the future of this symbolic grassland inhabitant focus on protecting large, contiguous grassland tracts. Maintaining a variety of vegetation heights and densities by mowing, light grazing, and prescribed burning is critical. Only portions of a site should be treated each year to protect patches of dense cover that are essential for successful nesting and brood rearing. Reducing woody encroachment and providing perch sites (fence posts, tree stumps, or rocks) are also important (Russell et al. 2016; Vickery et al. 2010).

MARBLED GODWIT

Limosa fedoa

Minnesota Seasonal Status: Regular breeding resident and migrant. The Marbled Godwit was uncommon during the MNBBA.
Migration: Short- to medium-distance migrant, wintering primarily along the coasts of southern California and Mexico.
Food: Primarily aquatic invertebrates, secured by probing.
Nest: Shallow nest depression on the ground lined with grasses.
Conservation Concern: Assigned a high PIF Continental Concern Score of 15/20; officially listed as *Special Concern* in Minnesota.

Photo by Karl Bardon.

Marbled Godwits are hard to miss in the large, remnant grasslands of western Minnesota. If you are lucky enough to encounter a breeding pair, they will leave a lasting impression. Reminiscing about a trip to Grant and Traverse Counties in 1879, Roberts (1919) wrote: "the great marbled Godwit was so abundant, so constant and insistent in its attentions to the traveler on the prairie, and so noisy that it became at times a nuisance. They were continually hovering about the team [of horses], perfectly fearless and nearly deafening us with their loud, harsh cries—'go-wit, go-wit.'"

Standing nearly 0.5 m tall, the most distinctive feature of this speckled brown bird is the long, slightly upturned bill. Ranging from 8–13 cm in length, the bill is usually bicolored: black at the tip and bright orange or pink at the base. The brown specks on the throat and upper chest grade into horizontal bars on the belly. In flight, cinnamon-colored wing linings are diagnostic. The females are a little heavier than the males and have a slightly longer bill. The Latin name *Limosa* refers to the species' muddy foraging habitats; the origin of the term *fedoa* is unclear but may refer to its probing habits (Sandrock and Prior 2014).

Marbled Godwits have a restricted breeding range, limited largely to the northern Great Plains in the United States and the Prairie Provinces of Canada. Two very small, isolated populations occur along the south end of James Bay in Ontario and Quebec, and on the Alaskan peninsula. The highest breeding densities occur in southeastern Alberta.

MINNESOTA BREEDING DISTRIBUTION

"Fifty years ago," Roberts wrote in 1932, the Marbled Godwit "was an abundant summer resident throughout all the prairie region and still earlier it undoubtedly nested to some extent in more open areas in the southeastern part of the state." Evidence that it nested widely includes Hatch's (1892) report of Godwits occasionally nesting as far east as the "vicinity of Minneapolis" and an 1871 report of confirmed nesting approximately 50 miles northwest of St. Paul (Roberts 1932). Confirmed nesting accounts with more detailed location information, however, were limited to Grant, Hennepin, Otter Tail, Polk, and Stearns Counties. Despite Roberts's assertion that the species was present throughout Minnesota's grasslands, Green and Janssen (1975) later noted there were no early records south of the Minnesota River.

The stately Godwit was an easy target for settlers in the 1800s and "a plump and toothsome morsel for the table" (Bent 1927). Hunting pressure, coupled with the extensive loss of grasslands, resulted in a significant population decline and the bird's disappearance from portions of its breeding range. By the early twentieth century, Minnesota's Godwit population was restricted to limited numbers in west-central Minnesota and the Red River valley. Numbers declined so rapidly throughout its range that Bent (1927) speculated it might soon "join the ranks of those that are gone but not forgotten." Yet, after the Migratory Bird Treaty Act was signed in 1918 and market hunting declined, a slow recovery was underway.

By 1975, Green and Janssen noted that the Marbled Godwit was once again well established as a breeding resident in northwestern Minnesota, with occasional reports of birds in the west-central and central regions of the state as far east as Stearns County. Janssen (1987) later identified 11 counties where nesting had been confirmed since 1970, stretching in a narrow band from Lac qui Parle County north to Kittson and Roseau Counties. He also mentioned a 1978 report of birds nesting in Brown County, but positive confirmation was lacking. By 1998, Hertzel and Janssen had documented one more county where nesting had been confirmed: McLeod, west of the Twin Cities.

The MBS documented a total of 286 Marbled Godwit breeding-season locations. Well-distributed throughout the Tallgrass Aspen Parklands Province, the birds were also found farther south on prairies along the Glacial Lake Agassiz beach ridges, in the upper Minnesota River valley in Lac qui Parle and Big Stone Counties, and in a region of grasslands centered in Stearns County.

More than a decade later, MNBBA participants documented only 183 Marbled Godwit records in 2.6 percent (124/4,747) of the surveyed atlas blocks. Nesting was confirmed in 21 blocks **(A)**. The birds were observed in 24 of Minnesota's 87 counties and were documented nesting in 11 counties. Four of the counties where breeding was confirmed were new since Hertzel and Janssen (1998): Chippewa, Clearwater, Pope, and Stearns.

The Marbled Godwit's MNBBA breeding distribution remained similar to that reported by the

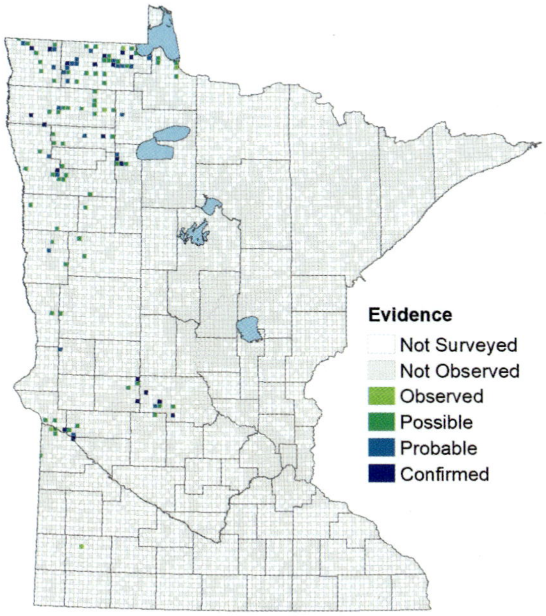

A. Breeding distribution of the Marbled Godwit in Minnesota based on the Breeding Bird Atlas (2009–13).

MBS, with records stretching from the Aspen Parklands, south to the Upper Minnesota River valley, and east to Stearns County. Overall abundance, however, appears to have declined in the northwest. The MNBBA found only one record in Red Lake County, and the number of records in Pennington, Clay, and Wilkin Counties was considerably fewer than the number reported by the MBS. The continued loss of grassland and wetland habitats, in part driven by high commodity prices in the first decade of the twenty-first century, may be contributing to some of the decline. Poor wetland conditions and continued expansion of the region's gravel industry may also be contributing factors.

Historically, the Marbled Godwit's continental breeding range stretched as far south as Iowa and Nebraska and as far east as Wisconsin (Gratto-Trevor 2020). Today, however, breeding populations are virtually gone from all three states.

BREEDING HABITAT

The Marbled Godwit depends on large tracts of native grassland embedded with shallow wetlands **(B)**. Suitable landscapes are relatively level, harbor few trees, and are rarely less than 100 ha in size (Granfors 2009; Dechant, Sondreal, et al., "Marbled Godwit," [1998] 2001a). Vegetation is usually less than 15 cm high; tall, dense cover is avoided. Hayfields, idle pastures, and recently grazed grasslands may be used if they are structurally similar to shortgrass prairies (Melcher et al. 2010). Nests are usually located at least 200 m from the edge of wetlands. Temporary or seasonal wetlands with short shoreline vegetation are preferred, although in drought years, larger, semipermanent wetlands are used (Dechant, Sondreal, et al., "Marbled Godwit," [1998] 2001a; Melcher et al. 2010).

POPULATION ABUNDANCE

In 2012, the North American population of Marbled Godwits was estimated at 173,500 breeding adults (Andres et al. 2012). PIF's (2021) recent update placed the number at 170,000. In Minnesota, a survey conducted by the USFWS several years prior to the MNBBA estimated the statewide population numbered approximately 400–500 breeding pairs (Russell, pers. comm.).

In the Prairie Potholes Bird Conservation Region, which is the core of the species' breeding range and encompasses approximately 70 percent of its breeding population, BBS data document a significant population decline averaging −0.99 percent per year from 1967 to 2019 **(C)**. Unfortunately, the small size of Minnesota's population results in a projected population trend that is not statistically robust. Nevertheless, although the trend line is relatively stable, a small decrease is evident in recent years (Sauer et al. 2020).

Habitat loss and degradation continue to be the primary factors impacting Marbled Godwit populations (Gratto-Trevor 2020). Local abundance can also vary widely from year to year in response to wetland conditions. Godwits depend on the availability of seasonal wetlands early in the breeding season for feeding. Adapted to the variable precipitation patterns encountered across the Great Plains, if drought conditions reduce seasonal wetland abundance early in the season in one area, Godwits will move to other more suitable sites (Russell, pers. comm.; Gratto-Trevor 2020).

B. Large grasslands for nesting and shallow wetlands for foraging provide the critical features required by the Marbled Godwit during the breeding season. Photo by Gerald J. Niemi.

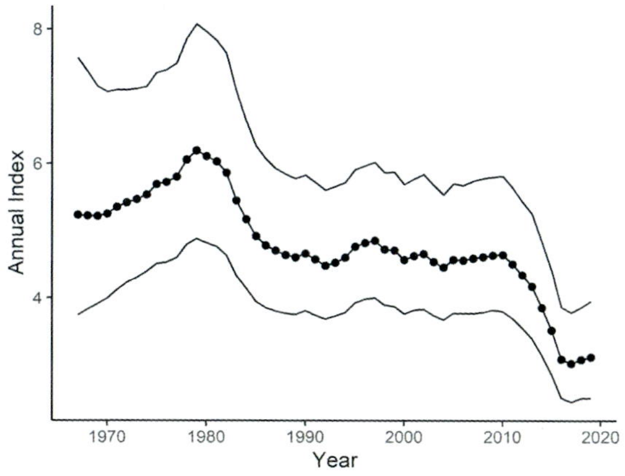

C. Breeding population trend and 95 percent credible interval (solid black lines) for the Marbled Godwit in the Prairie Potholes Bird Conservation Region based on the 1967–2019 Breeding Bird Survey (Sauer et al. 2020).

CONSERVATION

Since the Marbled Godwit is considered an indicator of quality grassland habitats, its frequent association with some of Minnesota's largest and highest-quality grasslands is a clear indication of its vulnerability. This narrow habitat tolerance, coupled with its very restricted breeding distribution and small population size, prompted the species' designation as a high conservation priority (PIF 2021). In Minnesota, the Marbled Godwit is officially listed as *Special Concern* (Coffin and Pfannmuller 1988).

Management recommendations focus on maintaining large habitat complexes that support quality grassland habitat and a diversity of wetlands; tracts larger than 100 ha are recommended (Dechant, Sondreal, et al., "Marbled Godwit," [1998] 2001a). Burning, mowing, and grazing are all appropriate management options if sequenced properly to provide the right conditions during the nesting season. Light to moderate grazing in the spring, for example, maintains low cover for the adults to forage in during the early summer months but provides denser cover later in the season for chicks (Russell et al. 2016).

Warmer temperatures might pose a risk to a species with such a restricted breeding distribution. Fortunately, under a 2°C warming scenario, NAS (2019) projected a 23 percent loss of the Godwit's current breeding distribution by 2050 and a gain of 15 percent, resulting in its classification as a species with low vulnerability. Some of the gains were actually predicted in western Minnesota. Nevertheless, given its small breeding range, the future of this impressive grassland inhabitant warrants close monitoring.

AMERICAN WOODCOCK

Scolopax minor

Photo by Mike Lentz.

Minnesota Seasonal Status: Regular breeding resident and migrant. The American Woodcock was common during the MNBBA.
Migration: Short-distance migrant wintering throughout the southeastern United States, along the Atlantic and Gulf coasts from New Jersey to Texas and north to Missouri and Tennessee. Eastern breeding populations in the United States, such as in New Jersey, Maryland, and Virginia, may be permanent residents.
Food: Chiefly consumes invertebrates, especially earthworms, which are extracted by probing into the ground or soft sediment, and occasionally seeds. Its long bill has a flexible upper mandible specialized for capturing food.
Nest: On the ground in a shallow depression of leaves and litter.
Conservation Concern: A game species; assigned a moderate PIF Continental Concern Score of 12/20.

The American Woodcock, a plump shorebird, is a popular and challenging Minnesota game bird. It is well-known for its bizarre looks and its unique sky dances during its spring courtship rituals. As Roberts (1932) described his experience: "on almost any evening in April and early May, the weird aerial nuptial performance of several pairs may be easily observed. This 'song of the Woodcock' . . . has always attracted the intense interest of ornithologists and bird-lovers and has elicited many lengthy attempts at describing it."

One is usually attracted to the courtship area by the buzzy *Peent* call of the male, emitted during crepuscular periods on the ground where its sky dances are initiated. The sky dance flight is a rapid, erratic spiraling of the male up into the sky, where melodious chirps can be heard along with the whistling caused by three modified outer primary feathers. The male then returns to the same area it started from and continues with its courtship behaviors. The description of its aerial "song" has given the Woodcock its alternative name *Timber-doodle.*

The American Woodcock's plumage is well camouflaged against the ground leaf litter where it spends most of its time resting, nesting, or foraging. It uses its long, stout bill to probe for earthworms or other creatures in moist soils where its borings can often reveal its recent presence. Its genus name, *Scolopax,* is Latin for "woodcock," while its species name, *minor,* distinguishes it from its European counterpart, *Scolopax rusticola.*

The American Woodcock is patchily distributed and found in low abundance throughout eastern North America from the Maritime Provinces in Canada to southeastern Manitoba, and south to eastern Texas. It most commonly breeds above 40 degrees North latitude in the upper midwestern and northeastern United States, and in southern Canada.

MINNESOTA BREEDING DISTRIBUTION

Roberts (1932) reported that the American Woodcock was a "summer resident, breeding throughout the state, but best represented in the counties bordering the Mississippi River south of Anoka," a city and county located just north of the Twin Cities. He reported that he had summer records throughout the entire state, except in the northeastern angle, now known as the Arrowhead region. In his 1936 edition of *The Birds of Minnesota,* Roberts makes special mention of one of his observers, H. F. Kendall, who reported seeing the bird "near Virginia, St. Louis County, on July 31 and August 11, 1932, suggesting that it nested there."

Roberts noted that already in the 1930s, "[t]he Woodcock has undoubtedly decreased in Minnesota, but it was never abundant enough to encourage sportsmen to make it a special object of pursuit, except, perhaps, in the bottom-lands of the Mississippi River." He mentioned that sportsmen had found particularly "wonderful shooting trips" south of Lake Pepin about 1872. No one else seems to have encountered such conditions in Minnesota.

Roberts (1932) reported only three records of nests with eggs, from Lake Minnetonka and Minneapolis, Hennepin County, as well as an "adult with three nearly grown young" observed by Surber in Pine County. Roberts (1936) added a nest found in the Fort Snelling Reservation in April 1932, and his young protégé Mr. Breckenridge banded three nestlings from the Fort Snelling nest. One of the nestlings was later recovered in Ocean Springs, Mississippi, in January 1933.

Green and Janssen (1975) briefly summarized the American Woodcock as a resident in "northeastern, north central, central, and east central regions and in a few adjacent counties to the southeast and northwest." They reported confirmed nesting in 18 counties, including all three counties in the Arrowhead region and in the northwest in Pennington County. In the forty years since Roberts (1932, 1936) wrote his book, the breeding distribution had become quite different from what he observed and described.

Marshall (1982) provided a review of the American Woodcock distribution in the state from, as he described it, the "scanty literature, personal reports, and the history of land use changes." Marshall said that in the late 1800s Woodcocks were found primarily in the Mississippi and Minnesota River bottoms above and below St. Paul, where hunters found them abundant. He points out that hunting magazines, such as *Sports Afield,* reported occasions when sportsmen were "shooting 40 birds in three hours." In addition, newspaper articles said that in 1862, 1864, and 1873 Woodcock were being sold in markets and restaurants.

Evadene Swanson (1940) quoted numerous reports of hunting of the American Woodcock in the period from 1850 to 1900. For example, one said that near Hastings, Minnesota, "in September 1835, two men shot 55 birds in one day," and another that "the average bag in that locality in 1879 was 20 woodcock in a morning." In the late nineteenth century, a report stated that "the woodcock is scarce enough where you find them." However, another report refers to the woodcock presence and abundance in the Minnesota River bottoms in 1897, when "one hunter got 11 in a half-day tramp."

Marshall (1982) surmised that indeed American Woodcock may have been largely confined to the Mississippi and Minnesota River floodplains during the early years because much of the forested habitat of Minnesota was mature and unsuitable for the species. In addition, the mortality of woodcock was high owing to hunting in spring, summer, and fall, especially from market hunting in the southern United States, which may have kept populations low. Indeed, concerns about population levels in the early 1900s led to enactment of federal and state laws that confined hunting seasons and set bag limits.

Marshall (1982) maintained that the changes in the distribution of American Woodcock in Minnesota were undoubtedly related to changes in land-use practices during the sixty years from the 1920s to the 1980s. The species' continued scarcity in southeastern Minnesota was associated with intensive agriculture and dairy farming, which left unsuitable habitat. But in the northern forested areas, logging and wildfires created the early stages of forest succession, which was good habitat for the woodcock.

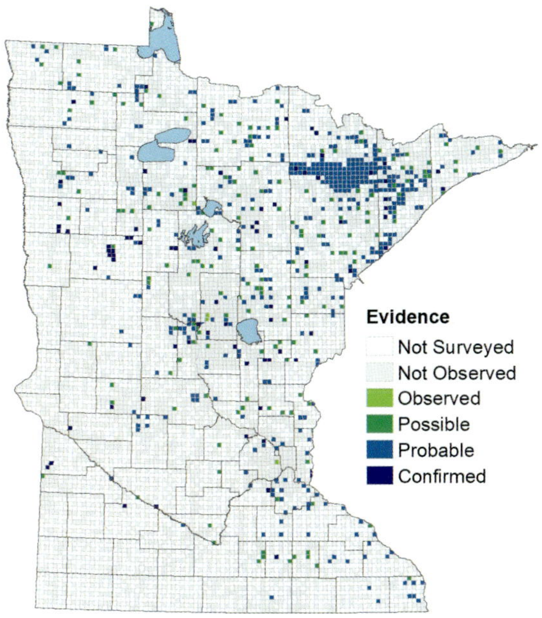

A. Breeding distribution of the American Woodcock in Minnesota based on the Breeding Bird Atlas (2009–13).

He went on to describe the expansion of the species' range into the northern and western parts of Minnesota. Woodcock were not seen until the mid-1960s in Itasca State Park, and Dr. Daniel Svedarsky, a professor at the University of Minnesota in Crookston, recorded nests and broods in a prairie area invaded by brush in Pennington County in 1975, 1976, and 1977 (Marshall 1982).

Janssen (1987) described the expansion of the American Woodcock's breeding distribution into the southwestern and western regions of Minnesota. He reported confirmed nesting since 1970 in 31 counties, including in the far western county of Big Stone and in the extreme southwest in Pipestone County. Hertzel and Janssen (1998) later identified 37 counties with confirmed nesting records since 1970. They included 10 additional counties not included by Janssen (1987): Becker, Brown, Grant, Itasca, Kandiyohi, Koochiching, Pennington, Redwood, Waseca, and Winona. They did not include confirmed nesting in Carlton, Kanabec, Pine, or Washington Counties, which had been listed by Janssen (1987).

The MBS began their inventories in the late 1980s and reported 89 breeding-season locations. Most locations are scattered in the north-central and northeastern part of the state. Notable locations include one in Houston County in extreme southeastern Minnesota, one on the border between Lincoln and Yellow Medicine Counties, two in Marshall County, and five in Kittson County in extreme northwestern Minnesota.

Superb efforts by the participants in the MNBBA resulted in a whopping 1,026 records of the American Woodcock, which were widely distributed in Minnesota except for the extreme southwestern region **(A)**. MNBBA observers reported records from 14.3 percent (696/4,853) of all blocks surveyed and 12.4 percent (289/2,337) of the priority blocks. Confirmed nesting was recorded in 79 blocks and in 33 counties. Confirmed nesting in counties not previously recorded by Hertzel and Janssen (1998) since 1970 included Crow Wing, Isanti, Kanabec, Lac qui Parle, Marshall, Morrison, Norman, Pine, Polk, Pope, Scott, Steele, Todd, and Washington Counties. One confirmed nest straddled two counties: Brown and Renville along the Minnesota River. The MNBBA did not record confirmed nesting from several counties previously reported but did report probable or possible nesting in most of these. The exceptions were Big Stone, Le Sueur, Pipestone, and Ramsey, where no nesting evidence was reported. The differences between the MNBBA results and Hertzel and Janssen's (1998) are partly due to the longer time period of coverage included in the earlier report but may also be due to variations in coverage, changes in habitats and landscapes, annual variation, and differences in the definitions of what constitutes a confirmed nesting record.

The extensive surveys in northern St. Louis County were completed by Steve Wilson, who dedicated considerable effort to a systematic count in that region. His specific interest was to determine the breadth of the species' distribution in a regional area. His results suggest that the American Woodcock's distribution and abundance are likely underestimated in Minnesota.

The landcover suitability map for the American Woodcock predicts that the species could be distributed throughout the state where suitable habitat exists, although it was expected to be most abundant in the north-central and northeastern portions **(B)**. Several linear areas, such as along the Minnesota River and a broad swath along the Mississippi River floodplain, are predicted to have moderately suitable areas.

In their review of the American Woodcock in North America, McAuley and colleagues (2020) provided little background on historical changes in the species' distribution. However, they cited Marshall's 1982 summary of northward expansions of breeding birds that colonized available habitats opened up by logging activity. Thereafter, few data were available until the late 1960s, when American Woodcock singing-ground surveys were initiated in 1968 across

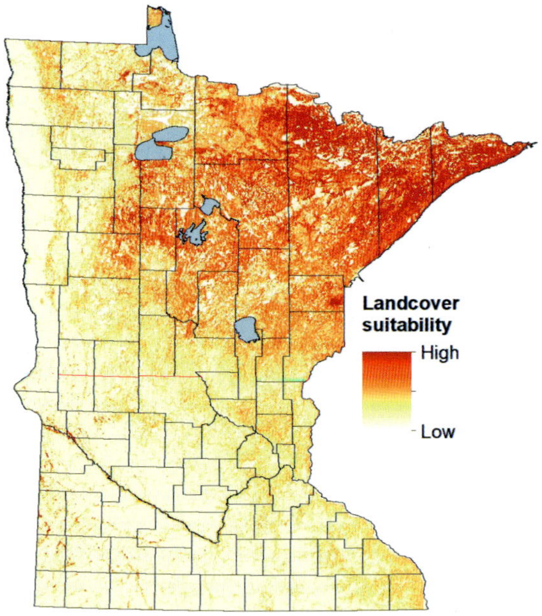

B. Landcover suitability map for the American Woodcock in Minnesota based on habitat, landscape context, and climate data gathered during the Minnesota Breeding Bird Atlas (2009–13) using the MaxEnt modeling approach.

much of the species' eastern and central range (Kelley et al. 2008).

In Wisconsin and Michigan, similar changes in the distribution of the American Woodcock have been identified (Cutright et al. 2006; S. D. Robbins 1991; Chartier et al. 2013). For instance, Chartier and colleagues (2013) noted that Woodcock were less abundant and more scattered throughout Michigan prior to European settlement, but currently the species is found throughout the state. Highest densities are in the northern Lower Peninsula and eastern Upper Peninsula. They attribute the increased abundance of the species to the proportion of land covered in young forests and the increased proportion of aspen–paper birch cover. The presettlement forests of Michigan had less than 5 percent aspen–paper birch. Today it is greater than 26 percent.

In summary, the historical reports of the American Woodcock in Minnesota suggest the species was largely confined to the Mississippi and Minnesota River floodplains in the 1800s. Excessive hunting of the species throughout the United States likely kept populations low until hunting regulations began to be implemented in the early twentieth century. As Marshall (1982) in Minnesota and Chartier and colleagues (2013) in Michigan suggest, the Woodcock population began to increase and expand after the major logging periods in the late 1800s and early 1900s. However, forest fires were also frequent, and regeneration of aspen occurred following these fires for hundreds of years prior to the nineteenth century. Two additional factors that limit our understanding of the American Woodcock range in Minnesota are the lack of observations or documents from early explorers, and the scant knowledge about the effect of the lack of adequate food resources, such as earthworms. Regarding the latter, Frelich and colleagues (2006) have documented that much of the forested areas of Minnesota that were glaciated lacked earthworms. It is only recently that this food resource has increased in many forested regions of central and northern Minnesota.

BREEDING HABITAT

The American Woodcock has been extensively studied in Minnesota and in many parts of its range. Nesting and foraging habitats are described as shrub thickets of alder and willow, recently cut and regenerating forests, abandoned farmland, edges of forest where shrub canopy is high, and young to mid-aged forest interspersed with openings (Roberts 1932; Gutzwiller et al. 1983; Gregg 1984; Shartell 2018; McAuley et al. 2020). Shrub canopy cover has been documented as high, ranging from 75 to 87 percent (Morgenweck 1977).

Kelley and colleagues (2008) divided the breeding habitat of the American Woodcock into four main categories. First, singing grounds where males perform their courtship activities included clear-cuts, natural openings, roads, pastures, cultivated fields, and reverting agricultural fields. They

C. Typical shrubby and often soggy breeding habitat of the American Woodcock in Minnesota. Photo by Gerald J. Niemi.

D. Microhabitat nesting site for the American Woodcock. Note the adult female Woodcock, incubating on the nest in the center left of the photo, is well camouflaged. Photo by Deborah Petersen.

noted that singing grounds were usually less than 109 m from diurnal cover areas. Second, nesting and brood-rearing habitat was primarily in young, second-growth hardwood stands (e.g., aspen). Brood habitats were characterized by dense hardwood cover with an abundance of earthworms. Third, diurnal habitat was described as more variable, including early successional growth and mature forest with a dense understory. They suggested that the presence of earthworms was critical. Fourth, nocturnal habitat was described as open areas, such as clear-cuts, abandoned agricultural fields, and pastures. Woodcock move into these areas at dusk.

Soil conditions such as damp, rich organic soils and a well-developed ground cover provide optimal conditions for foraging and nesting **(C, D,** and **E)**. American Woodcock avoid heavy clay soils and acidic bogs (e.g., the Red Lake Peatland and the Sax-Zim Bog area) that provide poor habitat for earthworms.

POPULATION ABUNDANCE

PIF (2021) estimated a global population of 3.5 million individuals, all of which occur in the United States and Canada. No population estimates were available for Minnesota. The most reliable measures of population trends for Woodcock are singing-ground surveys, which were initiated in 1968 (Kelley et al. 2008; McAuley et al. 2020). These consist of permanently established routes throughout the breeding range of the species. They are 6 km routes with 10 listening stations. Seamans and Rau (2019) report an overall decline of the American Woodcock from 1968 to

2019 in the Central Region of the United States of −0.89 percent per year, but during the same period the population has been increasing in Minnesota at a rate of 0.49 percent per year. Of the six US states and two Canadian provinces in the Central Region, only Manitoba and Minnesota recorded a positive trend from 1968 to 2019.

In the development of the American Woodcock Conservation Plan, Kelley and colleagues (2008) compared changes in the historic (1970–75) to current (2000–2004) number of singing males on the singing-ground surveys. They estimated that the "historic" number of displaying males was 3.07 million in 1970–74, compared with 2.24 million in 2000–2004. In Minnesota, they reported a decrease in vocalizing males from 276,204 in 1970–74 to 235,259 in 2000–2004, a loss of about 41,000.

CONSERVATION

The American Woodcock has a moderate conservation concern because of its declining population nationally and in Minnesota, where it has also been designated a *Species in Greatest Conservation Need* (MNDNR 2019a). The American Woodcock is a game species. In Minnesota, the estimated number of Woodcock hunters has varied from 9,430 to 13,310 from 2002 to 2019, while the success rate per active hunter was relatively consistent, from 2.3 birds to 3.6 birds from 2002–3 to 2019–20 (MNDNR 2015b; Davros and Dexter 2019). From 2004 to 2019–20, the estimated hunter harvest rates varied from a low of 24,980 in 2011–12 to a high of 46,867 in 2016–17.

Kelley and colleagues (2008) stated that, nationally, it is "widely believed that loss of early succession forest habitat is responsible for declines in woodcock recruitment and in overall population status." Therefore, their primary goal is "to half the decline in woodcock populations and return them to densities which provide adequate opportunity for utilization of the woodcock resource." They have outlined specific conservation plans for Bird Conservation Region (BCR) 11 (Prairie Potholes), which includes the northwestern and western regions of Minnesota, and BCR 12 (Boreal Hardwood Transition), which includes portions of northern Minnesota. In both regions, a major goal is to increase the amount of young, regenerating forests. In addition, in BCR 12 they recommended retention of management techniques that encourage establishment of young forest habitats and shrubland in riparian areas. The southeastern and the central regions of Minnesota are included in BCR 23 (Prairie Hardwood Transition). Because much of the land in BCR 23 is privately owned, the goals are to work with private forestland owners to improve habitat conditions for the Woodcock.

Climate change may be a looming factor that is detrimental to the future of the American Woodcock in the United States. Under a 2°C warming scenario, NAS (2019) projected moderate vulnerability for the species' breeding distribution in North America, including a predicted 36 percent range loss and a 20 percent gain northward into Canada by 2050. Its breeding range retractions were forecast in the western regions of Minnesota under the imminent 1.5°C warming scenario and the 2.0°C warming scenario by 2050.

Kelley and colleagues (2008) described a landscape-level approach to managing American Woodcock. They recommended management units of 200 to 400 ha, which would support approximately 500 Woodcock. Ideally the management units would be within 1.5 to 3.0 km of one another to support gene flow among the populations of Woodcock in these units. They also point out that landscape-level management can be difficult to implement under multiple ownership patterns such as occurs throughout much of Minnesota. Despite the overall recent increases observed in Minnesota, with the loss of and dynamic nature of forests and wetlands, plus the predicted alterations to Minnesota's climate, the future of the American Woodcock is indeed precarious.

E. LEFT: The modest nest of the American Woodcock with four eggs on the ground amid a thick leaf layer. **BELOW:** A precocial, one-day-old American Woodcock capable of walking but not able to fly. Photos by Deborah Petersen.

WILSON'S SNIPE

Gallinago delicata

Minnesota Seasonal Status: Regular breeding resident and migrant; overwinters near open water in southern Minnesota. The Wilson's Snipe was common during the MNBBA.
Migration: Short- to long-distance migrant that primarily winters from the southern United States to northern South America.
Food: Its long bill is used to probe wet soils for larval insects and other invertebrates.
Nest: Shallow ground depression lined with grasses.
Conservation Concern: A game species; assigned a moderate PIF Continental Concern Score of 9/20.

A haunting sound may be our first introduction to this mysterious bird. At one moment it seems overhead, then behind us, and then in front, constantly moving as we ponder its skyward origin: "One is both thrilled and puzzled when he hears it for the first time, for it seems like a disembodied sound, the sighing of some wandering spirit, until the author is discovered, a mere speck, sweeping across the sky" (Bent 1927).

The Wilson's Snipe's renowned courtship display, known as *winnowing*, begins as the bird slowly rises upward, flying in large circles. As it gains height, it begins to swoop back and forth, rising more rapidly until it climbs nearly 50 m. Then, suddenly, it dives and we hear the *hu-hu-hu 'winnow'* as air rushes over its spread outer tail feathers. As it comes out of the dive, still in midair, the sound stops and the entire performance begins again as the bird rises upward. More frequently performed by males, the display occurs throughout the day but is most common after sunset. The birds also have a variety of vocalizations, including a sharp whistle that sounds like *"jick, jick, jick, jick"* (H. Mueller 2020).

On the ground, the Snipe's cryptic, mottled-brown plumage blends perfectly with the dead grasses and forbs in its favored wetland habitats. When disturbed, however, it darts into the air with a fast, zigzag flight that attempts to divert those in pursuit. The Snipe's slender build, small head, and heavily barred breast distinguish it from the Woodcock's stockier build, big head, and plain buffy-orange breast. The top of the Snipe's head also has longitudinal stripes, compared to the Woodcock's traverse stripes. Both birds show stripes on their

Photo by Mike Lentz.

backs. *Gallinago delicata,* the Snipe's Latin name, means "a hen-like bird of delicate appearance."

For many years, biologists considered the North American Snipe a subspecies of the Common Snipe, a species found throughout Eurasia. Although Alexander Wilson, the prominent ornithologist for whom the bird is named, described how the birds differed in the early 1800s, they were not officially recognized as distinct species until 2002 (Banks et al. 2002).

The Wilson's Snipe's breeding range extends across much of Canada and the northern states from the Pacific Northwest, where it resides year-round, east to New England. It reaches some of its highest breeding densities in North Dakota, Manitoba, and Alberta.

MINNESOTA BREEDING DISTRIBUTION

Roberts's 1932 account of the Wilson's Snipe described it as nesting "rather uncommonly throughout Minnesota, usually choosing wet meadows or marshes of small extent and occasionally the open, bushy lowlands bordering tamarack swamps at some distance from water." Most of his narrative, however, describes the bird's aerial display rather than its distribution and abundance. At the time of his writing, nests had been found in only three counties, Hennepin, Polk, and Sherburne, while inferred nesting was reported from Anoka, Pine, and Kittson Counties. Together, they suggested a rather broad distribution, at least through central and northwestern Minnesota.

In 1975, Green and Janssen described the Snipe as a statewide resident that was most abundant in the north-central region and in adjacent counties to the east; it was "least numerous" in southeastern and southwestern Minnesota. Janssen (1987) included a distribution map that depicted the species' range as confined largely to northern and central Minnesota. He identified 11 counties where nesting had been confirmed since 1970; all but one (Goodhue) were north of the Minnesota River. Hertzel and Janssen (1998) later added five northern counties to the list.

By 2014, the MBS documented 687 breeding-season locations for the species. The majority of records were in the Tallgrass Aspen Parklands and Laurentian Mixed Forest Provinces. Beyond west-central Minnesota, records were sparse in the southern counties, but included several in the Prairie Coteau region of southwestern Minnesota. Notable during this time was a 2008 confirmed nesting report in Rock County (MOU 2021a).

MNBBA participants reported 1,728 Wilson's Snipe records in 23.5 percent of the surveyed atlas blocks **(A)**. Breeding was confirmed in only 17 blocks (0.4 percent) but deemed probable in 513 (10.7 percent). The birds were observed in 70 of Minnesota's 87 counties and were confirmed breeding in 13 counties, all north of the Minnesota River. Five of these counties were new to Hertzel and Janssen's (1998) list: Big Stone, Kanabec, Koochiching, Lake of the Woods, and Roseau.

The Wilson's Snipe was most abundant in northern Minnesota, from the Tallgrass Aspen Parklands Province east throughout the Laurentian Mixed Forest Province. Sparsely distributed across southern Minnesota, most records were located along the South Dakota border. The dense cluster of probable records in St. Louis County was contributed by MNBBA volunteer Steve Wilson, who was motivated

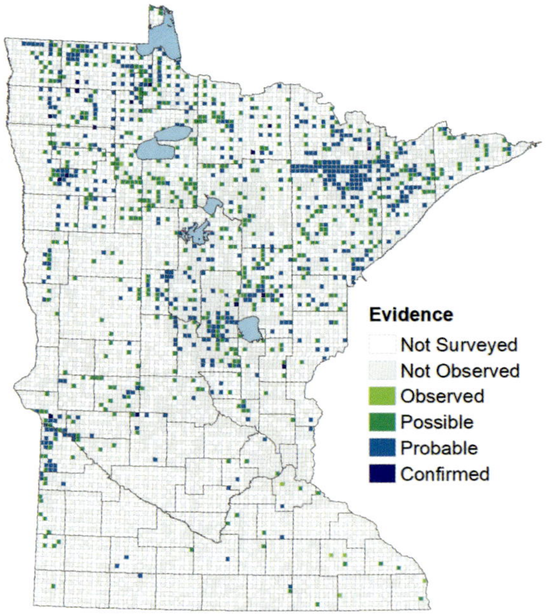

A. Breeding distribution of the Wilson's Snipe in Minnesota based on the Breeding Bird Atlas (2009–13).

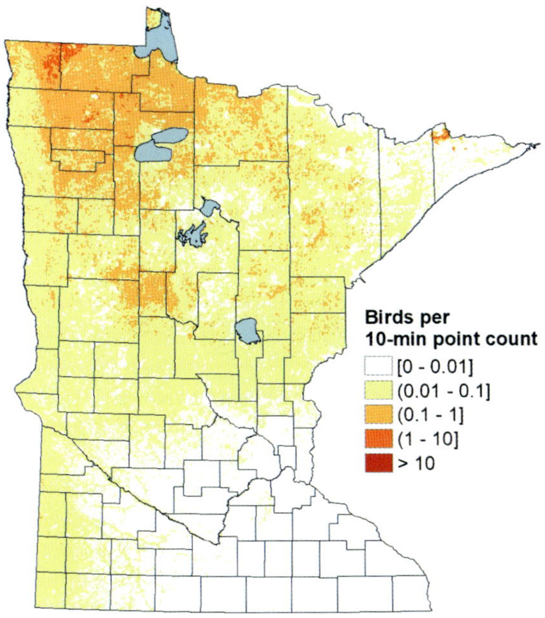

B. Predicted breeding distribution (birds detected per 10-minute point count) of the Wilson's Snipe in Minnesota based on habitat, landscape context, and climate data gathered during the Minnesota Breeding Bird Atlas (2009–13) using the Generalized Linear Modeling method without an adjustment for detectability.

to assess the Snipe's local distribution and abundance in the region.

The MNBBA probability map predicts that the Snipe's highest breeding densities occur in eastern Kittson and western Roseau Counties **(B)**. Moderate densities are found throughout the remainder of the Aspen Parklands and in pockets throughout the western Laurentian Mixed Forest Province, including in the southern Pine Moraines and Outwash Plains. The 2007 Ham Lake fire area north of Grand Marais provides especially suitable habitat. Given the number of MNBBA records collected in central St. Louis and Lake Counties, the model may underpredict the species' abundance in some regions.

It is difficult to assess if there have been any significant changes to the species' statewide status and distribution in the past one hundred years. East of Minnesota, the southern periphery of the species' breeding range is receding (e.g., Chartier et al. 2013; Cadman et al. 2007). Increased development and wetland drainage appear to be the primary culprits. West of Minnesota, however, the species is faring well throughout much of the Great Plains (Sauer et al. 2020).

BREEDING HABITAT

A shallow wetland inhabitant, the Wilson's Snipe has two basic requirements: soft organic soils that supply an abundance of invertebrates within probing distance and patches of clumped vegetation that are dense enough to provide cover but still provide a view of approaching predators. Tuck (1972) broadly described its habitat as sites with organic soils, mostly peatlands, which occur within the boreal forest region. Suitable habitat ranges from open bogs with little tree growth to alder–willow swamps, wet meadows, the riparian edges of rivers, open-water wetlands, and pastures **(C)**.

POPULATION ABUNDANCE

Solid estimates of the species' population size are lacking. Shorebird biologists use a low, conservative estimate of 2 million birds established by Morrison and his colleagues in 2001. Subsequent reviews noted that no new information was available to suggest any changes to this original, conservative estimate (Morrison et al. 2006; Andres et al. 2012; PIF 2021).

The BBS shows a surveywide, slowly increasing population trend for the species from 1966 to 2019 (Sauer et al. 2020). Although the long-term trend in Minnesota demonstrates wide fluctuations, the data document a statistically significant increase averaging 1.96 percent per year from 1967 to 2019 **(D)**. Wetland conditions are believed to be responsible for much of the annual variability (H. Mueller 2020).

CONSERVATION

Given its relatively small population and the vulnerability of its wetland habitats, the Snipe is considered a moderate conservation priority (PIF 2021). Warming temperatures pose another concern. Under a 2°C warming scenario, NAS (2019) projected a moderate vulnerability for the species' North American breeding distribution, with habitat losses spanning the southern periphery of its breeding range, including in Minnesota. Even a temperature rise of 1.5°C projects the species' statewide extirpation. Fortunately, the Snipe has a more extensive distribution in the state than NAS considered in their models. Nevertheless, its future may be quite precarious.

C. The soft organic soils of Minnesota's peatlands provide suitable habitat for the Wilson's Snipe across the state's northern forest region. Photo by Lee A. Pfannmuller.

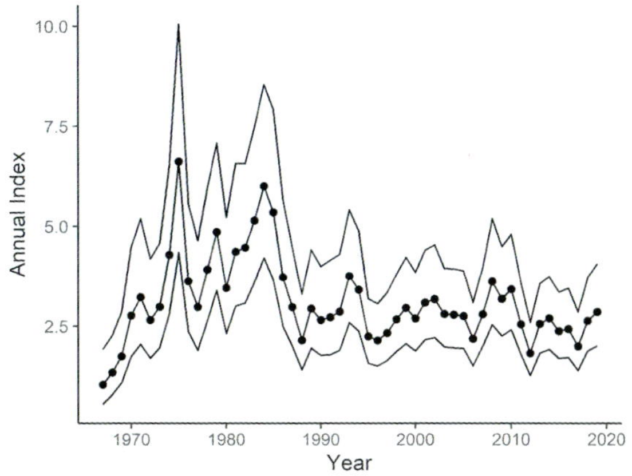

D. Breeding population trend and 95 percent credible interval (solid black lines) for the Wilson's Snipe in Minnesota based on the 1967–2019 Breeding Bird Survey (Sauer et al. 2020).

A century ago, hunting was the primary threat to the Snipe, a popular game bird. Individuals "shot hundreds, sometimes thousands of Snipe in a day, and tens of thousands in a single season" (Tuck 1972). Even when harvest regulations were enacted, compliance was initially poor. These unsustainable harvest levels, coupled with drought conditions on the breeding range and extended cold periods on the bird's wintering range in the 1930s, prompted the season's closure from 1941 through 1953.

Today, few hunters concentrate solely on Snipe but rather harvest the bird incidentally while they hunt waterfowl (K. Arnold 1994). The total US harvest in 2019 was approximately 92,700 birds (Raftovich et al. 2020), significantly lower than the 422,000 average annual harvest from 1964 to 1976 (E. Martin 1979). In Minnesota, the 2019 harvest was 1,764 birds (Davros and Dexter 2019).

The Snipe's future relies on the protection and restoration of wetland habitats, especially south of the boreal forest region, where a larger percentage of its critical habitat has been lost or severely degraded. Since the bird is responsive to the abundance of wetlands at the landscape level (Saunders et al. 2019), Russell and his colleagues (2016) recommend maintaining a large network of wetlands that provide open, shallow waters interspersed with mudflats in order to provide the high-quality nesting habitat required by the mysterious "winnower."

SPOTTED SANDPIPER

Actitis macularius

Minnesota Seasonal Status: Regular breeding resident and migrant. The Spotted Sandpiper was uncommon during the MNBBA.
Migration: Short- to long-distance migrant that winters from the southern United States through South America.
Food: Aquatic and terrestrial invertebrates secured by probing shallow waters, shorelines, and beaches.
Nest: Small scrape lined with dead grasses and usually located within a few hundred meters of water, where herbaceous vegetation can provide some cover.
Conservation Concern: Assigned a moderate PIF Continental Concern Score of 10/20.

Photo by Rebecca Field.

Spend summers along the shorelines of Minnesota's lakes and rivers and you will likely spot a small, light-colored bird with long legs foraging in the sand or among the rocks and constantly tipping its tail up and down. Once disturbed, it will take flight, engaging in another unusual performance as it flies just above the water or ground "on rapidly vibrating, down-curved wings, gliding a few feet now and then, and describing a wide-spread curve to an alighting place ahead" (Roberts 1932). The Spotted Sandpiper was referred to as a "Teeter-peep" or "Tip-up" in earlier days; these behaviors will help confirm its identity.

Smaller and slimmer than a Killdeer, the bird has plumage that is also diagnostic. Brown spots on a white breast and belly are unique to this shorebird during the breeding season. Also look for a white line above the eye, an orange bill with a black tip, and pale yellowish-pink legs. The head, back, and wings are sandy brown; in flight, a white wing stripe is evident. A common call is a long series of *weet, weet, weet* notes (J. M. Reed et al. 2020). *Actitis macularius,* the species' scientific name, translates to "spotted dweller on the shore."

The most intriguing aspect of the Spotted Sandpiper is its sexual role reversal. The female is larger, initiates most of the courtship activity, and establishes and defends the territory. The smaller male assumes the majority of incubation responsibilities and raises the young. This unusual reproductive behavior was largely uncovered during a long-term investigation conducted by Dr. Lewis Oring and his students in Itasca State Park and on Leech Lake from 1969 to 1991. Over the course of the investigation, more than 1,000 Sandpipers were banded and closely monitored. Their landmark study is now considered a classic in the field of behavioral biology (Oring et al. 1991).

Sparsely distributed across most of Canada and the United States, the Spotted Sandpiper is a solitary species. It is rare to sight more than a single bird or, at most, a single family.

MINNESOTA BREEDING DISTRIBUTION

In Minnesota, Roberts (1932) described the species as a common summer resident. "It is found," he wrote, "throughout the state without regard to any special environment. Any variety of shore-line or water-margin, even though it be only that of a stagnant pool in a quarry or brickyard, seems to be acceptable to the plebeian taste of this little Sandpiper." At the time, nesting records were available from Jackson County in the southwest, Hennepin, Sherburne, and Wright Counties in central Minnesota, and Cook, Itasca, Polk, and St. Louis Counties in the north.

Decades later, Green and Janssen (1975) and Janssen (1987) described the Spotted Sandpiper as a statewide breeding resident that was most abundant along the Mississippi River and in the northern lakes region. Janssen (1987) identified 25 counties where nesting had been confirmed since 1970; all but five were located north of the Minnesota River. Hertzel and Janssen (1998) later added five counties to the list, including three south of the Minnesota River. In contrast to these accounts, by 2014 nearly half of the 103 breeding-season locations identified by the MBS were located either along or south of the Minnesota River valley; only two records were reported along the Mississippi River valley south of the Twin Cities.

During the MNBBA, participants reported 603 Spotted Sandpiper detections in 10.4 percent (493/4,747) of the surveyed atlas blocks **(A)**. Breeding was confirmed in 44 blocks (0.9 percent) and deemed probable in 161 (3.4 percent). An unusually high number of MNBBA reports were collected in several west-central counties, owing largely to intensive survey efforts by one individual. Observed in 80 of Minnesota's 87 counties, the Sandpiper was confirmed breeding in 26 counties. Among the latter, eight counties were south of the Minnesota River. Thirteen of the counties where nesting was confirmed were additions to the list published by Hertzel and Janssen (1998).

Despite early reports of the Sandpiper's abundance along the Mississippi River, only a handful of MNBBA records were found along the river valley south of the Twin Cities. Its abundance in the northern lakes region, where lakeshore development has grown significantly, may have declined as well. Itasca County alone witnessed a 31 percent increase in lakeshore development in just six years, from 1992 through 1998 (D. Dempsey 2006). High water levels in some regions may also be contributing to declining numbers. Shorelines inundated by water no longer provide the narrow band of shoreline habitat that is critical for the Sandpiper (Robert Russell, pers. comm.). Although one may presume that the "teeter bird" was a common species in Roberts's time, during the MNBBA it was classified as uncommon.

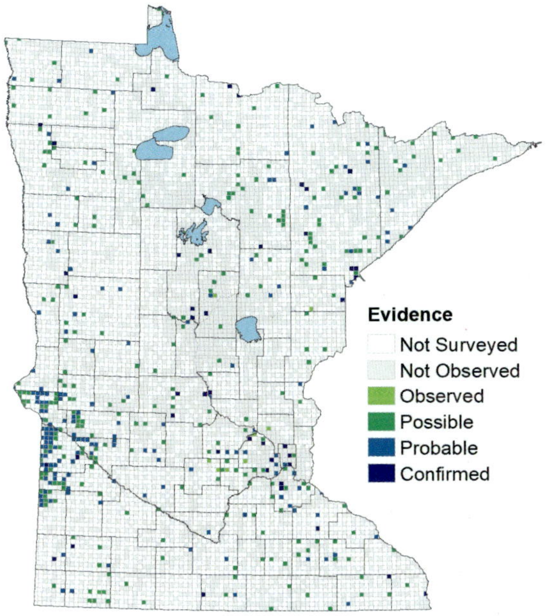

A. Breeding distribution of the Spotted Sandpiper in Minnesota based on the Breeding Bird Atlas (2009–13).

Across North America, the Spotted Sandpiper has not experienced any large-scale changes to its distribution in the past one hundred years (J. M. Reed et al. 2020). Many states and provinces, however, report a decline in abundance.

BREEDING HABITAT

During the summer, the Spotted Sandpiper selects habitats ranging from remote islands and the shorelines of lakes and rivers to gravel pits and artificial wetlands **(B)**. Breeding densities are especially high on islands surrounded by deep water because of the protection they provide from terrestrial predators (J. M. Reed et al. 2020). In recent years, the species has been a regular resident at many sewage treatment plants, where the steep, rocky shorelines do not seem to be a deterrent (Kim Eckert, pers. comm.). Regardless of which habitat is selected, three features are important: shorelines for foraging and displaying; open habitats for nesting; and patches of dense vegetation for brood cover (J. M. Reed et al. 2020). Open areas with low herbaceous cover are preferred for nesting. If open areas are not available close to the shoreline, the birds select suitable sites some distance away and then move the chicks to the shoreline once they have hatched.

POPULATION ABUNDANCE

Because shoreline habitats are not adequately sampled by the roadside BBS, biologists have instead examined migration data and consulted with species experts to estimate the Sandpiper's population size. The most recent North American population estimate is 660,000 birds (Andres et al. 2012; PIF 2021), which is significantly higher than previous estimates of 150,000 birds (Morrison et al. 2001; Morrison et al. 2006). The higher estimate was based in part on a study in Ontario that estimated the population in that province alone numbered in the hundreds of thousands (Ross et al. 2003). A statewide estimate for Minnesota is not available.

The Spotted Sandpiper's sparse distribution, coupled with the limitations of monitoring shoreline nesting birds, means that statistical analyses of population trends using BBS data are not very robust. With this caveat, BBS data across the survey region and in Minnesota suggest that Spotted Sandpiper populations are declining. Surveywide, numbers have declined at an average annual rate of −1.45 percent from 1966 to 2019; in Minnesota the average decline is considerably higher at −2.75 percent per year (Sauer et al. 2020). In contrast, Andres and his colleagues (2012) believed that the available data for the Spotted Sandpiper suggest a stable population trend.

The most important factor that appears to influence local populations is the "availability of safe nesting sites" (J. M. Reed et al. 2020). Islands or shorelines that are bounded by large expanses of low herbaceous cover provide the most protection from predators. Weather can also influence annual reproduction but is considered "secondary to predation."

CONSERVATION

The Spotted Sandpiper is ranked as a moderate conservation priority based on its relatively small population size and its declining abundance as suggested by BBS data (PIF 2021). Shorebird biologists, however, have classified it as a Species of Low Concern (US Shorebird Conservation Plan Partnership 2016). Regardless of its ranking, the species' dependence on shoreline habitats is cause for concern. Lake shorelines are a heavily used resource in Minnesota. Many lakeshore homeowners, in an effort to create a "clean" lakefront, eliminate natural habitats valuable for shoreline plants and animals. Initiatives that promote shoreline restoration are extremely important and can benefit species such as the Spotted Sandpiper. Dredge spoil islands, which provide suitable habitat along the Mississippi River, should also be managed to maintain the semi-open habitat favored by the species (Russell et al. 2016).

Warming temperatures may pose another challenge. NAS (2019) classified the Sandpiper as moderately vulnerable under a 2°C warming scenario when models predicted the loss of 27 percent of its continental breeding range by 2050 with a northward gain of 20 percent. As temperatures rise, the Sandpiper's Minnesota distribution is projected to gradually retreat northward.

Currently a well-distributed species in the state, the "Teeter Snipe" faces an uncertain future. Close monitoring is warranted, given continued pressure on its highly prized shoreline habitat and other potential threats. Luckily, the species' utilization of such modified habitats as gravel pits and sewage ponds demonstrates a level of adaptability that provides some flexibility.

B. Low vegetative cover for nesting, nestled between an open shoreline for foraging and denser cover for the safety of broods, provides the basic needs for the Spotted Sandpiper. Photo by Lee A. Pfannmuller.

SOLITARY SANDPIPER

Tringa solitaria

Minnesota Seasonal Status: Regular breeding resident and migrant restricted to northern Minnesota. The Solitary Sandpiper was very rare during the MNBBA.
Migration: Medium- to long-distance migrant that winters in Central and South America.
Food: Consumes a variety of arthropods, mollusks, and frogs gathered along the edges of small pools.
Nest: Selects old tree and shrub nests constructed by other species, such as the American Robin and the Cedar Waxwing.
Conservation Concern: Assigned a moderate PIF Continental Concern Score of 11/20.

Photo by Sparky Stensaas.

Few breeding species in Minnesota are as rare or as poorly known as the graceful Solitary Sandpiper. Although the bird was first described in 1813 by well-known ornithologist Alexander Wilson, ninety years passed before the first nest was discovered in 1903 in the remote Canadian wilderness. More than a century later, only a few field studies have been published that focus exclusively on the species' life history and habits.

A rare nester in the state, the Solitary Sandpiper is seen more regularly during migration as it travels to and from its boreal breeding grounds. Unlike many shorebirds that migrate in large groups, numbering from as few as twenty individuals to hundreds of birds, the Solitary Sandpiper, as its common and scientific names imply, travels alone or in very small groups of two to six birds (Janssen 1987). Also known as the "woodland tattler," it "frequently indulges in a peculiar tilting and nodding habit similar to that of the spotted sandpiper, but it is more deliberate and not so pronounced" (Bent 1929).

Diagnostic features in the field include its size (larger than a Spotted Sandpiper and smaller than a Lesser Yellowlegs), its brownish-gray breast, olive legs, speckled olive-brown back and wings, and a prominent eye-ring. In flight, it is easily distinguished from the Spotted Sandpiper by its dark rump and central tail feathers and by the absence of a white wing stripe.

The Solitary Sandpiper is largely a species of boreal Canada. The only known breeding records in the United States are in northern Minnesota and central Oregon.

MINNESOTA BREEDING DISTRIBUTION

Roberts (1932) knew the Solitary Sandpiper as a "common, sometimes an abundant, spring and fall migrant throughout the state." Its status as a breeding species, however, remained uncertain. Most of the summer records available to Roberts were from July and were classified as early fall migrants instead of breeding birds. Even a few late June records (June 26, 27, and 30) were considered potential fall migrants. In retrospect, repeated observations of a bird at the town of Biwabik, St. Louis County, in June 1908 and observations on June 26 and 27 in 1922 from the Popular River in Cook County are of particular interest. Both sites are now known to be within the species' potential breeding range, but it was uncertain at the time whether these were breeding birds or migrants. Roberts also included one questionable breeding record from Jackson County in 1901 by a respected observer, Mr. J. C. Knox. A pair of birds was observed for several days in early June on a "sparsely wooded island and point in the Des Moines River." Several days later, four newly hatched young were seen. Given the location, it seems far more likely that the birds were Spotted Sandpipers, a species relatively common in that region of the state.

The first well-documented breeding record of the Solitary Sandpiper was reported in July 1973. While driving on a county road along the Mississippi River in Aitkin County, Terry Savaloja observed a pair of Solitary Sandpipers flying overhead. When they landed along the shoreline, a half-grown downy young also appeared (Savaloja 1973). Then, in 1982, after spotting Solitary Sandpipers at two different wetlands in central Cook County the previous two summers, Ken and Molly Hoffman observed a pair of adults and a short-tailed chick in an old, drained beaver pond on June 30 (Hoffman and Hoffman 1982b). Nesting continued to be documented in the area during the summers of 1983, 1984, and 1987 (Green and Janssen 1984; Hendrickson and Eckert 1985; Wilson and Shedd 1988).

The Cook County site remained the second breeding confirmation for nearly thirty years, until the MBS documented fledged young in southern Lake of the Woods County in 2012. This record was included in the MNBBA. During the course of the MNBBA, eight additional Solitary Sandpiper records were validated (four *Probable* records and four *Possible* records), for a total of nine records.

Each MNBBA record was scrutinized carefully, as were nine other atlas records that were ultimately invalidated. Identifying which records represented late spring migrants, early fall migrants, or possible breeding records was challenging. Spring migrants have been observed in the state as late as early to mid-June and fall migrants as early as late June (Janssen 1987). As a result, two primary criteria for accepting Solitary Sandpiper breeding records from atlas participants were delineated. First, records during the first week of June were invalidated unless the observer included documentation noting that either a pair or an adult with young was observed. Second, the observations had to be in a suitable habitat in northern Minnesota. This analysis resulted in the validation of the nine MNBBA records men-

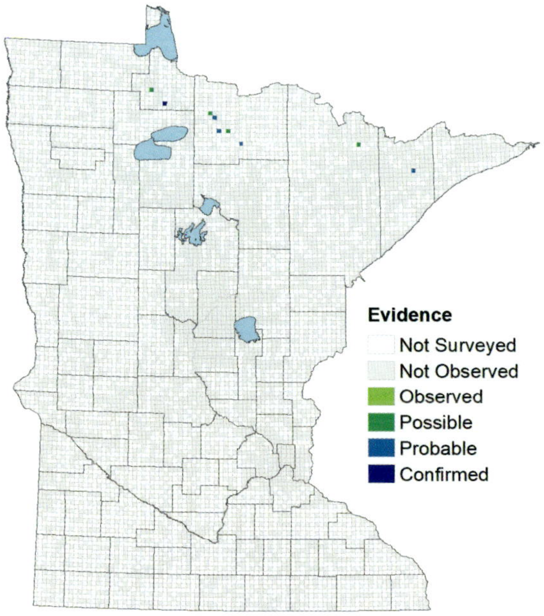

A. Breeding distribution of the Solitary Sandpiper in Minnesota based on the Breeding Bird Atlas (2009–13).

tioned above as well as the invalidation of nine other records **(A)**. All validated records are from the northern reaches of four northern counties: Koochiching, Lake, Lake of the Woods, and St. Louis.

Because the species occupies remote wetlands throughout the boreal forest region, little is known about any historical changes to its breeding distribution.

BREEDING HABITAT

The Solitary Sandpiper is found in peatlands with scattered open pools and ponds. More details are available to describe the marshes, rice paddies, and brackish coastal habitats it uses in the spring and fall than are available to describe its relatively inaccessible breeding habitat. Moskoff (2020) simply describes suitable nesting habitat as "freshwater lakes and ponds in areas of muskeg bogs and spruce trees" and "margins of shallow water near woodlands, mainly coniferous forest" **(B)**. Sites that provide 2.5–7.5 cm of water are preferred (Russell et al. 2016).

The best qualitative description of the Solitary Sandpiper's breeding habitat in Minnesota was provided by Ken and Molly Hoffman when they discovered the birds nesting in Cook County (Hoffman and Hoffman 1982b). All observations were made at "drained beaver ponds with extensive mud areas and standing dead trees." At least one of the sites was a large black spruce bog where the trees had died because of the beaver dam's high water; in its place were standing dead trees, a mix of marsh grasses, and dense thickets of alder brush. A mat of sphagnum was largely dead and appeared like a large mud flat bisected by two small rivulets meandering through the site.

POPULATION ABUNDANCE

Because it is so difficult to adequately survey breeding areas, population estimates for the Solitary Sandpiper are based on counts taken during migration. Recent estimates of the North American population are 190,000 individuals (Andres et al. 2012; PIF 2021), which are slightly higher than previous estimates of 150,000 (Morrison et al. 2001; Morrison et al. 2006).

Although biologists originally suspected the Sandpiper's population was declining (S. Brown et al. 2001), they now report that trends are quite variable from year to year, with uncertainty about the long-term trend (Andres et al. 2012). Data collected in the core BBS survey region suggest a relatively stable population from 1966 to 2019 (Sauer et al. 2020). The trend should be viewed with considerable caution, however, given the species' extremely low abundance in the region (an average of 0.09 birds per route). The statistical reliability of the data improves when BBS routes in the non-core region (i.e., remote areas of Alaska and Canada) are included in the analysis. Indeed, within Canada, they illustrate a significant increase averaging 2.34 percent per year from 1993 through 2019 (Sauer et al. 2020). Nevertheless, the species' relative abundance is still quite low (0.27 birds per route).

CONSERVATION

Because so little information is available on the Solitary Sandpiper, specific conservation measures have not been delineated. Historically, the species' boreal habitat has been considered secure, but the recent loss of habitat due to urbanization, agriculture, forestry, and energy development now raises concern, particularly in western Canada, the heart of the species' breeding range (NABCI Canada 2012).

A long-distance migrant, the Solitary Sandpiper, like many other migrant shorebirds, is still hunted during migration and on its wintering grounds in the Caribbean and in South America (J. Smith 2016b). Fortunately, because of its solitary nature, hunting is not considered to have impacted Solitary Sandpiper numbers to the same degree that it impacts species like the Red Knot, which congregates in sizable numbers at large stopover sites during migration and on its wintering grounds (Moskoff 2020).

Warming temperatures may also pose a challenge to this boreal nesting species. Audubon biologists (NAS 2019) classified the Sandpiper as moderately vulnerable under all three warming scenarios they examined, ranging from 1.5°C to 3°C. Although the models did not consider the species' presence in Minnesota, as temperatures rise and its breeding range retreats northward, the Sandpiper's occurrence in Minnesota would become increasingly unlikely.

Meanwhile, more intensive surveys in key counties could provide additional information about the species' habits and habitats and help unravel some of the mystery that embraces this poorly known inhabitant of our northern peatlands and boreal forests. Appropriately named, this solitary shorebird remains a rare and elusive member of Minnesota's avifauna.

B. A small pond in St. Louis County's boreal forest is typical of the northern nesting habitat selected by the Solitary Sandpiper. Photo by Gerald J. Niemi.

WILLET

Tringa semipalmata inornata

Minnesota Seasonal Status: Summer vagrant and regular migrant; bred in Minnesota from 1931 to 1933. The Willet was very rare during the MNBBA.
Migration: May be a short- to long-distance migrant depending on how far birds travel to their wintering grounds. Wintering birds can be found along the Atlantic coast from Virginia to Texas, along the Pacific coast of California, and along both coasts of Mexico south to northern South America. Birds in the West Indies are permanent residents.
Food: Feeds primarily on insects, crustaceans, and, along the coasts, on marine worms; may also take small fish and some vegetation.
Nest: Ground nest lined with grasses or pebbles; when placed within marshes, nests are hidden by surrounding vegetation.
Conservation Concern: Assigned a high PIF Continental Concern Score of 14/20.

Photo by Sparky Stensaas.

This large, stocky, grayish shorebird is inconspicuous while standing but, when alarmed, it becomes a different bird, loudly calling its territorial song, *pill-will-willet,* and flashing its boldly patterned black-and-white wings. John James Audubon (1827–38) captured this scene in the *Birds of North America*: "The sight of a Crow, a Turkey Buzzard, a quadruped of any kind, and more especially of a gunner, at once excites the greatest alarm; and, rising on wing, they fly above and around you at a considerable distance, vociferating their anger with great vehemence."

The Willet has two "disjunct breeding populations differing in ecology, in morphology, and subtly in vocalizations" (Lowther, Douglas, and Gratto-Trevor 2020). Inland populations belong to the subspecies *[T.] semipalmata inornata,* or Western Willet, and are found from Alberta east to Manitoba, south through the Dakotas to northern Nebraska, and west to northern California. Coastal populations belong to the subspecies *[T.] s. semipalmata,* or Eastern Willet, and are found locally along the Atlantic coast from the Maritime Provinces (Stewart et al. 2015) south to Florida, and west along the Gulf coast to Louisiana and Texas (Melvin 2006). They also occur in the West Indies, coastal Mexico, and locally in northern South America. The region of highest abundance is the northern Great Plains, where nearly 50 percent of the Willet population breeds (North Dakota Game and Fish Department 2016).

MINNESOTA BREEDING DISTRIBUTION

Roberts encountered this species during his first summer trip to the prairie in 1879, when he and a companion visited Grant and Traverse Counties on June 5–21 (Krosch 1987). His notes about the Willet included two entries in his diary. The first was on June 16 at Lake Traverse: "A pair of willets made much ado and evidently has a nest in a marsh along the lake"; and the second on June 20 near Herman, Grant County: "pair of Willets that seemed to have a nest" (Roberts 1991). A complete narrative and bird list of this trip were published in the *Bulletin of the Nuttall Ornithological Club* (Roberts and Benner 1880) and reprinted in *The Loon* (Roberts and Benner 1987). In the list, Roberts described the Willet as "common and breeding." Many years later, when he wrote his definitive book on Minnesota birds, he listed those two breeding observations plus another from Cantwell, who "states that in May, 1889, it was numerous in Lac qui Parle County and one collected May 16 contained an egg ready to be deposited" (Roberts 1932). Roberts concluded, however, that since about 1909, the species "has almost entirely disappeared from the state." He listed several migrant records from the 1920s.

The only modern breeding records in Minnesota were obtained in 1931, 1932, and 1933 in western Swift County. Documented because of the field perseverance and bird knowledge of two small-town women and an adventurous boy, the location was on the Pomme de Terre River "four miles west of Halloway." On July 26, 1931, "Mr. Peterson of Pipestone visited the locality where these birds were seen by Mrs. Wilson [July 5, 1931] and found there three Willets, one adult and two young. . . . A farmer boy living nearby said that he had found the nest at an earlier date on an island in the river. It contained two eggs only" (Roberts 1932). In 1932, "in the same locality where this species nested in 1931, two adult Willets and one downy young were found on June 24, 1932, by Mrs. C. E. Peterson and Mrs. E. O. Wilson; and again the next year, on June 2, Mrs. Peterson reported finding a pair at the same place" (Roberts 1936). This record was also documented in *The Flicker* (Risser 1932).

MOU (2021b) seasonal reports from 1961 to 2019 show the Willet as a regular migrant in Minnesota, encountered more frequently in the spring than in the fall. Most records are from the southern half of the state and the northwestern prairie counties, where spring flocks of 30 to 80 birds occasionally

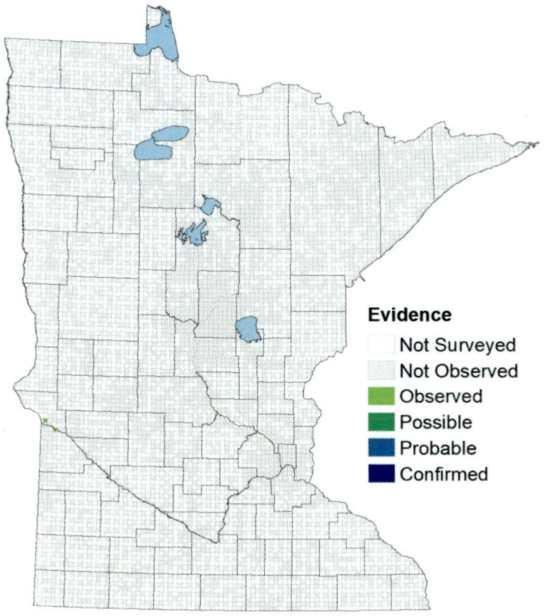

A. Breeding distribution of the Willet in Minnesota based on the Breeding Bird Atlas (2009–13).

have been seen (Bardon 2002a). Nevertheless, it can occur anywhere in the state where there is shorebird habitat, especially at Duluth, where it is regular in the spring (including three flocks of 15 to 25 birds) and occasional in the fall. Because spring migration for the species may end in mid-June and fall migration begins just weeks later, in early July, it is difficult to determine if summer observations are migrants or possibly nesting birds (Bardon 2002b). Using a cautious approach for dates, with a summer gap of June 10–30 between spring and fall migration, there are several periods when Willets were reported in appropriate nesting situations in prairie counties. These were from 1961 to 1966 (mostly Salt Lake, Lac qui Parle County); 1970 to 1980 (Marshall, Swift, Traverse, and Wilkin Counties); and 2002 to 2006 (Becker, Lac qui Parle, Marshall, and Traverse Counties). Breeding evidence, however, was never obtained. More investigation is still warranted in appropriate landscapes in Minnesota. There are persistent breeding populations in South Dakota, including in counties adjacent to Minnesota (Drilling, Sparks, et al. 2016; R. A. Peterson 1995; Tallman et al. 2002) and in North Dakota (Stewart 1975; North Dakota Game and Fish Department 2016). The literature from other states adjacent to Minnesota does not show evidence of nesting.

MNBBA participants reported just two possible records: one in 2011 and one in 2013 **(A)**. Both were in early June and thus could still be spring migrants. They were seen in appropriate habitat in Big Stone County and Lac qui Parle County, where many of the June records discussed above were obtained. About 20 miles east, in adjacent Swift County, is the locality where the birds nested from 1931 to 1933.

BREEDING HABITAT

Along the Atlantic Seaboard and Gulf coast, the Willet nests in coastal marshes, barrier islands and beaches, and sparsely vegetated sand dunes (Lowther, Douglas, and Gratto-Trevor 2020; Melvin 2006). In the interior, large "expanses of short, sparse grasslands, particularly native grasslands" provide critical nesting habitat for the species (North Dakota Game and Fish Department 2016) **(B)**. Shallow waters, with little shoreline vegetation, are preferred for foraging. Densely vegetated wetlands are avoided. Brackish or saline wetlands and alkali flats may also be used in the Great Plains and Great Basin (Lowther, Douglas, and Gratto-Trevor 2020).

POPULATION ABUNDANCE

Morrison and his colleagues (2001) provided a total population estimate of "at least" 250,000 with approximately 10 percent breeding in Canada. Their "rough estimate" for the eastern subspecies, including the Maritime Provinces, was 90,000 birds and, for the western subspecies, including the Canadian Prairie Provinces, was 160,000. This estimate was reviewed again in 2012 and 2021 but, given the absence of new information, no changes were made (Andres et al. 2012; PIF 2021).

Surveywide, BBS trend results from 1966 to 2019 show a steady, statistically significant decline averaging −0.62 percent per year (Sauer et al. 2020). The population trend in the Prairie Potholes Bird Conservation Region may be the best indicator of the status of the Western Willet population. Here BBS data show a significant decline averaging −0.66 percent per year from 1967 to 2019 **(C)**. The loss of wetlands and grasslands is likely the primary factor responsible for the regional decline.

B. Native grasslands interspersed with shallow wetlands are the preferred breeding habitat of the Willet in the Great Plains. Photo by Lee A. Pfannmuller.

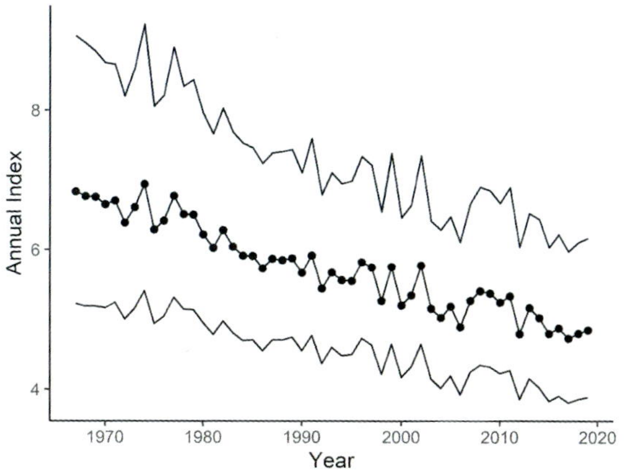

C. Breeding population trend and 95 percent credible interval (solid black lines) for the Willet in the Prairie Potholes Bird Conservation Region based on the 1967–2019 Breeding Bird Survey (Sauer et al. 2020).

The Eastern Willet population was almost extirpated from its range along the Atlantic coast by market hunters who sought its tasty flesh and eggs in the late nineteenth and early twentieth centuries. Protection was finally obtained under the Migratory Bird Treaty Act of 1918. Not until the mid- to late twentieth century did its recovery and recolonization take hold and produce a breeding population that appears to be stable (Schneider et al. 2010).

CONSERVATION

The species is considered a high conservation priority (PIF 2021); its small population size, declining population trend, and threats to its breeding and wintering habitat all contributed to its high ranking. Unfortunately, the availability of suitable habitat may not be the only challenge to this inconspicuous shorebird. As temperatures rise, this prairie pothole inhabitant may have to contend with drought conditions and more extreme rainfall events. Nevertheless, NAS (2019) deemed the Willet's North American breeding distribution was stable under all three warming scenarios they examined. As temperatures rise, available habitat in northwestern Minnesota was even projected to increase. Birders should definitely keep an eye out for this rare grassland inhabitant in the coming years.

WILSON'S PHALAROPE

Phalaropus tricolor

Minnesota Seasonal Status: Regular breeding resident and migrant. The Wilson's Phalarope was rare during the MNBBA.
Migration: Long-distance migrant that winters in South America.
Food: Feeds on aquatic invertebrates obtained by probing mud and, while swimming, by spinning in a circle to create a small whirlpool that brings prey items to the surface.
Nest: Shallow ground scrape lined with vegetation.
Conservation Concern: Assigned a moderate PIF Continental Concern Score of 10/20 and classified as a *Threatened* species in Minnesota.

Photo by Rebecca Field.

Slender and graceful, the Wilson's Phalarope was described by Roberts (1932) as "the most beautiful and attractive of the many Shore Birds that come to Minnesota." Bent (1927) similarly praised the bird when he reflected on a visit he made to a prairie wetland that was "teeming with bird life." Amid the ceaseless din of calling terns and gulls darting overhead, the persistent calls of the Killdeers, and the throngs of noisy blackbirds, "the almost silent, gentle, dainty, little phalaropes stand out in memory as charming features in the picture."

Like many shorebirds, the Wilson's Phalarope is known for its sexual role reversal. Once the female lays her eggs, her maternal duties are finished, and she proceeds to court another male and initiate a second nesting. The female is also larger and more colorful than the male. A little smaller than a Killdeer, the female has as her most prominent features a black stripe that extends from the base of the bill through the eye and down each side of the neck, a light blue-gray crown, a white chin and patch above the eye, and a rufous neck. A white belly contrasts with the gray and rufous wings; in flight, a white rump is prominent. The smaller male may show only a hint of the female's black and rufous neck colors. Common vocalizations include a short, nasal call and a deeper, often hollow-sounding call note (Colwell and Jehl 2020).

The Phalarope's small breeding range stretches from central British Columbia south to northern California and east through the Great Plains, including western Minnesota. Small localized populations are found in the Great Lakes region and in several other scattered localities. Populations are most abundant in the northern Great Plains.

MINNESOTA BREEDING DISTRIBUTION

Roberts (1932) described the Wilson's Phalarope as a summer resident restricted to southern and western Minnesota. Nesting was confirmed in Jackson County in the southwest, in McLeod and Hennepin Counties in central Minnesota, and in Polk County in the northwest. Although the species was considered common at Leech Lake in 1902 and 1903 and in Kittson County in 1929, no nests were found. Regarding its abundance, Roberts's (1932) own words are most apt: "Previous to about 1900 it was an abundant summer resident throughout the southern and western parts of the state, every large slough or shallow lake harboring many pairs during the nesting-season. Then, for some not very evident reason, it almost entirely disappeared over much of this region for fifteen or twenty years. . . . Just as it seemed that it was about to become a bird of the past it began to reappear and has steadily increased until now it is back again over much, if not all, of its former range, though it is still far short of the old-time abundance."

Forty years later, Green and Janssen (1975) described the Wilson's Phalarope as a summer resident throughout central and western Minnesota, yet "very scarce in many areas." The breeding range they delineated excluded south-central, southeastern, north-central, and northeastern Minnesota. The bird was considered most abundant in northwestern Minnesota; its status south of the Minnesota River was unclear. The only breeding record since 1900 was from Lac qui Parle County in 1948.

Janssen (1987) accompanied his updated status account with a more restricted range map. With the exception of Lac qui Parle and Lincoln Counties, where nesting was confirmed in 1978 and 1979, respectively, all of southern Minnesota was excluded. The species' breeding range now was largely confined to northwestern Minnesota, stretching as far east as Aitkin and Morrison Counties. A recent increase in the number of Phalarope reports in north-central Minnesota was attributed to an increase in local wild rice cultivation. Janssen identified eight counties where nesting had been confirmed since 1970: Aitkin, Big Stone, Lac qui Parle, Lincoln, Mahnomen, Marshall, Morrison, and Polk. Hertzel and Janssen (1998) later added three more counties to the list: Anoka, Lake of the Woods, and Wilkin.

Beginning in the late 1980s through 2014, the MBS identified 69 breeding-season locations for the Phalarope. In addition to known localities in Lac qui Parle and Lincoln Counties, breeding-season observations were documented south of the Minnesota River in Brown, Murray, and Nobles Counties. Records were also found in Anoka County and in several central counties, including Morrison, Todd, Stearns, and Douglas.

Far fewer breeding-season records were found by MNBBA participants, who reported only 31 Phalarope records in 28 surveyed atlas blocks. Breeding was confirmed in four blocks in widely dispersed

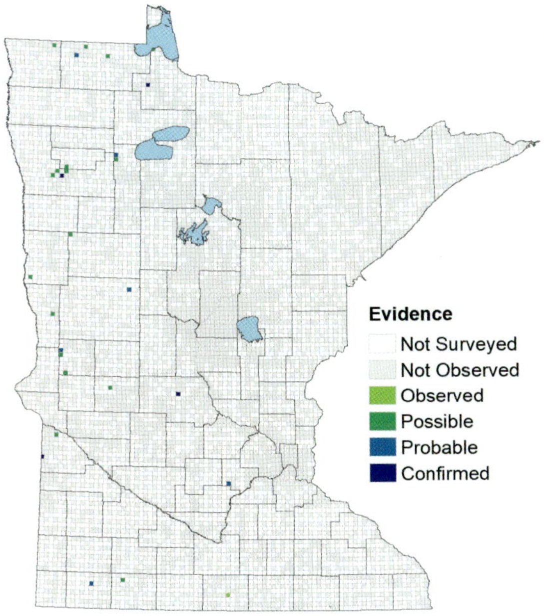

A. Breeding distribution of the Wilson's Phalarope in Minnesota based on the Breeding Bird Atlas (2009–13).

counties: Lac qui Parle, Lake of the Woods, Polk, and Stearns **(A)**. The bird was observed in 18 of Minnesota's 87 counties; far northwestern Minnesota was the core of the species' abundance, where 13 of the 28 blocks with reported detections occurred. Records were largely absent from the greater Twin Cities metropolitan region with the exception of one probable record in Carver County. Five records occurred south of the Minnesota River.

Unfortunately, this graceful shorebird remains a rare and sparsely distributed species in Minnesota; it has never regained its widespread distribution and abundance observed during the nineteenth century. Its abundance appears to have declined even further in the past twenty to thirty years. Similar changes have been noted throughout the eastern portion of its range, where it is largely a localized and sporadic breeder (Russell et al. 2016). In contrast, the Phalarope has expanded its breeding range in Alaska and across Canada, from British Columbia east to New Brunswick (Colwell and Jehl 2020; Cadman et al. 2007).

BREEDING HABITAT

Found in both alkaline and freshwater wetlands, Wilson's Phalaropes prefer shallow wetland basins with mixed patches of open water and patchy emergent vegetation **(B)**. Wild rice ponds and sewage treatment ponds may also be used. Overall, suitable sites provide open water, emergent vegetation, and an open shoreline (Dechant, Johnson, et al., "Wilson's Phalarope," [1999] 2002). Wetland complexes larger than 100 ha are preferred, and nesting usually occurs within 100 m of the wetland edge (Potter et al. 2007). Early in the nesting season, Phalaropes are more likely to nest in drier upland grasslands, moving to wetter meadows as the season becomes hotter and drier.

POPULATION ABUNDANCE

The most recent population estimates, based on counts conducted at migration stopovers and wintering sites, are consistent at 1.5 million birds (Andres et al. 2012; PIF 2021). Shorebird biologists believe that, in addition to a significant decline in numbers in the nineteenth century, a similar decline occurred later in the twentieth century, pointing to reduced counts at key staging areas in the 1990s and reduced numbers at important wintering sites in Bolivia and Argentina (Lesterhuis and Clay 2010).

Although the BBS roadside survey may not be a robust monitoring tool for wetland shorebirds, it does cover a significant portion of the species' breeding range. From 1966 to 2019, the BBS illustrates a reliable, nonsignificant decline averaging −0.58 percent per year across the entire survey region **(C)**. In Minnesota, BBS data are too sparse to provide statistically robust trend projections.

Lesterhuis and Clay (2010) summarized their concern regarding the species' status: "While its breeding range has considerably expanded in recent decades, the species no longer breeds at a number of former sites, and the population has not shown a marked increase in size despite the expansion in range." Many factors may be negatively impacting populations, but habitat loss and degradation are considered the most significant threats, especially in the northern Great Plains.

B. The Wilson's Phalarope is attracted to shallow alkaline wetland basins with emergent vegetation, such as the Salt Lake WMA in western Minnesota. Photo by Lee A. Pfannmuller.

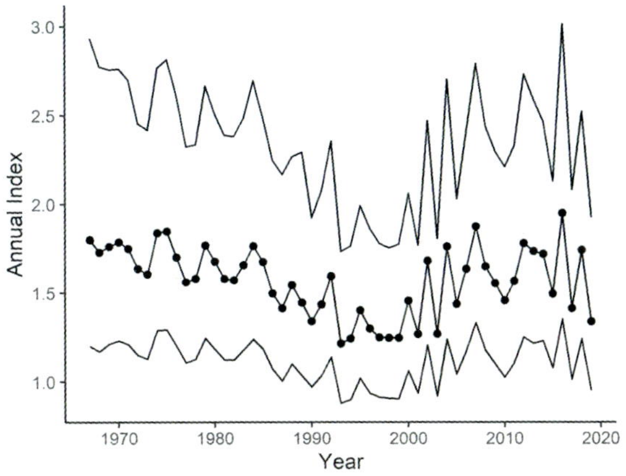

C. Surveywide breeding population trend and 95 percent credible interval (solid black lines) for the Wilson's Phalarope based on the 1967–2019 Breeding Bird Survey (Sauer et al. 2020).

CONSERVATION

In Minnesota, the Phalarope was upgraded from *Special Concern* status to *Threatened* in 1996, when MBS field surveys in northwestern Minnesota revealed the species was absent from many sites with suitable habitat (MNDNR 1995). Nationally, concern regarding its relatively small population size and threats to its breeding and wintering habitats led to its designation as a moderate conservation priority (PIF 2021).

Aggressive efforts to protect and restore Minnesota's wetlands and grasslands will benefit this rare shorebird. Shallow wetland basins in particular provide critical nesting habitat and should be a focus of such efforts. In northwestern Minnesota, impoundments designed to reduce spring flood levels can improve waterfowl habitat and incorporate measures that provide habitat for nesting and migrating shorebirds, including establishing nesting islands and beaches with sparse vegetation and gentle inclines (Dechant, Johnson, et al., "Wilson's Phalarope," [1999] 2002; MNDNR 2019m).

Unfortunately, warming temperatures may negatively impact the shallow wetlands that Phalaropes depend on during the breeding season and at major migratory stopovers. The species was deemed moderately vulnerable under a 2°C warming scenario; NAS (2019) projected a loss of 30 percent of the species' continental breeding range by 2050, as the Great Plains and Intermountain West become more arid and shallow seasonal wetlands disappear. Fortunately, the models predicted a 29 percent gain northward in central Canada and in northwestern Minnesota, making the graceful Phalarope one of the species that might be seen more often in the state as temperatures rise.

FRANKLIN'S GULL

Leucophaeus pipixcan

Photo by Karl Bardon.

Minnesota Seasonal Status: Regular breeding species and migrant. The Franklin's Gull was rare during the MNBBA.
Migration: Long-distance migrant that winters along the Pacific coast of South America.
Food: Feeds primarily on insects, both aquatic and terrestrial, and small fish. Often feeds in agricultural fields near nesting colonies.
Nest: Floating mat of emergent vegetation anchored to bulrushes or cattails; old muskrat houses are also used. A colonial nesting species.
Conservation Concern: Assigned a high PIF Continental Concern Score of 14/20; officially listed as *Special Concern* in Minnesota.

Small and elegant, the Franklin's Gull was a well-known and welcome sight among the early settlers who farmed Minnesota's rich prairie soils. Known as the "prairie pigeon" or "prairie dove," the Gulls arrived each spring as the ice was going out and the farmers were beginning their spring plowing. Job (1910) believed the birds were "one of the prettiest sights which the wide prairies can afford," describing how they would alight in the furrows immediately behind the plow and, "with graceful carriage, hurry about to pick up the worms and grubs which the plow has just unearthed."

The Franklin's Gull's black head, interrupted by arcs of white above and below the eyes, and its bright red legs and bill are diagnostic. A flush of salmon pink also appears on the breast early in the breeding season. The species may be confused with Bonaparte's Gull, a common migrant in Minnesota and an occasional summer visitor. Slightly smaller, the Bonaparte's Gull has a black bill and pink legs and lacks any white around the eyes. In flight, its upper wings are lighter than the darker, slate-gray mantle of the Franklin's Gull. Calling incessantly at breeding colonies, the Franklin's Gull's most common vocalization is a nasal-sounding *keeah*.

The breeding colonies of the Franklin's Gull, a species of North America's interior, are primarily restricted to the Canadian provinces of Alberta, Saskatchewan, and western Manitoba and the northern states of Montana, the Dakotas, and western Minnesota. Breeding densities are high wherever nesting colonies are located. The species is named in honor of Sir John Franklin, a British explorer "who perished while searching for the Northwest Passage" in 1847 (Sandrock and Prior 2014).

MINNESOTA BREEDING DISTRIBUTION

Roberts (1932) described the Franklin's Gull as an abundant summer resident, nesting "throughout western Minnesota in great colonies." Its fidelity to a given nest site, however, was quite low. "Unlike most birds that breed in colonies, they usually change nesting-sites from year to year; in successive years these may be miles apart, though still within the limits of the same big lake or slough. Occasionally, however, the birds arrive in force at the old locality and then, after a brief stay, all disappear to locate for that season at an entirely new lake." Roberts surmised that the abandonment was prompted by changes in water levels or food availability. Despite his comments on their wide distribution, the only documented nesting record was from Heron Lake in Jackson County, where at least 50,000 nests were reported in 1916 (Roberts 1932). Green and Janssen (1975) later reported that the site supported 100,000 nests in 1937.

In their updated account, Green and Janssen (1975) also noted confirmed nesting in five additional western counties in the early 1900s: Clay, Kandiyohi, Marshall, Todd, and Traverse. Since the mid-1960s, however, only two breeding colonies were active: Lake Osakis in Todd/Douglas Counties and Agassiz NWR in Marshall County. When Janssen (1987) identified colonies that had been active since 1970, Thief Lake WMA in Marshall County and Heron Lake in Jackson County were added to Lake Osakis and Agassiz NWR.

Given the bird's ephemeral nature, tracking the distribution and abundance of Minnesota's Franklin's Gull breeding population is difficult. In some years, "local populations may not breed at all" (Burger and Gochfeld 2020). Nevertheless, historical records suggest that nesting was either confirmed or attempted at approximately 14 sites in the state since the late 1800s. They stretched from Jackson County in the south, north to Marshall and Lake of the Woods Counties, and east to Crow Wing and Kandiyohi Counties (Pfannmuller 2014b). Most sites were used for just one to three years before they were abandoned. Only three wetland complexes have been used repeatedly in the past one hundred years: Agassiz NWR, Thief Lake WMA, and Heron Lake.

Although Heron Lake supported the largest known colony in the early 1900s, activity at the site has dramatically decreased. Intensive cultivation in the lake's small watershed means that spring and summer rains have an immediate impact on water levels, often drowning nests, eggs, and young. Conservation efforts in the watershed have improved, but the site is still used only intermittently. The most recent report was of 500 young in 1992 (Chaffee 1992).

Since nesting began at Agassiz NWR in the 1930s, reports show population numbers ranged from zero to at least 125,000 pairs in 1968 (USFWS 1968). The most recent published account reports 14,800 pairs in 2005 (USFWS 2005a). Then, for three consecutive years, from 2006 to 2008, the Gulls began nesting at

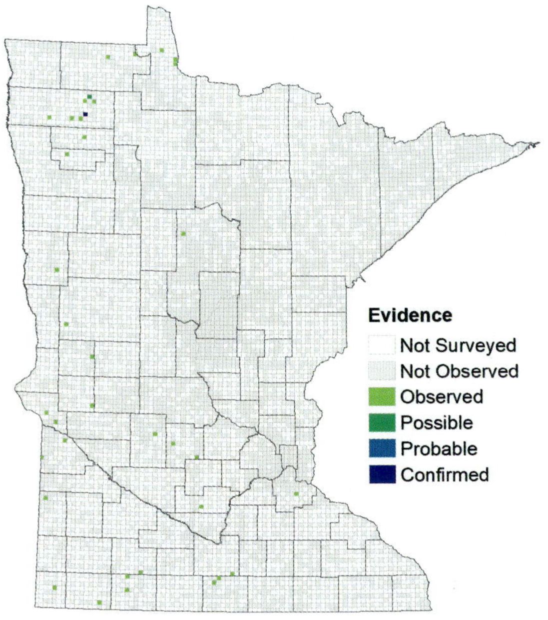

A. Breeding distribution of the Franklin's Gull in Minnesota based on the Breeding Bird Atlas (2009–13).

Agassiz in the spring but soon abandoned the refuge. High predation may have prompted the initial move in 2006 (Dokken 2008).

Nearby, at Thief Lake WMA, Janssen (1987) reported 5,000–10,000 nesting pairs in 1980; Burger and Gochfeld (2020) reported 2,000 pairs in 1994 and none in 2005–7. In 2008, once the Gulls abandoned Agassiz NWR, they relocated to Thief Lake. Unfortunately, heavy rains later prompted their abandonment of that site as well (Dokken 2008).

During the MNBBA, Agassiz NWR was the only site where breeding was confirmed. Participants reported 42 Franklin's Gull records from 37 surveyed atlas blocks and confirmed nesting in just one block **(A)**. A group of 60 adults and immatures was observed at Salt Lake in early July 2013, but the birds had probably dispersed from a nearby colony in South Dakota. Other observations were likely post-breeding birds, birds whose nests failed, or first-year birds. Biologists believe the species may not breed until two years of age (Burger and Gochfeld 2020).

The establishment of large, managed wetland complexes like Agassiz NWR has enabled the Franklin's Gull to recover from the Dust Bowl years, when numbers declined precipitously and entire colonies were lost and never reoccupied. Indeed, the species' expansion into many western states, especially in the Great Basin, is attributed to the creation of new state and federal wildlife refuges (Burger and Gochfeld 2020). But the days of the species' presumed abundance in western Minnesota have long since passed. Today, the continued presence of the Franklin's Gull in the state is restricted solely to the public lands that were so wisely set aside in Marshall County.

BREEDING HABITAT

Large, extensive wetland complexes, where stands of emergent vegetation are interspersed with open water, are the chosen habitat of this species **(B)**. Nests are usually placed at the interface of the emergent vegetation and the open water (Burger and Gochfeld 2020). Water levels are extremely important in determining whether a site is selected for nesting; levels that are too high or too low can initiate colony abandonment (Beyersbergen et al. 2004). Vorland (1984) noted that some of Minnesota's largest nesting colonies were reported in years following droughts or following managed drawdowns that served to rejuvenate emergent vegetation.

POPULATION ABUNDANCE

Establishing a population estimate for the Franklin's Gull is challenging. Most colonies are now located on public lands, so surveys by land managers are often the best source of information. Unfortunately, when resources are limited, waterbird surveys may not be conducted consistently. The most recent estimate places the North American population at approximately 830,000 breeding adults (PIF 2021). A statewide estimate is not available.

BBS data for the Franklin's Gull in Minnesota are unreliable but are statistically robust for the Prairie Potholes Bird Conservation Region, the core of the species' breeding range. For the entire survey period, from 1967 through 2019, these data show a nonsignificant population decline averaging −1.37 percent per year **(C)**. Despite the overall decline, in recent years numbers began a slow increase, likely in response to higher precipitation levels and improved wetland conditions. Indeed, unfavorable water levels are considered the primary threat to the Gull's population.

B. When water levels are suitable, this large wetland complex in Thief Lake WMA in Marshall County supports one of the last remaining nesting colonies of the Franklin's Gull in Minnesota. Photo by Lee A. Pfannmuller.

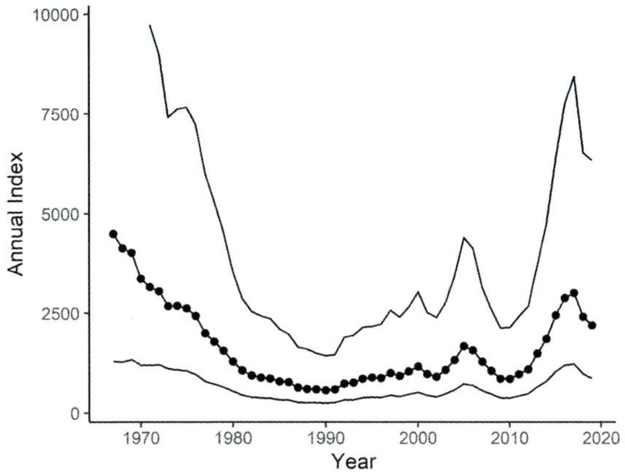

C. Breeding population trend and 95 percent credible interval (solid black lines) for the Franklin's Gull in the Prairie Potholes Bird Conservation Region based on the 1967–2019 Breeding Bird Survey (Sauer et al. 2020).

CONSERVATION

Like all colonial species that nest in large numbers and in localized areas, the Franklin's Gull is considered a vulnerable species. Coupled with its small population size and threats to its restricted breeding and wintering habitat, it is designated a high conservation priority (PIF 2021) and is officially listed as *Special Concern* in Minnesota (Minnesota Administrative Rules: Chapter 6134). Periodically monitoring primary nesting sites and identifying newly established colonies should be a priority and are critical to assessing the species' overall status.

Management recommendations for the species focus on maintaining stable water levels during the nesting season and reducing disturbance. Nevertheless, even these efforts may not be sufficient to address severe weather conditions such as drought and sudden, high rainfall events. The significance of Thief Lake WMA as an alternative nesting site in close proximity to Agassiz NWR illustrates the importance of maintaining large wetland complexes throughout the landscape.

Unfortunately, warming temperatures also pose a risk. Indeed, an analysis by NAS (2019) classified the Franklin's Gull as highly vulnerable to warming temperatures that ranged from an increase of 1.5° to 3.0°C. Although the models predicted a small northern expansion of its breeding range, it was not nearly sufficient to compensate for major losses farther south. Even under the lowest warming scenario of 1.5°C, the species is projected to disappear from 57 percent of its entire breeding range, including Minnesota. The future of this inland "seagull" does indeed seem precarious.

RING-BILLED GULL

Larus delawarensis

Minnesota Seasonal Status: Regular breeding resident and migrant; occasionally observed in winter. The Ring-billed Gull was uncommon during the MNBBA.
Migration: Short- to medium-distance migrant that winters along the United States coastline and in the lower Great Lakes, southern states, northern Mexico, and Caribbean.
Food: Opportunistic omnivore that employs a diversity of foraging strategies to secure insects, earthworms, fish, rodents, and refuse.
Nest: Ground nest placed in low-lying areas with sparse vegetation; a colonial nester that often nests with other species; colonies range in size from just a few pairs to thousands.
Conservation Concern: Assigned a low PIF Continental Concern Score of 6/20.

The most common gull observed in Minnesota is the Ring-billed Gull; its shrill and raucous calls and aerobatic antics are familiar to many. Although nesting sites are usually located near water, the gregarious and adaptable birds are frequently seen wherever there are ample feeding opportunities provided by people and/or refuse, including lakeshores, outdoor eateries, and garbage dumps. Nonbreeding birds may be seen wandering throughout Minnesota during the summer.

The species is aptly named; the adult's most diagnostic feature is the black ring near the tip of its yellow bill. One of three gull species that nest in Minnesota, the Ring-billed Gull is easily distinguished from the adult Franklin's Gull by the absence of a black head. But, for the novice, distinguishing it from a Herring Gull may be challenging. Although the latter species lacks a black ring-tipped bill when it is an adult, immature Herring Gulls and Ring-billed Gulls both have black bill tips for several years. Luckily, the Herring Gull is nearly 40 percent larger and, by its first full summer, can usually be identified by its pale pink legs compared to the Ring-billed Gull's yellow legs (P. J. Grant 1982).

Distributed across southern Canada, the Ring-billed Gull's breeding range also encompasses the northern states, from the mid–Pacific coast through the Great Lakes and east to upstate New York and Maine. Breeding densities are high wherever large breeding colonies are located.

Photo by Sparky Stensaas.

MINNESOTA BREEDING DISTRIBUTION

One hundred years ago, the Ring-billed Gull was only a common spring and fall migrant in the state (Roberts 1932). Nonbreeding adults and immature birds were occasionally observed during the summer, but there were no known breeding records. Elsewhere, the species was a breeding resident from British Columbia, east through the Great Lakes, and farther east to the north Atlantic coast. Nesting was documented in southeastern Wisconsin in 1860 and on Lake Michigan in 1879, 1881, and 1882 (Cutright et al. 2006). West and north of Minnesota, Bent (1921) observed two large colonies on Stump Lake in North Dakota in 1901 and several colonies on Manitoba's Lake Winnipegosis in 1913. These latter sites were likely the source of the nonbreeding Ring-billed Gulls observed in Minnesota.

Despite its broad distribution, the species declined precipitously beginning in the mid-1800s in response to widespread persecution, including the collection of eggs for food and the slaughter of adults for the millinery trade. By the early 1900s, not only had populations plummeted, but the species' breeding range was largely restricted to undisturbed sites in the "inland prairie regions" of Canada and the United States (Pollet et al. 2020). Passage of the 1918 Migratory Bird Treaty Act eventually ended the carnage and launched an unprecedented growth in population numbers.

The first undisputed account of Ring-billed Gulls nesting in Minnesota occurred in 1936 when a pair with two downy young was found on Lake Superior's rocky shoreline in Cook County (M. D. Thompson 1936b). Then, following a 1957 nesting attempt in the Duluth-Superior Harbor (Cohen 1958), reports began to accumulate from several sites across the state. In 1960, an egg was found on an island in Leech Lake (Dickerman and Lefebvre 1961), and in 1963 at least 100 nesting pairs were found on Hennepin Island on Mille Lacs Lake (Hiemenz 1964). Two large colonies were found on islands on the US side of Lake of the Woods in 1981 (Hirsch 1981), and three nests were found on Marsh Lake in Big Stone County in 1984 (Janssen 1987). Although nesting pairs were eventually found at a few other locations, the nesting attempts and colonies documented in the Duluth-Superior Harbor, Lake of the Woods, Mille Lacs Lake, Leech Lake, and Marsh Lake would grow to support the majority of the state's breeding population. By 1998, nesting had been confirmed in seven counties since 1970: Big Stone, Cass, Faribault, Hennepin, Lake of the Woods, Mille Lacs, and St. Louis (Hertzel and Janssen 1998).

During the MNBBA, 902 Ring-billed Gull records were reported in 14.0 percent (663/4,741) of the surveyed atlas blocks **(A)**. Most records were of nonbreeding individuals. Breeding was confirmed in 18 blocks spanning 12 counties across the northern two-thirds of the state. In addition to the established colonies noted above, a large colony (1,000–2,000 Gulls) was reported on Pelican Lake in Crow Wing County. Small numbers of nesting pairs were also found at a few other sites. To the seven nesting counties identified by Hertzel and Janssen (1998), the atlas added records in Carlton, Crow Wing, Douglas, Grant, and Hubbard Counties. The colonies on Marsh Lake and the Minnesota River were in blocks that straddled three counties: Big Stone, Chippewa, and Lac qui

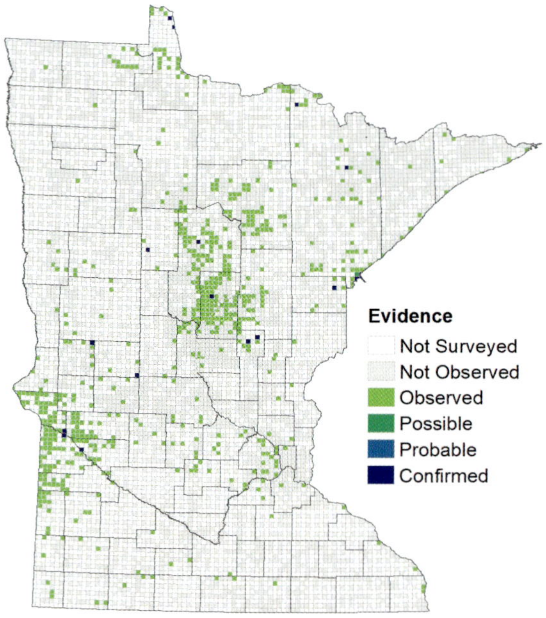

A. Breeding distribution of the Ring-billed Gull in Minnesota based on the Breeding Bird Atlas (2009–13).

B. Devoid of any vegetation, this level sand and gravel beach is typical of many Ring-billed Gull nesting colonies in the Great Lakes.
Photo by Gerald J. Niemi.

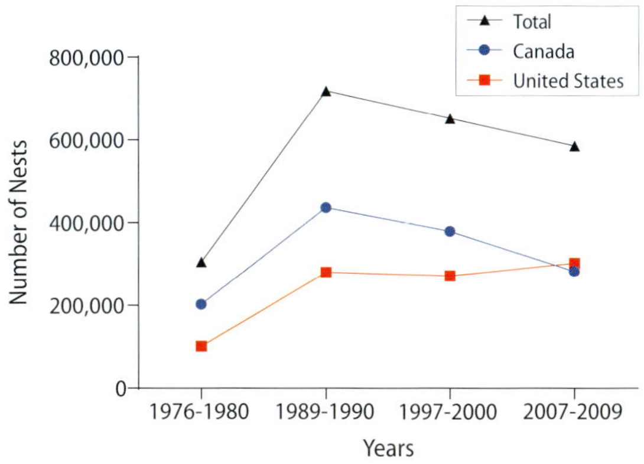

D. Population trend of Ring-billed Gulls on the North American Great Lakes (data from Morris et al. 2011, table 1).

Parle. Following the atlas, confirmed nesting reports were also reported in Morrison, Red Lake, and Scott Counties (Janssen 2019; MOU 2021b).

The Ring-billed Gull is now firmly established as a regular breeding resident in Minnesota. It not only has recovered from the persecution it experienced in the 1800s and early 1900s but also has reoccupied most of its former North American breeding range and expanded farther in the western United States and Canada, in the Maritime Province of Newfoundland and Labrador, and in the Great Lakes (Pollet et al. 2020).

BREEDING HABITAT

Although Ring-billed Gulls forage in a variety of open habitats, their nesting requirements are quite narrow. Colonies are usually located on low-lying, level areas with sparse vegetation **(B)**. Nesting substrates include sandbars, small pebbles, exposed rock, driftwood, and even concrete. Most sites are islands, either natural or man-made, the latter including dredge-spoil islands and rooftops. Peninsulas on large lakes may also be used. Protection from mammalian predators, however, is paramount. Nests are often placed near or beneath small plants that provide some protection from aerial predators (Pollet et al. 2020).

POPULATION ABUNDANCE

The North American Ring-billed Gull population is estimated at 3.7 million individuals (PIF 2021). Although there is no statewide survey that provides a one-time estimate of Minnesota's breeding population, Ring-billed Gull nest counts collected during a 2015 survey of American White Pelican and Double-crested Cormorant colonies (Hamilton and Cuthbert 2016) and a 2016 survey of the Duluth-Superior Harbor (Bracey, Ponshok, and Strand, pers. comm.) document at least 28,408 nesting pairs in the state. Nearly 46 percent of the pairs occupied just one site: Interstate Island in the Duluth-Superior Harbor.

The BBS is not the best tool for monitoring colonial waterbirds, but the surveywide population increase of the Ring-billed Gull is well-documented, with a statistically reliable, nonsignificant average increase of 0.76 percent per year from 1966 to 2019 **(C)**. Minnesota BBS data also suggest a long-term increase. Perhaps the most reliable data are collected from Great Lakes nesting colonies, which document a 92 percent increase from the late 1970s to the most recent survey completed in 2009, with the largest increase occurring between 1976 and 1990 **(D)**.

In the Great Lakes, the introduction of new food sources (rainbow smelt in 1922 and alewife in the 1930s) and cycles of low water levels that exposed more nesting areas contributed to the increase (Morris et al. 2011; Cutright et al. 2006). Some of the increase in the Duluth-Superior Harbor is attributed to cycles of high water levels that swamped low-lying colonies in the eastern Great Lakes and prompted the birds to explore sites farther west (Pollet et al. 2020). The construction of reservoirs in the Great Plains also contributed to the species' success, as did the Gull's omnivorous diet and adaptability to human-modified landscapes, including landfills and dredge-spoil islands. Numbers have increased so rapidly in some areas that efforts to control or discourage nesting populations may be permitted for airline safety or to limit agricultural damage (Pollet et al. 2020; Morris et al. 2011).

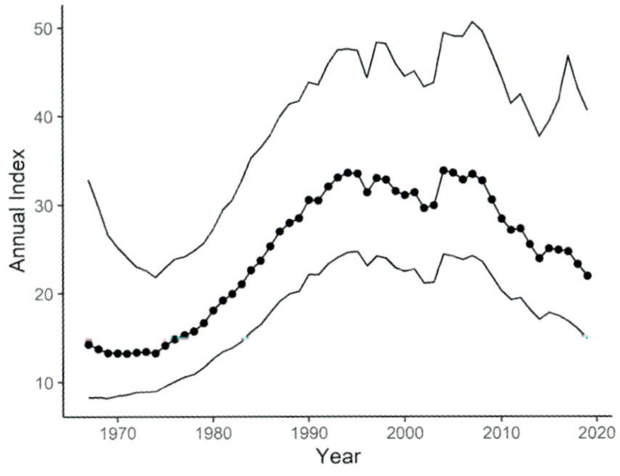

C. Surveywide breeding population trend and 95 percent credible interval (solid black lines) for the Ring-billed Gull based on the 1967–2019 Breeding Bird Survey (Sauer et al. 2020).

CONSERVATION

With an expanding and widespread population, the Ring-billed Gull is a low conservation priority (PIF 2021). Indeed, the species' population growth has had serious consequences for one of Minnesota's rarest colonial waterbirds, the Common Tern. The two species frequently utilize the same undisturbed and sparsely vegetated islands for nesting. Ring-billed Gulls often usurp quality habitat before the later-arriving Terns return each spring, prompting researchers to implement creative management techniques to discourage nesting Gulls in Tern colonies (e.g., Maxson, Mortensen, et al. 1996; Mortensen and Ringle 2007; Penning and Cuthbert 1993). To date, the efforts have been largely successful but require due diligence each summer.

Even warming temperatures are unlikely to be a major threat to the Ring-billed Gull. Its North American breeding range is projected to remain stable under a 2°C warming scenario, with a 23 percent loss by 2050 compensated by a 61 percent gain (NAS 2019). Losses are especially high throughout the eastern Great Lakes states, while gains are predicted in the northern Great Lakes and across Canada. Although climate conditions are projected to become unsuitable for Ring-billed Gulls in southwestern Minnesota, the large colonies in the northern counties may expand even further.

HERRING GULL

Larus argentatus

Minnesota Seasonal Status: Regular breeding resident and migrant; regular in the winter. The Herring Gull was uncommon during the MNBBA.
Migration: Partial migrant that winters along coastlines from Alaska and northeastern Canada south through Mexico and the Caribbean; also found in the lower Great Lakes and near open water in the southern United States.
Food: Consumes a variety of animals, including invertebrates, fish, mammals, and young birds, as well as trash and carrion.
Nest: Shallow nest placed on the ground and occasionally on cliff ledges; frequently nests in large colonies, often with other colonial nesting species, but may be found as isolated pairs or in small colonies numbering fewer than 10 pairs.
Conservation Concern: Assigned a moderate PIF Continental Concern Score of 11/20.

Photo by Karl Bardon.

The Herring Gull is among the most abundant and widely distributed gulls in North America. Although it occurs throughout Canada and the Great Lakes, its close association with densely populated areas along the north Atlantic coast means it is also one of the continent's best known "seagulls." The species' common name refers to its consumption of Atlantic herring, suggesting this opportunistic predator and scavenger has a more restricted diet than it actually does. Its Latin name, *Larus argentatus,* means a seabird ornamented in silver, a reference to its silver-gray wings.

The largest of three gulls that regularly breed in Minnesota, the Herring Gull may be mistaken for the smaller Ring-billed Gull. Adult Herring Gulls, however, lack the smaller species' black ring near the tip of the bill. Instead, they have a red spot near the tip of the bill's lower mandible and pink legs compared to the Ring-billed Gull's yellow legs. Unfortunately, both species display an array of tan-streaked plumages for several years until they attain the full adult plumage, which can make identification quite challenging. The Herring Gull's large vocal repertoire includes a long, drawn-out *mew* call and a higher-pitched *keow* call (Weseloh et al. 2020).

The Herring Gull is a circumpolar species; in North America, breeding densities are highest along the Atlantic coast, from Newfoundland south to the Delmarva Peninsula, with local concentrations at several inland sites, including the Great Lakes.

MINNESOTA BREEDING DISTRIBUTION

Roberts (1932) described the Herring Gull's breeding range as restricted to two large bodies of water: Lake of the Woods and the North Shore of Lake Superior. He speculated that it might also nest "elsewhere about the large northern lakes." Yet, at the time, nesting records were limited to three islands in Lake of the Woods (Gull Rock, Big Oak, and Fourblock) and to two islands along the North Shore. The latter included an island near Two Harbors, presumably Knife Island, and a site known as Two Islands in Taconite Harbor, now known as Bear Island and Gull Island (J. Green, pers. comm.).

Then, in 1945, a third North Shore colony was found near Beaver Bay (E. Jones 1945), and in 1966 birds were found nesting on Mille Lacs Lake (Huber 1966b). The first nesting in the Duluth-Superior Harbor was in 1973, when a single nest was found within a nesting colony of Ring-billed Gulls (Harris and Matteson 1975). By 1975, Green and Janssen described the species' breeding distribution as the rocky islands and shoreline cliffs that bordered the North Shore of Lake Superior, the border lakes from Lake of the Woods east through Cook County, Lake Vermilion, and occasionally Mille Lacs Lake. Large lakes often supported colonies numbering 100 or more pairs; small lakes usually supported only one to five pairs.

The largest colonies occurred along the North Shore. This was confirmed when Goodermote (1980, 1984, 1989) conducted a series of surveys from Knife Island to the Canadian border. From 1978 to 1979 he identified 37 rookeries supporting 6,334 nests. The estimate rose to 7,748 in 1984 and 9,037 in 1989. The largest colonies (more than 1,000 nests) were located on the Susie Islands, at Taconite Harbor, and at Silver Bay.

Farther west, Hirsch (1981) documented 332 nesting pairs on six islands in the Minnesota portion of Lake of the Woods, and Mortensen and Ringle (2007) found two nesting pairs on a small island in Leech Lake in 1983. By 1987, Janssen identified seven counties with confirmed nesting since 1970: Cass, Cook, Koochiching, Lake, Lake of the Woods, Mille Lacs, and St. Louis. The number remained unchanged when Hertzel and Janssen (1998) provided an update several years later. Meanwhile, the MBS identified 66 breeding-season locations confined largely to the Arrowhead region.

The 346 Herring Gull detections reported by MNBBA participants documented a similar distri-

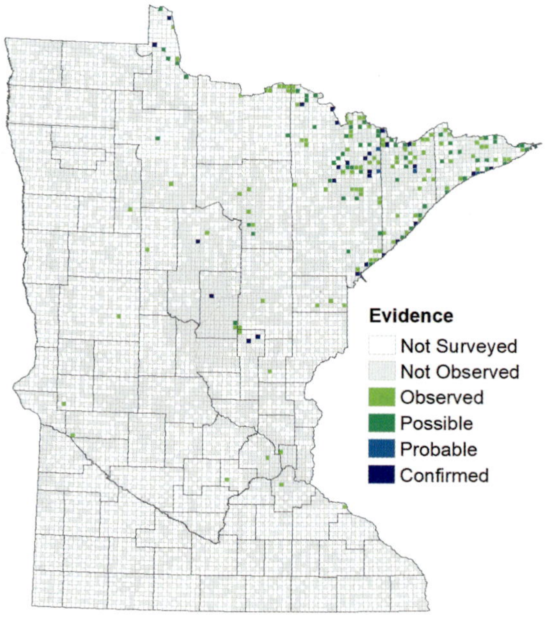

A. Breeding distribution of the Herring Gull in Minnesota based on the Breeding Bird Atlas (2009–13).

bution **(A)**. Found in 4.5 percent (212/4,741) of the surveyed atlas blocks, breeding was confirmed in 31 blocks and seven counties. Although they were not reported in Koochiching County, one pair was found nesting on Pelican Lake in Crow Wing County in 2010. There were also nesting reports of this adaptable species from rooftops in the city of Duluth: at the Miller Hill Mall in 2009 and 2010 (one pair); and on the Essentia Building in downtown Duluth in 2012 (several pairs). Observations of nonbreeding birds elsewhere in the state were uncommon.

Overall, the core of the species' Minnesota breeding range remained Lake Superior's North Shore and the border lakes, from Lake of the Woods east through Cook County, a distribution similar to that described by Roberts (1932). Elsewhere, the only major change to the Gull's breeding range was its southern extension along the Atlantic coast. Originally confined to Maine and Nova Scotia, it expanded south to the Carolinas between 1910 and the 1960s (Weseloh et al. 2020).

BREEDING HABITAT

In the Great Lakes, nesting Herring Gulls are associated with treeless natural islands, breakwaters, dredge-spoil islands, and rocky islets **(B)**. Well-drained sites are preferred. The most important requirement, however, is protection from predators. Colonies are most successful when vegetative cover is sparse, providing some cover from weather while still allowing visibility in all directions. Foraging may occur up to 100 km from the colony. Most feeding occurs in open water, but the species also feeds at inland sites, including landfills and agricultural facilities (Weseloh et al. 2020; Wires, Lewis, et al. 2010).

POPULATION ABUNDANCE

Weseloh and his colleagues (2020) provided a North American population estimate of 200,000 breeding pairs, with 120,000 pairs along the Atlantic coast, 70,000 pairs in the Great Lakes, and 7,500 pairs on a few well-known lakes elsewhere in Canada and Alaska. The authors cautioned, however, that Canada's boreal forest has "thousands of potentially suitable [nesting] lakes" that likely support "tens of thousands of [nesting] pairs." In Minnesota, Herring Gull nest counts collected during surveys of Minnesota's White Pelican and Cormorant colonies (Hamilton and Cuthbert 2016), during the Great Lakes colonial waterbird survey (Cuthbert and Wires 2013), and during surveys in the Duluth-Superior Harbor (Bracey, pers. comm.) generate a conservative statewide estimate of 4,390 nesting pairs. The only data missing are estimates from the inland lakes in northeastern Minnesota. This compares to 1989, when more than 9,000 pairs were found just along Minnesota's North Shore (Goodermote 1989).

B. Bare rocks around the perimeter of this small island along the shore of Lake Superior provide an ideal site for a Herring Gull nesting colony. Photo by Gerald J. Niemi.

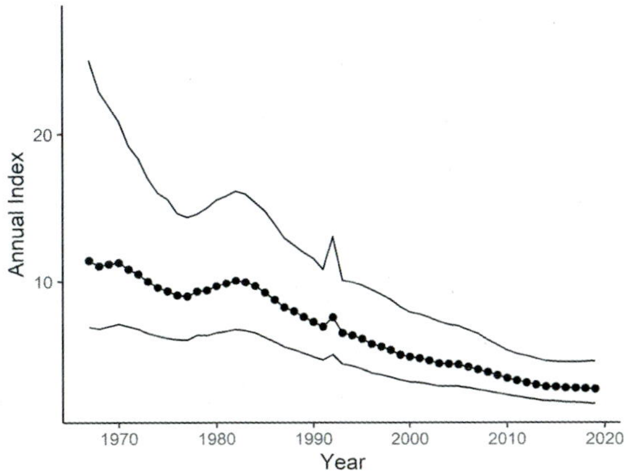

C. Surveywide breeding population trend and 95 percent credible interval (solid black lines) for the Herring Gull based on the 1967–2019 Breeding Bird Survey (Sauer et al. 2020).

Although it is not well suited to monitoring colonial waterbirds, the BBS documents a reliable and significant population decline across the survey region, averaging −2.68 percent per year from 1966 to 2019 **(C)**. BBS data also suggest a declining population in Minnesota (Sauer et al. 2020). Colonial waterbird surveys of nesting colonies in the United States portion of the Great Lakes, however, demonstrate an increasing population trend from 1977 to 2010 **(D)**; recent increases are due largely to increases observed on Lake Michigan (Cuthbert and Wires 2013).

Herring Gull populations have recovered from earlier times when the birds were shot for their feathers and their eggs were collected for food. Protection by the 1918 Migratory Bird Treaty Act helped spur their recovery, as did the availability of supplemental food sources, including waste at landfills and at coastal fishing ports. An exploding alewife population in the Great Lakes from the 1960s to the 1980s also contributed (Scharf and Shugart 1998). Some of the recent declines are attributed to a decrease in commercial fishing, an increase in predatory species, such as Great Black-backed Gulls and Bald Eagles, and improved waste management practices (Weseloh et al. 2020).

CONSERVATION

Despite the bird's widespread distribution, the Herring Gull's significant population decline in North America is largely responsible for its designation as a moderate conservation priority (PIF 2021). Nevertheless, at some sites in the Great Lakes where numbers are increasing, Gull numbers are controlled owing to property damage. Such efforts, however, are minimal and have limited impact (Cuthbert and Wires 2013).

Contaminants and human disturbance have been considered the species' primary threats (Wires, Lewis, et al. 2010). Organochlorines in particular cause mortality in embryos and chicks, retard chick growth, and cause deformities. Contaminant levels, however, have declined significantly since they were first measured in the mid-1900s (Weseloh et al. 2020).

Although rising sea levels may impact coastal habitats used by Herring Gulls, climate change models predict that the species has a low vulnerability under warming scenarios that ranged from 1.5° to 3.0°C (NAS 2019). Breeding range losses were compensated by gains across northern Canada. Unfortunately, this adaptable species is predicted to decline significantly in the Great Lakes region, including in Minnesota. The Gull's use of human refuse and human-modified habitats may help buffer some of the predicted loss. Nevertheless, given these projections, continued monitoring of the species' Great Lakes population will provide critical data in the coming years.

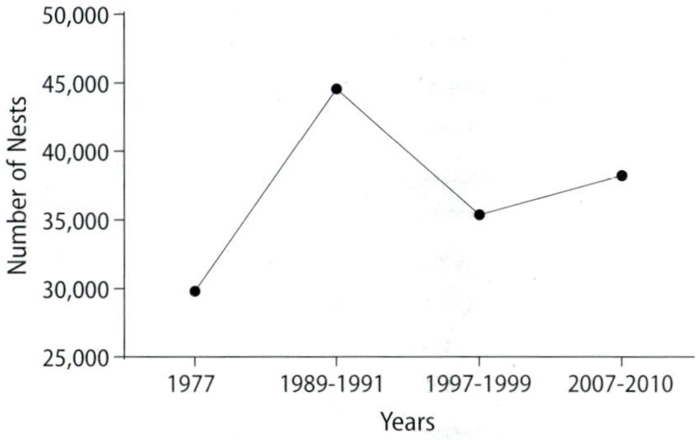

D. Population trend of Herring Gulls on the US portion of the Great Lakes from 1977 to 2007–10 (from Cuthbert and Wires 2013, table 10b).

CASPIAN TERN

Hydroprogne caspia

Minnesota Seasonal Status: Regular migrant, summer visitor, and very rare breeding species since 1969. The Caspian Tern was rare during the MNBBA.
Migration: Northern populations are short- to long-distance migrants that winter along the southern coasts of the United States, in coastal habitats in the Caribbean, and as far south as northern South America.
Food: Feeds primarily on fish taken by aerial dives.
Nest: Shallow ground depression often lined with vegetation, pebbles, or debris. May nest singly, in groups of just a few pairs, or in much larger colonies.
Conservation Concern: Assigned a moderate PIF Continental Concern Score of 10/20.

Photo by Michael Furtman.

One of Minnesota's rarest breeding species, the Caspian Tern has a large size, heavy white body, and short tail that may fool you into thinking it is a gull. But its remarkably long and pointed wings, its black cap, and its large, crimson-colored bill betray its true identity as Minnesota's rarest regularly occurring tern. Nearly twice the size of the Common Tern and Forster's Tern, the Caspian Tern lacks the long, forked tail of these two similar species, sporting instead a shorter tail with just a small notch in the center. If its size doesn't give it away, its raucous calls are also diagnostic. A raspy, low-pitched *kraa* or *rau* are common vocalizations.

In awe of the bird, Bent (1921) wrote that it was the "king of all the terns," deserving instead to be called the "Imperial Tern." Roberts (1932) considered it "a magnificent bird, handsome in form and coloring, and graceful and dashing in action." Chances are, when you spot a Caspian Tern you will see it flying overhead with a heavy, strong wingbeat. Its scientific name, *Hydroprogne caspia*, translates to "water swallow of the Caspian Sea," a reference to the bird's swallow-like wings and tail and the location where the type specimen was collected.

Global in distribution, the Caspian Tern breeds in widely separated locations in North America—along the Gulf coast, the north Atlantic coast, and the Pacific coast. Interior populations are scattered in the western states, the Great Lakes, and central Canada.

MINNESOTA BREEDING DISTRIBUTION

Roberts (1932) described the Caspian Tern as a regular spring and fall migrant in the late 1800s and early 1900s, usually passing through in small numbers and frequently spotted near many of Minnesota's largest lakes and rivers. There were occasional summer reports but no documented nesting records. Most summer records were from northern lakes, but the birds were occasionally seen at Lake Minnetonka in early June, and one was taken on June 10, 1920, in Le Sueur County. Roberts surmised that at least a few birds might nest somewhere in the state.

It was 1969, however, before the first Caspian Tern nesting attempt was documented on Leech Lake (Warner and Beimborn 1969). Two nesting pairs were found on Gull Island, a tiny site less than 0.4 ha in size, next to a large nesting colony of Common Terns. The Leech Lake nesting may have represented a southward expansion of the Manitoba breeding population, the nearest known colonies of which were located on Lake Winnipeg and Lake Winnipegosis, nearly 650 km north of Leech Lake (Wires and Cuthbert 2000). Warner and Beimborn (1969) originally reported that the nearest colonies were located in Lake Superior (Gravel Island) and Lake Michigan (Beaver Island). Recent accounts of the Tern's historical status in Wisconsin, however, do not mention these sites as active colonies in the 1960s (Cutright et al. 2006). Regardless of their origin, North American Caspian Terns began expanding throughout their breeding range in the 1960s, moving north along the Pacific coast, south along the Gulf coast, and south from central Canada's large freshwater lakes (Cuthbert and Wires 2020).

Unfortunately, Minnesota's first documented nesting on Leech Lake was not successful and stood as the solitary record for thirty-five years. Then, in 2004, one nest with three eggs was found on Hennepin Island in Mille Lacs Lake, within another well-known Common Tern colony (McDowell 2004). This nesting attempt was also unsuccessful.

In the interim, the Leech Lake Band of Ojibwe began closely monitoring waterbird nesting efforts on Leech Lake and actively managing the lake's islands to provide suitable nesting habitat (Mortensen and Ringle 2007). Efforts focused on the Common Tern, but staff closely tracked the presence and nesting efforts of other colonial waterbirds, including Caspian Terns.

Their records document the presence of nonbreeding, adult Caspian Terns on Leech Lake every year from 1990 through 2006, except in 1992. Numbers usually ranged from 10 to 100 adults but, in 1990, a record 1,200 birds were observed (Mortensen and Ringle 2007). The birds were believed to be late migrants, immatures, or adults that had unsuccessfully nested elsewhere. Then, in 2007, 13 pairs nested successfully on Little Pelican Island, just north of Gull Island, fledging a total of 11 chicks. By 2010, the colony had grown to 43 nests; in 2015, 130 nests were documented (Hamilton and Cuthbert 2016); and, in 2020, 358 nests were reported (S. Mortensen, pers. comm.).

Several management actions implemented by the Leech Lake staff encouraged the Tern's nesting efforts. Most important was the gradual clearing of woody vegetation on Little Pelican Island to create about one hectare of open beach habitat that was very attractive to both Common and Caspian Terns (Mortensen 2009). A buffer zone around the island was also established during the open-water season to reduce public disturbance, and an island stabi-

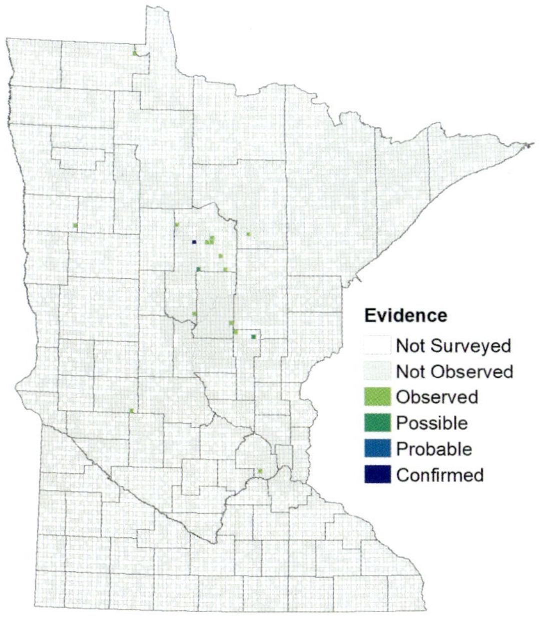

A. Breeding distribution of the Caspian Tern in Minnesota based on the Breeding Bird Atlas (2009–13).

lization project was initiated in 2000 to mitigate shoreline erosion (Mortensen and Ringle 2007).

During the MNBBA, Little Pelican Island remained the state's only active Caspian Tern nesting colony. Eight of the 17 atlas blocks where the species was observed were in the vicinity of the island; three more were located around the perimeter of Mille Lacs Lake where the species has nested in the past; and one was located on the shoreline of Lake of the Woods, where the birds were found nesting in 2017 (see below). Most observations were of single birds; in a few cases, two to three birds were reported **(A)**.

Since the MNBBA was completed, a second Minnesota nesting site was discovered on a remote island near the Northwest Angle in Lake of the Woods. When it was first visited on June 25, 2017, there were a total of 35 nests with eggs; on July 3, 60 nests and 98 adults were counted (K. Bardon and C. Herwig, pers. comm. 2017). Like many islands near the Angle, the site has supported nesting colonies of Common Terns, Ring-billed Gulls, American White Pelicans, and Double-crested Cormorants in past years.

BREEDING HABITAT

Caspian Terns are found in many aquatic habitats, ranging from coastal estuaries to barrier islands to alkaline and freshwater lakes. In the Great Plains, they are found nesting on islands in large freshwater or alkaline lakes. In the Great Lakes, they nest on remote, natural islands and, recently, on man-made habitats such as dredge spoil islands (Wires and Cuthbert 2000; Shuford and Craig 2002). All sites share three characteristics: they are remote, affording little access to ground predators or humans; they are close to an abundant fish resource; and they are sparsely vegetated **(B)**.

POPULATION ABUNDANCE

In the absence of any systematic monitoring program that tracks North American Caspian Tern breeding populations, Wires and Cuthbert (2000) contacted field biologists and natural resource managers to assess the Tern's population size and distribution. Their work generated a North American population estimate of 32,000 to 34,000 breeding pairs from the late 1980s to 1998 **(C)**. Approximately 21,000 pairs nested in the United States. This represented a significant increase above the 9,454 pairs estimated from the mid-1970s to the early 1980s (Spendelow and Patton 1988). Ten years later, PIF (2021) provided an updated estimate of 78,000 birds in North America and a global estimate of 530,000.

Numerous factors likely contributed to the recent population increase. Declines in predatory fish populations in the Great Lakes have resulted in a greater abundance of forage fish, providing a plentiful food supply for waterbirds like the Caspian Tern. The remote location of many nesting sites also affords protection, and the species' acceptance of artificial nesting sites, such as dredge spoil islands, has provided new nesting opportunities (Cuthbert and Wires 2020; Wires and Cuthbert 2000).

CONSERVATION

Like so many waterbirds, Caspian Tern populations suffered for years as eggs and adults were collected for food and feathers; their colonial nesting habits made them especially vulnerable. Despite recent population increases in North America, this colonial nesting species, with its small population, warrants continual vigilance. Because it is dependent on coastal environments throughout much of its range, rising water levels will certainly pose a threat in coming years. Under a 2°C warming scenario, however, small losses in the Caspian Tern's North American breeding range are predicted to be overshadowed by significant gains, especially throughout central Canada (NAS 2019). Small gains are even predicted in northern Minnesota.

In the interim, intensive efforts to protect known colonies and provide additional nesting sites are critically important. The efforts of the Leech Lake Band of Ojibwe to provide and maintain suitable habitat on Little Pelican Island have been crucial for Minnesota's Common Tern population. These efforts have provided the added benefit of attracting the most impressive tern of all, the Caspian Tern.

C. Approximate population sizes of North America's five Caspian Tern populations.

Population	Population Estimate	Years of Estimate
Great Lakes	6,416	1997–1998
Atlantic coast	142–144	1995–1997
Gulf coast	2,303	1996–1997
Central Canada	8,780–9,980	1989–1992
Pacific coast/Western region	14,846–14,982	1991–1998
Total	32,487–33,825	1989–1998

Source: Summarized from Wires and Cuthbert 2000, table 1.

B. An aerial view of Little Pelican Island on Leech Lake highlights the importance of sparsely vegetated, remote islands for the nesting Caspian Tern. Photo by Steve Mortensen.

BLACK TERN

Chlidonias niger

Minnesota Seasonal Status: Regular breeding resident and migrant. The Black Tern was uncommon during the MNBBA.
Migration: Medium- to long-distance migrant that winters along the coasts of Central America and northern South America.
Food: An aerial forager that either snatches flying insects in mid-air or skims the water surface, dipping its bill down to feed on aquatic invertebrates and fish.
Nest: Shallow platform of aquatic vegetation, usually built on a mat of floating vegetation or atop muskrat houses.
Conservation Concern: Assigned a moderate PIF Continental Concern Score of 12/20.

Years ago, there was a time when every prairie slough was teeming with ducks, rails, grebes, and gulls. Invariably, the Black Tern was among those species encountered. Not only were its black body and sooty gray wings unmistakable, it declared its presence with a persistent, metallic-sounding cry as it darted overhead. Bent (1921) described the bird as a "restless waif of the air, flitting about hither and thither with a wayward, desultory flight, light and buoyant as a butterfly." While feeding, its flight was likened to that of a flycatcher as it sallied in mid-air to nab flying insects, and to that of a swallow as it glided smoothly just above the water to capture emerging insects. Indeed, the scientific name, *Chlidonias,* means "like a swallow."

Even today, it is rare to come upon a prairie wetland in the summer and not immediately discern the presence or absence of Black Terns. Like other gulls and terns, Minnesota's smallest member of the Laridae family aggressively defends its nesting colony. As Roberts (1932) recounted, if the birds are present, the intruder is immediately harangued as "the whole colony comes screaming about," repeating their high-pitched *kreea* alarm call as they "take turns in dashing at the head of the intruder in furious protest against the unwelcome intrusion."

Global in distribution, the Black Tern has a breeding range in North America that stretches from British Columbia, south to northern California, and east to Ontario, southwestern Quebec, and the Great Lakes region. Breeding densities are highest in the northern Great Plains.

Photo by Gerald Hoekstra.

MINNESOTA BREEDING DISTRIBUTION

In the early 1900s, Roberts (1932) not only described the Black Tern as "an abundant summer resident" but also wrote that "no other water-bird is more generally and abundantly distributed in Minnesota." It could be found, he wrote, in suitable wetlands "in all parts of the state." Given the rarity of the species' marshy habitat in northeastern and north-central Minnesota, one has to question Roberts's broad assertion of its statewide abundance. Confirmed nesting records were available from only a handful of central and western counties: Hennepin (Minneapolis), Kittson, Morrison, and Sherburne Counties and Leech Lake. Inferred nesting was reported in Marshall County.

Green and Janssen (1975) later described the Tern's distribution as including all but the extreme northeastern region of the state, with no confirmed nesting reports from Lake or Cook Counties. Janssen (1987) added that the Tern was also absent from large portions of St. Louis and Koochiching Counties. He wrote that it was rare in southeastern Minnesota, in counties just west of the Mississippi River floodplain, and had declined in abundance throughout the river valley. He delineated 30 counties where nesting had been confirmed since 1970. Hertzel and Janssen (1998) later added an additional nine counties to the list.

Concern regarding the species' status prompted the MNDNR to conduct a volunteer, statewide survey from 1990 through 1995. Observed in 71 counties and confirmed breeding in 35 counties, the Tern was least abundant in the agricultural regions of western and southern Minnesota and from Koochiching County east to Cook County. Although there were a few scattered sightings in St. Louis and Lake Counties, nesting was confirmed only as far east as Itasca and Aitkin Counties (Baker and Hines 1996). Similar results were documented by the MBS, which identified 475 breeding-season locations from the late 1980s through 2014.

The extent of the Black Tern's breeding distribution documented by the MNBBA has changed little since Janssen's (1987) update. Participants reported 754 detections in 10.6 percent (503/4,751) of the surveyed blocks and confirmed nesting in 88 blocks (1.9 percent) **(A)**. Probable breeding evidence was reported in an additional 68 blocks (1.4 percent). Observed in 72 of Minnesota's 87 counties, breeding was confirmed in 38 counties. The latter included 15 counties not depicted in the map published by Hertzel and Janssen (1998). Counties lacking any breeding evidence were largely restricted to northeastern and southern Minnesota.

The MNBBA landcover suitability model predicts that Tern habitat is most suitable in central Minnesota, as far north as Itasca, southern Beltrami, and Clearwater Counties **(B)**. Scattered pockets outside of this area are distributed throughout the state and include wetlands in eastern Marshall County and the Swan Lake wetland complex in Nicollet County.

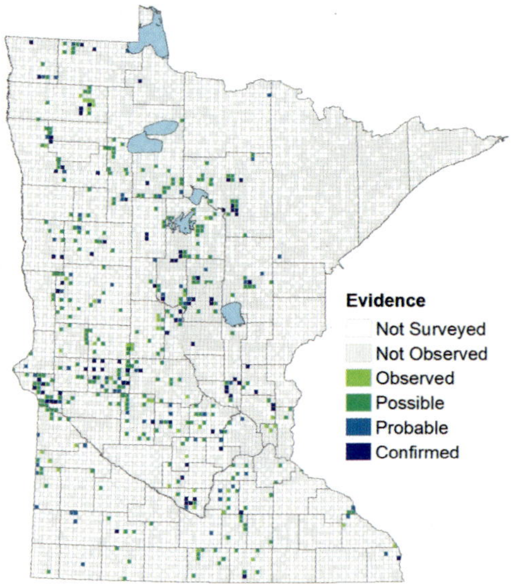

A. Breeding distribution of the Black Tern in Minnesota based on the Breeding Bird Atlas (2009–13).

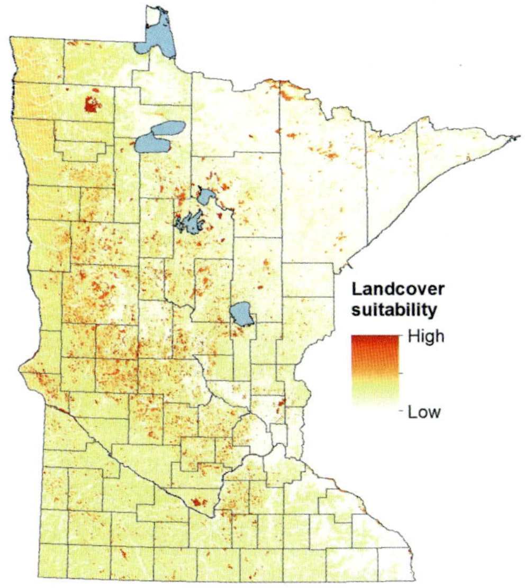

B. Landcover suitability map for the Black Tern in Minnesota based on habitat, landscape context, and climate data gathered during the Minnesota Breeding Bird Atlas (2009–13) using the MaxEnt modeling approach.

Notwithstanding Roberts's original assertion regarding the species' statewide occurrence, it appears that the overall distribution of the Black Tern has been remarkably stable over the past one hundred years. The same is true elsewhere within its breeding range. Although breeding colonies located in several states along the southern periphery of its breeding range, including Missouri, Kentucky, and Illinois, are no longer active, the extent of its core distribution in the northern Great Plains remains largely unchanged (Heath et al. 2020).

BREEDING HABITAT

The Black Tern is an inhabitant of shallow, productive wetlands with emergent and floating aquatic vegetation **(C)**. Small, isolated wetlands, large wetland complexes, the backwaters of large river floodplains, and shallow, protected bays of lakes, can each provide suitable habitat (Zimmerman, Dechant, et al. 2002; Wires, Lewis, et al. 2010). Terns also prefer sites that are embedded in landscapes with a high density of wetlands.

Some open water is important. In northwestern Minnesota, dense cattail stands were usually avoided. Sites with lower stem densities and higher species diversity were preferred (G. Brewer 1992). In Marshall County, nest sites had an average of 72 percent cover; nests were absent from sites with less than 40 percent cover. Higher cover density may reduce the impacts of wind and wave action on nests and help conceal the eggs and young chicks from predators (Maxson, Fieberg, and Riggs 2007).

POPULATION ABUNDANCE

As with population estimates of most marsh birds, those of the Black Tern have been speculative and quite variable over the years, with North American estimates ranging from 100,000 to 500,000 (Kushlan et al. 2002) to 300,000 to 750,000 (Wetlands International 2012). A more recent analysis by PIF (2021) resulted in a much higher North American estimate of 2.3 million individuals and a global population estimate of 5.1 million. Unfortunately, no statewide estimate is available.

Given the species' general abundance, BBS Black Tern data are statistically robust and demonstrated a nonsignificant surveywide population decline averaging −2.44 percent per year from 1966 to 2019 (Sauer et al. 2020). In Minnesota, the Tern has shown a far more dramatic, statistically significant decline, averaging −7.37 percent per year since 1967, for a cumulative loss of 98 percent by the year 2019 **(D)**!

Habitat loss and degradation are considered among the primary factors responsible for the decline. The loss of more than 50 percent of the wetlands in the United States has impacted all wetland-dependent species. Wetlands that remain, even those that are protected, are often subject to major water-level fluctuations that directly affect nesting success. Overgrowth of cattails also poses a challenge because the Terns rarely nest in sites with greater than 75 percent cover (Heath et al. 2020).

Despite these impacts on wetland habitats, banding studies in Wisconsin led Matteson and his colleagues (2012) to conclude that "the most compelling explanation for the large-scale population decline in Wisconsin relates to adult survival probability and natal recruitment into the breeding population." At the Horicon Marsh, "one of the largest remaining breeding colonies in the state, annual adult survival probability was estimated at 62% and nestling survival to recruitment age (2 yr) was < 2%." The authors suggested that the survival challenges may not originate on the breeding grounds but occur instead during migration and on the wintering grounds.

C. Dense cover provided by a variety of emergent vegetation supplies abundant nesting material and protects Black Tern nests from wind and predators in this shallow wetland in Traverse County. Photo by Lee A. Pfannmuller.

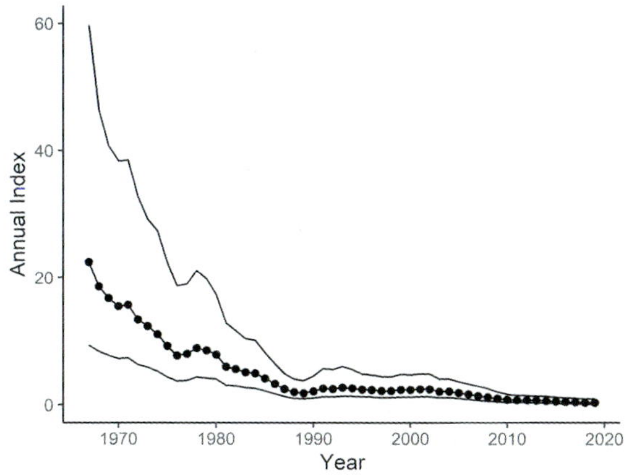

D. Breeding population trend and 95 percent credible interval (solid black lines) for the Black Tern in Minnesota based on the 1967–2019 Breeding Bird Survey (Sauer et al. 2020).

CONSERVATION

The Black Tern's declining numbers and dependence on wetland habitats have prompted its designation as a moderate conservation priority (PIF 2021). Until further research unravels the factors responsible for its decline and surveys provide a solid population estimate, protecting and restoring large wetland complexes remains critical. Specific management recommendations include managing water levels to promote emergent vegetative growth and maintaining stable water levels during the nesting season (e.g., Wires, Lewis et al. 2010; Soulliere et al. 2007; Zimmerman, Dechant, et al. 2002). When vegetation becomes too dense, disking, prescribed burning, and herbicide applications are recommended to reduce cover. Providing for healthy muskrat populations can also keep cattails at bay.

Fortunately, despite its dependence on prairie wetlands, the Black Tern may be less threatened by warming temperatures than many of the region's waterbirds. Under a 2°C warming scenario, NAS (2019) classified the species as climate stable with a projected loss of 20 percent of its continental breeding range by 2050 and a significant gain northward of 47 percent. A significant portion of the loss, however, is projected to occur in the Dakotas and western Minnesota. The tern's unsustainable rate of population decline could reduce its ability to readily adapt to these changing climatic conditions. As a result, Minnesota's emphasis on protecting and restoring productive wetlands may be vital to this little tern's future.

COMMON TERN

Sterna hirundo

Photo by Michael Furtman.

Minnesota Seasonal Status: Regular breeding resident and migrant. The Common Tern was rare during the MNBBA.
Migration: Medium- to long-distance migrant; 70 percent of breeding birds in central North America were found wintering along the coast of Peru, with the remaining 30 percent found elsewhere in northwestern South America, the Gulf of Mexico, and Central America (Bracey, Lisovski, et al. 2018).
Food: Aerial diver that feeds primarily on small fish.
Nest: Shallow scrape or natural depression on the ground lined with vegetation and debris. A colonial nesting species.
Conservation Concern: Assigned a moderate PIF Continental Concern Score of 12/20 and officially classified as Threatened in Minnesota.

One of three white terns that grace Minnesota's waters during the breeding season, the Common Tern is restricted to large bodies of water in northern Minnesota where rocky islets and small islands with sandy beaches provide suitable nesting habitat. Circumpolar in distribution, the bird is "an elegant and dainty creature" with "its spotless and delicate plumage and its buoyant, graceful flight" (Bent 1921). Its long, pointed wings and its aerobatic foraging maneuvers are reminiscent of the flight of a swallow, earning it the vernacular name of "sea swallow" along the East Coast. Indeed, the Latin term *hirundo* means "swallow."

Easy to distinguish from the much larger Caspian Tern, the Common Tern can be very difficult to distinguish from Minnesota's other white tern, the Forster's Tern. As adults, both species have a white underbody, a light gray back and tail, and a black cap. The Common Tern's black-tipped bill and feet are an intense orange-red color compared to a duller orange on the Forster's Tern. Both species have forked tails, but the Common Tern's is less deeply forked and has darker outer tail feathers. Vocalizations can help identification. Most of the Common Tern's nasal and burry calls are longer in duration than the Forster's Tern's short buzzy *kerr* call. Learning to recognize these subtle features takes practice. Luckily, habitat may be the best clue. Unlike the Common Tern, the Forster's Tern nests in shallow wetlands in western and southern Minnesota.

In North America, the Common Tern's breeding range includes Canada, from Alberta east to the Maritime Provinces, the Great Lakes states, the Atlantic coast south to South Carolina, and the Caribbean. Valued for its fine plumage, feathers, wings, and entire birds adorned women's hats during the late 1800s, leading to a precipitous population decline. Although passage of the 1918 Migratory Bird Treaty Act provided protection and enabled populations to recover, new threats are posing additional challenges. Today the Common Tern is one of Minnesota's least common species.

MINNESOTA BREEDING DISTRIBUTION

Although Roberts (1932) suspected there were additional Common Tern colonies in the state, only five were known in the early 1900s: Spirit Island and Hennepin Island on Mille Lacs Lake; Big Oak Island and Fourblock Island on Lake of the Woods; and a small boulder reef on Gull Lake in Cass County. Subsequently, there have been reports of the species occasionally nesting in small numbers at a few additional sites: Island Lake in St. Louis County (Hines 1993), Cotton Lake in Becker County (Green and Janssen 1975), the Shipwreck Islands on Kabetogama Lake (McKearnan 1986), and Baby Lake in Cass County (2014; MOU 2021a). But for decades now, the Common Tern has consistently nested in only four locations: the Duluth-Superior Harbor; Spirit and Hennepin Islands on Mille Lacs Lake; numerous islands in Lake of the Woods; and a small archipelago on Leech Lake. The Tern's history at each location follows.

Duluth-Superior Harbor

Since the first nesting pair was found on Minnesota Point in 1937, the Common Tern has nested in multiple locations throughout the harbor, periodically moving in response to encroaching vegetation or disturbance (Harris and Matteson 1975). The first major colony was established on Hearding Island, which was created in 1934. Terns nested on this dredge-spoil island for approximately ten years (1946–55). At its peak in 1953, the site supported 87 nesting pairs. But as vegetation succeeded to shrubs and young saplings, the site became less suitable. Although a few Terns continued to nest there for several years, a new dredge-spoil island, Barkers Island, was established in 1955 just offshore from Superior, Wisconsin. Within one year of its creation, 57 nests were found on the island, and 108 were found the following year (Finseth 1957; Cohen 1958).

Data regarding the Tern's nesting history in the harbor during the 1960s is limited. Cohen (1961) mentioned a large nesting colony on Minnesota Point but provided no details. The only information was a report of approximately 60 adults incubating at the Sky Harbor Airport, at the tip of Minnesota Point, in 1963 (Huber 1963a). The status of this colony during the 1970s and early 1980s is clearer. Ranging in size from 3 to 79 nesting pairs, the colony persisted, although the site was increasingly unsuitable due to encroaching vegetation and high rates of predation (Penning and Cuthbert 1993). The hazards posed by birds loafing on the runway finally prompted airport personnel to actively discourage nesting from 1986 through 1989. As early as 1971, the majority of the harbor's Terns actually began nesting at another disturbed dredge-spoil site, the Port Terminal. This became the harbor's main colony until many birds moved to a small island across from Wisconsin Point in 1987 and 1988.

The Common Tern's precarious situation in the harbor prompted the Minnesota and Wisconsin Departments of Natural Resources to acquire another dredge-spoil island solely for the Common Tern population. Inaccessible to the public, Interstate Island has been intensively managed to reduce vegetation and Ring-billed Gull predation. The site has supported between 68 and 302 Tern nesting pairs from 1985 to 2019 (Penning and Cuthbert 1993; Bracey, Chatterton, and Niemi 2016; Strand 2019, pers. comm.). Although the birds have nested at a variety of sites within the St. Louis Estuary in the past, nesting is now restricted to this one "highly, vulnerable site" (Bracey, Chatterton, and Niemi 2016).

Mille Lacs Lake

Spirit and Hennepin Islands comprise the Mille Lacs NWR, the smallest wildlife refuge in the United States (0.2 ha). Tiny rocky islands, the sites are located approximately 10 km from each other and 5 km from the lake's southern shoreline. Both have supported Common Tern colonies since at least 1886 (Roberts 1932). In 1915, 100 nests were reported on Hennepin Island and 150 nests on Spirit Island; in 1930, approximately 500 nesting pairs were observed on Hennepin Island and 80 on Spirit Island.

When Rice Lake NWR staff began monitoring the colonies in the 1960s, the numbers had declined significantly. The first survey on Hennepin Island in 1963 found 100 nests; by 1976 there were only 57. Regular surveys of Spirit Island began in 1976 when only 11 nests were found (J. Green, personal files).

A major factor in the Common Tern's decline was the appearance of the Ring-billed Gull on Hennepin Island in 1963 and on Spirit Island in 1970. The Gulls return to their breeding colonies earlier than the Terns, usurping what little habitat is available. Terns last nested on Spirit Island in 1998 but continue to nest on Hennepin Island. There, refuge staff is collaborating with the Mille Lacs Band of Ojibwe to protect the species from the growing Ring-billed Gull population, the recent appearance of Double-crested Cormorants, and the loss of habitat from shoreline erosion (McDowell 2011).

Lake of the Woods

Along the lake's southern shoreline, the Common Tern has nested for many years on a long sand spit known as Pine and Curry Island and, occasionally, on nearby Morris Point. Because the federally *Threatened* Piping Plover also nests here, field studies launched in the early 1980s enabled close monitoring of the Tern population.

Unfortunately, the Common Tern colony was at its peak long before the plover research began, with nearly 1,000 nests reported in 1932 (BMNH Records). Between 1979 and 1999, numbers dropped to approximately 200 pairs per year, ranging from a low of 25 nests in 1980 to a high of 485 in 1986 (Herwig, pers. comm.). Between 2001 and 2016, the colony's size declined further, with an average of only 77 nests per year; this included five years when the Terns failed to nest at all (Herwig, pers. comm.). Overall, when the lake's water levels are low, the Tern is successful, but when water levels are high, nest losses due to waves and storm events can be significant.

Population numbers are higher on islands near the Northwest Angle. Efforts by the MNDNR to monitor these remote sites have added immeasurably to understanding the significance of the region. Between 2001 and 2016, the average number of Common Tern nests was 213 per year (Herwig, pers. comm.). Combined with the yearly average during the same time period on Pine and Curry Island, the Minnesota portion of the lake supported approximately 290 Common Tern nests per year, with the Northwest Angle contributing more than 70 percent of the population. Water levels and gull depredation, however, are concerns here as well.

The Common Tern has always contended with the vagaries of weather and natural changes in water levels. As a result, it is prone to "boom and bust" years of productivity. Today, however, these challenges are compounded by other activities on Lake of the Woods. With dams on the lake's major outlets, the international Lake of the Woods Control Board is responsible for managing water levels to balance competing priorities, including recreation, hydropower generation, and flooding. Unfortunately, high levels maintained during the summer can exacerbate problems for the Common Tern. The loss of sandy beaches, especially along the lake's south shore, is another challenge. When sediments lost through shoreline erosion are not naturally replenished, critical Tern habitat declines. Indeed, nearly 1,500 m of Pine and Curry Island's sandy shoreline disappeared from 1985 to 2004 (Herb et al. 2004). High water levels likely contributed to the loss, but other factors may also be at play.

Leech Lake

An archipelago of three islands in southern Leech Lake has supported a Common Tern nesting colony for many years. Owned and managed by the Leech Lake band of Ojibwe, the archipelago includes Gull Island, Little Pelican Island, and Pelican Island. Until 1997, a small number of Terns also nested on Little Pipe Island, located closer to the lake's southern shoreline.

Gull Island supported the largest Tern colony from the 1920s through 1988. In 1933, approximately 2,000 to 3,000 birds were reported nesting (Roberts 1936). Numbers remained high through the early 1970s but then began a gradual decline. By 1988, only 142 nests were found (Mortensen and Ringle 2007). Competition with a growing nesting population of Ring-billed Gulls was the primary culprit. The Ring-billed Gulls first arrived in 1960, and the tiny island (0.16 ha) supported more than 1,000 breeding pairs by 1991 (Dickerman and Lefebvre 1961; Mortensen and Ringle 2007). Each spring, the Common Terns returned to a diminishing amount of suitable nesting habitat (J. H. Miller 1987; J. M. Reed et al. 1991). By 1989, they had abandoned Gull Island and moved north to Little Pelican Island.

Larger than Gull Island, Little Pelican Island presented its own challenges. In addition to the presence of mammalian and avian predators, encroaching vegetation, human disturbance, and a burgeoning population of Ring-billed Gulls, which also began nesting on the island in 1996, Double-crested Cormorants arrived in 1998. Working hard to address each challenge, the Leech Lake Band of Ojibwe has succeeded in maintaining a breeding population of 100–200 pairs of Terns since 2004 (Mortensen and Ringle 2007; Hamilton and Cuthbert 2016). Recently, some Terns and Ring-billed Gulls also began nesting on the largest island in the archipelago, Pelican Island.

MNBBA Results

Since Common Tern nesting populations are largely inaccessible, the MNBBA depended on data contributed by Dr. Francesca Cuthbert and her colleagues at the University of Minnesota. During the atlas, they conducted a statewide census of Double-crested Cormorant and American White Pelican colonies. Three of the Common Tern's major colonies co-occur with these species (Hamilton and Cuthbert 2016); data on the Duluth colony was provided by the University's Natural Resources Research Institute.

The data from MNBBA volunteers and the University provided Common Tern detections in 43 of the surveyed blocks. Breeding was confirmed in six blocks (the atlas block where Interstate Island is located in Duluth is not visible at this scale) **(A)**. In addition to documenting nesting at the four primary sites, approximately 10–20 nesting pairs were found in 2010 on Bird Island in Pelican Lake in Crow Wing County. Other significant observations included about 20 pairs on a sandbar between Upper

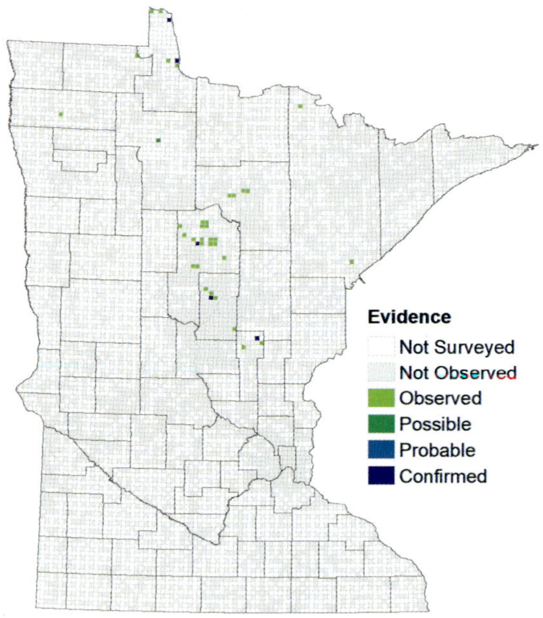

A. Breeding distribution of the Common Tern in Minnesota based on the Breeding Bird Atlas (2009–13).

and Lower Red Lake and a pair observed foraging in Voyageurs National Park. Unfortunately, no evidence of nesting was found at either site.

BREEDING HABITAT

Minnesota's Common Tern nesting colonies demonstrate the range of suitable habitats utilized by the species. Although all the colonies are located on islands in large bodies of water, the nesting substrates vary from sandy depressions amid rocky boulders to sand and gravel beaches and artificially created dredge-spoil islands **(B)**. Scattered herbaceous vegetation or sheltered nooks among rocks and boulders help provide cover for the young chicks. Some elevation helps protect the nests from wash-overs during storms and high winds and provides good visibility. An abundant supply of small fish is critical (Cuthbert et al. 2003).

As the species' history in the Duluth-Superior Harbor illustrates, encroaching vegetation is a deterrent to nesting colonies, eventually forcing their abandonment unless the vegetation is managed to prevent succession. Sites that are not protected from human disturbance and predators are also either avoided or used intermittently.

POPULATION ABUNDANCE

Jennifer Arnold and colleagues (2020) tabulated approximately 120,000 breeding pairs along the Atlantic coast, in the Great Lakes, in southern Manitoba, and in the Caribbean; more than 80 percent of the pairs occurred along the Atlantic coast and nearly 8 percent within the Great Lakes. Although population estimates are not available, they noted the birds "are dispersed on medium-sized lakes (area greater than 5 square km) throughout much of Canada east of the Rocky Mountains, including most of the boreal forest in addition to transition zones and prairie lakes to the south and west." PIF (2021) provided an updated estimate of 470,000 individuals in the United States and Canada and a global population estimate of 2.8 million. In Minnesota, estimates vary considerably, depending on how and when surveys were conducted. Nevertheless, data for the state's primary colonies were compiled to determine a statewide estimate for five time periods spanning from the 1930s to 2015 **(C)**. In the 1930s the statewide estimate ranged from 2,580 to 3,080 nesting pairs; in 2015 the population was approximately 930 pairs.

BBS population trends were unreliable but suggest a nonsignificant, surveywide decline averaging −0.85 percent per year from 1966 to 2019 (Sauer et al. 2020). The Great Lakes population in the United States, however, has been closely monitored since the 1970s and demonstrates an increase in population numbers from approximately 2,500 nesting pairs in 1977 to 4,200 nesting pairs during a three-year census window from 2007 to 2009 **(D)**. These estimates include the Duluth-Superior Harbor. The increase was attributed to the management and protection efforts underway at many colonies. By contrast, in Canada's portion of the Great Lakes, nesting Common Terns declined 41 percent from 1976 to 2009 (Morris et al. 2012), resulting in an overall decline across the Great Lakes region. Both the number of colony sites and the total number of breeding pairs have declined, with the steepest reductions occurring before 2000 (J. Arnold et al. 2020). A recent study by Bracey (2021) suggests that the number of years before juveniles in the lower Great Lakes reach maximum breeding potential (five years) may be contributing to the decline because juveniles reach sexual maturity earlier in other inland colonies and along the Atlantic coast.

The complicated history of the Common Tern in Minnesota is not unlike that in other states. Loss of habitat, disturbance, avian and mammalian predation, and competition with the Ring-billed Gull and Double-crested Cormorant are among the many factors challenging biologists throughout the species' breeding range (Cuthbert et al. 2003). Historically, the Tern's low site fidelity enabled it to adapt quickly to the ephemeral nature of its open beach habitat with natural conditions providing a revolving array of suitable habitats over time. In Minnesota, where islands with sandy or gravel beaches have always been a rare commodity, nature now requires significant assistance by resource managers to sustain the state's population.

CONSERVATION

The Common Tern is considered a high priority throughout the Great Lakes region. It was officially listed as *Special Concern* in Minnesota in 1984 (Coffin and Pfannmuller 1988) and upgraded to *Threatened* in 1995 (Minnesota Administrative Rules: Chapter 6134) when field studies documented declining population levels despite the intensive management

B. Interstate Island WMA in the Duluth-Superior Harbor is intensively managed to provide suitable nesting habitat for the Common Tern and to minimize interference by Ring-billed Gulls. Photo by Derek Hamilton, courtesy of USFWS/UMN.

C. Common Tern population numbers in Minnesota, 1930s–2015.

Site	1930s	1984	2004	2010	2015
Leech Lake					
Pelican Island					85[d]
Little Pelican Island		489[c]	186[d]	127[d]	78[d]
Gull Island	1,000–1,500[a]				
Mille Lacs Lake					
Hennepin Island	580[b]	47[c]	138[d]	216[d]	118[d]
Spirit Island		46[c]			0[d]
Lake of the Woods					
Pine and Curry Island	1,000[a]	139[c]	109 (all failed)[e]		228[h]
Northwest Angle		Not surveyed	250[d]	282[d]	295[h]
Duluth-Superior Harbor					
Interstate Island	First nest 1937		204[f]	238[f]	101[f]
Port Terminal		113[c]			
Sky Harbor		27[c]			
Other sites					
Pelican Lake				10–20[g]	
Kabetogama Lake		20[c]			
Unnamed site in Becker County					25[h]
Total	**2,580–3,080**	**881**	**887**	**873–883**	**930**

Note: Estimates of breeding populations vary considerably from year to year and site to site, particularly for counts conducted on the ground. Only recently have aerial counts by Cuthbert and colleagues provided some consistency between surveys. As a result, some of the numbers reported above are counts of breeding adults, while others are reports of nests conducted under different survey circumstances. Nest counts early in the season may be misleading if disturbance or storms cause some or a complete loss of nests. At best, these numbers represent initial breeding efforts only.

[a]MOU files (transcribed and provided by J. C. Green). [b]Roberts 1932. [c]McKearnan 1986. [d]Hamilton and Cuthbert 2016. [e]Haws 2004. [f]Data provided by Fred Strand, retired wildlife manager, Wisconsin DNR. [g]MNBBA data. [h]Data provided by C. Herwig, MNDNR (pers. comm.).

efforts that were underway. At the continental level, the Tern is considered a moderate conservation priority (PIF 2021).

The Common Tern has survived in Minnesota in large part owing to the remarkably creative and intensive management efforts of dedicated federal, state, and tribal resource personnel. Three of the state's four breeding populations (Mille Lacs Lake, Leech Lake, and Duluth-Superior Harbor) are actively managed to address numerous challenges, including vegetation encroachment, mammalian predation, beach erosion, and competition with Ring-billed Gulls and Double-crested Cormorants (Penning and Cuthbert 1993; Mortensen and Ringle 2007; Pfannmuller 2014a). Innovative management options continue to be pursued, including the possibility of establishing nesting habitat on two modified pontoon boats in Mille Lacs Lake (Wessel 2015).

Management has also been underway on Lake of the Woods, but the challenges are considerable in this large, remote lake. Large-scale loss of habitat on Pine and Curry Island is one of the greatest challenges, while islands in the Northwest Angle are so remote that fieldwork is costly and difficult. Loss of nests due to high water and wave action is a major threat and will require close cooperation with the Lake of the Woods Control Board to help ensure the species' future in this important region.

Because the Common Tern is a fish consumer, contaminants present another concern. Tissue analyses conducted on Terns in the Duluth-Superior Harbor in the late 1970s concluded that the concentrations of organochlorine compounds were high enough that potential impacts to reproductive success could not be excluded (Niemi, Davis, et al. 1986). Fortunately, their levels in the Great Lakes have been decreasing since the 1980s (Arnold et al. 2020). Elevated mercury levels, however, have been reported in Terns as recently as 2012 in Alberta and documented as a concern at several colonies on the East Coast (Arnold et al. 2020). In Lake Superior, a study of two Common Tern nesting colonies (Interstate Island in Duluth-Superior Harbor and Ashland Island in Chequamegon Bay, Wisconsin) demonstrated how colony location, foraging sites, and different stages of the breeding season impacted the level of the tern's exposure to mercury (Bracey, Etterson et al. 2021).

Warming temperatures may also pose a threat to the Common Tern. As major storm events become more common, they can cause considerable damage to ground-nesting waterbirds, while rising water levels may reduce available habitat. NAS (2019), however, classified the Tern with a low vulnerability to climate change under three different warming scenarios. Models predicted that major losses to the species' breeding range in western Canada were offset by even larger gains across the Arctic. Although the NAS models inaccurately depict the species' current distribution in Minnesota, impacts to the state's Common Tern population appear minimal.

As challenging as it has been for Minnesota biologists to protect local colonies, a recent genetic analysis of Common Tern breeding populations highlights the conservation significance of their efforts. Szczys and her colleagues (2017) provided evidence that the Common Tern population in the lower Great Lakes is the oldest population in North America. Their analysis also demonstrated a tenfold increase in the rate at which inland Terns have dispersed to coastal breeding habitats since the 1960s, despite the successful management efforts underway. The authors suggest that this movement is symptomatic of the decline in suitable habitat in the Great Lakes. Factors during the nonbreeding season, however, may also be contributing to the decline. In the coastal waters of Peru, where 70 percent of the inland Common Terns fitted with geolocators were found wintering, warming temperatures are negatively impacting prey availability (Bracey, Lisovski, et al. 2018). Clearly, the Common Tern's future in the Great Lakes is in jeopardy.

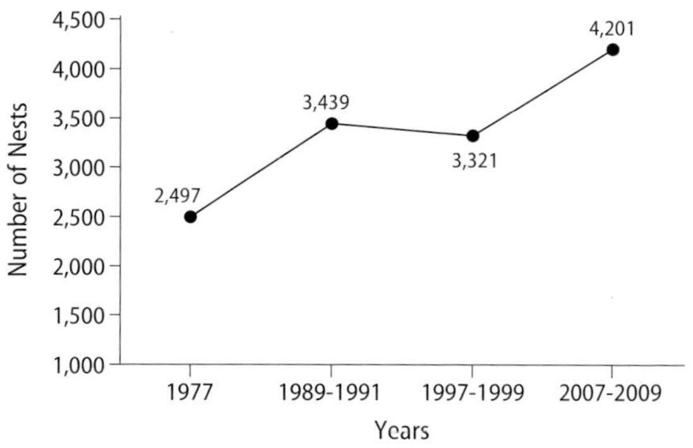

D. Population trend for the Common Tern in the US portion of the Great Lakes from 1977 to 2007–9 (from Cuthbert and Wires 2013, table 11c).

FORSTER'S TERN

Sterna forsteri

Minnesota Seasonal Status: Regular breeding resident and migrant. The Forster's Tern was uncommon during the MNBBA.
Migration: Great Plains populations are short- to medium-distance migrants that winter from the United States' Gulf coast to northern Central America.
Food: Aerial diver that feeds on small fish, insects, and crustaceans.
Nest: Platform of aquatic vegetation constructed on floating vegetation or old muskrat houses. A colonial nester.
Conservation Concern: Assigned a moderate PIF Continental Concern Score of 13/20; officially classified as *Special Concern* in Minnesota.

Photo by Mike Lentz.

A marshland inhabitant, the Forster's Tern is one of four nesting terns in Minnesota. It is easily distinguished from the larger Caspian Tern by its deeply forked tail and from the smaller Black Tern by its white plumage. Distinguishing it from the Common Tern, however, can be quite challenging. These two elegant species are so similar that early ornithologists did not identify the Forster's Tern as a separate species until 1858. Even the renowned ornithologists Alexander Wilson and John James Audubon were fooled.

Today, skilled observers rely on subtle differences in the birds' tails and outer wing feathers for correct identification. Familiarity with their vocalizations is also helpful. Both birds have a buzzy quality to their calls, but the differences are fairly distinct. Knowledge of their habitat and distribution, however, provides the most valuable clue. The Forster's Tern is a wetland species widely dispersed throughout western and central Minnesota. The rarer Common Tern is restricted to large bodies of water in northern Minnesota that provide rocky islets and sandy islands for nesting.

The Forster's Tern is an emblematic species of the Prairie Potholes Region; its most extensive population occurs in the northern Great Plains, from southern Alberta east to Manitoba and south across the Dakotas and western Minnesota. Populations also occur in the western Great Basin and along the Gulf coast from Texas to Alabama and south to northern Mexico.

MINNESOTA BREEDING DISTRIBUTION

In the early 1900s, Roberts (1932) described the Forster's Tern as a summer resident on the state's western "prairie sloughs and marshes." Nesting was confirmed at Heron Lake in Jackson County and at King Lake in Meeker County; breeding colonies in Kittson (Twin Lakes) and Renville (Hodgson Lake) Counties were also noted. Roberts's short account makes it difficult to assess the species' abundance in those early years. Although he commented that the Tern nested "in considerable numbers at Heron Lake" and "at similar places in the open country," he went on to note that it "commonly assembles in little colonies of a half-dozen to two dozen pairs."

Within a few years, the Terns were also found nesting at two sites in eastern Minnesota: about 25 pairs were discovered on Otter Lake in Ramsey County (Engstrom 1940); and a single nest was found in backwaters of the Minnesota River in Hennepin County (E. T. Mitchell 1941). The latter report also mentioned a nesting attempt that summer in Chisago County. Only Hennepin County continued to support small numbers of breeding birds beyond the 1960s (MOU 2021b), although other small colonies in the metropolitan region would later be identified.

When Green and Janssen (1975) and Janssen (1987) prepared their updated accounts, the Tern was described as a regular breeding species in western Minnesota, stretching east to the Twin Cities. Both accounts identified some of the larger colonies (i.e., hundreds of nesting pairs), including Heron Lake in Jackson County, Agassiz NWR in Marshall County, Lake Osakis in Todd/Douglas Counties, and Clearwater Lake in Wright County. Janssen's account (1987) delineated 20 counties with confirmed nesting since 1970. Hertzel and Janssen's (1998) update excluded Becker, Goodhue, Murray, and Redwood Counties from Janssen's original map and added three new counties: Kandiyohi, Otter Tail, and Washington.

Coincidentally, Guertin and Pfannmuller (1985) reported on the MNDNR's Nongame Wildlife Program's efforts to assess the status and distribution of Minnesota's colonial nesting waterbirds. Fifteen Forster's Tern colonies were reported active in at least one year from 1981 to 1983. The colonies were located in 13 counties, all within the species' primary breeding range. Ten of the colonies supported fewer than 50 nests each. The largest colony was located on Lake Osakis in Todd/Douglas Counties, where the population numbered approximately 1,000 pairs in 1983, but just one year later was reported to support only 200 pairs. These data, coupled with the experience of long-time field observers, led biologists to conclude that the species had declined significantly in the past fifty years. As a result, the Forster's Tern was officially listed as a state *Special Concern* species in 1984 (Coffin and Pfannmuller 1988).

The listing prompted a field investigation of the species in 1985 and 1986 (Cuthbert and Louis 1993). Fifteen of the 16 colonies selected for study supported nests in at least one of the two breeding seasons; approximately half of the sites supported fewer than 15 nesting pairs each. The total statewide population was estimated to number fewer than 1,000 pairs.

Nearly twenty-five years later, MNBBA participants reported 157 Forster's Tern records in 2.4 percent (116/4,737) of the surveyed atlas blocks **(A)**. Included were a small number of observations near

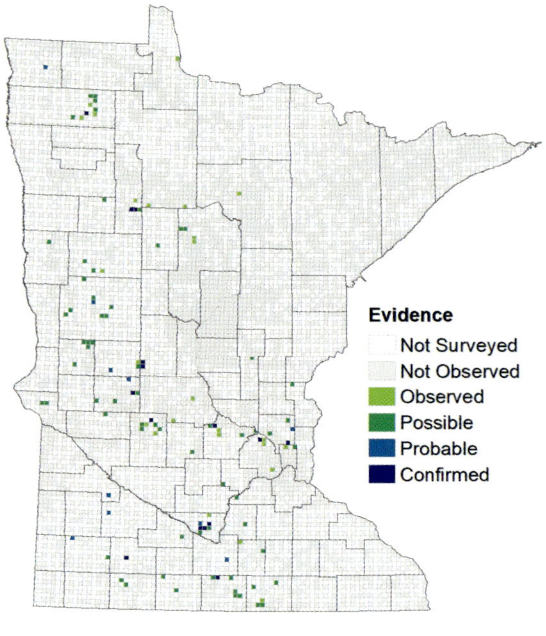

A. Breeding distribution of the Forster's Tern in Minnesota based on the Breeding Bird Atlas (2009–13).

B. A small shallow lake in Anoka County with an abundance of floating and emergent aquatic vegetation supported a small nesting colony of the Forster's Tern and Black Tern. Photo by Lee A. Pfannmuller.

Common Tern nesting colonies. Despite the similarity of the two tern species, the Forster's Tern records were all reported by experienced observers and retained. Breeding was confirmed in 15 blocks that represented 12 different breeding locations. Spanning western Minnesota and stretching as far east as the Twin Cities metropolitan region and Chisago County, the birds were reported from 42 of Minnesota's 87 counties and were reported nesting in 11 counties; probable breeding was reported in seven counties. The combined total of confirmed and probable nesting counties (18) is slightly larger than the 15 breeding locations reported in the 1980s. MNBBA data, however, did not allow an estimate of nesting pairs.

Distributional changes elsewhere in the breeding range of the Forster's Tern are not well-documented. In their comprehensive review of the species' status, McNicholl and his colleagues (2020) noted that the ephemeral nature of the Tern's nesting habitat "blurs distinction between changes in breeding distribution and long-term fluctuations."

BREEDING HABITAT

Freshwater and saline wetlands and the marshy shorelines of lakes and rivers are the domain of this marsh-nesting species. But in the interior Great Plains, the prairie potholes and sloughs are where the Forster's Tern is most at home. Here nesting colonies are found in extensive semipermanent or permanent wetlands where emergent vegetation is abundant and widely interspersed with open water **(B)**.

Unfortunately, changing water levels, high winds, and storms can easily render the Tern's nesting sites unsuitable. As a result, it is not uncommon for the birds to select other suitable sites at the beginning of the nesting season or to abandon sites midseason and attempt to renest elsewhere (Cuthbert and Louis 1993; Fraser 1997).

POPULATION ABUNDANCE

Population data are sparse for the Forster's Tern, whose nests are often located in large wetland complexes that are difficult to access. Nevertheless, biologists have provided North American estimates ranging from 47,000 to 51,000 breeding individuals (Kushlan et al. 2002) to a more recent estimate of 150,000 (PIF 2021). As noted earlier, in the 1980s, Minnesota supported fewer than 1,000 breeding pairs, or 2,000 individuals (Cuthbert and Louis 1993).

BBS monitoring data for the species lack statistical rigor in every state, Canadian province, and ecological region except the Gulf Coastal Prairie, where the resident population shows a stable trend in numbers (Sauer et al. 2020). Although trends were unreliable elsewhere, they suggest a survey-wide, nonsignificant population decline averaging −1.17 percent per year from 1966 to 2019. In the core of the species' breeding range in the Prairie Potholes Bird Conservation Region, however, the population demonstrates a small increasing trend that is most prevalent beginning in the late 1990s. Higher precipitation levels in recent years, combined with focused conservation efforts in the region, have likely improved wetland conditions for the species. Farther east, however, the Great Lakes Colonial Waterbird Survey documented a dramatic 77 percent decline in Forster's Tern numbers in the United States between censuses conducted in 1989–91 and 2007–09 **(C)**. Low water levels and the spread of the invasive, nonnative species Phragmites and narrow-leaved cattail, whose dense growth chokes wetlands and renders them uninhabitable by Terns, were considered responsible (Cuthbert and Wires 2013).

CONSERVATION

The Tern's small, declining population and the vulnerability of its wetland habitats to numerous threats, including drainage, changing water levels, invasive species, and pesticides, led to its designation as a moderate conservation priority (PIF 2021). Although it can quickly adapt to unsuitable conditions by relocating to other wetlands, the strategy is beneficial only if suitable sites are available.

The Tern's dependence on wetlands in the northern Great Plains is of particular concern in light of the projected impacts of rising temperatures (NABCI 2010). Nonetheless, under a 2°C warming scenario, NAS (2019) deemed the Forster's Tern breeding range to be stable with a loss of only 13 percent by 2050 and a gain of 43 percent in northern Canada and east across the Great Lakes. Minnesota would witness a loss of habitat in the south and potentially significant gains in the north.

Meanwhile, efforts to protect and restore wetland habitats are critical. It is especially important to ensure that large nesting colonies that have been active consistently over the years are protected and periodically monitored. These sites are the Agassiz NWR in Marshall County, Clearwater Lake in Wright County, Lake Osakis in Todd/Douglas Counties, North Heron Lake in Jackson County, and Swan Lake in Nicollet County. Together they composed 90 percent of the estimated state population in 1985 and 92 percent in 1986 (Cuthbert and Louis 1993). All but North Heron Lake were documented nesting sites during the MNBBA. Most sites are also important to many other wetland-dependent species, so periodic monitoring would serve multiple purposes.

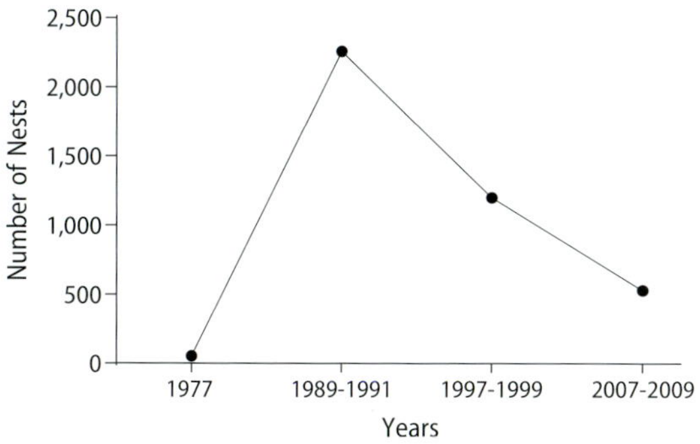

C. Forster's Tern population trend in the US portion of the Great Lakes from 1977 to 2007–9 (from Cuthbert and Wires 2013, table 13b).

COMMON LOON

Gavia immer

Minnesota Seasonal Status: Regular breeding resident and migrant; some birds linger in Minnesota through mid-December. The Common Loon was common during the MNBBA.
Migration: Short- to medium-distance migrant; winters are spent in the coastal waters of North America and northern Mexico.
Food: Primarily small fish, secured by diving and swimming under water.
Nest: Mound of grasses, twigs, and aquatic vegetation placed on the ground near the lakeshore and often on an island.
Conservation Concern: Assigned a moderate PIF Continental Concern Score of 10/20; officially recognized as Minnesota's state bird.

Photo by David Brislance.

Since the earliest beginnings, the Common Loon has enthralled North America's inhabitants. An omen of death, a bird with magical powers, a warrior spirit, and a symbol of good harvest; these are just a few of the qualities attributed to the Loon by Native American cultures as diverse as the Algonquin in Quebec and the Inuit in Alaska. Today, these oral legends are augmented by scores of written documents, ranging from technical research papers about the bird's life history to eloquent soliloquies about its unique vocalizations. Considered a keystone wilderness species, the Common Loon is now an indicator of the health and sustainability of an entire ecosystem. Yet, like the Ojibwe of northern Minnesota, visitors to the north woods still experience a spiritual connection with the bird. The Loon, Dunning (1985) wrote, "has the ability to awaken us to the greatest mysteries of existence—both the infiniteness of our own subconscious and the vastness of all we may never know about life on this earth."

Breeding adults display an elegant array of black-and-white feather patterns, which make the Common Loon an unmistakable species. The red color in their eyes is an adaptation for seeing well in the low-light conditions of deep water. Turning from a dull red-brown in winter to a brilliant deep red in summer, the eye color may also enhance courtship and territorial defense during the breeding season. As striking as the bird's appearance is, it is the Loon's vocal repertoire that leaves a lasting impression. From the mournful wail to the vibrant yodel and gleeful tremolo, we are entertained by an array of conversational pieces that arouse our senses, often in the darkness of night.

While the Common Loon is largely restricted to the boreal forests of Canada and Alaska, its breeding range also extends across the northern tier of the United States and includes both Iceland and Greenland. Minnesota supports the largest United States breeding population south of Alaska, but the centers of greatest abundance are found in Canada. The species' Latin name, *Gavia immer,* means a seabird colored like the black ash remains of a fire.

MINNESOTA BREEDING DISTRIBUTION

Although it is considered a species of Minnesota's northern forests, the Common Loon was found on lakes throughout the state in the late 1800s and early 1900s. Hatch (1892) referenced the species' abundance when he arrived in Minnesota in the late 1850s, noting that some of the Loon's favorite haunts near White Bear, Waseca, and Minnetonka were also sites where visitors were "tacitly allowed to shoot at them to their heart's content." Roberts (1932) also described the species as a common breeding resident throughout the state. Although less common in western Minnesota, confirmed nesting records were available from Jackson County in the southwest corner of the state, from the prairie–forest border counties of Otter Tail and Becker, and from Polk County in the far northwest. Farther east, in central Minnesota, confirmed records were available from Hennepin, McLeod, and Meeker Counties, while nesting was inferred in Isanti and Sherburne Counties and in Itasca State Park. Roberts noted that numbers had declined considerably in the forty years since Hatch first described the species as abundant, owing largely to the popularity of shooting the bird for mere sport.

The Common Loon's statewide breeding distribution retreated northward by the time that Green and Janssen prepared an updated account of its status in 1975. Largely absent from counties south of the Minnesota River, breeding pairs were confined to the forested counties of central and northern Minnesota by the early 1970s. Similar changes were noted in other states on the southern edge of the species' breeding range, including Illinois, Indiana, Massachusetts, Michigan, Ohio, and Pennsylvania (Paruk, Evers, et al. 2021). Little had changed when Janssen provided an updated account in 1987 and delineated 32 counties where nesting had been confirmed since 1970. The only county south of the Minnesota River with confirmed nesting was Scott in the western Twin Cities metropolitan region. Hertzel and Janssen (1998) would later add another six counties to the list, all north of the Minnesota River.

During the mid-1900s, increased recreational activity and shoreline development on Minnesota's northern lakes began to raise concern about the potential impacts on Minnesota's state bird. Significant die-offs of migrants congregating on Lake Michigan due to botulism poisoning during the falls of 1963, 1964, and 1965 added to the concern (Fay 1966). Then, in 1966, a field biologist spent four months canoeing in the Boundary Waters Canoe Area Wilderness (BWCAW) and failed to observe a single young chick (Ream 1976).

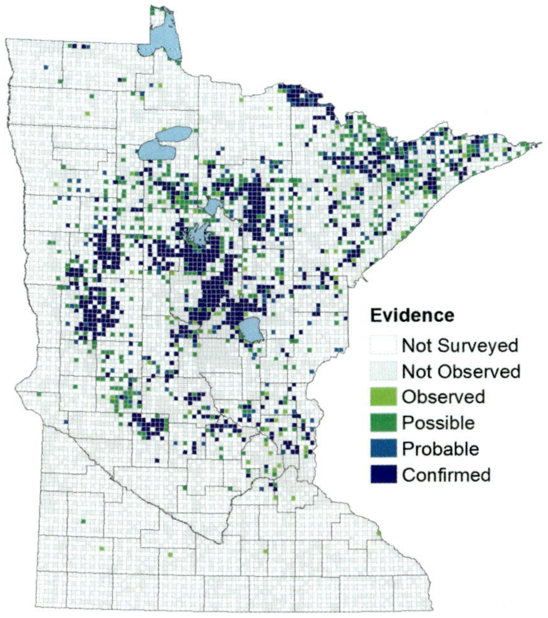

A. Breeding distribution of the Common Loon in Minnesota based on the Breeding Bird Atlas (2009–13).

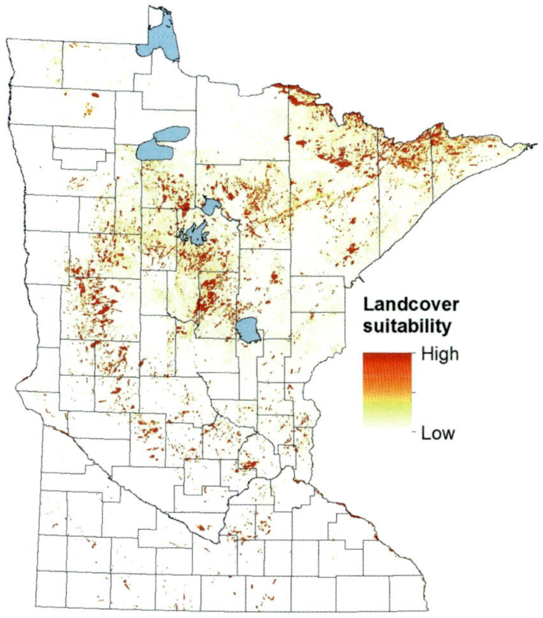

B. Landcover suitability map for the Common Loon in Minnesota based on habitat, landscape context, and climate data gathered during the Minnesota Breeding Bird Atlas (2009–13) using the MaxEnt modeling approach.

These concerns, coupled with widespread public interest in the species, prompted the initiation of numerous studies across the state. Many efforts focused on understanding the impacts of recreational activity on nest site selection and productivity (e.g., McIntyre 1975; Titus and VanDruff 1981; Tischler 2001). Other studies focused on assessing regional and statewide population numbers (e.g., Hirsch and Henderson 1980; Mooty and Goodermote 1985; Mooty and Perry 1988; Mathisen 1988; McIntyre 1978, 1988; Strong and Baker 2000). Today, the Common Loon is one of the most well-studied members of Minnesota's breeding avifauna. In-depth studies have been conducted in the BWCAW (S. T. Olson and Marshall 1952; Ream 1976; Titus and VanDruff 1981), Voyageurs National Park (Reiser 1988; Paruk, Nelson, et al. 2008), Itasca State Park (McIntyre 1975), and the Whitefish chain of lakes in north-central Minnesota (Valley 1987; Tischler 2001). Although one study demonstrated a local population decline over a 15-year interval from 1971 to 1986 (McIntyre 1988), the overwhelming conclusion of these diverse studies was that the Common Loon is an adaptable species that has maintained a stable and healthy population in Minnesota for several decades.

During the MNBBA, participants reported 2,856 Common Loon records in 30.9 percent (1,528/4,941) of the surveyed atlas blocks. Breeding was confirmed in 762 blocks (15.4 percent) **(A)**. The species was observed in 56 of Minnesota's 87 counties and was documented breeding in 41 counties; one block where breeding was documented straddled both Scott and Dakota Counties and was the only record located south of the Minnesota River. However, probable breeding records were recorded in two southern counties (Lyon and Sibley) and possible breeding records were recorded in four southern counties (Murray, Wabasha, Waseca, and Watonwan). Six of the counties where nesting was confirmed were not documented previously by Hertzel and Janssen (1998): Carlton, Dakota, Mille Lacs, Pine, Polk, and Swift. The MNBBA landcover suitability map predicts that sites with the highest suitability are concentrated in the Laurentian Mixed Forest Province and the northern Eastern Broadleaf Forest Province **(B)**. Suitable sites are also scattered in other areas, including the Aspen Parklands and along the prairie–forest border in west-central Minnesota.

Although the Common Loon has apparently retreated from the southern counties where it may have been a resident in the late 1800s and early 1900s, its distribution has remained relatively stable over the past fifty years. Comparison of the map of breeding-season locations identified by the MBS with the MNBBA distribution map actually suggests that the species has become more common throughout central and east-central Minnesota in the past twenty years.

As noted earlier, breeding populations also retreated along the southern periphery of its breeding range in other northern states in the face of residential and agricultural development. Reports in the late 1900s, however, suggest that recolonization of some of these areas may be underway in the northeastern United States (Paruk, Evers, et al. 2021), Ontario (Cadman et al. 2007), and Michigan (Chartier et al. 2013).

BREEDING HABITAT

The Common Loon was long considered a species of the deep, clear lakes of northern Minnesota. Such sites were often remote from human disturbance, dotted with islands, and bounded by an irregular, rocky shoreline **(C)**. Large clear-water lakes that averaged between 20 and 80 ha in size and that provided an abundance of small fish, suitable nesting sites, and a shallow, protected location to serve as a nursery for young chicks were deemed critical (McIntyre 1986). Large lakes seemed particularly important, so that the large-bodied Loon had enough room to take off from the water's surface and gain sufficient altitude to rise above the shoreline's trees. Because the species is dependent on its eyesight to capture prey, water clarity was considered equally important, while minimal human disturbance was deemed critical for reproductive success. However, as more field studies were conducted, biologists learned that the species' habitat preferences were more variable than was once thought.

C. In Minnesota's north country, the Common Loon typically nests along the protected shoreline of islands in large, deepwater lakes that provide abundant fish populations. Photo by Lee A. Pfannmuller.

Indeed, the land of 10,000 lakes proved to be the perfect testing ground to validate the species' dependence on large lakes. Several studies had noted that birds occasionally used lakes smaller than 20 ha in size. Data collected by the MNDNR Volunteer Loon Watcher Survey included five reports of the Common Loon successfully nesting on lakes as small as 5–8 ha (Perry and Woizeschke 2000). These observations prompted a more in-depth study of lake occupancy in Crow Wing County during the summers of 1988 and 1989. The results clearly demonstrated the importance of small lakes to Minnesota's Common Loon population. Nearly 35 percent of the lakes that were 4–10 ha in size had at least one Loon present (Perry and Woizeschke 2000). Another broader analysis conducted in 1989 throughout the state found that 30 percent of the lakes that were 4–20 ha in size were occupied by at least one Loon; statewide, those lakes supported 32 percent of Minnesota's Common Loon population (Strong and Baker 2000). Other studies have shown that Loons occupying small lakes usually defend territories that embrace multiple lakes, presumably to provide access to more foraging sites (S. T. Olson and Marshall 1952; Miller and Dring 1988; Piper, Paruk, et al. 1997).

Many of these smaller lakes are embedded in landscapes fragmented with scattered woodlands and farmlands and are often more eutrophic than those located farther north on the Canadian Shield. More than half of the nests found within the Whitefish chain of lakes, for example, were located on cattail hummocks along the lakes' shallow, marshy shorelines (Tischler 2001).

Although the Common Loon has proved to be more adaptable than originally thought, water clarity remains particularly important to this visual predator. The Minnesota Loon Monitoring Program has shown a significant correlation between water clarity and the number of adult Loons, the number of juvenile Loons, and the number of Loons per acre (Larson 2016).

POPULATION ABUNDANCE

Paruk, Evers, and colleagues (2021) recently provided a global breeding population estimate of 254,000 to 265,000 territorial pairs; the number of nonbreeding birds raised the estimate for the total adult population to approximately 611,000 to 636,000 birds. Of these birds, approximately 31,000 to 39,000 adults were estimated to reside in the United States. PIF (2021) projected a higher global estimate of 1.2 million adults. In Minnesota, a statistically robust statewide survey was conducted during the 1989 breeding season. The survey estimated the state's total adult population at 10,355 to 12,897 individuals (Strong and Baker 2000).

Following the 1989 survey, the MNDNR launched its volunteer Loon Monitoring Program in 1994, which focused on monitoring the species in six 100-lake regions, or index areas. Each area was selected to represent different regions of the species' breeding distribution. The program engages nearly 1,000 citizen scientists who visit each of the 600 lakes for one morning during a ten-day period in late June or early July every year and report the number of Loons observed.

From 1994 to 2020, the survey demonstrated that the Common Loon's population has remained relatively stable within these areas. Together, the regions average two adults per 40 ha of lake, varying from a low of approximately 0.5–1.3 adult Loons per 40 ha in the Kandiyohi index area to a high of 2.7–4.7 adult Loons per 40 ha in the Itasca index area. Overall, 67 percent of monitored lakes report Common Loon detections (Larson 2021). Minnesota's stable population trend is similar to that reported throughout much of the species' breeding range (Paruk, Evers, et al. 2021).

Although the BBS is not considered a reliable tool for monitoring Common Loon populations in some areas (Paruk, Evers, et al. 2021), the data are statistically reliable in Minnesota where the species is common, widespread, and encountered frequently enough along BBS routes to result in a statistically valid trend. These data demonstrate a nonsignificant increase averaging 0.79 percent per year from 1967 to 2019, with an even larger increase since 1993 **(D)**. Although the statewide population trends are different for the BBS and MNDNR surveys, the good news is that Minnesota's Common Loon population is stable, and more likely increasing. The BBS data may better reflect the species recolonization of portions of its former breeding range in the state, especially in east-central Minnesota, an area not included in the index areas surveyed by the MNDNR.

Historically, the Common Loon in the United States' portion of the Great Lakes region has experienced the greatest declines in both abundance and distribution (Paruk, Evers, et al. 2021). Indeed, the species is extirpated in several states where it formerly occurred (Illinois, Indiana, Iowa, and Ohio), and its breeding range has receded north since the late 1800s and early 1900s in Michigan, Wisconsin, and Minnesota. Nevertheless, the Great Lakes region now supports over 50 percent of the United States breeding population. As noted earlier, Michigan, Ontario, and Minnesota have recently documented small southern range expansions. Focused conservation efforts by resource agencies and local

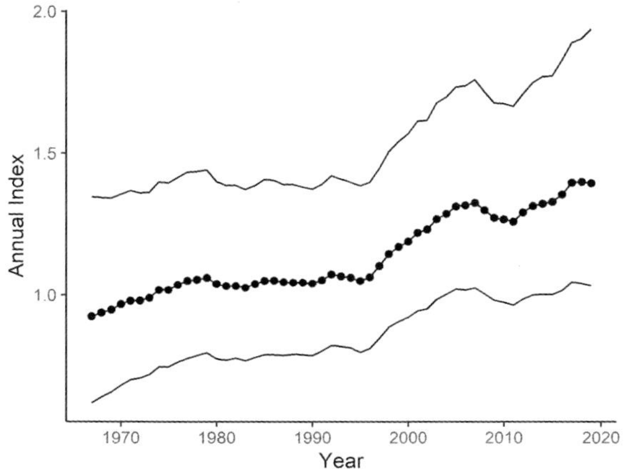

D. Breeding population trend and 95 percent credible interval (solid black lines) for the Common Loon in Minnesota based on the 1967–2019 Breeding Bird Survey (Sauer et al. 2020).

lake associations are surely one of the principal factors responsible for the population increases.

As hopeful as these trends are, there is cause for concern. A long-term investigation (1993–2019) in a large area of northern Wisconsin revealed that territorial occupancy may not be a suitable indicator of population sustainability. Indeed, even though the number of territorial pairs remained relatively stable during the 27-year study, the population experienced significant declines in chick mass, fledgling success, the number of non-breeders, and the total adult population (Piper, Grear, et al. 2020). The authors proposed that the most likely cause of the decline was a decrease in food abundance. Similar declines have been observed in Canada, especially in the eastern provinces, where the downward trend in productivity has been attributed to acid rain and mercury contamination (Tozer et al. 2013; Bianchini et al. 2020).

CONSERVATION

An iconic symbol of the northern wilderness, the Common Loon is considered a moderate conservation priority (PIF 2021). It is not federally listed but has been classified as a state *Threatened* species in Michigan and New Hamsphire (Paruk, Evers, et al. 2021). In 1961 it was legally designated Minnesota's state bird following twenty years of public discussion and debate (Strnad 1961).

Although shooting was considered a major mortality factor for the Common Loon nearly a century ago, largely because of its perceived threat to sport fishing, today's threats to the Common Loon are complex and wide-ranging. Mortality from botulism and other diseases, particularly in areas where large numbers of birds congregate during migration or in the winter, poses an ongoing risk. As recently as 2012, an estimated 3,000 Loons died from botulism poisoning on Lake Michigan, where birds congregate each fall before migrating to their wintering grounds along the Atlantic coast and the Gulf of Mexico (Taylor 2013). This was not a rare occurrence. Between 1999 and 2009, 24,000 Loons died from botulism on Lake Michigan (Porter 2013).

Water-based recreational activities and expanded shoreline development are also considered major challenges. Detailed studies in Minnesota and elsewhere have shown mixed results in this regard. Valley (1987), for example, showed a significant correlation between recreational disturbance and hatching success along the Whitefish chain of lakes, while most analyses in the BWCAW showed little impact on productivity. Indeed, recreational activity increased in the wilderness area 800–900 percent from 1950 to 1976, while the adult Loon population actually increased 35 percent (Titus and VanDruff 1981). Despite increases in lakeshore development, including an estimated 500 percent increase in seasonal housing units in the state's northern counties from 1990 to 2000 (Freshwater Society 2007), Minnesota's Loon population has remained relatively stable. A Wisconsin study, however, demonstrated that although the species may adapt to high levels of recreational disturbance, territorial Loons were either absent from lakes or not reproductively successful when the density of shoreline buildings exceeded 25 buildings per km (M. W. Meyer 2006).

Climate change may pose another threat. Warmer temperatures, for example, could negatively impact the coastal fisheries that Common Loons depend on during the nonbreeding season (Paruk, Evers, et al. 2021). Nevertheless, during the summer months, NAS (2019) predicted that under a 2°C warming scenario the species would lose only 16 percent of its North American breeding range by the year 2050. Because it would be compensated by a northward gain of 14 percent, the Loon's breeding distribution was considered to have a low vulnerability to climate change. Most of the loss, however, would occur in the northern United States, including throughout a significant portion of Minnesota.

A long-lived species high on the food chain, the Common Loon is also vulnerable to a host of environmental contaminants. One of the most thoroughly studied contaminants is mercury, which reaches northern lakes through atmospheric deposition of sulfur dioxide and nitrogen oxides and is present in all levels of the aquatic food chain. The Common Loon's fish-based diet can accumulate deadly levels of mercury that impact the bird's behavior and survival. Birds with high blood-mercury levels, for example, spend less time incubating their eggs (Paruk, Evers, et al. 2021). Fortunately, acid deposition has declined in Minnesota, but the accumulation of methylmercury in fish remains a potential threat (MPCA 2002). Equally dangerous is the ingestion of lead fishing sinkers, consumed when birds are searching lake bottoms for small pebbles to aid their digestion. One study in New England documented that lead poisoning was responsible for 44 percent of the dead Loons collected over a 14-year period (Sidor et al. 2003).

Finally, all birds that spend a portion of their lives on coastal waters are susceptible to the impacts of oil spills. The extent of the threat was magnified by the 2010 Deepwater Horizon oil spill in the Gulf of Mexico. Between April 22 and July 15, 2010, an estimated 795 million liters of oil created an oil slick that extended thousands of square kilometers and polluted nearly 1,770 km of shoreline. To combat the spill, another 3 to 3.8 million liters of toxic dispersant chemicals were also released into the Gulf waters (Weflen 2013; Bylander and Grumney 2015; Pallardy 2016). This was a major concern for Minnesota's Loon population, nearly 85 percent of which winters in the Gulf of Mexico (T. Kennedy 2016). Although adults return each spring to breed, juveniles remain on the coast for up to three years (Paruk, Evers, et al. 2021).

Approximately 900 Loons were killed directly by the Deepwater Horizon oil spill; the number of Minnesota birds that perished is unknown (Bylander and Grumney 2015; Myers 2018). Although it is a significant one-time loss, the long-term impact of directly ingesting oil and oil dispersants, or of ingesting other aquatic organisms impacted by these contaminants, remains unclear. Assessing these potential impacts is the focus of a comprehensive, long-term investigation conducted by the MNDNR, the University of Connecticut, and the US Geological Survey. The study has already detected traces of oil and oil dispersant in the blood, fat, feathers, and eggs of Minnesota Loons (Bylander and Grumney 2015). Luckily, the natural resources damage settlement for the oil spill, announced in the fall of 2018, provides $16 million for fish and wildlife rehabilitation. This includes an impressive $7.2 million for Minnesota's Loon population. Plans are underway to use the dollars to protect critical Loon breeding habitat on Minnesota lakes and to promote the use of non-toxic fishing tackle (Myers 2018). The hope is that such initiatives will help buffer the multiple threats facing Minnesota's most endearing wildlife species.

DOUBLE-CRESTED CORMORANT

Nannopterum auritum

Minnesota Seasonal Status: Regular breeding resident and migrant; occasionally observed during the winter, primarily in southern Minnesota. The Double-crested Cormorant was common during the MNBBA.
Migration: Most Cormorants that breed in Minnesota are short- to medium-distance migrants, wintering along the southern Atlantic coast, the Gulf coast to southern Mexico, and on inland lakes and reservoirs in the southeastern United States.
Food: Almost entirely small fish, obtained by diving and pursuing prey underwater.
Nest: Uses a variety of materials to construct a platform that may be placed on the ground, in trees or shrubs, or on mats of emergent wetland vegetation. A colonial nesting species.
Conservation Concern: Assigned a low PIF Continental Concern Score of 8/20.

From gluttonous and detestable to adaptable and extraordinary, the words used to describe the Double-crested Cormorant suggest an intriguing yet complicated history. Known as a supreme fish predator, the Cormorant has been both valued and vilified for centuries. In China, records document fishermen using Great Cormorants as early as 317 BCE to retrieve fish (Wires 2014). A constricting ring was often placed at the base of their throat, enabling the birds to consume small forage fish but preventing them from swallowing the larger, more valuable fish that were returned to the fisherman's boat. Although the practice has largely been replaced by modern fishing operations, centuries ago it was also employed in Japan and even spread as far as England, France, and Belgium "where it was done more for sport" (Wires 2014). In North America, on the other hand, Cormorants have long been viewed as competitors for the continent's valued fisheries resources. Persecuted and hunted since European explorers and settlers first arrived, the Double-crested Cormorant may be the only native bird that continues to be the target of so much ill-will centuries later.

Along the rivers and lakes where it nests and forages, Cormorants are easily identified by their unique behavior of resting on a pier or shoreline rock with their wings fully extended to either side. Although the function of the pose has been debated, the birds usually orient themselves to the sun or wind, leading most biologists to suggest that the behavior helps dry the water-soaked flight feathers. While on the water, the Cormorant's low-lying profile, upturned bill, and long, sinuous neck, are also diagnostic. In flight, the birds often fly in lines or in loosely organized, V-shaped flocks.

Photo by Rebecca Field.

Some of the Cormorant's most unique features, however, are visible only on close inspection. Indeed, when viewed "face to face" during the breeding season, the bird has been described as "an uncanny and rather fearsome object" (Roberts 1932). The small feather crests on its head, an orange gular pouch, a hook at the tip of the bill, brilliant blue-green eyes, and a blue mouth lining are among the striking features. The Cormorant's scientific name was recently changed from *Phalacrocorax auritis* to *Nannopterum auritum* when studies revealed its close relationship to the Flightless Cormorant (*Nannopterum harrisi*) (Kennedy and Spencer 2014). Although the word *Nannopterum* is of Greek origin and means "very small wings," the Double-crested Cormorant has normal-sized wings and is fully capable of flight. The Latin word *auritum* means "ears" and refers to the bird's rarely visible feather crests.

The Double-crested Cormorant can be found along both coastlines during the breeding season as well as at inland sites in the central Canadian provinces and the north-central United States, from Montana east through the Great Lakes. The core of its breeding population is centered in the northern Great Plains, including southern Minnesota.

MINNESOTA BREEDING DISTRIBUTION

The Cormorant certainly captured the attention of the region's Native Americans and early European explorers. Blackduck Lake, Crowduck Island, and Cormorant Point are just a few examples of Minnesota sites that clearly reference the species' abundance hundreds of years ago. The 1821 Cass expedition, for example, noted Cormorants on Lake Winnibigoshish (Schoolcraft 1821), and in 1832 Henry Schoolcraft recorded the species' presence on Leech Lake (Schoolcraft 1834). When Hatch wrote the first comprehensive account of Minnesota birds in 1892, he wrote: "Except when the water is frozen firmly, there is no time in the year when they may not be seen in almost every general section where the conditions are favorable to their habits of feeding."

Migrant flocks passing through the state in the spring and fall were often enormous in size. Roberts (1932) included a spring 1918 report from Game and Fish Commissioner Carlos Avery, referring to complaints he received of ten thousand or more Cormorants in the bottomlands of the Minnesota River. An excerpt from an April 1926 account of the species' migration north along the Mississippi River, south of La Crosse, also notes the sheer magnitude of their numbers (Grasset 1926): "The flight was so large that at times it was impossible to see the sunset sky through the mass. . . . The number of birds is variously estimated as from 100,000 to 1,000,000 birds."

Widely perceived as a menace to local fish populations, the Double-crested Cormorant was routinely molested and shot in nighttime roosts and nesting colonies during the 1800s and early 1900s. Permits to shoot the species were issued by the Minnesota Game and Fish Commission, but the species' quick demise suggests that many citizens simply took action into their own hands. Numbers had declined so precipitously by 1930 that Roberts (1932) identified only three active nesting colonies in the state that collectively supported just a few hundred birds: Gull Rock in Lake of the Woods, Crane Island in Lake Minnetonka, and Kawishiwi Lake in Lake County. Several large, well-known colonies had been abandoned, including three sites in southwestern Minnesota: Heron Lake and Loon Lake in Jackson County, and Lake Shetek in Murray County.

Despite the paucity of nesting birds, pillaging of the unprotected Cormorant continued. In 1948, the editor of the statewide bird journal *The Flicker,* Dr. Dwain Warner (1948), wrote:

> Information on the wholesale slaughter of this species on its rookeries and elsewhere in Minnesota during 1948 by organized groups and shooting parties has reached the editor's desk. Is this killing justifiable? Or are these people misinformed as to the true relationships of this species to fish and the fishing industries? Or is this unprotected bird being shot largely in the name of "sport" at times when the game seasons are closed?

The note ended with a call for more information about the species' status. Observers soon responded with reports of small colonies at widely scattered locations, including Rice Lake NWR in Aitkin County (Hofslund 1950b), Gull Rock in Lake of the Woods (Dobie 1956), and Shields Lake in Rice County (Hanlon 1956). Although the carnage continued and the species was not federally protected until 1972, populations in Minnesota and throughout North America were slowly rebounding.

Unfortunately, a new threat in the 1950s—the widespread use of pesticides—took its toll on many fish-eating species, including the Double-crested Cormorant. Contaminant levels were so high in the Great Lakes' fish populations that they likely led to the Cormorant's extirpation on Lake Superior and Lake Michigan (Dorr et al. 2020). But as contaminant levels gradually declined in the 1970s, this adaptable species once again bounced back. Populations not only stabilized but also began a slow and steady growth in numbers in the 1980s and early 1990s, followed by a period of rapid growth beginning in the late 1990s.

As populations were beginning to recover in the late 1970s, the MNDNR's Nongame Wildlife Program solicited reports of colonial waterbird nesting colonies across Minnesota that depicted the rapidly changing status of the Cormorant. In 1977, 12 active colonies were identified in eight counties (Henderson 1977). Just eight years later, 26 colonies were identified in 18 counties, including seven colonies in Lake of the Woods. One of the latter sites, Crowduck Island, supported the state's largest nesting colony, numbering 1,024 breeding pairs (Guertin and Pfannmuller 1985). During this same period, both Green and Janssen (1975) and Janssen (1987) described the species as a statewide breeding resident, common in all but the northeastern and north-central regions. By 1998, Hertzel and Janssen would identify confirmed nesting records from 28 Minnesota counties since 1970.

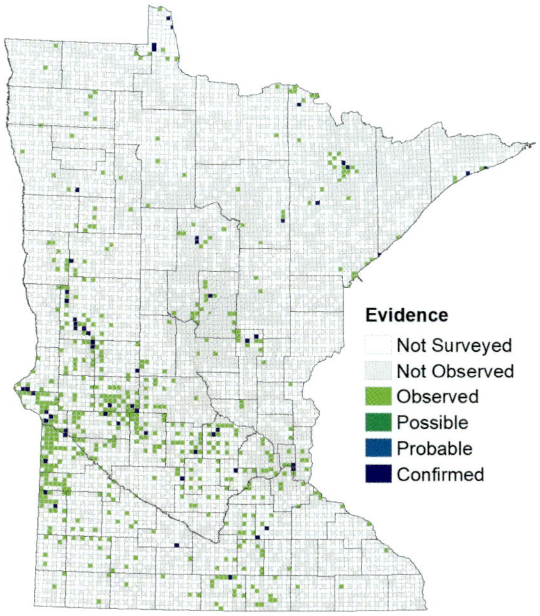

A. Breeding distribution of the Double-crested Cormorant in Minnesota based on the Breeding Bird Atlas (2009–13).

During the MNBBA, Cormorants were most numerous across central Minnesota, from the Twin Cities west to Lac qui Parle and Big Stone Counties. Participants reported 1,165 records from 12.7 percent (606/4,762) of the surveyed atlas blocks and confirmed breeding in 57 blocks **(A)**. Unless the birds were seen at the nesting colony, all records were coded as observed. Cormorants may travel up to 62 km to foraging sites (Dorr et al. 2020), so observations away from the nesting colony could not be assumed to indicate possible or probable breeding status. With reports in 77 of Minnesota's 87 counties and confirmed breeding in 30 counties, MNBBA records were augmented with data collected during a 2010 statewide survey of Double-crested Cormorant colonies (Wires, Cuthbert, and Hamilton 2011).

What Minnesotans have learned in the past one hundred years, humanity has known for millennia: the Double-crested Cormorant is an extraordinary and very adaptable species (Wires 2014). It has made a remarkable recovery from near decimation in the early 1900s, not only in Minnesota but throughout the Great Lakes and North America. Nevertheless, given the reports of its abundance in the early 1900s, particularly during migration, it certainly has not returned to its former level of abundance.

BREEDING HABITAT

A colonial species, Cormorants select a variety of sites for nesting, including islands, cliffs, coastlines, wetlands, riparian woodlands, and artificial structures, such as towers and bridges (Wires, Lewis, et al. 2010; Dorr et al. 2020). Sites that are free from mammalian predators and human disturbance, yet close to open waters with abundant fish populations, are most suitable.

Within each site, the birds may nest either on the ground **(B)** or in trees and shrubs **(C)**. The latter are usually alive when the site is first selected but die within a few years from the bird's excrement and activity (Dorr et al. 2020). When the original nesting trees die, Cormorants often remain at the site but begin nesting on the ground. On the ground, nesting substrates vary from large boulders and small,

B. After years of use, two small trees are all that remain on this bare island in Meeker County that supports a ground nesting colony of Double-crested Cormorants. Photo by USDA Wildlife Services-MN.

C. A dead tree on an island in Lake Johanna, Pope County, supports nests of the Double-crested Cormorant and Great Blue Heron. Photo by Lee A. Pfannmuller. Permission granted by the Scientific and Natural Areas Program, Minnesota Department of Natural Resources.

rocky islets to broad, flat beaches of bedrock, sand, or gravel. Occasionally, the birds nest on mats of emergent vegetation in large wetlands (D. Anderson and Ellis 1966).

Cormorants also demonstrate flexibility in selecting nesting partners. Although they nest in single-species colonies, the birds frequently co-nest with other colonial nesting waterbirds. Ground-nesting colonies are often shared with American White Pelicans, Common Terns, and Ring-billed Gulls and/or Herring Gulls; tree-nesting colonies are often shared with Great Blue Herons, Great Egrets, and/or Black-crowned Night-Herons.

POPULATION ABUNDANCE

PIF (2021) estimated the population in the continental United States and Canada at approximately 560,000 breeding adults; an earlier and more detailed analysis by the USFWS (2020a) provided a much larger estimate of 871,000 to 1.03 million birds. The Cormorant's population increase is well illustrated by data collected by the BBS. Across the survey region, from 1966 through 2019, the average annual increase was statistically significant at 2.55 percent (Sauer et al. 2020). The increase is evident in every Canadian province and in forty-three of the forty-nine continental states. In the Prairie Potholes Bird Conservation Region, where numbers are especially high, the average annual increase was 4.33 percent **(D)**. Although not statistically robust, the increase has been equally dramatic in Minnesota, where populations show an average increase of 5.76 percent per year from 1967 through 2019 (Sauer et al. 2020).

Roberts's (1932) account documented three nesting colonies, totaling approximately 125 breeding pairs. More than fifty years later, in 1984, 26 colonies had been documented and the statewide population was estimated to number approximately 3,900 to 6,500 nesting pairs (Henderson 1984). Beginning in 2004, the MNDNR initiated a periodic census of all Double-crested Cormorant and American White Pelican colonies in the state. Several emerging issues prompted the census; principal among them was the local population control that began in 2005 in response to concerns that growing Cormorant numbers were impacting local fisheries. Periodic monitoring efforts would help state biologists assess how these control efforts were affecting the statewide Cormorant population. Because Pelicans frequently nest in association with Cormorants, it was equally important to ensure that Cormorant control activities were not impacting co-nesting species. The census is conducted every five years; results from three of these surveys are available (Wires, Haws, et al. 2006; Wires, Cuthbert, and Hamilton 2011; Hamilton and Cuthbert 2016) and are displayed, along with estimates dating back to 1932, in **(E)**. Although colonies may move from year to year, population numbers have remained relatively stable from 2004 to 2015 with approximately 15,500–16,000 breeding pairs nesting at 36–42 colony sites. Significant culling operations that are underway in Minnesota, discussed below, have likely contributed to the recent leveling off of population numbers.

CONSERVATION

Due to its wide distribution and growing population, the Double-crested Cormorant is a low conservation priority (PIF 2021). Indeed, its successful recovery has led to growing public complaints and resource conflicts. Among the principal issues raised are concerns that Cormorant populations are impacting local fish populations, decimating fisheries' stocks at aquaculture facilities and natural ponds managed for bait fish production, negatively impacting vegetation at nesting colonies, and serving as a vector for diseases such as West Nile virus and Newcastle disease (MNDNR 2019e). At least one study in North Dakota documented the spread of Newcastle disease from Cormorants to domestic turkeys (Heckert et al. 1996).

In response to these concerns, the USFWS established a series of regulatory tools that enable private and public entities to address damage caused by the bird. The first tool is a depredation permit that enables private landowners to use many options to alleviate damage to their property caused by federally protected species, including Cormorants. Between 1999 and 2012, an average of 529 Cormorants was taken each year in Minnesota under authorized depredation permits. The numbers began increasing considerably beginning in 2008 **(F)**.

The second tool is a permit authorized under the federal Aquaculture Depredation Order that allows private, state, and federal fish-hatchery owners in thirteen states to take Cormorants that are "committing or about to commit depredations to aquaculture stocks." The Aquaculture Depredation Order was established in 1998 and updated in 2003; since 2003, an average of 1,713 birds was taken each year at Minnesota hatchery facilities **(F)**.

Finally, take was also authorized under the Public Resource Depredation Order. Established in 2003, the order authorizes federally recognized tribes, the US Department of Agriculture's Wildlife Services office, and state resource agencies "to reduce the occurrence and/or minimize the risk of, adverse impacts of Cormorants to public resources." Numerous requirements had to be met before actions could be

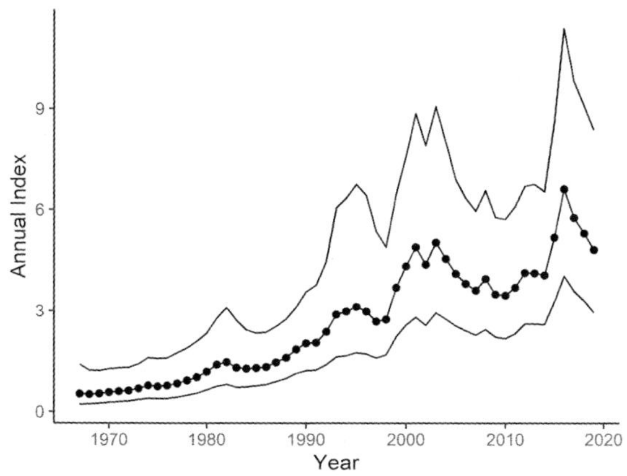

D. Breeding population trend and 95 percent credible interval (solid black lines) for the Double-crested Cormorant in the Prairie Potholes Bird Conservation Region based on the 1967–2019 Breeding Bird Survey (Sauer et al. 2020).

taken using the authorities granted by this order. In Minnesota, the Public Resource Depredation Order was used to control the Double-crested Cormorant on lakes where public fisheries were considered at risk. Between 2005 and 2012, a total of 19,694 Cormorants were taken under this authority **(F)**. The majority were taken at Leech Lake, where there were concerns about the lake's fisheries and about the Cormorant's impact on nesting Common Terns.

Collectively more than half a million birds (506,319) were taken across the species' range in the United States from 1999 to 2012 under the legal authorities mentioned above (USFWS 2014). More recent data is not available for Minnesota, but at the national level "from 2007 to 2018, an average of 51, 456 cormorants were taken each year (range 14,181 to 84, 410) under all available authorities within the 48 states" (USFWS 2020a). Most of the birds ("on average, 41%") were taken each year under the former Public Resource Depredation Order.

Although many factors—from environmental contaminants to disturbance to rising sea levels—have impacted and will continue to impact the distribution and abundance of the Double-crested Cormorant (Dorr et al. 2020; NAS 2019), one of the most significant has been and may continue to be conflicts with humans over fishery resources. Primarily a fish-eating species, the Cormorant is an opportunistic feeder. Most fish consumed are small, less than 15 cm in length (Dorr et al. 2020). Countless studies have examined the species' feeding habits and the wide variety of prey consumed. Losses due to Cormorant predation are clear and well-documented at confined aquaculture facilities (Wires and Cuthbert 2003). The species' impacts on natural fish populations, however, are less clear. A complex array of factors can affect local fisheries in a given year, and teasing out the contribution of one single factor, such as Cormorant predation, at the exclusion of all others is extremely difficult.

In Minnesota, an intensive study of the fish population in Leech Lake, where one of the largest Cormorant colonies resides, concluded that "predation pressure by the expanding Cormorant population could have been sufficient to impact walleye survival to older age classes" (MNDNR 2019e). A collaborative effort to reduce the Cormorant population on the lake and maintain it at a level of 500 breeding pairs was initiated in 2005, after extensive review and public input.

Unfortunately, the public is often quick to ascribe poor fishing success to the local presence of Cormorants. As a result, management operations have taken place not only at Leech Lake but also, using the authority granted by the Public Resource Depredation Order, at Potato Island on Lake Vermilion. Efforts to discourage nesting were also attempted for a season at Knife Island on Lake Superior. Control efforts were undertaken at Lake Waconia in Carver County and Wells Lake in Rice County under the authority of the depredation permits for animal damage on private property (Hamilton and Cuthbert 2016).

In 2014, the USFWS moved to extend the effective date of the Aquaculture and Public Resource Depredation orders for five years, through June 2019. In response to this proposed extension, a lawsuit was filed by the Public Employees for Environmental Responsibility, arguing that the USFWS had failed to examine the long-term consequences of Cormorant control. Upon hearing the case, a US District Court then rendered the order null and void until the USFWS could take measures to ensure that they are in full compliance with the National Environmental Policy Act. Subsequently, the USFWS issued a new rule that took effect February 21, 2021, authorizing the take of Cormorants under specific conditions and, when they are implemented, "within the limits and allocations set by the Service" (USFWS 2020c). The allocations are detailed in an accompanying Environmental Impact Statement (USFWS 2020a).

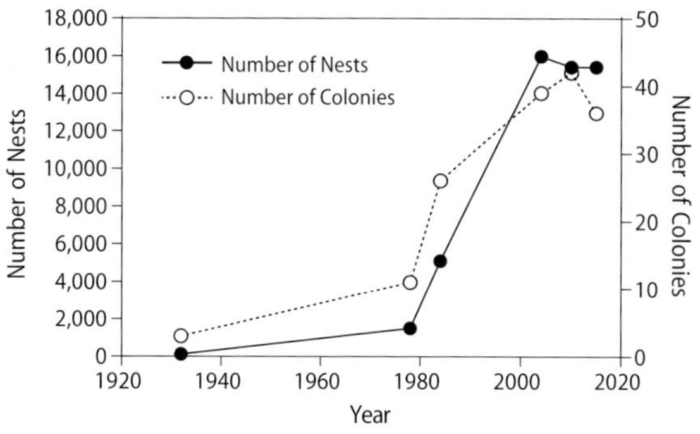

E. Double-crested Cormorant population numbers in Minnesota (number of nesting pairs and number of colonies) (compiled from Roberts 1932; Henderson 1978, 1984; Hamilton and Cuthbert 2016).

In reviewing the conservation challenges imposed by these sociopolitical conflicts, Wires and Cuthbert (2006) wrote: "Although managing fish-eating birds to benefit fishery yields may increase some fish populations, this approach does not resolve or address the underlying problems causing current fish population declines across the continent."

It is hoped that an informed and conservation-conscience citizenry in the twenty-first century will ensure that control efforts are undertaken only when absolutely necessary and that the Double-crested Cormorant, like all fish-eating birds, from loons to grebes to pelicans, will always continue to have a proper place in healthy ecosystems.

F. Double-crested Cormorant take in Minnesota with different regulatory tools from 1999 to 2012.

Regulatory Tool	1999	2000	2001	2002	2003	2004	2005	2006	2007	2008	2009	2010	2011	2012	Total
Depredation permits	24	20	9	13	9	25	50	50	0	374	868	1,700	2,398	1,859	7,399
Aquaculture Depredation Order					1,249	2,553	1,857	1,865	1,246	1,676	1,725	1,667	1,748	1,540	17,126
Public Resource Depredation Order							2,793	3,103	2,461	2,601	3,084	2,222	1,848	1,582	19,694
Total	**24**	**20**	**9**	**13**	**1,258**	**2,578**	**4,700**	**5,018**	**3,707**	**4,651**	**5,677**	**5,589**	**5,994**	**4,981**	**44,219**

Source: USFWS 2014.

AMERICAN WHITE PELICAN

Pelecanus erythrorhynchos

Minnesota Seasonal Status: Regular breeding resident and migrant. The American White Pelican was common during the MNBBA.
Migration: Short- to medium-distance migrant that winters on coastal waters in the southern United States and Mexico and, rarely, on inland lakes and rivers.
Food: Feeds primarily on rough fish caught by scooping prey into its pouch. Often employs a cooperative feeding strategy, in which a group of Pelicans drives fish into shallow waters where they are easier to catch.
Nest: Shallow ground depression; may be lined with vegetation. A colonial nesting species, Pelicans often nest with other waterbirds, particularly Ring-billed Gulls and Double-crested Cormorants.
Conservation Concern: Assigned a moderate PIF Continental Concern Score of 10/20 and officially listed as *Special Concern* in Minnesota.

In western Minnesota, the American White Pelican is a giant among the landscape's small grassland sparrows, secretive marsh birds, and nesting waterfowl. One of the largest birds in North America, it has a wingspan of 2.7 m. Roberts (1932) wrote that, while it is an "awkward and somewhat grotesque appearing bird" on land, "on the water it floats in a stately, dignified manner with the great bill and pouch held closely against the neck and breast." But in the air the Pelican excels, maintaining "the dignity of its position in the grandeur of its flight" (Bent 1922).

The adult's white plumage is offset by its black wing feathers, orange feet, and enormous orange bill. *Erythrorhynchos,* the species' Latin name, translates to "red beak," a somewhat inaccurate description of its most distinctive feature. The bill's lower mandible has a large, expandable gular pouch that serves several functions, including scooping up fish while foraging, delivering threats to other Pelicans, and regulating its body temperature. On hot summer days, the Pelican vibrates its vascularized pouch to cool down. A vertical plate, or horn, adorns the upper mandible. Once incubation is underway, the horn is shed and white plumes on the crown and back of the head are replaced with darker feathers.

The Pelican breeds in widely dispersed colonies that stretch from British Columbia east to Ontario and south through the western United States; a small nonmigratory population occurs in Mexico. The core of the breeding population is located in the northern Great Plains.

Photo by Rebecca Field.

MINNESOTA BREEDING DISTRIBUTION

The preponderance of geographical places that are named for the Pelican is testament to the species' presence in the state long before European settlers arrived (DiMatteo 2016). Lake Shetek in Murray County, for example, incorporates the original Ojibwe name for the Pelican, *shetek.* Then there are the Pelican River, Pelican Point in Heron Lake, Pelican Island in Leech Lake, six Pelican Lakes strewn across northern Minnesota, and the town of Pelican Rapids!

Although these place-names provide a clue to the species' abundance and distribution, DiMatteo (2016) found it difficult to accurately assess the bird's historical status. Early accounts mention its presence on Lake of the Woods, on the Red River, and throughout west-central and north-central Minnesota. It was described by many as a common breeding resident (e.g., Bullis 1892; Hatch 1874; Roberts 1932). Detailed descriptions of nesting colonies, however, were rare. Roberts (1932) identified only two sites with definitive nesting evidence: the Mustinka River in Grant County (eggs collected in 1878); and Big Sandy Lake in Aitkin County (two prefledglings shot in 1904). Other sites that may have supported breeding colonies included Heron Lake in Jackson County, Pelican Lake in Grant County, Lake Shetek in Murray County, and the Leech Lake vicinity. Roberts pointed out, however, that "these reports were more or less in the nature of rumors and, as no actual nests were found, were probably based on the presence of nonbreeding birds, spending the summer on various lakes."

Regardless of their breeding status, by the early 1900s, sport shooting and human disturbance had taken their toll and were responsible for major population declines in Minnesota and throughout the species' breeding range (Knopf and Evans 2020). Confirmed nesting reports were absent in the state from 1904 until 1968, when approximately 65 to 75 pairs were found nesting on an island in Marsh Lake in the Lac qui Parle game refuge (Breckenridge 1968a). In the years that followed, this small colony grew exponentially to support thousands of breeding pairs. Other colonies also gradually appeared, including on Lake of the Woods (Hirsch 1981), Leech Lake (Mortensen 2000), Pigeon Lake in Meeker County (Wiens 2003), and Minnesota Lake in Faribault County (Fall 1994b).

The Pelican's return to Minnesota is attributed to the growth of the Chase Lake colony in North Dakota. Established in the early 1900s, the site is now among the largest White Pelican colonies in North America. Because Pelicans do not breed until at least their third year, young birds may wander long distances from the colony until they reach sexual maturity (DiMatteo 2016). Although many will eventually return to their natal breeding colony, if the site is already at carrying capacity, they will explore other suitable sites.

The growing abundance of Pelicans was well-documented by the MNBBA. Aided by the addition of data from a statewide survey of American White Pelican and Double-crested Cormorant colonies (Wires, Cuthbert, and Hamilton 2011), 1,270 Pelican records were reported from 14.7 percent (697/4,749) of the surveyed atlas blocks. Breeding was confirmed in 13 blocks **(A)**; all records outside of known colonies were coded as observed. Statewide, the birds were reported in 67 of Minnesota's 87 counties, a sight that would have been unimaginable one hundred years ago!

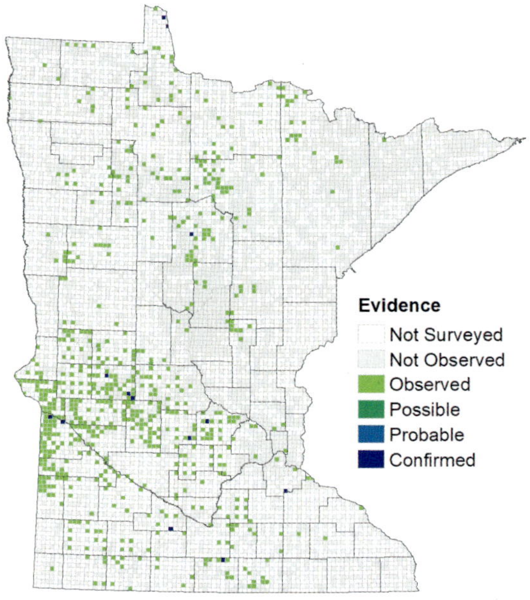

A. Breeding distribution of the American White Pelican in Minnesota based on the Breeding Bird Atlas (2009–13).

B. American White Pelicans nesting with Double-crested Cormorants on an island in Lake Johanna. Photo by Lee A. Pfannmuller. Permission granted by the Scientific and Natural Areas Program, Minnesota Department of Natural Resources.

BREEDING HABITAT

In the northern Great Plains, Pelican colonies are located on islands found on freshwater lakes, rivers, or impoundments. Mainland peninsulas are occasionally used, but reproductive success is usually lower owing largely to higher rates of nest predation. Nesting islands are typically flat, with little or no vegetation, and may be located 50 km or more from foraging sites **(B)**. The latter include a wide range of shallow water habitats that supply an abundance of forage fish. The low site tenacity demonstrated by Pelicans is an adaptation that enables them to relocate if a site becomes unsuitable due to changing water levels, disturbance by humans or predators, or changes in food availability (Knopf and Evans 2020).

POPULATION ABUNDANCE

Recent assessments of the size of the population in the United States and Canada generated estimates of 134,000 breeding adults by King and Anderson in 2005 and a significantly higher estimate of 410,000 birds by PIF in 2021. These numbers compare to a conservative estimate of 30,000 adults by Thompson in 1933. Surveywide, BBS data corroborate the increase, documenting a statistically reliable, nonsignificant increase averaging 3.08 percent per year from 1966 to 2019 **(C)**.

In Minnesota, biologists have tracked the Marsh Lake Pelican colony for decades, but the first statewide survey was launched in 2004 to monitor Cormorant and Pelican nesting colonies statewide. The most recent survey in 2015 tallied a total of 16,406 breeding Pelican pairs and nests at eight colony sites **(D)**, down from earlier years owing largely to water level fluctuations at Marsh Lake (Hamilton and Cuthbert 2016). Nevertheless, in any given year, Marsh Lake contributed 63–84 percent of the total statewide breeding population. Designated a Minnesota Stewardship Species by Audubon Minnesota (Pfannmuller 2012), the continental significance of Minnesota's White Pelican population is clear. Indeed, Marsh Lake is one of the largest colonies in North America, often vying for first or second place with North Dakota's Chase Lake colony. The latter supported only 7,100 nesting pairs in 2020 because of high water levels (MacPherson 2020).

Legal protection aided the species' recovery in the early 1900s, but efforts to protect and restore wetland habitats have been equally important. The rapid expansion of the aquaculture industry in the southeastern United States may also be improving the overwinter survival of birds that winter along the Gulf coast (King and Anderson 2005).

CONSERVATION

Minnesota has a unique obligation to responsibly manage and protect this spectacular species. The potential impacts of warming temperatures, however, are a concern. Not only is the number of wetlands predicted to decline, but more intense rain events may cause more frequent flooding of nesting sites (NABCI 2010). Nevertheless, under a 2°C warming scenario, NAS (2019) classified the Pelican as a climate-stable species when models predicted its current breeding range would decrease 18 percent by 2050 but expand by 56 percent in new locations. Although the species could disappear from southwestern Minnesota, it would increase in abundance farther north.

In the short term, protecting nesting colonies from disturbance is the most important management strategy. Pelicans are especially sensitive to disturbance early in the nesting season, when an entire colony may abandon its eggs and young. Recent efforts to legally control nesting Cormorant populations throughout the Great Lakes, including Minnesota, have also raised concerns about the unintended disturbance to neighboring Pelicans.

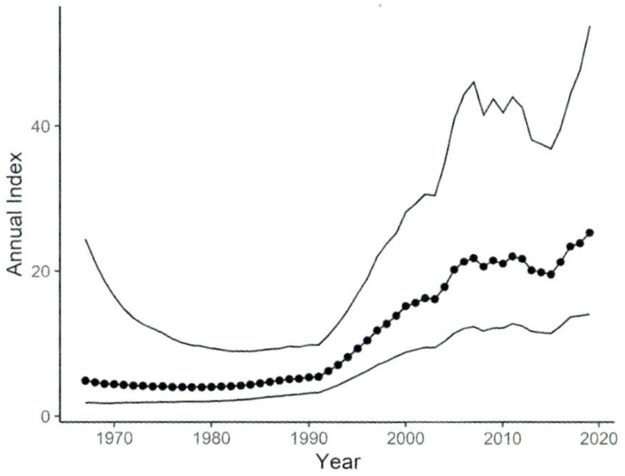

C. Surveywide breeding population trend and 95 percent credible interval (solid black lines) for the American White Pelican based on the 1967–2019 Breeding Bird Survey (Sauer et al. 2020).

Unfortunately, as Pelican numbers increase, more opportunities for conflict occur. An incident in Meeker County in 2011 is an example. When high water levels forced a nesting colony from an island in Minnesota Lake to the adjacent shore, a farmer destroyed more than 70 percent of the nests and nearly 3,000 adults abandoned the colony. The farmer was subsequently sentenced and fined $12,500 (US Department of Justice 2011). Some anglers have also expressed concern about the impact of Pelicans on local fish populations. Studies have repeatedly demonstrated, however, that this magnificent bird's diet is composed primarily of rough forage fish with relatively low economic value.

D. American White Pelican breeding colonies and nest counts in Minnesota, 2004/5–2015.

Site Name	County	2004/2005	2010	2011	2012	2015
Big Twin Lake	Martin	16	0	0	0	0
Goose Lake	Pope	0	NC	0	0	48
Hanska Lake	Brown	NA	3	0	0	P
Lake Hassel	Swift	19	0	0	0	0
Lake Johanna	Pope	97	735	1,203	1,904	1,966
Red Lake	Beltrami	340	0	0	0	0
Swartout Lake	Wright	49	913	11	176	378
Lake of the Woods[a]	Lake of the Woods	569	1,028	1,312	1,036	877
Leech Lake[a]	Cass	11	174	239	314	357
Marsh Lake[a]	Big Stone/Lac Qui Parle	13,178	11,233	16,581	15,544	10,289
Minnesota Lake[a]	Faribault	974	1,370	1,887	1,868	979
Pigeon Lake[a]	Meeker	357	543	1,273	1,181	1,512
Total		**15,610**	**15,999**	**22,506**	**22,023**	**16,406**

Source: Hamilton and Cuthbert 2016, adapted from table 3.
[a]Denotes sites with multiple islands and/or with an adjacent peninsula. NA: Not Applicable. NC: Not Counted. P: Present but not nesting.

AMERICAN BITTERN

Botaurus lentiginosus

Minnesota Seasonal Status: Regular breeding resident and migrant. The American Bittern was uncommon during the MNBBA.
Migration: Northern populations are short- to medium-distance migrants, wintering from the southern United States to Central America; southern birds may be year-round residents.
Food: Feeds on invertebrates, amphibians, small fish, and small mammals.
Nest: Platform constructed with emergent wetland vegetation and placed above standing water.
Conservation Concern: Assigned a moderate PIF Continental Concern Score of 12/20.

The American Bittern is an intriguing species. This large, solitary member of the heron family stands from 60 to 85 cm in height and has a wingspan averaging 92 to 115 cm. It should be easy to spot a bird this size as it moves among the cattails and wetland shrubs, but many have been fooled as a Bittern stands motionless just a few meters away. The bird's buffy-brown plumage, splashed with spots of beige and black (i.e., *lentiginosus*), blends perfectly with the surrounding vegetation. Distinctive features include a long, pointed bill; yellow-green legs; rufous-brown vertical stripes stretching from the white throat down to the belly; and a black line extending along each side of the neck.

But it is the male's unique *pump-er-lunk* call that captures one's attention and is the source of many of the bird's colorful colloquial names, including "thunder-pumper," "stake-driver," and "water-belcher." The performance begins with a series of bill-snapping clicks, followed by "abrupt contortions of the head and neck unpleasantly suggestive of those of a person afflicted by nausea" that enable the bird to inflate his esophagus and produce loud, guttural sounds reminiscent of a wooden pump handle (Brewster 1902). The genus name *Botaurus* is a reference to this deep, bellow-like call.

Those who spot an American Bittern may be treated to another unusual behavior. When the bird senses danger, it stands erect with its neck stretched upward and its bill pointed skyward. If a breeze begins to sway the surrounding vegetation, the Bittern sways in the same direction. And, if the danger it perceives moves, the bird adjusts its stance to keep its eyes focused on the potential threat (Bent 1926; C. Evans 1938).

Photo by Rebecca Field.

Found from British Columbia east to the Maritime Provinces and south through the northern United States, the American Bittern is most abundant during the breeding season in the northern Great Plains.

MINNESOTA BREEDING DISTRIBUTION

Both Hatch (1892) and Roberts (1932) devoted far more of their accounts on this secretive marsh bird to its unique behavior and "pump-handle" call than to their assessment of its statewide distribution. "The concealing coloration and the solitary, stealthy habits of the Bittern," Roberts wrote, "lead to its being comparatively little known, though it is an abundant bird throughout the state wherever suitable lowlands exist." Confirmed nesting records, however, were limited to nine counties north of a line between Lac qui Parle County in the west and Hennepin County in the east.

Green and Janssen (1975) also made brief note of the Bittern's statewide distribution, adding only that it was most abundant in central Minnesota. A few years later, however, Janssen (1987) wrote that it was confined to counties north of the Minnesota River, having declined "drastically" in the southern and western regions of the state in recent years. Indeed, just a few years earlier, the Bittern was officially listed as *Special Concern* in the state due to its absence from much of southern Minnesota (Coffin and Pfannmuller 1988). Janssen (1987) identified eight northern counties with confirmed nesting since 1970. Hertzel and Janssen (1998) later deleted one of the counties (Clearwater) but added three more, all north of the Twin Cities. By 2014, the MBS identified 353 breeding-season locations for the species, the vast majority located within the northern forests and aspen parklands.

Little had changed when MNBBA participants took to the field and reported 625 American Bittern records in 10.2 percent (484/4,746) of the surveyed atlas blocks. Once again, MNBBA records were most abundant in the Laurentian Mixed Forest and Tallgrass Aspen Parklands Provinces. Few were reported in central Minnesota, while records were entirely absent from the core of the Red River valley and in the Paleozoic Plateau of southeastern Minnesota. Records south of the Minnesota River were limited and most common in southwestern Minnesota. Breeding was confirmed in just seven blocks, while probable breeding was reported in 120 blocks **(A)**. Reported from 52 of Minnesota's 87 counties, two of the five counties where breeding was confirmed were additions to the list compiled by Hertzel and Janssen in 1998 (Koochiching and Lac qui Parle).

The MNBBA probability map identifies northern Minnesota as the core of the species distribution **(B)**. Although breeding densities are low across the region, higher densities are predicted in some of the expansive wetlands located in the northwest corner of the state. This suggests that the size and landscape

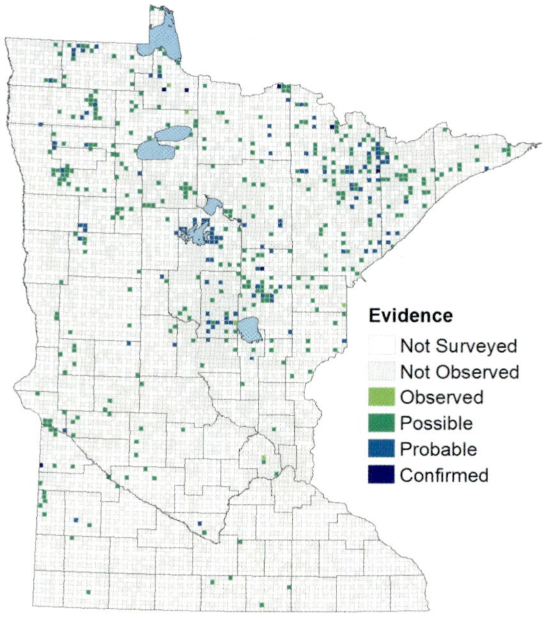

A. Breeding distribution of the American Bittern in Minnesota based on the Breeding Bird Atlas (2009–13).

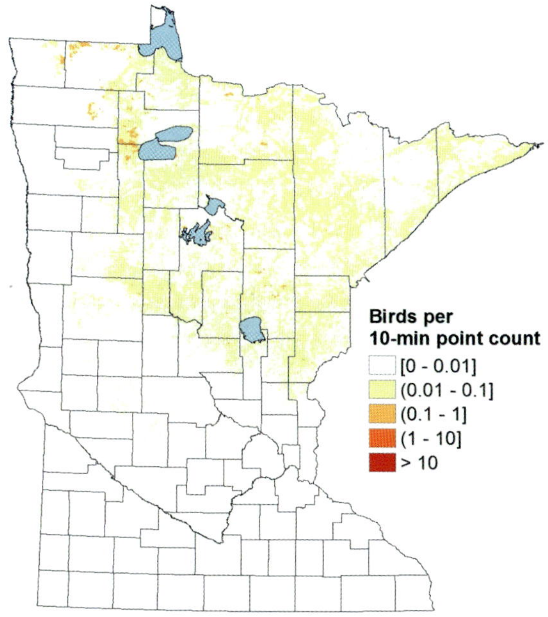

B. Predicted breeding distribution (birds detected per 10-minute point count) of the American Bittern in Minnesota based on habitat, landscape context, and climate data gathered during the Minnesota Breeding Bird Atlas (2009–13) using the Generalized Linear Modeling method without an adjustment for detectability.

context of these wetlands may be important criteria in determining habitat suitability. The model, however, appears to underpredict the species' abundance in west-central Minnesota.

Unlike one hundred years ago, the American Bittern no longer has a statewide distribution, and its center of abundance appears to have shifted north. A nearly identical change has occurred in Wisconsin. Most abundant in the southern regions of the state in the early 1900s, the species receded north to central Wisconsin by the mid-1900s. When the state's first breeding bird atlas was completed in 2000, the center of abundance again shifted farther north (Cutright et al. 2006). A similar retreat was documented in Michigan (Chartier et al. 2013). Although the loss of wetlands has contributed to the observed changes, the northern shift may be part of a much longer, historical change in the species' distribution. Lowther, Poole, and colleagues (2020) speculated that the Bittern may be adapted to northern climates and has moved north as the glaciers retreated, unable to tolerate the warmer southern environments.

BREEDING HABITAT

Bitterns utilize permanent and semipermanent wetlands dominated by dense stands of emergent vegetation, willow and alder shrublands, and wet meadows **(C)**. Grasslands are also used, but sites that have been recently mowed, heavily grazed, or burned are usually avoided. Suitable grasslands are usually within close proximity to wetlands (Svedarsky 1992; Dechant, Sondreal, et al., "American Bittern," 2004). The species' area-sensitivity was demonstrated by a Minnesota study that found Bitterns tended to avoid wetlands located less than 200 m from agricultural lands (Saunders et al. 2019).

In late summer, the Bittern confines itself to the dense cover and protection of permanent wetlands, where it undergoes a nearly complete molt that leaves it flightless. During this period, it requires deep, stable water levels that provide sufficient food resources and protection from predators (Azure 1998).

POPULATION ABUNDANCE

The BBS under-samples wetland habitats but provides the only range-wide population data for this solitary nesting marsh bird. Biologists have used those data to generate a North American population estimate of 2.5 million breeding adults (PIF 2021). No statewide estimate is available.

BBS trend data across the survey region lack statistical precision but suggest a relatively stable American Bittern population, with a nonsignificant decline of −0.74 percent per year from 1966 to 2019 (Sauer et al. 2020). Unfortunately, the species is not experiencing a significant population increase anywhere within its breeding range. The Prairie Potholes Bird Conservation Region, the center of its breeding distribution, is one of the few areas where BBS data are reliable but also illustrate an average, nonsignificant decline of −0.98 percent per year. The Great Lakes Marsh Monitoring Program documents a significant population decline of −2.7 percent per year from 1996 through 2013 across the southern Great Lakes (Tozer 2016). A similar downward trend in Bittern numbers is evident in Minnesota (Sauer et al. 2020), where extensive wetland drainage across the western and southern counties has likely contributed to the loss.

CONSERVATION

Designated a Minnesota *Special Concern* species in 1984, the American Bittern was delisted in 1995, when field surveys found a relatively large, well-distributed population throughout the state's northern forests (MNDNR 1995). Habitat loss and degradation, however, remain significant threats, leading to its designation as a moderate conservation priority across North America (PIF 2021).

Saunders and her colleagues (2019) recommended that wetland restoration and conservation efforts that are most likely to benefit American Bitterns are those located in the northern forest region "or within areas containing larger intact wetland complexes." Isolated wetlands embedded in agricultural landscapes are unlikely to provide suitable habitat. Large, public wetland complexes should also be managed to mimic natural water level fluctuations; complete wetland drawdowns prior to mid-August, when Bitterns are molting, can be detrimental (L. H. Frederickson and Reid 1986).

The importance of protecting wetlands in northern Minnesota was reinforced by recent climate change models (NAS 2019). Although the species' North American breeding distribution demonstrated a low vulnerability to warming temperatures under all three warming scenarios NAS examined, the species will gradually retreat even farther to the north. Indeed, under the 3°C warming scenario, by the year 2080 it is projected to disappear from all but the very northern reaches of the state. Unfortunately, the American Bittern's unique *pump-er-lunk* call may become a distant memory to many Minnesotans.

C. Large, diverse wetlands covered with tall stands of emergent vegetation and embedded within larger tracts of public land are critical for American Bitterns. Photo by Lee A. Pfannmuller.

LEAST BITTERN

Ixobrychus exilis

Minnesota Seasonal Status: Regular breeding resident and migrant. The Least Bittern was uncommon during the MNBBA.
Migration: Short- to medium-distance migrant that winters in the southeastern United States, Mexico, Central America, and the Caribbean; some southern populations may be nonmigratory.
Food: Small fish and aquatic invertebrates compose the primary food items.
Nest: Loosely woven platform of marsh vegetation placed slightly above the water surface.
Conservation Concern: Assigned a moderate PIF Continental Concern Score of 11/20.

Photo by Travis Bonovsky.

The Least Bittern is the smallest and "least" well-known member of the heron family. Less than half the size of its relative, the American Bittern, the Least Bittern has "quiet, retiring habits" (Bent 1926) that challenge those in pursuit of the bird. Even if one navigates its deepwater, wetland habitat, there is little guarantee of success. "The elusiveness and quiet disposition of the Least Bittern, combined with its slender form and perfect mimicry of its surroundings in color and action, cause it to be overlooked even where it is dwelling commonly" (Roberts 1932).

If you are lucky enough to catch a glimpse, the species' most distinctive features are the blackish-green crown and back, chestnut stripe behind the eye, striped neck, chestnut-colored wings, and yellow bill. A buffy-white line of feathers on both sides of the back is diagnostic. The female is just a little lighter in color than the male. The male's quiet *coo-coo-coo* call resembles a less musical Black-billed Cuckoo, but a variety of other quacking calls are also given by both sexes. The only similar species is the Green Heron, which is larger in size and darker in coloration, with a reddish-brown neck and a greenish-brown crown, back, and wings.

Like the American Bittern, the Least Bittern has the unusual habit of freezing in an upright stance when alarmed. The feathers are compressed against the body and the head and bill are pointed upward. Cryptically melting into the surrounding habitat, the bird may also sway back and forth as though it were a cattail or bulrush in the wind.

The Least Bittern differs from the taller herons and egrets that use their long legs to wade through shallow wetlands, searching for food. Instead, this small Bittern usually deftly maneuvers through the dense emergent vegetation by grasping shoots with its feet and then dipping down to snatch food morsels in the deeper waters below. Such behavior enables it to utilize deepwater sites that might otherwise seem unsuitable for such a small bird (Bent 1926; Poole, Lowther, et al. 2020).

The species' breeding distribution stretches from the coast of Canada's Maritime Provinces, south to the Caribbean and portions of Mexico and Central America, and west through the eastern United States. Absent through the Appalachian region, small, scattered populations also occur farther west. Owing to its secretive habits, its relative abundance across its breeding range is poorly known. The species' unusual scientific name adds to its mystique. The Swedish biologist who named it apparently was not a Greek language scholar, labeling it a little (*exilis*) mistletoe eater (*Ixobrychus*) (Sandrock and Prior 2014).

MINNESOTA BREEDING DISTRIBUTION

In the early 1900s, Roberts (1932) wrote that "this nimble little acrobat of the reeds is a common inhabitant of the more extensive sloughs and cane-fringed lakes of southern and western Minnesota." Although observations were apparently common from Heron Lake in Jackson County, north to the Canadian border, and east along the Minnesota River valley, confirmed nesting records were available from just six counties: Carver, Hennepin, Meeker, Marshall, Otter Tail, and Polk. Commenting on the lack of records from northern Minnesota, Roberts wrote, "Our files contain no records for the heavily forested portion of the state although suitable lowlands are not lacking."

Green and Janssen (1975) described the Least Bittern as a resident throughout much of the state, with the exception of northeastern Minnesota and the adjacent counties of Itasca and Koochiching. Janssen (1987) included a distribution map that excluded northeastern Minnesota and all or portions of the adjacent north-central counties, the Red River valley, and all of southeastern Minnesota with the exception of the Mississippi River valley. He noted that the number of observations in southern Minnesota was increasing, but it was unclear if this represented an increase in Least Bitterns or an increase in the number of observers. Seven counties where nesting had been confirmed since 1970 were delineated: Clearwater, Marshall, Nicollet, Polk, Pope, Stearns, and Wright. Hertzel and Janssen (1998) later deleted Clearwater from the list. By 2014, the MBS had documented 68 breeding-season locations. Surpris-

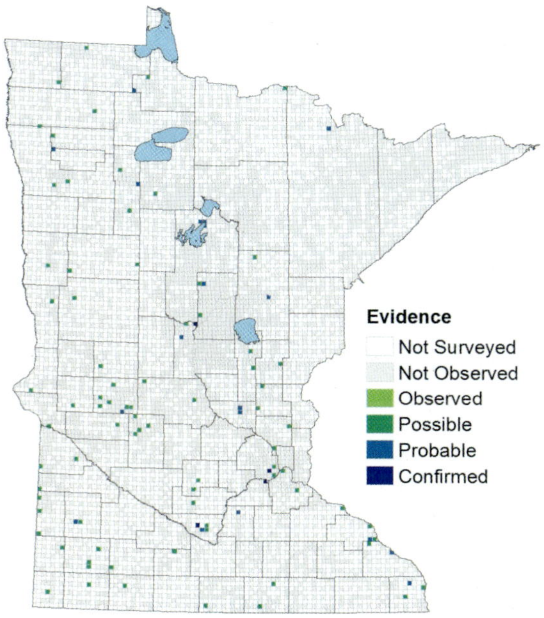

A. Breeding distribution of the Least Bittern in Minnesota based on the Breeding Bird Atlas (2009–13).

ingly, more than one-third of the records were from wetlands in southwestern Minnesota.

During the course of the MNBBA, participants reported Least Bitterns in 2.2 percent (103/4,741) of the surveyed atlas blocks. Breeding was confirmed in only four blocks but considered probable in 19 blocks **(A)**. Reported in 57 of Minnesota's 87 counties, the counties where breeding was confirmed included: Cass/Morrison (the block straddled both counties), Hennepin, Scott, and Nicollet. All but Nicollet were additions to the list published by Hertzel and Janssen in 1998. In the years following the MNBBA, through the 2019 breeding season, confirmed nesting records were also added from Jackson, St. Louis, Sherburne, Ramsey, and Washington counties (MOU 2021b), further expanding the species' known breeding range.

Although the MNBBA records were few in number, they were widely scattered throughout many regions of the state. The species remains rare throughout northeastern Minnesota, from Pine County north across the Arrowhead region and west to Beltrami, Itasca, and Koochiching Counties. In southeastern Minnesota, reports were restricted primarily to the Mississippi River valley. It is difficult to know if the paucity of Least Bittern records is an accurate representation of its abundance or a consequence of its secretive nature.

Roberts's (1932) broad assertions regarding the species' abundance in southern and western Minnesota in the late 1800s and early 1900s suggest that the small number of MNBBA records represents a decline in overall abundance. The significant loss of wetlands, especially in southern Minnesota, certainly supports that conclusion. Nevertheless, it seems that the overall extent of the Bittern's distribution in the state has changed very little. It is encouraging that, despite the extensive wetland loss in the agricultural region, the species still retains a presence there. Although there are sufficient wetlands to support a robust population of Least Bitterns in the northeastern and north-central counties, these habitats appear to be unsuitable to the species, whose Minnesota breeding distribution represents the northern periphery of its continental breeding range.

A comprehensive review of the species in 2009 makes no mention of historical change to its distribution anywhere within its breeding range (Poole, Lowther, et al. 2020), although the loss of wetlands has certainly resulted in a decline in abundance.

BREEDING HABITAT

Least Bitterns nest in freshwater and brackish wetlands with dense stands of emergent vegetation **(B)**. Clumps of wetland shrubs may also be present. Hemi-marshes, with an equal dispersion of vegetation and open water, are preferred. Drier wetland sites are usually avoided, as are sites with dry, muddy openings and/or shorter vegetation. In addition to natural wetland basins and floodplain wetlands, impoundments are also used (Poole, Lowther, et al. 2020).

POPULATION ABUNDANCE

Estimates of the size of the Least Bittern population are, at best, professional guesses. In 2001, biologists proposed a rough global estimate of only 128,000 breeding adults (Wetlands International 2012); PIF (2021) estimated a global population of 310,000, with approximately 130,000 in the United States and Canada. Statewide estimates are not available. The species' sparse distribution and low detectability on BBS routes result in a trend line that has low reliability, even surveywide. Nevertheless, BBS data are the only long-term data available and suggest a relatively stable population from 1966 through 2019 (Sauer et al. 2020).

Monitoring data are also collected by the Great Lakes Marsh Monitoring Program, coordinated by Bird Studies Canada. Begun in 1995, the binational program surveys marsh birds and frogs at more than 5,000 locations in the Great Lakes Basin. From 1996 until 2013, Least Bitterns were found to have a statistically significant annual decline in the region averaging −1.7 percent per year (Tozer 2016).

Although many long-time bird-watchers believe the species has declined over the years, others contend that knowledge of the species is too sparse to make any accurate assessment of its population trend (Poole, Lowther, et al. 2020). Certainly, the large-scale loss and degradation of wetlands throughout eastern North America have negatively impacted such a wetland-dependent species to some degree.

CONSERVATION

The Least Bittern's very small population size, its uncertain status, and its dependence on wetland habitats are primary factors influencing its designation as a moderate conservation priority (PIF 2021). Management recommendations tailored specifically to Least Bitterns are lacking. More in-depth study of the species' breeding biology and habitat requirements would help direct site-specific conservation efforts. Meanwhile, aggressive programs designed to protect and restore wetland habitats will provide the greatest benefits.

Fortunately, the impacts of climate change may be minimal. Classifying the species as climate stable under all three warming scenarios they examined, NAS (2019) predicted only minor losses along the southern edge of its breeding range and huge gains (up to 251 percent) across Canada. Losses in western Minnesota would be compensated by gains in the north. Although the Least Bittern remains vulnerable to rising water levels, this unique, secretive marsh bird may become a more familiar inhabitant in the years ahead.

B. Hemi-marshes that provide an equal abundance of emergent aquatic vegetation and open water are prime breeding habitats for the Least Bittern, which prefers deeper water than the American Bittern. Photo by Lee A. Pfannmuller.

GREAT BLUE HERON

Ardea herodias

Minnesota Seasonal Status: Regular breeding resident and migrant; overwintering birds are occasionally observed in southern Minnesota. The Great Blue Heron was common during the MNBBA.
Migration: Some populations are nonmigratory, but most northern birds winter from the southern states south to northern South America.
Food: Consumes a variety of small animals, including fish, amphibians, birds, and crustaceans.
Nest: Platform nest built of sticks and usually placed in a tree, 6–18 m above the ground. A colonial nesting species.
Conservation Concern: Assigned a low PIF Continental Concern Score of 8/20.

Photo by Rebecca Field.

With its abundance of wetlands, lakes, and rivers, Minnesota provides ideal habitat for one of the state's most conspicuous aquatic inhabitants, the Great Blue Heron. Whether quietly stalking prey in a secluded wetland or flying with its slow and steady wingbeats, the Great Blue Heron is a familiar sight statewide. Even in the densely populated Twin Cities, it is not uncommon to observe birds flying to and from their breeding colonies located along one of the region's major river valleys.

Similar in height to the Sandhill Crane, with a long, sturdy bill, an elongated neck, long legs, and blue-gray plumage, the Heron is often referred to as a "Blue Crane." Compared to the Crane's distinct red crown, the Heron's crown feathers are white, bordered by a wide black stripe above each eye. The Heron is also adorned with beautiful long plumes along its back, neck, and crown during the breeding season. This physical trait provided a valued resource for the millinery trade in the late 1800s and early 1900s and resulted in significant mortality.

Distributed across southern Canada and the United States, extending south into portions of Mexico and the Caribbean, the Great Blue Heron is most abundant from the Great Lakes states south to the Gulf coast (Vennesland and Butler 2020).

MINNESOTA BREEDING DISTRIBUTION

Roberts (1932) reported that this common, statewide resident nested primarily in the forests of eastern Minnesota. Only one of the 16–18 heronry locations he knew of at the time "lies west of the heavy forests" in Pope County. Because of its proximity to the Twin Cities, one of the best-known colonies was located on Crane Island in Lake Minnetonka. Discussed extensively by Roberts, the colony was active as early as 1876 and supported both Great Blue Herons and Double-crested Cormorants. Unfortunately, the lake's rising popularity led to a dramatic increase in shoreline development and recreational activity and the colony's eventual abandonment. There were also occasional reports of birds nesting in central and western Minnesota, including at Swan Lake in Nicollet County and Lake Shetek in Murray County. Yet, despite the species' abundance, nesting was confirmed in just seven counties: Benton, Crow Wing, Hennepin, Murray, Nicollet, Polk, and Rice.

Green and Janssen (1975) reiterated the Heron's statewide distribution but noted that nesting colonies were absent in southwestern Minnesota and colonies in the northwest were "small and scarce." Shortly thereafter, several comprehensive field assessments of the species' distribution and abundance began. Minnesota's national forests, for example, conducted forest-wide surveys that identified 12 Great Blue Heron colonies in the Chippewa National Forest (Mathisen and Richards 1978) and 14 colonies in the Superior National Forest (Siderits 1980). Most colonies supported fewer than 150 nests.

A few years later, Guertin and Pfannmuller (1985) summarized the efforts of the MNDNR Nongame Wildlife Program to assess the statewide status and distribution of colonial waterbirds. A total of 127 active Great Blue Heron colonies were identified in 46 counties from 1981 to 1983. Twenty-five of the colonies were mixed-species colonies, frequently including Double-crested Cormorants, Great Egrets, and Black-crowned Night-Herons. The majority of sites supported only one to 50 Great Blue Heron nests; two colonies supported 500 to 1,000 nests. The largest colony was located on Long Lake in Kandiyohi County and supported 1,186 Great Blue Heron nests in 1981.

During the MNBBA, 2,385 Great Blue Heron records were tallied in 34.1 percent of the surveyed atlas blocks (1,625/4,766). Breeding was confirmed in 90 blocks (1.9 percent); adjacent blocks with confirmed nesting often represented a single nesting colony (e.g., two blocks in Grant County represented one breeding locality, Egret Island on Pelican Lake). Detected in all 87 Minnesota counties, breeding was confirmed in 47 counties **(A)**. The list of breeding counties included 10 that were new to the list of 54 identified by Hertzel and Janssen in 1998, including two counties in southwestern Minnesota. Because Herons travel several kilometers from the nesting colony to their foraging territories (Vennesland and Butler 2020; Custer and Galli 2002), all but two atlas records of Herons observed away from a known colony were coded as observed.

The statewide distribution of Great Blue Herons has changed little, especially since the 1970s. The total number of colonies, however, appears to have declined from 127 reported in 1985 to approximately 80–85 confirmed during the MNBBA. Elsewhere, apart from the significant population decline witnessed during the plume trade, the species has reoccupied much of

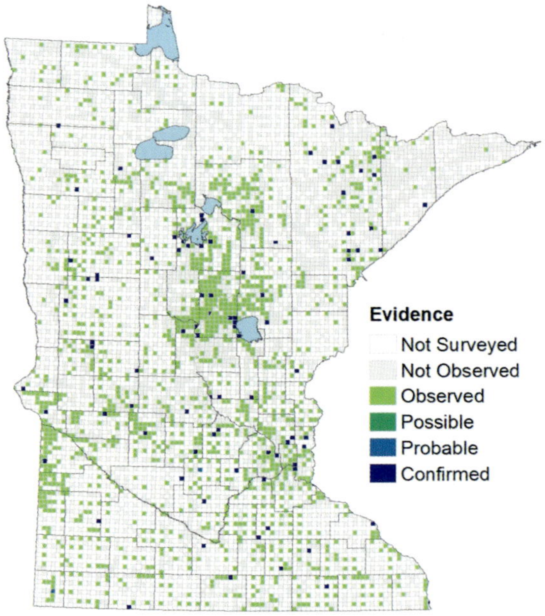

A. Breeding distribution of the Great Blue Heron in Minnesota based on the Breeding Bird Atlas (2009–13).

B. A small island on the Mississippi River, just north of downtown Minneapolis, provides the seclusion and protection needed by a small Great Blue Heron nesting colony. Photo by Lee A. Pfannmuller.

its former range, but likely at smaller numbers than in the past (Vennesland and Butler 2020).

BREEDING HABITAT

Nesting colonies are often located on forested islands in rivers and lakes, in trees bordering beaver ponds, and in remote lowland forests **(B)**. A primary criterion for a suitable colony site is its proximity to productive wetlands and shallow lakes for foraging. Sites that are difficult for mammalian predators to reach and that provide a buffer from other disturbances are favored. Tall trees for nesting are preferred, but the birds occasionally nest in shrubs, on the ground, or even on artificial structures (Vennesland and Butler 2020).

Nesting colonies may be viable for decades, but the Heron's acidic excrement stresses trees and their use of the canopy's outer limbs can cause physical damage that eventually leads to the colony's demise (Kerns and Howe 1967). Other threats to nesting trees include wind and, more recently, Dutch elm disease. Nevertheless, dead and dying trees are often preferred as nest sites until they fall or become so degraded they can no longer support the weight of the birds and their nests (B. A. Carlson 1995).

POPULATION ABUNDANCE

PIF (2021) estimated 620,000 birds residing in the United States and Canada, a number significantly higher than the estimate of 83,000 provided by Kushlan and his colleagues in 2002. Although the number still seems low, the species is sparsely distributed across its range. No statewide population estimate is available, but surveys of Minnesota's Double-crested Cormorant and American White Pelican colonies provide a clue. From 2004 to 2015, 26 Cormorant colonies were surveyed that also supported Great Blue Herons (Hamilton and Cuthbert 2016). Located primarily in central and southern Minnesota, the total number of Heron nests ranged between 1,358 and 1,741. This represents only a portion of Minnesota's population, but these sites supported the largest Heron colonies in the 1970s and 1980s.

Surveywide, the BBS population trend was unreliable owing to the species' low abundance but illustrates a dramatic population increase from 1966 through the mid-1990s followed by a somewhat slower decline. Similar trends occur throughout the Great Lakes region (Sauer et al. 2020). In Minnesota, BBS data reveal a reliable and significant downward trend from 1967 through 2019 **(C)**. Elsewhere, significant, long-term declines occur only in Alaska, Florida, and Nova Scotia.

Several factors may be contributing to the drop in numbers. Warner (1985), for example, documented a decline in Great Blue Herons nesting at the Pig's Eye Island Heron Rookery in response to the loss of elms from Dutch elm disease. Another Minnesota study demonstrated the detrimental effects that a local population of raccoons, as well as Great Horned Owls and American Crows, can have on a colony (Von Duyke 2006). Ontario biologists speculated that declines may be related to a documented decrease in amphibians throughout the Great Lakes (Cadman et al. 2007; Crewe, Timmermans, and Jones 2005). Finally, in a state that has lost nearly 50 percent of its wetlands, the availability of suitable foraging sites is likely another contributing factor.

CONSERVATION

A low conservation priority (PIF 2021), the Great Blue Heron faces few major threats. During the 1970s, many industrial contaminants, such as PCBs and dioxins, were a concern, reducing reproductive success and survival. Although these chemicals no longer present a risk, vigilance is warranted, especially for colonies located in industrial areas. Even warmer temperatures are projected to only further expand the species' distribution across northern Canada (NAS 2019). Nevertheless, the impacts of climate change on wetlands, including coastal habitats, need to be closely monitored.

In Minnesota, protection and monitoring efforts are best focused on large, mixed-species colonies that have persisted for decades. Several of these colonies are already protected as Scientific and Natural Areas, such as the Egret Island Rookery in Grant County and the Bruce Hitman Heron Rookery on Lake Johanna in Pope County. Both are closed to visitors during the nesting season. Establishing buffers around important colonies to reduce disturbance during the nesting season can also be an effective management strategy (Vennesland and Butler 2020).

Years ago, resource agencies were actively engaged in monitoring Minnesota's colonial waterbirds. Unfortunately, with a plethora of competing resource priorities, most efforts have been discontinued. Periodic surveys of Cormorant and Pelican colonies are now the only source of limited data on Great Blue Herons. At a minimum, these should continue. Because the effort includes some of the largest mixed-species heronries in the state, it provides an index that helps ensure that "one of the largest, most stately, and picturesque of Minnesota birds" remains a familiar sight (Roberts 1932).

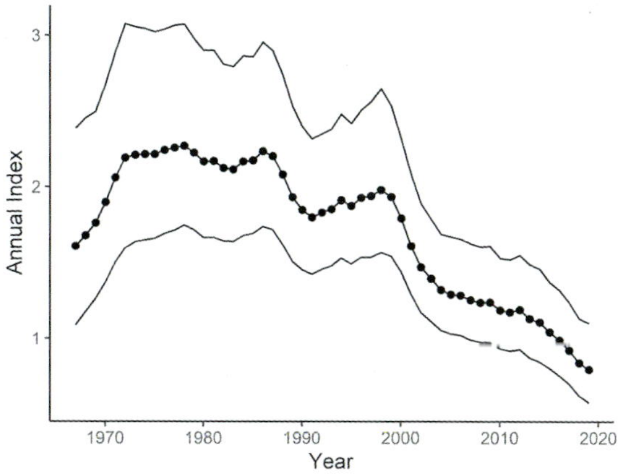

C. Breeding population trend and 95 percent credible interval (solid black lines) for the Great Blue Heron in Minnesota based on the 1967–2019 Breeding Bird Survey (Sauer et al. 2020).

GREAT EGRET

Ardea alba

Minnesota Seasonal Status: Regular breeding resident and migrant. The Great Egret was uncommon during the MNBBA.
Migration: Northern populations are migratory, but the distance traveled varies depending on wetland conditions and temperatures. Migrants are found in the southern states south through Central America and the Caribbean.
Food: Primarily small fish, but also consumes a variety of other vertebrates and invertebrates.
Nest: Usually a stick platform on tall trees and shrubs; occasionally nests on the ground. A colonial nesting species.
Conservation Concern: Assigned a low PIF Continental Concern Score of 6/20.

Photo by Jean Brislance.

Dazzling," "majestic," and "graceful" are just a few of the superlatives used to describe the Great Egret. Smaller than the Great Blue Heron, the Egret was eulogized by Bent (1926) when he wrote, the "long, graceful, flowing plumes, reaching far beyond its tail like a bridal train, and the exquisite purity of its snowy-white plumage make a picture of striking beauty." Present only during the breeding season, the plumes were famously prized as ornaments for women's hats. The largest and most common of Minnesota's three egret species, the Great Egret is easily identified by its yellow bill and its black legs and feet.

One of the species most valued by the early millinery trade, the Great Egret suffered carnage on a scale that is difficult to imagine and that nearly brought the bird to extinction. Because adults were shot at nesting colonies, the slaughter also resulted in the loss of two or three eggs or chicks per nesting pair. In 1902 alone, approximately 1.5 tons of Great Egret plumes were sold in London markets, a quantity that required at least 200,000 adults (Serratore 2018).

Luckily, the Egret was quick to recover after the passage of the 1918 Migratory Bird Treaty Act. Globally distributed, in North America this long-time symbol of the National Audubon Society occurs across the southeastern United States and north along the upper Mississippi River valley and the Atlantic coast. Scattered populations occur in southern Canada, the western states, throughout the Caribbean, and from Mexico south to Argentina and Chile. In North America, it reaches its highest breeding densities in the lower Mississippi River valley.

MINNESOTA BREEDING DISTRIBUTION

In the late 1800s, the Great Egret was an occasional straggler in the state (Roberts 1932). Records included sightings in Fillmore County, along the Minnesota River above Fort Snelling, at Heron Lake in Jackson County, and along the Red River valley, from Big Stone Lake north to Moorhead. Early summer records were believed to be "far-wandering adult birds"; late summer reports were likely juveniles who were known to disperse northward after fledging (McCrimmon et al. 2020). By the early 1900s, however, the summer stragglers had disappeared. It would take the birth of a national conservation movement to bring plume hunting to an end, and nearly twenty-five years of recovery before Great Egrets returned to Minnesota.

The first reports were in August 1932, when nine adults were observed at Sager Lake in Martin County (Roberts 1936), with additional reports in 1937 in Hennepin, Martin, and Murray Counties (Rysgaard 1937c). Then, during the following summer, "an acquaintance of Dr. G. H. Luedtke of Fairmont saw about 25 adult Egrets with nests and eggs at East Chain Lake" in Martin County (A. B. Erickson and Upson 1938). Although this was a second-hand account, a "well-authenticated" nesting report along the Mississippi River near Winona was investigated by Walter Breckenridge (1939) the following year. He found three Great Egret pairs nesting in a colony with Great Blue Herons, Double-crested Cormorants, and Black-crowned Night-Herons.

By 1975, Green and Janssen would report nesting in 17 counties south of a line that stretched from Grant County east to Anoka and Washington Counties. Included were reports from well-known, mixed-species colonies at Shields Lake in Rice County (Hanlon 1956) and Lake Johanna in Pope County (Tvrdik 1966). Summer observations of birds even farther north, including in Clearwater and Otter Tail Counties, led the authors to conclude that the Egret would continue to expand northward.

In the late 1970s, the MNDNR's Nongame Wildlife Program initiated an effort to identify all colonial waterbird nesting sites in the state by compiling information from field staff, published accounts, other resource agencies, and academic institutions. From 1981 to 1983, 22 active Great Egret colonies were identified in 21 counties located in central, southeastern, and northwestern Minnesota (Guertin and Pfannmuller 1985). All nesting sites were located within mixed-species colonies that included Great Blue Herons, Double-crested Cormorants, and Black-crowned Night-Herons. Only five colonies supported more than 50 nests. The two largest colonies were on Long Lake in Kandiyohi County (860 nests in 1981) and on Pelican Lake in Grant County (743 nests in 1981). A few years later, Janssen identified 29 counties where nesting had been confirmed since 1970, including Becker and Marshall (Janssen 1987). Hertzel and Janssen (1998) deleted Becker County from the list and added Cass County.

MNBBA participants reported 815 Great Egret records in 10.2 percent (483/4,744) of the surveyed atlas blocks. Breeding was confirmed in 24 blocks, at least two of which represented one breeding locality: Pelican Lake in Grant County **(A)**. With one

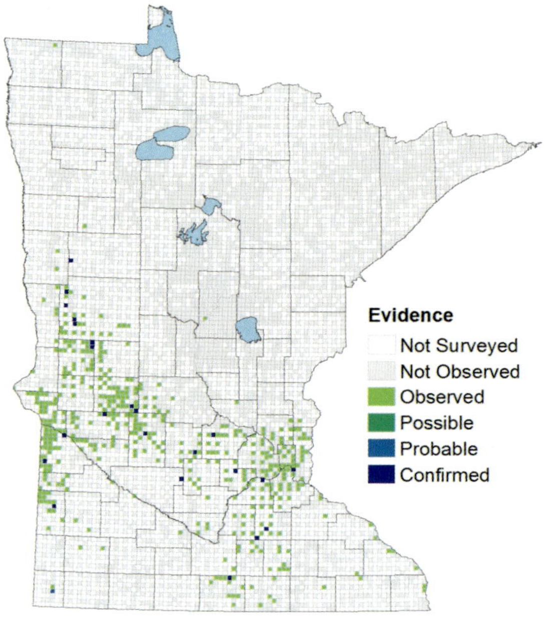

A. Breeding distribution of the Great Egret in Minnesota based on the Breeding Bird Atlas (2009–13).

B. A large, wooded island on Wells Lake in southern Rice County provides the perfect setting for this mixed-species rookery, which includes a nesting population of the Great Egret. Photo by Lee A. Pfannmuller.

well-documented exception in southwest Minnesota, unless Egrets were seen at the nesting colony, all detections were coded as observed.

Breeding was confirmed in 19 counties distributed across central and southern Minnesota. Five of the counties, however, were attributed to two colonies that were located in blocks that straddled multiple counties: Pig's Eye and Marsh Lake straddled three and two counties, respectively. Four counties were additions to Hertzel and Janssen's 1998 list: Becker, Carver, Lincoln, and Renville. Only one colony was reported along the Mississippi River south of the Twin Cities. However, observations were reported along the entire stretch of the river, suggesting that colonies may have been inaccessible to atlas observers.

The MNBBA results are similar to those reported by Guertin and Pfannmuller (1985). The latter reported 22 active colonies; the atlas documented approximately 20–23 breeding localities. Elsewhere, the species not only has reoccupied portions of its former breeding range but also, beginning in the late 1900s, has expanded farther north in New England, the Great Lakes, and central Canada, and west into Idaho and Oregon (McCrimmon et al. 2020).

BREEDING HABITAT

Like other members of the heron family, the Great Egret requires suitable nest sites located in a landscape with sufficient high-quality foraging habitat. Secluded woodlands with mature trees, surrounded by abundant shallow wetlands for foraging, are ideal. Protection from predators is also important, so islands and remote wetlands are preferred **(B)**. The majority of Great Egrets in the Upper Midwest nest in mixed-species colonies. Nests are frequently located in the same trees as those of other waterbirds, especially Great Blue Herons.

POPULATION ABUNDANCE

The dramatic impact of plume hunting on the Great Egret population is well-documented. Numbers plummeted nearly 95 percent and, by 1911, only 1,000–1,500 breeding pairs remained in the United States (R. Allen 1958). Today, the breeding population is thriving and is estimated to number at least 710,000 in the United States and Canada and 9.5 million globally (PIF 2021).

A statewide estimate is available from surveys conducted by Dr. Cuthbert and her colleagues at the University of Minnesota. Since 2004, they have periodically monitored Minnesota's American White Pelican and Double-crested Cormorant colonies. Because many of these colonies support other colonial waterbirds, including Great Egrets, their data were a significant contribution to the MNBBA. Between 2004 and 2015, the number of colonies that included Great Egrets ranged from 13 to 17, and the number of Great Egret nesting pairs ranged from 2,573 to 3,338 (Hamilton and Cuthbert 2016). Given that only 20–23 colonies were reported during the MNBBA, the colonies monitored by the University likely represent a significant portion of the statewide population.

Long-term monitoring data collected by the BBS also document a healthy population, increasing an average of 1.47 percent per year from 1966 to 2019 across the survey region (Sauer et al. 2020) **(C)**. Likewise, Minnesota BBS data illustrate a significant average increase of 3.33 percent per year.

Factors responsible for the species' increase are not well understood. Indeed, biologists have questioned why the Egret has not been as significantly impacted by the continuing loss and degradation of wetland habitats as other wetland-dependent species. Overall, the Egret demonstrates a high degree of adaptability and flexibility that has served it well during the past 150 years.

CONSERVATION

With steadily increasing populations, the Great Egret is not a conservation priority (PIF 2021). Even the species' response to warming temperatures is deemed to be positive, with range expansions projected across North America, including in northern Minnesota (NAS 2019). Yet, as a wetland-dependent species, increasingly arid conditions may threaten quality foraging sites while rises in sea level will likely impact coastal habitats.

In addition to protecting known colonies, management recommendations focus on reducing disturbances during the nesting season and establishing buffers of at least 100 m around each colony. Such measures could be more important at smaller colonies, which may be more sensitive to disturbance and more likely to be abandoned (J. P. Kelly et al. 2007).

At present, data collected incidentally at Cormorant colonies provide the only statewide population index of Minnesota's colonial waterbirds. Should this effort be discontinued, a small-scale monitoring program for the state's largest mixed-species colonies would provide a valuable population index for these important sentinels of our wetland communities, including one of the most striking waterbirds of all, the Great Egret.

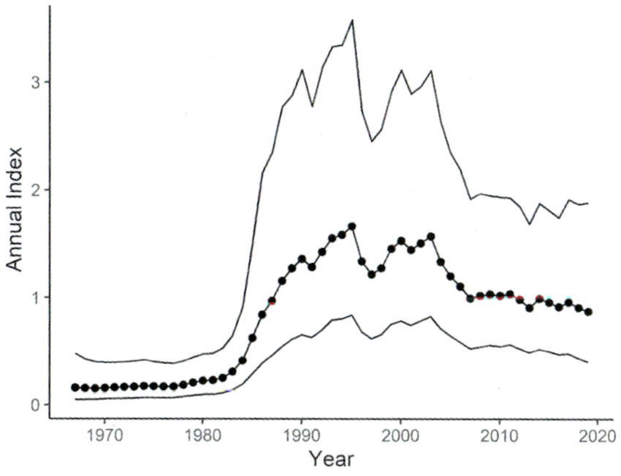

C. Surveywide breeding population trend and 95 percent credible interval (solid black lines) for the Great Egret based on the 1967–2019 Breeding Bird Survey (Sauer et al. 2020).

SNOWY EGRET

Egretta thula

Minnesota Seasonal Status: Regular summer visitant and migrant; formerly bred in Minnesota from 1971 to 1986; recently reported nesting in Grant County (2018). The Snowy Egret was very rare during the MNBBA.
Migration: Northern breeding birds winter from the southern coasts of the United States to South America; farther south, many breeding populations are permanent residents. During migration, wandering or dispersing birds can be found almost anywhere in the United States north to southern Canada.
Food: Feeds primarily on aquatic animals, such as fish, frogs, crayfish, worms, and insects in shallow waters. Foraging strategies are diverse and include foot-stirring, foot-raking, and wing-flicking; often forages in mixed-species flocks.
Nest: Consists of sticks or emergent vegetation constructed in a tree, in dense shrubs, or atop marshland vegetation placed within large multi-species colonies.
Conservation Concern: Assigned a low PIF Continental Concern Score of 7/20.

Smaller than Minnesota's more common Great Egret, the Snowy Egret has been described as "one of nature's daintiest and most exquisite creatures" (Bent 1926). During the breeding season, the fine, wispy plumes on its breast, lower back, and the back of its head add to its delicate appearance, while its bright yellow feet are a startling contrast to its black legs. Because Snowy Egrets use their feet to forage more than any other member of the heron family, biologists believe the dazzling yellow color may aid in capturing prey (Parsons and Master 2020).

Unfortunately, the beauty of this elegant bird was a liability in the late 1800s when feather-adorned women's hats were all the rage. Its breeding plumes, called aigrettes, were sold in 1886 for $32 an ounce, which was twice the price of gold (Allan 1974). The wanton killing of birds for fashion motivated high-society women from the nascent Audubon societies to take political action. Together with sportsmen, professional ornithologists, and humanitarians they helped pass the Lacey Act in 1900. Prohibiting the trade of wildlife taken in violation of state laws, the act essentially ended the New York millinery trade, especially once the protection of birds was further strengthened by the 1918 Migratory Bird Treaty Act.

Photo by Michael Furtman.

The primary breeding range of the Snowy Egret is along the Atlantic and Gulf coasts, with inland fingers to southern Kansas and up the Mississippi River to southeastern Missouri. In the west it breeds locally in many states from Arizona and California north to Oregon and Montana. Disjunct breeding populations are present in eastern South Dakota and in the Great Lakes at Green Bay, Lake Michigan, and on West Sister Island in western Lake Erie (Parsons and Master 2020; Rodewald et al. 2016). It is a permanent resident in many parts of Mexico, Central America, the Caribbean, and South America. In North America, the highest breeding densities are found in southern California and along the Gulf coast of Louisiana.

MINNESOTA BREEDING DISTRIBUTION

In summarizing the status of the species in Minnesota, Roberts (1932) wrote: "If the Snowy Egret ever occurred at all in Minnesota it was as a very rare straggler many years ago when Egrets were much more abundant than in recent years." Mounted specimens, shot in April 1884 at Heron Lake, were reportedly seen on display in the town by J. W. Preston in 1885. Their existence, however, was unknown by long-time Heron Lake resident Thomas Miller (Roberts 1932). The first documented records in the state were not reported until 1950 in Faribault County (Longley 1950) and 1951 at Lake Traverse, Traverse County (Enstrom and Bardon 1952).

Snowy Egret records slowly increased in the 1960s and 1970s, with nesting documented at three sites: first, Lake Johanna, Pope County, in 1971 (Hitman 1972); second, Big Stone NWR, Lac qui Parle County, in 1977, 1978, 1980, and 1981 (Ehlers 1977; Green 1979; Guertin and Pfannmuller 1985); and, third, Agassiz NWR, Marshall County, in 1986 (J. P. Mattsson 1986). Only a few (1–12) Snowy Egret nests were found at each site. The species has also been reported at the Egret Island SNA heron rookery in Grant County, but specific data are lacking. All these sites are large rookeries supporting Great Blue Herons, Great Egrets, Double-crested Cormorants, and/or Black-crowned Night-Herons. Often remote or difficult to access, these sites can be challenging to survey. Composed of hundreds or even thousands of birds, the colonies are chaotic and considerable patience is required to assess their approximate size and composition. Rare species like the Snowy Egret can easily be missed.

With one exception noted below, additional breeding records are lacking. Nevertheless, it is possible that the species still nests in small numbers. MOU records from 1985 to 2019 consistently show clusters of observations in counties with multispecies rookeries, such as Egret Island in Grant County and Long Lake in Kandiyohi County; the latter, however, no longer appears active. In recent years, the new North Ottawa Impoundment in Grant County has also been a source of summer records. Another cluster of reports in the Twin Cities region suggests possible breeding at sites like the Minnesota Valley NWR, Carlos Avery WMA, and the Pig's Eye Island Heron Rookery SNA. Unfortunately, since the early 1980s, statewide colonial waterbird surveys have been limited to three surveys of known Double-crested Cormorant and American White Pelican colonies conducted from 2004/2005 to 2015 (Hamilton and Cuthbert 2016). Although the surveys included most of the state's largest waterbird colonies, Snowy Egrets were not reported.

Observed regularly in Minnesota from mid-April through mid-October (MOU 2021b), many Snowy Egrets are migrants or dispersing or wandering birds from other colonies. Most records are from the Prairie Parkland and Eastern Broadleaf Forest Provinces. Except for a few vagrants from the Duluth estuary (2000, 2006, 2007), the only other reports from the Laurentian Mixed Forest Province are single vagrants in Hubbard (2008), Lake (1984), and northern St. Louis (1985) Counties. Most observations are during the spring, the prime time for bird-watching. There have been a couple of notable spring influxes of Snowy Egrets in recent years. One was in 2000, which was well-documented (Tustison 2001), and another was in 2004 (Budde, Lind, and Svingen 2004). In the fall, most birds are gone by the end of September.

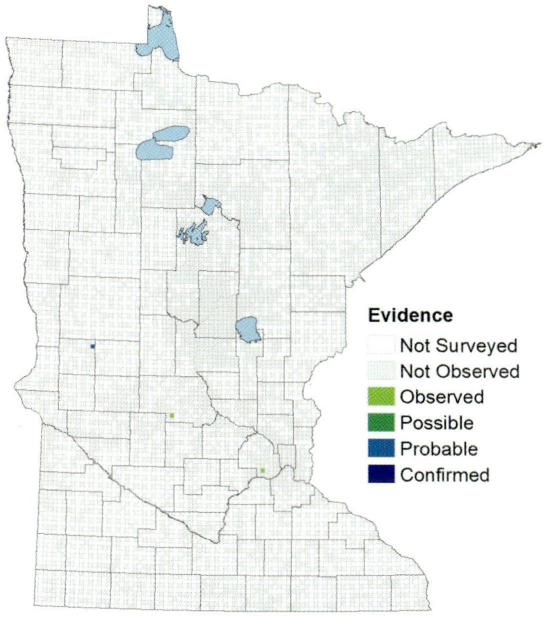

A. Breeding distribution of the Snowy Egret in Minnesota based on the Breeding Bird Atlas (2009–13).

During the MNBBA, Snowy Egrets were reported in three atlas blocks **(A)**. A probable observation from the Egret Island colony in Grant County was recorded with the following notes: "8 adults seen foraging in stream connecting Melby Lake and Pelican Lake; adults fly out to island after foraging where huge heronry exists but nest not visible at this distance; strongly suspect breeding with other herons and egrets." One possible MNBBA record was reported from the Paynesville sewage treatment plant, Stearns County ("single adult bird with plume"), and one observed record was reported from southern Hennepin County ("2 birds seen"). During the MNBBA, there were several summer observations of Snowy Egrets in the MOU database that were not submitted to the atlas but were from counties where known multispecies colonies exist: Big Stone, Lac qui Parle, and Kandiyohi. After the atlas, one MOU record from June 25, 2018, of three to four birds at Lac qui Parle WMA indicated possible nesting in the Marsh Lake colony and one record of an adult with five immatures on September 9, 2018, confirmed nesting in Grant County.

BREEDING HABITAT

With nests in both coastal and freshwater habitats, breeding sites range from upland islands in reservoirs, swamps, and river edges to coastal marshes and dredge-spoil islands. In the Upper Midwest, the mixed-species colonies that the Snowy Egret joins are usually on forested islands in large lakes or in stands of trees along edges of reservoirs and large marshes **(B)**. These sites provide the isolation that Egrets favor. Foraging habitat includes slow, shallow streams and river and lake shorelines, as well as impoundments and sewage treatment sites.

POPULATION ABUNDANCE

Approximately twenty years ago, the North American Snowy Egret population was estimated as at least 143,000 breeding adults (Kushlan et al. 2002). More recently, PIF (2021) estimated the continental population at 220,000 and the global population at 2.1 million. Population trends are stable to slightly increasing. BBS data demonstrate a statistically reliable, nonsignificant population increase averaging 0.87 percent per year from 1966 to 2019 throughout the survey region **(C)**.

Plume hunting had a huge impact on Snowy Egret populations in the late nineteenth century. The number of birds declined, and the species' range contracted. Once protective measures were enacted in the early 1900s, populations recovered and expanded north beyond their previous breeding range. More recently, however, the northern periphery of the species' breeding range has receded southward. Competition for nest sites with other colonial waterbirds, such as Cattle Egrets and Black-crowned Night-Herons, has been cited as one potential cause (Parsons and Master 2020).

CONSERVATION

Given its broad distribution and its recently increasing population, the Snowy Egret is considered a low conservation priority (PIF 2021). Threats on the breeding grounds include habitat loss from urban development, industrial agriculture, and wetland degradation from water management programs. In some areas, pesticides and other ecological contaminants have had a detrimental effect on hatching and fledgling survival (Parsons and Master 2020). Wintering habitat in the tropics may be at risk from the same threats. Although NAS (2019) ranked the species as climate stable under all three warming scenarios they examined, the loss of coastal habitats due to rising water levels is bound to present challenges. Considering these concerns, Kushlan et al. (2002) stressed the need to intensify monitoring efforts and expand efforts to identify key habitats that this elegant waterbird depends on throughout its life cycle.

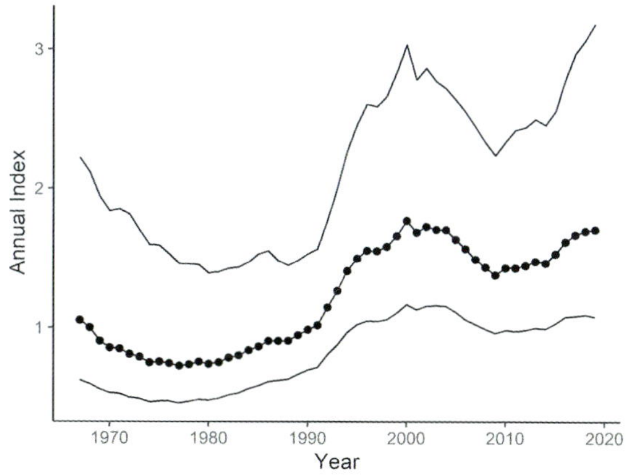

C. Surveywide breeding population trend and 95 percent credible interval (solid black lines) for the Snowy Egret based on the 1967–2019 Breeding Bird Survey (Sauer et al. 2020).

B. Nesting Snowy Egrets are attracted to large, mixed-species colonies like this well-known rookery on Pelican Lake in Grant County, which supports hundreds of Great Egrets as well as Great Blue Herons and Double-crested Cormorants.
Photo by Gary T. Seim. Permission granted by the Scientific and Natural Areas Program, Minnesota Department of Natural Resources.

LITTLE BLUE HERON

Egretta caerulea

Photo by Sparky Stensaas.

Minnesota Seasonal Status: Summer vagrant and migration visitant; formerly bred in Minnesota from 1971 to 1981. The Little Blue Heron was very rare during the MNBBA.
Migration: Most birds winter in the southeastern United States, the Caribbean, Mexico, and Central America, but banding records show some migration as far south as South America.
Food: Feeds primarily on small fish, amphibians, and crustaceans, including crabs and crayfish. It may also feed on grasshoppers and other insects in upland fields.
Nest: Platform nest of sticks placed in low shrubs and small trees. The species is a colonial nester and occupies small territories in multispecies colonies or nests in small single-species colonies. Some larger Little Blue Heron colonies are associated with Cattle Egrets, whose aggressive behavior may impact Heron nesting success.
Conservation Concern: Assigned a moderate PIF Continental Concern Score of 11/20.

The "shy and retiring" Little Blue Heron (Bent 1926) is actually closely related to the Snowy Egret and may occasionally hybridize with it (Rodgers and Smith 2020). Although they are nearly identical in size, there are obvious plumage and behavioral differences that distinguish these two very rare Minnesota species. The Little Blue Heron has an almost all-white plumage as a juvenile and maintains a mottled white appearance the following spring until it molts into the dark slate-blue plumage of an adult. It is a solitary forager that quietly stands or slowly stalks its prey, in contrast to the Snowy Egret, which is a social feeder with active, dancing foraging techniques.

Because the Little Blue Heron does not have the extravagant white plumes of the Snowy Egret, it was not a prime target of the unregulated plume hunting in the late 1800s and early 1900s that devastated so many colonial birds. But the habitat disturbance involved in that killing likely had some effect on Little Blue Heron populations. Since that time, its numbers have increased, and its range has expanded northward.

Found primarily in the southeastern United States, from Virginia south along the coast and in adjacent inland areas to Texas, the Little Blue Heron's breeding range extends north along the Mississippi River to the southern Ohio River and in the southern Great Plains to Kansas. In the East, it is a rare and local breeder up the coast to southern Maine, and in the West it can be found in isolated colonies in New Mexico and California. There are two disjunct north-central populations where the birds are present in small numbers. The first is in Ohio, where the birds have sporadically nested on West Sister Island in western Lake Erie since 1983 (Rodewald et al. 2016); the second is in northeastern South Dakota, where birds have bred at a few sites since 1980 (Drilling 2013; Drilling, Sparks, et al. 2016). In North America, the highest breeding densities are found along the Gulf coast of Louisiana and Texas.

South of the United States, local breeding populations can be found throughout the West Indies, coastal Mexico, Central America, and northern parts of South America. Some of these birds migrate south while others stay year-round (Rodgers and Smith 2020); the same variation in migratory status is seen in US coastal populations.

MINNESOTA BREEDING DISTRIBUTION

The Little Blue Heron was not known in Minnesota in 1932 at the time when Roberts was writing *The Birds of Minnesota*. The first observations were by William Longley in what is now called the Whitewater WMA in Winona County on May 21, 1953, and August 17, 1956 (Lupient 1953; Longley 1956b). There were other spring records before nesting was first documented in 1971 at the large multispecies rookery on Lake Johanna in Pope County (Hitman 1972; Green and Baumhofer 1972). Additional nesting evidence at the colony was reported through 1981 (Eckert 1973a; Carr 1974; Hiemenz 1981).

Breeding was subsequently reported at three other large multispecies colonies in central Minnesota: first, Pelican Lake in Grant County (1972–79; Eckert 1973b); second, Big Stone NWR in Lac qui Parle County (1976–81; Buer and Buer 1977; Eckert 1978a; Green 1979); and, third, Pig's Eye Island in Ramsey County (1979; Green 1980; Janssen 1987). Since these reports were published, there have been no additional reports of Little Blue Herons nesting in Minnesota. Guertin and Pfannmuller (1985) noted that three of the sites (Pelican Lake, Lake Johanna, and Big Stone NWR) were surveyed in 1984, but Little Blue Herons were not detected. Since then, statewide surveys of colonial waterbirds have been limited to three surveys of Double-crested Cormorant and American White Pelican colonies conducted from 2004/2005 to 2015 (Hamilton and Cuthbert 2016). Although somewhat limited in scope, these surveys included many of the state's largest mixed-species colonies, including three of the sites where the Little Blue Heron was previously documented nesting. Two of the sites were surveyed on the ground; one was surveyed by air. Nevertheless, none of the surveys detected this rare species.

During migration, when the birds are vagrants or dispersers in the state, most observations (about 80 percent) are in southern Minnesota. Sites in northern Minnesota where they have been observed are primarily in counties on the western margin of the state; a few records are along the North Shore (Duluth, 1985; Grand Portage, 2003).

During the MNBBA, only two Little Blue Heron records were reported **(A)**; both were coded as ob-

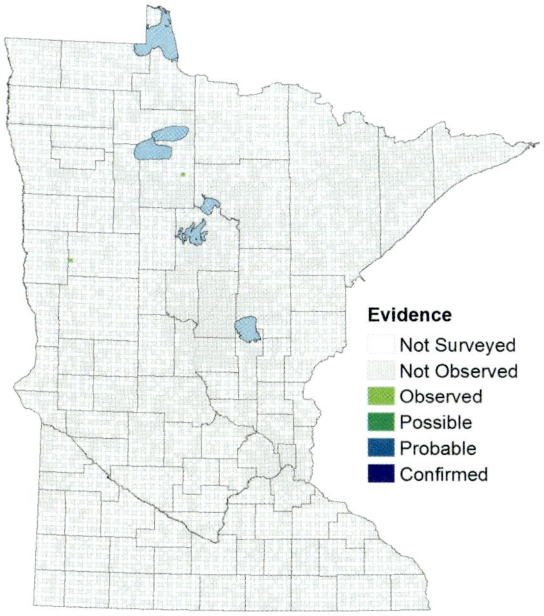

A. Breeding distribution of the Little Blue Heron in Minnesota based on the Breeding Bird Atlas (2009–13).

served records. The first was reported from Lake Park Township in Becker County by the USFWS on June 1, 2009. Given the early June date, it was probably a migrant. The second record is from Beltrami County on June 17, 2009.

BREEDING HABITAT

Colony sites are located in many wetland habitats, including marshes, swamps, lakes, ponds, and impoundments **(B)**. Most breeding is at freshwater inland sites but also occurs in coastal areas on dredge-spoil islands. Birds may forage in freshwater or saltwater habitats (Rodgers and Smith 2020).

POPULATION ABUNDANCE

As recently as 2002, the North American Waterbird Conservation Plan stated that there is "insufficient information" to provide any population figures (Kushlan et al. 2002). By 2010, the Upper Mississippi River/Great Lakes Waterbird Conservation Plan cited a North American population estimate of 200,000 to 300,000 breeding individuals and a population of approximately 2,200 breeding pairs within the Great Lakes Region south to northern Arkansas, northeastern Oklahoma, central Kentucky, and central Tennessee (Wires, Lewis, et al. 2010). PIF (2021) recently placed the continental population at 270,000 and the global population (North and South America) at 1.1 million. The Little Blue Heron is difficult to survey on the ground and from the air because it is usually a minor species in the large mixed-species colonies that it occupies.

Long-term monitoring data collected by the BBS demonstrate a statistically reliable, significant population decline across the survey region that averaged −1.25 percent per year from 1966 to 2019 **(C)**. Only two southeastern states, Kentucky and North Carolina, reported increasing populations. In their comprehensive review of the species, Rodgers and Smith (2020) suggest that the "alteration of foraging habitat and food availability" are the primary factors impacting populations.

CONSERVATION

Although the Little Blue Heron is now a protected species, populations are still at risk from a variety of habitat threats, such as adverse wetland conditions resulting from drainage and sedimentation, and habitat loss from development. Ecological contaminants are also a concern (Rodgers and Smith 2020). Although warming temperatures are likely to expand the species' North American breeding distribution farther north (NAS 2019), rising sea levels are bound to present challenges to the birds in coastal habitats. In light of these concerns, as well as its small population size and declining population trend, the species was designated a moderate conservation priority by PIF (2021).

In Minnesota, three of the colonies where it was found breeding are designated by the MNDNR as Scientific and Natural Areas: Egret Island (Pelican Lake), Bruce Hitman Heron Rookery (Lake Johanna), and Pig's Eye Island Heron Rookery (St. Paul). Although these sites are protected and visitors are not allowed during the nesting season, foraging sites located near breeding colonies and at migratory stopover sites need protection as well. As Rodgers and Smith (2020) noted, "continued population decline, despite protection of colony sites, suggests that preservation of important foraging habitat has not been successful."

The recent statewide surveys of American White Pelican and Double-crested Cormorant colonies conducted by Hamilton and Cuthbert (2016) have been critical in monitoring the most common species that nest in many of the state's largest colonies. Should these efforts cease, resource professionals should consider establishing another cost-effective monitoring tool. Although it is difficult to document the presence of rare species in these large colonies, it is important to know that they continue to support large and thriving populations of common species such as the Great Blue Heron, Great Egret, and Black-crowned Night-Heron. If the Little Blue Heron experiences another northern extension of its breeding range in the state, these large, multispecies colonies remain as important sites to attract the birds.

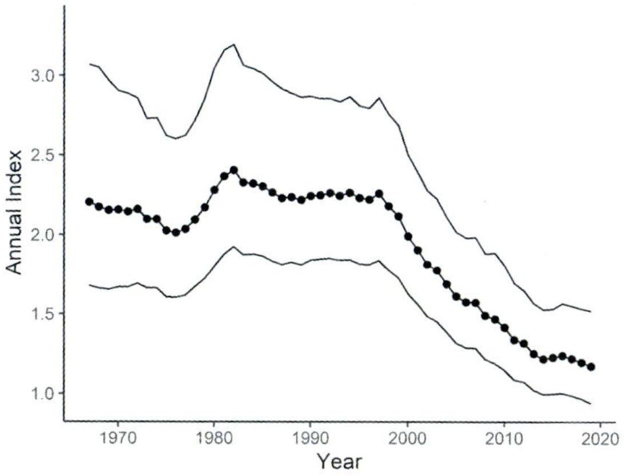

C. Surveywide breeding population trend and 95 percent credible interval (solid black lines) for the Little Blue Heron based on the 1967–2019 Breeding Bird Survey (Sauer et al. 2020).

B. Lake Johanna in Pope County is one of the few rookeries in the state where Little Blue Herons formerly nested. Densely vegetated in the 1970s, today the island is almost totally devoid of vegetation. Photo by Lee A. Pfannmuller. Permission granted by the Scientific and Natural Areas Program, Minnesota Department of Natural Resources.

CATTLE EGRET

Bubulcus ibis

Minnesota Seasonal Status: Summer visitant and regular migrant; formerly bred in Minnesota from 1971 to 1992. The Cattle Egret was very rare during the MNBBA.

Migration: Northern birds winter from the West Indies south to northern South America; southern populations are permanent residents. Juveniles disperse widely, in random directions, and rarely return to their natal colony to breed.

Food: Associated with foraging cattle, which flush the Egret's desired insect prey, such as grasshoppers and crickets. Small vertebrates are also taken. Cattle Egrets usually forage in flocks in upland pastures and along water courses.

Nest: Platform or shallow bowl constructed of sticks and placed in trees or shrubs; may use abandoned nests of other species. Cattle Egrets are colonial nesters.

Conservation Concern: Assigned a moderate PIF Continental Concern Score of 9/20.

Photo by Rebecca Field.

The Cattle Egret is an Old World species that arrived on the eastern coast of South America about 1877, presumably having flown across the Atlantic Ocean from Africa. Its dispersal tendencies are remarkable, and even now it is regularly seen at sea (Telfair 2020). Originally native to the tropical and subtropical areas of Africa and Asia, it expanded its range north to Spain and Portugal about the same time that it crossed the ocean to South America. It arrived in Florida in 1941, bred there in 1953, and rapidly spread across the continent. The enormous scale and speed of its expansion has been described as "among the greatest recorded for any avian species" (Telfair 2020).

Half the size of a Great Egret, the Cattle Egret is easily distinguished from all other herons that breed in Minnesota. During the height of the breeding season, its bill and legs become various shades of red, and its white plumage is highlighted with pale orange plumes on the head, lower back, and breast. Behaviorally, it is the only heron that forages in flocks in upland pastures. Indeed, its habit of feeding in livestock pastures gave rise to its genus name, *Bubulcus,* which translates to "pertaining to cattle."

In the Western Hemisphere, the Cattle Egret occurs from South America north to southern Canada. Nesting has been confirmed in all but four of the US contiguous states. Highest breeding densities are found in the core of its breeding range from Florida to Texas, with disjunct high population nodes in southern California and the Dakotas.

In his comprehensive review of the species, Telfair (2020) commented on the unique aspects of the Cattle Egret's life history that helped spur its remarkable expansion: "Apparent keys to the spread and success of the Cattle Egret are its dispersal tendencies, gregariousness, diet, foraging adaptability—especially as humans increasingly convert large areas of landscape to pasture for livestock production and to crop fields for rice agriculture—and its breeding adaptability and success."

MINNESOTA BREEDING DISTRIBUTION

First reported in Minnesota in June 1959 in Pope County (Hanna 1959), a series of nine more Cattle Egret records in the state were logged from 1961 to 1970. Although most of these reports were in southeastern counties and the Twin Cities, three were from western Minnesota. Single birds were observed during spring 1969 in Lac qui Parle and Nicollet Counties (MOU 2021b), while a flock of eight birds was observed on August 21, 1970, in northern Grant County (Moos 1971).

From 1971 to 1992, nesting was documented in five western counties at large mixed-species colonies: first, Lake Johanna, Pope County, in 1971 (Hitman 1972), 1984 (Hendrickson and Eckert 1985; Guertin and Pfannmuller 1985), and 1985 (Wilson and Shedd 1986); second, Pelican Lake, Grant County, in 1970 (Hitman 1972), 1971 (Eckert 1973a), 1973 (W. H. Wyatt 1973), 1985 (Wilson and Shedd 1986), and 1986 (Shedd and Wilson 1987); third, Long Lake, Kandiyohi County, in 1984 (Guertin and Pfannmuller 1985); fourth, Big Stone NWR, Lac qui Parle County, in 1977 (Eckert 1978a) and 1978 (Green 1979); and, fifth, Thief Lake WMA, Marshall County, in 1992 (Wiens 1993).

Although some birds have been observed every summer from 1993 through 2020 (MOU 2021a), nesting records ceased after 1992. Many of the summer observations have come from the same counties where nesting was documented. As with the Snowy Egret and Little Blue Heron, which were first found nesting in the state during this same time frame, it is not clear if the absence of nesting records is due to a lack of intensive colony surveys or to the fact that these species are no longer nesting in the state. Periodic surveys conducted since 2004/2005 at many colonies where the Cattle Egret bred have failed to document its presence but were not designed to survey rare species (Hamilton and Cuthbert 2016).

All five of the former Cattle Egret nesting sites in western Minnesota are near larger populations in the Dakotas, where they have nested since the 1970s (Tallman et al. 2002). In North Dakota, eight colonies surveyed in 2014–15 supported a mean of 22 breeding pairs each (Drilling 2015). In South Dakota, the species is considered a "fairly common breeder in the northeastern part of the state," including at Sand Lake NWR in Brown County (Tallman et al. 2002). The statewide population is estimated at 700 to 1,400 birds (Drilling, Sparks, et al. 2016). Dispersers in Minnesota most likely come from this Great Plains population.

As a migrant, the Cattle Egret is regularly seen from the Twin Cities northwestward to the prairie counties, where it is observed in summer. Many of the flocks in western counties in the fall are juveniles, probably wandering from the Dakotas (Ellwanger 1997). In the forested northern counties, scattered vagrants are found in the spring (six records) but in the fall (22 records) most birds are observed at vagrant hot spots along the North Shore of Lake Superior from mid-October to mid-November (MOU 2021b). These fall records fit the pattern of this Egret dispersing long distances after the breeding season.

During the MNBBA, eight observed Cattle Egret records were reported in seven atlas blocks ranging from Douglas County southeast to Winona

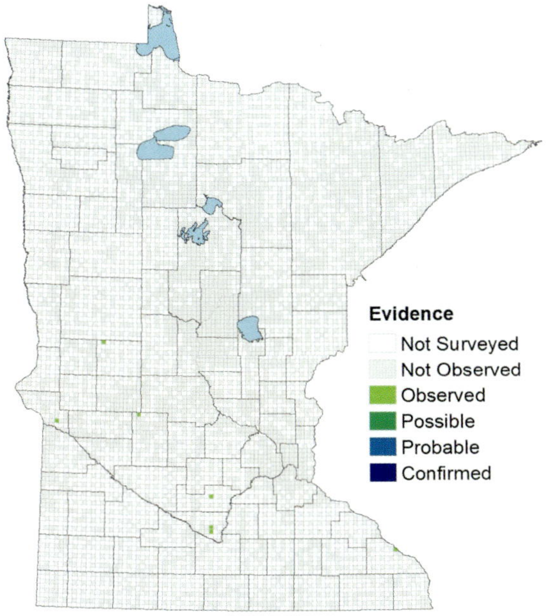

A. Breeding distribution of the Cattle Egret in Minnesota based on the Breeding Bird Atlas (2009–13).

County **(A)**. Although surveyors originally reported two possible records, lacking any documentation to support possible or probable nesting, they were changed to observed status. Like all herons and egrets, Cattle Egrets may forage 2.5 to 35 km from the nesting colony; two studies, however, found the majority of trips were within 15 km of the colony (Telfair 2020). Unless birds are seen at a nest, observations cannot be assumed to be evidence of breeding status. During the atlas period, at least 15 additional Cattle Egret breeding-season records were recorded in the MOU database (MOU 2021a) and their distribution was similar to the MNBBA records.

BREEDING HABITAT

The Cattle Egret nests in both coastal and upland habitats, usually in established mixed-species colonies of other herons. Sites include: "coastal barrier, marsh, and dredge-material islands; periphery and islands in reservoirs, lakes, quarries, and wetlands; swamps; riparian and upland woodlands and groves, with and without understory **(B)**. Upland sites may or may not be associated with water" (Telfair 2020).

POPULATION ABUNDANCE

Precise population estimates are lacking, but an earlier estimate placed the Texas population alone at 1.16 million (Kushlan et al. 2002). PIF (2021) estimated the global population to number 48 million and the continental population in the United States and Canada to number 2.8 million. Long-term monitoring data across the BBS survey region demonstrated a reliable and significant annual decline averaging −1.00 percent per year from 1966 to 2019 **(C)**.

The Cattle Egret reached a peak in its population growth in the United States about 1980. It then stabilized, and the growth rate lessened and even declined in many regions, especially in the Carolinas (Telfair 2020). Some of the decline is attributed to conversion of agricultural lands to residential development and to predation by raccoons. These factors are universal, however, and in some regions where Cattle Egrets are declining, other waterbird populations are increasing. PIF (2021) speculated that the decline in the number of cattle on US farms may be contributing to the downward trend of Cattle Egrets. Approximately 10 million fewer cattle were managed on farms in early 2021 compared to 1996 (USDA 2021). Nest availability also appears to be an important factor regulating populations. Cattle Egrets seek active heronries where other species have already initiated nesting (Telfair 2020).

CONSERVATION

Although the Cattle Egret is a non-native bird and its range expansion has been dramatic, it is not considered an "invasive species," according to a definition that means "likely to cause harm" (Avery and Tillman 2005). The Cattle Egret has demonstrated the ability to occupy landscapes already significantly altered by agricultural development. Yet, despite its adaptability, populations are declining, which has spurred its designation as a moderate conservation priority (PIF 2021).

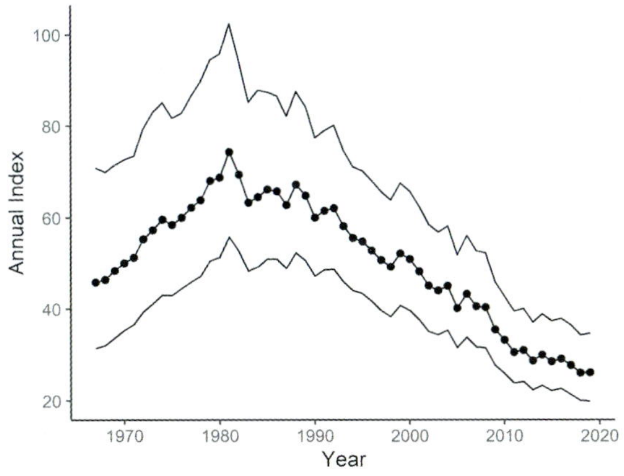

C. Surveywide breeding population trend and 95 percent credible interval (solid black lines) for the Cattle Egret based on the 1967–2019 Breeding Bird Survey (Sauer et al. 2020).

Given its habit of foraging in agricultural lands, pesticides might pose a significant threat to the species. Numerous studies, however, indicate that they have not had a detrimental effect (Telfair 2020). A recent assessment suggests that warmer temperatures also pose no risk and will only promote the species' expansion farther northward (NAS 2019). Large-scale changes in land use, however, such as agricultural practices that convert pastures to cropland, can adversely impact populations.

Overall, the rapid dispersal that the Cattle Egret experienced across North America in the latter half of the 1900s has largely abated. More persistent survey work is necessary to assess the Cattle Egret's current breeding status in Minnesota.

B. No longer active, the mixed-species rookery on Long Lake in Kandiyohi County was one of the few sites where Cattle Egrets were documented nesting in Minnesota. Photo by Lee A. Pfannmuller.

GREEN HERON

Butorides virescens

Minnesota Seasonal Status: Regular breeding resident and migrant. The Green Heron was uncommon during the MNBBA.
Migration: Northern populations are short- to long-distance migrants that winter in Florida, the Caribbean, and along the Gulf of Mexico south to northern South America; farther south populations are nonmigratory.
Food: Principally small fish, but aquatic and terrestrial invertebrates are also taken, as well as amphibians, reptiles, and small mammals.
Nest: Platform of small twigs, usually placed in a tree or shrub close to water. Typically a solitary nester, but some birds nest in small, loose colonies that rarely number more than 10 to 20 pairs.
Conservation Concern: Assigned a moderate PIF Continental Concern Score of 12/20.

Photo by Rebecca Field.

Canoe quietly through a wetland or stroll along the edge of a lake's shallow bay and you may encounter a Green Heron. Often perched on a small branch hanging low above the water, it sits motionless, waiting to surprise unsuspecting prey that swim beneath its vigilant post. If your arrival startles the bird, you may hear its raspy alarm call as it flushes and flies to a quieter spot. As the frightened bird flies away, it frequently leaves behind a trail of white excrement that is the origin of some of its most colorful colloquial names, such as "Shite-poke" and "Chalk-line."

The Green Heron's scientific name, *Butorides virescens*, literally translates to "a bittern-resembling, green bird." When it is crouched down low, the most prominent features of its stocky build are its long, dagger-like bill, its greenish-gray crown, a chestnut-colored face and neck, and a dark back and wings all tinged with hints of blue and green. When it raises its head, chestnut-and-white stripes along the length of the neck are visible as the bird assumes a streamlined form that may suggest an altogether different bird. Bent (1926) described this frequent transformation and concluded "this bird well deserves to be called 'rubber neck.'"

One of the most intriguing behaviors of the Green Heron is its use of bait, such as worms and mayflies, to attract small fish. When fish swim close to investigate the bait dangling from its bill, the Green Heron extends its neck and nabs the prey in a sudden flash. Even small twigs and leaves may be employed as lures. One writer recounted watching a bird drop "a small dried leaf onto the surface of the water, occasionally relocating it to a preferred location as the leaf began to drift away or sink" (McLeish 2014).

Ornithologists have long debated the taxonomic relationships among the small herons within the genus *Butorides*, especially between the Old World species, the Striated Heron (*B. striata*), and North America's Green Heron (*B. virescens*). Although they have, at times, been considered one and the same species (*B. striata*), the most recent checklist supplement still recognizes them as distinct species (Chesser, Billerman, et al. 2020).

During the breeding season, the Green Heron is primarily restricted to the eastern United States and the Pacific coast. Populations extend north into British Columbia on the West Coast and into southern Ontario, Quebec, New Brunswick, and Nova Scotia in the east. Nonmigratory populations occur along the southern coasts of the United States, throughout the Caribbean, and from Mexico to northern South America.

MINNESOTA BREEDING DISTRIBUTION

In the early 1930s, Roberts (1932) described the Green Heron as a species restricted to southern Minnesota and most common along the Mississippi River valley from the Twin Cities south to the Iowa border. Sherburne and Pine Counties were the northern limit of its breeding range in eastern Minnesota; in the west, it was absent north of Lincoln County. Occasional reports of birds collected in eastern North Dakota and Manitoba led Roberts to speculate that "it may be a straggler in northern Minnesota but thus far it has eluded observation." Confirmed nesting records were limited to Pine County and to three east-central counties: Carver, Hennepin, and Ramsey.

More than forty years later, Green and Janssen (1975) reported an increasing number of summer reports from northern counties, including as far north as Lake of the Woods, Marshall, and St. Louis Counties. Crow Wing and Pine Counties, however, defined the northern limit of breeding evidence. When Janssen published an updated account in 1987, the species had continued its northward expansion. Breeding was confirmed as far north as southern St. Louis County in the east and Clearwater County in the west. Summer reports farther north continued to increase, as did reports from southwestern Minnesota; records were missing, however, from the Red River valley. Janssen identified 15 counties where nesting had been confirmed since 1970. Hertzel and Janssen (1998) later added an additional 12 counties to the list.

From the late 1980s through 2014, the MBS documented 188 breeding-season locations sparsely distributed from Houston County in the southeast to Rock County in the southwest and north to the Canadian border in Roseau County. In the northern forests, records were found as far east as central St. Louis County.

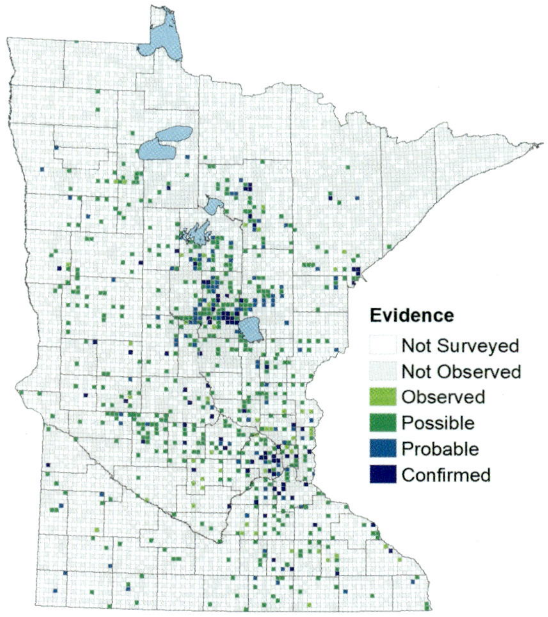

A. Breeding distribution of the Green Heron in Minnesota based on the Breeding Bird Atlas (2009–13).

The distribution of the 974 Green Heron records reported by MNBBA participants **(A)** is similar to that depicted by the MBS. The species was largely absent from the far northern reaches of the state and the Red River valley; the core of its breeding range extended from southeastern Minnesota north to Itasca, southern Beltrami, and Clearwater Counties. Records were most abundant in the Twin Cities metropolitan region and the Brainerd Lakes region, the two regions where search efforts were most intense. Reported from 15.4 percent (732/4,754) of the surveyed atlas blocks, breeding was confirmed in just 84 blocks (1.8 percent). Breeding was considered probable in an additional 110 blocks (2.3 percent). Observed in 81 of Minnesota's 87 counties, the Green Heron was absent only from Cook, Kittson, and Lake of the Woods Counties in the north and from Martin, Pipestone, and Traverse Counties in the west. Twenty of the 38 counties where breeding was confirmed were additions to the list published by Hertzel and Janssen (1998).

The landcover suitability model for the Green Heron **(B)** predicts that the most suitable habitat is located in east-central Minnesota north to the southern edges of the northern forest and west through the prairie–forest ecotone. In southern and western Minnesota, forested floodplains, especially the upper Minnesota River valley, also provide quality habitat opportunities.

The Green Heron has clearly expanded its range northward since Roberts first summarized its status nearly one hundred years ago. MNBBA breeding records in northern Itasca and St. Louis Counties are nearly 150 miles north of the Pine County record documented by Roberts in 1919. W. E. Davis and Kushlan (2020) also noted a northward range expansion across the middle of the continent and along the Pacific coast in the twentieth century.

BREEDING HABITAT

Ideal sites for this wetland inhabitant provide trees and shrubs for nesting and shallow waters for feeding. Riparian areas with dense thickets of willows and alders, or lined with trees, are preferred **(C)**. Dry upland nesting sites, some distance from suitable foraging areas, and islands covered with small trees and shrubs are frequently used as well. Wetland habitats range from marshes, shallow lakes, and small streams to reservoirs, stormwater retention ponds, and drainage ditches (W. E. Davis and Kushlan 2020). Foraging usually occurs in secluded areas close to the water's edge.

POPULATION ABUNDANCE

Given its largely solitary nesting habits and its secluded wetland habitats, it is difficult to assess the size and status of the Green Heron's population. The North American Waterbird Conservation Plan simply noted that insufficient data were available (Kushlan et al. 2002). More recently, PIF (2021) estimated the total population to number 1.2 million, with 770,000 breeding adults residing in the United States and Canada.

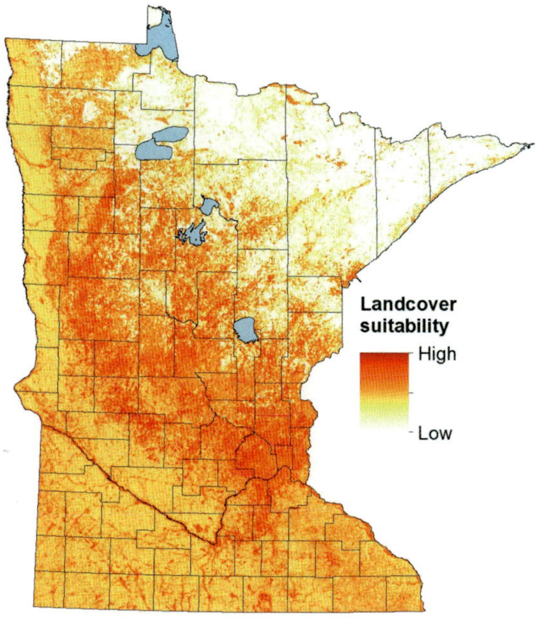

B. Landcover suitability map for the Green Heron in Minnesota based on habitat, landscape context, and climate data gathered during the Minnesota Breeding Bird Atlas (2009–13) using the MaxEnt modeling approach.

C. Small trees and shrubs lining a shallow backwater of the Mississippi River provide an ideal nesting habitat for the Green Heron. Photo by Lee A. Pfannmuller.

Despite its limitations for monitoring wetland birds, the BBS provides the only long-term data available for monitoring the species' population trend. Unfortunately, because the Green Heron is encountered so infrequently along most BBS routes, the statistical robustness of the trend data is compromised. This is true for Minnesota as well as for the entire BBS survey region. Despite these limitations, the data suggest a range-wide average decline of –1.33 percent per year from 1966 to 2019 (Sauer et al. 2020). Factors responsible for the widespread decline are not known, but undoubtedly include the large-scale loss and degradation of wetlands. BBS data in Minnesota also suggest a declining population.

CONSERVATION

Though it is ranked as a moderate conservation priority (PIF 2021), the dearth of information on the species' status and population size has been an impediment to understanding its true conservation status. Nevertheless, its use of a variety of altered habitats, such as reservoirs and stormwater retention ponds, provides a level of adaptability that helps guard against the loss and degradation of natural wetland habitats. Although warmer temperatures may prompt range expansions in many regions of North America, the Great Lakes states are one of the few regions where some losses are projected (NAS 2019). Indeed, Minnesota could witness some habitat losses in the south and some gains in the north as temperatures rise. In the meantime, efforts to protect and restore wetlands will help secure the future of this shy yet clever inhabitant.

BLACK-CROWNED NIGHT-HERON

Nycticorax nycticorax

Minnesota Seasonal Status: Regular breeding resident and migrant. The Black-crowned Night-Heron was uncommon during the MNBBA.
Migration: Northern populations are short- to medium-distance migrants that winter in the Caribbean and from the southern United States south through Central America.
Food: An opportunistic feeder that consumes anything available, occasionally including young in adjacent nests, carrion, and refuse.
Nest: Nests are constructed of small twigs and placed within trees or shrubs or constructed in wetlands with emergent vegetation. A colonial nester, the species frequently nests in mixed-species colonies.
Conservation Concern: Assigned a moderate PIF Continental Concern Score of 11/20.

Photo by Ashley J. Peters.

A handsome bird, the Black-crowned Night-Heron is aptly named for its prominent black crown and its crepuscular and nocturnal foraging habits. Although breeding pairs may forage during the day to meet the incessant demands of newly hatched young, the birds prefer to dine from dusk to sunrise. This habit is a clever adaptive strategy that reduces competition with other members of the heron family that feed in the same wetlands during the day (Hothem et al. 2020).

Stockier than most herons, with a shorter neck and shorter legs, the Black-crowned Night-Heron also has a heavier bill designed to grab, rather than stab, prey items. Its black cap, back, and shoulders contrast with its gray wings, white face, and white breast. During the breeding season, long white plumes on the back of the head add an element of elegance. The female is slightly smaller than the male and has fewer plumes. Young Night-Herons do not acquire the adult plumage until their second full summer. Heavily streaked as hatchlings, they retain some of the streaking the following summer and appear more mousy-gray than the adults.

The Night-Heron has a diverse vocabulary, with its most common call being a crow-like, guttural *quawk*. Indeed, the bird's genus and species name, *nycticorax,* is derived from the Greek word *nyktikorax*, which means "night raven." The young and adults also utter a squeal-like scream that has been described as rather "ghastly" (Gross 1923).

The Black-crowned Night-Heron breeds on every continent except Antarctica and Australia. In North America, it occurs wherever suitable wetlands are present but it is especially abundant in coastal areas and in the northern Great Plains. Resident populations also occur from Mexico through South America.

MINNESOTA BREEDING DISTRIBUTION

Roberts (1932) described the Black-crowned Night-Heron as a breeding resident throughout the "prairie and semiprairie" regions of the state. It was most abundant in southern and central Minnesota, with confirmed nesting records from seven counties: Grant, Hennepin, Jackson, Kandiyohi, McLeod, Murray, and Ramsey. Approximate numbers of breeding pairs or nests were given for sites in four of the counties: 200 nests at the Grant County colony (1924); "hundreds" of nests at Lake Shetek in Murray County (1921); 100 nests at the McLeod County colony; and perhaps 1,000 nesting pairs at Heron Lake in Jackson County in 1898, with "about one-fourth" as many in 1916. Roberts also mentioned established colonies in Pine and Meeker Counties but confirmed nesting records were not available.

Summer observations of Night-Herons in the west-central counties of Douglas and Otter Tail and farther north in Kittson, Marshall, and St. Louis suggested the birds might be nesting in those counties as well. Indeed, decades later, Green and Janssen (1975) confirmed the discovery of a colony in 1963 in the Agassiz NWR in Marshall County.

Guertin and Pfannmuller (1985) reported on the MNDNR Nongame Wildlife Program's efforts to assess the status and distribution of Minnesota's colonial waterbirds. Seventeen Black-crowned Night-Heron colonies were reported active in at least one year from 1981 to 1983. Ten of the colonies were mixed-species colonies; four of the colonies each supported more than 200 nesting pairs (Henderson 1984). With the exception of two colonies located in Marshall County, the remaining 15 were located in central and southwestern Minnesota.

Janssen's (1987) update delineated a similar distribution and identified 18 counties where nesting had been confirmed since 1970. Hertzel and Janssen (1998) added two more counties to the list. MBS field surveys documented 78 breeding-season observations, again demonstrating a similar pattern of distribution.

Survey efforts by MNBBA participants, complemented by surveys conducted in 2010 by Hamilton and Cuthbert (2016), resulted in the identification of only 66 Black-crowned Night-Heron records in 49 of the surveyed atlas blocks **(A)**. Restricted to central, south-central, and northwestern Minnesota, nesting was confirmed in ten blocks and deemed probable in three blocks. New counties with confirmed nesting records were Carver and Dakota. Numerous obser-

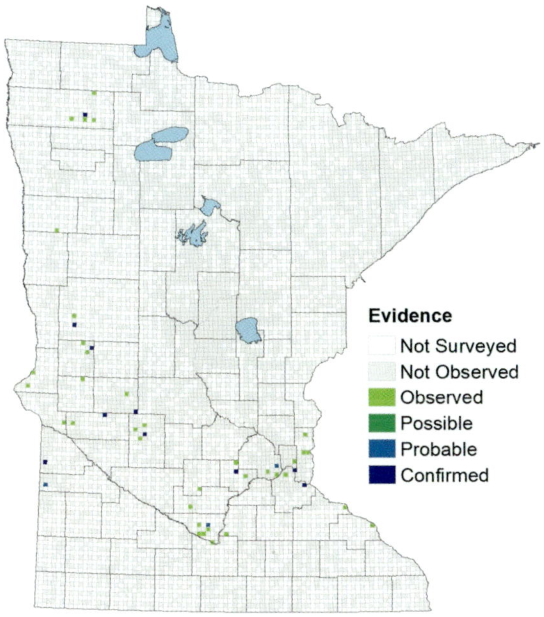

A. Breeding distribution of the Black-crowned Night-Heron in Minnesota based on the Breeding Bird Atlas (2009–13).

vations in Nicollet County strongly suggested nesting also occurred at Swan Lake and/or Middle Lake, two large and well-known waterbird nesting sites.

Overall, the statewide distribution of the Black-crowned Night-Heron appears similar to that reported by Roberts (1932), with the exception that colonies are no longer present in southwestern Minnesota south of Lac qui Parle County. Nevertheless, it seems that the species was far more abundant in the late 1800s and early 1900s, even in the 1980s, than it was during the MNBBA. The number of nests was estimated at five of the 10 colonies identified during the atlas and totaled only 79 nests. Even though the number of breeding pairs in the state is certainly larger, it is likely well below that reported decades ago.

Beyond Minnesota, few data are available to document any major changes to the species' distribution. The only notable change has been its expansion to sites "where irrigation and water impoundments provide suitable habitat" (Hothem et al. 2020).

BREEDING HABITAT

Night-Herons are found nesting in woodlots, riparian forests, and shrub thickets, either on the mainland or on islands. In the Prairie Pothole Region and farther west, the species frequently nests within dense stands of emergent vegetation in large wetland complexes **(B)**. Sites with minimal access are preferred. Mammalian predators and human disturbances are usually considerably reduced in such settings. Although the species may travel some distance to foraging areas, it prefers nesting sites that are close to shallow bodies of water with abundant prey.

POPULATION ABUNDANCE

The Black-crowned Night-Heron is a challenging species to monitor. Not only is it a nocturnal feeder but also nesting colonies located within large wetland complexes are difficult to access. When it nests in colonies with other tree-nesting waterbirds, the Night-Heron nests are often located in the subcanopy, where they are less visible during aerial surveys. In 2002, Kushlan and his colleagues simply suggested that the continental breeding population numbered more than 50,000 individuals. By 2021, PIF placed the population estimate for the United States and Canada at 420,000 birds with a global population estimate of 3 million. Although not statistically robust, BBS data signal a slowly declining population trend across the survey region, averaging −0.56 percent per year from 1966 to 2019 (Sauer et al. 2020).

Census data are also sparse in Minnesota. Adams and his colleagues (1973) estimated that in the 1940s the statewide population was a minimum of 6,060 nesting pairs based on the number of pairs reported at just two known colonies: Heron Lake in Jackson County (5,000 nests); and a site that "no longer exists" in Ramsey County (1,060 nests). Henderson (1984) estimated the statewide population at 1,858–3,850 nesting pairs in the early 1980s. Since then, only those Night-Heron colonies that are also occupied by Cormorants have been periodically surveyed. Among the colonies visited, five supported a minimum of 305 Night-Heron nesting pairs in 2004; four supported a minimum of 77 nesting pairs in 2010, and three supported a minimum of nine nesting pairs in 2015 (Hamilton and Cuthbert 2016).

Unfortunately, these latter surveys are the only data that provide even a glimpse of the species' current status in Minnesota. Despite caveats associated with the scope of the work, it appears that Minnesota's Black-crowned Night-Heron population has declined significantly. This is corroborated by Minnesota BBS data that suggest a sharp population decline from 1967 to 2019 (Sauer et al. 2020).

The status of several historically important Black-crowned Night-Heron nesting sites lends further credence to the decline. At least four important colonies are no longer active or their Night-Heron populations appear significantly reduced in size: first, Pig's Eye south of downtown St. Paul (population significantly reduced; Hamilton and Cuthbert 2016); second, Heron Lake in Jackson County (last observation of birds in 1995; Koster 1995); third, Agassiz NWR in Marshall County (fewer than 50 nesting pairs in 2016; Jordan Young, refuge biologist, pers. comm. 2017); and, fourth, Long Lake in Kandiyohi County (last known nesting report in 2010; Hamilton and Cuthbert 2016).

Factors responsible for the species' decline are not well understood. Habitat loss and degradation, coupled with human disturbance, are considered to be the major threats. Additional concerns include nest predation, interspecific competition for nesting sites, and bad weather. High pesticide contaminant levels were found in some populations in the 1970s and 1980s, but the impact on reproductive success appeared minimal (Hothem et al. 2020).

CONSERVATION

Given the species' uncertain status, threats to its breeding habitat, and its small continental population, it has been designated a moderate conservation priority (PIF 2021). Fortunately, warming temperatures may not pose a risk. NAS (2019) considered the Black-crowned Night-Heron's North American breeding distribution to be stable under the three warming scenarios they examined. Small losses projected in southern Minnesota will likely be offset by gains across the northern counties. Nevertheless, increasing water levels in coastal habitats and the Great Lakes, as well as more intense rainstorms in the Great Plains, could directly impact nesting colonies. Although wetland protection and restoration are important long-term conservation strategies, the heron's precarious status warrants a more accurate population assessment.

B. Extensive wetland complexes provide the Black-crowned Night-Heron with multiple nesting opportunities in western and southern Minnesota. Photo by Gary T. Seim.

YELLOW-CROWNED NIGHT-HERON

Nyctanassa violacea

Minnesota Seasonal Status: A summer and migratory visitant; documented breeding sporadically in Minnesota from 1955 to 1986. The Yellow-crowned Night-Heron was very rare during the MNBBA.
Migration: Northern birds winter primarily in southern Florida but have been reported in other coastal and inland areas south to Central America. Birds may wander following the breeding season.
Food: Primarily crustaceans (crabs on the coasts; crayfish inland). Other prey items are usually only a minor component of the diet, depending on the locality. Feeds during the day and night.
Nest: Stick platform placed in trees or bushes; usually only one nest per tree. The species nests in small colonies or as isolated pairs.
Conservation Concern: Assigned a moderate PIF Continental Concern Score of 12/20.

Photo by Kim Rewinkel.

The Yellow-crowned Night-Heron has an on-and-off record as a nesting species in Minnesota. It was not even observed in the state until 1955 and has been documented nesting irregularly in small numbers (usually one to four pairs) at three sites: LaCrescent (1955–86); Pig's Eye Lake, South St. Paul (1963; 1981–86); and Aitkin County rice paddies (1981–83). There has been no firm breeding evidence since 1986; subsequent records have been migrants, summer visitants, or a few fall wandering juveniles.

This heron is one of two Night-Herons that occur in Minnesota—and the rarest. Just a bit smaller than the Black-crowned Night-Heron, the slender Yellow-crowned Night-Heron is distinguished by its distinct head pattern. A broad, creamy-white crown, a bold white cheek below the eye, and long, creamy-white plumes at the back of the crown embellish the black head. The "dandelion yellow" forehead (Watts 2020), though difficult to discern, especially in poor light, is responsible for its common name. Unfortunately, juveniles of both Night-Heron species are difficult to distinguish from one another, both being heavily streaked with brown and white feathers their first year. Nevertheless, by their second year, young Yellow-crowned Night-Herons largely appear similar to an adult. All that is missing is the creamy-white color of the crown, which appears brownish gray until the fourth year (Watts 2020). The Latin name for the genus, *Nyctanassa,* means "queen of the night"; *violacea* suggests that its dark gray plumage sometimes appears violet.

In North America, breeding populations occur from southern coastal New England south to Florida and west to Texas. There are inland populations throughout the southeastern states and north along the Mississippi River to Missouri and southern Ohio. Scattered nesting colonies occur in the western Great Lakes states; the core of its breeding range is from east Texas to Georgia and Florida. The species is a permanent resident in many coastal areas from the southern United States to northern South America.

MINNESOTA BREEDING DISTRIBUTION

The first recorded observation and nesting record of the Yellow-crowned Night-Heron in Minnesota was in 1955 in the Mississippi River backwaters near La Crescent, Houston County, when a pair of adults and a nest with young were found by Brother Theodore Voelker (Theodore 1955). Four pairs nested in 1956 (Theodore 1956). According to MOU records, this small colony remained active through 1977 (MOU 2021b). Then, in 1963, two adults were found at a second site: Pig's Eye Lake, Ramsey County (Huber 1963b). Nesting was documented at this site in 1964, 1965, and 1966 (Huber 1964c, 1966a, 1966b).

The expansion of the Yellow-crowned Night-Heron into Minnesota was not unique. Indeed, Watts (2020) noted that the species experienced a "dramatic northward range expansion between 1925 and 1960, with 11 new state breeding records." This included the nearby states of Michigan (1960, first record; 1971, first bred; Chartier et al. 2013), Wisconsin (1941, first record; 1955, first bred; S. D. Robbins 1991), and Iowa (1953, first record; 1956, first bred; Kent and Dinsmore 1996). Small colonies continue to persist in these states.

In the 1960s and 1970s, this Night-Heron was infrequently observed away from the two known nesting sites. An injured bird was collected south of Clements, Redwood County, in 1966 and given to the University of Minnesota (MMNH specimen 21881; Sogaard 1966), and one adult was observed in 1972 in the large, mixed-species rookery on Lake Johanna, Pope County (Green and Baumhofer 1972). Birds were also observed in the rice paddies of Aitkin County in 1972, 1973, and 1976, but there was no nesting evidence. In 1978, an adult was seen building a nest and incubating from June 7 to July 18 at the Maplewood Nature Center, Ramsey County, but no young were produced (Gange 1978). Unfortunately, an intensive survey at the Pig's Eye rookery in 1973 did not find any Yellow-crowned Night-Herons (Adams et al. 1973).

Then, during the 1980s, probable breeding was confirmed in 1981 in Aitkin County's rice paddies when a pair was seen throughout the nesting season and a family of two adults and two young was photographed in August (W. Nelson 1981). They bred again in 1982 (W. Nelson 1982) and in 1983 (Green and Janssen 1984). Probable nesting was reported in 1982 at Thief Lake, Marshall County, when an adult was flushed from a Black-crowned

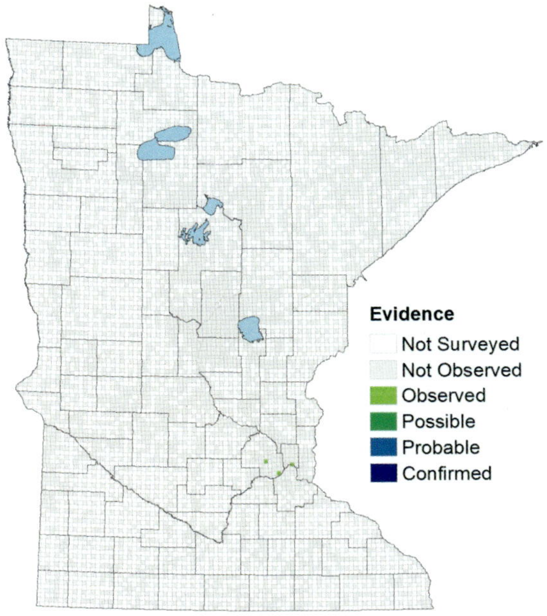

A. Breeding distribution of the Yellow-crowned Night-Heron in Minnesota based on the Breeding Bird Atlas (2009–13).

Night-Heron colony (Steva 1982). And, at Pig's Eye, nesting of a few pairs (2–5) was confirmed again in 1981 (S. Richardson 1981) while probable nesting was reported in 1984 (Warner 1984) and 1986 (Shedd and Wilson 1987). This colony, which is now a state protected site, is in an industrial area and has the unlovely name Pig's Eye Island Heron Rookery SNA. The name commemorates Pierre "Pig's Eye" Parrant, a person of ill-repute who was the first European settler to live within the city limits of Minnesota's capital, although not on this island, which he tried to buy.

From 1987 to 2020, there were no documented breeding records anywhere in the state. Nevertheless, the Yellow-crowned Night-Heron continued to be observed as a migrant and summer visitant almost every year from Otter Tail County and the Twin Cities southward through 2012; there were no summer reports from 2013 through 2017. Then, in 2018, one juvenile was reported at the Coon Rapids Dam in Hennepin County. Two years later, an "apparent juvenile" was reported at the North Ottawa Impoundment in Grant County, and one juvenile each was reported in Hennepin and Steele Counties (MOU 2021a, 2021b). Prior to 2013, most summer records were clustered in the Twin Cities counties of Hennepin, Ramsey, and Dakota. This may represent an increase in Twin Cities' bird-watchers or dispersing Night-Herons from undocumented nesting sites. Outside of the Twin Cities, there was a pronounced decline in the number of observations from 2005 to 2020.

MNBBA participants reported three adult Yellow-crowned Night-Heron records from three atlas blocks, all within the Twin Cities area **(A)**. One was on an urban pond in Minnetonka, where it was present from July 23 to August 20, 2009. A second was seen on June 21, 2009, flying west from the general direction of the Pig's Eye slough. The third was seen on July 11, 2012, from the Old Cedar Avenue bridge on the Minnesota Valley NWR. MOU (2021b) records also include a juvenile bird photographed near Long Meadow Lake in the Minnesota Valley NWR, September 8–15, 2012. These records from the NWR, as well as the observation near Pig's Eye, point to possible nesting birds in the Twin Cities.

In Minnesota, the known colony sites are rarely monitored. Aside from the nesting areas noted above, two accounts also indicate an additional breeding site in southwestern Minnesota, presumably Heron Lake (Beyersbergen et al. 2004; Watts 2020). An actual site location for these records could not be located. The species has not been found breeding in South Dakota (Drilling, Sparks, et al. 2016) or North Dakota (Drilling 2015).

BREEDING HABITAT

In coastal areas, the Yellow-crowned Night-Heron nests in forested wetlands, mangroves, cypress swamps, and bayous and occasionally on coastal islands and spoil banks. Inland, the species is found along forested rivers, streams, and in shrubby wetlands and forested uplands near riparian areas. It does not occupy open marshes. In Minnesota, it has also nested in abandoned rice paddies overgrown with shrubs and with an abundant supply of frogs (W. Nelson 1982) **(B)**. It is tolerant of suburban development for both nesting and feeding when water and a good food supply are close by.

POPULATION ABUNDANCE

The species' small and scattered colonies, usually well hidden by trees and located in sites that are often inaccessible, make it difficult to assess population size or trends with any accuracy. Wires, Lewis, and colleagues (2010) provided a general estimate of 50,000 to 100,000 breeding individuals in North America, while PIF (2021) recently estimated a global population of 400,000 birds with 130,000 residing in the United States and Canada. Although they have a low level of reliability, BBS data suggest a stable to slowly declining population throughout the survey region (Sauer et al. 2020).

CONSERVATION

The Yellow-crowned Night-Heron's small population is the primary factor that prompted its designation as a moderate conservation priority (PIF 2021). It is not state-listed in Minnesota, but a few states on the northern periphery of its breeding range have given it a "listed" status, including Wisconsin, where it is a *Threatened* species (Wisconsin DNR 2015). Rarely the focus of conservation or management actions, the Night-Heron is predicted to have a positive response to climate change (NAS 2019). As it gradually expands northward with rising temperatures, it may even become a regular breeding resident as far north as southern Minnesota. Rising sea levels, however, are likely to pose a challenge to the species along the Gulf coast.

B. Small trees and shrubs bordering wild rice paddies in Aitkin County are one of the few habitats where the Yellow-crowned Night-Heron was historically reported breeding in Minnesota. Photo by Lee A. Pfannmuller.

TURKEY VULTURE

Cathartes aura

Minnesota Seasonal Status: Regular summer resident and migrant; late migrants have been reported in early winter. The Turkey Vulture was common during the MNBBA.
Migration: Short- to long-distance migrant; winters in the southern United States, Mexico, Central America, and South America.
Food: Carrion, sometimes fruit; rarely a predator on vertebrates and insects.
Nest: On the ground, rock ledges, broken treetops, abandoned buildings, and in caves and hollow trees.
Conservation Concern: Assigned a low PIF Continental Concern Score of 5/20.

Photo by David Brislance.

The Turkey Vulture, commonly referred to as a "buzzard," is a unique species and the only vulture that occurs in Minnesota, except for the very rare Black Vulture (*Coragyps atratus*) that occasionally ventures into the state. It derives its common name from its red unfeathered head, dark body, and size that resemble those of a turkey. Its scientific genus name, *Cathartes,* is based on the Greek word for "purifier" or "cleaner," which refers to its beneficial scavenging behavior, while its species name, *aura,* is thought to be derived from the Mexican and South American name for vulture (Sandrock and Prior 2014).

Three subspecies of Turkey Vulture are commonly recognized, although up to six subspecies may be possible, depending on the authority (Kirk and Mossman 2020). In Minnesota, both *Cathartes aura septentrionalis* (the larger subspecies from eastern North America; Wied, 1939) and *C. a. meridionalis* (the smaller subspecies from western North America; Swann, 1921) can occur, and they are known to intergrade in the state.

Turkey Vultures may not be handsome birds up close but, as Chapman (1966) stated, "they soar overhead in graceful circles, perfect pictures of repose in motion." The species recently has been increasing in Minnesota. The birds are easily identified by their "teetering" flight pattern and V-shaped wing form with nary a wing flap required as they search for food by both sight and smell. Their services as scavengers are invaluable.

The Turkey Vulture is widely distributed throughout most of North America from southern Canada to southernmost South America. It is rare in the open regions of western North America and most abundant in Texas, the southeastern United States, and parts of California.

MINNESOTA BREEDING DISTRIBUTION

Historically, the Turkey Vulture has bred locally throughout the state. As of the 1930s, Roberts (1932) reported it "was formerly more common in Minnesota" but "still to be seen in fair numbers throughout the state, reaching the Canadian boundary from Kittson County in the west to Lake Superior in the east." He cited several accounts from the late 1800s that indicated the species was at least locally abundant throughout many parts of Minnesota. He described Turkey Vultures as "abundant; breeds" in east-central Minnesota in 1870; "very common" in the Red River valley in 1885; and "seen every hour in the day, sometimes several together" in Otter Tail County in 1893. A stronghold appears to have been southeastern Minnesota, along the Mississippi River and along the cliffs in Houston, Fillmore, and Winona Counties, and also along the St. Croix River.

Roberts noted several confirmed breeding records, including from Becker, Jackson, and Nicollet Counties. He included several nineteenth-century observations of nesting, which were simply attributed to Hatch in Hennepin and Traverse Counties, but there were no details in Hatch's 1892 summary *Notes on the Birds of Minnesota.*

Green and Janssen (1975) and Janssen (1987) stated that the primary breeding range was the forested regions of northern Minnesota and the Mississippi and St. Croix River valleys. D. H. Johnson (1982) identified nesting records in *The Flicker* and *The Loon* from Clearwater County "around" 1930 and in 1965, from Beltrami County in 1951, and from St. Louis County in 1956. Confirmed nesting records since 1970 have been documented from ten counties, including Beltrami and Clearwater Counties in the northwest; Aitkin, St. Louis, and Lake Counties in the northeast; Fillmore, Olmsted, and Wabasha Counties in the southeast; and Ramsey and Renville Counties (D. H. Johnson 1982; Hertzel and Janssen 1998).

MBS reported 214 records of breeding-season locations. These locations were distributed throughout the state, except for northwest Minnesota and south-central Minnesota. Observations in south-central Minnesota were largely confined to the Minnesota River valley.

The MNBBA documented the substantial population and distribution of the species in Minnesota, consistent with the increase that has occurred throughout much of North America over the past forty years. MNBBA records of the Turkey Vulture were carefully reviewed and scrutinized because, although the bird is easily observed flying overhead, determining whether these observations imply breeding evidence can be difficult. Many probable and possible records were converted to observations, especially if details were lacking.

Turkey Vulture records were reported for every county of Minnesota, including 29.8 percent (1,423/4,771) of all blocks surveyed and 37.7 percent (880/2,337) of priority blocks **(A)**. A total of 92 percent (1,313/1,423) of the records were observations. Confirmed nesting was reported from 28 blocks (16 priority blocks) in 23 counties. Probable nesting was recorded from 79 blocks (53 priority blocks).

Roberts (1932) did not provide insight on the reasons for the apparent decline of the Turkey Vulture

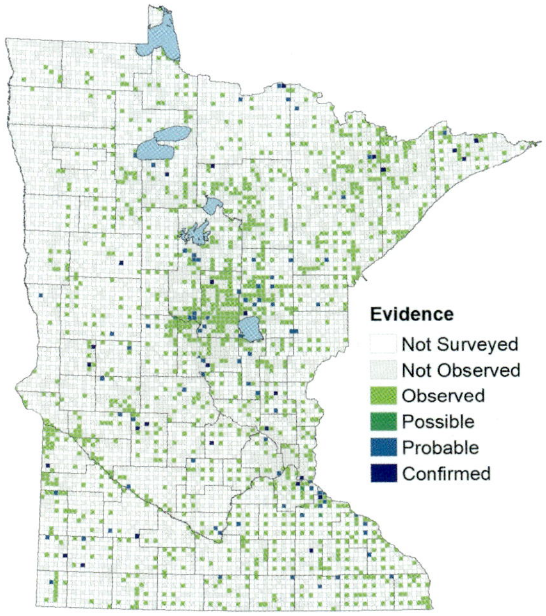

A. Breeding distribution of the Turkey Vulture in Minnesota based on the Breeding Bird Atlas (2009–13).

in Minnesota from the late 1800s to the 1930s. Kirk and Mossman (2020), in their review of the Turkey Vulture in North America, suggested that increases in human settlement, expanded agriculture, the species' avoidance of human disturbance, and human persecution all likely contributed to this decline. Beginning between 1920 to 1940 and continuing into the 1990s, the species' distribution began to expand northward in the Midwest and the Northeast and into eastern Canada. The observations during breeding bird atlases completed in Michigan, Ontario, and Wisconsin all strongly supported the substantial increases in the Turkey Vulture's distribution and population in this region (Cadman et al. 2007; Cutright et al. 2006; Chartier et al. 2013).

B. Besides forested areas, in Minnesota Turkey Vultures often use old farm structures, such as barns and sheds or even old houses, for nesting. Photo by Lee A. Pfannmuller.

The range and population expansion of the Turkey Vulture was largely attributed to increased road building resulting in a steady supply of carrion, increased deer populations contributing to increased carrion, openings created by logging activity, and warming temperatures, which facilitate movement and increased survival northward (Kirk and Mossman 2020).

BREEDING HABITAT

Turkey Vultures prefer a mix of open country and forested areas that provide a combination of both open foraging areas and isolated nesting opportunities, especially those away from human disturbance (Mossman and Hartman 1992; Kirk and Mossman 2020). They tend to avoid open cropland. They use large trees or cliffs free from disturbance for communal roosting, often located in hilly country that provides updrafts for flight. Turkey Vultures commonly are observed foraging near streams, lakes, and roads in search of carrion **(B)**.

POPULATION ABUNDANCE

PIF (2021) estimated a United States and Canadian population of 8.4 million breeding adults and a Minnesota population of 29,000 adults. Its global population is estimated at 28 million! Eleven years earlier, Farmer and colleagues (2008) had only suggested a North American population of greater than 2 million based on the combined four migration counts around the Gulf of Mexico. Hence, there is considerable uncertainty in estimating populations of the Turkey Vulture because of substantial changes in its population over time and its wanderings over large areas.

Population trends were unreliable for Minnesota based on the BBS, but the trend in the United States and Canada illustrated the substantial increase averaging 1.78 percent per year from 1966 to 2019 **(C)**. More specifically, annual counts during fall migration at Hawk Ridge in Duluth, Minnesota, also indicated a significant 3.25 percent per year increase from 1974 to 2019 **(D)**. Counts before 1980 at Hawk Ridge were fewer than 500 individuals, while counts after 1996 have consistently been more than 1,000. The highest count was 2,573 in the fall of 2018; the lowest count was 184 in 1973.

CONSERVATION

The Turkey Vulture's substantial breeding population, its population increase and range expansions, and its wide distribution in the Americas renders it of little conservation concern. As a carrion forager, the Turkey Vulture experienced eggshell thinning owing to increased loads of DDT, DDE, and other organochlorine compounds, but the extent to which populations were affected is unclear (Farmer et al. 2008). Habitat changes have continued, but shooting of Turkey Vultures has declined with educational programs that have emphasized the lack of harm this species has on humans and livestock.

The Turkey Vulture greatly benefits from human activities such as garbage dumps, livestock operations, fishing, and especially road networks and road kills, though vehicle collisions kill many vultures. It should be considered a valuable species that cleans up many of the unwanted items humans leave across the countryside. Given its wide distribution across North, Central, and South America, the Turkey Vulture is not predicted to be affected by climate change. Rather, the species' distribution will likely expand farther northward (NAS 2019).

Based on recent population increases, the future of the Turkey Vulture in Minnesota appears optimistic, so keen observers can continue to enjoy the graceful, effortless flight of this "master of the air."

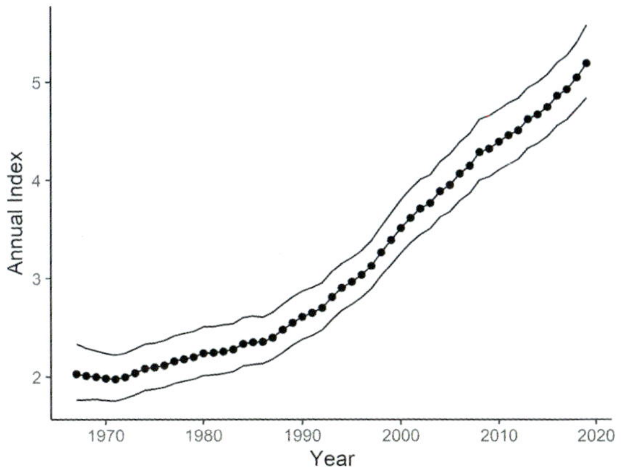

C. Surveywide breeding population trend and 95 percent credible interval (solid black lines) for the Turkey Vulture based on the 1967–2019 Breeding Bird Survey (Sauer et al. 2020).

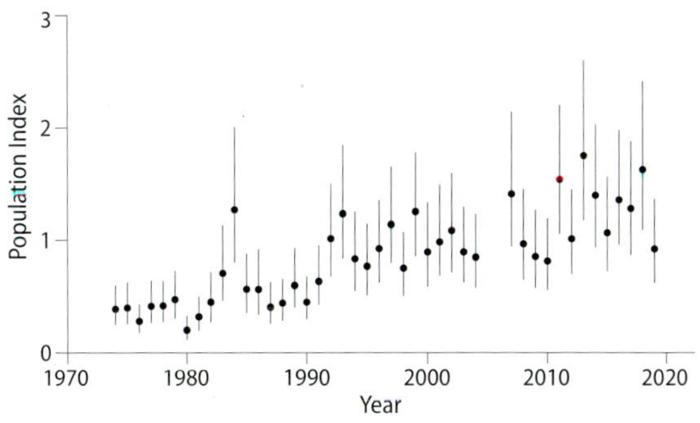

D. Fall migration counts and 95 percent credible interval (vertical lines) of Turkey Vultures at Hawk Ridge in Duluth show a significant increase in its population index from the early 1970s to 2019 as calculated by the Raptor Population Index (Oleyar et al. 2021).

OSPREY

Pandion haliaetus

Minnesota Seasonal Status: Regular breeding resident and migrant. Individuals have been recorded lingering into early winter in southern Minnesota, usually along the Mississippi River. The Osprey was uncommon during the MNBBA.
Migration: Short- to long-distance migrant that winters along the southern Atlantic coast and the Pacific coast, in Florida, along the Gulf coast, and in the Caribbean, coastal Mexico, and Central and South America (Martell, Henny, et al. 2001).
Food: Exclusively a fish-eating species, which often hovers before diving feet first into the water to grab a fish.
Nest: Natural or artificial structures, usually at the top of trees, rocky cliffs, towers, utility poles, channel markers, or poles erected specifically for Ospreys. Common features include proximity to water in good feeding areas, openness allowing easy access to nest, safety from predators, and sufficient width and stability to accommodate a large nest. Nests are often reused each year.
Conservation Concern: Assigned a low PIF Continental Concern Score of 7/20.

Photo by Mike Lentz.

With all the water in Minnesota, the Osprey is a well-known and spectacular fish-eating icon. The species has had a rocky breeding population history over the course of the past 150 years. Roberts (1932) stated it was a "summer resident, breeding throughout the wooded portion of the state. Formerly common now becoming scarce." He suggested its decline in the early 1900s was due to indiscriminate shooting of the birds, and, even though the species was eventually protected, isolated shooting still occurred. The DDT era of the 1950s and 1960s again greatly reduced its population. Today it has substantially recovered, though many threats still exist or are emerging.

The Osprey is a widely distributed species and occurs throughout the northern hemisphere as well as Australia and the South Pacific. Its scientific genus name, *Pandion,* derives from Pandion, the mythical king of Athens whose daughters were turned into birds. The name's relationship to the Osprey or its habits, however, is a mystery (Sandrock and Prior 2014). The scientific species name, *haliaetus,* is appropriately from Greek and Latin, meaning "sea eagle."

Smaller than eagles, the Osprey is one of the easier raptors to identify. Its dominant dark brown back, white belly, and white head with a prominent dark eye stripe quickly distinguish it from other large birds. In flight, it has long, narrow wings with black wrist marks and a silhouette often shaped like the letter M. It is one of the few raptors that hovers in flight as it searches for food over water. Its call is a series of loud *kyews* most often associated with nesting activity. It is found widely near lakes and rivers where suitable nesting sites are available. Ospreys have also adapted well to living in close association with people. As Bent (1937) stated, "security and a good food supply seem to be all that they require."

MINNESOTA BREEDING DISTRIBUTION

Roberts (1932) commented that the Osprey "has now largely vanished as a summer resident from . . . summer-resort lakes, though still fairly common in wilder regions where the gun and rifle are not quite so much in evidence." Regarding the Osprey's fishing behavior, he said, "There are few sights about our lakes so thrilling and spectacular as the plunge of this great bird from a height of forty or fifty feet." Further, he wrote, "when it sights a fish it pauses for a moment, then on set wings it plunges swiftly . . . if the strike is successful it emerges with the fish . . . with head pointed forward."

Roberts surprisingly documented few nests in his 1932 treatise on the birds of Minnesota. He reported confirmed nesting in Aitkin, Cook, Itasca, Lake, and St. Louis Counties and in Itasca State Park. All the nests were observations of young in the nest or just out of the nest, except Cook County, which was of a nest in a Great Blue Heron colony. In his 1936 revision, Roberts included one nest with eggs in Crow Wing County, and two nests in Lake of the Woods County, one of which he labeled as unusual because it was in a live spruce tree.

Besides illegal shooting, Roberts (1932, 1936) also attributed the decline of the Osprey to negative interactions with Bald Eagles. Evans (1982) pointed out that interspecific aggression occurs with Bald Eagles but cited evidence that actual predation was infrequent. Even though the population of Ospreys was not large after the 1930s, it declined further when the use of DDT began to enter the environment in the 1950s. DDT and its metabolite DDE disrupted calcium metabolism and resulted in thin eggshells, which broke upon incubation. This ultimately led to poor reproduction and substantially reduced the populations in Minnesota and throughout North America (Poole, Bierregaard, and Martell 2002).

In 1975, Green and Janssen reported that Ospreys were "resident in north central and northeastern Minnesota and adjacent Becker, Mille Lacs, and Pine Counties." Even though Ospreys had been observed in the Upper Mississippi National Wildlife and Fish Refuge in southeastern Minnesota, no nests had been located in that region of the state. Green and Janssen reported confirmed nesting in 15 counties. Besides the seven already reported by Roberts, they included Becker, Beltrami, Carlton, Cass, Clearwater, Hubbard, Mille Lacs, and Pine Counties.

D. H. Johnson (1982) stated that during the 1981 breeding season there were an estimated 160 "producing" osprey nests in the state. Producing nests were defined as those where at least one young survives to fledging age. He included nests from the Chippewa and Superior National Forests and an incomplete inventory of Crow Wing, Becker, Mahnomen, and Hubbard Counties. About 25 of the nests were on artificial structures. A majority of the nests documented in Minnesota, 70 percent (112 of 160), were in the Chippewa National Forest, owing to the diligent efforts of John Mathisen, the wildlife biologist on the Chippewa, who led the effort to document nesting. From 1968 to 1981, Johnson reported Osprey nesting data from the Chippewa National Forest, where the number of active nests varied from a low of 40 in 1968 to a high of 142 in 1980. Nesting success varied from a low of 13 percent in 1968 to a high of 65 percent in 1980. Johnson reported nesting from 17 counties; new nesting records were reported from Mahnomen County, and a nest with one chick was reported from Lake Bronson State Park in Kittson County.

Finally, Johnson pointed out a nest with two young that was found near Winona, Winona County, in 1981 but on the Wisconsin side of the Mississippi River. Its breeding presence in this region had not been noted for quite some time. Hatch (1892) said the Osprey was "almost universally distributed throughout the state." Roberts, as noted above, also said it was found throughout the state, and with the decline of the Bald Eagle "it might return once more to such former nesting-places as Lake Minnetonka, White Bear Lake, Mille Lacs, and along the bluffs of the Mississippi River." S. D. Robbins (1991) in Wisconsin surmised that "one might expect this bird to occur throughout Wisconsin each summer wherever suitable lake and river habitats exist. The nineteenth-century observers found it to be so."

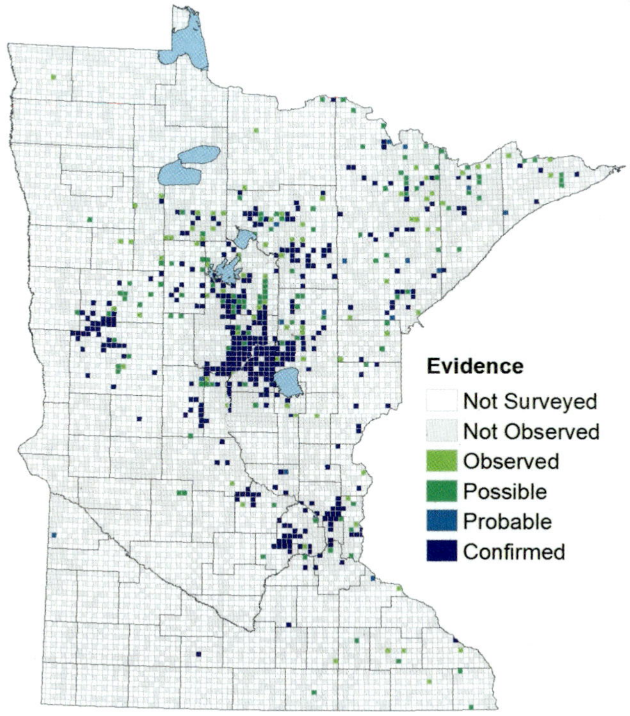

A. During the Minnesota Breeding Bird Atlas (2009–13), Ospreys were primarily found nesting near lakes and where many observers were located, such as in central Minnesota.

In 1983, Coffin and Pfannmuller (1988) stated that Henny had estimated the entire contiguous US population as only 8,000 pairs. In 1984 and 1985, Hennepin County Parks initiated a program to restore breeding populations of Osprey to the Twin Cities area (Gillette and Englund 1985; Martell, Englund, and Tordoff 2002).

In his 1987 update, Janssen reinforced that the Osprey breeding distribution in Minnesota was in the north-central and northeastern regions and west to Mahnomen County. He confirmed nesting in 15 counties since 1970 with most in northeastern and north-central Minnesota. Besides those previously reported, he extended confirmed nesting south to Carver and Kanabec Counties and west to Mahnomen County. By 1998, Hertzel and Janssen had added four more counties to the list since 1970, including a southern extension to Winona County and west to Otter Tail County, and nesting in Anoka and Todd Counties.

The MBS began their surveys in the late 1980s and have currently recorded 79 breeding-season locations. Their locations are primarily in north-central and northeastern Minnesota but extend to Kittson, Marshall, and Pennington Counties in the northwest and to Kandiyohi County in the west-central region and include one location in Freeborn County in the southern tier of counties. A statewide survey conducted by the MNDNR in 2004 found around 600 nesting pairs, concentrated in the north-central portion of the state.

The results of the intensive efforts by MNBBA participants illustrate the massive recovery by the Osprey since the 1950s and 1960s. The MNBBA reported 938 records, which included 12.9 percent (618/4,791) of the surveyed atlas blocks **(A)**. Confirmed nesting was identified in 397 blocks, probable in 15 blocks, and possible in another 120 blocks. This distribution represents the continued expansion of the population described by both Green and Janssen (1975) and Janssen (1987) to Otter Tail and Douglas Counties, and a possible nesting in Lac qui Parle County in western Minnesota. The MNBBA reported confirmed nesting in 34 counties, including southwest to Watonwan County, south to Le Sueur County, and southeast to Winona County.

The reintroduction program in the Twin Cities metropolitan area in the 1980s and 1990s greatly expanded breeding in the area and southward (Gillette and Englund 1985). The high density of nesting

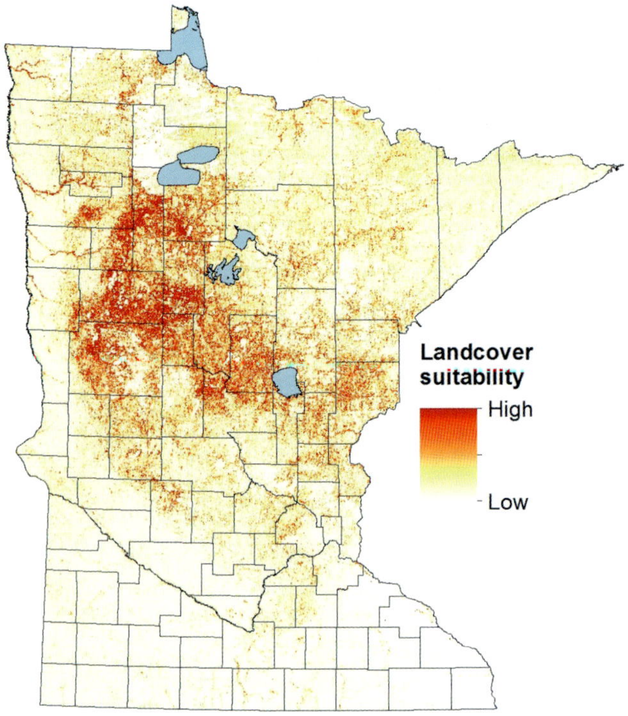

B. Landcover suitability map for the Osprey in Minnesota based on habitat, landscape context, and climate data gathered during the Minnesota Breeding Bird Atlas (2009–13) using the MaxEnt modeling approach.

records in the metropolitan area, in central Minnesota, and in western Becker and Otter Tail Counties is a testament to the recovery of the Osprey and to the dedication of MNBBA participants in those regions.

The landcover suitability map for the Osprey highlights the extensive availability of lakes in the central, west-central, and east-central regions of Minnesota **(B)**. In addition, several riverine systems are well represented.

Since the banning of DDT in 1972, Ospreys have again been expanding their range in Minnesota, presumably to some of their formerly occupied territories. As Roberts pointed out, shooting Ospreys had reduced their populations into the 1930s. Even with the passage of laws protecting migratory birds that occurred between the 1930s and the 1960s, Osprey populations were unlikely to be large because raptors were shot well into the 1950s, and then DDT further reduced its populations. Today the Osprey has substantially recovered, though many threats still exist or are emerging.

In their North American review of the Osprey, Bierregaard and colleagues (2020) point out that not only did the banning of DDT aid the species, but many management efforts were also important to its recovery. For example, reintroduction programs, construction of artificial nesting structures, and the creation of reservoir habitats have aided the expansion into formerly unoccupied areas. The recovery of Osprey populations is also well-documented in surrounding states and provinces, including Michigan (Chartier et al. 2013), Ontario (Cadman et al. 2007), and Wisconsin (Cutright et al. 2006).

BREEDING HABITAT

The Osprey is a water-loving species with numerous opportunities for nesting in Minnesota. It is found widely near lakes and rivers where suitable nesting sites exist. Ospreys have also adapted well to living in close association with people. They nest in many locations near water with accessible fish and nest sites free from predators. Poole, Bierregaard, and Martell (2002) found that Ospreys will commute up to 10 to 20 km for food in shallow water (0.5–2 m deep). Beaver activity can be beneficial to the Osprey if dead snags for nesting are created along with shallow ponds with available fish.

Natural nest sites include trees, cliffs, large boulders along shorelines, and especially sites on predator-free islands (Bierregaard et al. 2020). When in trees, Ospreys usually nest at the top of the tree. As previously noted, many artificial nest sites are also used, including towers, power poles, buoys, channel markers, duck blinds, and poles erected specifically for Ospreys **(C)**. K. L. Peterson and colleagues (2020) recently summarized nest use and nesting success in the restored population of Osprey in the Twin Cities metropolitan area. They reported no differences in Osprey nesting success between nests selected opportunistically and those on artificial structures. Moreover, collectively the birds have averaged 76 percent successful nests and 1.6 young per nest since 1995. Ground nesting is also known where mammalian predators are absent, such as on Gardiners Island, New York, where approximately 300 pairs historically nested. Primary predators are mammals (e.g., raccoon), but Bald Eagles and Great Horned Owls are known to prey on nestlings, and Great Horned Owls on adults.

POPULATION ABUNDANCE

Osprey populations have varied considerably over the past one hundred years in Minnesota. PIF (2021) estimated the United States and Canadian population at 400,000 adults, with a population in Minnesota of 2,700 individuals. They also estimated a global Osprey population of 1.2 million birds. Rosenberg, Kennedy, and colleagues (2016) calculated that the Osprey population in North America had increased by more than 200 percent from 1970 to 2014. Population estimates by Environment Canada (2011) were much broader and ranged from 50,000 to 500,000 adults. The MNBBA did not attempt to estimate the population of Ospreys in the state.

The BBS roadside counts are generally unreliable for detecting species like the Osprey. However, it was reliable for Florida, where trends for 79 BBS routes showed a significant increase in the Osprey population from 1966 to 2019, averaging 3.30 percent per year (Sauer et al. 2020). Based on 1,198 routes, the trends surveywide in the United States and Canada should be interpreted with caution but illustrate that the population was significantly increasing an average of 1.90 percent per year over the same time period. In Minnesota, the number of Osprey detections was relatively low from 1967 to the mid-1990s and has steadily increased. Its breeding population remains highly variable, and its overall increase from 1967 to 2019 is averaging 2.10 percent per year **(D)**.

The Osprey trend based on fall migration counts at Hawk Ridge in Duluth, Minnesota, indicated an increasing count from 1974 to the turn of the century but a decreasing count since then up to 2019 **(E)**. For

C. Ospreys nest in many locations near water, including trees, cliffs, and manmade structures, such as this artificial pole constructed near a power line structure. Photo by Gerald J. Niemi.

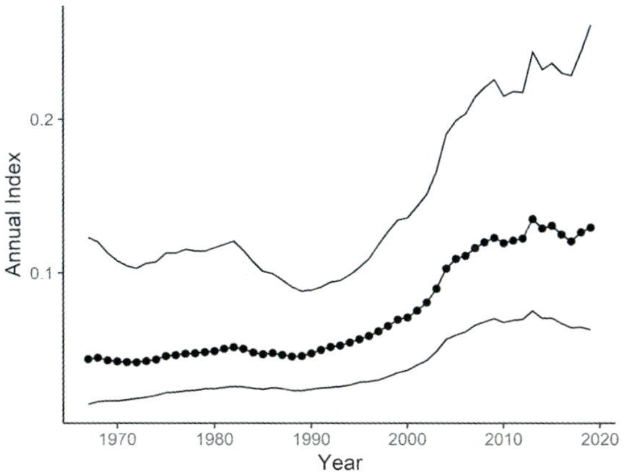

D. Breeding population trend and 95 percent credible interval (solid black lines) for the Osprey in Minnesota based on the 1967–2019 Breeding Bird Survey (Sauer et al. 2020). Note the Osprey's rapid but highly variable population increase.

example, fall counts at Hawk Ridge in the 1970s varied from 62 to 182 with the highest count ever, 539, in 2005. In contrast, from 2009 to 2020 the fall counts have dropped and ranged from 149 to 396.

CONSERVATION

At the current time, there is minimal concern regarding the population status of the Osprey. Its population has been increasing and has a very broad geographic range, which includes a substantial distribution across the globe. In Minnesota, Ospreys were officially listed as *Special Concern* in 1984 (Coffin and Pfannmuller 1988) but delisted in 1996 (MNDNR 1995). Current Osprey populations may be among the highest in more than one hundred years.

Poole, Bierregaard, and Martell (2002) and Bierregaard and colleagues (2020) identified many potential emerging problems for the Osprey, including land conversion of forests to agriculture with loss of nest trees and increased turbidity in foraging waters, continued shooting on wintering grounds and at fish farms, automobile collisions, and ingestion of lead, plastics, and chemicals, such as mercury and organochlorine compounds. Bierregaard and colleagues (2020) emphasized the many positive effects that management and conservation activities have had and can continue to have on Ospreys. These activities include the creation of reservoirs and waterfowl impoundments; protection and improvement of bodies of water, such as controlling water levels that are favorable to foraging Ospreys; reintroduction programs of Ospreys to new areas; and, most important, providing Ospreys with artificial nest sites, which has led to dramatic increases in Osprey numbers.

Under all three scenarios of climate change, the NAS (2019) projected a stable breeding range in North America for the Osprey and included range expansions primarily to the far north in Canada by 2050. Some larger patches of losses in its breeding range were forecast in eastern Virginia and North Carolina, around Lake Michigan, in central California, and scattered patches of losses throughout the continent. The Osprey winter range was also predicted to be stable under all three climate scenarios, with some winter range gains north and westward in the southeastern United States. Uncertainties associated with these predictions include how the species will respond to sea-level rise in coastal areas or find suitable habitat and food within its potential expanded range northward.

Roberts (1932) pointed out that "aside from man, the Osprey's chief enemy is the Bald Eagle." He further speculates that "it may be difficult or impossible for the Osprey to maintain itself when the Eagle is abundant." Bald Eagles have increased dramatically along with the Osprey. Bald Eagles are pirates: they rob Ospreys of their fish and prey upon Osprey nestlings. The interactions between these two species will be an interesting play to observe in the future, and the theater is here, in the outdoors of Minnesota.

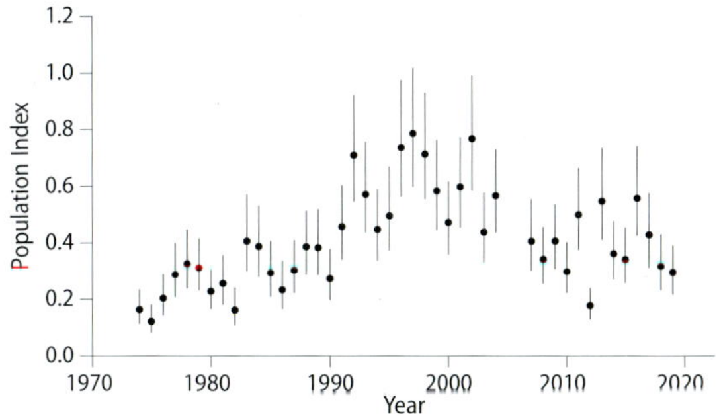

E. Fall migration counts and 95 percent credible interval (vertical lines) of Ospreys at Hawk Ridge in Duluth have an unusual pattern of an increasing population index from the early 1970s to the late 1990s but then a decline after the early 2000s, as calculated by the Raptor Population Index (Oleyar et al. 2021).

NORTHERN HARRIER

Circus hudsonius

Minnesota Seasonal Status: Regular breeding resident and migrant; occasionally observed during the winter months. The Northern Harrier was common during the MNBBA.
Migration: Partial to medium-distance migrant that winters throughout the central and southern United States south to Central America and the Caribbean.
Food: Plunges from low flight to capture prey on the ground; food consists primarily of small mammals and small birds.
Nest: Small ground depression or platform constructed of grasses and forbs; nests are located in upland fields or over very shallow water at the edge of a wetland.
Conservation Concern: Assigned a moderate PIF Continental Concern Score of 11/20.

Slowly gliding low over open fields and wetlands, the Northern Harrier peruses the landscape for potential prey. It has long wings and a long, banded tail, but the bird's most diagnostic feature is its white rump, clearly visible as it "veers from side to side, rarely pursuing a straight course" (Bent 1937). Indeed, its genus name, *Circus,* refers to its circular flight. The species is sexually dimorphic; the female is dark brown with heavy streaking on the rufous breast, while the smaller male is gray with speckled feathers dotting its white underside. The bird's disc-shaped head is unique among Minnesota's diurnal raptors. Similar to the head of an owl, its shape, as well as the stiff hairs that surround it and the large external ear openings, enhance the Harrier's ability to use sound as well as vision to capture prey.

Once known as the Marsh Hawk, the species has a broader habitat tolerance than its former name implied, occupying a diversity of open habitats dominated by dense vegetation. Unlike many hawks in the late 1800s and early 1900s, the Northern Harrier was considered a species largely beneficial to farmers. Roberts (1932) quoted Dr. A. K. Fisher (1893), who described it as "probably the most active and determined foe of meadow mice and ground squirrels, destroying greater numbers of these pests than any other species." Although it was not entirely immune to the farmer's wrath, Roberts believed it survived the early carnage better than most birds of prey.

Photo by Michael Furtman.

With a breeding distribution that stretches across Canada, the northern United States, and many western states, the Harrier is most abundant in the northern Great Plains and the central Canadian provinces. Formerly considered a global species, the Northern Harrier was recently recognized as a separate species, distinct from its near relative, the Hen Harrier (*Circus cyaneus*), which occurs in Eurasia (K. G. Smith, Wittenberg, et al. 2020).

MINNESOTA BREEDING DISTRIBUTION

In the late 1800s, Hatch (1892) considered the Northern Harrier as "undeniably the most abundant of the hawks that visit the state." Forty years later, Roberts (1932) commented that although numbers were reduced, it remained abundant. It was often the only hawk seen during extensive forays across the state, particularly in the open country of western and central Minnesota. At the time, confirmed nesting records were available from 12 counties stretching from Jackson and Lincoln Counties in the southwest, east to the Twin Cities, and northwest to Marshall County.

When Green and Janssen (1975) prepared an updated account of its status, the Northern Harrier was considered one of the three most common raptors in the state; the other two were the Red-tailed Hawk and the American Kestrel. Evidence, however, pointed to sharp population declines in some regions. Just a few years later, D. H. Johnson (1982) summarized the status of all raptors nesting in Minnesota and noted that from 1970 to 1981 there were no observations or nest records of Northern Harriers from 16 counties where they had been reported prior to 1970. Eleven of the 16 counties were south of the Minnesota River. Johnson believed the absence of records reflected a true population decline due to the "tremendous loss of habitat." By 1987, Janssen commented that the Harrier had virtually disappeared from southern Minnesota but remained common in the northwest and north-central regions. He identified 13 counties where nesting had been confirmed since 1970; all were located north of the Minnesota River valley (Janssen 1987).

Several years later, Hertzel and Janssen (1998) identified 17 counties with confirmed nesting since 1970. One county (Mahnomen) identified by Janssen (1987) was no longer listed, but the five counties added were all south of the Minnesota River, suggesting that local populations were slowly rebounding. Fieldwork by the MBS from the late 1980s through 2014 further confirmed the species' broadened distribution, with 473 breeding-season locations well spread across the state.

During the MNBBA, participants reported 1,417 Northern Harrier records in 21.9 percent (1,048/4,780) of the surveyed atlas blocks. Located in south-central Minnesota, Watonwan was the only county where the species was not reported. Breeding was confirmed in 54 blocks located in 35 counties **(A)**. The birds were more sparsely distributed in the densely forested counties of northern Minnesota, in the southeast, and along the Minnesota-Iowa border. Indeed, the Harrier's landcover suitability map predicts that northwestern Minnesota provides the most suitable habitat, with scattered pockets located

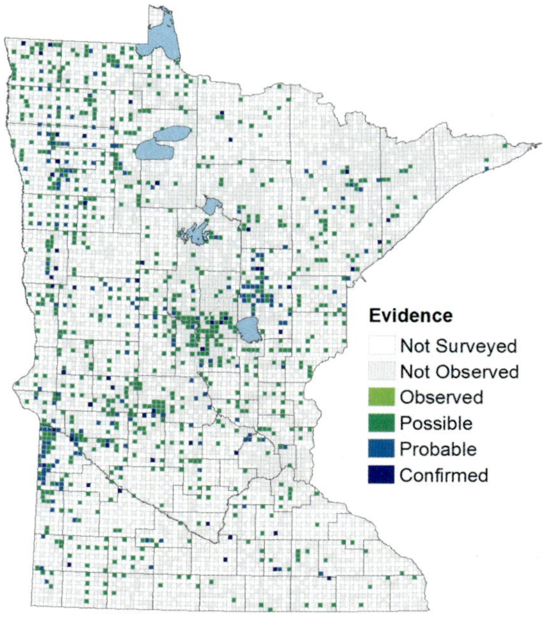

A. Breeding distribution of the Northern Harrier in Minnesota based on the Breeding Bird Atlas (2009–13).

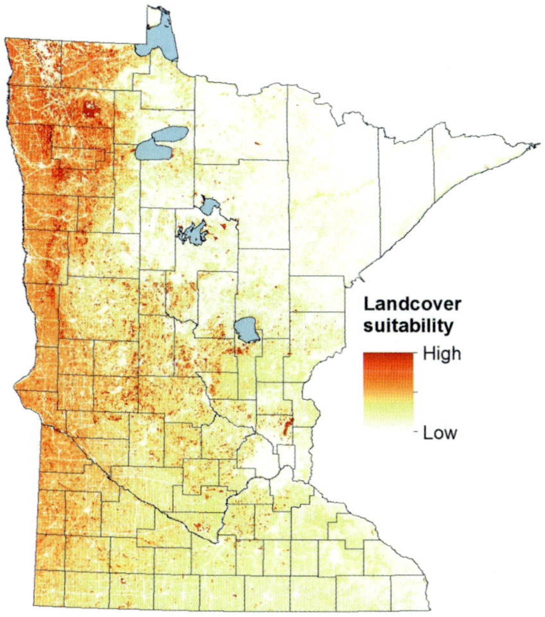

B. Landcover suitability map for the Northern Harrier in Minnesota based on habitat, landscape context, and climate data gathered during the Minnesota Breeding Bird Atlas (2009–13) using the MaxEnt modeling approach.

throughout the central and southwestern regions as well **(B)**.

Clearly the Northern Harrier is faring better than records suggested in the mid- to late 1900s and has reoccupied its former range south of the Minnesota River valley. It is doubtful, however, that the species has recovered to the former levels of abundance implied by Roberts in the early 1900s. Who would have imagined forty years ago that it would eventually become less abundant than the Bald Eagle, with only 1,417 MNBBA records compared to the Eagle's 2,113 records? Among all Minnesota raptors, the Northern Harrier fell to fifth place behind the Red-tailed Hawk, Bald Eagle, Turkey Vulture, and American Kestrel.

BREEDING HABITAT

The Northern Harrier occupies open upland and wetland habitats **(C)**, including disturbed sites such as croplands, lightly grazed pastures, and old fields, and native habitats such as wetlands and native prairies (K. G. Smith, Wittenberg, et al. 2020; Slater and Rock 2005). Primary habitat features include large, open landscapes dominated by dense vegetation and abundant residual cover (Dechant, Sondreal, et al., "Northern Harrier," [1998] 2002b). Woody vegetation becomes a deterrent when it covers more than 30 percent of a site (K. G. Smith, Wittenberg, et al. 2020). Even when a site appears entirely suitable, the ultimate predictor of Harrier occupancy may be the abundance of prey, particularly voles and other small rodents. Considered a nomadic species, local population densities and reproductive success are strongly tied to prey abundance, especially microtine voles (K. G. Smith, Wittenberg, et al. 2020).

POPULATION ABUNDANCE

PIF (2021) generated a population estimate of 820,000 breeding adults in the United States and Canada and a statewide estimate of 26,000. Given the Northern Harrier's relatively low abundance, BBS trends across the survey region and in Minnesota are unreliable owing to the small number of data points. The most statistically robust data come from two regions in the northern plains: the Prairie Potholes (which includes western Minnesota) and the Badlands Bird Conservation Regions. In the former, Northern Harriers show a significant decline averaging −1.03 percent per year **(D)**; in the latter, populations have been relatively stable. Fall migration counts at Hawk Ridge in Duluth, Minnesota, also show a relatively stable trend in Harrier numbers from 1974 through 2019. In contrast, the majority of migration count sites farther east report significant declines from 2009 to 2019 (Oleyar et al. 2021).

Habitat loss due to wetland drainage, intensive agricultural practices, and the reforestation of many farmlands in the northeastern United States is considered the principal factor responsible for the decline. During the mid-1900s, these impacts were compounded by the ingestion of organochloride pesticides in vertebrate prey items, particularly

C. Large, open landscapes of wet meadows and mesic grasslands are commonly used by the Northern Harrier in western Minnesota for foraging and nesting. Photo by Lee A. Pfannmuller.

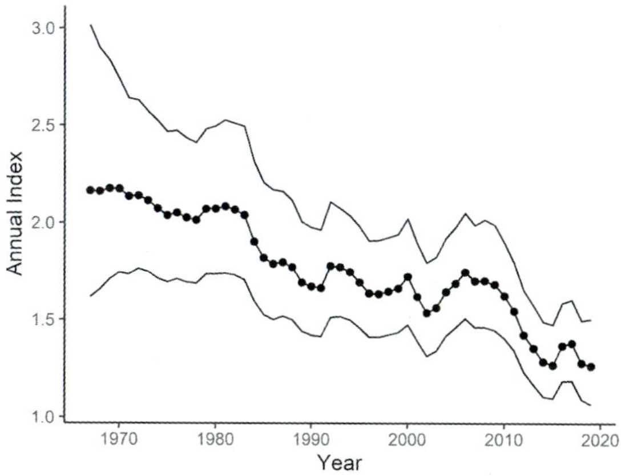

D. Breeding population trend and 95 percent credible interval (solid black lines) for the Northern Harrier for the Prairie Potholes Bird Conservation Region, based on the 1967–2019 Breeding Bird Survey (Sauer et al. 2020).

DDT and DDE. Lower reproductive success, due to a reduction in eggshell thickness and egg mass, was directly linked to declining numbers (K. G. Smith, Wittenberg, et al. 2020). In recent years, mercury was found in the feathers of Harriers migrating through Duluth. Fortunately, the levels were considered to pose a low risk (Keyel et al. 2020).

CONSERVATION

Given its small and declining population, and its dependence on wetland and grassland habitats, the Northern Harrier is considered a moderate conservation priority (PIF 2021). Management recommendations focus on protecting and restoring large tracts of land with moderate to dense herbaceous cover. Because residual ground cover is important for nesting, large tracts should always be managed in small units that allow some of the vegetation to be mowed or burned regularly while still leaving some units idle for three to five years (Dechant, Sondreal, et al., "Northern Harrier," [1998] 2002b; Slater and Rock 2005).

Habitat protection and restoration efforts are especially important given the potential consequences of warming temperatures, particularly in the Great Plains, the core of the species' breeding range. Although NAS (2019) classified the Harrier with a low vulnerability to climate change under all three warming scenarios they examined, the predicted losses in Minnesota are substantial. Under the 2°C warming scenario, nearly half of the species' breeding range in the state is lost; at 3°C it becomes restricted to far northwestern Minnesota. Fortunately, the Harrier's ability to recover from past assaults is noteworthy and may provide hope for its long-term adaptation to further challenges.

SHARP-SHINNED HAWK

Accipiter striatus

Minnesota Seasonal Status: Regular breeding resident and migrant; also frequently regular in winter, especially in southern Minnesota. The Sharp-shinned Hawk was uncommon during the MNBBA.
Migration: Short- to medium-distance migrant. Most Minnesota populations spend winters in the southern United States, the Caribbean, and Mexico.
Food: Predator that captures small birds, some small mammals, and large insects by surprise attacks on the wing or from a perch.
Nest: Placement is usually well hidden on the horizontal branches of a coniferous tree, close to the trunk and well below the canopy.
Conservation Concern: Assigned a low PIF Continental Concern Score of 7/20.

Photo by David Brislance.

The smallest of Minnesota's three "accipiter" raptors, the Sharp-shinned Hawk resembles a small Cooper's Hawk with a square tail and short, rounded wings. Adult males have a blue-gray crown, back, and upper wings with rufous barring on the breast. Adult females have brownish upperparts and are larger than males. Juveniles, like the one pictured here, are brownish above and whitish below with brown streaks. Commonly called "Sharpies," they are mostly silent except for a *kek-kek-kek* heard near the nest.

The unusual common name, Sharp-shinned, refers to the thin tarsus exposed below the short feathers on its upper leg. Its scientific species name, *striatus,* is Latin for "streaked" or "lined" and appropriately refers to its streaked upperparts and barred tail. Roberts (1932) unsympathetically commented that its decline in the early twentieth century was "not to be deplored" because they prey on other birds. Yet, a positive perspective on this predator is its ecological function of keeping small birds wary and healthy.

Sharp-shinned Hawks are widely but patchily distributed across the forested areas of North America, Central America, and South America. They are not common anywhere within their wide breeding range.

MINNESOTA BREEDING DISTRIBUTION

Historically, Roberts (1932) simply described the Sharp-shinned Hawk as a "summer resident, breeding sparingly throughout the state." He felt the species was formerly more abundant than in the early twentieth century, with declines possibly due to habitat loss and shooting. Roberts (1932) included two confirmed nests, each with six eggs, in Cass County (1919). He also reported an unusual 1921 nest discovered by Surber with three full-fledged young in Murray County found "in an oak tree in the midst of a nesting-colony of Night Herons."

More than forty years later, Green and Janssen (1975) highlighted that no summer records existed since the 1930s from a line west of Fillmore, Rice, Stearns, Becker, and Roseau Counties, roughly the least forested areas of Minnesota. They added that Sharpies were found in north-central, northeastern, central, and southeastern regions but were "generally scarce everywhere."

D. H. Johnson (1982) summarized that nine nesting records of the Sharpie existed for the state. Besides the nests described by Roberts, he added reports of a nest in St. Louis County in 1945, one in Winona County in 1953, two unpublished records of nests from Clearwater County, and two that he observed in Hubbard County. He pointed out that the low number of nests found is surprising only to those who have never tried to look for a Sharpie nest.

Janssen (1987) outlined a similar distribution to that described by Green and Janssen (1975), but excluded all of central, southern, and western Minnesota. Janssen (1987) and Hertzel and Janssen (1998) included confirmed nesting since 1970 in six counties: Beltrami, Cass, Clearwater, Cook, Hubbard, and Lake.

Another tribute to the secretive breeding behavior of the Sharp-shinned Hawk was provided by the MBS. Despite many sharp-eyed MBS observers, the survey has recorded only 20 breeding-season locations. The locations ranged from three in Morrison County of central Minnesota and northeastward to two in Cook County, including locations in Aitkin, Cass, Itasca, Lake, and St. Louis Counties.

Because Sharpies are difficult to identify, all MNBBA reports were carefully scrutinized. This resulted in 156 records of the Sharp-shinned Hawk, all heavily distributed in northeastern, northern, and north-central Minnesota **(A)**. Breeding evidence was obtained in only 3.1 percent (145/4,735) of the blocks surveyed. A total of 11 confirmed nesting records were included from Becker, Crow Wing, Cook, Itasca, Pine, and St. Louis Counties, and from one block that straddled two counties: Cass and Todd. Most records were of possible nesting, which were places where Sharp-shinned Hawks were observed in suitable nesting habitat.

The breeding distribution of the Sharp-shinned Hawk remains an enigma in Minnesota. Given that the confirmed breeding distribution in Wisconsin (Cutright et al. 2006) extends to the southern part of that state, it is reasonable to assume that the breeding distribution in Minnesota potentially includes southeastern Minnesota. Misidentification with the Cooper's Hawk and the secretive behavior of the Sharp-shinned Hawk, especially in the breeding season, render it difficult to precisely define the species' breeding distribution in the state.

In their review of the Sharp-shinned Hawk in North America, Bildstein and colleagues (2020)

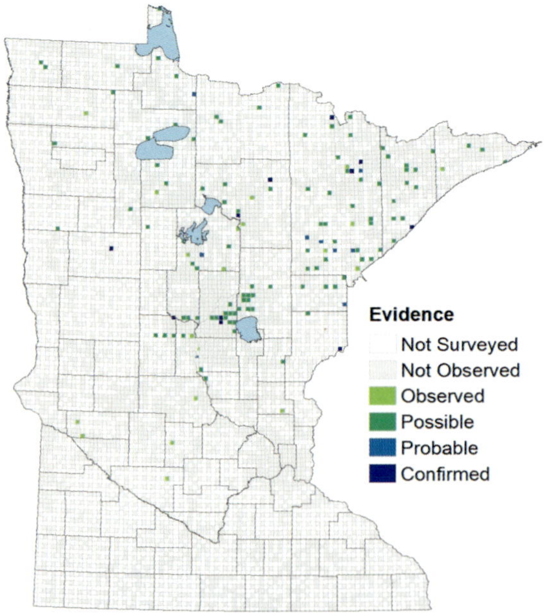

A. Breeding distribution of the Sharp-shinned Hawk in Minnesota based on the Breeding Bird Atlas (2009–13).

B. The breeding habitat of the Sharp-shinned Hawk in a mature pine stand in northern Minnesota has a well-developed canopy and an open understory. Photo by Gerald J. Niemi.

stated that the loss of eastern deciduous forests at the end of the nineteenth century and early twentieth century certainly affected the species' distribution, particularly in eastern North America. These changes also occurred in Minnesota and were observed by Roberts (1932). Continued reductions in forested habitats and ongoing shooting of the species likely led to further declines in Minnesota through the 1900s and were further exacerbated by DDT and its effects in the mid-1900s (D. L. Evans 1982). Bildstein and colleagues (2020) suggest the species "appears to be expanding its breeding range southward, continent-wide," but little evidence indicates that it has reestablished populations in southern Minnesota.

BREEDING HABITAT

Sharp-shinned Hawks nest in moderately aged to mature coniferous and mixed deciduous–coniferous forest stands (D. L. Evans 1982) that are frequently described as "dense stands with a well-developed canopy" (Platt 1976) **(B)**. Rosenfield and colleagues (1991) suggested that the species nests in dense, younger forests with more conifers compared with the Cooper's Hawk. Niemi and Hanowski (1992a) consistently found the species in dense black spruce forests of the Red Lake Peatlands in northern Minnesota.

POPULATION ABUNDANCE

PIF (2021) estimated a population of 410,000 adult Sharp-shinned Hawks in the United States and Canada and 4,500 adults in Minnesota. Rosenberg, Kennedy, and colleagues (2016) estimated an increase in the population of the Sharpie of 68 percent from 1970 to 2014.

At Hawk Ridge in Duluth, Sharpies are the second most abundant raptor counted (average over 13,000 per fall) and the most abundant raptor captured and banded (60,147 from 1972 to 2009) (D. L. Evans et al. 2012). Raptor counts at Hawk Ridge from 1974 to 2019 indicated a nonsignificant but declining trend and a stable recent trend from 2009 to 2019 **(C)**. Fall counts from 2017 to 2020 have been 22,163, 11,148, 18,947, and 13,593, respectively. The fall 2016 count of 22,602 was the largest ever at Hawk Ridge since regular counts began in 1972.

CONSERVATION

Overall, there are few current conservation concerns for the Sharp-shinned Hawk, but its overall population in North America is relatively small. The species appears secure, but populations are difficult to monitor because identification and detection are troublesome and primary breeding habitats are in remote locations.

The declines in the 1980s and 1990s were attributed to pesticides, natural population cycles, declining availability of Neotropical migrant songbirds, and longer cycles of spruce budworm infestations in boreal Canada, which resulted in lower availability of forest songbirds as a food source (Farmer et al. 2008). Keyel and colleagues (2020) found relatively high values of mercury in the feathers of Sharp-shinned Hawks captured at Hawk Ridge. They attributed these high values to the species' diet of birds because most raptors with mammalian diets had relatively low levels of mercury. The overall effect of mercury or pesticide levels on the species' population is unclear.

Bildstein and colleagues (2020) suggest that shooting still occurs to some degree but is not a likely factor affecting the species' overall population. In the past, thousands of Sharpies were shot at migration sites. For example, Sharpies were a popular shooting target at Hawk Ridge in Duluth owing to their habit of flying low and thus within range of shotguns at this site. Activities by the Duluth Bird Club led efforts to stop the shooting in the early 1950s (Bronoel 1954b).

With a 2°C warming scenario, NAS (2019) projected a moderate vulnerability for the Sharp-shinned Hawk breeding distribution in North America. In Minnesota, the species was predicted to be extirpated from the state under a 3°C warming scenario by 2080. Analyses of fall migration data from Hawk Ridge from 1974 to 2009 found that Sharpies were migrating on average four days later (Rosenfield et al. 2011); this was further supported by analysis through 2019 by Steiner and colleagues (2022). The species' wide distribution in North America may buffer it from the possible effects of a changing climate, but migratory short-stopping by Sharpies will result in fewer Sharpies being counted. "Short-stopping" refers to the species not migrating as far south because they prey on birds at bird feeders, which have increased, plus warmer winters have reduced the need for many birds to fly as far south in winter.

In summary, the future of the Sharp-shinned Hawk is complicated because climate change predicts a reduced breeding range in Minnesota, while Bildstein and colleagues (2020) suggest a possible expanding population southward. The species has suffered many setbacks in its population in the past from human persecution, habitat change, and exposure to chemicals. The secretive nature of this species is a potential advantage because it tends to avoid interactions with people, except during migration, but the resulting limited understanding of the species' status is a disadvantage. Management or conservation measures, therefore, could be difficult to implement in a timely manner.

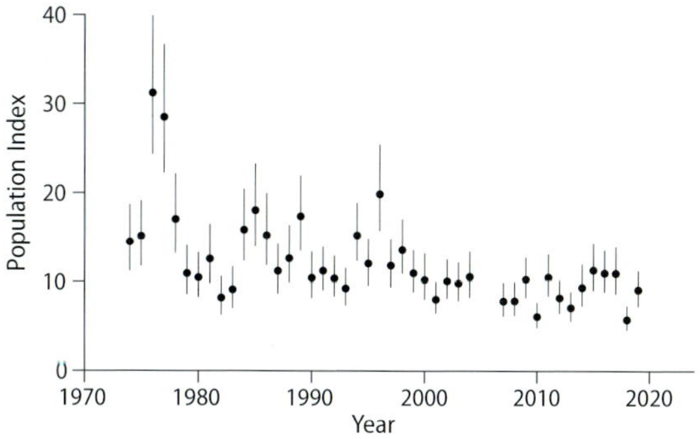

C. Fall migration counts and 95 percent credible interval (vertical lines) of Sharp-shinned Hawks at Hawk Ridge in Duluth show a nonsignificant but declining trend in its population index from 1974 to 2019, but the trend has stabilized from 2009 to 2019 as calculated by the Raptor Population Index (Oleyar et al. 2021).

COOPER'S HAWK

Accipiter cooperii

Minnesota Seasonal Status: Regular breeding resident, migrant, and regular species in the winter. Overwintering individuals are observed in southern Minnesota, including the Twin Cities. The Cooper's Hawk was uncommon during the MNBBA.
Migration: Primarily a short- to medium-distance migrant to the southern United States and Mexico but may be a permanent resident in southeastern Minnesota.
Food: Carnivore focused primarily on medium-sized birds, such as robins, jays, thrushes, and woodpeckers, or mammals, such as chipmunks, squirrels, and small rabbits, which it captures using brief perch-and-scan episodes in woodlands.
Nest: Placed in both large deciduous and coniferous trees, usually ranging from 8 to 15 m high. Most are placed on preexisting bases, such as squirrel nests, previously used nests of other species, or detritus in the fork of a tree.
Conservation Concern: Assigned a low PIF Continental Concern Score of 7/20.

Photo by Tom Muir.

The Cooper's Hawk is the quintessential woodland hawk and has had a highly variable population over the past 150 years (Rosenfield 2018; Rosenfield et al. 2020). It is a mid-sized accipiter, smaller than the Northern Goshawk but larger than the Sharp-shinned Hawk. It is a common summer resident in the southeast, central, and east-central portions of the state, but scarce in the southwest, western, and northeastern parts of Minnesota. It derives its common and scientific name from William C. Cooper, a New York zoologist and the father of James C. Cooper, for whom the former Cooper Ornithological Society is named. W. C. Cooper shot the type specimen in New York in 1828.

The Cooper's Hawk is among the most difficult raptors to identify because all three accipiters are similar in appearance and all have the *flap and glide* flight pattern. All of the adult accipiters have a blue-gray back, rufous barring, and short, broad wings. The Cooper's Hawk is best distinguished from the Sharp-shinned Hawk by its larger size and long tail, which is rounded at the tip. It has a dark, somewhat flat crown contrasting with a light-colored nape and buffy cheeks suggesting a beret. Distinguishing the Cooper's Hawk from the Northern Goshawk is based largely on its smaller size and thinner appearance, plus rufous or brown coloration on the breast, relatively longer tail, and lack of an eye line. It is the most common of the accipiters observed during the summer in Minnesota because it is found in urban areas and woodlots, although it can still be secretive and inconspicuous.

The Cooper's Hawk is widely distributed across North America, with primary populations in the Eastern Temperate Forest Biome and in California. Scattered breeding populations exist throughout the western United States and Canada as far north as Alberta and Saskatchewan. The species is not abundant anywhere.

MINNESOTA BREEDING DISTRIBUTION

Roberts (1932) reported that Cooper's Hawks were "breeding throughout the wooded portions of the state," but by the 1930s it was "by no means as common as formerly." He reported many confirmed nests from Minneapolis and nests with eggs in Anoka, Becker, Isanti, Marshall, and Polk Counties spanning the period from 1883 to 1928.

Roberts pointed out that the Cooper's Hawk best deserves "the name of Hen Hawk or Chicken Hawk" because of its extensive predation of poultry, game, and ducklings. He further stated that "if this and the other two species of the bird-killing trio [referring to the three accipiters, including the Sharp-shinned Hawk and Northern Goshawk] could be eliminated or held in check, much of the blame that now attaches to Hawks in general would disappear." Indeed, the Cooper's Hawk declined in abundance over broad areas owing to illegal shooting and, beginning after World War II, the use of pesticides, such as DDT, which led to reproductive failures.

Forty years later, Green and Janssen (1975) identified the Cooper's Hawk as a "resident throughout the state; very scarce on the prairie in the west-central, southwestern, and south-central regions." D. H. Johnson (1982), in his review of raptors nesting in Minnesota, identified 70 nesting records from 1883 to 1981. These records included locations in 23 counties, ranging from southeastern Minnesota, west to Brown and Grant Counties, northwest to Marshall and Polk Counties, and east to St. Louis County. He reported pre-1970 summer observations from northern St. Louis (1938), Lake (1953), and Cook (1967) Counties.

Several years later, Janssen (1987) labeled the species as a breeding resident throughout the state, especially in the southeastern, east-central, and central regions, but absent from northeastern Minnesota. He confirmed nesting in 14 counties since 1970 within the same range described by Johnson in 1982. Hertzel and Janssen (1998) added three counties with confirmed nesting records since 1970, including a range extension to extreme northwestern Minnesota in Kittson and Marshall Counties.

The MBS documented 90 breeding-season locations of the Cooper's Hawk. Their locations were scattered throughout southern and western Minnesota, northwest to Marshall County, and east to central St. Louis County. Locations were most common in northern Morrison County and southern Beltrami County, but none was located in east-central Minnesota in Aitkin, Carlton, or Pine Counties, or in Cook or Lake Counties.

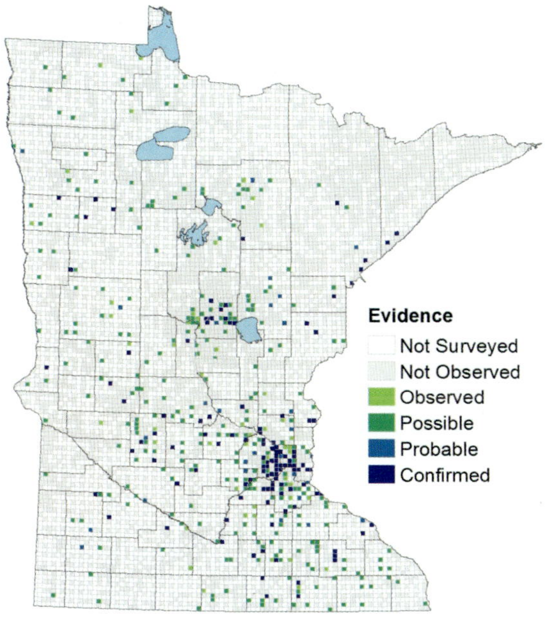

A. Breeding distribution of the Cooper's Hawk in Minnesota based on the Breeding Bird Atlas (2009–13).

The MNBBA documented 581 records of the Cooper's Hawk, which were widespread throughout the state. Breeding evidence was found in every county except the following eight: Cook, Cottonwood, Jackson, Koochiching, Lincoln, Pennington, Pipestone, and Wadena Counties, which are spread throughout the state **(A)**. Records were included from 9.8 percent (468/4,754) of all blocks. Confirmed nesting activity was reported from 42 counties, and these locations were also widespread throughout a substantial part of the state. The breeding records were densest in the metropolitan area of Hennepin and Ramsey Counties, and in Crow Wing County. These regions were well covered by participants during the MNBBA. Since the atlas period, numerous confirmed breeding records have been added, including, for example, from Cottonwood and Wadena Counties (MOU 2021a).

Evans (1982) summarized the rise and fall of the Cooper's Hawk where shooting and human persecution led to declines in the early 1900s until legislation, law enforcement, and educational programs led to improved conditions. In the mid-1900s, the effects of DDT—along with high nest predation, habitat loss, changes in forested habitats, and even excessive take by falconers—caused additional population declines. Today, with changes in poultry practices and the availability of alternate prey, such as Rock Pigeons, Mourning Doves, and European Starlings, conflicts with Cooper's Hawks are more isolated events in the United States. Shooting still occurs in some overwintering areas, such as in Mexico.

Rosenfield and colleagues (2020) stated that significant population increases and range expansions began to occur in the 1990s, when Cooper's Hawks began to colonize urban and suburban areas. MNBBA data also confirm the recolonization of the woodlots, riparian areas, and other suitable forested habitats in the western portion of the state where Green and Janssen (1975) formerly emphasized that it was scarce following the DDT era.

In summary, the Cooper's Hawk has made a strong population recovery in Minnesota since its decline and relatively low abundance throughout much of the twentieth century. Most reports mention problems in identification of the Cooper's Hawk, especially confusion with the smaller Sharp-shinned Hawk. Vigilant identification of this species is especially warranted.

BREEDING HABITAT

Habitat used by Cooper's Hawks for nesting was broadly described as extensive forests, woodlots of 4–8 ha, and occasionally isolated trees in more open areas (Rosenfield et al. 2020) **(B)**. Pine plantations were described as extensively used in southeastern Wisconsin (Rosenfield et al. 1991; Rosenfield 2018). Although Cooper's Hawks formerly avoided urban landscapes, they have "adapted remarkably well" to human disturbance and habitat fragmentation and are now frequently found nesting in suburban and urban areas (Rosenfield et al. 1991).

POPULATION ABUNDANCE

Based on the BBS, PIF (2021) included a United States population of 1,000,000 and a Minnesota population of 14,000. The BBS, however, was unreliable for estimating a population trend for the species. Fall migration counts at Hawk Ridge in Duluth indicated a highly variable population **(C)**. Counts at Hawk Ridge were very low from 1972 to 1990 (less than 80 per year) but increased substantially up to the turn of the century (103 to 319 per year). Since that time, counts have been lower (range 38 to 179), including low counts in 2006 (38), 2019 (53), and 2020 (52). Rosenberg, Kennedy, and colleagues (2016) calculated that the Cooper's Hawk population in the United States and Canada had increased by more than 200 percent from 1970 to 2014! Fall counts in the eastern United States, such as at Hawk Mountain, Pennsylvania, and Cape May, New Jersey, have also shown similar patterns of change since the early 1970s (Oleyar et al. 2021).

B. Cooper's Hawks are primarily found in mature deciduous forests, often with open understories, and in small woodlots such as this area in the western Twin Cities region. Photo by Gerald J. Niemi.

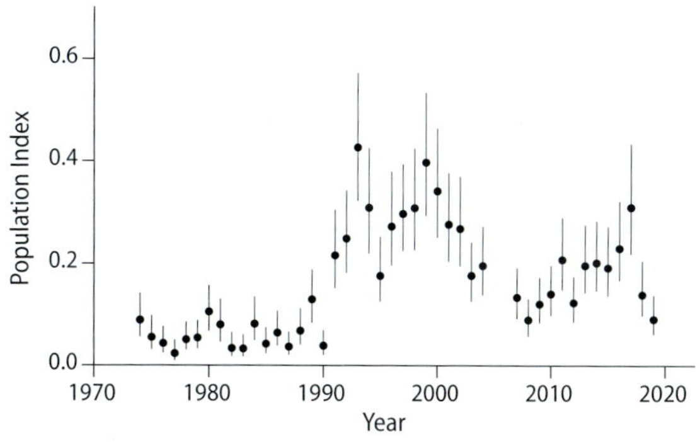

C. Fall migration counts and 95 percent credible interval (vertical lines) of Cooper's Hawks at Hawk Ridge in Duluth show a highly variable but insignificant trend in its population index from 1974 to 2019, as calculated by the Raptor Population Index (Oleyar et al. 2021).

CONSERVATION

The Cooper's Hawk was formerly extirpated, endangered, threatened, or a species of conservation concern in 15 US states. Rosenfield and colleagues (2020) state that the species "has recovered remarkably well from the two major eras that affected the populations of many raptors in North America: DDT and heavy direct persecution."

Current mortality factors include collisions with vehicles and man-made structures and nest predation by raccoons and Great Horned Owls. Degradation of habitat may alter the suitability of some nesting or foraging habitats, but the species' use of selectively logged forests, plantations, and its occupation of suburban areas suggests it is adapting well to many human-altered environments (Rosenfield 2018). Under all three of its warming scenarios, NAS (2019) projected stable breeding populations for the Cooper's Hawk across North America. The largest area where its range was forecast to decline with a 2°C warming scenario was the Prairie Coteau region of southwestern Minnesota and northwestern Iowa. Losses, however, were more than compensated for by northern range expansions.

NORTHERN GOSHAWK

Accipiter gentilis

Minnesota Seasonal Status: Regular species throughout the year, including migration and winter. The Northern Goshawk was rare during the MNBBA.
Migration: Often referred to as a partial migrant. Some birds remain in their breeding areas year-round, while others migrate; the proportion of the population that migrates may depend on winter conditions and prey availability.
Food: Renowned for its swift, aerial pursuit of prey, the Goshawk is an opportunistic feeder consuming a variety of birds and small mammals, including Ruffed Grouse, Red Squirrels, and Eastern Chipmunks.
Nest: Large, open platform constructed with twigs and lined with bark and fresh green foliage.
Conservation Concern: Assigned a moderate PIF Continental Concern Score of 10/20; officially listed as *Special Concern* in Minnesota.

Photo by Karl Bardon.

A Holarctic species, the Northern Goshawk has been a favorite among falconers for nearly 2,000 years; its aggressive pursuit of prey is extolled in Helen Macdonald's book, *H is for Hawk* (2014). Reflecting on their elusiveness, the author wrote, "Looking for goshawks is like looking for grace; it comes, but not often, and you don't get to say when or how." It is an apt description of the secrecy and rarity of this inhabitant of Minnesota's mature northern forests.

An accipiter, the Goshawk has short, rounded wings and a long tail superbly adapted for navigating the forest understory with supreme agility as it chases its woodland prey. The adult male has a beautiful gray plumage on its back and upper wings, with much lighter gray feathers underneath. Its most distinct features are a white stripe above the eye that separates a black crown from a black patch through the red eye. Nearly 25 percent larger in size, the female has plumage that is similar but somewhat browner on her head, back, and upper wings. The alarm call is a harsh and loud *ki ki ki* or *kree-ah*.

Frequently described in terms that highlight its fierceness, the Northern Goshawk was deemed by Bent (1937) to be the "largest, the handsomest, and the most dreaded of the *Accipiter* tribe." He quoted another admirer who wrote eloquently about the bird's physical attributes, describing it as a "fiery-hearted warrior" that ranked "second to none in martial beauty and fearlessness." All these qualities are on full display to any who venture too close to a Goshawk nest.

Sparsely distributed throughout its breeding range, in North America the Northern Goshawk is found across the boreal forest region from Alaska, across Canada, and south into the northern Great Lakes states, New England, the western United States, and a small region of northern Mexico. The common name Goshawk originates from an old English term that translates to *goose hawk*, a reference to its occasional pursuit of waterfowl. The scientific name for the species, *gentilis*, is a reference to its nobility.

MINNESOTA BREEDING DISTRIBUTION

On the southern periphery of its breeding range in Minnesota, the Northern Goshawk was not familiar to the early chroniclers of the state's bird fauna except as a winter visitor. Hatch (1892) described it as regular during the winter months except when the weather was very mild. He suspected that it bred in the state, but he had no documented evidence. The same year that his report on Minnesota birds was published, his suspicion would prove true. Frederick Nutter, a keen observer of birds, published a small note in the journal *Oologist* describing a Goshawk nest he found in Hennepin County, not far from Minneapolis (Nutter 1892).

Forty years later, Roberts's (1932) comprehensive account of the species included Nutter's record and three confirmed nesting attempts in Roseau County in 1926 and 1927. Then, in 1936, he reported on two active nests found in Carlton County and a deserted nest with three eggs found near the French River, along Lake Superior's North Shore (Roberts 1936). Nevertheless, like Hatch, he believed the species' presence was largely confined to the winter season and that it only rarely nested in the state. It is more likely, however, that this secretive denizen of the boreal forest had simply gone undetected by early naturalists, who spent far more time in the open landscapes of southern and western Minnesota than in the dense forested landscapes farther north.

By the early 1940s, more reports of Goshawks nesting in northern Minnesota began to accumulate. Marius Morse (1942) reported on three nesting attempts in Carlton (1934) and St. Louis Counties (1935, 1942). Two of these nests (1934 and 1935) were first reported by Roberts (1936) but lacked some of the clarifying details that Morse provides. Indeed, one of the two Carlton County nests that Roberts attributed to Morse (the nest observed in May 1935) was actually in St. Louis County, not Carlton County. Additional attempts in St. Louis County were documented in 1943, 1945, and 1946 (Lakela 1943, 1945; R. Olson 1946). When Green and Janssen published their updated account in 1975, they described the Northern Goshawk as a breeding resident in the far northern counties of the state, bounded by eastern Roseau County in the northwest and Carlton County in the south. Although Nutter's 1892 nesting report in Hennepin County remained the only documented nesting record south of this region, they noted a suspected nesting effort in Morrison County in 1938, just at the southern border of the northern forest.

Gullion (1981) and Parmelee (1982) both provided some insights on the ecology of the Northern Goshawk and additional accounts of nesting at the Cloquet Forestry Center in Carlton County and Itasca State Park, respectively. D. H. Johnson's (1982) review, however, provided the most comprehensive summary of the nesting distribution and population status of nesting Goshawks in Minnesota. Since Nutter's original record in 1892, a total of 61 Goshawk nests had been reported through the 1981 breeding season. He delineated a breeding range that extended from eastern Roseau County south to Clearwater, Hubbard, and Crow Wing Counties and east to Pine County and the Arrowhead region. Despite fluctuations in abundance, he concluded that the Minnesota population appeared relatively stable. When Janssen (1987) wrote his updated account of the species' status a few years later, little had changed; he delineated ten counties where the species was known to have bred since 1970.

Although knowledge of the species' true breeding status and distribution was increasing, the Northern Goshawk remained an elusive member of Minnesota's breeding bird community. The only opportunity to closely observe the species was during the annual fall migration at Hawk Ridge in Duluth, Minnesota. Most autumns, a few hundred Goshawks are tallied as they migrate south from their northern Minnesota and Canadian breeding grounds (Hawk Ridge Bird Observatory 2019). Several thousand, however, were observed during invasion years in the early 1970s, early 1980s, early 1990s, and early 2000s (Hofslund 1973; D. L. Evans et al. 2012). While the number of Goshawks is relatively small compared with the thousands of Broad-winged Hawks that pass over the ridge each year, it is still substantially higher than the number observed at the well-known mecca for hawk migration, Hawk Mountain in Pennsylvania, where an average annual count of Goshawks is approximately 70 birds (Hawk Mountain 2019).

Interest in the Northern Goshawk began to accelerate in the late 1980s and early 1990s. As timber harvest levels in Minnesota's northern forests began to rise, so did concern for those species that depended on large tracts of mature forest, including the Northern Goshawk. Numerous field studies were launched that not only examined the Minnesota population in depth but also placed it in context with populations throughout the western Great Lakes region. More intensive field inventories began in 1994 (e.g., Martell and Dick 1996; Hamady et al. 2003), followed by intensive studies of foraging and nesting habits (e.g., Boal et al. 2003, 2005a, 2006; Smithers 2001; Smithers et al. 2005). Efforts to better understand the origins of migrants that pass through Duluth each fall were also initiated (Hawks et al. 2007).

Although these studies significantly expanded knowledge of the Northern Goshawk, the species itself remained quite rare. When Hertzel and Janssen (1998) published updated maps of nesting records, they delineated just eight counties where nesting had been confirmed since 1970. They deleted three counties from the map Janssen published in 1987 (Itasca, Crow Wing, and Mahnomen) and added one (St. Louis). Meanwhile, MBS field biologists combed the northern counties, but by 2014 they had reported only 13 breeding-season locations. Because of continuing concerns regarding the Northern Goshawk's status in the state (MNDNR 2012), it was officially listed as *Special Concern* in 2013 (MNDNR 2013).

During the MNBBA, data on Goshawks were secured from two primary sources. First, data were gathered by MNBBA observers who reported 42 Northern Goshawk records from less than 1.0 percent (37/4,735) of the surveyed atlas blocks. Breeding was confirmed in five blocks. The birds were observed in ten of Minnesota's 87 counties and were confirmed breeding in three counties: Lake, St. Louis, and Roseau. The Roseau County record was an addition to the list of confirmed counties published by Hertzel and Janssen in 1998. Second, data were provided by the MNDNR Northern Goshawk Monitoring Program for the years 2009 to 2013. Because a limited number of Goshawks are still legally taken for falconry, the atlas illustrates only the highest level of breeding activity documented by the MNBBA and MNDNR records to the county level rather than to the block level **(A)**.

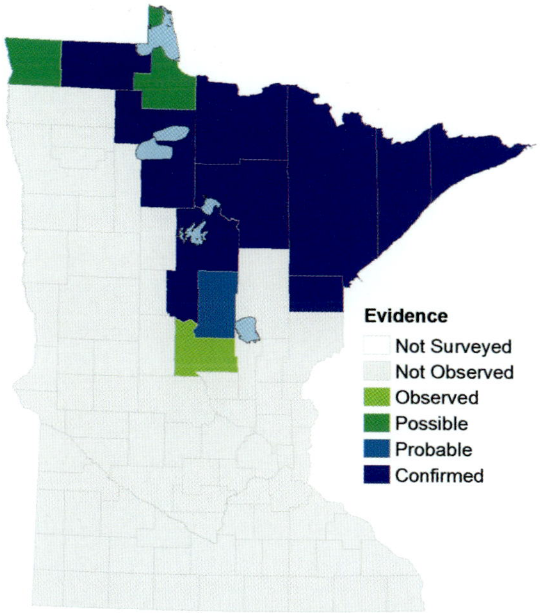

A. County-level breeding distribution of the Northern Goshawk in Minnesota based on the Breeding Bird Atlas (2009–13).

Remarkably, in the years following the MNBBA, Northern Goshawks were confirmed nesting in Hennepin County along Nine Mile Creek (Janssen 2019). Individual birds were first reported in the vicinity in May 2014 and, the following year, in May and June. Photos and additional documentation were submitted of a nest, adults, and two juveniles in 2016 (MOU 2021a). As of the fall of 2021, no additional Goshawk nesting reports have been submitted in the Twin Cities region.

Our understanding of the habits and distribution of the Goshawk have improved considerably in the past one hundred years. Although Roberts (1932) considered the species to be primarily a winter visitant, it likely has always been a rare, year-round resident of Minnesota's northern forests. Studies in the late 1990s in Wisconsin, however, actually documented a breeding range for the Northern Goshawk that extended across the northern two-thirds of the state, including several records south of the northern forest region (Rosenfield et al. 1998). The authors concluded that not only was there no evidence for a range contraction in the state, but the species might actually be expanding to the south. Whether or not Minnesota's Goshawk population is expanding southward in Minnesota, or reoccupying portions of its former breeding range, the recent report of nesting in Hennepin County, more than 130 years after Nutter's 1892 report, is intriguing.

Nevertheless, in their comprehensive review of the species, Squires and colleagues (2020) note that there are few data to document any historical changes in the Goshawk's distribution. They surmise that populations in the western and northern regions of its range have changed little but speculate that eastern populations might have been negatively impacted by the loss of a major prey item, the Passenger Pigeon, and by the loss of forest lands. As many agricultural lands in New England succeed back to mature forest, they believe that eastern populations may recover.

B. A tall aspen tree in a mature pine forest in northern Minnesota provides the perfect nesting opportunity for the Northern Goshawk. Photo by Gaea Crozier.

BREEDING HABITAT

With a broad distribution throughout North America, the Northern Goshawk has adapted well to a diversity of mature, forested habitats. In the western Great Lakes region, they utilize a variety of mixed deciduous–coniferous forest types. The vegetation composition of the forest is less important than the structural features, which invariably include large trees in a mature or old-growth stand with closed canopies (Squires et al. 2020) **(B)**.

In Minnesota, the birds select stands with high canopy closure (60–95 percent) for nesting (Boal et al. 2006). Even in stands dominated by conifers, 80 percent of nests were placed in aspen, followed by birch, white pine, red pine, and red oak (Boal et al. 2001). Nest trees were usually the tallest and largest (i.e., largest diameter at breast height) trees available in the nesting area. Stands were also characteristically multistoried, with relatively open flight paths available between the canopy and shrub or sapling layer, and between the shrub or sapling layer and the understory. Such structural layering is important for this forest predator, which secures its prey by aerial pursuit.

Habitat used for foraging is similar to that selected for nesting. In general, nesting stands had larger and taller canopy trees and fewer understory trees than foraging habitats (Boal et al. 2005a). Overall, young deciduous stands, lowland conifers, and open areas were used less than predicted based on their availability for both nesting and foraging, while mature forests were used more than predicted based on their availability (Boal et al. 2005a).

Northern Goshawks also require large tracts of forests. Individual birds in Minnesota, regardless of their sex, used areas that averaged approximately 2,600 ha. A mated pair, however, used an even larger area. The average size of 11 breeding territories ranged from 5,000 ha to 7,800 ha (Boal et al. 2001).

As noted earlier, studies have suggested that breeding birds in Minnesota (both males and females) remain in the vicinity of their nest site throughout the winter months (Boal et al. 2003). However, year-round residency may depend on prey availability.

POPULATION ABUNDANCE

PIF (2021) estimated the population of this global species at 210,000 breeding adults in the United States and Canada. Unfortunately, this estimate incorporates BBS data, which is not a robust monitoring tool for such a sparsely distributed, dense-forest inhabitant (Rosenberg, Kennedy, et al. 2016). The statewide estimate for Minnesota, 820 birds, is equally tenuous.

Given the inadequacies of the BBS protocol, raptor biologists have experimented with the use of conspecific calls to survey Goshawk populations with reasonably good success (e.g., J. T. Kimmel and Yahner 1990; Kennedy and Stahlecker 1993). A study in Minnesota further improved the methodology (Roberson 2001; Roberson et al. 2005). Using these call techniques as a foundation, Hargis and Woodbridge (2006) designed a regional survey approach, arguing that sampling smaller land units for such a widely dispersed species with large habitat require-

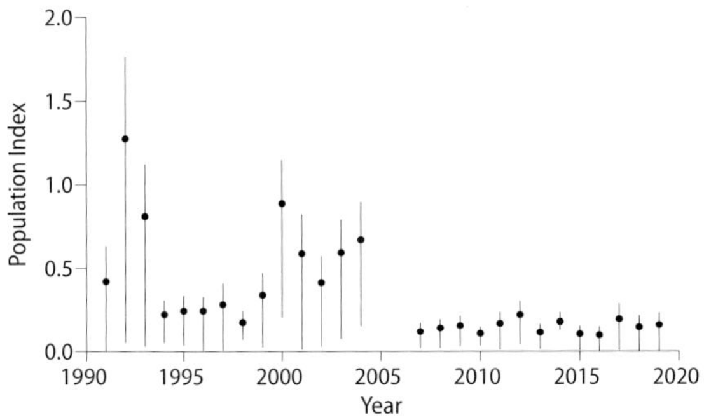

C. Fall migration counts and 95 percent credible interval (vertical lines) of Northern Goshawks at Hawk Ridge in Duluth show a long-term decline in numbers from 1991 to 2019, as calculated by the Raptor Population Index (Oleyar et al. 2021).

ments was problematic. In 2008, Bruggeman and his colleagues (2011) implemented the regional survey design proposed by Hargis and Woodbridge across the entire western Great Lakes region, an area that encompassed portions of Minnesota, Wisconsin, and Michigan. The results indicated that Goshawks occupied 27 percent of the potential Goshawk habitat in the region. To date, this is the only systematic survey conducted for Goshawks in Minnesota and across the entire Great Lakes region. Unfortunately, implementation of a species-specific inventory and monitoring protocol across such a large area for a rare and sparsely distributed species can be costly for resource-strapped natural resource agencies. Although the work conducted by Bruggeman and colleagues (2011) provides an important foundation of monitoring data, lack of funding has prevented additional surveys.

Meanwhile, conspecific calls were used to locate nesting Goshawks in suitable habitat on public lands in northern Minnesota beginning in the 1990s. The surveys documented an average of just 29 nesting attempts per year; approximately 79 percent of these occurred in territories that had been occupied the previous year. From 1991 through 2012, a cumulative total of 130 active territories were identified (MNDNR 2012).

In addition to the low number of nests, Goshawk nesting success and productivity were found to be lower in Minnesota than in other regions of North America. Among 26 nesting attempts that were closely monitored from 1998 to 2001, 1.14 young fledged per nesting attempt, and 1.85 young fledged per successful nest (Boal et al. 2005b). These rates were similar to two of three studies conducted elsewhere in the western Great Lakes region, and lower than one. Productivity rates throughout the western Great Lakes, however, were lower than the averages reported in 16 studies in western North America (Boal et al. 2006). More recent data from Minnesota document slightly lower reproductive success, with an average of 1.05 young per nesting attempt and 1.54 young per successful nest from 2003 to 2013 (Crozier, pers. comm.).

Fall migration counts conducted at sites throughout the United States provide another source of long-term data on raptor populations that can be used as an index of trends. Migration counts conducted at Hawk Ridge in Duluth from 1991 to 2019 showed a decline in the number of Northern Goshawks, with an average annual decrease of −5.73 percent **(C)**. These and data from other locations also illustrate the variability in migrant numbers. Irruptive years, when migrant numbers are unusually high, occur at approximately 10-year intervals. The cycles are believed to be connected to similar cycles in prey abundance in Canada, particularly for snowshoe hare and grouse (Squires et al. 2020). More recently, however, the size of the irruptions appears to have dampened. Milder winters and increased prey availability may be causing shorter migrations, with many birds remaining north of Duluth through the winter, or the change may represent a significant decline in Goshawks. Indeed, from 2009 to 2019, an examination of migration count sites across the United States and southern Canada revealed that Goshawk numbers were significantly declining at 10 of the 21 reporting sites; trends from the remaining 11 sites were nonsignificant (Oleyar et al. 2021).

Although a number of factors can influence populations and reproductive success in a given year, including inclement weather and low prey availability, habitat degradation and habitat fragmentation are now considered the primary threats to this forest predator. "The availability of large patches of mature forest preferred by the species is in decline regionally due to fragmented land ownership as well as fragmentation of historically large contiguous stands resulting from past and current forest management practices" (MNDNR 2012). Timber harvest in particular can be detrimental to local populations, resulting in nest disturbance, nest destruction, and loss of large tracts of mature forest (Squires et al. 2020). As noted earlier, some birds are still taken for falconry, but the sport is tightly regulated at both the state and federal level in order to minimize impact on population trends (Squires et al. 2020).

CONSERVATION

Although it has a small population size and moderate threats to its breeding population, the Northern Goshawk's wide distribution has resulted in its designation as a moderate conservation priority (PIF 2021). A petition filed by several western conservation organizations in 1991 to federally list the western population was declined in 1998 by the USFWS (1998b). However, the species has been designated a Sensitive Species in many national forests, including the Chippewa and Superior National Forests in Minnesota (USFS 2012). As noted earlier, the Northern Goshawk is officially listed as *Special Concern* in the state.

Because more than 40 percent of the Goshawk's range occurs in the boreal forests of Canada, attention to forest conservation and management issues in Canada is extremely important. Unfortunately, as with many boreal species, the impacts of climate change are a concern. NAS (2019) classified the species as moderately vulnerable under a 2°C warming scenario, predicting the loss of 36 percent of its southern breeding range by 2050 and only a 13 percent gain to the north. Even under a 1.5°C warming scenario, which is considered imminent by many scientists, the Goshawk is predicted to be extirpated from Minnesota. In light of these predictions, the apparent decline in Goshawk population numbers should be closely monitored.

The bird's sensitivity to forest management practices coupled with its dependence on large, mature forest tracts are further cause for concern. Recognizing the importance of forest management activities for the sustainability of the state's Goshawk population, the MNDNR developed a series of Northern Goshawk Management Considerations to help guide and inform forest landscape planning initiatives on state forest lands (MNDNR 2003b). These considerations acknowledge the importance of maintaining a forest landscape that includes a mosaic of different forest conditions, including a range of patch sizes, forest types, and stand ages, to provide sufficient habitat for Goshawks. Specific habitat considerations are outlined for breeding territories, nest areas, and post-fledging areas that address the need for large tracts of mature forest in and near known territories and the need to minimize disturbance during the nesting season. Similar management guidelines are also recommended in Wisconsin (Kreitinger et al. 2013).

Fortunately, Minnesota has a long-standing tradition of integrated, landscape-scale forest management that coordinates forest management across ownership boundaries and strives to maintain a diverse landscape for all forest-dependent species, including the "bold and rapacious" Northern Goshawk (Roberts 1932).

BALD EAGLE

Haliaeetus leucocephalus

Minnesota Seasonal Status: Regular breeding resident and migrant and formerly a winter visitant, the Bald Eagle is now regular in winter in southeastern Minnesota and other parts of Minnesota where open water is found, such as Lake Superior. The Bald Eagle was a common species during the MNBBA.
Migration: Short- to medium-distance migrant that winters in the lower forty-eight states and in coastal areas of Canada, Alaska, and Mexico. The Bald Eagle is primarily found along major rivers, lakes, and reservoirs and wherever there is suitable open water.
Food: Chiefly fish and carrion but also large birds, mammals such as muskrats and hares, reptiles, amphibians, and crustaceans; occasionally garbage.
Nest: Large trees with branches capable of holding the nest, cliffs, and on rare occasions on the ground. Across their range, pairs average 1.5 nests/pair within their territory and up to five nests have been reported.
Conservation Concern: Assigned a moderate PIF Continental Concern Score of 9/20. The Bald Eagle was formerly listed as *Endangered* in the United States except in Minnesota, Wisconsin, Michigan, Oregon, and Washington where it was *Threatened* (1978). The state of Minnesota also listed it as *Threatened* (1984). It has been delisted at both the federal and state level because of population increases, range expansion, and adaptation to human disturbance. The Bald Eagle is still protected under the federal Bald and Golden Eagle Protection Act and the Migratory Bird Treaty Act.

The Bald Eagle is a magnificent specimen. Regardless of how many times one sees an adult, the contrast of its white head and tail against its dark background plus its sheer size translates to pure reverence in any observer. It is no wonder that the Bald Eagle has the distinction of being our national bird. Forbush and May (1955), writing in the first person, capture this feeling well, along with conflicts about its behavior:

> I experienced the feeling of awe with which I watched this living embodiment of the emblem of American freedom ascending grandly into the autumnal sky. My emotions on that occasion were somewhat similar to those experienced when I first viewed in the near distance the great dome of the capitol at

Photo by David Brislance.

> Washington, for around both bird and building cluster many memories and traditions of a great country and a mighty people. Our Eagle may deserve some of the epithets that have been heaped upon him; he may be a robber, a skulker and a carrion feeder; nevertheless, he is a powerful and noble bird and a master of the air.

The "Bald" Eagle is not bald; the word refers to the white on its head that is reminiscent of baldness. Its scientific genus name, *Haliaeetus,* is Greek and translates to *salt eagle,* while its species name, *leucocephalus,* is also Greek and appropriately refers to the white feathers on its head. Bent (1937) described the voice of the Bald Eagle as "ridiculously weak and insignificant." He quoted the following from Hoffman: "The cry of the Eagle heard oftenest near its nest, is a high-pitched very metallic *kweek kuk kuk, kweek-a-kuk-kuk* with the quality of an unoiled caster."

The Bald Eagle is widely but patchily distributed across North America from Florida to the Maritime Provinces, north to Alaska, and south to California. The species is rare in open country lacking water. Most prominent population centers are found along coastal zones and inland lakes, reservoirs, and major river systems, such as the Pacific and Atlantic coasts, the upper Midwest in Minnesota, Michigan, and Wisconsin, and the Pacific Northwest.

MINNESOTA BREEDING DISTRIBUTION

Historically in Minnesota, Roberts (1932) reported that "records of forty to fifty years ago show that one or more pairs nested [on] about nearly every large lake and here and there along large rivers throughout the wooded parts of the state." In those earlier years, he stated the nesting Bald Eagle was even found occasionally in groves in the prairie regions near places with ample water. He cited such areas as the Des Moines River, Loon Lake and Heron Lake in Jackson County, Otter Tail County, and the Red River valley as previous nesting areas. He quoted observations by Elliott Coues (1878), an early ornithological explorer in Minnesota: "While steaming down the Red River from Moorhead to Pembina, we frequently saw Bald Eagles sailing overhead, and several nests were noticed on the tops of tall, isolated trees."

By the 1930s, Roberts reported that the "sight of an Eagle, except in some far-away place, has become a rare happening even during the seasons of migration." He said that the "Bald Eagle aerie can be found only in remote or protected regions." Roberts blamed the decline on the "settling of the state" and commented that the "rifle and gun have been the chief causes of its disappearance." Hatch (1892) described the apparent widespread, cruel treatment of the Bald Eagle where "many a cabin in the solitudes of the deep, dark forests, has its young eagle chained to its gable, or the convenient out-house." Referring to the citizens of Minneapolis, "it was no unusual thing to see individuals of the species chained, like a monkey, to a box or outhouse in different places in the city."

Although Roberts reported the wide distribution of the species, he described confirmed nests with young only at Lake Minnetonka, Hennepin County (1874), and Itasca State Park (1902 and 1917), plus "2 young" in Stearns County (1894). In his 1936 revision, Roberts added two confirmed nests, both with two young, detected by Nestor Heimenz in Crow Wing County, plus an "occupied nest" found by Marius Morse in Itasca County.

The status of nesting Bald Eagles was tenuous from the 1930s to the early 1960s. However, J. P. Mattsson (1988) provided an annotated list of historical Bald Eagle records from 1873 to 1959. These records represent a combination of observations of individual birds at specific locations and a variety of nesting data, such as whether a nest, a pair, or young was reported. In addition, pre-1976 Bald Eagle records in the MNDNR Heritage Database were concentrated in the Chippewa National Forest, the Superior National Forest, and the Boundary Waters Canoe Area Wilderness, with additional observations from Aitkin, Clearwater, and southern St. Louis Counties (M. Miller and Pfannmuller 1991).

In their 1975 summary, Green and Janssen described confirmed nesting in 17 counties and inferred nesting in two counties. These included a summary of all county nesting records known at that time in Minnesota. The authors stated that the Bald Eagle was primarily a resident in the northern regions of the state and that the only known recent active nesting in the southern half of Minnesota was in the Upper Mississippi NWR in Houston County. They emphasized that the area producing the most eaglets was the Chippewa National Forest.

D. H. Johnson (1982) summarized Bald Eagle nesting data from 1960 to 1981 and identified 24 counties in which nesting occurred, from Houston County in the extreme southeast to Marshall County in the northwest and most counties in northern and northeastern Minnesota. Except for the Houston County record, no nests were found in most of the western counties or south of a line from Chisago County to northern Morrison and Otter Tail Counties. Johnson summarized nesting activity in the Chippewa National Forest from 1963 to 1981. The number of known nests in 1963 was 48, with 20 identified as active. This increased to 199 known nests and 75 active nests in 1981. The number of successful nests increased from six to 60 during this same period. In 1973, when statewide surveys were initiated, 115 known, occupied breeding areas were located (Coffin and Pfannmuller 1988; M. Miller and Pfannmuller 1991), and this increased to 437 occupied breeding areas and 300 successful nests in 1990.

Janssen (1987) summarized a breeding distribution similar to that described by D. H. Johnson (1982) and slightly expanded the number of counties with confirmed nests to 25. This included confirmed nesting since 1970 in Anoka, Sherburne, and Stearns Counties on the southern periphery of the species' range. In 1998, Hertzel and Janssen would expand confirmed nesting of the Bald Eagle to 39 counties since 1970, with further extensions to the southwest as far as Brown, Lac qui Parle, Redwood, and Yellow Medicine Counties, which are all counties bordering the Minnesota River.

MBS, which began in the late 1980s, has recorded 188 breeding-season locations during its county surveys. Breeding locations have spanned a considerable portion of the state, but by far the most locations are in the north-central and northern regions from Cook to St. Louis County. Several locations were noted along the Minnesota River.

Given Roberts's pessimistic view of the Bald Eagle's status in the 1930s, he would be astonished with the species' spectacular recovery and the status of the Bald Eagle today. The MNBBA included a total of 2,113 records of the Bald Eagle. The species was recorded in every county except Lincoln and Pipestone in the extreme southwestern part of the state **(A)**. The Bald Eagle was reported in an astounding 29.3 percent

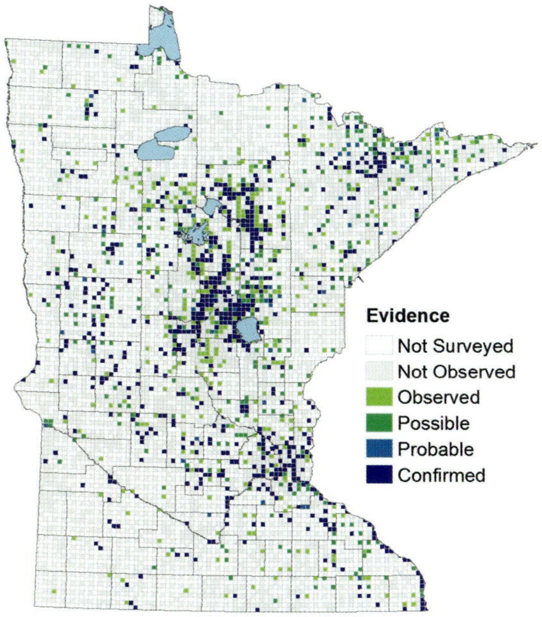

A. Breeding distribution of the Bald Eagle in Minnesota based on the Breeding Bird Atlas (2009–13).

(1,432/4,886) of surveyed blocks. Equally impressive was confirmed nesting in 671 blocks, or 47 percent of all blocks where the species was recorded.

The MNBBA results have further expanded and now document the general breeding distribution of the Bald Eagle in Minnesota as of 2013. These results provide evidence for recolonization of many areas in western, southwestern, southern, and southeastern Minnesota that likely were formerly occupied by the species. Many of these extensions were previously identified by M. Miller and Pfannmuller (1991), Hertzel and Janssen (1998), and the MNDNR (2019b); however, the extent of the Bald Eagle's range currently illustrated by the MNBBA is unprecedented in the documented history of Minnesota. Even though Roberts (1932) reported observations from Jackson County and the Red River valley, it is unknown how extensive Bald Eagle distribution was throughout the entire state in Roberts's time.

In his review of the Bald Eagle in North America, Buehler (2020) stated that the present distribution of the Bald Eagle is probably similar to its historical breeding distribution. The story of the recovery of the distribution and population of the Bald Eagle in Minnesota is also similar to stories that have been documented in Michigan, Wisconsin, and Ontario (Cadman et al. 2007; Cutright et al. 2006; Chartier et al. 2013).

BREEDING HABITAT

The Bald Eagle nests in large trees, often in the supercanopy and usually in old-growth or mature forests. In northern Minnesota, nest sites are usually within 0.8 km of water and are predominantly in white or red pines (Coffin and Pfannmuller 1988; Buehler 2020). In southeastern Minnesota, Mundahl and colleagues (2013) noted that of 53 nest trees that were used by the Bald Eagle, most were the tallest trees in the forest and were more than 50 m from water, and most successful nests were those with the surrounding forest containing basal area densities of greater than 20 square m/ha. This exemplifies the importance of maintaining large trees in the Minnesota landscape, especially in riparian areas close to large rivers and lakes.

Formerly, the Bald Eagle was sensitive to human disturbance and land use near nest sites, so buffer zones were created to protect these sites (Mathisen et al. 1977). Recently, many birds appear to have adapted and tolerate human disturbance near nests in urbanized landscapes and in rural areas near residential areas **(B** and **C)**.

POPULATION ABUNDANCE

Over the past 150 years, the Bald Eagle has had a highly variable population history in Minnesota. In the late 1800s, Roberts reported it as common throughout the wooded portions of the state, as well as scattered where nesting trees and adequate water and food were available. As human settlements and land conversion began to expand in the late 1800s, available habitat was reduced; the population decline was further exacerbated by increased shooting (Evans 1982). This led Roberts

B. Typical Bald Eagle nest in a large tree with the nest positioned slightly below the top, which allows easy access and a spectacular view of the surrounding area. This nest has been used for several years, is on the edge of a field near a small town, and is half a kilometer from a large lake. Photo by Gerald J. Niemi.

C. Another example of the nest of the Bald Eagle in an open wetland near excellent foraging areas in east-central Minnesota. Photo by Lee A. Pfannmuller.

(1932) to declare that the species was "formerly common, now much reduced in numbers and absent as a nesting bird from parts of the state where it once bred regularly."

The passage of the federal Migratory Bird Treaty Act of 1918 and, later, the Bald and Golden Eagle Protection Act of 1940 may have reduced illegal shooting, but the use of DDT beginning in the mid-1900s severely reduced the species' populations (Carson 1962). Based on the paucity of reports, populations were very low and scattered in northern forested regions near selected large lakes and rivers. The banning of DDT in 1972 marked the start of the recovery for the Bald Eagle in Minnesota. Fall migration counts from Hawk Ridge in Duluth clearly indicate the recovery of the population **(D)**. Migration counts at Hawk Ridge have increased an average of 6.5 percent per year from 1991 to 2019. Recent fall counts at Hawk Ridge from 2017 to 2020 have been 6,099, 4,603, 6,177, and 5,300 respectively.

The MNDNR (2006d) completed surveys of active Bald Eagle nests from 1973 to 2005 and documented the substantial increase in its nesting population in Minnesota. In 1973, the number of known active nests was slightly more than 100; at the end of the MNDNR surveys in 2005 the number of known active nests was 872. Based on a random survey of nests, they estimated that Minnesota had 1,312 total active nests in 2005.

PIF (2021) estimated a global population of 200,000 breeding adults, which included the United States and Canada, but the USFWS (2020b) estimated 316,700 Bald Eagles nesting in the lower 48 US states alone. This estimate was 4.4 times more than they estimated in 2009. Clearly, the PIF (2021) global estimate of Bald Eagles is highly conservative, and their populations are rapidly increasing. This is a strong testament to the power of conservation efforts and educational programs.

A limited number of pairs (less than 10) were reported in Mexico (Buehler 2020). Previously, Farmer and colleagues (2008) had estimated a conservative range of 100,000 to 1,000,000 Bald Eagles in North America. Environment Canada (2011) also indicated a conservative population range for Canada of 50,000 to 500,000 adults. Trend signals were mixed in the United States, based on Christmas Bird Counts (NAS 2018) and the BBS (Sauer et al. 2020). The BBS was not used here because it was deemed unreliable for all states and regions and even surveywide in the United States and Canada.

Density estimates are rare, but Mundahl and colleagues (2013) found a mean distance between nests of 1.52 km and densities ranging from 0.32–9.72 nests per 100 square km for 53 active nests in southeastern Minnesota.

CONSERVATION

The Bald Eagle was listed as *Threatened* when Minnesota's first endangered species list was created in 1984 (MNDNR 2019b). The Bald Eagle was listed as federally *Endangered* in 1978, but *Threatened* in Minnesota. The Northern States Bald Eagle Recovery Plan had a goal of 300 occupied breeding territories in Minnesota, which was achieved in 1987 (M. Miller and Pfannmuller 1991). In 1996, because of the increased number of occupied breeding areas, the Bald Eagle was moved down on Minnesota's state list to *Special Concern* status (MNDNR 1995). Finally, the results of the MNDNR surveys showing 872 active nests and estimating a total of 1,312 active nests in the state in 2005 helped justify the delisting of the Bald Eagle from the federal list of *Threatened* and *Endangered* species in 2007 (USFWS 2007b). The Bald Eagle was removed from the state list in 2013 (MNDNR 2013).

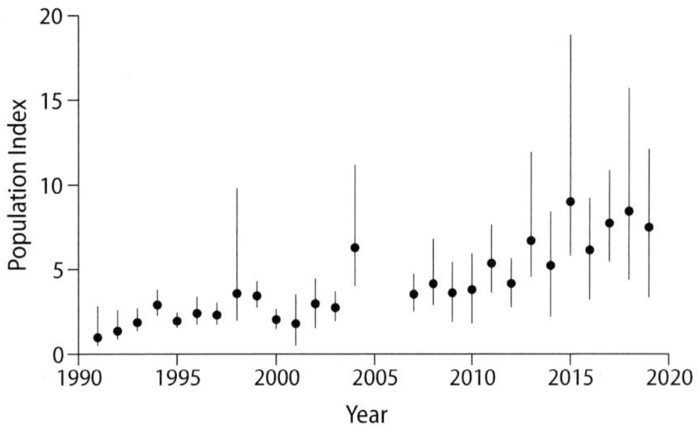

D. Fall migration counts and 95 percent credible interval (vertical lines) of Bald Eagles at Hawk Ridge in Duluth show a strong and significant increase in its population index from 1991 to 2019, as calculated by the Raptor Population Index (Oleyar et al. 2021).

Even though the Bald Eagle is no longer listed, it still has many possible threats, and most are related to human activity (MNDNR 2019b). DDT has been banned, but organophosphates, heavy metals, lead poisoning, and oil spills are continuing concerns. The Bald Eagle is still shot or intentionally poisoned. Collisions with vehicles have become a greater problem because they frequently feed on dead carcasses, especially deer, along roadsides. Many improvements have been made with electrical poles and power lines, but full implementation to retrofit this infrastructure has been slow, so raptors, including eagles, are still electrocuted. Finally, despite the tolerance of the Bald Eagle to human activity, protection of nesting, roosting, and wintering areas can still be a problem in some areas. Impacts of climate change on the Bald Eagle's breeding range were estimated as "low vulnerability" by NAS (2019) under all three climate scenarios they evaluated. Its winter range was predicted to expand northward, but some losses were projected in the southwestern United States. Education of the public is an ongoing activity and is a focus of environmental education around Minnesota. There is considerable management and conservation guidance available, such as that provided by the MNDNR (2019b) and the USFWS (2007c).

Lead poisoning is a specific concern for Bald Eagles in Minnesota, and researchers in the state have been on the cutting edge of unraveling these concerns (e.g., J. L. Kramer and Redig 1997; Cruz-Martinez et al. 2012; Bruggeman et al. 2018). For instance, Cruz-Martinez and colleagues (2012) found elevated lead levels in 334 of 1,277 Bald Eagles in the upper midwestern United States, including Minnesota. In addition, increased odds of elevated lead levels were associated with the season, primarily late fall and early winter periods (December and January), coinciding with periods following the deer hunting season. Lead bullets are used for killing deer, and Bald Eagles frequently feed on deer carcasses. Even though the Bald Eagle population appears healthy in the state, it is unknown how lead is affecting populations.

Finally, D. H. Johnson's (1982) insights are worth repeating. He stated that despite improvements in the Bald Eagle's population, the potential damage from "environmental contaminants, urbanization, lakeshore development, agricultural expansion, shooting, trapping, lead poisoning, and nest disturbances" is ongoing. Indeed, since 1982 the species seems to be adjusting reasonably well to human settlements, but lead poisoning and contaminants loom as continuing issues of concern.

RED-SHOULDERED HAWK

Buteo lineatus

Minnesota Seasonal Status: Breeding resident, migrant, and occasionally seen in winter in southern Minnesota. The Red-shouldered Hawk was uncommon during the MNBBA.
Migration: Eastern populations are short- to medium-distance migrants in the eastern United States as far south as Mexico, while Pacific coast populations are permanent residents.
Food: Hunts from an elevated perch for small mammals, amphibians, reptiles, small birds, and insects.
Nest: Stick nest in the crotch of a large deciduous tree or sometimes a coniferous tree.
Conservation Concern: Assigned a low PIF Continental Concern Score of 8/20. This species is officially listed as *Special Concern* in Minnesota.

The Red-shouldered Hawk is the third most common of the four "buteos" that nest in Minnesota; only the Swainson's Hawk is rarer. The species is largely found along floodplain forests of the Mississippi and St. Croix Rivers and in the Hardwood Hills region of west-central Minnesota with its forested lakes and wetlands.

Early in the nesting season, it is one of the noisiest raptors screaming its loud *KEE-aah* call, reminiscent of a Blue Jay. True to its name, the species has a red shoulder patch that is frequently difficult to see, especially in flight or if obscured by vegetation while on a perch. It is a medium-sized raptor, similar in size to the Broad-winged Hawk and smaller than the Red-tailed Hawk. Adults are rufous-colored underneath with fine white barring and narrow black-and-white bands on their tail. Juveniles are less distinctively colored but have narrow banding on the tail. Its scientific name, *lineatus,* is Latin and means "striped or decked out," referring to its banded tail and its underparts. Sandrock and Prior (2014) emphasize that "decked out" aptly describes the striking plumage of this bird.

Roberts (1932) apologized that "During fifty-odd years of bird work in Minnesota the writer has never seen the Red-shouldered Hawk alive, though it has been constantly looked for." He would certainly have seen this species today in Minnesota, which implies that it is more common today than in the late 1880s and early 1900s. Yet it is not as common as the Red-tailed or Broad-winged Hawk.

Photo by Gerald Hoekstra.

The Red-shouldered Hawk has two widely distributed but disjunct breeding populations. One is in the eastern United States, including as far northwest as Minnesota and northeast to Maine and to Quebec, Canada. The second is in California and north to Oregon. Its highest breeding densities are in Florida.

MINNESOTA BREEDING DISTRIBUTION

Historically, the Red-shouldered Hawk was apparently a rare bird species in Minnesota. Roberts (1932) stated it "undoubtedly breeds in the south-eastern part of the state." He devoted considerable text to purported observations of the species in the late 1800s and early 1900s but dismisses them as errors or mistakes in identification. For example, Currier stated he saw one on June 8, 1902, and several between May 22 and June 8, 1903, near Walker in Cass County (Currier 1904). Roberts also reported an observation by Surber near the Iowa border on September 22, 1920. Roberts's first observation was of a wing and tail brought to him in September 1922 by Frank Blair, a former superintendent of the State Game Farm near Mound in Hennepin County. The bird had apparently been shot at the farm. Later, he also accepted an observation by young "Mr. Walter Breckenridge who reported seeing one just above Marine on the St. Croix River, Washington County, on July 3, 1927." He was willing to accept this observation because Breckenridge was from Iowa and familiar with the species. Subsequently, in later years, additional observations were made in the same area, which suggested a nesting pair. Breckenridge would later go on to be a well-respected professor and director of the Bell Museum at the University of Minnesota.

In Roberts's revised treatise on the birds of Minnesota (1936), he reported the first confirmed nest with three eggs that he found in Wayzata, Hennepin County, on April 22, 1934. The nest was one used the previous year by a Cooper's Hawk. He also reported a nest with three eggs found by Nestor Heimenz in 1935 from St. Cloud in Stearns County. He concluded that the Red-shouldered Hawk should be regarded as "not an uncommon bird in the southeastern part of the state, and occasional farther north and west. All the records indicated that this species is undoubtedly extending its range northward and westward."

By 1975, Green and Janssen described the breeding distribution of the Red-shouldered Hawk from southeastern Minnesota and extending north to Leech Lake in Cass County, northwest to Tamarac NWR in Becker County, and south and west to Swan Lake in Nicollet County. They reported confirmed nesting in nine counties, ranging from Olmsted to Crow Wing, and inferred nesting from five counties, all in southeastern Minnesota.

A few years later, D. H. Johnson (1982) documented 39 nests of the Red-shouldered Hawk from 1935 to 1981, including northwest to Otter Tail and Hubbard Counties. He also stated that observations had been made west to Lac qui Parle County, northwest to Beltrami County, and northeast to Aitkin and Pine Counties. In contrast to the range extensions that had been reported beginning in the 1930s, D. H. Johnson (1982) felt "the species is having problems in Minnesota," but did not suggest any potential causes. Janssen (1987), however, suggested the Red-shouldered Hawk was "expanding northwestward in the state," noting breeding records as far northwest as Mahnomen County and northeast to Aitkin and Pine Counties. He included confirmed nesting from 11 counties since 1970. Hertzel and Janssen (1998) expanded the list to 18 counties with confirmed nesting records since 1970 but generally within the same breeding distribution described by D. H. Johnson (1982) and Janssen (1987).

Stucker and colleagues (2004) provided a review of the status of the Red-shouldered Hawk in Minnesota. They used reports from the MBS, the MNDNR Rare Features database, and other observations from 1980 to 2003 and reported a total of 460 Red-shouldered Hawk locations statewide among 39 counties, primarily in southeastern, east-central, and north-central Minnesota. They also included three locations in northeastern Minnesota, two in southeastern St. Louis County, and one in eastern Cook

County. Their records went well beyond the previously known breeding range and their significance is unclear, but they acknowledged that the species may nest at low densities at these locations and perhaps at other locations in northeastern Minnesota.

MBS recorded 334 breeding-season locations of the Red-shouldered Hawk. Their intensive surveys, which began in the late 1980s, revealed a substantial number of locations in west-central Minnesota, especially in the Hardwood Hills Ecological Subsection, the northwestern area of Morrison County (e.g., Camp Ripley Training Center), and along the Crow Wing, Minnesota, Mississippi, and St. Croix River floodplains.

The participants of the MNBBA documented 251 records of the Red-shouldered Hawk with a pattern of distribution similar to what had been previously reported **(A)**. These records showed a strong pattern from the southeast in Houston County, northwest to Norman and Polk Counties, and northeast to Itasca and Carlton Counties. Breeding evidence was recorded in 4.0 percent (188/4,742) of all surveyed blocks. Nesting was confirmed in 43 blocks within eighteen counties. The relatively low ratio of confirmed nesting compared with probable and possible nesting (43/131 or 1:3) likely reflects the secretive behavior of this species once nesting has begun (Dykstra et al. 2020).

The MNBBA records for the Red-shouldered Hawk had a strong association with the Eastern Broadleaf Forest Province and the southern portion of the Laurentian Mixed Forest Province. All of these records were highly associated with riparian areas of river, lake, and wetland ecosystems within the mixed deciduous–coniferous forests and, as previously noted, along the Mississippi, St. Croix, and Crow Wing Rivers and their tributaries. Detections were also high in the Twin Cities and Brainerd areas, where river floodplains exist but also where coverage was intensive. In 2020, a confirmed nest was also found in LeSueur County (MOU 2021a), where there have been no recent breeding records, but MBS recorded several Red-shouldered Hawk locations along the Minnesota River.

In their review of the Red-shouldered Hawk in North America, Dykstra and colleagues (2020) mentioned little about distributional changes in the northern portions of its eastern population, except for reported range expansions northward in Michigan. S. D. Robbins (1991) thought that Red-shouldered Hawks were probably never common in Wisconsin. He pointed out that observations of this species in the mid-1800s were probably misidentifications with the more common Broad-winged Hawk. Early reports in Wisconsin by Kumlien and Hollister ([1903] 1951) said it was by no means a common bird. Cutright and colleagues (2006), during their breeding bird atlas from 1995 to 2000, found that the species was widespread but uncommon throughout Wisconsin.

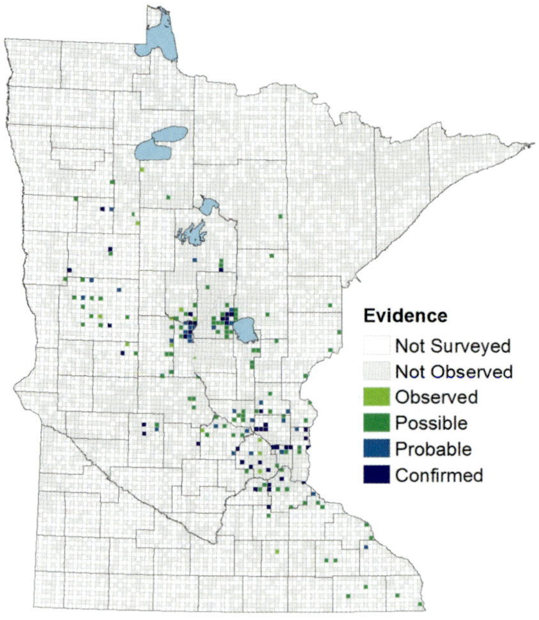

A. Breeding distribution of the Red-shouldered Hawk in Minnesota based on the Breeding Bird Atlas (2009–13).

In summary, the observations of Red-shouldered Hawks in northwestern Minnesota reported by Janssen (1987) and the extensive documentations by the MBS and the MNBBA in north-central Minnesota in comparison with those of Roberts (1932, 1936) and Green and Janssen (1975) suggest a range expansion in Minnesota beginning around the 1930s. Even though Currier's (1904) observations in 1902 and 1903 were from central Minnesota in Cass County, many of the previous reports have emphasized that the species has never been numerous and could have been missed early in the twentieth century. D. H. Johnson (1982) questioned this supposition because he felt that the species' extensive calling in early spring and summer render it difficult to overlook.

BREEDING HABITAT

The Red-shouldered Hawk is broadly described as nesting in bottomland hardwoods, riparian areas, flooded deciduous swamps, large woodlots, and suburban areas (Dykstra et al. 2020). S. P. Stucker and colleagues (2004) emphasized the species' use of upland deciduous forests, particularly in the central and north-central regions of the state, and especially the importance of wetland openings within and adjacent to forest habitats. However, they also found one-quarter of their extensive records of the species were from lowland forests, such as those dominated by silver maple floodplains, and to a lesser extent in green and black ash forests. Detailed studies in Cass and Morrison Counties by Henneman (2006) and Henneman and Andersen (2009) found Red-shouldered Hawks in large, mature deciduous forests without openings but also in mature deciduous forests with small forest openings from timber harvest (less than 5 ha). Many of these habitat characteristics were also summarized by the MNDNR (2019h) **(B)**.

Red-shouldered Hawk nest sites are generally in mature northern hardwood forests, and the birds use many species of deciduous trees and sometimes coniferous trees (Dykstra et al. 2020). Belleman (1998) found 38 nests in the Camp Ripley Military Reservation in central Minnesota. Nest trees were in aspen (61 percent), oak (34 percent), and birch (5 percent) and their average age was sixty years in aspen and sixty-seven years in oak. McLeod, Belleman, and colleagues (2000) found 58 nests, primarily high in mature upland hardwood stands with thick canopies and large-diameter trees in central Minnesota.

The effects of forest fragmentation on Red-shouldered Hawks are equivocal. In Iowa, Bednarz and Dinsmore (1982) found this species inhabits large woodlots closer to water and far from roads and buildings. They suggested that suitable breeding habitats should be greater than 100 ha in size. Henneman and Andersen (2009) in central Minnesota suggested that landscape context is an important consideration for Red-shouldered Hawks. Large, non-forested areas were detrimental to their populations, but small forest openings less than 5 ha may not affect their occupancy. S. P. Stucker and colleagues (2004) in southeastern Minnesota emphasized the importance of floodplain forests along major rivers for the Red-shouldered Hawk and, in central Minnesota, the need to manage large tracts of forests. The relatively large number of confirmed nests in Twin Cities suburbs found by the MNBBA indicated a tolerance for human disturbance (North and Faber 2014), but this number might also have been influenced by the extensive volunteer efforts in that region. Dykstra and colleagues (2020) reported that home ranges ranged in size from 90 to 200 ha.

POPULATION ABUNDANCE

Using BBS data, PIF (2021) estimated a United States and Canadian population of 1.8 million breeding adults and 1,500 in the state of Minnesota. The

B. Red-shouldered Hawks are especially fond of riparian forest habitats for foraging, cover, and nesting, such as those found along rivers, streams, wetlands, and ponds. Photo by Gerald J. Niemi.

species was too rare on point counts used in the MNBBA to adequately estimate a population in Minnesota. Farmer and colleagues (2008) estimated the continental population to be much lower than PIF's number, ranging from 10,000 to 100,000 based on raptor migration count sites. However, because PIF also estimated a greater than 200 percent increase in the species' population from 1970 to 2014, the discrepancy between these two estimates is partly due to the differences in time (Rosenberg, Kennedy, et al. 2016).

Population trend estimates can be difficult to determine for this species. The trends estimated from the BBS were unreliable for Minnesota because the mean number of birds detected per route per year was substantially less than one. The paucity of Red-shouldered Hawk detections in the BBS were also low for most other states and regions, but the overall trend surveywide from 1966 to 2019 illustrated an increasing population that averaged 2.33 percent growth per year **(C)**. Despite a sample size of 1,761 routes, the low number of detections suggests cautious interpretation of the trend.

Farmer and colleagues (2008) reported that the number of Red-shouldered Hawks on Christmas Bird Counts in the eastern and southeastern United States had significantly increased from 1974 to 2004. The Raptor Population Index from fall migration sites with enough observations in the eastern United States indicated a mixed pattern with some sites having a significant decline, many with stable trends, and a few with significantly increasing trends (Oleyar et al. 2021).

McLeod and Andersen (1998) noted a high density of 1 nest per square km at the Camp Ripley Training Center in central Minnesota, but estimates in other locations summarized by Dykstra and colleagues (2020) were much lower and ranged from 0.03 to 0.15 nests per square km.

CONSERVATION

As an official species of *Special Concern* in Minnesota, the Red-shouldered Hawk also remains a concern in many parts of its range outside of Minnesota. Conservation concerns emanated in the mid-1900s because many counts had indicated a slow decline in the north-central United States. This was likely due to problems with DDT exposure as well as habitat loss and fragmentation. Even though the species was less affected by DDT compared with many other raptors, the combined effects of DDT with habitat changes took their toll on populations of the Red-shouldered Hawk. In the early twentieth century, the species was shot, but legal protection and educational programs have reduced this source of mortality.

Cutting of large contiguous forest tracts and fragmentation of breeding habitats into small blocks of forest for agriculture or urbanization have been documented to reduce populations (Dykstra et al. 2020). These changes reduce effective nesting and foraging areas for Red-shouldered Hawks. As forests are fragmented into smaller units, they become more suitable for both the Red-tailed Hawk, its closest competitor, and the Great Horned Owl, one of its nest predators.

Following the 1984 listing of the Red-shouldered Hawk in Minnesota, the MNDNR (2019h) initiated a number of studies of the species. Among its management recommendations were retaining large amounts of mature deciduous forest, using thinning and light-selection cuts in forest cuts within these forest types, protecting wetland openings that are used for foraging, and maintaining a canopy closure of greater than 70 percent. The species is sensitive to human disturbance. Activity within 300 m of nest sites should be minimized during the breeding season of April 1 to June 30. Finally, maintaining previous nest sites is encouraged because Red-shouldered Hawks will reuse their nests.

The species' breeding range was predicted to be stable under all three climate change scenarios studied by NAS (2019). However, under the 2°C warming scenario a substantial part of the species' breeding range in northwestern Minnesota was predicted to be lost by 2050. This likely is because of extensive wildfires that may occur in that region. Any loss of its breeding range was gained by predictions of habitat increases north of Minnesota in central and eastern Canada. Similarly, its wintering range was projected to be stable under all three climate scenarios, with winter range expansions northward in the United States. Overall, the recent apparent range expansion of the Red-shouldered Hawk in Minnesota and its population increases in the southeastern United States are encouraging. These accounts indicate that it is unlikely that the species is reoccupying its former range in Minnesota, unless it was missed by early explorers in the 1800s or even by Roberts in the early 1900s. Therefore, the species expansion could be related to warming temperatures in the state; only time will tell if this is the case.

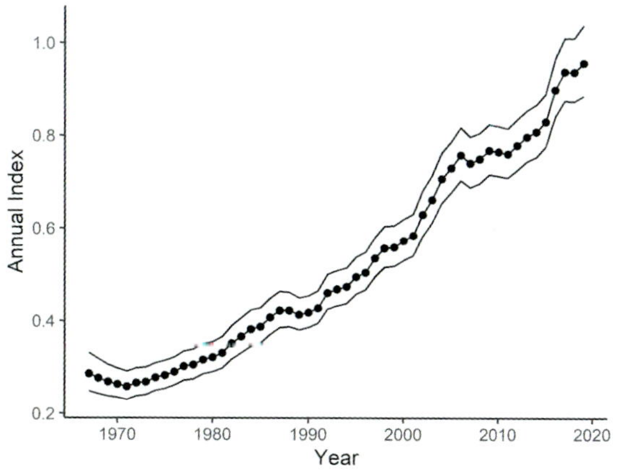

C. Surveywide breeding population trend and 95 percent credible interval (solid black lines) for the Red-shouldered Hawk based on the 1967–2019 Breeding Bird Survey (Sauer et al. 2020).

BROAD-WINGED HAWK

Buteo platypterus

Minnesota Seasonal Status: Regular summer resident and migrant. The Broad-winged Hawk was uncommon during the MNBBA.
Migration: Primarily a medium- to long-distance migrant to Central and South America, but over-wintering populations can be found sparingly along the Gulf coast and in southern Florida (Haines et al. 2003).
Food: A sit-and-wait predator that feeds on small mammals, amphibians, snakes, large insects, and birds, including nestlings.
Nest: Usually placed in the first main crotch of a deciduous tree or on a platform close to the bole of a coniferous tree.
Conservation Concern: Assigned a moderate PIF Continental Concern Score of 9/20.

The Broad-winged Hawk is the most common raptor in Minnesota's forests. Roberts (1932) specifically described it as "a common bird throughout the deciduous woodlands of the state and . . . equally well represented in the coniferous forests farther north." Even though it is found frequently in the contiguous forests of Minnesota, it can be secretive during the breeding season. Bent (1937) called them "gentle, retiring, quiet birds of the deep forest." They are, however, very conspicuous during fall migration. Broad-wings are well-known for their magnificent fall migrations at Hawk Ridge in Duluth, when tens of thousands can be observed in a single day. When not migrating, this species is most often observed sitting motionless on a tree branch or wire close to the road in forested areas where it searches for prey.

The Broad-winged Hawk is among the smallest of the "buteo" raptors, those with broad wings built for soaring (Goodrich et al. 2020). The species can occur in both a common light phase and a dark morph. The light phase is distinguished during flight by its short tail with two dark bands, one wide white band, and a second, less conspicuous narrow white band near the base, and rufous barring on the breast and wing linings. Dark morphs have a dark breast and wing coverts, silver lining on the wing feathers, and a dark tail with white band (Alderfer 2006), but the light phase is substantially more common. During the breeding season, the Broad-winged Hawk can be heard with a high-pitched whistle, described by

Photo by Laura Erickson.

Roger Tory Peterson (1947) as *P-wee-e-e-e-e-e* and suggestive of the Eastern Wood-Pewee, often given near the nest and while in flight. Beware, because the call is often imitated by Blue Jays.

Broad-wings are widely distributed in the contiguous forested regions of eastern and northeastern North America as far west as Texas, Missouri, and Minnesota, and northwest, but sparsely, to British Columbia. This species is rare in open country and not abundant anywhere within its breeding range. An endemic subspecies occurs year-round on several Caribbean islands.

MINNESOTA BREEDING DISTRIBUTION

In the late 1800s, Hatch (1892) described the Broad-wing as "fairly common from the borders of Iowa to Lake Superior. Rare in the northwestern sections of the state." Roberts (1932) documented nesting activity in Becker, Hennepin, Isanti, Kittson, Marshall, Meeker, Polk, and Stearns Counties, as well as in Itasca State Park.

More than forty years later, Green and Janssen (1975) noted there were "no summer records for the prairie in the west central, southwestern, and south-central regions, and the species is scarce in the southeastern region." Although they did not include nesting records, they stated it was the most numerous hawk in the northern forested regions. Several years later, Janssen (1987) reinforced its breeding distribution in the "heavier forested portions of the state where there are upland openings and wetlands." He confirmed nesting since 1970 in 23 Minnesota counties, including one nest in Brown County in southern Minnesota, and in the Twin Cities metropolitan area northwest to Kittson County and northeast to Cook County. In 1998, Hertzel and Janssen extended confirmed nesting since 1970 to five additional counties, including Scott and Winona Counties in the southeast, Kandiyohi in the west-central region, and Kittson and Lake of the Woods Counties in the northwest.

During intensive fieldwork, the MBS documented 233 breeding-season locations that spanned almost the entire state, from the southern tier of counties, such as Jackson (three locations), Martin (one), and Mower (one), to Roseau County (one) in the northwest to Cook County (more than 20) in the northeast. Locations were sparse in the southern and western regions of the state but included two locations in Ottertail County, and Janssen (2019) reported a confirmed nest there in 2017.

The MNBBA participants documented 959 records, which exemplified the wide distribution of the Broad-winged Hawk, ranging from southeastern to northwestern Minnesota and all the forested areas to the east **(A)**. Breeding evidence was recorded in 14.9 percent (708/4,761) of all surveyed blocks. Confirmed nesting was recorded from 89 blocks, and probable breeding evidence from 115 blocks. Records were relatively dense in the Twin Cities and Brainerd areas, where coverage was extensive, but also in the northern and northeastern portions of the state, where populations are most common.

The MNBBA probability map predicted that landcover suitability for the species was highest in the northeastern and north-central regions of

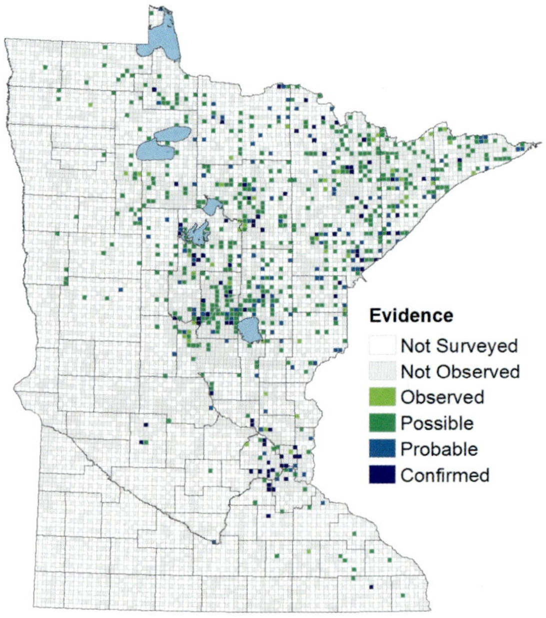

A. Breeding distribution of the Broad-winged Hawk in Minnesota based on the Breeding Bird Atlas (2009–13).

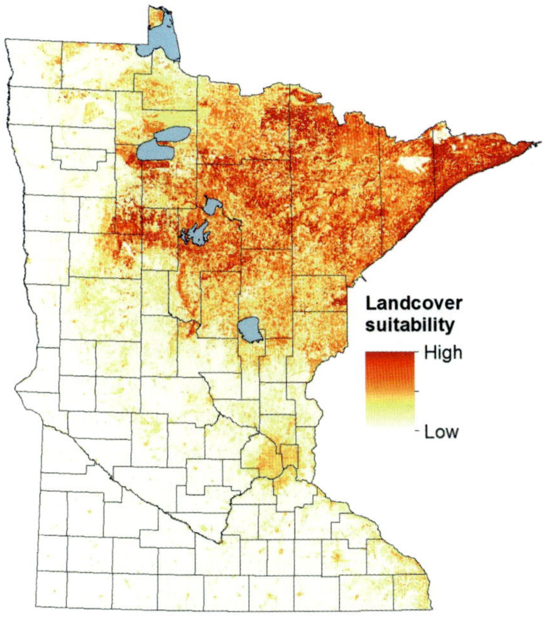

B. Landcover suitability map for the Broad-winged Hawk in Minnesota based on habitat, landscape context, and climate data gathered during the Minnesota Breeding Bird Atlas (2009–13) using the MaxEnt modeling approach.

Minnesota **(B)**. Lower suitability was predicted on the fringes of these regions, including in the northwest, the Twin Cities area, and southeastern Minnesota.

Roberts noted the species was present in deciduous and coniferous forests throughout the state, but nearly 50 percent of the forests of Minnesota have been lost since the mid-1800s. The forest-dwelling Broad-winged Hawk has undoubtedly disappeared or its population declined with the loss of forests in the southern and western portions of the state. Yet scattered populations still exist where some forests remain, especially in riparian habitats. In western Minnesota, observations were very rare in the Red River valley. A hint of the past in this area was described by Bent (1937), who cited J. W. Preston's observations from 1888: "In hidden retreats, where the tangled wilderness of lakes and forests guards in lonely silence the streams which feed the Red River of the North in Minnesota, I found the Broad-winged Hawk . . . breeding abundantly." The Red River valley on the border with North Dakota has been heavily converted to agricultural land.

In their review of the Broad-winged Hawk in North America, Goodrich and colleagues (2020) pointed out that the species has recently increased in the northeastern United States with reforestation. They suggested the species may be expanding westward into many forested areas in western states, including Arizona, Colorado, Idaho, Montana, New Mexico, and Wyoming in the 1980s. Cadman and colleagues (2007) in Ontario concluded that the Broad-wings' "relative abundance has probably not changed much since the 1800s, except in the Carolinian region where forest cover has declined greatly."

BREEDING HABITAT

Broad-winged Hawks have been described as nesting in deciduous and mixed deciduous–coniferous forests (Goodrich et al. 2020). Breeding territories of the species were frequently associated with openings and water nearby, such as ponds or wetlands (Keran 1978; Titus and Mosher 1981). On the basis of 29 nests from both Minnesota and Wisconsin, Keran (1978) stated that the species used managed forests, including oak and aspen from thirty-five to fifty years old in areas with an average density of 204 trees per ha. Niemi, Howe, and colleagues (2016) documented that Broad-wings used a wide variety of forested stands, including aspen, aspen–spruce–fir, jack pine, lowland hardwoods, oak, and white pine **(C)**.

POPULATION ABUNDANCE

Using BBS data, the global population was estimated as 1.9 million breeding adults with 980,000 in the United States and 64,000 individuals estimated in Minnesota (PIF 2021). Trend estimates from the BBS were unreliable throughout North America because of too few observations in all jurisdictions sampled. Despite the lack of detections, Rosenberg, Kennedy, and colleagues (2016) used data from the BBS and estimated that the breeding population in North America had increased by 53 percent since 1970. Fall migration counts at Hawk Ridge in Duluth from 1974 to 2019 revealed an insignificant and highly variable trend for the Broad-winged Hawk **(D)**. Annual fall counts at Hawk Ridge range from a total of 9,411 in 1998 to 160,703 in 2003.

C. The Broad-winged Hawk prefers deciduous and mixed deciduous–coniferous forests that are often near small ponds or wetlands where it forages for amphibians and small mammals. Photo by Gerald J. Niemi.

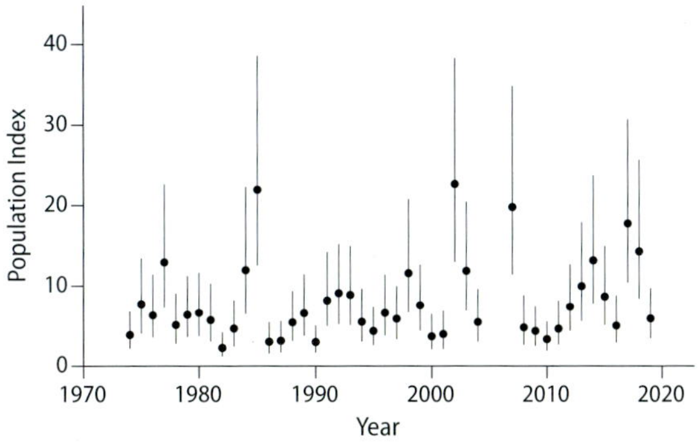

D. Fall migration counts and 95 percent credible interval (vertical lines) of Broad-winged Hawks at Hawk Ridge in Duluth have high variability in its population index from 1974 to 2019, as calculated by the Raptor Population Index (Oleyar et al. 2021).

CONSERVATION

Broad-wings avoided major population declines during the DDT era because of their reliance on amphibians and mammals for food, but increased use of DDT south of the United States may be a concern because of their greater consumption of insects in winter (Goodrich et al. 2020). Shooting, trapping, and collisions on roads are continued concerns on the hawk's wintering grounds. In the past, large-scale shooting of migrating Broad-winged Hawks commonly occurred at Hawk Ridge and other places in Minnesota up until the late 1940s, when laws were put in place to protect migratory birds.

With a 2°C warming scenario, NAS (2019) projected a low vulnerability for the Broad-winged Hawk's breeding distribution in North America. In Minnesota, it was predicted to have substantial loss of its breeding range in central to northwestern parts of the state under a 2°C warming scenario by 2050.

It is likely that loss of extensive forested habitat throughout its potential breeding range since the nineteenth century has reduced the overall population of the Broad-winged Hawk. However, its range appears to be expanding westward, and recent estimates suggest an increasing population in North America. In Minnesota, the species has had a widespread and solid population in the forested areas, but the predicted effects of climate change may substantially reduce available habitat for this iconic forest raptor.

SWAINSON'S HAWK

Buteo swainsoni

Minnesota Seasonal Status: Regular breeding resident and migrant. The Swainson's Hawk was rare during the MNBBA.
Migration: Long-distance migrant that spends winters in the Pampas of Argentina.
Food: Vertebrates, including mammals, birds, and reptiles, and large invertebrates, especially grasshoppers. Hunts by soaring or perching, or often on the ground.
Nest: Typically places nest in a grove of trees or occasionally in a solitary tree in open country, usually near the top.
Conservation Concern: Assigned a moderate PIF Continental Concern Score of 11/20.

Photo by Michael Furtman.

The Swainson's Hawk is a champion long-distance migrant among North American buteos, overwintering in southern South America. It was named after William Swainson, who first illustrated the species in an 1831 book, *Fauna Boreali-Americana,* which he co-authored with John Richardson during a British Arctic expedition to Canada.

Swainson's Hawks have the classic buteo shape: broad wings and short tails. They tend to be slimmer and have slightly longer wings, which Eckert (2017) states extend beyond the tail in a perched bird, a trait not found among other Minnesota buteos. The most diagnostic characteristics in flight are the contrasting white wing linings, dark flight feathers, and the V-shape of their wings as they soar. While sitting, most birds are light-bellied with dark or reddish-brown upperparts. However, as with many buteos, this can be highly variable, with light to dark plumages.

Roberts (1932) summarized comments by a well-known ornithologist from the 1800s, Dr. Elliott Coues. Coues was particularly interested in the species' diet; he said, "Those that I shot after midsummer all had their craws [stomach] stuffed with grasshoppers." And upon observing them foraging, he said, "It is amusing to see a Hawk catching grasshoppers, skipping about in an awkward way, and looking as if he were rather ashamed of being seen in such a performance."

The Swainson's Hawk is a raptor of the grassland biome in open country of western North America from northern Mexico to central Alberta and Saskatchewan. The highest densities are patchily distributed in eastern Colorado, western Texas, southeastern New Mexico, southeastern Idaho, and southern Alberta.

MINNESOTA BREEDING DISTRIBUTION

Roberts (1932) labeled the Swainson's Hawk a "summer resident throughout the western part of the state," but also noted observations in semi-open country east to the Mississippi River. He noted observations by A. W. Hewitt in 1898, who described the species as "Fairly common. Breeds" in Faribault County. He also cited observations from the late 1800s and early 1900s from adjacent Iowa that Swainson's Hawks were nesting in Winnebago County just south of Freeborn County in Minnesota and, as observed by Roberts, were breeding about to the Mississippi River. In northwestern Minnesota, he said, "Peabody and Langevin consider it quite common in the Red River Valley." He led a trip there in June 1928, where they observed 14 Swainson's Hawks from Polk County to the Canadian border. Roberts included four confirmed nesting records from Jackson (1901), Polk (1920 and 1921), and Traverse (1879) Counties.

Decades later, Green and Janssen (1975) described the Swainson's Hawk's summer distribution as "scarce except in the eastern part of their range" in Minnesota. They outlined a distribution confined to a narrow band along the western border with North and South Dakota, along the southern tier of counties bordering Iowa, and in the southeastern counties. Several years later, D. H. Johnson (1982) summarized 27 known nesting records, from the 1879 Traverse County record to a 1981 nesting record from Dakota County. These records included a mix of nests with eggs or young, but many were inferred from observations of nests without documentation. He included nesting records from 14 counties, including the farthest-north record of nesting, in Polk County from 1920 and 1921.

In 1987, Janssen described the Swainson's Hawk's distribution as the west-central and southwestern areas of the state and portions of south-central Minnesota north to southern Washington and Wright Counties. His breeding distribution map was similar to Green and Janssen (1975) but slightly wider along the border with the Dakotas. He emphasized that the species' breeding status north of Clay County was unknown, but there were summer observations throughout northwestern Minnesota. Collectively, Janssen (1987) and Hertzel and Janssen (1998) documented confirmed nests in 19 counties since 1970.

Martell, Willey, and Schladweiler (1998) added 16 nesting records from 1981 to 1996 that ranged from southwestern Minnesota, north to Kittson County, and southeast to Dakota and Fillmore County. In addition, they conducted intensive searches for nesting Swainson's Hawks in 1996–97 and recorded 28 locations. The locations were all in the southern portion of the state and ranged from Pipestone, Lac qui Parle, and Stevens Counties in the west to Washington and Winona County in the east; the latter is the easternmost county where the species has been recorded nesting in the state. Two of the counties, Stevens and Winona, were new county records not included in Hertzel and Janssen (1998).

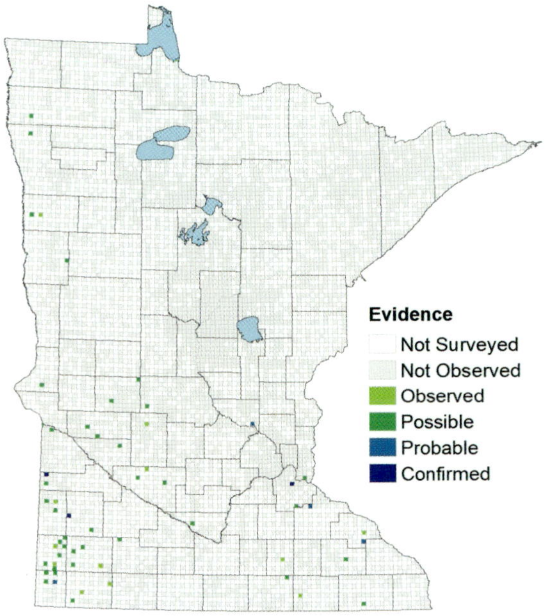

A. Breeding distribution of the Swainson's Hawk in Minnesota based on the Breeding Bird Atlas (2009–13).

MBS recorded 86 breeding-season locations for the Swainson's Hawk during county surveys. Their records were largely focused in southwestern Minnesota, especially in Lincoln, Lyon, Murray, Pipestone, and Rock Counties, and included a cluster in Dakota, Goodhue, and Rice Counties. The remaining locations were scattered in many counties, from Olmsted County in the southeast to southwestern Todd County, and northwest to Kittson, Polk, and Red Lake Counties.

The MNBBA observers identified 63 records of the Swainson's Hawk, with a breeding distribution similar to what had previously been described **(A)**. One extension was probable nesting in southern Sherburne County, as well as possible nesting evidence in Chippewa, Pope, Renville, Stearns, and Swift Counties in southwestern Minnesota. Swainson's Hawk was recorded in 1.2 percent of all surveyed blocks (58/4,740). Nesting was confirmed in three blocks, with one each in Dakota, Lyon, and Yellow Medicine Counties. MNBBA observers also recorded possible nesting north of Clay County in Norman, Polk, and Marshall Counties.

Bechard and colleagues (2020) reviewed the Swainson's Hawk in North America and did not identify any major breeding range changes. They described major population declines since the late 1800s and early 1900s and cited an 1882 observation by E. E. Thompson [Seton], who considered the Swainson's Hawk "abundant" as far east as Winnipeg, Manitoba. The decline of the species was largely attributed to human persecution because most settlers considered the species a "chicken hawk," but the decline was also attributed to decreases in habitat and food availability due to the plowing of grasslands.

In Minnesota's past, the Swainson's Hawk was largely confined to the open grasslands and prairie regions, where suitable food supplies and nesting trees were located. Their general distribution is likely similar today, but the overall population has declined substantially. In some areas, clearing of forested areas for agriculture since the 1930s in south-central, central, and southeastern Minnesota may have opened additional breeding habitat where recent nesting has or may have occurred (Green 1981).

BREEDING HABITAT

Bent (1937) simply described Swainson's Hawk as "a bird of the wide open spaces, prairies, plains, and deserts." Martell, Willey, and Schladweiler (1998) in Minnesota found Swainson's Hawk primarily nesting in shelterbelts, woodlots, and lone trees in or near agricultural areas. They found 11 nests in cottonwood, seven in ash, two in box elder, and one each in spruce and red pine. It typically nests in areas with scattered trees within grasslands, shrubland, or agricultural landscapes and often in a solitary tree along riparian areas, in shelterbelts, or on abandoned homesteads **(B)**.

POPULATION ABUNDANCE

PIF (2021) estimated a United States and Canadian breeding population of 820,000, with only 570 individuals in Minnesota. The BBS population trends indicated a significantly increasing population trend of 0.88 percent per year in the Prairie Potholes Bird Conservation Region from 1967 to 2019—one of the only regions that had enough data to reliably estimate a trend **(C)**. Rosenberg, Kennedy, and colleagues (2016) estimated a 35 percent increase in the species' population from 1966 to 2014.

B. Breeding habitat of the Swainson's Hawk is open country with scattered trees for nesting and perching, or woodlots and shelterbelts adjacent to prairies, grasslands, or agricultural landscapes. Photo by Gerald J. Niemi.

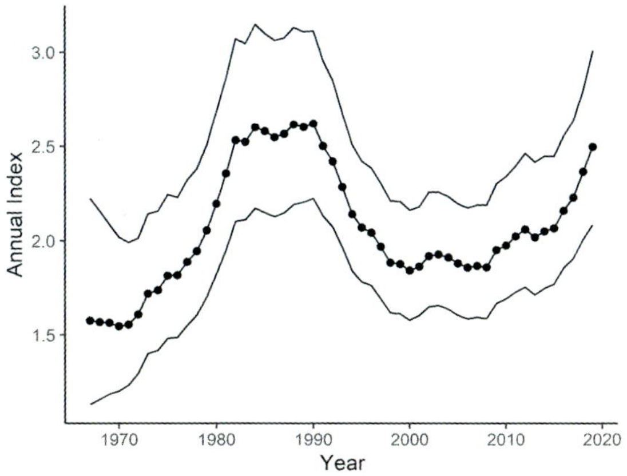

C. Breeding population trend and 95 percent credible interval (solid black lines) for the Swainson's Hawk in the Prairie Potholes Bird Conservation Region based on the 1967–2019 Breeding Bird Survey (Sauer et al. 2020).

CONSERVATION

Even though the population since the 1960s may be considered stable or potentially increasing in parts of western North America, Swainson's Hawk population levels have substantially declined since the 1800s and early 1900s. Primary factors in its long-term declines have been attributed to population declines in one of its primary prey species, Richardson's ground squirrel (*Spermophilus richardsonii*), and also to habitat changes in the North American grasslands and prairies (Bechard et al. 2020).

Many other factors contribute to mortality, including nest predation by American Crows and Great Horned Owls, electrocution from power lines, and potentially negative interactions with Red-tailed Hawks. Despite benefits derived from Swainson's Hawks' predation on mammals (e.g., ground squirrels and pocket gophers) and grasshoppers, shooting of the birds continues in North America, during migration in Mexico and Central America, and during winter in South America. During the DDT era, Swainson's Hawks were not known to be greatly affected, but organophosphate pesticides applied in the agricultural lands of Argentina reduced wintering populations in the 1990s. NAS (2019) in their analysis of climate changes predicted a stable breeding range in North America for the Swainson's Hawk under all three warming scenarios they examined through 2080. The future of the Swainson's Hawk in Minnesota suggests it will remain a rare species in the state because many factors appear to keep its North American and Minnesota population limited.

RED-TAILED HAWK

Buteo jamaicensis

Minnesota Seasonal Status: Regular breeding resident, migrant, and regular winter resident in the southern half of the state. The Red-tailed Hawk was common during the MNBBA.
Migration: Permanent resident to short-distance migrant. Migration patterns of Minnesota birds are complex and still largely unknown.
Food: Carnivore that primarily uses a sit-and-wait strategy to feed on small to medium-sized mammals, birds, amphibians, and reptiles.
Nest: Placed in trees on edges or within mixed deciduous–coniferous forests adjacent to openings of shrubland, grassland, or agricultural areas.
Conservation Concern: Assigned a low PIF Continental Concern Score of 6/20.

Photo by Mike Lentz.

The Red-tailed Hawk is one of Minnesota's most distinctive and common raptor species, especially in open country, along roadways, woodlands, and in openings in heavily forested areas. Roberts (1932) noted it was a "summer resident, breeding throughout the state," which is still true today but substantially less common in the heavily forested areas in the north. The Red-tailed Hawk is found in open country throughout the United States and to the tree line in northern Canada and Alaska. Highest densities are found in widely scattered patches in Kansas, North Dakota, Alberta, and California.

Red-tails are a polymorphic species with four subspecies in Minnesota and potentially 16 described in its North American population (Preston and Beane 2020). The four subspecies include the dark-phased Harlan's Hawk (*Buteo jamaicensis harlani*), the dark, often rufous-colored Western Red-tailed Hawk (*B. j. calurus*), and the light-phased Krider's Hawk (*B. j. kriderii*) (Green and Janssen 1975). The most commonly observed subspecies in Minnesota is the Eastern Red-tailed Hawk (*B. j. borealis*). The Red-tailed Hawk is in the genus *Buteo,* which refers to a large group of raptorial birds that tend to be "sit-and-wait" predators that circle and migrate on thermals with their relatively broad wings and short tails. The specific scientific name *jamaicensis* simply refers to Jamaica, where the first type specimen was collected.

Good identification marks of the adults include their rufous tail, terminal black bar on the tail, and, in most birds, a dark band across the breast. Juveniles have a banded tail and do not develop a red tail until they are more than a year old. The call of the Red-tailed Hawk is a distinctive two- to three-second call or scream—*kee-eeee-arrr,* which is often duplicated in inappropriate habitats in movies or on television.

MINNESOTA BREEDING DISTRIBUTION

In Minnesota, Roberts (1932) lamented the widespread disappearance of this species as a summer resident in Minnesota as follows: "During an automobile trip entirely across the state from Minneapolis to Pipestone County and back, August 31 to September 4, 1923, not a single Red-tail was seen. Only two were noted on a similar trip, June 18 to July 7, 1924, from Minneapolis west to the Dakota line, south along the wooded shores of Lakes Traverse and Big Stone to Lac qui Parle County and return." Roberts included confirmed nests with eggs or young from the following counties: Becker, Dakota, Goodhue, Hennepin, Isanti, Jackson, Kittson, Lac qui Parle, Lake, Polk, Sherburne, Traverse, and Wright. He also noted inferred nesting in Faribault, Nicollet, Stearns, Steele, and Wabasha Counties.

More than forty years later, Green and Janssen (1975) and Janssen (1987) labeled the Red-tail simply as a summer "resident throughout the state" and considered it among the most abundant raptors in most areas of the state, except in the forested areas. Janssen reported confirmed nesting in 38 counties since 1970, with a distribution similar to that presented by Roberts (1932). Later, Hertzel and Janssen (1998) expanded that to include an additional 13 counties since 1970, for a total of 51 counties. MBS recorded 506 breeding-season locations for the Red-tailed Hawk. Their locations were distributed throughout the state where MBS has completed inventories but were less numerous in the heavily forested regions of the northeast and east-central regions of Minnesota.

The participants in the MNBBA reported Red-tails in every county of the state and a total of 2,620 records **(A)**. In terms of records, the Red-tailed Hawk was the most abundant raptor during the MNBBA. They were recorded in 38.2 percent (1,852/4,848) of all surveyed blocks. The most frequent category recorded was possible nesting because, although the species is easily observed in an area, finding the actual nest can be challenging. Most confirmed or probable nesting records were identified in the Twin Cities metropolitan area, where volunteer efforts and coverage were extensive. Nesting was confirmed in 72 of the 87 Minnesota counties.

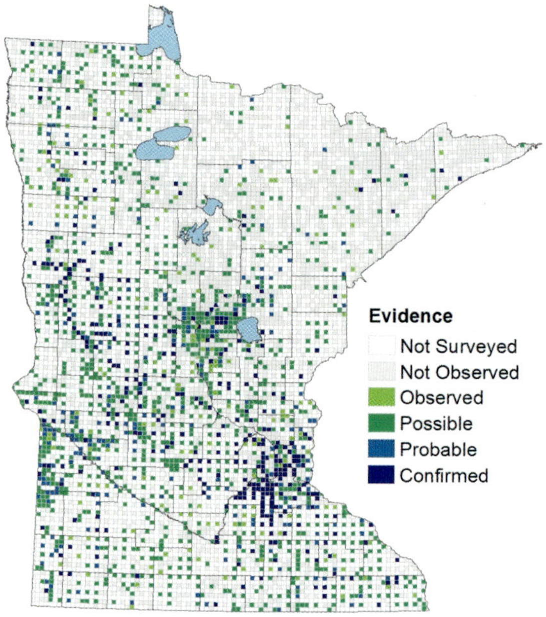

A. Breeding distribution of the Red-tailed Hawk in Minnesota based on the Breeding Bird Atlas (2009–13).

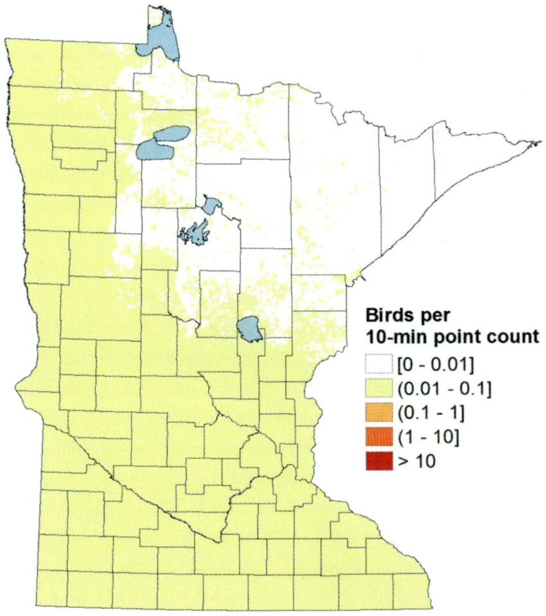

B. Predicted breeding distribution (birds detected per 10-minute point count) of the Red-tailed Hawk in Minnesota based on habitat, landscape context, and climate data gathered during the Minnesota Breeding Bird Atlas (2009–13) using the Generalized Linear Modeling method without an adjustment for detectability.

The probability map for the Red-tailed Hawk predicts a wide and generally uniform population distribution throughout much of western, southern, central, and northwestern regions of Minnesota **(B)**. Within this broad area, the species is found in open and semi-open country where it is found foraging, roosting, and nesting among scattered forest patches. Scattered breeding areas are also located in northeastern and northern Minnesota, where suitable openings exist in agricultural regions, farmsteads, and recently logged areas.

The population of Red-tailed Hawks is considered to be secure in Minnesota and in North America. Preston and Beane (2020) suggested that populations have expanded through much of North America during the mid- to late twentieth century in response to the widespread establishment of open, wooded parklands that have replaced grasslands and dense forests. The species adapts to urban, exurban, and agricultural development as long as adequate food supplies (primarily small to medium-sized mammals), perch sites, and trees or other structures for nesting are available.

Roberts (1932) bemoaned the loss of this species from his early years in Minnesota, when he observed it "in the vicinity of Minneapolis where now it is, in the summer, an object of interest." In recent years Red-tailed Hawks have been increasingly observed in the Twin Cities area, and the MNBBA provides solid evidence of their nesting throughout the region; this is a testament to its resilience and its adaptability to human-altered landscapes.

BREEDING HABITAT

Red-tailed Hawks are found in open and semi-open habitats, such as grasslands, prairies, wetlands, and agricultural lands adjacent to fragmented forests, woodlots, and contiguous forests (Green 1991; Preston and Beane 2020). They are often found in the same habitats as the Great Horned Owl, where competition for nest sites can be intense **(C)**. D. H. Johnson (1982) stated that Red-tails were most commonly associated with disturbed or fragmented forest habitats. He also noted the placement of their nests in or along the edge of openings, where they can gain easy access to nest sites. In the contiguous forested areas in the northern half of Minnesota, breeding habitat includes logged and burned-over areas (Green and Niemi 1978).

POPULATION ABUNDANCE

PIF (2021) estimated a global breeding population of 3.1 million adults, with 2.1 million in the United States, and a Minnesota breeding population of 42,000 adults. Using migration raptor count data, Farmer and colleagues (2008) provided a more conservative estimate of its global population of 100,000 to 2 million individuals, with 90 percent of the population in the United States and Canada (PIF 2021). The overall trend for BBS routes of all forms of Red-tailed Hawks censused in the United States averaged about 1.5 detections per route per year and indicated the population was increasing an average of 1.52 percent per year from 1966 to 2019. Similarly, surveywide in the United States and Canada, the BBS trend was significantly increasing at a rate of 1.29 percent per year **(D)**. Fall migration counts of Red-tails from Hawk Ridge in Duluth, however, indicated a significant decline from 1991 to 2019 of −2.08 percent per year (Oleyar et al. 2021). The fall 2020 count at Hawk Ridge of 4,720 was the lowest in thirty years.

C. Typical breeding habitat of the Red-tailed Hawk in Minnesota includes open habitats for foraging that are adjacent to woodlands with large trees for nesting. Photo by Gerald J. Niemi.

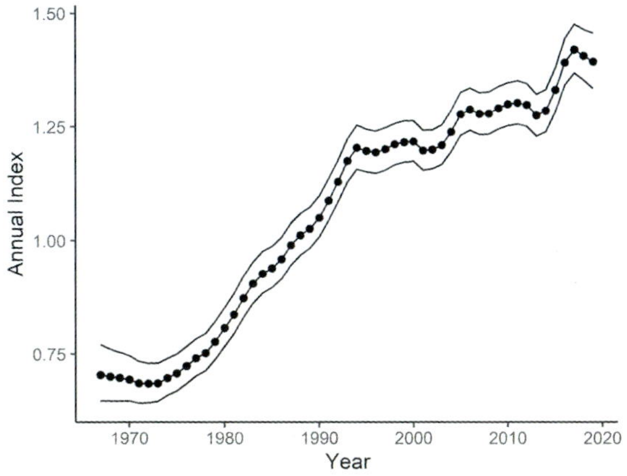

D. Surveywide breeding population trend and 95 percent credible interval (solid black lines) for the Red-tailed Hawk based on the 1967–2019 Breeding Bird Survey (Sauer et al. 2020).

CONSERVATION

The Red-tailed Hawk is of low conservation concern because it has a substantial North American population with a wide distribution and its overall breeding population is increasing. Many of the past threats, such as shooting, have subsided, but isolated events still occur. Educational programs and laws have helped, but they need continued attention and enforcement. Lead poisoning; pesticides from agricultural activities; collisions with cars, planes, wires, and wind turbines; and electrocution from power lines have continued to be sources of mortality. Because of its diet of small to medium-sized mammals, Red-tails were not greatly affected by DDT in the mid-twentieth century.

Great Horned Owls are known to prey upon Red-tailed Hawk nestlings, and corvids prey on their eggs. Bald Eagles will rob Red-tailed Hawks of food. Despite these threats, the species appears to be healthy at present. It was one of the most frequently killed raptors at wind farms in California. Its susceptibility to wind farm mortality will need to be monitored because the number of wind farms and turbines is expected to increase in the future.

Because of its wide and extensive southern distribution, the Red-tailed Hawk is not expected to be affected in the near future by climate change (NAS 2019). The future of the Red-tail in Minnesota appears promising, but recent declining migration counts at Hawk Ridge need continued monitoring. It should remain a strong avian representative of Minnesota's open landscapes for years to come.

EASTERN SCREECH-OWL

Megascops asio

Minnesota Seasonal Status: Regular, year-round resident. The Eastern Screech-Owl was rare during the MNBBA.
Migration: Sedentary; northern populations may move south during the winter, likely in response to food shortages.
Food: Nocturnal predator that consumes a wide array of invertebrates and small vertebrates; hunts from a perch below the tree canopy.
Nest: Cavity nester that uses natural cavities, abandoned woodpecker cavities, and artificial nest boxes.
Conservation Concern: Assigned a moderate PIF Continental Concern Score of 10/20.

Photo by Laura Erickson.

There is something quite endearing about a pint-sized owl with big yellow eyes. Minnesota has three owls that actually fit that description but, unlike the Northern Saw-whet Owl and Boreal Owl, only the Eastern Screech-Owl has ear tufts and is primarily restricted to the southern half of the state. Unlike other owls, the little Screech-Owl is dichromatic. The expression of the two color phases, reddish-brown and brownish-gray, appears to be independent of the season of the year, or the bird's sex or age, but may be related to local climatic conditions (Ritchison, Gehlbach, et al. 2020). Regardless of its hue, the mottled colors provide the perfect camouflage as it perches on a small branch or peeks out from the opening of a tree cavity. Like most raptors, the male is smaller than the female.

If startled, the Eastern Screech-Owl may stretch its body upward, compress its feathers, and close its eyes, assuming a posture that makes it look as though it were simply an extension of the tree. It may even turn sideways and raise one wing to cover its lower face (Gehlbach 1994). It is more common, however, to detect this nocturnal species when it begins vocalizing after dark. Although its name implies you will hear a hair-raising scream, screeches are given only when the bird is anxious, especially if young are threatened (Gehlbach 1994). Much more common is a soft, eerie-sounding trill known as the *tremolo* call, which is used primarily for courtship or as a way to maintain contact between the pair. Another frequently heard call is the *whinny* song. Shorter than the *tremolo,* it sounds like the soft whinny of a horse and is important for territorial defense (Gehlbach 1994). Both sexes vocalize.

Long considered the same species, the Western Screech-Owl and Eastern Screech-Owl were split into separate species in 1983. Since the birds are similar in appearance, their different vocalizations are among their most prominent distinctions. The Eastern Screech-Owl is largely an inhabitant of eastern North America, from the Great Plains east to the Atlantic coast and south to northeastern Mexico. Widely dispersed throughout its range, it is rarely present in high densities. The Western Screech-Owl (*Megascops kennicottii*) is restricted to states west of the Rocky Mountains, stretching south into central Mexico.

MINNESOTA BREEDING DISTRIBUTION

When Roberts (1932) prepared his account of the Eastern Screech-Owl, he described it as a common breeding resident throughout the state. Despite this assertion, at the time breeding records were available from only five counties: Grant, Hennepin, Pipestone, Polk, and Rice. Forty years later, Green and Janssen (1975) described a more limited distribution, noting the species "was absent from Cook, Lake, Koochiching, and Itasca counties." In St. Louis County there were only a few records in the southern third of the county. "The abundance of the species," they wrote, "has declined since the early 1960s." D. H. Johnson (1982) pointed to the Owl's spotty nesting activity; from 1883 to 1981, only 56 nest records had been documented in the state. The majority of these records were south of Grant and Morrison Counties in central Minnesota and were restricted to cities and small towns. Little was known about the species' status in rural areas.

Janssen (1987) delineated a distribution that excluded the northern third of Minnesota. He depicted the northern limit of the Eastern Screech-Owl's range as stretching from southwestern Pine County west to northern Wilkin County; south of this line, he identified seventeen counties where nesting had been confirmed since 1970. Hertzel and Janssen (1998) later added six counties to the list, all from Wadena County south. Because nocturnal owl surveys were not a focus of the MBS, field surveys documented only two Screech-Owl breeding-season locations, both in southwestern Minnesota.

During the MNBBA, observers reported 64 Eastern Screech-Owl detections from 1.2 percent (58/4,743) of the surveyed atlas blocks. Breeding was confirmed in 12 blocks and deemed probable in 11 blocks **(A)**. Detections were reported in 36 of Minnesota's 87 counties while breeding was confirmed in 14 counties, all from southern Morrison County southward. Three of the atlas blocks were breeding was confirmed were located in two counties each (Morrison/Stearns, Dakota/Hennepin, and Mower/Olmsted). Seven counties were additions to the list published by Hertzel and Janssen (1998). The

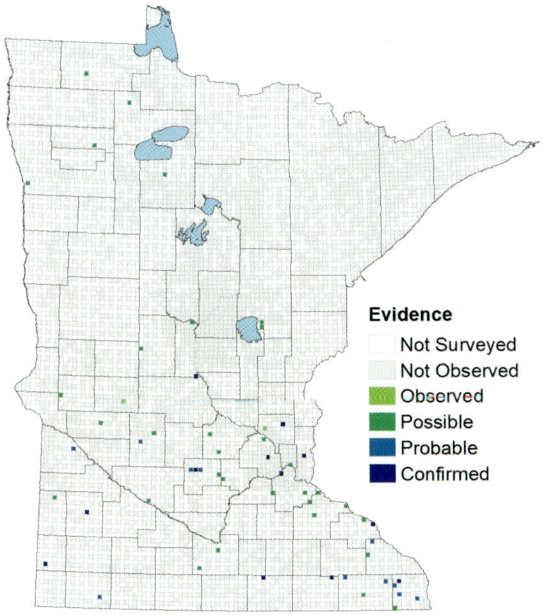

A. Breeding distribution of the Eastern Screech-Owl in Minnesota based on the Breeding Bird Atlas (2009–13).

majority of detections were located in the southern two-thirds of the state, from Mille Lacs Lake south to the Iowa border. The few northern records were restricted largely to northwestern Minnesota.

Overall, there have been few changes to the Eastern Screech-Owl's distribution in the state since Green and Janssen's account in 1975. It is less clear if its distribution has changed since the early 1900s. The owl's abundance, however, appears to have declined significantly. Areas where its nighttime *tremolo* was once common are now silent. Originally described by Roberts (1932) as "the commonest member of its family in Minnesota," during the MNBBA the Screech-Owl was far outnumbered by the Barred Owl, the Great Horned Owl, and the Northern Saw-whet Owl. Even the Great Gray Owl, which was not known to nest in Minnesota until 1935, outnumbered the Eastern Screech-Owl with 83 reports.

Elsewhere in North America, the species has expanded its breeding range, especially in Canada. Ritchison, Gehlbach, and colleagues (2020) suggested that urbanization, which is often accompanied by more tree plantings and warmer temperatures, may be facilitating the expansion. But, in his comprehensive review of Wisconsin birds, S. D. Robbins (1991) noted that the species' range had receded since the 1930s and 1940s in the far northern reaches of the state and in the central regions in the 1960s and 1970s. Given that cold winter weather can be a factor that limits the Eastern Screech-Owl's northern distribution, Cutright and his colleagues (2006) suggested that the species' presence in northern Wisconsin may vacillate in response to "mini-climatic" changes over long periods of time.

BREEDING HABITAT

The Eastern Screech-Owl requires woodlands with sufficient mature trees to provide nesting cavities. In Minnesota, the bird is found in oak woodlands and in farmlands and urban settings with large shade trees **(B)**. It avoids large tracts of dense, contiguous forest and areas with significant conifer cover (Danz et al. 2007). An open understory facilitates foraging and accommodates the species' swooping, U-shaped flight pattern, in which it makes a steep dive down from its perch and then flies at a level grade until it rises steeply to another perch or cavity (Ritchison, Gehlbach, et al. 2020).

The species fares best when mature woodlands are adjacent to forest edges or include gaps in the forest canopy that provide additional foraging opportunities (Danz et al. 2007). Where openings are limited, the habitat becomes more suitable for Barred Owls and Great Horned Owls.

POPULATION ABUNDANCE

PIF (2021) provided a population estimate of 500,000 breeding adults in the United States and Canada, with an additional 60,000 in Mexico. This is a decline from an earlier estimate of 800,000 (PIF Science Committee 2013). Because Screech-Owls are rarely detected on Minnesota BBS routes, a statewide population estimate is not available. Even projected population trends have little statistical reliability (Sauer et al. 2020).

B. In Anoka County, an oak woodland, with its mature trees and scattered openings, provides ideal foraging and nesting habitat for the Eastern Screech-Owl. Photo by Lee A. Pfannmuller.

Whether or not the Eastern Screech-Owl is declining is a matter of some debate. Limited BBS data indicates that the North American population declined 41 percent from 1970 to 2014 (Rosenberg, Kennedy, et al. 2016). Ritchison, Gehlbach, and colleagues (2020), however, suggested that biologists are confusing declines with population cycles; they believe the species has demonstrated an overall fairly stable population.

Anecdotal evidence in Minnesota suggests a statewide decline in numbers. In light of its adaptability to urban and suburban landscapes, one would expect that development would positively impact local populations. Studies in Texas, for example, demonstrated that populations residing in urban and suburban habitats were more stable and reproductively successful than populations residing in rural areas (Gehlbach 1994). On the other hand, increased populations of species that prey on adult and young Screech-Owls, including Barred Owls and Great Horned Owls, could result in local declines. As the MNBBA data illustrate, both predators were nearly twelve times more abundant than the Eastern Screech-Owl.

CONSERVATION

Although it has a large breeding range, the Eastern Screech-Owl's moderate conservation priority reflects concern regarding its apparent population decline and relatively small population size (PIF 2021). Its adaptability to human-dominated landscapes, however, led Ritchison, Gehlbach, and colleagues (2020) to conclude that "long-term negative human impacts on this species as a whole are doubtful." Given that very cold winters are one of the factors that regulate population numbers (Ritchison, Gehlbach, et al. 2020), warmer winters may facilitate the species' northward range expansion. Although NAS (2019) predicted that the Screech-Owl's breeding distribution was stable under all three warming scenarios examined, they projected a small northward range extension, including in northeastern Minnesota.

Until more is known about the Screech-Owl's status and the factors regulating populations, efforts to retain mature trees and nesting cavities in deciduous woodlots adjacent to open areas are important. Nest boxes are an effective tool for encouraging occupation of younger forests, where cavities may be lacking, but proper maintenance of nest boxes is essential.

GREAT HORNED OWL

Bubo virginianus

Minnesota Seasonal Status: Permanent, year-round resident. The Great Horned Owl was uncommon during the MNBBA.
Migration: Nonmigratory, but occasionally there are irruptions of northern birds into Minnesota during the winter months.
Food: Considered a "perch and pounce" predator, this species has one of the most diverse diets of any raptor but feeds largely on mammals and birds.
Nest: Primarily abandoned platform nests of hawks and crows; also nests in tree cavities, the hollows of standing stumps, and on human-built structures.
Conservation Concern: Assigned a low PIF Continental Concern Score of 8/20.

The Great Horned Owl is one of Minnesota's largest avian predators, with a reputation for its strength and ferocity. Considered by some as royalty among birds of prey, the "tiger owl" uses its powerful feet and talons to take prey that range in size from small rodents and songbirds to woodchucks, pheasants, Canada geese, and other hawks and owls. Only the smallest items are swallowed whole; large animals are torn apart using its sharp beak.

Were it not for the raucous calls of nearby crows attempting to drive away their potential predator, the Great Horned Owl might easily go undetected during the daylight hours. Perched near the trunk of a tall tree, the Great Horned Owl has mottled gray-brown plumage that serves as a seamless camouflage while it rests during the daytime hours. Two long feather tufts, sitting atop either side of the head, give rise to its common name. Other diagnostic features include a barred plumage pattern across the breast, a white throat patch, yellow eyes, and a facial disc that is often rufous in color. Although the plumage of both sexes is similar, like most owls the female is larger in size. The call, *whoo-hoo-hoo*, has been likened to the "sound of a distant foghorn, the far-away whistle of a locomotive, or the barking of a large dog in the distance" (Bent 1938). *Bubo,* the Latin term for the owl's genus, means "owl" and notably contains "the onomatopoetic *oo* sound of an owl call" (Sandrock and Prior 2014).

A highly adaptable species, the Great Horned Owl is found in every North American biome except the open tundra. It is sparsely distributed across North America, from the tree line south to the Gulf Coast and even farther south into Mexico and Central and South America.

Photo by Mike Lentz.

MINNESOTA BREEDING DISTRIBUTION

Roberts (1932) described this versatile predator as a common, permanent resident throughout the forested regions of the state, occurring even in riparian woodlands scattered across the prairie landscape. At the time of his writing, the Great Horned Owl was "much more common" than the Barred Owl, its population size augmented in part by an influx of northern migrants during the winter months. Confirmed nesting reports were available from seven counties, stretching from Houston County in the southeast to Roseau County in the far northwest.

Like Roberts, other accounts suggested that some Great Horned Owl populations might be migratory. In Minnesota, reports included accounts of summer residents in northern counties moving south during the winter as well as winter influxes of the much paler, arctic subspecies *B. v. subarcticus* into northern Minnesota (Green and Janssen 1975; Janssen 1987). Recently, however, Artuso and his colleagues (2020) concluded that most Great Horned Owls are permanent residents and not migrants. Dispersals, they note, are more accurately considered irruptions in response to declines in prey abundance. Once the owls move, it is unclear if the dispersing birds return the following spring. At least one study found that territorial owls were less likely to respond to food shortages than nonbreeding birds (Rohner 1997).

Although the annual movements of this species deserve more study, by 1987 the widely distributed Great Horned Owl had been confirmed nesting in 39 counties across the state since 1970 (Janssen 1987). By 1998, an additional 15 counties had been added to the list (Hertzel and Janssen 1998). In a review of the status and distribution of Minnesota raptors, D. H.

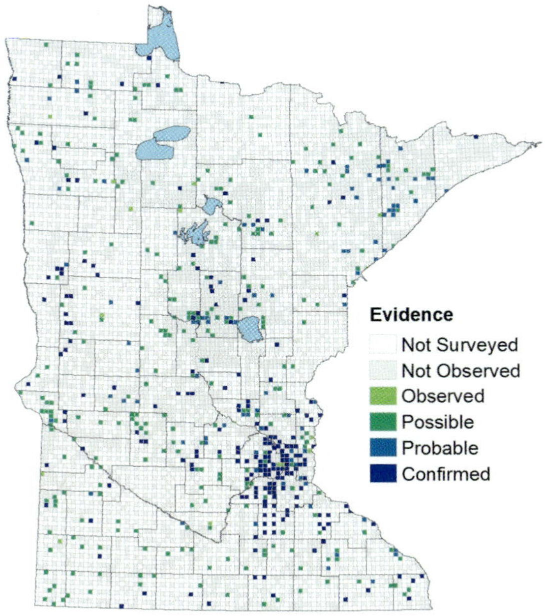

A. Breeding distribution of the Great Horned Owl in Minnesota based on the Breeding Bird Atlas (2009–13).

Johnson (1982) concluded that, although the species was uncommon in the northeastern forest region, it was still the most common breeding owl in the state. Since Johnson's report, field biologists with the MBS had documented 73 breeding-season locations for the species by 2014. The records were widely dispersed across all but the far northeastern counties.

MNBBA participants tallied 750 Great Horned Owl records in 12.2 percent (585/4,811) of the surveyed atlas blocks. Breeding was documented in 202 blocks (4.2 percent) and deemed probable in 91 blocks (1.9 percent) **(A)**. Owls were observed in all but three of Minnesota's 87 counties (Mahnomen, Mille Lacs, and Renville) and were confirmed breeding in 61 counties. Included were 18 counties not previously identified by Hertzel and Janssen (1998). The Great Horned Owl was most commonly reported in the Twin Cities metropolitan region, where its abundance may more accurately reflect a greater number of observers than a greater number of owls. The species is also more easily detected in the open habitats of residential areas compared with the dense forests of northern Minnesota.

Although considered "much more common" than the Barred Owl nearly one hundred years ago, during the MNBBA the Great Horned Owl and Barred Owl were nearly identical in abundance. A total of 750 records were reported for the Great Horned Owl and 756 for the Barred Owl. The Great Horned Owl was reported from 585 atlas blocks, and the Barred Owl from 558. Although both species are relatively common in the Twin Cities region, the Barred Owl is more frequently encountered in northern Minnesota, while the Great Horned Owl is more frequently encountered in western Minnesota.

Historically, the Great Horned Owl's wide distribution across North America has remained largely unchanged. Although many regions report a decline in abundance, given its low relative abundance and its nocturnal habits, it is difficult to assess the reliability of these long-term trends.

BREEDING HABITAT

Given the broad distribution of the Great Horned Owl, it is no wonder that the list of habitats used by the species is extensive. As Johnsgard (1988) commented, there is probably no other owl in North America that "lives in so many habitats and under so many climatic variations as the Great Horned Owl." The species demonstrates a preference, however, for open, second-growth deciduous or mixed deciduous–coniferous woodlands. Fragmented landscapes, with a mix of openings and scattered woodlots, are characteristic of most breeding territories (Artuso et al. 2020) **(B)**.

Johnsgard (1988) delineated three critical habitat features: suitable nest sites, roosting trees, and foraging areas. The Great Horned Owl is quite adaptable in utilizing a variety of potential nest sites. It prefers used nesting platforms of hawks and crows, but it also uses dead tree stumps, cliff ledges, and caves, and it may even nest on the ground. Roosting trees provide concealment and protection during the day and are typically conifer trees or deciduous trees that retain their leaves year-round. Foraging usually occurs over agricultural fields, wetlands, or grassland habitats, where scattered trees are available for perching as the owls keep a watchful eye for prey. The only intensive study conducted on Great Horned Owls in Minnesota documented that in the Anoka Sandplain, the birds utilized open fields and forest edges for foraging but preferred the landscape's upland oak forests for roosting and nesting (Fuller 1979).

B. The interspersion of tall, mature trees that are used for nesting, roosting, and hunting perches with open habitats for foraging is attractive to the Great Horned Owl. Photo by Lee A. Pfannmuller.

POPULATION ABUNDANCE

Given its nocturnal habits, its early nesting phenology, and its low population density, the Great Horned Owl is not well suited to the only long-term avian monitoring program conducted in North America, the BBS. Despite concerns regarding its application, biologists used these and other data to generate a population estimate of 3.8 million breeding adults in the United States and Canada, with an additional 1.9 million birds residing south of the United States (PIF 2021). Minnesota supports approximately 29,000 adults.

These same data suggest a relatively stable to slightly decreasing population trend across the BBS survey region from 1966 to 2019 (Sauer et al. 2020), with a cumulative decline of 27 percent from 1970 to 2014 (Rosenberg, Kennedy, et al. 2016). Minnesota data suggest a steeper decline averaging −1.38 percent per year. Nevertheless, the statistical reliability of both trends is compromised by the species' low abundance and nocturnal habits.

CONSERVATION

The Great Horned Owl's indiscriminating palate, coupled with powerful talons that can kill animals as large as waterfowl and grouse, did not endear it to either sportsmen or farmers over the years. Persecuted for decades, the species suffered especially great losses in agricultural regions, where the Owl could inflict considerable damage to farmstead poultry. In the early 1900s, Roberts (1932) used numerous reports from Minnesota game wardens, detailing the number of birds trapped and killed, as a measure of the species' abundance.

Although some illegal hunting may continue, today the Great Horned Owl is more revered than persecuted for its prowess. Deemed a low conservation priority (PIF 2021), this high-ranking predator faces one of its few threats from its potential exposure to the impacts of poisonous chemicals, especially agricultural pesticides that may accumulate in its prey (Artuso et al. 2020). Even a warmer climate, which poses a threat to many North American species, is predicted to have little impact on the remarkably adaptable Great Horned Owl (NAS 2019).

NORTHERN HAWK OWL

Surnia ulula

Photo by David Brislance.

Minnesota Seasonal Status: Regular but rare permanent resident. The Northern Hawk Owl was rare during the MNBBA.
Migration: Irruptive movements to more southern regions regularly occur and are thought to be responses to prey availability and climate.
Food: Primarily small mammals and birds, the latter especially in winter.
Nest: In cavities of decayed trees and in broken tops of trees, stumps, and vacant woodpecker holes.
Conservation Concern: Assigned a moderate PIF Continental Concern Score of 11/20.

The Northern Hawk Owl, as its name implies, was described by Roberts as "more like a Hawk than an Owl." It is hawk-like because it is active during the day and its appearance in flight is similar to a Cooper's Hawk, while its sitting profile is falcon-like. Distinguishing features include black borders on its facial disk, a long graduated tail, and barred underparts. It is most often seen perched on the top of a tree or stout bush. Bent (1937) described its flight from tree to tree as follows: "It commonly throws itself headlong downwards nearly to the ground, along which it skims toward the next tree, and on nearing its goal rises with a graceful aerial bound to the topmost perch offered."

The Northern Hawk Owl is one of the least-studied birds in North America. It is the only species in the genus *Surnia,* and its species name, *ulula,* refers to its wail or mournful call, which has been described as a trilling, rolling whistle *ulululululululul* . . . lasting up to 14 seconds (Alderfer 2006).

A Holarctic species, the Northern Hawk Owl occurs sparsely from the Maritime Provinces in eastern Canada across the boreal forest to Alaska. In the United States, its southern limit occurs in northern Minnesota, northern Wisconsin, and northern Michigan, and occasionally in the Rocky Mountains.

MINNESOTA BREEDING DISTRIBUTION

Roberts (1932) did not describe the Northern Hawk Owl as a nesting species in Minnesota. He said it was a "winter visitant from the north, chiefly in the northern and eastern parts of the state." He included several observations of potential nesting of the species in Minnesota from April to June but disregarded the observations as "not conclusive and may have resulted from errors in identification." The observations included secondhand information from a "reliable" source of a nest near Ada in Norman County in 1884; possible nesting in Roseau County in 1906, Cook County in 1923, Koochiching County in 1927, and Itasca County in 1928; and several late spring observations, one as far south as St. Paul on May 21, 1925.

Green and Janssen (1975) labeled the Northern Hawk Owl as a "casual summer resident." They noted that breeding had been reported from Norman, Roseau, and St. Louis Counties. The latter was described by Strnad (1963) as six individuals, including a possible nesting pair and four young, in July 1962. The report by Strnad was from the area now known as the Sax-Zim Bog and was instrumental in initiating the current attraction of the area to nature enthusiasts.

In 1982, D. H. Johnson summarized four nesting records for the Northern Hawk Owl and identified those records as accidental nesters. He included a nest with eggs in Roseau County from the summer of 1906 as a confirmed nest record, one which Roberts had questioned. Johnson also included the nesting record from the spring of 1884 in Ada, Minnesota, but, like Roberts, he had reservations because the habitat did not seem appropriate. Johnson cited ten additional summer observations of the species from 1905 to 1963. The other two nest records he cited were both from 1980 and are considered solid: the first Minnesota nesting record by Steve Loch in Aitkin County and the second by Martin Kehoe in Lake of the Woods County.

Janssen (1987) noted that nesting seemed to occur following major winter invasion years. He included the two nests reported by Johnson in Aitkin and Lake of the Woods Counties, but no new records were added of confirmed nests since 1970. Hertzel and Janssen (1998) would later add two additional confirmed county nesting records since 1970 in Lake and Roseau Counties. During its county surveys, the MBS recorded three breeding-season locations, two in St. Louis County and one in Lake County.

Grosshuesch (2008) summarized that the Northern Hawk Owl has been documented breeding after six of the past 11 winter irruptions and only twice in non-irruption years. He also stated that in the years before the major 2005–6 irruption (1962–2005) nesting had been confirmed at 14 sites, which included nine active nests and five sites with short-

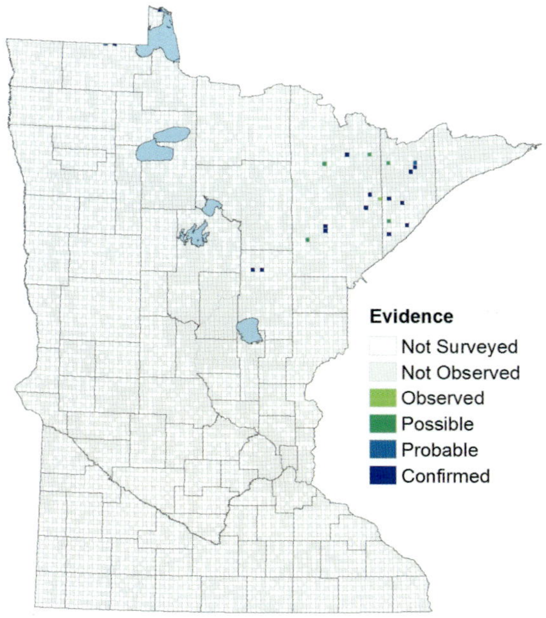

A. Breeding distribution of the Northern Hawk Owl in Minnesota based on the Breeding Bird Atlas (2009–13).

tailed young. These occurred in seven counties: Aitkin, Cook, Koochiching, Lake, Lake of the Woods, Roseau, and St. Louis. Cook, Koochiching, and St. Louis Counties were new records not included in Hertzel and Janssen (1998) since 1970.

The MNBBA participants identified 28 records of the Northern Hawk Owl, all of which come from five counties: Aitkin, Lake, Lake of the Woods, Roseau, and St. Louis **(A)**. Fifteen blocks in five counties had confirmed nests: two in Aitkin, six in Lake, one in Lake of the Woods, five in St. Louis, and one in Roseau. Records were included from 0.5 percent (23/4,735) of the surveyed atlas blocks.

Overall, the breeding status of the Northern Hawk Owl in Minnesota is elusive. The prevailing wisdom has been that breeding primarily occurs after invasion years, but nests have been found after noninvasive years. The MNBBA results confirmed nesting in five counties following noninvasive years. It remains a possibility that a small population in Minnesota may nest most years or every year in the remote forests and peatlands. Unfortunately, it would require a massive long-term search effort to definitively determine the regular breeding status of this species in the state.

Duncan and Duncan (2020) emphasized that few data were available to determine any historical changes in the distribution of the species. Furthermore, the authors suggested the Northern Hawk Owl may have historically nested in places like Minnesota, and recent nesting did not necessarily represent extensions of its breeding range.

BREEDING HABITAT

The most detailed study of Northern Hawk Owl nesting in Minnesota was completed by Grosshuesch (2008, 2009), who found nine Minnesota nesting sites in 2006 and an additional 41 nesting locations from Minnesota to Alberta between 1998 and 2006. Ninety-one percent of the nests were found in dead trees and 87 percent were found in open areas; no nests were greater than 40 m from an opening. The openings were primarily from forest fires and logging and contained large-boled trees for nesting sites **(B)**.

Duncan and Duncan (2020) also described the species' typical nesting habitat as moderately dense coniferous or mixed coniferous–deciduous forests bordering marshes or other areas cleared by fire or logging. Hannah and Hoyt (2004) suggested that burned-over sites are most suitable for up to eight years after a fire and before an extensive shrub layer develops.

POPULATION ABUNDANCE

PIF (2021) estimated a total US and Canadian population of 130,000 breeding adults, with approximately 49,000 in the United States, primarily in Alaska. They also surprisingly estimated a Minnesota population of 1,400 birds with a wide 95 percent confidence interval of 0–5,100. No trend estimates were available for Minnesota or any other regions. The species clearly has a very low population, and even an estimate of 1,000 seems high. Winter irruption years in Minnesota since the 1960s (and the number of estimated nonbreeding individuals observed) were: 1962–63 (47–125), 1977–78 (9), 1986–87 (9), 1990–91 (16), 1991–92 (159), 1995–96 (47), 1996–97 (100), 2000–2001 (190), 2001–2 (35), 2004–5 (475), and 2005–6 (139) (Grosshuesch 2008).

CONSERVATION

Duncan and Duncan (2020) summarized that the Northern Hawk Owl is not considered to be "at risk" in Canada and it is "globally secure." However, it is considered to be a "sensitive or vulnerable species" in the Canadian provinces of Alberta, Saskatchewan, and Quebec. Threats include vulnerability to shooting because it is diurnal, bold, and uses exposed perches. It has been documented as vulnerable to fur trapping and collisions with vehicles, trains, and transmission lines (Duncan and Duncan 2020).

The species benefits from forest fires and logging disturbances, especially if the latter include the availability of snags and appropriate perches. An example in Minnesota is well described by Wilson (1993). Sonerud (1997) suggested that perches at least 9 m high create an effective searching radius of at least 70 m, which can improve its hunting success. In Fennoscandia, the species has also used nest boxes.

With a 2°C warming scenario, NAS (2019) projected a low vulnerability for the Northern Hawk Owl breeding distribution in North America, including a 26 percent range loss and a 35 percent range gain primarily in the far north by 2050. In Minnesota, the species was predicted to be extirpated from the state even under our imminent 1.5°C warming scenario by 2050.

Without question, more data are needed on the status and ecology of the Northern Hawk Owl in Minnesota and throughout its range in North America. The species is listed as one of America's "100 most wanted" species to be observed by birders and wildlife enthusiasts, and Minnesota is the best place in the continental United States to see this species (Mlodinow and O'Brien 1996). The future of this secretive and unusual half owl–half hawk in Minnesota is difficult to predict. Its current status in the state is largely unknown, and its future viability is mostly dependent on lands to the north, where the bulk of its population resides.

B. Nesting habitats of the Northern Hawk Owl in northern Minnesota are typically in areas with large-boled dead trees and relatively close to openings that allow access to prey. This photo illustrates a recently burned area where a Northern Hawk Owl nested at the top of the tree in the center. Photo by David Grosshuesch.

BURROWING OWL

Athene cunicularia hypugaea

Minnesota Seasonal Status: Summer and migration vagrant; last successful breeding documented in 2007. The Burrowing Owl was very rare during the MNBBA.
Migration: Populations in Florida, the far southwestern United States, the Caribbean, Mexico, and South America are year-round residents; northern populations migrate to the southwestern states and Mexico.
Food: Primarily insects and small mammals.
Nest: Nests underground; western birds largely depend on the abandoned burrows of prairie dogs, Richardson's ground squirrels, and badgers; Florida birds dig their own burrows.
Conservation Concern: Assigned a moderate PIF Continental Concern Score of 11/20; officially listed as *Endangered* in Minnesota.

Photo by Gerald Hoekstra.

The scientific name of this "amusing little owl" (Roberts 1932) comes from the Greek word for Athena, the goddess of wisdom, whose favorite bird was the owl. In an ironic accident of taxonomic history, the Burrowing Owl, which occurs only in the Western Hemisphere, was named after a Greek goddess when it was described in 1822 by a European taxonomist. In North America, the Zuni of the desert southwest described it as the "priest of the prairie dogs" because of its use of the animal's empty burrows. Indeed, *cunicularia* means "burrower."

Well-camouflaged by its brown-and-white plumage, the Burrowing Owl's long legs may be its most distinctive feature, along with its round head and large yellow eyes. Nearly indistinguishable from each other, the sexes are similar in size, unlike most owls where the female is larger. Generally quiet birds, the most common vocalization during the breeding season is the male's soft and mellow *coo-coooo* call.

Although numerous subspecies are scattered from the Caribbean to South America, North America's Burrowing Owl population includes just two subspecies. The first, the Florida Burrowing Owl (*A. c. floridana*), was originally restricted to the prairies of central Florida but expanded as development created more open country. Breeding populations are now concentrated north and west of Lake Okeechobee and along Florida's southeastern coast.

Hundreds of miles away, the range of the Western Burrowing Owl (*A. c. hypugaea*) stretches from southern Alberta and Saskatchewan south to the Texas Panhandle, then west to the Great Basin and the interior valleys of California, Oregon, and Washington. Recently the population has declined, and the eastern margin of its range has receded westward. It no longer breeds regularly in Manitoba, Minnesota, and Iowa; in the Dakotas, it now is rarely found east of the Missouri River (Klute et al. 2003). The highest breeding densities occur in southeastern Colorado.

MINNESOTA BREEDING DISTRIBUTION

The paucity of written accounts describing Minnesota's avifauna in the 1800s obscures the early status of rare birds like the Burrowing Owl. Although Roberts (1932) characterized it as "a comparatively recent immigrant from the west," his perceptions may have been influenced by his failure to encounter the bird when visiting Grant and Traverse Counties in June 1879 (Roberts and Benner 1880). Just two years later, however, he discovered two pairs with young in Swift County (1881) and compiled records from other observers in Jackson County (sometime in the 1880s and in 1896); Lac qui Parle County (1893); Lincoln County (1897); and Marshall County (1897; nesting). "Thereafter," he wrote, "it increased steadily in numbers until by 1924 it was nesting commonly throughout Grant, Traverse, Pipestone, Lincoln, and Lac qui Parle counties and farther north in the Red River Valley."

By 2015, the Burrowing Owl could be considered as having breeding evidence from 27 western counties. These include, first, 21 counties identified by R. A. Grant (1965b), who did not provide a definition of "breeding" and included "all Minnesota records known" in 1964 by himself, R. Janssen, E. Strubbe, and N. Zeller; and, second, Sibley County, cited by Martell, Schladweiler, and Cuthbert 2001. The source of the latter record (1938–41) was simply identified as MNDNR files (Martell, pers. comm.); unfortunately, a recent search of MNDNR files could not locate additional details (Bob Dunlap, pers. comm). Other counties were added from published accounts: Rock and Cottonwood (Janssen 1987) and Watonwan (Svingen 2004); and from the field notebooks of N. M. Hiemenz (Kandiyohi County, August 7, 1943, "one seen near burrow" *fide* Tom McAlpin and Stearns County, May 30, 1946; "one seen fly out of burrow"; BMNH Records). There are also migration records (MOU 2021b) of vagrants from Duluth (1956, 1979, 1997) and Rochester (1965, 2006).

From the 1920s to the 1980s, anecdotal descriptions provide clues regarding changes in the species' abundance. In a letter to Roberts on August 12, 1924, Alfred Peterson of Pipestone County commented: "burrowing owls in a dozen known places—probably more numerous than usual" (A. Peterson 1924). Forty years later, R. A. Grant (1965b) found only four breeding pairs after traveling approximately 21,000 km (13,000 miles) across western Minnesota. He commented that "all who knew this species agreed that it was no longer as common as formerly." Overall, files maintained by Roberts and the MOU document that the number of breeding pairs each decade ranged from five to six in the 1930s and 1940s, to seven to eight in the 1950s, 1960s, and 1980s, and three in the 1970s. The slightly higher numbers in the 1960s and the 1980s may reflect special surveys in the 1960s (R. A. Grant 1965b) and public outreach efforts and research in the 1980s (Martell, Schladweiler, and Nibe 1990). The latter reference documented "13 successful nestings at eight sites" in Rock (four sites), Traverse (one site) Pipestone (two sites), and Yellow Medicine (one site) Counties from 1987 to 1990.

Coupled with three successful nestings in 1990 in Rock County (Martell, Schladweiler, and Nibe 1990), six incidental sightings of Burrowing Owls occurred in southwestern Minnesota during the 1990s. The number of records increased slightly from 2000 to 2015, with five incidental sightings, five unsuccessful breeding attempts **(A)**, and two successful nestings. The latter included a family of two adults and two juveniles observed August 8–18, 2007, near Edgerton in Pipestone County (Budde, Dunlap, et al. 2008). The second record was north of Glacial Ridge NWR in Polk County. At least two adults were observed at a burrow from April 17 to May 6, 2006 (Svingen 2006; Budde, Lind, and Marengo 2006). The next year, two adults and two juveniles were seen near the refuge from June 1 to July 28 (Kessen and Svingen 2008).

During the MNBBA, another pair was observed at a burrow near the refuge in 2010, but no young were seen **(B)**. Two incidental observations were reported that same year in Clay County (May 15) and Kittson County (May 18), both before the MNBBA safe dates. Following the atlas, the only documented report was in 2016, when an adult was observed at a burrow near Breckenridge, Wilkin County, from April 23 to May 23 (Budde, Kieser, et al. 2016).

BREEDING HABITAT

An inhabitant of open country, the species is found in the Great Plains in native prairies, "short-cropped" pastures, and roadside ditches that include deserted badger dens or colonies of prairie dogs or Richardson's ground squirrels (R. A. Grant 1965b; Klute et al. 2003; Roberts 1932) **(C)**. "Because of their preference for short grass, it is probable that both the burrowing owl and the Richardson's ground squirrel were associated with the coming and going of the bison in Minnesota" (Schladweiler 1985). The Florida subspecies is more eclectic in its choice of habitat and may nest on native grasslands or in a variety of developed sites, including cemeteries, golf courses, and suburban lots (Poulin, Todd, et al. 2020).

A. Unsuccessful Burrowing Owl breeding attempts in Minnesota, 2000–15.

Dates	County	Notes
May 6–June 2, 2002	Murray	Bird observed at burrow
May 22–June 2, 2004	Watonwan	Pair observed
April 30–May 14, 2005	Lac qui Parle	Bird observed at burrow
May 6–May 29, 2006	Norman	Pair observed
July 1–August 14, 2006	Stevens	Pair observed

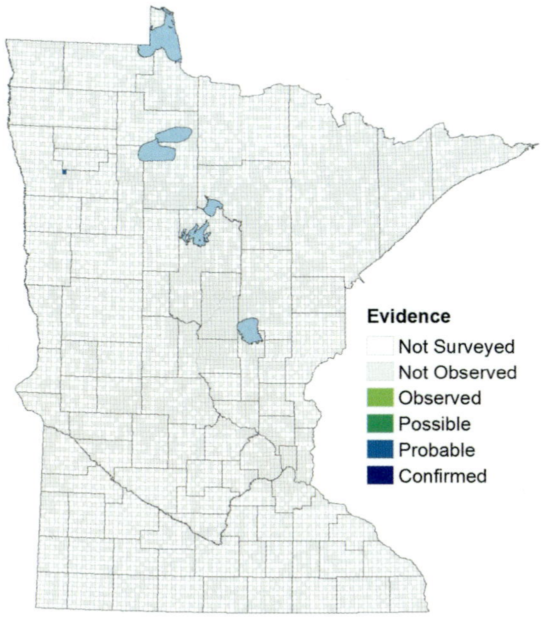

B. Breeding distribution of the Burrowing Owl in Minnesota based on the Breeding Bird Atlas (2009–13).

POPULATION ABUNDANCE

Approximately 18 million Burrowing Owls occur in the Western Hemisphere, with 990,000 birds residing in the United States and Canada (PIF 2021). Across the BBS survey region, data reveal an average annual population decrease of −0.73 percent from 1966 to 2019 (Sauer et al. 2020). Although the species' sparse distribution makes statistical precision difficult, no state or province has demonstrated a significant increase, while five, including North Dakota, have demonstrated significant declines.

Habitat loss is the primary factor responsible for the decline as large landscapes are converted from grasslands to intensive agriculture (Klute et al. 2003). This is particularly notable in the Prairie Potholes Bird Conservation Region, where BBS data suggest an especially large population decline. In North Dakota, the species has "disappeared from the eastern third of the state and is uncommon to rare in the best habitats north and east of the Missouri River" (R. K. Murphy et al. 2001).

Reductions in burrowing mammal populations, often an outcome of targeted control efforts, are also believed to contribute to the decline (Klute et al. 2003). Elsewhere, agricultural practices, such as irrigation and grazing, have actually benefited local populations (Poulin, Todd, et al. 2020).

CONSERVATION

Considered abundant in South America, the Burrowing Owl is listed as *Threatened* in Canada and Mexico (Klute et al. 2003) and *Endangered* in Minnesota (Coffin and Pfannmuller 1988). Its moderate conservation status in North America reflects its declining population and threats to its grassland habitat (PIF 2021).

An array of conservation actions has been employed in an effort to assist local populations, including the installment of nest boxes, vegetation management to maintain short cover, and captive breeding. In Minnesota, a reintroduction initiative from 1986 to 1990 released 105 young owls collected in South Dakota into four Minnesota counties (Pipestone, Rock, Traverse, and Yellow Medicine). The birds were kept in pens equipped with artificial nest burrows until they were ready to fledge. Unfortunately, once they were released, the birds were never relocated (Martell, Schladweiler, and Cuthbert 2001). Similar results followed a reintroduction effort in Manitoba (Poulin, Todd, et al. 2020). Other studies have documented low post-fledgling survival and large dispersal distances from the natal breeding site (COSEWIC 2017).

Fortunately, warmer temperatures do not pose a threat to the Burrowing Owl. NAS (2019) considered the species' breeding range stable under all three warming scenarios they examined; a small expansion is even predicted along Minnesota's border with the Dakotas. Meanwhile, biologists recommend carefully monitoring population trends for this unique member of the prairie avifauna (Poulin, Todd, et al. 2020).

C. Recently burned, this extensive grassland in Polk County is one of the few areas where the Burrowing Owl was documented nesting in Minnesota since the year 2000. Photo by Gary T. Seim.

BARRED OWL

Strix varia

Minnesota Seasonal Status: Permanent, year-round resident. The Barred Owl was uncommon during the MNBBA.
Migration: Sedentary; winter influxes of northern birds may be related to prey availability.
Food: Carnivorous, feeding primarily on small mammals, small birds, reptiles, amphibians, and invertebrates. Hunts from an elevated perch at night.
Nest: Secondary cavity nester that uses natural cavities or the hollowed-out tops of tree snags; also nests in old platform nests built by hawks and crows or squirrel nests.
Conservation Concern: Assigned a low PIF Continental Concern Score of 7/20.

Photo by Mike Lentz.

Unlike many nocturnal denizens of Minnesota's forests, Barred Owls proclaim their presence with a loud and unique vocal repertoire. "Strongly accented with a rhythmic swing and a wild, strenuous cadence" (Bent 1938), the calls are most common in the spring, during courtship, and in late summer and early fall, when young are dispersing. The most frequent vocalization is a low-sounding, extended call that translates to *who cooks for you? who cooks for you all?* Their vocal array also includes a duet performance by breeding pairs consisting of discordant cackles, hoots, and screams that suggest outbursts of "derisive laughter" (Bent 1938).

Lacking the Great Horned Owl's ear tufts, the smaller Barred Owl has a rounded head with large, dark brown eyes that suggest a rather sad and soulful expression. Its mottled brown plumage is distinguished by a band of horizontal barring beneath the facial disc and vertical barring on the chest and belly. Like most birds of prey, the female is larger than the male. The species' Latin name, *Strix varia,* actually refers to a screeching owl with a variegated plumage. The only other large owl in Minnesota with a rounded head is the Great Gray Owl. Nearly 25 cm larger in size and restricted to the northern forests, the Great Gray Owl's gray plumage is accented with yellow eyes and a white chin.

The Barred Owl is largely a species of the Eastern Deciduous Forest, stretching into western Canada and the Pacific Northwest; a small disjunct population also resides in Mexico. The species is sparsely distributed throughout its breeding range.

MINNESOTA BREEDING DISTRIBUTION

Long recognized as a common, year-round resident in Minnesota, the Barred Owl's preferred habitat was described by Roberts (1932) as "deep, dark forests"; he noted "it may occasionally be found in more open places and even in the sparse groves of timber bordering the prairie streams and lakes." Confirmed nesting records, however, were sparse and limited to four counties in southeastern Minnesota: Hennepin, Houston, Scott, and Wabasha.

Although Barred Owls were occasionally present in western Minnesota, forty years later nesting had yet to be documented in the region (Green and Janssen 1975). By 1987, Janssen identified 21 counties where nesting had been confirmed since 1970; only one record, in Brown County, occurred west of the northern and eastern forest regions. In 1998 Hertzel and Janssen deleted one of the counties identified by Janssen (Itasca) and added five new ones, including three in the Prairie Parkland: Blue Earth, Kandiyohi, and Nicollet. Among the 73 Barred Owl breeding-season observations collected by the MBS, only seven were within the Prairie Parkland.

Considered permanent residents, Barred Owls may disperse south during the winter months. Green and Janssen (1975) noted that winter influxes of birds occasionally occurred, especially in northern Minnesota, leading Janssen (1987) to suggest that the species might be "partially migratory." In his comprehensive review of North American owls, Johnsgard (1988) wrote that the northernmost Barred Owl populations "are probably semi-migratory, depending upon prey availability." Whether or not these are truly migratory movements—with the birds returning the following spring—or one-time dispersals, is unclear.

During the MNBBA, observers tallied 756 Barred Owl records in 11.6 percent (558/4,794) of the surveyed atlas blocks. Breeding was confirmed in 59 blocks (1.2 percent) and considered probable in 148 blocks (3.1 percent). The birds were reported in 66 of Minnesota's 87 counties and were confirmed breeding in 28 counties **(A)**. Of these, 15 were additions to the map published by Hertzel and Janssen (1998). The overwhelming majority of blocks with Barred Owls were reported in the Laurentian Mixed Forest and the Eastern Broadleaf Forest Provinces.

Overall, the species' distribution in Minnesota is remarkably similar to what it was nearly one hundred years ago. Elsewhere, Mazur and James (2020) note that the only major change in the twentieth century was the species' westward expansion into

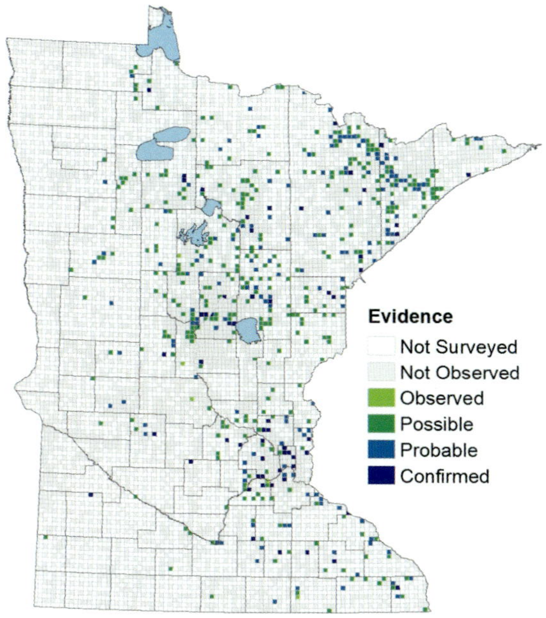

A. Breeding distribution of the Barred Owl in Minnesota based on the Breeding Bird Atlas (2009–13).

the boreal forests of Canada and the Pacific Northwest. Factors responsible for the expansion are believed to include fire suppression, increasing forest age, and the establishment of shelterbelts in Canada's Prairie Provinces. Each of these changes would provide more suitable habitat for the Barred Owl.

Although the species' distribution in Minnesota appears unchanged, its relative abundance has changed significantly. In Roberts's day, the Eastern Screech-Owl was considered the most abundant owl in the state. Today, the Screech-Owl is far outnumbered by both the Barred Owl and Great Horned Owl. The latter two species, nearly identical in abundance during the MNBBA, were nearly 12 times more common than the Screech-Owl.

BREEDING HABITAT

A mature forest species, Barred Owls can be found in both uplands and lowlands **(B)**. Older stands provide many important structural features, such as large snags and decaying trees that supply nesting cavities, and open understories that make foraging flights easier for this large predator. Dense canopies with persistent foliage (including both conifers and oaks) are ideal for daytime roosting and help provide protection from mobbing crows, winter winds, and cold temperatures. Open habitats adjacent to mature stands, such as forest clearings and wetlands, also may be used for foraging (Johnsgard 1988; Mazur and James 2020).

Several Minnesota studies shed additional light on the species' habitat preferences. Dunstan and Sample (1972) studied nesting pairs in Itasca County and found the birds preferred the interior of large forest tracts; fewer pairs were found in small woodlots or in young forest stands. Green (1995) classified the bird as an area-sensitive species. D. H. Johnson (1982) noted that, although Barred Owls are found in a variety of upland forests, nesting most frequently occurred in bottomland hardwoods.

The most in-depth Minnesota studies were conducted at the Cedar Creek Ecosystem Science Reserve in Anoka and Isanti Counties in the 1970s. Demonstrating some of the earliest applications of radio telemetry in field ornithology, the studies documented the Barred Owl's strong association with oak forests and mixed coniferous–deciduous forest uplands (Nicholls and Warner 1972; Fuller 1979).

POPULATION ABUNDANCE

PIF (2021) provided a population estimate of 3.5 million individuals in the United States and Canada and a statewide estimate of approximately 90,000 birds, or 2.6 percent of the continental population.

Unfortunately, the BBS is not a particularly effective tool for monitoring long-term population trends of owls. Their nocturnal habits, coupled with their low abundance and early nesting phenology, result in trend lines that often lack statistical precision. With these cautions in mind, the BBS data demonstrated a significant, surveywide population increase for Barred Owls, averaging 1.09 percent per year from 1966 to 2019 (Sauer et al. 2020). The BBS data from Minnesota lack even more statistical precision than those from the entire survey region but suggest the species may be increasing here as well. No state, province, or ecological region demonstrated a statistically significant decline.

Specific factors responsible for the population increase have not been closely examined but they appear to be varied. On the one hand, Barred Owls are increasing in areas where forests have matured and now provide increased habitat opportunities, such as New England. On the other hand, it is generally acknowledged that the Barred Owl has declined in regions where large tracts of mature and old-growth forest have disappeared, such as in portions of the eastern and southern United States (Johnsgard 1988). The latter may be offset by the expansion of the species' range into western Canada and the Pacific Northwest during the twentieth century (Johnsgard 1988; Mazur and James 2020).

CONSERVATION

The Barred Owl was persecuted for many years by farmers and sportsmen alike, but shooting is no longer a major source of mortality. In areas where habitat loss is a concern, management practices that increase rotation ages and emphasize the maintenance of snags and decaying trees are critical. Such practices are particularly important on public lands. Other than habitat loss, the primary concern for the Barred Owl may be the Great Horned Owl, the former being a prey item of the latter. Forest stands that support the more aggressive Great Horned Owl are avoided by Barred Owls.

Although a warming climate is considered a challenge for many species, NAS (2019) classified the Barred Owl as climate stable under all three warming scenarios they examined. Warmer temperatures, however, are predicted to result in numerous forest changes, including changes in tree species composition and forest disturbance regimes that could ultimately impact the distribution of many forest birds.

In Minnesota, the Barred Owl is largely considered a beneficial species with a secure future. However, conservationists have an entirely different view of the species in the Pacific Northwest, where, after decades of intensive conservation efforts to protect and restore the Spotted Owl's (*Strix occidentalis*) old-growth forest habitat, the larger and more aggressive Barred Owl is now considered the greatest threat to the future of this federally *Threatened* species (Welch 2009).

B. River corridors, with their mixture of upland hardwoods on the bluffs and lowland hardwoods in the floodplain, provide breeding habitat for the Barred Owl, even along the Mississippi River just south of downtown Minneapolis. Photo by Lee A. Pfannmuller.

GREAT GRAY OWL

Strix nebulosa

Minnesota Seasonal Status: Regular but local permanent resident in northern Minnesota. The Great Gray Owl was rare during the MNBBA.
Migration: Irruptive southward movements are frequently recorded.
Food: Primarily consumes small mammals, some birds, and medium-sized mammals.
Nest: Uses the tops of dead trees, old bird's nests, mistletoe clumps, and man-made platforms; sometimes nests on the ground.
Conservation Concern: Assigned a moderate PIF Continental Concern Score of 9/20.

Photo by David Brislance.

The Great Gray Owl is nicknamed the "Phantom of the Forest" because it is a secretive species that occurs in low numbers in dense coniferous forests and, despite its relatively large size, can be difficult to find (Nero 1980). Cadman and colleagues (2007) wrote, "Few birds capture the attention of even the most casual birders as dramatically as the Great Gray Owl. Whether because of its imposing size, almost human-like face, or apparently calm demeanor, this largest of North American owls is difficult not to notice and admire." Like the Boreal and Northern Hawk Owls, it is one of only three Minnesota breeding bird species that top the "100 most wanted" list for observation by birders and nature enthusiasts in the United States (Mlodinow and O'Brien 1996).

When observed, the Great Gray Owl is easy to identify. It is more than a half meter in height, with a substantial wingspan of a meter and a half. It has a large head, long tail, concentric rings around its yellow eyes, and lacks ear tufts. Eckert (2017) emphasizes that the white "bow tie" is a prominent field mark that can be seen from a distance or in poor light; in addition, it actually has a black "knot" in the middle. Its voice is seldom heard but consists of five to ten muffled *whoo* notes. Its scientific name, *nebulosa,* accentuates its mystique; in Latin it means "misty" or "cloudy," alluding to the bird's gray plumage and elusive nature.

The Great Gray Owl is a Holarctic species that occurs across northern North America and northern Eurasia. In North America it is widely distributed across the boreal forests of Canada from Quebec to Alaska and south in the Rocky Mountains to northern California and northwestern Wyoming. In the Midwest, the species is found in northern Minnesota, and breeding birds have also been found in Wisconsin and Michigan.

MINNESOTA BREEDING DISTRIBUTION

Great Gray Owls were known primarily as a winter visitor by Roberts (1932). In the 1936 edition of his treatise on Minnesota birds, he confirmed the first nesting record for Minnesota from Roseau County in April 1935, where a nest with three eggs was found by a trapper.

Forty years later, Green and Janssen (1975) described the Great Gray Owl as a "casual summer resident." They noted the 1935 nesting record and the 1970 summary of another nest in Roseau County, described in detail by Nero (1970). They also reported breeding-season observations from seven counties, including Aitkin, Clearwater, Cook, Itasca, Lake, Lake of the Woods, and St. Louis.

In 1982, D. H. Johnson summarized that there "are currently 49 nest records for the state." He included confirmed nests from nine counties: Aitkin, Beltrami, Cass, Itasca, Koochiching, Lake of the Woods, Pine, Roseau, and St. Louis. Based on the new breeding information, Janssen (1987) identified the Great Gray Owl as a summer resident and a "local permanent resident in the north." He reported confirmed nesting since 1970 in seven counties, including all of the counties listed by Johnson except Beltrami and Pine Counties. He noted that there had been sight observations of Great Gray Owls from Carlton, Cook, and Lake Counties. In 1998, Hertzel and Janssen added no new counties to the list of confirmed nesting since 1970.

In the late 1980s, the MBS began intensive coverage of counties in Minnesota. The MBS reported 15 breeding-season locations. The locations included five in Lake County; the rest were in Aitkin, southern Beltrami, Itasca, southeastern Pennington, Roseau, and St. Louis Counties.

The MNBBA reported the owl in 65 blocks from 11 counties, spread from Cook to Clearwater and Roseau Counties, and south to Aitkin County and northern Cass County **(A)**. The records included nine confirmed nesting blocks in four counties: one in Aitkin County, one in Cass County, three in

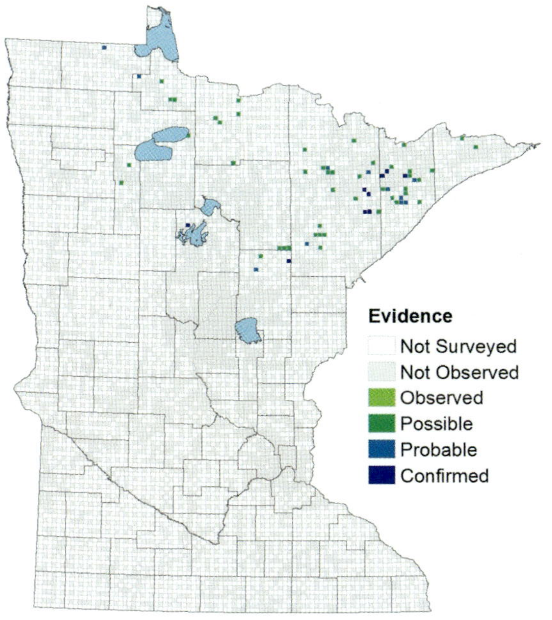

A. Breeding distribution of the Great Gray Owl in Minnesota based on the Breeding Bird Atlas (2009–13).

Lake County, and four in St. Louis County. Records in 45 of the 65 blocks were of possible nesting evidence. Records with some potential evidence of nesting were from all of the ten counties previously mentioned as having confirmed nesting, except that there was no evidence from Pine County. These data support Janssen's (1987) description of the Great Gray Owl as a local permanent resident in northern Minnesota.

In their review of the Great Gray Owl in North America, Bull and Duncan (2020) simply noted that there were no data on historical changes to its distribution. D. H. Johnson (1982) stated that "in all likelihood we have always had a regular, fluctuating, but until recently an undiscovered population of Great Grays in the state." Great Gray Owls nest in remote areas that were not well covered by bird-watchers in the 1800s or early 1900s. The increase in coverage that has occurred since the first nest was discovered in 1935 and especially in the 1970s is the best explanation for any apparent increase of the species' range southward into Minnesota.

BREEDING HABITAT

The Great Gray Owl prefers bogs, coniferous forests, and forest edges where suitable prey are found **(B)**. Great Gray Owls will readily nest in upland forests as long as they are near suitable foraging habitat, such as bogs or sedge meadows.

Suitable nest sites may often be a limiting factor for the species because it relies on finding a suitable flat surface, such as on dead snags, old nests of raptors or corvids, and man-made structures (Nero et al. 1974). Bohm (1988) placed 18 nest boxes in suitable habitat in Aitkin and St. Louis Counties during the winters of 1984–85 and 1986–87 and found that six of the 18 nest boxes were used. He pointed out that the configuration of many trees in habitats where the Great Gray Owl nests is not conducive to the construction of natural stick nests, so artificial nest structures can provide nesting opportunities.

POPULATION ABUNDANCE

PIF (2021) estimated a United States and Canadian population of 58,000 adults, with a US population of 11,000. They also estimated a Minnesota breeding population of 980 with a 95 percent confidence interval of 0–3,000. Most of the breeding population in the United States is found in the western Rocky Mountain states and Alaska. Like the estimate for the Northern Hawk Owl, the estimate seems high based on the evidence from the MNBBA.

The winter influx of Great Gray Owls and their potential cyclicity are legendary in Minnesota. Svingen and Lind (2005) provide an excellent review of the extraordinary numbers that occurred in 2004–5 and the progression of their chronological movements in the state. Using a network of randomized surveys, volunteers, and monthly surveys, they estimated that the total number of Great Gray Owls in Minnesota from fall 2004 through mid-May 2005 was about 5,225! Moreover, this was likely a conservative estimate and at least 13 times greater than previous records. They also summarized 11 previous irruptions that occurred in Minnesota from 1965 to 2002.

CONSERVATION

Conservation concerns for the Great Gray Owl are primarily due to its rarity in the region and the need to protect suitable nesting and foraging areas. Bull and Duncan (2020) emphasize that among conservation concerns is forest harvesting, which can reduce the availability of breeding habitat through removal of perch and nest trees. They suggest that forest harvesting be implemented in small, irregularly shaped patches with retention of suitable trees, which are essential for sighting prey and use as nest sites.

The use of artificial nest structures could potentially aid the breeding population of the Great Gray Owl in Minnesota and in North America. In Europe, Finland has a massive program of artificial nesting structures and many are used by this species (Sulkava and Huhtala 1997). For instance, in 1994 volunteers in Finland placed 22,691 owl nest boxes that were checked for occupancy, and much of this continues today! Unless implemented by volunteers, the magnitude of this effort is not realistic in most places. Maintaining suitable habitat through appropriate forest management is a more realistic option.

NAS (2019) projected a high vulnerability for the Great Gray Owl breeding range in North America with a 2°C warming scenario, including a 78 percent range loss and a 46 percent range gain primarily in the far north by 2050. In Minnesota, the species was predicted to be extirpated as a breeding species even under our current 1.5°C warming scenario.

Educational programs focused on the fact that the Owls are highly beneficial and primarily prey on voles and mice have reduced the illegal shooting of owls. However, this source of mortality has not been completely eliminated, and Great Grays are an easy target because they are so tame. Collisions with vehicles are also common because the Owls tend to forage in ditches near roads, especially during their winter movements to the south (Nero and Copland 1981).

Because of its remote breeding habitats, changes in the Phantom's population or distribution in Minnesota are a challenge to detect. The Great Gray Owl's future in North America is highly dependent on societal decisions regarding climate, forest management, and other land management, such as agriculture and peat or energy extraction. All of these activities can reduce the amount of suitable habitat for Great Gray Owls, especially within its primary breeding range north of Minnesota.

B. Typical breeding habitat of the Great Gray Owl consists of dense boreal forests composed of spruce, balsam fir, or tamarack trees for cover and edges or openings for hunting small mammals. Photo by Gerald J. Niemi.

LONG-EARED OWL

Asio otus

Minnesota Seasonal Status: Regular breeding resident, migrant, and likely winter resident. The Long-eared Owl was a rare species during the MNBBA.
Migration: Short- to medium-distance migrant that winters in the central and southern United States to Mexico.
Food: Consumes small mammals, especially voles, deer mice, and shrews, and occasionally birds, primarily captured below the canopy in open forests.
Nest: Stick nest in a tree made by another species, such as an American Crow, Common Raven, or another raptor, and occasionally on the ground or in a tree cavity.
Conservation Concern: Assigned a moderate PIF Continental Concern Score of 12/20.

Photo by Mike Lentz.

The Long-eared Owl is elusive to observe but, as Forbush and May (1955) suggested years ago "it is more common still than most of us realize." They elaborate: "It is so nearly nocturnal that it is seen rarely except when disturbed at its nest, or surprised in some retreat in the daytime. It seems to inhabit dense coniferous woods by preference, where it hides during daylight amid the dark foliage, though at night it often hunts in the open. . . . When in danger of being observed in such a retreat, they draw the feathers close to the body and stand erect so as to resemble the stub of a broken branch." Roberts (1932) remarked that this posture "presented a comical appearance."

Long-eared Owls resemble a smaller and thinner Great Horned Owl and, as its name suggests, it has conspicuous, long ear tufts that are closer together than in the Great Horned and not noticeable while the bird is in flight. Both species are similar, with a varied plumage: the upperparts with a mix of brown, black, gray, and white and the underparts boldly streaked and barred. It has yellow eyes and a rusty-colored facial disk. Its scientific name, *otus,* is derived from Greek, meaning "eared owl," and one of its former common names was Little Eared Owl (Sandrock and Prior 2014). Generally, the species is quiet, but in the breeding season its vocalization is a long series of 10 to more than 200 low, soft *hooo* notes; its alarm call has been described as a harsh bark, *"oooack oooack oooack"* (Alderfer 2006).

The Long-eared Owl is a widely distributed species that occurs across North America, Eurasia, and in parts of Africa, the Azores, and the Canary Islands. The species' breeding range in North America extends from the Maritime Provinces to British Columbia, south in the Rocky Mountains to Arizona and New Mexico, and the northern tier of US states from Colorado to southern Minnesota, and east to New York State.

MINNESOTA BREEDING DISTRIBUTION

Roberts (1932) described the Long-eared Owl as "common" and a generally distributed summer resident throughout the state. He cited evidence that it was a permanent resident in the southern portions of the state but only a summer resident in the northern forested areas. Confirmed nests were found from 1888 to 1922 in Hennepin, Isanti, Meeker, Polk, and Roseau Counties. Inferred nesting with only the nest observed was recorded in Goodhue and Stearns Counties.

In contrast with Roberts, Green and Janssen (1975) described a more restricted summer distribution. They wrote that Long-eared Owls were "resident throughout the wooded portions of the state" but not known to nest on the prairie or even in the wooded valleys, except in the northwestern region near Ada, Norman County, and Crookston, Polk County. D. H. Johnson (1982) summarized 51 nesting records from 1878 to 1981. He listed 22 counties with nesting records, including 15 not included by Roberts. The distribution of records ranged from northwestern Minnesota to the southeast in Goodhue County.

Janssen (1987) described the Long-eared Owl's distribution as the "heavily wooded portions of east-central, central, north-central, and portions of the northwestern regions." Areas along the Minnesota River floodplain from Nicollet County to Big Stone County were included in his distribution map. He reported there were no records of nesting from the prairies in the west-central, southwest, or south-central regions of Minnesota and its status in the southeastern portion is unknown. Collectively, Janssen and Hertzel and Janssen (1998) included eleven counties with confirmed nesting since 1970. MBS reported only one breeding-season location, but it

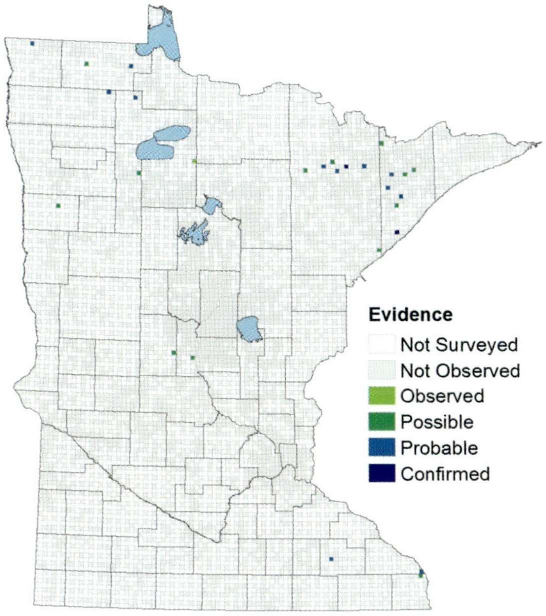

A. Breeding distribution of the Long-eared Owl in Minnesota based on the Breeding Bird Atlas (2009–13).

was from a county not previously mentioned, southeastern Douglas County.

Participants in the MNBBA reported 34 records within 28 of the surveyed blocks. The distribution of blocks was patchy, with a majority of the blocks (54 percent, 15/28) in Lake and St. Louis Counties **(A)**. Most of the rest were from northwestern Minnesota (Beltrami, Clearwater, Kittson, Marshall, Norman, and Roseau Counties). Single locations were in Morrison and Todd Counties in central Minnesota, and in Houston, Olmsted, and Winona Counties in southeastern Minnesota. Two confirmed nesting records were included from southern Lake County and north-central St. Louis County.

The Long-eared Owl, being nocturnal, cryptically plumaged, and not particularly vocal, can be difficult to observe or hear. In addition, vocalizations occur primarily from February to March and at night, when the number of observers is limited. D. H. Johnson (1982) underscored his opinion that this species is more abundant than current nest and sight records reveal because of its short spring vocalization period and its nocturnal habits. Eckert (1978b) noted that when the Long-eared Owl is foraging at dusk in open meadows, it may be misidentified as a Short-eared Owl.

In their review of the Long-eared Owl in North America, Marks and colleagues (2020) commented on the lack of data to assess historical changes in the species' distribution. The cumulative nesting records of the Long-eared Owl in Minnesota indicated it has a wide distribution within the state, ranging throughout the wooded portions, and having a scattered presence in more open landscapes, where forest patches, riparian forests, and coniferous patches such as pine plantations exist. The breeding distribution map presented by Janssen (1987) likely represents where the species was historically found and is still present today.

Unfortunately, as articulated by Neri and Mackentley (in Chartier et al. 2013) in their summary of the Long-eared Owl in Michigan, "a true understanding of the Long-eared Owl's status as a breeder is likely to remain elusive for the foreseeable future." This will probably remain true for Minnesota as well, unless targeted search efforts for breeding are completed, such as the winter effort carried out by P. Hertzel and Hertzel (2000) in southern Minnesota.

BREEDING HABITAT

Roberts (1932) described the haunts of the Long-eared Owl as "tamarack swamps, alder and willow thickets, and other low-lying wooded areas." Marks and colleagues (2020) termed its habitat as highly variable but emphasized "dense vegetation adjacent to grasslands or shrublands; also open forests." Niemi and Hanowski (1992b) summarized breeding habitats as coniferous and mixed coniferous–deciduous forests, especially near water **(B)**. They also noted red and jack pine plantations and windbreaks as important habitats in agricultural landscapes.

B. Breeding habitats for the Long-eared Owl can be highly variable but tend to be conifer or mixed deciduous–coniferous forests close to openings and often near water. Photo by Gerald J. Niemi.

POPULATION ABUNDANCE

PIF (2021) estimated a United States and Canadian population of 150,000 breeding adults in the United States and Canada and 9,900 in Minnesota. The BBS did not report on trends for this species because of insufficient data. However, Rosenberg, Kennedy, and colleagues (2016) estimated that the species has declined an average of 5.39 percent per year or a cumulative loss of 91 percent from 1970 to 2014. Declines in Minnesota are partly supported by fall banding at Hawk Ridge in Duluth. D. L. Evans and colleagues (2012) reported that a total of 2,243 Long-eared Owls were captured and banded from 1972 to 2009, with an average of 59 per year. The number captured was relatively consistent from 1974 to 2009, with the largest number (164) captured in 1978 and the lowest (9) in 1999. Prior to 1991, there were six years when more than 111 were captured, but since that time the number has been below 63.

Definitive reasons for the decline in Long-eared Owls are unknown. In his review of the Long-eared Owl, Holt (1997) stated that in most cases "there were insufficient data to convincingly conclude which factors influence population trends." He summarized the possibilities as habitat alteration, forest succession, urbanization, competition with Great Horned Owls, loss of habitat for prey species, rodenticides, shooting, and loss of riparian habitats and grasslands.

CONSERVATION

Given its large population decline, the Long-eared Owl has been designated as a Yellow Watch List species in North America by Rosenberg, Kennedy, and colleagues (2016). One area they identified as a focus for conservation was the Upper Mississippi River/Great Lakes Region. This region, especially from Iowa to western Ohio, is considered an important overwintering habitat and a target for conservation actions.

Under a 2°C warming scenario, the NAS (2019) projected a stable breeding range for the Long-eared Owl in North America with a 15 percent loss of its breeding range and a 31 percent gain northward by 2050. However, in Minnesota, breeding range retractions were predicted to occur in the south and southeastern regions. Habitat conservation efforts for Long-eared Owls have been suggested to focus on maintaining riparian woodlands and isolated tree groves. They have also been found using plantations, cemeteries, and shelterbelts, especially those with coniferous trees, which provide excellent cover for nesting and roosting. Most importantly, there is a dire need for more information on this secretive species.

SHORT-EARED OWL

Asio flammeus

Minnesota Seasonal Status: Regular breeding resident and migrant; occasionally observed during the winter months. The Short-eared Owl was rare during the MNBBA.
Migration: Northern populations are largely migratory, moving to southern states and northern Mexico in the winter.
Food: Feeds on a variety of small mammals and birds, but *Microtus* voles comprise a major portion of its diet.
Nest: A small scrape in the ground lined with grasses and downy feathers.
Conservation Concern: Assigned a moderate PIF Continental Concern Score of 12/20; officially classified as *Special Concern* in Minnesota.

Photo by Michael Furtman.

Owls are often thought of both as denizens of dense forests, and as quiet by day and active at night. The Short-eared Owl breaks that mold. An inhabitant of wet meadows and grasslands, this spectacular bird is as likely to be seen at dawn and dusk, and even in the middle of the day, as it is at night. Gliding low over the prairie's dewy grass at daybreak with slow and deliberate wingbeats, the Short-eared Owl might easily be mistaken for another grassland bird of prey, the Northern Harrier.

In flight, the Owl's shorter tail and the lack of the Harrier's characteristic white rump are sufficient to distinguish the two species. A dark wrist patch and black wing tips on the underside of the Short-eared Owl's wings are also diagnostic, while a buffy patch can be seen in the outer primaries on the upper wing. Despite its name, the head of this heavily streaked bird appears rounded because the short ear tufts are rarely visible. Dark feathers around the yellow eyes also impart a spectacled appearance. Although the Latin word *flammeus* suggests a flaming yellow or orange plumage, the color on the Owl's back and wings is a mottled tawny to dark brown. The birds are generally quiet, but during the breeding season the smaller male sings a low, soft *hoo* call that is repeated fifteen to twenty times, while both sexes include wing-claps in their courtship flights.

The Short-eared Owl's nomadic movements are a response to the abundance of local prey, especially *Microtus* voles. Birds that move south in the winter may find abundant food sources that entice them to stay for the coming breeding season. Alternatively, they may leave well-known breeding areas in the summer because food is limited and move to sites with more suitable prey populations (Wiggins et al. 2020).

Global in distribution, Short-eared Owls are found on every continent except Antarctica and Australia. In North America, their breeding range extends across Canada and the northern United States, where they occur from northern California east to northern New England. Populations reach their highest breeding densities in southern Alberta. To the south, the birds are also found in the Caribbean and South America.

MINNESOTA BREEDING DISTRIBUTION

Roberts (1932) described the Short-eared Owl as "an inhabitant of the marshes and low-lying fields throughout Minnesota." It was considerably more abundant on the western prairies, and less common but "generally distributed in the forested area wherever there [were] extensive marshes." He suspected the Owl migrated south during the winter yet still described it as a permanent resident. Nesting was confirmed in four counties: Grant, Jackson, Meeker, and Polk.

Green and Janssen (1975) described the Short-eared Owl as a summer resident "throughout most of the state," noting that it was absent from southeastern Minnesota and was most abundant in the west-central and northwestern counties. Numbers appeared to be declining owing to the loss of habitat. "If conditions are favorable," they wrote, "individuals may winter in the northwestern region."

In D. H. Johnson's (1982) review of the species status, he identified just 19 confirmed Short-eared Owl nest records since the late 1800s, which were scattered across 14 counties in western and central Minnesota; Aitkin County was the one outlier in eastern Minnesota. Summer observations were reported from an additional 16 counties, including both St. Louis (six records) and Lake (one record) Counties. The species' nomadic behavior was illustrated by an example in Lake of the Woods County. In the summer of 1980 it was relatively abundant; one year later it was "essentially nonexistent."

Janssen's updated account in 1987 identified only six counties with confirmed nesting since 1970: Beltrami, Chippewa, Hubbard, Lake of the Woods, Rock, and Roseau. Hertzel and Janssen (1998) later published an updated breeding map that eliminated the records from Chippewa, Rock, and Roseau Counties and replaced them with records from Aitkin, Lac qui Parle, and Marshall Counties.

By 2014, the MBS had identified 20 breeding-season locations for the Short-eared Owl. Apart from five records in Aitkin County, the majority were restricted to the far northwest, specifically Lake of the Woods, Marshall, Pennington, Red Lake, and Roseau Counties.

During the MNBBA, observers tallied a total of 47 Short-eared Owl records in only 41 (0.9 percent) of the surveyed atlas blocks **(A)**. Aside from a few

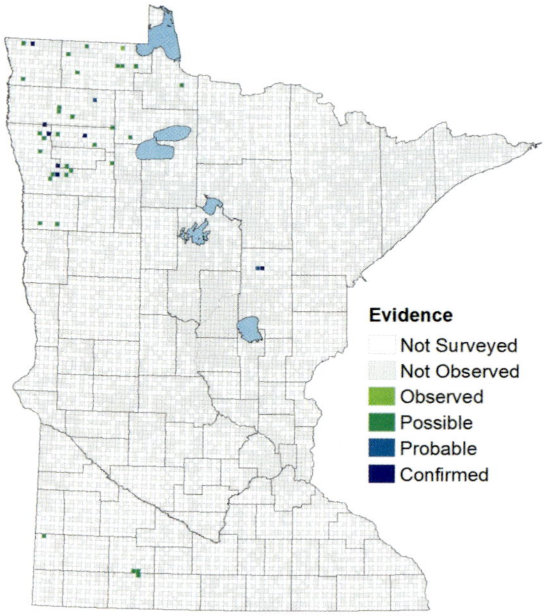

A. Breeding distribution of the Short-eared Owl in Minnesota based on the Breeding Bird Atlas (2009–13).

possible detections in southwestern Minnesota, the records were confined to Aitkin County and the northwest corner of the state, from Norman County north to the Canadian border and east to Lake of the Woods and Beltrami Counties. Breeding was confirmed in seven blocks and deemed probable in two blocks. The birds were reported from 13 of Minnesota's 87 counties and were confirmed breeding in five counties (one block straddled Marshall and Polk counties). Of the confirmed breeding records, three counties were additions to the list published by Hertzel and Janssen in 1998: Kittson, Pennington, and Polk.

The Short-eared Owl's breeding distribution in Minnesota clearly has receded over the past century. No longer is it a regular species beyond the large sedge meadows of Aitkin County, the Red Lake Peatlands, and the Aspen Parklands of northwest Minnesota. Elsewhere in North America, its breeding range has receded in New England, while its increasing presence in Florida is believed to represent colonization from Cuba (Wiggins et al. 2020).

BREEDING HABITAT

A species of the open country, the Short-eared Owl can be found in habitats ranging from tundra and coastal heathlands to sagebrush prairie, grasslands, and retired cropland (Wiggins 2004). In Minnesota, the species is primarily an inhabitant of grasslands, wet meadows, and open peatlands **(B)**; Conservation Reserve Program (CRP) grasslands are also occasionally used (Hanowski 1995). Agricultural lands, including hayfields and fallow stubble fields, may be suitable, but the birds avoid sites that are grazed, requiring some residual vegetation to help conceal the nest (Kantrud and Higgins 1992; Wiggins et al. 2020).

Large unfragmented landscapes are also important. In North Dakota, Short-eared Owls were rarely observed on sites that were smaller than 100 ha. Although they may use smaller tracts, they are often embedded in a much larger grassland landscape (Dechant, Sondreal, et al., "Short-Eared Owl," [1998] 2001b). The most important criteria governing habitat selection, however, may be prey densities: when food is abundant, smaller sites likely are more suitable.

POPULATION ABUNDANCE

PIF (2021) generated a population estimate of 600,000 breeding adults in the United States and Canada and a statewide estimate of 2,800 individuals. Two primary sources, the BBS and National Audubon's Christmas Bird Counts (CBC), suggest a long-term population decline. Although the data are sparse and somewhat unreliable for this rare species, declines have been reported by the BBS from 1966 to 2019 in Canada (averaging −4.42 percent per year) and in the Prairie Potholes Bird Conservation Region (averaging −3.36 percent per year) (Sauer et al. 2020). Overall, PIF estimated that the continental population declined approximately 65 percent from 1970 to 2014 (Rosenberg, Kennedy, et al. 2016). Unfortunately, given the species' low abundance, nomadic tendencies, and crepuscular habits, these trend data should be interpreted with caution. BBS data in Minnesota were too sparse to provide reliable trend information.

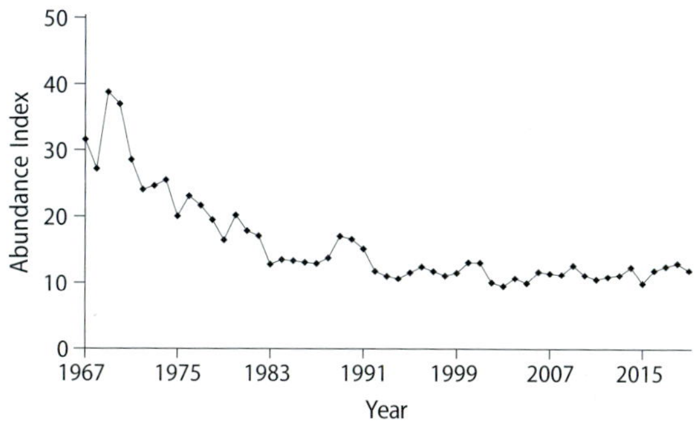

C. Surveywide population abundance indices for the Short-eared Owl based on the 1967–2019 Christmas Bird Counts (Meehan et al. 2020).

Christmas Bird Count (CBC) data provide another monitoring tool, focusing instead on the distribution and abundance of birds from mid-December through early January. CBC data collected from 1966 to 2019 also demonstrate a clear decline in numbers (Meehan et al. 2020) **(C)**. Indeed, using another metric, the number of Owls counted per party hour declined approximately 50 percent in the United States and 80 percent in Canada from 1966 to 2010 (Booms et al. 2014).

CONSERVATION

Considered a moderate conservation priority (PIF 2021), the Short-eared Owl has been designated a Common Bird in Steep Decline (Rosenberg, Kennedy, et al. 2016). Classified as a federal *Special Concern* species across Canada (COSEWIC 2008), it has been officially classified as *Special Concern* in Minnesota since 1984, owing to its declining abundance and restricted distribution (Coffin and Pfannmuller 1988).

An ecologically vulnerable species, the Short-eared Owl's dependence on large, unfragmented grasslands is the most significant threat to its status. Given its nomadic behavior, large-scale conservation efforts that focus on open grasslands, wet meadows, and fens across multiple landscapes are critical. In particular, it is essential that remaining tracts of grassland in the Aspen Parklands, the region most critical to the Short-eared Owl in Minnesota, are protected.

Although the species' future is strongly associated with the future of grasslands, the impact of warming temperatures should not be dismissed. NAS (2019) predicted that the Owl was moderately vulnerable under all three warming scenarios they examined. Breeding distribution losses ranged from 18 to 30 percent, while gains ranged from only 6 to 7 percent. While just a small loss is projected in northwestern Minnesota, the Short-eared Owl's future is clearly at stake on many fronts.

B. In northwestern Minnesota, large unfragmented grasslands and wet meadows that supply an abundance of small mammals for foraging and dense cover for nesting are critical for the Short-eared Owl. Photo by Lee A. Pfannmuller.

BOREAL OWL

Aegolius funereus

Minnesota Seasonal Status: Regular and possible permanent resident, migrant, or winter visitant. The Boreal Owl was very rare during the MNBBA.
Migration: Irruptive winter movements from northern breeding grounds, presumably due to food shortages or inclement weather.
Food: Primarily small mammals; also, birds and insects.
Nest: Secondary cavity nester that uses tree holes such as those previously excavated by woodpeckers.
Conservation Concern: Assigned a moderate PIF Continental Concern Score of 10/20.

Photo by David Brislance.

The Boreal Owl was primarily considered a wintering bird in Minnesota until the first nest was discovered in Minnesota and in the lower forty-eight states. The discovery was made in 1978 by Eckert and Savaloja (1979) along the Gunflint Trail, in extreme northeastern Minnesota. The Boreal Owl was previously known as Richardson's Owl in North America and is known as Tengmalm's Owl in Europe. The species has a Holarctic distribution across the boreal forest of northern North America, Europe, and Asia. Its distribution in North America extends from the Maritime Provinces in eastern Canada to Alaska and south in the Rocky Mountains to northern New Mexico.

The Boreal Owl is among only three species in Minnesota—along with the Great Gray and Northern Hawk Owls—on the "100 most wanted" list of birds desired for observation by birders and nature enthusiasts in North America (Mlodinow and O'Brien 1996). During periodic invasion years, this owl can attract people from all over the United States to Minnesota. Like many boreal species, it can be quite tame. Roberts (1932) noted a comment by P. O. Fryklund of Roseau in northwestern Minnesota that during an invasion year "it sought shelter inside barns and outhouses everywhere, one being caught in the basement of the Court House." During these invasions, many Boreal Owls perish from inclement weather and lack of food.

The Boreal is one of three small forest owls that nest in Minnesota, along with the Northern Saw-whet and Eastern Screech Owl. It is distinguished from the Screech Owl by its lack of ear tufts and from the Saw-whet by its whitish facial disk with a black border and pale bill. The species is best observed during invasion years in winter; or it can be heard, though rarely, in the solitude of the northern forests during March and April. Its primary "staccato" song is a trill lasting one to several seconds and frequently repeated after a short silence. The song is reminiscent of the winnowing of Wilson's Snipe. Eckert (2017) also cautions about the similarities of the territorial songs of the Boreal Owl and Northern Hawk Owl, both of which are rare nesting species in the northern boreal forests of the state.

MINNESOTA BREEDING DISTRIBUTION

Hatch (1892) in his "First Report of the State Zoologist" for Minnesota stated, "he would not be greatly surprised if the nest is found here ultimately." Roberts (1932), in his treatise on the birds of Minnesota, identified it only as a "rare winter visitant" and wrote that it "nests chiefly in the far north." The single potential record of breeding activity he noted was of an individual singing near Rose Lake in Cook County near the Canadian border (Schorger 1926). This record was later disputed by Green and Janssen (1975) because the details did "not preclude the possibility of confusion with the Saw-whet Owl."

Green and Janssen stated there were no summer records as of 1975 and noted only the potential misidentification of the species in 1926. The Boreal Owl was therefore unknown as a nesting species in Minnesota, and even in the lower forty-eight states, until the Eckert and Savaloja (1979) observation in May 1978. The nest hole was 0.3 m from the top of a 3.4 m black spruce stump and only 3 m from a parking area adjacent to the road. They also noted the nest was found after one of the largest winter invasions ever in the state, in 1977–78, when 66 individuals were recorded.

In 1987, Janssen reported three confirmed nests from Cook County, including the 1978 nesting record, another one reported by Eckert (1979), and one found by Matthiae (1982) in a waterfowl nest box in Cook County. By 1998, Hertzel and Janssen (1998) added confirmed nesting of the Boreal Owl in Lake, Roseau, and St. Louis Counties since 1970. The MBS detected one Boreal Owl breeding-season location in central Lake County.

The MNBBA reported seven records of the Boreal Owl in four blocks, all in central Lake County, where

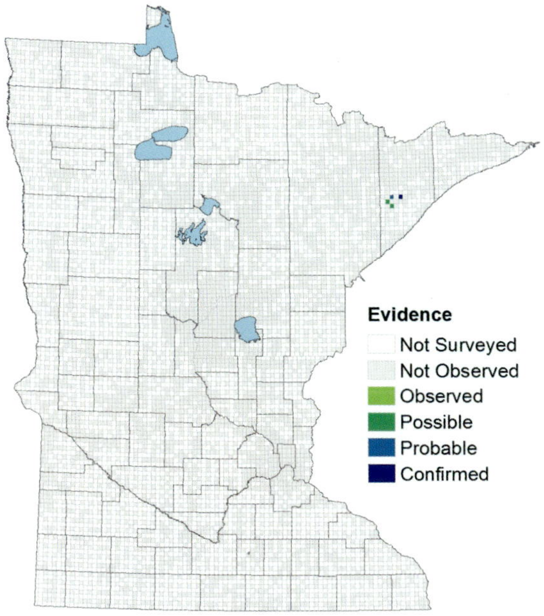

A. Breeding distribution of the Boreal Owl in Minnesota based on the Breeding Bird Atlas (2009–13).

two MNBBA regional coordinators of that area, Dave Grosshuesch and Jim Lind, diligently looked for evidence of Boreal Owl nesting. They included one confirmed, one probable, and two possible records of breeding evidence **(A)**.

In their review of the Boreal Owl in North America, Hayward and Hayward (2020) doubted that the new breeding records reported in the past forty years reflected a range expansion in North America. A range expansion is also unlikely in Minnesota; new records probably reflect increased coverage in the state by dedicated bird and nature enthusiasts. The species is small, secretive, and occurs in remote forests. Like all owls, it vocalizes at night in remote places in the north woods, where few people know its territorial song and fewer yet listen for owls on cold nights in March and April. D. H. Johnson (1982), in his review of raptors in Minnesota, also felt that further observations may reveal the Boreal Owl to be a regular nester in the northern regions of Minnesota.

BREEDING HABITAT

The Boreal Owl is a secondary cavity-nesting species that uses holes excavated primarily by woodpeckers, most notably Northern Flickers and Pileated Woodpeckers (Johnsgard 1988). Breeding habitat is broadly described as boreal forest characterized by black and white spruce, aspen, birch, and balsam fir (Hayward and Hayward 2020).

Lane and colleagues (2001) found that territorial Boreal Owls in northeastern Minnesota used poletimber- and sawtimber-sized trees in upland mixed forest stands; the latter was used more than expected given availability, but open brush and regenerating stands were used less than expected. Belmonte (2005) characterized the habitat near 31 Boreal Owl cavity trees in northeast Minnesota and found that upland mixed forests were more common within a 100 m radius of cavity trees, upland conifer stands more common at a 500 m radius, and lowland conifer stands more common at a 1,000 m radius compared with random sites **(B)**.

POPULATION ABUNDANCE

PIF (2021) estimated a United States and Canadian population of 500,000 adults and a global population of 1.54 million individuals, which includes its Holarctic distribution across the Northern Hemisphere. No estimates were available for Minnesota. Environment Canada (2011) reported a large increase in the population since the 1970s based on Christmas Bird Counts but assigned a low reliability to the estimate. They estimated the Canadian population was between 50,000 to 500,000 adults.

CONSERVATION

Conservation concerns for the Boreal Owl primarily emphasize its low overall population size, threats to its breeding and nonbreeding habitats, and the lack of information on its population trend. It is designated as a *Species in Greatest Conservation Need* by the MNDNR (2015b) because of its rarity, declining habitat, its need for a large home range with multiple habitats, and its restricted distribution in the state.

Hayward and Hayward (2020) suggest that disturbance, shooting, trapping, and pesticides do not have major impacts. They add that changes to forest conditions, such as those from logging, are likely to have the greatest effects on the species. To alleviate these impacts, they recommend maintaining large-diameter snags in clear-cuts and employing logging practices that minimize impacts on the Boreal Owl's primary prey populations, such as red-backed voles. Forestry practices that leave slash distributed across the logged site can help maintain small mammal populations and aid in their recolonization.

NAS (2019) projected a moderate vulnerability in the Boreal Owl breeding range in North America with a 2°C warming scenario, which included a 57 percent range loss and a 39 percent range gain, primarily in the far north of Canada and Alaska, by 2050. In Minnesota the species was predicted to be extirpated from the state under our current 1.5°C warming scenario by 2050.

Nest boxes have been effectively used to enhance nesting activity by the Boreal Owl in Fennoscandia (Korpimäki and Hakkarainen 2012) and in the Rocky Mountains (Hayward and Hayward 2020). Management costs for construction, placement, and routine monitoring can prohibit their establishment and long-term effectiveness. Even though the costs are high, Korpimäki and Hakkarainen (2012) reported that an average of 11,598 nest boxes for small owls were maintained in Finland from 1977 to 2009!

The future of the Boreal Owl in Minnesota is unclear, and it is still questionable whether it nests annually in the state. Furthermore, while Boreal Owls are clearly an irruptive species in certain years when many are observed, are they also a regular migrant in such small numbers and so elusive as to be virtually undetected? Unfortunately, its population is so sparse and difficult to detect that it may take years before anyone realizes it has disappeared from the state.

B. Boreal Owls prefer medium- to large-boled trees in mixed forest stands where they can find suitable nesting holes; conifers are especially used for song perches and cover. Photo by Gerald J. Niemi.

NORTHERN SAW-WHET OWL

Aegolius acadicus

Minnesota Seasonal Status: Regular breeding resident, migrant, and winter visitant. The Northern Saw-whet Owl was uncommon during the MNBBA.
Migration: Short-distance migrant; populations in northern regions move south in winter, while some individuals remain on their breeding grounds throughout the year. Irruptive fall movements are known and are thought to occur in four-year intervals.
Food: Primarily small mammals, especially deer mice and white-footed mice (*Peromyscus* spp.), voles, shrews, and occasionally bats; some evidence of small birds during migration and of insects. Hunts at night from a low perch either along forest edges or in forest openings.
Nest: Secondary cavity nester that uses old woodpecker holes, especially those excavated by the Northern Flicker and Pileated Woodpecker but will also use nest boxes.
Conservation Concern: Assigned a low PIF Continental Concern Score of 8/20.

The diminutive Northern Saw-whet Owl, described by Roberts (1932) as "not much larger than a plump House Sparrow," is the smallest owl that occurs in Minnesota. He emphasized that the combination of its "size and quiet, secretive, nocturnal habits cause it to be rather rarely seen and so regarded as less common than is really the case." The Saw-whet is best found when calling in the spring with its monotonous series of long, low whistles, variously described as *too-too-too-too* or *tink tank tenk tink,* with about two notes per second. This vocalization gives the species its unusual name of Saw-whet because it is reminiscent of a knife being sharpened on a whetstone or of water dripping.

Its scientific genus name *Aegolius* simply translates from Greek to "small owl," while its species name, *acadicus,* refers to where its type specimen was found in the Acadia region of eastern Canada. The plumage of the Saw-whet is reddish-brown above with reddish streaks on the breast, a reddish facial disk, and a lack of ear tufts. It is most similar to the Boreal Owl, but Saw-whets have a black bill (yellow in the Boreal) and lack the dark facial border found on the Boreal. Both species, however, have yellow eyes.

Photo by Laura Erickson.

The Northern Saw-whet Owl is widely distributed across the forested areas of the midwestern and northern United States, throughout the forested regions of Canada and southern Alaska, and south through the Cascades and Rockies to California, Arizona, and New Mexico. A resident population is also found in Mexico. The Northern Saw-whet Owl is sparsely distributed throughout its range; its exact distribution limits are not well known and vary annually.

MINNESOTA BREEDING DISTRIBUTION

Roberts (1932) stated that his "files contain enough records to show that it is fairly well represented throughout Minnesota at all times of the year," though "there is a partial withdrawal from the northern part of its range in the winter." Despite the species' broad distribution, Roberts included only three confirmed nesting observations in Hennepin, Kittson, and Sherburne Counties. One of the two nests he described in Hennepin County was "in the same stub as nests of the Arctic [Black-backed] and Pileated Woodpeckers."

Green and Janssen (1975) described the Saw-whet Owl as a "resident throughout the wooded portion of the state as far south as the Twin Cities area." They emphasized it was likely a resident in the southeastern portion of Minnesota, but there were no twentieth-century records from that region. In addition, "no old or modern summer records exist for the west central, southwestern, or south central regions." Besides the three counties mentioned by Roberts, the authors cited nesting records for Wadena County in west-central Minnesota.

D. H. Johnson (1982) summarized that 22 breeding records were known from 1881 to 1981 for 11 counties in Minnesota. Besides the four counties already mentioned, the additional seven included Becker, Beltrami, Cook, Crow Wing, Lake, Norman, and St. Louis. The distribution of these nesting records is consistent with Green and Janssen (1975) and ranges north from the Twin Cities to the extreme northwest and northeast counties in the state.

Following Johnson's summary, Janssen (1987) described the Northern Saw-whet Owl's distribution as "the more heavily wooded portions of the state from the southeast (very rare) to the northwest regions." He also reinforced that there were no nesting records or breeding-season observations from the south-central, southwest, or west-central prairie regions. Janssen included confirmed nesting in 15 counties since 1970, with eight new counties not previously confirmed: Aitkin, Cass, Itasca, Lake of the Woods, Marshall, Polk, Ramsey, and Red Lake. In a 1998 update, Hertzel and Janssen included confirmed nesting in 15 Minnesota counties since 1970. Their map did not include Aitkin, Itasca, or Red Lake, but

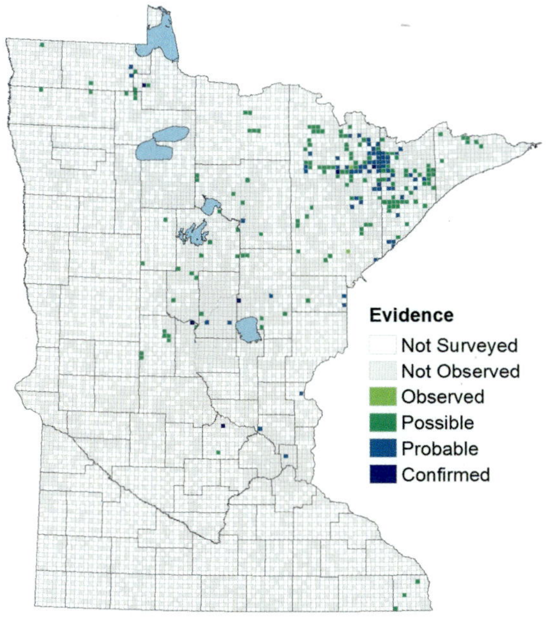

A. Breeding distribution of the Northern Saw-whet Owl in Minnesota based on the Breeding Bird Atlas (2009–13).

it added Hennepin, Hubbard, and Koochiching Counties as new county records.

Since the MBS began in the late 1980s, they have recorded 15 breeding-season locations for the Northern Saw-whet Owl. New potential breeding-season locations included one in southern Houston County in extreme southeastern Minnesota, one near the border between Pope and Stearns Counties, and one in central Pine County. The remaining locations were in Aitkin (one record), Becker (one), the border of Becker and Hubbard (one), Cass (two), Cook (three), Crow Wing (one), and northern St. Louis (three) Counties.

Participants in the MNBBA reported 329 records of the Northern Saw-whet Owl, including six confirmed blocks in Aitkin, Lake of the Woods, Morrison, St. Louis, and Wright Counties **(A)**. In addition, 69 blocks had probable nesting and 150 blocks had possible nesting. Breeding evidence was generally distributed in the wooded portions of the state, from three blocks in Houston County in the extreme southeast to one block in Kittson County in the extreme northwest.

In their review of the Northern Saw-whet Owl in North America, Rasmussen and colleagues (2020) simply stated that no data were available on historical changes to the distribution of this species. The combination of all these data suggest that the primary distribution of the Northern Saw-whet Owl in Minnesota is confined to the forested areas of the state. As previously noted, Roberts (1932) stated the species was breeding throughout the state. Since the Big Woods have largely been converted to agricultural and residential areas, it is logical that the species may be absent or very rare in this region today, but it was likely present prior to the many habitat changes that have occurred.

BREEDING HABITAT

The Northern Saw-whet Owl occurs in a wide variety of forested habitats with suitable nesting cavities (Rasmussen et al. 2020). Rasmussen and colleagues especially emphasized that the species breeds in mixed forest habitats of dense conifers (for roosting) and deciduous trees (for nesting and foraging). Using radio telemetry at the Cedar Creek Ecosystem Science Reserve, Forbes and Warner (1974) recorded the species in a combination of cedar swamps, where it was primarily found, as well as in oak upland and maple–basswood forests. The species typical breeding habitat in northern Minnesota is shown in the figure **(B)**.

POPULATION ABUNDANCE

PIF (2021) estimated the species' global population at 2 million individuals. No population estimates were available for Minnesota, and trend data from the BBS were lacking. Nocturnal species are not monitored effectively with diurnal roadside counts.

Environment Canada (2014b) provided a broader population estimate of the Northern Saw-whet Owl in Canada at 500,000 to 5 million adults. They suggested that the provincial owl surveys, combined with continental data from the Christmas Bird Counts, indicate a possible increase in the population in Canada; however, they emphasized that the reliability of this estimate was considered low. Rasmussen and colleagues (2020) also noted that population trend data are lacking for species like the Saw-whet because of their secretive behaviors.

The Northern Saw-whet Owl is one of the most frequently trapped raptors at Hawk Ridge in Duluth, which is among the most productive banding stations for the species in North America. D. L. Evans and colleagues (2012) reported a total of 20,676 Northern Saw-whet Owls, or an average of 544 captured and banded per year, from 1972 to 2009. This represented about 11 percent of the Northern Saw-whet Owls recorded at the US Geological Survey Bird Banding Laboratory.

CONSERVATION

Among the concerns regarding the Saw-whet is its estimated small global population size; however, it is considered to be secure because its range is widespread and large. Because of its small size and secretive habits, the Northern Saw-whet Owl has not been reported as susceptible to shooting or trapping.

With a 2°C warming scenario, NAS (2019) projected a moderate vulnerability for the Saw-whet's breeding distribution in North America, including a 46 percent range loss and a 37 percent range gain primarily in the far north by 2050. In Minnesota, the species was predicted to be extirpated from the state under a 3°C warming scenario by 2080. There are considerable uncertainties as to how the species would adapt to new winter or breeding habitats.

Besides climate change, concerns include the availability of suitable nesting cavities in mature and old-growth forests (Rasmussen et al. 2020). Young regenerating forests tend to be lacking in suitable foraging areas and available nesting trees.

The future of the Saw-whet appears precarious in Minnesota, but additional details on its ecology are highly desirable.

B. Northern Saw-whet Owls can often be found during the breeding season in mixed deciduous–coniferous forests with large trees and snags for nesting sites.
Photo by Gerald J. Niemi.

BELTED KINGFISHER

Megaceryle alcyon

Minnesota Seasonal Status: Regular breeding resident and migrant; occasionally observed during the winter along open streams and rivers in southern and central Minnesota. The Belted Kingfisher was common during the MNBBA.
Migration: Considered a partial migrant; birds can survive cold temperatures if open water is available. Most interior populations in Canada and Alaska migrate and winter from the southern United States to northern South America.
Food: Diver that feeds primarily on small fish and occasionally on other vertebrates and invertebrates.
Nest: Excavates a burrow in an earthen bank.
Conservation Concern: Assigned a moderate PIF Continental Concern Score of 10/20.

Photo by Mike Lentz.

The only member of the kingfisher family that is restricted to North America during the breeding season, the Belted Kingfisher is not easily mistaken for any other bird. Bent (1940) aptly described its unique silhouette: "Its long, heavy bill and its large head with its prominent crest, contrasting with its diminutive feet and its short tail, seem entirely out of balance and give it a top-heavy appearance. But its peculiar proportions and structure are beautifully adapted for the life it leads." The large head and beak aid the steep dives to capture aquatic prey seen below the water surface, while the small feet and tail enable it to efficiently excavate its nesting burrow.

The Belted Kingfisher's plumage is also unmistakable. Erect, blue-gray head feathers make the bird look as though it is having a bad-hair day but are, in fact, an ornamental, double-pointed crest. In addition to their blue-gray back and wings, both the male and female have a white neck collar bordered by another band of blue-gray across the upper breast. Unlike most birds, it is the female that is more ornately adorned with a second rusty-colored breast band that extends onto the sides of the belly.

Frequently seen perched on a dead branch extending over a small stream or lake, the Kingfisher gives a rattle call that may be the surest way to know that a bird is nearby. Prompted by the slightest disturbance, Bent (1940) likened the call to "the sound made by an old-fashioned policeman's or watchman's rattle," noting that "there is not much beauty in the voice of the kingfisher, but the loud rattling call always produces a thrill in the listener."

Found south of the tree line across Canada and Alaska and throughout all but the very southern reaches of the United States, the Belted Kingfisher is sparsely distributed throughout its breeding range.

MINNESOTA BREEDING DISTRIBUTION

The Belted Kingfisher is a common member of Minnesota's avifauna; Roberts (1932) remarked that its rattle call "is a common and familiar sound about the lakes and streams of Minnesota." Even in the early 1900s, "a few individuals are tempted to linger into December or even to pass the winter where there is open water, usually along rapid streams in the southern part of the state where fish can be had." As widespread as the species was, confirmed nesting reports were limited to three counties: Hennepin, Meeker, and Polk. Nesting was inferred in Goodhue County and the Mille Lacs region.

Little had changed when Green and Janssen (1975) wrote their updated account of the species forty years later. Janssen (1987) pointed out that the species was less common in the northwestern region and delineated 13 counties, widely spread across the state, where nesting had been confirmed since 1970. Hertzel and Janssen (1998) would later identify 18 counties with confirmed nesting since 1970 (they added six counties to Janssen's original list and deleted one: Clearwater). By 2014, MBS field staff reported 290 breeding-season observations for the Belted Kingfisher. Once again, these records were widely dispersed across the state, although they were more sparsely distributed in the northwestern counties.

MNBBA participants further affirmed the species' broad, statewide distribution. A total of 1,462 Kingfisher records were reported in 23.6 percent (1,122/4,761) of the surveyed atlas blocks. Breeding was confirmed in 110 (2.3 percent) of the blocks and was deemed probable in an additional 214 blocks (4.5 percent) **(A)**. The species was reported in all but one of Minnesota's 87 counties (Red Lake) and was confirmed nesting in 42 counties; 28 of the counties were additions to the list published by Hertzel and Janssen (1998). Common throughout the Laurentian Mixed Forest and Eastern Broadleaf Forest Provinces, Kingfishers were least common in the Red River valley and Aspen Parklands.

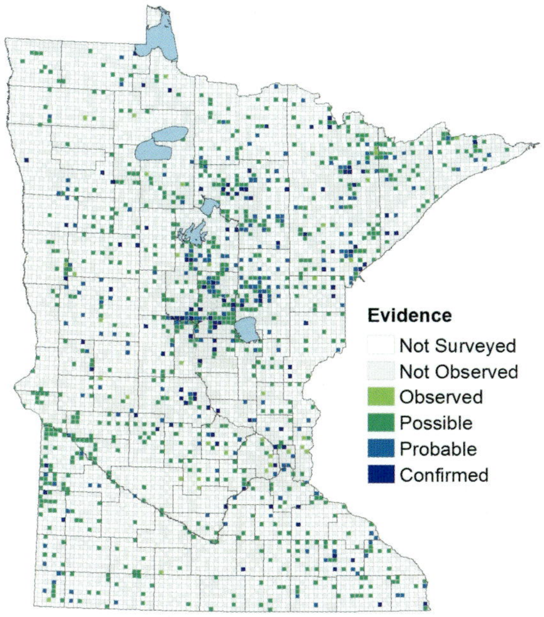

A. Breeding distribution of the Belted Kingfisher in Minnesota based on the Breeding Bird Atlas (2009–13).

Overall, it appears that the breeding distribution and abundance of the Belted Kingfisher has changed little since Roberts wrote the first comprehensive account of the species in 1932. However, warmer winter temperatures have increased the number of birds that linger here each winter. The same scenario is evident in many other states and provinces in the Upper Midwest. While numbers have declined in many regions, the species' widespread breeding distribution remains unchanged (e.g., Cadman et al. 2007 and Cutright et al. 2006).

BREEDING HABITAT

The most critical components of the Belted Kingfisher's habitat are bodies of water that support fish populations and earthen banks for nesting **(B)**. Ideally the two are located together, such as a cut stream bank, but frequently they are separated by some distance. Field studies have demonstrated that nest sites can be 2–3 km from the primary foraging area (Cornwell 1963; W. P. Mueller 2013a).

Although the body of water may be a lake, stream, river, or estuary, water clarity is important to enable this diving predator to sight prey as it flies or hovers overhead. Extensive beds of emergent and submerged vegetation seem to be a deterrent. One study suggested that stream riffles, which are an important source of prey, may be an important habitat feature (W. J. Davis 1982).

Most nesting burrows are excavated in cut banks along the shoreline or stream edge. If suitable sites are not available, then man-made banks are utilized, including ditches, road cuts, gravel pits, and dredge piles. Nesting banks that attract colonies of Bank Swallows are also used by Kingfishers. Occasionally old tree stumps or cavities in dead trees are used (J. F. Kelly et al. 2020).

POPULATION ABUNDANCE

PIF (2021) generated a population estimate of 1.8 million birds in the United States and Canada and a statewide estimate of 33,000 birds. Although its range of distribution is expansive, the Belted Kingfisher is a solitary species and is not present in high densities anywhere within its breeding range.

BBS roadside surveys are not well suited to monitoring riparian species but they provide the only long-term data for detecting population trends. The Kingfisher's low abundance also results in population trend lines that lack precision at all levels of analysis. This caveat should be kept in mind when considering that the BBS data document a surveywide decline averaging −0.88 percent per year from 1966 to 2019. In Minnesota, the rate of decline is a little higher, averaging −1.15 percent per year since 1967 (Sauer et al. 2020).

As of 2019, there is no state, province, or region in the United States or Canada where the Belted Kingfisher's population has demonstrated a statistically significant increase. Overall, populations are estimated to have declined 49 percent just from 1970 to 2014 (Rosenberg, Kennedy, et al. 2016). Factors responsible for the decline have not been examined closely. Return rates of banded fledglings are low, suggesting that mortality may be high for first-year birds; this estimate included a zero return rate for 46 fledglings banded in Minnesota (J. F. Kelly et al. 2020). Habitat loss and water-quality degradation are also important factors for this aquatic-dependent species.

B. Belted Kingfisher nest hole along the shoreline of a lake in the Boundary Waters Canoe Area Wilderness of northern Minnesota. Exposed banks near clear water provide essential breeding habitat for this Kingfisher. Photo by Jarad B. Niemi.

CONSERVATION

Despite its steady population decline and its designation as a moderate conservation priority (PIF 2021), little management attention has been directed at the Kingfisher. Conservation measures that will likely benefit the species are those that focus on improving water quality, including efforts to restore degraded and channelized streams, and management practices that ensure protection of riparian zones.

In addition to the need for clean water, the availability of suitable nesting sites may be a limiting factor. Wisconsin biologists commented on the importance of quarry and road cuts for nesting, particularly in northern counties, where Kingfishers frequently forage around lakes that lack eroded banks suitable for nesting (Cutright et al. 2006). Unfortunately, these man-made banks erode as they age and become less suitable for Kingfishers. Because similar challenges are present in Minnesota, protecting suitable nest sites can be an effective conservation tool.

Finally, as a fish-eating bird, the accumulation of environmental toxins in Kingfishers needs to be considered. One study that examined eggs collected before and after the use of DDT showed that the quality of the eggs decreased during the years of DDT use but concluded the species' reproductive success was not impaired (Fox 1974). Jeffrey Kelly and colleagues (2020) noted that the species "appears less susceptible to environmental contaminants than other fish-eating birds," possibly owing to its consumption of smaller fish that accumulate fewer toxins.

Although its declining population deserves close monitoring, in the short-term the future of the Belted Kingfisher seems secure in Minnesota. Even though the state is likely to witness the species' breeding range slowly recede eastward as temperatures rise (NAS 2019), climate change is not considered a major threat. The Kingfisher's biggest challenge may be its dependence on quality aquatic environments. Nearly 40 percent of Minnesota's rivers and lakes do not meet water quality standards (MPCA 2018).

RED-HEADED WOODPECKER

Melanerpes erythrocephalus

Minnesota Seasonal Status: Regular breeding resident and migrant; occasionally observed during the winter in southern and central Minnesota. The Red-headed Woodpecker was uncommon during the MNBBA.
Migration: Irregular or short-distance migrant; movements influenced by mast crop abundance.
Food: Omnivorous diet of seeds, arthropods, and fruits acquired by fly catching, ground foraging, bark gleaning, and bark drilling. Commonly caches food in bark crevices. Mast crops are important in the winter.
Nest: Excavates cavities in a dead tree or dead limb of a tree; may reuse its own cavities or those excavated by other woodpecker species.
Conservation Concern: Assigned a moderate PIF Continental Concern Score of 13/20.

The Red-headed Woodpecker seldom eludes observation. Its stunning red head, white belly, and black tail and wings, both with prominent white patches, easily identify the bird even with a quick glance. *Melanerpes erythrocephalus,* the species' scientific name, means a black creeper or woodpecker with a red head. The open country the species inhabits makes it even more conspicuous. A drive along western Minnesota's rural roads often reveals Red-headed Woodpeckers perched on fence posts and telephone poles, foraging on tasty morsels found on the road, or giving their hoarse-sounding *tchur* call. Unlike all other Minnesota woodpeckers, the sexes are identical.

Such a prominent bird is bound to be the subject of legends and tales. Important to many Native American tribes, it was a war symbol for the Cherokee, who carried its head into battle. And, in Longfellow's epic poem, *The Song of Hiawatha,* Hiawatha rewards the bird for helping defeat the evil magician, Pearl-Feather, by giving it a head of red feathers (Frei et al. 2020).

Well-known for its population fluctuations in response to locally abundant mast crops, the Red-headed Woodpecker is largely restricted to the eastern United States, from the Great Plains east to the Atlantic coast. The core of its range occurs in the Central Plains states of Iowa, Kansas, Nebraska, and South Dakota.

Photo by Mike Lentz.

MINNESOTA BREEDING DISTRIBUTION

Roberts (1932) described the Red-headed Woodpecker as a statewide summer resident, yet remarked that it was "most in evidence in the longer settled and cultivated portions of the state, extending its range into more heavily timbered and wilder regions only as they are cleared and inhabited." In the Red River valley, Roberts referenced Dr. Elliott Coues (1878), who reported that Red-headed Woodpeckers were common along the Red River at Pembina in June 1873. Unfortunately, when Roberts prepared his account of the species nearly fifty years later, "very few" had been seen in the region and, despite the Woodpecker's habit of alighting on roads, "it was absent from the numerous birds killed on the highways." Indeed, the few confirmed and inferred nesting records were restricted to the open landscapes of central and southern Minnesota in Goodhue, Hennepin, Isanti, Ramsey, and Rock Counties; only one nest was found farther north in Polk County. Perhaps the most significant change Roberts witnessed in his lifetime was an increasing number of Red-headed Woodpeckers overwintering in the state. Most reports were from southeastern Minnesota in localities where winter food was abundant, including at "corn-cribs, corn-shocks, chicken-yards" and feeding stations (Roberts 1932).

When Green and Janssen (1975) prepared their updated account, the species' summer distribution had changed very little, but winter observations were now reported regularly as far north as Mille Lacs and occasionally north to Wadena and southern St. Louis Counties. Janssen (1987) graphically depicted the species' breeding range as covering all of Minnesota except for most of the Arrowhead region and portions of Itasca, Koochiching, and Lake of the Woods Counties. He identified 34 counties where nesting had been confirmed since 1970. Six more counties were later added by Hertzel and Janssen (1998). By 2014, the MBS had reported 320 breeding-season locations. Concentrated in southwestern Minnesota, survey records were also scattered across southeastern and central Minnesota, with a few isolated records extending northwest into the Tallgrass Aspen Parklands Province.

A similar distribution was documented by MNBBA participants **(A)**. A total of 637 Red-headed Woodpecker detections were reported from 9.7 percent of the surveyed atlas blocks (463/4,776). Breeding was confirmed in 2.2 percent (104) of the blocks and considered probable in an additional 1.8 percent (85). Birds were reported in 82 of Minnesota's 87 counties and were confirmed breeding in 48 counties. Twenty-four of the counties were additions to the list published by Hertzel and Janssen (1998); most significant was the addition of four northwestern counties (Lake of the Woods, Marshall, Red Lake, and Roseau). Reports were common in southern Minnesota; records from farther north were concentrated in the Brainerd Lakes region and scattered through the northwestern parklands. Largely absent from the northeast and far north-central regions, the most notable record was in

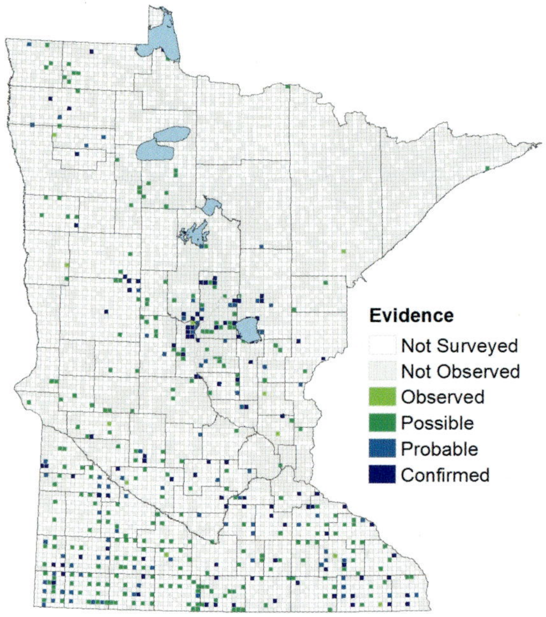

A. Breeding distribution of the Red-headed Woodpecker in Minnesota based on the Breeding Bird Atlas (2009–13).

B. In western Minnesota, old windbreaks with dead and dying trees are attractive to the Red-headed Woodpecker. Photo by Lee A. Pfannmuller.

Cook County. A single bird was observed in May 2010 on a farm near Grand Marais.

Overall, the breeding distribution of Minnesota's Red-headed Woodpecker population has experienced few changes in the past century. It may be more common in the Brainerd Lakes region, where impressive efforts by local atlas observers contributed to the large number of reports. Jack pine–oak barrens on the region's sandy outwash plains provide suitable habitat for the species, but extensive residential growth has also created a more open, fragmented landscape preferred by the Woodpeckers.

The species' distribution has remained largely intact across the eastern United States as well, but local changes in abundance are common. Periodic fluctuations in mast crops can change local habitat conditions over a short period of time, as can insect pests or fungal diseases that ravage trees. Dutch elm disease, for example, proved a boon to Iowa's Red-headed Woodpeckers in the mid-1900s, until the dead trees themselves became a rare resource (Frei et al. 2020).

BREEDING HABITAT

The Red-headed Woodpecker inhabits woodlands with a relatively open canopy, a sufficient number of dead trees or dead limbs for nesting, and an abundant supply of mast. Sites are often xeric with an open understory that facilitates ground foraging, as grasshoppers, beetles, seeds, and nuts are important components of the Woodpeckers' diet (Meyers 2010). In the Upper Midwest, oak savannas, prairie–forest edges, and lowland hardwoods are all typical habitats (Potter et al. 2007). In agricultural landscapes, shelterbelts and windbreaks provide many of the necessary amenities, as do golf courses, cemeteries, and other modified habitats **(B)**. The savanna-like conditions of recent forest burns are also suitable (Niemi 1978).

POPULATION ABUNDANCE

PIF (2021) estimated a population of 1.8 million individuals in the United States and Canada, with approximately 28,000 residing in Minnesota. Surveywide, BBS data demonstrated a significant population decline averaging −1.42 percent per year from 1966 to 2019 **(C)**. While the rate of the decline has lessened in recent years, the cumulative decline from 1970 to 2014 was estimated at 67 percent (Rosenberg, Kennedy, et al. 2016). The Woodpecker's status in Minnesota is more precarious. Although the BBS data are sparse and should be interpreted with caution, they indicate an average annual decline of −5.23 percent, resulting in a cumulative loss of 94 percent of Minnesota's Red-headed Woodpecker population from 1967 to 2019 (Sauer et al. 2020)!

In the early 1900s, declining numbers corresponded with the disappearance of a favorite food item, the Rocky Mountain Grasshopper; the loss of chestnut trees due to chestnut blight; and a decline in the abundance of mature beech forests (Frei et al. 2020). The Woodpecker's known susceptibility to car collisions (Howitz 2003) and nest site competition with a growing population of Starlings were and continue to be significant challenges. In the Midwest, however, habitat loss may be the largest threat. Oak savannas were once abundant in some of the most densely populated regions of Minnesota. In the Anoka Sand Plain alone, they composed nearly 54 percent of the habitat in the 1890s but less than 1 percent by 1990 (MNDNR 2006c). Loss of their preferred habitat may force the birds into more densely forested stands, where competition with similarly sized woodpeckers (i.e., Hairy and Red-bellied Woodpeckers) may reduce survival (Frei et al. 2020). In western Minnesota, the birds have adapted to the region's small woodlots and shelterbelts. Although these modified habitats provide an abundance of dead and decaying trees adjacent to open foraging areas, it is unknown if nesting success is comparable to that found in more natural open habitats, especially oak savannas.

CONSERVATION

Designated a moderate conservation priority (PIF 2021), the species has benefited from the efforts of Minnesota's Red-headed Woodpecker Recovery Project. Launched in 2007 by the Audubon Chapter of Minneapolis, the project has conducted extensive Woodpecker surveys, has prepared management guidelines that focus on retention of suitable nesting trees, and has been a catalyst for an in-depth study of the resident population at the Cedar Creek Ecosystem Science Reserve. Located in Anoka and Isanti Counties, the site supports the highest density of Red-headed Woodpeckers in the state.

Minnesota may provide more opportunities for this stunning Woodpecker in the future, but at the expense of habitat that becomes less suitable in the Central Plains. Models predict that much of the state's climate will become similar to that currently found in Nebraska and Kansas (Fitzpatrick and Dunn 2019), now the core of the species' breeding range. Indeed, under a 2°C warming scenario, the Woodpecker's highly vulnerable breeding distribution is projected to decrease 64 percent by 2050 and expand only 27 percent northward (NAS 2019). It may, however, move farther north than these models predict, since the expansion projected in Minnesota already lies well within its current breeding distribution.

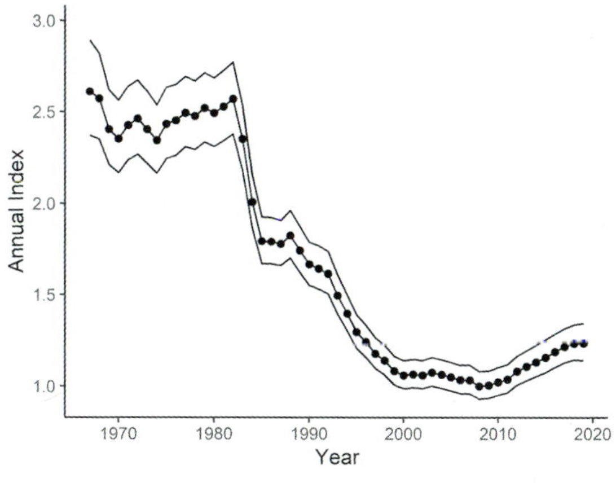

C. Surveywide breeding population trend and 95 percent credible interval (solid black lines) for the Red-headed Woodpecker based on the 1967–2019 Breeding Bird Survey (Sauer et al. 2020).

RED-BELLIED WOODPECKER

Melanerpes carolinus

Minnesota Seasonal Status: Permanent resident. The Red-bellied Woodpecker was common during the MNBBA.
Migration: Sedentary throughout most of its breeding range; northern birds may move south during harsh winters and in response to food availability.
Food: Generalist that feeds on arthropods, mast, fruit, seeds, and sap secured by hawking, foliage gleaning, and probing and excavating bark surfaces; eggs and nestlings of smaller birds are also consumed. A common visitor to bird feeders.
Nest: Primary cavity nester in snags or dead limbs in live trees.
Conservation Concern: Assigned a low PIF Continental Concern Score of 7/20.

The Red-bellied Woodpecker may be the most striking woodpecker in Minnesota. Most species are adorned in various patterns of black-and-white, but the Red-bellied's plumage is a regular, unbroken pattern on the back and wings that is often described as "zebra-like" in appearance. What really catches one's breath, however, is the intensity of the scarlet red color on the head. On the male, the red is on the top of the head and the back of the neck; on the female it is limited to the back of the neck. Some color also shows at the base of the bill. Despite its name, the red belly is evident only upon close inspection and is often just a faint tinge at the bottom of the bright white belly. The genus name, *Melanerpes,* translates to "black creeper," a reference to the bird's black outer tail feathers and its creeping movement along tree trunks.

The Woodpecker's loud call may be the best clue to its whereabouts. Its most common vocalization has been described as a rolling *churr* or *kwirr* (K. E. Miller et al. 2020). Its low pitch helps distinguish it from the similar but hoarser, higher-pitched call of the Red-headed Woodpecker. Another frequently heard call is a short, gruff-sounding *cha,* delivered singly or in a longer series of notes. Although the species' breeding range extends from the Central Plains east to the Atlantic coast, the highest breeding densities are found in the southeastern states, especially along the Gulf coast.

Photo by David Brislance.

MINNESOTA BREEDING DISTRIBUTION

Located on the northern periphery of the Red-bellied Woodpecker's breeding range, Minnesota is an excellent example of the species' northern range extension during the past 125 years. Although not always reliable, Hatch's 1892 publication *Notes on the Birds of Minnesota* was considered the first complete annotation of the state's avifauna. Red-bellied Woodpeckers, however, were not among the 304 birds Hatch described. The first confirmed record for the state came one year after his publication, when a male was shot in Houston County in 1893. The collector, Mr. Whit Harrison, reported that he had also seen a pair "several years previously" (Roberts 1932). Roberts first observed the species in June 1898, again in Houston County near the town of La Crescent.

Following the Mississippi River valley north, Red-bellied Woodpeckers were found in Goodhue County by 1900 and in Minneapolis by 1908. At the same time, they began spreading westward, evidenced by a 1907–8 winter record in Rochester and an observation in the fall of 1908 in Steele County. A nesting pair found in Red Wing provided the state's first confirmed breeding record in 1903. No doubt, as Roberts noted, the species had probably been breeding farther south in the state for many years. By the time Roberts prepared his 1932 treatise on Minnesota birds, Red-bellied Woodpeckers had become a common resident as far north as Red Wing, with a few pairs established in the Twin Cities. Although the metropolitan area was the northern limit of its range, he predicted it would be found farther north and west in coming years.

Indeed, in 1975, Green and Janssen's updated account reported that Red-bellied Woodpeckers, although still restricted largely to southeastern Minnesota, were now a permanent breeding resident as far north as Stearns County. Several years later, Janssen (1987) reported its range had expanded even farther, with confirmed nesting as far north as Aitkin County and as far west as Brown County. It was a common species south of a line that stretched from central Pine County in the east, west to southern Todd, Otter Tail, and Traverse Counties. He identified 12 counties, largely in the eastern half of the state, where nesting had been confirmed since 1970. By 1998 Hertzel and Janssen added another three counties to the list, all in southeastern and south-central Minnesota. Approximately fifteen years later, the MBS had documented breeding-season locations as far north as Itasca, Beltrami, and Clearwater Counties, across southwestern Minnesota, and west along the Minnesota River valley to Yellow Medicine County. The species was absent, however, from the Arrowhead region east of Duluth.

Continuing to mark its northward and westward expansion, MNBBA participants reported possible breeding evidence in five locations on the Canadian border in Lake of the Woods, Koochiching, and St. Louis Counties and confirmed breeding as far north as southern Beltrami and northwestern Itasca Counties. All total, 1,649 Red-bellied Woodpecker records were reported in 21.8 percent (1,034/4,748) of the surveyed atlas blocks. Breeding evidence was confirmed in 146 blocks (3.1 percent) and deemed probable in 200 blocks (4.2 percent) **(A)**. Detected in 80 of Minnesota's 87 counties, breeding was confirmed in 49

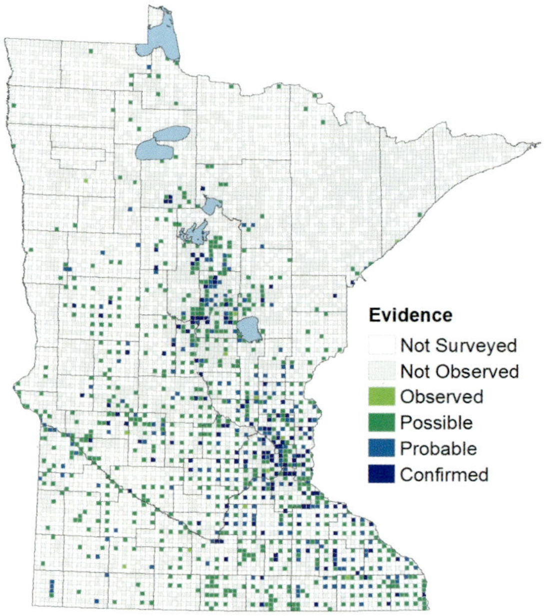

A. Breeding distribution of the Red-bellied Woodpecker in Minnesota based on the Breeding Bird Atlas (2009–13).

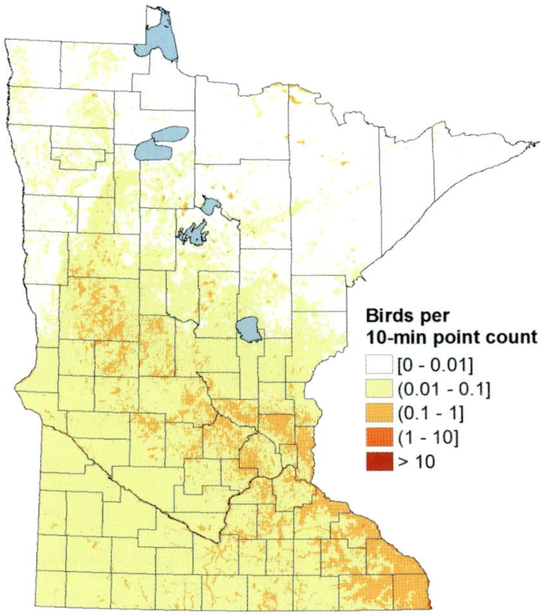

B. Predicted breeding distribution (birds detected per 10-minute point count) of the Red-bellied Woodpecker in Minnesota based on habitat, landscape context, and climate data gathered during the Minnesota Breeding Bird Atlas (2009–13) using the Generalized Linear Modeling method without an adjustment for detectability.

counties; 35 were additions to the list published by Hertzel and Janssen (1998). The species was most common in the Eastern Broadleaf Forest Province. Although it was encountered frequently from the Brainerd Lakes region north to Bemidji, it was absent from most of north-central and northeastern Minnesota. It was equally scarce in the Tallgrass Aspen Parklands Province and in the Red River valley.

The MNBBA probability map predicts that moderate breeding densities occur throughout the dissected river valleys of the Mississippi River and its tributaries in southeastern Minnesota and in scattered portions of east-central Minnesota west to the Hardwood Hills of west-central Minnesota **(B)**. The Blue Earth River and its tributaries in south-central Minnesota, as well as the entire Minnesota River valley, are also predicted to support moderate breeding densities. Elsewhere breeding pairs are either absent or predicted to be present in very low numbers.

The Red-bellied Woodpecker's Minnesota breeding range has expanded approximately 400 km to the northwest from the year 1903, when it first nested in Red Wing, to 2013, when it was reported nesting in Bemidji during the MNBBA. Similar range expansions have been observed elsewhere. In New England, its northward expansion has been facilitated by an aging forest landscape; in the Great Plains, its expansion has been facilitated by wooded floodplains and maturing tree plantings in urban areas. Some researchers have suggested that warmer winters and the increased interest in bird feeding have also been important factors facilitating north and westward movements (K. E. Miller et al. 2020).

BREEDING HABITAT

The Red-bellied Woodpecker's habitat is broadly described as mature hardwood forests that provide trees large enough for cavity excavation **(C)**. In the southern states, pine and mixed pine–hardwood forests are also used. Frequently found in mature floodplain forests, the species is equally at home in dry oak and mesic mixed upland forests and readily adapts to smaller woodlots in urban settings. Compared to other woodpeckers, Red-bellied Woodpeckers have been described as being more closely associated with high tree density and with a well-developed understory that has a high density of young trees and shrubs (Danz et al. 2007; K. E. Miller et al. 2020).

POPULATION ABUNDANCE

PIF (2021) provided a population estimate of 16 million Red-bellied Woodpeckers in the eastern United States and the very southern regions of eastern Canada; only 130,000 breeding adults are estimated to reside in Minnesota. As expected, the BBS shows a significant population increase throughout the species' range, with an average increase of 0.80 percent per year from 1966 to 2019 **(D)**. In Minnesota, the population is increasing at a much faster pace. Although the species is still uncommon enough along BBS routes to result in a trend line that lacks precision, it indicates an average annual increase of 4.39 percent since 1967 (Sauer et al. 2020). As high as this rate of increase is, it is exceeded in several northeastern states, including Connecticut (11.14 percent), New Jersey (5.53 percent), and New York (7.28 percent). The Florida peninsula is the only region where the species is declining.

As noted earlier, factors responsible for the species' population growth and northward expansion

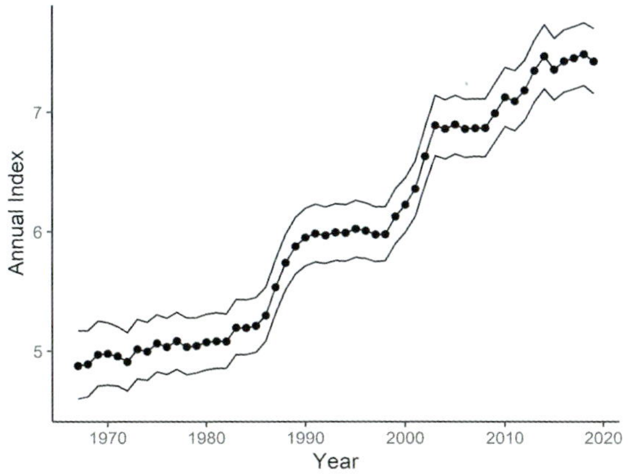

D. Surveywide breeding population trend and 95 percent credible interval (solid black lines) for the Red-bellied Woodpecker based on the 1967–2019 Breeding Bird Survey (Sauer et al. 2020).

include additional foraging opportunities provided by winter bird feeders and maturing forests in the northeastern United States. Warmer winter temperatures and increased winter survival may be equally important. Kirchman and Schneider (2014) noted that the species began its range expansion early in the twentieth century, but the pace was much faster after the 1950s, when winter temperatures began to increase.

CONSERVATION

With its rapidly expanding population and its adaptability to developed landscapes, the Red-bellied Woodpecker is not considered a conservation priority (PIF 2021), nor has it been the focus of targeted conservation efforts or management studies. Its positive response to warming temperatures certainly justifies its classification as a climate-stable species (NAS 2019). As long as forest management practices protect large trees that provide critical nesting cavities, the Red-bellied Woodpecker will likely continue to increase its presence in Minnesota.

C. The Red-bellied Woodpecker is common in mature floodplain forests along the Mississippi River. Photo by Lee A. Pfannmuller.

YELLOW-BELLIED SAPSUCKER

Sphyrapicus varius

Minnesota Seasonal Status: Regular breeding resident and migratory species in Minnesota, casual in winter. The Yellow-bellied Sapsucker was a common species during the MNBBA.
Migration: Short- to medium-distance migrant and the most migratory of the woodpeckers; its winter distribution includes the southeastern United States, Mexico, the Caribbean, and Central America.
Food: Sap and insects, especially gathered from its characteristic sap wells, a series of vertical holes around a tree trunk. The species also consumes fruit and is known to practice fly-catching.
Nest: Primary cavity excavator in trees.
Conservation Concern: Assigned a low PIF Continental Concern Score of 6/20.

Photo by David Brislance.

Known for its characteristic habit of drilling a series of holes, or sap wells, in the bark of trees, the Yellow-bellied Sapsucker is often thought of as a pest. Roberts (1932) even called them "the evildoers among the Woodpecker tribe" because of their girdling but acknowledged that sap wells "soon [become] the rallying-place of a varied and motley assemblage of interlopers, all intent upon getting a share of the spoils." The sapsucker creates the sap wells, a sugar source, for food, but they also serve an ecological role that provides benefits to other wildlife species. For instance, hummingbirds are known to place their nests near sap wells because they also use them as a natural food source. Wells attract insects, a protein source for the sapsucker, hummingbirds, and other insect-eating forest species. Moreover, the nest holes created by the sapsucker are later used by other birds and some mammals, such as flying squirrels (Walters et al. 2020). Many authors have also emphasized that tree girdling by the sapsucker rarely causes mortality to the tree.

The male Yellow-bellied Sapsucker is primarily a black-and-white bird distinguished from other medium-sized woodpeckers by its red fore-crown, chin, and throat, which is outlined in black. The female is similar, but its chin and throat are white. Its species name, *varius,* is Latin for "variegated" and refers to its mottled black-and-white plumage. Its "yellow-bellied" common name refers to the yellowish wash on its underparts. This species has a wide variety of vocalizations, but one of the most commonly heard, especially near the nest, is a cat-like mewing sound that slurs downward. Other calls are similar to those made by Blue Jays or the Red-shouldered Hawk. Its drumming is also distinctive and simply described by Roberts (1932) as "starts with three or four rapid notes and then slows down, putting variations of one, two, or three notes in equal time spaces." In contrast to other woodpeckers, one can actually count the notes of the sapsucker.

The Yellow-bellied Sapsucker is found across the midwestern and northeastern United States and Canada, southeast to northern Georgia in the Appalachian Mountains, and northwest to Alaska. Relatively high densities are widespread throughout the Yellow-bellied Sapsucker's breeding range, but those among the highest can be found in north-central and northeastern Minnesota.

MINNESOTA BREEDING DISTRIBUTION

The Yellow-bellied Sapsucker is a widely distributed species, especially in heavily forested regions and along major river systems. Roberts (1932) identified the species as a "common summer resident, breeding throughout the state." He documented a few nests, which varied in location from the Twin Cities area northwest to Itasca State Park and northeast to Lake County.

Forty years later, Green and Janssen (1975) stated the species was a summer resident throughout the forested parts of Minnesota but they found no modern summer records south of the Minnesota River and west of Rice County. However, several years later, Janssen (1987) and Hertzel and Janssen (1998) collectively documented nesting in 49 counties since 1970. The counties included virtually all of Minnesota, except the extreme southwestern counties of Lincoln, Murray, Nobles, Pipestone, and Rock.

The MBS recorded 1,165 breeding-season locations of the Yellow-bellied Sapsucker during their county surveys. They documented that the dominant distribution was in the northeastern and north-central regions of the state. Locations were sparse in the southern counties, such as Nobles, Pipestone, and Steele, which probably lacked suitable breeding habitats. However, the MBS included four breeding-season locations in Murray County and three in Rock County, plus multiple locations in Jackson, Lincoln, and Lyon Counties.

The MNBBA included 3,262 records of the Yellow-bellied Sapsucker and provided additional documentation on the species' extensive breeding distribution **(A)**. Confirmed breeding records were reported from 7.5 percent (354/4,750) of the surveyed blocks. These confirmed nesting records were most heavily reported from the northeastern and north-central regions, where breeding populations are highest, and in areas where search efforts were intense. Evidence of nesting was also concentrated along the Mississippi, St. Croix, and Minnesota River valleys.

Confirmed nesting was relatively high in many of the western counties, especially in the northwest

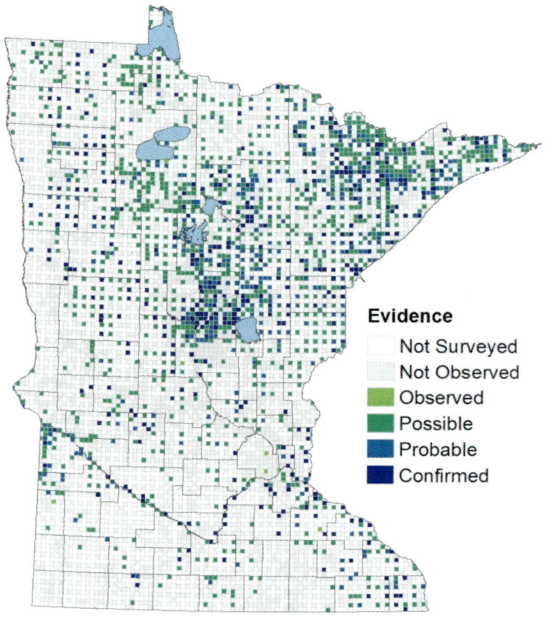

A. Breeding distribution of the Yellow-bellied Sapsucker in Minnesota based on the Breeding Bird Atlas (2009–13).

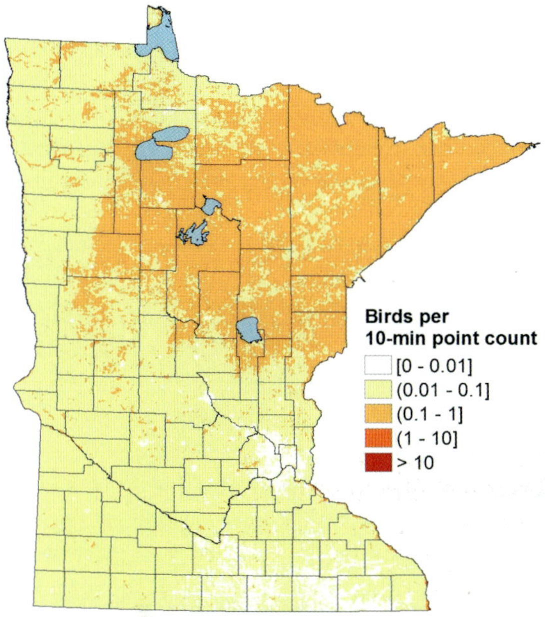

B. Predicted breeding distribution (birds detected per 10-minute point count) of the Yellow-bellied Sapsucker in Minnesota based on habitat, landscape context, and climate data gathered during the Minnesota Breeding Bird Atlas (2009–13) using the Generalized Linear Modeling method without an adjustment for detectability.

within the Tallgrass Aspen Parklands Province. A total of 14 new county records were added to those previously reported since 1970. Possible nesting was recorded from several counties in southern Minnesota, but no records were included from Nobles, Pipestone, Rock, Steele, and Wilkin Counties.

The probability map based on the MNBBA data indicated high use of the northeastern and north-central forests and lower densities in areas dominated by lowland coniferous forests **(B)**. Lower densities were predicted in western Minnesota and southern Minnesota, but larger populations were predicted in patches in the northwestern region, in the Hardwood Hills Subsection, and along the Mississippi River floodplain south of the Twin Cities. Since the MNBBA period, many confirmed breeding records have been documented from both Pipestone and Steele Counties and possible nesting from Nobles County (MOU 2021a). These were all areas where low populations were predicted to occur.

In their review of the Yellow-bellied Sapsucker in North America, Walters and colleagues (2020) indicated the species had retracted its range northward from northeastern Missouri. Roberts (1932) is mute on any changes to the breeding distribution of the species in Minnesota from the late 1800s to the early 1900s. The massive reduction of trees in the southern and western portions of Minnesota has undoubtedly resulted in a retraction of a substantial portion of the Yellow-bellied Sapsucker population from those areas. Yet scattered populations continue to exist in suitable woodlots of those regions. Aspen, one of the sapsucker's favorite trees, has increased in Minnesota because of its popularity with the forest products industry. Subsequently, populations of the sapsucker have increased in many northern forests. It is difficult, however, to determine the overall effect increased aspen has had on the population of the Yellow-bellied Sapsucker in the state.

BREEDING HABITAT

In Minnesota, the Yellow-bellied Sapsucker is found in a wide variety of forested habitats, including deciduous to mixed deciduous–coniferous forests, northern hardwood forests, and floodplain forests (Niemi and Hanowski 1992a; Green 1995). Deciduous cover types, including aspen–birch, lowland hardwoods, oak, and upland hardwood, were used most often (Niemi, Howe, et al. 2016) **(C)**. Nest sites are often found in trees with fruiting bodies of heartwood decay fungi (*Phellinus tremulae* and *Fomes fomentarius*) which are commonly found in aspen (Kilham 1971; Runde and Capen 1987). Heartrot allows for easier excavation of the nesting cavity.

POPULATION ABUNDANCE

PIF (2021) estimated a North American breeding population of 14 million breeding adults, with 410,000 in Minnesota. In Minnesota, the BBS trend indicated a significantly increasing population of the Sapsucker of 1.86 percent per year from 1967 to 2019 **(D)**. Rosenberg, Kennedy, and colleagues (2016) calculated an increase of 46 percent in the species' population based on the surveywide BBS trend in North America from 1970 to 2014. The Yellow-bellied Sapsucker was by far the most common woodpecker in forested areas of Minnesota and second in abundance to the Northern Flicker throughout Minnesota.

C. Yellow-bellied Sapsuckers are typically found in deciduous forests, especially in aspen, where they can easily excavate their cavities in trees with heart rot. Photo by Gerald J. Niemi.

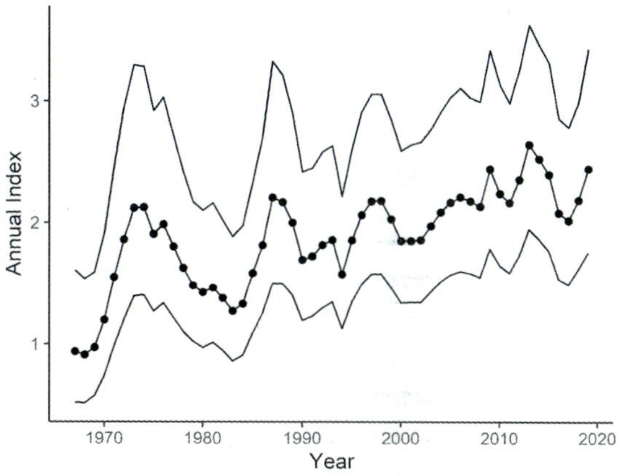

D. Breeding population trend and 95 percent credible interval (solid black lines) for the Yellow-bellied Sapsucker in Minnesota based on the 1967–2019 Breeding Bird Survey (Sauer et al. 2020).

CONSERVATION

The Yellow-bellied Sapsucker has attracted little management concern either in Minnesota or nationally. Rather than concern for how to increase the Yellow-bellied Sapsucker's population, the greatest concern is usually how to control its population. Many consider the Yellow-bellied Sapsucker a pest on trees because it girdles the bark of trees, especially those in orchards and backyards. Louise Lawrence (1967) estimated that each pair of sapsuckers killed only 1–2 trees per year by girdling, so the benefits of sap wells to other species may far outweigh the negatives (W. L. Foster and Tate 1966). Eberhardt (2000) suggested that even though some trees die as a result of sapsucker attacks, many animals may need sap for food. Thus, sapsucker activity could have far-reaching implications for many other wildlife species in the forest community.

Because the Yellow-bellied Sapsucker is protected by the federal Migratory Bird Treaty Act, it is illegal to shoot them. Regardless, shooting is considered an ineffective control method because transient birds can quickly replace local sapsucker populations.

With a 2°C warming scenario, NAS (2019) projected a moderate vulnerability for the Yellow-bellied Sapsucker's breeding distribution in North America, including a 50 percent range loss and a 51 percent range gain primarily in the far north by 2050. In Minnesota the species was predicted to be extirpated from the state under a 3°C warming scenario by 2080. It is unclear how quickly its essential breeding habitat, especially aspen, would recolonize the northern regions that are expected to be gained by the year 2050 or 2080.

AMERICAN THREE-TOED WOODPECKER

Picoides dorsalis

Minnesota Seasonal Status: Very rare permanent resident with some southern movements during the winter. The American Three-toed Woodpecker was rare during the MNBBA.
Migration: Southward movements in winter from northern Minnesota and Canada are regularly reported.
Food: Primarily a specialist on bark beetles in the family Scolytidae but also consumes larvae of wood-boring beetles gathered while pecking and scaling the bark of trees.
Nest: Excavates a hole in trees of many different species.
Conservation Concern: Assigned a low PIF Continental Concern Score of 8/20.

Photo by David Brislance.

A rare permanent resident of northern Minnesota, the American Three-toed Woodpecker is relatively tame and quiet compared with its closely related congener, the Black-backed Woodpecker (Janssen 1987). It is most often encountered in recently burned or beaver-altered habitats in heavily forested areas of the extreme northern portions of the state. Clues to its presence, like that of the Black-backed, are freshly scaled bark of trees on burned or dying trees (Tremblay, Leonard, and Imbeau 2020). Like many boreal birds, both the Black-backed and American Three-toed Woodpecker are "fearless birds, tame and unsuspicious, probably because of their unfamiliarity with man and his hostile intentions; both are less active than most woodpeckers," with the American Three-toed "being particularly quiet in its movements and sedentary in its habits" (Bent 1939).

The American Three-toed Woodpecker was formerly considered to be one species with a Holarctic distribution. Its common and scientific name has changed frequently over the years as more information has accumulated. For instance, it was formerly named the "Three-toed," "Ladder-backed," and "Northern Three-toed Woodpecker." Recent genetic evidence and its voice have separated it from *Picoides tridactylus* in Europe and Asia, so its new scientific name, *dorsalis,* refers to its mostly dark back with white barring. Any white coloration on its back quickly distinguishes it from the Black-backed Woodpecker.

Field identification of the species is primarily based on its medium size, black-and-white coloration, and stocky appearance. It has a white eye stripe, a back with white and black barring, white below, and heavily barred flanks with spotted primaries. Identification is greatly aided by habitat. It is found primarily in spruce forests. Like the Black-backed Woodpecker, it has three toes, with two in front and one behind, and the males have a yellow crown, the only woodpeckers with this unique anatomy and coloration. Its call, *pik* or *kik,* is higher in pitch, and its drumming is slower and shorter than the Black-backed Woodpecker.

The American Three-toed Woodpecker is the most northerly distributed woodpecker in North America. It is found across the boreal forests of Canada and Alaska and south in the Rocky Mountains to Arizona and New Mexico. It is found in low abundance throughout its breeding range.

MINNESOTA BREEDING DISTRIBUTION

Hatch included the American Three-toed Woodpecker in his 1892 report based on "rumor having never had the bird in my own hands for identification." Roberts's (1932) record of observations for this species was limited to ten localities and usually of single birds. He describes the species as frequently encountered only on the "Mesabi Iron Range . . . in the vicinity of Ely and Hibbing." Roberts did not document any nesting of this species in Minnesota but did report observations of a pair with one young male nearly fully grown in July 1902 in Itasca State Park. Roberts's most southerly record for this species was of an individual observed at Cambridge, Isanti County, during the winter of 1915.

More than forty years later, Green and Janssen (1975) identified the woodpecker's summer range in Minnesota as "within 50 miles of the Canadian border from the Northwest Angle to Cook County and in Itasca State Park." The first confirmed nesting record of the species was recorded by Eckert in 1981 along the Gunflint Trail in Cook County. Janssen (1987) suggested the summer records "indicate that it is an occasional breeder in extreme northern parts of the state." He noted the most southerly record in Minnesota was from Washington County in December 1981. Hertzel and Janssen (1998) added St. Louis County to the confirmed nesting records for the species since 1970.

The MBS recorded five breeding-season locations of the American Three-toed Woodpecker during their county inventories, but they had not yet reported coverage of northern Beltrami, Koochiching, or Lake of the Woods Counties. Their locations included two each in Cook and northern Lake Counties, and one in northern Itasca County.

The MNBBA documented 14 records in 11 blocks distributed from Lake of the Woods to Cook County **(A)**. This represented less than 1 percent of all blocks surveyed or priority blocks sampled in Minnesota. Nesting was confirmed in four blocks, including three in Lake County and one in southern Lake of the Woods County. In addition, MOU (2021a, 2021b) reported the first confirmed breeding of fledged young in Koochiching County (June 2010).

The historical status of the American Three-toed Woodpecker in Minnesota is unclear. It is unknown whether the species regularly nests in Minnesota. Since Hatch (1892) never observed or collected the species and Roberts (1932) had limited encounters, the lack of coverage in the northern coniferous forests in the mid-1800s and much of the twentieth century limits any conclusions. Historically, given that fire frequency and the breadth of forest fires were greater before European settlement in Minnesota (Heinselman 1973), it is possible that populations of this species were higher during this period because of their known use of areas affected by

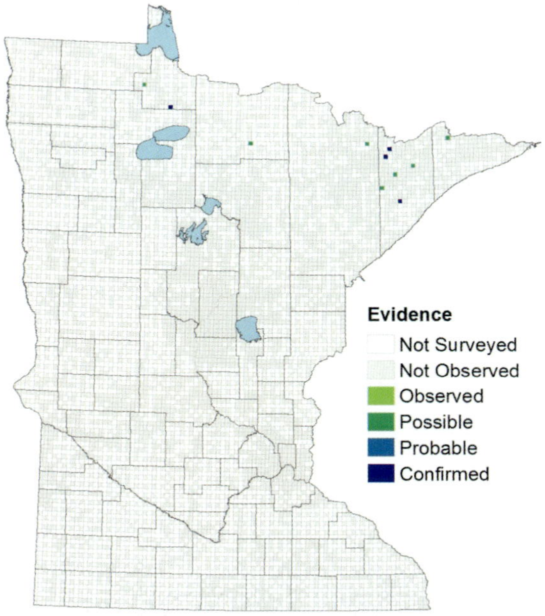

A. Breeding distribution of the American Three-toed Woodpecker in Minnesota based on the Breeding Bird Atlas (2009–13).

forest fires. Tremblay, Leonard, and Imbeau (2020), in their review of the American Three-toed Woodpecker in North America, stated that there were "no documented extirpations or range expansion for North America although this is difficult to determine given the species' low abundance and inconspicuous nature."

The American Three-toed Woodpecker was not listed by Cutright and colleagues (2006) in Wisconsin or by Chartier and colleagues (2013) in Michigan. Ontario reported no differences in the species' distribution from its first atlas, in 1981 to 1985, to its second atlas, in 2001 to 2005 (Cadman et al. 2007). In both Ontario atlases, Cadman and colleagues (1987, 2007) emphasized that the species can easily be overlooked because of its quiet nature and the difficulty of accessing its wet breeding habitat.

BREEDING HABITAT

Tremblay, Leonard, and Imbeau (2020) suggested that the American Three-toed Woodpecker prefers spruce and may prefer denser forests compared with the Black-backed Woodpecker. In Michigan, Vermont, Maine, and Canada, breeding habitat is associated with black spruce, tamarack, balsam fir, mixed-conifer forest, old-growth, and moist forests (Tremblay, Leonard, and Imbeau 2020). Occupied territories are commonly in areas of disturbance, mainly those associated with forest fire, wind, disease, or flooding, such as from beaver activity **(B)**.

Preferred breeding habitat was generally described by Roberts (1932) as older, dense conifer forests. Cadman and colleagues (1987) in Ontario emphasized coniferous woodlands that tended toward wet conditions found in bogs.

R. L. Hoover and Wills (1984) estimated a minimum habitat area of 400 ha to maintain a minimum viable population of 20 individuals. Goggans and colleagues (1988) found that each pair of American Three-toed Woodpeckers required more than 200 ha of montane, old-growth mixed conifer or lodgepole pine (*Pinus contorta*) forests. Based on these findings, both suggest a sensitivity to forest area size.

POPULATION ABUNDANCE

PIF (2021) estimated the United States and Canadian population at 1.6 million breeding adults but did not include an estimate for the state of Minnesota. Environment Canada (2014b) estimated a wide range of 500,000 to 5,000,000 adults. MNBBA did not include a population estimate because of too few detections.

The BBS indicated that trend estimates for all states, provinces, and surveywide were unreliable. Environment Canada (2014b) reinforced the lack of reliable data for this species in Canada; however, they also noted that the BBS and CBC combined suggested moderate increases in the population since 1970. In contrast, Rosenberg, Kennedy, and colleagues (2016) calculated a 27 percent decline in its North American population from 1970 to 2014. Generally, this species has a secretive behavior and minimal vocalizations, which render it more difficult to detect in most surveys.

CONSERVATION

Environment Canada (2014b) emphasized that trends in this species should be closely followed and that "timber harvesting, fragmentation, rapid cutting rotations, fire suppression, and salvage logging have all been shown to have negative influences on the species." Modeling studies on increased forest harvesting and management in Minnesota predicted a decline in this species if there was a substantial increase in logging in the most northern forest regions (Niemi and Hanowski 1992a).

In summary, a better understanding of this rare woodpecker will require a much greater focus on the management of the fire-dominated ecosystem that historically occurred naturally in the northern forests of Minnesota. The species' use of wet bog habitats, which are difficult to survey, and its sensitivity to forest patch size also needs consideration. Observers becoming more familiar with the species' vocalizations during the breeding season could also improve its detection. With continued fire suppression and the predicted reduction in coniferous forests in the state with climate change, the future of the American Three-toed Woodpecker is not likely to improve. NAS (2019) in their modeling of climate change did not recognize the existence of a breeding population in Minnesota but predicted moderate vulnerability with a 2°C warming scenario in the breeding range in North America by 2050.

B. Typical breeding habitat of the American Three-toed Woodpecker in Minnesota often consists of recently burned-over areas where they find a good supply of wood-boring beetle larvae. Photo by Gerald J. Niemi.

BLACK-BACKED WOODPECKER

Picoides arcticus

Minnesota Seasonal Status: Permanent resident and winter visitant. The Black-backed Woodpecker was uncommon during the MNBBA.
Migration: Southward movements from breeding areas in some years.
Food: Specialist on wood-boring beetles (*Cerambycidae*), which it extracts by scaling and pecking the bark of trees, especially those in areas affected by fire, wind, or other disturbances.
Nest: Excavates a hole in trees of many different species.
Conservation Concern: Assigned a low PIF Continental Concern Score of 8/20.

Photo by David Brislance.

The Black-backed Woodpecker was formerly known as the Black-backed Three-toed Woodpecker or Arctic Three-toed Woodpecker. Its scientific name, *arcticus,* is from the Greek term *arktikos* for "near the Great Bear" or *arktos,* "Ursa Major," which refers to its northern distribution. Roberts (1932) described it as a "permanent resident throughout the coniferous forests of the state; wanders southward in winter." In comparison, Janssen (1987) classified the species as a "rare to locally uncommon resident in the northeast and northcentral regions."

This species is closely related to the American Three-toed Woodpecker, which shares the unique three-toed anatomy and where the male sports a yellow cap. It is more common, noisier, and distinguished from the American Three-toed by its all-black back. The species' call note is often described as a sharp *pik* or *kyik,* sometimes given in rapid succession. For an observer with keen hearing abilities, the call note has been described as lower in pitch and its drum faster and longer than that of the American Three-toed Woodpecker. This species, like many bird species of the northern boreal forests, is rather tame and will often allow the observer a very close view. Major Bendire said they "are persistent drummers, rattling away for minutes at a time . . . especially active during the mating season, in April" when he "located more than one specimen by travelling in the direction of the sound when it was fully half a mile away" (in Bent 1939).

The Black-backed is found in patches across Canada from Labrador to British Columbia and Alaska, southward into the Rocky Mountains and the eastern portion of the Cascades to northern California. In the midwestern and northeastern United States, it is found in the extreme northern portions.

MINNESOTA BREEDING DISTRIBUTION

Roberts (1932) described the Black-backed Woodpecker as "common in summer only in the northern third of the state but has been found in limited numbers in the nesting season as far south as Lake Mille Lacs, Aitkin County." He noted nesting activity in Becker County and Itasca State Park and reported that many nests were found by Peabody and Warren on the Mesabi Iron Range in upper St. Louis County. He also described a "nest with eggs in a tamarack swamp near Lake Minnetonka" on May 9, 1891, as the "only breeding record south of the evergreen forests." Later, Green and Janssen (1975) labeled this record as suspect because it was not observed first-hand by Roberts.

Green and Janssen (1975) described the Black-backed Woodpecker's breeding range as extending as far west as Becker County and south to Crow Wing County. It was most numerous in Lake and Cook Counties. They included confirmed nesting in five counties: Becker, Clearwater, Cook, Lake, and St. Louis. Inferred nesting was indicated in Crow Wing County. Several years later, Janssen (1987) classified the species as a "rare to locally uncommon resident in the northeast and northcentral regions." He emphasized that Black-backed Woodpeckers were better represented in these areas in winter rather than summer, likely owing to movements in the fall and an exodus in the spring. Janssen (1987) identified confirmed nesting in eight counties since 1970: Beltrami, Clearwater, Cook, Hubbard, Itasca, Lake, Lake of the Woods, and St. Louis. In their updated account, Hertzel and Janssen (1998) recorded seven counties with confirmed nesting since 1970, excluding Hubbard County.

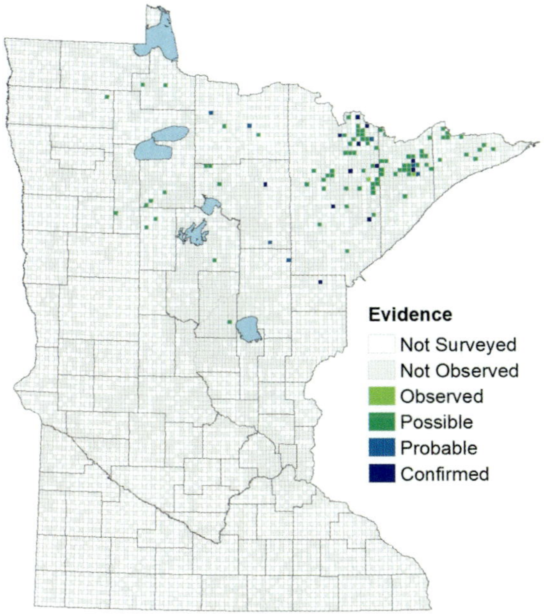

A. Breeding distribution of the Black-backed Woodpecker in Minnesota based on the Breeding Bird Atlas (2009–13).

The MBS recorded 139 breeding-season locations of the Black-backed Woodpecker during their county inventories. These locations were distributed from northeastern Becker County, Clearwater County, and northwestern Todd County to Cook County. Single locations were also noted in southern Aitkin, southern Pine, and northern Carlton Counties. The bulk of the observations were noted in Cook County and northern Lake and St. Louis Counties.

Participants in the MNBBA documented 158 records of the Black-backed Woodpecker. The records were distributed as far west as Marshall County, south to Aitkin and Crow Wing Counties, and northeast to Cook County **(A)**. The birds were detected in 2.3 percent (107/4,736) of all surveyed blocks and 1.8 percent (43/2,337) of priority blocks. Confirmed nesting was reported from 12 blocks in Carlton, Itasca, Lake, and St. Louis Counties. The confirmed nesting in Carlton County extended recent breeding observations southward beyond reports by Janssen (1987) and Hertzel and Janssen (1998). More than one hundred years ago Roberts (1932) "found this woodpecker rather common in Carlton County," including with "young that were nearly full grown." Overall, the species was most numerous in northern Cook, Lake, and St. Louis Counties and sparsely distributed in suitable coniferous habitat in many areas of northern and north-central Minnesota.

Tremblay, Dixon, and colleagues (2020) identified no known changes in the breeding range or population but did point out that it was formerly a common winter resident in southeastern Wisconsin. As for many rarer species that reside in the northern forested regions of Minnesota, coverage of these regions was sparse in the nineteenth century and well into the twentieth. Combined observations from records of the MOU, the MBS, and the MNBBA suggest the distribution of this species has not changed substantially over the past 150 years. Because the species has an affinity for naturally disturbed forests, recent fire suppression may have contributed to reduced populations in many areas. The Black-backed makes limited use of logged areas unless dead trees remain that have suitable wood-boring insect populations.

BREEDING HABITAT

Black-backed Woodpeckers are primarily found in mature conifer-associated forests and especially those affected by forest fire, insects, windthrow, or flooded areas created by beaver activity **(B)**. Niemi (1978) found the species relatively common in virgin, burned jack pine forests in the Boundary Waters Canoe Area Wilderness two to four years after the Little Sioux fire in northern Minnesota. It was relatively rare in the unburned forest surrounding these areas. Niemi, Howe, and colleagues (2016) significantly found it most common in black spruce–tamarack forests. It has not been commonly recorded in the Red Lake Peatland area, which has extensive lowland black spruce–tamarack forests, perhaps because of the lack of fire or insect outbreaks (Zlonis et al. 2017).

The importance of burned versus unburned forests to the Black-backed Woodpecker from a landscape perspective has been debated. For instance, Hutto (1995a) suggested that the Black-backed Woodpecker in the western United States was restricted to early post-fire coniferous forests and that populations were maintained by patches of recently burned forests. In contrast, Tremblay, Dixon, and colleagues (2020) provided a counterargument based on a landscape model that predicted burned forests provided only short-term pulses in the population, while long-term stability in the population was more dependent on older, unburned forests.

B. Black-backed Woodpeckers are attracted to recently burned areas such as the Ham Lake Fire in northeastern Minnesota. A pair was nesting in the tall dead jack pine on the right. Photo by Jarad B. Niemi.

POPULATION ABUNDANCE

PIF (2021) estimated a North American population of 1.7 million breeding birds and a Minnesota population of 2,600 individuals. Population estimates for Minnesota by the MNBBA were unreliable because of its low population. Breeding population trends for this species based on the BBS were also unreliable for Minnesota, for each Canadian province, and surveywide. Black-backed Woodpecker populations occur in many roadless areas not covered well by roadside counts like the BBS. It is likely that the North American population of this species is underestimated.

Density estimates are relatively uncommon. Niemi (1978) estimated three pairs per 40 ha in a recently burned, virgin jack pine forest in the Boundary Waters Canoe Area Wilderness of northern Minnesota. Overall populations in Minnesota forests are very low and difficult to detect but can be high locally in forests disturbed by fire, insects, wind, or other factors.

CONSERVATION

There are several conservation concerns for the Black-backed Woodpecker owing to habitat loss, habitat fragmentation, and the reduction of ecological processes that used to provide adequate habitat conditions for the species. The latter concern is primarily fire suppression, which minimizes the availability of burned trees and the beetles that colonize them, a major food source for the species. With a 2°C warming scenario, NAS (2019) projected a moderate vulnerability for the Black-backed Woodpecker's breeding distribution in North America, including a 45 percent range loss and a 32 percent range gain primarily in the far north by 2050. In Minnesota, the species was predicted to be extirpated from the state even under our imminent 1.5°C warming scenario by 2050.

The Black-backed Woodpecker is a unique species of the northern boreal forests of Minnesota. Its presence anytime of the year usually creates some excitement among the birding community and other wildlife observers. Its status, however, given climate change, fire suppression, and many unknown factors that may affect its population and distribution, renders its future as precarious in Minnesota.

DOWNY WOODPECKER

Dryobates pubescens

Minnesota Seasonal Status: Regular permanent resident. The Downy Woodpecker was common during the MNBBA.
Migration: Sedentary; occasional long-distance dispersals but no regular seasonal movements.
Food: Gleans and probes bark surfaces to secure arthropods, fruits, and seeds; also gleans from small twigs, shrubs, and plant stems. A common visitor to feeders.
Nest: Primary cavity excavator; cavity entrance usually located on the underside of a leaning limb of a dead or dying tree.
Conservation Concern: Assigned a low PIF Continental Concern Score of 7/20.

Photo by David Brislance.

The smallest of the nine woodpeckers that regularly breed in Minnesota, the Downy Woodpecker is likely the most endearing and well-known. Early ornithologists were effusive in elaborating on the bird's industrious qualities. Wilson (1832), for example, wrote that "the principal characteristics of this little bird are diligence, familiarity, [and] perseverance," while Audubon (1827–38) commented that it was "perhaps not surpassed by any of its tribe in its hardiness, industry, or vivacity." Such sentiments are no doubt shared by Minnesotans whose days are enriched by watching the little woodpecker visit their backyard feeders.

The male's crimson-red patch on the back of the head is absent in the female, though both are attired in a vivid pattern of black and white on the head, back, and wings. Despite its unmistakable plumage, distinguishing Downy Woodpeckers from their near look-alike, the larger Hairy Woodpecker, can be challenging. The most important features are the Downy Woodpecker's small size (nearly 6.5 cm smaller than the Hairy) and its dainty bill. The Hairy Woodpecker has a much larger and longer bill. Field guides also point to differences in the barring on the tail (present on the Downy; absent in the Hairy) and the partial black necklace on the upper breast (longer on the Hairy). In the field, listen for the Downy's whiny call, aptly described by Sampson (1934) as "a handful of staccato notes flung out in a rapid run," accelerating and declining in pitch until the end. Its *pik* call note is a little sharper than a similar note given by the Hairy (Roberts 1932).

Although the similarity of their appearances led biologists to believe that the two species were closely related, the emergence of DNA sequencing uncovered quite a different story. The two species actually evolved separately along two different branches of the woodpecker genetic tree (Weibel and Moore 2002a, 2002b). Even though they are not close cousins, their coevolution of similar plumages likely provides some evolutionary benefit. Weibel and Moore (2005) suggest that looking alike is a way to quickly distinguish possible competitors (i.e., their near look-alike) and restrict them from competing for limited resources in a shared environment, such as suitable nesting trees.

The source of the species' common name, Downy Woodpecker, is also an interesting account. An early naturalist, Mark Catesby, reportedly coined the common name based on the soft white feathers on the broad stripe down the middle of its back, compared to the more hair-like feathers on the back of the Hairy Woodpecker (Wilson 1832). Field observers, however, will not be able to detect any difference!

A common inhabitant of our cities, parks, and woodlands, the Downy Woodpecker is found from southern Alaska across Canada to the Atlantic coast, south to the Gulf coast, and throughout all but the extreme southwestern United States. Breeding densities are highest in southern New England and in the southeast.

MINNESOTA BREEDING DISTRIBUTION

Nearly one hundred years ago, Roberts (1932) described the Downy Woodpecker as a common, permanent resident throughout the state. In comparing its habits to that of the Hairy Woodpecker, he wrote, "It is not as partial to heavy timber as is the Hairy and is more frequently seen about orchards and wood-lots and in city streets and parks." Confirmed nesting reports were available from Hennepin, Houston, and Isanti Counties and from Leech Lake. Nesting was inferred in Goodhue County and the Mille Lacs region. Roberts mentioned the presence of two subspecies in the state, the Northern Downy (formerly *Picoides pubescens medianus*) and Nelson's Downy (formerly *P. p. nelsoni*), which are difficult to distinguish from each other. The Northern Downy was the common, statewide resident. Nelson's Downy was considered an occasional winter visitant from its Canadian breeding grounds.

Green and Janssen (1975) and Janssen (1987) also described the Woodpecker as a common resident statewide and mention the presence of two races or subspecies. Today ornithologists recognize seven Downy Woodpecker subspecies but have combined the Northern and Nelson's subspecies into one, *Dryobates pubescens medianus* (J. Jackson and Ouellet 2020). Janssen's 1987 account identified 34 Minnesota counties where nesting had been confirmed since 1970. Hertzel and Janssen (1998) added ten counties to the list, covering all regions of the state. The MBS documented a total of 699 breeding-season locations, again broadly distributed statewide.

MNBBA participants reported a total of 2,724 Downy Woodpecker records in 36.5 percent (1,730/4,745) of the surveyed atlas blocks and in 49.1 percent

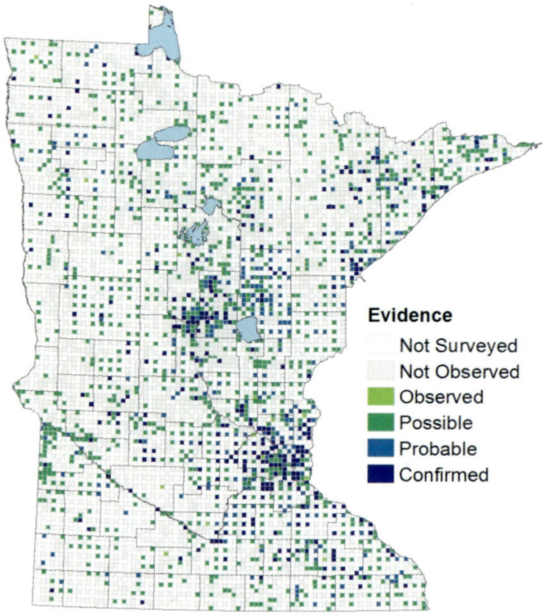

A. Breeding distribution of the Downy Woodpecker in Minnesota based on the Breeding Bird Atlas (2009–13).

(1,147/2,337) of the priority blocks **(A)**. The birds were observed in all 87 Minnesota counties while breeding was confirmed in 71 counties. Thirty-three of these counties were additions to the list published by Hertzel and Janssen (1998). Records were truly statewide in distribution, including from some of the most intensively cultivated regions of the state. Among the nine woodpecker species that inhabit Minnesota, the diminutive Downy was the third most commonly reported species during the MNBBA, exceeded only by the Northern Flicker and the Yellow-bellied Sapsucker.

MNBBA records certainly confirm that the Downy Woodpecker is as common and widely distributed as it was one hundred years ago. The same is true elsewhere within its breeding range where there are few data to suggest that there have been any major distributional changes (J. Jackson and Ouellet 2020).

BREEDING HABITAT

The Downy Woodpecker occupies a variety of open woodland habitats, including both upland and lowland forests and both deciduous and mixed forests **(B)**. Frequently found in forest edges and open areas with scattered trees, its dependence on mature trees for nesting and foraging precludes its use of the earliest successional stages of forest communities. Nevertheless, because of its small size, the species is able to excavate cavities in limbs as small as 10 cm in diameter. Ample habitat opportunities are thus available, including in many human-modified landscapes such as orchards, city parks, and residential neighborhoods. Even farm shelterbelts and small woodlots can provide sufficient habitat if mature trees are present (J. Jackson and Ouellet 2020).

POPULATION ABUNDANCE

BBS data generated a population estimate of 13 million Downy Woodpeckers in the United States and Canada, with approximately 260,000 residing in Minnesota (PIF 2021). From 1966 to 2019, the species demonstrated a statistically reliable and relatively stable population across the BBS survey region, punctuated by a significant increase in the late 1900s and a significant decrease beginning in 2003 **(C)**. The trend line is similar in Minnesota, but the species' low detectability along BBS routes in the state generates a less credible result (Sauer et al. 2020).

In northern Minnesota, detailed data collected in both the Chippewa and Superior National Forests also illustrate a recent decline in Downy Woodpecker numbers. When data for both forests are combined, they indicate a statistically significant, regional decline of –3.82 percent per year from 1995 to 2020, with a drop in numbers especially evident after 2005 (Grinde, Bednar, et al. 2020). The decrease may reflect broad land-use changes in northern Minnesota. The more open and fragmented forests preferred by the Downy Woodpecker are created by natural disturbances, such as windstorms and fire, and by timber management activities. Since bird monitoring began in Minnesota's national forests in the early 1990s, the level of timber harvest has consistently declined in both forests (Niemi, Howe, et al. 2016). On the one hand, an aging forest landscape might provide less suitable habitat for the Downy Woodpecker. On the other hand, an older forest should provide an abundance of aging aspen trees with heartrot that are suitable for excavating nest cavities. Potential competition with an expanding Red-bellied Woodpecker population may be another concern. The latter are known to "usurp and enlarge" Downy Woodpecker nests for their own use in areas where their ranges overlap (J. Jackson and Ouellet 2020). Indeed, given the Red-bellied Woodpecker's northern range expansion, coupled with its significantly increasing population across eastern North America, this competitor may be having a range-wide impact on Downy Woodpeckers. Additional field studies could shed more light on this and other factors affecting populations of this diminutive woodpecker.

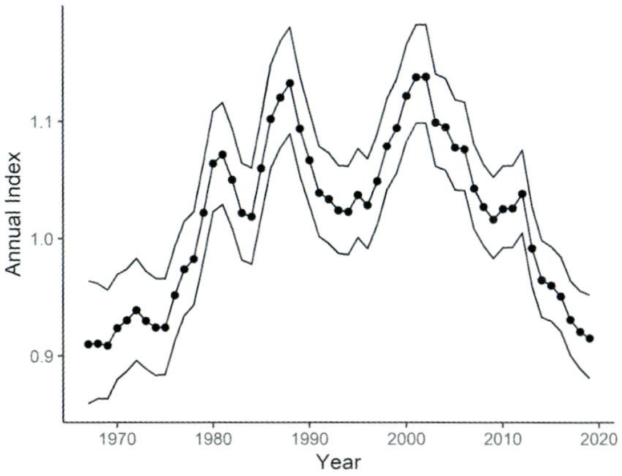

C. Surveywide breeding population trend and 95 percent credible interval (solid black lines) for the Downy Woodpecker based on the 1967–2019 Breeding Bird Survey (Sauer et al. 2020).

B. Breeding habitats selected by the Downy Woodpecker are diverse but typically include deciduous forest edges like these aspen trees bordering the edge of a mixed forest in northern Minnesota. Photo by Gerald J. Niemi.

CONSERVATION

Widely distributed with a long-term stable population, the Downy Woodpecker is not a conservation concern (PIF 2021). Even a comprehensive assessment of its potential response to climate change concluded that it was a climate-stable species whose breeding distribution gains in Canada would more than compensate for the small losses projected in the southwestern United States (NAS 2019).

Specific management recommendations are limited and primarily focused on snag retention. With respect to populations in Minnesota's extensive northern forests, forest management plans that ensure the availability of mature deciduous trees with heartrot fungus are important for the species' long-term sustainability (Danz et al. 2007). The soft interior of these trees is ideal for cavity excavation.

The sentiments expressed by many nineteenth-century naturalists ring true today. The little Downy Woodpecker, one of our most familiar year-round residents, is a "model of patient industry and perseverance," qualities that should serve it well into the future regardless of any uncertainties (Bent 1939).

HAIRY WOODPECKER

Dryobates villosus

Photo by David Brislance.

Minnesota Seasonal Status: Permanent resident. The Hairy Woodpecker was a common species during the MNBBA.
Migration: Minnesota breeding birds are sedentary; a subspecies that occurs in Canada migrates south into northern and central Minnesota in the fall.
Food: This bark driller and gleaner feeds largely on wood-boring beetles and bark beetles as well as ants and caterpillars and some fruits and seeds; a common visitor to feeders.
Nest: Primary cavity nester; cavity often placed in the trunk of a living tree with fungal heartrot.
Conservation Concern: Assigned a low PIF Continental Concern Score of 6/20.

Described as a "large edition of the Downy Woodpecker" (Bent 1939), the Hairy Woodpecker has a wide distribution across North America that is similar to that of the Downy. It also inhabits many of the same forest woodlands. When the two species are perched side by side, their physical differences are readily apparent. It can be challenging, however, if you catch only a quick glimpse of a single bird in the forest. Although the Hairy is nearly 50 percent longer than the Downy, without a comparison, size alone may be deceiving.

The most reliable distinction is the size of the bill. The Downy Woodpecker has a short, dainty bill that is approximately one-third the length of the head; the Hairy Woodpecker has a much longer and sturdier bill. The Hairy also lacks the dark bars on the Downy's white outer tail feathers. If you don't see the birds, their whiny calls are distinct; the Downy's call descends downward compared to the common one-pitch call of the Hairy. Their Latin names reference a more subtle distinction: the hairy-like (*villosus*) white plumage down the middle of the Hairy Woodpecker's back compared to the Downy's softer (*pubescens*) white plumage. Like the Downy, the only difference between the two sexes is the red on the back of the male's head.

Although the similarity of their physical appearance would suggest that the two species are closely related, recent genetic evidence suggests otherwise. Instead, the two species may be an example of convergent evolution: distantly related species that have evolved to look like each other over time (Weibel and Moore 2005).

MINNESOTA BREEDING DISTRIBUTION

The Hairy Woodpecker has long been recognized as a common, year-round resident throughout Minnesota. Not only is it common in the extensive forests of northern Minnesota but it can also be found wherever small woodlands are present, including across the agricultural landscape of western Minnesota. As Roberts (1932) noted nearly a century ago, it is even found residing in city parks and neighborhoods of some of the state's most densely populated areas. Nevertheless, he thought it might demonstrate "a preference for tamarack and spruce swamps and the wilder stretches of woodlands." Despite its wide distribution, confirmed nesting records were available from only two counties (Hennepin and Isanti) while inferred nesting records were reported from three counties (Aitkin, Goodhue, and Sherburne).

Neither Green and Janssen (1975) nor Janssen (1987) had much to add to Roberts's original account of the species' status and distribution. All three sources noted the presence of two subspecies in the state. Minnesota's breeding population is assigned to the eastern subspecies, *Dryobates villosus villosus*. A slightly larger and somewhat whiter subspecies, *D. v. septentrionalis,* occurs north and northwest of Minnesota. Known to wander during the winter months, birds identified as *D. v. septentrionalis* have been reported in northern and central Minnesota from early October through late April (Janssen 1987; Roberts 1932). Roberts even suggested it might be "an occasional summer resident in the extreme northern part of the state," based on two specimens collected in Marshall County in late June 1897 and 1928. The full extent and regularity of the northern subspecies' winter movements, however, are not well known.

Janssen's (1987) Hairy Woodpecker account noted that nesting had been confirmed in 31 counties scattered across the state since 1970. In 1998, Hertzel and Janssen added nine counties to the list. The MBS documented a similar statewide distribution for the species.

Only slightly less abundant during the MNBBA than the Downy Woodpecker, the Hairy Woodpecker had a nearly identical distribution across the state. While it is most abundant in the forests of eastern Minnesota, participants reported a total of 2,460 Hairy Woodpecker records in 35.4 percent (1,680/4,740) of the surveyed atlas blocks and in 44.5 percent (1,041/2,337) of the priority blocks **(A)**. They were reported in all 87 Minnesota counties and confirmed breeding in 62 counties. Thirty of the counties where breeding was confirmed were additions to the list published by Hertzel and Janssen in 1998. The Hairy Woodpecker was the fourth most commonly reported woodpecker, following the Northern Flicker, the Yellow-bellied Sapsucker, and the Downy Woodpecker.

Remarkably little has changed for the Hairy Woodpecker in Minnesota during the past one hundred years. The same is true throughout the species' breeding range. Although there have been no significant large-scale changes in its distribution, the loss of mature forests and the intensification of agricultural practices in the eastern United States have likely led to local declines, while in the Northeast, forest succession on abandoned farmland has created new habitat opportunities. In the Great Plains and arid southwestern states, tree plantings have

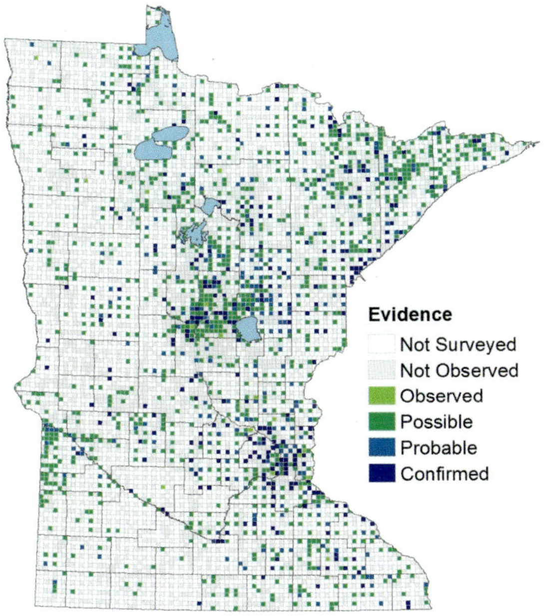

A. Breeding distribution of the Hairy Woodpecker in Minnesota based on the Breeding Bird Atlas (2009–13).

likely facilitated local population increases (J. Jackson et al. 2020).

BREEDING HABITAT

The Hairy Woodpecker is found anywhere that large trees prevail, including small woodlots, city parks, residential areas, cemeteries, and even recent burns (Danz et al. 2007) **(B)**. Generally mature deciduous and mixed deciduous forests (both upland and lowland) are preferred (Noon et al. 1979). Where timber harvests have been managed to retain snags, these residual trees may be used as perches or as foraging sites but are seldom used for nesting cavities (Loose and Anderson 1995). Although one study suggested the species required a minimum of 4 ha of wooded habitat for nesting (C. S. Robbins 1979), its sensitivity to the size of a forest stand and the composition of the forest landscape have not been well studied.

As a primary cavity nester, the limiting factor in habitat selection is the availability of suitable nest trees that are of sufficient size and have an adequate amount of decay for excavation. Hairy Woodpeckers prefer nesting in mature deciduous trees, frequently selecting a live tree with fungal heartrot. Their large, chisel-like bill provides them with some flexibility to select trees that are relatively healthy with harder wood (Runde and Capen 1987).

POPULATION ABUNDANCE

PIF (2021) generated a breeding population estimate of 8.9 million breeding adults; approximately 8.7 million reside in the United States and Canada, with the remainder in Mexico and Central America. Minnesota supports approximately 170,000 individuals. Although Hairy Woodpeckers are sparsely distributed throughout their breeding range, breeding densities found in portions of the Great Lakes region are among the highest in the BBS survey region (Sauer et al. 2020).

Because of their relatively low abundance, BBS population trends for the species are often unreliable. One of the few ecological regions where the trend line is statistically robust is the Boreal Hardwood Transition Bird Conservation Region, which includes the Laurentian Mixed Forest Province of northern Minnesota. From 1966 to 2019, Hairy Woodpeckers demonstrated an average annual increase of 1.32 percent in the region **(C)**. Indeed, the population in the United States and Canada is estimated to have increased 54 percent from 1970 to 2014 (Rosenberg, Kennedy, et al. 2016). Although Minnesota BBS data lack statistical precision, they also suggest an increasing population (Sauer et al. 2020) and are corroborated by data collected on Minnesota's two national forests from 1995 to 2020 (Grinde, Bednar, et al. 2020).

By and large, populations across New England, Canada, the Great Lakes, and the western states are increasing, while populations in many regions of the southeastern and central United States are declining (Sauer et al. 2020). Factors often cited as responsible for local declines are habitat fragmentation, loss of mature and old-growth forests, and competition for nesting cavities with European Starlings and House Sparrows. The latter species can be quite aggressive in securing a cavity once the Woodpecker has completed its excavation (J. Jackson et al. 2020; Bent 1939). Population increases, on the other hand, may be due to lower harvest levels on public lands and forest maturation.

B. A hardwood forest with an abundance of live, mature trees with some decay for easier nest excavation is home to multiple breeding pairs of the Hairy Woodpecker. Photo by Lee A. Pfannmuller.

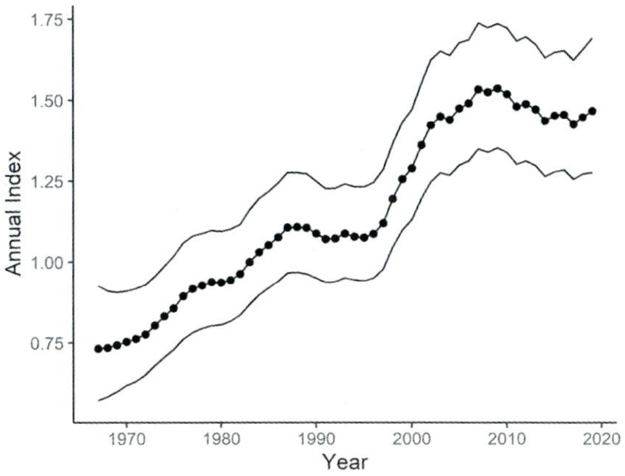

C. Breeding population trend and 95 percent credible interval (solid black lines) for the Hairy Woodpecker in the Boreal Hardwood Transition Bird Conservation Region based on the 1967–2019 Breeding Bird Survey (Sauer et al. 2020).

CONSERVATION

A widely distributed species with an increasing population, the Hairy Woodpecker is a low conservation priority (PIF 2021). Management recommendations focus on protecting forest stands with large, mature hardwood trees (Danz et al. 2007). Maintaining habitat corridors and large forest tracts also may be important (Hess et al. 2000). In the Pacific Northwest, researchers recommended managing approximately 45 percent of the landscape in stands that are older than forty years, and 30 percent of the landscape in stands that are older than sixty years to ensure that adequate nesting habitat is available (Ripper et al. 2007).

Warming temperatures may present some local challenges. NAS (2019) projected a low vulnerability for the species' North American breeding distribution under all three warming scenarios they examined; losses predicted across the Woodpecker's southern breeding range were largely compensated for by gains to the north. Unfortunately, as temperatures rise, the models predict the species' gradual disappearance across much of Minnesota by 2080. Eventually restricted to the northeastern counties under a 3°C warming scenario, the Hairy Woodpecker's ubiquitous presence in the state appears to be short-term.

NORTHERN FLICKER

Colaptes auratus

Minnesota Seasonal Status: Regular breeding resident and migrant; regularly observed during the winter, especially in southern Minnesota. The Northern Flicker was abundant during the MNBBA.
Migration: Variable, but most northern populations are short-distance migrants that winter in southern states.
Food: Ground forager that consumes arthropods, primarily ants; fruits are important in the fall and winter.
Nest: Excavates cavities in snags 2–5 m above the ground; occasionally uses an existing nest cavity.
Conservation Concern: Assigned a moderate PIF Continental Concern Score of 9/20.

Among the nine woodpecker species that nest in Minnesota, the handsome Northern Flicker is unique in many respects. Other than the confusing array of names it has been given over the years, it is our only woodpecker with brownish-gray feathers patterned with heavy black markings on its back, wings, and breast. A prominent black crescent adorns the upper breast and, in flight, its white rump is always conspicuous. It is exceeded in size only by the Pileated Woodpecker.

Plumage variations among Flickers have intrigued biologists for more than one hundred years. In the east, including Minnesota, the undersides of the tail and wings are yellow, as are the dorsal side of the flight feather shafts. The male and female have a red crescent on the back of the head, and the male has a black "mustache" at the base of the bill. Farther west, the yellow colors are replaced by salmon-red; the red crescent on the back of the head is absent, and the male's mustache is red. The two forms overlap and hybridize in a band that extends from southeastern Alaska to northwestern Texas (Wiebe and Moore 2020). Considered separate species for many years, the Red-shafted Flicker and Yellow-shafted Flicker are now considered one species, the Northern Flicker.

In addition to its unique plumage, the Northern Flicker is more likely to be seen foraging on the ground for its favorite food, ants, than excavating a tree trunk for larvae. Its loud and emphatic *wick-a-wick-a-wick-a* call might be mistaken for that of a Pileated Woodpecker, but the latter's call

Photo by David Brislance.

is delivered at a lower pitch and varies in tempo, unlike the higher-pitched and even delivery of the Flicker's call.

Widely distributed during the breeding season across the United States, Canada, and portions of Mexico, Central America, and the Caribbean, the Northern Flicker is absent only from some of the most arid regions of the Southwest. It is least abundant in the southeastern United States, from Louisiana and Arkansas east to the Atlantic coast.

MINNESOTA BREEDING DISTRIBUTION

Roberts's (1932) original treatise, *Birds of Minnesota,* included one account for the Northern Flicker (i.e., the Yellow-shafted Flicker) and one for the Red-shafted Flicker. Considered separate species at the time, his brief account of the Red-shafted Flicker was limited to a description of two specimens collected from Hennepin and Pipestone Counties, and written accounts of one bird observed in Clay County and one in Rice County. Upon closer examination, however, Roberts believed the specimens were actually hybrids between the Northern and Red-shafted species because the "pinkish-red" in the tail was "much paler than the typical Red-shafted Flicker." He does mention a 1925 report of a typical Red-shafted Flicker taken at Grafton, North Dakota, just 12 miles from Minnesota (Roberts 1932).

It was the Northern Flicker that Roberts devoted most of his attention to, describing it as a common breeding resident throughout the state. Confirmed and inferred nesting records were available from seven counties stretching from Jackson County east to Goodhue County, and north through Ramsey, Hennepin, McLeod, and Meeker Counties in central Minnesota, and Marshall County in the far northwest. Some of the most notable reports were of enormous flights of birds seen during migration, particularly during the fall. Roberts wrote of one flight he witnessed for several days in Itasca State Park in September 1921, when "the ground was covered with Flickers and Robins too numerous even to estimate." During his lifetime, the most notable change in the Flicker's distribution and abundance was an increase in the number of birds that attempted to spend the winter in southern Minnesota.

Green and Janssen's 1975 account of the species (then known as the Common Flicker) also described it as numerous and widespread throughout the state. Winter reports were now regular as far north as Duluth and Moorhead. When Janssen provided another update in 1987, the species' common name had changed back to Northern Flicker. Although its summer status remained unchanged, there was at least one report of an individual that spent the entire winter in Pennington County in the far northwest. Janssen identified 38 counties, covering all regions of the state, where Flickers had been confirmed nesting since 1970. Hertzel and Janssen (1998) later added six more counties to the list. Fieldwork conducted by the MBS further confirmed the species' abundance and statewide distribution.

The Northern Flicker was the most common woodpecker during the MNBBA, with nearly 1,200 more records than the second most common woodpecker, the Yellow-bellied Sapsucker. The only region of the state where detections were sparse was a narrow band of intensively cultivated lands along the Red River valley. Participants reported a total of 4,468 Northern Flicker records from 55.8 percent (2,663/4,772) of the surveyed atlas blocks and from 72.4 percent (1,691/2,337) of the priority blocks **(A)**. Breeding was confirmed in 6.8 percent (325) of the blocks and considered probable in another 10.5 percent (503). Detected in every county, breeding was confirmed in 70 counties; 31 were additions to the list Hertzel and Janssen published in 1998.

The MNBBA probability map predicted that the Flicker attains its highest breeding densities in the deciduous and mixed forests of northern Minnesota **(B)**. Although it was predicted to be less abundant in the northeastern forested region of Lake and Cook Counties, the Ham Lake fire of 2007 and other disturbances, such as wind and insect damage, have temporarily opened the forest canopy and provided

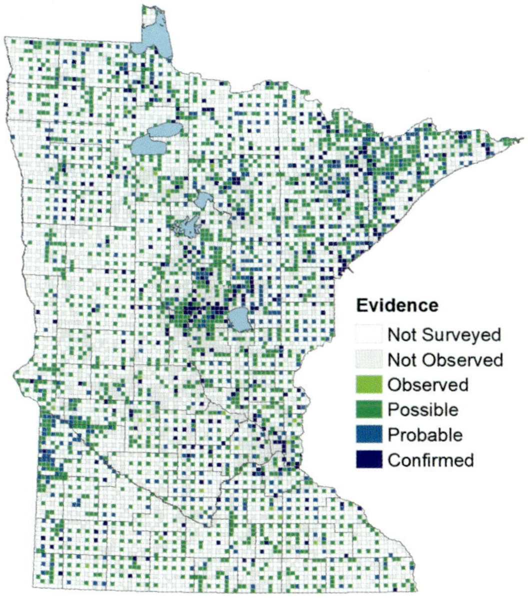

A. Breeding distribution of the Northern Flicker in Minnesota based on the Breeding Bird Atlas (2009–13).

more suitable habitat, as evidenced by the abundance of MNBBA records in the region.

As with many members of the woodpecker family, the overall distribution of the Northern Flicker has witnessed few changes over the past one hundred years in Minnesota and across North America. Local and regional changes in distribution and abundance usually reflect local habitat changes.

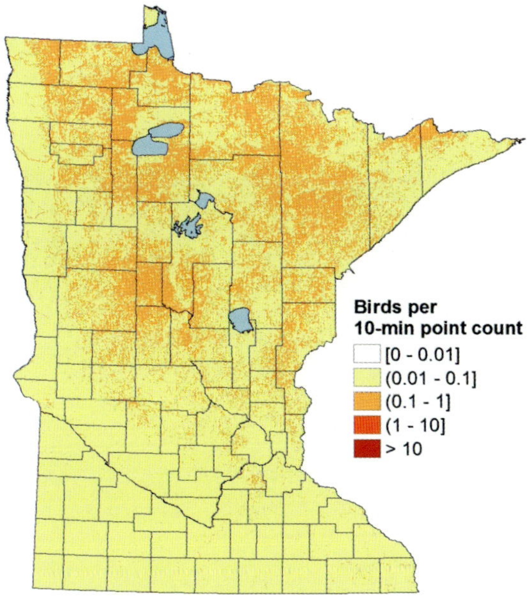

B. Predicted breeding distribution (birds detected per 10-minute point count) of the Northern Flicker in Minnesota based on habitat, landscape context, and climate data gathered during the Minnesota Breeding Bird Atlas (2009–13) using the Generalized Linear Modeling method without an adjustment for detectability.

BREEDING HABITAT

Flickers inhabit a variety of woodlands, including forest edges, farm woodlots, shelterbelts, riparian forests, young successional forests, and savannas **(C)**. Although it is generally an open woodland species, recent clear-cuts and other disturbances in northern counties also provide suitable habitat. Their exposed ground provides ideal foraging habitat, while the adjacent forest provides nest trees.

Nests are placed in dead snags or dead limbs of live trees. In northern forests, aspen trees are a favored nest tree because heartrot softens the interior and makes excavation easier (Wiebe and Moore 2020).

POPULATION ABUNDANCE

PIF (2021) provided a global population estimate of 12 million birds, with 11 million located in the United States and Canada. Approximately 200,000 individuals were estimated to reside in Minnesota. Given the species' statewide distribution and relative abundance, this estimate may be low.

The surveywide BBS trend shows a significant average population decline of –1.21 percent per year from 1966 to 2019 (Sauer et al. 2020). The decline is especially steep in the eastern and central United States, including Minnesota, where the rate of decline since 1967 averaged –2.52 percent per year **(D)**. Although the forest bird monitoring program conducted on the Chippewa and Superior National Forests from 1995 to 2020 documented an overall stable population trend for the species (Grinde, Bednar, et al. 2020), the data also reveal a decline on each forest in the past fifteen to twenty years.

Factors responsible for the species' decline are not well understood. Wiebe and Moore (2020) have proposed three hypotheses: competition for nesting cavities with European Starlings, which aggressively usurp nesting cavities from Flickers; a decline in suitable nest cavities due to habitat loss; and pesticide use in both agricultural fields and urban/suburban neighborhoods where Flickers forage. All three factors are likely more pronounced in the central and southern regions of Minnesota, where agricultural activity and suburban development are intense and where Starling populations are dense and widespread. Nevertheless, the pattern of the Flicker's population trends across Minnesota appears complex and deserves to be closely monitored.

CONSERVATION

Designated a moderate conservation priority owing to its declining population trend (PIF 2021), this common and widely distributed species de-

C. Flickers commonly inhabit open woodlands with mature trees and snags for nesting and an open understory for foraging. Photo by Lee A. Pfannmuller.

serves our attention for its most important role as a primary cavity excavator. Once abandoned, these cavities provide nesting habitat for a wide range of secondary cavity-nesting birds and roost sites for many birds and small mammals. Indeed, Wiebe and Moore (2020) note that the Northern Flicker is "a keystone species by excavating a large percentage of the nest cavities used by numerous secondary cavity-nesting species." As a result, forest management practices should retain dead and dying trees, especially near open foraging areas where pesticide use is minimal.

Fortunately, climate change is not a major concern for the Flicker. Under a 2°C warming scenario, NAS (2019) projects a low vulnerability for the species' North American and Minnesota breeding distribution. Only under a 3°C warming scenario are habitat losses predicted along the state's western border. In the interim, Minnesota's most common woodpecker merits more consideration for its critical role in forest communities.

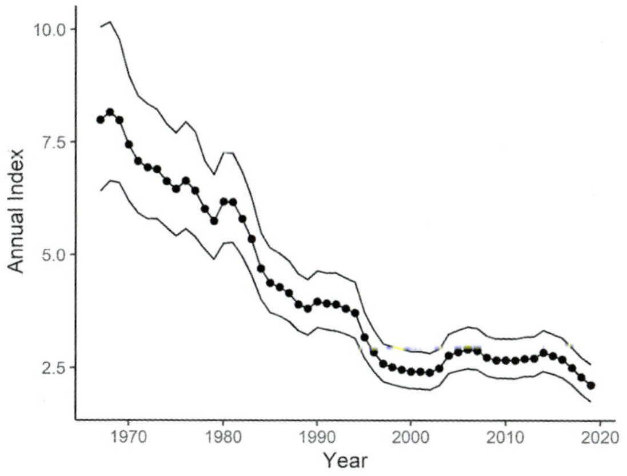

D. Breeding population trend and 95 percent credible interval (solid black lines) for the Northern Flicker in Minnesota based on the 1967–2019 Breeding Bird Survey (Sauer et al. 2020).

PILEATED WOODPECKER

Dryocopus pileatus

Minnesota Seasonal Status: Permanent breeding resident. The Pileated Woodpecker was common during the MNBBA.
Migration: Nonmigratory; movements reported during the fall are believed to represent dispersal of young birds or seasonal movements following the breeding season.
Food: Bark gleaner and excavator that feeds primarily on ants and larvae of wood-boring beetles. Large, oval excavations to extract food are diagnostic. Also consumes nuts and berries.
Nest: Primary cavity excavator; the nest cavity and its oblong nest entrance are usually excavated in a dead or dying tree approximately 5–20 m above the ground.
Conservation Concern: Assigned a low PIF Continental Concern Score of 7/20.

Count the Pileated Woodpecker among an elite group of birds that consistently spark excitement among all nature enthusiasts. Its impressive size, its powerful drumming, and its rather comical call each evoke images of a primeval forest inhabitant. Many vernacular names have been used over the years to describe this unique woodpecker, including "black log," "stump-breaker," and "laughing woodpecker." Harbingers of earlier times, one name is as appropriate today as it was one hundred years ago: "king of the woods" (Bent 1939).

Roberts (1932) aptly described the Pileated Woodpecker as "a magnificent bird, nearly the size of a Crow and nearly as black, but adorned with a flaming scarlet crest and showing, in flight, great white patches on the wings." Other characteristic features include a dark mustache that extends from the base of the bill down the throat. In the male, the mustache begins as red and merges into black; in the female, it is entirely black. The brilliant red crest extends from the forehead to the back of the head on the male; on the female, the forehead is gray or yellowish brown before grading into red. *Dryocopus pileatus,* the species' scientific name, refers to a crested (*pileatus*) woodcutter (*Dryocopus*).

A series of *wuk* notes that vary in intensity, speed, and pitch is one of the Pileated's most common vocalizations. It can sound similar to a Northern Flicker's call, but the Flicker's vocalization is more uniform in

Photo by David Brislance.

rhythm and tone and doesn't drop in pitch at the end. In flight, the Pileated demonstrates the undulating movement that is characteristic of most woodpeckers.

The Pileated Woodpecker is distributed throughout southern Canada, the eastern United States, and portions of the northern Rockies and Pacific Northwest, with some of the highest breeding densities found in the southeastern states.

MINNESOTA BREEDING DISTRIBUTION

Roberts (1932) speculated that the Pileated Woodpecker was common and widely distributed throughout Minnesota prior to European settlement. By the late 1800s, however, it had become "so rare as seldom to be seen except in remote places." Extensive forest clearing, coupled with rampant shooting for both sport and food, led to its early decline. However, as the public's attitude toward wildlife began to change, the Woodpecker's plight improved, and numbers slowly rebounded. By the early 1900s, Roberts wrote, "while seeing a Pileated is still a matter of special interest, it is by no means as rare an event as it was a decade or two ago." It was present even in isolated woodland groves of the western grasslands. Despite its large size and conspicuous behavior, Roberts had confirmed nesting reports from only two counties (Hennepin and Stearns), while inferred nesting reports were limited to Goodhue County and Itasca State Park.

Green and Janssen (1975) later reported that the species could be found in all forested regions of the state, including "the heavily timbered valleys and lakeshores of the prairie." It was rare only in the southwest, where habitat was limited. Janssen (1987) observed that it decreased in abundance as one moved from the forests of eastern Minnesota west to the prairie–forest transition and grasslands of western Minnesota. He delineated 21 counties where nesting had been confirmed since 1970; by 1998 Hertzel and Janssen added an additional nine counties to the list. By 2014, the MBS reported a total of 772 breeding-season locations, including nine in southwestern Minnesota.

During the MNBBA, observers reported 2,393 Pileated Woodpecker records from 31.3 percent (1,486/4,751) of the surveyed atlas blocks and confirmed breeding in 118 blocks (2.5 percent) **(A)**. The species was reported in 75 of Minnesota's 87 counties and was confirmed breeding in 37 counties. Fifteen of the counties were additions to the list published by Hertzel and Janssen in 1998. In addition to documenting seven possible and probable records in southwestern Minnesota, south of the Minnesota River valley, numerous records were documented from the Red River valley, including three confirmed nesting reports. Nevertheless, despite the large number of counties where the species was reported, it was sparsely distributed wherever the landscape was dominated by intensive cultivation and remains most abundant in the Eastern Broadleaf Forest and Laurentian Mixed Forest Provinces, much as it was nearly one hundred years ago.

The MNBBA probability map depicts the Pileated Woodpeckers' wide distribution throughout the state at albeit low population densities **(B)**. The far west-central and southwest regions of the state support even lower breeding densities, while scat-

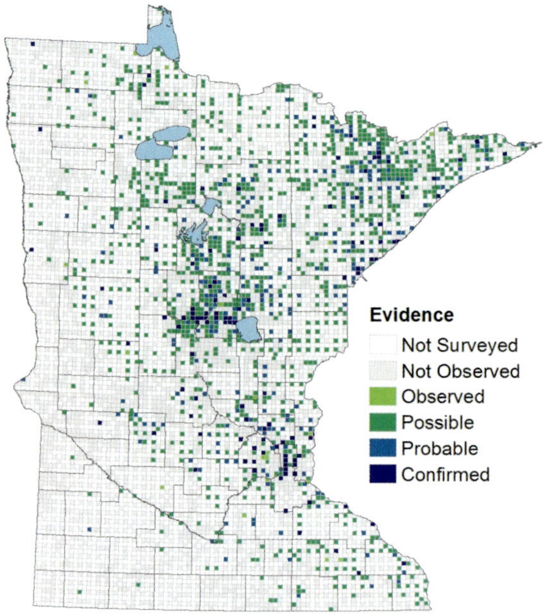

A. Breeding distribution of the Pileated Woodpecker in Minnesota based on the Breeding Bird Atlas (2009–13).

tered areas primarily in north-central and northeastern Minnesota are predicted to support slightly higher densities.

The population decline that Roberts (1932) chronicled in Minnesota in the late 1800s was witnessed throughout the species' range, as was its subsequent recovery (Bull and Jackson 2020). Indeed, populations continued to increase in numbers throughout the twentieth century. Its distribution, however, has remained relatively unchanged.

BREEDING HABITAT

The Pileated Woodpecker is an inhabitant of mature hardwood forests but can also be found in coniferous and mixed coniferous–deciduous stands (Green 1995). Although both upland and lowland forests are used, in eastern North America it often demonstrates a preference for floodplain forests that provide large trees, a greater density of stems, and a higher crown canopy **(C)**. If large trees are left standing on harvested sites, young regenerating forests also may harbor nesting birds (Bull and Jackson 2020).

Considered by some to be an area-sensitive species (e.g., Green 1995), during the MNBBA Pileated Woodpeckers were found in sites as diverse as an extensively forested block in central Cook County to an agricultural block in Clay County interrupted only by a narrow riparian corridor along the Buffalo River. Nesting was confirmed in both blocks. Biologists correctly caution that the mere presence of birds in small, isolated patches of habitat may not indicate the site is suitable for sustaining local populations. Nevertheless, the MNBBA documented four confirmed nesting reports in blocks that included narrow riparian corridors embedded within intensively cultivated landscapes. It appears that even these small ribbons of habitat can sometimes provide suitable habitat for this woodland species.

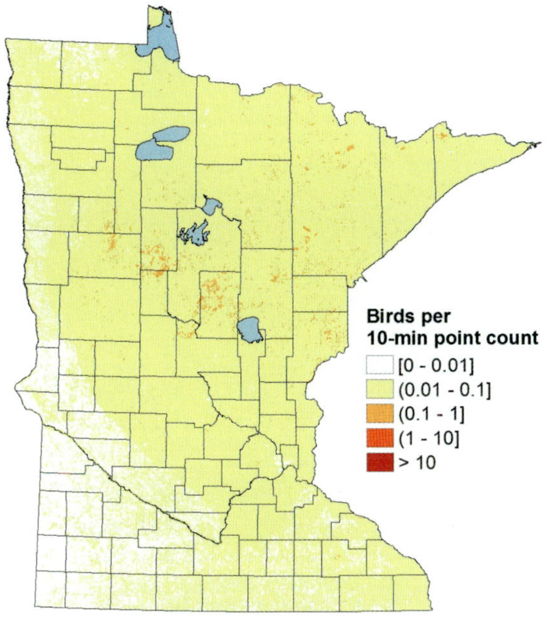

B. Predicted breeding distribution (birds detected per 10-minute point count) of the Pileated Woodpecker in Minnesota based on habitat, landscape context, and climate data gathered during the Minnesota Breeding Bird Atlas (2009–13) using the Generalized Linear Modeling method without an adjustment for detectability.

C. A large, mature tree in this hardwood forest reveals the tell-tale feeding evidence of its Pileated Woodpecker inhabitants. Photo by Lee A. Pfannmuller.

POPULATION ABUNDANCE

PIF (2021) analyzed BBS data to help generate a Pileated Woodpecker population estimate of 2.6 million birds in the United States and Canada. Approximately 55,000 adults are estimated to reside in Minnesota.

Since the BBS began in 1966, the Pileated Woodpecker population has been steadily increasing. Because of its relatively low abundance, many of the state and regional population trends lack statistical precision, including that across the entire survey region. Nevertheless, the data suggest a strong, range-wide increase averaging 1.28 percent per year from 1966 to 2019 (Sauer et al. 2020). BBS data in Minnesota suggest an even larger increase. Overall, biologists estimate that the Pileated Woodpecker's North American population increased 87 percent from 1970 to 2014 (Rosenberg, Kennedy, et al. 2016).

The spread of Dutch elm disease in the latter half of the twentieth century may be responsible for some of the Woodpecker's population growth. The dead trees are a rich source of carpenter ants, wood borers, and other wood-inhabiting invertebrates that provide a significant portion of the species' diet. Other important factors include forest management practices that emphasize retention of large trees and snags in riparian areas and the continued maturation of second-growth forests in the northeastern United States (Bull and Jackson 2020).

CONSERVATION

The most significant threats to the Pileated Woodpecker are the loss of mature forest habitats and "clean" forest management practices that eliminate large snags or live trees that provide essential foraging, roosting, and nesting habitat. Persecuted for many years, the Pileated Woodpecker is now considered a critical member of a healthy forest ecosystem. The large oval cavities it excavates provide nest sites and shelter for other birds and small mammals that depend on secondary cavities. As they chisel away at live and dead trees in search of tasty morsels, their work also accelerates the rate of forest decomposition and nutrient recycling and attracts other birds that feed on the exposed invertebrates (Bull and Jackson 2020).

The Pileated Woodpecker has shown a high degree of adaptability in utilizing a wide range of woodland habitats, a trait that has certainly aided its population recovery. Other than the loss and degradation of habitat, the species faces few significant threats. Even a warmer climate may be of little concern to this giant of the eastern forest landscape. Although small losses are projected along the western periphery of the species' range, including in Minnesota, NAS (2019) classified the species as climate stable under all three warming scenarios examined. In the near term, as long as sufficiently large trees and snags are available, this impressive species will continue to remain a common member of the state's avifauna.

AMERICAN KESTREL

Falco sparverius

Minnesota Seasonal Status: Regular breeding resident, migrant, and regular in winter in southern Minnesota. The American Kestrel was common during the MNBBA.
Migration: A short- to long-distance migrant, the species overwinters from the midwestern United States to Central and South America.
Food: Primarily a sit-and-wait predator on arthropods and small vertebrates, such as small mammals, birds, reptiles, and amphibians. The species will also hover while searching for prey.
Nest: An obligate, secondary cavity nester that relies on woodpecker-excavated or natural cavities, and on nest boxes.
Conservation Concern: Assigned a moderate PIF Continental Concern Score of 10/20.

Formerly known as a Sparrow Hawk, the American Kestrel is a familiar bird, often spotted on a wire or an available perch site in open country. Sandrock and Prior (2014) state the word "kestrel" derives from Middle French and Latin to mean "a kind of falcon" and "a noisy bell or rattle," respectively, the latter referring to its high-pitched call. Its scientific species name, *sparverius,* is Latin for "sparrow hawk," supposedly for its love of sparrows; however, it has a more diverse diet than just sparrows. The American Kestrel is the only kestrel in the western hemisphere, but 12 other kestrel species are found in other parts of the world.

The American Kestrel was described by Roberts (1932) as "arrayed in a varicolored dress of strikingly contrasted pattern which makes it one of the handsomest of its tribe." Both sexes have two conspicuous vertical facial markings, narrow pointed wings, and a relatively long tail. Males have bluish-gray wings, a rufous back with bars, and a rufous tail with a dark sub-terminal band. Females differ in having rufous wings and a rufous tail with dark barring. The voice, described as *killy-killy-killy-killy-killy,* is often sounded when excited or stressed from a predator or as aggression toward a competitor.

The American Kestrel's breeding range is widely distributed throughout North America, including Alaska to New Brunswick, and virtually all of the contiguous United States except southern Florida

Photo by Mike Lentz.

and along the Gulf coast. American Kestrels also nest throughout much of Central and South America. Seventeen subspecies are recognized, ranging from Alaska to Tierra del Fuego, Argentina. Patches of high density are scattered throughout the midwestern and western United States.

MINNESOTA BREEDING DISTRIBUTION

The American Kestrel is the smallest, most widespread, and most abundant falcon in Minnesota. Roberts (1932) said it was "breeding throughout the state; rather infrequent in the prairie groves, most numerous northeastward." He added, "Fifty years ago the Sparrow Hawk was an abundant summer bird throughout Minnesota. . . . In recent years it has disappeared to a considerable extent in the more thickly settled portion of the state." He speculated that the expansion of logging activity and forest fires had allowed American Kestrel populations to expand into the central and northeastern regions of the state. Roberts documented nesting activity from Minneapolis and throughout many counties in central and northern Minnesota.

Decades later, Green and Janssen (1975) described the kestrel as "one of the most numerous breeding hawks" but "least numerous in the heavily wooded portions of the state." They added that fire protection had reduced the availability of the burned-over areas that the species prefers in these forested regions. D. H. Johnson (1982) documented positive nesting from 1960 to 1981 in 29 counties, ranging from Winona County in the southeast, west to Lac qui Parle County, northwest to Pennington County, and northeast to Cook County.

A few years later, Janssen (1987) documented confirmed nesting in 26 counties since 1970. These were similarly distributed throughout the state but included Murray County in the extreme southwestern region of the state. Hertzel and Janssen (1998) later expanded the list to 40 counties with confirmed nesting since 1970.

The MBS has recorded 279 breeding-season locations of the American Kestrel during county surveys in the state. The locations were widely distributed statewide but were especially numerous in the southwest, across the southern tier of counties, and in Clay County, Clearwater County, and central St. Louis County.

The MNBBA illustrated the extensive distribution of the American Kestrel in Minnesota and established it as the most abundant falcon, with 1,782 records **(A)**. Reported from every county in the state, the species was recorded in 28.0 percent (1,352/4,831) of all blocks covered and in 36.4 percent (851/2,337) of the priority blocks. Confirmed nesting was reported in 57 counties. The only counties without confirmed or probable nesting were Becker, Houston, Hubbard, and Watonwan, but all four had possible nesting activity and all have had previously recorded breeding activity.

The probability map based on point counts gathered during the MNBBA identified lands with highest suitability to be the open areas of southern and western Minnesota **(B)**. Low landcover suitability was predicted for northeastern and southeastern forested areas and the Twin Cities metropolitan area.

In their review of the American Kestrel in North America, Smallwood and Bird (2020) emphasized that the species had increased greatly during the

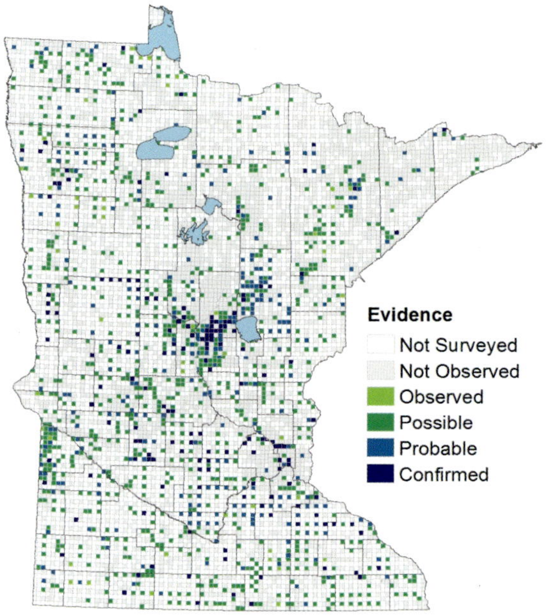

A. Breeding distribution of the American Kestrel in Minnesota based on the Breeding Bird Atlas (2009–13).

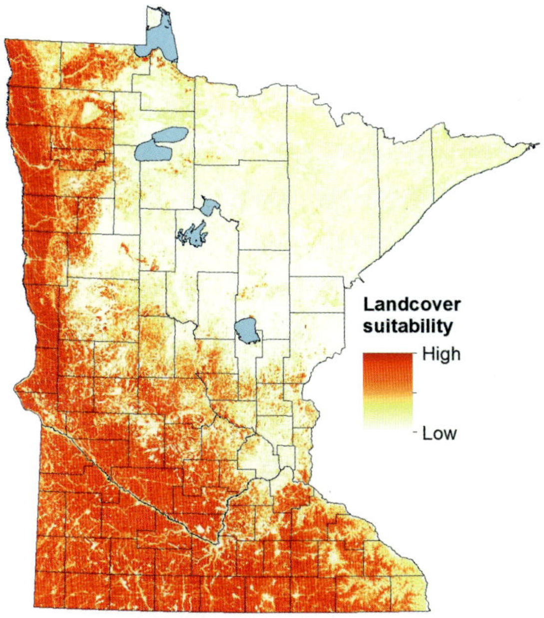

B. Landcover suitability map for the American Kestrel in Minnesota based on habitat, landscape context, and climate data gathered during the Minnesota Breeding Bird Atlas (2009–13) using the MaxEnt modeling approach.

eighteenth and nineteenth centuries in eastern North America with widespread deforestation and "horse-powered" agriculture. Both activities provided open areas for foraging as long as suitable nesting sites in trees were available. Subsequently, urban and suburban areas as well as reforestation rendered many locations again unsuitable for kestrels. Smallwood and Bird also suggest that suitable nest sites, available prey, and perch sites in open country may be limiting factors for the species.

In Minnesota, Roberts (1936) provided little insight into the reasons, besides human settlements, for the American Kestrel being "now much reduced in numbers." In Minnesota, the American Kestrel likely was historically common or even locally abundant in open areas that had suitable nest sites in trees, hunting perches, and available prey. These situations existed in many areas of the state, except the open prairies and grasslands and densely forested areas that lacked edges or openings. Intensive agriculture and settlement that reduced the availability of important nesting and foraging opportunities resulted in declining populations. These activities appeared to be especially prevalent beginning in the early 1900s, leading to the species' decline that Roberts lamented.

BREEDING HABITAT

In Minnesota, American Kestrels prefer open and partly open country with scattered trees, agricultural lands, some open urban areas, and recently logged or burned-over forests (Niemi and Hanowski 1992a). The combination of suitable habitat, nest sites, available prey, and perch sites is essential for successful nesting **(C)**.

POPULATION ABUNDANCE

The population estimate of the American Kestrel in the United States and Canada was 2.8 million individuals, with a global population of 9.2 million and a Minnesota population of 60,000 individuals (PIF 2021). Population trends for American Kestrels based on the BBS in Minnesota detected a significant decline of −2.85 percent per year from 1967 to 2019. Surveywide, from 1966 to 2019, the BBS also detected a decline of −1.41 percent per year, but Sauer and colleagues (2020) suggested caution in their interpretation. Migration counts from Hawk Ridge in Duluth, Minnesota, were mixed, with no significant change from 1974 to 2019. However, from 2009 to 2019, they significantly declined by 4.3 percent per year **(D)**.

CONSERVATION

Despite these recent population declines, there are currently few conservation actions being implemented because of the species' extensive breeding distribution throughout the Western Hemisphere. Nevertheless, the widespread decline in its populations is noteworthy and a cause for alarm. It has experienced considerable changes in population levels since European settlement in North America. The species expanded its population and distribution in the eighteenth and nineteenth centuries with widespread deforestation and conversion to agriculture, especially to pastureland. Recent declines have been largely attributed to "clean farming" and the subsequent lack of suitable nest or perch sites, and to the lack of available food (Smallwood and Bird 2020). The effects of current pesticide use on American Kestrels are unclear; indirect effects of pesticides due to their reduction in arthropod food resources may be greater.

C. American Kestrels are found in open landscapes where large trees or nest boxes are available for this hole-nesting bird, with perches suitable for locating prey.
Photo by Gerald J. Niemi.

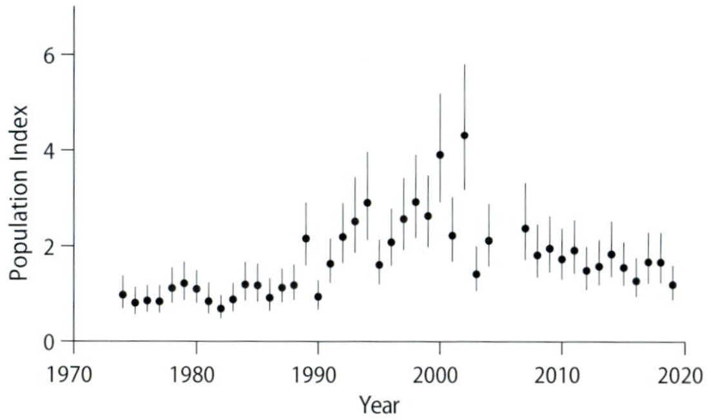

D. Fall migration counts and 95 percent credible intervals (vertical lines) of American Kestrels at Hawk Ridge in Duluth have a variable pattern in its population index. The trend was insignificant from 1974 to 2019, but significantly declining from 2009 to 2019. The index was calculated by the Raptor Population Index (Oleyar et al. 2021).

Shooting of American Kestrels, especially during migration periods, was likely a contributing factor to some reductions in the twentieth century until migratory bird laws were enacted and enforced. Effects of pesticides during the DDT era may have contributed to some reductions in populations but are not thought to be a major cause of population decline. The introduction of the European Starling to the United States in the late 1800s and its appearance in Minnesota in 1929 likely contributed to population declines because Starlings compete for nest boxes. More recently, predation by Cooper's Hawks has been reported, but its effect on populations is largely unknown.

From the perspective of climate change, NAS (2019) projected a stable breeding distribution in North America for all three climate change scenarios they examined. Under a 2°C warming scenario, only 7 percent of the Kestrel breeding range would be lost and 22 percent would be gained in North America by 2050. In Minnesota, no changes in its breeding range were predicted.

The future of the American Kestrel is of growing concern in North America and in Minnesota. Maintaining or restoring suitable nest and perch trees as well as considering the safe use of pesticides are likely to provide the best options for conserving the species. In any event, a major goal would be to avoid any further population declines of this beneficial and well-dressed falcon in Minnesota.

MERLIN

Falco columbarius

Minnesota Seasonal Status: Regular breeding resident and migrant; increasingly overwintering in the state. The Merlin was uncommon during the MNBBA.
Migration: Short- to long-distance migrant; individuals from Minnesota are thought to migrate across the western states and along coastal areas to Mexico, the Caribbean, and Central America, and as far south as Ecuador in South America.
Food: Predator of small to medium-sized birds, usually captured on the wing or off perches.
Nest: Site usually high in a coniferous tree within forested or open habitats of riparian areas of lakes and rivers and in urban areas; rarely on the ground, on a cliff, or in a cavity. Does not build its own nest, but uses old crow, raven, or hawk nests.
Conservation Concern: Assigned a low PIF Continental Concern Score of 7/20.

Photo by Michael Furtman.

The Merlin, formerly known as Pigeon Hawk, is a small, fast falcon that has experienced a massive range expansion and population increase since the 1960s, including extensive use of urban areas. The name Merlin derives from an old French name for the species, *esmerillon,* because it was originally named in Europe. Its scientific name, *columbarius,* is Latin for "pertaining to pigeons or doves," because its flight is similar to that of pigeons (Warkentin et al. 2020).

The Merlin is often referred to as "a miniature Peregrine Falcon" because much of its plumage has similarities with that of its larger congener (Roberts 1932). Adult males have a blue-gray back, whitish breast with brown streaking, rufous-colored leg feathers, and a dark tail with light gray banding and a white terminal band. Males are 25–30 percent smaller than females. Adult females have brownish upperparts, a whitish breast with brown streaking, and a dark tail with buffy bands and a white terminal band. Juveniles of both sexes resemble the female. Both sexes vocalize *kee-kee-kee,* which is usually elicited near the nest.

A Holarctic species, the Merlin breeds across northern North America, Europe, and Asia. In North America it is found north to the tree limit and the prairie parklands of the northern Great Plains. Its breeding distribution is expanding southward into the northeastern and midwestern states. It is sparsely distributed throughout its range.

MINNESOTA BREEDING DISTRIBUTION

Roberts (1932) labeled the species as "mainly a migrant, spring and fall," but he acknowledged that there were several summer records, evidence that it "breeds infrequently in the extreme northern part of the state." He noted observations of Merlins that indicated the possibility of nesting in many areas, specifically at Leech Lake and in Cook, Itasca, Marshall, and St. Louis Counties, but as of 1932 he had never reported a positive nesting in Minnesota. In Roberts's 1936 version of his treatise, three nests had been located in northern Cook County. One nest with four young was found on Gunflint Lake, and another with three young and one egg on Saganaga Lake. The third observation was thought suspicious; a nest was reportedly found on Winchell Lake, where "mice" were being brought to the nest. Merlins are known to feed primarily on birds, large insects, and bats. Small mammals are rarely reported as prey, although they have been in Sweden when small rodents are at high density (Warkentin et al. 2020). Four nests were also found on the north shore of Saganaga Lake, all with young, but these were located in Canada.

Decades later, Green and Janssen (1975) identified the Merlin as a summer resident "in the northeastern and north-central regions and adjacent counties in the northwestern region." They suggested it was a very scarce species, except along the Canadian border and possibly in Kittson and Roseau Counties. They reported confirmed nesting in Cook, Lake, and St. Louis Counties. Years later, D. H. Johnson (1982) summarized the nesting activity of Merlins: first, eight confirmed nesting records from 1935 to 1981 from Cook, Lake, and St. Louis Counties, including one record from Minnesota Point in Duluth in 1952; second, 33 summer observations from 1922 to 1981, and 35 records between 1952 and 1979 of "reported nesting" and "defending adults seen" from the same three counties; and, third, summer observations from Becker, Beltrami, Carlton, Cass, Hennepin, Itasca, Kittson, Lake of the Woods, Marshall, Morrison, Otter Tail, Roseau, and Stearns Counties. Possible nesting was also suggested for Agassiz NWR in Marshall County and for Wadena County.

Janssen, in 1987, summarized a breeding distribution of the Merlin similar to that described by D. H. Johnson (1982). He reported confirmed nesting in three counties since 1970, from northern Cook, Lake, and St. Louis. Hertzel and Janssen (1998) added a confirmed nesting in Lake of the Woods County since 1970. The MBS recorded 49 breeding-season locations for the Merlin. Most were located in northern Lake and St. Louis Counties, but locations were identified as far south as Carlton and Aitkin Counties, west to southern Beltrami County, and in the northwest, where there

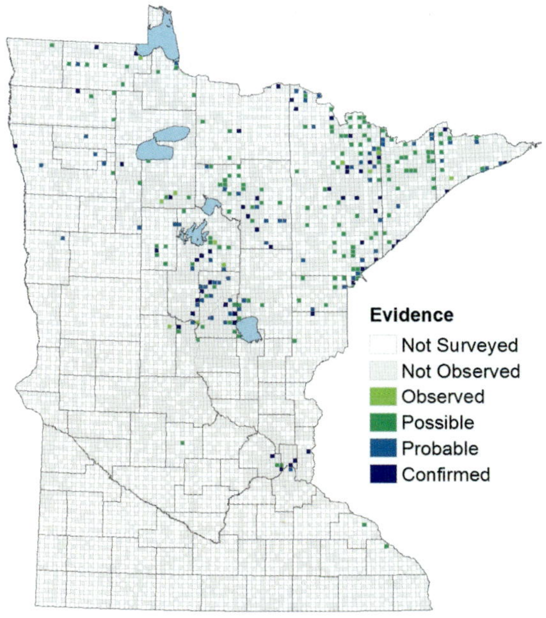

A. Breeding distribution of the Merlin in Minnesota based on the Breeding Bird Atlas (2009–13).

were single locations in Kittson and Roseau Counties. Confirmed nests were added for Pennington County in 2000 and in Red Lake County in 2005 (MOU 2021a).

The MNBBA reported 401 records of the Merlin and greatly expanded the potential breeding range of the species in Minnesota **(A)**. Records were located as far southeast as Winona County, southwest to Meeker County, and west to Clay and Polk Counties. A total of 56 blocks had confirmed nesting records, and 71 had probable nesting records. Merlins were detected in 6.1 percent (291/4,743) of all the surveyed blocks. Confirmed nesting was identified in 19 counties, ranging as far south as Dakota County and northwest to Marshall, Polk, and Roseau Counties. A confirmed breeding with fledged young was noted for Anoka County in 2017 (MOU 2021a).

Hatch (1892), Roberts (1932, 1936), and S. D. Robbins (1991) in Wisconsin all suggest that the Merlin was a rare species in Minnesota and Wisconsin in the late 1800s and early 1900s. Hatch (1892) stated, "I have never yet seen a specimen of them alive, much as I have been on the alert for them." In his detailed summary on the status of the Merlin, D. L. Evans (1982) documented that the "Prairie or Richardson's" subspecies suffered the most serious population declines during the DDT era in the mid-1900s. He also stated that the "Boreal" Merlins, the primary subspecies found in Minnesota, also declined to some extent but suggested they had stabilized by the early 1980s. Unfortunately, they are still being exposed to organochlorine compounds during migration and in overwintering areas, which is enough to affect reproductive success (Farmer et al. 2008).

Warkentin and colleagues (2020) emphasized that many data sets indicated Merlins had recovered from the widespread population declines of the 1960s and suggested either stable or increasing populations from the 1960s through the early 2000s. Merlins have made a remarkable recovery and expansion in their North American population through their use of urbanized landscapes. In Minnesota, Merlins have substantially increased over the past 100 to 150 years, especially through their colonization and use of old crow nests in urban areas. Similar increases have been reported in Michigan, Ontario, and Wisconsin.

BREEDING HABITAT

Merlins primarily use shorelines of lakes, rivers, and bogs; forest openings; and islands of larger lakes, especially those with large evergreen trees with suitable nest sites (D. H. Johnson 1982; Warkentin et al. 2020). More recently, urban and suburban woodlots or forested areas with large evergreen trees, food, and nest sites have proved to be suitable nesting areas **(B)**.

POPULATION ABUNDANCE

Population estimates by PIF (2021) were 1.6 million individuals in the United States and Canada and 12,000 in Minnesota. Rosenberg, Kennedy, and colleagues (2016) calculated that Merlin populations have increased by more than 200 percent from 1970 to 2014! Migration counts at Hawk Ridge, in Duluth, also increased significantly from 1974 to 2019 by an average of 6.18 percent per year, although the recent counts from 2009 to 2019 have been variable **(C)**. The Merlin's increase is thought to be primarily associated with the abatement of DDT use and the species' colonization of urban areas with their availability of nest sites and small to medium-sized birds for food.

B. Merlins typically nest in large coniferous trees where they use old raptor, American Crow, or Common Raven nests. These can be found in many locations along lakeshores, openings, or, as shown here, in urban areas. Photo by Lee A. Pfannmuller.

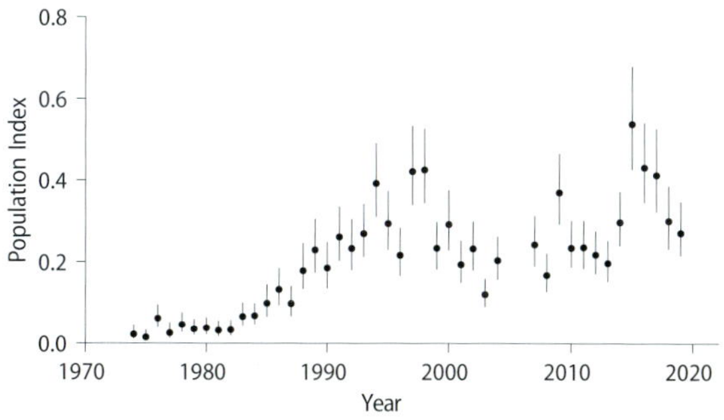

C. Fall migration counts and 95 percent credible interval (vertical lines) of Merlins at Hawk Ridge in Duluth show a consistent and significant increase in its population index from 1974 to 2019, as calculated by the Raptor Population Index (Oleyar et al. 2021).

CONSERVATION

The Merlin has had a highly variable population history over the past 100 to 150 years but currently its population is healthy. Roberts, based on his early experiences, would be surprised at the Merlin's current distribution and perhaps cautious. He warned that "if it were common it would be, despite its small size, a serious menace to bird-life, both wild and domesticated." Current evidence suggests this worry is unwarranted. Warkentin and colleagues (2020), in their summary of the foods taken by Merlins, listed no domesticated species but commonly found exotic species such as House Sparrows.

Overall causes of mortality are unclear. In the past, shooting was a problem and may continue to be in remote areas and on the wintering grounds in Central and South America (Warkentin et al. 2020). Information on mortality during migration due to collisions with windows, towers, or wind turbines is largely unknown. The potentially growing threat to Merlins may be from climate change. With a 2°C warming scenario, NAS (2019) projected a low vulnerability for the Merlin's breeding distribution in North America, including a 24 percent range loss and a 12 percent range gain primarily in the far north by 2050. In Minnesota, the species was predicted to be extirpated under the 3°C warming scenario by 2080. Yet, the salvation for the species' future may be its success in using urbanized landscapes, which may mitigate the effects of climate change.

PEREGRINE FALCON

Falco peregrinus

Minnesota Seasonal Status: Regular breeding resident, migrant, and winter visitant. The Peregrine Falcon was uncommon during the MNBBA.
Migration: A short- to long-distance migrant that winters along the North American coasts (Pacific, Atlantic, and Gulf); individuals can be found almost anywhere from the United States to South America. The species practices leap-frog migration, with the most northerly populations migrating the farthest, from northern North America to Argentina and Chile in southern South America.
Food: Primarily birds, small to large (e.g., waterfowl), occasionally mammals, and rarely amphibians, fish, and insects.
Nest: Cliffs, but more recently tall buildings, bridges, smokestacks, and other artificial structures.
Conservation Concern: Assigned a moderate PIF Continental Concern Score of 9/20. Formerly listed as a federal and a Minnesota *Endangered* species. The Peregrine Falcon was federally delisted in 1999 and downgraded by the state of Minnesota in 2013 to *Special Concern*.

Photo by Mike Lentz.

The Peregrine, whose name means "wanderer," is worldwide in distribution and found on all of the world's continents except Antarctica. It is thought to be the world's fastest bird. As Forbush and May (1955) eloquently described, "its plunge is so lightning-like that the bird seems to have been evolved out of a clear sky, and the sound of its rush is like that made by a rocket. The sudden arrowlike fall of this bird from far heights is one of the most impressive sights in bird life." Direct measurements of its stoops to attack prey have been recorded at more than 330 km per hr (200 miles per hour).

The Peregrine Falcon gained great notoriety during the 1960s and 1970s, when its population plunged owing to the use of persistent chemicals. It has made a remarkable recovery and is one of the grand success stories in conservation. This infamy has also bestowed on it the honor of being one of the most-studied bird species, with a bibliography that exceeds 2,000 primary scientific titles (C. M. White et al. 2020).

The Peregrine is a medium-sized raptor and, like all falcons, has distinctive, long, narrow, pointed wings and a tail of moderate length. Both sexes are similar, with a bluish-gray upperpart, barred breast, and dark facial mustaches with a light cheek, but females are larger than males. Juveniles are brownish instead of grayish. Peregrines have considerable geographic variation in their plumage from light-colored birds to dark morphs, while most in Minnesota are intermediate between these two extremes.

Besides its worldwide distribution, the Peregrine Falcon is widely found throughout North America, with its primary breeding distribution in extreme northern regions of Canada, in Alaska, and throughout western North America. It has been introduced in many places, including cities, across the United States and Canada. Populations are widely scattered throughout the species' breeding range.

MINNESOTA BREEDING DISTRIBUTION

Roberts's (1932) summary of the Peregrine Falcon, known then as the Duck Hawk, considered the species' nesting distribution to be the "Lake Superior region and along the bluffs of the Mississippi and St. Croix rivers." He further quipped that it "is probably about as well represented today as it ever was," but noted it was "nowhere a common bird." More specifically, Roberts summarized that a "half-dozen pairs nest along the north shore of Lake Superior, probably about the same number in the bluffs along the Mississippi River south of Red Wing, a fewer number along the upper St. Croix River, and an unknown but probably small number about cliff-bordered lakes and streams between the Canadian boundary and Lake Superior." He described only a few confirmed nests, ranging widely from near Lutsen in Cook County and Little Marais in Lake County along the North Shore of Lake Superior to Homer in Winona County. All observed nests had eggs, young, or both.

Green and Janssen (1975) listed the Peregrine Falcon as "formerly a resident" along the bluffs of the Mississippi River, the upper St. Croix River, the North Shore of Lake Superior, and the Boundary Waters Canoe Area Wilderness (BWCAW), with the last reported nesting of the species within these areas occurring in 1962, 1945, 1964, and 1964, respectively. The authors noted that "[f]rom 1965 through 1969 there were no summer reports, but in 1970 and 1971 there were observations of a single bird in two localities in the Twin Cities area."

D. H. Johnson (1982) summarized the nesting of the Peregrine Falcon reported by various authors from the mid-1930s to the 1980s. Besides the nesting records previously reported by Roberts (1932), Johnson included 13 nesting records ranging from 1939 to 1964. These records included Cook, Goodhue, Lake, Pine, St. Louis, Washington, and Winona Counties. Johnson noted the paucity of observations in the 1960s but mentioned two summer observations since the 1971 report by Green and Janssen. Both observations were in June 1981, in St. Louis and Nobles Counties.

Janssen (1987) reiterated much of what had been previously described by Green and Janssen and by D. H. Johnson. He reported on the reintroduction programs that were initiated in Cook and Wabasha Counties. The first successful reintroduction occurred in 1984 in Wabasha County. That program was started in 1982 at Weaver Dunes by Drs. Patrick Redig and Harrison "Bud" Tordoff, both from the University of Minnesota, in conjunction with the MNDNR Nongame Wildlife Program (Tordoff and Redig 1997, 1999; Midwest Peregrine Society 2017). The first reintroductions were unsuccessful because of predation by owls and raccoons. Later reintroductions were successful at buildings in Minneapolis and Rochester and on cliffs along the North Shore of Lake Superior. These initial humble efforts to reestablish Peregrine Falcon breeding in Minnesota have resulted in the extensive nesting we see in the state today.

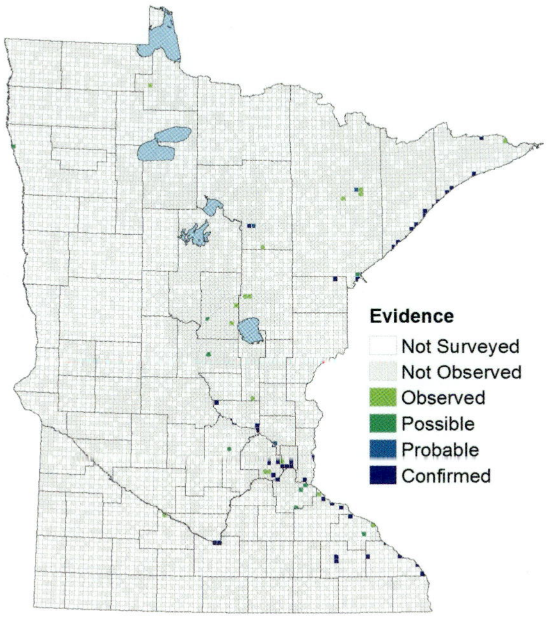

A. Breeding distribution of the Peregrine Falcon in Minnesota based on the Breeding Bird Atlas (2009–13).

As of 1998, Hertzel and Janssen reported confirmed nesting in 11 counties since 1970 (Dakota, Hennepin, Itasca, Lake, Olmsted, Ramsey, Sherburne, St. Louis, Washington, Winona, and Wright). Redig and colleagues (2021) summarized all Peregrine nesting activity since the reintroduction program began, which, in addition to the counties delineated by Hertzel and Janssen, now included nesting in Blue Earth, Carlton, Clay, Cook, Crow Wing, Goodhue, Houston, Stearns, and Wabasha Counties.

MNBBA participants reported 132 records for the Peregrine Falcon, primarily distributed on the North Shore of Lake Superior from St. Louis to Cook County and along the Mississippi River from Winona County north to Itasca County **(A)**. Many of the records were courtesy of the Midwest Peregrine Society. The species was detected in 1.7 percent (79/4,742) of the surveyed blocks and from a mere 0.9 percent (20/2,337) of the priority blocks. The general distribution of the many observations corresponded with the species' historical breeding distribution along the North Shore of Lake Superior and along the Mississippi River, especially in southeastern Minnesota.

Nesting was confirmed in 49 blocks in 17 counties. In addition to reports along the North Shore and the Mississippi River, other confirmed nesting included aeries in Cohasset, Itasca County; northern Cook County in the BWCAW; Cloquet in northern Carlton County; Washington County nesting along the St. Croix River; and Mankato, Blue Earth County. Probable nesting was identified in the iron mining region of central St. Louis County, and possible nesting was identified from East Grand Forks, Polk County, and Crow Wing and Morrison Counties in central Minnesota. Scattered observations were also noted from 17 blocks, including blocks not previously mentioned in Aitkin, Brown, Lake of the Woods, and Mille Lacs Counties.

The history of the decline in the Peregrine Falcon's population has been well-documented in Minnesota (Coffin and Pfannmuller 1988) and in North America (D. L. Evans 1982; C. M. White et al. 2020). For instance, Coffin and Pfannmuller summarized the fact that the use of organochlorine pesticides, especially DDT, after World War II caused the pesticides' accumulation in the food chain and eventual transfer to the Peregrine Falcon via its predation on birds. DDT was linked with eggshell thinning in the Peregrine Falcon and many other species of birds and resulted in poor or complete failure of nesting. In addition, habitat loss and increased human activity contributed to declines. D. H. Johnson (1982) highlighted that the Peregrine Falcon had begun declining in the late 1940s and early 1950s throughout the eastern United States; others noted similar declines throughout North America. C. M. White and colleagues (2020) identified the maximum extent of reduction in the species distribution, which occurred from 1972 to 1975.

Minnesota's active reintroduction program for Peregrine Falcons **(B)** was part of the recovery plan for the species in North America. This plan has led

B. Professor Patrick Redig and Jackie Fallon banding and gathering critical data on the health of a Peregrine Falcon chick at a nesting site in Minnesota. The Peregrine Falcon reintroduction and monitoring program is a highly successful bird species recovery program in Minnesota and in North America. Photo by Chet Ellingson, Midwest Peregrine Society.

C. Natural Peregrine Falcon nesting sites occur on the cliffs along the North Shore of Lake Superior. Photo from Midwest Peregrine Society.

to an active monitoring program and detailed reporting of aeries found in Minnesota from 1982 to 2020 (Fallon 2020) as well as in many states in the Midwest (Redig et al. 2021). Fallon (2020) reported 48 sites in Minnesota that were occupied by a pair of falcons and 122 fledglings at 46 territories. Fallon summarized that "on average, Minnesota is home to ~ 60+ occupied Peregrine territories, producing ~ 120–150 chicks annually." Furthermore, the estimated production in 2019 and 2020 averaged 2.67 birds per territory, which far exceeded the USFWS recovery goal of 1.0 young/successful territory.

BREEDING HABITAT

The Peregrine Falcon nests in high, inaccessible places that offer platforms for nesting and sweeping views of the surrounding landscape. These locations generally occur in high bluffs adjacent to large bodies of water **(C)**; but as observed in recovery programs, the species has accepted using many fabricated structures, such as buildings, towers, bridges, and smokestacks **(D, E)**. Roberts (1932) even commented on the potential preadaptation to such use when it was related to him "that during the grand parade in Washington at the inauguration of President Hoover, March 4, 1929, a Duck Hawk viewed the proceedings from various points along Pennsylvania Avenue."

The availability of food is also an important factor in the occupation of suitable breeding habitat. The species hunts in open areas such as shorelines, mudflats, and other spaces where birds congregate. Suitable places are found within human-dominated habitats, where pigeons, blackbirds, crows, starlings, shorebirds, or waterfowl can often be found in large groups **(F)**.

POPULATION ABUNDANCE

Roberts's (1932) summary suggested a small breeding population in Minnesota during the late 1800s and early 1900s, but the coverage of the state was minimal in remote areas, such as the North Shore of Lake Superior, the Canadian Shield in what is now the BWCAW and Voyageurs National Park, and the vast river networks of Minnesota. Coffin and Pfannmuller (1988) suggested that the Peregrine Falcon breeding population "once totaled perhaps 30 or 40 pairs."

PIF (2021) estimated the US and Canadian breeding population of the Peregrine Falcon as more than 72,000 adults, and a global population of 340,000; the latter including the massive area of North America, Europe, and Asia. They did not include an estimate for Minnesota; however, Fallon (2020) estimated an average of more than 60 occupied territories in Minnesota each year, which suggests a minimum of 120 Peregrine adults in the state. Environment Canada (2011) estimated a broad range, reflecting the uncertainty in population estimates, at 5,000 to 50,000 adults in Canada.

Population trends were unreliable for the Peregrine Falcon based on the BBS because of the species' rarity and the fact that the BBS roadside count method was not designed to adequately detect populations of many raptorial species, such as the Peregrine Falcon. The Raptor Population Index (Brandes et al. 2016; Crewe, Taylor, et al. 2016) was developed to assess population trends for migrating raptors at fall migration count sites, such as those at Hawk Ridge in Duluth. The count trend indicated a highly significant population increase that averaged 5.97 percent per year from 1974 to 2019 (Oleyar et al. 2021) **(G)**. From 2009 to 2020 Peregrine annual fall counts at Hawk Ridge have varied from 53 to 182.

CONSERVATION

Because of extremely low populations of the Peregrine Falcon in the 1950s and 1960s, primarily due to DDT and the closely related DDE, the species was listed as federally *Endangered* in 1970. Efforts to propagate and release falcons began in the late 1960s (C. M. White et al. 2020), but reintroductions were not initiated until 1974 and 1975 after a meeting of experts. However, the first Peregrine Falcon Eastern Population Recovery Plan was not published until 1979. After the Peregrine's introduction in many places, including cities, across the United States and Canada, established populations are now widely scattered throughout the species' current breeding range. The species was removed

D. A nest box used by Peregrine Falcons for many years at the top of the former Hotel Duluth, now Greysolon Plaza, in downtown Duluth, MN. Photo by Gerald J. Niemi.

from the federal endangered species list in 1999 (USFWS 1999a). Officially listed as *Endangered* in Minnesota in 1984 (Coffin and Pfannmuller 1988), it was downgraded to *Threatened* in 1996 (MNDNR 1995) and to *Special Concern* in 2013 (MNDNR 2013). Even though the recovery program has been highly successful, natural reproduction and full recovery of the population in North America still deserves to be closely monitored. Given its relatively small population, it has been designated a moderate conservation priority (PIF 2021). From the perspective of climate change, NAS (2019) considered the Peregrine Falcon's breeding range stable in North America with all three climate scenarios they examined. The winter range of the Peregrine was also predicted to be stable under all three warming scenarios and all predicted an expanding overwintering range in Minnesota.

Many issues are critical for the future conservation of viable populations of the Peregrine Falcon. These include protecting nest sites; providing education to avoid the illegal shooting of the Peregrine Falcon and all raptors; ensuring a suitable supply of birds that the species primarily forages upon; improving understanding of diseases and parasites; and the safe use of pesticides (C. M. White et al. 2020). The Peregrine Falcon has had a high-profile history in North America. Its continued existence is highly dependent on continued protection, education, and management, especially in monitoring its populations.

The future of this sleek, swift, and majestic species is substantially improved from the 1960s and 1970s, when nary a Peregrine Falcon could be found in Minnesota. Its recovery and return to something that resembles its recent historical distribution is a welcome sight and is an outstanding example of conservation efforts that can be effective when government agencies, universities, science, and the public work together with a vision and a realistic plan.

E. The Ford Avenue Bridge, located immediately north of Lock and Dam No. 1 on the Upper Mississippi River, connects the cities of Minneapolis and St. Paul and has provided nesting habitat for Peregrine Falcons for many years. The bridge's broad reinforced concrete arches and supporting structures first attracted nesting Peregrine Falcons in 1997. Since then, the bridge has been used by Peregrines during multiple breeding seasons, resulting in the production of 39 young through 2021 (Redig et al. 2021). Photo by Lee A. Pfannmuller.

F. A close-up view of a nest box high on a skyscraper of downtown Minneapolis with an adult Peregrine Falcon watching over three young Peregrines. The abundance of food resources even in this downtown location provides an ample supply for these healthy young chicks. Photo by Mark Martell.

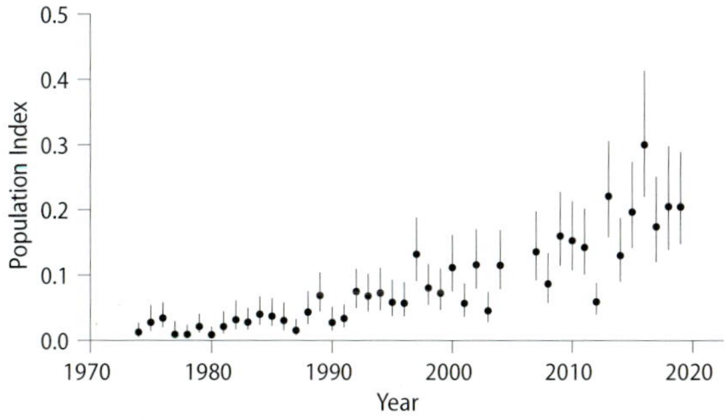

G. Fall migration counts and 95 percent credible interval (vertical lines) of Peregrine Falcons at Hawk Ridge in Duluth show a steady and significant increase in its population index from 1974 to 2019, as calculated by the Raptor Population Index (Oleyar et al. 2021).

GREAT CRESTED FLYCATCHER

Myiarchus crinitus

Minnesota Seasonal Status: Regular summer resident and migrant. The Great Crested Flycatcher was common during the MNBBA.
Migration: Medium- to long-distance migrant; most birds winter in Central America and northern South America.
Food: Aerial insectivore that also gleans arthropods from foliage and dives after prey items on the ground.
Nest: Unique among Minnesota flycatchers, the species is a secondary cavity nester that utilizes natural cavities, cavities excavated by woodpeckers, and an assortment of structures, including nest boxes, drainpipes, and buckets. Shed snakeskins may be used as nesting material, especially in southern latitudes.
Conservation Concern: Assigned a low PIF Continental Concern Score of 8/20.

In a family of flycatchers dominated by olive-gray plumages, where focusing on the fine intricacies of wing bars and eye-rings is necessary to identify many of the species, the Great Crested Flycatcher is a welcome reprieve. The size of a Cardinal, the bird has an olive-brown back, a gray neck and upper chest, a creamy yellow belly, and rufous feathers in its long tail and outer wing feathers. The dark olive-brown feathers on its crown are raised when excited to form a well-defined crest, true to its name. The Latin name, *Myiarchus*, refers to the species' fly-catching foraging habits; *crinitus* means long hair, a reference to its crest. Both sexes are alike.

Despite its distinct plumage, the secretive Great Crested Flycatcher is more frequently heard than seen. Its vocalizations have a unique raspy quality that, Audubon (1827–38) remarked, "transcends all others in shrillness." Most commonly heard is the loud *whee-eep, whee-eep* call, which serves as a contact call. At dawn or near sunset, the males may sing a more unique song that usually consists of three parts: one low-pitched whistle (*wheeyer*) followed by a pause and then another higher-pitched whistle (*wheeyer*), and ending with a low, burry note that is not always audible (*churr*) (Nice 1931b; W. J. Smith and Smith 1996).

The literature on Great Crested Flycatchers abounds with references to the bird's addition of

Photo by Gerald Hoekstra.

shed snakeskins to its nest. Several early naturalists believed they were employed as a tactic to scare away potential predators. But not all Great Crested Flycatchers use them, and some geographical variation is evident: more nests in southern latitudes contain snakeskins than nests in northern latitudes, which may reflect regional availability. K. E. Miller and Lanyon (2020) described one study that experimentally placed snakeskins in the nest boxes of Flycatchers nesting in Florida and found they reduced predation by small mammals, suggesting that in areas where snakes are abundant, the Flycatchers have adapted to use skins as a preventive measure. The authors point out, however, that other investigators maintain that snakeskins are just another example of the crumpled materials often incorporated into the Flycatcher's nest, including cellophane, wrappers, and onion skins.

Broadly distributed across the eastern United States, from the Great Plains east to the Atlantic coast, the Great Crested Flycatcher is also found in portions of Canada's Prairie Parklands and east along the southern regions of Ontario, Quebec, and the Maritime Provinces. Populations reach their highest breeding densities in the southeastern states.

MINNESOTA BREEDING DISTRIBUTION

Roberts (1932) considered the Great Crested Flycatcher a common summer resident throughout the state's forests and woodlands. He wrote, "Wherever there are woodlands throughout the length and breadth of the state, there the loud, rather harsh calls of the Crested Flycatcher may be heard." Even in western and southern Minnesota the species could be found within the scattered woodlots and shelterbelts. Although he considered it "one of the common birds of the heavy forests," confirmed nesting records were limited to six counties, located primarily in central Minnesota: Becker, Chisago, Hennepin, Otter Tail, Ramsey, and Sherburne. Nesting was inferred in Mille Lacs and Scott Counties.

Green and Janssen (1975) reported the species was a statewide resident except in Lake and Cook Counties, where it was scarce and where breeding had not yet been confirmed. Janssen (1987) also noted its scarcity in the northeast and added that it was rare in the far north-central counties as well. He included a summer distribution map that excluded Cook County, most of Lake, St. Louis, and Koochiching Counties, and the northeastern corner of Itasca County. The map identified 22 counties where nesting had been confirmed since 1970, stretching from Olmsted and Wabasha Counties in the southeast to Clearwater, Pennington, and Polk Counties in the northwest. Hertzel and Janssen (1998) later added six more counties to this list, including Brown and Murray Counties in the southwest. A total of 1,866 breeding-season locations documented by the MBS largely confirmed these earlier accounts but also included several records from Itasca County east through Cook County.

MNBBA participants further established the species' statewide distribution, tallying a total of 3,103 Great Crested Flycatcher records in 39.9 percent (1,896/4,751) of surveyed atlas blocks and in 55.2 percent (1,289/2,337) of the priority blocks. Breeding was confirmed in 115 blocks (2.4 percent) and identified as probable in 589 blocks (12.4 percent) **(A)**.

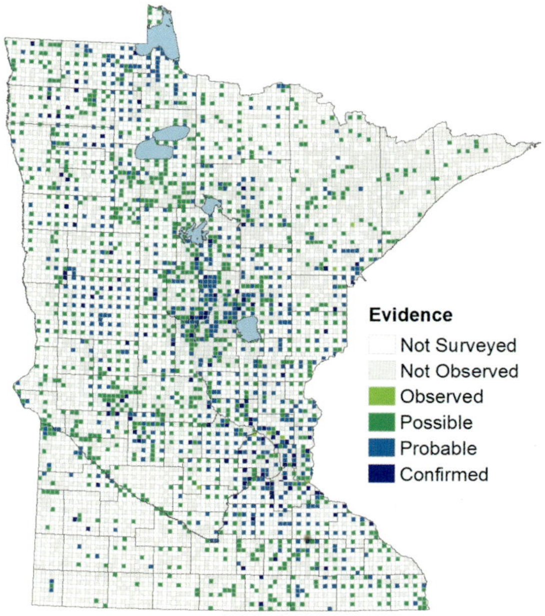

A. Breeding distribution of the Great Crested Flycatcher in Minnesota based on the Breeding Bird Atlas (2009–13).

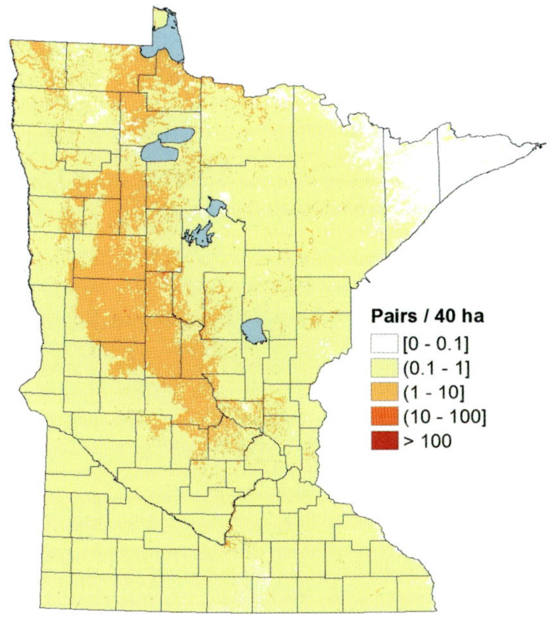

B. Predicted breeding distribution (pairs per 40 ha) of the Great Crested Flycatcher in Minnesota based on habitat, landscape context, and climate data gathered during the Minnesota Breeding Bird Atlas (2009–13) using the Generalized Linear Modeling method with an adjustment for detectability.

The species was found in all but one of Minnesota's 87 counties (Pipestone), with breeding confirmed in 47 counties. Of these, 24 were additions to the list published by Hertzel and Janssen in 1998. Overall, Great Crested Flycatchers were widely distributed throughout the state but were least abundant in the southwest and northeast regions.

The MNBBA probability map predicted that breeding densities are highest in the northern regions of the Eastern Broadleaf Forest Province and in northwestern Minnesota, east through northern Koochiching County **(B)**. The more open, parkland nature of forests in this ecological transition zone between the forests and grasslands provides ideal habitat for the Flycatcher.

The distribution and relative abundance of the Great Crested Flycatcher as described by Roberts in 1932 appears to have changed little in the intervening years. One can surmise, however, that the species is now more abundant along the western and southern edges of the state's northern forest region, where development has created a fragmented landscape of open woodlands favored by the species.

Elsewhere within its breeding range, it has expanded northwest into central Canada, where fire suppression has allowed trees in the prairie parklands to mature and provide nesting cavities (A. R. Smith 1996). Populations also appear to have expanded in eastern Canada, but it is unknown if the expansion is real or simply the result of an increased number of observers (K. E. Miller and Lanyon 2020).

BREEDING HABITAT

The Great Crested Flycatcher is primarily an inhabitant of hardwood forests or mixed stands of hardwoods and conifers. Although it is found in coniferous stands, it usually uses scattered deciduous trees, avoiding sites that are purely coniferous (Danz et al. 2007). Open or semi-open woodlands dominated by mature hardwoods are preferred, as are mesic to wet sites, including lowland hardwoods **(C)**. Natural crevices, knotholes, and broken limbs on live or dead trees provide their favored nesting cavities. If these are unavailable, abandoned woodpecker cavities or a variety of man-made structures are used.

Forest edges are frequently used. When the Flycatcher is found in the interior of a stand, the canopy is usually fairly open and broken. The interiors of closed-canopy forests are avoided. The species has adapted well to urban and suburban landscapes and is frequently found in neighborhoods and parks with mature shade trees (Danz et al. 2007; K. E. Miller and Lanyon 2020).

POPULATION ABUNDANCE

Relying on BBS data, PIF (2021) biologists generated a population estimate of 8.8 million breeding adults in the United States and Canada and a statewide estimate of 330,000. MNBBA point count data generated a considerably larger population estimate of 795,000 adults; the confidence interval, however, was quite large.

Despite many fluctuations, the Great Crested Flycatcher has demonstrated one of the most sta-

C. The Great Crested Flycatcher prefers open woodlands with mature trees that provide nesting cavities and foraging perches, like this oak savanna stand on the bluffs of the Mississippi River. Photo by Lee A. Pfannmuller.

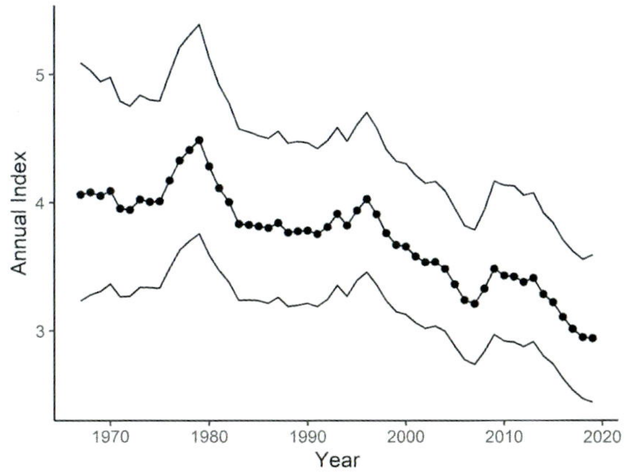

D. Breeding population trend and 95 percent credible interval (solid black lines) for the Great Crested Flycatcher in Minnesota based on the 1967–2019 Breeding Bird Survey (Sauer et al. 2020).

ble populations of any songbird in the BBS survey region. From 1966 to 2019, the average change in population numbers was only −0.10 percent per year (Sauer et al. 2020). This stable trend, however, masks wide variability across the continent. Populations are largely decreasing except in the southeastern states and in the Great Plains.

In Minnesota, Great Crested Flycatchers have demonstrated a significant decrease averaging −0.62 percent per year since 1967 **(D)**. Even at the state level, wide differences are visible from 1967 to 2019, with increasing populations in the west and south-central regions and declining populations in central and northern Minnesota (Sauer et al. 2020). This is further corroborated by evidence that the species has been slowly declining in the Chippewa National Forest from 1985 to 2020 (Grinde, Bednar, et al. 2020). Nest site availability and competition with other secondary cavity nesters are mentioned as possible factors regulating populations (K. E. Miller and Lanyon 2020).

CONSERVATION

Given its stable and relatively large population, the Great Crested Flycatcher is not a conservation priority (PIF 2021). Even under a 2°C warming scenario, NAS (2019) projected the species' North American breeding distribution had only a low vulnerability. By 2050, losses projected in western Minnesota would be partially compensated for by predicted gains in the northeast, where the species already is sparsely distributed. Meanwhile, management practices that create and maintain snags may be one of the most important conservation tools to ensure that the showy and vociferous Great Crested Flycatcher remains a common member of Minnesota's avifauna.

WESTERN KINGBIRD

Tyrannus verticalis

Minnesota Seasonal Status: Regular breeding species and migrant. The Western Kingbird was uncommon during the MNBBA.
Migration: Short- to medium-distance migrant that winters in southern Florida and along the Pacific coast of southern Mexico and Central America.
Food: Aerial insectivore that also gleans insects from foliage and dives after prey items on the ground.
Nest: Cup nest constructed in the upper third of a tree or shrub. Various structures are also used, including utility poles, fence posts, and building ledges.
Conservation Concern: Assigned a moderate PIF Continental Concern Score of 9/20.

A handsome bird as it perches on a telephone wire, the Western Kingbird has a habit of making its presence known. Like its relative the Eastern Kingbird, it can be a feisty character, "screaming and scrimmaging" when an intruder enters its territory or rising up in the air "claw to claw with a rival" (Bent 1942). Yet the Western Kingbird is considered by some to be a more social and "less quarrelsome" species than its eastern counterpart.

The Kingbird's gray head, lemon-yellow belly, and black tail, edged in white along the sides, are diagnostic. The only other large flycatcher with a yellow belly is the Great Crested Flycatcher, which is distinguished from the Kingbird by its rusty tail, browner head, and its preference for inhabiting woodlands, unlike the open country favored by the Western Kingbird. *Verticalis,* the species' scientific name, means "crown" and refers to the small, rarely visible patch of orange-red feathers on the top of its head.

Roberts (1932) described Western Kingbirds, then known as Arkansas Kingbirds, as "loquacious birds" that always "have something to talk about." Vocalizations include a sharp, low-pitched, wispy-sounding *kip* and a series of squeaky, warbling twitters. During the breeding season, the male also gives a "dawn song" that begins with a few sharp *kips* followed by a short, bubbly warble.

The Western Kingbird's breeding range is largely restricted to the western United States and southern Canada, stretching from British Columbia east to Ontario. Although breeding densities are high throughout the Central Plains, the highest densities are found in northwestern Texas and eastern New Mexico.

Photo by Gerald Hoekstra.

MINNESOTA BREEDING DISTRIBUTION

Roberts (1932) was the first to report Western Kingbirds in Minnesota in 1879, when he observed three pairs in Traverse County, right on the South Dakota border. Observations followed in 1889 in Lac qui Parle County and in 1893 in Pipestone and Otter Tail Counties. By 1898, Roberts considered the species common throughout southwestern Minnesota. The birds were already common in western Polk and Marshall Counties when they were first reported in the northwest region in 1912; farther east, a single bird was reported in St. Paul just three years later. After several Kingbirds were observed along Lake Pepin in 1927, Roberts predicted the species would soon become prominent throughout the "semi-prairie portion" of southeastern Minnesota.

By 1932, Roberts described the Kingbird as an abundant summer resident throughout Minnesota's western prairies. Nesting was confirmed in seven western counties (Grant, Jackson, Lac qui Parle, Marshall, Pipestone, Polk, and Traverse) and inferred in Murray County. Although the bird was "uncommon in the eastern half of the state," and largely absent from the northern forests, a nest with eggs was reported from Cass County in 1935 (Roberts 1936). Road surveys had suggested that Eastern Kingbirds were twice as common as Western Kingbirds, "but locally, here and there, the Western equaled or outnumbered the Eastern." Noting the species' eastward expansion, Roberts wrote, "Only one other species has come from the west in recent years and established itself over such a wide area in greater numbers, viz., [the] Brewer's Blackbird."

Thirty years later, in the late 1960s, Breckenridge (1968b, 1971) observed that both Eastern and Western Kingbirds were becoming scarce. Comparing survey data collected in 1968 and 1971 to similar surveys conducted in 1926, he detected a notable decline and speculated that agricultural pesticides were to blame. Green and Janssen (1975) also noted the Western Kingbird's decline in abundance. Nevertheless, its breeding range, which included all of western Minnesota stretching east across central and east-central Minnesota, had not changed significantly.

Janssen (1987) commented that "this species has a rather peculiar distribution in the state, and its absence from large areas which appear to be suitable for it are not easy to explain. Numbers fluctuate widely on the eastern edge of its range." He identified 20 counties with confirmed nesting since 1970. Hertzel and Janssen (1998) added seven more counties to the list. MBS field surveys were notable in finding only two of the Kingbird's 112 breeding-season locations in east-central Minnesota.

The MNBBA's 166 Western Kingbird detections documented a similar distribution but with more observations in east-central and northwestern Minnesota and fewer in the southwest **(A)**. Recorded in 3.0 percent of the surveyed atlas blocks (144/4,757), the records spanned 30 counties; breeding was confirmed in 20 blocks covering 11 counties. Three of

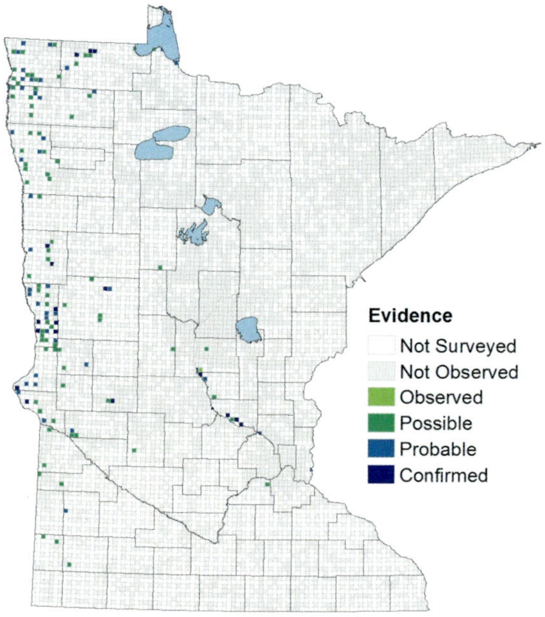

A. Breeding distribution of the Western Kingbird in Minnesota based on the Breeding Bird Atlas (2009–13).

B. Although human-modified habitats are often used, grasslands must include scattered trees for nesting and fences or trees for perching in order to attract the Western Kingbird. Photo by Lee A. Pfannmuller.

the 11 counties were new to the list published by Hertzel and Janssen (1998): Benton, Pope, and Roseau. The species was totally absent from southeastern Minnesota; Roberts's (1932) prediction that the Western Kingbird would become prominent in the region has never materialized. Overall during the atlas, the Eastern Kingbird outnumbered the Western Kingbird by more than twenty to one.

Since the early 1900s, the Western Kingbird's Minnesota distribution has experienced few changes, but its abundance has vacillated from the statewide declines originally observed by Breckenridge to the more recent regional changes observed in the past thirty to forty years. Factors responsible for the changes are not clear. In their comprehensive review of the species, Gamble and Bergin (2020) note that the species benefits from human activities. For example, when farmers convert grasslands and forests to agricultural fields, they also plant trees, construct buildings, and erect fence lines and utility poles, providing new nesting and perching opportunities for the birds (Gamble and Bergin 2020). Range expansions attributed to these factors since the late 1800s are documented in Texas, Oklahoma, Illinois, Arkansas, Manitoba, and Saskatchewan. Overall, the species is largely portrayed as one that is faring well. Most population fluctuations, they conclude, are local in nature and due largely to predation and weather events.

BREEDING HABITAT

The Western Kingbird inhabits open lands, including grasslands, cultivated fields, pastures and oak savannas, as well as open riparian woodlands **(B)**. Farm shelterbelts and urban areas are also suitable as long as they meet the species' primary requirements: open areas for foraging and scattered trees or tall structures for nesting and perching (Gamble and Bergin 2020).

The string of MNBBA records along the Mississippi River from Hennepin County to Morrison County suggests a strong correlation with riparian habitats. Although the juxtaposition of the river with adjacent open lands may have been an initial attraction for the birds, the nesting pairs along this corridor documented the species' adaptability to human-modified landscapes. Three of the nests were in electrical substations, two were on power poles, and one was placed in a line of trees bordering a parking lot.

POPULATION ABUNDANCE

PIF (2021) generated a population estimate of 30 million adults in the United States and Canada and a statewide estimate of 17,000. Surveywide, BBS data demonstrate a relatively stable but fluctuating population, decreasing an average of just −0.14 percent from 1966 to 2019 **(C)**. Indeed, between 1970 and 2014 the species actually experienced a cumulative increase of 20 percent (Rosenberg, Kennedy, et al. 2016) before beginning a steep downward trend in 2007. Although the Western Kingbird is too uncommon in Minnesota to generate a credible BBS trendline, the state lies on the eastern edge of a broad region that demonstrates significant population declines. Stretching from southern Manitoba, where numbers have declined an average of −3.18 percent per year from 1967 to 2019, south to Oklahoma, where the annual decline has averaged −1.93 percent, similar trends have been observed elsewhere in the Great Plains, including in South Dakota (−1.17 percent), Nebraska (−1.61 percent), and Kansas (−1.42 percent) (Sauer et al. 2020).

Factors responsible for the regional decline are unknown. Because Kingbirds often feed in heavily cultivated landscapes, pesticides may be a potential threat, as Breckenridge (1968b) speculated years ago. Although at least one study demonstrated that their direct impacts to the birds may be minimal (Gamble 1985), pesticide impacts to the abundance of local insect populations could reduce nesting success.

In Minnesota, the most significant land use change in western counties has been the intensification of farming operations. Small farms and their assortment of buildings, fences, and shelterbelts, all features that initially may have attracted Western Kingbirds, now are less common. No doubt pesticide use has also increased.

CONSERVATION

Concerns regarding the Western Kingbird's small winter range in Mexico and Central America helped secure its designation as a moderate conservation priority (PIF 2021). Management recommendations are limited to protecting and restoring riparian habitats and maintaining farm shelterbelts, which provide important nesting habitat (Gamble and Bergin 2020).

Warming temperatures, increased precipitation, and more severe storm events are all expected to place additional stress on grassland avifauna in the coming years. The Western Kingbird's adaptability may serve it well in this uncertain future, aided by its assessment as a climate-stable species (NAS 2019). Indeed, its breeding range is predicted to expand to the east, including in Minnesota. In the interim, studies should focus on factors responsible for the species' decline in the Great Plains.

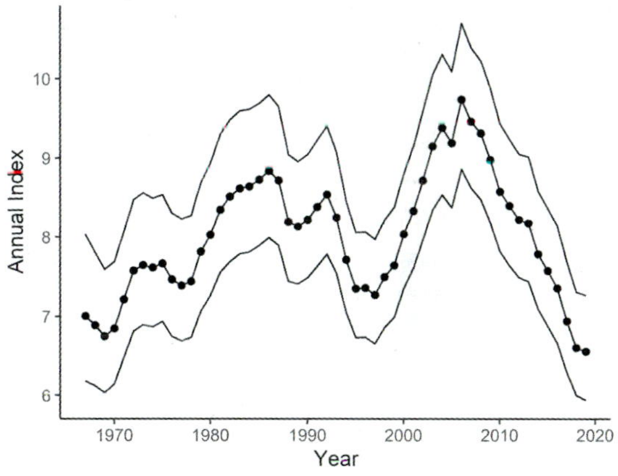

C. Surveywide breeding population trend and 95 percent credible interval (solid black lines) for the Western Kingbird based on the 1967–2019 Breeding Bird Survey (Sauer et al. 2020).

EASTERN KINGBIRD

Tyrannus tyrannus

Minnesota Seasonal Status: Regular breeding species and migrant. The Eastern Kingbird was common during the MNBBA.
Migration: Long-distance migrant that winters in South America.
Food: Aerial insectivore that also gleans insects from foliage, dives after prey items on the ground, and consumes some fruit as the summer progresses.
Nest: Cup nest usually placed within the upper third of a tree. Bits of trash, such as twine or plastic, may appear in the nest.
Conservation Concern: Assigned a moderate PIF Continental Concern Score of 11/20.

Photo by Jean Brislance.

The Eastern Kingbird is not easily overlooked. A prominent bird of the open country, it frequently perches along a fence line, atop a haystack, or on a lone tree amid a landscape of grasses. Ever watchful for territorial intruders, it strikes a rather formal pose in its charcoal gray and white attire. Hidden within the dark crown feathers, a small sliver of orange-red adds to its regal appearance but is rarely visible.

The Kingbird's scientific name literally translates to tyrant of tyrants, an apt description of its rather audacious behavior. In Bent's (1942) early account of the species, Tyler noted: "If we were limited to one adjective to suggest the kingbird's character as impressed on us by his behavior, I think most of us would use the word 'defiant'; if we were allowed one more, perhaps we should add 'fearless.'" Indeed, no bird is too big or too small to incite an offensive attack by the Kingbird. From tiny warblers to crows and red-tailed hawks, all have suffered from venturing within sight of the ever-watchful tyrant. One observer (Williams 1935) watched as a Kingbird even attacked a low-flying airplane!

Despite its belligerent reputation, the Eastern Kingbird is a welcome sight. Its faint, twittering call notes are often delivered as the bird sallies forth from its perch with its wings fluttering rapidly and its tail spread wide to prominently display the white tips. The notes have been variously described by different observers, but perhaps the best description was provided by Walter Faxon when he wrote, "He is trying to pronounce the word 'explicit,' but he is making a miserable, stuttering failure of it" (Bent 1942). Less often heard is a more musical dawn song given by the male during the breeding season, which is an uninterrupted series of more complex versions of the call notes (M. T. Murphy and Pyle 2020).

The Eastern Kingbird can be found throughout much of Canada, from the southern reaches of the Yukon and the Northwest Territories east to Newfoundland and southern Quebec. In the United States it is absent only from the southwestern states and along the Pacific coast. The core of the species' breeding range occurs in the Great Plains, from North Dakota south through Kansas.

MINNESOTA BREEDING DISTRIBUTION

As early as the late 1800s, Hatch (1892) described the Eastern Kingbird as a species he was "sure of seeing" wherever he traveled in the state. Forty years later, Roberts (1932) also described it as an abundant breeding species throughout Minnesota: "The Kingbird is such an abundant, noisy, and conspicuous fellow that it is probably one of the best-known birds in the state. It is everywhere from early May until September, in forests, in orchards, on the prairies, and in the outskirts of towns and cities, and is one of the most frequent 'post and wire' birds by the roadside." At the time, Roberts had breeding evidence (confirmed and inferred nesting reports) from 14 counties, including Houston County in the southeast, Jackson County in the southwest, Marshall and Polk Counties in the northwest, and Lake County in the northeast.

By the mid-1900s, however, changes were underway. Using data he and Roberts collected in 1926, Breckenridge (1968b, 1971) compared it to surveys conducted in 1968 and 1971 and found that the Eastern Kingbird was nearly 12 times more abundant in 1926; the Western Kingbird experienced similar declines. This led him to speculate that pesticide use was impacting these aerial insectivores in the agricultural regions of the state as increasingly intensive farming practices were implemented.

Although the Eastern Kingbird's numbers were declining, the species remained widely distributed throughout the state. Both Green and Janssen (1975) and Janssen (1987) noted that the bird could be found breeding in all regions of the state. Janssen included a statewide distribution map that identified 42 counties where nesting had been documented since 1970. Hertzel and Janssen (1998) later added 11 more counties to the list. Fieldwork conducted by the MBS also tallied more than 1,000 Eastern Kingbird breeding-season locations throughout much of the state.

With detections in all 87 counties and confirmed breeding records in 75 counties, MNBBA participants further confirmed the Kingbird's broad statewide distribution. Although reports came from all regions of the state, the species was least abundant in northeastern Minnesota, east of St. Louis County, and in the very northern regions of north-central Minnesota **(A)**. Nevertheless, breeding was confirmed in every county north of the Twin Cities except Clearwater, Cook, and Kanabec. A total of 3,672

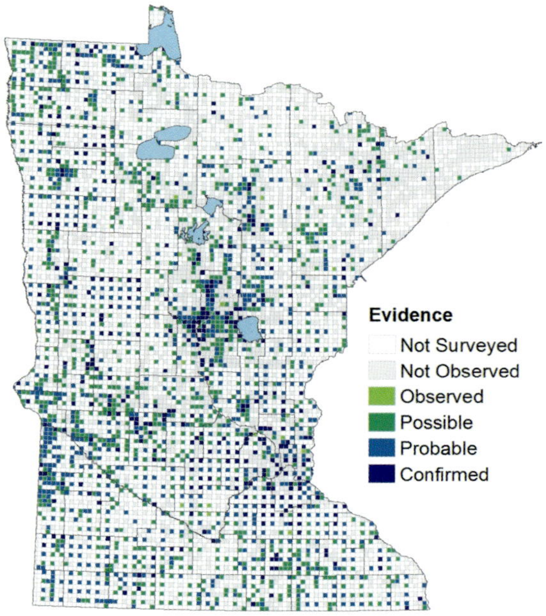

A. Breeding distribution of the Eastern Kingbird in Minnesota based on the Breeding Bird Atlas (2009–13).

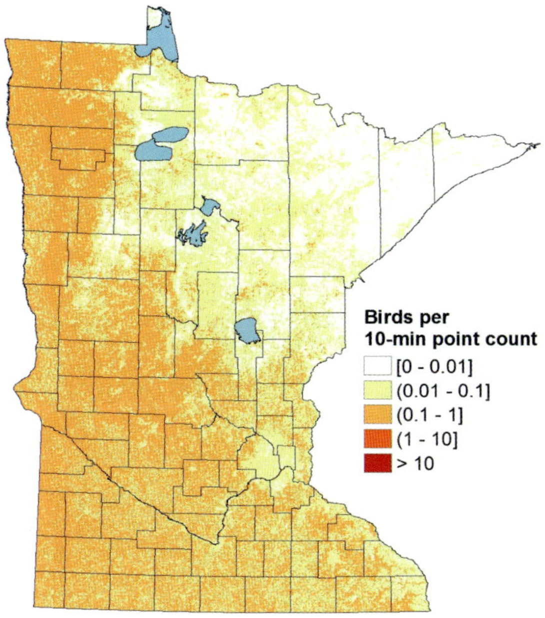

B. Predicted breeding distribution (birds detected per 10-minute point count) of the Eastern Kingbird in Minnesota based on habitat, landscape context, and climate data gathered during the Minnesota Breeding Bird Atlas (2009–13) using the Generalized Linear Modeling method without an adjustment for detectability.

Eastern Kingbird records were collected from 47.5 percent (2,274/4,791) of the surveyed atlas blocks. Breeding was confirmed in 352 blocks (7.3 percent). Twenty-eight of the 75 counties where breeding was confirmed were additions to the list published by Hertzel and Janssen in 1998.

The MNBBA probability model predicted breeding densities are lowest in the Laurentian Mixed Forest Province, especially in the Arrowhead Region, and in metropolitan regions such as the Twin Cities and Rochester **(B)**. Predicted breeding densities throughout western and southern Minnesota are quite uniform, with no areas of significantly higher abundance.

Since Hatch (1892) first wrote his account of the Eastern Kingbird nearly 150 years ago, the species' widespread distribution throughout the state has remained unchanged, despite an apparent decline in overall abundance. The same is true across North America. Although the loss of extensive forest habitat across the Eastern Deciduous Forest Biome and the planting of trees in the Great Plains likely provided some benefits to the species, no significant changes to its distribution have been noted in the past one hundred years (M. T. Murphy and Pyle 2020).

BREEDING HABITAT

The Eastern Kingbird is found in open landscapes that provide suitable trees for nesting and sufficient perches for foraging. Such general habitat requirements are met in a variety of places, ranging from pastures, shelterbelts, and grasslands to oak savannas, woodland edges, golf courses, and city parks **(C)**. Even in extensively forested landscapes, the species finds suitable habitat in small farmsteads, along forest edges, and in forest clear-cuts and burns with standing snags. Some studies, including in northern Minnesota and Wisconsin (Niemi, Howe, et al. 2016), have noted the species' affinity for sites associated with water and have reported high nesting densities in trees along the water's edge or in dead snags surrounded by water.

POPULATION ABUNDANCE

PIF (2021) generated a population estimate for Eastern Kingbirds of 26 million breeding adults in the United States and Canada and a statewide estimate of 420,000. A comparative analysis with MNBBA data was not conducted.

Despite their overall abundance, Eastern Kingbirds have been declining throughout most of their breeding range since the BBS began in 1966, with a cumulative loss of 38 percent from 1970 to 2014 (Rosenberg, Kennedy, et al. 2016). Unfortunately, no ecological region, province, or state has shown a significant increase since 1966. This includes Minnesota, where Kingbirds have experienced a statistically significant decline of −1.30 percent per year from 1967 to 2019. Indeed, since 1993, the rate of the decline in the state has more than doubled to −2.73 percent per year **(D)**.

Potential factors responsible for the decline include habitat loss and degradation; automobile collisions because the bird often nests and forages close to roads; and pesticide use, given its prevalence in agricultural lands and orchards. Some have suggested that population declines in the eastern United States may be the result of forest regeneration as abandoned farmlands succeed to forested habitats (M. T. Murphy and Pyle 2020). A study in New York dismissed the impact of forest regeneration and found, instead, that the percentage of the population's decline was nearly identical to the decline in agricultural lands used for pasture or grazing (M. T. Murphy 2001). Declines in aerial insect populations should also be considered.

C. The Eastern Kingbird is a common sight on farmsteads that have small trees for nesting and wooden fence lines bordering agricultural fields for foraging.
Photo by Lee A. Pfannmuller.

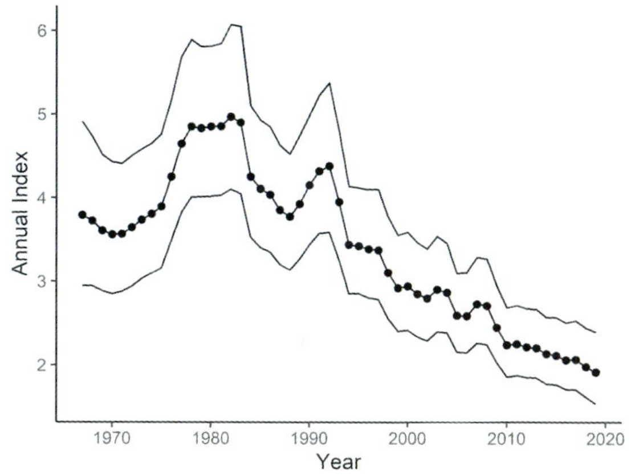

D. Breeding population trend and 95 percent credible interval (solid black lines) for the Eastern Kingbird in Minnesota based on the 1967–2019 Breeding Bird Survey (Sauer et al. 2020).

CONSERVATION

As with many aerial insectivores, the Eastern Kingbird's population decline is the principal factor responsible for its moderate conservation ranking (PIF 2021). Fortunately, the species' broad habitat requirements may provide some resilience as it confronts another challenge—climate change. Under a 2°C warming scenario, NAS (2019) projected a moderate vulnerability for the species' North American breeding distribution, with a range loss of 29 percent and a gain northward of 28 percent. Luckily, only range gains are predicted in Minnesota, specifically in the northeast.

Targeted recommendations for managing Kingbirds are limited. Based on field studies in New York, M. T. Murphy (2001) suggested that the protection of riparian habitats was a high priority because these were the only habitats where local populations were not declining. In addition to protecting riparian zones, he argued, land preservation efforts to protect small farms and open spaces are critical to prevent further decline of this emblematic open-country species.

OLIVE-SIDED FLYCATCHER

Contopus cooperi

Minnesota Seasonal Status: Regular breeding species and migrant. The Olive-sided Flycatcher was uncommon during the MNBBA.
Migration: Medium- to long-distance migrant that winters from southern Mexico to northern South America.
Food: Aerial insectivore that sallies from high perches to feed on flying insects.
Nest: Cup nest of twigs usually on a horizontal tree branch away from the trunk; coniferous trees are preferred, but deciduous trees are occasionally used.
Conservation Concern: Assigned a moderate PIF Continental Concern Score of 13/20.

This large, olive-gray flycatcher distinguishes itself from its congeners by being "a lover of high places" (Roberts 1932), selecting the tops of high trees, often dead snags, from which to announce its presence and hunt for its aerial prey. As it perches motionless, in wait for a potential morsel to fly by, one has ample time to note the gray flanks on either side of the chest that are separated by a thin line of white feathers. Many have described the plumage pattern as having the appearance of an open, unbuttoned vest. A white throat and belly are clearly visible, while small white feather tufts that extend behind the wings are more difficult to discern. Unlike many flycatchers, the Olive-sided Flycatcher lacks prominent wing bars. The genus name, *Contopus,* is a reference to the bird's relatively short legs; the species' name honors the nineteenth-century naturalist William Cooper.

Ornithologists have variously described the mnemonics of its common whistled song as *Come Right Here, Put Take Care,* or *Quip Peer-Peer,* but *Quick Three Beers* may be the long-time favorite. "Whatever the rendering," Roberts (1932) wrote, the listener "will hear a loud, penetrating, harsh utterance of several notes, so characteristic that, once learned, it can not be mistaken for anything else."

A northern forest species, the Olive-sided Flycatcher has a breeding range stretching from Alaska across boreal Canada, south along the coniferous forests of the Pacific Northwest and Rocky Mountains, and east across the Great Lakes states and New England. Sparsely distributed throughout its range, some of its highest breeding densities are found along the Pacific coast.

Photo by Dan Scolaro.

MINNESOTA BREEDING DISTRIBUTION

Roberts (1932) described the Olive-sided Flycatcher as a "common summer resident throughout the evergreen forests, where it breeds as far south as northern Isanti and Pine counties and as far west as Itasca Park and eastern Marshall County." Nevertheless, confirmed breeding records were limited to Cass and Itasca Counties and Cass Lake; breeding was inferred in Cook County. Roberts's enumeration of breeding records includes his own record from Isanti County on June 13, 1927, but the text specifies "bird seen, nest not located."

By 1975, Green and Janssen's updated account described the species' breeding range as "the northeastern and north central regions as far west as Kittson County and Many Point Lake, Becker County." Farther east, breeding evidence was found "as far south as Sturgeon Lake, Pine County." Because the Olive-sided Flycatcher is a late migrant that often lingers in southern Minnesota through early June, the authors discounted Roberts's 1927 breeding record in Isanti County. They considered it a late spring migrant rather than a summer resident. In addition to Robert's other breeding records from Cass, Cook, and Itasca Counties, they added an inferred nesting record from Carlton County. A few years later, Janssen (1987) identified three counties with confirmed nesting since 1970 (Aitkin, Clearwater, and Cook) but Hertzel and Janssen (1998) deleted the Clearwater County record.

Although the MBS had not completed surveys in north-central Minnesota, which provides prime Flycatcher habitat, the 145 breeding-season locations identified by 2014 documented a similar distribution to that delineated by earlier accounts. Observations were documented as far south as northern Pine County in the east and northern Wadena County in the west.

With two notable exceptions, the 217 Olive-sided Flycatcher records tallied during the MNBBA again reaffirmed earlier accounts of the species' distribution **(A)**. The species was reported from 3.5 percent (168/4,736) of the surveyed atlas blocks that spanned 18 northern counties. Breeding was confirmed only once, when a family group was observed in Koochiching County, resulting in a new county nesting record. Breeding was deemed probable in nine counties. Except for the two records discussed below, all were located within the Tallgrass Aspen Parklands and Laurentian Mixed Forest Provinces.

Two MNBBA records occurred in central Minnesota, in Todd and Sherburne Counties. Although several June records south of Mille Lacs Lake were invalidated as likely migrants, the 2013 probable record from Todd County was retained because it was based on two observations more than seven days apart in the same block and in the same general location—one in early July and one in early August. Satellite images of the site also revealed ideal habitat. A confirmed nesting record from the same block in 2010 was downgraded to possible because it lacked supporting documentation. Nevertheless, the three records suggest that the block is suitable for breeding Olive-sided Flycatchers. The second record was in Sherburne County, where multiple possible records were reported in a block over a three-year period, ranging from mid- to late June. Each observation was by skilled observers; four of the five records were from the MBS. Located within the Sherburne NWR, the block also appears to provide suitable habitat.

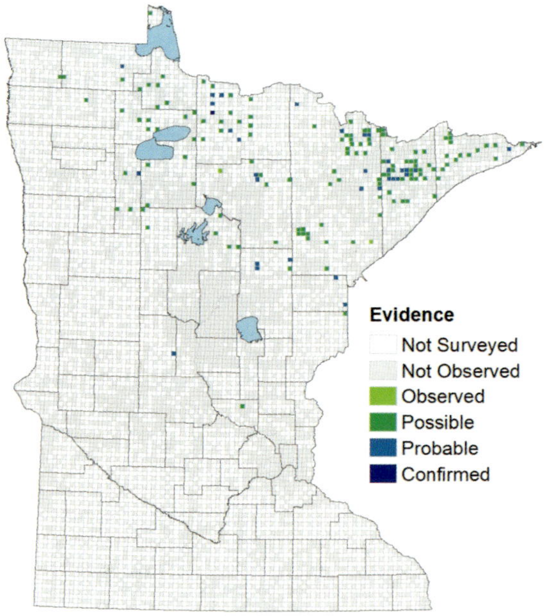

A. Breeding distribution of the Olive-sided Flycatcher in Minnesota based on the Breeding Bird Atlas (2009–13).

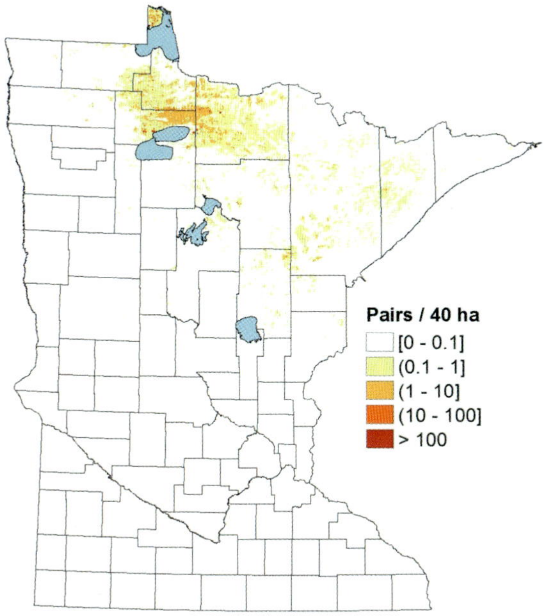

B. Predicted breeding distribution (pairs per 40 ha) of the Olive-sided Flycatcher in Minnesota based on habitat, landscape context, and climate data gathered during the Minnesota Breeding Bird Atlas (2009–13) using the Generalized Linear Modeling method with an adjustment for detectability.

The Flycatcher's probability map predicts that the highest breeding densities occur in the extensive wetlands north of Red Lake **(B)**. Although this map does not preclude the validity of the atlas records in Todd and Sherburne Counties, it does emphasize that records in central Minnesota require careful documentation.

Overall, the Olive-sided Flycatcher's current breeding distribution in Minnesota is remarkably similar to that described by Roberts in 1932. Elsewhere, it still occurs in small areas within the southern Appalachian Mountains but is largely extirpated from Maryland, North Carolina, Tennessee, Virginia, and West Virginia. Range contractions have also occurred in New England and Nova Scotia. By contrast, populations appear to have expanded in other Maritime Provinces and in northern California (Altman and Sallabanks 2020).

BREEDING HABITAT

The Olive-sided Flycatcher is found in coniferous forests with forest openings and edges **(C)**. These may be naturally located along streams, rivers, and wetlands, or they may be the result of fires, blowdowns, the establishment of beaver meadows, insect infestations, or timber-harvesting activities (Altman and Sallabanks 2020; Cutright et al. 2006; Kotliar 2007). The most important habitat features are an open canopy and the presence of tall trees or snags used for scouting insects, singing perches, and nesting. Open habitats, especially those near water, also provide a rich abundance of flying insects for this aerial insectivore.

POPULATION ABUNDANCE

The population of Olive-sided Flycatchers is estimated at 1.9 million individuals in the United States and Canada, with only 4,600 residing in Minnesota (PIF 2021). Although BBS trend data for the species lack some precision, they are considered statistically significant across the core survey region and demonstrate a large decline from 1966 to 2019, averaging −2.08 percent per year **(D)**. A large percentage of the Flycatcher's breeding range, however, occurs in northern Canada and Alaska, in areas located outside of the "core" North American region that the BBS sampled when it was first established. When recent data collected in these remote areas are combined with data from the "core" region, they provide a more biologically meaningful trendline that is also statistically significant with a very similar average annual decline of −2.02 percent from 1993 to 2019.

In Minnesota, the Flycatcher's downward BBS population trend lacks statistical precision but is corroborated by data collected in the Superior and Chippewa National Forests. A regional population trend generated by combining data from both forests indicates a significant decline, averaging −3.63 percent per year from 1995 to 2020 (Grinde, Bednar, et al. 2020).

Factors potentially contributing to the decline include reductions in the availability of snags in suitable habitats; decreased populations of beaver, resulting in fewer forest wetlands with snags; declines in the abundance of aerial insectivores; and habitat loss (Danz et al. 2007; Hutto 1995b; Nebel et al. 2010). On the breeding grounds, fire suppression has reduced one of the species' preferred habitats, post-fire stands. On the wintering grounds, an estimated 85 percent of the forests in the Andes, a region where winter densities of Olive-sided Flycatchers are especially high, have been lost or altered (Altman and Sallabanks 2020).

C. A low-lying meadow adjoining a mature coniferous forest with scattered tall trees that provide perches for singing and scouting insects is the type of breeding habitat the Olive-sided Flycatcher finds suitable. Photo by Gerald J. Niemi.

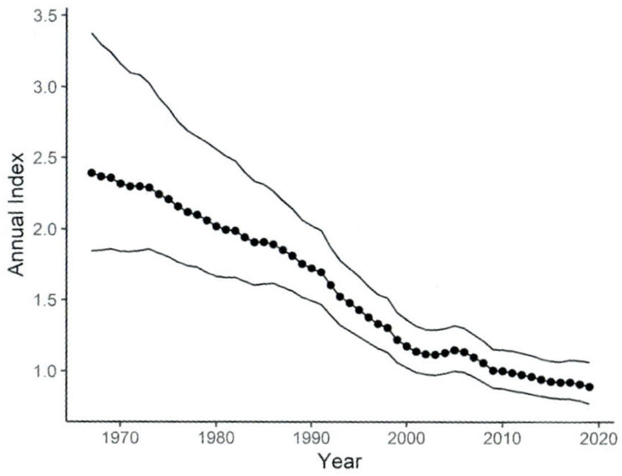

D. Surveywide breeding population trend and 95 percent credible interval (solid black lines) for the Olive-sided Flycatcher based on the 1967–2019 Breeding Bird Survey (Sauer et al. 2020).

CONSERVATION

Potential habitat threats, coupled with its population decline, prompted the Flycatcher's designation as a moderate conservation priority (PIF 2021). PIF also classified it as a Yellow Watch species that requires focused management attention (Rosenberg, Kennedy, et al. 2016).

Unfortunately, few studies have identified management practices critical for maintaining populations. Recommendations generally focus on retaining snags and tall trees that are adjacent to open areas, such as wetlands, burns, or harvested areas, and "maintaining natural patterns of forest fire that create suitable forest openings" (Niemi, Howe, et al. 2016).

Warming temperatures pose another concern. Not only will the extent of the boreal forest, where nearly 60 percent of the Olive-sided Flycatcher population resides, decrease but also forest-inhabiting aerial insectivores may be especially vulnerable as the synchrony between feeding young and the emergence of aerial insects is disrupted. Although NAS (2019) classified the species' breeding distribution as moderately vulnerable under a 2°C warming scenario, losing 34 percent of its southern breeding range and gaining only 16 percent to the north, the species is predicted to be extirpated from Minnesota even under an imminent 1.5°C warming scenario. Clearly, the future of this habitat specialist and proclaimed "lover of high places" (Roberts 1932) is fraught with challenges.

EASTERN WOOD-PEWEE

Contopus virens

Minnesota Seasonal Status: Regular summer resident and migrant. The Eastern Wood-Pewee was common during the MNBBA.
Migration: Long-distance migrant, overwintering in northwestern and western South America.
Food: Flying insects that are captured by sallies from a perch; occasionally gleans from foliage.
Nest: Cup nest placed on a branch of a tree or sapling in the subcanopy or canopy. The species is known to use a wide variety of deciduous and coniferous trees.
Conservation Concern: Assigned a moderate PIF Continental Concern Score of 10/20.

The Eastern Wood-Pewee is a common species in Minnesota woodlands, where it prefers open understories, especially oak forests that allow ample space to sally out and nab flying insects. Within these forests, one can hear its distinctive song, *pee-a-wee,* which appropriately provides the species with its common name. Bent (1942) described how its "leisurely notes, sliding smoothly and evenly as they change in pitch, give the impression of restfulness and peace, almost of sadness."

Its species name, *virens,* translates as "to be green" in Latin, in recognition of its overall olive-brown appearance, which camouflages it well in forest settings. The Eastern Wood-Pewee is very similar to its close relative the Western Wood-Pewee, which is an occasional species in western Minnesota. They are difficult to tell apart, except that their vocalizations are different and their ranges seldom overlap. The Eastern Wood-Pewee is found in the temperate and southern boreal forest in the east, while the Western is generally found from the Dakotas westward.

Identification of the Eastern Wood-Pewee can be challenging if the song is not heard. It is similar to the Eastern Phoebe, but larger, slimmer, not as dark, and does not bob its tail like the Phoebe. It is larger, slimmer, and has longer wings and tail than the *Empidonax* flycatchers like the Least Flycatcher. Pewees also lack a white eye-ring and have a slight crest. Both sexes are similar in appearance.

Eastern Wood-Pewees are found throughout the eastern United States and southern Canada as far west as southeastern Alberta, western Nebraska, and central Texas. Its highest densities have been found in southwestern Missouri and central Tennessee.

Photo by Travis Bonovsky.

MINNESOTA BREEDING DISTRIBUTION

In 1932, Roberts described the Eastern Wood-Pewee simply as "a common summer resident, breeding throughout the state." He included nesting records from Fillmore, Goodhue, Hennepin, Houston, Isanti, and Wabasha Counties, all counties in the southeastern portion of Minnesota. Farther north, the only nest records he reported were from Cass County and Itasca State Park. He commented that the "nest is a frail but beautiful structure" but "difficult to see from the ground as it rises but little above the limb on which it rests and has the entire outside covered with bits of lichen, bark, and spider-web, like the nest of a Hummingbird." Hence, confirmation of nesting can require a substantial effort when the nest is placed high in the canopy.

Green and Janssen (1975), more than forty years later, similarly described the Eastern Wood-Pewee's summer distribution, reporting that it is a "resident throughout the state" and "frequents groves on the prairie." Several years later, Janssen (1987) echoed the species' statewide distribution but noted the Eastern Wood-Pewee was least common in the southwest. He confirmed nesting since 1970 in 23 counties from Rock County in the southwest to Houston County in the southeast, northwest to Pennington County and northeast to St. Louis County. Hertzel and Janssen (1998) included confirmed nesting in five additional counties for a total of 27 counties but excluded Clearwater County, which was previously included by Janssen in 1987. The MBS recorded 1,932 breeding-season locations and, as of 2014, included every county in the state surveyed except Clay County in western Minnesota.

The MNBBA identified 3,018 breeding records for the Eastern Wood-Pewee from 40.3 percent of all surveyed blocks (1,911/4,746) and 55.2 percent of priority blocks (1,289/2,337) **(A)**. Nesting was confirmed in 43 counties, probable nesting in 38 counties, and possible nesting in an additional five counties. The only county with no records was Pipestone County in the southwestern corner of the state. The extensive distribution of the Eastern Wood-Pewee observed during the MNBBA and the number of confirmed nesting records are a strong tribute to the diligence of Minnesota's atlas observers, since documentation of nesting activity is not an easy task.

The probability map for the Eastern Wood-Pewee emphasizes the statewide distribution of the species but especially its presence in the broad ecotone between the coniferous forests in the northeast and the agricultural/grasslands of the western, southwestern, and southern portions of the state **(B)**. Highest densities were predicted in the Eastern Broadleaf Forest Province, and more specifically in the Hardwood Hills Subsection and in floodplain forests of the Mississippi River and many of its tributaries.

The overall breeding distribution of the Eastern Wood-Pewee likely has not changed substantially in Minnesota over the past one hundred-plus years. With almost a 50 percent loss of forests in the state, its breeding population has undoubtedly declined. The Watt and colleagues (2020) review of the Eastern Wood-Pewee in North America also described no major changes in this species' distribution and mostly indicated the same distribution that was

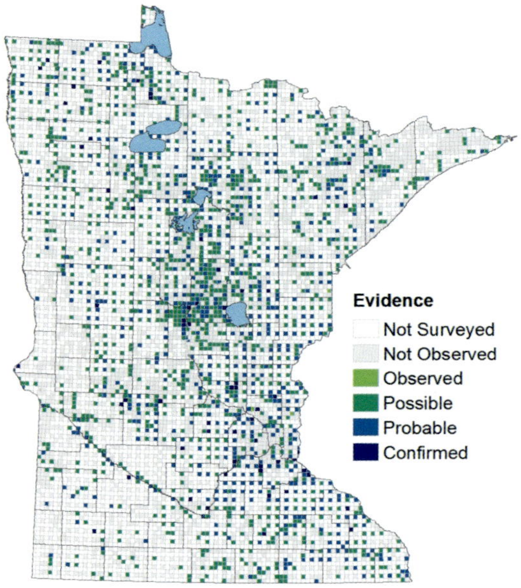

A. Breeding distribution of the Eastern Wood-Pewee in Minnesota based on the Breeding Bird Atlas (2009–13).

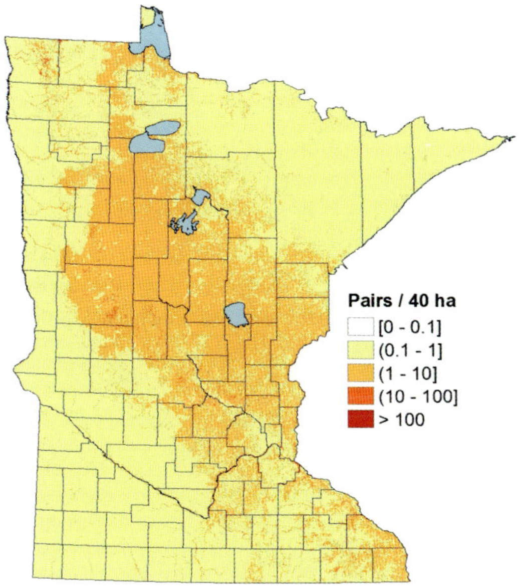

B. Predicted breeding distribution (pairs per 40 ha) of the Eastern Wood-Pewee in Minnesota based on habitat, landscape context, and climate data gathered during the Minnesota Breeding Bird Atlas (2009–13) using the Generalized Linear Modeling method with an adjustment for detectability.

described by Bendire in 1895 (cited in Bent 1942). However, they noted that the species has had a recent expansion westward in Kansas into forested and riparian corridors.

BREEDING HABITAT

Watt and colleagues (2020) concluded that the Eastern Wood-Pewee breeds in virtually every type of wooded community, including both deciduous and coniferous forest types. Roberts (1932) suggested the species has a "preference for oak woodlands but [occurs] commonly everywhere throughout the state." Extensive counts of this species in the national forests of northern Minnesota and Wisconsin during the breeding season indicated a significant preference for oak forest cover types but also substantial use of red pine, upland hardwoods, and beech–maple–birch forests cover types (Niemi, Howe, et al. 2016). Pewees avoid lowland coniferous forests and open habitats such as those recently logged. Forest habitats with open subcanopies that allow space to sally for flying insects are favored (Niemi and Hanowski 1984b, 1992a) **(C)**.

Many studies have emphasized that Eastern Wood-Pewees are not sensitive to the size of forest fragments (C. S. Robbins et al. 1989). For instance, Blake and Karr (1987) found the species in forest sizes that ranged from 1.8 to 600 ha, and Stauffer and Best (1980) observed its use of wooded riparian habitats as narrow as 20 m wide.

POPULATION ABUNDANCE

PIF (2021) estimated a US and Canadian population of 6.5 million breeding adults and a Minnesota population of 230,000. The MNBBA estimated a population of about 1.3 million breeding adults in Minnesota (95 percent confidence interval of 1.14–1.84 million). The MNBBA counts were based on a larger sample size in Minnesota than that used by PIF, and there were methodological differences with the counts made by PIF.

The BBS population trend for Minnesota was significantly decreasing from 1967 to the early 2000s, but it has stabilized and slightly increased since 2002. Overall, its trend from 1967 to 2019 is nonsignificant and averaged −0.27 percent per year **(D)**. The species' BBS trend surveywide is still significantly declining (Sauer et al. 2020). Rosenberg, Kennedy, and colleagues (2016) estimated a decline in the population of 44 percent from 1970 to 2014. In Minnesota's Chippewa and Superior National Forests, the species has been significantly declining from 1995 to 2020 at an average rate of −1.02 percent per year (Grinde, Bednar, et al. 2020). Factors responsible for the population declines are not clear but may be related to the overall loss of forest area throughout their range and declines in aerial insects, which are a critical food source during the breeding season (Nebel et al. 2010).

C. Breeding habitat of the Eastern Wood-Pewee in Minnesota generally consists of large trees with an open understory that allows ample room to forage for flying insects. Photo by Gerald J. Niemi.

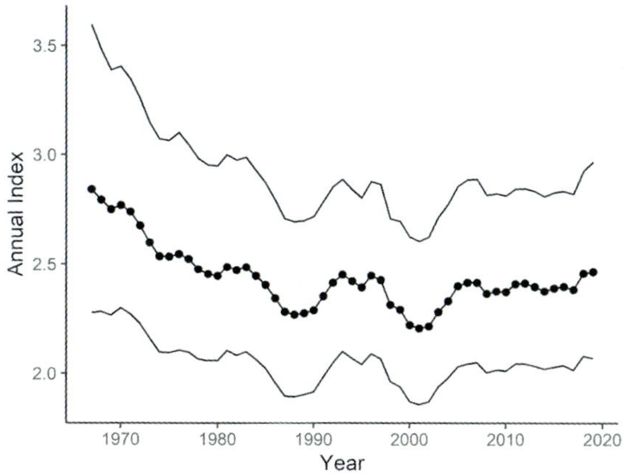

D. Breeding population trend and 95 percent credible interval (solid black lines) for the Eastern Wood-Pewee in Minnesota based on the 1967–2019 Breeding Bird Survey (Sauer et al. 2020).

CONSERVATION

The greatest concern for the Eastern Wood-Pewee is the large population decline of this species over the past forty-five-plus years. Loss of forested habitat in Minnesota over the past 150 years has certainly decreased its overall numbers. At present, however, the species' population is still relatively large and widespread in eastern North America.

Collisions with communication towers were considered a minor source of mortality. Watt and colleagues (2020) cited pesticides and subsequent reductions in insect prey as being possibly associated with reduced populations of the Eastern Wood-Pewee. Decalesta (1994) documented negative effects on its breeding population from high deer populations in the northeastern United States due to excessive browsing in the understory. The Eastern Wood-Pewee has been documented as tolerant of habitat fragmentation because of its use of edges and small forest fragments (Blake and Karr 1987; C. S. Robbins et al. 1989). NAS (2019) projected low vulnerability for the species' breeding range in North America with all three climate scenarios they examined through 2080. In Minnesota, breeding range retractions were predicted along the western regions, while improving conditions were predicted in the northeast. The status of the Eastern Wood-Pewee's wintering populations in South America is not well known; the species is difficult to identify on the wintering grounds because of its lack of vocalizations (Watt et al. 2020).

YELLOW-BELLIED FLYCATCHER

Empidonax flaviventris

Minnesota Seasonal Status: Regular breeding resident and migrant. The Yellow-bellied Flycatcher was uncommon during the MNBBA.
Migration: Medium-distance migrant; winters in southern Mexico and Central America.
Food: Flying insects that are caught in the air and gleaned from foliage.
Nest: On the ground in moss, roots, or thick vegetation.
Conservation Concern: Assigned a moderate PIF Continental Concern Score of 9/20.

Photo by David Brislance.

The Yellow-bellied Flycatcher is one of five closely related flycatchers found in Minnesota in the genus *Empidonax,* along with the Acadian, Alder, Least, and Willow Flycatchers. The Yellow-bellied Flycatcher is among the most distinct, with its yellowish eye-ring and pale yellow underparts that are the basis for its scientific species name, *flaviventris,* Latin for "yellow belly" (Sandrock and Prior 2014). The species is likely first detected by its voice; its song is referred to as a *che-bunk,* which is similar to the *che-bek* of the Least Flycatcher but with a softer ending. Its call notes, as described by Eckert (2017), are a "two-syllabled whistle (*che wee*) and an abrupt, metallic *killink!*"

The Yellow-bellied Flycatcher is the only flycatcher in Minnesota that is nearly completely found in coniferous forested habitat, especially lowland coniferous forests. To find this species in its "summer home," Bent (1942) provided the following advice: "the searcher must invade the moist, gloomy morass of some northern forest bog, beneath the shade of spruce and firs, and endure the attacks of hoards [*sic*] of black flies and mosquitos, to get even a glimpse of this woodland waif."

The Yellow-bellied Flycatcher is widely distributed across Canada from Labrador to British Columbia to the tree limit, and in the United States is restricted to the northern regions of the Upper Midwest and the northeastern New England states. Its highest densities are found in Labrador, Nova Scotia, and northern Minnesota.

MINNESOTA BREEDING DISTRIBUTION

The sum of Roberts's (1932) records indicate the Yellow-bellied Flycatcher was a fairly rare species in Minnesota. This may have been owing to a lack of coverage in appropriate habitat and difficulties in identification. He recorded the following observations, which he said, "show that it is a summer resident in northern Minnesota": Aitkin County (female in breeding condition in June 1897); Itasca State Park (behavior on July 9, 1917, indicated a nest or young nearby); Lake County (one collected on July 15, 1914); eastern Marshall County (brood of young being fed by adults on July 10, 1900); and St. Louis County (one collected on June 14, 1894). He summarized its distribution as "records for the northern part of the state are not numerous but show that it is widely distributed during the breeding-season from the eastern border of the Red River Valley to Lake Superior, as far south as Itasca Park and Aitkin County."

In 1975, Green and Janssen described a breeding distribution that was similar to what Roberts (1932) had reported. Their boundaries included northeastern and north-central Minnesota as far south as Pine County, southern Aitkin County, and west to Clearwater and Marshall Counties. They cited confirmed breeding records only from northern Lake County. Janssen (1987) emphasized that the species was "best represented in northern Cook, Lake, St. Louis, Lake of the Woods, Koochiching, and Itasca Counties." Janssen (1987) and Hertzel and Janssen (1998) both included three counties with confirmed nesting records since 1970: Aitkin, Cook, and Koochiching.

Up to 2014, the MBS has recorded 635 breeding-season locations during intensive county surveys. Locations were widely spread across northern Minnesota; the majority were from southern Beltrami County to Cook County. The spatial extent of their locations ranges south to Mille Lacs, northern Morrison, and Pine Counties, west to eastern Becker County, and northwest to Marshall, Pennington, and Roseau Counties.

The MNBBA provided a solid snapshot of the current breeding distribution with 889 records of the Yellow-bellied Flycatcher **(A)**. The species was recorded in 9.4 percent (444/4,735) of all surveyed blocks. Nesting was confirmed from five counties: Beltrami, Cass, Itasca, Lake, and St. Louis. In addition, probable breeding was identified in another eight counties: Aitkin, Carlton, Cook, Hubbard, Koochiching, Lake of the Woods, Pine, and Roseau. Possible nesting extended south to Kanabec and Mille Lacs Counties and west to Clearwater and eastern Marshall Counties.

The probability map for the Yellow-bellied Flycatcher predicts the primary distribution in the northeastern and north-central regions of the state **(B)**. Highest densities are predicted in the Red Lake Peatland north of Upper Red Lake and a broad region from southwestern St. Louis County to northeastern Lake and Cook Counties. The breeding distribution is slightly larger than what had been previously described, extending southward to Kanabec and Mille Lacs Counties. Recent studies have reinforced the substantial populations of the Yellow-bellied Flycatcher in the Agassiz Lowlands Subsection and in northeastern Minnesota (e.g., Niemi, Howe, et al. 2016; Zlonis et al. 2017).

Gross and Lowther (2020) suggested that the Yellow-bellied Flycatcher distribution has remained relatively stable in North America since the early twentieth century. They commented on the loss of

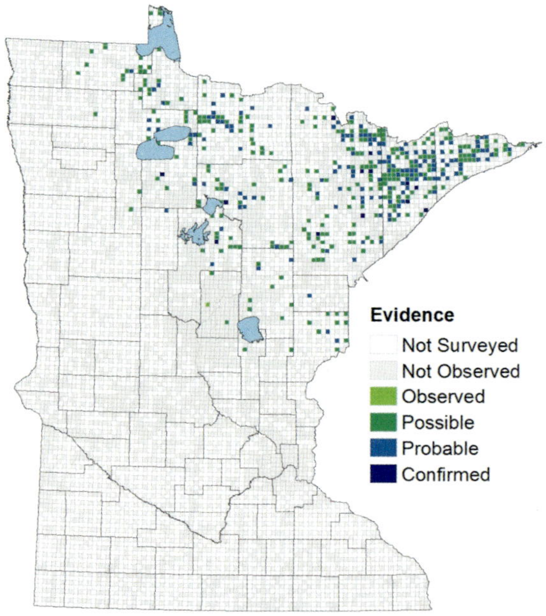

A. Breeding distribution of the Yellow-bellied Flycatcher in Minnesota based on the Breeding Bird Atlas (2009–13).

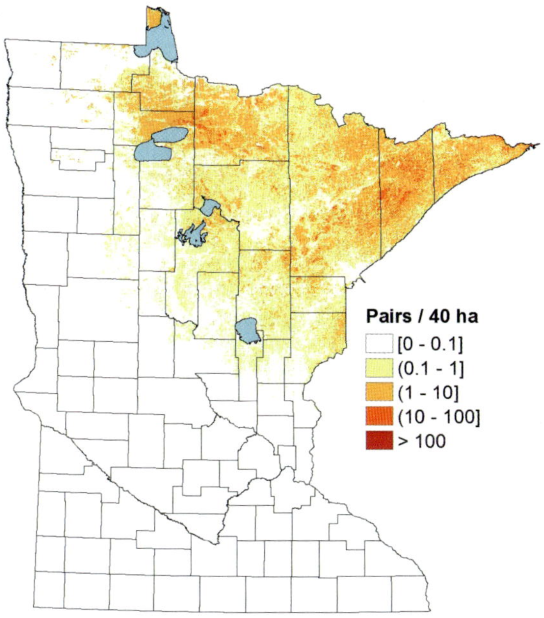

B. Predicted breeding distribution (pairs per 40 ha) of the Yellow-bellied Flycatcher in Minnesota based on habitat, landscape context, and climate data gathered during the Minnesota Breeding Bird Atlas (2009–13) using the Generalized Linear Modeling method with an adjustment for detectability.

breeding habitat in the southern fringes of the species' range in Canada and from southern Wisconsin. These changes have largely been due to logging, drainage of bogs, and agricultural development.

In Wisconsin, Kumlien and Hollister ([1903] 1951) found nests of the Yellow-bellied Flycatcher from 1860 to 1864 in Jefferson County, which is in the second tier of counties in southeastern Wisconsin. Roberts (1932) did not find the Yellow-bellied Flycatcher south of Aitkin County during the breeding season, and his documents indicate that he and others spent considerable time in the area of Mille Lacs and Pine Counties. Interestingly, Cadman and colleagues (2007) in Ontario questioned whether the training of atlas participants during the second atlas period (2001–5) resulted in increased familiarity with the songs and calls of the species and contributed to finding the species more widespread during their second breeding bird atlas compared with the first (1981–85). Moreover, Kumlien and Hollister stated that this species was "not as rare in Wisconsin as early writers have made us believe." It remains unknown if the lack of vocal identification skills confounded observations during the early years of settlement in central Minnesota. Alternatively, it seems unlikely that the species would have been missed during collection trips during the late 1800s or early 1900s because it is easy to identify in the hand. The lack of earlier observations of Yellow-bellied Flycatchers in more southerly regions of Minnesota remains an enigma.

BREEDING HABITAT

Yellow-bellied Flycatcher haunts have been broadly described as boreal conifer forest and peatlands, bogs, and muskeg (Gross and Lowther 2020). Roberts (1932) reported that all of the reports of this species were "found in dense spruce, tamarack, and arbor vitae swamps." These observations correspond well with recent quantitative studies in Minnesota, which found it significantly associated with black spruce–tamarack and mixed swamp conifer forest cover types as well as in upland aspen–spruce–fir and jack pine cover types (Niemi, Howe, et al. 2016) **(C)**.

POPULATION ABUNDANCE

PIF (2021) estimated a US and Canadian breeding population of 13 million adults but only 170,000 breeding adults in Minnesota. The MNBBA estimated a much larger breeding population of approximately 680,000 breeding adults (95 percent confidence interval of 567,000–1,683,000) in Minnesota. The discrepancy in these estimates is due partly to under-sampling of many roadless habitats in northern Minnesota where this species occurs that are not counted in the BBS roadside counts, as well as to methodological differences with PIF (2021).

Population trend estimates of the Yellow-bellied Flycatcher based on the BBS in Minnesota lacked enough data, but data were sufficient in the Boreal Hardwood Transition Bird Conservation Region, which includes northern Minnesota (Sauer et al. 2020). In this region, the Yellow-bellied Flycatcher population trend was stable at 1.22 percent per year from 1966 to 2019 **(D)**. However, BBS data from Ontario and throughout Canada from 1966 to 2019 indicated that the species was significantly increasing

C. Yellow-bellied Flycatchers are fond of coniferous forests, especially lowland forests, where they have adequate ground cover to hide their nests and a good supply of flying insects. Photo by Gerald J. Niemi.

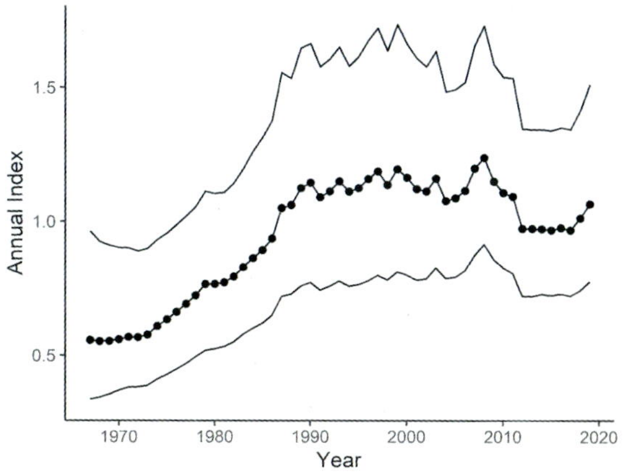

D. Breeding population trend and 95 percent credible interval (solid black lines) for the Yellow-bellied Flycatcher in the Boreal Hardwood Transition Bird Conservation Region from 1967 to 2019, based on the Breeding Bird Survey (Sauer et al. 2020).

at a rate that averaged 3.03 percent and 2.07 percent per year, respectively. In contrast, analysis by Ralston and colleagues (2015) found significant declines in this species' population within nine different regions of North America. Among the regions included was the Superior National Forest of northern Minnesota, where it has continued to significantly decline by an average of −2.04 percent per year from 1995 to 2020 (Grinde, Bednar, et al. 2020).

CONSERVATION

The lack of extensive coverage of many parts of northern Minnesota forests in the 1800s, early 1900s, and up to the 1970s prohibits statements about changes in the population or distribution of this species in Minnesota. Historical and recent declines in the coniferous forest of Minnesota (Jaakko Pöyry Consulting 1994) have likely contributed to overall declines in the population of this species over the past 150 years.

Gross and Lowther (2020) note that Yellow-bellied Flycatchers have been recorded at tower kills but impacts on the overall population are unknown. They also comment on the continued threats to this species from loss of forested habitat due to agricultural development in overwintering areas in Mexico.

Additional threats to the Yellow-bellied Flycatcher habitat in Minnesota include increased logging, especially in lowland coniferous forests; the loss of tamarack forests due to the eastern larch beetle in Minnesota; and climate change. NAS (2019) predicted a moderate vulnerability in its North American breeding range due to climate change, and extirpation from Minnesota is predicted with our current 1.5°C warming scenario. As with the moose and other boreal species, the continued existence of the Yellow-bellied Flycatcher in Minnesota is under imminent threat.

ACADIAN FLYCATCHER

Empidonax virescens

Minnesota Seasonal Status: Regular breeding resident and migrant. The Acadian Flycatcher was rare during the MNBBA.
Migration: Medium- to long-distance migrant that winters in southern Central America and northern South America.
Food: Aerial insectivore that snatches insects from undersides of leaves, catches them in mid-air, and gleans them from the foliage and ground.
Nest: Cup nest located in deciduous trees and shrubs.
Conservation Concern: Assigned a moderate PIF Continental Concern Score of 11/20; officially listed as *Special Concern* in Minnesota.

A relative newcomer to Minnesota, the Acadian Flycatcher is one of five *Empidonax* flycatchers that nest in the state. The species are so similar in appearance that some of North America's most skilled ornithologists have been fooled. Arthur C. Bent (1942), for example, reported that renowned ornithologist Alexander Wilson provided an excellent description of the Acadian Flycatcher in 1810, but his description of the bird's nest was clearly that of the Least Flycatcher, not the Acadian. The species also acquired its common name when a bird collected in Acadia, now Nova Scotia, was considered the type specimen for the species. Only years later did ornithologists discover that the Acadian Flycatcher does not occur within hundreds of miles of the locality. The Latin name for the species, *acadius,* was promptly changed to *virescens* (i.e., greenish color) but the vernacular name persisted (Bent 1942).

Fortunately, habitats and vocalizations help differentiate the species. Three of Minnesota's *Empidonax* species are woodland inhabitants: the Yellow-bellied Flycatcher, which is strictly a species of northern forests; the Least Flycatcher, which occurs statewide; and the Acadian Flycatcher, which is primarily restricted to southern and central woodlands. The most reliable way to distinguish the latter two species is by sound. The Least Flycatcher delivers an emphatic and unmistakable *che-bek.* The Acadian Flycatcher's song is similar but has a burry quality and sounds like an explosive hiccup or *peet-sah.*

Photo by Gerald Hoekstra.

Finer differences between the latter two species include the Acadian Flycatcher's slightly larger size (1.3 cm), a larger bill in proportion to its head, a bit of yellow on its belly, and a more greenish-appearing head and back. The head's sloping profile has a slightly raised crest, giving rise to one of its original names, the Small Green-crested Flycatcher (Bent 1942). Unfortunately, during the nonbreeding season, distinguishing the Acadian Flycatcher from the shrub-loving Alder and Willow Flycatchers is very challenging.

Confined to the eastern United States, the Acadian Flycatcher's breeding range stretches from central Minnesota east to southern New England and south to the Gulf coast and northern Florida. The highest breeding densities are found in a region that stretches from Louisiana northwest to Virginia and southern Pennsylvania.

MINNESOTA BREEDING DISTRIBUTION

Hatch's account of Minnesota birds in 1892 described the Acadian Flycatcher as a "fairly common" species. Although he noted its frequent confusion with the Traill's Flycatcher (now split into the Alder and Willow Flycatchers), he was quite certain of his identification. He also accepted reports of the species from the Root River in southeastern Minnesota and from the Red River valley in northwestern Minnesota.

Forty years later, Roberts (1932) summarily dismissed Hatch's account, writing that the "whole article is without foundation in fact." Although Hatch's report of birds along the Root River was at least plausible, the observer, Dr. Hvoslef, later notified Dr. Roberts that his identification was in error (Roberts 1932). Roberts himself, anticipating the Flycatcher would eventually appear in the state, spent many years collecting specimens of small *Empidonax* flycatchers, hoping to find evidence of the species' presence. His efforts, however, were to no avail.

In 1940, there was another report of Acadian Flycatchers singing along the Root River, "2½ miles east of Rushford," but with no additional details (Green and Janssen 1975). It was 1967 before Fred Lesher (1968) uncovered a nesting pair in Beaver Creek Valley State Park in Houston County. More reports soon followed. By 1987, Janssen detailed the species' expansion farther to the north and west and identified three counties where nesting had been confirmed: Houston, Goodhue, and Rice. By 1998, breeding was was also confirmed in Dakota, Hennepin, Ramsey, Scott, and Winona Counties (Hertzel and Janssen 1998).

MBS field surveys documented 54 Acadian Flycatcher breeding-season locations. Most records were found in the Eastern Broadleaf Forest Province, from Houston County northwest to Douglas County, and west along the Minnesota River valley to Nicollet County. Two records in southwestern Minnesota came from Lyon County and the border of Lincoln and Yellow Medicine Counties. The most remarkable MBS report, however, was from northern Kittson County. On June 19, 2012, a male was heard singing and calling in an aspen grove on the Caribou WMA. Given the unusual nature of the record, an audio recording was made and verified. It was also submitted as an MNBBA record.

Little had changed when MNBBA participants reported 34 Acadian Flycatcher records from 25 of the surveyed atlas blocks and confirmed breeding in four blocks **(A)**. The birds were reported in 15 of Minnesota's 87 counties and were confirmed breeding in four counties (Hennepin, Goodhue, and in a block that straddled Scott and Dakota). The majority of reports were from east-central and southeastern Minnesota, with records farther north and west in Brown, Kittson, Nicollet, Otter Tail, and Stearns Counties. Since the atlas, nesting has also been confirmed in Washington County (2019; MOU 2021a).

The Acadian Flycatcher remains a very rare species in Minnesota. Confined primarily to the far east-central and southeastern counties, it is also well established along the Minnesota River valley in Brown and Nicollet Counties. Three localities where the species has been consistently reported are Beaver Creek Valley State Park in Houston County; Murphy-Hanrehan Park Reserve in Scott

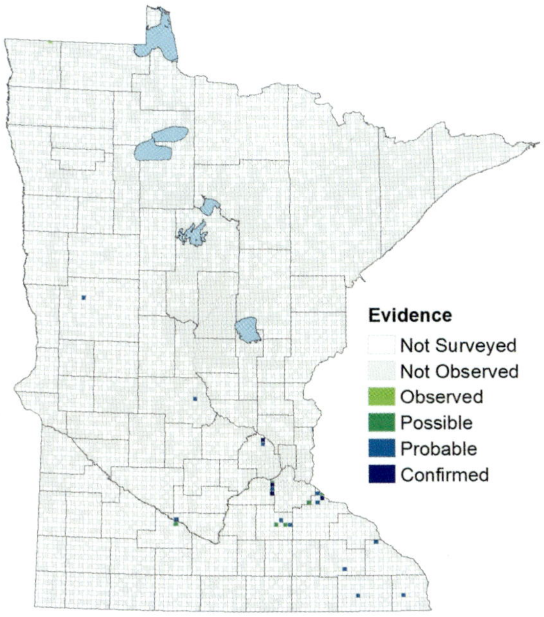

A. Breeding distribution of the Acadian Flycatcher in Minnesota based on the Breeding Bird Atlas (2009–13).

and Dakota Counties; and Seven Mile Creek Park in Nicollet County. Reports beyond these locations are rare and inconsistent from year to year. Although the Flycatcher was not reported from Seven Mile Creek during the MNBBA, other public lands where the species was reported included Elm Creek Park Reserve in Hennepin County, Nerstrand Big Woods State Park in Rice County, Maplewood State Park in Otter Tail County, and Falls Creek SNA in the far northeastern corner of Washington County. All of these sites support stands of mature hardwood forest.

Coincident with the species' appearance and range expansion in Minnesota, the Acadian Flycatcher's breeding range has expanded throughout the northeastern United States since the 1960s and 1970s. Records from the late 1800s and early 1900s document that it is reoccupying portions of its former breeding range. For example, it was known to breed in Massachusetts in 1888 but then disappeared until the 1960s. It was present in southern New York in the early 1900s but then disappeared until the early 1970s. Biologists believe the species may be responding to long-term changes in the availability of forest cover (M. Allen et al. 2020).

BREEDING HABITAT

The Acadian Flycatcher's breeding habitat has been broadly characterized as large, mature forest tracts. Key features often include an association with water, a lack of disturbance, a dense canopy, and a well-developed understory (M. Allen et al. 2020). In some areas of southern Minnesota, such as Beaver Creek Valley State Park, streamside ravines provide ideal habitat **(B)**. In other areas, upland hardwood forests, with scattered wetlands, are more typical of occupied sites, such as Murphy-Hanrehan and Elm Creek Park Reserves (Fall 1987; O. L. Johnson 1994).

Several studies have demonstrated the species' dependence on large, unfragmented forest tracts; in general, the abundance of Flycatchers increases with an increase in tract size (Temple 1988; C. S. Robbins 1980). One study, however, suggested that the species may be less sensitive to habitat fragmentation in narrow floodplain corridors that were at least 80 m in width (Chapa-Vargas and Robinson 2007).

B. Mature and undisturbed hardwood forests are preferred by the Acadian Flycatcher. Frequently the stands border small streams or larger rivers. Photo by Lee A. Pfannmuller.

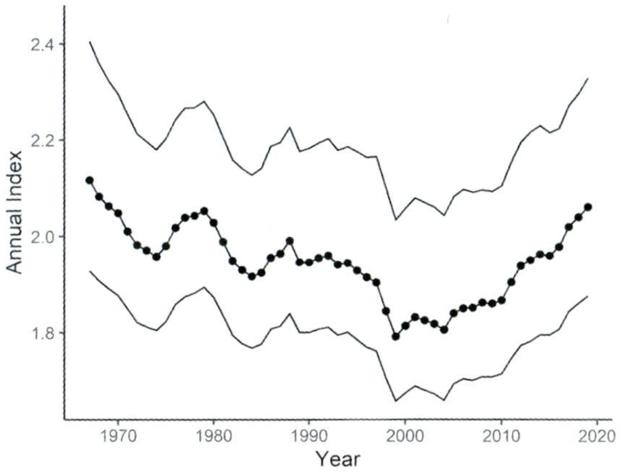

C. Surveywide breeding population trend and 95 percent credible interval (solid black lines) for the Acadian Flycatcher based on the 1967–2019 Breeding Bird Survey (Sauer et al. 2020).

POPULATION ABUNDANCE

PIF (2021) provided a population estimate of 5.2 million breeding adults in the United States and Canada. Across its breeding range, population numbers have been relatively stable from 1966 to 2019 **(C)**. Numbers slowly declined in the latter half of the twentieth century but began a slow increase beginning in 2000. Although a rough estimate of the species' population in Minnesota was given as 300 adults by Rosenberg (2004), the species' rarity on Minnesota BBS routes precludes a more precise estimate.

CONSERVATION

Concerns regarding the Acadian Flycatcher's small winter range, coupled with its dependence on intact, mature forests year-round, led to its assignment as a moderate conservation priority (PIF 2021). In Minnesota, the paucity of records and the declining abundance of large blocks of mature deciduous forest led to its 1996 classification as a state *Special Concern* species (MNDNR 1995). Fortunately, although climate change will result in a gradual loss of the species' breeding range in several states, from Iowa south to Texas and east to North Carolina, all three warming scenarios examined by NAS (2019) projected future gains in eastern Minnesota. Management recommendations focus on protecting mature deciduous forest tracts that are at least 10,000 ha in size, with as little internal disturbance as possible (M. Allen et al. 2020). Further opportunities to protect and manage forested, dissected river valleys and large upland forests with scattered wetlands should be explored. Such efforts are not only critical for the future of the Acadian Flycatcher but also for many other area-sensitive songbirds in the east-central and southeastern regions of the state.

ALDER FLYCATCHER

Empidonax alnorum

Minnesota Seasonal Status: Regular breeding species and migrant. The Alder Flycatcher was common during the MNBBA.
Migration: Long-distance migrant that winters in South America.
Food: Insects captured by fly-catching and foliage gleaning.
Nest: Cup nest placed in shrubs approximately 0.3 to 1.0 m from the ground.
Conservation Concern: Assigned a moderate PIF Continental Concern Score of 9/20.

Like all *Empidonax* flycatchers, the Alder Flycatcher sports an olive-green plumage above, a lighter belly below, two white wing bars, and a narrow white eye-ring, none of which can confirm its identification. Indeed, the species was lumped together with its look-alike, the Willow Flycatcher, until 1973; together they were known as Traill's Flycatcher (*Empidonax traillii*). Yet as early as 1895, ornithologist William Brewster recognized that the species sang two distinctly different songs. He designated the vocalist of one song a subspecies of the Traill's Flycatcher, the Alder Flycatcher (*E. t. traillii*). Regardless, confusion continued until detailed studies more than sixty years later provided evidence to officially divide the Traill's Flycatcher into two distinct species, the Alder Flycatcher and the Willow Flycatcher.

Today, the best clue for identifying the Alder Flycatcher remains its song, a two- or three-syllable buzzy *fee-beeur* or *fee-bee-o,* where the accent is on the second syllable (Lowther 2020a; Eckert 2008). The Latin name for the species, *alnorum,* refers to one of its preferred habitats, wet alder thickets. Unfortunately, habitat is not always a reliable field clue, as Willow Flycatchers may occupy similar habitats. With a more northerly distribution than the Willow Flycatcher, the Alder Flycatcher occurs across Alaska, boreal Canada, the Great Lakes states, and New England. Some of the highest breeding densities in North America occur in portions of northern Minnesota.

Because the two species were lumped together for so many years, much remains to be learned about the Alder Flycatcher's distribution and life history. Many early studies were conducted in regions where Willow Flycatchers prevail. Minnesota, however, provides opportunities to observe both species.

Photo by Michael Furtman.

The Willow has a more southern distribution in the state and the Alder a more northern distribution, but there is a broad region of overlap. The birds have even been observed within hearing distance of each other on the same site.

MINNESOTA BREEDING DISTRIBUTION

When Roberts published his comprehensive account of Minnesota birds in 1932, this inconspicuous little flycatcher was still commonly known as Traill's Flycatcher. Nevertheless, assuming Minnesota's population belonged to the northern subspecies (*E. t. trailli*), he referred to it as the Alder Flycatcher and described it as a common breeding resident throughout the state, especially in northern Minnesota. In fact, with one exception, Roberts's (1932) numerous depictions of the flycatcher's song were most similar to the song type of the Alder Flycatcher (*fee-bee-o* or *fee-beeur*) rather than that of the Willow Flycatcher (*fitz-bew*) (Zink and Fall 1981).

For years, biologists believed these songs simply represented regional differences in a widely distributed species. Field observers, however, began to note subtle regional differences in nest structures, habitats, and morphology (Aldrich 1953; Snyder 1953). These reports led Stein (1958, 1963) to conduct a series of extensive field studies. Although specimens he collected looked nearly identical, regardless of their origin, he found their vocalizations and general biology quite distinct. He depicted the Alder Flycatcher as a species of the northern boreal forest and the Willow Flycatcher as an open-country species of the Great Plains and prairies. By 1973, his carefully documented evidence led the American Ornithologists' Union, the official arbiter of bird nomenclature, to declare that the Traill's Flycatcher was indeed two different species.

Just two years later, Green and Janssen (1975) broadly described the Alder Flycatcher as a resident in northern Minnesota and the Willow Flycatcher as a resident in western and southern Minnesota. Considerable work was still necessary, they wrote, to sort out the finer details of each species' distribution. Meanwhile, S. Robbins (1974) was working to delineate the range of the two species in Wisconsin with more precision. The outcome was the identification of a broad region of overlap, extending from the southeast corner of the state northwest to Polk County, which borders east-central Minnesota. The entire area of overlap encompassed more than 20 of Wisconsin's 72 counties.

Zink and Fall (1981) soon followed S. Robbins's (1974) lead and identified a similar zone of sympatry in Minnesota. Roughly contiguous to that described by Robbins, the zone stretched from the east-central counties of Anoka and Ramsey, west to Pope County in west-central Minnesota, and north to Clay County. Included in their assessment was the identification of 11 sites where the two species co-occurred.

Although the Alder Flycatcher is still best described as a species of Minnesota's northern forests, it is occasionally reported farther south. Among the 1,314 breeding-season locations identified by the MBS, four were localities south of the Minnesota River in Brown, Dodge, Freeborn, and Mower Counties. By 1998, however, nesting had been confirmed in only seven northern counties: Aitkin, Beltrami, Cook, Lake, Ramsey, St. Louis, and Stearns (Hertzel and Janssen 1998).

The MNBBA's 2,501 Alder Flycatcher records extended the species' distribution even farther south of the Laurentian Mixed Forest and Tallgrass Aspen Parklands Provinces where it is most abundant. Additional records were found throughout east-central Minnesota and from nine counties south of the Minnesota River, stretching from Lincoln County in the west to Houston County in the east **(A)**. Detected in 29.0 percent (1,374/4,744) of the surveyed atlas blocks, breeding was confirmed in 26 blocks. The birds were reported in 56 of Minnesota's 87 counties. Twelve counties with confirmed nesting stretched from Hennepin County north to Canada. Of these, nine were additions to the list published by Hertzel and Janssen (1998). Nesting was deemed probable in an additional 31 counties.

Because flycatchers are late migrants and are difficult to identify when they are not singing, all atlas reports of Alder Flycatchers in southern Minnesota were carefully examined. At least nine reports were invalidated because they occurred in early June

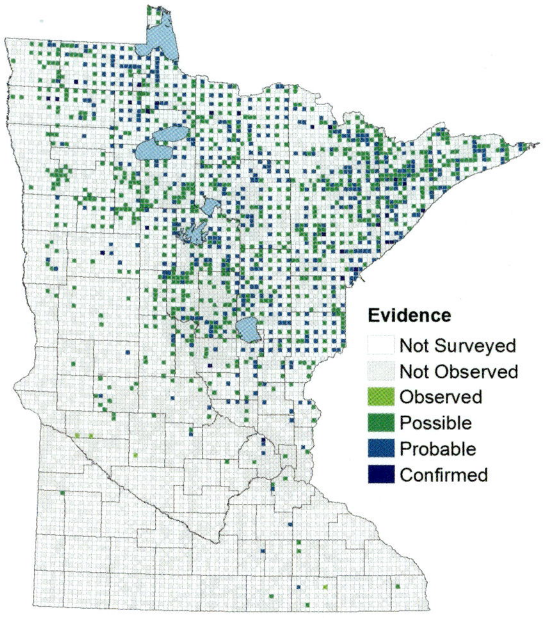

A. Breeding distribution of the Alder Flycatcher in Minnesota based on the Breeding Bird Atlas (2009–13).

and were considered to be late migrants. Accepted records were found during the second or third week of June or were accompanied by notes that supported their retention.

It is not clear if the Alder Flycatcher has actually extended its range southward in recent years or if it was simply overlooked by earlier observers. Regardless, its breeding status in southern Minnesota requires further study. Both MBS and MNBBA data also suggest that the range of overlap with the Willow Flycatcher may be wider than proposed by Zink and Fall (1981), extending north throughout the entire Red River valley and south into south-central and southeastern Minnesota.

The MNBBA probability map predicts low breeding densities throughout much of the state's Eastern Broadleaf Forest Province and largely moderate densities throughout the Laurentian Mixed Forest Province **(B)**. The shrubby wetlands of the Agassiz Lowlands Subsection, however, are predicted to provide some of the most suitable habitat for the species.

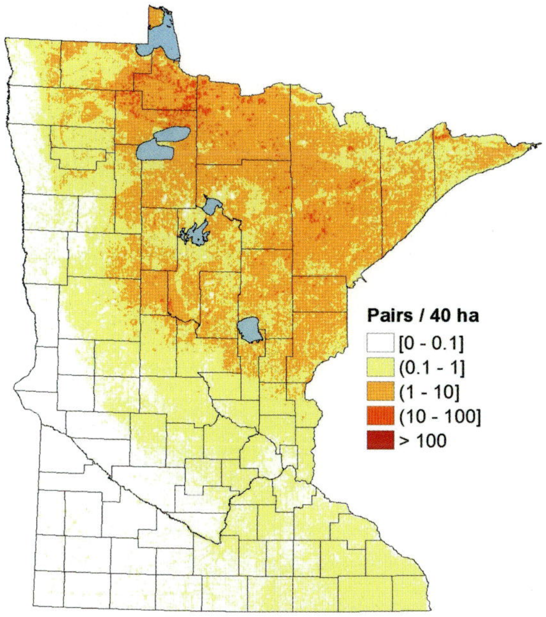

B. Predicted breeding distribution (pairs per 40 ha) of the Alder Flycatcher in Minnesota based on habitat, landscape context, and climate data gathered during the Minnesota Breeding Bird Atlas (2009–13) using the Generalized Linear Modeling method with an adjustment for detectability.

BREEDING HABITAT

In Minnesota, Zink and Fall (1981) described the species' typical habitat as "dense, boggy alder (*Alnus* spp.) and willow (*Salix* spp.) thickets, often near streams" **(C)**. It can also be found less commonly in dense shrubby upland habitats, especially in regenerating stands of young trees and shrubs following harvest or other disturbances (Lowther 2020a).

Willow Flycatchers, in contrast, are more likely to be found in scattered patches of small trees and tall shrubs embedded in open grassland habitats, as well as in low swales and along woodland edges (Zink and Fall 1981). When both species occur on the same site, the authors noted that the Alders were invariably found in "denser and taller vegetation with a wet boggy substrate, and Willow Flycatchers were either in more open vegetation (if very wet) or drier areas (if dense vegetation)." Other accounts have noted a similar separation of habitat preferences (Rodewald et al. 2016; Cutright et al. 2006).

POPULATION ABUNDANCE

PIF (2021) generated a population estimate of 120 million Alder Flycatchers in Canada and the United States, with approximately 1.1 million birds residing in Minnesota. MNBBA point count data generated a slightly higher estimate of 1.4 million birds (95 percent confidence interval of 1.2 to 2.0 million). The high density of breeding pairs in the Agassiz Lowlands Subsection, an area well-sampled by the MNBBA, likely contributed to this larger estimate.

The most recent analysis of BBS monitoring data combines data for the Alder and Willow Flycatchers. Across their breeding range, the Alder/Willow complex has demonstrated a significant, long-term population decline that averages −0.54 percent per year from 1966 to 2019 (Sauer et al. 2020). In Minnesota, the population has been relatively stable, with a notable decline in recent years **(D)**. Aside from changes in the availability of suitable habitat, the factors influencing these population trends are poorly understood and have received little attention.

C. Shrubby lowlands, often bordering a nearby stream, are the Alder Flycatcher's most common abode.
Photo by Lee A. Pfannmuller.

CONSERVATION

Deemed a moderate conservation priority given its declining population (PIF 2021), the Alder Flycatcher deserves considerably more research attention. Fortunately, under a 2°C warming scenario, the vulnerability of the North American breeding distribution of this signature species of Minnesota's northern shrub swamps is considered low (NAS 2019). Nevertheless, the same scenario projects the Flycatcher's disappearance from all but the extreme northeastern corner of the state by 2050; extensive losses are even projected under an imminent 1.5°C warming scenario. A future without the familiar *fee-bee-o* emanating from nearly every alder–willow shrubland in northern Minnesota is indeed difficult to imagine.

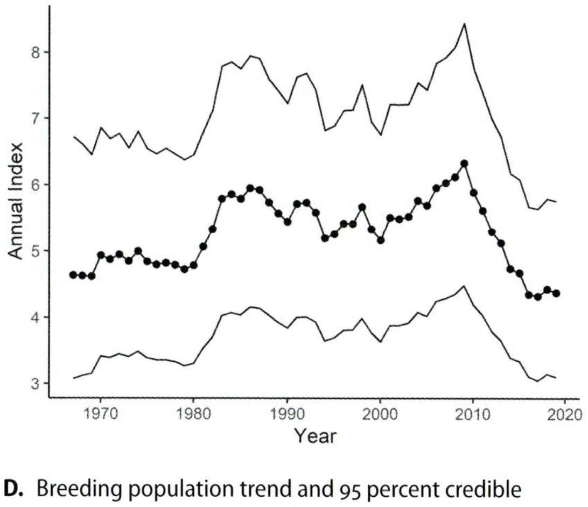

D. Breeding population trend and 95 percent credible interval (solid black lines) for the Alder/Willow Flycatcher complex in Minnesota based on the 1967–2019 Breeding Bird Survey (Sauer et al. 2020).

WILLOW FLYCATCHER

Empidonax traillii

Minnesota Seasonal Status: Regular breeding species and migrant. The Willow Flycatcher was uncommon during the MNBBA.
Migration: Medium- to long-distance migrant that winters from southern Mexico to northern South America.
Food: Insects secured by fly-catching and foliage gleaning.
Nest: Cup nest placed in a deciduous tree or shrub, usually within 2 m of the ground.
Conservation Concern: Assigned a moderate PIF Continental Concern Score of 10/20; the federally *Endangered* subspecies *E. t. extimus* is restricted to the southwestern United States.

In 1828, when Audubon described and named *Empidonax traillii* in recognition of his friend Thomas Traill, he believed he was identifying one broadly distributed species, the Traill's Flycatcher. His beautiful painting of the bird, perched on the branches of a sweet gum tree, was based on a pair he watched "in the skirts of the woods along the prairie lands of the Arkansas River" (Audubon 1827–38). He also possessed specimens from the Columbia River basin, which were collected by fellow ornithologist John Townsend. Today, both areas fall within the breeding range of the Willow Flycatcher and are classic examples of the species' primary habitat, riparian woodlands and shrublands.

Early on, however, the Traill's Flycatcher was known to sing two distinctly different songs. One song, *fitz-bew,* sounded like a two-syllable sneeze with the accent on the first syllable. A second song was described as a three-syllable *fee-bee-o,* with the accent on the second syllable. For years, biologists attributed the differences to regional variation in a widely dispersed species. It was not until observers reported hearing both songs at the same site that biologists began to take a closer look at this enigmatic little flycatcher.

Common across Canada and the northern United States, specimens looked nearly identical regardless of their origin. All had an olive-brown plumage on the back and wings with a lighter belly, two wing bars, and an indistinct white eye-ring. Researchers, however, began to note regional differences in the species' nest, habitat, and morphology (Aldrich 1953; Snyder 1953). Even in areas of sympatry, where both

Photo by Gerald Hoekstra.

song-types occurred, Stein (1958, 1963) found these differences remained intact and concluded that the Traill's Flycatcher was actually two different species. He described the Alder Flycatcher (*E. alnorum;* song type *fee-bee-o*) as a species of the northern boreal forest and the Willow Flycatcher (*E. traillii;* song type *fitz-bew*) as a species of the open country. Stein's work was officially adopted by the American Ornithologists' Union in 1973.

The southerly distributed Willow Flycatcher has two centers to its breeding range: the first stretches across the northeastern and north-central states south to the northern Appalachians and southern Ohio River valley. The second is in the Pacific Northwest, stretching southeast to Arizona and New Mexico. Areas with the highest breeding densities largely occur from central North Dakota west to the Pacific coast.

MINNESOTA BREEDING DISTRIBUTION

Roberts's (1932) treatise on Minnesota birds only included an account for *Empidonax traillii.* Commenting on its distribution and relative abundance, he wrote: "If looked for in the right situations, it will be found fairly common throughout the southern part of the state, even in the bushy lowlands of the prairies, and northward it becomes increasingly abundant until in the upper half of the state its harsh cry comes from every alder thicket and wooded swamp." Unbeknownst to Roberts, he was describing the combined breeding ranges of the Willow and Alder Flycatchers.

Following their official declaration as distinct species, Green and Janssen (1975) described the Willow Flycatcher as a breeding resident in southern and western Minnesota, while the Alder Flycatcher was a denizen of northern shrublands. A more in-depth assessment was conducted by Zink and Fall (1981). Combining their field observations with reports from others, they expanded the Willow Flycatcher's breeding range as far north as Marshall County in northwestern Minnesota and Aitkin County in north-central Minnesota. They also described a narrow zone of overlap where both species occurred that stretched from east-central Minnesota west to Pope County and north to Clay County. This zone was contiguous with one described several years earlier by S. Robbins (1974) in Wisconsin.

The MBS provided additional details. With 438 breeding-season locations, Willow Flycatchers were found as far north as Kittson County in northwestern Minnesota and Morrison County in central Minnesota. Despite its wide distribution, Hertzel and Janssen (1998) identified only seven counties where nesting had been confirmed: Anoka, Clay, Dakota, Hennepin, Hubbard, Murray, and Wilkin. Only two were located south of the Minnesota River.

Concentrated in the southern half of the state, the MNBBA's 592 Willow Flycatcher detections were reported from 8.3 percent (392/4,740) of the surveyed atlas blocks that were distributed across 65 of Minnesota's 87 counties **(A)**. Nesting was confirmed in 11 blocks spanning 11 counties and deemed probable in an additional 25 counties. Only one confirmed breeding location (Polk County) was in northern Minnesota. Eight of the 11 counties where breeding was confirmed were additions to the list published by Hertzel and Janssen (1998).

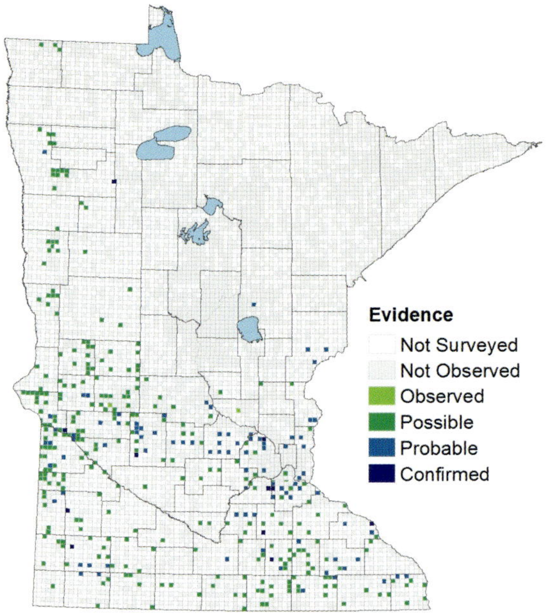

A. Breeding distribution of the Willow Flycatcher in Minnesota based on the Breeding Bird Atlas (2009–13).

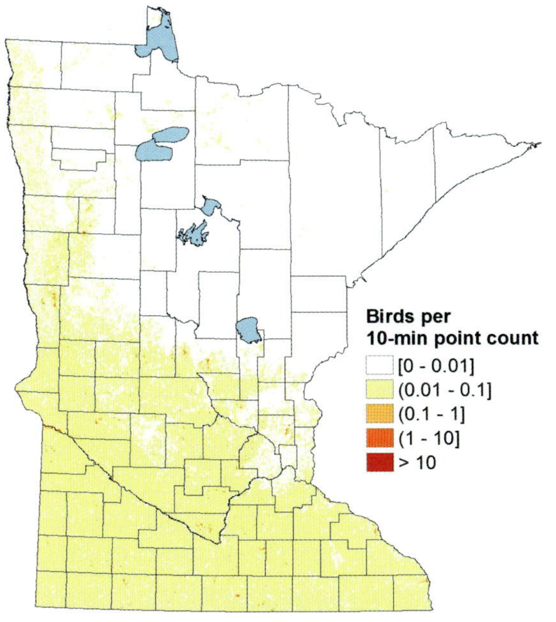

B. Predicted breeding distribution (birds detected per 10-minute point count) of the Willow Flycatcher in Minnesota based on habitat, landscape context, and climate data gathered during the Minnesota Breeding Bird Atlas (2009–13) using the Generalized Linear Modeling method without an adjustment for detectability.

Although the Willow Flycatcher was detected in six more counties than the Alder Flycatcher, it was more sparsely distributed, reported in 9.8 percent of the priority blocks compared to 37.3 percent for the Alder Flycatcher. The MNBBA data documented a similar distribution to that delineated by the MBS, with two exceptions: first, records extended farther north in east-central Minnesota, with probable nesting records in central Pine and southern Aitkin Counties; second, in west-central Minnesota, the birds were not restricted to the upper Minnesota River valley but were found in numerous blocks beyond the river's floodplain. Both MBS and MNBBA data suggest a broader zone of overlap where both species occur than that proposed by Zink and Fall (1981).

The MNBBA probability map predicts low breeding densities throughout southern Minnesota, with abundance decreasing north along the Red River valley **(B)**. The model suggests that Willow Flycatchers, although extremely rare, might be encountered farther east into Beltrami and Koochiching Counties.

Beyond Minnesota, eastern populations appear to be expanding to the south (e.g., in Ohio, Pennsylvania, and West Virginia) and to the north (e.g., in Ontario, New York, and Vermont). Although the factors responsible for these movements are not entirely clear, habitat changes at the landscape level are occasionally noted (Sedgwick 2020).

BREEDING HABITAT

A shrub specialist, the Willow Flycatcher is frequently associated with riparian habitats. Like the Alder Flycatcher, it is also found in dry, shrubby uplands, including young regenerating forests (Sedgwick 2020). Although it is closely associated with willow thickets, a wide variety of other shrubs are used. In the northern Great Plains, it has been described as an edge-adapted species found in riparian woodlands, low-lying prairies, and woodland edges. Farther east it is commonly associated with willow thickets, upland shrubs, and shrubby roadsides (Johnsgard 1986; Graber et al. 1974).

In Minnesota, Zink and Fall (1981) emphasized the species' habitat as scattered patches of small trees and tall shrubs embedded in open grasslands **(C)**, as well as low swales and woodland edges. When both Alder and Willow Flycatchers were present on the same site, the Alders were found in "denser and taller vegetation with a wet boggy substrate, and Willows were either in more open vegetation (if very wet) or drier areas (if dense vegetation)." Other accounts have noted a similar separation of habitat preferences (Rodewald et al. 2016; Cutright et al. 2006).

POPULATION ABUNDANCE

PIF (2021) generated a population estimate of 8.1 million adults in the United States and Canada, with approximately 48,000 in Minnesota. The species was too uncommon to develop a reliable statewide estimate with MNBBA data. The analysis of BBS data from 1966 to 2019 combines the data for the Willow and Alder Flycatchers and documents a significant, surveywide population decline, averaging −0.54 percent per year (Sauer et al. 2020). In Minnesota, the combined trend line for the two species illustrates a relatively stable population from 1967 to 2019 **(D)**. Despite the overall stability, a significant drop in numbers is evident in recent years.

C. The Willow Flycatcher is especially fond of upland shrub thickets in western and southern grasslands. Photo by Lee A. Pfannmuller.

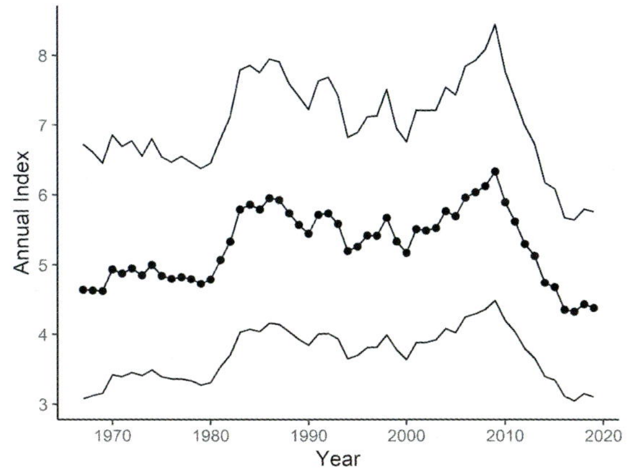

D. Breeding population trend and 95 percent credible interval (solid black lines) for the Alder/Willow Flycatcher complex in Minnesota based on the 1967–2019 Breeding Bird Survey (Sauer et al. 2020).

The loss of wetland habitats is likely the most important factor impacting Willow Flycatcher populations. Grazing, stream channelization, water diversions, invasive species, development, and more intensive agricultural practices all place these wetlands at risk. High rates of nest predation and nest parasitism by Brown-headed Cowbirds also negatively impact local populations (Sedgwick 2020). Together, these factors led to the listing of the Willow Flycatcher's southwestern subspecies (*E. t. extimus*) as federally *Endangered* (USFWS 2002).

CONSERVATION

With declining numbers, a relatively small population, and potential threats to its breeding habitat, the Willow Flycatcher has been assigned a moderate conservation score (PIF 2021). As temperatures rise, its breeding population is projected to gradually disappear from the southern portion of its range in the east-central states and expand northward across southeastern Canada. NAS (2019) projected range expansions across northern Minnesota under a 2°C warming scenario; should temperatures rise an additional degree, this Alder Flycatcher look-alike will disappear from most of the state's southern counties as it continues its trek northward.

Management recommendations are largely focused on efforts to protect the federally *Endangered* subspecies, *E. t. extimus,* which numbers fewer than 1,000 birds. Cowbird control to reduce nest parasitism has been the primary action taken (Sedgwick 2020). Elsewhere, populations of this easily overlooked Flycatcher are best served by measures that protect the shrubby riparian borders of wetlands and upland shrublands (Rodewald et al. 2016).

LEAST FLYCATCHER

Empidonax minimus

Minnesota Seasonal Status: Regular breeding species and migrant. The Least Flycatcher was common during the MNBBA.
Migration: Medium- to long-distance migrant; winters from Mexico to Central America.
Food: Almost exclusively flying insects; occasionally gleans from foliage.
Nest: Cup nest placed in the subcanopy or canopy; usually in deciduous tree species.
Conservation Concern: Assigned a moderate PIF Continental Concern Score of 11/20.

The Least Flycatcher is a common forest bird species that occurs in a variety of mature deciduous and mixed deciduous–coniferous forests throughout Minnesota. It is somewhat gregarious and tends to be found in breeding population clusters. It is one of five species of flycatchers in Minnesota in the genus *Empidonax*, which means "lord of the flycatchers" in Greek. Its Latin scientific name, *minimus,* emphasizes that it is the smallest member of the genus. As with others in this genus, key characteristics are its relatively small, chunky appearance, olive-green plumage, two white wing bars, and white eye-ring. It is almost impossible to distinguish from two other *Empidonax* flycatchers, the Alder and Willow Flycatcher, by plumage alone, but its habitat and vocalization are diagnostic. The Least Flycatcher was formerly called "chebek" in honor of its distinctive song, which is similar only to that of the Yellow-bellied Flycatcher.

Bent (1942) described it as the "smallest, earliest, tamest, smartest, bravest, noisiest, and most prominent member of its genus." Further, he stated, "He is a restless, active little fellow, far less sedate than our wood pewee, preferring to dart about among the foliage, or from one perch to another, in quest of his prey, rather than sit on one favorite perch and watch for it."

The Least Flycatcher is widely distributed and common across the northern United States and Canada to the tree limit, including south in the Appalachian Mountains to northern Georgia. Highest densities have been recorded in northern Minnesota and on the North Dakota border with Manitoba and Saskatchewan.

Photo by David Brislance.

MINNESOTA BREEDING DISTRIBUTION

According to Roberts in 1932, the Least Flycatcher was "one of the most common and widely distributed birds of the state." He reported nesting activity from Jackson and Fillmore Counties on the border with Iowa, to Kittson County in extreme northwestern Minnesota, and to Itasca County in northern Minnesota.

More than forty years later, Green and Janssen (1975) described the species' distribution in a similar fashion, simply stating that it is a resident throughout the state and "frequents groves on the prairies." A few years later, Janssen (1987) not only echoed this distribution but also stated that the species was "scarce to absent in many areas of the south-central and southwest regions." He reported confirmed nesting from 22 counties, including the northeastern counties of Cook, Lake, and St. Louis. Hertzel and Janssen (1998) added another four counties with confirmed nesting records since 1970. Collectively, the confirmed nesting records include the heavily forested areas of the northeast, northern, and north-central regions of the state. Records were scarce in the southern counties; Brown County was the only county with a confirmed nesting in these areas.

The MBS included 1,627 breeding-season locations in their detailed surveys of counties. The highest density of locations was in northeastern and north-central Minnesota, but locations were scattered throughout the state. For instance, locations were noted in the extreme southwestern counties of Lincoln (four locations), Murray (two), and Rock (two). However, numerous counties (such as Lyon, Nobles, Pipestone, and Watonwan) had no records.

The MNBBA further substantiated the widespread distribution of the species in the state with 3,774 breeding records and at least possible nesting from every county **(A)**. The participants of the MNBBA recorded the species in 43.3 percent (2,051/4,741) of the surveyed atlas blocks and 55.8 percent (1,305/2,337) of priority blocks. These included 75 confirmed nesting records, 60 of which came from priority blocks. A total of 31 counties had confirmed nesting records. Nesting records were more densely distributed in the northern forested portion of the state. However, possible or probable nesting was also quite substantial in many southwestern and western counties. Breeding records were sparse in most of the southern counties, but small populations can still be found in wooded areas of this region.

The probability distribution map for the Least Flycatcher highlighted potentially high densities in north-central and northwestern Minnesota and in many other areas of northern Minnesota **(B)**. These areas generally correspond with forested habitats that are dominated by deciduous trees or mixed deciduous–coniferous forests, such as the Tallgrass Aspen Parklands Province in the northwest, the Pine Moraines and Outwash Plains Subsection in

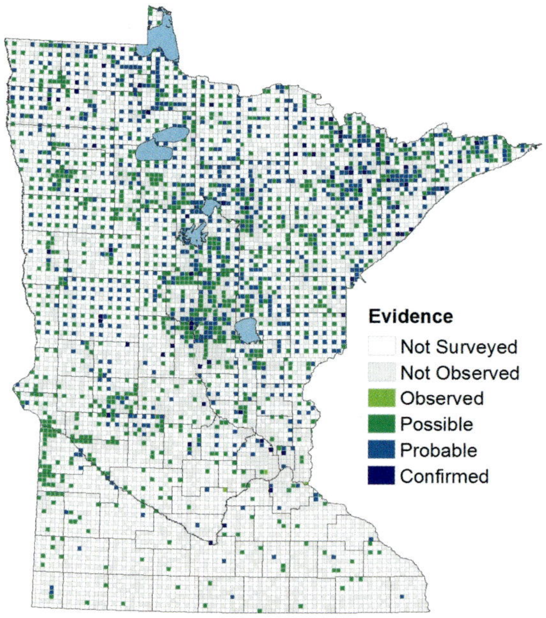

A. Breeding distribution of the Least Flycatcher in Minnesota based on the Breeding Bird Atlas (2009–13).

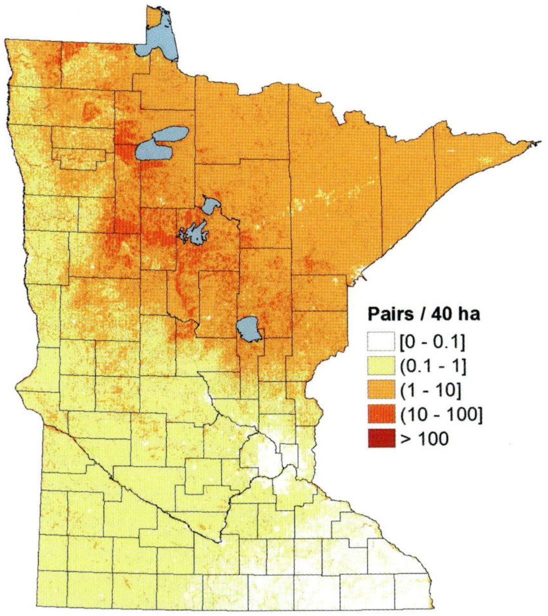

B. Predicted breeding distribution (pairs per 40 ha) of the Least Flycatcher in Minnesota based on habitat, landscape context, and climate data gathered during the Minnesota Breeding Bird Atlas (2009–13) using the Generalized Linear Modeling method with an adjustment for detectability.

west-central Minnesota, and the Western Superior Uplands Section in the east-central region.

In a summary of the Least Flycatcher in North America, Tarof and Briskie (2020) identified few changes in its distribution except for an expansion into Washington State since the late 1970s. However, S. D. Robbins (1991) observed that the Least Flycatcher has been moving northward in Wisconsin as mature deciduous forest cover has declined in the south. The low detection of the species in the southern portions of Minnesota renders Roberts's (1932) comment that it is the "most generally and evenly distributed member of its family" as no longer true. Furthermore, Roberts did not report any nests from northeastern Minnesota, even though currently some of the highest North American densities of Least Flycatchers occur in portions of that region. This was likely due to the lack of coverage in those regions during Roberts's time, but may be consistent with S. D. Robbins's (1991) comment that its distribution is moving northward in Minnesota as well.

BREEDING HABITAT

The Least Flycatcher occurs in a wide variety of forest types, including upland and lowland deciduous forests, mixed deciduous–coniferous upland forests, mixed lowland swamp conifers, and patches of forest in open agricultural and prairie groves (Bond 1957; Tarof and Briskie 2020). Forested habitats with open understories are especially important (Breckenridge 1956). Data gathered in four national forests of Minnesota and Wisconsin found a high frequency of occurrence and abundance in oak forests, followed by red pine, beech–maple–birch, upland hardwoods, and aspen–birch cover types (Niemi, Howe, et al. 2016) **(C)**.

POPULATION ABUNDANCE

PIF (2021) estimated a breeding population of 27 million adults in the United States and Canada, and a Minnesota population estimated at 990,000 breeding adults. The MNBBA, however, estimated a much higher population of 3.6 million adults in Minnesota (95 percent confidence interval of 3.2–5.9 million).

The BBS trends in Minnesota indicated a consistent and significant decline in the population from 1967 to 2019 averaging −1.16 percent per year **(D)**. Rosenberg, Kennedy, and colleagues (2016) estimated an overall population decline of −53 percent from 1970 to 2014. Tarof and Briskie (2020) summarized issues associated with declining population trends in the Least Flycatcher and focused on a combination of changes in forest conditions, including forest succession, fragmentation, and various forest disturbances. They also identified nest predation, brood parasitism, weather, deer densities, and adult mortality as contributing factors to lower nesting success in several detailed studies. Manolis and colleagues (2000) stressed that nesting failure in Minnesota was difficult to assess because of excessive nest heights constructed by the species.

CONSERVATION

Despite recent declines, the Least Flycatcher still has a large breeding population in North America, but its decline of over 50 percent is very substantial and a cause for concern. PIF (2021) lists it as a Common Species in Steep Decline. Matteson and colleagues (2009) labeled it a "regional stewardship species" because of its potential preference for large tracts of deciduous forest and its population decline in the boreal hardwood transition forests of northern Minnesota, Michigan, and Wisconsin. DellaSala and Rabe (1987) emphasized that Least Flycatchers were sensitive to disturbances and tended to be found in more interior portions of the forest. They stressed that management should include maintaining large contiguous forest tracts and minimizing disturbances.

C. Preferences for breeding habitat by the Least Flycatcher are mature forests with open understories that allow ample room to forage for flying insects. Photo by Debbie Petersen.

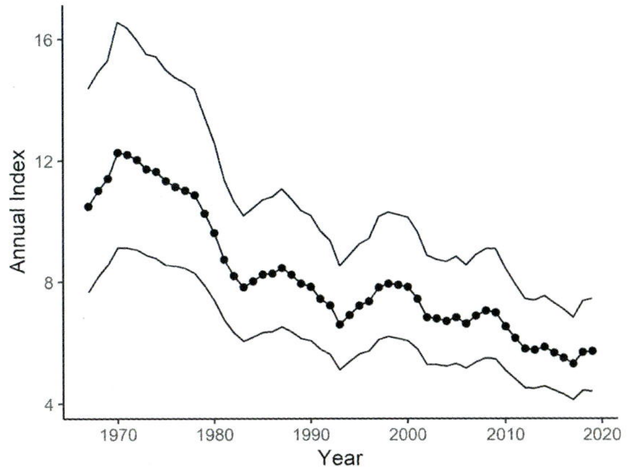

D. Breeding population trend and 95 percent credible interval (solid black lines) for the Least Flycatcher in Minnesota based on the 1967–2019 Breeding Bird Survey (Sauer et al. 2020).

With a 2°C warming scenario, NAS (2019) projected a moderate vulnerability for the Least Flycatcher's breeding distribution in North America, including a 37 percent range loss and a 32 percent range gain primarily in Canada and Alaska by 2050. In Minnesota, it was predicted to be extirpated from the state with a 3°C warming scenario by 2080. In contrast, it was projected to slightly expand its winter range northward in Texas and Florida under all three climate scenarios they examined.

The loss and fragmentation of Minnesota's forests over the past 150 years has reduced the distribution of Least Flycatchers in the southern and western portions of the state. Population declines were particularly noteworthy in the north-central region of Minnesota in the BBS and in the Chippewa National Forest, where the species has declined an average of −1.88 percent from 1995 to 2020 (Grinde, Bednar, et al. 2020). Land management and ownership in the north-central region is actively being parceled (fragmented) compared with northeastern Minnesota, where populations appear to be presently stable. Therefore, the impact of forest disturbances on the Least Flycatcher in Minnesota is consistent with more detailed studies of the species.

EASTERN PHOEBE

Sayornis phoebe

Minnesota Seasonal Status: Regular breeding species and migrant. The Eastern Phoebe was common during the MNBBA.
Migration: Short- to medium-distance migrant throughout most of its range; winters along the southern Atlantic coast, the Gulf coast, and in eastern Mexico. Populations in the southeastern states are permanent residents.
Food: Aerial insectivore that hawks flying insects.
Nest: Cup nest with a base of mud that is built up with leaves, grasses, and moss. Although natural sites such as rocky outcrops are still used, it is common to find nests built on bridges, culverts, or buildings.
Conservation Concern: Assigned a low PIF Continental Concern Score of 8/20.

When the first warm days of spring tease us in late March and early April, birders long for some visible sign that our summer songbirds have begun their northward journey. Although the waves of migrating vireos, thrushes, and warblers are still weeks away, our yearnings are satisfied when we are greeted by the raspy, namesake song of one of Minnesota's earliest migrants, the Eastern Phoebe. With little fanfare the birds suddenly appear, already settled into last year's favorite haunts and singing their familiar *phoe-be, phoe-be.* The song's rough, breathy quality is unlike the clear, sweet whistle of the Black-capped Chickadee's *fee-bee.*

Many early ornithologists commented on the Phoebe's amiable disposition. It so readily adapts to the presence of people that in 1804 it inadvertently became the first bird banded in North America by none other than John James Audubon. A nesting pair he observed in Pennsylvania was so docile that he decided to test his suspicion that the young would return to their natal site to breed. Having tied a silver thread to the leg of each nestling, when he visited the site the following spring he found it was reoccupied, presumably by the original, unmarked pair. But when he caught the adults in several nearby nests, he "had the pleasure of finding that two of them had the little ring on the leg" (Audubon 1827–38).

A familiar inhabitant of our farms, lakeside cabins, and riparian woodlands, the Phoebe has few distinctive features. A little larger than a House Sparrow, it is easily recognized as a flycatcher by its flat, slender bill and its erect, upright posture as it perches in the understory. Its brownish-gray head and back and its light undersides are rather plain, apart from a faint yellow cast that may be present on the belly and some light gray "smudges" on the sides of the breast. Eye-rings and wing bars are absent in the adults. The lack of notable plumage characteristics is actually a good clue to the species' identification, along with its constant tail bobbing. One field study concluded that this behavior actually served as an effective predator deterrent (Carder and Ritchison 2009).

Photo by David Brislance.

The Eastern Phoebe's breeding range extends from the Northwest Territories and Alberta southeast across much of the eastern United States. It is absent from many regions in the Deep South. Breeding densities are moderately high throughout the United States and southeastern Canada. The species' genus name, *Sayornis,* means "Say's bird," a reference to being named for Thomas Say, an entomologist on the nineteenth-century Long expedition up the Missouri and Platte Rivers where the type specimen was collected.

MINNESOTA BREEDING DISTRIBUTION

Like many of his contemporaries, Roberts (1932) described this common summer resident by commenting on the Phoebe's long-standing association with people: "Nearly everyone knows this plainly dressed little bird by one name or another, as it comes so familiarly about man's habitations and mingles its home activities so intimately with those of its host. There is scarcely a farm or dwelling in the country or a deserted shack in the forest that has not its pair of sociable, tail-wagging Phoebes." At the time, other common names included Bridge-Bird and Barn Pewee, referencing its frequent use of bridges and barns for nesting habitat.

Even the long list of nest records submitted by Roberts's friends and colleagues was a testament to the ease of documenting the breeding activity of a bird that often shared one's home as a nesting abode. Breeding evidence was available from 17 counties as well as from Cass Lake, Itasca State Park, and Leech Lake. The records were confined primarily to the forested regions of the state, including the Aspen Parklands Province, and stretched from Houston County in the southeast to Kittson and Roseau Counties in the northwest, Lake of the Woods in north-central Minnesota, and Carlton and St. Louis Counties in the northeast. The absence of records from the southwest and extreme northeast regions suggested that the species was relatively uncommon in the open grasslands and sparsely settled forest landscapes.

Green and Janssen (1975) also described the Phoebe as a statewide summer resident. Janssen (1987) noted that it was most common in northern Minnesota, decreasing in abundance as one moved south, and "scarce to absent in large portions of the south-central and southwestern regions south of the Minnesota River." He identified 42 counties where nesting had been confirmed since 1970, including three southwestern counties: Lac qui Parle, Pipestone, and Yellow Medicine. Hertzel and Janssen (1998) added 12 more counties to the list. More than half of the counties without confirmed nesting records (18) were south of the Minnesota River. Field surveys conducted by the MBS further documented the species' statewide distribution. Among the 300 breeding-season locations identified, multiple records were widely dispersed across the southwestern corner of the state.

During the MNBBA, observers reported 2,809 Eastern Phoebe records from 37.6 percent (1,789/4,763) of the surveyed atlas blocks and from 49.9 percent (1,167/2,337) of the priority blocks. Nesting was confirmed in 651 blocks **(A)**. There was at least one record from each of Minnesota's 87 counties and confirmed nesting records from 81 counties. The six counties lacking nesting records were in west-central and southwest Minnesota: Big Stone, Chippewa, Lyon, Nobles, Pipestone, and Traverse. Of the counties with confirmed nesting records, 29 were new to the list published by Hertzel and

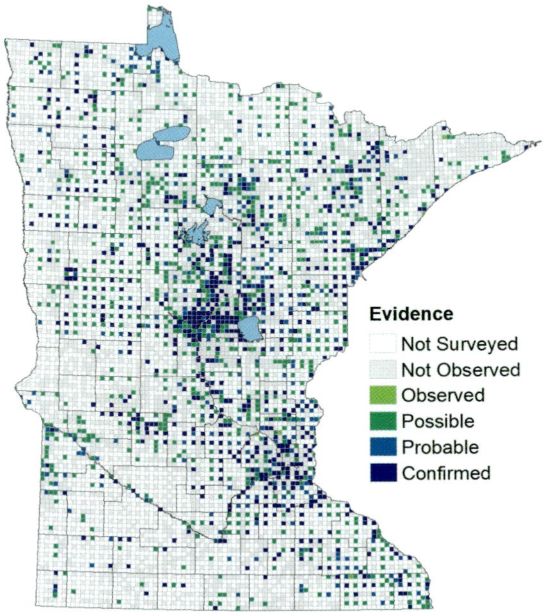

A. Breeding distribution of the Eastern Phoebe in Minnesota based on the Breeding Bird Atlas (2009–13).

Janssen (1998). Although widely distributed across the state, the Eastern Phoebe was least abundant in the Prairie Parkland Province.

The MNBBA probability map predicted the species' prominence in the northern reaches of the Eastern Broadleaf Forest Province and in the southern and western regions of the Laurentian Mixed Forest Province **(B)**. Breeding densities are also moderately high in the floodplain forests of southeastern Minnesota but quite low in the northeast and throughout much of the densely cultivated regions of the state.

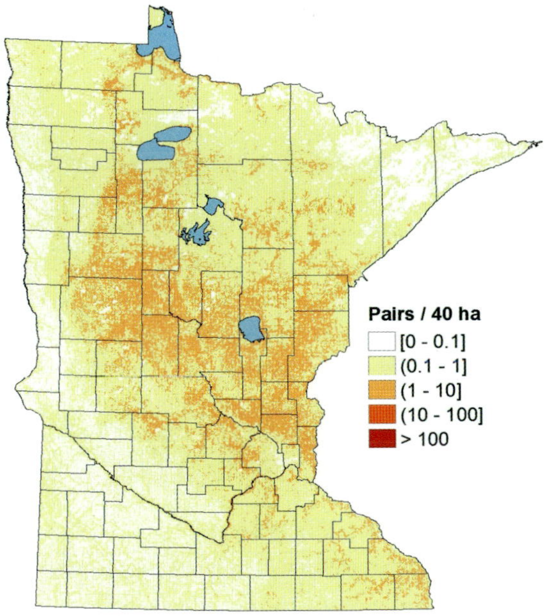

B. Predicted breeding distribution (pairs per 40 ha) of the Eastern Phoebe in Minnesota based on habitat, landscape context, and climate data gathered during the Minnesota Breeding Bird Atlas (2009–13) using the Generalized Linear Modeling method with an adjustment for detectability.

The Phoebe's distribution may have changed little in the past one hundred years, but it is likely more common across Minnesota's agricultural and northern forested landscapes than in earlier times. Historically the species' distribution and abundance were limited by the availability of natural nest sites, including rocky outcrops and ledges located near forested streams and lakeshores. But European settlement, and the subsequent construction of buildings and bridges, provided more nesting opportunities in previously unsuitable regions for this adaptable species. These changes also facilitated its expansion into the northern Great Plains and along the coastal regions of the Carolinas and Florida (Weeks 2020).

BREEDING HABITAT

The Eastern Phoebe occupies a wide range of habitats, from woodland edges, rural farmsteads, lake homes, and bridges to rocky ravines and outcrops **(C)**. As noted above, the limiting factor is the availability of suitable nest sites. Specifically, the birds need a structure, natural rock outcropping, or stream bank that provides some protection from predators and inclement weather. Suitable habitats usually share two other features: they are in the vicinity of water, and trees and woodlands are nearby (Weeks 2020). A diverse understory of small trees and shrubs seems particularly important, as the small branches are frequent perching sites used by this aerial insectivore to scout for prey.

POPULATION ABUNDANCE

PIF (2021) estimated a breeding population of approximately 35 million Eastern Phoebes in the United States and Canada, of which nearly 800,000 reside in Minnesota. The Phoebe is one of the few examples where the MNBBA point count data generated a slightly smaller estimate of 660,000 adults (95 percent confidence interval of 550,000–794,000). The point count sampling of many inaccessible sites, where development is negligible and which are less apt to provide suitable Phoebe habitat, likely contributes to the lower estimate.

Since the BBS began monitoring birds in the mid-1960s, Eastern Phoebe numbers have remained relatively stable, even increasing slightly. This is particularly true within the United States, where populations have demonstrated a significant increase from 1966 to 2019, while the species has declined in Canada over the same time period (Sauer et al. 2020). Although relatively stable, the long-term population trend in Minnesota shows a small decline that accelerates in recent years **(D)**. Although studies are limited, severe weather, especially during spring migration, may be a significant mortality factor (Weeks 2020).

C. Although the Eastern Phoebe frequently attaches its nest to our homes and cabins, along the Mississippi River it still uses rocky escarpments for nesting. Photo by Lee A. Pfannmuller.

CONSERVATION

Given its stable population, its relatively large population size, and its adaptability to human-modified landscapes, the Eastern Phoebe is not considered a conservation priority (PIF 2021). Although its North American breeding distribution is considered stable under a 2°C warming scenario, some losses are predicted along the western periphery of the Phoebe's breeding range, including in southeastern and central Minnesota (NAS 2019). Nevertheless, its adaptability may help mitigate such losses and enable the Phoebe to remain one of our most endearing avian neighbors.

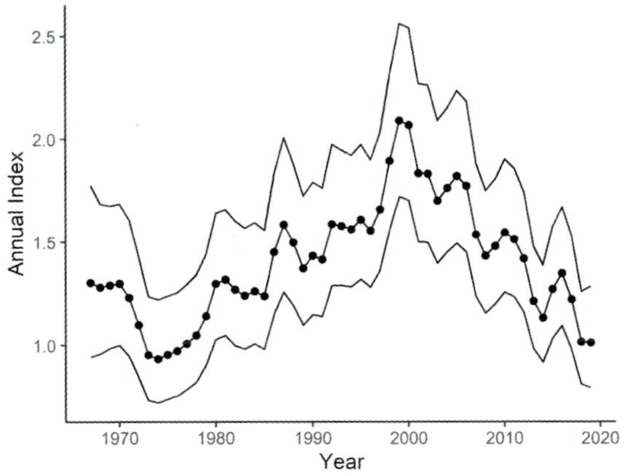

D. Breeding population trend and 95 percent credible interval (solid black lines) for the Eastern Phoebe in Minnesota based on the 1967–2019 Breeding Bird Survey (Sauer et al. 2020).

SAY'S PHOEBE

Sayornis saya

Photo by Andy Reago and Chrissy McClarren.

Minnesota Seasonal Status: Summer visitant, occasionally breeding in four western counties (Clay, Rock, Traverse, and Yellow Medicine), 2005, 2010–18. Occasional migrant, especially in western Minnesota or throughout the state. The Say's Phoebe was very rare during the MNBBA.
Migration: Short- to medium-distance migrant that winters in the southwestern United States and Mexico; populations in the Southwest are permanent residents.
Food: Insectivorous, feeding mostly on wild bees, wasps, winged ants, and many other terrestrial invertebrates.
Nest: Built in cavities, caves, and arroyo banks or on ledges or cliffs with overhead cover; also in mines, under bridges, and on eaves or other structures with similar overhangs.
Conservation Concern: Assigned a low PIF Continental Concern Score of 7/20.

Say's Phoebe is an extremely rare species in Minnesota. It is slightly larger than its close relative the Eastern Phoebe and is its counterpart in western North America. Both sexes are similar, with grayish-brown backs that are darker on the wings and head. Its tawny cinnamon belly and black tail are distinctive. Bent (1942) described its call note as quite different from the Eastern Phoebe, "a soft, plaintive *phee-eur,* rather sweet but somewhat melancholy in tone; it is often accompanied by a twitching of the tail and a raising of the crest." Furthermore, he terms its mating call as *"pit-tsee-ar,"* usually given repeatedly as it flutters in the air.

As the land west of the Mississippi River was explored, the Say's Phoebe was one of the New World species discovered by nineteenth-century naturalist Thomas Say, for whom this species is named. Say was the great-grandson of the pioneer naturalist John Bartram and was a self-taught entomologist. He was appointed as the chief zoologist to the Major Stephen H. Long expedition in 1819–20 to explore the central Great Plains from the Platte River west to the Rocky Mountains.

The Say's Phoebe is widely found throughout the western Great Plains and Great Basin with a disjunct population on the treeless tundra and shrub tundra in Alaska and the adjacent Yukon territory. The highest densities are found in New Mexico, southeastern Utah, and the state of Washington.

MINNESOTA BREEDING DISTRIBUTION

At the time that Roberts was gathering information for his 1932 book on Minnesota avifauna, there were no records of the Say's Phoebe in the state. The first documented observation (September 3, 1963) was of a migrant found in Winona County (Einhorn 1969). Subsequently, a specimen was discovered in the collection at the College of St. Benedict, Stearns County, presumably found on campus in May 1967 (Eckert 1968). Another migrant was observed (September 3, 1973) on a roadside power line in Rock County (Eckert 1974b). The next year, breeding territorial behavior, but no actual nest, was observed on many occasions from April 3 to July 14, 1974, at Blue Mounds State Park, Rock County (Eckert 1975b).

All subsequent twentieth-century records (15) were migrants in both spring and fall along the North Shore of Lake Superior (1975 Duluth, 1983 and 1997 Cook County) and in two southeastern counties (1988 Olmsted, 1998 Dakota). Additional observations were mostly during the spring on the western prairie edge in the following counties: Big Stone (two), Clay (three), Murray, Polk, Rock (two), and Wilkin. In the twenty-first century, because of the relatively large number of records from 2000 to 2013, the MOU in 2015 changed the classification of the Say's Phoebe from casual to regular (Tustison 2015b). These records included presumed migrants in widely scattered, nonprairie counties: Aitkin (October 25, 2005), Anoka (May 12–13, 2009), Clearwater (April 29, 2002), Cook (April 25, 2008), Fillmore (April 21–22, 2012), Kanabec (September 9, 2014), Lake (May 8, 2015), Sherburne (September 9, 2007), and St. Louis (McQuade Safe Harbor, October 7–11, 2012).

Beginning in the year 2000, an unprecedented number of records that could be categorized as either migrants or potential territorial birds were also reported from the prairie counties. A total of 12 spring records (April 29 through May 22) occurred in nine western border counties through 2014. Two counties at the western border had the most records: Big Stone (2001, 2006, 2013, 2014) and Clay (2004, 2005, 2010, 2013), which may reflect only that the Big Stone NWR and Felton Prairie are popular birding sites. However, a probable breeding pair was recorded May 15 to July 2, 2005, in a gravel pit in the Felton prairie complex.

Because the Say's Phoebe was assumed to not nest in Minnesota, it was not on the species list for data entry into the MNBBA and, therefore, it was impossible to enter observations in the MNBBA database. However, published records in *The Loon* and online MOU databases are summarized here. The most important finding was documentation of breeding in Florida Township, Yellow Medicine County, where a pair was observed from June 16 to July 26, 2010, by a dozen people at various times, and a nest with eggs was photographed on July 25, 2010 (Hertzel, Hertzel, and Dunlap 2011). From the description, the nest was found on the bridge over Florida Creek in atlas block T115R46a. Birds were observed the next year

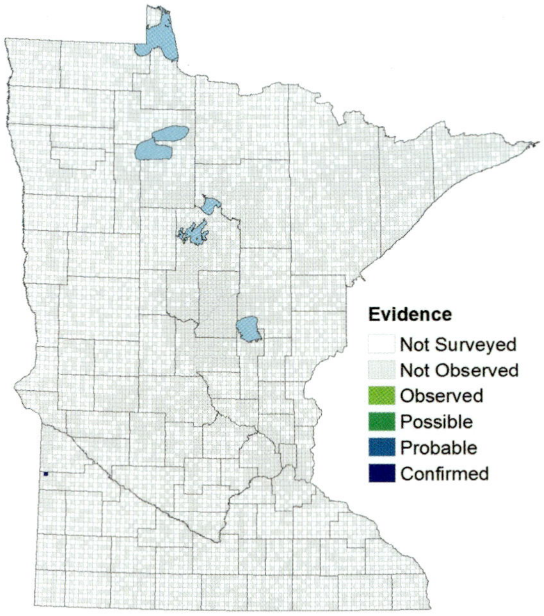

A. Breeding distribution of the Say's Phoebe in Minnesota based on the Breeding Bird Atlas (2009–13).

at the same site and "used the same nest as the previous year, producing a clutch of six" (Kessen and Svingen 2012). This information was converted into two MNBBA confirmed breeding records, one each for 2010 and 2011, and is displayed as one block in northwestern Yellow Medicine County **(A)**. Another record during the atlas years was from the Felton Prairie complex in Keene Township (atlas block T141R45d) of one bird on May 12 and May 22, 2013 (Budde, Dunlap, et al. 2013; Tustison 2013b).

B. The dry-mesic prairie in the Felton Prairie landscape of Clay County is one of a few Minnesota sites where the Say's Phoebe has been observed since the year 2000. Photo by Lee A. Pfannmuller.

Since the MNBBA was completed in 2013, additional breeding-season observations from some of the same locations have been submitted to MOU (2021a, 2021b). They include the following: Clay County, Felton Prairie complex (May 8–20, 2016; June 24 to August 8, 2017; June 16–24, 2018; June 9–29, 2019) and Yellow Medicine County (April 25, 2015). Three new records include two accounts of confirmed nesting in Traverse County, along County Road 2, northeast of Browns Valley (July 5 to August 10, 2018, and June 1–19, 2019) and reports of one bird seen at the Minnesota State University Science Center in Clay County (June 2–17, 2019) (MOU 2021a, 2021b).

In states that are adjacent to Minnesota, the Say's Phoebe is found breeding in North Dakota (mostly west of the Missouri River, Stewart 1975); South Dakota (found statewide but scarce in the eastern counties, Tallman et al. 2002; Drilling, Sparks, et al. 2016); Iowa (1960–72, nested, two northwestern counties, Kent and Dinsmore 1996; same area, first breeding bird atlas 1966–91, L. Jackson et al. 1996; but no records during the second breeding bird atlas, 2008–12); and the province of Manitoba (southwestern corner, Artuso et al. 2019).

In their review of the species in North America, Schukman and Wolf (2020) summarized various expansions and contractions in the historical breeding range of the Say's Phoebe. The species has expanded its breeding range northward in Alaska since 1960 to the foothills of the Arctic slope from the Brooks Range. Some changes were described in California, and expansion to the east in Manitoba from the mid-1980s and more records since 1950 in North Dakota were noted. The increase in records from Manitoba and North Dakota are consistent with the increase in observations of the Say's Phoebe in western Minnesota from the 1970s and 1980s.

BREEDING HABITAT

The Say's Phoebe adapts to human land use and structures in dry, open country within many landscape types: tundra, savanna, prairie, farmland, ranches, badlands, sagebrush flats, scrub, canyons, and cliffs **(B)**. It avoids forested areas and wet lowlands. Nest site availability may be a constraint on the species' use of a given habitat. Habitat occupancy at the eastern edge of its range in Iowa and Minnesota is sporadic and may be linked to climatic conditions of dry versus wet years (Kent and Dinsmore 1996).

POPULATION ABUNDANCE

PIF (2021) estimated the population of the Say's Phoebe at 5 million in the United States and Canada. Because of its rarity in Minnesota, no population estimates have been attempted. The BBS demonstrated that the Say's Phoebe population is stable or significantly increasing from 1966 to 2019 throughout all regions that have a suitable sample size (Sauer et al. 2020). For instance, surveywide the species is significantly increasing by an average of 0.78 percent per year from 1966 to 2019 **(C)**. Rosenberg, Kennedy, and colleagues (2016) estimated that the population increased by 40 percent from 1970 to 2014.

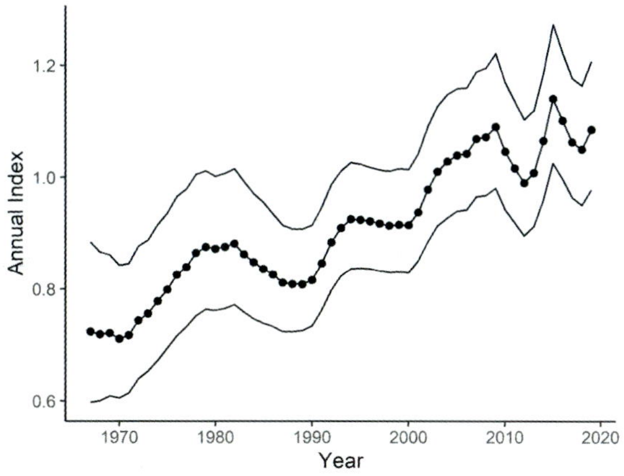

C. Surveywide breeding population trend and 95 percent credible interval (solid black lines) for the Say's Phoebe based on the 1967–2019 Breeding Bird Survey (Sauer et al. 2020).

CONSERVATION

The Say's Phoebe is of minimal conservation concern because its breeding population is increasing and it has a wide breeding distribution. Because the Say's Phoebe readily nests in human structures, its numbers may have increased historically with the creation of settlements in the West (Mollhoff 2016). Schukman and Wolf (2020) emphasized that no specific management guidelines have been suggested nor are there any known management efforts directed at this species. More intensive avian monitoring of the Felton Prairie complex and other suitable sites in western Minnesota would improve our understanding of this unique prairie species and its habits on the eastern periphery of its breeding range.

Fortunately, warmer temperatures are unlikely to be a detriment to the Say's Phoebe. NAS (2019) even predicted a small eastward expansion of the Phoebe's breeding range in the Dakotas with all three of the climate scenarios they examined. An increase in Minnesota records may likely follow and should encourage birders and wildlife enthusiasts to explore the western borders of the state for this handsome and distinctive member of the flycatcher family.

WHITE-EYED VIREO

Vireo griseus

Minnesota Seasonal Status: Summer vagrant; bred 1992, 2001; migratory visitant. The White-eyed Vireo was very rare during the MNBBA.
Migration: Short- to medium-distance migrant that winters along the southern Atlantic coast, the Gulf coast to Yucatan and Belize, and in Cuba and the Bahamas.
Food: An insectivore that feeds on caterpillars, moths, and butterflies gleaned from foliage.
Nest: Cup nest suspended in the fork of two twigs, well concealed, and low to the ground.
Conservation Concern: Assigned a low PIF Continental Concern Score of 8/20.

The White-eyed Vireo is one of the smallest North American vireos and is hard to see in the dense scrub that it inhabits. It is termed as "living in the thickets, down among the Catbirds, Thrashers, Sparrows, Chats, and Wrens" (Forbush and May 1955). A glimpse of either sex reveals a distinctive white eye, yellow spectacles, and two white wing bars, but more often it is the elaborate jumbled and ventriloquial song of the male that advertises its presence. The song is explosively loud and complex. Individuals "have repertoires of a dozen or more distinct songs, each comprising six to ten highly variable elements" (Hopp et al. 2020) and will frequently mimic the songs of other birds.

The White-eyed Vireo is found primarily in the southeastern states, from southern New England west to eastern Iowa and south to Texas and Florida. It is less common north of southern Missouri, the Ohio River, the Appalachian Mountains, and Virginia. Its highest densities are found along the coastal plain from southeast Texas to Florida, especially in Arkansas, Alabama, Louisiana, and Mississippi.

Photo by Larry Sirvio.

MINNESOTA BREEDING DISTRIBUTION

There were no records of the White-eyed Vireo included by Roberts (1932, 1936) from the late 1800s or early 1900s in Minnesota. The first observation was in Stearns County, July 31, 1941 (Hiemenz 1965). The species was not reported again until the 1960s, when there were two separate descriptions: May 16, 1964, Brown County (Janssen 2019); and May 23–24, 1965, Goodhue County (R. L. Huber 1965). In the 1970s, there were also two records: May 3, 1977, Washington County (Wojahn 1977); and May 25, 1977, Hennepin County (Fuller 1977).

In the decades after these early observations, the number of records steadily increased to 13 in the 1980s, 13 in the 1990s, and 23 from 2000 to 2013. Almost all records were vagrant migrants in the spring (61 percent) and the fall (10 percent); the rest were summer records. With a few exceptions, the migrant records were scattered throughout the southern third of the state, from Anoka County west to Swift County and south to the Iowa border. The spring migration was from the end of April and throughout May, and many of the birds found were singing males; in the fall, the birds were very late migrants, late October to late November. In these southern counties, it is difficult to know if they were displaced migrants or singing territorial birds. The few birds found in the northern two-thirds of the state included two spring records (1989, 2001) and one summer record (2004) from Clay County; one spring record from Otter Tail County (2000); and two remarkable fall records from Cook County: one at Lutsen, October 25–28, 2001 (Stephenson 2002) and another photographed in Grand Marais, November 8, 2015 (Tustison 2016).

There are 14 records for the White-eyed Vireo during the summer season, starting in the 1980s (1983, 1986, and 1989) through the 2013 season. Except for a bird banded at Buffalo River State Park, Clay County (Nielsen 2005), all incidental summer records were from the Twin Cities south (Brown, Houston, Martin, Pipestone, Ramsey, Rice, Scott,

Steele, Waseca, and Winona Counties). Most birds were recorded on only a few days in early June, but several were seen for a longer time: Martin County, June 6–July 1, 1986 (Sullivan 1986); Brown County, June 21–July 4, 1999 (Wiens 2000); Winona County, July 13–August 2002 (Eckert 2003); and Scott County, July 19–September 7, 2008 (Svingen 2009a). In addition, there are two breeding records that are well-documented: Houston County (Schumacher 1992) and Steele County (N. Swanson and Batt 2002).

During the MNBBA from 2009 to 2013, only one record was submitted. It was on June 15, 2010, from the MNDNR Natural Heritage database (T107R22a) **(A)**. It was a possible breeding observation by Steve Stucker, both seen and heard, on June 15, 2010, near Goose Lake in Waseca County. Subsequent to the MNBBA, in the MOU Seasonal Reports 2014–19, 12 migration records have been reported (eight in the spring and four in the fall) as well as one summer record (Houston County in 2019); all were south of the Twin Cities but with no evidence of breeding (MOU 2021b).

In a review of the status of the White-eyed Vireo, Hopp and colleagues (2020) present limited evidence that the species declined from the late 1800s to the mid-1930s in the eastern United States and that populations have increased in the central part of their breeding range since the mid-1970s. The latter is consistent with the increased observations in Minnesota since the 1970s.

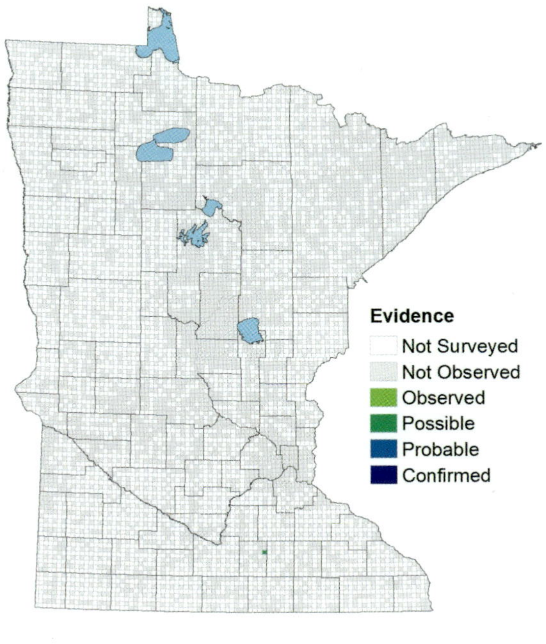

A. Breeding distribution of the White-eyed Vireo in Minnesota based on the Breeding Bird Atlas (2009–13).

B. The White-eyed Vireo prefers nesting in dense vegetation of shrubs or young trees, where it hides its nest close to the ground. Photo by Lee A. Pfannmuller.

In the states near or adjacent to Minnesota, the White-eyed Vireo breeds in eastern and southern Iowa (Kent and Dinsmore 1996). It has been expanding in Wisconsin since the 1960s. In their first breeding bird atlas from 1995 to 2000, Wisconsin identified five probable nesting detections but failed to find a nest (Cutright et al. 2006). Previously, a pair carrying food was observed in Dane County on several occasions in 1972 and a nest was found in 1988 in Waukesha County (S. D. Robbins 1991). In Michigan, the White-eyed Vireo was reported to be increasing in the 1960s and 1970s, but only a modest increase was detected in the species between its first breeding bird atlas in 1983–88 and its second atlas in 2002–8 (Reinoehl 2011). In South Dakota, the species is a casual spring migrant, accidental fall migrant, and accidental summer visitant in the east (Tallman et al. 2002), but was not detected during their second breeding bird atlas (Drilling, Sparks, et al. 2016). In North Dakota, it is a vagrant in the east (O'Connor 2000). Vagrant birds have been recorded in most states north of the regular breeding range to the Canadian border.

BREEDING HABITAT

The most common attribute of the White-eyed Vireo's habitat is the denseness of the vegetation. Its scrubby habitat can be found along forest edges, utility corridors, abandoned pastures, streamsides, and swamp thickets **(B)**. The vegetation can be brambles, briers, willow shrubs, wild plum, and dense saplings in early successional habitat (Hopp et al. 2020).

POPULATION ABUNDANCE

PIF (2021) estimated a population of 22 million White-eyed Vireos in the United States and Canada, but no estimates were available for the state of Minnesota. The BBS shows a significantly increasing population of 0.81 percent per year from 1966 to 2019 in the United States **(C)**. The only states where the White-eyed Vireo had a significant decline were in Maryland and West Virginia (Sauer et al. 2020). Overall, Rosenberg, Kennedy, and colleagues (2016) estimated that the species' population in the United States had increased by 33 percent from 1970 to 2014. Few data have been presented on reasons for the population expansion or increase in recent years, but the positive effects of climate change are suspected.

CONSERVATION

Because the White-eyed Vireo is extremely rare in Minnesota, there are no conservation plans for the species in the state. It has a substantial and increasing population and a relatively wide breeding range that should provide stability and little concern for its future. This will be an interesting species to watch over the coming years in Minnesota. With a 2°C warming scenario, NAS (2019) projected a stable breeding range for the White-eyed Vireo in North America, including only a 4 percent range loss and a 19 percent range gain along its entire western and northern breeding range by 2050. The species' range was predicted to increase in southeastern Minnesota even under our current 1.5°C warming scenario, as long as suitable habitat is available. Its winter range is projected to expand northward in the southeastern United States, from North Carolina to Texas, under all three climate scenarios that NAS (2019) examined.

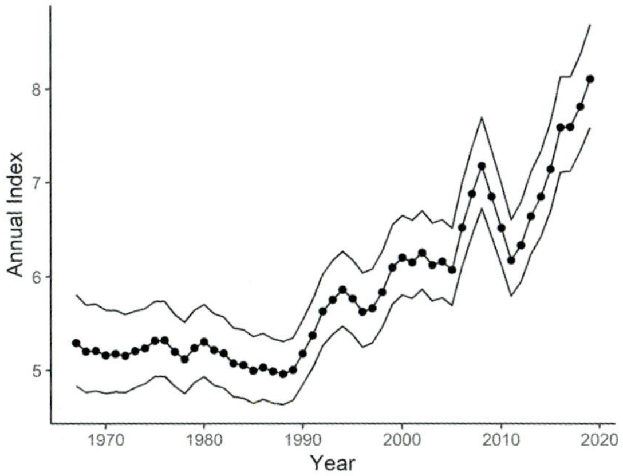

C. Breeding population trend and 95 percent credible interval (solid black lines) for the White-eyed Vireo in the United States based on the 1967–2019 Breeding Bird Survey (Sauer et al. 2020).

BELL'S VIREO

Vireo bellii

Minnesota Seasonal Status: Regular breeding resident and migrant. The Bell's Vireo was rare during the MNBBA.
Migration: Medium-distance migrant whose small winter range is largely restricted to the Pacific coast of Mexico and northern Central America.
Food: A foliage gleaner that feeds on arthropods, especially spiders and caterpillars.
Nest: Pendulant cup placed in the fork of a horizontal branch on a small tree or shrub.
Conservation Concern: Assigned a moderate PIF Continental Concern Score of 10/20; officially listed as *Special Concern* in Minnesota.

Small and inconspicuous, the Bell's Vireo has been described as a "timid, shy, retiring little bird" (Bent 1950). It is olive-gray on its back, wings, and head; one needs to look closely to observe its pale white eye-ring and its two white wing bars, only the lower one being prominent. A yellow wash may be visible on the sides of the light chest and underneath the tail. First described by Audubon in 1843, the Vireo was one of 14 new species he found on his last expedition, traveling along the Missouri River to western North Dakota (Kus et al. 2020; Souder 2004). The bird was named after one of his companions, J. G. Bell, a noted American taxidermist (Roberts 1932).

Once known as Bell's Greenlet, this little vireo makes its presence known with the male's unmistakable, raspy song. Roberts (1932) described it as a "low, sweet, but emphatic warble, easily recognized as coming from a Vireo." Nice (1931a) thought less of its sweetness but was impressed with its delivery: "There is no music in the Bell Vireo's refrain, but it possesses a quaint charm in its air of enthusiasm, in the rapid jumble of it all." Described by Nice as *whillowhee, whillowhee, wheé,* the song's paired structure is notable as the phrases often alternate endings with a rising and falling inflection.

Restricted to shrubby habitats in southern Minnesota, the Vireo's breeding range stretches from northern Mexico, across southern Texas, New Mexico, Arizona, and California, north into the Great Plains, and east to Indiana, Missouri, and Arkansas. In the United States, breeding densities are highest in southern Texas. The species is best known for its federally *Endangered* subspecies, *Vireo bellii pusillus,* the Least Bell's Vireo, which is confined to a stretch of the California coastline.

Photo by Larry Sirvio.

MINNESOTA BREEDING DISTRIBUTION

In the first comprehensive account of Minnesota's avifauna, Hatch (1892) described the Bell's Vireo as a "relatively common" summer resident, its favorite haunts being "our beautiful, sylvan lakes." Roberts (1932) later challenged that assertion: "The author's long acquaintance with Dr. Hatch leads him to believe that this is one of several misconceptions that found their way into his writings," noting that "no capable observer" had ever found any evidence to "substantiate the statement." Roberts provided only two records for the species: first, a nesting pair that he and his colleague William Kilgore found in May 1922 "in a dense poplar thicket on the west side of the Fort Snelling Reservation"; and, second, a well-documented sighting near Winona in August 1930 by Vernon Bailey and his wife, Florence.

More than twenty years passed before Brother Theodore published a remarkable account of six nesting pairs he found in 1952 near St. Mary's College in Winona (Theodore 1952). Additional records slowly accumulated, including a second nesting report in southeastern Houston County in 1961 (R. L. Huber 1961b). When Green and Janssen summarized the species' distribution in 1975, they described it as largely restricted to the Mississippi River corridor in southeastern Minnesota, where it was "very scarce except in Winona and Houston counties." An interesting summer record of a singing male in Rock County was also noted. Janssen's 1987 update reported breeding evidence in three new counties (Olmsted, Ramsey, and Wabasha) and a second summer record from southwestern Minnesota in Pipestone County.

Subsequently, the MBS reported 12 Bell's Vireo breeding-season locations, including records in Blue Earth, Dodge, Lyon, and Mower Counties, while other observers confirmed nesting in Blue Earth County at Minneopa State Park in 1990 and 1996 (MOU 2021b). By 1998, nesting had been confirmed in six counties since 1970: Blue Earth, Dakota, Olmsted, Ramsey, Wabasha, and Winona (Hertzel and Janssen 1998).

MNBBA records documented a small extension of the species' range north to Sherburne and Chisago Counties and west to Kandiyohi County. Although no records were collected in southwestern Minnesota, one possible breeding record was reported in northern Watonwan County in the south-central region. Spanning 15 counties, 32 records were collected from 25 atlas blocks **(A)**. Breeding was confirmed in three counties: Dakota, Hennepin, and Rice. A nest with eggs found in Rice County (2010) was a new county nesting record, while a successful nesting in Hennepin County (2010) was the first confirmed nesting in the county since Roberts's 1922 report (Doyle 2011).

Changes in the status of this little Vireo are difficult to determine. Given that Roberts (1932) found a nest in the Twin Cities years before it was reported elsewhere in the state, the species may have been present for some time in less accessible sites south along the Mississippi River. The same may have been true in southwestern Minnesota. In a landscape largely devoid of early field observers, the species likely expanded into the region from nearby populations in Iowa and South Dakota. As a result, the increase in records that began in the 1950s may simply represent better coverage by a larger cadre of skilled observers rather than a range expansion. This premise was suggested by J. Green (pers. comm.), who further suggested that the real range expansion occurred in the late 1900s when birds were more frequently reported not only in southeastern Minnesota, but also in the greater Twin Cities region and in scattered locations throughout south-central Minnesota.

On the other hand, Kus and her colleagues (2020) noted a "slight northern and eastern expansion" in North America during the 1900s, while

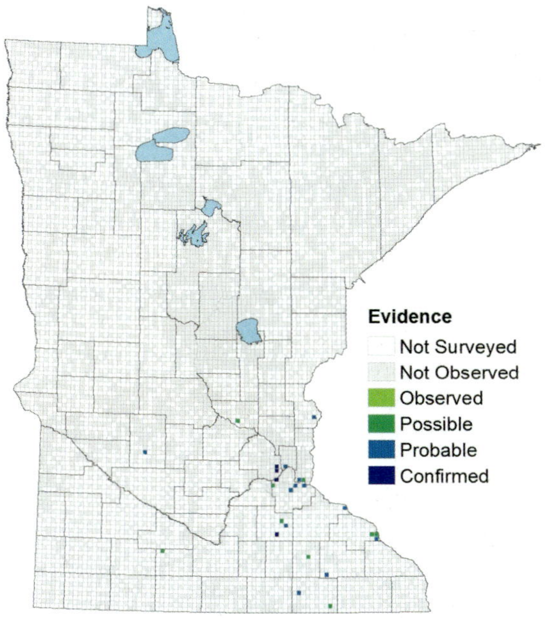

A. Breeding distribution of the Bell's Vireo in Minnesota based on the Breeding Bird Atlas (2009–13).

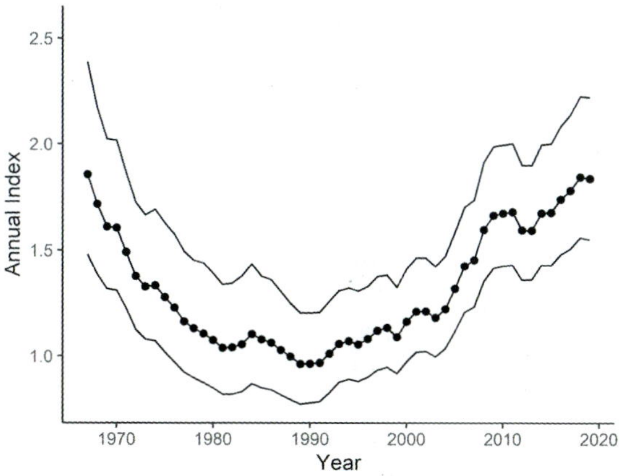

C. Surveywide breeding population trend and 95 percent credible interval (solid black lines) for the Bell's Vireo based on the 1967–2019 Breeding Bird Survey (Sauer et al. 2020).

Peterjohn (2001) noted its expansion into the Great Lakes region beginning in the 1920s. In the end, the Bell's Vireo's true history in Minnesota may remain a mystery.

BREEDING HABITAT

Dense, shrubby vegetation is the Vireo's primary domain **(B)**. In Wisconsin, the species was found in "shrubby patches of prairie and savanna habitat" and in "lowland shrub carr" (Cutright et al. 2006). In South Dakota, it was found in "shelterbelts and upland deciduous woodlands (34%), lowland woodlands and riparian areas (31%), shrublands (28%), and shrubby fields (9%)" (Drilling, Sparks, et al. 2016). Theodore (1952) found the Winona County nests in dense tangles of shrubs, grasses, and small trees. Although the species is often found near water, this association may be strongest in the arid southwest (Kus et al. 2020).

POPULATION ABUNDANCE

PIF (2021) generated a population estimate of 4.6 million birds in the United States and Canada with an additional 1.1 million residing in Mexico. Given the limited number of records in Minnesota, a statewide estimate is not available. Indeed, only 11 birds have been reported on Minnesota BBS routes from 1967 to 2019 (Pardieck et al. 2020).

Following a notable decline from approximately 1966 through 1990, Vireo numbers within the BBS survey region began increasing, demonstrating an overall stable population from 1966 to 2019 **(C)**. Although the population appears secure, in a few regional pockets population numbers have experienced a steep decline, especially along the lower Mississippi River valley. The principal factors cited are loss of habitat and high rates of cowbird parasitism (Kus et al. 2020).

CONSERVATION

Two factors are largely responsible for the designation of the Bell's Vireo as a moderate conservation priority (PIF 2021). First, its small winter range increases its vulnerability to small changes that can have a large impact on breeding populations. Second, flood control projects, reservoir construction, and intensive agricultural production threaten riparian corridors that provide critical breeding habitat (Kus et al. 2020). In Minnesota, the species' rarity and patchy distribution led to its *Special Concern* listing (MNDNR 2012).

Protecting and restoring riparian buffers to improve water quality can improve lowland habitat opportunities for the Bell's Vireo. In uplands, Budnik and his colleagues (2002) recommended improving habitat by increasing the density of large shrub patches (greater than 200 square m) and ensuring they are located away from elevated perches used by Brown-headed Cowbirds. Larger shrub patches also increase nest concealment.

The Vireo's preference for shrubby thickets, however, can conflict with grassland management priorities. Woody encroachment reduces a site's suitability for many grassland-dependent species, prompting managers to halt succession using controlled burns, mowing, and herbicide applications. As a result, biologists suggest that habitat improvements should focus on sites currently used by the Vireo and be augmented with the restoration of large grassland tracts that provide opportunities for a spectrum of grassland and shrub-dependent species.

In the future, the Bell's Vireo may actually become a more widespread member of the state's avifauna. Under all three warming scenarios examined, NAS (2019) projected a low vulnerability for the species' breeding distribution. Small losses in Mexico, the southwestern United States, and the Central Plains were compensated by larger gains in the north, including in southwestern Minnesota. Meanwhile, an awareness of the needs of shrubland species, particularly in the grassland habitats of southeastern Minnesota, is warranted.

B. A wetland in the floodplain of the Minnesota River that provides scattered shrubs and small trees is one of the few sites in the state where the Bell's Vireo nests. Photo by Lee A. Pfannmuller.

YELLOW-THROATED VIREO

Vireo flavifrons

Minnesota Seasonal Status: Regular breeding resident and migrant. The Yellow-throated Vireo was common during the MNBBA.
Migration: Medium- to long-distance migrant that winters in the Caribbean and from southeastern Mexico south to northern South America.
Food: Arthropods and some fruit and seeds secured by gleaning the foliage and bark of the mid- and upper tree canopies.
Nest: Tightly woven cup nest bound together with caterpillar silk and spider webs and placed in the canopy of deciduous trees.
Conservation Concern: Assigned a moderate PIF Continental Concern Score of 9/20.

Among the six vireos that regularly breed in Minnesota, each dressed in subtle shades of pale yellow, gray, olive-green, and white, the Yellow-throated Vireo is the most colorful of the lot. The yellow spectacles around each eye, the bright yellow throat and breast, and the white belly are enough to positively identify this woodland inhabitant. Two white wing bars are also prominent. Indeed, *Vireo flavifrons* means "vireo with a yellow front." Its plumage is similar to that of the Pine Warbler, but the latter has a much thinner bill, faint streaking on the breast, and is invariably found in mature conifer stands, which the Yellow-throated Vireo avoids.

The species' low, reedy song is comprised of many two- to three-syllable phrases that are separated from each other by nearly a full second. Its forlorn quality was described by Audubon (1827–38) as "measured and plaintive" notes that evoke an image of "a bird lost in the woods." Roberts (1932) described it as "a pleasing song, in form like that of the Red-eyed Vireo but richer and more musical in quality and with the phrases delivered at longer intervals."

Largely restricted to the Eastern Deciduous Forest Biome, the Yellow-throated Vireo's breeding range extends from the eastern Great Plains, where it is found along major river valleys, south to central Texas, and east to the Atlantic coast. To the north, it reaches into southern Manitoba, Ontario, and Quebec. While the vireo is sparsely distributed throughout its range, east-central Minnesota is one of the few areas where breeding densities are a little higher.

Photo by Gerald Hoekstra.

MINNESOTA BREEDING DISTRIBUTION

Roberts (1932) considered the Yellow-throated Vireo a common summer resident south and west of Minnesota's northern boreal forest, "its range extending northward through the Red River Valley in decreasing numbers as far as southern Manitoba." He speculated that the species' range might eventually extend into northeastern Minnesota as younger, deciduous forests replaced the mature conifers and created more opportunities for this deciduous-loving species. At the time, confirmed nesting records were available from Hennepin, Ramsey, and Wright Counties in east-central Minnesota, and from Mahnomen County in the northwest. Breeding was inferred in Isanti, Sherburne, and Waseca Counties and in the Mille Lacs region.

Green and Janssen's (1975) updated account identified confirmed nesting in seven additional counties stretching from Houston County in the southeast to Pennington County in the northwest. Farther east, breeding was inferred as far north as Pine and Crow Wing Counties, while summer observations were reported in Duluth. A few years later, Janssen's (1987) account included a distribution map for the species that excluded northeastern Minnesota, the adjacent counties of Koochiching and Itasca, much of southwestern Minnesota, and the southern Red River valley. Unlike Roberts's (1932) description, the Vireo was found to be more common in the northwestern corner of the state than in the southern Red River valley. Nesting had been confirmed in 12 counties since 1970; Hertzel and Janssen (1998) later added two additional counties. Although Itasca County was excluded from Janssen's map, he noted that the species continued to expand in the north with observations from Cass and southern Itasca Counties in north-central Minnesota and from Carlton and southern St. Louis Counties in the northeast.

By 2014, the 924 Yellow-throated Vireo breeding-season locations documented by the MBS found the species was common throughout the western half of the Laurentian Mixed Forest Province, including in Clearwater, southern Beltrami, and Itasca Counties. In addition to two reports in central St. Louis and Lake Counties, observations were scattered across the southwestern corner of the state.

The 1,457 records collected by MNBBA participants documented a similar distribution **(A)**. Reported from 21.4 percent of the surveyed atlas blocks (1,014/4,742), the birds were observed in all but seven of Minnesota's 87 counties: Cook and Lake Counties in the northeast; Wilkin County in the west; and Cottonwood, Jackson, Murray, and Pipestone Counties in the southwest. Nesting of this high-canopy-dwelling species was confirmed in just 3 percent (30) of the 1,014 atlas blocks where the Vireo was reported but was deemed probable in an additional 306 blocks. Fourteen of the 19 counties where nesting was confirmed were additions to the map published by Hertzel and Janssen (1998) and included several northeastern counties: Aitkin, Carlton, Pine, and St. Louis. Among Minnesota's vireos, the Yellow-throated ranked third in abundance behind the Red-eyed Vireo and Warbling Vireo.

The MNBBA probability distribution map predicts the areas of greatest abundance occur along the southern boundaries of the Laurentian Mixed

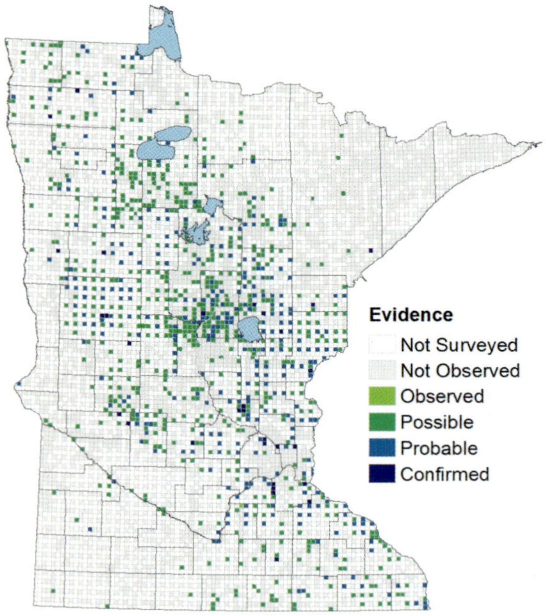

A. Breeding distribution of the Yellow-throated Vireo in Minnesota based on the Breeding Bird Atlas (2009–13).

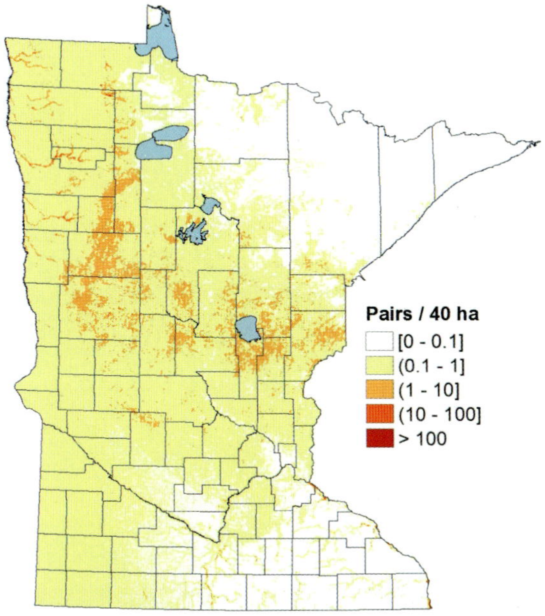

B. Predicted breeding distribution (pairs per 40 ha) of the Yellow-throated Vireo in Minnesota based on habitat, landscape context, and climate data gathered during the Minnesota Breeding Bird Atlas (2009–13) using the Generalized Linear Modeling method with an adjustment for detectability.

Forest Province and in the northern Eastern Broadleaf Forest Province **(B)**. The region's matrix of farmland, residential development, and forests provide a perfect landscape for this semi-open deciduous forest species. Floodplain forests, especially in northwestern Minnesota, are also predicted to support moderate breeding densities. On the other hand, the species was encountered far more frequently during the atlas in the southeast corner of the state than the model predicted.

As Roberts (1932) speculated nearly one hundred years ago, the Yellow-throated Vireo has continued to move farther into the Laurentian Mixed Forest Province, spurred by the loss of upland conifer cover and increased landscape fragmentation. Apart from similar changes elsewhere within its breeding range, there is little evidence of any large-scale change in the species' breeding distribution (Rodewald and James 2020).

BREEDING HABITAT

The Yellow-throated Vireo's primary habitat is mature, deciduous woodlands, including both uplands and floodplains **(C)**. Because it prefers open stands and forest edges, it also resides in parks and small towns where groves of mature shade trees are available. Dense forests are avoided unless there are scattered forest openings (Rodewald and James 2020). In national forests in northern Minnesota and Wisconsin, the species was strongly associated with oak forests (Niemi, Howe, et al. 2016).

Despite its preference for forest edges, several studies have pointed to the Vireo's area sensitivity. One study (Bond 1957) found it was 50 percent more common in forest stands larger than 32 ha; another demonstrated that numbers decreased in forests that were smaller than 100 ha (C. S. Robbins 1980). Yellow-throated Vireos may initially select large forest tracts but then selectively use the outer edges of those stands. Unfortunately, rates of cowbird parasitism and predation are often higher along edges (Rodewald and James 2020).

POPULATION ABUNDANCE

PIF (2021) used BBS data to project a population estimate of 4.7 million Yellow-throated Vireos in the eastern United States and southern Canada and a statewide estimate of 190,000, or 4.0 percent of the continental population. MNBBA data generated a larger estimate of 430,000 breeding adults with its more intensive sampling effort (95 percent confidence interval of 317,000–811,000).

Given the species' relatively low abundance, population trend data collected by the BBS lack statistical precision in most states, including Minnesota. Nevertheless, data from nearly every state and ecological region suggest either a stable or increasing population (Sauer et al. 2020). Range-wide, the Vireo's population was estimated to have increased 62 percent from 1970 to 2014 (Rosenberg, Kennedy, et al. 2016). Because the BBS's Prairie Hardwood Transition Bird Conservation Region encompasses the heart of the species' breeding range in Minnesota, it is a good surrogate for examining the species' population trend in the state. The data for the region are statistically robust and demonstrate a significant population increase, averaging 2.08 percent per year from 1966 to 2019 **(D)**.

C. The Yellow-throated Vireo is typically found in open stands of mature deciduous hardwoods or along the edges of stands where the tree density is much higher. Photo by Lee A. Pfannmuller.

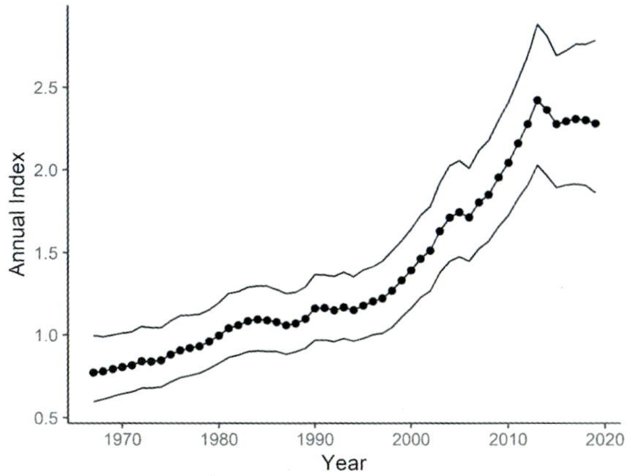

D. Breeding population trend and 95 percent credible interval (solid black lines) for the Yellow-throated Vireo in the Prairie Hardwood Transition Bird Conservation Region based on the 1967–2019 Breeding Bird Survey (Sauer et al. 2020).

The Vireo's increase is attributed in part to increased forest cover throughout the eastern United States, especially in the Northeast. Following two centuries of deforestation and intensive farming, a significant decline in agricultural land use has created a landscape that is increasingly forested and maturing over time (D. R. Foster et al. 2008). Sites that originally were cleared for small farms have since been abandoned, allowing them to succeed once again to forests.

CONSERVATION

Because it is largely restricted to the Eastern Deciduous Forest, the Yellow-throated Vireo has been assigned a moderate conservation ranking (PIF 2021). Unfortunately, few studies have examined the species' response to forest management practices. Nevertheless, retaining large blocks of mature deciduous forest, with scattered openings and a strong oak component, appears to be a key conservation measure.

Regarding the impacts of climate change, under a 2°C warming scenario NAS (2019) projected the loss of 19 percent of the Vireo's North American breeding distribution by the year 2050 with an equivalent expansion northward. Although the species' vulnerability to climate change was deemed low, the most extensive range losses are projected in the eastern Dakotas and western Minnesota, with only small gains in northeastern Minnesota.

Pesticides pose another threat. The species' disappearance from many small cities and towns in the northeastern United States in the early 1900s was attributed to extensive spraying programs to control Dutch elm disease and gypsy moths (Rodewald and James 2020). Such practices should be carefully monitored as we track the changes that are likely ahead for the Yellow-throated Vireo.

BLUE-HEADED VIREO

Vireo solitarius

Minnesota Seasonal Status: Regular summer resident and migrant. The Blue-headed Vireo was uncommon during the MNBBA.
Migration: Short- to medium-distance migrant; the species overwinters in the southeastern United States, Mexico, and Central America.
Food: Arthropods gleaned from foliage.
Nest: Cup nest suspended from fork of two twigs, usually at the end of a conifer shrub or sapling branch.
Conservation Concern: Assigned a low PIF Continental Concern Score of 7/20.

The Blue-headed Vireo is a handsome bird and one of the more distinctively colored vireos. It has classy white spectacles, a blue-gray head that contrasts with a bright olive back and white throat, and yellow-tinged wing bars. Bent (1950) said, "He appears to be a well-groomed aristocrat among birds." The sexes are very similar in plumage, but the male is slightly more colorful.

During the breeding season, the Blue-headed Vireo is the vireo most associated with coniferous trees and, even though its song is similar to those of the Red-eyed and Yellow-throated Vireos, it tends to be slower and richer in quality. Roberts (1932), in one of his first encounters with the species in 1902 at Itasca State Park, described it as "readily distinguished from either by its greater richness and volume of tone." Forbush and May (1955) say it "heralds its presence . . . by its wild sweet song, a charming cadence of the wooded wilderness. Its notes seem more spiritual and less commonplace than those of the familiar vireos of village and farmstead." Its Latin scientific name, *solitarius,* was provided by Alexander Wilson, who named the species in 1810 from a lone migrant he collected near Philadelphia.

The Blue-headed Vireo is solitary in the sense that it is found nesting away from most major population centers. It breeds across the northern coniferous forests from northeastern North America to British Columbia and south in the Appalachian Mountains to northeastern Mississippi. The highest breeding densities are found in Nova Scotia and north-central Ontario.

Photo by David Brislance.

MINNESOTA BREEDING DISTRIBUTION

Roberts (1932) knew the Blue-headed Vireo primarily from his time at Itasca State Park and described it overall as a "frequent summer resident of the coniferous forests." He reported nesting records from 1886 in Becker County, at Itasca State Park in 1919 and 1928, and one from Itasca County in 1930. Overall, he described its breeding distribution as south to Mille Lacs Lake, west to Itasca State Park, and northeast to Cook, Lake, and St. Louis Counties.

More than forty years later, Green and Janssen (1975) described a similar breeding distribution in northeastern and north-central Minnesota. They reported breeding observations from Carlton County but stressed that its western boundary was "poorly known" and likely extended only to Itasca State Park. Besides those reported by Roberts (1932), they reported confirmed nesting in Clearwater, Hubbard, Lake, and St. Louis Counties. Janssen (1987) reinforced the species' primary distribution in northeastern and north-central Minnesota and included confirmed nesting since 1970 in Aitkin, Cass, Clearwater, Cook, Crow Wing, Hubbard, Lake, and St. Louis Counties. He also noted summer observations from Beltrami and Roseau Counties and much farther south in Houston County. Later Hertzel and Janssen (1998) removed Clearwater County from the list of confirmed nesting since 1970.

During nest searches in Pine County in 1996, Pirjo and Dr. Markku Kuitunen of the country of Finland found a nest of the Blue-headed Vireo. This observation was gathered as part of a research project on the Least Flycatcher at the University of Minnesota Duluth. Unknown to them at the time, this is the most southerly confirmed nesting of the species in Minnesota.

The MBS included 355 breeding-season locations of the Blue-headed Vireo and further solidified its primary breeding distribution in northeastern and north-central regions from Clearwater to Cook County. However, they also included many breeding observation locations from Carlton County, at least five locations from Pine County, and one location in northern Mille Lacs County.

The MNBBA included 915 breeding-season records of the Blue-headed Vireo, including many from Cook and Lake Counties **(A)**. These records extended its possible breeding range to Kittson and Marshall Counties in the northwest as well as westward to Becker, Polk, and Wadena Counties. The species was recorded in 10.9 percent (514/4,734) of the surveyed atlas blocks. The MNBBA added confirmed nesting in Beltrami, Itasca, and Lake of the Woods Counties to those counties not included by Hertzel and Janssen (1998), plus evidence of probable nesting in Carlton, Clearwater, Koochiching, and Roseau Counties. It is likely that the Blue-headed Vireo nests

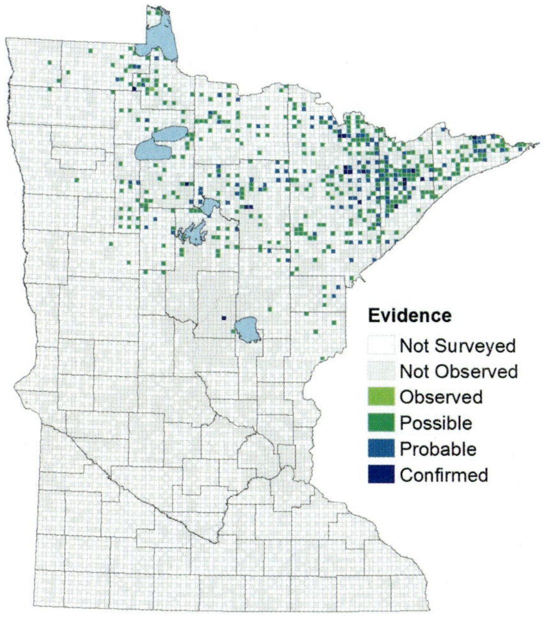

A. Breeding distribution of the Blue-headed Vireo in Minnesota based on the Breeding Bird Atlas (2009–13).

throughout the northern region of Minnesota and as far northwest as coniferous trees are present.

Roberts (1932) noted a June 4, 1873, observation by Coues from Pembina on the Red River, at the border of North Dakota and Minnesota. This date could represent a late migrant, but currently suitable habitats along the Red River are very rare or no longer exist in many of these western counties.

Cutright and colleagues (2006) cited observations in the early 1900s from southern areas of Wisconsin and range retractions northward. During the Wisconsin first and second breeding bird atlases, nesting was confirmed in Calumet, Jackson, and Waukesha Counties (Cutright et al. 2006; WSO 2021). Waukesha County is south of the Minnesota and Iowa border, while Calumet and Jackson Counties are equivalent in latitude to Winona County.

B. Blue-headed Vireos sing, nest, and forage in mature coniferous trees of northern Minnesota.
Photo by Gerald J. Niemi.

Perhaps the unusual observation in 1983 from Houston County cited by Janssen (1987) should signal further scrutiny along the southeastern counties in Minnesota. In their review of the Blue-headed Vireo in North America, Morton and James (2020) emphasized that the species has been recolonizing places where it formerly occurred in the nineteenth century. In these areas, reforestation of conifers has been implemented in the midwestern United States, such as in Wisconsin, Illinois, and Michigan, as well as in the province of Ontario.

BREEDING HABITAT

Blue-headed Vireo breeding haunts are described as evergreen forests with spruce, hemlock, and pine or deciduous growth with alder or willow in the understory (Morton and James 2020). In Minnesota, the species occurs in coniferous forests (Niemi and Pfannmuller 1979; Green 1995). Niemi, Howe, and colleagues (2016) noted observations in the Chippewa and Superior National Forests were in many forest cover types but most abundantly and most frequently in black spruce–tamarack forests. Additional cover types used included pine forests (jack, red, and white) and mixed swamp conifer forests composed of white cedar, spruce, and black ash **(B)**.

POPULATION ABUNDANCE

PIF (2021) estimated a US and Canadian population of 13 million breeding adults and a Minnesota population of 140,000. Estimates from the MNBBA for Minnesota were about three times greater at 400,000 breeding adults. Comparatively, Minnesota still has a relatively small portion of the species' global population.

The BBS trend in Minnesota was unreliable, but estimates for the Boreal Hardwood Transition Bird Conservation Region, which includes Minnesota, indicated a significant population growth of 2.97 percent per year from 1966 to 2019 **(C)**. Using data from the BBS, Rosenberg, Kennedy, and colleagues (2016) suggested an overall 200 percent increase in the breeding population from 1970 to 2014! In contrast, data from the Chippewa and Superior National Forests in northern Minnesota indicated a slight but insignificant decline in the population trend from 1995 to 2020 averaging −0.88 percent per year (Grinde, Bednar, et al. 2020).

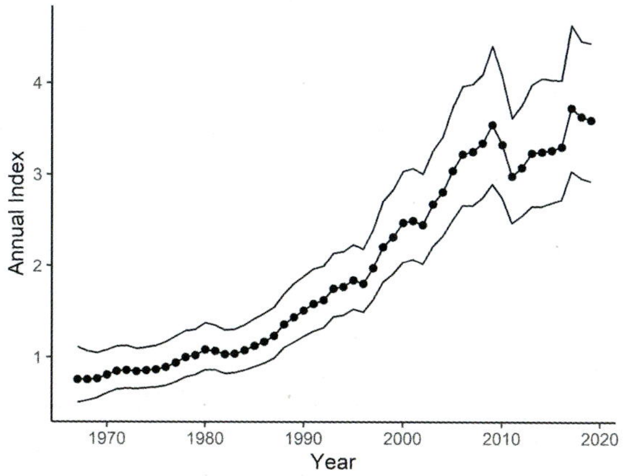

C. Breeding population trend and 95 percent credible interval (solid black lines) for the Blue-headed Vireo in the Boreal Hardwood Transition Bird Conservation Region based on the 1967–2019 Breeding Bird Survey (Sauer et al. 2020).

CONSERVATION

Given its current North American population size and its substantially increasing population throughout many parts of North America, there is minimal current conservation concern for the Blue-headed Vireo. Its declining population in northern Minnesota is of interest, especially since it is contradictory to trends in the broader regional area. Its population would be susceptible to substantial increases in logging activity of intermediate and mature coniferous forest cover types, and our warming climate is an emerging concern. Reductions in its population were predicted under moderate and high harvesting scenarios in Minnesota (Niemi and Hanowski 1992a). Regarding climate change, NAS (2019) projected a 67 percent range loss and a 41 percent range gain in North America, with most gains in the far north of Canada and Alaska by 2050. However, it was predicted to be extirpated from Minnesota under the 2°C warming scenario by 2050. Moderate winter range gains were projected for this Vireo in the southeastern United States under all three climate scenarios examined (NAS 2019).

Morton and James (2020) reported low numbers of Blue-headed Vireos in collisions with towers or buildings, suggesting that the species does not appear to be overly sensitive to mortality from human structures. The status of the Blue-headed Vireo breeding population in Minnesota deserves continued monitoring and scrutiny. It is a joy to watch this good-looking aristocrat as it solitarily gobbles insects among the spruce, fir, and pine of Minnesota's northern forests.

PHILADELPHIA VIREO

Vireo philadelphicus

Minnesota Seasonal Status: Regular breeding resident and migrant. The Philadelphia Vireo was rare during the MNBBA.
Migration: Medium-distance migrant that overwinters in southern Mexico and Central America.
Food: Small arthropods gleaned from foliage.
Nest: Pensile cup nest suspended in a fork, usually high in a tree.
Conservation Concern: Assigned a moderate PIF Continental Concern Score of 9/20.

Photo by Laura Erickson.

In Minnesota's northern forests, the Philadelphia Vireo is a rare and challenging species to identify. It may be more common than we are aware, but its song is virtually identical to that of the common Red-eyed Vireo. Detailed studies of these two species indicate they use similar breeding habitats and are known to be among the best examples of "adaptive interspecific territoriality," meaning they have adapted to co-existence because their territories do not overlap. Since territories are largely maintained by song, Philadelphia Vireos will imitate the songs of the Red-eyed Vireo to avoid physical contact, an important adaptation for the Philadelphia because it is smaller than the Red-eyed (Moskoff and Robinson 2020). Similarities in the songs of the two species are also supported by acoustic analyses. Field observations require visual confirmation for positive identification.

Forbush and May (1955) quote Bradford Torrey, who quipped, "It looks like one vireo and sings like another," a reference to its similarities in plumage to the Warbling Vireo and in song to the Red-eyed. The Philadelphia Vireo has a dark eyeline, a darker cap, and darker wing coverts compared with the Warbling Vireo, although consultation with a well-illustrated field guide is advisable. Its shorter bill, smaller size, and lack of a red eye distinguish it from the Red-eyed Vireo. Like the inappropriately named Connecticut, Tennessee, and Nashville Warblers, its scientific name *philadelphicus* has little association with Philadelphia, except that a migrant was collected and named there in 1851.

The Philadelphia Vireo has primarily been found in the northeastern forests of Minnesota, but verified observations from Becker County indicate that it should be looked for in other forested areas of the state. It is the most northern distributed vireo in North America and occurs across a narrow band in Canada from Labrador to northwestern Alberta. Its breeding range also extends southward into northern portions of Minnesota, Wisconsin, and Michigan. Populations are generally low. The highest densities have been observed in parts of Ontario and Quebec.

MINNESOTA BREEDING DISTRIBUTION

Roberts (1932) described "no positive breeding-records for Minnesota, but it seems probable that it will be found nesting in the northern part of the state." Even breeding observations appeared to be limited because he lists only Hibbing in St. Louis County and the border city of Pembina in North Dakota as locations where individuals were collected. The latter was from an observation in 1873 where he quotes Coues as saying the species "undoubtedly breeds about Pembina, in the heavy timber of the river-bottom."

More than forty years later, Green and Janssen (1975) emphasized its primary distribution in northeastern Minnesota as far south as Duluth. They found two breeding records with young being fed in eastern Becker County (1961) and a bird banded on June 16, 1966, in Little Falls, Morrison County (1966). Besides the "inferred" nesting in Becker County, they also list confirmed nesting in Lake and St. Louis Counties. A few years later, Janssen (1987) provided a more restrictive distribution, primarily in the northern tier of counties from Lake of the Woods County and northern Beltrami County east to northern Cook County. Janssen (1987) and Hertzel and Janssen (1998) only included confirmed nesting in Cook County since 1970. Both Green and Janssen (1975) and Janssen (1987) emphasized that the species was scarce throughout its breeding range, but identification is complicated because of its similarity to the substantially more common Red-eyed Vireo. The MBS included only 19 breeding-season locations; all were found in Cook, Lake, and St. Louis Counties, except for one location in central Becker County.

The MNBBA included a total of 32 records in 22 blocks, a paltry 0.5 percent of all blocks sampled. Similar to the observations by the MBS, most detections were in northeastern Minnesota, including Cook, Lake, and St. Louis Counties, plus one in northeastern Becker County **(A)**. There were three confirmed nesting observations: two in St. Louis County and one in Cook County. Probable nesting was also limited to one in St. Louis County; possible nesting observations were scattered throughout Cook County, southern and central Lake County, and northern St. Louis County, with one record in Becker County. Note that all MNBBA records of the Philadelphia Vireo were carefully reviewed before acceptance because of the notorious difficulty in identifying this species.

The difficulty in identifying the Philadelphia Vireo is well known. For instance, Moskoff and

Robinson (2020) emphasized that its "obscurity" is related to the overwhelming abundance of the Red-eyed Vireo. The Philadelphia Vireo is very easily overlooked even by experienced observers. The first nesting in Wisconsin was not detected until 1997 in Bayfield County, during Wisconsin's first breeding bird atlas (Cutright et al. 2006). In addition, another confirmed nesting was identified in Pierce County in 1999 near the St. Croix River, adjacent to Washington County in Minnesota. Because of these complications in detection, no information on any historical changes in its breeding distribution is available.

The Philadelphia Vireo is well represented by nesting activity in the southeastern region of Manitoba (Artuso et al. 2019) and by high relative abundance in the western regions of Ontario, areas that are adjacent and north of Cook and Lake Counties (Cadman et al. 2007). Observations of Vireos in June and early July should be scrutinized throughout the forested areas of Minnesota and pictures of any observations are important sources of confirmation.

BREEDING HABITAT

The Philadelphia Vireo breeds primarily in deciduous forests composed of aspen, birch, and ash in early to mid-successional stages (Erskine 1977; Rice 1978; Godfrey 1986) **(B)**. Holmes and colleagues (1986) in the northeastern United States found the species in mixed conifer and older forests as long as deciduous trees were present. The Philadelphia Vireo was also described as preferring woodland edges, aspen parklands, second growth on burned and logged areas, and shrubby riparian areas (Moskoff and Robinson 2020). Green and Niemi (1978) identified open woodlands, burned-over areas, and streamside willows and alders in the Superior National Forest of northeastern Minnesota as areas of habitat.

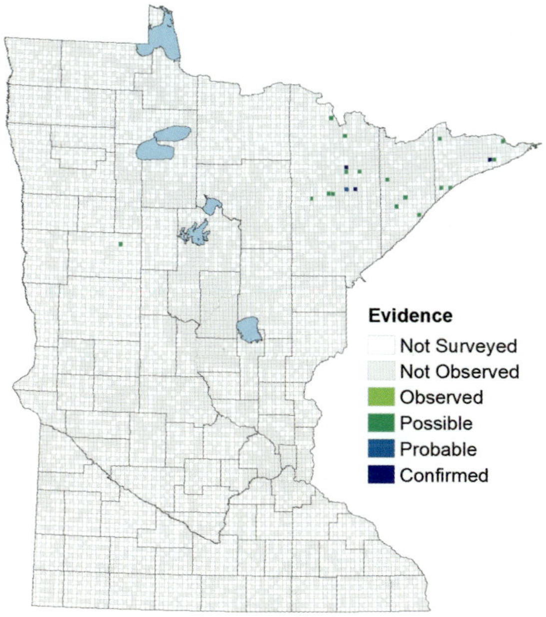

A. Breeding distribution of the Philadelphia Vireo in Minnesota based on the Breeding Bird Atlas (2009–13).

B. In northern Minnesota, the Philadelphia Vireo prefers to nest in early to middle-aged forests such as this aspen stand. Photo by Gerald J. Niemi.

POPULATION ABUNDANCE

PIF (2021) estimated a breeding population of 4 million breeding adults in the United States and Canada, but over 97 percent of the population was in Canada. They estimated a population of 3,200 adults in Minnesota. No population estimate was attempted by the MNBBA because of so few verified observations.

Two reliable trend estimates for the Philadelphia Vireo included the Boreal Hardwood Transition Bird Conservation Region, which included Minnesota, and surveywide in the BBS. These trends indicated a stable population with 1.21 percent growth per year from 1966 to 2019 in the Boreal Hardwood Transition Bird Conservation Region **(C)** and a significantly increasing population growth of 1.98 percent per year surveywide, also from 1966 to 2019 (Sauer et al. 2020). Rosenberg, Kennedy, and colleagues (2016) estimated an overall increase in the population of 87 percent from 1970 to 2014. Given the difficulty in the identification of this species, all of these estimates should be interpreted with caution.

CONSERVATION

The Philadelphia Vireo's primary breeding population exists in Canada where it is considered "unlikely to face significant habitat shortages in the near future" (Environment Canada 2014b). The species is listed as a *Species in Greatest Conservation Need* in Minnesota primarily because it is rare, has potentially declining or vulnerable habitat, and the causes for these declines are uncertain (MNDNR 2015b).

The Philadelphia Vireo was identified as susceptible to communication tower kills, with an estimated annual mortality of more than 38,000 adults or approximately 1 percent of its total population (Longcore et al. 2013). Moskoff and Robinson (2020) suggested some sensitivity to Brown-headed Cowbird nest parasitism and predation due to fragmentation of forests. However, they also advised that the species might be dependent on forest disturbance because of its potential disappearance from late-successional forests. Cutright and colleagues (2006) reported it was found nesting in Bayfield County of northern Wisconsin in aspen forests where the canopy was "opened by logging activity or natural disturbance."

Terborgh (1989) pointed out that the Philadelphia Vireo had a restricted winter distribution in southern Central America, but it seemed to accept disturbed habitats in its wintering grounds. With a 2°C warming scenario, NAS (2019) projected a moderate vulnerability for the Philadelphia Vireo's breeding distribution in North America, including a 63 percent range loss and a 42 percent range gain primarily in the far north of Canada and Alaska by 2050. In Minnesota, NAS (2019) considered the species extirpated under the 1.5°C warming scenario.

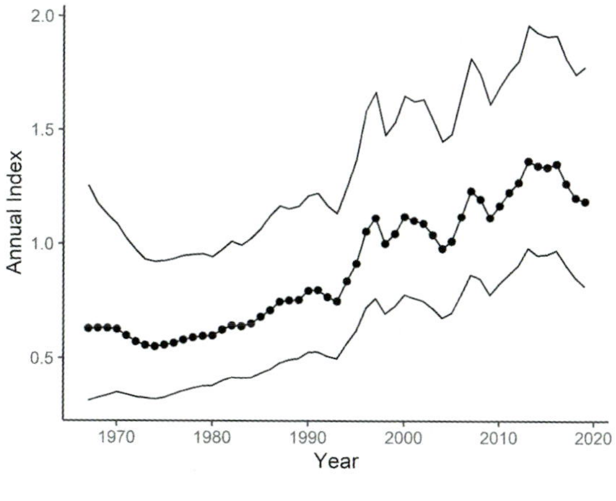

C. Breeding population trend and 95 percent credible interval (solid black lines) for the Philadelphia Vireo in the Boreal Hardwood Transition Bird Conservation Region based on the 1967–2019 Breeding Bird Survey (Sauer et al. 2020).

WARBLING VIREO

Vireo gilvus

Minnesota Seasonal Status: Regular breeding resident and migrant. The Warbling Vireo was common during the MNBBA.
Migration: Medium-distance migrant that spends winters in western Mexico and northern Central America.
Food: Foliage gleaner that feeds on arthropods.
Nest: A compact cup nest, usually placed high within a deciduous tree or shrub and some distance from the tree trunk.
Conservation Concern: Assigned a low PIF Continental Concern Score of 8/20.

For more than a century, naturalists have written enthusiastically about the Warbling Vireo's unique song yet struggled to delineate any other distinctive features. "The warbling vireo, if it were not for its song, would not be a notable species, for it is a little bird in leaf-green plumage, inconspicuous as it moves about among the foliage on the highest branches of its favorite elms and poplars" (Bent 1950). Such is the plight of the Warbling Vireo! But the species' lack of distinct markings actually provides one of the best identification clues. The Vireo is olive-gray on the back with a grayish-white belly that may show a hint of yellow, and its Latin name, *Vireo gilvus,* means "pale yellow vireo." The only notable feature is a faint, olive-gray eyeline. The male and female are identical.

Luckily, the Warbling Vireo's song is remarkable among vireos. Unlike the short, abrupt phrases that characterize the vocalizations of many vireo species that reside in Minnesota, the Warbling's Vireo's song is a long, musical warble that has been likened to the mnemonic *brigadier, brigadier, brigate.* It ends with a pronounced upward inflection that writer Pete Dunne described as sounding "like a happy drunk making a conversational point at a party" (CLO 2017h).

The Warbling Vireo is found in open woodlands from southeastern Alaska, east across western Canada, and across much of the northern and central United States. It is absent across a broad stretch of the southeastern United States. Farther west, the population extends south into western Mexico. It reaches some of its highest breeding densities in British Columbia and in the Rocky Mountain regions of western Montana and Colorado.

Photo by Michael Furtman.

MINNESOTA BREEDING DISTRIBUTION

Roberts (1932) described the Warbling Vireo as a common species throughout much of Minnesota in the late 1800s and early 1900s. It was "most abundant in open woodlands, preferring small groves of poplars and elms, in orchards or on tree-bordered country roads, and is one of the common birds of city parks and shaded streets." Even in western Minnesota, "every grove and tree-claim has its quota of Warbling Vireos." The Vireo was absent from northern Minnesota's forests for many years; the first report of it was in 1902, when a single bird was encountered in the town of Walker, on the west shore of Leech Lake. In the years that followed, reports began to document the species' gradual appearance in small settlements and cleared areas as far north as Crow Wing, Itasca, Lake of the Woods, and Pine Counties. It was only a matter of years, Roberts predicted, before the Warbling Vireo would become a common north country inhabitant. At the time of his publication, however, confirmed nesting reports were limited to just six counties south and west of the boreal forest region: Goodhue, Hennepin, Polk, Sherburne, Traverse, and Wright.

Indeed, more than forty years later, Green and Janssen (1975) documented summer records for the Warbling Vireo from nearly every county in the state. Nevertheless, it remained "quite scarce" in the far north-central and northeastern counties, where breeding records were virtually absent. Janssen (1987) later depicted the species as a breeding resident everywhere except the Arrowhead region north and east of Duluth and in much of Koochiching and eastern Itasca Counties. Noting that it was most abundant in east-central and southeastern Minnesota, he identified 22 counties where nesting had been confirmed since 1970. By 1998, Hertzel and Janssen identified 26 counties where nesting had been confirmed. The Vireo's broad distribution was further confirmed by the MBS, which documented 744 breeding-season locations for the species. Records were sparse in the northeast but included several records in St. Louis County and one in southern Cook County. Elsewhere the species was relatively common in areas that were surveyed.

MNBBA participants further confirmed the Warbling Vireo's broad distribution across all but the far northern regions of the Laurentian Mixed Forest Province **(A)**. It was absent from only two of Minnesota's 87 counties, Lake and Cook; there were a total of 2,409 Warbling Vireo records in 34.2 percent (1,624/4,754) of the surveyed atlas blocks and in 47.1 percent (1,101/2,337) of the priority blocks. Breeding was confirmed in 42 blocks (0.9 percent) and reported as probable in 525 blocks (11.0 percent). Eighteen of the 28 counties where breeding was confirmed were additions to the list published by Hertzel and Janssen (1998). Although the species was least common in the northeastern forests and north-central peatlands, it was reported throughout much of the remainder of the state, including the intensively cultivated regions of western and southern Minnesota.

The MNBBA probability map for the Warbling Vireo suggests that the core of the species' breeding range lies within western Minnesota, stretching from the northern edge of the Prairie Coteau in the southwest corner of the state to the Canadian border **(B)**. Small pockets of relatively high breeding densities are found scattered throughout the open, park-like landscape of the aspen parklands. Elsewhere, moderate densities are predicted along the

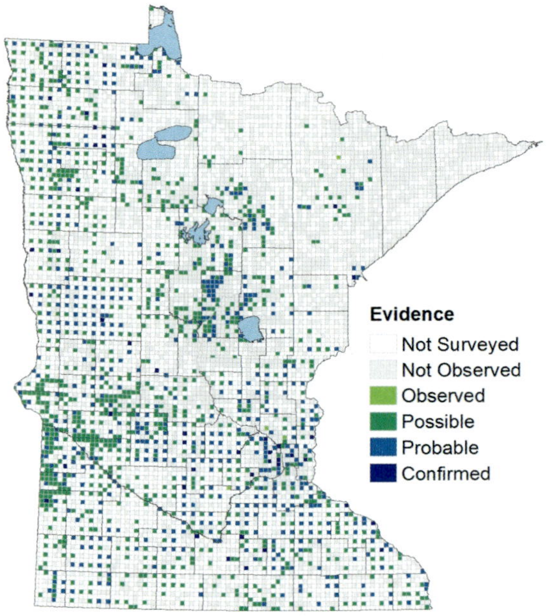

A. Breeding distribution of the Warbling Vireo in Minnesota based on the Breeding Bird Atlas (2009–13).

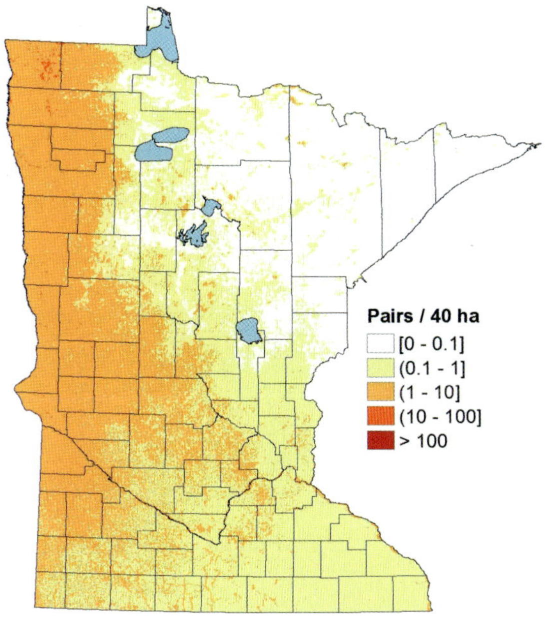

B. Predicted breeding distribution (pairs per 40 ha) of the Warbling Vireo in Minnesota based on habitat, landscape context, and climate data gathered during the Minnesota Breeding Bird Atlas (2009–13) using the Generalized Linear Modeling method with an adjustment for detectability.

Mississippi River south of the Twin Cities and in scattered areas west and southwest of the metropolitan region. Abundance rapidly declines moving east through the northern forested counties.

Roberts's (1932) prediction that the Warbling Vireo would become a common inhabitant of the north country was not entirely fulfilled. It has become more prevalent along the southern and western borders of the Laurentian Mixed Forest Province, especially in recreational communities near the cities of Brainerd and Grand Rapids. Nevertheless, it remains a rare species in the more densely forested landscape east and north of these cities.

Elsewhere there have been few large-scale changes in the species' distribution. This is attributed in part to the paucity of historic information on this obscure little bird (Gardali and Ballard 2020). The Vireo's expansion along the northern periphery of its breeding range in Quebec and Ontario has been facilitated largely by forest clearing in the late 1800s and 1900s. Other Great Lakes states, including Wisconsin (Cutright et al. 2006) and Michigan (Chartier et al. 2013), have seen little change in the species' historic distribution.

BREEDING HABITAT

The Warbling Vireo is strongly associated with mature deciduous trees, often in open, fragmented habitats bordering wetlands, lakes, and rivers. Open parklands within suburban neighborhoods or city parks provide ideal habitat **(C)**. Small woodlots, far from water, may also be used. The primary requirement is the presence of tall trees embedded in an otherwise semi-open canopy. The structure and composition of the understory vegetation seem to be less important. In Minnesota's northern forests, Warbling Vireos are commonly associated with small towns and cities; they were rarely encountered away from these areas in the Chippewa and Superior National Forests (Niemi, Howe, et al. 2016).

POPULATION ABUNDANCE

BBS data helped PIF (2021) to generate a population estimate of 52 million breeding adults in the United States and Canada, with an additional 1 million birds residing in Mexico. Minnesota is estimated to support approximately 490,000 birds. The much larger sample of MNBBA point counts generated a statewide estimate that was more than double the PIF estimate, at 1.2 million adults (95 percent confidence interval of 1.0–1.7 million).

From 1966 to 2019, Warbling Vireos have demonstrated a significant population increase averaging 0.64 percent per year across the BBS survey region (Sauer et al. 2020), with a cumulative increase of 55 percent from 1970 to 2014 (Rosenberg, Kennedy, et al. 2016). Range-wide, populations are declining in only five ecological regions, two of which cover Minnesota's Laurentian Mixed Forest and Eastern Broadleaf Forest Ecological Provinces. The decreases in these regions appear to be offset by a significant increase in the Prairie Potholes Region, which covers western Minnesota. This likely explains why Minnesota's Warbling Vireo population is not following the increasing trend observed at the continental scale but has maintained a stable trend line despite some small fluctuations **(D)**.

C. The Warbling Vireo is commonly attracted to park-like landscapes with mature deciduous trees located along lakeshores. Photo by Lee A. Pfannmuller.

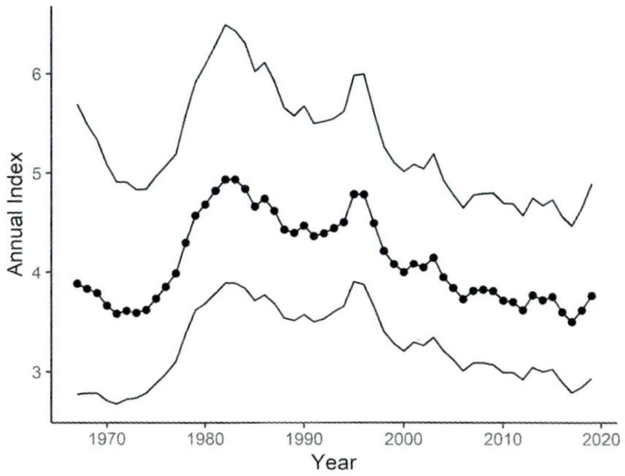

D. Breeding population trend and 95 percent credible interval (solid black lines) for the Warbling Vireo in Minnesota based on the 1967–2019 Breeding Bird Survey (Sauer et al. 2020).

Factors responsible for these regional differences are poorly understood. Population declines in the state's forested region may correlate with increased development, especially in east-central and southern Minnesota, where small, fragmented woodlots are replaced with more intensive land uses associated with agriculture and urbanization. Farther west, increases may be due to the maturation of farm shelterbelts and trees in small towns and cities. Range-wide, the Vireo's population increase is largely attributed to its adaptability to human-modified landscapes. Along the northern periphery of its continental breeding range, the landscape also became more suitable as mature coniferous forests were harvested and replaced with young, second-growth deciduous woodlands (Gardali and Ballard 2020).

CONSERVATION

With a large population dependent on open, fragmented woodlands, the Warbling Vireo is a low conservation priority (PIF 2021). Protecting riparian zones along wetlands, a management practice increasingly implemented to improve water quality, should benefit this relatively common species. Even warmer temperatures are predicted to pose little threat to the species, with habitat losses more than compensated for by gains; conditions may even improve for the species in eastern Minnesota (NAS 2019). In the near term, the future of the little Warbling Vireo, nearly inconspicuous in its pale olive-green plumage as it flits among the treetop canopies, remains quite secure.

RED-EYED VIREO

Vireo olivaceus

Minnesota Seasonal Status: Regular breeding resident and migrant. The Red-eyed Vireo was very abundant during the MNBBA.
Migration: Long-distance migrant that overwinters in northern South America.
Food: Insectivorous, feeding especially on caterpillars gleaned from foliage.
Nest: Pensile cup nest suspended from the end of a branch in a tree or shrub.
Conservation Concern: Assigned a low PIF Continental Concern Score of 6/20.

Photo by David Brislance.

The Red-eyed Vireo is one of the most abundant bird species in Minnesota forests. It is so common that most forests in northern Minnesota have a nesting pair in every five acres. Once an observer gets to know the song of the Red-eyed, it can be heard throughout Minnesota summers. In words the song is described as *"Here I am! Here I am! Don't you see me! Don't you hear me! Here I am!"* with each phrase rising in inflection. Bent (1950) claimed it has won the title of "Preacher" because "no other of our birds sings so persistently all day long, and because his long-continued . . . utterances, given in short, emphatic phrases, going on for hours, calls to mind a lengthy sermon."

This vireo indeed has a red iris, which provides its common name. Its genus, *Vireo,* translates in Latin "to be green," while its species name, *olivaceus,* is also Latin and aptly describes its olive-green upperparts. Distinguishing head features include a distinctive white eyeline with a black border on top and a gray crown. The species lacks wing bars, and both sexes are similar in plumage.

The Red-eyed Vireo is widely distributed in the eastern United States and the southern boreal forest across all the Canadian provinces to British Columbia. It is especially abundant in northern Minnesota, Wisconsin, and Ontario, as well as the New England states, western Pennsylvania, and northwestern Arkansas. Because of its ubiquity, it is a well-studied species with substantial information available on communication, breeding biology, foraging behavior, and migration (Cimprich et al. 2020).

MINNESOTA BREEDING DISTRIBUTION

Roberts (1932) described the Red-eyed Vireo as "an abundant summer resident throughout the state wherever there are trees" and as one of the "most generally distributed birds." He listed the confirmed county nesting records as not excessive in number but extensive in distribution across the state. He included nesting documentation from Wabasha and Goodhue Counties in southeastern Minnesota, from Hennepin and Sherburne Counties and Mille Lacs in the east-central part of the state, from Marshall County in the northwest, in Itasca, Lake, and St. Louis Counties in the northeast, and from Cass Lake, Leech Lake, and Itasca Park in the north-central region. He also thought that the species is a "common inhabitant of the natural groves and timber-claims of the western prairies."

Green and Janssen (1975), more than forty years later, described the Red-eyed Vireo simply as a "resident throughout the state" and "the most widespread and numerous vireo species." Janssen (1987) emphasized it was "a numerous resident throughout the heavily wooded areas of the state" and "least numerous in the southwest region." Janssen, along with Hertzel and Janssen (1998), documented and considerably expanded confirmed nesting since 1970 in counties ranging from Houston County in the extreme southeast to Murray County in the southwest, to Roseau County in the northwest, and to Cook County in the northeast.

The MBS illustrated the wide distribution of breeding locations throughout the state with a total of 4,253 records. No locations were included from many western Minnesota counties, such as Big Stone, Norman, Pipestone, and Wilkin. These were also counties without confirmed nesting records in Janssen (1987) or Hertzel and Janssen (1998). Breeding observations were also sparse in many counties in southwestern Minnesota, except along the Minnesota River and its forested floodplain.

The MNBBA included a whopping 7,380 records from every county in the state **(A)**. Breeding records were recorded in 61.8 percent (2,952/4,777) of blocks covered in the MNBBA. Many counties in the southwestern part of the state had few records. Most other counties had probable or confirmed nesting observations. An additional 17 counties had confirmed nesting records not previously cited by Janssen (1987) or Hertzel and Janssen (1998).

The MNBBA probability map illustrates the extensive distribution of the Red-eyed Vireo throughout the state, with the highest densities predicted in the heavily forested regions of northeastern and north-central Minnesota **(B)**. Outside of this region, densities are also relatively high in northwestern, western, south-central, and southeastern Minnesota, especially along forested river valleys. Red-eyed Vireos can be found almost anywhere with

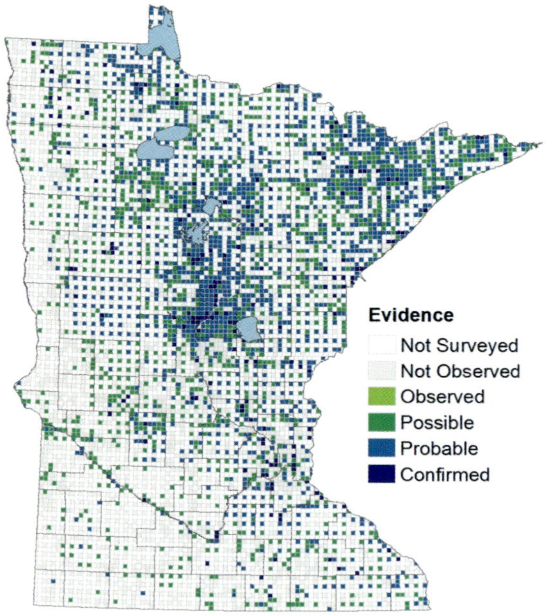

A. Breeding distribution of the Red-eyed Vireo in Minnesota based on the Breeding Bird Atlas (2009–13).

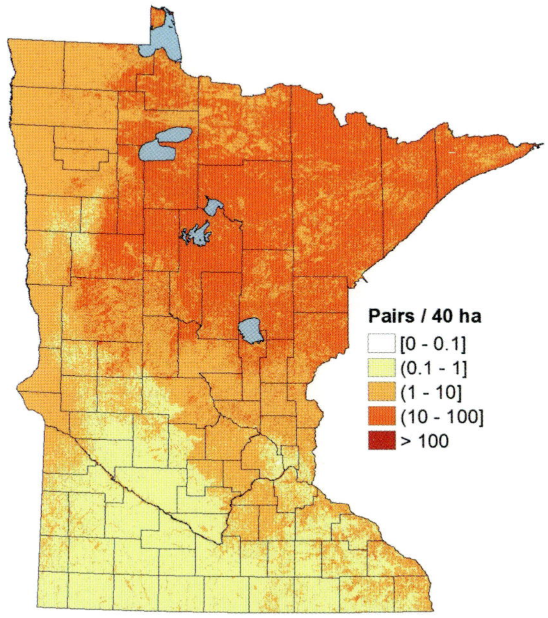

B. Predicted breeding distribution (pairs per 40 ha) of the Red-eyed Vireo in Minnesota based on habitat, landscape context, and climate data gathered during the Minnesota Breeding Bird Atlas (2009–13) using the Generalized Linear Modeling method with an adjustment for detectability.

suitable forested areas, including woodlots in agricultural and urbanized regions.

The Red-eyed Vireo has likely become less abundant in Minnesota compared with the mid-1850s because of the loss of more than 40 percent of forested area in the state. It is, however, still a very abundant species. Cimprich and colleagues (2020), in their review of the Red-eyed Vireo in North America, reported it was expanding in Oregon, Utah, and Newfoundland from the 1920s to 1940s, concurrent with the planting of "eastern" tree species in shelterbelts and other landscaping activities.

BREEDING HABITAT

Red-eyed Vireos occur in many upland forests, lowland deciduous forests, and mixed deciduous–coniferous forests as well as in residential areas, city parks, and cemeteries—wherever large trees grow (Cimprich et al. 2020). Niemi, Howe, and colleagues (2016) indicated wide use of most forest types but highest use of deciduous forest cover types, such as oak, aspen–birch, northern hardwoods, and lowland deciduous forests. Moderate use was also noted in many mixed deciduous–coniferous forests, such as aspen–fir–spruce, coniferous forests including pine (jack, red, and white) forest types, and regenerating forest cover types **(C)**.

C. S. Robbins and colleagues (1989) reported the Red-eyed Vireo was sensitive to forest fragmentation. They found that the species occurred in forest patches as small as 0.5 ha but was absent when those forest fragments were isolated. R. D. James (1976) found the species was generally absent or rare when understory shrubs were not present.

POPULATION ABUNDANCE

PIF (2021) estimated a US and Canadian population of 130 million breeding adults and a Minnesota population of 4.1 million adults. The MNBBA estimate for Minnesota was substantially higher, with 8.54 million breeding adults (95 percent confidence interval of 8.20–8.94 million). This higher estimate, however, is still only 4.5 percent of the Red-eyed Vireo's global population.

The BBS in Minnesota for the years 1967–2019 indicated a significantly increasing population trend of 0.48 percent per year **(D)**. On a broader scale, over the same period, Manitoba, Michigan, Ontario, Wisconsin, and surveywide all showed significant increases ranging from 0.57 to 1.67 percent per year. Rosenberg, Kennedy, and colleagues (2016) calculated that the population had increased by 43 percent from 1970 to 2014. In contrast with the statewide and other positive trends, in the Chippewa and Superior National Forests combined, the species population trend has been significantly declining from 1995 to 2020 by an average of −0.85 percent per year (Grinde, Bednar, et al. 2020). Given these changes in northern Minnesota, it will be imperative in the future to continue to watch these trends and potentially identify causes for changes in the species' populations at both regional and national scales.

In one of the few semiquantitative surveys completed in the early 1900s, Roberts (1932) stated, "in eight walks of a mile or two along different trails in

C. A typical forest habitat used by the ubiquitous Red-eyed Vireo. Many forests are suitable for the species. Photo by Gerald J. Niemi.

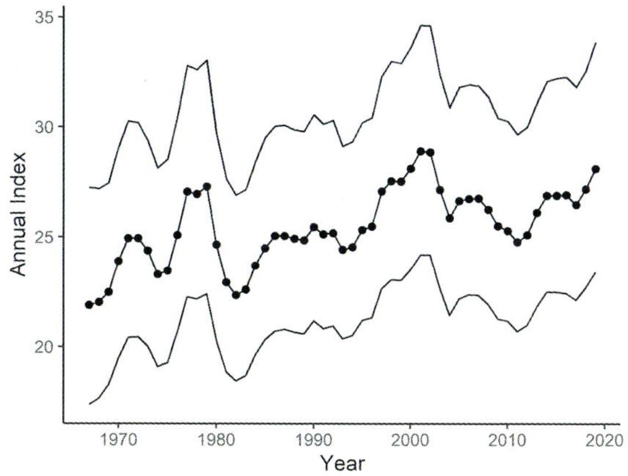

D. Breeding population trend and 95 percent credible interval (solid black lines) for the Red-eyed Vireo in Minnesota based on the 1967–2019 Breeding Bird Survey (Sauer et al. 2020).

Itasca Park, from July 5 to 16, 1919, over one hundred of these birds were checked." He probably meant observations of this species, and likely most were singing males.

CONSERVATION

From a conservation perspective, there are few concerns with the Red-eyed Vireo. The species has a wide geographic distribution and large breeding population and uses many forest and semi-forested habitats. Under a 2°C warming scenario, NAS (2019) projected low vulnerability for the Red-eyed Vireo's breeding distribution in North America with a 12 percent loss of its breeding range and a 17 percent gain by 2050. Minimal range retractions in Minnesota were predicted on its western borders under this same scenario.

Cimprich and colleagues (2020) reported some sensitivity to logging activity, but data from Minnesota suggest the species is tolerant of logging activity, especially if some live trees remain at the site (Grinde, Niemi, Adams, and Wigley 2017). However, population densities in regenerating forests are substantially lower than in intermediate and mature upland forests (Niemi, Howe, et al. 2016). The Red-eyed Vireo is frequently found dead at communication towers, but the overall impact of these towers on its population is believed to be minimal because of its large population. Ecological data from its wintering grounds in South America are minimal and deserve further study.

Despite its potential decline in some of Minnesota's northern forests, it remains a well-tracked, abundant, and easy-to-identify species. Therefore, we can hope that the population changes observed are part of a longer cycle that will rebound so Minnesotans can enjoy the Red-eyed Vireo well into the future, with its voracious appetite for caterpillars continuing to contribute to healthy forests in the state.

LOGGERHEAD SHRIKE

Lanius ludovicianus

Minnesota Seasonal Status: Regular breeding resident and migrant. The Loggerhead Shrike was rare during the MNBBA.
Migration: In the northern portion of its range, including Minnesota, the species is a short- to medium-distance migrant that winters in the southern United States and Mexico. Farther south, breeding populations are year-round residents.
Food: Primarily large arthropods, such as grasshoppers and crickets, as well as some small birds, reptiles, mammals, and carrion.
Nest: Cup nest placed in a dense and often thorny shrub or small tree.
Conservation Concern: Assigned a moderate PIF Continental Concern Score of 11/20 and officially listed as *Endangered* in Minnesota.

Photo by Matthew Jolley.

Chunky and conspicuous, the Loggerhead Shrike is the songbird family's version of a raptor. Smaller than a robin, this predatory bird can be seen sitting motionless on a telephone wire as it scans the open landscape for insects and small vertebrates. Lacking the strong feet and talons that enable hawks and owls to quickly snare prey from above and hold them while they feed, the Shrike depends on other clever foraging strategies. Larger vertebrate prey, for example, are often attacked from behind and paralyzed with a series of bites that sever the vertebrae. Once paralyzed, victims are impaled on thorns and barbed wire fences or wedged into tight spaces, where they are easily torn apart with the Shrike's sharply hooked bill. Smaller, non-toxic prey are usually consumed immediately, but larger items may be left for later consumption. Impaling and leaving behind noxious prey items, such as Monarch butterflies, also allows toxic chemicals to degrade prior to eating (Yosef 2020). This unique behavior is primarily exhibited by males and may also function as a method of food storage during breeding or as a courtship activity in which the male demonstrates his fitness to potential mates (Pruitt 2000).

Although a sun-dried frog impaled on a barbed wire fence is a sure sign that a Shrike is nearby, this rare bird is easy to identify. Its common name refers to the large size of its head relative to the rest of its body. Its gray body feathers contrast with its black wings, tail, and facial mask; the wings and tail also have conspicuous patches of white. As Bent (1950) noted, it has "little reputation in vocal performance." Harsh call notes are characteristic, as is a rarely heard, musical song delivered during the spring.

The Loggerhead Shrike can be confused with two other Minnesota birds, the Northern Mockingbird, which lacks the black mask and has an extremely long tail, and the Northern Shrike. The latter is only a winter resident in Minnesota, but both shrike species may be seen during migration. Subtle plumage differences, including the Northern's narrower face mask and barred belly feathers, help distinguish the two species from each other.

The species' Latin name, *Lanius ludovicianus,* translates to "butcher of the Louisiana Territory," a reference to where the first type specimen was collected; the species is also known as the butcherbird. Distributed across the southern half of the United States and south into Mexico, the Loggerhead Shrike's range extends northward into the Great Plains and Prairie Provinces and east through the grasslands of the Upper Midwest. It is also found in central Washington and central Oregon.

MINNESOTA BREEDING DISTRIBUTION

Roberts (1932) described the Loggerhead Shrike as an "abundant" summer resident throughout southern Minnesota, becoming "increasingly less common northward through the Red River Valley." Because the bird was frequently seen perched on telephone wires, Roberts was able to provide a metric of the Shrike's former abundance that present-day birders can easily relate to. During a 12-day automobile trip through southwestern Minnesota in June 1926, he counted 89 Loggerhead Shrikes as he traveled along nearly 500 miles of roadways. A similar trip two years later, which originated in Douglas County in west-central Minnesota and ended in Kittson County on the Canadian border, resulted in a count of 41 Shrikes. Today, birding enthusiasts would be delighted if they sighted just one Shrike! At the time, Roberts had compiled confirmed nesting records from nine counties (Fillmore, Hennepin, Martin, McLeod, Mower, Pipestone, Polk, Rock, and Traverse) and inferred nesting records from three counties (Marshall, Sherburne, and Stearns).

When Green and Janssen (1975) provided an updated account forty years later, little in their description suggested that the species had experienced any change in its status. Although they noted it was very scarce in the northeastern region and in adjacent counties in north-central Minnesota, they still described the Shrike as a statewide resident, with nesting records as far north as the town of Chisholm in St. Louis County.

But change was well under way by the early 1970s. The Shrike was added to the National Audubon Society's Blue List in 1972, an early-warning system designed to bring attention to declining species. And, just fifteen years after the North American BBS was launched in 1966, the Loggerhead Shrike had demonstrated a 50 percent decline in abundance (Coffin and Pfannmuller 1988). Indeed, these changes led to the official classification of the species as state *Threatened* in 1984. By the time Janssen's 1987 update was published, the Shrike was described as "a rare and very local resident in the state." Its distribution was narrowly restricted to western Minnesota, from Clay County south to the Iowa border, along the Minnesota River valley, and in eastern Minnesota, from Morrison County south. Breeding had been confirmed in 18 counties since 1970. Hertzel and Janssen (1998) later added five more counties to the list.

Classification as a state *Threatened* species spurred an increased emphasis on survey work and field investigations. Most significant was the initiation of a two-year assessment of the Shrike's statewide distribution, population dynamics, and habitat requirements from 1986 to 1988 (B. L. Brooks 1988). The statewide survey, conducted during the study's 1986 field season, identified 32 nesting pairs in 12 counties that ranged from Clay County in the northwest to Fillmore County in the southeast. Fifty-six percent of the pairs, however, were located in just three widely dispersed counties: Sherburne (eight pairs), Clay (six pairs), and Goodhue (four pairs). When Brooks compared her results with Loggerhead Shrike reports published by the MOU from 1975 to 1985, it appeared that little had changed in the past ten years (Brooks and Temple 1986).

In 1989, the MNDNR launched a Loggerhead Shrike monitoring program built on the outcomes of Brooks's field studies. Eight road transects were established in areas with the highest nesting concentrations. Conducted annually from 1989 through 1994, the monitoring indicated further declines in the species' abundance. Concern regarding the effectiveness of the survey, however, prompted the department to launch a new, two-year study (1995–96) designed to assess the most effective and efficient method for monitoring population trends (Eliason 1996). During the 1995 field season, 24 nesting attempts were observed in 13 counties, while in 1996 only 13 nesting attempts were observed in eight counties. Although three different survey techniques were tested, the study concluded that Loggerhead Shrike "numbers are so low and most of the birds so widely dispersed" that none of the methods was practical or cost-effective for monitoring populations.

In subsequent years, targeted survey work was limited. Surveys conducted by the MBS identified 33 breeding-season locations for the Loggerhead Shrike. Nearly one-third of the records, however, were restricted to Clay and Dakota Counties. In 2008, MNDNR staff conducted a survey in Dakota, Goodhue, and Rice Counties and in several counties in the Anoka Sandplain that had previously supported nesting pairs. Loggerhead Shrikes were observed at only seven locations (MNDNR 2012). The continued decrease in the number of reports led to the reclassification of the Shrike as state *Endangered* in 2013 (MNDNR 2013).

During the MNBBA, observers reported 50 Loggerhead Shrike records in just 33 of the surveyed atlas blocks. Breeding evidence was confirmed in 14 blocks; 50 percent of the blocks with confirmed breeding evidence were located in Dakota County **(A)**. The birds were observed in 18 of Minnesota's 87 counties and were confirmed breeding in seven counties: Anoka, Clay, Dakota, Goodhue, Olmsted, Wilkin, and Yellow Medicine. Wilkin and Yellow Medicine Counties were new additions to the list of counties with confirmed nesting published by Hertzel and Janssen (1998). Breeding was reported as possible in northwestern Polk County, where a family of birds was observed perched on a telephone wire. The young, however, were fully capable of flight, so the exact block where nesting occurred could not be determined. Confirmed breeding in the county had been documented just prior to initiation of the MNBBA in 2007 (MOU 2021a). An atlas observation from southern St. Louis County was particularly interesting. This was not the first time Shrikes had been reported in the area; summer reports from the county date back to at least 1966 (MOU 2021b). In addition to the records compiled by the atlas, one nesting record (fledged young) was reported to the MOU during the atlas period in Otter Tail County on August 7, 2011; unfortunately, details were insufficient to assign it to an atlas block (MOU 2021b). And, seven years following the atlas, nesting was confirmed in Kandiyohi County for the first time in 2020 (MOU 2021a).

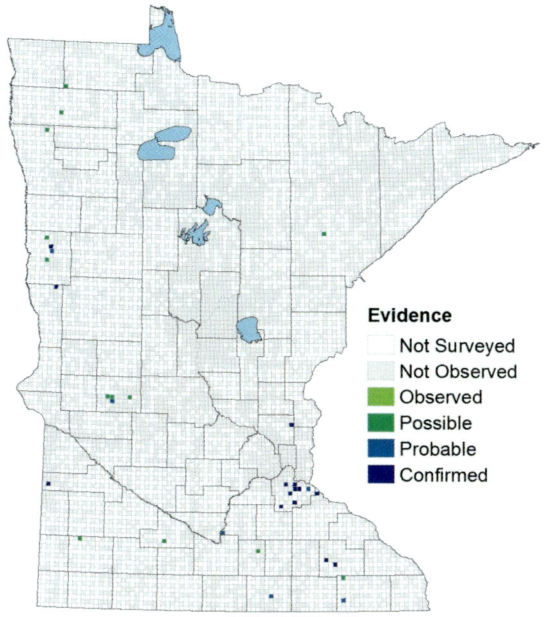

A. Breeding distribution of the Loggerhead Shrike in Minnesota based on the Breeding Bird Atlas (2009–13).

Although Loggerhead Shrikes continue to be reported from widely scattered sites throughout the state each year, the majority of confirmed nesting reports now remain restricted to Dakota County. This is a notable shift from the mid-1980s, when Brooks and Temple (1986) documented 12 of 32 nests from the central Minnesota counties of Benton, Morrison, and Sherburne. That same year, 1986, only two nests were reported in Dakota County. By 1995, the three central counties supported only one nest, while there were nine nest attempts in Dakota County (Eliason 1996). The Loggerhead Shrike clearly has become a very rare and vulnerable species in Minnesota.

Despite its rarity, an analysis of suitable habitat by the MNBBA illustrates that a large portion of western and southern Minnesota still provides habitat potentially conducive to supporting breeding Loggerhead Shrikes **(B)**. The reasons for their absence over much of the suitable landscape are not well understood (Yosef 2020).

Elsewhere in its breeding range, the species has experienced declines similar to those documented in Minnesota. As eastern forests were cleared in the late 1800s and early 1900s, small farm fields bordered by woody plantings provided ideal habitat for the Shrike. Similar range expansions occurred in the northern Great Plains and north into southern Canada as farm shelterbelts provided new nesting habitat (Cade and Woods 1997). Then, as agricultural practices intensified, the large, unbroken fields of wheat and corn became less suitable habitat for the

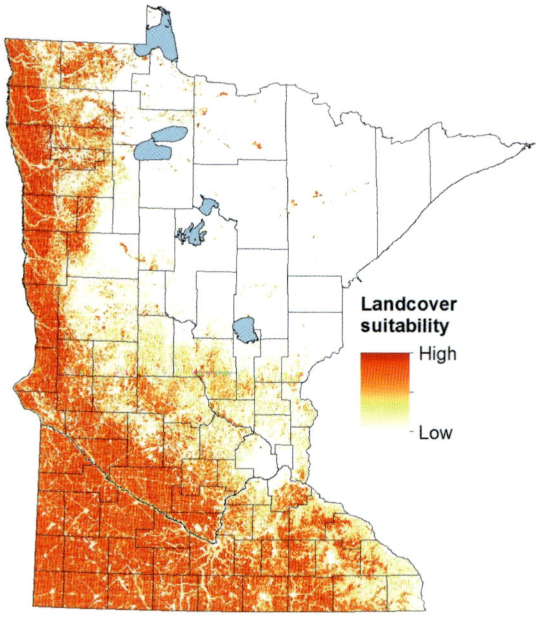

B. Landcover suitability map for the Loggerhead Shrike in Minnesota based on habitat, landscape context, and climate data gathered during the Minnesota Breeding Bird Atlas (2009–13) using the MaxEnt modeling approach.

Shrike. Declines were noted in many regions as early as the 1930s, 1940s, and 1950s (e.g., Cade and Woods 1997; Rodewald et al. 2016; Chartier et al. 2013). Not only did the species' abundance decline dramatically but also, earlier in the twentieth century, the overall extent of its breeding range receded "well south of its maximum known breeding distribution," including in southern Canada, New England (Cade and Woods 1997), Michigan (Chartier et al. 2013), and Wisconsin (Cutright et al. 2006).

BREEDING HABITAT

An inhabitant of open country, the Loggerhead Shrike occurs in a diversity of habitats, ranging from upland native prairie, oak savannas, and riparian areas to pastures, old orchards, golf courses, and cemeteries (Kaufman 1996; Yosef 2020). Across the Upper Midwest, its habitat has been broadly characterized as open, agricultural sites interspersed with grasslands (Brooks and Temple 1990b). In several states, the bird's abundance has been shown to positively correlate with the abundance of pasturelands (Yosef 2020). Because the Shrike is a "sit-and-wait" predator, the availability of perches, preferably less than 5 meters in height, is especially important. Indeed, a study in Florida demonstrated that reproductive success was "positively influenced by the density of hunting perches" (Yosef and Grubb 1993).

In Minnesota, the species occupies upland prairies, grasslands, and agricultural fields where perches are available to search for prey and scattered trees and shrubs are available for nesting **(C)**. Small farmsteads that still include hedgerows and fence lines are especially suitable, as are shelterbelts (Coffin and Pfannmuller 1988). Red cedar has been reported as the most frequently used nest tree (44 percent) in the state. A variety of other species that have a shrubby or bushy growth with thorns or spines are also used, including American plum, hawthorns, and Russian olive. Nests are usually placed in isolated trees within grassland habitats or within narrow hedgerows or windbreaks that are located adjacent to agricultural fields (Brooks and Temple 1990b).

In recent years, one of the most reliable sites for finding Loggerhead Shrikes in Minnesota has been a stretch of unpaved roads in central Dakota County. Fences and aboveground electrical wires provide the birds with perches; roads are bordered by croplands and pastures that supply small prey; and scattered homes and farm buildings are surrounded by small trees and shrubs that furnish nesting habitat **(D)**.

POPULATION ABUNDANCE

BBS data helped to generate a population estimate of 4.6 million birds in the United States and Canada with an additional 2.4 million residing in Mexico (PIF 2021). The same data provided an estimate of only 150 birds in Minnesota. MNBBA data were insufficient to generate a more precise estimate.

BBS data also clearly document the species' significant and steady population decline. From 1966 to 2019, the population declined an average of −2.56 percent per year within the BBS survey region **(E)**, with a cumulative decline of 74 percent from 1970 to 2014 (Rosenberg, Kennedy, et al. 2016). Numbers are too low in Minnesota to generate a statistically reliable population trend, but available data certainly suggest a precipitous decline. Nationwide, the only location where numbers are significantly increasing is in the Tamaulipan Brushlands in southern Texas (Sauer et al. 2020).

Given the Shrike's significant decline, biologists estimate that the amount of time before an additional 50 percent of the population is lost could be as little as twenty-four years (Rosenberg, Kennedy, et al. 2016). As noted earlier, in addition to a decline in numbers, the species' range has also contracted. Cade and Woods (1997) suggest that its distribution is retreating to a region they believe it may have occupied prior to European settlement.

Numerous hypotheses have been suggested to explain the Shrike's steady decline and range contractions, including habitat loss, low reproductive success, pesticide contamination, interspecific competition, and vehicle collisions (Yosef 2020; Pruitt 2000). The impact of pesticides may be particularly significant for a species that forages in agricultural fields that are the frequent target of chemical applications. For instance, the Shrike's initial decline was concurrent with the widespread use of organochlorine pesticides that began in the 1940s. Unfortunately, research

C. A rare, dry sand-gravel prairie protected on the Hastings Sand Coulee Scientific and Natural Area in Dakota County provides ideal habitat for the Loggerhead Shrike. The short grasses and herbs are excellent foraging habitat, while the scattered red cedar trees support nesting sites and perches.
Photo by Lee A. Pfannmuller.

D. A farmstead in rural Dakota County meets all the needs of the Loggerhead Shrike: a fence and overhead electrical wires for perching; a pasture bordered by scattered trees and shrubs; a newly planted cornfield adjacent to the pasture; and, in the distance, an unplowed knoll with grasses and scattered deciduous trees and red cedar. Photo by Lee A. Pfannmuller.

studies attempting to link the decline to pesticide use produced ambiguous results. In Illinois, for example, DDE levels in Shrike eggs collected from 1995 to 1996 were 79% lower than in eggs collected from 1971 to 1972 (Herkert 2004). Yet, when these chemicals were banned or their use was significantly restricted, Loggerhead Shrike numbers continued their downward trajectory while other species negatively impacted by organochlorines, such as the Peregrine Falcon and Double-crested Cormorant, began a slow recovery. Meanwhile, chemical companies continue to introduce new pesticides, further challenging biologists to untangle their respective roles in the Loggerhead Shrike's uncertain future. Indeed, direct mortality to the birds may not be the most significant impact of chemical applications. Instead, their ability to significantly reduce prey abundance may decrease overall Shrike survival (Yosef 2020).

Biologists also consider the loss and degradation of habitat to be major threats to Shrike populations. More intensive agricultural practices in the Great Plains and Upper Midwest and reforestation of areas in the Northeast have reduced the patchwork of grasslands, agricultural fields, and shrubs preferred by the species (Cade and Woods 1997; Pruitt 2000). In Minnesota, changes in agricultural practices have had a noticeable impact on habitat suitability. Conversion of pastures to row crops, the loss of hedgerows and windbreaks that provided nesting cover, and the absence of hunting perches or nesting trees in grassland habitats have all had local impacts on nesting Shrikes (Eliason 1996). Nevertheless, many biologists have noted that habitat that appears "visually suitable" for the birds remains unoccupied (Eliason 1996; Brooks and Temple 1990a; Pruitt 2000), suggesting that we lack a complete understanding of their habitat requirements or that other mortality factors are at play.

Overwinter survival may be a key factor negatively impacting Shrike populations that breed in the northern states and provinces. Not only do these migrating birds need to contend with habitat loss in southern states where they winter, but also they must compete for resources with those Shrikes that remain in the region as year-round residents (Brooks and Temple 1990a). Intraspecific competition among Shrikes may be exacerbated by fire ant infestations that have rendered many agricultural fields in the south uninhabitable, while the insecticides used to control the ants may pose another threat (Lymn and Temple 1991).

Vehicle collisions pose another concern. The bird's "sit-and-wait" foraging habit of perching on fences and power lines that border roads, and then swooping down low to capture prey, increases the chances of colliding with passing vehicles. Juvenile Shrikes may be especially susceptible to such accidents (Wiggins 2005). A study conducted in Virginia "estimated that 29 percent of Shrike mortality during the fall and winter could be attributed to such collisions" (Blumton 1989). Today, the species' vulnerability to these incidents may be even higher. As more acres are converted to croplands, roadsides are becoming the "last hunting haven for shrikes in many highly developed agricultural areas" (Lymn and Temple 1991).

The list of potential threats to the Shrike's survival is extensive. Predation, severe weather, disease, and potential competition with the American Kestrel, a species considered more tolerant of human-modified landscapes than the Shrike, are additional concerns for the species (Yosef 2020). Nevertheless, biologists agree that "despite a relatively intensive research and conservation effort in recent years, the factors responsible for the species' near range-wide declines are not yet clear" (Wiggins 2005).

CONSERVATION

Although it is a widely distributed species, the Loggerhead Shrike was assigned a moderate conservation priority because of its steady population decline and northern range contraction (PIF 2021). It was also designated a Common Bird in Steep Decline by the 2016 PIF Landbird Conservation Plan (Rosenberg, Kennedy, et al. 2016). As noted earlier, the Shrike has been classified as a state *Endangered* species in Minnesota since 2013.

Specific conservation plans and management recommendations are limited and challenging to carry out, in part because much of the currently occupied and suitable habitat occurs on private lands. Suggestions include leaving fence lines for Shrikes to use for perching and impaling their food, maintaining brush along fence lines, and leaving scattered trees in pastures and fields. Programs that provide financial incentives for farmers to retire highly erodible croplands and maintain them as permanent grass cover are beneficial. These recommendations are included in a brief landowner's guide for maintaining and encouraging Loggerhead Shrikes prepared by the MNDNR (1996).

Nevertheless, the Loggerhead Shrike has a precarious future in Minnesota. Although the species' breeding range was projected to remain reasonably stable under all three warming scenarios examined by NAS (2019), its status across the northern edge of its breeding range seems uncertain. As Pruitt (2000) warned, "Unless we make progress in elucidating underlying limiting factors to Loggerhead Shrike populations in these areas, regional extirpations seem likely." She believed the problem is particularly challenging because resource biologists "are reluctant to implement conservation efforts in the form of habitat improvement" given the perception that suitable, unoccupied habitat already remains available. Although interagency landscape initiatives designed to maintain a matrix of diverse agricultural lands interspersed with protected grassland habitats should be encouraged, the benefits of such efforts to the Loggerhead Shrike are clearly dependent on the cooperation and support of nearby private landowners.

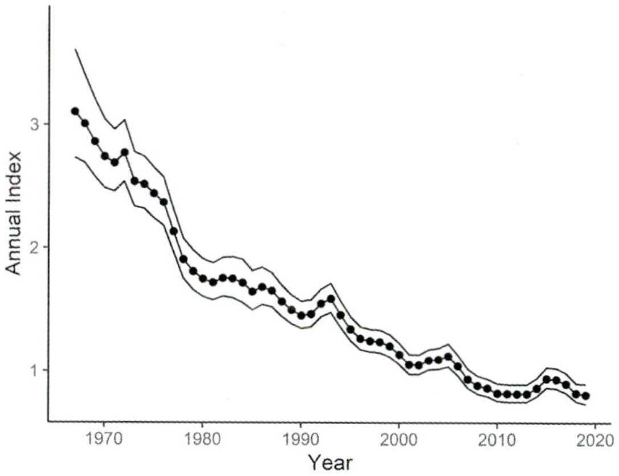

E. Surveywide breeding population trend and 95 percent credible interval (solid black lines) for the Loggerhead Shrike based on the 1967–2019 Breeding Bird Survey (Sauer et al. 2020).

CANADA JAY

Perisoreus canadensis

Minnesota Seasonal Status: Regular, permanent resident and occasional, erratic fall migrant. The Canada Jay was uncommon during the MNBBA.
Migration: Sedentary; irruptive movements southward have been noted in some years.
Food: Omnivorous, consuming arthropods, fruit, carrion, nestlings, and fungi.
Nest: Open nest, of low to moderate height and usually placed in a coniferous tree near the trunk.
Conservation Concern: Assigned a low PIF Continental Concern Score of 8/20.

Photo by David Brislance.

Few species have as many aliases as the Canada Jay. It was most recently officially called the Gray Jay, but also known as the Whiskey Jack, Whiskey John, Camp Robber, Moose Bird, Carrion Bird, and Lumberjack. These names were derived from their associations with lumberjack camps, moose, foraging on carrion, and even robbing campers of their food, such as stealing bacon off a frying pan. Forbush and May (1955) suggest that Whiskey Jack or John was anglicized from a Native American name, perhaps *wis-ka-tjon* or *Wiskedjak*. Its genus name, *Perisoreus,* was thought to translate from Greek for "around to heap up," which describes its behavior of caching and storing food (Sandrock and Prior 2014).

The Canada Jay is an unmistakable, fluffy gray bird with black at the back of the head and a white forehead; it has been described as being reminiscent of a large, overgrown chickadee. Both sexes are similar, but juveniles, who usually travel with the adults in summer, are a distinctive, sooty gray. Roberts (1932) stated that the Canada Jay "has a great variety of notes and calls" and that "[it] has been said that when one hears a strange noise in the north woods it is safe to attribute it to this species."

Canada Jays are found across Canada from Labrador to British Columbia and southward in mountainous regions and forests of western North America to New Mexico and California. In the midwestern and northeastern United States, the species is restricted to the northern regions. High densities have been found in widespread areas, including northern Alberta, British Columbia, Labrador, Manitoba, and Ontario.

MINNESOTA BREEDING DISTRIBUTION

Roberts (1932) reported the Canada Jay as a "common summer resident in the more northern evergreen forests of the state." He cited breeding activity at Mille Lacs, Itasca State Park, and in eastern Marshall County, and reported several observations from St. Louis County, including the only documented nest with eggs where he noted early nesting behavior: "February 22, 1898, building nest, snow still on the ground." Most of his reports were of family groups traveling together.

Years later, Green and Janssen (1975) described the Canada Jay's distribution as being in the northeastern and north-central portions of the state. They reported confirmed nesting from four counties: Cook, Itasca, Lake of the Woods, and St. Louis.

Janssen (1987) highlighted the fall irruptive movements of Canada Jays southward in Minnesota, such as one described by Roberts (1932) in 1929–30, when many wandered south and stayed throughout the winter in the Twin Cities area. Janssen reported additional movements, which occurred in 1965–66, 1974, 1976–77, and 1984, and one of the largest movements, in 1986 (Eckert 1987). The causes for these invasions are unknown, but lack of food is always one of the suspects. Janssen also reported confirmed nesting since 1970 in 11 counties, including seven that had not been previously reported. They were in Aitkin, Beltrami, Cass, Clearwater, Crow Wing, Hubbard, and Koochiching Counties. Hertzel and Janssen (1998) then added nesting in Lake County since 1970, confirming nesting in all of the counties in the northeast and north-central regions.

The MBS identified 369 breeding-season locations of the Canada Jay during their intensive county surveys. These locations substantiated the breeding locations previously identified in northeastern and north-central Minnesota. They also verified locations in northern Carlton County, west to northeastern Becker and Marshall Counties, and several locations in Roseau County.

The results of the MNBBA totaled 734 records, with the overall distribution similar to those previously described **(A)**. Canada Jays were identified in 8.1 percent (385/4,737) of all surveyed blocks. Confirmed nesting was reported from 96 blocks, with especially high concentrations from the northern tier of counties ranging from Lake of the Woods County to Cook County. Confirmed nesting in southeastern Roseau County represented the first nesting record for that county. One probable nesting observation was made of a pair in southern Crow Wing County, a region with extensive black spruce and tamarack bog habitat. The species was not observed in Carlton County, which is adjacent to the northwestern-most county in Wisconsin where it has been confirmed nesting (Cutright et al. 2006).

Strickland and Ouellet (2020), in their review of the Canada Jay in North America, suggested only minor changes in its historical distribution. They cited retractions northward in the species' range in Alberta, Michigan, Algonquin Park in Ontario, Quebec, and Wisconsin. Land clearing in Alberta in the early 1900s was mentioned as a cause. Roberts

(1932) identified "two full-fledged young" at Mille Lacs, which may represent the most southerly observation of potential nesting in Minnesota. Unfortunately, the location is vague. Both Aitkin and Crow Wing Counties, where nesting has been confirmed, abut Mille Lacs Lake, where suitable bog habitat exists. It is logical to conclude that there may have been some breeding range contractions in Minnesota in the southern and western margins, but documentation is lacking.

BREEDING HABITAT

Canada Jays strongly favor coniferous and mixed coniferous–deciduous forests where spruce is typically present (Strickland and Ouellet 2020). Extensive counts in northern Minnesota indicated a strong preference for mature black spruce–tamarack and jack pine forest cover types, but they were also detected frequently in regenerating conifers, mixed swamp conifers (commonly with white cedar and black ash), aspen-birch, and aspen–spruce–fir habitats (Niemi, Howe, et al. 2016) **(B)**. Nest site data for Minnesota were limited. The largest data set on nest sites is from Algonquin Provincial Park, Ontario, where Strickland and Ouellet (2020) reported that 652 nests were located from 1969 to 2010! The distribution of these nests was 57 percent in black spruce, 28 percent in balsam fir, and 14 percent in white spruce.

POPULATION ABUNDANCE

The most recent estimate of the Canada Jay population in the United States and Canada was 27 million breeding adults (PIF 2021), with a Minnesota population of 28,000 individuals. Population trends for the Canada Jay based on the surveywide BBS routes in North America from 1966 to 2019 indicated a stable trend, but the overall pattern since 2010 has shown a slight decline **(C)**. For instance, Rosenberg, Kennedy, and colleagues (2016) calculated a Canada Jay population decline in North America of 19 percent from 1970 to 2014. This was likely due to significantly declining populations in Alberta, the Northern Rockies region, and the Atlantic Northern Forest region. The species has also been declining by −1.47 percent per year in the Chippewa and Superior National Forests from 1995 to 2020 (Grinde, Bednar, et al. 2020).

Data from Christmas Bird Counts analyzed for 1959 to 1988 indicated a significantly increasing number observed surveywide, but significantly declining populations were detected in Colorado, Maine, and Quebec. Numbers observed in Minnesota increased by 3.6 percent per year based on 31 Christmas Bird Counts over the same period (Sauer, Schwartz, and Hoover 1996); however, a recent analysis by Menebroeker and colleagues (2016) analyzed five Christmas Bird Counts from northern Wisconsin and found a significantly declining count of Canada Jays beginning in the early 1990s.

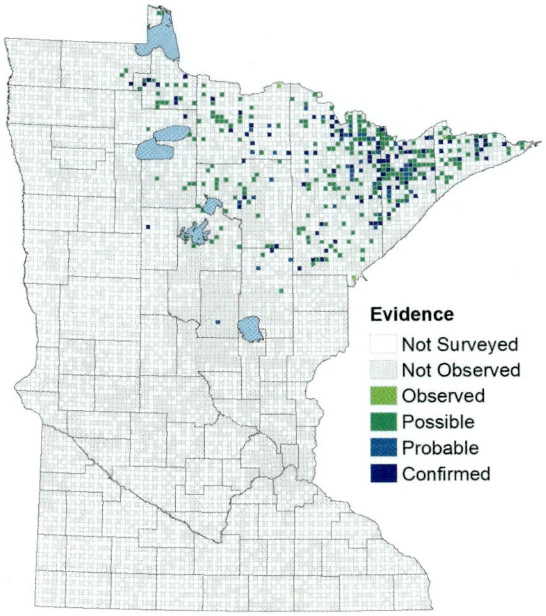

A. Breeding distribution of the Canada Jay in Minnesota based on the Breeding Bird Atlas (2009–13).

B. Canada Jays nest in many types of coniferous or mixed coniferous–deciduous forests, usually where some spruce trees are available, such as this habitat in northern Minnesota. Photo by Gerald J. Niemi.

CONSERVATION

Strickland and Ouellet (2020) emphasized that the primary threat to this species is climate change. They identified three major effects of these changes, the first of which is the northward advance of the boreal forest. Second, warmer and more variable fall, winter, and spring events (e.g., freeze-thaws) in Canada degrade perishable foods that are stored by Canada Jays as boli in the bark of trees to feed nestlings in the early spring when food is scarce; without this food source, survival of young is reduced (Waite and Strickland 2006; A. O. Sutton et al. 2021). Third, the irruption of the Rocky Mountain pine beetle (*Dendroctonus ponderosae*) in western North America has lowered the density of trees and subsequently decreased populations of the Canada Jay. There is current speculation that the Rocky Mountain pine beetle may move east across the Canadian boreal forest and potentially into Minnesota's coniferous forests.

NAS (2019) projected a moderate vulnerability for the Canada Jay's breeding range in North America with a 2°C warming scenario, including a 44 percent range loss and an 18 percent range gain, primarily in the far north of Canada and Alaska, by 2050. In Minnesota, the species was predicted to be extirpated from the state even under a 1.5°C warming scenario—an imminent threat because that is our current climate scenario.

The future of this species in Minnesota is dependent on the availability of coniferous forests. Unfortunately, that availability and the species' dependence on cold food storage in the form of boli are both projected to be affected by a warmer climate. Lower nest productivity in Algonquin Park at the southern edge of the species' range and at a similar latitude in Minnesota has already been noted (Strickland and Ouellet 2020). The long-term future of the Canada Jay in Minnesota is in jeopardy with climate change.

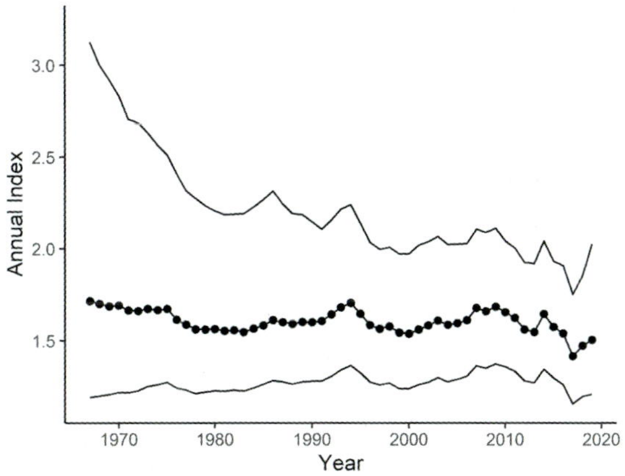

C. Surveywide breeding population trend and 95 percent credible interval (solid black lines) for the Canada Jay based on the 1967–2019 Breeding Bird Survey (Sauer et al. 2020).

BLUE JAY

Cyanocitta cristata

Minnesota Seasonal Status: Regular migrant and permanent resident. The Blue Jay was very abundant during the MNBBA.
Migration: Resident or short-distance migrant; 20 percent or less of the population may be migratory.
Food: Feeds primarily on arthropods and nuts; eggs and young birds are occasionally taken. A ground forager and foliage gleaner, the Blue Jay also catches flying insects in the air.
Nest: Cup nest in a deciduous or coniferous tree.
Conservation Concern: Assigned a moderate PIF Continental Concern Score of 9/20.

Photo by David Brislance.

Striking and unmistakable, the Blue Jay impressed many visitors to the New World as an exceptional bird. Audubon (1827–38) described it as a form of graceful, physical perfection, "arrayed by nature in a garb so resplendent." Explorer Jonathan Carver remarked, "Upon the whole this bird can scarcely be exceeded in beauty by any of the winged inhabitants of this or other climates" (Roberts 1932). Its dashing plumage, however, was not the only feature to garner attention. As Roberts noted, "Its intelligence and accomplishments also place it in the first rank." A socially complex species, the Blue Jay has a boisterous vocal repertoire known for its extensive diversity and mimicry, while its habit of caching acorns is considered a factor that aided the northern dispersal of oak trees following glaciation.

The species' Latin name, *Cyanocitta cristata,* can be translated to "a blue chattering bird with a crest," the latter being a prominent feature. Although the Blue Jay has a vast array of vocalizations, those most frequently heard are a loud, harsh *jay* or *jeer* call and a squeaky pump-handle call. The only vocalization akin to a song is known as the quiet and gentle "whisper song." Bent (1946) described it as "a potpourri of faint whistles and various low, sweet notes."

The Blue Jay is found in southern Canada from eastern British Columbia east to Newfoundland, and in the United States from the northern Great Plains, south to Texas, and east to the Atlantic coast; its highest breeding densities occur along the Gulf coast.

MINNESOTA BREEDING DISTRIBUTION

Roberts (1932) made two comments about the Blue Jay's status in Minnesota. First, he described it as a breeding species throughout the state. Less abundant in extensively forested landscapes, it was more common in "settled" areas, where it could take advantage of waste grains and cultivated fruits. Second, he provided evidence that many Blue Jays were permanent residents, noting that "if there is a winter southward movement, it is limited to only a portion of the birds." At the time, nesting evidence was available from eight counties stretching across southern, central, and northwestern Minnesota: Crow Wing, Goodhue, Hennepin, Isanti, Meeker, Polk, Sherburne, and Stearns.

In their updated account of the species' status, Green and Janssen (1975) again noted the Blue Jay's statewide distribution. They made no comment about its relative abundance in northern Minnesota during the summer but noted that it was less abundant in the region during the winter months. Although they considered the species to be both a permanent resident and a regular migrant, it was not clear if the birds present at a given site in the winter were the same birds that summered there or were instead migrants.

A few years later, Janssen (1987) identified 37 widely dispersed counties where nesting had been confirmed since 1970. Regarding the Jay's migratory status, he referenced studies that suggested many young birds in northern regions disperse from their natal sites, while adults in the north migrate but return the following spring. Hertzel and Janssen (1998) later confirmed nesting in 10 additional counties since 1970, while fieldwork conducted by the MBS also confirmed the Blue Jay's abundance in all but Minnesota's most intensively cultivated regions.

During the MNBBA, the Blue Jay was the tenth most common species, with a total of 6,321 records. It was reported in 64.2 percent (3,067/4,778) of the surveyed atlas blocks and was confirmed breeding in 422 blocks **(A)**. Observed in all 87 Minnesota counties, it was confirmed nesting in 69 counties. Twenty-eight of the counties with confirmed breeding were additions to the list published by Hertzel and Janssen in 1998. Blue Jays were present in nearly every priority block in the state (81.3 percent) except where croplands dominated. The MNBBA probability map predicts that moderate breeding densities are present nearly statewide, with scattered areas of lower abundance limited to western Minnesota **(B)**.

In Minnesota, the only significant change in the past one hundred years appears to be the Blue Jay's increased abundance in the northern forest region. At the continental scale, the species began expanding westward in both Canada and the United States in the 1940s, with the number of reports "dramatically increasing in the 1970s." Urbanization and the prevalence of bird feeders have likely aided this expansion (K. G. Smith, Tarvin, and Woolfenden 2020).

To this day, the true nature of the Blue Jay's migratory behavior remains somewhat elusive. K. G. Smith, Tarvin, and Woolfenden (2020) summarized what is undisputed: first, some individuals are present year-round; second, 20 percent or less of the population migrates; and, third, age does not appear to determine which birds migrate. The tendency for northern birds to migrate may also have subsided in response to the availability of food supplied by bird feeders (K. G. Smith, Tarvin, and Woolfenden 2020). Although this may influence populations residing in cities and towns, the number of Blue Jays observed

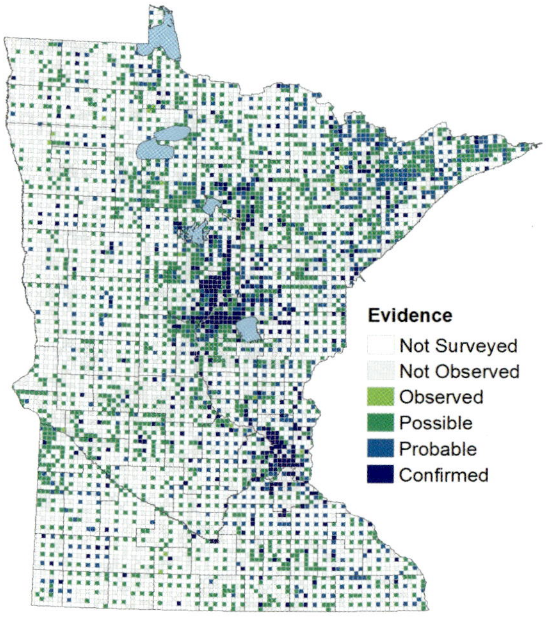

A. Breeding distribution of the Blue Jay in Minnesota based on the Breeding Bird Atlas (2009–13).

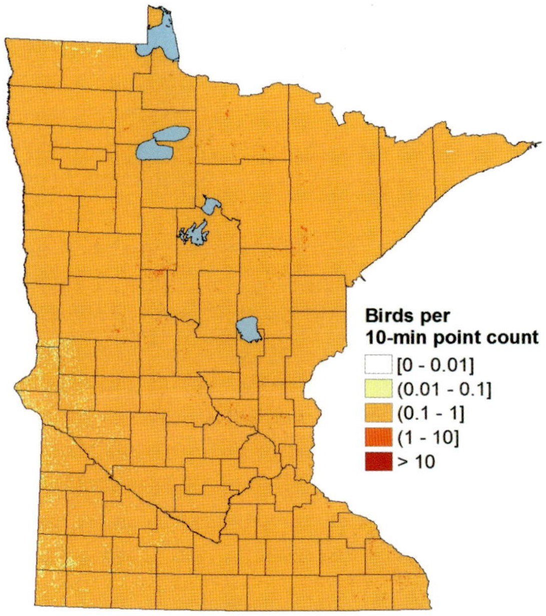

B. Predicted breeding distribution (birds detected per 10-minute point count) of the Blue Jay in Minnesota based on habitat, landscape context, and climate data gathered during the Minnesota Breeding Bird Atlas (2009–13) using the Generalized Linear Modeling method without an adjustment for detectability.

moving through Duluth each fall suggests that a significant number of northern birds still migrate south. From 2007 through 2016, the Hawk Ridge Bird Observatory documented an annual average of approximately 36,000 fall migrants (K. Bardon, pers. comm.).

BREEDING HABITAT

A woodland inhabitant, Blue Jays are found in upland and lowland forests, oak savannas, farmsteads, residential areas, and parklands **(C)**. Open woodlands and forest edges are preferred over dense forest stands (Kaufman 1996; K. G. Smith, Tarvin, and Woolfenden 2020). On national forest lands in northern Minnesota, Blue Jays reached some of their highest breeding densities in conifer-dominated stands, including mature lowland conifers in the Chippewa National Forest and mid-successional jack pine stands in the Superior National Forest (Niemi, Howe, et al. 2016).

Because acorns and seeds are common food items, forest stands with mast-producing trees and shrubs are often favored. Able to transport five or more acorns at a time, Jays eat and cache several thousand nuts each fall (Darley-Hill and Johnson 1981; W. C. Johnson and Adkisson 1985). The bird's affinity for caching acorns in the ground is considered an important factor responsible for the rapid dispersal of oaks following the latest glaciation. Some biologists also believe that the Jay's annual movements are best explained by its response to annual mast production, but studies have failed to demonstrate a correlation at the local or regional scale (K. G. Smith, Tarvin, and Woolfenden 2020).

POPULATION ABUNDANCE

PIF (2021) estimated a population of 17 million breeding adults in the United States and Canada and a Minnesota population of 480,000. Because the Blue Jay's vocalizations are not strongly tied to territory defense, a statewide population estimate was not generated with MNBBA data.

The BBS demonstrates a significant Blue Jay population decline across the survey region, averaging −0.59 percent per year from 1966 to 2019 (Sauer et al. 2020). Declines are prevalent in the central and southern states, while increases occur across Canada, with statistically significant increases occurring from Manitoba east through the Maritime Provinces. As with its northern neighbors, Minnesota's Blue Jay population shows a significant increase, averaging 0.48 percent per year **(D)**. Significant population increases were also observed in the Superior and Chippewa National Forests from 1995 to 2020 (Grinde, Bednar, et al. 2020).

Blue Jays were among the first species to be impacted by the West Nile virus outbreak that began in New York City in 1999, which resulted in widespread population declines throughout the eastern United States (K. G. Smith, Tarvin, and Woolfenden 2020). First detected in Minnesota in 2002, the impact was especially noticeable in the Twin Cities metropolitan region. The factors responsible for the Blue Jay's long-term decline, however, are less clear, particularly given their adaptability to residential areas and bird feeders. In Minnesota, an increase in mature forests as timber harvest declines in the northern counties is likely contributing to the state's increasing population (Grinde, Bednar, et al. 2020).

C. An oak savanna and adjacent woodlands in southeastern Minnesota provide the Blue Jay with mature trees for nesting and an abundance of mast for foraging. Photo by Lee A. Pfannmuller.

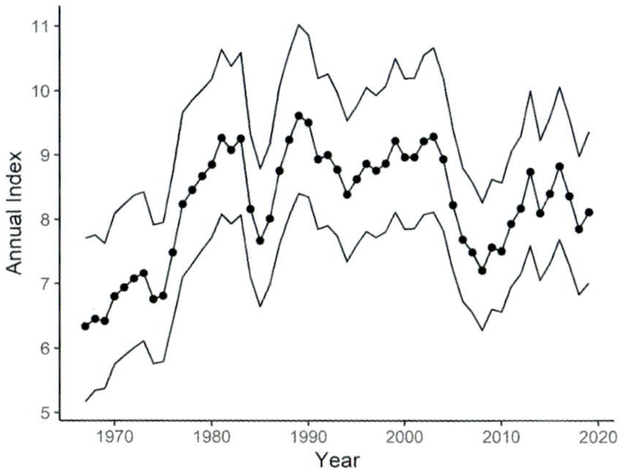

D. Breeding population trend and 95 percent credible interval (solid black lines) for the Blue Jay in Minnesota based on the 1967–2019 Breeding Bird Survey (Sauer et al. 2020).

CONSERVATION

Despite its decline, the Blue Jay remains an abundant and widespread species that has readily adapted to human-modified landscapes. It has a low conservation priority (PIF 2021); even its vulnerability to climate change is considered minimal. Models predict, however, that the western border of its range will recede eastward as temperatures rise, eventually extirpating the species from North Dakota and portions of northwestern Minnesota by 2080 (NAS 2019).

Unfortunately, the Blue Jay's reputation is not always that of an exalted member of the bird community. Its predation of songbird eggs and young has been cited as contributing to the decline of some migratory songbirds. Although evidence regarding their large-scale impact is not entirely clear, cameras have demonstrated that they can be a major nest predator at the local level (K. G. Smith, Tarvin, and Woolfenden 2020).

Described by Audubon (1827–38) as a study in contrasts, whose resplendent beauty belies its sometimes mischievous nest-robbing behavior, the Blue Jay will likely remain a major inhabitant of Minnesota's woodlands and forests long into the future.

BLACK-BILLED MAGPIE

Pica hudsonia

Minnesota Seasonal Status: Permanent resident. The Black-billed Magpie was an uncommon species during the MNBBA.
Migration: No regular migratory movements but individuals may disperse after the breeding season and during the winter.
Food: Carrion forms a major portion of the diet, as do grains, small mammals, ground-dwelling arthropods, and parasites gleaned from the hides of ungulates; primarily a ground forager.
Nest: Placed in a tree or tall shrub, the large, elaborate nest often includes a dome that covers the entire nest and provides one or more openings for the birds to enter.
Conservation Concern: Assigned a moderate PIF Continental Concern Score of 9/20.

If your first memorable encounter with the Black-billed Magpie was in the open country of the western Great Plains, you may be amazed to find it in the Aspen Parklands and mixed forests of northern Minnesota. Although one might think it a wayward wanderer, the bird has been well established as a permanent resident in the state since nesting was first documented in 1951.

Despite one's surprise, the Magpie's identity is unmistakable. Its striking appearance is marked by a black bill, head, chest, and back, and a white belly. *Pica,* the species' genus, refers to its black-and-white plumage. The wings and impressively long tail have an iridescent blue-green sheen that sparkles in the sun. Broad patches of white occur in the wing's upper scapulars, while the white outer primaries with black tips are obvious in flight. Like other corvids, males and females are identical and often announce their presence with a harsh, chattering alarm call.

The Black-billed Magpie has been a character in Native American lore for centuries, where its expressive personality may be depicted as anything from a loyal ally to a deceptive busybody. In some tribes, wearing a Magpie feather is a sign that you are bold and have little fear. The bird's inquisitive and voracious habits are legendary. On the Lewis and Clark expedition, Magpies entered the explorers' tents, snatched meat from their plates, and stole their dressed game (Baird et al. 1874).

Photo by Kathy Munsel.

A highly social species, Black-billed Magpies occur along the southern Pacific coastline of Alaska south across the southern Canadian provinces, from British Columbia east to Ontario, and in the United States from the Pacific Northwest east to Kansas, Nebraska, the Dakotas, and northern Minnesota.

MINNESOTA BREEDING DISTRIBUTION

The Black-billed Magpie has an interesting history in Minnesota. Roberts (1932) considered the bird a western "straggler," appearing only from late September through mid-February before returning west to its breeding grounds in the Dakotas. Often observed traveling together in small groups, Magpies were documented from 24 counties that were widely "distributed from the Canadian boundary on the north to the Iowa line on the south, and as far east as Pine and Goodhue Counties." The winter of 1921–22 was especially noteworthy. Roberts reported that a "remarkable invasion" occurred in many states, including Minnesota, where 51 Magpies were reported. This exceeded the cumulative number of records in all prior years. Nesting reports, however, were limited to two unconfirmed records from Lake of the Woods County (1920) and Marshall County (1921).

Twenty-five years later, A. B. Erickson (1957) provided a comprehensive update on the species' status. Records were now available from 43 counties dispersed across all but extreme northeast and southeast Minnesota, and the state's first confirmed nesting record was documented in 1951 in Beltrami County. Major fall invasions, averaging 20 to 50 Magpies once every ten years, continued; in between these years, birds were present in far fewer numbers. Summer observations remained scarce until 1971, when there were three confirmed nesting reports in Marshall County (Burger and Howe 1971).

In 1975, Green and Janssen classified the Magpie as a casual summer resident. Major fall invasions continued approximately every ten years, and nesting had been documented in two additional counties: Clay County (1972) and Roseau County (1973). Only a few more years passed before Magpies became firmly established as a nesting species in northwestern, north-central, and portions of northeastern Minnesota, prompting Janssen (1987) to upgrade their status to a regular, permanent resident. By 1998 nesting had been confirmed in seven counties since 1970: Aitkin, Kittson, Marshall, Pennington, Polk, Roseau, and St. Louis (Hertzel and Janssen 1998). By 2014, the MBS had documented the species' increasing presence in northern Minnesota, including a significant concentration of records in Clearwater County and southern Beltrami County that accounted for nearly one-third of the survey's 195 breeding-season observations.

During the MNBBA, Black-billed Magpies were most abundant in the Tallgrass Aspen Parklands Province and the far western boundaries of the Laurentian Mixed Forest Province but were regular as far east as St. Louis County and as far south as Aitkin County. There was also one observation in southern Stearns County. Although the overall limits of its breeding range were similar to that depicted by the MBS, its abundance in the Aspen Parklands

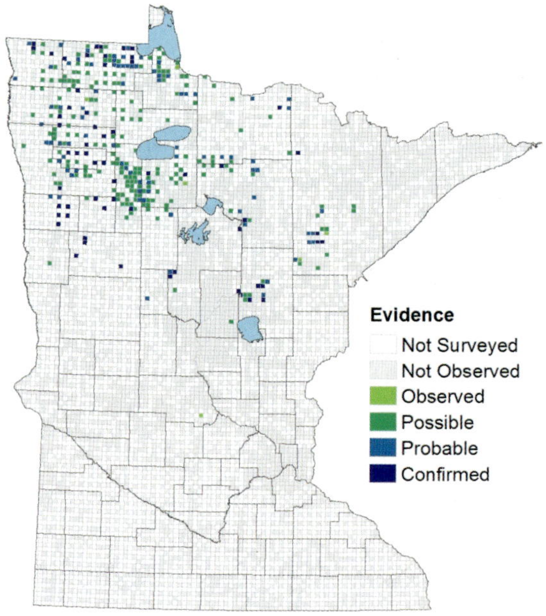

A. Breeding distribution of the Black-billed Magpie in Minnesota based on the Breeding Bird Atlas (2009–13).

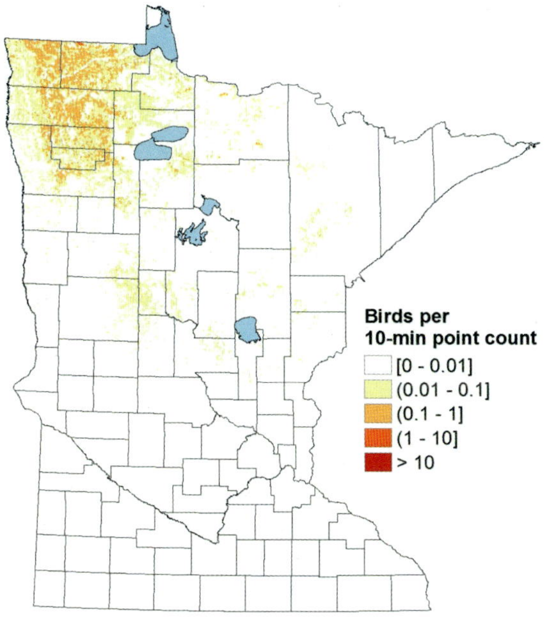

B. Predicted breeding distribution (birds detected per 10-minute point count) of the Black-billed Magpie in Minnesota based on habitat, landscape context, and climate data gathered during the Minnesota Breeding Bird Atlas (2009–13) using the General Linear Modeling method without an adjustment for detectability.

appears to have significantly increased. Participants reported 551 records of Black-billed Magpies from 7.2 percent (345/4,781) of the surveyed atlas blocks and confirmed breeding in 60 blocks **(A)**. The birds were observed in 22 of Minnesota's 87 counties and were confirmed breeding in 17 counties. Nine counties were additions to the list published by Hertzel and Janssen (1998). The probability map **(B)** closely mirrors the Black-billed Magpie's current distribution but also predicts that scattered pockets of suitable habitat may occur a little farther south into Ottertail, Todd, and Morrison Counties.

In a remarkably short time, the Black-billed Magpie has gone from being a rare straggler in the state to a permanent resident. This change is likely part of the species' recovery from earlier population declines and range contractions. Long associated with bison for the parasites they provided, the Magpie's range receded as this large ungulate was nearly exterminated (Trost 2020). At the same time, the bird's consumption of carrion, its opportunistic feeding on game birds and poultry, and its habit of picking parasites from livestock, often repeatedly pecking open wounds, led to its widescale persecution. In North Dakota alone, bounty was paid on approximately 10,500 Magpies each year from 1950 through 1956 (A. B. Erickson 1957). Despite these early setbacks, the bird has expanded its range in a number of areas, including in Minnesota and areas of Canada where development has opened the boreal forest (Trost 2020).

BREEDING HABITAT

In the Great Plains, the Black-billed Magpie is an open-country bird. During the breeding season, it selects riparian areas with dense thickets of trees and shrubs. Shelterbelts associated with farmsteads, particularly with livestock, also provide suitable habitat (Trost 2020). Many of the Magpies nesting in northern Minnesota are nesting and roosting close to small settlements or in thickets close to small farms **(C)**. Because it is a ground forager, open grasslands and fields are an important feature of their habitat.

POPULATION ABUNDANCE

PIF (2021) used BBS data to generate a population estimate of 6 million breeding adults in the United States and Canada and a statewide population of 35,000. Across the BBS survey region, Magpies have experienced a significant population decline, averaging −0.89 percent per year from 1966 to 2019; in the past thirty years, however, the data suggest a more stable trend **(D)**. The decline is most pronounced in the Central Plains, stretching from the Dakotas south to Kansas and west to Oregon and California (Sauer et al. 2020). Factors responsible for the downward trend are unknown, but a reduction in brood size is apparently widespread owing to starvation, related directly either to food availability or to disturbance at the nest site. Trost (2020) suggests that the birds' sensitivity to disturbance results from years of persecution. Predation on young is also high, averaging 24 to 30 percent (Trost 2020).

The continental decline contrasts strongly with the population trend in Minnesota. Although the species' relatively low abundance in the state compromises statistical analysis of the statewide trend, the species has clearly experienced a remarkable increase in numbers, especially in the past thirty years (Sauer et al. 2020). Its expansion into Minnesota may be a consequence of increased protection throughout the species' range. Populations in neighboring North and South Dakota, however, have experienced significant declines since 1966.

C. In northern Minnesota, the Black-billed Magpie is frequently found on small farmsteads with a grazed cattle pasture for foraging and a shelterbelt or shrubby thicket for nesting. Photo by Lee A. Pfannmuller.

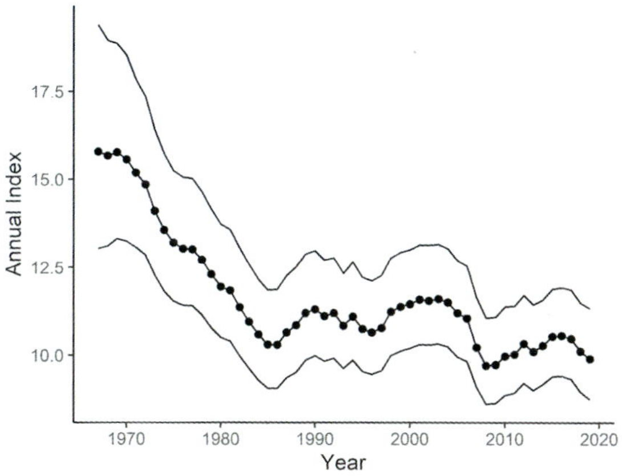

D. Surveywide breeding population trend and 95 percent credible interval (solid black lines) for the Black-billed Magpie based on the 1967–2019 Breeding Bird Survey (Sauer et al. 2020).

CONSERVATION

In the United States, Magpies are no longer persecuted as "vermin" and are protected by the Migratory Bird Treaty Act. Nevertheless, some prejudices remain, and the bird is still an unprotected species in Canada. Its long-term population decline is the primary factor prompting its designation as a moderate conservation priority (PIF 2021).

Warming temperatures may pose a significant challenge to the species. Despite its adaptability to human-modified landscapes, the Magpie's range is relatively restricted. Its preference for cooler climates in western regions increases its vulnerability. It is classified as moderately vulnerable under a 2°C warming scenario; NAS (2019) projected a 39 percent range loss by 2050 and only a 16 percent gain northward. Even under an imminent 1.5°C warming scenario, models predict the species will be extirpated from Minnesota, North Dakota, and much of Manitoba.

Yet, like many members of the Corvid family, the Black-billed Magpie may be more adaptable than the models predict. Opportunities to observe this remarkable bird in Minnesota continue to increase and, unlike many birds, only a brief glimpse of its striking plumage is necessary to positively identify it as a Black-billed Magpie!

AMERICAN CROW

Corvus brachyrhynchos

Minnesota Seasonal Status: Permanent resident and migrant. The American Crow was very abundant during the MNBBA.
Migration: Short-distance migrant to year-round resident. In Minnesota, American Crows are considered permanent residents, although a pulse of migrants each spring and fall may represent seasonal movements of breeding residents and/or of Canadian birds.
Food: Omnivorous ground forager; can be an economic pest when large roosts are located near agricultural crops.
Nest: Large, bulky bowl constructed primarily of dead twigs; usually placed high in a tree.
Conservation Concern: A game species in Minnesota; assigned a low PIF Continental Concern Score of 6/20.

Drive along any country road, interstate highway, or city street and you are bound to encounter American Crows. Usually viewed with passive indifference or outright hostility, the bird deserves our admiration just for its astonishing ability to adapt and thrive despite widespread persecution. Nearly two centuries ago, John James Audubon (1827–38) recognized that "scarcely one of his race would be left in the land, did he not employ all his ingenuity, and take advantage of all his experience, in counteracting the evil machinations of his enemies." Landscape changes that followed in the wake of European settlement were an added benefit. Open habitats, crop waste, and the persecution of many of the species' predators, including hawks and owls, all improved conditions for this cunning bird.

The American Crow's shiny black plumage and emphatically delivered *caw* call are its most prominent features and are usually sufficient to ensure positive identification. But in northern Minnesota, the resident Common Raven, another large black bird, presents some identification challenges. The latter species is nearly 17 cm larger, has a deeper, hoarser voice, and a thicker, chunkier bill. The Crow's Latin name, *Corvus brachyrhynchos,* actually translates to "short-billed raven." In flight, the raven's wedge-shaped tail is another diagnostic feature, differing from the Crow's nearly square, slightly rounded tail.

Photo by David Brislance.

An intelligent bird with strong family bonds, the American Crow is broadly distributed across North America. In the United States, it is absent only from large regions of the arid Southwest. An abundant species throughout much of its breeding range, it is common throughout Minnesota, especially in urban and agricultural settings.

MINNESOTA BREEDING DISTRIBUTION

Many early accounts of the American Crow focused more on its feeding habits and the damage it inflicted on corn and grain crops than on its abundance and distribution. Roberts's (1932) account was no different, discussing the relative merits of organized Crow hunts, bounties, and local trapping efforts. The species was, he noted, a common summer resident throughout the state, clearly having increased in abundance in recent decades. "Instead of diminishing in numbers it has thriven and multiplied as man has thriven and multiplied, and in the same places." At the time, nesting evidence was available from nine counties scattered across southern, central, and northern Minnesota: Aitkin, Goodhue, Hennepin (Minneapolis), Isanti, Kittson, Lincoln, Polk, Sherburne, and Stearns.

In the years that followed, the Crow remained a common breeding resident throughout the state (Green and Janssen 1975; Janssen 1987). Janssen (1987) identified 26 counties where breeding had been confirmed since 1970; Hertzel and Janssen (1998) added eight more counties to the list. The MBS identified 2,599 breeding-season locations in all corners of the state, with the exception of the most intensively cultivated counties, which were not the focus of survey efforts.

The MNBBA was further testament to the Crow's ubiquitous distribution. It was the sixth most commonly reported species during the atlas; observers reported 7,275 records in 73.0 percent (3,570/4,889) of the surveyed atlas blocks and confirmed breeding in 728 blocks **(A)**. Observations were reported in all 87 counties, and breeding was confirmed in 78 counties. Seven of the nine counties where breeding was not confirmed were in the far western regions of the state; two were in south-central and southeastern Minnesota. Breeding was most frequently confirmed in major metropolitan areas, including the Twin Cities, Brainerd, and Duluth. The only area where the species was relatively scarce was Koochiching County, where large peatlands dominate the landscape.

Although their breeding distribution in Minnesota remains similar to that described by Roberts nearly one hundred years ago, Crows are certainly more common in northern Minnesota during the winter months than they were in earlier times (Roberts 1932). Now it is common for Crows to be reported in northeastern and north-central Minnesota from December through February (MOU 2021b). Landfills and feedlots, even when the ground is covered in snow, provide a reliable food source throughout the coldest winter months. As average winter temperatures continue to rise, this ground forager also finds more opportunities to forage in natural habitats.

In addition to extending its winter distribution in Minnesota, the American Crow expanded north into Canada during the twentieth century, following settlement and the suppression of prairie fires. Range-wide, it was a rare inhabitant in cities and

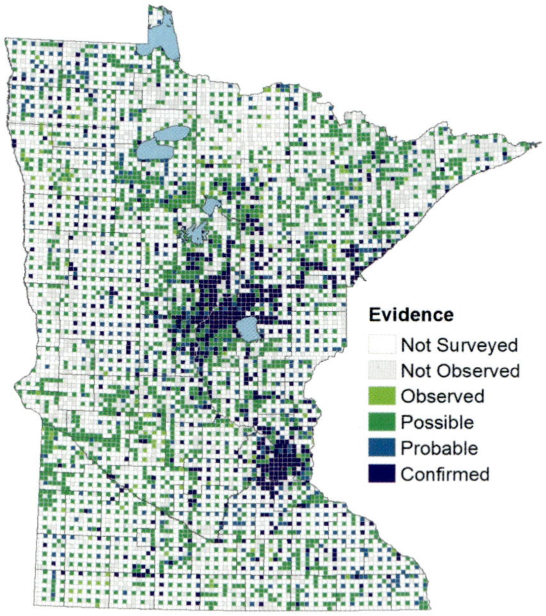

A. Breeding distribution of the American Crow in Minnesota based on the Breeding Bird Atlas (2009–13).

towns until the latter half of the 1900s, but tree plantings and the prohibition of hunting in residential areas facilitated its expansion. Considered uncommon throughout much of the northeastern United States in the late 1800s and early 1900s, it now is an abundant and widespread species throughout that region as well. The species' adaptation to human-modified environments certainly has aided its population growth and range expansion throughout North America (Verbeek and Caffrey 2020).

BREEDING HABITAT

Extremely adaptable, the American Crow is found in a wide range of natural and modified habitats, including woodlots, forest edges, riparian areas, city parks, cemeteries, and farmsteads **(B)**. Large landscapes devoid of trees, such as grasslands, are generally avoided. Suitable habitat needs to meet only two broad requirements: open areas for foraging on the ground, and scattered trees for nesting, roosting, and resting (Verbeek and Caffrey 2020).

Although Crows are also uncommon in extensively forested landscapes, they may be found in riparian areas along rivers and streams and in open areas, such as small settlements, campgrounds, and forest openings. In the four national forests in northern Wisconsin and Minnesota, Crows were most abundant in local towns and in habitats associated with water (Niemi, Howe, et al. 2016).

POPULATION ABUNDANCE

The US and Canadian population of American Crows is estimated at 28 million breeding adults, while Minnesota supports approximately 770,000 birds (PIF 2021). The Crow's complex behavior during the breeding season, moving between nighttime roosts and daytime territories, prohibited the generation of a reliable population estimate using MNBBA data. The PIF estimate seems low, however, given the species' statewide abundance.

Since the BBS began in 1966, data demonstrate a slowly increasing population across the survey region unil the year 2000, when numbers were significantly impacted by the outbreak of the West Nile Virus (WNV). Transmission rates were so high in the eastern United States that some local populations declined by nearly 70 percent (L. M. Reed et al. 2009). A small recovery that began in 2008 was short-lived before the population continued its downward trend a few years later. Reforestation of abandoned farmland in the northeastern states is cited as a possible cause (Verbeek and Caffrey 2020).

Farther west, in Minnesota, the decline was not as severe. Indeed, a decade after the WNV outbreak began in the United States, researchers discovered that the virus declined in virulence as it moved west (Koenig et al. 2010). Nonetheless, thousands of dead Crows were reported to the Minnesota Department of Health (2004) during the peak of the outbreak, halting the growth of Minnesota's increasing Crow population. And, as was observed across the BBS survey region, numbers continued to decline once the outbreak subsided **(C)**. Intensification of agricultural practices that eliminated pastures, windbreaks, and shelterbelts may be contributing to the downward trend as BBS data reveal a decline in Crow numbers across southwestern and south-central Minnesota (Sauer et al. 2020).

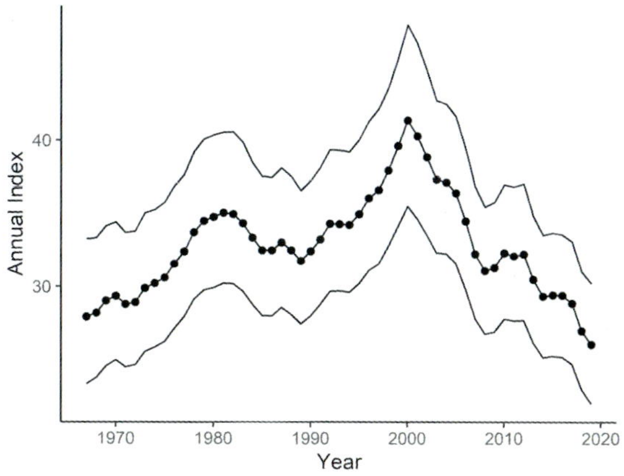

C. Breeding population trend and 95 percent credible interval (solid black lines) for the American Crow in Minnesota based on the 1967–2019 Breeding Bird Survey (Sauer et al. 2020).

CONSERVATION

Long the target of human persecution, the American Crow has survived and prospered despite many focused attempts to reduce local populations. Efforts have been most intense at winter roost sites, where Crows may congregate in numbers ranging from just a few hundred birds to 500,000 and larger (Gorenzel and Salmon 1992). Concerns range from crop depredation in agricultural landscapes to noise, defecation, and overall disturbance in urban areas.

Nevertheless, local efforts to reduce damage are largely unsuccessful. A 1922 report in Minnesota is typical, noting that the "enormous kill" by sportsmen's clubs did not reduce local Crow populations but instead resulted in the loss of "an immense number of useful Hawks, Owls and other large birds" (Roberts 1932). In Oklahoma, from 1934 to 1945, a total of 127 roosts were dynamited, resulting in the loss of more than 3.8 million birds. Despite the carnage, there was no noticeable reduction in damage to agricultural crops (H. G. Hanson 1946).

Such practices are largely of a bygone era. Today, when local control work is needed, efforts focus on techniques to exclude or frighten the birds before lethal control is initiated. Since 1989, it has also been legal to hunt Crows in Minnesota. During the 10 seasons from 2010 to 2019, the number of Crow hunters in Minnesota each year averaged 8,350 while the average annual harvest was approximately 63,100 birds (Davros and Dexter 2019).

With an abundant, widespread population that has prospered with the increase in human populations, the American Crow is poised to remain a plentiful member of the state's avifauna for many years to come. Even warming temperatures are not likely to have a major impact on this adaptable species (NAS 2019).

B. Like many habitats, a farm's woodlot bordering an open pasture provides the two essential requirements for the American Crow: open ground for foraging and trees for nesting and roosting. Photo by Lee A. Pfannmuller.

COMMON RAVEN

Corvus corax

Minnesota Seasonal Status: Permanent resident. The Raven was common during the MNBBA.
Migration: Sedentary, but movements into central and southern Minnesota are frequently noted, especially along the North Shore of Lake Superior.
Food: Omnivorous; forages on carrion, insects, grains, fruit, and garbage.
Nest: Highly variable; cliffs, trees, buildings, or wherever there is a suitable substrate.
Conservation Concern: Assigned a low PIF Continental Concern Score of 6/20.

Few bird species are as legendary as the Common Raven. Various folklore includes the two Ravens that attended to Odin, the chief god of the Norse. In Wales, the legendary Owein was accompanied by an army of Ravens. In Ireland, Cú Chulainn had two magic Ravens that warned him of impending danger. In the Bible, it was the Raven that Noah first sent forth from the Ark. The ebony and stately Raven gained notoriety forever in Edgar Allan Poe's poem "The Raven," and its famous quote, "Nevermore." Bernd Heinrich in his book *Ravens in Winter* suggests that the Common Raven is one of the world's most intelligent and fascinating birds because of its solitary and complex social behavior.

The Raven superficially looks like a large Crow, but beyond its larger size it can be distinguished from the Crow by its wedge-shaped tail, its prominent bill, and its habit of soaring or flying with shallower wingbeats than a Crow. Both sexes are similar. The Raven also has a variety of vocalizations; the one most commonly heard is a loud *kruuuk*.

The Raven is found widely distributed throughout the Northern Hemisphere in North America, Europe, and Asia. It has southward extensions in the Appalachian Mountains and in many western states, where its highest densities have been reported in Nevada, western Utah, and southern California.

Photo by David Brislance.

MINNESOTA BREEDING DISTRIBUTION

According to Roberts (1932), the Common Raven has had a highly variable and dubious distribution in the state. He summarized the species as "once a common, or even locally abundant, winter visitant in northern Minnesota as far south . . . as Otter Tail and Wadena counties." Most accounts by Roberts were of fall and winter observations, including those in the late 1800s in Hennepin, Pipestone, and Sherburne Counties. He also described it as common around lumbering camps, fishing stations, and "pineries" and even a nuisance in 1876–77 about hunting camps in Otter Tail and Wadena Counties. In 1895, it was labeled as being as "numerous in winter as Crows are now in the summer."

Roberts never recorded a nest of this species in the state, but forest rangers told him that it nested on cliffs in the boundary lakes of Saganaga and Gunflint as well as about Frear Lake in Cook County. Summer observations were so remarkably rare that he included sightings, such as at Cass Lake and Grand Rapids, Itasca County, in August 1916; three were seen in northern Itasca County on July 12, 1928; three were seen in eastern Pine County in July 1919; and one was seen near Eagle Mountain, Cook County, on July 6, 1922. In Roberts's 1936 revision, he summarized a letter from Warden C. E. Cooke, who had observed a nesting pair of Ravens on Drywood Island, Rainy Lake, St. Louis County, in which he was "quite sure" it was in 1929. He also reported an additional observation of a Raven 2 miles north of Duluth by Frank Craighead Jr. on July 15, 1935.

Roberts (1932, 1936) concluded, "the Raven is one of our disappearing birds and may be approaching extinction in the state." He never mentioned any reasons for its disappearance but recognized that many were shot or caught in wolf traps.

Green and Janssen (1975) described it as a resident in the north-central and northeastern regions. Most numerous in the Superior National Forest, it was not known in Itasca State Park until the 1960s, but it was found as far south as Mille Lacs and Pine Counties. They included confirmed nests in Carlton, Cook, Lake, and St. Louis Counties and inferred nesting in Koochiching County. They described its winter distribution as common in the northern regions in the 1800s and rare from about 1900 to the 1940s but increasing since the 1950s.

Janssen (1987) also commented on its range expansion since the 1950s, especially "westward and very slowly southward." He included confirmed nesting in eight counties since 1970: Cook, Koochiching, Lake, Lake of the Woods, eastern Marshall, Roseau, St. Louis, and Wadena. Hertzel and Janssen (1998) further documented its expansion with four additional nesting records confirmed since 1970 in Aitkin, Beltrami, Isanti, and Pine Counties.

The MBS documented extensive breeding-season locations throughout northern and north-central Minnesota. Surveyors detected 555 records extending westward to Becker, Clearwater, Kittson, Mahnomen, and Polk Counties. In addition, they recorded locations southwest to Todd County and south to Chisago and Mille Lacs Counties.

Breeding detections were extended even farther by the MNBBA, which included 2,162 records **(A)**. Confirmed nesting was recorded from 226 blocks, including confirmed county records from 16 new Minnesota counties that extended as far south as Washington. In total, this species was detected in 27.7 percent of all surveyed blocks in the state (1,318/4,763). Probable nesting was also documented from 127 blocks, including south to Chisago County and west to Otter Tail County; breeding-season observations were made in Cottonwood and Winona Counties. On June 24, 2010, a record submitted to the MOU (2021a) reported a young short-tailed Raven with an adult in Anoka County.

The Raven probability map emphasized a widespread distribution across northeastern, north-central, and northern Minnesota **(B)**. Slightly lower populations were predicted south to northern Isanti and Washington Counties, west to eastern Becker and northeastern Otter Tail Counties, and northwest to Kittson County. Pockets of its presence were also predicted in the Hardwood Hills Subsection in

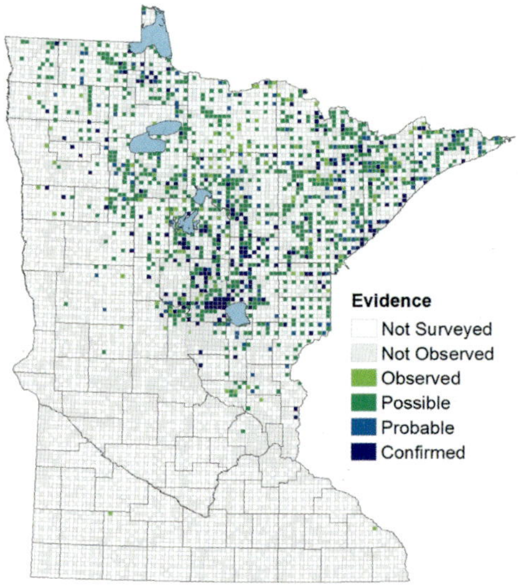

A. Breeding distribution of the Common Raven in Minnesota based on the Breeding Bird Atlas (2009–13).

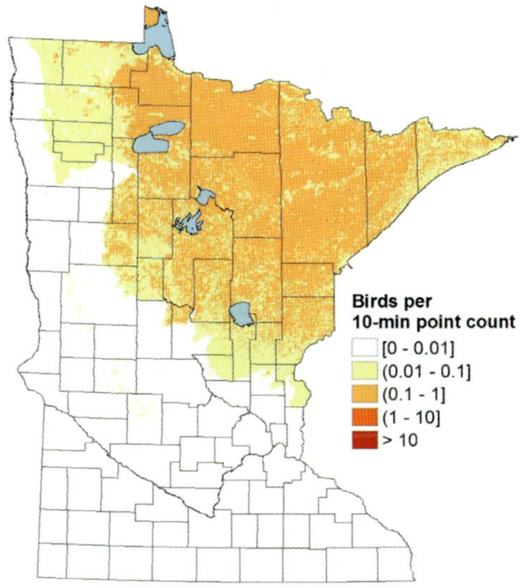

B. Predicted breeding distribution (birds detected per 10-minute point count) of the Common Raven in Minnesota based on habitat, landscape context, and climate data gathered during the Minnesota Breeding Bird Atlas (2009–13) using the Generalized Linear Modeling method without an adjustment for detectability.

western Minnesota, where possible and probable nesting records from the MNBBA existed.

The current distribution of the Common Raven represents a substantially broader breeding distribution than what has been suggested for the past 100–150 years. The species has made a remarkable recovery since Roberts's comment in 1932 that it is "approaching extinction." Declines in the late 1800s and early 1900s were likely associated with extensive human persecution via shooting, poisoning, trapping, conversion of vast areas of habitat to agricultural and urban landscapes, and possibly negative interactions with American Crows at carrion feeding sites. The Migratory Bird Treaty Act of 1918 greatly aided the species' recovery with reductions in human persecution as well as the loss of fear of humans with additional feeding (garbage and roadkill) and nesting opportunities (e.g., towers, poles, bridges, and abandoned buildings).

Boarman and Heinrich (2020), in their review of the Common Raven in North America, identified "shooting, poisons, baited traps, and disappearance of bison" as primary reasons for changes in its distribution, especially on the prairies. They point out that its disappearance in the New England states was largely attributed to deforestation, conversion of land to agriculture, and increases in American Crow populations.

BREEDING HABITAT

The Common Raven is a broadly distributed species found in many forest cover types, prairies, tundra, and grasslands. Many Raven populations prefer human settlements, while others prefer wilderness and avoid human settlements. In Minnesota, the species is largely confined to heavily forested areas and human settlements, especially in the northern portion of the state (Bruggers 1988). It will nest in a variety of locations, including trees and cliffs **(C, D)**.

POPULATION ABUNDANCE

The US and Canadian population of the Raven has been estimated as 8.3 million adults (PIF 2021), with a Minnesota population of 37,000 adults. It is Holarctic in its distribution, so its global population is substantially higher (29 million) when northern Europe and Asia are included. The MNBBA did not attempt a population estimate in Minnesota because individuals have large home ranges and wander extensively.

The BBS population trends for Minnesota indicated a significantly increasing population of 3.23 percent per year from 1967 to 2019 **(E)**. Grinde, Bednar, and colleagues (2020), however, reported a significant decline of –0.96 percent per year from 1995 to 2020 in the Chippewa and Superior National Forests of northern Minnesota. The discrepancy is likely due to the fact that the statewide BBS detects that Raven populations are expanding southward while the National Forest Monitoring Program, which more intensively covers the species' primary breeding range in northern Minnesota (Figure B), detects a slight decline.

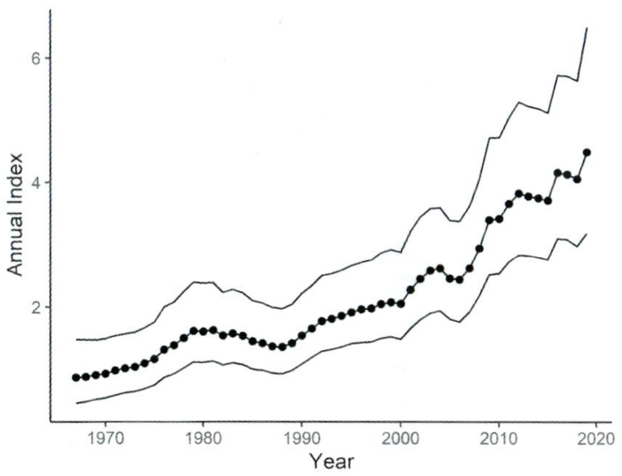

E. Breeding population trend and 95 percent credible interval (solid black lines) for the Common Raven in Minnesota based on the 1967–2019 Breeding Bird Survey (Sauer et al. 2020).

C. Common Ravens construct large nests in many forest settings, such as this mixed deciduous–coniferous forest in northeastern Minnesota. Photo by John C. Green.

D. Common Ravens also frequently use cliffs as a nesting site, as shown here along the North Shore of Lake Superior. Photo by John C. Green.

CONSERVATION

Despite the modest decline in northern Minneosta, there are few conservation concerns for the Raven because its overall population is increasing and it has a wide global distribution. The historical persecution of Ravens continues in some agricultural settings, where they consume economically valuable crops such as grains, almonds, and pistachios or prey upon game species and waterfowl eggs (Boarman and Heinrich 2020). However, in the United States the Common Raven is protected by the Migratory Bird Treaty Act of 1918.

Under a 2°C warming scenario, NAS (2019) projected low vulnerability for the Common Raven's breeding distribution in North America, with a 6 percent loss of its breeding range and a 10 percent gain in the far north in Canada and Alaska by 2050. In this scenario, breeding range retractions in Minnesota were predicted in southern and western regions as well as in scattered areas in the central part of the state.

The recovery of the Raven is a tribute to the propensity of our native species to recover if given the appropriate conditions. In the case of the Raven, we have simply stopped the indiscriminate shooting, poisoning, and trapping of this species and now can enjoy living with this intelligent and fascinating creature.

BLACK-CAPPED CHICKADEE

Poecile atricapillus

Minnesota Seasonal Status: Permanent resident. The Black-capped Chickadee was abundant during the MNBBA.
Migration: Nonmigratory, but erratic, short-distance movements southward can occur.
Food: Omnivore; largely insects, spiders, seeds, and berries gleaned from foliage or feeders.
Nest: Cavity nester in trees; often self-excavated in a natural cavity; occasionally a nest box or a cavity excavated by a woodpecker.
Conservation Concern: Assigned a low PIF Continental Concern Score of 7/20.

Photo by David Brislance.

The Black-capped Chickadee is one of the most well-known backyard birds in Minnesota; it is especially fond of the free banquets offered at feeders. Bent (1946) describes them as "cheerful little birds. When we watch a flock of them in winter they remind us of a group of happy, innocent little children playing in the snow." The species is widely distributed across the forested and urbanized areas of the midwestern and northern tier of US states and Canada and south in the Rockies to northern New Mexico. The highest densities are observed in the northeastern and upper midwestern portions of North America.

The plumage of the Black-capped Chickadee is among the most well-known of all songbirds. Both sexes are similar. They are small, compact birds with a black crown and forehead, black bib, white cheeks, gray back and wings, and whitish underparts. Its common vocalization is the well-known *chicka dee-dee* call, which is the origin of its common name. However, its territorial song, *feebee,* is less often heard because it is most frequently emitted in late winter and early spring, when fewer observers are present outdoors.

Black-capped Chickadees have a curious social system. During the breeding season, pairs are formed and territories are maintained as with most songbirds, but during the nonbreeding season, they will flock into groups and defend a home range. Groups are structured in what is termed "linear dominance hierarchies," in which males dominate females and older males generally dominate younger birds (Howitz 1981). Females select the dominant males because such pairings tend to have greater hatching success and better nestling survival. Studies show that dominant males have the best access to food, have larger territories, and have a higher annual survival.

MINNESOTA BREEDING DISTRIBUTION

The Black-capped Chickadee was described by Roberts (1932) as a "common permanent resident throughout the state." He included limited but wide-ranging confirmed nesting from many areas of the state, from the extreme northeastern regions in Cook County, to Houston County in the southeast, plus many counties throughout the forested regions of Minnesota.

Forty years later, Green and Janssen (1975) did not include a map of confirmed breeding observations but simply stated the Black-capped Chickadee was "breeding throughout the state." Janssen (1987) also described it as a "common to abundant permanent resident." He included a breeding distribution map with confirmed nesting documented in 44 of the 87 counties, all of which were widely distributed throughout Minnesota. A few years later, Hertzel and Janssen (1998) added another seven counties to the list of confirmed nesting in Minnesota since 1970. The MBS included 1,502 records of breeding-season locations, all of which were broadly distributed throughout the surveyed counties. However, no breeding locations were included from two western counties, Big Stone and Wilkin.

The MNBBA reported 5,646 breeding records from every county in the state and in 57.3 percent of the 4,756 blocks with coverage, including confirmed nesting in 603 blocks **(A)**. Nesting was confirmed in 90 percent (78/87) of the counties in the state. The only counties where nesting was not confirmed were Big Stone, Faribault, Grant, Lincoln, Martin, Redwood, Rock, Traverse, and Watonwan; all are located in the southwestern part of the state. Probable nesting was observed in all of these counties except three: Lincoln, Redwood, and Traverse. Confirmed nesting activity was particularly dense in areas with the most intense coverage, such as in the Twin Cities and Brainerd areas.

The MNBBA probability map exemplified the wide distribution of the Black-capped Chickadee throughout Minnesota **(B)**. The highest densities were predicted in the urban and suburban areas of the Twin Cities and other populated areas, such as the lakes region around Brainerd, and broadly in the forested areas. Lower densities were predicted in the southwestern and western portions of the state where there are fewer trees, as well as in many of the bog areas where trees are often too small and unsuitable for nest sites. Black-capped Chickadees also occur in riparian areas and woodlots throughout the state.

Roberts (1932) did not mention any changes to the species' population or distribution from the late 1800s to the early 1900s. Cutright and colleagues (2006), during their breeding bird atlas from 1995 to 2000, found the Black-capped Chickadee nesting throughout Wisconsin but suggested that it may have become more common than in previous years due to logging and opening of the forest in the northern portion of the state. In contrast, they also cited Kumlien and Hollister from the early 1900s, who stated that its status had not changed much since the previous century. In their review of the Black-capped Chickadee in North America, Foote and colleagues (2020) emphasized that the species has had "no substantial historical changes in distribution."

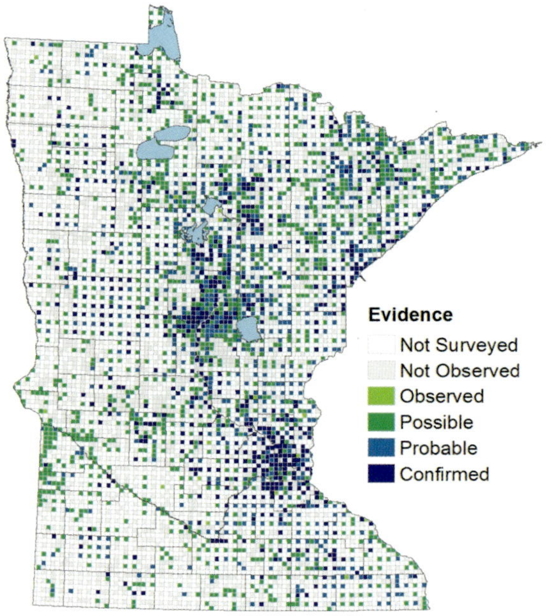

A. Breeding distribution of the Black-capped Chickadee in Minnesota based on the Breeding Bird Atlas (2009–13).

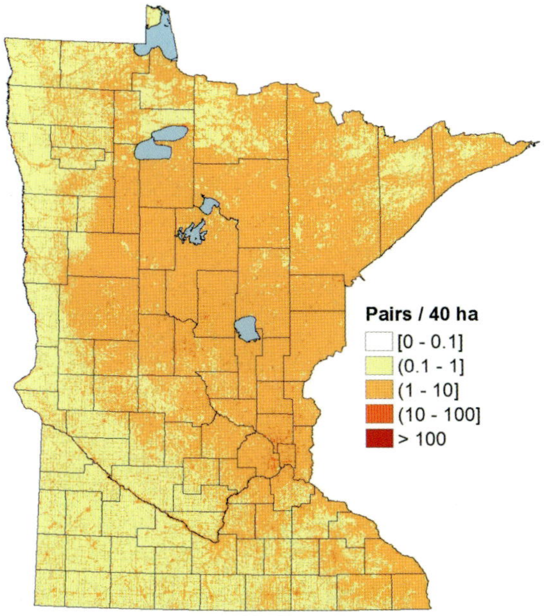

B. Predicted breeding distribution (pairs per 40 ha) of the Black-capped Chickadee in Minnesota based on habitat, landscape context, and climate data gathered during the Minnesota Breeding Bird Atlas (2009–13) using the Generalized Linear Modeling method with an adjustment for detectability.

Because this species nests in trees, it is difficult to comprehend that loss of forested habitat in Minnesota over the past 150 years has not had a negative effect on its abundance. Its commonness and the lack of comparable monitoring data from the 1800s or early 1900s render it impossible to gauge whether any change has occurred. The Black-capped Chickadee has adapted well to human-dominated landscapes, and its attraction to bird feeders may have mitigated any possible negative effects from the reduced number of trees in the state.

BREEDING HABITAT

The Black-capped Chickadee uses many habitats, including mixed deciduous–coniferous forests, open woods and parks, suburban areas, and woodland edges. It is highly tolerant of forest fragments (S. Smith 1991; Foote et al. 2020). Based on almost 5,000 observations in the heavily forested area in northern Minnesota and Wisconsin, Niemi, Howe, and colleagues (2016) indicated primary use of urban areas (particularly small towns) but equal use of all other forest types, except treeless open areas such as sedge meadows **(C)**. These forest cover types included deciduous forests, mixed deciduous–coniferous forests, lowland deciduous and coniferous forests, and recently logged forests, as long as some live and dead trees remain in these areas.

POPULATION ABUNDANCE

Black-capped Chickadees have been estimated to have a substantial US and Canadian breeding population of 43 million breeding adults (PIF 2021) and a Minnesota breeding population of 1.3 million. However, the MNBBA estimated a Minnesota breeding population of almost twice that size, approximately 2.0 million breeding adults (95 percent confidence interval of 1.80–2.35 million).

Although the BBS population trend for the Black-capped Chickadee in Minnesota has been highly variable, it demonstrates an overall, significantly increasing trend from 1967 to 2019 of 2.26 percent per year **(D)**. This significant population trend is also consistent with trends in Michigan, Ontario, and Wisconsin. Based on the BBS, Rosenberg, Kennedy, and colleagues (2016) estimated an overall 40 percent increase in its North American population from 1970 to 2014.

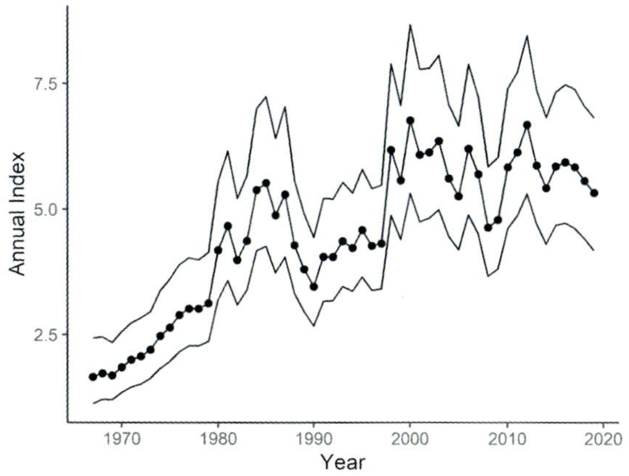

D. Breeding population trend and 95 percent credible interval (solid black lines) for the Black-capped Chickadee in Minnesota based on the 1967–2019 Breeding Bird Survey (Sauer et al. 2020).

CONSERVATION

Overall, the Black-capped Chickadee has received little attention directed toward conservation efforts because of its healthy and generally increasing population status. However, declining regional populations have been detected in some areas by the BBS, especially in the western United States and Canada. These areas are worthy of continued observation and identification of potential reasons for these declines. Black-capped Chickadee populations greatly benefit from food supplied by bird feeders, especially during the winter months (Desrochers et al. 1988; Brittingham and Temple 1988). Recent increases in the cost of birdseed and potential decreases in bird feeding may be related to some of the recent declines that have been observed.

The availability of suitable nesting cavities is an ongoing concern for any species that primarily excavates its own cavity. This is especially true for the Black-capped Chickadee, which does not have the anatomical apparatus of a woodpecker to excavate cavities in solid wood. Chickadees must rely on damaged trees or decaying wood to satisfactorily excavate a suitable cavity, so maintaining such trees (e.g., dead snags or partially rotted trees) is an important aspect of forest management. As identified by Foote and colleagues (2020), "overzealous" forest management that reduces or eliminates natural nest sites can be detrimental to the species.

Under both 1.5°C and 2°C warming scenarios, NAS (2019) projected a stable breeding range for the Black-capped Chickadee in North America. Its breeding range in Minnesota was predicted to have minimal changes. In addition, under both of these scenarios its winter range was projected to be stable. Milder winters could benefit this species by increasing overwinter survival, one of the major factors contributing to its mortality. Therefore, backyard bird-watchers and nature enthusiasts in most parts of Minnesota should continue to enjoy this energetic, chunky little bird for years to come.

C. Nesting habitat of the Black-capped Chickadee in Minnesota is typically mixed deciduous–coniferous forests, often in rural and urban areas with suitable nest trees. Photo by Gerald J. Niemi.

BOREAL CHICKADEE

Poecile hudsonicus

Minnesota Seasonal Status: Permanent resident. The Boreal Chickadee was uncommon during the MNBBA.
Migration: Sedentary; periodic irruptions south of its primary range occasionally occur.
Food: Omnivore; largely arthropods, seeds, and berries gleaned from foliage.
Nest: Cavity nester in trees; often self-excavates in a natural cavity but occasionally uses a hole excavated by another species, primarily woodpeckers.
Conservation Concern: Assigned a low PIF Continental Concern Score of 8/20.

One must look carefully for the Boreal Chickadee in the northern boreal forests of Minnesota at any season. It is often found associated with its near relative the Black-capped Chickadee, but is best distinguished by its brown cap and overall brownish-tan appearance. Its voice is easily distinguished from that of the Black-capped Chickadee. It has a very nasal quality and is described as *tseek-a-day-day*. It is a more secretive species compared with the Black-capped but, like many birds of the boreal region, it can be very tame and often allows very close views.

Roberts (1932) knew the species by its former name, the Hudsonian Chickadee. He said it was a "rather uncommon permanent resident in the extreme northern part of Minnesota," which is still true today. It can make irruptive movements during fall and winter into southern Minnesota counties and can even be locally common at bird feeders in some winters, especially along the North Shore of Lake Superior.

Genetically, the Boreal Chickadee is most closely related to the Gray-headed Chickadee (*Poecile cinctus*), found in boreal forests of both the Palearctic and Nearctic regions of the Northern Hemisphere, and the Chestnut-backed Chickadee (*P. rufescens*), a relative found in the western coastal forests of Canada and the United States (Ficken et al. 2020). The Boreal Chickadee is found throughout the boreal forest of Canada and Alaska, the northern portions of the Upper Midwest, and the northeastern United States. Densities observed were highest in Labrador; high densities may also occur in northern regions of Canada and in Alaska, but they are not well surveyed.

Photo by David Brislance.

MINNESOTA BREEDING DISTRIBUTION

Roberts (1932) included a relatively lengthy discussion on the historical status and nesting of the Boreal Chickadee in Minnesota from the late nineteenth century to the early twentieth century. He found "the records for Minnesota are not numerous and many observers who have passed much time in the field in the northern part of the state have not met with it." A specific comment by Roberts on its status in the extreme northern part of the state east of Ely, Minnesota, included the following: "Mr. Breckenridge, while spending nearly three weeks in May 1931; some eighteen miles east of Winton, in northern St. Louis County, found the Hudsonian Chickadee as common as the Black-capped Chickadee."

Roberts (1932) reported that the Rev. P. B. Peabody was the only individual fortunate enough to find a Boreal Chickadee nest, which he located near Hibbing in St. Louis County on May 24, 1901. He described observations by two of his colleagues, "Kilgore and Breckenridge," of adults feeding young out of a nest at Big Sandy Lake, Aitkin County, on June 30, 1929. He cited observations by Gustav Swanson in a locality not far north of Deer River, Itasca County, where two adults were carrying food and later, at the same locality, a young bird was being fed by the parents.

More than forty years later, Green and Janssen (1975) described the Boreal Chickadee distribution as primarily in northeastern and northern Minnesota. They included confirmed nesting in Lake County and St. Louis County and inferred nesting in Aitkin, Clearwater, and Itasca Counties. Later, Axelrod (1979), a student at the University of Minnesota, exquisitely described details of a Boreal Chickadee nest found by Bruce Fall in 1977 at La Salle Creek Bog in Hubbard County. Janssen (1987) suggested a more restricted distribution from that defined by Green and Janssen, which largely eliminated Cass County. Both Janssen (1987) and Hertzel and Janssen (1998) summarized confirmed nesting since 1970 in seven counties: Aitkin, Beltrami, Clearwater, Cook, Hubbard, Lake, and St. Louis.

The MBS included 74 records of breeding-season locations, which were broadly distributed from Itasca County to Cook County. The MBS listed multiple locations in Cook, Lake, Itasca, and St. Louis Counties as well as two in northern Aitkin County and one in northern Carlton County.

The MNBBA included 219 records of the Boreal Chickadee. The records encompassed a distribution similar to what was described by Janssen (1987), from Cook County to Roseau County and south to Aitkin County **(A)**. No observations of the species were recorded from Carlton, Clearwater, or Hubbard Counties, but confirmed nesting was reported from Cook, Itasca, Lake, Lake of the Woods, Koochiching, and St. Louis Counties. Boreal Chickadees were reported from 3.0 percent (141/4,733) of all blocks sampled. Probable nesting was recorded from 23 blocks as far west as Roseau and Beltrami Counties and south to northern Aitkin County.

The probability map based on the MNBBA point counts identified low densities for the Boreal Chick-

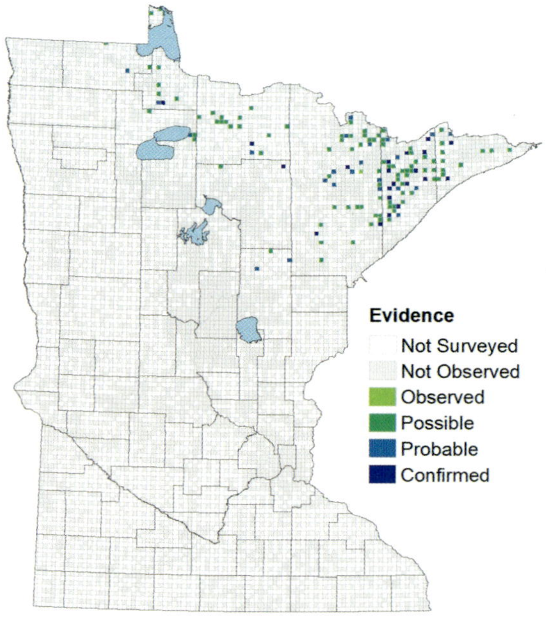

A. Breeding distribution of the Boreal Chickadee in Minnesota based on the Breeding Bird Atlas (2009–13).

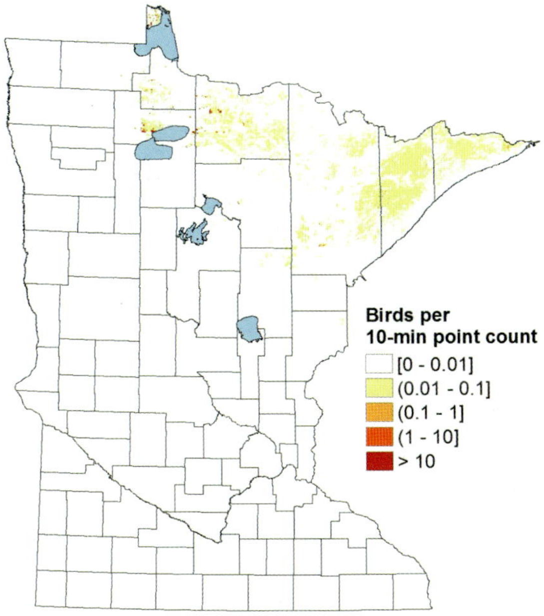

B. Predicted breeding distribution (birds detected per 10-minute point count) of the Boreal Chickadee in Minnesota based on habitat, landscape context, and climate data gathered during the Minnesota Breeding Bird Atlas (2009–13) using the Generalized Linear Modeling method without an adjustment for detectability.

adee throughout much of northeastern Minnesota and in the Northern Minnesota and Ontario Peatlands Section north and east of Upper Red Lake **(B)**.

Overall, the Boreal Chickadee is a regular nesting species throughout many northern and northeastern counties in Minnesota. However, its relatively low abundance, its remote forested habitats, and its lack of a territorial "advertising" song during the breeding season all make it difficult to observe on the breeding grounds, let alone to find nests. Ficken and colleagues (2020), in their review of the Boreal Chickadee in North America, do not identify any information on historical changes in its distribution. However, historical data on the species are lacking from the nineteenth century and even the early to mid-twentieth century. For instance, the first nest in Michigan was not found until 1954, and the first nest in Wisconsin was found in 1959.

BREEDING HABITAT

Throughout their range, Boreal Chickadees breed in boreal coniferous and mixed coniferous–deciduous woodlands (Ficken et al. 2020). In Minnesota, most breeding observations occur in the lowland coniferous forests composed of black spruce, tamarack, and white cedar (Green and Niemi 1978) **(C)**. The species is reported from many upland habitats, but usually when balsam fir, white spruce, or pine are present (Niemi and Hanowski 1992a). In the Agassiz Lowlands Subsection of northern Minnesota, Bednar and colleagues (2016) and Zlonis and colleagues (2017) found the species was primarily associated with mature lowland black spruce–tamarack bogs and the combination of lowland black spruce and white cedar evergreen forest.

POPULATION ABUNDANCE

Populations in the United States and Canada were estimated by PIF (2021) as 13 million breeding adults, with 17,000 in Minnesota. In contrast, Environment Canada (2014b) provided a broader, more conservative range of 5 to 50 million adults. Most population trend data for the Boreal Chickadee were unreliable because few or not enough counts have been completed in its primary breeding range: the vast roadless areas of Canada. However, the BBS counts in Alaska, Newfoundland, and the Yukon were reliable and indicated stable population counts since counts began in the 1960s to 2019 and in the more recent period from 1993 to 2019 (Sauer et al. 2020). In addition, BBS counts of the Boreal Chickadee in their "core and non-core" areas in the United States were reliable and stable from 1993 to 2019.

CONSERVATION

The MNDNR (2015b) listed the Boreal Chickadee as a *Species in Greatest Conservation Need* because of potential habitat loss, degradation, and fragmentation. Environment Canada (2014b) included the Boreal Chickadee as a species of conservation priority in several of its bird conservation regional strategies.

NAS (2019) projected a high vulnerability for the Boreal Chickadee's breeding range in North America with a 2°C warming scenario, which included a 51 percent range loss and a 21 percent range gain primarily in the far north by 2050. In Minnesota, the species was predicted to be extirpated from the state even under our current 1.5°C warming scenario.

Cadman and colleagues (1987) noted Boreal Chickadees have a preference for forests of poor quality that are not extensively harvested. In contrast, recently in Minnesota, there has been an increased interest in logging of coniferous forests, including black spruce (Bednar et al. 2016). Bednar and colleagues cautioned that extensive logging of mature black spruce–tamarack stands with high site indexes could negatively affect Boreal Chickadees. Ficken and colleagues (2020) reported on exploratory modeling in Ontario that suggests that the availability of old black spruce forests in winter may be critical to the persistence of local populations.

Gillespie and Kendeigh (1982) identified the Boreal Chickadee as a forest-interior species that avoided forest edges, a result consistent with studies by Zlonis and colleagues (2017) in northern Minnesota. Niemi and Hanowski (1992a), in their analysis of increased forest harvesting and management in Minnesota, predicted the amount of available habitat for Boreal Chickadees would decline under moderate (4.9 million cords per year) and intensive (7 million cords per year) harvest scenarios in Minnesota. However, these harvesting levels have never occurred in Minnesota.

In summary, we still know little about the nesting biology of the Boreal Chickadee. The imminent effects of climate change and future forest management of northern coniferous forests will be the two most important issues this species will face and decide whether it will remain as a breeding species in Minnesota's future.

C. Typical breeding habitat of the Boreal Chickadee in northern Minnesota: a mature black spruce forest with a ground cover of *Sphagnum* moss. Photo by Alexis Grinde.

TUFTED TITMOUSE

Baeolophus bicolor

Minnesota Seasonal Status: Permanent resident restricted primarily to southeastern Minnesota. The Tufted Titmouse was rare during the MNBBA.
Migration: Data suggest some birds, perhaps juveniles, occasionally move south in the fall.
Food: Omnivorous; forages in trees and on the ground. During the fall and winter, food items such as acorns and sunflower seeds may be cached under loose bark or in the cracks and furrows of tree limbs. A frequent visitor to bird feeders.
Nest: Secondary cavity nester; uses natural cavities, old woodpecker cavities, and nest boxes.
Conservation Concern: Assigned a low PIF Continental Concern Score of 7/20.

Photo by Mike Lentz.

It is hard not to be charmed and entertained by the lively, acrobatic Tufted Titmouse. "Dangling by one foot, they often take on a trapeze style envious of the best circus performers" as they search for morsels missed by others (Sorenson 2017). Titmice are equally inquisitive and "utterly fearless of human presence"; amusing accounts of them plucking hairs from the heads of startled bystanders to line their nest cavities are not uncommon (Bent 1946). One writer characterized the Titmouse as "a curious imp with mischievous tendencies," constantly scolding other Titmice, other birds, and anything perceived as a potential threat (Stevens 2018). Another writer simply concluded that, given the species' congenial nature, "we should probably ascribe to him without hesitation the first place in our hearts" (Bent 1946). Clearly, as Stevens (2018) noted, the pint-sized Tufted Titmouse has a big personality!

A little larger than a chickadee, the Titmouse is an easy bird to spot. Gray above and white below, the species has distinguishing features that include a thin flank of rufous feathers on each side of the belly, a patch of black just above the bill, and a conspicuous gray crest on the crown. Its scientific name, *Baeolophus bicolor,* means a "small-crested" bird of "two colors" (gray and white). The male and female have similar plumages. Although it has a remarkable array of vocalizations, the most familiar song is a loud *Peter, Peter, Peter* that may be sung incessantly for hours by a single male.

Largely restricted to the eastern United States, the Tufted Titmouse has a breeding range that extends from southeastern Minnesota south to eastern Texas and east to the Atlantic coast, where it can be found from Maine south through much of Florida. This little bundle of energy is abundant across the southeastern United States.

MINNESOTA BREEDING DISTRIBUTION

Roberts (1932) summarized the Tufted Titmouse's status in Minnesota as follows: "An infrequent straggler into the state from the south, mainly as a winter visitant thus far." The first record appeared in the spring of 1879, when a bird was collected in Minneapolis. Over the next fifty years, "nearly fifty" reports were widely dispersed across the state, including one record of a pair that spent much of the winter in the town of Fosston in Polk County. Most records were in the "late fall, winter, and early spring"; only three records were in the summer season. All three were reports of breeding pairs in the 1920s, restricted to Houston and Fillmore Counties in southeastern Minnesota. The observations prompted Roberts to predict that "this typically southern species is extending its range northward into Minnesota and, like the Cardinal, is appearing first as a cold weather visitant. Later it will, no doubt, become a summer resident and nesting species as has been the case with the Cardinal. The records from Houston and Fillmore counties suggest that this is already taking place."

The first documented nesting was actually much farther north, at the town of Marine on St. Croix, Washington County, in 1943 (Edgar 1943). Nesting reports from southeastern Minnesota soon followed. In the summer of 1946, Mary Lupient reported that two pairs of Titmice nested "last spring [1945] for the second time" in Rushford, Fillmore County, suggesting that the state's second confirmed nesting was in the summer of 1944 (Lupient 1946).

Thirty years later, Green and Janssen (1975) considered the Tufted Titmouse to be a permanent resident in the southeast, extending north to Washington County in the east and to Stearns County in the west. Winter records extended even farther north, occurring "accidentally as far north as Cook and Polk counties." By 1971 nesting was also reported in Ramsey County (Eckert 1971).

Janssen (1987) later reported that the Titmouse, known for its population fluctuations, had retreated southward. "From the 1940s to the mid-1970s this species expanded its breeding range northward in the state to where it was seen regularly in Hennepin County. Since the mid-1970s it has retreated southward from this area." Although it was still regularly observed in Houston and Fillmore Counties, it was no longer found during the summer at sites in the Greater Twin Cities Metropolitan Region, where it had previously been reported. Likewise, the 16 breeding-season locations identified by the MBS were restricted to Houston, Fillmore, and southern Olmsted Counties.

Tufted Timouse records collected during the MNBBA suggested that the bird may once again be moving northward **(A)**. Reported from 29 surveyed atlas blocks covering 14 counties, the species was found as far west as Brown County, along the

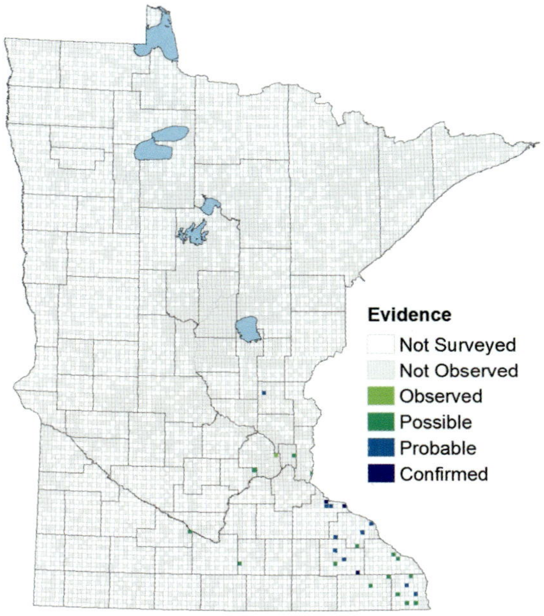

A. Breeding distribution of the Tufted Titmouse in Minnesota based on the Breeding Bird Atlas (2009–13).

Minnesota River valley, and as far north as Isanti County, in east-central Minnesota. Breeding was confirmed in three blocks located in Olmsted and Goodhue Counties and deemed probable in 10 additional blocks located in Isanti, Wabasha, and Houston Counties. Given Minnesota's increasingly warm winters, it is not surprising that the Tufted Titmouse is on the move again. Indeed, in 2015, breeding was once again confirmed in Hennepin County (MOU 2021a), the last confirmed nesting having occurred forty-six years earlier in 1969 (MOU 2021b).

Similar changes have been observed in Wisconsin. First sighted in the state in 1900, the species steadily increased in numbers and distribution through the 1960s. Populations became well established along the Mississippi and Wisconsin Rivers in southwestern Wisconsin and along the Chippewa River east to Eau Claire and Chippewa Falls. Beginning in the 1960s, however, numbers slowly declined and "bottomed out in 1980," only to begin a slow and gradual recovery in succeeding years (Cutright et al. 2006). As of 2019, Titmice have been observed as far north as Burnett County and confirmed breeding just to the south in Barron County, both located in northwestern Wisconsin (eBird 2021).

Although the Tufted Titmouse is well known for its regular population fluctuations, it is also widely cited as a species that has responded positively to a warming climate. Over a period of nearly one hundred years, the species has expanded the northern edge of its distribution nearly 500 miles in New England and 300 miles in the Great Lakes region (Bird Source 2016). Biologists believe the harsher winters in the Upper Midwest may be preventing the birds from expanding north as fast as they are in New England. Extreme cold likely pushes the birds south until there is a string of warm winters that again facilitate movements northward (Bird Source 2016).

Although a warmer climate is considered the primary factor responsible for the Tufted Titmouse's northern expansion, the popularity of backyard bird feeding has likely contributed to the observed changes. Forest succession on abandoned farmland in the northeastern United States is also cited as a contributing factor (Ritchison, Grubb, and Pravosudov 2020).

BREEDING HABITAT

Frequent inhabitants of mature deciduous forests **(B)**, Titmice are occasionally found in mixed coniferous and deciduous woodlands as well. Their general requirements appear to be stands of large trees with a relatively open understory and a dense canopy. Because mast is an important winter food item, oaks and other mast-producing species are important. These features are often found in lowland hardwood stands, which are a common haunt of the species. Although they are found in a variety of human-altered habitats, such as parks and orchards, they reach their highest breeding densities in large tracts of unfragmented forest (Cutright et al. 2006).

B. The Tufted Titmouse selects dense stands of deciduous trees where old woodpecker nesting cavities are available for use, such as this mature oak forest in the blufflands of southeastern Minnesota. Photo by Lee A. Pfannmuller.

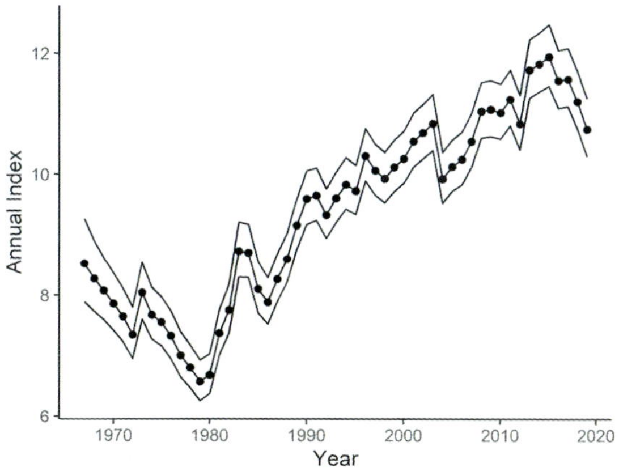

C. Surveywide breeding population trend and 95 percent credible interval (solid black lines) for the Tufted Titmouse based on the 1967–2019 Breeding Bird Survey (Sauer et al. 2020).

POPULATION ABUNDANCE

Using BBS data, PIF (2021) estimated that the small breeding range of the Tufted Titmouse supports approximately 12 million breeding adults. It is rarely reported on BBS routes in Minnesota; the statewide estimate is only 760 birds. Surveywide, BBS data clearly illustrate the species' steady and significant increase, averaging 0.44 percent per year from 1966 to 2019 **(C)**. The increases are most prevalent along the entire East Coast, in states across the Southeast Coastal Plain (Alabama, Georgia, and Mississippi), and inland across the Appalachian Mountains. Although data in Minnesota were insufficient to predict population changes, southeastern Minnesota is included within the Prairie Hardwood Transition Bird Conservation Region, where populations have experienced a nonsignificant increase of 0.81 percent per year (Sauer et al. 2020).

Noted earlier, factors responsible for the increase include an uptick in winter bird feeding as a recreational activity, the maturation of deciduous forests in the northeastern United States, and warming temperatures (Ritchison, Grubb, and Pravosudov 2020). At the local scale, factors that may limit populations include the availability of nesting cavities, annual variability in mast production, severe winters, and competition with other cavity-nesting species (Cutright et al. 2006).

CONSERVATION

The Tufted Titmouse is among the species experiencing a clear benefit from climate change (NAS 2019). It seems likely that residents north of Minnesota's far southeastern counties may be entertained by this lively and inquisitive little bird in the not-too-distant future, especially in areas with large tracts of mature deciduous forest that provide abundant cavities.

HORNED LARK

Eremophila alpestris

Minnesota Seasonal Status: Regular breeding species and migrant; regularly observed during the winter, especially in southern Minnesota. The Horned Lark was common during the MNBBA.
Migration: Short-distance migrant wintering in the central and southern United States. Birds present during Minnesota winters include northern migrants; it is not known if breeding residents remain through the winter.
Food: Omnivorous ground feeder that consumes seeds and arthropods.
Nest: Cup nest constructed in a small depression on the ground. Small articles, such as pebbles or corn-stalk fragments, are often placed beside the nest.
Conservation Concern: Assigned a moderate PIF Continental Concern Score of 9/20.

Photo by David Brislance.

In some of Minnesota's most barren habitats, including summer's corn and soybean fields or winter's snow-plowed, gravel roads, observers with keen ears and sharp eyes are rewarded with the sights and sounds of Horned Larks. The only member of the lark family native to North America, the bird has a light brown back that enables it to blend with the surrounding soil. Its most distinctive markings are located on the head, where patches of black separate yellow or white feathers located above and behind the eye and on the chin. Two small black feather tufts on the crown give rise to its common name. The female is simply a paler version of the male.

Like all members of the lark family, the Horned Lark has a beautiful song that is often delivered in flight. G. M. Sutton (1927) described it "as *tinkling*; and although it is sweet and delicate, it can hardly be called glorious." Roberts (1932) provided a detailed account of the aerial performance:

> Rising from the ground the bird mounts spi-rally, by successive leaps, upward and upward, until lost from sight. By listening carefully one discovers that he is pouring forth his short but musical warble again and again, hovering and circling meanwhile, a mere speck against the sky. After a few minutes, on set and nearly closed wings the singer drops head foremost, almost perpendicularly, and seems about to dash against the ground, when at the last instant he catches himself on outspread wings and jauntily alights or skims along a short distance over the prairie to drop down, perhaps, at the very spot from which the ascent was made.

A Holarctic species, the Horned Lark is found in the Western Hemisphere across Alaska, Canada, and nearly all of the continental United States south to central Mexico. Breeding densities are highest in the central plains states of Kansas and eastern Colorado. The genus name, *Eremophila,* refers to a species "fond of open spaces"; *alpestris* means "of the mountains," a likely reference to the species' montane habitats in portions of Europe and the Balkan Peninsula (Sandrock and Prior 2014; Beason 2020).

MINNESOTA BREEDING DISTRIBUTION

The species was known in the early 1900s as the Prairie Horned Lark; Roberts (1932) considered it a common summer resident in all but the heavily forested regions of the state. During his lifetime, its abundance declined rather dramatically in the southeastern counties in the early 1900s. Although common in western Minnesota, in "southeastern Minnesota there is not one pair today where there used to be fifty. Many miles may now be traveled without seeing a single bird over roads where they were once so common as to be called the 'Road Bird.'" The decline, he mused, might have been owing to the loss of suitable habitat and a scarcity of food. At the time, nesting evidence was reported in 11 widely dispersed counties: Anoka, Hennepin, Isanti, Martin, McLeod, Meeker, Pipestone, Polk, Ramsey, Stearns, and Wadena.

When Green and Janssen (1975) and Janssen (1987) provided updates on the species' status, little had changed. Since the bird was still scarce in the forests of northeastern and north-central Minnesota, Janssen described it as most common "on the western prairies and heavily cultivated regions in the south, central, and east-central regions." He delineated 16 counties where nesting had been confirmed since 1970; Hertzel and Janssen (1998) later added three more counties to the list. The MBS also documented the species' widespread occurrence in all but the Laurentian Mixed Forest Province of the state.

During the MNBBA, participants reported 2,327 Horned Lark records from 31.0 percent (1,480/4,772) of the surveyed atlas blocks. Breeding was confirmed in 79 blocks **(A)**. The birds were observed in 83 of Minnesota's 87 counties, being absent only from the northern counties of Carlton, Cook, Itasca, and Lake. Breeding was confirmed in 36 counties; 26 of the counties were additions to the list published by Hertzel and Janssen (1998). The species continued to be broadly distributed across the state's grassland and agricultural regions, including throughout southeastern Minnesota. Scattered records in the Laurentian Mixed Forest Province were largely confined to the more southern, developed regions of the province.

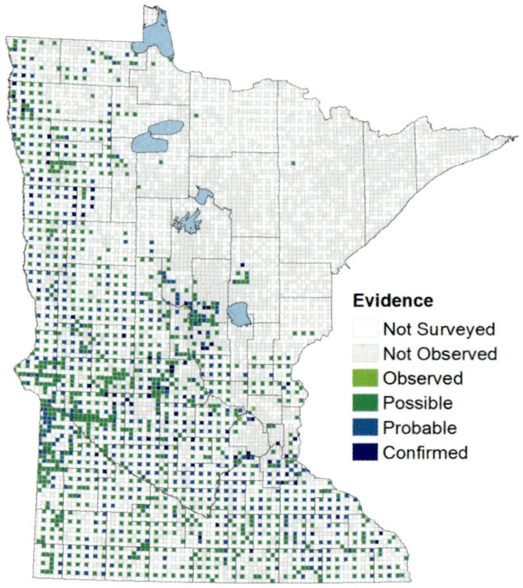

A. Breeding distribution of the Horned Lark in Minnesota based on the Breeding Bird Atlas (2009–13).

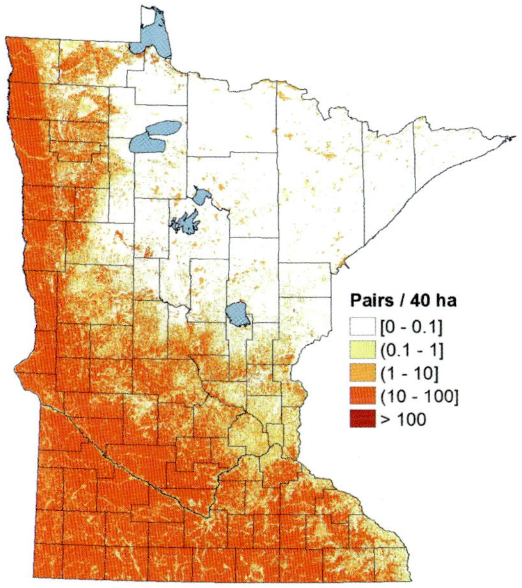

B. Predicted breeding distribution (pairs per 40 ha) of the Horned Lark in Minnesota based on habitat, landscape context, and climate data gathered during the Minnesota Breeding Bird Atlas (2009–13) using the Generalized Linear Modeling method with an adjustment for detectability.

The MNBBA probability map for the Horned Lark clearly reflects the relatively high breeding densities found throughout the cultivated landscapes of western and southern Minnesota and the species' sparse abundance across the northern forest **(B)**.

The Horned Lark's distribution in Minnesota is quite similar to that described by Roberts (1932). It is more difficult to assess the degree to which its abundance may have changed beyond noting its current abundance in southeastern Minnesota, a region where Roberts (1932) once highlighted its notable decline. Primarily a species of western North America, the Horned Lark expanded as the eastern deciduous forests were cleared in the late 1800s and early 1900s. But as farmlands began to be abandoned in the 1940s, left to succeed to young deciduous forests, or lost to suburban development, the species declined in abundance and its range contracted, especially in the northeastern states (Beason 2020).

BREEDING HABITAT

Bent (1942) aptly described the Horned Lark's breeding habitat when he wrote, "Here is a bird nesting where no bird has a right to nest, a bird in a niche that demands not vegetation but lack of it." Indeed, what most characterizes the Horned Lark's habitat is its relatively sparse vegetative cover, including cropland, roadside right-of-ways, pastures, and recently mowed haylands **(C)**. As the season progresses, habitat that is suitable early in the season may become unsuitable as the vegetation grows taller and denser. Occupied sites usually contain 10–37 percent bare ground, and breeding bird densities are often highest on sites with the lowest densities and heights of vegetation (Dinkins et al. [2000] 2002). Studies suggest that the Horned Lark is also an area-sensitive species, its density increasing as the size of the habitat patch increases (Ribic et al. 2009).

POPULATION ABUNDANCE

Using BBS data, PIF (2021) projected a population estimate of 100 million birds in the United States and Canada, with an additional 40 million residing in Mexico. Minnesota's population was estimated at 1.1 million birds. MNBBA data, however, generated a dramatically higher population estimate of 12.2 million breeding adults (95 percent confidence interval of 11.2–14.4 million). Different methods of calculating the estimate, combined with the MNBBA's more intensive sampling effort, are likely responsible for the variation. Given the prevalence of cropland habitat in Minnesota, the higher estimate is not unreasonable. Indeed, Pettingill's (1970) assessment more than fifty years ago is correct: the Horned Lark "may be the most common bird in the prairie country of North America."

Despite the species' abundance, Horned Lark populations across the BBS survey region have significantly declined an average of −1.91 percent per year from 1966 to 2019 (Sauer et al. 2020), with a cumulative decline of 65 percent from 1970 to 2014 (Rosenberg, Kennedy, et al. 2016). The species has shown a significant increase in only three states: Arkansas, Georgia, and North Carolina. Although population numbers in Minnesota have fluctuated, BBS data reveal a relatively stable population with a nonsignificant decline, averaging only −0.22 percent per year **(D)**.

As noted earlier, Horned Lark populations began declining in the northeast as farms were abandoned and succeeded to forest cover or were lost to residential development (Beason 2020). Today, however, the declines are prevalent throughout much of the species' range, including the western interior of the United States, a region where cropland loss from 1949 to 2002 was the lowest in the United States (Pfeffer et al. 2006). Perhaps, as with many grassland species, more intensive agricultural use of these lands, including pesticide use, may be contributing to the downward trend.

C. A young field of soybeans provides the open country and sparse cover that the Horned Lark favors for breeding habitat. Photo by Lee A. Pfannmuller.

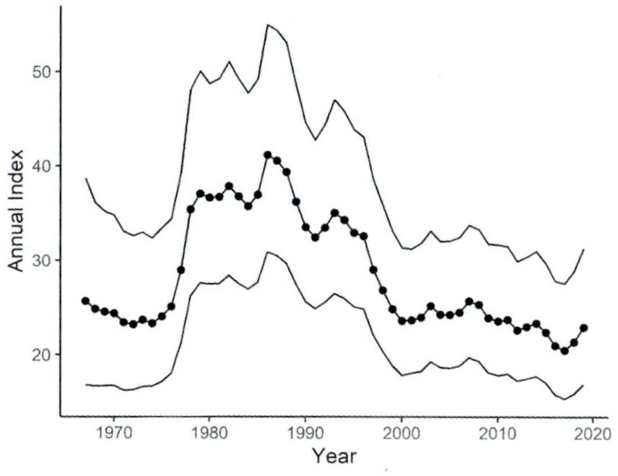

D. Breeding population trend and 95 percent credible interval (solid black lines) for the Horned Lark in Minnesota based on the 1967–2019 Breeding Bird Survey (Sauer et al. 2020).

CONSERVATION

The Horned Lark is designated a moderate conservation priority (PIF 2021) and a Common Bird in Steep Decline (Rosenberg, Kennedy, et al. 2016). Management recommendations for the species focus on establishing disturbances that create open and sparse vegetation, including burning to reduce woody vegetation and mowing or grazing to create shorter vegetation (Dinkins et al. [2000] 2002). Unfortunately, conservation programs designed to improve grassland habitats may not benefit the Lark. In many states it was entirely absent from Conservation Reserve Program lands. In other states, including Minnesota, the Lark used these sites more frequently when they were planted with native grasses rather than nonnative grasses and legumes. In general, Larks were nearly always more abundant in cropland than in grasslands (Dinkins et al. [2000] 2002).

Unlike many species at risk with a warming climate, the Horned Lark had a low vulnerability under all three warming scenarios examined by NAS (2019). Indeed, as its breeding distribution gradually recedes northward, it is predicted to become even more ubiquitous in Minnesota, expanding throughout the northern counties by 2080. Although it is important to understand the factors responsible for the bird's decline, the aerial performance of the Lark's faint but beautiful song should remain a feature of our landscape long into the future.

BANK SWALLOW

Riparia riparia

Minnesota Seasonal Status: Regular breeding species and migrant. The Bank Swallow was uncommon during the MNBBA.
Migration: Medium- to long-distance migrant that winters primarily in South America.
Food: Aerial insectivore.
Nest: Excavates a nesting burrow in exposed cliff faces, bluffs, and banks along rivers, lakes, and streams; readily adapts to man-made environments such as sand and gravel pits. Nests in colonies that range in size from 10–20 nesting pairs to several hundred pairs.
Conservation Concern: Assigned a moderate PIF Continental Concern Score of 11/20.

Photo by Karl Bardon.

Minnesota's smallest swallow, the Bank Swallow, is nearly as obscure as its close relative, the Northern Rough-winged Swallow. The Bank Swallow is brown above and white below; its signature trait is a dark upper breast band with a center point that dips into the white belly. As clear as this feature appears in field guides, when the birds are darting overhead it can be difficult to see unless they briefly come to rest on a small branch or at the entry to their nest burrows. Eckert (2005) offered some helpful tips to distinguish the bird from the Rough-winged Swallow and immature Tree Swallow, both of which show some color in the upper breast but lack the center streak to the upper belly. One suggestion is to look for the contrast between the Bank Swallow's pale lower back and rump and its darker tail. Calls can also aid identification. The Bank Swallow's bubbly, multi-syllable twitter can be distinguished from the short *brrrt* notes of the Northern Rough-winged Swallow and the more musical, liquid gurgle of the Tree Swallow.

The Bank Swallow's behavioral traits truly set it apart from other swallows in Minnesota. Like the Cliff Swallow, it is a colonial nesting species, but it has the unique habit of excavating its own nesting burrows. Indeed, the Latin translation of the species' scientific name, *Riparia riparia,* is "bank of stream," a reference to its common habit of dwelling along the bank of a stream or river. Using their bill, feet, and wings, the birds excavate a tunnel perpendicular to the bank face and approximately 60 cm deep. Their small, conical bill and strong leg muscles are adaptations specifically designed for the task (Garrison and Turner 2020).

Globally distributed, the Bank Swallow's North American breeding range extends from Alaska east to the Maritime Provinces and south across most of the northern and central United States. A small breeding population also occurs in northeastern Mexico. Breeding densities are highest in the Prairie Parklands of central Manitoba south through the northern Great Plains.

MINNESOTA BREEDING DISTRIBUTION

Roberts (1932) considered the Bank Swallow a common summer resident: "Wherever in the state there are exposed sandbanks, its nesting-holes are most certainly to be seen. It shows a preference for high, washed banks and deep cuts near water, but on the prairies and in the forests it may be found nesting by the roadside in banks no more than eighteen to twenty inches high." Breeding evidence was available from 12 counties covering nearly all regions of the state, including Goodhue and Lincoln Counties in the south, seven counties across central Minnesota, and Cass, Itasca, and Polk Counties in the north. Many birds were also reported nesting at Leech Lake.

Decades later, Green and Janssen (1975) noted that Bank Swallows were less abundant in "heavily forested areas where stream banks are scarce or absent." Janssen (1987) clarified that the bird was locally common in the northern forest region where sand and gravel pits were present. He identified 32 counties where nesting had been confirmed since 1970. In 1998, Hertzel and Janssen added three more counties to the list. Among the 277 breeding-season locations the MBS identified, Bank Swallows were especially common along the Minnesota River valley, in southwestern Minnesota, and along the Glacial Lake Agassiz beach ridges of northwestern Minnesota.

During the MNBBA, participants reported 809 Bank Swallow records from 13.2 percent (628/4,764) of the surveyed atlas blocks. Breeding was confirmed in 6.4 percent (305) of the blocks **(A)** and in all but

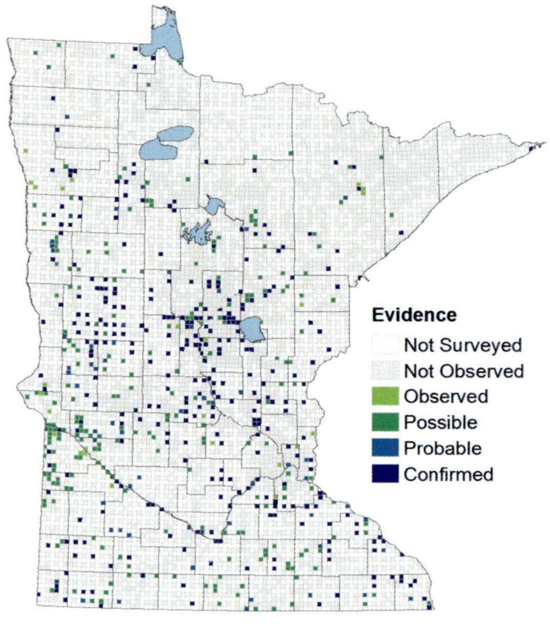

A. Breeding distribution of the Bank Swallow in Minnesota based on the Breeding Bird Atlas (2009–13).

B. An exposed bank in sandy soil provides the perfect nesting habitat for a Bank Swallow colony.
Photo by Debbie Petersen.

six counties (Dodge, Faribault, Freeborn, Kittson, Lake of the Woods, and Red Lake). Forty-seven of the counties were additions to the list published by Hertzel and Janssen (1998). Although the birds were reported from all 87 Minnesota counties, records were sparse north of a line from Crookston in northwestern Minnesota east to Duluth.

The distribution of Bank Swallows in Minnesota appears to have changed little during the past century. An adaptable species, it has found new opportunities in several man-made environments. Two notable examples are the cluster of MNBBA records in northern Clay County, an area along the Glacial Lake Agassiz beach ridges where gravel mining is common, and a small cluster of records in central St. Louis County, an area of taconite mining. In their comprehensive review of the species, Garrison and Turner (2020) noted no significant changes to the species' distribution other than local shifts resulting from changes in habitat.

BREEDING HABITAT

Historically, the Bank Swallow was restricted to sites where the erosive force of water cut away at stream banks and lakeshores. Today the species has broadened its preferences to include many man-made habitats, including sand and gravel mines and vertical road cuts **(B)**. Although many accounts suggest the species is always found near water, that association more likely characterizes natural nest sites and not the built environments it has adopted. Foraging habitats include a variety of open landscapes, including grasslands, agricultural fields, wetlands, open water, and occasionally woodlands (Garrison and Turner 2020).

Natural or man-made, colony sites are usually composed of alluvial soils that are easily excavated. Substrates with very fine grains allow the birds to excavate deeper burrows, which can result in higher reproductive success. Bent (1942) noted the species was "most numerous in the glaciated sections of the country where glacial deposits of sand and gravel abound."

POPULATION ABUNDANCE

PIF (2021) estimated a global population size of 29 million breeding adults with approximately 7.9 million residing in Canada and the United States. Their population estimate for Minnesota was 290,000 birds.

Unfortunately, this little Swallow has experienced a dramatic population decline. Throughout the BBS survey region, numbers declined an average of −3.69 percent per year from 1966 to 2019 (Sauer et al. 2020). From 1970 to 2014, the species experienced a cumulative decline of 89 percent (Rosenberg, Kennedy, et al. 2016). Indeed, it has not experienced a significant increase in any physiographic region, province, or state in North America. The significant downward trend has been even steeper in Minnesota, where the population has declined an average of −5.48 percent per year since 1967, for a cumulative decline of 95 percent through 2019 **(C)**!

Numerous factors may be responsible for the Bank Swallow's steep decline, but two appear especially important. First, disturbance and nest destruction at active sand and gravel mining operations certainly are responsible for some losses. In Wellington County, Ontario, 95 percent of all the aggregate mines that were operational in 2011 were occupied by colonies of Bank Swallows; work at 63 percent of the active pits destroyed 32 percent of the county's occupied burrows (COSEWIC 2013). Given the species' increasing dependence on such habitats, this level of annual loss likely has a significant impact on overall population numbers.

Second is the observation that many aerial insectivores, particularly flycatchers and swallows, have experienced significant population declines. As Robert Smith and his colleagues (2020) point out, the sheer scope of the decline and the number of species affected "suggests that there are powerful and broad-scale factors at work." The fact that declines are especially acute among species in the northeastern United States has led some to speculate that the declines reflect reductions in flying insect populations in this heavily industrialized region (Nebel et al. 2010).

CONSERVATION

Given its significant decline in numbers, the Bank Swallow has been designated a moderate conservation priority (PIF 2021) and recognized as a Common Bird in Steep Decline (Rosenberg, Kennedy, et al. 2016). It was officially listed as a federally *Threatened* species in Canada in 2013 (COSEWIC 2013), and, in 2016, Ontario published a detailed recovery strategy for the species. Principal among the recommendations outlined is the need to partner with the aggregate mining industry to develop and implement best-management practices that protect nesting colonies without negatively impacting the industry (Falconer et al. 2016).

Flood and erosion control projects pose another concern. Measures to stabilize eroding stream banks may reduce the availability of natural nest sites and should be closely evaluated for their potential impacts on the species. Fortunately, the negative impacts of warming temperatures appear minimal. NAS (2019) classified the species' breeding distribution as stable under all three warming scenarios they examined; minor range losses along the southern periphery of its breeding range were compensated by significant range expansions northward. Even northeastern Minnesota was predicted to provide more suitable habitat as temperatures rise, although suitable soils for nest excavation in this region could be a limiting factor.

Overall, resource professionals should consider the status and distribution of this industrious aerial excavator in their conservation work. Many of the smaller swallows, like the Bank Swallow, are easily overlooked, but as aerial insectivores, they provide an invaluable ecosystem service.

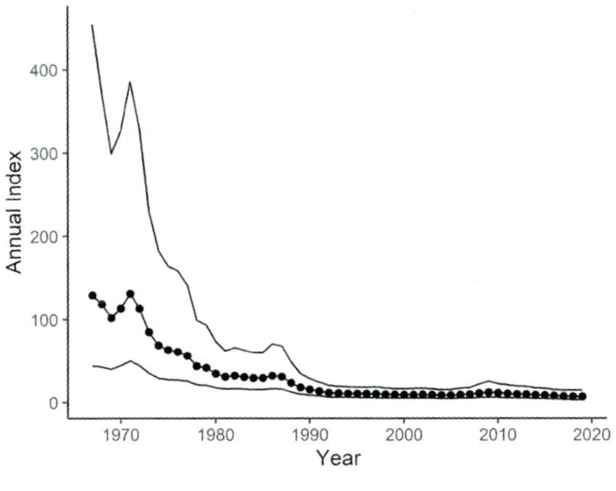

C. Breeding population trend and 95 percent credible interval (solid black lines) for the Bank Swallow in Minnesota based on the 1967–2019 Breeding Bird Survey (Sauer et al. 2020).

TREE SWALLOW

Tachycineta bicolor

Minnesota Seasonal Status: Regular breeding resident and migrant. The Tree Swallow was abundant during the MNBBA.
Migration: Short- to medium-distance migrant that overwinters in the southern United States, Mexico, Central America, and the Caribbean.
Food: Flying insects; rarely fruit, except during inclement weather and in winter.
Nest: Secondary cavity nester in trees, nest boxes, and occasionally holes in a variety of unusual locations.
Conservation Concern: Assigned a moderate PIF Continental Concern Score of 10/20.

The common swallow of fields, marshes, open water, and forest openings, the Tree Swallow is easily recognized by what its *bicolor* scientific name states: white below and steely, iridescent blue above. Some females, however, do retain the immature brown plumage above into their first spring and beyond. Its genus name, *Tachycineta,* is Greek for "fast mover" and aptly describes its swift, agile, and gliding flight while foraging for aerial insects. Tree Swallows are the first swallows to arrive in Minnesota in spring. They emit a variety of vocalizations; the most common is described as a prolonged chatter. Bent (1942) says it has "a pleasing, gentle quality . . . sometimes rippled into three or more syllables" that "always retains its delicacy."

The Tree Swallow has formerly been called the Water Swallow, Blue-backed Swallow, and White-bellied Swallow (Sandrock and Prior 2014). Frequently nesting near bodies of water, it is dependent on the availability of trees because it nests in natural tree cavities and old woodpecker holes. Several unusual nesting structures have also been recorded, such as in the eaves or holes of buildings, fence posts, fire hydrants, or even holes in the ground (Winkler et al. 2020). It readily accepts nest boxes; many placed to attract the Eastern Bluebird also benefit Tree Swallows.

The Tree Swallow is widely distributed and common to abundant across the United States and Canada. It is uncommon or absent from many parts of the southern United States and beyond the tree limit of northern Canada. Among the highest densities observed in North America are those in central and southwestern Minnesota and east-central Wisconsin.

Photo by Jean Brislance.

MINNESOTA BREEDING DISTRIBUTION

Nests have been found throughout the state, but Roberts (1932) stated the species was found less commonly in the prairie region, where he suggested it is replaced by the Barn Swallow. He emphasized that where there is a nest site, "there must be water not far away." He reported confirmed nests over widely scattered portions of the state, from Goodhue County in the southeast; Lake County in the northeast; McLeod, Grant, and Otter Tail Counties in the west; and Lake of the Woods County in the north.

Many years later, Green and Janssen (1975) simply stated that the Tree Swallow is a resident and "numerous in all regions." Janssen (1987) presented a map of confirmed nesting activity that included 49 counties in Minnesota since 1970. Hertzel and Janssen (1998) expanded the confirmed county nesting records since 1970 even further, to 60 of the 87 counties in Minnesota. The areas lacking confirmed nesting were primarily in the agricultural regions, where nesting structures or food are difficult to find. The MBS included 1,242 breeding-season locations from counties where they have completed their detailed inventories, including multiple locations from southern and west-central Minnesota.

The MNBBA included 4,820 records with confirmed nesting in every county and demonstrated that the Tree Swallow is one of the most widely distributed species in Minnesota **(A)**. It was recorded in 53.2 percent (2,541/4,779) of the blocks and confirmed nesting in 21.6 percent of those blocks. The MNBBA documented extensive breeding activity throughout much of southern Minnesota and especially in the central and west-central regions of the state.

The MNBBA probability map supported these observations, with highest densities predicted in the transition region between the forested areas of the northeast and the agricultural, open landscapes of the western and southern regions **(B)**. Higher densities were predicted along many of the large river floodplains such as the Minnesota and Mississippi Rivers.

The distribution of the Tree Swallow in Minnesota has likely not changed substantially from the past. Intensive agriculture and urbanization that has resulted in a landscape devoid of suitable nest sites, whether it be trees or nest boxes, have likely reduced local distributions and overall populations in the state. The placement and maintenance of nest boxes for the Eastern Bluebird have aided successful nesting for the Tree Swallow in otherwise unsuitable habitats. Because of the lack of detailed distributional data in Minnesota from the late 1800s and early 1900s, it is impossible to decipher specific population changes.

Winkler and colleagues (2020), in their review of the Tree Swallow in North America, primarily focus on changes in the species' distribution in the southern United States. Before the 1920s, the Tree Swallow was uncommon south of Maryland but is now a regular nesting species in South Carolina and Georgia, and westward to Texas and Oklahoma. Its expansion is continuing and likely due to land clearing, reintroduction of beaver, expansion of its diet, colonization of bluebird houses, and damming of streams.

BREEDING HABITAT

Tree Swallows are widely found in open areas with suitable nesting trees or nest boxes, and commonly near water (Niemi and Pfannmuller 1979; Mengelkoch et al. 2004; Winkler et al. 2020) **(C)**. These areas include meadows, agricultural fields, pastures,

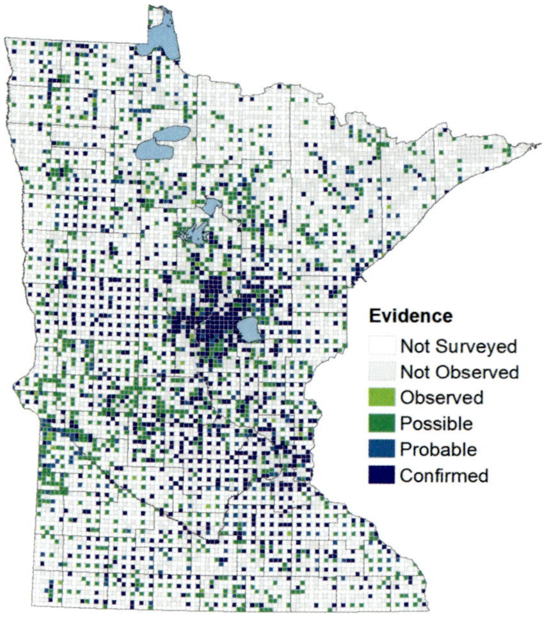

A. Breeding distribution of the Tree Swallow in Minnesota based on the Breeding Bird Atlas (2009–13).

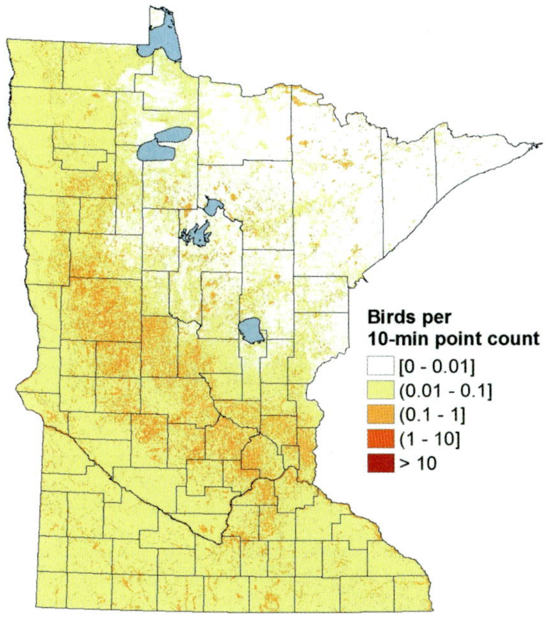

B. Predicted breeding distribution (birds detected per 10-minute point count) of the Tree Swallow in Minnesota based on habitat, landscape context, and climate data gathered during the Minnesota Breeding Bird Atlas (2009–13) using the Generalized Linear Modeling method without an adjustment for detectability.

wetlands, suburban areas, shorelines, and recently logged areas (Niemi and Hanowski 1984b; Niemi, Howe, et al. 2016). The species may be limited by the availability of suitable nest sites, such as those excavated by the Northern Flicker, Yellow-bellied Sapsucker, Hairy Woodpecker, and Downy Woodpecker.

POPULATION ABUNDANCE

Based on BBS data and other supporting information, PIF (2021) estimated the US and Canadian breeding population as 19 million adults, with 730,000 in Minnesota. The MNBBA did not estimate a breeding population because of the possibility of inflated estimates due to roadside bias in counting birds, where they are often found near nest boxes, on fence lines, and on telephone and power lines.

The BBS in Minnesota indicated a significantly increasing population of 1.05 percent per year from 1967 to 2019 **(D)**. However, over a similar time period, Tree Swallows have been significantly declining both surveywide and in Canada by an average of −0.67 percent and −1.22 percent per year, respectively. Rosenberg, Kennedy, and colleagues (2016) estimated that the population in North America has declined by −40 percent from 1970 to 2014.

Population declines in the Tree Swallow have been associated with reductions in the availability of aerial insects, their primary food source, and lack of suitable nest sites due to land use change, such as intensification of agricultural practices, land conversion to urbanized and industrial landscapes, and intensive clear-cutting of forests. Inclement weather events in spring, when the species initiates nesting, can also cause local declines (Hobson and Schieck 1999; Cadman et al. 2007; Nebel et al. 2010). Aerial insectivores, in general, have experienced declining populations in North America (Nebel et al. 2010). Among the common denominators are the loss and degradation of wetland habitats. Sediment contamination of many aquatic bodies, for example, has reduced populations of emerging, flying insects.

CONSERVATION

The Tree Swallow is still a common bird with a large population and wide breeding distribution, but its population decline over the past forty years supports moderate conservation concern. Several studies have documented the cocktail of contaminant compounds that both adults and young are exposed to from contaminated sediments in insect food sources (Ankley et al. 1992; P. D. Jones et al. 1993; Nichols et al. 1995). The overall effects of these contaminants on wildlife are not completely known, but Heinz and colleagues (2009) clearly showed that methylmercury has adverse effects.

C. Tree Swallows are usually found nesting where there is a suitable nesting tree or nest box close to water where they can forage on flying insects.
Photo by Gerald J. Niemi.

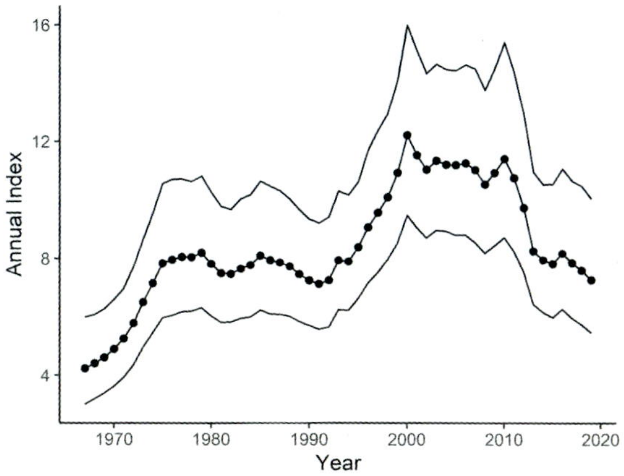

D. Breeding population trend and 95 percent credible interval (solid black lines) for the Tree Swallow in Minnesota based on the 1967–2019 Breeding Bird Survey (Sauer et al. 2020).

Reductions in available nest sites in open agricultural regions have been partly mitigated by nest box placement for the Eastern Bluebird, but the efforts still reflect a relatively small part of the Tree Swallow range. In forested areas, clear-cut forests provide limited nesting opportunities, but natural disturbance regimes such as fire, wind, and insect outbreaks leave many available nesting substrates with suitable cavities. In Minnesota, however, harvesting guidelines encourage leaving 6–12 trees per acre within a logging site (Minnesota Forest Resources Council 2005).

Under a 2°C warming scenario, NAS (2019) projected moderate vulnerability for the Tree Swallow's breeding distribution in North America, with a 28 percent loss of its breeding range and a 26 percent gain northward by 2050. Under this scenario, breeding-range retractions were predicted northeastward from southern and western regions of Minnesota. However, as Winkler and colleagues (2020) point out, Tree Swallows have been breeding in more southern regions in recent years. Dunn and Winkler (1999) found the species' average laying date was nine days earlier in North America from 1959 to 1981 owing to higher spring temperatures.

Many factors have negatively affected the ecology of the Tree Swallow over the past 150-plus years, such as habitat loss and changes, exposure to contaminants, changes in food supplies, and, most recently, warmer temperatures with more variable weather conditions. The overall future of this species is difficult to predict, but many areas of Minnesota would not be the same without this swift and agile bird scooping up flying insects.

NORTHERN ROUGH-WINGED SWALLOW

Stelgidopteryx serripennis

Minnesota Seasonal Status: Regular breeding resident and migrant. The Northern Rough-winged Swallow was uncommon during the MNBBA.
Migration: Short- to medium-distance migrant that winters in southern California and along the Gulf coast of the United States south through Central America; most populations along the southern border of the United States and farther south are year-round residents.
Food: Aerial insectivore.
Nest: Usually nests in burrows constructed by other species, including Belted Kingfishers, Bank Swallows, and small mammals. Breeding pairs may nest singly or in small groups consisting of a few pairs.
Conservation Concern: Assigned a moderate PIF Continental Concern Score of 10/20.

Photo by Jean Brislance.

The Northern Rough-winged Swallow's name is a direct reference to this little swallow's most unusual feature: small, stiff barbs along the outside of the first primary wing feather. The barbs are present on both the male and female; the male's barbs are longer and have a strongly curved hook at their tip, but the hook is either absent or only slightly curved on the female. Even the bird's scientific name emphasizes the uniqueness of this trait. *Stelgidopteryx* means "scraper wing" and *serripennis* means "saw feather." The purpose of these structural features remains somewhat of a mystery. In flight, the feathers may produce a shrill whistle (De Jong 2020); Lunk (1962) suggested the sound may serve a role in courtship or territorial defense.

Apart from this unique physical feature, the species' most infamous trait may be that it is frequently overlooked. Unlike Minnesota's other swallows, the Northern Rough-winged Swallow lacks distinctive plumage characteristics that enable quick identification, such as the Barn Swallow's deeply forked tail or the Cliff Swallow's rusty rump. Dull brown on the back and wings, with a light brown throat that grades into a dull white belly, the Rough-winged Swallow is easily confused with the Bank Swallow. Although it lacks the Bank Swallow's dark breast band, the absence of this feature can be difficult to discern while the birds are engaged in their aerial acrobatics. Look for the Rough-winged Swallow's square tail and listen for its low, hoarse call note, *brrr,* to help with identification.

Widely distributed from British Columbia, east across the southern edge of the Canadian provinces, and south across the United States and throughout much of Central America, the Northern Rough-winged Swallow is sparsely distributed throughout its breeding range.

MINNESOTA BREEDING DISTRIBUTION

The Swallow was described as a breeding resident throughout the state; Roberts's (1932) only caveat regarding the species' distribution was that he doubted it occurred in the state's Northwest Angle, remarking that the "conditions" there were not suited to the Swallow's needs. Presumably, he was considering the Angle's broad, flat landscape, where the vertical banks required for nesting are in short supply. There were, however, reliable reports from many northern regions, including the Red River valley, from Clay County north to Marshall County, and from Itasca State Park, Cass County, Itasca County, and the northwest corner of Kittson County. Confirmed and inferred nesting evidence was available from nine counties in southern and central Minnesota (Fillmore, Hennepin, Houston, McLeod, Pipestone, Rock, Sherburne, Wabasha, and Washington) as well as from Leech Lake and Itasca State Park. Roberts's brief account notes that, although the species was common, it was less abundant than the Bank Swallow.

Forty years later, Green and Janssen's (1975) updated account of the species' distribution noted that it was numerous in southern and central Minnesota and less common farther north, becoming scarce in the northeast. Janssen (1987) identified 20 counties, widely dispersed throughout the state, where nesting had been confirmed since 1970. Hertzel and Janssen (1998) later added another six counties to the list, all located from Aitkin County south. By 2014, fieldwork by the MBS documented 197 breeding-season locations. The species was sparsely distributed in northern Minnesota; more than half of the records were located from the Minnesota River valley south to the Iowa border.

Similar results were reported during the MNBBA. Although 615 Northern Rough-winged Swallow records were found in 83 of Minnesota's 87 counties, the majority of records were located in southern Minnesota, south of a line from Big Stone County in the west to Chisago County in the east **(A)**. The least common member of the swallow family, with the fewest total number of MNBBA records, the birds were detected in 9.8 percent (465/4,743) of the surveyed atlas blocks and in 13.1 percent (306/2,337) of the priority blocks. Breeding was confirmed in 110 blocks (2.3 percent) and 59 counties. Thirty-six of the counties were new to the list published by Hertzel and Janssen (1998); seven of the new counties were included because of blocks that straddled county lines. Despite their wide distribution, Roberts's (1932) original suspicion proved true; there were no records from the Northwest Angle!

There have been few changes in the distribution and relative abundance of the Northern Rough-winged Swallow in Minnesota over the past one hundred years. Elsewhere, records suggest the species may be expanding north in the Midwest and New England, and south along the Florida peninsula. Whether this represents a natural range expansion or is simply the result of better documentation by field observers is unknown (De Jong 2020). Anyone who has struggled to correctly identify swallows as they swiftly dart about appreciates the challenges of field identification. Even John James Audubon, in his first encounter with Northern Rough-winged Swallows, summarily dismissed them as Sand Swallows (now known as Bank Swallows) until, tired at the end of a long day, he sat down and shot a few and could clearly see their distinguishing features (Bent 1942).

BREEDING HABITAT

The Northern Rough-winged Swallow is a species of the open country. Although it is closely associated with water, its selection of habitat is likely influenced more by the availability of suitable nest sites. A burrow nester, the Swallow prefers vertical banks of sand, gravel, or clay. Whether they excavate their own nesting burrows or simply rely on burrows abandoned by other excavators, such as Kingfishers and Bank Swallows is still disputed (De Jong 2020). Regardless, suitable substrates for primary excavators are often adjacent to water bodies and found along rocky gorges, lake shorelines, and the banks of rivers and streams **(B)**. In agricultural regions they have been reported nesting in the steep banks of irrigation ditches (Dawson 1923).

An adaptable species, the Northern Rough-winged Swallow may nest in anything resembling a crevice along a vertical surface, including cracks in building walls, drainpipes in retaining walls, and quarry embankments (Bent 1942; Roberts 1932). Foraging usually occurs over water, open fields, and low brush (De Jong 2020).

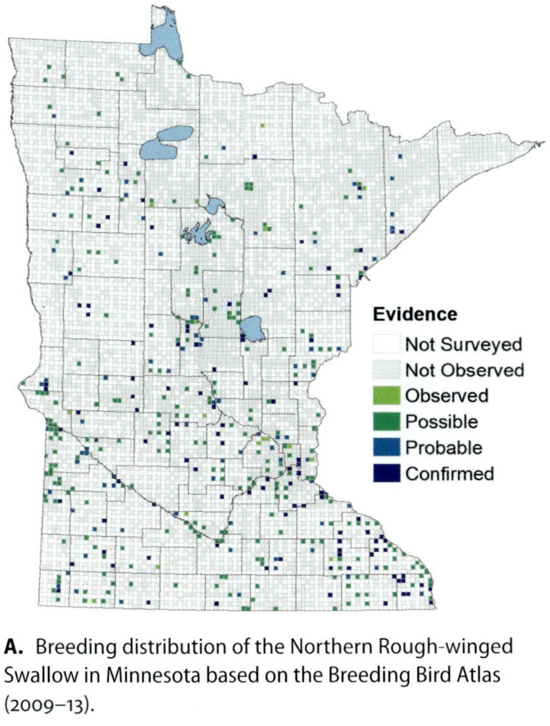

A. Breeding distribution of the Northern Rough-winged Swallow in Minnesota based on the Breeding Bird Atlas (2009–13).

B. Exposed stream banks often provide old burrows constructed by Bank Swallows or Belted Kingfishers that the Northern Rough-winged Swallow can use for nesting near an abundant food source. Photo by Lee A. Pfannmuller.

POPULATION ABUNDANCE

From Canada south through Central America, the Northern Rough-winged Swallow is estimated to number approximately 27 million breeding adults, of which 20 million reside in the United States and Canada (PIF 2021). Approximately 180,000 adults occur in Minnesota.

Across the BBS survey region, the population has experienced a nonsignificant decline, averaging only −0.22 percent per year from 1966 to 2019 **(C)**. Although the species has fared well in the southeastern United States, numbers are declining elsewhere. In Minnesota, population trends were somewhat unreliable owing to the species' sparse distribution along BBS survey routes, but they document a steeper and significant decline, averaging −1.91 percent per year, from 1967 to 2019 (Sauer et al. 2020).

Factors responsible for these long-term declines are not well known. The availability of suitable nest sites is considered a primary concern, yet the species has done well adapting to nesting opportunities in developed landscapes. A decline in prey items is another concern. The species is one of many aerial insectivores whose populations have declined in recent years. For many of these species, the declines became most notable in the 1980s and were particularly acute in the northeastern United States, leading some to speculate that industrialization may be negatively impacting insect populations (Nebel et al. 2010; R. Smith et al. 2020). Indeed, Bird Conservation Regions where Northern Rough-winged Swallow numbers have declined include the lower Great Lakes–St. Lawrence and Prairie Hardwoods Transition, which include such large metropolitan areas as Detroit, Cleveland, and Toronto. A steeper decline, however, has been documented in the Boreal Hardwood Transition Bird Conservation Region, which spans the northern Great Lakes and includes the extensively forested landscapes of northern Minnesota, Wisconsin, and Michigan (Sauer et al. 2020). Because the Swallow is an open-country species, often restricted to the riparian shorelines of rivers and lakes in forested landscapes, the increase in recreational homes along these shorelines may be contributing to the decrease in northern populations.

CONSERVATION

Management recommendations and conservation actions for this little Swallow are limited. Artificial nest burrows have been utilized successfully in some situations, resulting in occupancy and nest success as high as 80–90 percent (De Jong 2020; Lunk 1962). Fortunately, warmer temperatures are not a threat and are projected to expand the northern limits of the species' breeding range across northern Minnesota and southern Canada (NAS 2019). Nevertheless, the increasing attention directed at the loss of insect populations in general, and aerial insects in particular, warns against complacency (e.g., Sánchez-Bayo and Wyckhuys 2019). Designated a moderate conservation priority (PIF 2021), this rather plainly attired and relatively silent swallow definitely deserves careful monitoring in the years ahead.

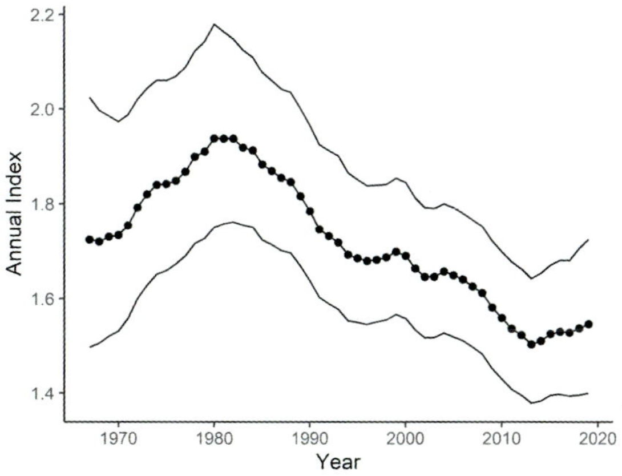

C. Surveywide breeding population trend and 95 percent credible interval (solid black lines) for the Northern Rough-winged Swallow based on the 1967–2019 Breeding Bird Survey (Sauer et al. 2020).

PURPLE MARTIN

Progne subis

Minnesota Seasonal Status: Regular breeding resident and migrant. The Purple Martin was uncommon during the MNBBA.
Migration: Long-distance migrant that winters in South America.
Food: Aerial insectivore.
Nest: Secondary cavity nester. Eastern populations nest almost entirely in multi-compartment, man-made nesting structures; western populations can still be found nesting singly in natural cavities.
Conservation Concern: Assigned a moderate PIF Continental Concern Score of 10/20; officially listed as *Special Concern* in Minnesota.

Photo by Mike Lentz.

For centuries, North Americans have had a love affair with the continent's largest swallow, the Purple Martin. From the hollow gourds that Native Americans hung to attract this cavity-nesting bird to their villages, to the elaborate condominiums that today's Martin enthusiasts construct, the species has always been a welcomed neighbor. "Those who have a word of criticism for it," Bent (1942) wrote, "are few and far between."

The male Purple Martin is distinguished by its glossy, bluish-black plumage. Swallows are notoriously difficult to identify as they dart overhead, but the male Martin is the only Minnesota swallow to show a dark belly as it sails by. The female, by contrast, has a lighter blue-black plumage on the back and a gray-brown plumage underneath. When she alights, a gray collar around her nape is also apparent. Their joyful, gurgling chatter is a medley of different vocalizations, including rattles, chirps, and clicks. Although early naturalists debated whether the Martin's vocal efforts were in fact melodious, they have always been pleasing to those who admire the bird.

Distributed across the eastern United States and the southern regions of central and eastern Canada, the Purple Martin also occurs along the Pacific coast, in the southwestern United States, and in Mexico. Breeding densities are highest in the southeastern states. The scientific name for the Martin's genus, *Progne,* is a reference to the legendary daughter of the king of Athens, Procne, who, according to Greek mythology, turned into a swallow.

MINNESOTA BREEDING DISTRIBUTION

Roberts (1932) considered the Purple Martin to be a common statewide summer resident. Even in the prairies it was found in "considerable numbers in the towns and about farmhouses." Although today we associate the species strictly with artificial nesting structures, in Roberts's day Martins were still found nesting in natural tree cavities, in cavities excavated by woodpeckers, on cliff ledges, and in rock crevices, especially in less settled regions of the state. Indeed, Roberts included an account of Martins nesting among large boulders on two tiny islands in Mille Lacs Lake, Spirit Island and Hennepin Island. Roberts described how the birds, first reported on the islands in 1886, were nesting "in the midst of the Common Tern colonies, with which birds they were at constant war, as the Terns no doubt ate the eggs and young of the Martins when they could reach them." Despite these challenges, Martins apparently persisted on the islands for more than fifty years. The last report was logged in 1938, when "many nests" were seen on Spirit Island (A. B. Erickson and Upson 1938). Roberts also compiled breeding evidence from 12 other counties and from Itasca State Park and Cass Lake. The counties stretched from Fillmore and Waseca Counties in the south to Lake of the Woods, Marshall, and Polk Counties in the far northwest.

Green and Janssen (1975) reported that Martins were a statewide breeding resident but were most common along the Mississippi River and about the towns and cities of central and western Minnesota. Janssen (1987) identified 36 widely dispersed counties where Martins had been confirmed nesting since 1970. Hertzel and Janssen (1998) added seven more counties to the list. The MBS reported 101 breeding-season locations restricted primarily to central and southern Minnesota. However, their focus on natural communities precluded an in-depth survey of small towns and recreational areas where the species is frequently encountered.

The 537 surveyed atlas blocks (11.2 percent) where MNBBA participants detected Purple Martins clearly demonstrate the species' association with the lakes region of central Minnesota **(A)**. The efforts of energetic atlas participants in Brainerd certainly contributed to the high number of records, but the region is also populated with many lakeshore residents who maintain Martin homes on their property. Indeed, the string of breeding records encompassing the shoreline of Mille Lacs Lake is an impressive demonstration of their enthusiasm for the species. Although the Martin was virtually absent from northeastern Minnesota and was sparsely distributed in the northwest and southern regions, the records still spanned 75 counties. Breeding was confirmed in 350 blocks (7.3 percent) that encompassed 68 counties; 32 of the counties were additions to the list published by Hertzel and Janssen (1998).

It is unclear if the absence of MNBBA records in northeastern Minnesota represents a range contraction or population decline. Roberts (1932) cited just a handful of records from the region that were restricted to single pairs nesting in natural cavities or woodpecker holes. Today, unlike central Minnesota's lakes region, the northern forest's extensive array of public lands precludes a concentration of lake cabins where Martin homes may be erected. In some areas of its breeding range, however, the species has receded southward, partic-

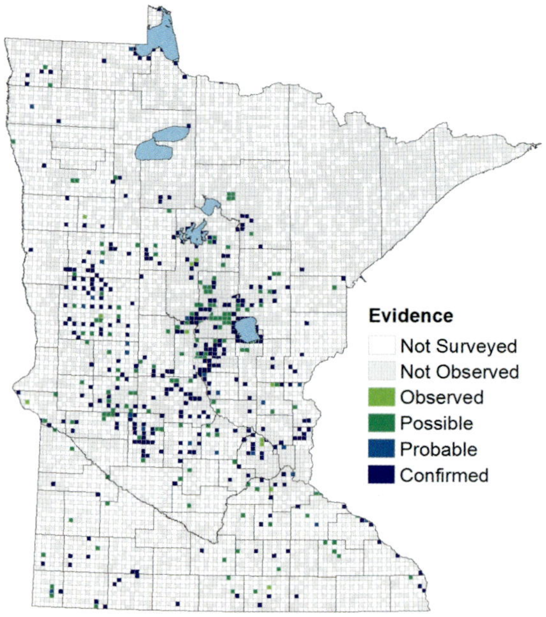

A. Breeding distribution of the Purple Martin in Minnesota based on the Breeding Bird Atlas (2009–13).

B. The enthusiasm of homeowners for establishing and properly managing Purple Martin houses may be the key to the species' long-term survival. Photo by Lee A. Pfannmuller.

ularly in northwestern Canada (C. R. Brown and Tarof 2020) and Ontario (Cadman et al. 2007). Factors responsible for the range contraction are unclear, but weather may be responsible. In the Great Plains, however, the installation of Martin houses has facilitated a small westward expansion (C. R. Brown and Tarof 2020).

BREEDING HABITAT

Prior to the large-scale use of Martin houses, the species was found along forest edges and in riparian areas that provided an abundance of dead snags with nesting cavities and aerial insects for food. However, early efforts to entice Martins to nest in hollowed-out nesting gourds were not uncommon in Native American villages (Sprunt 1942) and, in later years, on southern rice plantations (Forbush 1929). Because the species fiercely defends its breeding territory, nesting birds were encouraged as a means of protecting domestic poultry from hawks and eagles.

In Minnesota, nesting now occurs wherever suitable nesting structures are erected, from rural towns and lake cabins to urban and residential landscapes **(B)**. Guidelines usually recommend the structures be placed in open areas, 12 m from trees and within 9–36 m of housing (Purple Martin Conservation Association 2016). Nearby meadows, agricultural fields, marshes, and lakes provide foraging habitat.

POPULATION ABUNDANCE

PIF (2021) estimated the Purple Martin population in the United States and Canada at 8.7 million birds, with approximately 49,000 residing in Minnesota. Throughout the BBS survey region, the species has experienced a slow but significant decline, averaging −0.54 percent per year from 1966 to 2019 **(C)**. In Minnesota, the statistical precision of the BBS data is compromised by the species' low abundance, but the data indicate that Purple Martins have decreased at an alarming rate, averaging −4.80 percent per year, resulting in a cumulative decline of 92 percent from 1967 to 2019 (Sauer et al. 2020)! Populations have been plummeting across the northern and eastern portions of its breeding range, with similar declines in states and provinces across the Great Lakes, New England, the Southeastern Coastal Plain, and eastern Canada.

Many factors may be at play. Because the declines are steepest in regions where human population and industrial development densities are high, Nebel and colleagues (2010) postulated that atmospheric pollutants may be negatively impacting the species' sole diet item, aerial insects. Like other aerial insectivores, migrating Martins are also vulnerable during the spring if cold weather and precipitation impact aerial insect populations. Indeed, C. R. Brown and Tarof (2020) believe that adverse weather is greater than "all other sources of mortality combined."

Martins also compete with European Starlings and House Sparrows for nest cavities. In some areas, the availability of suitable cavities is a major factor limiting populations (C. R. Brown and Tarof 2020). Fewer homeowners erecting and maintaining Purple Martin nesting structures presents another concern. Kelly Applegate commented that "the martin house used to be an American tradition. Back in the 50s, you had your American Flag and your martin house in the front yard. But you just have a lack of interest from generations that are coming up now" (Mohr 2011).

Finally, aggregating the birds into large colonies in small spaces increases their vulnerability to nest predators and nest parasites. Both can significantly reduce reproductive success, particularly if sites are poorly managed (C. R. Brown and Tarof 2020).

CONSERVATION

The Purple Martin's population decline prompted its assignment as a moderate conservation priority (PIF 2021) and its official listing as *Special Concern* in Minnesota (Minnesota Administrative Rules: Chapter 6134). Fortunately, the impacts of climate change appear minimal because the species will experience greater gains than losses in its North American breeding distribution under all three warming scenarios examined by NAS (2019). The only loss of habitat, however, is projected to occur along the very northwestern edge of its breeding range, stretching from western Minnesota northwest to central Alberta.

Ultimately, the Purple Martin's future may really depend on dedicated bird enthusiasts. Having adapted to the use of artificial nest cavities, the species has become entirely dependent on the proper management of these structures. Websites and articles abound with recommendations on how to be a successful Purple Martin landlord, including installing predator guards and actively discouraging nest competitors (e.g., Hill 1993; Perry and Bloch 1987). Addressing these recommendations may be the best way to improve the future of this beloved Minnesota neighbor.

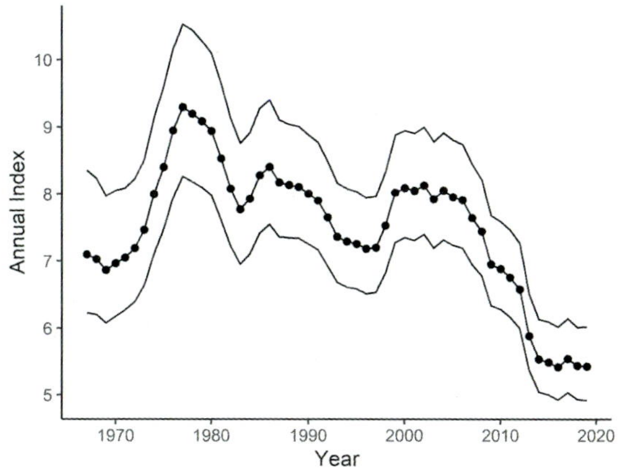

C. Surveywide breeding population trend and 95 percent credible interval (solid black lines) for the Purple Martin based on the 1967–2019 Breeding Bird Survey (Sauer et al. 2020).

BARN SWALLOW

Hirundo rustica

Minnesota Seasonal Status: Regular breeding species and migrant. The Barn Swallow was abundant during the MNBBA.
Migration: Medium- to long-distance migrant that winters in Central and South America.
Food: Aerial insectivore.
Nest: Cup nest made of mud and dried grass. Originally nested in caves and rocky crevices; today nearly all nests are located on man-made structures that provide a vertical wall with an overhanging eave or a flat surface. Nests singly or in colonies.
Conservation Concern: Assigned a low PIF Continental Concern Score of 8/20.

Photo by Karl Bardon.

Although many swallows challenge our identification skills, there is no mistaking the identity of the Barn Swallow. Its long, steely-blue tail is deeply forked, a trait unique among Minnesota's six breeding swallow species. The bird is boldly colored; its deep blue crown, back, upper wings, and tail contrast with its chestnut forehead and throat and its rufous chest and belly. The female is similar in appearance, though her outer tail feathers, or streamers, are slightly shorter; her chest and belly are also lighter. The Barn Swallow's song, described by Roberts (1932) as a "pleasing, twittering song," followed by mechanical gratings, can last for 20 seconds or longer and is given by both sexes in flight and while perched.

Given the species' global distribution and long association with humans, legends about the Barn Swallow abound. Many are about the bird's remarkable tail feathers. An Islamic legend relates how, in the Garden of Eden, the Swallow "offended the serpent Elbis, which retaliated by snapping a chunk out of the middle of the bird's tail" (Warren-Chadd and Taylor 2016). In North America, Native Americans attribute the Swallow's forked tail to an angry god who, upon discovering the bird had stolen fire from the gods to bring to native people, threw a fireball that singed the middle of its tail (M. B. Brown and Brown 2020). German legend has a more somber story of an abused wife who became a Swallow to escape her husband's beatings. Enraged, he pinned her down with a knife between her tail feathers. As she struggled to escape, her tail was split in two (Baars 2017). Legends aside, the length of the Swallow's tail and the extent to which the longest, outer tail feathers are symmetrical may be a signal of the bird's overall fitness, including both males and females. Although it appears important in mate selection, tail length is not a predictor of either reproductive success or annual survival (M. B. Brown and Brown 2020).

The Barn Swallow also figured prominently in the launch of the national conservation movement. Stories have been told for nearly a century about how the use of heron and egret feathers in the millinery trade incited outrage. But it was the slaughter of Barn Swallows solely for the purpose of using their cobalt blue feathers on women's hats that prompted George Bird Grinnell to write an editorial in his conservation magazine *Forests and Streams* in 1886 (CLO 2017b). Deploring the carnage, Grinnell asked his readers to sign a pledge to never wear clothing that featured wild bird feathers. He went on to propose the formation of the Audubon Society, an association dedicated to the protection of wild birds.

The Barn Swallow is found throughout Europe, Asia, northern Africa, and Mexico; North American breeding populations range from southern Alaska east to the Maritime Provinces and south throughout much of the United States and northern Mexico. Breeding densities are highest in the eastern and central states. Within this area, south-central Minnesota is one of the pockets of highest Barn Swallow density.

MINNESOTA BREEDING DISTRIBUTION

Roberts (1932) described this colorful Swallow as an abundant, statewide summer resident. Although he doubted it was present in the western grasslands prior to European settlement, for lack of suitable nesting sites, it had become "the commonest Swallow of the prairie country, finding suitable nesting-places about farmhouses and in the environs of towns." He went on to write, "Its good habits, beautiful plumage, and familiar and confiding ways have made it one of the best known and most loved birds of the rural districts." Although closely associated with human habitations in western Minnesota, Barn Swallows could still be found in crevices along the North Shore's steep escarpments in the early 1900s. At the time, Roberts reported confirmed nesting in nine counties (Hennepin, Isanti, Jackson, Kandiyohi, McLeod, Polk, St. Louis, Traverse, and Wabasha), while nesting was inferred in three counties (Itasca, Rock, and Sherburne). Nesting was also reported from Mille Lacs and Itasca State Park.

Years later, Green and Janssen (1975) simply noted that the Barn Swallow had become most common in the prairie region and "least numerous in heavily wooded areas." By the time Janssen wrote his updated account in 1987, the species was also common in central Minnesota. He identified 55 counties, widely dispersed across the state, where nesting had been confirmed since 1970. Hertzel and Janssen (1998) later added another nine counties to the list. By 2014, the 889 breeding-season locations documented by the MBS further demonstrated the species' wide distribution, particularly in agricultural regions of the state.

Indeed, during the MNBBA participants documented 4,863 Barn Swallow records in 55.4 percent (2,663/4,806) of the surveyed atlas blocks and in 70.9 percent (1,656/2,337) of the priority blocks. Breeding was confirmed in 1,022 blocks and in all 87 counties **(A)**. Although they were found statewide, the birds were least abundant in the eastern and northern regions of the Laurentian Mixed Forest Province. They were the most frequently reported member of the swallow family during the MNBBA, edging out the Tree Swallow by just 43 records.

The MNBBA probability map illustrates the species' predicted abundance throughout Minnesota **(B)**. Densities are lowest in the Laurentian Mixed Forest Province and are relatively low in

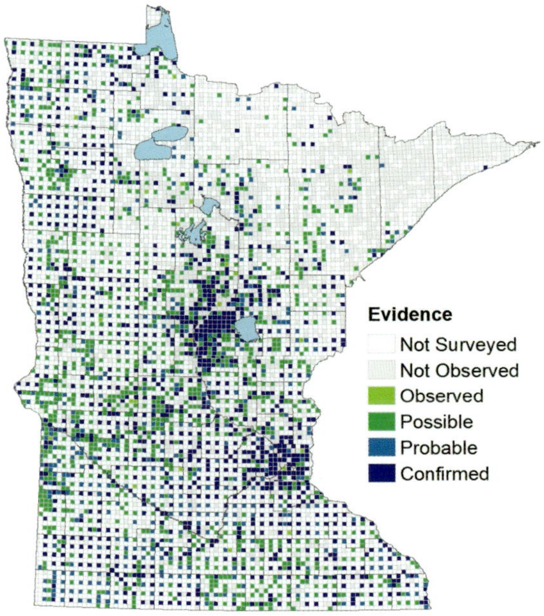

A. Breeding distribution of the Barn Swallow in Minnesota based on the Breeding Bird Atlas (2009–13).

the Tallgrass Aspen Parklands Province compared to higher breeding densities found throughout the Prairie Parkland and Eastern Broadleaf Forest Provinces. Scattered areas of higher abundance occur in southwestern and west-central Minnesota.

Since Roberts (1932) wrote his original account of the species' status, there have been no significant changes in the Barn Swallow's distribution. It is far more common, however, in the developed areas of the Laurentian Mixed Forest Province than it was one hundred years ago, when much of the region was sparsely settled. The Swallow's adaptation to human landscapes has enabled it to expand its distribution into many areas of the United States where it did not formerly occur. Perhaps most notable is its expansion south through the Atlantic and Gulf states in the latter half of the twentieth century (M. B. Brown and Brown 2020).

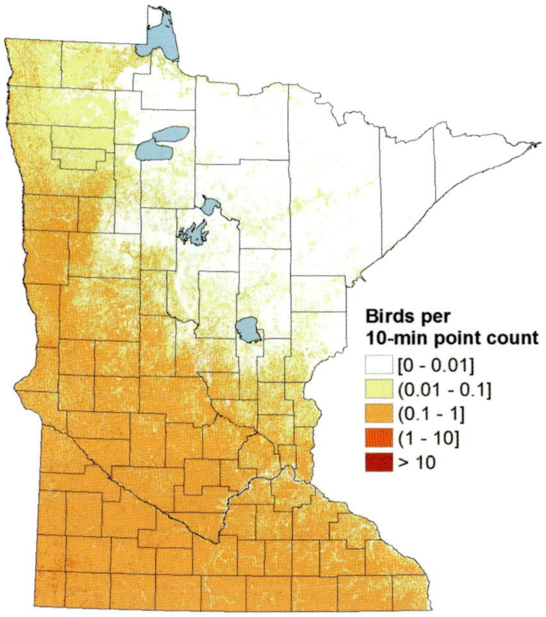

B. Predicted breeding distribution (birds detected per 10-minute point count) of the Barn Swallow in Minnesota based on habitat, landscape context, and climate data gathered during the Minnesota Breeding Bird Atlas (2009–13) using the Generalized Linear Modeling method without an adjustment for detectability.

BREEDING HABITAT

Although originally confined to caves, rocky ledges, and cut banks, today's Barn Swallow is almost entirely dependent on man-made structures for breeding **(C)**. The conversion is so complete that it is rare to observe the birds nesting in natural settings. While Barn Swallows are believed to have been largely solitary nesters (one to two pairs) in their natural habitat, today they often nest in small groups of two to 35 pairs (M. B. Brown and Brown 2020). Suitable nesting sites need to meet only three requirements: a nearby source of mud for nest construction, open areas for foraging, and buildings for nest placement (M. B. Brown and Brown 2020).

POPULATION ABUNDANCE

Approximately 47 million Barn Swallows reside in the United States and Canada, with nearly 1.4 million breeding adults in Minnesota (PIF 2021). Since the BBS began in the mid-1960s, Barn Swallow populations have experienced a significant decline throughout the continental survey region, averaging −0.55 percent per year from 1966 to 2019 (Sauer et al. 2020). The cumulative loss between the years 1970 and 2014 was 38 percent (Rosenberg, Kennedy, et al. 2016). The decline is widespread across southern Canada and the northern half of the United States. Population numbers in Minnesota have been relatively stable after experiencing a large increase and subsequent decline from 1967 through 2000 **(D)**.

An aerial insectivore, the Barn Swallow is one of many members of the foraging guild that is experiencing population declines. The drop in numbers is particularly acute in the northeastern United States. Nebel and his colleagues (2010) have suggested that the increase in atmospheric pollutants in this industrialized region and their negative impact on aerial insects may be responsible for the decline of aerial insectivores. M. B. Brown and Brown (2020) suggested that strong cold snaps in the northern regions during the spring, when Barn Swallows are just beginning their northward migration, may also regulate populations. Not only are these weather events physically difficult for the birds but also the availability of sufficient food resources is limited.

C. The Barn Swallow frequently selects buildings that provide some overhead protection and are located in or near fields and wetlands for foraging, to construct its nest. Photo by Gerald J. Niemi.

CONSERVATION

Although populations are declining, the Barn Swallow's global distribution, coupled with its adaptation to human-modified landscapes, is the primary reason it is not a conservation priority (PIF 2021). Even warmer temperatures do not appear to pose a threat (NAS 2019). As temperatures rise, breeding range expansions are predicted to exceed range losses, with no changes projected in Minnesota. Nevertheless, close monitoring is warranted, given recent studies regarding declining aerial insect populations (e.g., Sánchez-Bayo and Wyckhuys 2019). Overall, the Barn Swallow's adaptability has served it well. It remains one of the most abundant members of Minnesota's avifauna, ranking nineteenth among the 250 species reported during the MNBBA based on the total number of records.

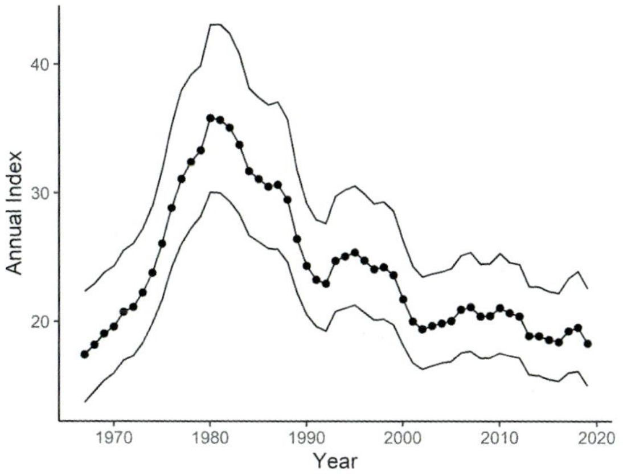

D. Breeding population trend and 95 percent credible interval (solid black lines) for the Barn Swallow in Minnesota based on the 1967–2019 Breeding Bird Survey (Sauer et al. 2020).

CLIFF SWALLOW

Petrochelidon pyrrhonota

Minnesota Seasonal Status: Regular breeding species and migrant. The Cliff Swallow was common during the MNBBA.
Migration: Long-distance migrant that winters in South America.
Food: Aerial insectivore.
Nest: Elaborate gourd-like structure constructed with mud and attached to a natural vertical surface, a man-made structure, or another Cliff Swallow nest. This highly social bird nests in large colonies.
Conservation Concern: Assigned a low PIF Continental Concern Score of 6/20.

The Cliff Swallow is well-known for its colonial habits and its propensity to nest under bridges or within barns and culverts. Indeed, it was often referred to as the Eave Swallow because of its habit of securing nests to the juncture of an overhanging roof and the supporting wall. But the name Cliff Swallow has endured and is an appropriate tribute to the species' original breeding range in the mountains of western North America, where vertical cliffs and rocky crevices provide ideal nesting habitat.

Petrochelidon pyrrhonota, the species' scientific name, translates to "red-backed rock swallow," a reference to the one feature that distinguishes it from all other Minnesota swallows: its pale-orange rump. Other characteristics of this handsome blue-gray bird include a square tail, a creamy yellow forehead, rufous patches on the sides of the head and throat, and creamy breast feathers that extend across the back of the neck. Both sexes are identical. Their call is a series of squeaky and harsh-sounding chatters that last a few seconds.

The Cliff Swallow nests in colonies that can number from a few hundred pairs to a few thousand. This gregariousness has led to the development of several adaptive behaviors, including a highly developed vocal system that allows adults to recognize the voice of their young, and the habit of parasitizing neighboring Swallow nests. Laying eggs in their neighbor's nest, or moving their own eggs into other nests, are strategies to increase reproductive success in colonies infested by the American swallow bug (*Oeciacus vicarius*). Swallow bugs feed on the blood of adults and young and can cause nest abandonment, low hatchling survival, premature fledging, and death (C. R. Brown et al. 2020).

Photo by Mike Lentz.

The third most abundant of the six swallows that nest in Minnesota, the Cliff Swallow's breeding distribution ranges across most of Canada, the United States, and central Mexico. Rare in the southeastern coastal states, it is abundant west of the Mississippi River.

MINNESOTA BREEDING DISTRIBUTION

Prior to European settlement, Cliff Swallows were limited to sites where natural cliffs and rocky escarpments provided suitable nesting habitat. Explorer G. W. Featherstonhaugh found the species nesting on granite escarpments along the upper reaches of the Minnesota River in 1834. Another observer reported it was abundant along the Blue Earth River about 1848, while Dr. Elliott Coues "found it was so abundant at Fort Pembina on the Red River of the North in 1873 as to be a nuisance" (Roberts 1932). Such natural nesting opportunities were fairly limited in the state.

As European immigrants began constructing homes, farms, and bridges, the structures created new nesting opportunities that enabled the Swallows to occupy areas that were previously uninhabitable. By the late 1800s, the species was abundant throughout Minnesota, although it was less common "in the heavily timbered regions of the state" (Roberts 1932). Breeding evidence was available from 11 widely dispersed counties, including a record in Lake County of a small colony nesting at Castle Danger, along Lake Superior's shoreline.

By the year 1900, however, the species "was rapidly disappearing," not only in Minnesota but in other states as well. The explanation, Roberts (1932) wrote, was two-fold: "first, the general destruction of its nests by the farmers, and, second, the occupation of its nests by the House Sparrow."

Fifty years later, populations had recovered to the point that Robinson and Beer (1955) found the species present "in moderate numbers in the north and northeastern portions of the state." They disputed several theories regarding its previous decline, including the negative impact of the invasive House Sparrow, concluding that all reasonable hypotheses were based on pure speculation. Nevertheless, the coincidence of the rapid expansion of the House Sparrow in Minnesota in the late 1800s with the subsequent decline of the Cliff Swallow in the early 1900s seems compelling.

Twenty years later, Green and Janssen (1975) considered the Cliff Swallow a statewide breeding resident, least common in the southwestern region and most abundant in the northwest and northeast regions, "especially in St. Louis and Marshall counties." In 1987, Janssen noted that the introduction of corrugated culverts for stream crossings in western Minnesota resulted in a tremendous increase in the region's breeding population. He identified 40 counties where nesting had been confirmed since 1970, while Hertzel and Janssen (1998) later added seven more counties to the list. Their confirmed nesting records were least abundant in south-central Minnesota. Yet by 2014, the 382 breeding-season locations collected by the MBS were largely concentrated in southwestern and south-central Minnesota and along the Glacial Lake Agassiz beach ridges in Clay and Wilkin Counties.

The 2,498 Cliff Swallow records collected by the MNBBA provide the most complete picture of the species' statewide distribution **(A)**. Reported from 29.8 percent (1,437/4,821) of the surveyed atlas blocks, the birds were regular throughout all but the Laurentian Mixed Forest Province. Here they were reported in only 20 percent of the priority blocks, compared to nearly 54–56 percent in each of the other three provinces. Breeding was confirmed in 919 blocks (19.1 percent) and in 86 of Minnesota's 87 counties. Thirty-nine of the counties were additions to Hertzel and Janssen's 1998 list of counties with confirmed nesting records. Cook County was the only county where Cliff Swallows were not observed.

In their comprehensive review of the species, C. R. Brown and colleagues (2020) discussed the Swallow's eastward range expansion, led by the growth and spread of settlements across the continent's interior. The successful introduction of House

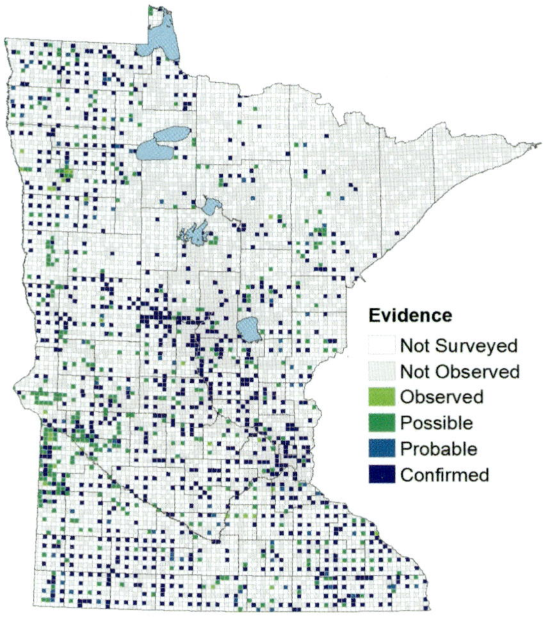

A. Breeding distribution of the Cliff Swallow in Minnesota based on the Breeding Bird Atlas (2009–13).

Sparrows in 1852 seems to have had its greatest impact in the northeastern states, the only region where Cliff Swallow populations still remain relatively small. Elsewhere, populations have increased in abundance and have expanded in distribution in the southeast and in the eastern Great Plains.

BREEDING HABITAT

Cliff Swallows still nest among natural cliffs and rocky escarpments, especially in the west, but the use of man-made structures is widespread and includes bridges, buildings, and culverts **(B)**. The availability of mud for nest construction is critical, and the birds may travel several kilometers to locate a suitable source. Nevertheless, many sites that appear ideal for a colony are not occupied, which suggests that the species' habitat requirements are not well understood (C. R. Brown et al. 2020).

POPULATION ABUNDANCE

PIF (2021) estimated a total of 78 million breeding adults in the United States and Canada, with an additional 5 million birds residing in Mexico. Minnesota supports approximately 2.7 million breeding adults. Since the BBS began in 1966, Cliff Swallow numbers have experienced a nonsignificant increase averaging 0.82 percent per year across the survey region (Sauer et al. 2020). In Minnesota, population numbers rose steeply from about 1967 through 1990 and have since leveled off, even demonstrating a small decline in recent years. Cumulatively, however, the state has witnessed a significant increase of 2.88 percent per year from 1967 to 2019 **(C)**.

The range-wide increase in Cliff Swallow numbers during the past thirty years largely reflects dramatic annual population increases in the central and southern United States, such as Iowa (13.81 percent), Missouri (14.96 percent), and Oklahoma (6.15 percent; Sauer et al. 2020). These increases are offset by declines throughout the northeastern states, southeastern Canada, and the Great Lakes, including northern Minnesota. This pattern is similar to that demonstrated by many aerial insectivores. Nebel and his colleagues suspect that the high density of industrial development across much of the northeast region may be negatively impacting aerial insect populations (Nebel et al. 2010). Weather can also be a significant cause of mortality in northern climates. Late spring cold snaps impact prey availability and, ultimately, adult survival (C. R. Brown et al. 2020). And, as noted above, nest site competition with House Sparrows may also impact Cliff Swallow populations, especially in the northeastern United States.

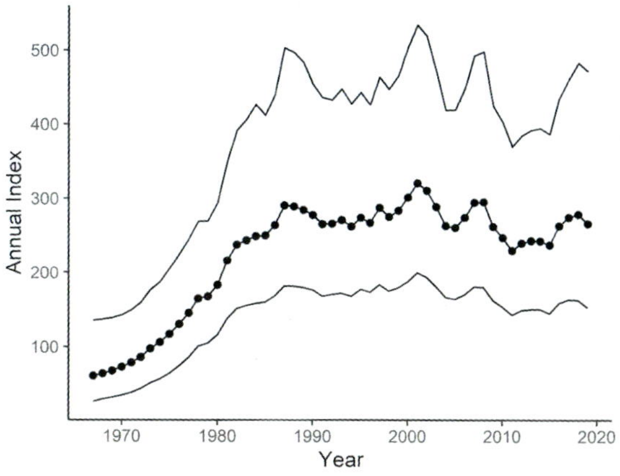

C. Breeding population trend and 95 percent credible interval (solid black lines) for the Cliff Swallow in Minnesota based on the 1967–2019 Breeding Bird Survey (Sauer et al. 2020).

B. An old wooden bridge crossing a stream on a rural road provides an ideal nesting site for a small Cliff Swallow colony. Photo by Lee A. Pfannmuller.

CONSERVATION

With its wide distribution and growing population, the Cliff Swallow is not considered a conservation priority (PIF 2021). Even climate change is projected to result in greater gains than losses in the species' North American breeding distribution as temperatures rise, with no changes projected in Minnesota (NAS 2019).

In some localities, particularly in the northeastern United States, management has focused on reducing the impacts of House Sparrows on Swallow nesting colonies. House Sparrows are either shot or trapped or Cliff Swallow nests are destroyed following the nesting season, preventing sparrows from using them as winter roosts or nesting chambers the following spring (C. R. Brown et al. 2020). Nest destruction usually has a minimal impact on Swallows because colonies frequently relocate in response to disturbance or swallow bug infestations.

Measures are occasionally taken to discourage the establishment of nesting colonies on major bridges that require maintenance or where road construction is underway. Yet, without due diligence, many efforts to dissuade the birds are unsuccessful. On the contrary, it is not uncommon for rural homeowners to actively attract nesting Swallows. As insect consumers, these industrious little birds provide an important ecological role. Groups of foraging Cliff Swallows from large colonies are especially adept at consuming large aggregations of swarming insects, including wasps, flies, and even grasshoppers (C. R. Brown et al. 2020).

RUBY-CROWNED KINGLET

Corthylio calendula

Minnesota Seasonal Status: Regular breeding resident, migrant, and occasional in winter. The Ruby-crowned Kinglet was uncommon during the MNBBA.
Migration: Short- to medium-distance migrant; winters in the southern and western United States, Mexico, and south to Guatemala.
Food: Small arthropods gleaned from foliage.
Nest: Pensile cup nest, often located high near the trunk of a coniferous tree; lays the largest clutch (up to 12) of any of North America's small songbirds.
Conservation Concern: Assigned a low PIF Continental Concern Score of 6/20.

The Ruby-crowned Kinglet has two distinctive and fascinating behavioral characteristics: its marvelous song and its nervous movements. Forbush and May (1955) describe "its extraordinary powers of song" and note that it "is really surprising that such a tiny creature should be capable of the strong and sustained notes." They explain that its larynx, the sound-producing organ, is not much bigger than a pinhead, and the muscles that move it are microscopic shreds of flesh. Yet, if humans had the same proportions, "we could converse with ease at a distance of a mile or more." Bent (1949) laments that those who have not heard it cannot appreciate what they have missed, especially since it is primarily heard in the conifer forests of the northern breeding grounds. He describes the song as a three- to four-note phrase repeated two to seven times, or as *"eee-tee-tee-tee-too-too-tu-tu-ti-ti-ta-tidaweet-tidaweet-tidaweet."*

Bent (1949) describes the motion of the Ruby-crowned as "more active, nervous, and irascible" compared with the Golden-crowned Kinglet. Forbush and May (1955) further observed the species "twitching their wings as they alight, and always too busy to pay attention to what may be going on around them." Its plumage is primarily olive-green above with dusky underparts, white eye-rings, and two white wing bars. The sexes are similar, but the male has a vermilion crown patch. Its scientific name, *calendula,* refers to the glowing nature of the ruby patch when it is erected, though it is frequently hidden. Bent emphasized, "What it lacks in color it makes up for in music."

Photo by David Brislance.

Ruby-crowned Kinglets are found in Canada from Labrador to British Columbia and in the western United States, especially in the Rocky Mountains. They also occur in the northern areas of the Upper Midwest and New England states. The highest densities are found in the coniferous forests of Labrador, the Colorado Rockies, and British Columbia.

MINNESOTA BREEDING DISTRIBUTION

Roberts (1932) never recorded the Ruby-crowned Kinglet as nesting in the state, although he cited Mr. Kendall, who "thought it may possibly be nesting in the vicinity of Hibbing, St. Louis County." Roberts also described several observations, including of singing males that span from late May to early August from Anoka, Itasca, Pine, and northern St. Louis Counties, but no definitive evidence of nesting.

The first nest found in Minnesota was in Stearns County in 1944, where it was found in a balsam fir (Hiemenz 1944). Green and Janssen (1975) described the Ruby-crowned Kinglet's primary breeding range as northeastern and north-central Minnesota, and west to Itasca State Park and eastern Marshall County. Summer observation records existed from as far south as Anoka and Dakota Counties. They remarked that it was "strange that the species with its distinctive song was completely overlooked as a summer resident in Minnesota until the 1940s." Besides Stearns County, nesting records were scarce and confirmed only in Cook County, but nesting was inferred in Crow Wing and southern St. Louis Counties.

Several years later, Janssen (1987) noted that the Ruby-crowned Kinglet was a scarce summer resident in the northeastern and north-central regions. Both Janssen and Hertzel and Janssen (1998) in their summary of confirmed nesting since 1970 included three counties: Beltrami, Cook, and St. Louis. As of 2014, the MBS documented 341 breeding-season locations of the Ruby-crowned Kinglet during their fieldwork. Its locations included a relatively extensive breeding distribution from southeastern Beltrami County and Itasca County and throughout much of Cook, Lake, and St. Louis Counties. In addition, they included breeding observation locations from eastern Marshall and Roseau Counties in extreme northwestern Minnesota.

The MNBBA participants documented 916 records of the Ruby-crowned Kinglet, most of which were sparsely distributed in the Laurentian Mixed Forest Province; a few records were in the Tallgrass Aspen Parklands Province **(A)**. The highest density of records was from the extreme northeastern portions of the state. Records were reported from 8.6 percent (405/4,736) of surveyed blocks. A total of 13 confirmed nesting records were reported from Cook, Lake, Itasca, Koochiching, Roseau, and St. Louis Counties. Probable nesting was observed in scattered counties throughout northern Minnesota, including Aitkin, Beltrami, Lake of the Woods, and Marshall, while possible nesting was identified in north-central Minnesota, including from Carlton, Cass, Hubbard, and Pine Counties. An observation of fledged young noted by Janssen (2019) on August 11, 2009, in Sherburne County was not included because of the lack of details and because it was beyond normal nesting dates.

The probability map of the Ruby-crowned Kinglet emphasized its distribution in northeastern and northern Minnesota **(B)**. The highest densities were

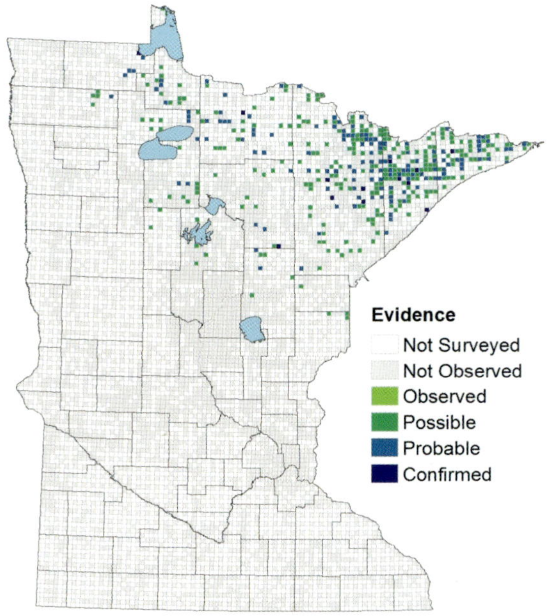

A. Breeding distribution of the Ruby-crowned Kinglet in Minnesota based on the Breeding Bird Atlas (2009–13).

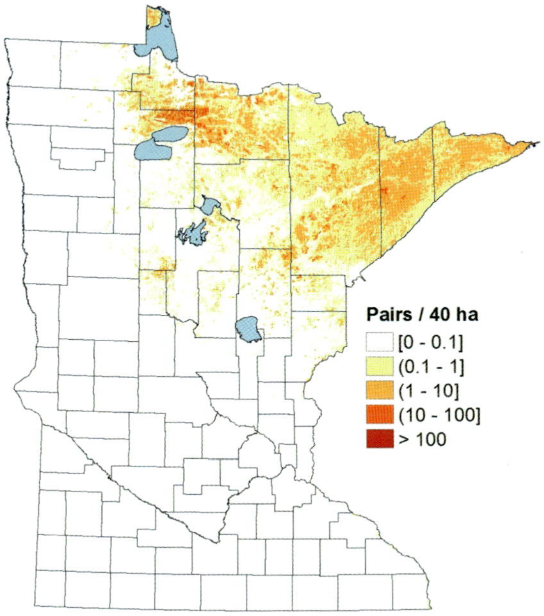

B. Predicted breeding distribution (pairs per 40 ha) of the Ruby-crowned Kinglet in Minnesota based on habitat, landscape context, and climate data gathered during the Minnesota Breeding Bird Atlas (2009–13) using the Generalized Linear Modeling method with an adjustment for detectability.

predicted in the Red Lake Peatland north of Upper Red Lake and in patchy areas in three ecological subsections of northeastern Minnesota: Tamarack Lowlands, Toimi Uplands, and Laurentian Uplands. These regions stretch from northeastern Aitkin County to Lake County.

In their review of the Ruby-crowned Kinglet in North America, D. L. Swanson, Ingold, and Wallace (2020) stated that nests were first found in New York and Vermont in the 1940s. They suggested that the finding of the nests was due not to a range expansion but to increased efforts by observers. This may also have been the case with the first nest found in Minnesota, which was fortuitously observed in an urban setting, in the gardens of a hospital in St. Cloud, Minnesota (Hiemenz 1944).

The lack of detections of the species and breeding evidence up to the 1940s is surprising but parallels observations in Wisconsin (Cutright et al. 2006). Wisconsin had no summer records until the 1940s, and the first nest, in northeastern Oneida County, was not found until 1964! A similar story was told by Chartier and colleagues (2013) for Michigan. The inability to find the nest can easily be forgiven because it is so well hidden high in a conifer and close to the trunk. But it is surely difficult to comprehend that early observers would have missed the distinctive and loud song of the Ruby-crowned Kinglet. The lack of breeding detections of the Ruby-crowned Kinglet in the 1800s and early 1900s will remain a mystery—or perhaps it indicates a recent southern range expansion?

BREEDING HABITAT

In Minnesota, breeding habitats of the Ruby-crowned Kinglet have been described as mature boreal forest, including spruce–balsam fir, black spruce–tamarack bogs, white cedar, and, to a lesser extent, pine forests (Green and Niemi 1978; Niemi and Pfannmuller 1979) **(C)**. Niemi, Howe, and colleagues (2016) most frequently recorded the highest densities in lowland black spruce–tamarack forests. This was also true in the Agassiz Lowlands Subsection, where the species preferred larger patches of evergreen forests and tended to occur in more stagnant, semi-open black spruce stands compared with the Golden-crowned Kinglet (Bednar et al. 2016; Zlonis et al. 2017). However, both species can be found in the same forest stand.

POPULATION ABUNDANCE

PIF (2021) estimated a breeding population of 100 million breeding adults in the United States and Canada, with an estimated 84,000 adults in Minnesota. The MNBBA estimated a substantially larger population, of more than 430,000 adults, but the population estimate was highly variable with a wide confidence interval. Still, the species' population in Minnesota was a small fraction of its estimated North American breeding population.

The BBS population trend for Minnesota from 1967 to 2019 was unreliable, but the Boreal Hardwood Transition Bird Conservation Region, which includes northern Minnesota, showed a stable population from 1966 to 2019. However, the recent trend in this region was significantly declining by −2.00 percent per year from 1993 to 2019 **(D)**. Trends in the Superior National Forest, where Ruby-crowned Kinglets are common, indicated a highly significant increase that has averaged 7.22 percent per year from 1995 to 2020 (Grinde, Bednar, et al. 2020).

C. Ruby-crowned Kinglets usually nest in mature coniferous forests composed of black or white spruce and balsam fir. Photo by Gerald J. Niemi.

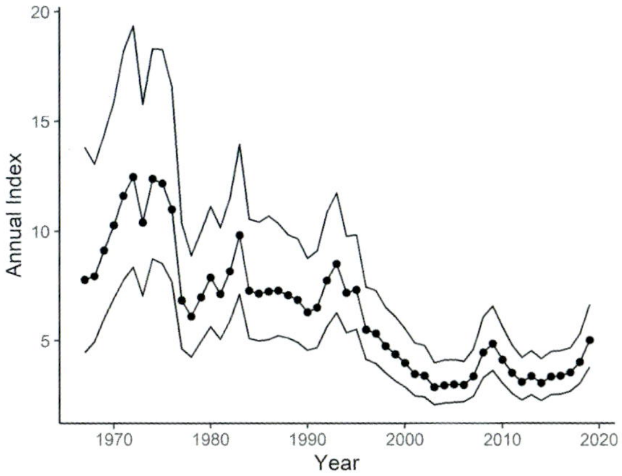

D. Breeding population trend and 95 percent credible interval (solid black lines) for the Ruby-crowned Kinglet in the Boreal Hardwood Transition Bird Conservation Region based on the 1967–2019 Breeding Bird Survey (Sauer et al. 2020).

CONSERVATION

At present, there are few national conservation concerns for this species because of its large breeding population and wide distribution, but climate change has raised new concerns about its long-term survival. The overall North American population of the Ruby-crowned Kinglet appears stable. D. L. Swanson, Ingold, and Wallace (2020) emphasized that the populations in some eastern North American regions were declining, while they were increasing in selected areas of western North America despite extensive logging. They also noted that factors associated with the declining populations were unknown. The increase of the Ruby-crowned Kinglet in northeastern Minnesota may be associated with the substantial decline in logging over the past fifteen years (Niemi, Howe, et al. 2016).

In a scenario similar to that of the Golden-crowned Kinglet, NAS (2019) projected a moderate vulnerability for the Ruby-crowned Kinglet breeding range in North America with a 2°C warming scenario. The projections included a 45 percent range loss and a 17 percent range gain in the far north by 2050. In Minnesota, the species was predicted to be extirpated even under a 1.5°C warming scenario; a scenario that is currently in existence. In addition, Longcore and colleagues (2013) identified the species as ranked second in mortality frequency at communication towers in the New England/Mid-Atlantic coastal region of North America. Climate change, migration hazards, and its adaptability to future habitat changes in coniferous forests will be critical to its long-term survival.

GOLDEN-CROWNED KINGLET

Regulus satrapa

Minnesota Seasonal Status: Regular breeding resident, migrant, and winter visitant. The Golden-crowned Kinglet was uncommon during the MNBBA.
Migration: Short-distance migrant that winters throughout much of the United States, including within much of its breeding area.
Food: Small arthropods, gleaned from foliage, especially in coniferous trees.
Nest: Pensile cup nest suspended from a fork in a coniferous tree.
Conservation Concern: Assigned a low PIF Continental Concern Score of 8/20.

Except for the hummingbird, the Golden-crowned Kinglet, about 10 cm in length, is among the smallest birds found in Minnesota. It is a tiny olive-green and gray bird with a black border on its crown, a white line above the eye, white wing bars, and whitish underparts. Forbush and May (1955) describe the male as "easily mistaken at times for a female, but whenever he becomes in the least excited the orange blazes forth," a reference to the central orange feathers of the yellow-gold crown. Further, they elaborate that the Golden-crowned Kinglet's "little nervous trick of flirting up the wing-tips, like a flash, will identify it at once." The combination of its twitchy behavioral pattern, especially while foraging, its small size, and head markings are diagnostic in distinguishing it from the New World Warblers.

Roberts (1932) describes its song as a "humble performance" in comparison to the "famed warble" of the Ruby-crowned Kinglet. Its song, he says is "a keen disappointment to hear . . . a series of fine, ascending, screeping sounds followed by two or three descending, full-toned, warbling notes which constitutes the only musical feature of the utterance." It is reminiscent of the high-frequency Brown Creeper song and requires acute hearing ability.

Both sexes of the Golden-crowned Kinglet have a golden crown and, in honor of the crown, the genus name of the species, *Regulus,* is Latin for "little king" and *satrapa* means "viceroy or prince." Both species of kinglets found in North America, the Golden-crowned and the Ruby-crowned Kinglet, look like small Neo-

Photo by David Brislance.

tropical warblers, but their evolutionary affinity is with Old World bird families. The Golden-crowned Kinglet has even been considered to be the same or a sister species of the Eurasian Goldcrest (*Regulus regulus*) (D. L. Swanson, Ingold, and Galati 2020).

The Golden-crowned Kinglet can be found across Canada from Labrador to British Columbia, south along the Pacific coast to California, and is patchily distributed in the Rocky Mountains. It occurs in northern areas of the Upper Midwest and the northeastern United States, and south in the Appalachian Mountains to northern Georgia. Its highest densities are found along the Pacific coast from California to British Columbia.

MINNESOTA BREEDING DISTRIBUTION

Roberts (1932) described the Golden-crowned Kinglet as locally common in the evergreen forests as far south as northern Isanti County and west to Itasca State Park. He listed "definite" nesting in Aitkin, Cass, Cook, Isanti, Itasca, and St. Louis Counties, as well as at Itasca State Park and at Mille Lacs.

More than forty years later, Green and Janssen (1975) reported that the species occurred in northern regions, sparingly in the eastern north-central regions, and as far west as Tamarac and Agassiz NWRs. Besides the counties noted by Roberts (1932), they reported confirmed nesting in Clearwater, Hubbard, and Mille Lacs Counties and inferred nesting from Koochiching, Lake, and Ramsey Counties. Several years later, Janssen (1987) outlined a more restricted distribution to northern Pine County and west to Clearwater and Roseau Counties; he acknowledged breeding observations from both Agassiz and Tamarac NWRs, and former breeding records as far south as Ramsey County. Hertzel and Janssen (1998) summarized confirmed nesting since 1970 in five counties: Aitkin, Beltrami, Clearwater, Lake, and St. Louis.

The MBS reported 350 breeding-season locations during its surveys of Minnesota's counties. Locations of extensive breeding observations ranged from Itasca County to Cook County, south to northern Mille Lacs and Pine Counties, southwest to Morrison and Douglas Counties, west to Becker County, and northwest to eastern Marshall and Roseau Counties. These observations expanded the breeding range outlined by Janssen (1987) to the south, southwest, and western portions of Minnesota.

The MNBBA reported 981 records for the Golden-crowned Kinglet, which were almost exclusively within the Laurentian Mixed Forest Province **(A)**. New confirmed nesting records were from Carlton, Crow Wing, Koochiching, and Pine Counties, and there were extensive records from both Lake and St. Louis Counties. Breeding records were recorded from 10.3 percent (489 blocks of 4,737) of the blocks surveyed in the state. Probable nesting records were from Becker, Lake of the Woods, Marshall, and Roseau Counties; possible nesting was found in Wadena County.

The probability map identified suitable habitat south to Chisago County and in the northwest to Kittson County, with patches in eastern Marshall, Red Lake, and the eastern portion of Becker and Otter Tail Counties **(B)**. A confirmed nesting record from Sherburne County (Janssen 2019) reinforces the availability of suitable habitat for this species farther south in Minnesota.

Whether the breeding distribution of the Golden-crowned Kinglet has changed over the past 150 years in Minnesota is unclear. Roberts (1932) states

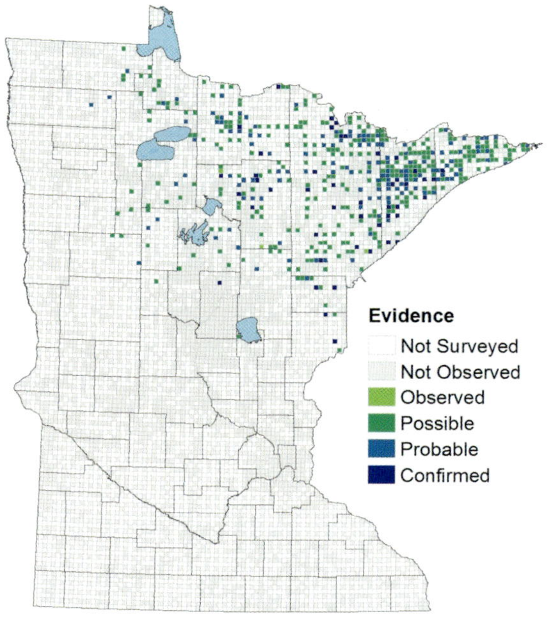

A. Breeding distribution of the Golden-crowned Kinglet in Minnesota based on the Breeding Bird Atlas (2009–13).

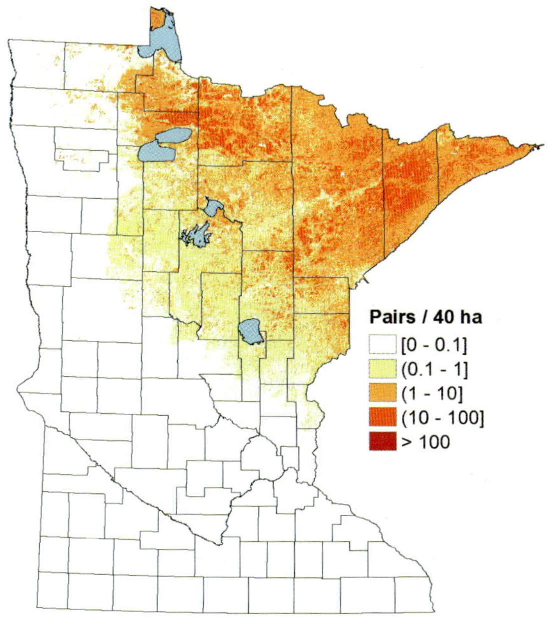

B. Predicted breeding distribution (pairs per 40 ha) of the Golden-crowned Kinglet in Minnesota based on habitat, landscape context, and climate data gathered during the Minnesota Breeding Bird Atlas (2009–13) using the Generalized Linear Modeling method with an adjustment for detectability.

little about the species' breeding distribution south of northern Isanti County. Besides the former breeding record in Ramsey County (prior to 1970) and the more recent record in Sherburne County, few breeding observations have been reported from areas south of Pine County.

D. L. Swanson, Ingold, and Galati (2020) reported range expansions in the eastern United States and southward extensions of its range in spruce plantations since the mid-1980s in Illinois, Indiana, and Ohio. Wisconsin (Cutright et al. 2006; WSO 2021) and Michigan (Chartier et al. 2013) have both reported nesting of Golden-crowned Kinglets in locations farther south than what has been observed in Minnesota, so nesting by the Golden-crowned Kinglet in more southerly locations of Minnesota should be scrutinized. This is especially true in early spring where spruce or coniferous forest patches exist. The species' high-frequency song can easily be missed.

BREEDING HABITAT

Galati (1991), in his book dedicated to the Golden-crowned Kinglet and with data primarily gathered from Itasca State Park, listed a wide range of conifer habitats used for nesting, including pure and mixed stands of balsam fir, white spruce, black spruce, tamarack, and both red and jack pines. Recent extensive coverage of the Agassiz Lowlands Subsection revealed high concentrations in both mature black spruce and white cedar cover types (Bednar et al. 2016) **(C)**. The Golden-crowned Kinglet and Ruby-crowned Kinglet occupied similar habitat types, but Golden-crowned Kinglets preferred more productive stands (Zlonis et al. 2017). In the Chippewa and Superior National Forests, Niemi, Howe, and colleagues (2016) indicated strongest preference for black spruce–tamarack forests, and moderate use of mature aspen–spruce–fir, mixed swamp conifers, and white pine forests.

POPULATION ABUNDANCE

PIF (2021) estimated a large population in the United States and Canada of 130 million breeding adults and 780,000 breeding adults in Minnesota. In contrast, MNBBA estimates were more than twice the PIF estimates, at 1.77 million breeding adults in Minnesota (95 percent confidence interval of 1.50–2.20 million). Among the differences in data used by the MNBBA and not incorporated in PIF estimates was the sampling of roadless areas, such as the Agassiz Lowlands Subsection **(B)**, many of which have high populations of Golden-crowned Kinglets.

There were too few detections to estimate a population trend in Minnesota, but the BBS estimate for the Boreal Hardwood Transition Bird Conservation Region, which includes northern Minnesota, was significantly increasing by 1.59 percent per year from 1966 to 2019 **(D)**. Golden-crowned Kinglet populations, however, in both the Chippewa and Superior National Forests of Minnesota have been stable (Grinde, Bednar, et al. 2020). In contrast, Rosenberg, Kennedy, and colleagues (2016) calculated an overall population decline in the United States and Canada of −25 percent from 1970 to 2014, primarily owing to declines in several regions of western North America (e.g., Oregon, Washington, and British Columbia) where mature coniferous forests have dwindled.

C. Typical breeding habitat of the Golden-crowned Kinglet in Minnesota consists of old-growth or mature conifer forest; the species is especially attracted to spruce, balsam fir, or pines. Photo by Gerald J. Niemi.

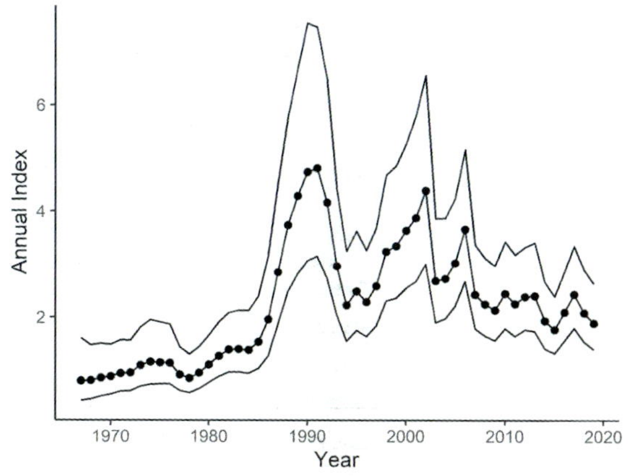

D. Breeding population trend and 95 percent credible interval (solid black lines) for the Golden-crowned Kinglet in the Boreal Hardwood Transition Bird Conservation Region based on the 1967–2019 Breeding Bird Survey (Sauer et al. 2020).

CONSERVATION

Currently there is low conservation concern for the Golden-crowned Kinglet because it has a broad breeding distribution and a potentially expanding population in some areas owing to reforestation and maturation of large conifers. Its overall population may be declining, so continued monitoring of its population is essential. In Minnesota, Niemi and Hanowski (1992a) predicted that Golden-crowned Kinglets would be negatively affected by loss of mature coniferous trees if harvest scenarios began to exceed 4 million cords per year. However, this level of logging has not been exceeded in more than twenty-five years, and current levels are substantially lower.

D. L. Swanson, Ingold, and Galati (2020) emphasized the species' affinity for breeding habitat in old-growth conifers of western North America. They presented evidence that the birds responded positively to insect outbreaks of spruce budworm and mountain pine beetle as well as the negative population effects of selective logging of conifer forests, especially if the thinning was relatively intense.

NAS (2019) projected a moderate vulnerability for the Golden-crowned Kinglet's breeding distribution in North America with a 2°C warming scenario, including a 43 percent range loss and a 36 percent range gain, primarily in the far north in Canada and Alaska, by 2050. In Minnesota, the species was predicted to be extirpated even under a 1.5°C warming scenario, a projection that currently exists.

The future in Minnesota for the energetic little Golden-crowned Kinglet is precarious because of the uncertainty of our northern coniferous forests and our current trajectory of climate change. Continuation of its expanding distribution and its ability to adapt to reforestation efforts in coniferous forests throughout its range will be key issues in its long-term survival.

CEDAR WAXWING

Bombycilla cedrorum

Minnesota Seasonal Status: Regular breeding resident and migrant and a regular winter visitor. The Cedar Waxwing was common during the MNBBA.
Migration: Considered a short-distance migrant, but winter movements are quite irregular because the birds respond largely to the availability of berries and fruits.
Food: Fleshy fruits and insects secured by foliage gleaning and fly-catching; cedar and mountain ash berries are common winter foods.
Nest: Cup nest placed in a deciduous or coniferous tree or shrub, generally 2–15 m high.
Conservation Concern: Assigned a low PIF Continental Concern Score of 6/20.

Photo by Mike Lentz.

From its berry-gorging appetite to its nomadic behavior and sleek plumage, the Cedar Waxwing, Bent (1950) wrote, impresses us because it is so "unlike most of the birds we know." With a palette of pale browns, grays, and yellows, the Waxwing's plumage is accented with a black eye mask outlined in white, a gray tail edged in bright yellow, and red, waxy secretions on its secondary wing feathers. The head crest adds a final element of distinction. Nearly identical in appearance, the female is slightly paler than the male. *Bombycilla,* the species' genus name, is a reference to the bird's silky plumage. The waxy wing tips, the source of its common name, are believed to serve a role in courtship. Older birds have more feathers with red on their tips than younger birds and consistently breed earlier and raise more young (Witmer et al. 2020). Although it is a member of the songbird family, the Cedar Waxwing's vocal repertoire is quite limited. At times barely audible, one of its most frequent vocalizations is a very high-pitched *bzeee* call note.

Because the Cedar Waxwing may be observed during the winter, it can be confused with the very similar Bohemian Waxwing, a common winter visitor to northern Minnesota. The latter is a chunkier bird, nearly twice the weight of the Cedar Waxwing, and is distinguished by its rusty undertail coverts and small patches of white and yellow in its wings.

Although it has many unique attributes, the Cedar Waxwing may be best known for its frugivorous diet and its occasional bouts of drunkenness when too many fermented berries are consumed. Insects provide additional nourishment during the breeding season, but fruits comprise more than 80 percent of their annual food consumption (Witmer et al. 2020). Unfortunately, weather patterns often impact fruit production, so Waxwings have adapted by foraging in nomadic flocks to improve their chances of locating a bountiful food supply. Their frugivorous diet also dictates a later breeding season. While most Minnesota songbirds breed from late May through early July, Cedar Waxwings usually breed from mid-June to mid-August.

Distributed across southern Canada and the northern half of the United States, the Cedar Waxwing reaches its highest breeding densities in the northeastern states and southeastern provinces.

MINNESOTA BREEDING DISTRIBUTION

In the early 1900s, Roberts (1932) considered the Cedar Waxwing a common, statewide breeding resident, noting it was "most abundant in the northern coniferous forests, where it is at home both on the uplands and in the spruce and tamarack swamps." Nesting evidence was available from 16 counties, ranging from Jackson and Goodhue Counties in southern Minnesota; to Anoka, Hennepin, and Sherburne Counties in east-central Minnesota; and north to numerous counties that span the Aspen Parklands and northern forest, including Marshall and Polk Counties east to Cook County.

Forty years later, Green and Janssen (1975) described the Cedar Waxwing as a resident in the forested regions of the state, but they were uncertain about its status in the western grasslands and agricultural counties. Although flocks of birds were observed in the region during the summer months, only two breeding records were documented from west-central and southwestern Minnesota: Roberts's 1898 record from Jackson County and a 1948 record from Lac qui Parle County. Yet only a few years later, Janssen (1987) would describe the species as not only numerous in the forests and open woodlands of eastern, central, and northwestern Minnesota but most numerous in the northwestern, east-central, and southwestern regions of the state. It remained quite rare, he commented, in west-central and southeastern Minnesota. Janssen also delineated 31 counties where nesting had been confirmed since 1970, most of which were located in northern and east-central Minnesota. By 1998, Hertzel and Janssen would add an additional 13 counties to the list. Meanwhile, MBS field biologists documented a total of 928 Cedar Waxwing breeding-season locations. The majority of records were in the northern forest counties, but the species was well distributed in all but the most intensively cultivated regions of the state.

The species' statewide distribution was confirmed by MNBBA participants, who reported 3,277 Cedar Waxwing detections from 44.7 percent of the surveyed atlas blocks and from 57.6 percent of the priority blocks **(A)**. Breeding was confirmed in 249 blocks (5.2 percent) and deemed probable in an additional 448 blocks (9.4 percent). The birds were reported in all 87 Minnesota counties and were confirmed breeding in 67 counties. Of the 20 counties where nesting was not confirmed, 11 were located in southern Minnesota, south of a line between the Twin Cities and Lac qui Parle County; most of those counties were in the southwestern corner of the state, where the species was more sparsely distributed.

Apart from Roberts's (1932) original statement that the Waxwing was common throughout the

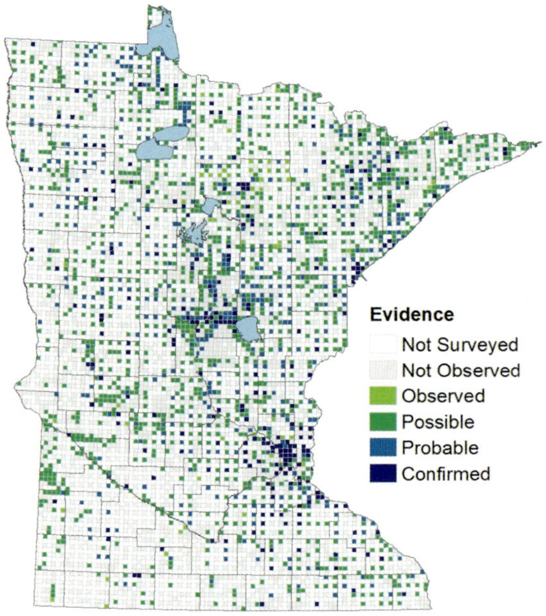

A. Breeding distribution of the Cedar Waxwing in Minnesota based on the Breeding Bird Atlas (2009–13).

state, yet most abundant in the northern forests, there is little in his account that sheds light on its relative abundance across the state. Although it remains common in the northern forest, it now appears almost equally common and widespread throughout the open woodlands that characterize the transitional forest belt that stretches from the southeast corner of the state through west-central and northwestern Minnesota. As described by later accounts, Waxwings remain least common in the Red River valley and, contrary to Janssen's (1987) assessment, in the southwestern corner of the state.

In the upper Midwest, the Cedar Waxwing's breeding distribution has shown no widespread changes. Historically, it was and remains a widely distributed species. Elsewhere within its breeding range, two distributional changes have been noted: first, in the past thirty years, the Cedar Waxwing has become more common in the northeastern United States and southeastern Canada; second, the southern periphery of the species' breeding range has expanded south in regions of Kentucky, southern Ohio, Maryland, and Virginia (Witmer et al. 2020).

BREEDING HABITAT

The availability of fruits and berries may be the best predictor of suitable habitat for the Cedar Waxwing. As long as sufficient food resources are available, the species can be found using upland and lowland woodlands, including deciduous, coniferous, and mixed forests. If scattered trees and shrubs are present, even old fields and grasslands may be used. Only the interior of forest stands are avoided, as the birds prefer open woodlands and edge habitats that provide more sunlight for fruiting plants (Witmer et al. 2020; Kaufman 1996; Ehrlich et al. 1988).

Although fruits and berries are the mainstay of the Cedar Waxwing's diet, insects are important during the summer months. Adept at fly-catching and hawking insects on the wing, the species often favors open riparian habitats that provide access to an abundant supply of emerging aquatic insects **(B)**. Human-modified landscapes that incorporate ornamental plantings of fruiting trees and shrubs also provide suitable habitats, including residential areas, cemeteries, orchards, parks, and golf courses.

POPULATION ABUNDANCE

PIF (2021) provided a population estimate of 64 million Cedar Waxwings in the United States and Canada, while Minnesota's population was estimated at approximately 1.5 million birds. Because of the species' non-territorial habits, a statewide estimate using MNBBA data was not generated.

Since long-term monitoring began in 1966, the Cedar Waxwing's population has remained quite stable, demonstrating a nonsignificant decrease averaging only −0.14 percent per year from 1966 to 2019 (Sauer et al. 2020). Overall, the Waxwing is increasing across much of the southern portion of its breeding range and is decreasing-to-stable in the northern regions. Indeed, in Minnesota, BBS data reveal a significant decline of −1.24 percent per year from 1967 to 2019 **(C)**; the decline is most pronounced since the early 2000s. In contrast, when data from the Chippewa and Superior National Forests are combined, Cedar Waxwings demonstrate a significantly increasing regional trend averaging 1.78 percent per year from 1995 through 2020 (Grinde, Bednar, et al. 2020).

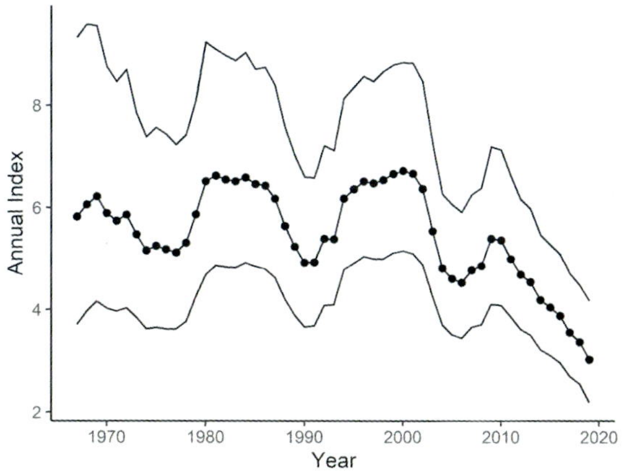

C. Breeding population trend and 95 percent credible interval (solid black lines) for the Cedar Waxwing in Minnesota based on the 1967–2019 Breeding Bird Survey (Sauer et al. 2020).

Unfortunately, detailed studies have not examined factors responsible for local population fluctuations. By and large, fluctuations are thought to reflect changes in food resources that are a consequence of changes in land use. Since the species is a frugivore, the cessation of DDT use to control insect infestations on fruit crops and ornamental plants may have contributed to population increases in some regions (Witmer et al. 2020).

CONSERVATION

Although population trends vary, the Cedar Waxwing's relatively stable and large population, coupled with its wide distribution, has made it a low conservation priority (PIF 2021). In some local situations, it even is considered an agricultural pest that damages fruit crops (Witmer et al. 2020).

NAS (2019) models regarding the Cedar Waxwing's vulnerability to climate change have deemed it to have a low vulnerability under all three warming scenarios examined. Losses to its North American breeding distribution in the United States are compensated by gains across northern Canada. Nevertheless, as temperature increases rise from 1.5°C to 3°C, the species' statewide distribution in Minnesota is projected to gradually recede northward until it is restricted to the far northeastern counties. Despite its general habitat requirements and its adaptability to human-modified landscapes, this handsome bird may eventually become a rare sight in southern Minnesota.

B. A small pond surrounded by native and ornamental plantings at Minnesota's Landscape Arboretum provides a perfect spot for nesting and foraging Cedar Waxwings. Photo by Lee A. Pfannmuller.

RED-BREASTED NUTHATCH

Sitta canadensis

Minnesota Seasonal Status: Permanent resident and migrant. The Red-breasted Nuthatch was common during the MNBBA.
Migration: Partial migrant. Northern populations appear to move south each winter, while southern populations demonstrate irruptive movements that may be a response to winter food shortages.
Food: Arthropods during the summer and a variety of seeds, including those from coniferous trees, in other seasons. Forages along small branches in the tree canopy and occasionally on the main trunk; also a common visitor to bird feeders.
Nest: Primary cavity excavator but occasionally uses natural cavities and old woodpecker holes. Resin is applied to the cavity entrance, presumably to reduce predation and competition with other cavity nesters; sometimes a small twig is used in the application.
Conservation Concern: Assigned a low PIF Continental Concern Score of 6/20.

A petite and more colorful version of the familiar White-breasted Nuthatch, the Red-breasted Nuthatch is an active little bird, continually scrambling along the high branches of the forest canopy searching for tasty morsels. Its distinctive black cap and black eyeline are diagnostic, as is the rufous-colored breast and belly. Both sexes are similar in appearance, but the female has a grayer crown, paler underparts, and a paler black eyeline.

The Red-breasted Nuthatch is most abundant in the conifer forests and mixed coniferous–deciduous forests of northern Minnesota. It gradually replaces the White-breasted Nuthatch in abundance as one moves north, especially in the Arrowhead Region of northeastern Minnesota. In areas of overlap, the calls of the two species may sound similar, but with a little experience the more nasal *yank-yank-yank* call of the Red-breasted Nuthatch will become familiar. Its voice has been likened to the sound of a "blast on a tiny tin trumpet" (Bent 1948).

Although it spends more time in the high distal branches of a tree than its trunk-loving relative, the Red-breasted Nuthatch displays all the acrobatic agility that is characteristic of the Sittidae family. Like all nuthatches, its common name derives from its habit of cramming nuts and seeds into bark crevices so

Photo by Laura Erickson.

that it can use its beak to hammer, or "hatch" them open. *Sitta,* the bird's genus, refers to birds that peck the bark of trees. Populations are fairly dependent on conifer seeds for their winter diet. When seed crops are poor, birds that are typically permanent residents are forced to move farther south to find food, resulting in periodic irruptive movements. Detailed studies of seasonal movements, however, are lacking.

This boreal species is found from southern Alaska east across southern Canada and south into the montane forests of the western United States, and east through the Great Lakes states, New England, and the Appalachian Mountains. The highest population densities occur in western South Dakota and portions of Washington, Oregon, and California.

MINNESOTA BREEDING DISTRIBUTION

In the early twentieth century, Roberts (1932) considered the Red-breasted Nuthatch to be a common summer resident of the "evergreen forests" where it is "breeding from northern Isanti County on the south to the eastern border of the Red River Valley on the west." At the time, nesting records were "wanting," but by 1936 nesting had finally been confirmed near Lake Winnibigoshish (1935) and in Cook County (1936) (Roberts 1936). Describing the species as a winter resident throughout the state, Roberts grouped it along with 12 other "half hardy" permanent residents that were present during the winter months in "much reduced numbers."

Decades later, Green and Janssen (1975) described the species as a "regular permanent resident; widely distributed in the northeastern and north central regions." They found the bird occasionally breeding as far south as the Twin Cities and identified seven counties where nesting was inferred and two widely separated counties where nesting was confirmed: Cook and Ramsey. Numbers, they noted, varied "markedly" from year to year, especially in the winter months. Janssen (1987) included a distribution map that delineated the species' breeding range as extending south to the Twin Cities and eastern Stearns County, and west to Clearwater and eastern Roseau Counties. Within this region, nesting had been confirmed in 13 counties since 1970. Hertzel and Janssen (1998) later deleted Washington County from Janssen's (1987) original map.

In the summers that followed, the birds were regularly reported south of the Twin Cities, including in Houston County, but nesting was never confirmed south of Dakota County (MOU 2021b). Although the majority of the 808 breeding-season locations identified by the MBS were located in the Laurentian Mixed Forest Province, the survey also documented records as far south as Winona and Houston Counties as well as farther west in eastern Douglas, Otter Tail, Becker, Mahnomen, and Polk Counties.

The 2,816 records collected by the MNBBA further expanded the southern and western boundaries of the Red-breasted Nuthatch's breeding distribution. Reported from 27.0 percent (1,281/4,750) of the surveyed atlas blocks, breeding was confirmed in 115 blocks **(A)**. The birds were reported in 49 counties and were confirmed breeding in 19 counties (including three counties straddled by one block). Ten of the counties were additions to Hertzel and Janssen's 1998 list, including Rice County in southern Minnesota and Polk and Becker Counties in northwestern Minnesota. The Polk County record was just northeast of Grand Forks, North Dakota. Nesting was also considered probable in Olmsted and Fillmore Counties. The birds remained most abundant in the Laurentian Mixed Forest Province but were scattered

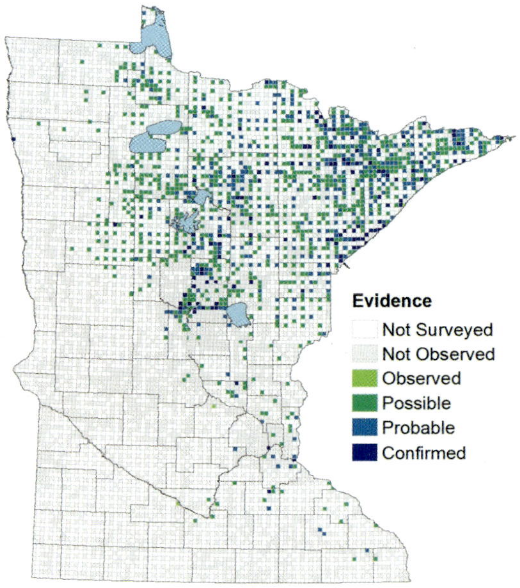

A. Breeding distribution of the Red-breasted Nuthatch in Minnesota based on the Breeding Bird Atlas (2009–13).

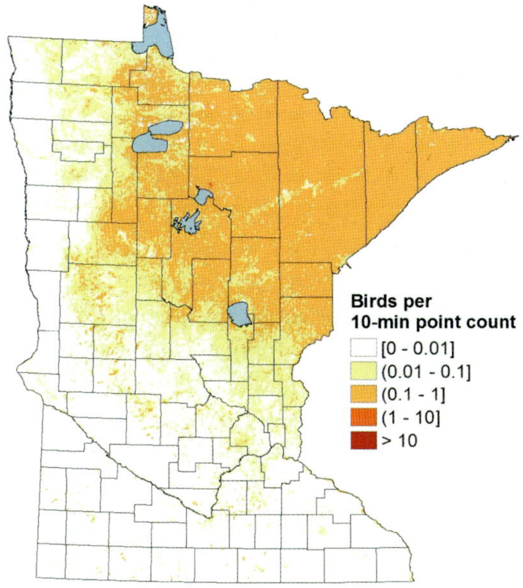

B. Predicted breeding distribution (birds detected per 10-minute point count) of the Red-breasted Nuthatch in Minnesota based on habitat, landscape context, and climate data gathered during the Minnesota Breeding Bird Atlas (2009–13) using the Generalized Linear Modeling method without an adjustment for detectability.

throughout the southern Eastern Broadleaf Forest Province and were found in suitable habitat as far west as Nicollet, Brown, and Blue Earth Counties in the Prairie Parkland Province. Clearly this species is finding suitable habitat in areas beyond what Roberts (1932) considered its primary breeding range nearly one hundred years ago.

The MNBBA probability map **(B)** predicts that the core of the species' breeding distribution is in the Laurentian Mixed Forest Province, with lower breeding densities projected throughout much of the northern Eastern Broadleaf Forest Province. Patches of moderate abundance occur as far west as Otter Tail and Douglas Counties and in sites scattered throughout southern Minnesota, including along the Mississippi River.

The southward expansion of Red-breasted Nuthatches in Minnesota is representative of a broader expansion observed elsewhere in North America (Ghalambor and Martin 2020). Ontario, for example, detected a notable expansion southward in the twenty years between their first atlas (1981–85) and their second atlas (2001–5), citing an increase in the number and age of conifer plantations as the primary factor responsible for the change (Cadman et al. 2007). Southern range expansions in the eastern United States, including in South Carolina and Georgia, have been attributed to habitat changes in the Appalachian Mountains (Renfrow 2005). The factors responsible for new breeding records in states south of Minnesota, including Iowa, Missouri, Kansas, and Illinois (Ghalambor and Martin 2020), are less clear, but backyard bird feeders, coupled with the planting of ornamental conifers, may be contributing to the change.

BREEDING HABITAT

An inhabitant of both upland and lowland coniferous forests, the Red-breasted Nuthatch may be found in deciduous woodlands if conifers are present (Danz et al. 2007; Green 1995). Trees that provide suitable opportunities for cavity excavation (e.g., snags, dead and decaying branches) are critical, as are abundant conifer cones for feeding (Ghalambor and Martin 2020) **(C)**. In northern Minnesota and Wisconsin, Niemi, Howe, and colleagues (2016) found that the birds strongly preferred mature upland pine stands but also demonstrated a wide tolerance for other forest cover types when scattered mature conifers were present.

POPULATION ABUNDANCE

The Red-breasted Nuthatch is twice as abundant in North America as the White-breasted Nuthatch. PIF (2021) generated an estimate of 20 million Red-breasted Nuthatches in the United States and Canada, compared to an estimate of only 10 million White-breasted Nuthatches. Statewide estimates for the two species, however, were nearly identical, with 340,000 for the smaller Nuthatch and 370,000 for its larger relative.

C. A subcanopy of conifers in a mature aspen forest provides nesting habitat for the Red-breasted Nuthatch in northern Minnesota. Photo by Gerald J. Niemi.

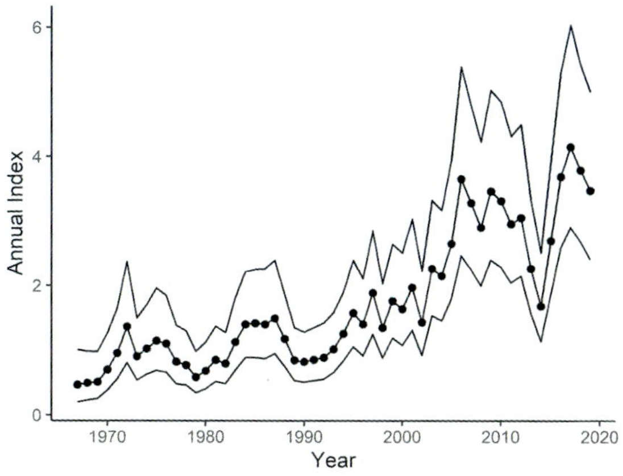

D. Breeding population trend and 95 percent credible interval (solid black lines) for the Red-breasted Nuthatch in Minnesota based on the 1967–2019 Breeding Bird Survey (Sauer et al. 2020).

BBS data reveal a slow but significant Red-breasted Nuthatch population increase throughout the survey region, averaging 0.81 percent per year from 1966 to 2019 (Sauer et al. 2020), with a cumulative increase of 86 percent from 1970 to 2014 (Rosenberg, Kennedy, et al. 2016). Amid some wide fluctuations that may reflect conifer seed crop abundance, populations in Minnesota have expanded at a much faster rate, averaging 3.94 percent per year from 1967 to 2019 **(D)**. This significant increase is corroborated by data collected in the Chippewa and Superior National Forests (Grinde, Bednar, et al. 2020). Niemi, Howe, and colleagues (2016) noted that a decrease in the level of timber harvest activities on the forests may be benefiting many species that, like the Nuthatch, prefer more mature forests.

CONSERVATION

Two factors affect the long-term sustainability of the Red-breasted Nuthatch: timber management activities in the boreal forest and the impacts of climate change on a species that depends on forests that dominate cooler regions. As long as timber management activities provide for diverse species composition and maintain mature snags, local management should adequately provide for the species. Warming temperatures, however, are a concern. NAS (2019) classified the species as moderately vulnerable under all three warming scenarios they examined, with projected losses to its continental breeding range exceeding projected gains. The situation is especially dire in Minnesota, where the species is nearly extirpated from the state under a 2°C warming scenario by 2050. In time, the White-breasted Nuthatch will likely replace its petite and more colorfully adorned relative throughout Minnesota's northern forests.

WHITE-BREASTED NUTHATCH

Sitta carolinensis

Minnesota Seasonal Status: Permanent resident. The White-breasted Nuthatch was common during the MNBBA.
Migration: Occasional reports of local movements of birds during the nonbreeding season may represent young of the year seeking areas with less dense breeding populations.
Food: Bark forager. Feeds primarily on arthropods during the summer; seeds are important at other times of the year and are frequently cached within crevices and furrows in tree bark. A common visitor to bird feeders.
Nest: Secondary cavity nester; nests are typically placed in old woodpecker holes or natural cavities. Occasionally excavates an existing hole to enlarge it but rarely excavates a new cavity by itself.
Conservation Concern: Assigned a low PIF Continental Concern Score of 6/20.

An acrobatic, handsome little bird, the White-breasted Nuthatch is a familiar and favorite among homeowners who enjoy feeding birds. A stocky bird dressed in black, white, and blue-gray colors, the Nuthatch's black crown extends down the back of the neck and is in sharp contrast to its white face, breast, and belly. The blue-gray feathers on the back and wing grade into the central tail feathers, while the outer tail feathers are marked with bold patches of white and black. A flush of chestnut can be seen in the lower belly and underneath the tail. The tiny, pointed bill is slightly upturned and a perfect instrument for probing the crevices and furrows of bark for wood borers, beetles, caterpillars, and seeds. The female differs only in having a paler cap on the crown and nape. Although described as "no great musician," the male has a simple song—*hah-hah-hah*—that is a welcome sound throughout the year (Bent 1948).

It is one of two nuthatch species in Minnesota (the other being the Red-breasted Nuthatch). The most unusual feature of these birds is their ability to move down a tree headfirst. The forests' other bark foragers, woodpeckers and Brown Creepers, move up a tree in search of food. If those species want to move down the tree trunk, they need to fly down before moving upward again; the agile little nuthatches

Photo by David Brislance.

simply turn to move in a different direction. Viewing the world upside down is a distinct feeding advantage because the birds can find prey that might not be visible to foragers who must approach in the opposite direction.

Sitta, the Nuthatch's genus name, is Greek for a bird that pecks the bark of trees. Its common name, however, is believed to have originated from its tendency to wedge food items, including acorns and nuts, into bark crevices. Once the food item is secured, the bird can hack or "hatch" out the tasty kernel within (Grubb and Pravosudov 2020).

Found throughout most of the United States and southern Canada, the White-breasted Nuthatch is scarce in the central Great Plains states, except in woodlots and riparian habitats, and largely absent from the Great Basin and Sonoran Desert. Farther south it can be found in southern Baja California and central Mexico. Some of the species' highest breeding densities in North America are found in the Upper Midwest, southern New England, and south of the Ohio River valley.

MINNESOTA BREEDING DISTRIBUTION

In the early twentieth century, Roberts (1932) wrote that the White-breasted Nuthatch was a common breeding resident "throughout the state wherever there is timber," but it was most abundant in southern Minnesota. It is also, he wrote, "one of our most familiar little birds, ever at home and busy among the shade trees of the city and the monarchs of the forest." Confirmed nest records were available from Hennepin, Meeker, Scott, and Sherburne Counties. Inferred nesting was reported in Marshall and Wadena Counties.

Many years later, Green and Janssen (1975) noted that the White-breasted Nuthatch was a permanent resident throughout the state but "very scarce and local during the breeding season in the northeastern region," where confirmed breeding reports were lacking. When Janssen provided an updated account of the species in 1987, nesting had been documented in St. Louis County (1974), in Lake County (1982), and in 27 other widely dispersed counties since 1970. Counties with confirmed nesting were most common in the Eastern Broadleaf Forest and the Tallgrass Aspen Parklands Provinces. Hertzel and Janssen (1998) later added four more counties to Janssen's (1987) list. The MBS further confirmed the species' abundance in all but the most intensively cultivated counties. Widely scattered records were also reported in the Arrowhead region of northeastern Minnesota, although they became increasingly scarce east of St. Louis County.

The MNBBA further expanded the breeding distribution of the White-breasted Nuthatch with detections in all 87 counties, including multiple records in the Arrowhead Region and throughout Minnesota's agricultural landscape **(A)**. With a total of 2,793 MNBBA records, the species was reported in 36.0 percent (1,707/4,744) of the surveyed atlas blocks. Breeding was confirmed in 209 blocks spanning 56 counties; probable breeding was reported in an additional 18 counties. Confirmed and probable records were least frequent in southwestern Minnesota; no confirmed or probable records were reported in Cook County. In the Arrowhead region and far north-central Minnesota, the Red-breasted Nuthatch surpasses the White-breasted Nuthatch in abundance.

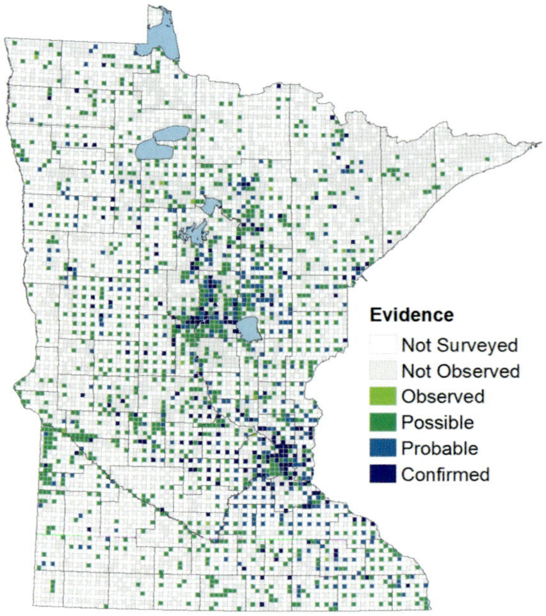

A. Breeding distribution of the White-breasted Nuthatch in Minnesota based on the Breeding Bird Atlas (2009–13).

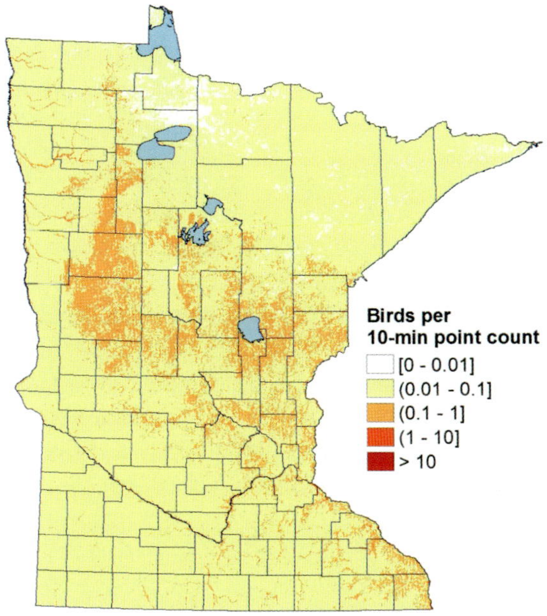

B. Predicted breeding distribution (birds detected per 10-minute point count) of the White-breasted Nuthatch in Minnesota based on habitat, landscape context, and climate data gathered during the Minnesota Breeding Bird Atlas (2009–13) using the Generalized Linear Modeling method without an adjustment for detectability.

The MNBBA probability map **(B)** predicts that the species is moderately abundant in the Eastern Broadleaf Forest Province and in the western and southern reaches of the Laurentian Mixed Forest Province. Scattered pockets of abundance are also predicted along major rivers, including the Mississippi River, the Minnesota River, and major drainages throughout southern and northwestern Minnesota.

Today the White-breasted Nuthatch remains a common species wherever woodlands are found. Least common in Minnesota's most heavily cultivated regions, as well as in the far northeastern and north-central counties, it likely is more abundant in the northern forests than it was a century ago. As the old-growth pine stands declined and the landscape became more developed, the forests became more fragmented and dominated by aspen–birch. These changes provided additional opportunities for the deciduous-loving White-breasted Nuthatch. Similar changes in the forest landscape and in the White-breasted Nuthatch's distribution have been noted in other Great Lakes states and provinces. Overall, however, few large-scale changes in distribution have been noted (Grubb and Pravosudov 2020).

BREEDING HABITAT

An inhabitant of mature forests, White-breasted Nuthatches prefer hardwood stands, often with closed canopies and open understories (Danz et al. 2007) **(C)**. Extensive monitoring in the national forests in northern Minnesota and Wisconsin demonstrated an overwhelming preference for upland oak stands followed by upland hardwoods and lowland hardwoods (Niemi, Howe, et al. 2016). Although mixed deciduous–coniferous stands are used, as the amount of conifers increases, White-breasted Nuthatches decrease in abundance and Red-breasted Nuthatches increase. Other suitable habitats include city parks, suburban woodlots, and riparian corridors that provide large, mature hardwood trees.

Overall, the White-breasted Nuthatch demonstrates some preference for woodland edges (Peck and James 1987). Although it is common in small woodlot settings, studies suggest that reproductive success and survivorship are higher in larger forest tracts (Galli et al. 1976; Doherty and Grubb 2002) and in stands that are not too isolated from one another (Grubb and Doherty 1999).

POPULATION ABUNDANCE

In the United States and Canada, the White-breasted Nuthatch's population numbers approximately 10 million breeding adults, with 370,000 birds residing in Minnesota (PIF 2021). Surveywide, BBS data document a significant population increase averaging 1.15 percent per year from 1966 to 2019 (Sauer et al. 2020). The rate of increase was especially high between the years 1970 and 2014, when biologists estimated a 124 percent growth in population numbers (Rosenberg, Kennedy, et al. 2016). In Minnesota, the Nuthatch population has experienced an even higher significant increase of 1.54 percent per year from 1967 to 2019 **(D)**.

C. The White-breasted Nuthatch is a common inhabitant of upland hardwood stands with open understories and mature trees. Photo by Gary T. Seim.

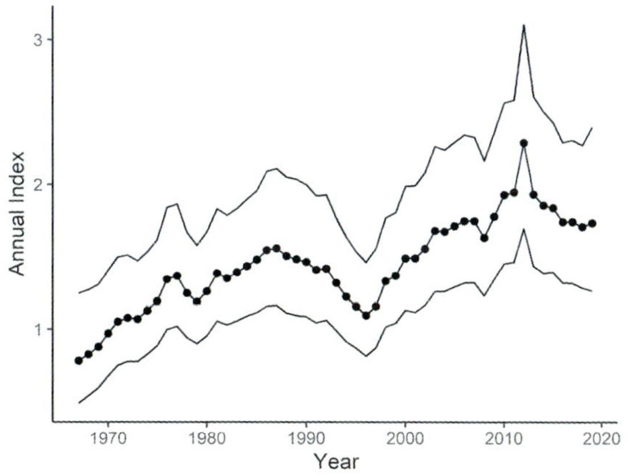

D. Breeding population trend and 95 percent credible interval (solid black lines) for the White-breasted Nuthatch in Minnesota based on the 1967–2019 Breeding Bird Survey (Sauer et al. 2020).

Factors responsible for the Nuthatch's population growth have not been delineated, but the increase is widespread throughout much of the eastern United States. The availability of mature shade trees in residential communities throughout North America and the ever-growing popularity of bird feeding are likely major contributors.

CONSERVATION

Although the White-breasted Nuthatch is faring quite well, management practices that retain dead and decaying mature trees are especially important for this cavity-nesting species. Its demonstrated preference for oak forests in northern Minnesota, which may reflect its dependence on acorns during the winter months, suggests that maintaining large, mature oaks in hardwood stands is also critical. In more urban/suburban and agricultural settings, protection of public lands that are sufficiently large (greater than 10 ha) and interconnected will also provide benefits.

An analysis by NAS (2019) predicted that the Nuthatch has a low vulnerability to climate change under all three warming scenarios examined. As temperatures rise, breeding range losses in southern states and the eastern Great Plains will be exceeded by projected gains to the north and west. At the same time, the Nuthatch is projected to gradually disappear from western and southern Minnesota while habitat conditions become more favorable in the northeast region, where it may eventually replace its retreating relative, the Red-breasted Nuthatch. Meanwhile, this handsome, acrobatic little bird is likely to remain a common and welcome visitor at bird feeders for many years.

BROWN CREEPER

Certhia americana

Minnesota Seasonal Status: Regular breeding resident and migrant; present throughout the winter, especially in southern Minnesota. The Brown Creeper was uncommon during the MNBBA.
Migration: Extent of migration is poorly known; most breeding populations are year-round residents, but birds that breed in the north may migrate south, overwintering in the United States and northern Mexico.
Food: Bark forager that consumes arthropods, systematically moving from the base of a tree upward.
Nest: One of the most unusual and difficult-to-locate bird's nests, it is usually wedged between loose bark and the trunk of a dead or dying tree.
Conservation Concern: Assigned a low PIF Continental Concern Score of 8/20.

Photo by David Brislance.

The Brown Creeper is aptly named. Many observers detect its presence when their eyes catch a bit of movement along a large tree trunk. Those who investigate further are rewarded by finding the diminutive Creeper intently searching for food as it scrambles up the tree. One writer compared it to "a fragment of detached bark that is defying the law of gravitation" (Bent 1948). Unlike nuthatches, which easily move headfirst down a tree trunk, once the Brown Creeper finishes its spiral journey upward, it takes flight and lands at the base of another tree to begin its upward journey once again. Appropriately, *Certhia americana* refers to a small, tree-dwelling bird in America.

Roberts (1932) remarked that "the little Creeper is an example of almost perfect adaptation to the special role that it plays in nature." Its mottled brown-and-white plumage on its back provides the ideal camouflage; its sharp, curved bill is a precision instrument for probing the bark's deep furrows for insects; its long, stiff tail helps balance its vertical position; and its large claws keep it clinging tightly to the tree's surface. Thin call notes, *tsit* or *zi-i-i-it,* may give the Creeper's presence away, but one must be within close range to hear them. The male's high-pitched, delicate song can also be difficult to hear but has been described as "one of the gems of bird music" (Bent 1948).

This forest inhabitant is found from southern Alaska, east across southern Canada, south through the Pacific Northwest and the Rocky Mountains, and east across the Great Lakes and New England. Populations also extend into Mexico and Central America. One rarely sees them in large numbers; it is more common to spot lone individuals, often widely dispersed from one another.

MINNESOTA BREEDING DISTRIBUTION

The Brown Creeper is a complicated bird. Roberts (1932) characterized it as a "half-hardy species" residing in "the northern evergreen forests as far south as Isanti County." Although he classified it as a permanent resident, he noted that it was less common in the region during the winter, when "continued cold weather and . . . a decreasing food supply take a toll." In southern Minnesota, he classified it as a "common migrant and uncommon winter resident." Given the difficulty of finding the species' well-hidden nest, it is not surprising that there were only two confirmed reports: one from central Minnesota, "probably southern Crow Wing County" (1870), and one from Cook County (1930). An inferred nesting report was available from Itasca County.

Years later, Green and Janssen (1975) considered the Brown Creeper a regular summer resident across north-central and northeastern Minnesota. Nesting reports in Hennepin and Goodhue Counties led them to speculate that the species was probably a resident "throughout much of the wooded portion of the state." Janssen's (1987) updated account identified new records from western Minnesota, including a breeding report from Brown County and summer observations from Cottonwood and Pope Counties. Farther east, numerous summer observations in the southeastern corner of the state, along with a breeding record in Winona County, led him to conclude that the species' breeding range extended south along the Mississippi River valley to Iowa. Since 1970, confirmed nesting records were available from eight widely dispersed counties (Aitkin, Anoka, Brown, Clearwater, Lake, Lake of the Woods, St. Louis, and Winona); Hertzel and Janssen (1998) added three more counties to the list (Cook, Dakota, and Polk).

Meanwhile, the MBS collected 250 Brown Creeper breeding-season observations. Most significant were records along the floodplain forests of the Minnesota River, west to Redwood County, and in several central Minnesota counties located largely west of the breeding range depicted by Janssen (1987): Kandiyohi, Morrison, Stearns, and Todd.

The MNBBA tallied 709 Brown Creeper detections from 10.1 percent (479/4,738) of the surveyed atlas blocks **(A)**. Their distribution spanned 33 counties that were located primarily east of a line from Fillmore County in the southeast to Kittson County in the northwest. Observers did an exceptional job locating 10 nests and confirming nesting in 28 blocks. Five of the 14 counties where breeding was confirmed were in the Twin Cities region; nine counties were additions to the 1998 list published by Hertzel and Janssen. The MNBBA failed to find Brown Creepers in the upper Minnesota River valley but did document more birds than the MBS in the lakes region of Cass, Crow Wing, and Itasca Counties and in the state's far northwestern corner. The core

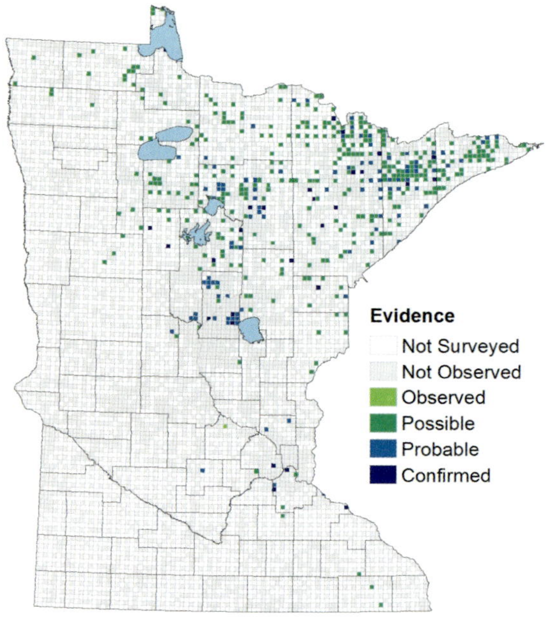

A. Breeding distribution of the Brown Creeper in Minnesota based on the Breeding Bird Atlas (2009–13).

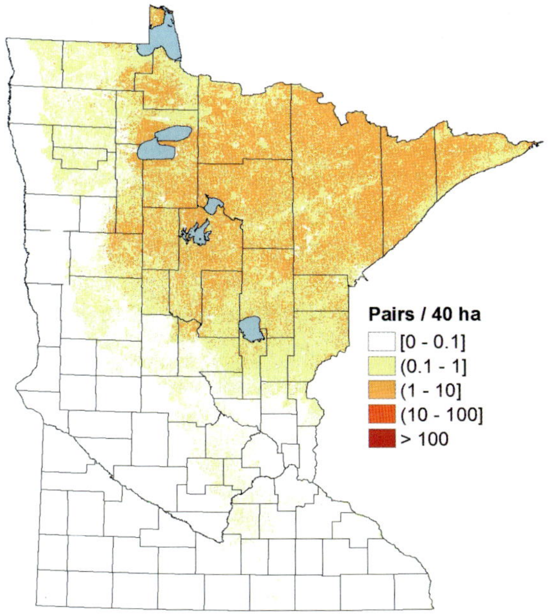

B. Predicted breeding distribution (pairs per 40 ha) of the Brown Creeper in Minnesota based on habitat, landscape context, and climate data gathered during the Minnesota Breeding Bird Atlas (2009–13) using the Generalized Linear Modeling method with an adjustment for detectability.

of the species' breeding range, however, remains in Cook, Lake, and northern St. Louis Counties.

The MNBBA probability map predicts that Brown Creepers may be found in moderate breeding densities across much of the Laurentian Mixed Forest Province **(B)**. Low densities are predicted in the Tallgrass Aspen Parklands, in portions of the Eastern Broadleaf Forest Province, and along the lower Mississippi River valley.

Accurately assessing changes in the inconspicuous Brown Creeper's distribution and abundance is difficult. Nevertheless, the absence of summer observations in southern Minnesota in the early 1900s suggests they were either absent or quite rare in abundance. Although they are still uncommon south of Crow Wing County, scattered locations, particularly floodplain forests along the lower Minnesota River and the Mississippi River, provide suitable habitat, as do mature forests in southeastern Minnesota. Elsewhere in its breeding range, the Creeper has expanded south along the lowlands of the mid-Atlantic states and in the Midwest, "especially in the bald-cypress and tupelo riparian forests along major rivers" (Poulin, D'Astous, et al. 2020).

BREEDING HABITAT

The Brown Creeper is closely associated with mature upland and lowland forests. Deciduous, coniferous, and mixed forest stands are all utilized **(C)**. In national forests in northern Minnesota and Wisconsin, the species' strongest association was with lowland hardwood stands, followed by white pine and aspen–birch–fir (Niemi, Howe, et al. 2016). Regardless of the cover type, a site's suitability is defined by the availability of large trees (greater than 30 cm diameter at breast height) for foraging and the presence of dead and dying trees with peeling bark for nesting (Poulin, D'Astous, et al. 2020).

POPULATION ABUNDANCE

Because the Brown Creeper's soft, high-pitched vocalizations are detected only at close range, the reliability of data collected by BBS roadside counts is relatively low. With this constraint in mind, PIF (2021) used BBS data to generate a population estimate of 9.5 million breeding adults in the United States and Canada and a Minnesota population of 160,000. MNBBA data, which sampled many off-road localities in northern Minnesota, generated a significantly higher estimate of 780,000 adults (95 percent confidence interval of 638,000–1.3 million).

Although BBS trend estimates for the Brown Creeper lack statistical reliability, they suggest a stable to slightly increasing population across the survey region from 1966 to 2019 (Sauer et al. 2020). A similar increase is evident in Minnesota, despite a notable downward trend that began after the 2010 breeding season. The songbird monitoring program in the Chippewa and Superior National Forests may be the best barometer of the species' population trend in Minnesota. From 1995 to 2020, the data reveal a stable population that is beginning to trend downward on both forests (Grinde, Bednar, et al. 2020).

Other than the loss of old-growth forests, there is little else to suggest what factors may be influencing large-scale population trends, either positive or negative. The birds are susceptible to cold winter temperatures, and thus difficult winters could affect survivorship and result in local population declines. Roberts (1932), for example, included an account by Mr. Daniel Smith of finding three Brown Creepers frozen but still clinging to the bark of trees during an "intensely cold" winter in 1912. Survivorship also may be impacted when prolonged wet or cold spells occur during the breeding season (Poulin, D'Astous, et al. 2020). Locally, timber management activities that reduce the availability of mature trees and snags also impact habitat suitability.

CONSERVATION

The species' reliance on large trees for foraging and nesting points to a major conservation concern: the protection and management of mature and old-growth forests. Fortunately, in Minnesota, old-growth forest management guidelines for state and federal lands help to ensure that these important stands are considered in forest management plans. Warmer temperatures, however, are a major challenge. The Brown Creeper is classified as moderately vulnerable under all three warming scenarios examined; NAS (2019) predicted that it would be nearly extirpated from Minnesota under the imminent warming scenario of 1.5°C. Under the 2°C scenario, eastern populations will retreat entirely to Canada. Collisions with buildings and towers are another threat. Identified as a "super collider" by T. Arnold and Zink (2011), several studies have documented the species' vulnerability to such hazards (Bracey 2011; Longcore et al. 2013; Loss et al. 2014).

Although Brown Creepers currently appear relatively secure in Minnesota, the challenges ahead appear formidable. In the interim, the next time you think you see a piece of bark moving on a tree trunk, take a closer look and enjoy the antics of a bird that is well adapted to a unique niche within the forest environment.

C. In a floodplain forest along the Mississippi River, the peeling, damaged bark on this mature deciduous tree provides an ideal nesting site for the Brown Creeper. Photo by Lee A. Pfannmuller.

BLUE-GRAY GNATCATCHER

Polioptila caerulea

Minnesota Seasonal Status: Regular breeding species and migrant. The Blue-gray Gnatcatcher was uncommon during the MNBBA.

Migration: Northern breeding birds are short- to medium-distance migrants that winter from the coast of the southern United States south through the Caribbean, Mexico, and northern Central America.

Food: Arthropods secured by foliage gleaning, sallying after flying insects, and hovering above the foliage. Gnats are consumed but are not a significant component of their diet.

Nest: Tightly woven cup nest placed in a deciduous tree at variable heights.

Conservation Concern: Assigned a low PIF Continental Concern Score of 7/20.

Photo by Mike Lentz.

"Perky" and "pixie-like" are among the many expressions used to capture the essence of the tiny Blue-gray Gnatcatcher. Smaller than a warbler and larger than a hummingbird, the species is a tree-top inhabitant. Its blue-gray plumage, white eye-ring, and black tail, bordered with white outer tail feathers, are diagnostic. During the breeding season, the male has a hint of black in the forehead and over the eye. *Polioptila caerulea,* the species' scientific name, is a direct reference to its blue-gray feathers. Described as a "little bird of intense activity," Bent (1949) noted that the Gnatcatcher's "most expressive feature . . . is its long, ever-active tail." Contributing nearly half of the bird's entire length, this unique attribute also provides a helpful clue to its identification.

The Gnatcatcher's high, wispy song and call notes can be easily overlooked, but the vocalizations are unlikely to be mistaken for any other bird. The nasal *twee* call note is actually more easily heard than the male's more complex, rambling song, which often includes phrases that mimic other birds (Kershner and Ellison 2020), or its simpler song of four to eight short, wheezy phrases. Its vocal mimicking led some early writers to nickname it the "little mockingbird."

Broadly distributed across the southern United States, north to the Great Lakes, southern New England, and extreme southeastern Canada, the species is largely absent from the Great Plains. It is also found year-round in Florida, Mexico, and northern Central America. The highest breeding densities are found in the south-central states, from southern Missouri and northern Arkansas east to North Carolina.

MINNESOTA BREEDING DISTRIBUTION

The first Minnesota record of a Blue-gray Gnatcatcher was a specimen collected in May 1877 in Minneapolis (Roberts 1932). Then, in 1895, a rather surprising account was published by Walton Mitchell, describing six Blue-gray Gnatcatcher nests found in the vicinity of St. Paul in Dakota County from 1892 to 1894. This account was all the more remarkable because nearly thirty years would pass until summer observations were again reported and additional breeding evidence was documented. The latter included an August 1924 report of "five individuals" from Frontenac in Goodhue County, a July 1930 report of "one adult and three small young" from northern Washington County, and a 1931 report of a nesting pair in Minneapolis (Roberts 1932). Although a specimen was collected as far west as Murray County in 1900, Roberts considered the species to be primarily restricted to the Mississippi River floodplain and its larger tributaries only as far north as Minneapolis. The Gnatcatcher, however, continued its trek northward, and by 1935 "a dozen or more" birds and an active nest were found in the St. Cloud area (Roberts 1936).

By 1975, Green and Janssen depicted the Gnatcatcher's breeding distribution as stretching from those counties bordering the Mississippi River north to southeastern Stearns County in east-central Minnesota and along the St. Croix River in Washington County. Spring and summer observations beyond these counties suggested that the breeding range might extend even farther north and west. Breeding had been confirmed in eight counties, from Stearns County in the north to Houston County in the south.

Janssen (1987) described the species' continued range expansion west along the Minnesota River valley and northwest along the Mississippi River valley. The species was now "well represented in Nicollet County and as far southwest as Lyon County (Garvin Park)." Since 1970, confirmed nesting records had been reported from 17 counties, including a record from Rock County in southwestern Minnesota and from Otter Tail County in west-central Minnesota. Hertzel and Janssen (1998) later added seven more counties to the list, further extending its breeding range north to Clay and Hubbard Counties. The 362 Blue-gray Gnatcatcher breeding-season locations identified by the MBS documented additional records in the southwest corner of the state and two records in the northwest Aspen Parklands. Farther east, records were found as far north as southern Aitkin, Crow Wing, and Pine Counties.

The MNBBA's 432 Blue-gray Gnatcatcher records outlined a distribution similar to that depicted by the MBS records. Reported from 6.3 percent (297/4,737) of the surveyed blocks **(A)**, breeding was confirmed in 57 blocks (1.2 percent) and deemed probable in an additional 70 blocks (1.5 percent). The birds were reported in 54 of Minnesota's 87 counties and were confirmed breeding in 28 counties, including 12 that were additions to the list published by Hertzel and Janssen (1998). New confirmed breeding records along the northern edge of the Gnatcatcher's breeding range included records in southern Cass, southern Crow Wing, and northern Kanabec Counties. Probable nesting was reported as far north as west-

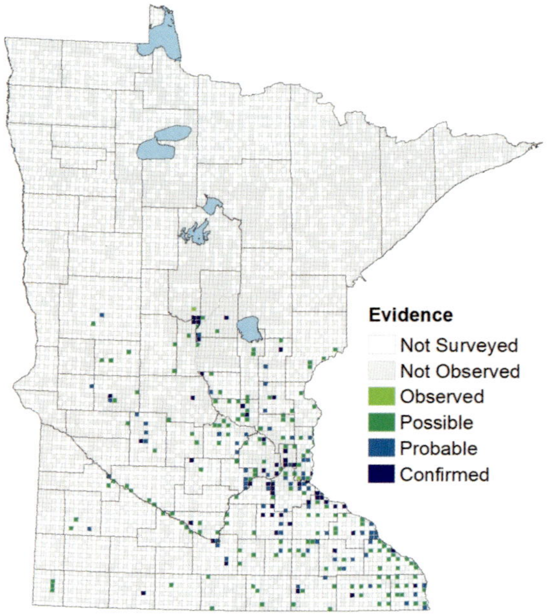

A. Breeding distribution of the Blue-gray Gnatcatcher in Minnesota based on the Breeding Bird Atlas (2009–13).

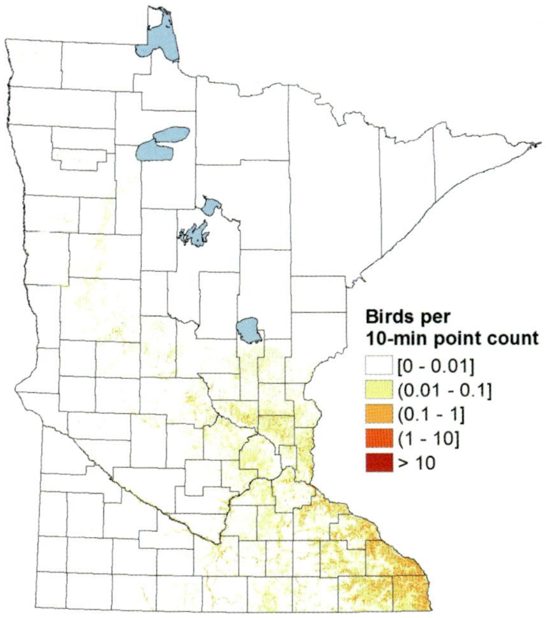

B. Predicted breeding distribution (birds detected per 10-minute point count) of the Blue-gray Gnatcatcher in Minnesota based on habitat, landscape context, and climate data gathered during the Minnesota Breeding Bird Atlas (2009–13) using the Generalized Linear Modeling method without an adjustment for detectability.

ern Polk County, when a pair was sighted in East Grand Forks in late May 2013 (the atlas block is too small to see on the MNBBA map). Now firmly established from southern Cass and Crow Wing Counties south through east-central and southeastern Minnesota, Gnatcatchers also occurred throughout the Minnesota River valley as far west as Yellow Medicine and Renville Counties and in scattered locations across south-central, southwestern, and west-central Minnesota.

The MNBBA probability map **(B)** predicts that the highest abundance of Blue-gray Gnatcatchers occurs along the river valleys of southeastern Minnesota, gradually decreasing north along the Mississippi and St. Croix Rivers, west across south-central Minnesota and the Minnesota River valley, and northwest along the prairie forest border. Wooded floodplains across south-central Minnesota provide stretches of suitable habitat, including the Blue Earth River drainage south of Mankato.

The Blue-gray Gnatcatcher has gradually expanded northward during the twentieth century across its breeding range. Attributing the expansion to a warming climate, Kershner and Ellison (2020) noted that its range has expanded nearly 314 km to the north just since the mid-1980s. Wisconsin's first breeding bird atlas, for example, documented a northward range extension of at least 50 km from the early 1990s through 2005 (Cutright et al. 2006). In South Dakota, the birds were first reported breeding in the state's southeast corner in 2004; when their second atlas was completed in 2012, breeding populations were reported in all corners of the state (Drilling, Sparks, et al. 2016). In Canada, Blue-gray Gnatcatchers have not yet expanded into Manitoba (Artuso et al. 2019), but they are well established in Ontario's southern shield and are continuing to expand north (Cadman et al. 2007).

BREEDING HABITAT

The Blue-gray Gnatcatcher inhabits a diversity of wooded habitats, from oak woodlands and chaparral in the southwestern United States to riparian forests and deciduous uplands in the East **(C)**. Along the northern periphery of its breeding range, it is frequently associated with riparian and lakeshore habitats (Kershner and Ellison 2020). Indeed, Mossman (1991) documented the species' preference for floodplain forests of silver maple–black ash and swamp white oak–black ash along the St. Croix River. It prefers the tops of tall trees in mature deciduous forests (Niemi and Hanowski 1992b); open understories and large trees are characteristic features of its northern habitat and may be more important than the composition of tree species.

Some studies have suggested that the Blue-gray Gnatcatcher is area sensitive, becoming more abundant in larger forest tracts (Niemi and Hanowski 1992b). Others, however, have suggested that features at the microhabitat level, including the presence of stream valleys and forest gaps, may be more limiting than the size of the forest stand (Kershner and Ellison 2020).

C. A restored oak savanna along the bluffs of the Mississippi River provides the large trees and open understory often preferred by the Blue-gray Gnatcatcher in Minnesota. Photo by Lee A. Pfannmuller.

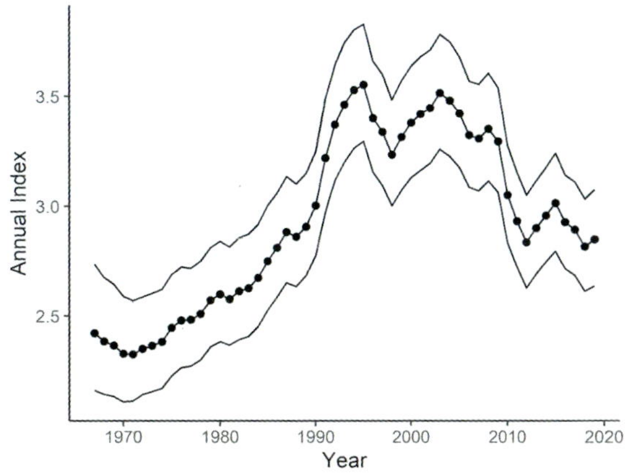

D. Surveywide breeding population trend and 95 percent credible interval (solid black lines) for the Blue-gray Gnatcatcher based on the 1967–2019 Breeding Bird Survey (Sauer et al. 2020).

POPULATION ABUNDANCE

Approximately 230 million Blue-gray Gnatcatchers reside in the United States and Canada, with an additional 30 million present south of the United States (PIF 2021). The estimate for Minnesota is 180,000 breeding adults. Numbers were too low to generate a reliable state estimate using MNBBA data.

Despite its range expansions and a significant population increase in the late twentieth century, the Blue-gray Gnatcatcher has demonstrated a relatively stable population trend. From 1966 to 2019, data document a small, significant increase averaging 0.30 percent per year throughout the BBS survey region **(D)**, with a cumulative increase of 27 percent from 1970 to 2014 (Rosenberg, Kennedy, et al. 2016). Because the Gnatcatcher is so rarely encountered on BBS routes in Minnesota, the data are not statistically robust. Nevertheless, the trend line is positive (Sauer et al. 2020).

CONSERVATION

With its large and expanding population, the Blue-gray Gnatcatcher is considered a low conservation priority (PIF 2021). Nevertheless, given its vulnerability to nest parasitism by Brown-headed Cowbirds and its potential status as an area-sensitive species, protection of large forest tracts, especially forest floodplains, is an important conservation measure. Although climate change is a potential threat to many North American songbirds, the Gnatcatcher is one species that has benefited from a warming climate. In future years, it is likely to become an even more familiar bird to Minnesotans, adding its high and wispy song to the morning chorus of songbirds even farther north within Minnesota's deciduous woodlands (NAS 2019).

HOUSE WREN

Troglodytes aedon

Minnesota Seasonal Status: Regular breeding species and migrant. The House Wren was abundant during the MNBBA.
Migration: Northern populations are short- to medium-distance migrants that winter in the southern United States and Mexico.
Food: Gleans terrestrial arthropods from trees, shrubs, and the ground.
Nest: Secondary cavity nester that utilizes an extraordinarily diverse range of "cavities," including natural tree hollows, used nests of other birds, nest boxes, flowerpots, and old boots.
Conservation Concern: Assigned a low PIF Continental Concern Score of 5/20.

Familiar to many backyard bird enthusiasts, the pint-sized House Wren is a bundle of energy, constantly flitting about with its upturned tail, its chattering alarm calls, and its bubbly song. The male's tireless determination is measured by the ceaseless delivery of his song. Long after the dawn chorus has ceased, he continues to belt out his cascade of shrill warbles and trills, persistent in his efforts to attract a mate and defend his piece of real estate. Many have attempted to phonetically describe the House Wren's complex song, but Bent (1948) captured its spirit when he characterized it as a "rather loud, hurried, strenuous, bubbling outpouring." The Chippewa named the Wren *O-du-ná-mis-sug-ud-da-we'-shi,* which translates to "a big noise for its little size" (Cooke 1884). A little smaller than a Black-capped Chickadee, the Wren's most notable physical feature is the barring on the wings, back, and tail; the sexes are identical.

Aggressiveness is another behavioral trait that many associate with the Wren. Competition for available nest sites can be intense among species that do not excavate their own cavities. House Wrens respond to this competition by pecking or removing the eggs and hatchlings of other cavity-nesting species nesting nearby, such as Eastern Bluebirds, Tree Swallows, and Black-capped Chickadees. Such behavior resulted in a firestorm of protest in the early 1900s led by illustrator and ornithologist, Althea Sherman. An avid backyard bird-watcher, Sherman's admiration for the Wren began to wane when, after attracting numerous pairs to nest boxes, she watched them systematically destroy the eggs and young of other endearing backyard inhabitants (Waters 2016). Her wide-ranging protests against the use of nest boxes may have dis-

Photo by Mike Lentz.

suaded some, but ornithologists understand that the Wren's behavior is a well-adapted survival tactic that has served the species quite well.

The House Wren's breeding range occupies "the broadest latitudinal range of any native passerine in the New World" (L. S. Johnson 2020). It can be found across southern Canada and the northern two-thirds of the United States, and from southern California and Arizona south to the West Indies, Mexico, Central America, and nearly all of South America. A total of 31 subspecies have been delineated. Arranged into five groups, one is known as the Northern House Wren, or *adeon*-type, and is restricted to North America. The species' scientific name, *Troglodytes aedon,* is of Greek origin and refers to "a hole-dwelling nightingale."

MINNESOTA BREEDING DISTRIBUTION

In Minnesota, Roberts (1932) considered the species to be an abundant, statewide summer resident. He wrote: "Not only is the House Wren a common bird among the dwellings of man but it is especially common throughout all the wild places of the state, particularly the desolate, burned-over areas of the north." Nesting evidence stretched from Fillmore and Goodhue Counties in the southeast, to Chisago, Grant, Hennepin, Isanti, Kandiyohi, and McLeod Counties in central Minnesota, and then north to Polk and Marshall Counties in the northwest and to Lake of the Woods County, St. Louis County, and Itasca State Park in northern Minnesota.

Little had changed when Green and Janssen (1975) and Janssen (1987) published their updated accounts of the species' status. The House Wren remained "widely distributed and numerous in all regions." Janssen (1987) identified 61 counties where nesting had been confirmed since 1970. Hertzel and Janssen (1998) later added another seven counties to the list. By 2014, the 1,600 breeding-season locations reported by the MBS provided more detail regarding the species' relative abundance. Although the House Wren was common throughout most of the surveyed regions of southern and western Minnesota, its abundance decreased considerably across the Laurentian Mixed Forest Province.

A similar distribution was depicted by the 4,750 House Wren records reported during the MNBBA **(A)**. Abundant across Minnesota's Eastern Broadleaf Forest, Tallgrass Aspen Parklands, and Prairie Parkland Provinces, it was also common along the southern and western edges of the Laurentian Mixed Forest Province, from Carlton County northwest to Lake of the Woods County. Farther east, it was quite rare except along the Iron Range in central St. Louis County. All told, the Wren was detected in 50.4 percent (2,402/4,769) of the surveyed atlas blocks and 68.7 percent (1,605/2,337) of the priority blocks. Breeding was confirmed in 547 blocks (11.5 percent) that covered all but three counties: Traverse, Lincoln, and Nobles. Several probable nesting records, however, were reported in each of these counties.

The MNBBA probability map for the House Wren predicts that the area of highest abundance largely occurs in a region that stretches from the Twin Cities, northwest along the prairie forest border, and north along the Glacial Lake Agassiz beach ridges and riparian corridors of northwestern Minnesota **(B)**. Fragmented forest landscapes throughout this region are particularly suitable for the species. Elsewhere, moderate breeding densities are predicted throughout most of southern and western Minnesota and along

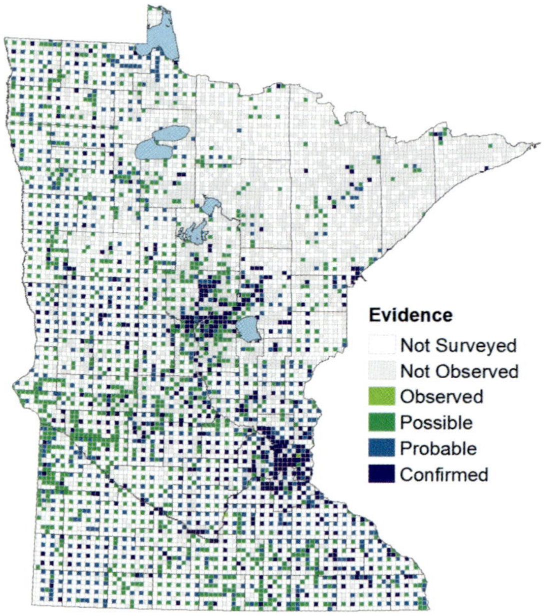

A. Breeding distribution of the House Wren in Minnesota based on the Breeding Bird Atlas (2009–13).

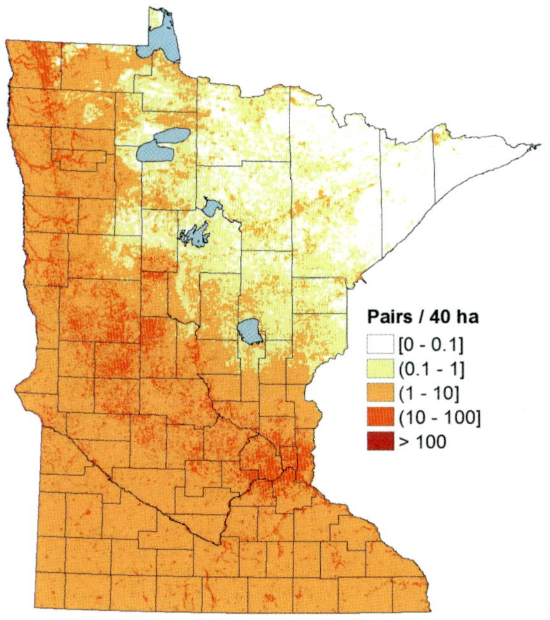

B. Predicted breeding distribution (pairs per 40 ha) of the House Wren in Minnesota based on habitat, landscape context, and climate data gathered during the Minnesota Breeding Bird Atlas (2009–13) using the Generalized Linear Modeling method with an adjustment for detectability.

the southern and western edges of the Laurentian Mixed Forest Province. Farther east, suitable habitat is restricted to residential communities, the Iron Range, and sites of recent disturbance.

Nearly one hundred years after Roberts (1932) described it as an abundant, statewide breeding resident, the House Wren remains a widely distributed species. Its sparse distribution in the northern counties, however, is likely a more recent phenomenon. During the late 1800s and early 1900s, the region's vast stands of pine, aspen, birch, fir, and spruce were harvested, leaving behind a landscape of slash that was prone to fires. As Roberts noted in 1932, the open country, where snags abounded, provided ideal habitat for the House Wren. Today harvested sites still provide suitable habitat but at a much smaller scale now that forest lands are responsibly managed. A similar scenario has been noted in northern Michigan. Early logging cleared much of the state's northern forest and likely resulted in a northern range extension of House Wrens until the mid-twentieth century, when second-growth forests dominated the landscape and provided less suitable habitat (R. Brewer et al. 1991).

Across the United States, the major change in the House Wren's (i.e., *adeon*-type) distribution has been its steady southward expansion as forests in the eastern states were cleared for agriculture and residential development. It became a common and widespread species in Ohio in the mid-1800s (Rodewald et al. 2016), reached West Virginia by the late 1800s, and South Carolina and Georgia by the 1940s and 1950s. Farther west, it reached central and southern Missouri by the 1960s and northern Alabama by the early 1970s. Along the leading line of expansion, the Wrens are usually present as disjunct populations. The birds first become established in residential areas and then expand outward from these centers of abundance (L. S. Johnson 2020).

BREEDING HABITAT

House Wrens occupy a range of semi-open habitats in the Upper Midwest, including the edges of upland and lowland deciduous forests, burns and clear-cuts with residual trees, farmsteads, parklands, and suburban lots (Danz et al. 2007). Their primary requirements are the presence of dead trees or dead tree limbs that provide suitable cavities, clearings or canopy gaps, and brushy tangles in the understory for feeding **(C)**. Nest boxes attract them to sites that lack natural cavities but only if other important features, such as suitable cover and food resources, are also available. Large contiguous forest landscapes

C. The House Wren is attracted to a wide variety of habitats, including forest openings with shrubby cover for feeding and dead or dying trees for nesting. Photo by Lee A. Pfannmuller.

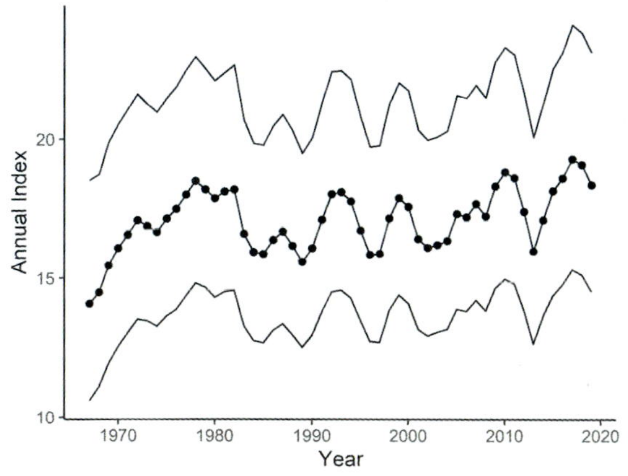

D. Breeding population trend and 95 percent credible interval (solid black lines) for the House Wren in Minnesota based on the 1967–2019 Breeding Bird Survey (Sauer et al. 2020).

are generally avoided, with the exception of recently harvested areas or burns.

POPULATION ABUNDANCE

BBS data were used by PIF (2021) to generate a population estimate for the Northern House Wren of 43 million breeding adults in the United States and Canada and a statewide estimate of 2 million for Minnesota. MNBBA data generated a state estimate that was more than twice as large, at 4.9 million adults (95 percent confidence interval of 4.5–5.4 million adults). The entire population, including all 31 subspecies, numbers approximately 190 million (PIF 2021).

Across the BBS survey region, House Wrens have experienced a small but significant increase averaging 0.49 percent per year from 1966 to 2019 (Sauer et al. 2020). Numbers in Minnesota demonstrate a similar trend line, with a significant annual increase of 0.52 percent **(D)**. Although a few studies have demonstrated a decline in Wren numbers following drought years and cold winter temperatures, other environmental parameters that impact survivorship have not been studied. Nest boxes can enhance local House Wren abundance, but not in all circumstances (L. S. Johnson 2020).

CONSERVATION

An abundant species that has adapted well to human-modified landscapes, the House Wren is not a conservation priority (PIF 2021). Unfortunately, under a 2°C warming scenario, NAS (2019) projected a moderate vulnerability for the species' continental breeding distribution, with a 30 percent range loss and only a 21 percent gain. Although no changes were predicted in Minnesota, when temperatures rise by 3°C, this familiar bundle of tireless energy is suddenly projected to recede from all but the north-central and northeastern counties by 2080.

WINTER WREN

Troglodytes hiemalis

Minnesota Seasonal Status: Regular breeding resident and migrant; occasionally found during the winter near open water in southern Minnesota. The Winter Wren was common during the MNBBA.
Migration: Northern populations are short-distance migrants that winter in the central and southern regions of the eastern United States; populations in southern New England and the Appalachian Mountains are permanent residents.
Food: Arthropods collected by gleaning foliage in the lower forest canopy and by probing the forest litter.
Nest: Domed nest placed within a natural cavity close to the ground, including the roots of upturned trees, under logs, within brush piles, or in rotten tree stumps or snags.
Conservation Concern: Assigned a low PIF Continental Concern Score of 8/20.

The Winter Wren, a denizen of our mature northern forests, is one of Minnesota's smallest songbirds, measuring just 10 cm in length. Not only does its small size make it difficult to observe but also one must bushwhack through dense lowland forests and scramble over mossy hummocks and downed logs to catch a fleeting glimpse of it. Dark brown with heavy barring, a short tail, and a plump body, the bird is well camouflaged as it hops among the decayed litter and woody debris of the forest floor—its favorite haunt—frequently stopping to bob its head or its entire body up and down.

Lucky for us, the Wren's most endearing quality is its magical song. Ounce for ounce, the little bird delivers a more robust and lengthy song than any other North American songbird. The Ojibwe named it *Ka-wa-miti-go-shi-que-na-go-mooch,* a term that honors its rambling melody, which may last 10–20 seconds (Kroodsma 2005). It left a lasting impression on Roberts (1932) who described it as a "wild ecstatic bit of woods melody." The listener, he wrote, is "filled with astonishment at the great power and endurance displayed by such a mite of a bird." Although many have sought to capture the nature of the song with words, hearing it just once imparts an unforgettable memory. "Dull indeed," Audubon (1827–38) wrote, "must be the ear that thrills not on hearing it."

The Winter Wren occurs across North America's boreal forest, stretching from British Columbia east to the Maritime Provinces and south across the northern Great Lakes states, New England, and the Appalachian ridge. The northern states and eastern provinces support the highest breeding densities. *Troglodytes* means "hole creeper," a reference to the species' nesting habits; *hiemalis* means "winter," a possible reference to its habit of occasionally overwintering in the northern states and provinces.

Photo by David Brislance.

MINNESOTA BREEDING DISTRIBUTION

Roberts (1932) considered the Winter Wren to be a regular summer resident across Minnesota's northern forests, stretching south to Pine and Aitkin Counties, and west to Cass and Lake of the Woods Counties. He wrote, "It is probable that this Wren breeds throughout this area, sparingly for the most part, but locally rather commonly." During a visit to Cook County in 1922, Thaddeus Surber reported to Roberts that he found the species "over the entire region wherever heavy timber occurs, and in the big swamps of spruce and white cedar, but we never found it in the more open second-growth." Breeding evidence was available from Cook, Itasca, and Lake of the Woods Counties, and near Cass Lake.

Decades later, Green and Janssen (1975) reported that the species was a summer resident primarily in north-central and northeastern Minnesota. Occasionally observed in Pine and Mille Lacs Counties, it had two summer records even farther south, in Anoka and Washington Counties. Janssen's 1987 update included observations in Goodhue, Hennepin, and Houston Counties, and a report of a family group in Chisago County, suggesting the birds might occasionally breed in east-central Minnesota and farther south along the Mississippi River. He identified eight counties where nesting had been confirmed since 1970: Aitkin, Cass, Chisago, Clearwater, Cook, Crow Wing, Lake, and St. Louis. Hertzel and Janssen (1998) deleted Clearwater County from the list and added Itasca County.

Meanwhile, the MBS documented 714 Winter Wren breeding-season locations. Although most records were restricted to northern Minnesota, a few sites were located in Fillmore, Houston, and Winona Counties in the southeast, in Nicollet and Redwood Counties along the Minnesota River, and in Chisago, Scott, and Washington Counties in east-central Minnesota. Several records also occurred in Marshall, Polk, and Roseau Counties and along the prairie–forest border of Mahnomen, Becker, and Otter Tail Counties.

MNBBA participants documented a more restricted breeding range than earlier accounts, with all but eight of the 1,837 detections occurring within the Laurentian Mixed Forest Province **(A)**. Winter Wrens were reported from 16.6 percent (786/4,741) of the surveyed atlas blocks and confirmed breeding was reported in only 2.7 percent (21) of the blocks where they were detected. They were found in 18 northern counties, including Marshall and Roseau, but records were absent south of Morrison and Pine Counties in the east and south of Clearwater County in the west. Although only a few atlas records oc-

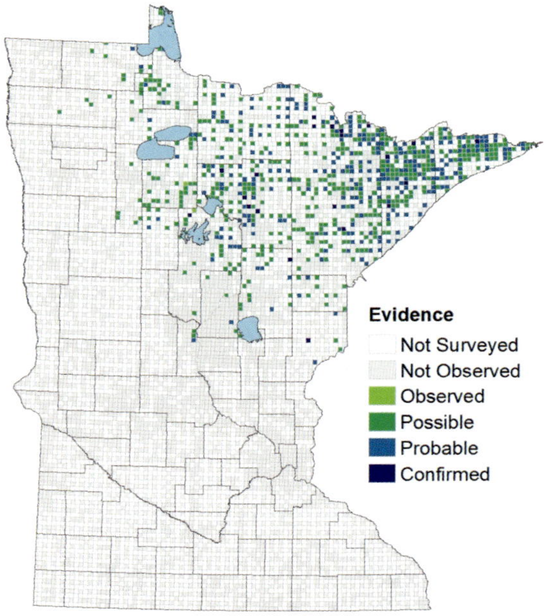

A. Breeding distribution of the Winter Wren in Minnesota based on the Breeding Bird Atlas (2009–13).

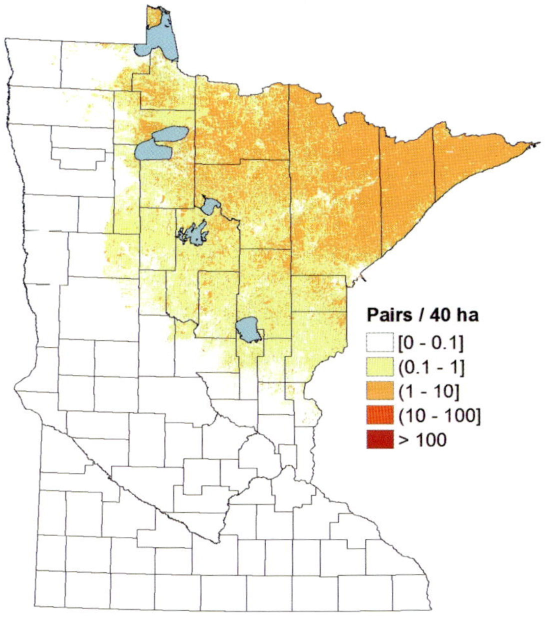

B. Predicted breeding distribution (pairs per 40 ha) of the Winter Wren in Minnesota based on habitat, landscape context, and climate data gathered during the Minnesota Breeding Bird Atlas (2009–13) using the Generalized Linear Modeling method with an adjustment for detectability.

curred along the southern perimeter of the northern forest, the MNBBA probability map predicted the Wren may occur in low breeding densities as far south as southern Chisago County and as far west as eastern Kittson, eastern Becker, and northeastern Otter Tail Counties **(B)**.

The species' status in southern Minnesota warrants further investigation. Because the Winter Wren's song is unmistakable, it is unlikely that its presence in southern Minnesota was missed by atlas participants. Nevertheless, during the atlas, the MOU (2021b) reported summer records in three west-central and southern counties that could not be attributed to atlas blocks: Douglas County (an adult carrying food in 2010); Fillmore County (July 18, 2011); and Steele County (June 12, 2013). The latter record could also be a late spring migrant.

In neighboring Wisconsin, scattered Winter Wren populations are present in remnant conifer stands, rocky ravines, and lowland forests as far south as Milwaukee in the east and Wyalusing State Park, south of Prairie du Chien in the west (Cutright et al. 2006). Both localities occur at latitudes that are approximately 30–35 miles south of the Minnesota/Iowa state border and suggest that additional surveys in southeastern Minnesota are warranted. Elsewhere, the species is believed to have expanded its range in southern New England in the late 1900s as mature forest cover increased following the abandonment of old farms (Hejl et al. 2020).

BREEDING HABITAT

In northern Minnesota and Wisconsin, Winter Wrens demonstrate a strong preference for mixed swamp conifer stands dominated by white cedar, balsam fir, and/or black spruce, followed by lowland hardwoods, black spruce–tamarack, and upland aspen–spruce–fir stands (Niemi, Howe, et al. 2016) **(C)**. In Minnesota's Agassiz Lowlands Subsection, the species reached its highest abundance in mature tamarack swamps (Bednar et al. 2016).

Water is a common feature of many sites and may include a small stream, lakeshore, or swampy wetland forest. The species' most important habitat requirement, however, is a dense understory littered with coarse woody debris, such as dead logs, slash piles, and upturned roots at the base of fallen trees. Characteristic of large (greater than 20 ha) mature forests, a diverse understory provides foraging opportunities, singing perches, and nest sites. When slash piles and other debris are present in recent clear-cuts or remain in pole-sized hardwood stands, they too can provide suitable habitat (Hejl et al. 2020).

POPULATION ABUNDANCE

PIF (2021) used BBS data to provide a population estimate of 11 million Winter Wrens in Canada and the United States and a Minnesota population numbering 180,000. MNBBA data, however, generated an estimate of 588,000 adults (95 percent confidence interval of 527,000–703,000). The MNBBA's sampling efforts in off-road areas of northeastern and north-central Minnesota, regions not surveyed by the BBS, likely contribute to a more accurate estimate.

The species has demonstrated a nearly stable but fluctuating population trend from 1966 to 2019 across the BBS survey region and in Minnesota **(D)**. Populations at both scales, however, suggest a strong

C. Downed logs and a dense understory in this lowland forest stand provide the Winter Wren with an array of foraging and nesting opportunities. Photo by Gerald J. Niemi.

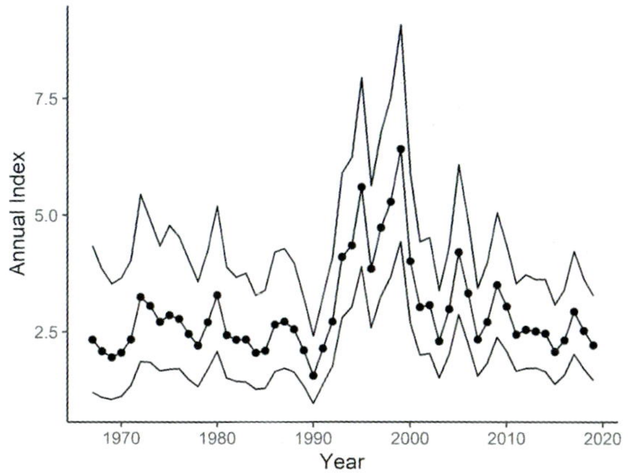

D. Breeding population trend and 95 percent credible interval (solid black lines) for the Winter Wren in Minnesota based on the 1967–2019 Breeding Bird Survey (Sauer et al. 2020).

downward trend in the past twenty years following an increase in numbers from 1990 to 2000. Indeed, in Minnesota, the Winter Wren has experienced a significant decline from 1995 to 2020 of −2.55 percent per year in the Chippewa National Forest and −0.99 percent per year in the Superior National Forest (Grinde, Bednar, et al. 2020).

Factors responsible for the decline are unknown. Harvest levels have decreased in Minnesota's national forests, resulting in an increase in mature forest cover that should benefit Winter Wrens (Niemi, Howe, et al. 2016). Yet declines in breeding populations often follow severe winters or prolonged cold spells (Green 1995; C. S. Robbins et al. 1986). These can be a significant hazard for short-distance migrants like the Winter Wren that winter in the southeastern United States.

CONSERVATION

We have several reasons to be cautious regarding the Winter Wren's future. Canada's boreal forests support over 90 percent of the Wren's population (PIF 2021). Untouched for years, these forests are now subject to expanded timber harvesting, hydroelectric development, and oil and gas exploration and development. Maintaining large tracts of mature and old-growth forest will provide the best insurance for Canada's Winter Wren populations.

Rising temperatures, however, are another concern. Under a 2°C warming scenario, NAS (2019) projected a moderate vulnerability for the species' continental breeding distribution, with a 56 percent range loss by 2050 and only a 47 percent gain northward. Included in this scenario is the species' extirpation from Minnesota, which is nearly complete even under the imminent likelihood of temperatures rising 1.5°C. Clearly, the future of this remarkable vocalist should not be taken for granted.

SEDGE WREN

Cistothorus stellaris

Minnesota Seasonal Status: Regular breeding resident and migrant. The Sedge Wren was common during the MNBBA.
Migration: Short- to medium-distance migrant that winters in the southeastern United States along the Atlantic coast, and west to Texas and northeastern Mexico.
Food: Gleans insects, spiders, and other arthropods from vegetation.
Nest: Globular nest of sedges and grasses; several extra dummy nests are made by the male in a territory.
Conservation Concern: Assigned a low PIF Continental Concern Score of 7/20.

The Sedge Wren, formerly called the Short-billed Marsh Wren, is an aptly named species because it is found nesting almost exclusively in habitats dominated by sedges or grasses with some low-lying shrubs. Bent (1948) suggests its habitat "is more of a meadow . . . for it shuns the wettest marshes where the long-billed marsh wren [Marsh Wren] loves to dwell among the tall, dense growths of cattails or bulrushes and where the water is a foot or more deep."

Identification of the Sedge Wren by sight can be challenging because it is "one of the shiest, most retiring, and elusive" species (Bent 1948). Even though the bird may be spotted singing on the top of a small bush or swaying atop a tall sedge or reed, "if we approach for a closer look he dives into the nearest and thickest cover, and we may not see him again" (Bent 1948). Therefore, it is best to know its song, which has been described as a chattering with two to five introductory notes followed by a trill. In words, *chip-chip-chip-chip, chip-chip-chip-chir-r-r-r-r-r,* it is reminiscent of striking two sticks or stones very rapidly together or rattling a bag of marbles (Bent 1948). In plumage, the sexes are similar. Their crown and back are streaked with white on a blackish and tawny background, with buffy underparts and a light whitish eyebrow.

Sedge Wrens in North America are found from the central United States south to Missouri and north to the western Great Lakes region, northwest to Manitoba, central Alberta, and east in the United States to upstate New York. The highest North American densities occur in central Minnesota and central North Dakota.

Photo by Sparky Stensaas.

MINNESOTA BREEDING DISTRIBUTION

Roberts (1932) stated the Sedge Wren, a secretive but vocal species, was found throughout the state but "most numerous" on the prairies, such as along the Red River valley in northwestern Minnesota. He recorded confirmed nests in Cass, Jackson, and Marshall Counties and in Minneapolis and inferred nesting in Aitkin and Lake of the Woods Counties.

Green and Janssen (1975) and Janssen (1987) both emphasized the Wren's distribution throughout the state, though they noted it was less numerous in the northeastern forested region. Janssen also suggested it was most numerous in the "western and central" regions. He listed confirmed nesting from five counties since 1970, including Clearwater, Hennepin, Le Sueur, Wadena, and Washington. Hertzel and Janssen (1998) would later add Brown, Cass, Kandiyohi, Marshall, Red Lake, and Stevens Counties to the list of confirmed nesting since 1970, but excluded Clearwater and Wadena Counties.

Until 2014, the MBS had included 1,893 breeding-season locations and reinforced the Sedge Wren's wide breeding distribution in Minnesota. They included locations from nearly every county in the state. Counties not included were primarily from southern Minnesota, including Houston, Renville, and Sibley.

The MNBBA participants reported a total of 3,555 records of the Sedge Wren. The distribution of records was statewide, but sparse in the extreme northeastern forested area, the extreme southeastern region, and along the Red River valley **(A)**. The MNBBA records included 42.6 percent (2,037/4,777) of the surveyed blocks and 56 confirmed nesting records. Participants reported new county nesting records from 27 counties, which included two records that straddled two counties (Big Stone/Lac qui Parle and Rice/Dakota). None of these four counties had previous confirmed nesting. Nesting was also confirmed in Aitkin County, the first record since Roberts reported an empty nest in 1893. Probable nesting was also recorded from every county in the state except Martin County in Minnesota's southern tier of counties. One probable nesting location was identified in Winona County, where the MBS had recorded breeding observations in three locations.

The MNBBA probability map emphasized potentially suitable habitat in northwestern Minnesota, the Tallgrass Aspen Parklands Province, and in central Minnesota **(B)**. Additional areas that are predicted to have higher densities include the Red Lake Peatland north of Upper Red Lake and the eastern edge of the Red River Valley Section. Relatively low populations are predicted in the heavily forested regions in north-central, northeastern, and southeastern Minnesota, as well as in the Twin Cities metropolitan area.

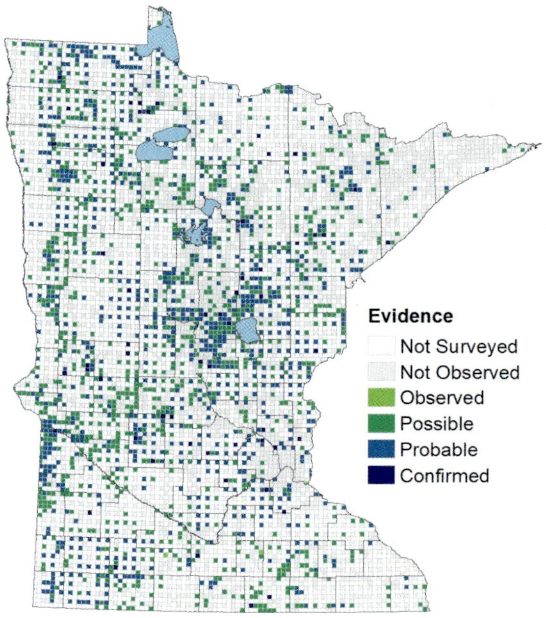

A. Breeding distribution of the Sedge Wren in Minnesota based on the Breeding Bird Atlas (2009–13).

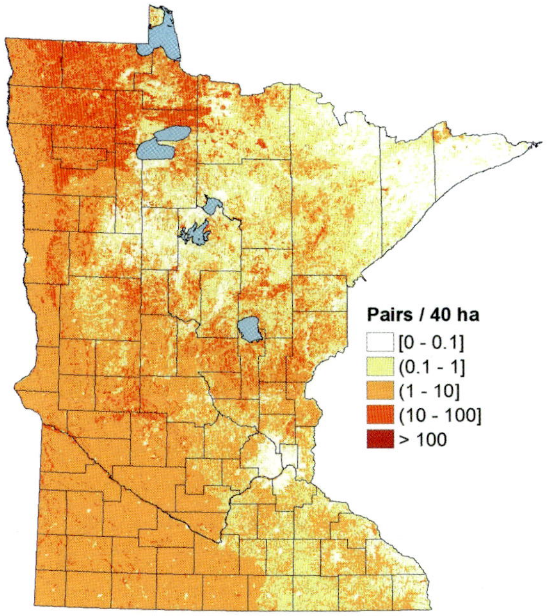

B. Predicted breeding distribution (pairs per 40 ha) of the Sedge Wren in Minnesota based on habitat, landscape context, and climate data gathered during the Minnesota Breeding Bird Atlas (2009–13) using the Generalized Linear Modeling method with an adjustment for detectability.

As Roberts and later authors have emphasized, this species is widely distributed throughout the state, and most identified its presence by its territorial song. Less attention has been directed to confirmation of nesting activity. The Sedge Wren is a secretive species; finding nests can be difficult because the birds make many decoy or dummy nests, plus sedge meadows can be difficult to traverse. Therefore, observations beyond song activity have been limited. McNair (1983) pointed out that finding nests may not always be an indicator of breeding activity because of the propensity for the species to make dummy nests. Because MNBBA rules state that confirmed nesting includes finding a nest, nest building, or carrying nesting material, the MNBBA data may overemphasize the species' breeding activity in Minnesota. Regardless, the MNBBA has greatly expanded records of potential nesting in the state; absolute evidence of "successful" nesting remains a challenging task and a subject of further research.

Herkert, Kroodsma, and Gibbs (2020), in their review of the Sedge Wren in North America, suggested the species has expanded eastward and northward following European settlement as forests were cleared for farming and pastureland. They also emphasized that such gains in potential habitat have been lost in many of these same areas owing to wetland drainage, more intensive agricultural practices, and reforestation in former grassland communities. These habitat changes have been mirrored in Minnesota, especially in the agricultural region of the state. The species' absence along the Red River valley is especially notable.

BREEDING HABITAT

Dechant, Sondreal, and colleagues ("Sedge Wren," [1998] 2002) and Pfannmuller (2012) emphasized mesic grasslands in western and southern Minnesota, describing tall, dense growths of grasses, sedges, and scattered small shrubs as being common features of suitable habitat. The Sedge Wren was also found abundantly in the open bogs, shrub swamps, and sedge fens in the Agassiz Lowlands Subsection of Minnesota (Nevers et al. 1981; Niemi and Hanowski 1984a; Bednar et al. 2016) **(C)**.

POPULATION ABUNDANCE

Based on the BBS, PIF (2021) estimated a US and Canadian breeding population of 5 million adults, with 1.6 million, or 32 percent, in Minnesota. In contrast, MNBBA estimated a population of 4.75 million birds in Minnesota (95 percent confidence interval of 4.10–6.72 million). The discrepancy between these estimates is due to the different databases and methodology used by PIF and MNBBA. The MNBBA was able to sample a greater proportion of roadless areas, and many of these areas, such as the Red Lake Peatland, have very high densities of Sedge Wrens (see https://mnbirdatlas.org/). Regardless, it is likely that the Sedge Wren population is higher in Minnesota than has been previously estimated.

The BBS trend estimates for Minnesota indicated a significantly increasing population of Sedge Wrens that averaged 1.42 percent per year from 1967 to 2019 **(D)**. This trend is consistent with North and South Dakota, which had significantly increasing populations over this same time period of 3.76 and 7.02 percent per year, respectively. As with many waterfowl, this increase is likely due to improved rainfall conditions.

C. Typical breeding habitat of Sedge Wrens consists of sedges and grasses where they can weave their dome-shaped nests. These habitats often have low shrubs and are usually in places where you get your feet wet. Photo by Gerald J. Niemi.

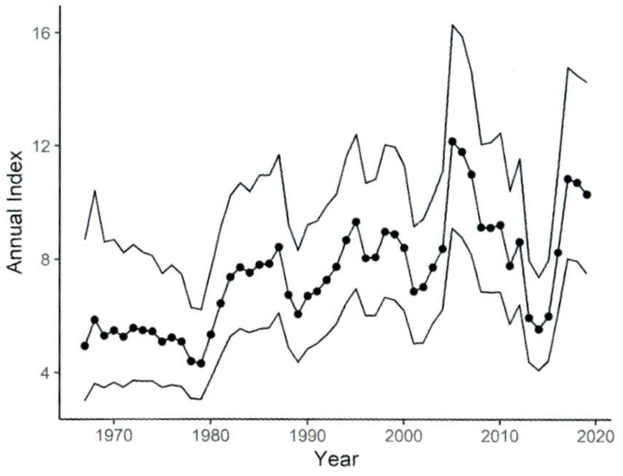

D. Breeding population trend and 95 percent credible interval (solid black lines) for the Sedge Wren in Minnesota based on the 1967–2019 Breeding Bird Survey (Sauer et al. 2020).

CONSERVATION

In North America there are few conservation concerns for the Sedge Wren because of its large breeding distribution and population. Minnesota, however, has a proportionately higher conservation responsibility because such a large percentage of its population occurs in the state. This responsibility has been acknowledged by designations by the MNDNR as a *Species in Greatest Conservation Need* and by Audubon Minnesota (Pfannmuller 2012) as a Stewardship Species.

Historically, degradation of habitat through draining of wetlands and conversion to agriculture and urbanization have been the major conservation concerns for this species. Wetland loss between 1950 and the mid-1970s was estimated at 1.9 million ha in the United States (Tiner 1984). The sedge and grass meadows that the species prefers are among the easiest to drain and most susceptible to wetland conversion.

Common factors that negatively affect nest success and population levels of the Sedge Wren include nest loss due to early harvesting of hay as well as inclement weather, predation, and strong intraspecific interactions with nest destruction by other Sedge Wrens (Picman and Picman 1980; Herkert, Kroodsma, and Gibbs 2020).

Under a 2°C warming scenario, NAS (2019) projected moderate vulnerability for the Sedge Wren's breeding range in North America, with a 35 percent loss of its breeding range and a 29 percent gain north and westward by 2050. Its breeding range was forecast to have a marginal retraction in southeastern Minnesota. The future effects of climate change and the availability of large sedge and grass meadows are key to conserving populations of this secretive but energetic, insectivorous Wren in Minnesota and across North America.

MARSH WREN

Cistothorus palustris

Minnesota Seasonal Status: Regular breeding resident, migrant, and occasional in winter. The Marsh Wren was common during the MNBBA.
Migration: Short- to medium-distance migrant that winters along the Atlantic and Pacific coasts, the southern United States, and Mexico.
Food: Invertebrates, primarily insects and spiders.
Nest: Globular, dome nest in cattails or bulrushes; many false or dummy nests (5 to 22) are usually built in a territory.
Conservation Concern: Assigned a low PIF Continental Concern Score of 7/20.

Photo by Gerald Hoekstra.

In contrast to its close relative the Sedge Wren, the Marsh Wren is a "real marsh" bird, inhabiting cattail swamps, bulrushes, oozy sloughs, and marshy borders of sluggish rivers, where it is at home with Bitterns, Soras, and Swamp Sparrows (Forbush and May 1955). Its scientific name, *palustris,* literally means *of the swamp* in Latin. Roberts (1932) described the species as "[at] home among the reeds, rushes, and cattails of sloughs where the water is several inches deep." He considered its ability to construct multiple dome nests a "remarkable feat" for such a small bird.

The Marsh Wren was formerly known as the Long-billed Marsh Wren, which is one way to distinguish it from the Sedge Wren, but bill length is a subtle characteristic and difficult to see when the bird is uncooperative among the cattails. Marsh Wrens also have a distinctive white line (supercilium) behind the eye that contrasts with a dark brown crown and a darkish back with prominent white stripes. Learning its song is the easiest path to identification. It is described as several well-spaced introductory notes, followed by a musical bubbling series of notes, and ending in a trill: *Chik chik chu-u-u-u-u-u-u-u-rrr* (Alderfer 2006). Welter (1935), who studied the species near Staples, Minnesota, for his dissertation at Cornell University in New York State, suggested its song sounds like a "sewing machine of the older sort being run rapidly." It can be delivered from the top of vegetation or while in flight.

The Marsh Wren has a complex taxonomy and history, with currently up to 14 subspecies that have been recognized. It has a widespread breeding distribution in northern North America, including northern portions of the Atlantic coast to the Pacific coast, throughout the southern Great Lakes region and the northern prairies, and patches throughout the western United States and Canada. Its highest densities have been found in North Dakota.

MINNESOTA BREEDING DISTRIBUTION

Aptly described by Roberts (1932) as a "summer resident throughout the state; common in the southern part and on the prairies to the Manitoba line, but more sparingly represented in the marshes of the northern forested area." Hatch (1892) emphasized its ubiquity when he wrote, "[That] it is generally distributed over the entire State there can be no doubt." Along the Red River valley in northwest Minnesota, he stated his collector, Mr. F. L. Washburn, found the species "breeding in large colonies." Roberts reported confirmed nests from 1879 to 1929 in Cass, Grant, Jackson, Kittson, Otter Tail, and Polk Counties and in Minneapolis. Inferred nesting was reported from Anoka County.

Green and Janssen (1975) described the Marsh Wren's breeding distribution throughout the state as scarce in the northeastern and adjacent north-central counties, where cattail marshes are uncommon. Janssen (1987) echoed this description and identified confirmed nesting in 13 counties since 1970. These counties were all widely scattered from the extreme northwest to the extreme southwest and northeast to southern St. Louis County. Hertzel and Janssen (1998) would later add six more counties with confirmed nesting records since 1970 but excluded Clearwater County.

The MBS provided a broader perspective on the breeding distribution of the Marsh Wren, but locations were supportive of the general descriptions previously noted. They included 893 records with higher concentrations west of Mower County in the south, west of Itasca County in the north-central region, and northwest to Roseau County. Scattered breeding-season observations were found northeast to southern St. Louis County; one location each in Carlton, Chisago, Cook, and Pine Counties; and one in Houston County in extreme southeastern Minnesota.

The MNBBA substantially filled in gaps to the breeding distribution but generally confirmed the Marsh Wren's overall breeding range in the state **(A)**. MNBBA participants reported 1,663 records in 21.4 percent (1,016/4,757) of all surveyed blocks. A majority of records were of possible nesting. Confirmed nesting was reported from 33 counties and records were noted from every county in the state except Cook County where MBS had noted one location previously. Marsh Wrens, however, were found at only one location in Carlton, Fillmore, and Lake Counties and only two locations each in Houston, Martin, Olmsted, and Winona Counties. Marsh Wren records were particularly numerous in Big Stone, Lac qui Parle, and Kandiyohi Counties where MNBBA participants were impressively industrious.

The probability distribution map accentuated the patchiness of the species' available habitat throughout Minnesota, especially small pockets in the southern and western portions of its range **(B)**. High densities were predicted at the Agassiz NWR in Marshall County, at Swan Lake in Nicollet County, and in many scattered areas in west-central, southwestern, and northwestern Minnesota.

The general distribution of this species in Minnesota appears to be largely intact, but the histor-

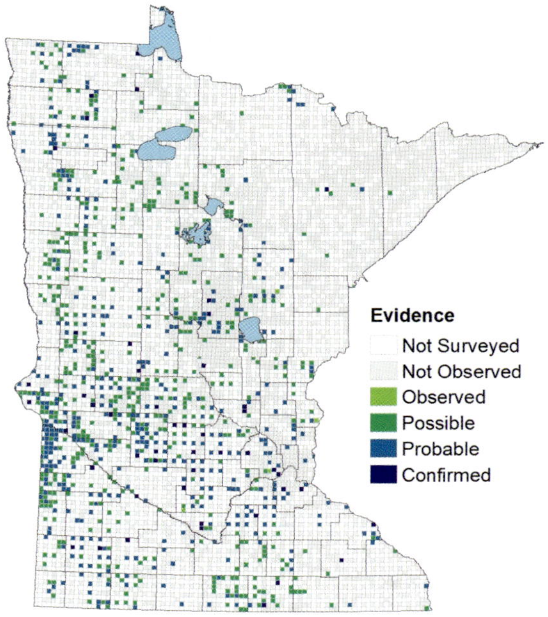

A. Breeding bird distribution of the Marsh Wren in Minnesota based on the Breeding Bird Atlas (2009–13).

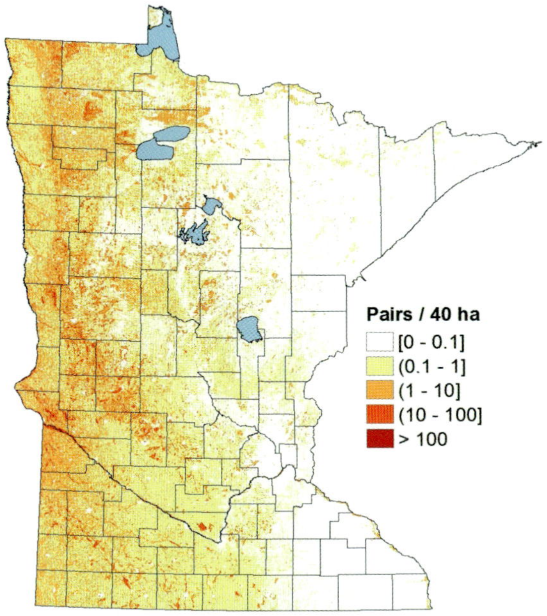

B. Predicted breeding distribution (pairs per 40 ha) of the Marsh Wren in Minnesota based on habitat, landscape context, and climate data gathered during the Minnesota Breeding Bird Atlas (2009–13) using the Generalized Linear Modeling method with an adjustment for detectability.

ical loss of wetland habitat due to agricultural, industrial, and residential development has clearly reduced breeding populations in the state. This is especially true in the southern and western portions of Minnesota. This is supported by previous atlas observations throughout many other parts of the species' range, including Michigan (R. Brewer et al. 1991; Chartier et al. 2013) and Ontario (Cadman et al. 2007). Fortunately, the Marsh Wren's breeding population is still widely distributed in Minnesota despite extensive habitat loss.

BREEDING HABITAT

Collectively, Welter (1935), Verner (1965), and Kroodsma and Verner (2020) emphasized cattail stands with scattered small patches of bulrush (*Scirpus* spp.) or phragmites (*Phragmites australis*) and cattail-sedge associations **(C)**. They also emphasized that the species prefers the exotic narrow-leaf cattail (*Typha angustifolia*) over the native broad-leaf cattail (*T. latifolia*). The narrow-leaf cattail is native to Eurasia and is invasive, plus it hybridizes with the native broad-leaf cattail. Both cattail species can displace native wetland plants and have increased dramatically since the early 1900s (Czarpata 2005). Marsh Wrens have been documented as more abundant within wetlands when the surrounding landscape contained more wetlands (Tozer et al. 2010; Panci et al. 2017; Saunders et al. 2019).

POPULATION ABUNDANCE

Based on the BBS, PIF (2021) estimated a breeding population of 11 million adults in the United States and Canada, with 460,000 in Minnesota. The MNBBA estimated a substantially higher breeding population of 2.35 million adults (95 percent confidence interval of 1.54–5.31 million). Several differences exist in the methods that PIF and MNBBA use to estimate populations. One of the primary differences is that the MNBBA is able to incorporate more accurate detection distances in their estimates and Marsh Wrens can be detected at short distances, which results in higher population estimates.

BBS trend estimates for Minnesota indicated a stable, albeit highly variable breeding population from 1967 to 2019 among the 61 routes where the species has been recorded **(D)**. In contrast, the Marsh Wren population has been significantly increasing in North Dakota (average of 3.53 percent per year), South Dakota (4.81 percent per year), and Manitoba (2.05 percent per year) from 1967 to 2019. Rosenberg, Kennedy, and colleagues (2016) estimated an overall increase in the population of 168 percent from 1970 to 2014!

CONSERVATION

Kroodsma and Verner (2020), in their review of the Marsh Wren in North America, emphasized minimal or no conservation concern for this species at the present time. It was formerly considered a *Species in Greatest Conservation Need* in Minnesota, but recent stable populations and increases elsewhere have eliminated concerns.

As noted above, Marsh Wren population concerns in the past were largely based on the extensive conversion of wetlands for agricultural, industrial, and residential development (Cadman et al. 1987; Chartier et al. 2013; Kroodsma and Verner 2020).

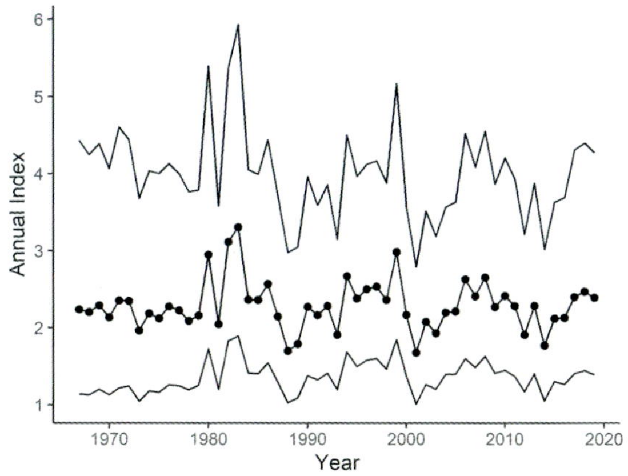

D. Breeding population trend and 95 percent credible interval (solid black lines) for the Marsh Wren in Minnesota based on the 1967–2019 Breeding Bird Survey (Sauer et al. 2020).

Fortunately, the species has responded positively to many wetland restoration efforts (Gutzwiller and Anderson 1987; Delphey and Dinsmore 1993). Gutzwiller and Anderson (1987) and Chartier and colleagues (2013) identified exotic giant cane (*Arundo donax*) and purple loosestrife (*Lythrum salicaria*), respectively, as having potentially negative effects on nesting of the species, though much additional research is necessary on this topic.

Under both 1.5°C and 2°C warming scenarios, NAS (2019) projected a stable breeding range for the Marsh Wren in North America. Substantial breeding range extensions northward into Canada were forecast under both scenarios. In Minnesota, under the 2°C warming scenario, slight breeding range losses were predicted in west-central and northwestern regions but with range gains in north-central and northern areas by 2050. Adaptation to new habitat created by a changing climate may be possible northward in Minnesota and Canada, as well as continued colonization of restored wetlands, but only time will tell.

C. The Marsh Wren, as its name implies, nests in cattail marshes where it weaves its dome-shaped nest; such marshes often contain shrubs and a variety of wetland plants. Photo by Gerald J. Niemi.

CAROLINA WREN

Thryothorus ludovicianus

Minnesota Seasonal Status: Summer visitant; breeds sporadically. In fall, winter, and spring a yearly visitant in the southeast; occasional elsewhere. The Carolina Wren was very rare during the MNBBA.
Migration: Permanent resident in its primary breeding range in the United States; in Minnesota it has not established a resident population.
Food: Mostly insects of many species, spiders, sometimes small frogs and lizards. Consumes berries and fruits in winter and frequents feeders for seed and suet.
Nest: Domed construction placed in a variety of cavities and openings; the hole for the nest site can be in natural sites, such as tree cavities, dense brush, and rocky nooks, and in human structures, such as outbuildings, mailboxes, flowerpots, and abandoned nest boxes.
Conservation Concern: Assigned a low PIF Continental Concern Score of 7/20.

Photo by Isaac Sanchez.

The Carolina Wren, despite its name, first appeared in the scientific avifauna of its day in 1790 with a taxonomic description of a specimen collected in Louisiana, near New Orleans. Thirty years later, John James Audubon, while descending the Mississippi River on a flatboat into Louisiana, described the essence of this favorite bird: "Carolina Wrens . . . exercising their Vocal powers all day—. . . During spring, these notes are heard from all parts of the plantations, the damp woods, the swamps, the sides of creeks and rivers, as well as from the barns, the stables and the piles of wood, within a few yards of the house. . . . So fond is this bird of the immediate neighbourhood of water, that it would be next to impossible to walk along the shore of any of the islands of the Mississippi, from the mouth of the Ohio to New Orleans, without observing several on each island" (Audubon 1841).

The Carolina Wren is slightly larger than the House Wren and both sexes are similar. It can be distinguished by its long, distinctive white eyebrow, reddish-brown upperparts, buffy underparts below, and white throat. It is an enthusiastic and loud singer. Alderfer (2006) describes the song as *chip mediator mediator mediator meep.*

Widely found in the southeastern United States from southeastern Texas east to Florida and northward to southern New England, then westward to the southern Great Lakes states, Iowa, and eastern Nebraska, the species has year-round populations in northeastern Mexico and the Yucatan. High densities occur throughout the southeastern United States but especially in northern Florida, southern Georgia, and southern Alabama.

MINNESOTA BREEDING DISTRIBUTION

The assessment that Roberts (1932) made about the presence of the Carolina Wren in Minnesota is related in his chapter "Some Changes in the Bird-Life of Minnesota in Recent Years," where he listed the wren among seven species from the south that were "rare fifty years ago" and "are becoming more frequent and spreading northward but are still nowhere common." The observations Roberts cited were primarily from the 1920s, but also included one from "Little Rock" (now Chaska) in 1848, and one by P. L. Hatch, who collected this species in 1868 in Minneapolis.

The confirmed breeding evidence Roberts gave was from the "outskirts" of St. Paul in the spring of 1927, when a pair built a nest in the eaves of a garage and successfully reared young. The records cited by Green and Janssen (1975) and Janssen (1987) were the same: "In the summer there are former records of family groups or nests from Houston, Fillmore, Ramsey, and Washington Counties. There are no recent breeding records" (Janssen 1987). The Houston County breeding record of August 2, 1947, is documented (M. D. Thompson 1947), and the 1927 St. Paul, Ramsey County, record above is from Roberts.

Although from 1960 through 1999, reports of the Carolina Wren have steadily increased and become yearly after 1985, they were mostly fall, winter, and spring visitants from the Twin Cities southeastward, with summer records from Dakota (1986), Hennepin (1967, 1997, 1998,) and Olmsted (1986, all summer 1999) Counties. There were also northern vagrants: Duluth (May 25, 1981), Crow Wing County (November 27, 1983), and a feeder in Lutsen, Cook County (1983). One unusual northern record from Otter Tail County was an overwintering bird at a feeder from November 21, 1999, to May 24, 2000 (Edwards and Edwards 2000). This record is interesting because the bird was fed all winter with copious amounts of mealworms and in the spring built a nest and laid five eggs that were infertile because she lacked a mate.

From 2000 to 2008, there was a substantial increase in Carolina Wren records in the MOU seasonal reports every year in fall, winter, and spring. These incidental records covered the state as far north as Lake and Beltrami Counties, and as far west as Clay and Yellow Medicine Counties; some winter birds were found at feeders. Summer reports were primarily from its usual range, the Twin Cities southeastward; however, there is one report from Clearwater County on June 11, 2005, with no details, and one in Brown County from July 10 to 29, 2006, in Flandrau State Park. This influx also produced the first actual breeding record since 1947, from Rochester, Olmsted County. The Rochester report was of an overwintering pair in which, on May 1, 2007, the incubating female was found in a garage sitting on six eggs, but it is not known if they hatched (Budde, Dunlap, et al. 2007; Kessen and Svingen 2008).

During the MNBBA, participants recorded four observations of the Carolina Wren; two in the north that appeared to be incidental visitants (Itasca County and southern Lake County along the Lake Superior shoreline) and two in the south in counties where this species has been regularly observed (Houston and Ramsey Counties) **(A)**. However, the MOU database revealed there were an additional 23 observations during the MNBBA time frame (2009–13) not submitted to the atlas. An examination of these observations revealed three with sufficient breeding evidence and locational information to be added to the MNBBA database as probable records: first, Rochester, Olmsted County (Block T106R14a: in suitable habitat, "singing throughout the day," one, August 20–22, 2012, then a pair seen August 24, 2012); second, Minneopa State Park, Blue Earth County (Block T108R27d: in suitable habitat, pair "along trail west of campground," August 4, 2013); and, third, Preston, Fillmore County (Block T102R10d: in suitable habitat, July 6–10, 2010, and again on July 3, 2012). Also during this time period, there were other summer incidental records from Dakota, Faribault, Hennepin, Ramsey, Rice, Steele, and Washington Counties. In 2012 there was an unprecedented winter influx of 31 individuals found in 18 counties.

After the atlas time period, 2014–19, the Carolina Wren has been reported regularly and in every season in 2016, 2017, and 2018, but sufficient evidence is lacking to regard it as a resident species with year-round records in an area. Reports are mostly incidental visitants in the southeast and vagrants elsewhere. Places like Rochester with decades of records should be checked carefully for their yearly presence and evidence of breeding. This species appears to be limited in the northern portions of its range by persistent snow cover and prolonged cold that are detrimental to its survival (Haggerty and Morton 2020); however, its northward expansion is likely to continue with impending climate change (see NAS 2019).

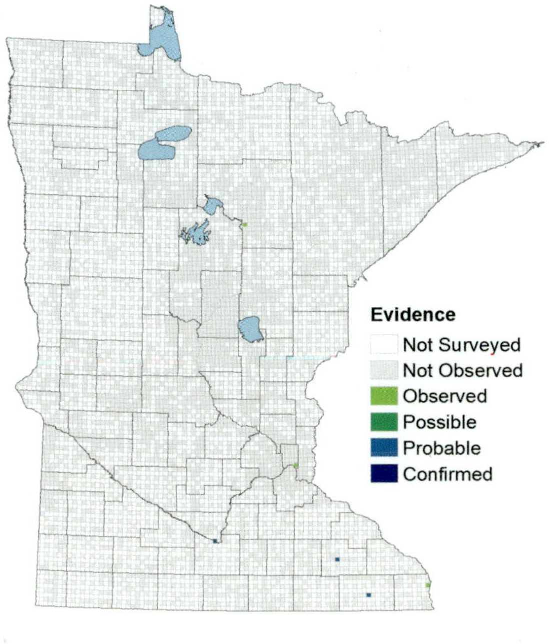

A. Breeding distribution of the Carolina Wren in Minnesota based on the Breeding Bird Atlas (2009–13).

B. Tangled vegetation and suitable nesting cavities found along the edge of a wet forest opening are typical features of the breeding habitat of the Carolina Wren. Photo by Lee A. Pfannmuller.

BREEDING HABITAT

Found in deciduous forests, the Carolina Wren nests along brushy edges, in the thick understory, or in disturbance gaps with tangled vegetation. Bottomland forests are often used, as are swamps, streams, and wetland margins with dense vegetation. The species has adapted well to human settlement and breeds in wooded suburbs and towns with areas of dense brushy vegetation, rocky crevices, or unused human structures or debris that provide suitable nesting cavities **(B)**.

POPULATION ABUNDANCE

PIF (2021) estimated the population of the Carolina Wren in the United States and Canada as 18 million birds, and no attempt has been made to estimate a population in Minnesota. The BBS surveywide population trend indicated a significant increase that averaged 1.32 percent per year from 1966 to 2019 **(C)**. It has been increasing in all states and provinces (Ontario) where it occurs in sufficient numbers. Rosenberg, Kennedy, and colleagues (2016) estimated that the Carolina Wren population had increased by 71 percent from 1970 to 2014.

CONSERVATION

The Carolina Wren is not considered a conservation concern because of its increasing population trend, its expanding breeding distribution, and its sedentary movement patterns that can avoid migration hazards. As winters become less harsh, it is also likely to continue expanding northward. For instance, NAS (2019) predicted a stable status for the Carolina Wren with all three of the climate warming scenarios they examined, including a modest breeding range expansion northward. An increasing population may also continue, given its association with built environments, especially if adequate cover, such as dense shrubbery, and food sources are available. As with most wildlife species, habitat degradation and ecosystem contamination can influence future populations. For instance, the Carolina Wren has been found to be susceptible to mercury exposure. High levels were detected in its blood due to its year-round foraging on spiders. Spiders are apex predators in the arthropod world and hence at a higher trophic level compared with other invertebrate prey. The high mercury loads in Carolina Wrens at contaminated sites were found to affect its nesting success compared to those at reference sites (A. Jackson et al. 2011).

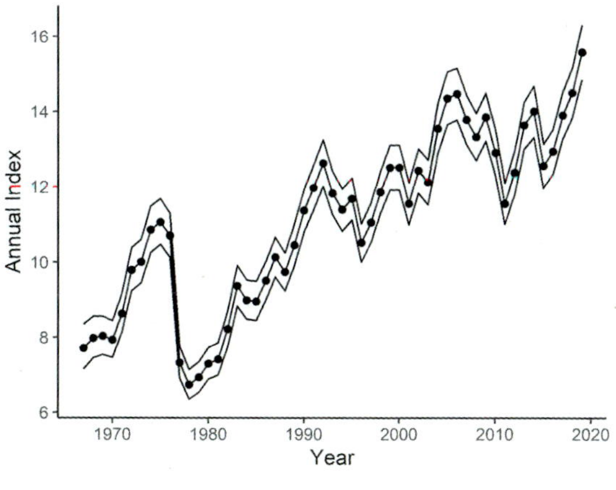

C. Surveywide breeding population trend and 95 percent credible interval (solid black lines) for the Carolina Wren based on the 1967–2019 Breeding Bird Survey (Sauer et al. 2020).

GRAY CATBIRD

Dumetella carolinensis

Minnesota Seasonal Status: Regular summer resident and migrant; occasionally lingers into the winter months at bird feeders. The Gray Catbird was abundant during the MNBBA.
Migration: Short- to medium-distance migrant that winters along the Atlantic and Gulf coasts of North America, Mexico, and Central America and in the Caribbean. Some coastal breeding populations in the United States are year-round residents.
Food: Ground forager and foliage gleaner that consumes both arthropods and fruit.
Nest: Cup nest, usually placed in shrubs.
Conservation Concern: Assigned a low PIF Continental Concern Score of 8/20.

Photo by David Brislance.

Smaller than a Robin but larger than a Bluebird, the Gray Catbird is aptly named for its dark gray plumage and raspy, catlike *mew* call. Its rather modest plumage is accented by a distinct black cap, black tail, and rusty undertail coverts hidden beneath the tail feathers. Commenting on its attire, Roberts (1932) wrote: "There is no more genteelly dressed or smarter appearing bird in all our woodlands than the Catbird. Somber and plain are its colors but its suit fits it to perfection—'tailor-made' it has been called." *Dumetella,* the scientific name for the Catbird's genus, is a derivative of the Latin term *dumetum* and translates to "thicket," an accurate description of the species' dense and often tangled shrubby habitat.

What the Catbird may lack in brilliant plumage it makes up for in its renowned vocal abilities. Roberts (1932) described it as one of "our finest songbirds," delivering "a forceful, varied, and richly musical medley." Audubon (1827–38) remarked that "he must have a dull ear indeed, and little relish for the simple melodies of nature, who can listen to them without delight." A member of the Mimidae bird family, the male Catbird is an exceptional mimic. Short notes, whines, gurgles, and chatter are intermixed with imitations of other bird songs. Even frogs and the sounds of inanimate objects may be heard. Sometimes continuing for up to 10 minutes, the songs can have a ventriloquial character to them, a result of the bird being able to produce sounds independently from both sides of its song box, the syrinx. "A good Catbird song needs no apology," one writer noted. "Phrase follows phrase in rapid succession, and snatches of all the bird songs in the neighborhood appear intermixed with occasional harsher notes, which are given with as much care and finish as the more melodious ones" (Bent 1948). Unlike the Brown Thrasher, another Minnesota mimic, the song phrases are rarely repeated.

Broadly distributed across southern Canada, the eastern United States, and the northern Great Plains, the Gray Catbird's breeding range also includes many western states, from eastern Washington south to northern New Mexico and Arizona. Some of the highest breeding densities can be found from east-central Minnesota east to Pennsylvania and southern New England.

MINNESOTA BREEDING DISTRIBUTION

In the early 1900s, Roberts (1932) considered the Gray Catbird a common summer resident statewide, though it was most abundant south of the northern forests. It could be found "in all the wooded portions of the state, including the scattered native groves and timber-claims of the prairie regions." Further north, "in the poplar groves and brush-lands of the northwestern counties it is equally common up to the Canadian boundary." Although it was "well represented" in the northern forest regions, it was not an abundant species. Breeding evidence was reported in Goodhue, Hennepin, Itasca, Marshall, Polk, Rock, and Stearns Counties.

Little had changed when Green and Janssen (1975) published their updated account of the species about forty years later. Although the Gray Catbird was still distributed across the state, they noted that in northern Minnesota it was largely confined to brushy openings in the region's extensively forested landscape. A few years later, Janssen (1987) identified 37 counties where nesting had been documented since 1970; most counties were located south of the northern forests. Hertzel and Janssen (1998) later added an additional 12 counties to the list. Then, from the late 1980s through 2014, the MBS documented a total of 1,430 breeding-season locations for the Gray Catbird. Observations were least numerous in the eastern half of the Laurentian Mixed Forest Province and in the intensively cultivated regions of the Red River valley and the upper Minnesota River valley.

Although records continued to be least abundant in the far north-central and northeastern counties, MNBBA detections were reported in all 87 Minnesota counties **(A)**. Participants reported 4,422 Gray Catbird records in 49.7 percent (2,361/4,752) of the surveyed atlas blocks and in 66.8 percent (1,562/2,337) of the priority blocks. Breeding was confirmed in 368 blocks distributed across 78 counties, including 32 that were additions to the list published by Hertzel and Janssen (1998). The Gray Catbird was most abundant in the Eastern Broadleaf Forest Province and in the southern half of the Laurentian Mixed Forest Province.

The Catbird's MNBBA probability model predicts that the highest breeding densities occur in the river valleys of southeastern Minnesota, just north of the Twin Cities metropolitan region in the Anoka Sand Plain Subsection, and north along the prairie–forest border extending to the Canadian border **(B)**. Else-

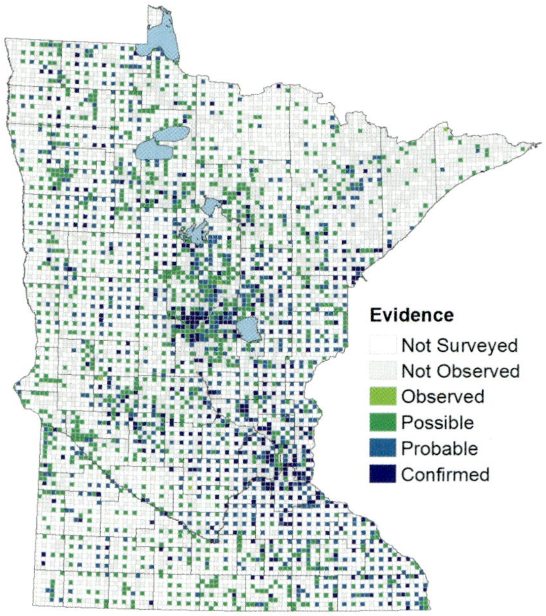

A. Breeding distribution of the Gray Catbird in Minnesota based on the Breeding Bird Atlas (2009–13).

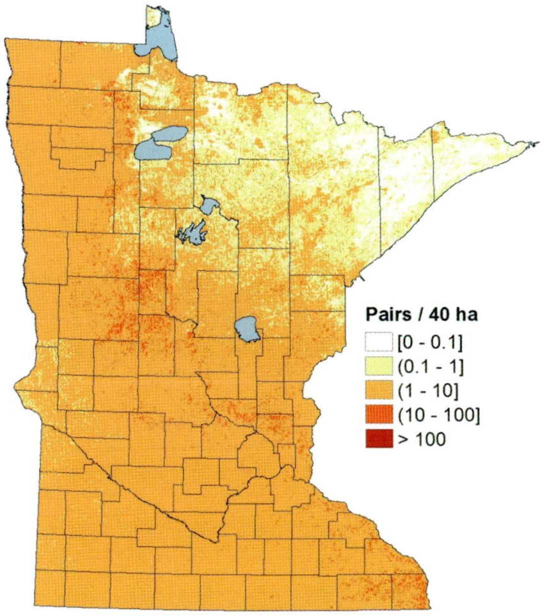

B. Predicted breeding distribution (pairs per 40 ha) of the Gray Catbird in Minnesota based on habitat, landscape context, and climate data gathered during the Minnesota Breeding Bird Atlas (2009–13) using the Generalized Linear Modeling method with an adjustment for detectability.

where the species is moderately abundant throughout the Prairie Parkland and Eastern Broadleaf Forest Provinces and throughout much of the southern and western regions of the Laurentian Mixed Forest Province. Densities are predicted to decrease moving east through the northern forests. One notable exception is the area of the Ham Lake Fire, where more than 75,000 acres burned in 2007 at the northern end of the Gunflint Trail.

With a similar distribution to that described by Roberts in 1932, the Gray Catbird remains a common to abundant species throughout much of the state. Although it is least common in the northeastern region, suitable habitat created by residential development in the western and southern portions of the Laurentian Mixed Forest Province likely provides more habitat for the Catbird than was available in the early 1900s. In their comprehensive review of the species, R. Smith and colleagues (2020) noted that no significant changes have been documented in the Gray Catbird's North American breeding distribution. Nonetheless, this shrub-loving species likely benefited from the wide-scale clearing of the eastern deciduous forest in the 1800s and early 1900s (Cadman et al. 2007).

BREEDING HABITAT

The Catbird accommodates its preference for dense shrubs and tangled vines in a wide range of habitats, from young deciduous forest stands, shrubby wetlands, riparian corridors, and forest clearings to fencerows, shelterbelts, power line corridors, and residential neighborhoods (Kaufman 1996; R. Smith et al. 2020) **(C)**. Coniferous stands are usually avoided. The species is strongly associated with highly fragmented landscapes (Keller and Yahner 2007; C. S. Robbins et al. 1989) and with high shrub densities wherever they occur (Lent 1990).

POPULATION ABUNDANCE

BBS data helped to generate a population estimate of 29 million breeding adults in the United States and Canada and a statewide estimate of 1.3 million (PIF 2021). The statewide estimate, which is derived from the more widely distributed MNBBA point counts, is significantly larger, at 3.2 million breeding adults (95 percent confidence interval of 3.0–3.7 million).

Although BBS data document a relatively stable population across the survey region, the trend line includes a major decline from 1966 through the late 1990s, followed by a significant increase through 2019 (Sauer et al. 2020). Major regional differences are also apparent, with increases in the western United States and Canada and decreases in the southeastern states, northern New England, and eastern Canada. The declines in the Northeast may be due in part to forest succession on abandoned farmsteads that formerly provided more of the open, shrubby habitat preferred by Catbirds. In Minnesota, the trend has also been relatively stable, with a small, nonsignificant increase averaging only 0.25 percent per year **(D)**. Indeed, the regional variability seen at the national level is also evident in

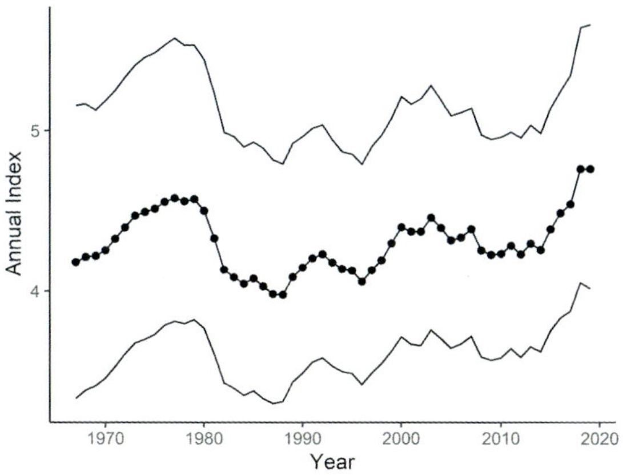

D. Breeding population trend and 95 percent credible interval (solid black lines) for the Gray Catbird in Minnesota based on the 1967–2019 Breeding Bird Survey (Sauer et al. 2020).

Minnesota, with numbers increasing in portions of southern and northwestern Minnesota but decreasing in the Arrowhead region and central Minnesota (Sauer et al. 2020). Factors that influence population levels have received little attention from researchers (R. Smith et al. 2020).

C. Deciduous shrubs and/or dense patches or stands of young saplings along forest edges are the defining feature of the Gray Catbird's breeding habitat. Photo by Lee A. Pfannmuller.

CONSERVATION

With a wide distribution and a large, stable population, the adaptable Gray Catbird has fared well in the face of increasing urbanization and development. Even within densely forested landscapes, timber management activities create favorable habitat opportunities. As a result, the species is a low conservation priority (PIF 2021).

Although intensive agricultural practices and dense urbanization would likely decrease the availability of shrubby habitats, the distribution and abundance of the Gray Catbird in Minnesota does not appear to be impacted by such changes. Indeed, the species was well distributed throughout some of the most intensively cultivated and urbanized regions of the state that still provide suitable habitat in shelterbelts, local parks, and along riparian corridors. Even warming temperatures are not a major threat. The species' North American breeding distribution was stable under all three warming scenarios that NAS (2019) examined, and no distributional changes were projected in Minnesota. Collisions with towers and vehicles do present some concern. The Gray Catbird was among the 10 most common species killed at communication towers in five of eight regions examined in North America (Longcore et al. 2013). Nevertheless, as long as habitat opportunities abound, the Gray Catbird is likely to remain an abundant breeding species in Minnesota.

BROWN THRASHER

Toxostoma rufum

Minnesota Seasonal Status: Regular breeding resident and migrant; occasionally observed during the winter. The Brown Thrasher was common during the MNBBA.
Migration: Northern populations are short-distance to partial migrants that winter in the southeastern United States; southern populations are year-round residents.
Food: Ground forager and occasional foliage gleaner whose diet consists primarily of arthropods but also includes berries, nuts, and waste grains.
Nest: Cup nest, usually placed within vines, shrubs, or small trees, generally less than 3 m above the ground.
Conservation Concern: Assigned a moderate PIF Continental Concern Score of 10/20.

The Brown Thrasher can be a rather shy and timid bird. More often heard then seen, the species spends much of its time on the ground, where it uses its long, slightly curved bill to sweep the forest floor as it searches for insects, seeds, and berries. *Toxostoma,* the species' genus name, translates to "a bow-shaped bill." If one happens upon a Thrasher feeding in a clearing, the bird is easily startled and quickly retreats to its favored habitat, the dense, tangled undergrowth of the nearest shrubbery. Luckily, even a brief glimpse provides enough time to identify this Robin-sized bird by its bright rufous plumage, heavily streaked breast, and long tail. Its short flight also reveals two white wing bars.

Like other members of the Mimidae bird family, the Brown Thrasher is revered as a superb songster. Many naturalists waxed eloquent about the fine songs of the Northern Mockingbird, the Brown Thrasher, and the Gray Catbird, often ranking the quality of their performances in that very order. Each bird's medley, however, has unique and pleasing qualities. The Brown Thrasher's song is distinguished by phrases that are usually sung twice. Once known as the Planting Bird, the mnemonic for the Thrasher's song is often described as *drop it, drop it; cover it up, cover it up; let it grow, let it grow; I'll pull it up, I'll pull it up.* Roberts (1932) remarked that "the great variety and richness of its notes, together with the seeming mimicry of the songs of other birds, has caused the Thrasher to be sometimes called the 'Mockingbird of the North.'"

Photo by David Brislance.

The Brown Thrasher is widely distributed across eastern North America, from southern Alberta east to southern Quebec and New Brunswick, and from the Great Plains states east to the Atlantic coast. The core of the species' breeding range is in the central Great Plains and the southeastern United States.

MINNESOTA BREEDING DISTRIBUTION

In the early 1900s, Roberts (1932) remarked that, although the Brown Thrasher was a common summer resident throughout the state, it was, like the Catbird, more abundant in southern regions than in the north. "In thickly forested regions it seeks the more open places and clearings, avoiding the heavy timber. Every patch of woodland out on the prairies has its pair or more of Thrashers." Breeding evidence was reported from Rock and Lac qui Parle Counties in southwestern Minnesota, from Goodhue and Hennepin Counties in eastern Minnesota, from Polk and Marshall Counties in the northwest, and from Cass Lake, Itasca County, and Itasca State Park in north-central Minnesota.

Many years later, Green and Janssen (1975) reiterated the species' statewide distribution and its absence from "large areas of dense forest." When present in the northern forests, it was found in forest openings created by timber management activities but appeared to be absent from natural forest openings. A few years later, however, Niemi (1978) documented that Brown Thrashers were common on the Little Sioux burn two to four years following the May 1971 fire that consumed nearly 15,000 acres.

Including a distribution map that covered the entire state, Janssen (1987) delineated 34 counties where nesting had been confirmed since 1970. Hertzel and Janssen (1998) later added another seven counties to the list, all within eastern and central Minnesota. Field biologists with the MBS documented 416 Brown Thrasher breeding-season locations by 2014. Again found statewide, the records were least common in the Laurentian Mixed Forest Province and most common in the southern half of the Prairie Parkland Province.

The 2,091 detections collected by MNBBA participants provided a more detailed illustration of the species' relative abundance across the state **(A)**. Although the Thrasher was observed in all 87 Minnesota counties, it was least abundant and more sparsely distributed throughout the northern Red River valley, the Tallgrass Aspen Parklands, and much of the Laurentian Mixed Forest Province. Reported from 29.5 percent (1,402/4,757) of the surveyed atlas blocks, breeding was confirmed in 166 blocks (3.5 percent) spanning 62 counties; 34 counties were additions to the list published by Hertzel and Janssen (1998). Breeding was also considered probable in 18 counties.

The MNBBA probability map predicted that the lowest breeding densities occur in the far northeastern and north-central counties of the Laurentian Mixed Forest Province, especially from Koochiching and Itasca Counties east to Cook County **(B)**. Moderate densities are predicted elsewhere throughout much of the state, with the exception of a few regions in eastern Minnesota, including the Anoka Sand Plain Subsection and the southeastern Blufflands Subsection, where the Thrasher is less abundant.

Over the past one hundred years, the Brown Thrasher's distribution appears relatively unchanged. It is likely more abundant along the southern edge of the Laurentian Mixed Forest Province

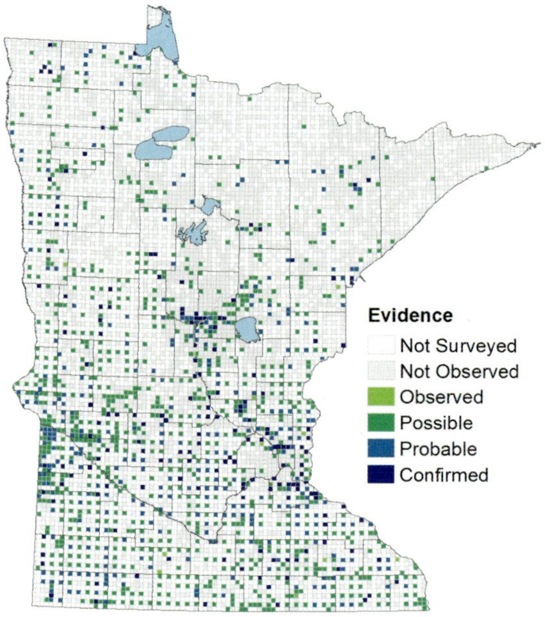

A. Breeding distribution of the Brown Thrasher in Minnesota based on the Breeding Bird Atlas (2009–13).

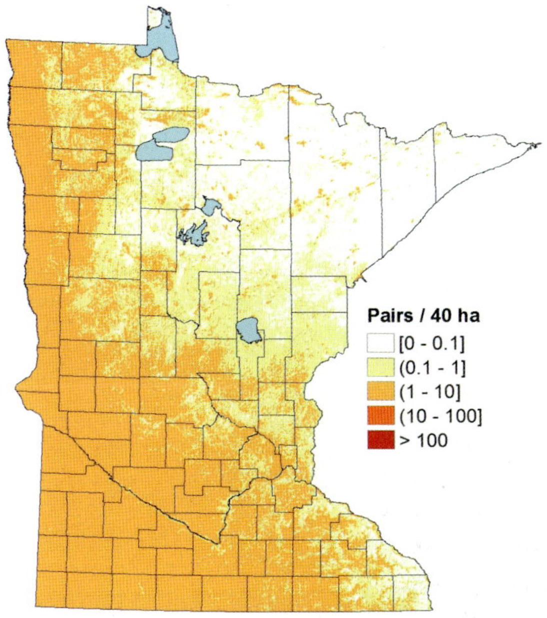

B. Predicted breeding distribution (pairs per 40 ha) of the Brown Thrasher in Minnesota based on habitat, landscape context, and climate data gathered during the Minnesota Breeding Bird Atlas (2009–13) using the Generalized Linear Modeling method with an adjustment for detectability.

than it was in the early 1900s. The landscape's extensive stands of mixed deciduous and coniferous forests have been replaced or fragemented by towns and cities, industrial development, and recreational opportunities, creating a more open landscape for this shrub-loving bird.

Historically the Brown Thrasher benefited from wide-scale forest clearing and settlement in the eastern deciduous forest in the late 1800s and early 1900s. Farther west, in the Great Plains, it was originally confined to shrubby riparian corridors, but woody plantings along fencerows and farmsteads provided new nesting opportunities. More recently, in the latter half of the twentieth century, Brown Thrashers have expanded into Canada's Maritime Provinces and into southern Florida (Cavitt and Haas 2020). Erskine (1992) suggested the expansion in eastern Canada may be attributed to warmer temperatures rather than to habitat changes.

BREEDING HABITAT

Few studies have examined the species' breeding-season requirements. Bent's (1948) description over seventy years ago still serves as a comprehensive description of the range of habitats Thrashers find suitable. He characterized the species as an inhabitant of the "rural, woodland and farming districts, living in bushy pastures, sproutlands, brier patches, tangles along fences, dry thickets, brushy hillsides, and the edges of woodlands, almost always far from human habitations." In the East, it was commonly associated with overgrown pastures, dry thickets, and forest edges **(C)**. In the West, it was associated with fencerows, farmstead shelterbelts, and "woody draws." In the more extensive forest landscapes of northern Minnesota, Brown Thrashers are found in the open, brushy clearings or dense, young growth of early successional stands (Niemi, Howe, et al. 2016).

POPULATION ABUNDANCE

BBS data helped generate a population estimate of 6.2 million breeding adults in the United States and Canada, with approximately 140,000 birds residing in Minnesota (PIF 2021). This is considerably lower than the estimate derived using MNBBA point count data. Analyses of this larger data set resulted in a statewide population estimate of nearly 1.1 million adults (95 percent confidence interval of 872,000–1.3 million).

Across the BBS survey region, the Brown Thrasher has demonstrated a significant, steady population decline, averaging −0.87 percent per year from 1966 to 2019 (Sauer et al. 2020). Range-wide, biologists estimated the population declined 37 percent from 1970 to 2014 (Rosenberg, Kennedy, et al. 2016). In Minnesota, the species has experienced a somewhat slower, nonsignificant drop in numbers averaging −0.54 percent per year from 1967 to 2019 **(D)**.

Maturation of shrubby habitats and young forests in eastern North America and more intensive agricultural practices that have eliminated fencerows and shelterbelts farther west are considered the principal factors responsible for the decline (Cavitt and Haas 2020). The loss of small family farms, for example, has certainly eliminated suitable habitat for Brown Thrashers throughout southern and western Minnesota. Between 2007 and 2012, the state lost 8 percent of its farms, a loss that was twice the national average (Pioneer Press 2014). Unlike its shrub-loving relative, the Gray Catbird, the Thrasher is less likely to inhabit suitable habitat in residential areas. A narrower habitat tolerance may help explain why its population is declining while that of the more cosmopolitan Catbird has remained relatively stable.

C. Scattered patches of dense shrubs and small trees bordering old fields, grasslands, or small streams are common features of the Brown Thrasher's breeding habitat in Minnesota. Photo by Lee A. Pfannmuller.

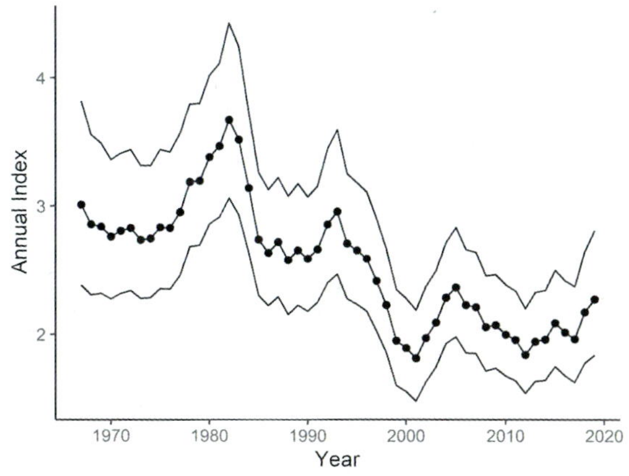

D. Breeding population trend and 95 percent credible interval (solid black lines) for the Brown Thrasher in Minnesota based on the 1967–2019 Breeding Bird Survey (Sauer et al. 2020).

Farther east, at least one study suggested the decline in Thrashers observed in New York might be related to the significant increase in the number of Northern Mockingbirds, a more aggressive bird that could possibly displace the Thrasher (Raynor 1976). Local food supplies may also regulate populations (Cavitt and Hass 2020).

CONSERVATION

The Brown Thrasher is ranked as a moderate conservation priority (PIF 2021); habitat loss is considered the most significant threat to its population. Additional concerns include its susceptibility to collisions with communication towers and, because of its ground-foraging habits, its susceptibility to car collisions (Cavitt and Haas 2020). Rising temperatures pose another major challenge. Under a 2°C warming scenario, the Brown Thrasher's highly vulnerable North American breeding distribution is projected to decline 63 percent by 2050 and increase north by only 23 percent (NAS 2019). Although its distribution will not suffer any losses in Minnesota, and may even expand in the northeast, populations should be closely monitored to ensure that this shrubland inhabitant, widely recognized for its diverse vocal repertoire, does not suffer any further declines.

NORTHERN MOCKINGBIRD

Mimus polyglottos

Minnesota Seasonal Status: Summer vagrant, occasionally breeds; occasional to rare spring and fall migrant; winter vagrant. The Northern Mockingbird was very rare during the MNBBA.
Migration: Permanent resident in its core range from Pennsylvania to Missouri, westward through the southern Great Plains states to California, and then south to the Gulf coast and Mexico. North of this region it is considered a partial migrant.
Food: Mainly an insectivore in the summer; switches to berries and fruits in the fall and winter.
Nest: Cup nest of dead twigs lined with grasses, leaves, and often trash; placed in dense vegetation, usually 1–3 m above the ground but sometimes higher, and occasionally on buildings.
Conservation Concern: Assigned a low PIF Continental Concern Score of 8/20.

Photo by Sparky Stensaas.

Most people are aware of the Mockingbird from its song, which is recalled in an old lullaby, "Hush, little baby, don't say a word, Papa's gonna buy you a mockingbird," and in a folk song, "Listen to the Mocking Bird." The melodic phrases of the Northern Mockingbird's song are not its only attributes, as shown by this description from a classic bird book: "The Mockingbird stands unrivaled. He is the king of song. . . . He equals and even excels the whole feathered choir. He improves upon most of the notes that he reproduces, adding also to his varied repertoire the crowing of chanticleer, the cackling of the hen, the barking of the house dog, the squeaking of the unoiled wheelbarrow, the postman's whistle. . . . He even imitates man's musical inventions" (T. G. Pearson 1917). Its species name, *polyglottus,* is aptly given for its multilingual capabilities, and its genus name, *Mimus,* for the imitator that it is.

Anyone who has spent time in the southern United States is likely familiar with its identification. Males and females are similar. It is a trim bird with grayish-brown upperparts, dark wings, and white-gray underparts. It has long wings, tail, and legs. In flight, its white wing patch, wing bars, and white edges of the tail are conspicuous.

The Northern Mockingbird's range spans the continent from New England to Texas, then westward, through New Mexico and Arizona to California. It also breeds in much of Mexico, the Bahamas, the Cayman Islands, and the Greater Antilles. Highest densities in the United States are found in southern and central Texas, southern Louisiana, southern Mississippi to central Georgia, and eastern New Mexico.

MINNESOTA BREEDING DISTRIBUTION

The Northern Mockingbird has been part of Minnesota's avifauna since the nineteenth century. Roberts (1932) cited records in 1883 (Otter Tail County), 1896 (Hennepin County), 1898 (Carver County), 1917 (Goodhue County), 1923 (McLeod County), and 1928 (Hennepin County), but he concluded that no "nests have been found in Minnesota thus far." In the 1936 edition of *The Birds of Minnesota,* he quoted an account by Sigurd F. Olson for the winter of 1932–33 in Ely, St. Louis County, where a mockingbird "appeared on December 26 1932 feeding on the berries of the Virginia creeper" on a neighbor's porch and remained there feeding as well on mountain ash trees nearby until "it froze to death, February 11th . . . following a night of about 40 [degrees below zero]." This episode illustrated why it is difficult for this species to become an established breeding resident in Minnesota.

The first nesting observation for the state was in 1968, when a pair successfully nested in a yard in Royalton, Morrison County, and returned in 1969 and nested again (Nagle 1969; R. L. Huber 1968; Russell 1969). Mockingbirds have nested six times since 1969, including the following: in 1977, Rock County (Eckert 1977a); in 1996, Hennepin County (Wiens 1997); and in 2002, Dakota County (J. P. Mattsson 2003a). Three other instances were reported by Hertzel (2019): in 2007, Hennepin County; and successful fledging of young in 2018 and 2019 in Fridley, Anoka County. Despite these nesting records from five counties, there is no evidence that a permanent population has been established in Minnesota.

The Northern Mockingbird occurs in the state yearly. Records from 1961 through 2019 are in the MOU seasonal reports (MOU 2021b). In the fifty-year time span from 1961 to 2019, vagrant birds have turned up in almost all counties in the state (84 out of 87). Most records are during the migratory seasons, when it is more abundant in the spring than the fall. During the summer season, migrants commonly linger well into June, and wandering vagrants can be found in late June and July. The number of spring migrants seen increased in the 1990s and became noticeably larger in the 2000s. Weather and climate change are likely associated with these trends. Increased numbers could lead to successful breeding and locational fidelity or to permanent residency for this species in Minnesota.

MNBBA participants reported records in eight blocks, all of which were coded as possible breeding evidence during the atlas years from 2009 to 2013. The records were widespread throughout the state and included two each in Cass County (one is on the western shoreline of Leech Lake; one occurs in a block that straddles a small portion of northwestern Morrison County, but the record was along the

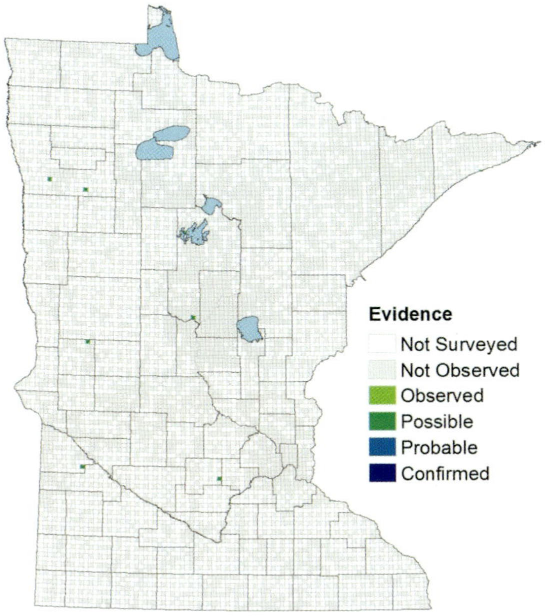

A. Breeding distribution of the Northern Mockingbird in Minnesota based on the Breeding Bird Atlas (2009–13).

southern shoreline of Sylvan Lake in Cass County) and Polk County, and one each in Cook (on the Lake Superior shoreline), Grant, Lac qui Parle, and McLeod Counties **(A)**. These records follow the pattern of incidental observations of birds lingering into late June. One record of a bird on July 12–13, 2013, in Grand Marais, Cook County, was probably a vagrant, as has commonly been observed on the North Shore. A query of the MOU seasonal reports during the atlas years and the atlas safe dates (June 1 to August 1) resulted in 15 additional summer records: 13 observations of lingering spring migrants (North Shore counties, seven; Twin Cities counties, three; singles in Waseca, Murray, and Goodhue Counties) and two July vagrants (Washington County, July 16, 2009; Blue Earth County, July 18, 2009).

Farnsworth and colleagues (2020) describe the species' gradual expansion northward throughout the nineteenth and twentieth centuries in North America, which they primarily attribute to a growing human population and the availability of suitable habitats and food. In states adjacent to Minnesota, the Northern Mockingbird is considered a permanent resident in Iowa, a rare summer resident in South Dakota (Tallman et al. 2002), and a rare but sporadic breeder in Wisconsin (Cutright et al. 2006).

BREEDING HABITAT

The Northern Mockingbird's habitat varies across the continent, but the species basically prefers open areas with sparse vegetation **(B)**. In the East, it frequents human-dominated landscapes, such as towns, farms, suburbs and other areas with parks, gardens, lawns, and thickets. In the West, it is found in desert scrub, chaparral, and residential and other developed areas.

POPULATION ABUNDANCE

Based on the BBS, PIF (2021) estimated a US and Canadian population of 34 million birds and a Minnesota population of 110 individuals. The BBS suggests a significant declining population surveywide, averaging −0.66 percent per year from 1966 to 2019 **(C)**. Fifteen eastern areas have significantly declining yearly trends; the largest declines are in Peninsular Florida (average of −2.15 percent per year), the Edwards Plateau in southern Texas (−1.92 percent per year), and New Jersey (−1.87 percent per year). There are also significant increasing trends in California (1.40 percent per year) and Utah (2.96 percent per year) in the West, and in New York (2.32 percent per year) and Massachusetts (1.88 percent per year) in the East. Overall, Rosenberg, Kennedy, and colleagues (2016) estimated a decline of −19 percent in the United States and Canada from 1970 to 2014.

B. The Northern Mockingbird is frequently found in open landscapes with scattered trees and shrubs, including very manicured sites like this small park in Jackson County. Photo by Lee A. Pfannmuller.

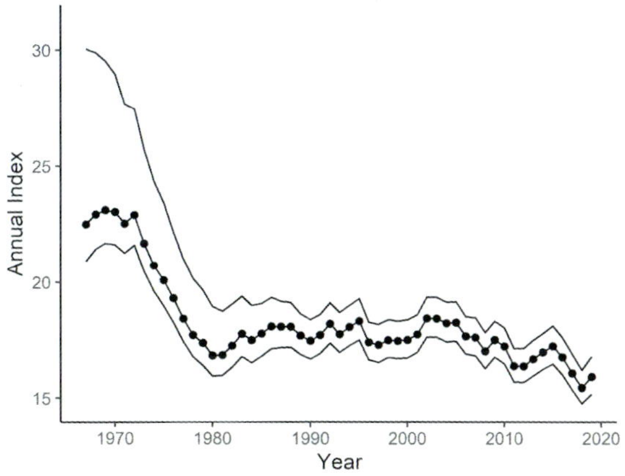

C. Surveywide breeding population trend and 95 percent credible interval (solid black lines) for the Northern Mockingbird based on the 1967–2019 Breeding Bird Survey (Sauer et al. 2020).

Because the population trends of the Northern Mockingbird have varied considerably across the United States, simple explanations of its overall decline are difficult to state. Farnsworth and colleagues (2020) suggested that, in the East, reforestation of former agricultural and pasturelands may be contributing to local declines, while in the Midwest the consolidation of farms into larger fields with fewer edges may be associated with less suitable habitat.

CONSERVATION

Given the species' wide breeding range and large population, plus its extensive use of human-dominated habitats, there is little conservation concern for the Northern Mockingbird in North America (Rosenberg, Kennedy, et al. 2016). However, because of its recent decline, continued monitoring is encouraged. Interestingly, the Migratory Bird Treaty Act of 1918 gave protection to this species, which was sorely needed because the Northern Mockingbird had been exploited in the caged-bird trade since the late 1700s. The species was kept as a pet because of its singing ability, and, by the end of the nineteenth century, its numbers in many northeastern areas were severely depleted. After protection in the early twentieth century, it took many decades for areas of local extinction to be repopulated by birds from the southern states (Farnsworth et al. 2020).

Northern Mockingbird populations are expected to increase northward in the future with a warming climate (NAS 2019), especially if suitable habitat is present. The availability of habitat remains a determining factor, but this species has adapted well to human environments. Projects like the MNBBA, designed to be repeated at ten- to twenty-year intervals, will help to untangle if and how these populations may change over time.

EUROPEAN STARLING

Sturnus vulgaris

Minnesota Seasonal Status: Regular breeding species and permanent resident. The European Starling was abundant during the MNBBA.
Migration: Largely considered a sedentary, permanent resident, but many birds do migrate. This is particularly true of northern populations, including those in the Great Lakes region and the Midwest. The migratory movements, however, are not well studied.
Food: Omnivorous, foraging in open areas such as urban lawns, agricultural pastures, and croplands; considered an economic pest.
Nest: An adaptable cavity nester that uses former woodpecker cavities, natural cavities, cliffs, burrows, nest boxes, and anything resembling a hole or cavity in any building or structure.
Conservation Concern: Introduced to North America in the late 1800s, this unregulated, non-native species is not a conservation concern. It causes damage to agricultural crops, poses safety concerns at airports, and aggressively outcompetes many native species for nesting cavities.

Photo by Gerald Hoekstra.

We have become so accustomed to considering the European Starling a pesky, undesirable species that we often give little thought to its rather remarkable story. In the early 1900s, many ornithologists were busy detailing the bird's rapid spread across North America, including its precise rate of travel, its directional movements, and the expansion of breeding populations in each newly established locality. As Bent noted in 1950, "The literature of American ornithology is flooded with references to its spectacular progress."

The European Starling is a common species throughout Eurasia; attempts to introduce it to North America began as early as 1872. At least six attempts were made in cities across the country, each successively failing before the species took hold in New York City's Central Park. In a project implemented by a group of Shakespeare enthusiasts who were committed to introducing each bird that the playwright mentioned in his plays to North America, 100 Starlings were released over a period of two years from 1890 to 1891 (CLO 2017e). The rest is history. By 1916, Starlings ranged from southern Maine to Virginia; by 1931, they were sighted at Churchill, Manitoba; and by 1942 the first flock was reported in California (Bent 1950). The clearing of the eastern deciduous forest coupled with the species' adaptable foraging and nesting habits propelled its success across North America.

A little smaller than a Robin, the European Starling is a stocky, black bird with a short tail. During the breeding season, the iridescent purple and green feathers on the head and chest sparkle in the sunlight, and the bill is bright yellow. During the nonbreeding season, the bill is black, and the browner plumage is covered in white spots that are the tips of new feathers that have emerged during the fall molt. The tips gradually wear away to reveal the bird's full iridescence by spring (Cabe 2020).

In a testament to the bird's diverse repertoire, observers have variously described the sounds and song of the Starling as ranging from "painful" and "disagreeable" to a "musical soliloquy." The song, in particular, is a thin and raspy warble that may last as long as a minute. Well-known for their mimicry of other bird songs, Starlings can fool listeners with imitations of at least 20 other common birds, including Killdeers, Flickers, Orioles, Wood Pewees, and Bluebirds (Bent 1950).

Global in distribution, the European Starling now occurs throughout North America, from southeastern Alaska east to Canada's Atlantic coastline and south throughout the United States and northern Mexico. Some of the densest breeding populations occur in the Corn Belt states of Illinois and Indiana.

MINNESOTA BREEDING DISTRIBUTION

When Roberts (1932) wrote a comprehensive account of the species in Minnesota, the Starling had just reached the southeastern corner of the state. The first report was submitted in 1931 by an observer in Houston County, who wrote: "A whole colony of these birds was found in a cupola of a barn on a farm in the eastern part of Fillmore County. The people who own the barn have been observing these birds for at least two years."

By 1933, a flock of 22 birds was observed just south of Minneapolis, and three birds were seen near Sturgeon Lake in Pine County (M. Morse 1933); by 1936, several nests were reported in St. Louis County (M. D. Thompson 1936a). In his 1936 revision of *The Birds of Minnesota*, Roberts wrote: "During the last four years the Starling has spread throughout Minnesota and westward into the Dakotas. It is nowhere as abundant, as yet, as farther east, but since it is now well established and nesting regularly, it will not be long before it will be an all too familiar bird about our cities and towns."

Roberts's prediction proved all too true. Forty years later, Green and Janssen (1975) noted that the species was already statewide in distribution by the end of the 1930s, although its abundance did not surge until the 1940s. Most common in Minnesota's open farm country, the species had even been reported in the Boundary Waters Canoe Area Wilderness. Janssen (1987) included a statewide distribution map that identified 34 widely dispersed counties where nesting had been confirmed since 1970. Hertzel and Janssen (1998) later added nine more counties to the list.

The MNBBA not only confirmed the species' abundance throughout Minnesota's farm country but, due to the efforts of atlas participants in the Twin Cities and Brainerd, the species' prominence in those areas of the state was documented as well **(A)**. A total of 4,055 European Starling records were reported from 46.5 percent of the surveyed atlas blocks and

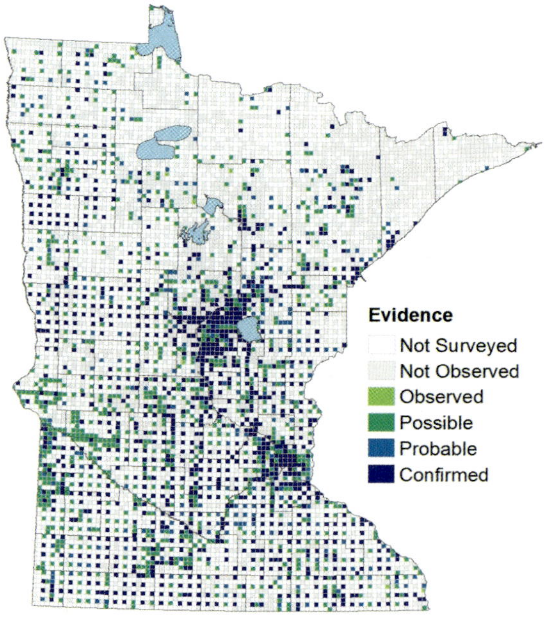

A. Breeding distribution of the European Starling in Minnesota based on the Breeding Bird Atlas (2009–13).

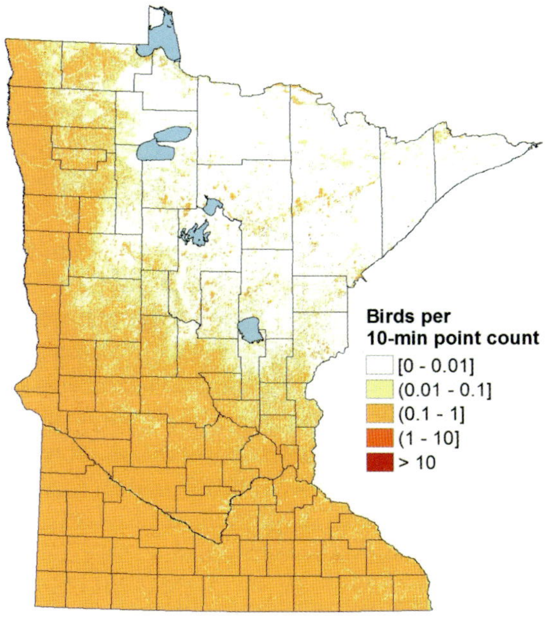

B. Predicted breeding distribution (birds detected per 10-minute point count) of the European Starling in Minnesota based on habitat, landscape context, and climate data gathered during the Minnesota Breeding Bird Atlas (2009–13) using the Generalized Linear Modeling method without an adjustment for detectability.

from 59.5 percent of the priority blocks. Breeding was confirmed in 1,026 blocks (21.2 percent). The species was least common in north-central and northeastern Minnesota. Regardless, it was reported from all 87 Minnesota counties and was confirmed nesting in all but one county, Lake of the Woods.

The MNBBA probability map predicted that moderate breeding densities occur throughout the settled and agricultural regions of southern and western Minnesota **(B)**. The species was predicted to be absent or present in low breeding densities throughout much of the Laurentian Mixed Forest Province, with the exception of settled and developed areas such as Duluth, Brainerd, Bemidji, International Falls, and the Iron Range. The area of predicted abundance in northern Cook County is the site of the 2007 Ham Lake Fire, where the species was indeed found breeding in 2009 and 2011.

Since the Starling's rapid spread across North America in the early 1900s, local populations have fluctuated in abundance, but the species' overall distribution has remained relatively unchanged. Yet, in his comprehensive review of the species, Cabe (2020) noted that Starlings may still be expanding slowly along the northern and southern peripheries of their breeding range.

BREEDING HABITAT

The European Starling has two primary habitat requirements: suitable nest cavities and open areas with short vegetation for foraging (Cabe 2020). Both requirements are easily met in a wide range of diverse habitats, especially in developed areas **(C)**. Buildings provide an endless supply of creative nesting opportunities, and farmland or residential areas provide an abundance of foraging opportunities. In Minnesota, the European Starling avoids dense forest cover and large areas of open vegetation, such as expansive peatland and grassland landscapes.

POPULATION ABUNDANCE

PIF (2021) estimated a global population of 250 million Starlings, with a population estimate of 93 million residing in the United States and Canada. Minnesota supports approximately 1.4 million breeding adults.

Despite its abundance, the European Starling has experienced a steady and statistically significant decline across the BBS survey region, averaging −1.30 percent per year from 1966 to 2019 (Sauer et al. 2020). Some of the largest declines have been in the northeastern United States and southeastern Canada. In Minnesota, the statistically significant decline has been slightly steeper, averaging −1.65 percent per year **(D)**. Surprisingly, "little is unambiguously known of the factors controlling Starling populations" (Cabe 2020). Nevertheless, in the northeastern states and provinces, the abandonment of farmsteads and pasturelands and their subsequent replacement by forest cover has likely contributed to the decline.

C. Farmsteads supply an abundance of food and nesting cavities for the non-native European Starling. Photo by Lee A. Pfannmuller.

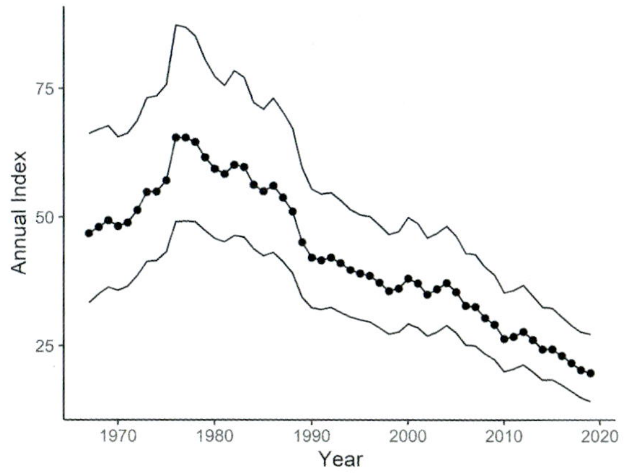

D. Breeding population trend and 95 percent credible interval (solid black lines) for the European Starling in Minnesota based on the 1967–2019 Breeding Bird Survey (Sauer et al. 2020).

CONSERVATION

As a nonnative species, the European Starling is not protected by the Migratory Bird Treaty Act. Management efforts have focused on limiting the economic damage it causes to crops, orchards, and livestock feed, albeit with limited long-term success. One study estimated the species' annual damage to US agriculture at $800 million (Pimentel et al. 2000). Control efforts have also been implemented on large, urban roosts, where the species' presence can create noise, odors, and safety concerns, including at airports (Linz et al. 2007). Although the Starling's aggressive competition for nesting cavities is often cited as a factor contributing to the decline of other cavity-nesting species, one study (Koenig 2003) failed to find any strong correlation at the continental level, with the potential exception of the Yellow-bellied Sapsucker.

This adaptable, aggressive species will likely remain a major component of Minnesota's avifauna long into the future. Even warmer temperatures are unlikely to provide a deterrent (NAS 2019). Years ago, Roberts (1932) foresaw a time in the future when, following its initial rapid growth and spread, the European Starling population would gradually decrease in numbers and its condition would be reduced to that of "innocuous desuetude." Indeed, despite its abundance, it does not elicit nearly the concern that it did when it first appeared. There is now a reluctant acceptance of its inevitable presence in the landscape.

EASTERN BLUEBIRD

Sialia sialis

Minnesota Seasonal Status: Regular breeding resident and migrant; occasionally present during the winter, especially in southern Minnesota. The Eastern Bluebird was common during the MNBBA.
Migration: Northern populations are partial, short- to medium-distance migrants that winter in the southern United States and Mexico; the extent of migration varies depending on weather and food resources. Farther south, birds are permanent residents.
Food: Omnivore; sits and waits on a perch to search for arthropods on the ground; berries are gleaned from vegetation.
Nest: Secondary cavity nester that uses natural tree cavities, old woodpecker cavities, and nest boxes.
Conservation Concern: Assigned a low PIF Continental Concern Score of 7/20.

Photo by Rebecca Field.

Few North American birds have garnered more adoration and concern than the Eastern Bluebird. Since colonists first set sight on the "blue robin," many have written eloquently of their love and admiration for this "sweet and amiable" bird (Chapman 1895). Even physical descriptions of it are poetic. Describing the rich colors of the male Bluebird, Bent (1949) wrote: "It has been said that he carries on his back the blue of heaven and the rich brown of the freshly turned earth on his breast." Eulogized less often, the paler female also dons blue above and rusty orange below.

Although some have described the species' song as sweet and pleasing, the Eastern Bluebird is not much of a songster when compared to other members of the thrush family. Soft and low in pitch, the entire song consists of three to eight notes sung in phrases of one to three notes each. Each phrase has a burry quality that is unmistakable. Naturalists have tried to describe the song with a variety of mnemonics, but Roberts (1932) described it best as a "soft, mellow warble."

Well-known for its population decline in the early to mid-1900s, followed by a rather extraordinary recovery, the Bluebird's acceptance of nest boxes not only helped to launch its recovery but also resulted in it becoming one of the most intensively studied songbirds in North America. Today it flourishes throughout eastern North America, where it is found from southern Manitoba south to Texas and east to the Atlantic coast. Breeding densities are especially high in the southeastern states. Resident populations are also found in Mexico and Central America.

MINNESOTA BREEDING DISTRIBUTION

Roberts (1932) described the Eastern Bluebird as a "common summer resident" throughout Minnesota. Considered most abundant in "well-settled" areas, it was not uncommon "in the desolate, burned-over areas and windfalls of the northern forests." Breeding was confirmed in nine counties, stretching north and northwest from Houston County in the southeast to Carlton, Cass, and Polk Counties. Although it was occasionally observed wintering as far north as Duluth, severe weather during the winter of 1895 demonstrated the Bluebird's vulnerability to cold temperatures. Roberts noted the species was virtually "exterminated" and nearly twelve to fifteen years passed before it returned to its former level of abundance. Apart from this dramatic decline in numbers, there is no hint in Roberts's account of the crisis that lay ahead.

In the mid-1900s a tsunami of events unfolded that led to the Bluebird's decline. As a secondary cavity nester, dependent on the availability of natural cavities in trees or wooden fence posts or on cavities previously excavated by woodpeckers, the Bluebird faced "a housing shortage" (Kraske 1981). First, fruit orchards, one of the Bluebird's favorite haunts, were more intensively managed as a cash crop. Dead and dying limbs were pruned, while potential nest cavities on live limbs were filled and painted with creosote to prevent further damage. Farming practices also changed. Dead trees were cut down, wooden fence lines were replaced with metal posts and wire, and shelterbelts and windbreaks were demolished to increase cropland acreage (Lindquist 1952; Kraske 1981; E. Smith 2018).

Aggressive species that can out-compete Bluebirds for nesting cavities, House Sparrows and European Starlings were considered even bigger culprits. Their introduction and rapid spread across North America posed a threat to local Bluebird populations, especially where nesting cavities were limited. Weather also played a major role, for the Bluebird is among the earliest songbirds to return each spring; the impact of late winter storms and freezing temperatures on local and regional Bluebird populations is well-documented (Sauer and Droege 1990).

Anecdotal accounts of population declines began as early as the 1920s. Dr. T. E. Musselman, a biologist in Illinois, was among the first to take action. Launching an investigation into the proper construction and placement of Bluebird next boxes, he established a small Bluebird trail, published his findings in ornithological journals, and appealed to others to initiate similar efforts (Lindquist 1952). In Minnesota, local bird clubs, individuals, and the MNDNR took note, directing their efforts to help local Bluebird populations.

By 1975, Green and Janssen described the species as a statewide summer resident but noted its absence from the northern forests and its scarcity in the western agricultural region. Janssen (1987) made specific mention of the construction of Bluebird trails and the expanding statewide population. Since 1970, nesting was confirmed in 70 of Minnesota's 87 counties. Hertzel and Janssen (1998) later added five more counties to this list, while the MBS further documented the species' presence in all but the most cultivated regions of the state.

The species' strong association with the open, fragmented forests that stretch from southeastern Minnesota north to the Aspen Parklands was clearly documented by the MNBBA **(A)**. A total of 3,313 Eastern Bluebird records were reported in 36.0 percent of the surveyed atlas blocks and in all 87 counties. Once again, impressive survey efforts in the Brainerd Lakes region documented the species' remarkable abundance in central Minnesota. Statewide, breeding was confirmed in 796 blocks (16.6 percent) spanning 76 counties; seven counties were additions to Hertzel and Janssen's 1998 publication. Ten of the 11 counties where breeding was not confirmed were located in Minnesota's agricultural region. Cook County was the only northern county where nesting was not reported; four probable records, however, were noted. Reflecting these outcomes, the MNBBA probability map **(B)** predicts moderate breeding densities throughout the central portion of the Eastern Broadleaf Forest Prov-

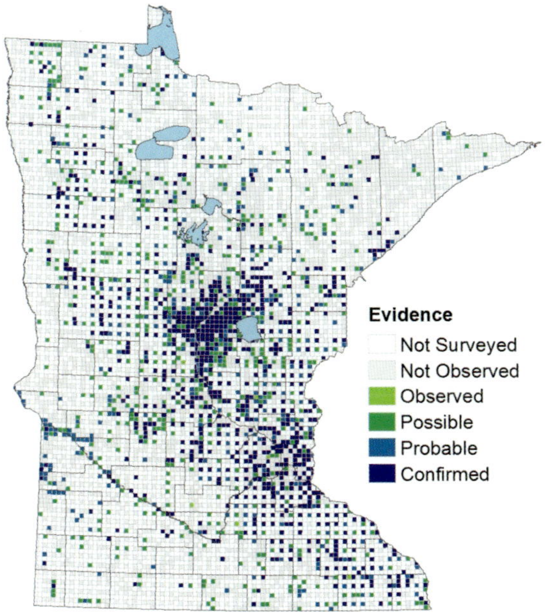

A. Breeding distribution of the Eastern Bluebird in Minnesota based on the Breeding Bird Atlas (2009–13).

ince and the southern and western perimeter of the Laurentian Mixed Forest Province. These landscapes are where open woodlands are prominent and recreational homeowners often provide nest boxes.

Gowaty and Plissner (2020) point out that the population of this open woodland inhabitant was likely much smaller before the eastern deciduous

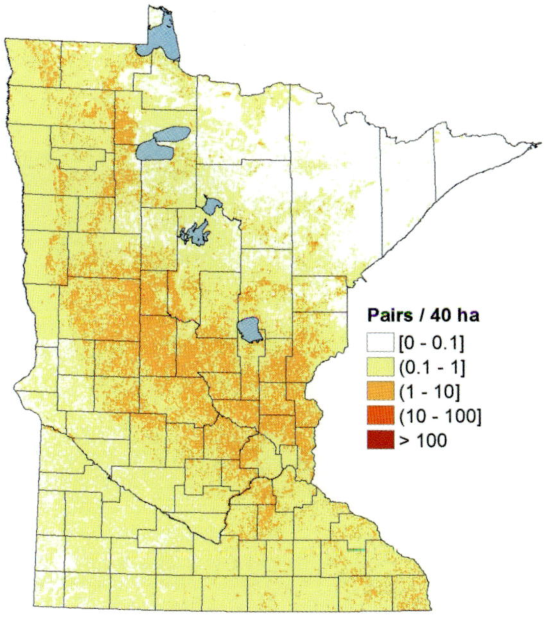

B. Predicted breeding distribution (pairs per 40 ha) of the Eastern Bluebird in Minnesota based on habitat, landscape context, and climate data gathered during the Minnesota Breeding Bird Atlas (2009–13) using the Generalized Linear Modeling method with an adjustment for detectability.

forest was cleared for agriculture, timber, and new frontier settlements. Indeed, "despite concerns about 20th century fluctuations in numbers, populations may have been more abundant in the 20th and early 21st centuries than in any other period of North American history since the Pleistocene."

BREEDING HABITAT

The Eastern Bluebird requires an open landscape with scattered trees, little to no understory, and sparse ground cover. Stands need to be old enough to provide sufficient snags and/or dead and dying limbs. Before settlers opened the eastern forests, the Bluebird was likely limited to scattered forest openings with residual trees, fire-maintained oak and pine savannas, and open pine stands (Gowaty and Plissner 2020). Following settlement, farmsteads, orchards, old fields, and forest edges provided new habitat opportunities **(C)**. Nest boxes further broadened the range of habitats utilized to include suburban lots, golf courses, and parks. In northern Minnesota, the species continues to be found in forest clearings with scattered trees and snags left behind by logging and fire (Niemi, Howe, et al. 2016).

POPULATION ABUNDANCE

Relying largely on BBS data, PIF (2021) established a population estimate of 23 million breeding adults, with 21 million residing in the United States and Canada; approximately 340,000 birds reside in Minnesota. MNBBA data generated a statewide estimate nearly twice as high, at 641,000 breeding adults (95 percent confidence interval of 502,000–827,000).

Across the BBS survey region, the Eastern Bluebird has demonstrated a strong, statistically significant population increase averaging 0.94 percent per year from 1966 to 2019 (Sauer et al. 2020). Although the rate of increase has slowed in recent years, the population experienced a 178 percent cumulative increase from 1970 to 2014 (Rosenberg, Kennedy, et al. 2016). The trend line is similar in Minnesota, where BBS data illustrate a significant population increase averaging 1.34 percent per year from 1967 to 2019 but a slower rate of growth beginning in the year 2000 followed by a small decline **(D)**.

Factors that influence Bluebird populations include nest site availability, winter food abundance, agricultural pesticides, nest competition with other species, and weather (Gowaty and Plissner 2020). In the northern Great Lakes states, a decline in forest harvest levels, corresponding with an increase in forest age, is likely contributing to slower rates of population growth observed throughout the region (Sauer et al. 2020).

C. Scattered mature trees in an old field provide nesting cavities for the Eastern Bluebird and plenty of open ground for foraging. Photo by Lee A. Pfannmuller.

CONSERVATION

A listed species in many states in the late 1900s, the Bluebird's robust population is a testament to its improved status and low conservation priority (PIF 2021). Even climate change is expected to have little impact on this treasured member of the thrush family. Minnesota is among the few states where changes are predicted. Under all three warming scenarios NAS (2019) examined, range retractions in western Minnesota are compensated by range expansions in south-central and northeastern counties.

Citizens devoted to the Bluebird's recovery deserve recognition for their hard work. In particular, the Minneapolis Chapter of the Audubon Society, in partnership with the MNDNR's Nongame Wildlife Program, promoted the construction and management of Bluebird houses and trails to help recover "one of the most agreeable" of our feathered friends (Audubon 1827–38).

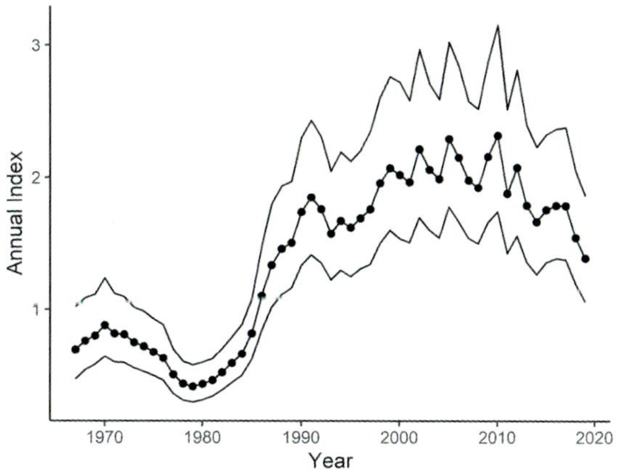

D. Breeding population trend and 95 percent credible interval (solid black lines) for the Eastern Bluebird in Minnesota based on the 1967–2019 Breeding Bird Survey (Sauer et al. 2020).

VEERY

Catharus fuscescens

Minnesota Seasonal Status: Regular breeding resident and migrant. The Veery was abundant during the MNBBA.
Migration: Long-distance migrant that winters primarily in south-central and southeastern Brazil.
Food: Arthropods and fruit, primarily secured on the ground.
Nest: Cup nest on the ground or slightly elevated in the base of a bush or small tree.
Conservation Concern: Assigned a moderate PIF Continental Concern Score of 12/20.

Photo by Gerald Hoekstra.

Like so many birds found in dense forests with impenetrable shrubbery, the Veery is more often heard than seen. Roberts (1932), who knew the species as the Willow Thrush, refers to it as a "wild, shy bird on its nesting-grounds." The Veery's song is an enigma because some have considered it "one of the strangest sounds in nature" (Bent 1949) while others described it as "one of the finest of our songsters" with "one of the sweetest sounds of the woodland" (Forbush and May 1955). The song is a flute-like, mournful descending spiral *veeerr veeerr veeerr,* while the distinctive call note is an abrupt descending *veer* (Alderfer 2006), both of which serve as a derivative for its common name. Forbush and May (1955) emphasized that "[t]o be fully appreciated, this song must be heard when one is alone in the deep woods, among the falling shades of the coming night. It breathes the spirit of the dying day. Sometimes at evening these thrush songs reply to one another like echoes in the moonlight."

Although a challenge to observe on the breeding grounds, the Veery is often easier to observe during migration. It is distinguished in plumage by its uniform reddish upperparts from crown to tail, the lack of an eye-ring, and relatively indistinct spotting on the upper breast, the combination of which is unique among the thrushes found nesting in Minnesota.

Although the Veery has a wide breeding distribution in North America, its highest global breeding densities occur in northern and north-central Minnesota, in eastern Ontario, and in western Quebec. The Veery distribution stretches south to northern Georgia in the Appalachian Mountains and west to British Columbia and Washington State.

MINNESOTA BREEDING DISTRIBUTION

Roberts (1932) described the Veery's distribution as "throughout the wooded areas of the state, increasing in abundance northward." He emphasized that conditions were generally unsuitable in the southwestern part of the state in the late nineteenth and early twentieth centuries, but that the Veery was common in selected locations, such as along the Minnesota River. By the early twentieth century, the species was no longer common south of the tamarack and spruce swamps of Isanti and Pine Counties. Roberts (1936) reported confirmed nesting from Cass, Crow Wing, Isanti, Itasca, Otter Tail, Sherburne, and Steele Counties, as well as at Itasca State Park, Mille Lacs, and Minneapolis.

Green and Janssen (1975) also present the Veery as a resident throughout the state but wrote that it is "very scarce and local south and west of a line between Washington, Stearns, and Otter Tail Counties." They emphasized that its status in southwestern Minnesota was poorly known and that in the southeastern region it had never been very numerous but was "largely restricted to a few areas, like Nerstrand Woods in Rice County, Forestville Woods in Fillmore County, and the Whitewater area of Wabasha and Winona counties."

Janssen (1987) said the Veery's distribution was well represented in the northern half of the state and in the groves on the prairies of the northwest region south to Otter Tail County. Its southern distribution extended to Anoka, Ramsey, and Washington Counties and it was breeding sparingly in riparian tributaries and heavy woods in the southeastern region. Further, the Veery was not known to breed in the southwest, but there were summer observations from Lyon County. He identified confirmed nesting since 1970 in 10 counties. Hertzel and Janssen (1998) added another six counties to the confirmed list since 1970, including a nesting record from Winona County in the southeast.

The MBS provided substantially greater detail, recording 2,277 breeding-season locations throughout the state. They confirmed the extensive distribution in the northern portion of the state, especially in the Laurentian Mixed Forest and Tallgrass Aspen Parklands Provinces, and in the northern regions of the Eastern Broadleaf Forest Province. Locations were scattered throughout the fringes of the southwest, including along the southern edges of the Minnesota River and into the southeastern counties of Fillmore, Houston, and Mower; one location was in Watonwan County.

MNBBA data exhibited a pattern similar to that observed by the MBS. It included 4,318 records that ranged south to Murray and Steele Counties and to where the species was previously recorded in Lyon, Mower, and Watonwan Counties **(A)**. Records were found in 42.2 percent (2,004/4,754) of surveyed blocks and 50.6 percent (1,182/2,338) of priority blocks. The species' extensive breeding presence throughout northern Minnesota was well-documented. Nesting was confirmed in 72 blocks, including new county records for Aitkin, Beltrami, Carlton, Kanabec, Kittson, Koochiching, Mahnomen, Pine, Rice, Roseau, Todd, and Washington Counties.

The probability map for the Veery in Minnesota suggests substantial densities throughout much of northeastern, northern, and north-central Minnesota, ranging from 1 to 10 pairs per 40 ha **(B)**. Much lower densities, ranging from 0.1 to 1 pair per 40 ha, were predicted for more southerly and western regions of the state. Higher densities were predicted for a few scattered areas of young regenerating aspen

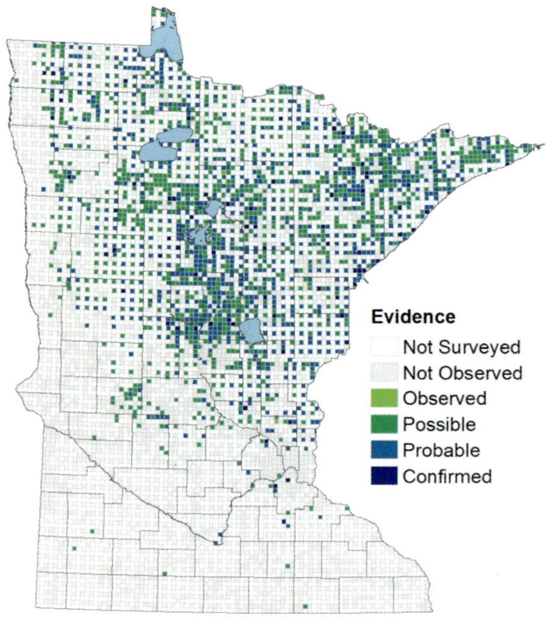

A. Breeding distribution of the Veery in Minnesota based on the Breeding Bird Atlas (2009–13).

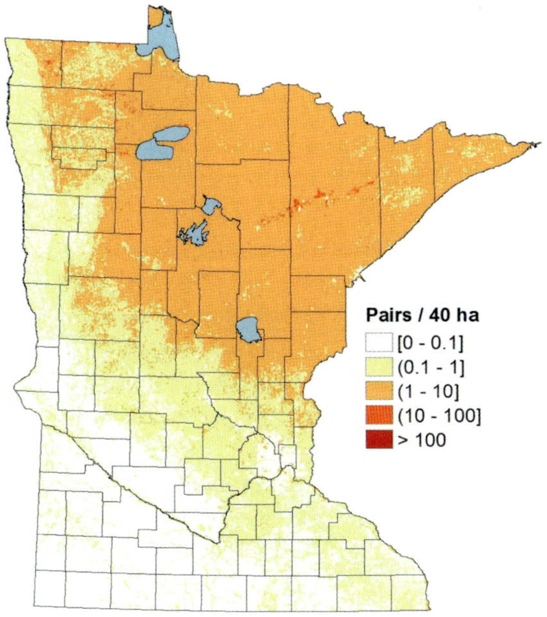

B. Predicted breeding distribution (pairs per 40 ha) of the Veery in Minnesota based on habitat, landscape context, and climate data gathered during the Minnesota Breeding Bird Atlas (2009–13) using the Generalized Linear Modeling method with an adjustment for detectability.

in the Iron Mining areas in northeastern Minnesota and in the Tallgrass Aspen Parklands Province.

The species' range described by Roberts in 1932 has undoubtedly been reduced by the conversion of forests and riparian woodlands to agricultural and urban uses since the mid-1800s. The breeding range, however, is larger than that reported by Janssen in 1987. Some of the expanded range is due to extensive coverage by the MBS and the MNBBA and not necessarily to any recent range expansion. In their review of the Veery in North America, Heckscher and colleagues (2020) mentioned no major changes in the breeding distribution of the species in North America.

BREEDING HABITAT

Roberts (1932) described the Veery as breeding in areas of "low, damp woodlands, preferring especially tamarack swamps, and thickets of poplar, willow, and alder, bordering streams and lakes." Niemi, Howe, and colleagues (2016) found that the Veery was most abundant in lowland hardwood and regenerating forests, but also common in aspen-birch, lowland shrub, and oak forest cover types. The common denominator is usually the presence of a dense shrub understory **(C)**.

POPULATION ABUNDANCE

Using BBS data, PIF (2021) estimated a US and Canadian breeding population of 11 million adults and a Minnesota population of 1.20 million. The MNBBA estimated a slightly higher breeding population of 1.96 million adults (95 percent confidence interval of 1.82–2.24 million). The PIF (2021) estimate would suggest that 11 percent of the global breeding population of the Veery occurs in Minnesota, while the MNBBA estimate would approach 18 percent.

The BBS indicated a nearly significant declining population trend in Minnesota of −0.41 percent per year from 1967 to 2019 **(D)**. This is consistent with trends across a broad region, including Michigan, Ontario, and Wisconsin, as well as throughout Canada and the United States. In contrast, the results of monitoring Veery populations in Minnesota's National Forests from 1995 to 2020 were conflicting (Grinde, Bednar, et al. 2020). In the Chippewa National Forest of north-central Minnesota, the species was significantly increasing, but it was stable in the Superior National Forest of northeastern Minnesota. Rosenberg, Kennedy, and colleagues (2016) estimated an overall global decline in the population of 40 percent from 1970 to 2014.

Declines in the Veery population have been attributed to forest fragmentation and degradation of its wintering grounds. C. S. Robbins and colleagues (1989) and Blake (1991) identified the Veery as area sensitive to forest fragmentation. Field evidence suggests that the Veery has lower reproductive success in small forest patches because of high predation of nests and high nest parasitism by Brown-headed Cowbirds. Heckscher and colleagues (2020) emphasized the vulnerability of Veery overwintering habitat and suggest it is the most important factor for effective conservation of the species. Habitat where the species overwinters in southern Brazil is under exceptionally rapid destruction from conversion to agriculture, development, and construction of hydroelectric dams.

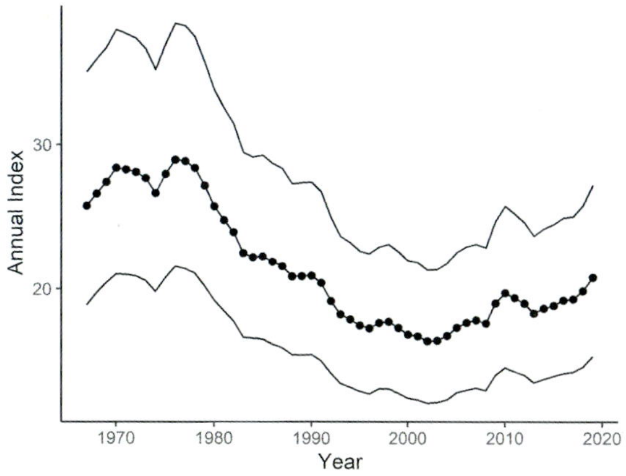

D. Breeding population trend and 95 percent credible interval (solid black lines) for the Veery in Minnesota based on the 1967–2019 Breeding Bird Survey (Sauer et al. 2020).

CONSERVATION

There are many concerns about the status of the Veery in North America. Its widespread decline in numbers and the vulnerability of its winter habitat, however, are the primary factors responsible for its moderate conservation ranking (PIF 2021). It has also been identified as a Minnesota Stewardship Species by Audubon Minnesota because the state supports a significant portion of the the species' breeding population, the highest of any US state (Pfannmuller 2012).

Under a 2°C warming scenario, NAS (2019) projected low vulnerability in its breeding range in North America, with a 43 percent loss of its breeding range and a 55 percent gain northward by 2050. However, substantial breeding range retractions were predicted for most of its Minnesota range, except in the extreme northeastern portion.

Even though the Veery is still a common to abundant species throughout a substantial part of Minnesota, its population and distribution face many threats. Minnesota has a key role to play in maintaining suitable landscapes and keeping its strange, sweet song echoing from our young forests.

C. Nesting habitats for the Veery in Minnesota are typically areas with a high density of shrubs and young trees that provide substantial cover for nest and foraging activity. Photo by Gerald J. Niemi.

SWAINSON'S THRUSH

Catharus ustulatus

Minnesota Seasonal Status: Regular breeding resident and migrant, accidental in winter. The Swainson's Thrush was an uncommon species during the MNBBA.
Migration: Medium- to long-distance migrant; overwinters in Mexico, Central America, and South America.
Food: Omnivore; arthropods and fruit, primarily gathered near the ground.
Nest: Cup nest in the understory of a shrub, sapling, or tree.
Conservation Concern: Assigned a moderate PIF Continental Concern Score of 10/20.

The Swainson's Thrush, formerly the Olive-backed Thrush, is one of the most secretive of our Minnesota breeding thrushes. Found in the most northerly forests of Minnesota, it is primarily heard rather than seen. Its song is similar to the flute-like quality of the Hermit Thrush but is less musical, lacks an introductory note, spirals upward in pitch, and has been described as *whip-poor-will-a-will-e-zee-zee-zee* (Bent 1949). Roberts (1932) cites Hoffmann's description of the song as "the voice of a thrush like a Veery's song inverted, going up instead of down the scale, but throatier, more gurgling, inferior in purity, richness, and suggestiveness to those of the three other common thrushes," referring to the Wood Thrush, Hermit Thrush, and Veery.

Identification of the Swainson's Thrush requires a careful examination of the bird. Look for a buffy-colored eye-ring and buffy or tawny cheeks that distinguish it from the similar Gray-cheeked Thrush, the latter a migratory species that breeds north of the Minnesota border. The upperparts from crown through the tail are an olive-brownish color and lack any hint of rust or red. This coloration provides its scientific name, *ustulatus,* which, derived from the Latin for "to burn slightly or to singe," refers to its charred or ash-like plumage (Sandrock and Prior 2014).

Although the Swainson's Thrush was thought to occur primarily in the northern forests of Minnesota, the intensive and astute MNBBA birders in the Brainerd area documented several records in central Minnesota. Range-wide, the species occurs throughout Canada and the western United States, with southern limits in the northeastern and upper midwestern United States. The highest densities are found in western North America from Oregon to British Columbia.

Photo by David Brislance.

MINNESOTA BREEDING DISTRIBUTION

Roberts (1932) described the Minnesota range of the Swainson's Thrush as "a summer resident in the northern evergreen forests as far south as Pine County and as far west as Itasca Park and Kittson County." He reported confirmed nesting in Cass, Cook, Itasca, and Kittson Counties; the latter was the first nest found in Minnesota in 1896. Roberts reported observations of the Swainson's Thrush in east-central Minnesota in the 1870s by Thomas M. Trippe as "Rather rare. Does not breed." However, a report from 1922 referred to the species as common in Cook County, while Roberts and his colleague William Kilgore also found it fairly common in northern Itasca County.

Years later, Green and Janssen (1975) described its breeding distribution as northeastern and north-central Minnesota, plus adjacent northwestern counties. They remarked that the species' present status in the northwestern counties and as far south as Pine County was unknown. They added confirmed nesting in Crow Wing, Lake, and St. Louis Counties to those previously noted by Roberts (1932). A few years later, Janssen (1987) suggested a more restricted breeding distribution, where the species was confined to the northernmost counties from Clearwater to northern Itasca, most of St. Louis, and all of Lake and Cook Counties. He suggested a northwestern distribution limited to eastern Roseau County. Janssen (1987) and Hertzel and Janssen (1998) only included confirmed nesting in Clearwater and Cook Counties since 1970.

The MBS included 362 breeding-season locations and emphasized the dominant distribution of the Swainson's Thrush in the eastern portion of St. Louis County and especially throughout Cook and Lake Counties. The MBS also included observations from Carlton, southern Beltrami, northern Itasca, eastern Marshall, and northern Roseau Counties.

The MNBBA included 859 records that further reinforced the species' major distribution in Cook, Lake, and northeastern St. Louis Counties, but also scattered probable and possible nesting in Aitkin, Beltrami, Hubbard, Kittson, Koochiching, Lake of the Woods, Marshall, and Roseau Counties **(A)**. Even though the breeding distribution was broad, it was patchy and appeared in only 7.6 percent (361/4,734) of all surveyed blocks. The most southerly observations were from north-central Minnesota, including southern Cass and Crow Wing Counties and southwestern Aitkin County. Nesting was also confirmed in Crow Wing and northern Cass Counties. Other than the seven confirmed nesting observations in Cook and Lake Counties, confirmed nesting records included one in central Itasca County and two in St. Louis County.

The probability map emphasizes the highest densities of the Swainson's Thrush from southwestern St. Louis County to extreme eastern Cook County **(B)**. High densities are also predicted throughout Koochiching County.

Breeding observations from the MBS and the MNBBA illustrated the breeding distribution in Minnesota as larger than previously outlined by Janssen (1987); but neither source included observations from Pine County, where Roberts (1932)

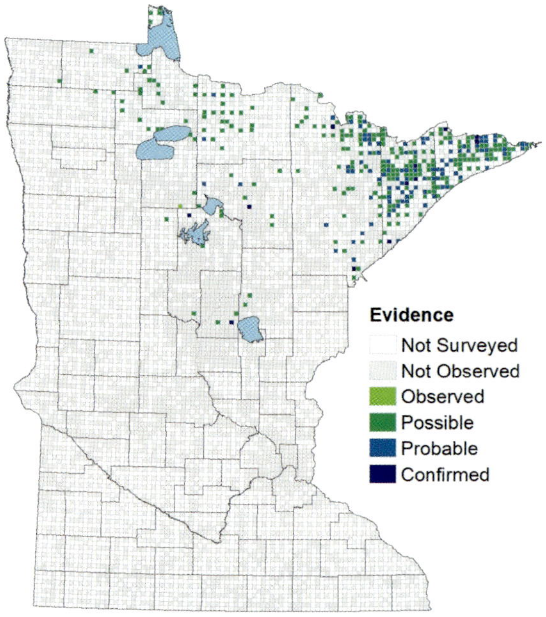

A. Breeding distribution of the Swainson's Thrush in Minnesota based on the Breeding Bird Atlas (2009–13).

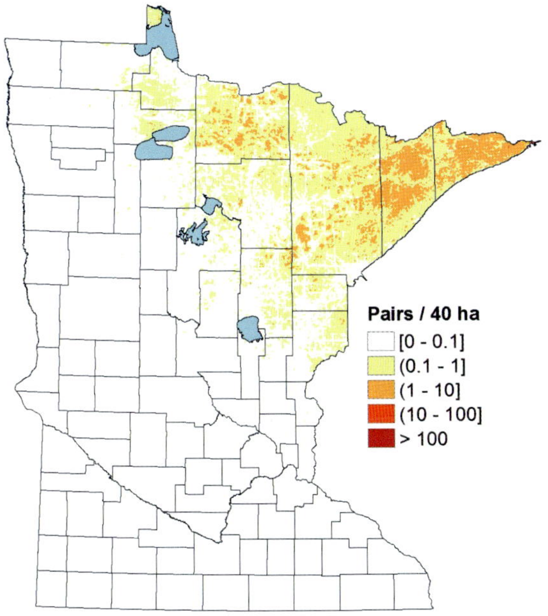

B. Predicted breeding distribution (pairs per 40 ha) of the Swainson's Thrush in Minnesota based on habitat, landscape context, and climate data gathered during the Minnesota Breeding Bird Atlas (2009–13) using the Generalized Linear Modeling method with an adjustment for detectability.

previously recorded the species. There still appeared to be scattered, though sparsely distributed, populations of the Swainson's Thrush in the northwestern counties of Kittson, Marshall, and Roseau as well as a few from the MNBBA observers in north-central Minnesota.

Mack and Yong (2020), in their North American review of the Swainson's Thrush, mention range contractions only in some coastal and mountainous areas of California and the loss of spruce forests in Massachusetts. In California, the range declines were attributed to loss of riparian habitats and cowbird parasitism. The species' status in Minnesota appears to be stable or potentially even more substantial than previously reported, except for the lack of observations in east-central Minnesota. However, the recent extensive coverage by the MBS and MNBBA and improved recognition of the Thrush's vocalizations may be associated with better documentation of its breeding range in Minnesota.

BREEDING HABITAT

Mack and Yong (2020) described the species' breeding habitat as strongly associated with coniferous, spruce–fir (*Picea* spp.–*Abies* sp.) forests, and mixed hardwood–conifer forests across Canada and the northern New England states. In Minnesota, the species is considered to be conifer-dependent but is also found in mixed coniferous–deciduous upland forests (Green and Niemi 1978; Niemi and Pfannmuller 1979). Habitat analyses by Niemi, Howe, and colleagues (2016) indicated a significant association with aspen–spruce–fir forest cover types but also found the species in mixed swamp conifers and jack pine forest cover types **(C)**. In the Agassiz Lowlands Subsection, the species was most often found in mixed lowland white cedar forests (Bednar et al. 2016).

POPULATION ABUNDANCE

Based primarily on the BBS data, PIF (2021) estimated a global population of 120 million breeding adults in the United States and Canada, with only 110,000 breeding adults in Minnesota. The MNBBA estimated a population twice that size, of 217,000 breeding adults, but the prediction was highly variable for a relatively small population in the state.

The BBS data indicated a nonsignificant trend in the Minnesota breeding population from 1967 to 2019, and the wide credible interval suggests considerable uncertainty in the trend estimate **(D)**. Similar nonsignificant trends over the same period were observed in the Canadian provinces of Manitoba and Ontario, but both provinces have recorded significantly declining populations from 1993 to 2019 (Sauer et al. 2020). Rosenberg, Kennedy, and colleagues (2016), based on the BBS data, estimated that the North American breeding population has declined by −32 percent from 1970 to 2014. A decline is consistent with population monitoring in the Superior National Forest of northeastern Minnesota, where a significant average decline of −3.88 percent per year was detected from 1995 to 2020 based on 74 forest stands sampled (Grinde, Bednar, et al. 2020).

Overall trend results for the Swainson's Thrush are equivocal. Ralston and colleagues (2015) also reported on the negative trends in the Superior National Forest but found stable or increasing trends in nine other areas studied, primarily in the northeastern United States. Many have reported potentially high mortality during migration for the Swainson's Thrush. For instance, Loss and colleagues (2014) identified a higher-than-average risk of mortality at residences. Bracey, Etterson, and colleagues (2016) found it to be the most common species found at window kills in residences at Minnesota Point in Duluth. Longcore and colleagues (2013) ranked the species as fifth in mortality at communication towers.

C. The Swainson's Thrush is usually found nesting in mature forested areas with coniferous trees or with a mix of coniferous and deciduous trees. Photo by Gerald J. Niemi.

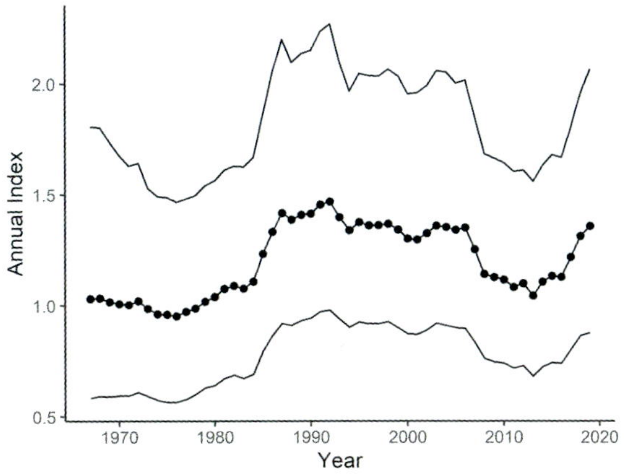

D. Breeding population trend and 95 percent credible interval (solid black lines) for the Swainson's Thrush in Minnesota based on the 1967–2019 Breeding Bird Survey (Sauer et al. 2020).

CONSERVATION

The Swainson's Thrush was considered a moderate conservation concern primarily because of its documented declining population, yet it has a large breeding population and wide breeding distribution in North America. Therefore, the species has not been a primary focus of conservation action in North America. NAS (2019), however, projected a moderate vulnerability for the Swainson's Thrush breeding distribution in North America with a 2°C warming scenario, including a 48 percent range loss and a 28 percent range gain in the far north by 2050. In Minnesota, the species was predicted to be extirpated even under our imminent 1.5°C warming scenario.

Mack and Yong (2020) cited highly variable population fluctuations of the Swainson's Thrush among different regions of North America and identified habitat loss, cowbird parasitism, lack of knowledge on its wintering grounds, and mortality during migration as causes for concern. The increase in urbanization as well as the number of communication towers and wind turbines could represent a growing concern for this long-distance migrant.

Even though the Swainson's Thrush is often missed by many observers, its melodic song is a cherished addition to the morning symphony in Minnesota's north woods for those with the ability to recognize it.

HERMIT THRUSH

Catharus guttatus

Minnesota Seasonal Status: Regular nesting species, migrant, and occasional winter visitor. The Hermit Thrush was common during the MNBBA.
Migration: Short- to medium-distance migrant, winters in eastern, southeastern, and southern areas of the midwestern United States to northeastern Mexico.
Food: Omnivore; insects, small invertebrates, and fruit primarily foraged on the ground.
Nest: Cup nest on the ground or slightly elevated in the base of a bush or small tree.
Conservation Concern: Assigned a low PIF Continental Concern Score of 6/20.

A medium-sized woodland thrush with a rusty, reddish tail and brown back is quickly identified as a Hermit Thrush. The larger Wood Thrush is reddish on its upperparts but not on the tail; the similarly sized Veery is reddish throughout its back and tail. Both Hermit Thrush sexes have a similar plumage.

Among the greatest traits of the Hermit Thrush are its vocal abilities. Bent (1949) said it is "the most gifted songster in North America." He elaborated, "The exquisite song of this modest bird . . . has captivated the affections of a host of bird lovers. Those who have been privileged to hear its song possess delightful memories of associations with the hermit; perhaps a wooded border of some mirrored lake or some fern-carpeted woodland; or again they may have heard the fluted notes ringing across some brilliant sunset scene." In words, the song has been described as *"oh, holy holy, ah, purity purity eeh, sweetly sweetly"* (CLO 2017a). Among the keys to identification is the introductory note that distinguishes it from the Swainson's Thrush and Wood Thrush. It has a characteristic call note, *tschup* or *eeee,* the latter similar to the Cedar Waxwing. Both can be learned with practice and time spent in the field.

The Hermit Thrush has a wide breeding distribution in the upper midwestern and northeastern United States and Canada, and south in the Appalachian Mountains to the state of Virginia. It is also found throughout the western United States and Canada. The highest densities occur in northern Wisconsin, northern Michigan, and northeastern Minnesota, plus in many isolated patches of Ontario, Nova Scotia, and northern Colorado.

Photo by Michael Furtman.

MINNESOTA BREEDING DISTRIBUTION

The Hermit Thrush was historically described as nesting in the northern evergreen forests of Minnesota, south to Isanti and Pine Counties and northwest to Itasca State Park and Roseau and Lake of the Woods Counties (Roberts 1932). Roberts reported confirmed nests in Cass, Itasca, Lake, and St. Louis Counties as well as at Itasca State Park and "Mille Lacs."

More than forty years later, Green and Janssen (1975) identified a similar distribution, but more precisely suggested that its southern distribution in Minnesota extended to the cities of Onamia in Mille Lacs County and Sturgeon Lake in Pine County. They reported additional counties with confirmed nesting records, including Carlton, Clearwater, Cook, Hubbard, Lake of the Woods, Pine, and Roseau Counties. Janssen (1987) depicted a similar breeding distribution and confirmed nesting in nine counties since 1970 in which Aitkin, Beltrami, Crow Wing, and Koochiching Counties were new records. Hertzel and Janssen (1998) updated the confirmed nesting records since 1970 to include the same counties identified by Janssen (1987), except that they excluded Clearwater County and added Roseau County.

The extensive coverage of the MBS recorded 1,178 breeding-season locations for the Hermit Thrush. These locations reinforced its primary breeding distribution in northeastern Minnesota but extended its southern breeding range to include Kanabec, Mille Lacs, Morrison, Pine, and Todd Counties. Western breeding locations included one each in Mahnomen and Otter Tail Counties, plus many locations in eastern Becker, northeastern Marshall, and Roseau Counties.

The MNBBA detected 2,691 records of the Hermit Thrush distributed extensively throughout the Laurentian Mixed Forest Province **(A)**. The breeding distribution was similar to the MBS, but no observations were included from Kanabec County, and only one possible record was reported in the far northeastern corner of Otter Tail County. The Hermit Thrush was recorded from 24.5 percent (1,162/4,751) of all blocks, including confirmed nesting in 1.4 percent (66 blocks). The only confirmed county nesting record not previously reported was from western Minnesota in Becker County, while probable nesting was also suggested for Mahnomen and Morrison Counties.

The probability map for the Hermit Thrush identified high densities in the Agassiz Lowlands Subsection of northern Minnesota **(B)**. Otherwise, the breeding population was evenly distributed throughout northeastern and north-central Minnesota but with patches of higher density noted in the Tamarack Lowlands Subsection of southwestern St. Louis County.

In their review of the Hermit Thrush in North America, Dellinger and colleagues (2020) commented on its potential range expansions in British Columbia, in some New England states, and in the southern Appalachian Mountains, but no changes in the midwestern United States. Michigan, Ontario, and Wisconsin have all reported southern range extensions from what had been previously reported (R. Brewer et al. 1991; Cadman et al. 1987; Cutright et al. 2006). These extensions appeared to be colonizations of coniferous patches, such as in spruce–fir forests and pine plantations.

Based on the results of the MBS and MNBBA, the current distribution of the Hermit Thrush in Minnesota appears to be larger than previously reported. The species is relatively easy to identify by sight and sound, making it unlikely to be missed in earlier periods. Based on these results, the Hermit Thrush should be looked for in suitable conifer patches in

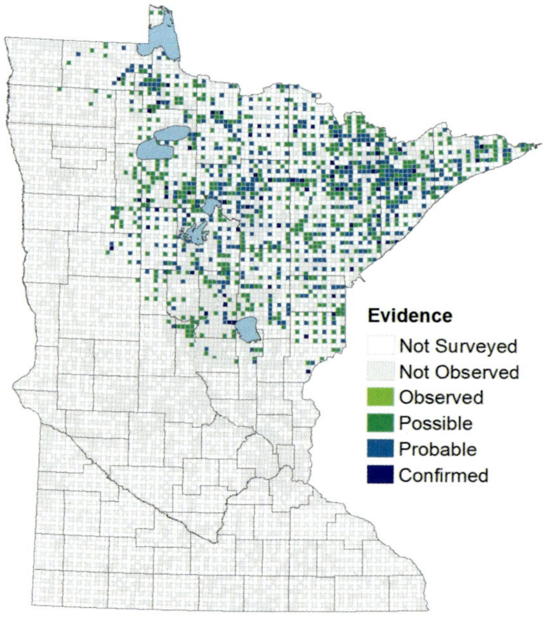

A. Breeding distribution of the Hermit Thrush in Minnesota based on the Breeding Bird Atlas (2009–13).

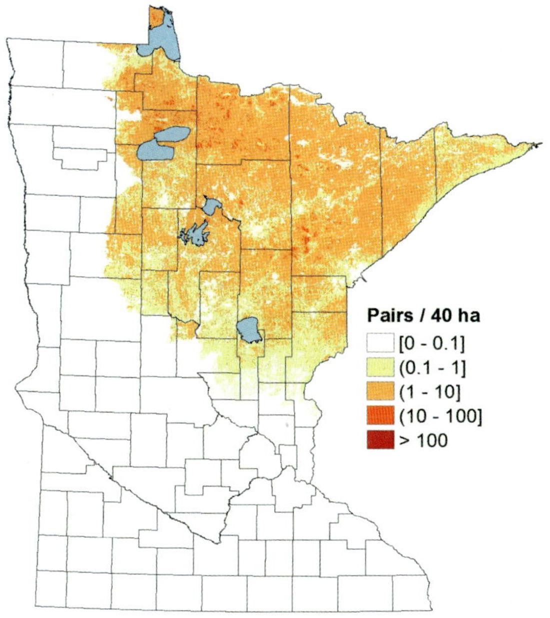

B. Predicted breeding distribution (pairs per 40 ha) of the Hermit Thrush in Minnesota based on habitat, landscape context, and climate data gathered during the Minnesota Breeding Bird Atlas (2009–13) using the Generalized Linear Modeling method with an adjustment for detectability.

central and southerly locations of Minnesota during the June breeding season.

BREEDING HABITAT

Across its substantial range, nesting areas have been broadly described as coniferous, deciduous, and mixed-forest types but with an affinity for conifers (Dellinger et al. 2020). The coniferous forest component has been emphasized in Minnesota by Niemi and Pfannmuller (1979), Green (1995), and Niemi and Hanowski (1992a, 1992b) **(C)**. Intensive coverage of the Agassiz Lowlands area north of Upper Red Lake indicated extensive use of black spruce–tamarack lowland and white cedar forests (Bednar et al. 2016). Niemi, Howe, and colleagues (2016) in the Chippewa and Superior National Forests found significantly higher use of black spruce–tamarack forests compared with all other forest types, although it was detected in almost every mature upland forest cover type.

C. The Hermit Thrush is found in many forest types, but highest densities during the nesting season in Minnesota are in forests dominated by black spruce and tamarack trees.
Photo by Gerald J. Niemi.

POPULATION ABUNDANCE

The US and Canadian population of the Hermit Thrush has been estimated at 72 million breeding adults (PIF 2021), with 330,000 in Minnesota. The MNBBA estimated a population that was about three times as high, or 986,000 breeding adults (95 percent confidence interval of 913,000–1,011,000). The higher estimate in Minnesota is likely due to methodological differences between PIF and the MNBBA and more thorough coverage of roadless areas, especially in the northern part of the state, where it is quite common.

The BBS in Minnesota indicated a significantly increasing breeding population averaging 1.05 percent per year from 1967 to 2019 **(D)**. This increasing trend contrasted with stable population trends in Michigan, Wisconsin (0.91 percent), and overall in Canada and the United States. Rosenberg, Kennedy, and colleagues (2016) calculated the North American population had increased by 35 percent from 1970 to 2014. Based on counts in Minnesota's two national forests, however, Hermit Thrush have been significantly declining from 1995 to 2020, and especially since 2010 (Grinde, Bednar, et al. 2020).

CONSERVATION

The Hermit Thrush currently is not considered a conservation concern in Minnesota nor elsewhere because of its large breeding population and its wide distribution in North America, but climate change is an emerging concern (NAS 2019). Optimal breeding habitat appears to be lowland coniferous forests in Minnesota; however, it is commonly found in many other forested habitat types.

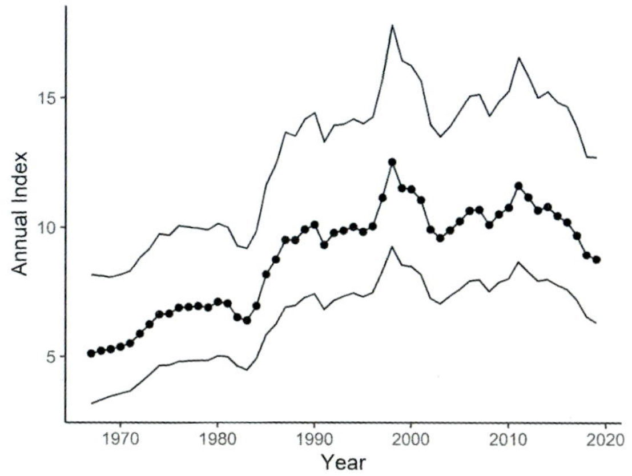

D. Breeding population trend and 95 percent credible interval (solid black lines) for the Hermit Thrush in Minnesota based on the 1967–2019 Breeding Bird Survey (Sauer et al. 2020).

NAS (2019) projected a moderate vulnerability for its breeding distribution in North America under a 2°C warming scenario, including a 49 percent range loss and an 18 percent range gain in the far north by 2050. In Minnesota, the species was predicted to be extirpated even under our imminent 1.5°C warming scenario by 2050. Besides climate change, future threats to the Hermit Thrush include effects of predation and inclement weather on nest success, diseases, and possibly hazards of migration. Flaspohler and colleagues (2001) in Michigan and Manolis and colleagues (2002) in Minnesota both found decreased nesting success near forest edges. Hanowski and Niemi (1995a) found the species to abandon selectively logged riparian areas several years after logging, but Schulte and Niemi (1998) found that residuals left in logged areas were key to maintaining its presence within these areas.

Dellinger and colleagues (2020) reported high mortality at buildings and residences, but neither Longcore and colleagues (2013) nor Loss and colleagues (2014) identified the Hermit Thrush as a high-risk species in tower collisions or window fatalities. Bracey (2011) detected five Hermit Thrush fatalities with window kills on Minnesota Point in Duluth over five migration seasons, lower than the 14 Swainson's Thrush fatalities.

Range-wide, the Hermit Thrush appears to be maintaining a stable population, but there are signs of a decline in northern Minnesota forests. Continued monitoring of its population and distribution are essential to retain confidence on its status. Climate change is likely the primary factor for whether Minnesotans will continue to hear the lovely, flute-like song of the Hermit Thrush in the years to come.

WOOD THRUSH

Hylocichla mustelina

Minnesota Seasonal Status: Regular breeding species and migrant. The Wood Thrush was uncommon during the MNBBA.
Migration: Medium-distance migrant that winters in southern Mexico and Central America.
Food: Ground forager that feeds primarily on soil invertebrates; fruits are common in the nonbreeding season.
Nest: Cup nest, usually placed in a deciduous tree or shrub, 2–5 m above the ground.
Conservation Concern: Assigned a high PIF Continental Concern Score of 14/20.

Minnesota is home to an abundance of woodland thrushes that grace the predawn and twilight hours with remarkable songs. But the "spiritual and appealing voices" of the Wood Thrush and Hermit Thrush are unparalleled (Roberts 1932). Because the deciduous-loving Wood Thrush has a more southern distribution than the boreal-inhabiting Hermit Thrush, early naturalists were more likely to encounter this species and write eloquently of its extraordinary, ethereal-sounding song. Thoreau ([1853] 1906) was among its many admirers, writing: "This is the only bird whose note affects me like music, affects the flow and tenor of my thought, my fancy and imagination."

Part of the song's appeal is the Thrush's ability to use its two voice boxes to sing different phrases simultaneously. Described as clear and flute-like, the song begins with a few short, low-pitched notes that are often inaudible, followed by variants of the phrase *ee-oh-lay,* and ending with a high trill. The song may be confused with that of the Hermit Thrush, but the latter invariably begins its song with a clear, sustained whistle. The Wood Thrush's *pit-pit-pit* call is also distinct.

When the bird forages for invertebrates on the forest floor, the Wood Thrush's rufous head and olive-brown back and wings blend with the browns and rusts of the forest litter. Its white eye-ring and the bold, dark spotting on the breast and belly are diagnostic. The sexes are indistinguishable.

Broadly distributed across the eastern United States and southeastern Canada, from the eastern Great Plains to the Atlantic coast, Wood Thrushes are most abundant in Mississippi and in the central Appalachian Mountains. The genus name, *Hylocichla*, comes from the Greek words *hyle*, which means "wood," and *kichle*, which means "thrush."

Photo by Gerald Hoekstra.

MINNESOTA BREEDING DISTRIBUTION

Roberts (1932) described the Wood Thrush as an abundant summer resident across Minnesota's southern deciduous woodlands, occurring north to the southern limit of the coniferous forests. It was also found in the floodplains of the Minnesota River, west to Chippewa County, and in scattered woodland groves in southwest Minnesota. Nesting evidence was available from Goodhue, Fillmore, Hennepin, Houston, Itasca, and Wabasha Counties.

During his lifetime, Roberts reported a northward expansion of the Thrush's distribution as the coniferous forests were harvested and replaced by hardwood stands more suitable to this deciduous forest inhabitant. When this expansion began was unclear. Early reports were dismissed by Roberts because they were based entirely on song identification. Given the similarity of the Wood Thrush's song to that of the Hermit Thrush, confusion with the latter could not be dismissed. Not until Roberts received credible accounts from trusted colleagues did he accept reports from the Mille Lacs region (1926) and from Pine (1919) and Itasca (1930) Counties; the latter included a confirmed nesting record.

By 1975, Green and Janssen reported the Wood Thrush was broadly distributed "except in the heavy coniferous forests adjacent to the Canadian border." It was most abundant in southeastern Minnesota but absent from "the forested central region" despite the apparent availability of suitable habitat. Nevertheless, they did note a new nesting record from Polk County and reports of territorial males along Lake Superior's North Shore. The authors believed these records represented an expansion in the knowledge of the Thrush's distribution rather than an expansion in its breeding range.

Janssen's 1987 update, however, made special note of the species' "widespread but local" presence "in maple basswood forests of the central and north-central regions." He extended the northern periphery of its breeding distribution to include southern Clearwater and Beltrami Counties, Itasca County, and the southern two-thirds of the Arrowhead Region. Records along the Minnesota River and in the southwest were considered evidence that the population was expanding westward. Yet, given Roberts's original account of the species' presence in this region, Wood Thrushes may have been reoccupying portions of their former range. Janssen identified 11 counties with confirmed nesting since 1970, including the northern counties of Clearwater, Crow Wing, and St. Louis. Hertzel and Janssen (1998) added seven counties to the list, including Becker, Beltrami, and Polk. The 422 breeding-season locations collected by the MBS also contained numerous records throughout the northern mixed forests, the eastern deciduous forests, and in prairie groves scattered across south-central Minnesota.

The breeding distribution documented by the MNBBA extended north and west of that described by Janssen (1987) **(A)**. The 585 sparsely distributed records were reported from 64 counties but in only 9.2 percent of the surveyed atlas blocks (434/4,738). Breeding was confirmed in 14 blocks (11 counties) and deemed probable in 96 blocks. Six counties with confirmed nesting were additions to Hertzel and Janssen's 1998 list, including Cass, Itasca, and Lake. Probable nesting was reported as far north as Koochiching, Lake of the Woods, and Roseau Counties.

Unlike earlier accounts, detections were most common in the Laurentian Mixed Forest Prov-

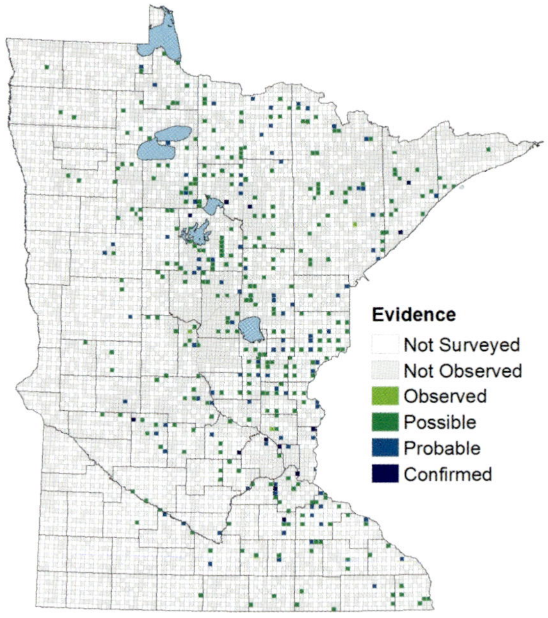

A. Breeding distribution of the Wood Thrush in Minnesota based on the Breeding Bird Atlas (2009–13).

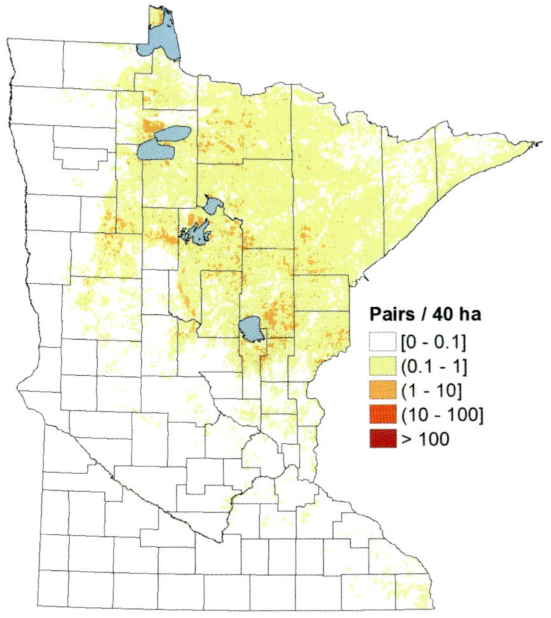

B. Predicted breeding distribution (pairs per 40 ha) of the Wood Thrush in Minnesota based on habitat, landscape context, and climate data gathered during the Minnesota Breeding Bird Atlas (2009–13) using the Generalized Linear Modeling method with an adjustment for detectability.

ince but also occurred as far west as Redwood and Renville Counties in the south, Grant County in west-central Minnesota, and Kittson, Marshall, and Roseau Counties in the northwest. Although the MNBBA probability map predicted moderate breeding densities along the southern and western boundaries of the Laurentian Mixed Forest Province **(B)**, even here the species was uncommon and sparsely distributed.

Clearly, Minnesota has witnessed a major shift in the Wood Thrush's distribution over the past century. Changes in the forest landscape, improved knowledge of the species' distribution, better song identification skills by birders, and reoccupation of its former breeding range have all likely contributed. What is undisputed is that this hardwood-loving species that once prospered in the state's southern mesic hardwoods has lost much of its former breeding range to agriculture and residential development. Nearby, in Wisconsin, the Wood Thrush has also expanded northward (S. D. Robbins 1991), but in Michigan its distribution "has not changed substantially" in nearly one hundred years (Chartier et al. 2013).

BREEDING HABITAT

Associated with deciduous and mixed forests, the Wood Thrush shows a strong preference for mature mesic uplands **(C)**. Forest structure may be more important than forest composition. F. C. James and colleagues (1984) highlighted the significance of "a well-shaded understory that contains at least a few small trees with low, exposed branches" for nesting, and "a fairly open forest floor with moist, decaying leafy litter" for foraging. Such conditions are frequently found in mature forests.

Although sites deemed most suitable for the Wood Thrush are at least 80 ha in size (Rosenberg, Hames, et al. 2003), it frequently nests in smaller stands and along woodland edges (e.g., Roth 1987; C. S. Robbins et al. 1989). Reproductive success, however, is low in such areas owing to high rates of predation and parasitism by Brown-headed Cowbirds (Weinberg and Roth 1998).

POPULATION ABUNDANCE

In the United States and Canada, PIF (2021) provided a Wood Thrush population estimate of 12 million breeding adults and a Minnesota estimate of 110,000. MNBBA data generated a higher estimate of 198,000 breeding adults (95 percent confidence interval of 155,000–254,000).

Minnesota and Ohio are the only two states where BBS data suggest an increasing population from 1967 to 2019 (Sauer et al. 2020). More intensive monitoring data from Minnesota's national forests corroborate this trend, demonstrating a combined regional increase of 1.63 percent per year from 1995 to 2020 (Grinde, Bednar, et al. 2020). This is in stark contrast to the Wood Thrush's population trend throughout much of the eastern United States and southeastern Canada, where the population experienced a significant, surveywide decline, averaging −1.30 percent per year from 1966 to 2019 **(D)**.

Significant predation of young and eggs in fragmented forest landscapes is considered the factor most responsible for the downward trend. Nests that

C. Typical breeding habitat of the Wood Thrush includes remnant Big Woods stands in east-central Minnesota with scattered small trees and shrubs in the understory and a relatively open forest floor. Photo by Lee A. Pfannmuller.

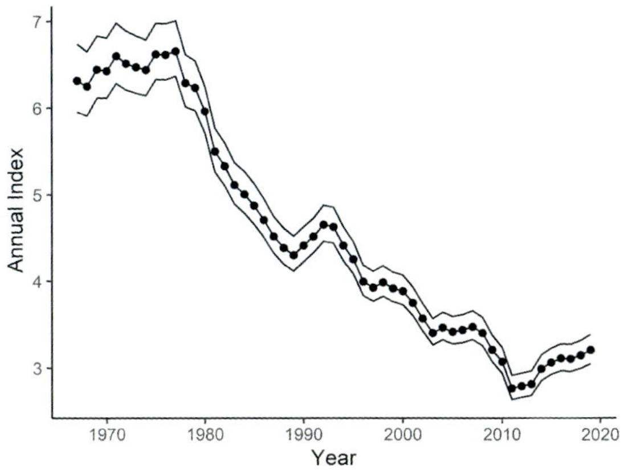

D. Surveywide breeding population trend and 95 percent credible interval (solid black lines) for the Wood Thrush based on the 1967–2019 Breeding Bird Survey (Sauer et al. 2020).

are not predated often sustain high levels of parasitism by cowbirds, further contributing to low reproductive success. On the wintering grounds, habitat loss may be impacting adult survival (M. Evans et al. 2020).

CONSERVATION

Given its widespread population decline, the Wood Thrush has garnered considerable attention and is considered a high conservation priority (PIF 2021). Biologists stress that conservation efforts need to focus on the species' entire life cycle, from its winter habitat in southern Mexico and Central America to its breeding habitat in the eastern United States and Canada. Ideally, large forest stands that are a minimum of 80 to 100 ha in size should be protected (Rosenberg, Hames, et al. 2003; M. Evans et al. 2020).

Unfortunately, even well-managed forests are not immune from the impact of invasive earthworms. When earthworm densities are high, they can severely reduce and even eliminate the forest's rich duff layer, rendering stands unsuitable for Wood Thrushes. Climate change raises additional concerns. Under a 2°C warming scenario, NAS (2019) projected a moderate vulnerability for the species' breeding distribution, with a 29 percent range loss by 2050 compensated by only a 15 percent gain. Conditions will likely deteriorate along the western periphery of the Thrush's breeding range in Minnesota but improve in the northeast region.

Clearly, vigilance is needed to ensure that this high-profile, beloved forest songster remains an important component of the state's avifauna. Its future requires close monitoring, more detailed information on factors impacting nesting success, and confidence that birders are accurately documenting its distribution.

AMERICAN ROBIN

Turdus migratorius

Minnesota Seasonal Status: Regular breeding resident and migrant statewide; regularly observed throughout the state during the winter, depending on weather conditions and food availability. The most frequently reported species, the American Robin was very abundant during the MNBBA.
Migration: Short- to medium-distance migrant; most of Minnesota's breeding Robins winter in the southern United States and Mexico.
Food: Ground forager and foliage gleaner that feeds on terrestrial invertebrates and fruit.
Nest: Cup nest placed in a tree or shrub 1.5–8 m above the ground. A variety of building structures may also be used, including windowsills, mailboxes, and fence posts.
Conservation Concern: Assigned a low PIF Continental Concern Score of 5/20.

Photo by Mike Lentz.

Awakening us with its cheery, predawn songs, amusing us as it cocks its head sideways searching for earthworms in our lawns, and delighting us when it chooses our backyard to nest, the American Robin is a beloved North American treasure. Nursery rhymes and children's books introduce us at an early age and our acquaintance continues for a lifetime because, no matter where you live in North America, you are bound to encounter the bird during at least one season of the year. Ornithologist Donald Kroodsma (2005) observed that: "So familiar is the Robin that it is the standard by which we know other birds. The bird in the bush is 'a little smaller than a Robin' or 'about the size of a Robin.'" The same is true for bird songs. "Learn to recognize the Robin's song," he noted, "and you've got the standard for the songs of other birds too." Indeed, Roger Tory Peterson's (1980) popular field guide describes the Scarlet Tanager's song as "suggesting a Robin with a sore throat"; the Rose-breasted Grosbeak's sweeter song is like a Robin that "has taken voice lessons"; and the Red-eyed Vireo's persistent singing is composed of "short abrupt phrases of Robin-like character."

Handsomely attired in a gray suit, black head, and rusty-orange breast, the Robin has finer details that are often overlooked. But take a closer look and you will see the broken white eye-ring, the finely streaked chin, and the white belly. In flight, white tips on the outer tail feathers are always apparent. The female is a little duller in coloration than the male, particularly on the head and, as even the most casual observer knows, the young are easily identified by their spotted breasts.

Like most members of the thrush family, the Robin sings its familiar *cheerily cheer-up* song both in the early morning hours just before sunrise and in the evening just after sunset. But its repertoire is considerably more diverse and includes a wide range of calls and notes delivered when the birds are disturbed, as well as a high-pitched, "ethereal" note or series of notes, known as the Whisper Song (Vanderhoff et al. 2020). Phonetically akin to *hisselly,* these notes may be interspersed between phrases of its common song or delivered at the end in song-like phrases (Bent 1949; Vanderhoff et al. 2020).

Named by early American settlers who thought the bird's orange breast resembled that of the much smaller, unrelated European Robin, the American Robin has a breeding distribution extending across most of North America, from the tree line south to the central interior of Mexico. Some of its highest breeding densities are found in southern Minnesota and Iowa east to Pennsylvania and southern New England.

MINNESOTA BREEDING DISTRIBUTION

In Minnesota, the American Robin has been considered an abundant breeding resident throughout the state since the late 1800s (Roberts 1932). Describing it as one of the "most beloved" of all Minnesota birds, Roberts commented that the Robin "has thoroughly adjusted itself to civilized conditions and has come to prefer our lawns, gardens, and orchards to the wilder places. It is, however, not unknown in the depths of the northern forests where it seems strangely out of place." The species' abundance in the western grasslands was less clear. All but one of the 17 counties where confirmed or inferred nesting reports were available were confined to the forested regions of the state and the northwest Aspen Parklands. Abundant during the summer's breeding season, even during the early 1900s some Robins were known to overwinter, especially in southern Minnesota.

In 1975, Green and Janssen reported that the species was a summer resident "throughout the state, from the prairie to the dense northern forests." Janssen (1987) included a distribution map in his updated account that identified 52 counties where nesting had been confirmed since 1970. By 1998, nesting had been documented in 20 additional counties (Hertzel and Janssen 1998). The MBS further confirmed the Robin's statewide abundance with a total of 2,466 breeding-season locations.

MNBBA participants reported a total of 10,227 American Robin records in 85.3 percent (4,213/4,940)

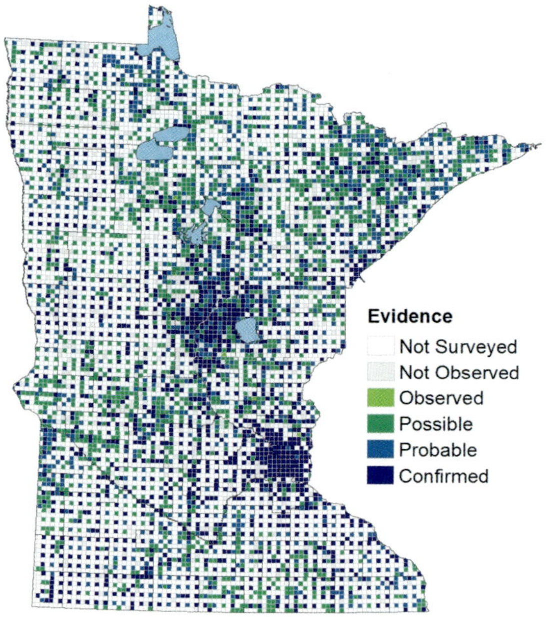

A. Breeding distribution of the American Robin in Minnesota based on the Breeding Bird Atlas (2009–13).

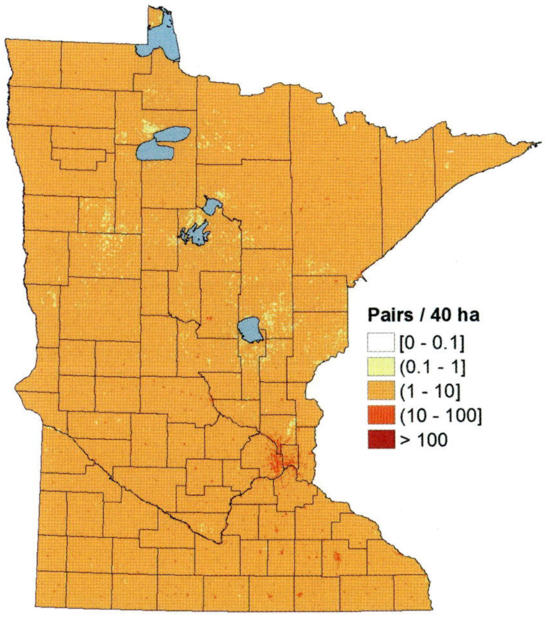

B. Predicted breeding distribution (pairs per 40 ha) of the American Robin in Minnesota based on habitat, landscape context, and climate data gathered during the Minnesota Breeding Bird Atlas (2009–13) using the Generalized Linear Modeling method with an adjustment for detectability.

of the surveyed atlas blocks and in 97.4 percent (2,276/2,337) of the priority blocks, leading all other species in the number of detections. Breeding was confirmed in 30.5 percent (1,509/4,940) of the blocks **(A)**. The birds were observed and confirmed breeding in all 87 counties and were distributed across all of Minnesota's ecological provinces. Intensive survey efforts by atlas participants in the Twin Cities and Brainerd Lakes region were especially effective at demonstrating the species' ubiquitous distribution.

The MNBBA probability map **(B)** depicts moderate densities of breeding birds are found, with few exceptions, statewide. Higher breeding densities are invariably associated with metropolitan areas and small communities, especially in southeastern Minnesota.

Roberts's simple statement nearly one hundred years ago that the Robin was abundant throughout the state clearly remains true today. And although one of his colleagues, T. Martin Trippe, stated "it avoids the tamarack regions," the Robin's MNBBA map makes it difficult to believe that the bird avoids anything other than open water! The same wide distribution and abundance is seen in states throughout the Upper Midwest (e.g., Cutright et al. 2006; Chartier et al. 2013; Rodewald et al. 2016). In many regions, the species' distribution has expanded with the establishment of tree plantings and parks, residential development, irrigation of crop lands, and, in many western states, with the introduction of earthworms (Vanderhoff et al. 2020).

BREEDING HABITAT

Originally a forest species, the American Robin is a habitat generalist that shows a preference for early successional forest stands created by timber harvest, windstorms, or fire. Today, developed landscapes—including residential lawns, city parks, golf courses, and cemeteries—provide ideal habitat **(C)**. Open ground for foraging and trees and shrubs for nesting are prime features of suitable habitats; the vegetative composition and age of a site are less critical (Danz et al. 2007). The National Forest Bird Monitoring Program in northern Wisconsin and Minnesota documented the species' utilization of virtually every habitat that was available, from mature upland stands of aspen–birch and lowland black spruce–tamarack to regenerating stands, open wetlands, and small towns (Niemi, Howe, et al. 2016). Although Robins used a diversity of habitats in the forests, they were most strongly associated with small residential areas.

POPULATION ABUNDANCE

BBS data helped to generate a population estimate of 370 million breeding adults in the United States and Canada, with approximately 6.9 million residing in Minnesota (PIF 2021). Unlike most estimates using MNBBA data, the MNBBA population estimate for the American Robin is considerably lower, at approximately 3.5 million adults (95 percent confidence interval of 3.4–3.7 million). This estimate likely reflects the species' lower abundance in off-road areas surveyed by the MNBBA.

Although the long-term population trend for Robins throughout the BBS survey region has been nearly stable since 1966, numbers increased markedly through the 1980s and then began declining

C. Among the wide variety of habitats where American Robins can be found are the manicured lawns and scattered trees and shrubs that are typical of residential neighborhoods. Photo by Lee A. Pfannmuller.

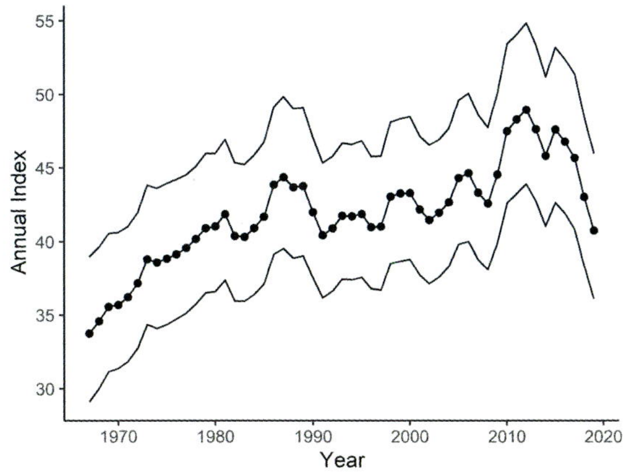

D. Breeding population trend and 95 percent credible interval (solid black lines) for the American Robin in Minnesota based on the 1967–2019 Breeding Bird Survey (Sauer et al. 2020).

in the early 2000s (Sauer et al. 2020). The stable trend also masks wide regional differences. From the northern Great Plains of Canada and the United States, east through Ohio, Kentucky, and Tennessee, Robin populations are increasing. Most declining populations are in the western states and provinces, the northern Great Lakes region, and along the north Atlantic coast. Factors responsible for these regional differences are unknown. Predation, disturbance, and inclement weather are considered the primary culprits impacting local breeding populations (Vanderhoff et al. 2020). Minnesota is among the states with a positive trend, with a significant average increase of 0.36 percent per year from 1967 to 2019 **(D)**.

CONSERVATION

Widespread and abundant, the adaptable American Robin is a low conservation priority (PIF 2021). It is not, however, without some challenges. Although distributional changes are not predicted in Minnesota, under a 2°C warming scenario the species is deemed moderately vulnerable to climate change, with a projected loss of 13 percent of its breeding range by 2050 and a gain of only 6 percent to the north (NAS 2019). Another concern is the Robin's susceptibility to pesticides and other contaminants. Notable declines in the 1950s and early 1960s stemming from DDT poisoning thrust it into the limelight following the publication of Rachel Carson's *Silent Spring* (Cutright et al. 2006). Populations did recover, but the Robin remains heavily dependent on earthworms found in residential lawns and parklands and on fruits that are taken from well-managed orchards (Vanderhoff et al. 2020). Given its continued contaminant exposure and its vulnerability to warming temperatures, the Robin's ubiquitous presence should not be taken for granted.

HOUSE SPARROW

Passer domesticus

Minnesota Seasonal Status: Regular permanent resident. The House Sparrow was a common species during the MNBBA.
Migration: Sedentary species with no migratory movements in North America.
Food: Ground forager that feeds primarily on grains but supplements its diet during the breeding season with arthropods; in urban settings, it readily consumes waste food items and frequently visits bird feeders.
Nest: Nest sites are quite variable, although the female prefers a site that is enclosed. Nest boxes, holes in buildings, rain gutters, dense vertical growths of ivy on buildings, and just about any hole in any man-made structure will suffice. The birds resort to constructing nests in trees and shrubs only when other options are limited or unavailable.
Conservation Concern: Introduced to North America in 1852, this unregulated non-native species is not a conservation concern; local populations are often managed to reduce competition for nesting cavities with native species.

Photo by Gerald Hoekstra.

Despised by many bird-lovers, the common House Sparrow has an adaptability and tolerance of human-modified landscapes that have enabled it to thrive throughout much of the Western Hemisphere. It is so abundant in our cities and towns that we hardly take notice of its aggressive behavior, social hierarchies, and curious nesting abodes.

In the 1800s European immigrants were zealous in their efforts to introduce the House Sparrow, originally native to Eurasia, to the United States, eager to bring familiar elements of their homeland back to America. Some also believed the bird would be a more effective predator of the insects that plagued their neighborhood's shade trees than our native species (Roberts 1932). As mistaken as this notion was, small flocks were introduced in at least one hundred cities in thirty-three different states and Canadian provinces in the latter half of the nineteenth century. The first effort was in 1850, when eight pairs were imported from England by the Brooklyn Institute in New York City. Although they did not fare well when released, a much larger number of birds was imported in 1852. Aided by numerous introductions across New England and Pennsylvania, and the additional release of several hundred pairs in New York City, the birds had successfully established themselves throughout the north-central United States by 1886. Introductions in Salt Lake City, San Francisco, and Texas also facilitated their western expansion (from Barrows 1889 as reported by Roberts 1932). By 1940, Kalmbach wrote: "At present the range of the English Sparrow in North America covers the entire continental United States except Alaska, all thickly settled parts of the contiguous Canadian Provinces, and similar areas in Mexico."

Multiple introductions certainly hastened the Sparrow's journey, but equally important was its dietary dependence on grains. Its progression west followed major roads and railways, where waste grain was dropped by vehicles, horses, and railroad cars loaded with grain or cattle. Cities and towns along these lines were often the first sites where small, local breeding populations became established. These then became the source populations for further expansion out to surrounding rural towns and farmsteads (Bent 1958).

In this sexually dimorphic species, the male House Sparrow is a handsome bird with a gray crown, rufous nape, silver cheeks, and a black bib. The underbelly is gray, and the wings and back are streaked with brown, black, and gray. The size of the male's black bib indicates his age and subsequent dominance over other males in the flock; this is an important social clue in a species that is very gregarious throughout the year (CLO 2017a). By contrast, the female House Sparrow is rather dull in appearance, with a grayish-brown belly, a light brown crown, and wings and back striped in brown, black, and gray.

Unrelated to North America's sparrows, which all have quite lovely, albeit short songs, the House Sparrow has a single *cheep*, vocalized primarily by the male, which is as complex as its song comes. Bent (1958) wrote: "Little can be said in favor of the [House] sparrow's voice."

Not long after their introduction, enthusiasm for the Sparrow waned considerably. Bent (1958) considered them "Dirty, soot-begrimed street gamins" and Roberts (1932) wrote: "on the whole it is an undesirable and harmful bird, not only doing comparatively little good itself but driving away and taking the rightful place of native birds of much greater value." Their aggressive behavior in forcing other small cavity-nesting birds like Purple Martins, House Wrens, Eastern Bluebirds, and Tree Swallows from nest boxes continues to vex many a homeowner to this day.

MINNESOTA BREEDING DISTRIBUTION

Roberts (1932) first observed a flock of House Sparrows in Minneapolis in the fall of 1876 only to discover later that a St. Paul resident had had a dozen birds brought from New York and released in his garden the previous summer, in June 1875. By 1877, the House Sparrow was established as a breeding bird in the Twin Cities. The species initially spread across the state by following the railroads, arriving in Fillmore County in 1886, in Duluth in 1887, along the Iron Range in 1892, and in the Red River valley about the same time.

By 1932, when Roberts published his treatise on Minnesota birds, the House Sparrow was already "an abundant permanent resident, breeding throughout the state." By 1975, Green and Janssen provided further clarity on its statewide distribution, noting its presence wherever there were towns or farms. It did not occur, however, in the extensively forested landscape of northern Minnesota. A few years later, Janssen (1987) identified 27 counties where the species had been confirmed nesting since 1970. Hertzel and Janssen (1998) later added an additional 11 counties to the list.

MNBBA participants tallied a total of 3,675 House Sparrow detections from 41.0 percent (1,979/4,831) of the surveyed atlas blocks and from 51.9 percent (1,212/2,337) of the priority blocks. Breeding was confirmed in 822 blocks (17.0 percent) **(A)**. House Sparrows were reported from every county except Cook. Breeding was confirmed in all but three counties: Cook, Koochiching, and Lake of the Woods. The species remains uncommon throughout the most extensively forested landscapes of northeastern and far north-central Minnesota.

The MNBBA probability distribution map predicts the highest breeding densities are associated with the largest human population centers in southern and western Minnesota **(B)**. The Twin Cities region, Rochester, Mankato, and St. Cloud are among the sites of high predicted House Sparrow densities. Breeding

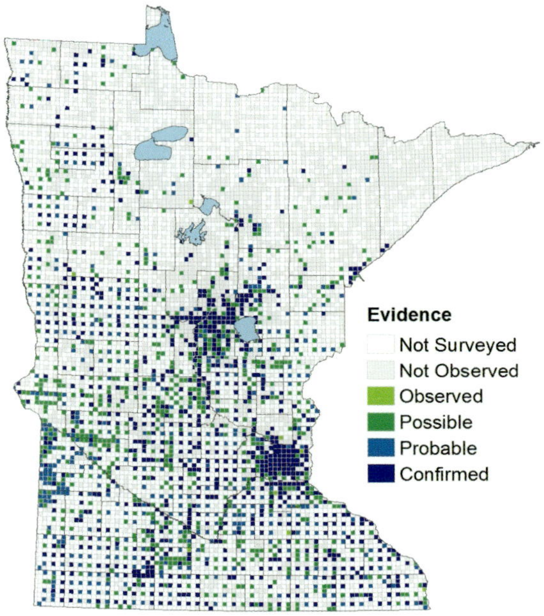

A. Breeding distribution of the House Sparrow in Minnesota based on the Breeding Bird Atlas (2009–13).

C. The House Sparrow is found in a variety of human-modified habitats; farmsteads provide ample nesting and foraging opportunities. Photo by Lee A. Pfannmuller.

densities are moderate throughout the southern half of Minnesota and north through the southern Red River valley. Densities are predicted to be extremely low throughout the Laurentian Mixed Forest Province, with the exception of the Iron Range in St. Louis and Itasca Counties and the cities and smaller communities that dot the region's landscape.

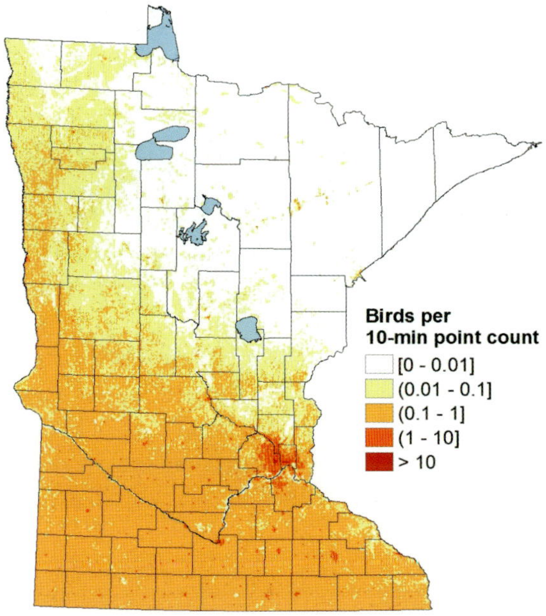

B. Predicted breeding distribution (birds detected per 10-minute point count) of the House Sparrow in Minnesota based on habitat, landscape context, and climate data gathered during the Minnesota Breeding Bird Atlas (2009–13) using the Generalized Linear Modeling method without an adjustment for detectability.

BREEDING HABITAT

The House Sparrow is largely an inhabitant of cities, towns, and farms. It is never found in natural settings, such as forests and grasslands, but can always be found near buildings, ranging from farmsteads in rural areas **(C)** to skyscrapers in the middle of major cities.

POPULATION ABUNDANCE

BBS data helped to generate a population estimate of 93 million birds in the United States and Canada; this represents approximately 13 percent of the estimated global population of 740 million birds (PIF 2021). Minnesota supports nearly 1.6 million House Sparrows.

At the beginning of the twentieth century, House Sparrows were considered the most abundant bird in the United States. Its numbers have declined markedly since then (Bent 1958). The decline was particularly noticeable in urban areas in the early 1900s, where it was attributed to the decrease in horses when motor vehicles became the favored mode of transportation. Even in rural areas, increased industrialization was considered a factor. Horse manure originally provided a "bountiful food supply" of waste grain along streets and highways and in farm pastures (Bent 1958). Although these changes occurred decades ago, the House Sparrow population has continued to decline. Since 1966, when the BBS began monitoring landbirds, the House Sparrow has experienced a significant long-term decline, averaging −2.95 percent per year from 1966 to 2019 (Sauer et al. 2020). The decline is evident throughout most of Canada and the United States. In Minnesota the decline has been nearly identical, with a significant annual decline averaging −3.01 percent **(D)**.

Recent declines are believed to be the result of more intensive agricultural practices. Increasing applications of pesticides reduce insect populations, which can lower food availability for nestlings, while more efficient farming practices reduce the availability of waste grain. Both are changes that are bound to impact House Sparrow numbers in agricultural landscapes (Lowther and Cink 2020).

CONSERVATION

As an introduced, non-native species, the House Sparrow is not a conservation concern and is not protected by the Federal Migratory Bird Treaty Act. More often, House Sparrows are the focus of management efforts to reduce their interference with nesting efforts by other species that have endeared themselves to local residents, such as House Wrens, Tree Swallows, and Eastern Bluebirds. Their aggressive takeover of nest boxes inhabited by these other species is legendary.

Warming temperatures, considered a threat for many North American birds, are unlikely to pose any problems for the House Sparrow (NAS 2019). Despite its decline, this adaptable species, which thrives in developed landscapes, is likely to remain a common breeding bird in Minnesota for many years.

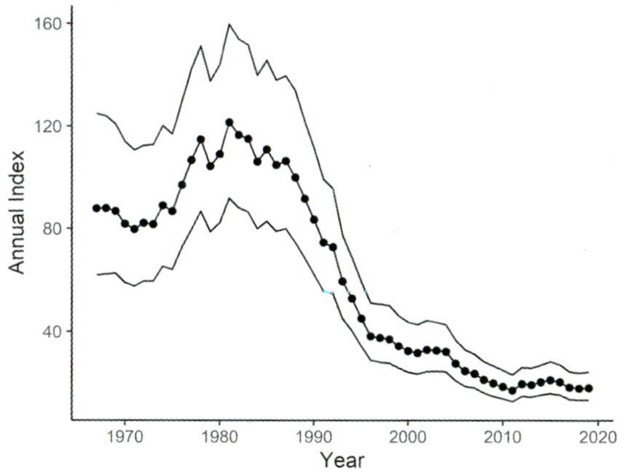

D. Breeding population trend and 95 percent credible interval (solid black lines) for the House Sparrow in Minnesota based on the 1967–2019 Breeding Bird Survey (Sauer et al. 2020).

SPRAGUE'S PIPIT

Anthus spragueii

Minnesota Seasonal Status: Summer visitant; bred in northwestern Minnesota from the late 1800s to the late 1900s. Migratory visitant in the northwestern counties; occasional elsewhere. The Sprague's Pipit was very rare during the MNBBA.
Migration: Short- to medium-distance migrant that winters in Mexico's Chihuahuan Desert and from southern Arizona east to southern Arkansas and western Louisiana.
Food: Ground forager that consumes primarily arthropods, including grasshoppers, crickets, beetles, and moths; also consumes seeds in the nonbreeding season.
Nest: Built on the ground near a clump of grass, often with a grass canopy.
Conservation Concern: Assigned a high PIF Continental Concern Score of 14/20; classified as *Endangered* in Minnesota. Classified as *Vulnerable* on the IUCN Red List (IUCN 2021).

Photo by Bradley Hacker.

The expedition to the Missouri River that John James Audubon organized in 1843 included Isaac Sprague as "Assistant Artist." On June 18, 1843, working out of Fort Union near the junction of the Missouri and Yellowstone Rivers, two hunters on the trip collected a "wee brown bird" that Audubon recognized as being new to the nation's avifauna. He named it Sprague's Missouri Lark. "On 24 June Sprague himself went out to search for the new pipits and found a nest, built of grass in a depression in the ground and containing five eggs" (Mearns and Mearns 1992).

A small, slender bird, Sprague's Pipit spends most of its time on the ground, where it "disappears completely in the prairie grass, walks or runs nimbly away without showing itself and, if flushed, flies quickly off" (Roberts 1932). Subtle features include its small head, thin bill, scaly brown plumage on the back and wings, and white outer tail feathers. Once in the air, however, the male Pipit fills the prairie sky with a remarkable aerial performance, flying "higher and higher in ascending circles" until he becomes nearly invisible (Roberts 1932). Silent as he flies into the wind, he suddenly stops flapping his wings, holds them horizontally as he glides backward, and delivers his flight song. Then, just as quickly, he flies forward and repeats the performance all over again (S. K. Davis et al. 2020). Some have written high praise of the Pipit's unearthly melody (e.g., Coues 1874), which is often likened to the sound of ringing bells, while others have described it as "rather monotonous and commonplace from constant repetition" (e.g., Roberts 1932). Other common names, such as "Missouri Skylark" and "Prairie Skylark" (T. G. Pearson 1917), are a tribute to this aerial display.

A grassland endemic whose range is restricted to Alberta, Manitoba, Saskatchewan, Montana, and North Dakota, the Sprague's Pipit also occurs locally and rarely in South Dakota and northwestern Minnesota. The core of its breeding range is in the grasslands of southeastern Alberta.

MINNESOTA BREEDING DISTRIBUTION

Formerly a resident in northwestern Minnesota, the Sprague's Pipit was found on the native grasslands of the Glacial Lake Agassiz lakebed. Roberts (1932) cites records that confirm its presence in the late 1800s in Kittson and Marshall Counties. He also assumed from the unpublished recollections of observers in Lac qui Parle and Pipestone Counties that the species was "formerly a summer resident throughout the western prairie region of Minnesota." Roberts's checklists (1923–31) in the University of Minnesota Archives, however, do not support that assumption. Later trips to northwestern Minnesota made by staff of the Museum of Natural History (now the Bell Museum) in 1928, 1929, and 1930 found the species in Kittson, Marshall, Pennington, Polk, and Red Lake Counties. Checklists from the 1928 expedition document that Sprague's Pipits were found in these counties almost every day from June 10 to 24. Roberts (1932) reported it to be "one of the common birds of the valley, its tinkling song being heard high overhead everywhere." These trips also produced the first state breeding records: Kittson County, June 16, 1929 (nest with six eggs); and Marshall County, northeast of Warren, July 1930 ("young birds awing but still being fed" [Roberts 1932]). The farthest south that birds were recorded on these trips, however, was northern Wilkin County.

In the following decades, only a few additional observations were reported, all within the northern Red River valley. On June 24, 1933, a nest was found with five eggs in Pennington County (Woolsey 1933); and on June 18, 1937, "a young pipit out of the nest but yet unable to fly more than a few feet" was found near Warren (Rysgaard 1937a). The only other summer records were in 1946 near Averill, Clay County, and Foxhome, Wilkin County (Gunderson 1946).

No additional information was available until the 1960s, when displaying birds were found at the Felton Prairie in June 1961 and June 1962 (R. L. Huber 1961c; Pieper 1963b). In 1962, two nests were also discovered between Felton and Ulen, Clay County, on May 30 and June 9 (R. L. Huber 1962b). Felton soon became the "usual spot" for observing the Sprague's

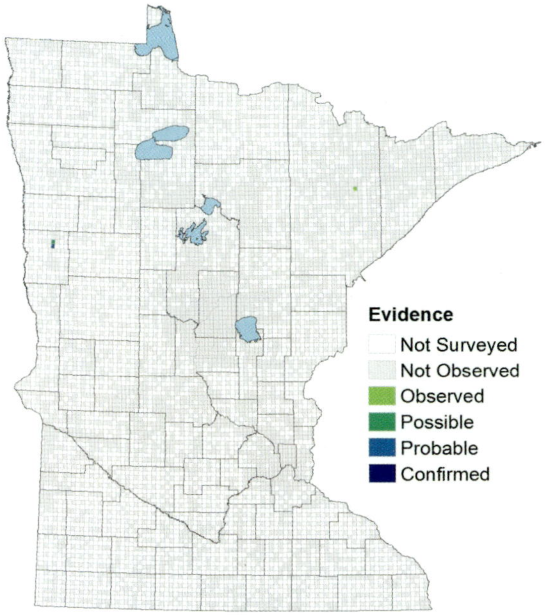

A. Breeding distribution of the Sprague's Pipit in Minnesota based on the Breeding Bird Atlas (2009–13).

Pipit. It was reported there almost every summer from 1961 to 1987, except 1977 and 1982. Unfortunately, little information was given on the number of birds present or where they were observed. This large prairie encompasses three townships in Clay County (Felton, Flowing, and Keene) and includes native grasslands, pastures, old gravel pits, and active gravel mines. Ownership is mixed and includes private lands and public units managed by the MNDNR, including the Bicentennial Unit of the Felton Prairie SNA, where one bird was heard calling in 1984 (Wyckoff 1985). After 1987 at Felton, there were only single records in 1992 (Svingen 1992) and 1994 (Wiens 1995), and then none until 2012.

The last nesting record for the state was in 1988 in Polk County in Tilden Township (Lambeth and Lambeth 1988). During the breeding season, occupancy was noted in only two other locations. In 1990 and 1991 several observations were made in Roseau County. The first were two displaying birds near Roseau, July 28–29, 1990 (Otnes 1990), followed by other observations that summer through September 1 (S. Carlson et al. 1991a). Again in 1991, two to three birds returned to the same site from April 27 to June 15, as reported by S. Carlson and colleagues (1991b), the MBS (S. L. Jones 2010), the MOU (2021b) and Wiens (1992). The MBS also reported a singing bird in Polk County in 1995 (S. P. Stucker 1996).

During the MNBBA, the Sprague's Pipit was reported in two locations **(A)**. The first was from Felton Prairie in 2012, where one bird was observed and heard displaying near the boundary of two adjacent blocks by multiple observers from June 8 to June 29 (one bird, two different locations; Chu 2013). A remarkable second record reported on June 27, 2013, by Steve Wilson was a "singing male, apparently plucked out of midair by one of a pair of Peregrine Falcons that arrived on the scene 30 minutes after the pipit was discovered." This observation was at a reclaimed taconite tailings basin (now an 11.6 square km grassland/wetland landscape) near Hoyt Lakes, St. Louis County (Tustison 2015b). All three MNBBA records are incidental observations and do not indicate a breeding population.

BREEDING HABITAT

The Sprague's Pipit breeds on dry grassland, preferentially on native prairie with a varied vegetative structure and a few shrubs **(B)**. It prefers an intermediate grass height compared to the surroundings and avoids areas with bare ground or lesser spikemoss ground cover. They "are likely influenced by the size of the grassland patches and the amount of grassland in the landscape" (S. L. Jones 2010). Croplands are avoided, and the birds are uncommon on non-native grasslands; they "prefer ungrazed to lightly grazed pastures" (A. R. Smith 1996).

POPULATION ABUNDANCE

The Sprague's Pipit's small population in the northern Great Plains numbers approximately 1.4 million birds (PIF 2021) and has been consistently declining for decades. Throughout the BBS survey region, BBS data document a significant decline, averaging −3.25 percent per year from 1966 to 2019 **(C)**. Indeed, between the years 1970 and 2014, the population experienced a cumulative decline of −75 percent (Rosenberg, Kennedy, et al. 2016). The most severe declines are in Canada, where the species is more abundant and where the population has experienced an average annual decline of −3.73 percent (Sauer et al. 2020).

B. Upland prairie remnants in northwestern Minnesota are one of the few habitats where the Sprague's Pipit may still be found. Photo by Lee A. Pfannmuller.

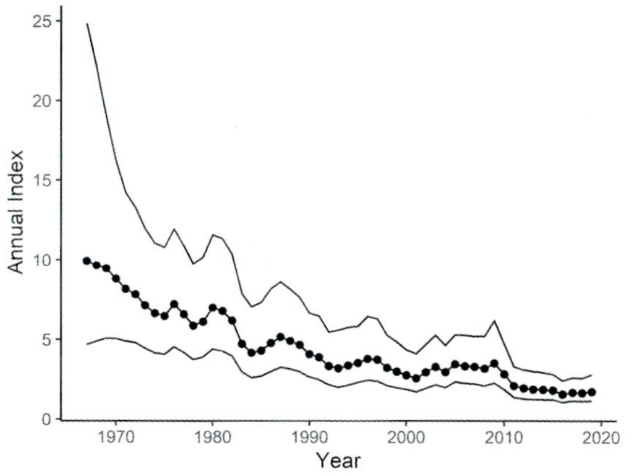

C. Surveywide breeding population trend and 95 percent credible interval (solid black lines) for the Sprague's Pipit based on the 1967–2019 Breeding Bird Survey (Sauer et al. 2020).

The limited range of the Sprague's Pipit has also contracted, especially in Canada, where it now occurs "several kilometers south" of where it was found in the late 1800s (S. K. Davis et al. 2020). In the United States, the range has contracted on the periphery "west and north in North Dakota and Minnesota and to the north in Montana. Data on South Dakota are inconclusive" (S. L. Jones 2010). Historically, much of the decline "occurred in the late 19th and early 20th centuries as the short- and mixed-grass prairies were converted to agriculture" (S. L. Jones 2010).

CONSERVATION

The species is *Threatened* in Canada (COSEWIC 2010); a petition for listing the Pipit under the US Endangered Species Act was submitted in 2008. Although the USFWS ruled "that the petition presented substantial information indicating that listing the Sprague's Pipit is warranted," it was placed on the candidate list because of higher listing priorities (S. L. Jones 2010). It is a high conservation priority in North America (PIF 2021) and was listed as *Endangered* in Minnesota in 1984 (Coffin and Pfannmuller 1988).

Habitat loss is a serious challenge for this grassland endemic, but climate change also poses a significant risk. It is moderately vulnerable under a 2°C warming scenario; NAS (2019) predicted it may lose 55 percent of its current breeding range by 2050 and gain 52 percent to the north. Projections are much grimmer under a 3°C warming scenario. The Prairie Skylark's future is clearly in jeopardy in North America.

EVENING GROSBEAK

Coccothraustes vespertinus

Minnesota Seasonal Status: Regular breeding resident, erratic migrant, and winter visitant in Minnesota. The Evening Grosbeak was uncommon during the MNBBA.
Migration: Permanent to semipermanent resident; nomadic.
Food: Invertebrates, including beetles, caterpillars such as spruce budworm, spiders, and aphids; also seeds and fruits of many trees (especially box elder), shrubs, and forbs, and commonly sunflower seeds at feeders.
Nest: Loose, saucer-like nest in trees or shrubs, most often in coniferous trees, but deciduous trees are also used.
Conservation Concern: Assigned a moderate PIF Continental Concern Score of 13/20. Classified as *Vulnerable* on the IUCN Red List (IUCN 2021).

Photo by David Brislance.

One of the most colorful and startling birds, the Evening Grosbeak is a large finch that can frequently be found during migration and in winter at bird feeders. Its gregarious behavior and appetite can quickly empty a bird feeder of its sunflower seeds. Unfortunately, the species has been declining in Minnesota and other parts of the eastern United States for many years. As described below, causes for the decline likely include many issues, such as changes in climate and breeding habitat, diseases, window and road kills, and changes in food supplies.

The vernacular name "Evening" Grosbeak is a misnomer. The original naming of the species by William Cooper in 1825 was derived from one of the first observations of the species by Major Delafield, a US agent for boundaries, who first met the bird in August 1823 near the Savannah River in north-central Minnesota. Delafield heard vocalizations of the bird but could not observe it until twilight, when it made its appearance close to his tent. His subsequent inference was "that this bird dwells in such dark retreats, and leaves them at the approach of night" (Bent 1968). However, it definitely does not vocalize or emerge only at night.

The species is one of the easiest to identify. Both sexes are husky birds with a large, conical bill. The male is a brightly colored yellow and black with a distinctive white wing patch and a bright yellow stripe over the eye. Female and immature birds are gray with black-and-white wings. Their call is a loud, distinctive *peer* that is often given in flight. Its genus name derives from the Greek *kokkus* for kernel and *thrauein* for shatter, thus "seed cracker." Its species name, *vespertinus,* maintains the evening quality of its common name, meaning "of or in the evening" in Latin (Sandrock and Prior 2014).

Evening Grosbeaks are patchily distributed in the Upper Midwest and in the northeastern and northwestern United States and Canada; extensions include southward into northern California and the Rocky Mountains to Arizona and New Mexico. The highest breeding densities are in the Northwest, including British Columbia and Washington State.

MINNESOTA BREEDING DISTRIBUTION

The Evening Grosbeak was described by Roberts (1932) as an enigmatic species that is "present throughout the year in the evergreen forests of Minnesota, most commonly and apparently regularly from Itasca and Koochiching counties eastward into Lake and Cook counties." He further pronounced it as "by no means evenly distributed throughout the area, occurring, usually a number together, at widely separated localities."

Roberts reported that no nests had been found, but he cited evidence of adults feeding young, observations of mated pairs, and birds captured with distinctive incubation patches. These observations ranged from June to August from 1900 to 1928 in Aitkin, Cook, Itasca, and Koochiching Counties.

More than forty years later, Green and Janssen (1975) emphasized that the Evening Grosbeak was a resident in the northeastern and north-central regions of Minnesota. They stated it was formerly "present only sparingly in the coniferous forest" but more numerous during the past twenty years. They also suggested evidence that its breeding range was expanding south and west in Minnesota, to where nesting was confirmed in Cass, Clearwater, and Crow Wing Counties, and to St. Louis County. Inferred nesting was included for Beltrami, Carlton, Cook, and Lake Counties.

Janssen in 1987 reinforced the range expansion of the Evening Grosbeak even farther westward in Minnesota. His breeding range map included its summer distribution reaching to eastern Becker County. Janssen updated confirmed nesting since 1970 and included seven counties: Aitkin, Becker, Clearwater, Crow Wing, Hubbard, Itasca, and St. Louis. In 1998, Hertzel and Janssen did not add any new confirmed county nesting records since 1970.

The MBS identified 114 breeding-season locations during their county surveys. They included two primary locations: one in northern Lake and St. Louis Counties, and a second concentration around the conjunction of eastern Becker, southeastern Clearwater, and northwestern Hubbard Counties. They also reported scattered locations in Aitkin, Carlton, Cook, southern Lake, Roseau, southern St. Louis, and Wadena Counties.

The MNBBA participants reported 247 records, which were strongly associated with the Laurentian Mixed Forest Province, especially in northeastern Minnesota **(A)**. A total of 165 blocks, or 3.5 percent of all blocks (165/4,734), included breeding-season records. Nesting was confirmed in 10 blocks, largely confined to Lake and St. Louis Counties, with one

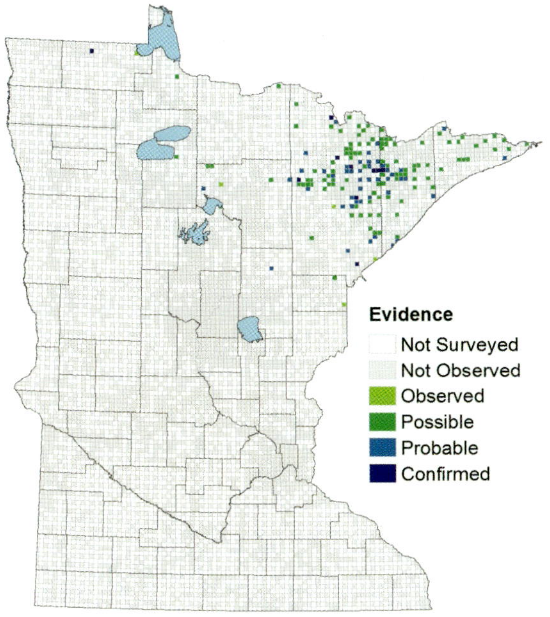

A. Breeding distribution of the Evening Grosbeak in Minnesota based on the Breeding Bird Atlas (2009–13).

in Roseau County. Probable and possible nesting behavior was also recorded in many northern counties of the state, most of which have been mentioned previously. Two exceptions were single observations of possible nesting in Koochiching and Lake of the Woods Counties in the extreme northern portions of Minnesota. In general, both of these counties receive scant coverage.

The probability map for the Evening Grosbeak predicts relatively low populations primarily confined to the northeastern and northern portions of Minnesota's forests **(B)**. Highest densities were predicted in north-central St. Louis County and in northern Cook County.

The lack of MNBBA records from the north-central counties, such as Becker, Cass, Clearwater, and Crow Wing, that were listed by Janssen (1987) and the MBS is symptomatic of the population decline and range retractions in North America over the past twenty years (Ralston et al. 2015; Niemi, Howe, et al. 2016). In their review of the Evening Grosbeak in North America, Gillihan and Byers (2020) describe the dynamic changes that have occurred in the distribution of this species in North America since the mid-1800s. Briefly, Evening Grosbeaks were a common species in the Rocky Mountains and western Canada until the mid-1800s and were considered rare east of the Rockies. They reported that minor irruptions began to occur in the 1850s in the Great Lakes, and major irruptions began to occur in the 1880s in eastern Canada and the northeastern states. The irruption in 1889–90 was particularly intense. The frequency and number of birds in irruptions began to increase in the early 1900s, which corresponded with the increase in breeding observations in Minnesota reported by Roberts.

As noted earlier, the first potential nesting observations were made by Roberts in the early 1900s. The first observation of adults with young in Wisconsin was not noted until the 1920s, and the first nest was found in 1964 (S. D. Robbins 1991). In Ontario, the first nest was not noted until 1944 (Cadman et al. 1987). All of these data appear to support the more "recent" arrival of this species to eastern North America.

This supposition has been questioned by Laura Erickson (2015), who, along with R. Brewer and colleagues (1991), noted that the type specimen of the Evening Grosbeak was collected in Sault Ste. Marie, Michigan, in April 1823. Erickson also pointed out that in 1940 James Bailley documented summer records of Evening Grosbeaks that formed an almost continuous belt from southeastern Manitoba to eastern Ontario. She also noted that the Ojibwe have a special name for the species, which implies they have possibly known the species for a considerable period. Adikwan (2019) writes that both the Pine Grosbeak and Evening Grosbeak were given the name *Zhingobiikens,* which means "little balsam bird," which comes from the Ojibwe name for balsam tree, *Zhingob*. Their association with balsam fir trees is likely because they frequently nest in coniferous trees, and balsam fir is the primary tree species attacked by spruce budworm, an important insect food source for Evening Grosbeaks.

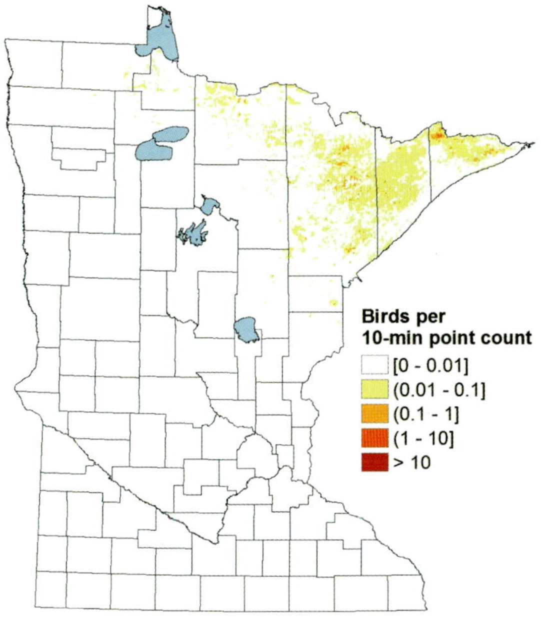

B. Predicted breeding distribution (birds detected per 10-minute point count) of the Evening Grosbeak in Minnesota based on habitat, landscape context, and climate data gathered during the Minnesota Breeding Bird Atlas (2009–13) using the Generalized Linear Modeling method without an adjustment for detectability.

Gillihan and Byers (2020) documented that the first breeding records in the New England region began to occur in the 1920s: 1926 in Vermont, 1937 in Massachusetts, 1946 in New York, 1962 in New Jersey, and 1983 in Connecticut. The number of wintering birds appeared to peak somewhere between the 1940s and the mid-1980s but has declined drastically since the mid-1980s. The increases in the Midwest and East have been attributed to the widespread planting of box elder trees as windbreaks and ornamentals, making their seeds available for consumption by Grosbeaks throughout the winter. Note that even though irruptions continued to occur in the species' western range, there was no similar expansion of its breeding range.

BREEDING HABITAT

Roberts (1932) stated that the Evening Grosbeak "has been found in the nesting-season in Minnesota only in the vicinity of dense spruce and arbor vitae swamps." Gillihan and Byers (2020) found that Evening Grosbeaks use coniferous forests and mixed coniferous–deciduous forests, while Takekawa and Garton (1984) emphasized their use of forests affected by spruce budworm, noting their economic value in these forests deriving from their insectivory.

Green (1995) suggested the species is conifer dependent but also has a food preference for ash (*Fraxinus* spp.). Niemi, Howe, and colleagues (2016), in extensive censuses in national forests of northern Minnesota and Wisconsin from May to early July, found Evening Grosbeaks most common in towns, but also in many coniferous cover types such as jack, red, and white pine, and mixed aspen–spruce–fir **(C)**. Their use of human-dominated environments scattered within the forests likely reflects post-breeding movements during summer into areas with suitable food sources, such as feeders and ornamental trees.

POPULATION ABUNDANCE

PIF (2021) estimated a US and Canadian population of 3.8 million breeding adults, with a Minnesota population of 21,000. Similarly, Environment Canada (2011) estimated a broader range of the adult population, between 500,000 and 5 million adults. The species was not abundant enough and was too variable to estimate its population size in Minnesota from the MNBBA.

C. In Minnesota, Evening Grosbeaks have a preference for coniferous forests, often mixed with deciduous trees such as black ash, aspen, and maple or, as shown, balsam fir that has been invaded by Eastern spruce budworm (*Christoneura fumiferana*). Photo by Gerald J. Niemi.

The BBS trend for the Evening Grosbeak in Minnesota was too unreliable. The BBS trend in the Boreal Hardwood Transition Bird Conservation Region, which includes northern Minnesota, showed a significant population decline from 1966 to 2019 averaging −5.02 percent per year **(D)**. Significant declines were also reported surveywide in the United States and Canada over the same time frame (−2.52 percent per year). Besides Minnesota, the species was too variable to reliably estimate a population trend in Manitoba, Michigan, Ontario, or Wisconsin. Rosenberg, Kennedy, and colleagues (2016) reported a drastic −94 percent decline in the population from 1970 to 2014. Population trends across North America indicated widespread declines in the population from 1966 to 2015 but also patches of increases in the population **(E)**.

Bonter and Harvey (2008) provided one of the few analyses of changes in winter populations for the Evening Grosbeak. They reported from Project FeederWatch (http://feederwatch.org) that the proportion of feeders reporting the species as present has declined by −50 percent, and mean flock sizes have declined by −27 percent from 1988 to 2006. Christmas bird counts conducted by the National Audubon Society since 1967 corroborate the downward trend (Meehan et al. 2022) **(F)**, resulting in fewer opportunities to observe the beauty of a winter flock of Evening Grosbeaks **(G)**.

In Minnesota, monitoring of breeding bird populations in the Superior National Forest in the northeastern corner of the state also showed a significant decline from 1995 to 2020, despite the species being highly variable from year to year among the 12 forest stands where it was reasonably detected (Grinde, Bednar, et al. 2020).

Environment Canada (2011) highlighted that the reasons for its population decline are not well known but suggested multiple causes, including large-scale forestry that has reduced the age structure of forests, diseases, and a reduced number of insect infestations (e.g., spruce budworm).

CONSERVATION

Because of the substantial decline in the population of Evening Grosbeaks over the past thirty years, the species has been designated a moderate conservation concern. As noted above, reasons for the decline have primarily focused on habitat, diseases, and the availability of food, but several other hypotheses have also been explored.

Habitat management to ensure an adequate availability of mature forests has been emphasized by McGarigal and McComb (1995) because of the species' breeding use of older forests. However, the evidence is equivocal. Gillihan and Byers (2020) reported that there was little evidence that logging or forest management had effects on Evening Grosbeak populations nor were there apparent effects of pesticide applications (e.g., of DDT) in western North America.

Bonter and Harvey (2008) suggested there are several diseases that may affect this species, including conjunctivitis (possibly caused by the bacterium *Mycoplasma gallisepticum*), salmonellosis (*Salmonella enterica* serotype Typhimurium), and West Nile virus. Although they noted that population changes are not necessarily related to diseases, they illustrated where the wintering population changes in this species have been most severe, such as in eastern and midwestern Canada and in the Rocky Mountains. The winter population declines have not been as severe in Minnesota in comparison with those areas. Klem (1989) reported that Evening Grosbeaks were the tenth most common bird killed at windows, which is often a hazard for species that frequent bird feeders.

Like many of the birds with northern Minnesota breeding distributions, the threat of a warming climate has consequences for the future of the species in the state. With a 2°C warming scenario, NAS (2019) projected a moderate vulnerability for the Evening Grosbeak's breeding distribution in North America, including a 55 percent range loss in the south and a 48 percent range gain primarily in the far north of Canada and in Alaska by 2050. In Minnesota, the species was predicted to be extirpated from the state in the 2°C warming scenario.

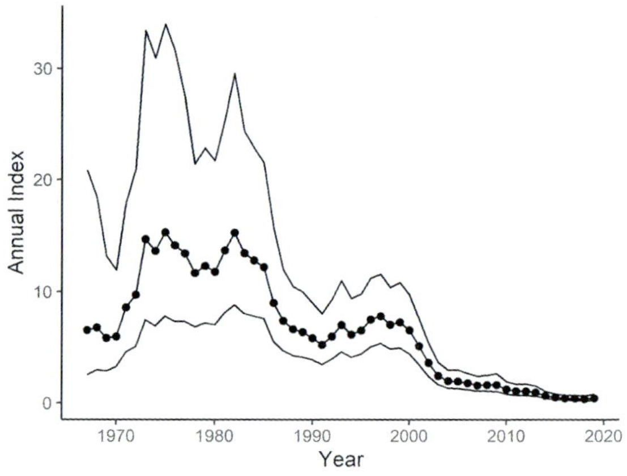

D. Breeding population trend and 95 percent credible interval (solid black lines) for the Evening Grosbeak in the Boreal Hardwood Transition Bird Conservation Region based on the 1967–2019 Breeding Bird Survey (Sauer et al. 2020).

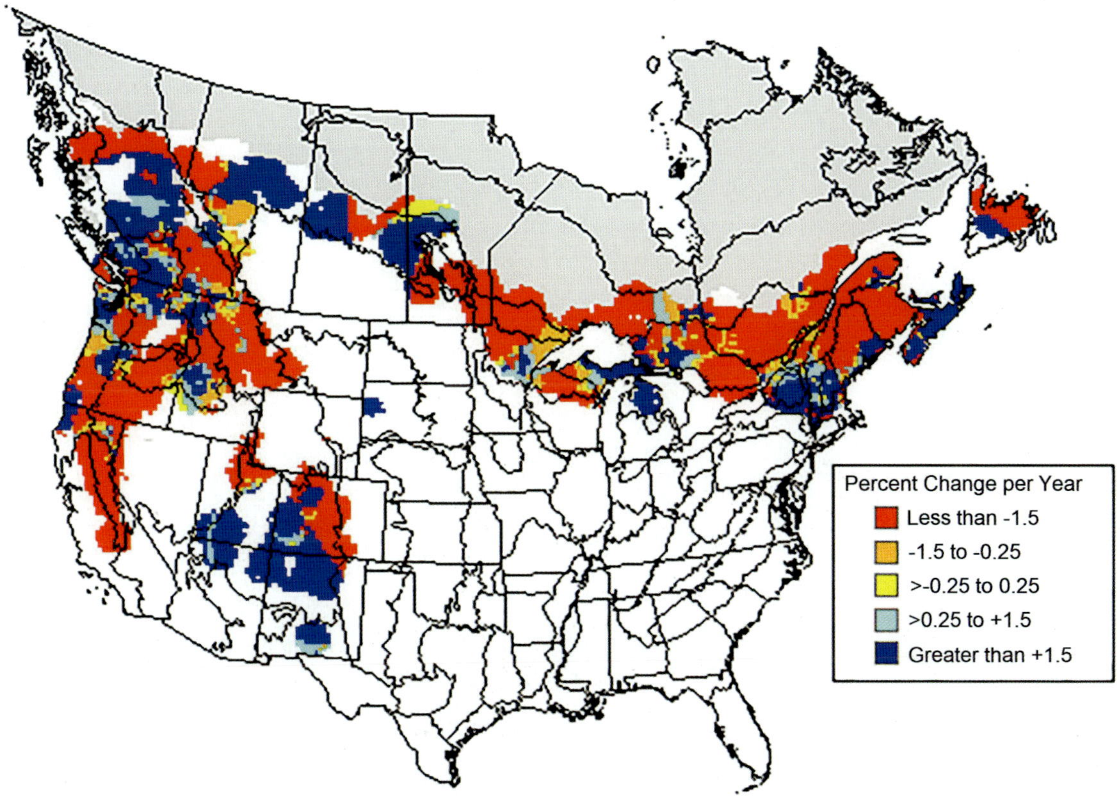

E. The pattern of population changes for the Evening Grosbeak in North America shows widespread population decline based on the 1966–2015 Breeding Bird Survey (Sauer et al. 2020).

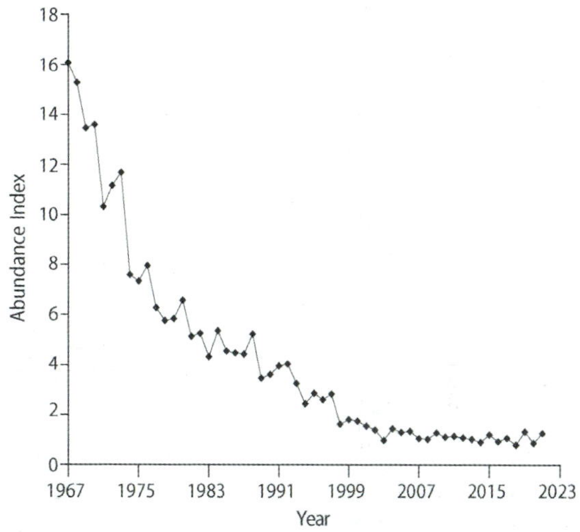

F. Surveywide population abundance indices for the Evening Grosbeak based on the 1967–2021 Christmas Bird Count (Meehan et al. 2022).

Laura Erickson (2015) noted that we do not know with any certainty what has caused the dramatic decline in the Evening Grosbeak's populations. She summarized contributing factors as, first, the reduction in spruce budworm populations; second, forestry practices that reduce the availability of seeds from maples, like box elder, which have been replaced by fast-growing softwood trees; third, exploitation of forest lands for mining, such as the Alberta tar sands; fourth, collisions with cars, because the species is drawn to roads to exploit salt and grit (2,000 were reportedly killed in a 16-km stretch in British Columbia [W. G. Smith 1981]); and, fifth, window-kill mortality, emphasized by Klem (1989).

The decline of the Evening Grosbeak is distressing to many who have enjoyed watching these gregarious and colorful birds gorge on seeds at bird feeders. While its decline is a combination of many factors, at this time its future remains a highly uncertain mystery.

G. Despite lower populations of Evening Grosbeaks in Minnesota, flocks of these colorful, gregarious birds can still be found in some places during winter, such as this group at the popular Sax-Zim Bog Welcome Center in northern Minnesota. Photo by Sparky Stensaas.

HOUSE FINCH

Haemorhous mexicanus

Minnesota Seasonal Status: Permanent resident, but some birds undergo a short, southward migration. The House Finch was common during the MNBBA.
Migration: The extent of seasonal movements is not well known.
Food: Ground forager and foliage gleaner that feeds almost entirely on plants, including seeds, buds, and fruit; a common visitor to bird feeders.
Nest: Cup nest placed in a variety of sites, including coniferous trees, hanging flower baskets, ivy, ledges, and buildings; occasionally uses old nests of other birds.
Conservation Concern: Assigned a low PIF Continental Concern Score of 6/20.

In 1932, Roberts added a footnote to his Purple Finch account that detailed all that was known about the House Finch. The only Minnesota record of this western species was a single bird shot in Minneapolis in 1876. The incident was documented by George G. Cantwell, a naturalist who spent several years in Minnesota in the late 1800s. Occurring far beyond its natural range, the bird was considered an anomaly. Today, nearly 150 years later, the adaptable House Finch can be found in developed landscapes throughout much of the state.

With its thick, conical bill, the male's bright red face and upper chest contrast with its brown back, wings, and heavily streaked belly. The Greek translation for its generic name, *Haemorhous*, is blood rump, a reference to the red rump that is visible when the bird is in flight. It is often confused with the male Purple Finch, but the latter bird lacks the male House Finch's heavy belly streaking and shows considerable red on the back of its head and along its entire back. The brown and heavily streaked females of the two species are more difficult to distinguish from each other, but the female House Finch lacks the distinctive brown cheek patch, bordered above and below by white, that is visible on the female Purple Finch.

Described by Bent (1968) as an "indefatigable songster," the House Finch's cheerful warble consists of a series of rapidly delivered syllables interspersed with buzzy slurs. Not only do songs vary among individuals but also one bird's song can include a wide repertoire of variations. Even females may occasionally sing a shorter, simpler song.

Photo by Mike Lentz.

Originally native to the southwestern United States and Mexico, the House Finch is now found across much of the United States and is continuing to expand north into Canada, especially in the western Provinces. Recent sightings have been reported as far north as southern Alaska (Badyaev et al. 2020). The species reaches its highest breeding densities in California.

MINNESOTA BREEDING DISTRIBUTION

The House Finch's rapid expansion across the United States is yet another story of the consequences of an unintended introduction and of a species' remarkable adaptability. It began in the early twentieth century, when there was a profitable trade in caged House Finches. Considered an economic pest in California's fruit orchards, the birds were not only trapped and sold locally but also shipped to pet dealers in the eastern United States, where they were advertised as "Hollywood Finches" or "Red-headed Linnets" (Elliott and Arbib 1953; Cant 1962). Unfortunately, due to the well-intended efforts of one private citizen who worked to end the illegal sale of songbirds, word spread among New York City pet dealers that their possession of Finches was illegal. Although the evidence is circumstantial, it seems that one or more of the dealers released the birds in the late 1930s to avoid prosecution. At the time, no one thought these West Coast birds would survive the cold, northern winters (Elliott and Arbib 1953). The rest, as the saying goes, is history.

Nearly one hundred years after the first report in 1876, House Finches were again reported in the Twin Cities in 1971 and 1972 (Freeland 1971; Arneson and Arneson 1972). The accounts, however, left open the possibility that the observers had seen the very similar Purple Finch. Several years passed before the first confirmed report was submitted in the fall of 1980. The description of two males, observed at a backyard feeder in Minneapolis, left no doubt that House Finches were seen (Bruggers and Bruggers 1981). Then, in December 1983, Breckenridge and his wife photographed a male at their feeder in Minneapolis (Breckenridge 1984). More reports soon followed from all corners of the state, including the first confirmed nesting in Rice County in 1989 (Janssen 1989). Janssen (1992) provided an excellent overview of the species' rapid range expansion in Minnesota. By the spring of 1992, he noted that House Finches had been observed in 80 of Minnesota's 87 counties. Counties lacking reports were confined to the northern regions of the state (Cook, Lake, Carlton, Itasca, Koochiching, Lake of the Woods, and Kittson). By 1998, nesting had been documented in 22 counties (Hertzel and Janssen 1998).

Fifteen years later, MNBBA participants reported 1,457 House Finch records from 18.2 percent of the surveyed atlas blocks (868/4,758), spanning all but three of Minnesota's 87 counties: Cook, Koochiching, and Pennington **(A)**. Breeding was confirmed in 251 blocks (5.3 percent) and 60 counties. Despite their broad distribution, records were concentrated in the Twin Cities metropolitan region. House Finches were least common in the Laurentian Mixed Forest Province, where they were found as far north as Lake of the Woods (confirmed breeding) and as

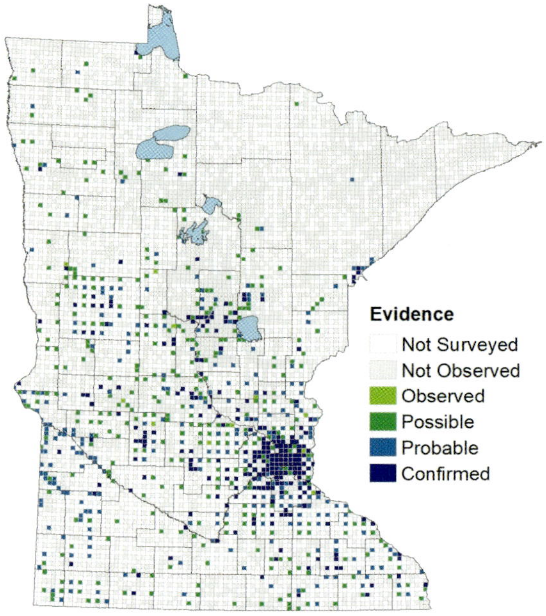

A. Breeding distribution of the House Finch in Minnesota based on the Breeding Bird Atlas (2009–13).

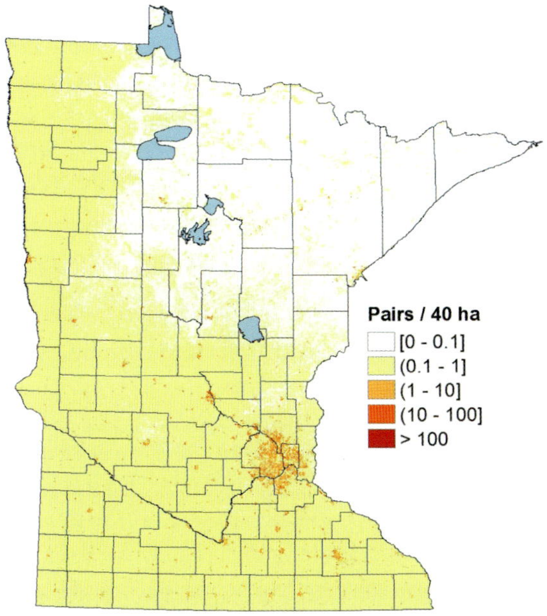

B. Predicted breeding distribution (pairs per 40 ha) of the House Finch in Minnesota based on habitat, landscape context, and climate data gathered during the Minnesota Breeding Bird Atlas (2009–13) using the Generalized Linear Modeling method with an adjustment for detectability.

far east as Two Harbors (probable breeding). The MNBBA probability model is a strong reflection of the species' current distribution, with moderate breeding densities restricted to the Twin Cities and smaller communities **(B)**.

Many species have undergone dramatic changes in the state since Roberts wrote his tome on Minnesota birds in 1932. Few, however, have so rapidly exploited suitable habitats across the state as the House Finch. From its first documented sighting in 1980 to the completion of the MNBBA thirty-three years later, the House Finch became a breeding resident in nearly all corners of the state!

Badyaev and colleagues (2020) provided a graphic timeline of the species' rapid westward expansion. In 1961 the House Finch was still largely confined to eastern New York; by 1981 it occupied much of New England and the Appalachian Mountains; and by 1990 it had reached the Great Plains. At the same time, native House Finch populations in the western United States expanded to the east, eventually merging with the introduced populations that were expanding westward. Badyaev and colleagues (2020) noted that "in just a few decades this species came to occupy one of the widest ecological ranges of any extant bird."

BREEDING HABITAT

In the southwestern United States and Mexico, the House Finch is associated with deserts, grasslands, shrublands, and other dry, open habitats. It is also found in areas as diverse as riparian draws, subalpine shrubs, and developed landscapes (Badyaev et al. 2020). In the East, however, the species is almost exclusively associated with urban areas, small towns, and farmsteads **(C)**. Weedy areas that provide food and trees or buildings that provide nesting sites are the only requirements. Although it avoids extensively forested landscapes, its eastern breeding habitats are "often associated with coniferous trees and evergreens" (Badyaev et al. 2020).

POPULATION ABUNDANCE

PIF (2021) projected a global population estimate of 40 million birds, with approximately 33 million residing in the United States and Canada. A total of 170,000 were estimated to occur in Minnesota. The more intensive sampling conducted by the MNBBA generated a statewide population twice the size of the PIF estimate at approximately 340,000 (95 percent confidence interval of 220,000–1.5 million).

Since population monitoring began in 1966, House Finches have actually demonstrated a highly variable population trend across the BBS survey region, with a nonsignificant decline averaging just −0.16 percent per year from 1966 to 2019 **(D)**. Their rapid expansion across eastern North America was offset by long-term population declines in California and Oregon, where approximately 30 percent of the North American population resides (Sauer et al. 2020; PIF 2021).

C. Restricted almost entirely to residential areas in the Midwest, the House Finch commonly makes its home in neighborhoods with a variety of nesting opportunities, including porch roofs, hanging flower baskets, and coniferous trees. Photo by Lee A. Pfannmuller.

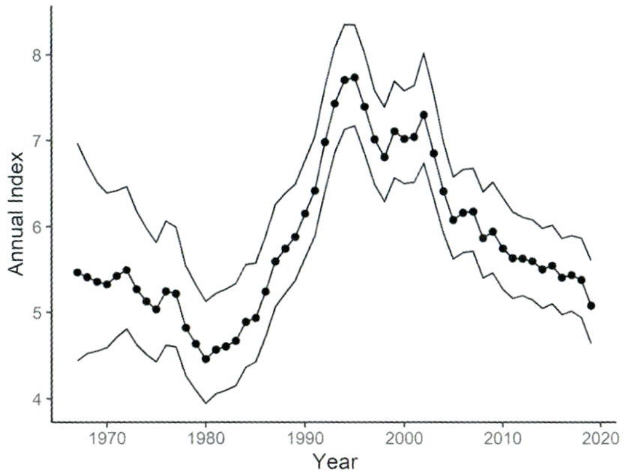

D. Surveywide breeding population trend and 95 percent credible interval (solid black lines) for the House Finch based on the 1967–2019 Breeding Bird Survey (Sauer et al. 2020).

The years of peak growth in eastern North America occurred from the early 1980s through the early 1990s. At that point, the species' expansion was halted by a new strain of *Mycoplasma gallisepticum*, a parasitic bacterium that affects poultry and pigeons. In House Finches, the bacterium causes red, swollen, and watery eyes that can crust over and result in vision loss. Birds with serious infections are more vulnerable to predators and are more likely to die owing to exposure and starvation (CLO 2017f). The outbreak initially resulted in the loss of millions of birds before some signs of disease resistance evolved.

In Minnesota, the House Finch's recent appearance and its low rate of detection along BBS routes result in a statistically unreliable trend estimate. Nevertheless, the species is increasing in every Bird Conservation Region and state east of the Great Plains (Sauer et al. 2020).

CONSERVATION

Given its rapid expansion and its adaptability to human development, the House Finch is not a conservation priority (PIF 2021). Even rising temperatures are projected to pose only small concerns under all three warming scenarios examined by NAS (2019). Overall, predicted losses to the species' breeding range are compensated for by even larger expansions northward, including in Minnesota (NAS 2019).

Although the House Finch's presence in the eastern United States resulted from a deliberate introduction, this native species might eventually have made its way eastward, albeit at a much slower rate without the aid of the pet trade. Unlike the European Starling and House Sparrow, the handsome male House Finch and his bubbly song have become welcome additions to the chorus of songbirds in residential areas throughout the state.

PURPLE FINCH

Haemorhous purpureus

Minnesota Seasonal Status: Regular breeding resident, migrant, and winter visitant. The Purple Finch was uncommon during the MNBBA.
Migration: Short-distance migrant that overwinters throughout the eastern, southeastern, and midwestern United States and southern regions of Nova Scotia, Quebec, and Ontario in Canada.
Food: Omnivorous, including buds, seeds, fruits, and arthropods from trees, shrubs, and on the ground.
Nest: Cup nest on a branch of a coniferous tree or a shrub.
Conservation Concern: Assigned a moderate PIF Continental Concern Score of 9/20.

The Purple Finch male is more raspberry-colored than purple, despite its common and scientific name, *purpureus*. It has similarities to the male House Finch, but its raspberry coloration is more extensive, especially on its breast and flanks. The female lacks the reddish color and can be confused with the Song Sparrow and female House Finch. She is best distinguished from both by two broad white stripes on the side of the head and heavy dark brown streaking. Both Purple Finch sexes have a chunky appearance, thick bill, and forked tail.

Roberts (1932) referred to this species as "a group of half-hardy" birds because most individuals migrate south of Minnesota but some will stay all winter. Therefore, some individuals can be common at bird feeders in winter, especially in the more southern regions or depending on the severity of the winter and the availability of food. The plumage of the male can be stunning against a snow-white background, rivaling any cardinal.

Purple Finches are one of the first species to sing in the spring. Forbush and May (1955) referred to it as "one of the most melodious of American finches. . . . He pours out his gushing, ecstatic warble from the top of some tree, . . . launches into the air, fluttering about with quivering wings in lowly emulation of the Skylark." Since the recent range expansion of the House Finch in Minnesota, there can be some confusion between the two species because their songs are similar. However, the House Finch is generally limited to urbanized habitats, while the Purple Finch retreats to northern forests during the breeding season.

The Purple Finch is found in low densities in the northeastern and upper midwestern United States,

Photo by David Brislance.

across Canada from the Maritime Provinces to the Pacific coast, north to northern British Columbia and northern Alberta, and south through California. Its highest densities are found in Nova Scotia and along the Pacific coast, especially in western Washington State.

MINNESOTA BREEDING DISTRIBUTION

The Purple Finch has historically been identified as "a common summer resident in the evergreen forests of the state as far south as northern Isanti and Pine counties and as far west as Itasca Park" (Roberts 1932). Primarily from 1900 to the 1930s, Roberts reported nesting activity in Aitkin, Cass, Cook, Isanti, Itasca, and Pine Counties, as well as near Cass Lake, Leech Lake, and Mille Lacs.

Green and Janssen (1975) reported a similar breeding distribution, primarily in northeastern and north-central Minnesota. They identified confirmed and inferred nesting from 15 counties in the northern and northeastern regions south to Isanti County, southwest to northeastern Stearns County, and west to Clearwater County. Janssen (1987) extended the breeding distribution farther south with a confirmed nesting in northern Anoka County, and northwest to Marshall County. He included a total of 10 counties with confirmed nesting since 1970; Hertzel and Janssen (1998) further expanded confirmed nesting since 1970 to include three additional counties.

The MBS included 438 breeding-season locations and indicated high concentrations in northeastern and north-central Minnesota. Scattered breeding-season locations also included Douglas, Otter Tail, and Todd Counties in west-central Minnesota.

The MNBBA recorded 979 breeding records that indicated a strong association of the Purple Finch with the Laurentian Mixed Forest Province **(A)**. Several observations were scattered in the Tallgrass Aspen Parklands Province and in the Eastern Broadleaf Forest Province. Confirmed nesting in the latter province was noted in Anoka and Washington Counties; the latter was the first confirmed nesting in Washington County.

A total of 692 blocks (14.6 percent) recorded confirmed, probable, or possible nesting activity. Besides Washington County, nesting was confirmed in 59 blocks, with new confirmed nesting in Carlton, Kittson, Morrison, Roseau, and Todd Counties. Probable nesting was also reported from Koochiching, Otter Tail, and Wadena Counties. Janssen (2019) added confirmed nests in Kanabec and Pennington Counties.

The breeding distribution of the Purple Finch in Minnesota as described by Roberts (1932) has not changed much since the late 1800s and early 1900s. Breeding populations have likely existed in the western and northwestern regions of Becker, Kittson, Marshall, and Otter Tail Counties where suitable forested areas exist. The Purple Finch sings earlier in the spring, such as in late April or in May, when few breeding bird counts have been initiated. Therefore, breeding activity is probably underestimated in many locations because song activity declines by June.

Wootton (2020), in his review of historical changes in the breeding distribution of the Purple Finch in North America, commented only on range contractions in northern Illinois, Indiana, and the southern Appalachian Mountains over the past eighty years. Changes in the breeding distribution

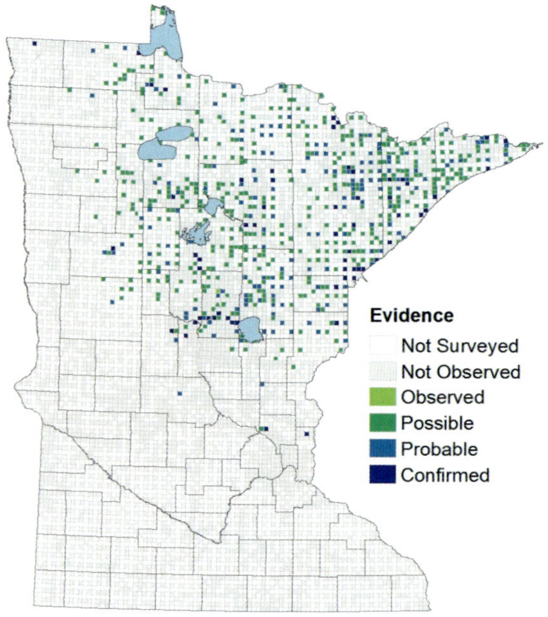

A. Breeding distribution of the Purple Finch in Minnesota based on the Breeding Bird Atlas (2009–13).

of the Purple Finch in neighboring states and provinces are equivocal. R. Brewer and colleagues (1991) suggested that the species had declined in Michigan since the early 1900s with the clearing of the forests, while Chartier and colleagues (2013) stated that it "appears that the distribution of Purple Finch has been consistent for 100 years" in Michigan. Previously, Cadman and colleagues (2007) found that "the overall distribution of the Purple Finch in Ontario does not appear to have changed much for as long as records have been kept."

Interestingly, Cutright and colleagues (2006) found confirmed nesting of Purple Finches in two southern Wisconsin counties, Sauk and Vernon. Vernon County is adjacent to Houston County, the most southeastern county in Minnesota. These confirmed observations suggest the species should be looked for in late April and May in suitable habitats of Minnesota's southeastern counties, such as among mature coniferous trees.

BREEDING HABITAT

The Purple Finch has generally been described as nesting in "moist or cool coniferous forests" and often in mixed coniferous–deciduous forest, conifers in parks, and residential areas (Niemi and Pfannmuller 1979) **(B)**. Niemi, Howe, and colleagues (2016) in the Chippewa and Superior National Forests found the species widely distributed in many habitats, including black spruce–tamarack bogs, aspen–spruce–fir, jack pine, edges of open wetlands, and frequently in small, rural towns. It was a widely distributed inhabitant of white cedar, tamarack, and black spruce forested bogs in the Agassiz Lowland Subsection, but usually in low abundance (Bednar et al. 2016).

POPULATION ABUNDANCE

PIF (2021) estimated the total population in the United States and Canada at 6.5 million breeding adults, with an estimated 73,000 adults in Minnesota. MNBBA estimates in Minnesota were substantially higher, at 374,000 adults (95 percent confidence interval of 292,000–504,000). The MNBBA estimate suggests that Minnesota has about 6 percent of the global breeding population of the Purple Finch. The higher estimate in Minnesota is likely due to methodological differences between PIF and the MNBBA and better coverage of the roadless regions in the northern portion of the state by the MNBBA. These roadless areas are not covered by the BBS routes that are used for estimating populations in North America.

The BBS trends for Minnesota were unreliable, but the Boreal Hardwood Transition Bird Conservation Region, which includes northern Minnesota, revealed a significant population decline that averaged −1.07 percent per year from 1966 to 2019 **(C)**. Rosenberg, Kennedy, and colleagues (2016) calculated a 47 percent decline in the species' global population from 1970 to 2014. In contrast, trends from the Chippewa and Superior National Forests combined in northern Minnesota indicated a significantly increasing trend from 1995 to 2020 (Grinde, Bednar, et al. 2020).

Several threats associated with population declines in the Purple Finch have been identified. Loss and colleagues (2014) stated that it was the most vulnerable species around residential homes, with a window collision risk 257 times greater than the average species. Interspecific competition with the House Finch has also been suggested because the House Finch tends to dominate Purple Finches at bird feeders, which results in lower winter survival rates (Wootton 2020). Cadman and colleagues (2007) suggested the recent reduction in spruce budworm populations, causing a loss of this food source for nesting finches, may be a factor in their decline.

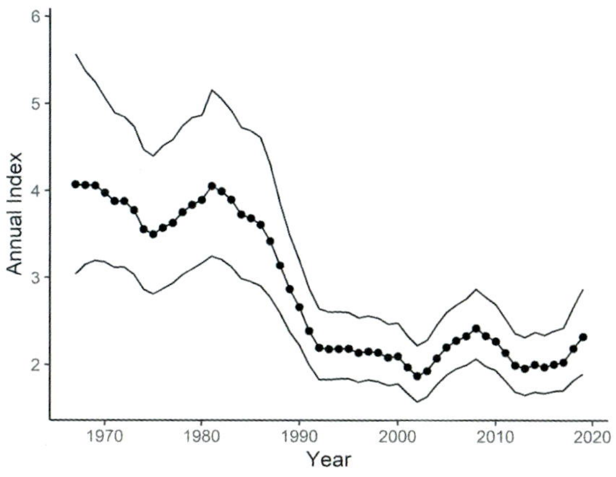

C. Breeding population trend and 95 percent credible interval (solid black lines) for the Purple Finch in the Boreal Hardwood Transition Bird Conservation Region based on the 1967–2019 Breeding Bird Survey (Sauer et al. 2020).

CONSERVATION

Overall, the population of the Purple Finch remains relatively large in North America, but many of the noted declines suggest continued monitoring and a close watch on population trends. Improvements in placement of bird feeders relative to windows can lower mortality for Purple Finches; controlling House Finches is not suggested, however, because they are a native species of North America.

With a 2°C warming scenario, NAS (2019) projected a low vulnerability for the Purple Finch's breeding range in North America. The projections included a 45 percent range loss but a 72 percent range gain into western Canada and Alaska by 2050. In Minnesota it was predicted to be extirpated under the 3°C warming scenario by 2080.

Therefore, the breeding distribution of the Purple Finch is likely to shift northward with climate change. Its winter use of feeders is compounded by mortality with window collisions and interactions with the House Finch. Unfortunately, the long-term future of the Purple Finch in Minnesota is a conundrum that cannot be viewed with much optimism.

B. Purple Finches in Minnesota prefer nesting habitat in mature forests with coniferous trees or a mixture of coniferous and deciduous trees. Photo by Gerald J. Niemi.

RED CROSSBILL

Loxia curvirostra

Photo by Gerald Hoekstra.

Minnesota Seasonal Status: Regular but rare and erratic summer resident, migrant, and winter visitor. The Red Crossbill was rare during the MNBBA.
Migration: Permanent resident, but highly nomadic.
Food: Seeds, primarily spruce and pine; also seeds of many species of deciduous trees and some invertebrates in summer.
Nest: Cup nest in coniferous trees. It is known to nest in any season of the year.
Conservation Concern: Assigned a low PIF Continental Concern Score of 8/20.

The Red Crossbill is one of the most unusual species in Minnesota. It is in the genus *Loxia,* meaning "crosswise"; its species name, *curvirostra,* is Latin for "bent bill." This unique and odd-looking bill structure is superb for removing seeds from conifer cones. The Red Crossbill is uncommon because it is a highly nomadic species that has been known to nest somewhere in any month of the year. Forbush and May (1955) refer to this behavior as "strange, erratic, and seemingly irresponsible." However, the availability of a suitable cone crop is the most important factor influencing when it nests.

The Red Crossbill is a medium-sized finch. The adult male has a brick red to a reddish-yellow plumage, with dark black-brownish flight feathers. In contrast, the female has a uniformly olive to grayish plumage, with a yellowish breast and rump. Red Crossbills have little plumage variation across their range but have considerable variation in body size, bill size and shape, and vocalizations. These variations have been considered as similar to the adaptive radiation of Darwin's finches in the Galapagos, with specialization on different species and cone sizes of conifers. Typical vocalizations include a two-note *kip-kip* call and a song of *too-tee too-tee too-tee* ending with a trill, which is often given from the top of a coniferous tree.

Currently there are 10 different "call types" or ecotypes that were previously referred to as subspecies in North America; each type is "characterized by distinct contact calls and other vocalizations" and appears "to be associated with, if not adapted to, particular conifer species" (Benkman and Young 2020). Regardless of their foraging adaptations, each type feeds on the seeds of multiple conifer species depending on their seasonal abundance. Four of the recognized types have been documented in Minnesota. Unfortunately, most of them cannot be easily distinguished by geographic range, voice, plumage, or morphology and, hence, require genetic evidence. In summary, Red Crossbills are a confusing species complex and only further study will reveal if additional ecotypes besides the Cassia Crossbill (*L. sinesciuris*) of southern Idaho will be elevated to species designation.

Red Crossbills are found throughout coniferous forests and mountainous regions of the Holarctic, including North America, Europe, and Asia. Their breeding distribution in North America is highly variable, with the highest recorded breeding densities in western mountainous regions of the United States and Canada, plus scattered populations in the upper Midwest, eastern Canada, and the northeastern United States.

MINNESOTA BREEDING DISTRIBUTION

Roberts (1932) stated that the Red Crossbill is a permanent resident of Minnesota's northern evergreen forests. He emphasized that "no nests have yet been reported from Minnesota though they certainly breed commonly in the northern part of the state." He described numerous observations of the species throughout the state from the late 1800s to the 1930s. For instance, in 1871, Thomas M. Trippe found the species "very abundant; breeds" in the east-central part of the state. In August 1879, Roberts found it common along the North Shore of Lake Superior from Duluth to Grand Marais, and in August 1922 the "Surbers found it generally distributed in Cook County," exclaiming crossbills to be present "literally in swarms" on the Caribou River.

Green and Janssen (1975) described the Red Crossbill as an irregular summer resident in the northern regions of Minnesota and its breeding distribution as basically the northern half of the state. They stated nesting had occurred as far south as Stillwater in Washington County and confirmed nesting from Clay County in 1967 (O. W. Johnson 1969). Inferred nesting was documented in Cook, Crow Wing, and St. Louis Counties.

Janssen (1987) highlighted the Red Crossbill's erratic nesting behavior from Clearwater and Pennington in the northwest to Cook and St. Louis in the northeast, and south to Ramsey in the metropolitan area of the Twin Cities. Further highlighting the species' seemingly vagabond nature, summer records existed from as far south as Cottonwood County in the southwest and Rice County in the south-central regions. He included confirmed nesting in Clearwater, Cook, Ramsey, and St. Louis Counties since 1970, while Hertzel and Janssen (1998) included only Cook, Ramsey, and St. Louis Counties since 1970 and eliminated Clearwater County. The MBS included 33 records of Red Crossbill breeding-season locations. The records were scattered between Clearwater County in the northwest, Otter Tail and Todd Counties in the west-central region, and Cook County in the northeast.

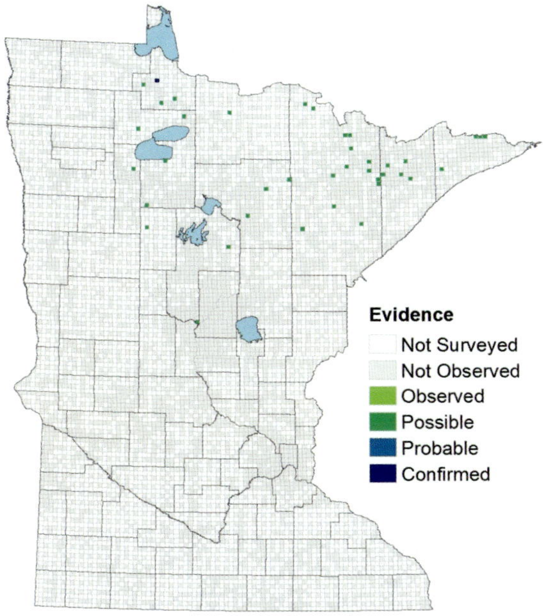

A. Breeding distribution of the Red Crossbill in Minnesota based on the Breeding Bird Atlas (2009–13).

The MNBBA included 44 records of the Red Crossbill that were distributed primarily in the northeastern and north-central regions of the state **(A)**. A total of 39 blocks had breeding records, including a new confirmed record for Lake of the Woods County. Possible nesting was also recorded in widely scattered locations, including four in Beltrami County, one record each in Hubbard and Koochiching Counties, three in Itasca County, 14 in St. Louis County, five in Lake County, and four in Cook County. One possible record also occurred in a block that straddled both Cass and Morrison Counties. The species had too few observations to produce a reliable probability distribution map. There were two recent confirmed records of fledged young: from Todd County in 2019 and St. Louis County in 2020 (MOU 2021a).

Changes in the historical distribution or abundance of the Red Crossbill are difficult to evaluate because of the lack of consistent data and its nomadic, erratic breeding behavior. In their review of this species in North America, Benkman and Young (2020) stated that there were no major changes in distribution in recent history, but regional changes in abundance have occurred. Historical nesting evidence in the southern portions of Minnesota is nonexistent. Historical data are similarly vague in Michigan (R. Brewer et al. 1991), in Ontario (Cadman et al. 2007), and in Wisconsin (Cutright et al. 2006). A definitive answer on whether the distribution or abundance of this species has changed in Minnesota remains elusive.

BREEDING HABITAT

The Red Crossbill is highly associated with mature coniferous forests, especially those dominated by spruce, pine, and hemlock (Benkman and Young 2020). Little specific data exists for Minnesota, but Green (1995) emphasized a seed preference for white and red pine. In the Chippewa and Superior National Forests, Niemi, Howe, and colleagues (2016) rarely recorded the species, but when it was observed in late May to early July, it was almost exclusively in mature coniferous habitats such as red pine, jack pine, and lowland black spruce–tamarack cover types **(B)**.

POPULATION ABUNDANCE

During their recent assessment on the conservation status of the Red Crossbill, PIF (2021) estimated a US and Canadian population of 9.6 million breeding adults, with a Minnesota population of 28,000. Detections were too few during the MNBBA to estimate a Minnesota breeding population. The BBS analyses for Minnesota and most other states or regions were too unreliable to estimate a trend from 1966 to 2019 (Sauer et al. 2020). However, the BBS trend in Canada was reasonable and indicated a stable population from 1966 to 2019 and especially from 1990 to 2019 **(C)**. Overall, Rosenberg, Kennedy, and colleagues (2016) estimated a population decline of –12 percent from 1970 to 2014 in the United States and Canada.

Reasons for a decline are likely associated with reductions or fragmentation of coniferous forests, especially old-growth trees. For instance, Benkman (1993) reported population reductions with the loss of old-growth forests in the northwestern United States because conifers greater than sixty years old produce the largest cone crops. Environment Canada (2014b) suggested that extensive loss of pine trees in British Columbia and Alberta due to the mountain pine beetle will likely influence future Red Crossbill abundance in the region.

B. Red Crossbill breeding habitats are mature coniferous forests, such as this pine forest in northern Minnesota. Crossbills can nest in any season of the year.
Photo by Gerald J. Niemi.

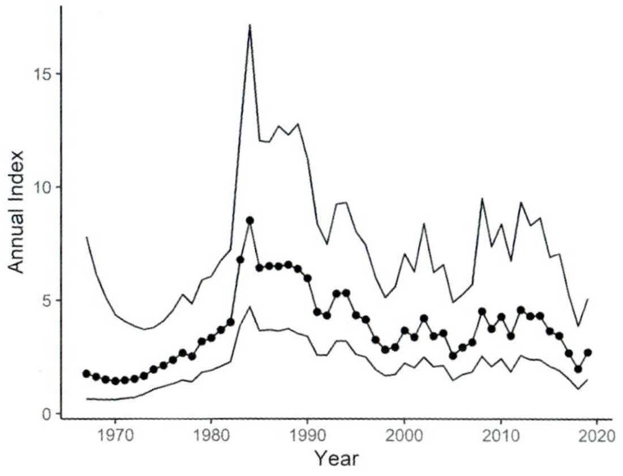

C. Breeding population trend and 95 percent credible interval (solid black lines) for the Red Crossbill in Canada based on the 1967–2019 Breeding Bird Survey (Sauer et al. 2020).

CONSERVATION

Despite apparent declines in its overall population in the United States and Canada, the Red Crossbill is of minimal conservation concern because it has a large breeding distribution and population across the entire Holarctic region. Benkman and Young (2020) stressed the need to maintain increased rotation times and to set aside core tracts of old-growth forest across the species' range in North America to adequately maintain populations of the various Red Crossbill ecotypes. NAS (2019) projected a moderate vulnerability for the Red Crossbill's breeding distribution in North America under a 2°C warming scenario, including a 40 percent range loss and a 20 percent range gain in the far north by 2050. In Minnesota they did not recognize a breeding population, so no predictions were noted.

Hahn (1995) emphasized that Red Crossbills have adapted to the extreme variability in its seed crop by evolving a nomadic wandering behavior, early onset of sexual maturity, and a long breeding season. The future of the Red Crossbill in Minnesota and throughout its range is dependent on the future of mature coniferous forests, especially with changes in climate and forest management. A better understanding of the genetics and interrelationships of its many ecotypes is a priority.

WHITE-WINGED CROSSBILL

Loxia leucoptera

Minnesota Seasonal Status: Regular but erratic breeding resident, migrant, and winter visitant. The White-winged Crossbill was rare during the MNBBA.
Migration: Permanent resident, highly nomadic.
Food: Seeds of spruce, tamarack, and occasionally balsam fir; invertebrates and seeds of many plant species in summer.
Nest: Cup nest, mostly in spruce tree away from trunk.
Conservation Concern: Assigned a low PIF Continental Concern Score of 6/20.

Photo by Mike Lentz.

The White-winged Crossbill is an appropriately named bird because, with its characteristic crossed mandibles and distinctive white wingbars, it is relatively easy to identify. Its scientific name in Greek means *leucos* (white) and *pteron* (wing). The species is a medium-sized finch. The male is red with a black tail and wings that include bold wingbars. The female is an olive-yellow with black streaks on its back and belly. They also have dark wings with white wingbars, but one wingbar is usually concealed. The species has been known to nest in every month of the year.

Like the Red Crossbill and many other birds, the White-winged Crossbill is best identified by its vocalizations. Its call note is a two- or three-note *cheet cheet,* usually heard in flight, while its song has been described as a "fine series of loud whistles, trills and twitters, suggesting that of Goldfinch" (Forbush and May 1955). The song can be heard from the top of a tree or as the birds fly about in large circles over the woods. S. D. Robbins (1991) points out the potential problem with its identification by stating, "Anyone who does not recognize the call note is bound to find the White-winged Crossbill rare." Nevertheless, the White-winged Crossbill is known to be almost motionless while feeding on cones on the tops of trees, and to vocalize only when leaving the tree where it was foraging. Hence, historically and today, the species may be easily overlooked.

The White-winged Crossbill has a more northerly distribution in North America compared with the Red Crossbill. Its breeding distribution is highly variable in Canada, the northern United States, and in scattered areas of the Rocky Mountains. Some of the highest recorded densities have been in northern Manitoba, southwestern Alberta, northeastern British Columbia, and Labrador. A subspecies (*L. l. bifasciata*) occurs throughout the Palearctic region from northern Scandinavia to Siberia.

MINNESOTA BREEDING DISTRIBUTION

Roberts (1932) documented its distribution as north and northeastward from northern Isanti County west to Wadena and Roseau Counties. In contrast to the Red Crossbill, Roberts never reported it during the summer in the southern counties of the state. He described no nests but stated that it "unquestionably breeds commonly in many localities." He included an observation by Thaddeus Surber of "parents feeding full-fledged young out of nest" on August 8, 1922, in Cook County. Several observers noted that it was "common in Itasca County"; "extremely abundant" from June 20 through July 1901, at Hibbing, St. Louis County; and "common on the Iron Range about Hibbing, Virginia, Biwabik, and Wolf" in the early 1900s.

More than forty years later, Green and Janssen (1975) still labeled it as "a very scarce resident in the northern coniferous forests." No nests had been found by the early 1970s, but fully fledged young being fed by adults were observed in Clearwater, Cook, and St. Louis Counties. After 1970, few data were added by Janssen (1987), except that the first summer record of the White-winged Crossbill in southern Minnesota was found at Luverne in Rock County on July 5, 1975. The first confirmed nest in the state was discovered by a young birder, Jacob Langeslag, in March 1996. Documented in detail by Little and Langeslag (1996), the nest was found, surprisingly, in Rice County in southeastern Minnesota. The second nesting record, also in 1996, was from Lincoln County (Schroeder 1996).

The MBS recorded breeding-season locations from six counties, including eastern Marshall and northwestern Hubbard Counties. The majority of observations were from Cook (seven locations), Itasca (more than 15), Lake (six), and St. Louis (six) Counties.

In contrast, the MNBBA had 27 records, including a new county record in Roseau County and only the third known nesting record for the state **(A)**. The MNBBA also recorded one probable nesting in Lake County and 25 possible nesting locations. These possible locations included three blocks in northern Beltrami County, five in Cook County, one in Itasca County, four in Koochiching County, seven in Lake County, two in Lake of the Woods County, two in

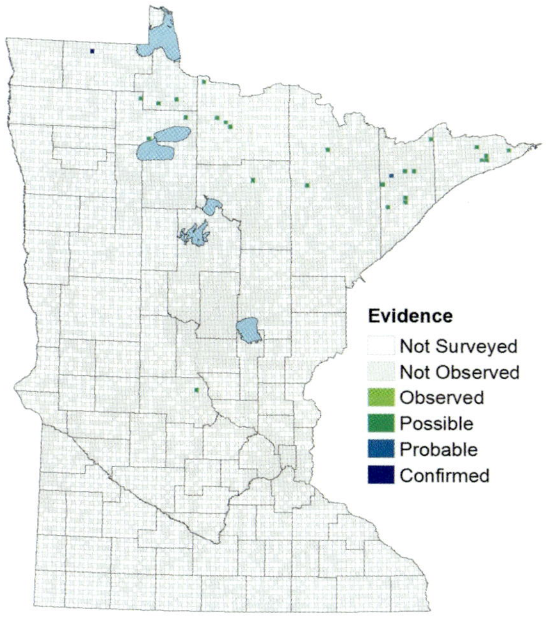

A. Breeding distribution of the White-winged Crossbill in Minnesota based on the Breeding Bird Atlas (2009–13).

St. Louis County, and one more southerly record in Stearns County. Another breeding confirmation of fledged young was recorded in St. Louis County on May 23, 2020 (MOU 2021a).

Benkman (2020), in his review of the White-winged Crossbill in North America, provided no perspective on historical changes to its North American distribution. This was mainly because of the lack of solid observational or nesting data. There is no basis to suggest that the distribution of this species has changed appreciably in Minnesota over the past 150 years.

S. D. Robbins (1991) cited one nesting record from Oconto County, Wisconsin, in 1894 and otherwise labeled the White-winged Crossbill as a "rare summer resident north." Cutright and colleagues (2006) had seven quads with confirmed nesting during the Wisconsin breeding bird atlas from 1995 to 2000, including one nest, the first in the twentieth century. Similar results were reported by R. Brewer and colleagues (1991) in Michigan, where there were two historical nesting records from 1891 and 1941 in the western Upper Peninsula. Michigan's atlas from 1983 to 1988 also revealed two nesting records from the western Upper Peninsula. Ontario's atlas reported that knowledge of the breeding distribution of the White-winged Crossbill was poorly known before its first atlas in 1981–85 (Cadman et al. 2007). They subsequently reported that the White-winged Crossbill had a broad distribution of observations from the northern regions of the province. During Ontario's second atlas, in 2001–5, they found the distribution had not changed, but substantially more evidence of breeding was identified. They attributed this to better identification skills by observers and to widespread and frequent irruptions that occurred during the second atlas period. Data from the second atlas in Ontario provided some evidence for problems with identification, especially by call notes while the species is in flight.

BREEDING HABITAT

In contrast with the Red Crossbill and its high association with pine, spruce, or hemlock, White-winged Crossbills are primarily associated with white and black spruce, tamarack, and balsam fir (Benkman 2020). In Minnesota, Green (1995) reinforced this by noting the species' preference for black and white spruce seeds. Data gathered in the Chippewa and Superior National Forests indicated a significant breeding habitat association with black spruce–tamarack forests and few observations outside this forest cover type (Niemi, Howe, et al. 2016) **(B)**.

POPULATION ABUNDANCE

PIF (2021) estimated a US and Canadian population of 40 million breeding adults and even estimated a Minnesota population of 8,800. The MNBBA did not venture an estimate because the species was too rare from 2009 to 2013. Sauer and colleagues (2020) suggested there were too few data to reliably estimate a population trend based on the BBS because of too few routes with detections or an unsatisfactory number of detections within routes. The species' population is also highly variable, and coverage is poor or nonexistent in its primary breeding range in high latitudes and high altitudinal regions. Despite these limitations, Rosenberg, Kennedy, and colleagues (2016) calculated that the population had increased by 163 percent from 1970 to 2014. This increase was supported by Environment Canada (2014b), which suggested that both the BBS and Christmas Bird Counts "show wide variation, but overall the population has shown an increase since about 1970."

B. Nesting habitat of the White-winged Crossbill is usually associated with large spruce trees, like these white spruce in northern Minnesota; the species is known to nest in any month of the year. Photo by Gerald J. Niemi.

CONSERVATION

The White-winged Crossbill is currently of low conservation concern because of its large and possibly increasing population and its wide Holarctic distribution across northern North America, Europe, and Asia. Environment Canada (2014b) cautions that "large-scale forestry that reduces the age-class distribution of spruce forests enough to limit cone production could be detrimental." Furthermore, Fowells (1965) noted that maximum cone production does not occur until trees are over sixty years of age.

Several researchers have specified the importance of conifer seed availability as a limitation on White-winged Crossbill populations (Pulliainen 1971; Bock and Lepthien 1976). When seed availability is low, the species is more nomadic, and mortality is greater (Benkman 2020). Benkman (1993) noted that broad-scale conservation in regions of the North American continent may be important because cone production during any year can be low in one or more regions. Thus, a coordinated network of suitable conservation regions across the continent may be essential for the long-term survival of this species.

Climate change is also a potential threat because of the White-winged Crossbill's northerly distribution. Under a 2°C warming scenario, NAS (2019) projected a moderate vulnerability for its breeding range in North America, with a 41 percent range loss and a 14 percent range gain primarily in the far north by 2050. In Minnesota it was predicted to be extirpated from the state even under our current 1.5°C warming scenario. Furthermore, it is questionable whether habitats in the far north would be colonized by small-coned conifers such as spruce, tamarack, or fir. Gambling on the presence of suitable habitat and appropriate food sources presents a risky future for this species if climate change continues.

PINE SISKIN

Spinus pinus

Minnesota Seasonal Status: Regular but highly variable permanent resident. The Pine Siskin was uncommon during the MNBBA.
Migration: Short- to medium-distance migrant; erratic annual movements as far south as the southern United States and Mexico.
Food: Highly variable diet, consisting of gleaning arthropods, seeds, and buds on conifers, and also on deciduous trees and shrubs, grasses, forbs, and a variety of weedy species. Commonly found at feeders, especially in winter or during migration.
Nest: Cup nest on outer branches of coniferous trees. Sometimes nesting as isolated pairs or frequently in small loose colonies.
Conservation Concern: Assigned a moderate PIF Continental Concern Score of 10/20.

Roberts (1932) whimsically referred to Pine Siskins as "possessed of a gipsy-like wanderlust that takes it now here, now there, and only irregularly and seemingly by chance back again to the same locality." Further, he said they are "wandering through the land, few or many together, winter and summer, settling for a season and rearing its family in most unexpected places." Pine Siskins are indeed an opportunistic nomadic species and are known to many Minnesotans as a highly variable species at feeders. They can be extremely common one year and gone the next. Like many finches, Pine Siskins are not rigidly tied to a reproductive schedule, but their annual cycle is likely related to the distribution and abundance of the seeds that constitute their primary diet.

Pine Siskins are small finches. Characteristic features include their highly streaked upperparts with greenish brown on a gray background, underparts gray streaked with white, notched tails, sharply pointed bill, and yellow in the wings and tail flight feathers. The sexes are similar, but older males tend to have brighter yellow coloration in the wings and tails. Bent (1968) describes the call notes as "weak, thin, lispy, buzzy, wheezy, and churring." He suggests a note during its undulating flight that sounds like *tit-a-tit*. Alderfer (2006) describes a common call note as a buzzy, rising *zreeeeee* and a song that is a "lengthy jumble of trills and whistles similar to that of the American Goldfinch."

Pine Siskins are widely distributed across the boreal forest in Canada to Alaska, much of the northern United States, and southward throughout the Rocky

Photo by Laura Erickson.

Mountains as far south as Mexico. The highest breeding densities have been recorded in the western Rocky Mountains, especially in British Columbia.

MINNESOTA BREEDING DISTRIBUTION

Roberts (1932) described the Pine Siskin as a common summer resident throughout the evergreen and pine forests during the breeding season; it undoubtedly nests over a wide area. He reported finding only one nest in a pine tree near Eveleth in St. Louis County in 1918, but he noted a July 1918 observation of "young birds out of the nest being fed by the parents in Pine County." Overall, he pronounced that it occurred over a substantial portion of the state but most commonly in the northern third.

Green and Janssen (1975) reinforced the vagabond nature of its nesting in the state but described its primary breeding distribution as being in the northeastern coniferous region. They note confirmed nesting from Itasca and St. Louis Counties as well as from Hennepin County in 1961 and 1966. Widespread inferred evidence of nesting was reported from Clearwater, Lac qui Parle, Lake, Pine, Rice, and Winona Counties. Most southern nesting generally occurred after large winter invasions. Janssen (1987) summarized confirmed nesting records since 1970 from 14 counties, ranging from Cook County in the northeast to Winona County in the southeast, and west to Stearns County in central Minnesota. He further points out a summer observation in June 1980 from Houston County, the most southeastern county of the state. In their summary, Hertzel and Janssen (1998) added Cass and Koochiching, for a total of 16 counties with confirmed nesting records since 1970.

The MBS reported 53 breeding-season locations that illustrated the wide and patchy nature of the species' breeding distribution across the northern half of the state. The most extensive locations were from Cook County, central St. Louis County, and eastern Becker/western Hubbard County.

The MNBBA included 252 records of the Pine Siskin that were also scattered across a substantial part of the state, except most of the southern and western portions of Minnesota **(A)**. The species was detected in only 5.3 percent (252/4,737) of all blocks surveyed, but they included almost 50 percent (43/87) of the counties. A total of 28 widely scattered blocks had confirmed nesting, including southwest to Kandiyohi County, south to Blue Earth, Rice, and Olmsted Counties, and northwest to Marshall County. Most records were from northeastern Minnesota, but also included confirmed records from several counties in the Twin Cities and in central Minnesota. The probability map predicted low abundance of the Pine Siskin throughout the northeastern and northern region of Minnesota **(B)**.

Dawson (2020), in his review of the Pine Siskin in North America, suggested that increased silviculture and establishment of conifers may have increased the normal range of this species farther south than its traditional range, in southward extensions of nesting in Illinois, Indiana, and Ohio. Nesting activity was likely augmented by the growth in availability of feeders. Similar extensions have been reported in Wisconsin (Cutright et al. 2006) and in Michigan (R. Brewer et al. 1991). Although there is an increased number of confirmed nesting reports in southern Minnesota, the extent to which this is a recent occurrence is unknown. The lack of better information on confirmed

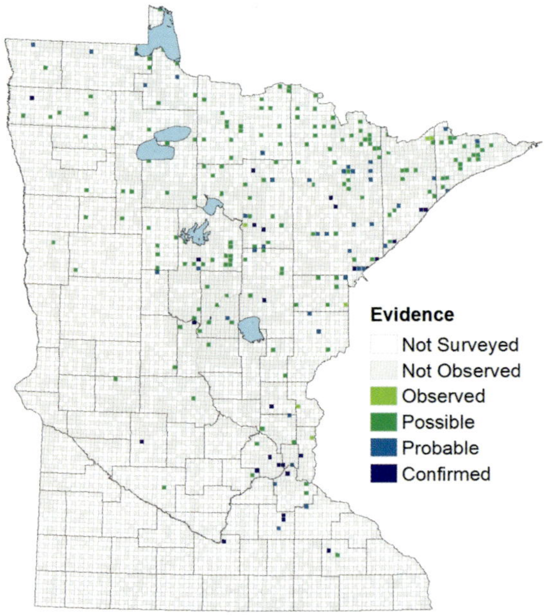

A. Breeding distribution of the Pine Siskin in Minnesota based on the Breeding Bird Atlas (2009–13).

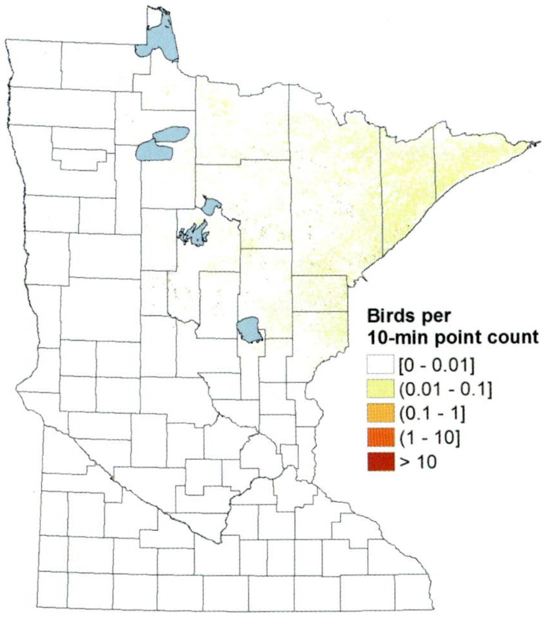

B. Predicted breeding distribution (birds detected per 10-minute point count) of the Pine Siskin in Minnesota based on habitat, landscape context, and climate data gathered during the Minnesota Breeding Bird Atlas (2009–13) using the Generalized Linear Modeling method without an adjustment for detectability.

historical nesting in southern Minnesota by Pine Siskins precludes definitive conclusions.

BREEDING HABITAT

Dawson (2020) described open coniferous forests, ornamental conifers in parks and cemeteries, and mixed coniferous–deciduous forests as primary nesting habitats for the Pine Siskin. Green (1995) emphasized a seed preference for birches, white cedar, and white spruce. Niemi, Howe, and colleagues (2016), in their analysis of Pine Siskin habitat preferences in national forests of Minnesota and northern Wisconsin, found the species widely distributed in many forest cover types but, interestingly, most frequently detected in small towns. Otherwise, the species was detected in aspen–spruce–fir, red and white pine, black spruce–tamarack, regenerating conifers, and mixed swamp conifer forest cover types **(C)**. Large coniferous, ornamental trees found in many urban areas, especially white or blue spruce, provide excellent nesting substrates.

POPULATION ABUNDANCE

The US and Canadian population of Pine Siskins was estimated at 45 million breeding adults (PIF 2021), but with only 18,000 in Minnesota. The species was too rare and variable to estimate a breeding population for Minnesota during the MNBBA.

The BBS population trend for Minnesota was also unreliable, but its surveywide trend, despite being highly variable, revealed a significant decline from 1966 to 2019 that averaged −2.19 percent per year **(D)**. Rosenberg, Kennedy, and colleagues (2016) calculated a −80 percent decline in the population from 1970 to 2014. BBS analyses indicated that the decline was widespread throughout the United States and Canada, including significant population declines in New Brunswick, Oregon, and Washington (Sauer et al. 2020).

Factors identified as potentially contributing to its decline included disease, inclement weather, reduction in coniferous trees, and nest predation (Dawson 2020). The most significant disease is salmonellosis (*Salmonella typhimurium*), a pathogen associated with bird feeders that can lead to death. Nest predation may be more serious because of the Siskin's gregarious, semicolonial nature in nesting, which provides greater opportunities for predators. Inclement weather, especially in the spring, when the species primarily nests, may also contribute to the decline. Unfortunately, all of these factors are difficult to quantify.

CONSERVATION

The steep population decline for the Pine Siskin renders it a conservation concern in North America. As stated, the primary factors associated with its decline appear to be diseases, the availability of coniferous trees, nest predation, and weather. Regarding disease, T. L. Rodgers (1991) recommended practices to reduce the spread of salmonellosis, including cleaning and disinfecting feeders at least yearly; ceasing feeding for 10 days if bird deaths are observed; and placing food in locations that will lessen the risk of the birds' contact with avian fecal material.

C. Breeding habitat in Minnesota largely consists of scattered coniferous trees, especially spruce, often in suburban and exurban settings. Photo by Gerald J. Niemi.

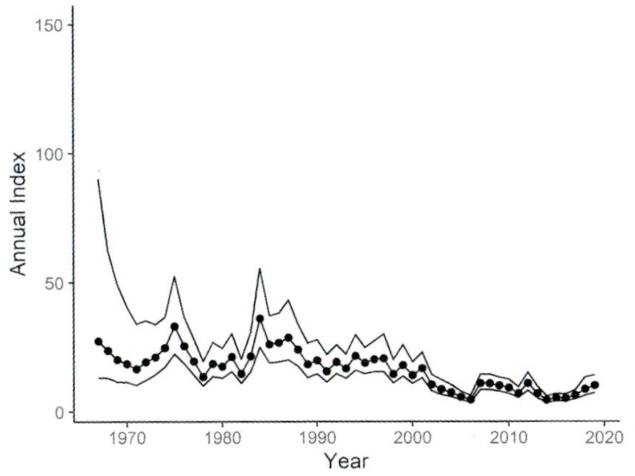

D. Surveywide breeding population trend and 95 percent credible interval (solid black lines) for the Pine Siskin based on the 1967–2019 Breeding Bird Survey (Sauer et al. 2020).

The availability of coniferous trees for nesting and foraging is complex. For instance, Niemi and Hanowski (1992b) estimated a decline in the species' population with moderate increases in forest harvesting, especially in reductions of coniferous forest cover types. However, logging in Minnesota has not reached the moderate levels of harvesting analyzed in that study. In addition, increases in urban conifers, such as in parks, cemeteries, and around homes, can counteract the loss of conifers in natural habitats for nesting and foraging. With a 2°C warming scenario, NAS (2019) projected a moderate vulnerability for the Pine Siskin's breeding range in North America, including a 47 percent range loss and a 28 percent range gain primarily in the far north by 2050. In Minnesota, the species' extirpation was projected even under the imminent 1.5°C warming scenario. NAS, however, began with an assumption that the Siskin's breeding range in the state is significantly smaller than the MNBBA and earlier resources have documented.

The future of the Pine Siskin in Minnesota and North America is difficult to predict. Its nomadic behavior makes it challenging to detect population trends and many factors affect its populations. Yet, it would be a shame to have any further reduction in these lively, energetic birds so commonly observed and enjoyed at feeders.

AMERICAN GOLDFINCH

Spinus tristis

Minnesota Seasonal Status: Regular breeding resident and migrant; regular in winter. The American Goldfinch was abundant during the MNBBA.
Migration: Northern populations are short- to medium-distance migrants that winter in the southern United States and northern Mexico; southern populations are year-round residents. Some birds are present year-round in Minnesota and are likely permanent residents.
Food: Foliage and ground gleaner that feeds exclusively on the seeds of annual plants, primarily composites. Insects are only consumed when encountered foraging for seeds. A common visitor to bird feeders.
Nest: Cup nest located in herbaceous plants, shrubs, or trees; deciduous shrubs are preferred.
Conservation Concern: Assigned a low PIF Continental Concern Score of 6/20.

When lingering winters suppress any hope that spring is on its way, the transformation of the male American Goldfinch into his radiant yellow and black nuptial dress brightens one's spirits. A welcome visitor to backyard bird feeders, the Goldfinch is one of our most beloved songbirds. "If the actions and expressions of birds are any indication of their dispositions and feelings," Roberts (1932) wrote, then "the handsome little Goldfinch is surely one of the most joyous and light-hearted of all our feathered throng. Sociable and genial among its kind, cheery and musical of voice, and gay and happy in demeanor," the bird has a "happy-go-lucky air." So endearing is the Goldfinch that it is the official state bird of Iowa and New Jersey and was the first bird to be proposed as Minnesota's state bird in 1940 by T. S. Roberts's colleague, William Kilgore (Strnad 1961).

The male's brilliant summer plumage needs little introduction. Its lemon-yellow body is accented with a black cap, while its black wings and tail feathers are interlaced with streaks of white. The short, conical bill is light orange in color. Lacking the male's black cap, the female has similar but much duller plumage. Carotenoid pigments obtained by their diet of composite seeds are responsible for their striking plumage. And the more colorful the male, the more attractive he is to females (McGraw and Middleton 2020).

Photo by David Brislance.

The Goldfinch's chattering song is a sweet and complex medley of warbling phrases that vary in length and composition. The Latin name for the species, *Spinus tristis,* means "little finch with a mournful song," since Linnaeus apparently interpreted the medley as rather sad and plaintive (Sandrock and Prior 2014). But its most recognizable vocalization is its flight call. "This is the song," Roberts (1932) wrote, that "it sings as it goes bounding along in easy, undulating flight, punctuating every upward curve with this joyful assurance of the good time it is having." Early naturalists described the call as *per-chic-o-ree;* today the phrase *po-ta-to-chip* is commonly used.

The American Goldfinch rarely initiates nest-building activities until July or early August. Timing coincides with the maturation of its primary food item, the seeds of annual plants. Members of the composite family, such as sunflowers, ragweed, and thistle, are especially favored. Not only are the seeds important food items for the young but also their feathery plumes are used for nest construction.

Widely distributed across southern Canada and the northern two-thirds of the United States, the Goldfinch is a common species throughout the fields and open woodlands of Minnesota. High breeding densities are found in large areas throughout its breeding range.

MINNESOTA BREEDING DISTRIBUTION

The American Goldfinch was a statewide resident even in the 1900s. Roberts (1932) wrote: "Over all the varied surface of Minnesota, even far out on the western prairies if there be so much as a vestige of arboreal growth, the Goldfinch is a common and often abundant summer resident." The species could also be seen throughout the winter months in southern Minnesota and as far north as Anoka County. Confirmed nesting records were available from 10 counties that stretched from the southeastern corner of the state north through the deciduous and mixed forest regions, including Cass, Crow Wing, Fillmore, Goodhue, Hennepin, Houston, Polk, Scott, Sherburne, and St. Louis counties.

Forty years later, Green and Janssen (1975) also commented on the species' statewide distribution, noting that it was most abundant in the south and least abundant in the heavily forested landscapes of northern Minnesota. Janssen (1987) delineated 21 widely dispersed counties where nesting had been confirmed since 1970; Hertzel and Janssen (1998) added 11 more counties to the list. The 2,303 breeding-season locations identified by the MBS further confirmed the species' broad distribution.

Indeed, during the MNBBA, the Goldfinch was detected in 61.9 percent (2,966/4,788) of the surveyed MNBBA blocks and in 80.2 percent (1,875/2,338) of the priority blocks **(A)**. Breeding was confirmed in 262 blocks (5.5 percent) and considered probable in an additional 1,257 blocks (26.3 percent). With 5,316 detections, the American Goldfinch was observed in all 87 Minnesota counties and was confirmed breeding in 56. Of these, 34 counties were additions to the list published by Hertzel and Janssen in 1998. Remarkable efforts by atlas volunteers in the Twin Cities and Brainerd Lakes region contributed to these impressive outcomes. Although the species was widespread across the state, it remained less common in portions of southwestern Minnesota and in the far northern regions of northeastern Minnesota. The American Goldfinch was the fourteenth most abundant species detected during the atlas.

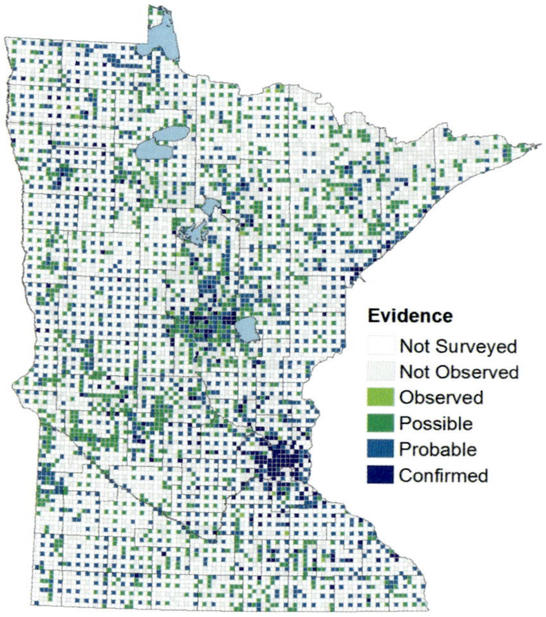

A. Breeding distribution of the American Goldfinch in Minnesota based on the Breeding Bird Atlas (2009–13).

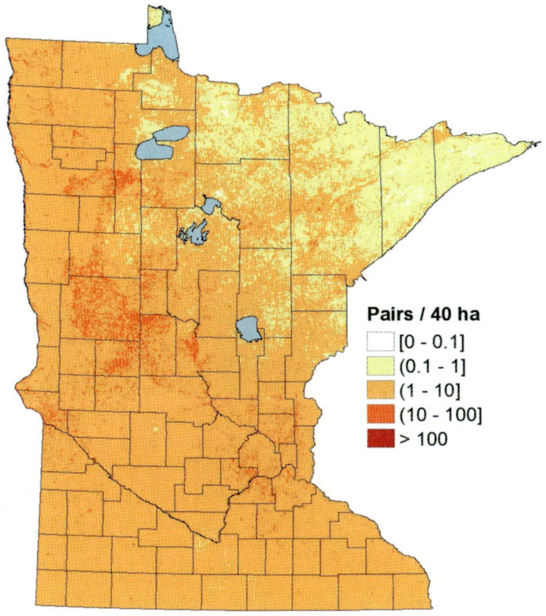

B. Predicted breeding distribution (pairs per 40 ha) of the American Goldfinch in Minnesota based on habitat, landscape context, and climate data gathered during the Minnesota Breeding Bird Atlas (2009–13) using the Generalized Linear Modeling method with an adjustment for detectability.

The MNBBA probability map suggests that the likelihood of encountering the American Goldfinch is highest in the Hardwood Hills and the Pine Moraines and Outwash Plains Subsections of northwestern Minnesota. Relatively high breeding densities are also predicted in other scattered regions of east-central and west-central Minnesota and in the northwest, where an association with riparian floodplains is evident **(B)**. Extensive peatlands in the north-central counties and the densely forested landscapes to the east are predicted to support the lowest breeding densities.

The broad distribution of the American Goldfinch remains unchanged since Roberts (1932) first described it as a statewide resident nearly one hundred years ago. In their comprehensive review of the species, McGraw and Middleton (2020) noted that there had been no major detectable changes to its historical North American breeding distribution.

BREEDING HABITAT

A species of open country with scattered trees and shrubs, the American Goldfinch is typically associated with old fields, young successional stands, roadsides, floodplains, brushy openings, forest edges, and prairie groves (Danz et al. 2007; McGraw and Middleton 2020; Bent 1968) **(C)**. The interiors of mature forests are avoided, but when forests are cleared for agriculture or small residential areas, the openings provide ideal habitat as long as scattered trees and shrubs are available. In northern Minnesota's two national forests, the American Goldfinch was most strongly associated with small towns (Niemi, Howe, et al. 2016).

A common feature of any habitat where the species is found is the presence of herbaceous plants of the composite family, especially thistles, the seeds of which are a preferred food item. Thistle plants are also a common nesting site. Their tight clusters of branches form an ideal little crotch for nest placement; the spines on all parts of the plant also provide a degree of predator protection (Pius 1948; Lewis 1952). A study in the 1940s found more than 700 Goldfinch nests in a 400-ha site in west St. Paul; the majority (591) were placed in the uppermost cluster of branches of thistle plants, primarily tall thistle (*Cirsium altissimum*) (Lewis 1952).

POPULATION ABUNDANCE

PIF (2021) used population data collected by the BBS to establish a population estimate of 44 million breeding adults in the United States and Canada; the statewide estimate for Minnesota was 2.3 million. MNBBA point count data, in contrast, provided more intensive statewide coverage and generated a significantly larger estimate of 4.4 million breeding adults (95 percent confidence interval of 3.9–4.9 million). Indeed, southern Minnesota is part of a broad region stretching across the upper Midwest and Great Lakes that supported some of the highest breeding densities of the American Goldfinch in North America from 2011 to 2015 (Sauer et al. 2020).

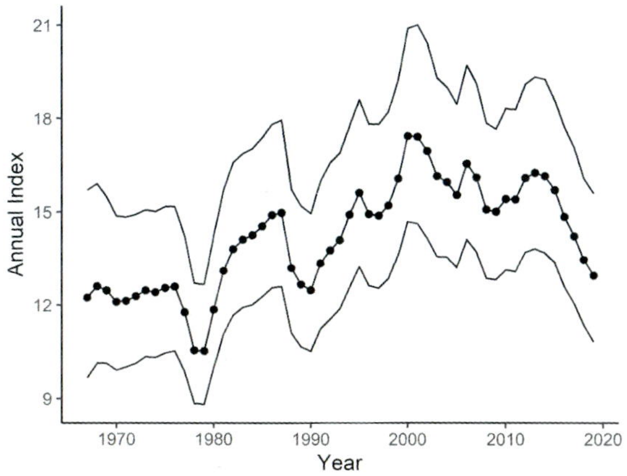

D. Breeding population trend and 95 percent credible interval (solid black lines) for the American Goldfinch in Minnesota based on the 1967–2019 Breeding Bird Survey (Sauer et al. 2020).

Throughout the BBS survey region, the American Goldfinch has experienced a slow, statistically significant population decline, averaging −0.60 percent per year from 1966 to 2019 (Sauer et al. 2020). By contrast, Minnesota's population has demonstrated a nearly stable population with a small, nonsignificant increase averaging just 0.11 percent per year **(D)**. Populations in the Chippewa and Superior National Forests also demonstrated nonsignificant increases from 1995 to 2020 (Grinde, Bednar, et al. 2020).

Factors that influence populations are variable. Overall, the species benefits from habitat changes that create open woodlands and fields. During the breeding season, predation is the most important factor impacting reproductive success, but mortality is believed to be highest during the winter (McGraw and Middleton 2020).

CONSERVATION

Despite its small continental population decline, the Goldfinch remains an abundant species and a low conservation priority (PIF 2021). Indeed, it is hard to imagine a future without this joyful, high-spirited bird. Nevertheless, NAS (2019) projected moderate vulnerability for the species' breeding distribution when predicted range declines consistently exceed predicted gains under all three warming scenarios examined. Even in Minnesota, the species gradually retreats eastward until, under the 3°C warming scenario, it is restricted to the northeast region. Such dire predictions for one of Minnesota's most common species are surely an incentive to aggressively address the challenges of climate change.

C. Old fields with abundant forbs and scattered small shrubs and trees are one of many habitats that attract the American Goldfinch. Photo by Lee A. Pfannmuller.

CHESTNUT-COLLARED LONGSPUR

Calcarius ornatus

Minnesota Seasonal Status: Summer resident only in Clay County; formerly bred in western counties from Jackson to Marshall. Migrant visitant along the western border and occasional elsewhere. The Chestnut-collared Longspur was very rare during the MNBBA.
Migration: Short- to medium-distance migrant that winters in the Chihuahua grasslands of north-central Mexico as well as in Arizona, New Mexico, western Oklahoma, and Texas.
Food: Consumes the seeds of grasses and forbs; also arthropods, especially grasshoppers, crickets, and beetles.
Nest: Slight depression on the ground composed of grasses; sometimes lined with feathers or animal hair.
Conservation Concern: Assigned a high PIF Continental Concern Score of 15/20; officially listed as an *Endangered* species in Minnesota. Classified as *Vulnerable* on the IUCN Red List (IUCN 2021).

Photo by Bradley Hacker.

The Chestnut-collared Longspur, a prairie grassland specialist, is one of seven endemic grassland species of the northern Great Plains. It was first collected in the spring of 1834 on the Platte River, western Nebraska, by John K. Townsend, a leading naturalist from Philadelphia. Townsend, along with Thomas Nuttall, was part of Nathaniel Wyeth's second expedition across the Rockies via South Pass to the Pacific Ocean. The first descriptions of Townsend's birds, written by Nuttall under Townsend's name, appeared in 1837 and included the Chestnut-collared Longspur (Mearns and Mearns 1992).

The plumages of the male and female Chestnut-collared Longspur are very different. As described by Roberts (1932), the females are "dull-colored, brownish birds; striped above and faintly so on the breast and sides." In contrast, the male is "black below, throat white or buffy; a chestnut collar on hindneck; top of head black with white and black stripes on sides; back striped; tail mostly white, middle feathers dark at tip and on margins." Further, "the male is conspicuous from its habit of perching on the tops of slender flowering plants . . . with which it sways back and forth in the breezes." The male has a distinctive two-syllable *kittle* call note and a rapid warble song usually given in flight or from a low perch.

The Chestnut-collared Longspur has a relatively small breeding distribution. Its core range includes southern Alberta, Saskatchewan, southwestern Manitoba, Montana, and the western Dakotas. Its highest breeding densities occur in southwestern North Dakota, in northwestern South Dakota, and in northeastern and north-central Montana.

MINNESOTA BREEDING DISTRIBUTION

Roberts (1932) first documented the species' Minnesota presence in 1879 when exploring the prairie in Grant and Traverse Counties with his friend Franklin Benner (Roberts and Benner 1880), where he found a nest and collected specimens (Krosch 1987). His description of the 1879 trip conveys the prairie scene before settlement and his surprise "to find the Chestnut-collared Longspur, previously known only from the plains farther west, an abundant breeding bird all over the dry upland prairies. In some sections it was so numerous as to outnumber all other birds together" (Roberts 1932). Additional county records during the nineteenth century by Roberts were from Jackson, Kittson, Lac qui Parle, Lincoln, Marshall, Otter Tail, Pipestone, and Redwood Counties.

In the early decades of the twentieth century, trips to western Minnesota by staff from the Museum of Natural History (now the Bell Museum) produced a few records from the northwestern counties and none from the southwest. By the 1920s, the native prairie of southwestern Minnesota was mostly settled, and only remnant populations of the Chestnut-collared Longspur were found on the uncultivated sandy beach ridges of Glacial Lake Agassiz that border the eastern Red River valley (Roberts 1932). From the 1930s to 1950s, a few more locations with breeding birds were found and documented, including in 1928 in Polk County (Roberts 1932), 1933 in Pennington County (Woolsey 1933); 1936 in Clay County (Wagner 1936); 1937 in Grant County (BMNH Records); 1937 in Marshall County (Rysgaard 1937d); 1942 in Wilkin County (Harrell 1942); 1949 in Clay County (M. K. Nelson 1949); and 1956 in Clay County (BMNH Records). From 1960 to 1995, incidental summer observations of mostly single birds have been published in the MOU Seasonal Reports from additional prairie counties: Pipestone, Yellow Medicine, Big Stone, and Norman. Breeding records were obtained from Traverse County in 1980 (Eckert 1980), and in 1984 a nest and about 30 birds were reported there (Wyckoff 1986b).

Since the 1960s, almost all Chestnut-collared Longspur nesting activity has been on a prairie and gravel mining complex near Felton, Clay County. It was discovered in 1960, when at least 25 displaying males were located (R. L. Huber 1960a). In later trips, a nest was documented on June 17, 1961 (Pieper 1963b) and another nest on June 9, 1962 (R. L. Huber 1962b). Ever since, yearly summer observations and population surveys have been reported on this remnant population. The first survey was conducted in 1980 and estimated 65 pairs (Eckert 1980). More structured surveys were conducted in 1984 and 1985, which estimated a population of 234–262 birds (Wyckoff 1986a).

Subsequently, seasonal reports in *The Loon* have indicated the species' presence or its probable nesting at Felton Prairie every year from 1986 to 2019, but no monitoring has occurred since the early 1980s. The breeding population has declined substantially,

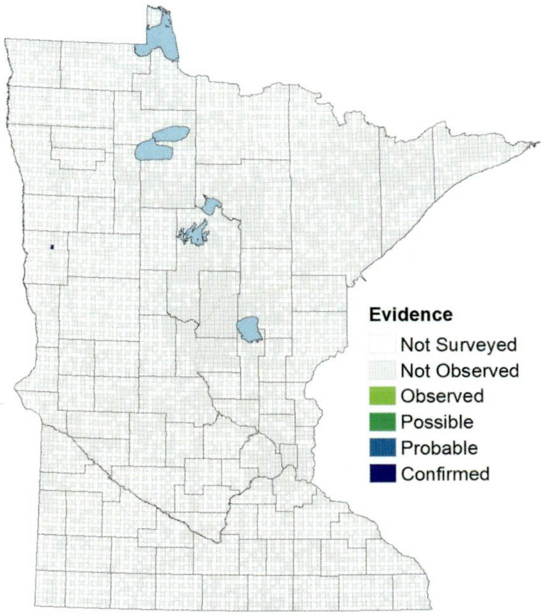

A. Breeding distribution of the Chestnut-collared Longspur in Minnesota based on the Breeding Bird Atlas (2009–13).

with only one to eight birds reported yearly from 2000 through 2019. Because the Felton Prairie is the only site in Minnesota where this species is reliably observed, it has been frequently visited. In 2014, the last breeding evidence, a juvenile, was reported and, despite frequent visits, only one to two birds were reported in 2017, 2018, and 2019.

During the MNBBA, all three records were on the Felton prairie complex in Clay County **(A)**. The records occurred on May 21, 2009 (confirmed, nest with 4 eggs); July 25, 2011 (possible); and June 14, 2012 (possible). All observations were made along 170th Street North in the southwest quarter of Keene Township. The MOU Query Sightings Database for 2009 to 2013 also included records for this location, which was visited by many observers each year (MOU 2021a). Observations indicated either the species' presence or gave a count of two to six birds each year. One observation in a location in the northwest quarter of Keene Township reported a female carrying food and four birds at that site on June 16, 2010.

In their review of historical changes in the distribution of the Chestnut-collared Longspur, Bleho and colleagues (2020) emphasized the substantially reduced distribution of this species along with the loss of native prairie grasslands. These changes in distribution and loss of populations in many parts of Minnesota were well-documented by Roberts (1932, 1936). Land-use changes have continued to the present. For instance, from 2009 to 2013 in the large Mississippi River Basin–Great Plains area, one-third of the area's land had been converted from natural habitat to cropland due to high crop prices, government incentives, and a growing global human population (Gage et al. 2016). Similar conversions of native prairie and grasslands to agricultural lands in the species' winter range have also likely occurred in the southern and southwestern United States and in central Mexico.

BREEDING HABITAT

The Chestnut-collared Longspur breeds only on the dry prairies of the northern Great Plains; the bulk of the population is in the Canadian provinces and three of the Great Plains states: Montana, North Dakota, and South Dakota. The species' preference is for short- to mixed-grass prairie on rolling landscapes with sparse, open vegetation **(B)**. This vegetative structure can be found on native dry prairies or on sites that have been recently grazed, mowed, or burned. The species will also use mowed pastures or fallow fields with stubble, but at much lower densities (Bleho et al. 2020). In a study in Felton Prairie, Wyckoff (1985) summarized that the Chestnut-collared Longspur preferred "well-drained sites away from trees and shrubs, in close-proximity to wet areas, where the vegetation was relatively short, sparse, and devoid of litter."

POPULATION ABUNDANCE

PIF (2021), using the BBS data, estimated the US and Canadian population as 3.1 million birds; no estimate was included for Minnesota. Overall, the BBS surveywide decline in the United States and Canada was significant and averaged −2.58 percent per year from 1966 to 2019 **(C)**. Significantly declining populations during the same period were found in Alberta (average of −5.37 percent per year), North Dakota (−2.96), Saskatchewan (−3.26), and South Dakota (−3.46) (Sauer et al. 2020). Rosenberg, Kennedy, and colleagues (2016) calculated a cumulative decline of −85 percent from 1966 to 2014.

B. The native, dry-mesic prairie habitats of Clay County provide one of the few nesting sites for the Chestnut-collared Longspur in Minnesota. Photo by Lee A. Pfannmuller.

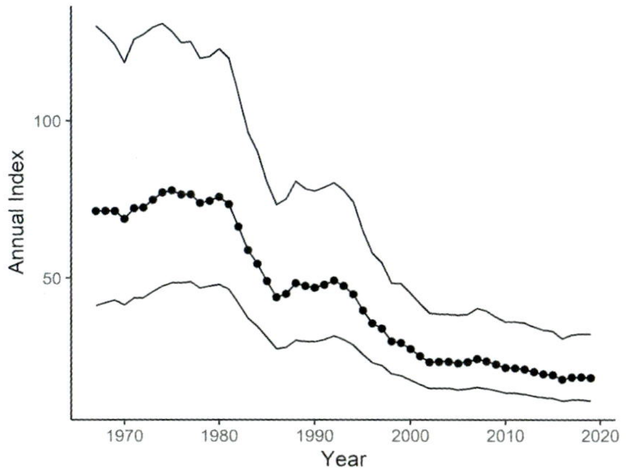

C. Surveywide breeding population trend and 95 percent credible interval (solid black lines) for the Chestnut-collared Longspur based on the 1967–2019 Breeding Bird Survey (Sauer et al. 2020).

The reason for the decline of the Chestnut-collared Longspur is undoubtedly associated with the loss of native prairie habitat across its breeding range. In Minnesota the native prairie has been reduced by nearly 99 percent since the mid-1800s, owing primarily to conversion to agricultural lands.

CONSERVATION

With a little more than 1 percent of the native Minnesota prairie remaining, the Chestnut-collared Longspur was designated in 1984 as *Endangered* in Minnesota (Coffin and Pfannmuller 1988). It was also recommended that the relict population on the Felton Prairie be "monitored annually for changes" (MNDNR 2019d). Unfortunately, monitoring has not occurred since the early 1980s, when a population of 132 males was recorded (Coffin and Pfannmuller 1988), nor have other populations been discovered. Presently, so few birds are observed that it is on the verge of being extirpated as a breeding species in the state.

Warming temperatures might hasten that outcome. Although NAS (2019) predicted that the Longspur may gain a small amount of breeding habitat in northwestern Minnesota under the 1.5°C and 2°C warming scenarios, all available habitat in the state would disappear as temperatures rise to 3°C. Indeed, the species would vanish from the northern Great Plains under the 3°C warming scenario. Only a tiny area located south of the Longspur's current breeding range was identified by NAS as "climate suitable" by the year 2080, leaving this colorful grassland endemic with a very uncertain future.

GRASSHOPPER SPARROW

Ammodramus savannarum

Minnesota Seasonal Status: Regular breeding resident and migrant. The Grasshopper Sparrow was uncommon during the MNBBA.
Migration: Short- to medium-distance migrant that winters from the southern United States through Central America and the Caribbean.
Food: Omnivorous ground forager that feeds on arthropods, especially grasshoppers, and seeds.
Nest: Cup nest with a partial dome, placed on the ground at the base of a clump of vegetation.
Conservation Concern: Assigned a moderate PIF Continental Concern Score of 12/20.

In the open grasslands, where windless days are memorable for their rarity, the little Grasshopper Sparrow and its weak, insect-like song are easily overlooked. Secretive and well-hidden, the species spends much of its time on the ground beneath a canopy of grasses and forbs. Even when startled, its first instinct is to run, only reluctantly taking to the air in a low, fluttering flight when the pursuer has approached too closely. Then, almost as soon as the flight has begun, the Sparrow dives back down into its grassy refuge. Appropriately, the species' Latin name, *Ammodramus savannarum*, translates to "sand runner of the meadows."

Flat-headed, with a short, notched tail, among the species' most distinctive features are its unstreaked buffy breast and flanks. Other traits include a striped crown, a streaked back, and, on close inspection, a buffy orange stripe between the eye and the bill. Although it is fond of grasshoppers, its common name refers to its short buzzy song. Describing it as *tsick, tsick, tsurrrr,* Roberts (1932) noted that it is "so thin and weak that even the grasshopper after which he is named might be ashamed to have it come from its fiddle." A longer song, often delivered in flight, is more musical and consists of a series of variable buzzy notes.

Found from the southern regions of central and eastern Canada, south to Texas, and east to the Atlantic coast, small breeding populations also occur in the northwestern United States and California. Small resident populations are found in Mexico, Central America, and the Caribbean. The highest breeding densities occur in the Great Plains, from North Dakota south through Nebraska.

Photo by Sparky Stensaas.

MINNESOTA BREEDING DISTRIBUTION

Roberts (1932) described the Grasshopper Sparrow as "one of the commonest small birds everywhere on the dry, upland prairies of Minnesota." Also found in "open woodlands," it even extended, albeit sparingly, into the northern forests "wherever there are extensive clearings and sparsely wooded areas." Confirmed nesting records, however, were limited to six counties in western and east-central Minnesota: Grant, Hennepin, Jackson, Polk, Sherburne, and Wilkin. Nesting was inferred in Ramsey County.

Green and Janssen (1975) described the species as a statewide summer resident but noted it was "generally scarce in the north central and northeastern regions." Janssen (1987) outlined a similar breeding distribution and identified seven counties with confirmed nesting since 1970: Brown, Clay, Cottonwood, Norman, Ramsey, Rock, and Wabasha. Hertzel and Janssen (1998) later added Aitkin, Kandiyohi, Sherburne, and Stevens Counties to the list. The 529 breeding-season locations identified by the MBS were largely confined to the Tallgrass Aspen Parklands, Prairie Parkland, and Eastern Broadleaf Forest Provinces.

Compared to data compiled by the MBS, the atlas documented more Grasshopper Sparrow reports farther east into the Laurentian Mixed Forest Province, including multiple records across Aitkin, Cass, and Crow Wing Counties **(A)**. Indeed, among the 23 counties with confirmed breeding was one record in central St. Louis County at a large, reclaimed taconite tailings basin. Nevertheless, the species remained absent throughout much of north-central and northeastern Minnesota. Seven of the nine counties where it was not detected were located in the northern half of the state.

Overall, MNBBA participants reported 761 Sparrow records from 9.8 percent (464/4,747) of the surveyed atlas blocks. Breeding was confirmed in 26 blocks and deemed probable in 121 blocks **(A)**. Seventeen of the 23 counties with confirmed nesting were new to the list published by Hertzel and Janssen (1998); most were located in central and southern Minnesota. Data submitted to the MOU (2021b) during the atlas confirmed nesting in three additional counties (Faribault, Hennepin, and Washington), but locational information was insufficient to assign the records to atlas blocks. The Grasshopper Sparrow is sparsely distributed across its breeding range; the MNBBA probability distribution map predicts that the highest breeding densities occur in the Prairie Coteau region of the southwest, the Glacial Lake Agassiz beach ridges in the northwest, and the Tallgrass Aspen Parklands Province **(B)**.

The Grasshopper Sparrow's distribution has changed little in the past century. More permanent openings in the western and southern regions of the Laurentian Mixed Forest Province have likely increased habitat for the species. Farther east, it is difficult to assess how much of a change in abundance or distribution has occurred since Roberts's time, but the smaller scale and lower frequency of forest disturbances has certainly reduced the amount of extensive clearings that may have once provided suitable habitat. Elsewhere, the loss of grassland habitat has led to major population declines. In many eastern states the species is present only as small, remnant populations (Vickery 2020).

BREEDING HABITAT

With the extensive clearing of eastern forests, this obligate grassland species was able to adapt to other dry, grassland habitats, including pastures, hayfields,

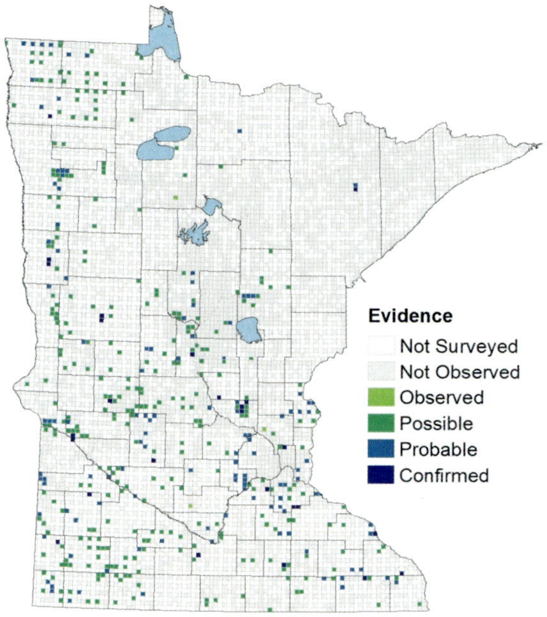

A. Breeding distribution of the Grasshopper Sparrow in Minnesota based on the Breeding Bird Atlas (2009–13).

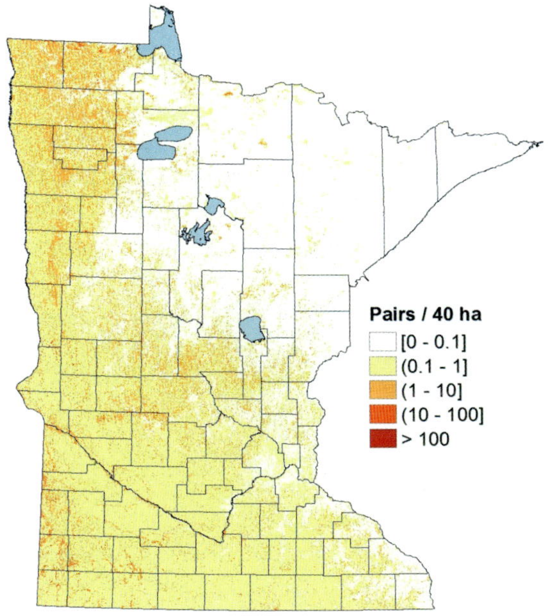

B. Predicted breeding distribution (pairs per 40 ha) of the Grasshopper Sparrow in Minnesota based on habitat, landscape context, and climate data gathered during the Minnesota Breeding Bird Atlas (2009–13) using the Generalized Linear Modeling method with an adjustment for detectability.

and, to a limited extent, croplands such as small grain fields (Bent 1968) **(C)**. Recently, the species has made use of planted cover provided by the Conservation Reserve Program (CRP) (Hanowski 1995; Dechant, Sondreal, et al., "Grasshopper Sparrow," [1998] 2002a). Important habitat features include short to moderately high vegetation, patches of bare ground (7–9 percent cover), tall, stiff-stemmed forbs that provide song perches, and a moderately deep litter layer. In the Upper Midwest, when shrub cover exceeds 4 percent, nesting densities decline (Dechant, Sondreal, et al., "Grasshopper Sparrow," [1998] 2002a; Elliot 2016). In contrast to Roberts's (1932) assertion that the species was abundant in "open woodlands," Elliot (2016) found the "presence of trees on the landscape is negatively correlated" with nesting densities.

Several studies have concluded that the Grasshopper Sparrow is an area-sensitive species, with higher breeding densities and/or higher reproductive success on larger tracts. This includes two recent studies conducted in Minnesota (Dunlap 2013; Elliot 2016). At least one study, however, has demonstrated regional variability in the bird's response (D. H. Johnson and Igl 2001).

POPULATION ABUNDANCE

PIF (2021) utilized BBS data to generate a population estimate of 33 million breeding adults in the United States and Canada, with an additional 1 million birds occurring farther south. The estimate for Minnesota was 150,000. Although MNBBA data generated an estimate of 948,000, its wide confidence interval suggests a low reliability in the projected number.

The most concerning aspect of the Grasshopper Sparrow's population is its steady decline. From 1966 to 2019 it experienced a significant decline, averaging −2.49 percent per year across the entire BBS survey region **(D)**. Although data collected along Minnesota BBS routes are sparse, the state's downward trend line is even steeper. Dropping an average of −6.44 percent per year from 1967 to 2019 (Sauer et al. 2020), the population has experienced a cumulative statewide loss of nearly −97 percent!

Habitat loss and degradation are the principal culprits. Not only have 50 percent of the Great Plains' grasslands been lost but also those that remain are increasingly fragmented, heavily grazed, subject to successional changes that reduce their suitability, and/or are located on poor or marginal soils (World Wildlife Fund 2018). Such losses are not confined to the species' breeding range. Studies in the Chihuahuan Desert of northern Mexico, a critically important area for wintering Grasshopper Sparrows, documented that 6 percent of the grasslands were converted to croplands each year from 2006 to 2011. At the present rate, all of the Desert's grasslands would be converted by 2025 (Pool et al. 2014).

CONSERVATION

Even with its broad distribution and large population, the Grasshopper Sparrow's dependence on grasslands and its significant population decline led to its assignment as a moderate conservation priority (PIF 2021). It is also designated a Common Bird in Steep Decline (Rosenberg, Kennedy, et al. 2016).

Unfortunately, this inconspicuous little sparrow faces many challenges. Although it has adapted well to nonnative grasslands, it is vulnerable to agricultural practices and farm policies that waver in response to commodity prices and overseas markets. When commodity prices are high, farmers are disinclined to enroll acres in set-aside programs, reducing the amount of suitable habitat. Harvesting grasslands for biofuel production, however, may provide benefits to the Grasshopper Sparrow and provide farmers with a cost-effective alternative when CRP payments decline. A Minnesota study demonstrated that as the percentage of grasslands harvested for biofuels increased, the shorter and sparser vegetation present the following spring resulted in an increase in Grasshopper Sparrow abundance (Dunlap 2013).

C. Restored prairies are among the many dry grassland habitats that the breeding Grasshopper Sparrow finds suitable. Photo by Lee A. Pfannmuller.

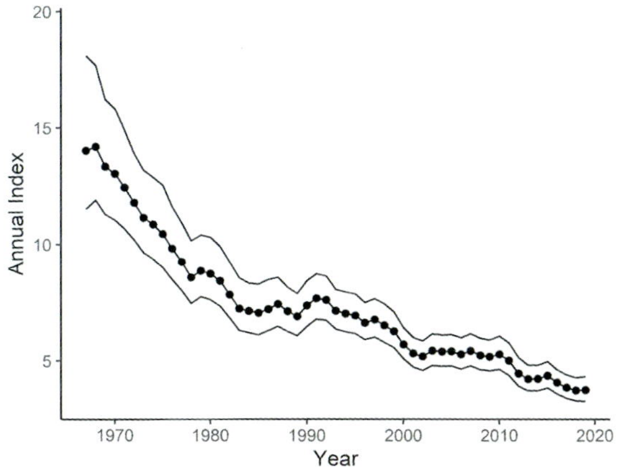

D. Surveywide breeding population trend and 95 percent credible interval (solid black lines) for the Grasshopper Sparrow based on the 1967–2019 Breeding Bird Survey (Sauer et al. 2020).

Warming temperatures may pose another challenge. Studies have highlighted their negative impact on grassland species, especially those that winter in the Chihuahuan Desert (Ruth 2015). Nevertheless, under a 2°C warming scenario, NAS (2019) predicted that by 2050 northern expansions of the Sparrow's breeding distribution (43 percent) would more than compensate for range losses farther south (20 percent) as long as suitable habitat is available. Significant expansions are also projected in northern Minnesota, where the species is already sparsely distributed.

Overall, the most important conservation measure is the protection and management of grasslands. Fortunately, the Grasshopper Sparrow's habitat requirements are well understood. Management that embraces a diversity of prescriptions and practices (e.g., grazing, mowing, and burning) provides a mosaic of cover conducive to an entire suite of grassland-dependent birds, including the imperiled little "sand runner" (Ruth 2015; Dechant, Sondreal, et al., "Grasshopper Sparrow," [1998] 2002a).

LARK SPARROW

Chondestes grammacus

Photo by Mike Lentz.

Minnesota Seasonal Status: Regular breeding resident and migrant. The Lark Sparrow was uncommon during the MNBBA.
Migration: Most birds are medium-distance migrants that winter in Mexico.
Food: Ground-foraging omnivore that feeds primarily on seeds and arthropods.
Nest: Cup nest placed in a small depression on the ground or in trees or shrubs less than 4 m high. It may reuse old nests of other species.
Conservation Concern: Assigned a moderate PIF Continental Concern Score of 10/20; officially listed as *Special Concern* in Minnesota.

Many grassland sparrows challenge us with their subtle brown plumages and their short and faintly delivered songs, but the Lark Sparrow's appearance is heralded by its distinctive facial markings and its lyrical melody. Bold chestnut and white stripes on the crown, and chestnut cheek patches bordered by black and white, are sufficient to correctly identify this species as it perches atop a small tree or tall forb. In flight, the white tips of the outer tail feathers are conspicuous. Its Latin name, *Chondestes grammacus,* means "a grain eater with lines," a reference to the bird's facial stripes.

The lark-like qualities of the Lark Sparrow's pleasing song were described by Roberts (1932) as "a series of runs and trills, liquid and clear, broken here and there by fine aspirate notes." Ridgway (1889) described it as "one continued gush of sprightly music; now gay, now melodious, and then tender beyond description,—the very expression of emotion." He believed it had "few, if any, rivals" among North American sparrows. The song is often heard late into the night; during courtship it may be delivered from the ground or while the male is in flight.

The Lark Sparrow is primarily a species of the western United States. Small numbers also breed in the southern Prairie Parklands of central Canada and occasionally as far east as Ohio and Tennessee. Year-round populations occur in California, Idaho, Texas, and northern Mexico. The species reaches its highest breeding densities in the Great Plains.

MINNESOTA BREEDING DISTRIBUTION

In the late 1800s and early 1900s, Roberts (1932) described the Lark Sparrow as a summer resident "south and west of the coniferous forests." Most abundant in southeastern Minnesota, it gradually decreased in numbers farther north through the Red River valley. Confirmed nesting records were available from Goodhue, Hennepin, Marshall, and Sherburne Counties; nesting was inferred in Isanti and Norman Counties. "Unfortunately," Roberts wrote, "it is one of the birds that has suffered the greatest diminution in numbers during the last score or so of years. The explanation of this decrease is not evident, but the fact remains that there is not one pair today where there were many twenty-five or thirty years ago."

Thirty years later, Hibbard (1965) lamented that much of the sandy savanna habitat where the species was found in Sherburne County was rapidly being converted to pine plantations. By 1975, Green and Janssen described the Lark Sparrow's distribution as largely confined to two regions: "one in the northwest and the other in the southeast." Despite scattered observations in the west-central and southwestern counties, breeding records were lacking. When reports were submitted from only four widely scattered counties in the summer of 1975, Eckert (1976) commented that "the range of this species continues to defy description."

Instead of the two breeding regions outlined by Green and Janssen (1975), Janssen (1987) delineated four "somewhat separate" breeding regions: southeast Minnesota bordering the Mississippi River; east-central and central Minnesota north to Washington and Sherburne Counties; the Minnesota River valley west to Chippewa and Yellow Medicine Counties; and the Red River valley. Nesting was confirmed in eight counties since 1970: Anoka, Clay, Dakota, Houston, Le Sueur, Polk, Renville, and Wabasha. Hertzel and Janssen (1998) added Otter Tail and Yellow Medicine Counties to the list. By 2014, the MBS documented just 61 breeding-season locations; all but one (Faribault County) were located within the breeding areas identified by Janssen (1987).

The 265 Lark Sparrow records collected during the MNBBA were more broadly distributed than those reported by the MBS. Several reports south of the Minnesota River included confirmed nesting in Blue Earth, Olmsted, and Watonwan Counties **(A)**. Reported in 3.4 percent (159/4,738) of the surveyed atlas blocks, the birds were detected in 51 of Minnesota's 87 counties. Breeding was confirmed in 53 blocks and deemed probable in 37 blocks. Among the 24 counties where nesting was confirmed, 17 were additions to the list published by Hertzel and Janssen in 1998.

The Lark Sparrow was largely absent from northern Minnesota, with two notable exceptions: one observed record of a bird seen by several individuals in Duluth on July 21, 2012; and one observed record in northern Cook County, reported by Robert Russell on June 23, 2009. The latter record was of three

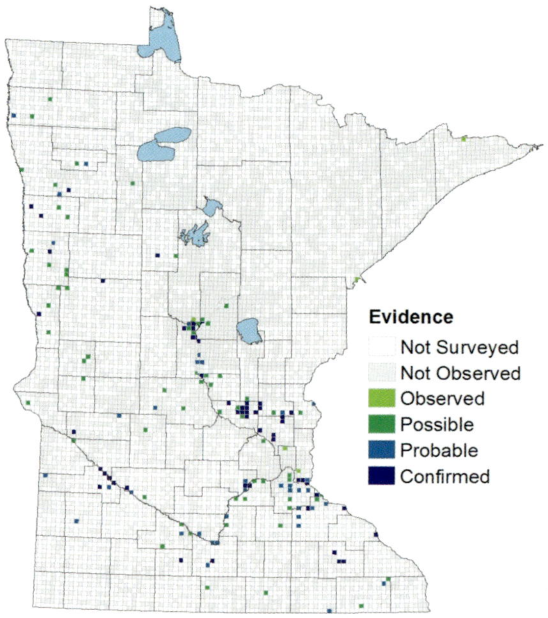

A. Breeding distribution of the Lark Sparrow in Minnesota based on the Breeding Bird Atlas (2009–13).

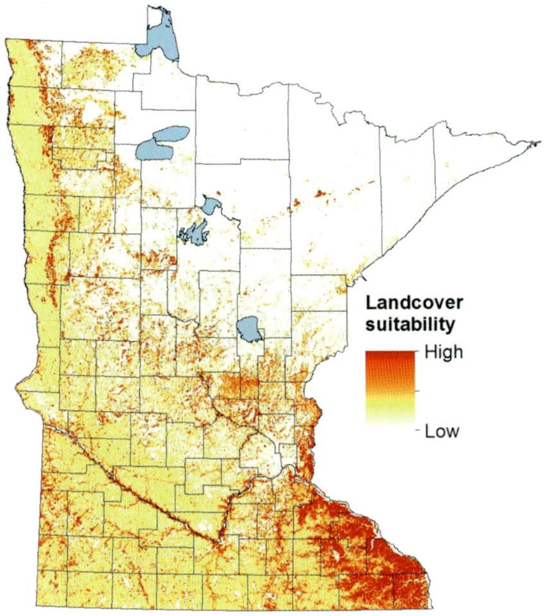

B. Landcover suitability map for the Lark Sparrow in Minnesota based on habitat, landscape context, and climate data gathered during the Minnesota Breeding Bird Atlas (2009–13) using the MaxEnt modeling approach.

singing males found along a section of the Old Gunflint Trail that was within the 2007 Ham Lake Fire area. Although the site consisted of at least 40 ha of "ideal habitat," the birds could not be relocated two days later. Russell (pers. comm.) noted that 2009 was a "big year for vagrant Lark Sparrows" in northern Minnesota and Wisconsin and speculated that the birds may have originated from a small breeding population in southern Manitoba.

Several areas with clusters of atlas records were located in ecological subsections characterized by the prominence of sandy outwash plains (i.e., the Pine Moraines and Outwash Plains and the Anoka Sand Plain). Dry grasslands, oaks, and jack pines grow in open, savanna-like communities on these soils, providing excellent habitat for the Lark Sparrow. In addition, Dakota and Goodhue Counties, the Minnesota River valley, and northwestern Minnesota are all areas where dry, sandy river terraces, granite outcrops, and gravel beach ridges provide suitable habitat. Each area is highlighted by the MNBBA landcover suitability map **(B)**, but the largest area of appropriate habitat is predicted to occur along the lower Mississippi River valley, despite the small number of atlas records south of Goodhue County.

It is difficult to assess the degree to which the Lark Sparrow's distribution has changed in the past century. Although its abundance has likely decreased, even in the Greater Twin Cities Metropolitan Region the species is still regularly reported at such sites as the Fish Creek Natural Area in Ramsey County and Crow Hanssen Park Reserve in Hennepin County (MOU 2021a). Unfortunately, Lark Sparrows have retreated from many eastern states, including New York, West Virginia, and Maryland, where they initially expanded as forests were cleared in the 1800s (J. Martin and Parrish 2020).

BREEDING HABITAT

The Lark Sparrow prefers structurally open habitats (J. Martin and Parrish 2020). In Minnesota, it is associated with oak savannas, dry grasslands, or pastures with small trees or shrubs scattered within or adjacent to open areas **(C)**. Sites with poor soils are ideal, as they provide the sparse ground cover and open bare areas that are used for foraging. Sandy soils are common, but sites that are gravelly or heavily grazed are also used. Abundance decreases as the amount of ground litter and the density of cover increase. Planted cover is rarely used unless it is managed to provide some scattered trees and shrubs and some open areas with sparse ground cover (Dechant, Sondreal, et al., "Lark Sparrow," [1999] 2002b).

POPULATION ABUNDANCE

The 2,300 Lark Sparrows that are estimated to reside in Minnesota are a very small percentage of the estimated 11 million breeding adults found across the United States and southern Canada (PIF 2021). Declining at an average rate of –1.18 percent per year, the population has experienced a significant decline across the BBS survey region from 1966 to 2019 **(D)**. Much of the decrease is attributed to the wide-scale loss of habitat in the eastern United States. Areas that were cleared for agriculture are now succeeding to forest, converting to residential development, or being more intensively cultivated. This loss has been minimized by significant increases in many western regions where higher precipitation in recent years may have improved habitat conditions and prey availability (J. Martin and Parrish 2020). In Minnesota, BBS data are too sparse to generate a reliable trend line (Sauer et al. 2020).

C. In southeastern Minnesota, the Lark Sparrow prefers open and sparsely vegetated sand dunes bordering the Mississippi River for nesting and foraging. Photo by Lee A. Pfannmuller.

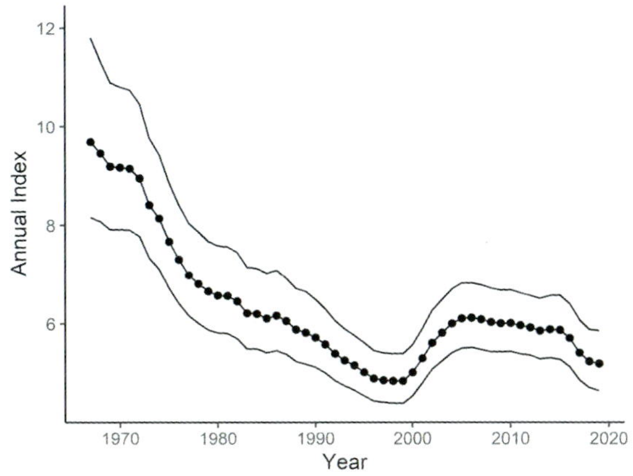

D. Surveywide breeding population trend and 95 percent credible interval (solid black lines) for the Lark Sparrow based on the 1967–2019 Breeding Bird Survey (Sauer et al. 2020).

CONSERVATION

Designated a moderate conservation priority (PIF 2021), the primary concerns regarding the Lark Sparrow's future are its declining population and threats to its breeding habitat. In Minnesota, it was officially classified *Special Concern* in 2013, owing to the small number of records documented by the MBS, its patchy distribution, and its rather narrow habitat requirements (MNDNR 2012). Fortunately, the Sparrow's breeding distribution is projected to remain stable under all three warming scenarios examined by NAS (2019). Indeed, major range expansions projected in Mexico are accompanied by small expansions along the eastern periphery of its breeding range, including in Minnesota.

Dechant, Sondreal, and colleagues ("Lark Sparrow," [1999] 2002b) recommended that grasslands be burned every five to eight years before Lark Sparrows return in the spring. Burning at moderate temperatures provides a matrix that includes unburned ground cover for nesting and more open areas with sparse cover for foraging. Removal of woody vegetation is a common grassland management practice, but retaining about 10 percent woody cover is critical for maintaining Lark Sparrows.

Although many grassland species have been the focus of intense studies, the Lark Sparrow has been largely overlooked. But its distinct plumage, its melodious song, and its unique habitat requirements make it one of the most interesting members of the sparrow family to inhabit Minnesota grasslands.

LARK BUNTING

Calamospiza melanocorys

Minnesota Seasonal Status: Summer vagrant; bred sporadically in the state from 1879 to 1964. Migrant visitant, particularly in western counties and the North Shore of Lake Superior. The Lark Bunting was very rare during the MNBBA.
Migration: Short- to medium-distance migrant that winters in Arizona, Texas, and the high plateau of northern Mexico in large wandering flocks, searching for the best winter food conditions.
Food: Ground forager and foliage gleaner that primarily consumes arthropods, especially grasshoppers, as well as waste grains and seeds of grasses and forbs.
Nest: Located on the ground in a shallow depression under or next to a shrub or a clump of grasses or forbs; small groups of buntings often nest close together.
Conservation Concern: Assigned a moderate PIF Continental Concern Score of 12/20.

The Lark Bunting has many old vernacular names, including White-winged Blackbird, Prairie Bobolink, and Buffalo-Bird. Some are suggestive of the species' grassland habitat, while others describe the distinguishing features of the male's breeding plumage: white wing coverts on an otherwise black bird. Roberts (1932) called the bunting "an especially interesting bird" and summarized Coues's (1874) description of the species as "one of the most singularly specialized of all the Sparrow tribe, pointing out that in some ways it is like a Lark, has the bill of a Grosbeak, the seasonal plumage changes of the Bobolink, notes and manner of singing like the Yellow-breasted Chat, [and the] gregarious habits of the Blackbirds." The species was first described in 1837 from specimens collected in 1833–34 by John Townsend, a member of the Wyeth expedition that traveled up the Platte River, across the Great Plains and Rocky Mountains, to the Pacific Ocean. Audubon (1827–38) used the specimens to illustrate the Lark Bunting in his classic *Birds of America.*

Striking in his nuptial plumage, during the nonbreeding season the male is a heavily streaked brown bird, similar to the more lightly streaked female. Regardless of the season, some of the distinguishing features for both sexes are a splash of white along the edge of the folded wing, a bluish-gray bill, and white-tipped tail feathers. Like many prairie birds, the male Bunting performs aerial song displays for courtship

Photo by Mike Lentz.

and territorial defense. Indeed, he is unique in possessing two different flight songs (Shane 2020). Saunders described the primary song as "a long mixture of series of short notes, two-note phrases or slurs. It is quite musical and pleasing" (Bent 1968). The aggressive flight song, however, contains harsher notes, is interrupted by more pauses, and is often delivered "communally with as many as 5 or 6 males displaying simultaneously" (Shane 2020).

A Great Plains endemic, the Lark Bunting has a range that spans from the southern Prairie Provinces of Canada south to the panhandles of Oklahoma and Texas; it is found west of the Missouri River and east of the Rocky Mountains. The core of its breeding range is in the northern high plains from central Montana south to northeastern Colorado. This species is nomadic, regularly changing its breeding locations depending on local site conditions.

MINNESOTA BREEDING DISTRIBUTION

The Lark Bunting was first discovered in Minnesota in 1879, when Roberts and Benner traveled to the prairies in Grant and Traverse Counties. A single male was collected near Herman in Grant County and a "pair was taken, the female having eggs about to be laid" in Traverse County, on the Prairie Coteau above Lake Traverse, where they were common (Roberts and Benner 1880). The next Minnesota records were migrants observed in Fillmore County by Dr. Hvoslef in 1893 and 1894 (Rustad 1951). Subsequently, the species' sporadic nature was demonstrated by the following observations in southwestern Minnesota: first, 1895, Jackson County, common; second, 1897, "common throughout the southwestern counties as far north as Traverse"; third, 1900, in all of the southwestern counties "over an area of at least ten thousand square miles"; fourth, 1901, counties unknown, "only a few birds seen, mostly transients"; fifth, 1927, considerable numbers in Rock and Pipestone Counties compared to 1929 when few were reported; sixth, 1936, small numbers in Lac qui Parle and Yellow Medicine Counties (Rysgaard 1936; Wilson 1937); and, seventh, 1964, Rock and Pipestone Counties, almost 1,000 juveniles.

Except as indicated, these records are from Roberts (1932) and B. W. Anderson and Getman (1965). The latter publication investigated the relationship of summer precipitation and habitat suitability and concluded, "there seemed to be a correlation between the occurrence of breeding Lark Buntings and below average precipitation." The last documented nesting of the Lark Bunting in Minnesota was in 1964. One can speculate that habitat change from intensive agriculture in southwestern Minnesota and/or the decline of the species' core population on the western plains was responsible for the decline in nesting records.

The Lark Bunting was never a regular summer resident; most of its breeding records occurred during years when the species was abundant in the state and precipitation was below normal, such as 1895–97, 1900, 1927, and 1936 (B. W. Anderson and Getman 1965). Roberts (1932) initially documented nesting in four counties: Traverse (1879); Lincoln (1897); Jackson (1898); and Pipestone (1927). Breeding records in four additional counties followed these from 1936 to 1964: first, Yellow Medicine in 1936 (Wilson 1937); second, Lac qui Parle in 1936 and 1964 (Rysgaard 1936; R. A. Grant 1965a; R. L. Huber 1964b); third, Big Stone in 1942 (Harrell 1942); and, fourth, Rock in 1964 (B. W. Anderson and Getman 1965).

Summer observations that lacked breeding evidence occurred primarily along the southwest edge of the state, from Clay County south to Rock County and east to Jackson and Cottonwood Counties, a slightly larger area than the nesting records. Most summer observations (35) were reported from 1960 to 1995. Subsequently, summer records were limited to a few early June migrants. One exception was a male observed in appropriate habitat at Glacial Ridge NWR, Polk County, from June 23 to July 1, 2007 (Kessen and Svingen 2008); a second observation on the refuge occurred on June 11, 2016 (Tustison 2016).

The MNBBA documented only one Lark Bunting record from a block in Jackson County, east of

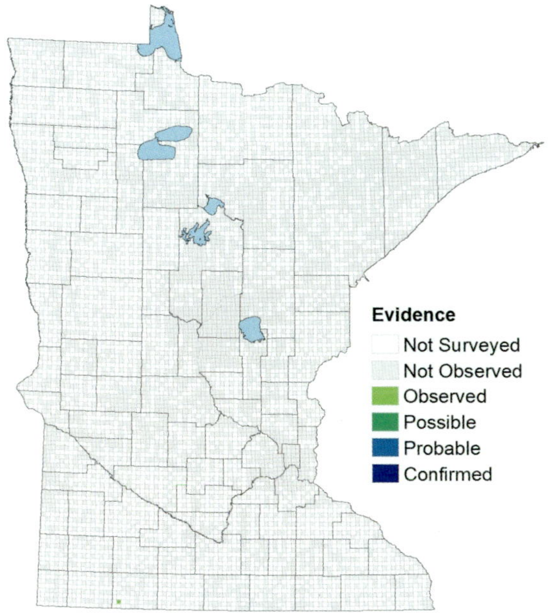

A. Breeding distribution of the Lark Bunting in Minnesota based on the Breeding Bird Atlas (2009–13).

Round Lake, on June 14, 2012 **(A)**. Although it was observed in appropriate habitat, it was never relocated by the block owner, John Schladweiler, or other bird-watchers (Schladweiler 2012). Since the atlas was completed, breeding-season observations have been reported in 2016 in Lac qui Parle, Lyon, Polk, and Wilkin Counties (MOU 2021a).

Except for the Dakotas, there is no breeding evidence for the Lark Bunting in adjacent states. It is accidental in Wisconsin (S. D. Robbins 1991) and casual to accidental in Iowa (Kent and Dinsmore 1996).

BREEDING HABITAT

The Lark Bunting is found in prairie grasslands, shrub-steppes, desert grasslands, and agricultural areas of the high plains **(B)**. Breeding grounds vary each year depending on the weather and its impact on prey abundance. Variation in precipitation and subsequent food availability influences breeding success, particularly "in seasonally unpredictable climates such as the Great Plains, where drought often affects food supplies over broad areas of the landscape" (Shane 2020). In Colorado, the Lark Bunting was more common in shortgrass prairie than in mid- and tall grasses but was also present in cropland and farm fields. "Buntings use shortgrass, if not heavily grazed and if interspersed with taller vegetation (sunflowers, ragweed, thistles)" (Kingery 1998).

POPULATION ABUNDANCE

The Lark Bunting population numbers approximately 12 million birds in the United States and Canada (PIF 2021). BBS data document a significant, surveywide decline from 1966 to 2019 that averaged −3.70 percent per year **(C)**. North Dakota witnessed an average annual decline of −6.83 percent, while the entire Prairie Pothole Region population has declined an average of −8.32 percent per year (Sauer et al. 2020). PIF estimated a cumulative population decline of −86 percent from 1970 to 2014 and projected another 50 percent of the population could be lost in just sixteen years (Rosenberg, Kennedy, et al. 2016). BBS data are the only data available to monitor long-term population trends. Nevertheless, their suitability for this nomadic species is a concern (Kingery 1998; Shane 2020).

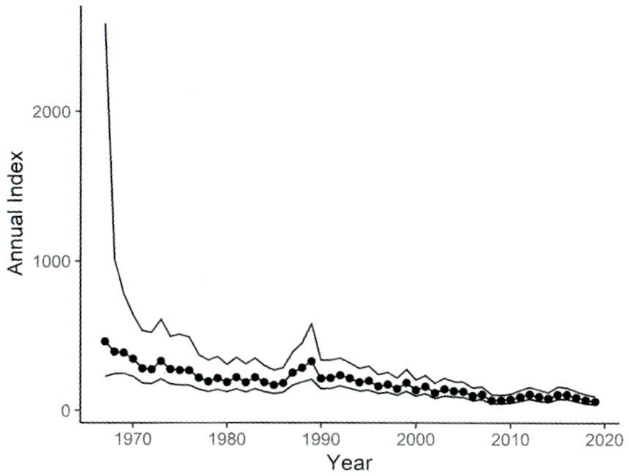

C. Surveywide breeding population trend and 95 percent credible interval (solid black lines) for the Lark Bunting based on the 1967–2019 Breeding Bird Survey (Sauer et al. 2020).

Factors responsible for the decline are unclear. Although much of the native grassland habitat within the species' breeding range is now tilled cropland, the birds do successfully nest in agricultural fields (Shane 2020). Loss and degradation of native grasslands in the winter range may be one factor influencing breeding populations. Indeed, Shane (2020) noted that "overwinter survival in Chihuahuan and Sonoran Deserts may be a major factor where survivorship may be limited by food availability."

CONSERVATION

The perilous status of the Lark Bunting's population, coupled with threats to its breeding and wintering habitat, were important considerations in assigning it a moderate conservation priority (PIF 2021). It is also designated a Common Bird in Steep Decline (Rosenberg, Kennedy, et al. 2016). Unfortunately, under a 2°C warming scenario, NAS (2019) projected a moderate vulnerability for the Bunting's breeding range, including a 27 percent range loss and a 12 percent gain by 2050. Because the small gains include eastward range expansions in the Dakotas, this very rare species might become a more regular sighting in western Minnesota.

Although planted cover provided by the Conservation Reserve Program has resulted in some local population increases in the Dakotas (Shane 2020), more field assessments of suitable habitat are needed to guide conservation efforts. Meanwhile, management recommendations are similar to those for many grassland species and include protecting large grassland tracts, delaying mowing until after the nesting season, using minimum tillage practices on croplands, and minimizing the use of persistent pesticides that degrade slowly (Shane 2020).

B. Typical breeding habitat of the Lark Bunting includes native short-grass prairie, sage brush prairies, and agricultural fields, such as grazed pastures, alfalfa fields, and hayfields. Photo by Lee A. Pfannmuller.

CHIPPING SPARROW

Spizella passerina

Minnesota Seasonal Status: Regular breeding resident and migrant; occasionally observed during the winter. The Chipping Sparrow was very abundant during the MNBBA.
Migration: Short- to medium-distance migrant that winters in the southern United States and northern Mexico; a year-round resident in Mexico and Central America.
Food: Omnivorous ground forager that feeds primarily on seeds supplemented with arthropods during the breeding season.
Nest: Cup nest placed at variable heights within a coniferous tree or shrub.
Conservation Concern: Assigned a moderate PIF Continental Concern Score of 9/20.

Photo by David Brislance.

Walk along any city street in the summer and you are bound to encounter the little Chipping Sparrow. You may hear it belting out its dry, monotone trill from the top of a nearby conifer or you may spot it running along the sidewalk, stopping to feed on seeds that have dropped from the boulevard's trees. So common is this species in urban settings that Roberts (1932) described it as the "most thoroughly domesticated of all our native sparrows."

In its breeding plumage the Chipping Sparrow is easily identified by its small size, rufous crown, black eyeline, and the creamy white line above the eye. Two faint wing bars are present, and the breast lacks any streaking. The male and female are indistinguishable. *Spizella passerina,* the species' scientific name, translates to "finch-like sparrow."

It is easy to forget that this urban dweller is actually a forest inhabitant. Although the bird is easily identified in cities and towns, in northern Minnesota birders can be led astray if they rely solely on its song for identification. Like the Chipping Sparrow, the Pine Warbler also gives a trill on a single note. But the sweetness of the Warbler's song and its slower delivery distinguish it from that of the Sparrow. Far more challenging is the similarity between the song of the Dark-eyed Junco and the Chipping Sparrow. Many a skilled birder has been confident they heard one of the two birds only to discover, upon spotting the songster, that it was indeed the other species. A rough rule of thumb is that the Junco's trill is a bit slower and a little more musical than the very rapid, mechanical trill of the Chipping Sparrow. In northern Minnesota, odds are that you heard a Chipping Sparrow because atlas records for the Sparrow outnumbered those of the Junco by ten to one. But it is always best to locate the vocalist and confirm its identity.

Broadly distributed across the United States and Canada, from eastern Alaska south through California and east to the Atlantic coast, the Chipping Sparrow is a common species throughout its breeding range. Small populations also occur in Mexico and Central America. Because of its adaptation to human-modified landscapes, it is one of the few species that has prospered despite intensive development.

MINNESOTA BREEDING DISTRIBUTION

In the early 1900s, Roberts (1932) considered the Chipping Sparrow a common breeding species throughout the state except for the northern Red River valley. Breeding evidence was reported in Crow Wing, Goodhue, Hennepin, Houston, Isanti, Lake of the Woods, Polk, Sherburne, and St. Louis, and Stearns Counties and from Cass Lake and Itasca State Park. Roberts wrote a charming account of the species' broad habitat preferences: "It dwells familiarly about our homes and builds its nest in the ornamental shrubbery, even in the vines and shrubs about our doorsteps. It is equally at home in wilder places and is an abundant bird in the spruce groves and open glades of the northern forests, where, like the Robin, it seems somewhat out of place to those who are wont to associate it with civilized surroundings."

Little has changed since the days of Roberts. Green and Janssen (1975) and Janssen (1987) each described the species as a common, statewide summer resident. Green and Janssen cited its abundance in the eastern and central regions of the state, while Janssen claimed it was most abundant in eastern and northern Minnesota. Janssen identified 45 counties where nesting had been confirmed since 1970. Hertzel and Janssen (1998) added another 11 counties to the list. The MBS reported a total of 973 breeding-season locations. Reports were common from all but the most intensively cultivated regions of the state, where survey efforts were minimal.

The Chipping Sparrow's statewide distribution was well-documented by the MNBBA. Participants reported 6,349 Chipping Sparrow records in 65.4 percent of the surveyed atlas blocks (3,135/4,792) and in 83.8 percent of the priority blocks (1,959/2,337). Breeding was confirmed in 13.7 percent (657) of the blocks **(A)**. Chipping Sparrows were the seventh most commonly detected species during the atlas and were reported from each of Minnesota's 87 counties. They were confirmed breeding in all but one county, Faribault, where breeding was probable. Detections were most prominent in the Twin Cities and Brainerd Lakes region, where atlas participants made an impressive field effort.

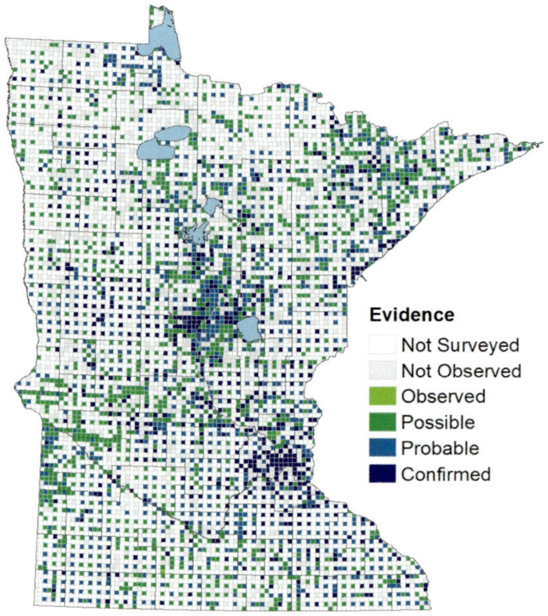

A. Breeding distribution of the Chipping Sparrow in Minnesota based on the Breeding Bird Atlas (2009–13).

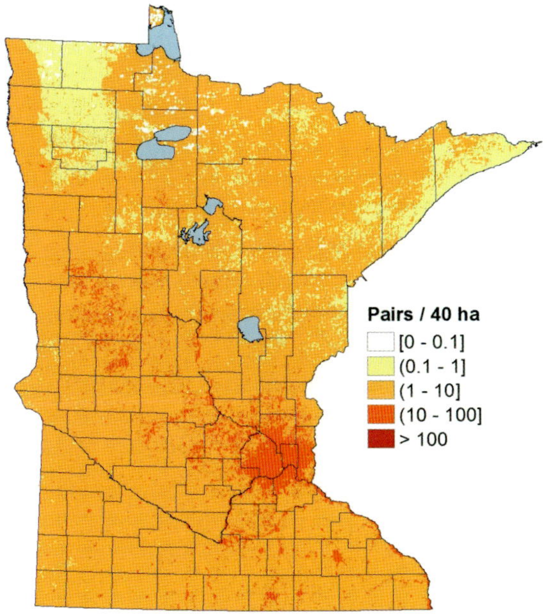

B. Predicted breeding distribution (pairs per 40 ha) of the Chipping Sparrow in Minnesota based on habitat, landscape context, and climate data gathered during the Minnesota Breeding Bird Atlas (2009–13) using the Generalized Linear Modeling method with an adjustment for detectability.

The Chipping Sparrow's predicted probability map indicates that its highest breeding densities are found in the Twin Cities metropolitan region and in smaller communities throughout the Eastern Broadleaf Forest Province **(B)**. Although it is found statewide, the lowest breeding densities are predicted in portions of southern Lake County, throughout much of Cook County, and in the Tallgrass Aspen Parklands Province in northwest Minnesota.

Atlas data confirm that the Chipping Sparrow remains at least as widespread today as it was nearly one hundred years ago, occurring even on farmsteads and in small communities throughout the Red River valley and the Upper Minnesota River valley. An inhabitant of open woodlands, the species surely benefited from the clearing of North America's eastern deciduous forest in the 1800s. Several accounts mention its decline following the introduction of the House Sparrow (Bent 1968; Cutright et al. 2006; Middleton 2020). In the Twin Cities the House Sparrow was well established by 1877, but Roberts (1932) makes no mention of the exotic species' impact on the Chipping Sparrow in Minnesota. Overall, the Sparrow is considered more abundant and widely distributed today than it was prior to European settlement (Middleton 2020).

BREEDING HABITAT

Found in a variety of forest habitats, the Chipping Sparrow shows a strong preference for upland conifer stands ranging in age from recent clear-cuts to mature forests **(C)**. It is also found in lowland conifers dominated by spruce and tamarack (Green 1995; Danz et al. 2007). Open understories, or the adjacency of open areas to forest stands, are important habitat components because of the Sparrow's ground-foraging habits. As a result, open woodlands and forest edges are especially suitable.

The open nature of city parks and housing subdivisions are perfectly suited to the Chipping Sparrow's habitat requirements. Even in the extensively forested landscapes of Minnesota and Wisconsin's national forests, the Chipping Sparrow was most frequently associated with small towns embedded within the forests' boundaries (Niemi, Howe, et al. 2016).

POPULATION ABUNDANCE

PIF (2021) estimated the Chipping Sparrow's global population at 240 million birds, with 230 million residing in the United States and Canada. Minnesota was estimated to support approximately 4.5 million breeding adults. A statewide population estimate generated using MNBBA point count data was identical at 4.5 million (95 percent confidence interval of 4.1–4.8 million).

Across southern Canada and the United States, BBS data reveal a long-term, significant population decline, averaging −0.63 percent per year from 1966 to 2019 (Sauer et al. 2020). Local population trends, however, vary widely. Populations in the western United States and Canada show strong declines, while populations in the central and southeastern United States are faring well; elsewhere populations are largely stable or slightly decreasing. In Minnesota, Chipping Sparrow numbers steadily increased from 1967 to 2006 before experiencing a rather sharp decline. Together, the increase and decrease averaged out to a nonsignificant trend from 1967 to 2019 **(D)**.

C. Mature coniferous trees are an important feature of the Chipping Sparrow's habitat in areas as diverse as lowland conifer bogs in northern Minnesota to a residential landscape with scattered mature conifers in the Twin Cities metropolitan region. Photo by Gerald J. Niemi.

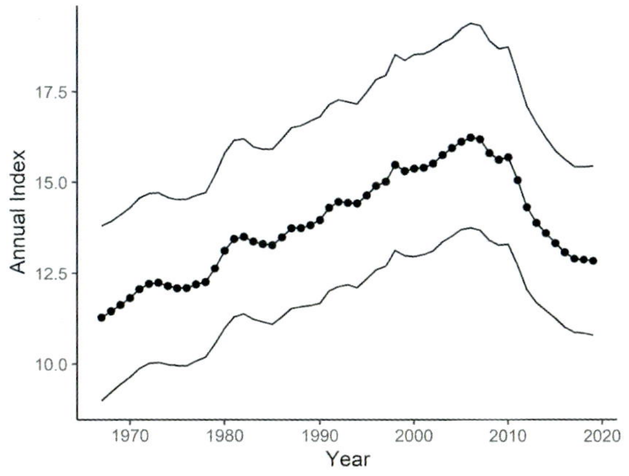

D. Breeding population trend and 95 percent credible interval (solid black lines) for the Chipping Sparrow in Minnesota based on the 1967–2019 Breeding Bird Survey (Sauer et al. 2020).

In Minnesota's two national forests, the Chipping Sparrow has demonstrated a significant annual decline, averaging −3.43 percent in the Chippewa National Forest and −4.15 percent in the Superior National Forest from 1995 to 2020 (Grinde, Bednar, et al. 2020). The decline may be attributed in part to the recent decrease in forest harvesting activities documented in both forests, especially given the Sparrow's high breeding densities in young regenerating conifer stands (Niemi, Howe, et al. 2016).

Other factors often cited as responsible for local population declines include Brown-headed Cowbird parasitism, competition in urban areas with the House Sparrow and the House Finch, forest succession that reduces forest openings, and intensive agricultural practices that eliminate shelterbelts and small woodlots. Factors cited for local population increases point largely to increased settlement and the landscape plantings that accompany such development (Middleton 2020).

CONSERVATION

Despite the Chipping Sparrow's broad distribution, large population, and adaptability to human-modified landscapes, its population decline prompted a moderate conservation ranking (PIF 2021). Rising temperatures present another concern. Under a 2°C warming scenario, NAS (2019) projected a moderate vulnerability for the species' breeding distribution, with a 26 percent northward range expansion minimally compensating for a 29 percent loss by 2050. With losses projected across western Minnesota, it is unclear if the region's scattered towns and farmsteads will be sufficient to maintain a resident population for this common avian neighbor.

CLAY-COLORED SPARROW

Spizella pallida

Minnesota Seasonal Status: Regular breeding resident and migrant. The Clay-colored Sparrow was common during the MNBBA.
Migration: Short- to medium-distance migrant that winters in portions of Texas, Baja California, and the highlands of northern and central Mexico.
Food: Omnivorous ground forager that consumes seeds and arthropods.
Nest: Cup nest, usually placed in a low shrub.
Conservation Concern: Assigned a moderate PIF Continental Concern Score of 10/20.

While warblers dazzle us with their striking plumages each spring, sparrows challenge us with their infinite shades of gray and brown. The Clay-colored Sparrow is one such example. Roberts (1932) described it as a "light-colored, plainly garbed, inconspicuous little bird." *Spizella pallida,* the species' scientific name, simply means "a little finch of pale color." As unadorned as the bird may appear, the following features distinguish it from other sparrows: a brown ear patch outlined in black; a white stripe that extends down from the base of the bill; and a white line above the eye. Other features include a light median crown stripe and the absence of any spots or streaking on the breast and belly.

The male's song may be the best clue to the presence of this shrub-loving species. Roberts (1932) described it as a low, insect-like buzz that is repeated two to three times, characterizing it as "hardly a song at all" and comparing it to "the performance of the cicada but rather more rasping and unattractive than the insect's fiddling." While it may not be a pleasing melody, it is unique and not easily confused with that of other species.

Confined to the northern interior of North America, the Sparrow's breeding range stretches from eastern British Columbia southeast to southern Ontario and Quebec, and south through the northern Great Plains and Great Lakes region. Small, disjunct populations also occur along Hudson Bay and James Bay. High breeding densities are found throughout Canada's Prairie Provinces and North Dakota.

MINNESOTA BREEDING DISTRIBUTION

Roberts's (1932) account of the Clay-colored Sparrow described it as an abundant summer resident. "Indeed, nowhere in the state, except in the deep unbroken forest, is the Clay-colored Sparrow without a goodly representation." It was "especially abundant in the poplar belt bordering the Red River Valley on the east." And, in northern Minnesota, it could be found wherever "there are clearings, old 'burns,' windfalls, or other openings." Nests were found in Fillmore, Goodhue, and Mower Counties in southeastern Minnesota; in Big Stone, Hennepin, Isanti, and Sherburne counties in central Minnesota; and in Kittson, Marshall, Norman, Pennington, and Polk Counties in the northwest. Nesting was also reported at Itasca State Park, Leech Lake, and Mille Lacs. Decades later, Green and Janssen (1975) simply noted that the Clay-colored Sparrow was a summer resident throughout the state but was "very scarce" in southern Minnesota. Janssen (1987) stated that it originally "bred sparingly across the southern part of the state," but nearly disappeared following the 1920s. Although reports beginning in the mid-1970s suggested that it was slowly returning to the region, Janssen excluded most of the southern two tiers of counties from the species' distribution map. He identified 21 counties where nesting had been confirmed since 1970. Hertzel and Janssen (1998) added another 11 counties to the list; none was south of the distribution limit identified by Janssen.

Photo by Jean Brislance.

MBS biologists documented 1,562 breeding-season locations by 2014, including many records in southwest and south-central Minnesota. Records from every county along the Iowa border, except Fillmore, supported Janssen's (1987) suggestion that the species was reoccupying portions of its former breeding range in the south. The species was still rare, however, in the southeast blufflands.

MNBBA participants reported the Clay-colored Sparrow in all 87 counties and confirmed breeding in 56 counties **(A)**. The latter included three new counties along the state's southern border: Fillmore, Freeborn, and Jackson. Breeding was deemed probable in 26 counties, and possible in the remaining five counties (Cook, Faribault, Houston, Martin, and Watonwan). A total of 3,872 detections were reported in 40.5 percent (1,933/4,767) of the surveyed atlas blocks. Records were most common in central and northwestern Minnesota.

The MNBBA probability map predicted that the Tallgrass Aspen Parklands Province supports the densest population of Clay-colored Sparrows, especially eastern Kittson County **(B)**. The prediction corroborates the Reverend P. B. Peabody's observation more than one hundred years ago that the species "is well-nigh the most abundant summer inhabitant" in Kittson County, occupying the region's "meadow brushland" (Roberts 1932). Elsewhere it is abundant along the Glacial Lake Agassiz beach ridges, in the Hardwood Hills and the Pine Moraines and Outwash Plains Subsections of west-central Minnesota, and in the shrubby wetlands north and south of Red Lake. Abundance is predicted to slowly decrease to the south but rapidly decreases across northern Minnesota. The exception is the former Ham Lake Fire area in northern Cook County. Even though Clay-colored Sparrows were not detected there, the site is predicted to provide suitable habitat.

Overall, the Clay-colored Sparrow's breeding distribution in Minnesota has experienced few changes in the past century. Although it was scarce in southern Minnesota in the mid-1900s, it now is a regular species throughout much of the region. Meanwhile, in the north, a loss of forest cover along the southern and western edges of the Laurentian Mixed Forest

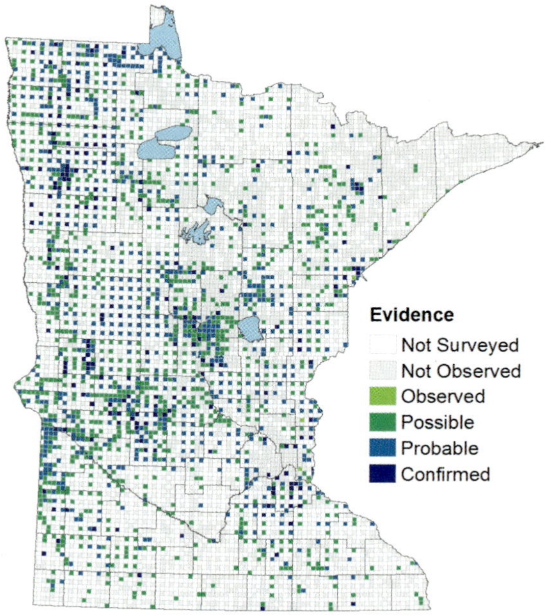

A. Breeding distribution of the Clay-colored Sparrow in Minnesota based on the Breeding Bird Atlas (2009–13).

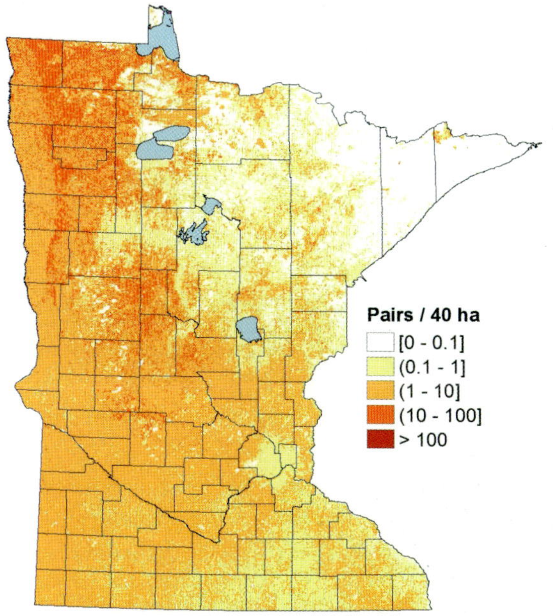

B. Predicted breeding distribution (pairs per 40 ha) of the Clay-colored Sparrow in Minnesota based on habitat, landscape context, and climate data gathered during the Minnesota Breeding Bird Atlas (2009–13) using the Generalized Linear Modeling method with an adjustment for detectability.

Province likely provides more habitat opportunities than in Roberts's day. Indeed, the Clay-colored Sparrow has gradually expanded its range to the north and east throughout North America as logging, farming, and development have opened up densely forested landscapes (T. A. Grant and Knapton 2020).

BREEDING HABITAT

Shrubs in an otherwise open landscape are the defining feature of the Clay-colored Sparrow's breeding habitat. The species occurs in dry shrubby uplands, such as forest clear-cuts and young conifer plantations, in dry-to-mesic grasslands interspersed with low shrubs, and in wet lowlands, such as alder–willow thickets **(C)**. Suitable foraging areas must be located nearby. Short, sparse vegetation is critical and includes cultivated fields, pastures, dry prairies, and restored grasslands (Dechant, Sondreal, et al., "Clay-colored Sparrow," [1998] 2002). Clay-colored Sparrows also utilize fields enrolled in the Conservation Reserve Program (CRP), though at lower breeding densities than native grasslands (T. A. Grant and Knapton 2020). Such sites are often planted with alfalfa and sweet clover, species that grow in dense clusters that are structurally similar to low shrubs.

Studies are equivocal regarding the species' sensitivity to patch size. At least one Minnesota study suggested that nest success is higher on large, contiguous grassland tracts (R. G. Johnson and Temple 1990). Winter and colleagues (2006), however, concluded that patch size may be less important than local habitat structure and landscape composition.

POPULATION ABUNDANCE

Based on BBS data, PIF (2021) generated a population estimate of 60 million breeding adults across the United States and Canada and approximately 1.9 million in Minnesota. The MNBBA-derived estimate for Minnesota was considerably higher, at 3.4 million (95 percent confidence interval of 3.1–3.8 million). The species' high abundance in the aspen parklands and some of the extensive wetlands surrounding Red Lake, regions well covered by the more intensive MNBBA surveys, likely contributed to the higher estimate.

Across their limited breeding range, Clay-colored Sparrows have experienced a statistically significant decline, averaging −0.99 percent per year from 1966 to 2019 (Sauer et al. 2020), with a cumulative decline of −43 percent from 1970 to 2014 (Rosenberg, Kennedy, et al. 2016). The BBS trend is similar in Minnesota, where the species has experienced a significant average annual decline of −0.86 percent, which is most pronounced since the late 1990s **(D)**.

The loss of native prairie and grassland habitats, on both the breeding and wintering grounds, is the most significant factor impacting Clay-colored Sparrows (T. A. Grant and Knapton 2020). Although much of the loss in North America occurred in the late 1800s and early 1900s, significant losses continue. Between 2008 and 2012, more than 7.3 million acres of previously uncultivated land in the United States were converted to cropland (DeVore 2018). Similar changes are occurring in the Chihuahuan Desert grasslands of northern Mexico, an important wintering area for Clay-colored Sparrows (T. A. Grant and Knapton 2020).

C. Just a few scattered shrubs in a low-lying grassland are sufficient to attract a Clay-colored Sparrow nesting pair. Photo by Lee A. Pfannmuller.

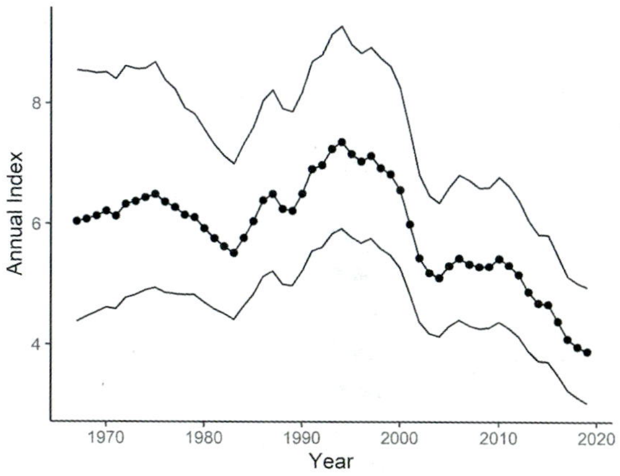

D. Breeding population trend and 95 percent credible interval (solid black lines) for the Clay-colored Sparrow in Minnesota based on the 1967–2019 Breeding Bird Survey (Sauer et al. 2020).

CONSERVATION

Given the loss of habitat, its restricted distribution, and its significant population decline, the Clay-colored Sparrow has been designated a moderate conservation priority (PIF 2021). Management recommendations focus on providing suitable shrub habitat without compromising a site's viability for other grassland species. Luckily, few shrubs are needed to attract this shrub-dependent species. One study found that as little as 3 percent shrub cover raised the probability of Clay-colored Sparrows being present to 69 percent; 20 percent shrub cover raised the probability to 95 percent (Madden et al. 1999). Allowing grasslands to remain idle long enough to encourage some shrubby cover, and protecting small patches when burning or mowing occurs, improves habitat (Dechant, Sondreal, et al., "Clay-colored Sparrow," [1998] 2002).

Another important measure includes protecting large grassland tracts. This practice may reduce the impacts of predation, which can be responsible for significant nest losses (Winter, Johnson, and Shaffer 2005). Promoting farm policies that ensure erodible croplands are taken out of production is equally important.

Compounding the loss of habitat are the impacts of climate change. The species is moderately vulnerable under a 2°C warming scenario. By 2050, NAS (2019) has projected a loss of 44 percent of the species' breeding distribution, including its extirpation from all but the northeastern corner of Minnesota; the loss would be compensated by a range-wide gain of only 19 percent. Under a 3°C warming scenario, the nearest population would be more than 600 km north of Minnesota. Clearly, this inconspicuous little sparrow has a precarious future ahead.

FIELD SPARROW

Spizella pusilla

Minnesota Seasonal Status: Regular breeding resident and migrant; occasionally observed during the winter. The Field Sparrow was common during the MNBBA.
Migration: Only those birds that occur in the most northern reaches of its breeding range, including Minnesota, are migratory and winter in the southern United States.
Food: Ground-foraging omnivore feeding largely on seeds and arthropods.
Nest: Cup nest located on or near the ground early in the season and in shrubs and small trees as the season progresses.
Conservation Concern: Assigned a moderate PIF Continental Concern Score of 12/20.

Nearly the size of its well-known relative the Chipping Sparrow, the small, plainly clad Field Sparrow is far less likely to be encountered where homes and people dominate the landscape. Fond of young successional habitats, such as old fields with shrubby edges, this little sparrow is easily overlooked. Bent (1968) described it as "a likeable, friendly little bird," characterized by its gentle nature. Lacking any boldly distinctive plumage features, its plain breast, rufous crown, rufous patch behind the eye, white eye-ring, and pink bill help differentiate it from other small sparrows. Its Latin name, *Spizella pusilla,* means "little finch."

During the breeding season, the male's unique song provides the best identification clue. Roberts (1932) wrote: "The plain appearance of this little bird is compensated for by a sweetly plaintive, appealing voice, and its song, though rather simple and not loud, is an attractive feature of the hills and dales where it makes its home." Sung from a low perch, the song begins with a series of long, melancholy notes, each slurring downward in pitch and then transitioning into a higher-pitched trill. Some have likened it to a bouncing rubber ball. Weygandt (1930) characterized it as "a little song, a trembling and lonely melody, minor certainly, but never uncheerful."

The Field Sparrow breeds from the northern Great Plains in eastern Montana south to central Texas, and east to southern New England and northern Florida. Some of the highest breeding densities occur in the mixed grass prairies of Nebraska and in the states of Kentucky, Missouri, and Tennessee.

Photo by Jean Brislance.

MINNESOTA BREEDING DISTRIBUTION

Roberts (1932) first encountered the Field Sparrow in northern Ramsey County in 1884. Like many southern species, it had slowly expanded its range northward following the Mississippi River, "becoming more generally distributed in Minnesota during the last forty or fifty years." By the late 1920s, it was established as a regular summer resident in the Twin Cities region and was "locally common" in the southern third of the state. It was most abundant in the southeastern counties, from Goodhue south to Houston, but records occurred as far north as Sherburne and Isanti Counties in the east (rarely to southern Pine County) and to Chippewa County in the west. It was an uncommon species in southwestern Minnesota, having first appeared in Pipestone County in 1923. Although there were records from localities farther north, Roberts questioned their authenticity and assigned them to "errors in identification." University staff had carefully searched for the species in central and northern Minnesota to no avail. At the time, nesting was confirmed in Fillmore, Goodhue, Hennepin, Houston, Isanti, Sherburne, and Wabasha Counties.

Little had changed when Green and Janssen (1975) published an updated account of the species' status. Summer records had been reported as far north as Otter Tail, Itasca, and southern St. Louis Counties, but nesting evidence was still confined to southeastern and east-central Minnesota. Janssen (1987) added that it was "well represented" along the Minnesota River valley as far as the South Dakota border, but it decreased in abundance in the south-central and southwestern regions. Although nesting had yet to be confirmed north of Morrison County, he noted an increasing number of observations in northern counties, including Aitkin, Hubbard, and Itasca. Since 1970, nesting had been confirmed in 19 counties. Hertzel and Janssen (1998) later added two more counties to the list.

The MBS reported a total of 570 breeding-season locations for the Field Sparrow. Although records in northern Minnesota remained rare, the species was reported at several locations in Hubbard County and as far north as southern Polk County. Records were common throughout southern Minnesota, including numerous records across the Prairie Coteau region of southwestern Minnesota.

A very similar distribution was documented by the MNBBA's 1,039 Field Sparrow records, with the exception of additional records extending farther to the north and east in the Laurentian Mixed Forest Province **(A)**. Reported in 11.3 percent (536/4,741) of the surveyed atlas blocks, the records spanned 73 of Minnesota's 87 counties. Sixteen of the 32 counties where nesting was confirmed had not been previously reported by Hertzel and Janssen (1998).

Probable breeding was reported as far north as southern Koochiching County and as far east as eastern Carlton County. Possible breeding records were found in northern Itasca and southern Beltrami Counties. The northern limit of confirmed breeding, however, occurred in Isanti County in the east, Morrison and Todd Counties in central Minnesota, and Otter Tail County in the west.

The MNBBA probability map predicts that moderate breeding densities occur primarily in the Blufflands and Rochester Plateau Subsections of south-

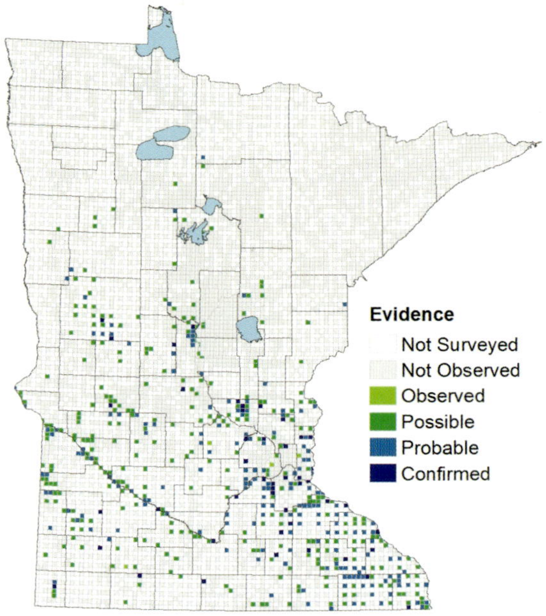

A. Breeding distribution of the Field Sparrow in Minnesota based on the Breeding Bird Atlas (2009–13).

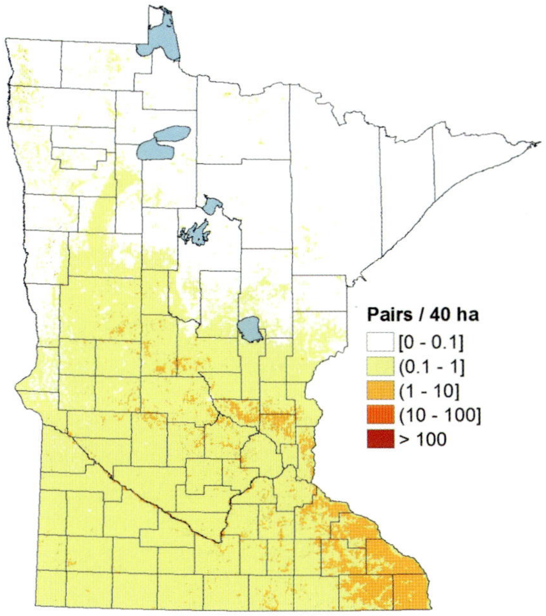

B. Predicted breeding distribution (pairs per 40 ha) of the Field Sparrow in Minnesota based on habitat, landscape context, and climate data gathered during the Minnesota Breeding Bird Atlas (2009–13) using the Generalized Linear Modeling method with an adjustment for detectability.

eastern Minnesota, the Anoka Sand Plain Subsection north of the Twin Cities, and along the entire length of the Minnesota River valley **(B)**; low densities are predicted elsewhere across the southern two-thirds of the state. Although the MNBBA did not detect the species in far northwestern Minnesota, the map suggests that some suitable habitat may be available. Indeed, the Manitoba breeding bird atlas (2010–14) found Field Sparrows in several blocks just north of the Minnesota border (Artuso et al. 2019).

Clearly, the Field Sparrow has continued its northward range expansion in Minnesota at a rather slow pace over the past one hundred years. The peatlands and dense forests of north-central and northeastern Minnesota, however, continue to provide a barrier to this species of open habitats. Elsewhere in its breeding range, few large-scale changes have been noted in its distribution (Carey et al. 2020).

BREEDING HABITAT

The Field Sparrow depends on successional habitats that include abandoned farm fields, hedgerows, shelterbelts, and woodland edges **(C)**. Shrub grasslands are generally preferred. When shrubby edges of woodlands are selected, the birds rarely venture more than a few meters into the woods or more than 12–15 m into the adjacent grasslands or fields (Dechant, Sondreal, et al., "Field Sparrow," [1999] 2002a).

Overall, the most important habitat components are dense grassy cover of moderate height and woody vegetation. Structural features appear to be more important than patch size, since sites smaller than 2 ha may be used. Sites with scattered patches of shrubs are preferred over continuous cover (Dechant, Sondreal, et al., "Field Sparrow," [1999] 2002a). At the landscape scale, one of the most consistent characteristics of the Field Sparrow's breeding habitat is its avoidance of human habitation (Carey et al. 2020).

POPULATION ABUNDANCE

BBS data generated a population estimate of 9.3 million breeding adults in the United States and southern Canada, with a Minnesota estimate of 100,000 (PIF 2021). MNBBA data, however, projected a statewide population of nearly 348,000 adults (95 percent confidence interval of 266,000–625,000).

Perhaps the most notable feature of the Field Sparrow is its precipitous population decline. From 1966 to 2019, BBS data documented a significant decline, averaging −2.18 percent per year across the survey region **(D)**. Since it experienced a cumulative decline of −62 percent between the years 1970 and 2014, PIF listed it as a Common Bird in Steep Decline (Rosenberg, Kennedy, et al. 2016). The decline is nearly range-wide; the only significant regions showing an increase are in the northern and central Great Plains. Because it is encountered infrequently along Minnesota BBS routes, the species' statewide population trend lacks statistical precision but suggests a declining population, dropping an average of −1.56 percent per year from 1967 to 2019 (Sauer et al. 2020).

C. Woodland edges adjacent to dense, grassy cover provide optimal nesting habitat for the Field Sparrow. Photo by Lee A. Pfannmuller.

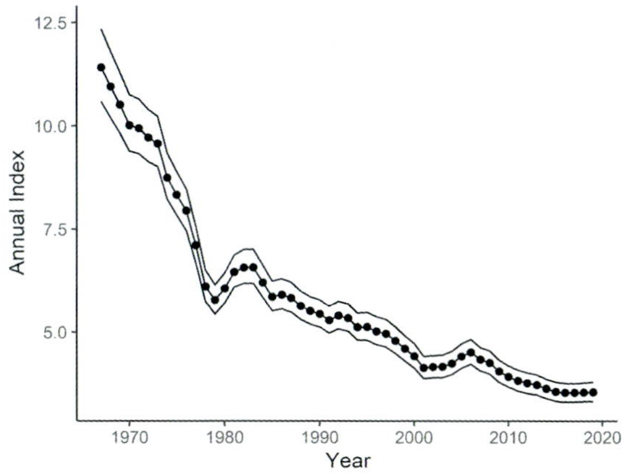

D. Surveywide breeding population trend and 95 percent credible interval (solid black lines) for the Field Sparrow based on the 1967–2019 Breeding Bird Survey (Sauer et al. 2020).

The loss and degradation of its ephemeral, successional habitat are considered the principal factors responsible for the Sparrow's decline. In New England, for example, the well-documented loss of farmland and its subsequent reforestation may have initially created increased opportunities for the species; but, as time passed, succession closed the forest canopy and eliminated suitable habitat. In contrast, shrub invasion of grassland sites that are not managed to prevent woody succession may be contributing to the Field Sparrow's increasing populations in the region of the Great Plains.

CONSERVATION

Although the Field Sparrow is still relatively abundant, it is designated a moderate conservation priority in light of its declining population (PIF 2021). Management recommendations focus on retaining some woody vegetation in grassland areas and conducting management burns on a rotational schedule. This allows the retention of woody vegetation and dense herbaceous cover for nesting birds in a portion of each managed area.

One of the most pleasing songs heard among the hedgerows and shelterbelts of southern and central Minnesota each summer, the Field Sparrow's sweet melody may soon be heard even farther north. Despite a projected loss of 47 percent of its North American breeding range under a 2°C warming scenario, NAS (2019) predicted a northward range expansion across Minnesota. Unfortunately, under all three warming scenarios they examined, losses to the species' breeding range were always significantly greater than the gains, gradually resulting in a very restricted breeding distribution for this likeable little sparrow.

DARK-EYED JUNCO

Junco hyemalis

Minnesota Seasonal Status: Regular breeding species and migrant; some individuals regularly overwinter in Minnesota. The Dark-eyed Junco was uncommon during the MNBBA.
Migration: Short- to medium-distance migrant that overwinters throughout the United States and parts of southern Canada south to northern Mexico.
Food: Seeds, arthropods, waste grain, and some fruit, primarily foraged from the ground.
Nest: Cup nest in highly variable locations, but usually on or near the ground in roots, moss, or grasses.
Conservation Concern: Assigned a low PIF Continental Concern Score of 8/20.

Photo by David Brislance.

The Dark-eyed Junco is known to most Minnesotans as a bird that frequents feeders during migration and in winter or is observed on the side of the road, where it flashes its two distinctive white outer tail feathers and darts away into the surrounding vegetation. The sexes are similar: they are grayish-black birds with white bellies. Younger birds and some females have more brownish plumages.

Dark-eyed Juncos were formerly known as Slate-colored Juncos, and many refer to them as "snowbirds," because they frequently begin to arrive in many parts of Minnesota when the snow begins to fall in autumn. Forbush and May (1955) suitably described a Dark-eyed Junco scene as: "A bleak gray day in the early winter—bare trees standing stark and black against a background of white snow—a cold wind sweeping across the drifted fields—and in a sheltered, brush-filled corner, a flock of lively little gray and white birds fluttering and twittering together."

Even though we know this species mostly during migration and winter, they are a nesting species in the bogs and coniferous forests of northern Minnesota. Here they can be heard in early spring singing their simple trill-like song, which is similar to that of the Chipping Sparrow but is considered more musical. The identification by song alone should be made with caution because Chipping Sparrows also nest in many of the same habitats.

There are five subspecies of juncos, all with distinct characteristics that provide considerable misery for taxonomists. The subspecies that occurs in Minnesota, *Junco h. hyemalis,* breeds throughout the mid-western and northeastern United States and Canada and west to Alaska. It integrates with *J. h. carolinensis* in the Appalachian Mountains, from southern New York State to West Virginia. The highest densities of the Minnesota subspecies are found in the western United States and Canada. All these subspecies have dark eyes but vary considerably in their plumage coloration. Currently, the only North American junco with a separate species status is the Yellow-eyed Junco (*J. phaeonotus*), found in the southwestern United States. Future research with these junco species and subspecies may reveal more taxonomic clarity.

MINNESOTA BREEDING DISTRIBUTION

Roberts (1932) emphasized a distribution in the northern coniferous forests. He described its southern extent from central Pine County northwest to Itasca State Park and eastern Marshall County. He further stated the species was "found evenly distributed, nesting alike among the rocks and fallen needles under the pines on the upland and in the sphagnum moss beneath the Labrador tea of the spruce and cedar swamps of the lowlands." He confirmed nesting in Aitkin, Marshall, Pine, and St. Louis Counties, plus in Itasca State Park.

Years later, Green and Janssen (1975) labeled the Junco's distribution as being primarily in the northeastern and north-central regions but scarce among the western and southern margins of these regions. They included additional confirmed nesting in Carlton, Clearwater, Cook, and Lake Counties, plus an observation of two young birds in northern Chisago County in 1950. Several years later, Janssen (1987) suggested a more restricted breeding distribution south to northern Carlton and Crow Wing Counties. He also noted an extensive summer record in Otter Tail County in 1977 from June 2 to July 12. Along with Hertzel and Janssen (1998), he included confirmed nesting from six counties since 1970: Aitkin, Beltrami, Cook, Koochiching, Lake, and Lake of the Woods.

The MBS included 100 breeding-season observations and reinforced the Junco's primary distribution as being in northeastern Minnesota, but included potential breeding locations in northern Pine, northern Cass, eastern Marshall, and northern Roseau Counties.

The MNBBA included 241 records, all of which were confined to the northern portions of the Laurentian Mixed Forest Province, especially the northern tier of counties from eastern Roseau and Beltrami Counties to Cook County **(A)**. Twelve confirmed nesting records were reported, with the only new county record being from northeastern Itasca County. Possible nesting was reported in Carlton, Cass, and Crow Wing Counties, at the southern extent of its breeding distribution in the state. Records were reported in only 3.3 percent (156/4,733) of the atlas blocks surveyed.

The MNBBA probability map identifies a patchy distribution of the Dark-eyed Junco throughout

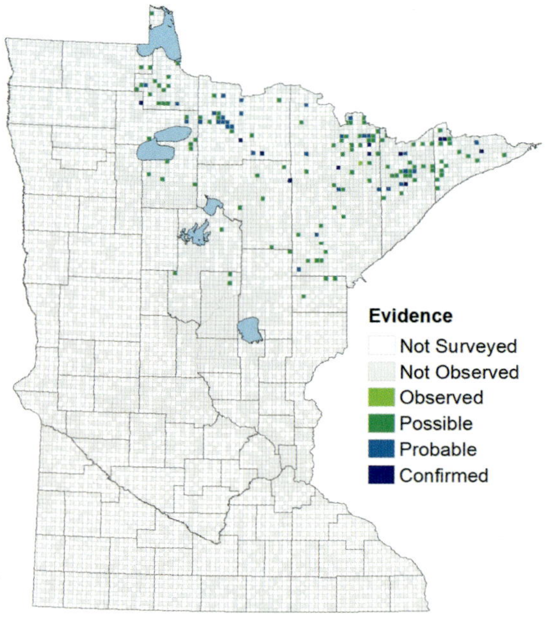

A. Breeding distribution of the Dark-eyed Junco in Minnesota based on the Breeding Bird Atlas (2009–13).

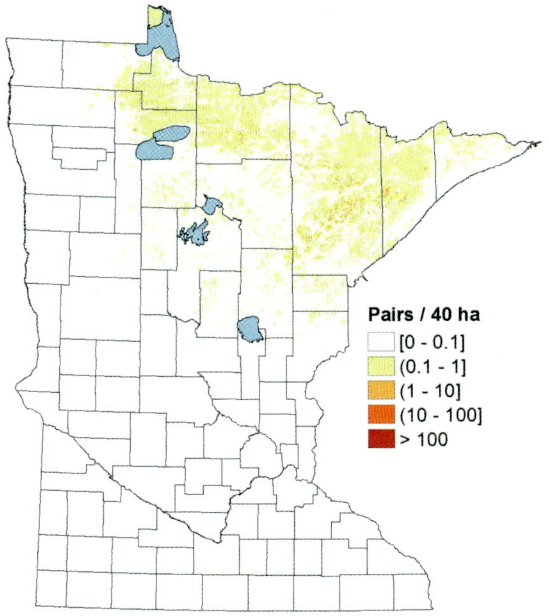

B. Predicted breeding distribution (pairs per 40 ha) of the Dark-eyed Junco in Minnesota based on habitat, landscape context, and climate data gathered during the Minnesota Breeding Bird Atlas (2009–13) using the Generalized Linear Modeling method with an adjustment for detectability.

northern and north-central Minnesota **(B)**. Concentrations were predicted in the lowland coniferous forests in the Agassiz Lowlands Subsection of northern Beltrami, western Koochiching, and parts of Lake of the Woods Counties. In addition, pockets of moderate populations were predicted in many areas of Lake and St. Louis Counties. Overall, the Junco is not predicted to be found in high densities anywhere in Minnesota.

Roberts (1932) provided few details on breeding distribution changes in Minnesota, though he documented Dark-eyed Juncos feeding their young in Pine County. In addition, the record of young birds in northern Chisago County in 1950 suggested some nesting in that region of the state (Green and Janssen 1975). The MNBBA recorded no detections in the western counties of Clearwater or Marshall, places where the species had previously been detected. The song of the Dark-eyed Junco can be difficult to separate from other species with a trill-type song (e.g., Chipping Sparrow), so misidentification can frequently occur. This difficulty may have played a role in the lack of detections in this region, but if the species is observed, misidentification is unlikely.

Nolan and colleagues (2020) provided little evidence of historical changes in the Dark-eyed Junco's breeding distribution in North America. In Minnesota, its historical record remains unclear. Southern and western portions of Minnesota were primarily prairie grasslands and deciduous forests, unsuitable habitat for the species. The limited information provided by Roberts (1932) suggests the species may have nested in more southerly (Pine County) and western (Marshall County) locations than it does today, especially where conifer forests existed, but its breeding distribution is unlikely to have been substantially more extensive in Minnesota.

BREEDING HABITAT

Across North America, this species breeds in boreal forests, pine and aspen forests, parklands, shrubby open areas, and even urban settings (Nolan et al. 2020). In Minnesota it is largely associated with both upland (e.g., spruce–fir and pine forests) and especially lowland coniferous forests dominated by black spruce or tamarack (Niemi and Hanowski 1992a, 1992b; Green 1995) **(C)**. Many sources report it in recently logged areas, but it is not often found in many of these habitats (Niemi and Hanowski 1984b; Schulte and Niemi 1998), with the exception of the Agassiz Lowlands Subsection (Zlonis et al. 2017).

POPULATION ABUNDANCE

PIF (2021), using data from the BBS, estimated the US and Canadian breeding population as 220 million adults, but only 22,000 in Minnesota. In contrast, the MNBBA estimated a Minnesota breeding population substantially higher, at 107,000 adults (95 percent confidence interval of 61,000–316,000). Regardless, Minnesota does not constitute a large proportion of the global population of Dark-eyed Juncos.

The BBS trend for Minnesota has too few observations to be reliable, but the trend for the Boreal Hardwood Transition Bird Conservation Region, which includes northern Minnesota, indicated a significantly declining trend that averaged −1.52 percent per year from 1966 to 2019 (Sauer et al. 2020) **(D)**. Declines were also significant but slightly less dramatic for the same period for all routes surveywide (average −0.70 percent per year). Rosenberg, Kennedy, and colleagues (2016) estimated an overall decline of −42 percent of the population from 1970 to 2014 in North America. In addition, Christmas Bird Counts surveywide indicated a decline that averaged −1.3 percent per year from 1959 to 1988 (Sauer, Schwartz, and Hoover 1996), a trend very similar to the BBS results.

C. Breeding habitat of the Dark-eyed Junco in Minnesota is typically mature upland or lowland coniferous forests where pairs hide their nests in the dense ground flora and moss. Photo by Gerald J. Niemi.

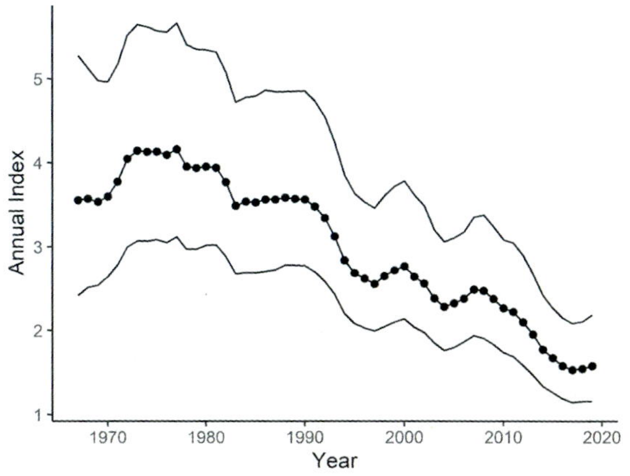

D. Breeding population trend and 95 percent credible interval (solid black lines) for the Dark-eyed Junco in the Boreal Hardwood Transition Bird Conservation Region based on the 1967–2019 Breeding Bird Survey (Sauer et al. 2020).

Documented reasons for the species' decline are unknown, but Nolan and colleagues (2020) identified nest predation as common, ranging annually from 20 to 80 percent in the Commonwealth of Virginia. Also, Longcore and colleagues (2013) listed the Dark-eyed Junco as one of the top 10 species killed at communication towers in the Prairie Potholes and Badlands Bird Conservation Regions in the midwestern United States.

CONSERVATION

The Dark-eyed Junco is considered to be of limited conservation concern in North America because of its wide breeding distribution and large overall population. However, NAS (2019) projected a moderate vulnerability for the Dark-eyed Junco's breeding range in North America with a 2°C warming scenario and extirpation from Minnesota in that warming scenario by 2050.

With its large population, the future of this lively snowbird as a migrant and overwintering species in Minnesota seems relatively secure, but its status as a nesting species is more precarious. Hardy Minnesotans who want to know more about its ecology will need to deal with its sparse Minnesota breeding population, a song that is difficult to separate from that of Chipping Sparrows, and a good supply of mosquitos and black flies where it nests.

WHITE-THROATED SPARROW

Zonotrichia albicollis

Minnesota Seasonal Status: Regular nesting species, migrant, and occasional in winter. The White-throated Sparrow was abundant during the MNBBA.
Migration: Short-distance migrant; winters in the southeast, northeast, and lower midwestern United States as far west as Arizona, as well as along the Pacific coast.
Food: Omnivorous, including arthropods, seeds, buds, fruit, and a variety of plant parts, primarily gathered from the ground.
Nest: Cup nest on the ground or slightly elevated and well concealed in vegetation.
Conservation Concern: Assigned a moderate PIF Continental Concern Score of 9/20.

Photo by David Brislance.

Roberts (1932) emphasizes, "There is no more welcome and pleasing sound in the spring than the clear, whistled song of the White-throated Sparrow." It is a familiar sound to many Minnesotans because its song, which seems to say *Old Sam Peabody, Peabody, Peabody,* or, as Canadians interpret it, *Oh Canada, Canada, Canada,* resonates throughout the bogs and forests in the northern parts of the state. The White-throat is also fond of feeders during migration.

Its sweet, clear whistled song has also spawned many vernacular names, including Old Sam Peabody, Canada Bird, Nightingale, Striped Bird, Whistle Bird, and Widow Bird. In plumage its white throat is diagnostic, which is emphasized by its scientific name *albicollis,* meaning "white neck." Besides the ever-present white throat, this relatively large sparrow has two distinct morphs. One has clear white and black stripes on the head with a diagnostic yellow eyebrow, while the other is less striking, with a tan and brown striped head and lacking the yellow eyebrow. The latter plumage was formerly thought to be of juvenile or immature birds, but more detailed research has indicated that both morphs are maintained in the population via a genetic process called assortative mating. Sexes of both morphs are identical.

The White-throated Sparrow as a breeding species is widely distributed from the eastern Canadian provinces and New England states to British Columbia, but in a narrow band along the ecotone between coniferous and deciduous forests. The highest reported densities are in northeastern Minnesota, north-central Ontario, and Labrador.

MINNESOTA BREEDING DISTRIBUTION

Historically, Roberts (1932) described the White-throated Sparrow as "a common summer resident in the evergreen forests from Isanti County northward." Its westward distribution included eastern Kittson and central Marshall Counties. Confirmed nesting was documented in the late 1800s and early 1900s in Cass, Cook, Isanti, Itasca, Lake, Lake of the Woods, Marshall, and St. Louis Counties, as well as in Itasca State Park and the Mille Lacs Lake area.

Green and Janssen in 1975 emphasized the White-throated Sparrow's primary breeding distribution in the northeastern and north-central regions of the state. They also suggested it had a marginal presence in the southern and western fringes of these areas. They added confirmed nests beyond those reported by Roberts for Anoka, Beltrami, Clearwater, Hubbard, and Pine Counties. Inferred nesting was also included for Carlton County. Janssen's (1987) findings reinforced the breeding distribution previously described. He confirmed nesting in 11 counties since 1970, while Hertzel and Janssen (1998) added two more counties for a total of 13.

The MBS recorded 1,805 breeding-season locations. Their observations reinforced the major range of the species in the northeastern and north-central parts of the state, but also included locations from northwestern, west-central, central, and east-central Minnesota, including the following counties: Becker, Chisago, Douglas, Mahnomen, Otter Tail, Polk, and Stearns.

The MNBBA included a total of 4,207 records with numerous detections in blocks from Aitkin County northwest to Beltrami and Lake of the Woods Counties, and all counties east to Cook County **(A)**. Records were detected in 29.6 percent (1,413/4,765) of the blocks covered in the state. Confirmed nesting was documented in 14 counties, including south to Anoka County. Probable nesting was also documented in Otter Tail County and Kittson County, and possible nesting in Becker, Benton, Mahnomen, Morrison, Polk, Stearns, and Wadena Counties. Many of these counties are on the western fringes of the breeding range of the White-throated Sparrow in Minnesota.

The MNBBA probability map predicted highest breeding densities of the White-throated Sparrow from southeastern Roseau County and east throughout the Arrowhead region of northeastern Minnesota **(B)**. Populations with lower densities were predicted in west-central Minnesota to Becker and Otter Tail Counties and south to Benton and Chisago Counties.

The overall changes in the breeding distribution and abundance of this species since the mid-1800s and early 1900s are difficult to assess. Falls and Kopachena (2020) in their North American summary on the breeding range of the species identified no major changes in its breeding range, except for slight extensions in British Columbia, New York, and Vermont. The collective evidence from Minnesota, Wisconsin, and Michigan suggests that the species' population has not changed much during the past hundred-plus years (R. Brewer et al. 1991; Chartier et al. 2013; Cutright et al. 2006; Roberts 1932).

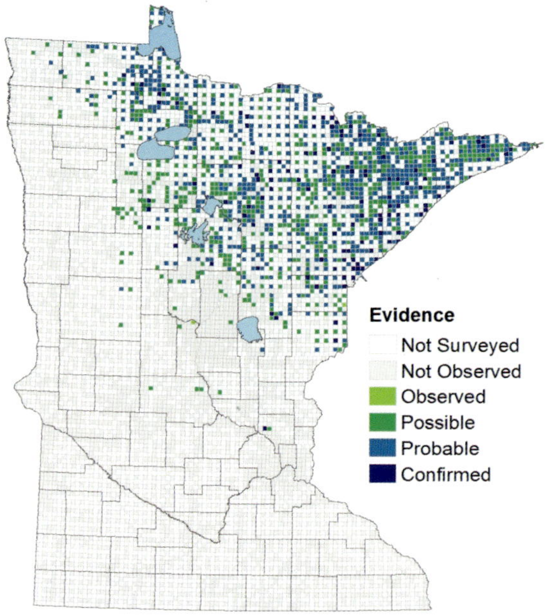

A. Breeding distribution of the White-throated Sparrow in Minnesota based on the Breeding Bird Atlas (2009–13).

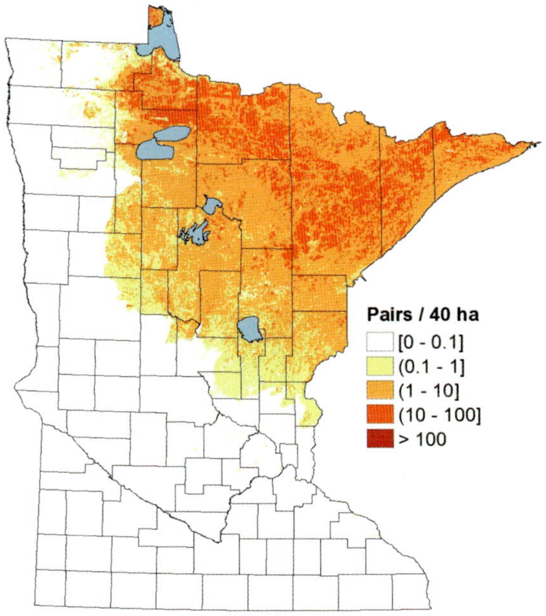

B. Predicted breeding distribution (pairs per 40 ha) of the White-throated Sparrow in Minnesota based on habitat, landscape context, and climate data gathered during the Minnesota Breeding Bird Atlas (2009–13) using the Generalized Linear Modeling method with an adjustment for detectability.

In Minnesota it is logical to assume that deforestation in the southern and western parts of the state may have resulted in a retraction of the species' range northward. However, Roberts (1932) makes no mention of the species' distribution including the southern or western parts of the state. Hence, this species may not have been a common species in these regions, or at best it was found marginally in forested habitats after forest fire or logging activities.

BREEDING HABITAT

The White-throated Sparrow is widely described as a habitat generalist in coniferous and mixed deciduous–coniferous forests with low, dense shrubby ground cover (Falls and Kopachena 2020). In Minnesota, it is most often found in lowland coniferous forests and in forests regenerating as a result of disturbances such as fire, logging, insects, and wind (Back 1979; Niemi and Pfannmuller 1979; Niemi and Hanowski 1984b, 1992a; Lind and Hanowski 2004) **(C)**. In the Chippewa and Superior National Forests, Niemi, Howe, and colleagues (2016) found White-throats significantly most abundant and most frequently recorded in black spruce–tamarack forests. However, they were also commonly found in aspen–spruce–fir, regenerating forests, lowland shrubs, and mixed swamp conifer forests. High densities were also found in black spruce–tamarack forests and regenerating lowland conifer forests in the Agassiz Lowland Subsection (Nevers et al. 1981; Niemi and Hanowski 1984b; Bednar et al. 2016).

POPULATION ABUNDANCE

PIF (2021) estimated a large US and Canadian breeding population of 160 million breeding adults, with 2 million in Minnesota. In contrast, the MNBBA estimated a larger breeding population in Minnesota of 2.82 million breeding adults (95 percent confidence interval of 2.70–3.00 million). Yet, the Minnesota population estimate would still represent less than 2 percent of the global population of this species. The substantial area of suitable habitat in Canada hosts the majority of its global breeding population.

The BBS trend estimated for Minnesota from 1967 to 2019 was insignificant, which suggests a stable breeding population **(D)**. However, the species' recent trend from 1993 to 2019 in Minnesota was significantly declining (average −0.68 percent per year). Many northeastern US states and New Brunswick and Ontario from 1966 to 2019 had significant negative trends, averaging from −1.08 percent per year for Ontario to −4.05 percent per year for Maine. The trends for this species were also significantly negative for all routes completed in Canada (average of −0.66 percent per year), in the United States (−2.21 percent), and surveywide (−0.74 percent) from 1966 to 2019 (Sauer et al. 2020). Rosenberg, Kennedy, and colleagues (2016) calculated an overall decline in the species' global breeding population of −29 percent from 1970 to 2014. Population trends from 1995 to 2020 in the Chippewa and Superior National Forests of northern Minnesota are stable and consistent with Minnesota's BBS trends (Grinde, Bednar, et al. 2020).

C. Although common in recently logged and burned-over forests, White-throated Sparrows achieve their highest densities in open lowland coniferous forests, where they find plenty of ground cover with food and nest sites.
Photo by Gerald J. Niemi.

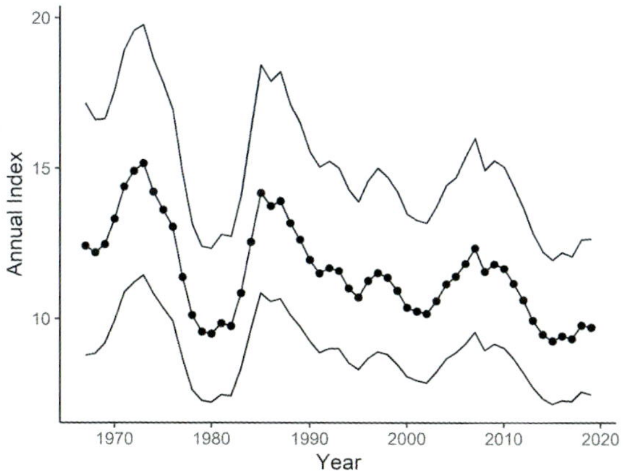

D. Breeding population trend and 95 percent credible interval (solid black lines) for the White-throated Sparrow in Minnesota based on the 1967–2019 Breeding Bird Survey (Sauer et al. 2020).

The causes for population declines are unknown, but a warming climate with subsequent movements northward in the breeding range or high mortality during migration have been implicated. For instance, Falls and Kopachena (2020) stated that the White-throated Sparrow is among the most vulnerable to building collisions, having been ranked in the top 3 of 164 species based on mortality studies.

CONSERVATION

Like many species with high estimated breeding populations and broad distributions in North America, the White-throated Sparrow would appear to have few current problems in maintaining its breeding population and distribution. Yet it is of moderate conservation concern because of its declining population, especially in the eastern portion of its range.

As with virtually every species with a northerly distribution in Minnesota, NAS (2019) projected a moderate vulnerability for the White-throated Sparrow's breeding range in North America with a 2°C warming scenario. They predicted a 56 percent range loss and a 34 percent range gain in North America and extirpation from Minnesota under the 2°C warming scenario.

The large decline of this abundant species over broad portions of its breeding range remains an issue to be watched in the future. Information that contributes to a better understanding of factors associated with its distribution and abundance on the breeding and wintering grounds as well as migratory movements will be especially useful. Further reductions in populations of the White-throat would not be welcome among those who enjoy hearing its familiar, clear whistled song wafting out of the Minnesota northern forests.

VESPER SPARROW

Pooecetes gramineus

Minnesota Seasonal Status: Regular breeding resident and migrant. The Vesper Sparrow was common during the MNBBA.
Migration: Partial migrant; northern breeding populations winter in the southern United States and Mexico.
Food: Ground-foraging omnivore that feeds on arthropods and seeds.
Nest: Ground nest placed at the base of a clump of vegetation, including forbs, shrubs, or small trees. A small ramp leading up to the nest and a canopy of bent-over vegetation may be present.
Conservation Concern: Assigned a moderate PIF Continental Concern Score of 11/20.

In a landscape dominated by croplands and grasslands as far as the eye can see, the Vesper Sparrow finds its home. Often perched on telephone wires and fences that line rural roads, this grayish-brown sparrow has only a few distinct features, including a white eye-ring and a lightly streaked breast. A small chestnut patch on the shoulder can be difficult to see but was the source of one of its early common names, the Bay-winged Bunting. When flushed, the white outer tail feathers provide an excellent clue to its identity, helping to distinguish it from an array of other grassland sparrows.

Its sweet song is easily confused with that of the Song Sparrow. Both usually begin with introductory notes followed by a series of twitters and trills. The Song Sparrow begins with several short, abrupt notes. The Vesper Sparrow begins with two to four long, clear notes that may be paired: the first two higher in pitch than the second two. It has been described as slower and having a "more definite form" than that of the Song Sparrow (Bent 1968).

It was the early naturalist and essayist John Burroughs who endowed the species with its common name, Vesper Sparrow, believing that "its singing was sweeter and more impressive towards evening" (Roberts 1932). But the Sparrow is equally vocal during nearly all hours of the day, beginning in earnest before sunrise, continuing but waning through the midday hours, and then peaking again at dusk. The Latin name, *Pooecetes gramineus,* simply translates to "grass dweller."

Photo by Gerald Hoekstra.

The species is found from eastern British Columbia south to northern Arizona and New Mexico, and east to New England and the mid-Atlantic coast. The core of its breeding range is largely west of the Mississippi River, with the highest densities in central Montana and southern Alberta.

MINNESOTA BREEDING DISTRIBUTION

Nearly a century ago, Roberts (1932) considered the Vesper Sparrow a common summer resident south of the densely forested northern landscape. Yet, even in northern Minnesota, "every clearing, old burned-over area, wind-fall, or cut-over region, has its considerable quota of Vesper Sparrows." It was, he wrote, "one of the commonest and most generally distributed of Minnesota birds." Breeding evidence was reported from Anoka, Crow Wing, Goodhue, Hennepin, Isanti, Lake of the Woods, Polk, and Sherburne Counties and from Itasca State Park, Cass Lake, and Leech Lake.

Green and Janssen (1975) described the Vesper Sparrow as a statewide resident that was least common in the northeast, where breeding evidence was absent from Lake and Cook Counties. Janssen (1987) noted the species was uncommon in the north-central region as well. His distribution map excluded the Arrowhead region and portions of several adjacent counties. He identified 20 counties where nesting had been confirmed since 1970; Hertzel and Janssen (1998) added another 11 counties to the list. Only Clearwater, Crow Wing, and Mille Lacs Counties were located in the northern forest region. By 2014, the MBS reported 490 Vesper Sparrow breeding-season locations. Records in the Laurentian Mixed Forest Province were largely limited to the province's western border. A few records, however, were found as far east as Itasca and St. Louis Counties.

MNBBA records expanded the species' distribution a little farther east with two confirmed breeding records in east-central St. Louis County, one probable nesting record in northern Lake County, and two probable records in Carlton County **(A)**. The St. Louis County records were from the town of Babbitt and from the reclaimed tailings basin of the old LTV Steel Mining Company. The Lake County record was in an area burned by the 2011 Pagami Creek Fire. Overall, 2,768 detections were reported from 33.6 percent (1,601/4,761) of the surveyed atlas blocks. Breeding was confirmed in 109 blocks (2.3 percent). The species was reported from every county except Cook and was confirmed breeding in 47 counties, including six in north-central and northeastern Minnesota: Cass, Clearwater, Crow Wing, Hubbard, St. Louis, and Wadena.

The MNBBA probability map predicted that the highest breeding densities occur in southwestern Minnesota and along the Red River valley **(B)**. Densities gradually decrease eastward, especially in northern Minnesota. The one exception is the open landscape of the Iron Range in St. Louis and Itasca Counties. In western Minnesota, floodplains are one of the few areas where lower densities are predicted.

The primary change to the Vesper Sparrow's distribution appears to be in northern Minnesota. During an era when the scale and frequency of disturbances in the forest landscape was much greater, Roberts (1932) noted that the species was common in openings created by timber harvest, burns, and windfalls. But large openings, such as the 92,000-acre Pagami Creek Fire in Lake County and the LTV

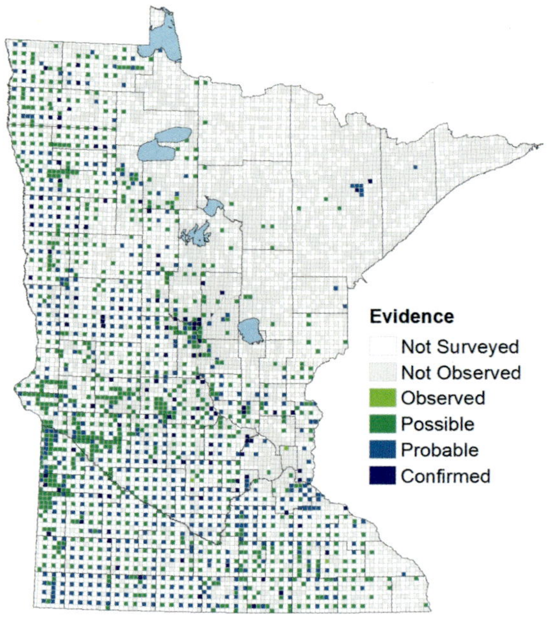

A. Breeding distribution of the Vesper Sparrow in Minnesota based on the Breeding Bird Atlas (2009–13).

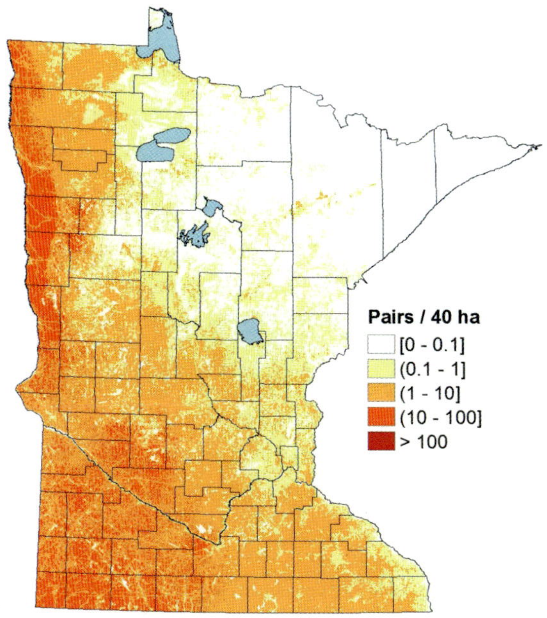

B. Predicted breeding distribution (pairs per 40 ha) of the Vesper Sparrow in Minnesota based on habitat, landscape context, and climate data gathered during the Minnesota Breeding Bird Atlas (2009–13) using the Generalized Linear Modeling method with an adjustment for detectability.

reclaimed tailings basin in St. Louis County, are now relatively rare, and the decline in the presence of the species is a likely response to the change.

Similar population fluctuations have been observed in New England and southeastern Canada. When forests were initially cleared, the Sparrow expanded into the regions, "becoming relatively numerous in many areas" where they had been quite uncommon earlier (S. L. Jones and Cornely 2020). Now, because many farmlands have been abandoned and vegetation has either succeeded to forest or has been cleared for development, populations are declining.

BREEDING HABITAT

Relatively dry, open sites with short, sparse vegetation, often interspersed with patches of bare ground, are the domain of the Vesper Sparrow. Habitats that meet these requirements range from native prairies, restored grasslands, and agricultural fields to woodland edges, shelterbelts, and jack pine barrens **(C)**. Conservation Reserve Program (CRP) lands are suitable, but studies have found that the Sparrow is more abundant in croplands than in CRP lands (Dechant, Dinkins, et al., "Vesper Sparrow," [2000] 2002).

Edges are a common feature of many sites, including fencerows, shelterbelts, and pastures bordered by small woodlots. Trees and shrubs bordering fields provide singing perches while fences, tall forbs, and crop residue serve the same purpose in more open landscapes (Dechant, Dinkins, et al., "Vesper Sparrow," [2000] 2002; S. L. Jones and Cornely 2020). Edges, however, have a cost associated with them. Cowbird parasitism is often higher when nests are located near field edges (Dechant, Dinkins, et al., "Vesper Sparrow," [2000] 2002).

POPULATION ABUNDANCE

Using BBS data, PIF (2021) generated a US and Canadian population estimate of 35 million breeding adults and a Minnesota population of 1.1 million. MNBBA data, however, generated a considerably larger statewide estimate of 3.5 million adults, which is likely owing to the more intensive sampling effort and the species' prevalence in the agricultural regions that dominate much of Minnesota's landscape (95 percent confidence interval of 3.3–3.8 million).

Unfortunately, BBS data also document a significant, surveywide population decline, averaging −0.84 percent per year from 1966 to 2019 (Sauer et al. 2020). In Minnesota, the decline has been steeper, averaging −1.55 percent per year since 1967 **(D)**. The only region where the Vesper Sparrow is significantly increasing is the Central Mixed Grass Prairie Bird Conservation Region, which stretches from Nebraska south to central Texas.

Several factors are responsible for the decline. As noted earlier, development and the wide-scale succession of old farms to forests have eliminated suitable habitat throughout the northeastern states and southeastern Canada. In the Midwest and Great Plains, increasingly intensive agricultural practices eliminate shelterbelts, hedgerows, and grassy field margins, increase chemical applications, and increase the frequency of mowing and haying, all of which reduce reproductive success. Finally, the loss

C. Typical Vesper Sparrow nesting habitat includes open areas with short vegetation for foraging and forbs, shrubs, or small trees for perching. A small windbreak bordering an agricultural field meets these requirements. Photo by Lee A. Pfannmuller.

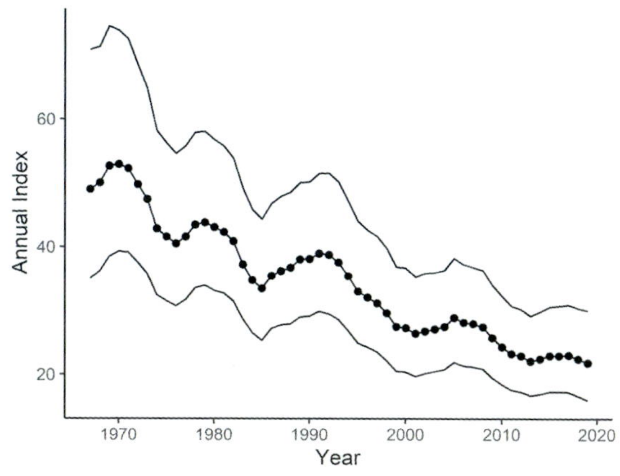

D. Breeding population trend and 95 percent credible interval (solid black lines) for the Vesper Sparrow in Minnesota based on the 1967–2019 Breeding Bird Survey (Sauer et al. 2020).

and overgrazing of wintering habitat is also contributing to the decline (S. L. Jones and Cornely 2020).

CONSERVATION

Despite its relatively large population and broad distribution, the Vesper Sparrow, given its population decline and threats to its breeding and wintering habitats, is considered a moderate conservation priority (PIF 2021). Warming temperatures pose another concern. NAS (2019) found the species moderately vulnerable under a 2°C warming scenario and predicted the loss of 46 percent of its breeding distribution by 2050, with a northward gain of only 39 percent. In Minnesota, as conditions deteriorate in the southern and western counties, the species is predicted to move to suitable habitat that becomes available in the north-central and northeastern counties. Eventually, under a 3°C warming scenario, models project the Vesper Sparrow will be restricted to the very eastern tip of Cook County.

As with all grassland species, habitat protection and restoration are key management recommendations. Mowing, grazing, and burning activities, however, should be delayed until after the breeding season to avoid destruction of nests and flightless young. Farmers can improve cropland habitat by retaining field edges created by hedgerows, shelterbelts, and fence lines. Breeding success can be enhanced by no-tillage agricultural practices or by a reduced number of till operations (S. L. Jones and Cornely 2020).

The Vesper Sparrow is one of the few species that successfully inhabits Minnesota's extensively cultivated landscapes. Its sweet musical trills are a tribute to a bird that can find what may appear the most inhospitable of habitats suitable. Nevertheless, its population decline and vulnerability to rising temperatures signal the need for close monitoring.

LECONTE'S SPARROW

Ammospiza leconteii

Minnesota Seasonal Status: Regular breeding resident and migrant. The LeConte's Sparrow was uncommon during the MNBBA.
Migration: Short-distance migrant that winters in the south-central United States.
Food: Omnivorous ground forager that consumes arthropods and seeds.
Nest: Cup nest located on or just above the ground.
Conservation Concern: Assigned a moderate PIF Continental Concern Score of 13/20.

Like several small grassland sparrows, the LeConte's Sparrow has been described as "so secretive, so nondescript, and so hard to observe that its presence often passes unnoticed" (Bent 1968). It is abundant in landscapes where wet grasses and sedges dominate, and access is often difficult. Indeed, between 1790, when the species was first described, and 1872, only four specimens were collected. Several more years would pass before the first nests were found in the early 1880s (Bent 1968). Well-hidden within dense clumps of dead grass, LeConte's Sparrow nests remained a rare find until a comprehensive study of grassland birds in northwestern Minnesota and southeastern North Dakota revealed a gold mine of 50 nests, providing a wealth of new information on this poorly studied species (Winter, Shaffer, et al. 2005).

Among the Sparrow's distinguishing features are a striped crown, a buffy stripe above the eye, a gray ear patch, a yellow-gold throat and breast, and brown streaks along the flanks. Its light, buffy plumage was described by Roberts (1932) as "having the general effect of a warm old-gold suffusion." The song is a strongly accented, buzzy trill that begins and ends with a soft chirp. Described by Swanson (Roberts 1936) as a gentle sneeze (*ta-zeee-tsa*), it is often heard during the evening hours. Unfortunately, the very short and high-pitched song can be easy to miss.

Named for Audubon's friend John LeConte, the species has a breeding range restricted largely to the Canadian provinces of Alberta, Saskatchewan, and Manitoba but that also dips south into the north-central United States. It is an uncommon species throughout its breeding range; the highest breeding densities are found in northwestern Minnesota.

Photo by Sparky Stensaas.

MINNESOTA BREEDING DISTRIBUTION

Roberts (1932) described the LeConte's Sparrow as a species of Minnesota's prairies and open woodlands but noted that it was "likely to occur wherever there are suitable marshes, wet meadows, or low-lying prairie with long, tangled grass and clumps of small bushes." He first encountered a pair in Minneapolis in 1877, occupying "a willow-grown bog beside a slough, which has long since disappeared." It was considered abundant in the Red River valley, where, in some places, it outnumbered "any other small marsh-dwelling species." Reports of nests with eggs were limited to Kittson County (1897, 1929); reports of young birds were noted in Hennepin, Marshall, Ramsey, and Sherburne Counties.

Apart from a later report that the species was common in the "extreme northern part of Koochiching County" (Roberts 1936), additional records were sparse. In 1959, Dwain Warner, the Bell Museum's curator of ornithology, reflected on the paucity of records. Among the thousands of observations on Minnesota birds that the museum had acquired since Roberts's 1932 and 1936 publications, there were only 12 summer records for the LeConte's Sparrow, stretching from Steele County in southeastern Minnesota northwest to Roseau County. Fortunately, more reports soon followed, including from a site near the town of Nimrod in Wadena County, where the birds were particularly common. Here David Parmelee and Richard Oehlenschlager (1973) located one nest with three young in 1972 and two nests, with five eggs each, in 1973. These were the first confirmed nesting records in the state since 1929.

Green and Janssen's (1975) account described the species as limited primarily to the northwestern counties, though it occurred "sparingly" in other northern and central regions. They had no recent records from southern Minnesota but noted a 1928 record of juveniles in Jackson County that was not included in Roberts (1932, 1936). Janssen (1987) described the Sparrow as "common and widespread" in central and northern Minnesota yet absent from heavily forested landscapes. It was found as far south as Lac qui Parle and Dakota Counties in the 1950s and 1960s but the southern limit of its current distribution was defined by Pine County in the east and Wilkin County in the west. Although two isolated records were reported from southwestern Minnesota in the early 1970s, since 1970 confirmed nesting records were limited to Aitkin, Beltrami, Marshall, and Wadena Counties. Hertzel and Janssen (1998) later added Becker and Red Lake Counties to the list. A similar distribution was documented by the MBS, which tallied 464 breeding-season locations by 2014, even though a large portion of north-central Minnesota had yet to be surveyed.

The MNBBA's 664 LeConte's Sparrow records included more reports from central Minnesota, including several along the Minnesota River valley **(A)**, suggesting that the species may be reoccupying the southern periphery of its former breeding range. Detected in 9.4 percent (445/4,752) of the surveyed atlas blocks that spanned 45 counties, breeding was confirmed in just four blocks in Aitkin, Kittson, and Roseau Counties. Nesting was also confirmed during the last year of the atlas in Clay and Marshall Coun-

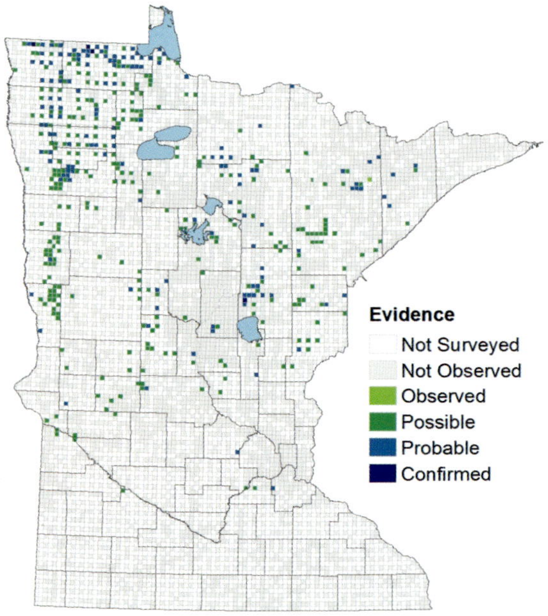

A. Breeding distribution of the LeConte's Sparrow in Minnesota based on the Breeding Bird Atlas (2009–13).

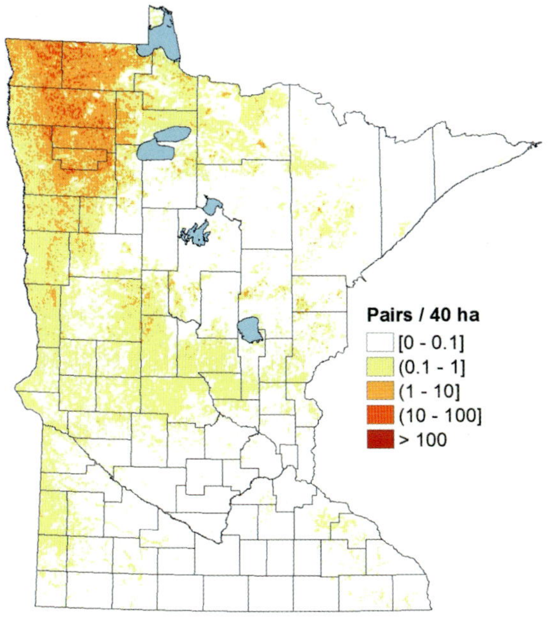

B. Predicted breeding distribution (pairs per 40 ha) of the LeConte's Sparrow in Minnesota based on habitat, landscape context, and climate data gathered during the Minnesota Breeding Bird Atlas (2009–13) using the Generalized Linear Modeling method with an adjustment for detectability.

ties, but locational data were insufficient to assign the records to atlas blocks (MOU 2021a). Probable breeding was documented by the atlas in 27 counties. Cook County was the only northern county where the species was not reported. Abundant in the Tallgrass Aspen Parklands Province, the species was also common in the grasslands of Clay and Wilkin Counties and in the peatlands of several northern counties, including Aitkin and St. Louis. The southern limit of its range was defined by Dakota County in the east and Redwood County in the west. The MNBBA probability map confirmed the importance of the Aspen Parklands but also predicted that potential habitat with low breeding densities may occur across central Minnesota and even sparingly farther south **(B)**.

Roberts (1932) provided no evidence of the Sparrow's occurrence south of the Minnesota River. With the exception of a few reports, it appears the species has always been restricted to the northern two-thirds of the state. Although Roberts believed it was largely absent from the northern forests, recent records likely reflect more intensive surveys of the region's wet meadows and peatlands rather than an expansion of the species' breeding range. Lowther's (2020c) review of the species also makes no mention of wide-scale changes in its North American distribution.

BREEDING HABITAT

Frequently found in low-lying, grassy meadows **(C)**, LeConte's Sparrows also use alkali fens, wet prairies, and open peatlands as well as drier uplands, including hay pastures, fallow fields, and planted cover (S. Cooper 1984; Dechant, Sondreal, et al., "Le Conte's Sparrow," [1998] 2002; Niemi 1985). Sites usually share four attributes: first, open and level terrain; second, relatively undisturbed grass and/or sedge cover; third, moderately tall, dense vegetation; and, fourth, abundant residual vegetation. Although the birds seem to "prefer areas free of shrubs and other woody vegetation," studies in Minnesota and Michigan also found them using sites with "scattered small willows" (Dechant, Sondreal, et al., "Le Conte's Sparrow," [1998] 2002). For example, Roberts (1932) first encountered the species in a "willow-grown bog"; Peabody (1901) noted its presence in "willow-dotted" meadows in the Red River valley; and Nevers and colleagues (1981) found high breeding densities in shrub-dominated habitats in the Red Lake Peatland.

POPULATION ABUNDANCE

PIF (2021) estimated the Sparrow's US and Canadian population to number 5.1 million breeding adults, with approximately 300,000 birds residing in Minnesota. Given the MNBBA's intensive sampling of many remote wetland areas, the statewide population estimate is considerably larger, at 603,000 birds (95 percent confidence interval of 429,000–2.2 million).

Although this frequent inhabitant of wet meadows may experience wide population fluctuations in response to changes in annual precipitation (Lowther 2020c; Winter, Shaffer, et al. 2005), since 1966 it has demonstrated only a small, nonsignificant decline, averaging −0.40 percent per year through 2019 **(D)**. Unfortunately, this overshadows a more dramatic drop in numbers documented across the BBS survey region since 1993 that averages −3.30 percent per year (Sauer et al. 2020). A similar BBS trend line is evident in Minnesota. Nevertheless, the factors that regulate population levels are poorly understood.

C. Although they are not restricted to mesic or wet habitats, LeConte's Sparrows are commonly found in large, low-lying meadows with scattered shrubs. In the Red Lake Peatland, shown here, Le Conte's Sparrow densities in the early 1980s ranged from 12–26 pairs per 10 ha (Niemi and Hanowski 1984a). Photo by Gerald J. Niemi.

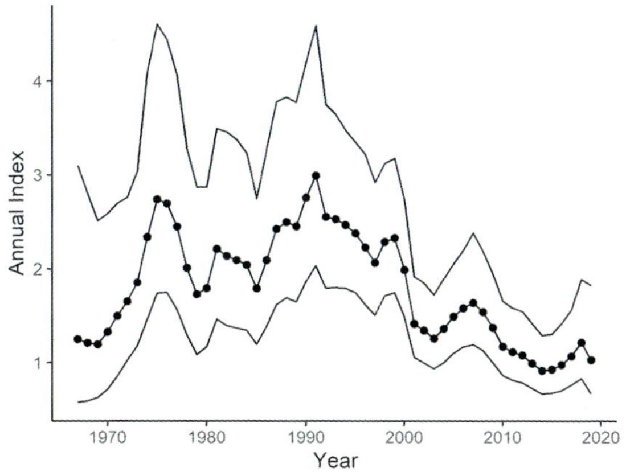

D. Surveywide breeding population trend and 95 percent credible interval (solid black lines) for the LeConte's Sparrow based on the 1967–2019 Breeding Bird Survey (Sauer et al. 2020).

CONSERVATION

The LeConte's Sparrow's recent population decline prompted its assignment as a moderate conservation priority (PIF 2021). It was also added to PIF's Yellow Watch List, a group of the most vulnerable species that require focused conservation attention (Rosenberg, Kennedy, et al. 2016). Habitat protection and restoration are essential. Although shrubs may be present in some suitable habitats, controlling succession on grassland tracts to prevent significant woody encroachment is especially critical. Burning or mowing every two to four years is recommended (Dechant, Sondreal, et al., "Le Conte's Sparrow," [1998] 2002). Because the birds depend on residual cover, large sites should be managed in units that allow at least half of each parcel to be maintained in older cover. In North Dakota, the birds were most abundant on sites two years after they were burned; unburned sites were avoided (Madden 1996; Madden et al. 1999).

Unfortunately, warming temperatures are expected to have a significant impact on species found in the northern Great Plains and Prairie Provinces (NABCI 2010). Under a 2°C warming scenario, the LeConte's Sparrow was deemed moderately vulnerable with a projected loss of 67 percent of its breeding distribution by 2050, compensated by only a 40 percent gain to the north (NAS 2019). The projected losses include the species' extirpation from Minnesota, with the nearest population located nearly 660 km to the north. Clearly, this unassuming little bird, with its very restricted breeding range, has an uncertain future.

NELSON'S SPARROW

Ammospiza nelsoni nelsoni

Minnesota Seasonal Status: Regular breeding resident and migrant. The Nelson's Sparrow was rare during the MNBBA.
Migration: Short-distance migrant that overwinters in marshes along the Atlantic and Gulf coasts.
Food: Insects, arthropods, mollusks, and seeds.
Nest: Cup nest on or near the ground, often partly or fully domed.
Conservation Concern: Assigned a moderate PIF Continental Concern Score of 12/20.

Photo by Gerald Hoekstra.

The Nelson's Sparrow has had a complex taxonomic history for more than one hundred years. Until 1995, it was considered a subspecies of the Sharp-tailed Sparrow (*Ammodramus caudacuta*) complex, a common name that now no longer exists. The complex has since been split into two sister species, the Nelson's Sparrow (*Ammodramus nelsoni*) and the Saltmarsh Sparrow (*A. maritima*), the latter found exclusively on the Atlantic coast (Shriver et al. 2020). Recent behavioral and genetic analyses have identified three subspecies of the Nelson's Sparrow. The subspecies that occurs in Minnesota (*Ammodramus n. nelsoni*) has a breeding range restricted from Minnesota and South Dakota to northwestern Alberta, with highest densities in North Dakota. The two other subspecies, *Ammodramus nelsoni alter* and *A. n. subvirgatus,* have breeding distributions along the coast of Hudson Bay to northwestern Quebec and along the northeast Atlantic coast, respectively (Shriver et al. 2020). Each of the subspecies differs in size and subtle plumage characteristics.

The Nelson's Sparrow was named after Edward Nelson, chief of the US Biological Survey from 1916 to 1927. It is a rare and shy bird found in wet, grassy meadows and marshes where the water is ankle- or knee-deep. As Roberts (1932) stated, "The nest of the Nelson's Sparrow has long been one of the special rarities—not so much on account of the scarceness of the bird as the difficulty in locating it and making sure of its identity."

Identification of the Nelson's Sparrow requires a careful look, with key features being a yellow-orange stripe above the eye, a brown stripe behind the eye, and a grayish stripe on top of the head and on the back of the neck. Consultation with a good field guide is highly recommended to distinguish the Nelson's from the LeConte's, Saltmarsh, Grasshopper, and other sparrows. As with visual identification, the song of the Nelson's Sparrow is distinctive if one is familiar with the other insect-like songs of other sparrows that often occur in the same habitat. Breckenridge (in Roberts 1932) said it has "a low initial note [that] slurs immediately into a high, wheezy, nasal buzz, which terminates in a low, short, grating *ur*." In words it has been described as *p-tshh-hhhhh-uk* (Alderfer 2006).

MINNESOTA BREEDING DISTRIBUTION

Roberts (1932) described the Nelson's Sparrow as a summer resident that breeds in northwestern Minnesota, but he emphasized the possibility of nesting throughout the southern portion of Minnesota as well. He documented many reports from the southern and western portions of the state but all, except for a June observation in Blue Earth County, were in April, May, July, and September. Therefore, most could have been migratory birds. June observations were also recorded in northwestern Minnesota in both Marshall and Kittson Counties, though only one nest with three eggs was confirmed in Kittson County. According to Breckenridge (in Roberts 1932), the nest was found in a very wet, boggy marsh and only after a considerable effort, on June 21, 1929.

Green and Janssen (1975) emphasized that the species resides only in the northwestern region of the state. They included confirmed nesting in Kittson County and inferred nesting in Clay, Mahnomen, and Marshall Counties. Janssen (1987) noted that until 1974 it was presumed restricted to northwestern Minnesota from Clay to Kittson Counties. However, he points out new potential nesting activity in Minnesota as follows: first, a nest found by Burt (1978) during a "week-long search for Yellow Rail nests" in late June at the McGregor Marsh in Aitkin County; second, observations during the breeding season in western Roseau and Pennington Counties by Eckert (1980) in surveys of prairie tracts; and, third, observations during the 1982 nesting season in Becker, Cass, and Polk Counties. Janssen (1987) and Hertzel and Janssen (1998) identified Aitkin and Marshall Counties as the only two counties where confirmed nests have been found since 1970.

The MBS recorded 115 locations of Nelson's Sparrow during its county surveys. The excellent MBS observers have greatly expanded the potential breeding range of the species in Minnesota, includ-

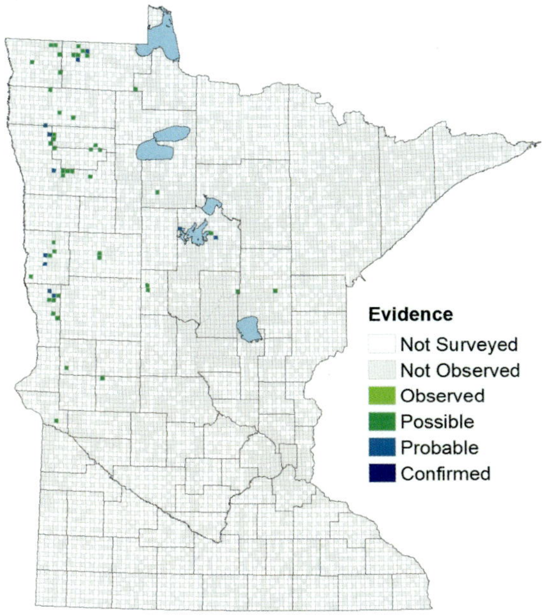

A. The breeding distribution of the Nelson's Sparrow in Minnesota based on the Breeding Bird Atlas emphasizes its primary distribution in the northwest and western parts of the state (2009–13).

ing a total of 16 counties with breeding-season locations. These locations ranged from Kittson and Roseau Counties in northwest Minnesota and south to Douglas, Lac qui Parle, Morrison, Otter Tail, Swift, Todd, and Wilkin Counties. They also recorded locations in north-central Minnesota, including several in the McGregor area, two locations in Itasca County, and multiple locations east of Leech Lake in Cass County. Scattered breeding observation locations were also identified in southern Beltrami, Hubbard, Red Lake, and Wadena Counties, as well as east to St. Louis County.

The MNBBA included breeding records from 58 blocks with a similar but not quite as extensive breeding range as indicated by the MBS **(A)**. Probable breeding was recorded in Roseau County in the northwest, the counties of Clay, Marshall, Polk, and Wilkin in the west, and Cass County to the east. Most of the remaining breeding observations were of possible nesting (81.0 percent, 47/58) in Aitkin, Becker, Beltrami, Big Stone, Crow Wing, Grant, Kittson, Lake of the Woods, Pennington, Pope, Red Lake, and Wadena Counties. No confirmed nests were found.

The collective evidence from these observations suggest the Nelson's Sparrow is more widely distributed in Minnesota than previously thought. Yet its distribution is sparse and localized in wet meadows and marshes—often in relatively large complexes that are difficult to access. Breckenridge's 1930 description of the nest he found was in a marsh in Kittson County "two square miles in extent" (in Roberts 1932). The description by Burt (1978) of the nest he found in the extensive McGregor Marsh of Aitkin County was in knee-deep water and hidden under a tuft of grass.

This species should be sought in large sedge and grass meadows throughout Minnesota. Visual identification can be a challenge in these habitats, and its faint song is frequently only vocalized at night. In their review of the Nelson's Sparrow in North America, Shriver and colleagues (2020) suggested that the species' overall historical distribution has not changed, but some local populations have been extirpated owing to habitat loss.

BREEDING HABITAT

The habitat preferences of the three subspecies of Nelson's Sparrow were described by Shriver and colleagues (2020) as all different and distinctive. They described the Minnesota subspecies as breeding in "freshwater marshes of the Prairie Pothole Region" where "cordgrass, sedges, and Phragmites are common." Hanowski and Niemi (1990) summarized the qualitative descriptions of breeding habitat for this species as bulrush wetlands, extensive marshy areas with cordgrass (*Spartina pectinata*), and wet boggy marshes.

Two of the most consistent places in Minnesota where the species has been found, the Agassiz NWR in Marshall County and McGregor Marsh in Aitkin County, were sampled for vegetation characteristics (Hanowski and Niemi 1990). Both areas are many square kilometers in extent **(B)**. The authors found the following mean values within territories of the species from both sites: vegetation height = 122 cm; water depth = 12 cm; density of grasses/sedges = 182 per square m; and density of forbs = 17 per square m.

POPULATION ABUNDANCE

PIF (2021) estimated a US and Canadian breeding population of all three subspecies as 1 million breeding adults, with a Minnesota population of 13,000. Due to its rarity, no reasonable population trends were available.

CONSERVATION

The MNDNR (2015b) listed the Nelson's Sparrow as a *Species in Greatest Conservation Need* because of its low population and vulnerability to declining wetland habitat. The species is ranked fourth by T. Arnold and Zink (2011) as a "super collider" with buildings. They documented 44 mortalities with buildings, or 25 times more collisions than the average bird. Given the rarity of this species, it is alarming that any individuals would even be detected colliding with buildings. Shriver and colleagues (2020) emphasized that the protection of extensive, undisturbed marshland habitat and interior grasslands is critical to maintaining stable populations.

NAS (2019) projected a moderate vulnerability for the Nelson's Sparrow's breeding range in North America under a 2°C warming scenario. Their models included a very small current breeding range for the species in Minnesota and extirpation by 2050 under the 2°C scenario.

The lack of data on this species' population and trends renders its overall status as precarious. More is known about its distribution in the state because of efforts by the MBS, the MNBBA, and several diligent bird-watchers. Its breeding distribution is larger than previously thought, but little is known about its nesting productivity. Fortunately, this secretive sparrow has persisted in the state, despite extensive habitat losses. It deserves more extensive study, but studies will not be easy, given its low detectability and challenging breeding habitat.

B. The Nelson's Sparrow prefers large, open wetlands with a variety of plant species such as sedges, grasses, cattails, bulrushes, and a diversity of forbs. Photo by Gerald J. Niemi.

HENSLOW'S SPARROW

Centronyx henslowii

Minnesota Seasonal Status: Regular breeding resident and migrant. The Henslow's Sparrow was uncommon during the MNBBA.
Migration: Short-distance migrant that winters in the southeastern United States.
Food: Omnivorous ground feeder that consumes arthropods, especially grasshoppers, crickets, and beetles, as well as seeds.
Nest: Cup nest located on or near the ground; ground litter may form a partial dome.
Conservation Concern: Assigned a high PIF Continental Concern Score of 15/20 and officially listed as *Endangered* in Minnesota.

Photo by Mike Lentz.

A small grassland sparrow, the Henslow's Sparrow is frequently described as a "shy, unobtrusive, and secretive little bird" (Bent 1968). Nearly identical words have been used to describe three other sparrows that regularly nest in Minnesota: the Grasshopper Sparrow, Nelson's Sparrow, and LeConte's Sparrow. All four are elusive, spend much of their time on the ground, are reluctant to fly, and have short, buzzy songs. All four inhabit sites dominated by grasses and/or sedges, are nearly identical in size, and have rather short, thin tails. Their taxonomic relationship has been debated for decades (e.g., Tordoff and Mengel 1951; Rising 1996). Once placed within the same genus, *Ammodramus,* they were recently split among three genera: *Centronyx, Ammodramus,* and *Ammospiza* (Chesser, Burns, et al. 2018).

Despite their similarities, Audubon (1827–38) recognized that the Henslow's Sparrow was a new species when he spotted it in Kentucky in 1820. "It was on the ground, amongst tall grass, and exhibited the usual habits of its tribe. Perceiving it to be different from any which I had seen, I immediately shot it," a standard practice in those days. The bird was named to honor his friend the Reverend J. S. Henslow, a minister and professor of botany at Cambridge University. It was often described as a "mouse with wings"; Audubon (1827–38) always found the species on the ground, where "it runs with rapidity, passing through the grass with the swiftness of a mouse." Even its flight was described as barely more than "respectable jumps over the grass" (N. Brown 1879).

Trying to distinguish this Sparrow in the field from the other three similar species makes one appreciate Audubon's remarkable skills. Subtle plumage variations, different habitat preferences, and unique songs, provide the essential clues. The rufous color on the back and wings of the Henslow's Sparrow, the olive-green tinge on the sides of its flat head, two black lines below the bill, and a finely streaked, buffy breast band are the best diagnostic features. This grassland species prefers sites with a very thick litter layer, a characteristic of grasslands that have not been recently burned. Its unique song, however, is the best identification clue. One of the shortest bird songs in North America, the simple *tse-slik* sounds to many listeners like a dry hiccup. "The song is so weak and fine," Roberts (1932) wrote, "that one is not quite sure, even when close by, that he has really heard anything."

The restricted breeding range of the Henslow's Sparrow spans the grasslands of Minnesota south to eastern Kansas and east across the Ohio River valley and southern Great Lakes to southern Ontario, western New York, and Pennsylvania. Small populations occur along the mid-Atlantic coast. Sparsely distributed throughout its breeding range, the highest densities occur in north-central Missouri.

MINNESOTA BREEDING DISTRIBUTION

Records of this little Sparrow were relatively sparse when Roberts published his 1932 treatise on Minnesota birds. He described it as a summer resident across southern Minnesota, stretching north to Grant and Isanti Counties. There were also two isolated records from the northwest region: a June 1898 record from Hallock, Kittson County; and a July 1879 record from Pembina, North Dakota, "just across the river from Minnesota." Given its rarity, the details of every report, 20 in total from 1880 to 1928, were provided. The records covered all corners of the state's grassland landscape, including Goodhue, Grant, Hennepin, Isanti, Jackson, Kittson, and Pipestone Counties. Nevertheless, only one confirmed nesting record was reported, in 1902 from Jackson County. The most detailed observation was from June 1928, when several birds were found "in the meadows along the south side of the Minnesota River Valley above Fort Snelling." Roberts remarked that "the bird is probably much more common and more widely distributed than these observations indicate, as its secretive habits and weak song make it difficult to find." When he published the second edition of his book in 1936, nesting had also been confirmed in Stearns County (Roberts 1936).

Despite Roberts's optimism about the species' abundance, breeding-season reports only trickled in during the next several decades. Through the 1960s, reports included a nest in Lac qui Parle County (Chambers 1947), a report of "several" birds in Mahnomen County (Janssen 1959), and one adult with a fledgling in Clay County (R. L. Huber 1961c). Generally, the pattern was for one or more pairs to appear at a site one year only to disappear in subsequent years.

The exception was in Winona County. Several grasslands with Henslow's Sparrows were visited by Brother Theodore Voelker in the 1950s. Voelker first reported three nests in 1953 (Herz 1954) and six nests that raised 20 young in 1955 (Guttman 1956). It is unclear if the site or sites he visited were in the same area that later became O. L. Kipp State Park in 1976, a park well-known for its Henslow's Sparrow population. Unfortunately, most of the early accounts submitted to the MOU simply refer to the "usual Winona County areas." Once the state park was established, nearly all Henslow's Sparrow records in Winona County make sole reference to the state park (MOU 2021b).

From the mid-1970s through the early 1990s, the majority of reports continued to come from southeastern Minnesota, again primarily from O. L. Kipp State Park. Gradually, however, reports began to accumulate from other sites. Eckert (1974a) found the species at Blue Mounds State Park in Rock County and referenced earlier reports by Kyllingstad in Pipestone County, while Fall and Eliason (1982) found a nest in Hennepin County. Other reports were submitted from several east-central and southeastern counties, including Dakota, Fillmore, Houston, Ramsey, Sherburne, Steele, and Wabasha (MOU 2021b). Farther west, reports came from Grant (Igl 1996), Norman (D. P. Johnson 1986), and Wilkin Counties (Millard 1995). Hanowski and Niemi (1983) even found several birds in the peatlands north of Waskish in Beltrami County.

During this period, a three-year (1987–89) field study was conducted at O. L. Kipp State Park by graduate student Lynelle Hanson (1994). Included was an assessment of 23 sites throughout the state where the birds had been reported since 1968. Unfortunately, one-third of the previously occupied sites no longer provided suitable habitat owing to urban development or more intensive agricultural practices. The majority of observations reported during the study again came from the state park, where 23 adults were found in 1987, 19 in 1988, and 22 in 1989.

Considered ephemeral in nature, Henslow's Sparrow populations often fluctuate considerably from year to year (Herkert, Vickery, and Kroodsma 2020). Indeed, their abundance at the state park in the late 1980s was followed by four consecutive years (1992–95) when they were largely absent from the site despite intensive field searches; the exception was three birds reported May 12, 1994 (Svingen 1998b). These lean years were followed by an abundance of reports beginning in 1996, when 21 birds were reported from five different locations (Svingen 1998b). In 2000 they were reported from 10 counties across the southern two-thirds of the state, and in 2005 they were reported from 32 counties, including 28 singing males at newly renamed Great River Bluffs State Park (formerly O. L. Kipp) **(A)**. The only regions lacking reports were north-central and northeastern Minnesota (MOU 2021b). Faber (2006) speculated that the unprecedented number of birds in Minnesota could be due to increased grassland management efforts and/or to an influx of birds from states farther south that were experiencing a significant drought. Meanwhile, by 2014, the MBS had documented 80 Henslow's Sparrow breeding-season locations, all south of a line from southern Otter Tail County in the west to southern Crow Wing County in the east.

Despite the abundance of reports, Henslow's Sparrow nests are exceedingly difficult to find and the number of confirmed nesting records in the state remained low. Hertzel and Janssen (1998) identified only two counties with nesting records since 1970: Hennepin and Winona. These remained the only two counties where breeding had been documented for nearly forty years, until the MNBBA began in 2009.

During the MNBBA, participants reported 185 Henslow's Sparrow records in 2.1 percent (99/4,740) of the surveyed atlas blocks. Breeding was confirmed in three blocks **(B)**. The birds were observed in 46 counties, which covered the southern half of the state and extended north through the Red River valley. Breeding was confirmed in Carver and Murray Counties and in a block in the Murphy-Hanrehan Park Reserve that straddled Scott and Dakota Counties. Submitted by Bruce Fall, notes that accompanied the Park Reserve record (2012) provided further details: "There are 7–8 HESP territories within the block in unburned restored prairie . . . plus another 16 territories in the adjacent block to the west." In a situation similar to what Lynelle Hanson (1994) found at Great River Bluffs State Park, the density of breeding pairs at the site suggests the species is "loosely colonial" (Herkert, Vickery, and Kroodsma 2020). The breeding records from Murphy-Hanrehan Park Reserve and from Carver County were observations of adults feeding young or carrying food; the Murray County record was of a juvenile sighted in a block where an adult male had been singing earlier in the season.

Despite the large number of counties where the Henslow's Sparrow was reported during the MNBBA, it remains a sparsely distributed and uncommon species in Minnesota. Years of abundance have been interrupted by years with very few records. Fortunately, as the number of birds has waxed and waned at Great River Bluffs State Park, Henslow's Sparrows have begun to establish a regular presence at several sites in the Twin Cities metropolitan region. In addition to the population reported at Murphy-Hanrehan Park Reserve, small Sparrow populations have been reported at Crow-Hassan

A. The dense, mesic grasslands that line the entrance road to Great River Bluffs State Park in Winona County have supported a small population of nesting Henslow's Sparrows since at least the 1970s. A deep litter layer, a thick cover of grasses and forbs, and tall, residual vegetation from the previous growing season are all apparent. Photo by Lee A. Pfannmuller.

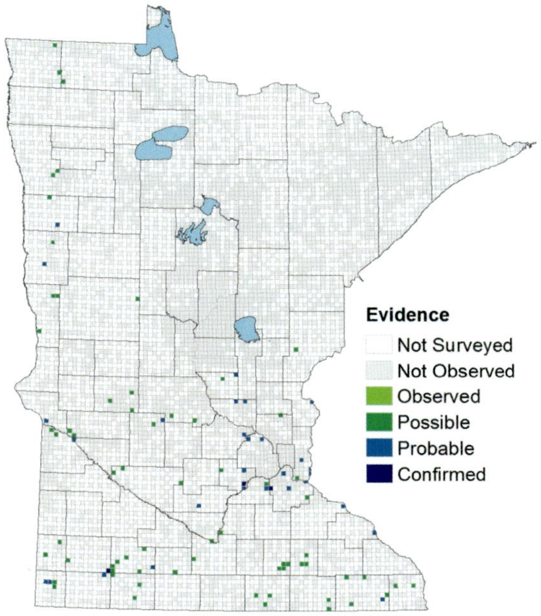

B. Breeding distribution of the Henslow's Sparrow in Minnesota based on the Breeding Bird Atlas (2009–13).

Park Reserve in Hennepin County, and at Afton State Park and Grey Cloud Dunes SNA in Washington County. The MNDNR and Three Rivers Park District have done an exceptional job managing all of these grasslands; the persistence of their Henslow's Sparrow populations will likely be determined by their continued management efforts.

In their comprehensive review of the species, Herkert, Vickery, and Kroodsma (2020) provided a detailed account of the historical changes in the Sparrow's breeding distribution. In the 1800s, there were two disjunct populations: one in the central grasslands of Nebraska and Kansas, and one in the coastal marshes of New England. As the eastern forests were cleared, populations increased and expanded north in the Great Plains and east across the northern Ohio River valley and northeastern states. However, by the mid-1900s, populations were declining, especially in the east. Field surveys of more than 1,000 grasslands from 1997 to 2000 failed to find any Henslow's Sparrows in New England; in New York, a small number of birds were concentrated in just two western counties (Shriver et al. 2005). Apart from a few remnant populations, the species has largely disappeared from the Atlantic coast and is now confined primarily to grasslands in the north-central United States.

BREEDING HABITAT

Historically, Henslow's Sparrows were found in coastal marshes, wet meadows, and weedy pastures along the Atlantic coast. Today, tallgrass prairies, hayfields, and wet meadows are their principal domain in the Midwest **(C)**. For many years, moisture was considered an important component of suitable breeding habitat, given the species' prevalence in coastal marshes and wet meadows. Recent studies in the Midwest have indicated that vegetation structure is a far stronger predictor of occupied habitat (Burhans 2002). Important components include a dense litter layer and standing residual vegetation (Herkert, Vickery, and Kroodsma 2020). The litter layer is used for nesting, foraging, and protection, and residual forbs are used for song perches. Woody cover is minimal to absent. These structural features are not only present on native grasslands and idle hayfields but are also available in planted grassland cover on Conservation Reserve Program (CRP) lands and reclaimed surface mines (Herkert [1998] 2002; T. R. Cooper 2012).

In Minnesota, Lynelle Hanson (1994) found that a dense litter layer, averaging 7.1 cm in depth, was an essential habitat component. It was important that

C. This large, restored prairie in Afton State Park, in Washington County, with its minimal woody vegetation, scattered forbs, and a well-developed litter layer, provides ideal habitat for the Henslow's Sparrow. Photo by Lee A. Pfannmuller.

the litter not be compressed to the ground but provide an open area underneath to allow easy mobility for the birds. This structure was a primary feature that differentiated sites used by the Sparrows from grasslands that were not used.

Some studies suggest that loose colonies are more frequently found on larger sites. Herkert ([1998] 2002) noted that, although individual breeding territories are quite small (0.18 to 1.0 ha), the birds "are more likely to be encountered, and densities may be higher, in large grassland areas than in small areas, and large grasslands may be needed to support persistent populations." Burhans (2002) summarized studies suggesting that, when population levels are high, the birds can be found using small tracts (3–20 ha) as well as larger sites, but the larger sites are occupied first; when population levels are low, only the larger sites are occupied. The 23 Minnesota sites that supported Henslow's Sparrows at least once from 1968 to 1987 ranged in size from just a few hectares to more than 100 ha (L. Hanson 1994). The patch of suitable habitat utilized by the species at Great River Bluffs State Park was 23 ha.

T. R. Cooper (2012) warned that most in-depth habitat studies have been conducted while the species has been experiencing a long-term population decline. When population levels are low, species are often restricted to the highest-quality habitat, providing a skewed view of suitable habitat (C. Smith 1992). Recent field studies in Illinois support this premise. Henslow's Sparrow numbers have experienced a measurable increase in the state since 2000 and the birds are now utilizing smaller, disturbed sites, habitat that is generally considered less desirable (Sauer et al. 2020; T. R. Cooper 2012).

POPULATION ABUNDANCE

BBS data generated a North American population estimate of 410,000 breeding adults within the Sparrow's limited breeding range in the United States and southeastern Canada. Minnesota supports approximately 9,600 adults, or 2.3 percent of the continental population (PIF 2021). Because the Henslow's Sparrow is so sparsely distributed, the BBS does not provide statistically reliable population trends. Nonetheless, it suggests a steady, range-wide decline, averaging −1.94 percent per year from 1966 to 2019, with a small increase from 1990 to 2010 **(D)**. Long-term population trends in Minnesota are difficult to discern with the species' low abundance, but numbers also appear to be declining (Sauer et al. 2020).

Nationally, population gains were observed in portions of the Central Hardwoods, Central Mixed Grass Prairie, and Eastern Tallgrass Prairie Bird Conservation Regions from approximately 1995 to 2005, while persistent decreases were observed farther east (Sauer et al. 2020). When the second state atlas was conducted in Ohio, Henslow's Sparrows were observed in 37 percent fewer priority blocks (Rodewald et al. 2016); a similar decline occurred during Pennsylvania's second atlas (Wilson et al. 2012). In New York, the birds were present in 348 blocks during the first atlas but in only 70 during the second atlas (New York State Department of Environmental Conservation 2006). In all three states, the first atlas was conducted in the 1980s and the second in the first decade of the twenty-first century.

Overall, the species' national decline is largely attributed to the loss and degradation of grassland habitat (T. R. Cooper 2012). Federal farm policies can significantly impact species dependent on the availability of undisturbed grasslands. Research biologists speculated that the Henslow's Sparrow was at a relatively high population level in 1966, when the BBS began, owing to the establishment of the Soil Bank Program in 1956. Designed to reduce farm commodity surpluses, at its peak the program retired 28.7 million acres in ten-year contracts in 1960. However, most contracts expired by 1969, which coincides with the start of the Sparrow's decline. Not until the new set-aside program, CRP, was initiated in 1985 did populations begin to slowly rebound (T. R. Cooper 2007).

Regionally, recent decreases in the northeastern United States are attributed to intensive agricultural practices that have converted pastures and hayfields to row crops, and to habitat loss from development. Inactive management of idle grasslands that results in woody encroachment also degrades otherwise suitable habitat (T. R. Cooper 2012). Factors responsible for population increases are less clear, other than increased enrollment of set-aside acres, such as CRP (T. R. Cooper 2007).

CONSERVATION

The plight of the Henslow's Sparrow has garnered the attention of many state and federal resource agencies and conservation organizations. Considered a high conservation priority by PIF (2021), the species was added to their Yellow Watch List, a group of North American birds that require focused care to prevent further declines (Rosenberg, Kennedy, et al. 2016). It is federally *Endangered* in Canada (COSEWIC 2011) and is classified as *Endangered* or *Threatened* in thirteen states (T. R. Cooper 2012). In Minnesota, the species was classified as *Special Concern* in 1984 and upgraded to *Endangered* in 1996.

In the United States, the USFWS included Henslow's Sparrow in their list of Migratory Nongame Birds of Management Concern in 1987 and 1995 (USFWS 1995). Pruitt (1996) subsequently prepared a comprehensive status assessment of the species as a means of evaluating whether it should also become a candidate for federal listing. Based on that assessment, the USFWS decided that "elevating Henslow's Sparrow to candidate status was not justified" (USFWS 1998a). In response, a formal petition to list the species was submitted in April 1998 by the Biodiversity Legal Foundation requesting that the USFWS re-evaluate their decision and list Henslow's Sparrow as a federally *Threatened* species.

Following additional review, the USFWS concluded that data were inadequate to justify changing their original assessment (USFWS 1998a). Indeed, their decision noted that the species' decline in abundance "may have stopped, and may even be reversing," a decision corroborated by the surveywide population increase documented by the federal BBS from 1990 to 2010 **(D)**. The Service further supported their decision with specific examples of areas where the Sparrow's numbers had significantly increased, including reclaimed strip-mining sites in Indiana and Ohio that supported over 2,000 territorial males.

The USFWS's ruling (USFWS 1998a) also included reports from three states (Illinois, Indiana, and Missouri) that noted that increases in Henslow's Sparrow numbers were due to the bird's use of new lands enrolled in CRP. Unfortunately, just a few years after their decision was published, federal changes to the conservation program became the most immediate threat to the species. When the program was reauthorized in 2014, several changes were enacted that could negatively impact Henslow's Sparrows and many other grassland-dependent species. Included among the changes was a lower enrollment cap (32 million acres decreased to 24 million acres by FY2018) and allowances for emergency harvesting and grazing without a reduction in payments. These changes also coincided with a period of high commodity prices, which reduced overall enrollment in the CRP (Stubbs 2014).

Recognizing the importance of CRP to the Henslow's Sparrow breeding population, as well as other threats that continued to pose concerns for the species, the USFWS developed a national conservation plan and an updated status assessment (T. R. Cooper 2012). One important outcome of this effort was the recognition of potential concerns about the species' wintering grounds, a small area that stretches from North Carolina south to central Florida and west to eastern Texas. The Sparrow's primary habitat during this season is longleaf pine (*Pinus palustris*) communities located within "savanna and pitcher plant (*Sarracenia* spp.) bog areas." Grassy power line corridors, hayfields, and "saline soil barrens" are used as well, but the birds show a strong preference for pine stands. Like the grassland habitats Henslow's Sparrows occupy during the breeding season, longleaf pine forests and savannas have also declined in abundance. Biologists estimate that "only 3 percent of the historic longleaf pine ecosystem" remains and "less than 10 percent of the historic" savannas remain (Pruitt 1996). And, like grasslands, these pine communities are fire-dependent. Unfortunately, prescribed burns are a costly and intensive management activity that can be challenging for resource-strapped agencies to implement.

Warming temperatures also pose serious challenges for the species. The Henslow's Sparrow was one of only six Minnesota breeding species whose breeding distribution was deemed highly vulnerable under all three warming scenarios analyzed by NAS (2019). Under a 2°C warming scenario, models predicted that 72 percent of the species' current North American breeding range would be lost by 2050 and compensated for by just a 10 percent gain to the north. Although the losses were widespread, including in southwestern Minnesota, the gains were limited to the northern regions of Minnesota, Wisconsin, and Michigan. Finally, habitat loss and climate change are not the only threats the Henslow's Sparrow has to contend with. Mortality at communication towers was identified as a major concern, comprising an estimated 1.6 percent of annual mortality (Longcore et al. 2013).

The key to maintaining suitable habitat for the Henslow's Sparrow focuses on protecting large grassland tracts (a minimum of 30 ha and preferably 100 ha or larger) and actively managing each site to control succession and woody encroachment (Herkert [1998] 2002). Grazing, burning, and mowing all improve habitat by reducing woody growth but also negatively impact the litter layer, a critical habitat component. As a result, it is important to protect large sites that enable portions of the tract to be actively managed in a given season while suitable habitat conditions prevail elsewhere on the tract (Herkert [1998] 2002; L. Hanson 1994). Because much of the species' habitat is on privately owned lands, it is particularly important that resource managers promote private land conservation programs such as CRP and CRP Enhancement. Collectively, such efforts should help to secure the future of Audubon's little "mouse with wings."

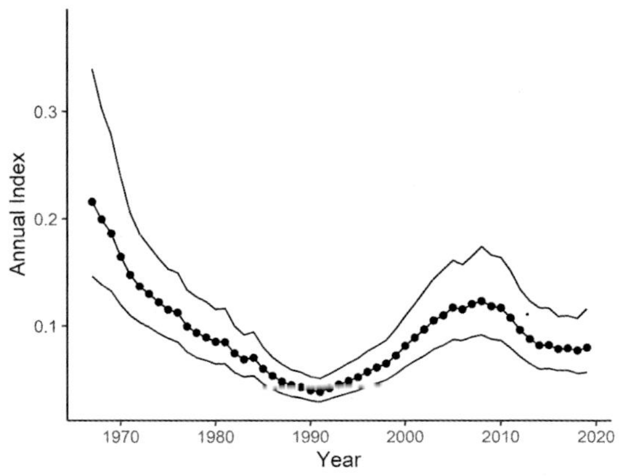

D. Surveywide breeding population trend and 95 percent credible interval (solid black lines) for the Henslow's Sparrow based on the 1967–2019 Breeding Bird Survey (Sauer et al. 2020).

SAVANNAH SPARROW

Passerculus sandwichensis

Minnesota Seasonal Status: Regular breeding resident and migrant; occasionally observed during the winter. The Savannah Sparrow was abundant during the MNBBA.
Migration: Short- to medium-distance migrant that winters in the southern United States, Mexico, and northern Central America. Small resident populations occur in southern California, Baja California, and Mexico.
Food: Omnivorous ground forager that feeds on a variety of seeds and arthropods.
Nest: Cup nest placed in a shallow ground depression, usually tucked within a mound of vegetation or well hidden by a canopy of forbs and grasses. A small tunnel underneath the grasses may lead to the nest.
Conservation Concern: Assigned a low PIF Continental Concern Score of 8/20.

Visit a hayfield or grazed pasture on a summer morning and you are bound to hear the faint, buzzy song of the Savannah Sparrow. Common in grassland habitats statewide, the species is easily overlooked, being far less conspicuous than many of its neighbors, such as Meadowlarks, Kingbirds, and Bobolinks. Small in size, with a short, notched tail, the Savannah Sparrow is brown above and light beige below, with streaking on the back, sides, and breast. The breast streaks may coalesce into a central spot. During the summer, a pale yellow line located just in front and above each eye is diagnostic.

Many early naturalists described the Savannah Sparrow's song as unmusical, insignificant, and shrill (Bent 1968). To those enamored with the dawn chorus, however, its pleasing and distinct quality is a true sign of the grassland's vitality. Although weakly delivered and easily masked by the wind, the three-part song is unmistakable. Several short notes serve as the introduction and are followed by a high, insect-like buzz and then a lower-pitched trill. Other than when it is perched on fence posts or telephone wires, this shy and elusive Sparrow usually keeps low to the ground, preferring to run than to take flight.

A widespread and complex species, the Savannah Sparrow displays considerable morphological variation across its breeding range. The most variable trait is the darkness of the plumage. Although

Photo by Gerald Hoekstra.

17 subspecies are currently recognized, Rising (2010) proposed that these be collapsed into four distinct species. Minnesota birds would remain as the nominate species, *Passerculus sandwichensis,* which translates to "little sparrow of Sandwich." Sandwich Bay on Unalaska Island in the Aleutians is where naturalist Johann Gremlin found and described the species in 1789. The species' common name is a reference to the locality of a type specimen collected in Savannah, Georgia, in 1811 by ornithologist Alexander Wilson (Rising 2010).

The Savannah Sparrow's breeding distribution stretches from Alaska east across Canada, south through most of California, and east across the northern Great Plains, the Great Lakes, New England, and the Appalachian Mountains. Small populations are present in Mexico. Breeding densities are highest in the northern Great Plains of the United States and in Canada's Prairie Provinces.

MINNESOTA BREEDING DISTRIBUTION

Roberts (1932) described this diminutive Sparrow as a common resident "throughout the state wherever conditions are suitable." On the western prairies, "it [was] especially abundant" but could also be found in northern Minnesota, where it was "confined to meadows adjoining lakes and marshy land and to old grass-grown clearings." Despite its abundance and widespread distribution, confirmed nesting records were confined to nine counties in western and east-central Minnesota: Grant, Hennepin, Itasca, Kittson, Otter Tail, Pennington, Polk, Ramsey, and Wilkin.

Green and Janssen (1975) and Janssen (1987) also described the species as a breeding resident throughout Minnesota wherever "low-lying grassy fields" were present. Janssen identified 13 counties with confirmed nesting since 1970; only one was located south of the Minnesota River (Olmsted). Hertzel and Janssen (1998) raised the number to 21, adding three in southeastern Minnesota (Dakota, Rice, and Winona). The MBS further confirmed the species' broad distribution, with a total of 1,465 breeding-season locations from all corners of the state.

During the MNBBA, participants reported 4,012 Savannah Sparrow records from 45.7 percent of the surveyed atlas blocks (2,189/4,789) and 62.8 percent of the priority blocks (1,469/2,338). Breeding was confirmed in 166 blocks (3.5 percent) but deemed probable in an additional 785 blocks (16.4 percent) **(A)**. Although the species was least abundant in the far north-central and northeastern reaches of the state, it was reported from all 87 counties. Breeding was confirmed in 53 counties, deemed probable in 32 counties, and considered possible in two counties.

The distribution of Savannah Sparrow records was especially uniform across northwestern Minnesota. This is the same region where the MNBBA probability map predicts the species is most abundant. High densities are also predicted across a relatively narrow band of central Minnesota, in scattered pockets in southeastern and southwestern Minnesota, and in the peatlands north of Red Lake **(B)**. Breeding densities are lowest in northeastern and north-central Minnesota and in the Twin Cities metropolitan region.

Long recognized as an inhabitant of Minnesota's "low-lying fields and prairies," the Savannah Sparrow was and remains a common species across all but the most heavily forested regions. The MNBBA, however,

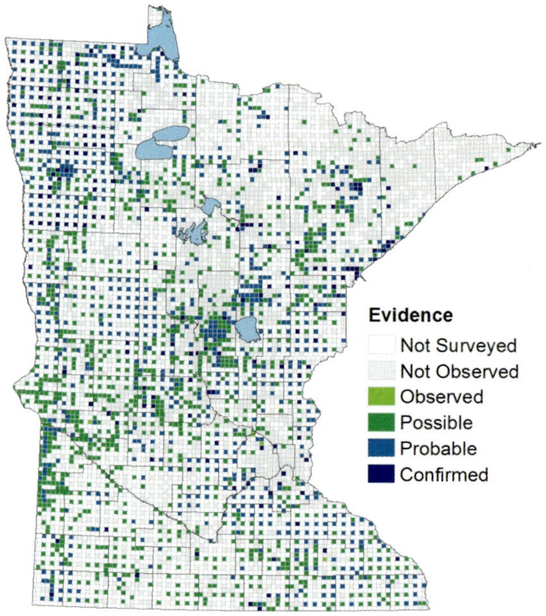

A. Breeding distribution of the Savannah Sparrow in Minnesota based on the Breeding Bird Atlas (2009–13).

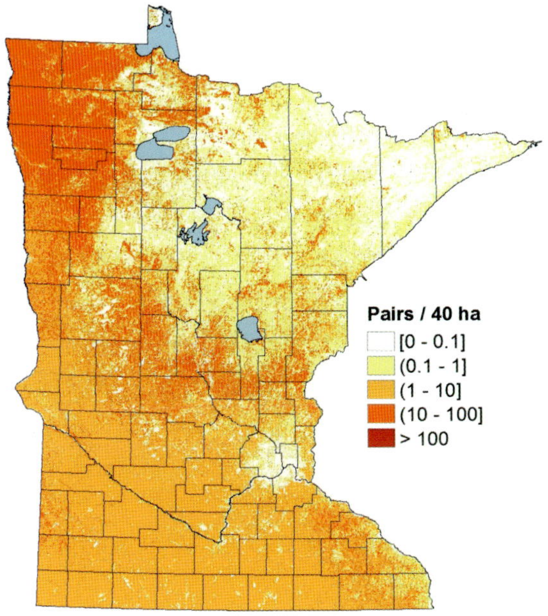

B. Predicted breeding distribution (pairs per 40 ha) of the Savannah Sparrow in Minnesota based on habitat, landscape context, and climate data gathered during the Minnesota Breeding Bird Atlas (2009–13) using the Generalized Linear Modeling method with an adjustment for detectability.

shed new light on the species' relative abundance, including the predicted significance of wet meadows in northwestern Minnesota as an area of particularly high breeding densities. Elsewhere, Wheelwright and Rising (2020) noted only two small changes to the Sparrow's national distribution: the apparent loss of one population in northwestern Arkansas and another at the mouth of the Colorado River.

BREEDING HABITAT

An inhabitant of open landscapes dominated by grasses and sedges, the Savannah Sparrow finds suitable habitat in a wide range of cover types. In the Upper Midwest and northern Great Plains, it is found in sites ranging from native prairies to lightly grazed pastures, alfalfa hayfields, and lands enrolled in the Conservation Reserve Program (CRP), to wet meadows and sedge fens **(C)**. Indeed, it dominated the avifauna associated with the "matted sedge and heath vegetation" of Minnesota's nutrient-poor, ombrotrophic peatlands (Niemi and Hanowski 1992a). In other parts of its range, it occupies tundra, salt marshes, and coastal dunes (D. A. Swanson [1998] 2002; Wheelwright and Rising 2020).

Important habitat features include the presence of grasses and sedges that are short to intermediate in height, an intermediate vegetation density, and a well-developed litter layer. Sites with extensive woody vegetation are avoided. Mesic sites are preferred over arid sites (D. A. Swanson [1998] 2002). Several studies have demonstrated that the species is more abundant on large tracts where rates of nest predation and cowbird parasitism were lower (R. G. Johnson and Temple 1986, 1990).

POPULATION ABUNDANCE

In the United States and Canada, PIF (2021) used BBS data to generate a population estimate of 170 million breeding adults; approximately 3.6 million adults reside in Minnesota. MNBBA data, which provided more extensive coverage of the state, generated a considerably larger population estimate of 6.9 million breeding adults (95 percent confidence interval of 6.1–7.5 million).

Throughout the BBS survey region, Savannah Sparrows experienced a significant decline, averaging −1.11 percent per year from 1966 to 2019 (Sauer et al. 2020), with a cumulative decline of −40 percent between 1970 and 2014 (Rosenberg, Kennedy, et al. 2016). The decline is widespread and includes all of eastern Canada and many states in the Upper Midwest and New England. In Minnesota, the Sparrow is declining at an even faster pace, averaging −1.98 percent per year from 1967 to 2019 and more than doubling to −4.13 percent per year since 1993 **(D)**.

As the eastern deciduous forest was cleared for agriculture in the 1800s, the Savannah Sparrow flourished. But as many farmlands were abandoned and succeeded to forest cover or were cleared for development, the availability of suitable sites declined. Agricultural practices also intensified, as largescale row crops were favored over smaller dairy farms and hayfields (Wheelwright and Rising 2020). The advent of conservation programs, particularly CRP, has provided improved habitat opportunities (e.g., Hanowski 1995), but the short-term contracts that dominate the program are susceptible to the vagaries of the commodity market. Other potential contributors to population declines include the Sparrow's sensitivity to disturbance at the nesting site and its exposure to agricultural pesticides (Wheelwright and Rising 2020).

C. The Savannah Sparrow finds a wide range of open habitats suitable for nesting, including large, mesic grasslands, such as this restored prairie in Brown County. Photo by Lee A. Pfannmuller.

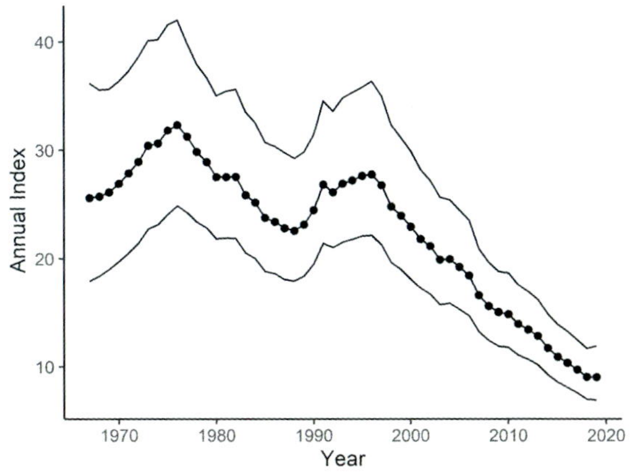

D. Breeding population trend and 95 percent credible interval (solid black lines) for the Savannah Sparrow in Minnesota based on the 1967–2019 Breeding Bird Survey (Sauer et al. 2020).

CONSERVATION

Although its population decline mirrors that of many grassland birds, the Savannah Sparrow's large population and broad distribution place it low on the list of conservation priorities (PIF 2021). Nevertheless, rising temperatures are a concern. Under a 2°C warming scenario, the Sparrow's North American breeding distribution was deemed moderately vulnerable, with a 31 percent decline by 2050, a northward gain of only 11 percent, and its near extirpation from Minnesota (NAS 2019).

Management should focus on protecting and restoring suitable habitats that are at least 50 ha in size and preferably larger than 100 ha. Grasses of intermediate height and density and a well-developed litter layer are critical. Although burning controls woody encroachment, higher breeding densities were found on sites that had not been burned for at least 12 months. The Savannah Sparrow may require at least two years of litter accumulation until a site is suitable for nesting (Tester and Marshall 1961; D. A. Swanson [1998] 2002). As a result, only a portion of larger sites (25–30 percent) should be burned in a given year. Light grazing is tolerated.

Despite its abundance and statewide distribution, the Savannah Sparrow deserves careful monitoring. Efforts to protect and restore its diverse grassland habitats, from native prairies to wet meadows, will benefit this sprightly and diminutive little songster and may help buffer some of the imminent threats of climate change.

SONG SPARROW

Melospiza melodia

Minnesota Seasonal Status: Regular breeding resident and migrant; occasional during the winter months throughout much of Minnesota. The Song Sparrow was very abundant during the MNBBA.
Migration: Short-distance migrant. Northern populations are migratory and winter in the southern United States; populations in the southern half of the breeding range are year-round residents.
Food: Omnivorous ground forager that consumes arthropods and seeds.
Nest: Compact cup nest; usually located on the ground but may be found in shrubs or small trees or attached to emergent vegetation in wetlands.
Conservation Concern: Assigned a low PIF Continental Concern Score of 8/20.

The Song Sparrow is an extraordinary little bird. Widely distributed from Alaska's remote Aleutian Islands south to Baja California and east to the Atlantic coast, the species displays more morphological variation than most North America species (Arcese et al. 2020). Birds in Alaska are one-third larger than and nearly twice as heavy as those on the East Coast, while birds that breed in the desert southwest are much paler than their darker counterparts in the Pacific Northwest (CLO 2017g). All told, 24 distinct subspecies have been identified based largely on differences in size and plumage.

Minnesota's Song Sparrows belong to the subspecies *Melospiza melodia melodia,* which ranges across the northern United States and southern Canada from the Great Plains east to the Atlantic coast. Their white underparts are heavily streaked and have a well-defined center breast spot; their rufous backs are streaked with black and gray, while their crown and head are streaked with rufous-brown and gray. Their short flights are characteristically accompanied by pumping their tails up and down.

The bird's vocal repertoire distinguishes it from all other birds and has endowed it with its common name. Even its Latin name, *Melospiza melodia,* translates to "melodious singing sparrow" (L. Erickson 1988). As variable as the songs are across the species' range, even from one individual, they have an unmistakable quality. Three strongly delivered introductory notes are followed by a trill and then a jumbled series of variable notes. Arthur Allen said: "The song is not clear, for it has quite a burr to it, and yet it is given with such obvious enthusiasm that you

Photo by David Brislance.

feel it is one of the pleasantest you have heard. 'Hip hip hooray boys, that spring is here again,' it seems to say, but its phrases can be as fickle as the weather" (A. Allen and Kellogg 1954). The enormous variability of songs and how each bird acquires its diverse repertoire has been the subject of extensive study. An early student of bird song, A. A. Saunders, wrote: "I have 885 records of the song and no two of them alike" (quoted in Bent 1968).

The Song Sparrow's intrigue does not stop here. This common little bird became the subject of one of the most renowned ornithological field studies ever conducted. Margaret Nice, an amateur ornithologist, undertook an intensive seven-year study of Song Sparrows in her Ohio backyard in the 1930s. Until then, most bird study was focused on collecting and describing museum specimens. Nice's detailed observations of the birds and their behavior in a natural setting launched an entirely new era of field ornithology (L. Erickson 1988; J. Smith 2016a).

Today the Song Sparrow remains a perfect study specimen for addressing a diversity of ornithological questions. Its presence in nearly every open and semi-open habitat in northern and central North America, coupled with its overall abundance, has resulted in a plethora of information about this remarkable little bird.

MINNESOTA BREEDING DISTRIBUTION

Since Roberts (1932) wrote his first comprehensive account of the Song Sparrow, it has been recognized as an abundant summer resident throughout the state. He wrote: "The Song Sparrow is among the most abundant and most generally distributed of Minnesota birds. Bushy meadows and the banks of lakes and streams are the chosen dwelling-places, but it is not confined to such surroundings and may be found almost anywhere about prairie groves, upland fields, clearings, and gardens, shunning only the deep shade of heavy timber." At the time, nesting was confirmed in 12 counties across central and northern Minnesota as well as at Itasca State Park and near Mille Lacs Lake, Cass Lake, and Leech Lake.

Decades later, both Green and Janssen (1975) and Janssen (1987) stated that the Song Sparrow was one of the "most numerous and evenly distributed breeding species in the state." Janssen (1987) identified 32 counties where nesting had been confirmed since 1970. Hertzel and Janssen (1998) later added 10 counties to the list. By 2014, field staff with the MBS documented 3,228 breeding-season locations for Song Sparrows, covering all regions and counties in the state.

MNBBA participants reported 9,019 Song Sparrow detections in 79.7 percent (3,828/4,804) of the surveyed atlas blocks and in 95.0 percent (2,222/2,338) of the priority blocks **(A)**. Breeding was confirmed in 587 (12.2 percent) surveyed blocks and considered probable in an additional 1,495 blocks (31.1 percent). The species was reported in all 87 Minnesota counties and was confirmed breeding in 80 counties, with the outstanding efforts of participants in the Brainerd Lakes area being particularly noteworthy. Song Sparrows were the fourth most common species detected during the atlas, behind the American Robin, Common Yellowthroat, and Red-winged Blackbird.

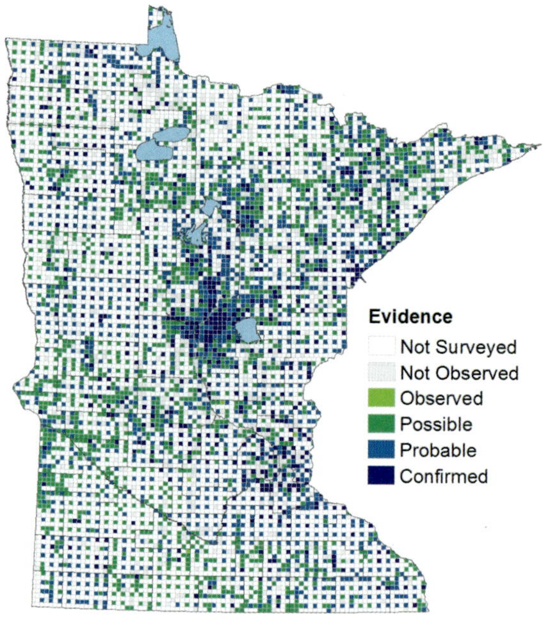

A. Breeding distribution of the Song Sparrow in Minnesota based on the Breeding Bird Atlas (2009–13).

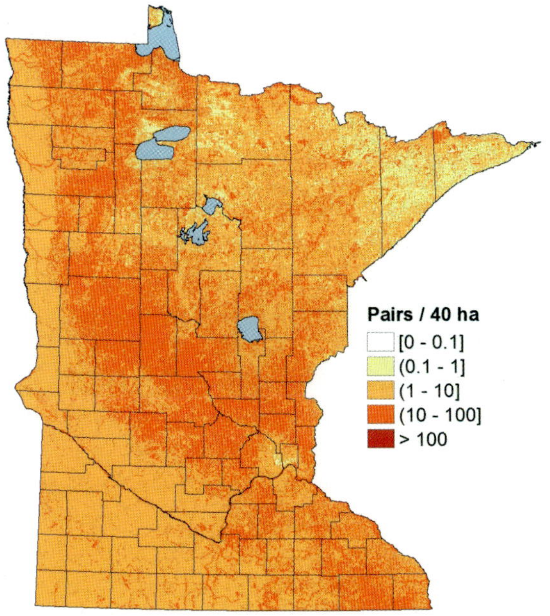

B. Predicted breeding distribution (pairs per 40 ha) of the Song Sparrow in Minnesota based on habitat, landscape context, and climate data gathered during the Minnesota Breeding Bird Atlas (2009–13) using the Generalized Linear Modeling method with an adjustment for detectability.

The MNBBA probability map reflects both the species' abundance and broad distribution throughout Minnesota **(B)**. High breeding densities are concentrated in a band that runs through the Eastern Broadleaf Forest Province northwest through the Tallgrass Aspen Parklands Province. The lowest breeding densities are predicted only for the core of the Twin Cities and portions of many of the densely forested counties. Overall, the state has few areas where one is not guaranteed to see the Song Sparrow and hear its melodious songs.

Historically, the distribution of the Song Sparrow has remained remarkably stable. Although it has always found suitable habitat in small wetlands and forest openings in the most heavily forested landscapes, the clearing of many such areas for agriculture and urban development has created even more habitat opportunities (Arcese et al. 2020). As the MNBBA results and the MNBBA probability map demonstrate, development is not a deterrent to the species.

BREEDING HABITAT

Song Sparrows are found in a wide range of open and semi-open habitats that support an abundance of scattered shrubs and small trees. Sites may be as diverse as desert scrub, arctic-alpine grasslands, temperate rain forests, floodplain forests, and tidal marshes (Arcese et al. 2020). An adaptable species, it can also be found in small towns and residential areas where small pockets of shrubby habitat are available.

In Minnesota, Song Sparrows are found in any small opening or clearing that provides the necessary cover, from small prairie groves to forest clearings to woodland edges (Danz et al. 2007). Regardless of this diversity, they are most frequently associated with shrubby riparian habitats along streams, rivers, lakeshores, and wetlands **(C)**.

POPULATION ABUNDANCE

Analyses using BBS data estimated the Song Sparrow population in the United States and Canada at approximately 130 million breeding adults, with a Minnesota statewide estimate of 5.3 million (PIF 2021). Data collected from the MNBBA's more intensive coverage of the state generated a statewide population estimate that was nearly twice as high, at 10.4 million breeding adults (95 percent confidence interval of 9.7–11.3 million).

Surveywide, the BBS has demonstrated a slow but significant decline in Song Sparrow numbers, averaging −0.59 percent per year from 1966 to 2019 (Sauer et al. 2020). Cumulatively, the loss from 1970 to 2014 was estimated at −25 percent (Rosenberg, Kennedy, et al. 2016). In Minnesota, Song Sparrows demonstrated a similar, but nonsignificant, downward trend, averaging −0.29 percent per year. Their rate of decline, however, has recently increased to a statistically significant average of −2.00 percent per year since 1993 **(D)**. Significant downward trends were also documented in the Chippewa and Superior National Forests, with a combined regional decline of −4.81 percent per year from 1995 to 2020 (Grinde, Bednar, et al. 2020).

Regionally, population trends vary widely. Increasing populations are seen throughout much

C. Although nesting Song Sparrows can be found in a wide range of habitats, shrubby riparian habitats are among the most common sites for finding them. Photo by Lee A. Pfannmuller.

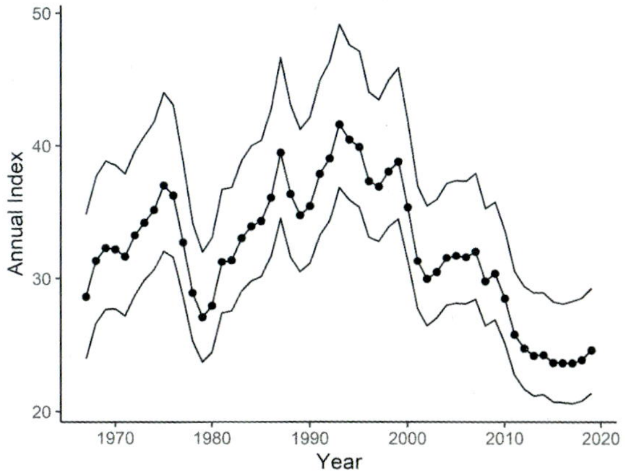

D. Breeding population trend and 95 percent credible interval (solid black lines) for the Song Sparrow in Minnesota based on the 1967–2019 Breeding Bird Survey (Sauer et al. 2020).

of the Great Plains, portions of the western Great Basin, and in several southwest and southeastern states. Declining populations are found throughout Canada, the Upper Midwest, the Northeast, and the Pacific Northwest. Local and regional trends are likely influenced by weather, food resources, nest predation, and brood parasitism rates. Habitat loss and degradation also impact local populations (Arcese et al. 2020).

CONSERVATION

Despite its slowly declining population, the Song Sparrow remains one of the most abundant and widely distributed songbirds in North America and is considered a low conservation priority (PIF 2021). Nevertheless, as temperatures rise, the species' North American breeding distribution will gradually shift northward. Indeed, under a 2°C warming scenario, NAS (2019) predicted a moderately vulnerable breeding distribution for the species, with a projected 27 percent range loss and a 17 percent gain in northern Canada and Alaska by the year 2050. Included among the projected losses was much of Minnesota's southern landscape.

It is easy to take common species, like the Song Sparrow, for granted. But the number of common, widely distributed species that are experiencing population declines is staggering. While many species are in greater peril, biologists estimate that in 75 years the current population of 130 million Song Sparrows will be reduced to 65 million (PIF 2021). And that doesn't account for the projected impacts of climate change. Lacking a continued focus on habitat conservation, many Minnesotans may face a future where even the melodious songs of the ubiquitous Song Sparrow become a distant memory.

LINCOLN'S SPARROW

Melospiza lincolnii

Photo by David Brislance.

Minnesota Seasonal Status: Regular breeding resident, migrant, and occasional in winter. The Lincoln's Sparrow was uncommon during the MNBBA.
Migration: Short- to medium-distance migrant; it overwinters in the southern United States, Mexico, and northern Central America.
Food: Arthropods and seeds or fruit primarily foraged from the ground or from low vegetation.
Nest: Cup nest on the ground or slightly elevated in sedge, moss, or shrubs.
Conservation Concern: Assigned a low PIF Continental Concern Score of 7/20.

The Lincoln's Sparrow is named after Thomas Lincoln, who was a member of Audubon's field crew and is credited with securing their first specimen of this elusive species during an expedition in Labrador. It can be a difficult species to find in the northern regions of Minnesota, but if one finds the right habitat, open bogs with scattered low bushes, and listens carefully for its song, then it is often quite common. As Forbush and May (1955) quipped, "It seems to dread discovery, else why should it steal so cautiously through the depths of leafy thickets. . . . Even when singing it seems careful to keep out of sight."

The song of the Lincoln's Sparrow has been described as a series of rapid bubbling notes that ends with a trill. Transcribed by Bent (1968) as "churr-churr-churr-wee-wee-wee-wah," songs "consist of 13 to 16 notes each, though the character of the song is such that one cannot be too sure of separating the notes and counting them exactly." Bent likened the quality of the song to that of a "house wren or purple finch." Field identification can also be challenging; the species may be mistaken for a Song Sparrow or an immature Swamp Sparrow, both of which can be found in the same habitat. Key characteristics include fine streaking on the breast, on a creamy buff background, that does not aggregate onto a central spot. It also has a grayish back and a narrow eye-ring. Both sexes are similar.

The Lincoln's Sparrow is widely distributed across Canada from the Maritime Provinces to British Columbia. In the United States, it is found in the northern portions of the upper Midwest, in the northeast, and scattered throughout the western states. The highest densities are found in British Columbia, northwestern Alberta, western Colorado, and Labrador.

MINNESOTA BREEDING DISTRIBUTION

Historically, the Lincoln's Sparrow was not known to nest in Minnesota until the mid-1940s. Roberts (1932) stated that "[i]t may nest in the spruce swamps of the northern part of the state, but thus far there is no entirely satisfactory record of the finding of the nest." He summarized its breeding status in Minnesota as a "problem as to its nesting in the state [that] remains to be definitely solved." The first documented nests of the Lincoln's Sparrow in Minnesota were reported by Lupient (1945), when three nests with five eggs each were found on June 9, 1945, by Brother Hubert in Cook County.

Green and Janssen (1975) summarized the species' primary breeding range as the northeastern region of the state as far south as northern Aitkin County and west to eastern Beltrami, Cass, and Lake of the Woods Counties. Besides Cook County, they included confirmed nesting in Lake County and as far south as Kelsey in St. Louis County. Janssen (1987) extended the potential breeding range south to northern Pine County and northwest to Roseau County, where a nest was found in 1976 (Nero 1982), the only county with confirmed nesting since 1970. He did not include Cass County in the breeding range. Hertzel and Janssen (1998) later added Cook County, for a total of only two confirmed counties since 1970. The MBS has recorded 316 breeding-season locations in Aitkin, Beltrami, Cook, Itasca, Lake, Roseau, and St. Louis Counties. In addition, two locations were recorded in northern Carlton County, one in Pine County, one in northeastern Crow Wing County, and two in Cass County.

The MNBBA included 518 breeding records found throughout most of the Laurentian Mixed Forest Ecological Province and in two blocks in the Tallgrass Aspen Parklands **(A)**. A total of 6.2 percent (292/4,735) of surveyed blocks recorded either confirmed, probable, or possible nesting. Besides Cook, Lake, Roseau, and St. Louis Counties with previously confirmed nests, new counties with confirmed nesting included Aitkin and Koochiching. Probable nesting was documented in Beltrami, Carlton, Cass, Clearwater, Hubbard, Itasca, Lake of the Woods, and Pine Counties. Overall, the current breeding distribution of the Lincoln's Sparrow ranges from Cook to Lake of the Woods Counties and possible nesting to western Minnesota in Becker, Marshall, and Roseau Counties and south to Aitkin, Mille Lacs, and Pine Counties.

The probability distribution of the Lincoln's Sparrow in Minnesota predicted its highest densities in the big bog areas north of upper Red Lake as well as the bogs in the Sax-Zim area of northeastern Minnesota **(B)**. Scattered populations were also predicted in potentially suitable habitats throughout northern Minnesota.

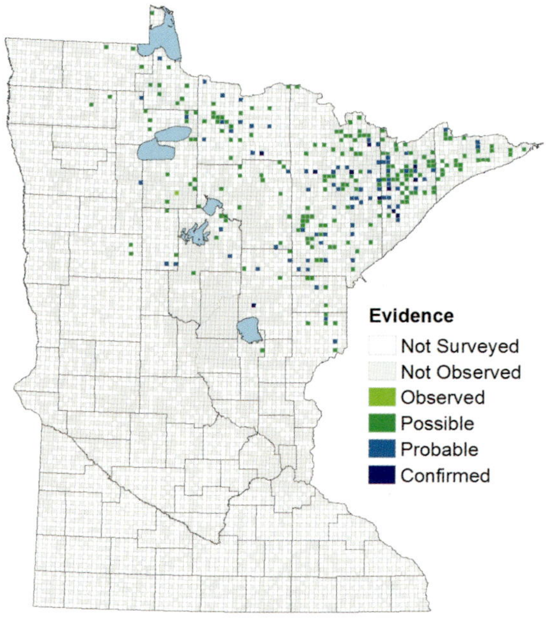

A. Breeding distribution of the Lincoln's Sparrow in Minnesota based on the Breeding Bird Atlas (2009–13).

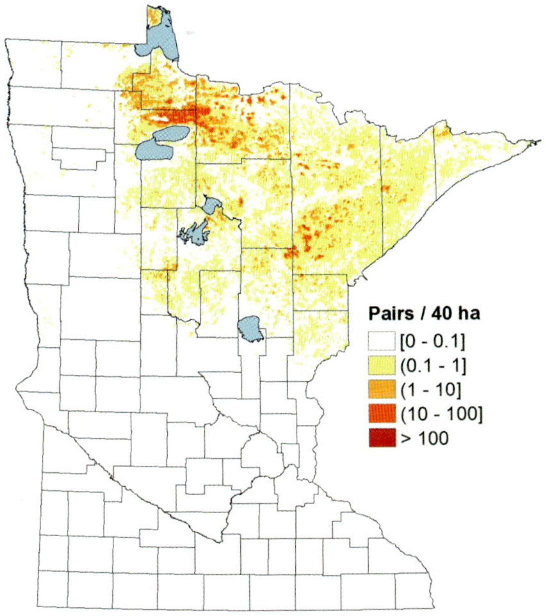

B. Predicted breeding distribution (pairs per 40 ha) of the Lincoln's Sparrow in Minnesota based on habitat, landscape context, and climate data gathered during the Minnesota Breeding Bird Atlas (2009–13) using the Generalized Linear Modeling method with an adjustment for detectability.

Data from both the MBS and the MNBBA showed a broader distribution for breeding-season locations of the Lincoln's Sparrow in Minnesota than what has previously been reported. Either the species has been overlooked or it has expanded its range since the early 1900s. Walkinshaw (1983) contemplated this same question in Michigan, where he also found the species more common than previously thought. He favored the expanded range concept because, he argued, there were "many good ornithologists 50–100 years ago and most of them were collectors." Both R. Brewer and colleagues (1991) and Granlund and colleagues (1994) also supported the idea that the species' range had expanded southward in Michigan.

Surprisingly, Ammon (2020), in her summary of the Lincoln's Sparrow in North America, stated that no historical changes in its distribution were known, but did not cite the changes discussed by Walkinshaw (1983). In contrast, Cutright and colleagues (2006) reported that the species appeared to have retracted northward since the early 1900s in Wisconsin because there were two nineteenth-century nest records from southeastern counties. Consistent with Michigan, the two Ontario breeding bird atlases of 1981–85 and 2001–5 also indicated an increase southward of the Lincoln's Sparrow (Cadman et al. 1987; 2007). They stated that the increase in the number of detections was real, but their regional coordinators pointed out that the increased number of detections may be a reflection of improved song recognition.

Roberts (1932) reported, "This Sparrow had been taken or satisfactorily identified in Minnesota in every month from June to September, but the localities are scattered from the southern to the northern counties and probably are explained by late spring and early fall migrants and stray birds not on their breeding-grounds." He admitted, however, that he had "never heard the song of the Lincoln's Sparrow." Unfortunately, whether its breeding range has recently expanded or whether the species had been previously overlooked remains unknown.

BREEDING HABITAT

The Lincoln's Sparrow nests in open bogs and sedge meadows with understories of moss and ericaceous shrubs, which frequently include scattered willows, bog birch, black spruce, or tamarack (Walkinshaw 1983; Niemi and Hanowski 1992a; Cutright et al. 2006) **(C)**. Two experimental studies in northern Minnesota clearly showed that the species responds positively to burns in comparison with logging activity (Schulte and Niemi 1998; Zlonis et al. 2019). Niemi, Howe, and colleagues (2016) in the Chippewa and Superior National Forests identified the species as significantly associated with sedge meadows, but it was also commonly found in open wetlands and black spruce–tamarack bogs, especially when tree densities were low.

POPULATION ABUNDANCE

Based on the BBS, PIF (2021) estimated a United States, Canada, and global population of 88 million adults, with 70,000 in Minnesota. The MNBBA estimated the Minnesota population as more than five times that size, or approximately 383,000 breeding adults; however, the confidence intervals were large, which indicates substantial uncertainty in this estimate. The large estimate for the global population was primarily due to the extensive habitat available

C. Lincoln's Sparrows are fond of open habitats with sedges, grasses, and scattered shrubs, often in wet or recently burned areas. Photo by Gerald J. Niemi.

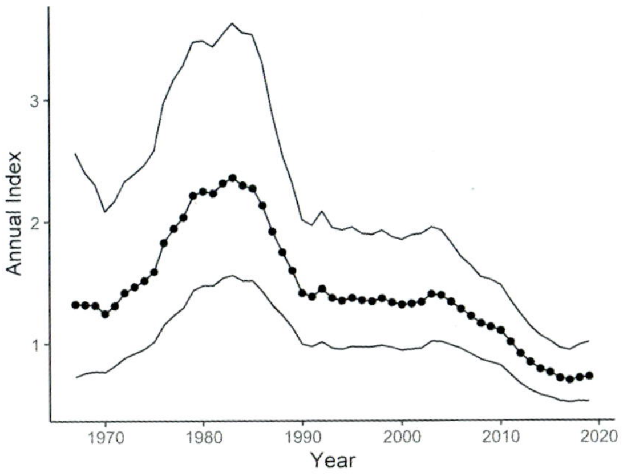

D. Breeding population trend and 95 percent credible interval (solid black lines) for the Lincoln's Sparrow in the Boreal Hardwood Transition Bird Conservation Region based on the 1967–2019 Breeding Bird Survey (Sauer et al. 2020).

across Canada to Alaska, in the Rocky Mountains, and in the western coastal region.

Detections of the Lincoln's Sparrow on the BBS were too scarce in Minnesota to estimate the species' trend. The trend for the Boreal Hardwood Transition Bird Conservation Region, which included northern Minnesota, was reliable and indicated a stable population from 1966 to 2019 **(D)**. However, the recent declining trend of −2.38 percent per year from 1993 to 2019 was significant. Rosenberg, Kennedy, and colleagues (2016) estimated an overall −25 percent decline in its North American population from 1970 to 2014. Unfortunately, few data exist on causes for the species' population decline. In contrast, Canada, with an estimated 86 percent of the breeding population, suggested that BBS trends were unreliable because routes cover only the southern portion of the species' breeding range (Environment Canada 2014b). They emphasized that few threats to the species are currently known.

CONSERVATION

With a 2°C warming scenario, NAS (2019) projected a moderate vulnerability for the breeding distribution of the Lincoln's Sparrow in North America by 2050, including a 48 percent range loss and an 11 percent range gain, primarily in the far north. In Minnesota the species was predicted to be extirpated even under our imminent 1.5°C warming scenario.

The status of the Lincoln's Sparrow is dependent on a better understanding of its breeding ecology in its northern range and its overwintering habitat. It appears to have expanded its range in Minnesota and elsewhere; however, its secretive behavior precludes a definitive statement on its status. Fortunately, it appears healthy in Canada, and Minnesotans are encouraged to learn more about this elusive sparrow.

SWAMP SPARROW

Melospiza georgiana

Minnesota Seasonal Status: Regular breeding resident and migrant; occasionally reported during the winter. The Swamp Sparrow was abundant during the MNBBA.
Migration: Short- to medium-distance migrant that winters from the northern regions of the eastern United States south through central Mexico.
Food: Omnivorous ground forager and foliage gleaner that consumes arthropods, seeds, and fruits.
Nest: Cup nest located just above the ground or water and supported by surrounding vegetation; occasionally placed on the ground.
Conservation Concern: Assigned a low PIF Continental Concern Score of 6/20.

A northern species restricted to wetland habitats, the Swamp Sparrow lives in obscurity compared to its near relative, the Song Sparrow. In addition to occupying many inaccessible habitats, the species spends much of its time on or just above the ground, where it nests and forages, well hidden beneath a canopy of cattails, ericaceous shrubs, or shrub thickets. Forbush (1929) described the species as follows: "The Swamp Sparrow is not a public character. He will never be popular or notorious. He is too retiring to be much in the public eye, and too fond of the impassable bog and morass to have much human company; and so he comes and goes unheralded and to most people unknown."

Slightly smaller than a Song Sparrow, the Swamp Sparrow can be distinguished from the former by its gray and unstreaked or faintly streaked breast compared to the heavily streaked breast of the Song Sparrow. Other features include a rufous crown divided by a lighter median stripe, a gray face and collar, and a dark eyeline. Its simple, emphatic song is a slow, musical trill consisting of several repeated notes, *weet-weet-weet-weet.*

The Swamp Sparrow's breeding range stretches across much of Canada and the northern United States, from the eastern Great Plains and Minnesota, east across the Great Lakes and New England. It reaches some of its highest breeding densities in Minnesota. It was given the species name *georgiana* by John Latham, who first discovered it during the nonbreeding season in Georgia in 1790.

Photo by Sparky Stensaas.

MINNESOTA BREEDING DISTRIBUTION

In the late 1800s and early 1900s, Roberts (1932) considered the Swamp Sparrow a common breeding species across the state, although he noted that it was less frequently encountered in the western grasslands, particularly along the Red River valley. Commenting on its favorite haunts, he wrote: "The Swamp Sparrow, as its name implies, makes its home chiefly in the wetter portions of bogs and marshes where the cattails, rushes, and rank grasses flourish. Sluggish meadow streams, bordered by low willows and alders, are much to its liking." At the time, nesting was confirmed in Hennepin, Jackson, Marshall, Otter Tail, and Sherburne Counties and inferred in Aitkin County and Itasca State Park.

Little had changed when Green and Janssen (1975) prepared an updated account of the species' status and commented that the Swamp Sparrow occurred statewide, wherever suitable wetlands were available. A few years later, Janssen (1987) identified 11 counties with confirmed nesting since 1970. All but two, Goodhue and Le Sueur, were north of the Minnesota River valley. He noted the species was particularly abundant in the "wooded swamps" of northern Minnesota. Five more counties were added by Hertzel and Janssen (1998), including Winona County in southeastern Minnesota. By 2014, the MBS had documented 2,249 Swamp Sparrow breeding-season locations. Observations were least numerous in the intensively cultivated regions of the Red River valley, where sampling was limited, in the upper Minnesota River valley, and in southeastern Minnesota.

The 5,057 Swamp Sparrow records reported by MNBBA participants delineated a breeding range very similar to that documented by the MBS **(A)**. With the bird found in 52.2 percent of the surveyed atlas blocks, breeding was confirmed in 152 blocks (3.2 percent) and deemed probable in an additional 945 blocks (19.8 percent). The birds were reported in all 87 Minnesota counties and were confirmed breeding in 39 counties, 37 of which were located north of the Minnesota River. Based on MNBBA records, the Swamp Sparrow was the third most abundant sparrow in Minnesota, following the Song Sparrow and Chipping Sparrow.

This shrubby, wetland-dependent species was predicted to reach its highest breeding densities in the extensive wetland landscape of the Agassiz Lowlands Subsection and the Tallgrass Aspen Parklands Province, located just north and west of Red Lake **(B)**. Interestingly, as early as 1900, Roberts (1932) commented on how abundant the species was in this region: "Along the Moose River, in western Beltrami County, where the Swamp Sparrow was abundant in July, 1900, the writer heard it singing frequently during the midnight hours on bright moonlight nights." Although there are small pockets of high abundance predicted throughout Minnesota, this expansive, largely inaccessible landscape provides ideal habitat. Miles of flat, poorly drained lake deposits, relics of Glacial Lake Agassiz, are covered by bogs, swamps, and fens that provide the wetlands and low vegetation that characterize most Swamp Sparrow habitats (MNDNR 2003a). Areas of lowest abundance correlate with high development, including the Twin Cities, Rochester, and St. Cloud.

In their comprehensive review of the species, Herbert and Mowbray (2020) noted that the Swamp Sparrow's breeding range appeared to expand north along the mid-Atlantic coast in the mid-1900s, from

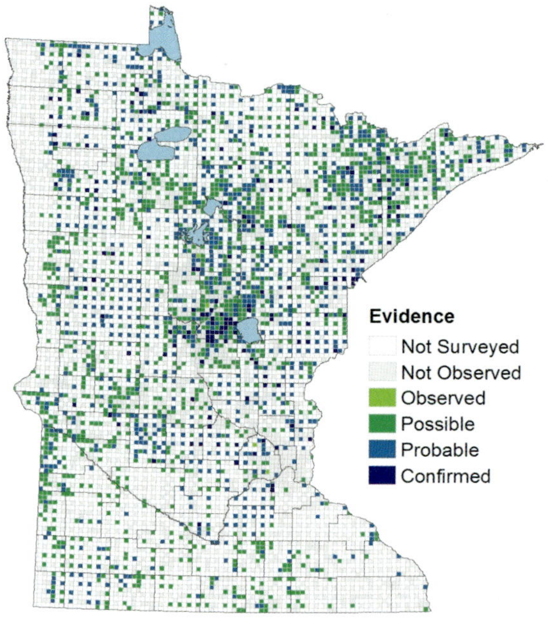

A. Breeding distribution of the Swamp Sparrow in Minnesota based on the Breeding Bird Atlas (2009–13).

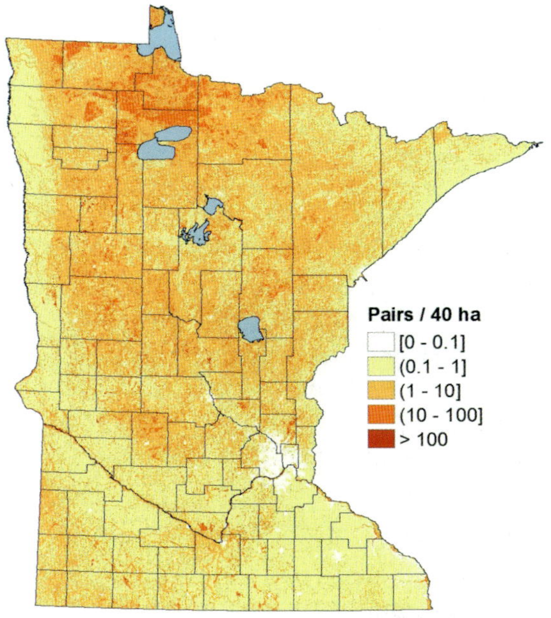

B. Predicted breeding distribution (pairs per 40 ha) of the Swamp Sparrow in Minnesota based on habitat, landscape context, and climate data gathered during the Minnesota Breeding Bird Atlas (2009–13) using the Generalized Linear Modeling method with an adjustment for detectability.

eastern Maryland north to the Hudson River, and south along the Chesapeake Bay. This may have been a true range expansion or a reflection of more intensive field surveys. Elsewhere, the Sparrow's extirpation from northeastern Missouri was also noted.

BREEDING HABITAT

Aptly named, the Swamp Sparrow resides in wetland habitats, ranging from freshwater wetlands in the interior to brackish and saltwater tidal marshes along the Atlantic coast. In Minnesota, the species is found in permanent and semipermanent wetlands dominated by cattails and bulrushes, in bogs and fens dominated by sedges and low ericaceous shrubs, and in shrub wetlands and riparian thickets dominated by alders and willows (Kaufman 1996; Danz et al. 2007) **(C)**.

Suitable sites usually share three features: shallow, standing water; low, dense cover that is usually less than 1.5 m in height; and elevated song perches (Reinert and Golet 1979). Shallow, standing water appears to be particularly important for foraging. Swamp Sparrows frequently feed on the ground in or adjacent to wet areas where they can glean aquatic seeds and invertebrates. They actually have longer leg bones than other members of the *Melospiza* genus. Herbert and Mowbray (2020) speculated that this allowed them to wade efficiently in water. Niemi (1985) noted that longer leg length, a feature shared by many shrub inhabitants, may instead be an adaptation to efficiently maneuver in habitats devoid of thick, sturdy branches.

POPULATION ABUNDANCE

BBS data helped generate a population estimate of 23 million breeding adults in the United States and Canada, with approximately 1.3 million residing in Minnesota, or 5.6 percent of the continental population (PIF 2021). This statewide estimate is significantly lower than the 3.7 million adults estimated using MNBBA data (95 percent confidence interval of 3.2–5.3 million). The higher estimate is reasonable, given that the core of the species' breeding range may be in the remote wetlands of northwestern Minnesota, an area inaccessible to the BBS roadside surveys but sampled by the MNBBA.

The Swamp Sparrow has demonstrated a slow, nonsignificant population increase across the BBS survey region, averaging 0.86 percent per year from 1966 to 2019 (Sauer et al. 2020), with a cumulative increase of 66 percent between 1970 and 2014 (Rosenberg, Kennedy, et al. 2016). In Minnesota, BBS data indicated a significant long-term increase averaging 1.80 percent per year **(D)**.

Factors influencing population trends are not well understood. Predation is believed to be a primary regulatory mechanism; also, because the species is dependent on wetland habitats and usually nests just above the water surface, local rainfall patterns likely have a major influence on annual reproductive success (Herbert and Mowbray 2020).

CONSERVATION

Broadly distributed, with a relatively large and increasing population, the Swamp Sparrow is not a conservation priority (PIF 2021). Nevertheless, approximately 80 percent of the species' North American population is confined to the boreal forest. As remote as this region may seem, between 1990 and 2000 nearly 7 million acres in three Canadian provinces (Ontario, Manitoba, and Saskatchewan) were impacted by forest clearing, road building, and other infrastructure developments associated with

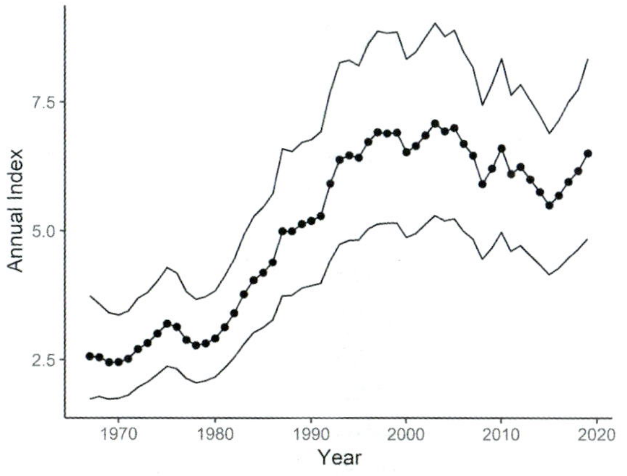

D. Breeding population trend and 95 percent credible interval (solid black lines) for the Swamp Sparrow in Minnesota based on the 1967–2019 Breeding Bird Survey (Sauer et al. 2020).

industries involved in natural resource extraction (Stanojevic et al. 2006a, 2006b).

Loss of habitat is not the only threat faced by species like the Swamp Sparrow that depend on this critical ecosystem. Warming temperatures are another concern. A recent analysis by NAS (2019) classified the Swamp Sparrow's breeding distribution as moderately vulnerable when temperatures rise 2°C. Models predicted the species could lose 35 percent of its current summer range, including throughout most of Minnesota, by the year 2050, while gaining 29 percent farther north in Canada. Such changes would force the birds to find suitable habitat across the Arctic tundra. Building and tower collisions have also been identified as a significant threat to the species (Loss et al. 2014), which was identified as a "super collider" by T. Arnold and Zink (2011).

Given the significance of Minnesota to the national population of this species, efforts to protect and restore wetland habitats are critical to ensuring that the Swamp Sparrow remains a significant component not only of the state's bird community, but also of North America's avifauna.

C. Water appears to be the universal feature of all Swamp Sparrow nesting habitats, from this dense cattail wetland, bordered by wetland shrubs, to fens, bogs, and riparian corridors. Photo by Lee A. Pfannmuller.

EASTERN TOWHEE

Pipilo erythrophthalmus

Minnesota Seasonal Status: Regular breeding resident and migrant; occasionally observed during the winter. The Eastern Towhee was uncommon during the MNBBA.
Migration: Northern populations are short-distance migrants that winter in the southeastern United States, although some birds remain in northern states during mild winters; southern populations are year-round residents.
Food: Omnivorous ground feeder and occasional foliage gleaner that consumes a variety of invertebrates, seeds, and fruits.
Nest: Cup nest placed within ground litter, occasionally in a low bush, or in a tangle of vegetation less than 2 m from the ground.
Conservation Concern: Assigned a moderate PIF Continental Concern Score of 11/20.

Compared to that of most members of the sparrow family that are distinguished by subtleties in their cryptic brown plumage, the Eastern Towhee's striking appearance seems rather brazen. Its black hood, back, and upper chest contrast sharply with its white belly and the deep chestnut feathers on its flanks. A small patch of white is visible in the wings and at the base of the outer tail feathers. The female Towhee is similar, but the male's black feathers are replaced with brown, providing an effective camouflage when she sits on her nest. The species' scientific name, *erythrophthalmus,* is Greek for "red eye," a reference to the red iris that is characteristic of northern populations. The genus name, *Pipilo,* is simply Latin for "to chirp."

One would think that a bird as boldly colored as the Eastern Towhee, and just a tad smaller than a Robin, would be easy to find. The species, however, "has a phenomenal ability to keep itself hidden from view" (Bent 1968). Those in pursuit of the bird will have a far easier time finding it if they learn its variable songs and calls. Although many mnemonics have been used to describe the song, *drink your tea* is the easiest to remember. The first two notes are quite distinct, the second being lower than the first, while the last syllable is a longer trill. The *chewink* call note is also diagnostic. Despite the Towhee's large repertoire, most vocalizations share a distinct metallic quality that is not easily confused with other species.

Photo by David Brislance.

Broadly distributed across the eastern United States, the Eastern Towhee's breeding range extends north into the southern regions of Canada's eastern provinces. It is most abundant in the southern Appalachian Mountains south to the Gulf coast states of Alabama, Georgia, Mississippi, and northern Florida.

MINNESOTA BREEDING DISTRIBUTION

A bird with many identities, the Eastern Towhee was known as the Red-eyed Towhee in the early 1900s and then the Rufous-sided Towhee until 1995, when the western subspecies, the Spotted Towhee, was separated into a distinct species. Other common names included Ground Robin, a reference to its ground-nesting and ground-foraging habits, and Chewink, the phonetic interpretation of its distinctive call note. Roberts (1932) knew the bird as the Red-eyed Towhee and described it as a statewide summer resident. Nesting was confirmed in six counties that stretched from Wabasha County in the southeast, north to Hennepin, Ramsey, and Sherburne Counties, and then north to Crow Wing and Marshall Counties and Itasca State Park. A nest with eggs was reported at Cass Lake, and nesting was inferred in Itasca County and near Mille Lacs Lake. The birds were also considered common "in the jack-pine woods and brush of Aitkin County."

Roberts noted, however, a significant decline in the Towhee's abundance in the "more thickly settled parts of Minnesota during the last twenty-five or thirty years." He attributed the decline, in part, to increasing development but believed other factors were at play. Plenty of suitable habitat still appeared to be available, leading him to suggest that "many stray cats and dogs, the inevitable accompaniment of civilization, may be a factor in the partial disappearance of this and certain other ground-nesting birds."

Accounts by Green and Janssen (1975) and Janssen (1987) described a statewide distribution nearly identical to that delineated by the nesting records enumerated by Roberts (1932). Found in a broad band extending from eastern Kittson and Lake of the Woods Counties in the northwest, southeast through east-central Minnesota, and south along the Mississippi River, the species was largely absent from northeastern and western Minnesota. Nesting had been confirmed in four counties since 1970: Anoka, Dakota, Hubbard, and Washington (Janssen 1987). Hertzel and Janssen (1998) added Brown, Crow Wing, Houston, and Scott Counties to the list.

The 280 breeding-season locations identified by the MBS documented some notable changes to the Towhee's distribution, including numerous records along the Minnesota River valley as far west as Yellow Medicine County, and several reports in the Prairie Coteau region of southwestern Minnesota. Elsewhere, records were concentrated in the Aspen Parklands, in southeastern Minnesota, and in the Northern Minnesota Drift and Lake Plains Section of north-central Minnesota, a region where dry, open woodlands that Towhees find suitable are common.

MNBBA participants reported 485 Eastern Towhee records in 5.5 percent (260/4,737) of the surveyed atlas blocks **(A)**. Breeding was confirmed in just 14 blocks (0.3 percent) and deemed probable in 91 blocks (1.9 percent). With the Towhee reported in 55 of Minnesota's 87 counties, breeding was documented in 11 counties; two of the blocks with confirmed breeding straddled two counties each

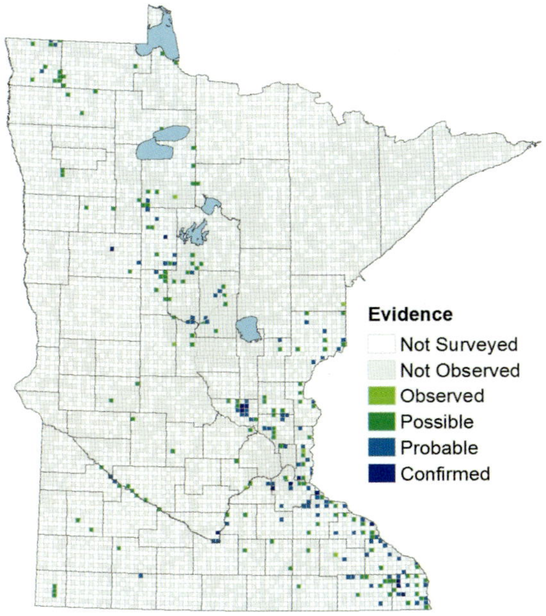

A. Breeding distribution of the Eastern Towhee in Minnesota based on the Breeding Bird Atlas (2009–13).

(Le Sueur/Nicollet and Dakota/Scott). Six counties were additions to the list published by Hertzel and Janssen (1998): Becker, Goodhue, Le Sueur, Nicollet, Sherburne, and Wabasha. The Towhee's MNBBA distribution was quite similar to that documented by the MBS, with the addition of numerous records in east-central Minnesota and the Twin Cities metropolitan region.

Roberts's (1932) observation that the Towhee was formerly abundant in suitable sites "throughout the state" is difficult to reconcile with the species' very limited distribution documented by the MNBBA. Although accounts are replete with references to the species' decline, including Roberts's, the true geographic scope of its original statewide distribution is unclear. One clue was provided by Roberts's mention of the Towhee's presence in the north woods, where "more open, small-growth areas" occurred, including along the "International Boundary." This suggests that conditions in the region may once have been suitable for Towhees following the large-scale clear-cuts and fires that were common in the late 1800s and early 1900s. Indeed, at least one study has demonstrated the species' positive response to fires (Rush et al. 2012). Beyond Minnesota, there have been no major historical changes to the species' breeding distribution, despite its widespread decline (Greenlaw 2020).

BREEDING HABITAT

Often characterized as an edge specialist, Eastern Towhees are found in old fields, shrubby thickets, brushy forest edges, and in open woodlands, including young successional forests **(B)**. Sites are typically dry or mesic with dense brushy cover and a well-developed litter layer that is used for nesting and foraging (Danz et al. 2007; Greenlaw 2020). Studies have shown a direct negative correlation between Towhee abundance and the percentage of forest canopy cover (C. S. Robbins et al. 1989). In national forests in Wisconsin and Minnesota, the species was strongly associated with stands that were recently logged or that were managed as open brush habitat (Niemi, Howe, et al. 2016).

POPULATION ABUNDANCE

PIF (2021) provided a population estimate of 29 million birds in the United States and Canada and a Minnesota estimate of 93,000. BBS data document a significant, surveywide population decline, with Towhee numbers decreasing an average of −1.40 percent per year from 1966 to 2019 **(C)**. Biologists estimate that the species experienced a cumulative decline of −43 percent from 1970 to 2014 (Rosenberg, Kennedy, et al. 2016). Some of the steepest annual declines (greater than 5 percent) have been reported in New England.

The Eastern Towhee is uncommon along Minnesota's BBS routes; statistical analyses of the statewide trend lack precision but suggest a relatively stable population (Sauer et al. 2020). Regional trends reveal a gradient across the state, with population numbers increasing in southeastern Minnesota and gradually transitioning to declining numbers in the northwestern counties. Indeed, in the Chippewa National Forest, which is located along the northwest extension of the Towhee's distribution in Minnesota, the birds have demonstrated a nonsignificant decline, averaging −0.86 percent per year from 1995 to 2020 (Grinde, Bednar, et al. 2020).

B. The Eastern Towhee prefers edge habitats, including overgrown fields with shrubby thickets and scattered small trees. Photo by Lee A. Pfannmuller.

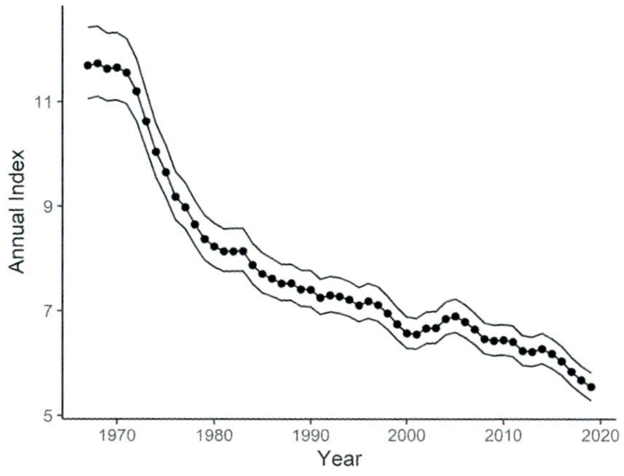

C. Surveywide breeding population trend and 95 percent credible interval (solid black lines) for the Eastern Towhee based on the 1967–2019 Breeding Bird Survey (Sauer et al. 2020).

Hagan (1993) concluded that large-scale changes in the forest landscape were largely responsible for the Towhee's decline. Greenlaw (2020) detailed that the "absence of midstage seral growth, poor understory development in forests, and thin litter accumulations on the ground" were the most important constraining factors. The declining population in Minnesota's Chippewa National Forest likely reflects a decrease in timber harvesting activities and the subsequent loss of early successional forests (Niemi, Howe, et al. 2016). And, as Roberts (1932) originally noted, the Towhee's ground-nesting habits make it especially prone to high rates of nest predation.

CONSERVATION

Although the Towhee remains a widespread species, its designation as a moderate conservation priority reflects concern regarding declining numbers and threats to its breeding habitat (PIF 2021). Unfortunately, under a 2°C warming scenario, NAS (2019) predicted the Towhee's North American breeding distribution to be highly vulnerable. By 2050, it would experience a 56 percent range loss across the southern states and only a 16 percent gain to the north. Under the same scenario, Minnesotans will likely witness the Towhee's disappearance from western regions of the state, where it is already sparsely distributed, while they observe its expansion to the northeast.

Focused conservation measures for the Towhee have not been delineated. Nevertheless, maintaining "an array of woody plant communities in mid-seral stages of secondary succession" will ensure the presence of the young shrubby forests that this unique member of the sparrow family relies on (Greenlaw 2020).

YELLOW-BREASTED CHAT

Icteria virens

Minnesota Seasonal Status: Summer visitant; occasionally breeds. Spring visitant; fall vagrant. The Yellow-breasted Chat was very rare during the MNBBA.
Migration: Medium-distance migrant that winters in Mexico and Central America.
Food: An omnivorous species that feeds on arthropods and many different berries and grapes. It forages in dense vegetation and can grip its food with its feet before eating.
Nest: Bulky cup nest placed near the ground or up to 3 m high in thick vegetation.
Conservation Concern: Assigned a moderate PIF Continental Concern Score of 11/20.

Photo by Jeremy Meyer.

The Yellow-breasted Chat's reputation as a "southern species" goes back to Mark Catesby's first description of this bird, collected in South Carolina in 1758; at the time it was only the third wood warbler described from North America (Green 2008). The classification of the chat as a warbler had long been questioned. Recently, because of its large size, its unusual vocal repertoire, behavior, and certain anatomical features, it was elevated to its own family, the Icteriidae (Chesser, Burns, et al. 2017). Its vocal repertoire has long been considered remarkable. Chapman ([1907] 1968) wrote (quoting Andrew Allison), "No description does justice to his singular medley of calls and whistles. Heard at night, when, especially if it be moonlight, the Chat often sings freely, the performance takes high rank among the songs of North American birds; not for its fluency or spiritual quality, but for its striking originality. . . . 'The voice of this bird is flexible to an almost unlimited degree. It has no notes suggesting its place among the Warblers.'"

The plumages of both sexes are similar. The Yellow-breasted Chat is the size of a thrush. Its back and long tail are an olive-green, which provides its scientific name, *virens,* Latin for "to be green." The upperparts provide a strong contrast to its bright yellow throat and breast blending into a white belly. It has distinct white spectacles with a black band over the upper bill and a white mustache stripe.

The breeding distribution of the Yellow-breasted Chat in the United States has two distinct sectors, which can be described as eastern and western. The eastern is found primarily in the southeastern United States, north to the southern Great Lakes states, and west to the Great Plains. The western population is in scattered nodes of suitable habitat from California and the Pacific Northwest, across the southern Prairie provinces, and then south from the Dakotas to Texas and the southwestern states; there are also breeding outliers in north-central Mexico. The species is found in highest abundance in the south-central states of Arkansas, Louisiana, Mississippi, and Alabama.

MINNESOTA BREEDING DISTRIBUTION

When Roberts (1932) compiled records for the Yellow-breasted Chat, he had little information and only a handful of observations throughout the whole state; he concluded that "the Chat is a very infrequent straggler from the south." These few records included a specimen from Heron Lake, Jackson County, sent to him on May 28, 1890, and a nest with eggs near Wilder, Jackson County, on July 1, 1898. Another record was a pair seen by Thaddeus Surber at Hokah, Fillmore County, on June 30, 1920. In 1934, there were two other observations: Mrs. C. E. Peterson banded two on August 26 and 27, at Madison, Lac qui Parle County, and Dr. C. C. Prosser found one near Lake Harriet in Minneapolis on May 14 (Roberts 1936).

In the decades that followed, more information about the Yellow-breasted Chat's distribution accumulated. In 1938, A. B. Erickson and Upson (1938) made a trip to Houston County in May and found the Chat "one of the most abundant warblers," and a nest was discovered there on June 13. The next information from the southeastern counties (1947 to 1955) was that the Chat was "very common" along the Mississippi River from Brownsville to Reno, Houston County (Bergerud 1947). Breeding evidence was obtained there in 1948 (Warner 1952) and in 1953 and 1955 by Brother Theodore Voelker (Herz 1954; Guttman 1956). Staff at the Minnesota Museum of Natural History (now the Bell Museum) heard or observed five individuals in Goodhue County in 1951 (Warner 1952). In the southwest corner of the state, at Ortonville, Big Stone County, Alma Chesley reported seeing Chats from 1923 to 1943, and nesting was confirmed in 1941 and 1942 (Chesley 1943). In Lac qui Parle County, Mrs. C. E. Peterson reported "banding 11 chats between 1934 to 1950" (Strnad 1959).

These records established the presence of a breeding population in southwestern and southeastern Minnesota. The population along the western border, which was reported for only a few years, has not persisted. The pattern of records in the southeast shows an increasing trend. In the decades from 1960 to 2016, Yellow-breasted Chat summer observations increased steadily in the southeast from 1960 to 1979, when the bird was seen in five years; from 1980 to 1999 it was seen in twelve years; and from 2000 to 2019 it was seen in fifteen years.

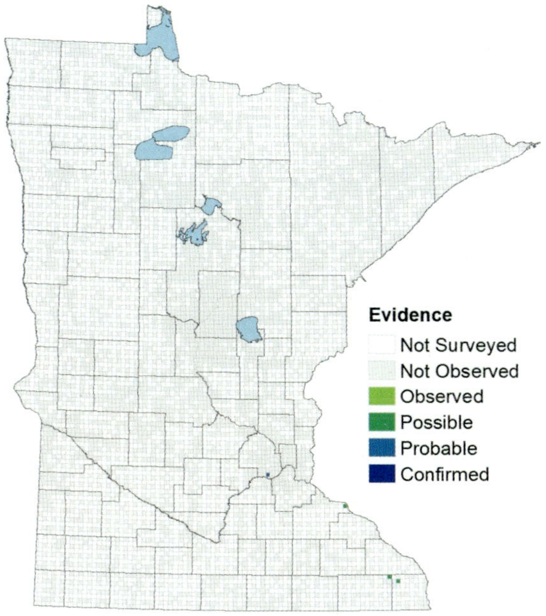

A. Breeding distribution of the Yellow-breasted Chat in Minnesota based on the Breeding Bird Atlas (2009–13).

Breeding evidence since 1960 included a nest with eggs at the Whitewater WMA, Winona County, in 1976 (Eckert 1977b), but no other breeding records until 2007, when a pair and an empty nest were seen at Black Dog Preserve, Minnesota Valley NWR (Kessen and Svingen 2008). After the MNBBA survey period, more breeding evidence was obtained in 2015 from Rock Dell Township, Olmsted County (Kessen et al. 2016). From the years 2000 to 2020, the Yellow-breasted Chat has been seen in the summer at several locations for one year and in multiple years at Great River Bluffs State Park, Winona County (five years); Afton State Park, Washington County (2017, 2018, 2019); and Cedar Rock SNA, Redwood County (2019 and 2020) (MOU 2021a, 2021b).

The MNBBA data include three possible records submitted during the safe nesting dates, all seen in the southeastern corner of the state in appropriate habitat in Goodhue County (June 10, 2009, in T112R13a), Houston County (July 4, 2012, in T104R7b), and Fillmore County (June 8, 2013, in T104R8a) **(A)**. Two Chats were also reported to the MOU in 2009 at Hyland Lake Park Reserve, Hennepin County, both heard about 500 m apart from each other. They were first found on June 2, 2009, by C. M. Brunell and last observed (both) by her on July 28, 2009.

Eckerle and Thompson (2020), in their North American review of the Yellow-breasted Chat, state that the species expanded northward into the northern tier of US states, such as Minnesota, and some Canadian provinces as forests were cleared in the nineteenth and twentieth centuries. As identified by Roberts (1932), the species began to appear in Minnesota in the late 1800s as the forests in the southern regions of the state were cleared and regenerated into shrub habitats. Some of this habitat has now been converted to agricultural lands, residential areas, or has ultimately matured into forests.

BREEDING HABITAT

The basic characteristic of the Yellow-breasted Chat's habitat is dense thickets and brushy tangles **(B)**. J. L. Dunn and Garrett (1997) described the habitat of the chat in the Eastern United States as "thickets, briar patches, willow clumps, multiflora rose brambles and other thorny clumps for breeding. Such thickets may either be in lowland floodplains or in upland regions." The chat is adapted to short-lived habitats, primarily shrubby ones, and abandons these areas once a tree canopy forms.

POPULATION ABUNDANCE

PIF (2021) gives the number of Yellow-breasted Chats in the United States and Canada as 15 million birds, but no estimate of the population in Minnesota was available. The BBS surveywide trend was significantly declining by an average of −0.72 percent per year from 1966 to 2019 **(C)**. Rosenberg, Kennedy, and colleagues (2016) calculated that its population had declined by 11 percent from 1970 to 2014. Eckerle and Thompson (2020) concluded that "densities vary considerably among different habitats for unknown reasons" and these changes can occur "independently of any apparent change in habitat." Population changes can be especially pronounced in species on the periphery of their range, such as the Chat is in Minnesota.

B. The open landscape of the Cedar Rock SNA in Redwood County with its sparse trees, dense shrubby thickets, and scattered rock pools provides ideal nesting habitat for the Yellow-breasted Chat. Photo by Lee A. Pfannmuller.

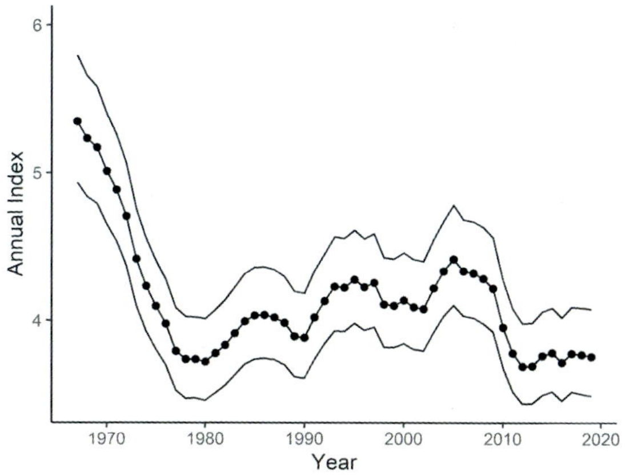

C. Surveywide breeding population trend and 95 percent credible interval (solid black lines) for the Yellow-breasted Chat based on the 1967–2019 Breeding Bird Survey (Sauer et al. 2020).

CONSERVATION

Although the Yellow-breasted Chat has a relatively large population and a wide geographic range, its declining population renders it of moderate conservation risk (PIF 2021). Because it breeds in patchy, short-lived habitats, management has been successful in many locations, as long as these ephemeral habitats are maintained over time. In the western United States, for example, protection of woody riprian corridors has "likely contributed to recent increases" (Eckerle and Thompson 2020). As soon as these successional patches are cleared, or succeed to mature forests, the species can no longer persist. Nevertheless, as noted above, one of the characteristics of the Yellow-breasted Chat is that population fluctuations often appear unrelated to changes in habitat (Eckerle and Thompson 2020). Indeed, the full range of factors regulating populations remains uncertain and further challenges management efforts.

Because the Yellow-breasted Chat has a more southerly US distribution, its breeding range has been identified as stable in response to climate change scenarios modeled by NAS (2019). Future climate conditions in Minnesota may even become more suitable to the species as states immediately to the south and west of Minnesota are predicted to support expanding populations as temperatures rise. Nevertheless, the Chat's dependence on the availability of patchy shrubby habitat may remain a limiting factor for this unique and eclectic species in Minnesota.

YELLOW-HEADED BLACKBIRD

Xanthocephalus xanthocephalus

Minnesota Seasonal Status: Regular breeding species and migrant; occasionally observed in southern Minnesota in early winter. The Yellow-headed Blackbird was common during the MNBBA.
Migration: Short- to medium-distance migrant wintering in the southern United States and Mexico.
Food: Primarily insectivorous during the breeding season, foraging on the ground or at the water's surface; seeds and waste grain are consumed the remainder of the year.
Nest: Cup nest secured to emergent, wetland vegetation; usually breeds in large colonies.
Conservation Concern: Assigned a moderate PIF Continental Concern Score of 9/20.

Sporting a bright yellow head and breast, the male Yellow-headed Blackbird is among a handful of birds whose physical description leaves no doubt as to its identity. Even its scientific name, *Xanthocephalus,* translates to "yellow head." Larger than its relative the Red-winged Blackbird, the male Yellow-headed Blackbird also has a conspicuous white wing patch and a small black mask immediately around its eyes and bill. The female is nearly 5 cm smaller than the male and weighs almost 50 percent less. The dull yellow on her dark brown plumage is confined to the upper breast and an area immediately above the eyes; she lacks any white in the wing. The male does not display the full adult plumage until his second spring.

Few attribute any musical quality to the bird's raucous vocalizations, and many writers decry their near vulgarities. One of its two songs borders on musical, beginning with several loud and fluid notes often followed by a trill. But the most frequently heard vocalization is a loud, raspy call that may begin with a few fluid notes but is followed by a rather buzzy, squawking call of *koh-koh-koh-waaaaa,* which is delivered with a rather dramatic performance. Taverner (1934) described it well when he wrote: "[the male] lowers his head as if to be violently ill, and disgorges a series of rough, angular consonants, jerkily and irregularly, with many contortions and writhings." Overall, the display gives "the general impression that the bird is in great agony" (Bent 1958).

Photo by Mike Lentz.

A gregarious species year-round, the Yellow-headed Blackbird nests in large, loose colonies that may include hundreds of individuals. During the post-breeding season and migration, it travels in flocks, stopping at staging areas shared with Red-winged Blackbirds, Brown-headed Cowbirds, Common Grackles, and Brewer's Blackbirds. The winter months are also spent foraging and roosting in large flocks numbering thousands of birds. Adult males, however, often travel in different flocks than the females and immature males (Twedt and Crawford 2020).

Restricted to the interior of western North America, the species' breeding range stretches from British Columbia and the Pacific Northwest, south to northern Arizona, New Mexico, and Texas, and east to the Great Lakes. A small resident population occurs in the far southwest extending into northern Baja California. The core of its breeding distribution occurs in central North Dakota.

MINNESOTA BREEDING DISTRIBUTION

In the early 1900s, Roberts (1932) described the Yellow-headed Blackbird as a common summer resident throughout the "open country but has invaded the forested regions where extensive reed-grown marshes exist." Indeed, nesting colonies were found as far north as Leech Lake in Cass County, at the mouth of the Rainy River in Lake of the Woods County, and near Duluth in St. Louis County. Confirmed nesting reports were available from eight counties: Big Stone, Goodhue, Hennepin, Jackson, Kittson, Marshall, Otter Tail, Polk, and Sherburne. Inferred nesting reports were available from Nobles County.

More than forty years later, little had changed. Green and Janssen (1975) described a nearly identical breeding distribution, further clarifying the species' absence from most of northeastern Minnesota, including Itasca and Koochiching Counties in north-central Minnesota and much of Pine and Aitkin Counties immediately to the south. Janssen (1987) reported that the species appeared to be expanding its range into the northeast region of the state. There were even June records from Cook County, but these were presumed to be summer vagrants. In 1998, Hertzel and Janssen identified 42 counties where Yellow-headed Blackbirds had been confirmed nesting since 1970, 10 more than Janssen had identified in 1987. Both publications included the northern counties of Aitkin, Itasca, and St. Louis. A similar distribution was documented by the MBS, although no breeding-season locations were found in the Arrowhead region, and records were sparse in southeastern Minnesota.

MNBBA records delineated a breeding distribution very similar to that documented by these earlier accounts **(A)**. Reported from 16.8 percent of the surveyed atlas blocks (800/4,753), the birds were reported from all but eight of Minnesota's 87 counties: four in northeastern Minnesota (Carlton, Koochiching, Lake, and Pine) and four in southeastern Minnesota (Dodge, Fillmore, Olmsted, and Ramsey). Breeding was confirmed in 110 blocks (2.3 percent) spanning 47 counties. Eighteen counties were additions to the list compiled by Hertzel and Janssen in 1998. The species was most frequently reported from the central and southern regions of the Prairie Parkland Province. Closely associated with emergent wetlands, the species' probability map also predicts that the Blackbird is

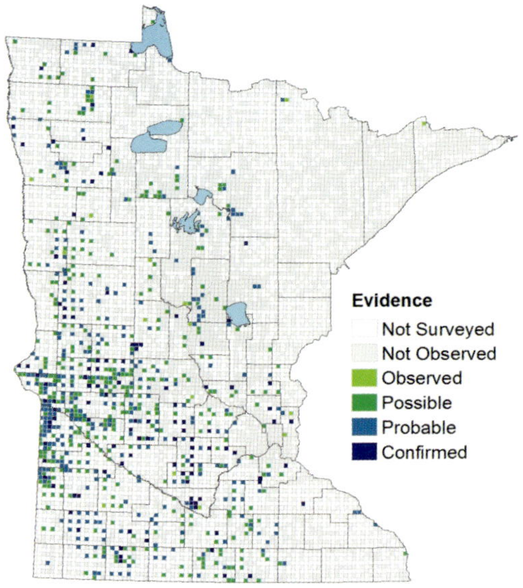

A. Breeding distribution of the Yellow-headed Blackbird in Minnesota based on the Breeding Bird Atlas (2009–13).

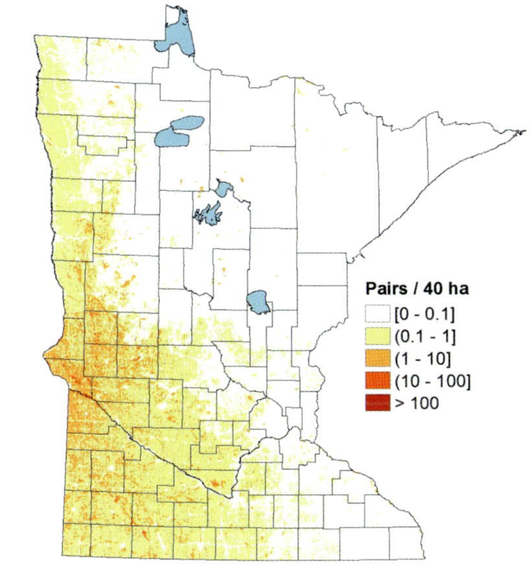

B. Predicted breeding distribution (pairs per 40 ha) of the Yellow-headed Blackbird in Minnesota based on habitat, landscape context, and climate data gathered during the Minnesota Breeding Bird Atlas (2009–13) using the Generalized Linear Modeling method with an adjustment for detectability.

most abundant in the west-central and southwestern regions of the Prairie Parkland Province, with densities declining rapidly farther to the north and east **(B)**.

Historically, few changes have been documented in the species' breeding range, with the exception that some accounts suggest it may be slowly spreading eastward. For example, breeding populations in the Great Lakes region were not documented until the mid-1900s, and the number of reports farther east and as far south as Florida has been increasing (Twedt and Crawford 2020). In Minnesota, the species' breeding range has changed little in the past one hundred years.

BREEDING HABITAT

The Yellow-headed Blackbird is primarily a species of deep freshwater marshes **(C)**. Productive, high-quality wetlands with dense stands of cattails or bulrushes, interspersed with channels of open water, are generally preferred. Sites that become dry during the nesting season are promptly deserted. One study found that territories were least likely to be established in wetlands where shoreline trees projected more than 30 degrees above the horizon (Orians 1980).

POPULATION ABUNDANCE

Using BBS data, PIF (2021) generated a population estimate of 11 million birds in the United States and western Canada, with a Minnesota population of approximately 250,000. The statewide estimate generated with MNBBA data was 549,000 breeding adults. The confidence interval for this highly colonial species, however, was quite large, lending a high degree of uncertainty to the estimate.

Across the BBS survey region, the Yellow-headed Blackbird has demonstrated a relatively stable population from 1966 to 2019, with a nonsignficiant increase averaging 0.71 percent per year (Sauer et al. 2020). Significant population increases are limited to the Prairie Potholes and Badlands Bird Conservation Regions, while significant decreases are restricted to the Coastal California and Prairie Hardwood Transition Bird Conservation Regions. Elsewhere, numbers are relatively stable. Contrary to the national trend, Yellow-headed Blackbird populations in Minnesota have declined significantly at an average rate of −2.93 percent per year from 1967 to 2019, for a cumulative decline of nearly −79 percent **(D)**. Increasing numbers through the early 1980s were followed by a rapid descent that averaged −6.08 per year from 1993 to 2019 (Sauer et al. 2020). The trendline is like that of so many of Minnesota's wetland-dependent species that occur in the western agricultural landscape: their populations are increasing in the core of the northern Great Plains but display a downward trajectory in the North Star state.

Factors responsible for Yellow-headed Blackbird population declines are often attributed to periodic droughts, loss and degradation of habitat, and, because of its consumption of waste grain, the impact of pesticides and herbicides. Many of the marshes selected for breeding are embedded in regions with intensive agriculture, and drift from aerial chemical applications may pose both direct and indirect impacts (Twedt and Crawford 2020; Kreitinger et al. 2013). Insecticide applications to wetlands near urban population centers may also pose a threat. One Minnesota study, however, found no evidence that larvicides used to control mosquito populations impacted breeding Yellow-headed Blackbirds. Instead, weather and predation "were probably more important influences" on the local population (Hanowski, Niemi, et

C. Unlike the Red-winged Blackbird, which selects a broad range of habitats, Yellow-headed Blackbirds nest in colonies that are largely restricted to large, deep wetlands interspersed with dense beds of emergent bulrushes and cattails. Photo by Lee A. Pfannmuller.

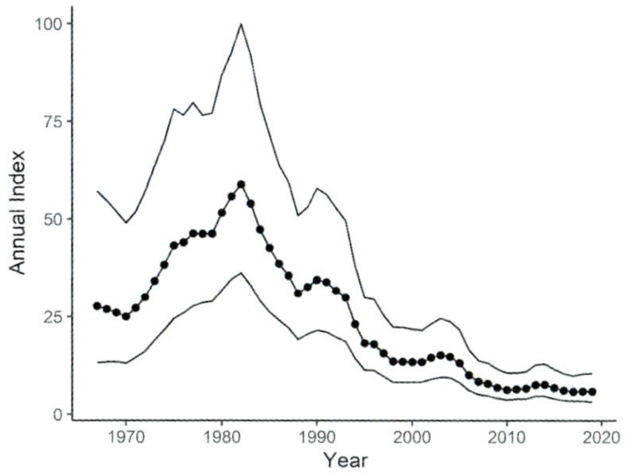

D. Breeding population trend and 95 percent credible interval (solid black lines) for the Yellow-headed Blackbird in Minnesota based on the 1967–2019 Breeding Bird Survey (Sauer et al. 2020).

al. 1997). Indeed, wetlands surrounded by intensive development or cultivation frequently receive heavy run-off during rain events, causing sudden rises in water levels that can kill emergent vegetation needed for nesting or, once the nesting season has begun, drown nests with eggs and young.

CONSERVATION

Like most blackbirds, the Yellow-headed Blackbird is often considered an agricultural pest and is the target of lethal control efforts by farmers growing grain crops. Although the damage they cause is usually minor compared to the damage that other blackbirds inflict (Dolbeer and Linz 2016), some birds are unintentionally killed when control efforts target large blackbird flocks.

In response to climate change, the Yellow-headed Blackbird's breeding distribution was assigned a low vulnerability under all three warming scenarios that NAS (2019) examined. Scattered range losses confined primarily to the northern Great Plains will be more than offset by range expansions in Canada, in the southwestern states, and in northern Mexico. In Minnesota, the species' breeding range is projected to gradually recede in the south but expand in the northwest.

Overall, the biggest concern for the Yellow-headed Blackbird may be the availability of suitable wetlands. Indeed, threats to its breeding and wintering habitats are largely responsible for its assignment of a moderate conservation priority (PIF 2021). Collaboration among resource agencies and conservation organizations is critical to effectively address this issue. Unlike Red-winged Blackbirds that can use sites as small as a roadside ditch, Yellow-headed Blackbirds require relatively large, productive wetland complexes. The protection of such sites is not only important for these emblematic wetland birds but for a host of other wetland-dependent species as well.

BOBOLINK

Dolichonyx oryzivorus

Minnesota Seasonal Status: Regular breeding species and migrant. The Bobolink was common during the MNBBA.
Migration: Long-distance migrant that winters in South America.
Food: Ground forager that consumes terrestrial arthropods, seeds, and small grains.
Nest: Ground nest, usually nestled at the base of grassland forbs.
Conservation Concern: Assigned a high PIF Continental Concern Score of 14/20.

Photo by Mike Lentz.

A quintessential grassland species with a unique plumage, an aerial display, and an unmistakable song, the Bobolink outrivals other members of the blackbird family on all accounts. It is the only North American songbird, for example, to essentially reverse its plumage. Many birds have a light underbelly and a dark back, but the male Bobolink is black on the underside, white on the lower back, rump, and shoulders, and creamy-yellow on the back of the head. The buffy-yellow female, however, might be mistaken as a grassland sparrow. Her dark crown stripes, her dark eyeline, and the absence of streaking on the back of her head help to aid identification. At the end of the breeding season, after undergoing a complete molt, the male assumes the female's buffy-yellow plumage until late winter, when he molts again into his spectacular nuptial attire. If plumage were not enough to distinguish this smallest member of the blackbird family, the Bobolink's migratory journey is one for the record books. From its breeding grounds in North America, it travels nearly 20,000 km to its wintering grounds in central South America.

Full of energy and complexity, the Bobolink's song is also remarkable. Bent (1958) characterized it as a "bubbling delirium of ecstatic music that flows from the gifted throat of the bird like sparkling champagne." Mathews (1921) called it "a mad, reckless song-fantasia, an outbreak of pent-up, irrepressible glee." Commonly sung in flight as the bird flutters its wings, the entire performance is a sight to behold. Roberts (1932) captured it best when he wrote, "An hour with the Bobolink in June is to the bird-lover an hour of long-remembered joy."

Distributed across southern Canada and the north-central and northeastern United States, the Bobolink is most abundant in the northern Great Plains, reaching its highest breeding densities in western North Dakota. Its unique Latin name, *Dolichonyx oryzivorus,* means "long-toed rice-eater." The long hind toe helps it perch on thin-stemmed grasses and forbs; "rice-eater" is a reference to a primary component of its winter diet.

MINNESOTA BREEDING DISTRIBUTION

Roberts (1932) described the Bobolink as a common summer resident found "throughout the state wherever there are suitable lowlands." Most common in the western grasslands, it was also present in small hayfields and pastures embedded across the forested landscape. Nesting was confirmed in eight widely scattered counties: Grant, Marshall, Pennington, Pipestone, and Polk Counties in western Minnesota; and Hennepin, Isanti, and Scott Counties in east-central Minnesota. Young were also observed in the Mille Lacs area. At the time, Roberts expressed concern that the Bobolink was declining in abundance and was being replaced by the more aggressive Brewer's Blackbird. The latter species, also dependent on open grasslands, was rapidly expanding east across the Great Lakes, New England, and the Maritime Provinces.

Both Green and Janssen (1975) and Janssen (1987) described the Bobolink's distribution in a manner similar to that of Roberts, pointing out that it was most abundant in western and northwestern Minnesota and least abundant in the north-central and northeastern regions. Janssen identified 15 counties, from St. Louis and Beltrami south to the Iowa border, where nesting had been confirmed since 1970. Hertzel and Janssen (1998) later added six more counties to the list. By 2014, the MBS reported a total of 1,653 breeding-season locations. Widespread throughout the Prairie Parkland and Tallgrass Aspen Parkland Provinces, many records were also reported in north-central Minnesota, from Clearwater County east to central St. Louis County.

MNBBA participants documented a slightly broader distribution. A total of 3,376 breeding records were collected in 37.6 percent (1,804/4,801) of the surveyed atlas blocks, including in some of the state's most cultivated landscapes, across southeastern Minnesota, and throughout the western and southern regions of the Laurentian Mixed Forest Province. Breeding was confirmed in 148 blocks (3.1 percent) and considered probable in an additional 794 blocks (16.5 percent) **(A)**. The birds were observed in all 87 Minnesota counties, including one probable record at a farmstead near Grand Marais, Cook County. Large hayfields scattered across central and southern St. Louis County also provide ideal habitat in an otherwise unsuitable landscape. The 61 counties where breeding was confirmed included 44 that were additions to those delineated by Hertzel and Janssen (1998). Reports remained uncommon in the Twin Cities and in far northeastern and north-central Minnesota. Despite Roberts's (1932) concern about its status nearly one hundred years ago, the Bobolink remains a common and widely distributed species. Compared to its 3,376 MNBBA records, only 728 records were reported for the Brewer's Blackbird.

The MNBBA probability map predicts that the core of the species' breeding distribution lies in the Red River valley and the Tallgrass Aspen Parklands Province of northwestern Minnesota **(B)**. Narrow bands of high abundance also stretch across central Minnesota and along the southwestern border

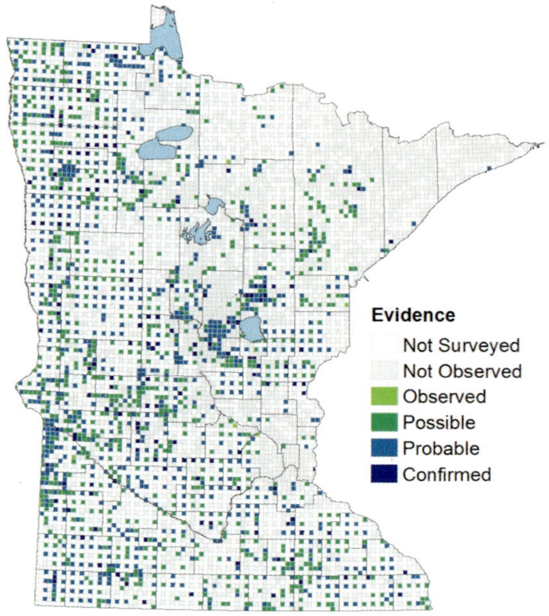

A. Breeding distribution of the Bobolink in Minnesota based on the Breeding Bird Atlas (2009–13).

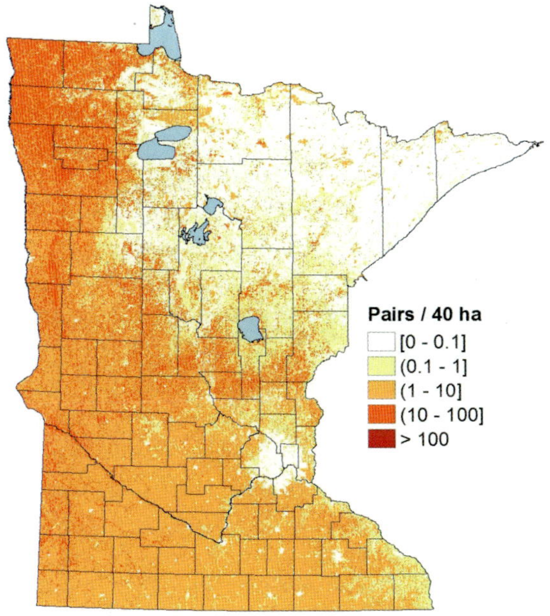

B. Predicted breeding distribution (pairs per 40 ha) of the Bobolink in Minnesota based on habitat, landscape context, and climate data gathered during the Minnesota Breeding Bird Atlas (2009–13) using the Generalized Linear Modeling method with an adjustment for detectability.

of the state. Present at moderate breeding densities across much of southern Minnesota, the species is predicted to be rare to absent throughout most of the northern forest.

Historically, Bobolinks expanded east as forests were cleared and expanded farther west as arid lands were irrigated and cultivated. Today, as some of the previously cleared areas in the east are abandoned and succeed to forest, the Bobolink's range is receding. Indeed, in the late 1900s, 18 northeastern states lost more than 5 million acres of hayfields, which were prime habitat for Bobolinks (Renfrew et al. 2020).

BREEDING HABITAT

Originally an inhabitant of mesic tallgrass prairies, the Bobolink now makes extensive use of hayfields, lightly grazed pastures, small grain fields, and planted cover, including Conservation Reserve Program (CRP) fields (Dechant, Sondreal, et al., "Bobolink," [1999] 2001) **(C)**. Madden and her colleagues (2000) suggested that the Bobolink's attraction to sites planted in non-native grasses, such as smooth brome and quack grass, results from their structural similarity to the native grass cover found in mesic prairies.

Suitable sites are characterized by moderate to tall grasses and forbs, moderate vegetation densities, moderately developed litter layers, and the absence of woody vegetation. Forbs are especially important for nesting cover. Bobolinks also demonstrate some area sensitivity, with higher nesting densities and higher reproductive success reported in larger habitat patches (e.g., D. H. Johnson and Igl 2001; R. G. Johnson and Temple 1986, 1990).

POPULATION ABUNDANCE

PIF (2021) estimated the Bobolink breeding population to number 10 million birds in the United States and Canada, with 1 million birds residing in Minnesota. The estimate derived using the more extensive MNBBA data is magnitudes higher, at 6.5 million (95 percent confidence interval of 5.8–7.6 million). Unfortunately, Bobolinks have experienced a significant population decline, averaging −1.54 percent per year across the BBS survey region from 1966 to 2019 (Sauer et al. 2020), with a cumulative loss of over −60 percent from 1970 to 2014 (Rosenberg, Kennedy, et al. 2016). Minnesota has witnessed a similar decline, averaging −1.62 percent per year **(D)**.

In addition to the loss of native grasslands, Renfrew and her colleagues (2020) identified three land use changes that threaten breeding populations: first, the decline in the number of acres maintained as hayfields in the northeastern states; second, an increase in the number of acres planted to alfalfa, a less-preferred grassland cover for Bobolinks; and, third, progressively earlier dates for mowing hayfields. Mowing occurs nearly two to three weeks earlier than occurred in the mid-1950s and now coincides with dates when young are still in the nest and unable to escape.

Challenges also face the Bobolink during its long migratory journey and on its wintering grounds. In the late 1800s and early 1900s, thousands of birds were killed during the fall migration and sent to eastern food markets or were slaughtered when they stopped to feed in rice fields in the southeastern United States (Bent 1958). Although United States rice cultivation has since moved west of the species' primary migratory corridor, the Bobolink is still considered a pest in South America's rice fields. Wintering birds and migrants are also trapped in the Caribbean and South America for sale in domestic and international pet markets (Renfrew et al. 2020).

C. A hayfield in St. Louis County with a rich abundance of forbs and tall grasses provides ideal nesting habitat for many Bobolink nesting pairs. Photo by Lee A. Pfannmuller.

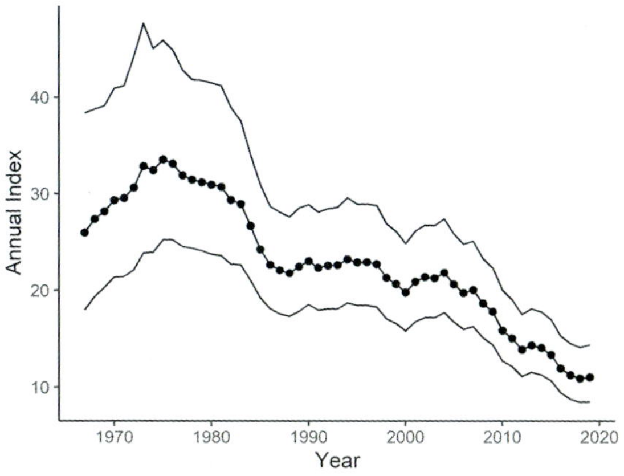

D. Breeding population trend and 95 percent credible interval (solid black lines) for the Bobolink in Minnesota based on the 1967–2019 Breeding Bird Survey (Sauer et al. 2020).

CONSERVATION

A high conservation priority (PIF 2021), the Bobolink was designated a Stewardship Species by Audubon Minnesota because such a significant portion of its total population (at least 10 percent) occurs in Minnesota (Pfannmuller 2012). Management recommendations focus on providing large tracts of suitable habitat, controlling the succession of woody vegetation, and protecting habitat from disturbance during the nesting season. CRP fields provide important breeding habitat and should be managed accordingly (Dechant, Sondreal, et al., "Bobolink," [1999] 2001). Unfortunately, most of these acres are not permanently protected and are subject to fluctuations in the commodity markets.

Given the Bobolink's preference for mesic conditions, warmer temperatures may pose additional challenges. Under a 2°C warming scenario, by the year 2050 NAS (2019) projected a moderate vulnerability for the Bobolink's North American breeding distribution, with a 58 percent range loss and a 43 percent gain northward. In Minnesota, its breeding range would recede from nearly all of the southern and central counties.

Minnesota has a special responsibility to provide for the future of this unique Stewardship Species. Aggressive efforts to protect and restore grassland habitats are critical to ensure that this grassland inhabitant, lauded for its joyful prairie song, remains a prominent voice in the early morning chorus, even with a warmer climate.

EASTERN MEADOWLARK

Sturnella magna

Minnesota Seasonal Status: Regular breeding species and migrant; occasionally observed in southern Minnesota during the winter. The Eastern Meadowlark was uncommon during the MNBBA.
Migration: Northern populations are short-distance migrants that winter in the southern United States; southern populations are year-round residents.
Food: Ground forager that feeds on insects and seeds.
Nest: Cup nest placed on the ground; often constructed with a canopy and tunnel entrance.
Conservation Concern: Assigned a moderate PIF Continental Concern Score of 11/20.

Although its name may suggest otherwise, the Eastern Meadowlark is not a lark but a member of the blackbird family. Its plump appearance and its bright yellow throat and belly, boldly marked with a dark V-shape on the breast, clearly identify it as a meadowlark. In flight, white outer tail feathers are visible. Unfortunately, its similarity to the Western Meadowlark is a source of continual frustration to most bird-watchers. Subtle plumage differences are frequently cited but challenging to identify in the field. Even Roberts (1932) remarked that, apart from its "voice and certain details in markings," the Eastern Meadowlark "is practically identical" to the Western Meadowlark.

Despite their similar appearance, the two species are distinguished by their markedly different songs and call notes. Bent (1958) quoted from Robert Ridgway (1877), who wrote: "We know of no two congeneric species, of any family of birds, more radically distinct in all their utterances than the eastern and western Meadow Larks." The Eastern Meadowlark's song consists of several clear, mournful whistles that are frequently rendered as *spring of the year, spring of the year,* compared to the Western Meadowlark's complex warble. The Eastern Meadowlark's call note is a higher pitched, buzzy note (*dzit* or *yert*); the Western Meadowlark's note is clearer and lower in pitch (*chuck*). Nevertheless, skeptics doubted that the birds were different species until Ridgway's (1902) analysis and delineation of the two species prompted the American Ornithological Society to declare them separate in 1908 (J. Allen et al. 1908).

Photo by Michael Furtman.

Subsequent studies have demonstrated that hybridization is "apparently rare" (Jaster et al. 2020).

Field biologists, however, recognized that in places like Minnesota, where the species ranges overlap, identification by song may not always be reliable (Eckert 2006; Dunlap 2019). Although the phenomenon is reported rarely, observers have noted meadowlarks singing intermediate songs that combine phrases from both species and, occasionally, a meadowlark that vocalizes both its repertoire and that of its sibling species. Such behavior is bound to baffle any field observer! The origin of this variability is not entirely understood. Hybridization or patterns of song learning may be responsible (Dunlap 2019).

The Eastern Meadowlark occurs across the eastern United States and southeastern Canada, extending west across the Central Plains states and south as far as Cuba and northern South America. Some of the highest breeding densities are found in Kansas, Missouri, and Oklahoma.

MINNESOTA BREEDING DISTRIBUTION

Roberts (1932) noted that the Eastern Meadowlark, with a summer distribution "confined to a limited area along the extreme eastern border of Minnesota," was slowly expanding westward. Nevertheless, it occurred regularly only from Mower County northwest to Le Sueur County, then east to western Hennepin County and north through Isanti County. From there it extended north to Duluth and Two Harbors. The species was also beginning to appear farther west along the Iron Range. Nesting evidence, however, was available only from Goodhue and Hennepin Counties.

Forty years later, as its breeding range continued to expand, Green and Janssen (1975) reported that it occurred north to Lake of the Woods County, west to Otter Tail County, and south to Blue Earth and Faribault Counties. The same distribution was delineated by Janssen (1987), who also identified five eastern counties with confirmed nesting since 1970: Aitkin, Anoka, Benton, Dakota, and St. Louis. Hertzel and Janssen (1998) later added Le Sueur, Wabasha, and Winona Counties to the list. Coincidentally, the MBS reported 256 breeding-season locations for the species. With the exception of a few records in southwestern Minnesota, its range had changed little since Green and Janssen's 1975 account.

The MNBBA's 730 Eastern Meadowlark detections not only documented records in southwestern Minnesota but included a small number of reports farther north, in the Red River valley and Aspen Parklands. Reported in 10.3 percent (490/4,749) of the surveyed atlas blocks, the records spanned 68 of Minnesota's 87 counties **(A)**. Breeding was confirmed in 37 blocks and deemed probable in 155 blocks. All but five of the 28 counties with confirmed breeding evidence were additions to the list provided by Hertzel and Janssen (1998). Nesting was newly confirmed north in Cass, Crow Wing, and Wadena Counties, while probable nesting was noted in Beltrami, Clearwater, Marshall, Polk, and Roseau Counties. The MNBBA probability map **(B)** predicted that the core of the species' breeding range lies in southeastern Minnesota, with scattered pockets of moderate abundance stretching north to Wadena County in north-central Minnesota and southern St. Louis County in the east.

Although the Eastern Meadowlark expanded north and west across Minnesota in the early 1900s, changes since the 1970s have been minimal. The scattered atlas reports from western Minnesota are not entirely new. Eckert (2010) noted that there were

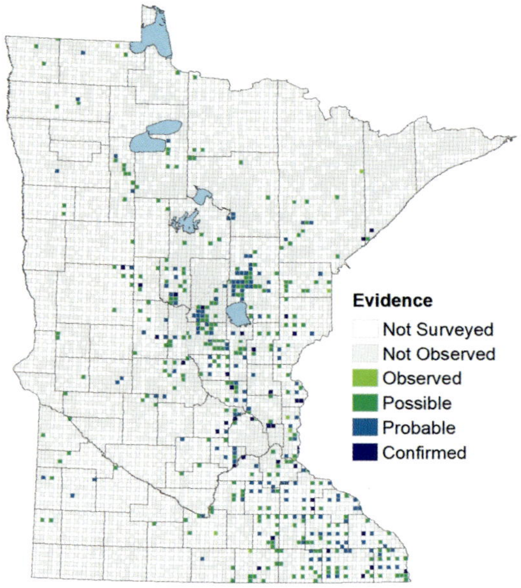

A. Breeding distribution of the Eastern Meadowlark in Minnesota based on the Breeding Bird Atlas (2009–13).

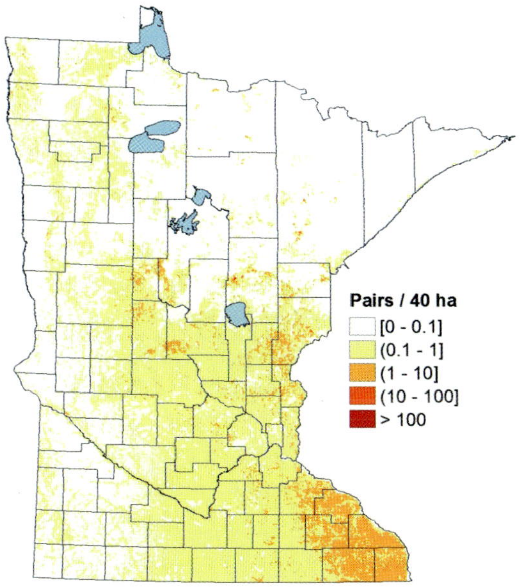

B. Predicted breeding distribution (pairs per 40 ha) of the Eastern Meadowlark in Minnesota based on habitat, landscape context, and climate data gathered during the Minnesota Breeding Bird Atlas (2009–13) using the Generalized Linear Modeling method with an adjustment for detectability.

places throughout the region "where there have been recent, credible sightings—and listenings—of Easterns in summer." Nevertheless, all the counties where the atlas documented nesting and all the counties where nesting was confirmed from 2014 to 2020 (MOU 2021a) were within the breeding range delineated by Green and Janssen (1975). Although both meadowlark species co-occur over a large portion of Minnesota, MNBBA data demonstrated that the Eastern Meadowlark is less likely to be found in western Minnesota than the Western Meadowlark is to be found in eastern Minnesota.

Beyond Minnesota, other changes in the species' distribution have been western expansions in Kansas and Texas and expansions in portions of Central America following forest clearing (Jaster et al. 2020).

BREEDING HABITAT

Originally dependent on native prairies, grassland openings, and oak savannas scattered across the Eastern Deciduous Forest, the species has adapted to other grass-dominated habitats, including weedy pastures, hayfields, golf courses, reclaimed surface mines, and restored grasslands **(C)**. Planted cover, including lands enrolled in the Conservation Reserve Program (CRP), also provides suitable habitat (Jaster et al. 2020; Hull [2000] 2002; Faanes 1981).

Sites are characterized by their significant grass cover, moderate forb density, low shrub cover, and the availability of song perches (Hull [2000] 2002). The latter include fence posts, telephone wires, shrubs, elevated mounds, and farm machinery (Bent 1958). In areas where both meadowlark species occur, Eastern Meadowlarks are more likely to be found in smaller, mesic to wet, low-lying sites that are embedded in more fragmented landscapes, while Western Meadowlarks prefer larger, drier upland sites (Longley 1955; Eckert 2006; Jaster et al. 2020).

Based on a review of the literature, Ribic and her colleagues (2009) concluded that the Eastern Meadowlark demonstrated a higher incidence of occurrence and greater breeding densities on larger tracts. Other studies, however, have suggested that the species' area sensitivity may vary regionally (Hull [2000] 2002).

POPULATION ABUNDANCE

PIF (2021) provided a global population estimate of 37 million birds, with approximately 24 million residing in the United States and Canada; Minnesota was estimated to support 93,000 individuals. The MNBBA-derived statewide estimate is 294,000 (95 percent confidence interval of 227,000–411,000), which is still a very small percentage of the entire population.

BBS data document a significant population decline for the Eastern Meadowlark across the BBS survey region that averaged −2.56 percent per year from 1966 to 2019 (Sauer et al. 2020). The decline is evident in nearly every state and province with the exception of a few regions on the western periphery of the species' breeding range, including the Prairie Potholes. In Minnesota, the Meadowlark's relatively low abundance on BBS routes results in an imprecise, but declining population trend (Sauer et al. 2020). Nevertheless, most of its Minnesota breeding range lies within the Prairie Hardwood Transition Bird Conservation Region, where a significant decline, averaging −3.26 percent per year, has been documented **(D)**.

One of the primary contributors to the species' decline is the loss of family farms, a former mainstay

C. Eastern Meadowlarks frequently nest in hayfields like this mesic meadow in the blufflands of southeastern Houston County. Photo by Lee A. Pfannmuller.

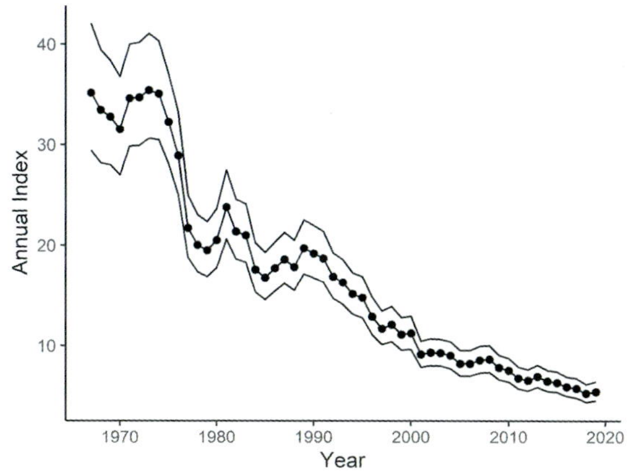

D. Breeding population trend and 95 percent credible interval (solid black lines) for the Eastern Meadowlark in the Prairie Hardwood Transition Bird Conservation Region of North America based on the 1967–2019 Breeding Bird Survey (Sauer et al. 2020).

of the agricultural landscape (CLO 2017c). Small farms, with their pastures and hay meadows, have been replaced by more intensive, large-scale agricultural operations devoted primarily to croplands. Mortality during severe winter storms also results in significant losses (Jaster et al. 2020). It is less clear why populations are stable or slightly increasing in a few localities.

CONSERVATION

Given its large and widespread population decline, the Eastern Meadowlark is a moderate conservation priority (PIF 2021) and is recognized as one of 24 common North American birds in steep decline (Rosenberg, Kennedy, et al. 2016). Its plight has also been highlighted by many state and federal resource agencies and conservation organizations.

Management guidelines recommend protecting large grassland tracts that provide a diversity of grasses and forbs with minimum shrub cover. Periodic disturbances created by burning, mowing, and light to moderate grazing are important to prevent woody encroachment. Federal farm policies that provide incentives to farmers to retire less productive farmlands are critical, as more than 95 percent of Eastern Meadowlarks in the United States are estimated to occur on private lands (NABCI 2011).

Although the species' vulnerability to warming temperatures is considered relatively low under a 2°C warming scenario, with minor expansions predicted in northeastern and south-central Minnesota (NAS 2019), its steep population decline requires close monitoring. Focused grassland conservation efforts are crucial if the Eastern Meadowlark is to remain "the outstanding and the most characteristic bird of the American farm" (Bent 1958).

WESTERN MEADOWLARK

Sturnella neglecta

Minnesota Seasonal Status: Regular breeding species and migrant; occasionally observed in southern Minnesota during the winter. The Western Meadowlark was common during the MNBBA.
Migration: Northern populations are short- to medium-distance migrants that winter in the southern United States and northern Mexico; southern populations are year-round residents.
Food: Ground forager that feeds on insects and seeds.
Nest: Cup nest placed on the ground; the structure often has a partially or fully domed roof.
Conservation Concern: Assigned a moderate PIF Continental Concern Score of 10/20.

Photo by Michael Furtman.

Nearly indistinguishable from the Eastern Meadowlark with its stocky silhouette, white outer tail feathers, and the black V-shaped collar on its yellow breast, the Western Meadowlark has a song that has been described as "its greatest charm" (Bent 1958). Native Americans and early explorers recognized that the western bird's introductory whistles and melodious warbles, unlike the simple, flutelike song of the Eastern Meadowlark, were unique. Its song has been described by many with glowing praise; Roberts (1932) wrote: "The song of the Western Meadowlark is to the treeless prairies what the songs of the Thrushes, Thrasher, and Rose-breasted Grosbeak are to the woodlands." The call notes of the two meadowlarks are also distinct. Chapman (1908) described the Western Meadowlark's call note as a "*chuck, chuck* followed by a wooden, rolling *b-r-r-r-r-r,* wholly unlike the sharp *dzit* or *yert* and metallic twitter of the eastern bird."

Unfortunately, relying on vocalizations for identification in areas where both species occur can be problematic. As both Eckert (2006) and Dunlap (2019) noted, not only are intermediate songs occasionally heard in Minnesota, but the same bird may deliver the songs and calls of both species! Dunlap (2019) concluded that either these birds are from rare hybrid pairings or they are learning each other's vocal repertoire. Regardless, such observations certainly add to the challenge of meadowlark identification.

The Western Meadowlark's breeding range stretches from north-central Mexico, across the western United States and southwestern Canada, east to southern Ontario and northwestern New York. Breeding densities are highest in the Great Plains. Audubon originally assigned the species its scientific name, *Sturnella neglecta,* lamenting that, even though the "celebrated explorers" Lewis and Clark were familiar with the bird, "no one has since taken the least notice of it" (Bent 1958). Indeed, the two meadowlark songsters were not officially recognized as separate species until 1908 (J. Allen et al. 1908).

MINNESOTA BREEDING DISTRIBUTION

Roberts (1932) described the Western Meadowlark as an abundant species throughout the western grasslands, extending east "into the open woodlands quite to the Wisconsin line, avoiding only the heavier forests." Breeding evidence was reported in Grant, Lyon, Martin, Otter Tail, Pipestone, and Polk Counties in the west and from Faribault, McLeod, Morrison, and Waseca Counties in central Minnesota.

When Green and Janssen (1975) provided an updated account, the Western Meadowlark had been reported in St. Louis County and was even locally abundant in some areas of north-central and northeastern Minnesota. Janssen (1987) described it as rare to absent in Lake and Cook Counties and casual to rare in the north-central region. He identified 19 counties where nesting had been confirmed since 1970, including five east-central counties (Anoka, Carver, Hennepin, Le Sueur, and Sherburne) and two southeastern counties (Olmsted and Wabasha). Hertzel and Janssen (1998) added three more counties to the list. By 2014, the 931 breeding-season locations identified by the MBS were scattered as far east as central St. Louis County in the north and southeastern Houston County in the south.

A similar distribution was documented by the MNBBA **(A)**. Although the 1,285 records spanned 76 of Minnesota's 87 counties, the species was most abundant in the Prairie Parkland and Tallgrass Aspen Parkland Provinces, with reports rapidly declining in abundance to the east. Reported from 17.4 percent (830/4,774) of the surveyed atlas blocks, breeding was confirmed in 53 blocks (1.1 percent) and deemed probable in 222 (4.7 percent). Fifteen of the 25 counties where breeding was confirmed were additions to the list compiled by Hertzel and Janssen (1998) and included extensions north to Kittson, Marshall, Polk, and Roseau Counties. The species' probability map closely reflects its MNBBA distribution, predicting moderate breeding densities along the entire western tier of counties but low densities immediately east in central and south-central Minnesota **(B)**.

With MNBBA reports in 830 blocks, the Western Meadowlark was nearly twice as abundant as the Eastern Meadowlark, which was reported in 490 blocks. Although their common names give a reasonable approximation of their statewide distributions, as Eckert (2006, 2010) noted, there are broad areas of overlap. During the MNBBA, these areas were located in a wide band of 58 counties that extended from southeastern Minnesota, across much of central Minnesota, and north to the Canadian border. Included were six counties where both species were confirmed breeding (Benton, Carver, Dakota, Goodhue, Otter Tail, and Stearns). Cook County was the only county where neither species was reported. Despite their broad distributional overlap, there is little evidence of interbreeding (S. K. Davis and Lanyon 2020). Nonetheless, as the MNBBA results confirm, it is best not to rely on the species' names to determine their likely presence anywhere in the state.

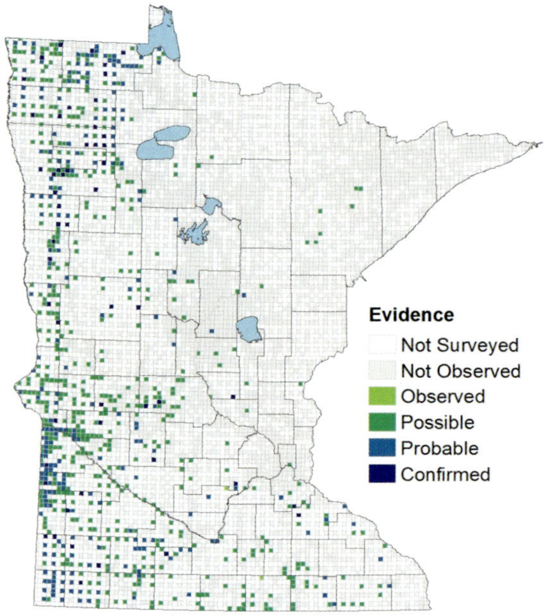

A. Breeding distribution of the Western Meadowlark in Minnesota based on the Breeding Bird Atlas (2009–13).

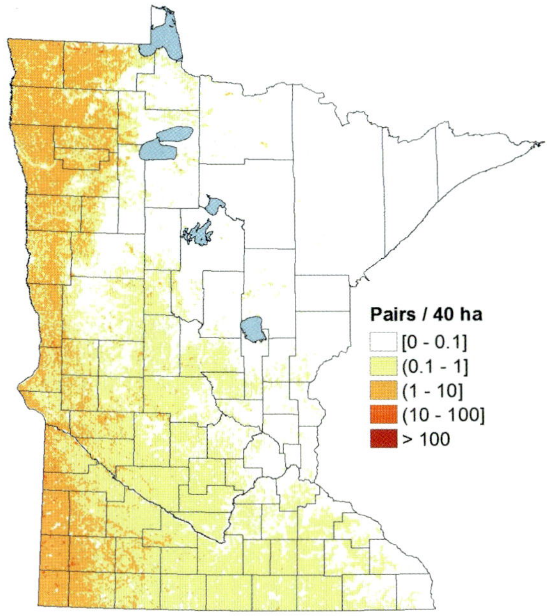

B. Predicted breeding distribution (pairs per 40 ha) of the Western Meadowlark in Minnesota based on habitat, landscape context, and climate data gathered during the Minnesota Breeding Bird Atlas (2009–13) using the Generalized Linear Modeling method with an adjustment for detectability.

The Western Meadowlark's Minnesota distribution has changed little in the past century, only moving slowly into the northeast region. Its continental breeding range has also expanded eastward from Missouri, Wisconsin, and Illinois to northwestern New York and southern Ontario (S. K. Davis and Lanyon 2020).

BREEDING HABITAT

Common in native grasslands, the Western Meadowlark displays a wide tolerance for other open habitats, including pastures, small grain fields, and hayfields **(C)**. Perennial cover planted on Conservation Reserve Program (CRP) fields also provides suitable habitat (e.g., Dechant, Sondreal, et al., "Western Meadowlark," [1999] 2002b; Hanowski 1995). In Minnesota, the number of meadowlarks (*Sturnella* spp.) observed on roadside counts increased with each incremental increase in the percentage of CRP acreage (Haroldson et al. 2006).

Western Meadowlarks are rarely found where the vegetation is very sparse or very tall. Sites with a moderate height and density of grasses and forbs are preferred (S. K. Davis and Lanyon 2020). Scattered trees and shrubs are tolerated and used as song perches. Where the ranges of Eastern and Western Meadowlarks overlap, the Western Meadowlark prefers larger, drier upland sites, while the Eastern Meadowlark prefers smaller, more fragmented, mesic to wet sites (Longley 1955; Eckert 2006).

Evidence that the Western Meadowlark is more successful on larger tracts is mixed. A North Dakota study demonstrated inconclusive results (D. H. Johnson and Igl 2001), while Ribic and her colleagues (2009) cited two studies that demonstrated higher breeding densities on larger grasslands. A Minnesota study demonstrated lower rates of nest predation and nest parasitism by Brown-headed Cowbirds on larger tracts (R. G. Johnson and Temple 1990).

POPULATION ABUNDANCE

PIF (2021) generated a population estimate of 95 million Western Meadowlarks in the United States and Canada with an additional 5 million birds in Mexico. Minnesota was estimated to support 440,000 birds. Estimates derived from MNBBA data were somewhat higher, at 540,000 breeding adults (95 percent confidence interval of 470,000–639,000).

Surveywide, BBS data indicate a significant population decline, averaging −0.88 percent per year from 1966 to 2019 (Sauer et al. 2020). Unlike many species whose populations include pockets of increases and decreases, Western Meadowlarks have not increased anywhere within the survey region except in Nevada and Wyoming. One of the steepest declines has occurred in Minnesota, which has experienced a significant loss of −6.74 percent per year from 1967 to 2019 **(D)** and a cumulative loss of −97 percent! Janssen (2000) brought attention to this decline when he examined the results from a BBS route located on the southern outskirts of the Twin Cities from 1968 to 2000. The route had a peak of 119 Meadowlarks in 1972 compared to only five birds in 2000. Although the growth of the Twin Cities' outer suburbs was certainly a major factor, when he examined 66 other Minnesota BBS routes where Western Meadowlarks had been reported, similar declines were noted. Losses due to

C. In western Minnesota, large, dry-mesic grasslands are the preferred habitat of the Western Meadowlark. Photo by Lee A. Pfannmuller.

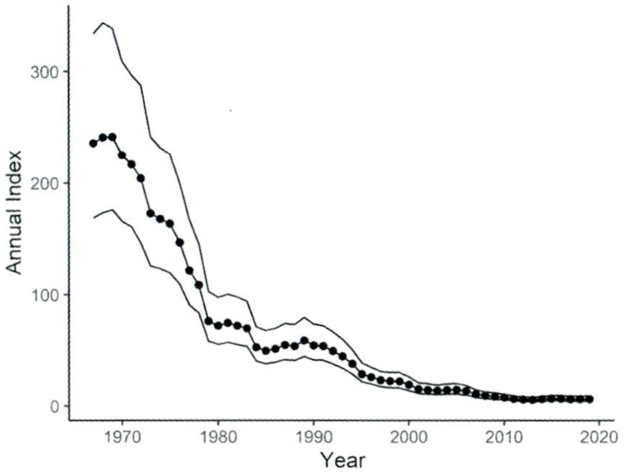

D. Breeding population trend and 95 percent credible interval (solid black lines) for the Western Meadowlark in Minnesota based on the 1967–2019 Breeding Bird Survey (Sauer et al. 2020).

residential development were coupled with losses due to more intensive agricultural practices, changes that are major threats to the species throughout its breeding range (S. K. Davis and Lanyon 2020).

CONSERVATION

The Meadowlark's declining population is the primary factor responsible for its designation as a moderate conservation priority, followed close behind by its threatened grassland habitats (PIF 2021). Ultimately, habitat protection and restoration are critical to the species' future. Active management by burning, mowing, and grazing is essential to maintain sites with minimal woody cover, while large grasslands should be managed to provide a mosaic of successional stages.

Federal farm programs that provide farmers with incentives to retire less-productive croplands are also important. The dense cover often planted on CRP acres may not provide the most ideal habitat for Western Meadowlarks, which prefer more moderate cover densities (Haroldson et al. 2006; Dechant, Sondreal, et al., "Western Meadowlark," [1999] 2002b), but CRP acres are adequate when native grasslands are not available.

NAS (2019) concluded that the species' North American breeding distribution has a low vulnerability to climate change under all three warming scenarios they examined. Small losses of habitat predicted in southwestern Minnesota should be more than compensated for by larger gains in the northwest. Nevertheless, given the Great Plains' projected vulnerability to rising temperatures (NABCI 2010), monitoring is warranted. Meanwhile, expanded grassland conservation efforts could help stem the Western Meadowlark's downward trajectory and return the "very spirit of the boundless prairie" (Bent 1958) to western Minnesota.

ORCHARD ORIOLE

Icterus spurius

Minnesota Seasonal Status: Regular breeding species and migrant. The Orchard Oriole was uncommon during the MNBBA.
Migration: Medium- to long-distance migrant, wintering from central Mexico to northern South America.
Food: Primarily a foliage gleaner feeding on terrestrial arthropods; also consumes some fruit and nectar.
Nest: Woven hanging basket placed within a tree or shrub; sometimes nests in small colonies, especially farther south.
Conservation Concern: Assigned a moderate PIF Continental Concern Score of 10/20.

The Orchard Oriole suffers from a bit of anonymity. Mention the word "oriole" to anyone who enjoys watching birds and you likely will hear them expound on the male's stunning colors and their enjoyment of a nearby nesting pair. They are, of course, referring to their love of the Baltimore Oriole; most are unaware that Minnesota is also home to the less well-known Orchard Oriole. Even the species' scientific name, *Icterus spurius,* is a reference to an initial confusion between the two species. The female Baltimore Oriole, a pale version of the brightly colored male, was believed to be the male Orchard Oriole. As a result, the latter was given the name Spurius, or False Baltimore (Wilson 1832).

Smaller than a Baltimore Oriole, the male Orchard Oriole is unmistakable in his deep black and rich chestnut-colored plumage. The juvenile male, however, does not acquire the adult plumage until his second winter season, looking instead like the adult female but with a black bib. The female may be difficult to distinguish from a female Baltimore Oriole, but the yellow-green plumage on her head, tail, and underparts lacks the orange tinge that is evident on the female Baltimore Oriole. The latter's head also appears darker as she ages, with black or dark brown feathers mixed within the orange.

Characterizing its song as "vivacious," Bent (1958) noted that it "is not as loud, nor as rich as that of the Baltimore oriole." Indeed, he believed it was quite different but "equally pleasing." Saunders (1935) described it as a "series of rather rapid, musical notes, exceedingly variable in time and pitch." Bubblier than the purer whistled phrases of the Baltimore Oriole, it begins with a few high-pitched notes followed by an array of lower-pitched notes and phrases.

Photo by Mike Lentz.

A species of open woodlands, the Orchard Oriole occupies a breeding range that stretches from the Great Plains east to the Atlantic coast. It occurs south through central Mexico and north to southern Saskatchewan, Manitoba, and Ontario. High breeding densities occur in scattered areas from North Dakota south to Kansas and in several coastal states, including Alabama, Louisiana, and Mississippi.

MINNESOTA BREEDING DISTRIBUTION

In the early 1900s, Roberts (1932) described the Orchard Oriole as "a comparatively uncommon summer resident throughout the southern part of the state." Considered most common in the riparian groves of western Minnesota, it was also present along the Mississippi River north into Stearns and Morrison Counties and north along the Red River valley. Relatively common in west-central Minnesota, it became increasingly rare going north toward the Canadian border. Nest records were available from Anoka, Goodhue, Hennepin, Jackson, Pipestone, and Traverse Counties. Nesting was inferred in Grant County.

Forty years later, Green and Janssen (1975) remarked that the species had largely retreated from the northern edge of its breeding range. There were no records north of Clay County since Roberts's publication, and it had become quite rare in the central region of the state: "[T]he species is scarce north of the Twin Cities in the east and north of the headwaters of the Minnesota River in the west." Reports in south-central Minnesota were also scarce. Similar changes were documented in other states along the northern periphery of its breeding range, including both Wisconsin (Cutright et al. 2006) and Michigan (R. Brewer et al. 1991). The decline was attributed to the loss of orchards and other agricultural lands in the face of increasing development (Scharf and Kren 2020).

A few years later, Janssen's (1987) updated account noted that the species was most numerous in southeastern and southwestern Minnesota and along the upper Minnesota River valley, east to Nicollet and Blue Earth Counties. It appeared to be moving north again in the Red River valley, with reports common as far north as Red Lake County. Since 1970, nesting had been confirmed in 16 counties, all south of Grant County in west-central Minnesota and Wright County in the east. Hertzel and Janssen (1998) later added five more counties to the list, including Wilkin County in northwestern Minnesota.

Contrary to Janssen's (1987) account, the 320 breeding-season locations identified by the MBS were sparse in southeastern Minnesota. Records were most abundant in the southwest and south-central counties, along the Upper Minnesota River valley, and in west-central Minnesota, becoming less common farther north along the Red River valley. Only two detections were in east-central Minnesota: one in Hennepin County and one in Washington County. Overall, the species' entire range appeared to have shifted westward.

In contrast, the MNBBA found the species to be relatively widespread across all of southern Minnesota, including the east-central and southeastern regions, and north throughout the Red River valley **(A)**. Impressive survey efforts in west-central Minnesota, from Big Stone County south to northern Lincoln County, also led to an abundance of records in that region. Overall, a total of 874 records were found in 13.2 percent (628/4,740) of the surveyed atlas blocks that spanned 69 of Minnesota's 87 counties. Breeding was confirmed in 60 blocks (1.3 percent) located in 40 counties, including four north of Wilkin County in northwestern Minnesota (Clay, Norman, Polk, and Roseau); probable breeding was documented in 19 additional counties. Twenty-eight counties with confirmed nesting were additions to the list published by Hertzel and Janssen (1998).

The MNBBA data reflect what states farther east have observed. In recent years, the Orchard Oriole has increased in abundance and reoccupied much of its former breeding range following population declines and range contractions (e.g., Chartier et al. 2011; Rodewald et al. 2016; Wilson et al. 2012). The species' current distribution in Minnesota is nearly identical to that described by Roberts nearly one hundred years ago. It also appears to be expanding its range farther north in east-central Minnesota, with confirmed breeding in Benton, Chisago, and Isanti

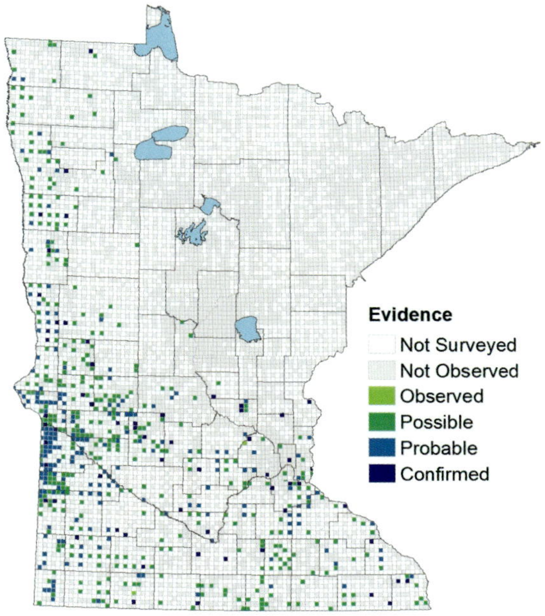

A. Breeding distribution of the Orchard Oriole in Minnesota based on the Breeding Bird Atlas (2009–13).

Counties, and is becoming a little more common in the northern Red River valley, with confirmed breeding in Roseau County and probable breeding in Kittson County, where Roberts described it as "a rare summer resident."

BREEDING HABITAT

The Orchard Oriole utilizes a range of open woodland habitats, including farm shelterbelts, shrublands, planted roadsides, and the margins of suburban areas with shade trees **(B)**. Suitable sites often border wetlands and streams. As its name implies, the species is widely associated with fruit orchards because the small, widely spaced trees provide ideal nesting sites (Bent 1958; Scharf and Kren 2020). Basic requirements include a relatively open landscape, the presence of trees that are usually smaller in diameter (7.6–15.2 cm) than those typically preferred by Baltimore Orioles, and a high diversity of trees close to the nest (Clawson 1980).

POPULATION ABUNDANCE

Using BBS data, PIF (2021) estimated a North American breeding population of 12 million adults; approximately 11 million birds reside in the United States and southern Canada (PIF 2021). The estimate for Minnesota is just 68,000 birds.

From 1966 to 2019, the species experienced a significant decline across the BBS survey region averaging −1.14 percent per year (Sauer et al. 2020), with a cumulative loss of −23 percent from 1970 to 2014 (Rosenberg, Kennedy, et al. 2016). Most losses occurred in southern states, while population increases were largely concentrated in the northern Great Plains and lower Great Lakes region. The species is still uncommon along Minnesota BBS routes, so even though the state population trend is increasing, the data is considered statistically unreliable. The core of the Oriole's breeding range in Minnesota, however, lies within the larger Prairie Potholes Bird Conservation Region, where the Orchard Oriole has demonstrated a significant increase averaging 2.52 percent per year **(C)**.

Although the ebbs and flows of the species' abundance and distribution have been widely noted, few have postulated factors that may be responsible for these changes. Because the Oriole is a frequent inhabitant of orchards, the impact of pesticide use must be considered, as well as the loss and degradation of riverine habitats in intensively cultivated landscapes. Nest parasitism by Brown-headed Cowbirds, particularly in southern states, is considered a major source of reduced nesting success (Scharf and Kren 2020). Less clear are the factors responsible for local population increases. Its adaptation to more residential environments has certainly created new habitat opportunities.

CONSERVATION

A moderate conservation priority, the primary concerns are the Orchard Oriole's declining population and threats to its breeding habitat, especially the loss and degradation of riparian areas (PIF 2021). Nevertheless, the species is still widely distributed throughout eastern North America and remains relatively common, particularly in regions south of Minnesota. Even a warmer climate may not pose the same problems for Orchard Orioles as it does for many temperate songbirds. Under a 2°C warming scenario, NAS (2019) projected a low vulnerability for the species' breeding distribution, including a 14 percent range loss in the central and southern Great Plains and a 27 percent gain across the northern periphery of its breeding range by 2050 (NAS 2019). As temperatures rise in Minnesota, models predict the species will gradually expand to the northeast.

In recent years, a great deal of emphasis has been placed on understanding and monitoring grassland species in Minnesota. Unfortunately, species like the Orchard Oriole, which are dependent on woodlands within grasslands, including riparian habitats, have been overlooked. This "spurious" Oriole could benefit from a more focused effort to ensure that its habitat requirements and management needs are of equal concern.

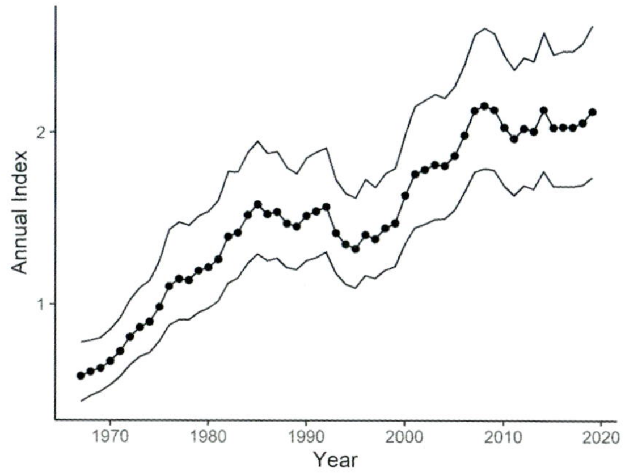

C. Breeding population trend and 95 percent credible interval (solid black lines) for the Orchard Oriole in the Prairie Potholes Bird Conservation Region based on the 1967–2019 Breeding Bird Survey (Sauer et al. 2020).

B. The Orchard Oriole is typically found in open, deciduous woodlands bordering small streams, like this one in west-central Minnesota. Photo by Lee A. Pfannmuller.

BALTIMORE ORIOLE

Icterus galbula

Minnesota Seasonal Status: Regular breeding species and migrant. The Baltimore Oriole was common during the MNBBA.
Migration: Medium- to long-distance migrant; most birds spend the winter in Central America or northern South America.
Food: Primarily a foliage gleaner feeding on terrestrial arthropods; also consumes fruit and nectar. A popular visitor to bird feeders, Orioles are attracted by oranges, nectar, and grape jelly.
Nest: Woven hanging basket, usually placed in a tall deciduous tree.
Conservation Concern: Assigned a moderate PIF Continental Concern Score of 10/20.

A mere glimpse of the male Baltimore Oriole—known as the "Golden Robin" in early days—takes your breath away. His stunning orange breast and belly contrast starkly with his deep black head and upper back. The colors are so unusual that he is easily identified even as he quickly flies across an opening to reach the highest branches of a nearby deciduous tree. As dramatic and distinct as the male is, the female's plumage is extremely variable. She may go unnoticed in her orange-yellow and olive-gray plumage or appear strikingly similar to the male, but a tad paler and with a more mottled head and upper back. Her plumage usually becomes more male-like with each successive molt. Juvenile males do not assume the full adult plumage until their second fall after fledging. *Icterus galbula*, the species' scientific name, means "a small yellow bird."

The male's distinctive song is a series of clear, flute-like whistles delivered in a succession of paired and single notes separated by short pauses. The quality of the song identifies the songster, not the song itself. Ornithologist Aretas A. Saunders remarked that "it is so exceedingly variable in form that one cannot pick out any one song, or even several songs, that could be said to be more typical of the species than others" (quoted in Bent 1958).

Considered a "marvelous piece of bird architecture" (Roberts 1932), the Oriole's nest is an intricately woven hanging basket constructed by the female. This "skillful artisan" (Bent 1958) completes the task in approximately one week, diligently weaving together an array of slender fibers. The species is

Photo by Mike Lentz.

equally renowned for its hybridization with the western Bullock's Oriole in a region of the Great Plains where the two species' ranges overlap. This zone of hybridization has been intensively studied and is one of the most cited examples of avian hybridization in the ornithological literature. The two birds were declared to be one species, the Northern Oriole, in 1983, but subsequent genetic studies confirmed that they are more closely related to other members of the blackbird family than to each other, resulting in their ultimate separation in 1995 (Rising and Flood 2020).

This woodland inhabitant was so named because its black and orange plumage resembled the colors of the coat of arms of the Baltimores, "the colonial proprietors of the Maryland colony" (Rising and Flood 2020). Found across southern Canada, from Alberta east to the southern Maritime Provinces, and throughout most of the eastern United States and the Great Plains, the Oriole's highest breeding densities are found in southern Nebraska and Kansas.

MINNESOTA BREEDING DISTRIBUTION

In the early 1900s, Roberts (1932) noted that the beloved Baltimore Oriole was a common summer resident across the state. It occurred throughout the deciduous forests of southeastern and central Minnesota and in scattered woodland groves in the western prairie and agricultural regions. As the mixed forests of the boreal landscape were cleared and replaced by extensive stands of aspen and birch, it also began spreading northward. Reports of nests, nests with eggs, or nests with young were compiled from nine counties—Goodhue, Hennepin, Isanti, Kittson, Morrison, Otter Tail, Polk, Waseca, and Wright—and from Itasca State Park.

Little has changed since Roberts's time. Green and Janssen (1975) noted that the species was scarce in the far northeastern counties of Cook and Lake but widely distributed throughout the remainder of the state. Janssen (1987) delineated 45 counties where nesting had been confirmed since 1970; Hertzel and Janssen (1998) added 10 more counties to the list. The largest region where nesting had not been confirmed stretched from Chisago County northeast to Cook County and west to Koochiching County. MBS field surveys documented a similar distribution.

During the MNBBA, the Baltimore Oriole was found in every county except Cook, and was confirmed nesting in 75 of Minnesota's 87 counties, including all but five counties in the Laurentian Mixed Forest Province (Carlton, Clearwater, Cook, Lake, and Koochiching) and seven counties in the Prairie Parkland Province (Big Stone, Cottonwood, Grant, Martin, Nobles, Pipestone, and Traverse). Twenty-six counties were additions to those delineated by Hertzel and Janssen (1998). MNBBA participants tallied a total of 3,423 Baltimore Oriole records in 43.2 percent (2,064/4,733) of the surveyed atlas blocks and confirmed breeding in 483 blocks **(A)**. Even in the intensively cultivated regions of western Minnesota, riparian woodlands, scattered woodlots, and shelterbelts were widely used by the species. The forests of northeastern and far north-central Minnesota remained the only area where the species was scarce.

The Oriole's probability map predicts moderate breeding densities throughout all but the more northern and eastern regions of the Laurentian Mixed Forest Province **(B)**. Areas of higher abundance are scattered throughout the state, including

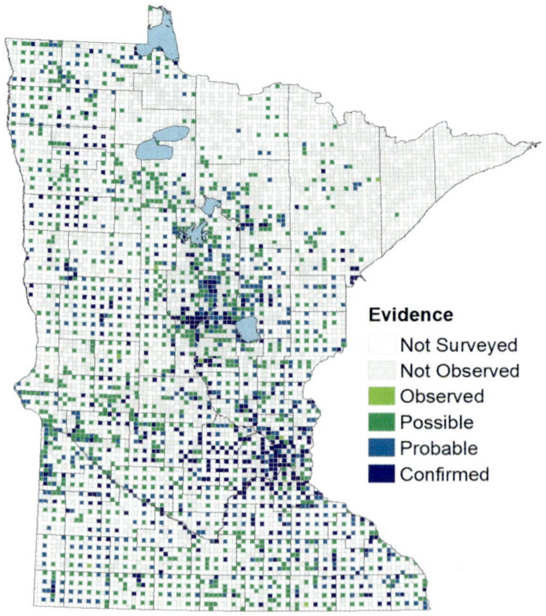

A. Breeding distribution of the Baltimore Oriole in Minnesota based on the Breeding Bird Atlas (2009–13).

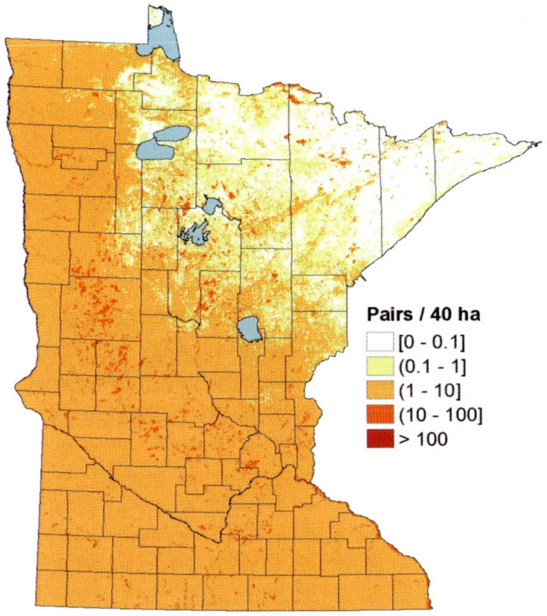

B. Predicted breeding distribution (pairs per 40 ha) of the Baltimore Oriole in Minnesota based on habitat, landscape context, and climate data gathered during the Minnesota Breeding Bird Atlas (2009–13) using the Generalized Linear Modeling method with an adjustment for detectability.

along the lower Mississippi River, along the prairie–forest border stretching from the Twin Cities northwest to Detroit Lakes, and along several riparian corridors in the Prairie Parkland Province. Although the species gradually declines in abundance across the northern forest landscape, pockets of moderate and high abundance are still found, including along the Mesabi Iron Range of central St. Louis County and southern Itasca County and along the Canadian border east of International Falls.

Historically, as more shelterbelts were planted across the Great Plains, Baltimore Orioles slowly expanded westward from the core of their breeding range in the eastern United States. They also moved northward as forests were cleared in the northern Great Lakes states and southern Canada (Rising and Flood 2020; Cadman et al. 2007). In regions where forest cover has increased, however, their abundance and distribution has declined. Such was the case in southern Ontario, which witnessed a decline in suitable habitat between the completion of its first atlas (1981–85) and its second atlas, twenty years later (Cadman et al. 2007). Elsewhere there have been few recent changes to the Oriole's distribution.

BREEDING HABITAT

The Baltimore Oriole inhabits open woodlands, including riparian corridors along streams and lakes, shelterbelts, forest edges, city parks, and residential areas **(C)**. Extensively forested landscapes are avoided. In northern Wisconsin and Minnesota, Orioles were most likely to be found in small towns and cities rather than in managed forest stands (Niemi, Howe, et al. 2016). Although they will use mixed woodlands, deciduous woods and trees are preferred.

POPULATION ABUNDANCE

PIF (2021) estimated the Oriole's population at 12 million breeding birds across southern Canada and the eastern United States. Approximately 470,000 birds, or 4.0 percent of the continental population, reside in Minnesota. The more extensive coverage conducted by the MNBBA generated a significantly higher statewide estimate of 3.2 million breeding birds (95 percent confidence interval of 2.5–4.4 million). Regardless of the estimate, Minnesota supports a relatively large percentage of the species' continental population and breeding range, prompting its designation as a Stewardship Species by Audubon Minnesota (Pfannmuller 2012).

Across the BBS survey region, the Oriole has demonstrated a statistically significant but slow, long-term population decline, averaging −0.84 percent per year from 1966 to 2019 (Sauer et al. 2020), with a cumulative decline of −42 percent from 1970 to 2014 (Rosenberg, Kennedy, et al. 2016). The population in Minnesota has been relatively stable, experiencing a rapid increase from 1967 to 1980, followed by a slow, long-term decline **(D)**. The factors responsible for the declining numbers are not well known. However, Orioles are susceptible, both directly and indirectly, to pesticides that are applied to shade trees to prevent or control infestations by gypsy moths or the spread of Dutch elm disease. Changing agricultural practices on the Oriole's wintering grounds also pose a threat to population numbers (Rising and Flood 2020).

C. Tall deciduous trees found in open woodlands or along a forest edge are the preferred nesting habitat for the Baltimore Oriole. Photo by Gerald J. Niemi.

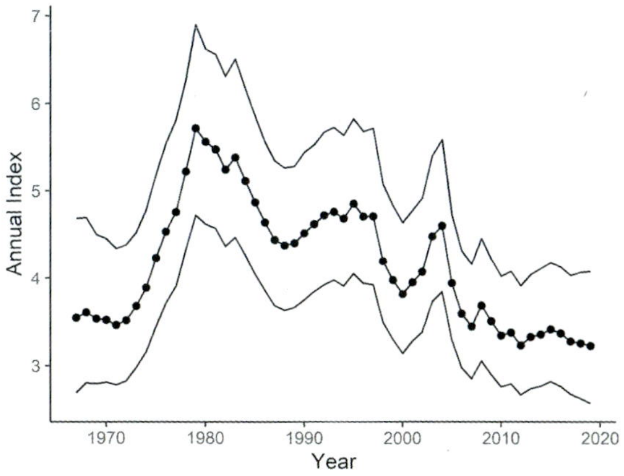

D. Breeding population trend and 95 percent credible interval (solid black lines) for the Baltimore Oriole in Minnesota based on the 1967–2019 Breeding Bird Survey (Sauer et al. 2020).

CONSERVATION

Despite its declining population, the species' adaptability to a variety of habitats has eased any immediate conservation concern for the species (PIF 2021). In years past, there was concern that the decline in elm trees from Dutch elm disease would negatively impact population numbers, but the species has readily accepted other large hardwood trees as suitable nest sites, including oaks, maples, and cottonwoods. The preservation of large trees in urban and rural areas, however, is an important conservation measure, coupled with reduced pesticide use. Even warmer temperatures pose few threats to Baltimore Orioles. Under all three warming scenarios analyzed by NAS (2019), the species' North American breeding distribution had a low vulnerability to rising temperatures. Small range losses along the western periphery of its breeding range were more than compensated for by gains to the north and southeast. In Minnesota, the species is projected to retain its statewide distribution, with habitat conditions actually improving throughout most of the state.

Overall, there is greater concern for the species' winter habitat in Central and South America than for its breeding habitat. Commonly associated with shade-grown coffee plantations, the trend to eliminate shade trees from many plantations could significantly impact wintering birds (Bock et al. 1993) and, ultimately, the number that return to brighten Minnesota's woodlands each spring.

RED-WINGED BLACKBIRD

Agelaius phoeniceus

Minnesota Seasonal Status: Regular breeding species and migrant; winter observations are most common in southern Minnesota. The Red-winged Blackbird was very abundant during the MNBBA.
Migration: Year-round resident or short-distance migrant. Most birds breeding in Canada and the northern United States spend winters in the southern states.
Food: Feeds primarily on terrestrial arthropods during the summer and waste grain and seeds during the winter.
Nest: Cup nest secured to standing vegetation.
Conservation Concern: Assigned a low PIF Continental Concern Score of 8/20.

From the largest wetland complexes to the smallest roadside ditches, Red-winged Blackbirds are virtually everywhere. One of the most abundant breeding birds in North America, this sleek black bird can be found as far north as Alaska and as far south as Costa Rica. Although the female may remain hidden in dense vegetation, the male Red-winged Blackbird commands our attention. Standing guard, conspicuously perched on a tall cattail, he sings his familiar, raspy *o-ka-lee* song. Flying from his perch to taunt us overhead, he sounds his *chack* alarm call as we near the territory he passionately defends. Naturalist Alexander Skutch wrote of one such encounter: "As I crossed one large meadow where several redwings apparently had nests, I had an escort of guardian males all the way; for as soon as I passed beyond the bounds of the domain of one of them and he dropped behind, another vigilant bird would take over, hover over me, and shriek down imprecations" (quoted in Bent 1958).

Like other members of its family, the Red-winged Blackbird is not intimidated by the size of any territorial intruder. Indeed, the larger the trespasser the more aggressive the response. During such encounters the beauty of the male Redwing is most apparent. The male hunches his shoulders forward, spreads his tail, and erects his crimson red shoulder feathers, or epaulets, to signal his indignation. Raised and puffed, the epaulets indicate aggression or courtship; lowered and hidden, they indicate submission or more subdued behavior. Smaller and more cryptically colored than the males, females

Photo by Joanna Eckles.

are heavily streaked in shades of brown, cream, and white. The species is highly gregarious during the nonbreeding season; its scientific name translates to gregarious (*Agelaius*) red (*phoeniceus*) bird.

Distributed across Canada, the United States, and portions of Mexico and Central America, the Red-winged Blackbird reaches some of its highest breeding densities from the northern Great Plains south to the Gulf coast of Louisiana and Texas. It was the third most commonly reported species during the MNBBA.

MINNESOTA BREEDING DISTRIBUTION

The Red-winged Blackbird was recognized as an abundant and widely distributed breeding species as early as 1892, when Hatch wrote the first comprehensive account of Minnesota birds. When Roberts published his more detailed account of the species in 1932, he reflected on the species' abundance: "There is, perhaps, no more abundant and widely distributed native bird in North America than the Red-winged Blackbird, and Minnesota, with its thousands of lakes, marshes, and sloughs, possesses an all too generous share."

Indeed, their abundance often wreaked havoc with farmers as the birds took a liking to young grain sprouts, ripened grain crops yet to be harvested, and ears of corn when the kernels were still soft and milky (Bent 1958). Late summer flocks were enormous in size. One observer wrote that, from a distance, they looked "like a dense cloud of black smoke," flying so close to one another that a single shot could take down 60 to 70 birds (Roberts 1932). Roberts himself noted that the birds that roost at some of the state's largest wetlands "are beyond computation," comprised of "millions and millions" of birds.

Despite the extensive loss of wetlands throughout much of Minnesota, the extent of the species' breeding distribution has remained essentially unchanged over the past one hundred years. Green and Janssen (1975) simply noted that it was a resident statewide, while Janssen (1987) described it as "one of the most common and evenly distributed breeding species in the state." By 1998 it had been confirmed breeding in 53 of Minnesota's 87 counties since the year 1970 (Hertzel and Janssen 1998). Fieldwork by the MBS further documented the species' abundance and wide distribution.

Likewise, MNBBA participants reported Red-winged Blackbirds in all corners of the state, with a total of 9,496 records in 77.9 percent (3,843/4,934) of the surveyed atlas blocks and in 90.1 percent (2,106/2,338) of the priority blocks. The species was outnumbered during the atlas only by the American Robin (10,227 records) and the Common Yellowthroat (9,609 records). Breeding was confirmed in 26.5 percent (1,307) of the surveyed blocks and in all 87 Minnesota counties **(A)**. Although the species was found statewide, it was least abundant in the far northeast and north-central regions. The extremely high density of records in the Brainerd Lakes region is a credit to the region's intensive volunteer efforts.

The MNBBA probability map predicts that the highest density of breeding birds occurs across the agricultural landscapes, wetlands, and fragmented forests south and west of the Laurentian Mixed For-

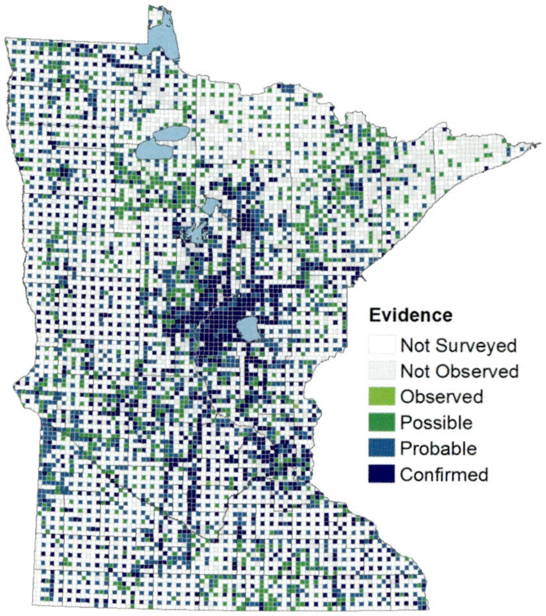

A. Breeding distribution of the Red-winged Blackbird in Minnesota based on the Breeding Bird Atlas (2009–13).

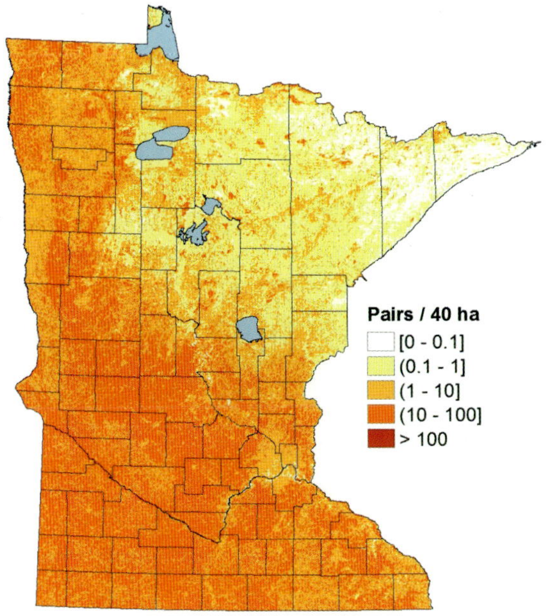

B. Predicted breeding distribution (pairs per 40 ha) of the Red-winged Blackbird in Minnesota based on habitat, landscape context, and climate data gathered during the Minnesota Breeding Bird Atlas (2009–13) using the Generalized Linear Modeling method with an adjustment for detectability.

est Province **(B)**. Suitable habitat is considerably less abundant throughout most of the northern forest, but small pockets that support higher breeding densities in the region still exist.

In their comprehensive review of the species, Yasukawa and Searcy (2020) make no mention of any historical changes to the species' distribution. Yet the extensive loss of wetlands has surely led to a decrease in abundance. In Minnesota, perhaps the most significant change has been the increasing number of birds observed during the winter months. Although large flocks of Redwings often remained in southern Minnesota during the winter in the early 1900s (Roberts 1932), over the years winter observations have gradually extended into northwestern, north-central, and northeastern Minnesota. Most northern records are of birds lingering into December and early January before moving south, but reports throughout January and February are becoming more common (MOU 2021b).

BREEDING HABITAT

A common wetland inhabitant, the Red-winged Blackbird is not particularly choosy, except to avoid the interior of dense forest stands. Water is usually present, but occasionally the bird will nest in drier sites, including pastures, hayfields, and low, wet meadows. Most bird enthusiasts associate the species with wetlands fringed with a dense growth of cattails, but this adaptable species can be found in sites as variable as a tiny roadside ditch with a few standing cattail stalks to large expansive wetlands that extend for several kilometers **(C)**.

POPULATION ABUNDANCE

PIF (2021) provided a population estimate of 170 million birds in the United States and Canada, with another 10 million birds residing in Mexico and Central America. An estimated 4.4 million birds reside in Minnesota. Population estimates using MNBBA data generated a significantly higher statewide estimate of 8.8 million breeding birds (95 percent confidence interval of 7.9–42.8 million). As staggering as these numbers seem, the northern Great Plains support even larger populations.

Despite its overall abundance, the species has experienced a steady decline. Throughout the BBS survey region, data indicate a statistically significant population decline, averaging −0.62 percent per year from 1966 to 2019 (Sauer et al. 2020). In Minnesota, the decline has been steeper, averaging −0.95 percent since 1967 **(D)**. Indeed, populations are declining at statistically significant levels in 20 of the 33 Bird Conservation Regions in the BBS survey area.

Regions with population declines have usually experienced significant losses in wetland habitat. In Minnesota, nearly 50 percent of the state's original wetlands have been lost (J. Anderson and Craig 1984). Although the species has shown a remarkable ability to adapt to a variety of nesting habitats, agricultural areas in the state that have been extensively tiled and tilled provide little suitable habitat. Such changes have accompanied the wide-scale conversion of small family farms, with their matrix of pastures and croplands, to large row crop enterprises. Yet, even in Minnesota's northern forest region, where wetland loss has been less extensive, Red-winged Blackbird populations demonstrated a small, nonsignificant decrease from 1995 through 2020 in the Chippewa and Superior National Forests (Grinde, Bednar, et al. 2020).

C. Red-winged Blackbirds are one of the most abundant inhabitants of cattail marshes across the Great Plains, but they are also found in a wide variety of other wetland and upland habitats. Photo by Lee A. Pfannmuller.

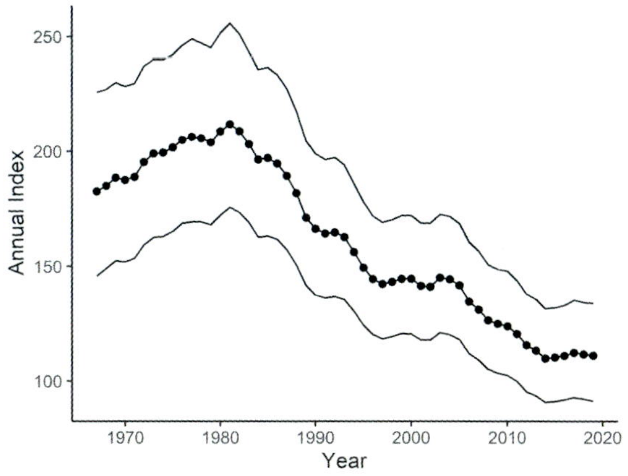

D. Breeding population trend and 95 percent credible interval (solid black lines) for the Red-winged Blackbird in Minnesota based on the 1967–2019 Breeding Bird Survey (Sauer et al. 2020).

CONSERVATION

Despite its population decline, the extremely abundant and widely distributed Red-winged Blackbird is considered an economic pest. When large winter roosts are located in agricultural areas, the damage inflicted can be substantial and economically devastating. Large roosts also pose a potential threat to aircraft and can be a local source of histoplasmosis, a respiratory infection caused by a fungus found in bird droppings. Because the USFWS permits the taking of birds that are causing damage to crops or property, or are creating a public health hazard, thousands of Red-winged Blackbirds are killed each year. Lethal control efforts, however, rarely have a significant impact on local populations. Instead, farmers are encouraged to employ a variety of nonlethal control measures, such as mechanical devices that scare the birds, or to alter their agricultural practices, including "switching to forage crops that are less vulnerable to bird predation" (Yasukawa and Searcy 2020).

Although many wetland species are vulnerable to the impacts of a warmer climate, rising temperatures are projected to cause very few negative impacts to the Red-winged Blackbird's breeding distribution, including none in Minnesota. Instead, significant range expansions are predicted, especially across Canada (NAS 2019). This abundant and adaptable member of Minnesota's avifauna is likely to remain one of the most common birds in the state far into the future.

BROWN-HEADED COWBIRD

Molothrus ater

Photo by Mike Lentz.

Minnesota Seasonal Status: Regular summer resident and migrant; occasionally observed during the winter in southern Minnesota. The Brown-headed Cowbird was abundant during the MNBBA.
Migration: Northern populations are short- to medium-distance migrants that winter in the southern United States and Mexico; southern populations are year-round residents.
Food: Ground forager consuming primarily seeds and arthropods.
Nest: Nest parasite, the female lays her eggs in the nests of other songbirds.
Conservation Concern: Assigned a low PIF Continental Concern Score of 7/20.

One of Minnesota's smallest blackbirds, the Brown-headed Cowbird has a notorious reputation as a nest parasite, imposing its family responsibilities on unsuspecting songbirds. Females search for an unattended host nest and deposit an egg before the host parent returns. To help ensure its success, one or more of the host's eggs are often removed from the nest before she retreats. Usually the first to hatch, the Cowbird chick may force the younger host chicks out of the nest, inadvertently smother them by its large size, or simply cause their starvation by its incessant feeding demands.

Among the more than 200 songbird species whose nests are parasitized, some address the intruder head-on. Robins and Catbirds, for example, remove Cowbird eggs from their nests; some birds may desert the parasitized nest; and others simply bury the eggs with additional nesting materials. The industrious Yellow Warbler constructs an entire new nest atop the invaded nest, often building a small high-rise of multiple nests if the nest is repeatedly parasitized. Needless to say, enough songbirds simply assume the responsibilities foisted upon them to make parasitism a successful strategy for the Cowbird, though it is one that can significantly impact the host species.

The Cowbird's Latin name translates to bronzed (*ater*) vagabond (*Molothrus*), while its common name refers to the brown head on the otherwise black male. The slightly smaller female lacks any distinctive features other than her plain gray-brown plumage. The male's signature vocalization is a series of high-pitched gurgling notes that sound like dripping water, followed by several slurred whistles. A high whistle is often heard while the birds are in flight.

Although the name Cowbird refers to the species' frequent association with cattle, in years past they were closely associated with the enormous herds of bison roaming the Great Plains. Here the birds feasted on insects residing on the bisons' hides and stirred up by their hooves. It was not until immigrants began clearing the deciduous forests that they began their eastward expansion (Lowther 2020b). Today the birds are found throughout the United States, southern Canada, and northern Mexico. Their highest breeding densities, however, remain in the Great Plains.

MINNESOTA BREEDING DISTRIBUTION

Roberts (1932) maligned the Brown-headed Cowbird for parasitizing species "generally of equal or greater value than itself" and described it as "all too abundant in every section of Minnesota." His sweeping assessment of the species' distribution, however, makes no distinction of its relative abundance in different regions of the state. Instead, he devoted most of his lengthy account to a detailed description of the species' unique reproductive strategy.

Green and Janssen (1975) briefly noted the Cowbird's statewide distribution, while Janssen (1987) described it as being most numerous in Minnesota's wooded regions, where nesting songbirds were abundant. This statement, however, is contrary to the species' origin in the western grasslands and to BBS data that illustrate higher numbers in western and southern Minnesota from 2011 to 2015 (Sauer et al. 2020). Janssen (1987) also described the species as widely dispersed across the state and identified 36 counties where the species had been confirmed parasitizing songbird nests since 1970. Hertzel and Janssen (1998) later added 16 more counties to the list of confirmations.

The MBS reported 2,142 breeding-season locations for Brown-headed Cowbirds. Their distribution map is the first to visually depict the species' relatively sparse distribution in the northeastern and north-central regions of the state. Indeed, their records include only one location in Cook County, while none were reported in Lake County.

The 5,170 Cowbird records collected by MNBBA participants provided further evidence of the species' relative abundance across Minnesota **(A)**. The species was reported from nearly every priority block south of a line from Roseau County south to Cass County and east to Carlton County. Farther east, its presence declined rapidly. It was found in

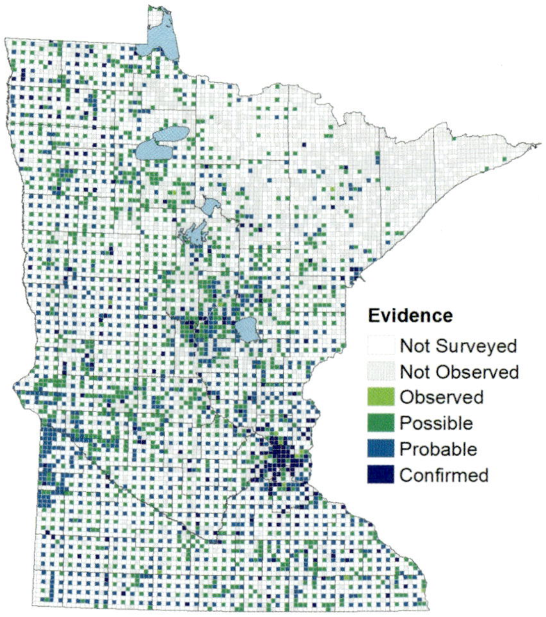

A. Breeding distribution of the Brown-headed Cowbird in Minnesota based on the Breeding Bird Atlas (2009–13).

only six priority blocks in Lake County and in two in Cook County. The bird was found in 55.6 percent (2,646/4,759) of the surveyed atlas blocks; breeding was confirmed in 5.0 percent (240) of the blocks. Observed in all 87 Minnesota counties, breeding was confirmed in 68 counties and deemed probable in 18 counties.

Although the Cowbird remains least abundant in Minnesota's northern forest, its broad habitat requirements have enabled it to exploit most of the state's landscapes. Similar expansions have occurred across North America. Cowbirds reached Nova Scotia, for example, in 1933, Prince Edward Island in 1953, Georgia in 1957, and Florida in the 1980s. A similar expansion occurred west to the Pacific coast, where they were found throughout California by 1900 and in Oregon and Washington by the 1950s (Lowther 2020b).

BREEDING HABITAT

Also known as the "Buffalo Bird," Cowbirds were originally associated with the Central Plains' largest herbivore, the bison. The open grasslands provided not only an abundance of prairie grass seeds, the bird's primary food source, but also a diversity of grasshoppers, beetles, and other insects that were flushed from the ground by the bison's movements (Bent 1958; Roberts 1932; Lowther 2020b).

While livestock replaced the role once served by bison, this adaptive species also made use of the open, fragmented landscapes that were left as the eastern forests were cleared. Today Cowbirds are associated not only with farmsteads but also with a wide variety of open habitats, ranging from forest edges, open woodlands, young forests, and grasslands to old fields, brushy thickets, and residential neighborhoods (Lowther 2020b) **(B)**. Fragmented landscapes provide open habitats for foraging while trees and shrubs provide perches for females to survey the nest-building activities of other birds (Lowther 2020b).

B. The Brown-headed Cowbird's preferred breeding habitat includes open grasslands for foraging and scattered trees and shrubs where females can search for nearby nests of other species to parasitize. Photo by Lee A. Pfannmuller.

Many studies conducted in the Midwest have repeatedly shown that Brown-headed Cowbird parasitism is lowest in regions with a high proportion of forest cover (e.g., Robinson et al. 1995; T. M. Donovan et al. 1997). Densely wooded habitats likely provide fewer feeding opportunities for the adults and increase the difficulty of searching for host nests. Fragmented landscapes, with multiple edges, were deemed more suitable, resulting in higher rates of parasitism (Wilcove et al. 1986; Paton 1994). Other studies, however, have demonstrated that this simple metric (i.e., high forest cover yields low Cowbird parasitism) is more complex, depending on the landscape context. For example, even in landscapes with a relatively high proportion of forest cover (30–50 percent), the interior of the forests may be as accessible to Cowbirds as the edges (Gustafson et al. 2002; Knutson, Niemi, et al. 2004).

POPULATION ABUNDANCE

The United States and Canada support a breeding population of approximately 130 million breeding birds. Minnesota supports approximately 2.4 percent of the continental population, or nearly 3.1 million birds (PIF 2021).

Although the Cowbird was found throughout eastern North America by the late 1800s, "it was not abundant" and was limited primarily to cultivated fields (Brittingham and Temple 1983). Its abundance began to rise sharply only in the first half of the twentieth century, when winter food supplies—especially the availability of waste grain in southern rice fields—and the number of cattle feedlots increased (Newton 2007). Since then, Cowbird numbers have actually declined significantly, averaging −0.51 percent per year across the BBS survey region from 1966 to 2019 (Sauer et al. 2020). The decline has been widespread, with the exception of several states in the northern Great Plains and along the southeastern Atlantic coast. Minnesota's population, however, has been relatively stable since the mid-1980s **(C)**.

Few studies have examined the factors responsible for these recent changes. A study in Missouri linked the species' decline to at least two factors: a more than 40 percent decline in the state's cattle production from 1975 to 2009, which reduced foraging habitat; and an 11 percent increase in forest cover between 1989 and 2010, which also reduced the amount of foraging habitat and increased the distances between foraging areas and host populations for breeding (Cox et al. 2012). Population declines in Minnesota are "likely associated with the decrease of small, family-owned farms that had dairy cows and horses" (Niemi, Lima, et al. 1995). Overall, the decline of Cowbirds has been regarded positively.

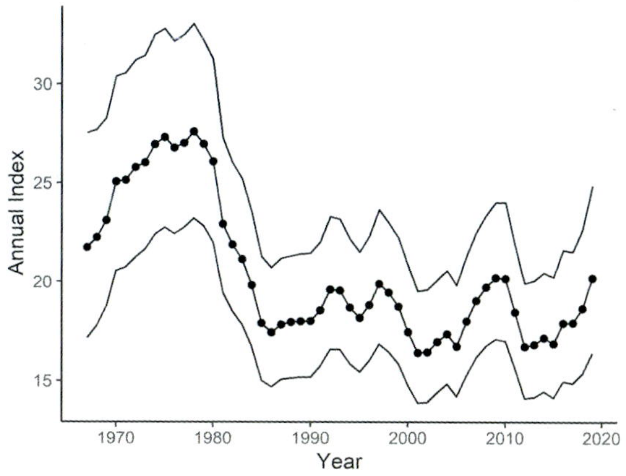

C. Breeding population trend and 95 percent credible interval (solid black lines) for the Brown-headed Cowbird in Minnesota based on the 1967–2019 Breeding Bird Survey (Sauer et al. 2020).

CONSERVATION

Conservation efforts are directed not at the Brown-headed Cowbird but instead at rare songbirds whose small and declining populations are negatively impacted by high rates of nest parasitism. Efforts to control Cowbirds at Kirtland's Warbler breeding sites in Michigan, for example, are well known. Although this federally *Endangered* species has recovered from the brink of extinction, the program to trap and remove Cowbirds that began in 1972 may continue indefinitely to ensure current Warbler population levels are sustained (Bocetti et al. 2020). Other vulnerable species for which Cowbird control measures have been implemented include the Black-capped Vireo (Grzybowski 2020), Least Bell's Vireo (Kus et al. 2020), and Southwestern Willow Flycatcher (Sedgwick 2020).

Despite its decline, the Brown-headed Cowbird is an abundant and widely distributed species. Even climate change will cause only minimal losses to its breeding distribution in a few southern states and northern Mexico. These losses are compensated for by range expansions across the Great Lakes states, including northeastern Minnesota, and Canada (NAS 2019). A low conservation priority (PIF 2021), the Cowbird will likely continue to be the focus of studies that investigate its complex relationship to the reproductive success of the songbirds that it frequently parasitizes.

BREWER'S BLACKBIRD

Euphagus cyanocephalus

Minnesota Seasonal Status: Regular summer resident and migrant; occasionally observed during the winter. The Brewer's Blackbird was uncommon during the MNBBA.
Migration: Populations in Canada and the north-central United States are short- to medium-distance migrants that winter in the southern United States and Mexico; farther west, populations are year-round residents.
Food: Omnivorous ground forager that primarily feeds on terrestrial insects, seeds, and waste grain.
Nest: Cup nest placed on the ground or in trees and shrubs; may breed as lone pairs or in small, loose colonies.
Conservation Concern: Assigned a moderate PIF Continental Concern Score of 10/20.

Many members of the blackbird family are well-known to Minnesota residents. From Red-winged Blackbirds and Bobolinks to Baltimore Orioles and Common Grackles, these species are abundant, widely dispersed, and have prominent characteristics that aid their identification. The most distinctive trait of the less common Brewer's Blackbird, however, is the absence of features that draw one's attention. Compared to a male Common Grackle, the Brewer's male is smaller in size, has a smaller, slimmer bill, and a shorter tail; in the sunlight, a purple sheen can be seen on his head and a green sheen on his body. Its plumage, coupled with its size and piercing yellow eye, help identify it as a Brewer's Blackbird. The female, by contrast, is a dull gray-brown bird; her eye is usually dark but occasionally lighter. The species' most common vocalizations include a hoarse *squee* and a very short, rushing gurgle.

The only bird that the Brewer's Blackbird may be confused with is the Rusty Blackbird, an uncommon migrant statewide and an extremely rare summer resident in northern Minnesota. Although it has a rusty plumage during the nonbreeding season, its summer plumage is difficult to distinguish from the Brewer's. Habitat is the best key: the Brewer's Blackbird is a species of open country, while the Rusty is a species of wet forests.

The Brewer's Blackbird is found from British Columbia south to northern Baja California and east

Photo by Larry Sirvio.

through the western states and Great Lakes region. Its Latin name translates to a species that eats well (*Euphagus*) and has a blue head (*cyanocephalus*). Its common name was given to it by Audubon in honor of his friend Dr. Thomas Brewer.

MINNESOTA BREEDING DISTRIBUTION

Prior to the 1900s, the Brewer's Blackbird was a western species. The eastern limit of its breeding range was Minnesota's Red River valley, where Roberts (1932) noted it was abundant "since the earliest records for that region." Because it is a species of open habitats, the Eastern Broadleaf Forest Province was considered an ecological boundary. But as small communities, farms, roads, and cities expanded during European settlement, the forested landscape opened and Brewer's Blackbirds began their eastward expansion.

Evidence was discovered that the birds were breeding in Minneapolis in 1914 and in Isanti and Sherburne Counties in 1915. When Roberts summarized the species' status in 1932, Brewer's Blackbirds were a common summer resident in all but Minnesota's most southern counties and the northern forests. Reports in 1928 from Pine and Carlton Counties, and along the Iron Range in St. Louis County, led him to speculate that "before many years it will be generally established in the coniferous region." At the time, confirmed nesting records were available from Douglas, Hennepin, Kittson, Marshall, and Polk Counties. Nesting was inferred in Isanti and Sherburne Counties.

The range expansion witnessed in Minnesota continued eastward. Taking advantage of the newly opened landscapes, Brewer's Blackbirds moved 1,200 km across the Great Lakes Region during the first half of the twentieth century. At the same time, in western Canada they moved 360 km north to the Northwest Territories (S. Martin 2020).

By 1975, Green and Janssen reported that the species had become a breeding resident throughout the state, except in the "heavily wooded portions of the northern regions, especially along the Canadian border from Lake of the Woods to the Pigeon River." Despite its scarcity in this region, scattered breeding records were available from Cook, Kittson, and St. Louis Counties. Brewer's Blackbirds were also scarce south of the Minnesota River. Janssen's (1987) distribution map for the species included all but the most southern tier of counties and identified 19 counties where nesting had been confirmed since 1970. Hertzel and Janssen (1998) added one more county to the list; all but two of the 20 counties (Dakota and Scott) were north of the Minnesota River. The 284 breeding-season locations identified by the MBS were largely restricted to the northern half of the state and to counties bordering the Minnesota River.

Although the distribution of MNBBA records for the Brewer's Blackbird was similar to that identified by the MBS, the 728 atlas records clearly demonstrated the species' prominence in northwestern Minnesota **(A)**. Indeed, 38 percent of the 539 blocks where the species was detected occurred in just 10 northwest counties stretching from Clay and Becker north to Kittson and Roseau. Scarce in the southwest, southeast, and far northeast corners of the

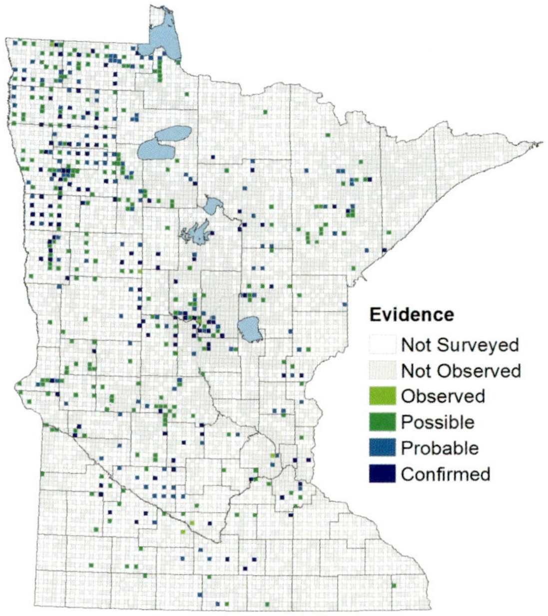

A. Breeding distribution of the Brewer's Blackbird in Minnesota based on the Breeding Bird Atlas (2009–13).

state, the species was encountered more frequently south of the Minnesota River than earlier accounts reported. Among the 44 counties where breeding was confirmed were six new counties south of the river: Blue Earth, Lyon, Redwood, Rice, Waseca, and Yellow Medicine. Nevertheless, 11 of the 15 counties where the Blackbird was not detected were also south of the river. Statewide, the species was found in 72 counties and in 11.2 percent of the surveyed blocks; breeding was confirmed in 2.7 percent of the blocks and deemed probable in an additional 3.0 percent. For a species fond of open landscapes, the Brewer's Blackbird's sparse distribution in southern Minnesota is somewhat surprising. Grassland landscapes in this region, however, are generally on much drier soils than the wet to mesic grasslands that the species seems to prefer farther north.

Despite Roberts's (1932) earlier prediction, further expansion of the species' breeding range beyond that observed in the first half of the 1900s has been limited, both in Minnesota and elsewhere (S. Martin 2020). In Michigan, for example, "the initial surge during the colonization phase" of the early 1900s "has waned" (Chartier et al. 2013), while a decline in the species' occurrence in Ontario suggested that "numbers may now be stabilizing" (Cadman et al. 2007).

BREEDING HABITAT

A species of open and semi-open habitats, the Brewer's Blackbird utilizes a range of sites from pastures, grasslands, and marshes to forest clear-cuts, burns, and plantations **(B)**. It is often closely associated with developed landscapes, including farmsteads, cemeteries, roadsides, and city parks. Suitable habitats share three features: short-grass foraging areas, perches where the males can watch for predators and monitor their mate's activities, and suitable nest sites (Horn 1968). Nests are frequently located near a water feature, such as an irrigation ditch, a pond, the edge of a wetland, or a shrub swamp (S. Martin 2020).

Some studies have noted that Brewer's Blackbirds displaced Common Grackles from grassland habitats as they expanded eastward, but not from marshy wetlands and suburban landscapes, where the Grackles appear to have a competitive advantage (Stepney 1979). In Minnesota, Roberts (1932) lamented the displacement of Bobolinks by the more aggressive Brewer's Blackbird: "Numerous meadows that were formerly the homes of many sweetly caroling Bobolinks are now occupied only by the harsh-voiced and otherwise undesirable Brewer's." Nevertheless, the Bobolink has prevailed; it was more than four times as abundant during the atlas than the Brewer's Blackbird.

POPULATION ABUNDANCE

PIF (2021) provided a population estimate of 23 million birds in the United States and Canada and a statewide estimate of 250,000. BBS data document a significant population decline, averaging −1.59 percent per year across the survey region from 1966 to 2019 (Sauer et al. 2020), with a cumulative loss of −61 percent from 1970 to 2014 (Rosenberg, Kennedy, et al. 2016). The downward trend in Minnesota's population is a little less steep, with a nonsignificant decline averaging −0.54 percent per year **(C)**.

Factors responsible for the species' decline are not well known. It is not considered an agricultural pest during the breeding season, since it feeds primarily on insects. However, during the nonbreeding season Brewer's Blackbirds associate with other blackbirds that can inflict significant crop damage.

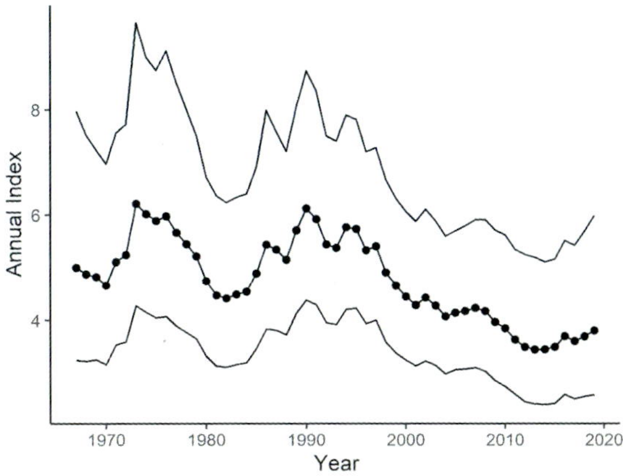

C. Breeding population trend and 95 percent credible interval (solid black lines) for the Brewer's Blackbird in Minnesota based on the 1967–2019 Breeding Bird Survey (Sauer et al. 2020).

As a result, they are often inadvertently taken when other species are the farmers' prime target. Like other blackbirds, Brewer's Blackbirds can be legally taken when they are causing, or are about to cause, crop depredation. It is also legal to take the species for the pet trade in Mexico, where many of the birds winter (S. Martin 2020).

Opening the vast landscape of the eastern deciduous forest initially facilitated the species' range expansion, but the loss of those open habitats to woody encroachment and development is likely contributing to their decline (Chartier et al. 2011). Pesticides may also pose a threat, given the species' prevalence in agricultural landscapes.

CONSERVATION

Although there are few concerns about threats to this habitat generalist, the Brewer's Blackbird is considered a moderate conservation priority given its significant population decline (PIF 2021). PIF has also identified it as one of 24 common species that have lost more than 50 percent of their population from 1970 to 2014 (Rosenberg, Kennedy, et al. 2016).

Minnesota's aspen parklands appear to be an important stronghold for the species, and its status in this region should be closely monitored. The potential impact of warming temperatures also warrants attention. The Blackbird's breeding distribution is considered moderately vulnerable under a 2°C warming scenario, with a projected 19 percent range loss compensated for by only an 8 percent gain northward by the year 2050 (NAS 2019). In Minnesota, the species' breeding range would gradually recede westward until, under a 3°C warming scenario, it is largely restricted to the northwestern counties, where the Aspen Parklands provide such critical habitat.

B. Wet meadows in northern Minnesota are one of the most common habitats used by nesting Brewer's Blackbirds. Photo by Lee A. Pfannmuller.

COMMON GRACKLE

Quiscalus quiscula

Photo by David Brislance.

Minnesota Seasonal Status: Regular summer resident and migrant; occasionally observed during the winter. The Common Grackle was abundant during the MNBBA.
Migration: Northern breeding populations, extending south throughout much of the central grasslands, are short-distance migrants that winter in the southern United States; southern populations are year-round residents.
Food: Omnivorous ground forager that feeds on seeds, waste grains, fruits, and arthropods; small vertebrates are consumed less frequently.
Nest: Cup nest, usually placed in coniferous trees and shrubs; less commonly found in deciduous trees and shrubs, emergent wetland vegetation, and in crevices and cavities. Often nests in loose colonies.
Conservation Concern: Assigned a moderate PIF Continental Concern Score of 9/20.

Proudly strutting across our lawns as it searches for edible morsels, the Common Grackle is one of the most adaptable and abundant members of the blackbird family, second only to the Red-winged Blackbird. Distributed across the eastern two-thirds of North America, the bird is unmistakable due to its large size, heavy bill, iridescent plumage, and long, strongly keeled tail (larger in males). In sunlight, the purple sheen on the male's head and the bronze sheen on its body are striking; the similar-looking female is a little smaller and her plumage less iridescent. The Grackle's vocalizations include a large repertoire of squeaks and croaks, but its most familiar song is one that sounds like a squeaky gate. The Latin name, *Quiscalus,* means "quail-like" and was assigned to the genus by Linnaeus. Although the Common Grackle does not look anything like a quail, Sandrock and Prior (2014) believed he might have been referencing the similar loud and strident calls made by both birds.

In Minnesota, the only species the Common Grackle might be mistaken for is the newly arrived Great-tailed Grackle. Rare and limited primarily to southwestern Minnesota, the latter species is considerably larger, with an exceptionally long tail that makes the Common Grackle's tail pale by comparison.

This very common species has a rather unique trait. The Common Grackle is one of many birds throughout the world that engages in what is known as "anting behavior." With wings spread, birds stand still over an anthill as the insects crawl over their body. Theories regarding the purpose of such behavior abound. One thought is that the formic acid secreted by the ants helps rid the birds of parasites. Other materials intentionally spread over their body while preening include the juice of lemons and limes and marigold flowers.

Unfortunately, the raucous and social Grackle may be best known as a consumer of agricultural crops and, to the dismay of many bird-lovers, the eggs and young of other songbirds. Roberts (1932) wrote, "It rifles nests of eggs and callow young to such an extent as to prevent most other species making their homes in the vicinity of a colony" of these nest raiders.

MINNESOTA BREEDING DISTRIBUTION

The Common Grackle, known to Roberts (1932) as the Bronzed Grackle, has long been a common summer resident of the state's open woodlands and agricultural landscapes. The conversion of Minnesota's western grasslands into productive agricultural fields provided the perfect setting for the Grackle to exploit its opportunities even further. By June 1926, when Roberts (1932) took a trip across southern Minnesota, he found them to be the most abundant bird along the entire route: "It was constantly in sight and even at that season was in flocks, sometimes covering several acres when feeding in pastures and fields. . . . Although the Grackle was known to have increased in numbers, its great abundance was a surprise."

The species was a common bird even in cities and parks. Confirmed nesting records were available from nine counties, including numerous records from Minneapolis in Hennepin County, as well as from Aitkin, Goodhue, and Isanti Counties in the eastern half of the state. Farther west, records stretched from Jackson and Pipestone Counties in the southwest and north to Grant, Marshall, and Polk Counties. Although Roberts did not specifically address their abundance in northeastern Minnesota, the species was probably scarce across the region's densely forested landscape.

Green and Janssen (1975) would later note its sparse distribution in the northern forested counties but wrote that it was not restricted entirely to agricultural lands or urban areas. They observed that the brushy riparian shorelines of rivers and lakes, often some distance from settled areas, also provided suitable habitat. Janssen (1987) noted the Grackle's abundance within windbreaks separating agricultural fields and in cities where conifer plantings were common. He reported a total of 41 widely dispersed counties where nesting had been confirmed since 1970. By 1998, Hertzel and Janssen would report that the species had been confirmed nesting in 48 counties. The MBS further confirmed its statewide distribution, including numerous breeding-season locations found throughout northeastern Minnesota.

During the MNBBA, participants reported 5,841 Common Grackle records in 58.1 percent (2,802/4,822) of the surveyed atlas blocks **(A)**. Breeding was confirmed in all 87 counties. With a few exceptions, Grackles were reported from virtually every priority block south of a line that extends from the Fargo-Moorhead region in northwestern Minnesota southeast to the Twin Cities. Although they were broadly distributed, north of this line occurrences were reported from nearly 50 percent fewer priority blocks.

Clearly, the extensive clearing and development that has occurred in Minnesota's northern forest has provided opportunities for the Common Grackle in the past one hundred years. Yet, as the species expanded into northeastern Minnesota, the most significant change in its North American distribution was its expansion westward. Facilitating this expansion was an increase in the number of shelterbelts, an ideal habitat for the species. As arid lands across the northern and central plains were irrigated and cultivated, shelterbelts became a more common feature across the landscape (Peer and Bollinger 2020).

BREEDING HABITAT

It is perhaps easier to describe what habitats the Common Grackle avoids than to list the wide variety of habitats it finds suitable. Although it is least abundant in densely forested landscapes, even here forest edges, clearings, riparian areas, and small settlements attract the birds to an otherwise unattractive landscape.

Overall, Grackles prefer any open or partially open landscape with scattered trees (Peer and Bollinger 2020). They are especially abundant in agricultural regions, where the combination of an abundant food source, open fields for foraging, and shelterbelts that provide nesting cover offer the ideal combination of features for successful reproduction **(B)**. Conifers are particularly attractive to Grackles. They are frequently found in loose colonies of up to 200 breeding pairs in conifer windbreaks and shelterbelts (Kaufman 1996; Peer and Bollinger 2020). One only needs to encounter the cacophony of sounds emitted by a gathering of nesting Grackles in western Minnesota's shelterbelts to appreciate the importance of this modified habitat to the birds.

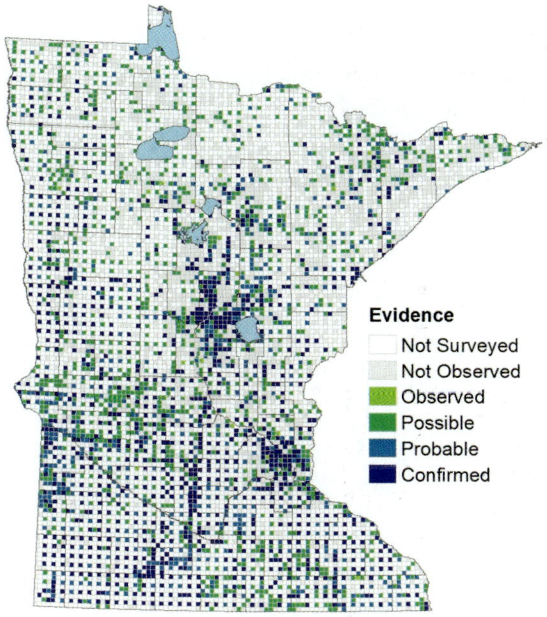

A. Breeding distribution of the Common Grackle in Minnesota based on the Breeding Bird Atlas (2009–13).

B. Farm shelterbelts planted with coniferous trees for nesting and adjacent to agricultural fields for foraging provide ideal nesting habitat for the Common Grackle throughout the agricultural landscape.
Photo by Lee A. Pfannmuller.

POPULATION ABUNDANCE

In the United States and Canada, Common Grackles have been estimated to number 67 million breeding adults, with approximately 3.7 million, or 5.5 percent of the continental population, residing in Minnesota (PIF 2021). Southwestern Minnesota actually supported some of the highest breeding densities of Grackles in the United States and southern Canada from 2011 to 2015 (Sauer et al. 2020).

Despite their abundance, Common Grackle populations have been declining quite steeply since wide-scale monitoring began in the 1960s. Across the survey region, BBS data illustrate a statistically significant decline of –1.47 percent per year from 1966 to 2019 (Sauer et al. 2020). This represents a cumulative decline of –54 percent from 1970 to 2014 (Rosenberg, Kennedy, et al. 2016). In Minnesota the decline has been equally steep and statistically significant, averaging –1.42 percent per year since 1967 **(C)**.

The species' decline may be due in large part to its overwhelming success. Common Grackles are considered a major agricultural pest. Along with other members of the blackbird family, Grackles are responsible for millions of dollars of damage to a variety of crops, including corn, sunflowers, grains, and fruit. Each year their damage to sunflowers alone was estimated at $7 million to $10 million (Kleingartner 2003). Damage by blackbirds is cited as one of the primary reasons North Dakota farmers have reduced sunflower production by more than 75 percent since the 1970s (Klosterman et al. 2013).

CONSERVATION

An abundant species, the Common Grackle is not a high conservation priority. Its moderate Continental Concern Score is influenced largely by its long-term population decline (PIF 2021). Warmer temperatures will cause only minimal changes to the species' breeding distribution, with northward range expansions more than compensating for losses in the Great Plains west of Minnesota and in the southeastern United States (NAS 2019).

The species' status as an economic pest means that management is focused on damage control rather than on conservation. Although they are a protected species under the Federal Migratory Bird Act, most blackbirds, including Common Grackles, can be taken legally without a permit when "they are causing serious injuries to agricultural or horticultural crops" (50 CFR §21.43—Depredation order for blackbirds, cowbirds, crows, grackles, and magpies). Indeed, an entire industry supports the farmer's need to control blackbirds. The success of these efforts over the years has impacted population numbers (Peer and Bollinger 2020). The Common Grackle's designation as a Common Bird in Steep Decline (Rosenberg, Kennedy, et al. 2016) is a call to closely monitor populations in the years ahead and not take this "common" species for granted.

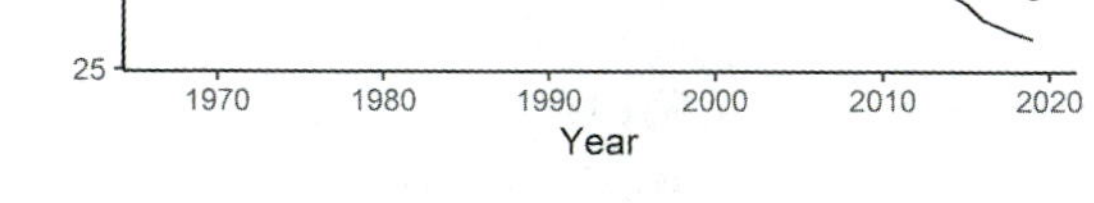

C. Breeding population trend and 95 percent credible interval (solid black lines) for the Common Grackle in Minnesota based on the 1967–2019 Breeding Bird Survey (Sauer et al. 2020).

GREAT-TAILED GRACKLE

Quiscalus mexicanus

Minnesota Seasonal Status: Regular summer resident and migrant. The Great-tailed Grackle was rare during the MNBBA.
Migration: Northern breeding populations are partial migrants that winter in the central and southern states; southern populations are year-round residents.
Food: Omnivorous ground forager; occasional fly-catching; and foliage gleaner. Arthropods compose the majority of the diet.
Nest: Cup nest, usually placed in trees and shrubs but occasionally in marsh vegetation. Usually nests in colonies numbering from just a few pairs to hundreds.
Conservation Concern: Assigned a low PIF Continental Concern Score of 5/20.

Found from Columbia north to southern Texas at the beginning of the twentieth century, the Great-tailed Grackle has been gradually expanding northward across the southwestern and west-central United States. Having arrived only recently in Minnesota, this impressive bird is still a rarity, limited primarily to the southwestern counties. If you are lucky enough to spot one, it will seem like a Common Grackle on steroids. A long-legged, slender bird with a large bill, the male is nearly 1.5 times larger than its smaller relative. Dark in appearance, with bright yellow eyes, he has black plumage with a purple iridescence. His most unique feature, however, is an extremely long and deeply keeled tail. The female is nearly 20 percent smaller than the male and is dark brown above and lighter below. Although she has an exceptionally long tail, it lacks the male's deep keel.

Loud and raucous, like other members of the Grackle genus this social bird has a variety of vocalizations that range from pleasant notes to loud and far-reaching whistles and squawks. Individual birds may even invent their own unique calls. The male's most frequently delivered territorial song is a clear and rapid *ki-ki-ki* (K. Johnson and Peer 2020). Because the bird's "loud, percussive, strident" calls are not unlike those of many quail species, Linnaeus is believed to have assigned the Grackle genus the Latin name *Quiscalus,* which means "quail-like" (Sandrock and Prior 2014).

Photo by Gerald Hoekstra.

By the spring of 2021, Great-tailed Grackles were found regularly throughout the western and central United States, stretching as far north as southern Washington in the west and central Minnesota in the east. Outliers occur both north and east of this primary range (eBird 2021).

MINNESOTA BREEDING DISTRIBUTION

The Great-tailed Grackle was still more than 1,000 miles away in southern Texas when Roberts (1932) wrote his treatise on Minnesota birds. It slowly began moving north in the early 1900s, and, by 1950, "a rapid range expansion was evident," with the birds moving along three major fronts: the Great Plains; the Rocky Mountain States; and the Pacific coast (Dinsmore and Dinsmore 1993). In the Great Plains, Oklahoma documented its first observation in 1953, Kansas in 1963, Missouri in 1972, Nebraska in 1976, and Iowa in 1983 (A. X. Hertzel and Hertzel 2001). It was only a matter of time before the birds were sighted in Minnesota.

The first observation was reported in June 1982 in Dakota County. An adult male was seen and heard by Paul Egeland at Black Dog Lake (Egeland 1983). An expert birder, Egeland first heard the bird and immediately recognized that it was a species he had never heard in Minnesota. Other experts, however, were hesitant to confirm the report because of the difficulty of distinguishing a Great-tailed Grackle from a Boat-tailed Grackle, the latter being restricted to marshes along the Gulf and Atlantic coasts. The two species are closely related and nearly identical in appearance. Indeed, for many years they were considered one species. Although details of the 1982 record were carefully documented, they were not considered sufficient to totally rule out the occurrence of a Boat-tailed Grackle.

Eleven years would pass before the species' presence in Minnesota was solidly confirmed. In April 1993, an adult male was seen near the town of Faribault in Rice County. While it remained in the vicinity from April 4 to April 10, birders were able to positively confirm its identity as a Great-tailed Grackle (K. Smith 1993; Glassel 1993). The bird's bright yellow eyes were markedly different from the duller yellow to brown eyes of the Boat-tailed Grackle; it also had a remarkably long tail, which is shorter in the Boat-tailed Grackle (K. Johnson and Peer 2020). More observations elsewhere in southern Minnesota soon followed, including the first state nesting record in the summer of 2000. Two nests were found in Jackson County in a wetland just 3 miles north of the Iowa border (Wiens 2001).

A. X. Hertzel and Hertzel (2001) provide an excellent summary of the species' expansion into the state through the spring of 2001. By 2008, the Grackle had been observed in 16 counties during the summer season; the summer season was defined using the Grackle's MNBBA summer safe dates (May 10 to July 20). The list included the original 1982 observation in Dakota County that was deemed a likely Great-tailed Grackle (Glassel 1993). Although the majority of counties were in southwestern Minnesota, the species was observed as far east as Chisago County (2005) in east-central Minnesota and Olmsted County (2006) in southeastern Minnesota (MOU 2021b). In 2004, the MOU Records Committee officially changed the status of the Great-tailed Grackle in Minnesota from Accidental to Regular (Eckert 2004).

Only a handful of records for this rare species were gathered during the MNBBA. Participants tallied just 11 records in eight atlas blocks. Breeding was confirmed in one block in southern Jackson County (2010), the site where the species has been found repeatedly over the years **(A)**. It was reported in five of Minnesota's 87 counties. Included are two possible records from Douglas County, where the species had not previously been reported; one probable and one possible record from Lyon County; a possible record from McLeod County; and an observed record in Lac qui Parle County.

In the years following the atlas (2014–2020), additional records are summarized by Janssen (2019) and the MOU (2021a, 2021b). Last reported nesting during the 2014 breeding season in Jackson County, the species has yet to be confirmed nesting elsewhere in the state. Overall, the speed of the Grackle's expansion appears to have abated. Its current distribution is still largely limited to the southwestern region, while a few scattered records occur across central Minnesota.

BREEDING HABITAT

The Great-tailed Grackle is a species of open country. Suitable habitats include brushy clear-

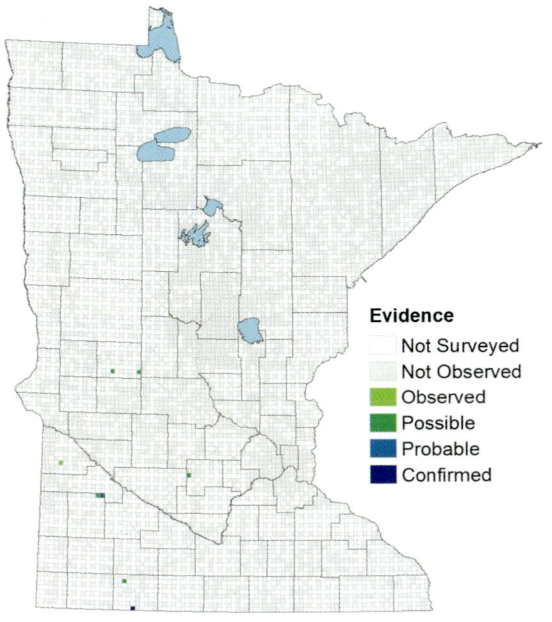

A. Breeding distribution of the Great-tailed Grackle in Minnesota based on the Breeding Bird Atlas (2009–13).

ings, open forests, and marshes. It is equally at home in agricultural and urban settings, including grazed pastures, cultivated fields, cemeteries, neighborhoods, and city parks. Only dense forests and arid lands are avoided. Primary habitat requirements are a water source, open country for foraging, and trees or shrubs for nesting and roosting. Lacking the latter, tall emergent wetland vegetation may be used (K. Johnson and Peer 2020; Kaufman 1996).

In Minnesota, A. X. Hertzel and Hertzel (2001) provided a brief description of the Jackson County nesting site. Here the species utilized a small wetland bordered by a fringe of emergent cattails for nesting and the surrounding pastures for foraging **(B)**. The two nests found in 2000 were anchored to cattails approximately 0.4 m above the water's surface. Although Great-tailed Grackles have adapted to nesting in trees and various man-made structures, they often select wetlands for nesting when they first colonize new sites (A. X. Hertzel and Hertzel 2001; Dinsmore and Dinsmore 1993).

POPULATION ABUNDANCE

PIF (2021) provided a total population estimate of 30 million breeding adults, with approximately 8.2 million residing in the United States and Canada. In Minnesota, the population likely numbers fewer than a few hundred individuals. The largest concentration of Great-tailed Grackles seen in the state was a group of 103 birds observed in Jackson County during the fall of 2008 (Janssen 2019). To the south, Iowa is estimated to support a population of 3,300 birds (PIF 2021). Predictably, within the United States, breeding densities are highest in the southwest, especially along the Texas-Mexico border.

BBS data indicate a surveywide population increase from 1966 to approximately 1985, followed by a gradual decline through 2019 **(C)**, with an average annual increase of 0.92 percent over the entire sampling period (Sauer et al. 2020). Some of the largest increases have been documented in California (11.06 percent per year), Colorado (8.71 percent per year), and Missouri (7.01 percent per year). A species that barely occurred beyond south Texas in 1900 had, by the end of the twentieth century, been documented nesting in at least fourteen states and had been observed in twenty-one states and three Canadian provinces (K. Johnson and Peer 2020). The Great-tailed Grackle's rapid expansion northward has been tied to an increase in irrigation in the arid southwest and central plains states, fire suppression that allowed the encroachment of woody vegetation on grasslands, and widespread tree planting and urban development (K. Johnson and Peer 2020). Factors responsible for its recent decrease in numbers are not well-known.

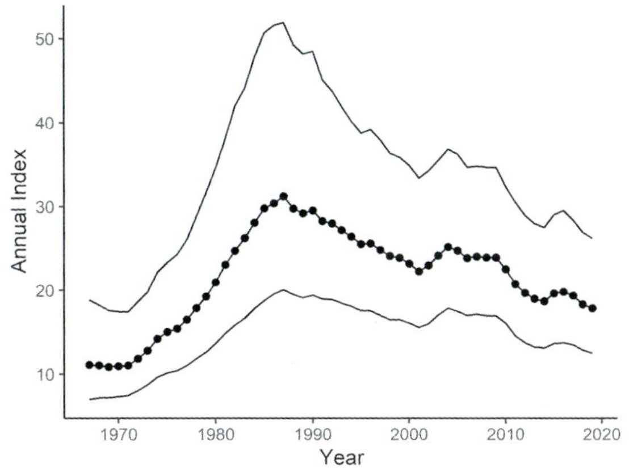

C. Surveywide breeding population trend and 95 percent credible interval (solid black lines) for the Great-tailed Grackle based on the 1967–2019 Breeding Bird Survey (Sauer et al. 2020).

B. A permanent wetland surrounded by a dense fringe of cattails, just north of the Iowa border, provides the only known nesting site for the Great-tailed Grackle in Minnesota. Photo by Gary T. Seim.

CONSERVATION

Highly adapted to the features provided by a settled landscape, the Great-tailed Grackle is not a conservation priority. A warming climate also poses little threat to this adaptable species (NAS 2019), only providing more favorable conditions as it treks northward. In many areas in the Southwest and in Central America, the Grackle is considered an agricultural pest. Its habit of foraging on emerging corn, sorghum, and citrus crops has led to the implementation of numerous measures to deter the birds from agricultural fields and orchards (K. Johnson and Peer 2020). Nevertheless, unless the current population decline continues, Minnesotans are likely to become much more familiar with Great-tailed Grackles in the coming years.

OVENBIRD

Seiurus aurocapilla

Minnesota Seasonal Status: Regular breeding resident and migrant. The Ovenbird was abundant during the MNBBA.
Migration: Short- to medium-distance migrant that winters along the southeastern Atlantic coast of the United States and in southern Mexico, Central America, and the Caribbean.
Food: Ground gleaner that feeds on arthropods.
Nest: Domed cup nest placed on the forest floor.
Conservation Concern: Assigned a moderate PIF Continental Concern Score of 9/20.

A common species throughout Minnesota's forests, the Ovenbird sings its signature song, *Teacher Teacher Teacher*, which is perhaps the loudest, most emphatic, and most easily identified warbler song in North America. Roberts (1932) described this "phantom bird of the woodland depths" as a bird "possessed of a voice and an exuberance of spirit during the courting-time that marks it among its fellows." Unbeknownst to many, the Ovenbird's vocal repertoire also includes an extended, complex song that some regarded as the bird's true "passion song" in the early 1900s (Bent 1953). The bird's delivery of the song is remarkable: the musical, lark-like warbles are usually given at twilight or dawn and are often part of an elaborate aerial display.

This sexually monomorphic understory inhabitant has distinguishing features such as olive-gray plumage on its upper body, an orange-rufous crown stripe bordered by black, a heavily streaked white breast, and a habit of walking on the ground, rather than hopping like most birds. As it strides along, its tail and body are moved in a fashion that appears as though the bird is bobbing up and down, giving rise to several other vernacular names, including Wood Wagtail. The species' Latin name, *Seiurus aurocapilla,* translates to "tail bobber with a golden crown," while its common name references the resemblance of its dome-shaped nest to an oven.

The Ovenbird's breeding distribution spans Canada's boreal forests, from northeastern British Columbia through the Maritime Provinces. In the United States, it is found from central Montana southeast to Arkansas, and east to the Appalachian Mountains and mid-Atlantic coast. Northern Minnesota, Wisconsin, and Michigan support the highest breeding densities.

Photo by David Brislance.

MINNESOTA BREEDING DISTRIBUTION

In the early 1900s, Roberts (1932) summarized the Ovenbird's status as a formerly abundant breeding bird throughout the forested regions of the state. However, he cautioned that it had recently become far less common in southern Minnesota, "where it has almost disappeared as a summer resident from localities where it was once abundant." Numbers appeared to be on the decline even in the extensive forests of northern Minnesota. In the prairie's scattered woodlands, the species was "much less frequent and chiefly a migrant." Nesting was confirmed in Aitkin, Becker, Cass, Hennepin, Marshall, and Stearns Counties and Itasca State Park, and inferred in Isanti and Wabasha Counties and at Cass Lake.

Reaffirming the species' distribution across Minnesota's forests, Green and Janssen (1975) noted recent reports from Norman County in the northwest, Pope County in west-central Minnesota, and Rice and Olmsted Counties in the south. Its status, they wrote, in the "groves and river bottoms on the prairie is uncertain." Janssen (1987) characterized the breeding range as occurring primarily in northern, east-central, and southeastern Minnesota. Scattered summer records from Brown, Chippewa, and Nicollet Counties along the Minnesota River valley, and from Nerstrand Big Woods State Park in Rice County and Camden State Park in Lyon County, suggested it was a "casual breeder" in some areas of south-central and southwestern Minnesota. Janssen identified 15 counties with confirmed nesting since 1970. Hertzel and Janssen (1998) added nine more counties to the list. The MBS documented a similar distribution, including several records in southwestern Minnesota and west along the Minnesota River valley to Yellow Medicine County.

Little had changed by the time MNBBA participants reported 5,755 Ovenbird detections from 45.7 percent (2,180/4,769) of the surveyed atlas blocks and confirmed breeding in 89 blocks (1.9 percent) **(A)**. The birds were detected in 75 of Minnesota's 87 counties and were confirmed nesting in 20 counties. Six counties with confirmed nesting were additions to the list published by Hertzel and Janssen (1998): Carlton, Crow Wing, Mille Lacs, Pine, Polk, and Stearns. Probable nesting was reported in 39 counties stretching from Houston and Fillmore Counties in the southeast north to the Canadian border. One of the most frequently reported birds during the MNBBA, the Ovenbird remained most abundant in the Laurentian Mixed Forest Province but was also present throughout the Eastern Broadleaf Forest and the Tallgrass Aspen Parklands Provinces. In western Minnesota, it was distributed along the entire length of the Minnesota River valley and in scattered pockets of suitable habitat in southwestern and south-central Minnesota.

Breeding densities are predicted to be highest in the northern forests, but the MNBBA probability map shows the birds are also well represented in the Hardwood Hills Subsection of the Eastern Broadleaf Forest Province in west-central Minnesota **(B)**. The only regions that appear totally inhospitable are the Twin Cities and much of the intensively cultivated landscapes of southwestern and western Minnesota.

The Ovenbird's distribution in the state has changed little in the past century. Although its status in the western prairies may have been uncertain from the 1800s through the mid-1900s, both the MBS and MNBBA demonstrate that suitable habitat is currently present in the floodplain forests of the

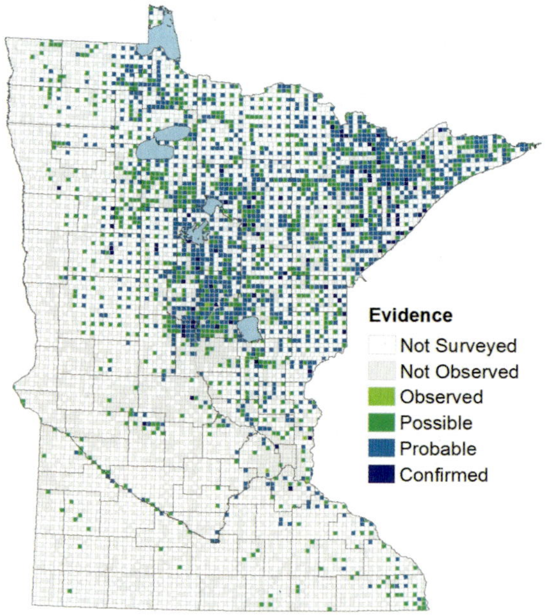

A. Breeding distribution of the Ovenbird in Minnesota based on the Breeding Bird Atlas (2009–13).

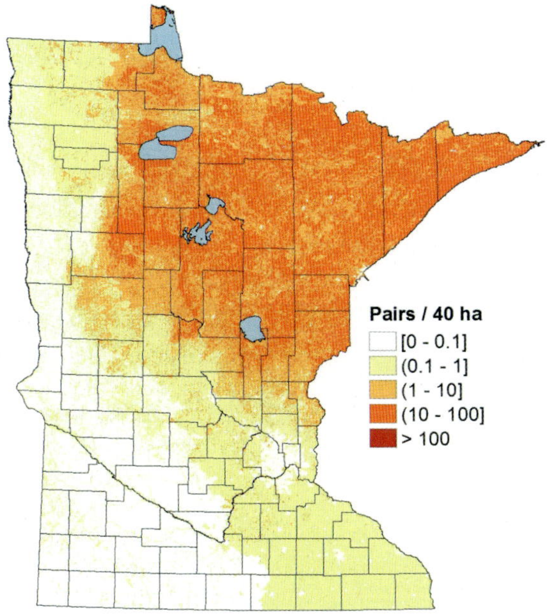

B. Predicted breeding distribution (pairs per 40 ha) of the Ovenbird in Minnesota based on habitat, landscape context, and climate data gathered during the Minnesota Breeding Bird Atlas (2009–13) using the Generalized Linear Modeling method with an adjustment for detectability.

Minnesota River valley and in woodlands scattered across the agricultural landscapes. The species' relative abundance has certainly declined in forested regions of the state where woodlands have given way to development and agriculture, especially in central and southeastern Minnesota. Nevertheless, the Ovenbird still remains present in these regions. Indeed, throughout North America, the breeding range of this northern forest species has remained remarkably stable (Porneluzi et al. 2020).

BREEDING HABITAT

The Ovenbird's habitat can be broadly described as large, mature forests with closed canopies, an open understory, and an intermediate density of shrubs (Danz et al. 2007). Deciduous, coniferous, and mixed forests all provide suitable habitat, as do both upland and lowland stands **(C)**. The forest's age and composition are less important than its structural features—even young, pole-sized forests are used if they meet the basic requirements of a closed canopy and an open understory. Sites with less ground cover and deep litter layers are favored by this ground nester and forager (Porneluzi et al. 2020).

Ovenbirds display some area sensitivity, although the size of the forest tract required may be quite variable, ranging from 100 to 885 ha (Porneluzi et al. 2020). The species can be found in small woodlands, but studies suggest that reproductive success is lowered by higher predation and nest parasitism rates. These negative impacts may be less evident when small stands are embedded within landscapes that are extensively forested compared with those found in landscapes fragmented by development and agriculture (Hanski et al. 1996).

POPULATION ABUNDANCE

The Ovenbird is one of the most common forest songbirds in North America. PIF (2021) generated a population estimate of 26 million breeding adults in the United States and Canada, with an estimated 1.5 million birds residing in Minnesota. The MNBBA's more extensive coverage of the state, including within the roadless Boundary Waters Canoe Area Wilderness where Ovenbirds are abundant, generated a significantly larger population estimate of 5.6 million adults (95 percent confidence interval of 5.4–5.8 million).

Since the BBS began in 1966, Ovenbirds have shown a stable population trend throughout the survey region (Sauer et al. 2020). Minnesota BBS data illustrate a similar trend with a nonsignificant, average annual increase of 0.26 percent from 1967 to 2019 **(D)**. A stable to increasing population in the Superior and Chippewa National Forests from 1995 to 2020 (Grinde, Bednar, et al. 2020) has been attributed to a steady decrease in logging activity, resulting in a greater proportion of older forest stands.

CONSERVATION

Given its abundance and stable population, the Ovenbird is not a high conservation priority. Nonetheless, the loss of suitable habitat on its breeding and wintering grounds remains a concern (PIF 2021). Management recommendations focus on maintaining large contiguous forest tracts, restoring connectivity, and managing a diversity of successional ages (Matteson et al. 2009). Forest management practices that retain only a minimal tree canopy, such as clearcuts, eliminate habitat for Ovenbirds until a canopy redevelops; practices that retain the stand's structural complexity, such as selective cuts, can continue to provide some suitable habitat.

C. Mature upland forests with a canopy of both deciduous and coniferous trees represent one of many forested habitats where the Ovenbird is common. Photo by Gerald J. Niemi.

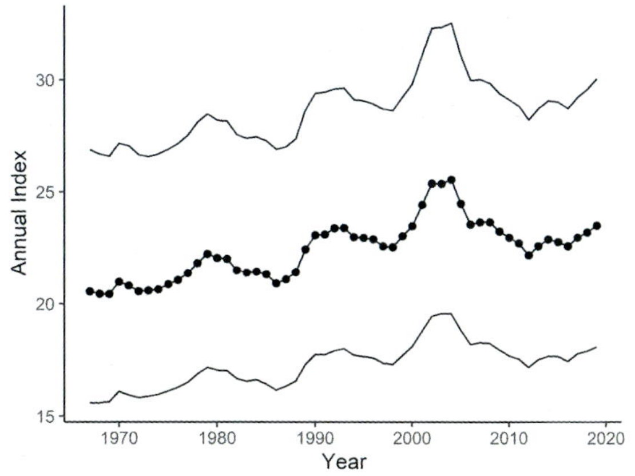

D. Breeding population trend and 95 percent credible interval (solid black lines) for the Ovenbird in Minnesota based on the 1967–2019 Breeding Bird Survey (Sauer et al. 2020).

Although the Ovenbird's population appears secure, it must contend with several human-induced threats. Studies, for example, have demonstrated that the species experiences a disproportionate level of annual mortality at communication towers (Longcore et al. 2013) and buildings (Loss et al. 2014). On Minnesota Point in Duluth, it makes up nearly 6 percent of the fatal window collisions (Bracey 2011). Another concern for this understory species is the impact of European earthworms on the forest litter layer. One study demonstrated a direct link between the presence of invasive European earthworms and lower Ovenbird breeding densities (Loss et al. 2012). Finally, under all three warming scenarios examined by NAS (2019), the species' breeding distribution demonstrated a moderate vulnerability to warming temperatures; models predicted that losses to its southern breeding range consistently exceeded northern expansions. In Minnesota, the Ovenbird's breeding distribution is projected to gradually recede northward until, under a 3°C warming scenario, it will be restricted to just a small area in northern Cook County.

It is hard to imagine a future without the Ovenbird's familiar song ringing out from the forest understory. Although this is an unlikely scenario in the near term, in the future the breeding range of the "teacher-bird" (Roberts 1932) may be largely restricted to forests many miles beyond our northern border.

LOUISIANA WATERTHRUSH

Parkesia motacilla

Minnesota Seasonal Status: Regular breeding resident and migrant. The Louisiana Waterthrush was rare during the MNBBA.
Migration: Medium-distance migrant that winters in the Caribbean and from portions of Mexico south to northern South America.
Food: Primarily aquatic invertebrates secured by probing along the stream edge or feeding from exposed rocks within the stream.
Nest: Cup nest placed in the roots of an upturned tree, a rocky crevice, or within small hollows along an earthen stream bank.
Conservation Concern: Assigned a moderate PIF Continental Concern Score of 12/20; officially classified as *Special Concern* in Minnesota.

Photo by Rebecca Field.

Among the largest of Minnesota's wood warblers, the Louisiana Waterthrush is often described as "thrush-like" in appearance. Lacking the brilliantly colored breeding plumage that characterizes many warblers, its brown-and-white attire is more typical of several forest-dwelling thrushes and helps it blend with its understory environment. Its most distinct features are a white stripe above each eye, a white throat, and a white chest and belly with heavy brown streaking. The male and female are indistinguishable. The Latin name for the genus, *Parkesia,* is a tribute to American ornithologist Kenneth Parkes; *motacilla* refers to the bird's tail bobbing.

The Louisiana Waterthrush and the Northern Waterthrush are so similar that many early ornithologists thought they were the same species. Both are closely associated with water, walk rather than hop, spend most of their time on the ground, and bob their tail and lower back up and down with each step. Subtle differences include the width and color of the line above the eye (whiter and wider on the Louisiana), the thin stripes on the chin and throat (absent on the Louisiana), and the brighter pink legs on the Louisiana. Many Northerns also have a pale yellow wash on their underparts, while a light buffy coloring may be evident on the Louisiana Waterthrush's flanks (Whitaker and Eaton 2020).

Fortunately, their songs are sufficiently distinct to aid identification. The Louisiana Waterthrush's song begins with several loud and clear slurred notes that precede a jumbled warble, unlike the short notes that precede the Northern Waterthrush's song. The more southerly distributed Louisiana Waterthrush also prefers forested stream valleys compared to the Northern Waterthrush's lakeshore and wetland habitats.

Restricted to the eastern United States and southern Ontario, the Louisiana Waterthrush is sparsely distributed throughout its breeding range. Population densities are highest in the Appalachian Mountains.

MINNESOTA BREEDING DISTRIBUTION

Documentation of the Louisiana Waterthrush's northward expansion into Minnesota began with the first state record in 1883 along the Root River in Fillmore County (Roberts 1932). By 1893 it was reported common in that area, and by 1898 it was found as far north as Red Wing in Goodhue County. Just four years later, in 1903, an adult male was taken in Hennepin County near Minnehaha Falls. It was not long before Roberts described it as a regular but uncommon breeding species in the Twin Cities region.

The range expansion continued north along the St. Croix River, where "hundreds" were reported as far north as Taylors Falls in 1908. When Thaddeus Surber visited southern Pine County ten years later, he reported Northern Waterthrushes along several tributaries of the St. Croix River: Crooked Creek, Sand Creek, and the mouth of the Tamarack River. Roberts (1932), however, assigned Surber's observations to the Louisiana Waterthrush (Green and Janssen 1975). Interestingly, both species are now known to occur in Pine County. The range expansion also continued west along the Minnesota River, reaching eastern Carver County in 1928. Regardless, the species remained most abundant "in the Mississippi River bottom-lands south of Red Wing" and along the St. Croix River (Roberts 1932). Confirmed or inferred nesting records were confined to Carver, Hennepin, Houston, and Washington Counties.

In the following years, the species was seldom observed beyond southeastern Minnesota and the St. Croix River. Most reports were from two localities: Beaver Creek Valley State Park in Houston County, and near the small town of Franconia in Chisago County (Longley 1973; MOU 2021b). Other reports were rare but documented the species' slow extension north along the Mississippi River to Stearns County (Eckert 1970) and west along the Minnesota River to Blue Earth, Nicollet, Scott, and Sibley Counties (MOU 2021b). By 1975, Green and Janssen added inferred and confirmed breeding records in Chisago, Stearns, and Winona Counties to the four counties already identified by Roberts (1932).

Roberts's decision to assign Thaddeus Surber's Pine County Waterthrush records to the Louisiana Waterthrush was reaffirmed by multiple records in the county in the early 1980s, including breeding confirmation in 1983 (Zumeta and Cincotta 1983). By 1987, Janssen identified five counties with confirmed nesting since 1970: Chisago, Pine, Houston, Olmsted, and Winona. Hertzel and Janssen (1998) added Fillmore and Washington Counties to the list. Meanwhile, the MBS documented 83 breeding-season locations, corroborating earlier accounts of the species' restricted distribution.

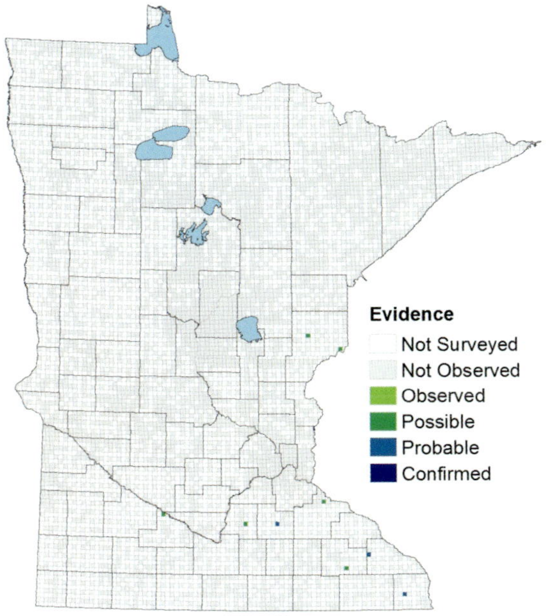

A. Breeding distribution of the Louisiana Waterthrush in Minnesota based on the Breeding Bird Atlas (2009–13).

MNBBA participants reported Louisiana Waterthrushes in just 10 blocks. Dispersed across 10 counties, breeding was confirmed in only one block located in northeastern Washington County along the St. Croix River **(A)**. (Because of its small size, the block is not visible on the MNBBA map.) Probable nesting was reported in Houston, Rice, and Winona Counties. The westernmost record was in the Minnesota River valley in a block that straddled Brown and Nicollet Counties; the most northern records were in Pine County.

In 1932, Roberts predicted that the Louisiana Waterthrush would continue its westward expansion in Minnesota, as so many other avian immigrants from the south had done. Although the Brown County/Nicollet County MNBBA record is nearly 140 river miles west of the 1928 Carver County record, there have been no other significant changes to the species' Minnesota breeding distribution since the mid-1900s. Indeed, the species has clearly declined in abundance. Elsewhere in the late twentieth century, populations expanded north in New England and Michigan. These movements may have been a response to forest maturation in areas that had been logged fifty to one hundred years earlier (B. J. Mattsson et al. 2020). Nevertheless, when Michigan conducted their second atlas (2002–8), there was a notable decline in Waterthrush numbers (Chartier et al. 2011).

BREEDING HABITAT

In Minnesota, the Louisiana Waterthrush occupies riparian forests ranging from mature, mesic forests bordering clear, cold-water streams **(B)** to mature bottomland hardwoods bordering the much larger Mississippi, St. Croix, and Minnesota Rivers (e.g., Eliason and Fall 1989; J. H. Stucker and Cuthbert 2000; Roberts 1932). Along cold-water streams, territories are frequently located beside steeply sloped ravines; along larger rivers, they are often near the mouths of tributaries. A few territories along the St. Croix River, however, were far from any nearby tributary, nestled on steep, rocky slopes at the river's edge (Eliason and Fall 1989). In Washington and Pine Counties, several reports were along creeks flowing through broad floodplain forests where the only relief was provided by a low-rising stream bank (Zumeta and Cincotta 1983; Eliason and Fall 1989).

Good water quality that supports a healthy community of aquatic invertebrates is essential. Exposed rocks provide foraging perches and are an important source of food, acting as "catches for vegetative debris" that harbor aquatic invertebrates (Surdick 1995). Along the stream bank, rocky crevices, upturned root clumps, and crannies within earthen banks provide hollows and cavities for nesting (B. J. Mattsson et al. 2020).

POPULATION ABUNDANCE

The Louisiana Waterthrush population is estimated to number just 450,000 breeding adults (PIF 2021). Although the statewide population is surely quite small, the species has never been recorded on any Minnesota BBS route. Indeed, given its sparse distribution and the inaccessibility of its riparian habitat, BBS data are statistically unreliable across North America.

B. The Louisiana Waterthrush may be found in small stream corridors surrounded by mature hardwoods in southeastern and east-central Minnesota. Photo by Lee A. Pfannmuller.

Nevertheless, they do suggest a slow population increase averaging 0.37 percent per year from 1966 to 2019 (Sauer et al. 2020), with a cumulative increase of 34 percent from 1970 to 2014 (Rosenberg, Kennedy, et al. 2016). In the core of the species' breeding range in the Appalachian Mountains and Piedmont Bird Conservation Regions, numbers show a significant upward trend from 1993 to 2019 (Sauer et al. 2020).

Most Louisiana Waterthrush accounts cite numerous factors that could negatively affect breeding populations, including forest management, habitat loss, acid rain, and water-quality degradation. Reforestation of areas previously cleared is the primary factor cited as potentially responsible for recent increases (B. J. Mattsson et al. 2020).

CONSERVATION

Given its small population and its dependence on mature riparian habitats and good water quality, the Louisiana Waterthrush has been designated a moderate conservation priority (PIF 2021). These same factors also prompted its 1984 classification as *Special Concern* in Minnesota (Coffin and Pfannmuller 1988).

J. H. Stucker and Cuthbert's (2000) in-depth study of the species concluded that the protection of contiguous, mature riparian forests adjacent to cold-water streams was critical to its future in southeastern Minnesota. Natural fluvial processes that create eroded stream banks are especially important for nesting. Activities that disturb the forest or stream's integrity "threaten the continued presence and productivity of Louisiana Waterthrushes." For example, when habitat improvement projects stabilize eroded stream banks, critical microhabitat features required for nesting may no longer be present. Riparian forest buffers that are at least 100 m wide and are mature enough to shade the entire stream channel are also recommended (Potter et al. 2007).

Despite its dependence on cool mesic forests and cold-water streams, warmer temperatures may have little impact on the Louisiana Waterthrush. Under all three warming scenarios examined, NAS (2019) projected a stable breeding distribution with minimal range losses along the southern edge of its breeding range and substantial gains farther north, including throughout southeastern and east-central Minnesota. Collisions with communication towers, however, do pose a hazard, resulting in an annual mortality of 1.4 percent of the Waterthrush's already small population (Longcore et al. 2013). Nevertheless, the availability of suitable habitat may be the largest concern that threatens this species' future in the North Star state.

NORTHERN WATERTHRUSH

Parkesia noveboracensis

Minnesota Seasonal Status: Regular breeding resident and migrant. The Northern Waterthrush was uncommon during the MNBBA.
Migration: Medium- to long-distance migrant that winters in the Caribbean and from central Mexico to northern South America.
Food: Aquatic invertebrates and terrestrial arthropods secured primarily by foraging amid leaf litter and in shallow pools.
Nest: Cup nest placed in a small cavity in the upturned roots of a downed tree, at the base of a clump of ferns, or under a riparian embankment.
Conservation Concern: Assigned a low PIF Continental Concern Score of 8/20.

The more common of Minnesota's two waterthrush species, the Northern Waterthrush is a distinctly northern, boreal species. Distributed across Alaska and Canada, its breeding range dips south into northern Idaho and western Montana, into the Great Lakes region and New England, and south to Virginia. It reaches its highest abundance on Newfoundland Island and in British Columbia. Aptly named, the species is "a large wood warbler disguised as a thrush and exhibiting an extreme fondness for water" (Forbush 1929). The name *noveboracensis* means "from New York," a reference to the locality of the type specimen.

So similar are the behaviors and appearance of the Northern Waterthrush and Louisiana Waterthrush that Roberts's (1932) account of the former species began with a description of their shared traits, notably their love of forests, their close association with water, and their unique bobbing gait and tail wagging. Their similarities are so striking that even early ornithologists such as Alexander Wilson considered them the same species (Whitaker and Eaton 2020).

Physically, the Northern Waterthrush differs from the Louisiana Waterthrush by having a narrower eyebrow and a lightly spotted throat; many birds also have a pale yellow wash on the chest, underbelly, and in the line above the eye. Unfortunately, some Northerns lack the yellow plumage, making them even more difficult to distinguish from Louisianas. Roberts (1932) described the Northern's song as "clear, liquid, and ringing, and delivered with a vehemence

Photo by David Brislance.

that gives it an exceptional carrying quality." It begins with several short, staccato notes, compared with the longer, slurred notes that launch the Louisiana's song, followed by a more rapid warble. A somewhat similar flight song may be heard at dusk.

Despite their similarities, the two species are not aggressive toward one another in regions where their breeding ranges overlap, including east-central Minnesota. Small but significant differences in their habitat and foraging behaviors are among the traits that delineate their unique niches within the forest environment.

MINNESOTA BREEDING DISTRIBUTION

The Northern Waterthrush was described by Roberts (1932) as locally common throughout the coniferous and mixed forests of northern Minnesota, occurring as far west as Itasca State Park and as far south as Aitkin County. Although a nest had yet to be found when he published the first edition of *The Birds of Minnesota,* there were three reports of inferred breeding in Itasca County from 1923 to 1930. It would be the summer of 1934 before the first nest was discovered, at Caribou Lake in Cook County (Roberts 1936).

At the time, two different subspecies were recognized: Grinnell's Waterthrush (*Seiurus noveboracensis notabilis*), which occurred north and west of northern Michigan; and the Northern Waterthrush (*S. n. noveboracensis*), which occurred in the eastern United States and Canada. Grinnell's Waterthrush occurred throughout Minnesota, although Roberts acknowledged some specimens had the yellower plumage of the eastern race. Today these two subspecies are no longer recognized, and the Waterthrushes, originally considered related to Ovenbirds (*Seiurus aurocapilla*), were recently placed in their own genus, *Parkesia* (Chesser, Banks, et al. 2010).

When Green and Janssen (1975) published their updated account, the species' primary breeding range remained north and east of a line between Duluth and Leech Lake. Scattered records outside of this region included reports as far south as Pine County and as far west as Clearwater County. Nesting, however, had been confirmed only from the Arrowhead counties of Cook, Lake, and St. Louis.

Janssen (1987) noted new summer reports from the Cedar Creek area in Anoka County, including an adult seen carrying food in 1979 (Green 1980). Finding Waterthrush nests, however, continued to be difficult. By 1987, no confirmed breeding records had been documented since 1970. Nesting was finally confirmed in St. Louis County in 1995 (Wiens 1996).

Meanwhile, MBS field surveys documented what others were observing. Although the majority of the survey's 328 breeding-season locations were from the Arrowhead region, an increasing number of reports were located west and south of the species' primary range. These included reports from Kittson and Marshall Counties in the northwest and farther south in Otter Tail, Douglas, Stearns, Isanti, Anoka, and Chisago Counties.

During the MNBBA, the southern extent of the Northern Waterthrush's distribution was approximately 80 km north of the most southern records detected by the MBS **(A)**. Otherwise, the 876 records outlined a similar distribution, stretching from eastern Kittson County south to Todd County and east to southern Pine County. Reported from 10.3 percent (489/4,736) of the surveyed atlas blocks, breeding was confirmed in 10 blocks spanning Aitkin, Cook, Itasca, Lake, and St. Louis Counties. Probable

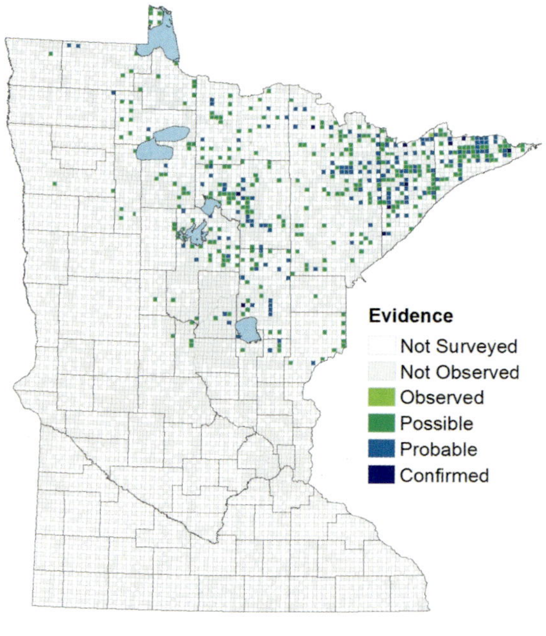

A. Breeding distribution of the Northern Waterthrush in Minnesota based on the Breeding Bird Atlas (2009–13).

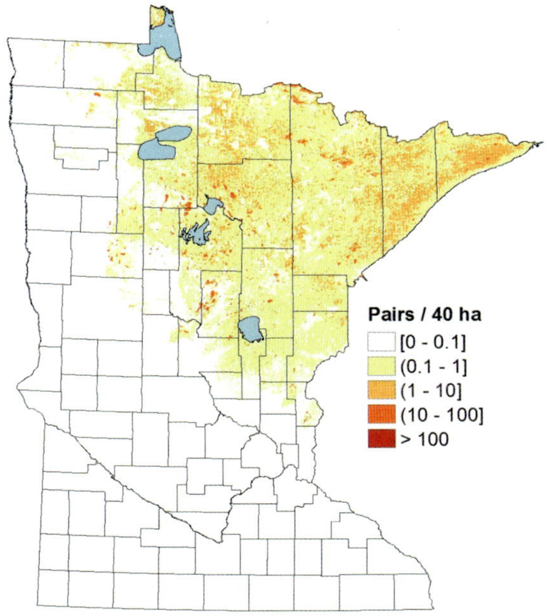

B. Predicted breeding distribution (pairs per 40 ha) of the Northern Waterthrush in Minnesota based on habitat, landscape context, and climate data gathered during the Minnesota Breeding Bird Atlas (2009–13) using the Generalized Linear Modeling method with an adjustment for detectability.

nesting was reported west to Marshall and Roseau Counties and south to Mille Lacs, Kanabec, and Pine Counties. The MNBBA probability map predicted varying breeding densities throughout the Laurentian Mixed Forest Province **(B)**. Moderate densities were most common in Cook and Lake Counties, with small pockets of higher densities farther west.

Although atlas records were absent south of Pine County, earlier MBS and MOU records clearly expanded the species' summer range farther south in the past few decades. Indeed, the species continues to be reported in Anoka County as recently as the summer of 2020 (MOU 2021b). The Waterthrush's presence in east-central Minnesota may fluctuate depending on water levels in wetlands, streams, and rivers, may signify a southern range extension, or, as proposed in Michigan, may simply represent a reoccupation of its former breeding range prior to European settlement (Chartier et al. 2011).

Nearly one hundred years ago, Roberts (1932) commented on the "wide stretch of country in Minnesota where there are no representatives of either species during the nesting season." Today, however, the ranges of the two waterthrushes overlap in a small region of east-central Minnesota, from southern Pine County south perhaps to Anoka, Chisago, and Isanti Counties (MOU 2021b).

Elsewhere, the species' distribution has had few changes (Whitaker and Eaton 2020). Recent atlas efforts in Wisconsin, however, have reported confirmed and probable breeding records from atlas blocks in western Wisconsin that are located at latitudes similar to the Twin Cities (eBird 2021). Although some Louisiana Waterthrushes may have been misidentified, the numerous reports raise the possibility of encountering breeding populations of Northern Waterthrushes even farther south in Minnesota.

BREEDING HABITAT

In the Great Lakes region and Canada, the Northern Waterthrush is associated with lowland hardwoods, conifer swamps, and alder and willow thickets. Where the soils are well drained, it is restricted to the riparian zones bordering streams, rivers, and lakes (Niemi, Howe, et al. 2016; Whitaker and Eaton 2020) **(C)**. Two of the most important habitat features are dense ground cover and the presence of surface water. Uprooted trees and jutting shoreline embankments provide small cavities or hollow areas for nesting, while downed logs along the shoreline are used to access shallow waters for foraging (Kaufman 1996; Whitaker and Eaton 2020).

Large forest tracts are another important Waterthrush requirement. One study concluded a minimum of 200 ha was needed, while the maximum probability of finding birds occurred when sites were larger than 3,000 ha (C. S. Robbins et al. 1989). Similarly, in southern Ontario, researchers found the species was more frequently encountered in extensively forested landscapes; its presence positively correlated with the proportion of forest cover within 2 km of a forest stand (Austen et al. 2001).

POPULATION ABUNDANCE

PIF (2021) generated a population estimate of 17 million breeding adults in the United States and Canada, with a statewide estimate of 43,000 birds.

C. Forested riparian shorelines in northern Minnesota are one of the most common habitats for finding nesting Northern Waterthrushes. Photo by Debbie Petersen.

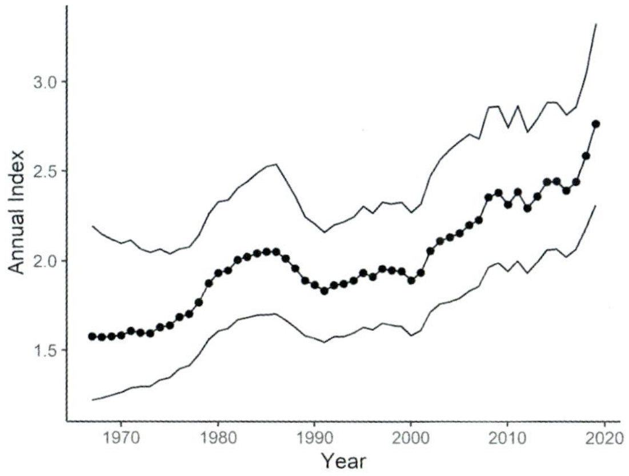

D. Surveywide breeding population trend and 95 percent credible interval (solid black lines) for the Northern Waterthrush based on the 1967–2019 Breeding Bird Survey (Sauer et al. 2020).

Estimates derived from the larger set of data collected by the MNBBA were highly variable but generated a median estimate that was orders of magnitude larger, at approximately 785,000 adults. Given the extensive availability of suitable aquatic habitats in Minnesota, the larger estimate is quite realistic.

Surveywide, BBS data demonstrated a significant increasing population trend, averaging 1.05 percent per year from 1966 to 2019 **(D)**. Because of the species' relatively low abundance in Minnesota, statistical analyses of BBS data lack precision but also suggest an increasing population (Sauer et al. 2020). Indeed, long-term monitoring of forest birds in the Chippewa and Superior National Forests from 1995 to 2020 demonstrate an increasing trend in both forests (Grinde, Bednar, et al. 2020).

CONSERVATION

Although the Northern Waterthrush is not a conservation priority (PIF 2021), challenges remain. For example, nearly 60 percent of its breeding population occurs in Canada's boreal forest, where landscape changes spurred by hydropower development, mineral extraction, and oil and gas development are a concern. Farther south, collisions with communication towers are a source of mortality (Whitaker and Eaton 2020). And, although climate change models project a low vulnerability for the species' North American breeding distribution under a 2°C warming scenario (NAS 2019), the species' future in Minnesota is not so bright. As temperatures rise, the Waterthrush will gradually withdraw northward until, under a 3°C warming scenario, models project its extirpation from Minnesota's northern landscape by 2080. Efforts to protect riparian zones around the region's lakes, streams, and rivers may be the most critical measure to help buffer such impacts.

GOLDEN-WINGED WARBLER

Vermivora chrysoptera

Minnesota Seasonal Status: Regular breeding resident and migrant. The Golden-winged Warbler was common during the MNBBA.
Migration: Medium- to long-distance migrant that winters in Central America and northern South America. Many of Minnesota's breeding birds overwinter in Honduras, Nicaragua, and Costa Rica (G. R. Kramer et al. 2017).
Food: Foliage gleaner that primarily consumes insects, spiders, and snails.
Nest: Cup nest located on or slightly above the ground and supported by stalks of surrounding vegetation.
Conservation Concern: Assigned a high PIF Continental Concern Score of 16/20.

Photo by Mike Lentz.

Minnesota provides critical habitat for many iconic and charismatic birds, including the Common Loon, Bald Eagle, and Trumpeter Swan. But it is the "dainty and diminutive" Golden-winged Warbler that may be the most extraordinary bird that graces our home state during its brief summer stay. The species is easily overlooked despite its striking plumage and unmistakable buzzy songs, and survived in relative obscurity until the late twentieth century. Even early ornithologists such as John James Audubon, Thomas Nuttall, and Alexander Wilson "had no idea where it bred" and thought it was simply a rare migrant passing through en route to its Canadian breeding grounds (Bent 1953). But the species' dramatic decline in the late 1900s catapulted it to new fame as a national symbol of the plight of Neotropical migrants.

With the species now teetering on the brink of extirpation throughout much of its breeding range, Minnesota's population of Golden-winged Warblers is, by contrast, faring quite well. Minnesota supports approximately 56 percent of the Warbler's entire population (PIF 2021), making it the state's premiere Stewardship Species (Pfannmuller 2012). The only songbird that comes even close is the Sedge Wren, with 17 percent of its global population residing in Minnesota. Clearly, the Golden-winged Warbler is in a class by itself and presents state resource managers with a unique responsibility for its conservation, management, and ultimate survival.

The species' common name provides the best identification clue for both the male and female. Prominent yellow-gold wing patches contrast sharply with their gray wings, back, and central tail feathers. Indeed, the Latin term *chrysoptera* means "gold-colored wing." The male is further distinguished by his black throat, black eye patch, and golden crown, each separated by patches of white. The female's yellow crown is tinged with olive, and her eye patch and throat are gray rather than black.

Unlike the brilliance of its plumage, the common song is rather inconspicuous and insect-like. Tyler described it as "composed of four notes, the first prolonged, and followed, after an almost imperceptible pause, by three shorter notes on a lower pitch. . . . [It] might be written *zeee, zer-zer-zer*" (Bent 1953). Like many warblers, the Golden-winged also has a second primary song that "sounds like a sibilant, rapid stutter followed by a lower buzzy note" (Confer et al. 2020).

Formerly found from southern Manitoba south through the Great Lakes states, and east to southern New England and the Appalachian Mountains, the species' breeding range is now divided into two primary population centers: a Great Lakes population that stretches from central Manitoba across the northern Great Lakes states to southern Quebec; and an Appalachian Mountain population that stretches from southeastern New York south to northeastern Georgia **(A)**.

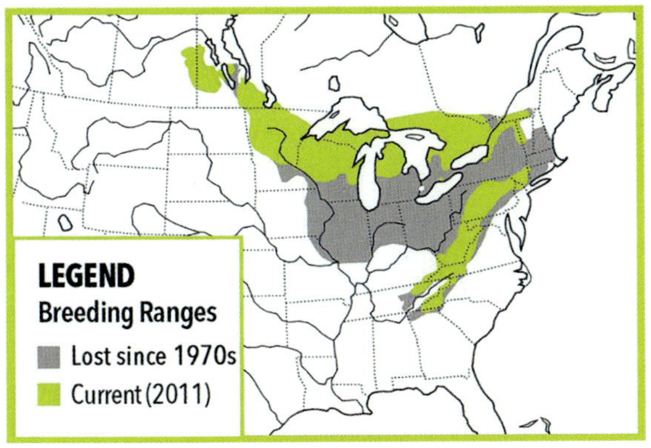

A. Former and current breeding range of the Golden-winged Warbler in North America. Reprinted from https://www.birds.cornell.edu/home/golden-winged-warbler-conservation-strategy-and-resources.

MINNESOTA BREEDING DISTRIBUTION

The Golden-winged Warbler was first documented nesting in Minnesota in 1878 in south Minneapolis (Svingen and Hertzel 2015). Roberts (1932) considered it a regular breeding species in the early 1900s from Lake Itasca east to Mille Lacs Lake and south across the Big Woods of Hennepin, Wright, and Stearns Counties. He also reported that it occurred, "though rarely," as far south as Fillmore County. The species always had a special charm for Roberts, due in part to his fond memories of first encountering it near the Twin Cities when he was a young man in the late 1800s. He wrote, "There was always, every springtime, a pair of these Warblers close by, and almost within reach of, the spray of the Laughing Waters [Minnehaha Falls], and several other pairs made their summer homes in the heavy timber south of Lake Calhoun."

These sites were gone by the early 1900s as the Twin Cities grew into a bustling metropolis. Roberts (1932) wrote, "It is now apparently a rare summer resident in what is left of the hardwood timber in the southern part of the state. All recent records in the nesting-season are from the coniferous forests or close to their southern border." Confirmed nesting records were limited to Becker and Hennepin Counties, while breeding was inferred in Mille Lacs and Sherburne Counties. Svingen and Hertzel (2015) provided additional details regarding the species' early presence in the state.

Years later, Green and Janssen (1975) discussed the species' distribution at length: "It is difficult to document any change in the range of the Golden-winged Warbler because ornithological exploration has been uneven, with the result that small local colonies may have been overlooked." They questioned Roberts's assertion that the species was a rare summer resident in the deciduous woodlands south of the Twin Cities. His sole evidence was a bird J. C. Hvoslef shot in Lanesboro in mid-June 1888, accompanied by a note that this was the fourth bird he had observed there. Svingen and Hertzel (2015) suggested that the June record may have been a late migrant rather than a breeding bird. Hvoslef's record aside, Green and Janssen (1975) noted two additional, albeit poorly documented, nesting records in southeastern Minnesota: the first in Winona County (1934) and the second in Olmsted County (1954).

Although its former status in the southeast is uncertain, changes appeared to be underway farther north. Other than brief references Roberts (1932) made to the species' presence in Otter Tail and Aitkin Counties in 1893, Cass County in 1903, Cass Lake in 1927, and in or near Itasca State Park intermittently from 1919 to 1929, there was little documentation from northern Minnesota in the early 1900s. Yet by the mid-1900s Golden-winged Warblers were frequently reported throughout north-central Minnesota. Indeed, Green and Janssen (1975) noted that in the twenty years prior to their publication, all but one summer specimen deposited in the collections at the Bell Museum were from north-central counties. The authors concluded that the species may have expanded northward and delineated a breeding range that extended from southern St. Louis County, northwest to Clearwater and Becker Counties, and southeast to Sherburne, Anoka, and Chisago Counties. In addition to Roberts's (1932) confirmed nesting reports in Becker and Hennepin Counties, they added Clearwater and Hubbard Counties. The expansion northward continued when Janssen (1987) reported on summer records from southern St. Louis, Lake, and Cook Counties in the northeast and from Itasca, southern Koochiching, Beltrami, and Lake of the Woods Counties in north-central Minnesota. His breeding distribution map, however, excluded much of northeastern Minnesota, east of a line from northern Lake of the Woods County southeast to southwestern Lake County. Since 1970, counties with confirmed nesting included Aitkin, Anoka, Clearwater, Crow Wing, Hubbard, Mille Lacs, and St. Louis. Hertzel and Janssen (1998) later added Becker County to the list.

By 2014, the MBS had documented 547 breeding-season locations for the Golden-winged Warbler. Although far north-central Minnesota had yet to be surveyed, the records extended the species' distribution farther west to include eastern Marshall and Roseau Counties in the northwest and portions of Douglas, Otter Tail, Stearns, and Todd Counties in central Minnesota.

MNBBA records further documented the species' broad distribution throughout the Laurentian Mixed Forest Province (except Cook County), the northern reaches of the Eastern Broadleaf Forest Province, and the far eastern region of the Tallgrass Aspen Parklands Province **(B)**. Although the extent of its distribution was similar to that delineated by the MBS, there were numerous possible and probable records along the Canadian border, including several as far west as north-central Roseau County, and a possible breeding record as far south as northern Hennepin County. Indeed, the species was especially abundant from Clearwater, southern Beltrami, and western Itasca Counites south to Sherburne County and east to Chisago County. Extensive documentation of Golden-winged Warblers in this region is due to the excellent work by local MNBBA participants and the existence of prime breeding habitat.

Overall, observers reported 1,329 Golden-winged Warbler detections in 16.9 percent (805/4,759) of the surveyed atlas blocks. Reported in 31 of Minnesota's 87 counties, breeding was confirmed in 14 counties, all within the species' primary range. Eight of the counties were additions to Hertzel and Janssen's (1998) list: Beltrami, Cass, Itasca, Kanabec, Lake, Morrison, Pine, and Sherburne. Since the atlas was completed, nesting has also been confirmed in Otter Tail (2015), Carlton (2018), and Lake of the Woods Counties (2019) (MOU 2021b).

The species' MNBBA probability distribution map predicted that the most suitable habitat for the Golden-winged Warbler in Minnesota, where it reaches its highest breeding densities, is found in the southern and western reaches of the Laurentian

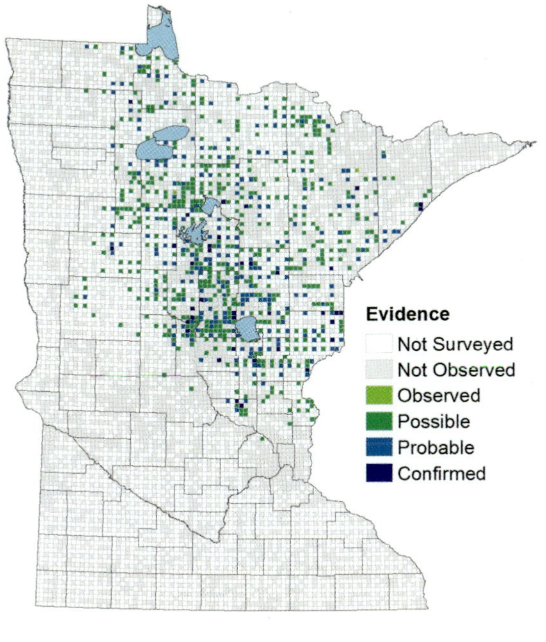

B. Breeding distribution of the Golden-winged Warbler in Minnesota based on the Breeding Bird Atlas (2009–13).

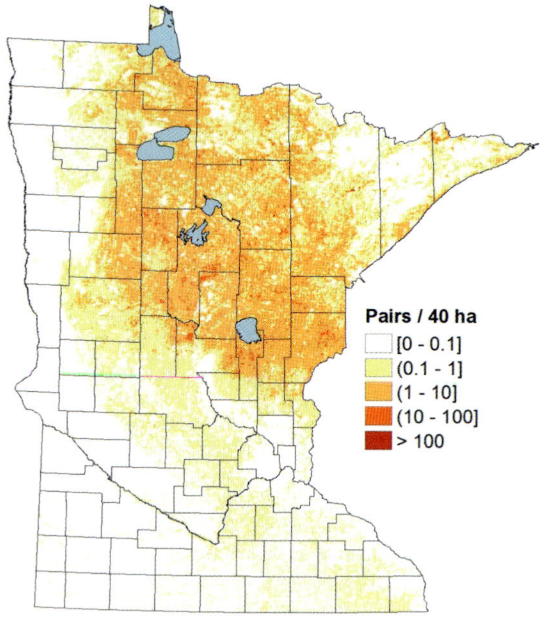

C. Predicted breeding distribution (pairs per 40 ha) of the Golden-winged Warbler in Minnesota based on habitat, landscape context, and climate data gathered during the Minnesota Breeding Bird Atlas (2009–13) using the Generalized Linear Modeling method with an adjustment for detectability.

Mixed Forest Province **(C)**. This relatively small region, especially the area from Hubbard, Cass, and Itasca Counties southeast to Mille Lacs and Pine Counties, supports the very core of the species' global population. In the far northeastern counties of Lake and Cook, the highest probability of occurrence was along the North Shore of Lake Superior and in the young growth of the 2007 Ham Lake Fire. Very low breeding densities were predicted in much of the Eastern Broadleaf Forest Province, where suitable habitat may be present but is likely occupied by its near relative, the Blue-winged Warbler.

Changes to the Golden-winged Warbler's breeding distribution in Minnesota are well-documented by a series of maps presented by Svingen and Hertzel (2015). Similar changes were witnessed throughout the species' breeding range. Logging of the eastern deciduous forest in the late 1800s and early 1900s created an abundance of new shrubland habitat for this original inhabitant of the Ohio River valley, southern New England, and Appalachian Mountains. Populations gradually pushed farther north into New England and the Upper Midwest. The Warbler first appeared in southern Ontario in the 1930s and was widespread throughout New England by the mid-1900s (Confer et al. 2020).

But, as human population centers continued to expand, fires were suppressed, and open woodlands gave way to intensive agricultural practices, the Warbler's range began to contract northward. Before the end of the twentieth century, Golden-winged Warblers were largely gone from southern Wisconsin, Illinois, Indiana, Ohio, and southern New England. In the next ten years the species' range further contracted in central Michigan, southern Ontario, New York, and throughout the Appalachian Mountains. Indeed, during the past "150 years, the range of the Golden-winged Warbler has geographically shifted more than that of almost any other avian species" (Golden-winged Warbler Working Group 2013). Habitat changes prompted much of the shift, but competition and hybridization with the Blue-winged Warbler also played a contributing role (Confer et al. 2020; Roth et al. 2012).

BREEDING HABITAT

Originally dubbed the "golden-winged swamp warbler" by Audubon (Helmberger 2012), the species was long considered an exclusive inhabitant of shrubby wetlands. But young, open upland stands also provide suitable habitat, especially when associated or intermixed with wetlands. Indeed, the species' use of early successional stands, particularly young aspen in the Great Lakes region, made it a poster child for protection and maintenance of young forest habitat beginning in the 1990s and overshadowed its equally important dependence on wetlands (Hanowski 2002).

Many studies have since demonstrated that Golden-winged Warblers utilize a variety of open upland and lowland habitats. Territories are often adjacent to a forest edge and include patches of forbs, grasses, shrubs, and scattered trees **(D)**. Patchiness and structural complexity of stands are particularly important (Golden-winged Warbler Working Group 2013). In northern Minnesota and Wisconsin, the species is strongly associated with lowland shrubs and sedge wetlands followed by open wetlands and regenerating stands (Niemi, Howe, et al. 2016). All open habitats, however, need to have scattered shrubs and trees to support Golden-winged Warblers. Where both Blue-winged Warblers and Golden-wings occur, the Golden-wings show a preference for wetter habitats (Will 1986).

Habitat selection, however, is more complex when examined throughout the nesting season. Until recently, assertions regarding habitat use were based largely on the location of singing males. When Streby and his colleagues (2012; 2015) used tiny radios to follow adults and young, they discovered the birds were also using adjacent forests. Females often nested in forest stands, and the young, once they fledged, spent significant time there, demonstrating that a matrix of cover types and ages, including shrubby wetland openings, are important elements of suitable habitat.

POPULATION ABUNDANCE

PIF (2021) generated a population estimate of 390,000 breeding adults in the United States and southeastern Canada and a statewide population of 220,000. When MNBBA data were used to estimate the size of the state population, the result was significantly larger, numbering 1.2 million; the estimate, however, had a large confidence interval. Nevertheless, the results suggest that both the state and national populations may be significantly larger than indicated by previous estimates. Regardless of its ultimate size, Minnesota is clearly the core of the species' North American breeding population, leading to its designation as a Minnesota Stewardship Species by Audubon Minnesota (Pfannmuller 2012).

Minnesota is also the only state where BBS data are robust enough to generate a reliable population trend. Since 1967, the population has remained fairly stable **(E)**. The trend is further corroborated by intensive monitoring data collected from the Chippewa and Superior National Forests, where the species has demonstrated relatively stable populations from 1995 through 2020 (Grinde, Bednar, et al. 2020). Unfortunately, this is in stark contrast

D. Shrubby wetland habitats near forest edges are a common nesting habitat for the Golden-winged Warbler in Minnesota. Photo by Lee A. Pfannmuller.

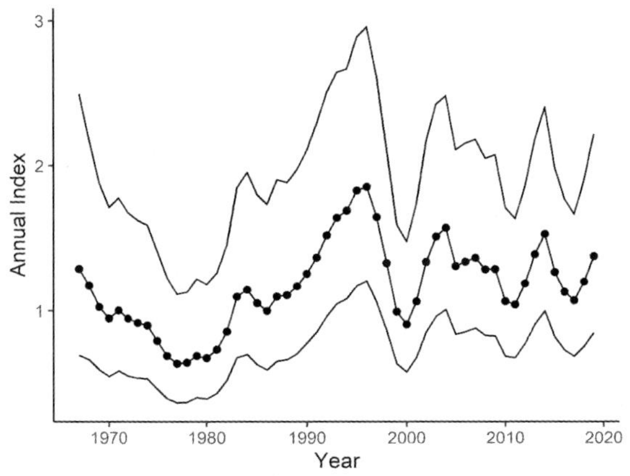

E. Breeding population trend and 95 percent credible interval (solid black lines) for the Golden-winged Warbler in Minnesota based on the 1967–2019 Breeding Bird Survey (Sauer et al. 2020).

to a declining population trend that averages −1.85 percent per year across the BBS survey region (Sauer et al. 2020). Indeed, there are no states or provinces where Golden-winged Warblers demonstrate a statistically significant increasing population trend. Among the twenty states and provinces that report Golden-winged Warblers along BBS routes, 10 reported statistically significant declining populations from 1966 to 2019, ranging from an average low of −2.39 percent per year in Wisconsin to a high of −7.33 percent per year in Tennessee (Sauer et al. 2020). Overall, biologists estimate that the population has declined more than 60 percent from 1970 to 2014 (Rosenberg, Kennedy, et al. 2016). Losses have been greatest within the Appalachian Mountains Bird Conservation Region, where numbers have declined by 98 percent (Roth et al. 2012).

Factors responsible for the species' range-wide decline are many. Although predation, cowbird parasitism, and competition and hybridization with the Blue-winged Warbler have added to the conservation challenges, widespread loss of both breeding and wintering habitat are considered the primary contributors to the decline (Confer et al. 2020).

CONSERVATION

When PIF began focusing conservation efforts on Neotropical songbirds in the early 1990s, the Golden-winged Warbler became one of the initiative's early mascots. The Golden-winged Warbler [GWW] Working Group was established in 1999 and launched the GWW Atlas Project. Led by the Cornell Lab of Ornithology, the atlas was designed to assess the distribution and habitat requirements of Golden-winged Warblers, Blue-Winged Warblers, and their hybrids. This and other research studies became the foundation for a major Status Assessment and Conservation Plan published in 2012 (Roth et al. 2012).

Coincidentally, PIF assigned the Warbler a high Continental Concern Score and placed it on their Red Watch List, a group of 19 species with restricted distributions and small, declining populations (Rosenberg, Kennedy, et al. 2016). In 2010, conservation organizations also petitioned the US Fish and Wildlife Service to list the species as *Endangered* or *Threatened* under the federal Endangered Species Act. Federal biologists are conducting a more thorough review to assess if listing is warranted (USFWS 2011).

Guidelines for managing breeding habitat have been developed by the Golden-winged Warbler Working Group (2013). Management guidelines for the Great Lakes region identified 16 focal areas where efforts should be concentrated, including north-central Minnesota and northwestern Wisconsin and the Aspen Parklands of northwestern Minnesota and southeastern Manitoba. The guidelines emphasize the importance of establishing habitat patches that have interspersed clumps of shrubs and/or saplings, an open canopy, and adjacency to forest stands.

Perhaps the most interesting, albeit challenging, aspect of the Golden-winged Warbler's life history and conservation is its relationship with its relative, the Blue-winged Warbler, a species found south of the Golden-winged Warbler's breeding range. Scientists believe the two species originated from a geographic split of a single population during the most recent glaciation nearly 10,000 years ago. Their time apart, however, was not long enough to allow sufficient differentiation to prevent successful interbreeding. So, when the Blue-winged Warbler began expanding north into the Golden-wing's breeding range in the 1800s, fertile, first-generation hybrids were produced. Known as Brewster's Warblers, the hybrids are encountered regularly in the area of overlap (Will 2009). Another variant, Lawrence's Warbler, results when the Brewster's Warbler mates with either a Golden-winged Warbler or Blue-winged Warbler. Volumes have been written over the decades about these hybrids and their parents. Indeed, D. H. Morse (1989) wrote that "they are probably the most intensively studied pair of hybridizing bird species in North America."

An equally important concern is the "genetic dilution" that results from hybridization (Confer et al. 2020). Even today, birds that appear to be "pure" representatives of either species carry genetic material from the other. In Minnesota, for example, one of 96 phenotypically "pure" Golden-winged Warblers that were sampled showed evidence of genetic interbreeding with Blue-winged Warblers (Vallender et al. 2009). The study's authors cautioned that it was only a matter of time before a larger percentage of the state's Golden-winged population becomes less "genetically pure." Indeed, recent genetic analyses suggest that the two species, if they can even be considered two separate species, have had a long history of successive episodes of isolation and sympatry. There appear to be only a half-dozen nuclear genes that clearly distinguish the two forms from each other (Toews et al. 2016).

Competition with the Blue-winged Warbler has further compounded problems because the two species utilize similar habitats and likely compete for available resources in areas of overlap. There is evidence that Golden-wings are often extirpated from an area within fifty years of when Blue-wings first arrive (Gill 1980). If Golden-winged Warblers keep retreating northward, this would not be a problem, but the amount of suitable habitat north of southern Manitoba seems limited at present.

Minnesota provides an excellent example of the interactions between the two species as the Blue-winged Warbler continues to expand its range northward since it first appeared in southeastern Minnesota in the late 1800s (Roberts 1932). Svingen and Hertzel (2015) reviewed how the statewide breeding range of each species has changed over time and provided a detailed account of all known hybrid records. Although many variables influence the number of reports received, their work strongly suggests that an increased rate of hybridization is occurring in the state.

Unfortunately, another challenge awaits the species. Under a 2°C warming scenario, NAS (2019) projected a moderate vulnerability for the Warbler's North American breeding distribution, losing 89 percent of its current breeding range, which will be compensated by a 121 percent range gain in northern Canada by the year 2050. Should temperatures rise 3°C, the species would experience a total loss of its current breeding distribution compensated by only a 61 percent gain northward by the year 2080. In Minnesota, even an imminent 1.5°C warming scenario is predicted to result in the species' disappearance from approximately half of its current breeding range in the state. Clearly climate change, regardless of its magnitude, will impact this signature species.

The Golden-winged Warbler presents resource managers with rather extraordinary and thought-provoking conservation challenges. Yet no matter how the future plays out, protecting the matrix of different forest cover types used by Minnesota's population is critical to ensuring that the state continues to provide suitable habitat for this stunning little Warbler that is so preciously abundant in our state.

BLUE-WINGED WARBLER

Vermivora cyanoptera

Minnesota Seasonal Status: Regular breeding species and migrant. The Blue-winged Warbler was uncommon during the MNBBA.
Migration: Medium-distance migrant that winters in the Caribbean, Mexico, and Central America.
Food: Insects and spiders secured by foliage gleaning and probing.
Nest: Cup nest placed on or near the ground at the base of a shrub or a tussock of herbaceous vegetation.
Conservation Concern: Assigned a moderate PIF Continental Concern Score of 13/20.

The Blue-winged Warbler is a relative newcomer to Minnesota, arriving in the 1880s as it gradually expanded its range north along the Mississippi River (Roberts 1932). The male's bright yellow crown and underparts, his bluish-gray wings with two white wing bars, and his thin black eyeline are diagnostic. With similar features, the female's plumage lacks the male's brilliance, and her wing bars are less prominent. The most common song is a two-part, buzzy vocalization, *beeeee bzzzzz,* the second half usually at a lower pitch than the first. An alternative song begins with a short note followed by a trill and low buzz, *wee-chi-chi-chi-chur*. After encountering the species, Roberts (1932) concluded that it "is so diminutive, so quiet and unobtrusive in its habits, and its song is so weak and insect-like that it may easily be overlooked."

The species was originally named the Blue-winged Yellow Warbler, and the first specimens collected in the 1700s were actually ascribed to the Pine Warbler, another wood warbler with blue wings and a yellow body. Although the misidentification was soon remedied, the Latin name for the species, *pinus,* remained for another two centuries until it was changed in 2010 to *cyanoptera*. Closely related to the Golden-winged Warbler, the two species hybridize in areas where their ranges overlap, producing fertile offspring.

Formerly an inhabitant of the Ozark Mountains east through the open woodlands of Tennessee, Kentucky, northern Alabama, and northern Georgia, Blue-winged Warblers have been expanding northward. They are now found throughout southern New England, southern Ontario, and southern regions of the Great Lakes states.

Photo by Mike Lentz.

MINNESOTA BREEDING DISTRIBUTION

The Warbler's first appearance in Minnesota occurred in May 1880, when Roberts (1932) found a male at Minnehaha Falls in Hennepin County. A few years later, in 1885, Dr. Hvoslef found the species near the town of Lanesboro in Fillmore County. In 1889 he found a family of four in the same area and the first nest in 1891. Reports were logged from another site in Fillmore County in 1916 (near the town of Preston), again in Hennepin County in 1919, in Winona County in 1922, and in Goodhue County in 1923. Detailing these accounts, Roberts summarized the species' distribution as restricted to "a narrow area along the southeastern border of the state."

Forty years later, Blue-winged Warblers were reported as a summer resident as far north as Dakota and Washington Counties by Green and Janssen (1975). Along with Roberts's record of confirmed nesting in Fillmore County, they added nesting confirmation in Goodhue, Olmsted, Washington, and Winona Counties. Reports of spring migrants were available farther north in Anoka, Sherburne, Stearns, and Wright Counties and to the west in Kandiyohi and Nicollet Counties, suggesting the bird might be more widely distributed during the breeding season. Coincident with the expansion was an increase in the number of reports of Brewster's and Lawrence's Warblers, both hybrids between the Blue-winged Warbler and Golden-winged Warbler. Most reports of hybrids were from Goodhue, Hennepin, and Winona Counties.

Janssen's 1987 update identified six counties where nesting had been confirmed since 1970: Hennepin, Houston, Olmsted, Scott, Washington, and Winona. Summer observations were now being reported along the Minnesota River valley in Scott and Nicollet Counties, and as far north as Stearns and Otter Tail Counties. Nesting was soon confirmed in Dakota and Fillmore Counties as well (Hertzel and Janssen 1998).

By 2014, the MBS documented 107 Blue-winged Warbler breeding-season locations. Although most records were restricted to the southeastern corner of the state, scattered reports were documented as far north as Morrison County in central Minnesota, and as far west as Lincoln County in southwestern Minnesota. The species was well represented along the lower Minnesota River valley from Shakopee to Mankato.

Additional northern records were documented when MNBBA participants reported 213 Blue-winged Warbler detections in 2.7 percent (129/4,737) of the surveyed atlas blocks. The most northerly reports came from southern Cass County and northern Mille Lacs County; the most westerly record was documented near the South Dakota border in Big Stone County. Breeding was confirmed in 11 blocks and included three records north of the Twin Cities in Anoka, Sherburne, and southern Pine Counties **(A)**. The birds were detected in 33 of Minnesota's 87 counties and were confirmed breeding in 10 counties; probable breeding was reported in an additional 13 counties. Six counties with confirmed breeding were new to the list published by Hertzel and Janssen (1998): Anoka, Carver, Pine, Rice, Sherburne, and Wright.

In their excellent review of the changing Minnesota distribution of the Blue-winged Warbler, the Golden-winged Warbler, and their hybrids, Svingen and Hertzel (2015) examined how the two species have come into increasingly frequent contact over the years. Their exhaustive review of both published and unpublished records found a total of 125 hybrid records in Minnesota since 1945; 108 of the records were from the southern half of the state. Unfortunately, as the Blue-winged Warbler continues to move northward, hybridization is likely to pose an

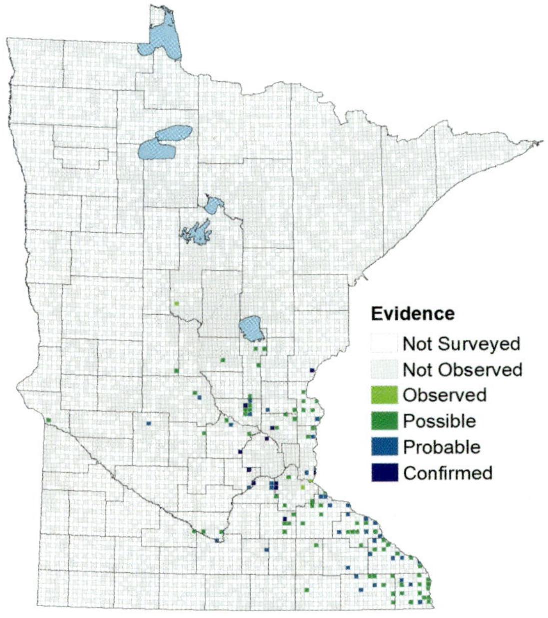

A. Breeding distribution of the Blue-winged Warbler in Minnesota based on the Breeding Bird Atlas (2009–13).

increasing threat to Minnesota's Golden-winged Warbler population (Vallender et al. 2009). Studies have shown that, within fifty years of Blue-winged Warblers occupying Golden-winged Warbler habitats, the latter species was likely to become locally extinct (Gill 1980). Svingen and Hertzel (2015) discuss the potential for a similar scenario in Minnesota.

The Blue-winged Warbler's gradual range expansion in Minnesota is simply a continuation of a much broader movement north that began in many eastern states in the late 1800s and early 1900s (Gill et al. 2020).

BREEDING HABITAT

Often described as a habitat generalist, the Blue-winged Warbler may be found in shrubby uplands, weedy fencerows, right-of ways, young successional growth, and forest edges **(B)**. Most sites are open, with scattered thickets of dense shrubs and young trees. Dense ground cover is common (Gill et al. 2020; Potter et al. 2007). J. L. Dunn and Garrett (1997) note that farther south the birds are frequently found on dry, brushy hillsides; farther north they are more common in mesic sites, including streamside thickets and lowland hardwoods. The species is not considered to be area sensitive (Potter et al. 2007).

Although the habitats of the Blue-winged Warbler and Golden-winged Warbler are similar in many respects, the Blue-winged Warbler tends to occupy a wider range of sites, tolerating stages of forest succession that are older than those tolerated by the Golden-winged Warbler (J. L. Dunn and Garrett 1997). In an area where the species' ranges overlapped in New York, Blue-winged Warblers occupied sites with less herbaceous cover and 25 percent more tree cover than Golden-winged Warblers (Gill et al. 2020).

POPULATION ABUNDANCE

Only 680,000 breeding adults are estimated to reside in the eastern United States and southeastern Canada, with Minnesota supporting approximately 23,000 birds (PIF 2021). Given its low abundance on BBS routes, the Warbler's population trend lacks statistical precision across the BBS survey region. Nevertheless, data suggest the population has declined an average of −0.67 percent per year from 1966 to 2019 **(C)**, with a cumulative decline of −22 percent between 1970 and 2014 (Rosenberg, Kennedy, et al. 2016). Although Blue-winged Warblers increased in number during the survey's early years, since 1996 they have sharply declined. Indeed, the species is largely declining throughout the core of its breeding distribution but stable to increasing along its northern periphery, including in the Boreal Hardwood Transition and the Prairie Hardwood Transition Bird Conservation Regions of the northern United States and southeastern Canada. This is evident in Minnesota as well, where the sparse BBS data suggest a nonsignificant increase averaging 0.94 percent per year (Sauer et al. 2020).

In the 1800s, when forests throughout the eastern deciduous forest biome were harvested, habitat opportunities for the Blue-winged Warbler were abundant. Favorable conditions likely continued as forest clearings gave way to small farmsteads. Today, such wide-scale opportunities are limited. Suitable habitat has often been replaced by development, intensive agricultural operations, or mature forests. The exception may be at the periphery of its breeding range, where the Warbler can continue to exploit sites not previously occupied. Unfortunately, in northern Minnesota, this brings it into direct contact with the Golden-winged Warbler.

B. Open woodlands with abundant shrubs and young trees are the Blue-winged Warbler's preferred habitat in southern Minnesota. Photo by Lee A. Pfannmuller.

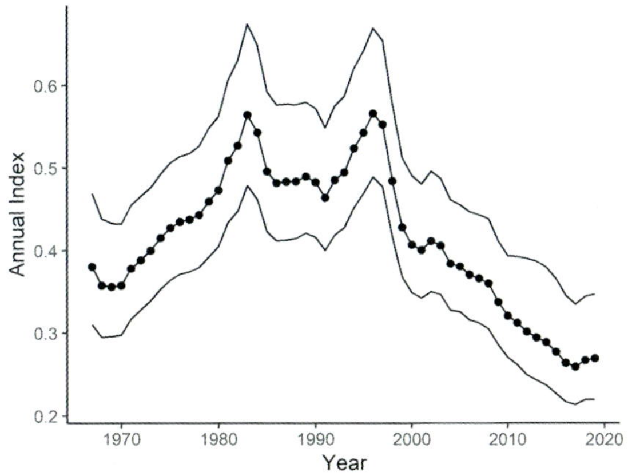

C. Surveywide breeding population trend and 95 percent credible interval (solid black lines) for the Blue-winged Warbler based on the 1967–2019 Breeding Bird Survey (Sauer et al. 2020).

CONSERVATION

In light of its restricted breeding range and small population size, the Blue-winged Warbler is a moderate conservation priority (PIF 2021). Nevertheless, much of the conservation focus is less on the Blue-winged Warbler than it is on the impacts of an expanding Blue-winged Warbler population on its near relative, the Golden-winged Warbler.

Despite its ability to capitalize on the establishment of new forest openings and edge habitats, the species may have some pressing challenges in the years ahead. NAS (2019) projected a moderate vulnerability for the species' North American breeding distribution under all three warming scenarios examined. As temperatures rise, it is predicted to expand northward into some of the same areas where the Golden-winged Warbler's breeding range is initially predicted to expand under the 1.5°C and 2°C scenarios, further exacerbating potential conflicts. Indeed, in Minnesota, the Blue-winged Warbler will likely replace the retreating Golden-winged. Other hazards also pose a risk to the species. For example, its small population is especially vulnerable to collisions with communication towers during migration, with annual mortality totaling nearly 4,000 birds (Longcore et al. 2013).

In the short-term, as we watch with some trepidation for the impacts that its northward journey will have on Minnesota's Golden-winged Warbler population, the Blue-winged Warbler's unique *bee-bzzz* song is likely to be heard at even more localities across the state.

BLACK-AND-WHITE WARBLER

Mniotilta varia

Minnesota Seasonal Status: Regular breeding resident and migrant. The Black-and-white Warbler was common during the MNBBA.
Migration: Short- to long-distance migrant that winters from the southeastern United States and Caribbean south to northern South America.
Food: Bark-forager and occasional foliage gleaner that consumes a variety of arthropods, especially caterpillars.
Nest: Cup nest placed on the ground at the base of a shrub or tree; occasionally nests in a stump cavity.
Conservation Concern: Assigned a moderate PIF Continental Concern Score of 11/20.

A common inhabitant of Minnesota's northern forests, the Black-and-white Warbler has an unmistakable zebra-like plumage. Often seen nimbly traversing tree trunks and large branches as it forages among the bark's furrows, the species is uniquely adapted for its unwarbler-like lifestyle. Each foot has an elongated hind claw, the legs are heavier and shorter than other warblers, and its thin bill curves slightly downward (J. L. Dunn and Garrett 1997). Behaving more like a nuthatch or creeper, the little Warbler "goes merrily along through life entirely oblivious of any inconsistency in its existence" (Roberts 1932). The Latin name *Mniotilta* translates to "moss-plucker," a reference to its bark foraging habits.

The primary difference between the sexes is that the black feathers in front of and behind the male's eyes are replaced with gray on the female, and her breast streaking is much lighter. Only during migration, when the similar-looking Blackpoll Warbler passes through the state, might a closer look be necessary. But the two species are easily distinguished by the Black-and-white Warbler's striped crown, which is visible year-round. The male Blackpoll Warbler has a black crown during the breeding season, and, like the female Blackpoll, a faintly streaked crown the remainder of the year.

Sounding like a squeaky wheel, the Black-and-white's primary song, *we-see, we-see, we-see,* is weak, high-pitched, and rhythmically delivered. Though it is easy to identify, like most warblers the species has an alternate song that is long and variable and call notes that can challenge identification when the

Photo by David Brislance.

bird is not visible. One early ornithologist wrote: "I know of no other warbler, except the Chat that can produce so great a variety of sounds" (Allison 1907).

This unique Warbler is found across the boreal forests of Canada, the Great Lakes region, and New England, south through the Appalachian Mountains, and southwest to central Texas. Some of its highest breeding densities occur in northern Minnesota.

MINNESOTA BREEDING DISTRIBUTION

Although the Black-and-white Warbler was formerly present across the forests of southern Minnesota, by the early 1900s Roberts (1932) found it largely restricted to Minnesota's northern landscape. "Since the cutting of the deciduous forests," he wrote, "it has become rare as a nesting bird south of the evergreen forests." Breeding evidence in the region was limited to a 1904 report of an adult feeding young in Wabasha County, and a note of a bird or pair "evidently with nest" that he observed in Minneapolis in 1875. In northern Minnesota, he considered the species to be regular but "nowhere exactly common." Confirmed nesting records were limited to the Mille Lacs region and Becker County. Nesting was inferred near Cass Lake and in Cook, Hennepin, Roseau, and Wabasha Counties.

Green and Janssen (1975) later described the species as a summer resident primarily in northeastern and north-central Minnesota. Farther south, it occurred east of the Mississippi River but was "quite scarce south and west of Mille Lacs Lake." In addition to Roberts's Becker County nesting record, nesting was now confirmed in Cook, Lake, and St. Louis Counties and, to the south, in Anoka County. More than fifty years had passed, however, since it had been observed in southeastern or northwestern Minnesota. But new reports soon led Janssen (1987) to suggest the Warbler's breeding range might extend south to northern Hennepin County and west to Roseau and Kittson Counties. He also identified confirmed nesting records since 1970 from Aitkin, Anoka, Cass, Clearwater, Cook, Itasca, Lake, and St. Louis Counties.

Then, in 1990, a pair was found nesting in Winona County, the first confirmed breeding record in southeastern Minnesota in more than seventy-five years (Wiens 1991). Nevertheless, despite extensive surveys of remnant forest tracts in the region by the MBS in the 1990s, no additional breeding records were reported. Yet, in northwestern Minnesota, the MBS found the species in numerous locations, from Todd and Otter Tail Counties north to Marshall, Kittson, and Roseau Counties.

The MNBBA's 3,423 Black-and-white Warbler records further expanded the species' occurrence in southern Minnesota **(A)**. There were not only scattered records throughout the Twin Cities Metropolitan Region but also two confirmed breeding records in southeastern Minnesota: a bird carrying nesting material in Fillmore County (May 10, 2012) and an adult feeding young in Goodhue County (August 20, 2010). Although several records in southern Minnesota were invalidated based on the date of the detection and/or insufficient supporting documentation, two records in southwest Minnesota were also retained: a possible record in Lyon County of two birds singing the first week of June 2011 in a block where Camden State Park occurs; and a possible record on the floodplain boundary of Brown and Nicollet County on June 22, 2009.

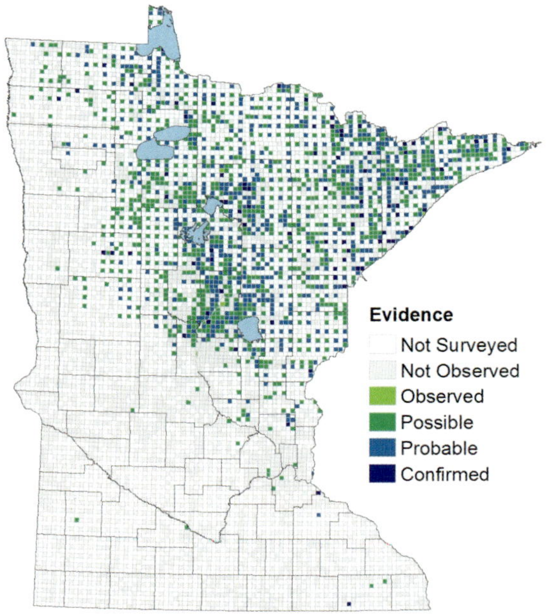

A. Breeding distribution of the Black-and-white Warbler in Minnesota based on the Breeding Bird Atlas (2009–13).

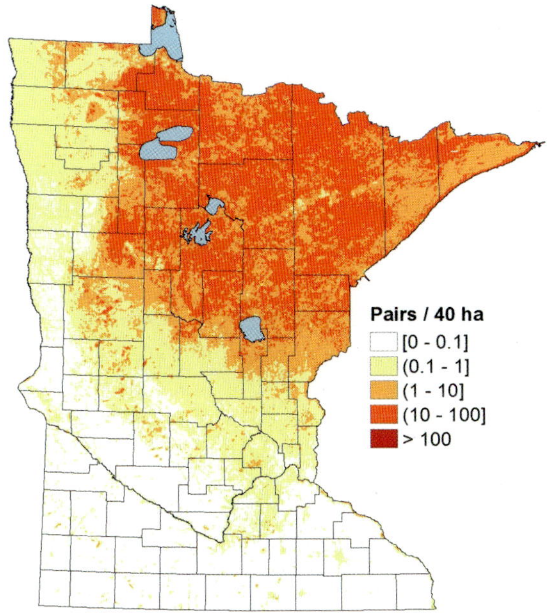

B. Predicted breeding distribution (pairs per 40 ha) of the Black-and-white Warbler in Minnesota based on habitat, landscape context, and climate data gathered during the Minnesota Breeding Bird Atlas (2009–13) using the General Linear Modeling method with an adjustment for detectability.

Overall, the Black-and-white Warbler was reported in 35.0 percent (1,662/4,748) of the surveyed atlas blocks. Breeding was confirmed in 83 blocks and deemed probable in 634 blocks. Twelve of the 19 counties with confirmed nesting were additions to the nine identified by Hertzel and Janssen (1998), including Roseau, Marshall, and Becker in northwestern Minnesota and Morrison County in central Minnesota. Nevertheless, the majority of records were from the Laurentian Mixed Forest Province; elsewhere records were widely scattered across the Tallgrass Aspen Parklands and in a few areas farther to the south and west. The MNBBA probability map **(B)** predicts moderately high breeding densities across nearly the entire Laurentian Mixed Forest Province.

Although the Black-and-white Warbler has not returned to the level of abundance in southern Minnesota suggested by Roberts (1932), it now is regularly observed in the northern regions of east-central Minnesota, including Sherburne, Isanti, Chisago, and northern Anoka Counties, with scattered records present even farther south. Records are also regularly reported from the northern Red River valley, where it was absent for more than fifty years (Green and Janssen 1975). Perhaps as forest remnants have matured, more suitable habitat is available. Elsewhere there have been no large-scale changes to the species' breeding distribution (Kricher 2020).

BREEDING HABITAT

The Black-and-white Warbler occupies deciduous and coniferous forests, including upland maple–basswood and aspen–fir stands and lowland hardwoods and black spruce–tamarack stands **(C)**. Although it generally prefers mid-successional to mature forests with a closed canopy, a well-developed understory, and low shrub density, it also occupies younger stands and lowland shrubs (Niemi, Howe, et al. 2016).

The Warbler may be an area-sensitive species. C. S. Robbins (1979, 1980) suggested that a minimum of 300 ha is needed to sustain a viable breeding population. Whitcomb and his colleagues (1981) reported the species was absent from stands smaller than 70 ha. Landscape context may be an important predictor of which stands are selected. Small stands may be more suitable when they are near larger forest tracts (Galli et al. 1976).

POPULATION ABUNDANCE

The United States and Canada support a population of approximately 18 million breeding adults, with 1.1 million, or 6.1 percent, residing in Minnesota. Predictably, the MNBBA's extensive coverage of northern Minnesota, which supports one of the highest breeding densities of Black-and-White Warblers in North America (Sauer et al. 2020), provided a significantly higher statewide estimate, numbering 5.6 million adults (95 percent confidence interval of 5.2–7.1 million).

Across the BBS survey region, the species' population trend has been relatively stable, with a small, nonsignificant decline, averaging −0.28 percent per year from 1966 to 2019 (Sauer et al. 2020). Numbers are also fairly stable in Minnesota but show a

C. A mid-successional white birch stand with a dense shrub understory is among the many habitats where the Black-and-white Warbler is found nesting. Photo by Gerald J. Niemi.

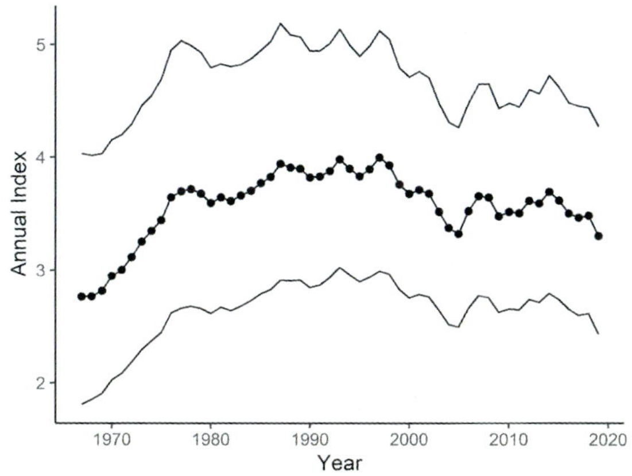

D. Breeding population trend and 95 percent credible interval (solid black lines) for the Black-and-white Warbler in Minnesota based on the 1967–2019 Breeding Bird Survey (Sauer et al. 2020).

very small, nonsignificant increase of 0.33 percent per year **(D)**. This increase is further supported by songbird monitoring data from the Superior and Chippewa National Forests, which are located within the core of the species' breeding distribution in the state. The data indicate a combined, regional population that has significantly increased an average of 1.75 percent per year from 1995 to 2020 (Grinde, Bednar, et al. 2020).

CONSERVATION

Given its slow, widespread decline and concerns regarding its summer and winter habitats, the Black-and-white Warbler was designated a moderate conservation priority (PIF 2021). Although little is known about the factors that regulate population numbers, the species warrants attention from a forest management perspective based on its potential area sensitivity. Its vulnerability to collisions with windows and communication towers is also a major concern (Kricher 2020; Longcore et al. 2013). Fortunately, the species' North American breeding distribution was projected to have a low vulnerability under a 2°C warming scenario, with a predicted range loss of 41 percent and a northward gain of 52 percent by 2050 (NAS 2019). Its plight in Minnesota, however, is not dissimilar to that predicted for many boreal songbirds. As temperatures rise, its breeding range gradually retreats to the northeast until, under a 3°C warming scenario it is extirpated from the state by 2080.

Luckily, the unmistakable Black-and-white Warbler is currently faring well in Minnesota. Populations are stable to increasing, and the Warbler appears to be slowly reoccupying portions of its former range to the south and west of the northern mixed forest as local habitat conditions improve.

PROTHONOTARY WARBLER

Protonotaria citrea

Minnesota Seasonal Status: Regular breeding resident and migrant. The Prothonotary Warbler was rare during the MNBBA.
Migration: Medium- to long-distance migrant that winters in Central America and northern South America.
Food: Bark and foliage gleaner that feeds primarily on arthropods.
Nest: Unique among eastern wood warblers as a secondary cavity nester, uses old Downy Woodpecker nesting cavities and natural tree cavities; also uses nest boxes.
Conservation Concern: Assigned a high PIF Continental Concern Score of 14/20.

Photo by Sparky Stensaas.

Along the backwaters of large rivers, few sights are as stunning as a Prothonotary Warbler. The sheer intensity of the male's yellow plumage, offset by its dark gray wings, can take one's breath away. Although a bit duller, the female's plumage is just as striking. Nevertheless, in dense bottomland forests the first clue to the Warbler's presence may be the male's simple, clear song. The upslurred and high-pitched syllables *peet, tsweet, tsweet* are repeated several times in a series that gradually increases in volume.

The Prothonotary was known for many years as the Golden Swamp-Warbler. The genus name comes from the Latin word *protonotarius,* which means "chief notary or scribe." In the Roman Catholic Church, protonotaries, clad in bright yellow robes to signify their status, managed the church's official business. The similarity of their robes to the little Swamp-Warbler's plumage gave rise to the bird's scientific nomenclature, *Protonotaria citrea,* which literally translates as "first notary of lemon colors." Early naturalists Bagg and Eliot (1937) lamented the inappropriate label and its etymology, writing, "Here, certainly, is a bothersome name fit only to be eschewed!" The name, however, has stuck and "has never detracted from its charm and beauty" (Bent 1953).

An inhabitant of the southeastern United States, extending north to the lower Great Lakes states and southern New England, the Prothonotary Warbler is sparsely distributed throughout most of its breeding range. Breeding densities are highest in southern Louisiana, eastern Virginia, and North Carolina.

MINNESOTA BREEDING DISTRIBUTION

The Prothonotary Warbler was first found near Minnesota in August 1874, when a specimen was collected "on the Wisconsin side of the Mississippi River opposite the mouth of the Root River" by Dr. Hvoslef of Lanesboro (Roberts 1932). Sixteen years would pass, however, before the species was found breeding in "limited numbers" across the river near the Minnesota town of La Crescent, in Houston County (Harrison 1890). Additional records from islands stretching along the Mississippi River, from Red Wing south to Lake Pepin, were published by C. B. Johnson (1898) of Red Wing and included an even earlier breeding record in 1889 in Goodhue County. Given these accounts, Roberts concluded that the species first appeared in Minnesota sometime between 1870 and 1890.

That same year, 1898, Roberts and a colleague surveyed the lower stretch of the Mississippi River. Traveling from Hastings to the Iowa border, they found the Warbler "increasing in numbers as we went southward. It was confined strictly to the bottom-lands and almost entirely to the wooded and periodically flooded islands." Eventually, outlying records were reported from Brainerd, Minneapolis, Rochester, and near the town of Preston in Fillmore County, but the floodplain forests south of Hastings were considered the species' primary breeding range. There was one notable exception. In 1915, "a small colony" of Prothonotary Warblers was found along the Rum River in Isanti County. It was present for many years, and Roberts was puzzled by how the birds found the locality, especially since they had bypassed "equally satisfactory and much more extensive nesting-grounds" along the St. Croix and Minnesota Rivers. In total, Roberts (1932) reported confirmed nesting in just three counties: Goodhue, Houston, and Isanti.

As the years passed, additional reports beyond the species' primary range were documented. Most notable was a 1945 report of a well-established breeding population along the St. Croix River, from Stillwater north to Taylors Falls (Hubert 1945). By 1975, Green and Janssen described the species as a summer resident in southeastern Minnesota and along the Mississippi (north to Stearns County), Rum, and St. Croix Rivers. Janssen (1987) later described it as a regular summer resident as far north as central Chisago County and as far west as Scott County. Even farther west was a 1984 nesting report along the Minnesota River valley in Brown County. He identified six counties with confirmed nesting since 1970: Brown, Hennepin, Houston, Ramsey, Wabasha, and Washington. Hertzel and Janssen (1998) subsequently eliminated Washington County and added Anoka, Dakota, Nicollet, Scott, Sherburne, and Winona Counties to the list. The distribution of the 56 breeding-season locations identified by the MBS differed little from earlier accounts. The exception was the absence of records along the Mississippi River north of the Twin Cities.

The majority of the 28 blocks where the species was reported during the MNBBA were restricted to sites along the Mississippi, Minnesota, and St. Croix Rivers **(A)**. The blocks spanned 19 counties stretch-

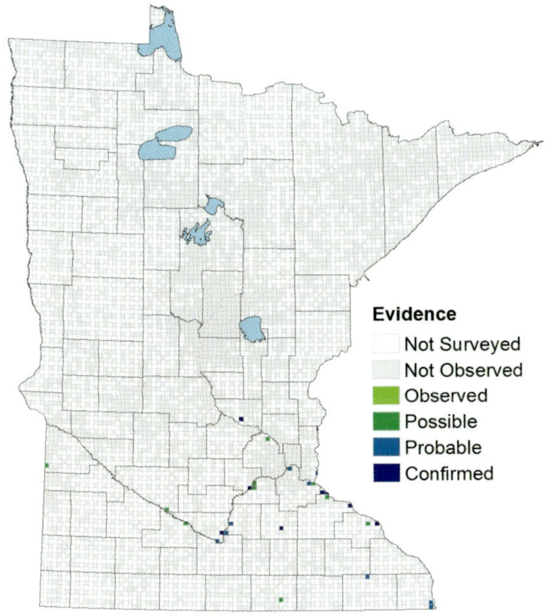

A. Breeding distribution of the Prothonotary Warbler in Minnesota based on the Breeding Bird Atlas (2009–13).

ing from the Twin Cities north to Sherburne County, west to Brown, Lac qui Parle, and Renville Counties, and south to Fillmore, Freeborn, and Houston Counties. Breeding was confirmed in nine blocks and nine counties; two blocks along the Minnesota River straddled two counties each (Carver/Scott and Nicollet/Le Sueur). Unfortunately, the county where nesting was confirmed in these blocks is not known. Among the remaining five counties, Goodhue, Rice, and Washington were additions to Hertzel and Janssen's list of confirmed nesting since 1970.

Extensions to the species' breeding range during the MNBBA included a possible record along the Lac qui Parle River in Lac qui Parle County, where a male was heard and seen in June 2010 and one was photographed in 2011. Farther south, there was a probable breeding record in northern Fillmore County when a bird was observed twice in 2010 along the Root River. Another possible record occurred along the Shell Rock River in Freeborn County. These latter reports demonstrate the occasional suitability of smaller streams as breeding habitat. Three years after the atlas was completed, nesting was also confirmed along the Minnesota River in Yellow Medicine County, lending further evidence to the species' establishment along the upper Minnesota River valley (MOU 2021a). Although the number of MNBBA records was less than the number detected by the MBS, it is difficult to know if that represents a real decrease in numbers or the inaccessibility of its prime habitat.

Roberts (1932) considered the Prothonotary Warbler to be the first of several southern species to settle in Minnesota. Although a few of these species—such as the Northern Cardinal and Red-bellied Woodpecker—now occur well beyond southeastern Minnesota, the Prothonotary's strong association with the floodplain forests of slow-moving, large rivers has likely limited its breeding opportunities elsewhere in the state. Aside from the northern range extension witnessed in the late nineteenth and early twentieth centuries in Minnesota, there have been few changes to the Warbler's continental distribution (Petit 2020).

BREEDING HABITAT

The Prothonotary Warbler inhabits seasonally flooded, lowland deciduous forests with suitable nesting cavities **(B)**. Such sites are typically within the floodplains of large rivers, but the riparian shorelines of smaller streams, lakes, and ponds are occasionally used (Petit 2020; Cutright et al. 2006). Primary features include low-lying, flat terrain, standing or slow-moving waters, a dense forest canopy (50–75 percent cover), and a sparse understory (Petit 2020). The species' preference for nesting over water is believed to be related to higher insect availability and to lower mammalian predation due to the inaccessibility of flooded stands (Petit and Petit 1996; J. P. Hoover 2006). Several studies have suggested that forests smaller than 100 ha in size and riparian edges less than 30 m wide are avoided (e.g., Whitcomb et al. 1981; Kahl et al. 1985).

B. The Prothonotary Warbler selects floodplain forests along waterways with slow-moving or standing water that are characterized by a sparse understory and a canopy of mature trees that supply nesting cavities. This site is a backwater channel of the Mississippi River. Photo by Gary T. Seim.

POPULATION ABUNDANCE

The population estimate of 2.1 million Prothonotary Warblers in the United States and Canada may underestimate its true population size (PIF 2021). BBS data were a primary source for the estimate. The survey is conducted along roads, and floodplain forests are not easily accessible even by secondary roads, where most BBS routes are conducted. Indeed, the bird has been reported only twice on Minnesota BBS routes, so a statewide estimate is not available (Sauer et al. 2020).

Although it lacks statistical reliability, the Warbler's BBS trend line suggests a population decline averaging −0.70 percent per year from 1966 to 2019 (Sauer et al. 2020), with a cumulative population loss of −38 percent from 1970 to 2014 (Rosenberg, Kennedy, et al. 2016). The losses are greatest in the southern states of Alabama, Arkansas, Florida, Georgia, and Louisiana.

Factors considered responsible for the decline are varied. The bird's narrow habitat requirements increase its vulnerability to habitat loss. Indeed, in the lower forty-eight states only 10 percent of the original bottomland forest remains (Dickson et al. 1995). J. P. Hoover (2009) specifically noted impacts that resulted from hydrological modifications of floodplain communities, while competition for nest cavities and brood parasitism by Brown-headed Cowbirds were estimated to impact 10–27 percent of nesting efforts (Petit 2020).

CONSERVATION

Given its declining population and habitat specificity, the Prothonotary Warbler has garnered the attention of many resource agencies and conservation organizations and was designated a high conservation priority (PIF 2021). Conservation measures focus on protecting large, unbroken stands of bottomland forests, particularly along the floodplains of major rivers. Within these stands, management practices that retain dead and dying trees with potential nest cavities are critical. Riparian buffers that are at least 90 m wide are also recommended (Potter et al. 2007). Fortunately, warmer temperatures are not considered a threat to the species; indeed, models project it will expand farther north across the eastern United States, including in eastern Minnesota (NAS 2019), provided suitable habitat is available.

Because of the remoteness of the Prothonotary Warbler's floodplain habitat, few Minnesotans have the opportunity to encounter this stunning and unique inhabitant of the backwaters of some of the state's largest rivers. Nevertheless, we can be content that its future appears secure as long as these vital waterways are protected.

TENNESSEE WARBLER

Leiothlypis peregrina

Minnesota Seasonal Status: Regular breeding resident and migrant. The Tennessee Warbler was uncommon during the MNBBA. It can be rare during the breeding season, but common to abundant during spruce budworm (*Choristoneura fumiferana*) outbreaks, of which it is a well-known specialist.
Migration: Medium- to long-distance migrant that winters in the Caribbean, Central America, and northwestern South America.
Food: Arthropods gleaned from foliage, especially spruce budworm caterpillars.
Nest: Cup nest on the ground, usually hidden near the base of a shrub or hidden in *Sphagnum* moss.
Conservation Concern: Assigned a moderate PIF Continental Concern Score of 10/20.

The Tennessee Warbler has little connection to Tennessee except as a stopover site en route to its northern breeding grounds or to its southern overwintering areas. The species was named and described by Alexander Wilson when he encountered it on the banks of the Cumberland River in Tennessee in 1811. Its scientific species name, *peregrina*, refers to its "wandering" migratory behavior. Recent evidence on its use of coffee plantations as wintering grounds suggests it would be more aptly named the "Coffee Warbler."

Roberts (1932) described its spring presence in Minnesota: "Insignificant in size and inconspicuous in garb, it made up for these shortcomings by numbers and incessant vocal effort." Indeed, Tennessee Warblers can be very abundant at times during spring migration, and the male is easiest to identify by its distinctive three-part song: *ticka ticka ticka, swit swit, sit-sit-sit-sit-sit-sit-sit*. It is loud, long, and usually repeated over and over again. Spring and summer characteristics of the male include bright olive-green above with a gray crown and nape, a white supercilium and underparts, black eye-stripe, and a faint white wing bar. Females are duller, with less contrast between the crown and back, but more yellowish throughout, especially in the breast.

Tennessee Warblers are found nesting in a narrow band across Canada from Labrador to British Columbia, including the northern portions of the upper Midwest and the northern New England states. The highest densities have been recorded

Photo by David Brislance.

in eastern Quebec and from northern Manitoba to northern British Columbia.

MINNESOTA BREEDING DISTRIBUTION

Roberts (1932) described the species historically as "undoubtedly breeding in the northern part of Minnesota, as it has been found in limited numbers in the nesting-season from Lake Superior west to eastern Marshall County and south to Cass Lake." He reported that no nests had been found in Minnesota during his tenure but included observations of birds during the breeding season in Cook, Lake, Marshall, and St. Louis Counties and at Cass Lake. Roberts dismissed a nest found by Hatch (1892) at Lake Minnetonka in June 1881 as an identification error.

More than forty years later, Green and Janssen (1975) described a similar distribution for the Tennessee Warbler as a breeding resident "in northeastern and north central regions as far south as Duluth and Itasca State Park." They emphasized that the species was scarce through most of this area except "along the Canadian border in Cook County, northern Lake County, and northern St. Louis County." They reported confirmed nesting only from Lake County but inferred nesting from Clearwater, Cook, and St. Louis Counties. Janssen (1987) modified the breeding distribution of Green and Janssen (1975) to exclude northern Aitkin County, most of Itasca County, and both Cass and Marshall Counties. He suggested the breeding range extended into northern Hubbard County. Janssen did not report any confirmed nests, but Hertzel and Janssen (1998) identified St. Louis County as the first county where nesting had been confirmed since 1970.

During extensive county surveys, the MBS recorded 163 breeding-season locations of the Tennessee Warbler. These locations included many in Cook County and northern St. Louis County and a few in western Lake County. Additional observations included one location in southern Clearwater County and two in southern Beltrami County.

The MNBBA included a total of 298 breeding-season records. Nesting was confirmed in St. Louis and Lake Counties, probable nesting in Cook and Roseau Counties, and possible nesting in Aitkin, Beltrami, Cass, Kittson, Koochiching, Itasca, Lake of the Woods, and Marshall Counties **(A)**. A total of 3.8 percent (178/4,735) of all blocks sampled included breeding evidence. The spatial pattern of the Tennessee Warblers' breeding distribution from the MNBBA was similar to that described by Roberts in 1932 and by Green and Janssen in 1975. The bulk of the records were located as previously described, largely in the northeastern counties of Cook, Lake, and St. Louis. The probability map illustrates this distribution well, with predictions highest in northeastern and northern Minnesota **(B)**.

Rimmer and McFarland (2020) reviewed the North American distribution of the Tennessee Warbler and concluded that no significant changes had been documented for this species. They underscored the fact that short-term fluctuations in local distributions were dictated by insect outbreaks, especially by the spruce budworm. This also appears to be the case in Minnesota, where Tennessee Warbler populations and their distribution are highly associated with the presence of spruce budworm outbreaks. Although Roberts (1932) did not mention spruce budworm, he reported "that it breeds sparingly in

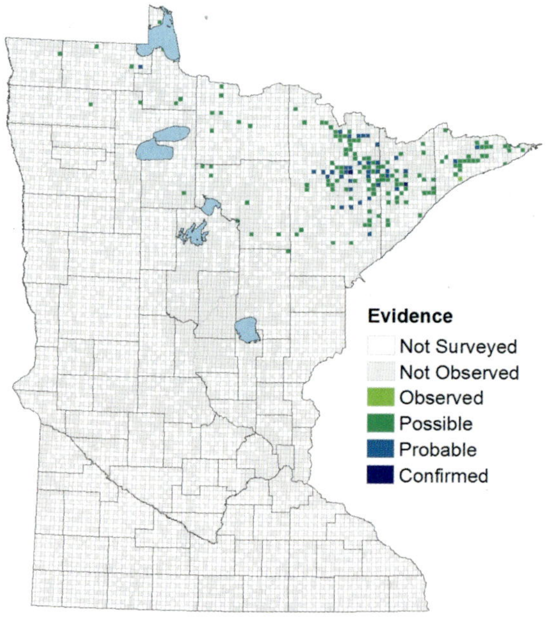

A. Breeding distribution of the Tennessee Warbler in Minnesota based on the Breeding Bird Atlas (2009–13).

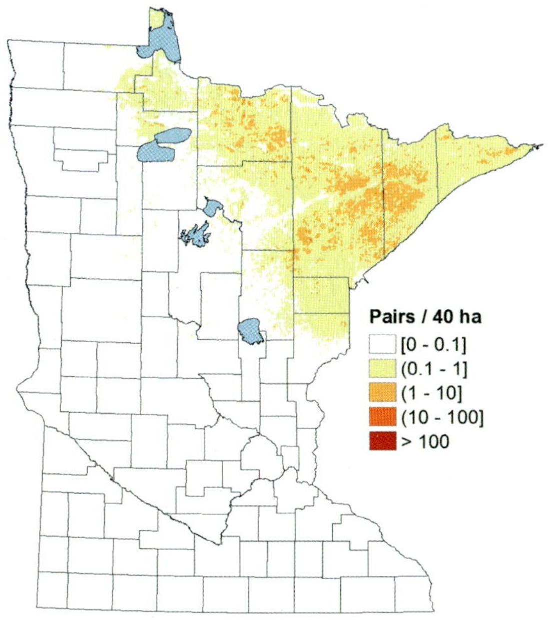

B. Predicted breeding distribution (pairs per 40 ha) of the Tennessee Warbler in Minnesota based on habitat, landscape context, and climate data gathered during the Minnesota Breeding Bird Atlas (2009–13) using the Generalized Linear Modeling method with an adjustment for detectability.

the spruce and arbor vitae swamps of the northern part of the state." Nearly one hundred years later, this statement is still true in Minnesota; however, during spruce budworm outbreaks the species can be locally common or abundant.

BREEDING HABITAT

Rimmer and McFarland (2020) summarized Tennessee Warbler habitat use as varying from early successional forests to mature forests in Ontario, all wooded habitats in southwestern James Bay, boreal spruce–fir forests in Maine, and boreal mixedwood forests in Saskatchewan. Niemi, Howe, and colleagues (2016) found that the Tennessee Warbler used many forested habitats, including regenerating conifers, aspen–birch, and lowland conifer forests **(C)**. In one of the few landscape studies in the region, Grinde (2015) observed that the Tennessee Warbler was more consistently found in a landscape matrix with small forest patches. Because the species' breeding population in Minnesota is closely related to the spruce budworm cycle, its habitat use aligns with the patchiness of these outbreaks.

POPULATION ABUNDANCE

Primarily using BBS data, PIF (2021) estimated the species' US and Canadian population at 110 million breeding adults, while Rosenberg, Kennedy, and colleagues (2016) calculated a 15 percent decline in its population from 1970 to 2014. PIF (2021) estimated a Minnesota breeding population of 36,000 adults. The MNBBA, however, estimated a much larger population of almost 204,000, with a 95 percent confidence interval of 147,000–324,000 breeding adults. This much larger estimate in Minnesota can be explained in three ways: first, spruce budworm populations were relatively high in northeastern Minnesota during the MNBBA count period from 2009 to 2013 (Grinde, Walton, et al. 2018); second, methodological differences exist between the PIF and the MNBBA estimates; and, third, the MNBBA had substantially better coverage of the northern portions of Minnesota.

The BBS population trends for Minnesota were unreliable because of too few observations. BBS surveywide trends indicated a stable trend from 1966 to 2019, but a significant negative trend averaging −2.33 percent per year from 1993 to 2019 **(D)**. Reasons for population reductions have most often been associated with the status of spruce budworm infestations, which can be highly variable over time.

CONSERVATION

Overall populations of the Tennessee Warbler in North America are healthy, but they fluctuate widely with available food supplies. The species is beneficial to healthy forests because they consume insects that are harmful to the growth of trees. For instance, a Tennessee Warbler can consume 65 to 100 caterpillars, or 6 grams, per day. Niemi (2010) calculated that this level of consumption for a 12-gram bird is equivalent to a 150-pound human eating 75 pounds of hamburger in a day! Holling (1978) provided

C. During spruce budworm outbreaks, Tennessee Warblers can be found in a variety of forested habitats in northern Minnesota, but especially in forests with coniferous trees. Photo by Gerald J. Niemi.

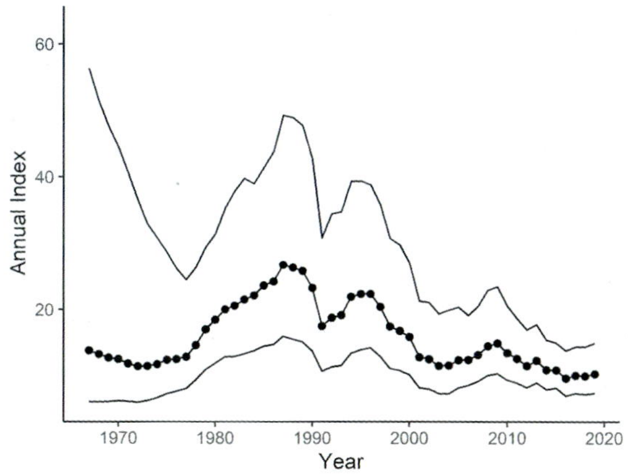

D. Surveywide breeding population trend and 95 percent credible interval (solid black lines) for the Tennessee Warbler based on the 1967–2019 Breeding Bird Survey (Sauer et al. 2020).

modeling evidence that the insectivory of species like the Tennessee Warbler could extend the time between spruce budworm cycles. Any decline in insectivorous bird species may result in more frequent spruce budworm outbreaks and thereby reduce the health of forest ecosystems.

The Tennessee Warbler may be susceptible to tower kills during migration. They have been reported as the fourth or fifth most frequently killed species at towers in many regions of the United States where it migrates (Longcore et al. 2013). Greenberg, Bichier, and colleagues (1997) raised concerns about overwintering populations in the tropics because of reductions in floristic diversity, pesticide applications, and conversions of shade coffee plantations to more intensively managed plantations.

NAS (2019) projected a moderate vulnerability for the Tennessee Warbler's breeding distribution in North America with a 2°C warming scenario, including a 50 percent range loss and a 25 percent range gain in the far north by 2050. In Minnesota, the species was predicted to be extirpated from the state under our current 1.5°C warming scenario. Whether breeding habitat in the far north in Canada and Alaska will be suitable for this species remains an open question.

The future status of the Tennessee Warbler in Minnesota is precarious. The species has a boom-and-bust population cycle on the breeding grounds that is dependent on the status of spruce budworm populations. The birds are also susceptible to tower kills during migration, and their overwintering situation in the tropics appears to be tenuous and certainly under-studied. The Tennessee Warbler's foraging is a natural insecticide, making it a valued member of Minnesota's avifauna and a contributor to the health of Minnesota's forests.

NASHVILLE WARBLER

Leiothlypis ruficapilla

Minnesota Seasonal Status: Regular breeding resident and migrant. The Nashville Warbler was an abundant species during the MNBBA.
Migration: Short- to medium-distance migrant that overwinters in southern Texas, Mexico, and south as far as Panama; casual winter visitor in the Caribbean.
Food: Gleans arthropods from foliage.
Nest: On the ground in thick foliage or moss hummocks.
Conservation Concern: Assigned a moderate PIF Continental Concern Score of 9/20.

Referred to as a "beautiful, bright and sprightly little bird" (Forbush and May 1955), the Nashville Warbler, like the Tennessee Warbler, has little association with Nashville, Tennessee, except that the type specimen was collected there during its migration. Its species name, *ruficapilla,* refers to the often hidden rufous-red crown of the male. The olive-green male sports a bright yellow breast that contrasts with its grayish head and neck. Both sexes have a distinctive white eye-ring, while the female is generally more subdued in coloration, has an olive-gray head, and retains an even fainter reddish crown. Like many warblers, the Nashville is much easier to identify by song as it flits through the vegetation. Its song, often sung on the top of a conifer during the breeding season, begins with several two-part notes followed by a trill, commonly described as *seebit seebit seebit ti ti ti ti ti ti.*

The Nashville Warbler is one of the most ubiquitous species found in the northern forests of Minnesota. It nests primarily in lowland coniferous forests consisting of black spruce and tamarack trees but is also commonly found in young forests created by logging, forest fire, and insect defoliation. It is an important Minnesota species because a high proportion of its North American breeding population occurs in the state. Fortunately, it appears to be highly adaptable, and its appetite for many harmful insects renders it a valuable bird for the health of Minnesota forests.

In North America, the Nashville Warbler occurs as two subspecies and in two separate breeding populations: one (*Leiothlypis ruficapilla ruficapilla*) in the northeastern and upper midwestern United States and Canada, and the other (*L. r. ridgwayi*) in the western United States and Canada. The highest densities are found in northeastern Minnesota but are also substantial in the Canadian provinces of Ontario and Quebec.

Photo by David Brislance.

MINNESOTA BREEDING DISTRIBUTION

The Nashville Warbler is historically described as breeding throughout the forested part of the state, especially in the tamarack and spruce swamps (Roberts 1932). During Roberts's time in Minnesota, from the late 1800s to the early 1900s, the species was "sparingly" found in Hennepin and adjacent counties "in and about the numerous large tamarack swamps, but since these have been largely cut off or drained, it has disappeared as a nesting bird." Roberts then reported the breeding range as confined almost entirely to the Canadian Zone, from northern Isanti County on the south, and as far west as the evergreens extended. He documented breeding activity at seven locations, and all were in tamarack and spruce swamps. They included Aitkin, Becker, and Itasca Counties, plus at Cass Lake, Leech Lake, Itasca State Park, and Mille Lacs.

More than forty years later, Green and Janssen (1975) emphasized that the Nashville Warbler's primary breeding distribution is in north-central and northeastern Minnesota. They added confirmed nesting records from as far south as Anoka County and reported confirmed nesting in Cass, Clearwater, Crow Wing, Lake, and St. Louis Counties that were not noted by Roberts (1932). They also noted that the Nashville Warbler uses tamarack stands in the southern and western regions of its breeding range. In 1987 Janssen reported confirmed nesting in 12 counties since 1970, of which Beltrami, Cook, Hubbard, Lake of the Woods, and Mille Lacs Counties were new records. Several years later, Hertzel and Janssen (1998) added confirmed nesting in Koochiching County since 1970.

The MBS recorded 2,078 breeding-season locations during their inventories of counties in the state. Their locations were dominated by those in northeastern and north-central Minnesota but included breeding observation locations throughout the northwestern counties, including Kittson, Mahnomen, Marshall, and Polk Counties. Observations were also made southwest to Douglas and Otter Tail Counties, and south to Anoka, Chisago, Stearns, and Washington Counties.

The MNBBA further documented not only the extensive distribution and abundance of the Nashville Warbler in the forested regions of Minnesota but also the struggle to confirm nesting activity **(A)**. The difficulty in finding nests of this species was a point noted by Roberts (1932) and by many regional breeding bird atlases. A total of 4,370 records of this species were gathered during the MNBBA. The ratio of blocks with confirmed nesting to all the other observations was only 7 percent (104 confirmed blocks to 1,529 blocks where the species was recorded). The MNBBA extended confirmed nesting of the Nashville Warbler to northwest Minnesota, including Marshall, Pennington, and Roseau Counties and south to Hennepin, Morrison, and Washington Counties. In addition, nesting was confirmed in the east-central region in Carlton and Pine Counties.

It is unknown whether the Nashville Warbler has historically declined in abundance, but Roberts (1932) noted its range had contracted from the southern portions of its former range in Minnesota due to the loss of forests, especially large tamarack swamps. The recent documentation of nesting in Hennepin, Morrison, and Washington Counties is likely due to a combination of, first, a recolonization of more southerly locations; second, the more liberal definition of confirmed nesting used in breeding bird atlas accounts compared with those used by the MOU; and, third, the expanded and more intensive coverage of nesting activity by the MNBBA. Lowther and Williams (2020), in their review of the Nashville Warbler in North America, provided little insight on changes

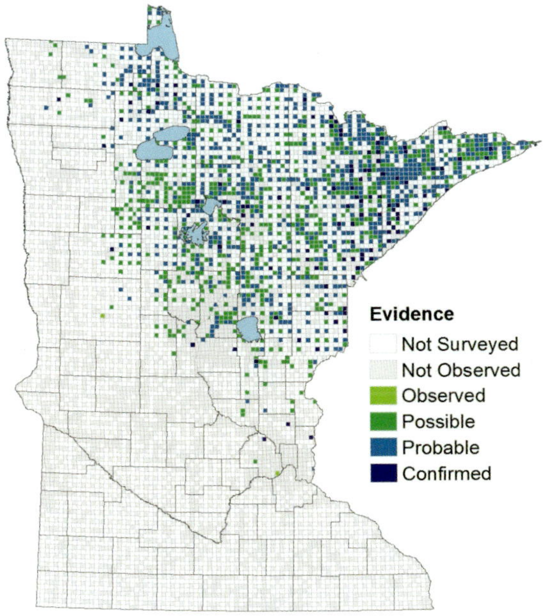

A. Breeding distribution of the Nashville Warbler in Minnesota based on the Breeding Bird Atlas (2009–13).

in the species' range except to emphasize that it responds to forest clearing and regrowth, so its distribution changes accordingly. Hence, it is possible that small populations of Nashvilles have always existed in these more southerly locations, but they are likely to be more variable and not present every year.

BREEDING HABITAT

In Minnesota, the Nashville Warbler is widely distributed in many different forest cover types and can be ubiquitous in early-successional, intermediate, and mature forests. Niemi, Howe, and colleagues (2016) reported its highest densities in black spruce–tamarack cover types in the Chippewa and Superior National Forests **(B)**. The Nashville Warbler was one of the most abundant species in the Agassiz Lowlands Subsection of northern Minnesota. It occurred in virtually all age classes of lowland coniferous forests, including white cedar, tamarack, and black spruce (Bednar et al. 2016).

POPULATION ABUNDANCE

PIF (2021) estimated a US and Canadian population of 40 million breeding adults and a Minnesota breeding population of 2.20 million. By comparison, the MNBBA estimated a Minnesota population of 9.21 million breeding adults (95 percent confidence interval of 8.81–9.80 million), or more than four times the PIF estimate. The BBS indicated that northeastern Minnesota had among the highest breeding densities in North America (Sauer et al. 2020). The reasons for discrepancies in population estimates are due to methodological differences in estimates between PIF and the MNBBA and the absence of BBS routes, used by PIF in establishing their estimates, in remote, roadless areas. In contrast, the MNBBA sampled these roadless areas where the Nashville Warbler breeding populations are very high. The BBS trends from 1967 to 2019 for the Nashville Warbler indicated a stable population in Minnesota but a recent significant decline, on average −1.82 percent per year from 1993 to 2019 **(C)**. However, in Minnesota's two national forests, the Chippewa and Superior, the species has been in-

B. Although Nashville Warblers are common in many young forests, they are most abundant in mature black spruce–tamarack lowland forests in Minnesota. Photo by Gerald J. Niemi.

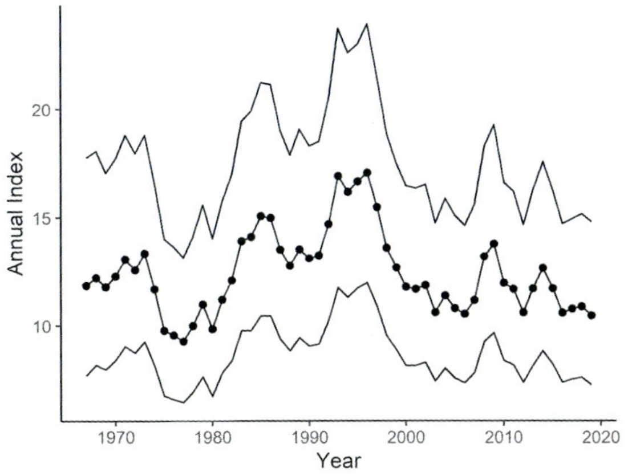

C. Breeding population trend and 95 percent credible interval (solid black lines) for the Nashville Warbler in Minnesota based on the 1967–2019 Breeding Bird Survey (Sauer et al. 2020).

creasing an average of 1.16 percent from 1995 to 2020 (Grinde, Bednar, et al. 2020).

CONSERVATION

Lowther and Williams (2020) suggested that the Nashville Warbler was of low priority for management because it readily adapts to second-growth and cutover areas. However, the high proportion of the breeding population that occurs in Minnesota was a factor identified by Audubon Minnesota (Pfannmuller 2012) in its selection of the Nashville Warbler as a Stewardship Species. As previously noted, this species may be more common today in Minnesota because it readily uses recently disturbed forests, such as those resulting from logging, forest fire, and wind. Schulte and Niemi (1998) found that Nashville Warblers were significantly more abundant in recently logged areas compared with similar areas that were recently burned.

Other factors that have been identified as affecting the species' populations include climate change, collision with man-made structures (towers and windows), and ground predators of nests (Lowther and Williams 2020). Loss and colleagues (2014) listed the species as vulnerable to collisions with windows and estimated that this species' risk was 22 times greater than the risk to an average bird.

With a 2°C warming scenario, NAS (2019) projected a moderate vulnerability for the Nashville Warbler's breeding distribution in North America, including a 53 percent range loss but a 78 percent range gain primarily in northern Canada by 2050. In Minnesota, the species was predicted to be extirpated from the state under the 2°C warming scenario by 2050. The Nashville Warbler is a highly beneficial insectivore that contributes to healthy forests; we hope this important species will be a part of our native avifauna for years to come.

CONNECTICUT WARBLER

Oporornis agilis

Minnesota Seasonal Status: Regular breeding resident and migrant. The Connecticut Warbler was uncommon during the MNBBA.
Migration: Long-distance migrant that winters in South America.
Food: Insectivore that gleans arthropods from low foliage and the ground.
Nest: Cup nest placed on or near the ground, often in a mossy hummock.
Conservation Concern: Assigned a moderate PIF Continental Concern Score of 13/20.

An inhabitant of lowland coniferous forests, the Connecticut Warbler is perhaps the most secretive and poorly known of Minnesota's wood warblers. If spotted, the male's gray hood, white eye-ring, pale yellow belly, and olive-green wings and back are diagnostic. The female's plumage is a little lighter; her white eye-ring helps distinguish her from the similar-looking female Mourning Warbler. Frequently, however, the only clue to the bird's presence is the male's song. Loud and emphatic, the phrase *chippy-chuppy, chippy-chuppy, chippy-chuppy* accurately depicts its rhythm and sound. University of Minnesota professor Ned Huff (1929) wrote: "In his summer home, the Connecticut Warbler is a shy and elusive bird, so secretive in his manner that he would rarely be seen here, even by those looking for him, were it not for his betraying song."

Although the bird was described in 1812 from a migrant specimen collected in Connecticut, it was 1883 before the first nest was discovered in Manitoba. The Warbler's distribution and habits, however, remained "clouded in mystery" for many more years (Bent 1953). Few comprehensive studies, for example, were undertaken in its remote breeding habitat. Only recently have biologists placed its three former relatives (Kentucky, MacGillivray's, and Mourning Warblers) into a separate genus (*Geothylpis).* This left the Connecticut Warbler as the sole representative of the genus *Oporornis* (Pitocchelli et al. 2020). The Greek name *Oporornis* means "autumn bird," which suggests that this rare boreal species is more frequently observed during the fall migration than at other times of the year (Sandrock and Prior 2014).

During the summer, the Warbler is sparsely distributed in a narrow band of boreal forest that

Photo by Sparky Stensaas.

stretches from eastern British Columbia to western Quebec and south to northern Minnesota, Wisconsin, and Michigan. The northern Great Lakes states comprise only a small percentage of the Warbler's breeding range.

MINNESOTA BREEDING DISTRIBUTION

Roberts (1932) detailed the years of ornithological exploration that followed the Connecticut Warbler's discovery in 1812 until its documentation as a Minnesota nesting species in 1929. It was an uncommon migrant in the state, so summer records were limited. The first specimen was collected in July 1879, along the Red River at Pembina, North Dakota, a location west of its known breeding range. The second was collected in July 1886 in St. Louis County. Reports of singing birds in Aitkin, Isanti, and Marshall Counties and in Itasca State Park soon followed. Adults carrying food were observed in Isanti County and Itasca State Park, but the discovery of a nest eluded observers until 1929, when Professor Huff found two nests in a black spruce–tamarack bog in Aitkin County. The species, Roberts (1932) concluded, "so far as discovered, makes its summer home only in cold tamarack and spruce swamps of typical *Canadian Zone* character. Such places are numerous and wide-spread in northern Minnesota."

Decades later, Green and Janssen (1975) described the species as a rare summer resident in northeastern and north-central Minnesota. The exception was an area stretching "from Koochiching County northwestward," where it was "plentiful" in lowland black spruce–tamarack bogs and in upland jack pine and aspen stands. In addition to Huff's nesting report in Aitkin County, nesting records were now available from Hubbard and Lake Counties. Consistently reported near Cambridge in Isanti County from 1914 to 1917, recent reports from the site were lacking.

Janssen's (1987) update identified four counties with confirmed nesting since 1970: Aitkin, Beltrami, Hubbard, and Lake. In 1998, Hertzel and Janssen added Itasca County to the list. Meanwhile, MBS surveyors identified 248 breeding-season locations that extended the species' summer range west to eastern Kittson County.

The breeding distribution delineated by the 263 MNBBA records was nearly identical to that outlined by earlier accounts **(A)**. Detected in 3.4 percent (161/4,734) of the surveyed atlas blocks, the species occurred in 16 northern counties stretching west to eastern Kittson County and south to central Pine County. Breeding was confirmed in just two blocks: one each in Aitkin and St. Louis Counties. Probable breeding was reported in seven counties: Beltrami, Carlton, Cass, Itasca, Koochiching, Lake, and Lake of the Woods. The MNBBA probability map **(B)** predicts moderately high breeding densities in small, scattered areas across Aitkin, Itasca, Lake, and St. Louis Counties and throughout the Agassiz Lowlands Subsection, north and east of Red Lake. The subsection's flat, poorly drained landscape is dominated by peatland communities, including black spruce–tamarack bogs, the Connecticut Warbler's preferred habitat.

Other than the loss of the small population in Isanti County, the statewide distribution of this rare Warbler is remarkably unchanged over the past one hundred years. The same is true throughout its very restricted breeding range (Pitocchelli et al. 2020).

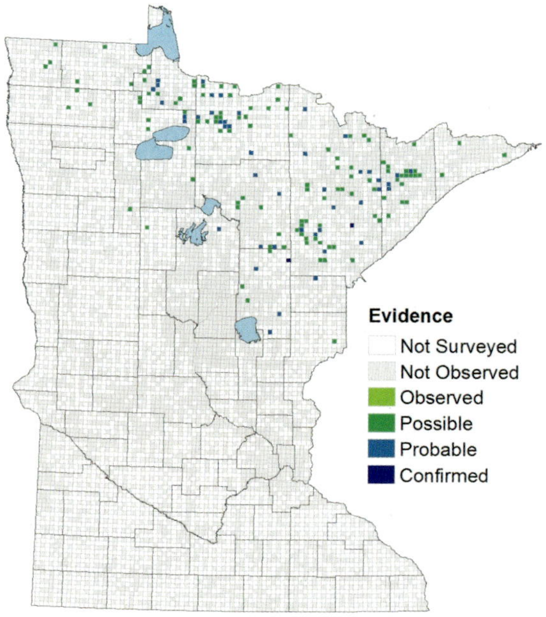

A. Breeding distribution of the Connecticut Warbler in Minnesota based on the Breeding Bird Atlas (2009–13).

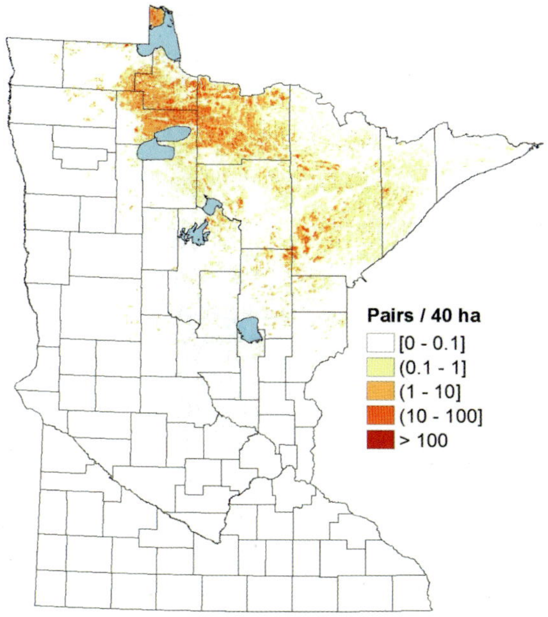

B. Predicted breeding distribution (pairs per 40 ha) of the Connecticut Warbler in Minnesota based on habitat, landscape context, and climate data gathered during the Minnesota Breeding Bird Atlas (2009–13) using the Generalized Linear Modeling method with an adjustment for detectability.

BREEDING HABITAT

The mature, lowland coniferous forests inhabited by Connecticut Warblers are usually composed of scattered black spruce and tamarack trees with a low understory of ericaceous shrubs underlain by a mat of sphagnum moss **(C)**. Occasionally the birds are found in upland forests, primarily pine stands, including jack pine barrens and mixed aspen–jack pine woodlands (Niemi, Howe, et al. 2016; Pitocchelli et al. 2020).

An in-depth assessment of the species' breeding habitat in Minnesota concluded that the Warbler relies more on landscape characteristics for selecting suitable habitat than on features at the stand level (Lapin et al. 2013). Most important was the availability of large contiguous stands of lowland conifers surrounded by upland conifers. This supported an earlier observation that the species may be area sensitive (Green 1995). However, in the Agassiz Lowlands Subsection, habitat features at both the stand and landscape level were important predictors of suitable habitat. The proportion of stagnant black spruce–tamarack forests figured prominently at both levels (Zlonis et al. 2017).

POPULATION ABUNDANCE

PIF (2021) estimated a US and Canadian breeding population of 1.8 million breeding adults with a Minnesota population of 47,000. This is considerably lower than the MNBBA estimate of 561,000 breeding adults, which includes data from remote, roadless areas not surveyed by BBS routes. The confidence interval for the latter estimate, however, was quite large.

Although most BBS trend data for this boreal species lack statistical reliability, they do suggest widespread population declines from 1966 to 2019. Even when data from BBS routes established in remote areas of Canada and Alaska are added to data from the core survey region, the trend lines remain unreliable but still show declining numbers. Unfortunately, in most states and provinces the rate of the decline also notably increased from 1993 to 2019. In Minnesota, reliable data are available from the Chippewa and Superior National Forests, where the combined regional trend indicates a significant decrease averaging −8.52 percent per year from 1995 to 2021 **(D)**. Grinde, Siebers, and Bednar (2021) reported that the Warbler had experienced "one of the most consistent declines of any species in the [National Forest] monitoring program." Indeed, 2018 was the first year since 1995 when no Connecticut Warblers were reported in either forest. Only two birds were detected in 2019 and 2020, and only one in 2021. Wisconsin reported similar results when targeted field surveys in 2021 discovered only one active breeding site (Brady et al. 2021).

The species' alarming decline is often attributed to nest predation and habitat loss (Niemi, Howe, et al. 2016), especially on the breeding grounds (Hallworth et al. 2021). The latter may be less important in Minnesota, where inaccessible peatlands are less impacted than most upland forests. Other concerns include climate change, loss of winter habitat, and building and tower collisions (T. Arnold and Zink 2011; Longcore et al. 2013; Pitocchelli et al. 2020).

C. A sparse overstory of stunted black spruce trees found in the peatlands of northern Beltrami County is one of the few habitats where the Connecticut Warbler may be found nesting. Photo by Lee A. Pfannmuller.

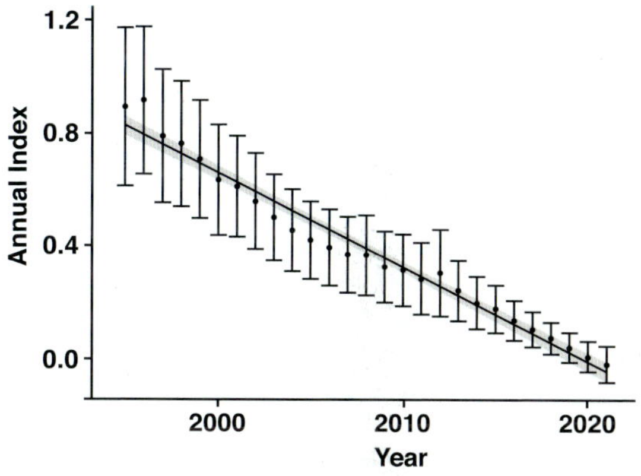

D. Combined regional population trend graph for the Connecticut Warbler within the Chippewa and Superior National Forests, from 1995 to 2021 (Grinde, Siebers, and Bednar 2021).

CONSERVATION

PIF (2021) ranked the Connecticut Warbler as a moderate conservation priority, given its restricted breeding range and its small and declining population, and placed the species on its Yellow Watch List, a list of North American birds that require action to stem population declines (Rosenberg, Kennedy, et al. 2016). The MNDNR (2015b) identified it as a *Species in Greatest Conservation Need,* and the USFS (2012) designated it a Sensitive Species in Minnesota's national forests.

Kudell-Ekstrum (2002) noted the absence of management guidelines for the species. Niemi, Howe, and colleagues (2016), however, stressed the need to protect large tracts of black spruce–tamarack peatlands, especially sites that are adjacent to upland coniferous forests, which also provide suitable habitat. In order to maintain "as much black spruce on the landscape as possible," Grinde, Bednar, and colleagues (2020) recommended limiting the harvest of black spruce stands, especially on non-productive sites.

Minnesota's vast peatlands are renowned for their pristine beauty, unique avifauna, and their critical role as carbon reservoirs. Unfortunately, they are especially sensitive to climate change; as temperatures rise, the abundance of lowland conifers in these remarkable wetlands is predicted to decrease (Handler et al. 2014). In Minnesota, even a rise of 1.5°C is predicted to reduce the acreage of lowland conifers required by the Connecticut Warbler and result in the species' statewide extirpation and the loss of 85 percent of its continental breeding range (NAS 2019). The nearest suitable habitat could be almost 700 km to the north, near Hudson Bay. Given the scope of these changes, the departure of the elusive Connecticut Warbler may be inevitable. Indeed, the exodus may already have begun.

MOURNING WARBLER

Geothlypis philadelphia

Minnesota Seasonal Status: Regular breeding resident and migrant. The Mourning Warbler was a common species during the MNBBA.
Migration: Medium- to long-distance migrant, overwinters in Central America and northwestern South America.
Food: Arthropods gleaned from foliage.
Nest: Cup nest on or near the ground in a well-concealed location.
Conservation Concern: Assigned a moderate PIF Continental Concern Score of 12/20.

The Mourning Warbler was named for the black bib of the male, which was reminiscent of someone in "mourning" attire. In contrast, its scientific name *philadelphia* was simply where it was first observed near the city of Philadelphia by Alexander Wilson, who officially described the species. The Mourning Warbler can be a secretive, inconspicuous species, especially during migration. Wilson never saw another one and Audubon saw very few. It is, however, an abundant summer species in the young forests and forest edges of Minnesota.

Like many warblers, the Mourning Warbler can be difficult to observe in the shrubs and near the ground of regenerating forests, but its *chirry chirry chirry, chorry chorry* song draws an observer's attention to its presence. "It is a cheerful song, not at all consistent with the bird's somber name," wrote Griscom and Sprunt (1957). Visually, the male Mourning Warbler is distinguished by its gray hood, black bib, greenish back, and yellow breast, while the female has similar but more diluted coloration and lacks the black bib.

The Mourning Warbler is found nesting across the northeastern and upper Midwest and throughout Canada from Newfoundland to northeastern British Columbia. Its highest densities in North America are found in northeastern Minnesota and parts of Ontario.

MINNESOTA BREEDING DISTRIBUTION

Roberts (1932) described the species as well distributed in northern and north-central forested areas and their fringes south to Isanti County and westward to Marshall and Polk Counties. He reported nesting activity from Aitkin, Carlton, Cass,

Photo by David Brislance.

Cook, Isanti, and St. Louis Counties, as well as from east-central Minnesota and the Mille Lacs area. He described accounts of the Mourning Warbler by T. Martin Trippe in the summer of 1870 as a "common nesting bird" in the region from Carlton County west to Otter Tail Lake. Roberts emphasized that this information was of special interest because it "supplied a missing chapter in the life history of a bird about which little was known up to that time." He added that Dr. Elliott Coues, a famous ornithologist and one of the founding members of the American Ornithologists' Union, found Mourning Warblers in 1873 "breeding abundantly along the Red River, between Dakota and Minnesota."

Roberts in 1880 found possible evidence of a "pair, evidently with young," in Sherburne County and commonly observed them in "burned-over" areas of Stearns County in 1881. He also identified numerous accounts of the commonness of the species in Carlton County and its abundance "in the vicinity of Virginia, Hibbing, Biwabik, and elsewhere on the Mesabi Iron Range of northern St. Louis County." He suggested that it "is apparently more abundant in the extreme northern part of the state than it is in the central counties."

Green and Janssen (1975) included a similar distribution with confirmed nesting in Clearwater, Hubbard, and Lake Counties, and inferred nesting in Itasca, Marshall, Otter Tail, and Roseau Counties not reported by Roberts (1932). They emphasized the presence of singing males in Anoka, Hennepin, and Washington Counties in the Twin Cities area. Several years later, Janssen (1987) documented confirmed nesting in seven counties since 1970, including Aitkin, Clearwater, Cook, Lake, Pine, St. Louis, and Washington Counties. Hertzel and Janssen (1998) would later add Anoka and Dakota Counties to the list of confirmed counties where nesting has been observed since 1970.

The MBS included 744 breeding-season locations during their counts that were distributed throughout most of the areas previously described. They also recorded breeding observations in Fillmore, Hennepin, Sibley, Stearns, and Wright Counties, as well as records that straddled Renville/Nicollet and Scott/Dakota Counties.

The extensive coverage of the MNBBA included 1,988 records and verified the broad breeding distribution in the state. Confirmed nesting not previously reported included Becker, Beltrami, Lake of the Woods, and Morrison Counties, as well as in Murphy-Hanrehan Park Reserve on the border of Scott/Dakota Counties noted by MBS and MOU (2021a) **(A)**. As is the case for many species in which nests are difficult to find, only 6 percent of all blocks where the species was observed had confirmed nesting (60 confirmed out of 975 total blocks).

The Minnesota probability map emphasized its highest populations in northeastern Minnesota, with populations west to Clearwater County and south to Pine County. Isolated populations likely exist in the western and southern fringes of its major breeding range in Minnesota, including south to the Twin Cities metropolitan area **(B)**. Highest densities were predicted in Cook, Lake, and St. Louis Counties, eastern Koochiching County, eastern Pine County, and northeastern Itasca County.

Pitocchelli (2020) makes little mention of historical range contractions of the Mourning Warbler in North America. Yet, over the past 150 years, the Mourning Warbler's range has contracted, especially

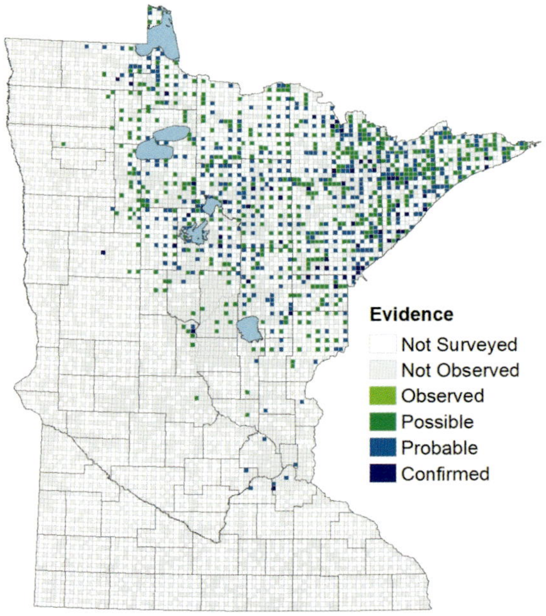

A. Breeding distribution of the Mourning Warbler in Minnesota based on the Breeding Bird Atlas (2009–13).

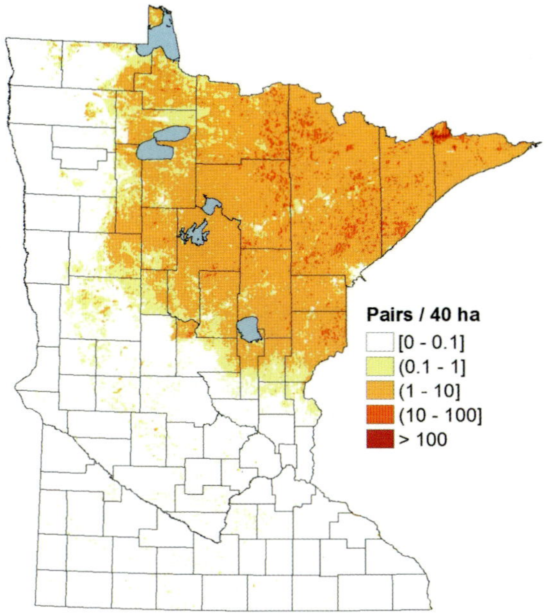

B. Predicted breeding distribution (pairs per 40 ha) of the Mourning Warbler based on habitat, landscape context, and climate data gathered during the Minnesota Breeding Bird Atlas (2009–13) using the Generalized Linear Modeling method with an adjustment for detectability.

in the southern and western fringes of Minnesota. Compared with Wisconsin (Cutright et al. 2006; WSO 2021), the species appears to have a more restricted distribution in Minnesota. Confirmed nesting and breeding observations in Wisconsin extended much farther south into the Iowa Driftless region and into Grant County, its most southeastern county. Roberts (1932) makes no mention of breeding activity in southeastern Minnesota. In an extensive effort of nest searching from 1996 to 1998 in southeastern Minnesota by Knutson, Niemi, and colleagues (2004), no observations out of 975 nests found were of this species.

BREEDING HABITAT

The Mourning Warbler is broadly described as nesting in upland early successional cover types resulting from forest disturbances such as logging, fire, and wind; tree fall gaps; second growth; and edges with thick understories (Pitocchelli 2020). Niemi, Howe, and colleagues (2016) indicated a significant habitat association with recently cut areas, such as open, dry uplands and regenerating forest cover types **(C)**. The species was also found commonly but less abundantly in aspen–birch, aspen–spruce–fir, and red pine cover types. In a comparison of 46 equally distributed sites between burned and logged areas of similar age, Schulte and Niemi (1998) found Mourning Warblers more abundant in logged areas. Similarly, Zlonis and Niemi (2014) found the species more common in managed second-growth forests compared with adjacent wilderness forests, even though the latter included a history of natural disturbances.

POPULATION ABUNDANCE

PIF (2021) estimated a US and Canadian breeding population of 14 million breeding adults and a Minnesota population of 760,000. Using MNBBA data, Minnesota's breeding population was estimated at 1.91 million breeding adults (95 percent confidence interval of 1.72–2.25 million), which would represent about 8 percent of the global breeding population. The majority of the breeding population occurs in Canada.

Population trends from the BBS for the Mourning Warbler indicated a significant decline in Minnesota from 1967 to 2019, averaging −0.84 percent per year **(D)**. Overall, these decreases in the population are consistent with surveywide counts over a similar time frame (averaging −1.60 percent per year) and are a little more severe in Canada (averaging −1.71 percent per year). Rosenberg, Kennedy, and colleagues (2016) calculated a −45 percent decline in the North American breeding population of Mourning Warblers from 1970 to 2014.

The reasons for the decline of the species are unclear and not well studied. Collisions with windows at low-rise buildings (4–11 stories high) during migration are high compared with many other species (Loss et al. 2014), and climate change, including breeding shifts to the north, may be a factor.

CONSERVATION

The Mourning Warbler's widespread population declines in North America are a cause for some conservation concern. In Minnesota, it is still a common bird species and locally abundant in many suitable habitats. Numerous studies in Minnesota

C. In Minnesota, Mourning Warblers are almost exclusively found nesting in shrubby, upland forests, such as those that have been recently logged or burned. Photo by Gerald J. Niemi.

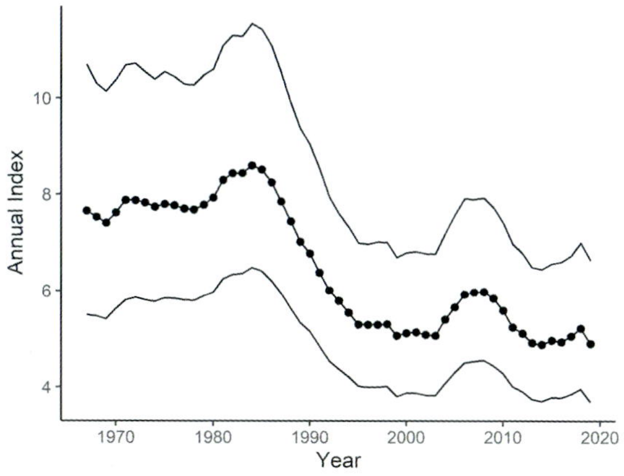

D. Breeding population trend and 95 percent credible interval (solid black lines) for the Mourning Warbler in Minnesota based on the 1967–2019 Breeding Bird Survey (Sauer et al. 2020).

have shown positive population responses following logging activity (Niemi and Hanowski 1984a; Hanowski, Danz, et al. 2003; Hanowski, Danz, et al. 2005; Atwell et al. 2008) and forest fire (Niemi and Probst 1990).

Maintaining an active silvicultural program to provide a constant supply of early successional vegetation has been proposed in the northeastern United States (Hunter et al. 2001). Logging activity in the state of Minnesota has been declining since 2005, especially in Minnesota's national forests (Niemi, Howe, et al. 2016). In contrast, over the past 150 years, Minnesota has seen extensive changes in its habitat profile of age classes and forest cover types. Young, early successional forests are more prevalent today compared with the original vegetation. It is possible that the population decline of Mourning Warblers represents a return to more normal levels following higher populations in the mid-1900s.

Additional threats to the species include changes on its wintering grounds and collisions during migration with windows, towers, and possibly wind turbines. With a 2°C warming scenario, NAS (2019) projected a moderate vulnerability for the Mourning Warbler's breeding range in North America, including a 56 percent range loss and a 41 percent range gain in northern Canada and Alaska by 2050. In Minnesota, the species was predicted to be extirpated under the 3°C warming scenario by 2080.

The recent decline of the Mourning Warbler is odd, given that there appears to be no shortage of breeding habitat in Minnesota. Continued scrutiny of its breeding population and changes in the future are essential to decipher elements associated with these fluctuations. The continued insectivory of the Mourning Warbler is an important component of healthy Minnesota forests.

KENTUCKY WARBLER

Geothlypis formosa

Photo by Jeremy Meyer.

Minnesota Seasonal Status: Casual, rare summer resident south, visitant elsewhere; occasional spring and fall migrant in southern half of Minnesota. The Kentucky Warbler was very rare during the MNBBA.
Migration: Medium- to long-distance migrant that winters from the Caribbean coast of southern Mexico through Central America to the northern areas of Columbia and Venezuela. It may also be found in winter in Bermuda and some of the Caribbean islands.
Food: Feeds on many types of insects, spiders, and some berries that are gleaned from vegetation.
Nest: Cup nest placed on or near the ground in dense understory vegetation, often at the base of a shrub, hummock, or low branch of a small tree.
Conservation Concern: Assigned a high PIF Continental Concern Score of 14/20.

In 1811, Alexander Wilson "discovered this handsome warbler and named it for the State in which he found it most abundant. The name is not inappropriate, for Kentucky is not far from the center of its abundance in the breeding season" (Bent 1953). The Kentucky Warbler is noted for its persistent, ringing song during the breeding season. Ridgway (1889) described its behavior as: "Considering its great abundance, the nest of this species is extraordinarily difficult to find . . . and he [the writer] has come to the conclusion that the female must slyly leave the nest at the approach of the intruder and run beneath the herbage until a considerable distance from the nest, when joined by her mate, the pair by their evident anxiety mislead the collector as to its location." It is secretive and difficult to see as it hops or runs through the dense undergrowth; when spotted the "tail is often held partly cocked upward and is sometimes flicked nervously" (J. L. Dunn and Garrett 1997). This behavior gave it the vernacular name of Kentucky Wagtail (T. G. Pearson 1917).

The Kentucky Warbler male has a distinctive plumage, especially with the black mask below its eye and large yellow eye stripes that look like spectacles. The female has a subdued mask that is usually reduced in intensity or mottled with a grayish tone. Otherwise, the male and female are similar, with olive-green upperparts, brilliant yellow underparts, and a short tail. In appropriate habitat of moist deciduous forests, the species is more often heard than seen. Its loud song consists of three to eight identical, up-slurred *chuurup* notes.

The breeding range of the Kentucky Warbler is the southeastern United States, from southern Pennsylvania west to Iowa, and then south to east Texas with fingers along the wooded river valleys in Nebraska, Kansas, and Oklahoma. The core of its range, where many of its highest densities occur, is along the Ohio River and the southern Mississippi River drainage.

MINNESOTA BREEDING DISTRIBUTION

The Kentucky Warbler was not part of Minnesota's known avifauna when Roberts (1932) wrote his opus. The first sighting was a persistent singing male near St. Mary's College, Winona County, in July 1952 (Theodore 1952) followed by one in Hennepin County (1956) and three other migrants in the 1950s. Noteworthy records after 1960 include Hennepin County, May 23, 1961 (road-kill specimen given to the Bell Museum; H. Huber 1961); singing male, below the Coon Rapids dam in Hennepin/Anoka Counties in fall 1963 (D. Pearson 1963), summer 1964 (D. Pearson 1964), and spring to summer 1965 (D. Pearson 1965; R. L. Huber 1966b); and one found on board an ore boat on Lake Superior, July 24, 1964 (by Captain J. P. Perkins *fide* JCG; Green and Janssen 1975).

The suspicion that the Kentucky Warbler might breed in Minnesota was deemed likely by observations in Nicollet County at Seven Mile Creek in 1992 (Filter 1993) and finally confirmed at the site in 1994 (Fall 1994a). From 1965 until breeding was confirmed in 1994, MOU seasonal reports included nine observations during the summer months that indicated probable breeding in the southern third of the state in Anoka, Lac qui Parle, Renville, Rice, and Washington Counties (MOU 2021b). There was also one observation record farther north: Itasca State Park, Hubbard County, June 4, 1978 (McKenzie 1978).

After the early 1990s, the number of Kentucky Warbler records in the MOU seasonal reports increased substantially and were reported every year, 1996–2008, during spring and summer but rarely in the fall (MOU 2021b). These observations were mostly clustered in two counties along the Minnesota River: Dakota/Scott (primarily Murphy-Hanrehan Park Reserve, ten years) and Blue Earth (primarily Williams Nature Center, nine years) with additional summer reports in several other counties, including LeSueur, Olmsted, and Rice.

From 2009 to 2013 participants in the MNBBA reported three new state locations for the Kentucky Warbler **(A)**, all within the distribution described above. The first record, coded possible, was on June 4, 2009, in Lake Louise State Park, in Mower County. Another record, coded possible, was de-

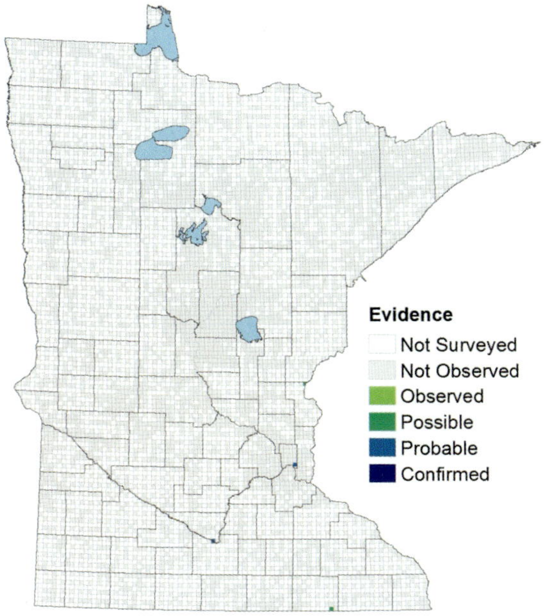

A. Breeding distribution of the Kentucky Warbler in Minnesota based on the Breeding Bird Atlas (2009–13).

tected on June 17, 2009, along the St. Croix River in the very northeastern corner of Chisago County. The third record, coded probable, was of a bird seen June 11–18, 2013, by many observers in Battle Creek Regional Park, Ramsey County. Finally, the MOU (2021a) database revealed that during the atlas the species was also reported during the breeding season at a fourth location, the Williams Nature Center in Blue Earth County, just south of the Minnesota River (T108R27d). Coded as probable, a bird was reported by several observers from May 29 to June 11, 2009, including a report on June 3 of two males singing to each other. The species was also observed near the center in May and June 2011, and in May 2013; previous reports were available from 2000 to 2008.

After the atlas survey years, the Kentucky Warbler has continued to be found at several locations, but at only two for extended periods: Sakatah Lake State Park, LeSueur County (May 31 to June 8, 2016) and Brown County (near New Ulm, May 31 to June 5, 2018) (MOU 2021a). Although there has not been confirmed breeding evidence since 1994, many records indicated a regular but rare breeding population, primarily along the Minnesota River valley.

In states adjacent to Minnesota, the Kentucky Warbler has had a breeding population in the southwestern portion of Wisconsin since 1924 (Grant County) and other areas near the confluence of the Wisconsin and Mississippi Rivers (S. D. Robbins 1991). In Iowa, many observations were made from 1960 to 1995, including a nesting record (1975) in Allamakee County bordering Minnesota (Kent and Dinsmore 1996).

McDonald (2020), in a review of the Kentucky Warbler in North America, cited recent evidence that the species has been expanding slowly northward, but cautions that this may represent reoccupation of the species' former breeding range.

BREEDING HABITAT

The Kentucky Warbler is a southern deciduous forest species with a specialized preferred habitat **(B)**. It is found in "deep shaded woods with dense, humid thickets, bottomlands near creeks and rivers, ravines in upland deciduous woods, and edges of swamps" (Kaufman 1996). McDonald (2020) described the "essential" habitat as "[w]ell-developed ground cover for ground nesting, and a thick understory." C. S. Robbins and colleagues (1989) classified it as a forest interior and area-sensitive species, while McDonald (2020) emphasized that riparian corridor width is important.

POPULATION ABUNDANCE

Based on the BBS, PIF (2021) estimated the population of the Kentucky Warbler at 2.6 million birds in the United States, but with no estimate in Minnesota because it is so rare. The BBS shows a significant, declining trend in the United States for the period from 1966 to 2019 that averaged −0.69 percent per year **(C)**. A summary of reasons given for the decline in the New York atlas is typical of many comments: "threats include loss and fragmentation of mature, moist deciduous forest, primarily from development

B. The southern deciduous forests where the Kentucky Warbler nests are characterized by dense tree canopies, thick understories, and a well-developed ground layer that helps conceal the warbler's nest. Photo by Gary T. Seim.

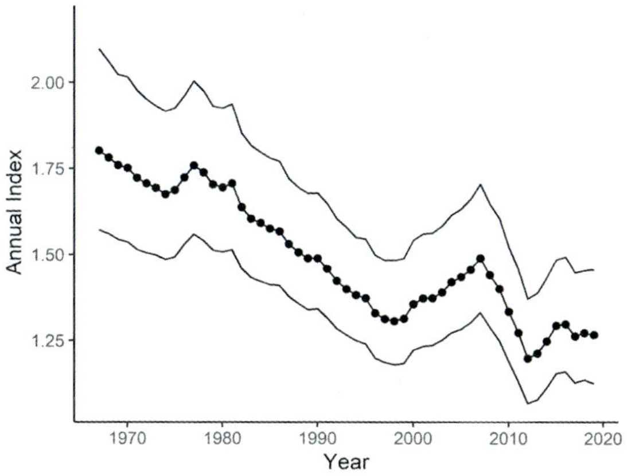

C. Breeding population trend and 95 percent credible interval (solid black lines) for the Kentucky Warbler in the United States based on the 1967–2019 Breeding Bird Survey (Sauer et al. 2020).

as well as loss of dense shrubby forest understory, the result of over-browsing by white-tailed deer and lack of disturbance in closed-canopy forests" (McGowan and Corwin 2008). Loss and colleagues (2014) found it was vulnerable to building collisions. They calculated that the species' risk of fatal collisions was 20 times greater with low-rise buildings and 14 times greater with high-rise buildings compared with the average bird.

CONSERVATION

The Kentucky Warbler has been identified as a species of high conservation concern because of its recent population declines and potential threats to its nonbreeding population. For the same reasons, PIF (2021) placed it on its Yellow Watch list, which includes a group of North American birds that are of continental concern. McDonald (2020) points out that it is listed as a "Bird of Conservation Concern" in many Bird Conservation Regions, USFWS regions, and by state agencies in approximately half the states where it occurs. However, it tends to be grouped with other forest-dwelling Neotropical migrants and, therefore, lacks specific management actions.

With potential movements northward, especially with our changing climate (NAS 2019), the Kentucky Warbler may become a more common species in Minnesota if suitable habitat is available. The warbler is a skulking species and notoriously hard to spot in its habitat of dense vegetation. It is a persistent and loud singer known for its high site fidelity, which, along with some observational determination, may produce more nesting evidence. Yet the Kentucky Warbler is not faring well in its current range and southern Minnesota may provide limited habitat opportunities.

COMMON YELLOWTHROAT

Geothlypis trichas

Minnesota Seasonal Status: Regular breeding resident and migrant. The Common Yellowthroat was very abundant during the MNBBA.
Migration: Short- to medium-distance migrant throughout most of its range, wintering in the southern United States, the Caribbean, Mexico, and Central America. Resident populations occur in the southeastern states, southern California, and Mexico.
Food: Arthropods secured primarily by foliage gleaning and ground foraging.
Nest: Cup nest, usually located on or just above the ground; the structure may include a slight overhang or roof. In very wet habitats the nests are usually placed higher to prevent flooding.
Conservation Concern: Assigned a moderate PIF Continental Concern Score of 9/20.

Competing for our attention in a region that supports one of the richest diversities of wood warblers in North America, the Common Yellowthroat may be a hands-down favorite. Abundant statewide, it is an easy species to find as it flits about in its brushy habitats, persistently scolding us with its raspy *tchat* call note. Unlike many of our northern warblers, who dash from tree-top to tree-top in their canopy abodes, this little masked robber is easily seen just a few feet above the ground. Indeed, it is Minnesota's most abundant warbler species. During the breeding season, the male's black facial mask, bordered by white above and bright yellow below, is unmistakable. The olive-gray female can be more difficult to identify because she lacks the black facial mask. Luckily, she is the only olive-green warbler with a yellow throat and white belly that occupies brushy sites during the summer.

The male's signature song, *witchity-witchity-witch*, is easily recognized and can be heard from afar. He also delivers a warbling flight song that Roberts (1932) whimsically described as a "real love-song . . . an outburst of the pent-up intensity of the full nuptial feeling." *Geothlypis trichas*, the species' scientific name, roughly translates to "a ground bird with a shrill song."

Despite its widespread distribution and abundance, few long-term studies of the species have been conducted. One of the few exceptions is a study undertaken by the late Dr. Pershing B. Hofslund

Photo by Michael Furtman.

(1959). A professor at the University of Minnesota Duluth from 1949 to 1982, Dr. Hofslund studied the breeding behavior of Yellowthroat populations in Michigan and Minnesota, including on the University's campus.

The Common Yellowthroat's breeding range stretches from southeastern Alaska through the boreal forests and parklands of Canada. Although its distribution is spotty across the southwest, it is also found throughout the United States and south into Mexico. High breeding densities occur in southeastern Canada and in the northern United States, from the Great Plains east across the Ohio River valley and New England. Within this core area, the highest breeding densities are attained in central Minnesota.

MINNESOTA BREEDING DISTRIBUTION

Since Hatch wrote his account of the Common Yellowthroat's status in Minnesota in 1892, the species has been known to be widespread throughout all corners of the state. Even forty years later, Roberts (1932) wrote: "Rare indeed would it be to find a bushy meadow by brook or lake-side, anywhere within the confines of our state, where the cheery and forcible ditty—*witch-i-te witch-i-te witch-i-te*—of the beautiful and animated little male Northern Yellow-throat would not greet the late spring and early summer visitor." At the time of his writing, nesting had been confirmed or inferred in 10 counties across western and central Minnesota: Cass, Douglas, Goodhue, Grant, Hennepin, Isanti, Jackson, Otter Tail, Pennington, and Scott.

Little had changed forty years later, when Green and Janssen (1975) noted that the Yellowthroat was the most "widespread and numerous warbler species in the state." Janssen (1987) added that it was also one of Minnesota's "most evenly distributed breeding species." He identified 28 widely dispersed counties where the species had been documented nesting since 1970. Hertzel and Janssen (1998) later added five more counties to the list. By 2014, the 4,616 breeding-season locations identified by the MBS exceeded the number of records tallied by the survey for any other breeding bird in the state.

MNBBA participants reported a remarkable total of 9,609 Common Yellowthroat records in 80.6 percent of the surveyed atlas blocks and in 95.5 percent of the priority blocks **(A)**. Breeding was confirmed in only 291 blocks (6.0 percent), demonstrating that documenting breeding activity even for a conspicuous, shrub-nesting warbler can be challenging. Probable breeding was documented in an additional 1,934 blocks (40.2 percent). The birds were observed in all 87 Minnesota counties and were confirmed breeding in 68 counties compared to 33 documented by Hertzel and Janssen (1998). The Common Yellowthroat was the second most abundant species during the MNBBA, second only to the American Robin, with 10,227 records. Among warblers, the Yellowthroat far outnumbered the second and third most abundant species: the Ovenbird with 5,755 records and the Yellow Warbler with 5,333 records.

The MNBBA probability map confirms the abundance of the species across Minnesota **(B)**. High breeding densities are predicted throughout the state; even the intensely cultivated regions of the Red River valley and the upper Minnesota

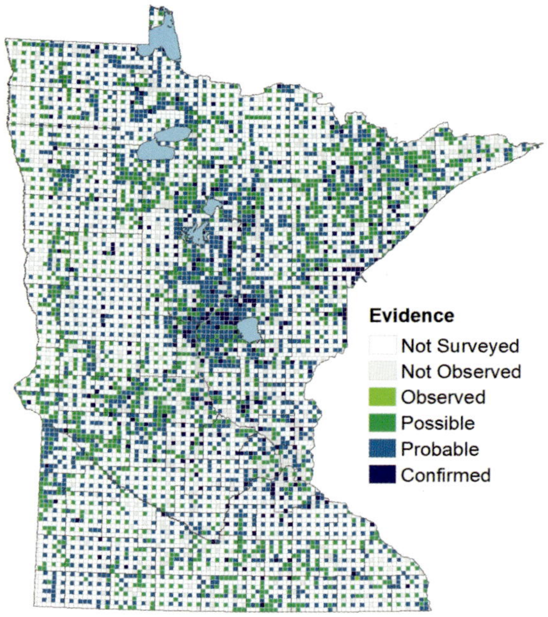

A. Breeding distribution of the Common Yellowthroat in Minnesota based on the Breeding Bird Atlas (2009–13).

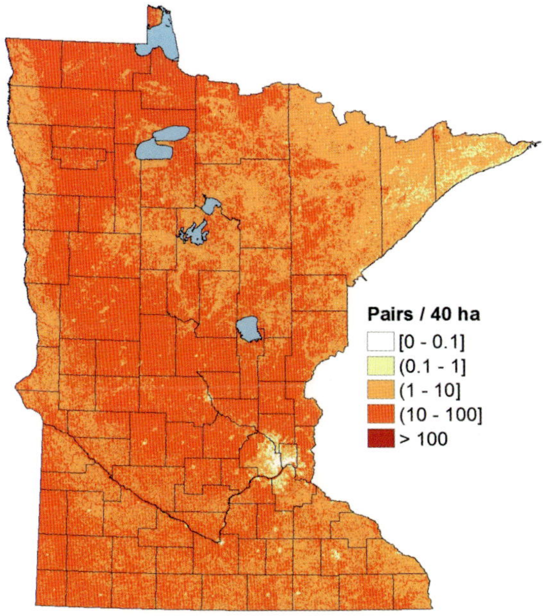

B. Predicted breeding distribution (pairs per 40 ha) of the Common Yellowthroat in Minnesota based on habitat, landscape context, and climate data gathered during the Minnesota Breeding Bird Atlas (2009–13) using the Generalized Linear Modeling method with an adjustment for detectability.

River valley support moderate breeding densities. The only large region where breeding densities are predicted to be quite low is the core of the Twin Cities metropolitan region, where suitable habitat is less abundant.

Today the Common Yellowthroat appears as widely distributed and as exceedingly common as it did one hundred years ago. It is as likely to be found in an extensive prairie wetland in southwestern Minnesota as it is in a tiny, low-lying opening in the forests of northeastern Minnesota. Despite its abundance, the significant loss of wetland habitat, especially in the southern and western regions of the state, has no doubt negatively impacted abundance levels. Nevertheless, the little Yellowthroat still finds suitable habitats across Minnesota's agricultural landscape. The same is true throughout its breeding range, where few changes to its overall distribution have been noted (Guzy and Ritchison 2020).

BREEDING HABITAT

One reason for the Common Yellowthroat's success is that it is able to occupy a wide range of habitats. Although typically associated with marshes and shrubby wetlands **(C)**, it also occupies nearly any patch of dense, low-lying vegetation, from wet prairies and weedy fields to the shrubby understory of open bogs. Even stands of upland shrubs and young regenerating conifers are suitable (Danz et al. 2007; Niemi, Howe, et al. 2016).

POPULATION ABUNDANCE

PIF (2021) generated a population estimate of 76 million breeding adults in the United States and Canada; an additional 1 million birds reside in Mexico. Minnesota was estimated to support approximately 7.1 percent of the continental population, or nearly 5.4 million birds. MNBBA data, however, generated a much higher statewide estimate of 14.7 million adults (95 percent confidence interval of 14.2–15.4 million). The MNBBA's coverage of many inaccessible wetlands not monitored by the BBS, including far north-central and northeastern Minnesota, likely contributes to a larger and more accurate population assessment.

The importance of Minnesota to the Common Yellowthroat population is evident from its abundance along BBS routes. In central Minnesota, where breeding densities are especially high, it is common to detect 50 or more Yellowthroats along each 25-mile route. Statewide, an average of 34 detections is reported per route, a number higher than that found in any other state (Sauer et al. 2020).

BBS data, however, reveal a slow, but statistically significant, surveywide decline, averaging −0.56 percent per year from 1966 through 2019 (Sauer et al. 2020). In Minnesota, population numbers initially increased from 1967 through the early 1980s but have been steadily declining ever since **(D)**. Indeed, from 1993 to 2019, the state population experienced an average decline that is equivalent to that seen at the national level since 1966, −0.56 percent per year. Data from the Chippewa and Superior National Forests corroborate this downward trend (Grinde, Bednar, et al. 2020).

C. Common Yellowthroats are abundant in a variety of wetland habitats, including this large wetland complex at the Carlos Avery WMA in Anoka County. Photo by Lee A. Pfannmuller.

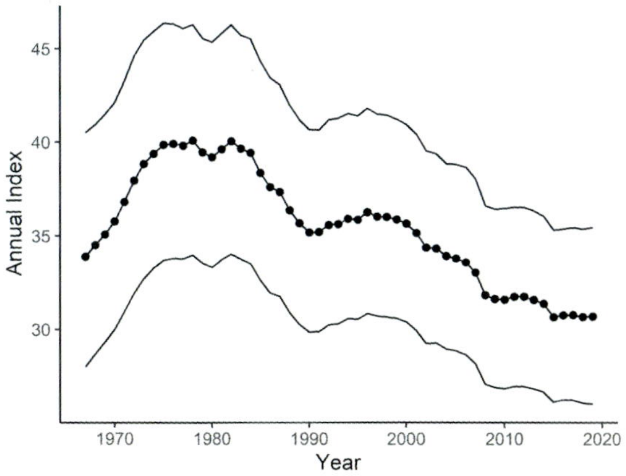

D. Breeding population trend and 95 percent credible interval (solid black lines) for the Common Yellowthroat in Minnesota based on the 1967–2019 Breeding Bird Survey (Sauer et al. 2020).

Nest predation, parasitism by Brown-headed Cowbirds, collisions with communication towers and buildings, and weather all impact local populations (Guzy and Ritchison 2020). But habitat loss and degradation are certainly the most important factors affecting this abundant little warbler. Although it is a habitat generalist, the Common Yellowthroat is most strongly associated with shrubby and herbaceous wetlands, habitats that have been drained or filled for decades for agricultural production and urban/suburban development.

CONSERVATION

Despite the downward trend in population numbers, the Common Yellowthroat continues to be a widespread and abundant species and is not considered a high conservation priority (PIF 2021). Even its vulnerability to climate change is thought to be minimal, with no distributional changes predicted in Minnesota (NAS 2019). During migration, however, nighttime collisions with communication towers and buildings may be a significant source of mortality (Bracey 2011; Loss et al. 2014). Loss and his colleagues estimated that the Common Yellowthroat is 10.9 times more vulnerable to building collisions than species with an average risk. Field studies of these hazards are now leading to practical management recommendations to help reduce bird fatalities.

The most important conservation measures that will benefit the Common Yellowthroat are those that target the protection and restoration of wetlands. In the grasslands region of the state, where wetland losses have been greatest, management work designed to protect and restore both wetlands and grasslands will help ensure that Minnesota remains a stronghold for the Common Yellowthroat's North American population.

HOODED WARBLER

Setophaga citrina

Minnesota Seasonal Status: Regular breeding species and migrant. The Hooded Warbler was rare during the MNBBA.
Migration: Medium-distance migrant that spends winters in portions of the Caribbean, southern and eastern Mexico, Central America, and the very northern tip of South America.
Food: Consumes insects and small spiders obtained by foliage gleaning, ground gleaning, and by hovering or sallying out to catch insects in midair.
Nest: Cup nest, usually placed in shrubs less than 1.5 m above the ground.
Conservation Concern: Assigned a moderate PIF Continental Concern Score of 9/20; officially listed as *Special Concern* in Minnesota.

Photo by Ivan Gabaldon.

One of the newest additions to Minnesota's breeding avifauna, the aptly named Hooded Warbler was first observed in the state in 1942 and first documented breeding in 1984. Frank Chapman ([1907] 1968) gave high praise to this exquisite little warbler when he wrote, "Its beauty of plumage, charm of voice, and gentleness of demeanor, make it indeed not only a lovely, but a truly loveable bird." He went on to note that the natural habitat of the species, the forest understory, added considerably to its charm because it was one of the few warblers that one did not have to strain to find in the highest reaches of the forest canopy.

As it quickly flits about the understory's small saplings and shrubs, the male Hooded Warbler is instantly recognizable with its distinctive black hood and its bright yellow forehead, cheeks, and belly. Females have a similar plumage, but the amount of black on the head varies from none to a faint outline of a partial hood. Like the American Redstart, the Hooded Warbler's white outer tail feathers are displayed when it rapidly fans its tail in and out. This behavior is thought to startle insects, causing them to flush from the foliage and be quickly snapped up with the Warbler's flycatching behavior. Indeed, its scientific name translates to lemon-colored (*citrina*) moth eater (*Setophaga*). Yet, like all warblers in the *Setophaga* genus, the Hooded Warbler's diet is considerably more diverse than its Latin name implies.

Vocally, the Warbler's short and clear song is an easily identified tune of *weeta-weeta-weet-tee-o*. The only song it might be confused with is that of the Magnolia Warbler, which is similarly described as *weta, weta, WETA,* the last note being rather loud and accented. Chiver and her colleagues (2020) note that the Magnolia's "song is often faster, of higher frequency, and not as loud as the Hooded's." Clearly, highly discriminating ears are helpful!

A species largely restricted to the eastern deciduous forests south of the Great Lakes, the Hooded Warbler has slowly been expanding its range northward. Along with several other southeastern warblers and vireos, it is also expanding to the west, particularly across the southwestern United States and into California. The species reaches its highest breeding densities from the Appalachian Mountains southwest to Arkansas, Louisiana, Mississippi, and eastern Texas.

MINNESOTA BREEDING DISTRIBUTION

When Roberts was actively documenting Minnesota's avifauna in the late 1800s and early 1900s, the Hooded Warbler had yet to be observed in the state. The first report of its presence would occur ten years after he published his two-volume treatise, *The Birds of Minnesota,* in 1932. In an area now known as Roberts Bird Sanctuary, located on the north end of Lake Harriet in Minneapolis, an adult male was observed and heard singing on May 17, 1942 (Davidson 1942). Twenty years would pass before the second report was logged. On June 2, 1962, an adult male was caught in a mist net at a banding station at Lake Demontreville in Washington County (Olyphant 1962). Then, from 1969 through 1973, two additional reports were documented from Hennepin County (Arneson 1972; Stanley 1973) as well as one new report each from Ramsey County (Peake 1970) and Mille Lacs County (Goldberg 1973). All six records, from 1942 through 1973, were of individual birds, including four adult males and two adult females. Three of the records were in the spring season, two in the summer, and one in the fall. When Green and Janssen published their account of the species in 1975, the Warbler's status was classified as accidental.

Beginning in the spring of 1975, the Hooded Warbler was documented somewhere in the state, during the spring or summer season, every year except 1986 (MOU 2021b). From 1975 through 1983, observations of single individuals continued to be reported in Hennepin County, from the Roberts Bird Sanctuary and Theodore Wirth Park, and in Anoka and Scott Counties. Additional outstate reports were logged from Clearwater, Crow Wing, Le Sueur, Olmsted (fall record), Otter Tail, Rice, and Sherburne Counties. Finally, forty-two years after the Hooded Warbler was first observed in the state, a pair nested successfully in 1984 at Murphy-Hanrehan Park Reserve, which straddles the Scott-Dakota county line on the southern edge of the Twin Cities metropolitan region (Fall 1984).

Encompassing a matrix of woodlands and abandoned farmland, the reserve includes a large, unfragmented tract of mature hardwoods. Approximately 3 square kilometers in size, the tract is a rare resource in a densely populated region that supports 3.6 million people (Fall 1989). Since the first pair was reported nesting in 1984, Murphy-Hanrehan has become the core of the state's breeding population, supporting as many as 35 territorial males during the 2001 breeding season (MOU 2021b).

During the 1980s and 1990s, as Hooded Warbler numbers continued to build in Dakota and Scott Counties, an increasing number of reports were logged from the Twin Cities metropolitan region, including in Anoka and Carver Counties, and from several southeastern counties, including Fillmore, Freeborn, Goodhue, and Winona. Scattered reports were also documented as far as Pipestone County in the spring of 1992, Cook County in the fall of 1995, and St. Louis County in the spring of 1999 (MOU 2021b). Most surprising was a well-documented report of a nesting pair at Camp Ripley in Morrison County during the 1996 breeding season (Merrill 1996), two summers after a male was first heard singing and was observed on the facility grounds (Merrill 1994).

In the following decade, during the years leading up to the initiation of the MNBBA (from 2000 to 2008), more spring reports came from the northwest region, including Beltrami and Kittson Counties, and the southwest region, including Lac qui Parle, Lyon, and Murray Counties. There was a summer report from Pennington County in 2008 and one from Pine County in 2002 (MOU 2021b). Nesting was confirmed in Anoka County in 2003. The report followed several seasons where birds had been observed at multiple sites in the county, including at Linwood Lake, where nesting was eventually confirmed (Bardon 2003). Nevertheless, despite all the widely scattered reports of birds since 1942, Murphy-Hanrehan remained the one site where the birds were most consistently documented during the summer months.

The Hooded Warbler's range expansion into Minnesota coincided with its expansion in the neighboring states of Wisconsin and Iowa. In Wisconsin, at least one to ten birds were reported nearly every year since 1939, the first year that records were published in *The Passenger Pigeon,* the state's ornithological field journal (S. D. Robbins 1991). Most records were confined to eastern counties near Madison and Milwaukee and were males that were briefly present from May 5 to May 25. The first nesting record was documented in 1975 in Manitowoc County in southeastern Wisconsin. Five more breeding records were reported in three southeastern counties by June 1980 (S. D. Robbins 1991). Then, by the late 1990s, during Wisconsin's first breeding bird atlas (1995–2000), the species was confirmed nesting in 13 counties, with probable or possible breeding in 15 additional counties, all within the southern two-thirds of the state (Cutright et al. 2006).

The species was reported as common in the floodplain forests of southeastern Iowa in the late 1800s, when two nests were found in Lee County, bordering the Mississippi River (Currier 1895). But that may have been a rare and localized occurrence. Between 1900 and 1967, there were only four records from four counties bordering the Mississippi River (Kent and Dinsmore 1996). Beginning in 1968, the number of reports slowly but steadily increased. When the first breeding bird atlas was conducted in Iowa (1985–90), the Hooded Warbler was reported in seven blocks as far west as the outskirts of Des Moines, including one confirmed nesting report just west of Iowa City. During the second atlas (2008–12), the species was reported in 14 blocks, including four where nesting was confirmed (Iowa Ornithologists' Union 2019).

Recent range expansions and increasing populations have been noted in other localities as well, including Michigan (Chartier et al. 2013), Ontario (Cadman et al. 2007), and in the southwestern United States and California (Patten and Marantz 1996).

Reported from nine counties during the MNBBA, the Hooded Warbler was found in only 15 surveyed atlas blocks. Breeding was confirmed in Dakota, Scott, and Washington Counties and deemed probable in Chisago and Hennepin Counties. Possible breeding activity was reported in Mower County, and the birds were observed in Morrison, Sherburne, and St. Louis Counties **(A)**. Three of the nesting localities were on public lands: Afton State Park in Washington County, Lebanon Hills Regional Park in Dakota County, and Murphy-Hanrehan Park Reserve in Dakota and Scott Counties.

In addition to these atlas records, a number of observers reported the Hooded Warbler at three other localities during the five breeding seasons that spanned the MNBBA (2009–13). Details were not specific enough to assign them to a given atlas block but included an observation at the Rapids Lake unit of the Minnesota Valley NWR in Carver County (July 1, 2009); and observations at the Hyland Lake Park Reserve (June 4, 2011) and Cedar Lake (June 7, 2011) in Hennepin County. Following the MNBBA, a pair was also reported nesting in 2015 at the Woodland Trails Park in Elk River, Sherburne County, where the species was observed in 2009 during the MNBBA (MOU 2021b).

Although the Hooded Warbler has become a regular breeding resident and migrant in Minnesota over a period of nearly seventy-five years, it remains relatively rare. While records are scattered across all regions of the state, its primary breeding range remains restricted to the broader Twin Cities metropolitan area, bounded by Dakota and Scott Counties in the south, Washington County in the east, and Anoka and Sherburne Counties in the north. Further expansion outward from this core is likely in the coming years. Nevertheless, for a species that has expanded its range from the core of its distribution in the southeastern United States, it is interesting that so few reports occur south of the Twin Cities, particularly from some of the larger forest tracts on state forest lands and in state parks in southeastern Minnesota. MBS surveys in this region in the 1990s documented only two Hooded Warbler records: one at Murphy-Hanrehan Park Reserve in the Twin Cities and one at Lake Louise State Park in Mower County. Now, more than twenty years later, a closer look across the southeastern region may be warranted.

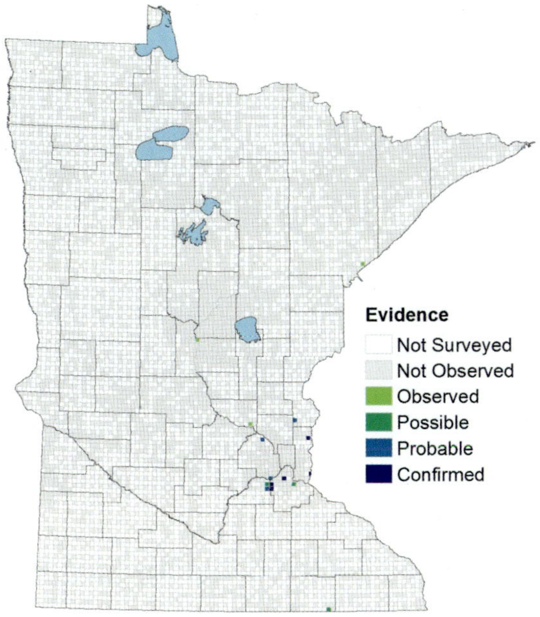

A. Breeding distribution of the Hooded Warbler in Minnesota based on the Breeding Bird Atlas (2009–13).

BREEDING HABITAT

The Hooded Warbler's breeding habitat is broadly described as mature forest interspersed with canopy gaps that allow a dense growth of small saplings and shrubs (Chiver et al. 2020) **(B)**. In the core of its breeding range in the southeastern United States, it is frequently associated with mesic to wet lowland forests dominated by cypress and tupelo with an understory dominated by palmetto and cane. Farther north, the species is found in mesic to dry upland forests. These include large tracts of mature hardwoods, selectively harvested stands of conifers and hardwoods, and former pastures with a sparse tree layer and a dense growth of shrubs and young trees (J. L. Dunn and Garrett 1997; Chiver et al. 2020; Danz et al. 2007; Bielefeldt and Rosenfield 2001).

The composition and age of the canopy appears less important than the structure and composition of the understory. Here a dense growth of shrubs and/or young saplings, vines, or semiwoody herbaceous plants is critical (Bielefeldt and Rosenfield 2001). The small woody stems provide support for the Warbler's nests, which are usually less than 1.5 m above the ground, and provide foraging habitat. Most foraging is restricted to an area less than 3 m above the ground (J. L. Dunn and Garrett 1997; Chiver et al. 2020; Bielefeldt and Rosenfield 2001).

In extensive forest tracts, pockets of dense shrubs occur where canopy gaps are created by natural disturbances—such as lightning, tree falls, and windstorms—or by selective harvesting. Trails and roads may also simulate such conditions, allowing enough light to penetrate to encourage a dense understory. Some have referred to the Hooded Warbler as a "gap specialist." But such conditions may be simulated in a variety of modified habitats, including selectively harvested conifer plantations and old pastures

B. A mature hardwood forest with a shrubby understory within the Murphy-Hanrehan Park Reserve is the core of the Hooded Warbler's distribution in Minnesota.
Photo by Lee A. Pfannmuller.

with scattered trees and shrubs (Chiver et al. 2020; Bielefeldt and Rosenfield 2001).

The literature on whether the species requires large forest tracts is conflicting. Many studies conducted farther east suggest that large forests are required. C. S. Robbins (1979), for example, found the Hooded Warbler absent from sites less than 485 ha in size, while in Illinois sites larger than 600 ha were required (Blake and Karr 1984). Another study found that the most significant factors influencing abundance were the height of the canopy, the percentage of forest within 2 km, and the foliage density in the understory (C. S. Robbins et al. 1989). Forest area, however, was not an important predictor of abundance. Other studies have found that, despite the species' apparent preference for large forest tracts, the Warbler successfully nests in tracts smaller than 5 ha (Norris et al. 2000; Eaton 1988), particularly when the small fragments are located near much larger tracts of continuous forest (Norris and Stutchbury 2002; Rush and Stutchbury 2008). Often subject to high rates of brood parasitism by Brown-headed Cowbirds, at least one study has shown that rates of parasitism are higher on forest tracts less than 30 ha in size than on tracts greater than 150 ha (Chiver et al. 2020).

The most comprehensive description of the Hooded Warbler's breeding habitat in Minnesota was provided by Fall (1989), who has closely monitored the local breeding population at Murphy-Hanrehan Park Reserve since nesting was first documented at the site in 1984 (Fall 1984). Fall's examination of old aerial photos shows the presence of a mature forest within the Reserve as early as 1937. The mixed hardwood stand is dominated by oak and elm species but includes many other trees, such as sugar maple, basswood, and bigtooth aspen. The understory is dense and includes oak, elm, and ironwood saplings and a diversity of shrubs, including dogwoods, prickly ash, and hazel. Fall speculated that the size and maturity of this forest tract were the major reasons a small, loose colony of Hooded Warblers was originally attracted to the area.

POPULATION ABUNDANCE

BBS data have generated a North American population estimate of 5.2 million birds in the eastern United States and southeastern Canada (PIF 2021). Given the number of birds reported each year at Murphy-Hanrehan and at a handful of other sites in the state, the statewide estimate for Minnesota is likely only a few hundred birds.

Surveywide, the BBS reveals that the Hooded Warbler has experienced a significant increase, averaging 1.09 percent per year from 1966 to 2019 **(C)**. Since 1970, biologists estimate that the continental population experienced a cumulative increase of 103 percent just through 2014 (Rosenberg, Kennedy, et al. 2016). Florida is the only state that has witnessed a statistically significant decline since 1966. Declines, however, are seen in the coastal regions of other states, including Texas and Louisiana (Sauer et al. 2020).

Factors responsible for this increase are largely unknown. On the breeding grounds, predation and brood parasitism are the most significant factors impacting local productivity, and both are exacerbated by landscape fragmentation. T. M. Donovan and Flather (2002) found that the greatest proportion of the Hooded Warbler's population occurs in relatively unfragmented landscapes, where productivity is presumably higher. As long as a sufficient amount of large, unfragmented forests remains available, population increases may continue.

CONSERVATION

Although Hooded Warbler populations are increasing throughout most of the species' breeding range, threats to both breeding and nonbreeding habitats remain primary concerns. This is especially troublesome in many regions of Central America, where the loss of mature forests has been significant (Chiver et al. 2020). As a result, PIF (2021) considers the Hooded Warbler to be a moderate conservation priority. In Minnesota, the species was officially listed as *Special Concern* in 1996 because of its small and restricted breeding population (MNDNR 1995).

Although research is somewhat conflicting, the Hooded Warbler should benefit from conservation measures that ensure the protection and management of large, continuous blocks of mature forest that are at least several hundred hectares in size. The long-standing population that resides at Murphy-Hanrehan Park Reserve occupies a site that is at least 300 ha. Disturbances, natural or prescribed, that maintain small openings with dense shrub growth are essential. Management guidelines in Wisconsin recommend maintaining forest tracts that are a minimum of 100 ha, with a dense understory of shrubs and saplings and small (0.03–1 ha) canopy gaps. The guidelines note that high deer densities can negatively impact the suitability of sites by reducing growth in the understory through browsing (Wisconsin DNR 2013).

Nest predation, nest parasitism, and collisions with communication towers are other threats that impact Hooded Warbler populations (Chiver et al. 2020; Longcore et al. 2013). The species' association with more mesic and wet forests in the core of its breeding range in the southeastern United States may also place it at risk as temperatures rise and drier climates prevail. Fortunately, climate change models predicted that the species' North American breeding distribution had a low vulnerability under a 2°C warming scenario, with a 16 percent range loss and a 17 percent gain to the north by the year 2050 (NAS 2019). Further expansions throughout southeastern Minnesota were also predicted.

In the short term, the Hooded Warbler appears to be a species on the move. An increasing population trend and a northward range expansion have established this once elusive species as a regular breeding resident in Minnesota. Biologists and birders alike should continue to keep an eye out for the species in suitable forest habitats throughout the state.

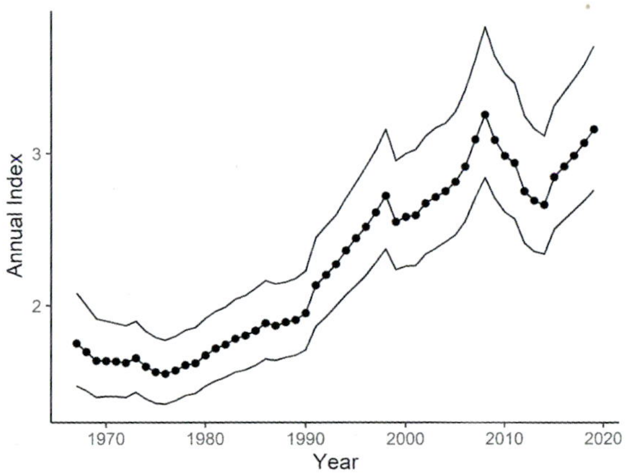

C. Surveywide breeding population trend and 95 percent credible interval (solid black lines) for the Hooded Warbler based on the 1967–2019 Breeding Bird Survey (Sauer et al. 2020).

AMERICAN REDSTART

Setophaga ruticilla

Minnesota Seasonal Status: Regular breeding resident and migrant. The American Redstart was abundant during the MNBBA.
Migration: Medium- to long-distance migrant that winters in the Caribbean and from southern Mexico to northern South America.
Food: Insectivore that secures prey items by hovering, gleaning, and fly-catching; also flushes prey by flashing its wings and tail.
Nest: Cup nest placed in a deciduous shrub or tree at variable heights.
Conservation Concern: Assigned a moderate PIF Continental Concern Score of 10/20.

Photo by David Brislance.

The male American Redstart is a sight to behold. In Cuba, where many Redstarts spend the winter, his extraordinary black and orange plumage has earned him the name *Candelita*, or "little torch" (Roberts 1932). The female's plumage is a subdued version of the male's stunning colors. His jet-black feathers are replaced with an olive-green wash while his flaming orange-red feathers, prominent on the sides of the chest and in the wings and tail, are replaced with bright yellow. Unlike other warblers, the male Redstart resembles the adult female until the end of his second summer **(A)**. Small patches of black feathers may appear very slowly beginning in the fall after hatching, but it is only after his first breeding season that he attains the full black and orange plumage of the adult male.

The species name, *ruticilla*, is Latin for red (*rutilus*) tail (*cilla*). Redstart, the vernacular name, is considered "a corruption of the Old German word *rothstert*," which also translates to "red tail" (Terres 1980). The anglicized word became "redstart" and was applied to the Eurasian Common Redstart (*Phoenicurus phoenicurus*), a small bird with a gray crown and back and an orange breast and tail. Later, the name was given to the American Redstart because it was thought to resemble the unrelated Eurasian species.

The Redstart, Roberts (1932) wrote, is "imbued with a spirit of unceasing activity"; its "tiny little body is ever in a state of fluttering agitation." It is known for its habit of dropping its wings and fanning its tail to flash their bright colors. A communication signal with other Redstarts, the flashing behavior also helps startle and flush insect prey from foliage (Sherry et al. 2020). Because the species forages most frequently at "low to mid-levels" in the forest (Dunn and Garrett 1997), the patient observer has ample opportunities to witness the "little torch" in action.

Like many warblers, the Redstart has a large vocal repertoire that can confuse even experienced listeners. Its weak, high-pitched warble lacks any outstanding features "except its extreme variation" (Bent 1953). A characteristic of its most common song is the abrupt, accented ending that sounds "sneeze-like." Each male also has three or four additional songs that lack the deliberate ending of the primary song (Sherry et al. 2020). Because melodies that lack the accented ending may differ among individual males, the repertoire of a local population of Redstarts will have "considerably more songs" than any one individual (Morse 1989), further confounding the challenge of song identification. The female, Roberts (1932) wrote, "is also a singer, though perhaps without as full a repertoire as her consort." Her weak and infrequent songs are often sung near the nest (Sherry et al. 2020).

The American Redstart is one of the most intensively studied members of the wood warbler family (Dunn and Garrett 1997). Sherry and his colleagues (2020) noted that the species' "broad geographic range and abundance, conspicuous foraging and communication behaviors, and accessibly low nests" contribute to the strong interest among researchers. Many unique aspects of its life history have also raised interesting evolutionary and behavioral questions.

For example, several morphological features that enable members of the flycatcher family to catch flying insects in midair are also present in the aerial-foraging Redstart (Morse 1987; Roberts 1932; Sherry et al. 2020). Like that of flycatchers, the Redstart's bill is wide and flat at the base, which allows it to sweep up insects efficiently. Also like those of flycatchers, its wings and tail have large surface areas

A. The plumage of a female American Redstart is more subtle than that of the adult male; first-year males look similar to the female. Photo by David Brislance.

that enable quick aerobatic maneuvers. And its stiff, bristle-like feathers at the base of the beak are a feature shared with many aerial insectivores, even though their precise function has been debated. Unlike the flycatcher, which perches, "patiently waiting for his prey to come near," the little Redstart "is continually in action, dashing here and there after flying insects or perhaps snapping up larva dangling in midair at the end of its long silken fiber" (Bent 1953). Nevertheless, given the Redstart's physical similarities with flycatchers, even renowned ornithologist Alexander Wilson considered the "little torch" to be a member of the flycatcher family (Roberts 1932).

Keast and his colleagues (1995) tackled the question of the Redstart's evolutionary convergence with flycatchers. They conducted an exhaustive examination of the physical differences between several wood warblers (including the Redstart) and several flycatchers (including the Least Flycatcher and Eastern Wood Pewee). The Redstart, they concluded, "is a unique adaptive and ecomorphological type among North American passerines." It has many of the physical features shared by all wood warblers, including their long, tapering wings that are designed for long-distance migratory flights and a long leg bone that is designed for hopping and clinging. But the Redstart is not a flycatcher, nor is it evolving to become a flycatcher. It simply has evolved similar physical features that enable it to take advantage of a unique foraging niche among wood warblers.

Another unusual aspect of the Redstart is the delayed maturation of the male's plumage. Biologists are keen to understand why the male never acquires his nuptial plumage until the end of his first breeding season. What is the evolutionary advantage of this strategy? Different theories abound (Sherry et al. 2020). One theory proposes that mimicking the female's cryptic plumage serves as an anti-predator defense, lowering mortality in yearling males. Another theory proposes that maintaining the female's plumage allows young males to avoid a molt during their first winter. This may reduce foraging competition with older males to find dietary sources of the carotenoid pigments that develop their reddish-orange feathers. Research studies have yielded many clues to this evolutionary conundrum, but, thus far, "the mystery remains" unsolved (Morse 1989).

The intriguing little Redstart is widely distributed across Canada's boreal forest and is absent only from portions of southern Alberta and Saskatchewan. In the United States, the species is found in the northern Rockies and east of the Great Plains. It

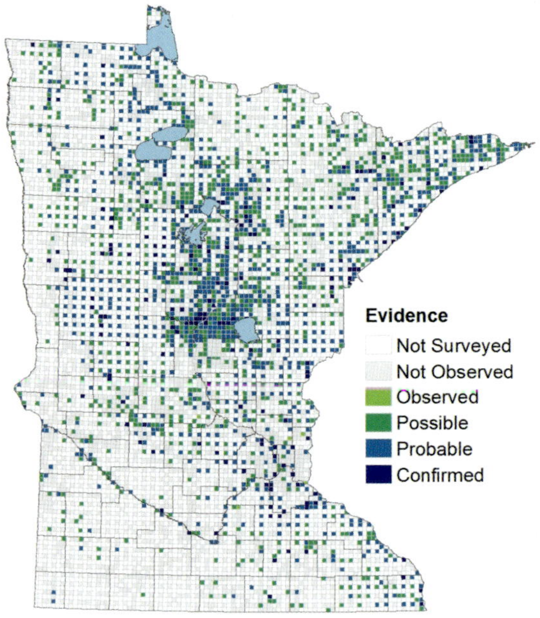

B. Breeding distribution of the American Redstart in Minnesota based on the Breeding Bird Atlas (2009–13).

is absent from the southern coastal plain, including most of Florida, and from portions of Kentucky and Tennessee. High breeding densities occur in British Columbia and in portions of several northern states and eastern provinces, including Minnesota.

MINNESOTA BREEDING DISTRIBUTION

Roberts (1932) described the American Redstart as an abundant and widely distributed species across "the heavily timbered regions" of Minnesota. "In some localities, as on the wooded islands of the Mississippi River bottom-lands where they border the state on the southeast, it frequently outnumbers all other small birds put together." It was also found in small woodland groves on the prairie, though in considerably fewer numbers. Confirmed breeding records were available from Cook, Goodhue, Hennepin, and Otter Tail Counties and in the Mille Lacs region; breeding was inferred in Cass and Ramsey Counties and from the Cass Lake and Leech Lake regions.

Forty years later, Green and Janssen (1975) also described the Redstart as a statewide breeding resident but characterized it as "very scarce south of the Minnesota River, from Blue Earth and Faribault counties westward." Janssen (1987) included a distribution map that eliminated the southwestern corner of the state, from Lac qui Parle County southeast to western Freeborn County, stating it was scarce to entirely absent from the region. He identified 27 counties with confirmed nesting since 1970. Hertzel and Janssen (1998) later added five counties to the list, including Brown County in south-central Minnesota.

By 2014, the MBS had documented 1,522 American Redstart breeding-season locations. Common throughout the Laurentian Mixed Forest and Eastern Broadleaf Forest Provinces, the species was also found in the floodplains of the Minnesota River valley west to Yellow Medicine and Renville Counties. Notable were scattered records across southwestern and south-central Minnesota.

MNBBA participants documented a somewhat broader distribution than that delineated by the MBS **(B)**. Reported from 45.3 percent (2,150/4,748) of the surveyed atlas blocks, the records covered all but two of Minnesota's 87 counties: Nobles and Martin. Breeding was confirmed in 169 blocks (3.6 percent) spanning 53 counties and deemed probable in an additional 937 blocks (19.7 percent). Twenty-six of the counties where breeding was confirmed were additions to the list published by Hertzel and Janssen (1998). Common throughout all but the Prairie Parkland Province, the American Redstart was still found in floodplain forests along the entire length of the upper Minnesota River valley, in small woodlots in the Red River valley, and in southwestern Minnesota. Confirmed nesting records, however, were sparse in the southwestern counties. Once again, the impressive efforts by atlas participants in the Brainerd Lakes region were

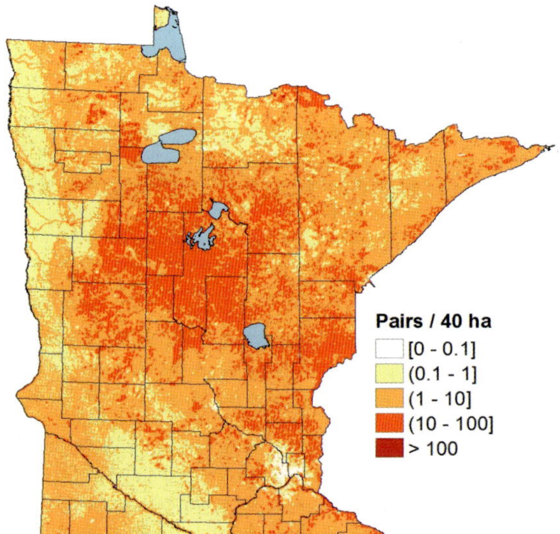

C. Predicted breeding distribution (pairs per 40 ha) of the American Redstart in Minnesota based on habitat, landscape context, and climate data gathered during the Minnesota Breeding Bird Atlas (2009–13) using the Generalized Linear Modeling method with an adjustment for detectability.

D. Mesic hardwood stands with a moderate to dense shrub layer are among the habitats where the American Redstart is most abundant. Photo by Gerald J. Niemi.

apparent. Overall, the American Redstart was the fifth most abundantly reported warbler during the MNBBA, with 4,069 detections.

The MNBBA probability map predicts that the highest breeding densities for this statewide resident occur along the Mississippi River valley and its tributaries in southeastern Minnesota and throughout north-central Minnesota, primarily south of Red Lake and Koochiching County **(C)**. Some of the lowest breeding densities are predicted in the Twin Cities, in the more intensively cultivated western counties, and in the bogs and lowland conifer forests of northern Minnesota.

The status of the American Redstart in Minnesota has changed little since the days of Roberts (1932). As prairie groves were converted to productive croplands, suitable habitat was likely at a premium in the 1900s, when Green and Janssen (1975) and Janssen (1987) noted its scarcity in southwestern Minnesota. Although it remains an uncommon species in the region, small patches of suitable habitat are now present.

Few major shifts in the Redstart's breeding distribution have been documented elsewhere within its breeding range other than localized extirpations due largely to habitat loss. The loss of bottomland timber, for example, was cited as a major factor for the species' absence from the Florida panhandle (Sherry et al. 2020).

BREEDING HABITAT

Redstarts are common in mid-age, mesic or moist, deciduous or mixed deciduous woodlands, with a diverse and abundant shrub/sapling layer **(D)**; water also is a common feature **(E)**. An open understory beneath the forest canopy and above the shrub/sapling layer is an important structural feature because it allows room for this fly-catching warbler to aerially pursue flying insects (Bond 1957; Sherry 1979). Although studies in national forests in Minnesota and Wisconsin demonstrated that oak uplands and lowland hardwoods were preferred, the species was found in many other habitats (Niemi, Howe, et al. 2016).

Evidence that the American Redstart is an area-sensitive species is equivocal. Some studies concluded that it prefers tracts larger than 4,000 ha (e.g., Ambuel and Temple 1983; C. S. Robbins et al. 1989); other studies have suggested that forest edges may be favored (DeGraaf 1985; Chasko and Gates 1982). The results suggest that further work is needed to delineate the important features of quality breeding habitat.

POPULATION ABUNDANCE

PIF (2021) generated a population estimate of 42 million breeding adults in the United States and Canada, with approximately 1.3 million birds residing in Minnesota. The statewide population estimate derived using the extensive set of MNBBA data was significantly larger at 7.7 million birds (95 percent confidence interval of 6.9–9.2 million).

American Redstart population numbers have been relatively stable across the BBS survey region from 1966 to 2019 (Sauer et al. 2020), experiencing a cumulative decline of only −12 percent from 1970 to 2014 (Rosenberg, Kennedy, et al. 2016). Different regions, however, reveal widely divergent trends. For example, populations in the northeast have experienced some of the largest declines, while some of the greatest increases have occurred in the Upper Midwest (Sauer et al. 2020). In Minnesota, populations have experienced a significant increase, averaging 0.55 percent per year **(F)**. Data from the Chippewa and Superior National Forests corroborate this upward trend (Grinde, Bednar, et al. 2020).

E. Although they can occur in many deciduous and deciduous–coniferous habitats, American Redstarts are frequently found in riparian forest stands bordering streams, rivers, and lakes. Photo by Lee A. Pfannmuller.

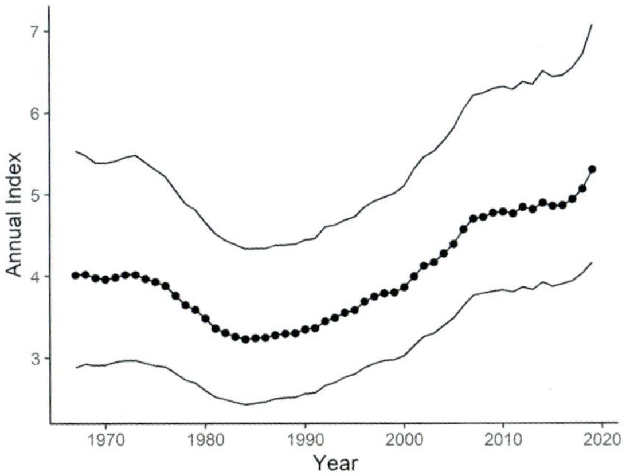

F. Breeding population trend and 95 percent credible interval (solid black lines) for the American Redstart in Minnesota based on the 1967–2019 Breeding Bird Survey (Sauer et al. 2020).

Factors that regulate populations are varied and include forest succession, cowbird nest parasitism, nest predation, food availability, and conditions on the wintering grounds. The Redstart's large population, for example, occupies a winter range that is only 63 percent the size of its breeding range, potentially limiting food availability and diminishing the bird's physical condition prior to its northward migration (Sherry et al. 2020).

CONSERVATION

With its relatively stable population trend, this widespread and abundant species is not a high conservation priority (PIF 2021). Only a few management recommendations have been identified. In mature forest stands, for example, selective thinning of the understory trees is recommended to maintain openness below the canopy for foraging. The forest canopy, however, should remain largely closed (Sallabanks 1993; Bushman and Therres 1988).

Warming temperatures will have some impact on the species. Its breeding distribution was projected to have a low vulnerability under a 2°C warming scenario, with losses along the southern periphery of its breeding range largely compensated by northward gains (NAS 2019). Although its breeding range will gradually recede from most of western Minnesota as temperatures rise, the diversity of woodland habitats that the Redstart finds suitable is a likely guarantee that it will continue to be one of Minnesota's most abundant warblers for many years.

CAPE MAY WARBLER

Setophaga tigrina

Minnesota Seasonal Status: Regular breeding resident and migrant; has occasionally been observed lingering into early winter at bird feeders in Minnesota. The Cape May Warbler was uncommon during the MNBBA.
Migration: Medium-distance migrant overwintering in the Caribbean, southern Mexico, and Central America.
Food: Invertebrates including insects (especially caterpillars such as spruce budworm, of which it is a specialist), spiders, fruit, and nectar during the winter. Forages primarily while perched but also known for hawking, hovering, or fly-catching for prey.
Nest: Cup nest, usually high in a coniferous tree and close to the trunk.
Conservation Concern: Assigned a moderate PIF Continental Concern Score of 12/20.

Photo by David Brislance.

The Cape May Warbler is poorly named because it has little to do with Cape May, New Jersey, except as a migrant. The type specimen of the species was collected in 1831 in Cape May County, where it was described by Alexander Wilson. Interestingly, neither Wilson nor Audubon ever saw this warbler alive. The Cape May Warbler was one of the prominent spruce-woods warblers studied by Robert MacArthur (1958) in his development of ecological niche theory that ultimately became one of the foundational studies in the advancement of quantitative ecology (Mayr 1982).

Roberts (1932) called it a "beautiful little bird" and "unusually attractive." Griscom and Sprunt (1957) were more explicit: "gaudily colored warblers there certainly are, but in any contest for the most handsome warbler, the male Cape May must surely be given consideration." The breeding plumage of the male, with its yellow head and chestnut-colored cheeks, is relatively easy to identify. Its white wing patch and black streaking on the yellow breast also aid in identification. The species' scientific name, *tigrina,* is Latin and refers to the "tiger-like" streaks on its breast and sides. The female is similarly colored but lacks the distinctive chestnut cheeks.

The song of the Cape May is relatively distinctive but requires practice, experience, and excellent auditory capabilities to confidently identify it. Common vocalizations are described as a high pitched *seet-seet-seet-seet-seet* or a two-part *seetee seetee seetee seetee seetee* (Alderfer 2006). Regardless of the season, the Cape May Warbler can be difficult to observe on its breeding grounds, where it occurs high on a spruce, especially white spruce, or balsam fir in upland forests infested with spruce budworm. The species is found patchily distributed across Canada from Labrador to British Columbia and in the northern portions of the Upper Midwest and the northeastern United States. Its highest densities have been observed in eastern Quebec and central Ontario.

MINNESOTA BREEDING DISTRIBUTION

Regarding the Cape May Warbler, Roberts (1932) quizzically stated that "if nesting at all in Minnesota it must be in very limited numbers and in restricted localities." His only report of potential nesting activity was an observation by Thaddeus Surber on June 19, 1919, who observed three or more pairs in Pine County that, he said, had "actions" indicating a nest was nearby.

Years later, Green and Janssen (1975) reported that Cape May Warblers were primarily resident in the northern part of the northeastern region of Minnesota. Observations during the breeding season had been noted as far south as southern St. Louis County and west to Itasca State Park. They identified confirmed nesting in northern Lake County and inferred nesting in Clearwater, Cook, and Hubbard Counties.

A few years later, Janssen (1987) described the Cape May Warbler's distribution as the northern part of northeastern Minnesota in "normal" years but stressed that its numbers can fluctuate widely with spruce budworm outbreaks. During infestations, it can be found in scattered locations in northeastern and north-central regions of the state. Janssen did not identify any confirmed nesting since 1970 but outlined its breeding distribution from Cook County to Clearwater County, northwest to eastern Roseau County, and southwest to northern Hubbard County; the southern boundary was defined by northern Itasca and southern St. Louis Counties. Hertzel and Janssen (1998) included only St. Louis County with a confirmed nesting record since 1970. The MBS recorded 155 breeding-season locations for the Cape May Warbler. The majority of their locations were clustered in northern St. Louis and western Lake Counties.

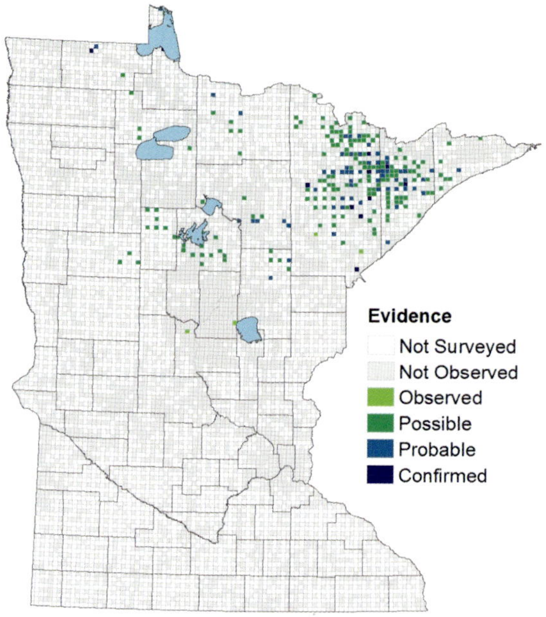

A. Breeding distribution of the Cape May Warbler in Minnesota based on the Breeding Bird Atlas (2009–13).

The MNBBA reinforced that the major breeding distribution of the species is in the northeastern counties of St. Louis, Lake, and Cook but included an observation as far south as Morrison County and possible nesting west to Becker County **(A)**. The Cape May Warbler was confirmed nesting in 11 blocks within Lake, Lake of the Woods, Roseau, and St. Louis Counties. There were 59 blocks (1.2 percent of surveyed blocks) with probable nesting, which included the four previously named counties plus Aitkin, Cass (block straddling Cass/Itasca), Cook, Itasca, and Koochiching. Possible nesting was reported from an additional 169 blocks. An observation of tailless young was reported by MOU (2021a) in Cook County in July 2012.

Baltz and Latta (2020), in their review of the Cape May Warbler in North America, primarily attributed historical changes in the distribution of the species to responses in budworm outbreaks. Because of the lack of observations of the species in the 1800s and early 1900s, the historical status of the Cape May Warbler in Minnesota is difficult to assess. The atlases from Michigan, Ontario, and Wisconsin all note the difficulty in censusing the species and in the identification of its nesting behavior. The Cape May Warbler's song is high, weak, and usually delivered from high atop a coniferous tree, often in remote areas. Its nest is hidden high in a coniferous tree near the trunk where, as noted by Roberts (1932), the nest is "invisible from the ground." Thus, the documentation by Cottrille (1962) in Lake County near Ely is impressive; her party was able to identify "eight small young three or four days old" that were "three and one-half feet from the top of a 23.5 foot black spruce." They discovered the Cape May nest after eight days of searching in an area with a spruce budworm outbreak. In summary, the status of this species begs for additional research, but it is challenging to study a species that is rare and highly variable in its population and distribution.

B. Cape May Warblers are highly associated with coniferous forests of balsam fir and spruce, especially if the trees are infested with spruce budworm. Photo by Gerald J. Niemi.

BREEDING HABITAT

The haunts of the Cape May Warbler are generally described as medium- to old-aged coniferous habitats with spruce and balsam fir (Baltz and Latta 2020) **(B)**. Niemi, Howe, and colleagues (2016) recorded a majority of observations in northeastern and north-central Minnesota in aspen–spruce–fir, aspen–birch, jack pine, coniferous swamps, and white pine stands. Most of these stands were of mixed types with varying densities of white spruce and balsam fir trees.

POPULATION ABUNDANCE

The US and Canadian population estimate of the Cape May Warbler by PIF (2021) was 7 million breeding adults, with an estimated 76 percent decline from 1970 to 2014 (Rosenberg, Kennedy, et al. 2016). About 96 percent of the Cape May Warbler breeding population occurs in Canada. Minnesota estimates by the MNBBA were approximately 852,000 breeding adults (95 percent confidence interval of 606,000–2,012,000), while PIF (2021) estimated a Minnesota population of only 73,000. During the MNBBA, Cape May Warbler populations were high because of several outbreaks of spruce budworm, especially in Minnesota's Superior National Forest, where most of the state's population occurs (Grinde, Bednar, et al. 2020). The BBS trends in all areas were deemed unreliable; however, the BBS trend in Minnesota did indicate a general tendency toward increasing from 1967 to the mid-1990s, followed by a slight decrease to the year 2000, and then an increase to 2017 **(C)**. The recent increase is consistent with the period of recent spruce budworm outbreaks in northeastern Minnesota. Population fluctuations in Cape May Warblers are not well known but are likely a combination of its dependence on spruce budworm cycles, high collision mortality during migration with windows and towers, and the possibilities of overwintering mortality, which unfortunately are not well known.

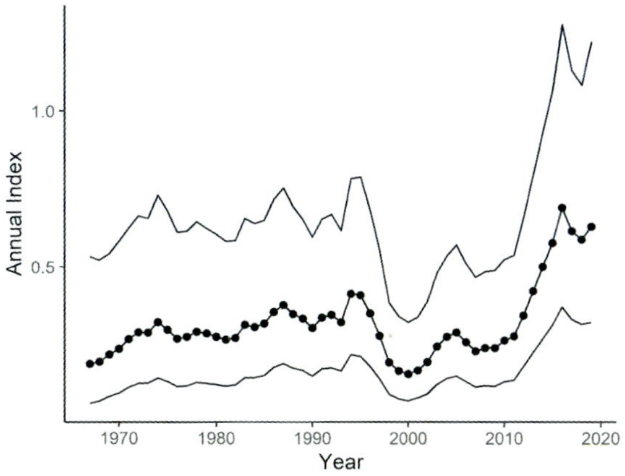

C. Breeding population trend and 95 percent credible interval (solid black lines) for the Cape May Warbler in Minnesota based on the 1967–2019 Breeding Bird Survey (Sauer et al. 2020).

CONSERVATION

Among the greatest concerns for the Cape May in North America are its large population decline since the early 1970s (Rosenberg, Kennedy, et al. 2016) and climate change. With a 2°C warming scenario, NAS (2019) projected a high vulnerability for the Cape May's breeding distribution in North America, including an 83 percent range loss and a 43 percent range gain by 2050. Although their models depicted a current breeding range in northeastern Minnesota that was extremely small, a rise of only 1.5°C would be sufficient to eliminate the Warbler from Minnesota and a large portion of the Canadian boreal forest. The drastic loss of breeding habitat would require a rapid expansion of coniferous trees northward, but this is unlikely to occur in a short time frame. Cape May Warblers prefer tall coniferous trees, not the stunted black spruce that often occur near the Arctic tundra. Its future does not look positive, given that its population is already declining.

Cape May Warblers have high mortality during migration. Loss and colleagues (2014) stated it has a high risk of collisions with low-rise buildings. They calculated its risk of mortality from building collisions as 16.7 times more than the average risk for a bird. Longcore and colleagues (2013) estimated their annual mortality at communication towers was about 1 million per year. Mortality from collisions was of most concern in peninsular Florida because a substantial part of the North American population migrates from Florida to overwinter in the Caribbean.

Because of threats throughout its life cycle, from breeding to migration to wintering, its future is dubious and definitely dependent on multinational actions to address these threats. Minnesota can help in the survival of this attractive and beneficial warbler by maintaining healthy coniferous forests.

CERULEAN WARBLER

Setophaga cerulea

Photo by Rebecca Field.

Minnesota Seasonal Status: Regular breeding resident and migrant. The Cerulean Warbler was rare during the MNBBA.
Migration: Long-distance migrant that winters in northern South America.
Food: Foliage gleaner that feeds largely on small insects.
Nest: Cup nest, usually placed high in the canopy of a deciduous tree.
Conservation Concern: Assigned a high PIF Continental Concern Score of 15/20; officially classified as *Special Concern* in Minnesota.

Few birds can claim to be the poster child of a major environmental campaign to halt mountaintop mining or the focus of a novel's subplot by a renowned writer (*Freedom* by Jonathan Franzen). It is an impressive resume for a tiny warbler that few have seen as it flits high within the forest canopy. Unfortunately, the Cerulean Warbler also has a small population that is declining at an alarming rate (Sauer et al. 2020).

Described by Bent (1953) as the "heavenly blue wood warbler," the species is aptly named for the cerulean feathers that don the male's head, back, and wings. Two white wing bars are prominent, as are a thin blue breast band and stripes along the flanks of the white belly. The female's plumage is pale blue-green above with a light yellow wash below. She lacks the male's breast band, but a soft yellow eyeline aids her identification. Weak, high-pitched, and in three parts, the male's buzzy, *zee zee zee zizizizi zeeet* song sounds like the phrase *Just a little sneeze* (Bent 1953).

Early ornithologists knew little about this canopy dweller, which was first described in 1831. Even Audubon's description of its nest was said to be "entirely erroneous" (Bent 1953). Nearly fifty years passed before the first Cerulean Warbler nest was discovered and properly described in New York. Today many details regarding its life history are still poorly known, a testament to the difficulty of studying a species that inhabits the high canopy of mature deciduous forests (Buehler et al. 2020). The Warbler is found from Minnesota south to Arkansas and east through the Appalachian Mountains; its breeding densities are highest in the mountains of West Virginia.

MINNESOTA BREEDING DISTRIBUTION

The Cerulean Warbler is another southern species that arrived in Minnesota by following the Mississippi River north in the late 1800s. Although Hatch (1892) claimed to have collected the first specimen in 1869, presumably in Minneapolis, his account sheds some doubt regarding the validity of his statements. More reliable were accounts of a specimen collected by Dr. Hvoslef in Fillmore County in 1885 and a specimen collected by Mr. Gleason in Carver County in 1898 (Roberts 1932). Reports slowly increased, with observations in Hennepin and Goodhue Counties in 1919; Houston, McLeod, and Olmsted Counties in 1924; and Meeker County in 1930. Although Roberts (1932) described the Cerulean Warbler as "a rare summer resident" breeding in southeastern Minnesota, the only breeding evidence was a 1929 report of three immature birds observed in Minneapolis.

In his 1936 edition of *The Birds of Minnesota,* Roberts provided a lengthy update on the species. Most remarkable were new reports of two small populations. The first was from Linwood Lake in Anoka County. Discovered in 1933, the birds were present "in large numbers" in 1934, and by summer's end four nests had been located (Prosser 1934). Nestor Hiemenz found another small population in 1934 in Stearns County. In a letter to Roberts, Hiemenz wrote that as "many as fifteen to twenty singing males could be located in a single afternoon's walk" (Roberts 1936).

Reports from several northern counties came in the mid-1960s: Mille Lacs (D. Bolduc 1966); Crow Wing (N. Roberts 1966); and Mahnomen (Kelleher 1967). By 1975, Green and Janssen described the species' breeding range as occurring "within a wide belt on both sides of the Mississippi River as far north as Mahnomen County." The Warbler was also spreading west, with reports from Douglas County in west-central Minnesota and Nicollet County in south-central Minnesota. Janssen's 1987 update noted a 1978 nesting record at Maplewood State Park, Otter Tail County, and a 1985 report of a pair seen at Flandrau State Park, Brown County. Since 1970, nesting had been confirmed in seven counties: Anoka, Goodhue, Hennepin, Nicollet, Olmsted, Otter Tail, and Scott. Hertzel and Janssen (1998) added Brown, Dakota, Ramsey, and Winona to the list. Meanwhile, MBS biologists documented 138 Cerulean Warbler breeding-season locations. Confined primarily to the Eastern Broadleaf Forest Province, records were found as far north as Becker, Crow Wing, and Pine Counties.

As national attention began to focus on the species' plummeting population, the Cerulean Warbler Atlas Project was launched. From 1997 to 2000, field observers across the eastern United States assessed the extent of the Warbler's breeding distribution. In Minnesota, 103 singing males were reported at 57 sites. Of these, nine sites had five or more singing males. Murphy-Hanrehan Park Reserve in Scott and Dakota Counties reported the highest number, with 20 males (Rosenberg, Barker, and Rohrbaugh 2000).

A few years later, MNBBA participants tallied just 45 Cerulean Warbler reports from 33 atlas blocks widely dispersed across 26 of Minnesota's 87 counties. Breeding was confirmed in five blocks that encompassed six counties: Brown/Renville, Goodhue, Le Sueur/Nicollet, and Winona **(A)**. Probable nesting was reported from 13 counties, including two where Cerulean Warblers had not previously been reported: Isanti and Lyon. Following the atlas, nesting was reported for the first time in Houston County (2016; MOU 2021b).

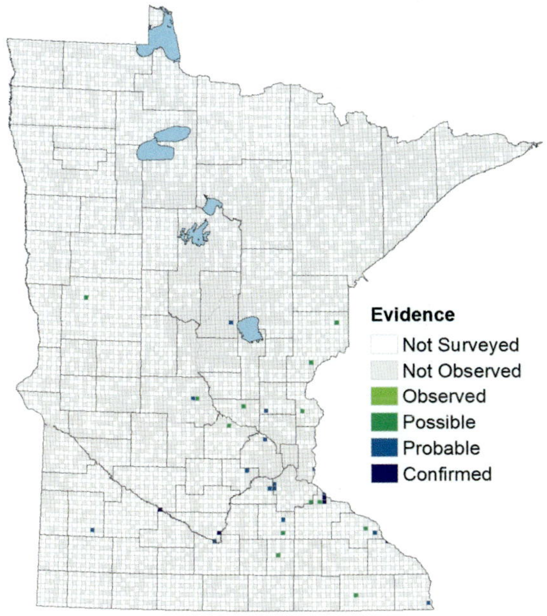

A. Breeding distribution of the Cerulean Warbler in Minnesota based on the Breeding Bird Atlas (2009–13).

Although the Warbler's Minnesota breeding distribution appears unchanged since the 1980s, the small number of MNBBA records certainly suggests a decline in abundance. The bird's thin, wispy song and its canopy-dwelling habits can challenge field observers. Nevertheless, because both the MNBBA and the Cerulean Warbler Atlas Project (conducted almost ten years earlier) used volunteers and paid surveyors to conduct their field assessments, the decrease in detections between the two projects may accurately portray the species' decline. Indeed, throughout the Warbler's breeding range there have been few changes in the extent of its distribution but significant declines in overall abundance.

BREEDING HABITAT

Rosenberg, Barker, and Rohrbaugh (2000) characterized the Warbler's breeding habitat as "distinctly bimodal," with birds evenly distributed between upland and lowland forests. In southeastern Minnesota, the species is found along the slopes of mesic deciduous forests bordering small streams and in floodplain forests along the Mississippi and Minnesota rivers. In central Minnesota, it often occupies drier hardwood forests that are interspersed with small wetlands (MNDNR 2019c) **(B)**. Regardless of the type of forest selected, suitable habitats include a structurally diverse forest canopy with tall, mature deciduous trees and scattered openings (Potter et al. 2007). The species is often associated with these canopy gaps.

The Warbler's dependence on large forest tracts may vary geographically (e.g., Thogmartin et al. 2004; P. B. Wood et al. 2006). Some studies have shown that smaller tracts are suitable in the north, while larger tracts are required in the south (Buehler et al. 2020). In the Upper Midwest, however, the Cerulean Warbler Atlas Project found 65 percent of all records were in forests larger than 400 ha; less than 10 percent were in forests smaller than 40 ha (Rosenberg, Barker, and Rohrbaugh 2000). Even when forest tracts are highly suitable, the species' distribution may be localized and patchy (Potter et al. 2007).

POPULATION ABUNDANCE

PIF (2021) provided a population estimate of only 530,000 adults in the United States and Canada, while BBS data document a significant decline in numbers across the survey region, decreasing an average of −1.84 percent per year from the mid-1960s to 2019 **(C)**. Cumulatively, the population declined −72 percent from 1970 to 2014, one of the steepest declines of any North American songbird (Rosenberg, Kennedy, et al. 2016). Unfortunately, given the species' low abundance, the trend line lacks statistical reliability. Statewide data are not available.

The loss and degradation of large, mature forest tracts is the most important factor impacting the Cerulean Warbler. In the core of its breeding range, one of the greatest threats is mountaintop coal mining, which removes an entire summit or ridge to expose buried seams of coal. The mature hardwood forests on these mountains provide ideal breeding habitat (Parr 2007). Habitat loss is also a concern on the wintering grounds, where the conversion of the species' montane forests to agriculture is occurring at a rapid pace (Terborgh 1989; C. S. Robbins et al. 1992).

B. In southeastern Minnesota, Cerulean Warblers are typically found in mature, mesic hardwood stands that have scattered canopy openings, a diverse understory, and are often associated with small streams.
Photo by Lee A. Pfannmuller.

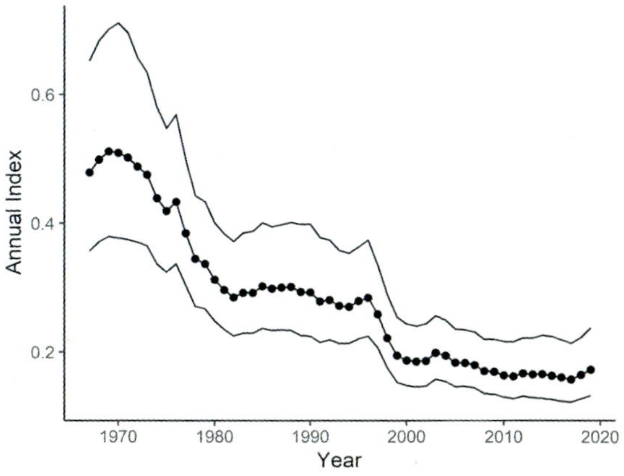

C. Surveywide breeding population trend and 95 percent credible interval (solid black lines) for the Cerulean Warbler based on the 1967–2019 Breeding Bird Survey (Sauer et al. 2020).

CONSERVATION

The Cerulean Warbler is widely recognized as a species in serious trouble. PIF (2021) designated it a high conservation priority and placed it on their Yellow Watch List, a group of North American songbirds in need of management attention to stabilize current population numbers (Rosenberg, Kennedy, et al. 2016). The USFWS (2007a) also prepared a national conservation plan. In Minnesota, the species is officially listed as *Special Concern* (MNDNR 1995).

Protection of large, mature deciduous forest tracts is critical to the Cerulean Warbler's future. Biologists advocate for maintaining forests that are at least 700 ha in size (Knutson, Butcher, et al. 2001). Uneven-aged management is recommended to retain large, mature trees and structural diversity in the understory and forest canopy (Hamel and Rosenberg 2007). Warming temperatures, however, may imperil sites that currently provide suitable habitat. Under a 2°C warming scenario, NAS (2019) projected the species was moderately vulnerable, losing 79 percent of its current breeding distribution by 2050. Fortunately, losses may be compensated by an 89 percent range expansion north into the Great Lakes region and southeastern Canada, including very small gains in eastern Minnesota.

Few Minnesotans have come to know and appreciate this charming little warbler as it flits high within the forest canopy. Lacking concerted conservation efforts, it may be even more difficult to do so in the years ahead.

NORTHERN PARULA

Setophaga americana

Photo by David Brislance.

Minnesota Seasonal Status: Regular breeding resident and migrant. The Northern Parula was a common species during the MNBBA.
Migration: Short- to medium-distance migrant that winters in southern Florida, the Caribbean, Mexico, and northern Central America.
Food: Arthropods gleaned from foliage.
Nest: Typically in a cup hollowed out within epiphytic hanging moss at variable heights in trees.
Conservation Concern: Assigned a low PIF Continental Concern Score of 8/20.

The Northern Parula, often referred to as "dainty" because it is one of the smallest warblers, is a common species in our northern forests, especially where hanging *Usnea* moss, or "old man's beard," is found. Its common name, Parula, is derived from the Latin genus *Parus* and the diminutive ending *ula* and translates to "little titmouse." Many ornithologists feel the term "Parula" is inappropriate for such a delicate warbler. It has often been referred to as the Blue Yellow-backed Warbler.

Roberts (1932) described its plumage as "conspicuously beautiful" with a "refined touch in coloration" that is "unexcelled among its fellows." The male is a handsome bird, with its yellow back, throat, and breast on a background of gray, distinct white eye crescents, black lores, and two white wing bars. The female is similar, but its coloration is slightly more diluted. They are both, indeed, sparkling birds, especially when seen in optimal light conditions. Like many birds of the forests, the species is most often identified by the male's song. It is a rising buzzy trill, described as *tse-e-e-e-ep,* ending abruptly in a slightly lower *tship.*

The Northern Parula is found exclusively in the forests of eastern North America with two broad populations, one in the northeastern and upper midwestern United States and Canada and the other in the southeastern United States. The highest densities occur in Nova Scotia and in northwestern Florida.

MINNESOTA BREEDING DISTRIBUTION

Roberts (1932) stated the Northern Parula was "a summer resident throughout the northern coniferous forests, where it breeds as far south as northern Isanti County and as far west as Itasca Park and eastern Polk County." He noted nesting activity in Becker, Itasca, and Pine Counties, as well as at Cass Lake and Itasca State Park. Descriptions within these areas indicated it was a common species. A field party visiting northern Itasca County in June 1923 noted "it may be detected in almost every suitable locality by the sharp-eyed and quick-eared observer." The latter suggested the species was easily overlooked because it can be difficult to find birds foraging or singing in the trees and because observers are potentially unaware of its song.

Green and Janssen (1975) described its breeding distribution as primarily in the northeastern and north-central regions of the state. They emphasized it was sparse in the northern portions of the north-central region and concurred with Roberts that its western distribution is limited to central Polk County. Confirmed nesting beyond those areas reported by Roberts included Clearwater, Hubbard, Lake, and St. Louis Counties, with inferred nesting in Cass, Cook, and Crow Wing Counties. Several years later, Janssen (1987) documented confirmed nests in five counties since 1970: Cass, Clearwater, Cook, Hubbard, and Lake. He also emphasized that the species' distribution in the northwestern portion of the state was not well known. In 1998, Hertzel and Janssen added no additional confirmed nesting records for this species since 1970.

The MBS included 524 breeding-season locations and largely confirmed the breeding distribution previously described. They also confirmed the sparseness of the species in the northwest and included no locations from that region.

The MNBBA included 1,388 records and confirmed the difficulty in finding nests of this species. Only 16 blocks were identified with confirmed nesting: a meager 2.4 percent (16/674) of all the blocks where breeding evidence was detected, which was 14.3 percent (674/4,727) **(A)**. These numbers are partly owing to the patchy nature of the Northern Parula's breeding habitat and the difficulty of definitively observing the species in a nest placed high in a tree covered in moss. The MNBBA confirmed nesting of the Northern Parula in five counties: Aitkin (first confirmed record), Cook, Lake, Itasca, and St. Louis Counties. Its distribution was very limited in the northwest and western regions, confirming Janssen's suspicions.

The predicted probability map based on MNBBA point counts indicates that the primary distribution of the Northern Parula is in the extreme northeastern regions of Minnesota, especially in Cook County **(B)**. Smaller, isolated populations were also predicted across much of northern and north-central Minnesota as far west as Becker and Otter Tail Counties and northwest to eastern Roseau County.

Roberts described the breeding distribution as being as far south as Isanti County and as far west as Polk

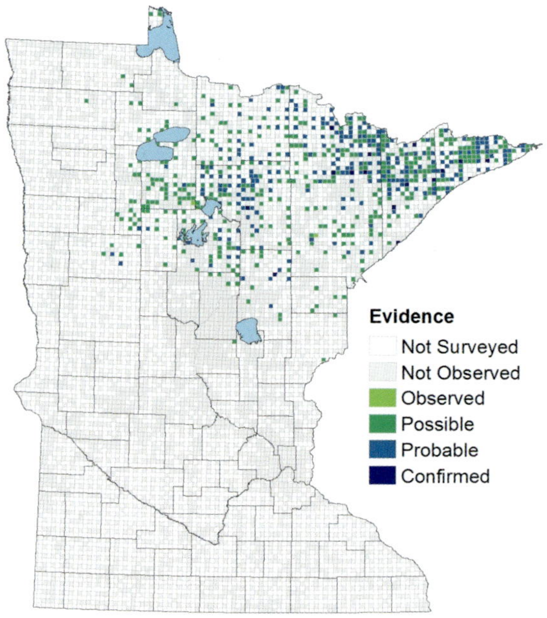

A. Breeding distribution of the Northern Parula in Minnesota based on the Breeding Bird Atlas (2009–13).

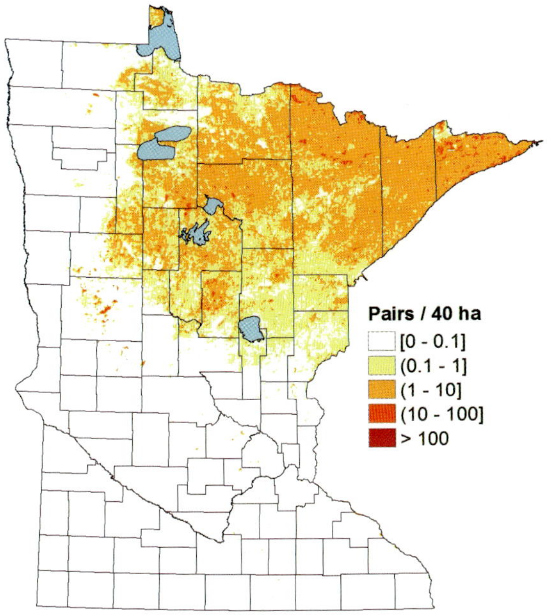

B. Predicted breeding distribution (pairs per 40 ha) of the Northern Parula in Minnesota based on habitat, landscape context, and climate data gathered during the Minnesota Breeding Bird Atlas (2009–13) using the Generalized Linear Modeling method with an adjustment for detectability.

County. It appears the breeding range of this species has retracted slightly since the early 1900s, but it was still confirmed nesting recently in Clearwater County (Janssen 2019). Retractions in western Minnesota are likely due to loss of forested habitat, especially riparian floodplain forests with suitable hanging moss for nesting. Moldenhauer and Regelski (2020), in their review of the Northern Parula in North America, state that this species had largely been extirpated or reduced in numbers in several northeastern states. These range contractions were purportedly due to air pollution and the limited growth of epiphytes that the species uses for nesting. They also point out that habitat loss in southern Michigan and Wisconsin, primarily due to agricultural and urbanization expansion, had severely reduced available habitat in those states. However, Chartier and colleagues (2013) indicated the species had increased and expanded its range in Michigan from the first atlas of 1983–88 to the second atlas of 2002–8.

Overall, it is difficult to draw firm conclusions regarding changes in the distribution of the Northern Parula in Minnesota. Roberts (1932) only reported the species as a migrant in southern Minnesota. Therefore, its breeding distribution is likely similar to that of the past, but the ability of early explorers to identify the species and its patchy nesting requirements in epiphytic moss may have rendered it easily overlooked. In Ontario, Cadman and colleagues (1987) also suggested that the species is easily overlooked because of its quiet song, especially when heard among a chorus of other actively singing birds.

BREEDING HABITAT

The Northern Parula nests in tall, mature coniferous forests, primarily with spruce, hemlock, and fir, where moss is abundant, such as in riparian, floodplain, and wet areas (Moldenhauer and Regelski 2020). Since hemlock is very rare in Minnesota, tamarack is also frequently used. In Minnesota, Collins (1981) suggested the species uses undisturbed mixed forests of deciduous-coniferous trees with more than 75 percent canopy cover. Niemi, Howe, and colleagues (2016) indicated a significantly higher use of mature aspen–spruce–fir and mixed swamp conifer forests (e.g., white cedar, black ash, and spruce) compared with other forest cover types. These habitats are frequently found in riparian areas adjacent to creeks, streams, rivers, wetlands, ponds, or lakes **(C)**. Lowland hardwood forests were also used. Blake and Karr (1987) in Illinois found the species only in forests at least 65 ha in size, while C. S. Robbins and colleagues (1989) reported the species was rarely found in forests of less than 100 ha. Niemi and Hanowski (1992b) classified the species as area sensitive for their studies assessing the effects of increased forest harvesting and management in Minnesota.

POPULATION ABUNDANCE

PIF (2021) estimated a US and Canadian global population of 18 million breeding adults, while the MNBBA estimated a breeding population of 1.07 million adults in Minnesota (95 percent confidence interval of 0.92–1.50 million). This was much higher than the PIF (2021) estimate of only 350,000 breeding adults in the state. The discrepancy between these two estimates is likely because of methodological differences between PIF and the MNBBA and the sampling of many roadless areas by the MNBBA that are not sampled by the BBS; the latter is used by PIF in their population estimates.

C. The Northern Parula prefers large coniferous trees with hanging moss, which it uses to construct its nest. These habitats are often associated with wet areas. Photo by Gerald J. Niemi.

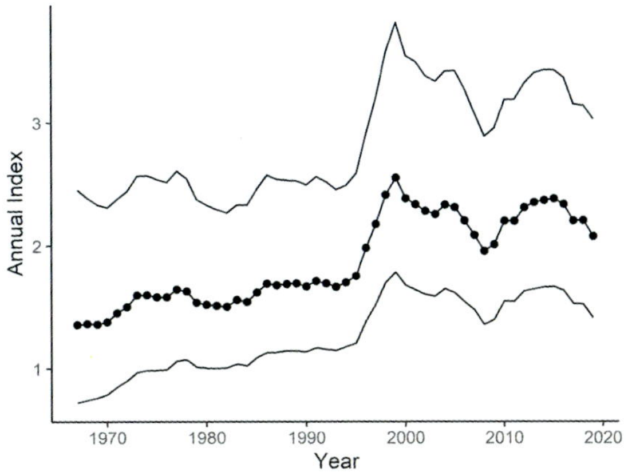

D. Breeding population trend and 95 percent credible interval (solid black lines) for the Northern Parula in Minnesota based on the 1967–2019 Breeding Bird Survey (Sauer et al. 2020).

The population trend for the Northern Parula based on the BBS indicated a stable population in Minnesota from 1967 to 2019 **(D)**. Trend estimates for sample routes completed in the United States, in Canada, and surveywide were all significantly increasing.

CONSERVATION

The Northern Parula has a relatively wide breeding distribution, a large North American breeding population, and currently an increasing population trend, so it is presently of minimal conservation concern. This species, as do all long-distance migrants, faces continued threats from obstructions along migratory pathways, such as towers, windows, and other manufactured structures, as well as diminishing suitable stopover habitat between breeding and wintering areas. Longcore and colleagues (2013) identified the Northern Parula as the seventh most frequently killed bird at communication towers in the Peninsular Florida region. Climate change poses yet another threat. NAS (2019), projected a moderate vulnerability for the Northern Parula's breeding distribution in North America under all three of its warming scenarios, including extirpation from Minnesota even with our current 1.5°C warming climate.

Maintaining mature forests along riparian corridors and in bottomland areas and encouraging conditions that allow epiphytic mosses to grow are important needs for this species (Moldenhauer and Regelski 2020). The Northern Parula has also been documented as susceptible to excessive habitat fragmentation. Based on this collective information, the future of the Northern Parula appears stable, but considerable uncertainty exists with regard to climate change.

MAGNOLIA WARBLER

Setophaga magnolia

Minnesota Seasonal Status: Regular breeding resident and migrant. The Magnolia Warbler was common during the MNBBA.
Migration: Short- to medium-distance migrant; overwinters from the southern United States to Mexico, south in Central America to Panama, and throughout the Caribbean.
Food: Arthropods gleaned from foliage.
Nest: Cup nest near the trunk in low to mid-canopy of coniferous trees, especially balsam fir and spruce.
Conservation Concern: Assigned a low PIF Continental Concern Score of 8/20.

Roberts (1932) emphasized that the more appropriate name for the Magnolia Warbler would be "Black and Yellow Warbler" in honor of its plumage or, as Griscom and Sprunt (1957) suggested, "Balsam or Spruce Warbler" in honor of its breeding habitat. Yet it received its scientific and common name from Alexander Wilson, who collected it in a magnolia tree during its migration in Mississippi, long before it was known as a breeding denizen of the boreal forests in the north.

In breeding plumage, the Magnolia Warbler is a brilliant specimen. Roberts (1932) admired its spring beauty among the "galaxy of Wood Warblers" when he said no member "shines more brightly in the hazy greens of the coming leafage than this little fellow in brilliant yellow and black with trimmings of purest white." The male Magnolia Warbler has a mix of colors: a gray crown, black back and cheeks, white wing bars, and black-and-white tail, all contrasting with its yellow underparts with black streaking. The female is similar, but more diluted in coloration. The male has a loud warble song characterized by three to four two-part notes, often described as *weeto weeto weeto wee,* with the last note highly variable, or, as Bent (1953) termed it, as *"Pretty, Pretty, Rachel."*

The Magnolia Warbler can be found throughout the upper midwestern and northeastern United States and Canada and south in the Appalachian Mountains to Virginia. It is also patchily distributed in the central and western provinces of Canada. The Magnolia reaches some of its highest global breeding densities in the extensive conifer forests of northern Minnesota. High densities are also found in parts of Ontario, Quebec, Labrador, and British Columbia.

Photo by David Brislance.

MINNESOTA BREEDING DISTRIBUTION

Historically described as a common summer resident in the northern evergreen forests, the Magnolia Warbler was found as far south as northern Mille Lacs County and west to Itasca State Park and eastern Marshall County (Roberts 1932). Roberts included nesting observations at Cass Lake (with one cowbird egg), Itasca State Park, Itasca County, and St. Louis County. The paucity of nesting observations was likely a consequence of limited coverage in the late 1800s and early 1900s in the northern coniferous forests. Roberts stated the Magnolia Warbler was "nowhere really abundant in Minnesota during the summer months" but considered it strictly a species of the *"Canadian Zone."*

More than forty years later, Green and Janssen (1975) reported confirmed nesting in Aitkin, Cass, Cook, Itasca, Lake, and St. Louis Counties and inferred nesting in Clearwater and Hubbard Counties. They emphasized that the Magnolia Warbler was most numerous in the eastern parts of its range. A few years later, Janssen (1987) included confirmed nesting from six counties since 1970: Aitkin, Beltrami, Clearwater, Cook, Hubbard, and Lake Counties. He also defined the breeding range as south to Carlton, southern Crow Wing, and Hubbard Counties and west to central Roseau County. Hertzel and Janssen in 1998 added no new confirmed nesting records to Janssen's list since 1970.

Observations by the MBS included 619 records of breeding-season locations for the Magnolia Warbler. The MBS included locations from southeastern Roseau County, eastern Becker County, and northern Aitkin and Carlton Counties, but the bulk of breeding observations were found in northern St. Louis and Lake Counties and throughout Cook County.

The MNBBA reported 1,683 records and clearly emphasized that the major distribution of the Magnolia Warbler was in the Laurentian Mixed Forest Ecological Province. Magnolias were especially common to abundant in eastern and northern St. Louis County, northern Lake County, and Cook County **(A)**. Magnolias were recorded in 14.3 percent of all blocks sampled (676 of 4,733) but only confirmed nesting in 3.4 percent (23 of 676) of the blocks where the species was detected. Observations of probable nesting were found in western Roseau County and northern Lake of the Woods County, as well as possible nesting in northern Marshall, eastern Mahnomen, central Becker, and northern Pine Counties. Collectively, these records slightly expand the breeding range of this species to the northwest and western portions of Minnesota from that presented by Janssen (1987). No new confirmed nests

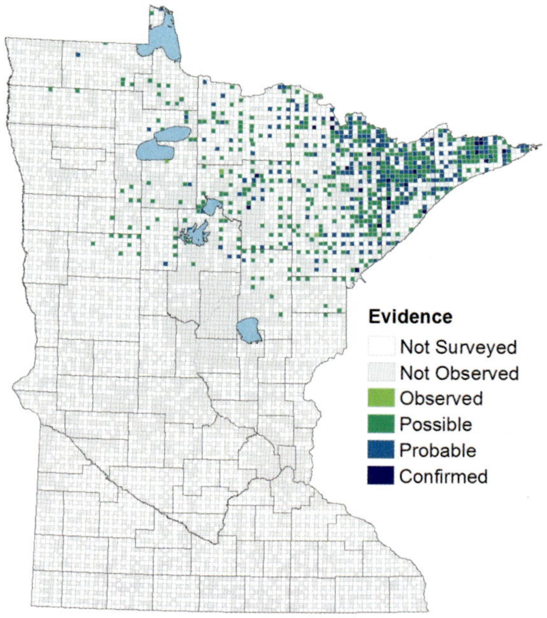

A. Breeding distribution of the Magnolia Warbler in Minnesota based on the Breeding Bird Atlas (2009–13).

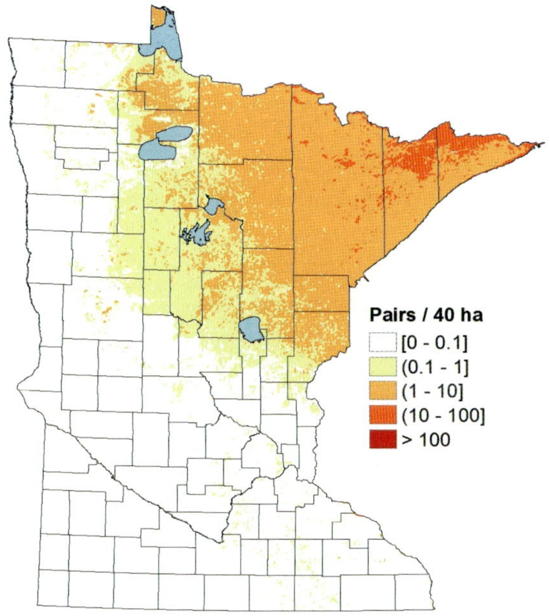

B. Predicted breeding distribution (pairs per 40 ha) of the Magnolia Warbler in Minnesota based on habitat, landscape context, and climate data gathered during the Minnesota Breeding Bird Atlas (2009–13) using the Generalized Linear Modeling method with an adjustment for detectability.

of Magnolias were located, which indicates how difficult it is to find nests among dense, young coniferous trees. Few observers have the patience or stealthy skills necessary to discover this bird's nest.

The probability map **(B)** also emphasizes that the highest densities occur in the northeastern region of the state, plus scattered high-density areas in St. Louis County. Low densities were predicted farther south, west, and to the northwest beyond where observations currently exist in Minnesota. However, the MBS and MNBBA included breeding evidence as far west as Becker, Clearwater, and Roseau Counties and the Wisconsin Breeding Bird Atlas I and II reported observations or probable breeding evidence for the Magnolia Warbler as far south as the extreme southwestern part of the state in Grant County (Cutright et al. 2006; WSO 2021). The Magnolia Warbler should be looked for in areas of dense, young conifers during the breeding season in many areas on the periphery of its current range.

In their review of the Magnolia Warbler in North America, E. H. Dunn and Hall (2020) identified range extensions southward in Ontario from 1985 to 2005. They also reported on expansions in several New England states in the early 1900s as forests regenerated following clearing in the 1800s. Some range contractions may have occurred in the western fringes of the species' range in Minnesota because balsam fir and spruce have become less common in these areas.

BREEDING HABITAT

The Magnolia Warbler occupies very distinctive breeding habitats, which are among the most easily recognized of Minnesota's warblers: young, dense coniferous tree species usually dominated by balsam fir, young white spruce, and less often jack pine (Collins et al. 1982; Niemi and Hanowski 1992a). Minnesota breeding habitats also included pole-sized and mature mixed aspen–spruce–fir forest cover types, especially characterized by balsam fir or white spruce in the understory (Niemi, Howe, et al. 2016) **(C)**. Additional habitats commonly used were 10- to 20-year-old upland red pine forests, dense jack pine forests, and lowland coniferous forests (Green and Niemi 1978; Niemi and Pfannmuller 1979).

POPULATION ABUNDANCE

PIF (2021) estimated a global population of 39 million breeding adults, with a Minnesota population of 340,000. MNBBA estimates for Minnesota's breeding population were about 1.43 million (95 percent confidence interval of 1.28–1.70 million), or a relatively small proportion of its overall population (3.7 percent). The population is vastly higher in Canada, where it is broadly estimated to be between 5 and 50 million adults (Environment Canada 2011).

The Magnolia Warbler's BBS population trend in Minnesota from 1967 to 2019 showed a stable population trend **(D)**, but its surveywide trend was increasing by 0.8% per year from 1966 to 2019 (Sauer et al. 2020). However, a significant decline has been observed in the Superior National Forest of northeastern Minnesota from 1995 to 2020 (Grinde, Bednar, et al. 2020). These equivocal results suggest that monitoring programs are important to maintain to detect population changes at both large and regional scales. For instance, BBS data from Michigan and Ontario indicated a stable population from 1966 to 2019, but a recent decline from 1993 to 2019 (Sauer et al. 2020).

C. Typical breeding habitat of the Magnolia Warbler in Minnesota is very distinctive. It consists of forests with young to medium-aged coniferous trees, usually balsam fir, white spruce, or occasionally pines. Photo by Gerald J. Niemi.

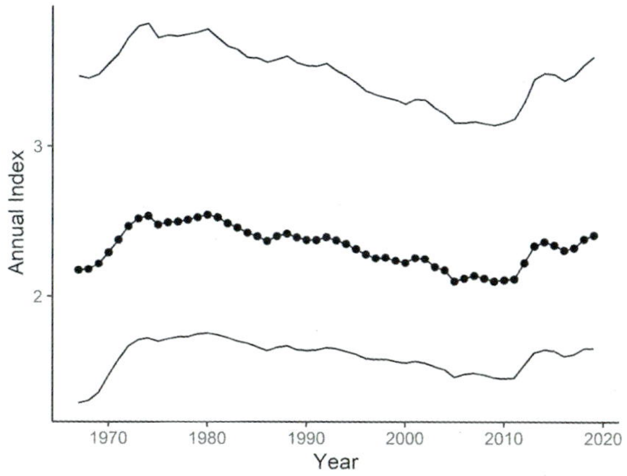

D. Breeding population trend and 95 percent credible interval (solid black lines) for the Magnolia Warbler in Minnesota based on the 1967–2019 Breeding Bird Survey (Sauer et al. 2020).

CONSERVATION

The Magnolia Warbler is of minimal conservation concern in North America due to its wide breeding geographic range, its substantial breeding population, and the lack of widespread threats to its current population. However, several possible or emerging threats exist, such as their susceptibility to mortality at communication towers (E. H. Dunn and Hall 2020; Longcore et al. 2013) and climate change.

Forest management issues may also be an emerging concern with mixed effects. Hobson and Bayne (2000) found forest clearing negatively affected its population, but the effect was short-term if regeneration included an adequate coniferous component. C. W. Pearson and Niemi (2000) emphasized that management for mixed stands of aspen, spruce, and fir was an effective strategy. Intermediate stages of red pine plantations prior to thinning can also provide breeding habitat for the species, but this is temporary if the site is thinned and it matures. Niemi, Howe, and colleagues (2016) suggest that forest management focused on fuel reduction and clear-cuts that remove young balsam fir and white spruce would have negative effects on Magnolia Warbler population levels.

With both 1.5°C and 2°C warming scenarios, NAS (2019) projected a moderate vulnerability for the Magnolia Warbler in North America, including a 52 percent range loss and a 45 percent range gain under the 1.5°C scenario. Moreover, in Minnesota the species was predicted to be extirpated from the state under both of these scenarios. Therefore, climate change may be one of the most important factors that will negatively affect this brilliantly plumaged and effective insectivore in Minnesota's future.

BAY-BREASTED WARBLER

Setophaga castanea

Minnesota Seasonal Status: Regular breeding resident and migrant. The Bay-breasted Warbler was rare during the MNBBA.
Migration: Medium- to long-distance migrant; the species overwinters from Panama to northern South America.
Food: Arthropods, especially caterpillars, all gleaned from trees, although the species will occasionally practice fly-catching.
Nest: Cup nest placed on a horizontal branch of a coniferous tree, highly variable in height.
Conservation Concern: Assigned a moderate PIF Continental Concern Score of 11/20.

Photo by Laura Erickson.

The Bay-breasted Warbler, the rarest of the trio of spruce budworm specialists in Minnesota (along with the Tennessee and Cape May Warblers), was one of the species studied intensively by MacArthur (1958) in his now-famous studies on the population ecology of warblers. Griscom and Sprunt (1957) claim it is an "unusually conspicuous" warbler and "attracts attention by reason of its rather large size." One might dispute that in Minnesota because of its rarity, but the keen observer can find it during migration or in spruce budworm outbreak areas.

Distinguishing characteristics of the male are its black face, chestnut crown, throat, and sides, cream-colored neck patch, and two white wing bars. The female is more subdued in coloration, with mottled black on its cheeks, less extensive chestnut on its crown, throat, and sides, and two white wing bars. Its common name derives from its reddish-brown breast, and its scientific name, *castanea,* is Latin for "chestnut." In the fall, adults and juveniles are challenging to tell apart from several other confusing fall warblers. Roberts (1932) stated they are "clothed in a dress of dull yellow, olive, and gray" and "they resemble so closely their travelling companions, the juvenile Black-polls as to be almost indistinguishable."

The song of the species can also be challenging, especially without excellent hearing. It is a high, thin, squeaking song, typically with five to ten lisping notes similar to the Cape May and Black-and-white Warblers. Described by Venier and colleagues (2020) as *see-atzee-atzee-atzee-atzee,* or *setzy-seetzy-seetzy-seetzy-see.*

Bay-breasted Warblers are found patchily distributed across Canada from the Maritime Provinces to British Columbia and in the northern portions of the Upper Midwest and northeastern United States. Its highest densities have been observed in eastern Quebec.

MINNESOTA BREEDING DISTRIBUTION

Roberts (1932) described the Bay-breasted Warbler as "a rare summer resident in the northern evergreen forest" of Minnesota, a status that remains the same today. Roberts did not report that any nests were found but documented breeding observations of pairs or specimens collected from Cass Lake and Itasca State Park, and from Cook, Lake, and Sherburne Counties. The observation from Cass Lake was from July 25, 1916, so it may have been an early fall migrant, but the observations from Sherburne County were from July 9–20, 1925. Regarding the latter observation, Roberts stated, "several seen in a tamarack swamp and both male and female taken in breeding condition."

More than forty years later, Green and Janssen (1975) described the Bay-breasted Warbler's distribution as primarily Cook, Lake, and northern St. Louis Counties. However, they proposed a potential distribution as far west as Roseau and Clearwater Counties and south to northern Hubbard and southern Itasca Counties. They included confirmed nesting only from northern Lake County. A few years later, Janssen (1987) restricted the warbler's primary breeding range to the northern portions of Beltrami, Cook, Lake, and St. Louis Counties, but did include all of Koochiching and Lake of the Woods Counties as well as west to eastern Roseau County. Janssen (1987) and Hertzel and Janssen (1998) both identified confirmed nesting in Cook and Lake Counties since 1970. The MBS has recorded 16 breeding-season locations during their intensive surveys of Minnesota counties. All of their locations were found in Cook, Lake, and St. Louis Counties.

The MNBBA documented 46 breeding records for the Bay-breasted Warbler. They included breeding evidence from 30 blocks, including four confirmed breeding records from St. Louis and Lake Counties **(A)**. Probable nesting was recorded from Cook County, and possible nesting in northern Aitkin County and Lake of the Woods County. These records represented only 0.6 percent (30 of 4,733) of the blocks covered during the MNBBA. The lack of nesting records in Minnesota reflected a combination of this species' genuine rarity, the difficulty in its identification, and a possible decline in its population.

With the exception of Roberts's (1932) reporting of possible Bay-breasted Warblers nesting in Sherburne County, this species' distribution does not appear to have changed in Minnesota for more than one hundred years. Its breeding distribution is closely tied to spruce budworm outbreaks. In their review of the Bay-breasted Warbler in North America, Venier and colleagues (2020) simply stated that

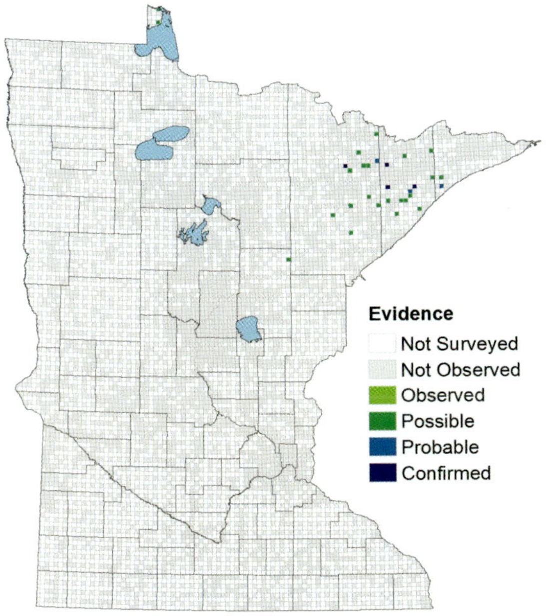

A. Breeding distribution of the Bay-breasted Warbler in Minnesota based on the Breeding Bird Atlas (2009–13).

there is "No information" available on historical changes to its distribution.

BREEDING HABITAT

The Bay-breasted Warbler is found in intermediate to mature spruce and balsam fir forests, most frequently in those infested with spruce budworm (Mayasich and Niemi 2003) **(B)**. Venier and colleagues (2020) also reported it from young mixed-wood stands during spruce budworm outbreaks and occasionally in bogs and swamps. Green (1995) emphasized the conifer dependency of this species in Minnesota forests and suggested their use of riparian, lowland coniferous forest, which is a habitat that is seldom studied.

POPULATION ABUNDANCE

PIF (2021) estimated a global population of 9.9 million breeding adults in the United States and Canada, but only 100,000 breeding adults were found in the United States. Of their 100,000 breeding adults, Minnesota was estimated to have 18,000 breeding birds (PIF 2021). The species was too rare in Minnesota to estimate a population based on the MNBBA point counts.

Data from the BBS were unreliable for estimating a population trend for the Bay-breasted Warbler in Minnesota. Reliable data were available for surveywide BBS routes from 1966 to 2019 and indicated a stable population **(C)**. Previously, Rosenberg, Kennedy, and colleagues (2016) had estimated a 9 percent decline in the US and Canadian population from 1970 to 2014. The decline was thought to be related to smaller populations of spruce budworm, a cyclical species that may or may not return to population levels found in the 1970s.

CONSERVATION

The MNDNR identified the Bay-breasted Warbler as a *Species in Greatest Conservation Need* based on habitat loss and its narrow thermal preferences (MNDNR 2015b). This renders the species' future in Minnesota as tenuous because of climate change and the predicted reduction of spruce–fir coniferous forests. NAS (2019) projected a high vulnerability for the Bay-breasted Warbler's breeding distribution in North America with a 2°C warming scenario, including an 84 percent range loss and a 43 percent range gain by 2050. They did not include Minnesota in its current breeding distribution. The species will need to colonize appropriate forests farther north in Canada and possibly in Alaska. However, during the second Ontario breeding bird atlas (2001–5), there was a substantial decline in the population of Bay-breasted Warblers in the Hudson Bay Lowlands of northern Ontario (Cadman et al. 2007). Whether the species can adapt to and thrive in habitats of northern Canada or Alaska is unclear at present.

Concerns for this species in Canada, where the bulk of its breeding population is found, are related to the warblers' dependence on mature forested areas and spruce budworm outbreaks (Environment Canada 2011). Winter populations may be affected because of the species' restricted wintering range and its use of mature forests that are subject to deforestation and development (Venier et al. 2020). Bay-breasted Warblers may also be negatively affected by short-rotation forestry practices and those that do not favor spruce or balsam fir (Environment Canada 2011).

B. Like those of the Cape May Warbler, the Bay-breasted Warbler's typical breeding habitats are coniferous forests composed of balsam fir and spruce. Populations can be very high if forests are infested with spruce budworm. Photo by Gerald J. Niemi.

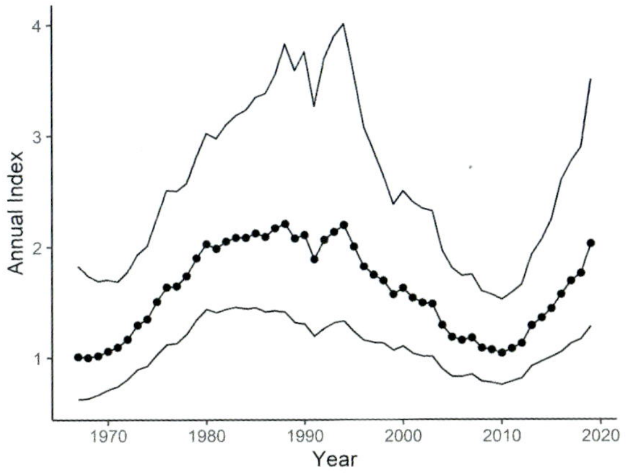

C. Surveywide breeding population trend and 95 percent credible interval (solid black lines) for the Bay-breasted Warbler based on the 1967–2019 Breeding Bird Survey (Sauer et al. 2020).

Collisions with communication towers have been identified as a serious concern for the Bay-breasted Warbler. T. Arnold and Zink (2011) identified this species as a "super collider" and ranked it number one as the species most frequently killed at tower sites. They estimated that its risk is 236 times greater than the average bird species. Longcore and colleagues (2013) supported the estimate of this risk and further estimated that 5.5 percent of the Bay-breasted Warbler population is killed annually at communication towers. Its susceptibility to tower-kill mortality is widespread in the United States. For instance, it is ranked second among all species in the combined Appalachian Mountains and Piedmont Bird Conservation Regions, sixth in the combined Lower Great Lakes/St. Lawrence, Atlantic Northern Forests, and New England/Mid-Atlantic Coasts Bird Conservation Regions, seventh in the Central Hardwoods Bird Conservation Region, and eighth in the combined Eastern Tallgrass Prairie, Prairie Hardwood Transition, and Boreal Hardwood Transition Bird Conservation Regions.

Given the information reported here, the future of the Bay-breasted Warbler is very questionable. During the breeding season, it is likely a slave to spruce budworm outbreaks and their inherent variability. Its adaptability to the impacts of climate change is uncertain. During the migration season, it faces substantial obstacles that are all projected to increase in the future, while the stability of its wintering grounds is largely unknown but certainly an additional concern.

BLACKBURNIAN WARBLER

Setophaga fusca

Minnesota Seasonal Status: Regular breeding resident and migrant. The Blackburnian Warbler was common during the MNBBA.
Migration: Medium- to long-distance migrant, winters in southern Central America and northwestern South America.
Food: Arthropods gleaned from foliage, especially coniferous trees.
Nest: Cup nest, primarily in coniferous trees of variable height and usually high in the canopy.
Conservation Concern: Assigned a moderate PIF Continental Concern Score of 9/20.

Photo by Mike Lentz.

No Minnesota bird species is more of an icon of the beautiful towering pines and spruce than the Blackburnian Warbler. Finding this little songster in the summer, when it sings from the pinnacle of a tall spruce tree, can strain one's neck. It is much easier to look for it during migration, when it often occurs at a reasonable height. Griscom and Sprunt (1957) stressed that "one never forgets his first sight of a Blackburnian" and to see a male singing in bright sunlight at the top of a conifer "is a red-letter experience." The song of the male is less stimulating than its plumage. It is a high-pitched series of *see see see see* notes followed by vibrant *tsee tsee tsee* notes.

Roberts (1932) eloquently described the experience of observing a male Blackburnian Warbler:

> The sight of this richly colored little bird amid the temperate surroundings of our northern clime always arouses a feeling that in some way he has wandered far from the luxuriant setting to which he seems naturally to belong. As he flashes in the sunlight, his brilliantly beautiful orange throat contrasted so effectively with the glossy black and white of the rest of the plumage, he never fails to elicit the astonishment and enthusiastic admiration of the ornithological tyro; and, indeed, it is probable that the veteran bird-lover rarely, if ever, flags in appreciation of his exceptional charm.

The Blackburnian belongs to the insect-devouring genus *Setophaga*; its scientific species name is *fusca*, an uninspiring term for "blackish." Its common name has a strange history: it was named after a British woman, Anna Blackburne, in the late 1700s (Griscom and Sprunt 1957). Sandrock and Prior (2014) suggest it was named for both Anna and Ashton Blackburne, two siblings who were amateur ornithologists and bird collectors. Ashton collected birds in the eastern United States and sent them back to Anna, who managed a bird museum (Gruson 1972).

The Blackburnian Warbler is primarily found in the northeastern and upper midwestern regions of the United States and the eastern and central provinces of Canada. It is also found south in the Appalachian Mountains as far as northern Georgia. The highest densities have been observed in northeastern Minnesota, northern Wisconsin, central Ontario, and northern New York.

MINNESOTA BREEDING DISTRIBUTION

Roberts (1932) described the Blackburnian Warbler as found "evenly distributed and generally common" in evergreen forests as far south as northern Isanti County, as well as in northern Sherburne, west to northeastern Otter Tail, eastern Becker, and Polk Counties. He reported confirmed nesting in Becker, Cass, Cook, Itasca, and Lake Counties, and in Itasca State Park and Mille Lacs.

By 1975, Green and Janssen emphasized that the species' primary distribution was in the northeastern and north-central regions of Minnesota. Besides confirmed nests reported by Roberts (1932), they included confirmed nesting in Beltrami, Clearwater, Hubbard, Mille Lacs, Pine, and St. Louis Counties. They highlighted that this species was found sparingly in the eastern fringes of the northwestern counties and the northern portions of counties in the east-central regions. Several years later, Janssen (1987) and Hertzel and Janssen (1998) identified seven counties where nesting had been confirmed since 1970, including Aitkin, Clearwater, Cook, Crow Wing, Hubbard, Lake, and St. Louis.

The MBS recorded 526 breeding-season locations, most of which were in areas previously identified. However, they also found the Blackburnian Warbler in north-central and eastern Roseau, northeastern Marshall, southeastern Mahnomen, northeastern Becker, northeastern Wadena, northern Mille Lacs, Kanabec, and eastern Chisago Counties, plus a stray observation in northeastern Douglas County.

The MNBBA reinforced the previously described breeding distribution, with nearly all 1,622 records reported from the Laurentian Mixed Forest Province **(A)**. The exceptions were observations in northwestern Minnesota, including possible nesting in central Roseau County and probable nesting in northeastern Marshall County. As with many secretive nesting species, only 4.6 percent of block records reported confirmed nesting. The MNBBA added extensive records of probable and possible breeding activity of the species to poorly sampled counties in extreme northern Minnesota, such as Beltrami, Koochiching, and Lake of the Woods.

Current evidence suggests that the Blackburnian Warbler has experienced a northward contraction all along the southern, western, and northwestern fringes of its breeding range in Minnesota. Isolated populations appear to exist in some areas where patches of large coniferous trees are found but certainly not in an "even distribution," as suggested by Roberts (1932). Patches with confirmed nesting still existed in many parts of central and southern Wisconsin, as far south as Walworth County (Cutright et al. 2006) and with possible nesting in Grant and Richland Counties, during the second Wisconsin breeding bird atlas (WSO 2021). These results may be due to the greater presence of hemlock in Wisconsin, which the species uses along with white spruce and pines.

In his review of the Blackburnian Warbler in North America, D. H. Morse (2020) cited some evidence for range expansion into western Canada and in New York State over the past one hundred years. He suggested the changes in New York were a result of recent reforestation in the Appalachian Plateau.

The Blackburnian Warbler's presence during the breeding season can be easily overlooked because its high-frequency, insect-like song is most often emitted from the tallest of coniferous trees, where observations are challenging. Given its more southerly presence in Wisconsin, this species warrants more attention in areas of Minnesota with tall coniferous pines or spruce by observers with good hearing and visual skills.

BREEDING HABITAT

Roberts (1932) aptly described the Blackburnian Warbler's habitat as "most commonly in heavy mixed timber, keeping constantly in the topmost branches of the tallest trees," adding that "towering pines and spruces are its special favorites." D. H. Morse (2020) considers its habitat to be "coniferous and mixed coniferous–deciduous forests, and especially in mature forests." Data from Minnesota found it most abundant in white pine forests, but also common in mature aspen–spruce–fir, jack pine, and red pine forest cover types (Niemi, Howe, et al. 2016) **(B)**.

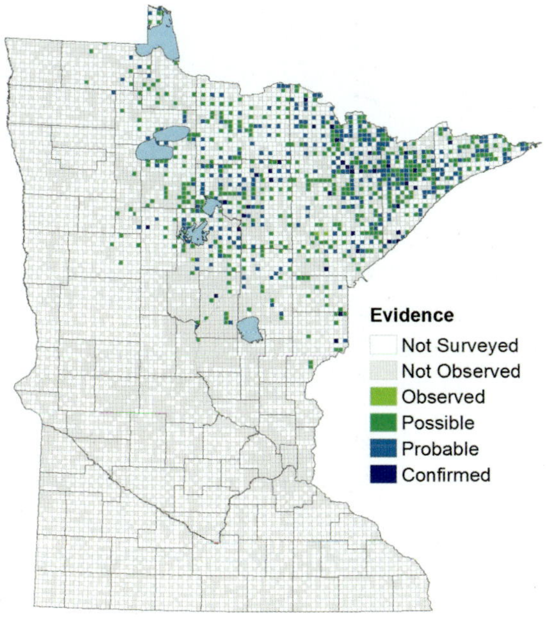

A. Breeding distribution of the Blackburnian Warbler in Minnesota based on the Breeding Bird Atlas (2009–13).

B. The Blackburnian Warbler is attracted to large coniferous trees, like these giant white pines, for nesting habitat. Photo by Gerald J. Niemi.

Blackburnians are often found in mixed coniferous–deciduous forest stands due to the presence of a few tall coniferous trees, especially white pine or white spruce. Even though these trees may represent a small proportion of the forest stand, they are essential for the Blackburnian Warbler's presence in these forest types.

POPULATION ABUNDANCE

PIF (2021) estimated a US and Canadian global population of 13 million breeding adults and estimated that 7.2 percent (about 960,000 adults) occurred in Minnesota. However, the MNBBA estimated a breeding population almost three times that size, at 2.76 million breeding adults, but the estimate was highly variable with wide confidence intervals and should be interpreted with caution.

The BBS trends in Minnesota from 1967 to 2019 indicated a stable population; however, the trajectory since the early 1990s illustrates a downward pattern **(C)**. The species' trend was stable when all routes in Canada, the United States, or surveywide were included in the analyses from 1966 to 2019, but an average decline of −0.57 percent per year was identified in the United States from 1993 to 2019 (Sauer et al. 2020). Rosenberg, Kennedy, and colleagues (2016) calculated an overall increase of 10 percent in the Blackburnian Warbler population from 1970 to 2014. Based on a sample of 233 forest stands, population trends in Minnesota's two national forests, the Chippewa and Superior, revealed a stable trend from 1995 to 2020 (Grinde, Bednar, et al. 2020). However, since 2010 the trend has been declining, especially in the Superior National Forest.

CONSERVATION

The Blackburnian Warbler is considered of minimal conservation concern in North America. Environment Canada (2011) stated there were few known threats on its breeding grounds. However, they emphasized that, first, short logging rotations will decrease the age of coniferous forests below optimum levels; second, the species is sensitive to habitat fragmentation (e.g., Hobson and Bayne 2000); and, third, deforestation on wintering grounds in northern South America could be detrimental (Terborgh 1989). Howe (1979) found the Blackburnian Warbler on small islands (less than 1 ha) on Burntside Lake in northern St. Louis County. Where large pines were present, the species may be tolerant of reduced patch sizes.

NAS (2019) projected a moderate vulnerability for the Blackburnian Warbler in North America with a 2°C warming scenario, including a 70 percent range loss and a 42 percent range gain in northern Canada by 2050. In Minnesota, it was projected to be extirpated under the 2°C warming scenario by 2050. Longcore and colleagues (2013) listed the Blackburnian Warbler as susceptible to tower kills.

In Minnesota, a key issue to sustaining this warbler is maintenance of large coniferous trees, especially white pine and white spruce, but climate change adds considerable uncertainty to its future. Regardless, reforestation of these large coniferous trees in many marginal areas of the western and southern fringes of its current range could provide benefits to this brilliantly plumaged species.

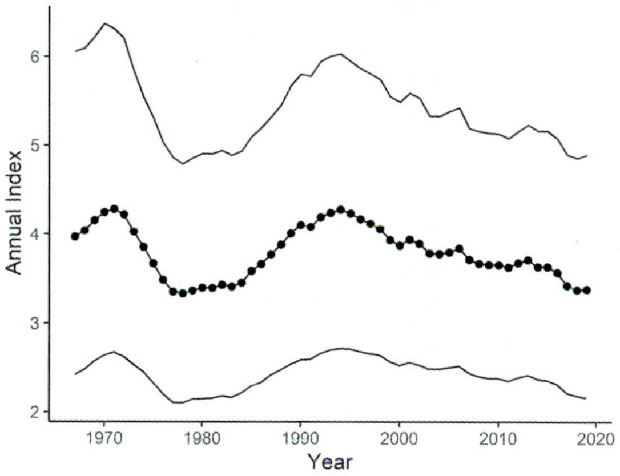

C. Breeding population trend and 95 percent credible interval (solid black lines) for the Blackburnian Warbler in Minnesota based on the 1967–2019 Breeding Bird Survey (Sauer et al. 2020).

YELLOW WARBLER

Setophaga petechia

Minnesota Seasonal Status: Regular breeding species and migrant. The Yellow Warbler was abundant during the MNBBA.
Migration: Medium- to long-distance migrant that winters from Mexico south to central South America.
Food: Arthropods obtained by foliage gleaning, hover-gleaning, and sallying from a perch out to catch insects in mid-air.
Nest: Cup nest placed in a small tree or shrub, usually located less than 2 m above the ground.
Conservation Concern: Assigned a low PIF Continental Concern Score of 8/20.

Photo by Mike Lentz.

For years, writers have attempted to capture with words the intense and striking plumage of the Yellow Warbler. Bent (1953) described it as "a flash of brilliant sunshine," while Chapman ([1907] 1968) wrote, "In his plumes dwells the gold of the sun." The male's chestnut breast streaks add a striking adornment to an otherwise intense palette of deep yellow. Even the grayish wings show shades of yellow, and the outer tail feathers display yellow on the inner webs. A bit duller in her brilliance, the female has yellow plumage with hints of green, while the chestnut breast streaks are either absent or only faintly visible. *Setophaga petechia,* the species' scientific name, means "a moth eater with red streaks."

Unfortunately, if the warbler is only heard and not seen, correctly identifying it can challenge those unfamiliar with its diverse song repertoire. A familiar mnemonic for the male's most characteristic song is *sweet-sweet-sweeter-than-sweet,* but many other variations exist. Ornithologist A. A. Saunders noted: "The songs are quite variable in form, so much so that it is the quality, rather than the form, that makes the song recognizable. This quality is difficult to describe, yet . . . after a little familiarity, is easily recognized" (Bent 1953).

One of the primary challenges for field surveyors is that the songs are quite similar to those of the Chestnut-sided Warbler. In areas where both species occur, it can be difficult to discriminate between the two. In fact, in areas of co-occurrence, Chestnut-sided Warblers are known to countersing with nearby Yellow Warblers just as they would with other Chestnut-sided Warblers (J. L. Dunn and Garrett 1997).

A shrubland inhabitant, the widely distributed Yellow Warbler displays more morphological variation across its breeding range than any other wood warbler (Lowther, Celada, et al. 2020). A total of 43 subspecies are recognized, which fall into three subgroups: first, the northern Yellow Warbler, which occupies Canada, the continental United States, and central Mexico; second, the Golden Warbler, found in the Caribbean; and, third, the Mangrove Warbler, which occurs in Central America. Only the northern Yellow Warbler is migratory. In the United States and Canada, breeding populations reach their highest densities in the northern Great Plains, but the species is abundant in many other locations as well.

MINNESOTA BREEDING DISTRIBUTION

In the early twentieth century, Roberts (1932) described the Yellow Warbler as one of the state's most abundant and widely distributed birds. Although it was inexplicably absent or rare in scattered areas, such as Itasca State Park, he considered it very common throughout most of the state. "A sparse little thicket of hazel or wild-rose, beside some tiny lakelet far out on the prairie-ocean, is all the invitation it requires to tarry awhile and establish its summer home. Everywhere it adapts itself most readily and cheerfully to the environment, however diversified, and in consequence forms a very considerable portion of the avian population in almost all localities."

Roberts believed it was a close contest between the Common Yellowthroat and the Yellow Warbler as to which species was the most abundant wood warbler in Minnesota, but he gave the advantage to the Yellow Warbler. Confirmed nesting records were available from 16 counties. The records ranged from Jackson, Lincoln, and Pipestone Counties in the southwest, north to Becker, Grant, Marshall, Otter Tail, Pennington, Polk, and Wilkin Counties, east to Goodhue, Hennepin, and Isanti Counties, and farther north to Itasca, Lake of the Woods, and St. Louis Counties. Nesting reports were also available from Mille Lacs, Leech Lake, and Big Stone Lake in west-central Minnesota.

Green and Janssen (1975) and Janssen (1987) also noted the Yellow Warbler's statewide distribution but stated it was least abundant in the heavily forested landscape of north-central and northeastern Minnesota. Janssen (1987) identified 40 counties, covering all regions of the state, where Yellow Warblers had been confirmed nesting since 1970. Hertzel and Janssen (1998) later added 10 more counties to the list of nesting records since 1970. By 2014, the MBS had documented 1,979 breeding-season locations, which were common throughout most of the state. Numerous records were also documented through the surveyed portions of north-central Minnesota and St. Louis County. Scattered records were found in Cook and Lake Counties.

MNBBA participants further documented the species' statewide distribution, reporting a total of 5,333 Yellow Warbler detections from 58.0 percent (2,762/4,765) of the surveyed atlas blocks and from 71.9 percent (1,681/2,337) of the priority blocks. Breeding was confirmed in 290 (6.1 percent) surveyed blocks and deemed probable in an additional 1,095 blocks (23.0 percent) **(A)**. The birds were observed in all 87 Minnesota counties, and breeding was confirmed in 65 counties. Twenty-one counties were additions to the list published by Hertzel and Janssen (1998). Although the species was found statewide, records were least abundant in Cook, Koochiching, and Lake Counties in northern Minnesota and in several southwestern counties. Once again, intensive search efforts in the Brainerd Lakes region documented the species' ubiquitous presence in Crow Wing and southern Cass Counties.

The MNBBA probability map predicts that the highest breeding densities occur in a broad band of central Minnesota that stretches from the outskirts of the Twin Cities north through the Tallgrass Aspen Parklands Province **(B)**. Elsewhere, high densities

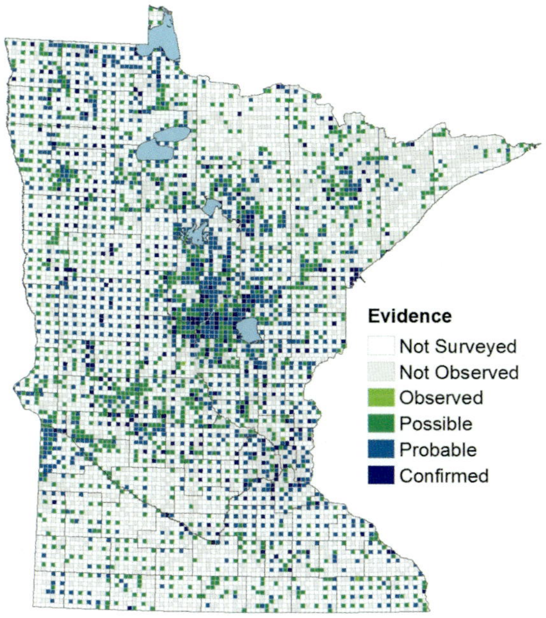

A. Breeding distribution of the Yellow Warbler in Minnesota based on the Breeding Bird Atlas (2009–13).

are predicted along riparian corridors throughout the Red River valley, along the length of the Minnesota River, and in scattered pockets throughout the state. Densities are lowest in the center of the Twin Cities metropolitan region and in northern Minnesota from Koochiching County east. The 2007 Ham Lake Fire in northern Cook County stands out as an area in the extensively forested landscape of Cook County that can support young forest and shrubland species like the Yellow Warbler.

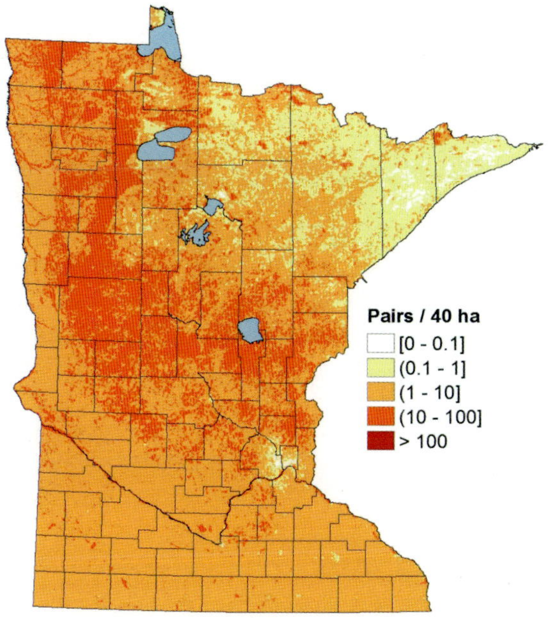

B. Predicted breeding distribution (pairs per 40 ha) of the Yellow Warbler in Minnesota based on habitat, landscape context, and climate data gathered during the Minnesota Breeding Bird Atlas (2009–13) using the Generalized Linear Modeling method with an adjustment for detectability.

Although it is difficult to know if the Yellow Warbler was ever as abundant as Roberts (1932) originally believed, among wood warblers it lagged behind the Common Yellowthroat and Ovenbird during the MNBBA. Nearly 4,300 more Common Yellowthroat detections were reported than were reported for the Yellow Warbler. The Ovenbird was just slightly more common than the Yellow Warbler, with 422 more detections. The species' statewide distribution, however, appears relatively unchanged. Indeed, no large-scale changes have been documented throughout the Yellow Warbler's North American breeding range (Lowther, Celada, et al. 2020).

BREEDING HABITAT

The Yellow Warbler can be found breeding in brushy riparian thickets, young successional forests, orchards, shelterbelts, abandoned fields, idle pastures, and any other upland open area with shrub thickets (Lowther, Celada, et al. 2020; J. L. Dunn and Garrett 1997) **(C)**. Nevertheless, it is most closely associated with wet, riparian thickets of willow and alder. Even the smallest wet area among an otherwise developed landscape will invariably support a pair of Yellow Warblers if willows and alders are present. In northern Minnesota, where many species specialize in young successional upland forests, including the Chestnut-sided Warbler, Mourning Warbler, and White-throated Sparrow, the Yellow Warbler is more closely associated with wetland habitats than young upland forests or shrub thickets (Niemi, Howe, et al. 2016).

POPULATION ABUNDANCE

PIF (2021) generated a global population estimate of 97 million breeding Yellow Warblers and a population of 93 million in the United States and Canada. Minnesota was estimated to support approximately 1.7 million birds. MNBBA data, with its extensive coverage of inaccessible wetlands, generated a much higher estimate of 8.1 million breeding adults (95 percent confidence interval of 7.4–9.0 million).

Across the BBS survey region, the Yellow Warbler has experienced a steady and significant decline that has been most pronounced since the late 1990s (Sauer et al. 2020). In contrast, population numbers in Minnesota were stable until the early 1990s, when they began a slow, nonsignificant increase averaging 0.43 percent per year from 1993 through 2019 **(D)**. The factors most frequently cited as responsible for local population declines are the loss of wetlands, riparian buffers, or other shrubby habitats due to development or more intensive agricultural practices. Other contributors include nest parasitism by Brown-headed Cowbirds and collisions with communication towers and buildings during migration. Population increases are often attributed to the species' adaptability to a variety of habitats (Lowther, Celada, et al. 2020; Cutright et al. 2006).

C. A dense alder thicket in Marshall County provides typical breeding habitat for the Yellow Warbler.
Photo by Lee A. Pfannmuller.

CONSERVATION

The Yellow Warbler has yet to raise the concern of ornithologists and has been assigned a relatively low conservation ranking (PIF 2021). Warmer temperatures, however, are projected to be a moderate threat to the species under a 2°C warming scenario, with a predicted range loss of 27 percent compensated by a northern gain of only 18 percent by the year 2050. Surprisingly, as temperatures rise in Minnesota, the distribution of this familiar wood warbler will gradually recede northward until, under a 3°C warming scenario, its dazzling, lemon-yellow plumage may disappear from all but the state's far northeastern corner.

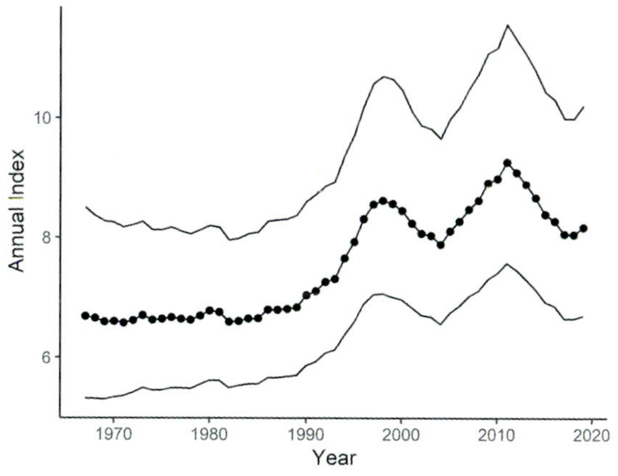

D. Breeding population trend and 95 percent credible interval (solid black lines) for the Yellow Warbler in Minnesota based on the 1967–2019 Breeding Bird Survey (Sauer et al. 2020).

CHESTNUT-SIDED WARBLER

Setophaga pensylvanica

Photo by David Brislance.

Minnesota Seasonal Status: Regular breeding resident and migrant. The Chestnut-sided Warbler was common during the MNBBA.
Migration: Medium- to long-distance migrant that winters from southern Mexico to Panama.
Food: Insects, spiders, and other arthropods, gleaned from foliage.
Nest: Cup nest, usually low or close to the ground in deciduous shrubs or saplings.
Conservation Concern: Assigned a moderate PIF Continental Concern Score of 12/20.

The Chestnut-sided Warbler is one of the most symbolic species of young forests in Minnesota's northern regions. Yet it can also be found in forest edges or in forest openings of mature forests, such as those created by wind, insects, or beaver activity. It obtains some of its highest North American breeding densities in Minnesota's north woods.

With prominent chestnut coloration on its sides, the male is appropriately named. It also has a brilliant yellow crown, black facial markings, a streaked yellow-green back, and a white belly that render it easy to identify. Its loud song is reminiscent of the words *pleased pleased pleased to meetcha,* with an emphasis on the *meetcha* (Curson et al. 1994). The female's similar coloring is more subdued in intensity, with the chestnut sides not as prominent.

The Chestnut-sided Warbler's scientific name, *pensylvanica,* is only indicative of where the type specimen was collected, near Philadelphia. Even though the species is found nesting in Pennsylvania, higher densities are found farther north. Many early explorers thought this species to be very rare. Even the famous John James Audubon observed only one in his entire life, despite his wanderings in America. He likely did not spend enough time in the regions where the species is most common, such as the upper midwestern and northeastern regions of the United States or Canada. The species is also found more sparingly in the Appalachian Mountains to northern Georgia.

MINNESOTA BREEDING DISTRIBUTION

In the late 1800s, Roberts (1932) described its distribution as "everywhere throughout the more open wooded regions of Minnesota and along the hazel-covered banks and rolling bluffs of the many water-courses that traverse the semiprairie region of the southeastern part of the state." He commented that "[i]t shunned only the dense unbroken forest and the open prairies of the west." However, by the early 1900s, Roberts lamented that the species had largely disappeared from the southern portions of its range, though "occasional pairs may still be found here and there."

Roberts reported nesting in Aitkin, Cass, Cook, Goodhue, Hennepin, Houston, Isanti, Itasca, Polk, Roseau, Stearns, and St. Louis Counties, as well as in Itasca State Park and the Mille Lacs Lake region. He emphasized that its breeding distribution from the early 1900s to the 1930s was common in forested areas northward from northern Isanti and Pine Counties.

Green and Janssen (1975) reported that no nests of the species had been found from the Twin Cities metropolitan area southward since 1940, but isolated summer observations existed from Dakota, Rice, and Winona Counties in southeastern Minnesota. Besides the counties reported by Roberts, Green and Janssen added confirmed nesting in Anoka, Beltrami, Clearwater, Crow Wing, Hubbard, Lake, and Mille Lacs Counties, plus inferred nesting from Wadena County.

Janssen (1987) included confirmed nesting since 1970 in nine counties. He highlighted summer observations from several counties in southern Minnesota, including Cottonwood, Dakota, Rice, and Winona. Janssen also suggested that nesting was again occurring in the Twin Cities area because 15–20 individuals were observed in Scott County in June 1986. Later, Hertzel and Janssen (1998) reported confirmed nesting in Rice and Scott Counties in the southern metropolitan area. In addition, they confirmed nesting since 1970 in Anoka, Becker, and Roseau Counties. In 2017, a nest was discovered in Washington County (MOU 2021a).

As of 2014, the MBS recorded 1,290 breeding-season locations and greatly expanded the Chestnut-sided Warbler's reported contemporary breeding range in Minnesota. The MBS included breeding observation locations in southern Minnesota and locations northwest to Clay, Kittson, and Marshall Counties.

In a testament to the species' abundance, the MNBBA participants included 3,781 breeding records of the Chestnut-sided Warbler! The distribution of records further expanded the breeding range southward, with the most extreme points being the possible nesting in Fillmore County and confirmed nesting in both Dakota and Goodhue Counties **(A)**. This was the first nesting record for Dakota County and the first in Goodhue County since Roberts (1932) reported it in 1925. Nesting was also confirmed for the first time in Kanabec, Koochiching, Lake of the Woods, Mahnomen, Morrison, Pine, and Sherburne Counties. MNBBA participants also included probable nesting in Benton, Clay, Kittson, Marshall, Pennington, and Todd Counties. Nesting was confirmed for 2.2 percent (103/4,747) of all blocks and for 6.2 percent of blocks where the species was reported.

Green and Janssen (1975), Janssen (1987), and Hertzel and Janssen (1998) collectively provided evidence for the species' recolonization into southern and western portions of the state since the early 1900s. Further evidence of an expansion into western Minnesota was a confirmed nest in Kandiyohi County in 2018 (MOU 2021a). However, observers need to be wary of identifications only by song because of confusion with the song of the Yellow Warbler, the similarity of which was also discussed by Roberts (1932). This is especially true in the western

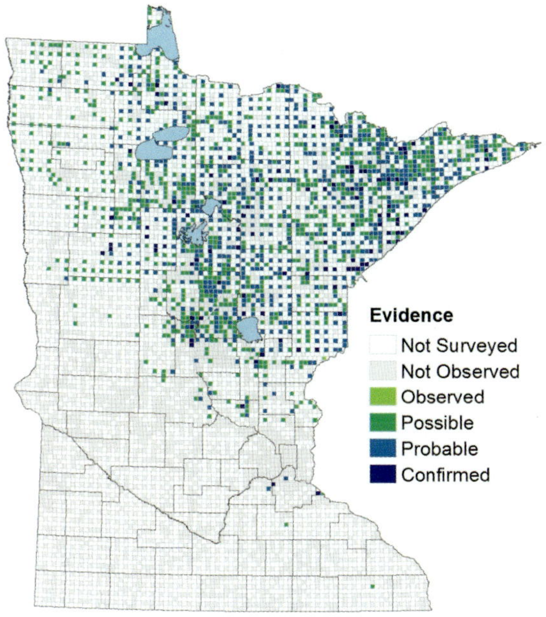

A. Breeding distribution of the Chestnut-sided Warbler in Minnesota based on the Breeding Bird Atlas (2009–13).

and southern areas of Minnesota, where the Yellow Warbler is more common. The decline of the Chestnut-sided Warbler's range since the early 1900s has also been documented for Wisconsin, Illinois, and Missouri (Byers et al. 2020). They emphasized that it had greatly benefited and expanded its range owing to logging activity and regeneration of second-growth forests that began in the early 1800s. As forests were later converted to agriculture and urban areas, the species' range began to decline, as observed by Roberts in the early 1900s.

In Minnesota, the species appears to be recolonizing areas formerly occupied in the west, southwest, and south, possibly because reforestation and protection efforts have increased. The species should be looked for in any suitable habitats in the southern and southwestern portions of Minnesota. Careful documentation that includes visual observations of the species is important to clarify any confusion with the Yellow Warbler song.

BREEDING HABITAT

The Chestnut-sided Warbler is highly associated with young upland forests, forest openings in deciduous and mixed deciduous–coniferous forests, and open habitats with shrubs. The species is especially associated with forested areas recently altered by logging, forest fires, or wind (Niemi and Pfannmuller 1979; Niemi and Hanowski 1984b; Niemi, Howe, et al. 2016). Schulte and Niemi (1998) found the species significantly more abundant in recently logged areas compared with burned areas **(B)**.

POPULATION ABUNDANCE

The US and Canadian breeding population was estimated at 18 million breeding adults, with 1.5 million adults estimated in Minnesota (PIF 2021). The MNBBA estimated Minnesota's population at 6.14 million breeding adults (95 percent confidence interval of 5.77–7.55 million). The MNBBA data suggest that the PIF (2021) estimates are conservative in Minnesota. Among the reasons for the discrepancy in estimates include the differences in methodology between PIF and the MNBBA and the lack of coverage of the northern, largely roadless areas of Minnesota that are not included in estimates by PIF (2021). The roadless areas, where densities of this warbler are high, were included in the MNBBA estimates.

In Minnesota, the BBS indicated a stable population trend from 1967 to 2019 **(C)**. This stable pattern was consistent with BBS trends in Manitoba, Michigan, and Wisconsin, but in Ontario Chestnut-sided Warblers had a significant declining trend over the same period of time (Sauer et al. 2020). A significant declining trend was also found in Minnesota's Superior National Forest from 1995 to 2020 (Grinde, Bednar, et al. 2020). A major threat to the species appears to be the loss of young forests due to forest succession and loss of forests due to agriculture and human settlements, especially in the southern portions of the species' breeding range in North America.

CONSERVATION

Concerns about the status of the Chestnut-sided Warbler have emerged because of the species' declining population not only over a broad area of North America but also in northern Minnesota. Pfannmuller (2012) identified the species as a Stewardship Species because a high proportion of its

B. Chestnut-sided Warblers are attracted to shrubby habitats in open areas of Minnesota, such as edges of forests or those created by disturbances like forest fires, logging, wind, tree fall gaps, or beaver activity. Photo by Gerald J. Niemi.

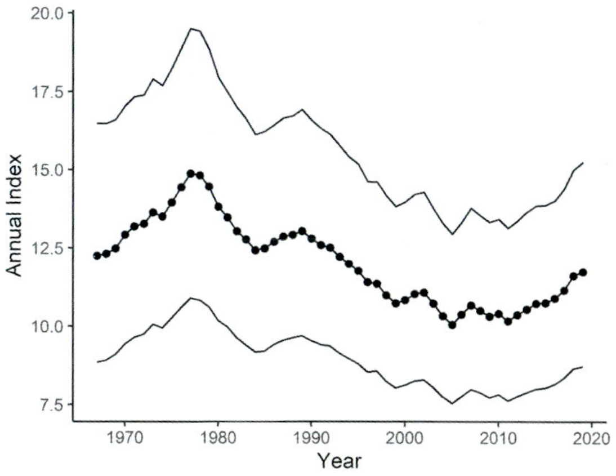

C. Breeding population trend and 95 percent credible interval (solid black lines) for the Chestnut-sided Warbler in Minnesota based on the 1967–2019 Breeding Bird Survey (Sauer et al. 2020).

population (more than 6 percent) was estimated to occur in Minnesota.

The Chestnut-sided Warbler responds positively to logging of forests and the subsequent regeneration of early successional shrub habitats (Hobson and Bayne 2000; Chizinski et al. 2011). In Minnesota, the total wood harvest began to decline in 1999 and has declined more steeply since 2005 (Niemi, Howe, et al. 2016). Zlonis and Niemi (2014) observed that the species was equally abundant in managed, logged forests and in wilderness areas without logging, presumably due to the availability of sufficient natural disturbances.

With a 2°C warming scenario, NAS (2019) projected a moderate vulnerability for the Chestnut-sided Warbler's breeding range in North America. In Minnesota, the species was predicted to be extirpated under the 3°C warming scenario by 2080. It is counterintuitive that the breeding population of the Chestnut-sided Warbler would be declining throughout many parts of its range or contracting northward with climate change, including in Minnesota, when its range may be expanding southward. Even though the population is still relatively large, the magnitude of the Chestnut-sided Warbler's population decline is alarming. Besides habitat loss and changes, Longcore and colleagues (2013) found that more than 1 percent of the population suffers mortality at communication towers.

Causes for the decline of the Chestnut-sided Warbler are worthy of additional study and continued monitoring. Programs such as the BBS volunteer counts (Sauer et al. 2020) and more detailed monitoring in core distribution areas, such as in the Chippewa and Superior National Forests (Grinde, Bednar, et al. 2020), are critical to determining the future of this species both in Minnesota and in North America.

BLACK-THROATED BLUE WARBLER

Setophaga caerulescens

Minnesota Seasonal Status: Regular breeding resident and migrant. The Black-throated Blue Warbler was uncommon during the MNBBA.
Migration: Medium-distance migrant, primarily overwinters in the Caribbean and in a very small area in southern Mexico and northern Central America.
Food: Insects, spiders, and other arthropods gleaned from deciduous shrubs and trees.
Nest: Cup nest, usually low or close to the ground in deciduous or coniferous shrubs or saplings.
Conservation Concern: Assigned a moderate PIF Continental Concern Score of 10/20.

The male Black-throated Blue Warbler is stunningly handsome and appropriately named, with its slate blue upperparts and black throat and sides. The female is quite different, with a plumage that is camouflaged by its greenish upperparts and yellow underparts. Both sexes have a distinctive white patch at the base of the primary feathers. Its scientific species' name, *caerulescens,* derives from Latin for the "bluish" plumage of the male. The song of the Black-throated Blue is a distinctive series of "three to four slowly drawled notes in a peculiar husky voice, the last note commonly slurred upward" (Griscom and Sprunt 1957). It is similar to the song of the Cerulean Warbler and has been described in words as *I am lazy.*

Chapman ([1907] 1968) relates a habitat description from Gerald Thayer in the eastern United States, where the species is more common, as "a bird of the ampler deciduous undergrowth in deep, moist woods—mixed virgin timber or very old second growth. . . . [I]ts movements . . . [are] more deliberate than many of its relatives." This description of its habitat and behavior also applies in Minnesota, where it is most commonly found in wooded hillsides along the North Shore of Lake Superior. It has also been described by many as one of the tamest of the warblers, which fortunately often allows a gratifying close look.

The Black-throated Blue is found in the northeastern and upper midwestern parts of North America, from Nova Scotia to western Ontario and Minnesota and south in the Appalachian Mountains to northern Georgia. Its highest densities are found scattered in Ontario, Nova Scotia, and New York State, as well as in western North Carolina and eastern Tennessee.

Photo by Mike Lentz.

MINNESOTA BREEDING DISTRIBUTION

Described as one of the rarest breeding warblers in the state by Roberts (1932), the Black-throated Blue Warbler was found as far south as Pine County and west to Itasca State Park. Nesting was reported from Cass Lake and Cook County. In his 1936 revision, Roberts added three additional breeding observations at Loon Lake in Cook County, at Lake Winnibigoshish in north-central Minnesota, and at Lake Vermilion in St. Louis County.

Green and Janssen (1975) emphasized its breeding distribution as northeastern and north-central Minnesota but noted that it was "scarce throughout most of this range except in Cook County." They included confirmed nesting in Beltrami, Cass, and Cook Counties and inferred nesting in Clearwater and northern St. Louis Counties. Several years later, Janssen (1987) suggested a greatly restricted breeding range, limited to Cook County and eastern Lake County. He stated the Black-throated Blue Warbler was "casual in the north-central region of Beltrami, Cass, and Clearwater Counties." Both Janssen (1987) and Hertzel and Janssen (1998) cite only confirmed nesting in Cook and Lake Counties since 1970.

The MBS recorded 288 breeding-season locations for the Black-throated Blue, which were not only extensive along the North Shore of Lake Superior from southeastern St. Louis to Lake and Cook Counties but also relatively common in the northern portions of these counties. Additional potential breeding locations were reported from eastern Becker County, Cass County near Leech Lake, northern Mille Lacs County, and western Itasca County.

The MNBBA reported 317 records, most of which were also concentrated in northern Cook, Lake, and St. Louis Counties, plus western Itasca County **(A)**. The latter area roughly corresponds to the eastern region of the Chippewa Plains Subsection, an area where rich, mesic hardwood and mixed forests predominate. Only two confirmed nesting records were reported, one each in Cook and Lake Counties. Probable nesting activity was recorded in southern Aitkin, northern Cass, western Itasca, and St. Louis Counties, and possible nesting in Beltrami and Koochiching Counties. Breeding observations were surprisingly sparse along the North Shore of Lake Superior compared with those from the MBS, except for in southeastern Lake County and Cook County. The species was recorded in only 3.0 percent of the priority blocks (69/2,337), the same percentage as in all blocks.

The probability map for the Black-throated Blue Warbler predicted the species to be almost exclusively found in Cook and eastern Lake Counties **(B)**. Some possible habitat is predicted to exist in other portions of the northeast, the north-central region, and east-central Minnesota.

Overall, this species continues to be a relatively rare warbler in Minnesota, except for the extreme northeastern portion of the state and in western Itasca County. In their review of the Black-throated Blue Warbler in North America, Holmes and colleagues (2020) stated that the breeding habitat for this species shrank considerably with the clearing of mixed deciduous/coniferous forests in the northeastern United States in the seventeenth and

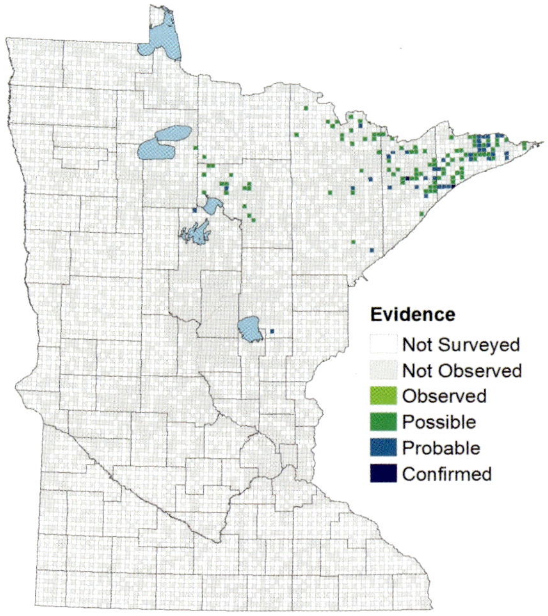

A. Breeding distribution of the Black-throated Blue Warbler in Minnesota based on the Breeding Bird Atlas (2009–13).

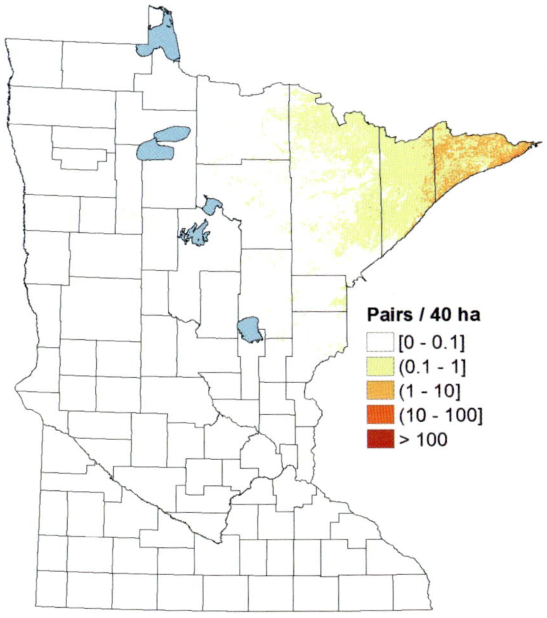

B. Predicted breeding distribution (pairs per 40 ha) of the Black-throated Blue Warbler in Minnesota based on habitat, landscape context, and climate data gathered during the Minnesota Breeding Bird Atlas (2009–13) using the Generalized Linear Modeling method with an adjustment for detectability.

eighteenth centuries. However, with the subsequent abandonment of farmland in the New England states, the species has increased again in the late nineteenth and twentieth centuries. The authors felt that the breeding habitat of the Black-throated Blue Warbler was more extensive now in the early twenty-first century than it was 100 to 150 years ago. There is little evidence that these changes have occurred in Minnesota because of the paucity of observations reported by Roberts from more southern locations in the state. However, the species' distribution may have retracted to some extent because the MNBBA did not record it from as far west as Clearwater County, nor were there any observations from Pine County, where Roberts had reported it previously.

BREEDING HABITAT

The breeding habitat of the Black-throated Blue Warbler has been described as undisturbed upland deciduous or mixed deciduous–coniferous forests (Holmes et al. 1996; Holmes et al. 2020). Niemi, Howe, and colleagues (2016) indicated the species prefers mature aspen–spruce–fir and mature upland hardwoods consisting of maple and basswood trees more than sixty years of age. The species avoided young or regenerating forests as well as most lowland forest cover types. The species' habitat is usually characterized by thick understories of saplings and shrubs, often with scattered gaps **(C)**.

POPULATION ABUNDANCE

The Canadian and US breeding population was estimated as 2.4 million breeding adults, with a Minnesota population of 6,500 (PIF 2021). However, the MNBBA estimated a substantially higher Minnesota breeding population of 78,000 adults. This discrepancy was likely due to different methods in estimating populations between PIF and the MNBBA, and the lack of coverage by the BBS routes in northern Cook, Lake, and St. Louis Counties upon which the PIF estimates were based (Sauer et al. 2020). These roadless areas were included in estimates by the MNBBA.

Population trend estimates from BBS counts in Minnesota are unreliable, but estimates from a larger region, the Boreal Hardwood Transition Bird Conservation Region, which includes northern Minnesota, indicated a significantly increasing population of 1.26 percent per year from 1966 to 2019 **(D)**. Recent estimates by Rosenberg, Kennedy, and colleagues (2016) from 1970 to 2014 revealed an overall increase in the species' population of 163 percent, which is also consistent with a recently expanding range.

CONSERVATION

The Black-throated Blue Warbler has a moderate conservation concern score. Its population appears to be increasing and its distribution expanding; however, C. S. Robbins and colleagues (1989) had identified the species as area sensitive, occurring mainly in forest tracts larger than 100 ha. Yet Holmes and colleagues (2020) pointed out that it uses stands with group-selection cuts and does not avoid transmission line corridors. In Minnesota,

C. The Black-throated Blue Warbler is fond of nesting in mature deciduous or mixed deciduous–coniferous forests with a shrubby understory, such as in canopy openings. Photo by Gerald J. Niemi.

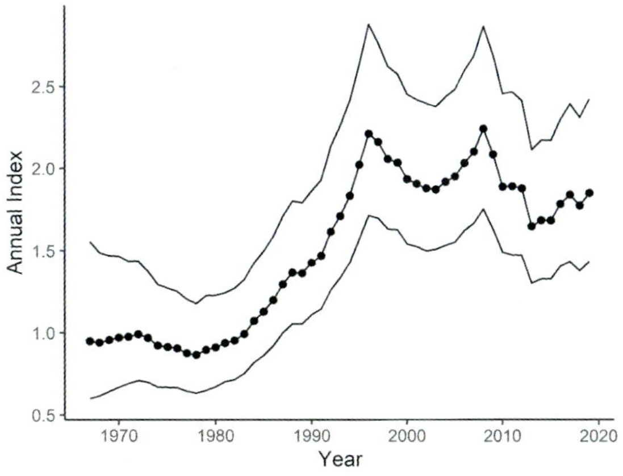

D. Breeding population trend and 95 percent credible interval (solid black lines) for the Black-throated Blue Warbler in the Boreal Hardwood Transition Bird Conservation Region based on the 1967–2019 Breeding Bird Survey (Sauer et al. 2020).

Grinde (2015) found that the Black-throated Blue Warbler experienced increased local extinction in small forest patches, which provides some support for Robbins's perspective.

NAS (2019) projected a moderate vulnerability for the Warbler in North America under a 2°C warming scenario, including a 64 percent range loss and a 41 percent range gain in eastern Canada by 2050. The species was predicted to be extirpated from Minnesota under the 2°C warming scenario by 2050. Holmes and colleagues (2020) emphasized it may have "benefited from climate change because of its ability to adapt to changes in breeding-season phenology," but warned that droughts in overwintering habitats can depress food abundance and, therefore, lower survivorship and physiological condition required for migration.

T. Arnold and Zink (2011) labeled the Black-throated Blue Warbler as a "super collider" because of its high incidence of mortality near communication towers. They estimated it was 96 times more at risk of tower kills compared with the average bird. Longcore and colleagues (2013) identify it as one of the most susceptible species to mortality at communication towers, with 4.9 percent of its population killed annually. Moreover, Loss and colleagues (2014) rated the species as highly susceptible to mortality due to collisions with windows.

The Black-throated Blue Warbler remains an uncommon species in Minnesota, except for portions of northeastern Minnesota. Even though its overall breeding population appears to be increasing, its status deserves to be watched over the coming years, especially with regard to overwintering habitat, climate change, mortality at windows and towers, and potential sensitivity to forest habitat fragmentation.

PALM WARBLER

Setophaga palmarum

Minnesota Seasonal Status: Regular breeding resident and migrant. The Palm Warbler was uncommon during the MNBBA.
Migration: Short- to medium-distance migrant that overwinters along the Atlantic and Pacific coasts, in the southern United States, south to Mexico and Central America, and in the Caribbean.
Food: Arthropods gleaned from the ground and foliage.
Nest: Cup nest on the ground, typically in *Sphagnum* moss.
Conservation Concern: Assigned a moderate PIF Continental Concern Score of 9/20.

Photo by David Brislance.

During the breeding season, the Palm Warbler is a secretive species restricted to Minnesota's peatland bogs in the northern regions of the state. Its song is inconspicuous, a weak, humble series of buzzy *tsee tsee tsee tsee* notes, often rising in the middle and reminiscent of an insect or of the song of the Yellow-rumped Warbler. In plumage and behavior, both sexes are similar and distinctive. The combination of its chestnut cap, yellow underparts with stripes on the breast and sides, olive-green upperparts, and especially its tail wagging render it a cinch to identify. Although the Palm Warbler is elusive during the breeding season, it can be quite conspicuous and abundant during migration or on the wintering grounds, where it can be found in many open areas, often with the Yellow-rumped Warbler.

The common name, Palm Warbler, is highly inappropriate. Griscom and Sprunt (1957) remarked that they amused themselves by "keeping an informal list of the people who have actually seen a Palm Warbler in a palm tree." Its common name and scientific species name were derived from the place it was first named in Santo Domingo, Dominican Republic, where it can be found overwintering. Roberts (1932) referred to it as the Yellow Red-poll Warbler, and Forbush and May (1955) called it the "little yellow wagtail."

The Palm Warbler is patchily distributed from far eastern Canada to northern Alberta, including northern Minnesota, the Upper Peninsula of Michigan, northern Wisconsin, and northern Maine.

MINNESOTA BREEDING DISTRIBUTION

Roberts (1932) called the Palm Warbler a "rare summer resident" in the northern part of the state as far south as northern Aitkin County and west to the "evergreen groves of eastern Marshall County." He described it as rare primarily because of the lack of observations, or at least consistent observations, during his field trips to many of the northern counties. For instance, only one specimen was collected—the only one he observed during a number of days around Lake Vermilion in northern St. Louis County. He described similar experiences with limited or a lack of observations during prime breeding periods at Lake Itasca, Cass Lake, Burntside Lake (St. Louis County), and Lake Vernon (Cook County). He summarized nesting activity in Aitkin, Cook, Marshall, and St. Louis Counties, as well as at Cass Lake.

In 1975, Green and Janssen emphasized the Palm Warbler's breeding distribution as the northeastern and north-central regions and in adjacent Marshall and Roseau Counties. They described it as scarce throughout most its range, except where "extensive areas of open or parklike tamarack-black spruce bogs" were present. Nesting was confirmed only in Aitkin and St. Louis Counties. Several years later, Janssen (1987) illustrated a more restricted breeding distribution from west-central St. Louis County to eastern Roseau and Lake of the Woods Counties. Janssen reported only confirmed nesting since 1970 in Beltrami County, but Hertzel and Janssen (1998) later included only confirmed nesting since 1970 in Koochiching County and did not include Beltrami County.

The MBS recorded 249 breeding-season locations and greatly expanded the breeding distribution reported by Janssen (1987). They detected the Palm Warbler from northern Carlton and Pine Counties, throughout Aitkin County, eastern Marshall County, and throughout northern St. Louis, Lake, and Cook Counties.

The MNBBA included a total of 356 records, which were all within the Laurentian Mixed Forest Ecological Province **(A)**. The species was recorded in only 4.0 percent (191/4,736) of all blocks surveyed. Confirmed nesting was recorded in only 8 percent (16/191) of

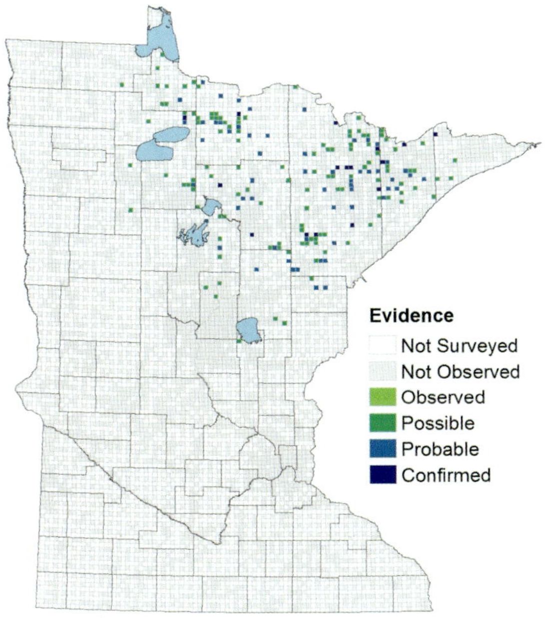

A. Breeding distribution of the Palm Warbler in Minnesota based on the Breeding Bird Atlas (2009–13).

the blocks where it was recorded and in a total of six counties: Beltrami, Cook, Itasca, Koochiching, Lake, and St. Louis. Probable nesting was also identified in Aitkin, Carlton, Cass, and Lake of the Woods Counties, plus possible nesting in Clearwater, Crow Wing, Mille Lacs, and Roseau Counties. Two reports of fledged young were confirmed in Glacier Ridge NWR in Polk County in 2018, the first nests ever reported in that county (MOU 2021a).

The distribution of the Palm Warbler identified by MBS and the MNBBA expanded the breeding range identified by Roberts in the late 1800s and early 1900s. It is unlikely that the breeding range has expanded in Minnesota; it is more likely that many previous efforts have overlooked the species. The Palm Warbler has a weak song that can be difficult to identify. Roberts (1932) aptly described its song as "one of the humblest of the group," "a simple, rather weak trill." For instance, the many records of the Palm Warbler by the MBS are likely due, in part, to their focus on rare habitats such as open peatlands and their experienced field staff.

Interestingly, Cutright and colleagues (2006) found a similar result for the Palm Warbler during the first breeding bird atlas in Wisconsin. Underscoring the dearth of information about its presence during the summer in Wisconsin, they quoted Kumlien and Hollister ([1903] 1951), who went so far as to declare that "there is no evidence that it was ever a summer resident in the state, even in the most northern counties." However, the Wisconsin breeding bird atlas found confirmed nesting throughout the northern tier of counties and as far south as Portage County. This species can be easily overlooked, and finding nests is a very challenging task that requires enduring patience.

BREEDING HABITAT

Palm Warbler nesting habitat is one of the easiest to identify among Minnesota's warblers **(B)**. Most modern descriptions emphasize open peatlands with scattered, low-statured tamarack or black spruce trees, commonly with dense shrubs (1–2 m), and ground cover of *Sphagnum* moss (Welsh 1971; Erskine 1977; W. H. Wilson 2020). Extensive surveys have confirmed its use of stagnant black spruce–tamarack bog forest cover types in the Agassiz Lowland Subsection of northwest Minnesota (Zlonis et al. 2017). Grinde (2015) found that Palm Warblers were more prone to local extinction in a landscape matrix with small patch sizes of suitable habitat.

POPULATION ABUNDANCE

PIF (2021) estimated a North American population of 13 million breeding adults and a Minnesota population of only 43,000. MNBBA estimates suggested a much larger breeding population of 620,000 adults, but the estimate was highly variable and should be interpreted with caution. Environment Canada (2011), with a large breeding population, estimated a variable population range from 5 to 50 million adults. They stated the Palm Warbler was abundant and widespread throughout Canada's boreal forest east of the Rocky Mountains.

Despite a suggestion by W. H. Wilson (2020) that BBS routes "are generally not adequate to determine trends" for this species, BBS routes within both core and non-core regions have been deemed reliable for Canada and surveywide in the United States and Canada. BBS found significant positive trends of 3.90 percent per year in Canada and a 3.87 percent per year increase surveywide from 1993 to 2019 using core and non-core routes (Sauer et al. 2020).

Any hints on population declines are scarce. W. H. Wilson (2020) suggested that high mortality has been documented at several tower sites and that peat harvesting in the boreal forests may contribute to declines. Longcore and colleagues (2013) documented the Palm Warbler's sensitivity to tower collisions by ranking it as the seventh most frequently killed bird species in the "Southeastern Coastal Plain and others" which incorporated a large area, including four Bird Conservation Regions. They also ranked the Palm Warbler as eighth in the Peninsular Florida Bird Conservation Region. Environment Canada (2011) stated there were few known threats either on its breeding or wintering grounds. They did acknowledge the loss of bog habitat in Quebec from the peat industry as a possible concern (see Desrochers et al. 1998).

CONSERVATION

Despite the moderate PIF conservation score, Environment Canada (2011) labeled the Palm Warbler as secure and showing a moderate increase in its breeding population since about 1970. The positive trend results from BBS in core and non-core areas from 1993 to 2019 is consistent with this interpretation.

NAS (2019) projected a high vulnerability for the Warbler's breeding distribution in North America with a 2°C warming scenario. The projections included a 58 percent range loss and a 26 percent range gain primarily in the far north by 2050. In Minnesota, the species was predicted to be extirpated from the state under our current 1.5°C warming scenario. It is likely to be among the first species to be extirpated from Minnesota due to climate change.

Because the Warbler's primary breeding population exists in Canada, the most fruitful conservation efforts are likely to be focused there during the breeding season. However, a better understanding of its migratory routes, stopover habitat, wintering grounds, how it responds to climate change, and its susceptibility to tower kills will be critical for its long-term viability.

B. Breeding habitat of the Palm Warbler in Minnesota is primarily in bogs. These habitats consist of widely scattered and stunted black spruce or tamarack trees, low shrubs, and a well-developed carpet of moss. Photo by Gerald J. Niemi.

PINE WARBLER

Setophaga pinus

Minnesota Seasonal Status: Regular breeding resident and migrant. The Pine Warbler was an uncommon species during the MNBBA.
Migration: Short-distance migrant; primarily overwinters in the southeastern United States.
Food: Arthropods gleaned from foliage, pine seeds, and infrequently seeds at feeders.
Nest: Cup nest in outer branches of the canopy in pine trees.
Conservation Concern: Assigned a low PIF Continental Concern Score of 7/20.

Photo by Jeremy Meyer.

The Pine Warbler is one of the most suitably named species. Its scientific species name, *pinus,* reflects its fondness for pine trees. Roberts (1932) emphasized that it "spends its entire existence, with the exception of the time consumed in migration, among pine trees."

During the breeding season, the male Pine Warbler is usually found first by its trill-like song emitted from the top of what else but a pine tree. The song is reminiscent of that of the Dark-eyed Junco or especially the Chipping Sparrow, which are both found in the same habitat and easily confused with the Pine Warbler. Griscom and Sprunt (1957) emphasized that its song "is really a series of notes uttered in rapid succession" and "sweeter, slower, softer and more pleasing" than that of the Chipping Sparrow. The male Pine Warbler is a relatively large, greenish-backed warbler with a bright yellow breast, faint streaking on its sides, two distinct white wing bars, and white spots on its tail. The female has the same coloration, which is more diluted in intensity.

The Pine Warbler is found throughout much of eastern North America, with a majority of its population in the southeastern United States and along the Atlantic coast. Northern populations occur from Nova Scotia and Maine in the northeast to Minnesota and western Ontario. The highest densities occur in the southeastern United States.

MINNESOTA BREEDING DISTRIBUTION

Historically in Minnesota, the Pine Warbler was distributed in coniferous forests as far south as northern Isanti County and west to Itasca State Park (Roberts 1932). Roberts reported nesting activity in Cass, Isanti, and St. Louis Counties, and in Itasca State Park and near Leech Lake. He noted it was one of the most common warblers throughout the coniferous region of northern Minnesota.

Green and Janssen (1975) emphasized that, since the early twentieth century, Pine Warblers were primarily found in the northeastern and north-central regions of the state, especially in Pine County and from Mille Lacs Lake to Red Lake. In addition to Roberts (1932), Green and Janssen (1975) reported confirmed nests in Beltrami and Clearwater Counties and inferred nesting in Hubbard, Lake, Lake of the Woods, and Pine Counties. Janssen (1987) suggested a more restricted range in northeastern Minnesota, describing the species as "sparsely represented or absent" in Cook, Lake, and St. Louis Counties. He included confirmed nesting in Clearwater, Crow Wing, and Pine Counties since 1970. In addition, he extended the breeding range southward to the Twin Cities metropolitan area, in Ramsey and Washington Counties, and to the central region of the state, in Morrison and Todd Counties. Several years later, however, Hertzel and Janssen (1998) excluded nesting in Clearwater County and added St. Louis County to the list of confirmed nesting counties in the state since 1970.

Since 2014, the MBS had recorded 477 breeding-season locations during their intensive county surveys. Their records clearly indicated extensive locations in northern Cook, northern Lake, and St. Louis Counties, and from southeastern Roseau County, eastern Mahnomen County, Becker County, northeastern Otter Tail County, and south to Anoka County.

The MNBBA found breeding records of the Pine Warbler throughout the Laurentian Mixed Forest Province and several records in the central portion of the Eastern Broadleaf Forest Province **(A)**. The most southerly evidence of possible breeding was in Washington County, while probable nesting was found south to Anoka and Sherburne Counties, northwest to Roseau County, and northeast to Cook County. Confirmed nesting records in Aitkin, Becker, Carlton, Hubbard, Itasca, Lake, Lake of the Woods, and Wadena Counties were additions to the list of counties where nesting had been confirmed since 1970 by Hertzel and Janssen (1998). Pine Warbler records represented 12.5 percent (591/4,737) of all blocks surveyed. Nesting was confirmed for only 22 blocks, a powerful indication of the difficulty of finding nests or observations of nesting behavior high in the canopy of pine trees.

The MNBBA's probability map for the Pine Warbler indicated a patchy pattern throughout north-central and northeastern Minnesota **(B)**. The highest probabilities predicted were in the major pineries of the state, especially in the north-central region, such as in the Chippewa National Forest. Densities were predicted to be much lower in the more deciduous forested regions, such as southern Aitkin and St. Louis Counties, Carlton and Pine Counties, and along the North Shore of Lake Superior.

Rodewald and colleagues (2020), in their review of the Pine Warbler in North America, commented on the many local extirpations and range expansions in North America where pines have been lost or have been introduced. Schulte and colleagues (2005) found similar contemporary distributions of the Pine Warbler in Wisconsin when compared with pre-1850 distributions based on landscape modeling. They point out that pine plantations established in the southern part of the state have provided new habitats in that region of Wisconsin.

The situation in Minnesota is less clear because the acreage of mature pine forests in the state has declined (Jaakko Pöyry Consulting 1994). Presum-

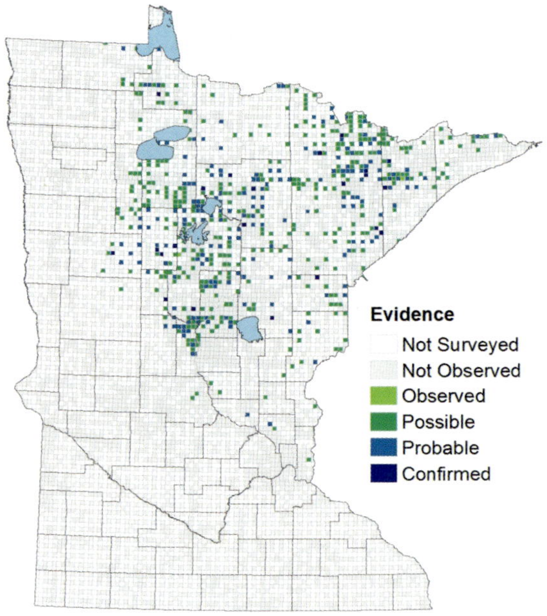

A. Breeding distribution of the Pine Warbler in Minnesota based on the Breeding Bird Atlas (2009–13).

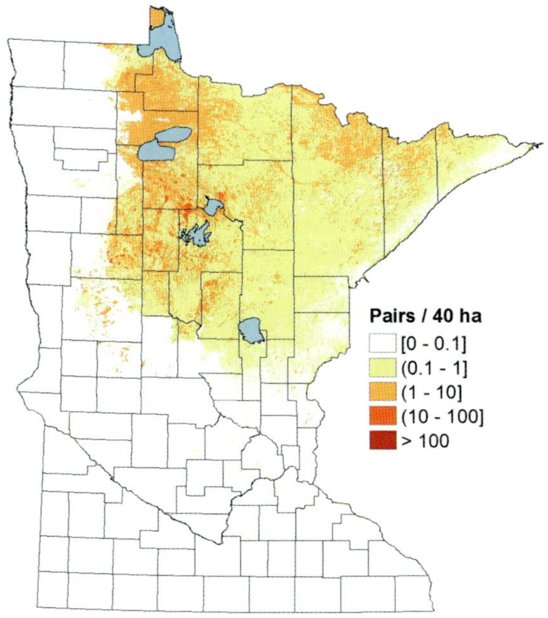

B. Predicted breeding distribution (pairs per 40 ha) of the Pine Warbler in Minnesota based on habitat, landscape context, and climate data gathered during the Minnesota Breeding Bird Atlas (2009–13) using the Generalized Linear Modeling method with an adjustment for detectability.

ably, the population and the range of the Pine Warbler contracted after the major logging eras of the late nineteenth and early twentieth centuries. This assumption agrees with Roberts's (1932) conclusion that the species was primarily distributed in the evergreen forests from northern Isanti to Itasca State Park. Consistent with the comments by Rodewald and colleagues (2020) and Schulte and colleagues (2005), the MNBBA data illustrates a potential expansion of the population of the Pine Warbler in the state in recent years with both western and southward breeding observations. Confirmed nesting by Pine Warblers in Beltrami (2017), Clearwater (2017), Ramsey (2020), and Stearns (2008) Counties further supports these recent expansions (MOU 2021a).

BREEDING HABITAT

Pine Warblers are strongly associated with large pine trees. Rodewald and colleagues (2020) stated that this species is known to have nested in 15 different species of pine across its breeding range. Roberts (1932) frequently mentions jack pine as a favorite tree of the species, while Janssen (1987) stated that "it is more frequently found in large stands of White Pine." Analysis of the species' use of forest cover types in the Chippewa and Superior National Forests in Minnesota indicated significant use of red pine and white pine forests with a lower preference for jack pine forests (Niemi, Howe, et al. 2016) **(C)**.

POPULATION ABUNDANCE

PIF (2021) estimated that the total US and Canadian population of the Pine Warbler was 13 million breeding adults, but only 170,000 in Minnesota. In comparison, the MNBBA estimated a Minnesota breeding population of about 840,000 breeding adults (95 percent confidence interval of 446,000–1,600,000), or about 6 percent of the North American population.

Trend estimates from the BBS were unreliable for Minnesota, but the surveywide results, based on more than 1,400 routes from 1967 to 2019, showed a significantly increasing trend that averaged 1.29 percent per year **(D)**. Rosenberg, Kennedy, and colleagues (2016) calculated a 68 percent increase in the population from 1970 to 2014. These trends were similar to the significantly increasing trends observed in both the Chippewa (1.65 percent per year) and Superior (4.90 percent per year) National Forests from 1995 to 2020 (Grinde, Bednar, et al. 2020). Collectively, these trend data provide solid evidence for an increasing population of Pine Warblers and further support the observed range expansions of the species in Minnesota.

CONSERVATION

Currently there is minimal conservation concern for the Pine Warbler because it has a substantial and increasing North American breeding population. Because of their strong association with pine, the species is susceptible to losses of pine forests within its range. It has also been reported as sensitive to fragmentation of these forests (C. S. Robbins et al. 1989) and appears to make minimal use of young pine plantations (Schroeder 1985). Atwell and colleagues (2008) found the species insensitive to selective logging in red pine forests in north-central Minnesota if the canopy was left intact. Rodewald and colleagues (2020) suggested that Pine Warblers may not be vulnerable to tower kills because they migrate via one long sustained flight from breeding grounds to overwintering areas, which helps them avoid or minimize such collisions.

C. The typical breeding habitat of the Pine Warbler demonstrates its fondness for large pine trees in Minnesota. Photo by Gerald J. Niemi.

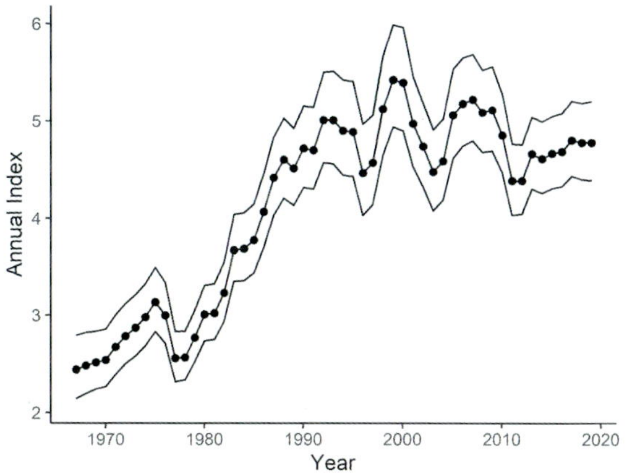

D. Surveywide breeding population trend and 95 percent credible interval (solid black lines) for the Pine Warbler based on the 1967–2019 Breeding Bird Survey (Sauer et al. 2020).

Under a 2°C warming scenario, NAS (2019) projected moderate vulnerability for the Pine Warbler's breeding distribution in North America with a 38 percent loss of its breeding range and a 20 percent gain northward by 2050. Its breeding range retractions in Minnesota were predicted in the western regions, while some gains were forecast in the southeast and northeastward into Canada.

Besides climate change, there appear to be few conservation concerns at present. Whether the species is now recovering from extensive loss of pine forests since the late nineteenth century and well into the twentieth is not completely clear. However, several lines of evidence suggest that the current maturation of pine trees and pine plantations may now be providing better breeding habitat conditions and facilitating an expansion or re-expansion into former ranges. From a management perspective, populations of Pine Warblers can serve as a strong indicator of the health of mature pine forests by virtue of their almost exclusive use of this forest cover type (Niemi, Hanowski, et al. 1997).

YELLOW-RUMPED WARBLER

Setophaga coronata

Photo by Rebecca Field.

Minnesota Seasonal Status: Regular breeding resident, migrant, and occasional in winter in southern Minnesota at feeders; most depart by late January. The Yellow-rumped Warbler was common during the MNBBA.
Migration: Short- to medium-distance migrant that overwinters in the southern United States, along the Atlantic, Gulf, and Pacific coasts, and in the Caribbean, Mexico, and Central America.
Food: Insects, other invertebrates, and occasionally berries gleaned from foliage. The species also commonly practices fly-catching.
Nest: Cup nest on a horizontal branch in coniferous trees, at highly variable heights.
Conservation Concern: Assigned a low PIF Continental Concern Score of 6/20.

The Yellow-rumped Warbler, formerly the Myrtle Warbler, is one of the most commonly observed warblers in Minnesota, but is primarily seen during migration, when it can be found almost anywhere in the state. It is the first warbler to arrive in the spring and the last to leave in the fall. It can be quite abundant during its migration through the state, but during the breeding season, it becomes far more secretive and vanishes into the northern coniferous forests to nest.

Roberts (1932) reminisced about his early encounters with this species when he was a boy of 16:

> When on a chilly fifteenth of April in 1874 . . . feelings of wonder and of delight were aroused that have never been repeated. Whenever this Warbler is encountered in numbers, there still is brought up from memory a vivid picture of the then village-like Minneapolis, filled with myriads of restless, twittering little birds, gay in their blue, black, and white, set off by four showy, golden patches. It seemed certain that this could never have happened before, such strangers are we at home until our senses are awakened to the things about us.

The Yellow-rumped Warbler's scientific name *coronata* refers to its yellow crown, while its common name refers to the easily recognized and appropriately named yellow on its rump. These two plumage patches add to the two yellow patches on its sides, all against a backdrop of bluish, black, and white that renders it a stunning plumage. Both the males and females have this pattern, although the female is more diluted in coloration. On the breeding grounds, the Yellow-rumped Warbler can best be found by its song, which is not easy to recognize. Roberts (1932) labeled it a weak effort with a "peculiar sliding intonation" described as *weech-weech-weech-weech.* It has also been described as a flat, slow trill, variable in intensity and usually rising (or sometimes falling) at the end (Curson et al. 1994). Its call note is its most diagnostic vocalization. It is a loud and sharp *chek* that once learned is easy to recognize.

In the past, the Yellow-rumped was classified as a distinct species from its western counterpart, the Audubon's Warbler, but recent classification by taxonomic experts combine the two into one species. When they were split, the eastern species was called the Myrtle Warbler in honor of its berry consumption on wax myrtle (*Myrica cerifera*) or northern bayberry (*Myrica pensylvanica*), particularly during migration or on the wintering grounds. In Minnesota, it is definitely a carnivorous insect consumer, but during inclement weather it is known to resort to berries. The Yellow-rumped is a widely distributed species in northeastern, upper midwestern, and western North America. Its highest densities are found in many parts of western North America and in Labrador.

MINNESOTA BREEDING DISTRIBUTION

Roberts (1932) stated that the Yellow-rumped Warbler's distribution was primarily north of Minnesota but that it was detected in the evergreen forests as far south as Cass Lake and west to eastern Marshall County. He found it "not generally common as a breeding bird" and reported very limited nesting activity. For instance, observations in Itasca, Marshall, and St. Louis Counties were of fledged young, young with adults, or feeding young out of the nest. Another observation in St. Louis County was of a "female with egg about ready to be laid," presumably of a bird that was collected.

Forty years later, Green and Janssen (1975) underscored the Yellow-rumped Warbler's distribution in the northeastern and north-central regions of the state. They included confirmed nesting in Clearwater, Cook, Itasca, Lake, and St. Louis Counties. The boundaries of their estimated breeding range extended west to eastern Becker and Mahnomen Counties, southwest to Cass, Crow Wing, and Wadena Counties, and south to northern Mille Lacs and Pine Counties. Several years later, Janssen (1987) slightly constricted the breeding range boundary in the southwestern portion to exclude Becker County and limited the boundary at Hubbard and Crow Wing Counties. Janssen in 1987 and Hertzel and Janssen in 1998 updated confirmed nesting since 1970 to include Aitkin, Beltrami, Clearwater, Cook, Itasca, Lake, Marshall, and St. Louis Counties.

The MBS recorded 650 breeding-season locations, many of which were found in the northeastern and north-central counties previously mentioned. They had breeding observation locations from northwestern Minnesota in Marshall and Roseau Counties, west to eastern Becker County, and south to northern Mille Lacs, Pine, northeastern Todd, and eastern Wadena Counties.

The MNBBA reported 1,970 records of the Yellow-rumped Warbler, which were primarily distrib-

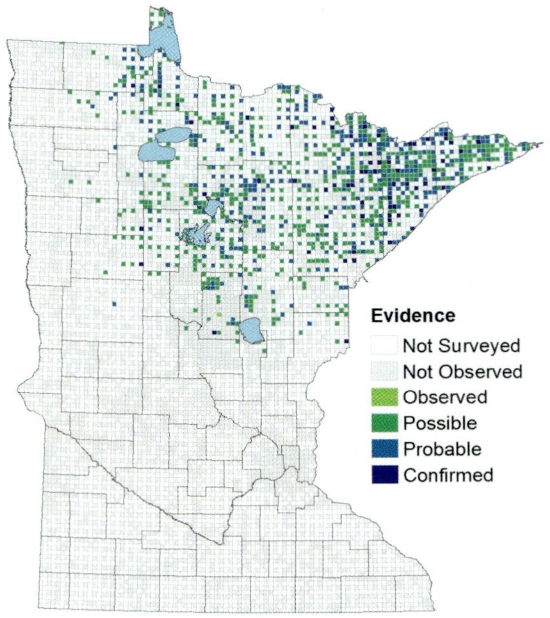

A. Breeding distribution of the Yellow-rumped Warbler in Minnesota based on the Breeding Bird Atlas (2009–13).

uted in the Laurentian Mixed Forest Province but included many records in the Tallgrass Aspen Parklands Province and in the Hardwood Hills Subsection **(A)**. Confirmed nesting records beyond those previously reported included Becker, Crow Wing, Koochiching, Lake of the Woods, Pine, and Roseau Counties. A probable nesting record extends the potential breeding range west to Otter Tail County. The MNBBA included breeding observations for 20.3 percent (964/4,748) of all blocks. As with many other secretive nesting species, only 6.7 percent (65/964) of all recorded blocks for this species had confirmed nesting activity, with the vast majority in Cook, Lake, and St. Louis Counties.

Data from both the MBS and MNBBA expanded the potential breeding distribution to the west, southwest, and south in Minnesota from that reported by Roberts and subsequent compilers through 1997. In their review of the Yellow-rumped Warbler in North America, P. D. Hunt and Flaspohler (2020) reported southward range expansions in New York, Pennsylvania, and West Virginia where reforestation and the creation of pine plantations had occurred. On April 27, 2020, a Yellow-rumped was detected carrying nesting material in Washington County (MOU 2021a), which indicates the most southerly potential nesting activity in Minnesota.

It is impossible to know whether Yellow-rumped Warbler breeding activity had been missed by Roberts and others during earlier times in Minnesota, such as on the fringes of its current breeding range in the northwest, western, or north-central regions where both the MBS and MNBBA observed this species. Alternatively, do these observations represent recent breeding range expansions in Minnesota? The Yellow-rumped Warbler is moderately difficult to identify by song but relatively easy to identify by sight. Given that coniferous trees and forests have contracted to the north with logging and forest clearing, an expansion south is counterintuitive. Observers in Minnesota should look for this species in more southern localities of Minnesota where substantial patches of coniferous trees or pine plantations occur.

BREEDING HABITAT

Yellow-rumped Warblers are considered a habitat generalist and found in many types of coniferous forest: mixed coniferous–deciduous forests, deciduous forests with scattered coniferous trees, and even young forests (Collins et al. 1982; P. D. Hunt and Flaspohler 2020; Cutright et al. 2006). Habitat data compiled during counts in Minnesota's Chippewa and Superior National Forests (Niemi, Howe, et al. 2016) documented its use of many habitat types, but the species was most abundant in mature black spruce–tamarack lowlands **(B)**. Other forest cover types used during the nesting season included jack, red, and white pine; regenerating conifer; and mixed swamp conifer.

POPULATION ABUNDANCE

PIF (2021) estimated a US and Canadian global breeding population of 170 million breeding adults, which makes the Yellow-rumped Warbler the most abundant warbler species in North America by far. They also estimated a Minnesota population of 350,000, but the MNBBA's estimate was substantially higher, with 2.65 million breeding adults (95 percent confidence interval of 2.34–3.75 million). Like many species that occur in the northern portions of Minnesota, especially in roadless areas, PIF tends to have more conservative breeding population estimates because of methodological differences, and they use the BBS roadside counts for their estimates. In contrast, the MNBBA sampled roadless areas where Yellow-rumps are very common and thus obtained a less conservative population estimate for the state.

B. Yellow-rumped Warblers commonly nest in a wide range of coniferous forest habitats in Minnesota. Photo by Gerald J. Niemi.

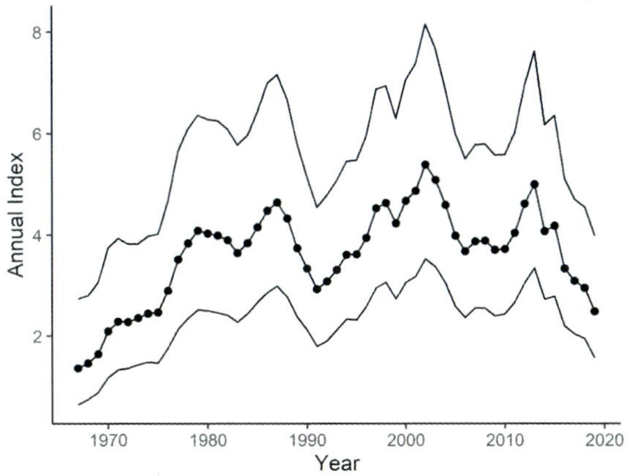

C. Breeding population trend and 95 percent credible interval (solid black lines) for the Yellow-rumped Warbler in Minnesota based on the 1967–2019 Breeding Bird Survey (Sauer et al. 2020).

Despite considerable variability in the Yellow-rumped Warbler population, the BBS estimated a stable population in Minnesota from 1967 to 2019 **(C)**. Rosenberg, Kennedy, and colleagues (2016) previously had calculated a stable population in North America from 1970 to 2014.

CONSERVATION

The Yellow-rumped Warbler has minimal conservation concerns primarily because it has a substantial and stable population; in addition, it has a wide breeding distribution in North America. P. D. Hunt and Flaspohler (2020) point out that any forest management directed toward increased coniferous forest tree species would benefit this species.

Longcore and colleagues (2013) identified the Yellow-rumped Warbler as one of the most frequently killed species at communication towers, where it was ranked fourth among all species in the Southeastern Coastal Plain (a large area that included four Bird Conservation Regions), and fifth in the Eastern Tallgrass Prairie Bird Conservation Region. NAS (2019) projected a moderate vulnerability for the Yellow-rumped in North America and extirpation from Minnesota with a 2°C warming scenario by 2050.

Overall, the status of the Yellow-rumped Warbler in Minnesota appears to be stable and possibly expanding, but projections of potential extirpation from the state as a breeding species due to a changing climate dampen that optimism. Only time will tell if Minnesotans can enjoy this colorful and voracious insectivore for years to come.

YELLOW-THROATED WARBLER

Setophaga dominica

Minnesota Seasonal Status: Summer visitant; sporadic breeding in five separate years. Regular spring migrant south, vagrant north; fall vagrant south and north. The Yellow-throated Warbler was very rare during the MNBBA.
Migration: Northern and interior populations are medium-distance migrants and winter in the Caribbean islands and along the Gulf coast in Mexico to Costa Rica. Populations in Florida and along the southern Atlantic coast are permanent residents.
Food: Insectivorous, feeding on lepidopteran larvae, flies, spiders, and many other insects. Forages by creeping along tree trunks and large branches and probing into bark crevices, needle clusters, and pine cones.
Nest: Cup nest lined with vegetation of moss, often in a clump of Spanish moss. Placed in the top of a pine, sycamore, cypress, or other canopy tree, usually 9–18 m high, but can be from 1 to 36 m high.
Conservation Concern: Assigned a moderate PIF Continental Concern Score of 9/20.

The Yellow-throated Warbler became known in 1766, when the type specimen was collected on the island of Dominica where it winters—hence, the roots of its species name. It was primarily known from the cypress swamps and piney woods of the deep South, where William Brewster (1877), a noted ornithologist from New England, studied its habits in Georgia and concluded that "nearly all the authors who have written on the Yellow-throated Warbler from personal observation compare his movements along the branches to those of the Black-and-white Creeper." Further, "At first I was inclined to the same opinion, but . . . I became convinced of the error of my earlier impressions. . . . They do, indeed, spend much of their time searching the larger branches for food, but it is much more in the manner of the Pine Warbler."

The male Yellow-throated Warbler is identified by a combination of several characteristics: a bright yellow throat, a white belly with dark streaking on the flanks, a black triangular patch below the eye, two distinct white wing bars, and a white eye-stripe. The female is similar, but has less black on its forehead and thinner side streaking. Its song has been likened to the Indigo Bunting and Louisiana Waterthrush, described as *tew-ew tew-ew tew tew tew tew wi* (Alderfer 2006).

Photo by Rebecca Field.

The Yellow-throated Warbler is mostly found in the southeastern United States. It is bounded in the north by states along the Ohio River valley from Pennsylvania to Illinois, south from Missouri to the Gulf states, west to eastern Texas, and from Florida up the Atlantic coast to New Jersey. It also breeds in the Bahamas, where it is found year-round. The highest densities are found in the southern Appalachian Mountains and along the southern Atlantic coast.

MINNESOTA BREEDING DISTRIBUTION

The Yellow-throated Warbler was not known as part of the state's avifauna until after Roberts (1932) and Green and Janssen (1975) were published. The first documented record was a bird found on May 6, 1980, in a wooded area in Roseville, Ramsey County (Soulen 1980). Records continued over the rest of that decade and increased markedly from 1991 to 2000.

During the 1990s through 2000, many of the published records were from Kandiyohi County, where the first observation occurred near Willmar in 1993. From 1994 to 2000, a singing male or a pair was observed in Sibley State Park from mid-May to early summer every year. The first evidence of nesting was reported on May 29, 1997, when the pair was observed at the nest, which was 9–11 m high in a large cottonwood along Lake Andrew. The nest was later abandoned, collected, and given to the Bell Museum of Natural History (R. Frederickson 1998). The remaining breeding-season records from 1994 to 2000 were in the Twin Cities metropolitan area, including Anoka, Hennepin, Dakota, Washington, and Wright Counties; elsewhere records were also reported in Rice and Marshall Counties. The Washington County record included multiple dates in one location of a warbler seen in William O'Brien State Park, Washington County, from June to August 2000 (Eckert 2001; Zumeta 2001).

Fewer summer records were reported early in the twenty-first century, but the first Minnesota successful breeding record (and the second breeding record) of the Yellow-throated Warbler was obtained in 2001. This warbler was seen at the Acacia Park Cemetery in Mendota Heights, Dakota County, from June 11 to July 1, 2001. The nest was discovered on June 21 about 10 m up in a Norway spruce; three young fledged on June 30 (J. P. Mattsson 2001).

During the MNBBA, the third breeding record and second successful nesting was obtained in 2013 near the Nature Store parking lot, Whitewater State Park, Winona County. A singing male was sighted by many on May 10, 2013, with subsequent reports continuing to July 11, when an adult was seen with three begging individuals **(A)**. The last day the family group was reported was July 27 (MOU 2021a). That same summer another observation was made: an adult on June 26, 2013, at Carley State Park along the north fork of the Whitewater River in Wabasha County, about 12 km away from Whitewater State Park (both referenced in Kessen and Svingen 2014).

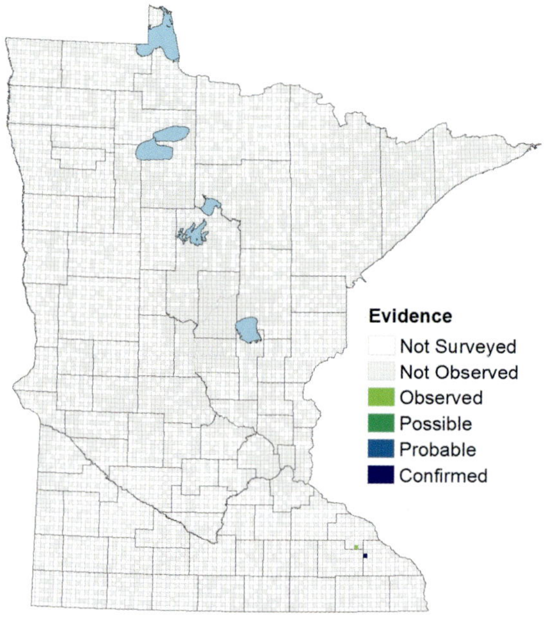

A. Breeding distribution of the Yellow-throated Warbler in Minnesota based on the Breeding Bird Atlas (2009–13).

From 2014 to summer 2020, two more confirmed breeding records were documented. First, in 2016, a pair was found in Gold Medal Park, "a small highly manicured city park with mowed lawn and scattered ornamental trees" in downtown Minneapolis (Fall 2016). The male was first seen on May 12; on May 13 two adults were seen. Fledglings were first seen on June 23. Adults were seen again the following year but only from April 22 to 25. Second, from May 19 to August 2, 2020, breeding was documented, including observations of an adult feeding young, at Vadnais Lake, Ramsey County (MOU 2021a, 2021b).

Three other summer observations were reported during this period. First, from 2015 to 2018, a singing male was observed again at the Whitewater State Park Nature Store, and from May 4–13, 2019, an adult was photographed; neither report included any evidence of breeding. Second, a singing male was heard June 5–8, 2018, at Crystal Springs SNA, Washington County. And third, a singing male was heard and seen June 26–27, 2018, June 2–8, 2019, and April 30–June 20, 2020, along the Reno-Hillside Road, Houston County (MOU 2021a, 2021b). It is not clear whether these summer records indicate an expanding breeding population of the Yellow-throated Warbler or whether they represent displaced migrants that persistently sing even in subsequent years at the same location.

Apart from the five (1997, 2001, 2013, 2016, 2020) documented breeding records, most observations in the twenty-first century have been single observations of Warblers in spring or summer in the extensively birded Twin Cities area (eight records) and in southeastern Minnesota (Dodge and Olmsted Counties).

Vagrants have increased north of the latitude of the Twin Cities from 2000 to 2020, including records from Otter Tail County (October 25–30, 2003); Morrison County (May 15, 2005); St. Louis County (September 9, 2008, May 18, 2009, and May 20, 2019); Crow Wing County (October 21–22, 2011); Isanti County (May 18, 2020); and Chisago County (May 30, 2020).

In states adjacent to Minnesota (Michigan, Wisconsin, and Iowa), the published records show that summer populations have been sporadic from the late 1800s. All have a few confirmed nesting records. However, during the last two decades, as in Minnesota, the number of observations has increased and the locations have expanded.

Only time will tell if the Minnesota presence of the Yellow-throated Warbler grows and leads to a sustainable population. Its status from the MOU has been designated as "casual" since 1988. What is known is that the distribution of the Yellow-throated Warbler has changed over time. The species disappeared from the northern periphery of its range in the late nineteenth and early twentieth centuries for reasons that are not well understood. However, after about 1940, its populations increased sufficiently to reclaim much of that range and may have even expanded farther northward in some areas (McKay and Hall 2020).

BREEDING HABITAT

In the southern United States, the Yellow-throated Warbler can be found in lowland swamps with cypress, live oak, and tupelo, or in mixed pine–deciduous forests and upland pine, especially loblolly pine. Both Spanish moss and super-canopy trees are important components. In the North, the species is found in the tributary valleys of the big northern rivers, often in sycamore bottomlands but also in upland woodlands, where it prefers forests with clumps of mature pines. Big super-canopy trees are a component in both habitats (J. L. Dunn and Garrett 1997). The structure of the forest vegetation, an open understory with tall canopy trees, seems more important than the actual species of trees in the forest. The presence of the species at Gold Medal Park in Minneapolis consists of an open understory with scattered trees **(B)**.

POPULATION ABUNDANCE

PIF (2021) estimated the population of the Yellow-throated Warbler in the United States at 2.0 million birds. The population trend from the BBS data revealed significant increases in Kentucky (average of 1.43 percent per year), North Carolina (1.66 percent per year) and West Virginia (5.59 percent per year), where trend estimates were deemed reliable, from 1967 to 2019. Population trends were also reliably identified in Georgia, South Carolina, and Tennessee, but trends were insignificant over this same time period (Sauer et al. 2020). Rosenberg, Kennedy, and colleagues (2016) estimated that the Yellow-throated Warbler population had increased by 50 percent from 1970 to 2014.

CONSERVATION

The Yellow-throated Warbler is of moderate conservation concern in the United States because of its relatively small population size, the limited data about its population and distribution, and threats to its breeding habitat. It does not have a formal national conservation status, and its population appears to be locally increasing. Two states at the northern edge of its range have included this species on their state lists: *Threatened* in Michigan (2009) and *Endangered* in Wisconsin (2015). At the western edge of its range, Kansas has designated its status as a nongame species in need of conservation. None of these states has a formal management plan. Regarding climate change, NAS (2019) predicted high vulnerability in the United States due to projected losses of its breeding range in the southeast but with expansions northward. Therefore, Minnesotans may look forward to more encounters with this stunning warbler.

B. In 2016, the Yellow-throated Warbler was confirmed nesting in Gold Medal Park, Minneapolis. Photo by Lee A. Pfannmuller.

BLACK-THROATED GREEN WARBLER

Setophaga virens

Minnesota Seasonal Status: Regular breeding resident and migrant. The Black-throated Green Warbler was a common species during the MNBBA.
Migration: Medium- to long-distance migrant, overwintering in the Caribbean, Central America, and northern South America.
Food: Arthropods, gleaned from foliage, especially caterpillars on conifers.
Nest: Cup nest, usually in conifers, often low but at variable heights.
Conservation Concern: Assigned a moderate PIF Continental Concern Score of 9/20.

Photo by David Brislance.

Aptly named, the Black-throated Green Warbler is indeed a colorful warbler with a green back and white belly, but its distinguishing features are its black throat and upper breast, two white wing bars, and its bright yellow neck and face. The sexes are similar in plumage, but the male is more brightly colored. During the breeding season, it is best identified with its distinctive and easily recognized buzzy song, described as *zee zee zee zo zee*.

Roberts (1932) describes one of his first encounters with the song of the Black-throated Green Warbler near Buffalo Lake in Wright County as "none is more distinct or enjoyable than the recollection of the exciting mystery that surrounded the origin of a certain sleepy, gliding song—sweet, but with a touch of melancholy—that came constantly from the tops of those great elms and basswoods. . . . This song fell upon interested but as yet untutored ears, and many were the hours of strenuous watching. . . . Two or three days elapsed before the tiny singer . . . was located . . . and the identity of the singer settled for once and all."

The Black-throated Green is primarily found in northeastern and upper midwestern portions of the United States and Canada, especially from Labrador to northern Alberta and south in the Appalachian Mountains to Alabama. The highest densities occur in the Upper Peninsula of Michigan, Labrador, and Nova Scotia.

MINNESOTA BREEDING DISTRIBUTION

Historically, the Black-throated Green Warbler was found by Roberts (1932) in the northern coniferous forests from northern Isanti County and west to Itasca State Park. He also described it as formerly "a summer resident in considerable numbers in the heavy deciduous forest ('Big Woods') of Wright County in 1875." Occasional pairs were found in the early 1930s in the heavy forests south of the coniferous region and in limited numbers around Lake Minnetonka in Hennepin County. He only reported nesting behavior from "Mille Lacs" and Itasca State Park.

Green and Janssen (1975) outlined a breeding distribution in the northeastern and north-central regions and added confirmed nesting from Clearwater, Cook, Lake, and Mille Lacs Counties, plus inferred nesting from Hubbard County. Janssen (1987) provided a similar distribution but cited observations south to northern Anoka County at Cedar Creek and as far west as Tamarac NWR in Becker County. He noted confirmed nesting in five counties since 1970, including Beltrami, Clearwater, Cook, Crow Wing, and Lake; in addition, he emphasized that the species' distribution in the northwestern region of the state was unclear. By 1998, Hertzel and Janssen had added confirmed nesting in Aitkin County since 1970.

By 2014, the MBS had identified 881 breeding-season locations of the Black-throated Green Warbler. The distribution of these locations emphasized the primary breeding distribution in the northeastern and north-central forests. However, the MBS included several breeding-season detections from northwestern Minnesota, including Roseau County and eastern Marshall County, west to Mahnomen and Becker Counties, and southwest to Douglas and Todd Counties.

The MNBBA included 1,741 records and reinforced the breeding distribution identified by the MBS. Nearly all the records were from the Laurentian Mixed Forest Province but they included scattered observations from the Tallgrass Aspen Parklands and from the extreme northern portion of the Eastern Broadleaf Forest Province **(A)**. Nesting was confirmed in 35 blocks (less than 1 percent of all blocks) and in only 4.4 percent (35/795) of the blocks where the species was detected. Additional counties with confirmed nesting from the MNBBA and not previously included in Hertzel and Janssen (1998) included Carlton, Cass, Itasca, and St. Louis. The MNBBA also identified probable nesting in Becker, Hubbard, Koochiching, Lake of the Woods, and Roseau Counties, plus possible nesting in Marshall County and in southeastern Pine County.

The probability map indicated the species' wide distribution in the northeastern and north-central regions, with especially high concentrations predicted in Cook and southeastern Lake Counties. Higher densities were also predicted in several patches farther west and south, including in portions of Beltrami, Carlton, Cass, Clearwater, Hubbard, Itasca, and Koochiching Counties **(B)**.

In their review of the Black-throated Green Warbler in North America, D. H. Morse and Poole (2020) suggest that the species might have expanded its range westward in recent years in Canada but state that the farthest northwestern breeding detections in the United States occurred in northwestern Minnesota. Morse and Poole also report an unusual disjunct breeding population that was found in the Ozarks and a recent expansion into the Ouachita Mountains of northwestern and west-central Arkansas.

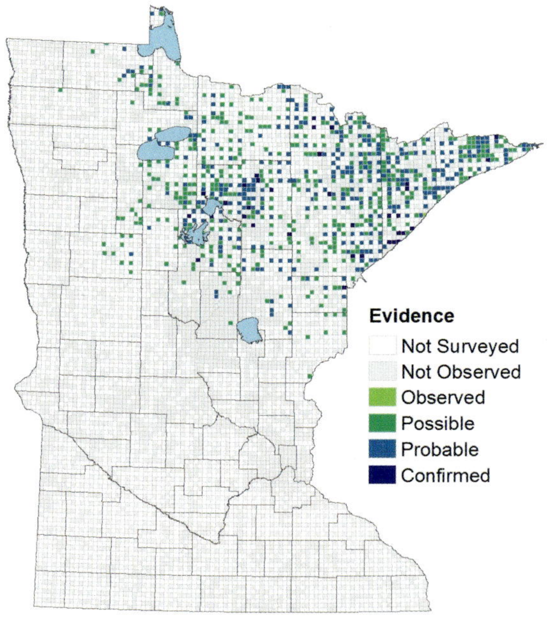

A. Breeding distribution of the Black-throated Green Warbler in Minnesota based on the Breeding Bird Atlas (2009–13).

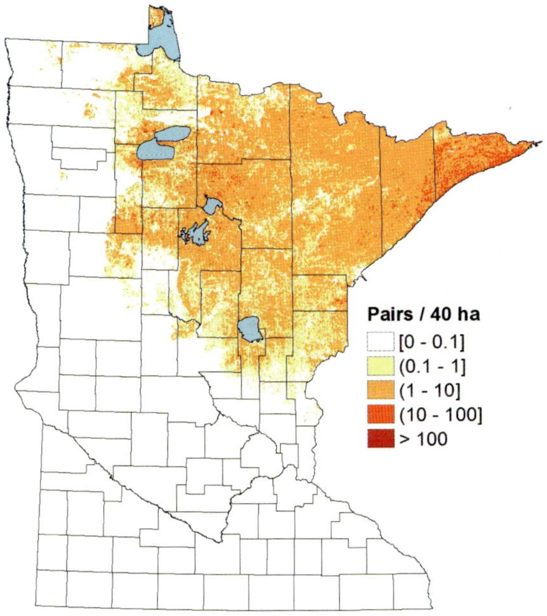

B. Predicted breeding distribution (pairs per 40 ha) of the Black-throated Green Warbler in Minnesota based on habitat, landscape context, and climate data gathered during the Minnesota Breeding Bird Atlas (2009–13) using the Generalized Linear Modeling method with an adjustment for detectability.

In Minnesota, Roberts's (1932) reports of the Black-throated Green Warbler in the Big Woods Subsection in the late 1800s documents that the species' range had contracted substantially with the opening of the forests for agriculture and urbanization. However, results from MBS and the MNBBA suggest that the Black-throated Green has been expanding its distribution westward and to the south in Minnesota in recent years. These findings are supported by similar documentation of southward distributional shifts by Cutright and colleagues (2006) in Wisconsin and by Chartier and colleagues (2013) in Michigan, based on their breeding bird atlases.

BREEDING HABITAT

Collins (1983) described the Black-throated Green Warbler as a characteristic inhabitant of the boreal coniferous forest and the transitional forests between coniferous and deciduous forests. In the Chippewa and Superior National Forests, Niemi, Howe, and colleagues (2016) supported the species' preference for many mixed types, such as aspen–spruce–fir, mature mixed swamp conifers that include white cedar, and mixed upland hardwood forests **(C)** but noted that it was also commonly found in mature red pine and white pine. The species tends to avoid lowland coniferous forest composed of black spruce and tamarack and uplands with jack pine. Several investigators have suggested it is sensitive to forest patch size (D. H. Morse and Poole 2020; Cutright et al. 2006), and, in Minnesota, Grinde, Niemi, Sturtevant, and colleagues (2017) suggested that Black-throated Green populations might have weather effects mitigated by large patches of forest.

POPULATION ABUNDANCE

PIF (2020) estimated a breeding population of 9.2 million breeding adults in the United States and Canada, with about 270,000 in Minnesota. The MNBBA, however, estimated a much larger population of the Black-throated Green Warbler in Minnesota, 1.43 million breeding adults (95 percent confidence interval of 1.26–1.94 million). The MNBBA estimate would indicate that more than 15 percent of the species' North American breeding population is found in Minnesota. The higher estimate by the MNBBA may be due both to better coverage of roadless areas in the state, such as the Boundary Waters Canoe Area Wilderness, where the species is prominent, and to methodological differences between PIF and the MNBBA. PIF (2021) primarily uses roadside counts from the BBS and, therefore, did not include data from areas where densities of this species can be high.

The BBS trend estimates for the Black-throated Green in Minnesota were nonsignificant from 1967 to 2019, which indicated a stable population **(D)**. Surveywide, the trend was also stable in North America. Rosenberg, Kennedy, and colleagues (2016), however, estimated a 41 percent increase in the breeding population from 1970 to 2014. These increases are partly supported by significantly increasing population trends that have been detected in both the Chippewa (1.43 percent per year) and Superior (1.73 percent per year) National Forests of northern Minnesota from 1995 to 2020 (Grinde, Bednar, et al. 2020).

C. The breeding habitat of the Black-throated Green Warbler is mixed deciduous–coniferous forests, a common habitat type in northern Minnesota, but it usually hides its nest in coniferous trees. Photo by Gerald J. Niemi.

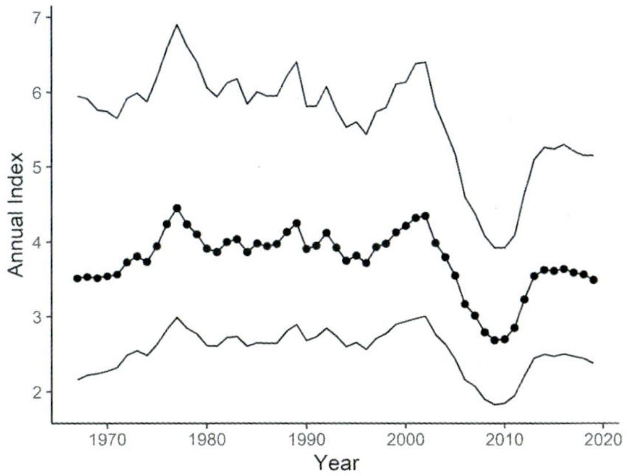

D. Breeding population trend and 95 percent credible interval (solid black lines) for the Black-throated Green Warbler in Minnesota based on the 1967–2019 Breeding Bird Survey (Sauer et al. 2020).

CONSERVATION

At present, there is only moderate conservation concern for the Black-throated Green Warbler because it has a large breeding population, a positive or stable population trend in most parts of its range, and a substantial breeding range in North America. Because the species shows a low affinity for younger forests (less than twenty years old) (Flaspohler et al. 2002; Niemi, Howe, et al. 2016), extensive logging or short-rotation forestry would have a negative effect on its overall population (Niemi and Hanowski 1992a). In addition, D. H. Morse and Poole (2020) reported that it showed population declines in areas sprayed with DDT and with applications of fenitrothion in spruce budworm control areas.

As with many of the species with northern boreal affinities, NAS (2019) projected a high vulnerability for the Warbler's breeding distribution in North America, including a 65 percent range loss but only a 20 percent range gain in northeastern Canada with a 2°C warming scenario. In Minnesota, the species was predicted to be extirpated under the 3°C warming scenario by 2080. The Black-throated Green Warbler is a colorful and welcome member of Minnesota's rich diversity of insect-loving warblers. Its presence is a strong symbol of Minnesota's ecotonal forests, where the eastern deciduous and boreal forest biomes meet. Despite a potential recolonization of its range west and south in Minnesota, this species will be one to watch and to listen for its buzzy territorial song as the climate scenario unfolds in Minnesota's future.

CANADA WARBLER

Cardellina canadensis

Minnesota Seasonal Status: Regular breeding resident and migrant. The Canada Warbler was common during the MNBBA.
Migration: Long-distance migrant that winters in northern South America.
Food: Arthropods secured by foliage gleaning, hover gleaning, fly-catching, and ground foraging.
Nest: Cup nest located on or near the ground in dense undergrowth.
Conservation Concern: Assigned a high PIF Continental Concern Score of 14/20.

A species of Minnesota's northern forests, the Canada Warbler spends much of its day among the understory's shrubs and young saplings. Elegantly attired, the male is easily identified: when sunlight filters through the canopy, "one may see it, glowing like a jewel, the black necklace standing out vividly against the brilliance of the yellow breast" (Griscom and Sprunt 1957). Even the paler female wears a faint necklace. Yellow lores and a white eye-ring impart a spectacled appearance to both sexes.

As the bird is a late spring arrival, the Canada Warbler's voice may be absent from the morning chorus until early June, when the woods suddenly come alive with the male's territorial proclamations. His emphatic song begins with a short chirp followed by a medley of rapidly delivered notes and short phrases that defy an easy mnemonic translation. Saunders remarked that "there seems to be no general rule about the form of the song, except that two notes in succession are rarely on the same pitch" (Bent 1953).

Critical of the species' common name, Roberts (1932) wrote: "This charming little bird, with its beautiful necklace of black pendants, is surely deserving of a less prosaic name than that with which it has been inflicted." The species, he argued, was no more Canadian than many other boreal-nesting songbirds. He and others proposed a more descriptive title, the Necklaced Warbler. Needless to say, the name Canada Warbler has endured.

Found across the boreal forest biome, south to the northern forests of the Great Lakes, New England, and the Appalachian Mountains, some of the Warbler's highest breeding densities are found in northeastern Minnesota.

Photo by Michael Furtman.

MINNESOTA BREEDING DISTRIBUTION

Roberts (1932) reported the Canada Warbler from the Canadian border south to northeastern Pine and southern Mille Lacs Counties and west to Itasca State Park and Lake of the Woods County, describing it as "most numerous in the far-northern part of the state." Areas of abundance included the North Shore of Lake Superior, the Iron Range in St. Louis County, and near Cass Lake, Leech Lake, and Lake Vermilion. In Cook County, only the Mourning Warbler was reported to be more abundant. Despite spending several summers at Itasca State Park, Roberts found just one male, in June 1917. Breeding evidence was limited to Cass and Lake of the Woods Counties and at Mille Lacs and Cass Lake.

Maps included in the Canada Warbler accounts by Green and Janssen (1975) and Janssen (1987) illustrated a similar breeding range to that described by Roberts while noting uncertainty about the extent of the species' distribution in the northwest and east-central regions. Indeed, two late June records in Anoka County in 1978 and 1980 led Janssen (1987) to suggest that the species might be a casual nester farther south than his distribution map indicated.

In 1998 Hertzel and Janssen identified four counties with confirmed nesting since 1970: Clearwater, Cook, Hubbard, and Lake. Coincidentally, in 1997 and 1998, two territorial males were found at the Boot Lake SNA in Anoka County, and in 1998 a pair, with a nest of five young, was found at the Falls Creek SNA in Washington County (Bardon 1998). Canada Warblers continued to be found in both counties during the 1999, 2000, and 2003 breeding seasons (MOU 2021b).

Meanwhile, the MBS tallied 447 breeding-season locations. While these were concentrated in the Arrowhead region, scattered records occurred as far west as southern Clearwater, eastern Mahnomen, and eastern Becker Counties and south to northern Mille Lacs and central Pine Counties.

MNBBA participants documented a similar distribution **(A)**. Among the 1,059 detections, however, were three possible breeding records in northwestern Minnesota from Marshall and Roseau Counties and in a block straddling Marshall and Pennington Counties. A singing male was also observed in Murphy-Hanrehan Park Reserve in Scott and Dakota Counties on June 14, 2009. Although this may have been a late migrant, the earlier nesting records from Anoka and Washington Counties, coupled with suitable habitat at the Park Reserve, prompted retaining the record's possible status. Overall, Canada Warblers were reported in 10.3 percent (486/4,736) of the surveyed atlas blocks and confirmed breeding in 33 blocks. Reported from 18 counties, breeding was confirmed in six counties and deemed probable in four counties. Counties with new confirmed breeding records since 1970 were Beltrami, Cass, Itasca, and St. Louis. Probable nesting was reported from three counties where nesting had yet to be confirmed: Aitkin, Carlton, and Pine. The MNBBA probability map predicts the highest breeding densities are primarily limited to Cook County and pockets of northern Lake and St. Louis Counties. Moderate breeding densities decline rapidly south and west of eastern Koochiching, northern Itasca, and eastern Pine Counties **(B)**.

The core of the Canada Warbler's distribution in Minnesota has changed very little since Roberts published his original account (1932). Even the areas of high abundance are nearly identical nearly one hundred years later, while occurrences beyond the Lau-

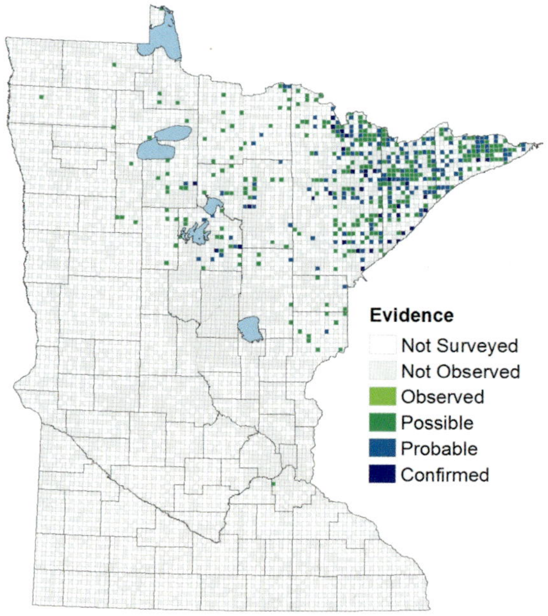

A. Breeding distribution of the Canada Warbler in Minnesota based on the Breeding Bird Atlas (2009–13).

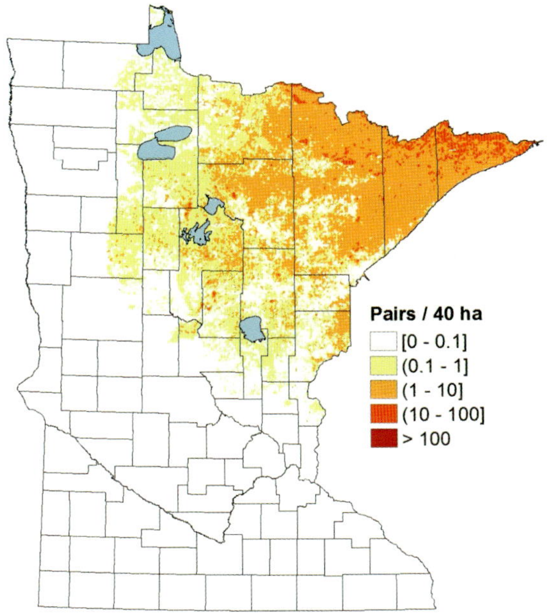

B. Predicted breeding distribution (pairs per 40 ha) of the Canada Warbler in Minnesota based on habitat, landscape context, and climate data gathered during the Minnesota Breeding Bird Atlas (2009–13) using the Generalized Linear Modeling method with an adjustment for detectability.

rentian Mixed Forest Province continue to be rare. Elsewhere, few changes in the species' continental distribution have been noted (Reitsma et al. 2020).

BREEDING HABITAT

The Canada Warbler is an inhabitant of mesic forests, including uplands of mature northern hardwoods and lowlands of white cedar and alder (J. L. Dunn and Garrett 1997; Cadman et al. 2007). In northern Minnesota, it is most strongly associated with mixed upland forests followed by lowland conifers (Grinde and Niemi 2016) **(C)**.

Shrubs and moist conditions are common features of suitable habitat. Sites with mountain maple (*Acer spicatum*) shrubs are typical, as are sites located near wetlands, small creeks, and streams. Mature stands with canopy gaps that promote a dense growth of shrubs and young saplings are preferred. Such gaps are created by selective timber harvests or by natural disturbances, such as windstorms, tree falls, or insect outbreaks (Grinde and Niemi 2016). A rich, uneven ground layer, covered with mossy hummocks, ferns, root masses, and woody debris is also important (Goodnow and Reitsma 2011; Lambert and Faccio 2005). Because the species is sensitive to forest fragmentation, forest stands located within or near extensively forested landscapes are more suitable (Grinde and Niemi 2016).

POPULATION ABUNDANCE

PIF (2021) estimated a US and Canadian population of 2.6 million birds, with 210,000 residing in Minnesota, or 8.1 percent of the continental population. MNBBA sampling of the roadless Boundary Waters Canoe Area Wilderness, which is not surveyed by the BBS yet supports high Canada Warbler breeding densities, contributed to a significantly higher statewide estimate of 1.1 million (95 percent confidence interval of 866,000–1.9 million).

Surveywide, BBS data suggest a downward population trend from 1966 to 2019; the decline is especially acute in New England and the Maritime Provinces (Sauer et al. 2020). Biologists have calculated a cumulative population loss of −62 percent from 1970 to 2014 (Rosenberg, Kennedy, et al. 2016). Minnesota BBS data, however, suggest a recently increasing population **(D)**, which is corroborated by a significant population increase in the Chippewa National Forest (3.14 percent per year) and a stable population in the Superior National Forest from 1995 to 2020. Disturbances that enhance understory growth, such as forest thinning operations, windstorms, and insect outbreaks, were believed to be associated with the increase (Grinde, Bednar, et al. 2020).

Factors responsible for the continental decline are largely unknown but likely include habitat loss. Grinde and Niemi (2016) suggested that declines may result from forest management practices that alter natural disturbance regimes. Harvesting upland hardwood stands before they reach full maturity (more than eighty years) stops natural processes that improve the stands' suitability for Canada Warblers. Overbrowsing by deer also hampers the growth of a well-developed understory, an important component of the Warbler's habitat (Cutright et al. 2006). Collision mortality may be contributing to the decline. Loss and colleagues (2014), for example, concluded the Warbler was 18 times more likely to collide with buildings than most species.

C. The Canada Warbler is an inhabitant of mature, mesic mixed forests with a well-developed understory of shrubs and small trees. Photo by Gerald J. Niemi.

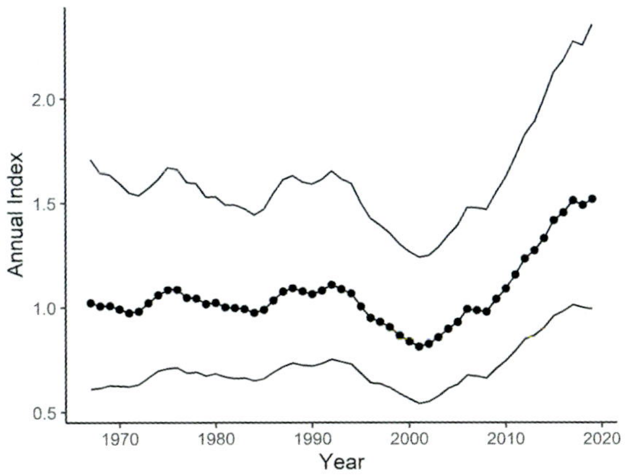

D. Breeding population trend and 95 percent credible interval (solid black lines) for the Canada Warbler in Minnesota based on the 1967–2019 Breeding Bird Survey (Sauer et al. 2020).

CONSERVATION

A high conservation priority due to its small population, declining numbers, and threats to its breeding and wintering habitats (PIF 2021), the Canada Warbler was added to PIF's Yellow Watch List, a group of species that warrant focused management attention (Rosenberg, Kennedy, et al. 2016). It is designated federally *Threatened* in Canada, with particularly steep population declines in southeastern Canada and the Maritime Provinces, where the majority of the nation's population resides (COSEWIC 2017). The loss of wintering habitat was highlighted as a major threat by Environment Canada (2011), which noted that forests in the northern Andes, the core of the Warbler's wintering range, are "the most threatened in the world."

Unfortunately, under a 2°C warming scenario, NAS (2019) projected a moderate vulnerability for the species' breeding distribution, with a range loss of 64 percent compensated for by a nearly equal gain farther north by the year 2050. As with several boreal species, the same scenario predicts the species' extirpation from Minnesota.

Management guidelines focus on protecting and managing large, mature (more than eighty years) mesic forest stands with a dense understory, a structurally complex forest floor, and a semi-open canopy (Lambert and Faccio 2005). Forest management plans and timber harvest activities need to be sensitive to the species' habitat requirements. In the short term at least, Minnesota plays a vital role in the future of the "Necklaced Warbler."

WILSON'S WARBLER

Cardellina pusilla

Minnesota Seasonal Status: Regular, rare summer resident; uncommon spring and fall migrant statewide. The Wilson's Warbler was rare during the MNBBA.
Migration: Short- to medium-distance migrant that winters in Mexico and Central America; some individuals winter in the United States along the Gulf coast in Louisiana and in southern California.
Food: Primarily consumes insects and spiders; forages by gleaning, sallying, and hover gleaning from the ground to the canopy.
Nest: Cup nest built on or near the ground in a tussock of moss or sedge or at the base of a shrub, usually from the ground up to 1 m.
Conservation Concern: Assigned a moderate PIF Continental Concern Score of 10/20.

Photo by David Brislance.

As its scientific name indicates (*pusilla* means "small"), the Wilson's Warbler is the smallest of the wood warblers in Minnesota, but it was named after one of the giants of American ornithology, Alexander Wilson. He was the first to publish a description of this species in 1811 of a bird he collected migrating through New Jersey. He originally named it the Green Black-capped Flycatcher. Modern taxonomists have divided the eastern and western populations into three subspecies, which many believe will ultimately be divided into two species.

Both sexes of the Wilson's Warbler are similar, except the female lacks the distinct black cap on the top of the head that is found in the male. Otherwise, they have olive-green upperparts contrasted with bright lemony yellow underparts. In the shrubby habitats of Minnesota's far north, it is most likely first detected by its song, which Griscom and Sprunt (1957) described as *chee chee chee chee chee che-che-che-ch-ch-wit-wit,* with the pitch falling in the middle and rising at the end.

The Wilson's Warbler ranges across the northern forests and taiga regions of Canada and Alaska. Of its three subspecies, the one in Minnesota, *Cardellina pusilla pusilla,* breeds across the boreal coniferous forest and taiga from Newfoundland and the Maritime Provinces westward to the Canadian Mackenzie District in the Northwest Territories; small outlying populations occur in northern New England and the Great Lakes states. The two other subspecies are *C. p. pileolata,* which breeds from Alaska through interior British Columbia and the Rocky Mountains, and *C. p. chryseola,* which breeds on the Pacific coast from southwestern British Columbia south to southern California. Highest densities of the Wilson's Warbler occur along the Pacific coast, in British Columbia, and in Newfoundland.

MINNESOTA BREEDING DISTRIBUTION

The first record of the Wilson's Warbler in Minnesota was in 1900, when Roberts and two companions took a trip (June 27 to July 11) to Mud Lake and Thief Lake, Marshall County, where they observed two individuals. In 1901, Dart went back to the big marshes in northwestern Marshall County and saw a pair "several times . . . on the edge of a spruce and tamarack swamp near Mud Lake" (Roberts 1932). Roberts concluded that the Wilson's Warbler was a "spring and fall migrant throughout the state, probably nesting in very limited numbers in the extreme northern part," although firm evidence of breeding had not been found. Subsequently, further indication of resident Wilson's Warblers in Marshall County was a checklist of the "Birds of Mud Lake National Wildlife Refuge" (Carlsen and Harris 1955) that designated the Wilson's Warbler as a "Common migrant. Possibly breeds."

Breeding season observations were nonexistent until July 13, 1975, when Fall (1977) published an observation at Itasca State Park, Hubbard County, and July 26 and 29, 1975, when birds were heard along Blesner Creek, Cook County Highway 7 (Eckert 1976). Interestingly, decades later one bird was heard on June 16, 2007, in the same place along shrubby Blesner Creek (J. C. Green, personal files).

The first evidence of confirmed nesting for the state was recorded on June 29, 1980, in a small colony along Hog Creek in Lake County (Wilson 1980). From the mid-1970s, there have been almost yearly observations during midsummer in appropriate shrubby habitats in Cook, Lake, and St. Louis Counties, as well as some observations in Clearwater, Pennington, and Lake of the Woods Counties. After the North American BBS was established in Minnesota in 1967, one to four birds were reported in seven years from 1980 to 1992 on northern routes. When new routes were created in 1993, especially several in swampy/brushy habitat in the north (i.e., the Jordan, Lude, and Red Lake BBS routes), from four to ten birds were reported yearly from 1993 to 2015. It is not known if this increase was due to better sampling in appropriate habitat or was an actual increase in the Wilson's Warbler population.

The MNBBA participants reported 76 records of the Wilson's Warbler. The survey work done for the MNBBA, plus other project surveys in northern Minnesota during the atlas years, increased the knowledge of the distribution of the Wilson's Warbler. The species was recorded from 1.2 percent (57/4,734) of the surveyed blocks **(A)**. The MNBBA produced two

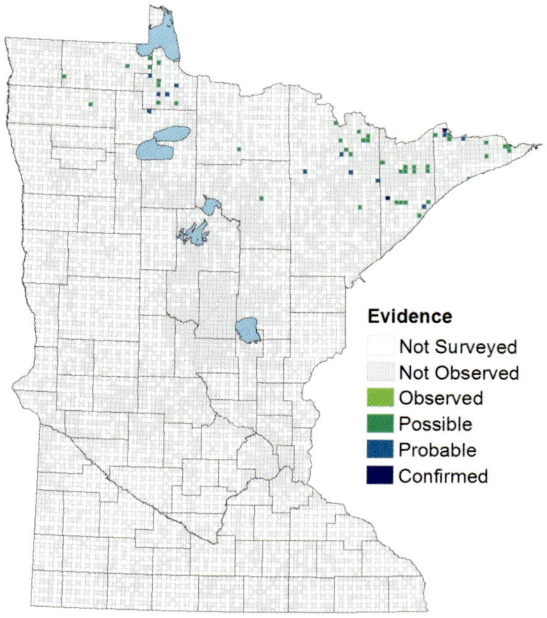

A. Breeding distribution of the Wilson's Warbler in Minnesota based on the Breeding Bird Atlas (2009–13).

confirmed records: one in Lake County and another, the first documented nesting in Cook County. Probable records occurred in 12 atlas blocks. A cluster of records in the peatlands and shrublands of the Agassiz Lowlands Subsection south of Lake of the Woods was mostly due to targeted field surveys. The other areas that produced most of the records were in Cook, Lake, and north and central St. Louis Counties. Additional possible breeding evidence was recorded from Itasca, Koochiching, Marshall, and Roseau Counties. Because this warbler is a very late spring migrant (through mid-June), it is difficult to know whether these birds were residents or migrants.

In Minnesota, the Wilson's Warbler appears to have a sustained population along the northern border from Cook County west to Roseau County. Because of the lack of breeding observations up until the 1970s, no definitive conclusions can be drawn on its historical status in Minnesota. Roberts's comment that the species is a migrant and probably nesting in limited numbers in northern Minnesota likely remains true today. Ammon and Gilbert's (2020) review of the Wilson's Warbler in North America provided few clues on the historical status of the species, except for extirpations in southern California. The majority of its breeding population occurs in remote areas of North America, so new knowledge will continue to be elusive or sluggish in progress.

BREEDING HABITAT

Across its wide range, Wilson's Warblers breed in a variety of landscape conditions but always in dense ground cover in wet, shrubby habitat along streams, lakes, beaver ponds, bogs, and regenerating openings in forests (Ammon and Gilbert 2020) **(B)**. The dense shrubbery can be deciduous thickets of willow, alder, and other understory shrubs, or stunted conifers in bogs with thick ericaceous species. In Minnesota, breeding birds were found primarily along rivers, streams, and creeks and in peatland landscapes. In Cook County, just south of the Canadian border, potential breeding birds have also been observed in large, regenerating burns (e.g., Zlonis et al. 2019).

POPULATION ABUNDANCE

Based on the BBS, PIF (2021) estimated a US and Canadian population of 81 million birds and a Minnesota population of 23,000 birds. Because of too few detections, the MNBBA did not attempt to make a prediction of the species' population.

The BBS also had too few detections in Minnesota to estimate a population trend, but the surveywide trend indicated an average decline of −1.70 percent per year from 1966 to 2019 **(C)**. Most of the declining populations were in the west, such as in the Canadian province of British Columbia and in the US states of California, Oregon, and Washington (Sauer et al. 2020). The causes of its decline are primarily identified as the extensive loss of riparian habitat (Ammon and Gilbert 2020).

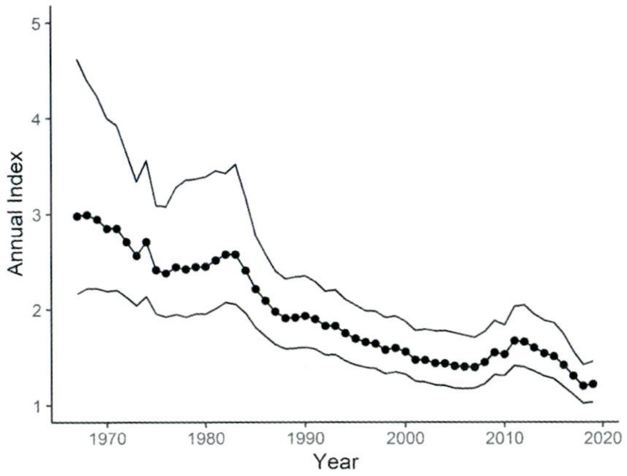

C. Surveywide breeding population trend and 95 percent credible interval (solid black lines) for the Wilson's Warbler based on the 1967–2019 Breeding Bird Survey (Sauer et al. 2020).

CONSERVATION

PIF (2021) designated the Wilson's Warbler as a Common Bird in Steep Decline primarily because of its population decline in western North America, where several western states have also identified it as a conservation priority (Ammon and Gilbert 2020). In Minnesota, its status is unclear because breeding observations have become more prevalent in recent years, likely because of more focused field efforts. NAS (2019) projected a moderate vulnerability for the warbler in North America with a 2°C warming scenario but did not include Minnesota within its current breeding range.

Besides loss of riparian habitats in California, recent research has shed light on another reason for the substantial decline in the western subspecies *C. p. chryseola,* which breeds along the Pacific coast. The use of genetic markers from feathers of the three subspecies to map their wintering ranges showed that each subspecies has a distinct wintering area (McBride 2016). More research on the life-cycle connectivity between breeding, migration, and wintering areas for different subspecies is needed to understand this species' status and the causes of recent declines in subspecies populations. Gathering these data will be challenging because most of the breeding populations of Wilson's Warbler occur in remote northern landscapes that are difficult to access and traverse. Continued field efforts in Minnesota will help to untangle its status in the southern periphery of its breeding range.

B. Shrubby riparian areas are the most typical habitats to find nesting Wilson's Warblers in northern Minnesota.
Photo by Gerald J. Niemi.

SUMMER TANAGER

Piranga rubra

Minnesota Seasonal Status: Summer visitant in the south, where there have been several unsuccessful pairings with Scarlet Tanagers; summer vagrant north. Migration records occur statewide but are most common in the spring and in southern Minnesota. The Summer Tanager was very rare during the MNBBA.
Migration: A medium- to long-distance migrant that winters in the Neotropics from central Mexico to northern South America.
Food: The Summer Tanager is noted for feeding on bees and wasps, including their larvae by tearing open the nest. A variety of other arthropods are eaten, as well as berries and fruits in season.
Nest: Loosely constructed shallow cup placed 3–10 m high, often in an oak, pine, or cottonwood.
Conservation Concern: Assigned a moderate PIF Continental Concern Score of 9/20.

Photo by Travis Bonovsky.

The Summer Tanager is a quintessential red bird and found more commonly in the South. It was first discovered by Mark Catesby (1754) in South Carolina and illustrated as plate 56 in volume 1 of his book *The Natural History of Carolina, Florida and the Bahama Islands*. From that illustration, it was described in 1758 as "*Fringilla rubra* Linnaeus." Plate 56, labeled as "The 'summer red-bird' and the 'western plane tree,'" is an iconic work of art and today is widely available as a fine-art print. This handsome illustration shows the Red Tanager superimposed on an artfully arranged cluster of leaves and fruit of the "plane tree" (American Sycamore, or *Platanus occidentalis*), another uniquely southeastern species. In the South, the species was called the "summer red-bird" to differentiate it from the "winter red-bird," the Northern Cardinal, which is most conspicuous in winter.

The species' scientific name, *rubra,* emphasizes the all-red color of the male, while the female is a yellowish-orange with greenish wings. Both sexes of the Summer Tanager lack the darker wings of the Scarlet Tanager. The song of the male is very similar to that of the American Robin and is not hoarse like that of the Scarlet Tanager.

The Summer Tanager has two widespread subspecies. The eastern nominative race, *Piranga rubra rubra,* is found throughout the Southeast, from southern Pennsylvania and Ohio west to southern Iowa, then south from Missouri to central Texas and eastward to Florida. The Western race, *P. r. cooperi,* occurs from northern Mexico and western Texas through New Mexico and Arizona, with local populations in southern Nevada and California. The highest breeding densities are found from southeast Oklahoma through northwest Arkansas to southeastern Missouri.

MINNESOTA BREEDING DISTRIBUTION

Roberts (1932) reported only two records of the Summer Tanager in Minnesota. One was a specimen collected in Pipestone County in late May 1891, and the other was a bird seen and well identified at Lake Pepin in Goodhue County on May 14, 1926. Green and Janssen (1975) found only seven more reports for the next forty years and concluded it was a casual spring migrant. The first record after 1926 was on June 8, 1952, when a bird was seen carrying food in Frontenac, Goodhue County (Lupient 1952). Four of the other early records were from the 1960s: one was seen in Winona County on June 14, 1963 (Shira 1964); two were observed in Duluth May 12–16, 1964 (male) (Green 1964) and May 19–24, 1965 (female) (Green 1965); and one was reported in 1968 in St. Paul (M. E. Olson 1968).

By far, most observations of the Summer Tanager have been during spring and fall migration. In the 1970s, 10 records (eight spring and two fall) were published in *The Loon,* but the number of records showed an obvious increase during the 1980s (19 records) and the 1990s (from 1991 to 1998 the species was reported every year in the spring and most years in the fall). Bardon (1999) summarized all previous records from 1963 to 1999. He found that there were two recent years with statewide large influxes, mostly immature and adult males, in 1995 (20 birds) and 1999 (21 birds, mostly spring, two fall).

Three of the four records submitted to the MNBBA fit the pattern of a species wandering north of its normal breeding range **(A)**: a possible record in Rice County, June 4, 2011; an observed record in Aitkin County, June 6, 2013, at a feeder; and a probable record in Blue Earth County, June 8 and 19, 2011. Another probable record submitted to the MNBBA in Afton State Park in southeastern Washington County was reported on July 9, 2009, and again on July 18 when two males were heard; the report on July 25, 2009, was presumably the same bird(s) (Kessen and Svingen 2010). Tanagers may persist at Afton State Park, as shown by subsequent observations in 2014 (Kessen et al. 2015), 2015 (Kessen et al. 2016), 2016 (June 5–21, six observations; MOU 2021a), and 2017 (June 1–August 4; multiple observations; MOU 2021a); however, no breeding evidence has been reported. The MBS also reported multiple observations in suitable habitat at

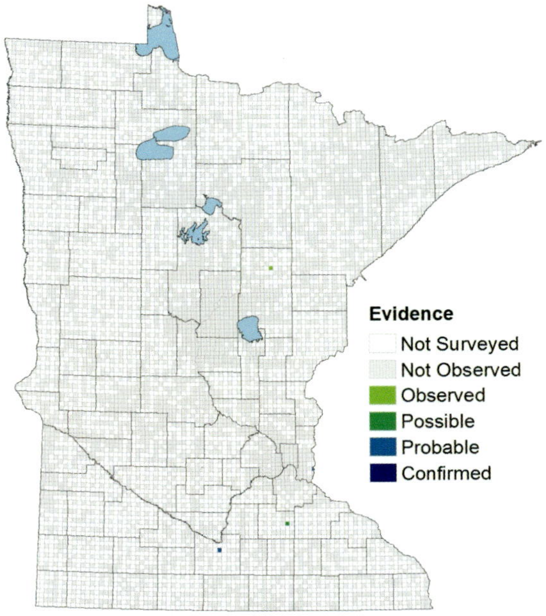

A. Breeding distribution of the Summer Tanager in Minnesota based on the Breeding Bird Atlas (2009–13).

Kilen Woods State Park in Jackson County on June 19 and 26, 2007 (Kessen and Svingen 2008).

Breeding information regarding hybrid tanagers have fewer but similarly detailed records. In 2003, a male Summer Tanager was observed paired to a female Scarlet Tanager in Murphy-Hanrehan Park Reserve, Scott County (Fall 2003) and in Lebanon Hills Regional Park, Dakota County (J. P. Mattsson 2003b). At both sites, a female Scarlet Tanager and male Summer Tanager were subsequently observed feeding a fledgling Brown-headed Cowbird. Then, in 2005 in Lebanon Hills Regional Park, Mattsson again found a male Summer Tanager paired with a female Scarlet Tanager, and his detailed observations "documented evidence of successful nesting and the production of at least one hybrid tanager" and one cowbird nestling. Having observed a female Summer Tanager earlier in the summer near the male Summer Tanager's territory, he also concluded "that production of pure Summer Tanager offspring is a realistic possibility in Minnesota" (J. P. Mattsson 2007). In Kandiyohi County from June 9 to 15, 2018, a first-year male Summer Tanager and an adult female Summer Tanager were seen by many observers at the Mount Tom parking lot, Sibley State Park; some observations indicated courtship behavior (MOU 2021a). Finally, from May 20 to June 18, 2020, a singing male persisted at Murphy-Hanrehan Park Reserve, Scott County but no other bird was seen (MOU 2021a).

The origin of Summer Tanager migrants is uncertain. The species is known to breed only sparingly in southern Iowa (L. Jackson et al. 1996). In two states, Wisconsin (S. D. Robbins 1991) and South Dakota (Tallman et al. 2002), the Summer Tanager is a visitant in spring and a vagrant during fall migration. Earlier records from the nineteenth and early twentieth centuries indicated that it might once have bred in both states.

In a review of the status of the Summer Tanager in North America, W. D. Robinson (2020) suggested that its distribution is declining "along edges of [its] range in most areas" of the eastern United States and in some regions of the west. Habitat destruction is implicated in the west, but factors responsible for changes along the northern edge of its eastern range are unknown. Expanding populations in California, Nevada, and New Mexico have been hypothesized as due to warmer and wetter summers in those regions. In Minnesota, there has certainly been an increase in observations of Summer Tanagers but this also coincides with the increase in observers over the past one hundred years. Increasing temperatures would logically provide increased opportunities for Summer Tanagers to colonize Minnesota forests.

BREEDING HABITAT

The eastern population breeds in deciduous forests, especially oak–hickory or oak–pine woods with canopy gaps, and in deciduous bottomlands and parks **(B)**. Where the Summer Tanager is sympatric with the Scarlet Tanager, the former "prefers sites with shorter average tree height, less canopy cover and fewer tree species" and the latter "tends to be most numerous in closed-canopy forest, where the trees are taller" (W. D.

B. The Summer Tanager is fond of nesting in open oak and oak–pine forest habitats, such as this area in southern Minnesota. Photo by Lee A. Pfannmuller.

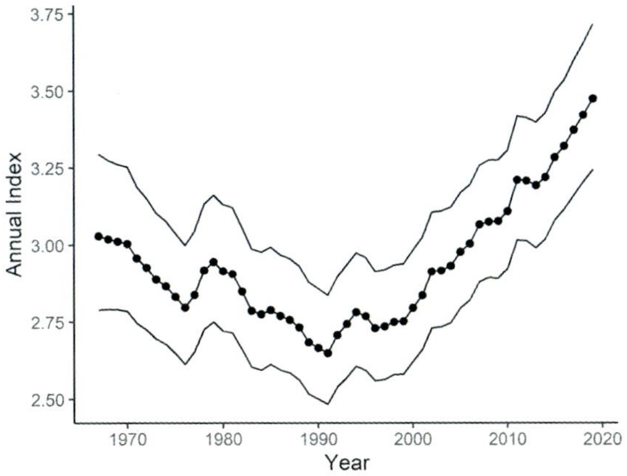

C. Breeding population trend and 95 percent credible interval (solid black lines) for the Summer Tanager in the United States based on the 1967–2019 Breeding Bird Survey (Sauer et al. 2020).

Robinson 2020). The western population breeds in cottonwood–willow riparian woodlands.

POPULATION ABUNDANCE

PIF (2021), using BBS data, estimated the US population at 11 million birds; owing to its rarity, no estimates for Minnesota were included. Trend analysis from the BBS shows that the breeding population in the United States significantly increased an average of 0.24 percent per year from 1966 to 2019 **(C)**. Rosenberg, Kennedy, and colleagues (2016) calculated that the overall population of the Summer Tanager in the United States had increased by 9 percent from 1970 to 2014. Across its breeding range, the Summer Tanager's population trends vary considerably. Nearly all states west of the Mississippi River demonstrate increasing trendlines, but in the east there is little consistency. W. D. Robinson (2020) identified habitat change, such as the loss of riparian forests, as a primary cause of population declines.

CONSERVATION

At present there is limited conservation concern for the Summer Tanager in Minnesota and nationally. The species generally is not specifically mentioned in management plans, except when grouped with other long-distance Neotropical migrants (W. D. Robinson 2020). NAS (2019) predicted that the species' range is stable in the United States with all three warming scenarios they examined, projecting range gains northward. As with many species that have major distributions south of Minnesota, the Summer Tanager should be looked for in suitable habitats in Minnesota, especially in the southern and central forested regions. This beautiful red bird is a treat to observe and would be a welcome addition to the avifauna of Minnesota.

SCARLET TANAGER

Piranga olivacea

Minnesota Seasonal Status: Regular breeding resident and migrant. The Scarlet Tanager was common during the MNBBA.
Migration: Medium- to long-distance migrant that winters from Panama through northwestern South America.
Food: During summer, arthropods are secured by gleaning foliage and hawking flying insects from a perch; fruit is commonly taken at other times of the year.
Nest: Loosely woven cup nest, frequently placed in a mature deciduous tree on a horizontal branch 6–9 m above the ground.
Conservation Concern: Assigned a moderate PIF Continental Concern Score of 12/20.

Photo by Gerald Hoekstra.

"Flaming," "brilliant," and "blindingly gorgeous" are just a few of the descriptions that attempt to portray the splendor of the male Scarlet Tanager. The striking contrast between the intense, deep red of the bird's head, back, and belly and the jet-black wings and tail is simply stunning. The female, who must be less conspicuous while sitting on the nest, has a more subdued yellow-green plumage with grayish-brown wings and tail. Males acquire this same plumage during the nonbreeding season following a late summer molt. Thus the Latin name, *Piranga olivacea,* a tanager of olive-green color.

As many writers have noted, the Scarlet Tanager is a "bird of contradictions" (Bent 1958). Roberts (1932) addressed this very point: "Although the male Scarlet Tanager is the most brilliantly hued bird in this northern country he is not the most easily discovered. Bright birds in their chosen haunts are frequently as inconspicuous as are dull-colored birds in theirs."

The Tanager's song represents another contradiction. One might expect that such a splendid-looking bird would have an equally splendid song. But far duller birds deliver more resplendent melodies than the Tanager. Nevertheless, what it lacks in musical talent it compensates for with an unmistakable burry quality to its vocalizations. Easy to misidentify, the Tanager's song resembles that of the American Robin, Rose-breasted Grosbeak, and Red-eyed Vireo, all of whom sing various versions of the phrases *Cheery-Up, Cheery-O.* When the male Scarlet Tanager sings this melody, however, it has a unique raspy quality. The female sings a similar song, though it sounds much softer. Another good identification clue is the Tanager's unique call, an emphatic sounding *chick-burr*.

The Scarlet Tanager is a bird of the eastern deciduous forest, with a breeding range extending across southeastern Canada south to Arkansas in the west and the Carolinas in the east. The highest breeding densities are found in the Appalachian Mountains.

MINNESOTA BREEDING DISTRIBUTION

Roberts (1932) described the Scarlet Tanager as a "fairly common summer resident in all suitable places and is quite as numerous in the evergreen forests, as far north as Itasca Park, as it is in the deciduous woodlands of the south. Farther north it decreases in numbers but is found quite to the Canadian boundary." The species was found in smaller numbers in woodland groves bordering the prairie's lakes and streams. Confirmed or inferred breeding evidence was reported from Goodhue, Hennepin, Isanti, Itasca, Jackson, Sherburne, and Wabasha Counties and Cass Lake.

Years later, Green and Janssen (1975) commented that the species was most abundant in the state's central woodlands and least abundant in the southern and western prairies. It could be found north as far as Beltrami, Cook, and Norman Counties. Janssen (1987) stated it was abundant in southeast, central, and east-central Minnesota. Other than the floodplain forests of the upper Minnesota River valley, his distribution map for the Tanager excluded only the Prairie Parkland Province. He identified 12 counties where nesting had been confirmed since 1970; Hertzel and Janssen (1998) later identified 15 counties. Two counties delineated earlier by Janssen were not included: Clearwater and Winona. The distribution of the 1,149 breeding-season locations identified by the MBS mirrored the breeding distribution identified by Janssen (1987) but also included numerous records in the southwestern corner of the state.

A similar distribution was delineated by the 1,470 Scarlet Tanager records collected by MNBBA participants **(A)**. Reported from 20.6 percent (979/4,743) of the surveyed blocks, nesting was confirmed in only 34 blocks, but probable nesting was reported in 239 blocks. Tanagers were detected in 75 of Minnesota's 87 counties and breeding was confirmed in 24 counties. Sixteen counties where the species had not previously been reported nesting since 1970 were all within the species' primary breeding range.

Scarlet Tanagers were most abundant in north-central Minnesota, specifically within the Northern Minnesota Drift and Lake Plains Section of the Laurentian Mixed Forest Province. The section's diverse topography supports a mosaic of mesic hardwood forests and drier forests of pine and boreal hardwoods (quaking aspen and paper birch) that provide ideal habitat for the Tanager (MNDNR 2003a). The MNBBA probability map reflects the importance of this entire ecological region to the species. Moderate breeding densities are also predicted farther south in portions of east-central and southeastern Minnesota **(B)**.

The Scarlet Tanager's distribution in Minnesota has not changed notably since Roberts described it as common across the state's deciduous and coniferous forests and sparsely distributed in riparian woodlands bordering the western lakes and streams. In his comprehensive review of the species, Mowbray (2020b) noted only a few, small historical changes in its continental distribution, including an expansion westward in southern Manitoba and southeastern Saskatchewan and eastward in South Carolina.

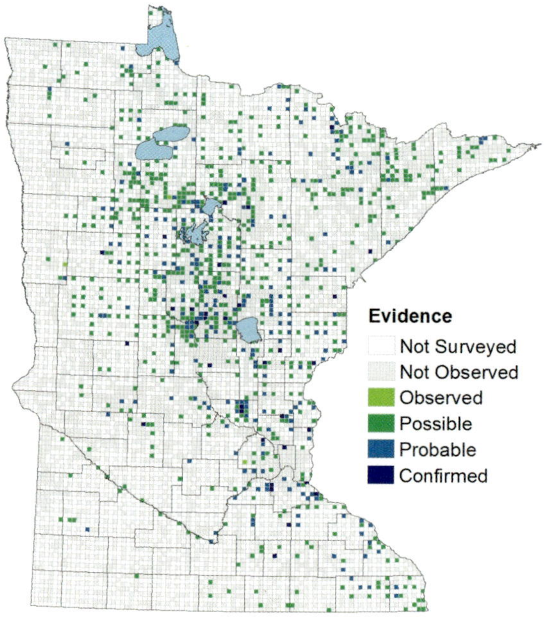

A. Breeding distribution of the Scarlet Tanager in Minnesota based on the Breeding Bird Atlas (2009–13).

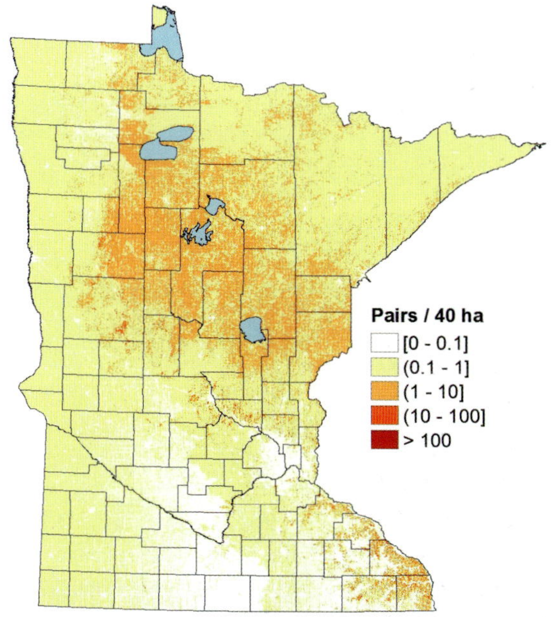

B. Predicted breeding distribution (pairs per 40 ha) of the Scarlet Tanager in Minnesota based on habitat, landscape context, and climate data gathered during the Minnesota Breeding Bird Atlas (2009–13) using the Generalized Linear Modeling method with an adjustment for detectability.

BREEDING HABITAT

The Scarlet Tanager is an inhabitant of mature upland forests. Deciduous forests are predominant, but mixed deciduous–coniferous and coniferous forests are also used **(C)**. Occasional use of residential woodlands with large shade trees is reported, including parks and cemeteries (Mowbray 2020b; Danz et al. 2007). In northern Minnesota and Wisconsin, oak forests were the primary habitat utilized, but the birds were also found in other cover types, including beech–maple–birch, upland hardwoods, hemlock, red pine, and white pine (Niemi, Howe, et al. 2016).

Several studies have documented the Tanager's sensitivity to the size of a forest tract and its negative response to a fragmented forest landscape where stands are small and isolated from one another (Mowbray 2020b). The degree of sensitivity varies geographically. In the Midwest and Atlantic coast regions, where the forest landscape is highly fragmented, Scarlet Tanagers require larger tracts than in the extensive forest landscapes of the northern Great Lakes and New England (Rosenberg, Lowe, and Dhondt 1999).

POPULATION ABUNDANCE

PIF (2021) provided a United States and Canadian population estimate of 2.6 million Scarlet Tanagers, with a statewide population of 49,000. Given the species' broad statewide distribution, the estimate seems particularly low. Indeed, the extensive coverage of MNBBA point count data generated a significantly larger estimate of 852,000 individuals (95 percent confidence interval of 663,000–1.1 million).

From 1966 to 2019, BBS data demonstrated a fluctuating but relatively stable population across the BBS survey region **(D)**, with a cumulative decline of only −7 percent from 1970 to 2014 (Rosenberg, Kennedy, et al. 2016). The trend line for Minnesota lacks statistical precision but also suggests a somewhat stable population with numbers slowly declining in recent years, while the BBS regional trend map illustrates a gradient of stable to increasing populations in southern Minnesota transitioning to declining populations in northern Minnesota (Sauer et al. 2020). Indeed, forest bird monitoring conducted in the Chippewa and Superior National Forests from 1995 to 2020 demonstrates a downward trend in numbers that is most pronounced since 2005 (Grinde, Bednar, et al. 2020).

Habitat loss, cowbird parasitism, and forest fragmentation are frequently cited as factors that negatively impact Tanager populations. Yet in Minnesota's national forests, Brown-headed Cowbirds have declined and timber harvest levels have decreased. Forest fragmentation is also less common in northern Minnesota, where extensive forest tracts still dominate much of the landscape. Nevertheless, some of the species' largest population declines are occurring in northern states, while farther south many populations are stable or increasing. Some authors have pointed to suspected high mortality during migration and on the wintering grounds, but these factors should be impacting populations across the species' breeding range (Mowbray 2020b). Clearly, much remains to be learned about the factors influencing Tanager population numbers.

C. Large stands of mature deciduous forests are common nesting habitat for the Scarlet Tanager. Photo by Lee A. Pfannmuller.

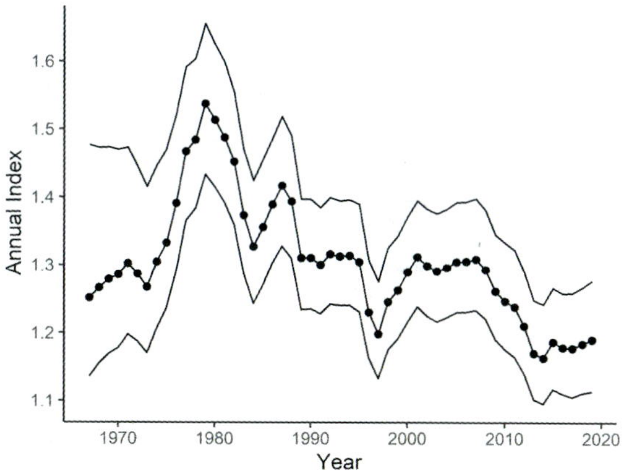

D. Surveywide breeding population trend and 95 percent credible interval (solid black lines) for the Scarlet Tanager based on the 1967–2019 Breeding Bird Survey (Sauer et al. 2020).

CONSERVATION

Although the Scarlet Tanager's rate of population decline is lower than that documented for many long-distance migrants, biologists remain concerned about the species' relatively small population size and threats to its breeding and wintering habitats. As a result, it has been assigned a moderate conservation priority (PIF 2021).

Regarding the Tanager's potential response to climate change, NAS (2019) projected a moderately vulnerable North American breeding distribution under a 2°C warming scenario. By the year 2050, the species would experience a 47 percent range loss compensated by only an 18 percent gain in southeastern Canada. In Minnesota, models predict the western edge of its breeding range will gradually recede eastward until, under a 3°C warming scenario, the Scarlet Tanager would disappear from all but the northeastern corner of the state. Collisions with cars, residential buildings, and communication towers are another well-documented concern (Bracey 2011; Longcore et al. 2013; Longley 1981).

Management recommendations stress the importance of managing large forest tracts where disturbances to the forest canopy are limited. Forest management practices that maintain upland deciduous forests with a strong oak component and a moderate density of large trees are especially important (Niemi, Howe, et al. 2016). In highly fragmented landscapes, managers should focus on protecting remaining forest tracts and on connecting small, isolated woodlots with forested corridors (Rosenberg, Lowe, and Dhondt 1999; Rosenberg, Rohrbaugh, et al. 1999). It would be sad, indeed, if such a stunning bird were to become only a distant memory.

NORTHERN CARDINAL

Cardinalis cardinalis

Minnesota Seasonal Status: Permanent resident. The Northern Cardinal was common during the MNBBA.
Migration: Sedentary; reported movements may be dispersing juveniles.
Food: Opportunistic omnivore, foraging on the ground or in low vegetation for a variety of berries, seeds, and arthropods; common at bird feeders.
Nest: Cup nest placed in small trees and shrubs, 1–2 m above the ground.
Conservation Concern: Assigned a low PIF Continental Concern Score of 5/20.

Photo by Jean Brislance.

Minnesotans lucky enough to enjoy Northern Cardinals at their backyard bird feeders never tire of their presence. In the coldest days of January, nothing is more pleasant than watching the bright red male at the feeder, framed by a backdrop of falling snow. In the bleak days of March, they are one of the first birds to remind us that warmer days are ahead with their springtime song. And, in late spring and summer, we watch and listen as the male and female court, sing duets, and settle in our backyard to nest, eventually introducing the feeder's bounty to their newly fledged young. Bent (1968) wrote the following tribute: "In the cardinal we have a rare combination of good qualities, brilliant plumage, a rich and pleasing voice, beneficial food habits, and devotion to its mate and family."

The male Northern Cardinal needs no introduction with his bright red feathers, red crest, red bill, and black mask. It is said that the Cardinal was given its name because its red plumage was similar to the red robes and birettas worn by the cardinals of the Roman Catholic Church. Even the female is immediately recognizable, with traces of red throughout her gray-brown plumage. Like many red birds, when Cardinals undergo a molt, their consumption of red and orange berries impacts the color of their new feathers. Although the carotenoids in these foods produce a yellow pigment, the birds carry an enzyme that converts the yellow to red (Goldman 2016). In fact, males and females that are brighter in color are more reproductively successful and more attentive parents (Halkin et al. 2021). The relationship between the Cardinal's color and its reproductive success, however, may be "more relaxed" for Cardinals living in urban areas (T. M. Jones et al. 2010).

Much has been written of the male's diverse song repertoire, "a rich variety of flutelike whistles that enthrall anyone who listens" (Kroodsma 2005). But few know the female Cardinal is a skilled vocalist with a similar repertoire. Although her voice may be softer and heard less often, she counter-sings with her mate during courtship, sings to ward off female intruders, and sings as she sits on the nest incubating eggs and brooding the young (Halkin et al. 2021).

The Northern Cardinal is found across the eastern and central United States, south into Central America, and north into southern Ontario, Quebec, New Brunswick, and Nova Scotia. Populations are also found in southern Arizona, New Mexico, and California. The core of its distribution lies in the southeastern United States, west to central Texas.

MINNESOTA BREEDING DISTRIBUTION

Our year-round enjoyment of the Northern Cardinal, especially in southern Minnesota, is a relatively recent phenomenon. Originally found farther south, it was only referred to as the "Northern" Cardinal because it was the most northerly representative of its genus, *Cardinalis.* The first record of its presence in Minnesota was in the fall of 1875 in Minneapolis (Roberts 1932). This was followed by observations in Sherburne (1887), Kandiyohi (1894), Fillmore (1898), and Martin (winter of 1913–14) Counties. Many of the first arrivals were single males that appeared in the fall, winter, or spring months, only to disappear come summer.

Indeed, fifty years would pass before the species was first confirmed nesting in Minnesota in 1925 in Steele County. A few years later, Roberts (1932) described the Cardinal as a breeding resident largely confined to the southeastern region of the state. It was most common from the town of Red Wing south, but also bred in smaller numbers as far north as Minneapolis and as far west as Owatonna. Elsewhere, records were limited to scattered reports of single birds observed during the nonbreeding season. These included a male found in Red Lake Falls, Red Lake County, nearly 300 miles north of any previous record. The bird remained in town throughout the entire winter of 1931–32, feeding on dried grapes and berries.

By the 1960s, Green and Janssen (1975) reported that the species had expanded north to Morrison County and west to Lac qui Parle County. Three inferred breeding records occurred north of Mille Lacs Lake in Aitkin, Cass, and Crow Wing Counties and one to the west in Clay County. Summer observations occurred as far north as Red Lake and Duluth.

As the Cardinal continued its trek northward, Janssen (1987) reported observations in Cook, Kittson, Koochiching, Lake, Pennington, and Roseau Counties. Summer observations were also increasing in the southwest region of the state. Nevertheless, despite these new reports, the core of the Cardinal's distribution remained "in the east-central and southeast regions in the Mississippi River Valley and its tributaries." Since 1970, nesting had been confirmed in 19 counties; all but one (St. Louis County) were located south of Morrison County. Hertzel and Janssen (1998) later added six counties to Janssen's list, all in the southern two-thirds of the state. Among the 605 breeding-season locations identified by the MBS, many were located along the Upper Minnesota River valley and across the southwest corner of the state. The most northern MBS records were in southern Itasca and southern Mahnomen Counties.

As noted by Roberts (1932) and described in more detail by Janssen (1987), the Cardinal's range expansion is initiated by lone individuals visiting areas beyond the species' breeding range in late fall, winter, and spring. Although some birds retreat to their breeding range come summer, others remain and establish new breeding territories. An analysis of banding records by Dow and Scott (1971) suggested that these "pioneers" are young birds, often

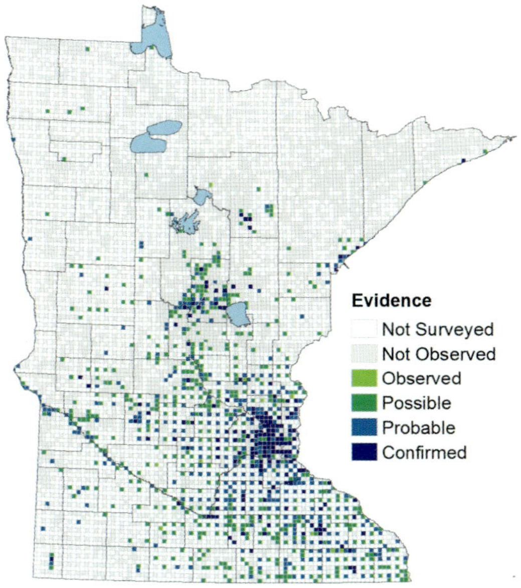

A. Breeding distribution of the Northern Cardinal in Minnesota based on the Breeding Bird Atlas (2009–13).

first-year hatchlings, which are responding to dispersal pressure.

MNBBA participants found the Northern Cardinal in 79 of Minnesota's 87 counties and reported 2,166 records in 22.1 percent of the surveyed atlas blocks **(A)**. Nesting was confirmed in 39 counties. Although it remained rare across northern Minnesota, nesting was confirmed near the Lake Superior shoreline in Cook County. Other counties with new nesting records north of the Twin Cities included Aitkin, Crow Wing, Isanti, Itasca, and Mille Lacs. New breeding records also occurred as far west as Otter Tail and Stevens Counties in west-central Minnesota and Rock County in the far southwest. Overall, the Northern Cardinal is now well established in the east-central, central, south-central, and southeastern regions of the state. Although widespread, it remains most abundant from Brainerd south to the Twin Cities, throughout southeastern Minnesota, and along the Minnesota River valley.

The Northern Cardinal's highest breeding densities are predicted to occur within the core of the Twin Cities metropolitan region and along the Mississippi River south of the Twin Cities **(B)**. Moderate densities are predicted throughout southeastern and east-central Minnesota, while low densities are found elsewhere across the southern two-thirds of the state.

Along the northern periphery of its breeding range, the Cardinal has also moved farther north in New England, in other Great Lakes States, and into the very southern tier of the eastern Canadian provinces. It also expanded west, often following major river valleys, including the Missouri, Platte, Niobrara, and Colorado. Its steady progression northward is widely recognized to be a response to warming temperatures, to fragmentation of extensively forested landscapes, especially in the Great Lakes region, and to the popularity of winter bird feeding (Halkin et al. 2021).

BREEDING HABITAT

Throughout most of its range, the Northern Cardinal can be found in semi-open habitats, including woodland edges, shrubby thickets in overgrown fields or forest clearings, farm shelterbelts, hedgerows, city parks, and suburban landscapes **(C)**. Its primary requirements are patches of dense, thick vegetation that provide sufficient cover to conceal the nest and perches for singing (Halkin et al. 2021).

C. Urban parks, with their patches of scattered trees and shrubs that provide singing perches and nesting sites, are a common setting for the Northern Cardinal.
Photo by Lee A. Pfannmuller.

POPULATION ABUNDANCE

PIF (2021) estimated the Northern Cardinal's population in the United States and Canada to number 120 million breeding adults, with an additional 10 million birds in Mexico, and approximately 530,000 birds in Minnesota. This number is somewhat lower than the statewide estimate of 624,000 adults generated with MNBBA data (95 percent confidence interval of 518,000–789,000). One would expect these estimates to be similar, given the species' association with developed landscapes and its absence in off-road areas sampled by the MNBBA.

The Northern Cardinal's population trend from 1966 to 2019 is either relatively stable or increasing in every state except Alabama, Georgia, South Carolina, and Tennessee (Sauer et al. 2020). Across the BBS survey region, the population has experienced a cumulative increase of 17 percent between 1970 and 2014 (Rosenberg, Kennedy, et al. 2016). Northern Cardinal numbers in Minnesota demonstrate a strong, statistically significant increase averaging 2.79 percent per year from 1967 to 2019 **(D)**.

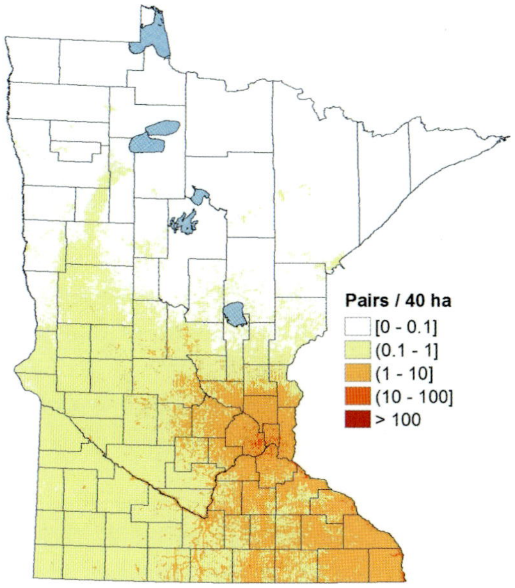

B. Predicted breeding distribution (pairs per 40 ha) of the Northern Cardinal in Minnesota based on habitat, landscape context, and climate data gathered during the Minnesota Breeding Bird Atlas (2009–13) using the Generalized Linear Modeling method with an adjustment for detectability.

CONSERVATION

An abundant, widely dispersed species with an increasing population trend, the Northern Cardinal is a low conservation priority (PIF 2021). Adapted to a wide variety of modified landscapes, it is among a suite of species that has benefited from disturbance and moderate winters. Even warmer temperatures will continue to improve habitat conditions farther north so that Minnesotans throughout the state should see even more of the Cardinal in coming years (NAS 2019).

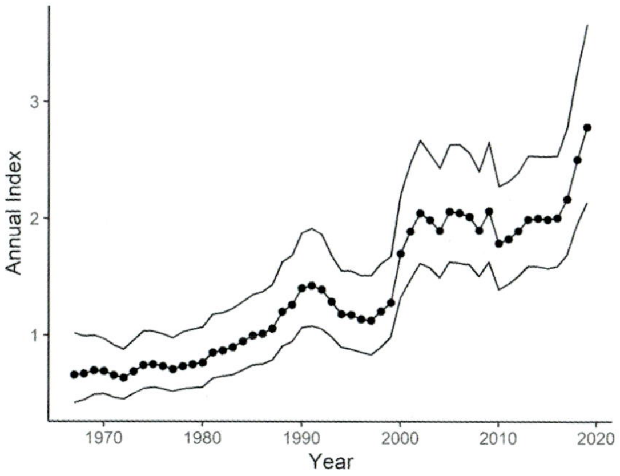

D. Breeding population trend and 95 percent credible interval (solid black lines) for the Northern Cardinal in Minnesota based on the 1967–2019 Breeding Bird Survey (Sauer et al. 2020).

ROSE-BREASTED GROSBEAK

Pheucticus ludovicianus

Minnesota Seasonal Status: Regular breeding species and migrant. The Rose-breasted Grosbeak was abundant during the MNBBA.
Migration: Medium- to long-distance migrant that overwinters in Mexico and Central America, rarely in the southern United States and Cuba.
Food: Omnivorous, consuming invertebrates (especially beetles), fruit, and seeds, gathered in many ways.
Nest: Loose open cup in vertical branch or crotch of shrub, sapling, or small tree.
Conservation Concern: Assigned a moderate PIF Continental Concern Score of 11/20.

Photo by Michael Furtman.

As Forbush and May (1955) declared: "The Rose-breasted Grosbeak is an admirable bird. It is beautiful, tuneful and useful, which from a human standpoint is about all that could be desired." The male has a startling appearance, "[g]arbed in a richly contrasting suit of black and white, with a resplendent rose-red breast-shield" (Roberts 1932), which is reminiscent of formal attire. The female plumage is much different from that of the male. It looks like a large sparrow, a combination of brown, tan, and white streaks, with a distinct, broad white eyeline and two white wing bars. It is well camouflaged for sitting on the nest. The large bill of both sexes is also diagnostic.

Roberts (1932) described the Rose-breasted Grosbeak as a "sweet-voiced singer." Its song is a "loud, sweet, pure-toned warble" that is similar to that of the American Robin but much faster, and to that of the Scarlet Tanager but not as hoarse. If you hear a Robin that you believe may be on a stimulant, then it may be this Grosbeak. The male is one of the few known to sing while on the nest, and its song has a ventriloquistic quality that may confuse potential predators (Forbush and May 1955).

The Rose-breasted Grosbeak was also named the "Potato-bug Bird" because of its appetite for consuming harmful potato beetles. The species is in the same family as the Blue Grosbeak but is only distantly related to two other species with "grosbeak" in their name that are found in Minnesota, the Evening and Pine Grosbeaks, both of which are in different families. The Rose-breasted and Blue Grosbeak are in the same family as tanagers and cardinals, all with Neotropical origins, while the Evening and Pine Grosbeaks have Palearctic origins. The Rose-breasted is a relatively common species in Minnesota and can be heard in late spring and summer singing from a forest edge or in young to intermediate-aged forests. It is primarily found nesting in the northeastern and midwestern United States and in Canada as far northwest as northern Alberta. The highest densities are found in the upper midwestern United States, including Michigan, Minnesota, and Wisconsin.

MINNESOTA BREEDING DISTRIBUTION

The Rose-breasted Grosbeak was historically widely distributed across the state; Roberts (1932) described it as abundant in the southern deciduous forests and less abundant in the north. He also noted it was common in prairie groves and along lakes and streams. Nesting was confirmed from the late 1800s to 1932 from Crow Wing, Goodhue, Houston, Isanti, Itasca, Polk, Rock, Stearns, and Wabasha Counties, plus Itasca State Park and in the cities of Minneapolis and St. Paul.

Green and Janssen (1975) simply described the species as a "resident throughout the state" and "most numerous in deciduous forests of the central and southern regions." Janssen (1987) similarly emphasized the breeding distribution throughout the wooded portions of the state. He included confirmed nesting since 1970 from 33 Minnesota counties that range from nearly every corner of the state. Counties with confirmed nesting were concentrated in the southern and central parts of the state, though these areas were also where the highest concentrations of bird observers were located. Hertzel and Janssen (1998) further expanded the total number of counties with confirmed nesting to 44, with most in the southern half of the state. The MBS breeding-season locations indicated the species was present in every county sampled, except from Wilkin County north along the Minnesota-North Dakota border.

The MNBBA clearly illustrates the Rose-breasted Grosbeak's statewide distribution, with potential nesting in every county of the state **(A)**. This includes possible nesting in five priority blocks in Wilkin County and confirmed or probable nesting in all counties bordering North Dakota. Confirmed nesting was reported in 187 blocks, while observed to confirmed nesting was reported from 48.3 percent (2,292/4,746) of all blocks sampled. The atlas added confirmed nesting in 27 counties not previously reported since 1970 by Hertzel and Janssen (1998).

Previous descriptions of the Rose-breasted Grosbeak had emphasized that it primarily occurred in the central and southern regions of the state. However, when coverage included the whole state and incorporated climate, habitat, landscape potential, and detectability, the heart of the species' distribution was predicted to be in the north-central regions and along the major river valleys **(B)**. The species was particularly common in regions dominated by deciduous forests. The extreme northwestern counties near the boundary between Kittson and Roseau Counties were predicted to have the highest densities.

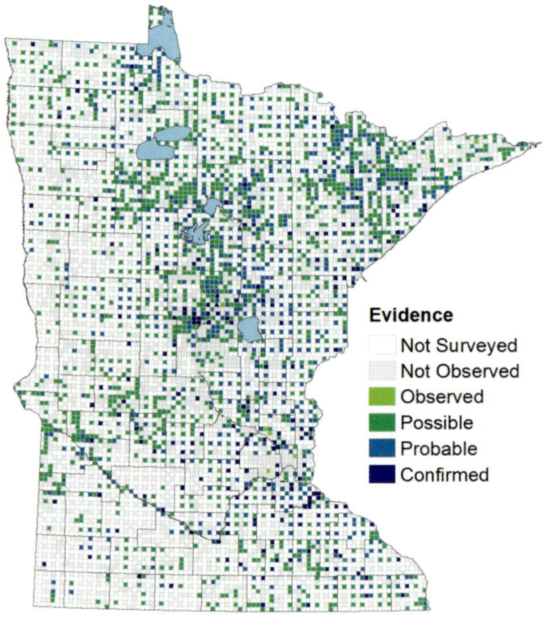

A. Breeding distribution of the Rose-breasted Grosbeak in Minnesota based on the Breeding Bird Atlas (2009–13).

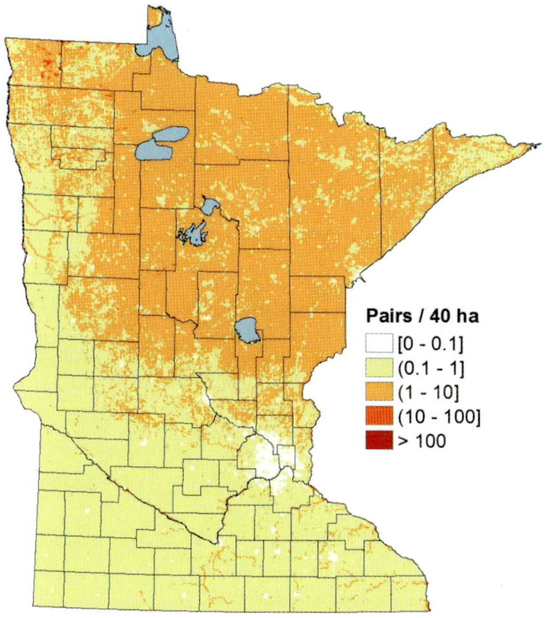

B. Predicted breeding distribution (pairs per 40 ha) of the Rose-breasted Grosbeak in Minnesota based on habitat, landscape context, and climate data gathered during the Minnesota Breeding Bird Atlas (2009–13) using the Generalized Linear Modeling method with an adjustment for detectability.

In their review of the species, Wyatt and Francis (2020) describe many range contractions and extensions in their summary of historical changes in the breeding distribution of the Rose-breasted Grosbeak that have occurred since the late 1800s. In particular, they emphasized that urbanization and agriculture appear to have produced mixed effects on the species. If the conversion of forests is not too extensive and forest fragments are still present but the canopy is open, then the changes may benefit the species. However, complete clearing without maintaining suitable forest habitat, such as has occurred in a substantial part of southern and western Minnesota, will greatly limit available habitat to a few woodlots or riparian forests along lakes, streams, and rivers. In contrast, the Rose-breasted Grosbeak has adapted to many urban environments where suitable habitat remains and is tolerant of several successional stages following forest cutting, especially as shrubs and saplings mature.

BREEDING HABITAT

Rose-breasted Grosbeaks use many habitats, including deciduous, mixed wooded uplands and lowlands, ecotones at the edge of riparian areas, second-growth woodlands, parks, gardens, and orchards, though most of these descriptions are from the eastern United States (Wyatt and Francis 2020). Probst and colleagues (1992) recorded it as most abundant in mid-successional stages of aspen regeneration. In the Chippewa and Superior National Forests, the highest densities were found in lowland shrub, regenerating forests, and upland hardwoods, but it was still present in most aspen–birch forests, mixed aspen–spruce–fir forests, and jack and red pine forests (Niemi, Howe, et al. 2016) **(C)**.

POPULATION ABUNDANCE

Based on the BBS, PIF (2021) estimated a US and Canadian breeding population of 4.7 million breeding adults and a Minnesota population of 430,000 adults. The MNBBA estimated a Minnesota breeding population at least three times that, at 1.41 million (95 percent confidence interval of 1.25–2.32 million).

The BBS in Minnesota revealed a stable population trend for the species from 1967 to 2019; however, the trend pattern is nearly significant and suggests a long-term decline beginning in the late 1970s **(D)**. Rosenberg, Kennedy, and colleagues (2016) calculated a −30 percent decline in the North American population of the species from 1970 to 2014. The BBS trends for North America also indicated widespread declines from 1966 to 2015 throughout the eastern United States and in much of the upper Midwest, including both northern and north-central Minnesota. These declining trends are consistent with breeding bird monitoring in the Chippewa and Superior National Forests, where Rose-breasted Grosbeaks are declining an average of −0.75 percent per year in a trend that has been especially steep from 2006 to 2020 (Grinde, Bednar, et al. 2020).

Reasons for the declining populations in the Rose-breasted Grosbeak are not well-known. As it is a forest-dwelling species, loss of forest area can be detrimental to its population. Its sensitivity to forest fragmentation is unclear. However, Grinde, Niemi, Adams, and Wigley (2017) found a positive response of the species to forest harvesting in north-central Minnesota when it occurred within a large forested area with minimal fragmentation from roads or any form of human development. The species also appears to be vulnerable to window kills. Loss and colleagues (2014) found the species to be sensitive to collisions with both low- and high-rise buildings, where it ranked thirteenth and seventeenth among species with a greater risk factor of 24 and 16, respectively, relative to an "average" species risk. Wyatt and Francis (2020) do not mention any concerns with overwintering habitat for the species.

C. One of the Rose-breasted Grosbeak's most preferred breeding habitats in Minnesota is young to mid-aged dense aspen forests, where it builds its bulky nest, which is often quite easy to spot. Photo by Gerald J. Niemi.

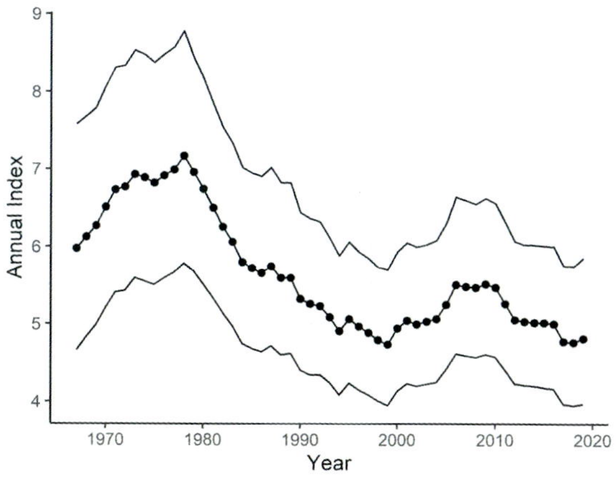

D. Breeding population trend and 95 percent credible interval (solid black lines) for the Rose-breasted Grosbeak in Minnesota based on the 1967–2019 Breeding Bird Survey (Sauer et al. 2020).

CONSERVATION

Even though the Rose-breasted Grosbeak uses many different habitat types for nesting and has a relatively wide breeding distribution, its recent declining trend in North America deserves scrutiny. Pfannmuller (2012) listed the species as a Stewardship Species in Minnesota because Minnesota is important to its overall population. Based on PIF (2021), Minnesota is estimated to account for about 9 percent of its global breeding population or, using the MNBBA results, the proportion would be about 30 percent.

Under a 2°C warming scenario, NAS (2019) projected low vulnerability for the Rose-breasted's breeding distribution in North America with a 45 percent loss of its breeding range and a 48 percent gain northward into Canada by 2050. Its breeding range retractions in Minnesota were predicted in the western and northwestern regions. Fortunately, the species is common in Minnesota with a relatively large population. Continued vigilance is warranted to better understand what may be affecting populations of the Rose-breasted Grosbeak.

BLUE GROSBEAK

Passerina caerulea

Minnesota Seasonal Status: Regular breeding resident and migrant. The Blue Grosbeak was rare during the MNBBA.
Migration: Medium-distance migrant that winters in southern Mexico, Central America, and the Caribbean.
Food: Omnivore that feeds on insects and seeds and forages on the ground and on foliage.
Nest: Cup nest placed in small trees or shrubs.
Conservation Concern: Assigned a low PIF Continental Concern Score of 8/20.

Photo by Laura Erickson.

A relative newcomer to Minnesota, the male Blue Grosbeak may be mistaken for an impressively large Indigo Bunting. Nearly 4 cm larger than the Bunting, the Grosbeak can be easily identified by its two chestnut-colored wing bars. Its gray bill is also noticeably larger and thicker and is surrounded by a small patch of black feathers. As with the male Bunting, the blue plumage is apparent only when the right light conditions allow the sun's blue wavelengths to be scattered; otherwise it appears nearly black. The cinnamon-brown female also has chestnut-colored wing bars and an occasional hint of blue on some of her feathers. First-year males display varying amounts of blue and brown.

The Blue Grosbeak is an uncommon bird throughout much of its range. Its song has been described as a burry, musical warble, somewhat similar to that of the Purple Finch, "though not so quick and energetic" (Nehrling 1896). Audubon (1827–38) compared its "sweet" song to that of the Bobolink and described its "shy and bashful" behavior as just the opposite of its spirited and neighborly relative, the Northern Cardinal. Much remains to be known regarding many aspects of this poorly studied species' life history, including its breeding biology, migratory behavior, and site fidelity (Lowther and Ingold 2020).

The Blue Grosbeak is distributed across the southern United States, from central California east to the Mid-Atlantic states. During the twentieth century, it has gradually expanded into the northern Great Plains. Migrant breeding populations also extend south into northern Mexico, while resident populations are present in southern Mexico and Central America. In the United States, the highest breeding densities are found in eastern North Carolina and in southern Georgia and Mississippi.

MINNESOTA BREEDING DISTRIBUTION

The Blue Grosbeak was first observed in Rock County on June 6, 1961, near the town of Beaver Creek (B. W. Anderson 1962), just miles from the borders of South Dakota and Iowa, states where the species was also relatively new (Bent 1968; Lovell 1951; L. Jackson et al. 1996). Two females and a male were seen along the side of a gravel road. A dead male, the apparent victim of a car collision, was found just a little farther away.

Following this initial sighting, Grosbeaks were consistently observed in the same vicinity, becoming a regular, yet rare summer resident. By 1975, Green and Janssen reported they had expanded north and east into two adjacent counties, Murray and Nobles. A pair was even observed much farther east in Dakota County on May 17, 1972 (J. Dempsey 1980). The only breeding evidence, however, was "an observation of two adults and three immatures near Beaver Creek, Rock County, on September 4, 1972" (Green and Janssen 1975). Breeding was also inferred in 1974, when a female was observed carrying nest materials at Blue Mounds State Park in Rock County (Eckert 1975a). Then, on July 16, 1976, a pair was found tending a nest in the park that contained "one newly hatched young, one egg, and one cowbird egg," becoming the state's first confirmed nesting record (Eckert 1977b).

In the coming years Blue Grosbeaks slowly expanded from their core breeding area in the southwestern corner of the state. By 1976 they were reported in Pipestone County, in Renville County in 1989, and in Lincoln County in 1991. Before the MNBBA began in 2009 they had been reported in 14 southwestern counties stretching north to Chippewa, Lac qui Parle, and Renville and east to Brown, Cottonwood, and Jackson (MOU 2021b). Elsewhere, scattered summer records were noted in Clay County in July 1988 (A. Bolduc 1988); again in Dakota County in 1999 and 2005 (both observations were of immature birds, MOU 2021b); and in Scott County in 2008 (MOU 2021b).

MNBBA participants documented a similar distribution. Sixty-four Blue Grosbeak records were reported in 1.1 percent of the surveyed atlas blocks (52/4,740) that spanned 19 of Minnesota's 87 counties. Most records were in the southwestern counties of Murray, Nobles, Pipestone, and Rock, the same counties where Hertzel and Janssen (1998) had previously reported confirmed nesting. Scattered reports were also noted farther to the north and east **(A)**. Breeding was confirmed in four blocks in three new counties: Dakota, Lyon, and Renville. Although one of the four blocks straddled Renville and Redwood Counties, accompanying notes documented that the birds were nesting in Renville County. In the southwest and south-central regions, Blue Grosbeaks were reported for the first time in McLeod, Martin, and Watonowan Counties, and in a block straddling Blue Earth/Nicollet Counties, expanding the species' core breeding range a little farther east. Other records of interest were probable breeding in Clay and Yellow Medicine Counties and possible breeding in Goodhue and Stevens Counties. Indeed, during the atlas, breeding was confirmed by other observers in Clay County (2012, MOU 2021a). In 2016 breeding was also confirmed in Yellow Medicine County, while Hennepin and Scott Counties were added to the list in 2017 (MOU 2021a).

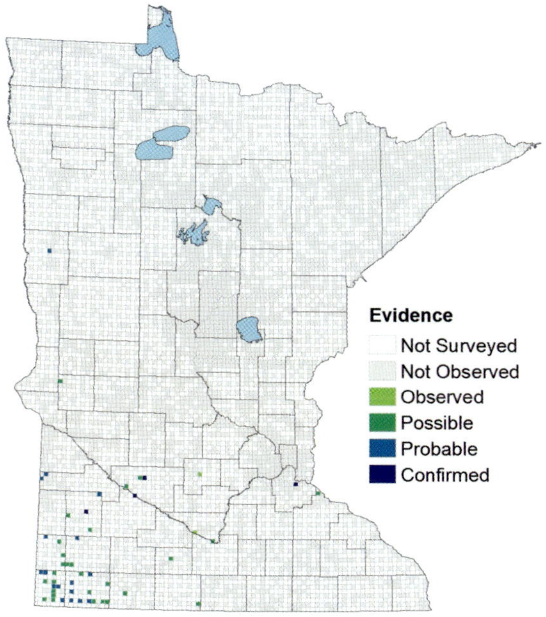

A. Breeding distribution of the Blue Grosbeak in Minnesota based on the Breeding Bird Atlas (2009–13).

The Blue Grosbeak's expansion into the northern Great Plains began in the early 1900s (Lowther and Ingold 2020). Rudolph Anderson (1907) considered it a hypothetical species in Iowa until it was first observed in 1932 (L. Jackson et al. 1996). Nesting was confirmed in 1964 (L. Jackson et al. 1996), and Blue Grosbeaks are now common in western Iowa along the Missouri River (Iowa Ornithologists' Union 2019). In South Dakota, the earliest report was in 1911 on the Pine Ridge Reservation in the southwestern corner of the state (Lovell 1951). Today, the species is found across the southern half of the state and in scattered locations farther north, again especially along the Missouri River (Drilling, Sparks, et al. 2016). In North Dakota, the Grosbeak was first observed in 1967 near the South Dakota border in Emmons County (Igl 1995) and was confirmed nesting in 1979 in Burleigh County, immediately north of Emmons County. The species is still uncommon in the state but, like its distribution in Iowa and South Dakota, most observations are near the Missouri River, where shrubby riparian habitats are common (Igl 1995).

The Blue Grosbeak is an enigma. As William Youngworth (1958) noted years earlier, its expanding distribution is just the opposite of what has been observed for decades for many grassland species whose populations have declined. The factors responsible for its northward movements are poorly understood.

BREEDING HABITAT

The Grosbeak inhabits a variety of semi-open habitats, including woodland edges, overgrown fields, shrubby thickets along streams or in grasslands, and hedgerows **(B)**. Some of the primary habitat requirements include low canopy cover, low shrub density, scattered trees, and heavy ground cover (Lowther and Ingold 2020).

In Minnesota, the birds have been observed in farm shelterbelts (R. L. Huber 1964a), farm woodlots (McIntosh 1964), gravel pits (Wallestad 2016), and shrubby riparian thickets (Eckert 1975a). A breeding territory in Dakota County was described by Hertzel (2012) as "an overgrown, weedy embankment" bordering a landfill. "A few cedar trees were scattered about the top of the dike and a wooded, marshy area was directly to the west." Wallestad (2016) found a strong association with gravel pits that were located near riparian habitats that bordered drainage ditches, ponds, small streams, and even large rivers. Indeed, common features in many descriptions are small trees or shrubs, disturbed habitats, and proximity to water or wetlands.

POPULATION ABUNDANCE

Biologists estimate that the Blue Grosbeak population numbers approximately 35 million breeding adults throughout the Western Hemisphere, with a population estimate of 21 million in the United States and Canada. Minnesota supports approximately 16,000 birds (PIF 2021). MNBBA data were insufficient to generate a reliable state estimate for comparison.

B. Scattered small trees and shrubs in a disturbed grassland are typical of many Blue Grosbeak nesting sites.
Photo by Lee A. Pfannmuller.

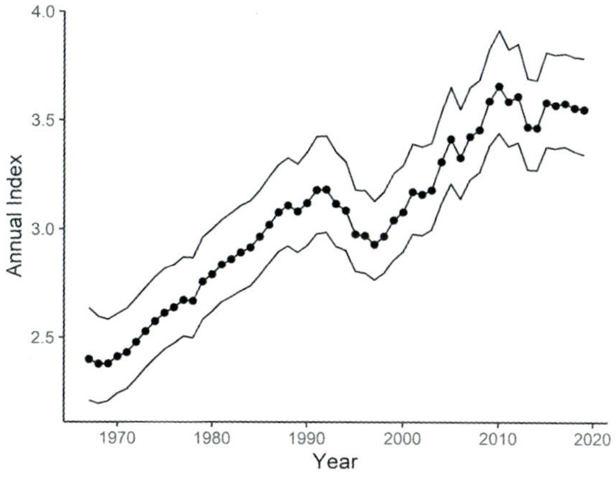

C. Surveywide breeding population trend and 95 percent credible interval (solid black lines) for the Blue Grosbeak based on the 1967–2019 Breeding Bird Survey (Sauer et al. 2020).

The Blue Grosbeak has demonstrated a steady and significant population increase averaging 0.73 percent per year across the BBS survey region from the mid-1960s through 2019 **(C)**. Biologists estimate that the North American population increased 45 percent from 1970 to 2014 (Rosenberg, Kennedy, et al. 2016). The species is not detected frequently enough in Minnesota to generate a statistically robust trend line, but BBS data clearly reflect the species' increasing presence in the state (Sauer et al. 2020).

Although factors responsible for the Grosbeak's expanding distribution and increasing population are not well understood, Ned Johnson (1993) postulated that the changes were due to warming temperatures. He suggested that a suite of southern species expanding north, including the Blue Grosbeak, was responding to increases in annual precipitation and humidity, climate variables that characterized the regions from which they were expanding. More than a century earlier, Coues (1874) attributed the bird's northern expansion to clearing of the eastern deciduous forest.

CONSERVATION

Although it is rarely abundant anywhere, the Blue Grosbeak was assigned a low conservation score (PIF 2021). Regarding climate change, NAS (2019) ranked its breeding distribution as stable under all three warming scenarios they examined. Small range losses in the south were compensated for by gains across the northern periphery of its breeding range. While the models did not include Minnesota within the species' current breeding distribution, warmer temperatures are likely to spur this newcomer's continued expansion into southern and western Minnesota.

INDIGO BUNTING

Passerina cyanea

Minnesota Seasonal Status: Regular breeding resident and migrant. The Indigo Bunting was common during the MNBBA.
Migration: Short- to long-distance migrant that winters in southern Florida, the Caribbean, and in southern Mexico south to northern South America.
Food: Omnivorous diet of arthropods and seeds. Foraging occurs primarily on the ground but also on the foliage of shrubs and small trees.
Nest: Cup nest, usually located within 1 m of the ground and placed within a shrub, small tree, or tall forb.
Conservation Concern: Assigned a moderate PIF Continental Concern Score of 9/20.

The male Indigo Bunting's azure blue plumage is stunning when seen in the bright sunlight. Even Audubon (1827–38) noted, "When the male of the species . . . is in full plumage, the richness of his apparel cannot fail to attract and please the eye of any observer." One would think it would be easy to spot such a striking bird in its open woodland habitat, but such is not always the case. The Indigo Bunting is a small bird, nearly the same size as the petite Chipping Sparrow. Unless the lighting is just right, it may not appear blue at all but rather black. All birds with blue feathers actually lack any blue pigment. Like the blue sky on a sunny day, when the sunlight reaches tiny structures in the Bunting's black feathers, the light is scattered. Because short blue waves are scattered the most, we see a magnificent indigo color when we admire the Bunting through our binoculars.

Although the male is often conspicuously perched atop a tree branch, singing his heart out, the much duller female usually remains well hidden. A pale, rusty brown above with faint breast streaks, there may be just a hint of light blue in her tail. Males do not acquire the full "blue" plumage until their second breeding season, donning a very mottled blue and brown plumage in their first full year.

A persistent songster well into late summer, the male Indigo Bunting has a diverse repertoire. The most familiar song rendition, where most syllables are delivered in well-articulated pairs, sounds like *sweet, sweet—where, where—here, here—see it, see it.* Roberts (1932) described it as "sweet-toned but rather characterless and is delivered in a lazy, indolent fashion." The mnemonic, however, only serves as a foundation. Each male may embellish it by incorporating more than 100 unique notes or song phrases (Payne 2020).

Photo by David Brislance.

A common species wherever open woodlands and early successional habitats are found, the Indigo Bunting has a breeding range stretching across southeastern Canada, south throughout the eastern United States, and west to the eastern Great Plains. Scattered populations are also present in the southwest United States.

MINNESOTA BREEDING DISTRIBUTION

Roberts (1932) provides a lengthy description of the Indigo Bunting's distribution in Minnesota in the late 1800s and early 1900s. Found in nearly all corners of the state, it was most abundant in the southeast, especially in counties bordering the Mississippi River north to the town of Red Wing. "Farther north it becomes less frequent but has been found along the Red River as far north as Norman and Marshall counties; at Leech Lake, Cass County; in Itasca County; along the North Shore of Lake Superior from Duluth to Grand Marais; and in the interior of Cook County." Although it had yet to be found in Kittson and Roseau Counties, along the Canadian border, documentation of its breeding status in Winnipeg suggested it may be a rare resident in these counties as well. The Bunting could also be found in small woodland groves that dotted the southwestern prairie region. Within the northern forests, it was "much less frequent" and largely restricted to "clearings and burnt-over areas," although Roberts never found it at Itasca State Park despite spending several summers there. Nesting was confirmed or inferred in Fillmore, Goodhue, Hennepin, Houston, Lake, Lake of the Woods, and Wabasha Counties.

Years later, both Green and Janssen (1975) and Janssen (1987) simply noted that the Indigo Bunting occurred throughout the state but was "most numerous in the southeastern, south-central, east-central, and central regions" (Green and Janssen 1975). Janssen identified 18 counties where nesting had been confirmed since 1970. Hertzel and Janssen (1998) later added four more counties to the list.

By 2014, the MBS reported 913 Indigo Bunting breeding-season observations. Although the majority of records were in the Eastern Broadleaf Forest Province and in the southern counties of the Laurentian Mixed Forest Province, a large number were reported along the Minnesota River valley and throughout the southwest corner of the state, where earlier observers reported the Bunting as relatively uncommon. The planting and maturation of farm shelterbelts in the southwest likely contributed to the species' abundance in the region.

The species was found in all 87 Minnesota counties during the MNBBA; participants reported 2,590 Indigo Bunting records in 32.9 percent (1,563/4,756) of the surveyed atlas blocks and in 45.7 percent (1,068/2,337) of the priority blocks **(A)**. Breeding was confirmed in 67 blocks and 38 counties; it was deemed probable in 40 counties. The species remained most abundant in the Eastern Broadleaf Forest Province and in the southern and western regions of the Laurentian Mixed Forest Province. Once again, volunteers in the Brainerd Lakes region did an admirable job documenting the species' local abundance.

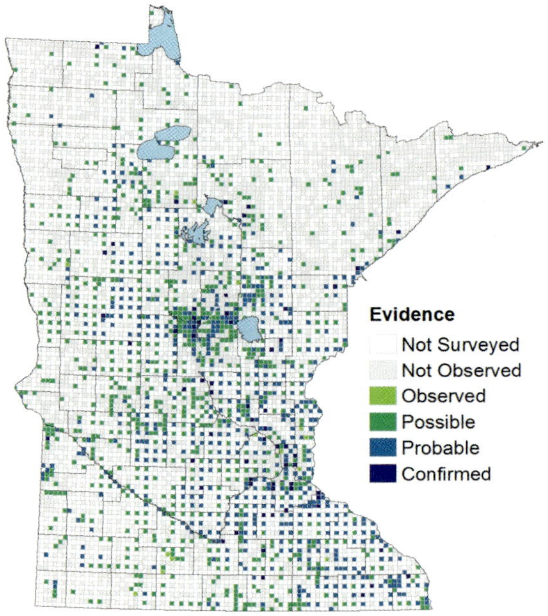

A. Breeding distribution of the Indigo Bunting in Minnesota based on the Breeding Bird Atlas (2009–13).

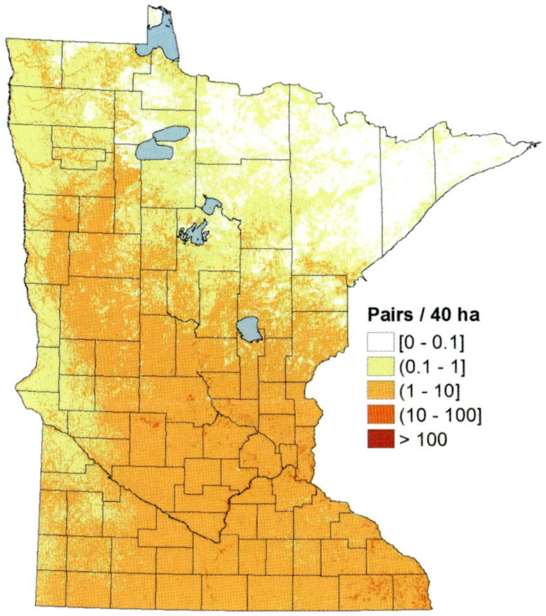

B. Predicted breeding distribution (pairs per 40 ha) of the Indigo Bunting in Minnesota based on habitat, landscape context, and climate data gathered during the Minnesota Breeding Bird Atlas (2009–13) using the Generalized Linear Modeling method with an adjustment for detectability.

The MNBBA probability map predicts the highest breeding densities would occur in scattered pockets in southeastern, south-central, and east-central Minnesota, including along the lower Minnesota River valley **(B)**. Rare to absent across northeastern and far north-central Minnesota, moderate breeding densities are predicted throughout much of central and southern Minnesota and along tributaries of the Red River in the northwest.

Roberts's (1932) detailed delineation of the Bunting's distribution provides a good foundation for comparison nearly one hundred years later. Still found in all corners of the state, Indigo Buntings are no longer most abundant only in the southeast but throughout the Eastern Broadleaf Forest Province, from Aitkin, Crow Wing, and Otter Tail Counties south. In the northwest they have expanded beyond Marshall and Norman Counties north to the Canadian border and are widespread throughout the southern two-thirds of the Prairie Parkland Province. Uncommon throughout the northern forests in Roberts's time, they are likely more widespread now that development has fragmented the forest landscape from Brainerd north to Duluth, Grand Rapids, and Bemidji. Beyond Minnesota, the species expanded north into Canada, particularly into the Maritime Provinces in the 1970s, west into California, Arizona, and New Mexico beginning early in the twentieth century, and south along the Florida peninsula from the early to mid-1900s (Payne 2020).

BREEDING HABITAT

Indigo Buntings are found in a wide variety of open woodland and shrubby habitats, ranging from overgrown fields, roadsides, and brushy pastures to disturbed forests, aspen clear-cuts, forest openings, and woodland edges **(C)**. The variety of habitats they use during the breeding season is so diverse that it is easier to identify where they are not found: mature forests, intensively cultivated lands, and densely developed urban and suburban areas. Suitable habitat tends to be more xeric then mesic, although the Buntings can be found in shrubby lowlands (Payne 2020). The most important habitat features appear to be sufficient perches for the territorial males and a high density of shrubs or tall forbs for nesting.

POPULATION ABUNDANCE

PIF (2021) generated a population estimate of 77 million breeding adults in the United States and Canada, with a statewide estimate of 630,000 adults. MNBBA data, however, provided a significantly higher population estimate of 1.85 million adults (95 percent confidence interval of 1.6–3.3 million).

The Indigo Bunting's population trend is well monitored by the BBS, which covers the species' entire breeding range. Since the survey began in 1966, the population has experienced a significant decline, averaging −0.62 percent per year through 2019 (Sauer et al. 2020), with a cumulative decline of −25 percent from 1970 to 2014 (Rosenberg, Kennedy, et al. 2016). Declines have been most pronounced in the southeastern states. Although Minnesota's population has fluctuated considerably over the years, numbers show a long-term, stable population from 1967 through 2019 **(D)**.

Indigo Buntings were likely a less common species when European settlers first arrived. The mature deciduous forests of eastern North America provided fewer nesting opportunities for this denizen of open woodlands. But as forests were cleared, the open habitat proved ideal for the species. Even as recently as the early twentieth century, population densities increased significantly in the heart of the Midwest. A long-term study in Illinois, from 1909 to 1958, documented a 47 percent increase in Bunting numbers (Graber and Graber 1963). Declines documented since 1966 may be due to more intensive agricultural practices that eliminate edge habitats, to increasing urbanization, and, in the northeast, to the successional loss of old fields to mature forests (Payne 2020).

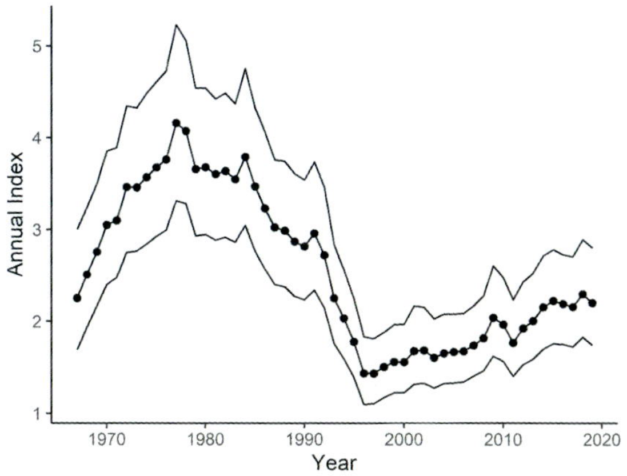

D. Breeding population trend and 95 percent credible interval (solid black lines) for the Indigo Bunting in Minnesota based on the 1967–2019 Breeding Bird Survey (Sauer et al. 2020).

CONSERVATION

Because populations have been steadily declining in the past fifty years, the Indigo Bunting has been assigned a moderate conservation ranking (PIF 2021) despite its widespread distribution and abundance. Threats to the population are considered relatively low, including its vulnerability to warming temperatures. Under a 2°C warming scenario, models predict a 14 percent range loss compensated for by a 15 percent gain by 2050, including gains in northeastern Minnesota (NAS 2019). In the near future, the status of this little "blue" gem appears quite secure.

C. A site located in southeastern Minnesota includes scattered trees and saplings that provide song perches for the Indigo Bunting and small shrubs and forbs that provide abundant nesting opportunities. Photo by Lee A. Pfannmuller.

DICKCISSEL

Spiza americana

Minnesota Seasonal Status: Regular summer resident and migrant. The Dickcissel was common during the MNBBA.
Migration: Long-distance migrant that winters primarily in central Venezuela.
Food: Omnivorous diet of arthropods and seeds captured while foraging on the ground and in low vegetation.
Nest: Cup nest located near the ground in dense grasses, forbs, or low shrubs.
Conservation Concern: Assigned a moderate PIF Continental Concern Score of 11/20.

Photo by Mike Lentz.

The Dickcissel is a species full of surprises. At first glance, the male may be mistaken for a meadowlark. His black V-shaped throat patch and his bright yellow chest are not unlike the yellow underparts and prominent black crescent on the breasts of the Eastern and Western Meadowlarks. However, the Dickcissel is significantly smaller in size, and its yellow plumage is restricted to small areas on the face and upper chest, while the belly is largely gray. Similar in many respects, the female lacks the male's black throat, and her yellow breast is much paler. Both sexes also have a rufous shoulder patch, the male's being more extensive. Although ornithologists have written numerous mnemonics to describe the male's simple song, the rendition of *Dick! Dick! chisl, chisl, chisl* has prevailed.

But it is the "little meadowlark's" erratic nesting behavior that provides the greatest surprise. Roberts (1932) wrote, "There is always a pleasant excitement as to whether or not [the] Dickcissel can be found and where. When it leaves in the fall, after perhaps a season of abundance and intimate acquaintance, it may be lost almost entirely from that locality for no one knows how long." Indeed, its nesting history is marked by periodic irruptions when birds disperse hundreds of miles from their core breeding range in the Central Plains. Here, 70 percent of the Dickcissel's population occurs in Iowa, Kansas, Missouri, Nebraska, and Oklahoma. Beyond this core, Temple (2020) identified a much larger area, stretching from the Great Plains east across the Eastern Deciduous Forest, where the species usually breeds in lower numbers. Prompted by weather and food availability, seasonal shifts in distribution and abundance are common throughout both areas. Roberts (1932) described the outcome as an annual "game of 'hide and seek.'" In some years, the birds may even nest in the southern portion of their range early in the season and then move north to initiate a second nesting or to renest if earlier attempts failed (Basili et al. 1997).

MINNESOTA BREEDING DISTRIBUTION

Historically, the state's most irruptive summer resident was considered a breeding species south and west of the northern forests (Roberts 1932). Although Dickcissels were "less frequent northward in the Red River Valley," summer records stretched as far as the Canadian border. At the time, confirmed nesting records were reported in Goodhue, Hennepin, and Polk Counties, while reports of fledged young were noted in McLeod and Wright Counties. Regardless of their erratic behavior, the birds were always present somewhere in the state each year. Roberts made special note of an irruption that occurred in 1925, when reports describing a virtual "deluge" of Dickcissels came from all corners of the state. Then, one year later, they were nearly absent from the entire eastern half of Minnesota. Examining records dating back to 1874, he noted that there had been years of high abundance approximately eight to ten times over a period of fifty-five years.

Green and Janssen (1975) later described the species' distribution as primarily restricted to southern Minnesota north to Traverse County in the west and to Chisago County in the east. Even in years of high abundance, Janssen (1987) noted it rarely extended north of southern Crow Wing and Pine Counties. He identified only four counties where nesting had been confirmed since 1970: Martin, Murray, Rock, and Wabasha. Hertzel and Janssen (1998) later added Aitkin, Sherburne, and Winona Counties to the list. Breeding-season locations identified by the MBS were concentrated south of the Minnesota River but also extended north into Wilkin, Clay, and Norman Counties. Although the species was found as far east as St. Louis County, it was rare across northern Minnesota.

During a major irruption that occurred during the 1988 breeding season, Janssen (1988) noted the relationship between the summer's extreme drought and the Dickcissel's abundance, a connection that had been identified by others as a major factor prompting large shifts in distribution (Emlen and Wiens 1965; Sealy 1976). Then, during the summer of 2012, in the middle of the MNBBA, the largest Dickcissel irruption ever reported in the state occurred (Svingen and Hertzel 2013). Not only was it evident in Minnesota, but Dickcissels were abundant across the Upper Midwest and southern Canada (eBird 2012). In their comprehensive review of the irruption, Svingen and Hertzel (2013) documented reports from every Minnesota county except Cook. Documentation included records from the MNBBA as well as reports to eBird and the MOU. Once again, drought conditions coincided with peak numbers of Dickcissels moving northward.

Over the course of the MNBBA, participants reported 1,672 Dickcissel records in 20.4 percent (973/4,774) of the surveyed atlas blocks. Although the records spanned 78 of Minnesota's 87 counties, the species was most abundant in southern Minnesota **(A)**. The nine counties where the species was not reported were all in the northern half of the state. Breeding was confirmed in 30 counties, stretching as far north as St. Louis County in the east, Crow Wing

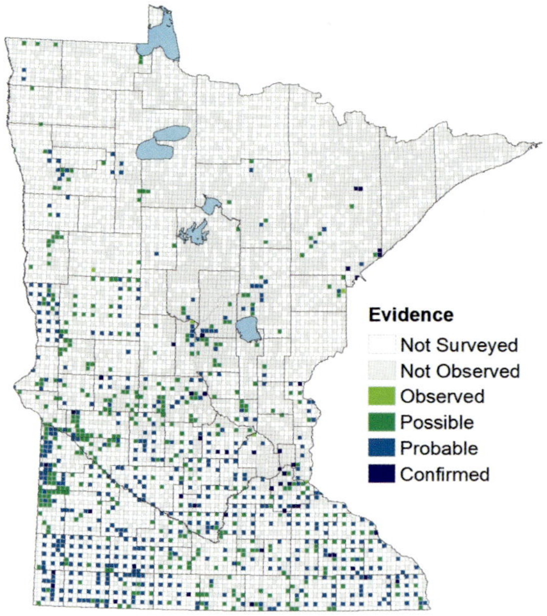

A. Breeding distribution of the Dickcissel in Minnesota based on the Breeding Bird Atlas (2009–13).

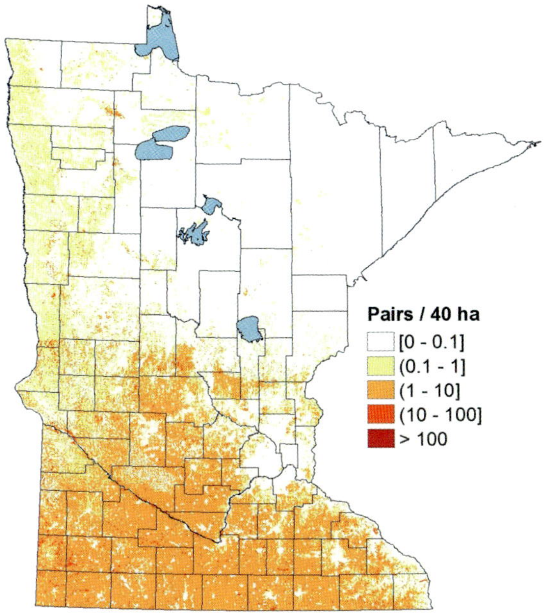

B. Predicted breeding distribution (pairs per 40 ha) of the Dickcissel in Minnesota based on habitat, landscape context, and climate data gathered during the Minnesota Breeding Bird Atlas (2009–13) using the Generalized Linear Modeling method with an adjustment for detectability.

and Todd Counties in central Minnesota, and Pope County in the west. Breeding was deemed probable in an additional 41 counties, with records north to Clearwater, Itasca, Kittson, and Polk. Fifty-one percent of the Dickcissel detections were collected during the 2012 irruption, including 65 percent of the confirmed breeding records.

The MNBBA probability map predicts the occurrence of moderately high breeding densities across southern Minnesota, especially south of the Minnesota River valley **(B)**. Farther north, the probability of detecting Dickcissels is highest in the Red River valley, although their predicted densities are quite low in all but a few areas.

Historically, the most significant change in the Dickcissel's distribution was its retreat from New England and the Mid-Atlantic states. Common in the region in the nineteenth century, it is only sporadically reported there today. Clearing the eastern forests initially created more habitat opportunities for the birds, but the development that followed resulted in a significant loss of suitable habitat (Temple 2020).

BREEDING HABITAT

Dickcissels can be found in grassland habitats ranging from hayfields and lightly grazed pastures to native prairies and restored grasslands **(C)**. The planted cover found in Conservation Reserve Program (CRP) fields is also suitable (Dechant, Sondreal, et al., "Dickcissel," [1999] 2002a). Important habitat features include dense cover, moderate to tall vegetation, moderately deep litter, and elevated song perches (Temple 2020). A high abundance of forbs is particularly important because they provide song perches, nesting cover, and insect prey (Dechant, Sondreal, et al., "Dickcissel," [1999] 2002a). Some shrub cover is tolerated. Although the birds use small grasslands, studies have documented lower reproductive success on sites smaller than 10 ha (Temple 2020).

POPULATION ABUNDANCE

PIF (2021) estimated a US and Canadian breeding population of 28 million adults, with approximately 240,000 birds in Minnesota. Although MNBBA data generated a significantly higher statewide estimate of 1.2 million adults, data from the 2012 irruption was included.

Across the BBS survey region, the Dickcissel has demonstrated a long-term, significant population decline, averaging −0.59 percent per year from 1966 to 2019. The decline was steepest in the first fifteen years of the survey but appears more stable since the 1980s (Sauer et al. 2020). Temple (2020), however, estimated that current population numbers are at least two-thirds of those seen in the late 1960s. In Minnesota, where population numbers may vary significantly from year to year, the species has experienced a significant decline, averaging −3.33 percent per year since 1967 **(D)**.

Intensive agricultural practices that convert grasslands and hayfields to row crops have certainly decreased the availability of suitable habitat for Dickcissels, while planted cover on lands enrolled in the CRP has created new habitat. Unfortunately, many suitable sites, such as hayfields and roadsides, are mowed before the young have fledged. This is a particular problem for the Dickcissel, which is a late nesting species (Dechant, Sondreal, et al., "Dickcissel," [1999] 2002a).

C. The Dickcissel is attracted to open fields of dense herbaceous vegetation with scattered trees and shrubs. Photo by Lee A. Pfannmuller.

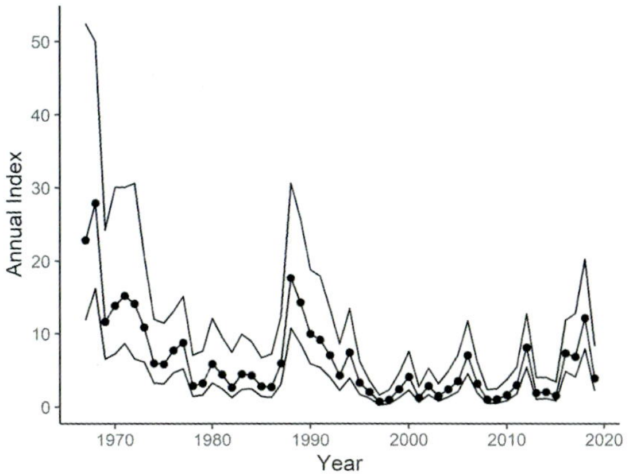

D. Breeding population trend and 95 percent credible interval (solid black lines) for the Dickcissel in Minnesota based on the 1967–2019 Breeding Bird Survey (Sauer et al. 2020).

A major threat also occurs in Venezuela, the core of the species' winter range. Here Dickcissels are considered a major agricultural pest, feeding on rice and sorghum fields. Although most farmers now use nonlethal methods to scare the birds away, lethal methods are still employed by some and can result in significant mortality (Temple 2020).

CONSERVATION

In the late 1970s, some biologists were predicting the species' demise, but its adaptability to restored grasslands, hayfields, and pastures has resulted in a much brighter future. Its moderate conservation ranking largely reflects the significant threats to the population on the wintering grounds (PIF 2021). Fortunately, NAS (2019) considered the Dickcissel's breeding distribution to be stable under all three climate change scenarios they examined. Indeed, the species may be among the biggest avian beneficiaries of a warmer climate; even with a warming scenario of 3°C, its breeding range may double in size across the eastern United States!

Perhaps the most important key to the "little meadowlark's" future is to ensure the future of farm assistance programs like CRP and the Environmental Quality Incentives Program. These provide farmers with technical assistance and financial resources to return environmentally sensitive agricultural lands to vegetative cover that is attractive to species like the Dickcissel. Although enrollment in these programs can fluctuate with changes in commodity markets, in an era where the loss of grassland cover continues at an unprecedented rate, they remain vital to the future of many grassland birds (DeVore 2018).

To understand and appreciate the world around us, people create categories. For the purposes of this book, Minnesota's breeding birds have been placed into two primary categories: first, bird species that were reported during the MNBBA as either confirmed, probable, or possible breeders or that were simply observed at least once during the 2009–13 breeding seasons; second, bird species that were not reported during the MNBBA and have never been regular breeders or no longer are regular breeders in the state. Here we honor the history of the latter group of 19 species that have disappeared from Minnesota or that have arrived from other places, have adapted to Minnesota's landscape, and now have a history of intermittent breeding in the state. A diverse group, these birds can be further placed into four categories: first, birds that have become extinct or extirpated (e.g., Passenger Pigeon or Swallow-tailed Kite, respectively); second, species that originally did not occur in the United States but were intentionally introduced by people (e.g., Chukar); third, vagrant species that have sporadically nested but have not established a sustaining population (e.g., Snowy Plover); and, fourth, species that occur naturally in other parts of the continent and have bred with a sister species native to Minnesota (e.g., Scissor-tailed Flycatcher). Although "Current Breeding Species" includes a few species that could be categorized as non-regular because they have similar attributes to those described above, credible breeding-season records for them were collected during the years of the MNBBA (e.g., Northern Bobwhite, Burrowing Owl, Mute Swan, Kentucky Warbler, and Summer Tanager).

The majority of species in "Former and Incidental Nesting Species" are considered vagrants, birds that are observed in locations outside their normal range. Vagrancy is a common phenomenon in the bird world. In a state like Minnesota, with many accomplished birders, the total number of species on the state's checklist includes many vagrants that are not part of the state's regular avifauna. For example, in Minnesota the MOU 2020 checklist gives a total of 446 bird species that have been seen in the state. This book identifies 250 (56 percent) of them as breeding species. This number is fluid and can change yearly. Happily, vagrants and their numbers are fun to list, as birders are well aware. Vagrancy is a biological concept that is interesting to contemplate and study. It encompasses many factors that cause a species to navigate or to be blown to a place beyond its normal range. Sometimes this migration results in adaptation to a new range.

Nomenclature definitions used in the species accounts are necessary for understanding their status. Most of the terms used in "Former and Incidental Nesting Species" are the same as those used in "Current Breeding Species," and they are described in chapter 3, "Reading the Species Accounts." When describing less common species, most bird lists use a status term to differentiate frequency of yearly occurrence. Those used here follow those used in the MOU 2020 checklist, which is available online at moumn.org. The terms are: Regular, Casual, Accidental, Extirpated, and Extinct. Two of these terms are used frequently in the following species accounts: Casual records have been accepted by the Minnesota Ornithologists Union's Record Committee (MOURC) in four to eight of the last ten years; and Accidental records have been accepted by the MOURC in none to two of the last ten years. Because these categories are sensitive to the number of records accepted by MOURC, they change decade by decade.

Finally, most of the accounts that follow are brief summaries of each species' seasonal status, conservation ranking, national distribution, and breeding history in Minnesota. The exception is the Swallow-tailed Kite, an iconic species with a rich historical record dating back to 1832. Like the Passenger Pigeon, it is a former breeding species in Minnesota. Unlike the Passenger Pigeon, its story is not widely known, even though it remains a breeding species in the southeastern United States, southern Mexico, Central America, and South America.

II. FORMER AND INCIDENTAL NESTING SPECIES

This diorama painting by artist Francis Lee Jaques of Swallow-tailed Kites, flying and perched over Minnesota's lower Mississippi River valley, was completed in 1946. It stood in the foyer of the former Bell Museum at the University of Minnesota and is currently on display in the Minnesota Journey Galleries in the new Bell Museum. Photo provided by Bell Museum.

CHUKAR

Alectoris chukar

Minnesota Seasonal Status: Formerly a permanent resident that was introduced in Minnesota in the 1930s and 1940s; extirpated since the 1970s; on rare occasions escaped game birds may be seen.
Conservation Concern: Assigned a low PIF Continental Concern Score of 5/20. Primarily managed as a game species in some areas of the western United States.

The Chukar is a medium-sized partridge that is native to Eurasia, primarily from Greece to Mongolia and China. It was first introduced to North America in Illinois in 1893. Christensen (2020) documented that roughly 795,000 Chukars were released in forty-one US states and 10,600 birds were released in six Canadian provinces between 1931 and 1970. Today the species primarily thrives in the Great Basin of the western United States from Nevada and Utah north to eastern Oregon, western Idaho, and eastern Washington. The species prefers the steep Rocky Mountain terrain and is a popular upland game bird in many western states.

MINNESOTA BREEDING DISTRIBUTION

R. L. Huber (1960b) stated that the Chukar was first released in Minnesota on September 3, 1937, in Martin County, and that the species "appeared to take hold, at first, in Stearns, Benton, Sherburne, Wright, Hennepin, Anoka, Chisago, Pine, Itasca, St. Louis, and Lake Counties." Subsequently, Green and Janssen (1975) documented that the birds were released in all but two of Minnesota's counties. Releases apparently continued from 1937 to 1947 (Janssen 1987), but the species persisted only in the open-pit iron ore mines in Ely, St. Louis County, because of suitable breeding habitat and artificial winter feeding by the town's residents (Green and Janssen 1975). The last observation of a brood was in 1966 (Green and Janssen 1975), and the last remnant flock was maintained until the early 1970s (Janssen 1987). The last record of the species based on MOU seasonal reports was in 1979 (MOU 2021b); however, rare observations, which are likely escapees from game farms, can still occur. The Minnesota Ornithological Records Committee dropped the species from the state list because it was an introduced species that maintained only a semi-wild population for a short time and then died out (Janssen 1987).

PASSENGER PIGEON

Ectopistes migratorius

Minnesota Seasonal Status: Former regular summer resident in the nineteenth century. Extinct.

The story of the violent, rapid extinction of the Passenger Pigeon, once the most numerous land bird in North America, is well known: from the early nineteenth century, when clouds of migrants darkened the sky, to 1902, when the last bird in the wild was credibly documented, to 1914, when the last bird died in a zoo. This extinction joined seven other species that have not survived in North America since the beginning of colonial settlement: Great Auk (*Pinguinus impennis*), Eskimo Curlew (*Numenius borealis*), Labrador Duck (*Camptorhynchus labradorius*), Carolina Parakeet (*Conuropsis carolinensis*), Bachman's Warbler (*Vermivora bachmanii*), Ivory-billed Woodpecker (*Campephilus principalis*; the USFWS proposed officially listing the species as extinct in September 2021; the proposal was controversial and preparation of the final proposal has been delayed [USFWS 2022]), and Imperial Woodpecker (*Campephilus imperialis*; critically endangered and presumed extinct [IUCN 2021]). The Passenger Pigeon was unique because of its vast numbers, estimated at 3–5 billion birds, and the swiftness of its decline facilitated by plenty of guns, commercial hunting, railroad transport, telegraph communication, and a ready urban market (Blockstein 2020).

The Passenger Pigeon's gregarious breeding behavior was unlike that of any other North American land bird (Blockstein 2020) and had far-reaching impacts on Eastern deciduous forests with mast-bearing trees, especially oaks and American beech. By nesting in immense flocks that numbered millions of birds, the species was able to exploit the huge abundance of the mast crops that varied yearly in time and space. The enormity of the nesting spectacle created forest disturbances that had ecological effects at a landscape scale (Coffin and Pfannmuller 1988). The core of its range was the Great Lakes states east to New York and southern New England.

The biological, ecological, historical, and cultural aspects of the Passenger Pigeon's demise are well covered in two books: Schorger (1955) and Greenberg (2014).

MINNESOTA BREEDING DISTRIBUTION

In Minnesota, the Passenger Pigeon's history can be summarized by a quote from Roberts (1932): "An abundant summer resident throughout the state, breeding both irregularly in large colonies and regularly in isolated pairs in all wooded regions. The large spring flocks ceased arriving about 1880 and after that date [the species] diminished rapidly in numbers." An old nesting record from June 1890, from the Mississippi River bottomlands, about 16 miles south of Minneapolis, was recently reprinted (Emerson [1905] 2020). The last known nest was collected on June 21, 1895, in Oak Park, Minneapolis, Hennepin County, and the account was published in *The Oologist* in 1897 (reprinted: O. V. Jones 2017). After that there are no other records. Eight specimens from Minnesota are preserved in the Bell Museum (Green and Janssen 1975).

KING RAIL

Rallus elegans

Minnesota Seasonal Status: Summer visitant; incidental breeding 1945–76 (eight times), 2017; rare summer resident in the nineteenth century. Vagrant migrant.
Conservation Concern: In Minnesota, first classified in 1984 as *Special Concern*, reclassified as *Endangered* in 1996. Federally, it is designated by the USFWS as a *Focal Species* (T. R. Cooper 2008). Assigned a high PIF Continental Concern Score of 15/20.

The King Rail's major distribution is on the Atlantic and Gulf coasts with scattered records inland to the eastern Great Plains. Across its range, the King Rail's breeding habitat includes many wetland types. In coastal areas, the species is found in brackish marsh, marsh–swamp, and domestic rice fields. In inland freshwater systems, King Rails accept a variety of wet prairie marshes, sloughs, and small potholes, which formerly dominated the prairie landscape. The edge between open water and vegetation, created by variable hydrology and climate, appears to create favorable nesting conditions (Pickens and Meanley 2020). Rails are notoriously secretive and are heard more often than seen; most species nest in inaccessible dense wetlands.

MINNESOTA BREEDING DISTRIBUTION

Roberts (1932) called the King Rail a summer resident, common in the southern part of the state. He quoted an observer, A. Hewitt, who said "that in 1898 he discovered nine nests in Faribault County and that a pair was to be found nesting in almost every prairie slough." By the 1920s, it was essentially extirpated from most of the prairie and was a very rare summer resident. Janssen (1987) called the King Rail a casual migrant and summer visitant.

According to all records from Roberts (1932), Janssen (1987), and the MOU (2021b), the King Rail range was entirely in the southern half of Minnesota. There are only a few reliable records of visitants or vagrants in northern counties: Marshall (1968), Cass (1936), and Duluth, St. Louis County (Reynolds 1974; Eckert 1984a). The bird observed in 1984 called constantly for eight days from cattails along Mud Lake, a backwater of the St. Louis River. The confirmed county breeding records from 1945 to 1976 are clustered in these regions: Twin Cities (Anoka, Hennepin, and Wright Counties), southeast (Houston County), and southwest (Faribault, Jackson, Martin, and Watonwan Counties). In these areas, the last confirmed breeding was from LaCrescent, Houston County, in 1976 (Janssen 1987).

The MOU observations in *The Loon* (MOU 2021b) of all migrants, summer visitants, vagrants, and resident breeding records delineate a pattern of decline after the 1980s. An examination of all records by decade demonstrates the downward trend: 1960s (ten), 1970s (nine), 1980s (two), 1990s (three), 2000s (zero), and 2010s (two). The three records from the early 1990s are: spring migrant (1990, Dakota County), fall migrant (1991, Swift County), and summer visitant (1992, Hennepin County [Eckert 1992; Risen 1992; Neubeck 1992]). After this decade, there are no records for over twenty years, until 2014, when a spring migrant was found in Goodhue County on May 27, 2014 (Tustison 2014a, 2014b).

This gap in records coincides with the accelerated landscape change that started in the 1960s, when intensive row crop agriculture of corn and soybeans displaced pasture and prairie in the Minnesota River Prairie Subsection of the Prairie Parkland Province. The Minnesota southeastern prairie had already undergone drastic habitat change between pre-settlement times (70–95 percent prairie, 20–25 percent wetland) and the late-twentieth-century agricultural landscape (82 percent row crops, 9 percent pasture, 3 percent wetland) (MNDNR 2006c).

After a forty-year gap, a breeding record for the King Rail was obtained in 2017 from Grant County in the Minnesota River Prairie Subsection. The record was on the North Ottawa Impoundment of the Bois de Sioux Watershed, a 3-square-mile area in a 74-square-mile watershed, finished in 2016. The management objectives were flood control and improved water quality, along with farming and wildlife habitat. In 2017, the King Rail (one or two birds) was seen by about 130 observers from June 15 to August 19 (North Ottawa Township, T129N, R44W). On August 12, 2017, a juvenile was spotted crossing the road in the same area where the adults had been seen the day before. It was estimated to be two-thirds or three-quarters grown by Kim Eckert. Documentation of these records included Tustison (2017), Chu (2018 [date of observation in this article is an error; it should be 2017]), Kessen and Svingen (2018), and Budde, Kieser, and colleagues (2018).

With a warming climate, NAS (2019) predicted a modest increase in the potential breeding range of the King Rail in Minnesota. Primary increases were projected along the Mississippi River corridor under all three scenarios they examined and into southern and east-central Minnesota with a 3°C warming scenario by 2080.

WHOOPING CRANE

Grus americana

Minnesota Seasonal Status: Former summer resident, nineteenth century. Accidental migrant.
Conservation Concern: Officially designated as *Endangered* in 1970 under the US Endangered Species Act, and in 1978 under Canadian federal laws. Listed as endangered on the IUCN Red List (2021). The first International Recovery Plan was approved in 1980 and the Third Revision in 2007 (CWS and USFWS 2007). Assigned a PIF Continental Concern Score of 16/20 and designated a Yellow Watch List species.

The Whooping Crane is the tallest bird and one of the most critically endangered species in North America. Its name is derived from its bugling call. Its historic breeding range and migration range extended from the Canadian prairie provinces south through the north-central Great Plains and included a nonmigratory population in southwestern Louisiana. Never common, by the end of the nineteenth century the entire population was estimated at only a few hundred cranes; by 1941 only a single migratory flock remained (15 birds counted), and the Louisiana population was extirpated (Urbanek and Lewis 2020). Unregulated shooting and habitat destruction during the nineteenth century were the primary causes of decline. Over the decades since then, the only wild self-sustaining population, which breeds at the Wood Buffalo National Park in Canada and winters at Aransas NWR in Gulf coastal Texas, has increased to more than 300 birds. PIF (2021) estimated the global wild population as 370 individuals. There are also several captive and experimental populations numbering up to 200 birds.

These increases have been assisted through multiple efforts by government agencies like the United States–Canadian Whooping Crane Recovery Team, which is responsible for monitoring and management. Additionally, other public and private organizations have programs for captive breeding, experimental reintroductions, and public outreach (Urbanek and Lewis 2020). These include the Rocky Mountain Experiment (failed), Florida Experimental Nonessential Population (introductions stopped in 2008), Eastern Migratory Population Experiment (aka "Operation Migration," discontinued in 2015), the USGS captive breeding program at Patuxent (funding stopped in 2018), and the Louisiana repopulation program (started in 2011). The professional North American Crane Working Group, dedicated to the conservation of cranes and their habitats in North America, was founded in 1988.

MINNESOTA BREEDING DISTRIBUTION

In Minnesota, there is little evidence about the Whooping Crane in the nineteenth century. Roberts (1936) concluded that it was a former "summer resident throughout the prairie regions." The available evidence was sparse, but it appeared that the Whooping Crane was rare except perhaps during peak migration. Roberts gave dates for spring migration from March 20 to April 23. One nest record from the prairie near Elbow Lake, Grant County, on May 21, 1876, was documented in a taxidermist's notebook published in *The Auk* (Deane 1923). Of the six specimens from that era (Green and Janssen 1975), four have locations and dates in the Bell Museum files: three are from Lac qui Parle County—no. 3813 (April 21, 1874), no. 3814 (April 10, 1893), no. 9812 (April 10, 1895)—and one is from Meeker County—no. 8482 (September 17, 1878). One intriguing bit of evidence for Minnesota breeding is from the diary notes of a nineteenth-century taxidermist about the "great marsh beyond Thief Lake," where he saw a pair with eggs on June 19, 1889 (in Nute 1945). The last record known to Roberts was a pair on April 23, 1917, shot near Badger, Roseau County; he stated that one of these birds is "now in the University Museum collection."

In the decades after the shooting of this unfortunate pair, there have been just three credible records up until the year 2000: November 7, 1951, Rice Lake NWR, Aitkin County (Dundas 1968); October 11, 1990, Gatzke, Marshall County; and October 21–28, 1990, Burnham Creek WMA, Polk County (Hedemark and Svedarsky 1990). Being careful to consider the wildness of each bird, there have been three acceptable records after the year 2000. The first was on April 9, 2007, at the Pembina Trail Preserve, Polk County, when a small migrating flock of seven Whooping Cranes flew directly overhead from two observers at about 9:15 a.m. (Hier 2007; Svingen 2007b, 2008). In 2010 there were two records, both in Marshall County: May 21, 2010, Agassiz NWR (Knutsen 2010; Tustison 2011a); and May 30, 2010, five miles west of Newfolden (Tustison 2011a, 2011b).

Since about 2005, Whooping Cranes have increasingly been seen in Minnesota from the experimental re-introduction projects at the International Crane Foundation in Wisconsin. The first one reported in the MOU sightings database was in 2005; then other records came from six more counties through 2014. Many more observations were reported from 2018 to 2021 from nine mostly eastern counties. Since the birds have not yet been reproducing successfully in the wild, they are not from a sustainable population and are not yet a part of Minnesota's avifauna.

Unfortunately, with a warming climate, NAS (2019) predicted high vulnerability for the Whooping Crane under all three scenarios they examined, including the 1.5°C warming scenario considered imminent by climate scientists. Therefore, based on climate, there appears to be a low probability of a breeding population being established in Minnesota's future.

BLACK-NECKED STILT

Himantopus mexicanus

Minnesota Seasonal Status: Summer visitant, two nest records (2014, 2022). Migrant, recent spring visitant (2000–18), fall vagrant.

Conservation Concern: Populations appear stable, increasing in the United States in the last forty years at the edges of its range. Hawaiian subspecies is both federally and state listed as *Endangered*. Assigned a low PIF Continental Concern Score of 8/20.

The Black-necked Stilt has a wide geographic range in the Western Hemisphere. It breeds in North America, the West Indies, Central and South America, and has one population in the Hawaiian Islands. In the United States, breeding is primarily in the interior west and along the Gulf and Atlantic coasts. Often nesting in close association with American Avocets, the species is found in scattered areas with favorable habitat like salt and sewage ponds, shallow wetlands, or flooded fields. Its US western range is from Montana, the Dakotas (sparse), south to New Mexico, and east to the lower Mississippi River, and then eastward along coasts to Florida and north to New Jersey. It wanders far beyond breeding sites and is sometimes transported by storms (Robinson et al. 2020). Northern populations are migratory.

MINNESOTA BREEDING DISTRIBUTION

In Minnesota, the Black-necked Stilt was included in Roberts (1932, 1936) because there were old records in the adjacent states of Iowa and North Dakota. His status summary for Minnesota was, "so far as is known it has not been reported from this state," and he went on to dismiss its presence on P. L. Hatch's list (1892) as "enthusiastic and imaginative" writing. Subsequent books about Minnesota birds did not list this species (Green and Janssen 1975; Janssen 1987). The first documented records for the state were in 1989 in two counties: first, an individual observed at Roseau River WMA, Roseau County, on April 23, 1989 (Provost 1989), and April 24–25, 1989 (S. Wood 1989); and second, a pair observed at several locations south of the town of Alberta, Stevens County, July 12–19, 1989 (Buer and Buer 1989; Winkelman 1989). In 1992 there was a record in Hennepin County (Krych 1992), but no other observations were reported until the decades of 2000 and 2010.

The decade of 2000 began with a record on May 24, 2000, in Martin County and another on May 25, 2003, in Wright County. The next year (2004) saw a dramatic increase in records: 1–3 birds in four different counties from April 14 to May 18; Svingen (2005a) predicted nesting "in the near future." In four following years (2005, 2007, 2008, 2010) during spring migration (April and May), there were incidental observations of single birds in four other counties. Skipping a year, the next three years (2012–14) produced spring observations in three counties per year. These records are all from the yearly compilations in the MOU Seasonal Reports (MOU 2021b). Excepting Roseau and Marshall Counties, all observations of Stilts have been in the southern half of Minnesota, scattered in 20 counties.

Finally, in 2014, there was a report of nesting attempted at the wastewater treatment ponds in Herman, Grant County. A pair was first observed on May 24, a nest found on May 27, and then a complete clutch of four eggs was recorded on June 1 (Hertzel 2015). Unfortunately, heavy rains on June 15 appeared to have washed out the nest and no birds were seen on June 16 (MOU-net posting). With this breeding evidence, Minnesota now joined adjacent states that have incidental first nesting records (Hertzel 2015): North Dakota (1993), South Dakota (1994), Wisconsin (1999), and Iowa (2013). These records do not represent a sustained population, although subsequent nesting evidence is occasionally reported.

Additional records from 2015 to 2018 were mostly during the spring migration period (mid-April to late May), with some summer visitants (June through August), and two late fall observations (October 16–18, 2016, Blue Earth County, MOU 2021b; October 14, 2018, Grant County, eBird confirmed). In July 2016, two family groups were found in flooded weedy fields along a township road south of Fish Lake in Jackson County (July 21–28); they may have nested locally. In 2017, observations indicated persistent birds in Grant County at the North Ottawa Impoundment. Two were present from June 15 to August 17 (Tustison 2017), and the behavior of this pair provided evidence of breeding: "Copulation was observed and they defended suspected nest area from intrusion by Great Egret and Ring-billed Gulls, but no eggs or young were ever seen" (Kessen and Svingen 2018). In 2019 and 2020, 10 new records were submitted to the MOU Records Committee; eight of these have been accepted by the Records Committee (MOURC 2022).

During the next two years, there was a significant increase in the number of sightings; 32 records were submitted to the MOURC in 2021 and 19 in 2022. Records in the MOU sightings database for 2022 included six observations of a female Black-necked Stilt at a wetland in northern Big Stone County from June 26 through July 31. Then, on August 3 and August 5, observers found and photographed two adults and two young at the same location. The record was accepted by the MOURC (Tustison 2022) and becomes the first successful nesting confirmed in Minnesota.

NAS (2019) does not include a current breeding population in Minnesota, but all three of their climate models predict a stable breeding range in North America. In addition, all three climate warming scenarios they examined include breeding range gains in the Dakotas, and the 3°C scenario predicts range gains in western Minnesota by 2080.

SNOWY PLOVER

Charadrius nivosus

Minnesota Seasonal Status: Summer vagrant, one breeding record. Vagrant spring migrant.
Conservation Concern: The Pacific coast population of the Snowy Plover was listed under the Endangered Species Act as *Threatened* in 1993; identified critical habitat was published by rule in 2012 (USFWS 2012). Its populations had declined from habitat degradation and disturbance and its range had contracted southward from Washington and Oregon. Remaining populations are considered a Bird of Conservation Concern by the USFWS (2008). PIF has also assigned it a high Continental Concern Score of 15/20 and designated it a Yellow Watch List species.

The distribution of the Snowy Plover is primarily on the Pacific and Gulf coasts and inland in the southern Great Plains as far north as Nebraska. It nests on the ground in a variety of bare habitat types: sandy beaches on the coasts and barren edges of inland saline lakes or rivers (Page et al. 2020). It also uses man-made habitats, such as wastewater ponds, reservoirs, dredge spoils, and salt evaporation ponds. All these habitats may be ephemeral, depending on weather and management regimes. The species winters on the Gulf and Pacific coasts in the United States and Mexico. There are year-round populations in the Bahamas and the Caribbean islands. Vagrants occur north to the southern prairie provinces and east to the Mississippi River drainage to Ohio (Sibley 2014).

The Snowy Plover is one of the rarest shorebirds in North America, with an estimated population of 24,000 birds in the United States and Canada, and an additional 7,000 birds in Mexico (PIF 2021). Two areas in the United States that have the highest numbers of birds, each with more than 5,000 individuals, are the Great Salt Lake in Utah and the Salt Plains NWR in Oklahoma. Populations in other states and areas are much smaller. For all of South Dakota, North Dakota, and Saskatchewan, the population was given as about 30 birds (Thomas et al. 2012). The Plover has bred in both Dakotas, where it is now categorized as Rare/Accidental on state checklists. In North Dakota it was documented as breeding in 2007 at the Lost Lake NWR, a small, 960-acre refuge near the Missouri River (Aron et al. 2008). The first modern breeding record in South Dakota was also in 2007, when successfully nesting birds were found at Stone Lake, Sully County, east of the Missouri River (R. D. Olson 2008). The expansion of this Plover's range into the northeastern Great Plains along the Missouri River drainage is recent (K. L. Hunt et al. 2013).

MINNESOTA BREEDING DISTRIBUTION

Although Roberts (1932) briefly cited one record in 1929 in Goodhue County, the details were determined to be too sketchy to be credible. In Minnesota, the first acceptable record was not until 1976, when a Plover was seen at the sewage lagoons in Marshall, Lyon County (Kyllingstad 1976). There have been 10 acceptable records since then in both the spring and summer months. The five spring migration records range from April 24 to May 30 and include records from two western counties: Agassiz NWR, Marshall County, in 1987 (J. P. Mattsson 1987) and 2009 (Svingen 2009b); and Big Stone NWR, Lac qui Parle County, in 1981 (Berber 1981; Eckert 1982b). There is another spring migrant record at the Albany sewage plant in Stearns County, June 4–5, 2004 (Svingen 2004), and a record along Lake Superior in Duluth, May 30–31, 2007 (Zierman 2007; Svingen 2007b).

The summer observations are all in northwestern counties of Minnesota. Two records are from Lake of the Woods County: one during July 11–18, 1983, at Pine/Curry Island (Wiens 1982), and the other near the lake at Zippel Bay (Hiemenz 1983). Other summer records are near the border with the Dakotas: Polk County, August 16, 1997 (Hertzel 1998); Clay County, June 30, 1986 (Eckert 1986a); and Grant County, June 8–9, 2017, North Ottawa Impoundment (Kessen and Svingen 2018). There is one breeding record in 2006 from Big Stone NWR, Lac qui Parle County. The birds were first found on July 1; then on July 8 the adults were seen with two downy chicks; the birds remained in the refuge through August 27 (Chu 2007; Svingen 2007a).

Although NAS (2019) did not include Minnesota in the Plover's current breeding range, all three of the climate scenarios they examined included expansions northward in the United States, but not as far north as Minnesota. Given these predictions, there is an increasing probability that in the future the Snowy Plover could colonize even more areas in Minnesota.

LONG-BILLED CURLEW

Numenius americanus

Minnesota Seasonal Status: A former (presumed nineteenth-century) summer resident. Vagrant migrant.
Conservation Concern: In the United States, it is not federally listed in the Endangered Species Act but is listed in several states as a *Species of Concern*. In Canada, it was listed in 2004 as a species of *Special Concern* under the Species at Risk Act. Assigned a moderate PIF Continental Concern Score of 12/20.

The Long-billed Curlew is a Great Plains resident, nesting west of the Missouri River from the southern Canadian provinces to Nevada and New Mexico. It winters on the southern Atlantic coast, sparingly on the US Gulf coast east of Louisiana, and in southern California. The major part of its winter range is in northern Mexico, along both coasts to Central America (Dugger and Dugger 2020). As a grassland-endemic species, it uses a variety of foraging and nesting habitats in diverse landscapes of the northern plains. A map in the USFWS status assessment of 2009 for this species shows that its historic breeding range included southeastern Wisconsin, Illinois, and western Indiana (Fellows and Jones 2009, Figure 1.2). Rampant market hunting in the nineteenth century is mentioned as a cause of its demise in the Midwest. The 2009 status assessment concludes: "Long-billed Curlew populations are lower than historically, and their range continues to contract. We believe that high levels of concern . . . are warranted, particularly in the shortgrass and mixed-grass prairies of the western Great Plains."

MINNESOTA BREEDING DISTRIBUTION

In *The Birds of Minnesota,* Roberts (1932) stated that "according to early accounts it was formerly a summer resident, breeding . . . south and west of the heavy timber." Roberts reported no Minnesota records of nests or dependent young and he had only weak second-hand evidence of breeding. The strongest evidence was from May 24–30, 1883: "pair near Loon Lake in southern Jackson County whose actions indicated that they had a nest near by," as recounted by J. W. Preston, and "female taken by Albert Lano on April 24, 1891, in Lac qui Parle County, contained an egg ready for the shell." He also had information in 1893 from Dr. W. C. Portman of Jackson County that "it had been very abundant on the high prairie about Jackson and that he had found the nest and eggs there." All the other records from *The Birds of Minnesota* are during migration, both spring and fall, with the last one dated in 1897. In Wisconsin, S. D. Robbins (1991) stated that in the early settlement period of the mid-nineteenth century it was a common breeding species in the southern counties; the last date for nesting was 1859.

The second edition of *The Birds of Minnesota* (Roberts 1936) adds migration records of small flocks, four or five birds, from McLeod County on May 14, 1934, and Meeker County on May 26, 1934, from information provided by J. P. Jensen. A letter from Jensen in 1936 to Roberts gives details from a trip on July 5, 1920, about nesting evidence south of Ada, Norman County, where "an area of considerable size of primitive prairie had been left here unplowed." He related seeing two pairs: one was in a "courtship act" and the other "acted as if there were young near by" (archived Bell Museum files). Other records from the 1930s are in the MOU files: October 28, 1936, Stearns County, and April 24, 1937, Douglas County.

No other records occurred for twenty years, until May 24, 1958, when a Long-billed Curlew was seen on Minnesota Point in Duluth (R. L. Huber 1959a). Since then, starting in the 1960s, there have been three to five vagrant records per decade through the 2000s. Twenty observations in this period were of migrants, 90 percent in the spring. With four exceptions, they all occurred south of Wilkin, Stearns, and Anoka Counties. In addition to the Duluth record, there are two northern records in Aitkin County: April 20, 1972 (Blanich 1972) and April 16–18, 2005 (Svingen 2005b). The fourth northern record is from Ely, St. Louis County, on May 21, 1987 (Schon 1988). Of the two summer vagrant records since the 1960s, one is from July 19 to 28, 2002, in Anoka County (Eckert 2003), and the other was on June 26, 1989, when a bird was seen flying at Blue Mounds State Park, Rock County (Hertzel 1990). As of 2022, the last documented record of a Long-billed Curlew in Minnesota was May 17, 2008, in Big Stone County (MOU 2021a).

Climate models examined by NAS (2019) did not include a current breeding range in Minnesota for the Curlew, but under the 1.5°C and 2°C warming scenarios, breeding range gains were predicted in northwestern Minnesota.

LITTLE GULL

Hydrocoloeus minutus

Minnesota Seasonal Status: Summer vagrant, one nest record (1986). Regular migrant in spring and fall.

Conservation Concern: Assigned a moderate PIF Continental Concern Score of 13/20.

The Little Gull is one of two species of small gulls that have found their way across the Atlantic Ocean from the Old World and become migrants and incidental breeders in the New World; the other species is the Black-headed Gull, abundant in Europe, extremely rare in America, and accidental in Minnesota. The Little Gull, the smallest gull in the world, is common in Europe and ranges throughout the Palearctic. The first documented occurrence in North America was about 1819 on the Saskatchewan River in the Canadian Arctic as recorded by Franklin's expedition (1819–22) to the Coppermine River (Ewins and Weseloh 2020). The North American breeding range is mainly in wetlands in the Great Lakes–St. Lawrence River drainage and in the lowlands of Hudson and James Bay in Manitoba and Ontario.

The first North American breeding record was on Lake Ontario in 1962 (Scott 1963); subsequently there have been 67 records with breeding evidence from 1962 to 1989, all in the Great Lakes drainage or closely adjacent, but none thereafter (Ewins and Weseloh 2020). Dates of confirmed nesting in the Upper Great Lakes region are Michigan, six (1976–80); Wisconsin, seven (1972–80); Manitoba, one (1981); and Minnesota, one (1986). These attempts have had low breeding success. In winter, Little Gulls occur in the lower Great Lakes and on the Atlantic coast from the Maritime Provinces to North Carolina, including offshore waters; vagrants have been reported in Texas as well as on the Pacific coast. Migration and dispersal from North American breeding areas have not been studied and can only be inferred from post-nesting migration, staging, and wintering sites. Flocks on the North American Atlantic coast may have Eurasian origins.

MINNESOTA BREEDING DISTRIBUTION

The first confirmed Minnesota record of the Little Gull was on May 27, 1973, when three adults and three immatures were observed in a large multispecies flock of gulls and terns (about 1,600 birds of 11 species) resting during stormy weather at the Sky Harbor Airport runway, Minnesota Point, Duluth, and continuing through May 31 (Green 1974). The next year, on June 17, 1974, a smaller mixed flock in the Duluth-Superior Harbor contained five immatures (Green 1974). An earlier published first record (1972) was considered not acceptable by the MOU Records Committee. For the next ten years, Little Gulls were seen almost yearly, mostly in spring at Duluth when flocks of Bonaparte's Gulls migrate.

In 1984 and 1985 single Little Gulls were spotted near Heron Lake, Jackson County. At Heron Lake the next year during May and June, a pair of Little Gulls was found in the Franklin's Gull colony; after a search, a nest with four eggs was located on June 2, 1986, but they were predated before they hatched (Schladweiler 1986).

After the first records in 1973 and 1974, Little Gulls continued to be very rare yearly migrants. The MOU Seasonal Reports document their migration (MOU 2021b). Most records are one or two birds, predominantly adults, and are more frequent in the spring (66 percent) than in the fall. The bulk of the spring migration is early to late May with extreme dates of April 26, 2008, to June 17, 1974. Most of the fall migration is mid-September through mid-October with extreme early dates of August 21, 2000 (Budde 2001), and August 24, 1987, at Lower Red Lake and late migrants on November 8, 1977, and November 4–14, 1995, both at Mille Lacs Lake.

The geographical pattern of spring versus fall migration is striking. In the spring, two-thirds of the records are from the Duluth-Superior Harbor, where Bonaparte's Gulls are a common migrant. In the fall, many of the records are from Minnesota's big inland lakes, where Bonaparte's Gulls congregate. These lakes, from northwest to southeast, are Lake of the Woods, Upper and Lower Red Lake, Lake Winnibigoshish, and Mille Lacs Lake. Some of these fall migrants are in juvenile plumage, as documented in an article about the Little Gull in *The Loon* (Svingen 2000); other records in the Seasonal Report from 2000 to 2006 were also identified as juveniles. Two summer records may be related to regional breeding activity: August 14–19, 1984, in Worthington, Nobles County (near Heron Lake in Minnesota and Spirit Lake, Iowa), and July 23–24, 1993, one adult photographed with Bonaparte's Gulls on Rainy Lake, Koochiching County (south of breeding sites on the Canadian Shield in Manitoba and Ontario).

NAS (2019) did not include a current breeding range in Minnesota for the Little Gull, but under the imminent 1.5°C and the 2°C warming scenarios breeding range gains were predicted in the western arm of Lake Superior.

WHITE-FACED IBIS

Plegadis chihi

Minnesota Seasonal Status: Summer visitant, bred in the 1890s. Regular migrant, uncommon in spring and rare in fall.
Conservation Concern: Assigned a PIF Continental Concern Score of 7/20. Great Basin population given a Species of Management Concern status by the USFWS.

The White-faced Ibis is one of three *Plegadis* Ibises in the New World: the Glossy Ibis is found in the Old World and the US Atlantic-Caribbean area; the Puna Ibis in South America; and the White-faced Ibis in the United States, Mexico, and southern South America. The White-faced Ibis distribution in the west is from California, Idaho, and Montana east to the central Great Plains and southern Texas, and Louisiana. Great Basin colony sizes range from hundreds to thousands of pairs (Ryder and Manry 2020). Northern populations migrate to the Gulf coast. Glossy Ibis and White-faced Ibis are closely related and difficult to tell apart; however, the Glossy Ibis is rare in Minnesota.

The White-faced Ibis nests in dense cattails or bulrushes in freshwater marshes or in trees or shrubs over water. Colony sites often shift from year to year, depending on water levels. The birds are nomadic and wander widely to take advantage of recent water-level changes and food availability (Kaufman 1996). This wandering adaptation means that populations can respond to changes in weather cycles, like drought. These movements make population numbers difficult to estimate; however, trends in the twentieth century include declines in the 1960s and 1970s from DDT pesticide contamination and then population recovery (Great Basin Bird Observatory 2010).

The breeding populations in the eastern Great Plains represent a range expansion that began in the latter half of the twentieth century. This is not a re-population of the historic nineteenth-century breeding range, as is commonly thought. Actual breeding in the 1800s is documented by only three breeding records (two in Minnesota), and all "available information suggests that the species was scarce or absent in the region until the middle twentieth century" (Jorgensen and Dinsmore 2005). The first modern record in Iowa was in 1977, and the first breeding record in 1986 (Kent and Dinsmore 1996); in South Dakota the first record was in 1962, and the first nesting in 1978 (Tallman et al. 2002); and in North Dakota the first nesting was in 1978 (Drilling 2015).

MINNESOTA BREEDING DISTRIBUTION

In Minnesota, the nineteenth-century breeding records were from Heron Lake, Jackson County, where eggs were collected in 1894 and again on June 22 and July 2, 1895. This was a small group of four nesting pairs in a colony of Black-crowned Night Herons, where "two fine male Ibises were taken" (Peabody 1896). Roberts (1932) mentioned this record and gave other citations for where it was documented (Peabody 1895a, 1895b). He listed other old records from Meeker County (one shot, October 1, 1891), Heron Lake (two shot, November 1893), and Kandiyohi County (two shot, October 12, 1907). Another potential specimen was located by Hertzel (2006) at the Redwood County Historical Society building in Redwood Falls. It was dated around 1900 in Three Lakes Township; however, because of the poor condition of the specimen, a definitive identification as a White-faced Ibis could not be made.

After that time, there were no other records until the mid-twentieth century. The first was on October 6, 1956, from Weaver on the Mississippi River (Longley 1956a). (Longley used "Glossy Ibis" in the title because the species was called "White-faced Glossy Ibis" at that time.) Three records (Green and Janssen 1975) were obtained from the 1960s: in May 1962 at Frontenac on the Mississippi River, Goodhue County, and in Lac qui Parle County (R. L. Huber 1962a); and in May 1967 near Okabena, Jackson County (Janssen 1967).

A regular yearly seasonal compilation of records was started by the MOU in 1970 (MOU 2021b). The results show a gradual increase in the species' numbers: 1970s (four credible records); 1980s (six confirmed records); and 1990s (eight confirmed records). The proportions of these records are spring, 58 percent; summer, 22 percent; and fall, 20 percent. Since the year 2000, MOU records have become more numerous, with annual observations of one to five birds becoming more common. In the 2010s, migration records dramatically increased, with birds sometimes seen in flocks of 20 to 30 birds of *Plegadis* spp. The summer records seem to be straggling migrants; there is no indication of breeding. All these records are in 55 counties in the unforested ecoregions; one exception is a vagrant at Duluth, May 3, 2015 (Janssen 2019). Because the Glossy Ibis is a very rare bird with only six state records (1991–2017), these Ibises are presumed to be predominantly White-faced Ibis.

Even though this Ibis was not included as a current breeding species in Minnesota by NAS (2019), all three of their climate models predict the possibility of small breeding range gains in northwestern Minnesota. Breeding range gains in the Dakotas, especially prominent under all three scenarios, suggest a continuation of future observations in Minnesota.

SWALLOW-TAILED KITE

Elanoides forficatus

Minnesota Seasonal Status: Former summer resident in the nineteenth century, last breeding in 1904. Vagrant in spring, summer, and fall.
Conservation Concern: No current federal listing; designated as *Endangered* in South Carolina and *Threatened* in Texas. The Swallow-tailed Kite Conservation Alliance (PIF 2021) estimated a global population of 260,000 breeding adults and a United States population of 7,500, which is about 3 percent of the global population. Assigned a moderate PIF Continental Concern Score of 12/20.

The Swallow-tailed Kite was first described by Mark Catesby in 1731 and was given its scientific name by Carl Linnaeus in 1758. Two subspecies of the Swallow-tailed Kite live in the Western Hemisphere. The North American one, *Elanoides forficatus forficatus*, breeds in coastal states from South Carolina to Texas and in Central America. It winters in South America in the range of the South American subspecies, *E. f. yetapa*, which breeds as far south as the Pantanal in Paraguay and Brazil.

This Kite's striking black-and-white plumage and its forked tail are distinctive, as is its manner of gliding flight, by which it soars effortlessly for miles, feeding on the wing. The Swallow-tailed Kite is a social raptor: the birds breed in clusters, are gregarious while feeding and during migration, and congregate in large pre-migration roosts of hundreds to thousands of birds (Strand 2016). In late summer, the birds head for their wintering areas in Brazil from the southern tip of Florida, starting a 16,000 km round-trip journey that crosses the Caribbean (mostly via Cuba) to the Yucatán and south to Central and South America (Avian Research and Conservation Institute 2018).

Historically, the US distribution of the Swallow-tailed Kite was along the Atlantic and Gulf coasts from the Carolinas to Texas and up the Mississippi valley to far northwestern Minnesota. This range encompassed eighteen states and included the lower Ohio River valley (K. D. Meyer 2020) and

Minnesota's adjacent states, excepting North Dakota. Now the breeding range is only on the Florida peninsula, with a few habitat enclaves from South Carolina to Texas. The Kite is a vagrant species in the eastern part of the country northward along the East Coast to Massachusetts and northwest to Minnesota at the edge of the Mississippi River drainage.

These changes represented "the disappearance of the Swallow-tailed Kite from three-fourths of its US breeding range between 1880 and 1910," as reported by the South Carolina Department of Natural Resources (2015). The authors went on to state that it "was one of the most dramatic range contractions of any bird before the post-World War II decline of the peregrine falcon." Habitat change as the continent was settled is an obvious cause, especially landscape-level destruction by drainage and logging (R. Brown et al. 1997), probably combined with the Kite's unique social breeding biology and demography. In Minnesota as elsewhere, human persecution also probably played a part (Kessen and Hertzel 1999) through unregulated shooting of raptors at migration hotspots through the 1930s (Bronoel 1954b).

Today, many Minnesotans are acquainted with the Swallow-tailed Kite only through the beautiful diorama at the Bell Museum created by the great wildlife artist Francis Lee Jaques in 1946. His dramatic painting depicts several Kites perched and flying over a high bluff above the Mississippi River valley (see page 587).

MINNESOTA BREEDING DISTRIBUTION

In Minnesota, the details of the Swallow-tailed Kite's distribution in the nineteenth century are better known than in other states along the Mississippi River drainage because of the sightings that Thomas Sadler Roberts compiled for *The Birds of Minnesota* (1932). He was diligent in gathering written notes from a cadre of sport hunters, collectors, explorers, local naturalists, and a few professional ornithologists, as well as making his own field notes. Roberts's compilation, preserved in the University of Minnesota Library's archives, gives many nineteenth-century occurrences (1871 to 1907) in counties near the Upper Mississippi River.

Noteworthy for Minnesotans, the first records for this species were obtained by territorial agent Henry Rowe Schoolcraft as he returned from his discovery of the headwaters of the Mississippi River. The first record was on July 14, 1832, on the Mississippi River at the juncture with the Manomin Lake outlet in Beltrami County. In his exploration narrative, Schoolcraft wrote, "We also observed the *falco furcatus,* or swallow tailed hawk, a species not heretofore, thought to inhabit the continent, so far north" (Mason 1958). The second record was on July 19, 1832, as his party descended the Crow Wing Chain of Lakes from Leech Lake. At the outlet of the eighth Crow Wing Lake his narrative continues: "Soon after entering this channel, one of my men fired at, and brought down, a fork-tailed hawk, a species which had before been noticed on the wing, but we had now an opportunity of closer scrutiny."

Roberts knew this spectacular Kite as a boy in the 1870s and covered it extensively in his monumental life's work, *The Birds of Minnesota*. His joy in observing this striking bird is captured in this quote: "This beautiful bird . . . once was a common summer resident of all the heavy woodland of Minnesota. It was not an infrequent experience in the writer's boyhood days at Minneapolis to see a half-dozen or so in company, making their way northward in graceful flight in the early spring and circling overhead as they drifted back southward in the late summer and early fall" (Roberts 1932).

In the book, he documented the nest records from the nineteenth century: three in Becker County (one in 1883, two in 1886); one in Hennepin County (1887); and one record from Stearns County of a pair feeding dependent young (1904). Specimens collected during the 1880s in Minnesota are now in museum collections: Harvard (MCZ no. 324976), American Museum of Natural History (AMNH no. 8067), and Bell Museum (MMNH no. 8719). Also, three specimens from the twentieth century are in the Bell Museum: Cass County (September 14, 1919); Fillmore County (August 18, 1949); and Washington County (April 29, 1966).

The observational records that Roberts compiled included several of his own from his 1874–1900 journals (Roberts 1991). The journal notes were detailed and included sightings of multiple birds. On July 8, 1875, on a trip to Buffalo Lake, Wright County, he stated: "we saw quite a number of Swallow-tailed Kites. . . . The Kites were continually soaring in circles over the woods"; and in late July 1875, at Lake Minnetonka, he noted: "Saw a number of Swallow-tailed Hawks" (Roberts 1991). A map of these nineteenth-century records outlines a breeding distribution in the shape of a big oval in the central part of the state, from the "Big Woods" northwestward to the headwaters of the Mississippi River. The west margin was bounded by the wooded lakes of Otter Tail and Becker Counties. The eastern margin was along the Mississippi River drainage in Mille Lacs, Aitkin, and Cass Counties, and includes these ecological subsections: Big Woods, Mille Lacs Uplands, Hardwood Hills, and Pine Moraines/Outwash Plains. During spring migration, birds entered Minnesota along the southern border in May and remained through August to early September.

By 1932, Roberts concluded there were "only a few records since 1900," and the species had "become very rare and [was] apparently approaching extinction in Minnesota." His summary of its current status was, "Formerly a summer resident." The evidence of decline in his museum files consisted of six records for the first decade of the 1900s from four counties (Jackson, Aitkin, Stearns, Clearwater); two in the 1910s (Cass 1914, "skin collected"; and Hennepin 1916 in Wayzata); and two in the 1920s (Sherburne, McLeod). The last known breeding record was in 1904 (Stearns County), although a few years later Francis Lee Jaques reported that birds "were observed for several years in succession at Aitkin, Aitkin County, about 1907" (Roberts 1932). Roberts went on to note that "an occasional pair may still make some remote place in Minnesota its summer home, but it is feared that this wonderful bird is a thing of the past here in the north."

In his reminiscences about those early years, Roberts recalled that "the swallow-tailed Kite was fairly common at that time [1874] and nested in the forest about Lake Minnetonka. During migration little parties of a half-dozen or so might be seen sweeping along in graceful flight, and never ceased to be a joy. The entire disappearance of this beautiful bird, from our state, is difficult to explain" (Roberts 1944).

There is about a twenty-five-year gap, from 1923 to 1949, with only two records: a specimen was taken from Fillmore County (Willis 1949), and a bird was seen by several people on the west shore of Mille Lacs Lake, Crow Wing County (W. Lawrence 1949). In the 1950s, only one record was described (Graham 1956); other observations mentioned in *The Flicker* are not credible. The first modern record was at Gun Club Lake, below Fort Snelling, on April 25, 1961 (Berg 1961). Then, in 1966, a dead bird was found in Washington County and became specimen MMNH no. 23152 (photo in *The Loon,* Green 1967). In the 1970s, three records occurred in places within its historical range: Cedar Creek, Anoka County (Maxson 1976), and within or near Itasca State Park (Hubbard and Clearwater Counties, respectively) (Parmelee 1977). The absence of records in the 1980s was followed by an exciting observation from May 15 to 27, 1999, at Lower Sakatah Lake, Rice County, where a Kite was observed feeding over the lake (Krienke 1999; W. Swanson 1999) and photographed (Hertzel 1999).

Since 1999, there have been six observations: August 11–15, 2002 (Scott and Dakota Counties); August

18, 2003 (Rice County); May 4, 2007 (Anoka County); August 22–28, 2008 (Watonwan County); August 16, 2020 (Pine County); and September 6, 2020 (Carlton County). The vagrancy of the Swallow-tailed Kite is now widespread throughout the eastern United States (Sibley 2014; K. D. Meyer 2020; eBird 2020).

When the Swallow-tailed Kite was resident in Minnesota, its habitat was described as "breeding throughout the heavily timbered areas of the state" from Hennepin County to Itasca State Park and occasionally farther north (Roberts 1932). This is part of the Eastern Broadleaf Forest Province, starting with the "Big Woods" and containing many lakes, both big and small, and the big river system of the Mississippi. The landscape's forested habitat, with openings of lakes, swamps, marshes, and the prairie parklands, is well suited for aerial foragers that require big trees for nesting (K. D. Meyer 2020). Interestingly, there are no records of this Kite as a resident in the Mississippi River bottomlands southeast of the Twin Cities. Perhaps the ecological landscape is not appropriate, or they occurred in lower densities and were simply missed, but the latter possibility seems unlikely because they are so easily identified.

Regarding climate change, a northward expansion of the Kite is not expected. Even with the warmest 3°C climate scenario analyzed by NAS (2019), a minimally projected breeding range gain northward would be limited to the southeastern region of the United States by 2080. With the exception of visits by sporadic vagrants, the Bell Museum diorama will remain a testament to the species' former status in the state.

BARN OWL

Tyto alba

Minnesota Seasonal Status: Sporadic permanent resident; confirmed breeding in seven years from 1891 to 2021; visitant in all four seasons.
Conservation Concern: Globally secure. Assigned a moderate PIF Continental Concern Score of 9/20.

The Barn Owl is truly a cosmopolitan species, with populations on all continents except Antarctica, and is now considered the most widespread landbird in the world. The Latin name literally means "white owl," a name given when this owl was first scientifically described in 1769. However, because of its open rural habitat and tolerance of human settlements, where it was welcomed as a rodent catcher, this owl has had a much longer folk history. Taxonomists

Barn Owl.

have established three major lineages—one in western Eurasia and Africa, one in southeast Asia and Australasia, and one in the Americas—and created 28 subspecies. The Barn Owl occurs in temperate and tropical zones but is not found in the boreal regions like Scandinavia or the Canadian Taiga.

The Barn Owl's global success is the result of two attributes: first, its remarkable hunting skills, including exceptionally keen hearing, strong flight, and nocturnal navigation; and second, its use of a wide variety of habitats and nest sites. Open country habitats include farmland, grasslands, deserts, and marshlands. Nest sites can be in various cavities such as cliffs, caves, riverbanks, trees, or nest boxes, and in structures such as barn lofts, attics, and church steeples (Marti et al. 2020).

In North America, the Barn Owl is a permanent resident where it occurs. Its range is predominantly south of the Canadian provinces and excludes parts of the northern border states. Breeding locally throughout the prairies and Midwest, birds in the region may migrate south during the winter because of their low tolerance of cold weather. The account of the first nesting in the Prairie Provinces near Winnipeg, Manitoba, in 1994 emphasizes why: "Low fat reserves, a slightly lower metabolic rate than for other owls, and feathers with a lesser insulative value are some reasons why Barn Owls are poorly adapted to cold climates" (Nero 1995).

Regardless, northern states adjacent to Minnesota all have scattered records of nesting throughout the twentieth century. In Wisconsin, where its status is "rare summer and winter resident," there are about 60 records of breeding from 1899 to 1999 (Cutright et al. 2006). In Iowa, its status is "rare resident," with 16 nesting records from 1960 to 1995 (Kent and Dinsmore 1996). In contrast, in South Dakota, it is considered a "rare migrant and summer resident," with nest records from 1931 to 1990 (Tallman et al. 2002), and in North Dakota it is described as "rare, irregular and local" (Stewart 1975).

MINNESOTA BREEDING DISTRIBUTION

In Minnesota, Roberts (1932) gave the Barn Owl status as "Occurs irregularly and infrequently throughout the year, chiefly in the southern part of the state." This statement also described its occurrence throughout the twentieth century; both Green and Janssen (1975) and Janssen (1987) gave it a casual resident status. The MOU changed the Barn Owl status from accidental to casual in 2019 (Tustison 2019). Classified as a permanent resident because there are continuous yearly records, the Barn Owl is at the northern limit of its range in Minnesota. The Bell Museum contains eight catalogued specimens, representative of all seasons, from the years 1916, 1917, 1922, 1955, 1980, 1990, 1995, and 2006. Although these Barn Owl records are from many seasons, no summer observations were published by the MOU from 1924 to 1956.

Starting in 1960, using the MOURC database of "acceptable" records and records published in *The Loon,* eight summer observations from seven counties in the southern and western ecological provinces were recorded. Two are especially noteworthy: Olmstead County, at an abandoned barn, 1973–76 (Fowler 1980); and Houston County, 2014–19, audio recordings of wild owls visiting a captive Barn Owl (Bloem et al. 2018). During the atlas project, 2009–13, there were no submitted observations; there was one record of a Barn Owl found dead in Anoka County on May 5, 2013 (Tustison 2015a). Other seasonal records after 1960 include 20 observations: ten spring, seven fall, and three winter. The winter records are all from St. Louis County: two found dead near Duluth in February 1960 (Kuyava 1960) and in January 1984 (Kuester 1984), and one along St. Louis County Highway 7 in the Sax-Zim Bog, about January 1 to 12, 2020, when it died (Eckert 2020). These winter records are the only observations from the Laurentian Mixed Forest Province in any season. It is difficult to know if this increase in records represents changes in the abundance of this species or changes in other variables, such as fewer barns, weather, climate, or observers. The Barn Owl is difficult to find; it is strictly nocturnal, nests in obscure places on private land, and can have vocalizations similar to other owls.

Only seven documented breeding records have occurred in more than a century. The nesting records can be grouped into three time clusters: first, records from Roberts (1932) in Rice County (1891) and Martin County (1923–24); second, published records in Martin County in 1963 (M. Nelson et al. 1964) and from Dakota County in 1990 (Janssen 1990a) and 1991 (S. Carlson et al. 1991b); and third, recent reports of nesting in Becker County (October 18, 2018, 2 young banded; R. Frederickson 2019), in Ottertail County (October 17, 2019, 2 downy young; MOU 2021a), and in Wabasha County (October 5–14, 2021, 5 owlets; Tustison 2021). There are two older published breeding records that lack details, are not in appropriate habitat, and are from the northern coniferous–hardwood forest. These records were "accepted" by the MOU; they are from Hubbard County, 1944 (Longley 1944), and Lutsen, Cook County, 1953 (W. Brown 1961). They should be disregarded.

Regarding climate change, the Barn Owl is considered to have a stable breeding range in North America with modest expansions northward under both a 1.5°and 2°C warming scenario by 2050 (NAS 2019).

SCISSOR-TAILED FLYCATCHER

Tyrannus forficatus

Minnesota Seasonal Status: Vagrant in spring, summer, and fall. One hybrid nesting attempt failed.
Conservation Concern: Its population is secure throughout its range. Assigned a moderate PIF Continental Score of 11/20.

The Scissor-tailed Flycatcher nests in open country in the southern Great Plains from Texas north to southern Nebraska; it winters in Central America. Its habitat includes savannas, fields, pastures, towns, and landscaped areas such as golf courses. The nest is typically placed in isolated trees; the Flycatcher feeds mostly on grasshoppers. There are vagrant records throughout the United States extending into southern Canada (Sibley 2014).

MINNESOTA BREEDING DISTRIBUTION

The vagrant records of the Scissor-tailed Flycatcher in Minnesota go back to Roberts (1932), where he cited two letters describing observations from the early twentieth century: about 1912 in spring, Kandiyohi County; and July 18, 1930, Jackson County. The quarterly magazine of the MOU, *The Flicker,* began in 1929, but it was not until 1957 that the first record of this flycatcher was published: a bird found near a beach on Lake of the Woods at Williams on July 3, 1957 (F. Kelly 1958). One other observation exists from that time: June 5, 1959, Anoka County (Lupient 1959). One specimen (MMNH no. 14562) was taken in Carlton County on October 23, 1958 (Green and Janssen 1975).

In every decade (1960–2020) since the first published record, there have been five to eight yearly records of this vagrant flycatcher in the MOURC database. Like some other flycatcher species, the Scissor-tailed Flycatcher is prone to long-distance vagrancy. It is very easy to identify and perches often in the open. The MOU Records Committee data system contains 95 records; the MOU 2020 checklist classifies this species as casual. These records, from 42 counties, are scattered in all regions of the state but especially in the eastern third (14) and central third (16) of the counties. The western third, which is closer to this flycatcher's Great Plains breeding area, has records in 12 counties. Forty percent of the 95 observations are from June and July. During the atlas years (2009–13), no records were submitted to the atlas database. However, during the atlas safe dates, there are a few records in the MOURC data system that were accepted as credible reports: Dodge County (July 31, 2009; Svingen 2010), Sibley County (June 26, 2009; Svingen 2010), Cottonwood County (August 2, 2012; Tustison 2013a), and Cook County (June 3, 2013; Tustison 2013b). The latter record is obviously a migrant. For Scissor-tailed Flycatcher vagrants, the North Shore of Lake Superior is an unsurpassed migrant trap, with 34 records from St. Louis (17), Lake (8), and Cook (9) Counties. Almost all vagrant migrant records are from May 7 to October 28; the exception is from Rice County, April 12–13, 2015 (Tustison 2015b).

An attempted nesting by a female Scissor-tailed Flycatcher with a Western Kingbird was observed from June 4 to July 17, 1997, south of Elk River, Sherburne County. The female built several nests, one of which was visited by the Western Kingbird. About five hours of observations of these birds were made by several people, with exceptional details given in the published account (Fall 1997). In 2017, one adult Scissor-tailed Flycatcher summered in Mower County from June 25 to July 19, 2017, and was seen by many birdwatchers (Kessen and Svingen 2018; Tustison 2017). One observer commented that the "bird was ranging widely but seemed to have a nest in the cell tower near I-90" (MOU 2021b).

All of the climate change scenarios examined by NAS (2019) indicated northward expansions of the Scissor-tailed Flycatcher's breeding range, but not as far as Minnesota, by 2080. However, with a warming climate, there are likely to be more observations in Minnesota's future.

WESTERN WOOD-PEWEE

Contopus sordidulus

Minnesota Seasonal Status: Incidental summer resident, mostly hybrid nesting with Eastern Wood-Pewee. Vagrant migrant.
Conservation Concern: High risk from the destruction of wintering range in tropical forests. Assigned a moderate PIF Continental Concern Score of 12/20 and identified as a Common Bird in Steep Decline.

The Western Wood-Pewee and Eastern Wood-Pewee are sibling species, both widespread and common in their respective ranges. In *Birds of the World,* the range map of the Western Wood-Pewee shows that the closest its breeding range gets to Minnesota is along the far western border of North and South Dakota; the map for the Eastern Wood-Pewee breeding range covers the eastern half of the Dakotas (Bemis and Rising 2020). The Eastern Wood-Pewee is resident throughout Minnesota. There is no specimen of the Western Wood-Pewee for Minnesota and therefore no direct genetic evidence for hybridization. The two species look alike and cannot be distinguished in the field by their plumage. The best way to make an identification is by vocalizations, which are usually diagnostic.

MINNESOTA BREEDING DISTRIBUTION

All seven records of the Western Wood Pewee that have been accepted by the MOU Records Committee were identified by voice by an observer with previous experience of the calls of both species. The first observation was on September 26, 1971, in Cottonwood County (Egeland 1972). The other observations during migration were on August 16, 1984, in Duluth (Eckert 1984b) and on August 25, 1985, in Clearwater County (Lafond 1986; Eckert 1986b).

Several summer records that have detailed published field observations are clustered along the far northwestern corner of the state. The first was in Pelan Pioneer Park, Roseau County, and involved a nesting pair observed May 29 to July 8, 1977; the male was identified by voice as a Western Wood-Pewee (Paulson 1977). The nest was successful and three young fledged (Eckert 1978a). At that same site on June 10 and June 25, 1978, another nest was found, and a probable identification was given as a Western Wood-Pewee (breeding was not confirmed; Green 1979). Another bird, identified as a Western Wood-Pewee, was carefully described on June 21, 1992, at Hayes State Park, Roseau County, and remained in the area for several days (Bardon 1992; Powell 1993; Eckert 1993). Another summer record, not from western Minnesota, was a vagrant at Kelsey, in the Sax-Zim Bog, St. Louis County, on June 22, 1989 (Eckert 1989b; Eckert 1990a).

A second confirmed nesting from June 21 to July 21, 2007, near Jackson, Jackson County, was well-documented and accepted by the MOU, although detailed descriptions were never published (Svingen 2008; Kessen and Svingen 2008). At Gary Pines Wayside in Norman County, on June 19–22, 2009, a bird identified as a Western Wood-Pewee was observed by many experienced birders, but the details were not sufficient to be accepted by the MOU Records Committee (Svingen 2010).

A recent hybrid pairing at Buffalo River State Park, Clay County, from June 3 to July 26, 2017, received much attention; seven forms with detailed observations, some with sonograms and photos, were submitted to the MOU Records Committee. Details of one observation were published in *The Loon* as a Note of Interest (Tustison 2018). In accepting this record, the MOU Records Committee stated, "On 8 August 2017, an unsuccessful attempt was made to capture this bird for DNA analysis to determine genetically whether or not the bird might be a hybrid with Eastern Wood-Pewee" because, although its vocalizations were typical for a Western Wood-Pewee, one was a "potentially questionable, atypical vocalization" (Tustison 2017). One photo, June 5, 2017, was published and the record was accepted (Kessen and Svingen 2018).

Even though the breeding range of this Pewee does not include Minnesota, all three climate warming scenarios examined by NAS (2019) predicted retractions of its breeding range westward. This would indicate an even lower probability of encountering the species anywhere in Minnesota's future.

ROCK WREN

Salpinctes obsoletus

Minnesota Seasonal Status: Summer vagrant; two breeding records (1984, 2004). Vagrant migrant.
Conservation Concern: Assigned a moderate PIF Continental Concern Score of 11/20.

The Rock Wren is one of three resident large wrens of the West, each in a separate genus, with common names that fit the land: rock, canyon, and cactus. The Rock Wren's northern populations are migratory; the Canyon Wren and Cactus Wren are year-round residents. The description of the type specimen, collected near the junction of Plum Creek with the South Platte River in Colorado, was published in 1823 (AOU 1998), during the time that the West was explored by official government expeditions. The exploration of the Platte River was part of Stephen H. Long's expedition to the Rocky Mountains, authorized by Secretary of War John C. Calhoun, to document the tribes, natural resources, and physical geography of the area; the expedition had Thomas Say and Titian R. Peale as scientific naturalists. The Rock Wren was the first of the three iconic western wrens to be officially added to the avifauna of the United States.

The breeding range of the Rock Wren is from southern British Columbia, Alberta, Saskatchewan, and (rarely) Manitoba, south through the Dakotas (Missouri River west), western Nebraska, and Kansas to central Texas and west to California. It winters from California, Arizona, New Mexico, and the Texas panhandle south to Mexico and, rarely, Honduras. As a wandering stray, the Rock Wren is found extensively in the East and Midwest, particularly the central states from Minnesota south to Louisiana.

Rock Wren.

In states near Minnesota, it breeds regularly in the western Dakotas; formerly in Iowa, near Sioux City (Kent and Dinsmore 1996); and is a vagrant in Wisconsin (two records) and Michigan (eight records).

The habitat and behavior of the Rock Wren are distinctive, although its breeding biology has been little studied. It inhabits rocky slopes and canyons with crevices, talus debris, boulder piles, and quarries, and gravel pits with jumbled debris. It often calls from the top of piles while bouncing up and down on its short legs (Lowther, Kroodsma, and Farley 2020).

MINNESOTA BREEDING DISTRIBUTION

The Rock Wren has been recorded in Minnesota for almost one hundred years but is found only as vagrant individuals during spring and fall migration (with two exceptions). The first record was of a bird collected in a rock quarry near Pipestone, Rock County, on May 13, 1922 (Roberts 1932). This bird became the only state specimen (MMNH no. 6269). Spring migration has a few more records than the fall. Migration timing in the spring is from mid-April to early June and in the fall from late September to late October. There are 31 migrant records of vagrants in 23 separate years, from 1922 to 2020, scattered throughout the state in 15 separate counties, including the four corner counties: Kittson (2015) to Rock (1922, 2006) to Houston (1999) to Cook (1966, 2005, 2008). As would be expected, many of the records (12) are in the western prairie counties. Other clusters are in places that bird-watchers frequent: eight in the Twin Cities; seven from North Shore and Duluth; and single incidental records from Houston, Blue Earth, Rice, and Beltrami Counties. The 2020 MOU checklist classifies the Rock Wren as accidental.

The most interesting records are the two with breeding evidence. The first one, in 1984 at Bemidji, Beltrami County, has been mostly overlooked or dismissed as not credible. However, the circumstances, behavior, and observer familiarity with the species were well described and published (Schmierer 1984). Two birds were found using a rubble pile in the North Bemidji rail yard and bringing nesting material there; they were also foraging for insects in an open boxcar. After observing (May 27 to June 17, 1984) these activities and one copulation, Al Schmierer wrote: "one might speculate at the possibility of their arrival by rail to Bemidji. Perhaps they had attempted to nest in a boxcar further west and were carried east, a trip of perhaps a day." Unfortunately, there was much disturbance in the rail yard, especially on June 5. The Wrens were hard to find after that and disappeared completely after June 17. This observation is similar to two vagrant breeding events from Churchill, Manitoba, with suspected railroad transport (Crosby and Beckett 1957; Seutin and Chartier 1989).

The other breeding record was more expected: near the western border with North Dakota at the Felton Prairie Complex in Clay County. Many people saw the birds from May 8 to July 17, 2004, as the Wrens frequented a rock pile in one of the gravel pits that dot this prairie, which is on an old Glacial Lake Agassiz beach ridge. Careful watching of the Wrens' activities around the rock pile on July 17 was rewarded with observations of two adults and four chicks estimated to be at least 29 days old (Chu 2005; Svingen 2004). Other details of this record were also published (Stensaas 2005). At least one bird returned to the same site in 2005 and was observed from May 10 into August (MOU 2021b).

The vulnerability of the Rock Wren's North American breeding distribution to climate change ranges from low to moderate, depending on which warming scenario NAS (2019) examined. Only very small expansions to its breeding range are projected, and none would bring its distribution any closer to Minnesota.

BEWICK'S WREN

Thryomanes bewickii

Minnesota Seasonal Status: Summer vagrant; formerly bred in the 1920s and again in the 1950s. Vagrant migrant.

Conservation Concern: Listed as *Endangered, Threatened,* or a *Watch* species in 16 states formerly in the eastern portion of its breeding range, including Wisconsin; assigned a moderate PIF Continental Concern Score of 11/20.

The Bewick's Wren was first collected by John James Audubon in 1821 in Louisiana and named after his British engraver, Thomas Bewick. In the southern Appalachians and the southern Midwest, it was the common garden wren. Now it is almost gone east of the Mississippi River but remains widespread in the west from Texas and Missouri to the California coast and northward to British Columbia. The most common explanation for the drastic decline in the east is competition from the nest-destroying and expanding House Wren (Kennedy and White 2020). Bewick's Wren is usually a permanent resident where it breeds. In Iowa and Wisconsin, where it once bred, it now occurs only as a sporadic visitor; the decline in Wisconsin was notable after the 1960s (S. D. Robbins 1991).

MINNESOTA BREEDING DISTRIBUTION

Although it has bred in Minnesota several times, the Bewick's Wren was always at the edge of or beyond its regular breeding range. From the 1920s through the 1960s, there were three to eight records per decade (28 in total), while there were only 11 records from the 1970s through 2020; the last record was in 2015. With two exceptions of birds that were considered vagrants (see below), observations have been confined to the south half of the Eastern Broadleaf Forest Province. The records also show that the Bewick's Wren in Minnesota went from being a resident in the 1920s, to a casual visitant from 1930 to 1990, to an accidental vagrant now.

When T. S. Roberts wrote the 1932 edition of *The Birds of Minnesota,* he had only a few sight records, mostly from Fillmore County. The most important were from Thaddeus Surber, who investigated the valley of the Root River in July and August 1920 and reported these observations: "Rushford, July 14, six; Whalen, July 20, ten; Preston, Aug. 13, two." Roberts also had a note from Mary L. Mills near Preston, who "states that a pair nested in one of her Wren boxes in 1923 and 1924" (Roberts 1932). There are 18 other single records from the 1920s through the 1950s in 10 southeastern counties from Houston and Freeborn north to Anoka and Sherburne, suggesting that there may have been a small and geographically limited breeding population in this region during those years.

Years later, both Green and Janssen (1975) and Janssen (1987) gave the status of this wren as "Casual." By 1987 Janssen downgraded its status to a "summer visitant," although it was considered a summer resident in the early 1900s. All summer records were from the east-central and southeastern regions of the state.

Actual nesting evidence was scarce. It was obtained in Fillmore County in 1923 and 1924 as published in Roberts (1932). Documentation of breeding also occurred in the 1950s: Fort Snelling, Hennepin County, April 23 to July 8, 1950, nest found May 30 (Hofstead 1950); and Anoka, Anoka County, April 15 to July 12, 1959, nest found May 10 (Wagner 1959). The 1951 Seasonal Report column in *The Flicker* gives a hint of what happened to this wren in the 1950s: "Bewick's wrens started nesting at 617 Westwood Ave., Minneapolis, but unfortunately after two weeks they were driven away by house wrens" (Lupient 1951).

As mentioned above, two vagrant records from outside the southeastern quarter of the state have been reviewed and accepted by the MOU Records Committee. The first was in Duluth, a noted vagrant hotspot, on September 8, 1945, by Olga Lakela (1953). The second was on April 8, 1976, near Deerwood, Crow Wing County, and is well described, but the location indicates it was a displaced migrant (Blanich 1976).

Records after the 1960s are rare, but the southeastern distribution remains the same: Sherburne County (three records), Twin Cities (two records), and single records in Freeborn, Olmsted, and Winona Counties. The records per decade vary from two to five: two in the 1970s, five in the 1980s, and three in the 1990s. Almost all observations were from spring and summer (early April through mid-July), but without breeding evidence except for a persistent singing male at Sherburne NWR from May 21 to July 17, 1998 (Eckert 1998b; Wiens 1999). Interestingly, there are sporadic older records from Sherburne County: near Zimmerman (Futcher and Pieper 1952; Bystrom 1983) and Clear Lake (Bratlie 1990). There were no acceptable records after 1998 until 2015, when one bird was found on the Hastings Sand Coulee SNA, Dakota County, June 1–9 (Tustison 2015b; A. D. Smith 2016).

As temperatures rise, this resident of the Southwest, West Coast, and northern Mexico is projected to have a stable North American breeding distribution under all three warming scenarios examined by NAS (2019). Although its breeding range is predicted to gradually expand northward, it only moves as close as western Nebraska under the 3°C warming scenario by 2080, barely increasing its chances of observation in Minnesota.

MOUNTAIN BLUEBIRD

Sialia currucoides

Minnesota Seasonal Status: Summer vagrant; one breeding record (1990–91); five other pairings with Eastern Bluebird. Regular migrant, spring and fall; accidental in winter.

Conservation Concern: Population secure. Variable trends in different landscape conditions. Assigned a moderate PIF Continental Concern Score of 11/20.

The Mountain Bluebird is a western species that breeds at high elevations and latitudes in open country of western North America. It is found in open habitats where the ground is covered with short grasses and short vegetation and where there are suitable cavity-nesting sites such as meadows, the prairie–

Mountain Bluebird.

forest edge, aspen parklands, and the early stages of burned or logged forested areas (L. S. Johnson and Dawson 2020). The species rarely wanders eastward to southern New England (Sibley 2014). It overlaps the range of the Eastern Bluebird along the eastern Great Plains (Peterson 2008). Directly west of Minnesota the overlap occurs in the Dakotas, principally west of the Missouri River and in the North Dakota Turtle Mountains (Stewart 1975; Tallman et al. 2002). It sometimes pairs with the Eastern Bluebird where their ranges overlap (Kaufman 1996).

MINNESOTA BREEDING DISTRIBUTION

The first record of a Mountain Bluebird in Minnesota was a male with a female on April 5–7, 1935, south of St. Cloud in Sherburne County (Hiemenz et al. 1935) and was included in the second edition of *The Birds of Minnesota* (Roberts 1936). The pair was collected and put on display in the museum of the State Teachers' College, St. Cloud. There is one specimen in the Bell Museum: MMNH no. 25706 (Green and Janssen 1975). A wintering bird was reported in Duluth between February 2–14, 1943, and again March 14–16, 1943 (Swedenborg 1947; Lakela 1953, respectively). Until 1970, when this Bluebird became a regular casual migrant, there were four other records (MOU 2021b) of late March migrants in the following counties: Lac qui Parle (1948), Beltrami (1949), Mower (1955), and Clay (1967).

In the decades from the 1970s to the 2010s, one to ten Mountain Bluebirds have been recorded during migration, essentially every year. From the MOU Seasonal Reports (MOU 2021b) there are slightly more records, 57 percent, in the spring than in the fall. The bulk of the spring migration is March 10 to April 25; the fall migration is October 23 to December 15. These birds are most often reported near the Twin Cities, along the North Shore, and in counties near the Dakota borders. There are winter reports from seven counties, including Cook in four years. There are summer records in six years (1935, 1986, 1990, 1991, 1994, and 1997), but none in the atlas years (2009–13) and none thereafter (2014–20).

Five summer records of hybrid nesting pairs, all males mated to Eastern Bluebird females, have been recorded, along with one record of a pure Mountain Bluebird nesting pair (1990–91). The first observation of a hybrid pair was reported on June 11, 1986, when a male Mountain Bluebird was verified nesting with a female Eastern Bluebird in a nest box east of Aitkin, Aitkin County. Five eggs were laid in the box, but only two hatched, and the young left the box about July 19 (Steblay 1986). In June 1990, a participant in the Bluebird Recovery Program reported a pair of Mountain Bluebirds nesting in a box near Florian, Marshall County; on June 14, 1990, the owner of the bluebird box confirmed that the pair had six young in the box. Two days later, birders from the Twin Cities arrived and photographed the male and the young in the nest, which fledged later that day (Janssen 1990b). The next year (1991) this pair returned and produced six young (Wiens 1992). This is the only confirmed record of a pure Mountain Bluebird nesting pair.

Another northwestern Minnesota record was of a male Mountain Bluebird mated to an Eastern Bluebird that nested in eastern Polk County, April 22 to May 16, 1990 (S. Carlson et al. 1990; Wiens 1991). The next year on June 29, 1991, a male mated with a female Eastern Bluebird was reported on a bluebird trail in Kittson County just south of the Manitoba border; the pair fledged four chicks. The owner of the bluebird trail said that a pure pair of Mountain Bluebirds nested in an adjacent box in 1990 (Bardon 1991b). The next record was on April 22, 1994, when a brilliant blue male "apparent Mountain/Eastern hybrid" was flushed from Highway 210 in Todd County; it was with a female Eastern Bluebird (Millard 1994). On June 4, 1994, the nest box they were using was discovered and it contained five eggs (Millard 1994). The last state record was reported on May 5 to August 1, 1997, from Bemidji, Beltrami County, where a Mountain Bluebird male nested with an Eastern Bluebird female (Benson et al. 1997, 1998). After this cluster of nesting records (1986–97), there has been no other breeding evidence in subsequent decades through 2020.

Since there has been no nesting evidence of Mountain Bluebirds in Minnesota for more than twenty years, predictions of its breeding range by NAS (2019) indicate further retractions westward with all three climate warming scenarios they examined. Therefore, this retraction would reduce the future probability of colonization and nesting activity in Minnesota.

THICK-BILLED LONGSPUR

Rhynchophanes mccownii

Minnesota Seasonal Status: Former summer resident in the nineteenth century. Vagrant migrant.
Conservation Concern: In the United States it has no designated federal status; in Canada federally designated as *Special Concern* in 2006 (Environment Canada 2014a). In Minnesota it is considered extirpated as a summer resident and does not have an official status. Assigned a high PIF Continental Concern Score of 15/20 and placed on the Yellow Watch List.

Formerly known as McCown's Longspur, the Thick-billed Longspur is a resident of the dry upland prairie of the northern Great Plains. The species' new name is a translation of the name of its genus; the name was changed by the AOU in 2020. First discovered in 1851 by an army captain, John P. McCown, in western Texas, it winters in flocks from Oklahoma and west Texas south to northern Mexico. The grassland habitat of this endemic species of the short-grass prairie was originally a patchwork of different vegetation types, heights, and extent that was created by drought and the dominant grazing mammals: bison, prairie-dogs, and elk (Askins et al. 2007). In the nineteenth century, market hunting, exploration, and settlement severely disrupted these ecological processes. After 1900 the Thick-billed Longspur's breeding range contracted severely, and now the bird is found only in two disjunct populations centered in Colorado and Montana/southern Canada (With 2020). This Longspur will nest in agricultural open landscapes like overgrazed pasture and plowed fields that mimic the original grassland mosaic. Conservation practices like fire management and the federally subsidized Conservation Reserve Program favor tall domestic or native grasses, but these do not replicate the natural-grazing mosaic structure (Askins et al. 2007). Analyses of data from the BBS and Audubon Christmas Bird Counts since the early 1960s show a continuing but yearly variable decline on both breeding and wintering ranges (With 2020).

MINNESOTA BREEDING DISTRIBUTION

In Minnesota, the Thick-billed Longspur has not been a summer resident since the nineteenth century. All that is known about this species as a breeding resident was covered in *The Birds of Minnesota*. Roberts (1932) defined its status as: "Once a common summer resident on the high, dry prairie along the southwestern border of the state from Pipestone County to Big Stone County and more sparingly northward along the Red River." The first record was on May 16, 1889, in Lac qui Parle County. In the 1890s, breeding evidence was well-documented from Pipestone, Lincoln, and Lac qui Parle Counties, mostly by A. D. Brown, an egg collector who published in the periodical *Ornithology and Oology* and described information about its breeding presence. Roberts (1932) summarized Brown's writings as: "Apparently it occupied only the high ridge of Minnesota—the *Coteau des Prairies*. It visited only in dry seasons—when very dry it was most abundant, and in wet seasons it was entirely absent. Up to 1891 it bred on the higher and drier parts of the prairie, but after that nests were found only in the high parts of wheat fields." The incursion of this Longspur breeding in Minnesota was short-lived and confined to the southwestern corner of the state. Reports farther north along the border were fragmentary. It was not reported after 1899.

Since 1900 there have been four credible published records, all vagrants. The first, on October 19, 1935, of two males in a flock with other migrants in Aitkin County, was considered so significant it was published in *The Auk* (Peterson and Peterson 1936); but much later the MOU expressed doubt (Eckert 1982b). The next record, a photographed male, was on June 11, 1982, in Grand Marais Harbor and was probably a displaced spring migrant (Eckert 1982a). Two other vagrants were both seen in the Duluth-Superior Harbor at the Erie Pier dredge disposal site: October 19–20, 1986, bird photographed (Janssen 1986a); and a male, May 8, 2000 (Benson 2000; Eckert 2000).

NAS (2019) indicated a status of high vulnerability of this Longspur in all three of the climate warming scenarios they examined. Even an imminent climate scenario, an increase of 1.5°C, predicted a 98 percent loss in its North American breeding range; an increase of 2°C predicted extinction. The outlook for a return of this extirpated species to Minnesota appears dire.

BAIRD'S SPARROW

Centronyx bairdii

Minnesota Seasonal Status: Summer vagrant. Formerly sporadic in the Red River valley; two nest records (1930, 1937).
Conservation Concern: A petition to officially list the species as federally *Threatened* in the United States was denied (USFWS 1999b); officially listed as *Endangered* in Minnesota. Listed as *Special Concern* in Canada in 2012. PIF assigned it a high PIF Continental Concern Score of 15/20 and added it to its Yellow Watch List.

The Baird's Sparrow has been part of Minnesota's avifauna since the 1890s but remains an enigma owing to the sporadic nature of its occurrences, the challenge of distinguishing the bird from other grassland sparrows, and the difficulty in confirming nesting. It was discovered by Audubon in 1843 in North Dakota near Old Fort Union, where the Missouri River crosses from Dakota to Montana. He named the collected specimen after his young assistant, the naturalist Spencer Fullerton Baird, who went on to have a long, distinguished career at the Smithsonian Institute (1848–87). This secretive bird of dense grasslands was not recorded again for almost thirty years (M. T. Green et al. 2020). It is the rarest of the Northern Great Plains endemic species and it is found from the Canadian Prairie Provinces south to Montana, North Dakota, and northern South Dakota. Its wintering grounds are mostly in the north-central Mexican Chihuahuan Desert grasslands, with northern enclaves in west Texas and southeastern New Mexico. The Baird's Sparrow has a sibling species, the Henslow's Sparrow *(Centronyx henslowii)*, which occurs in the north-central United States.

Its breeding habitat is the mixed-grass and fescue northern prairies, especially in Canada (Alberta, Saskatchewan), where it is most abundant. Its habitat occupancy is nomadic, which likely evolved with prairie landscape disturbances created by drought, fire, and bison herds. This leads to annual population fluctuations ranging from none to many birds at any given site. When preferred habitat conditions are found, it may nest in loose colonies (Kaufman 1996). Recent studies have shown that it will also use agricultural lands, hayfields, and pastures when the grassland structures are suitable (M. T. Green et al. 2020).

MINNESOTA BREEDING DISTRIBUTION

In Minnesota, the first specimens were collected in the 1890s from Kittson and Marshall Counties. Roberts (1932) described its presence as "known in Minnesota up to the present time only from the Red River Valley" and gave records in 1886, 1887, 1898, and 1900. It was not found again until the Museum of Natural History (now the Bell Museum) made field trips to northwestern Minnesota in 1928, 1929, and 1930. Birds were observed in Marshall, Pennington, Polk, Traverse, and Wilkin Counties. Only one nest was discovered during these trips: Pennington County on July 22, 1930 (Roberts 1932). No other Baird's Sparrows were found away from these Red River valley counties. The only other published Minnesota nest record was one near Euclid, Polk County, on June 20, 1937 (Rysgaard 1937a, 1937b). However, a note in the Museum files from a field trip in 1932 states: "set of eggs in Dr. Langevin's collection, labelled Baird's Sparrow" was marked 6-20-1917, Polk County (MMNH Records), becoming the third nesting record. In the MOU magazine, *The Flicker,* there are very few observations after the 1920s and 1930s: one record in 1942 and three in the 1950s, including two from Buffalo River State Park (Pieper 1954; R. L. Huber 1959b).

In 1961, the Avifaunal Club visited a native prairie on a Glacial Lake Agassiz beach ridge known as the Norcross Beach (about 4–5 miles east and 1–3 miles south of Felton, Clay County), which they had discovered in 1960. On July 4, 1961, they visited again and observed five singing Baird's Sparrows (R. L. Huber 1961c; Pieper 1963b). The Felton prairie complex was also searched in 1961 by Dennis Raveling, who observed many singing Baird's Sparrows and collected five on July 15 (Museum field report, MMNH Records).

Through the 1960s, 1970s, 1980s, and 1990s, Felton Prairie was the place to see three northern plains endemic species on the eastern margin of their range: Baird's Sparrow, Sprague's Pipit, and Chestnut-collared Longspur. Records compiled by the MOU Records Committee give yearly records for the Clay County prairie complex: Baird's Sparrow was recorded during three years in the 1960s, seven years in the 1970s, three years in the 1980s, and three years in the 1990s. After June 6, 1997, there are no more records from Felton. Other county records are Wilkin County (1954, 1959, 1976, 1977), Roseau County (1991; Bardon 1991a), and Polk County (2002; J. Richardson 2003; Eckert 2003; Budde and

Svingen 2003). Outside its range in the Red River valley, there are two vagrant records: Crow Wing County (1986; Risen 1986) and Douglas County (2016; Kessen and Svingen 2017).

In adjacent states, the Baird's Sparrow breeds in the Dakotas (Stewart 1975; Tallman et al. 2002) and is a casual visitant in Wisconsin (S. D. Robbins 1991). Minnesota books give the Baird's Sparrow's status as "Casual. Spring migrant; former summer resident" (Green and Janssen 1975) and "Casual. Summer resident only in Clay County since about 1960; formerly more widespread" (Janssen 1987). The MOU now classifies the species as accidental. The Baird's Sparrow has probably been extirpated as a breeding resident in Minnesota since the 1960s.

Climate change may have a devastating impact on the Baird's Sparrow. The species' North American breeding distribution was projected to be highly vulnerable under all three warming scenarios that NAS (2019) examined. As temperatures rise, its breeding range gradually contracts until, under the 3°C warming scenario in 2080, projections are that the species will be extinct. Unfortunately, the opportunities to observe this species, even in the heart of its breeding range in Montana, are disappearing far too quickly.

RUSTY BLACKBIRD

Euphagus carolinus

Minnesota Seasonal Status: Summer vagrant, formerly bred (1981, 1982, 1986). Regular migrant, winter visitant.
Conservation Concern: No federal status in the United States. Listed as *Special Concern* in Canada in 2006, a USFWS Bird of Conservation Concern in 2008, and a Common Bird in Steep Decline in North America by PIF in 2016. The IUCN Red List (2021) has classified the species as vulnerable. Assigned a moderate PIF Continental Concern Score of 12/20.

The Rusty Blackbird, a New World blackbird discovered in the Carolinas, was once called the "Rusty Grackle." The emphasis on "rusty" is certainly fitting because of its color change to a rusty-flecked winter plumage and its song, which sounds like a rusty creaking hinge. Its black plumage in summer is similar to its generic sibling, Brewer's Blackbird (*Euphagus cyanocephalus*), leading to difficult identification problems in Minnesota, where their distributions overlap. Their plumages are similar, but differences in breeding habitat, migration paths, and timing help in identification. The Rusty Blackbird breeds primarily in the Alaska and Canadian taiga south through the Boreal Forest to Labrador, northern New England, and the Adirondacks. The Brewer's Blackbird stronghold is open landscapes in the West, and it breeds from California north to British Columbia then east to Manitoba and the western Great Lakes states. The winter ranges of these two blackbirds overlap in the southern states.

When it bred in Minnesota, the Rusty Blackbird was very rare and found near the Canadian border in conifer forests and muskegs. The Brewer's Blackbird occurs in open country with fields, farms, and parks and is widespread in much of the state. During the MNBBA, there were no Rusty Blackbird records, and the Brewer's Blackbird was ranked as uncommon. Recent fall migration counts along Hawk Ridge and the Duluth lakeshore in 2007–11 documented the Rusty Blackbird as an abundant migrant from mid-September through mid-November (Bardon 2012; average yearly totals = 13,000, Karl Bardon, personal data).

Despite high migration counts, the Rusty Blackbird is in trouble in parts of its distribution. A study based on trend analysis from nearly forty years of BBS data (1966–96) and Audubon Christmas Counts (1958–88) determined a decline of about 90 percent (Greenberg and Droege 1999). Hypothesized causes are varied and unsatisfactory; subsequent research and monitoring has been undertaken by the International Rusty Blackbird Working Group.

MINNESOTA BREEDING DISTRIBUTION

In his compilation of records through the 1930s, Roberts (1932) described the Rusty Blackbird as: "An abundant spring and fall migrant through the state. Found occasionally in the winter. . . . Does not nest in the state as far as known." The MOU considered Rusty Blackbird records reported in *The Flicker* in the 1940s and 1950s to be fall migrants and winter residents; this was supported by regular observations of a few birds in winter flocks of Red-winged Blackbirds in the Minnesota River valley (MOU 2021b). In the 1960s, when the MOU began regularly publishing a Seasonal Report, Rusty Blackbirds were considered migrants and winter visitants, as the birds had become very rare and irregular in the winter months. This changed when summer observations were obtained in Cook County near the Canadian border in the 1970s: in 1973, Mineral Center (Carr 1973); 1975, near Saganaga Lake (Niemi 1975); 1976, Grand Portage (Eckert 1977b); and 1979, Fiddle Creek Road (Green 1980).

The first breeding records for the Rusty Blackbird were obtained in 1981, also in Cook County. At Swamp Lake, off the Gunflint Trail, on June 29, 1981, Ken and Molly Hoffman followed a pair and three begging long-tailed juveniles along the bog edge of the lake (Hoffman and Hoffman 1981). The next year they found a pair carrying food along the South Brule River, searched, and finally saw the nest with four nestlings on June 12, 1982 (Hoffman and Hoffman 1982a). That same year, on June 17, several Rusty Blackbirds were found by Steve Wilson while canoeing Hog Creek in Lake County, and one pair was carefully observed carrying food to begging young (S. G. Wilson 1982). The only other breeding record for the state was in Cook County in 1986: a reported nesting on McDonald Lake (Shedd and Wilson 1987). These sightings changed the status of the species to "migrant and summer resident; casual in winter" (Janssen 1987).

However, in the thirty years from 1989 to 2018, the MOU Seasonal Report has listed only four single observations of Rusty Blackbirds during the breeding season (1999, 2006, 2015, 2017; MOU 2021b), all without details. The most recent (June 18, 2018) unpublished breeding report was along the Gunflint Trail, Cook County: two birds, carrying food, T63N R1E, sw. ¼, vicinity of the Brule River (MOU 2021a). Credible summer observations of rare breeding species should include details of plumage, location, behavior, and habitat. Since the flurry of observations from 1973 to 1986, the only detailed summer record (June 16, 1992) was near Maude Lake, St. Louis County (T65N R16W): three territorial pairs and three flyovers were recorded by University of Minnesota Duluth Biology Department field staff when conducting a yearly breeding bird monitoring program for the Superior National Forest (1991–2017, Natural Resources Research Institute data). The Rusty Blackbird continues to be a regular migrant and winter visitant in Minnesota, but summer resident status has not been warranted for decades.

Climate change will have a significant impact on the Rusty Blackbird's breeding distribution. Even though NAS (2019) included a small section of Cook County within the species' current distribution, under a rather imminent 1.5°C warming scenario, its breeding range is predicted to retract at least 450 km north of Minnesota. By 2080, under a 3°C warming scenario, projections limit it to the Arctic region. Its future in Minnesota is likely to be only that of a spring and fall migrant.

ACKNOWLEDGMENTS

< Male Pileated Woodpecker. Photo by David Brislance.

The list of contributors to the Minnesota Breeding Bird Atlas is indeed impressive, and it testifies to the scope and diversity of the individuals and institutions involved. These contributors provided the necessary funds, designed the field sampling, organized the participants, collected the data, and painstakingly reviewed each atlas detection. They analyzed the data, designed the website, edited the manuscript, and handled everything in between. The talents and skills necessary to undertake an initiative of this magnitude extend far beyond the obvious need to identify birds! We sincerely appreciate every element of work that went into this remarkable initiative.

We are immensely grateful to all the following organizations, agencies, committees, and individuals who so generously contributed to the establishment and ultimate success of Minnesota's first Breeding Bird Atlas. The Minnesota Breeding Bird Atlas has worked hard to acknowledge everyone who contributed to the success of the atlas. We take full responsibility for any errors of omission and offer our apologies.

Also, we would like to thank some very important people behind the scenes. We will never find enough appropriate words to acknowledge our appreciation to our spouses for their patience, support, and assistance (especially proofreading and traveling to all corners of the state taking photos of each species' habitat). The personal sacrifices they have made while waiting for us to complete all the work associated with finalizing the atlas, including establishing the website and preparing this book, are remarkable. Gary Seim, Bonnie Niemi, and John Green: we sincerely thank you and owe you a debt of gratitude that we can never adequately repay.

We know this book would never have been finished without the voluntary efforts of our technical editor, Kim Rewinkel. Kim's exceptional mastery of the English language, her attention to detail, and her perseverance in reading and comprehending the University of Minnesota Press guidelines were simply extraordinary. We are so grateful that she was also willing to persist through these years of effort—from the website to the completion of this book. We know that the fuel that drove us and Kim was our love of birds.

Project Partners and Financial Supporters

The **Minnesota Environment and Natural Resources Trust Fund,** which provided significant funding support as recommended by the Legislative-Citizen Commission on Minnesota Resources (LCCMR).

Audubon Minnesota, which served the vital role as the Atlas's Program Manager from 2008 to 2017.

Natural Resources Research Institute (NRRI), University of Minnesota, whose staff was instrumental in the design and implementation of the MNBBA point count data component and in all statistical analyses of these data. We also wish to thank the Institute for supporting Dr. Gerald J. Niemi during preparation of parts of this book, Dr. Alexis Grinde for assuming a major role in managing and administering many of Dr. Niemi's research projects, and Dr. Matt Etterson for his support and scientific consultation. Other projects administered by NRRI that provided support for Dr. Niemi and helped contribute to the atlas's assessment of the status of many of Minnesota's avifauna included support received from the US Forest Service—Chippewa and Superior National Forests and the Northern Research Station; the US Environmental Protection Agency, including the Office of Research and Development and the Great Lakes National Program Office; UPM Blandin; the National Council for Stream and Air Improvement Inc.; and the Aitkin County Land Department.

United States Fish and Wildlife Service for providing critical technical and financial support.

Minnesota Ornithologists' Union for providing financial support, sharing bird records, and designing and managing the program's website from 2009 to 2014. The organization's records and publications also contributed significantly to the preparation of historical accounts for each species.

Minnesota Department of Natural Resources for providing financial and technical support and sharing many critical sources of avian data, especially for game species.

Bird Conservation Minnesota for launching the Minnesota Breeding Bird Atlas and contributing to its design and implementation.

Bell Museum of Minnesota for providing support during a critical juncture in the atlas project's development.

Organizational Committees and MNBBA Data Reviewers

Steering Committee. Dave Cahlander, Francesca Cuthbert, Bill George, Janet Green, Peter Harris, Carrol Henderson, Greg Hoch, Ann Kessen, Scott Lanyon (Chair), Mark Martell, John Moriarty, Gerald Niemi, Cynthia Osmundson, Bill Penning, Bonnie Sample, Mary Shedd, Valerie Slocum, and Tom Will.

Technical Committee. David Anderson, Rich Baker, Paul Bolstad, Nick Danz, Diane Granfors, Janet Green, Anthony Hertzel, Robert Janssen, Andy Jenks, Doug Johnson, Melinda Knutson, Robert Russell, Steve Stucker, Wayne Thogmartin, and Ron Windingstad (Chair).

MNBBA Data Reviewers. Carol Carter, Steve Cordts, Francesca Cuthbert, Bruce Fall, Janet Green, Anthony Hertzel, Ann Kessen, Jeff Lawrence, Mark Martell, Gerald Niemi, Lee Pfannmuller, Steve Stucker, and Steve Weston.

Project Staff

Program Managers. Mark Martell (2008–15) and Lee Pfannmuller (2015–17).

Research Coordinator. Gerald Niemi (2009–17).

Program Coordinator. Bonnie Sample (2008–14).

Program Liaison. Kristin Hall.

MNBBA Website Managers (2009–14). David Cahlander and Anthony Hertzel.

Final Website Development (2014–17). George Host (Project Coordinator), Alexis Liljenquist (Data Entry), Ashley Peters (Species Photographs), Jane Reed (Designer), Kim Rewinkel (Administrative Support and Development of Literature Cited Sections), Nicolas Walton (Data Analysis and Website Assistance), and Norman Will (Programmer).

Cornell Website Manager. Jeff Gerbracht.

Financial Manager. Susan Swanson.

Administrative Support. Carol Carter, Harriet Liedtke, Linnea Rowse, Emily Sames, George Skinner, and Steve Weston.

Field Survey Staff. Ryan Atwell, Josh Bailley, Josh Bednar, Lars Benson, Annie Bracey, Brandon Breen, Carol Carter, Jessica

Chatterton, Allison Coffman, Shawn Conrad, Nate Dennis, Brian Dirks, Paul Dolan-Linne, Tom Erickson, Randy Frederickson, Alexis Grinde, David Grosshuesch, Matti Hakkila, Eric Hall, Terry Hams, Ted Keyel, Carly Lapin, Jim Lind, Gerald Niemi, Diana Ostrowski, Debbie Petersen, Anna Peterson, Lee Pfannmuller, John Schaldweiler, Heidi Seeland, Jackie Smith, Mark Stensaas, Kyle Te Poel, Steve Weston, Katie Winkler, and Ed Zlonis.

Website Copy Editors. Mary Keirstead and Molly McBeath.

Book Production. Wilsted & Taylor Publishing Services: Christine Taylor (Project Manager); LeRoy Wilsted (Production Assistant); Nancy Evans (Copy Editor); Nancy Koerner, Laurel Muller, and Michael Starkman (Designers); Jennifer Brown, Sophia Fox, and Melody Lacina (Proofreaders); and Evan Winslow Smith (Color).

Technical Support. Rebecca Toov, University of Minnesota Archives, for assistance in working with T. S. Roberts's files and the Minnesota Ornithologists' Union files; and the Minnesota Supercomputing Institute (MSI; www.msi.umn.edu), University of Minnesota, for providing access to their computer cluster for analyses of MNBBA data.

Technical Reviewers. Many individuals helped with the species accounts, providing technical information and editing comments. We would especially like to thank Christine Herwig, who provided substantial assistance and updated information for the Common Tern and Piping Plover accounts, and Jeff Lawrence and Danelle Larson, who examined and provided input on all of the waterfowl accounts and the Sandhill Crane. Bob Dunlap and Mark Martell also provided assistance with investigating Burrowing Owl historical records. The accounts for all of these species were immeasurably improved by their generous contributions. Others who provided a critical review included John Green, Bonnie Niemi, Gary Seim, and Dave Zumeta. John Moriarty also provided us with many important historical documents and books.

In addition to these contributions, the following individuals provided comments on individual species accounts (noted below) and on the overall MNBBA website design and navigation:

Annie Bracey (American Kestrel, American Three-toed Woodpecker, Bay-breasted Warbler, Black-backed Woodpecker, Black-capped Chickadee, Boreal Chickadee, Broad-winged Hawk, Cape May Warbler, Common Tern, Dark-eyed Junco, Hermit Thrush, Lincoln's Sparrow, Merlin, Nelson's Sparrow, Red-shouldered Hawk, Rose-breasted Grosbeak, Swainson's Thrush, Tennessee Warbler, Veery, White-throated Sparrow); Gaea Crozier (Canada Jay, Northern Goshawk); Francesca Cuthbert (Black-crowned Night-Heron, Piping Plover, Ring-billed Gull); Kim Eckert (Common Raven, Gray Partridge, Greater Prairie-Chicken, Sharp-tailed Grouse, Spotted Sandpiper, Wilson's Warbler, Wood Thrush); Matt Etterson (Loggerhead Shrike, Sedge Wren); Alexis Grinde (Canada Warbler, Chestnut-sided Warbler); David Grosshuesch (Great Gray Owl, Northern Hawk Owl); Kristin Hall (Chimney Swift, Virginia Rail); Douglas Johnson (Gray Partridge, Ring-necked Pheasant); Mike Larson (American Woodcock, Ruffed Grouse); Jeff Lawrence (Mallard, Ring-necked Duck, Trumpeter Swan); Steve Lewis (Common Loon, Franklin's Gull); Jim Lind (Common Raven, Pine Siskin, White-winged Crossbill); Mark Martell (Bald Eagle, Osprey, Peregrine Falcon); Ashley Peters (Eastern Bluebird, Pileated Woodpecker); Robert Rosenfield (Cooper's Hawk, Sharp-shinned Hawk); Robert Russell (Gray Partridge, Marbled Godwit, Spotted Sandpiper, Upland Sandpiper, Willet, Wilson's Phalarope); Charlotte Roy (Sharp-tailed Grouse, Spruce Grouse); Bonnie Sample (Common Nighthawk, Grasshopper Sparrow); John Schladweiler (Burrowing Owl, Swainson's Hawk); Sparky Stensaas (Great Gray Owl, Northern Hawk Owl); Tom Will (Blue-winged Warbler, Golden-winged Warbler); Steve Wilson (Boreal Owl, Eastern Whip-poor-will); and Edmund Zlonis (Golden-crowned Kinglet, Wilson's Warbler, Yellow-bellied Flycatcher).

Photographers for the Book. The following individuals and organizations contributed photographs of the 250 species observed during the MNBBA and photographs of their breeding habitats. We would like to thank each of them for their very generous contributions that embellish each species account. We would also like to extend our appreciation to Jean Brislance for graciously allowing us to use countless species photographs taken by his late father, David Brislance.

Karl Bardon, Travis Bonovsky, Cynthia Bridge, David Brislance, Jean Brislance, Courtney Celley/USFWS, Gaea Crozier, Francesca Cuthbert, Joanna Eckles, Chet Ellingson, Laura Erickson, Rebecca Field, Michael Furtman, Ivan Gabaldon, John Green, Alexis Grinde, David Grosshuesch, Bradley Hacker, Derek Hamilton, Gerald Hoekstra, Terry Johnson, Matthew Jolley, Linda Kramer (minnemom), Mike Lentz, Gregory Lis, Mark Martell, Stephen Maxson, Jeremy Meyer, Midwest Peregrine Society, Steve Mortensen, Tom Muir, Kathy Munsel, Gerald Niemi, Jarad Niemi,

Ashley Peters, Debbie Petersen, Lee Pfannmuller, Andy Reago and Chrissy McClarren, Kim Rewinkel, Isaac Sanchez, Ryan Sanderson, Dan Scolaro, Gary Seim, Larry Sirvio, Chris Sloan, Paul Sparks, Sparky Stensaas, Dan Svedarsky, and US Department of Agriculture (USDA) Wildlife Services, Minnesota.

Volunteers

Regional Coordinators. Minnesota is a large state with a great diversity of bird habitat. Within the state there are four major ecoregions, each having distinct environmental conditions and natural resource communities: prairie landscapes are characteristic of western Minnesota, aspen parklands are found in the northwestern region, coniferous forests are widespread throughout the north country, and, in southeastern Minnesota, deciduous hardwood forests are most common.

To help us effectively organize a survey project of this size with this much landscape variation, it took local knowledge of both birds and their habitat to support effective survey coverage and accurate reporting. The Minnesota Breeding Bird Atlas created 32 administrative regions within the state to help make this happen. Each region had a Regional Coordinator to provide local leadership for the atlas. These volunteer Coordinators were essential to the success of the atlas. Regional Coordinators helped recruit volunteer surveyors, answer surveyor questions related to species identification and data recording, tracked project progress, and reviewed reported observations:

Carol Carter, Ted Dick, Joel Dunnette, Randy Frederickson, Janet Green, Carl Greiner, Lee Grim, Dave Grosshuesch, Christine Herwig, Kate Kelnberger, Ann Kessen, Jim Lind, Suzi Maki, Norma Malinowski, Pete Mattson, Matt Mecklenburg, Warren Nelson, Mike North, Ken and Pam Perry, Lee Pfannmuller, Jeremy Ridlbauer, Bonnie Sample, Joel Schmidt, Beth Siverhus, Sparky Stensaas, Bill Stjern, Bill Tefft, Stacey Trapp, Bill Unzen, John Voz, Steve Weston, and Bob Williams.

Atlas Participants. The following participants each generously contributed their time and expertise to the success of the atlas:

David Adair, Sabin Adams, Michael Ager, Roxie Aho, Pam Albin, Ray Amundson, Bart Anderson, Glen and Ellie Anderson, Jacob Anderson, Renner Anderson, Tom Anderson, Gordon Andersson, Cindy Angerhofer, Sarah Archbold, Sarah Archbold and Kofi Wing, John Arenz, Charles Argue, Melis Arik, Brett Arne, Don Arnosti, Gary Ash, Wiley Askew, Sandy Aubol, Jackie Augustine, Gustave Axelson, Rob Baden, Matt Baer, Jerold Bahls, Wes Bailey, Joshua Bailly, Darrellyn Barrett, Kelly Barrett, Dave Bartkey, Grace Bartlett, Al Batt, Jared Beck, Helen Becker, Karine Becker, Josh Bednar, Elaine Beitelspacher, Tom Bell, Dave Benson, Steve Benson, Dedrick Benz, Frank Berdan, Jean Beyer, Emily Bjerre, Bill Blackburn, Jo and Steve Blanich, Michael Blassey, Karla Bloem, Milton Blomberg, Suzanne Blue, Rachel Blume, Debby Bocnuk, Karen de Boer, Richard Bohannon, Brad Bolduan, Jason Bolish, Brad Bond, Jerry Bonkoski, Lauren Borer, Cheryl Boyes, Annie Bracey, Gary Bradshaw, Nancy Braker, Brandon Breen, John Brenneman, Kenneth Brennen, Denise Bricher, Wayne Brininger, Judd Brink, Ann Brogger, Sandy Brogren, Bill Brown, Steven Brown, Glenn Browne, O. William Bruins, Mary Kay Bryan, Amy Bryant, Carole Brysky, Paul Budde, Holly Burd, Cheryl Burgmeier, Robert Burmaster, Thomas Bursch, Corey Butler, Mike Butterfield, Bob Bystrom, David Cahlander, Lee Carlson, Pete Carlson, Anita Carpenter, Carol Carter, Jeffrey Carter, Adriana Casillas, Katy Cass, Ann Caturia, Neil Chartier, Jessica Chatterton, Robert and Audrey Chonko, Clay Christensen, Sally Christenson, Judy Chucker, Joel Claus, Dick Clayton, Judith Clayton, Laura Coble, Lewis Coffey, Allison Coffman, Jon Cole, Don and Cheryl Collins, Ahndree Conn, Shawn Conrad, Thomas Cooper, Pam Coy, Stuart Crane, Donna Crider, Christina Crowther, Gaea Crozier, Cheryl Culbreth, Val Cunningham, Noel Cutright, John Cyrus, Lisa Czaia, Steven Daigle, Geoff Davidge, Stephanie Davidson, Tim Dawson, Priscilla Dean, Lowell Deede, Nathan Dennis, Marcel Derosier, Wendy DeVault, Pat DeWenter, Nancy Dietz, Maria Digatono, Herb Dingmann, Brian Dirks, Morris Dodge, Rachel Doherty, Dale Doudrick, John Douglas, Jessica Dowler, Diana Doyle, Ryan Drum, Maria and Steve Duane, Matt Dufort, Peter Dullea, Bob Dunlap, Bill and Carol Dunn, Joel Dunnette, Roger Eberhart, Joanna Eckles, Cindy Edwardson, Paul Egeland, Claudia Egelhoff, Kathy Eich, John Ellis, Carol Eloranta, Ely Field Station, Marilyn Emerson, Norm Emerson, Deanne Endrizzi, Lars Erdal, Laura Erickson, Tom Erickson, Thomas Ericsson, Ronald Erpelding, Matthew Etterson, Andrew Evans, Audrey Evers, Steve Falkowski, Bruce Fall, Susan Fall, Sharon Fandel, Kathryn Farion, Greg Feinberg, Lisa Fennie, Kathy Fielden, Barbara Finkelson, Jim Fitzpatrick, Paul Flicek, Andy Forbes, Jason Frank, Jenna Frank, Randy Frederickson, Pamela Freeman, Travis

Freiwald, Merrill Frydendall, Cathy Gagliardi, Jim Gallagher, Jim Gay, Stephen and Patricia Monson Geerts, John Geissler, Bill George, Gloria Gervais, Richard Gibson, Glennie Gilleen, Jenn Gillen, Larry Gillette, Thomas Gilligan, Lynn Glesne, Kevin Gohman, Malcolm Gold, Mackenzie Goldthwait, Diane Granfors, Janet Green, Carl Greiner, Terry Grier, Lee Grim, Dave Grosshuesch, Lissa Grover, Diane Gulbrandson, Paul Gunderson, David Gustafson, Dan and Pam Guynn, Kyle Hafemeister, Roger Haglund, Jennie Hakes, Matti Hakkila, Andrew Halbritter, Joel Halbritter, Eric Hall, Kristin Hall, Maya Hamady, Terry Hams, Allen Hansen, Clifford Hansen, Kenneth Hanson, Kimberly Hanson, Jon Harder, Doug Harr, Bettie Harriman, Kenneth A. Harris, Peter Harris, Katherine Haws, Seth Hayden, Jan Hayne, Michelle Hed, Dale Hedlund, Melissa Hein, Brian Heins, Chad Heins, Robert Heise, Ed Heit, Drew Hempel, Mike Hendrickson, Dan Hertel, Anthony Hertzel, Christine Herwig, Jeanine Hess, John Hiebert, Tony Hilmoe, Sauli Hirsimaki, Ashley Hitt, Pete Hoeger, Gerald Hoekstra, Steven Hogg, Bill Hohengarten, Bob Holtz, Natalie Hoogshagen, Rebecca Horton, Steve Houdek, Marshall Howe, Cynthia Huener, Gary Huschle, Ann Hutchinson, Johnnie Hyde, C. J. Jacobsen, Gerald Jacobson, Timothy Jacobson, Gregory Jahner III, Robert Janssen, Vicky Jaskierski, Karen Jenkins, Ronald Johannsen, David Johnson, Douglas Johnson, Jay and Kirsten Johnson, Katy Johnson, Linda Johnson, Paul Johnson, Rex Johnson, Ruth Johnson, Jeanie Joppru, Brian Jungels, Endel Kallas, Laura Kaplan, Sharon Kaufman, Meagan Keefe, Kate Kelnberger, Ann Kessen, Ted Keyel, Memuna Khan, Doug Kibbe, James Kiehne, Don Kienholz, Douglas Kieser, Mary Klees, David Kleis, Gregg Knutsen, Suzanne Kosko, Michael Koutnik, Jan and Larry Kraemer, Paul Kreiss, Dean Krepela, Sherri Krier, Judy Krish, Susan Kroll, Scott Krych, Margaret Kuchenreuther, Dee Kuder, Ron Laack, Robin La Fortune, Andrea Lambrecht, Barb Lang, Wendy Langley, Carly Lapin, Jill Larsen, Cheryl Larson, Chris Larson, Katherine Lasee, Tony Lau, Linda Laumb, Steven Lawrenz, Susan Leaf, Mary Lee, Michael Lee, Mike Lehrke, Jolene Leland, Bruce Lenning, Gene Letty, Steve Lewis, Harriet Liedtke, Susan Light, Madeleine Linck, Jim Lind, Jason Linkert, Kelly Linton, Susan Lippolt, Perry Loegering, Dodie Logue, Scott Loss, Ellen Lowery, Bryan Lueth, Wade Lund, Jennifer Lust, Larry Lyrenmann, Toni Lyrenmann, Mike Magnuson, Thomas Maiello, Mike Majeski, Suzie Maki, Sarah Malick, Norma Malinowski, Thomas Malone, Craig Mandel, Jim Manolis, William Marengo, Lorrene Maroney, Guy Marshall, Mark Martell, Mark Martin, Becky Marty, Peter Mattson, Jim Mattsson, Laura May, Joseph McCartin, Joan McKearnan, Gretchen Mehmel, Andrew Mehus, Kenmund Mehus, Scott Mehus, Nicole Menard, Tammy Mercer, Bob Merrill, Cody Messerschmidt, Tyler Messerschmidt, Scott Meyer, Mike Meyman, Nathan Mielke, Gabe Miller, Mary Miller, Molly Miller, Richard Miller, Tony Mitzel, Cory Mlodik, Kirk Mona, Kent Montgomery, Andres Morantes, Matt Morin, Steve Mortensen, Susan Morton, Michael Moul, Thomas Mowbray, Mary Munn, Bruce Munson, Carmelita Nelson, Colleen Nelson, Dave Nelson, Deborah Nelson, Warren Nelson, Peter Nichols, Gerald Niemi, Collin Nienhaus, Kathy Nordgaard, Mike North, Jonathan Norum, Timothy Notch, Rochelle Nygaard, Andrew Nyhus, Donna Oglesby, Bridget Olson, John O'Reilly, Earl Orf, Don Orke, Alan Orr, Joan Osgood, Nancy Osvold, Nick Palaia, Barbara Pardo, Andy Paulios, Rebecca Pearson, Larry Pedersen, Bill Penning, Ethan Perry, Ken and Pam Perry, Kurt Peters, Bailey Petersen, Debbie Petersen, Anna Peterson, Jim Peterson, Lori Petrauski, Lee Pfannmuller, Dick Phillips, Matthew Pierce, Jeanne Pietig, Rick Pike, Bill Pinegar, Susan Plankis, Ryan Plantz, Jenni Poliseno, Julia Ponder, Nate Popkin, Adele Porter, Katie Pratt, Tom Prestby, Jerry Pruett, Catherine Rackliffe, Keith Raitz, Don Rakstad, Rachel Ramadhyani, Red Lake Wildlife Management Area, Ron and Maureen Refsnider, Lisa Reid, Chris Reimer, Margaret Rheude, Brenna Richards, Jeremy Ridlbauer, Kim and Cindy Risen, Kathy Rivers, William Robb, Peg Robertsen, Bob Roederer, Deb Rogers, Mariah Rogers, Gary Ronning, Connie Rose, Kristi Rosenquist, Amber Roth, Joe Rudich, Ken Rueter, Carl Runke, Lorna Rupp, Ann Russ, Robert Russell, James Ryan, Bonnie Sample, Dave Sample, Mary Savina, Dan Schad, Eileen Schantz-Hansen, Kim Schaubschlager, Jenny Schaust, Matthew Schaut, Karen Schik, John Schladweiler, Aanan Schlief, Allison Schmidt, Joel Schmidt, Mike Schrage, Robert Schroeder, Roger Schroeder, Jon Schubbe, Jen Schultz, Bill Schuna, Carl Schwartz, Peter Scofield, Dorene Scriven, Blaine Seeliger, Gary Seim, Andrew Sell, Julian Sellers, Stewart Shaw, Mary Shedd, Anna Side, Robert Sime, Tad Simons, Casey Simonson, Larry Sirvio, Beth Siverhus, Neil Skoog, Bridget Skrukrud, Bruce Smisek, Daniel Smith, Drew Smith, Kathleen Smith, Kevin Smith, Riley Smith, Paul Smithson, Jennifer Sobiech-Pariseau, Kenton Spading, Richard Specht, Cari Sprague, Paul Spreitzer, Wally Stadick, Colleen Staeger, Alisha Stalker, James Standefer, Siah St. Clair, Michael Steffes, Cheryl Steinmueller, Sparky

Stensaas, Pamela Stevenson, Sid Stivland, Bill Stjern, David Stocke, Nancy Stokes, Ashley Storms, Arthur Straub, Ben Stubbs, Steve Stucker, Paul Suchanek, Sue Suchy, Superior National Forest Staff, Scott Swanson, Michael Sweet, Lisa Swelland, Mark Tacke, Dan Tallman, Jeanne Tanamachi, Stephanie Tangen, John Tautin, David Taylor, Terry Taylor, Bill Tefft, Kyle Te Poel, Susan Tertell, Adam Thompson, Molly Thompson, Rolf Thompson, Gary Tischer, Matt Toenies, Adam Tome, Cindy Torkelson, Howard Towle, Stacey Trapp, Renee Tressler, Cathy Troendle, Tom Tustison, Bill Unzen, Carol Urness, Sara Vacek, Ken Vail, Aurial Van Der Laar, Luke Vaneps, Amber Beth Van Ningen, Diane Vejtruba, Jean Vertefeuille-Cutler, Jeff Vetsch, Jennifer Vieth, John Voz, Donald Vrchota, Lance Vrieze, Linda Wadsworth, Brenda Wagner, Mark Wagner, Beth Walters, Lucas Wandrie, Helen Wang, Bob and Amy Wangerin, Susan Wasson, Suzanne Weber, Leif Wedul, Garrett Wee, Howard Weinberg, Denis Weis, Jan Welsh, Rick West, Steve Weston, Mary Westra, Linda Whyte, Jane Wickland, Ben Wieland, Terry Wiens, Dennis Wiesenborn, Fern Wilcox, Wildlife Staff of Three Rivers Park District, Larry Wilebski, Krista Wilkowske, Tom Will, Bob Williams, Donna Williams, Jim Williams, Steve Wilson, Suzanne Winckler, Lucas Windingstad, Ron Windingstad, Sylvia Winkelman, Daniel Wohlfert, Kevin Woizeschke, Chris Wood, Christina Wood, Quentin Yoerger, Jeff Zajac, John Zakelj, Mark Zembryki, Jean Zettervall, Nancy Zimmerman, Rhonda Zimmerman, Roy Zimmerman, Ed Zlonis, and Sally Zodrow.

Photographers. During the course of the atlas, volunteers also posted images of birds they observed while they were conducting atlas fieldwork on the website www.mnbba.org. We would like to thank the following individuals for those contributions:

Terence Brashear, David Brislance, Carole Brysky, David Cahlander, Shawn Conrad, John Cyrus, Bob Dunlap, Dianne Endrizzi, Dan Guynn, Jr., Eric Hall, Katie Haws, Mike Hendrickson, Jeanine Hess, Marshall Howe, Linda Johnson, Jan Kraemer, Larry Kraemer, Dee Kuder, Robin LaFortune, Jim Lind, Ellen Lowery, Tammy Mercer, Wayne Michalski, Teri Miller, Cody Neu, Peter Nichols, Andrew Nyhus, Ronald Refsnider, John Schladweiler, Larry Sirvio, Beth Siverhus, Siah St. Clair, Johnnie Stowers, Ben Stubbs, Renee Tressler, Ben Wieland, Larry Wilebski, Ben Wilson, and Shawn Zierman.

Graphics Assistants. Gretchen Seim, Kristen Seim, Nick Walton, and Ed Zlonis.

Institutions and Organizations That Contributed Data to the MNBBA Database

(Specific data contributed are detailed in Appendix B.)

Audubon Minnesota; Cornell Lab of Ornithology; Golden-winged Warbler Working Group; Grand Portage National Monument; Hawk Ridge Bird Observatory; Hormel Nature Center; Midwest Peregrine Society; Minnesota Department of Natural Resources; Minnesota National Guard; Minnesota Ornithologists' Union; Minnesota Prairie-Chicken Society; The Nature Conservancy; Red Lake Wildlife Management Area; Twin Cities Osprey Project; United States Department of Agriculture: US Forest Service; United States Department of the Interior: US Fish and Wildlife Service; US Geological Society, Voyageurs National Park; and University of Minnesota Wildlife Rehabilitation Center.

Institutions and Organizations That Contributed Data and Graphics for the Species Accounts

Hawk Count (www.hawkcount.org).

Hawk Ridge Bird Observatory (www.hawkridge.org).

Minnesota Department of Natural Resources, the Division of Ecological and Water Resources and the Division of Fish and Wildlife (www.dnr.state.mn.us).

National Audubon Society, Christmas Bird Count (CBC) Data. These data are provided by National Audubon and through the generous efforts of Bird Studies Canada and countless volunteers across the Western Hemisphere (www.christmasbirdcount.org).

North American Waterfowl Breeding Population and Habitat Survey (https://catalog.data.gov/dataset/waterfowl-breeding-population-and-habitat-survey).

US Forest Service, Department of Agriculture, for allowing their forest bird monitoring program data on the Chippewa and Superior National Forests to be included in the MNBBA.

US Geological Survey, Patuxent Wildlife Research Center, North American Breeding Bird Survey (www.pwrc.usgs.gov/bbs/). We especially wish to thank Dr. John Sauer, who generously provided us with the most recent Breeding Bird Survey data through 2019, and the thousands of volunteers who each year gather data on thousands of BBS routes throughout North America that provide critical monitoring data on North America's avifauna.

APPENDIX A

Point Count Sampling, Data Review Process, and Data Analysis Methods Used by the Minnesota Breeding Bird Atlas

Data were compiled and summarized by Nicholas G. Walton and Edmund J. Zlonis.

Chapter 1 provides a general overview of the US Public Land Survey System used to establish a sampling grid for the MNBBA and a brief discussion of the three primary means of collecting MNBBA data: first, survey volunteers and paid survey staff; second, staff from the University of Minnesota who conducted point counts in nearly every MNBBA atlas block; and third, data contributed by agencies and organizations that were collected within the five-year sample period.

The information included in the following pages is for those readers interested in more details about the point count protocol employed by the atlas, how survey staff were trained, how the data collected by volunteers was reviewed for quality control, and, finally, the methods employed in the analysis of the data collected.

Point Counts

POINT COUNT METHODOLOGY AND TRAINING

The basis for the systematic point count protocol is a 10-minute observation/listening period with detections divided into one-minute intervals. The basic technique has been developed and used across the globe for thirty years or more (e.g., Hanowski and Niemi 1995b; Hanowski, Lind, et al. 2005; Howe et al. 1997; Niemi, Howe, et al. 2016; Ralph et al. 1995) and has been incorporated into many standard protocols across North America (e.g., Knutson, Danz, et al. 2008; Matsuoka et al. 2014). Extensive analysis on detection of breeding birds has been undertaken and scrutinized for many years (Etterson et al. 2009; Sólymos et al. 2013). The point count protocol not only was an effective technique for gathering standardized data in a time-efficient fashion but also allowed consistent data to be gathered from the entire state of Minnesota.

During a 10-minute point count, all birds seen or heard are recorded (e.g., unlimited distance), but distances from the point are estimated to allow habitat-specific data to be used (e.g., observations within 25, 50, 100 m, and greater than 100 m of the census point) and to determine species-specific effective sampling distance. This is a critical process to aid the calculation of species densities within specific habitat cover types and population sizes in Minnesota. Point counts were completed from the last week of May (in southern Minnesota) to the second week of July (northern Minnesota), but most counts were completed in June. Point counts were gathered from approximately 0.5 hour before to 4 hours after sunrise on days with little wind (less than 15 km/hr) and little or no precipitation. Extensions from sunrise to 6 hours were allowed for counts in western Minnesota owing to the high proportion of windy days and based on the experience of ornithologists in that region. All points were located and marked with a global positioning system (GPS), cover type of vegetation was visually estimated and mapped, and pictures were taken in two directions perpendicular to the road. In addition, incidental breeding evidence was recorded using the MNBBA breeding evidence codes. All data and pictures were entered into an online database. In cases where the point count was not conducted from a road, four pictures were taken in each of the four cardinal directions (north, south, east, and west).

Prior to the field season, recordings of Minnesota bird songs were provided as a learning tool, and all observers were required to pass an identification test of bird songs for 86 species. In our previous efforts, a standard for number of correct responses was established by giving the test to observers who were trained in identifying birds by sound and who had four to five years of field experience. This was done to identify test recordings that were not good representations of songs heard in northern Minnesota or Wisconsin. Based on the results of trained observers, the standard for passing was set at 85 percent correct responses. Songs on the tape were grouped by habitat (e.g., upland deciduous, lowland coniferous) to simulate field cues that would aid in song identification.

Observer field training was also conducted in late May. Observers completed simultaneous practice counts at several points used in the monitoring program. Data were compiled for each observer. The species list and numbers of individuals recorded on these counts by each observer were compared to those of an experienced observer. Deviations in the number counted or species missed were noted and discussed in the field. In addition to field training and testing, all observers were required to have a hearing test to ensure that their hearing was within normal ranges, as established by audiologists, for all frequencies (125 to 8000 hertz).

POINT COUNT SELECTION AND SAMPLING DESIGN

Within each priority block of a township the objective was to sample three 10-minute point counts per township. Sample points were identified using a two-stage process. First, geographic information system (GIS) analysis was used to identify candidate points for all townships before entering the field. Second, once in the field, protocols were followed to determine if the point needed to be moved based on field level observations.

The first point selected within the township was a random point. A random coordinate was selected within the northeast quadrant of each township or an alternative township if the northeast quadrant was too small (e.g., less than 8 square miles). The closest accessible point to a secondary road from the random point was selected for sampling. The second and third points to be sampled within the township were identified within the most abundant and second most abundant habitat in the township. The National Land Cover Data (NLCD) (1999–2003) and Minnesota Gap Analysis Project (GAP) data for the entire state of Minnesota, with all primary habitat types identified at a scale of 30 square meters resolution (Homer et al. 2004), were used to identify primary habitats. An expanded spiral search pattern from the random point was used to locate the nearest (straight line distance) points in each of the dominant and subdominant cover types that were within 50 m of an accessible road. Landcover was filtered so that a point was only considered within the target habitat if 40 percent of the land cover pixels in a 100 m radius represented that cover type. Points were a minimum of 250 m apart to avoid overlap among the point counts. Note that the primary and secondary habitat types sampled were based on the entire township, but we sampled within the priority block. If the primary or secondary habitat did not occur within the priority block or was not accessible, the closest points within the primary and secondary habitats adjacent to the priority block, regardless of direction, were selected. In most cases this was not an issue, except for agricultural areas where wetlands were often the secondary habitat as well as rare. In these cases, it was sometimes necessary to travel reasonably long distances to sample secondary habitat types. Additional details regarding point count selection are available on the MNBBA website, including rules for moving points in the field.

POINT COUNT SAMPLING OF ACCESSIBLE AND INACCESSIBLE TOWNSHIPS

The sample design for priority blocks within townships that were accessible by roads was used throughout a majority of Minnesota. Overall, depending on the density of the road network, between two to five townships (6 to 15 point counts) could be covered in a morning. However, many townships are not accessible by roads and, therefore, required an alternative sampling strategy.

An inaccessible block was defined as a priority block within a township with few or no public roads. This was a common situation in the Boundary Waters Canoe Area Wilderness (BWCAW), the Red Lake Peatlands (RLP), and several other townships in Minnesota. In these situations, access to the block was achieved via water (canoe or boat), trails (all-terrain vehicle or non-motorized bicycle), or, in the case of the RLP, MNDNR's *Marsh Master*.® The most convenient access point within a priority block was identified. The random point was selected from the closest foot-accessible location where observers could initiate their morning survey (e.g., along a trail or portage). Dominant and subdominant habitat types were defined based on the NLCD database and were selected a priori from the access point with the same distance restrictions as described above. These locations were often relocated based on ground reconnaissance, and considerable latitude was given for the use of portages or other access venues (e.g., canoe, boat, or bicycle). In most cases, inaccessible blocks were covered with two individuals and only one priority block was covered in one day. In many cases, these one-day visits were the only time many of these priority blocks were covered, especially if they were in remote locations.

Data Review

During the initial years of the atlas, a technical team of knowledgeable field ornithologists (Bruce Fall, Anthony Hertzel, Ann Kessen, and Steve Stucker) reviewed many of the records collected by field participants to ensure that data protocols were being followed and, when necessary, proper documentation was provided. Regional coordinators were also responsible for reviewing data being entered by participants in their regions, and a few species experts were asked to examine portions of the data being collected.

Then, at the end of the 2013 breeding season, a final, comprehensive review was conducted of all 381,000 MNBBA records to ensure data were as accurate as possible. Issues addressed included the correct use of breeding evidence codes, proper verification of species records outside of their known breeding range, and, ultimately, correct species identification. Identifying many of Minnesota's common species in the field is usually not a problem, but some species can create vexing identification challenges. Even experienced birders, for example, can confuse the rapid, high-pitched trill of a Dark-eyed Junco with that of the common Chipping Sparrow, be challenged to distinguish the song of a rare Philadelphia Vireo from that of the widespread Red-eyed Vireo, or question whether the quick glimpse of a female Sharp-shinned Hawk was really a male Cooper's Hawk. When questions arose, the original observers were often contacted to see if they remembered additional details to help decide on the record's final outcome. When in doubt, records were invalidated.

Draft distribution maps created from the MNBBA detections for each species were also reviewed multiple times to identify records beyond the species' known distribution. Each of these outliers, and any notes that accompanied them or comments that arose from contacting the observer, were carefully considered until they were certified for inclusion in the final MNBBA database. Conducted primarily by the book's authors, the final stage of the review process took nearly six months.

Data Analysis

HABITAT ASSOCIATIONS

We calculated a quantitative value on the strength of a bird species' association to a habitat type using the point count data gathered across Minnesota (Appendix D). We used a value called the percent perfect indicator (PPI) (Dufrêne and Legendre 1997). The PPI is a score for a bird species relative to a habitat type and varies from 0 to 100, where 0 means that individuals of the species were never detected in that habitat and 100 means that all individuals of the species were always found in that habitat type. A value of 100 is seldom obtained. The PPI is calculated for each species for each habitat type and is the relative abundance of the species in the habitat multiplied by the detection frequency for the species in that habitat × 100. The combination of relative abundance and the consistency of a species' use of a habitat (frequency) is a strong indicator of the affinity of the species to that habitat type. We used 13 habitat types derived from the LANDFIRE Existing Vegetation Types (Rollins et al. 2006) and the National Wetlands Inventory (http://www.fws.gov/wetlands/) landcover data sets. In the calculation of the PPI, it was essential not to include habitats with a small number of records because their relative importance is then overemphasized for the species. To this end, we excluded habitats that were dominant in less than 1 percent of the point counts. This resulted in many less common though ecologically important habitats being removed or combined with a similar habitat type for this statewide analysis (e.g., marsh and wet meadow were combined). In these cases, we often report less common but important habitat types in the *Breeding Habitat* description within each species account.

SPECIES DISTRIBUTION MODELS AND POPULATION ESTIMATES

Among our goals was to produce useful species distribution models for as many of Minnesota's breeding birds as possible. This goal is important because, even with hundreds of volunteer observers and thousands of point counts, it is impossible to cover all areas in the state. Species distribution models allow us to fill in areas that were not sampled. Ideally, these models could also be used to estimate statewide species populations. However, reliable population estimates require large sample sizes, and these estimates are poor for species not typically detected from territorial behavior. We chose three modeling strategies to maximize the number of useful models. In total, we created models for 118 species. The modeled results for the species that had adequate sample sizes and acceptable statistical models are included within each respective species account.

In addition to the MNBBA point count data, we used one supplementary point count data set to generate distribution models and population estimates: the Minnesota National Forest Bird Monitoring Program (NFB). NFB is a long-term breeding bird monitoring program implemented in the Chippewa and Superior National Forests of Minnesota. Even though the program was initiated in 1991 and was still ongoing in 2020, we included 961 off-road point count locations sampled during the same time frame as the atlas period, from 2009 to 2013. All point count locations were randomly selected from forest stands and in cover types that have been undersampled in Minnesota. The bird sampling protocols used for point counts in the NFB were compatible with those used in the MNBBA (Hanowski, Lind, et al. 2005; Etterson et al. 2009; Niemi, Howe, et al. 2016). Prior to incorporation of these data in the final models, their inclusion in the overall results was evaluated and determined to have no bias on the resulting models, but the larger sample sizes improved the reliability of the models for many of the less common species.

Breeding Bird Probability Distribution Modeling Strategy 1

This analysis closely followed Ball and colleagues (2016) and used a Poisson Generalized Linear Model (GLM) (McCullagh and Nelder 1989) with a log link. Detection probability and detection distance were accounted for using QPAD (a software program used to calculate species density; see Sólymos et al. 2013). This avenue allowed us to model density (i.e., pairs per 40 ha [100 acres]) by estimating a species' count-specific offset based on the species' effective detection radius (Sólymos et al. 2013). We applied this approach to 69 species that were reliably sampled by point counts; that were detected as singing birds from a minimum of 75 point counts; and that had at least 50 percent of their records noted as being of singing birds. Modeling was conducted in R (R Core Team 2016).

We used a forward "branching" variable selection based on a Bayesian information criterion (Schwarz 1978) to construct models from a suite of 44 candidate covariates. For instance, variables were grouped into similar types (land use/landcover, disturbance, landcover structure, landscape metrics, and climate), and forward selection was applied to one group at a time. After applying forward selection to the first group of candidate covariates, all variables found to improve the model from that group became the starting point (i.e., null model) for the selection process applied to the second group of candidate covariates. This process continued until all variable groups were evaluated.

We used bootstrap aggregation (Breiman 1996; Efron 2014) to estimate the variability in the selection process and for population estimates. We iterated this selection process 240 times to produce 240 models and predictions per species. The first run used the full data set (minus a 10 percent holdout data set), and all subsequent iterations used a random selection of the data with replacement. The random selection was stratified by year and region to address spatial and temporal issues.

The 240 models per species were used to produce statewide population estimates and predicted species distribution maps. The median population estimate (i.e., a single model) was used as our estimate for the Minnesota state population with 2.5 percent and 97.5 percent quantiles as our 95 percent confidence interval (CI). For the predictive species distribution maps, the median prediction for each cell was used. It was possible that the model that produced the median prediction for a given cell might not be the model that produced the median prediction for an adjacent cell. This bootstrap method has a smoothing effect on the predicted distribution and reduces the probability of extreme predictions.

The Minnesota Supercomputing Institute (MSI; www.msi.umn.edu) at the University of Minnesota provided access to their computer cluster and greatly reduced the amount of time necessary to run these analyses.

Breeding Bird Probability Distribution Modeling Strategy 2

For species recorded on at least 75 point counts that did not meet the requirements for strategy 1 (e.g., territorial behavior), we used the same method but without the QPAD offset for a species-specific distance estimation. This resulted in predictions in units of expected birds per 10-minute point count rather than by unit area. Birds in this group were reliably sampled by point counts. These included non-passerines or passerines that were primarily detected by means other than singing (e.g., Killdeer, Ruby-throated Hummingbird, and Blue Jay). Because we did not account for detectability in this case, some species in this group had variability that was too high to model with a GLM as described above. In those cases, we used a negative binomial GLM. Otherwise, the method was very similar to that described in strategy 1 above. We applied strategy 2 to 28 species.

Breeding Bird Probability Distribution Modeling Strategy 3

We used MaxEnt (Phillips et al. 2006; Elith et al. 2011) to model the distributions of species that did not meet the criteria for strategies 1 or 2 but had at least 20 geo-referenced records (point count and/or volunteer data). This strategy resulted in predictions that range from best to worst habitat. While this does not allow numeric estimates, it does allow us to estimate where many rare or elusive species (e.g., Common Nighthawk, Sandhill Crane, and Black Tern) are more or less likely to be breeding. Our MaxEnt analysis was similar to that used by Zlonis and colleagues (2017). We used a suite of 36 potential covariates derived from the same set of variables referenced in strategy 1 above. Model development and selection were implemented using the ENMeval package (Muscarella et al. 2014) in program R (R Core Team 2016). To simplify interpretation, we limited MaxEnt analyses to linear and quadratic transformations of covariates (Merow et al. 2013). We also applied five-fold cross-validation (using 80 percent of the data) and averaged models from these five partitions. To account for bias in volunteer reporting locations (e.g., more likely to be located near cities and roads), we used a targeted background approach where the environmental characteristics available to each species were characterized from the locations of volunteer observations as opposed to random locations. The top model was determined using Akaike Information Criterion corrected for small sample size (AICc) (Burnham and Anderson 2002), and raw MaxEnt output was interpreted as an index of habitat suitability (Merow et al. 2013; Merow and Silander 2014). This method was applied to 21 species.

APPENDIX B

Organizations Contributing Data to the Minnesota Breeding Bird Atlas

The Minnesota Breeding Bird Atlas was designed to provide a comprehensive assessment of the status of Minnesota's breeding birds during the years 2009–13. We accomplished this in three ways: first, by enlisting hundreds of volunteers who contributed data for atlas blocks throughout the state and augmenting their efforts with paid field staff to sample remote and poorly sampled regions; second, by establishing a network of point counts that were surveyed by trained field staff in order to provide a solid, statistical basis for analysis of relative abundance; and, third, by soliciting data from other resource agencies and conservation organizations, placing a particular emphasis on rare species or species that were not well surveyed by the point counts or standard atlas procedures (e.g., aerial waterbird surveys and nocturnal owl surveys). Only data that were collected within the 2009–13 atlas period and that could be attributed accurately to a given atlas block were solicited.

We wish to acknowledge the organizations and agencies listed in the following table who generously contributed additional data to the Minnesota Breeding Bird Atlas. We sincerely appreciate their willingness to share data they collected for the purposes of other important field initiatives. The following list is not only our way of acknowledging their contributions but also, when the atlas is repeated in future years, a resource to ensure that the comparison of atlas efforts is as complete as possible. The specific data sets contributed by each organization are detailed. When available, specific language and/or the observers who made these contributions also are acknowledged.

ORGANIZATIONS CONTRIBUTING DATA TO THE MINNESOTA BREEDING BIRD ATLAS

ORGANIZATION	DATA CATEGORY	DATA SET	SPECIFIC ACKNOWLEDGMENTS
AUDUBON MINNESOTA			
	Chimney Swifts	Chimney Swift Site 2013 Phase I	
CORNELL LAB OF ORNITHOLOGY			
	General Nest Records	Cornell Nest Watch Project 2009–13	Data contributed by David Bonter; we wish to thank the volunteers who collected the nesting data for this study through The Cornell Laboratory of Ornithology's Nest Box Network, The Birdhouse Network, NestWatch, the Smithsonian Migratory Bird Research Center's Neighborhood NestWatch, and various state nest monitoring projects that have contributed their historic data.
		Nest Watch 09	
		Nest Watch 10	
GOLDEN-WINGED WARBLER WORKING GROUP			
	Golden-winged Warblers		Data contributed by Amber Roth.
GRAND PORTAGE NATIONAL MONUMENT			
	General Records		Data contributed by Amy Seitz of the Grand Portage Reservation.
HAWK RIDGE BIRD OBSERVATORY			
	Owls	Great Lakes Owl Surveys	Data contributed by Dave Grosshuesch, who wishes to acknowledge Gerald Niemi and Ron Regal, University of Minnesota Duluth, for their support and analysis and the Minnesota Department of Natural Resources for funding.
HORMEL NATURE CENTER			
	General Records	Hormel NC Bird Surveys, 2009–13	Data contributed by Richard Smaby.
MIDWEST PEREGRINE SOCIETY			
	Peregrine Falcon		
MINNESOTA DEPARTMENT OF NATURAL RESOURCES			
	American Woodcock	Volunteer Woodcock Banding	Data contributed by Earl Johnson.
	Common Loons	Loon Watcher Survey	Data from the Loon Watcher Survey, Minnesota Loon Monitoring Program, Shallow Lakes Program, and Sensitive Lakeshore Program were contributed by Nongame Staff in the Brainerd Office, Pam Perry and Kevin Woizeschke.
		Minnesota Loon Monitoring Program (MLMP)	
	General Records	Shallow Lakes Program	
		Sensitive Lakeshore Program	
		All Bird Surveys	Data contributed by the Red Lake Wildlife Management Area.

ORGANIZATIONS CONTRIBUTING DATA TO THE MINNESOTA BREEDING BIRD ATLAS, *continued*

ORGANIZATION	DATA CATEGORY	DATA SET	SPECIFIC ACKNOWLEDGMENTS
Minnesota Department of Natural Resources *(continued)*			
	Grassland Birds	Grassland Bird Distribution and Habitat Surveys, Southwestern Minnesota	Data contributed by Lisa Harn.
		Grassland Surveys: Lincoln County Grassland Surveys: Yellow Medicine County	Data from Lincoln and Yellow Medicine Counties contributed by Lisa Gelvin-Innvaer.
		SPICE (Sustaining Prairies in a Changing Environment) Surveys, 2009–13	Data contributed by Daren Carlson.
	Northern Goshawk	Northern Goshawk Surveys	Data contributed by Gaea Crozier, who wishes to acknowledge the Chippewa National Forest, Superior National Forest, and MNDNR staff and contractors for providing data.
	Prairie Grouse	Prairie Chicken Survey	
		Sharp-tailed Grouse Dancing Ground Survey	
	Rare Species	Natural Heritage Information System	
	Ring-necked Duck	Ring-necked Duck Surveys	Data contributed by Christine Herwig, who wishes to acknowledge the Wetland Wildlife Populations and Research Group for their field efforts.
	Ruffed Grouse	Ruffed Grouse Drumming Count Survey	
	Sandhill Cranes	Sandhill Crane Surveys, Northwestern Minnesota	Data contributed by Jeff Lawrence, who wishes to acknowledge the Minnesota Department of Natural Resources and the US Fish and Wildlife Service, Webless Migratory Gamebird Grant Program, for their support.
	Scientific and Natural Areas Bird Surveys	Scientific and Natural Areas Avian Inventory Data	Data contributed by Katherine Haws.
	Trumpeter Swans	Trumpeter Swan Reports and Surveys	Data contributed by Christine Herwig.
MINNESOTA NATIONAL GUARD			
	Camp Ripley	2009 Camp Ripley Songbird Point Count Survey	Nancy Dietz and Brian Dirks helped enter data and coordinated the entry of Camp Ripley data by the following individuals: Bill Brown, Mary Lee, Chris Larson, Jason Linkert, Wade Lund, Laura May, Mike Meyman, Timothy Notch, Adam Thompson, Matt Toenies, and Beth Walters.
		2010 Ruffed Grouse Drumming Survey	
		2010 Camp Ripley Bluebird Nest-box Monitoring	

ORGANIZATION	DATA CATEGORY	DATA SET	SPECIFIC ACKNOWLEDGMENTS
Minnesota National Guard *(continued)*			
		2010 Camp Ripley Call-broadcast Survey	
		2010 Camp Ripley Wood Duck Nest-box Monitoring	
		2011 Camp Ripley Songbird Survey	
		2011 Camp Ripley Owl Survey	
		Camp Ripley Breeding Songbird Survey—AHATS	
MINNESOTA ORNITHOLOGISTS' UNION			
	All Birds		Data contributed by the Minnesota Ornithologists' Union; we specifically wish to acknowledge the efforts of David Cahlander in providing these data.
MINNESOTA PRAIRIE CHICKEN SOCIETY			
	Greater Prairie Chickens and Sharp-tailed Grouse	2010–11 Inventory Data	Data collection by the Minnesota Prairie Chicken Society, which also provided the data.
THE NATURE CONSERVANCY			
	Grassland Birds	Death of a Meadowlark 2013	Data contributed by Marissa Ahlering, who wishes to acknowledge the US Fish and Wildlife Service for providing funding support for the fieldwork.
THREE RIVERS PARK DISTRICT			
	All birds		Data contributed by wildlife staff from regional parks across the District.
TWIN CITIES OSPREY PROJECT			
	Osprey	Three Rivers Park District Osprey Project	
		Allete Powerline Surveys	
UNITED STATES DEPARTMENT OF THE INTERIOR, US GEOLOGICAL SURVEY			
	Grassland Birds	Conservation Reserve Program (CRP) Breeding Birds in Grant County	Data contributed by Lawrence D. Igl, who wishes to acknowledge the following field observers for the Conservation Reserve Program (CRP) and Waterfowl Production Area (WPA) field surveys: Lawrence D. Igl (2009–13), Jeremiah C. Slagter (2009), Kyle L. McLean (2010), Brian J. Chepulis (2011–12), and Joseph Orr (2013).
		Waterfowl Protection Area (WPA) Breeding Birds in Big Stone County	
		WPA Breeding Birds in Clay County	

ORGANIZATIONS CONTRIBUTING DATA TO THE MINNESOTA BREEDING BIRD ATLAS, ***continued***

ORGANIZATION	DATA CATEGORY	DATA SET	SPECIFIC ACKNOWLEDGMENTS
United States Department of the Interior, US Geological Survey *(continued)*			
		WPA Breeding Birds in Cottonwood County	
		WPA Breeding Birds in Lac qui Parle County	
	Common Loons	DNR Special Permit 16508_COLO	Data contributed by Kevin P. Kenow, who wishes to acknowledge Carrol Henderson, Minnesota Department of Natural Resources, for funding support, and Steven Houdek, Luke Fara, and Pete Boma, US Geological Survey, Upper Midwest Environmental Sciences Center, for field support.
		2011 BandingData_COLO	
UNITED STATES FISH AND WILDLIFE SERVICE			
Agassiz National Wildlife Refuge (NWR)	General Bird Surveys	Agassiz Black Tern Survey	Data contributed by Gregg Knutsen.
		Agassiz Breeding Bird Point Count Survey	
		Agassiz Brood Surveys	
		Agassiz Colonial Nesting Waterbird Surveys	
		Agassiz Bald Eagle Aerial Survey	
		Agassiz–Webster Creek Bird Survey	
Bloomington Field Office	Bald Eagles	Incidental Bald Eagle nest sightings	
Division of Migratory Bird Management	American Woodcock	American Woodcock Singing-ground Survey	Data contributed by Becky Rau.
Glacial Ridge NWR	General Bird Surveys	Glacial Ridge NWR Bird Surveys	
Habitat and Population Evaluation Team (HAPET)	Waterfowl and Yellow Rails	HAPET 4 square mile waterfowl productivity survey	
		Student Career Experience Program (SCEP) Yellow Rail Surveys	Data contributed by Anna Sidie-Slettedahl.
Litchfield Wetland Management District (WMD)	General Bird Surveys	Litchfield WMD Bird Surveys	Data contributed by Bridget Olson.
Morris WMD	Grassland Birds	USFWS WPA Grassland Point Count Surveys	Data contributed by Sara Vacek.

ORGANIZATION	DATA CATEGORY	DATA SET	SPECIFIC ACKNOWLEDGMENTS
United States Fish and Wildlife Service *(continued)*			
Rydell NWR	Eastern Bluebirds and Wood Ducks	Rydell NWR Blue Bird Box Survey	Data contributed by Jessica Dowler.
		Rydell NWR Wood Duck Nest Box Survey	
Upper Mississippi River (UMR) NWR	Great Blue Herons (GBH) and Bald Eagles	UMR NWR GBH Nest Survey	Data contributed by Brian Stemper.
		UMR NWR Bald Eagle Survey	
UNITED STATES FOREST SERVICE, SUPERIOR NATIONAL FOREST			
	Forest Birds	Monitoring Avian Productivity and Survivorship (MAPS) data Superior National Forest	Data contributed by Dave Grosshuesch, who wishes to acknowledge the US Forest Service for providing funding support for the project.
UNIVERSITY OF MINNESOTA			
Department of Fisheries, Wildlife, and Conservation Biology	Waterbirds	American White Pelican and Double-crested Cormorant Survey Data	Data contributed by Linda Wires, US Fish and Wildlife Service, and Francesca Cuthbert, University of Minnesota, who wish to acknowledge the US Fish and Wildlife Service and Minnesota Department of Natural Resources for providing funding support for the project.
Minnesota Cooperative Fish and Wildlife Research Unit	Forest Birds	Forest Songbird Study	
		Golden-winged Warbler Study in Tamarac NWR	
Natural Resources Research Institute	Forest Birds	Bird Point Count Data Sets: National Forests (2009–13), Agassiz Lowlands (2013), Aitkin County (2013), Northeastern Minnesota roadless areas (2010–13)	Data contributed by Gerald Niemi, Natural Resources Research Institute, University of Minnesota Duluth.
VOYAGEURS NATIONAL PARK			
	General Birds	Common Loon Productivity Surveys	Data contributed by Steve Windels, who wishes to acknowledge Greg Grabas, Paul Watton, and John Brett with the Canadian Wildlife Service–Ontario, for their field efforts.
		Colonial Waterbird Surveys	
		Forest Breeding Bird Surveys	
		Marsh Monitoring Program	
WILDLIFE REHABILITATION CENTER			
	General Birds	Avian Nursery Admits–2011	Data contributed by Tami Vogel.
		Avian Nursery Admits–2013	

APPENDIX C

Minnesota Breeding Bird Atlas Data Summary

Data were compiled and summarized by Nicholas G. Walton.

The table shows the number of species records for all blocks and priority blocks, plus the percentages of the total blocks or priority blocks by breeding evidence codes (see chapter 1, Table 1.2) for 250 bird species detected during the Minnesota Breeding Bird Atlas (2009–13). The last two columns represent the number of points in which a species was detected on all possible point counts sampled (All, n = 6,919) and on the first randomly selected point count not associated with a primary or secondary habitat (Random, n = 2,314), excluding any revisits to the point.

MINNESOTA BREEDING BIRD ATLAS DATA SUMMARY

	ALL BLOCKS										PRIORITY BLOCKS (N = 2,337)										POINT COUNTS	
	Observed		Possible		Probable		Confirmed		Total		Observed		Possible		Probable		Confirmed		Total		All	Random
Species	Blocks	%	Blocks	%	Blocks	%	Blocks	%	Blocks	Records	Blocks	%	Blocks	%	Blocks	%	Blocks	%	Blocks	Records		
Canada Goose	245	9	955	36	129	5	1,289	49	2,618	4,877	195	14	445	32	69	5	684	49	1,393	2,862	760	260
Mute Swan	1	50	1	50	0	0	0	0	2	4	0	0	1	100	0	0	0	0	1	3	0	0
Trumpeter Swan	67	9	281	36	180	23	244	32	772	1,173	43	13	106	31	82	24	110	32	341	540	17	3
Wood Duck	51	3	662	38	263	15	762	44	1,738	2,828	36	4	311	34	98	11	473	52	918	1,529	103	37
Blue-winged Teal	29	3	504	52	275	28	159	16	967	1,524	21	4	248	49	143	28	99	19	511	826	105	22
Northern Shoveler	16	7	141	58	76	31	11	5	244	314	9	8	68	62	29	26	4	4	110	142	11	3
Gadwall	11	5	93	46	69	34	28	14	201	280	6	7	45	49	29	32	12	13	92	127	14	7
American Wigeon	6	12	25	52	14	29	3	6	48	55	5	24	10	48	3	14	3	14	21	25	0	0
Mallard	188	7	1,110	40	440	16	1,028	37	2,766	5,229	165	11	533	36	230	16	533	36	1,461	2,875	523	158
American Black Duck	10	16	32	52	8	13	12	19	62	73	4	15	10	37	5	19	8	30	27	36	0	0
Northern Pintail	10	22	21	47	7	16	7	16	45	55	4	17	13	57	4	17	2	9	23	28	5	2
Green-winged Teal	11	8	87	65	26	20	9	7	133	154	8	11	43	61	12	17	7	10	70	84	6	0
Canvasback	6	8	29	38	18	23	24	31	77	94	4	11	13	34	11	29	10	26	38	52	5	2
Redhead	8	4	84	41	88	43	26	13	206	274	5	5	31	31	48	48	16	16	100	144	12	3
Ring-necked Duck	11	2	310	52	154	26	126	21	601	820	7	3	127	46	79	29	64	23	277	402	15	5
Lesser Scaup	7	15	27	59	12	26	0	0	46	49	5	17	17	59	7	24	0	0	29	32	2	0
Bufflehead	6	26	2	9	2	9	13	57	23	27	3	27	2	18	0	0	6	55	11	11	0	0
Common Goldeneye	6	2	149	46	25	8	142	44	322	472	3	2	59	43	14	10	62	45	138	220	7	4
Hooded Merganser	23	3	391	44	98	11	386	43	898	1,180	16	3	202	42	44	9	216	45	478	617	14	5
Common Merganser	21	7	152	52	33	11	88	30	294	397	8	6	74	55	16	12	37	27	135	209	14	8
Red-breasted Merganser	3	11	14	52	5	19	5	19	27	36	2	22	4	44	2	22	1	11	9	11	0	0
Ruddy Duck	9	5	103	52	61	31	26	13	199	257	7	7	50	48	32	30	16	15	105	145	11	1
Northern Bobwhite	0	0	1	25	0	0	3	75	4	4	0	0	0	0	0	0	1	100	1	1	0	0
Wild Turkey	19	2	465	49	130	14	338	36	952	1,384	13	2	276	51	73	13	181	33	543	851	161	47
Ruffed Grouse	5	0	389	31	651	51	227	18	1,272	2,239	3	0	232	38	255	42	120	20	610	1,114	158	49
Spruce Grouse	0	0	28	44	14	22	21	33	63	87	0	0	13	39	10	30	10	30	33	50	2	1
Sharp-tailed Grouse	1	0	67	20	253	75	16	5	337	566	1	1	42	33	71	56	13	10	127	204	22	7
Greater Prairie-Chicken	0	0	13	9	119	84	10	7	142	394	0	0	6	15	27	69	6	15	39	106	5	0
Gray Partridge	0	0	42	49	32	37	12	14	86	96	0	0	22	47	20	43	5	11	47	54	2	2
Ring-necked Pheasant	7	0	947	62	380	25	200	13	1,534	2,977	4	0	627	59	292	27	139	13	1,062	2,194	1,625	508
Pied-billed Grebe	10	1	401	56	115	16	190	27	716	1,043	6	2	205	53	53	14	120	31	384	605	99	26
Horned Grebe	0	0	3	75	0	0	1	25	4	4	0	0	0	0	0	0	1	100	1	1	0	0
Red-necked Grebe	2	1	77	39	32	16	86	44	197	296	1	1	36	38	16	17	42	44	95	158	4	3
Eared Grebe	0	0	4	33	3	25	5	42	12	20	0	0	3	43	2	29	2	29	7	7	0	0

	ALL BLOCKS										PRIORITY BLOCKS (N = 2,337)										POINT COUNTS	
	Observed		Possible		Probable		Confirmed		Total		Observed		Possible		Probable		Confirmed		Total		All	Random
Species	Blocks	%	Blocks	%	Blocks	%	Blocks	%	Blocks	Records	Blocks	%	Blocks	%	Blocks	%	Blocks	%	Blocks	Records		
Western Grebe	5	9	24	45	17	32	7	13	53	68	2	6	17	52	13	39	1	3	33	43	4	4
Clark's Grebe	0	0	3	100	0	0	0	0	3	3	0	0	0	0	0	0	0	0	0	0	0	0
Rock Pigeon	37	2	970	53	399	22	418	23	1,824	2,896	22	2	587	53	315	29	179	16	1,103	1,897	350	114
Eurasian Collared-Dove	2	2	46	46	40	40	11	11	99	133	0	0	22	38	30	52	6	10	58	85	0	0
Mourning Dove	17	1	1,439	49	999	34	493	17	2,948	6,125	10	1	695	39	754	43	314	18	1,773	4,117	2,163	701
Yellow-billed Cuckoo	3	2	108	83	18	14	1	1	130	160	2	3	66	84	11	14	0	0	79	99	53	16
Black-billed Cuckoo	0	0	534	85	62	10	29	5	625	739	0	0	358	86	42	10	18	4	418	488	206	66
Common Nighthawk	25	11	152	65	43	18	15	6	235	274	18	17	59	56	25	24	4	4	106	126	8	3
Eastern Whip-poor-will	1	1	94	61	54	35	6	4	155	214	1	2	33	52	26	41	3	5	63	79	1	1
Chimney Swift	36	6	344	56	150	25	82	13	612	836	19	5	206	59	87	25	39	11	351	493	72	26
Ruby-throated Hummingbird	26	2	775	68	231	20	114	10	1,146	1,575	13	2	532	69	170	22	54	7	769	1,069	122	47
Virginia Rail	0	0	207	67	86	28	14	5	307	377	0	0	96	66	43	29	7	5	146	177	7	2
Sora	2	0	474	74	155	24	13	2	644	839	1	0	254	71	94	26	9	3	358	479	128	50
Common Gallinule	0	0	4	33	1	8	7	58	12	35	0	0	1	25	0	0	3	75	4	9	0	0
American Coot	8	2	215	62	41	12	85	24	349	490	6	3	104	58	21	12	49	27	180	265	17	5
Yellow Rail	0	0	32	42	44	58	0	0	76	84	0	0	9	38	15	62	0	0	24	25	7	4
Sandhill Crane	50	4	515	43	331	28	290	24	1,186	1,943	29	5	260	42	189	30	145	23	623	1,146	335	122
American Avocet	0	0	2	50	1	25	1	25	4	4	0	0	2	50	1	25	1	25	4	4	0	0
Killdeer	17	1	1,225	51	520	22	618	26	2,380	4,190	5	0	722	46	392	25	442	28	1,561	3,049	1,500	492
Piping Plover	0	0	0	0	1	50	1	50	2	8	0	0	0	0	0	0	0	0	0	0	0	0
Upland Sandpiper	4	2	177	71	50	20	20	8	251	349	3	2	93	68	30	22	10	7	136	189	67	22
Marbled Godwit	5	4	72	58	26	21	21	17	124	183	4	7	30	52	11	19	13	22	58	93	10	3
American Woodcock	4	1	187	27	426	61	79	11	696	1,026	1	0	92	32	164	57	32	11	289	417	10	3
Wilson's Snipe	8	1	584	52	513	46	17	2	1,122	1,728	4	1	319	50	296	47	14	2	633	1,023	325	107
Spotted Sandpiper	8	2	280	57	161	33	44	9	493	603	5	2	157	58	87	32	21	8	270	326	7	2
Solitary Sandpiper	0	0	3	33	4	44	2	22	9	9	0	0	2	33	2	33	2	33	6	6	2	0
Willet	2	100	0	0	0	0	0	0	2	2	2	100	0	0	0	0	0	0	2	2	0	0
Wilson's Phalarope	1	4	17	61	6	21	4	14	28	31	1	10	5	50	3	30	1	10	10	11	0	0
Franklin's Gull	35	95	1	3	0	0	1	3	37	42	19	95	0	0	0	0	1	5	20	22	1	0
Ring-billed Gull	644	97	1	0	0	0	18	3	663	902	271	98	0	0	0	0	5	2	276	402	62	19
Herring Gull	112	53	65	31	4	2	31	15	212	346	50	50	43	43	2	2	6	6	101	161	13	6
Caspian Tern	14	82	2	12	0	0	1	6	17	17	5	100	0	0	0	0	0	0	5	5	0	0
Black Tern	45	9	302	60	68	14	88	17	503	754	22	9	124	51	48	20	51	21	245	415	46	11

	ALL BLOCKS										PRIORITY BLOCKS (N = 2,337)										POINT COUNTS	
	Observed		Possible		Probable		Confirmed		Total		Observed		Possible		Probable		Confirmed		Total		All	Random
Species	Blocks	%	Blocks	%	Blocks	%	Blocks	%	Blocks	Records	Blocks	%	Blocks	%	Blocks	%	Blocks	%	Blocks	Records		
Common Tern	36	84	1	2	0	0	6	14	43	64	13	100	0	0	0	0	0	0	13	17	2	0
Forster's Tern	25	22	66	57	10	9	15	13	116	157	11	20	32	57	8	14	5	9	56	81	1	0
Common Loon	85	6	491	32	190	12	762	50	1,528	2,856	54	8	254	38	113	17	245	37	666	1,383	349	126
Double-crested Cormorant	549	91	0	0	0	0	57	9	606	1,165	302	96	0	0	0	0	14	4	316	532	107	34
American White Pelican	684	98	0	0	0	0	13	2	697	1,270	394	100	0	0	0	0	1	0	395	737	126	45
American Bittern	5	1	352	73	120	25	7	1	484	625	4	2	181	71	66	26	4	2	255	325	62	16
Least Bittern	0	0	80	78	19	18	4	4	103	133	0	0	43	88	5	10	1	2	49	60	0	0
Great Blue Heron	1,533	94	0	0	2	0	90	6	1,625	2,385	887	96	0	0	2	0	31	3	920	1,413	225	74
Great Egret	458	95	0	0	1	0	24	5	483	815	268	97	0	0	1	0	7	3	276	484	75	24
Snowy Egret	2	67	0	0	1	33	0	0	3	3	2	100	0	0	0	0	0	0	2	2	0	0
Little Blue Heron	2	100	0	0	0	0	0	0	2	2	1	100	0	0	0	0	0	0	1	1	0	0
Cattle Egret	7	100	0	0	0	0	0	0	7	8	5	100	0	0	0	0	0	0	5	6	0	0
Green Heron	48	7	490	67	110	15	84	11	732	974	33	8	261	66	62	16	42	11	398	542	42	12
Black-crowned Night-Heron	36	73	0	0	3	6	10	20	49	66	21	72	0	0	3	10	5	17	29	40	0	0
Yellow-crowned Night-Heron	3	100	0	0	0	0	0	0	3	3	2	100	0	0	0	0	0	0	2	2	0	0
Turkey Vulture	1,313	92	3	0	79	6	28	2	1,423	1,911	808	92	3	0	53	6	16	2	880	1,256	85	20
Osprey	86	14	120	19	15	2	397	64	618	938	39	16	68	28	10	4	129	52	246	374	15	4
Northern Harrier	0	0	814	78	180	17	54	5	1,048	1,417	0	0	506	79	101	16	32	5	639	885	123	39
Sharp-shinned Hawk	20	14	105	72	9	6	11	8	145	156	11	13	59	71	6	7	7	8	83	91	4	1
Cooper's Hawk	50	11	266	57	32	7	120	26	468	581	33	12	167	61	20	7	52	19	272	336	18	7
Northern Goshawk	11	30	20	54	1	3	5	14	37	42	8	38	11	52	0	0	2	10	21	25	1	0
Bald Eagle	406	28	309	22	46	3	671	47	1,432	2,113	196	30	203	31	29	4	232	35	660	1,005	48	15
Red-shouldered Hawk	14	7	101	54	30	16	43	23	188	251	9	9	55	56	17	17	17	17	98	131	9	0
Broad-winged Hawk	48	6	504	67	115	15	89	12	756	959	28	7	271	64	81	19	41	10	421	561	43	10
Swainson's Hawk	12	21	39	67	4	7	3	5	58	63	12	29	24	59	3	7	2	5	41	44	4	3
Red-tailed Hawk	194	10	1,105	60	228	12	325	18	1,852	2,620	134	12	658	58	163	14	170	15	1,125	1,693	170	56
Eastern Screech-Owl	2	3	33	57	11	19	12	21	58	64	1	4	16	57	6	21	5	18	28	30	2	1
Great Horned Owl	18	3	274	47	91	16	202	35	585	750	11	4	136	48	43	15	96	34	286	370	18	5
Northern Hawk Owl	1	4	5	22	2	9	15	65	23	28	1	8	3	25	1	8	7	58	12	15	0	0
Burrowing Owl	0	0	0	0	1	100	0	0	1	1	0	0	0	0	0	0	0	0	0	0	0	0
Barred Owl	5	1	346	62	148	27	59	11	558	756	3	1	154	58	80	30	27	10	264	374	16	5
Great Gray Owl	0	0	45	69	11	17	9	14	65	83	0	0	14	64	6	27	2	9	22	32	0	0
Long-eared Owl	1	4	13	46	12	43	2	7	28	34	0	0	9	69	3	23	1	8	13	16	0	0

	ALL BLOCKS										PRIORITY BLOCKS (N = 2,337)										POINT COUNTS	
	Observed		Possible		Probable		Confirmed		Total		Observed		Possible		Probable		Confirmed		Total		All	Random
Species	Blocks	%	Blocks	%	Blocks	%	Blocks	%	Blocks	Records	Blocks	%	Blocks	%	Blocks	%	Blocks	%	Blocks	Records		
Short-eared Owl	1	2	31	76	2	5	7	17	41	47	1	4	18	75	1	4	4	17	24	29	7	3
Boreal Owl	0	0	2	50	1	25	1	25	4	7	0	0	0	0	0	0	0	0	0	0	0	0
Northern Saw-whet Owl	2	1	150	66	69	30	6	3	227	329	0	0	47	57	35	43	0	0	82	122	0	0
Belted Kingfisher	37	3	761	68	214	19	110	10	1,122	1,462	22	3	428	64	140	21	81	12	671	908	87	29
Red-headed Woodpecker	11	2	263	57	85	18	104	22	463	637	4	1	175	57	64	21	66	21	309	434	42	11
Red-bellied Woodpecker	12	1	676	65	200	19	146	14	1,034	1,649	7	1	453	66	139	20	88	13	687	1,118	330	111
Yellow-bellied Sapsucker	8	0	1,040	58	390	22	354	20	1,792	3,262	4	0	635	56	250	22	245	22	1,134	2,033	623	190
American Three-toed Woodpecker	0	0	7	64	0	0	4	36	11	14	0	0	3	75	0	0	1	25	4	7	0	0
Black-backed Woodpecker	2	2	86	80	7	7	12	11	107	158	1	2	31	72	3	7	8	19	43	63	2	1
Downy Woodpecker	14	1	1,132	65	283	16	301	17	1,730	2,724	7	1	734	64	206	18	200	17	1,147	1,840	304	102
Hairy Woodpecker	18	1	1,138	68	255	15	269	16	1,680	2,460	10	1	662	64	197	19	172	17	1,041	1,522	198	62
Northern Flicker	24	1	1,811	68	503	19	325	12	2,663	4,468	12	1	1,084	64	379	22	216	13	1,691	3,004	705	229
Pileated Woodpecker	19	1	1,103	74	246	17	118	8	1,486	2,393	9	1	640	73	171	19	62	7	882	1,401	302	103
American Kestrel	46	3	877	65	273	20	156	12	1,352	1,782	39	5	521	61	185	22	106	12	851	1,176	72	21
Merlin	12	4	152	52	71	24	56	19	291	401	5	3	96	63	27	18	25	16	153	207	12	4
Peregrine Falcon	17	22	9	11	4	5	49	62	79	132	11	55	3	15	1	5	5	25	20	29	1	1
Great Crested Flycatcher	7	0	1,185	62	589	31	115	6	1,896	3,103	5	0	747	58	459	36	78	6	1,289	2,182	896	270
Western Kingbird	2	1	80	56	42	29	20	14	144	166	0	0	40	52	29	38	8	10	77	90	7	4
Eastern Kingbird	15	1	1,123	49	784	34	352	15	2,274	3,672	10	1	591	42	580	41	241	17	1,422	2,435	503	140
Olive-sided Flycatcher	2	1	133	79	32	19	1	1	168	217	1	1	66	73	22	24	1	1	90	116	26	7
Eastern Wood-Pewee	3	0	1,228	64	620	32	60	3	1,911	3,018	2	0	748	58	505	39	34	3	1,289	2,098	789	238
Yellow-bellied Flycatcher	1	0	282	64	148	33	13	3	444	889	0	0	127	56	91	40	7	3	225	440	129	45
Acadian Flycatcher	1	4	6	24	14	56	4	16	25	34	0	0	0	0	6	75	2	25	8	12	0	0
Alder Flycatcher	4	0	827	60	517	38	26	2	1,374	2,501	1	0	437	50	410	47	23	3	871	1,586	716	240
Willow Flycatcher	2	1	290	74	89	23	11	3	392	592	2	1	152	66	67	29	8	3	229	332	62	17
Least Flycatcher	3	0	1,150	56	823	40	75	4	2,051	3,774	1	0	615	47	629	48	60	5	1,305	2,511	817	279
Eastern Phoebe	6	0	835	47	297	17	651	36	1,789	2,809	1	0	548	47	221	19	397	34	1,167	1,959	426	138
Say's Phoebe	0	0	0	0	0	0	1	100	1	2	0	0	0	0	0	0	1	100	1	2	0	0

	ALL BLOCKS										PRIORITY BLOCKS (N = 2,337)										POINT COUNTS	
	Observed		Possible		Probable		Confirmed		Total		Observed		Possible		Probable		Confirmed		Total		All	Random
Species	Blocks	%	Blocks	%	Blocks	%	Blocks	%	Blocks	Records	Blocks	%	Blocks	%	Blocks	%	Blocks	%	Blocks	Records		
White-eyed Vireo	0	0	1	100	0	0	0	0	1	1	0	0	1	100	0	0	0	0	1	1	1	1
Bell's Vireo	0	0	9	36	12	48	4	16	25	32	0	0	4	36	6	55	1	9	11	15	1	0
Yellow-throated Vireo	1	0	677	67	306	30	30	3	1,014	1,457	0	0	414	61	248	36	21	3	683	1,011	168	53
Blue-headed Vireo	1	0	375	73	120	23	18	4	514	915	0	0	202	65	98	32	9	3	309	531	86	24
Philadelphia Vireo	0	0	18	82	1	5	3	14	22	32	0	0	10	100	0	0	0	0	10	14	0	0
Warbling Vireo	4	0	1,053	65	525	32	42	3	1,624	2,409	3	0	629	57	435	40	34	3	1,101	1,726	483	158
Red-eyed Vireo	2	0	1,288	44	1,479	50	183	6	2,952	7,380	1	0	581	34	1,003	59	111	7	1,696	4,397	2,609	861
Loggerhead Shrike	0	0	13	39	6	18	14	42	33	50	0	0	5	38	3	23	5	38	13	19	1	0
Canada Jay	3	1	242	63	44	11	96	25	385	734	2	1	105	54	23	12	65	33	195	358	50	21
Blue Jay	17	1	1,906	62	722	24	422	14	3,067	6,321	8	0	1,111	58	542	29	239	13	1,900	4,003	1,850	596
Black-billed Magpie	6	2	198	57	81	23	60	17	345	551	6	3	96	48	52	26	45	23	199	346	58	24
American Crow	136	4	2,198	62	508	14	728	20	3,570	7,275	71	3	1,213	59	408	20	369	18	2,061	4,651	3,030	988
Common Raven	120	9	845	64	127	10	226	17	1,318	2,162	73	10	437	60	104	14	119	16	733	1,270	358	113
Black-capped Chickadee	19	1	1,407	52	698	26	603	22	2,727	5,646	8	0	743	44	521	31	407	24	1,679	3,686	1,287	410
Boreal Chickadee	1	1	96	68	23	16	21	15	141	219	1	1	42	56	18	24	14	19	75	120	11	5
Tufted Titmouse	1	3	15	52	10	34	3	10	29	36	1	5	10	48	8	38	2	10	21	27	1	1
Horned Lark	4	0	984	66	413	28	79	5	1,480	2,327	1	0	707	62	366	32	64	6	1,138	1,906	1,396	518
Bank Swallow	26	4	238	38	59	9	305	49	628	809	16	5	100	28	36	10	199	57	351	474	47	20
Tree Swallow	30	1	1,167	46	310	12	1,034	41	2,541	4,820	20	1	590	39	220	14	691	45	1,521	3,021	720	229
Northern Rough-winged Swallow	17	4	254	55	84	18	110	24	465	615	13	4	163	53	57	19	73	24	306	430	43	10
Purple Martin	11	2	153	28	23	4	350	65	537	878	9	3	67	25	17	6	172	65	265	418	29	9
Barn Swallow	24	1	1,176	44	441	17	1,022	38	2,663	4,863	11	1	596	36	335	20	714	43	1,656	3,417	1,323	418
Cliff Swallow	28	2	407	28	83	6	919	64	1,437	2,498	18	2	197	22	58	6	643	70	916	1,628	236	62
Ruby-crowned Kinglet	0	0	258	64	134	33	13	3	405	916	0	0	129	57	87	39	9	4	225	461	83	28
Golden-crowned Kinglet	3	1	306	63	133	27	47	10	489	981	1	0	166	56	92	31	37	12	296	572	132	44
Cedar Waxwing	56	3	1,374	65	448	21	249	12	2,127	3,277	38	3	814	61	338	25	155	12	1,345	2,171	934	311
Red-breasted Nuthatch	3	0	814	64	349	27	115	9	1,281	2,816	0	0	419	57	246	34	66	9	731	1,661	635	189
White-breasted Nuthatch	12	1	1,136	67	350	21	209	12	1,707	2,793	8	1	692	63	263	24	128	12	1,091	1,822	446	146
Brown Creeper	1	0	354	74	96	20	28	6	479	709	0	0	194	72	61	23	16	6	271	400	92	34
Blue-gray Gnatcatcher	2	1	168	57	70	24	57	19	297	432	1	0	116	58	51	26	32	16	200	291	77	28
House Wren	3	0	1,170	49	682	28	547	23	2,402	4,750	2	0	687	43	571	36	345	21	1,605	3,405	1,596	524

	ALL BLOCKS										PRIORITY BLOCKS (N = 2,337)										POINT COUNTS	
	Observed		Possible		Probable		Confirmed		Total		Observed		Possible		Probable		Confirmed		Total		All	Random
Species	Blocks	%	Blocks	%	Blocks	%	Blocks	%	Blocks	Records	Blocks	%	Blocks	%	Blocks	%	Blocks	%	Blocks	Records		
Winter Wren	2	0	499	63	264	34	21	3	786	1,837	2	0	238	55	180	41	14	3	434	1,038	277	78
Sedge Wren	3	0	1,082	53	896	44	56	3	2,037	3,555	1	0	557	45	635	51	42	3	1,235	2,204	821	238
Marsh Wren	4	0	551	54	413	41	48	5	1,016	1,663	1	0	275	46	280	47	36	6	592	985	235	54
Carolina Wren	4	57	0	0	3	43	0	0	7	11	3	75	0	0	1	25	0	0	4	6	0	0
Gray Catbird	4	0	1,265	54	724	31	368	16	2,361	4,422	4	0	730	47	569	36	259	17	1,562	3,086	902	312
Brown Thrasher	12	1	913	65	311	22	166	12	1,402	2,091	4	0	615	62	255	26	119	12	993	1,536	408	131
Northern Mockingbird	0	0	8	100	0	0	0	0	8	9	0	0	5	100	0	0	0	0	5	5	1	0
European Starling	11	0	970	43	245	11	1,026	46	2,252	4,055	7	1	478	34	204	15	701	50	1,390	2,830	665	225
Eastern Bluebird	8	0	592	34	326	19	796	46	1,722	3,313	3	0	326	32	209	20	485	47	1,023	1,910	350	131
Veery	4	0	1,053	53	875	44	72	4	2,004	4,318	2	0	483	41	656	55	41	3	1,182	2,706	1,462	478
Swainson's Thrush	1	0	249	69	99	27	12	3	361	859	0	0	125	62	68	34	7	4	200	451	122	34
Hermit Thrush	1	0	659	57	436	38	66	6	1,162	2,691	1	0	315	48	304	46	37	6	657	1,493	476	157
Wood Thrush	3	1	321	74	96	22	14	3	434	585	1	0	208	72	74	26	7	2	290	393	97	29
American Robin	4	0	1,644	39	1,056	25	1,509	36	4,213	10,227	0	0	597	26	690	30	989	43	2,276	6,404	4,039	1,353
House Sparrow	8	0	738	37	411	21	822	42	1,979	3,675	4	0	361	30	322	27	525	43	1,212	2,416	417	145
Sprague's Pipit	1	33	1	33	1	33	0	0	3	4	0	0	0	0	0	0	0	0	0	0	0	0
Evening Grosbeak	5	3	105	64	45	27	10	6	165	247	4	4	58	62	29	31	3	3	94	141	21	7
House Finch	12	1	356	41	249	29	251	29	868	1,457	4	1	238	44	169	31	134	25	545	890	110	48
Purple Finch	1	0	469	68	164	24	59	9	693	979	0	0	293	64	125	27	37	8	455	640	120	39
Red Crossbill	0	0	38	97	0	0	1	3	39	44	0	0	17	100	0	0	0	0	17	17	2	1
White-winged Crossbill	0	0	25	93	1	4	1	4	27	28	0	0	9	100	0	0	0	0	9	9	2	2
Pine Siskin	6	2	176	70	42	17	28	11	252	318	3	2	105	71	29	20	11	7	148	193	41	12
American Goldfinch	23	1	1,424	48	1,257	42	262	9	2,966	5,316	16	1	725	39	988	53	146	8	1,875	3,505	2,597	872
Chestnut-collared Longspur	0	0	0	0	0	0	1	100	1	3	0	0	0	0	0	0	0	0	0	0	0	0
Grasshopper Sparrow	3	1	314	68	121	26	26	6	464	761	2	1	187	68	78	28	10	4	277	390	102	37
Lark Sparrow	5	3	64	40	37	23	53	33	159	265	3	3	45	44	24	23	31	30	103	155	4	1
Lark Bunting	1	100	0	0	0	0	0	0	1	1	1	100	0	0	0	0	0	0	1	1	0	0
Chipping Sparrow	2	0	1,450	46	1,026	33	657	21	3,135	6,349	0	0	719	37	777	40	463	24	1,959	4,377	1,818	661
Clay-colored Sparrow	5	0	1,052	54	689	36	187	10	1,933	3,872	2	0	598	48	511	41	135	11	1,246	2,509	1,144	374
Field Sparrow	4	1	310	58	176	33	46	9	536	1,039	2	1	197	56	126	36	25	7	350	652	202	64
Dark-eyed Junco	1	1	112	72	31	20	12	8	156	241	1	1	62	70	20	22	6	7	89	149	34	4
White-throated Sparrow	2	0	585	41	711	50	115	8	1,413	4,207	1	0	217	29	444	60	84	11	746	2,337	1,294	417

	ALL BLOCKS										PRIORITY BLOCKS (N = 2,337)										POINT COUNTS	
	Observed		Possible		Probable		Confirmed		Total		Observed		Possible		Probable		Confirmed		Total		All	Random
Species	Blocks	%	Blocks	%	Blocks	%	Blocks	%	Blocks	Records	Blocks	%	Blocks	%	Blocks	%	Blocks	%	Blocks	Records		
Vesper Sparrow	5	0	958	60	529	33	109	7	1,601	2,768	3	0	643	53	482	40	85	7	1,213	2,263	1,356	491
LeConte's Sparrow	1	0	297	67	143	32	4	1	445	664	1	0	166	60	107	39	2	1	276	417	145	41
Nelson's Sparrow	0	0	47	81	11	19	0	0	58	68	0	0	19	95	1	5	0	0	20	25	1	0
Henslow's Sparrow	0	0	67	68	29	29	3	3	99	185	0	0	20	54	16	43	1	3	37	79	0	0
Savannah Sparrow	4	0	1,234	56	785	36	166	8	2,189	4,012	4	0	698	48	636	43	131	9	1,469	2,771	1,508	470
Song Sparrow	6	0	1,740	45	1,495	39	587	15	3,828	9,019	3	0	696	31	1,118	50	405	18	2,222	5,856	3,263	1,066
Lincoln's Sparrow	1	0	204	70	77	26	11	4	293	518	1	1	97	58	60	36	9	5	167	281	50	17
Swamp Sparrow	4	0	1,388	56	945	38	152	6	2,489	5,057	1	0	635	44	714	49	107	7	1,457	3,030	856	265
Eastern Towhee	6	2	149	57	91	35	14	5	260	485	3	2	96	57	61	37	7	4	167	270	40	20
Yellow-breasted Chat	0	0	3	75	1	25	0	0	4	5	0	0	2	100	0	0	0	0	2	2	0	0
Yellow-headed Blackbird	16	2	372	46	302	38	110	14	800	1,426	11	2	189	39	195	41	85	18	480	939	248	73
Bobolink	5	0	857	48	794	44	148	8	1,804	3,376	3	0	486	41	585	49	108	9	1,182	2,167	1,072	326
Eastern Meadowlark	8	2	290	59	155	32	37	8	490	730	6	2	175	54	112	35	30	9	323	500	127	38
Western Meadowlark	2	0	553	67	222	27	53	6	830	1,385	2	0	324	61	162	31	40	8	528	904	445	148
Orchard Oriole	2	0	370	59	196	31	60	10	628	874	2	1	209	55	130	34	37	10	378	516	54	15
Baltimore Oriole	9	0	1,050	51	522	25	483	23	2,064	3,423	5	0	662	49	382	28	314	23	1,363	2,379	646	211
Red-winged Blackbird	5	0	1,053	27	1,478	38	1,307	34	3,843	9,496	1	0	331	16	930	44	844	40	2,106	6,000	3,421	1,127
Brown-headed Cowbird	15	1	1,369	52	1,022	39	240	9	2,646	5,170	2	0	853	47	816	45	145	8	1,816	3,804	2,594	875
Brewer's Blackbird	6	1	265	49	142	26	126	23	539	728	4	1	145	40	112	31	103	28	364	511	98	32
Common Grackle	19	1	1,194	43	562	20	1,027	37	2,802	5,841	12	1	571	34	363	22	742	44	1,688	3,968	1,833	631
Great-tailed Grackle	1	12	5	62	1	12	1	12	8	11	0	0	2	67	1	33	0	0	3	3	0	0
Ovenbird	4	0	951	44	1,136	52	89	4	2,180	5,755	0	0	413	34	745	61	59	5	1,217	3,320	1,923	615
Louisiana Waterthrush	0	0	6	60	3	30	1	10	10	12	0	0	4	100	0	0	0	0	4	4	1	1
Northern Waterthrush	1	0	326	67	152	31	10	2	489	876	1	0	168	58	113	39	7	2	289	534	111	36
Golden-winged Warbler	2	0	534	66	232	29	37	5	805	1,329	2	0	328	61	188	35	22	4	540	896	302	98
Blue-winged Warbler	3	2	78	60	37	29	11	9	129	213	1	1	56	64	26	30	5	6	88	138	49	19
Black-and-white Warbler	4	0	941	57	634	38	83	5	1,662	3,423	0	0	462	47	453	46	61	6	976	2,165	1,005	292
Prothonotary Warbler	0	0	10	36	9	32	9	32	28	40	0	0	3	23	4	31	6	46	13	20	1	0
Tennessee Warbler	1	1	140	78	36	20	2	1	179	298	1	1	64	71	24	27	1	1	90	128	48	17

	ALL BLOCKS										PRIORITY BLOCKS (N = 2,337)										POINT COUNTS	
	Observed		Possible		Probable		Confirmed		Total		Observed		Possible		Probable		Confirmed		Total		All	Random
Species	Blocks	%	Blocks	%	Blocks	%	Blocks	%	Blocks	Records	Blocks	%	Blocks	%	Blocks	%	Blocks	%	Blocks	Records		
Nashville Warbler	2	0	670	44	753	49	104	7	1,529	4,370	2	0	289	34	492	57	78	9	861	2,402	1,309	416
Connecticut Warbler	0	0	113	70	46	29	2	1	161	263	0	0	55	65	27	32	2	2	84	134	46	13
Mourning Warbler	0	0	511	52	404	41	60	6	975	1,988	0	0	272	43	313	50	44	7	629	1,288	501	161
Kentucky Warbler	0	0	2	50	2	50	0	0	4	10	0	0	0	0	1	100	0	0	1	3	0	0
Common Yellowthroat	2	0	1,654	43	1,934	50	291	7	3,881	9,609	0	0	663	30	1,367	61	202	9	2,232	6,001	3,686	1,168
Hooded Warbler	3	20	3	20	4	27	5	33	15	26	1	20	0	0	1	20	3	60	5	10	0	0
American Redstart	5	0	1,039	48	937	44	169	8	2,150	4,069	2	0	592	44	642	48	114	8	1,350	2,719	1,107	389
Cape May Warbler	4	2	169	70	59	24	11	5	243	454	3	2	88	61	48	33	5	3	144	233	87	23
Cerulean Warbler	0	0	13	39	15	45	5	15	33	45	0	0	7	41	9	53	1	6	17	25	2	0
Northern Parula	2	0	420	62	238	35	16	2	676	1,388	2	1	211	53	174	44	9	2	396	824	270	89
Magnolia Warbler	3	0	415	61	235	35	23	3	676	1,683	2	1	207	53	171	43	14	4	394	951	323	110
Bay-breasted Warbler	0	0	23	77	3	10	4	13	30	46	0	0	15	79	2	11	2	11	19	21	7	2
Blackburnian Warbler	3	0	462	59	285	36	36	5	786	1,622	2	0	241	51	210	44	22	5	475	986	361	124
Yellow Warbler	4	0	1,373	50	1,095	40	290	10	2,762	5,333	2	0	722	43	762	45	195	12	1,681	3,417	1,447	438
Chestnut-sided Warbler	1	0	887	53	675	41	103	6	1,666	3,781	0	0	420	42	508	51	72	7	1,000	2,395	1,271	420
Black-throated Blue Warbler	0	0	100	72	37	27	2	1	139	317	0	0	44	64	24	35	1	1	69	144	25	10
Palm Warbler	0	0	118	62	57	30	16	8	191	356	0	0	61	55	40	36	9	8	110	183	16	10
Pine Warbler	2	0	383	65	184	31	22	4	591	988	0	0	198	58	130	38	13	4	341	570	121	33
Yellow-rumped Warbler	2	0	557	58	340	35	65	7	964	1,970	2	0	279	50	238	42	44	8	563	1,226	439	130
Yellow-throated Warbler	1	50	0	0	0	0	1	50	2	2	0	0	0	0	0	0	0	0	0	0	1	0
Black-throated Green Warbler	1	0	439	55	320	40	35	4	795	1,741	0	0	229	48	227	48	21	4	477	1,052	363	122
Canada Warbler	0	0	291	60	162	33	33	7	486	1,059	0	0	130	49	118	44	20	7	268	579	195	66
Wilson's Warbler	0	0	43	75	12	21	2	4	57	76	0	0	16	55	11	38	2	7	29	37	4	2
Summer Tanager	1	25	1	25	2	50	0	0	4	5	1	50	0	0	1	50	0	0	2	3	1	0
Scarlet Tanager	3	0	703	72	239	24	34	3	979	1,470	1	0	423	69	171	28	21	3	616	913	181	53
Northern Cardinal	10	1	532	51	333	32	177	17	1,052	2,166	4	1	305	48	233	37	90	14	632	1,397	481	152
Rose-breasted Grosbeak	10	0	1,406	61	689	30	187	8	2,292	4,097	6	0	801	55	529	36	131	9	1,467	2,656	962	299
Blue Grosbeak	2	4	29	56	17	33	4	8	52	64	0	0	19	58	12	36	2	6	33	37	3	2
Indigo Bunting	14	1	887	57	595	38	67	4	1,563	2,590	8	1	545	51	464	43	51	5	1,068	1,862	644	214
Dickcissel	5	1	431	44	492	51	45	5	973	1,672	2	0	216	35	381	61	27	4	626	1,107	113	36

APPENDIX D

Percent Perfect Indicator Values on Bird Habitat Affinity

Data were compiled and summarized by Nicholas G. Walton.

The table shows percent perfect indicator (PPI) values for 124 bird species recorded in 13 habitat types during 6,919 point counts randomly gathered in townships throughout Minnesota (2009–13). PPI values measure the strength of an association of a species with the habitat type and can range from 0 to 100; 0 means the species was never detected in the habitat type and 100 indicates all detected individuals of the species were always found in the habitat type. PPI values are calculated by a combination of number of detections and the frequency that a species is detected in a habitat type (Dufrêne and Legendre 1997).

PERCENT PERFECT INDICATOR VALUES ON BIRD HABITAT AFFINITY

Species	Bog	Cropland	Developed	Lowland Coniferous Forest	Marsh and Wet Meadow	Northern Hardwoods	Northern Mixed Forests	Oak Forest	Pine–Oak Barrens	Pine Forest	Shrub Wetland	Upland Coniferous Forest	Upland Grassland
Wild Turkey	0	0	0	0	0	0	0	3	0	0	0	0	1
Ruffed Grouse	1	0	0	1	0	0	2	0	1	0	0	3	0
Ring-necked Pheasant	0	14	0	0	4	0	0	1	0	0	0	0	10
Rock Pigeon	0	2	0	0	0	0	0	1	0	0	0	0	0
Mourning Dove	0	8	6	0	4	1	0	6	1	0	1	0	7
Black-billed Cuckoo	0	0	0	0	1	0	0	0	0	0	1	0	1
Chimney Swift	0	0	11	0	0	0	0	0	0	0	0	0	0
Ruby-throated Hummingbird	0	0	0	0	0	1	1	1	0	0	0	0	0
Sora	0	0	0	0	1	0	0	0	0	0	0	0	1
Sandhill Crane	0	0	0	0	2	0	0	0	1	0	2	0	1
Killdeer	0	16	1	0	2	0	0	0	0	0	0	0	5
Upland Sandpiper	0	1	0	0	0	0	0	0	0	0	0	0	0
Wilson's Snipe	1	0	0	1	2	0	0	0	0	0	6	0	1
American Bittern	0	0	0	0	1	0	0	0	0	0	1	1	0
Red-headed Woodpecker	0	0	0	0	0	0	0	0	0	0	0	0	0
Red-bellied Woodpecker	0	0	0	0	0	0	0	26	0	0	0	0	1
Yellow-bellied Sapsucker	2	0	0	4	0	2	5	2	1	3	0	7	0
Downy Woodpecker	0	0	1	0	0	1	1	2	1	0	0	0	0
Hairy Woodpecker	1	0	0	1	0	0	1	0	0	0	0	2	0
Northern Flicker	2	0	0	1	1	1	2	1	1	1	0	3	1
Pileated Woodpecker	2	0	0	1	0	1	2	1	0	1	0	3	0
American Kestrel	0	0	0	0	0	0	0	0	0	0	0	0	0
Great Crested Flycatcher	0	1	0	1	1	3	1	3	6	1	1	0	1
Eastern Kingbird	0	2	1	0	1	0	0	0	0	0	0	0	6
Olive-sided Flycatcher	5	0	0	0	0	0	0	0	0	0	0	0	0
Eastern Wood-Pewee	0	0	0	1	0	3	1	9	4	13	0	1	1
Yellow-bellied Flycatcher	22	0	0	2	0	0	0	0	0	0	0	5	0
Alder Flycatcher	6	0	0	4	2	1	1	0	1	0	17	1	0
Willow Flycatcher	0	0	0	0	1	0	0	1	0	0	0	0	0
Least Flycatcher	1	0	0	2	1	2	4	0	2	12	2	5	0
Eastern Phoebe	0	0	1	0	0	1	0	4	2	0	0	0	1
Yellow-throated Vireo	0	0	0	0	0	2	0	1	0	2	0	0	0
Blue-headed Vireo	5	0	0	2	0	0	0	0	0	3	0	4	0
Warbling Vireo	0	2	0	0	2	0	0	0	0	0	0	0	4
Red-eyed Vireo	5	0	0	8	1	9	11	3	7	15	4	9	1

Species	Bog	Cropland	Developed	Lowland Coniferous Forest	Marsh and Wet Meadow	Northern Hardwoods	Northern Mixed Forests	Oak Forest	Pine–Oak Barrens	Pine Forest	Shrub Wetland	Upland Coniferous Forest	Upland Grassland
Canada Jay	7	0	0	1	0	0	0	0	0	0	0	4	0
Blue Jay	7	1	1	5	0	3	5	5	4	7	2	9	1
Black-billed Magpie	0	0	0	0	0	0	0	0	0	0	0	0	1
American Crow	1	4	5	3	1	4	3	8	6	3	2	1	4
Common Raven	5	0	0	2	1	0	2	0	0	6	0	4	0
Black-capped Chickadee	3	0	4	3	0	3	3	4	4	5	1	4	1
Boreal Chickadee	5	0	0	0	0	0	0	0	0	0	0	0	0
Horned Lark	0	39	0	0	0	0	0	0	0	0	0	0	1
Bank Swallow	0	0	0	0	0	0	0	0	0	0	0	0	0
Tree Swallow	0	0	2	0	2	0	0	1	0	0	0	0	1
Northern Rough-winged Swallow	0	0	0	0	0	0	0	2	0	0	0	0	0
Purple Martin	0	0	1	0	0	0	0	0	0	0	0	0	0
Barn Swallow	0	2	2	0	1	0	0	1	0	0	0	0	2
Cliff Swallow	0	0	0	0	2	0	0	0	0	0	0	0	0
Ruby-crowned Kinglet	14	0	0	1	0	0	0	0	0	0	0	4	0
Golden-crowned Kinglet	13	0	0	1	0	0	0	0	0	1	0	9	0
Cedar Waxwing	1	0	4	1	1	1	1	1	3	3	1	1	1
Red-breasted Nuthatch	6	0	0	4	0	1	4	0	1	21	0	11	0
White-breasted Nuthatch	0	0	0	0	1	2	1	8	0	1	0	0	0
Brown Creeper	1	0	0	1	0	0	1	0	0	14	0	3	0
Blue-gray Gnatcatcher	0	0	0	0	0	0	0	21	0	0	0	0	0
House Wren	0	5	9	0	1	1	0	14	1	0	0	0	5
Winter Wren	14	0	0	4	0	0	3	0	0	1	0	14	0
Sedge Wren	0	1	0	0	15	0	0	0	0	0	8	0	4
Marsh Wren	0	0	0	0	18	0	0	0	0	0	1	0	1
Gray Catbird	0	1	1	0	5	1	0	13	2	0	3	0	2
Brown Thrasher	0	2	1	0	0	0	0	0	0	0	0	0	2
European Starling	0	3	8	0	0	0	0	0	0	0	0	0	2
Eastern Bluebird	0	1	0	0	1	0	0	2	0	0	0	0	2
Veery	4	0	0	5	2	5	9	0	4	6	13	4	0
Swainson's Thrush	10	0	0	1	0	0	0	0	0	0	0	7	0
Hermit Thrush	13	0	0	6	0	1	3	0	0	12	1	9	0
Wood Thrush	0	0	0	1	0	1	1	0	0	1	0	0	0
American Robin	4	6	19	4	2	2	4	6	2	2	3	5	5
House Sparrow	0	1	33	0	0	0	0	0	0	0	0	0	0
Evening Grosbeak	0	0	0	0	0	0	0	0	0	0	0	4	0

PERCENT PERFECT INDICATOR VALUES ON BIRD HABITAT AFFINITY, *continued*

Species	Bog	Cropland	Developed	Lowland Coniferous Forest	Marsh and Wet Meadow	Northern Hardwoods	Northern Mixed Forests	Oak Forest	Pine–Oak Barrens	Pine Forest	Shrub Wetland	Upland Coniferous Forest	Upland Grassland
House Finch	0	0	27	0	0	0	0	0	0	0	0	0	0
Purple Finch	1	0	0	1	0	0	1	0	1	1	0	1	0
Pine Siskin	1	0	0	0	0	0	0	0	0	0	0	0	0
American Goldfinch	1	3	8	1	2	2	1	9	2	0	3	0	4
Grasshopper Sparrow	0	0	0	0	0	0	0	0	0	0	0	0	2
Chipping Sparrow	1	2	15	1	0	1	1	7	5	6	0	1	3
Clay-colored Sparrow	0	3	0	0	2	0	0	0	2	0	2	0	13
Field Sparrow	0	0	0	0	0	0	0	8	0	0	0	0	2
Dark-eyed Junco	1	0	0	0	0	0	0	0	0	0	0	1	0
White-throated Sparrow	20	0	0	13	0	0	8	0	0	1	6	17	0
Vesper Sparrow	0	26	0	0	0	0	0	0	0	0	0	0	3
LeConte's Sparrow	0	0	0	0	0	0	0	0	0	0	1	0	3
Savannah Sparrow	0	10	0	0	1	0	0	1	0	0	0	0	15
Song Sparrow	1	5	2	2	7	3	2	6	3	0	7	1	8
Lincoln's Sparrow	4	0	0	0	0	0	0	0	0	0	0	0	0
Swamp Sparrow	4	0	0	2	16	0	0	0	0	0	15	0	1
Eastern Towhee	0	0	0	0	0	0	0	4	0	0	0	0	0
Yellow-headed Blackbird	0	1	0	0	8	0	0	0	0	0	0	0	1
Bobolink	0	7	0	0	1	0	0	0	0	0	0	0	16
Eastern Meadowlark	0	0	0	0	0	0	0	0	0	0	0	0	3
Western Meadowlark	0	4	0	0	1	0	0	0	0	0	0	0	6
Baltimore Oriole	0	1	1	0	1	1	0	4	2	0	0	0	3
Red-winged Blackbird	0	12	3	0	27	1	0	1	0	0	4	0	13
Brown-headed Cowbird	0	8	7	0	3	1	0	11	3	0	1	0	7
Brewer's Blackbird	0	1	0	0	0	0	0	0	0	0	0	0	0
Common Grackle	0	9	14	0	2	0	0	0	0	0	0	0	4
Ovenbird	4	0	0	8	0	8	14	1	6	20	1	14	0
Northern Waterthrush	6	0	0	1	0	0	0	0	0	0	0	3	0
Golden-winged Warbler	0	0	0	1	1	2	2	0	1	2	3	0	0
Black-and-white Warbler	6	0	0	6	0	3	7	0	3	9	3	7	0
Tennessee Warbler	2	0	0	1	0	0	0	0	0	0	0	3	0
Nashville Warbler	20	0	0	10	0	1	8	0	1	7	4	16	0
Connecticut Warbler	7	0	0	1	0	0	0	0	0	0	1	0	0
Mourning Warbler	1	0	0	5	0	1	6	0	0	5	1	8	0
Common Yellowthroat	5	4	1	5	16	4	2	2	1	1	16	1	9
American Redstart	0	0	0	1	1	5	4	7	6	4	2	1	1
Cape May Warbler	1	0	0	1	0	0	0	0	0	0	0	7	0

Species	Bog	Cropland	Developed	Lowland Coniferous Forest	Marsh and Wet Meadow	Northern Hardwoods	Northern Mixed Forests	Oak Forest	Pine–Oak Barrens	Pine Forest	Shrub Wetland	Upland Coniferous Forest	Upland Grassland
Northern Parula	2	0	0	2	0	0	3	0	0	4	0	10	0
Magnolia Warbler	4	0	0	1	0	0	3	0	0	3	0	14	0
Blackburnian Warbler	5	0	0	2	0	0	3	0	1	15	0	12	0
Yellow Warbler	0	1	1	0	13	3	0	1	2	0	8	0	3
Chestnut-sided Warbler	3	0	0	6	1	3	9	0	3	9	4	7	0
Black-throated Blue Warbler	0	0	0	0	0	0	1	0	0	0	0	2	0
Palm Warbler	6	0	0	0	0	0	0	0	0	0	0	0	0
Pine Warbler	0	0	0	0	0	0	0	0	0	48	0	2	0
Yellow-rumped Warbler	12	0	0	3	0	0	2	0	1	11	0	10	0
Black-throated Green Warbler	4	0	0	2	0	0	4	0	0	17	0	8	0
Canada Warbler	2	0	0	1	0	0	4	0	0	1	0	11	0
Scarlet Tanager	0	0	0	1	0	1	1	4	1	10	0	0	0
Northern Cardinal	0	0	15	0	0	0	0	18	0	0	0	0	1
Rose-breasted Grosbeak	2	0	0	3	1	1	3	3	2	5	4	4	1
Indigo Bunting	0	1	1	0	0	1	0	22	2	0	0	0	1
Dickcissel	0	1	1	0	0	0	0	0	0	0	0	0	1

APPENDIX E

Interpreting Historical Nesting Records

As noted in chapter 3, "Reading the Species Accounts," under Minnesota Breeding Distribution, the criteria that define confirmed nesting records for birds can vary within organizations over time, or among organizations or field survey protocols at a single point in time. These differences can be challenging when examining the nesting or breeding status of a species over a longtime horizon. In Minnesota, the first definitive and comprehensive accounts of the breeding status of birds are found in Roberts's 1932 publication *The Birds of Minnesota* and the updated edition published in 1936. Roberts, however, used no specific criteria to judge reports other than his vast knowledge and field experience. Each species account in his volumes includes a section titled "Minnesota Nesting." This is a description of "nesting" reports Roberts compiled for each species prior to the book's publication in 1932; new nesting reports of significance were added in appendices to the 1936 edition. The reports include his own records, those of his colleagues at the Museum of Natural History (now the Bell Museum), and those he received from field observers across the state. The records vary from accounts that clearly meet today's criteria for confirmed nesting (such as "nest with fresh eggs" or "adult feeding young in nest") to accounts that today's ornithologists would never deem acceptable (such as "many adults present") and many other descriptions in between (e.g., "fledged young able to fly"). We could not duplicate these valuable entries in the space of this volume but used our best judgment to describe the extent of the species' breeding distribution at the time of Roberts's publications. For rare or uncommon species, each of his records was often replicated in the text; for common and abundant species we often outlined the boundaries of the species' Minnesota breeding range by summarizing those county breeding records that we accepted as confirmed nesting. Occasionally we separated county records with the strongest evidence (i.e., confirmed nesting) from those records that we believed inferred nesting or breeding; at other times we referred to the sum of all records for a given county as simply providing breeding evidence. Although our summaries of Roberts's records vary, they were guided by criteria outlined in subsequent Minnesota documents (i.e., Green and Janssen 1975, Janssen 1987, Hertzel and Janssen 1998, Janssen 2019, and the specific guidelines used in the MNBBA [chapter 1, Table 1.2]). Because each of these later volumes defined its own criteria for nesting, our species accounts always deferred to those interpretations when they were cited. Readers interested in the specific details of the historical record should always consult the original sources.

APPENDIX F

Predicted Species Extirpations due to Warming Temperatures

The table shows the predicted extirpation of Minnesota breeding birds under three different warming scenarios (NAS 2019). The three global warming scenarios were based on the 2014 Intergovernmental Panel on Climate Change Fifth Assessment Report. The climate scenarios included first, a 1.5°C warming above preindustrial levels, which is considered imminent between 2030 and 2052 if we do nothing to mitigate climate change; second, a 2°C warming, which could happen as soon as 2050; and third, a 3°C warming, which could occur by 2080 (Wilsey, Bateman, et al. 2019). Data were extracted from the NAS interactive web page titled "Survival by Degrees: 389 Bird Species on the Brink" (https://www.audubon.org/climate/survivalbydegrees).

A species was designated extirpated at a given warming scenario when the NAS (2019) models portrayed the absence of any current or newly available "thermal breeding range" for the species in Minnesota. Even when a negligible amount of suitable habitat remained, the species was not designated extirpated. The text in the species accounts, however, may note that the species is "virtually" or "nearly" extirpated when those conditions exist. Nevertheless, we consider the precision of the NAS analysis to be uncertain, especially at increasingly small spatial scales. There were several species that NAS did not model as occurring in Minnesota, such as the Horned Grebe and American Three-toed Woodpecker. Although some of these species are unlikely to remain in the state as temperatures rise, the impacts of a warmer climate on their continental distribution were discussed in the text.

PREDICTED SPECIES EXTIRPATIONS DUE TO WARMING TEMPERATURES

Species	1.5°C	2°C	3°C
Trumpeter Swan			Extirpated
American Wigeon			Extirpated
Lesser Scaup			Extirpated
Bufflehead			Extirpated
Common Goldeneye			Extirpated
Ruffed Grouse			Extirpated
Spruce Grouse	Extirpated	Extirpated	Extirpated
Red-necked Grebe			Extirpated
Yellow Rail	Extirpated	Extirpated	Extirpated
Wilson's Snipe	Extirpated	Extirpated	Extirpated
Franklin's Gull			Extirpated
Sharp-shinned Hawk			Extirpated
Northern Goshawk		Extirpated	Extirpated
Northern Hawk Owl	Extirpated	Extirpated	Extirpated
Great Gray Owl	Extirpated	Extirpated	Extirpated
Boreal Owl	Extirpated	Extirpated	Extirpated
Northern Saw-whet Owl			Extirpated
Yellow-bellied Sapsucker			Extirpated
Black-backed Woodpecker	Extirpated	Extirpated	Extirpated
Merlin			Extirpated
Olive-sided Flycatcher	Extirpated	Extirpated	Extirpated
Yellow-bellied Flycatcher	Extirpated	Extirpated	Extirpated
Alder Flycatcher			Extirpated
Least Flycatcher			Extirpated
Blue-headed Vireo		Extirpated	Extirpated
Philadelphia Vireo	Extirpated	Extirpated	Extirpated
Canada Jay	Extirpated	Extirpated	Extirpated
Black-billed Magpie	Extirpated	Extirpated	Extirpated
Boreal Chickadee	Extirpated	Extirpated	Extirpated
Ruby-crowned Kinglet	Extirpated	Extirpated	Extirpated
Golden-crowned Kinglet	Extirpated	Extirpated	Extirpated
Red-breasted Nuthatch			Extirpated
Brown Creeper		Extirpated	Extirpated
Winter Wren		Extirpated	Extirpated
Swainson's Thrush	Extirpated	Extirpated	Extirpated

PREDICTED SPECIES EXTIRPATIONS DUE TO WARMING TEMPERATURES, *continued*

Species	1.5°C	2°C	3°C
Hermit Thrush	Extirpated	Extirpated	Extirpated
Sprague's Pipit	Extirpated	Extirpated	Extirpated
Evening Grosbeak		Extirpated	Extirpated
Purple Finch			Extirpated
White-winged Crossbill	Extirpated	Extirpated	Extirpated
Pine Siskin	Extirpated	Extirpated	Extirpated
Chestnut-collared Longspur			Extirpated
Clay-colored Sparrow			Extirpated
Dark-eyed Junco		Extirpated	Extirpated
White-throated Sparrow		Extirpated	Extirpated
LeConte's Sparrow		Extirpated	Extirpated
Nelson's Sparrow		Extirpated	Extirpated
Savannah Sparrow			Extirpated
Lincoln's Sparrow	Extirpated	Extirpated	Extirpated
Northern Waterthrush			Extirpated
Golden-winged Warbler			Extirpated
Black-and-White Warbler			Extirpated
Tennessee Warbler	Extirpated	Extirpated	Extirpated
Nashville Warbler		Extirpated	Extirpated
Connecticut Warbler	Extirpated	Extirpated	Extirpated
Mourning Warbler			Extirpated
Cape May Warbler	Extirpated	Extirpated	Extirpated
Northern Parula	Extirpated	Extirpated	Extirpated
Magnolia Warbler	Extirpated	Extirpated	Extirpated
Blackburnian Warbler		Extirpated	Extirpated
Chestnut-sided Warbler			Extirpated
Black-throated Blue Warbler		Extirpated	Extirpated
Palm Warbler	Extirpated	Extirpated	Extirpated
Yellow-rumped Warbler		Extirpated	Extirpated
Black-throated Green Warbler			Extirpated
Canada Warbler		Extirpated	Extirpated
Total	27	41	66

BIBLIOGRAPHY

Abraham, Jason. 2014. "A Search for Secrets of the Spruce Grouse." *Minnesota Conservation Volunteer* 77 (March/April): 18–25.

Ackerman, Joshua T., C. Alex Hartman, Mark P. Herzog, John Y. Takekawa, Julie A. Robinson, Lewis W. Oring, Joseph P. Skorupa, and Ruth Boettcher. 2020. "American Avocet (*Recurvirostra americana*)." Version 1.0. In Poole 2020. doi:10.2173/bow.ameavo.01.

Adams, A. Michael, James Bunn, Bruce S. Davis, Ann L. Jones, and L. W. Odne. 1973. "A Study of the Pig's Eye Lake Heron Colonies." *Loon* 45: 32–45.

Adikwan, Ogimaagiizhig Odoodeman (Charles Grolla). 2019. *Binesi-Dibaajimowinan: Ojibwe Bird Stories Unique to Northern Minnesota*. Cass Lake, MN: Leech Lake Tribal College, LLTC Design and Print Shop.

Alderfer, Jonathan. 2006. *Complete Birds of North America*. Washington, DC: National Geographic Society.

Aldrich, John W. 1953. "Habits and Habitat Differences in Two Races of Traill's Flycatcher." *Wilson Bulletin* 65: 8–11.

Alisauskas, Ray T., and Todd W. Arnold. 1994. "American Coot." In Tacha and Braun 1994, 127–43.

Allan, Thomas B. 1974. *Vanishing Wildlife of North America*. Washington, DC: National Geographic Society.

Allen, Arthur A., and Peter P. Kellogg. 1954. *Songbirds of North America in Color, Sound and Story*. Ithaca: Cornell University Press. 10-inch 33⅓ RPM record.

Allen, Arthur W. 1984. *Habitat Suitability Index Models: Gray Partridge*. Biological Report 82(10.73). Washington, DC: US Fish and Wildlife Service.

Allen, J. A., Charles W. Richmond, William Brewster, Jonathan Dwight Jr., C. Hart Merriam, Robert Ridgway, and Witmer Stone. 1908. "Fourteenth Supplement to the American Ornithologists' Union *Check-List of North American Birds*." *Auk* 25: 343–99.

Allen, Michael C., Megan M. Napoli, James Sheehan, Terry L. Master, Peter Pyle, Donald R. Whitehead, and Terry Taylor. 2020. "Acadian Flycatcher (*Empidonax virescens*)." Version 1.0. In Rodewald 2020. doi:10.2173/bow.acafly.01.

Allen, R. P. 1958. *A Progress Report on the Wading Bird Survey*. Tavernier, FL: National Audubon Society.

Allison, Andrew. 1907. "Notes on the Spring Birds of Tishomingo County, Mississippi." *Auk* 24: 12–25.

Altman, Bob, and Rex Sallabanks. 2020. "Olive-sided Flycatcher (*Contopus cooperi*)." Version 1.0. In Poole 2020. doi:10.2173/bowolsfly.01.

Alvo, Robert, and Michel Robert. 1999. *COSEWIC Status Report on the Yellow Rail* Coturnicops noveboracensis *in Canada*. Ottawa: Committee on the Status of Endangered Wildlife in Canada.

Ambuel, Bruce, and Stanley A. Temple. 1983. "Area-Dependent Changes in Bird Communities and Vegetation of Southern Wisconsin Forests." *Ecology* 64: 1057–68.

Ammon, Elisabeth M. 2020. "Lincoln's Sparrow (*Melospiza lincolnii*)." Version 1.0. In Poole and Gill 2020. doi:10.2173/bow.linspa.01.

Ammon, Elisabeth M., and William M. Gilbert. 2020. "Wilson's Warbler (*Cardellina pusilla*)." Version 1.0. In Rodewald 2020. doi:10.2173/bow.wlswar.01.

Anders, Angela D., and Eric Post. 2006. "Distribution-wide Effects of Climate on Population Densities of a Declining Migratory Landbird." *Journal of Animal Ecology* 75: 221–27.

Anderson, Bertin W. 1962. "First Blue Grosbeak Specimen for Minnesota." *Flicker* 34: 99.

Anderson, Bertin W., and S. Peter Getman. 1965. "Lark Bunting Occurrence in Southwestern Minnesota." *Loon* 37: 63–69.

Anderson, Bryce. 1979. "Observations from Pine/Curry Islands–Lake of the Woods." *Loon* 51: 144–45.

Anderson, Daniel W., and John W. Ellis. 1966. "Cormorant Nesting in Northwestern Minnesota." *Loon* 38: 5–8.

Anderson, Jeffrey P., and William J. Craig. 1984. "Growing Energy Crops on Minnesota's Wetlands: The Land Use Perspective." Pub. No. CURA 84-3. Minneapolis: University of Minnesota Center for Urban and Regional Affairs.

Anderson, Leon Joseph. 1973. "Habitat Use, Behavior, Territoriality, and Movements of the Male Canada Spruce Grouse of Northern Minnesota." Master's thesis, University of Minnesota.

Anderson, Rudolph Martin. 1907. "The Birds of Iowa." *Proceedings of the Davenport Academy of Sciences* 11: 125–417.

Andres, Brad A., Paul A. Smith, R. I. Guy Morrison, Cheri L. Gratto-Trevor, Stephen C. Brown, and Christian A. Friis. 2012. "Population Estimates of North American Shorebirds, 2012." *Wader Study Group Bulletin* 119: 178–94.

Angell, LeRoy. 1961. "Sandhill Cranes." *Flicker* 33: 125.

Ankley, Gerald T., Keith Lodge, Daniel J. Call, Mary D. Balcer, Larry T. Brooke, Phillip M. Cook, Russell G. Kreis Jr., et al. 1992. "Integrated Assessment of Contaminated Sediments in the Lower Fox River and Green Bay, Wisconsin." *Ecotoxicology and Environmental Safety* 23: 46–63.

Anteau, Michael J., and Alan D. Afton. 2006. "Diet Shifts of Lesser Scaup Are Consistent with the Spring Condition Hypothesis." *Canadian Journal of Zoology* 84: 779–86.

Anteau, Michael J., Jean-Michel DeVink, David N. Koons, Jane E. Austin, Christine M. Custer, and Alan D. Afton. 2020. "Lesser Scaup (*Aythya affinis*)." Version 1.0. In Poole 2020. doi:10.2173/bow.lessca.01.

AOU (American Ornithologists' Union). 1985. "Thirty-fifth Supplement to the American Ornithologists' Union *Check-List of North American Birds*." *Auk* 102: 680–86.

———. 1998. *Check-list of North American Birds*. 7th ed. Washington, DC: American Ornithologists' Union.

Arcese, Peter, Mark K. Sogge, Amy B. Marr, and Michael A. Patten. 2020. "Song Sparrow (*Melospiza melodia*)." Version 1.0. In Poole and Gill 2020. doi:10.2173/bow.sonspa.01.

Arneson, Mary. 1972. "Hooded Warbler in Minneapolis." *Loon* 44: 58–59.

Arneson, Mary, and Ruth E. Arneson. 1972. "Possible House Finch in Minneapolis." *Loon* 44: 53–54.

Arnold, Jennifer M., Stephen A. Oswald, Ian C. T. Nisbet, Peter Pyle, and Michael A. Patten. 2020. "Common Tern (*Sterna hirundo*)." Version 1.0. In Billerman 2020. doi:10.2173/bow.comter.01.

Arnold, Keith A. 1994. "Common Snipe." In Tacha and Braun 1994, 117–25.

Arnold, Todd W., and Robert M. Zink. 2011. "Collision Mortality Has No Discernible Effect on Population Trends of North American Birds." *PLoS One* 6: e24708. doi:10.1371/journal.pone.0024708.

Aron, Carol, Paul Van Ningten, Michael Rabenberg, Cheryl Jacobs, Gregory Pavelka, and Ricky D. Olson. 2008. "Breeding by the Snowy Plover in North Dakota and South Dakota." *Prairie Naturalist* 40: 51–52.

Artuso, Christian, Andrew R. Couturier, Ken D. De Smet, Rudolf F. Koes, Denis Lepage, Jon McCracken, Randall D. Mooi, and Peter Taylor, eds. 2019. *The Atlas of the Breeding Birds of Manitoba, 2010–2014*. Winnipeg, Manitoba: Bird Studies Canada. Accessed August 16, 2019. https://www.birdatlas.mb.ca/mbdata/maps.jsp?lang=en&lang=en.

Artuso, Christian, C. Stuart Houston, Dwight G. Smith, and Christoph Rohner. 2020. "Great Horned Owl (*Bubo virginianus*)." Version 1.0. In Poole 2020. doi:10.2173/bow.grhowl.01.

Askins, Robert A., Felipe Chávez-Ramirez, Brenda C. Dale, Carola A. Haas, James R. Herkert, Fritz L. Knopf, and Peter D. Vickery. 2007. "Conservation of Grassland Birds in North America: Understanding Ecological Processes in Different Regions: 'Report of the AOU Committee on Conservation.'" *Ornithological Monographs* 64: iii–46. doi:10.2307/40166905.

Atwell, Ryan C., Lisa A. Schulte, and Brian J. Palik. 2008. "Songbird Response to Experimental Retention Harvesting in Red Pine (*Pinus resinosa*) Forests." *Forest Ecology and Management* 255: 3621–31.

Audubon, John James. 1827–38. *Birds of America*. https://www.audubon.org/birds-of-america.

———. 1841. *Birds of North America*. Vol. 3. New York: J. J. Audubon; Philadelphia: J. B. Chevalier.

Audubon Minnesota. 2016. *Chimney Swift—Artificial Tower Locations Map*. Accessed April 21, 2016. http://www.arcgis.com/home/webmap/viewer.html?webmap=47cf28dde45d463fa5f940591dea58f7.

Austen, Madeline J. W., Charles M. Francis, Dawn M. Burke, and Michael S. W. Bradstreet. 2001. "Landscape Context and Fragmentation Effects on Forest Birds in Southern Ontario." *Condor* 103: 701–14.

Avery, Michael L., and Eric A. Tillman. 2005. "Alien Birds in North America—Challenges for Wildlife Managers." In *Proceedings of the 11th Wildlife Damage Management Conference*, edited by Dale L. Nolte and Kathleen A. Fagerstone, 82–89. Fort Collins: National Wildlife Research Center.

Avian Research and Conservation Institute. 2018. "Swallow-tailed Kite Breeding Biology, Demography and Habitat Management." ARCI website. http://www.arcinst.org/swallow-tailed-kite.

Axelrod, Miriam. 1979. "Observations on a Boreal Chickadee Nest." *Loon* 51: 135–40.

Azure, David A. 1998. "Aspects of American Bittern Ecology in Northwest Minnesota." Master's thesis, University of North Dakota, Grand Forks.

Baars, Els. 2017. "Why the Barn Swallow Has a Long Fork in Its Tail (001)." *Tails of Nature* (blog), March 26. https://talesofnature.com/2017/03/26/why-the-barn-Swallow-has-a-long-fork-in-its-tail-001/.

Back, Gary N. 1979. "Avian Communities and Management Guidelines of the Aspen-Birch Forest." In DeGraaf and Evans 1979, 67–79.

Badyaev, Alexander V., Virginia Belloni, and Geoffrey E. Hill. 2020. "House Finch (*Haemorhous mexicanus*)." Version 1.0. In Poole 2020. doi:10.2173/bow.houfin.01.

Bagg, Aaron Clark, and Samuel Atkins Eliot Jr. 1937. *Birds of the Connecticut Valley in Massachusetts*. Northampton, MA: Hampshire Bookshop.

Baird, Spencer F., Thomas M. Brewer, and Robert Ridgway. 1874. *A History of North American Birds: Land Birds*. 2 vols. Boston: Little, Brown.

Baker, Richard J., and Jeff Hines. 1996. *Black Tern Sightings in Minnesota 1990–1995: Final Report*. US Fish and Wildlife Service. http://files.dnr.state.mn.us/eco/nongame/projects/consgrant_reports/1996/1996_baker_hines.pdf.

Baldassarre, Guy A. 2014. *Ducks, Geese, and Swans of North America*. 2 vols. Baltimore, MD: Johns Hopkins University Press.

Ball, Jeffrey R., Péter Sólymos, Fiona K. A. Schmiegelow, Samuel Hache, Jim Schieck, and Erin Bayne. 2016. "Regional Habitat Needs of a Nationally Listed Species, Canada Warbler (*Cardellina canadensis),* in Alberta, Canada." *Avian Conservation and Ecology* 11: 10.

Baltz, Michael E., and Steven C. Latta. 2020. "Cape May Warbler (*Setophaga tigrina*)." Version 1.0. In Rodewald 2020. doi:10.2173/bow.camwar.01.

Banks, Richard C., Carla Cicero, Jon L. Dunn, Andrew W. Kratter, Pamela C. Rasmussen, J. Van Remsen Jr., James D. Rising, et al. 2002. "Forty-third Supplement to the American Ornithologists' Union *Check-list of North American Birds*." *Auk* 119: 897–906.

———. 2004. "Forty-fifth Supplement to the American Ornithologists' Union *Check-List of North American Birds*." *Auk* 121: 985–95.

Bannor, Brett K., and Erik Kiviat. 2020. "Common Gallinule (*Gallinula galeata*)." Version 1.0. In Poole and Gill 2020. doi:10.2173/bow.comgal1.01.

Bardon, Karl J. 1991a. "Baird's Sparrow in Roseau County." *Loon* 63: 284–85.

———. 1991b. "Nesting Mountain Bluebirds." *Loon* 63: 207–8.

———. 1992. "Western Wood-Pewee in Roseau County." *Loon* 64: 179.

———. 1997. "Clark's Grebe Nest at Lake Osakis, Todd County." *Loon* 69: 225–26.

———. 1998. "Breeding of Canada Warblers in Southern Minnesota." *Loon* 70: 169–70.

———. 1999. "An Influx of Summer Tanagers." *Loon* 71: 216–20.

———. 2002a. *High Counts of Birds in Minnesota: A Compilation of the Highest Single Count Totals for Regular Species*. Minnesota Ornithologist's Union Occasional Papers, no. 4. Minneapolis: Minnesota Ornithologists' Union.

———. 2002b. "Northbound or Southbound?—The Enigma of Summer Shorebird Migration." *Loon* 74: 65–82.

———. 2003. "Hooded Warbler Nest in Anoka County." *Loon* 75: 235–36.

———. 2012. "Fall Diurnal Migration of Passerines and Other Non-raptors at Hawk Ridge and Lake Superior, Duluth (2007–2011)." *Loon* 84: 8–20.

Barrows, Walter B. 1889. *The English Sparrow (Passer domesticus) in North America, Especially in Its Relations to Agriculture*. US Department of Agriculture Division of Economic Ornithology and Mammalogy Bulletin 1. Washington, DC: Government Printing Office.

Basili, Gianfranco D., S. L. Brown, Elmer J. Finck, Dan L. Reinking, K. L. Steigman, Stanley A. Temple, and John L. Zimmerman. 1997. "Breeding Biology of Dickcissels across Their Range and over Time." In *Continental-Scale Ecology and Conservation of the Dickcissel*, edited by Gianfranco D. Basili, 108–51. PhD diss., University of Wisconsin-Madison.

Bateman, Brooke L., Chad Wilsey, Lotem Taylor, Joanna Wu, Geoffrey S. LeBaron, and Gary Langham. 2019. "North American Birds Require Mitigation and Adaptation to Reduce Vulnerability to Climate Change." *Conservation Science and Practice* 2 (8): e242. doi:10.1111/csp2.242.

BBC (British Broadcasting Corporation). 2019. "Winged Warriers: Pigeons in the First World War." https://www.historyextra.com/period/first-world-war/winged-warriors-pigeons-in-the-first-world-war/.

Beason, Robert C. 2020. "Horned Lark (*Eremophila alpestris*)." Version 1.0. In Billerman 2020. doi:10.2173/bow.horlar.01.

Bechard, Marc J., C. Stuart Houston, Jose H. Sarasola, and A. Sidney England. 2020. "Swainson's Hawk (*Buteo swainsoni*)." Version 1.0. In Poole 2020. doi:10.2173/bow.swahaw.01.

Bednar, Josh D., Edmund J. Zlonis, Hannah G. Panci, Ron Moen, and Gerald J. Niemi. 2016. "Development of Habitat Models and Habitat Maps for Breeding Bird Species in the Agassiz Lowlands Subsection, Minnesota, USA." Minnesota Department of Natural Resources Report T-39-R-1/F12AF00328. Natural Resources Research Institute Technical Report NRRI/TR-2015-32.

Bednarz, James C., and James J. Dinsmore. 1982. "Nest-Sites and Habitat of Red-shouldered and Red-tailed Hawks in Iowa." *Wilson Bulletin* 94: 31–45.

Bell, Ronald. 1985. "Bufflehead Brood at Agassiz National Wildlife Refuge." *Loon* 57: 136.

Belleman, B. Ann. 1998. "Red-shouldered Hawk Breeding Ecology and Habitat Use in Central Minnesota." Master's thesis, University of Minnesota.

Bellrose, Frank C. 1976. "The Comeback of the Wood Duck." *Wildlife Society Bulletin* 4: 107–10.

———. 1980. *Ducks, Geese and Swans of North America*. 3rd ed. Harrisburg, PA: Stackpole Books.

Bellrose, Frank C., and Daniel J. Holm. 1994. *Ecology and Management of the Wood Duck*. Harrisburg, PA: Stackpole Books.

Belmonte, Lisa Rose. 2005. "Home Range and Habitat Characteristics of Boreal Owls in Northeastern Minnesota." Master's thesis, University of Minnesota.

Bemis, Carrie, and James D. Rising. 2020. "Western Wood-Pewee (*Contopus sordidulus)*. Version 1.0. In Poole and Gill 2020. doi:10.2173/bow.wewpew.01.

Bendire, Charles Emil. 1895. "Life Histories of North American Birds." *U.S. National Museum Special Bulletin* 3. Washington, DC.

Benkman, Craig W. 1993. "Logging, Conifers, and the Conservation of Crossbills." *Conservation Biology* 7: 473–79.

———. 2020. "White-winged Crossbill (*Loxia leucoptera*)." Version 1.0. In Billerman 2020. doi:10.2173/bow.whwcro.01.

Benkman, Craig W., and Matthew A. Young. 2020. "Red Crossbill (*Loxia curvirostra*)." Version 1.0. In Billerman et al. 2020. doi:10.2173/bow.redcro.01.

Benson, Dave. 2000. "McCown's Longspur at 40th Avenue West in Duluth." *Loon* 72: 246–47.

Benson, Dave, Paul Budde, Wally Swanson, and Tom Tustison. 1997. "The Spring Season (1 March to 31 May 1997)." *Loon* 69: 186–211.

———. 1998. "The Fall Season (1 August to 30 November 1997)." *Loon* 70: 82–105.

Bent, Arthur Cleveland. 1919. *Life Histories of North American Diving Birds (Order Pygopodes)*. US National Museum Bulletin 107. Washington, DC: Smithsonian Institution.

———. 1921. *Life Histories of North American Gulls and Terns (Order Longipennes)*. US National Museum Bulletin 113. Washington, DC: Smithsonian Institution.

———. 1922. *Life Histories of North American Petrels and Pelicans and Their Allies*. US National Museum Bulletin 121. Washington, DC: Smithsonian Institution.

———. 1923. *Life Histories of North American Wild Fowl (Order Anseres). Part I*. US National Museum Bulletin 126. Washington, DC: Smithsonian Institution.

———. 1925. *Life Histories of North American Wild Fowl: Order Anseres, Part II*. US National Museum Bulletin 130. Washington, DC: Smithsonian Institution.
———. 1926. *Life Histories of North American Marsh Birds (Orders Odontoglossae, Herodiones, and Paludicolae)*. US National Museum Bulletin 135. Washington, DC: Smithsonian Institution.
———. 1927. *Life Histories of Familiar North American Shore Birds (Order Limicolae). Part I*. US National Museum Bulletin 142. Washington, DC: Smithsonian Institution.
———. 1929. *Life Histories of North American Shore Birds (Order Limicolae). Part II*. US National Museum Bulletin 146. Washington, DC: Smithsonian Institution.
———. 1932. *Life Histories of North American Gallinaceous Birds (Order Galliformes and Columbiformes)*. US National Museum Bulletin 162. Washington, DC: Smithsonian Institution.
———. 1937. *Life Histories of North American Birds of Prey, Part 1 (Order Falconiformes)*. US National Museum Bulletin 167. Washington, DC: Smithsonian Institution.
———. 1938. *Life Histories of North American Birds of Prey, Part 2 (Order Falconiformes and Strigiformes)*. US National Museum Bulletin 170. Washington, DC: Smithsonian Institution.
———. 1939. *Life Histories of North American Woodpeckers (Order Piciformes)*. US National Museum Bulletin 174. Washington, DC: Smithsonian Institution.
———. 1940. *Life Histories of North American Cuckoos, Goatsuckers, Hummingbirds, and Their Allies (Orders Psittaciformes, Cuculiformes, Trogoniformes, Coraciiformes, Caprimulgiformes, and Micropodiiformes)*. US National Museum Bulletin 176. Washington, DC: Smithsonian Institution.
———. 1942. *Life Histories of North American Flycatchers, Larks, Swallows, and Their Allies: Order Passeriformes*. US National Museum Bulletin 179. Washington, DC: Smithsonian Institution.
———. 1946. *Life Histories of North American Jays, Crows, and Titmice: Order Passeriformes*. US National Museum Bulletin 191. Washington, DC: Smithsonian Institution.
———. 1948. *Life Histories of North American Nuthatches, Wrens, Thrashers, and Their Allies*. US National Museum Bulletin 195. Washington, DC: Smithsonian Institution.
———. 1949. *Life Histories of North American Thrushes, Kinglets, and Their Allies (Order Passeriformes)*. US National Museum Bulletin 196. Washington, DC: Smithsonian Institution.
———. 1950. *Life Histories of North American Wagtails, Shrikes, Vireos, and Their Allies: Order Passeriformes*. US National Museum Bulletin 197. Washington, DC: Smithsonian Institution.
———. 1953. *Life Histories of North American Wood Warblers (Order Passeriformes)*. US National Museum Bulletin 203. Washington, DC: Smithsonian Institution.
———. 1958. *Life Histories of North American Blackbirds, Orioles, Tanagers, and Allies: Order Passeriformes*. US National Museum Bulletin 211. Washington, DC: Smithsonian Institution.
———. 1968. *Life Histories of North American Cardinals, Grosbeaks, Buntings, Towhees, Finches, Sparrows, and Allies (Order Passeriformes)*. Parts 1, 2, and 3. US National Museum Bulletin 237. Washington, DC: Smithsonian Institution.
Berber, Robert B. 1981. "Snowy Plover at Big Stone NWR." *Loon* 53: 220–21.
Berg, William E. 1961. "Report of Swallow-tailed Kite sighting." *Loon* 33: 57.
———. 1990. "Sharp-tailed Grouse Management Problems in the Great Lakes States: Does the Sharp-tail Have a Future?" *Loon* 4: 42–45.
———. 1997. *The Sharp-tailed Grouse in Minnesota*. Minnesota Wildlife Report, no. 10. St. Paul: Minnesota Department of Natural Resources.
Bergerud, Arthur T. 1988. "Population Ecology of North American Grouse." In *Theory and Synthesis. Vol. 2 of Adaptive Strategies and Population Ecology of Northern Grouse*, edited by Arthur T. Bergerud and Michael W. Gratson, 578–685. Minneapolis: University of Minnesota Press.
Bergerud, Tom. 1947. "Observations in Lake Pepin Region and Southeastern Minnesota." *Flicker* 19: 79.
Beyersbergen, Gerard W., Neal D. Niemuth, and Michael R. Norton. 2004. *Northern Prairie and Parkland Waterbird Conservation Plan: A Plan Associated with the Waterbird Conservation for the Americas Initiative*. Denver, CO: Prairie Pothole Joint Venture.
Bianchini, Kristin, Douglas C. Tozer, Robert Alvo, Satyendra P. Bhavsar, and Mark L. Mallory. 2020. "Drivers of Declines in Common Loon (*Gavia immer*) Productivity in Ontario, Canada." *Science of the Total Environment* 738: 139724.
Bielefeldt, John, and Robert N. Rosenfield. 2001. "Nest Site Habitats and Breeding Biology of Hooded Warblers in Southeastern Wisconsin." *Passenger Pigeon* 63: 242–52.
Bierregaard, Richard O., Alan F. Poole, Mark S. Martell, Peter Pyle, and Michael A. Patten. 2020. "Osprey (*Pandion haliaetus*)." Version 1.0. In Rodewald 2020. doi:10.2173/bow.osprey.01.
Bildstein, Keith L., Kenneth D. Meyer, Clayton M. White, Jeffrey S. Marks, and Guy M. Kirwan. 2020. "Sharp-shinned Hawk (*Accipiter striatus*)." In Billerman et al. 2020. doi:10.2173/bow.shshaw.01.
Billerman, Shawn M., ed. 2020. *Birds of the World*. Ithaca, NY: Cornell Laboratory of Ornithology. https://birdsoftheworld.org/bow.
Billerman, Shawn M., B. K. Keeney, Paul G. Rodewald, and T. S. Schulenberg, eds. 2020. *Birds of the World*. Ithaca, NY: Cornell Laboratory of Ornithology. https://birdsoftheworld.org/bow.
Bird Source. 2016. "CBC Data Shows Tufted Titmouse Range Expansion." Accessed January 13, 2019. http://www.birdsource.org/Features/Tuti/.
Bird Studies Canada and NABCI. 2014. Bird Conservation Regions. Published by Bird Studies Canada on behalf of the North American Bird Conservaton Initiative. Accessed February 29, 2020. http://www.birdscanada.org/research/gislab/index.jsp?targetpg=bcr.
Blake, John G. 1991. "Nested Subsets and the Distribution of Birds on Isolated Woodlots." *Conservation Biology* 5: 58–66.
Blake, John G., and James R. Karr. 1984. "Species Composition of Bird Communities and the Conservation Benefit of Large versus Small Forests." *Biological Conservation* 30: 173–87.
———. 1987. "Breeding Birds of Isolated Woodlots: Area and Habitat Relationships." *Ecology* 68: 1724–34. doi:10.2307/1939864.
Blancher, Peter J., Kenneth V. Rosenberg, Arvind O. Panjabi, Bob Altman, Andrew R. Couturier, Wayne E. Thogmartin, and the Partners in Flight Science Committee. 2013. "Handbook to the Partners in Flight Population Estimates Database, Version 2.0." PIF Technical Series, no. 6. https://www.partnersinflight.org/wp-content/uploads/2018/11/PIF-Technical-Series-06-–-Handbook-to-PIF-Pop-Est-DB-2013.pdf.
Blanich, Jo. 1972. "Long billed Curlew Observation." *Loon* 44: 94.
———. 1976. "Bewick's Wren in Crow Wing County." *Loon* 46: 79.
Bleho, Barbara, Kevin Ellison, Dorothy P. Hill, and Lorne K. Gould. 2020. "Chestnut-collared Longspur (*Calcarius ornatus*)." Version 1.0. In Rodewald 2020. doi:10.2173/bow.chclon.01.

Blockstein, David E. 2020. "Passenger Pigeon (*Ecotopistes migratorius*)." Version 1.0. In Poole and Gill 2020. doi:10.2173/bow.paspig.01.

Bloem, Karla A., Marjon Savelsberg, and Rose A. Yoakum. 2018. "Recent Barn Owl Records in Minnesota." *Loon* 90: 39–41.

Blumton, Arlene K. 1989. "Factors Affecting Loggerhead Shrike Mortality in Virginia." Master's thesis, Virginia Polytechnic Institute and State University, Blacksburg.

BMNH (Bell Museum of Natural History) Records. Collection ua-00876. University of Minnesota, University Archives.

Boal, Clint W., David E. Andersen, and Patricia L. Kennedy. 2001. "Home Range and Habitat Use of Northern Goshawks (*Accipter gentilis*) in Minnesota." Final report to the Minnesota Department of Natural Resources, Nongame Wildlife Program. http://files.dnr.state.mn.us/eco/nongame/projects/consgrant_reports/2001/2001_boal_etal.pdf.

———. 2003. "Home Range and Residency Status of Northern Goshawks Breeding in Minnesota." *Condor* 105: 811–16.

———. 2005a. "Foraging and Nesting Habitat of Breeding Male Northern Goshawks in the Laurentian Mixed Forest Province, Minnesota." *Journal of Wildlife Management* 69: 1516–27.

———. 2005b. "Productivity and Mortality of Northern Goshawks in Minnesota." *Journal of Raptor Research* 39: 222–28.

Boal, Clint W., David E. Andersen, Patricia L. Kennedy, and Aimee M. Roberson. 2006. "Northern Goshawk Ecology in the Western Great Lakes Region." *Studies in Avian Biology* 31: 128–36.

Boarman, William I., and Bernd Heinrich. 2020. "Common Raven (*Corvus corax*)." Version 1.0. In Billerman 2020. doi:10.2173/bow.comrav.01.

Bocetti, Carol I., Deahn M. Donner, and Harold F. Mayfield. 2020. "Kirtland's Warbler *(Setophaga kirtlandii*)." Version 1.0. In Poole 2020. doi:10.2173/bow.kirwar.01.

Bock, Carl E., and Larry W. Lepthien. 1976. "Synchronous Eruptions of Boreal Seed-eating Birds." *American Naturalist* 110: 559–71.

Bock, Carl E., Victoria A. Saab, Terrell D. Rich, and David S. Dobkin. 1993. "Effects of Livestock Grazing on Neotropical Landbirds in Western North America." In *Status and Management of Neotropical Migratory Birds,* edited by Deborah M. Finch and Peter W. Stangel, 296–309. Fort Collins, CO: US Department of Agriculture, Forest Service, Rocky Mountain Forest and Range Experiment Station.

Boe, Janet S. 1992a. "A Survey for Breeding Horned Grebes in Minnesota." *Loon* 64: 75–78.

———. 1992b. "Wetland Selection by Eared Grebes, *Podiceps nigricollis,* in Minnesota." *Canadian Field-Naturalist* 106: 480–88.

Boe, Janet S., and Gary L. Nuechterlein. 1990. *Colony Site and Nest Site Selection by Eared Grebes in Minnesota.* Final report to the Minnesota Department of Natural Resources, Nongame Wildlife Program. http://files.dnr.state.mn.us/eco/nongame/projects/consgrant_reports/1990/1990_boe.pdf.

Boe, Janet S., and Kristie Prahl. 1987. "Colony Sites of Eared Grebes in Minnesota in 1986." *Loon* 59: 14–16.

Bohm, Robert T. 1988. "Use of Nestboxes by Great Gray Owls in Northern Minnesota." *Loon* 60: 121–23.

Boissoneault, Lorraine. 2021. "Piping Plovers: Despite New Challenges, the Birds Make Their Comeback." June 23. https://www.greatlakesnow.org/2021/06/piping-plovers-recovery-population/.

Bolduc, Alison. 1988. "Blue Grosbeak at the Felton Prairie." *Loon* 60: 129–30.

Bolduc, Donald. 1966. "Northerly Record for the Cerulean Warbler." *Loon* 38: 73.

Bond, Richard R. 1957. "Ecological Distribution of Breeding Birds in the Upland Forests of Southern Wisconsin." *Ecological Monographs* 27: 351–84.

Bonter, David N., and Michael G. Harvey. 2008. "Winter Survey Data Reveal Rangewide Decline in Evening Grosbeak Populations." *Condor* 110: 376–81. doi:10.1525/cond.2008.8463.

Booms, Travis L., Geoffrey L. Holroyd, Marcel A. Gahbauer, Helen E. Trefry, David A. Wiggins, Denver W. Holt, James A. Johnson, et al. 2014. "Assessing the Status and Conservation Priorities of the Short-Eared Owl in North America." *Journal of Wildlife Management* 78: 772–78.

Boyce, Mark S. 1980. "Possible Competitive Displacement between Common Nighthawks and Bats in Southwestern Oregon." *Murrelet* 61: 35.

Bracey, Annie M. 2011. "Window Related Avian Mortality at a Migration Corridor." Master's thesis, University of Minnesota Duluth.

———. 2021. "Ecology and Conservation of Common Terns (*Sterna hirundo*) Breeding in the North American Great Lakes Region." PhD diss., University of Minnesota.

Bracey, Annie M., Jessica Chatterton, and Gerald J. Niemi. 2016. "St. Louis River AOC R2R Support Projects: Ecological Monitoring and Assessment (CR#6403)." Final Report to Minnesota Pollution Control Agency. Natural Resources Research Institute, University of Minnesota Duluth.

Bracey, Annie M., Matthew A. Etterson, Gerald J. Niemi, and Richard F. Green. 2016. "Variation in Bird-Window Collision Mortality and Scavenging Rates within an Urban Landscape." *Wilson Journal of Ornithology* 128: 355–67.

Bracey, Annie M., Matthew A. Etterson, Frederick C. Strand, Sumner W. Matteson, Gerald J. Niemi, Francesca J. Cuthbert, and Joel C. Hoffman. 2021. "Foraging Ecology Differentiates Life Stages and Mercury Exposure in Common Terns (*Sterna hirundo*)." *Health and Ecological Risk Assessment* 17: 398–410. doi:10.1002/ieam.4341.

Bracey, Annie M., Simeon Lisovski, David Moore, Ann McKellar, Elizabeth Craig, Sumner Matteson, Fred Strand, et al. 2018. "Migratory Routes and Wintering Locations of Declining Inland North American Common Terns." *Auk* 135: 385–99.

Brady, Ryan S., Richard A. Staffen, Nicholas M. Anich, and Carly N. Lapin. 2021. "Status and Trend of the Connecticut Warbler in Wisconsin: Steep Decline Demands Urgent Conservation Action." *Passenger Pigeon* 83, nos. 2–3: 115–26.

Brandes, David, David Oleyar, Steve Hoffman, and Laurie Goodrich. 2016. *The Raptor Population Index, 2016: Regional Trend Summaries and Conservation Assessments.* Accessed November 1, 2019. http://rpi-project.org/2016.

Bratlie, Byron. 1990. "A Bewick's Wren in Sherburne County." *Loon* 62: 154.

Breckenridge, Walter John. 1939. "More Minnesota American Egret News." *Flicker* 11: 13.

———. 1949. *A Century of Minnesota Wild Life.* 1st ed. Minneapolis: Minnesota Museum of Natural History, University of Minnesota.

———. 1956. "Measurements of the Habitat Niche of the Least Flycatcher." *Wilson Bulletin* 68: 47–51.

———. 1968a. "Minnesota Nesting of White Pelicans." *Loon* 40: 100.

———. 1968b. "Where Are Our Kingbirds?" *Loon* 40: 56–57.

———. 1971. "Our Disappearing Kingbirds." *Loon* 43: 89–90.

———. 1984. "House Finch in Hennepin County." *Loon* 56: 64.

———. 2009. *My Life in Natural History.* Minneapolis: Bell Museum of Natural History, University of Minnesota.

Breiman, Leo. 1996. "Bagging Predictors." *Machine Learning* 24: 123–40.
Bremer, Paul E. 1966. "Possible Sandhill Crane Breeding." *Loon* 38: 139.
Brennan, Leonard A., Fidel Hernandez, and Damon Williford. 2020. "Northern Bobwhite (*Colinus virginianus*)." Version 1.0. In Poole 2020. doi:10.2173/bow.norbob.01.
Brewer, Gwenda L. 1992. "1991 Summary Report: Location of Breeding Colonies and Evaluation of Critical Nesting Habitat for the Black Tern (*Chlidonias niger*) in Northwestern Minnesota: Kittson and Roseau Counties." Report to the Minnesota Department of Natural Resources.
Brewer, Richard, Gail A. McPeek, and Raymond J. Adams Jr. 1991. *The Atlas of Breeding Birds of Michigan*. East Lansing: Michigan State University Press.
Brewster, William. 1877. "The Yellow-throated Warbler (*Dendroica dominica*)." *Bulletin of the Nuttall Ornithological Club* 2: 102–6. https://www.jstor.org/stable/24723308.
———. 1891. "A Study of Florida Gallinules, with Some Notes on a Nest Found at Cambridge, Massachusetts." *Auk* 8: 1–7.
———. 1895. "Notes on Certain Flycatchers of the Genus *Empidonax*." *Auk* 12: 157–63.
———. 1902. "Voices of a New England Marsh." *Bird-Lore* 4: 43–56.
Brigham, R. Mark, Janet Ng, Ray G. Poulin, and Scott D. Grindal. 2020. "Common Nighthawk (*Chordeiles minor*)." Version 1.0. In Poole 2020. doi:10.2173/bow.comnig.01.
Brisbin, I. Lehr, Jr., and Thomas B. Mowbray. 2020. "American Coot (*Fulica americana*)." Version 1.0. In Poole and Gill 2020. doi:10.2173/bow.y00475.01.
Brittingham, Margaret Clark, and Stanley A. Temple. 1983. "Have Cowbirds Caused Forest Songbirds to Decline?" *BioScience* 33: 31–35.
———. 1988. "Impacts of Supplemental Feeding on Survival Rates of Black-capped Chickadees." *Ecology* 69: 581–89.
Bronoel, Joel K. 1948. "Harbor Island, St. Louis Bay, Duluth." *Flicker* 20: 107.
———. 1953. "Harbor Island Census—1953." *Flicker* 25: 111.
———. 1954a. "Census of Harbor Island in St. Louis Bay—1954." *Flicker* 26: 118.
———. 1954b. "Duluth Club Participates in Hawk Conservation." *Flicker* 26: 94–95.
———. 1955. "Census of Harbor Island in St. Louis Bay—1955." *Flicker* 27: 174.
Brooks, Bonnie L. 1988. "The Breeding Distribution, Population Dynamics, and Habitat Availability and Suitability of an Upper Midwest Loggerhead Shrike Population." Master's thesis, University of Wisconsin–Madison.
Brooks, Bonnie L., and Stanley A. Temple. 1986. "The Breeding Distribution of the Loggerhead Shrike in Minnesota: A Preliminary Report." *Loon* 58: 151–54.
———. 1990a. "Dynamics of a Loggerhead Shrike Population in Minnesota." *Wilson Bulletin* 102: 441–50.
———. 1990b. "Habitat Availability and Suitability for Loggerhead Shrikes in the Upper Midwest." *American Midland Naturalist* 123: 75–83.
Brooks, William. 2006. "Red-necked Grebe." In Cutright et al. 2006, 136–37.
Brown, Charles R., Mary B. Brown, Peter Pyle, and Michael A. Patten. 2020. "Cliff Swallow (*Petrochelidon pyrrhonota*)." Version 1.0. In Rodewald 2020. doi:10.2173/bow.cliswa.01.
Brown, Charles R., and Scott Tarof. 2020. "Purple Martin (*Progne subis*)." Version 1.0. In Poole 2020. doi:10.2173/bow.purmar.01.
Brown, Mary B., and Charles R. Brown. 2020. "Barn Swallow (*Hirundo rustica*)." Version 1.0. In Rodewald 2020. doi:10.2173/bow.barswa.01.
Brown, Nathan Clifford. 1879. "A List of Birds Observed at Coosada, Central Alabama." *Bulletin of the Nuttall Ornithological Club* 4: 7–13.
Brown, Raymond E., J. Howard Williamson, and Dan B. Boone. 1997. "Swallow-tailed Kite Nesting in Texas: Past and Present." *Southwestern Naturalist* 42: 103–5.
Brown, Stephen, Catherine Hickey, Brian Harrington, and Robert Gill, eds. 2001. *The U.S. Shorebird Conservation Plan*. 2nd ed. Manomet, MA: Manomet Center for Conservation Sciences.
Brown, Woodward H. 1961. "Barn Owl Record for 1953." *Flicker* 33: 23.
Brua, Robert B. 2020. "Ruddy Duck (*Oxyura jamaicensis*)." Version 1.0. In Rodewald 2020. doi:10.2173/bow.rudduc.01.
Bruggeman, Jason E., David E. Andersen, and James E. Woodford. 2011. "Northern Goshawk Monitoring in the Western Great Lakes Bioregion." *Journal of Raptor Research* 45: 290–303.
Bruggeman, Jason E., William T. Route, Patrick T. Redig, and Rebecca L. Key. 2018. "Patterns and Trends in Lead (Pb) Concentrations in Bald Eagle (*Haliaeetus leucocephalus*) Nestlings from the Western Great Lakes Region." *Ecotoxicology* 27: 605–18.
Bruggers, David J. 1988. "The Behavior and Ecology of the Common Raven in Northeastern Minnesota." PhD diss., University of Minnesota.
Bruggers, Denise, and David J. Bruggers. 1981. "House Finch in Minnesota." *Loon* 53: 109.
Budde, Paul E. 2001. "The Fall Season (1 August to 30 November 2000)." *Loon* 73: 78–107.
Budde, Paul E., Robert M. Dunlap, Doug W. Kieser, James W. Lind, and William C. Marengo. 2008. "The Fall Season, 1 August through 30 November 2007." *Loon* 80: 56–93.
———. 2013. "The Spring Season, 1 March through 31 May 2013." *Loon* 85: 165–98.
Budde, Paul E., Robert M. Dunlap, Doug W. Kieser, James W. Lind, William C. Marengo, and Peder H. Svingen. 2007. "The Spring Season, 1 March–31 May 2007." *Loon* 79: 197–229.
Budde, Paul E., Doug W. Kieser, William C. Marengo, and Brian T. Smith. 2018. "The 2017 Fall Season: 1 August to 30 November 2017." *Loon* 90: 42–74.
Budde, Paul E., Doug W. Kieser, Brian T. Smith, and William C. Marengo. 2016. "The Spring Season, 1 March through 31 May 2016." *Loon* 88: 150–184.
Budde, Paul E., James W. Lind, and William C. Marengo. 2006. "The Spring Season, 1 March through 31 May 2006." *Loon* 78: 190–223.
Budde, Paul E., James W. Lind, and Peder H. Svingen. 2004. "The Spring Season (1 March through 31 May 2004)." *Loon* 76: 193.
Budde, Paul E., and Peder H. Svingen. 2003. "The Fall Season (1 August to 30 November 2002)." *Loon* 75: 77–105.
Budnik, Joel M., Frank R. Thompson III, and Mark R. Ryan. 2002. "Effect of Habitat Characteristics on the Probability of Parasitism and Predation of Bell's Vireo Nests." *Journal of Wildlife Management* 66: 232–39.
Buehler, David A. 2020. "Bald Eagle (*Haliaeetus leucocephalus*)." Version 1.0. In Poole and Gill 2020. doi:10.2173/bow.baleag.01.
Buehler, David A., Paul B. Hamel, and Than Boves. 2020. "Cerulean Warbler (*Setophaga cerulea*)." Version 1.0. In Poole 2020. doi:10.2173/bow.cerwar.01.
Buer, Chuck, and Micki Buer. 1977. "Little Blue Herons in Lac qui Parle County." *Loon* 49: 48–49.
Buer, Micki, and Chuck Buer. 1989. "Black-necked Stilts in Stevens County." *Loon* 61: 139.
Bull, Evelyn L., and James R. Duncan. 2020. "Great Gray Owl (*Strix nebulosa*)." Version 1.0. In Billerman 2020. doi:10.2173/bow.grgowl.01.
Bull, Evelyn L., and Jerome A. Jackson. 2020. "Pileated Woodpecker (*Dryocopus pileatus*)." Version 1.0. In Poole 2020. doi:10.2173/bow.pilwoo.01.

Bullis, Otto. 1892. "Ornithology." *Annual Report of the Minnesota State Horticultural Society* 20: 360–64.

Burger, Joanna. 1971. "The Western Grebe in Minnesota." *Loon* 43: 4–9.

Burger, Joanna, and Michael Gochfeld. 2020. "Franklin's Gull (*Leucophaeus pipixcan*)." Version 1.0. In Poole 2020. doi:10.2173/bow.fragul.01.

Burger, Joanna, and Marshall Howe. 1971. "Black-billed Magpie: New Breeding Records in Minnesota." *Loon* 43: 78–79.

Burhans, Dirk E. 2002. "Conservation Assessment: Henslow's Sparrow *Ammodramus henslowii*." General Technical Report NC-226. St. Paul, MN: US Department of Agriculture, Forest Service, North Central Research Station.

Burnham, Kenneth P., and David R. Anderson. 2002. *Model Selection and Multimodel Inference: A Practical Information-Theoretic Approach*. 2nd ed. New York: Springer-Verlag.

Burt, William. 1978. "A Sharp-tailed Sparrow Nest." *Loon* 50: 48–49.

Bushman, Ellen S., and Glenn D. Therres. 1988. *Habitat Management Guidelines for Forest Interior Breeding Birds of Coastal Maryland*. Maryland Department of Natural Resources Wildlife Technical Publication 88-1.

Butler, Christopher J., Jeffrey B. Tibbits, and Katrina Hucks. 2013. "Status of 10 Bird Species of Conservation Concern in U.S. Fish and Wildlife Service Region 6." Final Report. Edmond: University of Central Oklahoma.

Byers, Bruce E., Michael Richardson, and Daniel W. Brauning. 2020. "Chestnut-sided Warbler (*Setophaga pensylvanica*)." Version 1.0. In Poole 2020. doi:10.2173/bow.chswar.01.

Bylander, C. B., and Ray Grumney. 2015. "Loons under Watch." *Star Tribune,* October 30. http://stmedia.startribune.com/documents/1OUTFEATURE103015gr_Loon+page_online.pdf.

Bystrom, Bob. 1983. "Bewick's Wren in Sherburne County." *Loon* 55: 121.

Cabe, Paul R. 2020. "European Starling (*Sturnus vulgaris*)." Version 1.0. In Billerman 2020. doi:10.2173/bow.eursta.01.

Cade, Thomas J., and Christopher P. Woods. 1997. "Changes in Distribution and Abundance of the Loggerhead Shrike." *Conservation Biology* 11: 21–31.

Cadman, Michael D., Paul F. J. Eagles, and Frederick M. Helleiner, eds. 1987. *Atlas of the Breeding Birds of Ontario*. Waterloo, Ontario: University of Waterloo Press.

Cadman, Michael D., Donald A. Sutherland, Gregor G. Beck, Denis Lepage, and Andrew R. Couturier, eds. 2007. *The Atlas of the Breeding Birds of Ontario, 2001–2005*. Toronto: Bird Studies Canada, Environment Canada, Ontario Field Ornithologists, Ontario Ministry of Natural Resources, and Ontario Nature.

Cant, G. 1962. "The House Finch in New York State." *Kingbird* 12: 68–72.

Carder, Michelle L., and Gary Ritchison. 2009. "Tail Pumping by Eastern Phoebes: An Honest, Persistent Predator-Deterrent Signal?" *Journal of Field Ornithology* 80: 163–70.

Carey, Michael, Dirk E. Burhans, and Douglas A. Nelson. 2020. "Field Sparrow (*Spizella pusilla*)." Version 1.0. In Poole 2020. doi:10.2173/bow.fiespa.01.

Carlsen, J. C., and Stanley W. Harris. 1955. "Birds of Mud Lake National Wildlife Refuge." *Flicker* 27: 138–47.

Carlson, Becky A. 1995. "Nest Site Characteristics of Great Blue Herons (*Ardea herodias*) in Northeast Ohio." *Ohio Journal of Science* 95: 312–15.

Carlson, C. Edward. 1942. "The Prairie Chicken in Minnesota: Cyclic Die-Offs Complicate Its Management." *Minnesota Conservation Volunteer* 4 (May): 45–49.

Carlson, Steve, Oscar Johnson, Kim Risen, and Dick Ruhme. 1990. "The Spring Season (1 March to 31 May 1990)." *Loon* 62: 207–25.

———. 1991a. "The Fall Season: 1 August to 30 November 1990." *Loon* 63: 115–32.

———. 1991b. "The Spring Season (1 March to 31 May 1991)." *Loon* 63: 247–67.

Carlyle, William J. 1997. "The Decline of Summerfallow on the Canadian Prairies." *Canadian Geographer* 41: 267–80.

Carr, Marjorie M. 1973. "Rusty Blackbirds." *Loon* 45: 134–35.

———. 1974. "The 1973 Summer Season." *Loon* 46: 16–27.

Carroll, John P., Philip J. K. McGowan, and Guy M. Kirwan. 2020. "Gray Partridge (*Perdix perdix*)." Version 1.0. In Billerman 2020. doi:10.2173/bow.grypar.01.

Carson, Rachel. 1962. *Silent Spring*. New York: Houghton Mifflin.

Carter, Carol W., and Laurence N. Gillette. 2002. "Common Nighthawk Population Decline in the Seven County Metropolitan Region." *Loon* 74: 3–7.

Catesby, Mark. 1754. *The Natural History of Carolina, Florida, and the Bahama Islands: Containing the Figures of Birds, Beasts, Fishes, Serpents, Insects and Plants: Particularly the Forest-Trees, Shrubs, and Other Plants, not Hitherto Described, or Very Incorrectly Figured by Authors. Together with Their Descriptions in English and French. To Which Are Added, Observations on the Air, Soil, and Waters: With Remarks upon Agriculture, Grain, Pulse, Roots, Etc. To the Whole Is Prefixed a New and Correct Map of the Countries Treated Of*. London: for C. Marsh and T. Wilcox.

Cavitt, John F., and Carola A. Haas. 2020. "Brown Thrasher (*Toxostoma rufum*)." Version 1.0. In Poole 2020. doi:10.2173/bow.brnthr.01.

Chaffee, Cindy. 1992. "Heron Lake Colonial Waterbird Nesting Survey and Sago Pondweed Study." Final report to the Minnesota Department of Natural Resources.

Chambers, Amy. 1944. "Minnesota Nesting Record, 1943." *Flicker* 16: 10–14.

———. 1947. "Minnesota Nesting Records, 1947." *Loon* 19: 90–95.

Chapa-Vargas, Leonardo, and Scott K. Robinson. 2007. "Nesting Success of Acadian Flycatchers (*Empidonax virescens*) in Floodplain Forest Corridors." *Auk* 124: 1267–80.

Chapman, Frank M. 1895. *Handbook of Birds of Eastern North America*. New York: D. Appleton.

———. (1907) 1968. *The Warblers of North America*. Unabridged and unaltered replication. New York: Dover.

———. 1908. *Camps and Cruises of an Ornithologist*. New York: D. Appleton.

———. 1966. *Handbook of Birds of Eastern North America*. 2nd ed. New York: Dover.

Chartier, Allen T., Jennifer J. Baldy, and John M. Brenneman, eds. 2013. *Michigan Breeding Bird Atlas II*. Kalamazoo, MI: Kalamazoo Nature Center.

Chasko, Gregory G., and J. Edward Gates. 1982. "Avian Habitat Suitability along a Transmission-Line Corridor in an Oak-Hickory Forest Region." *Wildlife Monographs* 82: 3–41.

Chesley, Alma H. 1943. "Nesting of the Yellow-breasted Chat." *Flicker* 15: 52–53.

Chesness, Robert A. 1964. "Minnesota's Bobwhite Quail." *Loon* 36: 60–61.

Chesser, R. Terry, Richard C. Banks, F. Keith Barker, Carla Cicero, Jon L. Dunn, Andrew W. Kratter, Irby J. Lovette, et al. 2010. "Fifty-first Supplement to the American Ornithologists' Union *Check-List of North American Birds*." *Auk* 127: 726–44.

———. 2011. "Fifty-second Supplement to the American Ornithologists' Union *Check-list of North American Birds*." *Auk* 128: 600–613. doi:10.1525/auk.2011.128.3.600.

Chesser, R. Terry, Shawn M. Billerman, Kevin J. Burns, Carla Cicero, Jon L. Dunn, Blanca E. Hernández-Baños, Andrew W. Kratter, et al. 2021. "Sixty-second Supplement to the American Ornithological Society's *Check-list of North American Birds*." *Auk* 138: 1–18. doi:10.1093/ornithology/ukab037.

Chesser, R. Terry, Shawn M. Billerman, Kevin J. Burns, Carla Cicero, Jon L. Dunn, Andrew W. Kratter, Irby J. Lovette, et al. 2020. "Sixty-first Supplement to the American Ornithological Society's *Check-list of North American Birds*." *Auk* 137: ukaa030. doi:10.1093/auk/ukaa030.

Chesser, R. Terry, Kevin J. Burns, Carla Cicero, Jon L. Dunn, Andrew W. Kratter, Irby J. Lovette, Pamela C. Rasmussen, et al. 2017. "Fifty-eighth Supplement to the American Ornithologists' Union *Check-list of North American Birds*." *Auk* 134: 751–73.

Chesser, R. Terry, Kevin J. Burns, Carla Cicero, Jon L. Dunn, Andrew W. Kratter, Irby J. Lovette, Pamela C. Rasmussen, et al. 2018. "Fifty-ninth Supplement to the American Ornithological Society's *Check-list of North American Birds*." *Auk* 135: 798–813.

Chiver, Ioana, Lesley J. Evans Ogden, and Bridget J. Stutchbury. 2020. "Hooded Warbler (*Setophaga citrina*)." Version 1.0. In Rodewald 2020. doi:10.2173/bow.hoowar.01.

Chizinski, Christopher J., Anna Peterson, JoAnn M. Hanowski, Charles R. Blinn, Bruce Vondracek, and Gerald J. Niemi. 2011. "Breeding Bird Response to Partially Harvested Riparian Management Zones." *Forest Ecology and Management* 261: 1892–1900.

Christensen, Glen C. 2020. "Chukar (*Alectoris chukar*)." Version 1.0. In Poole and Gill 2020. doi:10.2173/bow.chukar.01.

Chu, Philip C. 2005. "First Confirmed Breeding of the Rock Wren in Minnesota." *Loon* 77: 59–62.

———. 2007. "First Breeding of the Snowy Plover in Minnesota." *Loon* 79: 194–96.

———. 2013. "Sprague's Pipit in Clay County." *Loon* 85: 43–44.

———. 2018. "King Rail in Grant County." *Loon* 90: 186–87.

Cimprich, David A., Frank R. Moore, and Michael P. Guilfoyle. 2020. "Red-eyed Vireo (*Vireo olivaceus*)." Version 1. In Rodewald 2020. doi:10.2173/bo2.reevir1.01.

Cink, Calvin L., Peter Pyle, and Michael A. Patten. 2020. "Eastern Whip-poor-will (*Antrostomus vociferus*)." Version 1.0. In Rodewald 2020. doi:10.2173/bow.whip-p1.01.

Clark, Robert G., Joseph P. Fleskes, Karla L. Guyn, David A. Haukos, Jane E. Austin, and Michael R. Miller. 2020. "Northern Pintail (*Anas acuta*)." Version 1.0. In Billerman 2020. doi:10.2173/bow.norpin.01.

Clark, William R., and Todd R. Bogenschutz. 1999. "Grassland Habitat and Reproductive Success of Ring-necked Pheasants in Northern Iowa." *Journal of Field Ornithology* 70: 380–92.

Clawson, Sharon D. 1980. "Comparative Ecology of the Northern Oriole (*Icterus galbula*) and the Orchard Oriole (*Icterus spurius*) in Western Nebraska." Master's thesis, University of Nebraska–Lincoln.

CLO (Cornell Laboratory of Ornithology). 2017a. *All About Birds*. https://www.allaboutbirds.org.

———. 2017b. "Barn Swallow." Retrieved from *All About Birds*. https://www.allaboutbirds.org/guide/Barn_Swallow/overview.

———. 2017c. "Eastern Meadowlark." Retrieved from *All About Birds*. https://www.allaboutbirds.org/guide/Eastern_Meadowlark/id.

———. 2017d. "Eurasian Collared-Dove." Retrieved from *All About Birds*. https://www.allaboutbirds.org/guide/Eurasian_Collared-Dove.

———. 2017e. "European Starling." Retrieved from: *All About Birds*. https://www.allaboutbirds.org/guide/European_Starling.

———. 2017f. "House Finch." Retrieved from *All About Birds*. https://www.allaboutbirds.org/guide/House_Finch/lifehistory.

———. 2017g. "Song Sparrow." Retrieved from *All About Birds*. https://www.allaboutbirds.org/guide/Song_Sparrow/id.

———. 2017h. "Warbling Vireo." Retrieved from *All About Birds*. https://www.allaboutbirds.org/guide/Warbling_Vireo/id.

Coffin, Barbara A., and Lee A. Pfannmuller, eds. 1988. *Minnesota's Endangered Flora and Fauna*. Minneapolis: University of Minnesota Press.

Cohen, Robert R. 1958. "Common Tern Colony on Barker's Island, St. Louis Bay." *Loon* 30: 39.

———. 1961. "Second Year Study, Harbor Island, Duluth." *Loon* 33: 111–13.

Collins, Scott L. 1981. "A Comparison of Nest-site and Perch-site Vegetation Structure for Seven Species of Warblers." *Wilson Bulletin* 93: 542–47.

———. 1983. "Geographic Variation in Habitat Structure of the Black-throated Green Warbler (*Dendroica virens*)." *Auk* 100: 382–89.

Collins, Scott L., Frances C. James, and Paul G. Risser. 1982. "Habitat Relationships of Wood Warblers (*Parulidae*) in Northern Central Minnesota." *Oikos* 39: 50–58.

Colwell, Mark A., and Joseph R. Jehl Jr. 2020. "Wilson's Phalarope (*Phalaropus tricolor*)." Version 1.0. In Poole and Gill 2020. doi:10.2173/bow.wilpha.01.

Confer, John L., Patricia Hartman, and Amber Roth. 2020. "Golden-winged Warbler (*Vermivora chrysoptera*)." Version 1.0. In Poole 2020. doi:10.2173/bow.gowwar.01.

Connelly, John W., Michael W. Gratson, and Kerry P. Reese. 2020. "Sharp-tailed Grouse (*Tympanuchus phasianellus*)." Version 1.0. In Poole and Gill 2020. doi:10.2173/bow.shtgro.01.

Conway, Courtney J. 2020. "Virginia Rail (*Rallus limicola*)." Version 1.0. In Poole and Gill 2020. doi:10.2173/bow.virrai.01.

Conway, Courtney J., and William R. Eddleman. 1994. "Virginia Rail." In Tacha and Braun 1994, 193–206.

Cooke, W[ells] W. 1884. "Bird Nomenclature of the Chippewa Indians." *Auk* 1: 242–50.

Cooper, Sam. 1984. "Habitat and Size of the LeConte's Sparrow's Territory." *Loon* 56: 162–65.

Cooper, Thomas R., ed. 2007. *Henslow's Sparrow Conservation Action Plan Workshop Summary: March 7–8, 2007*. Unpublished report.

———. 2008. *King Rail Conservation Plan*. Version 1. Fort Snelling, MN: US Fish and Wildlife Service.

———. 2012. *Status Assessment and Conservation Plan for the Henslow's Sparrow (*Ammodramus henslowii*)*. Version 1.0. Bloomington, MN: US Fish and Wildlife Service. https://www.fws.gov/midwest/es/soc/birds/pdf/HESP_SAandConsPlan11Oct2012.pdf.

Cordts, Steve. 2005. "2005 Minnesota Waterfowl Breeding Population Survey." In *Status of Wildlife Populations, Fall 2005*, edited by Margaret H. Dexter, 109–39. Unpublished report. St. Paul: Minnesota Department of Natural Resources, Division of Fish and Wildlife.

———. 2019. "2019 Waterfowl Breeding Population Survey, Minnesota." Unpublished Report. St. Paul: State of Minnesota, Department of Natural Resources, Division of Fish and Wildlife.

Cornicelli, Lou, Michelle Carstensen, Gino D'Angelo, Michael A. Larson, and Jeffrey S. Lawrence, eds. 2017. *Summaries of Wildlife Research Findings 2015*. St. Paul: Minnesota Department of Natural Resources.

Cornwell, George W. 1963. "Observations on the Breeding Biology and Behavior of a Nesting Population of Belted Kingfishers." *Condor* 65: 426–31.

COSEWIC (Committee on the Status of Endangered Wildlife in Canada). 2008. *COSEWIC Assessment and Update Status Report on the Short-eared Owl* Asio flammeus *in Canada*. Ottawa: COSEWIC.

———. 2009a. *Assessment and Status Report on the Yellow Rail* Coturnicops noveboracensis *in Canada*. Ottawa: COSEWIC.

———. 2009b. *COSEWIC Assessment and Status Report on the Horned Grebe* Podiceps auratus, *Western Population and Magdalen Islands Pupulation, in Canada*. Ottawa: COSEWIC.

———. 2010. *Assessment and Status Report on the Sprague's Pipit* Anthus spragueii *in Canada*. Ottawa: COSEWIC.

———. 2011. *COSEWIC Assessment and Status Report on the Henslow's Sparrow* Ammodramus henslowii *in Canada*. Ottawa: COSEWIC.

———. 2013. *Assessment and Status Report on the Bank Swallow* Riparia riparia *in Canada*. Ottawa: COSEWIC.

———. 2017. *Assessment and Status Report on the Canada Warbler Cardellina canadensis in Canada*. Ottawa: COSEWIC.

Cottrille, Betty Darling. 1962. "A Search for Nesting Cape May Warblers." *Flicker* 34: 38–40.

Coues, Elliott. 1874. *Birds of the Northwest: A Handbook of the Ornithology of the Region Drained by the Missouri River and Its Tributaries*. US Geological Survey of the Territories, no. 3. Washington, DC: Government Printing Office.

———. 1878. *Field Notes on Birds Observed in Dakota and Montana along the Forty-ninth Parallel during the Seasons of 1873 and 1874*. Author's ed. Washington, DC: US Geological and Geographical Survey. https://pubs.usgs.gov/unnumbered/70159117/report.pdf.

Cox, W. Andrew, Frank R. Thompson III, Brian Root, and John Faaborg. 2012. "Declining Brown-Headed Cowbird (*Molothrus ater*) Populations Are Associated with Landscape-Specific Reductions in Brood Parasitism and Increases in Songbird Productivity." *PLoS One* 7: e47591. doi:10.1371/journal.pone.0047591.

Craik, Shawn, John Pearce, and Rodger D. Titman. 2020. "Red-breasted Merganser (*Mergus serrator*)." Version 1.0. In Billerman 2020. doi:10.2173/bow.rebmer.01.

Crewe, Tara L., Philip Taylor, Denis Lepage, Laurie Goodrich, J. Brown, and J. Sodergren. 2016. *The Raptor Population Index, 2016 Analysis Methods and Trend Results*. Accessed November 1, 2019. http://rpi-project.org/2016/.

Crewe, Tara L., Steven T. A. Timmermans, and Kathy E. Jones. 2005. *The Marsh Monitoring Program Annual Report, 1995–2003: Annual Indices and Trends in Bird Abundance and Amphibian Occurrence in the Great Lakes Basin*. Port Rowan, Ontario: Bird Studies Canada.

Crosby, John, and Eva Beckett. 1957. "Rock Wren at Churchill, Manitoba." *Canadian Field-Naturalist* 71: 82–83.

Cruz-Martinez, Luis, Patrick T. Redig, and John Deen. 2012. "Lead from Spent Ammunition: A Source of Exposure and Poisoning in Bald Eagles." *Human-Wildlife Interactions* 6: 94–104.

Cullen, Sean A., Joseph R. Jehl Jr., and Gary L. Nuechterlein. 2020. "Eared Grebe (*Podiceps nigricollis*)." Version 1.0. In Billerman 2020. doi:10.2173/bow.eargre.01.

Cummins, J. F., and D. F. Grigal. 1981. *Legend to Map—Soils and Land Surfaces of Minnesota, 1980*. Soil Series No. 110, Miscellaneous Publication 11. St. Paul: University of Minnesota Department of Soil Science.

Currier, Edmonde S. 1895. "The Hooded Warbler." *Iowa Ornithologist* 1: 67–70.

———. 1904. "Summer Birds of the Leech Lake Region, Minnesota." *Auk* 21 (1): 29–44.

Curson, Jon, David Quinn, and David Beadle. 1994. *Warblers of the Americas: An Identification Guide*. New York: Houghton Mifflin.

Custer, Christine M., and Joan Galli. 2002. "Feeding Habitat Selection by Great Blue Herons and Great Egrets Nesting in East Central Minnesota." *Waterbirds* 25: 115–24.

Cuthbert, Francesca J. 2016. "Rebounding Plovers Are a Hopeful Sign." *Star Tribune*, July 11. https://www.pressreader.com/usa/star-tribune/20160711/281904477509832.

Cuthbert, Francesca J., and Mei-Yao Louis. 1993. "The Forster's Tern in Minnesota: Status, Distribution, and Reproductive Success." *Wilson Bulletin* 105: 184–87.

Cuthbert, Francesca J., and Sarah P. Saunders. 2017. "Piping Plover Breeding Biology and Management in the Great Lakes, 2017." Report to the US Fish and Wildlife Service. East Lansing, MI: US Fish and Wildlife Service, East Lansing Field Office.

Cuthbert, Francesca J., and Linda R. Wires. 2013. "The Fourth Decadal U.S. Great Lakes Colonial Waterbird Survey (2007–2010): Results and Recommendations to Improve the Scientific Basis for Conservation and Management." Final report to the US Fish and Wildlife Service.

———. 2020. "Caspian Tern (*Hydroprogne caspia*)." Version 1.0. In Billerman 2020. doi:10.2173/bow.caster1.01.

Cuthbert, Francesca J., Linda R. Wires, and Kristina Timmerman. 2003. *Status Assessment and Conservation Recommendations for the Common Tern (*Sterna hirundo*) in the Great Lakes*. Fort Snelling, MN: US Fish and Wildlife Service.

Cutright, Noel, Bettie R. Harriman, and Robert W. Howe, eds. 2006. *Atlas of the Breeding Birds of Wisconsin*. Waukesha: Wisconsin Society of Ornithology.

CWS and USFWS (Canadian Wildlife Service and US Fish and Wildlife Service). 2007. *International Recovery Plan for the Whooping Crane*. Ottawa: Recovery of Nationally Endangered Wildlife (RENEW), and Albuquerque, NM: US Fish and Wildlife Service.

Czarpata, Elizabeth J. 2005. *Invasive Plants of the Upper Midwest*. Madison: University of Wisconsin Press.

Danz, Nicholas P., Gerald J. Niemi, James W. Lind, and JoAnn M. Hanowski. 2007. *Birds of Western Great Lakes Forests*. http://www.nrri.umn.edu/mnbirds.

Darley-Hill, Susan, and W. Carter Johnson. 1981. "Acorn Dispersal by the Blue Jay (*Cyanocitta cristata*)." *Oecologia* 50: 231–32.

Davidson, Frances S. 1942. "The Hooded Warbler in Minneapolis." *Flicker* 14: 41–43.

Davis, Brian. 2019. "The Scoop on Duck Bills." *Ducks Unlimited* (website). https://www.ducks.org/conservation/waterfowl-research-science/the-scoop-on-duck-bills.

Davis, George. 1978. "Bufflehead Brood in Marshall County." *Loon* 50: 213–14.

Davis, Stephen K., and Wesley E. Lanyon. 2020. "Western Meadowlark (*Sturnella neglecta*)." Version 1.0. In Poole 2020. doi:10.2173/bow.wesmea.01.

Davis, Stephen K., Mark B. Robbins, and Brenda C. Dale. 2020. "Sprague's Pipit (*Anthus spragueii*)." Version 1.0. In Poole 2020. doi:10.2173/bow.sprpip.01.

Davis, William E., Jr., and James A. Kushlan. 2020. "Green Heron (*Butorides virescens*)." Version 1.0. In Poole and Gill 2020. doi:10.2173/bow.grnher.01.

Davis, William James. 1982. "Territory Size in *Megaceryle alcyon* along a Stream Habitat." *Auk* 99: 353–62.

Davros, Nicole M. 2015. "2015 Minnesota August Roadside Survey." Madelia: Minnesota Department of Natural Resources. http://files.dnr.state.mn.us/publications/wildlife/population2015/1-farmland-wildlife.pdf.

Davros, Nicole, and Margaret Dexter. 2019. "2019 Small Game Hunter Mail Survey." Minnesota

Department of Natural Resources, Wildlife Research Unit. https://files.dnr.state.mn.us/wildlife/research/smallgame/survey_2019.pdf.
Dawson, William Leon. 1923. *The Birds of California: A Complete, Scientific and Popular Account of the 580 Species and Subspecies of Birds Found in the State*. Vols. 1 and 2. San Francisco: South Moulton Company.
Dawson, William R. 2020. "Pine Siskin (*Spinus pinus*)." Version 1.0. In Poole 2020. doi:10.2173/bow.pinsis.01.
Deane, Ruthven. 1923. "Extracts from the Field Notes of George B. Sennett." *Auk* 40: 626–33.
Decalesta, David S. 1994. "Effect of White-tailed Deer on Songbirds within Managed Forests in Pennsylvania." *Journal of Wildlife Management* 58: 711–18.
Dechant, Jill A., Meghan F. Dinkins, Douglas H. Johnson, Lawrence D. Igl, Christopher M. Goldade, and Betty R. Euliss. (2000) 2002. *Effects of Management Practices on Grassland Birds: Vesper Sparrow*. Jamestown, ND: Northern Prairie Wildlife Research Center. http://digitalcommons.unl.edu/usgsnpwrc/117.
Dechant, Jill A., Meghan F. Dinkins, Douglas H. Johnson, Lawrence D. Igl, Christopher M. Goldade, Barry D. Parkin, and Betty R. Euliss. (1999) 2002. *Effects of Management Practices on Grassland Birds: Upland Sandpiper*. Rev. ed. Jamestown, ND: Northern Prairie Wildlife Research Center.
Dechant, Jill A., Douglas H. Johnson, Lawrence D. Igl, Christopher M. Goldade, Amy L. Zimmerman, and Betty R. Euliss. (1999) 2002. *Effects of Management Practices on Grassland Birds: Wilson's Phalarope*. Rev. ed. Jamestown, ND: Northern Prairie Wildlife Research Center.
Dechant, Jill A., Marriah L. Sondreal, Douglas H. Johnson, Lawrence D. Igl, and Christopher M. Goldade. (1998) 2002. *Effects of Management Practices on Grassland Birds: Clay-colored Sparrow*. Jamestown, ND: Northern Prairie Wildlife Research Center. http://digitalcommons.unl.edu/cgi/viewcontent.cgi?article=1126&context=usgsnpwrc.
Dechant, Jill A., Marriah L. Sondreal, Douglas H. Johnson, Lawrence D. Igl, Christopher M. Goldade, Melvin P. Nenneman, and Betty R. Euliss. (1998) 2001a. *Effects of Management Practices on Grassland Birds: Marbled Godwit*. Rev. ed. Jamestown, ND: Northern Prairie Wildlife Research Center.
———. (1998) 2001b. *Effects of Management Practices on Grassland Birds: Short-Eared Owl*. Jamestown, ND: Northern Prairie Wildlife Research Center. http://digitalcommons.unl.edu/usgsnpwrc/136.
———. (1998) 2002a. *Effects of Management Practices on Grassland Birds: Grasshopper Sparrow*. Jamestown, ND: Northern Prairie Wildlife Research Center. http://digitalcommons.unl.edu/usgsnpwrc/147.
———. (1998) 2002b. *Effects of Management Practices on Grassland Birds: Northern Harrier*. Rev. ed. Jamestown, ND: Northern Prairie Wildlife Research Center.
Dechant, Jill A., Marriah L. Sondreal, Douglas H. Johnson, Lawrence D. Igl, Christopher M. Goldade, Barry D. Parkin, and Betty R. Euliss. (1998) 2002. *Effects of Management Practices on Grassland Birds: Sedge Wren*. Jamestown, ND: Northern Prairie Wildlife Research Center.
———. (1999) 2002a. *Effects of Management Practices on Grassland Birds: Field Sparrow*. Rev. ed. Jamestown, ND: Northern Prairie Wildlife Research Center. http://digitalcommons.unl.edu/cgi/viewcontent.cgi?article=1135&context=usgsnpwrc.
———. (1999) 2002b. *Effects of Management Practices on Grassland Birds: Lark Sparrow*. Rev. ed. Jamestown, ND: Northern Prairie Wildlife Research Center. http://digitalcommons.unl.edu/cgi/viewcontent.cgi?article=1138&context=usgsnpwrc.
Dechant, Jill A., Marriah L. Sondreal, Douglas H. Johnson, Lawrence D. Igl, Christopher M. Goldade, Amy L. Zimmerman, and Betty R. Euliss. (1998) 2002. *Effects of Management Practices on Grassland Birds: Le Conte's Sparrow*. Rev. ed. Jamestown, ND: Northern Prairie Wildlife Research Center. http://digitalcommons.unl.edu/cgi/viewcontent.cgi?article=1139&context=usgsnpwrc.
———. (1999) 2001. *Effects of Management Practices on Grassland Birds: Bobolink*. Rev. ed. Jamestown, ND: Northern Prairie Wildlife Research Center.
———. (1999) 2002a. *Effects of Management Practices on Grassland Birds: Dickcissel*. Rev. ed. Jamestown, ND: Northern Prairie Wildlife Research Center. http://digitalcommons.unl.edu/cgi/viewcontent.cgi?article=1133&context=usgsnpwrc.
———. (1999) 2002b. *Effects of Management Practices on Grassland Birds: Western Meadowlark*. Rev. ed. Jamestown, ND: Northern Prairie Wildlife Research Center.
———. 2004. *Effects of Management Practices on Grassland Birds: American Bittern*. Jamestown, ND: Northern Prairie Wildlife Research Center.
Degernes, Laurel A., and R. K. Frank. 1991. "Causes of Mortality in Trumpeter Swans *Cygnus buccinator* in Minnesota 1986–1989." *Wildfowl Supplement* 1: 352–55.
DeGraaf, Richard M. 1985. "Breeding Bird Assemblages in New England Northern Hardwoods." In *The Impact of Timber Management Practices on Nongame Birds in Vermont*, edited by Ronald J. Regan and David E. Capen, 5–22. Montpelier: Vermont Fish and Wildlife Department.
DeGraaf, Richard M., and Keith E. Evans, eds. 1979. *Workshop Proceedings on Management of Northeastern and North Central Forests for Nongame Birds*. US Department of Agriculture, Forest Service, General Technical Report NC-51. St. Paul, MN: US Department of Agriculture, USDA Forest Service, North Central Forest Experiment Station.
DeGraaf, Richard M., Virgil E. Scott, Robert H. Hamre, Liz Ernst, and Stanley H. Anderson. 1991. *Forest and Rangeland Birds of the United States: Natural History and Habitat Use*. Agriculture Handbook, no. 688. Washington, DC: US Department of Agriculture, Forest Service.
De Jong, Michael J. 2020. "Northern Rough-winged Swallow *(Stelgidopteryx serripennis)*." Version 1.0. In Poole and Gill 2020. doi:10.2173/nrwswa.01.
DellaSala, Dominick A., and Dale L. Rabe. 1987. "Response of Least Flycatchers *Empidonax minimus* to Forest Disturbances." *Biological Conservation* 41: 291–99.
Dellinger, Rachel, Petra Bohall Wood, Peter W. Jones, and Therese M. Donovan. 2020. "Hermit Thrush *(Catharus guttatus)*." Version 1.0. In Poole 2020. doi:10.2173/bow.herthr.01.
Delphey, Philip J., and James J. Dinsmore. 1993. "Breeding Bird Communities of Recently Restored and Natural Prairie Potholes." *Wetlands* 13: 200–206.
Dempsey, Dave. 2006. *Minnesota Calling: Conservation Facts, Trends and Challenges*. http://www.belwin.org/media/50yearvision/minnesota_calling.pdf.
Dempsey, Joanne. 1980. "1972 Record of Blue Grosbeaks in Dakota County." *Loon* 52: 192.
De Smet, Ken D. 1983. "Breeding Ecology and Productivity of Red-necked Grebes in Turtle Mountain Provincial Park, Manitoba." Master's thesis, University of North Dakota, Grand Forks.
———. 1987. "Organochlorines, Predators and Reproductive Success of the Red-necked Grebe in Southern Manitoba." *Condor* 89: 460–67.

Desrochers, André, Susan J. Hannon, and Kelly E. Nordin. 1988. "Winter Survival and Territory Acquisition in a Northern Population of Black-capped Chickadees." *Auk* 105: 727–36.

Desrochers, André, Line Rochefort, and Jean-Pierre L. Savard. 1998. "Avian Recolonization of Eastern Canadian Bogs after Peat Mining." *Canadian Journal of Zoology* 76: 989–97.

Dessecker, Daniel R., Gary W. Norman, and Scot J. Williamson. 2006. *Ruffed Grouse Conservation Plan*. Association of Fish and Wildlife Agencies, Resident Game Bird Working Group. https://www.ruffedgrousesociety.org/UserFiles/File/RG_ConservationPlan.pdf.

DeVore, Brian. 2018. *Wildly Successful Farming: Sustainability and the New Agricultural Land Ethic*. Madison: University of Wisconsin Press.

Dexter, Margaret. 2017. "2017 Small Game Hunter Mail Survey." Minnesota Department of Natural Resources, Wildlife Research Unit. https://files.dnr.state.mn.us/wildlife/research/smallgame/2017.pdf.

Dickerman, Robert W., and Eugene Lefebvre. 1961. "Ring-billed Gull Nesting in Minnesota." *Loon* 33: 23.

Dickson, James G., Frank R. Thompson III, Richard N. Conner, and Kathleen E. Franzreb. 1995. "Silviculture in Central and Southeastern Oak Pine Forests." In *Ecology and Management of Neotropical Migratory Birds: A Synthesis and Review of Critical Issues,* edited by Thomas E. Martin and Deborah M. Finch, 245–66. New York: Oxford University Press.

DiMatteo, Jon J. 1992. "Status, Production and Migration of Greater Sandhill Cranes on Agassiz National Wildlife Refuge, Minnesota." In *Proceedings 1988 North American Crane Workshop, Feb. 22–24, 1988. Lake Wales, Florida,* edited by Don A. Wood, 13–20. State of Florida Game and Fresh Water Fish Commission Nongame Wildlife Program Technical Report 12. Tallahassee: State of Florida Game and Freshwater Commission Nongame Wildlife Program.

———. 2016. "Population Dynamics and Management Implications for American White Pelicans *(Pelecanus erythrorhynchos)* Breeding at Marsh Lake, Lac qui Parle Wildlife Management Area, Minnesota." PhD diss., North Dakota State University.

Dinkins, Meghan F., Amy L. Zimmerman, Jill A. Dechant, Barry D. Parkin, Douglas H. Johnson, Lawrence D. Igl, Christopher M. Goldade, and Betty R. Euliss. (2000) 2002. *Effects of Management Practices on Grassland Birds: Horned Lark*. Rev. ed. Jamestown, ND: Northern Prairie Wildlife Research Center.

Dinsmore, James J., and Stephen J. Dinsmore. 1993. "Range Expansion of the Great-tailed Grackle in the 1900s." *Journal of the Iowa Academy of Science* 100: 54–59.

Dobie, John. 1956. "Cormorants at Lake of the Woods." *Flicker* 28: 108–9.

Doherty, Paul F., Jr., and Thomas C. Grubb Jr. 2002. "Survivorship of Permanent-Resident Birds in a Fragmented Forested Landscape." *Ecology* 83: 844–57.

Dokken, Brad. 2008. "June Rains Put Damper on Gull Project at Agassiz Wildlife Refuge." *Grand Forks Herald,* July 13. https://www.grandforksherald.com/sports/outdoors/2077121-june-rains-put-damper-gull-project-agassiz-wildlife-refuge.

———. 2012. "Agassiz National Wildlife Refuge Celebrates 75th Anniversary." *Grand Forks Herald,* June 22. http://www.grandforksherald.com/content/agassiz-national-wildlife-refuge-celebrates-75th-anniversary.

Dolbeer, Richard A., and George M. Linz. 2016. "Blackbirds." Wildlife Damage Management Technical Series. US Department of Agriculture, Animal and Plant Health Inspection Service, Wildlife Services. http://digitalcommons.unl.edu/nwrcwdmts/1.

Donovan, Edward. 1820. *The Natural History of British Birds; or, A Selection of the Most Rare, Beautiful, and Interesting Birds Which Inhabit This Country: The Descriptions from the Systema Naturæ of Linnæus; with General Observations, Either Original, or Collected from the Latest and Most Esteemed English Ornithologists; and Embellished with Figures, Drawn, Engraved, and Coloured from the Original Specimens*. London.

Donovan, Therese M., and Curtis H. Flather. 2002. "Relationships among North American Songbird Trends, Habitat Fragmentation, and Landscape Occupancy." *Ecological Applications* 12: 364–74.

Donovan, Therese M., Peter W. Jones, Elizabeth M. Annand, and Frank R. Thompson III. 1997. "Variation in Local-Scale Edge Effects: Mechanisms and Landscape Context." *Ecology* 78: 2064–75.

Dorr, Brian S., Jeremy J. Hatch, and Datlaf Vaughn Weseloh. 2020. "Double-crested Cormorant *(Phalacrocorax auritus)*." Version 1.0. In Poole 2020. doi:10.2173/bow.doccor.01.

Dow, Douglas D., and David M. Scott. 1971. "Dispersal and Range Expansion by the Cardinal: An Analysis of Banding Records." *Canadian Journal of Zoology* 49: 185–98.

Doyle, Diana. 2011. "Successful Bell's Vireo Nest in Hennepin County." *Loon* 83: 42–43.

Drewien, Rod C., Carroll D. Littlefield, Lawrence H. Walkinshaw, and C. E. Brown. 1975. "Conservation Committee Report on Status of Sandhill Cranes." *Wilson Bulletin* 87: 297–302.

Drilling, Nancy E. 2013. *South Dakota 2012 Colonial Waterbird Survey*. Technical Report SC-Colony SD-06. Brighton, CO: Rocky Mountain Bird Observatory.

———. 2015. *North Dakota Colonial and Semi-Colonial Waterbird Inventory, 2014–2015*. Final Report. Technical Report SC-NDCWB-02. Brighton, CO: Bird Conservancy of the Rockies.

Drilling, Nancy E., Robert A. Sparks, Brittany J. Woiderski, and Jason P. Beason. 2016. *South Dakota Breeding Bird Atlas II: Final Report*. Technical Report M-SDBBA2-07. Brighton, CO: Rocky Mountain Bird Observatory.

Drilling, Nancy E., Rodger D. Titman, and Frank McKinney. 2020. "Mallard (*Anas platyrhynchos*)." Version 1. In Billerman 2020. doi:10.2173/bow.mallar3.01.

Dubovsky, J. A. 2019. *Status and Harvests of Sandhill Cranes: Mid-Continent, Rocky Mountain, Lower Colorado River Valley and Eastern Populations*. Administrative Report. Lakewood, CO: US Fish and Wildlife Service.

Dubowy, Paul J., Carles Carboneras, and Guy M. Kirwan. 2020. "Northern Shoveler (*Anas clypeata*)." Version 1.0. In Billerman 2020. doi:10.2173/bow.norsho.01.

Dufrêne, Marc, and Pierre Legendre. 1997. "Species Assemblages and Indicator Species: The Need for a Flexible Asymmetrical Approach." *Ecological Monographs* 67: 345–66.

Dugger, Bruce D., and Katie M. Dugger. 2020. "Long-billed Curlew (*Numenius americanus*)." Version 1.0. In Poole and Gill 2020. doi:10.2173/bow.lobcur.01.

Dugger, Bruce D., Katie M. Dugger, and Leigh H. Fredrickson. 2020. "Hooded Merganser (*Lophodytes cucullatus*)." Version 1.0. In Poole 2020. doi:10.2173/bow.hoomer.01.

Duncan, David C. 1987. "Nest-Site Distribution and Overland Brood Movements of Northern Pintails in Alberta." *Journal of Wildlife Management* 51: 716–23.

Duncan, James R., and Patricia A. Duncan. 2020. "Northern Hawk Owl (*Surnia ulula*)." Version 1.0. In Billerman 2020. doi:10.2173/bow.nohowl.01.

Dundas, Les. 1968. "1951 Whooping Crane Record." *Loon* 40: 21.
Dunlap, Robert M. 2013. "Responses of Songbirds and Small Mammals to Harvests of Native Grasslands for Biofuels in Western Minnesota." Master's thesis, University of Minnesota.
———. 2019. "An Individual Meadowlark Giving the Primary Calls of Both Species." *Loon* 90: 143–48.
Dunn, Erica H., and George A. Hall. 2020. "Magnolia Warbler (*Setophaga magnolia*)." Version 1.0. In Rodewald 2020. doi:10.2173/bow.magwar.01.
Dunn, Jon L., and Kimball L. Garrett. 1997. *A Field Guide to Warblers of North America*. Boston: Houghton Mifflin.
Dunn, Peter O., and David W. Winkler. 1999. "Climate Change Has Affected the Breeding Date of Tree Swallows throughout North America." *Proceedings of the Royal Society of London, Series B: Biological Sciences* 266: 2487–90.
Dunning, Joan. 1985. *The Loon: Voice of the Wilderness*. Dublin, NH: Yankee Publishing.
Dunstan, Thomas C., and Steve D. Sample. 1972. "Biology of Barred Owls in Minnesota." *Loon* 44: 111–15.
Dybas, Cheryl Lyn. 2018. "Historical Sign of Chesapeake Winter, the Canvasback, Still Brightens the Bay." *National Geographic* (blog), March 17. https://blog.nationalgeographic.org/2018/03/17/historical-sign-of-chesapeake-winter-the-canvasback-still-brightens-the-bay.
Dyke, Steve R., Sandra K. Johnson, and Patrick T. Isakson. 2015. *North Dakota State Wildlife Action Plan*. Bismarck: North Dakota Game and Fish Department.
Dykstra, Cheryl R., Jeffrey L. Hays, and Scott T. Crocoll. 2020. "Red-shouldered Hawk (*Buteo lineatus*)." Version 1.0. In Poole 2020. doi:10.2173/bow.reshaw.01.000.
Eadie, John M., Mark L. Mallory, and Harry G. Lumsden. 2020. "Common Goldeneye (*Bucephala clangula*)." Version 1.0. In Billerman 2020. doi:10.2173/bow.comgol.01.
Eaton, Stephen W. 1988. "Hooded Warbler." In *The Atlas of Breeding Birds in New York State,* edited by Robert F. Andrle and James R. Carroll, 418. Ithaca: Cornell University Press.
Eberhardt, Laurie S. 2000. "Use and Selection of Sap Trees by Yellow-bellied Sapsuckers." *Auk* 117: 41–51.
eBird. 2012. "Dickcissels Push Northward." Ithaca: Cornell Lab of Ornithology. http://ebird.org/content/ebird/news/dickcissel2012/.
———. 2020. *eBird: An Online Database of Bird Distribution and Abundance*. Black-billed Cuckoo for Dec–Feb, Past 10 Years. http://www.ebird.org/map/bkbcuc.
———. 2021. *eBird: An Online Database of Bird Distribution and Abundance*. http://www.ebird.org.
Eckberg, Daniel J. 2010. "Icon under Fire: The Giant Canada Geese of Rochester, Minnesota." Senior thesis, University of Minnesota.
Eckerle, Kevin P., and Charles F. Thompson. 2020. "Yellow-breasted Chat (*Icteria virens*)." Version 1.0. In Poole and Gill 2020. doi:10.2173/bow.yebcha.01.
Eckert, Kim R. 1968. "Say's Phoebe and Ground Dove Specimens Found." *Loon* 40: 18–19.
———. 1970. "Louisiana Waterthrush in Stearns Co." *Loon* 42: 147–48.
———. 1971. "The Summer Season (June 1 to July 31, 1971)." *Loon* 43: 114–19.
———. 1973a. "Cattle Egrets and Little Blue Herons in Grant Co. Heronry." *Loon* 45: 21–22.
———. 1973b. "The 1972 Fall Season—Aug. 1 to Nov. 30." *Loon* 45: 11–15.
———. 1974a. "Henslow's Sparrow and Bell's Vireo in Southwest Minnesota." *Loon* 46: 122–23.
———. 1974b. "Say's Phoebe and Barn Owl Observed in Rock County." *Loon* 46: 37–39.
———. 1975a. "Possible Blue Grosbeak Nesting at Blue Mounds State Park." *Loon* 47: 47–48.
———. 1975b. "Possible Say's Phoebe Nesting at Blue Mounds State Park." *Loon* 47: 13.
———. 1976. "The Summer Season (June 1–July 31, 1975)." *Loon* 48: 14–28.
———. 1977a. "Second Minnesota Nesting of the Mockingbird." *Loon* 49: 229–30.
———. 1977b. "The Summer Season (June 1–July 31, 1976)." *Loon* 49: 29–40.
———. 1978a. "The Summer Season (June 1–July 31, 1977)." *Loon* 50: 9–24.
———. 1978b. "Unusual Diurnal Hunting Behavior of a Long-eared Owl." *Loon* 50: 122.
———. 1979. "Another Boreal Owl Nest Found in Cook County." *Loon* 51: 198–99.
———. 1980. "A Survey of Birds on Selected Prairie Tracts." *Loon* 52: 170–77.
———. 1981. "First Minnesota Nesting Record of Northern Three-toed Woodpecker." *Loon* 53: 221–23.
———. 1982a. "First 20th Century Record of McCown's Longspur in Minnesota." *Loon* 54: 195.
———. 1982b. "Proceedings of the Minnesota Ornithological Records Committee." *Loon* 54: 42–46.
———. 1984a. "A King Rail in Duluth." *Loon* 56: 190–91.
———. 1984b. "Western Wood-Pewee in Duluth." *Loon* 56: 263–64.
———. 1986a. "Minnesota's Fifth Snowy Plover Record." *Loon* 58: 142–43.
———. 1986b. "Proceedings of the Minnesota Ornithological Records Committee." *Loon* 58: 43–45.
———. 1987. "A Documented Invasion of Gray Jays in Duluth, Fall 1986." *Loon* 59: 41–44.
———. 1989a. "Identification and Status of Clark's Grebe in Minnesota." *Loon* 61: 99–108.
———. 1989b. "Western Wood-Pewee in the Sax-Zim Bog." *Loon* 61: 147–48.
———. 1990a. "Proceedings of the Minnesota Ornithological Records Committee." *Loon* 62: 12–14.
———. 1990b. "Proceedings of the Minnesota Ornithological Records Committee." *Loon* 62: 134–36.
———. 1991. "Proceedings of the Minnesota Ornithological Records Committee." *Loon* 63: 154–55.
———. 1992. "Proceedings of the Minnesota Ornithological Records Committee." *Loon* 64: 138–41.
———. 1993. "Proceedings of the Minnesota Ornithological Records Committee." *Loon* 65: 33–36.
———. 1998a. "A First State Record: Eurasian Collared-Dove." *Loon* 70: 199–200.
———. 1998b. "Proceedings of the Minnesota Ornithological Records Committee." *Loon* 70: 155–59.
———. 2000. "Proceedings of the Minnesota Ornithological Records Committee." *Loon* 72: 163–68.
———. 2001. "Proceedings of the Minnesota Ornithological Records Committee." *Loon* 73: 35–38.
———. 2003. "Proceedings of the Minnesota Ornithologists' Union Records Committee." *Loon* 75: 41–46.
———. 2004. "Proceedings of the Minnesota Ornithologists' Union Records Committee." *Loon* 76: 39–45.
———. 2005. "Birding by Hindsight: A Second Look at Swallows." *Loon* 77: 111–14.
———. 2006. "Birding by Hindsight: A Second Look at Meadowlarks." *Loon* 78: 111–14.
———. 2008. "Birding by Hindsight: A Second Look at Flycatchers." *Loon* 80: 226–30.
———. 2010. "Birding by Hindsight: A Second Look at Second Looks (Part Two)." *Loon* 82: 138–42.

———. 2017. *Birding by Hindsight: A Second Look at Bird Identification*. Wrenshall, MN: Stone Ridge Press.

———. 2020. "Barn Owl in St. Louis County." *Loon* 92: 86–87.

Eckert, Kim R., and Micki Buer. 1991. "Possible Clark's Grebe Breeding Record in Minnesota." *Loon* 63: 220–25.

Eckert, Kim R., and Terry L. Savaloja. 1979. "First Documented Nesting of the Boreal Owl South of Canada." *American Birds* 33: 135–37.

Edgar, Marjorie. 1943. "Tufted Titmouse Nesting in Minnesota." *Flicker* 15: 35.

Edwards, Brandon P. M., Adam C. Smith, Teegan D. S. Docherty, Marcel A. Gahbauer, Caitlyn R. Gillespie, Alexis R. Grinde, Taylor Harmer, et al. 2023. "Point Count Offsets for Estimating Population Sizes of North American Landbirds; NA-POPS." *Ibis* 165: 482–503. doi:10.1111/ibi.13169.

Edwards, Eddy, and Judy Edwards. 2000. "Carolina Wren Visits Otter Tail County." *Loon* 72: 243–44.

Efron, Bradley. 2014. "Estimation and Accuracy after Model Selection." *Journal of the American Statistical Association* 109: 991–1007.

Egeland, Paul. 1966. "Common Gallinule near Grand Rapids." *Loon* 38: 109.

———. 1972. "A New Minnesota Species." *Loon* 44: 115–16.

———. 1983. "A Probable Great-tailed Grackle in Dakota County." *Loon* 55: 83–84.

Ehlers, Bradley D. 1977. "Rookery Survey on Big Stone National Wildlife Refuge." *Loon* 49: 185–86.

Ehrlich, Paul R., David S. Dobkin, and Darryl Wheye. 1988. *The Birder's Handbook. A Field Guide to the Natural History of North American Birds*. New York: Simon & Schuster.

Einhorn, C. M. 1969. "Say's Phoebe Seen in Minnesota." *Loon* 41: 10–11.

Eliason, Bonita C. 1996. "Statewide Survey and Habitat Protection for the Loggerhead Shrike in Minnesota. Final Report to US Fish and Wildlife Service Partnerships for Wildlife Program." Minnesota Department of Natural Resources, Natural Heritage and Nongame Research Program. https://files.dnr.state.mn.us/eco/nongame/projects/consgrant_reports/1996/1996_eliason.pdf.

Eliason, Bonita C., and Bruce A. Fall. 1989. "Louisiana Waterthrushes in Washington County: Results of the 1988 Minnesota County Biological Survey Work." *Loon* 61: 34–37.

Elith, Jane, Steven J. Phillips, Trevor Hastie, Miroslav Dudík, Yung En Chee, and Colin J. Yates. 2011. "A Statistical Explanation of MaxEnt for Ecologists." *Diversity and Distributions* 17: 43–57.

Elliot, Lisa H. 2016. "Grasshopper Sparrow Distribution, Habitat Associations, and Use as an Indicator Species for Grassland Birds in Southwestern Minnesota." Master's thesis, University of Minnesota.

Elliott, John J., and Robert S. Arbib Jr. 1953. "Origin and Status of the House Finch in the Eastern United States." *Auk* 70: 31–37.

Elliott-Smith, Elise, Mark Bidwell, Amanda E. Holland, and Susan M. Haig. 2015. *Data from the 2011 International Piping Plover Census*. Data Series 922. Reston, VA: US Geological Survey. https://pubs.usgs.gov/ds/0922/pdf/ds922.pdf.

Elliott-Smith, Elise, and Susan M. Haig. 2020. "Piping Plover (*Charadrius melodus*)." Version 1.0. In Poole 2020. doi:10.2173/bow.pipplo.01.

Elliott-Smith, Elise, Susan M. Haig, and Brandi M. Powers. 2009. *Data from the 2006 International Piping Plover Census*. Data Series 426. Reston, VA: US Geological Survey. https://pubs.usgs.gov/ds/426/pdf/ds426.pdf.

Ellis, John H. 1965. "Common Gallinule at Agassiz Refuge." *Loon* 37: 83.

Ellison, Walter G., ed. 2010. *Second Atlas of the Breeding Birds of Maryland and the District of Columbia*. Baltimore, MD: Johns Hopkins University Press.

Ellwanger, Lane. 1997. "Summary of Cattle Egret Sightings in Big Stone County." *Loon* 69: 46.

Emerson, W. Otto. [1905] 2020. "Passenger Pigeon Nesting in Minnesota." *Loon* 92: 40–41. Reprint from *The Warbler* 1: 74–76.

Emlen, John T., and John A. Wiens. 1965. "The Dickcissel Invasion of 1964 in Southern Wisconsin." *Passenger Pigeon* 27: 51–59.

English, Philina A., Joseph J. Nocera, Bruce A. Pond, and David J. Green. 2017. "Habitat and Food Supply across Multiple Spatial Scales Influence the Distribution and Abundance of a Nocturnal Aerial Insectivore." *Landscape Ecology* 32: 343–59.

Engstrom, Hugh R. 1940. "1940 Minnesota Nesting Records." *Flicker* 12: 46–52.

Enstrom, Oscar, and Jack Bardon. 1952. "Snowy Egret Seen at Lake Traverse." *Flicker* 24: 86.

Environment Canada. 2011. *Status of Birds in Canada*. https://wildlife-species.canada.ca/bird-status/index-eng.aspx?sY=2011&sL=e.

———. 2014a. *Management Plan for the McCown's Longspur (*Rhynchophanes mccownii*) in Canada. Species at Risk Act* Management Plan Series. Ottawa: Environment Canada.

———. 2014b. *North American Breeding Bird Survey–Canadian Trends Website*. Data version 2012. Gatineau, Quebec: Environment Canada.

Erickson, Arnold B. 1957. "Distribution of the American Magpie in Minn." *Flicker* 29: 95–99.

Erickson, Arnold B., and Walter H. Petraborg. 1952. "Pinnated Grouse—Prince of the Prairie." *Minnesota Conservation Volunteer* 15 (May/June): 6–11.

Erickson, Arnold B., and Robert Upson. 1938. "1938 Minnesota Nesting Records." *Flicker* 10 (3/4): 4–13.

Erickson, Laura. 1988. "Margaret Morse Nice's Song Sparrows." For the Birds Radio Program. Original air date May 11. http://www.lauraerickson.com/radio/program/10548/margaret-morse-nices-song-sparrows/.

———. 2015. "Whatever Happened to Evening Grosbeaks?" *Minnesota Birding* 52 (6): 1–3. https://moumn.org/newsletter/issues/2015-11.pdf.

Erskine, Anthony J. 1977. *Birds in Boreal Canada: Communities, Densities, and Adaptations*. Canadian Wildlife Service Report Series, no. 41. Ottawa: Canadian Wildlife Service.

———. 1992. *Atlas of Breeding Birds in the Maritime Provinces*. Halifax, Nova Scotia: Nova Scotia Museum.

Etterson, Matthew A., Gerald J. Niemi, and Nicholas P. Danz. 2009. "Estimating the Effects of Detection Heterogeneity and Overdispersion on Trends Estimated from Avian Point Counts." *Ecological Applications* 19: 2049–66.

Evans, Charles. 1938. "A Bittern Uses Its Coloration." *Flicker* 10: 5.

Evans, David L. 1982. *Status Report on Twelve Raptors*. Special Scientific Report–Wildlife 238. Washington, DC: US Fish and Wildlife Service.

Evans, David L., Gerald J. Niemi, and Matthew A. Etterson. 2012. "Autumn Raptor Banding at Hawk Ridge, Duluth, Minnesota U.S.A., 1972–2009: An Overview." *Journal of Raptor Research* 46: 36–49.

Evans, Melissa, Elizabeth Gow, Roland R. Roth, Mark S. Johnson, and Todd J. Underwood. 2020. "Wood Thrush (*Hylocichla mustelina*)." Version 1.0. In Poole 2020. doi:10.2173/bow.woothr.01.

Ewins, Peter J., and Datlaf Vaughn Weseloh. 2020. "Little Gull (*Hydrocoloeus minutus*)." Version 1.0. In Billerman 2020. doi:10.2173/bow.litgul.01.

Faanes, Craig A. 1981. *Birds of the St. Croix River Valley: Minnesota and Wisconsin*. North

American Fauna, no. 73. Washington, DC: US Fish and Wildlife Service.
Faber, Raymond A. 2006. "Henslow's Sparrow Studies at Great River Bluffs State Park during the 2005–2006 Field Seasons through 30 June 2006." Annual Plan T-62911. Minnesota Department of Natural Resources.
Faber, Raymond A., and Joseph J. Hickey. 1973. "Eggshell Thinning, Chlorinated Hydrocarbons, and Mercury in Inland Aquatic Bird Eggs, 1969 and 1970." *Pesticides Monitoring Journal* 7: 27–36.
Falconer, Myles, Kristyn Richardson, Audrey Heagy, Douglas C. Tozer, Becky Stewart, Jon McCracken, and Ron Reid. 2016. *Recovery Strategy for the Bank Swallow (*Riparia riparia*) in Ontario*. Ontario Recovery Strategy Series. Peterborough, Ontario: Ontario Ministry of Natural Resources and Forestry.
Fall, Bruce A. 1977. "Early Summer Warbler Records and Boreal Chickadee Nest near Itasca Park." *Loon* 49: 198–201.
———. 1984. "Male and Female Hooded Warbler, Murphy-Hanrahan Park, Scott County, June 1984." *Loon* 56: 278.
———. 1987. "Acadian Flycatcher Breeding Range Extension in Minnesota." *Loon* 59: 117–21.
———. 1989. "Additional Hooded Warbler Breeding Records from Minnesota." *Loon* 61: 4–9.
———. 1994a. "First Confirmed Kentucky Warbler Breeding Record in Minnesota." *Loon* 66: 115–17.
———. 1994b. "Newly Discovered American White Pelican Colonies in Southern Minnesota, Minnesota Lake." *Loon* 66: 111–12.
———. 1997. "Scissor-tailed Flycatcher Paired with Western Kingbird." *Loon* 69: 183–85.
———. 2003. "Summer Tanager / Scarlet Tanager Mixed-Species Nesting." *Loon* 75: 221–25.
———. 2016. "Third Successful Breeding of Yellow-throated Warbler in Minnesota." *Loon* 88: 84–85.
Fall, Bruce A., and Roger D. Eliason. 1982. "Henslow's Sparrow Nest, Hennepin County." *Loon* 54: 192.
Fall, Bruce A., Howard C. Towle, and Paul E. Budde. 2016. "The 2015–2016 Winter Season: 1 December 2015 through 29 February 2016." *Loon* 88: 97–120.
Fallon, Jacquelyn M. 2020. "Minnesota 2020 Annual Peregrine Falcon Report." Duluth, MN: Midwest Peregrine Society. https://midwestperegrine.umn.edu/?vw=annual&opt2=MN_2018.
Falls, J. Bruce, and Jeffrey G. Kopachena. 2020. "White-throated Sparrow (*Zonotrichia albicollis*)." Version 1.0. In Poole 2020. doi:10.2173/bow.whtspa.01.
Farmer, Christopher J., Laurie J. Goodrich, Ernesto Ruelas Inzunza, and Jeff P. Smith. 2008. "Conservation Status of North America's Birds of Prey." In *State of North America's Birds of Prey,* edited by Keith L. Bildstein, Jeff P. Smith, Ernesto Ruelas Inzunza, and Richard R. Veit, 303–420. Series in Ornithology, no. 3. Cambridge, MA: Nuttall Ornithological Club and American Ornithologists' Union.
Farnsworth, George, Gustavo Adolfo Londono, Judit Ungvari Martin, Kim C. Derrickson, and Randall Breitwisch. 2020. "Northern Mockingbird (*Mimus polyglottos*)." Version 1.0. In Poole 2020. doi:10.2173/bow.normoc01.
Farrell, Claire E., Scott Wilson, and Greg Mitchell. 2017. "Assessing the Relative Use of Clearcuts, Burned Stands, and Wetlands as Breeding Habitat for Two Declining Aerial Insectivores in the Boreal Forest." *Forest Ecology and Management* 386: 62–70.
Fay, L. D. 1966. "Type E Botulism in Great Lakes Water Birds." *Transactions of the North American Wildlife and Natural Resources Conference* 31: 139–49.
Featherstonhaugh, George. 1847. "A Canoe Voyage up the Minnay Sotor." By George William Featherstonhaugh 1835. https://digicoll.library.wisc.edu/WIReader/Contents/Voyage.html.
Fellows, Suzanne D., and Stephanie L. Jones. 2009. *Status Assessment and Conservation Action Plan for the Long-billed Curlew (*Numenius americanus*)*. Biological Technical Publication FWS/BTP-R6012–2009. Washington, DC: US Fish and Wildlife Service, Biological Technical Publication.
Ficken, Millicent S., Margaret A. McLaren, and Jack P. Hailman. 2020. "Boreal Chickadee (*Poecile hudsonicus*)." Version 1.0. In Poole and Gill 2020. doi:10.2173/bow.borchi2.01.
Filter, Lawrence W. 1993. "Kentucky Warbler in Nicollet County." *Loon* 65: 45–46.
Finseth, Ole A. 1957. "A New Common Tern Colony." *Loon* 29: 39.
———. 1958. "Ornithological History of Minnesota." *Loon* 30: 9–13.
Fisher, Albert K. 1893. *The Hawks and Owls of the United States in Their Relation to Agriculture*. Washington, DC: US Department of Agriculture, Division of Ornithology and Mammalogy.
Fitzpatrick, Matthew C., and Robert R. Dunn. 2019. "Contemporary Climatic Analogs for 540 North American Urban Areas in the Late 21st Century." *Nature Communications* 10: 614. doi:10.1038/s41467-019-08540-3.
Flaspohler, David J., Casey J. Fisher Hukins, Brian R. Bub, and Peter J. Van Dusen. 2002. "Temporal Patterns in Aquatic and Avian Communities Following Selective Logging in the Upper Great Lakes Region." *Forest Science* 48: 339–49.
Flaspohler, David J., Stanley A. Temple, and Robert N. Rosenfield. 2001. "Species-Specific Edge Effects on Nest Success and Breeding Bird Density in a Forested Landscape." *Ecological Applications* 11: 32–46.
Foote, Jennifer R., Daniel J. Mennill, Laurene M. Ratcliffe, and Susan M. Smith. 2020. "Black-capped Chickadee (*Poecile atricapillus*)." Version 1.0. In Poole 2020. doi:10.2173/bow.bkcchi.01.
Forbes, James E., and Dwain W. Warner. 1974. "Behavior of a Radio-tagged Saw-whet Owl." *Auk* 91: 783–95.
Forbush, Edward Howe. 1912. *A History of the Game Birds, Wild-Fowl, and Shore Birds of Massachusetts and Adjacent States*. Massachusetts State Board of Agriculture. Boston: Wright & Potter.
———. 1929. *Birds of Massachusetts and Other New England States*. Vol. 3, *Land Birds from Sparrows to Thrushes*. Norwood, MA: Berwick and Smith.
Forbush, Edward Howe, and John Bichard May. 1955. *A Natural History of American Birds of Eastern and Central North America*. New York: Bramhall House.
Forsyth, Douglas J., Pamela A. Martin, Ken D. De Smet, and Morley E. Riske. 1994. "Organochlorine Contaminants and Eggshell Thinning in Grebes from Prairie Canada." *Environmental Pollution* 85: 51–58.
Foster, David R., Brian Donahue, David B. Kittredge, Glenn Motzkin, Brian Hall, Billie L. Turner, and Elizabeth Chilton. 2008. "New England's Forest Landscape: Ecological Legacies and Conservation Patterns Shaped by Agrarian History." In *Agrarian Landscapes in Transition: Comparisons of Long-Term Ecological and Cultural Change,* edited by Charles L. Redman and David R. Foster, 44–88. New York: Oxford University Press.
Foster, William L., and James L. Tate Jr. 1966. "The Activities and Coactions of Animals at Sapsucker Trees." *Living Bird* 5: 87–113.
Fowells, Harry Ardell. 1965. *Silvics of Forest Trees of the United States*. US Department of Agri-

culture Handbook, no. 271. Washington, DC: United States Department of Agriculture.
Fowler, Joan. 1980. "The Birds of Olmsted County." *Loon* 52: 23–35.
Fox, Glen A. 1974. "Changes in Eggshell Quality of Belted Kingfishers Nesting in Ontario." *Canadian Field-Naturalist* 88: 358–59.
Franson, J. Christian, Scott P. Hansen, and John H. Schulz. 2009. "Ingested Shot and Tissue Lead Concentrations in Mourning Doves." In *Ingestion of Lead from Spent Ammunition: Implications for Wildlife and Humans,* edited by Richard T. Watson, Mark Fuller, Mark Pokras, and W. Grainger Hunt, 175–86. Boise, ID: The Peregrine Fund. https://www.peregrinefund.org/subsites/conference-lead/2008PbConf_Proceedings.htm. doi:10.4080/ilsa.2009.0202.
Franzen, Jonathan. 2010. *Freedom: A Novel.* New York: Farrar, Straus and Giroux.
Fraser, Gail. 1997. "Feeding and Nesting Behaviors of the Forster's Tern on Lake Osakis, MN." Master's thesis, North Dakota State University.
Frederickson, Leigh H., and Frederic A. Reid. 1986. "Wetland and Riparian Habitats: A Nongame Management Overview." In *Management of Nongame Wildlife in the Midwest: A Developing Art,* edited by James B. Hale, Louis B. Best, and Richard L. Clawson, 59–96. Proceedings of a symposium held at the 47th Midwest Fish and Wildlife Conference, Grand Rapids, Michigan, December 17, 1985. Chelsea, MI: BookCrafters.
Frederickson, Randy. 1998. "Minnesota's First Yellow-throated Warbler Nest." *Loon* 70: 230–32.
———. 2019. "A Successful Barn Owl Nest in Becker County." *Loon* 91: 177–78.
Freeland, David B. 1971. "House Finch Record for Minnesota." *Loon* 43: 127.
Frei, Barbara, Kimberly G. Smith, James H. Withgott, Paul G. Rodewald, Peter Pyle, and Michael A. Patten. 2020. "Red-headed Woodpecker (*Melanerpes erythrocephalus*)." Version 1. In Rodewald 2020. doi:10.2173/bow.rehwoo.01.
Frelich, Lee E., Cindy M. Hale, Stefan Scheu, Andrew R. Holdsworth, Liam Heneghan, Patrick J. Bohlen, and Peter B. Reich. 2006. "Earthworm Invasion into Previously Earthworm-Free Temperate and Boreal Forests." *Biological Invasions* 8: 1235–45.
Freshwater Society. 2007. "Critical Water Issues: Shoreland Standards and Land Use." http://freshwater.org/wp-content/uploads/joomla/PDFs/critical-water/shorelandstandardsandlanduse.pdf.
Fronczak, David L. 2014. "Distribution, Migration Chronology, and Survival Rates of Eastern Population Sandhill Cranes." Master's thesis, University of Minnesota.
Fronczak, David L., David E. Andersen, Everett E. Hanna, and Thomas R. Cooper. 2015. "Annual Survival Rate Estimate of Satellite Transmitter-marked Eastern Population Greater Sandhill Cranes." *Journal of Fish and Wildlife Management* 6: 464–71.
Fujisaki, Ikuko, Elise V. Pearlstine, and Frank J. Mazzotti. 2010. "The Rapid Spread of Invasive Eurasian Collared Doves *Streptopelia decaocto* in the Continental USA Follows Human-altered Habitats." *Ibis* 152: 622–32.
Fuller, Mark R. 1979. "Spatiotemporal Ecology of Four Sympatric Raptor Species." PhD diss., University of Minnesota.
Fuller, Pepper. 1977. "White-eyed Vireo in Wayzata." *Loon* 49: 174.
Furtman, Michael. 1999. *Ruffed Grouse: Woodland Drummer.* Mechanicsburg, PA: Stackpole Books.
Futcher, John, and William Pieper. 1952. "Bird Notes on Sherburne County." *Flicker* 24: 161.
Gage, Anne M., Sarah K. Olimb, and Jeff Nelson. 2016. "Plowprint: Tracking Cumulative Cropland Expansion to Target Grassland Conservation." *Great Plains Research* 26: 107–16.
Galati, Robert. 1991. *Golden-crowned Kinglets: Treetop Nesters of the North Woods.* Ames: Iowa State Press.
Galli, Anne E., Charles F. Leck, and Richard T. T. Forman. 1976. "Avian Distribution Patterns in Forest Islands of Different Sizes in Central New Jersey." *Auk* 93: 356–64.
Gamble, Lawrence R. 1985. "Nesting Ecology of Western Kingbirds (*Tyrannus verticalis Say*) in the Trans-Pecos Area of Texas Relative to Pesticidal Contamination." Master's thesis, Texas A&I University, Kingsville, Texas.
Gamble, Lawrence R., and Timothy M. Bergin. 2020. "Western Kingbird (*Tyrannus verticalis*)." Version 1.0. In Poole 2020. doi:10.2173/bow.weskin.01.
Gange, Megan. 1978. "Northern-most Nest of a Yellow-crowned Night Heron." *Loon* 50: 214–15.
Gardali, Thomas, and Grant Ballard. 2020. "Warbling Vireo (*Vireo gilvus*)." Version 1.0. In Poole and Gill 2020. doi:10.2173/bow.warvir.01.
Garrison, Barrett A., and Angela Turner. 2020. "Bank Swallow (*Riparia riparia*)." Version 1.0. In Billerman 2020. doi:10.2173/bow.banswa.01.
Gates, Charles M., ed. 1933. *Five Fur Traders of the Northwest: Being the Narrative of Peter Pond and the Diaries of John Macdonell, Archibald N. McLeod, Hugh Faries, and Thomas Connor.* Minneapolis: University of Minnesota Press.
Gauthier, Gilles. 2020. "Bufflehead (*Bucephala albeola*)." Version 1.0. In Poole 2020. doi:10.2173/bow.buffle.01.
Gauthier, Gilles, and James N. M. Smith. 1987. "Territorial Behaviour, Nest-Site Availability, and Breeding Density in Buffleheads." *Journal of Animal Ecology* 56: 171–84.
Gehlbach, Frederick R. 1994. *The Eastern Screech Owl: Life History, Ecology, and Behavior in the Suburbs and Countryside.* College Station: Texas A&M University Press.
Gerber, Brian D., James F. Dwyer, Stephen A. Nesbitt, Rod C. Drewien, Carol D. Littlefield, Thomas C. Tacha, and Paul A. Vohs. 2020. "Sandhill Crane (*Antigone canadensis*)." Version 1.0. In Poole 2020. doi:10.2173/bow.sancra.01.
Ghalambor, Cameron K., and Thomas E. Martin. 2020. "Red-breasted Nuthatch (*Sitta canadensis*)." Version 1.0. In Poole and Gill 2020. doi:10.2173/bow.rebnut.01.
Gill, Frank B. 1980. "Historical Aspects of Hybridization between Blue-winged and Golden-winged Warblers." *Auk* 97: 1–18.
Gill, Frank B., Ronald A. Canterbury, and John L. Confer. 2020. "Blue-winged Warbler (*Vermivora cyanoptera*)." Version 1.0. In Poole and Gill 2020. doi:10.2173/bow.buwwar.01.
Gillespie, Walter L., and S. Charles Kendeigh. 1982. "Breeding Bird Populations in Northern Manitoba." *Canadian Field-Naturalist* 96: 272–81.
Gillette, Laurence N. 1974. "Status of the Trumpeter Swan in Minnesota." *Loon* 46: 169.
———. 1976. "More on Ruffed Grouse in Hennepin County." *Loon* 48: 122.
———. 1986. "Trumpeter Swans Nest in the Chippewa National Forest." *Loon* 58: 194.
Gillette, Laurence N., and Judy Voigt Englund. 1985. "The Hennepin County Park Reserve District's Osprey Reintroduction Project." *Loon* 57: 52–58.
Gillihan, Scott W., and Bruce E. Byers. 2020. "Evening Grosbeak (*Coccothraustes vespertinus*)." Version 1.0. In Poole and Gill 2020. doi:10.2173/bow.evegro.01.
Giudice, John H., and John T. Ratti. 2020. "Ring-necked Pheasant (*Phasianus colchicus*)." Version 1.0. In Billerman 2020. doi:10.2173/bow.rinphe.01.

Glassel, Raymond A. 1993. "A Minnesota Great-tailed Grackle with Notes on the Species' Range Expansion in the Midwest." *Loon* 65: 148–50.

Godfrey, W. Earl. 1986. *The Birds of Canada*. Rev. ed. Ottawa: National Museums of Canada.

Goggans, Rebecca, Rita D. Dixon, and L. Claire Seminara. 1988. *Habitat Use by Three-toed and Black-backed Woodpeckers*. Technical Report 87-3-02. Oregon Department of Fish and Wildlife, Nongame Wildlife Program, USDA Deschutes National Forest. https://digital.osl.state.or.us/islandora/object/osl:18053.

Goldade, Christopher M., Jill A. Dechant, Douglas H. Johnson, Amy L. Zimmerman, Brent E. Jamison, James O. Church, and Betty R. Euliss. 2002. *Effects of Management Practices on Wetland Birds: Yellow Rail*. Jamestown, ND: Northern Prairie Wildlife Research Center. doi:10.3133/70159826.

Goldberg, M. G. 1973. "Hooded Warbler Netted." *Loon* 45: 65.

Golden-winged Warbler Working Group. 2013. *Best Management Practices for Golden-winged Warbler Habitats in the Great Lakes Region: A Guide for Land Managers and Landowners*. Cornell Lab of Ornithology. http://www.gwwa.org/resources/GWWA-GLRegionalGuide_130808_lo-res.pdf.

Goldman, Jason G. 2016. "Why Some Birds Have Red Feathers." *Audubon,* May 23. https://www.audubon.org/news/why-some-birds-have-red-feathers.

Goodermote, Donald L. 1980. "Herring Gull Nest Counts on the North Shore of Lake Superior." *Loon* 52: 15–17.

———. 1984. *A Survey of Nesting Herring Gulls along the North Shore of Lake Superior from Knife River to the Pigeon River, during the Period 1977–1984*. Minnesota Department of Natural Resources, Nongame Wildlife Program. https://www.dnr.state.mn.us/eco/nongame/projects/research_reports/abstracts/birds/goodermote1984.html.

———. 1989. *A Survey of Nesting Herring Gulls on Selected Sites, along the North Shore of Lake Superior from Knife River to the Pigeon River—1989*. Minnesota Department of Natural Resources, Nongame Wildlife Program. https://www.dnr.state.mn.us/eco/nongame/projects/research_reports/abstracts/birds/goodermote1989.html.

Goodnow, Marissa L., and Leonard R. Reitsma. 2011. "Nest-site Selection in the Canada Warbler (*Wilsonia canadensis*) in Central New Hampshire." *Canadian Journal of Zoology* 89: 1172–77.

Goodrich, Laurie J., Scott T. Crocoll, and Stanley E. Senner. 2020. "Broad-winged Hawk (*Buteo platypterus*)." Version 1.0. In Poole 2020. doi:10.2173/bow.brwhaw.01.

Goodwin, Arthur B. 1957. "A Study of Ring-necked Duck Nesting in the Pothole Region of Mahnomen County, Minn." *Flicker* 29: 22–29.

Gorenzel, W. Paul, and Terrell P. Salmon. 1992. "Urban Crow Roosts in California." In *Proceedings of the Fifteenth Vertebrate Pest Conference 1992,* edited by John E. Borrecco and Rex E. Marsh, 97–102. Davis: University of California, Davis.

Gowaty, Patricia A., and Jonathan H. Plissner. 2020. "Eastern Bluebird (*Sialia sialis*)." Version 1.0. In Poole 2020. doi:10.2173/bow.easblu.01.

Graber, Richard R., and Jean W. Graber. 1963. "A Comparative Study of Bird Populations in Illinois, 1906–1909 and 1956–1958." *Illinois Natural History Survey Bulletin* 28: 385–528. doi:10.21900/j.inhs.v28.169.

Graber, Richard R., Jean W. Graber, and Ethelyn L. Kirk. 1974. *Illinois Birds: Tyrannidae*. Biological Notes, no. 86. Urbana: Illinois Natural History Survey.

Graham, Mrs. Sterling. 1956. "Swallow-tailed Kite Again Recorded in Minnesota." *Flicker* 28: 41.

Granfors, Diane. 2009. *Marbled Godwit Expert Model*. US Fish and Wildlife Service, Habitat and Population Evaluation Team. http://www.fws.gov/midwest/hapet/MarbledGodwitModel.html.

Granlund, James, Gail A. McPeek, and Raymond John Adams, eds. 1994. *The Birds of Michigan*. Bloomington and Indianapolis: Indiana University Press.

Grant, Peter J. 1982. *Gulls: A Guide to Identification*. Vermillion, SD: Buteo Books.

Grant, R. A. 1965a. "Breeding Lark Buntings and American Avocets in Lac qui Parle County." *Loon* 37: 82–83.

———. 1965b. "The Burrowing Owl in Minnesota." *Loon* 37: 2–17.

Grant, Todd A., and Richard W. Knapton. 2020. "Clay-colored Sparrow (*Spizella pallida*)." Version 1.0. In Poole 2020. doi:10.2173/bow.clcspa.01.

Grasset [Grassett], Frank G. 1926. "An Unusual Flight of Cormorants." *Wilson Bulletin* 38: 234–35.

Gratto-Trevor, Cheri L. 2020. "Marbled Godwit (*Limosa fedoa*)." Version 1.0. In Poole and Gill 2020. doi:10.2173/bow.margod.01.

Great Basin Bird Observatory. 2010. "White-faced Ibis." In *Nevada Comprehensive Bird Conservation Plan,* Spp-20. Version 1.0. Reno, NV: Great Basin Bird Observatory. http://www.gbboinfo.org/secure/Species%20Accounts/White-faced_Ibis.pdf.

Green, Janet C. 1964. "Male Summer Tanager in Duluth." *Loon* 36: 63.

———. 1965. "First Summer Tanager Banded in Minnesota." *Loon* 37: 113.

———. 1966. "First Common Gallinule Record for Northeastern Minnesota." *Loon* 38: 72.

———. 1967. Photo of Bell Museum of Natural History specimen #23152, Swallow-tailed Kite found dead April 29, 1966, beside State Highway 95. *Loon* 67: 67.

———. 1974. "Arctic Tern and Little Gull as Migrants in Duluth." *Loon* 46: 52–57.

———. 1975. "Mute Swans in Western Lake Superior." *Loon* 47: 42–43.

———. 1979. "The Summer Season (June 1–July 31, 1978)." *Loon* 51: 28–44.

———. 1980. "Summer Season (June 1–July 31, 1979)." *Loon* 52: 122–39.

———. 1981. "Summer Season (June 1–July 31, 1980)." *Loon* 53: 132–47.

———. 1991. "A Landscape Classification for Breeding Birds in Minnesota: An Approach to Describing Regional Biodiversity." *Loon* 63: 80–91.

———. 1995. *Birds and Forests: A Management and Conservation Guide*. St. Paul: Minnesota Department of Natural Resources.

———. 2008. "Taxonomic Chronology of Wood-Warblers." *Minnesota Birding* 45 (1): 8–9.

Green, Janet C., and James S. Baumhofer. 1972. "The 1972 Summer Season." *Loon* 44: 105–11.

Green, Janet C., and Robert B. Janssen. 1975. *Minnesota Birds: Where, When, and How Many*. Minneapolis: University of Minnesota Press.

———. 1984. "The Summer Season (June 1–July 31, 1983)." *Loon* 56: 51–63.

Green, Janet C., and Gerald J. Niemi. 1978. *Birds of the Superior National Forest*. Washington, DC: US Department of Agriculture, Forest Service.

Green, Michael T., Peter E. Lowther, Stephanie L. Jones, Stephen K. Davis, and Brenda C. Dale. 2020. "Baird's Sparrow (*Centronyx bairdii*)." Version 1.0. In Poole and Gill 2020. doi:10.2173/bow.baispa.01.

Greenberg, Joel. 2014. *A Feathered River across the Sky: The Passenger Pigeon's Flight to Extinction*. New York: Bloomsbury.

Greenberg, Russell, Peter Bichier, Andrea Cruz Angon, and Robert Reitsma. 1997. "Bird Popu-

lations in Shade and Sun Coffee Plantations in Central Guatemala." *Conservation Biology* 11: 448–59.
Greenberg, Russell, and Sam Droege. 1999. "On the Decline of the Rusty Blackbird and the Use of Ornithological Literature to Document Long-Term Population Trends." *Conservation Biology* 13: 553–59.
Greenlaw, Jon S. 2020. "Eastern Towhee (*Pipilo erythrophthalmus*)." Version 1.0. In Rodewald 2020. doi:10.2173/bow.eastow.01.
Gregg, Larry. 1984. *Population Ecology of Woodcock in Wisconsin*. Technical Bulletin No. 144. Madison, WI: Department of Natural Resources.
Gregg, Larry, Brian Heeringa, and Daniel Eklund. 2004. *Conservation Assessment for Spruce Grouse (*Falcipennis canadensis*)*. Park Falls, WI: US Department of Agriculture, Forest Service, Eastern Region.
Greij, Eldon D. 1994. "Common Moorhen." In Tacha and Braun 1994, 145–57.
Grewe, Al. 1968. "Sandhill Cranes in Morrison County." *Loon* 30: 159.
Grinde, Alexis R. 2015. "Spatio-temporal Ecology of Forest Birds." PhD diss., University of Minnesota.
Grinde, Alexis, Josh Bednar, Steve Kolbe, and Reid Seibers. 2020. "Minnesota National Forest Breeding Bird Monitoring Program Annual Report 1995–2020." Natural Resources Research Institute Technical Report NRRI/TRC-2020/34.
Grinde, Alexis R., and Gerald J. Niemi. 2016. "Influence of Landscape, Habitat, and Species Co-occurrence on Occupancy Dynamics of Canada Warblers." *Condor* 118: 513–31.
Grinde, Alexis R., Gerald J. Niemi, Cheryl Adams, and T. Bently Wigley. 2017. "The Important Role of Young Forests in Forest Biodiversity Management." *Forestry Source* 22 (2): 12–13.
Grinde, Alexis R., Gerald J. Niemi, Brian R. Sturtevant, Hannah Panci, Wayne Thogmartin, and Peter Wolter. 2017. "Importance of Scale, Land Cover, and Weather on the Abundance of Bird Species in a Managed Forest." *Forest Ecology and Management* 405: 295–308.
Grinde, Alexis R., Reid Siebers, and Josh Bednar. 2021. *Minnesota National Forest Breeding Bird Monitoring Program Annual Report, 1995–2021*. Natural Resources Research Institute, University of Minnesota, Duluth, Technical Report NRRI/TR-2021/35.
Grinde, Alexis R., Nicholas G. Walton, Joshua D. Bednar, Alexis L. Liljenquist, and Steve R. Kolbe. 2018. "Minnesota National Forest Breeding Bird Monitoring Program Annual Report 1995–2018." Natural Resources Research Institute Technical Report NRRI/TR-2018/47.
Griscom, Ludlow, and Alexander Sprunt Jr. 1957. *The Warblers of America*. New York: The Devin-Adair Company.
Gross, Alfred O. 1923. "The Black-crowned Night Heron (*Nycticorax nycticorax naevius*) of Sandy Neck." *Auk* 40: 1–30, 191–214.
Gross, Douglas A., and Peter E. Lowther. 2020. "Yellow-bellied Flycatcher (*Empidonax flaviventris*)." Version 1.0. In Poole 2020. doi:10.2173/bow.yebfly.01.
Grosshuesch, David A. 2008. "Presence of Breeding Northern Hawk Owls (*Surnia ulula*) in Minnesota, 2006." *Loon* 80: 180–87.
———. 2009. "Breeding Biology and Nest-Site Characteristics of Northern Hawk Owls in the Southern Boreal Forest Region of North America." Master's thesis, University of Minnesota.
Groves, Deborah J. 2017. *The 2015 North American Trumpeter Swan Survey: A Cooperative North American Survey*. Juneau, AK: US Fish and Wildlife Survey. https://www.trumpeterswansociety.org/file_download/16524876-ce39-48cd-ae86-cb0b82756773.
Grubb, Thomas C., Jr., and Paul F. Doherty, Jr. 1999. "On Home-range Gap-crossing." *Auk* 116: 618–28.
Grubb, Thomas C., Jr., and Vladimir V. Pravosudov. 2020. "White-breasted Nuthatch (*Sitta carolinensis*)." Version 1.0. In Poole 2020. doi:10.2173/bow.whbnut.01.
Gruson, Edward S. 1972. *Words for Birds: A Lexicon of North American Birds with Biographical Notes*. New York: Quadrangle Books.
Grzybowski, Joseph A. 2020. "Black-capped Vireo (*Vireo atricappilla*)." Version 1.0. In Poole and Gill 2020. doi:10.2173/bow.bkcvir1.01.
Guertin, David S., and Lee A. Pfannmuller. 1985. "Colonial Waterbirds in Minnesota: An Update of Their Distribution and Abundance." *Loon* 57: 67–78.
Gullion, Gordon W. 1954. "The Reproductive Cycle of American Coots in California." *Auk* 71: 366–412.
———. 1967. "Selection and Use of Drumming Sites by Male Ruffed Grouse." *Auk* 84: 87–112.
———. 1970. "Factors Influencing Ruffed Grouse Populations." *Transactions of the North American Wildlife Conference* 35: 93–105.
———. 1981. "The Impact of Goshawk Predation upon Ruffed Grouse." *Loon* 53: 82–84.
———. 1984. *Grouse of the North Shore*. Oshkosh, WI: Willow Creek Press.
Gullion, Gordon W., and Alvin A. Alm. 1983. "Forest Management and Ruffed Grouse Populations in a Minnesota Coniferous Forest." *Journal of Forestry* 81: 529–36.
Gullion, Gordon W., and William H. Marshall. 1968. "Survival of Ruffed Grouse in a Boreal Forest." *Living Bird* 7: 117–67.
Gunderson, Dan, Elizabeth Dunbar, and Jiwon Choi. 2019. "A Look at Minnesota Farming in 7 Charts." *MPR News*, April 11. https://www.mprnews.org/story/2019/04/11/ag-census-2017-minnsota-snapshot.
Gunderson, Harvey L. 1946. "Chestnut-collared Longspur and Sprague's Pipit." *Flicker* 18: 65.
Gustafson, Eric J., Melinda G. Knutson, Gerald J. Niemi, and Mary A. Friberg. 2002. "Evaluation of Spatial Models to Predict Vulnerability of Forest Birds to Brood Parasitism by Cowbirds." *Ecological Applications* 12: 412–26.
Guttman, Burton S. 1956. "Minnesota Nesting Records, 1955." *Flicker* 28: 133–37.
Gutzwiller, Kevin J., and Stanley H. Anderson. 1987. *Habitat Suitability Index Models: Marsh Wren*. US Fish and Wildlife Service Biological Report 82(10.139). Washington, DC: US Fish and Wildlife Service, National Ecology Center.
Gutzwiller, Kevin J., Kevin R. Kinsley, Gerald L. Storm, Walter M. Tzilkowski, and James S. Wakeley. 1983. "Relative Value of Vegetation Structure and Species Composition for Identifying American Woodcock Breeding Habitat." *Journal of Wildlife Management* 47: 535–40.
Guzy, Michael J., and Gary Ritchison. 2020. "Common Yellowthroat (*Geothlypis trichas*)." Version 1.0. In Rodewald 2020. doi:10.2173/bow.comyel.01.
Haas, George Hartmann. 1974. "Habitat Selection, Reproduction, and Movements in Female Spruce Grouse." PhD diss., University of Minnesota.
Hagan, John M., III. 1993. "Decline of the Rufous-sided Towhee in the Eastern United States." *Auk* 110: 863–74.
Haggerty, Thomas M., and Eugene S. Morton. 2020. "Carolina Wren (*Thryothorus ludovicianus*)." Version 1.0. In Poole 2020. doi:10.2173/bow.carwre.01.
Hahn, Thomas P. 1995. "Integration of Photoperiodic and Food Cues to Time Changes in Reproductive Physiology by an Opportunistic Breeder, the Red Crossbill, *Loxia curvirostra* (Aves: Carduelinae)." *Journal of Experimental Zoology* 272: 213–26.

Haig, Susan M., D. Brown, W. Harrison, Ross Lock, Lee A. Pfannmuller, Ed Pike, and Mark Ryan. 1988. *Recovery Plan for the Great Lakes/Northern Great Plains Populations of the Piping Plover (*Charadrius melodus*)*. US Fish and Wildlife Service Report.

Haines, Aaron M., Mike J. McGrady, Mark S. Martell, B. James Dayton, M. Blake Henke, and William S. Seegar. 2003. "Migration Routes and Wintering Locations of Broad-winged Hawks Tracked by Satellite Telemetry." *Wilson Bulletin* 115: 166–69.

Halkin, Sylvia L., Daniel P. Shustack, M. Susan DeVries, Jodie M. Jawor, and Susan U. Linville. 2021. "Northern Cardinal (*Cardinalis cardinalis*)." Version 2.0. In Rodewald and Keeney 2021. doi:10.2173/bow.norcar.02.

Hallworth, Michael T., Erin Bayne, Emily McKinnon, Oliver Love, Junior A. Tremblay, Bruno Drolet, Jacques Ibarzabal, et al. 2021. "Habitat Loss on the Breeding Grounds Is a Major Contributor to Population Declines in a Long-distance Migratory Songbird." Proceedings of the Royal Society B 288 (1949): 20203164. doi:10.1098/rspb.2020.3164.

Hamady, Maya, Julie Adams, Tom Engel, and Susan Backe. 2003. "Spatially Explicit Goshawk Surveys on the Kabetogama and Washington/Sturgeon River State Forest." Minnesota Department of Natural Resources, Nongame Wildlife Program.

Hamel, Paul B., and Kenneth V. Rosenberg. 2007. "Developing Management Guidelines for Cerulean Warbler Breeding Habitat." In *Proceedings of the 15th Central Hardwood Forest Conference,* edited by David S. Buckley and Wayne K. Clatterbuck, 364–74. General Technical Report SRS-101. US Department of Agriculture, Forest Service, Southern Research Station.

Hamilton, Derek, and Francesca J. Cuthbert. 2016. "Assessing Distribution, Abundance, and Population Change in the American White Pelican and Double-crested Cormorant in Minnesota: Comparison to Three Census Periods, 2004/05, 2010, and 2015." Final report to the Minnesota Department of Natural Resources.

Hammell, Gord S. 1973. "The Ecology of the Lesser Scaup (*Aythya affinis* Eyton) in Southwestern Manitoba." Master's thesis, University of Guelph, Ontario.

Handler, Stephen, Matthew J. Duveneck, Louis Iverson, Emily Peters, Robert M. Scheller, Kirk R. Wythers, Leslie Brandt, et al. 2014. "Minnesota Forest Ecosystem Vulnerability Assessment and Synthesis: A Report from the Northwoods Climate Change Response Framework Project." General Technical Report NRS-133. Newtown Square, PA: US Department of Agriculture, Forest Service, Northern Research Station.

Hanlon, Robert W. 1956. "Heron Island—General Shields Lake." *Flicker* 28: 130–32.

Hanna, Darrell M. 1959. "Possible Sight Record of the Cattle Egret in Minnesota." *Flicker* 31: 103–4.

Hannah, Kevin C., and Jeff S. Hoyt. 2004. "Northern Hawk Owls and Recent Burns: Does Burn Age Matter?" *Condor* 106: 420–23.

Hanowski, JoAnn M. 1995. "Breeding Bird Composition and Species Relative Abundance Patterns on Conservation Reserve Program (CRP) Land in Western Minnesota." *Loon* 67: 12–16.

———. 2002. "Habitats and Landscapes Used by Breeding Golden-winged Warblers in Western Great Lakes Forests." *Loon* 74: 127–33.

Hanowski, JoAnn M., Donald P. Christian, and Gerald J. Niemi. 2000. "Landscape Requirements of Prairie Sharp-tailed Grouse *Tympanuchus phasianellus campestris* in Minnesota, USA." *Wildlife Biology* 6: 257–63.

Hanowski, JoAnn M., Nicholas P. Danz, James W. Lind, and Gerald J. Niemi. 2003. "Breeding Bird Response to Riparian Forest Harvest and Harvest Equipment." *Forest Ecology and Management* 174: 315–28.

———. 2005. "Breeding Bird Response to Varying Amounts of Basal Area Retention in Riparian Buffers." *Journal of Wildlife Management* 69: 689–98.

Hanowski, JoAnn M., James W. Lind, Nicholas P. Danz, and Tim Jones. 2005. "Regional Breeding Bird Monitoring in Western Great Lakes National Forests." In *Bird Conservation Implementation and Integration in the Americas: Proceedings of the Third International Partners in Flight Conference,* edited by C. John Ralph and Terrell D. Rich, 974–81. US Department of Agriculture, Forest Service, General Technical Report PSW-GTR-191. Albany, CA: US Department of Agriculture, Forest Service, Pacific Southwest Research Station.

Hanowski, JoAnn M., and Gerald J. Niemi. 1983. "Henslow's Sparrow in Red Lake Peatland." *Loon* 55: 120–21.

———. 1985. *Habitat Characteristics for Bird Species of Special Concern and Their Importance in Habitat Reclamation*. Final report to Minnesota Department of Natural Resources.

———. 1990. "An Approach for Quantifying Habitat Characteristics for Rare Wetland Birds." In *Ecosystem Management: Rare Species and Significant Habitats,* edited by Richard S. Mitchell, Charles J. Sheviak, and Donald Joseph Leopold, 51–56. Albany: University of the State of New York, New York State Museum.

———. 1995a. "A Comparison of On- and Off-Road Bird Counts: Do You Need to Go Off Road to Count Birds Accurately?" *Journal of Field Ornithology* 66: 469–83.

———. 1995b. "Experimental Design Considerations for Establishing an Off-road, Habitat Specific Bird Monitoring Program Using Point Counts." In Ralph et al. 1995, 145–50.

Hanowski, JoAnn M., Gerald J. Niemi, Ann R. Lima, and Ronald R. Regal. 1997. "Response of Breeding Birds to Mosquito Control Treatments of Wetlands." *Wetlands* 17 (4): 485–92.

Hanski, Ilpo K., Timothy J. Fenske, and Gerald J. Niemi. 1996. "Lack of Edge Effect in Nesting Success of Breeding Birds in Managed Forest Landscapes." *Auk* 113: 578–85.

Hanson, H. G. 1946. "Crow Center of the United States." *Oklahoma Game and Fish News* 2: 4–7.

Hanson, Harold C. 1953. "Muskeg as Sharp-tailed Grouse Habitat." *Wilson Bulletin* 65: 235–41.

———. 1965. *The Giant Canada Goose*. Carbondale: Southern Illinois University Press.

Hanson, Lynelle G. 1994. "The Henslow's Sparrow (*Ammodramus henslowii*) of Minnesota: Population Status and Breeding Habitat Analysis." Master's thesis, Central Michigan University.

Haramis, G. Michael, Dennis G. Jorde, Stephen A. Macko, and Jane L. Walker. 2001. "Stable-Isotope Analysis of Canvasback Winter Diet in Upper Chesapeake Bay." *Auk* 118: 1008–17.

Hargis, Christina D., and Brian Woodbridge. 2006. "A Design for Monitoring Northern Goshawks at the Bioregional Scale." *Studies in Avian Biology* 31: 274–87.

Haroldson, Kurt J., Richard O. Kimmel, Michael R. Riggs, and Alfred H. Berner. 2006. "Association of Ring-Necked Pheasant, Gray Partridge, and Meadowlark Abundance to Conservation Reserve Program Grasslands." *Journal of Wildlife Management* 70: 1276–84.

Harrell, Byron E. 1942. "Minnesota Nesting Records, 1942." *Flicker* 14: 33–40.

Harris, James T., and Sumner W. Matteson. 1975. "Gull and Terns Nesting at Duluth." *Loon* 47: 70–77.

Harrison, Whit. 1890. "The Prothonotary Warbler." *Oologist* 7: 228–29.

Hart, John Fraser, and Susy Svatek Ziegler. 2008. *Landscapes of Minnesota: A Geography*. St. Paul: Minnesota Historical Society Press.
Haseltine, Susan D., Gary H. Heinz, William L. Reichel, and John F. Moore. 1981. "Organochlorine and Metal Residues in Eggs of Waterfowl Nesting on Islands in Lake Michigan off Door County, Wisconsin, 1977–1978." *Pesticides Monitoring Journal* 15: 90–97.
Hatch, Philo L. 1874. "Report on the Birds of Minnesota." *Bulletin of the Minnesota Academy of Natural Sciences* 1: 43–68.
———. 1892. *First Report: Accompanied with Notes on the Birds of Minnesota*. The Geological and Natural History Survey of Minnesota Zoological Series. Minneapolis: Harrison & Smith Printers.
Hawk Mountain. 2019. "Record Highs, Lows, Averages (by Species), 1934–2011." *Raptorpedia*. Retrieved from *Raptorpedia*: http://www.hawkmountain.org/raptorpedia/record-highs-lows/page.aspx?id=349.
Hawk Ridge Bird Observatory. 2019. Hawk Ridge Bird Observatory [website]. https://www.hawkridge.org.
Hawks, Shawn E., David L. Evans, and James W. Grier. 2007. "Determining the Sources of Northern Goshawks Passing through Duluth, Minnesota." *Loon* 79: 187–93.
Haws, Katherine V. 2004. "Piping Plover and Common Tern Investigations: Lake of the Woods, 2003–2004." Minnesota Department of Natural Resources.
———. 2008. "Piping Plover and Common Tern Investigations, Lake of the Woods, 2004–2007." Final report to the Minnesota Department of Natural Resources, Nongame Wildlife Program.
Hayward, Gregory D., and Patricia H. Hayward. 2020. "Boreal Owl (*Aegolius funereus*)." Version 1.0. In Billerman 2020. doi:10.2173/bow.borowl.01.
Heath, Shane R., Erica H. Dunn, and David J. Agro. 2020. "Black Tern (*Chlidonias niger*)." Version 1.0. In Billerman 2020. doi:10.2173/bow.blkter.01.
Heckert, Robert A., Michael S. Collins, Ruth J. Manvell, Ian Strong, James E. Pearson, and Dennis J. Alexander. 1996. "Comparison of Newcastle Disease Virus Isolated from Cormorants in Canada and the USA in 1975, 1990 and 1992." *Canadian Journal of Veterinary Research* 60: 50–54.
Heckscher, Christopher M., Louis R. Bevier, Alan F. Poole, William Moskoff, Peter Pyle, and Michael A. Patten. 2020. "Veery (*Catharus fuscescens*)." Version 1.0. In Rodewald 2020. doi:10.2173/bow.veery.01.
Hedemark, Michael, and W. Daniel Svedarsky. 1990. "A Whooping Crane in Minnesota." *Loon* 62: 177–81.
Heidel, Kathy. 1983. "Bufflehead and Ducklings Found in Carver Park Reserve." *Loon* 55: 123–24.
Heinselman, Miron L. 1973. "Fire in the Virgin Forests of the Boundary Waters Canoe Area, Minnesota." *Journal of Quaternary Research* 3: 329–82.
Heinz, Gary H., Susan D. Haseltine, William L. Reichel, and Gary L. Hensler. 1983. "Relationships of Environmental Contaminants to Reproductive Success in Red-breasted Mergansers *Mergus serrator* from Lake Michigan." *Environmental Pollution* 32: 211–32.
Heinz, Gary H., David J. Hoffman, Jon D. Klimstra, Katherine R. Stebbins, Shannon L. Kondrad, and Carol A. Erwin. 2009. "Species Differences in the Sensitivity of Avian Embryos to Methylmercury." *Archives of Environmental Contamination and Toxicology* 56: 129–38.
Hejl, Sallie J., Jennifer A. Holmes, and Donald E. Kroodsma. 2020. "Winter Wren (*Troglodytes hiemalis*)." Version 1.0. In Billerman 2020. doi:10.2173/bow.winwre3.01.
Helmberger, Marshall. 2012. "The Ways of Warblers." *Minnesota Conservation Volunteer* (March/April): 16–26.
Henderson, Carrol L. 1977. "Minnesota Colonial Bird Nesting Site Inventory." St. Paul: Minnesota Department of Natural Resources. https://files.dnr.state.mn.us/eco/nongame/projects/consgrant_reports/1977/1977_colonial_birds.pdf.
———. 1978. "Minnesota Colonial Waterbird Nesting Site Inventory." Minnesota Department of Natural Resources. https://files.dnr.state.mn.us/eco/nongame/projects/consgrant_reports/1978/1978_colonial_birds.pdf.
———. 1979a. "Last Call for Cranes?" *Minnesota Conservation Volunteer* 42 (May/June): 48–57.
———. 1979b. *Minnesota Sandhill Crane Report*. Minnesota Department of Natural Resources. https://files.dnr.state.mn.us/eco/nongame/projects/consgrant_reports/1979/1979_sandhillcrane.pdf.
———. 1980. *The Mourning Dove in Minnesota*. St. Paul: Minnesota Department of Natural Resources. https://www.leg.mn.gov/docs/pre2003/other/810700.pdf.
———. 1984. "Colonial Waterbirds in Minnesota." Report by the Minnesota Department of Natural Resources, Nongame Wildlife Program. http://files.dnr.state.mn.us/eco/nongame/projects/consgrant_reports/1984/1984_colonial_birds.pdf.
———. 1994. *Landscaping for Wildlife*. Minnesota Department of Natural Resources.
———. 2016. "The Resilience of Sandhill Cranes." *Minnesota Conservation Volunteer* 79 (March/April): 8–19.
———. 2017. "Visions of Swans." *Minnesota Conservation Volunteer* 80 (March/April): 8–21.
Hendrickson, Mike. 1987. "A Clark's Grebe in Duluth." *Loon* 59: 210–11.
Hendrickson, Mike, and Kim R. Eckert. 1985. "The Summer Season, June 1 to July 31, 1984." *Loon* 57: 39–49.
———. 1991. "A Record Movement of Common Nighthawks." *Loon* 63: 68–69.
Henneman, Carlene. 2006. "Habitat Associations of Red-shouldered Hawks in Central Minnesota Landscapes." Master's thesis, University of Minnesota.
Henneman, Carlene, and David E. Andersen. 2009. "Occupancy Models of Nesting-Season Habitat Associations of Red-shouldered Hawks in Central Minnesota." *Journal of Wildlife Management* 73: 1316–24.
Henny, Charles J., and Norman E. Holgerson. 1974. "Range Expansion and Population Increase of the Gadwall in Eastern North America." *Wildfowl* 25: 95–101.
Hepp, Gary R., and Frank C. Bellrose. 2020. "Wood Duck (*Aix sponsa*)." Version 1.0. In Poole 2020. doi:10.2173/bow.wooduc.01.
Herb, William, Omid Mohseni, and Heinz Stefan. 2004. *Lake of the Woods Shoreline Erosion: Sensitivity to Lake Level and Wind, and Potential Erosion Control Strategies*. St. Anthony Falls Laboratory Project Report No. 465.
Herbert, John A., and Thomas B. Mowbray. 2020. "Swamp Sparrow (*Melospiza georgiana*)." Version 1.0. In Rodewald 2020. doi:10.2173/bow.swaspa.01.
Herkert, James R. (1998) 2002. *Effects of Management Practices on Grassland Birds: Henslow's Sparrow*. Rev. ed. Jamestown, ND: Northern Prairie Wildlife Research Center.
———. 2004. "Organochlorine Pesticides Are Not Implicated in the Decline of the Loggerhead Shrike." *Condor* 106: 702–5.
Herkert, James R., Donald E. Kroodsma, and James P. Gibbs. 2020. "Sedge Wren (*Cistothorus platensis*)." Version 1.0. In Poole and Gill 2020. doi:10.2173/bow.sedwre.01.

Herkert, James R., Peter D. Vickery, and Donald E. Kroodsma. 2020. "Henslow's Sparrow (*Centronyx henslowii*)." Version 1.0. In Rodewald 2020. doi:10.2173/bow.henspa.01.

Herr, Andrea M. 1992. "Assessing Spatial Patterns of Greater Sandhill Crane Nesting Habitat Using GIS and Remote Sensing." Master's thesis, University of Minnesota.

Hertzel, Anthony X. 1990. "Summer Record for a Long-billed Curlew." *Loon* 62: 58–59.

———. 1998. "Snowy Plover in Polk County." *Loon* 70: 177–78.

———. 1999. "Swallow-tailed Kite, 16 May 1999, Lower Sakatah Lake, Rice County." *Loon* 71: cover.

———. 2003. "The Northern Bobwhite in Minnesota." *Loon* 75: 3–7.

———. 2006. "An Old Ibis Specimen from Redwood County." *Loon* 78: 224–26.

———. 2012. "Blue Grosbeak Nesting in Dakota County." *Loon* 84: 53–54.

———. 2015. "First Minnesota Nesting Record of the Black-necked Stilt." *Loon* 87: 39–40.

———. 2019. "Nesting Northern Mockingbirds." *Loon* 91: 32–35.

Hertzel, Anthony X., and Paul Hertzel. 2001. "Great-tailed Grackles Nesting in Minnesota." *Loon* 73: 108–14.

Hertzel, Anthony X., T. Benjamin Hertzel, and Robert M. Dunlap. 2011. "Say's Phoebe (*Sayornis saya*) Nesting in Minnesota." *Loon* 83: 6–9.

Hertzel, Anthony X., and Robert B. Janssen. 1998. *County Nesting Records of Minnesota Birds*. Minnesota Ornithologists' Union Occasional Papers, no. 2. Minneapolis: The Minnesota Ornithologists' Union.

Hertzel, Paul, and Anthony X. Hertzel. 2000. "Locating Wintering Long-eared Owls in Minnesota." *Loon* 72: 4–9.

Herwig, Christine. 2014. "Perils for Plovers." *Nature Notes* (e-newsletter). Minnesota Scientific and Natural Areas. http://content.govdelivery.com/accounts/MNDNR/bulletins/e1fbb8.

Herz, Josephine D. 1954. "Minnesota Nesting Season—1953." *Flicker* 26: 66–73.

Hess, Gene K., Richard L. West, Maurice V. Barnhill III, and Lorraine M. Fleming. 2000. *Birds of Delaware*. Pittsburgh: University of Pittsburgh Press.

Hibbard, Edmund A. 1965. "Comments on the Distribution and Abundance of the Lark Bunting and Other Prairie Fringillids in Minnesota and North Dakota." *Loon* 37: 70–72.

Hiemenz, Nestor M. 1944. "The Eastern Ruby-crowned Kinglet Nesting in Minnesota." *Flicker* 16: 80–81.

———. 1964. "Ring-billed Gull Nesting in Minnesota." *Loon* 36: 133.

———. 1965. "Unusual Minnesota Records." *Loon* 37: 52–53.

———. 1981. "Little Blue Heron Nests at Lake Johanna." *Loon* 53: 224.

———. 1983. "Another Snowy Plover at Lake of the Woods." *Loon* 55: 177.

Hiemenz, Nestor M., Richard D. Voth, and George W. Friedrich. 1935. "Mountain Blubirds [*sic*] in Minnesota." *Auk* 52: 456–57.

Hier, Ross H. 2007. "Whooping Cranes in Polk County." *Loon* 79: 240–41.

Hill, James R., III. 1993. "Management Tips for Purple Martin Landlords." *Purple Martin Update* 4 (2): 26–27.

Hinds, David S., and William A. Calder. 1971. "Tracheal Dead Space in the Respiration of Birds." *Evolution* 25 (2): 429–40.

Hines, Jeff. 1993. "Establishing Common Tern Habitat." *Loon* 65: 31–32.

Hirsch, Katherine V. 1981. "New Colonies in Lake of the Woods." *Loon* 54: 37–39.

Hirsch, Katherine V., and Carrol L. Henderson. 1980. "Common Loon Report: Results of an Observation Card Program for Common Loons in Minnesota—1980." Minnesota Department of Natural Resources. https://www.leg.state.mn.us/docs/pre2003/other/820020.pdf.

Hitman, Bruce A. 1972. "Cattle Egret, Little Blue Heron, and Possible Snowy Egret Nestings at the Lake Johanna Rookery, Pope County, Minnesota." *Loon* 44: 36–43.

Hobson, Keith A., and Erin Bayne. 2000. "Effects of Forest Fragmentation by Agriculture on Avian Communities in the Southern Boreal Mixedwoods of Western Canada." *Wilson Bulletin* 112: 373–87.

Hobson, Keith A., and Jim Schieck. 1999. "Changes in Bird Communities in Boreal Mixedwood Forest: Harvest and Wildfire Effects over 30 Years." *Ecological Applications* 9: 849–63.

Hoch, Greg. 2011. "The Bird That Followed the Plow." *Minnesota Conservation Volunteer* 74 (September/October): 40–45.

Hoffman, Ken, and Molly Hoffman. 1981. "Rusty Blackbird Family Group at Swamp Lake, Cook County." *Loon* 53: 223.

———. 1982a. "Minnesota's First Rusty Blackbird Nest." *Loon* 54: 141–43.

———. 1982b. "Second Breeding Record for the Solitary Sandpiper South of Canada." *Loon* 54: 144–47.

Hoffman, Richard W., and Allan E. Thomas. 2007. *Columbian Sharp-tailed Grouse (*Tympanuchus phasianellus columbianus*): A Technical Conservation Assessment*. US Department of Agriculture, Forest Service, Rocky Mountain Region.

Hofslund, Pershing B. 1950a. "Census Review, Harbor Island, Duluth, Minnesota." *Flicker* 22: 128–29.

———. 1950b. "Cormorant–Great Blue Heron Rookery." *Flicker* 22: 128.

———. 1952. "Sixth Annual Harbor Island Census." *Flicker* 24: 164.

———. 1959. "A Life History Study of the Yellowthroat, *Geothlypis trichas*." *Proceedings of the Minnesota Academy of Science* 27: 144–74.

———. 1973. "An Invasion of Goshawks." *Raptor Research* 7: 107–8.

Hofstead, Russell. 1950. "Bewick's Wren Nesting at Fort Snelling." *Flicker* 22: 108.

Holling, Crawford S. 1978. *Adaptive Environmental Assessment and Management*. New York: John Wiley & Sons.

Holmes, Richard T., Sara A. Kaiser, Nicholas L. Rodenhouse, T. Scott Sillett, Michael S. Webster, Peter Pyle, and Michael A. Patten. 2020. "Black-throated Blue Warbler (*Setophaga caerulescens*)." Version 1.0. In Rodewald 2020. doi:10.2173/bow.btbwar.01.

Holmes, Richard T., Peter P. Marra, and Thomas W. Sherry. 1996. "Habitat-specific Demography of Breeding Black-throated Blue Warblers (*Dendroica caerulescens*): Implications for Population Dynamics." *Journal of Animal Ecology* 65: 183–95.

Holmes, Richard T., Thomas W. Sherry, and Franklin W. Sturges. 1986. "Bird Community Dynamics in a Temperate Deciduous Forest: Long-term Trends at Hubbard Brook." *Ecological Monographs* 56: 201–20.

Holt, Denver W. 1997. "The Long-eared Owl (*Asio otus*) and Forest Management: A Review of the Literature." *Journal of Raptor Research* 31: 175–86.

Homer, Collin G., Chengquan Huang, Limin Yang, Bruce K. Wylie, and Michael Coan. 2004. "Development of a 2001 National Land-Cover Database for the United States." *Photogrammetric Engineering and Remote Sensing* 70: 829–40.

Hoover, Jeffrey P. 2006. "Water Depth Influences Nest Predation for a Wetland-Dependent Bird

in Fragmented Bottomland Forests." *Biological Conservation* 127: 37–45.
———. 2009. "Prothonotary Warblers as Indicators of Hydrological Conditions in Bottomland Forests." In *Proceedings of the Fourth International Partners in Flight Conference: Tundra to Tropics: Connecting Bird, Habitats and People, 13–16 February 2008, McAllen, Texas,* edited by Terrell D. Rich, Coro Arizmendi, Dean W. Demarest, and Craig Thompson, 128–37. Partners in Flight. https://tethys.pnnl.gov/publications/proceedings-fourth-international-partners-flight-conference-tundra-tropics.
Hoover, Robert L., and Dale L. Wills. 1984. *Managing Forested Lands for Wildlife*. Denver: Colorado Division of Wildlife and US Department of Agriculture, Forest Service, Rocky Mountain Region.
Hopp, Steven L., Alice Kirby, and Carol A. Boone. 2020. "White-eyed Vireo (*Vireo griseus*)." Version 1.0. In Billerman et al. 2020. doi:10.2173/bow.whevir.01.
Horn, Henry S. 1968. "The Adaptive Significance of Colonial Nesting in the Brewer's Blackbird (*Euphagus cyanocephalus*)." *Ecology* 49: 682–94.
Hothem, Roger L., Brianne E. Brussee, William E. Davis Jr., Albert Martinez-Vilalta, Anna Motis, and Guy M. Kirwan. 2020. "Black-crowned Night-Heron (*Nycticorax nycticorax*)." Version 1.0. In Billerman 2020. doi:10.2173/bow.bcnher.01.
Houston, C. Stuart, Cameron Jackson, and Daniel E. Bowen Jr. 2020. "Upland Sandpiper (*Bartramia longicauda*)." Version 1.0. In Poole 2020. doi:10.2173/bow.uplsan.01.
Howe, Robert W. 1979. "Distribution and Behavior of Birds on Small Islands in Northern Minnesota." *Journal of Biogeography* 6: 379–90.
Howe, Robert W., Gerald J. Niemi, Stephen J. Lewis, and Daniel A. Welsh. 1997. "A Standard Method for Monitoring Songbird Populations in the Great Lakes Region." *Passenger Pigeon* 59: 183–94.
Howitz, James L. 1981. "A Population Study of the Black-capped Chickadee." PhD diss., University of Minnesota.
———. 2003. "The Red-headed Woodpecker in Minnesota." *Loon* 75: 123–30.
Huber, Harding. 1961. "Kentucky Warbler Specimen for Minnesota." *Flicker* 33: 91.
Huber, Ronald L. 1959a. "Whimbrel and Jaeger on Minnesota Point, St. Louis County." *Flicker* 31: 138.
———. 1959b. "Yellow Rails in Becker and Mahnomen Counties, Minnesota." *Flicker* 31: 82–83.
———. 1960a. "Chestnut-collared Longspurs in Clay County, Minnesota." *Flicker* 32: 99–100.
———. 1960b. "The Chukar in Minnesota." *Flicker* 32: 64.
———. 1961a. "Breeding Range Extension of Eared and Western Grebes." *Flicker* 33: 93–94.
———. 1961b. "Nesting Bell's Vireos Banded in Houston County." *Flicker* 33: 109–10.
———. 1961c. "The Nesting Season." *Flicker* 33: 71–78.
———. 1962a. "The Spring Season." *Flicker* 34: 50–51, 54–58.
———. 1962b. "The Summer Season." *Flicker* 34: 80–85.
———. 1963a. "The Breeding Season." *Flicker* 35: 85–94.
———. 1963b. "Yellow-crowned Night Herons Expanding Range in Minnesota." *Flicker* 35: 102–3.
———. 1964a. "Blue Grosbeaks and Lark Bunting in Rock County." *Loon* 36: 26–28.
———. 1964b. "The Summer Season." *Loon* 36: 83–91.
———. 1964c. "Yellow-crowned Night Heron Nesting in South Saint Paul." *Loon* 36: 134–35.
———. 1965. "The Spring Season." *Loon* 37: 136–44.
———. 1966a. "The Summer Season." *Loon* 38: 8–24.
———. 1966b. "The Summer Season." *Loon* 38: 118–29.
———. 1968. "The Summer Season." *Loon* 40: 118–30.
Hubert, B. 1945. "Prothonotary Warblers." *Flicker* 17: 91.
Huff, Ned L. 1929. "The Nest and Habits of the Connecticut Warbler in Minnesota." *Auk* 46: 455–65.
Hughes, Janice M. 2020a. "Black-billed Cuckoo (*Coccyzus erythropthalmus*)." Version 1.0. In Rodewald 2020. doi:10.2173/bow.bkbcuc.01.
———. 2020b. "Yellow-billed Cuckoo (*Coccyzus americanus*)." Version 1.0. In Rodewald 2020. doi:10.2173/bow.yebcuc.01.
Hull, Scott D. (2000) 2002. *Effects of Management Practices on Grassland Birds: Eastern Meadowlark*. Rev. ed. Jamestown, ND: Northern Prairie Wildlife Research Center.
Hunt, Kelsi L., Nuray Taygan, Daniel H. Catlin, Jay H. Felio, and James D. Fraser. 2013. "Demography of Snowy Plovers (*Charadrius nivosus*) on the Missouri River." *Waterbirds* 36: 220–24.
Hunt, Pamela D., and David J. Flaspohler. 2020. "Yellow-rumped Warbler (*Setophaga coronata*)." Version 1.0. In Rodewald 2020. doi:10.2173/bow.yerwar.01.
Hunter, William C., David A. Buehler, Ronald A. Canterbury, John L. Confer, and Paul B. Hamel. 2001. "Conservation of Disturbance-Dependent Birds in Eastern North America." *Wildlife Society Bulletin* 29: 440–55.
Hutto, Richard L. 1995a. "Composition of Bird Communities Following Stand-Replacement Fires in Northern Rocky Mountain (U.S.A.) Conifer Forests." *Conservation Biology* 9: 1041–58.
———. 1995b. "USFS Northern Region Songbird Monitoring Program: Distribution and Habitat Relationships." USFS Contract R1-95-05. Report to US Department of Agriculture, Forest Service, Region 1.
Igl, Lawrence D. 1995. "A Noteworthy Record and the Breeding Distribution of the Blue Grosbeak in North Dakota." *Prairie Naturalist* 27: 205–10.
———. 1996. "LeConte's, Sharp-tailed, and Henslow's Sparrow in Grant County." *Loon* 68: 127–28.
Iowa Ornithologists' Union. 2019. "Species Accounts." *Iowa Breeding Bird Atlas II*. http://bba.iowabirds.org/SpeciesAccounts.aspx.
IUCN (International Union for Conservation of Nature). 2021. *The IUCN Red List of Threatened Species*. Version 2021-1. Accessed April 28, 2021. https://www.iucnredlist.org.
Jaakko Pöyry Consulting, Inc. 1994. "Final Generic Environmental Impact Statement Study on Timber Harvesting and Forest Management in Minnesota." Tarrytown, NY: Jaakko Pöyry Consulting, Inc. https://conservancy.umn.edu/handle/11299/169527.
Jackson, Allyson K., David C. Evers, Matthew A. Etterson, Anne M. Condon, Sarah B. Folsom, Jennifer Detweiler, John Schmerfeld, et al. 2011. "Mercury Exposure Affects the Reproductive Success of a Free-Living Terrestrial Songbird, the Carolina Wren (*Thryothorus ludovicianus*)." *Auk* 128: 759–69.
Jackson, Bette J., and Jerome A. Jackson. 2020. "Killdeer (*Charadrius vociferus*)." Version 1.0. In Poole and Gill 2020. doi:10.2173/bow.killde.01.
Jackson, Jerome A., and Henri R. Ouellet. 2020. "Downy Woodpecker (*Dryobates pubescens*)." Version 1.0. In Rodewald 2020. doi:10.2173/bow.dowwoo.01.
Jackson, Jerome A., Henri R. Ouellet, and Bette J. Jackson. 2020. "Hairy Woodpecker (*Dryobates*

villosus)." Version 1.0. In Rodewald 2020. doi:10.2173/bow.haiwoo.01.
Jackson, Laura Spess, Carol A. Thompson, and James J. Dinsmore. 1996. *The Iowa Breeding Bird Atlas*. Iowa City: University of Iowa Press.
James, Frances C., Richard F. Johnston, Noel O. Wamer, Gerald J. Niemi, and William J. Boecklen. 1984. "The Grinnellian Niche of the Wood Thrush." *American Naturalist* 124: 17–47.
James, Ross D. 1976. "Foraging Behavior and Habitat Selection of Three Species of Vireos in Southern Ontario." *Wilson Bulletin* 88: 62–75.
Janssen, Robert B. 1959. "Henslow's Sparrow Extends Range in Minnesota." *Flicker* 31: 102.
———. 1967. "White-faced Ibis at Heron Lake." *Loon* 39: 108.
———. 1986a. "A McCown's Longspur in Duluth." *Loon* 58: 198.
———. ed. 1986b. "Minnesota Records of Clark's Grebe." *Loon* 58: 109–10.
———. 1987. *Birds in Minnesota*. Minneapolis: University of Minnesota Press.
———. 1988. "A Major Invasion of Dickcissels (1988)." *Loon* 60: 143–44.
———. 1989. "Minnesota's First House Finch Nest." *Loon* 61: 93.
———. 1990a. "Barn Owl Nests in Dakota County." *Loon* 62: 231.
———. 1990b. "Minnesota's First Mountain Bluebird Nest." *Loon* 62: 160–61.
———. 1992. "The House Finch in Minnesota." *Loon* 64: 79–81.
———. 1994. "Possible Breeding for Clark's Grebes in Minnesota." *Loon* 66: 208.
———. 2000. "The Western Meadowlark in Minnesota: Does It Have a Future?" *Loon* 72: 127–32.
———. 2019. *Birds in Minnesota*. Rev. ed. Minneapolis: University of Minnesota Press.
Jaster, Levi A., William E. Jensen, and Wesley E. Lanyon. 2020. "Eastern Meadowlark (*Sturnella magna*)." Version 1.0. In Poole 2020. doi:10.2173/bow.easmea.01.
Jensen, Jack R. 1959. "Sandhill Cranes in Roseau County." *Loon* 31: 138.
Jessen, Robert, and Carrol L. Henderson. 1979. "Canvasback Breeding in Minnesota." *Loon* 51: 46.
Job, Herbert K. 1910. "The Franklin's Gull." The National Association of Audubon Societies, Educational Leaflet 44. *Bird-Lore* 12: 124–27.
Johnsgard, Paul A. 1983. *The Grouse of the World*. Lincoln: University of Nebraska Press.
———. 1986. *Birds of the Rocky Mountains with Particular Reference to National Parks in the Northern Rocky Mountain Region*. Boulder: Colorado Associated University Press.
———. 1988. *North American Owls: Biology and Natural History*. Washington, DC: Smithsonian Institute Press.
———. 1995. *This Fragile Land: A Natural History of the Nebraska Sandhills*. Lincoln: University of Nebraska Press.
———. 2010. *Waterfowl of North America*. Rev. ed. Lincoln: University of Nebraska-Lincoln Libraries.
Johnson, C. B. 1898. "Prothonotary Warbler." *Oologist* 15: 53–54.
Johnson, David H. 1982. "Raptors of Minnesota—Nesting Distribution and Population Status." *Loon* 54: 73–104.
Johnson, David H., and Lawrence D. Igl. 2001. "Area Requirements of Grassland Birds: A Regional Perspective." *Auk* 118: 24–34.
Johnson, Douglas P. 1986. "Henslow's Sparrow in Norman County." *Loon* 58: 130.
Johnson, Jay M. 1976. "Distribution of Sandhill Cranes in Minnesota." In *Proceedings of the International Crane Workshop*, edited by James C. Lewis, 59–68. Stillwater: Oklahoma State University. https://www.savingcranes.org/proceedings-of-the-international-crane-workshop-1975/.
Johnson, Jeff A., Michael A. Schroeder, and Leslie A. Robb. 2020. "Greater Prairie-Chicken (*Tympanuchus cupido*)." Version 1.0. In Poole 2020. doi:10.2173/bow.grpchi.01.
Johnson, Kevin. 2020. "Green-winged Teal (*Anas crecca*)." Version 1.0. In Billerman 2020. doi:10.2173/bow.gnwtea.01.
Johnson, Kristine, and Brian D. Peer. 2020. "Great-tailed Grackle (*Quiscalus mexicanus*)." Version 1.0. In Poole and Gill 2020. doi:10.2173/bow.grtgra.01.
Johnson, L. Scott. 2020. "House Wren (*Troglodytes aedon*)." Version 1.0. In Poole 2020. doi:10.2173/bow.houwre.01.
Johnson, L. Scott, and Russell D. Dawson. 2020. "Mountain Bluebird (*Sialia currucoides*)." Version 1.0. In Rodewald 2020. doi:10.2173/bow.moublu.01.
Johnson, Ned K. 1993. "Pioneering and Natural Expansion of Breeding Distributions in Western North American Birds." In *A Century of Avifaunal Change in Western North America*, edited by Joseph R. Jehl Jr. and Ned K. Johnson, 27–44. Studies in Avian Biology, no. 15. Lawrence, KS: Cooper Ornithological Society.
Johnson, Oscar L. 1994. "Unusual Summer Observations in Hennepin County." *Loon* 66: 150–51.
Johnson, Oscar W. 1969. "Red Crossbill Breeding in Minnesota." *Auk* 86: 352–53.
Johnson, Richard G., and Stanley A. Temple. 1986. "Assessing Habitat Quality for Birds Nesting in Fragmented Tallgrass Prairies." In *Wildlife 2000: Modeling Habitat Relationships of Terrestrial Vertebrates*, edited by Jared Verner, Michael L. Morrison, and C. John Ralph, 245–49. Madison: University of Wisconsin Press.
———. 1990. "Nest Predation and Brood Parasitism of Tallgrass Prairie Birds." *Journal of Wildlife Management* 54: 106–11.
Johnson, W. Carter, and Curtis S. Adkisson. 1985. "Dispersal of Beech Nuts by Blue Jays in Fragmented Landscapes." *American Midland Naturalist* 113: 319–24.
Jones, Evelyn. 1945. "Herring Gulls Nesting at Beaver Bay." *Flicker* 17: 92–93.
Jones, Oliver V. 2017. "The Passenger Pigeon." *Loon* 89: 142. Reprint, *Oologist* 14: 14.
Jones, Paul D., John P. Giesy, John L. Newsted, David A. Verbrugge, Donald L. Beaver, Gerald T. Ankley, Donald E. Tillitt, et al. 1993. "2,3,7,8-Tetrachlorodibenzo-p-dioxin Equivalents in Tissues of Birds at Green Bay, Wisconsin, USA." *Archives of Environmental Contamination and Toxicology* 24: 345–54.
Jones, Stephanie L. 2010. *Sprague's Pipit (*Anthus spragueii*) Conservation Plan*. Washington, DC: US Fish and Wildlife Service.
Jones, Stephanie L., and John E. Cornely. 2020. "Vesper Sparrow (*Pooecetes gramineus*)." Version 1.0. In Poole and Gill 2020. doi:10.2173/bow.vesspa.01.
Jones, Todd M., Amanda D. Rodewald, and Daniel P. Shustack. 2010. "Variation in Plumage Coloration of Northern Cardinals in Urbanizing Landscapes." *Wilson Journal of Ornithology* 122: 326–33.
Jorgensen, Joel G., and Stephen J. Dinsmore. 2005. "An Assessment of the Status of White-faced Ibis (*Plegadis chihi*) in the Great Plains." *North American Birds* 59: 376–81.
Kahl, Richard B., Thomas S. Baskett, Jonathan A. Ellis, and James N. Burroughs. 1985. *Characteristics of Summer Habitats of Selected Nongame Birds in Missouri*. Research Bulletin no. 1056. Columbia: University of Missouri-Columbia, Agricultural Experiment Station.
Kalmbach, Edwin Richard. 1940. *Economic Status of the English Sparrow in the United States*. US Department of Agriculture Technical Bulletin No. 711. Washington, DC: US Department of Agriculture.

Kaminski, Richard M., and Milton W. Weller. 1992. "Breeding Habitats of Nearctic Waterfowl." In *Ecology and Management of Breeding Waterfowl,* edited by Bruce D. J. Batt, Alan D. Afton, Michael G. Anderson, C. Davison Ankney, Douglass H. Johnson, John A. Kadlec, and Gary L. Krapu, 568–89. Minneapolis: University of Minnesota Press.

Kantrud, Harold A., and Kenneth F. Higgins. 1992. "Nest and Nest Site Characteristics of Some Ground-Nesting, Non-Passerine Birds of Northern Grasslands." *Prairie Naturalist* 24: 67–84.

Kantrud, Harold A., and Robert E. Stewart. 1977. "Use of Natural Basin Wetlands by Breeding Waterfowl in North Dakota." *Journal of Wildlife Management* 41: 243–53.

Kaufman, Ken. 1996. *Lives of North American Birds*. Boston: Houghton Mifflin.

Keast, Allen, Laura Pearce, and Sari Saunders. 1995. "How Convergent Is the American Redstart (*Setophaga ruticilla*, Parulinae) with Flycatchers (Tyrannidae) in Morphology and Feeding Behavior?" *Auk* 112: 310–25.

Keating, William H. 1824. *Narrative of an Expedition to the Source of St. Peter's River, Lake Winnepeek, Lake of the Woods, &c., Performed in the Year 1823, by Order of the Hon. J. C. Calhoun, Secretary of War, under the Command of Stephen H. Long*. Vol. 2. Philadelphia: Carey & Lea.

Kelleher, Kevin. 1967. "Cerulean Warblers Nesting in Mahnomen County." *Loon* 39: 108–9.

Keller, Gregory S., and Richard H. Yahner. 2007. "Seasonal Forest-Patch Use by Birds in Fragmented Landscapes of South-Central Pennsylvania." *Wilson Journal of Ornithology* 119: 410–18.

Kelley, James, Scot Williamson, and Thomas R. Cooper, eds. 2008. *American Woodcock Conservation Plan: A Summary of and Recommendations for Woodcock Conservation in North America*. Wildlife Management Institute. https://timberdoodle.org/sites/default/files/woodcockPlan_0.pdf.

Kelly, Frank. 1958. "Scissor-tailed Flycatcher (*Muscivora forficata*) in Northern Minnesota." *Flicker* 30: 37.

Kelly, Jeffrey F., Eli S. Bridge, and Michael J. Hamas. 2020. "Belted Kingfisher (*Megaceryle alcyon*)." Version 1.0. In Poole 2020. doi:10.2173/bow.belkin1.01.

Kelly, John P., Katie Etienne, Cheryl Strong, Mark McCaustland, and Michael L. Parkes. 2007. "Status, Trends, and Implications for the Conservation of Heron and Egret Nesting Colonies in the San Francisco Bay Area." *Waterbirds* 30: 455–78.

Kennedy, E. Dale, and Douglas W. White. 2020. "Bewick's Wren (*Thryomanes bewickii*)." Version 1.0. In Poole 2020. doi:10.2173/bow.bewwre.01.

Kennedy, Martyn, and Hamish G. Spencer. 2014. "Classification of the Cormorants of the World." *Molecular Phylogenetics and Evolution*. 79: 249–57. doi:10.1016/j.ympev.2014.06.020.

Kennedy, Patricia L., and Dale W. Stahlecker. 1993. "Responsiveness of Nesting Northern Goshawks to Taped Broadcasts of 3 Conspecific Calls." *Journal of Wildlife Management* 57: 249–57.

Kennedy, Tony. 2016. "New Push to Protect Our Iconic Loons." *Star Tribune*, September 4. http://www.startribune.com/minnesota-on-the-cusp-of-a-new-era-of-loon-conservation/392254151/.

Kent, Thomas H., and James J. Dinsmore. 1996. *Birds in Iowa*. Iowa City and Ames: Privately printed.

Keran, Doug. 1978. "Nest Site Selection by the Broad-winged Hawk in North Central Minnesota and Wisconsin." *Journal of Raptor Research* 12: 15–20.

Kerns, J. M., and J. F. Howe. 1967. "Factors Determining Great Blue Heron Rookery Movement." *Journal of the Minnesota Academy of Science* 34: 80–83.

Kershner, Eric L., and Walter G. Ellison. 2020. "Blue-gray Gnatcatcher (*Polioptila caerulea*)." Version 1.0. In Poole 2020. doi:10.2173/bow.buggna.01.

Kessen, Ann E., and Anthony X. Hertzel. 1999. "The Historical Record of the Swallow-tailed Kite in Minnesota." *Loon* 71: 178–81.

Kessen, Ann E., and Peder H. Svingen. 2007. "The Summer Season, 1 June through 31 July 2006." *Loon* 79: 15–44.

———. 2008. "The Summer Season, 1 June through 31 July 2007." *Loon* 80: 16–40.

———. 2010. "The Summer Season, 1 June through 31 July 2009." *Loon* 82: 7–34.

———. 2012. "The Summer Season, 1 June through 31 July 2011." *Loon* 84: 20–46.

———. 2014. "The Summer Season: 1 June through 31 July 2013." *Loon* 86: 24–50.

———. 2017. "The 2016 Summer Season: 1 June to 31 July 2016." *Loon* 89: 9–35.

———. 2018. "The 2017 Summer Season: 1 June to 31 July 2017." *Loon* 90: 6–31.

Kessen, Ann E., Peder H. Svingen, and Paul E. Budde. 2015. "The 2014 Summer Season, 1 June through 31 July 2014." *Loon* 87: 9–32.

———. 2016. "The 2015 Summer Season, 1 June through 31 July 2015." *Loon* 88: 6–31.

Keyel, Edward R., Matthew A. Etterson, Gerald J. Niemi, David C. Evers, Christopher R. DeSorbo, Joel C. Hoffman, John W. Nichols, et al. 2020. "Feather Mercury Increases with Feeding at Higher Trophic Levels in Two Species of Migrant Raptors, Merlin (*Falco columbarius*) and Sharp-shinned Hawk (*Accipiter striatus*)." *Condor* 122: 1–17. doi:10.1093/condor/duz069.

Kilham, Lawrence. 1971. "Reproductive Behavior of Yellow-bellied Sapsuckers. I. Preference for Nesting in *Fomes*-Infected Aspens and Nest Hole Interrelations with Flying Squirrels, Raccoons, and Other Animals." *Wilson Bulletin* 83: 159–71.

Kimmel, J. Timothy, and Richard H. Yahner. 1990. "Response of Northern Goshawks to Taped Conspecific and Great Horned Owl Calls." *Journal of Raptor Research* 24: 107–12.

Kimmel, Richard O. 1988. "Potential Impacts of Ring-necked Pheasants on Other Game Birds." In *Pheasants: Symptoms of Wildlife Problems on Agricultural Lands,* edited by Diana L. Hallett, William R. Edwards, and George V. Burger, 253–65. Bloomington, IN: North Central Section of The Wildlife Society.

King, D. Tommy, and Daniel W. Anderson. 2005. "Recent Population Status of the American White Pelican: A Continental Perspective." Special publication, *Waterbirds* 28 (1): 48–54.

Kingery, Hugh E. 1998. "Lark Bunting." In *Colorado Breeding Bird Atlas*, edited by Hugh E. Kingery, 468–69. Denver: Colorado Bird Atlas Partnership and Colorado Division of Wildlife.

Kirchman, Jeremy J., and Kathryn J. Schneider. 2014. "Range Expansion and the Breakdown of Bergmann's Rule in Red-bellied Woodpeckers (*Melanerpes carolinus*)." *Wilson Journal of Ornithology* 126: 236–48.

Kirk, David A., and Michael J. Mossman. 2020. "Turkey Vulture (*Cathartes aura*)." Version 1.0. In Poole and Gill 2020. doi:10.2173/bow.turvul.01.

Kittelson, Steve. 1996. "The Return of a Giant in the Bird World to Minnesota." *Loon* 68: 81–85.

Kleingartner, Larry. 2003. "Sunflower Losses to Blackbirds: An Economic Burden." In *Management of North American Blackbirds,* edited by George M. Linz, 13–14. Fort Collins, CO: National Wildlife Research Center.

Klem, Daniel, Jr. 1989. "Bird-Window Collisions." *Wilson Bulletin* 101: 606–20.

Klett, Albert T., Terry L. Shaffer, and Douglass H. Johnson. 1988. "Duck Nest Success in the Prairie Pothole Region." *Journal of Wildlife Management* 52: 431–40.

Klosterman, Megan E., George M. Linz, Anthony A. Slowik, and H. Jeffrey Homan. 2013. "Comparisons between Blackbird Damage to Corn and Sunflower in North Dakota." *Crop Protection* 53: 1–5.

Klute, David S., Loren W. Ayers, Michael T. Green, William H. Howe, Stephanie L. Jones, Jill A. Shaffer, Steven R. Sheffield, et al. 2003. *Status Assessment and Conservation Plan for the Western Burrowing Owl in the United States*. US Fish and Wildlife Service Biological Technical Publication FWS/BTP-R6001–2003. Washington, DC: US Fish and Wildlife Service.

Knight, Elly C., and Erin M. Bayne. 2017. *Habitat Selection at Different Scales for a Declining Aerial Insectivorous Bird as Determined by Autonomous Recording Technology*. Edmonton, Alberta: Bioacoustic Unit, University of Alberta and Alberta Biodiversity Monitoring Institute. doi:10.13140/RG.2.2.22277.04327.

Knight, Elly C., Janet W. Ng, Caitlin E. Mader, R. Mark Brigham, and Erin M. Bayne. 2018. "An Inordinate Fondness for Beetles: First Description of Common Nighthawk (*Chordeiles minor*) Diet in the Boreal Biome." *Wilson Journal of Ornithology* 130 (2): 525–31.

Knopf, Fritz L., and Roger M. Evans. 2020. "American White Pelican (*Pelecanus erythrorhynchos*)." Version 1.0. In Poole 2020. doi:10.2173/bow.amwpel.01.

Knutsen, Gregg. 2010. "Whooping Crane at Agassiz National Wildlife Refuge." *Loon* 82: 193.

Knutson, Melinda G., Greg Butcher, Jane Fitzgerald, and Julie Shieldcastle. 2001. *Partners in Flight Bird Conservation Plan for The Upper Great Lakes Plain (Physiographic Area 16)*. La Crosse, WI: US Geological Survey, Upper Midwest Environmental Sciences Center in cooperation with Partners in Flight.

Knutson, Melinda G., Nick P. Danz, Todd W. Sutherland, and Brian R. Gray. 2008. *Landbird Monitoring Protocol for the U.S. Fish & Wildlife Service, Midwest and Northeast Regions*. Version 1.0. US Fish and Wildlife Service Biological Monitoring Team Technical Report BMT-2008-01. LaCrosse, WI: US Fish and Wildlife Service.

Knutson, Melinda G., Gerald J. Niemi, Wesley E. Newton, and Mary A. Friberg. 2004. "Avian Nest Success in Midwestern Forests Fragmented by Agriculture." *Condor* 106: 116–30.

Koenig, Walter D. 2003. "European Starlings and Their Effect on Native Cavity-Nesting Birds." *Conservation Biology* 17: 1134–40.

Koenig, Walter D., Wesley M. Hochachka, Benjamin Zuckerberg, and Janis L. Dickinson. 2010. "Ecological Determinants of American Crow Mortality due to West Nile Virus during Its North American Sweep." *Oecologia* 163: 903–9.

Kolbe, Stephen R. 2018. "The Common Nighthawk in Northern Minnesota." Master's thesis, University of Minnesota.

Kopischke, Earl D., and Roger N. Johnson. 1973. "Current Status of Wild Turkeys in Southeastern Minnesota." *Loon* 45: 49–50.

Korpimäki, Erkki, and Harri Hakkarainen. 2012. *The Boreal Owl: Ecology, Behaviour and Conservation of a Forest-Dwelling Predator*. Cambridge: Cambridge University Press.

Korschgen, Carl E., Kevin P. Kenow, William L. Green, Douglas H. Johnson, Michael D. Samuel, and Louis Sileo. 1996. "Survival of Radiomarked Canvasback Ducklings in Northwestern Minnesota." *Journal of Wildlife Management* 60: 120–32.

Koster, Susan K. 1995. "1995 Heron Lake Colonial Waterbird Survey." Report to the Minnesota Department of Natural Resources.

Kotliar, Natasha B. 2007. "Olive-sided Flycatcher (*Contopus cooperi*): A Technical Conservation Assessment." Report to US Department of Agriculture, Forest Service, Rocky Mountain Region.

Kramer, Gunnar R., Henry M. Streby, Sean M. Peterson, Justin A. Lehman, David A. Buehler, Petra B. Wood, Darin J. McNeil, et al. 2017. "Nonbreeding Isolation and Population-specific Migration Patterns among Three Populations of Golden-winged Warblers." *Condor* 119: 108–21.

Kramer, Janet L., and Patrick T. Redig. 1997. "Sixteen Years of Lead Poisoning in Eagles, 1980–95: An Epizootiologic View." *Journal of Raptor Research* 31: 327–32.

Krapu, Gary L., David A. Brandt, Kenneth L. Jones, and Douglas H. Johnson. 2011. "Geographic Distribution of the Mid-Continent Population of Sandhill Cranes and Related Management Applications." *Wildlife Monographs* 175: 1–38.

Kraske, Robert. 1981. "Whatever Happened to Bluebirds?" *Minnesota Volunteer* (January/February): 42–48.

Kreitinger, Kim, Yoyi Steele, and Andy Paulios, eds. 2013. *The Wisconsin All-Bird Conservation Plan*. Version 2.0. Madison: Wisconsin Department of Natural Resources, Wisconsin Bird Conservation Initiative. http://www.wisconsinbirds.org/.

Kricher, John C. 2020. "Black-and-white Warbler (*Mniotilta varia*)." Version 1.0. In Poole 2020. doi:10.2173/bow.bawwar.01.

Krienke, Cindy. 1999. "A Swallow-tailed Kite in Southern Minnesota." *Loon* 71: 176–77.

Kroodsma, Donald E. 2005. *The Singing Life of Birds: The Art and Science of Listening to Birdsong*. New York: Houghton Mifflin.

Kroodsma, Donald E., and Jared Verner. 2020. "Marsh Wren (*Cistothorus palustris*)." Version 1.0. In Poole 2020. doi:10.2173/bow.marwre.01.

Krosch, Penelope. 1987. "Three New Species for Minnesota: Roberts and Benner Trip to Western Minnesota." *Loon* 59: 64–66.

Krych, Scott. 1992. "Minnesota's Third Record of Black-necked Stilt." *Loon* 64: 166–67.

Kudell-Ekstrum, Janet. 2002. *Conservation Assessment for Connecticut Warbler (*Oporornis agilis*)*. US Department of Agriculture, Forest Service, Eastern Region.

Kuester, Lance. 1984. "Common Barn-Owl." *Loon* 56: 195, photograph.

Kumlien, Ludwig, and Ned Hollister. (1903) 1951. *The Birds of Wisconsin. Bulletin of the Wisconsin Natural History Society* 3, nos. 1–3; Milwaukee: Milwaukee Public Museum. Reprinted in installments by the Wisconsin Society for Ornithology in *The Passenger Pigeon*, vols. 10–13 (1948–1951), with annotations by A. W. Schorger. Reprinted as a separate volume in 1951 by the Wisconsin Society for Ornithology, Madison.

Kus, Barbara, Steven L. Hopp, R. Roy Johnson, and Bryan T. Brown. 2020. "Bell's Vireo (*Vireo bellii*)." Version 1.0. In Poole 2020. doi:10.2173/bow.belvir.01.

Kushlan, James A., Melanie J. Steinkamp, Katharine C. Parsons, Jack Capp, Martin Acosta Cruz, Malcolm Coulter, Ian Davidson, et al. 2002. *Waterbird Conservation for the Americas: The North American Waterbird Conservation Plan, Version 1*. Washington, DC: Waterbird Conservation for the Americas. https://iwjv.org/resource/north-american-waterbird-conservation-plan.

Kuyava, Gary C. 1960. "Barn Owl at Duluth." *Flicker* 32: 101.

Kyllingstad, Henry. 1976. "Snowy Plover at Marshall." *Loon* 48: 115.

Lafond, Kenneth. 1986. "Western Wood-Pewee in Clearwater County." *Loon* 58: 50.

Lakela, Olga. 1940. "The Status of Piping Plovers (*Charadrius melodus*) on Minnesota Point." *Flicker* 12: 34.

———. 1943. "More Summer Goshawks in Minnesota." *Flicker* 15: 38.

———. 1944. "American Avocet in Duluth." *Flicker* 16: 55.

———. 1945. "Goshawk Nesting in St. Louis County." *Flicker* 17: 89.

———. 1946. "Additional Observations on the Nesting of Piping Plovers in Duluth." *Flicker* 18: 11.

———. 1953. "A Check-list of Birds from Northeastern Minnesota Based on Observations Made Mostly in the Duluth Area." *Flicker* 25: 30–35.

Lambert, J. Daniel, and Steven D. Faccio. 2005. *Canada Warbler Population Status, Habitat Use, and Stewardship Guidelines for Northeastern Forests*. VINS Technical Report 05-4. Woodstock: Vermont Institute of Natural Science.

Lambeth, David O., and Sharon O. Lambeth. 1988. "Sprague's Pipits Nest in Polk County." *Loon* 60: 104–8.

Lane, William H., David E. Andersen, and Thomas H. Nicholls. 2001. "Distribution, Abundance, and Habitat Use of Singing Male Boreal Owls in Northeast Minnesota." *Journal of Raptor Research* 35: 130–40.

Lapin, Carly N., Matthew A. Etterson, and Gerald J. Niemi. 2013. "Occurrence of Connecticut Warbler Increases with Coniferous Forest Patch Size." *Condor* 115: 168–77.

LaPorte, Nicholas, Robert W. Storer, and Gary L. Nuechterlein. 2020a. "Clark's Grebe (*Aechmophorus clarkii*)." Version 1.0. In Rodewald 2020. doi:10.2173/clagre.01.

———. 2020b. "Western Grebe (*Aechmophorus occidentalis*)." Version 1.0. In Rodewald 2020. doi:10.2173/bow.wesgre.01.

Larson, Krista. 2016. Minnesota Loon Monitoring Program. 22nd Anniversary Report: 1994–2015. https://files.dnr.state.mn.us/eco/nongame/projects/mlmp_2015long.pdf.

———. 2021. *Results of the Minnesota Loon Monitoring Program (MLMP), 1994–2020*. Final report. Minnesota Department of Natural Resources, Nongame Wildlife Program.

Lawrence, Jeffrey S., and John H. Giudice. 2013. "2013 Ring-necked Duck Breeding Pair Survey." In *Status of Wildlife Populations, Fall 2013, Wetland Wildlife,* edited by Margaret H. Dexter, 141–57. Unpublished report. St. Paul: Minnesota Department of Natural Resources, Division of Fish and Wildlife.

Lawrence, Jeffrey S., John H. Giudice, and Robert G. Wright. 2016. *2016 Northwest Minnesota Sandhill Crane Breeding Ground Survey*. Minnesota Department of Natural Resources.

Lawrence, Louise de Kiriline. 1967. *A Comparative Life-History Study of Four Species of Woodpeckers*. American Ornithologists' Union Ornithological Monograph, no. 5. Anchorage, KY: American Ornithologist's Union.

Lawrence, Wynne. 1949. "Swallow-tailed Kite near Garrison, Minnesota." *Flicker* 21: 89–90.

Lee, Forrest B., Robert L. Jessen, N. J. Ordal, R. I. Benson, J. P. Lindmeier, and L. L. Johnson. 1964. *Waterfowl in Minnesota*. Minnesota Department of Conservation Division of Fish and Game Technical Bulletin No. 7.

Lent, Richard A. 1990. "Relationships among Environmental Factors, Phenotypic Characteristics, and Fitness Components in the Gray Catbird (*Dumetella carolinensis*)." PhD diss., State University of New York at Stony Brook.

Leopold, Aldo. 1931. *Report on a Game Survey of the North Central States*. Madison, WI: Sporting Arms and Ammunition Manufacturers' Institute.

———. 1949. *A Sand County Almanac: And Sketches Here and There*. London: Oxford University Press.

Leopold, Aldo, Miles D. Pirnie, and William Rowan. 1944. Introduction to *The Canvasback on a Prairie Marsh*, by H. A. Hochbaum, xi–xii. Washington, DC: American Wildlife Institute.

LeSchack, Craig R., S. Keith McKnight, and Gary R. Hepp. 2020. "Gadwall (*Mareca strepera*)." Version 1.0. In Billerman 2020. doi:10.2173/bow.gadwal.01.

Lesher, Fred. 1968. "Acadian Flycatcher Nest in Southeastern Minnesota." *Loon* 40: 4–6.

Lesterhuis, Arne J., and Rob P. Clay. 2010. *Conservation Plan for Wilson's Phalarope (*Phalaropus tricolor*)*. Version 1.1. Manomet, MA: Manomet Center for Conservation Sciences.

Leston, Lionel, and Theodore A. Bookhout. 2020. "Yellow Rail (*Coturnicops noveboracensis*)." Version 1.0. In Poole 2020. doi:10.2173/bow.yelrai.01.

Lever, Christopher. 2005. *Naturalised Birds of the World*. London: T. & A. D. Poyser.

Lewis, Hubert. 1952. "Thistle-Nesting Goldfinches." *Flicker* 24: 105–10.

Lind, James W., and JoAnn M. Hanowski. 2004. "Population Declines in a Boreal Forest Denizen, the White-throated Sparrow." *Loon* 76: 115–22.

Lindquist, Willis. 1952. "He Brought the Bluebirds Back." *Flicker* 24: 6–11.

Link, William A., John R. Sauer, and Daniel K. Niven. 2020. "Model Selection for the North American Breeding Bird Survey." *Ecological Applications* 30: e02137. doi:10.1002/eap.2137.

Linz, George M., H. Jeffrey Homan, Shannon M. Gaulker, Linda B. Penry, and William J. Bleier. 2007. "European Starlings: A Review of an Invasive Species with Far-Reaching Impacts." In *Managing Vertebrate Invasive Species: Proceedings of an International Symposium*, edited by Gary W. Witmer, Will C. Pitt, and Kathleen A. Fagerstone, 378–86. USDA/APHIS/WS, Fort Collins, CO: National Wildlife Research Center.

Little, Jon, and Jacob Langeslag. 1996. "White-winged Crossbill Nest in Rice County." *Loon* 68: 86–89.

Longcore, Jerry R., Daniel G. McAuley, Gary R. Hepp, and Judith M. Rhymer. 2020. "American Black Duck (*Anas rubripes*)." Version 1.0. In Poole and Gill 2020. doi:10.2173/bow.ambduc.01.

Longcore, Travis, Catherine Rich, Pierre Mineau, Beau MacDonald, Daniel G. Bert, Lauren M. Sullivan, Erin Mutrie, et al. 2013. "Avian Mortality at Communication Towers in the United States and Canada: Which Species, How Many, and Where?" *Biological Conservation* 158: 410–19.

Longley, William H. 1944. "Minnesota Nesting Data 1944." *Flicker* 16: 71–76.

———. 1950. "Snowy Egret in Faribault County, Minnesota." *Flicker* 22: 130.

———. 1955. "Meadowlark Habitat." *Loon* 27: 133.

———. 1956a. "A Glossy Ibis at Weaver, Minnesota." *Flicker* 28: 167.

———. 1956b. "A Second Little Blue Heron in Minnesota." *Flicker* 28: 165.

———. 1970. "Sandhill Cranes at the Carlos Avery Wildlife Area." *Loon* 42: 124–28.

———. [Langley, misspelled in source]. 1973. "The Louisiana Waterthrush Still Nests in Chisago County." *Loon* 45: 95–96.

———. 1981. "Car-Killed Birds." *Loon* 53: 169–70.

Longley, William H., and Byron E. Harrell. 1946. "Minnesota Nesting Records, 1946." *Flicker* 18: 79–84.

Loose, Steven S., and Stanley H. Anderson. 1995. "Woodpecker Habitat Use in the Forests of Southeast Wyoming." *Journal of Field Ornithology* 66: 503–14.

Lord, John, and D. J. Munns, eds. 1970. *Atlas of Breeding Birds of the West Midlands*. Illustrated by T. K. Beck. London: Collins.

Loss, Scott R., Gerald J. Niemi, and Robert B. Blair. 2012. "Invasions of Non-Native Earthworms Relative to Population Declines of Ground-Nesting Songbirds across a Regional Extent in Northern Hardwood Forests of North America." *Landscape Ecology* 27: 683–96.

Loss, Scott R., Tom Will, Sara S. Loss, and Peter P. Marra. 2014. "Bird-Building Collisions in the United States: Estimates of Annual Mortality and Species Vulnerability." *Condor* 116: 8–23.

Lovell, Harvey B. 1951. "Blue Grosbeak in the Badlands of South Dakota." *Condor* 53: 257.

Lowther, Peter E. 2020a. "Alder Flycatcher (*Empidonax alnorum*)." Version 1.0. In Poole and Gill 2020. doi:10.2173/bow.aldfly.01.

———. 2020b. "Brown-headed Cowbird (*Molothrus ater*)." Version 1.0. In Poole and Gill 2020. doi:10.2173/bow.bnhcow.01.

———. 2020c. "LeConte's Sparrow (*Ammospiza leconteii*)." Version 1.0. In Rodewald 2020. doi:10.2173/bow.lecspa.01.

Lowther, Peter E., Claudio Celada, Nedra K. Klein, Christopher C. Rimmer, and David A. Spector. 2020. "Yellow Warbler (*Setophaga petechia*)." Version 1.0. In Poole and Gill 2020. doi:10.2173/bow.yelwar.01.

Lowther, Peter E., and Calvin L. Cink. 2020. "House Sparrow (*Passer domesticus*)." Version 1.0. In Billerman 2020. doi:10.2173/bow.houspa.01.

Lowther, Peter E., Hector D. Douglas III, and Cheri L. Gratto-Trevor. 2020. "Willet (*Tringa semipalmata*)." Version 1.0. In Poole and Gill 2020. doi:10.2173/bow.willet1.01.

Lowther, Peter E., and James L. Ingold. 2020. "Blue Grosbeak (*Passerina caerulea*)." Version 1.0. In Poole 2020. doi:10.2173/bow.blugrb1.01.

Lowther, Peter E., and Richard F. Johnston. 2020. "Rock Pigeon (*Columba livia*)." Version 1.0. In Billerman 2020. doi:10.2173/bow.rocpig.01.

Lowther, Peter E., Donald E. Kroodsma, and Greg H. Farley. 2020. "Rock Wren (*Salpinctes obsoletus*)." Version 1.0. In Poole and Gill 2020. doi:10.2173/bow.rocwre.01.

Lowther, Peter E., Alan F. Poole, James P. Gibbs, Scott M. Melvin, and Frederic A. Reid. 2020. "American Bittern (*Botaurus lentiginosus*)." Version 1.0. In Poole 2020. doi:10.2173/bow.amebit.01.

Lowther, Peter E., and Janet McI. Williams. 2020. "Nashville Warbler (*Leiothlypis ruficapilla*)." Version 1.0. In Poole 2020. doi:10.2173/bow.naswar.01.

Lunk, William Allan. 1962. "The Rough-winged Swallow, *Stelgidopteryx ruficollis (Vieillot)*: A Study Based on Its Breeding Biology in Michigan." *Bulletin of the Nuttall Ornithological Club* 4: 1–155.

Lupient, Mary. 1945. "Minnesota Nesting Records, 1945." *Flicker* 17: 82–88.

———. 1946. "Tufted Titmice in Minnesota." *Loon* 18: 42.

———. 1951. "Seasonal Report." *Flicker* 23: 50–53.

———. 1952. "Seasonal Report." *Flicker* 24: 119–24.

———. 1953. "Seasonal Report." *Flicker* 25: 147–49.

———. 1959. "Seasonal Report." *Flicker* 31: 73–76.

Lymn, Nadine, and Stanley A. Temple. 1991. "Land-use Changes in the Gulf Coast Region: Links to Declines in Midwestern Loggerhead Shrike Populations." *Passenger Pigeon* 53: 315–25.

Lyons, Tim. 2020a. "Minnesota's Wild Turkey Harvest—Fall 2019, Spring 2020." Minnesota Department of Natural Resources.

———. 2020b. "2020 Minnesota August Roadside Survey." Minnesota Department of Natural Resources.

———. 2021. "2021 Minnesota August Roadside Survey." Farmland Wildlife Populations and Research Group. Minnesota Department of Natural Resources. https://files.dnr.state.mn.us/recreation/hunting/pheasant/roadside_survey.pdf?20220624-09.

MacArthur, Robert H. 1958. "Population Ecology of Some Warblers of Northeastern Coniferous Forests." *Ecology* 39: 599–619.

Macdonald, Helen. 2014. *H Is for Hawk*. New York: Grove Press.

Mack, Diane Evans, and Wang Yong. 2020. "Swainson's Thrush (*Catharus ustulatus*)." Version 1.0. In Poole and Gill 2020. doi:10.2173/bow.swathr.01.

MacPherson, James. 2020. "Pelican Nests at North Dakota Refuge Down Due to High Water." *US News*, August 28. https://www.usnews.com/news/us/articles/2020-08-28/pelican-nests-at-north-dakota-refuge-down-due-to-high-water.

Madden, Elizabeth M. 1996. "Passerine Communities and Bird-Habitat Relationships on Prescribe-burned, Mixed Grass Prairie in North Dakota." Master's thesis, Montana State University.

Madden, Elizabeth M., Andrew J. Hansen, and Robert K. Murphy. 1999. "Influence of Prescribed Fire History on Habitat and Abundance of Passerine Birds in Northern Mixed-grass Prairie." *Canadian Field-Naturalist* 113: 627–40.

Madden, Elizabeth M., Robert K. Murphy, Andrew J. Hansen, and Leigh Murray. 2000. "Models for Guiding Management of Prairie Bird Habitat in Northwestern North Dakota." *American Midland Naturalist* 144: 377–92.

Magnus, Lester T. 1955. "Sandhill Cranes on the Roseau River Refuge and Public Hunting Grounds, Roseau, Minnesota." *Loon* 27: 93.

Mallory, Mark L., Donald K. McNicol, and Patrick J. Weatherhead. 1994. "Habitat Quality and Reproductive Effort of Common Goldeneyes Nesting near Sudbury, Canada." *Journal of Wildlife Management* 58: 552–60.

Manolis, James C., David E. Andersen, and Francesca J. Cuthbert. 2000. "Uncertain Nest Fates in Songbird Studies and Variation in Mayfield Estimation." *Auk* 117: 615–26.

———. 2002. "Edge Effect on Nesting Success of Ground Nesting Birds near Regenerating Clearcuts in a Forest-Dominated Landscape." *Auk* 119: 955–70.

Marengo, William. 2000. "American Avocets Possibly Breeding in Traverse County." *Loon* 72: 239–40.

Marks, Jeffrey S., Dave L. Evans, and Denver W. Holt. 2020. "Long-eared Owl (*Asio otus*)." Version 1.0. In Billerman 2020. doi:10.2173/bow.loeowl.01.

Marshall, William H. 1982. "Minnesota Woodcock." *Loon* 54: 203–11.

Martell, Mark S., and Ted Dick. 1996. "Nesting Habitat Characteristics of the Northern Goshawk (*Accipter gentilis*) in Minnesota." Final report to the Minnesota Department of Natural Resources, Nongame Wildlife Program.

Martell, Mark S., Judy Voigt Englund, and Harrison B. Tordoff. 2002. "An Urban Osprey Population Established by Translocation." *Journal of Raptor Research* 36: 91–96.

Martell, Mark S., Charles J. Henny, Peter E. Nye, and Matthew J. Solensky. 2001. "Fall Migration Routes, Timing, and Wintering Sites of North American Ospreys as Determined by Satellite Telemetry." *Condor* 103: 715–24.

Martell, Mark S., John Schladweiler, and Francesca J. Cuthbert. 2001. "Status and Attempted Reintroduction of Burrowing Owls in Minnesota, U.S.A." *Journal of Raptor Research* 35: 331–36.

Martell, Mark S., John Schladweiler, and Jill Nibe. 1990. "Minnesota Burrowing Owl Conservation, 1990 Report." The Raptor Center, University of Minnesota and Minnesota Department of Natural Resources, Nongame Program. https://files.dnr.state.mn.us/eco/nongame/projects/consgrant_reports/1990/1990_martell_etal.pdf.

Martell, Mark S., Serena Willey, and John Schladweiler. 1998. "Nesting and Migration of Swainson's Hawks in Minnesota." *Loon* 70: 72–81.
Marti, Carl D., Alan F. Poole, Louis R. Bevier, David Christie, Guy M. Kirwan, and Jeffrey S. Marks. 2020. "Barn Owl (*Tyto alba*)." Version 1.0. Billerman 2020. doi:10.2173/bow.brnowl.01.
Martin, Elwood M. 1979. *Hunting and Harvest Trends for Migratory Game Birds Other Than Waterfowl: 1964–76*. Special Scientific Report—Wildlife 218. Washington, DC: US Fish and Wildlife Service.
Martin, John W., and Jimmie R. Parrish. 2020. "Lark Sparrow (*Chondestes grammacus*)." Version 1.0. In Poole and Gill 2020. doi:10.2173/bow.larspa.01.
Martin, Kristen A. 2012. "Habitat Suitability of the Yellow Rail in South-Central Manitoba." Master's thesis, University of Manitoba, Winnipeg.
Martin, Stephen G. 2020. "Brewer's Blackbird (*Euphagus cyanocephalus*)." Version 1.0. In Poole and Gill 2020. doi:10.2173/bow.brebla.01.
Martin, Thomas E. 1981. "Limitation in Small Habitat Islands: Chance or Competition?" *Auk* 98: 715–34.
Marzilli, V. 1989. "Up on the Roof." *Maine Fish and Wildlife* 31: 25–29.
Mason, Phillip P., ed. 1958. *Schoolcraft's Expedition to Lake Itasca: The Discovery of the Source of the Mississippi*. East Lansing: Michigan State University Press.
Mathews, F. Schuyler. 1921. *Fieldbook of Wild Birds and Their Music*. Rev. ed. New York: G. P. Putnam and Sons.
Mathisen, John E. 1988. *Report on Results of Monitoring of Common Loons Done in 1988 on the Chippewa National Forest*. Cass Lake, MN: Chippewa National Forest.
Mathisen, John E., and Ann Richards. 1978. "Status of Great Blue Herons on the Chippewa National Forest." *Loon* 50: 104–6.
Mathisen, John E., David J. Sorenson, L. D. Frenzel, and Thomas C. Dunstan. 1977. "A Management Strategy for Bald Eagles." *Transactions of the North American Wildlife and Natural Resources Conference* 42: 86–92.
Matsuoka, Steven M., C. Lisa Mahon, Colleen M. Handel, Péter Sólymos, Erin M. Bayne, Patricia C. Fontaine, and C. John Ralph. 2014. "Reviving Common Standards in Point-Count Surveys for Broad Inference across Studies." *Condor* 116: 599–608.
Matteson, Sumner W., Timothy A. Andryk, and John Wetzel. 1988. "Wisconsin Trumpeter Swan Recovery Plan." *Passenger Pigeon* 50: 118–30.
Matteson, Sumner W., Kimberly Kreitinger, G. Bartelt, Gregory Butcher, D. Sample, and Tom Will. 2009. *Partners in Flight Bird Conservation Plan for the Boreal Hardwood Transition, Bird Conservation Region 12—U.S. Portion*. Version 1.0. Partners in Flight. www.partnersinflight.org.
Matteson, Sumner W., Michael J. Mossman, and David A. Shealer. 2012. "Population Decline of Black Terns in Wisconsin: A 30-Year Perspective." *Waterbirds* 35: 185–93.
Matthiae, Thomas M. 1982. "A Nesting Boreal Owl in Minnesota." *Loon* 54: 212–14.
Mattsson, Brady J., Terry L. Master, Robert S. Mulvihill, and W. Douglas Robinson. 2020. "Louisiana Waterthrush (*Parkesia motacilla*)." Version 1.0. In Poole 2020. doi:10.2173/bow.louwat.01.
Mattsson, James P. 1986. "Snowy Egret Nest at Agassiz National Wildlife Refuge." *Loon* 58: 134–35.
———. 1987. "A Snowy Plover at Agassiz, NWR, Marshall County." *Loon* 59: 155.
———. 1988. *Annotated Historical Records of Bald Eagles from the Northern United States*. Biological Report No. 88(10). Washington, DC: US Fish and Wildlife Service.
———. 2001. "First Successful Minnesota Breeding Record of the Yellow-throated Warbler." *Loon* 73: 236–40.
———. 2003a. "Northern Mockingbirds Nest in Dakota County." *Loon* 75: 116.
———. 2003b. "Summer Tanager and Scarlet Tanager: Interspecific Pairing in Minnesota." *Loon* 75: 226–30.
———. 2007. "A Nesting Summer Tanager X Scarlet Tanager in Minnesota." *Loon* 79: 7–10.
Maxson, Stephen J. 1976. "1974 Report of a Swallow-tailed Kite." *Loon* 48: 182–83.
———. 1989. "Patterns of Activity and Home Range of Hens." In *Ruffed Grouse,* edited by Sally Atwater and Judith Schnell, 118–29. Harrisburg, PA: Stackpole Books.
Maxson, Stephen J., John R. Fieberg, and Michael R. Riggs. 2007. "Black Tern Nest Habitat Selection and Factors Affecting Nest Success in Northwestern Minnesota." *Waterbirds* 30: 1–9.
———. 2008. "Sandhill Cranc Ncst Habitat Sclcction and Factors Affecting Nest Success in Northwestern Minnesota." In *Proceedings of the Tenth North American Crane Workshop, Feb. 7–10, 2006, Zacatecas City, Zacatecas, Mexico*, edited by Martin J. Folk and Steve A. Nesbitt, 90–97. North American Crane Working Group.
Maxson, Stephen J., Steve A. Mortensen, Donald L. Goodermote, and Christopher S. Lapp. 1996. "Success and Failure of Ring-billed Gull Deterrents at Common Tern and Piping Plover Colonies in Minnesota." *Colonial Waterbirds* 19: 242–47.
Maxson, Stephen J., Jodie L. Provost, and George H. Davis. 1990. "Fall Migration of Sandhill Cranes in Northwestern Minnesota, 1988–89." *Loon* 62: 14–19.
Maxson, Stephen J., and Michael R. Riggs. 1996. "Habitat Use and Nest Success of Overwater Nesting Ducks in Westcentral Minnesota." *Journal of Wildlife Management* 60: 108–19.
Mayasich, Joseph, and Gerald J. Niemi. 2003. *Conservation Assessment for Bay-breasted Warbler (*Dendroica castanea*).* Report to US Department of Agriculture, Forest Service, Eastern Region. Natural Resources Research Institute Technical Report NRRI/TR-2003/17.
Mayr, Ernst. 1982. *The Growth of Biological Thought: Diversity, Evolution, and Inheritance*. Cambridge: Belknap Press of Harvard University Press.
Mazur, Kurt M., and Paul C. James. 2020. "Barred Owl (*Strix varia*)." Version 1.0. In Poole and Gill 2020. doi:10.2173/bow.brdowl.01.
McAuley, Daniel G., Daniel M. Keppie, and R. Montague Whiting Jr. 2020. "American Woodcock (*Scolopax minor*)." Version 1.0. In Poole 2020. doi:10.2173/bow.amewoo.01.
McBride, Abby. 2016. "Tracing Feathers: Where Did That Bird Come From? Ask Its Feather." *Living Bird* 35: 14–16.
McCrimmon, Donald A., Jr., John C. Ogden, G. Thomas Bancroft, Albert Martínez-Vilalta, Anna Motis, Guy M. Kirwan, and Peter F. D. Boesman. 2020. "Great Egret (*Ardea alba*)." Version 1.0. In Billerman 2020. doi:10.2173/bow.greegr.01.
McCullagh, P., and John A. Nelder. 1989. *Generalized Linear Models*. 2nd ed. New York: CRC Press.
McDonald, Mary Victoria. 2020. "Kentucky Warbler (*Geothlypis formosa*)." Version 1.0. In Poole 2020. doi:10.2173/bow.kenwar.01.
McDowell, Michelle. 2004. *Mille Lacs National Wildlife Refuge 2004 Report*. Annual Report from Ricc Lakc National Wildlife Refuge. US Fish and Wildlife Service.
———. 2011. *Mille Lacs National Wildlife Refuge 2011 Report*. Annual Report from Rice Lake National Wildlife Refuge. US Fish and Wildlife Service.
McGarigal, Kevin, and William C. McComb. 1995. "Relationships between Landscape Structure

and Breeding Birds in the Oregon Coast Range." *Ecological Monographs* 65: 235–60.
McGowan, Kevin J., and Kimberley Corwin, eds. 2008. *The Second Atlas of Breeding Birds in New York State*. Ithaca: Comstock Publishing Associates.
McGraw, Kevin J., and Alex L. Middleton. 2020. "American Goldfinch (*Spinus tristis*)." Version 1.0. In Rodewald 2020. doi:10.2173/bow.amegfi.01.
McIntosh, Jean. 1964. "Blue Grosbeak Observation." *Loon* 36: 103–4.
McIntosh, Michael. 1997. *Traveler's Tales: The Wanderings of a Bird Hunter and Sometime Fly Fisherman*. Camden, ME: Down East Books.
McIntyre, Judy W. 1975. "Biology and Behavior of the Common Loon (*Gavia immer*) with Reference to Its Adaptability in a Man-Altered Environment." PhD diss., University of Minnesota.
———. 1978. "The Common Loon: Part III: Population in Itasca State Park Minnesota 1957–1976." *Loon* 50: 38–44.
———. 1986. "Common Loon." In *Audubon Wildlife Report 1986*, edited by Roger L. di Silvestro, 679–95. New York: National Audubon Society.
———. 1988. "The Minnesota Report: A 15-Year Survey Comparison." In *Papers from the 1987 Conference on Loon Research and Management*, edited by Paul I. V. Strong, 118–30. Meredith, NH: North American Loon Fund.
McKay, Bailey, and George A. Hall. 2020. "Yellow-throated Warbler (*Setophaga dominica*)." Version 1.0. In Poole 2020. doi:10.2173/bow.yetwar.01.
McKearnan, Joan Elizabeth. 1986. "Status and Breeding Success of Common Terns in Minnesota." Master's thesis, University of Minnesota.
McKenzie, Ann. 1978. "Kentucky Warbler at Itasca Park." *Loon* 50: 169–70.
McLeish, Todd. 2014. "Green Herons: Birds That Bait." *Northland Woods: The Outside Story*, June 23. http://northernwoodlands.org/outside_story/article/green-herons.
McLeod, Mary Anne, and David E. Andersen. 1998. "Red-shouldered Hawk Broadcast Surveys: Factors Affecting Detection of Responses and Population Trends." *Journal of Wildlife Management* 62: 1385–97.
McLeod, Mary Anne, B. Ann Belleman, David E. Andersen, and Gary W. Oehlert. 2000. "Red-shouldered Hawk Nest Site Selection in North-Central Minnesota." *Wilson Bulletin* 112: 203–13.
McNair, Douglas B. 1983. "The Significance of Breeding Season Records of Sedge Wrens in the Southeast States." *Oriole* 48: 49–52.
McNicholl, Martin K., Peter E. Lowther, and John A. Hall. 2020. "Forster's Tern (*Sterna forsteri*)." Version 1.0. In Poole and Gill 2020. doi:10.2173/bow.forter.01.
McRoberts, Jon T., Mark C. Wallace, and Stephen W. Eaton. 2020. "Wild Turkey (*Meleagris gallopavo*)." Version 1.0. In Poole 2020. doi:10.2173/bow.wiltur.01.
Mearns, Barbara, and Richard Mearns. 1992. *Audubon to Xantus: The Lives of Those Commemorated in North American Bird Names*. San Diego: Academic Press.
Meehan, T. D., G. S. LeBaron, K. Dale, A. Krump, N. L. Michel, and C. B. Wilsey. 2020. "Abundance Trends of Birds Wintering in the USA and Canada, from Audubon Christmas Bird Counts, 1966–2019." Version 3.0. New York: National Audubon Society.
———. 2022. "Trends in Relative Abundance for Birds Wintering in the Continental USA and Canada: Audubon Christmas Bird Counts, 1966–2021." Version 4.0. New York: National Audubon Society.
Melcher, Cynthia P., Adrian Farmer, and Guillermo Fernández. 2010. *Conservation Plan for the Marbled Godwit (*Limosa fedoa*)*. Version 1.2. Manomet, MA: Manomet Center for Conservation Sciences.
Melvin, Scott M., and James P. Gibbs. 2020. "Sora (*Porzana carolina*)." Version 1.0. In Poole 2020. doi:10.2173/bow.sora.01.
Melvin, Stefani. 2006. "Willet." *The Texas Breeding Bird Atlas*. College Station and Corpus Christi: Texas A&M University System. Retrieved from the Texas Breeding Bird Atlas: http://txtbba.tamu.edu/species-accounts/willet/.
Menebroeker, Ryan C., Paula Spaeth Anich, Derek H. Ogle, and Nicholas M. Anich. 2016. "Recent Declines in Gray Jays on Christmas Bird Counts in Northern Wisconsin." *Passenger Pigeon* 78: 161–69.
Mengelkoch, Jean M., Gerald J. Niemi, and Ronald R. Regal. 2004. "Diet of the Nestling Tree Swallow." *Condor* 106: 423–29.
Merow, Cory, and John A. Silander Jr. 2014. "A Comparison of Maxlike and Maxent for Modelling Species Distributions." *Methods in Ecology and Evolution* 5: 215–25.
Merow, Cory, Matthew J. Smith, and John A. Silander Jr. 2013. "A Practical Guide to MaxEnt for Modeling Species' Distributions: What It Does, and Why Inputs and Settings Matter." *Ecography* 36: 1058–69.
Merrill, Samuel B. 1994. "The Birds of Camp Ripley." *Loon* 66: 117–26.
———. 1996. "Hooded Warbler Nest in Morrison County." *Loon* 68: 245–46.
Messinger, Lindsey, and Nicole M. Davros. 2018. *2018 Minnesota August Roadside Survey*. Madelia, MN: Minnesota Department of Natural Resources.
Mettler, Bonnie J. 1977. "Factors Contributing to the Increase of the Gray Partridge in Minnesota." *Loon* 49: 205–10.
Meyer, Kenneth D. 2020. "Swallow-tailed Kite (*Elanoides forficatus*)." Version 1.0. In Poole and Gill 2020. doi:10.2173/bow.swtkit.01.
Meyer, Michael W. 2006. "Evaluating the Impact of Multiple Stressors on Common Loon Population Demographics: An Integrated Laboratory and Field Approach." Final report for EPA Grant No. R829085. Wisconsin Department of Natural Resources. https://cfpub.epa.gov/ncer_abstracts/index.cfm/fuseaction/display.highlight/abstract/1916/report/F.
Meyers, Chet. 2010. 2010 Final Report. "Red-headed Woodpecker Recovery Project." http://www.redheadrecovery.org.
Michigan DNR (Department of Natural Resources). 2012. *Mute Swans—Invading Michigan's Waters*. https://www.trumpeterswansociety.org/file_download/inline/4dbdb442-ab6a-4560-be4d-005d76321137.
Middleton, Alex L. 2020. "Chipping Sparrow (*Spizella passerina*)." Version 1.0. In Poole and Gill 2020. doi:10.2173/bow.chispa.01.
Midwest Peregrine Society. 2017. *Midwest Peregrine Society* [website]. http://midwestperegrine.umn.edu/.
Millais, John G. 1913. *British Diving Ducks*. 2 vols. London: Longman's Green & Co.
Millard, Steve. 1981. "Bobwhite in Rock County." *Loon* 53: 167–68.
———. 1994. "Hybrid Bluebird in Todd County." *Loon* 66: 149–50.
———. 1995. "Henslow's Sparrow in Wilkin County." *Loon* 67: 186.
Miller, E., and T. Dring. 1988. "Territorial Defense of Multiple Lakes by Common Loons: A Preliminary Report." In *Papers from the 1987 Conference on Loon Research and Management*, edited by Paul I. V. Strong, 1–14. Meredith, NH: North American Loon Fund.
Miller, Jacob H. 1987. "A Study of Aggressive Interactions in a Mixed Larid Breeding Colony in Minnesota." PhD diss., University of Minnesota.

Miller, Karl E., and Wesley E. Lanyon. 2020. "Great Crested Flycatcher (*Myiarchus crinitus*)." Version 1.0. In Poole 2020. doi:10.2173/bow.grcfly.01.

Miller, Karl E., David L. Leonard Jr., Clifford E. Shackelford, Raymond E. Brown, and Richard N. Conner. 2020. "Red-bellied Woodpecker (*Melanerpes carolinus*)." Version 1.0. In Rodewald 2020. doi:10.2173/bow.rebwoo.01.

Miller, Mary, and Lee A. Pfannmuller. 1991. "Bald Eagle Population in Minnesota." *Loon* 63: 268–71.

Miller, Michael R., David Duncan, Karla Guyn, P. Flint, and Jane Austin, eds. 2001. *The Northern Pintail in North America: The Problem and a Prescription for Recovery, Part 1. Proceedings of the Northern Pintail Workshop, 23–25 March 2001.* Sacramento, CA.

Mini, Anne E., Erin R. Harrington, Emily Rucker, Bruce D. Dugger, and Thomas B. Mowbray. 2020. "American Wigeon (*Anas americana*)." Version 1.0. In Poole 2020. doi:10.2173/bow.amewig.01.

Minnesota Department of Health. 2004. "West Nile Virus in Minnesota: 2003 Update." *Disease Control Newsletter* 32 (3): 21–24. http://www.health.state.mn.us/divs/idepc/newsletters/dcn/2004/0403dcn.pdf.

Minnesota Forest Resources Council. 2005. *Sustaining Minnesota Forest Resources: Voluntary Site-Level Forest Management Guidelines for Landowners, Loggers and Resource Managers.* St. Paul: Minnesota Forest Resources Council.

Minnesota Interagency Climate Adaptation Team. 2013. *Adapting to Climate Change in Minnesota: 2013 Report of the Interagency Climate Adaptation Team.* https://glslcities.org/wp-content/uploads/2015/09/Adapting_to_Climate_Change_in_Minnesota_-_2013_Report_of_the_Interagency_Climate_Adaptation_Team_2013.pdf.

Minnesota Prairie Plan Working Group. 2011. *Minnesota Prairie Conservation Plan.* Minneapolis: Minnesota Prairie Plan Working Group.

Mitchell, Carl D., and Michael W. Eichholz. 2020. "Trumpeter Swan (*Cygnus buccinator*)." Version 1.0. In Rodewald 2020. doi:10.2173/bow.truswa.01.

Mitchell, Christine A., Thomas W. Custer, and Phillip J. Zwank. 1994. "Herbivory on Shoalgrass by Wintering Redheads in Texas." *Journal of Wildlife Management* 58: 131–41.

Mitchell, Earl T. 1941. "Nesting of Forster's Tern in Hennepin County." *Flicker* 13: 19.

Mitchell, Walton. 1895. "Breeding of the Blue-gray Gnatcatcher in Minnesota." *Auk* 12: 185.

Mladenoff, David J., Gerald J. Niemi, and Mark A. White. 1997. "Effects of Changing Landscape Pattern and USGS Land Cover Data Variability on Ecoregion Discrimination across a Forest-Agriculture Gradient." *Landscape Ecology* 12: 379–96.

Mlodinow, Steven G., and Michael O'Brien. 1996. *America's 100 Most Wanted Birds: Finding the Rarest Regularly Occurring Birds in the Lower 48 States.* Helena and Billings, MT: Falcon Press Publishing Co.

MNDNR (Minnesota Department of Natural Resources). 1990. *Sharp-tailed Grouse Management Plan.* 2: 1–15. St. Paul: Minnesota Department of Natural Resources.

———. 1995. *Statement of Need and Reasonableness: In the Matter of Proposed Amendment of Minnesota Rules, Chapter 6134: Endangered and Threatened Species.* State of Minnesota, Department of Natural Resources, Division of Fish and Wildlife.

———. 1996. "Landowners Guide for Maintaining and Encouraging Loggerhead Shrikes." Nongame Wildlife Program. https://files.dnr.state.mn.us/eco/rsg/shrikeflyer.pdf.

———. 2000. "Trumpeter Swans Make Transition to Wild." *Minnesota Conservation Volunteer* 63 (September/October): 62–63.

———. 2003a. *Field Guide to the Native Plant Communities of Minnesota: The Laurentian Mixed Forest Province.* St. Paul: Minnesota Department of Natural Resources, Ecological Land Classification Program, Minnesota County Biological Survey, and Natural Heritage and Nongame Research Program.

———. 2003b. *Northern Goshawk Management Considerations.* https://files.dnr.state.mn.us/natural_resources/ets/goshawk_mgmt%20guideline.pdf.

———. 2005a. *Field Guide to the Native Plant Communities of Minnesota: The Eastern Broadleaf Forest Province.* St. Paul: Minnesota Department of Natural Resources, Ecological Land Classification Program, Minnesota County Biological Survey, and Natural Heritage and Nongame Research Program.

———. 2005b. *Field Guide to the Native Plant Communities of Minnesota: The Prairie Parkland and Tallgrass Aspen Parklands Provinces.* St. Paul: Minnesota Department of Natural Resources, Ecological Land Classification Program, Minnesota County Biological Survey, and Natural Heritage and Nongame Research Program.

———. 2006a. *Long Range Duck Recovery Plan.* https://files.dnr.state.mn.us/recreation/hunting/waterfowl/duckplan_042106.pdf.

———. 2006b. *Long Range Plan for the Wild Turkey in Minnesota.* https://files.dnr.state.mn.us/recreation/hunting/turkey/long_range_turkey_plan_2007.pdf.

———. 2006c. *Tomorrow's Habitat for the Wild and Rare: An Action Plan for Minnesota Wildlife, Comprehensive Wildlife Conservation Strategy.* St. Paul: Minnesota Department of Natural Resources, Division of Ecological Services.

———. 2006d. *2005 Minnesota Bald Eagle Surveys.* https://files.dnr.state.mn.us/eco/nongame/projects/eagle_report_2005.pdf.

———. 2011. "Minnesota Trumpeter Swan Population Tops 5,500." News release, January 20.

———. 2012. *Statement of Need and Reasonableness: In the Matter of Proposed Amendment to and Repeal of Rules Governing Minnesota's List of Endangered, Threatened, and Special Concern Species in Minnesota Rules, Chapter 6134: Endangered and Threatened Species.* State of Minnesota, Department of Natural Resources, Division of Ecological and Water Resources.

———. 2013. *Minnesota's List of Endangered, Threatened, and Special Concern Species.* St. Paul: Minnesota Department of Natural Resources, Division of Ecological and Water Resources.

———. 2015a. *Feasibility Study for the Restoration of Wild Northern Bobwhite in Minnesota.* https://files.dnr.state.mn.us/aboutdnr/reports/legislative/2015-bobwhite-feasability-draft.pdf.

———. 2015b. *Minnesota's Wildlife Action Plan, 2015–2025.* St. Paul: Minnesota Department of Natural Resources, Division of Ecological and Water Resources. http://www.dnr.state.mn.us/mnwap/index.html.

———. 2015c. *Minnesota Waterfowl Hunting Regulations Handbook.* https://www.leg.mn.gov/docs/2015/other/150906.pdf.

———. 2015d. *2015 Minnesota Hunting and Trapping Regulations Handbook.* St. Paul: Minnesota Department of Natural Resources.

———. 2016. "Trumpeter Swans Continue to Be a Conservation Success Story." Minnesota Department of Natural Resources news release. February 11.

———. 2019a. "American Woodcock—*Scolopax minor*." http://www.dnr.state.mn.us/forestlegacy/koochiching_washington/americanwoodcock.html.

———. 2019b. "Bald Eagle (*Haliaeetus leucocephalus*)." *Rare Species Guide*. http://www.dnr.state.mn.us/rsg/profile.html?action=elementDetail&selectedElement=ABNKC1001.

———. 2019c. "Cerulean Warbler (*Dendroica cerulean*)." *Rare Species Guide*. http://www.dnr.state.mn.us/rsg/profile.html?action=elementDetail&selectedElement=ABPBX03240.

———. 2019d. "Chestnut-collared Longspur (*Calcarius ornatus*)." *Rare Species Guide*. http://www.dnr.state.mn.us/rsg/profile.html?action=elementDetail&selectedElement=ABPBXA6040.

———. 2019e. "Double-crested Cormorant." http://www.dnr.state.mn.us/birds/doublecrestedcormorant.html.

———. 2019f. "Ecological Classification System." https://www.dnr.state.mn.us/ecs.

———. 2019g. "Grouse Hunting." http://dnr.state.mn.us/hunting/grouse/index.html.

———. 2019h. "Red-shouldered Hawk (*Buteo lineatus*)." *Rare Species Guide*. http://www.dnr.state.mn.us/rsg/profile.html?action=elementDetail&selectedElement=ABNKC1903.

———. 2019i. "Ring-necked Pheasant (*Phasianus colchicus*)." http://www.dnr.state.mn.us/birds/ringneckedpheasant.html.

———. 2019j. "Ruffed Grouse (*Bonasa umbellus*)." http://www.dnr.state.mn.us/birds/ruffedgrouse.html.

———. 2019k. "Sharp-tailed Grouse (*Tympanuchus phasianellus*)." http://www.dnr.state.mn.us/birds/sharptailedgrouse.html.

———. 2019l. "Trumpeter Swan (*Cygnus buccinator*)." *Rare Species Guide*. http://www.dnr.state.mn.us/rsg/profile.html?action=elementDetail&selectedElement=ABNJB02030.

———. 2019m. "Wilson's Phalarope (*Phalaropus tricolor*)." *Rare Species Guide*. http://www.dnr.state.mn.us/rsg/profile.html?action=elementDetail&selectedElement=ABNNF20010.

———. 2019n. "Wood Duck (*Aix sponsa*)." http://www.dnr.state.mn.us/birds/woodduck.html.

———. 2019o. "Yellow Rail (*Coturnicops noveboracensis*)." *Rare Species Guide*. http://www.dnr.state.mn.us/rsg/profile.html?action=elementDetail&selectedElement=ABNME01010.

———. 2020a. "About the Bird Maps." https://www.dnr.state.mn.us/eco/mcbs/about_bird_maps.html.

———. 2020b. "Invasive Species in Minnesota." http://www.dnr.state.mn.us/invasives.

———. 2020c. "Minnesota Biological Survey." https://www.dnr.state.mn.us/mbs/index.html.

———. 2021. "Pheasant Management." https://www.dnr.state.mn.us/wildlife/pheasant/index.html.

Mohr, Elizabeth. 2011. "As the Purple Martin Population Dwindles in Minnesota, a Team Works to Attract and Study the Birds." *Pioneer Press*, July 27. http://www.twincities.com/2011/07/27/as-the-purple-martin-population-dwindles-in-minnesota-a-team-works-to-attract-and-study-the-birds/.

Moldenhauer, Ralph R., and Daniel J. Regelski. 2020. "Northern Parula (*Setophaga americana*)." Version 1.0. In Poole 2020. doi:10.2173/bow.norpar.01.

Mollhoff, Wayne J. 2016. *The Second Nebraska Breeding Bird Atlas*. Bulletin of the University State Museum, no. 29. Lincoln: University of Nebraska State Museum.

Moos, Louis M. 1971. "Cattle Egret Observation." *Loon* 43: 96.

Mooty, Jack J., and Donald L. Goodermote. 1985. "Common Loon Numbers in the Knife Lake Area—Boundary Waters Canoe Area Wilderness." *Loon* 57: 12–14.

Mooty, Jack J., and Pamela Skoog Perry. 1988. *Common Loon Numbers in North-Central Minnesota*. St. Paul: Minnesota Department of Natural Resources.

Morgenweck, Ralph O. 1977. "Diurnal High Use Areas of Hatching-Year Female American Woodcock." In *Proceedings of the Sixth Woodcock Symposium, Held at Fredericton, New Brunswick, October 4, 5, and 6*, edited by Daniel M. Keppie and Ray B. Owen Jr., 155–60. Fredericton: New Brunswick Department of Natural Resources.

Morris, Ralph D., Cynthia Pekarik, and David J. Moore. 2012. "Current Status and Abundance Trends of Common Terns Breeding at Known Coastal and Inland Nesting Regions in Canada." *Waterbirds* 35: 194–207.

Morris, Ralph D., D. Vaughn Weseloh, Linda R. Wires, Cynthia Pekarik, Francesca J. Cuthbert, and David J. Moore. 2011. "Population Trends of Ring-billed Gulls Breeding on the North American Great Lakes, 1976 to 2009." *Waterbirds* 34: 202–12.

Morrison, R. I. Guy, Robert E. Gill Jr., Brian A. Harrington, Susan Skagen, Gary W. Page, Cheri L. Gratto-Trevor, and Susan M. Haig. 2001. *Estimates of Shorebird Populations in North America*. Canadian Wildlife Service Occasional Papers, no. 104. Ottawa: Canadian Wildlife Service.

Morrison, R. I. Guy, Brian J. McCaffery, Robert E. Gill, Susan K. Skagen, Stephanie L. Jones, Gary W. Page, Cheri L. Gratto-Trevor, et al. 2006. "Population Estimates of North American Shorebirds, 2006." *Wader Study Group Bulletin* 111: 67–85.

Morse, Douglass H. 1989. *American Warblers: An Ecological and Behavioral Perspective*. Cambridge: Harvard University Press.

———. 2020. "Blackburnian Warbler (*Setophaga fusca*)." Version 1.0. In Poole 2020. doi:10.2173/bow.bkbwar.01.

Morse, Douglass H., and Alan F. Poole. 2020. "Black-throated Green Warbler (*Setophaga virens*)." Version 1.0. In Rodewald 2020. doi:10.2173/bow.btnwar.01.

Morse, Marius. 1933. "The Starlings Arrive." *Flicker* 5: 16–17.

———. 1942. "Breeding Goshawks." *Flicker* 14: 30.

Mortensen, Steve. 2000. "American White Pelican Nesting on Leech Lake, Cass County." *Loon* 72: 122.

———. 2009. "Successful Nesting of Caspian Terns in Minnesota." *Loon* 81: 156–57.

Mortensen, Steve, and John Ringle. 2007. "Changes in Colonial Waterbird Populations on Leech Lake." *Loon* 79: 130–42.

Morton, Eugene S., and Ross D. James. 2020. "Blue-headed Vireo (*Vireo solitarius*)." Version 1.0. In Poole 2020. doi:10.2173/bow.buhvir.01.

Moskoff, William. 2020. "Solitary Sandpiper (*Tringa solitaria*)." Version 1.0. In Poole 2020. doi:10.2173/bow.solsan.01.

Moskoff, William, and Scott K. Robinson. 2020. "Philadelphia Vireo (*Vireo philadelphicus*)." Version 1.0. In Poole 2020. doi:10.2173/bow.phivir.01.

Mossman, Michael J. 1991. "Breeding Birds of the St. Croix River, Wisconsin and Minnesota." *Passenger Pigeon* 53: 39–77.

Mossman, Michael J., and Lisa M. Hartman. 1992. "Turkey Vulture Nest Records from Wisconsin." *Passenger Pigeon* 54: 31–41.

MOU (Minnesota Ornithologists' Union). 2020. "Checklist of the Birds of Minnesota, 2020." https://moumn.org/checklist.pdf.

———. 2021a. *Review Reported Birds: Query Sightings Database*. Accessed April 27, 2021. http://moumn.org/avian/query.php.

———. 2021b. *Seasonal Report Archive for* The Loon. Accessed April 27, 2021. http://moumn.org/loon/reports/.

MOURC (Minnesota Ornithologists' Union Records Committee). 2022. Database accessed August 16, 2022. https://moumn.org/cgi-bin/rqd.pl?sort=date&op=all.

Mowbray, Thomas B. 2020a. "Canvasback (*Aythya valisineria*)." Version 1.0. In Poole and Gill 2020. doi:10.2173/bow.canvas.01.

———. 2020b. "Scarlet Tanager (*Piranga olivacea*)." Version 1.0. In Poole and Gill 2020. doi:10.2173/bow.scatan.01.

Mowbray, Thomas B., Craig R. Ely, James S. Sedinger, and Robert E. Trost. 2020. "Canada Goose (*Branta canadensis*)." Version 1.0. In Rodewald 2020. doi:10.2173/bow.cangoo.01.

MPCA (Minnesota Pollution Control Agency). 2018. "Minnesota's Impaired Waters List." . https://gisdata.mn.gov/dataset/env-impaired-water-2018.

———. 2002. "Acid Rain in Minnesota." Air Quality Fact Sheet no. 1.11, January.

Mueller, Helmut. 2020. "Wilson's Snipe (*Gallinago delicata*)." Version 1.0. In Poole and Gill 2020. doi:10.2173/bow.wilsni1.01.

Mueller, William P. 2013a. "Belted Kingfisher (*Ceryle alcyon*)." In Kreitinger et al. 2013. http://www.wisconsinbirds.org/plan/species/beki.htm.

———. 2013b. "Red-necked Grebe (*Podiceps grisegena*)." In Kreitinger et al. 2013. http://www.wisconsinbirds.org/plan/species/rngr.htm.

Muller, Martin J., and Robert W. Storer. 2020. "Pied-billed Grebe (*Podilymbus podiceps*)." Version 1.0. In Poole and Gill 2020. doi:10.2173/bow.pibgre.01.

Mundahl, Neal D., Anthony G. Bilyeu, and Lisa Maas. 2013. "Bald Eagle Nesting Habitats in the Upper Mississippi River National Wildlife and Fish Refuge." *Journal of Fish and Wildlife Management* 4: 362–76.

Murphy, Michael T. 2001. "Source-Sink Dynamics of a Declining Eastern Kingbird Population and the Value of Sink Habitats." *Conservation Biology* 15: 737–48.

Murphy, Michael T., and Peter Pyle. 2020. "Eastern Kingbird (*Tyrannus tyrannus*)." Version 1.0. In Rodewald 2020. doi:10.2173/bow.easkin.01.

Murphy, Robert K., Kristin W. Hasselblad, Chris D. Grondahl, John G. Sidle, Ron E. Martin, and David W. Freed. 2001. "Status of the Burrowing Owl in North Dakota." *Journal of Raptor Research* 35: 322–30.

Muscarella, Robert, Peter J. Galante, Mariano Soley-Guardia, Robert A. Boria, Jamie M. Kass, María Uriarte, and Robert P. Anderson. 2014. "ENMeval: An R Package for Conducting Spatially Independent Evaluations and Estimating Optimal Model Complexity for Maxent Ecological Niche Models." *Methods in Ecology and Evolution* 5: 1198–1205.

Myers, John. 2018. "Minnesota Loon Conservation Efforts Get $7.2M from Deepwater Horizon Oil Spill Settlement." *Pioneer Press*, October 9. https://www.twincities.com/2018/10/09/minnesota-loon-conservation-efforts-get-7-2m-from-deepwater-horizon-oil-spill-settlement/.

NABCI (North American Bird Conservation Initiative). 2010. *The State of the Birds: 2010 Report on Climate Change, United States of America*. Washington, DC: US Department of the Interior.

———. 2011. *The State of the Birds: 2011 Report on Public Lands and Waters*. Washington, DC: US Department of the Interior.

———. 2016. *The State of North America's Birds 2016*. Ottawa, Ontario: Environment and Climate Change Canada. www.stateofthebirds.org.

NABCI Canada (North American Bird Conservation Initiative Canada). 2012. *The State of Canada's Birds, 2012*. Ottawa: Environment Canada.

Nagle, Leo. 1969. "Mockingbirds Nesting at Royalton, Minnesota in 1968." *Loon* 41: 128.

NAS (National Audubon Society). 2018. *Christmas Bird Count Database*. Accessed August 6, 2019. http://netapp.audubon.org/CBCObservation/Historical/ResultsBySpecies.aspx?1.

———. 2019. "Survival by Degrees: 389 Bird Species on the Brink." https://www.audubon.org/climate/survivalbydegrees.

Natural Resources Conservation Service. 1999. *Ruby-throated Hummingbird (*Archilochus colubris*)*. Fish and Wildlife Habitat Management Leaflet, no. 14. https://efotg.sc.egov.usda.gov/references/public/SC/Ruby-throated_Hummingbird.pdf.

Nebel, Silke, Alex Mills, Jon D. McCracken, and Philip D. Taylor. 2010. "Declines of Aerial Insectivores in North America Follow a Geographic Gradient." *Avian Conservation and Ecology—Écologie et conservation des oiseaux* 5: 1. doi:10.5751/ACE-00391-050201.

Nehrling, Henry. 1896. *Our Native Birds of Song and Beauty, Being a Complete History of All the Songbirds, Flycatchers, Hummingbirds, Swifts, Goatsuckers, Woodpeckers, Kingfishers, Trogons, Cuckoos, and Parrots, of North America*. Vol. 2. Milwaukee: George Brumder.

Nelson, Martin K. 1949. "A Nest of the Chestnut-collared Longspur in Clay County." *Flicker* 21: 64.

Nelson, Maynard, Earl Kopischke, and Robert Chesness. 1964. "A Family of Barn Owls." *Loon* 36: 6–7.

Nelson, Warren. 1981. "Probable Nesting of Yellow-crowned Night Heron in Aitkin County." *Loon* 53: 232.

———. 1982. "Yellow-crowned Night Herons Nest Again in Aitkin County." *Loon* 54: 250–51.

———. 1991. "The Yellow Rails of McGregor Marsh." *Loon* 63: 92–97.

Nero, Robert W. 1970. "Great Gray Owls Nesting near Roseau." *Loon* 42: 88–93.

———. 1980. *The Great Gray Owl: Phantom of the Northern Forest*. Washington, DC: Smithsonian Institute Press.

———. 1982. "Post-Copulatory Display in the Lark Bunting and Other Species." *Wilson Bulletin* 94: 585–90.

———. 1995. "Barn Owl Nesting in Manitoba." *Blue Jay* 53: 159–66.

Nero, Robert W., and Herbert W. R. Copland. 1981. "High Mortality of Great Gray Owls in Manitoba—Winter 1980–81." *Blue Jay* 39: 158–65.

Nero, Robert W., Spencer G. Sealy, and Herbert W. R. Copland. 1974. "Great Gray Owls Occupy Artificial Nest." *Loon* 46: 161–65.

Nesbitt, Stephen A. 1975. "Feather Staining in Florida Sandhill Cranes." *Florida Field Naturalist* 3: 28–30.

Neubeck, Peter. 1992. "King Rail Record." *Loon* 64: 170.

Nevers, Marybeth, JoAnn M. Hanowski, and Gerald J. Niemi. 1981. "Breeding Bird Census in the Red Lake Peatlands of Northern Minnesota." *American Birds* 35: 99–100.

Newton, Ian. 2007. *The Migration Ecology of Birds*. London: Academic Press.

New York State Department of Environmental Conservation. 2006. *New York State Breeding Bird Atlas*. http://www.dec.ny.gov/animals/7312.html.

Ng, Janet. 2009. "Habitat Use and Home Range Characteristics of Common Nighthawks (*Chordeiles minor*) in Mixed-grass Prairie." Master's thesis, University of Regina, Regina, SK.

Nice, Margaret M. 1931a. *The Birds of Oklahoma*. Vol. 3. Rev. ed. Norman: University of Oklahoma Press.

———. 1931b. "Notes on the Twilight Songs of the Scissor-tailed and Crested Flycatchers." *Auk* 48: 123–25.

Nicholls, Thomas H., and Dwain W. Warner. 1972. "Barred Owl Habitat Use as Determined by Radiotelemetry." *Journal of Wildlife Management* 36: 213–24.

Nichols, John W., Christen P. Larsen, Michael E. McDonald, Gerald J. Niemi, and Gerald T.

Ankley. 1995. "A Bioenergetics-Based Model for Accumulation of Polychlorinated Biphenyls by Nestling Tree Swallows, *Tachycineta bicolor*." *Environmental Science and Technology* 29: 604–12.

Nielsen, Gary E. 2005. "White-eyed Vireo Banded in Clay County." *Loon* 77: 116.

Niemi, Gerald J. 1975. "Second Summer Observation of a Rusty Blackbird." *Loon* 47: 144–45.

———. 1978. "Breeding Birds in Burned and Unburned Areas in Northern Minnesota." *Loon* 50: 73–84.

———. 1985. "Patterns of Morphological Evolution in Bird Genera of New World and Old World Peatlands." *Ecology* 66: 1215–28.

———. 2010. "Opportunistic Tennessee Warblers." *Minnesota Conservation Volunteer* 73 (March/April): 40–47.

Niemi, Gerald J., and Thomas E. Davis. 1979. "Notes on the Nesting Ecology of the Piping Plover." *Loon* 51: 74–79.

Niemi, Gerald J., Thomas E. Davis, Gilman D. Veith, and Barbara Vieux. 1986. "Organochlorine Chemical Residues in Herring Gulls, Ring-billed Gulls, and Common Terns of Western Lake Superior." *Archives of Environmental Contamination and Toxicology* 15: 313–20.

Niemi, Gerald J., and JoAnn M. Hanowski. 1984a. "Effects of a Transmission Line on Bird Populations in the Red Lake Peatland, Northern Minnesota." *Auk* 101: 487–98.

———. 1984b. "Relationships of Breeding Birds to Habitat Characteristics in Logged Areas." *Journal of Wildlife Management* 48: 438–43.

———. 1992a. "Bird Populations." In *The Patterned Peatlands of Minnesota,* edited by H. E. Wright Jr., Barbara A. Coffin, and Norman E. Aaseng, 111–29. Minneapolis: University of Minnesota Press.

———. 1992b. "Detailed Species Descriptions—Forest Birds." In *Forest Wildlife: A Technical Paper Prepared for a Generic Environmental Impact Statement on Timber Harvesting and Forest Management in Minnesota,* compiled by Jaakko Pöyry Consulting, Inc. St. Paul, MN: Jaakko Pöyry Consulting.

Niemi, Gerald J., JoAnn M. Hanowski, Ann R. Lima, and David J. Mladenoff. 1998. "Biogeographic Patterns of Breeding Birds in Minnesota." *Loon* 70: 3–11.

Niemi, Gerald J., JoAnn M. Hanowski, Ann R. Lima, Tom Nicholls, and Norm Weiland. 1997. "A Critical Analysis on the Use of Indicator Species in Management." *Journal of Wildlife Management* 61: 1240–52.

Niemi, Gerald J., Robert W. Howe, Brian R. Sturtevant, Linda R. Parker, Alexis R. Grinde, Nicholas P. Danz, Mark D. Nelson, et al. 2016. *Analysis of Long-term Forest Bird Monitoring Data from National Forests of the Western Great Lakes Region.* US Department of Agriculture, Forest Service, General Technical Report NRS-159. Newtown Square, PA: US Department of Agriculture, Forest Service, Northern Research Station.

Niemi, Gerald J., Ann Lima, JoAnn Hanowski, and Lee Pfannmuller. 1995. "Recent Trends of Breeding Birds in Minnesota and Minnesota Forested Regions, 1966–1993." Loon 67: 191–201.

Niemi, Gerald J., and Lee A. Pfannmuller. 1979. "Avian Communities: Approaches to Describing Their Habitat Associations." In DeGraaf and Evans 1979, 154–79.

Niemi, Gerald J., and John R. Probst. 1990. "Wildlife and Fire in the Upper Midwest." In *Management of Dynamic Ecosystems,* edited by James M. Sweeney, 31–46. West Lafayette, IN: The Wildlife Society, North Central Section.

Niemuth, Neal D. 2005. *Prairie Pothole Joint Venture: 2005 Implementation Plan: Section IV—Waterbird Plan.* http://ppjv.org/assets/pdf/11_Waterbird_Plan.pdf.

Niemuth, Neal D., and Mark S. Boyce. 2000. "Enhanced Avian Diversity in Wisconsin Pine Barrens through Aggregated Timber Harvest: A Simulation." *Transactions of the North American Wildlife and Natural Resources Conference* 65: 2015.

Nocera, Joseph J., Jules M. Blais, David V. Beresford, Leah K. Finity, Christopher Grooms, Lynda E. Kimpe, Kurt Kyser, et al. 2012. "Historical Pesticide Applications Coincided with an Altered Diet of Aerially Foraging Insectivorous Chimney Swifts." *Proceedings of the Royal Society B: Biological Sciences* 279: 3114–20.

Nolan, Val, Jr., Ellen D. Ketterson, Dan A. Cristol, Christopher M. Rogers, Ethan D. Clotfelter, Russ C. Titus, Sara J. Schoech, et al. 2020. "Dark-eyed Junco (*Junco hyemalis*)." Version 1.0. In Poole and Gill 2020. doi:10.2173/bow.daejun.01.

Noon, Barry R., Verner P. Bingman, and J. Paige Noon. 1979. "The Effects of Changes in Habitat on Northern Hardwood Forest Bird Communities." In DeGraaf and Evans 1979, 33–48.

Nord, W. H., and Walter John Breckenridge. 1941. "Record Flight of Sandhill Cranes." *Flicker* 13: 2–4.

Norris, D. Ryan, and Bridget J. M. Stutchbury. 2002. "Sexual Differences in Gap-Crossing Ability of a Forest Songbird in a Fragmented Landscape Revealed through Radiotracking." *Auk* 119: 528–32.

Norris, D. Ryan, Bridget J. M. Stutchbury, and Trevor E. Pitcher. 2000. "The Spatial Response of Male Hooded Warblers to Edges in Isolated Fragments." *Condor* 102: 595–600.

North, Michael R., and William E. Faber. 2014. "The Effect of Twin Cities Metropolitan Area Development on Regional Breeding Bird Distributions." *Loon* 86: 111–26.

North American Waterbird Conservation Plan. 2006. "Conservation Status and Distribution of Solitary-Nesting Waterbird Species (A Species-level Categorization Relative to All Waterbirds and Derived within the Spatial Context of the NAWCP Area). Accessed June 15, 2016. https://mnbirdatlas.org/wp-content/uploads/2017/09/Conservation-Status-and-Distribution-of-Solitary-Nesting-Waterbird-Species-2006.pdf.

North American Waterfowl Management Plan Committee. 2004. *North American Waterfowl Management Plan 2004. Implementation Framework: Strengthening the Biological Foundation.* Canadian Wildlife Service, US Fish and Wildlife Service, Secretaria de Medio Ambiente y Recursos Naturales. https://nawmp.wetlandnetwork.ca/Media/Content/files/impfr-en-k.pdf.

———. 2012. *North American Waterfowl Management Plan 2012: People Conserving Waterfowl and Wetlands.* US Department of the Interior, Environment Canada, and Environment and Natural Resources, Mexico. https://nawmp.org/document/2012-nawmp-plan-people-conserving-waterfowl-and-wetlands.

———. 2014. *Revised Objectives: An Addendum to the 2012 North American Waterfowl Management Plan.* Washington, DC: US Fish and Wildlife Service. https://nawmp.org/sites/default/files/2017-12/NAWMP_Revised_Objectives_North_American_Waterfowl_Management_Plan_Final_9-22-14.pdf.

North Dakota Game and Fish Department. 2016. "Willet." *North Dakota Shorebirds and Water Birds.* https://gf.nd.gov/wildlife/id/shorebirds/willet.

Nuechterlein, Gary L. 1975. "Nesting Ecology of Western Grebes on the Delta Marsh, Manitoba." Master's thesis, Colorado State University.

Nute, Grace Lee. 1945. "A Bird-loving Taxidermist of the 1890's." *Minnesota Conservation Volunteer* 8 (May/June): 28–34.

Nutter, Frank H. 1892. "The Specimens I Didn't Get." *Oologist* 9: 67–69.

O'Connor, Robert. 2000. *Birding the Fargo-Moorhead Area*. Moorhead: Minnesota State University Moorhead and Audubon Society of Fargo-Moorhead.

Ojakangas, Richard W., and Charles L. Matsch. 1982. *Minnesota's Geology*. Minneapolis: University of Minnesota Press.

Oleyar, David, D. Ethier, Laurie Goodrich, David Brandes, R. Smith, J. Brown, and J. Sodergren. 2021. *The Raptor Population Index: 2019 Analyses and Assessments*. https://rpi-project.org/2019/.

Olson, Manley E. 1968. "Summer Tanager Seen in St. Paul." *Loon* 40: 59.

Olson, Ricky D. 2008. "First Documented Snowy Plover Breeding in South Dakota." *South Dakota Bird Notes* 60: 92–94.

Olson, Ross. 1946. "Nesting Birds of Prey." *Flicker* 18: 16–17.

Olson, Sigurd T., and William H. Marshall. 1952. *The Common Loon in Minnesota*. Minnesota Museum of Natural History Occasional Papers, no. 5. Minneapolis: Minnesota Museum of Natural History.

Olyphant, Murray, Jr. 1962. "Hooded Warbler Banded in Washington County." *Loon* 34: 130.

Orians, Gordon H. 1980. *Some Adaptations of Marsh Nesting Blackbirds*. Princeton, NJ: Princeton University Press.

Oring, Lewis W. 1978. "Red-breasted Mergansers Breeding in Cass County." *Loon* 50: 122–23.

Oring, Lewis W., J. Michael Reed, Mark A. Colwell, David B. Lank, and Stephen J. Maxson. 1991. "Factors Regulating Annual Mating Success and Reproductive Success in Spotted Sandpipers (*Actitis macularia*)." *Behavioral Ecology and Sociobiology* 28: 433–42.

Otis, David L., John H. Schulz, David Miller, Ralph E. Mirarchi, and Thomas S. Baskett. 2020. "Mourning Dove (*Zenaida macroura*)." Version 1.0. In Poole 2020. doi:10.2173/bow.moudov.01.

Otnes, Mark. 1990. "Sprague's Pipit in Roseau County." *Loon* 62: 167.

Page, Gary W., Lynne E. Stenzel, John S. Warriner, Jane C. Warriner, and Peter W. Paton. 2020. "Snowy Plover (*Charadrius nivosus*)." Version 1.0. In Poole 2020. doi:10.2173/bow.snoplo5.01.

Pallardy, Richard. 2016. "Deepwater Horizon Oil Spill of 2010." In *Encyclopedia Britannica*. Retrieved from *Encyclopedia Britannica*: https://www.britannica.com/event/Deepwater-Horizon-oil-spill-of-2010.

Palmer, Ralph S., ed. 1976. *Handbook of North American Birds*. Vol. 2, *Waterfowl, Part 1*. New Haven, CT: Yale University Press.

Panci, Hannah G., Gerald J. Niemi, Ronald R. Regal, Douglas C. Tozer, Thomas M. Gehring, Robert W. Howe, and Christopher J. Norment. 2017. "Influence of Local, Landscape, and Regional Variables on Sedge and Marsh Wren Occurrence in Great Lakes Coastal Wetlands." *Wetlands* 37: 447–59.

Pardieck, Keith L., David J. Ziolkowski Jr., Michael Lutmerding, Veronica I. Aponte, and Mary-Anne R. Hudson. 2020. *North American Breeding Bird Survey Dataset, 1966–2019*. Geological Survey Data Release. Laurel, MD: US Geological Survey, Patuxent Wildlife Research Center. doi:10.5066/P9J6QUF6.

Parmelee, David F. 1977. "Recent Sightings of the Swallow-tailed Kite at Itasca State Park." *Loon* 49: 181.

———. 1982. "Itasca Goshawks." *Loon* 54: 71–72.

Parmelee, David F., and Richard J. Oehlenschlager. 1973. "Status of LeConte's Sparrow in Minnesota." *Loon* 45: 131–32.

Parr, Michael J. 2007. "Saving the Cerulean Warbler." *Birding* (September/October): 24–27. http://aba.org/birding/v39n5p24.pdf.

Parsons, Katharine C., and Terry L. Master. 2020. "Snowy Egret (*Egretta thula*)." Version 1.0. In Poole and Gill 2020. doi:10.2173/bow.snoegr.01.

Partch, Max C. 1970. "Prairie Chicken Exodus: Notes on the Prairie Chicken in Central Minnesota." *Loon* 42: 5–19.

———. 1973. "A History of Minnesota's Prairie Chickens." In *The Prairie Chicken in Minnesota*, edited by W. Daniel Svedarsky and Terrance J. Wolfe, 15–29. Crookston: University of Minnesota Crookston.

Paruk, James D., David C. Evers, Judith W. McIntyre, Jack F. Barr, Jay Mager, and Walter H. Piper. 2021. "Common Loon (*Gavia immer*)." Version 2.0. In Rodewald and Keeney 2021. doi:10.2173.bow.comloo.02.

Paruk, James D., A. Nelson, Lori Siegel, Jennifer Fox, Steve Windels, David Evers, and Cory Counard MacNulty. 2008. "Effects of 2000 Rule Curve on the Reproductive Success of the Common Loon (*Gavia immer*) in Voyageurs National Park." Report to Voyageurs National Park. Gorham, ME: BioDiversity Research Institute.

Paton, Peter W. C. 1994. "The Effect of Edge on Avian Nest Success: How Strong Is the Evidence?" *Conservation Biology* 8: 17–26.

Patten, Michael A., and Curtis A. Marantz. 1996. "Implications of Vagrant Southeastern Vireos and Warblers in California." *Auk* 113: 911–23.

Paulson, Dennis. 1977. "Western Wood Pewee, Roseau County." *Loon* 49: 169–70.

Payne, Robert B. 2020. "Indigo Bunting (*Passerina cyanea*)." Version 1.0. In Poole 2020. doi:10.2173/bow.indbun.01.

Peabody, Putnam Burton. 1895a. "Dragging for Bobolinks." *Oologist* 12: 123–26.

———. 1895b. "Glossy Ibis at Heron Lake." *Nidiologist* 2: 116–17.

———. 1896. "White-faced Glossy Ibis Breeding in Minnesota." *Auk* 13: 79.

———. 1901. "Nesting Habits of Le Conte's Sparrow." *Auk* 18: 128–34.

———. 1922. "Haunts and Breeding Habits of the Yellow Rail." *Journal of the Museum of Comparative Oology* 2: 33–44.

Peake, John S. 1970. "Hooded Warbler Observation." *Loon* 42: 36.

Pearce, John, Mark L. Mallory, and Karen Metz. 2020. "Common Merganser (*Mergus merganser*)." Version 1.0. In Billerman 2020. doi:10.2173/bow.commer.01.

Pearson, Carol W., and Gerald J. Niemi. 2000. "Effects of Within-Stand Habitat and Landscape Patterns on Avian Distribution and Abundance in Northern Minnesota." In *Disturbance in Boreal Forest Ecosystems: Human Impacts and Natural Processes,* edited by Susan G. Conrad, 81–95. Proceedings of the Annual Meeting of the International Boreal Forest Research Association, Duluth, MN, August 4–7, 1997. US Forest Service General Technical Report NC-209. US Department of Agriculture, Forest Service, North Central Research Station.

Pearson, David. 1963. "Kentucky Warbler Seen in Hennepin County." *Flicker* 35: 139.

———. 1964. "Observations of a Kentucky Warbler." *Loon* 36: 94–96.

———. 1965. "Kentucky Warbler Again Seen in Anoka County." *Loon* 37: 114.

Pearson, T. Gilbert, ed. 1917. *Birds of America*. Vol. 3, Nature Lovers Library. New York: The University Society.

Peck, George K., and Ross D. James. 1987. *Breeding Birds of Ontario: Nidiology and Distribution*. Vol. 2, *Passerines*. Toronto: The Royal Ontario Museum.

Peer, Brian D., and Eric K. Bollinger. 2020. "Common Grackle (*Quiscalus quiscula*)." Version 1.0. In Poole and Gill 2020. doi:10.2173/bow.comgra.01.

Penning, William L., and Francesca J. Cuthbert. 1993. "The History of Colonial Waterbird Management in the Duluth-Superior Harbor, 1937–1990." *Loon* 65: 163–74.

Perring, Franklyn H., and S. Max Walters. 1962. *Atlas of the British Flora*. London: Thomas Nelson and Sons.

Perry, Pamela Skoog, and Scott D. Bloch. 1987. "A Survey of Purple Martins Nesting in Central Minnesota." *Loon* 59: 93–98.

Perry, Pamela Skoog, and Kevin Woizeschke. 2000. "A Survey of Common Loons on Small Lakes in Central Minnesota." In Svingen and Hertzel 2000, 1–4.

Peterjohn, Bruce G. 2001. *The Birds of Ohio with Breeding Bird Atlas Maps*. Wooster, OH: Wooster Book Company.

Peters, Albert S. [1918] 2018. "First Minnesota Yellow Rail Eggs." *Loon* 77: 77. Reprint, *Oologist* 35: 28.

Peterson, Alfred. 1924. Letter to Thomas S. Roberts, August 12, 1924. Minnesota Ornithology Correspondence, Incoming (Box 69, Folder 666, pp. 110–12). University of Minnesota Libraries, University Archives. http://purl.umn.edu/185060.

Peterson, Kenneth L., Steven Hogg, and Judy Voigt Englund. 2020. "Nest-Site Selection and Nesting Success in a Restored Population of Ospreys (*Pandion haliaetus*) in the Minneapolis–St. Paul, Minnesota (USA) Area." *Journal of Raptor Research* 54: 245–54. doi:10.3356/0892-1016-54.3.245.

Peterson, Richard A. 1995. *The South Dakota Breeding Bird Atlas*. Aberdeen: The South Dakota Ornithologists' Union.

Peterson, Roger Tory. 1947. *Field Guide to the Birds*. Boston: Houghton Mifflin.

———. 1980. *A Field Guide to the Birds East of the Rockies*. 4th ed. Boston: Houghton Mifflin.

———. 2008. *Peterson Field Guide to Birds of North America*. Boston: Houghton Mifflin.

Peterson, Theodore, and Mrs. Peterson. 1936. "Smith's and McCown's Longspur Seen in Minnesota." *Auk* 53: 342.

Petit, Lisa J. 2020. "Prothonotary Warbler (*Protonotaria citrea*)." Version 1.0. In Poole and Gill 2020. doi:10.2173/bow.prowar.01.

Petit, Lisa J., and Daniel R. Petit. 1996. "Factors Governing Habitat Selection by Prothonotary Warblers: Field Tests of the Fretwell-Lucas Models." *Ecological Monographs* 66: 367–87.

Pettingill, Olin Sewall, Jr. 1970. *Ornithology in Laboratory and Field*. Minneapolis: Burgess.

Pfannmuller, Lee A. 2012. *Stewardship Birds of Minnesota: Our Global Responsibility*. St. Paul: Audubon Minnesota.

———. 2014a. "Common Tern Minnesota Conservation Plan." In *Blueprint for Minnesota Bird Conservation*. Audubon Minnesota. https://mn.audubon.org/sites/default/files/common_tern_minnesota_conservation_plan_10-16-2014_0.pdf.

———. 2014b. "Franklin's Gull Minnesota Conservation Plan." In *Blueprint for Minnesota Bird Conservation*. Audubon Minnesota. https://mn.audubon.org/sites/default/files/franklin_gull_conservation_plan_10-21-2014.pdf.

Pfannmuller, Lee, Gerald Niemi, Janet Green, Bonnie Sample, Nicholas Walton, Ed Zlonis, Terry Brown, et al. 2017. *The First Minnesota Breeding Bird Atlas (2009–2013)*. https://mnbirdatlas.org.

Pfeffer, Max J., Joe D. Francis, and Zev Ross. 2006. "Fifty Years of Farmland Change: Urbanization, Population Growth and the Changing Farm Economy." In *Population Change and Rural Society,* edited by William A. Kandel and David L. Brown, 103–26. Dordrecht, The Netherlands: Springer.

Phillips, Steven J., Robert P. Anderson, and Robert E. Schapire. 2006. "Maximum Entropy Modeling of Species Geographic Distributions." *Ecological Modelling* 190: 231–59.

Pickens, Bradley A., and Brooke Meanley. 2020. "King Rail (*Rallus elegans*)." Version 1.0. In Rodewald 2020. doi:10.2173/bow.kinrai4.01.

Picman, Jaroslav, and Anna K. Picman. 1980. "Destruction of Nests by the Short-billed Marsh Wren." *Condor* 82: 176–79.

Pieper, William R. 1954. "A Western Minnesota Field Trip." *Flicker* 26: 84–85.

———. 1963a. "Preliminary Observations on the Yellow Rail in a Becker County Marsh." *Flicker* 35: 28–29.

———. 1963b. "Sprague's Pipit in the Felton Area." *Flicker* 35: 68–70.

PIF (Partners in Flight). 2019. Avian Conservation Assessment Database Archives: Global Assessment Scores (Version 2019) [data set]. https://pif.birdconservancy.org/wp-content/uploads/2021/03/PopEsts_ContGlobal_3-01-19.xlsx.

———. 2021. *Partners in Flight Databases*. Version 2021. https://www.partnersinflight.org/what-we-do/science/databases.

PIF Science Committee. 2013. *Population Estimates Database*. Version 2013. https://pif.birdconservancy.org/population-estimates-database-archive/.

Pimentel, David, Lori Lach, Rodolfo Zuniga, and Doug Morrison. 2000. "Environmental and Economic Costs of Nonindigenous Species in the United States." *BioScience* 50: 53–65.

Pioneer Press. 2014. "Minnesota Loses 6K Farms between 2007 and 2012, USDA Census Shows." *Pioneer Press,* March 3. http://www.twincities.com/2014/03/03/minnesota-loses-6k-farms-between-2007-and-2012-usda-census-shows/.

Piper, Walter H., Jason Grear, Brian Hoover, Elaina Lomery, and Linda M. Grenzer. 2020. "Plunging Floater Survival Causes Cryptic Population Decline in the Common Loon." *Ornithological Applications* 122: duaa044. doi:10.1093/condor/duaa044.

Piper, Walter H., James D. Paruk, David C. Evers, Michael W. Meyer, Keren B. Tischler, Margaret Klich, and Jerry J. Hartigan. 1997. "Local Movements of Color-Marked Common Loons." *Journal of Wildlife Management* 61: 1253–61.

Pitocchelli, Jay. 2020. "Mourning Warbler (*Geothlypis philadelphia*)." Version 1.0. In Rodewald 2020. doi:10.2173/bow.mouwar.01.

Pitocchelli, Jay, Julie Jones, David Jones, and Julie Bouchie. 2020. "Connecticut Warbler (*Oporornis agilis*)." Version 1.0. In Poole 2020. doi:10.2173/bow.conwar.01.

Pius, B. 1948. "Goldfinches in St. Paul." *Flicker* 20: 47–51.

Platt, Joseph B. 1976. "Sharp-shinned Hawk Nesting and Nest Site Selection in Utah." *Condor* 78: 102–3.

Poling, Trisha D., and Steven E. Hayslette. 2006. "Dietary Overlap and Foraging Competition between Mourning Doves and Eurasian Collared-Doves." *Journal of Wildlife Management* 70: 998–1004.

Pollet, Ingrid L., Dave Shutler, John W. Chardine, and John P. Ryder. 2020. "Ring-billed Gull (*Larus delawarensis*)." Version 1.0. In Poole 2020. doi:10.2173/bow.ribgul.01.

Pool, Duane B., Arvind O. Panjabi, Alberto Macias-Duarte, and Deanna M. Solhjem. 2014. "Rapid Expansion of Croplands in Chihuahua, Mexico Threatens Declining North American Grassland Bird Species." *Biological Conservation* 170: 274–81.

Poole, Alan F., ed. 2020. *Birds of the World*. Ithaca, NY: Cornell Laboratory of Ornithology. https://birdsoftheworld.org/bow.

Poole, Alan F., Rob O. Bierregaard, and Mark S. Martell. 2002. "Osprey (*Pandion haliaetus*)." *The Birds of North America*, edited by Alan F. Poole. Ithaca, NY: Cornell Laboratory of Ornithology.

Poole, Alan F., and Frank B. Gill, eds. 2020. *Birds of the World*. Ithaca, NY: Cornell Laboratory of Ornithology. https://birdsoftheworld.org/bow.

Poole, Alan F., Peter E. Lowther, James P. Gibbs, Frederic A. Reid, and Scott M. Melvin. 2020. "Least Bittern *(Ixobrychus exilis)*." Version 1.0. In Poole 2020. doi:10.2173/bow.leabit.01.

Porneluzi, Paul A., Michele A. Van Horn, and Therese M. Donovan. 2020. "Ovenbird *(Seiurus aurocapilla)*." Version 1.0. In Poole 2020. doi:10.2173/bow.ovenbi1.01.

Porter, Adele. 2013. "Flying with the Loons." *Minnesota Conservation Volunteer* 76 (January/February): 8–19.

Potter, Brad A., Greg J. Soulliere, Dave N. Ewert, Melinda G. Knutson, Wayne E. Thogmartin, John S. Castrale, and Mike J. Roell. 2007. *Upper Mississippi River and Great Lakes Region Joint Venture Landbird Habitat Conservation Strategy*. Fort Snelling, MN: US Fish and Wildlife Service.

Poulin, Jean-François, Émilie D'Astous, Marc-André Villard, Sallie J. Hejl, Karen R. Newlon, Mary E. McFadzen, Jock S. Young, et al. 2020. "Brown Creeper (*Certhia americana*)." Version 1.0. In Poole 2020. doi:10.2173/bow.brncre.01.

Poulin, Ray G., L. Danielle Todd, Elizabeth A. Haug, Brian A. Millsap, and Mark S. Martell. 2020. "Burrowing Owl (*Athene cunicularia*)." Version 1.0. In Poole 2020. doi:10.2173/bow.burowl.01.

Powell, Abby N. 1993. "Bird Surveys in Minnesota's Northwestern State Parks." *Loon* 65: 4–11.

Preble, Norman A. 1957. "Nesting Habits of the Yellow-billed Cuckoo." *American Midland Naturalist* 57: 474–82.

Preston, Charles R., and Ronald D. Beane. 2020. "Red-tailed Hawk (*Buteo jamaicensis*)." Version 1.0. In Poole 2020. doi:10.2173/bow.rethaw.01.

Price, E. W., and Francesca J. Cuthbert. 2002. "Piping Plover Habitat and Recovery Assessment for St. Louis River Estuary, Minnesota/Wisconsin." Report submitted to Twin Cities Field Office of the US Fish and Wildlife Service.

Price, Jeff, Sam Droege, and Amy Price. 1995. *The Summer Atlas of North American Birds*. San Diego: Academic Press.

Probst, John R., Don S. Rakstad, and David J. Rugg. 1992. "Breeding Bird Communities in Regenerating and Mature Broadleaf Forests in the USA Lake States." *Forest Ecology and Management* 49: 43–60.

Prose, Bart L. 1987. *Habitat Suitability Index Models: Plains Sharp-tailed Grouse*. Biological Report 82(10.142). Washington, DC: US Fish and Wildlife Service, National Ecology Center.

Prosser, C. C. 1934. "The Cerulean Warbler." *Flicker* 6: 63–65.

Provost, Jodie L. 1991. "Biology of Sandhill Cranes on Breeding and Fall Staging Areas in Northwestern Minnesota." Master's thesis, University of Minnesota.

Provost, Thomas. 1989. "Minnesota's First Black-necked Stilt." *Loon* 61: 197.

Pruitt, Lori. 1996. *Henslow's Sparrow Status Assessment*. June. Accessed June 1, 2023. https://www.partnersinflight.org/wp-content/uploads/2017/03/PIF-Henslows-Sparrow-Status-Assessment.pdf.

———. 2000. *Loggerhead Shrike Status Assessment*. Fort Snelling, MN: US Fish and Wildlife Service.

Pulliainen, Erkki. 1971. "Winter Nutrition of Crossbills (*Loxia curvirostra* and *L. leucoptera*) in Northeastern Lapland in 1969." *Annales Zoologici Fennici* 8: 326–29.

Purple Martin Conservation Association. 2016. "Location." https://www.purplemartin.org/purple-martins/attracting/44/location/.

Raftovich, Robert V., Kathleen K. Fleming, S. C. Chandler, and C. M. Cain. 2020. *Migratory Bird Hunting Activity and Harvest during the 2018–19 and 2019–20 Hunting Seasons*. Laurel, MD: US Fish and Wildlife Service.

Ralph, C. John, John R. Sauer, and Sam Droege. 1995. *Monitoring Bird Populations by Point Counts*. US Department of Agriculture, Forest Service, General Technical Report PSW-GTR-149. Albany, CA: US Department of Agriculture, Forest Service, Pacific Southwest Research Station.

Ralston, Joel, David I. King, William V. DeLuca, Gerald J. Niemi, Michale J. Glennon, Judith C. Scarl, and J. Daniel Lambert. 2015. "Analysis of Combined Data Sets Yields Trend Estimates for Vulnerable Spruce-Fir Birds in Northern United States." *Biological Conservation* 187: 270–78.

Rasmussen, Justin Lee, Spencer G. Sealy, and Richard J. Cannings. 2020. "Northern Saw-whet Owl (*Aegolius acadicus*)." Version 1.0. In Poole 2020. doi:10.2173/bow.nswowl.01.

Rave, David P., Michael C. Zicus, John R. Fieberg, Lucas Savoy, and Kevin Regan. 2014. "Trends in Eggshell Thickness and Mercury in Common Goldeneye and Hooded Merganser Eggs." *Wildlife Society Bulletin* 38: 9–13.

Raynor, Gilbert S. 1976. "Decrease in Breeding Brown Thrashers and Rufous-sided Towhees on Central Long Island." *Kingbird* 26: 190–93.

R Core Team. 2016. *R: A Language and Environment for Statistical Computing*. Vienna, Austria: R Foundation for Statistical Computing. https://www.R-project.org/.

Ream, Catherine H. 1976. "Loon Productivity, Human Disturbance, and Pesticide Residues in Northern Minnesota." *Wilson Bulletin* 88: 427–32.

Redig, Patrick T., John S. Castrale, and Amber Burnette. 2021. "Midwest Peregrine Falcon Restoration, 2021 Report." Midwest Peregrine Society.

Reed, J. Michael, Lewis W. Oring, and Julie A. R. Alberico. 1991. "Common Terns at Minnesota's Largest Colony: Breeding Site Shift, Population Decline, and Reproductive Success." *Loon* 63: 12–16.

Reed, J. Michael, Lewis W. Oring, and Elizabeth M. Gray. 2020. "Spotted Sandpiper (*Actitis macularius*)." Version 1.0. In Poole 2020. doi:10.2173/bow.sposan.01.

Reed, Lisa M., Michael A. Johansson, Nicholas Panella, Robert McLean, Terry Creekmore, Rose Puelle, and Nicholas Komar. 2009. "Declining Mortality in American Crow (*Corvus brachyrhynchos*) Following Natural West Nile Virus Infection." *Avian Diseases* 53: 458–61.

Reif, Charles B. 1941. "Minnesota Nesting Record, 1941." *Flicker* 13: 27–31.

Reinert, Steven E., and Francis C. Golet. 1979. "Breeding Ecology of the Swamp Sparrow in a Southern Rhode Island Peatland." *Transactions of the Northeast Section of The Wildlife Society* 36: 1–13.

Reinoehl, Jack. 2011. "White-eyed Vireo (*Vireo griseus*)." In *Michigan Breeding Bird Atlas II*, edited by Allen T. Chartier, Jennifer J. Baldy, and John Brenneman. Kalamazoo, MI: Kalamazoo Nature Center. https://naturecenter.org/Conservation/Michigan-Breeding-Bird-Atlas-ll.

Reiser, M. Hildegard. 1988. "Effects of Regulated Lake Levels on the Reproductive Success, Distribution and Abundance of the Aquatic Bird Community in Voyageurs National Park, Minnesota." PhD diss., Northern Arizona University, Flagstaff.

Reitsma, Len, Marissa Goodnow, Michael T. Hallworth, and Courtney J. Conway. 2020. "Canada Warbler (*Cardellina canadensis*)." Version 2.0. In Poole 2020. doi:10.2173/bow.canwar.01.

Renfrew, Rosalind B., ed. 2013. *The Second Atlas of the Breeding Birds of Vermont, 2003–2007.* Vermont Center for Ecostudies—Vermont Atlas of Life. http://val.vtecostudies.org/projects/vermont-breeding-bird-atlas/breeding-bird-species. Accessed June 2017.

Renfrew, Rosalind B., Allan M. Strong, Noah G. Perlut, Stephen G. Martin, and Thomas A. Gavin. 2020. "Bobolink (*Dolichonyx oryzivorus*)." Version 1.0. In Rodewald 2020. doi:10.2173/bow.boboli.01.

Renfrow, Frank. 2005. "Range Expansion and Habitat Selection in Breeding Populations of Red-breasted Nuthatch (*Sitta canadensis*) in the Southeastern United States." *North American Birds* 59: 516–23.

Reynolds, Dave. 1974. "King Rail in Duluth." *Loon* 46: 121–22.

Reynolds, Ronald E., Terry L. Shaffer, Randy W. Renner, Wesley E. Newton, and Bruce D. J. Batt. 2001. "Impact of the Conservation Reserve Program on Duck Recruitment in the US Prairie Pothole Region." *Journal of Wildlife Management* 65: 765–80.

Ribic, Christine A., Rolf R. Koford, James R. Herkert, Douglas H. Johnson, Neal D. Niemuth, David E. Naugle, Kristel K. Bakker, et al. 2009. "Area Sensitivity in North American Grassland Birds: Patterns and Processes." *Auk* 126: 233–44.

Rice, Jake. 1978. "Ecological Relationships of Two Interspecifically Territorial Vireos." *Ecology* 59: 526–38.

Richardson, John. 2003. "Baird's Sparrows, Polk County." *Loon* 75: 52–53.

Richardson, Susan. 1981. "Yellow-crowned Night Heron's Nest at Pig's Eye." *Loon* 53: 171.

Ridgway, Robert. 1877. "Ornithology." Vol. 4, part 3. In *Report of the Geological Exploration of the Fortieth Parallel,* by Clarence King, 303–643, 652–62. Washington, DC: Government Printing Office.

———. 1889. *The Ornithology of Illinois*. Springfield, IL: H. W. Rokker.

———. 1902. *The Birds of North and Middle America: A Descriptive Catalogue of the Higher Groups, Genera, Species, and Subspecies of Birds Known to Occur in North America, from the Arctic Lands to the Isthmus of Panama, the West Indies and Other Islands of the Caribbean Sea, and the Galapagos Archipelago*. Part 2. US National Museum Bulletin 50. Washington, DC: Government Printing Office.

Rimmer, Christopher C., and Kent P. McFarland. 2020. "Tennessee Warbler (*Leiothlypis peregrina*)." Version 1.0. In Poole 2020. doi:10.2173/bow.tenwar.01.

Ripper, Dana, James C. Bednarz, and Daniel E. Varland. 2007. "Landscape Use by Hairy Woodpeckers in Managed Forests of Northwestern Washington." *Journal of Wildlife Management* 71: 2612–23.

Risen, Kim W. 1986. "Baird's Sparrow in Crow Wing County." *Loon* 58: 131–32.

———. 1992. "Unusual Feeding Behavior Observed for King Rail." *Loon* 64: 162–63.

———. 1993. "Yellow Rails Recorded from Unusual Locations in Minnesota." *Loon* 65: 196–99.

Rising, James D. 1996. *A Guide to the Identification and Natural History of the Sparrows of the United States and Canada*. New York: Academic Press.

———. 2010. "The Many Savannah Sparrows." *Birding* 42 (November): 44–55.

Rising, James D., and Nancy J. Flood. 2020. "Baltimore Oriole (*Icterus galbula*)." Version 1.0. In Rodewald 2020. doi:10.2173/bow.balori.01.

Riske, Morley E. 1976. "Environmental and Human Impacts on Grebes Breeding in Central Alberta." PhD diss., University of Calgary.

Risser, Alden. 1932. "The 1932 Nesting Season." *Flicker* 4: 19–32.

Ritchison, Gary, Frederick R. Gehlbach, Peter Pyle, and Michael A. Patten. 2020. "Eastern Screech-Owl (*Megascops asio*)." Version 1.0. In Rodewald 2020. doi:10.2173/bow.easowl1.01.

Ritchison, Gary, Thomas C. Grubb Jr., and Vladimir V. Pravosudov. 2020. "Tufted Titmouse (*Baeolophus bicolor*)." Version 1.0. In Rodewald 2020. doi:10.2173/bow.tuftit.01.

Robbins, Chandler S. 1979. "Effect of Forest Fragmentation on Bird Populations." In DeGraaf and Evans 1979, 198–212.

———. 1980. "Effect of Forest Fragmentation on Breeding Bird Populations in the Piedmont of the Mid-Atlantic Region." In *Proceedings of the 1979 Mid-Atlantic Natural History Symposium: Bird Populations, A Litmus Test of the Environment,* edited by J. F. Lunch, 31–36.

Robbins, Chandler S., Danny Bystrack, and Paul H. Geissler. 1986. *The Breeding Bird Survey: Its First Fifteen Years, 1965–1979*. US Fish and Wildlife Service Resource Publication, no. 157. Washington, DC: US Fish and Wildlife Service.

Robbins, Chandler S., Deanna K. Dawson, and Barbara A. Dowell. 1989. "Habitat Area Requirements of Breeding Forest Birds of the Middle Atlantic States." *Wildlife Monographs* 103: 3–34.

Robbins, Chandler S., John W. Fitzpatrick, and Paul B. Hamel. 1992. "A Warbler in Trouble: *Dendroica cerulea*." In *Ecology and Conservation of Neotropical Migrant Landbirds,* edited by John M. Hagan III and David W. Johnston, 549–62. Washington, DC: Smithsonian Institute Press.

Robbins, Samuel. 1974. "The Willow and Alder Flycatchers in Wisconsin: A Preliminary Description of Summer Range." *Passenger Pigeon* 36: 147–52.

Robbins, Samuel D., Jr. 1991. *Wisconsin Birdlife: Population and Distribution Past and Present.* Madison: University of Wisconsin Press.

Roberson, Aimee M. 2001. "Evaluating and Developing Survey Techniques Using Broadcast Conspecific Calls for Northern Goshawks in Minnesota." Master's thesis, University of Minnesota.

Roberson, Aimee M., David E. Andersen, and Patricia L. Kennedy. 2005. "Do Breeding Phase and Detection Distance Influence the Effective Area Surveyed for Northern Goshawks?" *Journal of Wildlife Management* 69: 1240–50.

Roberts, Nancy. 1966. "Summer Observations of Cerulean Warbler and Lincoln's Sparrow in Crow Wing County." *Loon* 38: 142.

Roberts, Thomas S. 1919. *Water Birds of Minnesota: Past and Present.* Extracted from *Biennial Report of the State Game and Fish Commission of Minnesota, for the Biennial Period Ending July 31, 1918*. Minneapolis, MN.

———. 1932. *The Birds of Minnesota*. 2 vols. Minneapolis: University of Minnesota Press.

———. 1936. *The Birds of Minnesota*. 2nd ed. 2 vols. Minneapolis: University of Minnesota Press.

———. 1944. "Early Rambles of a Bird Lover in Minnesota." *Flicker* 16: 39–46.

———. 1991. *Shotgun and Stethoscope: The Journals of Thomas Sadler Roberts*. Edited and transcribed by Penelope Krosch. Minneapolis: James Ford Bell Museum of Natural History, University of Minnesota.

———. *See also* Thomas Sadler Roberts Field Checklists, 1923–31.

Roberts, Thomas S., and Franklin Benner. 1880. "A Contribution to the Ornithology of Minnesota." *Bulletin of the Nuttall Ornithological Club* 5: 11–20.

———. 1987. "A Contribution to the Ornithology of Minnesota." *Loon* 59: 66–72. Reprint of Roberts and Benner 1880.

Robinson, Gerald, and James R. Beer. 1955. "Survey of Cliff Swallow Colonies in Minnesota." *Flicker* 27: 148–52.

Robinson, Julie A., J. Michael Reed, Joseph P. Skorupa, and Lewis W. Oring. 2020. "Black-necked Stilt (*Himantopus mexicanus*)." Version 1.0. In Poole and Gill 2020. doi:10.2173/bow.bknsti.01.

Robinson, Scott K., Frank R. Thompson III, Therese M. Donovan, Donald R. Whitehead, and John Faaborg. 1995. "Regional Forest Fragmentation and the Nesting Success of Migratory Birds." *Science* 267: 1987–90.

Robinson, W. Douglas. 2020. "Summer Tanager (*Piranga rubra*)." Version 1.0. In Poole 2020. doi:10.2173/bow.sumtan.01.

Robinson, William L. 1980. *Fool Hen: The Spruce Grouse on the Yellow Dog Plains*. Madison: University of Wisconsin Press.

Rodewald, Paul G., ed. 2020. *Birds of the World*. Ithaca, NY: Cornell Laboratory of Ornithology. https://birdsoftheworld.org/bow.

Rodewald, Paul G., and Ross D. James. 2020. "Yellow-throated Vireo (*Vireo flavifrons*)." Version 1.0. In Poole 2020. doi:10.2173/bow.yetvir.01.

Rodewald, Paul G., and B. K. Keeney, eds. 2021. *Birds of the World*. Ithaca, NY: Cornell Laboratory of Ornithology. https://birdsoftheworld.org/bow.

Rodewald, Paul G., Matthew B. Shumar, Aaron T. Boone, David L. Slager, and Jim McCormac, eds. 2016. *The Second Atlas of the Breeding Birds in Ohio*. University Park: Pennsylvania State University Press.

Rodewald, Paul G., James H. Withgott, and Kimberly G. Smith. 2020. "Pine Warbler (*Setophaga pinus*)." Version 1.0. In Poole 2020. doi:10.2173/bow.pinwar.01.

Rodgers, James A., Jr., and Henry T. Smith. 2020. "Little Blue Heron (*Egretta caerulea*)." Version 1.0. In Poole 2020. doi:10.2173/bow.libher.01.

Rodgers, Tommie L. 1991. "Pine Siskins Die from Salmonellosis." *Migrant* 62: 105–6.

Rohner, Christoph. 1997. "Non-Territorial 'Floaters' in Great Horned Owls: Space Use during a Cyclic Peak of Snowshoe Hares." *Animal Behaviour* 53: 901–12.

Rohwer, Frank C., William P. Johnson, and Elizabeth R. Loos. 2020. "Blue-winged Teal (*Spatula discors*)." Version 1.0. In Poole and Gill 2020. doi:10.2173/bow.buwtea.01.

Rollins, Matthew G., Robert E. Keane, Zhiliang Zhu, and James P. Menakis. 2006. "An Overview of the LANDFIRE Prototype Project." In *The LANDFIRE Prototype Project: Nationally Consistent and Locally Relevant Geospatial Data for Wildland Fire Management*, edited by Matthew G. Rollins and Christine K. Frame, 5–43. US Department of Agriculture, Forest Service, General Technical Report RMRS-GTR-175. Fort Collins, CO: US Department of Agriculture, Forest Service, Rocky Mountain Research Station.

Rosenberg, Kenneth V. 2004. *Partners in Flight Continental Priorities and Objectives Defined at the State and Bird Conservation Region Levels: Minnesota*. Ithaca: Cornell Lab of Ornithology.

Rosenberg, Kenneth V., Sara E. Barker, and Ronald W. Rohrbaugh. 2000. *An Atlas of Cerulean Warbler Populations*. Ithaca: Cornell Lab of Ornithology.

Rosenberg, Kenneth V., Adriaan M. Dokter, Peter J. Blancher, John R. Sauer, Adam C. Smith, Paul A. Smith, Jessica C. Stanton, et al. 2019. "Decline of the North American Avifauna." *Science* 366: 120–24. doi:10.1126/science.aaw1313.

Rosenberg, Kenneth V., Ralph S. Hames, Ronald W. Rohrbaugh Jr., Sara Barker Swarthout, James D. Lowe, and André A. Dhondt. 2003. *A Land Manager's Guide to Improving Habitat for Forest Thrushes*. [Ithaca]: Cornell Lab of Ornithology. http://com-bbimages.s3.amazonaws.com/bbimages/clo/pdf/thrushguide.pdf.

Rosenberg, Kenneth V., Judith A. Kennedy, Randy Dettmers, Robert P. Ford, Debra Reynolds, John D. Alexander, Carol J. Beardmore, et al. 2016. *Partners in Flight Landbird Conservation Plan: 2016 Revision for Canada and Continental United States*. PIF Science Committee. https://partnersinflight.org/resources/the-plan/.

Rosenberg, Kenneth V., James D. Lowe, and André A. Dhondt. 1999. "Effects of Forest Fragmentation on Breeding Tanagers: A Continental Perspective." *Conservation Biology* 13: 568–83.

Rosenberg, Kenneth V., Ronald W. Rohrbaugh Jr., Sara E. Barker, James D. Lowe, Ralph S. Hames, and André A. Dhondt. 1999. *A Land Manager's Guide to Improving Habitat for Scarlet Tanagers and Other Forest-interior Birds*. [Ithaca]: Cornell Lab of Ornithology. http://com-bbimages.s3.amazonaws.com/bbimages/clo/pdf/tanager.pdf.

Rosenfield, Robert N. 2018. *The Cooper's Hawk: Breeding Ecology and Natural History of a Winged Huntsman*. Blaine, WA: Hancock House.

Rosenfield, Robert N., John Bielefeldt, Robert K. Anderson, and Joseph M. Papp. 1991. "Status Reports: Accipiters." In *Proceedings of the Midwest Raptor Management Symposium and Workshop*, edited by Beth Biron Pendleton and Diane L. Krahe, 42–49. Scientific and Technical Series, no. 15. Washington, DC: National Wildlife Federation.

Rosenfield, Robert N., John Bielefeldt, Dale R. Trexel, and Thomas C. J. Doolittle. 1998. "Breeding Distribution and Nest-Site Habitat of Northern Goshawks in Wisconsin." *Journal of Raptor Research* 32: 189–94.

Rosenfield, Robert N., Dan Lamers, David L. Evans, Molly Evans, and Jenna A. Cava. 2011. "Shift to Later Timing by Autumnal Migrating Sharp-shinned Hawks." *Wilson Journal of Ornithology* 123: 154–58.

Rosenfield, Robert N., Kristin K. Madden, John Bielefeldt, and Odette E. Curtis. 2020. "Cooper's Hawk (*Accipiter cooperii*)." Version 1.0. In Rodewald 2020. doi:10.2173/bow.coohaw.01.

Ross, Drew. 1998. "Gaining Ground: A Swan's Song." *National Parks* 72: 34–39.

Ross, R. Kenyon, Ken Abraham, Rob Clay, B. Collins, J. Iron, R. Janes, D. McLachlin, et al. 2003. *Ontario Shorebird Conservation Plan*. Downsview, ON: Environment Canada.

Roth, Amber M., Ronald W. Rohrbaugh, Thomas C. Will, and David A. Buehler, eds. 2012. *Golden-winged Warbler Status Review and Conservation Plan*. http://www.gwwa.org/plan.html.

Roth, Roland R. 1987. "Assessment of Habitat Quality for Wood Thrush in a Residential Area." In *Integrating Man and Nature in the Metropolitan Environment: Proceedings of a National Symposium on Urban Wildlife, 4–7 November 1986, Chevy Chase, Maryland*, edited by Lowell W. Adams and Daniel L. Leedy, 139–49. Columbia, MD: National Institute for Urban Wildlife.

Roy, Charlotte L. 2013. "Investigation of Trematodes and Faucet Snails Responsible for Lesser Scaup Die-Offs." In *Summaries of Wildlife Research Findings 2013*, edited by Lou Cornicelli, Michelle Carstensen, Marrett D. Grund, Michael A. Larson, and Jeffrey S. Lawrence, 159–65. St. Paul: Minnesota Department of Natural Resources.

———. 2016a. "2016 Minnesota Prairie-Chicken Population Survey." Minnesota Department of Natural Resources, Forest Wildlife Populations and Research Group. https://files.dnr.state.mn.us/recreation/hunting/prairiechicken/2016-survey.pdf.

———. 2016b. "2016 Minnesota Spring Grouse Surveys." Minnesota Department of Natural

Resources, Forest Wildlife Populations and Research Group. https://files.dnr.state.mn.us/recreation/hunting/grouse/grouse_survey_report16.pdf.
———. 2018. "2018 Minnesota Spring Grouse Surveys." Minnesota Department of Natural Resources, Forest Wildlife Populations and Research Group. https://files.dnr.state.mn.us/recreation/hunting/grouse/grouse_survey_report18.pdf.
———. 2020a. "2020 Minnesota Prairie-Chicken Harvest Survey." Minnesota Department of Natural Resources, Forest Wildlife Populations and Research Group. https://files.dnr.state.mn.us/recreation/hunting/prairiechicken/2020-harvest.pdf.
———. 2020b. "2020 Minnesota Prairie-Chicken Population Survey." Minnesota Department of Natural Resources, Forest Wildlife Populations and Research Group. https://files.dnr.state.mn.us/recreation/hunting/prairiechicken/2020-survey.pdf.
———. 2020c. "2020 Minnesota Spring Grouse Surveys." Minnesota Department of Natural Resources, Forest Wildlife Populations and Research Group. https://files.dnr.state.mn.us/wildlife/grouse/reports/spring_survey_2020.pdf.
Roy, Charlotte L., and Andrew J. Gregory. 2019. "Landscape and Population Genetics Reveal Long Distance Sharp-tailed Grouse (*Tympanuchus phasianellus*) Movements and a Recent Bottleneck in Minnesota." *Conservation Genetics* 20: 259–73.
Roy, Charlotte L., John Giudice, and Chris Scharenbroich. 2017. "Monitoring Spruce Grouse in Minnesota: A Pilot Study (2014–2016)." In Cornicelli et al. 2017, 69–92.
———. 2020a. "Evaluation of Cantus-call and Pellet Surveys for Spruce Grouse (*Falcipennis canadensis canace*) at the Southern Extent of Their Range." *Journal of Field Ornithology* 91: 44–63. doi:10.1111/jofo.12323.
———. 2020b. "2020 Minnesota Spruce Grouse Survey." Minnesota Department of Natural Resources.
Roy, Charlotte L., Christine M. Herwig, William L. Hohman, and Robert T. Eberhardt. 2020. "Ring-necked Duck (*Aythya collaris*)." Version 1.0. In Poole 2020. doi:10.2173/bow.rinduc.01.
Roy, Charlotte L., and Lindsey Shartell. 2017. "Sharp-tailed Grouse Responses to Fall Prescribed Fire and Mowing." In Cornicelli et al. 2017, 36–47.
Runde, Douglas E., and David E. Capen. 1987. "Characteristics of Northern Hardwood Trees Used by Cavity-Nesting Birds." *Journal of Wildlife Management* 51: 217–23.
Rusch, Donald H., Stephen Destefano, Michael C. Reynolds, and David Lauten. 2020. "Ruffed Grouse (*Bonasa umbellus*)." Version 1.0. In Poole and Gill 2020. doi:10.2173/bow.rufgro.01.
Rush, Scott A., Nathan Klaus, Tim Keyes, John Petrick, and Robert Cooper. 2012. "Fire Severity Has Mixed Benefits to Breeding Bird Species in the Southern Appalachians." *Forest Ecology and Management* 263: 94–100.
Rush, Scott A., and Bridget J. M. Stutchbury. 2008. "Survival of Fledgling Hooded Warblers (*Wilsonia citrina*) in Small and Large Forest Fragments." *Auk* 125: 183–91.
Russell, Robert P., Jr. 1969. "The Summer Season." *Loon* 41: 105–19.
Russell, Robert P., Katherine E. Koch, and Stephen J. Lewis. 2016. *Upper Mississippi Valley/Great Lakes Regional Shorebird Conservation Plan*. Version 2.0. Bloomington, MN: US Fish and Wildlife Service, Division of Migratory Birds.
Rustad, Orwin. A. 1951. "The Lark Bunting in Rice County." *Flicker* 23: 18–19.
Ruth, Janet M. 2015. "Status Assessment and Conservation Plan for the Grasshopper Sparrow (*Ammodramus savannarum*)." Version 1.0. Lakewood, CO: US Fish and Wildlife Service.
Ryder, Ronald A., and David E. Manry. 2020. "White-faced Ibis (*Plegadis chihi*)." Version 1.0. In Poole and Gill 2020. doi:10.2173/bow.whfibi.01.
Ryser, Fred A., Jr. 1985. *Birds of the Great Basin: A Natural History*. Reno: University of Nevada Press.
Rysgaard, George N. 1936. "The 1936 Nesting Season." *Flicker* 8: 26–45.
———. 1937a. "Afoot in Northern Minnesota." *Flicker* 9 (3/4): 1–5.
———. 1937b. "American Avocets." *Flicker* 9 (3/4): 6.
———, ed. 1937c. "American Egrets." *Flicker* 9 (3/4): 6.
———. 1937d. "1937 Minnesota Nesting Records." *Flicker* 9 (3/4): 7–12.
Sallabanks, Rex. 1993. "*Setophaga ruticilla* (American Redstart)." Element Stewardship Abstract Report. Arlington, VA: The Nature Conservancy, Conservation Science Division, in association with the Network of Natural Heritage Programs and Conservation Data Centers.
Sampson, Elizabeth. 1934. "Downy Goes Visiting." *Bird-Lore* 36: 356–60.
Sánchez-Bayo, Francisco, and Kris A. G. Wyckhuys. 2019. "Worldwide Decline of the Entomofauna: A Review of Its Drivers." *Biological Conservation* 232: 8–27.
Sandrock, James, and Jean C. Prior. 2014. *The Scientific Nomenclature of Birds in the Upper Midwest*. Iowa City: University of Iowa Press.
Sauer, John R., and Sam Droege. 1990. "Recent Population Trends of the Eastern Bluebird." *Wilson Bulletin* 102: 239–52.
Sauer, John R., William A. Link, and James E. Hines. 2020. *The North American Breeding Bird Survey, Analysis Results 1966–2019: US Geological Survey Data Release*. doi:10.5066/P96A7675.
Sauer, John R., S. Schwartz, and B. Hoover. 1996. *The Christmas Bird Count Home Page*. Version 95.1. Laurel, MD: US Geological Survey Patuxent Wildlife Research Center.
Saunders, Aretas Andrews. 1935. *A Guide to Bird Songs: Descriptions and Diagrams of the Songs and Singing Habits of the Land Birds of Northeastern United States*. New York: D. Appleton-Century.
Saunders, Sarah P., Kristin A. L. Hall, Nina Hill, and Nicole L. Michel. 2019. "Multiscale Effects of Wetland Availability and Matrix Composition on Wetland Breeding Birds in Minnesota, USA." *Condor: Ornithological Applications 121 (3)*. doi:10.1093/condor/duz024.
Savaloja [Savoloja, misspelled in source], Terry. 1973. "Evidence of Solitary Sandpiper Breeding." *Loon* 45: 96.
———. 1974. "New Yellow Rail Area." Loon 46: 41–42.
Scharf, William C., and Josef Kren. 2020. "Orchard Oriole (*Icterus spurius*)." Version 1.0. In Poole 2020. doi:10.2173/bow.orcori.01.
Scharf, William C., and Gary W. Shugart. 1998. "Distribution and Abundance of Gull, Tern, and Cormorant Nesting Colonies of the Great Lakes, 1989 and 1990." No. 1, *Gale Gleason Environmental Institute Publication,* edited by William W. Bowerman and Amy S. Roe. Sault Ste. Marie, MI: Lake Superior State University Press.
Schimpf, Ann, and Lee A. Pfannmuller. 1984. "Will the Piping Plover Survive?" *Minnesota Volunteer* 47 (May/June): 50–55.
Schladweiler, John. 1985. "The Burrowing Owl—An Endangered Species on Minnesota's Western Grasslands." *Blazing Star* (newsletter). Minnesota Department of Natural Resources,

Nongame Wildlife Program, Natural Heritage Program, and Scientific & Natural Areas Program.
———. 1986. "First State Nesting Record for the Little Gull." *Loon* 58: 166–70.
———. 2012. "Lark Bunting in Jackson County." *Loon* 84: 153–54.
———. 2016. "Greater Prairie-Chickens in Pipestone County." *Loon* 88: 186.
Schmierer, Al. 1984. "A Pair of Rock Wrens at Bemidji." *Loon* 56: 192–93.
Schneider, Todd M., Giff Beaton, Timothy S. Keyes, and Nathan A. Klaus, eds. 2010. *The Breeding Bird Atlas of Georgia*. Athens: University of Georgia Press.
Schon, Steven. 1988. "Long-billed Curlew Visits Ely." *Loon* 60: 89–90.
Schoolcraft, Henry Rowe. 1821. *Narrative Journal of Travels through the Northwestern Regions of the United States: Extending from Detroit through the Great Chain of American Lakes to the Sources of the Mississippi River, Performed as a Member of the Expedition under Governor Cass. In the Year 1820*. Albany: E. & E. Hosford.
———. 1834. *Narrative of an Expedition through the Upper Mississippi to Itasca Lake, the Actual Source of This River; Embracing an Exploratory Trip through the St. Croix and Burtwood (or Broule) Rivers, in 1832, under the Direction of Henry R. Schoolcraft*. New York: Harper & Brothers.
———. 1854. *Summary Narrative of an Exploratory Expedition to the Sources of the Mississippi River, in 1820: Resumed and Completed, by the Discovery of Its Origin in Itasca Lake, in 1832*. Philadelphia: Lippincott, Grambo & Co.
Schorger, Arlie W. 1926. "Richardson's Owl (*Cryptoglaux funereal richardsoni*) in Cook County, Minnesota." *Auk* 43: 544.
———. 1942. "The Wild Turkey in Early Wisconsin." *Wilson Bulletin* 54: 173–82.
———. 1945. "The Ruffed Grouse in Early Wisconsin." *Transactions of the Wisconsin Academy of Sciences, Arts and Letters* 37: 35–90.
———. 1955. *The Passenger Pigeon: Its Natural History and Extinction*. Madison: University of Wisconsin Press.
Schoumatoff, Alex. 2014. "500,000 Cranes Are Headed for Nebraska in One of Earth's Greatest Migrations." *Smithsonian,* March. https://www.smithsonianmag.com/science-nature/500000-cranes-are-headed-nebraska-one-earths-greatest-migrations-180949816/.
Schroeder, Michael A., Erik J. Blomberg, David A. Boag, Peter Pyle, and Michael A. Patten. 2020. "Spruce Grouse (*Falcipennis canadensis*)." Version 1.0. In Rodewald 2020. doi:10.2173/bow.sprgro.01.
Schroeder, Richard L. 1985. *Habitat Suitability Index Models: Pine Warbler*. Report FWS/OBS-82/10.28. Washington, DC: US Fish and Wildlife Service, Division of Biological Services, Western Energy and Land Use Team.
Schroeder, Roger. 1996. "Possible White-winged Crossbill Nesting." *Loon* 68: 243–44.
Schukman, John M., and Blair O. Wolf. 2020. "Say's Phoebe (*Sayornis saya*)." Version 1.0. In Rodewald 2020. doi:10.2173/bow.saypho.01.
Schulte, Lisa A., and Gerald J. Niemi. 1998. "Bird Communities of Early-Successional Burned and Logged Forest." *Journal of Wildlife Management* 62: 1418–29.
Schulte, Lisa A., Anna M. Pidgeon, and David J. Mladenoff. 2005. "One Hundred Fifty Years of Change in Forest Bird Breeding Habitat: Estimates of Species Distributions." *Conservation Biology* 19: 1945–56.
Schulz, John H., Paul I. Padding, and Joshua J. Millspaugh. 2006. "Will Mourning Dove Crippling Rates Increase with Nontoxic-Shot Regulations?" *Wildlife Society Bulletin* 34: 861–65.
Schumacher, Carol. 1992. "Probable Minnesota White-eyed Vireo Nesting Record, Houston County, 1991." *Loon* 64: 50–52.
Schummer, Michael L., Alan D. Afton, Shannon S. Badzinski, Scott A. Petrie, Glenn Olsen, Kevin Jacobs, Mark Mitchell, et al. 2013. "Using Satellite Telemetry to Evaluate the Effectiveness of the Waterfowl Breeding Population and Habitat Survey for Counting Lesser and Greater Scaup in North America." 6th North American Duck Symposium, Memphis, Tennessee.
Schwarz, Gideon. 1978. "Estimating the Dimension of a Model." *Annals of Statistics* 6: 461–64.
Scott, George A. 1963. "First Nesting of the Little Gull (*Larus minutus*) in Ontario and in the New World." *Auk* 80: 548–49.
Sealy, Spencer G. 1976. "The 1973 Dickcissel Invasion of Southern Manitoba." *Canadian Field-Naturalist* 90: 464–66.
Seamans, Mark E. 2020. *Mourning Dove Population Status, 2020*. Washington, DC: US Fish and Wildlife Service, Division of Migratory Bird Management.
Seamans, Mark E., and Rebecca D. Rau. 2019. "American Woodcock Population Status, 2019." Laurel, MD: US Fish and Wildlife Service.
Sedgwick, James A. 2020. "Willow Flycatcher (*Empidonax traillii*)." Version 1.0. In Poole and Gill 2020. doi:10.2173/bow.wilfly.01.
Seeley, Mark W. 2015. *Minnesota Weather Almanac*. 2nd ed. St. Paul: Minnesota Historical Society Press.
Serratore, Angela. 2018. "Keeping Feathers off Hats–And on Birds." *Smithsonian*, May 15. https://www.smithsonianmag.com/history/migratory-bird-act-anniversary-keeping-feathers-off-hats-180969077/.
Seutin, Gilles, and B. Chartier. 1989. "The Rock Wren, *Salpinctes obsoletus,* Breeding at Churchill, Manitoba." *Canadian Field-Naturalist* 103: 416–17.
Shane, Thomas G. 2020. "Lark Bunting (*Calamospiza melanocorys*)." Version 1.0. In Poole and Gill 2020. doi:10.2173/bow.larbun.01.
Sharrock, J. T. R. 1976. Atlas of Breeding Birds in Britain and Ireland. Calton, Staffordshire: T. and A. D. Poyser for the British Trust for Ornithology and Irish Wildbird Conservancy.
Shartell, Lindsey. 2018. "Use of Managed Forest Wildlife Openings by American Woodcock." Minnesota Department of Natural Resources. https://files.dnr.state.mn.us/wildlife/research/summaries/2017/upland/2017upland004.pdf.
Shedd, Mary, and Steven G. Wilson. 1987. "The Summer Season, June 1 to July 1, 1986." *Loon* 59: 19–35.
Sherry, Thomas W. 1979. "Competitive Interactions and Adaptive Strategies of American Redstarts and Least Flycatchers in a Northern Hardwoods Forest." *Auk* 96: 265–83.
Sherry, Thomas W., Richard T. Holmes, Peter Pyle, and Michael A. Patten. 2020. "American Redstart (*Setophaga ruticilla*)." Version 1.0. In Rodewald 2020. doi:10.2173/bow.amered.01.
Shields, William M., and Keith L. Bildstein. 1979. "Bird versus Bats: Behavioral Interactions at a Localized Food Source." *Ecology* 60: 468–74.
Shira, A. F. 1964. "Summer Tanager Seen in Winona County." *Loon* 36: 31.
Shriver, W. Gregory, Thomas P. Hodgman, and Alan R. Hanson. 2020. "Nelson's Sparrow (*Ammospiza nelsoni*)." Version 1.0. In Rodewald 2020. doi:10.2173/bow.nstspa.01.
Shriver, W. Gregory, Andrea L. Jones, Peter D. Vickery, Andrew Weik, and Jeffery Wells. 2005. "The Distribution and Abundance of Obligate Grassland Birds Breeding in New England and New York." US Department of Agriculture, Forest Service, General Technical Report PSW-GTR-191.
Shuford, W. David, and David P. Craig. 2002. *Status Assessment and Conservation Recommendations for the Caspian Tern (*Sterna Caspia*)*

in North America. Portland, OR: US Fish and Wildlife Service.
Sibley, David Allen. 2014. *The Sibley Guide to Birds*. 2nd ed. New York: Alfred A. Knopf.
Siderits, Karl. 1980. "1980 Bald Eagle, Osprey and Blue Heron Nesting, Superior National Forest." *Loon* 52: 193.
Sidor, Inga F., Mark A. Pokras, Andrew R. Major, Robert H. Poppenga, Kate M. Taylor, and Rose M. Miconi. 2003. "Mortality of Common Loons in New England, 1987 to 2000." *Journal of Wildlife Diseases* 39: 306–15.
Sivets, Jerry. 1975. "Ruffed Grouse in Hennepin County." *Loon* 47: 143.
Slater, Gary L., and Christine Rock. 2005. *Northern Harrier (*Circus cyaneus*): A Technical Conservation Assessment*. Prepared for the US Department of Agriculture, Forest Service, Rocky Mountain Region, Species Conservation Project. Mount Vernon, WA: Ecostudies Institute. https://www.fs.usda.gov/Internet/FSE_DOCUMENTS/stelprdb5182007.pdf.
Small, Robert J., James C. Holzwart, and Donald H. Rusch. 1991. "Predation and Hunting Mortality of Ruffed Grouse in Central Wisconsin." *Journal of Wildlife Management* 55: 512–20.
Smallwood, John A., and David M. Bird. 2020. "American Kestrel (*Falco sparverius*)." Version 1.0. In Poole and Gill 2020. doi:10.2173/bow.amekes.01.
Smith, Alan R. 1996. *Atlas of Saskatchewan Birds*. Regina: Saskatchewan Natural History Society.
Smith, Andrew D. 2016. "Bewick's Wren in Dakota County." *Loon* 88: 40–41.
Smith, Charles R. 1992. "Henslow's Sparrow (*Ammodramus henslowii*)." In *Migratory Nongame Birds of Management Concern in the Northeast,* edited by Kathryn J. Schneider and Diane M. Pence, 315–30. Newton Corner, MA: US Fish and Wildlife Service.
Smith, Elizabeth Zimmerman. 2018. "A Draft History of Bluebirds and Bluebirding." *Sialis.org*. Retrieved from http://www.sialis.org/history.htm.
Smith, Joe. 2016a. "The Battles of Song Sparrows: How a Scientific Outsider Changed How We Study Birds." Birds & Birding (blog). *Nature: Cool Green Science,* August 8. http://blog.nature.org/science/2016/08/08/battles-song-sparrows-scientific-outsider-changed-study-birds-ornithology-margaret-morse-nice/.
———. 2016b. "Shorebirds Are Still Being Hunted, Often Legally." Birds & Birding (blog). *Nature: Cool Green Science,* March 28. http://blog.nature.org/science/2016/03/28/shorebirds-are-still-being-hunted-often-legally/.
Smith, Kevin. 1993. "A Minnesota Great-tailed Grackle." *Loon* 65: 148.
Smith, Kimberly G., Keith A. Tarvin, and Glen E. Woolfenden. 2020. "Blue Jay (*Cyanocitta cristata*)." Version 1.0. In Poole 2020. doi:10.2173/bow.blujay.01.
Smith, Kimberly G., Sara Ress Wittenberg, R. Bruce Macwhirter, and Keith L. Bildstein. 2020. "Northern Harrier (*Circus hudsonius*)." Version 1.0. In Rodewald 2020. doi:10.2173/bow.norhar2.01.
Smith, P. William. 1987. "The Eurasian Collared-Dove Arrives in the Americas." *American Birds* 41: 1371–79.
Smith, P. William, and Herbert W. Kale II. 1986. "Eurasian Collared-Doves Collected in Florida." *Florida Field Naturalist* 14: 104–7.
Smith, Robert J., Margret I. Hatch, David A. Cimprich, and Frank R. Moore. 2020. "Gray Catbird (*Dumetella carolinensis*)." Version 1.0. In Poole 2020. doi:10.2173/bow.grycat.01.
Smith, Susan M. 1991. *The Black-capped Chickadee: Behavioral Ecology and Natural History*. Ithaca: Comstock Publishing of Cornell University Press.
Smith, W. G. 1981. "Observations on a Large Highway Kill of Evening Grosbeaks in British Columbia." *Syesis* 14: 163.
Smith, W. John, and Anne Marie Smith. 1996. "Vocal Signalling of the Great Crested Flycatcher, *Myiarchus crinitus* (Aves, Tyrannidae)." *Ethology* 102: 705–23.
Smithers, Brett L. 2001. "Northern Goshawk Food Habits in Minnesota: An Analysis Using Time-Lapse Recording Systems." Master's thesis, Texas Tech University.
Smithers, Brett L., Clint W. Boal, and David E. Andersen. 2005. "Northern Goshawk Diet in Minnesota: An Analysis Using Video Recording Systems." *Journal of Raptor Research* 39: 264–73.
Snyder, Lester Lynne. 1953. "On Eastern Empidonaces with Particular Reference to Variation in *E. traillii*." *Contributions of the Royal Ontario Museum of Zoology and Palaeontology* 35: 1–26.
Sogaard, James. 1966. "Yellow-crowned Night Heron Record." *Loon* 38: 108.
Sólymos, Péter, Steven M. Matsuoka, Erin M. Bayne, Subhash R. Lele, Patricia Fontaine, Steve G. Cumming, Diana Stralberg, et al. 2013. "Calibrating Indices of Avian Density from Non-standardized Survey Data: Making the Most of a Messy Situation." *Methods in Ecology and Evolution* 4: 1047–58.
Sólymos, Péter, Steven M. Matsuoka, Steven G. Cumming, Diana Stralberg, Patricia Fontaine, Fiona K. A. Schmiegelow, Samantha J. Song, et al. 2018. "Evaluating Time-removal Models for Estimating Availability of Boreal Birds during Point Count Surveys: Sample Size Requirements and Model Complexity." *Condor* 120: 765–86.
Sonerud, Geir A. 1997. "Hawk Owls in Fennoscandia: Population Fluctuations, Effects of Modern Forestry, and Recommendations on Improving Foraging Habitats." *Journal of Raptor Research* 31: 167–74.
Sorenson, Sharon. 2017. "For the Birds: The Secret Life of Titmice." *Courier & Press*, December 9. https://www.courierpress.com/story/life/columnists/sharon-sorenson/2017/12/10/birds-secret-life-titmice/925378001/.
Souder, William. 2004. *Under a Wild Sky: John James Audubon and the Making of* The Birds of North America. New York: North Point.
Soulen, Thomas K. 1980. "Minnesota's First Yellow-throated Warbler." *Loon* 52: 111–12.
Soulliere, Greg J., Brad A. Potter, Dan J. Holm, Diane A. Granfors, Mike J. Monfils, Steve J. Lewis, and Wayne E. Thogmartin. 2007. *Upper Mississippi River and Great Lakes Region Joint Venture Waterbird Habitat Conservation Strategy*. Fort Snelling, MN: US Fish and Wildlife Service. http://www.uppermissgreatlakesjv.org/docs/UMRGLR_JV_WaterbirdHCS.pdf.
Sousa, Patrick J. 1985. "Habitat Suitability Index Models: Gadwall (Breeding)." Biological Report 82(10.100). Washington, DC: US Fish and Wildlife Service.
South Carolina DNR (Department of Natural Resources). 2015. "Swallow-tailed Kite (*Elanoides forficatus*)." *South Carolina State Wildlife Action Plan (SWAP) 2015 Supplemental Volume: Species of Greatest Conservation Need (Priority Species)*. http://www.dnr.sc.gov/swap/species2015.html.
Southwick, Hiram C. 1956. "Avocets in Southwestern Minnesota." *Flicker* 28: 168.
Spaulding, Allen W., Karen E. Mock, Michael A. Schroeder, and Kenneth I. Warheit. 2006. "Recency, Range Expansion, and Unsorted Lineages: Implications for Interpreting Neutral Genetic Variation in the Sharp-tailed Grouse (*Tympanuchus phasianellus*)." *Molecular Ecology* 15: 2317–32.
Spendelow, Jeffrey A., and Stephen R. Patton. 1988. *National Atlas of Coastal Waterbird Colonies in the Contiguous United States:*

1976–1982. Biological Report 88(5). Washington, DC: US Fish and Wildlife Service.
Spiker, Chas. J. 1929. "The Hungarian Partridge in Northwest Iowa." *Wilson Bulletin* 41: 24–29.
Sprunt, Alexander, Jr. 1942. "Purple Martin." In Bent 1942, 489–509.
———. 1970. *South Carolina Bird Life*. Columbia: University of South Carolina Press.
Squires, John R., Richard T. Reynolds, Jaume Orta, and Jeffrey S. Marks. 2020. "Northern Goshawk (*Accipiter gentilis*)." Version 1.0. In Billerman 2020. doi:10.2173/bow.norgos.01.
Stanley, Evelyn. 1973. "Notes on the Hooded Warbler in Minneapolis." *Loon* 45: 133–34.
Stanley, Greg. 2019. "Where Are All the Pheasants? Habitat Loss Is Tied to More Farming." *Star Tribune*, October 13. https://www.startribune.com/where-are-all-the-pheasants/562922162/.
Stanojevic, Zoran, Peter Lee, and Jeannette D. Gysbers. 2006a. *Recent Anthropogenic Changes within the Boreal Plains Ecozone of Saskatchewan and Manitoba: Interim Report (A Global Forest Watch Canada Report)*. Edmonton, Alberta: Global Forest Watch Canada.
———. 2006b. *Recent Anthropogenic Changes within the Northern Boreal, Southern Taiga, and Hudson Plains Ecozones of Quebec (A Global Forest Watch Canada Report)*. Edmonton, Alberta: Global Forest Watch Canada.
Stanton, Jessica C., Peter Blancher, Kenneth V. Rosenberg, Arvind O. Panjabi, and Wayne E. Thogmartin. 2019. "Estimating Uncertainty of North American Landbird Population Sizes." *Avian Conservation and Ecology* 14: 4. doi:10.5751/ACE-01331-140104.
Stauffer, Dean F., and Louis B. Best. 1980. "Habitat Selection by Birds of Riparian Communities: Evaluating Effects of Habitat Alterations." *Journal of Wildlife Management* 44: 1–15.
Steblay, Jon. 1986. "Mountain Bluebird Paired with an Eastern Bluebird." *Loon* 58: 194–96.
Stedman, Stephen J. 2020. "Horned Grebe (*Podiceps auritus*)." Version 1.0. In Billerman 2020. doi:10.2173/bow.horgre.01.
Steeves, Tanner K., Shannon B. Kearney-McGee, Margaret A. Rubega, Calvin L. Cink, and Charles T. Collins. 2020. "Chimney Swift (*Chaetura pelagica*)." Version 1.0. In Poole 2020. doi:10.2173/bow.chiswi.01.
Stein, Robert Carrington. 1958. "The Behavioral, Ecological and Morphological Characteristics of Two Populations of the Alder Flycatcher, *Empidonax traillii* (Audubon)." New York State Museum and Science Service Bulletin, no. 371. Albany: University of the State of New York, State Education Department.
———. 1963. "Isolating Mechanisms between Populations of Traill's Flycatchers." *Proceedings of the American Philosophical Society* 107: 21–50.
Steiner, Ryan, Gerald Niemi, Frank Nicoletti, Mary Jane Evans, and Edmund Zlonis. 2022. "Changes in Survey Effort Can Influence Conclusions about Migration Phenology." *Journal of Raptor Research* 56 (2): 171–79.
Stenlund, Milton H., and Lester T. Magnus. 1951. "The Spruce Hen Comes Back." *Conservation Volunteer* 14 (November/December): 20–24.
Stensaas, Sparky. 2005. "Further Notes on Minnesota's First Nesting Rock Wrens." *Loon* 77: 186.
Stephenson, Jeff J. 2002. "Late White-eyed Vireo in Cook County." *Loon* 74: 121–22.
Stepney, Philip H. R. 1979. "Competitive and Ecological Overlap between Brewer's Blackbird and the Common Grackle, with Consideration of Associated Foraging Species." PhD diss., University of Toronto.
Steva, Shelley. 1982. "Yellow-crowned Night Heron at Thief Lake." *Loon* 54: 244–45.
Stevens, Bryan. 2018. "Tufted Titmouse: A Small Songbird with a Fierce, Gigantic Personality." *Bristol Herald Courier,* March 18. https://www.heraldcourier.com/community/tufted-titmouse-a-small-songbird-with-a-fierce-gigantic-personality/article_f89170ac-2972-11e8-9034-1b5bb7d84bb8.html.
Stewart, Rebecca L. M., Katherine A. Bredin, Andrew R. Coutuier, Andrew G. Horn, Denis Lepage, Scott Makepeace, Philip D. Taylor, et al., eds. 2015. *Second Atlas of Breeding Birds of the Maritime Provinces*. Sackville: Bird Studies Canada, Environment Canada, Natural History Society of Prince Edward Island, Nature New Brunswick, New Brunswick Department of Natural Resources, Nova Scotia Bird Society, Nova Scotia Department of Natural Resources, Prince Edward Island Department of Agriculture and Forestry.
Stewart, Robert E. 1975. *Breeding Birds of North Dakota*. Fargo, ND: Tri-College Center for Environmental Studies.
Stokes, Donald, and Lillian Stokes. 1989. *The Hummingbird Book: The Complete Guide to Attracting, Identifying, and Enjoying Hummingbirds*. Boston: Little, Brown.
Stout, Bonnie E., and Gary L. Nuechterlein. 2020. "Red-necked Grebe (*Podiceps grisegena*)." Version 1.0. In Billerman 2020. doi:10.2173/bow.rengre.01.
Strand, Ginger. 2016. "A Safe Place to Land." *Nature Conservancy Magazine*, April 1. https://www.nature.org/en-us/explore/magazine/magazine-articles/aprilmay-2016-issue-a-safe-place-to-land/.
Streby, Henry M., John P. Loegering, and David E. Andersen. 2012. "Spot-mapping Underestimates Song-territory Size and Use of Mature Forest by Breeding Golden-winged Warblers in Minnesota, USA." *Wildlife Society Bulletin* 36: 40–46.
Streby, Henry M., Sean M. Peterson, Gunnar R. Kramer, and David E. Andersen. 2015. "Post-independence Fledgling Ecology in a Migratory Songbird: Implications for Breeding-grounds Conservation." *Animal Conservation* 18: 228–35.
Strickland, Dan, and Henri R. Ouellet. 2020. "Canada Jay (*Perisoreus canadensis*)." Version 1.0. In Rodewald 2020. doi:10.2173/bow.gryjay.01.
Strnad, Forest V. 1959. "The Minnesota Bird Banders." *Flicker* 31: 112–14.
———. 1961. "The President's Page." *Flicker* 33: 29–31.
———. 1963. "Hawk-Owls in St. Louis County, Minnesota." *Flicker* 35: 129–34.
Strong, Paul I. V., and Richard J. Baker. 2000. "An Estimate of Minnesota's Summer Population of Adult Common Loons." In Svingen and Hertzel 2000, 5–13.
Strubbe, Ernest H. 1965. "Common Gallinule Record for Stevens County." *Loon* 37: 106.
Stubbs, Megan. 2014. *Conservation Reserve Program (CRP): Status and Issues*. Congressional Research Service Report R42783. https://crsreports.congress.gov/product/details?prodcode=R42783.
Stucker, Jennifer H., and Francesca J. Cuthbert. 2000. "Biodiversity of Southeastern Minnesota Forested Streams: Relationships between Trout Habitat Improvement Practices, Riparian Communities and the Louisiana Waterthrush." Report to the Natural Heritage and Nongame Wildlife Program. St. Paul: University of Minnesota, Department of Fisheries and Wildlife.
Stucker, Steven P. 1996. "Sprague's Pipit in Polk County, 31 May–1 June 1995." *Loon* 68: 67–68.
———. 2004. "Notes on the Distribution and Identification of the Cackling Goose (*Branta hutchinsii*) and Canada Goose (*Branta canadensis*) in Minnesota." *Loon* 76: 225–28.
Stucker, Steven P., Pamela Skoog Perry, and Kevin R. Woizeschke. 2004. "Notes on the

Breeding Distribution of Red-shouldered Hawks in Minnesota." *Loon* 76: 3–8.
Sugden, Lawson G. 1973. *Feeding Ecology of Pintail, Gadwall, American Wigeon and Lesser Scaup Ducklings in Southern Alberta*. Canadian Wildlife Service Report, Series 24. Ottawa: Canadian Wildlife Service.
Sulkava, Seppo, and Kauko Huhtala. 1997. "The Great Gray Owl (*Strix nebulosa*) in the Changing Forest Environment of Northern Europe." *Journal of Raptor Research* 31: 151–59.
Sullivan, Paul. 1986. "White-eyed Vireo in Martin County." *Loon* 58: 139.
Surdick, Jim. 1995. "The Louisiana Waterthrush in Southeast Minnesota." *Loon* 67: 201–4.
Sutton, Alex O., Dan Strickland, Nikole E. Freeman, and D. Ryan Norris. 2021. "Climate-driven Carry-over Effects Negatively Influence Population Growth Rate in a Food-caching Boreal Passerine." *Global Change Biology* 27: 983–92.
Sutton, George Miksch. 1927. "Flocking, Mating, and Nest-building Habits of the Prairie Horned Lark." *Wilson Bulletin* 34: 131–41.
Svedarsky, W. Daniel. 1977. "Roadside Nesting by Prairie Grouse in Northwest Minnesota." *Prairie Naturalist* 9: 41–42.
———. 1979. "Spring and Summer Ecology of Female Greater Prairie Chickens in Northwestern Minnesota." PhD diss., University of North Dakota, Grand Forks.
———. 1992. "Observations on Nesting of the American Bittern in Northwest Minnesota." *Prairie Naturalist* 24: 241–50.
Svedarsky, W. Daniel, Richard J. Oehlenschlager, and Tod D. Tonsager. 1982. "A Remnant Flock of Greater Prairie Chickens in North Central Minnesota." *Loon* 54: 5–13.
Svedarsky, W. Daniel, John E. Toepfer, Ronald L. Westemeier, and Robert J. Robel. 2003. *Effects of Management Practices on Grassland Birds: Greater Prairie-Chicken*. Jamestown, ND: Northern Prairie Wildlife Research Center. https://pubs.er.usgs.gov/publication/70159589.
Svedarsky, W. Daniel, and Gerald L. Van Amburg. 1996. *Integrated Management of the Greater Prairie-chicken and Livestock on the Sheyenne National Grasslands*. Special Report to the North Dakota Game and Fish Department and US Forest Service. Bismarck, ND: North Dakota Game and Fish Department.
Svedarsky, W. Daniel, Ronald L. Westemeier, Robert J. Robel, Sharron Gough, and John E. Toepher. 2000. "Status and Management of the Greater Prairie-chicken *Tympanuchus cupido pinnatus* in North America." *Wildlife Biology* 6: 277–84.
Svedarsky, W. Daniel, Terrance J. Wolfe, and John E. Toepfer. 1997. *The Greater Prairie-Chicken in Minnesota*. Minnesota Wildlife Report 11. St. Paul: Minnesota Department of Natural Resources. https://files.dnr.state.mn.us/recreation/hunting/prairiechicken/pc_mn_97.pdf.
Svingen, Peder H. 1991. "Notes on a Clark's Grebe at Thielke Lake." *Loon* 63: 194–96.
———. 1992. "Sprague's Pipit in Clay County." *Loon* 64: 231.
———. 1998a. "The American Avocet in Minnesota." *Loon* 70: 11–20.
———. 1998b. "Recent Status of the Henslow's Sparrow in Minnesota." *Loon* 70: 153–54.
———. 2000. "The Little Gull in Minnesota." *Loon* 72: 10–13.
———. 2004. "Proceedings of the Minnesota Ornithologists' Union Records Committee." *Loon* 76: 152–61.
———. 2005a. "An Influx of Black-necked Stilts." *Loon* 77: 49.
———. 2005b. "Proceedings of the Minnesota Ornithologists' Union Records Committee." *Loon* 77: 169–74.
———. 2006. "Burrowing Owls in Polk County." *Loon* 78: 233.
———. 2007a. "Proceedings of the Minnesota Ornithologists' Union Records Committee." *Loon* 79: 50–57.
———. 2007b. "Proceedings of the Minnesota Ornithologists' Union Records Committee." *Loon* 79: 123–29.
———. 2008. "Proceedings of the Minnesota Ornithologists' Union Records Committee." *Loon* 80: 5–13.
———. 2009a. "Proceedings of the Minnesota Ornithologists' Union Records Committee." *Loon* 81: 7–12.
———. 2009b. "Proceedings of the Minnesota Ornithologists' Union Records Committee." *Loon* 81: 115–19.
———. 2010. "Proceedings of the Minnesota Ornithologists' Union Records Committee." *Loon* 82: 3–6.
Svingen, Peder H., and Anthony X. Hertzel, eds. 2000. *The Common Loon: Population Status and Fall Migration in Minnesota*. Minnesota Ornithologist's Union Occasional Papers, no. 3. Minneapolis: Minnesota Ornithologists' Union.
———. 2013. "The 2012 Dickcissel Irruption in Minnesota." *Loon* 85: 35–42.
———. 2015. "Dynamic Distribution of Blue-winged Warbler (*Vermivora cyanoptera*), Golden-winged Warbler (*V. chrysoptera*), and Their Hybrids in Minnesota." *Loon* 87: 147–55.
Svingen, Peder H., and Karla A. Kinstler. 2002. "First Minnesota Nesting Record of the Eurasian Collared-Dove." *Loon* 74: 7–12.
Svingen, Peder H., and James W. Lind. 2005. "The 2004–2005 Influx of Northern Owls, Part II: Great Gray Owl." *Loon* 77: 194–208.
Swanson, David A. (1998) 2002. *Effects of Management Practices on Grassland Birds: Savannah Sparrow*. Rev. ed. Jamestown, ND: Northern Prairie Wildlife Research Center. https://pubs.usgs.gov/unnumbered/70159923/report.pdf.
Swanson, David L., James L. Ingold, and Robert Galati. 2020. "Golden-crowned Kinglet (*Regulus satrapa*)." Version 1.0. In Poole 2020. doi:10.2173/bow.gockin.01.
Swanson, David L., James L. Ingold, and George E. Wallace. 2020. "Ruby-crowned Kinglet (*Regulus calendula*)." Version 1.0. In Poole 2020. doi:10.2173/bow.ruckin.01.
Swanson, Evadene B. 1940. "The Use and Conservation of Minnesota Wildlife 1850–1900." Master's thesis, University of Minnesota. Transcribed and edited by Nancy Hertzel, Anthony X. Hertzel, and Carrol L. Henderson. 2007. Minnesota Department of Natural Resources. https://www.lrl.mn.gov/docs/2007/other/070554.pdf.
Swanson, George A., Gary L. Krapu, James C. Bartonek, Jerome R. Serie, and Douglas H. Johnson. 1974. "Advantages in Mathematically Weighting Waterfowl Food Habits Data." *Journal of Wildlife Management* 38: 302–7.
Swanson, Gustav. 1929. "The Quail in Minnesota." *Flicker* 1: 33–34.
Swanson, Norma, and Al Batt. 2002. "White-eyed Vireo Nesting in Steele County." *Loon* 74: 40–41.
Swanson, Wally. 1999. "A Swallow-tailed Kite." *Loon* 71: 177–78.
Swedenborg, E. D. 1947. "Ten Years of Winter Birds." *Flicker* 19: 14–18.
Szczys, Patricia, Stephen A. Oswald, and Jennifer M. Arnold. 2017. "Conservation Implications of Long-distance Migration Routes: Regional Metapopulation Structure, Asymmetrical Dispersal, and Population Declines." *Biological Conservation* 209: 263–72.
Szuba, Kandyd J., and James F. Bendell. 1983. "Population Densities and Habitats of Spruce Grouse in Ontario." In *Resources and Dynamics of the Boreal Zone: Proceedings of a*

Conference Held at Thunder Bay, Ontario, August 1982, edited by Ross W. Wein, Roderick R. Riewe, and Ian R. Methven, 199–213. Ottawa: Association of Canadian Universities for Northern Studies.

Tacha, Martha C., and Thomas C. Tacha. 1985. *Status and Distribution of Sandhill Cranes in Minnesota*. Final Project Report. Minnesota Department of Natural Resources. https://files.dnr.state.mn.us/eco/nongame/projects/consgrant_reports/1985/1985_tacha.pdf.

Tacha, Thomas C., and Clait E. Braun, eds. 1994. *Migratory Shore and Upland Game Bird Management in North America*. Washington, DC: International Association of Fish and Wildlife Agencies.

Takekawa, John Y., and Edward O. Garton. 1984. "How Much Is an Evening Grosbeak Worth?" *Journal of Forestry* 82: 426–28.

Tallman, Dan Allen, David L. Swanson, and Jeffrey S. Palmer. 2002. *Birds of South Dakota*. 3rd ed. Aberdeen: South Dakota Ornithologists' Union.

Tarof, Scott A., and James V. Briskie. 2020. "Least Flycatcher (*Empidonax minimus*)." Version 1.0. In Poole 2020. doi:10.2173/bow.leafly.01.

Taverner, Percy A. 1934. *Birds of Canada*. Ottawa: National Museum of Canada.

Taylor, Peter. 2018. "Bufflehead." In *The Atlas of the Breeding Birds of Manitoba, 2010–2014*, edited by Christian Artuso, Andrew R. Couturier, Ken D. De Smet, Rudolf F. Koes, Denis Lepage, Jon McCracken, Randall D. Mooi, et al. Winnipeg, Manitoba: Bird Studies Canada. http://www.birdatlas.mb.ca/accounts/speciesaccount.jsp?sp=BUFF&lang=en.

Taylor, Rick. 2013. "Great Lakes Loons Dying in Record Numbers from Botulism Outbreak Spurred by Ecological Disturbance." *Ann Arbor News*, August 13. http://www.annarbor.com/passions-pursuits/great-lakes-loons-dying-in-record-numbers-from-botulism-outbreak-spurred-by-ecological-disturbance/.

Telfair, Raymond C., II. 2020. "Cattle Egret (*Bubulcus ibis*)." Version 1.0. In Billerman 2020. doi:10.2173/bow.categr.01.

Temple, Stanley A. 1988. "When Is a Bird's Habitat Not Habitat?" *Passenger Pigeon* 50: 37–41.

———. 2020. "Dickcissel (*Spiza americana*)." Version 1.0. In Poole and Gill 2020. doi:10.2173/bow.dickci.01.

Terborgh, John. 1989. *Where Have All the Birds Gone? Essays on the Biology and Conservation of Birds That Migrate to the American Tropics*. Princeton, NJ: Princeton University Press.

Terres, John K. 1980. *The Audubon Society Encyclopedia of North American Birds*. New York: Alfred A. Knopf.

Tester, John R. 1995. *Minnesota's Natural Heritage: An Ecological Perspective*. Minneapolis: University of Minnesota Press.

Tester, John R., Susan M. Galatowitsch, Rebecca A. Montgomery, and John J. Moriarty. 2020. *Minnesota's Natural Heritage*. 2nd ed. Minneapolis and London: University of Minnesota Press.

Tester, John R., and William H. Marshall. 1961. "A Study of Certain Plant and Animal Interrelations on a Native Prairie of Northwestern Minnesota." Minnesota Museum of Natural History, Occasional Paper No. 8. Minneapolis: University of Minnesota Press.

Theodore, I. 1952. "Bell's Vireo in Winona County." *Flicker* 24: 142–44.

——— [L., misspelled in source]. 1955. "Yellow-crowned Night Heron Nests in Southeast Minnesota." *Flicker* 27: 171–72.

———. 1956. "Yellow-crowned Night Herons Return to Houston County." *Flicker* 28: 169.

Thogmartin, Wayne E. 2010. "Sensitivity Analysis of North American Bird Population Estimates." *Ecological Modelling* 221: 173–77. doi:10.1016/j.ecolmodel.2009.09.013.

Thogmartin, Wayne E., John R. Sauer, and Melinda G. Knutson. 2004. "A Hierarchical Spatial Model of Avian Abundance with Application to Cerulean Warblers." *Ecological Applications* 14: 1766–79.

Thomas, Susan M., James E. Lyons, Brad A. Andres, Elise Elliott-Smith, Eduardo Palacios, John F. Cavitt, J. Andrew Royle, et al. 2012. "Population Size of Snowy Plovers Breeding in North America." *Waterbirds* 35: 1–14.

Thomas Sadler Roberts Field Checklists, 1923–1931. Bell Museum of Natural History Records. Collection ua-00876. University of Minnesota, University Archives. https://archives.lib.umn.edu/repositories/14/archival_objects/345125.

Thompson, Ben H. 1933. *History and Present Status of the Breeding Colonies of the White Pelican (*Pelecanus erythrorhynchos*) in the United States*. Occasional Paper No. 1. US Department of the Interior, National Park Service.

Thompson, Milton D. 1936a. "The Immigration of the Starling in Northeastern Minnesota." *Flicker* 8: 49.

———. 1936b. "Minnesota Nesting Records of the Ring-billed Gull." *Flicker* 8: 50.

———. 1947. "Field Trip in Houston County." *Flicker* 19: 70–71.

Thoreau, Henry David. [1853] 1906. *The Writings of Henry David Thoreau: Journal V, March 5–November 30, 1853*. Edited by Bradford Torrey. Boston: Houghton Mifflin.

Tiner, Ralph W., Jr. 1984. *Wetlands of the United States: Current Status and Recent Trends*. Washington, DC: US Fish and Wildlife Service, National Wetlands Inventory.

Tischler, Keren B. 2001. "Pilot Study of Common Loon (*Gavia immer*) and Watercraft Interactions on Minnesota Lakes." Final report to the Minnesota Department of Natural Resources. https://files.dnr.state.mn.us/eco/nongame/projects/consgrant_reports/2001/2001_tischler.pdf.

Titus, James R., and Larry W. VanDruff. 1981. "Response of the Common Loon to Recreational Pressure in the Boundary Waters Canoe Area, Northeastern Minnesota." *Wildlife Monographs* 79: 3–59.

Titus, Kimberly, and James A. Mosher. 1981. "Nest-site Habitat Selected by Woodland Hawks in the Central Appalachians." *Auk* 98: 270–81.

Toepfer, John E. 2007. "Status and Management of the Greater Prairie-Chicken in Wisconsin—2006." *Passenger Pigeon* 69: 258–89.

Toews, David P. L., Scott A. Taylor, Rachel Vallender, Alan Brelsford, Bronwyn G. Butcher, Philipp W. Messer, and Irby J. Lovette. 2016. "Plumage Genes and Little Else Distinguish the Genomes of Hybridizing Warblers." *Current Biology* 26: 2313–18.

Tordoff, Harrison B., and Robert M. Mengel. 1951. "The Occurrence and Possible Significance of a Spring Molt in Leconte's Sparrow." *Auk* 68: 519–22.

Tordoff, Harrison B., and Patrick T. Redig. 1997. "Midwest Peregrine Falcon Demography, 1982–1995." *Journal of Raptor Research* 31: 339–46.

———. 1999. "Close Inbreeding in Peregrine Falcons in Midwestern United States." *Journal of Raptor Research* 33: 326–28.

Townsend, Charles W. 1926. "Virginia Rail." In Bent 1926, 292–301.

Tozer, Douglas C. 2016. "Marsh Bird Occupancy Dynamics, Trends, and Conservation in the Southern Great Lakes Basin: 1996 to 2013." *Journal of Great Lakes Research* 42: 136–45.

Tozer, Douglas C., C. Myles Falconer, and Debbie S. Badzinski. 2013. "Common Loon Reproductive Success in Canada: The West Is Best but Not for Long." *Avian Conservation and Ecology* 8: 1. doi:10.5751/ACE-00569-080101.

Tozer, Douglas C., Erica Nol, and Kenneth F. Abraham. 2010. "Effects of Local and Landscape-scale Habitat Variables on Abundance and Reproductive Success of Wetland Birds." *Wetlands Ecology and Management* 18: 679–93.

Tremblay, Junior A., Rita D. Dixon, Victoria A. Saab, Peter Pyle, and Michael A. Patten. 2020. "Black-backed Woodpecker (*Picoides arcticus*)." Version 1.0. In Rodewald 2020. doi:10.2173/bow.bkbwoo1.01.

Tremblay, Junior A., David L. Leonard Jr., and Louis Imbeau. 2020. "American Three-toed Woodpecker (*Picoides dorsalis*)." Version 1.0. In Rodewald 2020. doi:10.2173/b.attwoo1.02.

Trost, Charles H. 2020. "Black-billed Magpie (*Pica hudsonia*)." Version 1.0. In Billerman 2020. doi:10.2173/bow.bkbmag1.01.

Trottier, Garry C. 1992. *Conservation of Canadian Prairie Grasslands: A Landowner's Guide.* Ottawa: Canadian Wildlife Service.

Tuck, Leslie M. 1972. "The Snipes: A Study of the Genus *Capella*." Canadian Wildlife Service Monograph, Series 5. Ottawa: Canadian Wildlife Service.

Turner, Robert E. 1965. "Will Trumpeter Swans Return to Minnesota?" *Loon* 37: 122–26.

Tustison, Thomas A. 2001. "Influx of Snowy Egrets into Minnesota: Spring 2000 Migratory Movement." *Loon* 73: 3–7.

———. 2011a. "Proceedings of the Minnesota Ornithologists' Union Records Committee." *Loon* 83: 3–5.

———. 2011b. "Proceedings of the Minnesota Ornithologists' Union Records Commitee." *Loon* 83: 99–103.

———. 2013a. "Proceedings of the Minnesota Ornithologists' Union Records Committee." *Loon* 85: 3–7.

———. 2013b. "Proceedings of the Minnesota Ornithologists' Union Records Committee." *Loon* 85: 99–103.

———. 2014a. "King Rail in Goodhue County." *Loon* 86: 209.

———. 2014b. "Proceedings of the Minnesota Ornithologists' Union Records Committee." *Loon* 86: 107–10.

———. 2015a. "Proceedings of the Minnesota Ornithologists' Union Records Committee." *Loon* 87: 4–8.

———. 2015b. "Proceedings of the Minnesota Ornithologists' Union Records Committee." *Loon* 87: 95–97.

———. 2016. "Proceedings of the Minnesota Ornithologists' Union Records Committee." *Loon* 88: 91–93.

———. 2017. "Proceeding of the Minnesota Ornithologists' Union Records Committee." *Loon* 89: 87–91.

———. 2018. "Western Wood-Pewee in Clay County." *Loon* 90: 34–35.

———. 2019. "Proceedings of the Minnesota Ornithologists' Union Records Committee." *Loon* 91: 89–92.

———. 2021. "Proceedings of the Minnesota Ornithologists' Union Records Committee." *Loon* 93: 113.

———. 2022. "Proceedings of the Minnesota Ornithologists' Union Records Committee." *Loon* 94: 115–19.

Tvrdik, Genevieve M. 1966. "Observations of a Rookery Containing Great Blue Herons, Common Egrets, Black-crowned Night Herons, and Double-crested Cormorants." *Loon* 38: 106–7.

Twedt, Daniel J., and Richard D. Crawford. 2020. "Yellow-headed Blackbird (*Xanthocephalus xanthocephalus*)." Version 1.0. In Poole and Gill 2020. doi:10.2173/yehbla.01.

University of Minnesota Institute on the Environment. 2008. "Minnesota Statewide Conservation and Preservation Plan." https://www.lccmr.leg.mn/documents/scpp/final_plan/scpp_final_report.pdf.

Urbanek, Richard P., and James C. Lewis. 2020. "Whooping Crane (*Grus americana*)." Version 1.0. In Poole 2020. doi:10.2173/bow.whocra.01.

USDA (US Department of Agriculture). 2017. "CRP Enrollment and Rental Payments by State, 1986–2018." Excel document available at US Department of Agriculture Conservation Reserve Program Statistics. https://www.fsa.usda.gov/programs-and-services/conservation-programs/reports-and-statistics/conservation-reserve-program-statistics/index.

———. 2020. "The Conservation Reserve Program: A 35 Year History." https://www.fsa.usda.gov/Assets/USDA-FSA-Public/usdafiles/Conservation/PDF/35_YEARS_CRP_B.pdf.

———. 2021. "United States Cattle Inventory Down Slightly." News release, January 29. https://www.nass.usda.gov/Newsroom/archive/2021/01-29-2021.php.

US Department of Justice. 2011. "Minnesota Lake Man Charged with Destroying Pelican Nests and Killing Chicks." News release, September 15. https://www.justice.gov/archive/usao/mn/press/sep029.pdf.

USFS (US Forest Service). 2012. "Regional Forester Sensitive Animal Species for the Eastern Region." https://www.fs.usda.gov/Internet/FSE_DOCUMENTS/stelprdb5384459.pdf.

USFWS (US Fish and Wildlife Service). 1968. *Agassiz National Wildlife Refuge Narrative Report.* https://catalog.data.gov/dataset/agassiz-national-wildlife-refuge-narrative-report-calendar-year-1968.

———. 1995. *Migratory Nongame Birds of Management Concern in the United States: The 1995 List.* Prepared by the Office of Migratory Bird Management. Washington, DC: US Fish and Wildlife Service. Accessed June 1, 2023. https://ecos.fws.gov/ServCat/DownloadFile/101289?Reference=60929.

———. 1998a. "Endangered and Threatened Wildlife and Plants; 90-Day Finding for a Petition to List the Henslow's Sparrow as Threatened." *Federal Register* 63 (174): 48162–64.

———. 1998b. "Endangered and Threatened Wildlife and Plants; Notice of 12-Month Finding on a Petition to List the Northern Goshawk in the Contiguous United States West of the 100th Meridian." *Federal Register* 63 (124): 35183–84.

———. 1999a. "Endangered and Threatened Wildlife and Plants; Final Rule to Remove the American Peregrine Falcon from the Federal List of Endangered and Threatened Wildlife, and to Remove the Similarity of Appearance Provision for Free-Flying Peregrines in the Conterminous United States." *Federal Register* 64 (164): 46542–58.

———. 1999b. "Endangered and Threatened Wildlife and Plants; 90-Day Finding for a Petition to List the Baird's Sparrow as Threatened with Critical Habitat." *Federal Register* 64 (98): 27747–49.

———. 2002. *Southwestern Willow Flycatcher Recovery Plan.* Albuquerque, NM: US Fish and Wildlife Service, Region 2.

———. 2005a. *Agassiz National Wildlife Refuge Colonial Nesting Bird Surveys, 2005.* Unpublished report.

———. 2005b. *Mourning Dove National Strategic Harvest Management Plan.* Washington, DC: US Department of the Interior.

———. 2007a. "A Conservation Action Plan for the Cerulean Warbler (*Dendroica cerulean*)." https://partnersinflight.org/wp-content/uploads/2017/03/CeruleanWarbler-Focal-Species-Plan.pdf.

———. 2007b. "Endangered and Threatened Wildlife and Plants; Removing the Bald Eagle in the Lower 48 States from the List of Endangered and Threatened Wildlife; Final Rule; Endangered and Threatened Wildlife and Plants; Draft Post—Delisting and Monitoring

Plan for the Bald Eagle (*Haliaeetus leucocephalus*) and Proposed Information Collection; Notice." *Federal Register* 72 (130): 37346–72.
———. 2007c. "National Bald Eagle Management Guidelines." https://www.fws.gov/northeast/ecologicalservices/eaglenationalguide.html.
———. 2008. *Birds of Conservation Concern 2008*. Arlington, VA: US Fish and Wildlife Service, Division of Migratory Bird Management.
———. 2011. "Endangered and Threatened Wildlife and Plants; 90-Day Finding on a Petition to List the Golden-winged Warbler as Endangered or Threatened." *Federal Register* 76 (106): 31920–26.
———. 2012. "Endangered and Threatened Wildlife and Plants; Revised Designation of Critical Habitat for the Pacific Coast Population of the Western Snowy Plover." *Federal Register* 77 (118): 36727–36869.
———. 2014. *Draft Environmental Assessment: Management of Double-crested Cormorants under 50 CFR 21.47 and 21.48*. Arlington, VA: US Fish and Wildlife Service, Division of Migratory Bird Management. https://peer.org/wp-content/uploads/attachments/4_3_14_FWS_DEA.pdf.
———. 2016a. *Mid-winter Waterfowl Survey*. US Fish and Wildlife Service, Division of Migratory Bird Management. https://migbirdapps.fws.gov/mbdc/databases/mwi/aboutmwi_allflyways.htm.
———. 2016b. *Waterfowl Breeding Population and Habitat Survey: Retrievable Population Data, 1955–2015*. US Fish and Wildlife Service, Division of Migratory Bird Management, Patuxent, Maryland. https://migbirdapps.fws.gov/mbdc/databases/mas/maydb.asp?opt=maydb. Accessed November 10, 2016.
———. 2016c. *Trends in Duck Breeding Populations 1955–2015*. Washington, DC: US Fish and Wildlife Service, Division of Migratory Bird Management. https://ecos.fws.gov/ServCat/DownloadFile/204890.
———. 2019. *Waterfowl Population Status, 2019*. Washington, DC: US Fish and Wildlife Service, Division of Migratory Bird Management. https://www.fws.gov/media/waterfowl-population-status-2019.
———. 2020a. *Final Environmental Impact Statement: Management of Conflicts Associated with Double-crested Cormorants*. Falls Church, VA: US Fish and Wildlife Service, Division of Migratory Bird Management. https://cdxapps.epa.gov/cdx-enepa-II/public/action/eis/details?eisId=313501.
———. 2020b. "Final Report: Bald Eagle Population Size: 2020 Update." Washington, DC: US Fish and Wildlife Service, Division of Migratory Bird Management.
———. 2020c. "Migratory Bird Permits; Management of Conflicts Associated with Double-Crested Cormorants (*Phalacrocorax auritus*) throughout the United States." *Federal Register* 85 (249): 85535–56.
———. 2021. "Endangered Species." Endangered Species Database. Accessed April 28, 2021. https://www.fws.gov/endangered.
———. 2022. "Endangered and Threatened Wildlife and Plants: 6-Month Extension of Final Determination on the Proposed Removal of the Ivory-Billed Woodpecker from the List of Endangered and Threatened Wildlife." *Federal Register* 87 (129): 40477–478.
US Shorebird Conservation Plan Partnership. 2016. *US Shorebirds of Conservation Concern—2016*. http://www.shorebirdplan.org/wp-content/uploads/2016/08/Shorebirds-Conservation-Concern-2016.pdf.
Vallender, Rachel, Steven L. Van Wilgenburg, Lesley P. Bulluck, Amber Roth, Ron Canterbury, Jeffrey Larkin, R. Monica Fowlds, et al. 2009. "Extensive Rangewide Mitochondrial Introgression Indicates Substantial Cryptic Hybridization in the Golden-winged Warbler (*Vermivora chrysoptera*)." *Avian Conservation and Ecology* 4: 4. http://www.ace-eco.org/vol4/iss2/art4/.
Valley, Paul J. 1987. "Common Loon Productivity and Nesting Requirements on the Whitefish Chain of Lakes in North-Central Minnesota." *Loon* 59: 3–11.
Vanderhoff, Natasha, Peter Pyle, Michael A. Patten, Rex Sallabanks, and Frances C. James. 2020. "American Robin (*Turdus migratorius*)." Version 1.0. In Rodewald 2020. doi:10.2173/bow.amerob.01.
Venier, Lisa A., Steve Holmes, and Janet McI. Williams. 2020. "Bay-breasted Warbler (*Setophaga castanea*)." Version 1.0. In Poole 2020. doi:10.2173/bow.babwar.01.
Vennesland, Ross G., and Robert W. Butler. 2020. "Great Blue Heron (*Ardea herodias*)." Version 1.0. In Poole 2020. doi:10.2173/bow.grbher3.01.
Verbeek, Nicolaas A., and Carolee Caffrey. 2020. "American Crow (*Corvus brachyrhynchos*)." Version 1.0. In Poole and Gill 2020. doi:10.2173/bow.amecro.01.
Verner, Jared. 1965. "Breeding Biology of the Long-billed Marsh Wren." *Condor* 67: 6–30.
Vest, Josh L., Michael R. Conover, Clay Perschon, John Luft, and Jeffery Hall. 2009. "Trace Element Concentrations in Wintering Waterfowl from the Great Salt Lake, Utah." *Archives of Environmental Contamination and Toxicology* 56: 302–16.
Vickery, Peter D. 2020. "Grasshopper Sparrow (*Ammodramus savannarum*)." Version 1.0. In Poole and Gill 2020. doi:10.2173/bow.graspa.01.
Vickery, Peter D., Daniel E. Blanco, and Bernabé López-Lanús. 2010. *Conservation Plan for the Upland Sandpiper (*Bartramia longicauda*)*. Version 1.1. Manomet, MA: Manomet Center for Conservation Sciences.
Von Duyke, Andrew. 2006. "Summary Report: Monitoring, Research, and Management Activities at the Peltier Island Heronry during 2004 and 2005 Field Seasons." Final report to the US Fish and Wildlife Service, State Wildlife Grants Program. http://files.dnr.state.mn.us/eco/nongame/projects/consgrant_reports/2006/swg_2006_von_duyke.pdf.
Vorland, Jeannine. 1984. "Franklin's Gull (*Larus pipixcan*) Breeding Colonies and Associated Habitats in Minnesota, 1984." Unpublished progress report to the Minnesota Department of Natural Resources, Nongame Wildlife Program.
Wagner, Jerome. 1959. "Bewick's Wren Nesting in Anoka County." *Flicker* 31: 96–97.
Wagner, Lulu E. 1936. Letter to Thomas S. Roberts, June 14, 1936. Minnesota Ornithology Correspondence, Incoming (Box 70, Folder 693, pp. 176–77). University of Minnesota Libraries, University Archives. http://purl.umn.edu/185033.
Waite, Thomas A., and Dan Strickland. 2006. "Climate Change and the Demographic Demise of a Hoarding Bird Living on the Edge." *Proceedings of the Royal Society of London, Series B: Biological Sciences* 273: 2809–13.
Walkinshaw, Lawrence H. 1949. *The Sandhill Cranes*. Institute of Science Bulletin, no. 29. Bloomfield Hills, MI: Cranbrook.
———. 1983. "The Lincoln's Sparrow in Michigan." *Jack-Pine Warbler* 61: 75–81.
Wallestad, Josh. 2016. "Hey, Minnesota Birders, Go Find a Blue Grosbeak!" *Minnesota Birding* 53 (6): 1–4.
Walters, Eric L., Edward H. Miller, and Peter E. Lowther. 2020. "Yellow-bellied Sapsucker (*Sphyrapicus varius*)." Version 1.0. In Poole and Gill 2020. doi:10.2173/bow.yebsap.01.
Warkentin, Ian G., Navjot S. Sodhi, Richard H. M. Espie, Alan F. Poole, Lynn W. Oliphant, and

Paul C. James. 2020. "Merlin (*Falco columbarius*)." Version 1.0. In Billerman 2020. doi:10.2173/bow.merlin.01.

Warner, Dwain W. 1948. "Destruction of Double-crested Cormorants." *Flicker* 20: 99.

———. 1952. "Yellow-breasted Chat in Southeastern Minnesota." *Flicker* 24: 88–89.

———. 1959. "Breeding Distribution of Henslow's and LeConte's Sparrows." *Flicker* 31: 153.

———. 1984. "Pig's Eye Heron Rookery Census—1984." Report to the Minnesota Department of Natural Resources. https://files.dnr.state.mn.us/eco/nongame/projects/consgrant_reports/1984/1984_pigs_eye.pdf.

———. 1985. "Pig's Eye Heron Rookery Report for 1985." Report to the Metropolitan Airports Commission and the Minnesota Department of Natural Resources. http://files.dnr.state.mn.us/eco/nongame/projects/consgrant_reports/1985/1985_pigs_eye.pdf.

Warner, Dwain W., and Donald A. Beimborn. 1969. "First Caspian Tern Nesting in Minnesota." *Loon* 41: 83–84.

Warren-Chadd, Rachel, and Marianne Taylor. 2016. *Birds: Myth, Lore, and Legend*. London: Bloomsbury.

Washburn, Lowell. 2007. "Undisputed King of Ducks." *Iowa Outdoors* September/October: 40–45.

———. 2014. "Confessions of a Coot Shooter." *Globe Gazette,* November 9. http://globegazette.com/features/confessions-of-a-coot-shooter/article_6f081f47-63ff-59f9-9358-136baf177803.html.

Waters, Hannah. 2016. "The House Wren: The Unjustly Accused." The Sketch (editorial). *Audubon,* May 20. http://www.audubon.org/news/the-house-wren-unjustly-accused.

Watt, Doris J., John P. McCarty, Sarah W. Kendrick, Felicity L. Newell, and Peter Pyle. 2020. "Eastern Wood-Pewee (*Contopus virens*)." Version 1.0. In Rodewald 2020. doi:10.2173/bow.eawpew.01.

Watts, Bryan D. 2020. "Yellow-crowned Night-Heron (*Nyctanassa violacea*)." Version 1.0. In Poole 2020. doi:10.2173/bow.ycnher.01.

Wechsler, Chuck. 1985. *The Pheasant in Minnesota*. St. Paul: Minnesota Department of Natural Resources, Section of Wildlife.

Weeks, Harmon P., Jr. 2020. "Eastern Phoebe (*Sayornis phoebe*)." Version 1.0. In Poole 2020. doi:10.2173/bow.easpho.01.

Weflen, Kathleen. 2013. "Travels with *Gavia immer*." *Minnesota Conservation Volunteer* 76 (January/February): 2–3.

Weibel, Amy C., and William S. Moore. 2002a. "Molecular Phylogeny of a Cosmopolitan Group of Woodpeckers (Genus *Picoides*) Based on *COI* and *cyt b* Mitochondrial Gene Sequences." *Molecular Phylogenetics and Evolution* 22: 65–75.

———. 2002b. "A Test of a Mitochondrial Gene-Based Phylogeny of Woodpeckers (Genus *Picoides*) Using an Independent Nuclear Gene, β-Fibrinogen Intron 7." *Molecular Phylogenetics and Evolution* 22: 247–57.

———. 2005. "Plumage Convergence in *Picoides* Woodpeckers Based on a Molecular Phylogeny, with Emphasis on Convergence in Downy and Hairy Woodpeckers." *Condor* 107: 797–809.

Weidensaul, Scott, Tara Rodden Robinson, Robert R. Sargent, Martha B. Sargent, and Theodore J. Zenzal. 2020. "Ruby-throated Hummingbird (*Archilochus colubris*)." Version 1.0. In Rodewald 2020. doi:10.2173/bow.rthhum.01.

Weinberg, Howard J., and Roland R. Roth. 1998. "Forest Area and Habitat Quality for Nesting Wood Thrushes." *Auk* 115: 879–89.

Welch, Craig. 2009. "The Spotted Owl's New Nemesis." *Smithsonian,* January. http://www.smithsonianmag.com/science-nature/the-spotted-owls-new-nemesis-131610387/.

Welsh, Daniel A. 1971. "Breeding and Territoriality of the Palm Warbler in a Nova Scotia Bog." *Canadian Field-Naturalist* 85: 31–37.

Welter, Wilfred A. 1935. "The Natural History of the Long-billed Marsh Wren." *Wilson Bulletin* 47: 3–34.

Weseloh, D. V., Craig E. Herbert, Mark L. Mallory, Alan F. Poole, Julie C. Ellis, Peter Pyle, and Michael A. Patten. 2020. "Herring Gull (*Larus argentatus*)." Version 1.0. In Billerman 2020. doi:10.2173/bow.hergul.01.

Wessel, Ann. 2015. "Why an Island in Mille Lacs Is Vital to the Common Tern." *St. Cloud Times,* June 12. http://www.sctimes.com/story/sports/outdoors/2015/06/12/island-mille-lacs-vital-common-tern/71120816/.

Wetlands International. 2012. *Waterbird Population Estimates*. http://wpe.wetlands.org/downloads/downloads.

Weygandt, Cornelius. 1930. *The Wissahickon Hills: Memories of Leisure Hours Out of Doors in an Old Countryside*. Philadelphia: University of Pennsylvania Press.

Wheelwright, Nathaniel T., and James D. Rising. 2020. "Savannah Sparrow (*Passerculus sandwichensis*)." Version 1.0. In Poole 2020. doi:10.2173/bow.savspa.01.

Whitaker, Darroch M., and Stephen W. Eaton. 2020. "Northern Waterthrush (*Parkesia noveboracensis*)." Version 1.0. In Poole 2020. doi:10.2173/bow.norwat.01.

Whitcomb, R. F., Chandler S. Robbins, J. F. Lynch, B. L. Whitcomb, M. K. Klimkiewicz, and D. Bystrak. 1981. "Effects of Forest Fragmentation on Avifauna of the Eastern Deciduous Forest." In *Forest Island Dynamics in Man-dominated Landscapes,* edited by Robert L. Burgess and David M. Sharpe, 125–206. New York: Springer-Verlag.

White, Clayton M., Nancy J. Clum, Tom J. Cade, and W. Grainger Hunt. 2020. "Peregrine Falcon (*Falco peregrinus*)." Version 1.0. In Billerman 2020. doi:10.2173/bow.perfal.01.

White, Donald H., and Eugene Cromartie. 1977. "Residues of Environmental Pollutants and Shell Thinning in Merganser Eggs." *Wilson Bulletin* 89: 532–42.

Wiebe, Karen L., and William S. Moore. 2020. "Northern Flicker (*Colaptes auratus*)." Version 1.0. In Rodewald 2020. doi:10.2173/bow.norfli.01.

Wiens, Terry P. 1982. "Snowy Plover at Lake of the Woods." *Loon* 54: 242.

———. 1986. "Nest Site Tenacity and Mate Retention in the Piping Plover (*Charadrius melodus*)." Master's thesis, University of Minnesota.

———. 1989. "The Summer Season, June 1 to July 31, 1988." *Loon* 61: 18–33.

———. 1991. "The Summer Season (1 June to 31 July 1990)." *Loon* 63: 43–59.

———. 1992. "The Summer Season (1 June to 31 July 1991)." *Loon* 64: 18–36.

———. 1993. "The Summer Season: 1 June to 31 July 1992." *Loon* 65: 14–31.

———. 1995. "The Summer Season: 1 June to 31 July 1994." *Loon* 67: 17–37.

———. 1996. "The Summer Season (1 June to 31 July 1995)." *Loon* 68: 37–66.

———. 1997. "The Summer Season (1 June to 31 July 1996)." *Loon* 69: 11–28.

———. 1999. "The Summer Season (1 June to 31 July 1998)." *Loon* 71: 14–35.

———. 2000. "The Summer Season (1 June to 31 July 1999)." *Loon* 72: 14–36.

———. 2001. "The Summer Season (1 June to 31 July 2000)." *Loon* 73: 10–34.

———. 2003. "The Summer Season (1 June through 31 July 2002)." *Loon* 75: 15–41.

———. 2004. "The Summer Season: 1 June to 31 July 2003." *Loon* 76: 14–39.

Wiens, Terry P., and Francesca J. Cuthbert. 1984. "Status and Reproductive Success of the Piping Plover in Lake of the Woods." *Loon* 56: 106–9.

Wiggins, David A. 2004. *Short-eared Owl (*Asio flammeus*): A Technical Conservation Assessment.* US Department of Agriculture, Forest Service, Rocky Mountain Region. https://www.fs.usda.gov/Internet/FSE_DOCUMENTS/stelprdb5182042.pdf.

———. 2005. *Loggerhead Shrike (*Lanius ludovicianus*): A Technical Conservation Assessment.* USDA Forest Service, Rocky Mountain Region. Accessed May 20, 2023. http://www.fs.fed.us/r2/projects/scp/assessments/loggerheadshrike.pdf.

Wiggins, David A., Denver W. Holt, and Shawne M. Leasure. 2020. "Short-eared Owl (*Asio flammeus*)." Version 1.0. In Billerman 2020. doi:10.2173/bow.sheowl.01.

Wilcove, David S., Charles H. McLellan, and Andrew P. Dobson. 1986. "Habitat Fragmentation in the Temperate Zone." In *Conservation Biology: The Science of Scarcity and Diversity,* edited by Michael E. Soule, 237–56. Sunderland, MA: Sinauer Associates.

Will, Thomas C. 1986. "The Behavioral Ecology of Species Replacement: Blue-winged and Golden-winged Warblers in Michigan." PhD diss., University of Michigan.

———. 2009. "Conservation Column." *Minnesota Birding* 46 (6): 1–4.

Will, Tom, Jessica C. Stanton, Kenneth V. Rosenberg, Arvind O. Panjabi, Alaine F. Camfield, Allison E. Shaw, Wayne E. Thogmartin, et al. 2019. "Handbook to the Partners in Flight Population Estimates Database, Version 3.0." PIF Technical Publication No 7. pif.birdconservancy.org/popest.handbook.pdf.

Williams, John R. 1935. "Kingbird (*Tyrannus tyrannus*) Attacks Airplane." *Auk* 52: 89.

Williamson, Scot J., Dan M. Keppie, Robert Davison, Dave Budeau, Suzanne Carrière, Dale Rabe, and Michael A. Schroeder. 2008. *Spruce Grouse Continental Conservation Plan.* Washington, DC: Association of Fish and Wildlife Agencies.

Willis, Franklin. 1949. "Swallow-tailed Kite in Fillmore County." *Flicker* 21: 71–72.

Wilsey, Chad, Brooke Bateman, Lotem Taylor, Joanna X. Wu, Geoffrey LeBaron, Robyn Shepherd, Chloe Koseff, et al. 2019. *Survival by Degrees: 389 Bird Species on the Brink.* New York: National Audubon Society.

Wilsey, Chad, Lotem Taylor, Brooke Bateman, Caitlin Jensen, Nicole Michel, Arvind Panjabi, and Gary Langham. 2019. "Climate Policy Action Needed to Reduce Vulnerability of Conservation-reliant Grassland Birds in North America." *Conservation Science and Practice* 1 (4): e.21. doi:10.1111/csp2.21.

Wilson, Alexander. 1810. *American Ornithology; or, The Natural History of Birds of the United States.* Vol. 2. Philadelphia: Bradford and Inskeep.

———. 1832. *American Ornithology; or The Natural History of the Birds of the United States.* London: Cassel, Petter & Galpin.

Wilson, Andrew M., Daniel W. Brauning, and Robert S. Mulvill, eds. 2012. *Second Atlas of Breeding Birds in Pennsylvania.* University Park: Pennsylvania State University Press.

Wilson, N. O. 1937. "Lark Bunting Invasion." *Journal of Minnesota Ornithology* 1: 9–10.

Wilson, Scott, Eric M. Anderson, Amy S. G. Wilson, Douglas F. Bertram, and Peter Arcese. 2013. "Citizen Science Reveals an Extensive Shift in Winter Distribution of Migratory Western Grebes." *PLoS One* 8: e65408. doi:10.1371/journal.pone.0065408.

Wilson, Steven G. 1980. "Minnesota's First Breeding Wilson's Warbler." *Loon* 52: 182–83.

———. 1982. "Rusty Blackbirds—Lake County." *Loon* 54: 246–47.

———. 1985. "Summer Distribution of Whip-poor-wills in Minnesota." *Loon* 57: 6–8.

———. 1993. "Nesting Northern Hawk Owls." *Loon* 65: 58–60.

Wilson, Steven G., and Mary Shedd. 1986. "The Summer Season (June 1 to July 31, 1985)." *Loon* 58: 21–38.

———. 1988. "The Summer Season (June 1 to July 31, 1987)." *Loon* 60: 16–30.

Wilson, W. Herbert, Jr. 2020. "Palm Warbler (*Setophaga palmarum*)." Version 1.0. In Poole 2020. doi:10.2173/bow.palwar.01.

Winkelman, Gerald. 1989. "Black-necked Stilts, Stevens County." *Loon* 61: 139–40.

Winkler, David W., Kelly K. Hallinger, Daniel R. Ardia, Raleigh J. Robertson, Bridget J. Stutchbury, and Robert R. Cohen. 2020. "Tree Swallow (*Tachycineta bicolor*)." Version 1.0. In Poole 2020. doi:10.2173/bow.treswa.01.

Winter, Maiken, Douglas H. Johnson, Jill A. Dechant, Therese M. Donovan, and W. Daniel Svedarsky. 2001. *Evaluation of the Bird Conservation Area Concept in the Northern Tallgrass Prairie: Annual Report 2001.* Jamestown, ND: Northern Prairie Wildlife Research Center.

Winter, Maiken, Douglas H. Johnson, and Jill A. Shaffer. 2005. "Variability in Vegetation Effects on Density and Nesting Success of Grassland Birds." *Journal of Wildlife Management* 69: 185–97.

Winter, Maiken, Douglas H. Johnson, Jill A. Shaffer, Therese M. Donovan, and W. Daniel Svedarsky. 2006. "Patch Size and Landscape Effects on Density and Nesting Success of Grassland Birds." *Journal of Wildlife Management* 70: 158–72.

Winter, Maiken, Jill A. Shaffer, Douglas H. Johnson, Therese M. Donovan, W. Daniel Svedarksy, Peter W. Jones, and Betty R. Euliss. 2005. "Habitat and Nesting of Le Conte's Sparrows in the Northern Tallgrass Prairie." *Journal of Field Ornithology* 76: 61–71.

Wires, Linda R. 2014. *The Double-crested Cormorant: Plight of a Feathered Pariah.* New Haven, CT: Yale University Press.

Wires, Linda R., and Francesca J. Cuthbert. 2000. "Trends in Caspian Tern Numbers and Distribution in North America: A Review." *Waterbirds* 23: 388–404.

———. 2003. "Fish-eating Bird Predation at Aquaculture Facilities in Minnesota: A First Step towards Bridging the Information Gap." Final report to Minnesota Sea Grant. Sea Grant Document Number MINNU-T-03-001. https://repository.library.noaa.gov/view/noaa/38036.

———. 2006. "Historic Populations of the Double-crested Cormorant (*Phalacrocorax auritus*): Implications for Conservation and Management in the 21st Century." *Waterbirds* 29: 9–37.

Wires, Linda R., Francesca J. Cuthbert, and D. Hamilton. 2011. *The American White Pelican and Double-crested Cormorant in Minnesota in 2010: Distribution, Abundance and Population Change.* Final report to the Minnesota Department of Natural Resources, Nongame Wildlife Program. https://files.dnr.state.mn.us/eco/nongame/projects/consgrant_reports/2011/swg_2011_wires_etal.pdf.

Wires, Linda R., Katherine V. Haws, Francesca J. Cuthbert, Nancy E. Drilling, and Alexander C. Smith. 2006. "The Double-crested Cormorant and American White Pelican in Minnesota: First Statewide Breeding Census." *Loon* 78: 63–73.

Wires, Linda R., Stephen J. Lewis, Gregory J. Soulliere, Sumner W. Matteson, D. V. "Chip" Weseloh, Robert P. Russell, and Francesca J. Cuthbert. 2010. *Upper Mississippi Valley / Great Lakes Waterbird Conservation Plan: A Plan Associated with the Waterbird Conservation for the Americas Initiative.* Final report. Fort Snelling, MN: US Fish and Wildlife Service.

Wisconsin DNR (Department of Natural Resources). 2013. *Wisconsin Hooded Warbler*

Species Guidance. Madison, WI: Bureau of Natural Heritage Conservation, Wisconsin Department of Natural Resources. PUB-ER-686. Last updated October 8, 2018.

———. 2015. *2015–2025 Wisconsin Wildlife Action Plan*. Madison, WI. http://dnr.wi.gov/topic/WildlifeHabitat/ActionPlan.html.

With, Kimberly A. 2020. "McCown's Longspur (*Rhynchophanes mccownii*)." Version 1.0. In Poole 2020. doi:10.2173/bow.mcclon.01.

Witmer, Mark C., D. James Mountjoy, and Lang Elliot. 2020. "Cedar Waxwing (*Bombycilla cedrorum*)." Version 1.0. In Poole 2020. doi:10.2173/bow.cedwax.01.

Wojahn, Barbara. 1977. "White-eyed Vireo Confirmed for Minnesota." *Loon* 49: 174.

Wolfe, Terry. 1984. "Summary of 1980–1984 Minnesota Prairie Chicken Inventory Data by County." *Loon* 56: 202.

Wood, Petra Bohall, Scott B. Bosworth, and Randy Dettmers. 2006. "Cerulean Warbler Abundance and Occurrence Relative to Large-Scale Edge and Habitat Characteristics." *Condor* 108: 154–65.

Wood, Stan. 1989. "Minnesota's First Black-necked Stilt." *Loon* 61: 197.

Woodin, Marc C., and Thomas C. Michot. 2020. "Redhead (*Aythya americana*)." Version 1.0. In Poole and Gill 2020. doi:10.2173/bow.redhea.01.

Woolsey, Ralph. 1933. "The 1933 Nesting Season." *Flicker* 5: 26–47.

Wootton, J. Timothy. 2020. "Purple Finch (*Haemorhous purpureus*)." Version 1.0. In Poole and Gill 2020. doi:10.2173/bow.purfin.01.

World Wildlife Fund. 2018. *The Plowprint Report: 2018*. https://www.worldwildlife.org/publications/2018-plowprint-report.

WSO (Wisconsin Society for Ornithology). 2021. "Wisconsin Breeding Bird Atlas II: Atlas Results to Date." https://wsobirds.org/atlas-results-to-date.

Wyatt, Valerie E., and Charles M. Francis. 2020. "Rose-breasted Grosbeak (*Pheucticus ludovicianus*)." Version 1.0. In Poole and Gill 2020. doi:10.2173/bow.robgro.01.

Wyatt, William H. 1973. "Cattle Egrets Nesting at Pelican Lake." *Loon* 45: 100.

Wyckoff, Ann Marie. 1985. *Population Assessment* Calcarius ornatus *(Chestnut-collared Longspur), Felton Prairie, Minnesota*. Minnesota Department of Natural Resources, Natural Heritage and Nongame Research Program.

———. 1986a. "Longspurs Breed in Traverse County." *Loon* 58: 51.

———. 1986b. "A Relict Population of Chestnut-collared Longspurs in Western Minnesota." *Loon* 58: 3–11.

Yasukawa, Ken, and William A. Searcy. 2020. "Red-winged Blackbird (*Agelaius phoeniceus*)." Version 1.0. In Rodewald 2020. doi:10.2173/bow.rewbla.01.

Yosef, Reuven. 2020. "Loggerhead Shrike (*Lanius ludovicianus*)." Version 1.0. In Poole and Gill 2020. doi:10.2173/bow.logshr.01.

Yosef, Reuven, and Thomas C. Grubb, Jr. 1993. "Effect of Vegetation Height on Hunting Behavior and Diet of Loggerhead Shrikes." *Condor* 95: 127–31.

Young, Matt. 2017. "Gadwalls: The Wonder Duck." *Ducks Unlimited*. http://www.ducks.org/conservation/waterfowl-research-science/gadwalls-the-wonder-duck.

Youngworth, William. 1958. "The Blue Grosbeak in Western Iowa: A Summary." *Iowa Bird Life* 28: 57–59.

Zenzal, Theodore J., and Frank R. Moore. 2016. "Stopover Biology of Ruby-throated Hummingbirds (*Archilochus colubris*) during Autumn Migration." *Auk* 133: 237–50.

Zicus, Michael C., Mark A. Briggs, and Richard M. Pace III. 1988. "DDE, PCB, and Mercury Residues in Minnesota Common Goldeneye and Hooded Merganser Eggs, 1981." *Canadian Journal of Zoology* 66: 1871–76.

Zicus, Michael C., and Steven K. Hennes. 1988. "Cavity Nesting Waterfowl in Minnesota." *Wildfowl* 39: 115–23.

Zicus, Michael C., Steven K. Hennes, and Michael R. Riggs. 1995. "Common Goldeneye Nest Attendance Patterns." *Condor* 97: 461–72.

Zicus, Michael C., David P. Rave, John R. Fieberg, John H. Giudice, and Robert G. Right. 2008. "Distribution and Abundance of Minnesota-Breeding Ring-necked Ducks *Aythya collaris*." *Wildfowl* 58: 31–45.

Zierman, Shawn. 2007. "Snowy Plover at Park Point, Duluth." *Loon* 79: 240.

Zimmerman, Amy L., Jill A. Dechant, Douglas A. Johnson, Christopher M. Goldade, Brent E. Jamison, and Betty R. Euliss. 2002. *Effects of Management Practices on Wetland Birds: Black Tern*. Jamestown, ND: Northern Prairie Wildlife Research Center. https://pubs.usgs.gov/unnumbered/70159831/report.pdf.

Zimmerman, Amy L., Brent E. Jamison, Jill A. Dechant, Douglas H. Johnson, Christopher M. Goldade, James O. Church, and Betty R. Euliss. 2002a. *Effects of Management Practices on Wetland Birds: Sora*. Jamestown, ND: Northern Prairie Wildlife Research Center. https://pubs.er.usgs.gov/publication/70159922/report.pdf.

———. 2002b. *Effects of Management Practices on Wetland Birds: Virginia Rail*. Jamestown, ND: Northern Prairie Wildlife Research Center. https://pubs.usgs.gov/unnumbered/70159833/report.pdf.

Zink, Robert M., and Bruce A. Fall. 1981. "Breeding Distribution, Song and Habitat of the Alder Flycatcher and Willow Flycatcher in Minnesota." *Loon* 53: 208–14.

Zlonis, Edmund J., and Gerald J. Niemi. 2014. "Avian Communities of Managed and Wilderness Hemiboreal Forests." *Forest Ecology and Management* 328: 26–34.

Zlonis, Edmund J., Hannah G. Panci, Josh D. Bednar, Maya Hamady, and Gerald J. Niemi. 2017. "Habitats and Landscapes Associated with Bird Species in a Lowland Conifer-Dominated Ecosystem." *Avian Conservation and Ecology* 12: 7. doi:10.5751/ACE-00954-120107.

Zlonis, Edmund J., Nicholas G. Walton, Brian R. Sturtevant, Peter T. Wolter, and Gerald J. Niemi. 2019. "Burn Severity and Heterogeneity Mediate Avian Response to Wildfire in a Hemiboreal Forest." *Forest Ecology and Management* 439: 70–80. doi:10.1016/j.foreco.2019.02.043.

Zumeta, David C. 2001. "Yellow-throated Warbler in Washington County." *Loon* 73: 62–63.

Zumeta, David C., and Mary Ann Cincotta. 1982. "A Spruce Grouse Record from Pine County." *Loon* 54: 200–202.

———. 1983. "Louisiana Waterthrush in Pine County." *Loon* 55: 89.

Lee A. Pfannmuller worked at the Minnesota Department of Natural Resources (DNR) for thirty years developing extensive expertise in field ecology, forest bird management, endangered species policy, and biodiversity conservation. She is coeditor of *Minnesota's Endangered Flora and Fauna* (Minnesota, 1988), the first comprehensive assessment of the state's endangered species. As director of the DNR's Division of Ecological Resources, Pfannmuller was responsible for directing programs focused on Minnesota's rare resources, native plant communities, lakes and rivers, and invasive species. Following her retirement from the DNR, she spent several years preparing an *Implementation Blueprint for Minnesota Bird Conservation* for Audubon Minnesota and served as Chair of the Bell Museum Advisory Board from 2014 through 2022.

Gerald J. Niemi is a retired Professor and currently Emeritus in the Department of Biology, University of Minnesota–Duluth (UM-D). He helped establish the Natural Resources Research Institute (NRRI) at UM-D and was named Director of the Center for Water and the Environment in 1989, a position he was affiliated with until 2019. In his various roles at NRRI, he established an ornithological research program focused on how birds in Minnesota are affected by the use of natural resources and natural processes such as logging, forest fire, energy development, pollution, and tourism. Niemi is a Fellow of the American Ornithological Society and has authored or coauthored more than 160 peer-reviewed publications and more than 100 reports and popular articles.

Janet C. Green is an avid bird-watcher and conservationist based in Duluth, Minnesota. She has served on several citizen advisory committees with public agencies, including the Pollution Control Board, Environmental Quality Board, Minnesota Forest Resources Council, and the Minnesota DNR. With the DNR's Division of Ecological Resources, Green published a book in 1995, *Birds and Forests: A Management and Conservation Guide*. She is also coauthor of the annotated checklist *Minnesota Birds: Where, When, and How Many* (Minnesota, 1975) and *Birds of the Superior National Forest*. Green cofounded Duluth's Hawk Ridge Nature Reserve in 1972 and served on the advisory board of Audubon Minnesota.

Kim Rewinkel moved to Minnesota in 2013 after taking a position at the University of Minnesota–Duluth's Natural Resources Research Institute. At NRRI she edited manuscripts and reports for publication and served as a research administrator for projects in the Center for Water and the Environment. She also conducted field surveys for the Great Lakes Coastal Wetland Monitoring Project in Minnesota and Wisconsin, including the Apostle Islands on Lake Superior. Rewinkel now lives in Anchorage, Alaska, and works in grant administration and proposal development at Kansas State University.

Francesca J. Cuthbert is Professor in the Department of Fisheries, Wildlife and Conservation Biology at the University of Minnesota–Twin Cities. Her research focuses on the recovery of the endangered Great Lakes Piping Plover population, biology and management of Double-Crested Cormorants in North America, and colonial waterbird population dynamics and conservation in the Great Lakes and Asia. Cuthbert serves on the editorial board of the *Journal of Great Lakes Research,* and her research appears frequently in *Journal of Wildlife Management, Journal of Applied Ecology,* and other publications. In 2021 she was awarded the Ralph W. Schreiber Conservation Award by the American Ornithological Society.

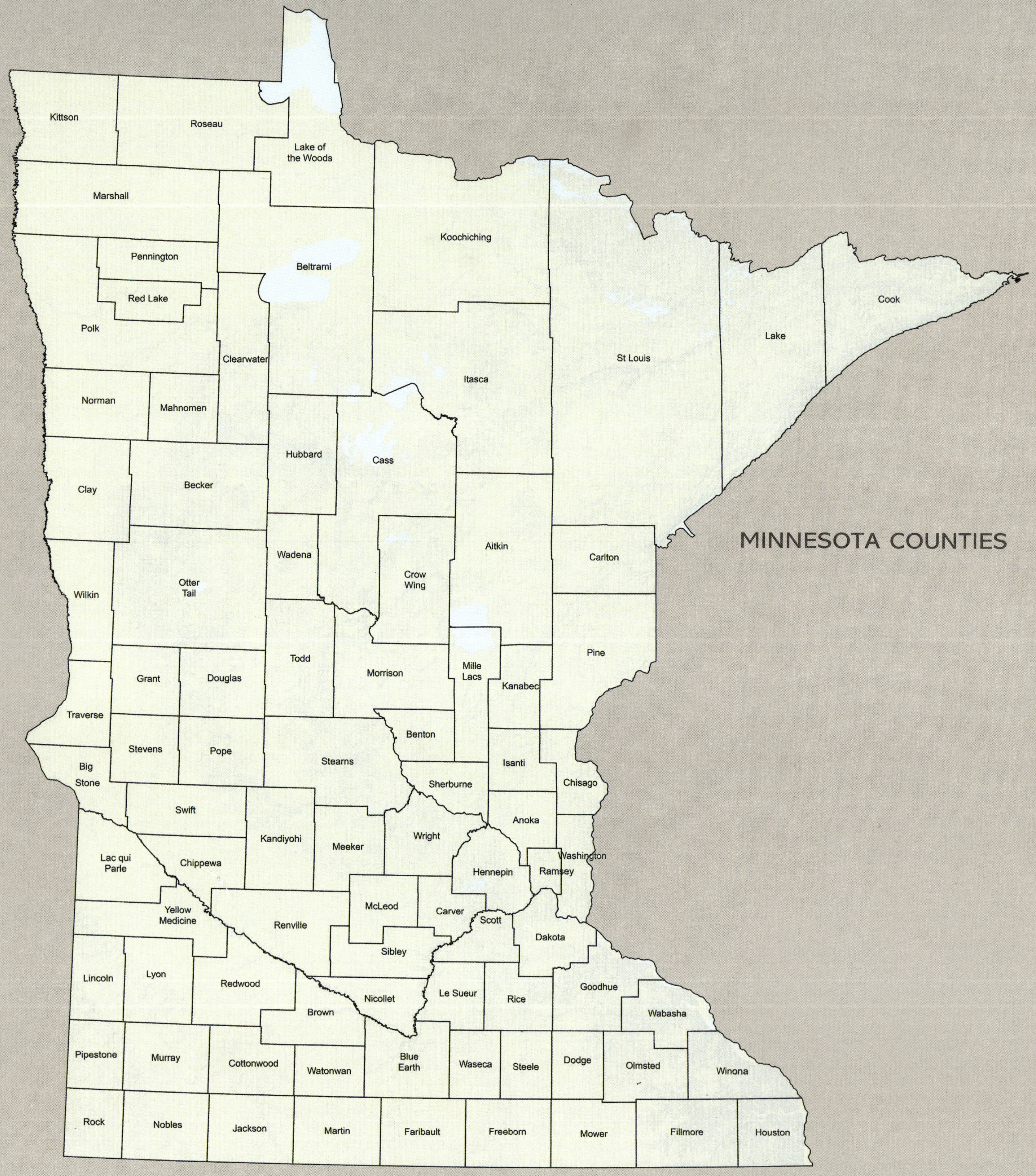
MINNESOTA COUNTIES
Kittson
Roseau
Lake of the Woods
Marshall
Pennington
Red Lake
Polk
Beltrami
Koochiching
Cook
Lake
St Louis
Itasca
Clearwater
Norman
Mahnomen
Hubbard
Cass
Clay
Becker
Aitkin
Carlton
Wadena
Otter Tail
Wilkin
Crow Wing
Todd
Morrison
Mille Lacs
Kanabec
Pine
Grant
Douglas
Traverse
Benton
Stevens
Pope
Stearns
Isanti
Big Stone
Sherburne
Chisago
Swift
Anoka
Kandiyohi
Meeker
Wright
Washington
Lac qui Parle
Chippewa
Hennepin
Ramsey
McLeod
Carver
Yellow Medicine
Renville
Scott
Dakota
Sibley
Lincoln
Lyon
Redwood
Goodhue
Le Sueur
Nicollet
Rice
Brown
Wabasha
Pipestone
Murray
Cottonwood
Watonwan
Blue Earth
Waseca
Steele
Dodge
Olmsted
Winona
Rock
Nobles
Jackson
Martin
Faribault
Freeborn
Mower
Fillmore
Houston